# INDIAN CHRONOLOGY

(SOLAR, LUNAR AND PLANETARY)

# INDIAN CHRONOLOGY

**(SOLAR, LUNAR AND PLANETARY)**

**A PRACTICAL GUIDE**
**TO THE**
**INTERPRETATION AND VARIFICATION**
**OF**
**TITHIS, NAKSHATRAS, HOROSCOPES**
**AND**
**OTHER INDIAN TIME-RECORDS**
**B.C. 1 TO A.D. 2000**

**L. D. SWAMIKANNU PILLAI**

GYAN

*Published by*

Gyan Publishing House
5, Ansari Road
Daryaganj, New Delhi-110002
Phone: 011-47034999, 9811692060
E-mail: books@gyanbooks.com

*Distribution Network*
gyanbooks.com
India, USA, Canada, UK, Australia, France



ISBN : 978-81-212-3603-4 (PB)
First Published, Madras, 1911

2nd Impression 2020

*Printed at:* Gyan Press, Delhi.



**Indian Chronology**
Author: L. D. Swamikannu Pillai

# PREFACE.

THIS book, which was announced to the public under various names *, is now issued under a title which restricts it to its original scope, namely, as a **Manual of Indian Chronology**, designed to present in one view and in a usable form all the material ordinarily required by investigators in that field of research.

The bulk of the tables, as well as of the letter-press, is devoted to a detailed exhibition of the *decursus* of Indian Chronology, solar, lunar and planetary, with eclipses, week-days, English calendar months and dates, *adhika* and *kshaya masas*, and mean places of the Sun and major planets, from B.C. 1 to A.D. 2000. The portion of the work, called Table X, will, it is hoped, supply a long-felt want among Epigraphists and Archæologists, official as well as non-official, in India, inasmuch as it will enable them to verify on the spot and *in less than a minute in any case*, the exact English equivalent of any Indian date occurring in any inscription as well as to test its genuineness. The instructions set opposite the Eye-Table at the end of the book are alone sufficient for any person, however previously unacquainted with the subject, to calculate quite accurately the ending moment of a *tithi* or *nakshatra*.

The work is intended, however, for the use of a much wider public than the scientific few who are interested in Indian chronological research. The same principles, whereby we are enabled to re-call into the light of history the inscriptions that are being unearthed at the rate of a thousand every year by the various Indian Epigraphical departments, continue to guide the religious observances and civil usages of millions of people in India at the present day, and no one interested in the social and religious life of this vast continent can fail to take at least some slight account of a system, so ancient and so thoroughly ingrained in the habits of the people. The Indian calendar is generally considered either too abstruse for popular, or too inaccurate for scientific, exposition, but neither of these objections is more than a fallacy. A good deal has been done in the past to put Indian chronology on its proper footing, namely, as the faithful handmaid of Indian history, and there is no reason why the torch so worthily carried in their day by the late Professor Kielhorn, the present Dr. Fleet and other workers in the field, should not be kept bright and burning in the hands of their successors, provided they are equipped with the necessary knowledge of data and principles.

Lastly, there is hardly any person, holding a position of consequence, public or private, in India, who is not occasionally confronted with the citation of an Indian date. To Magistrates, Judges, Lawyers, Missionaries, Business Men of every description, as well as to Historians, it is of interest to know the meaning of chronological terms in daily use among the people, although all may not, equally with the epigraphist and the archæologist, feel inclined to enter upon a regular study of Indian chronology. To those with scanty leisure, Tables XII and XVIII, forming a luni-solar and planetary Ephemeris for the 80 years A.D. 1840 to A.D. 1920, together with the simple instructions as to their use, contained in the Second and Fourth Parts of the book headed "Use of the Tables" and "Planetary Chronology" as well as in the Eye-Table, will be found no less handy than serviceable. The publications ordinarily known by the name of *Jantri* are unscientific, and few of them are up to date, whereas the ephemeris portion of the present work will enable any one to understand and decide, with reference to first principles, any question of current Indian chronology for seventy years past and some ten years to come. A table (XV) has also been inserted to link together the Indian and Muhammadan calendars.

* "*Tithis, Nakshatras and other Indian dates B.C. 1 to A.D. 2000*" (*Indian Review* for May 1910); "*Indian Astronomy and Chronology with a Criticism on Indian Astrology*" (Advt. page of the *Journal of the South Indian Association*, Madras), etc.

The book necessarily contains a vast amount of figures and tables, but the perusal of no part of it will requisition a greater knowledge of mathematics than is implied in the four simple rules of Arithmetic. Decimals have been used freely in order to ensure accuracy, but not only have tables whereby the decimal parts of a day, degree, etc., can be converted into *ghatikas*, *palas*, hours, minutes and seconds, been freely inserted, but the more important tables for the calculation of anomalies and equations for *tithis* and *nakshatras* (Table IX) and the Nakshatra Table itself (Table XI) have been twice given, once in the form of decimals and again as days, *ghatikas* and *palas*.

It was originally intended to limit this compilation to solar and lunar chronology, but the frequent occurrence, in chronological records, in Indian literature as well as in daily Indian life, of allusions to the mean or actual places of planets, induced the writer to add a considerable number of tables on this interesting, though somewhat difficult, subject. The tables of anomalies and equations, used for planets, are based mainly on those of *Vavilala Kuchinna*, as reproduced in Warren's *Kala Sankalita*. The computation of the geocentric longitudes of planets for the years A.D. 1840 to A.D. 1920, contained in Table XVIII, entailed an enormous number of detailed calculations, but the results will, it is hoped, be found sufficiently accurate for practical purposes. By "practical purposes" the author means solely the purposes of Indian chronology, and he wishes it to be distinctly understood that he does not believe in astrology any more than in the need for, or efficacy of, the fasts, festivals and donations mentioned in Chapter XVI, "Notes on tithis in relation to festivals", which find a place in this book, merely and solely because of the frequent references in inscriptions to Hindu fasts and feasts.

In conclusion, the author wishes to state that, while he has spared no pains to acquaint himself with the works of previous writers on the subject, he has endeavoured to work out details upon his own method which is explained in Part III—"Construction of Tables". To one honoured name in the field of Indian astronomy and chronology, that of **Professor Hermann Jacobi**, he wishes, though an utter stranger, to express special indebtedness, since to him the author owes his first knowledge of the subject as well as numerous illustrations serving to test the correctness of his method.

Lastly, he must thank the *Premier Press*, of Messrs. Hoe & Co., for the skill with which they have executed a work, which, though packed with figures, has been reproduced with a high degree of neatness consistently with economy. The author would have been glad to have inserted the proper diacritical marks throughout the text, but this would have entailed further delay in publishing a work, which, in the endeavour to make it accurate and reliable in more important respects, has already been kept too long in the press. Such marks have, therefore, been sparingly used, except in the Index, which gives the proper marks for every Sanskrit word occurring in the text, and to which reference should be made in case of doubt. The Index also explains the meaning of terms not occurring in the text, such as *Dakshinayana, Uttarayana, Aparapaksha*, etc.

MADRAS,
*25th March 1911.*

L.D.S.

# CONTENTS.

## PART I.—Relations between Indian Astronomy and Indian Chronology.



## PART II.—Use of the Tables.



## PART III.—Construction of Tables.



## PART IV.—Planets and Planetary Chronology. ... 

# TABLES.

# ADDENDA.

## Section 6 of Text and TABLE VI.

### Surya Siddhanta Constants for Centuries from 0 Kaliyuga to 3101 K.Y. (B.C. 3102 to B.C. 1.)

| Year B.C. | Kaliyuga. | Commencement of Solar Year. Month and date. | Commencement of Solar Year. Fraction of day. | First New Moon in Solar Year. | ☾'s anomaly at commencement of Solar Year. |
|---|---|---|---|---|---|
| **3102 B.C.** | 0 | 15F | ·57924 | 2·17013 | 4·71797 |
| **3101** | 1 | 15F | ·83800 | 20·80902 | 11·76692 |
| **3001** | 101 | 16F | ·71365 | 24·27066 | 0·24311 |
| **2901** | 201 | 17F | ·58980 | 27·73230 | 16·27388 |
| **2801** | 301 | 18F | ·46494 | 1·66335 | 4·75007 |
| **2701** | 401 | 19F | ·34059 | 5·12499 | 20·78084 |
| **2601** | 501 | 20F | ·21624 | 8·58663 | 9·25703 |
| **2501** | 601 | 21F | ·09189 | 12·04827 | 25·28780 |
| **2401** | 701 | 21F | ·96754 | 15·50991 | 13·76399 |
| **2301** | 801 | 22F | ·84319 | 18·97155 | 2·24018 |
| **2201** | 901 | 23F | ·71883 | 22·43319 | 18·27095 |
| **2101** | 1001 | 24F | ·59448 | 25·89483 | 6·74714 |
| **2001** | 1101 | 25F | ·47013 | 29·35647 | 22·77791 |
| **1901** | 1201 | 26F | ·34578 | 3·28752 | 11·25410 |
| **1801** | 1301 | 27F | ·22143 | 6·74916 | 27·28487 |
| **1701** | 1401 | 28F | ·09708 | 10·21080 | 15·76106 |
| **1601** | 1501 | 28F | ·97273 | 13·67244 | 4·23725 |
| **1501 B.C.** | 1601 | 29F | ·84837 | 17·13408 | 20·26802 |
| **1401** | 1701 | 1M | ·72402 | 20·59572 | 8·74421 |
| **1301** | 1801 | 2M | ·59967 | 24·05736 | 24·77498 |
| **1201** | 1901 | 3M | ·47532 | 27·51900 | 13·25117 |
| **1101** | 2001 | 4M | ·35097 | 1·45005 | 1·72736 |
| **1001** | 2101 | 5M | ·22662 | 4·91169 | 17·75813 |
| **901** | 2201 | 6M | ·10226 | 8·37393 | 6·23432 |
| **801** | 2301 | 6M | ·97791 | 11·83557 | 22·26509 |
| **701** | 2401 | 7M | ·85356 | 15·29721 | 10·74128 |
| **601** | 2501 | 8M | ·72921 | 18·75885 | 26·77205 |
| **501** | 2601 | 9M | ·60486 | 22·22049 | 15·24824 |
| **401** | 2701 | 10M | ·48051 | 25·68213 | 3·72443 |
| **301** | 2801 | 11M | ·35615 | 29·14377 | 19·75520 |
| **201** | 2901 | 12M | ·23180 | 3·07482 | 8·23139 |
| **101** | 3001 | 13M | ·10745 | 6·53646 | 24·26216 |
| **1** | 3101 | 13M | ·98310 | 9·99810 | 12·73885 |

*N.B.*—Odd years in centuries B.C. should be converted into positive years before using Table VII. See instructions in *N.B.* to Table XXII.

It will be seen that the *Solar Year* 0 K.Y. commenced on Feb. 15·57924. If to this be added the Sodhya of 2·1707 days we have, Feb. 15·5792 + 2·1707 = Feb. 17·75, or midnight between 17th and 18th Feb. B.C. 3102, which was the moment of commencement of the Era, 0 Kaliyuga, according to Surya Siddhanta. See foot-note on p. (3) of the Text.

---

## TABLE XVII.

### Mean Places of Major Planets at Commencement of each Century from B.C. 1601 to B.C. 1 (K.Y. 1501 to K.Y. 3101.)

| | | Mars. | Mercury. | Jupiter. | Venus. | Saturn. |
|---|---|---|---|---|---|---|
| **B.C. 1601** | K.Y. 1501 | 6·3122 | 73·0963 | 207·2519 | 359·5156 | 320·0274 |
| **1501** | 1601 | 66·8392 | 149·1143 | 1·9965 | 198·5007 | 101·9348 |
| **1401** | 1701 | 127·3662 | 225·1323 | 156·7411 | 37·4855 | 243·8412 |
| **1301** | 1801 | 187·8932 | 301·1503 | 311·4857 | 236·4709 | 25·7481 |
| **1201** | 1901 | 248·4202 | 17·1683 | 106·2303 | 75·4560 | 167·6550 |
| **1101** | 2001 | 308·9472 | 93·1863 | 260·9749 | 274·4411 | 309·5619 |
| **1001** | 2101 | 9·4742 | 169·2043 | 55·7195 | 113·4262 | 91·4688 |
| **901** | 2201 | 70·0012 | 245·2223 | 210·4641 | 312·4113 | 233·3757 |
| **801** | 2301 | 130·5282 | 321·2403 | 5·2087 | 151·3964 | 15·2826 |
| **701** | 2401 | 191·0552 | 37·2583 | 159·9533 | 350·3815 | 157·1895 |
| **601** | 2501 | 251·5822 | 113·2763 | 314·6979 | 189·3666 | 299·0964 |
| **501** | 2601 | 312·1092 | 189·2943 | 109·4425 | 28·3517 | 81·0033 |
| **401** | 2701 | 12·6362 | 265·3123 | 264·1871 | 227·3368 | 222·9102 |
| **301** | 2801 | 73·1632 | 341·3303 | 58·9317 | 66·3219 | 4·8171 |
| **201** | 2901 | 133·6902 | 57·3483 | 213·6763 | 265·3070 | 146·7240 |
| **101** | 3001 | 194·2172 | 133·3663 | 8·4209 | 104·2921 | 288·6309 |
| **1** | 3101 | 254·7442 | 209·3843 | 163·1655 | 303·2772 | 70·5378 |

## EXPLANATION.

With the aid of these two tables it is possible to calculate solar dates, tithis and *nakshatras* for any date between B.C. 3102 and B.C. 1 and also any horoscope from B.C. 1601 to B.C. 1. The places of planets' nodes for any of these years can be easily deduced from the places given in Table XVII in the body of the book for the centuries 1 B.C. to A.D. 2000.

It was thought unnecessary to carry the record of the mean places of planets further back than B.C. 1601, first because it is extremely doubtful whether planetary places were noted in India even roughly before that date, and secondly, because any person wishing to ascertain the mean places of planets at an earlier date can easily do so by noting in the above table the increase of mean sidereal longitude of any planet for 100, 200, 300............etc., years, and deducting this increase from the mean place for B.C. 1601.

We will now present the reader with the working of two interesting problems in B.C. dates: our examples are taken from Dr. Fleet's article " The day on which Buddha died" in the *Journal of the Royal Asiatic Society* for 1909.

I. *The day on which Buddha died* :—B.C. 483, *Karttika Sukla* 8.

| | | Commencement of Solar year. | First New Moon in Solar year. | | Moon's Anom. at commencement of Solar Year. | |
|---|---|---|---|---|---|---|
| (Contents p. 4) | B.C. 501 | March ... 9·60 | 22·22 | | 15·25 | |
| (Table VII, p. 7) | Add for 18 years. | ·66 | 10·66 | | 16·66 | |
| | B.C. 483 | March ... 10·26 | 32·88 | | 31·91 | |
| | | | 29·53 | (Syn. Mon.) | 27·55 | (1 Anom. month) |
| | | | 3·35 | | 4·36 | |
| | | | | | 3·35 | |
| | | | | | 7·71 | (Moon's Anomaly at First New-moon in Solar year B.C. 483.) |
| Table VIII (Addition for Karttika Sukla 8 in a year in which some previous month was *adhika*. * is the same as that for Sukla 8 of the next month Margasira). | | | 214·59 | | 21·71 | |
| | | | 217·91 | | 29·42 | |
| | | | | | 27·55 | (1 Anom. month) |
| | | | | | 1·87 | |
| Sun's Equation for 218 days (Eye-Table) | | −·13 | | | −·13 | (Sun's Equation) |
| Moon's Equation for 1·74 days (Eye-Table) | | −·15 | | | 1·74 | |
| | | Sum −·28 | −·28 | | | |
| | | | 217·66 | | | |
| Add commencement of Solar year ... ... | | ... March | 10·26 | | | |
| | | " | 227·92 | | | |

By Table VIII (p. 11), 227 days from 1 March=13 Oct., and ·92 day =(by Eye-Table) 22 hours.

∴ On Dr. Fleet's hypothesis. the tithi on which Buddha died ended on 13 Oct. B. C. 483, at 22 hrs. after Sunrise =(according to our reckoning) 4 A.M. on 14th Oct.

II. *Required the day of the Summer Solstice in the year B.C. 483*, (*i.e.*, the commencement of the Buddhist *vassa* in that year.)

The Summer Solstice happens when the sun is at 90° Tropical Longitude. Table XVII—A. gives the Sun's sidereal longitude for every complete day of the Solar year, while the table in Sec. **284**, p. (103) of the text gives us the difference between Indian sidereal and true tropical longitude for various epochs A.D. which we can apply, with sign reversed, to epochs B.C. Our epoch B.C. 483 preceded A.D. 520 by just 1003 years.

For 1000 years the movement of the precession is (by the Table in Sec. **284**) 16·5 degrees.

∴ 90° Tropical longitude in B.C. 483 corresponded to 90°+16·5°=106·5° Sidereal longitude.

By Table XVII−A, after 111 days complete, Sun's sidereal longitude is 107·26°−1·10°=106·16°. For the balance ·34 of a degree (106·5°—106·16°=34°) we may add (by Table XVII−C) ·33 of a day.

∴ In B.C. 483 the sun attained 90° Tropical or 106·5° Sidereal longitude at 111·33 days of the Solar year.

Since the year commenced on March 10·26, we have, as the day of the Summer Solstice, March 10·26+111·33= March 121·59=(by Table VIII) 29th June. This agrees with Prof. Jacobi's calculation, cited by Dr. Fleet *loc. cit.*

* The first New-moon in Solar year B.C. 483 being at 3·35 days, we learn from Table II, p. 2, that this was a year in which either Jyeshtha or Ashadha was *Adhika*. In fact Jyeshtha was *Adhika*.

## TABLE XI—B. Shortest Interval in Days, Ghatikas and Palas

### NAKSHATRAS.

| Order. | Names of Nakshatras. | Order | Ordinarily Vaisakha I (d. g. p.) | Order | Ordinarily Jyeshtha II (d. g. p.) | Order | Ordinarily Ashadha III (d. g. p.) | Order | Ordinarily Sravana IV (d. g. p.) | Order | Ordinarily Bhadrapada V (d. g. p.) |
|---|---|---|---|---|---|---|---|---|---|---|---|
| 27 | Revati ... ... ... ... ... | 3 | 0 59 21 | 5 | 0 48 15 | 7 | 0 37 9 | 9 | 0 26 2 | 11 | 0 14 56 |
| 1 | Asvini ... ... ... ... ... | 4 | 2 0 4 | 6 | 1 48 58 | 8 | 1 37 51 | 10 | 1 26 45 | 12 | 1 15 39 |
| 2 | Bharani ... ... ... ... ... | 5 | 3 0 47 | 7 | 2 49 41 | 9 | 2 38 34 | 11 | 2 27 28 | 13 | 2 16 22 |
| 3 | Krittika (*Tamil* Kiruttigai) ... ... ... | 6 | 4 1 30 | 8 | 3 50 24 | 10 | 3 39 17 | 12 | 3 28 11 | 14 | 3 17 5 |
| 4 | Rohini ... ... ... ... ... | 7 | 5 2 13 | 9 | 4 51 7 | 11 | 4 40 0 | 13 | 4 28 54 | 15 | 4 17 48 |
| 5 | Mirgasira (*Tamil* Mirugasiram) ... ... | 8 | 6 2 56 | 10 | 5 51 50 | 12 | 5 40 43 | 14 | 5 29 37 | 16 | 5 18 31 |
| 6 | Ardra (*Tamil* Arudra or Tiruvadirai) ... ... | 9 | 7 3 39 | 11 | 6 52 33 | 13 | 6 41 26 | 15 | 6 30 20 | 17 | 6 19 14 |
| 7 | Punarvasu ... ... ... ... | 10 | 8 4 22 | 12 | 7 53 16 | 14 | 7 42 9 | 16 | 7 31 3 | 18 | 7 19 57 |
| 8 | Pushya (*Tamil* Pusam) ... ... ... | 11 | 9 5 4 | 13 | 8 53 59 | 15 | 8 42 52 | 17 | 8 31 46 | 19 | 8 20 40 |
| 9 | Aslesha (*Tamil* Ayilyam) ... ... ... | 12 | 10 5 47 | 14 | 9 54 42 | 16 | 9 43 35 | 18 | 9 32 29 | 20 | 9 21 23 |
| 10 | Magha (*Tamil* Magham) ... ... ... | 13 | 11 6 30 | 15 | 10 55 25 | 17 | 10 44 18 | 19 | 10 33 12 | 21 | 10 22 6 |
| 11 | Purva Phalguni (*Tamil* Puram) ... ... | 14 | 12 7 13 | 16 | 11 56 8 | 18 | 11 45 1 | 20 | 11 33 55 | 22 | 11 22 49 |
| 12 | Uttara Phalguni (*Tamil* Uttiram) ... ... | 15 | 13 7 56 | 17 | 12 56 51 | 19 | 12 45 44 | 21 | 12 34 38 | 23 | 12 23 32 |
| 13 | Hasta (*Tamil* Hastam) ... ... ... | 16 | 14 8 39 | 18 | 13 57 34 | 20 | 13 46 27 | 22 | 13 35 21 | 24 | 13 24 14 |
| 14 | Chitra (*Tamil* Chittirai) ... ... ... | 17 | 15 9 22 | 19 | 14 58 17 | 21 | 14 47 10 | 23 | 14 36 4 | 25 | 14 24 58 |
| 15 | Svati ... ... ... ... | 18 | 16 10 5 | 20 | 15 59 0 | 22 | 15 47 53 | 24 | 15 36 47 | 26 | 15 25 41 |
| 16 | Visakha (*Tamil* Visakam) ... ... ... | 19 | 17 10 47 | 21 | 16 59 43 | 23 | 16 48 36 | 25 | 16 37 30 | 27 | 16 26 24 |
| 17 | Anuradha (*Tamil* Anusham) ... ... | 20 | 18 11 30 | 22 | 18 0 26 | 24 | 17 49 19 | 26 | 17 30 13 | 1 | 17 27 7 |
| 18 | Jyeshta (*Tamil* Kettai) ... ... ... | 21 | 19 12 13 | 23 | 19 1 9 | 25 | 18 50 2 | 27 | 18 38 56 | 2 | 18 27 50 |
| 19 | Mula (*Tamil* Mulam) ... ... ... | 22 | 20 12 56 | 24 | 20 1 52 | 26 | 19 50 45 | 1 | 19 39 39 | 3 | 19 28 33 |
| 20 | Purva Ashadha (*Tamil* Puradam) ... ... | 23 | 21 13 39 | 25 | 21 2 35 | 27 | 20 51 28 | 2 | 20 40 22 | 4 | 20 29 16 |
| 21 | Uttara Ashadha (*Tamil* Uttiradam) ... ... | 24 | 22 14 22 | 26 | 22 3 18 | 1 | 21 52 11 | 3 | 21 41 5 | 5 | 21 29 59 |
| 22 | Sravana (*Tamil* Tiruvonam) ... ... | 25 | 23 15 5 | 27 | 23 4 1 | 2 | 22 52 54 | 4 | 22 41 48 | 6 | 22 30 42 |
| 23 | Sravishta or Danishta (*Tamil* Avittam) ... | 26 | 24 15 48 | 1 | 24 4 44 | 3 | 23 53 37 | 5 | 23 42 31 | 7 | 23 31 25 |
| 24 | Satabhisaj or Sataraka (*Tamil* Sadayam) ... | 27 | 25 16 30 | 2 | 25 5 27 | 4 | 24 54 20 | 6 | 24 43 14 | 8 | 24 32 8 |
| 25 | Purva Bhadrapada (*Tamil* Purattadi) ... | 1 | 26 17 14 | 3 | 26 6 10 | 5 | 25 55 3 | 7 | 25 43 57 | 9 | 25 32 51 |
| 26 | Uttara Bhadrapada (*Tamil* Uttirattadi) ... | 2 | 27 17 57 | 4 | 27 6 53 | 6 | 26 55 46 | 8 | 26 44 40 | 10 | 26 33 34 |
| 27 | Revati ... ... ... ... | 3 | 28 18 40 | 5 | 28 7 36 | 7 | 27 56 29 | 9 | 27 45 23 | 11 | 27 34 17 |
| | | 4 | 29 19 23 | 6 | 29 8 19 | 8 | 28 57 12 | 10 | 28 46 6 | 12 | 28 35 0 |

### YOGAS.

| Order | Names of Yogas. | Order | I (d. g. p.) | Order | II (d. g. p.) | Order | III (d. g. p.) | Order | IV (d. g. p.) | Order | V (d. g. p.) |
|---|---|---|---|---|---|---|---|---|---|---|---|
| 27 | Vaidhriti ... ... ... ... | 5 | 0 53 57 | 9 | 0 33 17 | 13 | 0 12 37 | 18 | 0 48 26 | 22 | 0 27 46 |
| 1 | Vishkamba ... ... ... ... | 6 | 1 50 26 | 10 | 1 29 46 | 14 | 1 9 6 | 19 | 1 44 56 | 23 | 1 24 16 |
| 2 | Priti ... ... ... ... | 7 | 2 46 55 | 11 | 2 26 16 | 15 | 2 5 35 | 20 | 2 41 25 | 24 | 2 20 45 |
| 3 | Ayushmat ... ... ... ... | 8 | 3 43 25 | 12 | 3 22 45 | 16 | 3 2 5 | 21 | 3 37 54 | 25 | 3 17 15 |
| 4 | Saubhagya ... ... ... ... | 9 | 4 39 54 | 13 | 4 19 14 | 17 | 3 58 34 | 22 | 4 34 24 | 26 | 4 13 44 |
| 5 | Sobhana ... ... ... ... | 10 | 5 36 23 | 14 | 5 15 43 | 18 | 4 55 4 | 23 | 5 30 53 | 27 | 5 10 14 |
| 6 | Atiganda ... ... ... ... | 11 | 6 32 53 | 15 | 6 12 13 | 19 | 5 51 33 | 24 | 6 27 23 | 1 | 6 6 43 |
| 7 | Sukarman ... ... ... ... | 12 | 7 29 22 | 16 | 7 8 42 | 20 | 6 48 2 | 25 | 7 23 52 | 2 | 7 3 13 |
| 8 | Dhriti ... ... ... ... | 13 | 8 25 51 | 17 | 8 5 12 | 21 | 7 44 32 | 26 | 8 20 21 | 3 | 7 59 41 |
| 9 | Sula ... ... ... ... | 14 | 9 22 21 | 18 | 9 1 41 | 22 | 8 41 1 | 27 | 9 16 50 | 4 | 8 56 12 |
| 10 | Ganda ... ... ... ... | 15 | 10 18 50 | 19 | 9 58 10 | 23 | 9 37 31 | 1 | 10 13 20 | 5 | 9 52 40 |
| 11 | Vriddhi ... ... ... ... | 16 | 11 15 20 | 20 | 10 54 40 | 24 | 10 34 0 | 2 | 11 9 49 | 6 | 10 49 10 |
| 12 | Dhruva ... ... ... ... | 17 | 12 11 49 | 21 | 11 51 9 | 25 | 11 30 29 | 3 | 12 6 19 | 7 | 11 45 39 |
| 13 | Vyaghata ... ... ... ... | 18 | 13 8 18 | 22 | 12 47 38 | 26 | 12 26 58 | 4 | 13 2 48 | 8 | 12 42 9 |
| 14 | Harshana ... ... ... ... | 19 | 14 4 48 | 23 | 13 44 8 | 27 | 13 23 28 | 5 | 13 59 17 | 9 | 13 38 38 |
| 15 | Vajra ... ... ... ... | 20 | 15 1 17 | 24 | 14 40 36 | 1 | 14 19 57 | 6 | 14 55 47 | 10 | 14 35 8 |
| 16 | Siddhi ... ... ... ... | 21 | 15 57 46 | 25 | 15 37 6 | 2 | 15 16 27 | 7 | 15 52 16 | 11 | 15 31 36 |
| 17 | Vyatipata ... ... ... ... | 22 | 16 54 16 | 26 | 16 33 35 | 3 | 16 12 56 | 8 | 16 48 46 | 12 | 16 28 7 |
| 18 | Variyas ... ... ... ... | 23 | 17 50 45 | 27 | 17 30 5 | 4 | 17 9 25 | 9 | 17 45 15 | 13 | 17 24 35 |
| 19 | Parigha ... ... ... ... | 24 | 18 47 15 | 1 | 18 26 34 | 5 | 18 5 55 | 10 | 18 41 44 | 14 | 18 21 5 |
| 20 | Siva ... ... ... ... | 25 | 19 43 44 | 2 | 19 23 4 | 6 | 19 2 24 | 11 | 19 38 13 | 15 | 19 17 34 |
| 21 | Siddha ... ... ... ... | 26 | 20 40 13 | 3 | 20 19 33 | 7 | 19 58 53 | 12 | 20 34 43 | 16 | 20 14 4 |
| 22 | Sadhya ... ... ... ... | 27 | 21 36 43 | 4 | 21 16 3 | 8 | 20 55 22 | 13 | 21 31 12 | 17 | 21 10 32 |
| 23 | Subha ... ... ... ... | 1 | 22 33 12 | 5 | 22 12 31 | 9 | 21 51 52 | 14 | 22 27 42 | 18 | 22 7 3 |
| 24 | Sukla ... ... ... ... | 2 | 23 29 41 | 6 | 23 9 1 | 10 | 22 48 21 | 15 | 23 24 11 | 19 | 23 3 31 |
| 25 | Brahman ... ... ... ... | 3 | 24 26 11 | 7 | 24 5 30 | 11 | 23 44 51 | 16 | 24 20 40 | 20 | 24 0 2 |
| 26 | Indra ... ... ... ... | 4 | 25 22 40 | 8 | 25 2 0 | 12 | 24 41 20 | 17 | 25 17 10 | 21 | 24 56 30 |
| 27 | Vaidhriti ... ... ... ... | 5 | 26 19 9 | 9 | 25 58 29 | 13 | 25 37 50 | 18 | 26 13 39 | 22 | 25 52 59 |
| | | 6 | 27 15 39 | 10 | 26 54 59 | 14 | 26 34 19 | 19 | 27 10 8 | 23 | 26 49 29 |
| | | 7 | 28 12 8 | 11 | 27 51 28 | 15 | 27 30 48 | 20 | 28 6 38 | 24 | 27 45 58 |
| | | 8 | 29 8 37 | 12 | 28 47 58 | 16 | 28 27 18 | 21 | 29 3 7 | 25 | 28 42 27 |
| | | | | | | 17 | 29 23 47 | | | | |

## from New Moon to ending Moment of each Nakshatra and Yoga.

### NAKSHATRAS.

| Order | Ordinarily Asvina VI d. g. p. | Order | Ordinarily Karttika VII d. g. p. | Order | Ordinarily Margasira VIII d. g. p. | Order | Ordinarily Pausha IX d. g. p. | Order | Ordinarily Magha X d. g. p. | Order | Ordinarily Phalguna XI d. g. p. | Order | Chaitra, when no Adhika Masa. XII d. g. p. | Order | Chaitra when there is Adhika Masa. XIII d. g. p. |
|---|---|---|---|---|---|---|---|---|---|---|---|---|---|---|---|
| 13 | 0 3 50 | 16 | 0 53 26 | 18 | 0 42 20 | 20 | 0 31 14 | 22 | 0 20 8 | 24 | 0 9 1 | 27 | 0 58 38 | 2 | 0 47 31 |
| 14 | 1 4 33 | 17 | 1 54 9 | 19 | 1 43 3 | 21 | 1 31 57 | 23 | 1 20 50 | 25 | 1 9 44 | 1 | 1 59 21 | 3 | 1 48 14 |
| 15 | 2 5 16 | 18 | 2 54 52 | 20 | 2 43 46 | 22 | 2 32 40 | 24 | 2 21 34 | 26 | 2 10 27 | 2 | 3 0 3 | 4 | 2 48 57 |
| 16 | 3 5 59 | 19 | 3 55 35 | 21 | 3 44 29 | 23 | 3 33 22 | 25 | 3 22 16 | 27 | 3 11 10 | 3 | 4 0 46 | 5 | 3 49 40 |
| 17 | 4 6 42 | 20 | 4 56 18 | 22 | 4 45 12 | 24 | 4 34 5 | 26 | 4 22 59 | 1 | 4 11 53 | 4 | 5 1 29 | 6 | 4 50 23 |
| 18 | 5 7 25 | 21 | 5 57 1 | 23 | 5 45 54 | 25 | 5 34 48 | 27 | 5 23 42 | 2 | 5 12 35 | 5 | 6 2 12 | 7 | 5 51 6 |
| 19 | 6 8 8 | 22 | 6 57 44 | 24 | 6 46 37 | 26 | 6 35 31 | 1 | 6 24 25 | 3 | 6 13 19 | 6 | 7 2 55 | 8 | 6 51 49 |
| 20 | 7 8 51 | 23 | 7 58 27 | 25 | 7 47 20 | 27 | 7 36 14 | 2 | 7 25 8 | 4 | 7 14 1 | 7 | 8 3 38 | 9 | 7 52 32 |
| 21 | 8 9 34 | 24 | 8 59 10 | 26 | 8 48 3 | 1 | 8 37 57 | 3 | 8 25 50 | 5 | 8 14 44 | 8 | 9 4 21 | 10 | 8 53 14 |
| 22 | 9 10 16 | 25 | 9 59 53 | 27 | 9 48 46 | 2 | 9 38 40 | 4 | 9 26 34 | 6 | 9 15 27 | 9 | 10 5 3 | 11 | 9 53 57 |
| 23 | 10 10 59 | 26 | 11 0 36 | 1 | 10 49 29 | 3 | 10 39 22 | 5 | 10 27 16 | 7 | 10 16 10 | 10 | 11 5 46 | 12 | 10 54 40 |
| 24 | 11 11 42 | 27 | 12 1 18 | 2 | 11 50 12 | 4 | 11 40 5 | 6 | 11 27 59 | 8 | 11 16 53 | 11 | 12 6 29 | 13 | 11 55 23 |
| 25 | 12 12 25 | 1 | 13 2 1 | 3 | 12 50 54 | 5 | 12 40 48 | 7 | 12 28 42 | 9 | 12 17 35 | 12 | 13 7 12 | 14 | 12 56 6 |
| 26 | 13 13 8 | 2 | 14 2 44 | 4 | 13 51 37 | 6 | 13 41 31 | 8 | 13 28 25 | 10 | 13 18 19 | 13 | 14 7 55 | 15 | 13 56 49 |
| 27 | 14 13 51 | 3 | 15 3 27 | 5 | 14 52 20 | 7 | 14 42 14 | 9 | 14 29 8 | 11 | 14 19 1 | 14 | 15 8 38 | 16 | 14 57 32 |
| 1 | 15 14 34 | 4 | 16 4 10 | 6 | 15 53 3 | 8 | 15 42 57 | 10 | 15 29 50 | 12 | 15 19 44 | 15 | 16 9 21 | 17 | 15 58 14 |
| 2 | 16 15 16 | 5 | 17 4 53 | 7 | 16 53 46 | 9 | 16 43 40 | 11 | 16 30 34 | 13 | 16 20 27 | 16 | 17 10 3 | 18 | 16 58 57 |
| 3 | 17 15 59 | 6 | 18 5 36 | 8 | 17 54 29 | 10 | 17 44 22 | 12 | 17 31 16 | 14 | 17 21 10 | 17 | 18 11 46 | 19 | 17 59 40 |
| 4 | 18 16 42 | 7 | 19 6 18 | 9 | 18 55 12 | 11 | 18 45 5 | 13 | 18 31 59 | 15 | 18 22 53 | 18 | 19 12 29 | 20 | 19 0 23 |
| 5 | 19 17 25 | 8 | 20 7 1 | 10 | 19 55 54 | 12 | 19 45 48 | 14 | 19 32 42 | 16 | 19 23 35 | 19 | 20 13 12 | 21 | 20 1 6 |
| 6 | 20 18 8 | 9 | 21 7 44 | 11 | 20 56 37 | 13 | 20 46 31 | 15 | 20 33 25 | 17 | 20 24 19 | 20 | 21 13 55 | 22 | 21 1 49 |
| 7 | 21 18 51 | 10 | 22 8 27 | 12 | 21 57 20 | 14 | 21 47 14 | 16 | 21 34 8 | 18 | 21 25 1 | 21 | 22 14 38 | 23 | 22 2 32 |
| 8 | 22 19 34 | 11 | 23 9 10 | 13 | 22 58 3 | 15 | 22 47 57 | 17 | 22 34 50 | 19 | 22 25 44 | 22 | 23 15 21 | 24 | 23 3 14 |
| 9 | 23 20 16 | 12 | 24 9 53 | 14 | 23 58 46 | 16 | 23 48 40 | 18 | 23 35 34 | 20 | 23 26 27 | 23 | 24 16 3 | 25 | 24 3 57 |
| 10 | 24 20 59 | 13 | 25 10 36 | 15 | 24 59 29 | 17 | 24 49 22 | 19 | 24 36 16 | 21 | 24 27 10 | 24 | 25 16 46 | 26 | 25 4 40 |
| 11 | 25 21 42 | 14 | 26 11 18 | 16 | 26 0 12 | 18 | 25 50 6 | 20 | 25 36 59 | 22 | 25 27 53 | 25 | 26 17 29 | 27 | 26 5 23 |
| 12 | 26 22 25 | 15 | 27 12 1 | 17 | 27 0 54 | 19 | 26 59 48 | 21 | 26 37 42 | 23 | 26 28 35 | 26 | 27 18 12 | 1 | 27 6 6 |
| 13 | 27 23 8 | 16 | 28 12 44 | 18 | 28 1 37 | 20 | 27 51 31 | 22 | 27 38 25 | 24 | 27 29 19 | 27 | 28 18 55 | 2 | 28 6 49 |
| 14 | 28 23 51 | 17 | 29 13 27 | 19 | 29 2 20 | 21 | 28 52 14 | 23 | 28 39 8 | 25 | 28 30 1 | 1 | 29 19 38 | 3 | 29 7 32 |
| 15 | 29 24 34 | | | | | | | | | 26 | 29 30 44 | | | | |

### YOGAS.

| Order | VI d. g. p. | Order | VII d. g. p. | Order | VIII d. g. p. | Order | IX d. g. p. | Order | X d. g. p. | Order | XI d. g. p. | Order | XII d. g. p. | Order | XIII d. g. p. |
|---|---|---|---|---|---|---|---|---|---|---|---|---|---|---|---|
| 26 | 0 7 6 | 4 | 0 42 56 | 8 | 0 22 16 | 12 | 0 1 36 | 17 | 0 37 26 | 21 | 0 16 46 | 26 | 0 52 35 | 3 | 0 31 55 |
| 27 | 1 3 36 | 5 | 1 39 25 | 9 | 1 18 46 | 13 | 0 58 6 | 18 | 1 33 55 | 22 | 1 13 15 | 27 | 1 49 5 | 4 | 1 28 25 |
| 1 | 2 0 5 | 6 | 2 35 55 | 10 | 2 15 15 | 14 | 1 54 35 | 19 | 2 30 24 | 23 | 2 9 45 | 1 | 2 45 34 | 5 | 2 24 54 |
| 2 | 2 56 35 | 7 | 3 32 24 | 11 | 3 11 44 | 15 | 2 51 5 | 20 | 3 26 54 | 24 | 3 6 14 | 2 | 3 42 3 | 6 | 3 21 24 |
| 3 | 3 53 4 | 8 | 4 28 54 | 12 | 4 8 14 | 16 | 3 47 34 | 21 | 4 23 23 | 25 | 4 2 43 | 3 | 4 38 33 | 7 | 4 17 53 |
| 4 | 4 49 33 | 9 | 5 25 23 | 13 | 5 4 43 | 17 | 4 44 3 | 22 | 5 19 53 | 26 | 4 59 13 | 4 | 5 35 2 | 8 | 5 14 22 |
| 5 | 5 46 3 | 10 | 6 21 52 | 14 | 6 1 12 | 18 | 5 40 32 | 23 | 6 16 22 | 27 | 5 55 42 | 5 | 6 31 32 | 9 | 6 10 52 |
| 6 | 6 42 32 | 11 | 7 18 21 | 15 | 6 57 42 | 19 | 6 37 2 | 24 | 7 12 51 | 1 | 6 52 12 | 6 | 7 28 1 | 10 | 7 7 21 |
| 7 | 7 39 1 | 12 | 8 14 51 | 16 | 7 54 11 | 20 | 7 33 31 | 25 | 8 9 21 | 2 | 7 48 41 | 7 | 8 24 30 | 11 | 8 3 50 |
| 8 | 8 35 31 | 13 | 9 11 20 | 17 | 8 50 40 | 21 | 8 30 1 | 26 | 9 5 50 | 3 | 8 45 10 | 8 | 9 20 58 | 12 | 9 0 20 |
| 9 | 9 32 0 | 14 | 10 7 50 | 18 | 9 47 10 | 22 | 9 26 30 | 27 | 10 2 20 | 4 | 9 41 39 | 9 | 10 17 29 | 13 | 9 56 49 |
| 10 | 10 28 30 | 15 | 11 4 19 | 19 | 10 43 39 | 23 | 10 22 59 | 1 | 10 58 49 | 5 | 10 38 9 | 10 | 11 13 57 | 14 | 10 53 18 |
| 11 | 11 24 59 | 16 | 12 0 48 | 20 | 11 40 9 | 24 | 11 19 29 | 2 | 11 55 18 | 6 | 11 34 38 | 11 | 12 10 28 | 15 | 11 49 48 |
| 12 | 12 21 28 | 17 | 12 57 18 | 21 | 12 36 38 | 25 | 12 15 58 | 3 | 12 51 47 | 7 | 12 31 8 | 12 | 13 6 56 | 16 | 12 46 17 |
| 13 | 13 17 58 | 18 | 13 53 47 | 22 | 13 33 7 | 26 | 13 12 27 | 4 | 13 48 17 | 8 | 13 27 37 | 13 | 14 3 27 | 17 | 13 42 47 |
| 14 | 14 14 27 | 19 | 14 50 16 | 23 | 14 29 36 | 27 | 14 8 57 | 5 | 14 44 46 | 9 | 14 24 6 | 14 | 14 59 55 | 18 | 14 39 16 |
| 15 | 15 10 57 | 20 | 15 46 46 | 24 | 15 26 6 | 1 | 15 5 26 | 6 | 15 41 16 | 10 | 15 20 36 | 15 | 15 56 25 | 19 | 15 35 45 |
| 16 | 16 7 25 | 21 | 16 43 15 | 25 | 16 22 35 | 2 | 16 1 55 | 7 | 16 37 45 | 11 | 16 17 5 | 16 | 16 52 53 | 20 | 16 32 15 |
| 17 | 17 3 55 | 22 | 17 39 45 | 26 | 17 19 5 | 3 | 16 58 25 | 8 | 17 34 14 | 12 | 17 13 35 | 17 | 17 49 24 | 21 | 17 28 44 |
| 18 | 18 0 24 | 23 | 18 36 14 | 27 | 18 15 34 | 4 | 17 54 54 | 9 | 18 30 44 | 13 | 18 10 4 | 18 | 18 45 52 | 22 | 18 25 13 |
| 19 | 18 56 54 | 24 | 19 32 43 | 1 | 19 12 3 | 5 | 18 51 24 | 10 | 19 27 13 | 14 | 19 6 33 | 19 | 19 42 23 | 23 | 19 21 43 |
| 20 | 19 53 23 | 25 | 20 29 13 | 2 | 20 8 33 | 6 | 19 47 53 | 11 | 20 23 42 | 15 | 20 3 2 | 20 | 20 38 51 | 24 | 20 18 12 |
| 21 | 20 49 53 | 26 | 21 25 42 | 3 | 21 5 2 | 7 | 20 44 22 | 12 | 21 20 12 | 16 | 20 59 32 | 21 | 21 35 22 | 25 | 21 14 42 |
| 22 | 21 46 22 | 27 | 22 22 12 | 4 | 22 1 32 | 8 | 21 40 51 | 13 | 22 16 41 | 17 | 21 56 1 | 22 | 22 31 50 | 26 | 22 11 11 |
| 23 | 22 42 51 | 1 | 23 18 41 | 5 | 22 58 1 | 9 | 22 37 21 | 14 | 23 13 10 | 18 | 22 52 31 | 23 | 23 28 20 | 27 | 23 7 40 |
| 24 | 23 39 21 | 2 | 24 15 10 | 6 | 23 54 30 | 10 | 23 33 50 | 15 | 24 9 40 | 19 | 23 49 0 | 24 | 24 24 49 | 1 | 24 4 10 |
| 25 | 24 35 50 | 3 | 25 11 40 | 7 | 24 51 0 | 11 | 24 30 20 | 16 | 25 6 9 | 20 | 24 45 29 | 25 | 25 21 19 | 2 | 25 0 39 |
| 26 | 25 32 19 | 4 | 26 8 9 | 8 | 25 47 29 | 12 | 25 26 49 | 17 | 26 2 38 | 21 | 25 41 59 | 26 | 26 17 48 | 3 | 25 57 9 |
| 27 | 26 28 48 | 5 | 27 4 38 | 9 | 26 43 58 | 13 | 26 23 18 | 18 | 26 59 8 | 22 | 26 38 28 | 27 | 27 14 18 | 4 | 26 53 38 |
| 1 | 27 25 18 | 6 | 28 1 7 | 10 | 27 40 28 | 14 | 27 19 48 | 19 | 27 55 37 | 23 | 27 34 57 | 1 | 28 10 47 | 5 | 27 50 8 |
| 2 | 28 21 47 | 7 | 28 57 37 | 11 | 28 36 57 | 15 | 28 16 17 | 20 | 28 52 7 | 24 | 28 31 27 | 2 | 29 7 16 | 6 | 28 46 37 |
| 3 | 29 18 17 | | | | | 16 | 39 12 47 | | | 25 | 29 27 56 | | | | |

## TABLE XI--C. Annual Correction (additive) for Nakshatras and Yogas A.D. 1840--A.D. 1920.

| A. D. | Nakshatras. d. | gh. | p. | Yogas. d. | gh. | p. |
|---|---|---|---|---|---|---|
| 1840 | 0 | 40 | 22 | 1 | 15 | 7 |
| 1841 | 1 | 29 | 15 | 2 | 46 | 5 |
| 1842 | 0 | 11 | 35 | 0 | 10 | 25 |
| 1843 | 0 | 54 | 28 | 1 | 41 | 2[illegible] |
| 1844 | 1 | 43 | 21 | 3 | 12 | 20 |
| 1845 | 0 | 19 | 42 | 0 | 36 | 41 |
| 1846 | 1 | 8 | 35 | 2 | 7 | 38 |
| 1847 | 1 | 57 | 28 | 3 | 38 | 36 |
| 1848 | 0 | 33 | 49 | 1 | 2 | 57 |
| 1849 | 1 | 22 | 42 | 2 | 33 | 54 |
| 1850 | 2 | 11 | 35 | [illegible] | 4 | 52 |
| 1851 | 0 | 47 | 56 | 1 | 29 | 12 |
| 1852 | 1 | 36 | 49 | 3 | 0 | 10 |
| 1853 | 0 | 13 | 9 | 0 | 24 | 30 |
| 1854 | 1 | 2 | 3 | 1 | 55 | 28 |
| 1855 | 1 | 50 | 56 | 3 | 26 | 26 |
| 1856 | 0 | 27 | 16 | 0 | 50 | 46 |
| 1857 | 2 | 16 | 9 | 2 | 21 | 44 |
| 1858 | 2 | 5 | 2 | 3 | 52 | 41 |
| 1859 | 0 | 40 | 23 | 1 | 16 | 47 |

| A. D. | Nakshatras. d. | gh. | p. | Yogas. d. | gh. | p. |
|---|---|---|---|---|---|---|
| 1860 | 1 | 30 | 16 | 2 | 47 | 59 |
| 1861 | 0 | 6 | 37 | 0 | 12 | 19 |
| 1862 | 0 | 55 | 30 | 1 | 43 | 17 |
| 1863 | 1 | 44 | 23 | 3 | 1[illegible] | 15 |
| 1864 | 0 | 20 | 44 | 0 | 38 | 35 |
| 1865 | 1 | 9 | 37 | 2 | 9 | 31 |
| 1866 | 1 | 58 | 30 | 3 | 40 | 30 |
| 1867 | 0 | 34 | 50 | 1 | 4 | 51 |
| 1868 | 1 | 23 | 43 | 2 | 35 | 48 |
| 1869 | 0 | 0 | 4 | 0 | 0 | 9 |
| 1870 | 0 | 48 | 57 | 1 | 31 | 6 |
| 1871 | 1 | 37 | 50 | 3 | 2 | 4 |
| 1872 | 0 | 14 | 11 | 0 | 26 | 24 |
| 1873 | 1 | 3 | 4 | 1 | 57 | 22 |
| 1874 | 1 | 51 | 57 | 3 | 28 | 20 |
| 1875 | 0 | 20 | 18 | 0 | 52 | 40 |
| 1876 | 1 | 17 | 11 | 2 | 23 | 38 |
| 1877 | 2 | 6 | 4 | 3 | 54 | 35 |
| 1878 | 1 | 42 | 24 | 1 | 18 | 56 |
| 1879 | 1 | 31 | 18 | 2 | 49 | 54 |

| A. D. | Nakshatras. d. | gh. | p. | Yogas. d. | gh. | p. |
|---|---|---|---|---|---|---|
| 1880 | 0 | 7 | 38 | 0 | 14 | 14 |
| 1881 | 0 | 56 | 32 | 1 | 45 | 11 |
| 1882 | 1 | 45 | 24 | 3 | 16 | 9 |
| 1883 | 0 | 21 | 45 | 0 | 40 | 29 |
| 1884 | 1 | 10 | 38 | 2 | 11 | 27 |
| 1885 | 1 | 59 | 31 | 3 | 42 | 24 |
| 1886 | 0 | 35 | 52 | 1 | 6 | 45 |
| 1887 | 1 | 24 | 45 | 2 | 37 | 43 |
| 1888 | 0 | 1 | 6 | 0 | 2 | 3 |
| 1889 | 0 | 49 | 59 | 1 | 33 | 1 |
| 1890 | 1 | 38 | 52 | 3 | 3 | 58 |
| 1891 | 0 | 15 | 12 | 0 | 28 | 19 |
| 1892 | 1 | 4 | 5 | 1 | 59 | 17 |
| 1893 | 1 | 52 | 58 | 3 | 30 | 14 |
| 1894 | 0 | 23 | 20 | 0 | 54 | 34 |
| 1895 | 1 | 18 | 12 | 2 | 25 | 32 |
| 1896 | 2 | 7 | 5 | 3 | 56 | 30 |
| 1897 | 0 | 43 | 26 | 1 | 20 | 50 |
| 1898 | 1 | 32 | 19 | 2 | 51 | 48 |
| 1899 | 0 | 8 | 40 | 0 | 16 | 8 |

| A. D. | Nakshatras. d. | gh. | p. | Yogas. d. | gh. | p. |
|---|---|---|---|---|---|---|
| 1900 | 0 | 57 | 33 | 1 | 47 | 6 |
| 1901 | 1 | 46 | 26 | 3 | 18 | 3 |
| 1902 | 0 | 22 | 46 | 0 | 42 | 24 |
| 1903 | 1 | 11 | 39 | 2 | 13 | 21 |
| 1904 | 2 | 0 | 32 | 3 | 44 | 19 |
| 1905 | 0 | 36 | 53 | 1 | 8 | 39 |
| 1906 | 1 | 25 | 46 | 2 | 49 | 37 |
| 1907 | 0 | 2 | 7 | 0 | 3 | 57 |
| 1908 | 0 | 51 | 0 | 1 | 34 | 55 |
| 1909 | 1 | 39 | 53 | 3 | 5 | 52 |
| 1910 | 0 | 16 | 14 | 0 | 30 | 13 |
| 1911 | 1 | 5 | 7 | 2 | 1 | 11 |
| 1912 | 1 | 54 | 0 | 3 | 32 | 8 |
| 1913 | 0 | 30 | 21 | 0 | 56 | 29 |
| 1914 | 1 | 19 | 14 | 2 | 27 | 27 |
| 1915 | 2 | 8 | 7 | 3 | 58 | 24 |
| 1916 | 0 | 44 | 27 | 1 | 22 | 45 |
| 1917 | 1 | 33 | 20 | 2 | 53 | 42 |
| 1918 | 0 | 9 | 41 | 0 | 18 | 3 |
| 1919 | 0 | 58 | 34 | 1 | 49 | 0 |

### *EXPLANATION.*

In the *N.B.* to Sec. **263,** p. (95) of the text additive tables for Nakshatras were promised with the next edition. The convenience of having additive formulæ for the current epoch A.D. 1840 to A.D. 1920 is so great that the necessary tables are presented to the reader in the above addenda as Tables XI—B. and XI—C. Examples will now be given, showing how these tables should be used.

*I. Required the ending moment of Nakshatra Krittika for the month of July 1910* (*vide* p. 232 of Tables).

| | | d. | gh. | p. | ☾'s Anom. d. | gh. | p. | |
|---|---|---|---|---|---|---|---|---|
| (Table XII. p. 15) Jyeshtha New Moon. | June | 7 | 39 | 7 | 2 | 30 | 40 | |
| (Table XI—B. previous page) Shortest interval from Jyeshtha New Moon to No. 3 Krittika Nakshatra. | | 26 | 6 | 10 | 26 | 6 | 10 | |
| *Add* Nakshatra correction for 1910 (XI—C. *Supra.*) | | 0 | 16 | 14 | 0 | 16 | 14 | |
| | June | 34 | 1 | 31 | 28 | 53 | 4 | |
| (Table IX-*l*, p. 23.) Nakshatra Eqn. for Anom. of 1 d. 19 gh. 47 p. | | | − 6 | 20 | 27 | 33 | 17 | (1 Anom. Month.) |
| | June | 33 | 55 | 11 | 1 | 19 | 47 | |

Krittika Nakshatra ended on June 33, *i.e.*, on 3 July 1910, at 55 ghatikas 11 palas after mean Lanka Sunrise. If we want local time, we should use Table XIII.

*II. Required the ending moment of Yoga Vyatipata (No. 17) in July 1910.*

| | | d. | gh. | p. | ☉'s Anom. d. | gh. | p. | ☾'s Anom. d. | gh. | p. |
|---|---|---|---|---|---|---|---|---|---|---|
| (Table XII, p. 151.) Ashadha New Moon. | July | 7 | 10 | 57 | 84 | 58 | 29 | 4 | 29 | 13 |
| (Table XI—B, previous page) Shortest interval from Ashadha New Moon to No. 17 Vyatipata Yoga. | | 3 | 58 | 34 | 3 | 58 | 34 | 3 | 58 | 34 |
| (XI—C. *Supra*) *Add* Yoga correction for A. D. 1910. | | 0 | 30 | 13 | 0 | 30 | 13 | 0 | 30 | 13 |
| (Table IX-*i*, p. 20*a*) Sun's Yoga Eqn. for 89 d. 27 gh. + 1 gh. 25 p. | | 11 | 39 | 44 | 89 | 27 | 16 | 8 | 58 | 0 |
| „ p. 20*b*. Moon's Yoga Eqn. for 8 d. 59 gh. 25 p. − 19 gh. 40 p. | | | | | | | | | | |
| Sum − 18 gh. 15 p. | | − | 18 | 15 | (☉'s Yoga Eqn.) | | | + | 1 | 25 |
| | July | 11 | 21 | 29 | | | | 8 | 59 | 25 |

*i. e.*, No. 17 Yoga Vyatipata ended at 21 gh. 29 p. after mean Lanka Sunrise on 11 July 1910.

*Addendum.* **TABLES XI—D and XI—E. Shortest Interval in days and decimals of a day from new moon to ending moment of each Nakshatra and each Yoga.**

## EXPLANATION.

Tables XI–D and XI–E, which are printed on the next two pages, are the additive tables promised in N.B. to S. 263, p. (95) of the Text. One of the intervals in Table XI–D, as well as the result of the application of Table XI–E, should be *added* to the moment of new moon in order to ascertain the ending moment of any *nakshatra* or *yoga*. Though the tables are, as a whole, additive, it should be remembered that in using Table XI–E, the correction corresponding to the decimal portion of the argument should be *subtracted* from the correction corresponding to the integral number of days after which the first new moon appears in each solar year. Where great accuracy is not essential, it is unnecessary to calculate the correction corresponding to the decimal portion of the argument. The argument for Table XI–E is always the day of the First New Moon in the Solar Year. When the First New Moon is less than a day old in the Solar Year, prefix 0 to the argument. Thus, for A.D. 1915 (First New Moon in Solar Year, 0·9855) the Nakshatra correction is 2·2089 – (·0733 + ·0004) = 2·1352 ; and the Yoga correction is 4·11036 – (·13641 + ·00076) = 4·07319.

### *EXAMPLES.*

The examples on the previous page will first be worked out with the aid of Tables XI–D and XI–E on the next two pages.

*I.—Required the ending moment of Nakshatra Krittika for the month of July 1910.* (see Tables, p. 232.)

| | Month. | Day. | Fraction of day. | ☾'s Anom. days. |
|---|---|---|---|---|
| (Table X, p. 126) Jyeshtha New Moon, A.D. 1910 ... | June | 7 | ·65 | 0·535 |
| | | | | + 1·976 |
| (Table–XI D, next page). Shortest interval from Jyeshtha New Moon to No. 3. Krittika Nakshatra ... | | + 26 | ·10 | + 26·10 |
| | | | | 28·61 |
| | | | | - 27·55 (1 Anom. Month.) |
| | | | | 1·06 |
| (Table XI–E). The first New Moon in the Solar Year being, according to Table X, at 25·91 days, this is the argument for Table XI–E (next page). The Nakshatra correction for A.D. 1910–11 is, therefore by Table XI–E, 0·34 *minus* ·07= ... ... ... ... | | + | ·27 | + ·27 |
| | June | 34 | ·02 | 1·33 |
| Table (IX–*j*). Nakshatra equation corresponding to ☾'s Anom. of 1·33 days ... ... ... | | | – ·11 | |
| | June | 33 | ·91 | |

*i.e.*, (by Eye-table) the Nakshatra Krittika in **July** 1910 ended on June 33, that is, on 3 July, at 55 ghatikas after mean sunrise.

*II.—Required the ending moment of Yoga Vyatipata (No. 17) in July 1910.* (see Tables, p. 232.)

| | days | ☉'s Anom. days. | ☾'s Anom. days. |
|---|---|---|---|
| (Table X, p. 126.) Ashadha New Moon. A.D. 1910. ... | July 7·18 | 25·91 | 0·53 |
| | | +59·06 | +3·95 |
| (Table XI–D next page). Shortest interval from Ashadha New Moon to No. 17 Vyatipata Yoga ... ... | + 3·98 | + 3·98 | + 3·98 |
| (Table XI–E). Applying, as before, the argument for A.D. 1910–11, *i.e.*, 25·91 to the Yoga correction table in XI–E, we have 0·63 *minus* ·13= ... ... | + ·50 | + ·50 | + ·50 |
| | July 11·66 | 89·45 or 89d. 27gh. | 8·96 or 8 days 57½ gh. |
| | =July 11, 39½gh. | | +1½gh. [☉'s yoga equation by Table IX–(*i*)] |
| Table IX–(*i*) : ☉'s Yoga equation for 89d. 27 gh.= +1½ gh.<br>Table IX–(*i*) : ☾'s Yoga equation for 8d. 59 gh.= –19⅔ gh.<br>–18⅛ gh. | –18⅛gh. | | 8 days, 59 gh. |

Ending moment of No. 17 Vyatipata Yoga, July 11, 21⅓gh. after mean (Lanka) sunrise.

*N. B.*—The difference between this result and that arrived at on the last page is due to the fact that in the present working we have used only two decimal places for mean ending moments and anomalies.

We see that the method is applicable to Nakshtras and Yogas in any year, B.C. or A.D., not merely in the years A.D. 1840–A.D. 1920, as in Tables XI–B and XI–C on the three previous pages.

*Continued on page 12.*

*Addendum.* **TABLE XI—D. Shortest Interval in Days, from New-Moon to ending**

**NAKSHATRAS.**

| Order | Names of Nakshatras. | Order | Ordinarily Vaisakha I | Order | Ordinarily Jyeshtha II | Order | Ordinarily Ashadha III | Order | Ordinarily Sravana IV | Order | Ordinarily Bhadrapada V | Order | Ordinarily Asvina VI | Order | Ordinarily Karttika VII | Order | Ordinarily Margasira VIII |
|---|---|---|---|---|---|---|---|---|---|---|---|---|---|---|---|---|---|
| 27 | Revati ... ... | 3 | 0·9892 | 5 | 0·8041 | 7 | 0·6190 | 9 | 0·4339 | 11 | 0·2498 | 13 | 0·0638 | 16 | 0·8906 | 18 | 0·7055 |
| 1 | Asvini ... ... | 4 | 2·0011 | 6 | 1·8160 | 8 | 1·6309 | 10 | 1·4458 | 12 | 1·2608 | 14 | 1·0757 | 17 | 1·9025 | 19 | 1·7174 |
| 2 | Bharani ... ... | 5 | 3·0130 | 7 | 2·8279 | 9 | 2·6428 | 11 | 2·4578 | 13 | 2·2727 | 15 | 2·0376 | 18 | 2·9144 | 20 | 2·7293 |
| 3 | Krittika ... ... | 6 | 4·0249 | 8 | 3·3398 | 10 | 3·6548 | 12 | 3·4697 | 14 | 3·2846 | 16 | 3·0995 | 19 | 3·9263 | 21 | 3·7412 |
| 4 | Rohini ... ... | 7 | 5·0368 | 9 | 4·8518 | 11 | 4·6567 | 13 | 4·4816 | 15 | 4·2965 | 17 | 4·1114 | 20 | 4·9382 | 22 | 4·7532 |
| 5 | Mirgasira ... | 8 | 6·0488 | 10 | 5·8637 | 12 | 5·6786 | 14 | 5·4935 | 16 | 5·3084 | 18 | 5·1233 | 21 | 5·9502 | 23 | 5·7651 |
| 6 | Ardra ... ... | 9 | 7·0607 | 11 | 6·8756 | 13 | 6·6905 | 15 | 6·5054 | 17 | 6·3203 | 19 | 6·1352 | 22 | 6·9621 | 24 | 6·7770 |
| 7 | Punarvasu ... | 10 | 8·0726 | 12 | 7·8875 | 14 | 7·7024 | 16 | 7·5173 | 18 | 7·3322 | 20 | 7·1472 | 23 | 7·9740 | 25 | 7·7889 |
| 8 | Pushya ... ... | 11 | 9·0845 | 13 | 8·8991 | 15 | 8·7143 | 17 | 8·5292 | 19 | 8·3442 | 21 | 8·1591 | 24 | 8·9859 | 26 | 8·8008 |
| 9 | Aslesha ... ... | 12 | 10·0964 | 14 | 9·9113 | 16 | 9·7262 | 18 | 9·5412 | 20 | 9·3561 | 22 | 9·1710 | 25 | 9·9978 | 27 | 9·8127 |
| 0 | Magha ... ... | 13 | 11·1083 | 15 | 10·9232 | 17 | 10·7382 | 19 | 10·5531 | 21 | 10·3680 | 23 | 10·1829 | 26 | 11·0097 | 1 | 10·8246 |
| 11 | Purva Phalguni ... | 14 | 12·1202 | 16 | 11·9352 | 18 | 11·7501 | 20 | 11·5650 | 22 | 11·3799 | 24 | 11·1948 | 27 | 12·0216 | 2 | 11·8366 |
| 12 | Uttara Phalguni ... | 15 | 13·1322 | 17 | 12·9471 | 19 | 12·7620 | 21 | 12·5769 | 23 | 12·3918 | 25 | 12·2067 | 1 | 13·0336 | 3 | 12·8485 |
| 13 | Hasta ... ... | 16 | 14·1441 | 18 | 13·9590 | 20 | 13·7739 | 22 | 13·5888 | 24 | 13·4037 | 26 | 13·2186 | 2 | 14·0455 | 4 | 13·8604 |
| 14 | Chitra ... .. | 17 | 15·1560 | 19 | 14·9709 | 21 | 14·7858 | 23 | 14·6007 | 25 | 14·4156 | 27 | 14·2306 | 3 | 15·0574 | 5 | 14·8723 |
| 15 | Svati ... ... | 18 | 16·1679 | 20 | 15·9828 | 22 | 15·7977 | 24 | 15·6126 | 26 | 15·4276 | 1 | 15·2425 | 4 | 16·0693 | 6 | 15·8842 |
| 16 | Visakha ... ... | 19 | 17·1798 | 21 | 16·9947 | 23 | 16·8096 | 25 | 16·6246 | 27 | 16·4395 | 2 | 16·2544 | 5 | 17·0812 | 7 | 16·8961 |
| 17 | Anuradha ... | 20 | 18·1917 | 22 | 18·0066 | 24 | 17·8216 | 26 | 17·6365 | 1 | 17·4514 | 3 | 17·2663 | 6 | 18·0931 | 8 | 17·9080 |
| 18 | Jyeshtha ... ... | 21 | 19·2036 | 23 | 19·0186 | 25 | 18·8335 | 27 | 18·6484 | 2 | 18·4633 | 4 | 18·2782 | 7 | 19·1050 | 9 | 18·9199 |
| 19 | Mula ... ... | 22 | 20·2156 | 24 | 20·0305 | 26 | 19·8454 | 1 | 19·6603 | 3 | 19·4752 | 5 | 19·2901 | 8 | 20·1169 | 10 | 19·9319 |
| 20 | Purva Ashadha ... | 23 | 21·2275 | 25 | 21·0424 | 27 | 20·8573 | 2 | 20·6722 | 4 | 20·4871 | 6 | 20·3020 | 9 | 21·1289 | 11 | 20·9438 |
| 21 | Uttara Ashadha ... | 24 | 22·2394 | 26 | 22·0543 | 1 | 21·8692 | 3 | 21·6841 | 5 | 21·4990 | 7 | 21·3139 | 10 | 22·1408 | 12 | 21·9557 |
| 22 | Sravana ... ... | 25 | 23·2513 | 27 | 23·0662 | 2 | 22·8811 | 4 | 22·6960 | 6 | 22·5109 | 8 | 22·3259 | 11 | 23·1527 | 13 | 22·9676 |
| 23 | Sravishta ... | 26 | 24·2632 | 1 | 24·0781 | 3 | 23·8930 | 5 | 23·7079 | 7 | 23·5229 | 9 | 23·3378 | 12 | 24·1646 | 14 | 23·9795 |
| 24 | Satabhisaj ... | 27 | 25·2751 | 2 | 25·0900 | 4 | 24·9049 | 6 | 24·7199 | 8 | 24·5348 | 10 | 24·3497 | 13 | 25·1765 | 15 | 24·9914 |
| 25 | Purva Bhadrapada . | 1 | 26·2870 | 3 | 26·1019 | 5 | 25·9169 | 7 | 25·7318 | 9 | 25·5467 | 11 | 25·3616 | 14 | 26·1884 | 16 | 26·0033 |
| 26 | Uttara Bhadrapada. | 2 | 27·2990 | 4 | 27·1139 | 6 | 26·9288 | 8 | 26·7437 | 10 | 26·5586 | 12 | 26·3735 | 15 | 27·2003 | 17 | 27·0153 |
| 27 | Revati ... ... | 3 | 28·3109 | 5 | 28·1258 | 7 | 27·9407 | 9 | 27·7556 | 11 | 27·5705 | 13 | 27·3854 | 16 | 28·2123 | 18 | 28·0272 |
| | | 4 | 29·3228 | 6 | 29·1377 | 8 | 23·9526 | 10 | 28·7675 | 12 | 28·5824 | 14 | 28·3973 | 17 | 29·2242 | 19 | 29·0391 |
| | | | | | | | | | | | | 15 | 29·4093 | | | | |

**YOGAS.**

| Order | Names of Yogas. | Order | I | Order | II | Order | III | Order | IV | Order | V | Order | VI | Order | VII | Order | VIII |
|---|---|---|---|---|---|---|---|---|---|---|---|---|---|---|---|---|---|
| 27 | Vaidhriti ... | 5 | 0·8992 | 9 | 0·5548 | 13 | 0·2104 | 18 | 0·8075 | 22 | 0·4631 | 26 | 0·1187 | 4 | 0·7157 | 8 | 0·3713 |
| 1 | Vishkamba ... | 6 | 1·8407 | 10 | 1·4963 | 14 | 1·1519 | 19 | 1·7490 | 23 | 1·4016 | 27 | 1·0601 | 5 | 1·6572 | 9 | 1·3128 |
| 2 | Priti ... | 7 | 2·7822 | 11 | 2·4378 | 15 | 2·0934 | 20 | 2·6905 | 24 | 2·3460 | 1 | 2·0016 | 6 | 2·5987 | 10 | 2·2543 |
| 3 | Ayushmat ... | 8 | 3·7237 | 12 | 3·3793 | 16 | 3·0349 | 21 | 3·6319 | 25 | 3·2875 | 2 | 2·9431 | 7 | 3·5402 | 11 | 3·1958 |
| 4 | Saubhagya ... | 9 | 4·6652 | 13 | 4·3208 | 17 | 3·9763 | 22 | 4·5734 | 26 | 4·2290 | 3 | 3·8846 | 8 | 4·4817 | 12 | 4·1373 |
| 5 | Sobhana ... | 10 | 5·6067 | 14 | 5·2622 | 18 | 4·9178 | 23 | 5·5149 | 27 | 5·1705 | 4 | 4·8261 | 9 | 5·4232 | 13 | 5·0788 |
| 6 | Atiganda ... | 11 | 6·5482 | 15 | 6·2037 | 19 | 5·8593 | 24 | 6·4561 | 1 | 6·1120 | 5 | 5·7676 | 10 | 6·3647 | 14 | 6·0203 |
| 7 | Sukarman ... | 12 | 7·4896 | 16 | 7·1452 | 20 | 6·8008 | 25 | 7·3979 | 2 | 7·0535 | 6 | 6·7091 | 11 | 7·3062 | 15 | 6·9618 |
| 8 | Dhriti ... | 13 | 8·4311 | 17 | 8·0867 | 21 | 7·7423 | 26 | 8·3394 | 3 | 7·9950 | 7 | 7·6506 | 12 | 8·2477 | 16 | 7·9032 |
| 9 | Sula ... | 14 | 9·3726 | 18 | 9·0282 | 22 | 8·6838 | 27 | 9·2809 | 4 | 8·9365 | 8 | 8·5921 | 13 | 9·1891 | 17 | 8·8447 |
| 10 | Ganda ... | 15 | 10·3141 | 19 | 9·9697 | 23 | 9·6253 | 1 | 10·2224 | 5 | 9·8780 | 9 | 9·5335 | 14 | 10·1306 | 18 | 9·7862 |
| 11 | Vriddhi ... | 16 | 11·2556 | 20 | 10·9112 | 24 | 10·5668 | 2 | 11·1639 | 6 | 10·8195 | 10 | 10·4750 | 15 | 11·0721 | 19 | 10·7277 |
| 12 | Dhruva ... | 17 | 12·1971 | 21 | 11·8527 | 25 | 11·5083 | 3 | 12·1053 | 7 | 11·7609 | 11 | 11·4165 | 16 | 12·0136 | 20 | 11·6692 |
| 13 | Vyaghata ... | 18 | 13·1386 | 22 | 12·7942 | 26 | 12·4498 | 4 | 13·0468 | 8 | 12·7024 | 12 | 12·3580 | 17 | 12·9551 | 21 | 12·6107 |
| 14 | Harshana ... | 19 | 14·0801 | 23 | 13·7357 | 27 | 13·3912 | 5 | 13·9883 | 9 | 13·6439 | 13 | 13·2995 | 18 | 13·8966 | 22 | 13·5522 |
| 15 | Vajra ... | 20 | 15·0216 | 24 | 14·6771 | 1 | 14·3327 | 6 | 14·9298 | 10 | 14·5854 | 14 | 14·2410 | 19 | 14·8381 | 23 | 14·4937 |
| 16 | Siddhi ... | 21 | 15·9630 | 25 | 15·6186 | 2 | 15·2742 | 7 | 15·8713 | 11 | 15·5269 | 15 | 15·1825 | 20 | 15·7796 | 24 | 15·4352 |
| 17 | Vyatipata ... | 22 | 16·9045 | 26 | 16·5601 | 3 | 16·2157 | 8 | 16·8128 | 12 | 16·4684 | 16 | 16·1240 | 21 | 16·7211 | 25 | 16·3766 |
| 18 | Variyas ... | 23 | 17·8460 | 27 | 17·5016 | 4 | 17·1572 | 9 | 17·7543 | 13 | 17·4099 | 17 | 17·0655 | 22 | 17·6625 | 26 | 17·3181 |
| 19 | Parigha ... | 24 | 18·7875 | 1 | 18·4431 | 5 | 18·0987 | 10 | 18·6958 | 14 | 18·3514 | 18 | 18·0070 | 23 | 18·6040 | 27 | 18·2596 |
| 20 | Siva ... | 25 | 19·7290 | 2 | 19·3846 | 6 | 19·0402 | 11 | 19·6373 | 15 | 19·2929 | 19 | 18·9484 | 24 | 19·5455 | 1 | 19·2011 |
| 21 | Siddha ... | 26 | 20·6705 | 3 | 20·3261 | 7 | 19·9817 | 12 | 20·5788 | 16 | 20·2343 | 20 | 19·8899 | 25 | 20·4870 | 2 | 20·1426 |
| 22 | Sadhya ... | 27 | 21·6120 | 4 | 21·2676 | 8 | 20·9232 | 13 | 21·5202 | 17 | 21·1758 | 21 | 20·8314 | 26 | 21·4285 | 3 | 21·0841 |
| 23 | Subha ... | 1 | 22·5535 | 5 | 22·2091 | 9 | 21·8647 | 14 | 22·4617 | 18 | 22·1173 | 22 | 21·7729 | 27 | 22·3700 | 4 | 22·0256 |
| 24 | Sukla ... | 2 | 23·4950 | 6 | 23·1506 | 10 | 22·8061 | 15 | 23·4032 | 19 | 23·0588 | 23 | 22·7144 | 1 | 23·3115 | 5 | 22·9671 |
| 25 | Brahman ... | 3 | 24·4365 | 7 | 24·0920 | 11 | 23·7476 | 16 | 24·3447 | 20 | 24·0003 | 24 | 23·6559 | 2 | 24·2530 | 6 | 23·9086 |
| 26 | Indra ... | 4 | 25·3779 | 8 | 25·0335 | 12 | 24·6891 | 17 | 25·2862 | 21 | 24·9418 | 25 | 24·5974 | 3 | 25·1945 | 7 | 24·8501 |
| 27 | Vaidhriti ... | 5 | 26·3194 | 9 | 25·9750 | 13 | 25·6306 | 18 | 26·2277 | 22 | 25·8833 | 26 | 25·5389 | 4 | 26·1360 | 8 | 25·7915 |
| | | 6 | 27·2609 | 10 | 26·9165 | 14 | 26·5721 | 19 | 27·1692 | 23 | 26·8248 | 27 | 26·4804 | 5 | 27·0774 | 9 | 26·7330 |
| | | 7 | 28·2024 | 11 | 27·8580 | 15 | 27·5136 | 20 | 28·1107 | 24 | 27·7663 | 1 | 27·4219 | 6 | 28·0189 | 10 | 27·6745 |
| | | 8 | 29·1439 | 12 | 28·7995 | 16 | 28·4551 | 21 | 29·0522 | 25 | 28·7078 | 2 | 28·3633 | 7 | 28·9604 | 11 | 28·6160 |
| | | | | | | 17 | 29·3966 | | | | | 3 | 29·3048 | | | | |

## Moment of each Nakshatra and Yoga.

### NAKSHATRAS.

| Order | Ordinarily Pausha IX | Order | Ordinarily Magha X | Order | Ordinarily Phalguna XI | Order | Chaitra, when no Adhika Masa. XII | Order | Chaitra when there is Adhika Masa. XIII |
|---|---|---|---|---|---|---|---|---|---|
| **20** | 0·5204 | **22** | 0·3353 | **24** | 0·1502 | **27** | 0·9771 | **2** | 0·7920 |
| **21** | 1·5323 | **23** | 1·3472 | **25** | 1·1622 | **1** | 1·9890 | **3** | 1·8039 |
| **22** | 2·5442 | **24** | 2·3592 | **26** | 2·1741 | **2** | 3·0009 | **4** | 2·8158 |
| **23** | 3·5562 | **25** | 3·3711 | **27** | 3·1860 | **3** | 4·0128 | **5** | 3·8277 |
| **24** | 4·5681 | **26** | 4·3830 | **1** | 4·1979 | **4** | 5·0247 | **6** | 4·8396 |
| **25** | 5·5800 | **27** | 5·3949 | **2** | 5·2098 | **5** | 6·0366 | **7** | 5·8516 |
| **26** | 6·5919 | **1** | 6·4068 | **3** | 6·2217 | **6** | 7·0486 | **8** | 6·8635 |
| **27** | 7·6038 | **2** | 7·4187 | **4** | 7·2336 | **7** | 8·0605 | **9** | 7·8754 |
| **1** | 8·6157 | **3** | 8·4306 | **5** | 8·2456 | **8** | 9·0724 | **10** | 8·8873 |
| **2** | 9·6276 | **4** | 9·4426 | **6** | 9·2575 | **9** | 10·0813 | **11** | 9·8992 |
| **3** | 10·6396 | **5** | 10·4545 | **7** | 10·2694 | **10** | 11·0962 | **12** | 10·9111 |
| **4** | 11·6515 | **6** | 11·4664 | **8** | 11·2813 | **11** | 12·1081 | **13** | 11·9230 |
| **5** | 12·6634 | **7** | 12·4783 | **9** | 12·2932 | **12** | 13·1200 | **14** | 12·9350 |
| **6** | 13·6753 | **8** | 13·4902 | **10** | 13·3051 | **13** | 14·1320 | **15** | 13·9469 |
| **7** | 14·6872 | **9** | 14·5021 | **11** | 14·3170 | **14** | 15·1439 | **16** | 14·9588 |
| **8** | 15·6991 | **10** | 15·5140 | **12** | 15·3290 | **15** | 16·1558 | **17** | 15·9707 |
| **9** | 16·7110 | **11** | 16·5260 | **13** | 16·3409 | **16** | 17·1677 | **18** | 16·9826 |
| **10** | 17·7230 | **12** | 17·5379 | **14** | 17·3528 | **17** | 18·1796 | **19** | 17·9945 |
| **11** | 18·7349 | **13** | 18·5498 | **15** | 18·3647 | **18** | 19·1915 | **20** | 19·0064 |
| **12** | 19·7468 | **14** | 19·5617 | **16** | 19·3766 | **19** | 20·2034 | **21** | 20·0183 |
| **13** | 20·7587 | **15** | 20·5736 | **17** | 20·3885 | **20** | 21·2153 | **22** | 21·0303 |
| **14** | 21·7706 | **16** | 21·5855 | **18** | 21·4004 | **21** | 22·2273 | **23** | 22·0422 |
| **15** | 22·7825 | **17** | 22·5974 | **19** | 22·4124 | **22** | 23·2392 | **24** | 23·0541 |
| **16** | 23·7944 | **18** | 23·6094 | **20** | 23·4243 | **23** | 24·2511 | **25** | 24·0660 |
| **17** | 24·8064 | **19** | 24·6213 | **21** | 24·4362 | **24** | 25·2630 | **26** | 25·0779 |
| **18** | 25·8183 | **20** | 25·6332 | **22** | 25·4481 | **25** | 26·2749 | **27** | 26·0898 |
| **19** | 26·8302 | **21** | 26·6451 | **23** | 26·4600 | **26** | 27·2868 | **1** | 27·1017 |
| **20** | 27·8421 | **22** | 27·6570 | **24** | 27·4719 | **27** | 28·2987 | **2** | 28·1137 |
| **21** | 28·8540 | **23** | 28·6689 | **25** | 28·4838 | **1** | 29·3107 | **3** | 29·1256 |
| | | | | **26** | 29·4957 | | | | |

### YOGAS.

| Order | IX | Order | X | Order | XI | Order | XII | Order | XIII |
|---|---|---|---|---|---|---|---|---|---|
| **12** | 0·0269 | **17** | 0·6240 | **21** | 0·2796 | **26** | 0·8767 | **3** | 0·5323 |
| **13** | 0·9684 | **18** | 1·5655 | **22** | 1·2211 | **27** | 1·8182 | **4** | 1·4737 |
| **14** | 1·9099 | **19** | 2·5070 | **23** | 2·1626 | **1** | 2·7596 | **5** | 2·4152 |
| **15** | 2·8514 | **20** | 3·4485 | **24** | 3·1041 | **2** | 3·7011 | **6** | 3·3567 |
| **16** | 3·7929 | **21** | 4·3900 | **25** | 4·0455 | **3** | 4·6426 | **7** | 4·2932 |
| **17** | 4·7344 | **22** | 5·3314 | **26** | 4·9870 | **4** | 5·5841 | **8** | 5·2397 |
| **18** | 5·6759 | **23** | 6·2729 | **27** | 5·9285 | **5** | 6·5256 | **9** | 6·1812 |
| **19** | 6·6173 | **24** | 7·2144 | **1** | 6·8700 | **6** | 7·4671 | **10** | 7·1227 |
| **20** | 7·5588 | **25** | 8·1559 | **2** | 7·8115 | **7** | 8·4086 | **11** | 8·0642 |
| **21** | 8·5003 | **26** | 9·0974 | **3** | 8·7530 | **8** | 9·3501 | **12** | 9·0057 |
| **22** | 9·4418 | **27** | 10·0389 | **4** | 9·6945 | **9** | 10·2916 | **13** | 9·9472 |
| **23** | 10·3833 | **1** | 10·9804 | **5** | 10·6360 | **10** | 11·2331 | **14** | 10·8886 |
| **24** | 11·3248 | **2** | 11·9219 | **6** | 11·5775 | **11** | 12·1745 | **15** | 11·8301 |
| **25** | 12·2663 | **3** | 12·8634 | **7** | 12·5190 | **12** | 13·1160 | **16** | 12·7716 |
| **26** | 13·2078 | **4** | 13·8048 | **8** | 13·4604 | **13** | 14·0575 | **17** | 13·7131 |
| **27** | 14·1493 | **5** | 14·7463 | **9** | 14·4019 | **14** | 14·9990 | **18** | 14·6546 |
| **1** | 15·0907 | **6** | 15·6878 | **10** | 15·3434 | **15** | 15·9405 | **19** | 15·5961 |
| **2** | 16·0322 | **7** | 16·6293 | **11** | 16·2849 | **16** | 16·8820 | **20** | 16·5376 |
| **3** | 16·9737 | **8** | 17·5708 | **12** | 17·2264 | **17** | 17·8235 | **21** | 17·4791 |
| **4** | 17·9152 | **9** | 18·5123 | **13** | 18·1679 | **18** | 18·7650 | **22** | 18·4206 |
| **5** | 18·8567 | **10** | 19·4538 | **14** | 19·1094 | **19** | 19·7065 | **23** | 19·3621 |
| **6** | 19·7982 | **11** | 20·3953 | **15** | 20·0509 | **20** | 20·6479 | **24** | 20·3035 |
| **7** | 20·7397 | **12** | 21·3368 | **16** | 20·9924 | **21** | 21·5894 | **25** | 21·2450 |
| **8** | 21·6812 | **13** | 22·2783 | **17** | 21·9338 | **22** | 22·5309 | **26** | 22·1865 |
| **9** | 22·6227 | **14** | 23·2197 | **18** | 22·8753 | **23** | 23·4724 | **27** | 23·1280 |
| **10** | 23·5642 | **15** | 24·1612 | **19** | 23·8168 | **24** | 24·4134 | **1** | 24·0695 |
| **11** | 24·5056 | **16** | 25·1027 | **20** | 24·7583 | **25** | 25·3554 | **2** | 25·0110 |
| **12** | 25·4471 | **17** | 26·0442 | **21** | 25·6998 | **26** | 26·2969 | **3** | 25·9525 |
| **13** | 26·3886 | **18** | 26·9857 | **22** | 26·6413 | **27** | 27·2384 | **4** | 26·8940 |
| **14** | 27·3301 | **19** | 27·9272 | **23** | 27·5828 | **1** | 28·1799 | **5** | 27·8355 |
| **15** | 28·2716 | **20** | 28·8687 | **24** | 28·5243 | **2** | 29·1214 | **6** | 28·7769 |
| **16** | 29·2131 | | | **25** | 29·4658 | | | | |

## TABLE XI—E. ANNUAL CORRECTION.

**Argument**:—*Date of appearance of 1st New-Moon in each Solar Year according to Table X.*

The correction corresponding to the decimal portion of the argument should be subtracted from the Nakshatra or Yoga correction corresponding to the integral portion; thus the Nakshatra correction corresponding to an argument 28·53= 0·11 *minus* ·04=0·07.

### NAKSHATRAS.

| Arg. | Corrn. | Arg. | Corrn. | Arg. | Corrn. | Arg. | Corrn. | Arg. | Corrn. |
|---|---|---|---|---|---|---|---|---|---|
| **0** | 2·20890 | **·01** | ·00075 | **·30** | ·02244 | **·59** | ·04413 | **·88** | ·06582 |
| **1** | 2·13410 | **·02** | ·00150 | **·31** | ·02319 | **·60** | ·04488 | **·89** | ·06657 |
| **2** | 2·05930 | **·03** | ·00224 | **·32** | ·02394 | **·61** | ·04563 | **·90** | ·06732 |
| **3** | 1·98450 | **·04** | ·00299 | **·33** | ·02468 | **·62** | ·04638 | **·91** | ·06807 |
| **4** | 1·90970 | **·05** | ·00374 | **·34** | ·02543 | **·63** | ·04712 | **·92** | ·06882 |
| **5** | 1·83490 | **·06** | ·00449 | **·35** | ·02618 | **·64** | ·04787 | **·93** | ·06956 |
| **6** | 1·76010 | **·07** | ·00524 | **·36** | ·02693 | **·65** | ·04862 | **·94** | ·07031 |
| **7** | 1·68531 | **·08** | ·00598 | **·37** | ·02768 | **·66** | ·04937 | **·95** | ·07106 |
| **8** | 1·61049 | **·09** | ·00673 | **·38** | ·02842 | **·67** | ·05012 | **·96** | ·07181 |
| **9** | 1·53569 | **·10** | ·00748 | **·39** | ·02917 | **·68** | ·05086 | **·97** | ·07256 |
| **10** | 1·46089 | **·11** | ·00823 | **·40** | ·02992 | **·69** | ·05161 | **·98** | ·07330 |
| **11** | 1·38609 | **·12** | ·00898 | **·41** | ·03067 | **·70** | ·05236 | **·99** | ·07405 |
| **12** | 1·31129 | **·13** | ·00972 | **·42** | ·03142 | **·71** | ·05311 | | |
| **13** | 1·23649 | **·14** | ·01047 | **·43** | ·03216 | **·72** | ·05386 | | |
| **14** | 1·16169 | **·15** | ·01122 | **·44** | ·03291 | **·73** | ·05460 | | |
| **15** | 1·08689 | **·16** | ·01197 | **·45** | ·03366 | **·74** | ·05535 | | |
| **16** | 1·01209 | **·17** | ·01272 | **·46** | ·03441 | **·75** | ·05610 | | |
| **17** | 0·93729 | **·18** | ·01346 | **·47** | ·03516 | **·76** | ·05685 | | |
| **18** | 0.86249 | **·19** | ·01421 | **·48** | ·03590 | **·77** | ·05760 | | |
| **19** | 0·78769 | **·20** | ·01496 | **·49** | ·03665 | **·78** | ·05834 | | |
| **20** | 0·71288 | **·21** | ·01571 | **·50** | ·03740 | **·79** | ·05909 | | |
| **21** | 0·63808 | **·22** | ·01646 | **·51** | ·03815 | **·80** | ·05984 | | |
| **22** | 0·56328 | **·23** | ·01720 | **·52** | ·03890 | **·81** | ·06059 | | |
| **23** | 0·48848 | **·24** | ·01795 | **·53** | ·03964 | **·82** | ·06134 | | |
| **24** | 0·41368 | **·25** | ·01870 | **·54** | ·04039 | **·83** | ·06208 | | |
| **25** | 0·33888 | **·26** | ·01945 | **·55** | ·04114 | **·84** | ·06283 | | |
| **26** | 0·26408 | **·27** | ·02020 | **·56** | ·04189 | **·85** | ·06358 | | |
| **27** | 0·18928 | **·28** | ·02094 | **·57** | ·04264 | **·86** | ·06433 | | |
| **28** | 0·11448 | **·29** | ·02169 | **·58** | ·04338 | **·87** | ·06508 | | |
| **29** | 0·03968 | | | | | | | | |

### YOGAS.

| Arg. | Corrn. | Arg. | Corrn. | Arg. | Corrn. | Arg. | Corrn. | Arg. | Corrn. |
|---|---|---|---|---|---|---|---|---|---|
| **0** | 4·11036 | **·01** | ·00139 | **·30** | ·04176 | **·59** | ·08212 | **·88** | ·12249 |
| **1** | 3·97117 | **·02** | ·00278 | **·31** | ·04315 | **·60** | ·08351 | **·89** | ·12388 |
| **2** | 3·83198 | **·03** | ·00417 | **·32** | ·04454 | **·61** | ·08491 | **·90** | ·12527 |
| **3** | 3·69279 | **·04** | ·00557 | **·33** | ·04593 | **·62** | ·08630 | **·91** | ·12666 |
| **4** | 3·55360 | **·05** | ·00696 | **·34** | ·04732 | **·63** | ·08769 | **·92** | ·12805 |
| **5** | 3·41441 | **·06** | ·00835 | **·35** | ·04872 | **·64** | ·08908 | **·93** | ·12945 |
| **6** | 3·27522 | **·07** | ·00974 | **·36** | ·05011 | **·65** | ·09047 | **·94** | ·13084 |
| **7** | 3·13603 | **·08** | ·01113 | **·37** | ·05150 | **·66** | ·09186 | **·95** | ·13223 |
| **8** | 2·99684 | **·09** | ·01253 | **·38** | ·05289 | **·67** | ·09326 | **·96** | ·13362 |
| **9** | 2·85765 | **·10** | ·01392 | **·39** | ·05428 | **·68** | ·09465 | **·97** | ·13601 |
| **10** | 2·71846 | **·11** | ·01530 | **·40** | ·05568 | **·69** | ·09604 | **·98** | ·13641 |
| **11** | 2·57927 | **·12** | ·01670 | **·41** | ·05707 | **·70** | ·09743 | **·99** | ·13780 |
| **12** | 2·44008 | **·13** | ·01809 | **·42** | ·05846 | **·71** | ·09882 | | |
| **13** | 2·30089 | **·14** | ·01949 | **·43** | ·05985 | **·72** | ·10022 | | |
| **14** | 2·16170 | **·15** | ·02088 | **·44** | ·06124 | **·73** | ·10161 | | |
| **15** | 2·02251 | **·16** | ·02227 | **·45** | ·06263 | **·74** | ·10300 | | |
| **16** | 1·88332 | **·17** | ·02366 | **·46** | ·06493 | **·75** | ·10439 | | |
| **17** | 1·74413 | **·18** | ·02505 | **·47** | ·06542 | **·76** | ·10578 | | |
| **18** | 1·60494 | **·19** | ·02645 | **·48** | ·06681 | **·77** | ·10798 | | |
| **19** | 1·46575 | **·20** | ·02784 | **·49** | ·06820 | **·78** | ·20851 | | |
| **20** | 1·32656 | **·21** | ·02923 | **·50** | ·06959 | **·79** | ·10996 | | |
| **21** | 1·18737 | **·22** | ·03063 | **·51** | ·07098 | **·80** | ·11135 | | |
| **22** | 1·04818 | **·23** | ·03291 | **·52** | ·07238 | **·81** | ·11274 | | |
| **23** | 0·90899 | **·24** | ·03340 | **·53** | ·07377 | **·82** | ·11413 | | |
| **24** | 0·76980 | **·25** | ·03480 | **·54** | ·07516 | **·83** | ·11553 | | |
| **25** | 0·63061 | **·26** | ·03619 | **·55** | ·07655 | **·84** | ·11692 | | |
| **26** | 0·49142 | **·27** | ·03758 | **·56** | ·07795 | **·85** | ·11831 | | |
| **27** | 0·35223 | **·28** | ·03897 | **·57** | ·07934 | **·86** | ·11970 | | |
| **28** | 0·21304 | **·29** | ·04036 | **·58** | ·08073 | **·87** | ·12109 | | |
| **29** | 0·07385 | | | | | | | | |

*Addendum.* *Continued from page 9.*

*III.—Required the ending moment of No. 17 Anuradha Nakshatra in June 484 A.D.* (*see* Tables, p. 232).

| | | days. | ☾'s Anom. days. |
|---|---|---|---|
| (Table X, p. 48). Ashadha New Moon, A.D. 484 ... | June | 9·67 | 4·50 |
| | | | + 3·95 |
| (Table XI-D, overleaf) Shortest Interval from Ashadha New Moon to No. 17 Anuradha Nakshatra ... ... | | +10·74 | +10·74 |
| (Table XI-E, overleaf) Nak. correction for A.D. 484-85 is that corresponding to 24·39=0·41 *minus* ·02= ... | | + ·38 | + ·38 |
| *i.e.* | June | 20·79 | 19·57 |
| (Eye-Table) ☾'s tithi equation for Anom. of 19·57 days, reduced to Nakshatra Equation, is ... ... ... | | + ·38 | |
| | June | 21·17 | |

Our Nakshatra Anuradha (No. 17) ended on 21 June, A.D. 484, at ·17, *i.e.*, 10 ghat. after mean sunrise.

*N.B.*—In using the Eye-Table for Nakshatra and Yoga equations, we should first of all find the equation as if we were investigating a tithi, and then reduce the tithi equation to a Nakshatra or Yoga equation according to the scale furnished by the Eye-Table. We give another example to illustrate this.

*IV.—Required the ending moment of Yoga Sukla (No. 24) in June 484 A.D.* (*see* Tables, p. 232).

| | | days. | ⊙'s Anom. days. | ☾'s Anom. days. | |
|---|---|---|---|---|---|
| (Table X, p. 48). Ashadha New Moon, A.D. 484 ... | | June 9·67 | 24·39 | 4·50 | |
| | | | +59·06 | + 3·95 | |
| (Table XI-D, overleaf). Shortest interval from Ashadha New Moon to No. 24 Yoga Sukla ... ... ... | | + 10·57 | +10·57 | +10·57 | |
| (Table XI-E, overleaf). Yoga correction for A.D. 484-85 is that corresponding to an argument of 24·39 days, *i.e.* 0·77 *minus* ·05 ... ... ... = | | + 0·72 | +0·72 | + 0·72 | |
| | | June 20·96 | 94·74 | +19·74 | |
| (Eye-Table) Sun's tithi equation for 94·74 days, is —·04d. This with sign reversed (for Yoga), and reduced to scale of Yoga equation, becomes ... ... | +·03d. | | | + 0·03 | (⊙'s Yoga Equation.) |
| | | | | 19·77 | |
| (Eye-Table). Moon's tithi eqn. for 10·77 days is +·41d. This, reduced to scale of Yoga equation, becomes | +·35d. | | | | |
| Sum of Sun's and Moon's Yoga equations ... | +·38 | + ·38 | | | |

Ending moment of our Yoga No. 24, Sukla. ... ... June 21·34=20½gh. after mean sunrise on 21 June, A.D. 484.

# INDIAN CHRONOLOGY

## (SOLAR, LUNAR AND PLANETARY)

## B.C. 1 to A.D. 2000.

## PART I.

## Relations between Indian Astronomy and Indian Chronology.

*N.B.*—Decimal parts of a day may be converted into *ghatikas* and *palas* (*naligais* and *vinadis*) by means of Table XIX, and into *hours* and *minutes* by means of Table XX.

### CHAPTER I.

#### THE SYNODICAL MONTH AND THE INDIAN SOLAR YEAR.

**1. Astronomical Constants.**—A regular treatise on astronomy always closes with a chapter on astronomical constants, *viz.*, those elements of calculation which are of use in predicting astronomical events. A book on astronomical computation, like the present one, must begin, not end, with astronomical constants.

**2. First Astronomical Constant.—The Moon's Synodical Month or Lunation.**—The first astronomical constant we have to know in Hindu astronomy is the moon's synodical month of 29½ days. The exact length of this period, according to the *Sûrya Siddhânta*,* which we shall always follow, except when otherwise stated, is 29·530587946 days. The ancient Hindu astronomers purposely calculated this period to what we should now call nine places of decimals in order that there might be no error even after thousands of years. The period fixed by modern astronomy does not differ from the above in the first six decimal places, and as ·0000008, the actual difference, is $\frac{7}{100}$ of a second, it follows that the difference between European and Indian astronomy in 5,000 years or 61,844 synodical months may amount to 4,260 seconds or a little over 1 hour and 11 minutes. We shall see, when we come to compute new moons, that, in practice, the difference between the European and the Indian computation of new moons is very much less. [*Vide* Sec. **118**, Rule (3) *infra.*]

**3. What is a Synodical Month?**—A synodical month or lunation is the interval between one new moon and another. You are apt to imagine that this is the period in which the moon travels once round the earth. It is not so. It is the period in which the moon gains one complete revolution over the apparent or visible motion of the sun: and as this is a most essential fact, which you should thoroughly understand in order to work any problem in Indian dates intelligently, we will spend some time in considering it.

The moment of new moon is the moment when sun and moon have the same longitude, *i.e.*, are at the same distance measured from a fixed point in the heavens. When once

* The Surya Siddhanta, the best known system of Indian astronomy, is believed to have been current in its present form since the 11th Century A.D. and is the standard for all India. Numerous other Siddhantas are extant, but only one of them, the First Arya Siddhanta, is referred to in the present work, in addition to the Surya Siddhanta.

this moment is past, the moon resumes her journey at the rate of, say, 13 degrees a day and the sun resumes his at the rate of one degree a day. It follows that the moon gains twelve degrees over the sun in a day and therefore she gains 360 degrees in about 30 days. All these figures of days are approximate, but they will enable us to picture to ourselves what takes place. The synodical month, then, is the period during which the moon gains 360 degrees over the sun, and its exact length is 29·530587946 days.

**4. The Solar Year.**—The next astronomical constant which we have to study is the solar year, the length of which, according to the *Sûrya Siddhânta*, is 365·258756484 days. You have probably heard that the length of the year, according to modern astronomy, is 365·2422408 days, that the Julian calendar made out the year to be 365·25 days, *i.e.*, .00776 of a day in excess of the correct figure; that to rectify this error, Pope Gregory in 1582 ordered the dropping out of 10 days and the British Parliament in 1752 * similarly dropped out 11 days, and that to avoid a recurrence of the error, we now drop out three leap years out of every hundred. You may be inclined to wonder that the Hindu astronomers adopted for their solar year a period which is not less than one ghaṭika in excess of the modern astronomical year. The year of 365·2422408 days is, however, a *tropical* year, whereas the Hindu astronomical year is an *anomalistic* † year, and we should really compare the Hindu year with the modern *anomalistic* year, which we seldom hear of in practice, but the correct length of which is 365·2596296 days, being an excess of ·001 day over the Hindu year. This no doubt makes a difference of several days in the course of four or five thousand years, but as the Hindu year is essentially lunar and not solar, not much practical inconvenience is caused by the difference.

Perhaps you would like to know the practical difference between a tropical and an anomalistic year. A tropical year is that which brings the *seasons* round at the same time of the year, whereas an anomalistic year is that which brings back the sun's *anomaly*, *i.e.*, the rate at which he moves round the earth. This rate varies according as the sun is near to, or removed from, his perigee, *i.e.*, the point when he is nearest the earth, and once a year the sun returns, as it were, to his old pace. Now this varying pace of the sun is very important for calculating the moment of new moon, as well as for calculating the moment of sunrise, and this is probably why the Hindu astronomers reckon the course of the sun by the anomalistic, instead of by the tropical, year.

**5. The Decursus of the Hindu Solar Reckoning.**—Hindu astronomers reckon the present chronology from the midnight between 17th and 18th February 3102 B.C., which is commonly called the beginning of *Kaliyuga*. On the morning of 18th February 3101 B.C., that is one year later, one complete Hindu solar year had run out by 6·13 a.m., *i.e.*, at ·25875 of the day. As however the Hindu day is always reckoned from sunrise, mean sunrise for the whole of India being at 6 a.m., the first year is reckoned to have been completed at 13 minutes (or exactly ·00876 of a day) Indian time of the day, on 18th February 3101 B.C. At this moment the year 1 of Hindu chronology began. You probably think it

* This was the occasion when the NEW STYLE (more fully explained in Ch. XXVI, Secs. **231** to **241**, "*Vara* or Week-day ") was introduced.

† Strictly speaking, a *sidereal* year, but the designation *anomalistic* year is more suitable at this stage in order to lead up to the sun's *anomaly*.

was the year 2 which began on 18th February 3101 B.C., and not the year 1 ; but the Hindus generally reckon completed or expired years, and not current years, as the European calendar does ; and this is another point which you should thoroughly understand. The first year of the Hindu chronology, which began on 18th February 3102 B.C., is, according to Hindu reckoning, the year 0. By adding 3101 to an English calendar year A.D. you can always arrive at the corresponding (expired) year of Kaliyuga. Thus the present year 1910 A.D. is K.Y. 5011. For a B.C. year, the K.Y. equivalent is obtained by subtracting it from 3102, not 3101.

**6. Correspondence between A.D. and K.Y. years.**—Tables VI, VII, and X exhibit in a very simple and intelligible manner the correspondence between A.D. and K.Y. years. Advantage is there taken of the fact that a century year A.D. corresponds to a century year K.Y. increased by 1. Thus 100 A.D. corresponds to 3201 K.Y., 1900 A.D. to 5001 K.Y., and so on.

If you reckon 365·258756484 days for every Hindu year, you will find that the commencement of 3101 K.Y. fell on 16th March, 1 B.C., at ·15379 of the day. The Hindu astronomers, however, have to drop 2·1706944 days out of this reckoning, because the Indian Solar Year 0 Kaliyuga really began 2·1707 days before the moment above assigned for its commencement. This dropping out or correction is called *Sodhya*,* and we shall meet with it again.

It thus happens that 3101 K.Y. commenced in 1 B.C. on 13th March at ·9831 of the day after mean sunrise (6 a.m.). From this point Tables VI, VII, and X will carry us on regularly through every year up to 2000 A.D.

## CHAPTER II.

### FIXING THE MOMENT OF MEAN NEW MOON.

**7.** At the first moment of K.Y. 0, according to the Sûrya Siddhânta, the sun and moon had the same mean longitude, that is, the moon was *new* at that instant. At the first moment of 1 K.Y., 365·258756484 days would have passed, that is 12 synodical months, and in addition 10·891701134 days. That is, at the first moment of K.Y. 1, the moon was not new as she was at the first moment of K.Y. 0, but she was 10·89170 days old: and the first mean new moon in K.Y. 1 occurred on 29·530587946 *minus* 10·891701134 days = 18·638886812 days after the commencement of K.Y. 1. In this way in every Hindu year the first mean new moon would occur 18·63889 days later than in the previous year. In the year K.Y. 2 a mean new moon occurred 2 × 18·63889 days = 37·27778 days later than the commencement of the year, but as this period exceeds a synodical month, the *first* mean new moon in K.Y. 2 really occurred 37·27778 *less* 29·53059, that is, 7·74719 days after the commencement of the year K.Y. 2. The day of occurrence of the first mean new moon in each successive year after 0 K.Y. is given in Table VII, and we see from that table that in 100 K.Y. the first mean new moon occurred 3·46164 days after the commencement of the year. We follow the table

* The *Sodhya* should, properly speaking, be applied to every year, including year 0 Kaliyuga. The point should be borne in mind when calculating solar dates and tithis in any year. B.C. The nature of the correction called *Sodhya* is explained in Sec. 66.

through 200, 300, etc., years, till we find that in 3000 K.Y. the first mean new moon occurred 15·25746 days after the commencement of the solar year. For 3101 K.Y. (or 1 B.C.) we reckon the time of occurrence of the first mean new moon thus :—

| | Days. |
|---|---|
| For 3000 years ... ... ... ... ... ... | 15·25746 |
| For 100 years ... ... ... ... ... ... | 3·46164 |
| For 1 year ... ... ... ... ... ... | 18·63889 |
| | 37·35799 |
| *Deduct* 1 completed synodical month ... | 29·53059 |
| Remainder ... | 7·82740 |
| *Add* the *Sodhya* which causes the solar year to commence 2·1707 days earlier than it does by mean computation (*vide* Sec. 66 *infra*) ... ... | + 2·1707 |
| Occurrence of first mean new moon in K.Y. 3101 (1 B.C.) ... ... | 9·99810 days. |

That is, in 1 B.C., the first mean new moon of the Hindu solar year occurred almost exactly 10 days (properly 10 days less ·0019 of a day) after the commencement of the solar year.

**8.** From 1 B.C. we can follow the occurrence of the first mean new moon in each successive year according to the Tables VI and VII, the figures in which have merely to be added up suitably for each year. Thus if we require the date of the first mean new moon in the current solar year K.Y. 5011, we proceed according to the tables :—

For K.Y. 5001, 16·70809 days. (Tab. VI).
For 10 years, 9·20534 „ (Tab. VII).

For K.Y. 5011, 25·91343 „

That is, the first mean new moon in the year A.D. 1910–11, K.Y. 5011, occurs 25·91343 days after the commencement of the solar year, and as the solar year itself commences, as we may see from other columns of the same tables, on Ap. 12·62041
+·58756

*i.e.*, on April 13·20797

it follows that the addition of 25·91343 days, or 39·12140 days from 1 April, in other words 9·12140 days in May 1910, will give us the time of occurrence of the first mean new moon in K.Y. 5011. We may, if we like, convert the decimal places into *ghatikas* and *palas* or *hours* and *minutes*, according to Table XIX or XX, and we shall have as the result, 9th May 1910, 7 ghaṭikas and 17 palas or 2 hours and 55 minutes. This is the time of occurrence of *mean* new moon. About the *actual* ending moment of this or any other new moon tithi, we shall learn presently.

*N.B.*—For the meaning of *ghatika* and *pala*, See Sec. **130** *infra*.

## CHAPTER III.

### SUN'S AND MOON'S ANOMALY AND ACTUAL MOMENT OF NEW MOON.

**9.** We computed the time of occurrence of the first mean new moon in K.Y. 5011, and we arrived at the result, 7 ghaṭikas and 17 palas after mean sunrise on 9th May 1910. If now we look into a Panchângam, computed from the English Nautical Almanac, like Mr. Srauti's Tiruvâdi Panchângam or Messrs. Râghava Châri & Son's Nungumbaukam Panchângam, we shall find the time of occurrence of the new moon in May 1910 to be 9th May, 13 ghaṭikas and 2 palas after sunrise. It is encouraging that we are so near the Panchângam result, but we are not near enough. The Panchângams in question are calculated for the latitude and longitude of Kumbakonam and Madras, respectively, whereas our calculation is for mean sunrise (6 a.m.) at the Equator on the meridian of Ujain, which is accepted by Hindu astronomers as a starting point in the same way as Greenwich is accepted by modern English astronomers. It is known that Madras time is 18 minutes or 45 palas in advance of Ujain time, and, further, sunrise at Madras on the 9th May 1910 is noted in the Panchangam as occurring 12 minutes or 30 palas before 6 a.m. These two circumstances account for a difference of 75 palas or 1 ghaṭika and 15 palas between our mean time and the Panchangam time, but there is still a difference of 4 ghaṭikas and 30 palas or ·0750 of a day to be accounted for.

**10.** We account for it thus. Although mean new moon occurs every 29·53059 days, the actual new moon does not recur at exactly this interval. About the time when the moon is due to become new, she may be fast, or she may be slow, and we have to take this circumstance into consideration. Hindu astronomers have devised a table of the moon's anomaly which enables us to calculate exactly by how many degrees the moon is in advance of, or behind, her mean position at new moon or at any other time. The original tables of the moon's anomaly and equation of the centre, as these constants are called, will be found in Professor Jacobi's standard article on the subject in Vol. I of the *Epigraphia Indica*. For the sake of practical convenience, we may convert the degrees into time and further calculate the anomaly for each ·001 of the equation, and the result will be our Table IX. This table shows us exactly what time should be added to or deducted from mean time in order to arrive at the actual moment of occurrence of new moon.

**11.** To use this table, you should know the moon's anomaly at the time you are dealing with, and the moon's anomaly is determined for centuries and odd years in the same way as the *mean* moment of the first new moon is determined for every solar year. The moon completes an anomalistic month, that is, returns to a particular pace round the earth in 27·5546 days, and in a solar year the moon's anomaly increases by 365·25875 less 13 × 27·5546 = 7·04896 days as shown in Table VII, Column 4. In 100 years the anomaly increases by 16·03078 days and we calculate, as before, the anomaly for the commencement of the solar year K.Y. 3101 (B.C. 1).

| | Days. |
|---|---|
| For 3000 years ... ... ... ... ... ... ... | 12·49527 |
| For 100 years ... ... ... ... ... ... ... | 16·03078 |
| For 1 year ... ... ... ... ... ... ... | 7·04896 |
| For 3101 K.Y. ... ... ... ... ... ... ... | 35·57501 |
| *Deduct* one completed anomalistic month ... | 27·5546 |
| | 8·02041 |

*Add* the moon's anomaly at the first instant of K.Y. 0, which, according to the Sûrya Siddhânta, was 90° from perigee or 6·8886 days, and *deduct* Sodhya 2·1707 days, 4·7179

4·7179 12·73831 days.

**12.** The result, 12·7383 days, is just what we find entered in Table VI as the moon's anomaly at the commencement of K.Y. 3101 (B.C. 1). From this point we go on, year by year and century by century, till we reach 1600 A.D., when a correction, or *bija*, introduced by the

Hindu Astronomer Ganesa Daivajna, comes into operation. The effect of this *bija* is to raise the annual increase of the moon's anom. from 7·04896 d. to 7·04898 d. or by 2 seconds of time every year, and also to diminish the length of the anomalistic month from 27·5546 to 27·55459797 or ·000002 d., *i.e.*, ⅕ of a second every month. As a result of this correction, the anomaly for the commencement of K.Y. 5011 (1910-11 A.D.) is 2·1759 days. As the first new moon in the same year occurs (*vide* Section 8 *supra*) on the 25·9134th day after the commencement of the year, the anomaly for the time of the first new moon is 25·9134

+ 2·1759

28·0893

*Deduct* completed anomalistic month—27·5546

·5347 of a day.

**13.** As new moon is the moment when the moon has gained exactly 360° over the sun, we have, in determining this moment as well as the moon's anomaly at this moment, to take account of the sun's pace as well as the moon's. The sun's pace depends on his position in the anomalistic year, and that we know (Sec. 4) is the Hindu solar year. We also know that at the time of occurrence of the first new moon in K.Y. 5011 the sun had advanced 25·9134 days in the anomalistic year. We simply look for the corresponding equation in the table of sun's anomaly (Table IX-*c*) and we find it to be midway between +·145 and +·144. We put down +·1445 as the equation we require and *add* thereto ·5347, the moon's anomaly already found. Net moon's anomaly, ·6792.

We look in the table of moon's anomaly (Table IX-*a*) against ·679 of a day and we find the equation to be between —·059 and —·060. The actual anomaly entered in the table is ·677, and as our anomaly is ·679, we take a suitable equation by proportional parts, *i.e.*, ·0592.

We now *add* the total of sun's and moon's equations (+·1445 —·0592), or +·0853 to the mean time already found, 9·1213* d., and obtain 9·2066, *i.e.*, 12 ghaṭikas 24 palas after mean sunrise on 9th May 1910. For the present we might be satisfied with this result, which agrees with sufficient closeness with that (13 ghaṭikas 2 palas) of the Nautical Almanac and the Panchângams based thereon; for if we add the effect of longitude and local sunrise, already adverted to in Section 9 *supra*, our local time for the ending moment of the new moon tithi will be 13 ghaṭikas 39 palas after sunrise.

**14.** The reasons for the various processes whereby we arrive at the actual ending moment of any tithi will be clear to us when we come to Chapter XXIII "Theory of anomalies and equations," but one word on the subject may perhaps be usefully said before we close this chapter. The reader will observe that each anomaly table in our Table IX is divided into four equal portions, two having additive equations and two subtractive. Each equation, as given in the table, consists of three decimal places, the first two being given in vertical columns and the third in a horizontal column. The horizontal column is not necessary when all we want to know is how many hours or how many ghaṭikas after sunrise a

* The reader should note that two fractions of the day are usually quoted for the moment of occurrence of any astronomical event, *tithi*, *nakshatra*, *yoga*, *karana*, etc. The first fraction (in this case ·9134) is part of a period of time reckoned from the beginning of the solar year, while the second fraction (in this case ·1213) is the time of the day, reckoned from mean Lanka sunrise, 6 a.m. The difference between the two fractions is (as will be seen from Sec. 8 *supra*) the fraction of day marking the commencement of the solar year, one of the data entered in Table X.

new moon or other tithi occurred, and consequently, the horizontal place of the equation, namely, its third decimal place, is omitted in the Eye-table at the end of the book. On the other hand, when such a course is necessary for very accurate work, a *fourth* figure may be added to the third shown in the horizontal column by noting the difference between two successive anomalies and taking a proportional part of the difference between the corresponding equations. Thus, supposing we want the equation for a ☾'s anomaly of 7·25 days: the anomalies and equations next to those we want are—

| | | | | |
|---|---|---|---|---|
| ☾'s An. 7·077 d. | ... | ☾'s An. 7·275 d. | ... | Diff. ·198 d. |
| Eqn. – ·413 d. | ... | Eq. – ·414 d. | .. | Diff. ·001 d. |

Inasmuch as the difference in *anomaly* between 7·077 and 7·25 is ·173 which is $\frac{·173}{·198}$ =·9 of the tabular difference, we can *add* 9 to the lower *equation* and take ·4139 as the equation corresponding to ☾'s anomaly, 7·25 days.

We should note that sometimes the *lesser anomaly has the higher equation*. Thus, supposing we require the equation for ☾'s An. 20·375 d. We note the following as the nearest anomalies :

| | | | | |
|---|---|---|---|---|
| ☾'s An. 20·210 d. | ... | ☾'s An. 20·577 d. | ... | Diff. ·297 d. |
| Eqn. +·414 d. | ... | Eqn. +·413 d. | ... | Diff. ·001 d. |

Our anomaly is lower than 20·577 by ·202, which is $\frac{·202}{·297}$= ·7 of the difference. We therefore *add* 7 as a fourth figure to the equation +·413 and take +·4137 as the equation we require.

In all cases we should add to, or subtract from, the equation *corresponding* to the anomaly from which we took the difference. In this case, if we had taken the difference from 20·280, it would have been ·095 and the proportional part $\frac{·095}{·297}$=·3, which we should *deduct* from the fourth place of ·414. Thus our equation would be +·414 *minus* ·0003= +·4137, the same as before.

The same observations apply to solar anomalies.

**15.** For ordinary results, we do not require a fourth place in the equation, and we can take the nearest three place equation that we can find in Table IX and thus, by means of a simple sum in addition or subtraction, discover, without a moment's trouble, the actual from the mean moment of a tithi or nakshatra. Even then we shall be far more accurate than by any of the rough-and-ready methods now in use.

## CHAPTER IV.

### THE SOLAR MONTHS.

**16.** The solar months of the Indian calendar are named in Table II. The reader will note that the Tamil names of months are practically the same as the Bengal names, but that the first Bengal month *Vaisâkha* gives the name to the second Tamil month and so on, the last Bengal month *Chaitra* being the first Tamil month *Chittirai*.

Also, the Malayalam names of months are generally the same as the signs of the zodiac : the first two are called *Medam* and *Edavam* instead of *Mesham* and *Rishabham*.

But whether in the Tamil country or in Malabar or in Bengal, the measure of the solar months is the same. Like the solar year, each solar month ends at a *fraction of the day*, that is, at the moment when the next sankrânti takes place. For purposes of computation, the *sankrânti*, as well as the month to which it gives its name, is reckoned from the very moment at which the previous month ends. But in practice in the Tamil country, when a sankrânti takes place after sunset, the next month begins next day; and when the sankrânti occurs before sunset, *that* is the first day of the next month, and the old month loses a day. We shall have occasion, when dealing with the use of the tables, to note some of the practical consequences of the working of this rule, and we shall mention at the same time the corresponding rules in other parts of India. (Secs. **144**, **145** *infra.*)

**17.** Note that the solar months in the Indian calendar, which have 30 days each, are placed at fairly regular intervals, and that the months with 31 days each, as also those with 29 days each, are arranged continuously. Thus the series is:

| | |
|---|---|
| 30 | 30 |
| 31 | 29 |
| 31 | 29 |
| 31 | 29 |
| 31 | 29 |
| | 29 |
| | 30 |

which is fairly symmetrical. Owing, however, to the working of the rule about *sankrântis* before and after sunset, a month may have occasionally an extra day, and solar months with 32 days are not infrequent, as we may see from Table XII. Any consequent confusion can always be avoided if you invariably determine *the fraction of day* at which a month ends and the next begins.

---

## CHAPTER V.

### THE LUNAR MONTHS IN RELATION TO SOLAR MONTHS

**18.** The lunar months are the *doors* to the Indian calendar, but the solar months are the *hinges* on which the doors move. Every lunar month takes its name in Bengal from the solar month *in* which it occurs, and in the Tamil country from the *next* solar month after that in which it occurs.

**19.** Thus the lunar Vaiśâkha *must* begin sometime in the solar Vaiśâkha (Bengal) or sometime in the solar *Chittirai* (Tamil). The lunar month being only 29·53 days in length, there may be two new moons between the beginning and end of a solar month, and in this case *both* receive the same name, the first being called *Adhika* and the second *Nija* or true. The second is called 'true' because it immediately precedes a sankrânti. *Adhika* months occur ordinarily once in three years, as may be seen from Tables X and XII.

**20.** More rarely, that is about once or twice a century, a lunar month may begin and end without a solar month beginning between, and then some lunar month must be suppressed as there is no hinge on which this particular door can turn. The suppressed lunar month is said to be *kshaya** or in defect.

**21.** We will now take three examples illustrating the different kinds of lunar years: (1) a year consisting of twelve lunar months, corresponding to as many solar months; (2) a year consisting of thirteen lunar months, which will include an *adhika, i.e.*, an extra or *intercalary* lunar month; (3) a year of thirteen lunar months, exhibiting two intercalary or *adhika* lunar months and one suppressed or *kshaya* lunar month.

---

* Beginners find great difficulty in understanding *adhika* and *kshaya* months, because the idea is so utterly unlike anything in any other calendar; but the examples given on the next and following pages will enable the reader to cross safely over this *pons asinorum* of the Indian calendar.

## SCHEME OF MONTHS IN K.Y. 5011, A.D. 1910-11.

| Lunar Months ... | A.D. Date and Fraction of Day of mean New Moon. | Day and Fraction of Day of Solar Year when mean New Moon occurs, also Sun's Anomaly for New Moon. | Day and Fraction of day of Solar Year when Sankranti occurs: names of Solar Months commencing at each sankranti. S.M. = SOLAR MONTH | Moon's Anomaly at moment of each mean New Moon. (To the ☾'s An. at first new moon of Solar Year add 1·976 days for each lunar month) |
|---|---|---|---|---|
| | **A.D. 1910.** Indian Solar Year commences Ap. 13·2080. (Tab. X, p. 126.) | | **K.Y. 5011.** Mesha sankranti Vaisakha S.M. ; Chittirai and Medam S.M. ; solar year begins. | |
| 1. **Vaisakha** ... | May 9·1214. | 25·9134 (Tab. X.) | 30·9353. Vrishabha sank. Jyeshta S.M. ; Vaikasi or Edavam S.M. | 0·534 (Table X.) |
| 2. **Jyeshta** ... | June 7·6520. | 55·4440 | 62·3555. Mithuna sank. Ashada S.M. ; Ani S.M. | 2·510 |
| 3. **Ashada** ... | July 7·1826. | 84·9746 | 94·0003. Karkata sank. Sravana S.M. ; Adi S.M. | 4·486 |
| 4. **Sravana** ... | Aug. 5·7132. | 114·5052 | 125·4755. Simha sank. Bhadrapada S.M. ; Avani S.M. | 6·462 |
| 5. **Bhadrapada.** | Sep. 4·2437. | 144·0357 | 156·4942. Kanya sank. Asvina S.M. ; Purattasi S.M. | 8·438 |
| 6. **Asvina** ... | Oct. 3·7743. | 173·5663 | 186·9355. Tula sank. Kartika S.M. (Beng.); Aippasi S.M. | 10·414 |
| 7. **Karttika** ... | Nov. 2·3049. | 203·0969 | 216·8289. Vrischika sank. Margasira S.M ; Kartigai S.M. | 12·390 |
| 8. **Margasira** ... | Dec. 1·8355. | 232·6275 | 246·3192. Dhanus sank. Pausha S.M. ; Margali S.M. | 14·366 |
| 9. **Pausha** ... | Dec. 31·3661. | 262·1581 | 275·6369. Makara sank. Magha S.M. ; Tai S.M. | 16·342 |
| | **A.D. 1911.** | | | |
| 10. **Magha** ... | Jan. 29·8967. | 291·6887 | 305·0850. Kumbha sank. Phalguna S.M. ; Masi S.M. | 18·318 |
| 11. **Phalguna** ... | Feb. 28·4273. | 321·2193 | 334·9053. Mina sank. Chaitra S.M. ; Panguni S.M. | 20 294 |
| 12. **Chaitra** ... | Mar. 29·9579. | 350·7499 | 365·2587. Mesha sank. Vaisakha S.M. ; Chittirai S.M. | 22·270 |

*N.B.*—The main point to which the reader's attention should be directed is that the lunar month following a new moon has its name determined by its occurring *before* a particular sankranti. Thus, a lunar month commencing at any time between 0 day of the solar year and the 30·9353rd day is called **Vaisakha**; similarly, a lunar month commencing at any time between 156·4942 days and 186·9355 days of the solar year is called **Asvina** and so forth. The lunar month commencing before the Mesha Sankranti is called **Chaitra**

**22.** For the first we shall select the current Indian year K.Y. 5011, A.D. 1910-11; for the second we shall select last year, K.Y. 5010, A.D. 1909-10; and for the third, because there has been no suppressed month since A.D. 1822, and there will be none again till A.D. 1963, we will select the very first year of our chronology, K.Y. 3101, B.C. 1.

**23.** The solar year K.Y. 5011, A.D. 1910, which is a normal year, opens, as every year does, with the *Mesha Sankrânti* or entrance of the sun into the Indian *first point of Aries*, from which our celestial longitudes are reckoned. We have already seen how this moment is determined, namely, by the successive addition of 365·25875 days since the first moment of the year 0 *Kaliyuga*, less the *sodhya* of 2·1707 days. The *Mesha Sankrânti* determines all kinds of solar years in use in India, and it also determines directly the lunar year, since the first lunar month *Chaitra* is defined to be that whose commencement precedes the Mesha Sankrânti and the first day of the lunar year is that on which *Chaitra Sukla Pratipadâ* or the first tithi of the bright fortnight of Chaitra was current at sunrise. In our scheme of months in Table X and elsewhere the reader will observe that the lunar month *Chaitra* stands last, but this is only for purposes of computation, and, after all, the lunar month *Chaitra* begins in a previous solar year, and so it stands last among the lunar months of that year.

The moment of the *Mesha Sankrânti* marks the commencement of the solar month *Vaisâkhâ* in Bengal, of the solar month *Chittirai* in the Tamil country, and of the solar month *Medam* in Malabar, Travancore, and Cochin.

Each month begins and ends with a sankrânti, and the second and other sankrântis are named in the order of the signs of the zodiac, *Vrishabha*, *Mithuna*, *etc.*

**24.** There is no difficulty about the lunar months in A.D. 1910-11, since the new moons and sankrântis occur *alternately*, each *door* having its own *hinge* to turn on. If you like to calculate the actual moment of occurrence of each new moon, it will be a useful exercise for you to do so, and you can use the anomalies noted in columns 2 and 4.

**SCHEME OF MONTHS IN THE YEAR A.D 1909-10, K.Y. 5010.**

| Lunar Months ... | A.D. Date and Fraction of Day of mean New Moon. | Day and Fraction of Day of Solar Year when mean New Moon occurs, also Sun's Anomaly for New Moon. | Day and Fraction of Day of Solar Year when sankranti occurs; names of Solar Months commencing at each sankranti. S.M.=SOLAR MONTH. | Moon's Anomaly at moment of each mean New Moon. [See this col. on p.(9).] | Sum of Sun's and Moon's equations (by Table IX *a* to *c*). |
|---|---|---|---|---|---|
| | **A.D. 1909.** Indian Solar Year commences. Ap. 12·2492 (Tab. X). | | **K.Y. 5010.** Mesha sankranti Vaisakha S.M.; Chittirai or Medam S.M. Solar Year begins. | | |
| **1. Vaisakha** ... | Ap. 20·2238 | 7·2745 (Tab. X). | 30·9353. Vrishabha sank. Jyeshta S.M.; Vaikasi or Edavam S.M. | 2·401(Tab.X) | |
| **2. Jyeshta** ... | May 19·7543. | 36·8051. | 62·3555. Mithuna sank. Ashada S. M.; Ani S.M. | 4·377. | |

**SCHEME OF MONTHS IN THE YEAR A.D 1909–10, K.Y. 5010.**

| Lunar Months ... | A.D. Date and Fraction of Day of mean New Moon. | Day and Fraction of Day of Solar Year when mean New Moon occurs, also Sun's Anomaly for New Moon. | Day and Fraction of Day of Solar Year when sankranti occurs, names of Solar Months commencing at each sankranti. S.M.=SOLAR MONTH. | Moon's Anomaly at moment of each mean New Moon. | Sum of Sun's and Moon's equation (by Table IX *a* to *e*). |
|---|---|---|---|---|---|
| | **A.D. 1909.** | | **K.Y. 5010.** | | |
| 3. **Ashada** ... | June 18·2849 | 66·3357 | 94·0003. Karkata sankranti Sravana S.M.; Adi S.M. | 6·353 | |
| 4. } **Sravana** | July 17·8155 | 95·8663 | 125·4755. Simha sank. Bhadrapada S.M.; Avani S.M. | 8·329 | |
| 5. } | Aug. 16·3461 | 125·3970<br>— ·4470<br>*124·9500 | | 10·305<br>— ·1253(☉'s Eq.)<br>10·180 | — ·1253<br>— ·3217<br>— ·4470 |
| 6. **Bhadrapada.** | Sep. 14·8767 | 154·9275 | 156·4942. Kanya sank. Asvina S.M.; Purattasi S.M. | 12·281 | |
| 7. **Asvina** ... | Oct. 14·4073 | 184·4581 | 186·9355. Tula sank. Karttika S.M. (Beng.); Aippasi S.M. (Tam.). | 14·257 | |
| 8. **Karttika** ... | Nov. 12·9379 | 213·9886 | 216·8289. Vrischaka sank. Margasira S.M.; Kartigai S.M. (Tam.). | 16·233 | |
| 9. **Margasira** ... | Dec. 12·4685 | 243·5192 | 246·3192. Dhanus sank. Pausha S.M.; Margali S.M. | 18·209 | |
| | **A.D. 1910.** | | | | |
| 10. **Pausha** ... | Jan. 10·9990 | 273·0498 | 275·6369. Makara sank. Magha S.M.; Tai. S.M. | 20·185 | |
| 11. **Magha** ... | Feb. 9·5296 | 302·5804 | 305·0850. Kumbha sank. Phalguna S.M.; Masi S.M. | 22·161 | |
| 12. **Phalguna** ... | Mar. 11·0602 | 332·1110 | 334·9053. Mina sank. Chaitra S.M.; Panguni S.M. | 24·137 | |
| 13. **Chaitra** ... | Ap. 9·5908 | 361·6416 | 365·2587. Mesha sank. Vaisakha S.M.; Chittirai S.M. | 26·113 | |

**25.** In this year, as in 1910-11, the lunar months, up to Śrâvana, alternate with the solar months; but between the *Sankrântis* due at 94·30003 d. and 125·4755 d. of the solar year, we have two mean new moons at 95·8663 d. and 125·3970 d., respectively. We may satisfy ourselves by computation of the anomalies, as shown above, that the second new moon really precedes the sankrânti at 125·4755 d. In such a case both the new moons receive the same name (here *Srâvana*), and the first is called *Adhika Srâvana*, the second *Nija Srâvana*. After this, the numerical order of the lunar months is disturbed, for the 6th new moon is called Bhâdrapada, whereas in 1910-11, the 5th is Bhâdrapada, and the remaining new moons, including Bhâdrapada, alternate with the remaining sankrântis, the total number of new moons for the year being 13 instead of 12.

We now pass to the third example.

* Actual moment of occurrence of new moon after allowing for solar and lunar a omaly.

## SCHEME OF MONTHS FOR 1 B.C., K.Y. 3101.

| Lunar Months ... | Date of mean New Moon by Christian era and fraction of day. | Day of Solar Year when mean New Moon occurs; also Sun's anomaly. | Day of Solar Year when sankranti occurs, also dates of commencement and end of Solar Months. S.M.=Solar Month. | Moon's anomaly at moment of New Moon. | Sun's and Moon's equations (by Table IX *a* to *c*). |
|---|---|---|---|---|---|
| | **1. B.C.**<br>Indian Solar Year commences Mar. 13·9831 (Table X, page **24**). | | **K.Y. 3101.**<br>Mesha sankranti Vaisakha S.M.; Chittirai or Madam S.M. | | |
| **1. Vaisakha ...** | Mar. 23·98 | 9·99810 | 30·93523.<br>Vrishabha sank. Jyeshta S. M.; Vaikasi or Edavam S.M. | 22·736 | |
| **2. Jyeshta ...** | Apr. 22·51 | 39·52869 | 62·3555.<br>Mithuna sank. Ashada S.M.; Ani S.M. | 24·712 | |
| **3. Ashada ...** | May 22·04 | 69·05928 | 94·0003.<br>Karkata sank. Sravana S.M.; Adi S.M. | 26·688 | |
| **4. Sravana ...** | June 20·57 | 98·58986 | 125·4755.<br>Simha sank. Bhadrapada S.M.; Avani S.M. | 1·109 | |
| **5. Bhadrapada.** | July 20·10 | 128·12045 | 156·4942.<br>Kanya sank. Asvina S.M.; Purattasi S.M. | 3·085 | |
| **6. 7. } Asvina** | Aug. 18·63<br>Sep. 17·16 | 157·65104<br>187·18163<br>— ·5844<br>*186·5972 | 186·9355.<br>Tula sank. Karttika S.M.; Aippasi S.M. | 5·061<br>7·037<br>— ·1724<br>6·865 | —·1724<br>—·4120<br>—·5844 |
| **8. Karttika ...** | Oct. 16·69 | 216·71221<br>— ·5143<br>*216.1979 | 216·8289.<br>Vrischika sank. Margasira S.M; Kartigai (Tam.) S.M. | 9·013<br>—·1288<br>8·884 | —·1288<br>—·3855<br>—·5143 |
| **9. Margasira ...** | Nov. 15·23 | 246·24280<br>— ·3212<br>*245·9216 | 246·3192.<br>Dhanus sank. Pausha S.M.; Margali S.M. | 10·989<br>— ·0522<br>10·937 | —·0522<br>—·2690<br>—·3212 |
| **Pausha ...** | **(Kshaya)** | | 275·6369.<br>Makara sank. Magha S.M.; Tai S.M. | | |
| **10. Magha ...** | Dec. 14·176<br>**1 A.D.** | 275·77339<br>— ·0414<br>*275·7319 | 305·0850.<br>Kumbha sank. Phalguna S.M.; Masi S.M. | 2·965<br>+ ·0391<br>13·004 | +·0391<br>—·0805<br>—·0414 |
| **11. Phalguna ...** | Jan. 13·29 | 305·30398<br>+ ·2501<br>305·5540 | 334·9053.<br>Mina sank. Chaitra S.M.; Panguni S.M. | 14·941<br>+ ·1193<br>15·060 | +·1193<br>+·1308<br>+·2501 |
| **12. 13. } Chaitra** | Feb. 11·82<br>Mar. 13·35 | 334·83457<br>*+·4710<br>335·3055<br>364·36515 | 365·2587.<br>Mesha sank. Vaisakha S.M.; Chittrai or Medam S.M. | 16·917<br>+ ·1685<br>17·085<br>18·893 | +·1685<br>+·3025<br>+·4710 |

* Day and fraction of day, marking exact moment of actual New Moon, after allowing for anomaly.

**26** In this year there is nothing specially worthy of note till we reach *Asvina:* but for that month there is an *Adhika Asvina* just as we had an *Adhika Srâvana* in A.D. 1909. There is a peculiarity about *Nija Asvina* in the year B.C.1, which we must note carefully. The mean *Nija Asvina* at 187·1816 days is not before the sankrânti at 186·9355 days, but the actual new moon, after calculation of anomalies, is found to be before 187·1816 days and is therefore *Nija Asvina*, not Kârttika. Generally speaking, when a *nija* or *adhika* month is within ·6 day on either side of a sankrânti, calculation, or at least a consideration, of the anomalies is necessary before we can determine its true character.

**27.** Passing Âśvina, in the year 1 B.C., we find that each of the subsequent mean new moons is within ·6 day of a sankrânti, and we must therefore calculate the anomalies in order to be quite sure whether each new moon is on this or that side of a sankrânti. Having made these calculations, we find that between two sankrântis at 246·3192 d. and 275·6369 d. there is no new moon, although there is a new moon just under six ghaṭikas (·095 d.) *after* the second sankrânti. Had this new moon preceded the sankrânti ever so little, instead of following it, or had our method been inaccurate at this stage to the extent of 2 hours, the consequences we are about to state would not have followed. As it is, there is no new moon between the two sankrântis at 246·3192 d. and 275·6369 d., and therefore the lunar month which has no hinge to turn on is shut; this fact is expressed by saying that *Pausha* lunar month is *kshaya* or suppressed.

**28.** How do we know that the lunar month to be suppressed is *Pausha* and not any other? Because that is the first new moon which we are unable to place *before* a sankrânti. We find that *Mâgha* and *Phâlguna* are each followed by a sankrânti, though a long way off, and between the sankrântis at 334·9053 d. and 365·2587 d. there are again two new moons, that is to say Chaitras, the first of these being *adhika* and the second *nija*. Here also we note that appearances are deceptive, for the mean new moon at 334·8346 is apparently a Phâlguna, but the calculation of the anomalies shows it to be an *adhika Chaitra*.

**29.** Generally, (1) a *kshaya* month is preceded and followed, though not immediately, by an *adhika* month; (2) there are, as a rule, only one or two *kshaya* months in a century; (3) the *kshaya* months must be calculated from the mean moment by means of anomalies before we can prove them to be *kshaya;* and (4) only one of three months, Mârgasira, Pausha, and Mâgha, can be *kshaya*, because these are the three lunar months which turn on solar months of 29 days each.

The following table enables us to know an *adhika* month or a *kshaya* month by mere inspection of the date of occurrence of the first new moon in a solar year :—

| Names of Lunar Months. | Ending moments of Solar Months. | Periods of Lunar Months. | Limits of Adhika and Kshaya Months. |
|---|---|---|---|
| | | | The lunar month noted in column I will be an *adhika* month if first New Moon in Solar Year occurs |
| **Vaisakha** ... | 30·93528 d. | 29·53059 d. | before 1·40469 d. |
| **Jyeshta** ... | 62·35555 d. | 59·06117 d. | „ 3·29438 d. |

| Names of Lunar Months. | Ending Days of Solar Months. | Periods of Lunar Months. | Limits of Adhika and Kshaya Months. |
|---|---|---|---|
| | | | The lunar month noted in column I will be *adhika* if first New Moon in Solar Year occurs |
| **Ashada** ... | 94·00028 d. | 88·59176 d. | before 5·40851 d. |
| **Sravana** ... | 125·47555 d. | 118·12235 d. | ,, 7·35320 d. |
| **Bhadrapada** ... | 156·49417 d. | 147·65293 d. | ,, 8·84122 d. |
| **Asvina** ... | 186·93555 d. | 177·18353 d. | ,, 9·75201 d. |
| **Karttika** ... | 216·82888 d. | 206·71411 d. | ,, 10·11475 d. |
| | | | The lunar month noted in column I will be *kshaya* if first New moon in Solar year occurs |
| **Margasira** ... | 246·31916 d. | 236·24470 d. | after 10·07446 d. and<br>before 10·11475 d. |
| **Pausha** ... | 275·63694 d. | 265·77529 d. | after 9·86164 d. and<br>before 10·07447 d. |
| **Magha** ... | 305·08499 d. | 295·30588 d. | after 9·77910 d. and<br>before 9·86164 d. |
| | | | The lunar month noted in column I will be *adhika* if first new moon in solar year occurs |
| **Phalguna** ... | 334·90527 d. | 324·83647 d. | after 9·77912 d. and<br>before 10·06880 d. |
| **Chaitra** ... | 365·25875 d. | 354·36705 d. | after 10·06880 d. and<br>before 10·89170 d. |

*N.B.*—All the figures in column 4 are obtained by subtracting the corresponding figures in column 3 from those in column 2.

**30.** We see that if the first new moon in a solar year occurs before 1·40469 days (the difference between 30·93528 and 29·53059) of the solar year, there will be a second new moon before the end of the first solar month. Reasoning in the same manner, we see that if the first new moon in a solar year occurs after 1·40469 days, but before 3·29438 days of the solar year, there will be two new moons between the commencement of the first and second solar months, that is, there will be an *Adhika* as well as a *Nija Jyeshtha*. Similar reasoning will enable us to connect the possibility of occurrence of the several *adhika* and *kshaya* months with the occurrence of the first new moon before or between the days mentioned in the fourth column. It will be a useful exercise for the reader to try and reason out for himself each line of this table. For determining *mean* intercalations and *mean* suppression, the table can be used as it stands; and taken with suitable anomalies, it is a safe and reliable guide for ascertaining true *adhika* and true *kshaya* months. We believe that this is the first time that so simple a method has been announced for recognizing *adhika* and *kshaya* months.

**TITHIS.**

**31.** Tithis are in use over the whole of India for religious purposes and over the greater part of it for civil purposes also. To understand *tithis* thoroughly is to have mastered the system of Hindu chronology. Hence " Tithis " take the lead in the second title of this work.

**32.** A lunation or synodical month is divided into thirty tithis or lunar days of equal mean length. The names of tithis are familiar to all Hindus, but they are given for convenience of reference on the same page as Table II. The first fifteen tithis, corresponding to the bright half of the month, are called *Sukla paksha ;* and the second fifteen are called *Krishna paksha* or *Bahula paksha*. The last or 30th tithi is new moon or Amâvâsyâ, and it is called sometimes by the name of the month of which it marks the end, and sometimes by the name of the following month. Thus the moment of Amâvâsyâ, which marks the beginning of Vaiśâkha and which in this work is invariably called the "Vaiśâkha new moon" is the same as the ending moment of the 30th tithi of Chaitra. In an inscription* "Vaiśâkha Amâvâsyâ" often means the Amâvâsyâ at the *end* of Vaiśâkha, but for convenience of computation, the first new moon in the solar year is in this work called the Vaiśâkha new moon, while the 30th tithi of Vaiśâkha is called the Jyeshṭha new moon, and so on with the rest of the lunar months. (See remarks by Dr. Thibout, *Indian Antiq.*, Ap. 1895, p. 88.)

**33.** To find the mean ending moment of a particular *tithi*, all you have to do is to add the corresponding duration in days, according to either Table II or the Eye-table, to the moment of the previous new moon.

Thus, to find the mean ending moment of the 18th tithi of Mâgha lunar month, called *krishna* or *badi* (contraction for *bahula divasa*) *Tritiya* in K.Y. 5010, A.D. 1909-10, we proceed as follows :—

| | | |
|---|---|---|
| Mâgha new moon, A.D. 1910 (Table X) | ... February | 9·5296 d. |
| Duration of 18 tithis (Table II or Eye-table) | ... | 17·7183 d. |
| Mâgha 3 *badi*, *i.e.*, end of 18th tithi | ... February | 27·2479 d. |

The *mean* tithi in question ended at ·2479 of the 27th day of February, A.D. 1910.

**34.** For the *actual* ending moment of the tithi, we add the equations of sun's and moon's anomalies.

| | Anomaly at new moon. ⊙ | ☾ | | |
|---|---|---|---|---|
| | 302·58 d. | 22·161 d. | A.D. 1910, February | 27·2479 d. |
| Add for 18 tithis | 17·72 d. | 17·718 d. | ⊙Eq. for 320·30 d = +·1491 d. | |
| | | | ☾Eq. for 12·473 d = —·1332 d. | |
| | 320·30 d. | 39·879 d. | | |
| Deduct | | 27·555 (1 anom. month.) | +·0159 d. | +·0159 d. |
| | | 12·324 d. | | 27·2638 d. |
| | | +·1491 (⊙'s equation.) | | |
| | | 12·473 | | |

27·2638 d. or, in ghaṭikas and palas, 15 ghaṭikas 50 palas after mean sunrise on 27th February 1910. This is the absolute ending moment of Magha *badi* 3, in K.Y. 5010.

## CHAPTER VI.

### NAKSHATRAS.

**35.** The system of quoting dates by Nakshatras is as old as that of quoting by tithis and has prevailed in India from equally ancient times. There are twenty-seven nakshatras or lunar mansions through which the moon passes in her monthly journey

* That is, on the *amanta* system or the system of reckoning lunar months from New Moon to New Moon. There is also another system, prevalent in certain parts of India, of reckoning the lunar months from Full Moon to Full Moon, hence called the *purnimanta* system. On this system, which in all probability was more widely current formerly than it is now, the New Moon is at the middle of a lunar month and the two halves of each month are properly designated its two *pakshas* or wings, one on either side of new moon. Also, the new moon at the end of the amanta Chaitra lunar month, for instance, would, on the *purnimanta* system, be properly called the "Vaisakha new moon." As explained elsewhere, it is sometimes a difficult question whether a lunar month, referred to in an inscription, is to be understood in the *amanta* or in the *purnimanta* sense. See Sections **107** to **111** *infra*.

through the stars. For purposes of astronomical computation, the moon is supposed to spend an equal amount of time in each of the 27 nakshatras ; and as the total period of the moon's journey through the stars occupies 27·32166 days, it follows that the mean duration of each nakshatra is $\frac{27\cdot32166}{27}=1\cdot01191$ days, *i.e.*, 1 day and nearly 18 minutes. Thus, if we know the ending moment of a particular nakshatra, all we have to do to find out the ending moment of the next nakshatra, or, for that matter, of any other nakshatra, is to add as many times 1·0119 days as there are nakshatras between the ending moment which we know and the ending moment which we wish to find out. Having found this ending moment, we have next to add or subtract the moon's equation of time according to her anomaly at the ending moment of the nakshatra. We should remember that the sun's anomaly and the sun's equation of time do not enter into the calculation of the ending moment of nakshatras, because we are now concerned with the moon's own journey among the stars and not with the space gained by her over the sun.

**36.** When all we wish to know is the nakshatra for a particular tithi we may ascertain this directly from the mean tithi. The mean tithi, as we know, gives us the moon's longitude *minus* the sun's longitude. If to any tithi, therefore, we add the sun's longitude, we shall get the moon's longitude and this will enable us to find out the nakshatra corresponding to the longitude. The sun's longitude for nakshatras for every whole day of the solar year is given in the last column of the calendar (Table VIII) ; while the sun's longitude for decimals of a day is given in Table V. (The process of determining the sun's longitude for the purpose of nakshatras consists, briefly, in making 360° equal to 29·53059 day-spaces, and allowing for the *sodhya* of 2·1707 days. This conversion is necessary in order that we may add the moon's *less* the sun's longitude, which is calculated at the rate of 360° for 29·53059 days, to the sun's own longitude). To give an example, supposing we wish to find the nakshatra corresponding to the 27th day of February, A.D. 1910, which we found was the 18th tithi of lunar Mâgha, ending at ·2487 of the day, after sunrise, we proceed to determine the sun's longitude for the nakshatra. From the tithi we know that the sun's longitude for solar anomaly for the tithi is 320·30 days. The corresponding sun's longitude for nakshatra is as follows :—

| | | |
|---|---|---|
| For 320 days (according to Table VIII) | 25·6960 | days. |
| For ·30 day (according to Table V) | ·0242 | " |
| Add moon's *less* the sun's longitude, *i.e.*, the equivalent of 18 tithis in days | 17·7183 | " |
| | 43·4385 | |
| Deduct the equivalent of 360° of the moon's longitude | 29·5306 | |
| | 13·9079 | |

We look in Table III for the nakshatra equivalent of 13·9079 days in lunation space. We find this to be number 13, whose longitude extends from 13·12470 to 14·21843. Now, No. 13 nakshatra is called *Hasta*, which is therefore the nakshatra current at the ending moment of our tithi.

**37.** When we wish to find the *current* nakshatra corresponding to a tithi, the above method may be varied by converting the tithi into degrees at the rate of 12° for each tithi and adding the sun's longitude from Table XVII A and C, as in the example in Sec. **287.**

**38.** When we wish to find the *ending moment* of a nakshatra, the following method is preferable. This method, which is very simple, is indicated in Table XI which is provided in ghaṭikas and palas as well as in decimals of a day. To use this table, we should know the number of the nakshatra and the number of the new moon which the nakshatra follows, *i.e.*, whether it is the 1st, 2nd, 3rd, or 4th new moon, and so on, and then under the new moon in question we look out the "interval" opposite the number of the given nakshatra. The next step is to deduct from the interval so found the equivalent (according to the Correction Table in Table XI) of the time of occurrence of the first new moon in the Solar Year (given in Table X). The result, added to the date of the New Moon, gives us the mean ending moment of the nakshatra. If we wish to have the absolute ending moment, we take the equation for the nakshatra (Table IX *j* to *l*) corresponding to the moon's anomaly, and if we wish further to have the absolute ending moment in local time, we add or subtract the corrections according to Table XIII for the given latitude and longitude.

**39.** Suppose we want the ending moment of Nakshatra Mûla in Lunar Kârttika A.D. 1910, we proceed as follows:—

| | | Days. | ☾'s Anomaly Days. |
|---|---|---|---|
| First New Moon in Solar Year (Table X) ... | | 25·9134 | 0·535 |
| Add for Kârttika New Moon ( „ ) ... | | 177·1835 | 11·856 |
| Add commencement of Solar Year ( „ ) ... | | April 13·2080 | |
| Days from April 1 A.D. 1909 ... | | 216·3049 | |
| "Interval" corresponding to No. 19 "Mûla" Nakshatra under Kârttika (No. 7 New Moon)—Table XI ... ... | Days. 6·1353 | | |

Deduct equivalent of 25·9134 according to the correction table in Table XI.

| | | Days. |
|---|---|---|
| Equivalent of | 25 | =1·8700 |
| „ | ·91 | = ·0681 |
| „ | ·0034 | = ·0002 |
| „ | 25·9134 | =1·9383 |

| | | | |
|---|---|---|---|
| | −1·9383 | | |
| | + 4·1970 | + 4·1970 | + 4·197 |
| Days from April 1 ... | | 220·5019 | 16·588 |
| Nakshatra equation (Table IX-*k*) for anomaly of 16·588 days is +·2465 ... ... ... | | + ·2465 | |
| | | 220·7484 days from Ap. 1. | |

That is (by Tables VIII and XIX) 44 ghaṭikas 54 palas after mean sunrise on 6th November, A.D. 1910.

*N.B.*—For the years A.D. 1840 to A.D. 1919 we can get the ending moment of any Nakshatra yet more quickly from Tables XI and XII in the following manner :—

| | d. | gh. | p. | days | gh. | palas. | | d. | gh. | p. |
|---|---|---|---|---|---|---|---|---|---|---|
| (Table XII) Kârttika New Moon 1910, November | | | | 2 | 18 | 18 | ☾'s Anom. | 12 | 23 | 28 |
| Nak. interval No. 19 (Table XI—A, col. VII) | 6 | 8 | 7 | | | | | | | |
| Deduct for Nak. 1910-11 (Table XII, last col.) | —1 | 56 | 18 | | | | | | | |
| | 4 | 11 | 49 | 4 | 11 | 49 | | 4 | 11 | 49 |
| | | | | 6 | 30 | 7 | | 16 | 35 | 17 |
| Nak. Equation for ☾'s Anom. of 16 d. 35 gh. 17 p. (Table IX-*l*) + | | | | | 14 | 50 | | | | |
| | | | | 6 | 44 | 57 | | | | |

The ending moment is Nov. 6, 44 ghaṭikas 57 palas after Ujain sunrise, which only differs by 3 palas from the result above arrived at.

**42**. If we want this in local time, say at Tanjore (Lat. 11°) we have to add as correction, according to Table XIII [Lat. 11° (Tanjore)—208th day of Hinḍu Solar Year, on 6th Nov., when the Solar Year commences on 13 April] + 994 seconds of time = 41 palas.

44 ghaṭikas 54 palas + 41 palas = **45** ghatikas **35** palas, after sunrise at Tanjore on **6th November, A.D. 1910.**

This is the absolute ending moment of "Mûla" Nakshatra in Lunar Kârttika, A.D. 1910 (Kaliyuga 5011), according to the *Surya Siddhânta*. According to Mr. Srauti's Panchângam, which is based on the Greenwich Nautical Almanac, the absolute ending moment is 52 ghaṭikas 24 palas and according to the Kanjanur "Panchângam (No. 28)" it is 47 ghaṭikas 16 palas.

**43**. Suppose that in the same month we required to know the ending moment of No. 13 "Hasta" Nakshatra, instead of No. 19 "Mûla", we would have to take the "Interval" in Table XI for No. 13 and deduct therefrom the equivalent of 25·9134, *i.e.*, 1·9383, *which is the same for all months and all nakshatras throughout the Solar Year* 1910 *A.D.* Against No. 13 under the 7th New Moon in Table XI there are two "Intervals", *i.e.*, 0·0638 and 27·3854. We take the second interval as we cannot deduct 1·9383 from the first interval.

| Nakshatra Interval. | Days. | ☾'s Anomaly. |
|---|---|---|
| | | 0·535 |
| 27·3854 | 216·3049 | 11·856 |
| −1·9383 (Annual Nak. Corrn.) | | |
| 25·4471 | + 25·4471 | + 25·447 |
| | 241·7520 | 37·838 |
| | | −27·555 (1 an. month.) |
| | | 10·283 |
| Nakshatra equation for ☾'s anom. of 10·283 days (Tab. IX—*J*.) | −·2915 | |
| | 241·4605 days from 1st April, *i.e.*, | |

(by Tables VIII and XIX) 27 ghatikas 38 palas after *mean* sunrise on 27th November, A.D. 1910.

**44.** Lastly, if we want the ending moment of No. 15 " Svâti " Nakshatra in the same month, we find two " intervals " for No. 15 in Table XI, from each of which 1·9383 can be deducted. In other words, there will be two " Svâti's " in Lunar Kârttika 1910; one about 2·08 *minus* 1·94 or at ·14 of a day after new moon and the other at 29·41 *minus* 1·94 or 27·47 days after new moon. The equation for the mean anomaly of the first Svâti will be that for an anomaly of 0·53 + 11·86 + ·14 = 12·53 days ; in this case the equation is − ·118. The absolute ending moment of the first Svâti will be ·305 + ·14 − ·118 = ·327 or 19¾ ghaṭikas after *mean* sunrise on the new moon day, *i.e.*, the 216th from 1st April, or 2nd November, A.D. 1910.

The equation for the ☾'s mean anomaly at the ending moment of the second Svâti will be that for (0·53 + 11·86 + 27·47 − 27·55) = 12·31 days ; that is, by Table IX-*j*, the equation will be − ·138.

The absolute ending moment of the second Svâti will then be

$$\begin{array}{rr} & 216{\cdot}3049 \\ 27{\cdot}47 - {\cdot}138 = & +27{\cdot}332 \\ \hline & 243{\cdot}6369 \end{array}$$ days from April 1,

*i.e.*, (by Tables VIII and XIX) 38¼ ghaṭikas on 29th November, A.D. 1910.

This is midway between Mr. Srauti's time (36¾ ghaṭikas) and the " Kanjanur " time (41 ghaṭikas) for the nakshatra in question.

*N.B.*—We have not corrected our result for local time.

**45.** Besides the system of the division of the moon's path among the stars into 27 equal nakshatra spaces, there are two others, called Garga's system and Brahma's system, by which the sidereal month of 27·32166 days is divided into 27 convenient but *unequal* stages. Our Table III gives the collective duration of the nakshatras according to the unequal systems, and the ending moment of each nakshatra according to either of these systems can be found by adding the corresponding collective duration to the ending moment of the last Nakshatra Revatî, which should be first determined. We have also given in Table III the ..s of the deities presiding over the several nakshatras.

**46.** It is sometimes found that an extra Nakshatra *Abhijit* is interpolated between Nos. 21 and 22. Its duration is shown in Table III in the column dealing with " Brahma's System ".

## CHAPTER VII.

### YOGAS.

**47.** A yoga is the time during which the sun and moon together accomplish 13 degrees 20 minutes of space. There are thus 27 yogas which together make 360°. The names of all the yogas are given in Table III, from which it will be seen that the collective duration of 27 yogas is 25·42022 days.

**48.** Unlike tithis and nakshatras, the yogas do not represent the stages of the actual motion of any heavenly body or of a set of heavenly bodies. They are thus the result of a mathematical rather than of an astronomical conception. If we draw a line

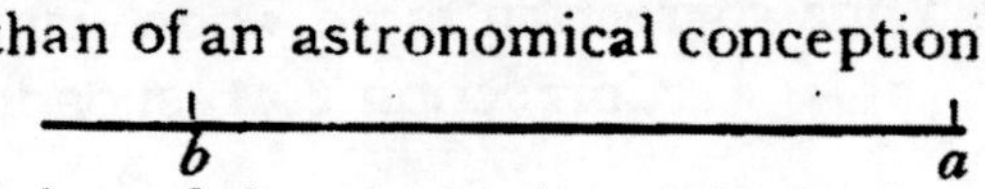


where *a* represents the motion of the moon in a given time and *b* the motion of the sun in the same time, it follows that *a*—*b* is the moon's elongation or the space gained by the moon over the sun in the given time. If we regard separately the spaces travelled by the sun and the moon in this time, we might say that together they have done $a+b$. This $a+b$, then, is the yoga, while $a-b$ is the tithi. To derive the yoga from the tithi, we make use of the formula

$$a-b+2b=\text{Yoga}.$$

just as we derived the nakshatra from the tithi by means of the formula $a-b+b$. (Sec. **36** *supra.*)

**49**. Supposing we want to determine the yoga current at the ending moment of the tithi for 27th February 1910, we proceed as follows. We have found *supra* (Sec. **33**) that the 18th tithi ended at ·2479 day on 27th February, 1910, and that the corresponding solar day was 320·30 days, while the corresponding moon's anomaly was 12·324 days.

In accordance with the formula $a-b+2b$, we add twice the sun's longitude to the tithi.

The sun's longitude for 320·30 days is calculated as follows:—

| | | |
|---|---|---|
| For 320 days (last column of Table VIII) | ... | 25·6960 |
| For ·30 day (Table V, sun's longitude) | ... | ·0242 |
| | | 25·7202 |
| Multiply this by 2 | ... ... | 51·4404 |
| Deduct completed lunation | ... ... | 29·5306 |
| | | 21·9098 |
| Add equivalent of 18 tithis (Table II) | ... | 17·7183 |
| | | 39·6281 |
| Deduct completed lunation | ... ... | 29·5306 |
| | | 10·0975 |
| The yoga corresponding to this lunation space is the 10th, i.e., *Ganda*, (Table III) whose space ends at | ... ... ... | 10·9372 |

This, therefore, was the yoga current at the ending moment of our tithi.

*N.B.*—Or as in Sec. **37**, the sun's longitude may be calculated directly from table XVII A and C thus :—

| | | |
|---|---|---|
| Sun's longitude for 320 days | ... ... ... ... | 313·2534° |
| Do. ·30 day | ... ... ... ... | ·2957 |
| | | 313·5491° |
| Twice ☉'s longitude | ... ... ... ... ... | = 627·0982° |
| Add tithi × 12 | ... ... ... ... ... ... | = 216° |
| | | 843·0982 |
| Deduct completed circles | ... ... ... ... ... | 720° |
| ☾'s Yoga longitude | ... ... ... ... ... | 123·0982 |

From the Eye-table we see that the current Yoga was Ganda.

### Easy Method for Yogas.

**51**. A very simple method of determining the ending moment of any Yoga is indicated in Table XI which is to be used in precisely the same manner for Yogas as for Nakshatras. We proceed to show how Table XI should be used for Yogas for any year, any lunar month, and any yoga. We take the solar year, A.D. 1910-11, the Lunar month Kārttika, and Yoga "Ayushmat".

For the year 1910-11 A.D., the correction corresponding to 25·9134 (first new moon in solar year) according to the Yoga correction table, is determined as follows :—

| | | | |
|---|---|---|---|
| Correction corresponding to | 25 | = | 3·4797 |
| " " | ·91 | = | ·1267 |
| " " | ·0034 | = | ·0005 |
| | | | 3·6069 |

In Lunar Kârttika, A.D. 1910, there are two yoga " intervals " for No. 3 " Ayushmat ", from each of which 3·6537 can be deducted. Let us calculate the mean and absolute ending moments of each of these " Ayushmats ".

| | | | Days. | ☾'s Anom. Days. |
|---|---|---|---|---|
| First New Moon in solar year | ... ... ... | | 25·9134 | 0·535 |
| Add for Kârttika New Moon (Table X) | ... ... | | 177·1835 | 11·856 |
| | | | 203·0969 | 12·391 |
| First No. 3. Yoga : Interval | ... | 3·8846 | | |
| Correction | ... | —3·6069 | | |
| Net Interval | ... ... ... | + ·2777 | + ·2777 | + ·278 |
| | | | 203·3746 | 12·669 |
| Commencement of solar year | ... ... | | Ap. 13·2080 | or 12., 40gh., 10 p. ⊙'sEqn. + 7gh. 55p. |
| | | | Ap. 216·5826 or 216d. 34gh. 57p. | 12d., 48gh. 5p. |

(1) ⊙'s Yoga Equation for Anom. of 203·3746 days, *i.e.*, 203d., 22gh. is, by Table IX—*i*, + 7gh., 55p.
(2) ☾'s Yoga Equation for Anom. of 12d., 48gh. 5p. is, by Table IX-*i*, – 5gh. 10p.
(3) Sum of ⊙'s + ☾'s Equations = 2gh. 45p. + 2gh. 45p.

216d., 37gh. 42p. from 1 April, *i.e.* (by Tables VIII and XIX) 37 ghaṭikas 42 palas after mean sunrise on 2 November, A.D. 1910 (Lanka time).

This is the absolute ending moment (according to Sûrya Siddhânta) of the first Ayushmat Yoga in the lunar month of Kârttika, Kaliyuga 5011 (expired).

**52.** For the second " Âyushmat " yoga of the same lunar month, we proceed as follows :—

| | | Days | ☾'s Anom. Days. |
|---|---|---|---|
| Kârttika New Moon ... ... | | 203·0969 | 12·381 |
| Net " interval " for No. 3 yoga :— From 29·3048 | | | |
| Deduct yoga correction for the year | —3·6069 | | |
| | 25·6979 | 25·6979 | 25·698 |
| | | 228·7948 | 38·079 |
| | | or 228 d., 47gh. 41p. | 27·555 |
| Commencement of solar year April 13·2080 | | or 13d., 12gh. 29p. | 10·524 or |
| | | 242d., 0gh. 10p. | 10d., 31gh. 30p. |
| (1) ⊙'s Yoga Eqn. (Table IX–*i*) for 228d., 48gh. is + 5gh. 12p. | | | Add. ⊙'s Eqn. + 5gh. 12p. |
| | | | 10d., 36gh. 42p. |

242d., 0gh., 10p.

(2) By Table IX–*i*, ☾'s Yoga Eqn., for 10d., 36gh., 42p. is—15gh., 8p.

(3) Sum of ☉'s and ☾'s Yoga Equations = —9gh., 56p. — 9gh. 56p.

241d., 50gh. 14p. from 1 April,

*i.e.*, (by Tables VIII and XIX) 50 ghaṭikas 14 palas after mean sunrise on 27 November, A.D. 1910.

**53**. For the years A.D. 1840 to A.D. 1919 yogas may be calculated very readily by means of Tables IX–*i*, XI–A, and XII. Thus, for "Vyatîpâta" yoga in April, A.D. 1910:—

| | d. gh. p. | d. gh. p. | ☉'s An. d. gh. p. | ☾'s An. d. gh. p. |
|---|---|---|---|---|
| Table XII, Chaitra new moon, A.D. 1910, April | | 9 35 27 | 361 38 30 | 26 6 48 |
| (Table XI–A, col. XIII, (Chaitra follg. Adhika masa) Interval for No. 17 "Vyatîpâta" yoga | 17 49 25 | | | |
| Deduct Yoga correction for the year (Table XII) | 1 0 45 | | | |
| | 16 48 40 + | 16 48 40 | 16 48 40 | 16 48 40 |
| Tab. IX–*i* ☉'s Yoga eqn. for 13d. 11g.—8gh. 25p. | | 26 24 7 | 378 27 10 | 42 55 28 |
| | | | * 365 15 30 | † 27 33 20 |
| ,, ☾'s Eqn. for 15d. 13gh. 43p. + 7gh. 40p. | | | 13 11 40 | 15 22 8 |
| Sum of ☉'s and ☾'s Eqns. —0gh. 45p. | | — 0 45 | | — 8 25 (☉'s Eqn.) |
| | April | 26 23 22 | | 15 13 43 |

That is, "Vyatîpâta" yoga in April 1910 ended at 23 ghaṭikas, 22 palas after mean sunrise on 26 April.

**54**. Yogas in their astronomical sense are not very much in use at present, but certain astrological *yogas* (Amritayoga, Siddhayoga, Maranayoga) are much more in use, being based on combinations of Nakshatras with certain week-days.

Apart from inscriptions in the Vikrama Era, where yogas are freely cited, one hears of the *yoga* proper in such auspicious combinations as the following :—

> *Kapilâ Shashthî*, i.e., the combination of Bhâdrapada Krishna Shasṭhi (in a year in which either Bhâdrapada or some previous lunar month is *adhika*) with Nakshatra *Rohinî* and Yoga *Vyatîpâta* on a Tuesday. The combination is proverbially rare and supposed to be very difficult to foresee, but it can be easily predicted by means of the present method. Look in Table X for Bhâdrapada mean new moon (in a year having this or some previous month as *adhika*) falling on a Wednesday, mean ending moment being less

* The deduction of the period of an entire solar year is made here.

† The deduction of the period of an entire anomalistic lunar month is made here.

than ·50, and calculate *Rohini* Nakshatra and *Vyatîpâta* Yoga for that month. Of course the tithi, the nakshatra, or the yoga *may* fall on Monday or Wednesday, and then there will be no *Kapilâshashti*.

*N.B.*—In A.D.17, A.D.193, A.D.396, A.D.440............A.D. 1513 these were *Kapilashashtis* : none between A.D. 1840 and A.D. 2000. Let the student try A.D. 1736 and A.D. 1787.

*Ardhodaya yoga*, i.e., a combination on a *Sunday* by day-time, of Nakshatra *Rohinî* Yoga *Vyatîpâta*, and *Mâgha* amâvâsyâ (*i.e.*, the amâvâsyâ at the *end* of lunar Mâgha).

*Champa Shashthî*, i.e., a combination on a *Sunday* or *Tuesday* of *Mârgasîra* śukla Shashṭi with Nakshatra *Satabhishaj* and Yoga *Vaidhriti*.

*Mahâmahâ Varunî*, i.e., a combination of *Phâlguna* krishna trayodaśî with *Saturday* and Nakshatra *Satabhishaj* and Yoga *Subha*.

*Dasaharî*, i.e., a combination in the *forenoon* of *Jyeshtha* śukla daśamî with *Wednesday* (or *Tuesday*) and Nakshatra *Hasta* and Yoga *Vyatîpâta*.

*Mahâsivarâtri*, i.e., Mâgha krishna chaturdaśî combining at *midnight* with *Sunday* or *Tuesday* and Yoga *Siva*.

*N.B.*—See further Chapter XVI below.

## CHAPTER VIII.

### KARANAS, AHAS AND TYAJYAM.

#### Karanas.

**55**. The list of *karanas* is given in Table III. Every tithi is divided into two *karanas*, and the ending moment of the second of every set of two *karanas* coincides with the ending moment of a tithi. For the other *karana* all we have to do is to add half a tithi, ·49217 day, to the last mean tithi, as well as to the corresponding solar and lunar anomalies and calculate from these data the absolute ending moment of the *karana*.

Thus on 27 February 1910 the mean tithi (18th) ended at ... ·2479 day.
For the 37th *karana* (*Bava*) add ... ... ... ... ·4922

·7401 day

| | days. | | days. |
|---|---|---|---|
| Sun's anomaly | 320·30 | Moon's anomaly | 12·324 |
| | ·49 | | ·492 |
| | 320·79 ; ⊙'s Eqn. +·1501. | | 12·816+·150=12·966 |

☾'s Equation—·0840.
Sum of the equations +·1501—·0840 = ... ... ... ... + ·0661

·8062 day

*i.e.*, the *karana Bava* ended at 48 ghaṭikas 22 palas on 27 February 1910.

**56**. Properly speaking, two karanas ought to be shown in the panchângam, corresponding to every tithi: but the panchângams in practice mark only that karaṇa whose ending moment is 30 ghaṭikas or less from sunrise. If the tithi itself ends at 30 ghaṭikas or less from sunrise, the same ending moment is entered in the panchângams against both tithi and karana. In the above case, *i.e.*, 27 February 1910, the 36th karana *Vishti* (called *Bhadra* in the Tamil country) will be found marked in the panchângams as having the same ending moment as the tithi.

### Ahas and Tyajyam.

**57.** Having discussed the five component parts of a panchângam proper, *viz.*, *vâra, tithi, nakshatra, yoga* and *karana*, it remains to notice two other entries usually found in a line with the first five, *viz.*, the *ahas* and the *tyâjyam.*

**58.** The *ahas* or duration of daylight is simply the normal period of daylight in India, *viz.*, twelve hours or 30 ghaṭikas, with the corrections for sunrise and sunset, deduced according to the rules to be presently laid down (Chapters IX and XI). Thus, if on a particular day the sun rose at 5-25 a.m. and set at 6-20 p.m. the *ahas* would be 12 hours $+35' + 20' =$ 12 hours 55 minutes, or 32 ghaṭikas 18 palas. (Tables XX and XIX.)

**59.** The *tyajyam* (lit., portion to be abandoned for purposes of business) is a definite portion of the total duration of each *nakshatra*, which portion is considered inauspicious. If the *tyâjyam*, always reckoned from the first moment of each nakshatra, ends by day, it is called a day-tyâjyam, and if it ends by night, it is called a night-tyâjyam, and in the latter case it is usual to deduct the *ahas* and state that the *tyâjyam* ends so many ghaṭikas and palas after sunset.

**60.** The following are the fractional parts which are *tyâjyam* for the several nakshatras :—

**Tyajyam.**

| No. | Nakshatra | Tyajyam |
|---|---|---|
| 1 | Asvini (five-sixths) ... ... ... | $\frac{5}{6}$ |
| 2 | Bharani (two-fifths) ... ... ... | $\frac{2}{5}$ |
| 3 | Krittika (one-half) ... ... ... (**Tam.** Kiruttigai) | $\frac{1}{2}$ |
| 4 | Rohini (two-thirds) ... ... ... | $\frac{2}{3}$ |
| 5 | Mrigasira (seven-thirtieths) ... ... (**Tam.** Mirugasiram) | $\frac{7}{30}$ |
| 6 | Ardra (seven-twentieths) ... ... (**Tam.** Arudra or Tiruvadirai) | $\frac{7}{20}$ |
| 7 | Punarvasu (one-half) ... ... ... | $\frac{1}{2}$ |
| 8 | Pushya (one-third) ... ... ... (**Tam.** Pusam) | $\frac{1}{3}$ |
| 9 | Aslesha (eight-fifteenths) ... ... (**Tam.** Ayilyam) | $\frac{8}{15}$ |
| 10 | Magha (one-half) ... ... ... (**Tam.** Magham) | $\frac{1}{2}$ |
| 11 | Purva Phalguni (one-third) ... ... (**Tam.** Puram) | $\frac{1}{3}$ |
| 12 | Uttara Phalguni (three-tenths) ... ... (**Tam.** Uttaram) | $\frac{3}{10}$ |
| 13 | Hasta (eleven-thirtieths) ... ... (**Tam.** Hastam) | $\frac{11}{30}$ |
| 14 | Chitra (fourteen-fifteenths) ... ... (**Tam.** Chittirai) | $\frac{14}{15}$ |
| 15 | Svati (seven-thirtieths) ... ... | $\frac{7}{30}$ |
| 16 | Visakha (seven-thirtieths) ... ... (**Tam.** Visakam) | $\frac{7}{30}$ |
| 17 | Anuradha (one-sixth) ... ... ... (**Tam.** Anusham) | $\frac{1}{6}$ |
| 18 | Jyeshtha (seven-thirtieths) ... ... (**Tam.** Kettai) | $\frac{7}{30}$ |
| 19 | Mula (one-third) ... ... ... (**Tam.** Mulam) | $\frac{1}{3}$ |
| 20 | Purva Ashadha (two-fifths) ... ... (**Tam.** Puradam) | $\frac{2}{5}$ |
| 21 | Uttara Ashadha (one-third) ... ... (**Tam.** Uttiradam) | $\frac{1}{3}$ |
| 22 | Sravana (one-sixth) ... ... (**Tam.** Tiruvonam) | $\frac{1}{6}$ |
| 23 | Sravishtha or Dhanishtha (one-sixth) ... (**Tam.** Avittam) | $\frac{1}{6}$ |
| 24 | Satabhisaj or Satataraka (three-tenths) ... (**Tam.** Sadayam) | $\frac{3}{10}$ |
| 25 | Purva Bhadrapada (four-fifteenths) ... (**Tam.** Purattadi) | $\frac{4}{15}$ |
| 26 | Uttara Bhadrapada (two-fifths) ... (**Tam.** Uttirattadi) | $\frac{2}{5}$ |
| 27 | Revati (one-half) ... ... ... | $\frac{1}{2}$ |

*N.B.*—Authorities differ as to the *tyajyam* for certain of the nakshatras: for instance, Mr. Srauti has informed the author that his tyajyams for nakshatras Nos. 13, 14, and 19 are $\frac{7}{30}$, $\frac{1}{3}$, and $\frac{14}{15}$ (instead of $\frac{11}{30}$, $\frac{14}{15}$, and $\frac{1}{3}$) respectively.

**61.** The method of working out the *tyâjyam* is simple. We take the total duration of each nakshatra from the ending moment of the previous nakshatra up to its own ending moment and cut off the *tyâjyam.*

Thus, according to the table in section **175** on 12 July, 1910, Nakshatra Uttara Phalgunî ended at Madras at 51 ghaṭikas 50 palas.

On 13 July 1910, the next Nakshatra Hasta ended in the same latitude at 56 ghaṭikas 30 palas.

The total duration of the Nakshatra Hasta was, therefore, 64 ghaṭikas 40 palas.

The *tyâjyam* for Hasta is $\frac{11}{30}$

Now, $\frac{11}{30}$ of 64 ghaṭikas 40 palas = 11 × (2 ghaṭikas 9 palas) = 23 ghaṭikas 39 palas.

∴ The tyâjyam for Nakshatra Hasta ended on
12 July 1910, 51 ghaṭikas 50 palas
+ 23 ghaṭikas 39 palas

13 July 1910, 15 ghaṭikas 29 palas, which was a day *tyâjyam*.

Again, on 20 July 1910, Nakshatra Mûla ended at 31 ghaṭikas 56 palas.

On 21 July 1910, Nakshatra Pûrva Ashâdhâ ended at 35 ghaṭikas 36 palas.

∴ Total duration of Pûrva Ashâdhâ was 63 ghaṭikas 40 palas,

of which the *tyâjyam* was $\frac{2}{5}$, *i.e.*, 2 × (12 ghaṭikas 44 palas) = 25 ghaṭikas 28 palas.

∴ Tyâjyam for Mûla ended on 20 July 1910 at 57 ghaṭikas 24 palas;

which, being night-time, we deduct the *ahas*, 31 ghaṭikas 27 palas (arrived at in accordance with subsequent rules regarding sunrise and sunset);

∴ Tyâjyam ended at 25 ghaṭikas 57 palas after sunset on 20 July 1910, and was a night *tyâjyam*.

Our *tyâjyam* will, of course, differ from that entered in other panchângams exactly in proportion to the difference between our ending moment and that of other panchângams for Nakshatras.

## CHAPTER IX.

### SUNRISE.

**62.** As explained in sections **130, 131**, Part II "Use of the Tables", all Indian time is reckoned, in the first instance, from Lankâ sunrise, and corrections are then applied so as to arrive at the time reckoned from local sunrise. These corrections are three in number :—

(1) A correction for terrestrial longitude;
(2) A correction for equation of time;
(3) A correction for the sun's tropical longitude.

**63**. To understand the nature of these corrections, we shall give in popular language the astronomical theory of local time, as applied by Indian Astronomers.

(1) CORRECTION FOR TERRESTRIAL LONGITUDE.—If the sun moved uniformly along the celestial equator, all places on earth would have sunrise at exactly 6 a.m., local time. On this supposition local time for any moment of the day could be deduced from Lankâ time by adding or subtracting four times as many

minutes of time as there are degrees of longitude between Lankâ (or Ujjain) and the given place. The longitude of Ujjain being 76° East of Greenwich, any place whose longitude is more than 76° from Greenwich has a positive correction, and any place whose longitude is less than 76° has a negative correction for terrestrial longitude. For example : the longitude of Madras being 80½° East of Greenwich, *i.e.*, 4½° more than Ujjain, local time at Madras is obtained from Lankâ time by adding $4\frac{1}{2}° \times 4' = 18$ minutes of time.

*N.B.*—A meridian of terrestrial longitude makes a revolution of 360° in 24 hours, that is, it revolves 1° in $24 \times 60 \div 360 = 4$ minutes. This interval of 4 minutes may be called the *time-difference* corresponding to a degree of terrestrial longitude.

The TIME–DIFFERENCE or correction for terrestrial longitude for nearly 200 important places in India (their importance being gauged by their population according to the Census Report of 1901) is given in Table XIII-A, pages 189, 190. The correction is given in seconds of time which can be converted into *palas* or *vinadis* by simply dividing the correction by 24.

(2) CORRECTION FOR EQUATION OF TIME.—The sun does not move uniformly throughout the year, as we know from Table IX-*c* and Table XVII-A. Hence, for each day of the solar year, we have to apply a correction which depends on the day's equation of the centre. The correction for each day of the solar year on account of the equation of time is given in Table XIII and it is expressed, like the first correction, in seconds of time.

(3) CORRECTION FOR THE SUN'S TROPICAL LONGITUDE.—The sun does not move along the equator, as in our first supposition, but along the ecliptic ; and therefore it is only at the points where the ecliptic cuts the equator, *i.e.*, at the moments of vernal and autumnal equinox that the first supposition holds good. At other times of the year, a correction has to be applied, depending on the sun's tropical longitude, *i.e.*, on his distance from either equinox measured along the ecliptic.

**64**. Now, how are we to determine the sun's tropical longitude ? The longitude that is given by the day of the Indian solar year is a sidereal longitude, *i.e.*, the longitude of the sun measured from a fixed point of the ecliptic, which fixed point was the true vernal equinox about A.D. 500 or Kaliyuga 3600. Every year since then, the vernal equinox has been moving further and further away from that fixed point, and the sun's tropical longitude for any year subsequent to 3600 Kaliyuga must be found by adding to his sidereal longitude 3 degrees for every subsequent period of 200 years.* In this portion of our treatise degrees are always converted into days, and for the purpose of Sûrya Siddhânta tithis, the tropical longitude for any day in any year is determined *by adding to the day of the solar year* 1 *day for every 64 years expired since* 3740 *Kaliyuga* (*A.D.* 639). Other Siddhântas give the precession as 1 day for every 60 years, which is more correct than 1 day for every 64 years (*see* Sec. **277**).

[**65**. By reckoning 3740 Kaliyuga and not 3600 Kaliyuga as the date of coincidence of the sidereal and tropical longitude, we eliminate the Sodhya of 2·1707 days (Surya Siddhanta) ; otherwise we would have to deduct the Sodhya from the day of the solar year in order to arrive at the sidereal longitude and then add to the sidereal longitude so arrived at as many days as there were periods of 64 years since 3600 Kaliyuga. ($2{\cdot}1707 \times 64 = 139{\cdot}9248$, which is nearly 140.)

* This is the correction for precession of equinoxes. The addition to be made to Indian sidereal longitudes on account of the precession of equinoxes is given in degrees in sec. **284** under "Platnetary Chronology" for various epochs from A.D. 520 to A.D. 2051.

**66.** The present is perhaps the best occasion in a popular work of this description for explaining what is meant exactly by the *Sodhya* when applied to the sun's sidereal longitude. The sun's equation of the centre, according to Table XVII-A on 0 day of the solar year, is +2·1378 degrees, which, by Table VII (b) is the sun's motion for 2·17 days.

In other words, when the mean sun is at 0 day of the solar year, the actual sun has already done 2·17 days journey of his annual course. Therefore, whenever we wish to express in days the sidereal longitude of the true or actual sun, *i.e.*, his distance from the position he occupied on 0 day of the mean solar year, we deduct the Sodhya of 2·1707 days from the day of the solar year we are at.

**67.** For the year A.D. 1919 the tropical longitude is found by adding to the day of the Solar Year $\frac{1919-639}{64}=\frac{1280}{64}$, *i.e.*, exactly 20 days, and this addition holds good for 64 years, *i.e.*, for 32 years before and 32 years after A.D. 1919 or from A.D. 1887 to A.D. 1951. In Table XIII the correction for Tropical longitude which applies to the year A.D. 1919 is assumed to apply to the whole period of 80 years from A.D. 1840 instead of only to the period of 32 years from A.D. 1887. There is no great error in this assumption.

**68.** For 30 important places in India, including one for every degree of N. Latitude between 8° and 35°, and also Calcutta and Bombay which are dealt with separately at pp. 183 to 188, the total of all the three corrections is given in Table XIII for each day of the solar year. Further, the same table gives the equivalence of English and Indian solar dates for all solar years which commence on the 13th April; for other years we must make a *deduction* or *addition* according as the Indian solar year in question commenced *after* or *before* 13th April.

**69.** With these elementary notions on Time, we are ready to work out quite accurately (according to the Siddhântas) any problem in local time. We shall work out, by way of illustrations, the somewhat formidable array of problems which are presented in Professor Jacobi's learned article in Vol. II of the *Epigraphia Indica*.

(i) Kaliyuga 4128 ; 4 Bhâdrapada (Bengal solar month); place RATNAGIRI, 17° N. Lat.)

Correction (1) for terrestrial longitude; – 34 vinadis. (We do not use this correction.)

Correction (2), equation of time: for 4 Bhâdrapada, *i.e.*, the 129th day of solar year, in 17° lat. + 425 seconds.

Correction (3), tropical longitude for Kaliyuga 4128, 4 Bhâdrapada: the number of days to be added to 129 is $\frac{4128-3740}{64}=\frac{388}{64}=6$. Now, 129 + 6 being 135, it follows that

correction (3) is that for the 135th day, 17° N. Lat., *i.e.*, + 746 seconds.

Total of the corrections (2) and (3) is + 425 + 746 = + 1171 seconds $=\frac{1171}{24}$ or 49 vinadis.

*N.B.*—Corrections (2) + (3) are sufficient for determining local sunrise: in this case the result " + 49 vinadis " means that on the day in question the sun rose at Ratnagiri 49 vinadis or 20 minutes before 6 a.m. [Result arrived at by Professor Jacobi, 50 vinadis which is also 20 minutes of time.]

(ii) Kaliyuga 4325, 4 Mârgaśira (Bengal solar month), SRINAGAR, latitude 34°.

Correction (1) : – 8 vinadis. (We do not use this correction.)

Correction (2): + 437 seconds (220th day.)

Correction (3): For $220 + \frac{4325-3740}{64} = 220 + \frac{585}{64} = 220 + 9 = 229$ days,
the correction is – 2248 seconds.

Total of corrections (2) and (3): + 437 – 2248 sec. = – 1811 seconds = – 75 vinadis.
[Professor Jacobi arrives at the result – 74 vinadis.]

(iii) Kaliyuga 4128; 7 Jyeshṭha (Bengal solar month); ALIGARH, latitude 28°.

Correction (1), + 14 vinadis. (We do not use this correction.)

Correction (2), – 291 seconds (37th day).
Correction (3), + 2576 seconds (43rd day).
} Total of corrections (2) + (3): + 2285" = 95 vinadis (94 vinadis according to Professor Jacobi).

(iv) Kaliyuga 3585; Âshâdha *sukla* 12 (lunar month), ERAN, 24° N. Lat.
See PART III, sec. **261**, p. (92) of the text.

**70**. The reader should note carefully when the total of all *three* corrections should be used and when the total of corrections (2) and (3) will suffice. When we merely want to know by how many seconds *local* sunrise preceded or followed 6 a.m., we should simply add up the second and third corrections, for here there is no need to consider the effect of terrestrial longitude. If the total correction is positive, sunrise precedes 6 a.m.; if the total correction is negative, sunrise is later than 6 a.m. The first three examples above illustrate this aspect of the problem.

When we want to know how many seconds should be added to or deducted from the ending moment of a *tithi* or *nakshatra*, of which we already know the Lankâ time, *i.e.*, the time relatively to Lankâ sunrise, we should sum up all *three* corrections, and if the sum is positive we should add it to Lankâ time, while if the sum is negative it should be subtracted from Lankâ time.

For examples, see the tables for determining the local Madras time of the ending moments of 30 tithis and 27 nakshatras in sections **167** and **175**, Part II *infra*.

## CHAPTER X.

### LAGHNA.

**71**. Closely analogous to the calculation of sunrise is that of the *laghna*, i.e., the portion of the ecliptic which appears at the eastern horizon at a *particular moment of the day*. At other times we might require to know the moment of the day when a *particular portion of the ecliptic* will be *laghna*. An example of the first kind of problem is :—

" What is the laghna at the moment of a person's birth " ?

*N.B.*—The portion of the ecliptic which is *laghna* or " cut " at the horizon at any given moment is ordinarily referred to as " the *laghna* ".

An example of the second problem is :—

" At what moment of a day, suitable for marriage, will a particular sign of the zodiac be laghna, so that the hour as well as the day most suitable for marriage may be determined " ?

The best exposition of the Laghna is that given by Professor Jacobi at pp. 189, 190 of the *Indian Antiquary* for 1900, Vol. XXIX. A slight adaptation of Professor Jacobi's rules will be necessary to suit our Sunrise Table XIII.

**72.** The Rules are :— .

(1) Calculate the time-difference for the interval of space between the sun and the portion of the ecliptic which is laghna.

(2) Apply the necessary corrections for tropical longitude.

**73.** *Problem I.*—At how many *ghatikas* and *palas* after sunrise was the 165th degree of the ecliptic (*i.e.*, the 15th degree of the sign *Kanya vide* Eye-table—) *laghna* on the 6th Jyeshṭha (Bengal solar month) in Kaliyuga 4000, Latitude 20° North ? By Table VIII, the end of 6th solar *Jyeshtha* was ·60 + 30·93 + 6 = 37·53 days from the first day of the solar year.

(*N.B.*—We add ·60 on account of the commencement of the solar year according to Table X—*Surya Siddhanta*).

Now, the sidereal longitude corresponding to any day of the solar year is found from Tables XVII-A and XVII-C (Tables, p. 207 208) thus:—

| | |
|---|---|
| ☉'s Long. for 37 days of the solar year (Table XVII–A). | 34·32° |
| ☉'s Long. for ·53 day (Table XVII-C) ... ... | ·52 |
| | 34·84° |

For the present purpose we take the nearest whole degree, *i.e.*, 35 degrees.

The interval in space between 35° and 165° = 130°.

Now 360° of the ecliptic rise above the horizon in 1 day or 60 ghaṭikas.

$$\therefore 130° \quad \ldots \quad \ldots \quad \ldots \quad \frac{130 \times 60}{360} = 21\frac{2}{3} \text{ ghatikas}$$

$$= 21 \text{ ghaṭikas } 40 \text{ palas.}$$

In other words, the 165th degree of the ecliptic was *laghna* about 21 ghaṭikas 40 palas after sunrise on the day in question.

**74.** Next, we proceed to apply the corrections for tropical longitude.

The difference between the tropical and sidereal longitude of the sun for the year 4000 Kaliyuga $= \frac{4000-3740}{64} = 4$ days. Now 37·53 days + 4 days = 41·53 days.

The correction for the tropical longitude of the sun is that pertaining to the 41st day in Table XIII, p. 165, under latitude 20°, *i.e.*, + 1865 seconds.

For the correction for tropical longitude of the laghna, *i.e.*, sidereal 165° of the ecliptic in Kaliyuga 4000 we have to determine the day when the sun accomplishes 165° of the ecliptic, which is determined as follows from Tables XVII-A and XVII-C :—

| | | | |
|---|---|---|---|
| (Table XVII-A) | 169 days | = | 164·43° |
| (Table XVII-C) | ·58 day | = | ·57° |
| | 169·58 days | = | 165° |

We add 4 days to 169·58 days for tropical longitude in the year 4000 Kaliyuga. Total 173·58 or 174 days.

The correction for tropical longitude corresponding (in Kaliyuga 4000) to sidereal 165°, or tropical 174 days, is, + 151 seconds.

We add this correction *with the sign reversed* to the first :—

+ 1865 − 151 = 1714 seconds = + 72 palas.

Adding this correction to the approximation first arrived at, we have

| | |
|---|---|
| 21 ghaṭikas | 40 palas. |
| + 1 ghaṭika | 12 palas. |
| 22 ghaṭikas | 52 palas. |

This was the moment (according to Professor Jacobi, *loc. cit.*) when the 165th degree of the ecliptic was *laghna* on the day in question.

*N.B.—Firstly,* we had to convert 165° of the ecliptic into days in order to apply Table XIII, which expresses the various portions of the ecliptic as days of the solar year.

*Secondly,* we reversed the sign of the correction in the case of the *laghna* because what we wanted was the difference between the corrections on account of the tropical longitude of the sun and of the *laghna* respectively.

**75**. *Second Problem.*—" What portion of the ecliptic was *laghna* on the same day at 20 ghaṭikas after sunrise " ?

We have seen that the Solar 6th Jyeshṭha corresponds to 35° sidereal longitude.

Now, a time-difference of 20 ghaṭikas answers to an interval in space of $20 \times \frac{360}{60} =$ 120 degrees.

∴ The portion of the ecliptic which was laghna at 20 ghaṭikas after sunrise was (*roughly*) 120° + 35° = 155° or the 5th degree of the 6th sign *Kanya* (*vide* Eye-table for the longitude corresponding to the different Rāśes).

This is enough for all ordinary calculations of laghna.

If we wish to be quite accurate, we proceed as follows. The correction for tropical longitude of the sun on the day in question was, as we have seen, + 1865 seconds.

The correction for the tropical longitude of the *laghna*, i.e., 155° of the ecliptic must be found by applying Tables XVII-A and XVII-C.

| | | |
|---|---|---|
| (Table XVII-A) | 159 days = | 154·57° |
| (Table XVII-C) | ·44 day = | ·43° |
| | 159·44 days = | 155·00° |
| Add for tropical longitude | 4 days | |
| | 163·44 days. | |

The correction for tropical longitude pertaining to the 20th degree of latitude and the 163rd day (in Kaliyuga 4000) is (by Table XIII) + 333 seconds.

We add this *with sign reversed* to the first correction :

+ 1865 − 333 = + 1532 = + 64 palas.

In other words, on the day in question, the 155th degree of the ecliptic was laghna at 20 ghaṭikas *plus* 64 palas.

We want, however, the degree of the ecliptic which was *laghna* at exactly 20 ghaṭikas after sunrise.

**76.** The annexed small table adapted from Prof. Jacobi's Table of *udayâsus* in Vol. I, *Ep. Ind.*, enables us to accomplish this stage of the problem.

We learn from this table that in the 20th degree of latitude the 6th sign of the zodiac (Kanyâ, with which we are concerned) takes 257 seconds to rise *one* degree above the horizon.

$\therefore$ 64 palas or 1532 seconds are taken to rise $\frac{1532}{257} = 5{\cdot}9611^\circ$

or (by Table XX) 5 degrees 57 minutes 40 seconds,

or nearly 5 degrees 58 minutes.

In other words, we should deduct 5 degrees 58 minutes from 155° in order to know which portion of the ecliptic was *laghna* at exactly 20 ghaṭikas after sunrise on the day in question.

The portion in question was 155° *minus* 5 degrees 58 minutes = 149 degrees 2 minutes, which is exactly the result arrived at by Professor Jacobi.

**TABLE.**

**Time (in seconds) taken by a Degree of the Ecliptic to rise above the Horizon.**

DEGREES OF TERRESTRIAL LATITUDE FROM 8° TO 35°.

| Signs of the Zodiac. | Degrees of Ecliptic. | | | 8° | 9° | 10° | 11° | 12° | 13° | 14° | 15° | 16° | 17° | 18° | 19° | 20° | 21° |
|---|---|---|---|---|---|---|---|---|---|---|---|---|---|---|---|---|---|
| I | 0°— 30° | XII | 330°—360° | 210 | 208 | 206 | 204 | 202 | 201 | 199 | 197 | 195 | 194 | 192 | 190 | 188 | 186 |
| II | 30 — 60 | XI | 300 —330 | 229 | 227 | 226 | 224 | 223 | 221 | 220 | 218 | 217 | 215 | 214 | 212 | 211 | 206 |
| III | 60 — 90 | X | 270 —300 | 253 | 253 | 252 | 252 | 251 | 251 | 250 | 250 | 249 | 248 | 248 | 247 | 246 | 249 |
| IV | 90 —120 | IX | 240 —270 | 263 | 263 | 264 | 264 | 265 | 266 | 266 | 267 | 267 | 268 | 269 | 269 | 270 | 270 |
| V | 120 —150 | VIII | 210 —240 | 250 | 251 | 253 | 254 | 256 | 257 | 259 | 260 | 262 | 263 | 265 | 266 | 268 | 269 |
| VI | 150 —180 | VII | 180 —210 | ... | ... | 239 | 241 | 243 | 244 | 246 | 248 | 250 | 252 | 254 | 255 | 257 | 259 |

| Signs of the Zodiac. | Degrees of Ecliptic. | | | 22° | 23° | 24° | 25° | 26° | 27° | 28° | 29° | 30° | 31° | 32° | 33° | 34° | 35° |
|---|---|---|---|---|---|---|---|---|---|---|---|---|---|---|---|---|---|
| I | 0°— 30° | XII | 330°—360° | 184 | 182 | 180 | 178 | 176 | 174 | 172 | 170 | 168 | 165 | 163 | 161 | 159 | 157 |
| II | 30 — 60 | XI | 300 —330 | 208 | 206 | 204 | 203 | 201 | 199 | 197 | 196 | 194 | 192 | 190 | 188 | 186 | 184 |
| III | 60 — 90 | X | 270 —300 | 245 | 244 | 244 | 243 | 242 | 242 | 241 | 240 | 239 | 238 | 238 | 237 | 237 | 236 |
| IV | 90 —120 | IX | 240 —270 | 271 | 273 | 273 | 274 | 274 | 275 | 275 | 276 | 277 | 277 | 278 | 278 | 279 | 280 |
| V | 120 —150 | VIII | 210 —240 | 270 | 272 | 274 | 276 | 278 | 279 | 281 | 283 | 285 | 287 | 289 | 291 | 293 | 295 |
| VI | 150 —180 | VI | 180 —210 | 261 | 263 | 265 | 267 | 269 | 271 | 273 | 275 | 278 | 280 | 282 | 284 | 286 | 288 |

# CHAPTER XI.

## SUNSET.

**77.** The considerations to which we have introduced the reader in dealing with Sunrise and Laghna will enable him to deal also with *sunset,* which is only the supplement (in the language of geometry) of *sunrise.*

Supposing the sun is at a point of the ecliptic whose tropical longitude is 100°, then we have to consider at sunrise how long, relatively to the mean sun on the equator, the mean sun on the ecliptic takes to rise 100°, and at sunset we have to consider how long, relatively to the mean sun on the equator, the mean sun on the ecliptic takes to fall 80°. [*see* Sec. **63** (3)]

**78.** Accordingly, to determine the moment of sunset by local time, we should take the sum of (1) the correction for equation of time, which is the same or nearly the same as for sunrise, and (2) the correction for tropical longitude, which will be the supplement of the tropical longitude for sunrise. The supplement of an angle is the difference between it and 180°. In our sunrise tables, degrees are expressed as days and by Tables XVII-A and XVII-C, 180° = 184·80 days or in round figures 185 days.

**79.** What we have to do, therefore, to determine the second correction for sunset is to find the difference between the day of the solar year under consideration and 185 days, and take the correction for tropical longitude for the day so arrived at. If the solar day we are considering exceeds 185 days, the sign of the tropical longitude for the difference will remain unchanged: otherwise the sign should be reversed. This correction should then be added to the equation of time in order to determine the interval between sunset and 6 p.m. local time, while the total so arrived at should be added again to correction (1) adverted to in sec. **63**, *i.e.*, the correction for terrestrial longitude, in order to arrive at the interval between 6 p.m. *Lankâ* time and *local* sunset.

**80**. In the absence of any other standard authorities on the subject, we may compare our calculations of sunset for A.D. 1910-11 with those given in the Kumbakonam Mutt Panchângam, edited by Mr. Viśvanâtha Srauti, B.A.

**Latitude 11° N. (Tanjore).**

| Date by European Calendar as well as by Tamil Solar Year *Sâdhârana*, A.D. 1910. | Day of Solar Year according to Table XIII. | Equation of Time in Seconds by Tab. XIII under Lat. 11° for the day of Solar Year entered in col. 2. | Supplement to Trop. longitude in Days (*i.e.*, difference between col. 2 increased by 20 and 185 days). | Correction for Trop. Longitude acc. to Tab. XIII for the Supplemental day arrived at in column 4 (*sign to be reversed if necessary*). | Sum of Corrections arrived at in columns 3 and 5. | Moment of Sunset according to col. 6. | Sunset according to Mr. Srauti's Panchangam for 1910. |
|---|---|---|---|---|---|---|---|
| 1 | 2 | 3 | 4 | 5 | 6 | 7 | 8 |
| | | | | | | P.M. | P.M. |
| Chittirai 1 (Ap. 13) ... ... | 1 | -435″ | 164th day | - 22″ | - 457″= - 7′ | 6-7 | 6-7 |
| Vaikasi 2 (May 15) ... ... | 32 | -352″ | 133rd day | - 312″ | - 664″= -11′ | 6-11 | 6-11 |
| Ani 3 (Je. 16) ... ... | 65 | -147″ | 100th day | - 986″ | -1133″= -19′ | 6-19 | 6 19 |
| Adi 3 (Jl. 18) ... ... | 97 | +163″ | 68th day | -1469″ | -1306″= - 22′ | 6-22 | 6-23 |
| Avani 4 (Au. 19) ... ... | 129 | +411″ | 36th day | -1168″ | - 757″= -13′ | 6-13 | 6-14 |
| Purattasi 5 (S. 20) ... ... | 161 | +515″ | 4th day | - 144″ | + 371″= + 6′ | 5-54 | 5-55 |
| Aippasi 6, (O. 22) ... ... | 192 | +494″ | 27th day (212 - 185 = 27) | + 969″ | +1463″= +24′ | 5-36 | 5-36 |
| Kartigai 8 (N. 23) ... ... | 224 | +353″ | 59th day | +1523″ | +1876″= +31′ | 5-29 | 5-30 |
| Margali 3 (D. 17) ... ... | 249 | +140″ | 84th day | +1278″ | +1418″= +23′ | 5-37 | 5-37 |
| Tai 21 (F. 3) A.D. 1911 ... | 296 | -268″ | 133rd day | + 312″ | + 44″= + 1′ | 5-59 | 6-1 |
| Panguni 2 (Mr. 15) A.D. 1911 ... | 337 | -421″ | 172nd day | + 12″ | - 409″= - 7′ | 6-7 | 6-8 |

**81.** We observe (1) that sufficiently accurate results can be obtained by taking whole days instead of days and fractions of days in columns 2 and 4 *supra*; in other words, it is not necessary to calculate the day of the solar year with exactitude as we did for Problem I under "**Laghna**", (2) that a *positive* correction in column 6 re-appears in column 7 as a *deduction* from 6 p.m., while a *negative* correction in column 6 re-appears in column 7 as an addition to 6 p m.

Supposing a birth took place 2 hours after sunset on 13 April, 1910, the moment of birth would be 6 p.m. + 2 hours + 7′ = 8-7 p.m. In this case sunset was 6-7 p.m. On the other hand, if a birth took place 2 hours after sunset on 23 November, 1910, then the correction for sunset being + 31′, the local time at which the birth took place would be 6 p.m. + 2 hrs. - 31′ = 7-29 p.m. In this case sunset was at 5-29 p.m.

## CHAPTER XII.

### ECLIPSES.

**82**. Eclipses are a most valuable and unerring historical record, and eclipses are freely cited in Indian historical documents. Accordingly, the date of occurrence of solar and lunar eclipses has been indicated in Tables X and XII by means of the symbol ● for solar eclipses and ⊙ for lunar eclipses. A solar eclipse can occur only at new moon and a lunar eclipse can occur only at full moon, and as full moons are not specially indicated in Tables X and XII, every lunar eclipse is shown merely as occurring in a particular lunar month. The actual date of a lunar eclipse can be found by working out the exact ending moment of the full-moon tithi, *i.e.*, *panchadasi* or *paurnami*. For this, of course, it will be necessary to add the collective duration of 15 tithis, *i.e.*, 14·76 days (Eye-table), to the mean moment of new moon and also to the moon's and sun's anomaly at that moment and add to, or subtract from, the mean moment of full moon the sum of the sun's and moon's equations. (See examples worked out in sections **93** and **94**.)

**83**. We may preface our observations on eclipses by the following passages from a well-known French work, first published in the 18th century, and still regarded as a standard authority on chronology, " *L'Art de vérifier les Dates* ;—

" The magnitude of an eclipse is indicated in digits. A digit is the twelfth part of the diameter of a heavenly body : accordingly, an eclipse of four digits is one where the third part of the diameter of a heavenly body is in shadow ; if one half of the diameter is in shadow, the eclipse is said to be of six digits ; and if the whole of the diameter is in shadow, the eclipse is said to be of twelve digits. Further, in a total eclipse, if the part which is least in shadow is removed from the edge of the shadow by as much as two digits, these two digits are added to the twelve, and the eclipse is said to be one of fourteen digits...............

**84**. " The duration of an eclipse of the moon is proportional to its magnitude ; but it does not depend on this factor alone. We may say, in general, that a total eclipse of the moon will last at least 3½ hours, and at most 4 hours and a few minutes. A partial eclipse above 6 digits may last as much as 3¼ hours or as little as 2½ hours ; it seldom goes beyond those limits. An eclipse of between three and six digits lasts between two and three hours. The duration corresponding to 2 digits is 1½ hours ; to 1 digit, 1 hour ; to half a digit, ¾ of an hour. The duration of total obscurity in a total eclipse is from 1½ to 2 hours........

**85**. " The Moon has no light of her own, but reflects that received by her from the Sun. Consequently, when in an eclipse the Earth passes between the Moon and the Sun, and cuts off the rays of the Sun from the Moon, the latter can shed no more light on Paris than on Pekin or on Constantinople. An eclipse of the Moon, therefore, commences at all points of the Earth at the same instant and lasts everywhere for the same length of time. All that we need see is whether the time of a lunar eclipse falls between local sunset and sunrise, as otherwise the eclipse will not be visible.

**86**. " On the other hand, the Sun is a self-luminous body: the Moon is very much smaller than the Earth : she is, in fact, a mere point in comparison with the Sun. From these well-known facts it follows that in a solar eclipse, the Moon passing between the Earth and the Sun, cannot cut off the rays of the Sun from all parts of the Earth, at the same instant. She casts her shadow in succession on different parts of the Earth, and even the successive passage of the Moon's shadow covers during a single eclipse only a small total extent of the Earth's surface. Thus, journeying from West to East, a solar eclipse may first strike Spain, then France, then Germany, and then in succession the eastern countries as far as the farthest East of Asia, but the whole time Africa may remain untouched by the Moon's shadow ".

**87.** The extent of the earth's surface, affected by a solar eclipse, is therefore a matter of calculation : and an exact calculation is by no means easy to introduce into a popular work of this kind.

The book entitled *L'Art de vérifier les Dates* gives (1) *all* lunar eclipses and (2) *all* solar eclipses visible anywhere in the civilized world known to history, marking, by a series of ingenious conventions, the path of each solar eclipse, the regions where it is visible and the latitudes where it is central (*a*) under the fifth meridian, *i.e.*, the meridian of Paris, (*b*) under the meridian where it is noon at the moment of true conjunction, and (*c*) under the 155th meridian. Other useful details, given under each eclipse, in the same valuable work, need not be reproduced here, as nothing short of a translation of the whole Chapter on Eclipses would give the reader an adequate idea of its contents.

**88.** We have given in Tables X and XII *all* lunar eclipses, and it is easy for the reader to calculate the ending moment of the full moon tithi, which will enable him to know whether the eclipse could have been seen in India at the place under consideration. We have given *all* solar eclipses, as given in the French work, from A.D. 1 to A.D. 300, and after the latter date we have, as a rule, not mentioned eclipses which, according to the same authority, could not have been visible in India. From A.D. 1900 to A.D. 2000 all solar eclipses are marked.

**89.** The points where the moon's orbit crosses the ecliptic are called the *nodes* and in order that an eclipse may occur, the opposition at the time of a lunar eclipse, or the conjunction at the time of a solar eclipse, must happen at or near a node. The sun's distance from either node is therefore an important factor in determining whether an eclipse is or is not possible at a particular New or Full Moon. Eclipse-Table X-A (pp. 130, 131) for Centuries and Odd Years will enable the reader to calculate the Sun's distance from either node at any moment from B.C. 1 to A.D. 2000, and by applying the following canons (adapted from Professor Jacobi's article in Vol. 1 of *Epigraphia Indica*) he may determine whether a solar or a lunar eclipse is certain, doubtful, or impossible at a particular New or Full Moon respectively.

**Solar and Lunar Eclipses as determined by Sun's distance from node at New and Full Moon.**

IF SUN'S DISTANCE FROM NODE AT NEW MOON IS

| | | |
|---|---|---|
| Between 0 and 15·598 days, or between 157·714 and 173·312 days, | A SOLAR ECLIPSE IS | **Certain.** |
| Between 15·771 and 18·198 days or between 157·541 and 155·114 days, | | **Doubtful.** |
| Between 18·371 and 154·939 days, | | **Impossible.** |

IF SUN'S DISTANCE FROM NODE AT FULL MOON IS

| | A LUNAR ECLIPSE IS |
|---|---|
| Between 0 and 10·052 days, or between 163·260 and 173·312 days, | **Certain.** |
| Between 10·225 and 13·345 days, or between 163·087 and 159·967 days, | **Doubtful.** |
| Between 13·518 and 159·794 days, | **Impossible.** |

**90.** We shall now apply the canons to some actual cases, and see how far they will be of use to us in practice.

Opening the first page of Table X, we see a succession of solar eclipses at the Jyeshṭha new moon in the years A.D. 12, A.D. 13, and A.D. 14, and we see a similar succession of lunar eclipses in the same lunar month in the years A.D. 20, A.D. 21, and A.D. 22. We also see that in Lunar Jyeshṭha, A.D. 12, as well as in Lunar Jyeshṭha A.D. 22, there was a lunar as well as a solar eclipse. We shall see if our canons will help us in verifying these eclipses. The reader should note how an eclipse may be located with the help of the secular distance of the sun from the node (*i.e.*, Distance from node for Centuries and Odd Years) as given in Table X–A and the head line entries of Table X.

| ☉'s Distance from Node Table X–A | (1) A.D. 12 days | (2) A.D. 13 days | (3) A.D. 14 days | (4) A.D. 20 days | (5) A.D. 21 days | (6) A.D. 22 days. |
|---|---|---|---|---|---|---|
| For century 1 B.C. ... | 62·91 | 62·91 | 62·91 | 62·91 | 62·91 | 62·91 |
| For odd year ... ... | 50·30 | 68·93 | 87·56 | 26·06 | 44·69 | 63·33 |
| Up to first New Moon in Solar Year ... ... | 26·95 | 16·06 | 5·17 | 28·41 | 17·52 | 6·62 |
| Up to Jyeshṭha New Moon | 29·53 | 29·53 | 29·53 | 29·53 | 29·53 | 29·53 |
| Up to Jyeshṭha Full Moon | ... | ... | ... | 14·76 | 14·76 | ... |
| | 169·69 | 177·43 | 185·17 | 161·67 | 169·41 | 162·39 |
| Deduct 173·31 days ... | (7) | * 173·31 | * 173·31 | ... | ... | (8) |
| | | 4·12 | 11·86 | | | |
| Up to Jyeshṭha Full Moon. | 14·76 | ... | ... | ... | ... | 14·76 |
| | 184·45 | | | | | 177·15 |
| Deduct ... ... | * 173·31 | ... | ... | ... | ... | * 173·31 |
| | 11·14 (1) | | | | | 3·84 (6) |

We can now classify these results.

**SOLAR ECLIPSE:** *Certain*, Results at foot of columns (2), (3), and also those in columns (1) and (6) marked (7) and (8) respectively.

**LUNAR ECLIPSE:** *Certain*, Results at foot of columns (5) and (6).
*Doubtful*, Results at foot of columns (1) and (4).

* The period of conjoint revolution of the sun's and moon's nodes is 346·6246 days and the half revolution is 173·3123167 days. This period should therefore be deducted whenever the sun's distance from a node exceeds 173·31 days.

**91.** We shall next see to what extent the canons will be of use to us in predicting eclipses. We may take as an example the solar year A.D. 1910. *

We first of all find out the distance of the sun from the node at the first new moon in the solar year.

| | |
|---|---|
| ☉'s distance from node for century beginning A.D. 1900 ... ... | 120·62 |
| " " " for odd year 10 ... .. .... ... | 13·03 |
| " " " at first new moon in solar year A.D. 1910 ... | 25·91 |
| | 159·56 |

**92.** As regards solar eclipses, we have to see whether 159·56 by itself, or the same increased by 29·53, or any multiple of 29·53 entered in the head lines of Table X, will bring us within the limits of *certain* or even *doubtful* solar eclipses.

We see at once that the figure 159·56 is itself within the limits of a *certain* solar eclipse. The date is the first new moon in the solar year, *i.e.*, 9th May, 1910. This eclipse was invisible in India.

The addition to 159·56 of 29·53 gives us 189·09, which by deduction of 173·31 is equivalent to 15·78, which is beyond the limits of certainty. There was in fact no solar eclipse on the corresponding date, Jyeshṭha new moon, 7th June, 1910.

The addition to 159·56 of 177·18 (Kârttika new moon) will give us 336·74, which by deduction of 173·31 leaves 163·43, which is within the limits of certainty. In fact, there was a solar eclipse (invisible in India) on *Kârttika* New Moon, *i.e.* (according to Table X or XII) 2nd November, 1910.

The addition of no other multiple of 29·53 to 159·56 will give us a figure within "certain" or even "doubtful" limits for solar eclipses.

**93.** As regards lunar eclipses, our starting figure 159·56 should be increased by the day equivalent of the full moon tithi, 14·76 days.

$$159{\cdot}56 + 14{\cdot}76 = 174{\cdot}32 \text{ days.}$$
$$\text{Deduct } 173{\cdot}31$$
$$1{\cdot}01 \text{ days.}$$

We see that in the first lunar month of the solar year A.D. 1910 there must be a lunar eclipse.

The actual ending moment of the full moon tithi works out as follows :—

| | | Days. | Anomaly of first new moon. |
|---|---|---|---|
| First new moon in solar year | | 25·9134 | 0·535 |
| | | 14·7653 | 14·765 |
| ☉'s Equation for 40·67 | +·1140 | 40·6787 | 15·300 |
| ☾'s Equation for 15·414 | +·1650 | | +·114 (☉'s Eqn.) |
| | +·2790 | + ·2790 | 15·414 |
| Commencement of Solar Year, Ap. | | 13·2080 | |
| | | 54·1657 | |
| *Add* total correction for latitude 13° N. on 40th day of Solar Year (Table XIII) | | ·0286 | |
| | | 54·1943 | |

* This portion of the text was written before the eclipse in question happened. The force of the illustration is of course not lessened by its referring to an event now past and it has therefore been retained.

It thus appears that the centre of the eclipse occurred at Madras at ·1943 of a day, *i.e.*, at 11 ghaṭikas 39 palas after sunrise. It is evident that a lunar eclipse at such an hour would be invisible in India.

**94.** The ⊙'s distance from node at the first full moon of the solar year A.D. 1910 being 1·01, it is evident that an addition of 177·18 days, which will take us to the Kârttika Full Moon, will give us 178·19, which by deduction of 173·31 gives us 4·88 days, which is within the limits of a "certain" lunar eclipse. Let us calculate the actual ending moment of this Full Moon tithi.

| | Days of Solar year. | ☾'s Anomaly. Days. |
|---|---|---|
| First Full Moon in Solar Year A.D. 1910 (sec. 93) ... | 40·6787 | 15·300 |
| *Add* for Kârttika Full Moon ... | 177·1835 | 11·856 |
| | 217·8622 | 27·156 |
| ⊙'s Equation for 217·86 days = — ·1263 | | —·1263 (⊙'s Eqn.) |
| ☾'s Equation for 27·029 days = + ·0453 | —·0810 | 27·0297 |
| — ·0810 | 217·7812 | |
| Commencement of Solar Year 1910, Ap. 13 | ·2080 | |
| | 217·9892 | |
| Total correction (by Table XIII) for 217th day, | +·0111 | |
| | 218·0003 | |

∴ the centre of the eclipse at Madras was on the 218th day, reckoned from April 13, *i.e.*, (by Table VIII) 17th November, 1910, at ·0003 of the day, *i.e.*, 1 pala or half a minute after sunrise on 17th November, 1910.

This very nearly agrees with the prediction in the Nautical Almanac reduced to the longitude of Madras. The eclipse was visible at Madras.

No other multiple of 29·53, added to 1·01, will give us a lunar eclipse in the Solar Year A.D. 1910 within the limits of certainty.

**95.** A third use of the canons is to enable us to detect omissions from or incorrect entries in lists of eclipses.

The book entitled *L'Art de vérifier les Dates* states that there was no eclipse in the year A.D. 555.

But seeing that the ⊙'s distance * from node at the first new moon in the solar year A.D. 554—55 was 163·66 days, and that the year was one with an *adhika* month, it is plain from the head-lines of Table X that the addition to 163·66 of 354·37 gives us 518·03 days, which, deducting 346·62 leaves 171·41 days, which indicates that *certainly* there was a solar eclipse at the corresponding New Moon, *i.e.*, on Chaitra New-Moon day, 8th March A.D. 555. As a matter of fact, it appears from OPPOLZER's *Canon der Finsternisse*, that there was a solar eclipse on 8th March, A.D. 555.

*23·76
139·68
0·22
163·66

**96**. Let us next take a case in which the canons will show us that the *Art de vérifier* is right and other lists wrong. The French book mentions a Lunar Eclipse (penumbral) on 6th May A.D. 692, while other lists do not mention it.

| |
|---|
| *154·79 |
| 154·53 |
| 3·24 |
| 312·56 |
| 29·53 |
| 14·76 |
| 356·85 |
| —*346·62 |
| 10·23 |

We find that the ☉'s distance* from node at the full moon following Jyeshṭha New Moon (21 Ap. A.D. 692) was, as shown in the margin, 10·23 days, which makes it almost certain that there was then a Lunar Eclipse. The French chronological work is, therefore, right.

---

## CHAPTER XIII.

### JOVIAN CYCLES, YEARS, AND MONTHS.

[*Note.*—For the calculation of the mean period of a Jovian year, see sec. **230** *infra.*]

**97**. A Jupiter's cycle of 60 Jovian years is often met with in Indian chronology as a mode of reckonng time. Each of the years in the cycle receives a name corresponding to the name of one of the years in Table I, but while the so-called Jupiter's cycle in Table I, as used to this day in Southern India, is merely a cycle of 60 Indian solar years, the Jupiter's cycle, properly so called, which has been in use in Northern India and which is given in Table XIV, is a cycle of real Jovian years, calculated by means of the multiplication tables in Table V and the following table for centuries.

| Kaliyuga. | Christian Era. | Jupiter's Samvat (with bija). | Kaliyuga. | A.D. | Jupiter's Samvat (with bija). |
|---|---|---|---|---|---|
| 3001 | B.C. 101 | 36·1117 | 4101 | 1000 | 8·9817 |
| 3101 | B.C. 1 | 17·2817 | 4201 | 1100 | 50·1517 |
| 3201 | A.D. 100 | 58·4517 | 4301 | 1200 | 31·3217 |
| 3301 | 200 | 39·6217 | 4401 | 1300 | 12·4917 |
| 3401 | 300 | 20·7917 | 4501 | 1400 | 53·6617 |
| 3501 | 400 | 1·9617 | 4601 | 1500 | 34·8317 |
| 3601 | 500 | 43·1317 | 4701 | 1600 | 15·9917 |
| 3701 | 600 | 24·3017 | 4801 | 1700 | 57·1717 |
| 3801 | 700 | 5·4717 | 4901 | 1800 | 38·3417 |
| 3901 | 800 | 46·6417 | 5001 | 1900 | 19·5117 |
| 4001 | 900 | 27·8117 | 5101 | 2000 | 0·6817 |

**98**. We shall take an example of a Jupiter's Ṣamvat and show how it is calculated. The Date of a Sanskrit Manuscript is given by its author as "Saka 1396, (Vikrama) Samvat 1531, year *Subhakrit*, Kārttika Sudi 9, after Dakshināyana on Wednesday, *Dhanishtha* Nakshatra, *Vriddhi* yoga, *Kaulava* karana, the moon in *Kumbha*" (*Indian Antiquary*, XIX, p. 27). For the Nakshatra, Yoga, Karana and Zodiacal Sign in this case, see sec. **266**, *infra*.

We first calculate the tithi in the usual way :—

By Table X the Kārttika New Moon of Vikrama 1531 (A.D. 1474), was (2) Monday, October 10·85

☾'s Equation + ·20
☉'s Equation – ·15

+ ·05

| | | ☉'s Anomaly. | ☾'s Anomaly. |
|---|---|---|---|
| | | 20·27 | 4·559 |
| | | 177·18 | 11·856 |
| Add for 9 tithis | 8·86 | 8·86 | 8· 86 |
| Equation | + ·05 | | 25·275 |
| (4) Wednesday, Oct. | 19·76 | 206·31 | ☉'s Eqn.— ·15 |
| | | | 25·12 |

We have now to find the equivalent in Jupiter's cycle of the 206·31 days of the solar year A.D. 1474.

| | | |
|---|---|---|
| Equivalent of A.D. 1400 (by above table) | | 53·6617 |
| " | " 74 years (Table V) Jupiter's Samvat | 14·8658 |
| " | " 200 days ( do. ) | ·5540 |
| " | " 6 days ( do. ) | ·0166 |
| | | 69·0981 |
| | Deduct completed cycle | 60 |
| | | 9·0981 |

That is, No. 9 of Jupiter's cycle (according to Table XIV) *i.e.*, *Subhakrit* had expired on the day in question of the solar year.

**99**. The year of Jupiter's cycle, ordinarily quoted in N. India, is that which was completed at the *beginning* of the solar year, and in this case inasmuch as only ·098 of the Jovian year, *i.e.*, only 36 days, had elapsed since the beginning of Subhakrit up to the 206th day of the solar year, the previous Jovian year, *Plava*, must have been current at the beginning of the solar year A.D. 1474. When the Jovian year quoted is only that which had been completed at the beginning of the solar year, all we have to do is to calculate the Jovian year for the *integral* solar year we are dealing with or, which is much simpler, read off the corresponding year of Jupiter's cycle from Table XIV.

**100**. Altogether, we see there are three meanings which might be attached to the citation of a year of Jupiter's cycle :—

(1) A year of the southern cycle (Table I), which is merely a solar year with a Jovian *name*.

(2) A year of the northern cycle (Table XIV), which is the Jovian year actually completed at the *beginning* of a solar year.

(3) A year of the northern cycle (Table XIV), being the Jovian year actually completed at the actual *moment* we are dealing with.

The context alone must enable us to conjecture which of these meanings is to be attached to a citation of a Jovian year.

**101**. Also, where the second of the three modes of citation is followed, it is evident that if any other Jovian year has had its beginning *as well as its end* in the course of a solar year, that Jovian year will be *kshaya* or suppressed.

Hence it is that certain Jovian years are actually suppressed in the northern system of reckoning. The suppressed years are shown by blanks (......) in Table XIV.

For instance, the Jovian equivalent of A.D. 1514 (solar year) is 34·8317
+ 14·1638
48·9955

And the Jovian equivalent the next year, of A.D. 1515* (solar year) is 34·8317 + 15·1755 = 50·0072. That is, the 48th Jovian year, **Vrisha** was current † at the beginning of Solar Year A.D. 1514* and the 50th Jovian year (**Subhanu**), was current † at the beginning of Solar Year A.D. 1515*. Therefore the 49th Jovian year **Chitrabhanu** which was not used in this particular cycle of 60 years, was *kshaya* or suppressed. Of course there would be no suppression of a Jovian year if the third mode of citation was always used: but it is evident that the third mode, which requires an exact knowledge of the actual beginning and end of a Jovian year and which might be used by an author, would not be in *popular* use.

**102**. *Fourth Mode of Citation* (4). There is a fourth mode of citation of Jovian years which we often find in Malayâlam inscriptions, and that is to cite, instead of one year out of every sixty, one year out of every twelve. In twelve Jovian years the planet Jupiter passes through the twelve signs of the zodiac and each sign corresponds to one of the twelve years of Jupiter's revolution. In the above case Jupiter having completed 9 Jovian years ‡ would be described as being in the 10th sign of the zodiac, i.e., *Makara* (Eye-table.)

**103**. *Fifth Mode of Citation* (5). In this the 12 Jovian years receive the names of the 12 lunar months, *Kârttika* being the first and *Asvina* the 12th. The Jovian years § (properly, months) are distinguished from lunar months by the prefix *Mahâ*. Thus, Mahâ-Kârttika, Mahâ-Mârgaśîra, etc. In the example in sec. **98**, Jupiter was in Mahâ-Śrâvana.

**104**. *Sixth Mode of Citation* (6). This is the heliacal rising system, in which the Jovian cycle begins with the heliacal rising of Jupiter, *i.e.*, his re-appearance after a conjunction with the sun. The system is exhaustively dealt with by Mr. Dikshit in his introduction to Vol. III of Dr. Fleet's *Corpus Inscriptionum Indicarum*, and all the citations of this kind, hitherto found in inscriptions, are there collected.

**105**. Readers in Southern India will be mainly interested in the 1st and 4th modes of citation, of which the latter is otherwise called the "mean-sign system"§ and which is met with in Travancore inscriptions, *e.g.*, those discussed by Dr. Schram in *Indian Antiquary* Vol. XXV (1896), pp. 9—11; and by Professor Kielhorn at pp. 53, 174. *Ibid*.

We shall cite an example: "Kollam 389, Jupiter in Kumbha, and the Sun 18 days old in Mîna, THURSDAY, Pushya Nakshatra, 10th Lunar day".

Kollam 389 is equivalent, as we shall see in sec. **112** *infra* to A.D. 389 + 825 = A.D. 1214 (Indian Solar Year).

We might calculate the day of the solar year in one of two ways, either by taking the equivalent of the solar date from Table VIII or the Eye-table or by calculating the solar date of the tithi. The first is the simpler method and it gives us 335 + 18 = 353 days.

* For convenience of reference, we use the expressions "Solar Year A.D. 1514" to mean "the Indian Solar Year which began in A.D. 1514, *i.e.*, Saka 1436=Vikrama 1571=Kaliyuga 4615, which began in March, A.D. 1514".

† The Jovian year *current* at a particular moment is, as the reader will have observed from the example already given, the Jovian year actually *completed* at that moment, omitting fractions. As no other Jovian year is ever referred to in the northern system, we need not be careful to distinguish between *expired* and *current* Jovian years, but may use the expressions as synonymous.

‡ "Jovian year" does not mean the period of revolution of Jupiter's orbit (which is a little less than 12 solar years) but one-twelfth of that period, which is roughly equivalent to a solar year.

§ In Part IV. of this book on Planetary Chronology, the reader will find a direct method of calculating either the mean or the actual sign or rasi occupied by Jupiter at any moment. (See Sec. **288**).

| The Jovian equivalent of 1214 A.D. is 31·3217 + 14·1638 = | 45·4855 | (Sec. 98.) |
|---|---|---|
| " " 350 days is (35 days × 10) = | ·9694 | Table V. |
| " " 3 " | ·0083 | do. |
| | 46·4632 | |

Dividing the total equivalent by 12, we have as remainder 10·4632, *i.e.*, Jupiter has left the 10th sign and is in the 11th, *i.e.*, Kumbha (4th mode of citation). Or, we can calculate Jupiter's mean sign for 353 days of A.D. 1214 direct from Table XVII (pp . 202, 203), thus : 220·10° + 64·86° + 29·33° = 314·29°, *i.e.*, (by Eye-Table) Jupiter was in *Kumbha.*

**106.** The equivalent in the Northern Cycle of the 46th Jovian year is (by Table XIV) *Pramathin* (3rd mode of citation), while in the Southern Cycle (1st mode of citation) the Jovian name for the solar year A.D. 1214 is (by Table I) *Bhava.* The Jovian year (Northern System) *current* at the beginning of the solar year A.D. 1214 was No. 45 *Bahudhanya,* which would be cited in cases where the 2nd mode of citation was in use.

## CHAPTER XIV.

### THE VARTTAMANA, PURNIMANTA AND KARTTIKADI SYSTEMS OF CITATION.

**107.** The ordinary or normal mode of citing an Indian date is by means of *gata* or 'expired' years (already explained, sec. **5** *supra*), *amanta* months (or months beginning with the first tithi after new moon or *amâvâsyâ*), and *Chaitrâdi* years (or lunar years beginning with *Chaitra*). The lunar year can begin with Chaitra only when the solar year begins with Mesha Sankrânti. Therefore a reckoning by Chaitrâdi lunar years is the same as reckoning by Meshâdi Solar Years.

**108.** Exceptional modes of citation, used at certain epochs of history, or still used in certain parts of India, are :—

*Varttamâna*, or current years, opposed to *gata* (expired); *Pûrnimânta* months, beginning with the tithi after full moon; and *Kârttikâdi* lunar years (or "southern" years) beginning with the lunar month *Kârttika.*

**109.** In the *Pûrnimânta* system, each month beginning with a full moon is named after the next *amânta* month, but takes in the dark fortnight preceding each new moon. It is to this system that we owe the name *paksha,* lit. a wing, the dark and bright fortnights being the wings on either side of a new moon.

**110.** In the *Kârttikâdi* lunar year system, the year beginning with lunar Kârttika takes the number of the Meshâdi solar year then running, and has in common with it the five months, Lunar Kârttika to Lunar Phâlguna.

*Canons.*

I. A *dark* fortnight in *n*th Amânta month is in $(n+1)$th *Pûrnimânta* month.

II. Any month or tithi in *N*th year, expired, is in $(N+1)$th year *current.*

Chaitra
Jyeshtha
Ashadha
Sravana
Bhadrapada
Asvina

III. Any of the months named in the margin in *N*th year Chaitrâdi, expired, is ascribed to the

(*a*) $(N-1)$th year *Kârttikâdi, expired,* or to

(*b*) *N*th year, *Kârttikâdi, current.*

*N.B.*—Conversely, any of the above months in *n*th year *Karttikadi*, expired, is ascribed to $(n+1)$th year *Chaitradi*, expired.

For clearness' sake, we will present these three rules under three aspects, one of which, when we are in doubt, every problem before us must assume.

*First Aspect.*—If we have before us a dark fortnight or *bahula* tithi of the *n*th month in the *N*th year, known to be an expired year, but do not know whether the system is *amânta* or *pûrnimânta*, we must search for the tithi both in the *n*th and the $(n-1)$th month in our ordinary (Amânta) tables.

*Second Aspect.*—If we have before us a dark fortnight or *bahula* tithi of the *n*th month in the *N*th year, and do not know whether the month is *amânta* or *pûrnimânta*, nor whether the year is expired or current, we must, in using our tables, look for the tithi—

(1) in the *n*th month, *N*th year, expired;

(2) in the $(n-1)$th month, *N*th year, expired;

(3) in the *n*th month, $(N-1)$th year, expired; and

(4) in the $(n-1)$th month, $(N-1)$th year, expired.

*Third Aspect.*—If we have before us a dark fortnight or *bahula* tithi of the *n*th month (from Chaitra to Âśvina, inclusive) in the *N*th year and do not know whether the month is *amânta* or *pûrnimânta*, nor whether the year is expired or current, *Chaitrâdi* or *Kârttikâdi*, we must, in using our tables, look for the tithi—

(1) in the *n*th month, *N*th year, expired;

(2) in the $(n-1)$th month, *N*th year, expired;

(3) in the *n*th month, $(N-1)$th year, expired;

(4) in the $(n-1)$th month, $(N-1)$th year, expired;

(5) in the *n*th month $(N+1)$th year, expired; and

(6) in the $(n-1)$th month $(N+1)$th year expired.

*N.B.*—For bright fortnight or *Sukla* dates in any month, no search need be made in any case in the $(n-1)$th month; and whether for bright or dark fortnight dates in the months of Karttika to Phalguna inclusive, no search need be made in any case in $(N+1)$th year, *expired*.

Readers who prefer Professor Kielhorn's mode of stating the same rules will find it in ***Ind. Antiq.***, Vol. XIX, p. 22 or at p. 406 of ***Ep. Ind.***, Vol. I. Any rules must be somewhat perplexing at first sight and should be carefully reasoned out by those interested in such investigations.

**111.** The Lunar year beginning with lunar Chaitra of any solar year *n* is itself $(n+1)$. Thus, Chaitra 1265 may mean:—

(*a*) Lunar Chaitra at the end of current solar 1264, *i.e.*, at the end of expired solar 1263.

(*b*) Do. do. do. expired solar 1264 and before the beginning of solar 1265. (expired)

(*c*) Do. do. do. do. do. 1265 and do. 1266 do.

This ambiguity is independent of the difference between Chaitrâdi and Kârttikâdi systems, and is peculiar to the lunar month Chaitra.

## CHAPTER XV.

### THE ERAS NOW IN USE OR MET WITH IN INSCRIPTIONS.

**112.** In the following Summary of Indian Eras, adapted from Cunningham and other authorities, we give in each case the formula whereby a year of each era, "expired" or "current" as the case may be, may be obtained from the A.D. year, also noting where the era is or was in use, and whether it is amânta or pûrnimânta, Chaitrâdi or Kârttikâdi and so forth.

(1) *Kaliyuga* (expired year=3102 *minus* B.C. year; A.D. year+3101). Used all over India as a solar (Meshâdi) and luni-solar (Chaitrâdi) year. The years in this era, with A.D., Śaka, and Vikrama equivalents, are regularly given in our Table X from B.C. 1 to A.D. 2000; and in Table XII from A.D. 1840 to A.D. 1920. For the years B.C. 3102 to B.C. 1, K.Y. 0 to K.Y. 3101, see Addendum, p. (114.)

(2) *Saptarshi* (current year =A.D. year+76). Mostly *current* and mostly *pûrnimânta* and *Chaitrâdi*. At present in use in Kashmir and formerly in Multan and elsewhere. Quoted generally in cycles of 100 years. Thus A.D. 1910=Saptarshi 1986, current, quoted only as "86 Saptarshi, current."

(3) *Vikrama Era* (expired year=A.D. year +57). Extensively used at present in Guzerat and all over Northern India, except Bengal: *Kârttikâdi* and *Amânta* in Guzerat, hence called *Southern Vikrama; Chaitrâdi* and *Pûrnimânta* in N. India, hence called *Northern Vikrama.* Other names, "Malava Era"; and also simply "Samvat". The Vikrama years, expired, from B.C. 1 to A.D. 2000 are regularly shown in our Table X, and from A.D. 1840 to A.D. 1920 in our Table XII.

(4) *Saka Era* (expired year =A.D. year *minus* 78). Extensively used in India both in the past, and at present. *Meshâdi* (for solar year) and *Chaitrâdi* (for luni-solar year): generally *amânta*, but *pûrnimânta* in Northern India. Śaka equivalents of Kaliyuga and Vikrama years, expired, from B.C. 1 to A.D. 2000, are regularly given in our Table X, and from A.D. 1840 to A.D. 1920 in our Table XII.

(5) *Kollam Era* (current year=A.D. year *minus* 825). Used in Malabar, Cochin, and Travancore since A.D. 825. Begins in N. Malabar with solar *Kanya* and in S. Malabar with *Chingam* (Simha); (For memory aids, see sec. **138** *infra*); always *solar* and *current.* For years from A.D. 1840 to A.D. 1920, Kollam equivalents are regularly given in our Table XII. For previous years, the following century equivalents will be of use:—

| A.D. | Kollam | A.D. | Kollam |
|---|---|---|---|
| 900= | Kollam 75 | 1500= | Kollam 675 |
| 1000= | „ 175 | 1600= | „ 775 |
| 1100= | „ 274 | 1700= | „ 875 |
| 1200= | „ 375 | 1800= | „ 975 |
| 1300= | „ 475 | 1900= | „ 1075 |
| 1400= | „ 575 | 2000= | „ 1175 |

(6) *Bengali San* (or *Sen*) (current year=A.D. year *minus* 593). *Meshâdi, current solar year.* Now in use in Bengal. For A.D. 1840 to A.D. 1920, B. San equivalents are regularly given in our Table XII. For previous years the following century equivalents will be of use:—

| A.D. | B.San | A.D. | B.San | A.D. | B.San |
|---|---|---|---|---|---|
| A.D. 600= | B.San 7 | A.D. 1100= | B.San 507 | A.D. 1600= | B.San 1007 |
| „ 700= | „ 107 | „ 1200= | „ 607 | „ 1700= | „ 1107 |
| „ 800= | „ 207 | „ 1300= | „ 707 | „ 1800= | „ 1207 |
| „ 900= | „ 307 | „ 1400= | „ 807 | „ 1900= | „ 1307 |
| „ 1000= | „ 407 | „ 1500= | „ 907 | „ 2000= | „ 1407 |

(7) *Chedi* or *Kalachuri Era* (current year=A.D. year *minus* 247), *current, Asvinâdi, pûrnimânta.*

| A.D. 300=Chedi 53 current. | A.D. 800=Chedi 553 current. |
|---|---|
| " 400= " 153 " | " 900= " 653 " |
| " 500= " 253 " | " 1000= " 753 " |
| " 600= " 353 " | " 1100= " 853 " |
| " 700= " 453 " | " 1200= " 953 " |

Not now in use, but was in use under the Kalachuri kings in Western and Central India.

(8) *Gupta* (current year=A.D. year *minus* 319).

| A.D. 400=Gupta 81 current. | A.D. 600=Gupta 281 current. |
|---|---|
| " 500= " 181 " | " 700= " 381 " |

Not now in use. Was in use in Central India and Nepal in the centuries above indicated as *current, Chaitrâdi, pûrnimânta.*

(9) *Valâbhi Era* (current year=A.D. year *minus* 318). *Current, Kârttikâdi, amânta,* or *pûrnimânta.*

| A.D. 400=Valâbhi 82 current. | A.D. 900= Valâbhi 582 current. |
|---|---|
| " 500= " 182 " | " 1000= " 682 " |
| " 600= " 282 " | " 1100= " 782 " |
| " 700= " 382 " | " 1200= " 882 " |
| " 800= " 482 " | " 1300= " 982 " |

Not now in use. Was in use in Kathiavâd and neighbourhood in centuries above indicated.

(10) *Vilâyati year* (current year=A.D. year *minus* 592). *Kanyâdi, solar, current.*

Now in use in Orissa. Months begin invariably *on* the day of Sankrânti instead of the next or the third day after [see Sec. **144** (1) *infra*].

This era bears to Bengali San pretty much the same relation that the Valâbhi bore to the Gupta era.

(11) *Amli year* (current year=A.D. year *minus* 592). *Luni-solar, current.* Lunar year changes on *Bhâdrapada sukla 12th,* i.e., some days before, or some days after, Kanyâ Sankrânti. Otherwise the same as Vilâyati. Current in Orissa. Probably bears the same relation to Vilâyati that the luni-solar year in the Telugu country bears to the solar year of the Tamil country (*i.e.,* only New Year Day is different in each case).

(12) *Fasali year* (luni-solar) in Bengal, and N.W. India (current year=A.D. year *minus* 592); *pûrnimânta and lunar Asvinâdi* (*i.e.,* it begins on pûrnimânta Âśvina, bahula pratipada); luni-solar year for Bengal, just as the Bengal San is the solar year for the same tract of country. Like all luni-solar years, the *Fasali* takes the number of the next Solar San. Thus, A.D. 1900 was, as we have seen, Bengal San 1307 current, but the luni-solar *Fasali* beginning on *Asvina krishna pratipada of* A.D. 1900 takes the number of the next Bengal San, *i.e.,* 1308 current. Hence the formula for Bengal *Fasali* is "current A.D. year *minus* 592".

(13) *Dakhan Fasli* (current year = A.D. year *minus* 593) was up to A.D. 1556 (when its number was 963, *vide* Table XV, p. 195) a Hijra year, but from that year forward it has been kept as a solar year (identical in number with Bengal San), beginning, in parts of Bombay, when the Sun enters the Nakshatra *Mrigasira,* i.e.* (now) about 7th or 8th June, The months, their periods of beginning, and number of days, are the same as in the Muhammadan calendar.

(14) The *Madras Fasli* year is an *agricultural, solar, tropical* year, beginning on 1 July and having no divisions of its own into months and days. Its formula is "current Fasli = A.D. year *minus* 590". The agricultural or revenue year from 1 July 1910 to 30 June 1911 might be expressed quite as well by the expression "Revenue Year 1910–11" as by the expression "Fasli 1320". The former expression would convey better sense to most people, and the expression "Fasli year", the use of which a blind tradition has consecrated in Madras, serves no other purpose than to confuse any one who is not a village karnam. It is a chronological anomaly and will no doubt gradually drop out of use.

(15) The Mahratta *Sûr-San, Shahûr-San* or *Arabi-San* (current year = A.D. year *minus* 599) was extensively used under the Mahratta supremacy, and is still occasionally met with. Begins, like the Dakhan Fasali, when the Sun enters the Nakshatra Mrigaśira.*

(16) *Harshakâla Era* (current year = A.D. year *minus* 606). Was current in Muttra (*Mathura*) in Kanauj, Nepal. Harsha 94 current = A.D. 700, and so on.

| | |
|---|---|
| (17) *Nevar Era* (used in Nepal from A.D. 878 to A.D. 1768), *expired, Kârttikâdi, Amânta.* Formula "Nevar expired = A.D. year *minus* 879". | Nevar 21, expired = A.D. 900.<br>Nevar 821, expired = A.D. 1700. |

(18) *Chalukya Era* (expired year = A.D. year *minus* 1076). This was in use only from A.D. 1079 to A.D. 1162. It followed the Śaka months and pakshas. Chalukya 24 expired = A.D. 1100.

(19) *Lakshmana Sena Era :* (expired year = A.D. year *minus* 1108, according to present day almanacs). Now in use in Tirhut and Mithila, but along with the Śaka or Vikrama Era.

Dr. Kielhorn concludes that the formula for this era was (from A.D. 1194 to A.D. 1551). "Expired Lakshmana Sena year = A.D. year *minus* 1119". According to this formula A.D. 1200 would be equivalent to Lak. Sena 81 expired. The years of this era were *expired, Kârttikâdi, amânta.*

(20) *Mahratta Raja Saka Era* (current year = A.D. year *minus* 1673) was established on the accession of Śivâji: the number of the year used to change every *Jyeshtha sukla trayodasî*, which was the tithi or date of Śivâji's accession. Otherwise same as southern luni-solar amânta Śaka years.

(21) *Tarikh-i-Ilahi* (= mighty or divine era), dating from the accession of Akbar (14th February 1556), fell into disuse *temp.* Shah Jehan. Days and months were solar, without intercalations.

---

* By Eye-Table, the Sun enters Nakshatra Mrigasira when his longitude is 53° 20′. The number of days corresponding to this (by Table XVII-A) = 56. When the solar year commences (as it did in A.D. 1911) on 13th April, the sun's entry into Mrigasira Nakshatra would be on 56 + 13 = 69 days from 1st April, *i.e.*, (Table VIII) on 8 June.

## CHAPTER XVI.

### NOTES ON TITHIS IN RELATION TO FESTIVALS.

**113. Day on which a festival is celebrated.**—As a *tithi* generally covers a portion of two days, it sometimes happens that though for civil purposes the tithi of a day is that which is current at sunrise, yet for religious purposes a tithi may have to be celebrated on the previous day when it begins. When a tithi is appointed for the celebration of a feast or fast, to be kept at forenoon, midday, late afternoon, midnight, etc., it is obvious that the feast or fast must be observed on the day when the tithi covers the prescribed part of the day. The day for such purposes is divided, first of all, into five portions *between sunrise and sunset.*

(1) *Prâtahkâla,* or early forenoon, 6 ghaṭikas from sunrise.
(2) *Samjava,* or forenoon, 6 to 12 ghaṭikas from sunrise.
(3) *Madhyâhna,* or midday, 12 to 18 ghaṭikas from sunrise.
(4) *Aparâhna,* or afternoon, 18 to 24 ghaṭikas from sunrise.
(5) *Sâyâhna,* or late afternoon, 24 to 30 ghaṭikas from sunrise.

(*a*) The 4 ghaṭikas before sunrise are called *arunodaya* or rise of dawn.
(*b*) The 6 ghaṭikas after sunset are called *pradosha* or evening.
(*c*) The 2 ghaṭikas in the middle of the night are called *nisitha,* midnight.
(*d*) A tithi is *pûrvaviddha* when it commences more than 4 ghaṭikas before sunset of one day and ends before sunset of the following day; a festival on such a tithi is celebrated on the first day of the tithi, not on the second.

*N.B.*—**Tithi dvayam**:—Two tithis meeting, *i.e.*, one commencing, and the other ending, between 18 and 24 ghatikas after Sunrise, when a similar meeting does not take place next day.

**114. Festivals connected with Nakshatras as well as Tithis.**—On the Nakshatra Śravishta (No. 23) in Lunar Śrâvana, the ceremony of Upakarma (Âvani Avittam) is celebrated. Our Table XI shows that this festival would ordinarily fall about full moon in Śrâvana. In the same manner, all festivals associated with particular *nakshatras* may be equally well connected with particular *tithis.* In Southern India, however, it is usual to connect *nakshatra* festivals with certain solar months in which they are celebrated; e.g., *Panguni Uttaram* (i.e., *Uttara Phalgunî* Nakshatra in the Solar month called *Panguni*), *Chittirai Mûlam* (i.e., *Mûla* Nakshatra in the Solar month called *Chittirai*); *Avani Avittam* (*vide supra*), etc. *Tithi* festivals are also often connected in Southern India with Solar Months; and if a festival falls, for instance, on *sukla* panchamî in a particular Solar Month, and there are two Śukla panchamîs in that month, the first is called *Sûnya tithi* and the second only is celebrated.

**115.** Most Hindu festivals are associated in the first place with *tithis*, and the following notes for which we are chiefly indebted to Prof. Kielhorn's article in Vol. XXVI of the *Indian Antiquary*, pages 177—187, will be of interest to Hindu readers as well as to others. Prof. Kielhorn described the Hindu festivals in order of the lunar months: by way of variation, we will discuss them in the order of tithis. Our arrangement will enable the reader to locate at once an *ashtamî* or an *ekâdasî*, although the month is, as usual, suppressed in such citations.

### PRATIPADA (TITHI I).

The **Chaitra** Sukla *pratipadâ*, i.e., that which precedes the Mesha Sankrânti, is the beginning of the Hindu Lunar year, New Year's Day (Lunar) being that on which the pratipadâ is current at sunrise. (When there is an *Adhika Chaitra*, that begins the year). The tithi is therefore called *Vatsarârambha* (commencement of the year). It is also *Navarâtrârambha*, there being another *Navarâtrârambha* on *Asvina sukla pratipadâ*.

**Karttika** Śukla 1 is *Balipratipadâ* or *Bali pûja* and it is pûrvaviddha as to time.

**Bhadrapada** bahula 1 is *Mahâlayârambha*.

**Phalguna** bahula 1 is *Vasantotsava*.

### DVITIYA (TITHI II).

**Ashada** Śukla 2 is *Ratha-yâtra dvitîyâ* or *Râma-rathotsava*.

**Karttika** Śukla 2 is *Yama dvitîyâ* or *Bhrâtri-dvitîya* (because sisters make presents to brothers), and the time is afternoon.

The bahula dvitiya in **Ashada, Sravana, Bhadrapada** and **Asvina** is called *Asûnyasayana-vrata* and the fast is broken at moon-rise.

### TRITIYA (TITHI III).

**Chaitra** Śukla 3 is *Gaurî-tritîyâ ;* also *Matsa-jayantî ;* (afternoon); also *Manvâdi* (forenoon).

**Vaisakha** Śukla 3 is *Kalpâdi* (forenoon); *Tretâyugâdi* (forenoon); *Akshaya-tritîyâ* (especial, when combined with Wednesday and Nak. Rohinî; time, forenoon); also *Parasurâma jayanti*.

**Jyeshtha** Śukla 3 is *Rambhâ-tritîyâ*, when Bhavâni is worshipped at *purvaviddha*.

**Sravana** Sukla 3 is *Madhu-sravâ* in Guzerat.

**Sravana** bahula 3 is *Kajjali-tritîyâ*.

**Bhadrapada** Sukla 3 is *Varâha-jayanti* (afternoon); *Haritâlikâ*, when Parvati is worshipped; *Manvâdi* (forenoon). The tithi is by some called *Sivâ*.

**Phalguna** bahula 3 is *Kalpâdi* (afternoon).

### CHATURTHI (TITHI-IV).

The Sukla Chaturthî in **every month** is called *Ganesa Chaturthî* or *Vinâyaka Chaturthî*, the chief being *Mâgha Chaturthî* (*Ganesa jayantî*). It is celebrated at midday. *Tila Chaturthî* is another name for **Magha** Śukla *chaturthî*. It is performed in the evening *Kunda Chaturthî* is another name for the same festival.

**Bhadrapada** Sukla Chaturthî is special when it falls on Sunday or Tuesday.

Similarly, the bahula Chaturthî in **every month** is *Sankashtachaturthî* and is a fast day for people in difficulties: the fast is broken at moonrise which is the time. This Chaturthî is called *Angâraka Chaturthî* if it falls on Tuesday, and continues till moonrise.

**Sravana** bahula Chaturthî is called, par excellence, **Bahula** *Chaturthî* and cows are then worshipped.

### PANCHAMI (TITHI V).

**Chaitra** Sukla 5 is *Sri Panchamî* according to some : it is also *Kalpâdi* (forenoon).

**Sravana** Śukla 5 is *Nâga Panchamî* and snakes are then worshipped. If tithi begins within 6 ghaṭikas after Sunrise of one day and ends within 6 ghaṭikas after Sunrise on the next day, the tithi is celebrated on the former day and that is Nâga Panchamî.

**Bhadrapada** Sukla 5 is *Rishi Panchamî* (midday).

**Asvina** Sukla 5 is *Lalitâ Panchamî* or *Upânga-lalitâ-vrata*, when Durgâ is worshipped in the afternoon.

**Margasira** Śukla 5 is *Nâgapûja* or *Nâgapanchamî*.

**Magha** Śukla 5 is *Vasanta panchamî*, when Rati and Kâma are worshipped in forenoon. *Sripanchamî* is another name.

**Phalguna** bahula 5 is *Ranga panchamî*, when colours are thrown about.

### SHASHTHI (TITHI VI).

**Sravana** Śukla 6 is *Kalki-Jayantî* (sunset), the last avatâra of Vishnu.

**Sravana** bahula 6 is *Hala-Shashthî.*

**Bhadrapada** Śukla 6 is *Sûrya-Shashthî* or *Skanda-Shashthî.*

**Bhadrapada** bahula 6 is Chandra *Shashthî.* It is called *KAPILA-SHASHTHI* when it combines on **Tuesday** with Nak. Rohinî and Yoga Vyatîpâta and the Sun is in Hasta. (See Sec. **54**).

**Karttika** Śukla 6 is special for feeding of Brahmans when it falls on Tuesday.

**Margasira** Śukla 6 is *Skanda-Shashthî* or *Mahâ-Shashthî; Champâ-Shashthî,* when Siva is worshipped as Khandoba. This tithi is special when it falls on Sunday or Tuesday and combines with Nakshatra Satabhishaj and Yoga Vaidhriti, or either of the two.

[*N.B.*—Our Table XI shows that Nak. Satabhishaj (No. 24) may concur with the tithi in a year in which some month before Margasira is 'adhika'; but Yoga Vaidhriti (No. 27) can never concur with this tithi.]

### SAPTAMI (TITHI VII).

A *Saptamî* on Tuesday, combined with Nak. Revati (that is, by Tab. XI, **bahula** 7 in Âshâdha or **Sukla** 7 in Pausha or Mâgha) is very auspicious. A *Sukla saptami,* falling on Sunday, is called *VIJAYA*, and is special for donations. A *Sukla saptami,* joined with the 1st quarter of Nak. Hasta, is called *BHADRA.* A *Sukla saptami,* coinciding with a Sankrânti is called *MAHAJAYA*, which, for making donations, is superior even to an eclipse.

**Vaisakha** Śukla 7 is *Gangâ-Saptamî* or *Gangotpatti* (birth of Gangâ—midday).

**Sravana** bahula 7 is *Sîtalâ-Saptamî* or *Sîtalâ-Vrata*, time *purvaviddha.*

**Bhadrapada** Sukla 7 is called by some *Aparâjitâ.*

**Asvina** Śukla 7 : about this tithi *Sarasvatî* is worshipped under Nakshatra Mûla.

**Karttika** Śukla 7 is *Kalpâdi* (forenoon).

**Margasira** Śukla 7 is *Sûryavrata.*

**Magha** Śukla 7 is *Ratha-Saptamî* or *Mahâ-Saptamî* (time, *arunodhaya*); *Manvâdi* (forenoon).

### ASHTAMI (TITHI VIII).

An *Ashtamî*, falling on Wednesday, is special and receives the name of *Budhâshtamî.* The *Sukla-Ashtamî* in **every month** is sacred to Durgâ or Annapuranî; while the *Bahula-Ashtamî* in **every month,** called *Kâlâshtamî,* celebrated at purvaviddha, is sacred to Krishna.

**Chaitra** Śukla 8 is *Bhavâni-utpatti:* when joined with Wednesday and Nak. Punarvasu, bathing on this tithi is special.

**Sravana** bahula 8: *Janmâshtamî, Krishnâshtamî* or *KRISHNA JAYANTI* (midnight); special when combined with Nak. Rohinî; less so, when joined only to *Monday* or *Wednesday; Manvâdi* (afternoon).

**Bhadrapada** Sukla 8: *Jyeshthâ-Gaurî-pûjana-vrata,* when moon is in Nak. *Jyeshthâ.*

**Bhadrapada** bahula 8 : *Mahalakshmî vrata* (purvaviddhâ) ; *Ashtaka-srâddha.*

**Asvina** Śukla 8 is *Mahâshtamî* and is special, when joined to Tuesday.

**Karttika** Śukla 8 is *Gopâshtamî*, when cows are worshipped.

**Do.** bahula 8 is *Krishnâshtamî*, *Kâla-Bhairavâshtamî*, or *Kâla-Bhairava-jayantî.*

**Margasira** bahula 8 is *Ashtaka-srâddhâ* in the afternoon. The same is the case with bahula 8 in **Pausha, Magha** and **Phalguna.**

**Pausha** Śukla 8 is special when joined to Wednesday and Nak. Bharani (Rohini or Ârdhra, according to some).

**Magha** Śukla 8 is *Bhîshmâshtamî* and is celebrated at midday.

**Magha** bahula 8 is Birth of Sîtâ.

### NAVAMI (Tithi IX).

**Chaitra** Śukla 9 is *Râma-navamî* or *Râma-jayantî*, at midday.

**Bhadrapada** do. is *Aduhkha-navamî.*

**Asvina** do. is *Mahâ-navamî* or *Durgâ-navamî* : Manvâdi (forenoon).

**Karttika** do. is *Kretayugâdi* (forenoon).

**Margasira** do. is *Kalpâdi* (forenoon).

**Magha** bahula 9 is *Râmadâsa Navamî.*

### DASAMI (Tithi X).

**Jyeshtha** Śukla 10 is *Dasa-harâ* (expiating ten sins) (see Sec. **54**) or *Gangâvatâra.*

**Ashada** do. is *Manvâdi* (forenoon).

**Asvina** do. is *Vijayadasamî* or *DASARA* (afternoon ; special with Nak. Sravana) ; *Buddha-jayantî.*

### EKADASI (Tithi XI).

Every *Ekâdasî* is sacred, like every *Amâvâsyâ*, and receives a special name. It is called **Vijaya** when joined with the Nak. *Punarvasu.* The following are the names of the 24 Ekâdasîs :—

| *Month Sukla.* | *Bahula.* | *Month Sukla.* | *Bahula.* |
|---|---|---|---|
| **Chaitra.**—*Kâmadâ Ekâdasî.* | *Varûthinî ekâdasî.* | **Bhadrapada.**-*Vishnu parivartanotsava* or Parivartinî Ekâdasî. (Vishnu turning on his side : called **Vishnu srinkhala,** when 11th and 12th tithis meet in Nakshatra *Sravana.* | *Indirâ Ekâdasî.* |
| **Vaisakha.**—*Mohinî do.* | *Aparâ do.* | | |
| **Jyeshtha.**—*Nirjalâ do.* | *Yoginî do.* | | |
| **Ashada.**—*Vishnusayanotsava ; Sayani* or *Vishnusayani Ekâdasî* (*i.e.*, Vishnu going to sleep). | *Kâmadâ* or *Kâmikâ Ekâdasî.* | | |
| **Sravana.**—*Putradâ Ekâdasî.* | *Ajâ Ekâdasî.* | | |

7

| *Month Sukla.* | *Bahula.* | *Month Sukla.* | *Bahula.* |
|---|---|---|---|
| **Asvina.**—*Pâsânkusâ Ekâdasî.* | *Ramâ Ekâdasî.* | **Pausha.**—*Putradâ Ekâdasî* or *MUKKOTTI* or *VAIKUNTHA EKADASI* *Manvâdi* (forenoon). | *Shattilâ Ekâdasî.* |
| **Karttika.**—*Prabodhini Ekâdasî* (Waking of Vishnu), *Bhishmapanchaka vrata* commences. | *Utpatti Ekâdâsi.* | **Magha.**—*Jayâ Ekâdasî.* | *Vijayâ Ekâdasî.* |
| **Margasira.**—*Mokshadâ Ekâdasî.* | *Saphalâ Ekâdâsî.* | **Phalguna.**—*Amalakî Ekâdasî.* | *Pâpamochinî Ekâdasî.* |

### DVADASI (Tithi XII).

The *Dvadasi* is called *Mahâ-dvâdasî* in the following circumstances :—

11th Tithi current at sunrise on two successive days : the next Dvâdaśî is called **Unmilani.**

12th Do. do. do. do. : the Dvâdaśî is called **Vanjuli.**

12th Tithi, to be followed by a full moon or a new moon tithi, current at two sunrises —**Pakshavardhini.**

12th Tithi, joined with Nakshatra *Pushya,* is **Jaya** [*i.e.,* by Tab. XI, Bahula 12 in Bhâd ; or Sukla 12 in Phâlg. or Chait.]

Do. do. do. do. do. *Sravana* is **Vijaya** [*i.e.,* by Tab. XI, Śukla 12 in Bhâd., Asv. ; Bahula 12 in Phâlg.]

Do. do. do. do. do. *Punarvasu* is **Jayanti** [*i.e.,* by Tab. XI, Śukla 12 in Phâlg. ; Bahula 12 in Śrâvana].

Do. do. do. do. do. *Rohini* is **Papanasini** [*i.e.,* by Tab. XI, Śukla 12 in Pausha, Mâgha ; Bahula 12 in Âshâḍha].

**Vaisakha** Śukla 12, joined with Nak. *Hasta,* and Jupiter and Mars in *Simha,* Sun in *Mesha,* is *VYATIPATA.*

**Ashada** Śukla 12 is the commencement of *Châturmâsya vrata.*

**Sravana** do. is *Vishnoh-pavitrâ-ropanam.*

**Bhadrapada** do. is *Vâmana jayantî* (midday) ; called **Sravana Dvadasi,** when joined with Nak. *Sravana,* and specially, when further joined with *Wednesday.*

**Asvina** bahula 12 is *Govatsa Dvâdasî* (evening).

**Karttika** Śukla 12 (1) end of *Châturmâsya vrata,* which began on same tithi in Âshâḍha.

(2) *Prabodhotsava* or *Utthâna dvâdasî* (preparation for waking Vishnu).

(3) *Tulasi vivâha* (marriage of Vishnu with the Tulasi plant).

(4) *Manvâdi* (forenoon).

**Magha** Śukla 12: *Bhîshma dvâdasî.*

Do. bahula 12: **Tila dvadasi** or **Vijaya,** when combined with Nak. *Sravana.*

[*N.B.*—This can only happen when a month previous to *Magha* is Adhika—See Table XI.]

**TRAYODASI (Tithi XIII).**

**Chaitra** Śukla 13: *Madana trayodasî* or *Ananga Pûjana vrata* (purvaviddha); god of Love worshipped.

**Bhadrapada** bahula 13: (1) *Kaliyugâdi* (afternoon).

(2) **Magha trayodasi**, when combined with Nak. *Magha.*

(3) *GAJACHAYA*, when joined to Nak. *Magha* and Sun in *Hasta.*

**Asvina** bahula 13: *Dhana trayodasî,* when money lenders worship money.

**Magha** Sukla 13: *Kalpâdi* (forenoon).

**Phalguna** bahula 13: (1) **Varuni**, when joined with Nak. *Satabhishaj.*

(2) **Mahavaruni,** do. do. do. + *Saturday.*

(3) *MAHA-MAHA-VARUNI* when joined with Nak. *Satabhishaj* + *Saturday* + Yoga *Subha.*

[*N.B.*—Combinations (1) to (3) can happen only when Phalguni or some previous month is Adhika—See Table XI.]

**CHATURDASI (Tithi XIV).**

Bahula Chaturdasî in **every month** is *Sivarâtri.*

**Vaisakha** Sukla 14: *Narasimha jayantî* (sunset): special when joined to Nak. **Svati** + *Saturday.*

**Sravana** Śukla 14: *Varalakshmî vrata.*

**Bhadrapada** Śukla 14: *Ananta chaturdasî.*

**Asvina** bahula 14: *Naraka chaturdasî* (moonrise), fasting to avoid *naraka* or hell. Dîpavali (see *Asvina amâvâsyâ* below), may fall on this tithi if joined to Nak. Svâti.

**Karttika** Śukla 14: *Vaikunta chaturdasî* (midnight).

**Margasira** Śukla 14: *Pâshâna chaturdasî.*

**Magha** bahula 14: *MAHA SIVARATRI* (midnight, when Nak. Sravana is current).

Special, when, combined with *Sunday* or *Tuesday* and Yoga *Siva.*

[N.B.—*Yoga Siva can combine with Sivaratri only when some month previous to Magha is Adhika*· See Table XI.]

**SUKLA PANCHADASI (Tithi XV): Sukla 15 or Full Moon** (*Pûrnimâ*).

A Śukla 15 or Pûrnimâ is called *Somavatî* when it falls on Monday and is special for donations.

It is called **Chudamani** when it is *further* joined with a lunar eclipse.

Most pûrnimâs receive special names, which are given below :—

**Chaitra** pûrnimâ : (1) *Manvâdi* (forenoon).

(2) *Hanumat-jayantî.*

(3) Special for bathing, when combined with Sunday, Thursday or Saturday.

**Vaisakha** pûrnimâ : *Kûrma jayantî* (late afternoon).

**Jyeshtha** pûrnimâ : (1) *Manvâdî* (forenoon).

(2) *Vata pûrnimâ* or *Vata Sâvitrî* (purvaviddhâ).

(3) **Maha-jyaishthi,** when Moon and Jupiter are in Nak. *Jyeshthâ* and Sun in Rohinî.

**Ashada** pûrnimâ : (1) *Manvâdi* (forenoon).

(2) *Sivasayanotsava,* or *Kokilâ-vrata* or *Vyâsapûjâ.*

**Sravana** pûrnimâ : (1) *Rig-yajuh-Srâvanî* (*i.e.*, Srâvana for followers of Rig and Yajur Vedas).

(2) *Rakshabandhana* (tying a string round the arm) or *Raki-pûrnimâ* or *Nâralî pûrnimâ* (throwing cocoanuts into the sea).

(3) *Hayagriva jayantî.*

**Bhadrapada** pûrnimâ : *Praushthapadî pûrnimâ* or *srâddha.*

**Asvina** pûrnimâ : (1) *Kojâgarî pûrnimâ* or *Kojâgara vrata* (midnight). Lakshmi and Indra worshipped ; games of chance.

(2) *Navânna pûrnimâ,* when new grain is cooked.

**Karttika** pûrnimâ : (1) *Manvâdi* (forenoon).

(2) *Châturmâsya vrata* ends.

(3) *Tripurî pûrnimâ* or *tripurotsava.*

(4) Special, when joined to Nak. **Krittika.**

(5) **Maha karttiki,** when joined to. Nak. *Rohini*; or when Moon and Jupiter are both in Nak. *Krittika.*

(6) *Padmakayoga,* when Moon is in Nak. *Krittika* and Sun in Nak. *Visâkha.*

(7) *KRITTIKA* festival in Southern India.

**Margasira** pûrnimâ : (1) *Datta treya* or *Datta-jayantî* (evening).

(2) Special for donations of salt, when joined to Nak. **Mrigasira.**

(3) **Ardra** festival in honour of Śiva at Chidambaram and elsewhere. Siva, in his incarnation of Nateśa, is said to have been born on Mârgaśira Pûrnimâ under the Nakshatra *Ardra:* the combination could take place only when some month previous to Mârgaśira was *adhika*. (See Table XI).

**Magha** pûrnimâ: **Maha-Maghi,** when Moon and Jupiter are both in Nak. **Magha.**

**Phalguna** pûrnimâ: (1) *Manvâdi* (forenoon).

(2) *Holikâ* or *Hutâsani pûrnimâ* (evening).

(3) *Kâman Pandigai* (destruction of Cupid by Śiva) in South India..

**Bahula Panchadasi** (Tithi XV): **AMAVASYA—New Moon.**

A Solar Eclipse on Sunday is **Chudamani**, and is special for donations.

Amâvâsyâ at the end of **Sravana** and beginning of **Bhadrapada:** *Pithori* or *Kusotpâtini.*

Do. do. **Bhadrapada** do. **Asvina**: *Sarvapitri* or *MAHALAYA ÂMAVASYA:* special when Sun and Moon are both in Nak. *Hasta.*

Do. do. **Asvina** do. **Karttika** is *DIPAVALI*, with previous and following tithis: that on which Nak. is *Svâti* being the special day.

Do. do. **Pausha** do. **Magha** (1) **ARDHODAYA,** when joined with *Sunday* in day time + Nak. *Sravana* + Yoga Vyatîpâta.

[*N.B.*—This can only happen when some month previous to Magha is Adikha. See Sec. **54**].

(2) **Mahodaya** when any one of these special features is wanting.

Do. do. **Magha** do. **Phalguna**: (1) Dvâparayugâdi (afternoon).

(2) Special for Srâddhas, when joined with Nak. *Satabhishaj* or Nak. *Dhanishtha.*

Do. do. **Phalguna** do. **Chaitra**: *Manvâdi* (afternoon).

## CHAPTER XVII.

### MODERN EQUATIONS FOR LUNAR INEQUALITIES.

**116.** In compliance with a desire expressed by many persons interested in the *Drig-ganita* system, it has been thought advisable to include in the present work three very simple tables (XVI-A, XVI-B and XVI-C) for correcting the ending moments of tithis worked out according to the *Sûrya-Siddhânta,* so as to bring them within 2 ghaṭikas of *Drig-ganita* or *Nautical Almanac* results, further approximation not being possible within the limits of a popular work. Seeing that Siddhânta and Drig-ganita panchângams occasionally exhibit a divergence of 17 ghaṭikas (7 hours) in regard to tithis intermediate between new-moon and new-moon, the approximations now made possible will go a great way towards leading to a better understanding between the followers of the two systems.

**117.** Modern Astronomers use five principal equations in order to determine the moon's actual, from her mean, elongation (or tithi). These are (1) the sun's equation of the centre, (2) the moon's equation of the centre, (3) the moon's evection, (4) the moon's variation, (5) the moon's annual equation. Of these (5) is already included in the sun's equation and both are given together in Tables IX-*c* and IX-*h*, (4) depends only on the tithi and is given in the last column of Table XVI-A, while (2) and (3) are given respectively in Tables XVI-B and XVI-C.

**118.** The example given below will show clearly how Table XVI is to be used, but it will be useful to bear in mind the following rules :—

Rule (1) Table IX gives *equated anomalies* (as explained in sec. **251**) for *mean* tithis, *i.e.*, each anomaly given in Table IX-*a*, *b* and *g*, for the moon, contains the equation which has to be separated before applying the equation to the *mean* tithi, whereas the anomaly given in Tables XVI–B and XVI–C is a *mean* anomaly for the *actual* moment of the tithi. The principle of the method is to work until the equation obtained from the *equated* anomaly of Tables IX–*a*, *b* and *g*, is equal to the equation obtained from the *mean* anomaly in Tables XVI–B and XVI–C. Accordingly, Table XVI enables us to ascertain an ending moment not *per saltum*, but gradually and by successive approximations, using the equated anomalies of Tables IX *a*, *b* and *g*, as our base.

Rule (2). Tables XVI-B and XVI-C apply, as they stand, only to the first quarters (0 d. 0 gh. 0 p. to 6 d. 53 gh. 20 p.) of the anomalistic month, when the equations are **negative**. For the 2nd quarter (6 d. 53 gh. 20 p. to 13 d. 46 gh. 40 p.) the equation is **negative**, and the argument should be obtained by deducting the given anomaly *from* 13 days 46 ghat. 40 palas. For the third quarter (13 days 46 gh. 40 p. to 20 d. 40 gh. 0 p.) the equation is **positive** and the argument should be obtained by deducting 13 d. 46 ghat. 40 palas *from* the given anomaly. For the fourth quarter (20 d. 40 gh. 0 p. to 27 d. 33 gh. 17 p.) the equation is **positive** and the argument is obtained by deducting the given anomaly *from* 27 days 33 ghat. 17 palas.

Rule (3). On account of differences between Indian and European astronomical constants, 45 palas have to be added to the moon's mean elongation or tithi and 14 ghaṭikas to the moon's mean anomaly. These corrections are applicable from the year A.D. 1800 to the year A.D. 2000.

*Example.*

**119**. To find the ending moment of *bahula Shashthi* (21st tithi), **Nija Sravana** A.D. 1909.

| | d. | gh. | pal. | ☉'s Anomaly. d. | gh. | pal. | ☾'s Anomaly. d. | gh. | pal. |
|---|---|---|---|---|---|---|---|---|---|
| According to the Ephemeris for A.D. 1909 (Table XII), | | | | | | | | | |
| Mean new-moon of Nija Śrâvana was on Aug. ... | 16 | 20 | 46 | 125 | 23 | 49 | 10 | 18 | 19 |
| Add equivalent of 21 tithis (*bahula shashthi*) ... | 20 | 40 | 17 | 20 | 40 | 17 | 20 | 40 | 17 |
| Add (for diff. between Ind. and Eur. constants)... | | | + 45 | | | | | + 14 | |
| | 37 | 1 | 48 | 146 | 4 | 6 | 31 | 12 | 36 |
| ☉'s Eqn. for 146 d. 4 gh. (Table IX-*h*)—9gh. 40p. | | | | Ded. 1 an. month. | | | 27 | 33 | 17 |
| ☾'s Variation for Shashṭhi (Table XVI-A)—1gh. 43p. | | | | | | | 3 | 39 | 19 |
| | | | | | | | — | 11 | 23 |
| —11gh. 23p. (**A**) | | | | | | | | | |
| | | | | | | | 3 | 27 | 56 |
| ☾'s Anom. for 3 days 27gh. 56p. by Table IX-*g* is —16gh. 40p. (**B**) | | | | | | | — | 16 | 40 |
| | | | | | | (**C**) | 3 | 11 | 16 |

| | d. | gh. | p. |
|---|---|---|---|
| The argument for Table XVI-B is the given moon's anomaly (C):— | 3 | 11 | 16 |
| Do. Table XVI-C is 2 (☾'s—☉'s long.)—☾'s Anom. | | | |
| Now, 2 (☾'s—☉'s long.) for 6 tithis is given in Table XVI-A as either 11d. 7gh. 19p. or 38d. 34gh 36p. | 11 | 7 | 19 |
| The given ☾'s anom. has to be deducted from either of these data | 3 | 11 | 16 |
| | 7 | 50 | 3 |
| Since the last anomaly is in the second quarter, we deduct it from [*vide* Rule (2) *supra.*] | 13 | 46 | 40 |
| | 5 | 56 | 37 (**D**) |

This is the argument for Table XVI-C.

The equation acc. to Table XVI-B. for 3d. 11gh. 16p. is —20gh. 33p.

Do. do. Table XVI-C. for 5d. 56gh. 37p. is — 6gh. 27p.

Total... —27gh. 0p. (**E**)

We compare the last equation (E) with (B) found *supra* from Table IX-*g* (*i.e.*, 16gh. 40p). Clearly, our Sûrya Siddhânta equation is too small and we must try again with—27gh. 0p. (E) in lieu of 16gh. 40p. (B). That is, we must diminish the argument for Table XVI-B by 10gh.20p and increase the argument for Table XVI-C by the same amount.

Our new arguments for Table XVI-B will be 3d. 0gh. 56p. and its eqn.—19gh. 38palas.

Do. do. Table XVI-C will be 6d. 6gh. 57p. and its eqn.— 6gh. 30palas.

Total... —26gh 8p. (**F**)

This time we began with —27 gh. (E) and ended with —26 gh. 8 p. (F) which shows that our last argument for Table XVI-B must be increased by 1 gh. and our last argument for Table XVI-C correspondingly diminished.

Our new argument for Table XVI-B is 3d. 1gh. 56p. and its equation —19gh. 43p.

Do. do. Table XVI-C is 6d. 5gh. 57p. — 6gh. 30p.

—26gh. 13p. (**G**)

We began with – 26gh. 8p. and ended with – 26gh. 13p. (G).

We may be satisfied with this result. Our final equation will be – 26gh. 13p.

| | | |
|---|---|---|
| (1) Sun's equation of the centre and ☾'s annual eqn. + Variation | | —11gh. 23p. |
| (2) Moon's do. do. and ☾'s evection | | —26gh. 13p. |
| | | —37gh. 36p. |

This being deducted from our mean tithi, we have

| | | |
|---|---|---|
| mean tithi or elongation | Aug. 37. | 1gh. 48p. |
| | | – 37gh. 36p. |
| | Aug. 36. | 24gh. 12p. |
| To which *add* for local time (Table XIII); 1651″ = | | 1gh. 9p. (Lat. 13°, 146th day). |
| | Aug. 36 or Sep. 5, | 25gh. 21p. |

which is about 2 ghaṭikas in excess of the time given in Drig-ganita panchângams for the year *Saumya* (A.D. 1909—10), whereas the Siddhânta panchângams give 12 or 13 ghaṭikas more as the ending moment of bahula Shashṭhi in Nija Śrâvana of that year.

**120.** Ordinarily, the greatest difference between Siddhânta and *Drig-ganita* panchângams occurs from the 6th to the 10th tithis. For the reason of this divergence, the reader is referred to the article on the *Moon* in the *Encyclopedia Britannica,* where we are told that ancient astronomers were unacquainted with the evection and variation, as perturbing elements of the lunar theory.

**121.** If the reader wishes to know how *Drig-ganita* panchângams obtain their tithis from the *Nautical Almanac,* we may state the process in a very few lines:—

Taking the above example, we look out in the *Nautical Almanac* the ⊙'s and ☾'s longitude at Greenwich mean noon on two or three neighbouring days, say 4th, 5th and 6th September 1909.

| (1) September 1909. | (2) ☾'s long. at noon. | | | (3) ☾'s long. at noon. | | | (4) ☾'s elongation. | | | (5) ☾'s elngation for 24 hrs. | | |
|---|---|---|---|---|---|---|---|---|---|---|---|---|
| | ° | ′ | ″ | ° | ′ | ″ | ° | ′ | ″ | ° | ′ | ″ |
| 4 | 41 | 14 | 39·3 | 161 | 17 | 19·1 | 240 | 57 | 20·2 | | | |
| 5 | 55 | 25 | 16·2 | 162 | 15 | 29·9 | 253 | 9 | 46·3 | 12 | 12 | 26·1 |
| 6 | 69 | 11 | 33·6 | 163 | 13 | 42·8 | 265 | 57 | 50·8 | 12 | 48 | 4·5 |

Columns 2 and 3 of this table are extracted from the *Nautical Almanac* for 1909. Column 4, Moon's elongation, is of course column 2 *less* column 3, and the difference between the elongations of two successive days is entered in column 5. The tithi we want is Bahula Shashṭhi, *i.e.*, the 21st tithi which ends when the Moon's elongation is 21 × 12° = 252°. This, we find, was sometime before Greenwich noon on 5 September 1909. Since the Moon's exact elongation at Greenwich noon on 5th September 1909 was 253°.9′46·3″ which is 1° 9′ 46·3″ (= 1·1628°, by Table XIX) more than we require, we have to deduct the time taken by the Moon to effect this elongation. This last time may be determined by working out a proportion either to the elongation of the previous 24 hours (12° 12′ 26·1″ = 12·2072° by Table XIX) or to the elongation of the next 24 hours (12°48′ 4·5″ = 12·8011° by Table XIX). The result is (1) $\frac{1 \cdot 1628}{12 \cdot 2072}$ = ·0953 day, or (2) $\frac{1 \cdot 1628}{12 \cdot 8011}$ = ·0908 day.

If we deduct the first result from Greenwich noon, we have September 4·9048, to which we add ·4605 day for the difference between Greenwich astronomical and Ujjain Lanka time, and again add 1651 seconds or 27½ minutes or ·0190 day for the difference between Ujjain time and Madras time (Table XIII; lat. 13°, 146th day). Net result, September 4·9048 + ·4605 + ·0190 = September 5·3843 or 23 ghat. 4 palas on 5th September 1909.

On the other hand, if we deduct the second result, ·0908 day from Greenwich noon, 5th September 1909, we have September 4·9092 to which we add, as before,

+ ·4605

+ ·0190

and also (for Madras standard time) 9 minutes + ·0062, and

we have, as the final result, September 5·3949

or (by Table XIX) **23** ghaṭikas **40** palas, which differs by only 13 palas from the result given in the Nungumbaukam Panchângam.

Our working shows that a difference of about 15 palas might result according as we took the proportional part from the elongation of the previous or of the following 24 hours.

**122.** This is very much in favour of the *Drig-ganita* method, even if exact astronomical accuracy is wanted, and that method only postulates the possession of a Nautical Almanac. In the absence of a Nautical Almanac, or for years long past or a long way ahead, we can obtain a very fair approximation to Nautical Almanac results by means of the modern equations in our Table XVI.

## CHAPTER XVIII.

### THE MUHAMMADAN YEAR.

**123.** The Muhammadan Era, commonly called *Hejra* or *Hijra*, is dated from Friday, 16 July, 622 A.D., the day of the Flight of the Prophet.

The Muhammadan calendar is not luni-solar like the Indian Calendar, but strictly lunar. The length of the synodical month according to the Muhammadan calendar is 29 days 12 hours 44 minutes. The 3 seconds reckoned in addition in the Hindu calendar are neglected in the Muhammadan. The months are alternately 29 and 30 days long, which, again, if carried out uniformly, would result in an error of 44 minutes per month, which, in 360 months, or 12 lunar years, would amount to $\frac{360 \times 44}{60 \times 24} = 11$ days. These 11 days are accordingly added to the last month of the calendar in 11 out of every 30 years, and the year containing an extra day is, from analogy, sometimes called a leap year.

**124.** Every 30 years form a cycle in the Muhammadan calendar. The years which have 355 days, instead of 354, are in some countries the 2nd, 5th, 7th, 10th, 13th, 15th, 18th, 21st, 24th, 26th and 29th. Elsewhere, the 2nd, 5th, 8th, 10th, 13th, 15th, 19th, 21st, 24th, 27th and 29th are kept as leap years.

It will be noticed that the difference of usage is as regards the 8th, 19th and 27th years in each cycle.

Our Muhammadan Calendar (Table XV) assumes that the 8th, 19th and 27th years in each cycle are leap years, and in paragraph 4 of the explanation at page 197 of the Tables we have shown clearly how Table XV should be used by those who observe the 7th, 18th and 26th years in each cycle as leap years.

**125.** All the books that deal with the Muhammadan Calendar mention the above difference of usage, but they do not give the reason of it. The reason is, however, not far to seek.

If we write down the accumulated error which remains at the end of each year by reason of the neglect to take account of 44 minutes, which we adverted to in the first section of this chapter, we obtain the following result under the system of reckoning the 8th, 19th and 27th as leap years :—

| Year. | Error. | How error is corrected by leap year. | Year. | Error. | How error is corrected by leap year. |
|---|---|---|---|---|---|
| 1st year, | 8 hrs. 48′ | | 16th year, | *Nil.* | |
| 2nd year, | 17 hrs. 36′ | 1st day added by leap year. | 17th year, | 5 hrs. 36′ | |
| 3rd year, | 2 hrs. 24′ | | 18th year, | 14 hrs. 24′ | |
| 4th year, | 11 hrs. 12′ | | 19th year, | 23 hrs. 12′ | 7th day added by leap year. |
| 5th year, | 20 hrs. 0′ | 2nd day added by leap year. | 20th year, | 8 hrs. 0′ | |
| 6th year, | 4 hrs. 48′ | | 21st year, | 16 hrs. 48′ | 8th day added by leap year. |
| 7th year, | 13 hrs. 36′ | | 22nd year, | 1 hr. 36′ | |
| 8th year, | 22 hrs. 24′ | 3rd day added by leap year. | 23rd year, | 10 hrs. 24′ | |
| 9th year, | 7 hrs. 12′ | | 24th year, | 19 hrs. 12′ | 9th day added by leap year. |
| 10th year, | 16 hrs. 0′ | 4th day added by leap year. | 25th year, | 4 hrs. 0′ | |
| 11th year, | 0 hrs. 48′ | | 26th year, | 12 hrs. 48′ | |
| 12th year, | 9 hrs. 36′ | | 27th year, | 21 hrs. 36′ | 10th day added by leap year. |
| 13th year, | 18 hrs. 24′ | 5th day added by leap year. | 28th year, | 6 hrs. 24′ | |
| 14th year, | 3 hrs. 12′ | | 29th year, | 15 hrs. 12′ | 11th day added by leap year. |
| 15th year, | 12 hrs. 0′ | 6th day added by leap year. | 30th year, | *Nil.* | |

We see that by observing the 8th, 19th and 27th years as leap years, a day is added by means of a leap year in order to avoid the error accumulating to more than *1 day* at the end of the following year.

Those that observe the 7th, 18th and 26th as leap years add a day by means of a leap year in order to avoid the error accumulating to more than *half a day* at the end of the following year. This in fact is what is done by both the schools at the end of the 15th year.

**126.** The months and the number of days in each are as follows :—

| Month. | No. of days. | Day of Commencement reckoning from beginning of year. | Month. | No. of days. | Day of Commencement reckoning from beginning of year. |
|---|---|---|---|---|---|
| 1. Muharram ... | 30 | ... | 7. Rajab ... | 30 | 177 |
| 2. Safar ... | 29 | 30 | 8. Shaban ... | 29 | 207 |
| 3. Rabi-ul-awwal ... | 30 | 59 | 9. Ramzan ... | 30 | 236 |
| 4. Rabi-ul-akhir or Rabi-us-sani ... | 29 | 89 | 10. Shawwal ... | 29 | 266 |
| | | | 11. Zilkada ... | 30 | 295 |
| 5. Jumadal-awwal ... | 30 | 118 | 12. Zil-hijja ... / Do. (when intercalary). | 29 / 30 | 325 |
| 6. Jumadal-akhir or Jumad-us-sani ... | 29 | 148 | | | |

All readers are familiar with the names of two of the Muhammadan months, *Muharam* and *Ramzan*, and all are equally familiar with the fact (due to the strictly lunar character of the year) that Muharram is liable to occur at any part of the solar year.

**127.** The Muhammadan day is reckoned from sunset to sunset, and the first day of Muharram is that on which the moon is for the first time visible at sunset after new moon. This is called the heliacal rising of the moon. Our readers are familiar with the Muhammadan usage of observing the heliacal rising of the moon from eminences of any sort, and passing word as soon as she is seen, so as to give notice that Muharram has commenced.

To the reader who is by this time quite at home among tithis and their ending moments, we would quote the following rule from Messrs. SEWELL & DIKSHIT's *Indian Calendar :—*

"It is well to note that where the first tithi of a month ends not less than 5 ghatikas before sunset, the heliacal rising of the moon will probably take place on the same evening ; but where the first tithi ends 5 ghatikas or more after sunset, the heliacal rising will probably not take place till the following evening. When the first tithi ends within these two periods between 5 ghatikas before and 5 ghatikas after sunset, the day of the heliacal rising can only be ascertained by elaborate calculations ".

Where so much depends upon sunset and moonrise, it is obivous that places with different longitudes and latitudes will often not agree in regard to the exact date of commencement of the Muhammadan year.

*N.B.*—Further directions for the use of our Table XV relating to the Muhammadan year are given as a concluding note to that table, *vide* Tables, p. 197.

---

Before commencing Part II of our work, which gives special hints with examples regarding the use of each of the more important tables, we take the opportunity to insert a note on Broken Periods :—

**Note on Broken Periods.**—It is very important that the reader, dealing with the Indian or any other calendar, should know thoroughly how to reckon broken periods. It is generally a question of addition or subtraction, but unless one goes through the process with reflection, there is always a danger of adding or subtracting 1 too much.

To take the simplest case, how many days are there from Friday in one week to Tuesday in the next ? Most people would use their fingers in such a case and answer '4' or '5' according as they began the reckoning with Friday or with Saturday. Now, such uncertainty is fatal to exact reckoning, and we should therefore follow certain rules whereby we may avoid all uncertainty.

First of all, we must always note the meaning of the question with reference to the *next sub-division of the broken period :* that is, in this case whether the meaning is from a particular hour on Friday to the same hour on Tuesday. This makes the meaning quite clear. From 6 a.m. on Friday to 6 a.m. on Tuesday following, there are 4 days: and the proper way to answer the question is to convert Friday and Tuesday into week-day figures 6 and 3, and to add a whole period (7) to 3 before deducting 6. Thus, 7+3=10. From 10 take 6, and we have the answer, 4 days. Some exercise is necessary in order to do even these simple week-day problems correctly.

The rule about *adding a whole period* when you have taken to reckon from a fraction of one period to a fraction of the next is very important. How many days are there from 23rd October in one year to 15th March in the next, the second being a leap year. If we refer to Table VIII we shall find that 23rd October is 296 days from 1st January, and 15th March (where we have had to pass 29th February) is 75 days. Add 365 to 75. Total 440. Deduct 296. Answer, 144 days.

If the 1st of a Tamil month be the 15th July, what will be the 7th of the Tamil month ? In such cases we must *make the same addition to or the same deduction from the one side as we make in regard to the other.* In the example before us, the addition to be made is 6, and the answer, 21st July.

If the 1st of a Tamil month be the 15th July, what will be the 29th of the Tamil month. We add 28 to each side. Therefore the answer is 15+28=43rd July, from which we deduct the whole of the completed period (July=) 31 days. Final answer, 43 *minus* 31=12th August.

---

# PART II.—USE OF THE TABLES.

## CHAPTER XIX.

### USE OF EPHEMERIS OR TABLE XII (Solar Dates).

**128.** The main interest of the present work to the general Indian reader will lie in its enabling him to discover for himself the chief data given in Panchângams for any year between B.C. 1 and A.D. 2000. Panchângams in India are strictly confined to the year to which they relate, and the greatest difficulty is sometimes felt in verifying an Indian date only two or three years' old, while forecasting a date a year or two hence for the purpose of a marriage or any other future event is a matter of equal difficulty. By the use of the present work, future and past events can be dated according to the rigour of the Indian calendar and that with the greatest ease.

**129.** For the use of the general Indian reader, a special table called Table XII has been inserted in the present work, which gives solar dates, new moons and lunar and solar anomalies for the eighty years ending with A.D. 1919, *i.e.*, for 70 years before 1910 and 10 years after that date. Before instructing the reader how to look up dates in this special table, we shall give a few brief hints as to Indian time.

**130.** Indian time is kept in *ghatikas* (Tam. *naligais*) or sixtieth parts of a day, each *ghatika* being equivalent to 24 minutes of English time ($\frac{2}{5}$ of an hour), and in *palas* (Tam. *vinadis*) or sixtieth parts of a ghaṭika, each *pala* being equivalent to 24 seconds of English time. The *ghatikas* and *palas* are reckoned, not from midnight to midnight as in English time, but from sunrise to sunrise. Now, as the moment of sunrise depends on the latitude and longitude of each place, there should, strictly speaking, be as many *panchângams* as there are places in India. Indian astronomers get over this difficulty by calculating time in the first place according to one central latitude and longitude, and then applying the necessary corrections in order to deduce the time for other places. The central latitude is the Equator and the central longitude is that of Ujjain (75° 46′ 6″ East of Greenwich) where there was an ancient observatory. To combine the central latitude and the central longitude, they imagined an island called Lankâ in the Indian Ocean, situated on the equator and having the same longitude as Ujjain. This Lankâ is, of course, not Ceylon.

**131.** The standard time in the Indian calendar is therefore called Lankâ time, and 6 a.m. at Lankâ is the mean sunrise from which all times are, in the first instance, calculated. The time marked in Table XII is invariably Lankâ time, *i.e.*, time reckoned from 6 a.m. at Lankâ. The correction to be made at the present epoch (A.D. 1840 to A.D. 1920) in order to deduce the actual time from local sunrise at thirty of the most important places in India is given in seconds of time in the column "Total Correction" in Table XIII. In no calendar that we are aware of is the correction worked out for so many places in India in a shape ready for instant use.

**132.** We are now ready to use Table XII, and we will first learn how to look up solar dates in that table. Solar dates, *i.e.*, dates based solely on solar months, are in use in the Tamil and Malayalam tracts of Southern India (*i.e.*, in the Tamil Districts, in Malabar, in Travancore, and in Cochin) and in Bengal: elsewhere lunar dates are chiefly in use.

**133.** A solar date is thus cited :—

(Tamil) *Sâdhârana* varusham, Âni 15.

(Malayalam) *Kollam* 1086, Dhanus 20.

(Bengal) *San* 1316, Mâgha 28.

**134.** TAMIL. SOLAR DATES.—The year *Sâdhârana* is one of 60 years comprised in Jupiter's cycle. The revolutions of the cycle from A.D. 967 to A.D. 1926 are clearly indicated with corresponding English years in Table I. The reader will find that in the current cycle *Sâdhârana* is A.D 1910, and this fact is also indicated in Table XII under A.D. 1910.

**135.** Looking down the columns in Table XII under A.D. 1910, we find the following entries against *Ani* :—

3 Je. 14. 32 6 Jl. 15.

These abbreviated entries give us the following useful information :—

The month of *Ani* in A.D. 1910 begins on June 14, the day of the week being 3, *i.e.*, Tuesday.

The same month of Âni contains 32 days, the 32nd day being July 15, the corresponding day of the week being 6, *i.e.*, Friday.

Now, to deduce the 15th day of *Ani*, the first being given, is an easy matter : for we have only to add 14 to "3 June 14" and we get "17 June 28."

Now "17" for the day of the week is obviously 3, for we must cast out sevens, *i.e.*, whole weeks, when the figure indicating the day of the week exceeds 7.

We thus answer: "*Sâdhârana*, Âni 15" is Tuesday, 28 June, A.D. 1910.

**136.** Tamil years may also be quoted by *Kaliyuga* or *Saka* eras, and the years belonging to these eras are also indicated opposite the A.D. year in Table XII. We there see that A.D. 1910 corresponds to Kaliyuga 5011 and Saka 1832.

**137.** The reader should now note a peculiarity in the citation of Indian years, namely, that as a rule the years are "expired", not "current". To understand the difference between these expressions, we go back to the first year of Kaliyuga, which began on 18th February 3102 B.C. We are apt to call that the beginning of the first year: but Hindu reckoning prefers to call it the beginning of the year 0, and what it calls the year 1 is really what we should call the second year, *i.e.*, the "expired" year 1, and so the difference of 1 runs throughout the reckoning between "expired" (*gata*) and "current" (*vartiamâna*) years. In the present work, following the bulk of the authorities, all purely Indian eras (except the Kollam Ându) are expressed in expired years, and A.D. and Kollam years are alone expressed as current years. When the reader wants a "current" year for an Indian era (except Kollam), he should add 1 to the figure given in this work.

**138.** MALAYALAM DATES.—We have just stated that the Kollam or Malayâlam era reckons current not expired years: and we find from Table XII that Kollam Ându 1086, cited in our example, began in South Malabar in the month of CHINGAM (Tam. *Avani*) A.D. 1910 and in North Malabar in the next month KANNI (Tam. *Purattâsi*). To help the memory, we may recollect that "*Simha*" (the sign of the zodiac from which *Chingam* is derived) and "*South Malabar*" begin with "**S**.". In fact "**S–M**" may stand equally for "Singa**M**" and "**S**outh **M**alabar"; while both "**N**. Malabar" and "Ka**nn**i" contain a strong "**n**" (unlike the soft nasal *n* in Chingam.)

**139.** We find from Table XII that "Kollam 1086, Dhanus" begins on Thursday, 15 December 1910, and ends (the number of days in Dhanus being 30) on Friday, 13 January 1911. Now between 20 Dhanus and 30 Dhanus the difference is 10, and so we deduct 10 from 13 January 1911 and get 3 January 1911 as the equivalent of "Kollam 1086, Dhanus 20". To find the day of the week, we deduct 10 from 14 (next higher multiple of 7) *plus* 6, that is from 20, and we get 10, which is equivalent to 3 Tuesday.

### Bengal Solar Dates.

**140.** Our specimen date is San 1316, Mâgha 28.

We find from Table XII, p. 151, that San 1316 (expired) is the Solar year corresponding to A.D. 1910–11; and from Table XII, p. 136, that the 1st of Solar Mâgha falls on Sunday, 15 Jan. 1911; while the 29th of Solar Mâgha (the last day of the month) is Sunday February 12. Obviously the 28th Mâgha is Saturday, February 11, A.D. 1911.

*N.B.*—Our Table XII treats the B. San years as *expired*. It will be seen, however, from Sec. **112** (6) that B. San years are usually cited as *current*. To convert the *expired* years in Table XII into *current*, add, as usual, 1.

**141.** The reader will note that each Bengal Solar month has the same name as the Lunar month commencing during that Solar month; whereas in the Tamil country each Solar month is named after the Lunar month which began *before* the Solar month in question. This is an important practical difference to be borne in mind in calculating Bengal and Tamil Solar dates. For instance Tamil *Kârtigai* is the next month to Bengal Solar Kârttika.

*N.B.*—All Solar months were originally named from the corresponding Lunar months; and each Lunar month was in turn named from the nakshatra in which the moon ordinarily is at Full Moon in that month. See Table XI.

### Commencement of the Solar Month or Sankranti.

**142.** The astronomical commencement of a solar month is the moment of *Saṅkrânti*, i.e., the moment when the sun enters a sign of the Zodiac. To ascertain the moment of Sankrânti, we must add up two figures, namely, (1) the ghaṭikas and palas (and if we like the day of the English month) entered in Table XII against "Commencement of the Solar Year", and (2) the days, ghaṭikas and palas entered in the Eye-Table under the month of which we wish to know the Sankrânti or moment of commencement.

Thus, in the case of the three months which we selected as examples, the Sankrântis are as follows :—

| | Arya Siddhanta. | Surya Siddhanta. |
|---|---|---|
| Commencement of Solar Year corresponding to A.D. 1910–11. | 13 April 1910, 6 ghaṭ. 3 palas. | 12 ghat. 13 palas. |
| Sankrânti or moment of commencement of Tamil *Ani.* | 62 days 21 ghaṭ. 20 palas. | |
| Do. do. Malayâlam *Dhanus* | 246 days 19 ghaṭ. 9 palas. | |
| Do. do. Bengal Solar *Mâgha* | | 275 days 38 ghat. 13 palas |

Sankrânti of Tamil Âni, A.D. 1910 : 62 days 27 ghaṭ. 23 palas. ... ... (1)
or 75 days (counting from 1 April 1910): 27 ghaṭ. 23 palas ... (2)

Sankrânti of Mal. *Dhanus*, A.D. 1910 : 246 days 25 ghaṭ. 12 palas ... ... (3)
or 259 days (counting from 1 April 1910) : 25 ghaṭ. 12 palas ... (4)

Sankrânti of Beng. Solar *Mâgha* A.D. 1910–11 : 275 days 50 ghaṭ. 44 palas ... (5)
or 288 days (counting from 1 April) 50 ghaṭ. 44 palas. ... (6)

**143.** Results (2), (4) and (6) enable us to express the three Sankrântis in terms of the English Calendar : for by reference to Table VIII we find that—

(1) *Ani* Sankrânti, 75 days from 1 April 1910, is 14 June 1910, 27 g. 23 p.

(2) *Dhanus* Sankrânti, 259 days from 1 April 1910, is 15 Dec. 1910, 25 g. 12 p.

(3) *Mâgha* Sankrânti, 278 days from 1 April 1910, is 13 Jan. 1911, 50 g. 44 p.

These results agree with the commencement of Tamil *Ani* and Malayâlam *Dhanus* for 1910 as entered in Table XII, but the Sankrânti of Bengal Solar Mâgha, as above determined (13 January 1911), is not the commencement of that month as entered in Table XII, p. 136 (15 January 1911).

**144.** To account for this difference, we must observe that though the moment of a Solar Sankrânti is the same all over India for the same Siddhânta, the commencement of the corresponding month for *civil* purposes is not necessarily the same all over India. The following rules on the subject are adopted from Messrs. **Sewell & Dikshit's** " Indian Calendar ".

(1) In Orissa the Solar month of the *Amli* and *Vilâyati* eras begins on the day of Sankrânti, at whatever moment of the day the latter may happen. Accordingly in Orissa, the three Solar months above named would begin on the Sankrânti days, *i.e.*, the 14th June 1910, the 15th December 1910 and 13th January 1911 respectively.

(2) In Bengal, when the fraction of the day at which the Sankrânti happens does not exceed 45 ghaṭikas, the Solar month begins on the next day : and when the Sankrânti occurs after 45 ghaṭikas, the Solar month begins on the next day but one. Accordingly the Bengal months corresponding to *Ani* A.D. 1910 (Bengal Solar *Ashâdha*) and to Malayâlam *Dhanus* A.D. 1910 (Bengal Solar *Pausha*) will commence on 15 June 1910 and 16 December 1910 respectively and not on 14 June and 15 December 1910 (the days of the Sankrânti), while Bengal Solar *Mâgha* A.D. 1910–11 will not commence till 15 January 1911 (*i.e.*, 2 days after the Sankrânti).

(3) In Southern India, in the Tamil country, when the fraction of the day at which the Sankrânti occurs does not exceed 30 ghaṭikas, the Solar month begins on the same day. That is why the beginning of Tamil *Ani* A.D. 1910 coincides with the day of the Sankrânti, 14 June 1910. When the Sankrânti occurs after 30 ghaṭikas, the Solar month begins on the

next day. This is why the Tamil month *Tai*, corresponding to Bengal Solar *Mâgha*, A.D. 1910–11 is shown in Table XII as having commenced on 14 January 1911, *i.e.*, the day after the Sankrânti.

**145**. The authors of the "Indian Calendar" also state that in Malabar, when the Sankrânti occurs after 18 ghaṭikas, the Solar month commences on the next day. If this were so, Malayâlam *Dhanus* in A.D. 1910–11 would commence on 16th December 1910. As a matter of fact, however, all the Malayâlam Panchângams show *Dhanus*, A.D. 1910–11, as commencing on the same day as the Tamil *Mârgali*, i.e., on 15 December 1910. This fact imports an element of doubt into the so-called "Malabar rule" laid down by Messrs. SEWELL & DIKSHIT; and till the contrary is shown to be the case, we must assume that the Tamil rule is followed in Malabar.

**146**. The Bengal rule laid down by the same authorities is practically applied in Table XII, page 136 (Ephemeris for Bengal). Whatever may be the rules as regards the commencement of a civil month *in any part of India*, the data given in Table XII and the Eye-Table will enable any one to apply his own *civil* rule with certainty. The dates entered in Table XII for the commencement and end of Solar months apply in strictness only to Southern India and to the *Arya Siddhânta*. For the *Sûrya Siddhânta* and for tracts where a different *civil* rule from that in force in Southern India is followed, suitable corrections may have to be applied, as in Table XII, page 136 (Ephemeris for Bengal).

**147**. To illustrate the difference between Sankrântis according to Ârya and Sûrya Siddhântas, we will give a practical example from a case which recently came under the author's notice. A person produced a Tamil horoscope which contained among other data the following note on the date of his birth :—

"A.D. 1856, June 28, *Ani* 16."

A reference to the Tamil panchângams of the year A.D. 1856 showed that according to *all* of them, 28 June 1856 A.D., corresponded to 17th, not to 16th *Ani*.

The apparent discrepancy was explained by the sole fact that the person, though now residing in Southern India, was born at *Belgaum* where the Sûrya Siddhânta is followed, and where presumably the horoscope was cast by a Tamil Astrologer.

The Solar year corresponding to A.D. 1856 commenced on 11 April A.D. 1856, at 14 ghaṭ. 6 palas according to Sûrya Siddhânta (Table X) and at 7 ghaṭ. 55 palas according to Ârya Siddhânta. (Table XII.)

For *Ani* Sankrânti by the Tables II and XIX, we have to add 62 days 21 ghaṭ. 20 palas according to Sûrya Siddhânta, and 62 days 19 ghaṭ. 34 palas according to Ârya Siddhânta.

∴ According to Sûrya Siddhânta, *Ani* Sankranti occurred on the 73rd day (reckoned from 1 April 1856) at 35 ghaṭ. 26 palas (*i.e.*, after 30 ghaṭ.)

According to Ârya Siddhânta, *Ani* Sankrânti occurred on the 73rd day (reckoned from 1 April 1856) at 27 ghaṭ. 29 palas (*i.e.*, before 30 ghaṭikas).

**148**. In accordance with the *civil* rules for Solar months current in the Tamil country (Sec. **144** *supra*) the month of *Ani* in A.D. 1856 began (according to Sûrya Siddhânta) on the 74th day reckoned from 1 April 1856, *i.e.*, on the 13th June 1856 (*vide* Table VIII), and the 16th *Ani* was 13 + 15 = 28th June 1856.

According to *Arya Siddhânta*, the month of *Ani* in A.D. 1856 began on the 73rd day reckoned from April 1856, *i.e.*, on 12th June 1856 (*vide* Tab. VIII) and the 17th of *Ani* was 12 + 16 = 28th June 1856.

**149.** This example shows that there could easily be a difference of 1 day between Solar dates in one part of the country and another.

It also shows that such differences cannot be accounted for by the ordinary *Jantris* and *Panchângams* but must be explained by recourse to the fundamental principles of the Indian Calendar.

---

## CHAPTER XX.

### USE OF EPHEMERIS OR TABLE XII for Lunar Tithis.

**150.** Even in those parts of the Indian Continent where the Solar Calendar is used for *civil* purposes, lunar tithis are observed for *religious* purposes: elsewhere the civil as well as the religious calendar is *lunar* and is regulated by tithis, instead of by days of the month.

**151.** Like the lunar calendar of the Jews, and unlike that of the Muhammadans, the Hindu calendar may be described as luni-solar since all its periods of time, though regulated by the movements of the moon, are made to fit in with divisions of the Solar Year. A lunar tithi is not counted at all unless the Sun rises upon it. A Tithi, Nakshatra, or Yoga may begin, or end, at any moment of the day, but the Tithi, Nakshatra, or Yoga pertaining to a day is that which is current *at sunrise.* A lunar month takes its name from the next following *Solar* Sankrânti : if there are two lunar months entitled to derive their name from a single Solar Sankrânti, they both receive the same name and one, the first, is called *adhika.* On the other hand, if a new moon is followed by two Sankrântis before it is followed by another new moon, the lunar month which would ordinarily have been derived from the second of the two Sankrântis is suppressed and is said to be *Kshaya*, that is, in defect. Lastly the lunar year, beginning in the month of *Solar* Chaitra, which is itself the last month of the *Solar* year, is named after the *next Solar Year.* These considerations suffice to prove that the Hindu lunar year is properly a luni-solar year.

**152.** Lunar periods of time are characterized by an element of certainty or rather of palpable evidence, not found in solar periods, and in another sense the actual moments of lunar phases are marked by much greater uncertainty than Solar Sankrântis. We will explain each of these propositions.

**153.** *Palpable evidence of lunar periods.*—We cannot visibly perceive in the heavens the fact that the sun has completed any definite stage in his annual course: but new moon, when the moon's longitude is the same as the sun's, is a patent fact: so are full moon and each quarter of the lunar month. There cannot be a difference of a whole day between the moment of new moon in one part of the country, and the same moment in another part, just as we saw there might be in the commencement of a solar month. If an inscription states that a certain tithi fell on a Monday, and by calculating backwards we trace the tithi to a Saturday we may be quite certain that there is some error in the inscription.

**154.** On the other hand, there are greater fluctuations in the moon's than in the sun's movements, in the sense that the actual time of new moon may be as much as 14 hours before or 14 hours after the time of *mean* new moon. Two calculations are therefore necessary to determine the time at which the moon reaches any particular stage of her course: we must first ascertain the *mean* time, which is simply the expected time, taking an average over very long periods; and then we must calculate the *actual* time by making a correction to the mean time according to the moon's and sun's *anomaly* at the particular moment. Both these operations can be performed very easily and very accurately, that is to say, in exact accordance with the Siddhântas, by means of the tables in the present work. The correction is called a correction for the sun's and moon's *equation of the centre.* There are also one or two other minor corrections recognized in modern astronomy, but the Siddhântas do not make them. (See sections **121** to **127** *supra.*)

**155.** In the case of new moons from A.D. 1840 to A.D. 1920, it was thought best not to give the reader even the slight trouble of calculating the anomaly and the equation of the centre; and the exact moment of each new moon during these eighty years as well as the *exact* day of its occurrence, where it differs from the mean date, is entered in separate columns, in Table XII "General Ephemeris." Here, however, the day and fraction of a day marking the occurrence of *mean* new moon, and the sun's and moon's anomalies at the moment of mean new moon are also given, as they are necessary for the calculation of tithis between one new moon and another.

**156.** Supposing now the reader wishes to ascertain the exact time of new moon in July 1910. He finds the following entries:—

(Tab. XII) A.D. 1910.

| Date and fraction of day of **Ashadha** new moon. | Actual moment of new moon. |
|---|---|
| (**5**) July 7, 10 ghaṭikas 57 palas; | July **6, 49 ghatikas 55 palas.** |

*Moon's and Sun's anomaly in days, ghatikas and palas.*

Sun's anomaly: 84 days 58 ghaṭikas 29 palas.

Moon's anomaly: 4 days 29 ghaṭikas 13 palas.

**157.** As regards the *exact* moment of occurrence of new moon in July 1910, there is no difficulty at all; for it is Wednesday, 6th July 1910 (the **date** of mean moon as well as the *week-day* has to be diminished by one, by reason of the anomalies) and the time of day is **49 ghatikas 55 palas.**

**158.** This, as explained already, is mean Lankâ time. If now you wish to know the *true local time* of the *actual* moment of new moon in July 1910, and if you happen to be at one of the 30 places named in Table XIII, all you have to do is to apply to Lankâ time the correction indicated in the column "Total correction" for the day of the solar year entered under the new moon date in Table XII.

**159**. Supposing you are at Tanjore, the correction for the 84th day is + 2478 seconds of time, *i.e.*, 103 palas or + 1 ghaṭika 43 palas. Adding this to Lankâ time, 49 ghaṭikas 55 palas, we have **51** ghaṭikas **38** palas, exact time at Tanjore according to Sûrya Siddhânta. If now you wish to know how far this time agrees with modern astronomical computation, you have merely to look at the time of new moon according to the Tiruvâdi Almanac (Mr. Srauti's Panchângam) which is calculated from the English Nautical Almanac and is reduced to the latitude of Tanjore, and you find there "6th July 1910, **52** ghaṭikas **10** palas", showing a difference of 32 palas. This difference of 32 palas is due to lunar acceleration which has changed since the date of the Sûrya Siddhânta. (*Vide* Sec. **2**.)

**160**. If you are at Madras (lat. 13°) you will find the correction according to Table XIII to be + 2200 seconds of time, *i.e.*, + 92 palas or + 1 ghaṭika 32 palas. Adding this to mean Lankâ time, you get (49 ghat. 57 palas + 1 ghat. 32 palas =) 51 ghat. 29 palas. The time according to the "College" Panchângam (Mr. Raghavachari's) is 52 ghaṭikas 33 palas, showing a difference of 1 ghaṭika 4 palas. The increased difference, as compared with Mr. Srauti's Panchângam, is mainly due to two facts, (1) The correction for Madras local time is less than that for Tanjore local time by 11 palas, (2) Madras standard time, which is shown in the College Panchângam, is nine minutes or 22½ palas in advance of Madras Local time. There is still, however, an unexplained difference of a few palas between our tithi and that of the College Panchângam.

**161**. Next, suppose the reader wishes to know the ending moment of the 8th tithi or *ashtamî* in the bright fortnight following the new moon of 6th July 1910. For this we have to calculate first of all the *mean* ending moment of the 8th tithi, and then its *actual* ending moment. The mean ending moment is given by adding the collective duration of 8 tithis according to the Eye-table to the *mean* moment of new moon.

| | | |
|---|---|---|
| (Tab. XII) Mean moment of new moon | ... | A.D. 1910, 7 July 10 ghat. 57 palas. |
| *Add* collective duration of 8 tithis according to the Eye-table. | ... | 7 days 52 ghat. 29 palas. |
| | | A.D. 1910 15 July 3 ghat. 26 palas. |

**162**. This, then, is the *mean* ending moment of the 8th tithi or *ashtamî*. For the *actual* ending moment of the same tithi, we first of all add the collective duration of 8 tithis to the moon's and sun's anomaly noted in Table XII under the new moon of 7 July 1910.

| | Moon's anomaly | | | Sun's anomaly | | |
|---|---|---|---|---|---|---|
| | 4 days | 29 ghat. | 13 palas | 84 days | 58 ghat. | 29 palas. |
| Add collective duration of 8 tithis. | 7 days | 52 ghat. | 29 palas | 7 days | 52 ghat. | 29 palas. |
| | 12 days | 21 ghat. | 42 palas | 92 days | 50 ghat. | 58 palas. |
| Deduct ☉'s Eqn. | | —2 ghat. | 20 palas | | | |
| Net ☾'s anom. | 12 days | 19 ghat. | 22 palas | | | |

**163.** According to the Eye-table we find the Sun's Equation for an anomaly of 92 days, 51 ghat. to be ***minus*** 2 ghat. 20 palas, and we deduct this equation from the moon's anomaly already found. The Equation of the centre for the moon's net anomaly of 12 days 19 ghat. 22 palas is, by the Eye-table, ***minus*** 9 ghaṭikas.

**164.** The sum of moon's and sun's Equations :

| | |
|---|---|
| – 2 ghat. | 20 palas. |
| – 9 ghat. | |
| – 11 ghat. | 20 palas. |

**165.** If 11 ghaṭikas be deducted from the mean ending moment of the 8th tithi, we obtain, as the ***actual*** ending moment of *sukla ashtamî*, 14 July A.D. 1910, 52 ghaṭikas after sunrise.

**166.** This is Lankâ time. If now we wish to know the ending moment of the tithi in true local time, say at Madras (Lat. 13°), all we have to do is to apply the figure entered in the column " Total correction " under Madras (Lat. 13°) against the 92nd day of the Solar Year in Tab. XIII. The correction being + 2095 seconds or + 87 palas, that is + 1 ghat. 27 palas, the ending moment of **Sukla ashtami**, 14 July 1910, at Madras, is (52 ghat. + 1 ghat. 27palas=) 53 ghat. 27 palas. The "College" Panchângam gives 52 ghat. 33 palas.

**167.** We will now show the reader how to calculate successive tithis from the Eye-table and Table XII combined. The duration of 1 tithi, according to the Eye-table, is 59 ghaṭikas 4 palas, that is, 1 day less 1 ghaṭika *plus* 4 palas. Bearing this in mind, we shall proceed to calculate the successive tithis after new moon, 7 July 1910 ; the mean time of which is 10 ghat. 57 palas, or nearly 11 ghaṭikas.

## Details for determining the ending moments (in Madras Time) of 30 tithis, Lunar Ashadha, A.D. 1910.

| A.D. 1910 Tithi | July. | Mean ending moment of tithi. | ⊙'s Anom. | ⊙'s Equation. | ☾'s Anom. at mean end of tithi. | Add or subtract ⊙'s Equation. | ☾'s Equation. | Sum of ⊙'s & ☾'s Equations. | Actual end of tithi, Lanka time. | Correction for Madras time. | Ending moment of tithi, Madras time. | Ending moment, "College" Panchangam. | Ending moment Kanjanur Panchangam | Difference between Surya Siddhanta and Naut. almanac. | Full name of tithi. |
|---|---|---|---|---|---|---|---|---|---|---|---|---|---|---|---|
| | | gh. p. | d. gh. | gh. p. | d. gh. p. | d. gh. p. | gh. p. | gh. p. | July gh.p. | gh. p. | July gh.p. | July gh. p. | July gh. p. | gh.p. | |
| Amavasya Sukla paksha. | 7 | 10,57 | 84,58 | −0, 50 | 4 29,13 | 4 28,23 | −20,12 | −21,2 | 6 49,55 | +1,31 | 6 51,26 | 6 52,33 | 6 53,11 | −1,7 | |
| | | | | | | | | | | | | | | | *Sukla Paksha.* |
| 1 | 8 | 10, 1 | 85,57 | −1, 0 | 5 28,17 | 5 27,17 | −22,48 | −23,48 | 7 46,13 | +1,31 | 7 47,44 | 7 47, 7 | 7 49,41 | +0,37 | Pratipada. |
| 2 | 9 | 9, 5 | 86,56 | −1,12 | 6 27,21 | 6 26, 9 | −24,23 | −25,35 | 8 43,20 | +1,30 | 8 45, 0 | 8 42,52 | 8 47,16 | +2, 8 | Dvitiya. |
| 3 | 10 | 8, 9 | 87,55 | −1,23 | 7 26,25 | 7 25, 2 | −24,47 | −26,10 | 9 41,59 | +1,29 | 9 43,28 | 9 40,10 | 9 45,58 | +3,18 | Tritiya. |
| 4 | 11 | 7,13 | 88,54 | −1,34 | 8 25,29 | 8 24,55 | −23,59 | −25,33 | 10 41,40 | +1,29 | 10 43, 9 | 10 39,13 | 10 45,55 | +3,56 | Chaturthi. |
| 5 | 12 | 6,17 | 89,53 | −1,45 | 9 24,33 | 9 22,48 | −21,55 | −23,40 | 11 42,37 | +1,28 | 11 44, 5 | 11 40,10 | 11 47, 7 | +3,55 | Panchami. |
| 6 | 13 | 5,21 | 90,52 | −1,56 | 10 23,37 | 10 21,41 | −18,36 | −20,32 | 12 44,49 | +1,27 | 12 46,16 | 12 43, 1 | 12 46,36 | +3,15 | Shashti. |
| 7 | 14 | 4,25 | 91,51 | −2, 7 | 11 22,41 | 11 20,34 | −14,11 | −16,18 | 13 48, 7 | +1,27 | 13 49,34 | 13 47,25 | 13 52,56 | +2, 9 | Saptami. |
| 8 | 15 | 3,29 | 92,50 | −2,17 | 12 21,45 | 12 19,28 | − 8,52 | −11, 9 | 14 52,20 | +1,26 | 14 53,46 | 14 52,59 | 14 57,23 | +0,47 | Ashtami. |
| 9 | 16 | 2,33 | 93,49 | −2,28 | 13 20,49 | 13 18,21 | − 2,58 | − 5,26 | 15 57, 7 | +1,26 | 15 58,33 | 15 59,12 | 15 60, 0 | −0,39 | Navami. |
| | | | | | | | | | | | | Navami | 16 2,16 | | |
| 10 | 17 | 1,37 | 94,48 | −2,39 | 14 19,53 | 14 17,14 | + 3,11 | + 0,32 | 16 60, 0 | | 16 60, 0 | 16 60, 0 | | | Dasami |
| | | | | | | | | Dasami | 17 2, 9 | +1,25 | 17 3,34 | 17 5,22 | 17 7,11 | −1,48 | Dasami. |
| 11 | 18 | 0,41 | 95,47 | −2,50 | 15 18,57 | 15 16, 7 | + 9, 4 | + 6,14 | 18 6,55 | +1,24 | 18 8,19 | 18 10,56 | 18 11,46 | −2,37 | Ekadasi. |
| 12 | 18 | 59,46 | 96,46 | −3, 0 | 16 18, 1 | 16 15, 1 | +14,22 | +11,22 | 19 11, 7 | +1,24 | 19 12,31 | 19 15,24 | 19 15,44 | −2,53 | Dvadasi. |
| 13 | 19 | 58,49 | 97,45 | −3,11 | 17 17, 5 | 17 13,54 | +18,45 | +15,34 | 20 14,23 | +1,23 | 20 15,46 | 20 18,35 | 20 18,44 | −2,49 | Trayodasi. |
| 14 | 20 | 57,53 | 98,44 | −3,21 | 18 16, 9 | 18 12,48 | +22, 1 | +18,40 | 21 16,33 | +1,23 | 21 17,56 | 21 20,10 | 21 20,34 | −2,14 | Chaturdasi. |
| 15 | 21 | 56,57 | 99,43 | −3,32 | 19 15,18 | 19 11,41 | +24, 4 | +20,32 | 22 17,29 | +1,22 | 22 18,51 | 22 20,24 | 22 21,31 | −1,33 | Paurnami. |
| *Krishna Paksha* | | | | | | | | | | | | | | | *Krishna Paksha* |
| 1 | 22 | 56, 1 | 100,42 | −3,42 | 20 14,17 | 20 10,35 | +24,48 | +21, 6 | 23 17, 7 | +1,22 | 23 18,29 | 23 17,10 | 23 20,34 | +1,19 | Pratipada. |
| 2 | 23 | 55, 5 | 101,41 | −3,52 | 21 13,21 | 21 9,29 | +24,21 | +20,29 | 24 15,34 | +1,21 | 24 16,55 | 24 16,53 | 24 18,43 | +0, 2 | Dvitiya. |
| 3 | 24 | 54, 9 | 102,40 | −4, 2 | 22 12,25 | 22 8,23 | +22,43 | +18,41 | 25 12,50 | +1,21 | 25 14,11 | 25 13,41 | 25 15,46 | +0,30 | Tritiya. |
| 4 | 25 | 53,13 | 103,39 | −4,12 | 23 11,22 | 23 7,17 | +20, 4 | +15,52 | 26 9, 5 | +1,20 | 26 10,25 | 26 9,44 | 26 11,49 | +0,41 | Chaturthi. |
| 5 | 26 | 52,17 | 104,38 | −4,22 | 24 10,33 | 24 6,11 | +16,34 | +12,12 | 27 4,29 | +1,20 | 27 5,49 | 27 5,18 | 27 7,11 | +0,31 | Panchami. |
| 6 | 27 | 51,21 | 105,37 | −4,32 | 25 9,37 | 25 5, 5 | +12,23 | + 7,51 | ... 59,12 | +1,19 | 28 0,31 | 28 0,29 | 28 1,47 | +0, 2 | Shashti. |
| 7 | 28 | 50,25 | 106,36 | −4,42 | 26 8,41 | 26 3,59 | + 7,40 | + 2,58 | 28 53,23 | +1,19 | ... 54,42 | ... 55,22 | ... 54, 8 | −0,40 | Saptami. |
| 8 | 29 | 49,29 | 107,35 | −4,52 | 27 7,45 | 27 2,53 | + 2,40 | − 2,12 | 29 47,17 | +1,19 | 29 48,36 | 29 50, 5 | 29 49,48 | −1,29 | Ashtami. |
| 9 | 30 | 48,33 | 108,34 | −5, 2 | 0 33,32 | 0 28,30 | − 2,29 | − 7,31 | 30 41, 2 | +1,18 | 30 42,20 | 30 44,37 | 30 43,44 | −2,17 | Navami. |
| 10 | 31 | 47,37 | 109,33 | −5,11 | 1 32,36 | 1 27,25 | − 7,31 | −12,42 | 31 34,55 | +1,18 | 31 36,13 | 31 39, 6 | 31 37,49 | −2,53 | Dasami. |
| | Aug. | | | | | | | | Aug. | | Aug | Aug. | Aug. | | |
| 11 | 1 | 46,41 | 110,32 | −5,21 | 2 31,40 | 2 26,19 | −12,14 | −17,35 | 1 29, 6 | +1,18 | 1 30,34 | 1 33,34 | 1 32,10 | −3,10 | Ekadasi. |
| 12 | 2 | 45,45 | 111,31 | −5,30 | 3 30,44 | 3 25,14 | −16,27 | −21,57 | 2 22,48 | +1,17 | 2 25, 5 | 2 28,11 | 2 27, 3 | −3, 6 | Dvadasi. |
| 13 | 3 | 44,49 | 112,30 | −5,39 | 4 29,48 | 4 24, 9 | −19,58 | −25,37 | 3 19,12 | +1,17 | 3 20,29 | 3 23,14 | 3 22,38 | −2,45 | Trayodasi. |
| 14 | 4 | 43,53 | 113,29 | −5,49 | 5 28,52 | 5 23, 3 | −22,39 | −28,28 | 4 15,25 | +1,17 | 4 16,42 | 4 18,57 | 4 19, 4 | −2,15 | Chaturdasi. |
| 15 Amavasya | 5 | 42,57 | 114,28 | −5,58 | 6 27,56 | 6 22, 2 | −24,19 | −30,17 | 5 12,40 | +1,16 | 5 13,56 | 5 15,24 | 5 16,33 | −1,28 | Amavasya. |

**168**. The reader is presented in the above table with a full decursus of tithis for a whole lunar month, calculated in accordance with the Sûrya Siddhânta; and he is also furnished in the same table with the ending moments of the same tithis, as arrived at by two standard Panchângams current in Southern India, namely, (1) the Kanjanûr Panchângam of Annâvaiyangâr, commonly called "No. 28 Panchângam", and (2) the Nungambaukum Panchângam by Mr. Râghavachâri and his son, commonly called "The College Panchângam." The former of these panchângams is typical of the class known as "Vâkya Panchângams".

**169**. A *vâkyam* (meaning in Tamil a sentence) is simply a phrase or series of phrases, employed in accordance with a very ancient "Transnumeration" table, called "Kadapayadi", closely analogous to the present writer's Transnumeration Table, published at page 37 of his "Secret of Memory". It is generally believed that the vâkya process is based on the Ârya Siddhânta, but this is by no means well established.

Only solar dates in Southern India seem to follow the Ârya Siddhânta, but for lunar tithis and nakshatras, that Siddhânta, as proved by the above, as well as by the next, table, seems to have been given up long ago, even in Southern India, in favour of the more accurate Sûrya Siddhânta. If the Kanjanûr Panchângam had followed the Ârya Siddhânta, all mean tithis, including *Amâvâsyâs*, would have occurred according to that Panchângam 3 ghaṭikas 42 palas later than by the Sûrya Siddhânta, whereas the above table, and more especially the next table of nakshatras, shows a much smaller difference as a rule between that panchângam and the results arrived at on strict Sûrya Siddhânta principles.

**170**. The Nungumbaukum or "College" Panchângam belongs to the class of what are called in Southern India *Drig-ganita* panchângams, *i.e.*, those in which computation is checked by observation, the observation being understood to be that carried on at a modern standard observatory, like those at Greenwich, Paris and New York.

**171**. Practically, all the *Drig-ganita* panchângams are based on the Greenwich Nautical Almanac, which is published some three years in advance for each year. It is not to be supposed that the moon's and sun's places given in the Nautical Almanac are those observed then and there, for they are also calculated on the best available data, which include several elements of the lunar theory, neglected in the Siddhântas.

*N.B.*—For the manner in which Drig-ganita tithis are computed, see Section **121** *Supra*.

**172**. The reader will observe that in our specimen table the difference between Sûrya Siddhânta tithis and what we may call Nautical Almanac tithis, does not exceed 2 ghaṭikas at new moon or 4 ghaṭikas at other times. (The extreme difference between the two systems may amount at times to as much as 17 ghaṭikas, or 7 hours.) This no doubt is a disadvantage of the Sûrya Siddhânta system for strictly astronomical purposes (*i.e.*, on the assumption that Hindu Astronomy *must* agree with the most accurate results of European Astronomy); but if it is remembered that a Nautical Almanac panchângam cannot be constructed unless one has in hand the Nautical Almanac of the particular year, whereas a Siddhânta panchângam can be constructed in a few hours' time for any year, past, present or future, without any other materials than those furnished by Tables VI, VII, VIII and IX of this work, it will be seen that a panchângam on the purely Indian system is a great convenience.

Moreover, the Indian panchângam is not a thing of to-day or yesterday, but has been constructed on the lines now followed for at least 1500 years, and no person, unacquainted with the system, can hope to understand the thousands of inscriptions scattered all over India, on the proper reading of which the reconstruction of scientific Indian history largely depends.

**173**. *Adhika* or *Trisparsa* Tithis and *Kshaya* Tithis: A tithi which begins on one day, is current for the whole of the next day and ends in the morning of the third day is called an *Adhika* or *Trisparsa* tithi. *Trisparsa* means "touching three days." An example is furnished in the above table by *sukla navami* of the Kanjanûr Panchângam, and *sukla dasamî* of our own and the College Panchângam. The fact that a tithi is current for the whole of a day is indicated in the panchângams by entering '60 ghat' as its ending moment, *i.e.*, it does not come to an end all that day. On the other hand, a tithi which begins and ends *between* one sunrise and another is *kshaya* or in defect and is suppressed. Such a tithi is *krishna saptamî* by all three Panchângams in the above table.

*N.B.*—If the reader wishes to apply to the Siddhanta results, the corrections necessary to bring them up to the Nautical Almanac standard, he is enabled to do so in Chapter XVII of Part I of this work, Sections **116** to **122**.

## CHAPTER XXI.

### Use of Tables XII and XI—A.

(NAKSHATRAS.)

**174**. We next suppose the reader to be desirous of calculating the *nakshatras* in order for a particular lunar month. The longest interval between each mean new moon and the mean ending moment of each of the 27 nakshatras following new moon is given in Table XI-A (pp. 134, 135) in days, ghaṭikas and palas. The interval is subject to a single correction which holds good for the whole of a Solar Year. For the years A.D. 1840 to A.D. 1920 the corrections for Nakshatras are shown in the last column of Table XII "Deduct for Nakshatras". The deduction for the solar year A.D. 1910–11 is 1 day 56 ghaṭikas 18 palas. Taking, for example, Lunar Âshâdha in this year, we find at p. 134 that the first Nakshatra for Âshâḍha from which the deduction can be made is No. 7 Punarvasu, and we proceed to determine the mean ending moment of Punarvasu Nakshatra.

| | | d. | gh. | p. |
|---|---|---|---|---|
| (Table XI-A, p. 134.) Interval for Punarvasu, No. 7 Nak. in "Âshâḍha" | | 2 | 49 | 41 |
| (Table XII, p. 151.) *Deduct* for the year 1910–11 | | —1 | 56 | 18 |
| | | 0 | 53 | 23 |
| *Add* mean ending moment of Âshâḍha New moon. (Table XII, p.151.) | July | 7 | 10 | 57 |
| Mean ending moment of No. 7 Punarvasu Nakshatra | July | 8 | 4 | 20 |

**175**. Starting from this point, the following Table illustrates the whole of the processes for determining the ending moment of every nakshatra in the series of 27 from 6th July to 2nd August A.D. 1910. As in the last table, a comparison is also instituted between the ending moments of nakshatras, recorded in the "No. 28" and "College" Panchângams respectively, and those arrived at in the present table on the principles of Sûrya Siddhânta. The reader will note that the agreement among the different panchângams is, in this particular month, closer in regard to nakshatras than in regard to tithis. However, an extreme variation of 5 ghaṭikas has to be looked for even under nakshatras between Sûrya Siddhânta and Nautical Almanac results.

**Table showing mean and actual ending moment (Lanka time and Madras local time) for each Nakshatra from 6th July 1910 to 2nd August 1910.**

| Name of Nakshatra. | English month and day. | Mean ending moment of Nakshatra. | Moon's mean anomaly | Nakshatra equation (Tab. IX-I). | Actual ending moment of Nakshatra, Lanka time. | Correction for Madras (Tab. XIII). | Actual ending moment of Nakshatra, Madras time. | Kanjanur No. 28 Panchangam. | College Panchangam. | Surya Siddhanta compared with Nautical Almanac. |
|---|---|---|---|---|---|---|---|---|---|---|
| A.D. 1910 | M. d. | gh. p. | d. gh. p. | gh. p. | M. d. gh. p. | gh. p. | M d. gh. p. | gh. p. | gh. p | gh. p. |
| Punarvasu ... | July 8 | 4 20 | 5 12 25 | -20 35 | July 7 43 45 | +1 30 | July 7 45 15 | 46 0 | 42 50 | +2 25 |
| Pushya ... ... | July 9 | 5 3 | 6 13 8 | -22 20 | July 8 42 43 | +1 30 | July 8 44 13 | 45 23 | 40 27 | +3 46 |
| Aslesha ... ... | July 10 | 5 46 | 7 13 51 | -22 59 | July 9 42 47 | +1 29 | July 9 44 16 | 45 54 | 39 36 | +4 40 |
| Magha ... ... | July 11 | 6 29 | 8 14 34 | -22 23 | July 10 44 6 | +1 29 | July 10 45 35 | 47 38 | 40 28 | +5 7 |
| Purva Phalguni ... | July 12 | 7 12 | 9 15 17 | -20 35 | July 11 46 37 | +1 28 | July 11 48 5 | 50 36 | 43 13 | +4 52 |
| Uttara Phalguni .. | July 13 | 7 55 | 10 16 0 | -17 33 | July 12 50 22 | +1 28 | July 12 51 50 | 54 47 | 47 47 | +4 3 |
| Hasta ... ... | July 14 | 8 28 | 11 16 43 | -13 25 | July 13 55 3 | +1 27 | July 13 56 30 | 52 57 | 53 48 | +2 42 |
| Chitra ... ... | July 15 | 9 11 | 12 17 26 | - 8 22 | July 15 0 49 | +1 26 | July *15 2 15 | 5 56 | 0 56 | +1 19 |
| Svati ... ... | July 16 | 9 54 | 13 18 9 | - 2 45 | July 16 7 9 | +1 25 | July 16 8 34 | 12 16 | 8 33 | +0 1 |
| Visakha ... ... | July 17 | 10 37 | 14 18 53 | + 3 5 | July 17 13 42 | +1 25 | July 17 15 7 | 18 40 | 16 9 | -1 2 |
| Anuradha ... | July 18 | 11 20 | 15 19 35 | + 8 42 | July 18 20 2 | +1 24 | July 18 21 26 | 24 41 | 23 1 | -1 35 |
| Jyeshtha ... | July 19 | 12 3 | 16 20 18 | +13 42 | July 19 25 45 | +1 23 | July 19 27 8 | 30 3 | 28 53 | -1 45 |
| Mula ... ... | July 20 | 12 46 | 17 21 1 | +17 47 | July 20 30 33 | +1 23 | July 20 31 56 | 34 26 | 33 15 | -1 19 |
| Purva Ashada ... | July 21 | 13 29 | 18 21 44 | +20 45 | July 21 34 14 | +1 22 | July 21 35 36 | 37 17 | 36 11 | -0 35 |
| Uttara Ashada ... | July 22 | 14 12 | 19 22 27 | +22 29 | July 22 36 41 | +1 21 | July 22 38 2 | 39 43 | 37 42 | +0 20 |
| Sravana ... ... | July 23 | 14 55 | 20 23 10 | +22 58 | July 23 37 53 | +1 21 | July 23 39 14 | 40 29 | 37 59 | +1 15 |
| Sravishta ... | July 24 | 15 38 | 21 23 53 | +22 15 | July 24 37 53 | +1 20 | July 24 39 13 | 40 5 | 37 13 | +2 0 |
| Satabhisaj ... | July 25 | 16 21 | 22 24 36 | +20 26 | July 25 36 47 | +1 20 | July 25 38 7 | 38 47 | 35 38 | +2 29 |
| Purva Bhadrapada ... | July 26 | 17 4 | 23 25 19 | +17 39 | July 26 34 43 | +1 20 | July 26 36 3 | 36 15 | 33 25 | +2 38 |
| Uttara Bhadrapada ... | July 27 | 17 47 | 24 26 2 | +14 5 | July 27 31 52 | +1 19 | July 27 33 11 | 33 16 | 30 44 | +2 27 |
| Revati ... ... | July 28 | 18 30 | 25 26 45 | + 9 54 | July 28 28 24 | +1 19 | July 28 29 43 | 29 34 | 27 45 | +1 58 |
| Asvini ... ... | July 29 | 19 13 | 26 27 28 | + 5 14 | July 29 24 27 | +1 19 | July 29 25 46 | 25 35 | 24 27 | +1 19 |
| Bharani ... ... | July 30 | 19 56 | 27 28 11 | + 0 25 | July 30 20 21 | +1 18 | July 30 21 39 | 21 32 | 21 0 | +0 39 |
| Krittika ... ... | July 31 | 20 39 | 0 45 37 | - 3 40 | July 31 16 59 | +1 18 | July 31 18 17 | 17 30 | 17 29 | +0 48 |
| Rohini ... ... | Aug. 1 | 21 22 | 1 56 20 | - 9 9 | Aug. 1 12 13 | +1 18 | Aug. 1 13 31 | 13 43 | 13 55 | -0 24 |
| Mrigasira ... | Aug. 2 | 22 5 | 2 57 3 | -13 24 | Aug. 2 8 41 | +1 17 | Aug. 2 9 58 | 10 20 | 10 27 | -0 29 |
| Ardra ... ... | Aug. 3 | 22 48 | 3 57 46 | -17 8 | Aug. 3 5 40 | +1 17 | Aug. 3 6 57 | 7 36 | 7 19 | -0 22 |

## CHAPTER XXII.

### USE OF TABLE X.

**176.** Table X, which covers more than a hundred and thirty pages of the present work is intended to serve the epigraphist and the historian in somewhat the same manner as Table XII is intended to serve the general reader. If Table X is used in conjunction with the Eye-Table, nothing else is needed for the determination of the ending moments of tithis, correct to two places of decimals of a day, for any of the 2000 years between B.C. 1 and A.D. 2000. The manner of using Table X will be obvious to the reader who knows how to use Table XII.

* *Chitra* Nakshatra extends from 56 gh. 30 p. on 13th July 1910 to 2 gh. 15 p. on 15th July 1910. No *Nakshatra* comes to an end on 14th July 1910, and this fact is indicated in the panchangas by entering *Chitra* as adhika Nakshatra and showing it against *July 14, 60 gh. no palas* as well as against July 15th, 2 gh. 15 p. The first entry means simply that on 14th July 1910, *Chitra* was current for the whole of the civil day.

**177.** Suppose for example that the date corresponding to "*Saka Samvat 999, Phâlguna Sukla 3*", is to be ascertained, we proceed as follows :—

Śaka 999 expired

| This is how the two head-lines of Table X should be used for ascertaining ⊙'s and ☾'s anomalies. | | Days of S. Year. | ☾'s Anom. | |
|---|---|---|---|---|
| | First New Moon in Solar Year ... | 3·28 | 2·67 | |
| | *Add* collective days up to Phalguna New Moon, and remember there was an adhika month earlier in the year. ... ... | 324·84 | 21·74 | |
| | Collective duration of 3 tithis (by Eye-Table) ... ... | 2·95 | 2·95 | |
| | | 331·07 | 27·36 | +·165 (⊙'s Eqn.)=27·53 |

| | | |
|---|---|---|
| Sun's Eqn. (by Eye-Table) for | 331·07 days= | +·165 day |
| Moon's do. do. for | 27·53 days= | +·002 day |
| Sum of ⊙'s and ☾'s Equations. | | +·167 day |
| (Table X) Phalguna New Moon, Saka 999, A.D. 1078 | (4) Feb. | 14·78 } **Mean tithi.** |
| (Eye-Table) Duration of 3 tithis | (2) | ·2·95 } |
| *Add* mean tithi to sum of ⊙'s and ☾'s Eqn. | (7)* Feb. | 17·90 |

The ending moment of the tithi is therefore ·90 of a day, *i.e.*, (according to Eye-Table 54 ghat. on Saturday, 17 February A.D. 1078 ; which, so far as ghaṭikas are concerned, is the result arrived at by Dr. Fleet by applying Prof. Chhatre's table—*Ind. Antiq.*, Vol. XVII, p. 162.

**178.** In the above process the *mean ending moment* of the tithi is given in the *two* lines bracketed "**Mean tithi**", and *all the rest* of the process is directed to ascertaining the sum of the sun's and moon's equations of the centre.

**179.** A further simplification could have been effected by using Table VIII for ascertaining the sun's and moon's anomaly, thus :—

| | ⊙'s An. days. | ☾'s An. days. |
|---|---|---|
| (Table X) Śaka 999 expired. Anomaly of sun and moon at first new moon in Solar Year ... ... ... | 3·28 | 2·67 |
| (Table VIII) Phâlguna śukla 3, increased by 1 month (see Sec. **220** *infra*) because of an adhika month in the year : i.e., *Chaitra sukla* 3. | 327·79 | 24·69 |
| | 331·07 | 27·36 |
| | ⊙'s Eqn. | + ·165 |
| | | 27·53 |

We thus obtain the sun's and moon's anomalies by the additionof *two*, instead of *three*, set of figures, and we may proceed as before to determine the equations for these anomalies.

## Solar Dates by Table X.

**180.** Supposing the Kumbha Sankrânti for the same Śaka Samvat 999, expired, was wanted : we proceed as follows :—

| | |
|---|---|
| Commencement of solar year, Śaka 999 : | Mr. 23·6638 |
| Kumbha Sankrânti (by Eye-Table) | 305·0850 |
| | Mr. 328·7488, *i.e.*, 44 ghat. 56 p. (Tab. XIX.) |

* Since the sum of the several fractions of a day which go to make up the ending moment of the tithi exceeds unity, we increase the week-day by 1.

Now. by Table VIII (see Sec. **216** *infra*) 328 days from Mr. 1, A.D. 1077 is **22 Ja. 1078**.

By Table IV (see Sec. **231** to **241** *infra*) the week day of 22 Ja. 1078 was 3 + 6 + 6 + 22 = 37, which divided by 7 leaves remainder 2, *i.e.*, **Monday**. This result agrees with that arrived at by Dr. Fleet *loc. cit.* except as regards the moment of Sankrânti, which he puts down as about 47 ghat. 52 palas.

**181**. To take another example, suppose we want the third Panguni of the Viśvâvasu year which was current in or after the Śaka year 1347 (expired.)

By Table X, Śaka 1347 (exp.) was A.D. 1425.

By Table I, the Viśvâvasu of that epoch was A.D. 1425.

By Table X, the commencement of Śaka 1347 (exp.) was Mr. 26·7111

By the Eye-Table, the addition for the commencement of Panguni = 334·9053

361·6164 days.

By Table VIII, 361 days from 1 Mr. A.D. 1425 was 24th Feb. A.D. 1426.

Now, since ·6164 of a day exceeds ·5, that is, goes beyond sunset, therefore, by the rule in force in South India, [*vide* Sec. **144** (3)] the first Panguni in the year in question was 25th Feb. 1426.

Therefore the 3rd Panguni (Śaka 1347, Viśvâvasu) was 27th Feb. 1426.

And by Table IV the week day of 27th Feb. 1426 was 6 + 4 + 2 + 27 = 39, which, divided by 7, leaves remainder 4 (i.e.) *Wednesday*. This agrees with the result arrived at by Professor Kielhorn at page 83 of Appendix to Vol. VII, *Epigraphia Indica*.

**Nakshatras concurrent with Tithis.**

**182**. A tithi is often cited along with a Nakshatra ending on the same day. In such cases the *ending moment* of the Nakshatra can be most easily determined by Table XI. (For merely finding the *concurrent* Nakshatra, we may proceed much more expeditiously as in Section **185** or Section **186**.)

Thus in the case dealt with in sections **177** to **179**, the Nakshatra correction for Śaka 999 (A.D.1077) was (Tab. XI) that corresponding to argument 3·279, *i.e.*, ·2394 + ·0059 = ·2453. Deducting this from the interval for Revatî in the 12th lunar month (the year 999 Śaka contained an adhika month, and therefore Phâlguna was the 12th month), we have, as net interval for Revatî, 3·1860 *minus* ·2453 = 2·9407. Add this to New moon, Feb. 14·78 we have Feb. 17·72. ☾'s Anom. 2·67 + 21·74 + 2·94 = 27·35. Eqn. for Nak. (Tab. IX-*k*) = + ·016. Absolute ending moment of Nakshatra, Feb. 17·72 + ·016 or Feb. 17·74, *i.e.*, 44 ghaṭikas after sunrise on 17 Feb. A.D. 1078. This agrees with Mr. Dikshit's calculation, cited by Dr. Fleet at p. 162 of *Ind. Ant.*, Vol. XVII.

**183**. In South Indian inscriptions a solar date, a lunar tithi and the corresponding *nakshatra* are often cited together.

The following example shows how the citation should be verified.

*Ep. Indica*, Supplt. to Vol. VII, p. 182. "Śaka 1106: on the day of Satabhîshaj, which was the 14th *tithi* of the first fortnight and a Wednesday, the 26th solar day of the month of Simha".

Now by Table X, Saka 1106 commenced on March 24·3508, A.D. 1184.

By Eye-Table, Simha begins and Karkata ends 125·4755 days after commencement of solar year.

∴ Simha Sankrânti of Śaka 1106 was on 149·8263 days of the solar year.

That is, by the rule in Southern India already adverted to, the 1st day of Simha was the 150th day reckoned from 1st March 1184.

∴ the 26th day of Simha would be 150 + 25 or the 175th day from 1st March which by Table VIII, was 22 August.

In order that the 26th day might be the 14th tithi of the first fortnight, the new moon must have occurred 13 days before, *i.e.*, about 9 August.

In Table X we find Bhâdrapada new moon on Aug. 8·56 of A.D. 1184 (Wednesday).

By Eye-Table, 14 tithis = 13·78 days.

∴ the mean 14th tithi ended on August 22·34, A.D. 1184.

The day of the week was Wednesday: since the new moon was on Wednesday, the addition of 14 days or 2 whole weeks to the day of the week would still give us the same day, Wednesday.

**184.** If we want the actual ending moment of the 14th tithi, we proceed as follows:—

| | | ☾'s Anom. |
|---|---|---|
| First new moon in solar year, A.D. 1184 | 19·0909 | 1·191 days. |
| By Table VIII, 14th tithi of Śukla Bhâdrapada | 131·9033 | 21·685 |
| | 150·9942 | 22·876 |
| By Tab. IX, ☉'s Eqn. for 150·99 days = —·1672 | | —·1672 |
| „ „ ☾'s Eqn. for 22·709 days = + ·3550 | | |
| | | 22·709 |
| Sum = + ·1878 | + ·1878 | |
| | 151·1820 | |
| *Add* commencement of solar year, Mr. | 24·3508 | |
| | 175·5328 | days from 1st Mr. A.D. 1184. |

By Table XIX ·5328 day = 31 ghaṭikas 58 palas.

The actual ending moment of Bhâdrapada śukla 14, Saka 1106, was 31 ghat. 58 palas after mean sunrise on Wednesday, 22 Aug., A.D. 1184.

**185.** For the Nakshatra corresponding to this tithi we proceed as follows:—

| | Lunation space. |
|---|---|
| By Table VIII (last column) Sun's longitude for Nak. on 151st day = | 12·0321 |
| „ V (last column) do. do. ·18 day = | ·0145 |
| *Add* 14 tithis (by Table II) | 13·7809 |
| | 25·8275 |

By Table III, the lunation space last arrived at, 25·8275 corresponds to No. 24 Satabhishaj whose ending space is 26·2494

Our inscription is therefore correct in all respects.

**186.** An equally simple method of ascertaining the Nakshatra concurrent with the above tithi is the following, where, however, we use degrees of sun's and moon's longitude.

Ending moment of tithi in days of solar year was 150·99 + ·19 (Sum of eqns.) = 151·18 days.
By Tab. XVII-A & XVII-C., ☉'s Long. and Eqn. for 151·18 days = 146·69° – 2·04° + ·18 = 144·83°
Add Moon's Elongation, *i.e.*, No. of tithi × 12° = 14 × 12° = 168·00°

312·83°

By Eye-Table the longitude is that of Nakshatra *Satabhishaj*.

---

## CHAPTER XXIII.

### THEORY OF ANOMALIES AND EQUATIONS OF THE CENTRE.

### CONSTRUCTION OF TABLE IX.

**187.** The uses of Table IX, to which we come after discussing Tables XI and X, will be sufficiently obvious from examples already worked out. We therefore give in this place in popular language a theory of anomalies and their equations.

**188.** From the fact that the orbit of the moon as well as that of the earth is elliptical, not circular, it follows that the motions of these bodies cannot be uniform from day to day or from hour to hour. This irregularity is called the *eccentricity* of the orbit and the correction to be applied on this account is called the equation of the centre.

**189.** The following extract from Prof. Jacobi's Table in Vol. I of *Epigraphia Indica* will serve to introduce the reader to the general theory of Solar and Lunar anomalies, and it will also show how the material furnished by the Siddhântas has been worked into Table IX of the present work :—

Surya Siddhanta.

| ☾'s Eqn. +; ☉'s Eqn. – | | | | ☾'s Eqn. –; ☉'s Eqn. + | | | | Moon's Equation of the centre. | | | Sun's Eqn. of the centre. | | |
|---|---|---|---|---|---|---|---|---|---|---|---|---|---|
| Deg. | Min. | Deg. | Min. | Deg. | Min. | Deg. | Min. | Deg. | Min. | Sec. | Deg. | Min. | Sec. |
| 0 | 0 | 180 | 0 | 180 | 0 | 360 | 0 | 0 | 0 | 0 | 0 | 0 | 0 |
| 30 | 0 | 150 | 0 | 210 | 0 | 330 | 0 | 2 | 32 | 0 | 1 | 6 | 3 |
| 60 | 0 | 120 | 0 | 240 | 0 | 300 | 0 | 4 | 22 | 30 | 1 | 53 | 26 |
| 90 | 0 | 90 | 0 | 270 | 0 | 270 | 0 | 5 | 2 | 46 | 2 | 10 | 31 |

**190.** Let us try to interpret in detail the meaning of this table. We are supposed to measure the moon's rate of progress, beginning from perigee, the point when she is nearest the earth, and at every step we must distinguish the moon's mean position, *i.e.*, the position which she would have attained at a uniform rate of motion equal to the mean, and the actual position which she attains on account of the eccentricity of her orbit.

**191.** The mean and actual positions are the same at 0° or 360°, *i.e.*, at perigee and at 180°, *i.e.*, at apogee.

When the moon's mean position is 30° from perigee, her actual position has advanced by 2 degrees 32 minutes.

When her mean position ought to be 60°, we find her actually at 64° 22′ 30″ from perigee.

When her mean position ought to be 90°, that is half way between perigee and apogee, she is actually 95° 2′ 46″ from perigee.

From this point she begins to move more slowly, though her actual position is still in advance of the mean.

At 120° from perigee, she is 4° 22′ in advance of the mean position, that is, exactly as she was at mean 60°.

At 150° from perigee she is only 2° 32′ in advance of the mean position.

From 180° onwards she begins to slow down, and when she ought to be 210° from perigee, or 30° from apogee, we find she has reached only 210° *minus* 2° 32′ or 207° 28′. When she ought to be 270° from perigee, she is only 270° *minus* 5° 2′ 46″ or 264° 57′ 14″.

From 270° onwards she begins to move quicker, though she is still behind her mean position. At 300° she is behind by 4° 22′ 30″ and at 330° she is behind her mean position by only 2° 32′ and at 360° or at perigee she is even with her mean position.

**192.** In like manner we might trace the Sun's mean and actual positions from perigee through apogee back to perigee, using the figures in the last column of the above table, from which we see that the maximum equation of the centre for the sun is 2° 10′ 31″.

**193.** In our tables (except under Planets' Tables XVII, XVIII) we do not refer to the sun's or moon's position by degrees, but by days, which is more readily intelligible and handier for purposes of calculation.

**194.** Our Table IX with its numerous divisions (*a*) to (*l*) is simply the result of a careful expansion of the smaller tables from which the figures in Section **196** have been extracted, and we shall see presently how far our Table IX agrees with the original.

We saw that when the moon is actually 32° 32′ from perigee, she is 2° 32′ in advance of her mean position.

We turn the first of these figures into days with the help of Subsidiary Table VII (*a*) and for turning 2° 32′ into days we use Subsidiary Table VI (*a*).

| | | | | |
|---|---|---|---|---|
| By Sub. Table VII (*a*) | 30° | = | 2·2962 days. | |
| | 2° | = | ·1531 day. | |
| | 32′ | = | ·0408 day. | |
| | 32° 32′ | = | 2·4901 days. | |
| By Sub. Table VI (*a*) | 2° | = | ·1641 day. | |
| | 32′ | = | ·0437 day. | |
| | 2° 32′ | = | ·2078 day. | ... (1) |

**195.** We now turn to Table IX (*a*), and look up the equation for an anomaly of 2·4901 days. We there find,

| ☾'s Anomaly. | Equation. |
|---|---|
| 2·479 days | ·207 day. |
| 2·493 days | ·208 day. |
| ·014 day | ·001 day. |

$$(\cdot490 \textit{ minus } \cdot479=)\ \cdot011 \text{ day}; \ \frac{\cdot011}{\cdot014}\times\ \cdot001=\cdot008 \text{ day}.$$

∴ The equation for 2·490 days = ·2078 day. . . . . . . . . . . . . (2)

Our result (2) according to Table IX (*a*) is in exact agreement with result (1) deduced from the Siddhânta Table, which shows the anomaly and the equation of the centre in degrees minutes and seconds.

**196.** The reader will naturally ask why we turned the *anomaly* 32° 32′ into days by means of Table VII (*a*) and the *equation* 2° 32′ by means of Table VI (*a*). We proceed to explain.

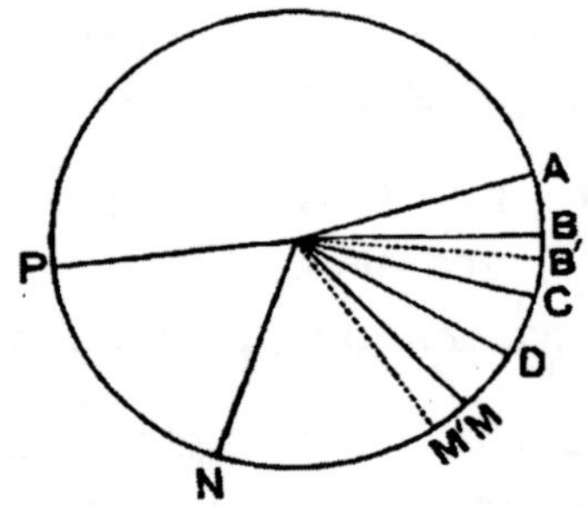


**197.** Let us suppose the sun and the moon to move in the same plane and in circular orbits, describing equal spaces in equal times. Let AB, BC, CD be the mean spaces described by the sun in three successive tithis, and let AM, MN, NP be the corresponding mean spaces described by the moon in the same tithis.

**198.** If there were no irregularity or eccentricity in the sun's and moon's motions, the first tithi would be the time taken by the moon to gain 12°, that is, the space BM over the sun; similarly the second and third tithis would be the periods in which the moon gains (MN *minus* BC) and (NP *minus* CD) over the sun.

**199.** But owing to the eccentricity of their orbits we will suppose the sun to be at B′ (actual position) when he ought to be at B (mean position) and the moon to be at M′ (actual position) when she ought to be at M (mean position).

**200.** Then in the period of a mean tithi (·9843 day) the moon gains over the sun the space B′M′ but our tithi is the period during which the moon actually gains over the sun 12°, *i.e.*, we must cut off from B′M′ a space equal to BM and determine the time during which that space is gained by the moon. Our problem would be solved if we knew the time during which B′M′ *minus* BM was gained by the moon.

**201.** Now B′M′ *minus* BM = MM′ *minus* BB′. The time during which MM′ − BB is gained, is evidently the time during which MM′ is gained *minus* the time during which BB′ is gained. The times during which MM′ and BB′ are respectively gained are obtained by turning MM′ and BB′ into days according to Sub. Table VI (*a*). This is the reason why in the Tables for the moon's and also the sun's equation of the centre we turn the equation into days invariably by Table VI (*a*). For Nakshatra and Yoga equations other scales which it was unnecessary to give in detail, were used for converting degrees into days.

**202.** As the equation for the sun as well as the equation for the moon is sometimes positive and sometimes negative, and for *tithis* we have to take the *difference* between the two equations, some confusion would result from our having to change signs so often. Therefore the sun's equations are tabulated, as in the extract given above, with the signs *reversed.* That is why the sun's eccentricity, starting from perigee, is shown in the table in Sec. **189** as negative whereas it is really positive.

**203.** The reader will also observe that the moon's equation in the above table is, as it should be, positive between perigee and apogee, whereas in Table IX (*a*) it is negative for the same period. The reason is that when the moon does in a given time *more* than the mean space, this is equivalent to a given space being done in *less* than mean time. This is why in Table IX, which derives equations of time from equations of space, the signs of space-equations are reversed. In the case of the sun's equations, a double reversion of signs takes place, first a reversion in order to make the operation of combining the sun's and moon's equations always an addition, and secondly a reversion in order to derive equations of time from equations of space. As a final result, the sun's equation of time is shown in Table IX (*c*) as positive after perigee and negative after apogee. A third reversion takes place in calculating sunrise, where we have to derive equations of space from equations of time. (*Vide* Sec. **255** *infra.*)

**204.** In calculating tithis, all we have to do, is to sum up the equations of the sun and the moon according to Table IX, whether they are positive or negative. Before taking the moon's equation, however, we in practice add to or deduct from the moon's anomaly the sun's equation and we determine the equations for the net moon's anomaly. The reason is that when the effect of the Sun's anomaly is to diminish or increase the mean time that would be necessary for a tithi, it is necessary to take the moon's anomalies for the altered mean time.

**205.** Conversely, in calculating *Yogas*, where we have to take sum of the proper anomalistic equations of the sun and the moon, our Yoga equations in Table IX (*i*) [p. 20 (*a*),] have had to be suitably altered as to their signs.

**206.** For *Nakshatras*, we have to convert the moon's eccentricity, in other words her equation of the centre, into days at the rate of the moon's sidereal motion, *i.e.*, 360° for 27·32166 days and this has accordingly been done in Table IX (*j*), (*k*), and (*l*).

**207.** The sun's anomaly in Tables IX (*c*), (*f*) and (*h*) is expressed, not in days of the *anomalistic* year, but as days of the *solar* year, and this is done for convenience of use, as explained in Part III, Sec. **253**.

**208.** For instance, in the above table, the sun's equation for an anomaly of 270° is entered as 2° 10′ 31.″

Now by Table VII (*b*) 270° = 202·92 + 71·02 = 273·94 days.
And by Table VI (*a*) 2° = ·1641 day.
10′ = ·0137 day.
31″ = ·0007 day.

2° 10′ 31″ = ·1785 day.

Since the solar year always begins with a mean anomaly of 102·0802 days, an anomaly of 273·94 days of the anomalistic year really belongs to 273·94 *minus* 102·08, *i.e.*, the 171·86th day of the solar year.

Accordingly we find in Table IX (*c*) the equation ·1785 day set down against the 171·86th day of the solar year.

**209.** The length of the modern anomalistic year is 365·2596 days while that of the Hindu solar year is 365·2587 days. There is no practical error in adopting, as the Indian Siddhântis have done, an identical period for the anomalistic and solar (sidereal) years. The modern anomalistic lunar month is, like the Hindu anomalistic month, 27·5546 days. In converting degrees of anomaly into days we, therefore, put—

Sun's anomaly : [Sub. Table VII (*b*)] 360° = 365·258756484 days.

Moon's anomaly : [Sub. Table VII (*a*)] 360° = 27·45459999 days.

**210.** The above theory of anomalies and equations is subject to an important variation in practice, which we alluded to briefly in Sec. **208**, but which it is now necessary to dwell on specially. Supposing a mean tithi is accomplished at A days of the solar year, when the Moon's anomaly is B days, then, if the ☉'s equation for A days is—*a*, this means that the tithi would be accomplished (supposing for the moment that the ☾'s anomaly had no influence) in *A—a* days of the solar year, at a moment when the moon's anomaly would be *B—a*. Now we proceed to determine the influence of this moon's anomaly *B—a* and find the moon's equation to be, let us suppose, —*b*. We then put down, as the actual ending moment of the tithi, *A—a—b* days of the solar year.

**211.** Strictly speaking, we ought to take (1) the sun's equation for $A + b + a$ days of the solar year, where *b* is the equation (positive or negative) of the moon for anomaly B, and *a* is the ☉'s equation, positive or negative, for A days of the solar year, (2) the moon's equation for anomaly of $B + a$ days, and then add the equations so found.

**212.** But in practice this refinement is not necessary for the sun's equation since the maximum value of $b + a$ is (·4138 + ·1784 =) ·5922 day, and the maximum variation of the sun's equation for this period is about ·0018 day or 6 palas only. We may note, however, that in the test example, worked by Prof. Jacobi for Âshâḍha Śukla 12, K.Y. 3585, and by ourselves in sections **228** and **259** to **262** *infra*, a difference of 4 palas does occur between his method and Mr. Dikshit's, and the learned Professor rightly surmises in a footnote (*Ep. Ind.* Vol. I, p. 430) that the difference must be due to an abridgment in the Hindu method. We now see what the abridgment consists in.

## CHAPTER XXIV.

### USE OF TABLE VIII.

**213.** Table VIII is a very comprehensive as well as a very useful table. It is designed to convey some very useful information by itself and taken with Tables VI and VII, it enables us to take the first and most important step in calculating tithis, that is, to ascertain the exact mean ending moment of a tithi in any year, past, present or future.

Let us interpret the table, taking the entries for a particular day, say the 340th day of the solar year, A.D. 1908 (Kaliyuga 5009).

**214. Column 1.** " Week day 4", *i.e.*, if the first day of the year was 1, the 340th day of the year would be the 4th day of the week, *i.e.*, an addition of 3; the first day of the year being 4, Wednesday, the 340th day would be 3 + 4 = Saturday.

**215. Column 2.** " Days reckoned from Jan. 1 : 5, 6 Decr." and there is the following footnote : " When two dates are given in any of these columns, use the first in a leap year : otherwise use the second ".

The meaning of these entries is that the 340th day of the English Calendar year, reckoned from 1 January, is 5 December if it was a leap year, otherwise it would be the 6 December. By the expression " if it was a leap year ", we mean, " if in the course of reckoning we have had to pass a 29th February ".

The entry in this column is meant to be of use only with reference to the English Calendar. If we want to know, for instance, how many days there are between 15 March and 23 Oct. we take the corresponding figures from col. 2 of Table VIII, *i.e.*, 74 and 296, and we know the interval to be 222 days. We should always remember that the meaning of such problems and their solution is : " From a particular hour on 15th March *to the same hour* on 23rd Oct. there are 222 whole days". (*Vide* note prefixed to Sec. **128**.)

**216. Column 3.** The third column means : " The 3rd Feb. is the 340th day reckoned from 1 March ". This column is useful for ascertaining the A.D. equivalent of dates in Indian Solar Years which began in March, *i.e.*, up to and inclusive of expired Kaliyuga 4853, A.D. 1752, see the examples in Sections **142, 180** and elsewhere, and rules in Sec. **144**.

**217. Column 4.** The fourth column means: " The 5th March is the 340th day reckoned from 1st April, if we have passed 29th February in our reckoning : otherwise the 340th day reckoned from 1st April is the 6th March of the following year ". This column is useful in ascertaining the A.D. equivalent of days in solar years, subsequent to A.D. 1752. See examples worked out in Chapters VI, VII and elsewhere.

**218. Column 5.** The fifth column means :—

" The 340th day of the Hindu Solar Year is, *more or less*, the 6th Panguni (Tamil), the 6th Meenam (Malabar) or the 6th Chaitra (Bengal Solar) ". We say " *more or less* ", because if we want to know exactly what day in the solar year corresponds to 6th Panguni, 6th Meenam, or 6th solar Chaitra, we should first of all determine the beginning of these months, according to (1) the moment of Sankranti, and (2) the rule of practice followed in the respective provinces. See examples worked out in Sec. **181** and elsewhere.

In the present case, supposing we are concerned with the Solar Year 5011 Kaliyuga, A.D. 1910 the year began as follows:—

| | Surya Siddhanta. | |
|---|---|---|
| (Tab. VI) A.D. 1900 ... | Ap. 12·6204 | |
| (Tab. VII) Add for 10 years. | ·5876 | |
| | Ap. 13·2080 | |
| Moment of Solar *Chaitra* Sankrânti according to Sûrya Siddhânta which is followed in Bengal (by Tab. VI.) | 334·9053 days. | |
| | 335·1133 days of Solar year. | |

| | Arya Siddhanta. | |
|---|---|---|
| | Ap. 12·5139 | |
| | ·5868 | |
| | Ap. 13·1007 | |
| Moment of *Panguni or Meenam* Sankrânti, according to Ârya Siddhânta, which is followed in the Tamil country and in Malabar (by Tab. II.) | 334·9200 days. | |
| | 335·0207 days of Hindu Solar year. | |

According to the rule followed in Malabar and the Tamil country, the month begins on the same day on which the Sankrânti occurs, if the Sankrânti occurred before ·50 of a day. [Sec. **144** (3)]. Therefore the 1st Panguni or 1st Meenam in the year, expired Kaliyuga 5011, was the 335th day of the Solar year and the 340th day would be the 6th Panguni or 6th Meenam.

In Bengal the month begins the day after Sankrânti, if the Sankrânti is before ·75 of the day. Therefore in Bengal the first Chaitra in the Solar year K.Y. 5011 (expired) will be the 336th day of the Solar year and the 340th day of the Solar year will only be the 5th Chaitra.

**219**. **Column 6.** The 6th column is the central column of the whole Table. The figure 340 in the present case is the guide in using the other columns.

**220**. **Column 7.** The 7th column should be read with the 8th. The two columns mean; "In a year where there is no *adhika mâsa*, the 1*st* tithi of the *dark* fortnight (indicated by *dark* figures) of Chaitra, which is called *bahula prathamî* or *badi* 1, ends on the 340th day of the Lunar Year at ·5861 of the day". If we find out the beginning of the lunar year, which we can do from Tables VI and VII, or Tab X, all we have to do is to add to it the ending moment of a tithi, as given in Table VIII, and then we know the day and fraction of day of the Solar Year or of the English Calendar year, when the mean tithi ends. Examples of this process are given throughout the work. If there has been an *adhika mâsa* during the year, which we can ascertain from Table X or XII, the tithi is advanced one month.

**221**. Thus, supposing we want the 3rd *bahula* tithi of Mâgha in expired K. Y. 5010 (A.D. 1909) when there was an adhika Śrâvana, we proceed as follows:—

| | | |
|---|---|---|
| Commencement of Solar year, A.D. 1909 | Ap. 12·9492 | (1) |
| First new moon in Solar year. | 7·2745 | (2) |
| 18th Mâgha (for which we take from Table VIII the figures against the 18th of the next lunar month, *Phâlguna*) ... ... ... | 313·0242 | (3) |
| (1) + (2) + (3), omitting 12 Ap. | 321·2479 (Solar Year). | |
| (1) + (2) + (3), including 12 Ap. | 333·2479 (days reckoned from 1 Ap.) | |

Our mean tithi, *Mâgha bahula tritiyâ*, ended on the 321st day of the Solar Year at ·2479 of the day or on the 333rd day of the English Calendar year, reckoned from 1st April (*i.e.*, 27 Feb. 1910) at ·2479 of the day.

**222. Column 8.** The 8th column, as we have just seen, gives the order of tithis in each lunar month, beginning with the bright fortnight (light figures) and ending with the dark fortnight (heavy figures).

**223. Column 9.** The 9th column gives the moon's anomaly at the ending moment of each tithi. The figure in this column should be added to the moon's anomaly at the moment of the first new moon in the solar year. The moon's anomaly at the first new moon in the solar year is given in the appropriate column of Table X. Tables VI and VII combined give only the moon's anomaly *at the commencement of the solar year* and if we are using Tables VI and VII to determine the moon's anomaly, we should take care to add the interval *between the commencement of the solar year* and *the first new moon in the solar year*. Thus, suppose we want the exact ending moment of the 12th tithi of the bright fortnight of Âshâḍha in the year A.D. 484 (K.Y. 3585 expired) and we wish to use Tables VI, VII, and VIII; we proceed as follows :—

| | Commencement of Solar year. | First new moon in solar year. | Moon's anomaly. |
|---|---|---|---|
| (Table VI) A.D. 400 | Mr. 17·4857 | 23·8447 d. | 21·748 d. |
| (Table V) *Add* for 84 years. | ·7355 | 0·5453 | 13·466 |
| | Mr. 18·2212 | 24·3900 | 35·214 |
| *Add interval between commencement of solar year and first new moon in solar year* | .. | | 24·390 |
| | | | 59·604 |
| *Deduct* two whole anomalistic months (Table II) | ... | | 55·109 |
| | | | 4·495 |
| *Add* (from Table VIII), for ending moment of Âshâḍha *Sudi* 12 (*no* Adhika masa) | ... | 70·8734 | 15·764 |
| | | 95·2634 | 20·259 |
| ⊙'s Eqn. for 95·26 days (Table IX-*c*) : | —·0455 * | | —·045 |
| ☾'s Eqn. for 20·214 days (Table IX-*b*) : | +·4138 | | 20·214 |
| | +·3683 | +·3683 | |
| *Add* commencement of solar year A.D. 484 | March 18 | ·2212 | |
| | March | 113·8529 days | |

reckoned from 1 Mr. A. D. 484, etc., as in Sec. **260** *infra*.

**224. Column 10.** The tenth and last column gives the sun's sidereal longitude for Nakshatras and Yogas: that is, it gives the sun's longitude by putting 365·25875 days = 29·5306.

* If we re-calculated the ⊙'s eqn. for ⊙'s anom. of 95·26+·37 (☾'s +⊙'s eqns.), we shall find it to be—·0466 which is the ⊙'s eqn. adopted by Prof. Jacobi in Vol. I, *Ep. Ind.* We have noted however in Sec. **212** *supra*, as well as in Sec. **260** *infra*, that this is a refinement seldom required in practice.

The method of using this column is explained in the chapters of the Text, headed "Nakshatras" and "Yogas" and also under "Use of the Tables", *vide* Sections **36**, **49**, **185**, etc.

This column gives the sun's longitude for whole days: for fractions of days, when necessary, we should add the equivalent from Table V (last column). The use of this column is not necessary if we prefer to take the Sun's longitude in degrees from Tables XVII-A and XVII-C. See examples in Secs. **186**, **287** and elsewhere.

---

## CHAPTER XXV.

### USE OF TABLES VII, VI AND TABLE V—LUNAR CYCLES.

**225**. The uses of these important tables will have become familiar from the explanation of Table VIII, contained in Sec. **223** and elsewhere. They enable us in fact to use Table VIII, whether for the Sûrya Siddhânta or for the Ârya Siddhânta, and whether for Solar or for Luni-Solar dates. Instead of giving constants, as Prof. Jacobi has done, for K.Y. 3100, 3200, etc., we have given constants for K.Y. 3101, 3201, etc., being the equivalents of A.D. century years. We thereby arrive at a very simple, and at the same time handy, method of determining the A.D. month, day and fraction of day, *marking the commencement of any Hindu solar year*. We believe it is the first time that this method has been used for this particular purpose.

**226**. At the end of Table VII are given constants for any period of 100, any period of 200, any period of 300 years, etc., up to a period of 3000 years. These figures lead us to a knowledge of the principal lunar cycles in the Indian Calendar, that is, periods of years after which new moons happen on the same day and more or less at the same hour of the Indian solar year. Thus we find from Table VII that, according to the Sûrya Siddhânta, a new moon after 19 complete S. Yrs. recurs on the same day of the S.Y. but 5 hrs. 24′ earlier.

| | | | | | | | |
|---|---|---|---|---|---|---|---|
| Do. | 122 | do. | do. | do. | do. | 2 ,, | 8′ later. |
| Do. | 385 | do. | do. | do. | do. | | 55′ ,, |
| Do. | 648 | do. | do. | do. | do. | | 17′ earlier. |
| Do. | 1315 | do. | do. | do. | do. | 6 ,, | ,, |
| Do. | 2329 | do. | do. | do. | do. | | 4½′ later. |

*N.B.*—2300, and 1300 years are the two most important lunar cycles according to Dr. Grattan Guinness, but he takes for comparison the *tropical* year and the Synodical month.

**227**. Under moon's anomaly the most important cycle is that of 43 years, as after 43 years, the anomaly increases by only ·00458 of a day, *i.e.*, 6 minutes and 3½ seconds.

**228**. After 46 years, mean new moons occur just 1 day 2 minutes and 3 seconds later. This is a useful fact to remember.

### TABLE V.

**229**. The exact method of using Table V for determining the *ending moments* of Nakshatras and Yogas is not explained, because the first two-thirds of this table are hardly necessary if Table XI is used. The third portion of Tab. V "Sun's Longitude for Nakshatras and Yogas" may be used in combination with the last column of Tab. VIII for determining the nakshatra *concurrent* with a tithi. Numerous examples of this method are given in

Secs. **36**, **40** and elsewhere, but our readers will probably prefer to use the alternative method provided by Tables XVII–A and XVII–C.

**230**. *Multiplication Table for Jupiter's Samvatsara.*—According to the Sûrya Siddhânta, Jupiter makes 364,220 revolutions (without *bîja* or correction) and 364,212 revolutions (with *bîja* or correction) in a yuga of 4,320,000 years. This means that the mean period of revolution of Jupiter is, with *bîja*, 4332·41581277 days, and without *bîja*, 4332·32065235 days. Each of these periods is nearly 12 years and a Jovian month or $\frac{1}{12}$ of the period of Jupiter's revolution is very nearly equal to a solar year. The Jovian month is therefore called in ordinary language a Jovian year and there are 1·0117 Jovian years in a solar year. Also a mean solar day = ·00277 of a Jovian year.

The multiplication table of Jupiter's samvatsaras is merely the multiplication of each of these quantities from 1 to 99. For further particulars regarding Jovian years, see sections **97** to **106** *supra*.

---

## CHAPTER XXVI.

### USE OF TABLE IV (VARA OR WEEK-DAY), TABLES III, II AND I.

**231**. The *vâra* or week-day is almost invariably quoted in Indian dates. The fact that week-days are the same in European and Indian reckoning (*e.g.*, a Monday in a date a thousand years ago corresponded, as it does to-day, to an Indian *Soma-vâra*, Tamil *Tingal*) is a striking proof of the common origin of the Indo-European mode of reckoning the week. This identity becomes all the more striking when we consider that everything else in the two reckonings (year, month, day of month, hour of day, etc.) is different. The week-day is, therefore, an important link between the two systems and it is well that we have an easy and at the same time a thoroughly accurate and reliable mode of identifying the week-day of any date, however remote, in the past. The ordinary rule laid down in Indian works on astronomical computation is to count the days from the beginning of Kaliyuga (18 Feb., 3102 B.C.) and cast off sevens—a truly formidable operation which few of our readers will venture upon.

*N.B.*—The number of days from the beginning of Kaliyuga up to any moment under consideration is called the *Ahargana*. This is constantly alluded to in Hindu works on astronomical computation and we have therefore given an *Ahargana* Table (Table XXII, last page of this work), but the reader will have no occasion to use it unless to verify allusions.

**232**. Table IV of the present work supplies an easy method of verifying the day of the week of the European Calendar. An equally simple, if not a simpler, method might be devised for discovering the week-day of any date of an Indian solar year, but inasmuch as all operations in the present work are directed towards ascertaining the A.D. or B.C. date corresponding to an Indian date, it will be enough if we are able to verify week-days, through the B.C. or A.D. equivalents of Indian dates.

**233**. The reader will observe in Table IV three lines in heavy type, consisting of the figures

**1, 2, 3, 4, 5, 6, 7, or 0.**

These are called co-efficients. Each century, year of a century and month of a year has its co-efficient which is shown in Table IV, and all we have to do is to add up the co-efficients for the component elements of a date, the day of the month being itself an additional co-efficient. Thus if we want to know the week day of 15 July 1910 we proceed as follows :

| | |
|---|---|
| (Table IV) co-effiicent of 1900 | **2** |
| " " of 1910 | **5** |
| " " of July | **5** |
| Day of month | 15 |
| Total | 27 |

Dividing 27 by 7, we have as remainder 6, which is equivalent to Friday, the 6th day of the week.

The 15th July 1910 was, therefore, a **Friday**.

It will simplify the operation if we cast off sevens in the very act of summing up the figures. Thus 2 + 5 being 7, we might neglect the first two figures, and 15 being 2 × 7 + 1, we need only add 5 + 1 which is 6 or Friday.

**234.** This method is applicable to any date, A.D. or B.C., **Old Style** or **New Style**, but the student should first understand the negative character of B.C. dates and also the difference between **Old Style** and **New Style**.

**235. B.C. dates.** All B.C. dates are negative. B.C. 44 is really the year "*minus* 44" in relation to A.D. 1. In determining week-days of B.C. dates, we should first of all convert the dates into positive figures : that is, for the century, we should take the next previous century increased by 1, and for the odd year we should take 101 *minus* the odd year we are dealing with.

Thus, supposing we want the week-day of 18 February 3102 B.C., the first day of Kaliyuga we procced as follows :—

| | |
|---|---|
| (Table IV). Co-efficient of 3201 B.C. (the century preceding 3102).......... | **3** |
| " " 101 *minus* 2 (the odd year before us) = 99........ | **4** |
| " " February (in an ordinary year)......................... | **2** |
| Day of month | 18 |
| Total | 27 |

Now, 27 divided by 7 leaves 6, *i.e.*, **Friday** which, according to all accounts, was the first week-day of Kaliyuga.

**236. Old Style.**—At present every fourth year A.D. is a leap year, but century years 1700, 1800, 1900, 2000 A.D. are leap years only if the first two figures are divsible by **4.**

This rule about century years was adopted, in most European countries (except Russia which still follows the Old Style) under the authority of a decree of Pope Gregory XIII, dated 1582, and in English speaking countries, under the authority of an Act of the British

Parliament, dated 1752. It was ordered by the same Act of Parliament (in order to correct the principal error of the Old Style,) that the day following 2 September 1752 (Wednesday) should be called the "14 September 1752" (Thursday) not the "3 September" (Thursday). This is the famous "dropping of 11 days" by Act of Parliament. **Wednesday, 2 September,** A.D. **1752** was therefore the last day of the Old Style in English-speaking countries and **Thursday, 14 September,** A.D. **1752** is the first day of the English New Style.

For Old Style dates, *i.e.*, dates down to, and inclusive of, 2 Sep. A.D. 1752, the reader should use the co-efficients of centuries appearing under "Old Style" in Table IV.

**237**. The reader will note that 1600 and 1700 appear under both Old Style and New Style. The reason is that in most European countries, except the United Kingdom and Russia, the New Style came into use in 1582, whereas 1600 and 1700 were Old Style in England.

**238**. Coming down to the co-efficients of odd years, we note that odd years of centuries have the same co-efficient, whether the style be Old or New, A.D. or B.C : only a B.C. odd year should first be deducted from 101 so as to render it positive. (*vide* Sec. **235**.)

**239**. Lastly, under co-efficients of months, we notice that the co-efficient of January in a leap year is 5, while in an ordinary year it is 6; and likewise February has 2 for its ordinary co-efficient, and 1 for its co-efficient in leap-years. The co-efficients of the other months do not change for leap-year.

**240**. In ordinary years, the week-day of the 1st of January is the co-efficient of the year. Thus the co-efficient of the year A.D. 1910 is 7 or 0, and the week-day of 1st January A.D. 1910 is also 7 or 0, *i.e.*, **Sunday**. The reason is that in ordinary years, the week-day of 1st January is the co-efficient of the year *plus* 6 + 1; and the addition of 7, *i.e.*, of a whole week, does not of course change the week-day.

In leap-years the week-day of 1st January is the co-efficient of the year *plus* 5 + 1; *i.e.*, in leap-years the week-day of 1 January is 1 less than the co-efficient of the year.

**241**. The student who has read the author's "Secret of Memory" will be able to dispense altogether with the use of Table IV for verifying week-days. For particulars of this interesting method, see "**Secret of Memory** *" Chapter **XIV**, p. 108.

### Tables III, II, and I.

**242**. The uses of these Tables will be self-evident. Tables II and III will be constantly handled by the reader, and their principal contents have, therefore, been included in a condensed form in the Eye-Table at the end of the book.

**243**. The "Limits of Adhika and Kshaya months" which form the middle portion of Table II present the whole of this intricate subject in a veritable nutshell. For explanation, See Sections **18** to **30**. When mean intercalations are required, that is, when Adhika months have to be determined without reference to anomalies, we should apply the "Limits of Adhika months" just as they are.

---

* Published by Higginbotham and Co., Madras—Price Rs. 2.

# PART III.—CONSTRUCTION OF THE TABLES.

[This part is intended to be of use in criticizing the method employed in the present work as well as in suggesting further improvements. It will be of immediate interest to those readers who are acquainted with one or other of the existing methods connected with the well-known names of JACOBI, KIELHORN, SCHRAM, SEWELL AND DIKSHIT, CHHATRE, etc., besides the older and less known names of WARREN and JERVIS. The general reader who makes his first acquaintance with the subject in the pages of this work will find the present part thoroughly intelligible and exceedingly interesting, provided he has mastered the first two Parts.]

## CHAPTER XXVII.

### TABLES FOR CONVERTING SPACE INTO TIME.

**244**. The method of calculating and verifying Indian dates, presented in this work, is intended to be of service to the general reader as well as to the scientific expert, the epigraphist, the archæologist, and the historian. The method is perfectly simple and at the same time absolutely correct according to the Siddhântas, so that it has become possible for the first time to dispense with all manner of approximations and rough and ready methods which, however valuable in the hands of an expert, are apt to mislead and confuse, more often than they assist, the general reader. The principal device by which this combination of extreme simplicity with accuracy and absolute fidelity to the original authorities, has been accomplished, is the reduction of all quantities required for calculation to whole days and fractions of a day.

**245**. The civil day (with its multiple, the week,) is the one measure of time that is common to European and Indian reckoning, everything else (year, month, ghaṭikas, palas hours, minutes, seconds,) being different in the two systems. Accordingly, the civil day and decimals of a day have been adopted throughout this work for expressing all manner of Indian dates as well as for working out ending moments of *tithis*, etc., and also for verifying the correspondence of English and Indian dates. Any decimal of a day can be converted readily into Indian *ghatikas* and *palas* (Tamil, *naligais* and *vinâdis*) or English hours, minutes, and seconds by means of Tables XIX and XX. To assist the reader in very exact computation, the fractions expressing minutes have been carried far enough to show the recurring places. If, for instance, we wish to know how many hours, minutes, and seconds are equivalent to ·40490 of a day, we turn to Table XX and find that ·40486 of a day is equal to 9 hours 43 minutes. The remainder of the decimal fraction is ·00004 which, the same table informs us, is between 3 and 5 seconds. So the answer is, 9 hours 43 minutes 4 seconds. The same decimal fraction is equivalent in Indian time (as we may see from Table XIX) to 24 ghaṭikas 18 palas. Ghaṭikas and hours, as fractions of a day, are also shown in the Eye-table at the end of the book.

**246**. The reader will notice that in the first three parts of the present work as well as in the connected tables (except the Subsidiary Tables VI-*a*, VII-*a*, and VII-*b*, which are intended mainly for purposes of comparison between the present tables and those of previous writers on the same subject,) measures of time alone are used, and that measures of space, *i.e.* (*degrees* of celestial longitude, *degrees* of mean anomaly of the sun and moon, etc.) have been altogether excluded. This is the principle known as Largeteau's method, which was first applied to Indian astronomical computation by Professor Jacobi in 1888. Messrs. Sewell and Dikshit have applied the same principle in their "*Indian Calendar*" (1896).

**247.** The present method is founded on Largeteau's principle, but differs essentially from it as well as all previous applications of it in one important respect. Instead of using Largeteau's method to discover how much *space* has been accomplished at a particular moment of time, the present writer has used the method of day-spaces or space days to discover the *moment of time* at which a particular extent of space has been accomplished. Thus, instead of determining the expired portion of a tithi, corresponding to a given moment of time (*i.e.*, generally, to mean sunrise on a particular day,) as is done by Messrs. Jacobi and Sewell, and then calculating the unexpired portion of the tithi by means of successive approximations, the present writer investigates, directly and once for all, the *ending moment of a tithi*, the very thing required by Indian usage.

**248.** By setting this object steadily in view, the author has been enabled to reduce to two or three very simple and easy steps, Messrs. Sewell and Dikshit's method, which covers a page and a half (pp. 81, 82) of their "Indian Calendar"; likewise he has considerably abridged Professor Jacobi's process, which consists, in the first place, of an approximation, on Largeteau's method, and thirteen or fourteen *subsequent* steps (*a*) to (*m*), as expounded in Volumes I and II of "*Epigraphia Indica*". As regards Mr. Dikshit's own *very accurate* but *very tedious* method, covering *several pages* of the introduction to Dr. Fleet's "*Gupta Inscriptions*", the present writer has been successful in arriving at *absolutely the same result* as Mr. Dikshit in two lines of working. (See Sections **223, 259**.)

**249.** The principles upon which space was converted into time for the purposes of the present work are set forth in the following paragraphs.

The principal measure of space, the distance of the moon from the sun, was converted into days in the ratio of 29·530587946 days to 360° in the case of the Sûrya Siddhânta and 29·5305925 days to 360° in the case of the Ârya Siddhânta. Subsidiary Table VI-A is the conversion Table for Sûrya Siddhânta.

The increase of the moon's age, according to the *Sûrya Siddhânta*, for each solar year is according to the above rate of conversion, 10·891701134 days. Instead of reckoning the increase of the moon's age, however, the present method reckons directly the *retardation in the date of appearance of the first new moon* in each solar year, for which purpose it is, of course, necessary to deduct 10·891701134 days from 29·530587946 days: result, 18·638886812 days. This, then, is the number of days by which the appearance of the first new moon is retarded each year, and the first thing to do every year is to calculate the interval of retardation for that year. The interval (if we take the retardation for one year) will, *ipso facto*, be the date of appearance of the first or *Vaisâkha* new moon in solar year 1 of Kaliyuga (expired). From this date all other mean new moons for that or any subsequent solar year may be found by the successive addition of multiples of 29·53059 days; and the mean ending moment of every *tithi* is given by the addition of the *tithi* equivalent in days (according to the Eye-Table) to the date of mean new moon. Precisely the same method was followed for the Ârya Siddhânta, Lalla's corrections being introduced at the appropriate date.

**250**. The mean anomaly, in the case of the moon as well as that of the sun, was reckoned from perigee, as in Professor Jacobi's article in the *Indian Antiquary* (1888), and not from apogee, as in his articles in Volumes I and II of *Epigraphia Indica*. For the purpose of Table IX, the moon's mean anomaly was converted into days in the ratio of 27·554599899 days to 360° in the case of the Sûrya Siddhânta, and of 27·554566936 days to 360° in the case of the Ârya Siddhânta. The increase of the moon's mean anomaly for a single solar year is thus :—

Sûrya Siddhânta, 7·048957797 days.
Ârya Siddhânta, 7·049310381 days.

The anomaly of the moon at the first moment of Kaliyuga was taken as 90° from perigee, that being the figure according to all the authorities. Subsidiary Table VII (*a*) is the conversion table for moon's anomaly according to the Sûrya Siddhânta. From the year A.D. 1600, the corrected period of the anomalistic month (27·55459797 days) has been adopted for Sûrya Siddhânta calculations.

**251**. The moon's mean anomaly, as entered in Table IX, corresponds to the *tithi* or space accomplished, while the equation is the addition to or deduction from the tithi, to be made in order to arrive at the *time* or *ending moment* of the tithi. Consequently, the equation in degrees was in every case added to or deducted from the mean anomaly, and the result, converted into days at the rate of 27·5546 days to 360°, is entered as the anomaly in Table IX.

**252**. For the purpose of Table IX–*c*, *f*, and *h*, the sun's mean anomaly, reckoned from perigee, was converted into days at the rate of 365·25875 days to 360° [Subsidiary Tab. VII (*b*)] and the sun's equation of the centre was converted at the same rate as the distance of moon from sun (29·53059 days to 360°), because the principal use of the sun's equation of the centre is to correct the mean into the true distance of the moon from the sun. The moon's equation of the centre in Table IX–*a*, *b*, *d*, *e*, *g*, was converted at the same rate, and for the same reason, from degrees into decimals of a day.

*N.B.*—In the Nakshatra and Yoga Equation Tables, IX–*i*, *j*, *k*, *l*, equations in degrees were converted into days at the rate of 27·32167 days for 360° in the case of Nakshatras, and of 25·4202 days for 360° in the case of Yogas.

**253**. The actual sun's anomaly, as entered in Table IX, is the result of a series of transformations which had to be carefully executed for each of the 680 stages of the anomaly. According to the Sûrya Siddhânta, the mean anomaly of the sun at the commencement of each solar year, neglecting the slow motion of the perigee which is allowed for in that Siddhânta but not in the others, is 102° 45′ or, in solar days, 104·25093 days. This amount has therefore to be added to the number of days expired in the solar year at a given moment in order that the sun's anomaly for that moment may be correctly expressed in days. From the sum, 2·1707 days have to be deducted for Sodhya. Thus the equation entered in Table IX (*c*) against the 182nd day of the solar year is the equation which really belongs to the following mean anomaly ; *viz.*, 182 days *plus* 104·25093 days *minus* 2·1707 days, that is, an anomaly of 284·0802 days : the latter figure, which is not required in practice, has wholly disappeared from Table IX (*c*), and the equation is simply entered against the day of the solar year for which, in practice, it is likely to be required.

**254**. For *nakshatras*, the following formula was used: (moon's longitude *minus* sun's longitude) *plus* sun's longitude = moon's longitude for *nakshatras*. As moon's *minus* sun's longitude is expressed according to the ratio 360° = 29·53059 days, the sun's longitude is expressed in the same ratio. Solar days are thus expressed in terms of the synodical month and the result of the reduction, after allowing for śodhya, is exhibited for whole days in the last column of Table VIII, "Sun's longitude for Nakshatras". *Fractions* of solar days can be converted into lunation-longitude by means of the third multiplication table in Table V. The first two multiplication tables in Table V were originally intended to be of use in determining the ending moment of a yoga or nakshatra, but in practice, either Table XI or Tables XVII-A and XVII-C will be found much easier and handier than Table V for the investigation of all kinds of problems connected with nakshatras and yogas. Table XI is referred to in Secs. **263** to **265** *infra* and Tables XVII-A and XVII-C in Sec. **266** *infra*.

**255**. The moment of *sunrise* for any latitude and longitude in India can be ascertained by means of Table XIII, based on the rules and table of *asus* given by Professor Jacobi in Volume I of the *Epigraphia Indica*. Professor Jacobi has himself given detailed tables for sunrise in Volume II of the same publication, but the results achieved by means of those tables can, it is believed, be more easily arrived at by the present Table XIII. For the purpose of determining the equation of time for each day of the solar year, the sun's equation of the centre, according to Table IX (*c*), was used, with the sign changed : likewise the *asus* given in Professor Jacobi's table and reproduced in Sec. **76** *supra*, had to be suitably modified.

Thus (according to Professor Jacobi's table) in the10th degree of Northern latitude 30 degrees of Sign 1 of the Zodiac take 1544 *asus* or 1544 × 4 seconds of time to rise;

or in lunation-longitude,

$$\left(\frac{30}{360}\right) \times 29{\cdot}53059, \text{ or } 2{\cdot}46088 \text{ units of space take } 1544 \times 4 \text{ seconds of time to rise;}$$

$$\therefore 1 \text{ unit of space takes } \frac{1544 \times 4}{2{\cdot}46088} = 1544 \times 1{\cdot}62 = 2510 \text{ seconds.}$$

This 2510, then, is the factor by which each day's equation of the centre according to Table IX (*c*) should be multipled (so long as the sun is in the first Sign) in order to give that day's equation of time.

And generally, all the *asus* in Professor Jacobi's table were multiplied by $\frac{4}{2{\cdot}46088} = 1{\cdot}62$, and the factors thus obtained were multiplied again by each day's equation according to Table IX (*c*) in the present work : the result was each day's equation of time in seconds as entered in Table XIII of this work. The same result could of course have been arrived at directly from Table XVII-A, "Sun's Equation in degrees for each day of Solar Year".

**256**. For using the tables in *Epigraphia Indica*, Vol. II, the sun's longitude (first sidereal and then tropical), corresponding to a given day of the solar year, has to be first determined from special tables in Vol. I, *Epigraphia Indica* ; then the *vinâdis* given under each degree of latitude and for each Sign of the Zodiac in Vol. II, *ibid.*, have to be multiplied by the equation of the centre, which in turn has to be calculated from the sun's mean anomaly, applied with proportional parts to Special Table XXIV-B in Vol. I. *Epigraphia Indica*: only then can the moment of sunrise for a given day in the solar year be ascertained.

**257.** In the present Table XIII on the other hand, the moment of sunrise for any day of any solar year is obtained by simply adding a figure in the column "Tropical Longitude" to the figure opposite the given day of the solar year in the column "Equation of Time"; and the result is the correction, in seconds of time, to be applied to mean sunrise at Lankâ (6 a.m.) to determine the local sunrise for the given latitude.

**258.** Column 3 under each degree of latitude in Table XIII is a further step towards simplification, for it gives, for the 80 years ending 1920, the *total* correction to be applied to mean Lankâ time in order to arrive at the local time, for each of 30 important places, including Calcutta, Madras and Bombay. In this column the difference in time corresponding to the longitude of each place, measured from Ujjain, is added to the correction for sunrise for each day of the solar year and for each degree of latitude.

*Example.—Ashâdha Sukla* 12, *Kaliyuga* 3585 (*expired*) A.D. 484.

**259.** A single test problem, the same as that selected by Mr. Dikshit in his Introduction to *Dr. Fleet's Gupta Inscriptions* as well as by Professor Jacobi in Vol. I of *Epigraphia Indica*, will suffice to demonstrate the absolute reliability and extreme simplicity of the above processes as carried out in the present work. The problem is to determine the ending moment of *Ashâdha sukla dvâdasî* in *Kaliyuga* 3585 (expired), A.D. 484. The ending moment has to be determined (*a*) for Lankâ and (*b*) for the latitude and longitude of Eran (Lat. 24°, Long. 78° 15'). Lastly, the problem has to be worked out, first according to the *Sûrya Siddhânta*, and then according to the *First Arya Siddhânta*.

**SURYA SIDDHANTA.**

| References to Tables. | Days of Solar Year. | Moon's mean Anomaly in days. |
|---|---|---|
| (Table X): Kaliyuga 3585; A.D. 484; * March 18·2212. | | |
| First New Moon in Solar year ... | 24·3900 | 4·494 |
| (Table VIII): Âshâḍha śukla 12 ... | 70·8734 | 15·764 |
| | 95·2634 | 20·258 |
| | | (☉'s Eqn.)—·0455 |
| Table IX (*c*): ☉'s Eqn. for 95·26 days of Solar Year = | —·0455 | |
| Table IX (*b*): ☾'s Eqn. for anom. of 20·212 days = | +·4138 | 20·212 |
| | +·3683 + ·3683 | |
| * English month, day, and fraction of day marking | 95·6317 | |
| commencement of solar year ... | *March 18·2212 | |
| | 113·8529 | |

Our result is: the tithi ended at ·8529 of a day, *i.e.*, at 51 ghaṭikas 11 palas on the 113th day of the English Calendar, counting from 1st March.

The reader will be pleased to note that this absolutely correct result for the ending moment of a tithi is obtained by *simply adding* up six or seven figures from Tables VIII, IX, and X, and that absolutely no other process is required for any tithi in any year.

**260.** Now, by Table VIII, the 113th day of the English Calendar, counting from 1st March, is 21st June ;

by Table XIX, ·8529 of a day = 51 ghaṭikas 11 palas ;
by Table IV we arrive at the week-day as follows :—

| | | |
|---|---|---|
| Co-efficient of 400 A.D. | ... | 2 |
| Co-efficient of odd year 84 | ... | 0 |
| Co-efficient of June | ... | 3 |
| Day of the month | ... | 21 |
| | | 26 |

Since 26, divided by 7, leaves remainder 5 = THURSDAY, the final answer is THURSDAY, 21st JUNE, A.D. 484, 51 GHATIKAS 11 PALAS after mean sunrise at Lankâ.

NOTE.—This is the absolutely correct ending moment, according to Mr. Dikshit ; but Professor Jacobi arrives at a result which is 4 palas short. To arrive at the latter result, ☾'s + ☉'s Eqn. must be added as in foot-note to p. (83) to ☉'s Anom. before ascertaining ☉'s Eqn., just as ☉'s Eqn. is in the actual working *supra*, added to ☾'s anom. before ascertaining ☾'s Eqn. This extra step, however, is a nicety seldom required in practice, since the error on this account can never exceed 6 palas. See Sec. **212** *supra*.

TO DETERMINE THE ENDING MOMENT OF THE ABOVE TITHI IN TRUE LOCAL TIME AT ERAN.

**261.** We first of all find the sun's sidereal longitude for 95·85 days, for which purpose we deduct the *sodhya*, 2·17 days. Remainder, 94 days nearly.

We turn to Table XIII and bring down the entry under Latitude 24° corresponding to the 94th day of the Solar Year, for ☉'s trop.* long. and the entry corresponding to the 95th day for the equation of time.

(1) Equation of time (95th day) : + 148 seconds of time.

(2) ☉'s tropical longitude (94th day) : +2647 seconds of time.

(3) To these figures from Table XIII we add the time-difference for the longitude of Eran (+2·53 degrees Ujjain longitude), namely, +2·53 × 240 or +608 seconds of time. Total : +148 + 2647 + 608 = +3403 seconds of time.

Now 3403 seconds of time, divided by 60, are 56 minutes 43 seconds or ·0393 of a day. Adding this to the mean Lankâ time already arrived at, *viz.*, ·8529, we obtain, as TRUE LOCAL TIME at ERAN for Âshâḍha Sukla 12, 3585 Kaliyuga, ·8922 of a day or 53 GHATIKAS 32 PALAS which is exactly the same as Mr. Dikshit's result.

**ACCORDING TO 1st ARYA SIDDHANTA.**

**262.** The simplest way of arriving at the ending moment of the above tithi according to the first Ârya Siddhânta—and this was apparently the method adopted by Mr. Dikshit—is to deduct from the ending moment, already arrived at, the difference between the *Sodhyas* of the Sûrya and Ârya Siddhântas (2·1707 days *minus* 2·1476 days = ·0231 day), this being the

* NOTE.—In the solar year, 3585 K.Y. we make no correction for the difference between ☉'s sidereal and tropical longitude because in 3600 K.Y., *i.e.*, only 15 years later, the ☉'s sidereal coincided with the ☉'s tropical longitude.

only difference between the Sûrya and Ârya Calendars at the epoch 3585 K.Y. We thus obtain ·8530 day less ·0231 day = ·8299 day= 49 GHATIKAS 48 PALAS (Table XIX). This is exactly Mr. Dikshit's result for the mean Lankâ time of the ending moment of the tithi according to the 1st Ârya Siddhânta. We may make the same deduction for true local time at Eran according to the same Siddhânta, and obtain Mr. Dikshit's result for this also.

NOTE.—There was also a difference of ·001 of a day, *i.e.*, $1\frac{1}{3}$ minutes or 4 palas between the commencement of the solar year, K.Y. 3585, according to the Surya and Arya Siddhantas (*vide* Tables VI and VII in the present work); but apparently Mr. Dikshit neglected this trivial difference.

## CHAPTER XXVIII.

### CONSTRUCTION OF TABLE FOR NAKSHATRAS AND YOGAS.

**263.** Table XI contains a very easy method of calculating directly the absolute ending moment of a *nakshatra* or *yoga.*

The interval between new moon and any particular nakshatra $n$ may be expressed by the formula $\frac{29{\cdot}5306-A}{q}+n\times d.$

where $A$ is the sun's longitude at new moon, expressed in terms of a lunation as in the last column of Table VIII, $n$ is the numerical order of the given nakshatra, counting from Revatî, $d$ is the mean duration of a nakshatra in days, and $q$ is the ratio of the length of the synodical to that of the sidereal month.

The above expression may be expanded into

$$\left\{29{\cdot}5306-\left(f+al-2{\cdot}1707\right)\times m\right\}\times\frac{1}{q}+n\times d,$$

where $f$ is the moment of occurrence of the first New Moon in a Solar Year, as given in Table X, $a$ is the number of lunations completed since the commencement of the Solar Year, $l$ is the period of a lunation in days (*i.e.*, 29·5306 days); 2·1707 days are the *sodhya*, $m$ is the factor for converting Sun's longitude into lunation space, according to Table V (last column); $\frac{1}{q}$ is the factor used in the 1st column of Table V $\left(i.e., \frac{1}{1{\cdot}0808}\right)$ and $n\times d$ is the collective duration in days of Nakshatras counted from Revatî up to the end of the given Nakshatra.

We may expand the above expression again as follows:

$$29{\cdot}5306\times\frac{1}{q}+2{\cdot}1707\times m\times\frac{1}{q}+n\times d-f\times m\times\frac{1}{q}-al\times m\times\frac{1}{q}.$$

Now all these expressions connected by + or —, except, $f\times m\times\frac{1}{q}$ may be calculated *once for all* for every *Nakshatra* in the lunar year, for

$$29{\cdot}5306\times\frac{1}{q}=27{\cdot}32167\ ;$$

$$2{\cdot}1707\times m\times\frac{1}{q}=\ {\cdot}16237\ ;$$

And the expression $({\cdot}16237+n\times d-al\times m\times\frac{1}{q})$, when calculated for each Nakshatra becomes one of the figures given in Table XI under "Interval between New Moon and each Nakshatra".

There remains $f\times m\times\frac{1}{q}$, whose maximum value is $29{\cdot}5306\times m\times\frac{1}{q}=2{\cdot}20891$ (the difference between 29·530587946 and 27·321674163, the lengths of the synodical and sidereal months).

The values of $f \times m \times \frac{1}{q}$, less than the maximum, may be calculated by means of the Nakshatra correction Table in Table XI.

Accordingly all that we have to do, in order to know the mean ending moment of any Nakshatra, is to take the corresponding figure from the "Interval" Table XI and deduct the value of $f \times m \times \frac{1}{q}$ as obtained from the "correction" table in Table XI.

*N.B.*—The correction tables might have been made additive instead of subtractive, by (1) subtracting all the quantities in the correction tables from 2·20890 in the case of Nakshatras and from 4·11036 in the case of Yogas, and (2) subtracting 2·20890 and 4·11036 respectively from all the intervals in the body of Table XI. This will be attended to in a future edition of the work, should a second edition be called for.

**264.** Thus, supposing the mean ending moment of *Múla* Nakshatra in the month of (lunar) *Kârttika* in A.D. 1910, is required :—

| | |
|---|---|
| The mean ending moment of *Kârttika* New Moon is (by Table X) November | 2·30 |
| The integer and first two decimal places of the "Interval" for "*Kârttika* Múla" (Table XI) | 6·13 |
| | Nov. 8·43 |
| Deduct $f \times m \times \frac{1}{q}$, *i.e.*, the equivalent of 25·91 according to the correction table in Table XI ; 1·87 + ·07 = 1·94. | −1·94 |
| | Nov. 6·49 |

This is the mean ending moment of the Nakshatra *Múla* in Lunar Kârttika, A.D. 1910.

If we want the absolute ending moment, we calculate the anomaly :

| | | Days. |
|---|---|---|
| Moon's mean anomaly at commencement of Solar Year | ... | 0·535 |
| *Add* increase of anomaly up to Kârttika New Moon | ... | 11·866 |
| *Add* "Interval" less $f \times m \times \frac{1}{q}$ | ... | 4·19 |
| | | 16·59 |
| Corresponding Equation for Nakshatras (Table IX–*k*) | ... | +·246 |

Adding this equation to the mean ending moment, Nov. 6·49, we have Nov. 6·74 or (by the Eye-Table) 44½ ghaṭikas after mean sunrise on Nov. 6, A.D. 1910. We could, of course, if we had liked, have calculated the absolute ending moment to four places of decimals instead of to two.

**265.** *Yogas* are worked in the same manner, by using the corresponding portions of Table XI, *i.e.*, by deducting from the "Yoga Interval" (for the given lunar month and given Yoga) the correction for the Solar Year according to the Yoga correction table in Table XI.

**266.** When a *Yoga* and *Nakshatra* are cited with the *tithi* as in the example in Sec. **98** *supra*, it will generally be enough to ascertain the nakshatra and the yoga which were concurrent with the tithi and this can most easily be done as follows :—

Sun's anomaly at ending moment of tithi, 203·31 days.

| | |
|---|---|
| By Table XVII–A, the ⊙'s longitude corresponding to 203 days is | 197·94° |
| " XVII–C, " " " ·31 day is | ·31° |
| | 198·25° |
| The tithi being the 9th, the moon's elongation was 9 × 12° = | 108·00° |
| Total | 306·25° |

This (by Eye-Table) is the longitude for Nakshatra *Dhanishtha.*

For. *Yoga* we add 2 × Sun's Longitude to Moon's Elongation.

That is, the yoga longitude is 108·00° + 2 × 198·25° = 108·00° + 396·50° = 504·50°

Deducting 360° which is one complete revolution, we have 504·50° − 360° = 144·50°, which (by Eye-Table) is the Longitude of No. 10 *Ganda* Yoga.

**267**. When we wish to find the ending moment of a Nakshatra or Yoga, we should apply Table XI as follows :—

In our example, the mean tithi ended at 8·86 days after *Kârttika New Moon.* The age of the first new moon in the solar year being 20·27 days, the corresponding Nakshatra correction is (by Table XI) 1·50 + ·02 = 1·52 = (by Eye-Table) 1 day 31 ghaṭikas and the Yoga correction (by Table XI) 2·78 + ·04 = 2·82 = 2 days 49 ghaṭikas.

From Table XI-A under col. VII Kârttika we obtain—

| | **Nakshatra.** | | | | **Yoga.** | | |
|---|---|---|---|---|---|---|---|
| | d. | gh. | p. | | d. | gh. | p. |
| No. 23 Dhanishṭha | 10 | 10 | 59 | No. 11 Vriddhi | 11 | 24 | 59 |
| *Deduct* correction | 1 | 31 | | | 2 | 49 | |
| | 8 | 40 | | | 8 | 36 | |

*N.B.*—The coincidence of a particular *Yoga* with a particular *Nakshatra* happens only once a year, and we could, by means of this coincidence, discover the month and the tithi even if these were not quoted.

This is the Tithi interval, 8 d. 51½ gh—11½ gh. This is the Tithi interval 8 d. 51½ gh.—15½ gh.

We can, with the last line of corrections, use for the Nakshatra and Yoga the ☉'s and ☾'s Anomalies already found (Sec. **98**) for the tithi, *i.e.,* ☉'s Anom. 206 days 18½ gh.; ☾'s Anom. 25 days 16 gh.

Our Nakshatra Eqn. will be that for ☾'s Anom. of 25 d. 16 gh. *less* 11½ gh., or for 25 d. 4½ gh.: Eqn. (Tab. IX-*l*) is + 11½ gh.

Our Yoga ☉'s Anom. will be 206 d. 18½ gh. *less* 15½ gh. or 206 d. 3 gh., for which ☉'s Yoga Eqn. (Tab. IX-*i*) is + 7 gh. 42 p.

Our Yoga ☾'s Anom. will be 25 d. 16 gh.—15½ gh. = 25 d. ½gh. for which ☾'s Yoga Eqn. (Tab. IX-*i*) is + 10 gh. 55 p.

The sum of ☉'s and ☾'s Eqns. for Yoga is + 7 gh. 42 p. + 10 gh. 55 p. = + 18 gh. 37 p.

These equations can be added to the mean ending moments of Nakshatra and Yoga.

The equation being added to Nakshatra mean ending moment, we have, as the absolute ending moment of the Nakshatra, Oct. 19·76, *i.e.,* 42½ gh.—11½ gh. + 11½ gh. = 42½ gh. on 19 Oct. A.D. 1474.

The Yoga ended at 42½ gh. — 15½ gh. + 18½ gh. = 45½ gh., when tithi also ended (*vide* Sec. **98**).

Now Dhanishṭha Nakshatra ends when moon's longitude is 306·7° (Eye-Table) and since the moon is said to have been in Kumbha râśi, her longitude must have been between 300° and 330° (Eye-Table).

Since the event happened between 300 and 306·7 degrees of moon's longitude, the exact hour is defined as being less than 30 ghaṭikas before the end of the Nakshatra (since by Table XVII-D the moon's long. increases by 6·7° in ·50 of a day).

*N.B.*—With reference to line 5 of Sec. **98**, the Tithi being the 18th, the *karana* (*vide* list in Table III) must have been *Kaulava* and must have ended at the same moment as the 9th tithi (for the reason explained in Sec. **55** *supra*).

# CHAPTER XXIX.

## INVESTIGATION OF ADHIKA AND KSHAYA MONTHS.

**268.** *Adhika* months are the cream of the Indian Calendar, while *kshaya* months are its *crème de la crème.* Figures of speech apart, it is certainly true that the success or failure of any computer in deducing *adhika* and *kshaya* months is the measure of his success or failure in dealing, as a whole, with the Indian Calendar. How far the present method satisfies this ordeal, will be for competent judges to decide.

**269.** Two in dependent English lists of *adhika* and *kshaya* months are at present in existence, the first by Prof. **Chhatre** (reproduced in *Ind. Antiq.*, Vol. XXIII, pp. 105–108), and the second by Messrs. **Sewell** and **Dikshit**. The lists in **Patell's** *Chronology* and **Cunningham's** *Indian Eras* are obviously copied, without check, from Chhatre whose reputation was and is, sufficient to justify such a procedure.

Messrs. Sewell and Dikshit certainly exercised an independent and erudite judgment in revising Prof. Chhatre's list, and they declare, in several parts of their work, the indubitable superiority of their list to that of *Chhatre.*

**270.** For the purposes of Table X in the present work, it was necessary to weigh carefully the merits of Mr. Chhatre's and Mr. Dikshit's calculations wherever they differed. The palm must no doubt be awarded, as a general rule, to Mr. Dikshit, but he seems to have failed, by oversight, to take notice of the *kshaya* months in A.D. 507 and A.D. 751 which are investigated below.

### Adhika Months.

### A.D. 629.

**271.** In regard to this year Messrs. Sewell and Dikshit are at some pains to explain (under *Additions and Corrections*, p. 150 of "Indian Calendar") that Chhatre's entry of Adhika Kârttika is wrong and that their own entry of Adhika Âśvina is correct. Let us verify this statement, and in so doing, note the comparative brevity and simplicity of our method.

| | Days. | ☉'s and ☾'s Eqns. | ☾'s Anom. | |
|---|---|---|---|---|
| First New moon in Solar Year A.D. 629, | 10·2145 | | 20·456 | |
| Add for 7th New moon | 177·1835 | −·172 | 11·856 | |
| | | −·342 | | |
| | 187·3980 | | 32·312 | |
| | −·514 | −·514 | −·172 | ☉'s Eqn. |
| 7th or Tulâ Sankrânti 186·9355: | 186·884 | | 32·140 | |
| | | | 27·555 | |
| | | | 4·585 | |

The 7th new moon was therefore *Asvina,* and obviously the 6th new moon also was an *Asvina*; so that the Adhika month was, as stated by Messrs. Sewell and Dikshit, Âśvina, not as stated by Chattre, Kârttika.

### A.D. 979.

**272.** In each of the next two cases Mr. Chhatre was out of reckoning by a very small margin. For A.D. 979 Messrs. Sewell and Dikshit have given *Srávana* as the Adhika month, whereas *Bhâdrapada* is the Adhika month in that year according to Prof. Chhatre's list. The calculations must in all these cases be made by the *Sûrya Siddhânta* and the foot-note to p. (9) of the text should be applied.

| | Days. | ☾'s Anomaly |
|---|---|---|
| First New moon in Solar Year A.D. 979, | 7·5649 | 5·023 |
| Add for 5th New moon | 118·1223 | 7·904 |
| | 125·6872 | 12·927 |
| Sum of ⊙'s and ☾'s Eqns. : −·1259−·1005= | −·2264 | −·126 (⊙'s Eqn.) |
| 5th or Simha Sankrânti 125·4755 : | 125·4608 | 12·801 |

The Adhika month was *Srâvana* by ·0147 day, *i.e.*, nearly one ghaṭika, and in this instance Messrs. Sewell and Dikshit are right. Our Table X shows Adhika Srâvana.

### A.D. 1199.

| | Days. | ☾'s Anom. |
|---|---|---|
| First New moon in Solar Year | 3·3683 | 8·539 |
| 3rd New moon in „ | 59·0612 | 3·952 |
| | 62·4295 | 12·491 |
| | | +·055 (⊙'s Eqn.) |
| | | 12·546 |
| Sum of ⊙'s + ☾'s Eqns. + ·0555 − ·1261 = − ·0706 | − ·0706 | |
| 3rd or Mithuna Sankrânti 62·3555 : | 62·3589 | |

The Adhika month, 3rd new moon, was *Ashâdha,* as stated by Messrs. Sewell and Dikshit, not *Jyeshtha,* as determined by Chhatre. Our Table X shows Adhika Âshâḍha.

It happens that this very interesting Adhika month, the 3rd lunar month in the year A.D. 1199—1200 is quoted in a South Indian inscription, discussed by Dr. Fleet at p. 156 of the *Indian Antiquary,* Vol. XIX (1890). In the inscription itself the Adhika month is identified as *Ashâdha,* not as *Jyeshtha.* Of course the margin of difference is so small (5 minutes), that a calculator might, without being convicted of error, place an *adhika* month of this description on either side of the Sankrânti. If we did not know that the difference was so small, we should be at a loss to account for the discrepancy between the statement in the inscription and Prof. Chhatre's list. As it is, a more striking instance cannot be imagined than this actual one, to demonstrate the value and importance to an epigraphist of a knowledge of how to calculate tithis accurately.

## Kshaya Months.

### A.D. 507.

| | Days. | ⊙'s and ☾'s Eqns. | | ☾'s Anom. | |
|---|---|---|---|---|---|
| First New moon in Solar Year | 10·1256 | —·129 | | 14·587 | |
| Add for 8th New moon in Solar Year | 206·7141 | —·064 | | 13·832 | |
| (1) 8th or Vrischika Sankrânti 216·8289 : | 216·8397 | —·193 | =216·6467 | 28·419 | |
| | | | | 27·555 | |
| | | | | 0·864 | |
| | | | | — ·129 | (⊙'s eqn.) |
| | | | | 0·735 | |
| | | +·039 | | 0·864 | |
| Add for 10th New moon | 59·0612 | —·355 | | 3·952 | |
| (2) 10th or Makara Sankrânti 275·637 : | 275·9009 | —·316 | =275·5849 | 4·816 | |
| | | | | + ·039 | (⊙'s eqn.) |
| | | | | 4·855 | |
| | | +·120 | | | |
| | | | | 4·816 | |
| Add for 11th New moon | 29·5306 | —·412 | | 1·976 | |
| (3) 11th or Kumbha Sankrânti 305·085 : | 305·4315 | —·292 | =305·1395 | 6·792 | |
| | | | | + ·120 | (⊙'s eqn.) |
| | | | | 6·908 | |

**273**. From (1) it follows that the 8th new moon was Kârttika and it is apparent from the time of occurrence of that new moon (1 ghaṭika before the end of the solar month) that the 7th new moon also must have been a *Kârttika.*

From (2) it follows that the 10th new moon was Pausha. [Foot-note to p. (9) of Text.]

From (3) it follows that the 11th new moon was Phâlguna. [ Do. do. ]

∴ The new moon between Pausha and Phâlguna, viz., *Mâgha,* was *kshaya* in this year.

It is also evident from the time of occurrence of this Phâlguna that the 12th new moon also must have been a Phâlguna.

Against this year Messrs. Sewell and Dikshit have noted only "Adhika Phâlguna", whereas Chhatre mentions also a *kshaya* month, Pausha. We have just seen that the kshaya month was really *Mâgha*, which is accordingly noted in our Table X.

Nor would the case have been different under the Ârya Siddhânta, since, as we remarked in the foot-note to Sec. **161** *supra,* the new moons under the Ârya Siddhânta at this epoch occurred only ·0231 of a day before the Sûrya Siddhânta new moons, and this would have made no difference as regards *adhika* and *kshaya* months in the present case.

## A.D. 751.

**274**. Only an *Adhika* Chaitra without a *kshaya* month is noted in this year by Messrs. Sewell and Dikshit, whereas Prof. Chhatre notes *Adhika* Kårttika and *Adhika* Chaitra besides Mårgasira *kshaya*. Let us work for the different new moons :—

| | Days. | ⊙'s and ☾'s Eqns. | ☾'s Anom. |
|---|---|---|---|
| First New moon in Solar Year A.D. 751: | 10·3034 | −·172 | 26·321 |
| Add for 7th New moon | 177·1835 | −·304 | 11·856 |
| (1) Beginning of 7th Solar month 186·9355 : | 187·4869 | −·476=187·011 | 38·177−·172 (⊙'s Eqn.)−27·55 =10·450 |
| | | | 27·555 |
| | | −·128 | 10·622 |
| Add for 8th New moon | 29·5306 | −·133 | 1·976 |
| (2) Beginning of 8th Solar month 216·828 : | 217·0175 | −·261=216·756 | 12·598−·128 (⊙'s Eqn)=12·470 |
| | | −·052 | |
| Add for 9th New moon | 29·5306 | +·077 | 1·976 |
| (3) Beginning of 9th Solar month 246·3192 : | 246·5481 | +·025=246·573 | 14·574−·052 (⊙'s Eqn.)=14·522 |
| | | +·040 | |
| Add for 10th New moon | 29·5306 | +·335 | 1·976 |
| (4) Beginning of 10th Solar month 275·6369 : | 276·0787 | +·375=276·454 | 16·550+·040 (⊙'s Eqn.)=16·590 |
| | | +·120 | |
| Add for 11th New moon | 29·5306 | +·385 | 1·976 |
| | 305·6093 | +·505 | 18·526 |
| (5) Beginning of 11th Solar month 305·0850 : | 305·6093 | +·505=306·114 | 18·526+·120 (⊙'s Eqn.)=18·664 |
| | | +·169 | |
| Add for 12th New moon | 29·5306 | +·412 | 1·976 |
| (6) Beginning of 12th Solar month 334·9053 : | 335·1399 | +·581=335·721 | 20·502+·169 (⊙'s Eqn.)=20·671 |
| | | +·176 | |
| Add for 13th New moon | 29·5306 | +·357 | 1·976 |
| (7) End of 12th Solar month 365·25875 : | 364·6705 | +·533=365·203 | 22·478+·176 (⊙'s Eqn.)=22·654 |

*N.B.*—The above is a convenient and concise method of working for *Adhika* and *Kshaya* months.

From (1) and (2) it follows [applying the rule in foot-note to p. (9) of the Text], that the 7th and 8th new moons were both Kårttika, the first of them being *Adhika*.

From (3) it follows that the 9th new moon was *Pausha*.

∴ The new moon between Kårttika and Pausha, viz., *Mârgasira* was *kshaya* and it is accordingly noted as such in our Table X.

From (4) it follows that the 10th new moon was *Mâgha*.

From (5) it follows that the 11th new moon was *Phâlguna*.

From (6) and (7) it follows that the 12th and 13th new moons were both *Chaitra* the first of them being *adhika*.

It would appear that by oversight Messrs. Sewell and Dikshit failed to take account of these facts, notwithstanding that Prof. Chhatre had mentioned them.

### A.D. 1963.

**275**. Messrs. Sewell and Dikshit, after noting that the last *Kshaya* month before our time was in A.D. 1822, go on to remark: "We are led to suppose that there will be no suppressed month till at earliest A.D. 1944, and possibly not till A.D. 1963".

There is no reason why the matter should be treated as one for conjecture, since anybody familiar with the present method can calculate that the next *Kshaya* month will be in A.D. 1963, as we have indeed noted in Table X. There will be another in A.D. 1982, as noted in the same Table.

---

## CHAPTER XXX.

### CONSTRUCTION OF PLANETARY TABLES XVII AND XVIII.

**276**. The mean sidereal periods, used in Tables XVII and XVIII, are those of modern European astronomy adjusted to the Indian sidereal year. The following table compares the sidereal period of each planet, ordinarily given in Indian astronomical works, with that adopted in this work as well as with the results of modern astronomy.

| Planet. | Indian Sidereal period. | Modern Astronomy (Encycl. Brit.) | Figure adopted in this work. |
|---|---|---|---|
| | Days. | Days. | Days. |
| Mars | 686·99749 | 686·979645 | 686·98814 |
| Mercury | 87·969702 | 87·969258 | 87·96939 |
| Jupiter | 4332·3206 | 4332·584821 | 4332·92322 |
| Venus | 224·698568 | 224·700786 | 224·70169 |
| Saturn | 10765·7730 | 10759·219817 | 10761·30664 |

**277**. In the first place it was thought unnecessary to observe in regard to planetary sidereal periods the same scrupulous adherence to Indian authorities which is incumbent in the case of the solar year and the moon's synodical month. In the next place, the difference in the length of the sidereal year between Indian and modern astronomy results in a slight displacement of the starting point of Indian celestial longitude, which displacement amounts to 7′ 6·26″

in 50 years and should be added to the precession, amounting, according to modern astronomy, to 41′ 52·27″ for fifty years. The total difference between Indian sidereal and modern tropical longitudes is thus 48′ 58·53″ for 50 years or 59″ per annum, while Bhâskara's estimate of the precession, 59·9007″ per annum, is only slightly larger.

**278**. If the slight annual displacement of the zero point of Indian longitudes, which is a practical postulate of Indian astronomy, however unrecognized in theory, is applied to the sidereal places of planets, their mean sidereal periods will have to be altered as shown above. In this manner alone will it be possible to apply to the planets the same precession as is applied to the sun, for the purpose of converting sidereal into tropical longitude.

*N.B.*—By taking this slight liberty with the mean sidereal periods of planets, the exact agreement of the mean place of any planet with its place in modern astronomy is secured, and the serious divergences between the two systems, commented on in Whitney's notes on the *Surya Siddhanta* have been effectually avoided. As a result, the place, whether mean or actual, assigned to any planet in Tables XVII and XVIII, may not tally exactly with Indian calculations, but the difference will generally be found to be very slight, while there is an obvious advantage in having for every possible epoch a mean place for each planet, identical with that assigned to it by modern astronomy.

**279**. The planetary anomalies and annual equations, which were used for Tables XVII and XVIII of this work, were taken from *Warren's Kâla Sankalita* and are ascribed by that author to a local Telugu Astronomer of the time, called *Vavilâla Kuchinna*. The tables are no doubt old-fashioned, but they are handy and sufficiently accurate for the purposes of Indian horoscope-chronology. This is the main reason for resuscitating them into present-day use.

**280**. The figures as to longitudes of apses and nodes and the greatest apparent latitudes of planets (Table XVII) are taken, partly from Warren and partly from the Siddhânta Siromani.

## PART IV—PLANETS AND PLANETARY CHRONOLOGY.

**281**. Indian astronomy reckons nine planets, spoken of collectively as *navagrahas*, namely the Sun, Moon, Mars, Mercury, Jupiter, Venus, Saturn, Rahu and Ketu. It will be seen that the first seven are named in the order of the days of the week, and this is always the order in which they are referred to in Indian astronomy. Rahu is another name for the moon's ascending node, one of the points at which the moon's orbit cuts that of the Sun (the ecliptic), the other point, or the descending node, being called Ketu. An eclipse cannot happen except when the moon is at either node and the sun is at the same or the opposite node; hence the popular legend which depicts *Rahu* as a dragon swallowing up the moon or the sun at the time of an eclipse. All the planets, except Rahu and Ketu, move like the Sun, from West to East in their apparent path round the earth. Rahu moves in the opposite direction, and Ketu is always assigned a position 180° from that of Rahu.

**282**. All the data necessary for calculating the geocentric longitude and latitude of the five planets, Mars, Mercury, Jupiter, Venus and Saturn, are given at pages 200 to 206 of the Tables (Table XVII).

**283.** The longitude of any planet is its distance, measured in degrees, from an arbitrary point in the ecliptic, which point may perhaps best be defined as the 0 point of the Sun's mean longitude. The Sun's *mean* longitude is always 0 at a certain part of the Indian Solar Year, namely, at 2·1707 days from the commencement of each Indian Solar Year: while *at the moment of commencement* of the Indian Solar Year, the Sun's *actual* longitude is 0. The difference between the Sun's mean and actual longitude at any time is his equation of the centre, and this difference is due to his varying pace at different times of the year as he journeys round the earth. The Sun's mean longitude and equation for every complete day of the Solar Year are given in Table XVII-A (pp. 207, 208), and these are most important data in Indian astronomy. Table XVII-C (p. 209) gives the increase of the Sun's mean longitude for hundredth parts of a day.

**284.** The Sun's mean longitude at any moment is ascertained by deducting 2·1707 from the number of days and fraction of day elapsed since the commencement of the Indian Solar Year, and applying to the result the Table of Sun's longitude (Table VII-B, p. 8). The difference between the Sun's longitude thus calculated and the longitude for the same moment given in any European *Nautical Almanac,* is the true Indian precession, and this difference should always be added to the longitude of a planet given in this work in order to ascertain the corresponding longitude according to the *Nautical Almanac.* The following table gives the precession to be added to Indian celestial longitudes at various epochs since A.D. 520 (when the European and Indian longitudes coincided), in order to arrive at the European celestial longitude. The rate of precession adopted for this table is 1 degree for every 61¼ years, or 59″ per annum, the rate arrived at in Sec. **277** *supra.*

(The European longitude is called the *tropical* longitude of a planet from its being regulated by the tropical year, whereas the Indian longitude is generally called a *sidereal* longitude from its being regulated by the sidereal year.)

| A.D. | Kaliyuga. | Indian Precession in Degrees. | A.D. | Kaliyuga. | Indian Precession in Degrees. | A.D. | Kaliyuga. | Indian Precession in Degrees. |
|---|---|---|---|---|---|---|---|---|
| **520** | 3621 | 0·0 | **1040** | 4141 | 8·5 | **1561** | 4662 | 17·0 |
| **551** | 3652 | 0·5 | **1071** | 4172 | 9·0 | **1591** | 4692 | 17·5 |
| **581** | 3682 | 1·0 | **1102** | 4203 | 9·5 | **1622** | 4723 | 18·0 |
| **611** | 3712 | 1·5 | **1132** | 4233 | 10·0 | **1653** | 4754 | 18·5 |
| **642** | 3743 | 2·0 | **1163** | 4264 | 10·5 | **1684** | 4785 | 19·0 |
| **673** | 3774 | 2·5 | **1194** | 4295 | 11·0 | **1715** | 4816 | 19·5 |
| **704** | 3805 | 3·0 | **1225** | 4326 | 11·5 | **1745** | 4846 | 20·0 |
| **735** | 3836 | 3·5 | **1255** | 4356 | 12·0 | **1775** | 4876 | 20·5 |
| **765** | 3866 | 4·0 | **1285** | 4386 | 12·5 | **1806** | 4907 | 21·0 |
| **796** | 3897 | 4·5 | **1316** | 4417 | 13·0 | **1837** | 4938 | 21·5 |
| **826** | 3927 | 5·0 | **1347** | 4448 | 13·5 | **1867** | 4968 | 22·0 |
| **856** | 3957 | 5·5 | **1377** | 4478 | 14·0 | **1898** | 4999 | 22·5 |
| **887** | 3988 | 6·0 | **1408** | 4509 | 14·5 | **1929** | 5030 | 23·0 |
| **918** | 4019 | 6·5 | **1439** | 4540 | 15·0 | **1960** | 5061 | 23·5 |
| **949** | 4050 | 7·0 | **1469** | 4570 | 15·5 | **1990** | 5091 | 24·0 |
| **980** | 4081 | 7·5 | **1500** | 4601 | 16·0 | **2020** | 5121 | 24·5 |
| **1010** | 4111 | 8·0 | **1531** | 4632 | 16·5 | **2051** | 5152 | 25·0 |

**285.** We will now explain briefly how the planetary tables at pages 200 to 206 are to be used in practice.

The mean longitude of a planet at any moment is obtained by adding together its mean longitude at the commencement of the century, the mean motion for the given odd year, and the mean motion for the given day. Thus for 202·54 days of the Indian Solar Year, A.D. 961 (Kaliyuga 4062) we have :—

| | Degrees. |
|---|---|
| Mars' mean longitude at commencement of K.Y. 4001, | 79·4873 |
| Mar's mean longitude for 61 years, | 155·72 |
| " " " 202 days, | 105·85 |
| " " " ·54 day, | ·2830 |
| *Answer* | 341·34° |

**286**. This is the mean longitude of the planet in his path round the Sun, though not his longitude, as he appears to us from the earth. To obtain this, which is called the geocentric longitude, and which is the longitude usually cited in Indian chronology, we have to perform a fairly tedious sum. So far as the geocentric longitudes of the planets from A.D. 1840 to A.D. 1919 are concerned, the reader is saved the trouble of doing any such sum, because the geocentric longitudes, as finally determined for Lankâ Sunrise on every tenth day in each solar year from A.D. 1840 to A.D. 1919, are given in a handy form at pages 210 to 225 (Table XVIII), so that all he has to do to ascertain the position of the planets in a horoscope, say, for April 13, A.D. 1911, is to look up the particular day of the Solar Year (in this instance 0 day against A.D. 1911 on p. 224) and set down the positions of the planets as follows :—

Sun, 357° + ·4 + 2·7 = 360·1 or 0·1° in Mesha ;
Mars, 298·4°, *i.e.*, 28·4° in Makara ;
Mercury, 16·3° in Mesha ;
Jupiter, 198·8°, *i.e.*, 18·8° in Tulâ ;
Venus, 31·8°, *i.e.*, 1·8° in Vrishabha ;
Saturn, 14·3° in Mesha ;
Rahu (Table XVII-B), A.D. 1911 : 21·90° in Mesha.

(*N.B.*—Only *mean* places, and not actuals, are calculated for Rahu and Ketu).

Ketu (*vide* Sec. **281**) 21·90° + or − 180° = 201·90,° *i.e.*, 21·90° in Kanyâ. The result for Ketu is the same, whether 180° are added to, or deducted from, the longitude of Rahu.

**287**. The place of the moon, the only remaining planet, can be deduced either from the tithi or from the Nakshatra. The latter method is illustrated in Sec. **292** *infra*. The former, which is easier, is shown below :—

We first of all calculate the ending moment of the tithi, as usual.

| | d. gh. | ☉'s Anom. d. gh. | ☾'s Anom d. gh. |
|---|---|---|---|
| (Tab. XII, p. 151) New Moon, March 1911 ... | Mar. 29 57½ | 350 45 | 22 16¼ |
| (Eye-Table) Collective duration of 15 tithis ... | 14 46 | 14 46 | 14 46 |
| | Mar. 44 43½ | 365 31 | 37 2¼ |
| Deduct 1 anomalistic year and month respectively | | 365 15 | 27 33⅓ |
| | | 0 16 | 9 29 |
| (Eye-table) ☉'s Eqn. = + 10½ gh ; ☾'s Eqn. = − 21 gh. | | | ☉'s Eqn. + 10½ |
| Sum of ☉'s and ☾'s Eqns. ... | − 10½ | | 9 39½ |
| | Mar. 44 33 = 33 gh. on Ap. 13. | | |

This was the ending moment of the tithi, when the moon's elongation or distance from the sun was $15 \times 12° = 180°$. For the moon's longitude at any moment = moon's elongation *plus* Sun's Longitude. (Secs. **36**, **254**.)

Therefore, if we find the Sun's longitude at ending moment of tithi and add it to the moon's elongation, 180°, we shall obtain the moon's longitude, or which is the same thing, the Nakshatra current at ending moment of tithi.

☉'s Longitude at Sunrise = 0·1° (see above).

Do. at ending moment of tithi = 0·1° *plus* increase of longitude for 33 ghat. or ·55 of a day (Eye-Table.) This increase, by Table XVII-C is ·54°.

∴ ☉'s Longitude at ending moment of tithi = 0·1° + ·54° = 0·64°.

∴ Moon's do. do. do. = 180° + 0·64° = 180·64°.

The corresponding Nakshatra (by Eye-Table) was *Chitrâ*, and the **Râśi** *Tulâ*.

We want however the moon's longitude for sunrise on the same day which was 33 ghaṭikas or ·55 day (Eye-Table) earlier.

The increase of ☾'s Long. for ·55 day is roughly (Tab. XVII-D) 7·33°.

∴ ☾'s Long. at Sunrise 13 Ap. A.D. 1911 will be roughly 180·64° − 7·33° = 173·31° which will be **23·3°** in Kanyâ. The Nakshatra is *Hasta*. (Eye-Table.)

*N.B.*—In strictness we should have ascertained the total duration of the Nakshatra and the unexpired part of the Nakshatra at sunrise. If this was ·50, Table XVII-D would give us 6·6° which we should deduct from 180·64°. Result 174·0° which is *Chitra* Nakshatra.

**288.** For years before A.D. 1840 the true geocentric longitudes of planets will have to be calculated in the following manner :—

The planets are divided into two classes: *Inferior* (Mercury and Venus) and *Superior* (Mars, Jupiter and Saturn.) Three quantities, which we may call ***a, b, v***, in the case of the inferior planets, Mercury and Venus, and A, B, N, in the case of the superior planets, Mars, Jupiter and Saturn, have to be determined first of all, before working out the actual geocentric longitude of any planet. Having determined these three quantities, the further procedure is the same, whether the planet is superior or inferior. We give below the calculations (1) for an *inferior* planet, Mercury, and (2) for a *superior* planet, Mars, for 202·54 days of the Solar Year A.D. 961 (*i.e.*, 9 ghaṭikas after sunrise on 11th October A.D. 961.)

I. *True Geocentric Longitude of Mercury for 202.54 days of the Indian Solar Year A.D. 961-62.*

*Time and English date. 9 ghatikas after sunrise on 11th October A.D. 961.*

## MERCURY (Inferior Planet).

(1)—*a*. Mean Long. of Mercury *minus* Mean Long. of Sun.

| Mean Long. of Mercury. | | Mean Long. of Sun. | |
|---|---|---|---|
| For A.D. 900 (p. 201) | 173·55° | 202 days (XVII-A) | 196·95° |
| 61 years „ | 100·37° | ·54 day (XVII-C) | ·53° |
| 202 days (p. 202) | 106·65° | | 197·48° |
| ·54 day | 2·21° | | |
| | 382·78° | | |
| *Deduct* | 360° | | |
| | 22·78° | | |

382·78°
−197·48°
185·30° *a*

(2)—*b*. Long. of Mercury's Apsis *minus* Sun's mean Long.
Long. of Mercury's Apsis for A.D. 900 (p. 201)* ... 220·44°
Sun's mean Long. (as above): −197·48°
22·96 *b*

(3)—*v*. Mean Long. of Sun ... 197·48° *v*

(4) Mercury's annual equation for *a* (p. 202): − 3·0° (4)

(5) Take half of (4) with reversed sign, and add to *b* ... 22·96
+ 1·5
24·5 (5)

(6) Mercury's anomalistic equation for (5) (p. 202): + 1·9° (6)

(7) Take half of (6), with reversed sign, and add to (5) ... 24·5
− 0·9
23·6 (7)

(8) Anomalistic equation of (7) (p. 202): + 1·8 (8)

(9) Take (8), with reversed sign, and add to *a* ... 185·3
− 1·8
183·5 (9)

(10) Annual equation of (9) (p. 202): − 2·0 (10)

(11) Add (8) and (10): +1·8° − 2·0° = − 0·2 (11)

(12) Add *v* and (11): 197·48° − 0·2° = 197·3° (12)

This last result **197·3°** is the true **geocentric place of Mercury** for the moment in question.

* The apses and nodes of planets, except the moon, have such slow motions that their places need not be calculated for odd years and days.

**289.** II. To find the *latitude* of Mercury (*i.e.*, his distance above or below the ecliptic for the same moment.)

(13) Mercury's mean longitude ... 22·78°

*minus* Longitude of Mercury's node (p. 201): −20·72

2·06° (13)

(14) Take (8) above, with sign reversed, and add to (13): − 1·8

0·3° (14)

(15) Sine of (14) according to Table XVII-F (p. 209) ... 18 (15)

(16) Equation corresponding to (9) *supra*, by Table XVII-E.

Equation for 183·75° = 2173.

" " ·2° = ·05 × 5 = ·25.

∴ Equation for 183·5° = 2173 *minus* ·25 = 2172·75 ... 2173 (16)

(17) Multiply (15) by 120, as a constant for Mercury, and divide the product by (16)

$$\frac{120 \times 18}{2173} = \frac{2160}{2173} = \cdot 9.$$

Now by Table XVII-F, 225 is the sine of 3·75°

$$\therefore \quad \cdot 9 \text{ is the sine of } \frac{3\cdot 75 \times \cdot 9}{225} = \frac{3\cdot 375}{225} = \cdot 015°.$$

***N.B.*—The corresponding constants for the other planets are the sines of their greatest apparent latitudes, given at foot of Table XVII-E.**

The latitude of Mercury is ·015° and inasmuch as (**14**) is less than 180°, *i.e.*, above the ecliptic, the latitude ·015° is North.

## MARS (Superior Planet).

**290.** I. *To find the Geocentric Longitude of Mars (Superior Planet) for 202.54 days in Solar Year A.D. 961-62.*

*Time and English date. 9 ghatikas after sunrise on 11th October A.D. 961.*

(1) A. Mean longitude of Sun *minus* Mars' mean longitude.

Mean longitude of Sun (already found) 197·48° or 197·48° + 360° = 557·48°

Mean longitude of Mars

| | | |
|---|---|---|
| For A.D. 900, | 79·49° | (Table XVII, p. 200.) See also p. (104) *supra*. |
| " 61 years, | 155·72° | |
| " 202 days, | 105·85° | |
| " ·54 day, | ·28° | |
| | 341·34° | |

−341·34°

216·14° A

(2) B. Long. of Mars' Apsis *minus* mean long. of Mars.

| | | |
|---|---|---|
| Long. of Mars' Apsis [*vide* foot-note on p. (106)] (p. 200) 130·04° or | 490·04° | |
| Mean longitude of Mars ... ... ... | − 341·34° | |
| | 148·70° | B |
| (3) N. Mean longitude of Mars ... ... ... | 341·34° | N |
| (4) Mars' annual equation for A (p. 200) ... ... ... | − 39·0° | (4) |
| (5) Take half of (4) with reversed sign, and add to B ... | 148·70° | |
| | + 19·5 | |
| | 168·2° | (5) |
| (6) Mars' anomalistic equation for (5) (p. 201) ... ... | + 2·4° | (6) |
| (7) Take half of (6), with reversed sign, and add to (5) ... | 168·2° | |
| | − 1·2° | |
| | 167·0° | (7) |
| (8) Anomalistic equation of (7) (p. 201) ... ... | + 2·7 | (8) |
| (9) Take (8) with reversed sign, and add to A ... ... | 216·14° | |
| | —2·7 | |
| | 213·4° | (9) |
| (10) Annual equation of (9) (p. 200) ... ... ... | − 38·0° | (10) |
| (11) Add (8) and (10) : + 2·7° − 38·0° ... ... ... | − 35·3° | (11) |
| (12) Add N and (11) : 341·34° − 35·3° ... ... ... | 306·0° | (12) |

The true **Geocentric Longitude** of Mars at the moment in question was **306·0°**

**291.** II. To find the latitude of Mars, *i.e.*, his distance above or below the Ecliptic for the same moment.

| | | |
|---|---|---|
| (13) Mars' true geocentric longitude ... ... | 306·0° | |
| *minus* Longitude of Mars' node for A.D. 900, (Tab. XVII, p. 200) [*vide* foot-note on p. (106).] | − 40·07 | |
| | 265·93° | (13) |
| (14) Take (10) above for Mars, with sign reversed, and add to (13): | + 38·0 | |
| | 303·9° | (14) |
| (15) Sine of (14) according to Table XVII-F ... ... | 2267 | (15) |
| (16) Equation for (9) above, by Table XVII-E ... ... | 2011 | (16) |

(17) Multiply (15) by 90, as a constant for Mars, and divide by (16)

$$\frac{90 \times 2267}{2011} = \frac{2040·30}{2011} = 101.$$

*N.B.*—The corresponding constants for the other planets are the sines of their greatest apparent latitudes, given at foot of Table XVII-E.

Now by Table XVII-F the sine 101 belongs to an angle between 0° and 3·75°
225 is the sine of 3·75°

$$\therefore 101 \quad " \quad \frac{101 \times 3{\cdot}75}{225} = 1{\cdot}68°$$

∴ The **latitude** of Mars sought is **1·7**° which is South since (14) is above 180°.

**292**. *To find the longitude and latitude of the moon for the above moment.*

We should first of all find the moon's true longitude which we may calculate either from the tithi direct, with the aid of Table XVII-A and XVII-C (as we did in Sec. **287**), or from the Nakshatra. We shall follow the latter method, although the former is easier.

The last New moon before our date (11 Oct. A.D. 961) was that on September 12·74 (Table X, p. 73). This was Âśvina New moon. The tithi for our date is that corresponding to the interval between Oct. 11·02 and Sep. 12·74, or between 41·02 and 12·74, *i.e*, 28·28 days. This, by Eye-Table, is less than the interval for 29 tithis, which is 28·55 days.

As our Nakshatra must fall on the same day as the Tithi, we must take a Nakshatra in the lunar month of Âśvina, whose interval, after correction, shall be near 28·55 days.

The Nakshatra correction for A.D. 961 is (by Tab. XI) that for 26·47 days, *i.e.*, 1·95 (correction for 26 days) + ·03 (correction for ·47 day) = 1·98, and 2·855 + 1·98 = 30·53. This is 1 day more than a lunar month and an interval of 30·53 days from Âśvina New moon takes us to 1 day from Kârttika New moon. The nearest interval to this in Table XI is 1·0757 days under Kârttika (col. VII), for which the Nakshatra is No. 14 *Chitrâ*.

The interval for No. 14 Chitrâ is 1·0757 days from Kârttika New moon,
*i.e.*, 1·08 + 2953 or 30·61 days from Âśvina New moon.
*Deduct* correction for A.D. 961 —1·98

Actual interval of No. 14 Nakshatra is 28·63 days from Âśvina New moon.

∴ Time of mean ending moment of No. 14 Nakshatra was Sep. 12·74 + 28·63
= Sep. 41·37 = Oct. 11·37.

The ☾'s anomaly for the ending moment of the Nakshatra: 7·09 (Anomaly of First New moon in solar year A.D. 961, *vide* Tab. X, p. 72) + 9·88 (addition for ☾'s Anom. at Âśvina New moon: Table X, head line) + 28·63 (actual interval from Âśvina New moon to No. 14 Nakshatra) = 45·60 days, from which deduct 1 anom. month, 27·55. Remainder, 18·05 days.

The Nakshatra eqn. for 18·05 days is + ·333 (Table IX-*k*). Nak. No. 14 Chitrâ ended therefore on Oct. 11·37 + ·33 = Oct. 11·70 or 42 ghaṭikas after sunrise (Eye-Table). The moment for which we want the moon's true place is earlier than this by ·55 day, which (by Table XVII-D) corresponds to 7·33 degrees of moon's longitude. (See, however, *N.B.* to Section **287**.)

At the ending moment of No. 14 Nak. the **moon's longitude** is 186·7° (Eye-Table); the true place of moon at the moment now in question (9 ghaṭikas. after sunrise on 11 Oct. A.D. 169) was, therefore, **186·7° − 7·3° = 179·4°.**

**293.** *Now to find the Moon's latitude at the moment sought.*

(1) Find the place of Moon's node, *i.e.*, Rahu.

By Table XVII-B place of Rahu for A.D. 900 is ... 147·65

By same table Rahu's motion for 61 years, 100·52°

*N.B.*—Remember that Rahu's mean motion for odd years and days should be subtracted from the mean place for century—*vide* Sec. **295** (4).

| | | |
|---|---|---|
| 200 days, | 10·60° | |
| ·54 day, | ·03 | |
| 61 yrs. 200·54 days | =111·13 | —111·13° |
| | | 36·52° (1) |

(2) Moon's true place *minus* place of her node ... ... 179·4°

—36·5

142·9 (2)

(3) Sine of (2) by Table XVII-F ... ... 2094 (3)

(4) Multiply (3) by $\frac{270}{3438}$, as a constant for the moon : $\frac{2094 \times 270}{3438} = 164.$

By Table XVII-F, 225 is the sine of 3·75°

∴ 164 ,, $\frac{164 \times 3{\cdot}75}{225} = 2{\cdot}73°$

Inasmuch as (2) is less than 180°, the **latitude** sought is **2·73°** North.

**294.** The reader should note—

(*a*) that for Venus the procedure is the same as for Mercury ;

(*b*) that for Jupiter and Saturn, the procedure is the same as for Mars ;

(*c*) that for Mercury and Venus, the geocentric longitude has for its basis the Sun's mean longitude and not that of the planet ; whereas the geocentric places of Mars, Jupiter and Saturn are based on their respective mean longitudes ;

(*d*) that we may always subtract a larger number of degrees from a smaller by adding 360° to the smaller number, as we did for (1) ***a*** on p. (106), for (1) A on p. (107) and for (2) B on p. (108) ;

(*e*) that the apsis of any planet except the moon does not change materially in position during a century, and that therefore we need not calculate the change of apsis for odd years or odd days ; [*vide* foot note on p. (106)].

(*f*) that after calculating ***a***, ***b*** and ***v***, and A, B and N, respectively, the procedure is the same for all planets, whether inferior or superior ;

(*g*) that the mean longitudes of planets for decimals of a day may be obtained from the longitude for odd days by prefixing decimals to the latter. Only we should remember that a mean revolution of Mercury is completed in 87 days and a mean revolution of Venus in 225 days, [*vide* Sec. **276**] and the increase of Mercury's longitude for ·89 day is not 0·422°, but 3·6422°. Similarly, the increase of Venus' longitude for ·231 day is not ·01009°, but ·37009° ;

(*h*) that the latitudes of planets are not referred to in Indian usage except in what is known as a planetary fight in astrology ;

(*i*) that the greatest apparent latitudes of the several planets and their corresponding Hindu sines are given at foot of Table XVII-E.

**295**. In using Table XVIII, *Geocentric* (*Indian*) *Longitudes of Sun and Major planets at Ujjain Sunrise for every ten days from A. D. 1840 to A. D. 1920*, the reader will do well to bear in mind the following instructions :—

(1) Table XVIII is intended to be of use in constructing or verifying the horoscopes of persons now living or to be born in the next nine years.

(2) The elements of a horoscope are the *râsi* positions of laghna, Sun, Moon, Mars, Mercury, Jupiter, Venus, Saturn, Rahu and Ketu.

(3) Of these, the positions of Sun,* Mars, Mercury, Jupiter, Venus and Saturn are given in Table XVIII.

(4) The *laghna* may be ascertained with the aid of the rules given in sections **71** to **76** *supra*. The position of Rahu for any day of the solar year may be ascertained from Table XVII-B. In using this table remember that the signs of Rahu are counted backwards and therefore the longitude of Rahu for odd years and days should be deducted from its longitude for centuries in order to arrive at its position at any moment. *Ketu* is always 180° from Rahu. It does not matter whether for Ketu we add or subtract 180° from Rahu's place : the result will always be the same.

(5) The position of the moon should be determined from her Nakshatra. It would have been no use to have given the position of the moon for every ten days of the year in Table XVIII, since her position shifts by 13° every day and is liable to considerable perturbations on account of her anomaly. Numerous examples are given in the text, showing how the moon's longitude may be ascertained either from the Tithi direct or from the ending momemt of the Nakshatra. [See especially Sections **287** and **292**.]

(6) For planets other than Moon, Rahu and Ketu, Table XVIII gives precise information for every ten days, beginning each year with what is called 0 day of the solar year. The 0 day of the solar year is the day on which the solar year commences, as shown in Table X, not as shown in Table XII (Ephemeris). The English date shown in Table X in the 4th column headed " Month and day A.D. ", is the 0 day of each solar year. For example, the 0 day of the solar year A.D. 1911-12 is April 13.

(7) English dates corresponding to days of the solar year are given most exactly in Table XVIII and not merely by approximation. The Tamil dates shown in the head-lines of Table XVIII are approximate : the exact Tamil date corresponding to any English date can be easily ascertained from Table XII.

---

* To the mean longitude of the Sun given in the second head-line of Table XVIII should be added the equation given in the third head-line, and the correction for ⊙'s long. given under each particular year. See example in Sec. **286**.

(8) For dates *between* every 10th day which is shown in Table XVIII, take proportional parts.

Thus, position of Mercury on 23 May 1910 is 44·1°
Do. do. do. 2 June 1910 is 36·6°

Difference for 10 days = 7·5°

$\therefore$ Difference for 6 days $= \frac{6}{10} \times 7{\cdot}5 = 4{\cdot}55°$

$\therefore$ Position of Mercury at Ujjain sunrise on 29 May 1910 is 44·1° — 4·5° = 39·6°.

*N.B.*—Except in A.D. 1905, A.D. 1909 and A.D. 1913, the actual places of Mercury from A.D. 1886 to A.D. 1919 are practically identical with those for A.D. 1840 to 1873. In other words, Mercury's actual places follow in general a cycle of 46 years. Compare Sec. **228** regarding the moon's cycle of 46 years.

(9) The interval between the 360th day of each solar year and the 0 day of next solar year is either 5 or 6 days according as the former Indian solar year is one of 365 or 366 days. To ascertain this point, refer to Table X: if the fraction of day at which a year commences increases beyond unity in the next year, the year with the former fraction is one of 366 days; where there is no increase of fraction, the length of the year in integral days is 365.

Thus, fraction of day for commencement of A.D. 1909-10 is ·9492
Do. do. do. 1910-11 is ·2080
Do. do. do. 1911-12 is ·4667

We infer that A.D. 1909-10 (Indian Solar Year) is a year of 366 days and that A.D. 1910-11 and A.D. 1911-12 are years of 365 days each. Of course an Indian Solar Year with 366 days will not, except by chance, correspond with an English leap year.

The difference in the longitude of any planet between the 360th day of one solar year and 0 day of the next solar year should be divided by 5 or 6 (according as the year is one of 365 or 366 days) in order to get the difference of longitude for *one day* during that interval, and the difference for one day should be multiplied by as many days as we want.

(10) Two English dates are marked in Table XVIII (just as in Table VIII, *vide* Sec. **215**), when there is a possibility of a year with Feb. 29 alternating with a year in which February has only 28 days. Thus under 330th day on p. 225, " March 6, 7 " is entered, which means that if before reaching either of these dates we have passed " February 29 ", March 6 is the equivalent of the 330th day of the Solar Year, otherwise March 7 is the 330th day.

(11) The positions marked in Table XVIII are all for Ujjain sunrise and any given moment for which we wish to verify or construct a horoscope should first be reduced to Ujjain time by adding or deducting the total correction for the place, day of the solar year and latitude shown in Table XIII or Table XIII-A. Of course the reader will remember that corrections shown in Table XIII are intended to enable him to obtain local time *from Ujjain time*, and that to obtain Ujjain time *from local time*, the signs of the corrections must be reversed.

(12) It is assumed that the reader knows the ordinary facts about the apparent motions of planets, *i.e.*, they are occasionally stationary or nearly so, and often in retrogression; that any planet is in conjunction with another whose longitude is the same; and in opposition when the distance between the longitudes or the two planets is 180°.

(13) The houses or signs in which the planets are situated according to Table XVII show only *occupation* (in the language of astrology). *Ownership* or *lordship* of houses by planets is purely an astrological convention and has nothing to do with astronomy.

(14) The entries at foot of Table XVIII show the limits or *ending* points of signs or *râsis* and *nakshatras,* while the Eye-Table gives the longitude at *commencement* of each Râśi or Nakshatra. Both sets of entries will be found very useful, and examples of their use are given throughout this work : see for example Sections **49, 186, 266, 287** ; foot-note to p. (45) of the text, etc.

(15) In order to keep the present work within its proper limits, it has been considered essential that the information furnished by it shall be such as will be directly useful for chronological investigation. That it may serve indirectly the purposes of Indian astrology is from the nature of the case, inevitable, but *judicial* astrology, or the interpretation of planetary phenomena for astrological purposes, is entirely beyond the author's scope : and it has been finally decided not to include in this book any criticism of judicial astrology.

**Construction of Indian Horoscopes.—Navamsas.**

**296**. Indian usage has, for a long time past, adopted a well-known method of exhibiting the places of planets at particular moments. This is the familiar horoscope, which, relieved of its astrological excrescence, has a scientific value as a chronological record; since a genuine horoscope may, by the exercise of a little ingenuity, be always referred to its date, however remote that date may be, and even though there may not be a single figure indicative of an era or a year. It is for this chronological reason that horoscopes are introduced into this work.

Having ascertained the longitudes of planets and the lagna, which is merely a device for indicating the time of day without the use of figures (see Sections **71** to **76**), we may exhibit them in the form of a horoscope, as shown on the next page. This is a horoscope for the moment already examined, *viz.*, sunrise on 13 April 1911, being 0 day of the Indian Solar Year A.D. 1911-12, Kaliyuga 5012.

**297**. It might seem at first sight that the first of these tables, which distributes the planets among 12 signs of 30 degrees each, cannot serve the purposes of an accurate chronological record. It is a fact, however, that no other horoscope within perhaps 5000 years will be exactly like this one: and given such a horoscope as this, one may, without further indications, though after some trouble, discover the year; and once the year is discovered, the month, day and hour can be identified in a few minutes.

**298**. In practice, it is usual for the astrologer or chronologer to frame a more detailed record, and as Indian tradition demands that figures should, as far as practicable, be dispensed with in chronological records, the details of degrees in a horoscope are given by means of a Navâmśa scheme such as that shown on the next page. In the Navâmśa scheme, each sign of 30° is sub-divided into 9 Navâmśas, whence their name, and 12 such Navâmśas constitute a full horoscope. Table XXI compares the Râśis and Navâmśas, and shows [illegible]ow, given a Râśi and the corresponding Navâmśa, the exact position of a planet c[illegible]cated within three degrees. For instance, if we know that Śukra is in Râśi **Vrishabha** (II) and Navâmśa **Makara** (X) we can, from the Navâmśa Table XXI, find that Śukra must be between 30° and 33·3°; this is a fairly close approximation to the exact longitude.

Other divisions of the Râśis into *dvâdasâmsas* (12 divisions of 2½ degrees each), *trimsâmsas* (30 divisions of a degree each) and *shashthiamsas* (60 divisions of half a degree each), will be self-evident.

| | | | |
|---|---|---|---|
| **XII**<br>**Mina**<br>OR<br>**Pisces.** | **I**<br>**Mesha** OR **Aries**<br>Lagna<br>Sun or Ravi<br>Mercury or Budha<br>Saturn or Sani<br>Rahu. | **II**<br>**Vrishabha**<br>OR<br>**Taurus**<br>Venus or<br>Sukra. | **III**<br>**Mithuna**<br>OR<br>**Gemini.** |
| **XI**<br>**Kumbha**<br>OR<br>**Aquarius.** | | | **IV**<br>**Kataka**<br>OR<br>**Cancer.** |
| **X**<br>**Makara**<br>OR<br>**Capricornus**<br>Mars or Kuja. | RASI. | | **V**<br>**Simha**<br>OR<br>**Leo.** |
| **IX**<br>**Dhanus**<br>OR<br>**Sagittarius.** | **VIII**<br>**Vrischika**<br>OR<br>**Scorpio.** | **VII**<br>**Tula**<br>OR<br>**Libra**<br>Jupiter or Guru<br>Ketu. | **VI**<br>**Kanya**<br>OR<br>**Virgo**<br>Moon or<br>Chandra. |

| | | | |
|---|---|---|---|
| **XII**<br>**Mina**<br>Guru. | **I**<br>**Mesha**<br>Lagna<br>Ravi. | **II**<br>**Vrishabha.** | **III**<br>**Mithuna.** |
| **XI**<br>**Kumbha.** | | | **IV**<br>**Kataka**<br>Chandra. |
| **X**<br>**Makara**<br>Sukra. | NAVAMSA. | | **V**<br>**Simha**<br>Budha<br>Sani. |
| **IX**<br>**Dhanus.** | **VIII**<br>**Vrischika.** | **VII**<br>**Tula**<br>Rahu. | **VI**<br>**Kanya**<br>Kuja. |

# TABLES.

## CORRIGENDA.

Text, p. (15), Sec. 32, last line, *for* Thibout, *read* Thibaut.

,, p. (28)—(31) and elsewhere, *for* laghna, *read* lagna.

,, p. (43), Sec. 112 (1). *for* "addendum, p. (114)", *read* "addenda, p. 4 of 'contents'."

,, p. (43), Sec. 112 (5). *Kollam Era. for* "A.D. year *minus* 825", *read* "A.D. year *minus* 824" and increase all the Kollam years shown against century years A D. by 1. Thus K. year 76 began in A.D. 900, K. year 1076 began in A.D. 1900, and so on.

,, p. (80), Sec. 212, *for* the citation **228**, *read* **223**.

---

| | | | | | | | | |
|---|---|---|---|---|---|---|---|---|
| TABLE IX, | p. 22, line 2, | from top | | | *for* | 3·777, | *read* | 13·777. |
| ,, X, | p. 30 | A.D. | 126 | ☾'s An. | *for* | 17·176 | *read* | 17·186. |
| ,, | p. 34 | A.D. | 210 | ,, | ,, | 3·543 | ,, | 3·643. |
| ,, | p. 36 | A.D. | 249 | ,, | ,, | 10·213 | ,, | 19·213. |
| ,, | p. 38 | A.D. | 281 | ,, | ,, | 3·623 | ,, | 2·620. |
| ,, | ,, | A.D. | 310 | ,, | ,, | 1·159 | ,, | 21·159. |
| ,, | p. 46 | A.D. | 454 | ,, | ,, | 14·381 | ,, | 15·381. |
| ,, | p. 52 | A.D. | 565 | ,, | ,, | 27·727 | ,, | 0·172. |
| ,, | ,, | A.D. | 566 | ,, | ,, | 23·894 | ,, | 23·884. |
| ,, | ,, | A.D. | 568 | ,, | ,, | 18·166 | ,, | 18·176. |
| ,, | p. 56 | A.D. | 628 | ,, | ,, | 23·985 | ,, | 23·995. |
| ,, | ,, | A.D. | 649 | ,, | ,, | 14·234 | ,, | 14·224. |
| ,, | ,, | A.D. | 652 | ,, | ,, | 4·694 | ,, | 4·674. |
| ,, | p. 60 | A.D. | 729 | ,, | ,, | 12·159 | ,, | 12·139. |
| ,, | p. 62 | A.D. | 773 | ,, | ,, | 12·489 | ,, | 12·469. |
| ,, | p. 72 | A.D. | 961 | ,, | ,, | 7·101 | ,, | 7·091. |
| ,, | p. 76 | A.D. | 544 | ,, | ,, | 10·277 | ,, | 10·287. |
| ,, | p. 77 | A.D. | 731 | ,, | ,, | 6·674 | ,, | 6·684. |
| ,, | p. 78 | A.D. | 854 | ,, | ,, | 8·170 | ,, | 8·710. |
| ,, | ,, | A.D. | 927 | ,, | ,, | 1·985 | ,, | 1·977. |

,, XI, p. 132, under V Bhadrapada against Yoga No. **5**, *for* 13·9813, *read* 13·9883.

,, XI-A, p. 134, under V Bhadrapada against Yoga No. **5**, *for* 13,58,53 *read* 13,59,18.

,, ,, p. 135, under VI Asvina against Yoga No. 16, *for* 20,13,28, *read* 20,14,4.

,, XII, To all the Kollam (Malayalam) years entered in col. 1 of Table XII add 1: thus, the Kollam year which began on 17 August 1911 is not K. 1086 as shown in Table XII, but K. 1087.

,, XVII, p. 200, against Odd Year 34, *for* 27·37, *read* 27·77.

,, ,, ,, 35 ,, 219·81 ,, 219·18.

,, p. 201, against Mar's Eqn. **11·0**, *for* 352·2, *read* 252·2.

,, ,, p. 203, line 4 from bottom, *for* 439·0, *read* 339·0.

,, ,, ,, ,, line 5 from bottom, *for* 032·6, *read* 332·6.

# TABLE I.

**Corresponding A.D. years—(Centuries A.D. in heavy type).**

(*N.B.*—The Indian year begins in March or April of the A.D. year.)

| Names of years in Jupiter's Cycle. (S. India.) | | | | | | | | | | | | | | | | |
|---|---|---|---|---|---|---|---|---|---|---|---|---|---|---|---|---|
| **1** Prabhava ... | **9** 67 | **10** 27 | **10** 87 | **11** 47 | **12** 07 | **12** 67 | **13** 27 | **13** 87 | **14** 47 | **15** 07 | **15** 67 | **16** 27 | **16** 87 | **17** 47 | **18** 07 | **18** 67 |
| **2** Vibhava ... | 68 | 28 | 88 | 48 | 08 | 68 | 28 | 88 | 48 | 08 | 68 | 28 | 88 | 48 | 08 | 68 |
| **3** Sukla ... | 69 | 29 | 89 | 49 | 09 | 69 | 29 | 89 | 49 | 09 | 69 | 29 | 89 | 49 | 09 | 69 |
| **4** Pramodhuta ... | 70 | 30 | 90 | 50 | 10 | 70 | 30 | 90 | 50 | 10 | 70 | 30 | 90 | 50 | 10 | 70 |
| **5** Prajapati ... | 71 | 31 | 91 | 51 | 11 | 71 | 31 | 91 | 51 | 11 | 71 | 31 | 91 | 51 | 11 | 71 |
| **6** Angirasa ... | 72 | 32 | 92 | 52 | 12 | 72 | 32 | 92 | 52 | 12 | 72 | 32 | 92 | 52 | 12 | 72 |
| **7** Srimukha ... | 73 | 33 | 93 | 53 | 13 | 73 | 33 | 93 | 53 | 13 | 73 | 33 | 93 | 53 | 13 | 73 |
| **8** Bhava ... | 74 | 34 | 94 | 54 | 14 | 74 | 34 | 94 | 54 | 14 | 74 | 34 | 94 | 54 | 14 | 74 |
| **9** Yuva ... | 75 | 35 | 95 | 55 | 15 | 75 | 35 | 95 | 55 | 15 | 75 | 35 | 95 | 55 | 15 | 75 |
| **10** Dhatri ... | 76 | 36 | 96 | 56 | 16 | 76 | 36 | 96 | 56 | 16 | 76 | 36 | 96 | 56 | 16 | 76 |
| **11** Isvara ... | 77 | 37 | 97 | 57 | 17 | 77 | 37 | 97 | 57 | 17 | 77 | 37 | 97 | 57 | 17 | 77 |
| **12** Bahudhanya ... | 78 | 38 | 98 | 58 | 18 | 78 | 38 | 98 | 58 | 18 | 78 | 38 | 98 | 58 | 18 | 78 |
| **13** Pramathi ... | 79 | 39 | 99 | 59 | 19 | 79 | 39 | 99 | 59 | 19 | 79 | 39 | 99 | 59 | 19 | 79 |
| **14** Vikrama ... | 80 | 40 | **11** 00 | 60 | 20 | 80 | 40 | **14** 00 | 60 | 20 | 80 | 40 | **17** 00 | 60 | 20 | 80 |
| **15** Vrisha ... | 81 | 41 | 01 | 61 | 21 | 81 | 41 | 01 | 61 | 21 | 81 | 41 | 01 | 61 | 21 | 81 |
| **16** Chitrabhanu... | 82 | 42 | 02 | 62 | 22 | 82 | 42 | 02 | 62 | 22 | 82 | 42 | 02 | 62 | 22 | 82 |
| **17** Subhanu ... | 83 | 43 | 03 | 63 | 23 | 83 | 43 | 03 | 63 | 23 | 83 | 43 | 03 | 63 | 23 | 83 |
| **18** Tarana ... | 84 | 44 | 04 | 64 | 24 | 84 | 44 | 04 | 64 | 24 | 84 | 44 | 04 | 64 | 24 | 84 |
| **19** Parthiva ... | 85 | 45 | 05 | 65 | 25 | 85 | 45 | 05 | 65 | 25 | 85 | 45 | 05 | 65 | 25 | 85 |
| **20** Vyaya ... | 86 | 46 | 06 | 66 | 26 | 86 | 46 | 06 | 66 | 26 | 86 | 46 | 06 | 66 | 26 | 86 |
| **21** Sarvajit ... | 87 | 47 | 07 | 67 | 27 | 87 | 47 | 07 | 67 | 27 | 87 | 47 | 07 | 67 | 27 | 87 |
| **22** Sarvadhari ... | 88 | 48 | 08 | 68 | 28 | 88 | 48 | 08 | 68 | 28 | 88 | 48 | 08 | 68 | 28 | 88 |
| **23** Virodhi ... | 89 | 49 | 09 | 69 | 29 | 89 | 49 | 09 | 69 | 29 | 89 | 49 | 09 | 69 | 29 | 89 |
| **24** Vikrita ... | 90 | 50 | 10 | 70 | 30 | 90 | 50 | 10 | 70 | 30 | 90 | 50 | 10 | 70 | 30 | 90 |
| **25** Khara ... | 91 | 51 | 11 | 71 | 31 | 91 | 51 | 11 | 71 | 31 | 91 | 51 | 11 | 71 | 31 | 91 |
| **26** Nandana ... | 92 | 52 | 12 | 72 | 32 | 92 | 52 | 12 | 72 | 32 | 92 | 52 | 12 | 72 | 32 | 92 |
| **27** Vijaya ... | 93 | 53 | 13 | 73 | 33 | 93 | 53 | 13 | 73 | 33 | 93 | 53 | 13 | 73 | 33 | 93 |
| **28** Jaya ... | 94 | 54 | 14 | 74 | 34 | 94 | 54 | 14 | 74 | 34 | 94 | 54 | 14 | 74 | 34 | 94 |
| **29** Manmatha ... | 95 | 55 | 15 | 75 | 35 | 95 | 55 | 15 | 75 | 35 | 95 | 55 | 15 | 75 | 35 | 95 |
| **30** Durmukha ... | 96 | 56 | 16 | 76 | 36 | 96 | 56 | 16 | 76 | 36 | 96 | 56 | 16 | 76 | 36 | 96 |
| **31** Hemalamba ... | 97 | 57 | 17 | 77 | 37 | 97 | 57 | 17 | 77 | 37 | 97 | 57 | 17 | 77 | 37 | 97 |
| **32** Vilamba ... | 98 | 58 | 18 | 78 | 38 | 98 | 58 | 18 | 78 | 38 | 98 | 58 | 18 | 78 | 38 | 98 |
| **33** Vikari ... | 99 | 59 | 19 | 79 | 39 | 99 | 59 | 19 | 79 | 39 | 99 | 59 | 19 | 79 | 39 | 99 |
| **34** Sarvari ... | **10** 00 | 60 | 20 | 80 | 40 | **13** 00 | 60 | 20 | 80 | 40 | **16** 00 | 60 | 20 | 80 | 40 | **19** 00 |
| **35** Plava ... | 01 | 61 | 21 | 81 | 41 | 01 | 61 | 21 | 81 | 41 | 01 | 61 | 21 | 81 | 41 | 01 |
| **36** Subhakrit ... | 02 | 62 | 22 | 82 | 42 | 02 | 62 | 22 | 82 | 42 | 02 | 62 | 22 | 82 | 42 | 02 |
| **37** Sobhana ... (Sobhakrit) | 03 | 63 | 23 | 83 | 43 | 03 | 63 | 23 | 83 | 43 | 03 | 63 | 23 | 83 | 43 | 03 |
| **38** Krodhi ... | 04 | 64 | 24 | 84 | 44 | 04 | 64 | 24 | 84 | 44 | 04 | 64 | 24 | 84 | 44 | 04 |
| **39** Visvavasu ... | 05 | 65 | 25 | 85 | 45 | 05 | 65 | 25 | 85 | 45 | 05 | 65 | 25 | 85 | 45 | 05 |
| **40** Parabhava ... | 06 | 66 | 26 | 86 | 46 | 06 | 66 | 26 | 86 | 46 | 06 | 66 | 26 | 86 | 46 | 06 |
| **41** Plavanga ... | 07 | 67 | 27 | 87 | 47 | 07 | 67 | 27 | 87 | 47 | 07 | 67 | 27 | 87 | 47 | 07 |
| **42** Kilaka ... | 08 | 68 | 28 | 88 | 48 | 08 | 68 | 28 | 88 | 48 | 08 | 68 | 28 | 88 | 48 | 08 |
| **43** Saumya ... | 09 | 69 | 29 | 89 | 49 | 09 | 69 | 29 | 89 | 49 | 09 | 69 | 29 | 89 | 49 | 09 |
| **44** Sadharana ... | 10 | 70 | 30 | 90 | 50 | 10 | 70 | 30 | 90 | 50 | 10 | 70 | 30 | 90 | 50 | 10 |
| **45** Virodhakrit ... | 11 | 71 | 31 | 91 | 51 | 11 | 71 | 31 | 91 | 51 | 11 | 71 | 31 | 91 | 51 | 11 |
| **46** Paridhavi ... | 12 | 72 | 32 | 92 | 52 | 12 | 72 | 32 | 92 | 52 | 12 | 72 | 32 | 92 | 52 | 12 |
| **47** Pramadicha ... | 13 | 73 | 33 | 93 | 53 | 13 | 73 | 33 | 93 | 53 | 13 | 73 | 33 | 93 | 53 | 13 |
| **48** Ananda ... | 14 | 74 | 34 | 94 | 54 | 14 | 74 | 34 | 94 | 54 | 14 | 74 | 34 | 94 | 54 | 14 |
| **49** Rakshasa ... | 15 | 75 | 35 | 95 | 55 | 15 | 75 | 35 | 95 | 55 | 15 | 75 | 35 | 95 | 55 | 15 |
| **50** Anala (Nala)... | 16 | 76 | 36 | 96 | 56 | 16 | 76 | 36 | 96 | 56 | 16 | 76 | 36 | 96 | 56 | 16 |
| **51** Pingala ... | 17 | 77 | 37 | 97 | 57 | 17 | 77 | 37 | 97 | 57 | 17 | 77 | 37 | 97 | 57 | 17 |
| **52** Kalayukta ... | 18 | 78 | 38 | 98 | 58 | 18 | 78 | 38 | 98 | 58 | 18 | 78 | 38 | 98 | 58 | 18 |
| **53** Siddharthi ... | 19 | 79 | 39 | 99 | 59 | 19 | 79 | 39 | 99 | 59 | 19 | 79 | 39 | 99 | 59 | 19 |
| **54** Raudra ... | 20 | 80 | 40 | **12** 00 | 60 | 20 | 80 | 40 | **15** 00 | 60 | 20 | 80 | 40 | **18** 00 | 60 | 20 |
| **55** Durmati ... | 21 | 81 | 41 | 01 | 61 | 21 | 81 | 41 | 01 | 61 | 21 | 81 | 41 | 01 | 61 | 21 |
| **56** Dundubhi ... | 22 | 82 | 42 | 02 | 62 | 22 | 82 | 42 | 02 | 62 | 22 | 82 | 42 | 02 | 62 | 22 |
| **57** Rudhirodgari... | 23 | 83 | 43 | 03 | 63 | 23 | 83 | 43 | 03 | 63 | 23 | 83 | 43 | 03 | 63 | 23 |
| **58** Raktaksha ... | 24 | 84 | 44 | 04 | 64 | 24 | 84 | 44 | 04 | 64 | 24 | 84 | 44 | 04 | 64 | 24 |
| **59** Krodhana ... | 25 | 85 | 45 | 05 | 65 | 25 | 85 | 45 | 05 | 65 | 25 | 85 | 45 | 05 | 65 | 25 |
| **60** Kshaya (Akshaya) | 26 | 86 | 46 | 06 | 66 | 26 | 86 | 46 | 06 | 66 | 26 | 86 | 46 | 06 | 66 | 26 |

# TABLE II.

| Signs of Zodiac; also names of Malabar Months. | Lunar Months; also Bengal Solar Months. | Tamil Solar Months. | Surya Siddhanta. Number of days in each Solar Month. | Moment of Sankranti; or No. of days up to beginning of each Solar Month. | Moment of each New Moon; or No. of days up to beginning of each Lunar Month. | Increase of ☾'s anomaly up to beginning of each Lunar Month. | — | No. of days in Anomalistic Months. | Arya Siddhanta. No. of days in each Solar Month. | Moment of Sankranti; or No. of days up to beginning of each Solar Month. |
|---|---|---|---|---|---|---|---|---|---|---|
| Mesha (**Mal.** Medam) | Vaisakha | Chittirai | 30·93528 | ... | ... | ... | **1** | 27·5546 | 30·92500000 | ... |
| Vrishabha (**Mal.** Edavam) | Jyeshta | Vaikasi | 31·42028 | 30·93528 | 29·53059 | 1·976 | **2** | 55·1092 | 31·40111111 | 30·925000 |
| Mithuna | Ashada | Ani | 31·64472 | 62·35555 | 59·06117 | 3·952 | **3** | 82·6638 | 31·60722222 | 62·3261111 |
| Karkata | Sravana | Adi | 31·47528 | 94·00028 | 88·59176 | 5·928 | **4** | 110·2184 | 31·46777777 | 93·9333333 |
| Simha | Bhadrapada | Avani | 31·01861 | 125·47555 | 118·12235 | 7·904 | **5** | 137·7730 | 31·03472222 | 125·4011111 |
| Kanya | Asvina | Purattasi | 30·44138 | 156·49417 | 147·65293 | 9·880 | **6** | 165·3276 | 30·45666666 | 156·4358333 |
| Tula | Kartika | Aippasi | 29·89333 | 186·93555 | 177·18353 | 11·856 | **7** | 192·8822 | 29·90333333 | 186·8925000 |
| Vrischika | Margasira | Kartikai | 29·49027 | 216·82888 | 206·71411 | 13·832 | **8** | 220·4368 | 29·50861111 | 216·7958333 |
| Dhanus | Pausha | Margali | 29·31777 | 246·31916 | 236·24470 | 15·808 | **9** | 247·9914 | 29·35055555 | 246·3044444 |
| Makara | Magha | Tai | 29·44805 | 275·63694 | 265·77529 | 17·784 | **10** | 275·5460 | 29·45666666 | 275·6550000 |
| Kumbha | Phalguna | Masi | 29·82027 | 305·08499 | 295·30588 | 19·760 | **11** | 303·1006 | 29·80833333 | 305·1116666 |
| Mina | Chaitra | Panguni | 30·35348 | 334·90527 | 324·83647 | 21·736 | **12** | 330·6552 | 30·33876157 | 334·9200000 |
| | | | | 365·25875 | 354·36705 | 23·712 | **13** | 358·2098 | | 365·2586805 |
| | | | | | 383·89764 | 25·688 | | | | |

**Limits of Adhika months.**

| | Before. | | Before. | | After. | Before. |
|---|---|---|---|---|---|---|
| **Vaisakha** | 1·40469 | **Bhadrapada** | 8·84122 | ... | ... | ... |
| **Jyeshta** | 3·29438 | **Asvina** | 9·75201 | **Phalguna** | 9·77912 | 10·06880 |
| **Ashada** | 5·40851 | **Kartika** | 10·11475 | **Chaitra** | 10·06880 | 10·89170 |
| **Sravana** | 7·35320 | ... | ... | ... | ... | ... |

**Limits of Kshaya months.**

| | After. | Before. |
|---|---|---|
| **Margasira** | 10·07446 | 10·11475 |
| **Pausha** | 9·86164 | 10·07447 |
| **Magha** | 9·77910 | 9·86164 |
| ... | ... | ... |

**Tithis.**

| Duration of Tithis. | | | Names of Tithis. | Duration of Tithis. | | | Names of Tithis. | Duration of Tithis. | | | Names of Tithis. |
|---|---|---|---|---|---|---|---|---|---|---|---|
| days. | | | | days. | | | | days. | | | |
| 0·98435 | **1** | Sukla | Pratipada | 10·82788 | **11** | Sukla | Ekadasi | 20·67141 | **6** | Krishna | Shashti |
| 1·96870 | **2** | Do. | Dvitiya | 11·81223 | **12** | Do. | Dvadasi | 21·65576 | **7** | Do. | Saptami |
| 2·95306 | **3** | Do. | Tritiya | 12·79659 | **13** | Do. | Trayodasi | 22·64012 | **8** | Do. | Ashtami |
| 3·93741 | **4** | Do. | Chathurthi | 13·78094 | **14** | Do. | Chaturdasi | 23·62447 | **9** | Do. | Navami |
| 4·92176 | **5** | Do. | Panchami | 14·76529 | **15** | Paurnami | | 24·60882 | **10** | Do. | Dasami |
| 5·90612 | **6** | Do. | Shashti | 15·74965 | **1** | Krishna | Pratama | 25·59318 | **11** | Do. | Ekadasi |
| 6·89047 | **7** | Do. | Saptami | 16·73400 | **2** | Do. | Dvitiya | 26·57753 | **12** | Do. | Dvadasi |
| 7·87482 | **8** | Do. | Ashtami | 17·71835 | **3** | Do. | Tritiya | 27·56188 | **13** | Do. | Trayodasi |
| 8·85918 | **9** | Do. | Navami | 18·70270 | **4** | Do. | Chathurthi | 28·54623 | **14** | Do. | Chaturdasi |
| 9·84353 | **10** | Do. | Dasami | 19·68706 | **5** | Do. | Panchami | 29·53059 | **15** | Amavasya | |

# TABLE III.

## Nakshatras.

| | Order and names of Nakshatras. | Collective duration of Nakshatras (equal space). | Deities presiding over Nakshatras. | Duration of Nakshatras. (Garga.) | Duration of Nakshatras. (Brahma.) |
|---|---|---|---|---|---|
| | | days. | | days. | days. |
| 1 | Asvini | 1·01191 | Asvin | 1·01191 | 1·0 |
| 2 | Bharani | 2·02383 | Yama | 1·51787 | 1·5 |
| 3 | Krittika (**Tam.** Kiruttigai) | 3·03574 | Agni | 2·52978 | 2·5 |
| 4 | Rohini | 4·04765 | Prajapati | 4·04765 | 4·0 |
| 5 | Mrigasira (**Tam.** Mirugasiram) | 5·05957 | Soma | 5·05956 | 5·0 |
| 6 | Ardhra (**Tam.** Arudra or Tiruvadirai) | 6·07148 | Rudra | 5·56552 | 5·5 |
| 7 | Punarvasu | 7·08340 | Aditi | 7·08339 | 7·0 |
| 8 | Pushya (**Tam.** Pusam) | 8·09531 | Brihaspati | 8·09531 | 8·0 |
| 9 | Aslesha (**Tam.** Ayilyam) | 9·10722 | Sarpah | 8·60126 | 8·5 |
| 10 | Magha (**Tam.** Magham) | 10·11914 | Pitarah | 9·61318 | 9·5 |
| 11 | Purva Phalguni (**Tam.** Puram) | 11·13105 | Bhaga | 10·62510 | 10·5 |
| 12 | Uttara Phalguni (**Tam.** Uttiram) | 12·14297 | Aryaman | 12·14297 | 12·0 |
| 13 | Hasta (**Tam.** Hastam) | 13·15488 | Savitri | 13·15488 | 13·0 |
| 14 | Chittra (**Tam.** Chittirai) | 14·16679 | Tvashtri | 14·16679 | 14·0 |
| 15 | Svati | 15·17871 | Vayu | 14·67275 | 14·5 |
| 16 | Visakha (**Tam.** Visakam) | 16·19062 | Indragni | 16·19062 | 16·0 |
| 17 | Anuradha (**Tam.** Anusham) | 17·20273 | Mitra | 17·20253 | 17·0 |
| 18 | Jyeshta (**Tam.** Kettai) | 18·21445 | Indra | 17·70849 | 17·5 |
| 19 | Mula (**Tam.** Mulam) | 19·22636 | Nirriti | 18·72040 | 18·5 |
| 20 | Purva Ashada (**Tam.** Puradam) | 20·23828 | Apah | 19·73232 | 19·5 |
| 21 | Uttara Ashada (**Tam.** Uttiradam) | 21·25019 | Visvadeva | 21·25019 | 21·0 |
| | | | Brahma | | 21·3217 (Abhijit) |
| 22 | Sravana (**Tam.** Tiruvonam) | 22·26210 | Vishnu | 22·26210 | 22·3217 |
| 23 | Sravishta or Danishta (**Tam.** Avittam) | 23·27402 | Vasava | 23·27401 | 23·3217 |
| 24 | Satabhisaj or Sataraka (**Tam.** Sadayam) | 24·28593 | Varuna | 23·77997 | 23·8217 |
| 25 | Purva Bhadrapada (**Tam.** Purattadi) | 25·29785 | Aja Ekapad | 24·79188 | 24·8217 |
| 26 | Uttara Bhadrapada (**Tam.** Uttirattadi) | 26·30976 | Ahi Budhnya | 26·30975 | 26·3217 |
| 27 | Revati | 27·32167 | Pushan | 27·32167 | 27·3217 |

## Yogas

| | Nakshatras and Yogas in lunation space. | | Names of Yogas. | Collective duration of Yogas. |
|---|---|---|---|---|
| | days. | | | days. |
| 1 | 1·09372 | 1 | Vishkamba. | 0·94149 |
| 2 | 2·18745 | 2 | Priti ... | 1·88298 |
| 3 | 3·28118 | 3 | Ayushmat ... | 2·82447 |
| 4 | 4·37490 | 4 | Saubhagya. | 3·76596 |
| 5 | 5·46863 | 5 | Sobhana ... | 4·70745 |
| 6 | 6·56235 | 6 | Atiganda ... | 5·64894 |
| 7 | 7·65608 | 7 | Sukarman ... | 6·59043 |
| 8 | 8·74980 | 8 | Dhriti ... | 7·53192 |
| 9 | 9·84353 | 9 | Sula ... | 8·47341 |
| 10 | 10·93725 | 10 | Ganda ... | 9·41489 |
| 11 | 12·03098 | 11 | Vriddhi ... | 10·35638 |
| 12 | 13·12470 | 12 | Dhruva ... | 11·29787 |
| 13 | 14·21843 | 13 | Vyaghata ... | 12·23936 |
| 14 | 15·31216 | 14 | Harshana ... | 13·18085 |
| 15 | 16·40588 | 15 | Vajra ... | 14·12234 |
| 16 | 17·49961 | 16 | Siddhi ... | 15·06383 |
| 17 | 18·59333 | 17 | Vyatipata ... | 16·00532 |
| 18 | 19·68706 | 18 | Variyas ... | 16·94681 |
| 19 | 20·78078 | 19 | Parigha ... | 17·88830 |
| 20 | 21·87451 | 20 | Siva ... | 18·82979 |
| 21 | 22·96823 | 21 | Siddha ... | 19·77128 |
| 22 | 24·06196 | 22 | Sadhya ... | 20·71277 |
| 23 | 25·15569 | 23 | Subha ... | 21·65426 |
| 24 | 26·24941 | 24 | Sukla ... | 22·59575 |
| 25 | 27·34314 | 25 | Brahman ... | 23·53724 |
| 26 | 28·43686 | 26 | Indra ... | 24·47873 |
| 27 | 29·53059 | 27 | Vaidhriti ... | 25·42022 |

## Karanas.

| | | |
|---|---|---|
| Kimstughna | ... 1, | |
| Bava | ... 2, 9, 16, 23, 30, 37, 44, 51, | 58=Sakuni. |
| Balava | ... 3, 10, 17, 24, 31, 38, 45, 52, | 59=Naga. |
| Kaulava | ... 4, 11, 18, 25, 32, 39, 46, 53, | 60=Chatushpada. |
| Taitila | ... 5, 12, 19, 26, 33, 40, 47, 54, | |
| Gara | ... 6, 13, 20, 27, 34, 41, 48, 55, | |
| Banija | ... 7, 14, 21, 28, 35, 42, 49, 56, | |
| Vishti (or Bhadra.) | ... 8, 15, 22, 29, 36, 43, 50, 57, | |

## TABLE IV.

### Perpetual Almanac for European Calendar.

**Centuries.**

| | | 1 | 2 | 3 | 4 | 5 | 6 | 7 or 0 | Week-day begin- |
|---|---|---|---|---|---|---|---|---|---|
| **B.C.** | ... | 3001 | 3101 | 3201 | | | | | ning from Sun- |
| | | 2301 | 2401 | 2501 | 2601 | 2701 | 2801 | 2901 | day which is 1. |
| | | 1601 | 1701 | 1801 | 1901 | 2001 | 2101 | 2201 | |
| | | 901 | 1001 | 1101 | 1201 | 1301 | 1401 | 1501 | |
| | | 201 | 301 | 401 | 501 | 601 | 701 | 801 | |
| | | | | | | | 1 | 101 | |
| **A.D.** | ... | 500 | 400 | 300 | 200 | 100 | 0 | | |
| **Old Style** | ... | 1200 | 1100 | 1000 | 900 | 800 | 700 | 600 | |
| | | | | 1700 | 1600 | 1500 | 1400 | 1300 | |
| **A.D.** | ... | 1600 | 1900 | | 1800 | | 1700 | | |
| **New Style** | ... | 2000 | 2300 | | 2200 | | 2100 | | |

**Odd Years.**

| 1 | 2 | 3 | 4 | 5 | 6 | 7 or 0 | Week-day begin- |
|---|---|---|---|---|---|---|---|
| 1 | 2 | 3 | ... | 4 | 5 | 6 | ning from Sun- |
| 7 | ... | 8 | 9 | 10 | 11 | ... | day which is 1. |
| 12 | 13 | 14 | 15 | ... | 16 | 17 | |
| 18 | 19 | ... | 20 | 21 | 22 | 23 | |
| ... | 24 | 25 | 26 | 27 | ... | 28 | |
| 29 | 30 | 31 | ... | 32 | 33 | 34 | |
| 35 | ... | 36 | 37 | 38 | 39 | ... | |
| 40 | 41 | 42 | 43 | ... | 44 | 45 | |
| 46 | 47 | ... | 48 | 49 | 50 | 51 | |
| ... | 52 | 53 | 54 | 55 | ... | 56 | |
| 57 | 58 | 59 | ... | 60 | 61 | 62 | |
| 63 | ... | 64 | 65 | 66 | 67 | ... | |
| 68 | 69 | 70 | 71 | ... | 72 | 73 | |
| 74 | 75 | ... | 76 | 77 | 78 | 79 | |
| ... | 80 | 81 | 82 | 83 | ... | 84 | |
| 85 | 86 | 87 | ... | 88 | 89 | 90 | |
| 91 | ... | 92 | 93 | 94 | 95 | ... | |
| 96 | 97 | 98 | 99 | ... | ... | ... | |

**Months.**

| | | 1 | 2 | 3 | 4 | 5 | 6 | 7 or 0 | Week-day begin- |
|---|---|---|---|---|---|---|---|---|---|
| **Ordinary years.** | | Aug. | Feb.<br>Mar.<br>Nov. | Jun. | Sep.<br>Dec. | Ap.<br>Jul. | Jan.<br>Oct. | May. | ning from Sun-<br>day which is 1. |
| **Leap years** | ... | Feb. | ... | ... | ... | Jan. | ... | ... | ... ... |

# TABLE V.

**Multip. Table for 1/1·0808. Mean time corresponding to Nakshatra-space.**

| | | | | | |
|---|---|---|---|---|---|
| **1** | 0·92520 | **·34** | ·31457 | **·67** | ·61988 |
| **2** | 1·85040 | **·35** | ·32382 | **·68** | ·62913 |
| **3** | 2·77560 | **·36** | ·33307 | **·69** | ·63839 |
| **4** | 3·70080 | **·37** | ·34232 | **·70** | ·64764 |
| **5** | 4·62599 | **·38** | ·35157 | **·71** | ·65689 |
| **6** | 5·55119 | **·39** | ·36083 | **·72** | ·66614 |
| **7** | 6·47639 | **·40** | ·37008 | **·73** | ·67539 |
| **8** | 7·40159 | **·41** | ·37933 | **·74** | ·68465 |
| **9** | 8·32679 | **·42** | ·38858 | **·75** | ·69390 |
| **10** | 9·27199 | **·43** | ·39783 | **·76** | ·70315 |
| **11** | 10·17719 | **·44** | ·40709 | **·77** | ·71240 |
| **12** | 11·10239 | **·45** | ·41634 | **·78** | ·72165 |
| **13** | 12·02759 | **·46** | ·42559 | **·79** | ·73091 |
| **14** | 12·95279 | **·47** | ·43484 | **·80** | ·74016 |
| **15** | 13·87799 | **·48** | ·44409 | **·81** | ·74941 |
| **16** | 14·80318 | **·49** | ·45335 | **·82** | ·75866 |
| **17** | 15·72838 | **·50** | ·46260 | **·83** | ·76791 |
| **18** | 16·65358 | **·51** | ·47185 | **·84** | ·77717 |
| **19** | 17·57878 | **·52** | ·48110 | **·85** | ·78642 |
| **20** | 18·50398 | **·53** | ·49035 | **·86** | ·79567 |
| **21** | 19·42918 | **·54** | ·49961 | **·87** | ·80492 |
| **22** | 20·35438 | **·55** | ·50886 | **·88** | ·81417 |
| **23** | 21·27958 | **·56** | ·51811 | **·89** | ·82343 |
| **24** | 22·20478 | **·57** | ·52736 | **·90** | ·83268 |
| **25** | 23·12998 | **·58** | ·53661 | **·91** | ·84193 |
| **26** | 24·05518 | **·59** | ·54587 | **·92** | ·85118 |
| **27** | 24·98037 | **·60** | ·55512 | **·93** | ·86043 |
| **28** | 25·90557 | **·61** | ·56437 | **·94** | ·86969 |
| **29** | 26·83077 | **·62** | ·57362 | **·95** | ·87894 |
| **·30** | ·27756 | **·63** | ·58287 | **·96** | ·88819 |
| **·31** | ·28681 | **·64** | ·59213 | **·97** | ·89744 |
| **·32** | ·29606 | **·65** | ·60138 | **·98** | ·90669 |
| **·33** | ·30531 | **·66** | ·61063 | **·99** | ·91595 |

**Multip. Table for 1/1·1616. Mean time corresponding to Yoga-space.**

| | | | | | |
|---|---|---|---|---|---|
| **1** | 0·86081 | **·34** | ·29267 | **·67** | ·57674 |
| **2** | 1·72162 | **·35** | ·30128 | **·68** | ·58535 |
| **3** | 2·58243 | **·36** | ·30989 | **·69** | ·59396 |
| **4** | 3·44324 | **·37** | ·31850 | **·70** | ·60257 |
| **5** | 4·30405 | **·38** | ·32711 | **·71** | ·61117 |
| **6** | 5·16486 | **·39** | ·33572 | **·72** | ·61978 |
| **7** | 6·02567 | **·40** | ·34432 | **·73** | ·62839 |
| **8** | 6·88648 | **·41** | ·35293 | **·74** | ·63700 |
| **9** | 7·74729 | **·42** | ·36154 | **·75** | ·64561 |
| **10** | 8·60810 | **·43** | ·37015 | **·76** | ·65421 |
| **11** | 9·46891 | **·44** | ·37876 | **·77** | ·66282 |
| **12** | 10·32972 | **·45** | ·38736 | **·78** | ·67143 |
| **13** | 11·19053 | **·46** | ·39597 | **·79** | ·68004 |
| **14** | 12·05134 | **·47** | ·40458 | **·80** | ·68866 |
| **15** | 12·91215 | **·48** | ·41319 | **·81** | ·69725 |
| **16** | 13·77295 | **·49** | ·42180 | **·82** | ·70586 |
| **17** | 14·63376 | **·50** | ·43040 | **·83** | ·71448 |
| **18** | 15·49457 | **·51** | ·43901 | **·84** | ·72309 |
| **19** | 16·35538 | **·52** | ·44762 | **·85** | ·73169 |
| **20** | 17·21619 | **·53** | ·45623 | **·86** | ·74030 |
| **21** | 18·07700 | **·54** | ·46484 | **·87** | ·74890 |
| **22** | 18·93781 | **·55** | ·47344 | **·88** | ·75751 |
| **23** | 19·79862 | **·56** | ·48205 | **·89** | ·76612 |
| **24** | 20·65943 | **·57** | ·49066 | **·90** | ·77474 |
| **25** | 21·52024 | **·58** | ·49927 | **·91** | ·78345 |
| **26** | 22·38105 | **·59** | 50788 | **·92** | ·79193 |
| **27** | 23·24186 | **·60** | ·51649 | **·93** | ·80055 |
| **28** | 24·10267 | **·61** | ·52509 | **·94** | ·80916 |
| **29** | 24·96348 | **·62** | ·53370 | **·95** | ·81777 |
| **·30** | ·25823 | **·63** | ·54231 | **·96** | ·82638 |
| **·31** | ·26685 | **·64** | ·55092 | **·97** | ·83498 |
| **·32** | ·27546 | **·65** | ·55953 | **·98** | ·84359 |
| **·33** | ·28407 | **·66** | ·56830 | **·99** | ·85220 |

**Sun's Longitude for Nakshatras and Yogas.**

| | | | | | |
|---|---|---|---|---|---|
| **·01** | ·0008085 | **·34** | ·0274885 | **·67** | ·0541684 |
| **·02** | ·0016170 | **·35** | ·0282970 | **·68** | ·0549769 |
| **·03** | ·0024254 | **·36** | ·0291054 | **·69** | ·0557854 |
| **·04** | ·0032339 | **·37** | ·0299139 | **·70** | ·0565939 |
| **·05** | ·0040424 | **·38** | ·0307224 | **·71** | ·0574024 |
| **·06** | ·0048509 | **·39** | ·0315309 | **·72** | ·0582108 |
| **·07** | ·0056594 | **·40** | ·0323394 | **·73** | ·0590193 |
| **·08** | ·0064679 | **·41** | ·0331478 | **·74** | ·0598278 |
| **·09** | ·0072764 | **·42** | ·0339563 | **·75** | ·0606363 |
| **·10** | ·0080848 | **·43** | ·0347648 | **·76** | ·0614448 |
| **·11** | ·0088933 | **·44** | ·0355733 | **·77** | ·0622533 |
| **·12** | ·0097018 | **·45** | ·0363818 | **·78** | ·0630617 |
| **13** | ·0105103 | **·46** | ·0371903 | **·79** | ·0638702 |
| **·14** | ·0113188 | **·47** | ·0379987 | **·80** | ·0646787 |
| **·15** | ·0121273 | **·48** | ·0388072 | **·81** | ·0654872 |
| **·16** | ·0129357 | **·49** | ·0396157 | **·82** | ·0662957 |
| **·17** | ·0137442 | **·50** | ·0404242 | **·83** | ·0671042 |
| **·18** | ·0145527 | **·51** | ·0412327 | **·84** | ·0679127 |
| **·19** | ·0153612 | **·52** | ·0420412 | **·85** | ·0687211 |
| **·20** | ·0161697 | **·53** | ·0428496 | **·86** | ·0695296 |
| **·21** | ·0169782 | **·54** | ·0436581 | **·87** | ·0703381 |
| **·22** | ·0177867 | **·55** | ·0444666 | **·88** | ·0714066 |
| **·23** | ·0185951 | **·56** | ·0452751 | **·89** | ·0719551 |
| **·24** | ·0194036 | **·57** | ·0460836 | **·90** | ·0727636 |
| **·25** | ·0202121 | **·58** | ·0468921 | **·91** | ·0735720 |
| **·26** | ·0210206 | **·59** | ·0477005 | **·92** | ·0743805 |
| **·27** | ·0218291 | **·60** | ·0485090 | **·93** | ·0751890 |
| **·28** | ·0226376 | **·61** | ·0493175 | **·94** | ·0759975 |
| **·29** | ·0234460 | **·62** | ·0501260 | **·95** | ·0768060 |
| **·30** | ·0242545 | **·63** | ·0509345 | **·96** | ·0776145 |
| **·31** | ·0250630 | **·64** | ·0517429 | **·97** | ·0784229 |
| **·32** | ·0258715 | **·65** | ·0525515 | **·98** | ·0792314 |
| **·33** | ·0266800 | **·66** | ·0533599 | **·99** | ·0800399 |

## Multiplication Tables for Jupiter's Samvatsaras.

| Jupiter's years in Solar years. | | Increase of Jupiter's year for Solar days. | Jupiter's years in Solar years. | | Increase of Jupiter's year for Solar days. | Jupiter's years in Solar years. | | Increase of Jupiter's year for Solar days. | Jupiter's years in Solar years. | | Increase of Jupiter's year for Solar days. | Jupiter's years in Solar years. | | Increase of Jupiter's year for Solar days. | Jupiter's years in Solar years. | | Increase of Jupiter's year for Solar days. |
|---|---|---|---|---|---|---|---|---|---|---|---|---|---|---|---|---|---|
| 1·0117 | **1** | ·00277 | 18·2106 | **18** | ·04986 | 35·4095 | **35** | ·09694 | 52·6084 | **52** | ·14403 | 9·8073 | **69** | ·19112 | 27·0062 | **86** | ·23820 |
| 2·0234 | **2** | ·00554 | 19·2223 | **19** | ·05263 | 36·4212 | **36** | ·09971 | 53·6201 | **53** | ·14680 | 10·8190 | **70** | ·19389 | 28·0179 | **87** | ·24097 |
| 3·0351 | **3** | ·00831 | 20·2340 | **20** | ·05540 | 37·4329 | **37** | ·10248 | 54·6318 | **54** | ·14957 | 11·8307 | **71** | ·19666 | 29·0296 | **88** | ·24374 |
| 4·0468 | **4** | ·01108 | 21·2457 | **21** | ·05817 | 38·4446 | **38** | ·10525 | 55·6435 | **55** | ·15234 | 12·8424 | **72** | ·19943 | 30·0413 | **89** | ·24651 |
| 5·0585 | **5** | ·01385 | 22·2574 | **22** | ·06093 | 39·4563 | **39** | ·10802 | 56·6552 | **56** | ·15520 | 13·8541 | **73** | ·20220 | 31·0530 | **90** | ·24928 |
| 6·0702 | **6** | ·01662 | 23·2691 | **23** | ·06371 | 40·4680 | **40** | ·11079 | 57·6669 | **57** | ·15788 | 14·8658 | **74** | ·20497 | 32·0647 | **91** | ·25205 |
| 7·0819 | **7** | ·01939 | 24·2808 | **24** | ·06647 | 41·4797 | **41** | ·11356 | 58·6786 | **58** | ·16065 | 15·8775 | **75** | ·20774 | 33·0764 | **92** | ·25482 |
| 8·0936 | **8** | ·02216 | 25·2925 | **25** | ·06924 | 42·4914 | **42** | ·11633 | 59·6903 | **59** | ·16342 | 16·8892 | **76** | ·21051 | 34·0881 | **93** | ·25759 |
| 9·1053 | **9** | ·02493 | 26·3042 | **26** | ·07201 | 43·5031 | **43** | ·11910 | 0·7020 | **60** | ·16619 | 17·9009 | **77** | ·21[illegible]28 | 35·0998 | **94** | ·26036 |
| 10·1170 | **10** | ·02770 | 27·3159 | **27** | ·07478 | 44·5148 | **44** | ·12187 | 1·7137 | **61** | ·16896 | 18·9126 | **78** | ·21605 | 36·1115 | **95** | ·26313 |
| 11·1287 | **11** | ·03047 | 28·3276 | **28** | ·07755 | 45·5265 | **45** | ·12464 | 2·7254 | **62** | ·17173 | 19·9243 | **79** | ·21881 | 37·1232 | **96** | ·26590 |
| 12·1404 | **12** | ·03324 | 29·3393 | **29** | ·08032 | 46·5382 | **46** | ·12741 | 3·7371 | **63** | ·17456 | 20·9360 | **80** | ·22158 | 38·1349 | **97** | ·26867 |
| 13·1521 | **13** | ·03601 | 30·3510 | **30** | ·08309 | 47·5499 | **47** | ·13018 | 4·7488 | **64** | ·17727 | 21·9477 | **81** | ·22435 | 39·1466 | **98** | ·27144 |
| 14·1638 | **14** | ·03878 | 31·3627 | **31** | ·08586 | 48·5616 | **48** | ·13295 | 5·7605 | **65** | ·18004 | 22·9594 | **82** | ·22712 | 40·1583 | **99** | ·27421 |
| 15·1755 | **15** | ·04155 | 32·3744 | **32** | ·08863 | 49·5733 | **49** | ·13572 | 6·7722 | **66** | ·18281 | 23·9711 | **83** | ·22989 | | | |
| 16·1872 | **16** | ·04432 | 33·3861 | **33** | ·09140 | 50·5850 | **50** | ·13849 | 7·7839 | **67** | ·18558 | 24·9828 | **84** | ·23266 | | | |
| 17·1989 | **17** | ·04709 | 34·3978 | **34** | ·09417 | 51·5967 | **51** | ·14126 | 8·7956 | **68** | ·18835 | 25·9945 | **85** | ·23543 | | | |

# TABLE VI.

## Surya Siddhanta Constants for Centuries.

| Year A.D. | Kaliyuga. | Commencement of Solar Year. Month and Date. | Commencement of Solar Year. Fraction of day. | First new moon in Solar Year. | ☾'s anomaly at commencement of Solar Year. |
|---|---|---|---|---|---|
| 1 B.C. | 3101 | 13M | ·98310 | 9·99810 | 12·73835 |
| 100 A.D. | 3201 | 14M | ·85875 | 13·45974 | 1·21454 |
| 200 | 3301 | 15M | ·73440 | 16·92138 | 17·24532 |
| 300 | 3401 | 16M | ·61004 | 20·38302 | 5·72151 |
| 400 | 3501 | 17M | ·48569 | 23·84466 | 21·75228 |
| 500 | 3601 | 18M | ·36134 | 27·30630 | 10·22845 |
| 600 | 3701 | 19M | ·23698 | 1·23735 | 26·25922 |
| 700 | 3801 | 20M | ·11264 | 4·69899 | 14·73539 |
| 800 | 3901 | 20M | ·98828 | 8·16063 | 3·21156 |
| 900 | 4001 | 21M | ·86393 | 11·62227 | 19·24288 |
| 1000 | 4101 | 22M | ·73958 | 15·08391 | 7·71850 |
| 1100 | 4201 | 23M | ·61523 | 18·54556 | 23·74927 |
| 1200 | 4301 | 24M | ·49088 | 22·00720 | 12·22544 |
| 1300 | 4401 | 25M | ·36652 | 25·46884 | 0·70161 |
| 1400 | 4501 | 26M | ·24217 | 28·93048 | 16·73238 |
| 1500 | 4601 | 27M | ·11782 | 2·86153 | 5·20855 |
| 1600 | 4701 | 27M | ·99347 | 6·32317 | 21·35937 |
| 1700 | 4801 | 28M | ·86912 | 9·78481 | 9·83810 |
| 1800 | 4901 | 10Ap | ·74477 | 13·24645 | 25·87144 |
| 1900 | 5001 | 12Ap | ·62041 | 16·70809 | 14·35017 |

## Arya Siddhanta Constants for Centuries.

| Year A.D. | Kaliyuga. | Commencement of Solar Year. Month and Date. | Commencement of Solar Year. Fraction of day. | First new moon in Solar Year. | ☾'s anomaly at commencement of Solar Year. |
|---|---|---|---|---|---|
| 1 B.C. | 3101 | 14M | ·020832 | 9·97497 | 12·44500 |
| 100 A.D. | 3201 | 14M | ·888887 | 13·43662 | 0·91097 |
| 200 | 3301 | 15M | ·756943 | 16·89826 | 16·93155 |
| 300 | 3401 | 16M | ·624998 | 20·35990 | 5·39752 |
| 400 | 3501 | 17M | ·493054 | 23·82154 | 21·41810 |
| 500 | 3601 | 18M | ·361109 | 27·28318 | 9·88407 |
| 600 | 3701 | 19M | ·229165 | 1·22789 | 25·95007 |
| 700 | 3801 | 20M | ·097220 | 4·70320 | 14·46145 |
| 800 | 3901 | 20M | ·965276 | 8·17851 | 2·97285 |
| 900 | 4001 | 21M | ·833331 | 11·65383 | 19·03883 |
| 1000 | 4101 | 22M | ·701387 | 15·12914 | 7·55022 |
| 1100 | 4201 | 23M | ·569442 | 18·60446 | 23·61622 |
| 1200 | 4301 | 24M | ·437498 | 22·07977 | 12·12760 |
| 1300 | 4401 | 25M | ·305553 | 25·55508 | 0·63900 |
| 1400 | 4501 | 26M | ·173609 | 29·03039 | 16·70498 |
| 1500 | 4601 | 27M | ·041664 | 2·97511 | 5·21637 |
| 1600 | 4701 | 27M | ·909720 | 6·45043 | 21·28237 |
| 1700 | 4801 | 28M | ·777775 | 9·92574 | 9·79375 |
| 1800 | 4901 | 10Ap | ·645831 | 13·40105 | 25·85975 |
| 1900 | 5001 | 12Ap | ·513886 | 16·87636 | 14·37114 |

## Subsidiary Table VI (a) Difference between Moon's and Sun's Longitude in days (360° = 29·5305880 days).

| Degrees. | Day. | Minutes of degree. | Fraction of day. | Minutes of degree. | Fraction of day. | Minutes of degree. | Fraction of day. | Seconds of degree. | Fraction of day. | Seconds of degree. | Fraction of day. | Seconds of degree. | Fraction of day. |
|---|---|---|---|---|---|---|---|---|---|---|---|---|---|
| 1 | ·0820294 | 1 | ·0013671 | 23 | ·0314447 | 45 | ·0615220 | 1 | ·0000228 | 23 | ·0005241 | 45 | ·0010254 |
| 2 | ·1640588 | 2 | ·0027343 | 24 | ·0328118 | 46 | ·0628892 | 2 | ·0000456 | 24 | ·0005469 | 46 | ·0010481 |
| 3 | ·2460882 | 3 | ·0041015 | 25 | ·0341789 | 47 | ·0642564 | 3 | ·0000683 | 25 | ·0005696 | 47 | ·0010707 |
| 4 | ·3281176 | 4 | ·0054687 | 26 | ·0355461 | 48 | ·0656235 | 4 | ·0000911 | 26 | ·0005924 | 48 | ·0010937 |
| 5 | ·4101471 | 5 | ·0068358 | 27 | ·0369132 | 49 | ·0669907 | 5 | ·0001139 | 27 | ·0006152 | 49 | ·0011165 |
| 6 | ·4921765 | 6 | ·0082030 | 28 | ·0382804 | 50 | ·0683578 | 6 | ·0001367 | 28 | ·0006380 | 50 | ·0011393 |
| 7 | ·5742059 | 7 | ·0095700 | 29 | ·0396475 | 51 | ·0697250 | 7 | ·0001595 | 29 | ·0006608 | 51 | ·0011621 |
| 8 | ·6562353 | 8 | ·0109373 | 30 | ·0410147 | 52 | ·0710921 | 8 | ·0001823 | 30 | ·0006836 | 52 | ·0011849 |
| 9 | ·7382647 | 9 | ·0123044 | 31 | ·0423819 | 53 | ·0724593 | 9 | ·0002051 | 31 | ·0007064 | 53 | ·0012076 |
| 10 | ·8202941 | 10 | ·0136716 | 32 | ·0437490 | 54 | ·0738265 | 10 | ·0002278 | 32 | ·0007291 | 54 | ·0012304 |
| 20 | 1·6405882 | 11 | ·0150387 | 33 | ·0451162 | 55 | ·0751936 | 11 | ·0002506 | 33 | ·0007519 | 55 | ·0012532 |
| 30 | 2·4608823 | 12 | ·0164059 | 34 | ·0464833 | 56 | ·0765608 | 12 | ·0002734 | 34 | ·0007747 | 56 | ·0012760 |
| 40 | 3·2811764 | 13 | ·0177730 | 35 | ·0478505 | 57 | ·0779279 | 13 | ·0002962 | 35 | ·0007975 | 57 | ·0012988 |
| 50 | 4·1014705 | 14 | ·0191402 | 36 | ·0492176 | 58 | ·0792951 | 14 | ·0003190 | 36 | ·0001203 | 58 | ·0013216 |
| 60 | 4·9217647 | 15 | ·0205073 | 37 | ·0505848 | 59 | ·0806622 | 15 | ·0003418 | 37 | ·0008434 | 59 | ·0013444 |
| 70 | 5·7420588 | 16 | ·0218745 | 38 | ·0519520 | | | 16 | ·0003646 | 38 | ·0008659 | 60 | ·0013671 |
| 80 | 6·5623529 | 17 | ·0232417 | 39 | ·0533191 | | | 17 | ·0003874 | 39 | ·0008886 | | |
| 90 | 7·3826470 | 18 | ·0246088 | 40 | ·0546863 | | | 18 | ·0004101 | 40 | ·0009114 | | |
| 100 | 8·2029411 | 19 | ·0259760 | 41 | ·0560534 | | | 19 | ·0004329 | 41 | ·0009342 | | |
| 200 | 16·4058822 | 20 | ·0273431 | 42 | ·0574206 | | | 20 | ·0004557 | 42 | ·0009570 | | |
| 300 | 24·6088233 | 21 | ·0287103 | 43 | ·0587877 | | | 21 | ·0004785 | 43 | ·0009798 | | |
| 360 | 29·5305880 | 22 | ·0300775 | 44 | ·0601549 | | | 22 | ·0005013 | 44 | ·0010026 | | |

# TABLE VII.

## Surya Siddhanta—Odd years.

| | Fraction of day, marking commencement of Solar Year. | First new moon in Solar Year. | ☾'s anomaly for beginning of Solar Year. | | Fraction of day, marking commencement of Solar Year. | First new moon in Solar Year. | ☾'s anomaly for beginning of Solar Year. | | Fraction of day, marking commencement of Solar Year. | First new moon in Solar Year. | ☾'s anomaly for beginning of Solar Year. |
|---|---|---|---|---|---|---|---|---|---|---|---|
| **1** | ·25876 | 18·63889 | 7·04896 | **38** | ·83274 | 29·07418 | 19·86899 | **75** | 1·40673 | 9·97888 | 5·13443 |
| **2** | ·51751 | 7·74719 | 14·09791 | **39** | 1·09150 | 18·18247 | 26·91795 | **76** | ·66549 | 28·61776 | 12·18339 |
| **3** | ·77627 | 26·38607 | 21·14687 | **40** | ·35026 | 7·29077 | 6·41231 | **77** | ·92424 | 17·72606 | 19·23235 |
| **4** | ·03502 | 15·49437 | 0·64123 | **41** | ·60901 | 25·92966 | 13·46127 | **78** | 1·18300 | 6·83436 | 26·28130 |
| **5** | ·29378 | 4·60267 | 7·69019 | **42** | ·86777 | 15·03796 | 20·51022 | **79** | 1·44176 | 25·47325 | 5·77566 |
| **6** | ·55254 | 23·24156 | 14·73915 | **43** | 1·12652 | 4·14626 | 0·00458 | **80** | ·70051 | 14·58155 | 12·82462 |
| **7** | ·81129 | 12·34986 | 21·78810 | **44** | ·38528 | 22·78515 | 7·05354 | **81** | ·95927 | 3·68985 | 19·87358 |
| **8** | ·07005 | 1·45815 | 1·28246 | **45** | ·64404 | 11·89344 | 14·10250 | **82** | 1·21802 | 22·32873 | 26·92253 |
| **9** | ·32881 | 20·09704 | 8·33142 | **46** | ·90279 | 1·00174 | 21·15146 | **83** | 1·47678 | 11·43703 | 6·41689 |
| **10** | ·58756 | 9·20534 | 15·38038 | **47** | 1·16155 | 19·64063 | 0·64581 | **84** | ·73554 | 0·54533 | 13·46585 |
| **11** | ·84632 | 27·84423 | 22·42933 | **48** | ·42031 | 8·74893 | 7·69477 | **85** | ·99430 | 19·18422 | 20·51481 |
| **12** | ·10508 | 16·95253 | 1·92369 | **49** | ·67906 | 27·38782 | 14·74373 | **86** | 1·25305 | 8·29252 | 0·00917 |
| **13** | ·36383 | 6·06082 | 8·97265 | **50** | ·93782 | 16·49611 | 21·79269 | **87** | 1·51181 | 26·93140 | 7·05812 |
| **14** | ·62259 | 24·69971 | 16·02161 | **51** | 1·19658 | 5·60441 | 1·28704 | **88** | ·77056 | 16·03970 | 14·10708 |
| **15** | ·88135 | 13·80801 | 23·07057 | **52** | ·45533 | 24·24330 | 8·33600 | **89** | 1·02932 | 5·14800 | 21·15604 |
| **16** | ·14010 | 2·91631 | 2·56492 | **53** | ·71409 | 13·35160 | 15·38496 | **90** | 1·28808 | 23·78689 | 0·65040 |
| **17** | ·39886 | 21·55520 | 9·61388 | **54** | ·97284 | 2·45990 | 22·43392 | **91** | 1·54683 | 12·89519 | 7·69935 |
| **18** | ·65761 | 10·66250 | 16·66284 | **55** | 1·23160 | 21·09878 | 1·92828 | **92** | ·80559 | 2·00349 | 14·74831 |
| **19** | ·91637 | 29·30238 | 23·71180 | **56** | ·49036 | 10·20708 | 8·97723 | **93** | 1·06434 | 20·64237 | 21·79727 |
| **20** | ·17513 | 18·41068 | 3·20615 | **57** | ·74911 | 28·84597 | 16·02619 | **94** | 1·32310 | 9·75067 | 1·29163 |
| **21** | ·43388 | 7·51898 | 10·25511 | **58** | 1·00787 | 17·95420 | 23·07515 | **95** | 1·58186 | 28·38956 | 8·34059 |
| **22** | ·69264 | 26·15787 | 17·30407 | **59** | 1·26663 | 7·06257 | 2·56951 | **96** | ·84061 | 17·49786 | 15·38954 |
| **23** | ·95140 | 15·26616 | 24·35303 | **60** | ·52538 | 25·70146 | 9·61846 | **97** | 1·09937 | 6·60616 | 22·43850 |
| **24** | ·21015 | 4·37446 | 3·84739 | **61** | ·78414 | 14·80976 | 16·66742 | **98** | 1·35813 | 25·24504 | 1·93286 |
| **25** | ·46891 | 23·01335 | 10·89634 | **62** | 1·04290 | 3·91805 | 23·71638 | **99** | 1·61688 | 14·35334 | 8·98182 |
| **26** | ·72767 | 12·12165 | 17·94530 | **63** | 1·30165 | 22·55694 | 3·21074 | **100** | ·87565 | 3·46164 | 16·03078 |
| **27** | ·98642 | 1·22995 | 24·99426 | **64** | ·56041 | 11·66524 | 10·25970 | **200** | ·75130 | 6·92328 | 4·50696 |
| **28** | ·24518 | 19·86884 | 4·48862 | **65** | ·81917 | 0·77354 | 17·30865 | **300** | ·62694 | 10·38492 | 20·53775 |
| **29** | ·50393 | 8·97713 | 11·53757 | **66** | 1·07792 | 19·41242 | 24·35761 | **400** | ·50259 | 13·84657 | 9·01393 |
| **30** | ·76269 | 27·61602 | 18·58653 | **67** | 1·33668 | 8·52073 | 3·85197 | **500** | ·37824 | 17·30821 | 25·04471 |
| **31** | 1·02145 | 16·72432 | 25·63549 | **68** | ·59543 | 27·15961 | 10·90093 | **600** | ·25389 | 20·76985 | 13·52089 |
| **32** | ·28020 | 5·83262 | 5·12985 | **69** | ·85419 | 16·26791 | 17·94988 | **700** | ·12954 | 24·23149 | 1·99708 |
| **33** | ·53896 | 24·47151 | 12·17881 | **70** | 1·11295 | 5·37621 | 24·99884 | **800** | ·00519 | 27·69312 | 18·02786 |
| **34** | ·79772 | 13·57981 | 19·22776 | **71** | 1·37170 | 24·01509 | 4·49320 | **900** | ·88084 | 1·62418 | 6·50404 |
| **35** | 1·05647 | 2·68810 | 26·27672 | **72** | ·63046 | 13·12339 | 11·54216 | **1000** | ·75648 | 5·08582 | 22·53482 |
| **36** | ·31523 | 21·32699 | 5·77108 | **73** | ·88922 | 2·23169 | 18·59112 | **2000** | ·51297 | 10·17164 | 17·51504 |
| **37** | ·57399 | 10·43529 | 12·82004 | **74** | 1·14797 | 20·87058 | 25·64007 | **3000** | ·26945 | 15·25746 | 12·49527 |

# SUBSIDIARY TABLE VII (a)

## Moon's Anomaly in days (360° = 27·5545999 days).

| Degrees. | Days. | Minutes of degree. | Fraction of day. | Minutes of degree. | Fraction of day. | Minutes of degree. | Fraction of day. | Seconds of degree. | Fraction of day. | Seconds of degree. | Fraction of day. | Seconds of degree. | Fraction of day. |
|---|---|---|---|---|---|---|---|---|---|---|---|---|---|
| **1** | ·0765405 | **1** | ·0012757 | **23** | ·0293405 | **45** | ·0574054 | **1** | ·0000215 | **23** | ·0004890 | **45** | ·0009568 |
| **2** | ·1530811 | **2** | ·0025513 | **24** | ·0306162 | **46** | ·0586811 | **2** | ·0000425 | **24** | ·0005103 | **46** | ·0009780 |
| **3** | ·2296217 | **3** | ·0038270 | **25** | ·0318919 | **47** | ·0599568 | **3** | ·0000638 | **25** | ·0005315 | **47** | ·0009993 |
| **4** | ·3061622 | **4** | ·0051027 | **26** | ·0331676 | **48** | ·0612324 | **4** | ·0000850 | **26** | ·0005528 | **48** | ·0010205 |
| **5** | ·3827028 | **5** | ·0063784 | **27** | ·0344432 | **49** | ·0625081 | **5** | ·0001063 | **27** | ·0005740 | **49** | ·0010418 |
| **6** | ·4592433 | **6** | ·0076540 | **28** | ·0357189 | **50** | ·0637838 | **6** | ·0001276 | **28** | ·0005953 | **50** | ·0010631 |
| **7** | ·5357839 | **7** | ·0089297 | **29** | ·0369946 | **51** | ·0650595 | **7** | ·0001488 | **29** | ·0006166 | **51** | ·0010843 |
| **8** | ·6123244 | **8** | ·0102054 | **30** | ·0382703 | **52** | ·0663351 | **8** | ·0001701 | **30** | ·0006378 | **52** | ·0011056 |
| **9** | ·6888650 | **9** | ·0114811 | **31** | ·0395459 | **53** | ·0676108 | **9** | ·0001913 | **31** | ·000659 | **53** | ·0011268 |
| **10** | ·7654055 | **10** | ·0127568 | **32** | ·0408216 | **54** | ·0688865 | **10** | ·0002126 | **32** | ·0006804 | **54** | ·0011481 |
| **20** | 1·5308111 | **11** | ·0140324 | **33** | ·0420973 | **55** | ·0701622 | **11** | ·0002339 | **33** | ·0007016 | **55** | ·0011694 |
| **30** | 2·2962166 | **12** | ·0153081 | **34** | ·0433730 | **56** | ·0714878 | **12** | ·0002551 | **34** | ·0007229 | **56** | ·0011906 |
| **40** | 3·0616222 | **13** | ·0165838 | **35** | ·0446486 | **57** | ·0727135 | **13** | ·0002764 | **35** | ·0007441 | **57** | ·0012119 |
| **50** | 3·8270277 | **14** | ·0178595 | **36** | ·0459243 | **58** | ·0739892 | **14** | ·0002977 | **36** | ·0007654 | **58** | ·3012331 |
| **60** | 4·5924333 | **15** | ·0191351 | **37** | ·0472000 | **59** | ·0752649 | **15** | ·0003189 | **37** | ·0007867 | **59** | ·0012544 |
| **70** | 5·3578388 | **16** | ·0204108 | **38** | ·0484757 | **60** | ·0765405 | **16** | ·0003402 | **38** | ·0008079 | **60** | ·0012757 |
| **80** | 6·1232444 | **17** | ·0216865 | **39** | ·0497514 | | | **17** | ·0003614 | **39** | ·0008292 | | |
| **90** | 6·8886499 | **18** | ·0229622 | **40** | ·0510270 | | | **18** | ·0003827 | **40** | ·0008504 | | |
| **100** | 7·6540555 | **19** | ·0242378 | **41** | ·0523027 | | | **19** | ·0004040 | **41** | ·0008717 | | |
| **200** | 15·3081110 | **20** | ·0255135 | **42** | ·0535784 | | | **20** | ·0004252 | **42** | ·0008930 | | |
| **300** | 22·9621665 | **21** | ·0267892 | **43** | ·0548541 | | | **21** | ·0004465 | **43** | ·0009142 | | |
| **360** | 27·5545999 | **22** | ·0280649 | **44** | ·0561297 | | | **22** | ·0004677 | **44** | ·0009355 | | |

## TABLE VII—*contd.*

### Arya Siddhanta—Odd years.

| | Fraction of day, marking commencement of Solar Year. | First new moon in Solar Year. | ☾'s anomaly for beginning of Solar Year. | | Fraction of day, marking commencement of Solar Year. | First new moon in Solar Year. | ☾'s anomaly for beginning of Solar Year. | | Fraction of day, marking commencement of Solar Year. | First new moon in Solar Year. | ☾'s anomaly for beginning of Solar Year. |
|---|---|---|---|---|---|---|---|---|---|---|---|
| **1** | ·25868 | 18·63903 | 7·04931 | **38** | ·82986 | 29·07937 | 19·88238 | **75** | 1·40104 | 9·98913 | 5·16084 |
| **2** | ·51736 | 7·74745 | 14·09862 | **39** | 1·08854 | 18·18780 | 26·93169 | **76** | ·65972 | 28·62815 | 12·21015 |
| **3** | ·77604 | 26·38648 | 21·14793 | **40** | ·34722 | 7·29624 | 6·42640 | **77** | ·91840 | 17·73659 | 19·25946 |
| **4** | ·03472 | 15·49492 | 0·64264 | **41** | ·60590 | 25·93526 | 13·47571 | **78** | 1·17708 | 6·84502 | 26·30877 |
| **5** | ·29340 | 4·60335 | 7·69195 | **42** | ·86458 | 15·04370 | 20·52502 | **79** | 1·43576 | 25·48405 | 5·80348 |
| **6** | ·55208 | 23·24238 | 14·74126 | **43** | 1·12326 | 4·15214 | 0·01973 | **80** | ·69444 | 14·59249 | 12·85279 |
| **7** | ·81076 | 12·35082 | 21·79057 | **44** | ·38194 | 22·79116 | 7·06904 | **81** | ·95312 | 3·70092 | 19·90210 |
| **8** | ·06944 | 1·45924 | 1·28528 | **45** | ·64062 | 11·89959 | 14·11835 | **82** | 1·21180 | 22·33994 | 26·95141 |
| **9** | ·32812 | 20·09827 | 8·33459 | **46** | ·89930 | 1·00803 | 21·16766 | **83** | 1·47048 | 11·44838 | 6·44612 |
| **10** | ·58680 | 9·20671 | 15·38390 | **47** | 1·15799 | 19·64705 | 0·66237 | **84** | ·72917 | 0·55681 | 13·49543 |
| **11** | ·84549 | 27·84573 | 22·43321 | **48** | ·41667 | 8·75549 | 7·71168 | **85** | ·98785 | 19·19584 | 20·54474 |
| **12** | ·10417 | 16·95417 | 1·92792 | **49** | ·67535 | 27·39452 | 14·76099 | **86** | 1·24653 | 8·30428 | 0·03945 |
| **13** | ·36285 | 6·06260 | 8·97723 | **50** | ·93403 | 16·50294 | 21·81030 | **87** | 1·50521 | 26·94329 | 7·08876 |
| **14** | ·62153 | 24·70162 | 16·02654 | **51** | 1·19271 | 5·61138 | 1·30500 | **88** | ·76389 | 16·05173 | 14·13807 |
| **15** | ·88021 | 13·81006 | 23·07585 | **52** | ·45139 | 24·25041 | 8·35432 | **89** | 1·02257 | 5·16017 | 21·18738 |
| **16** | ·13889 | 2·91850 | 2·57056 | **53** | ·71007 | 13·35884 | 15·40363 | **90** | 1·28125 | 23·79919 | 0·68209 |
| **17** | ·39757 | 21·55752 | 9·61987 | **54** | ·96875 | 2·46728 | 22·45294 | **91** | 1·53993 | 2·90763 | 7·73140 |
| **18** | ·65625 | 10·66596 | 16·66918 | **55** | 1·22743 | 21·10630 | 1·94764 | **92** | ·79861 | 2·01607 | 14·78071 |
| **19** | ·91493 | 29·30498 | 23·71849 | **56** | ·48611 | 10·21474 | 8·99695 | **93** | 1·05729 | 20·65508 | 21·83002 |
| **20** | ·17361 | 18·41341 | 3·21320 | **57** | ·74479 | 28·85376 | 16·04627 | **94** | 1·31597 | 9·76352 | 1·32474 |
| **21** | ·43229 | 7·52185 | 10·26251 | **58** | 1·00347 | 17·96219 | 23·09558 | **95** | 1·57465 | 28·40254 | 8·37405 |
| **22** | ·69097 | 26·16088 | 17·31182 | **59** | 1·26215 | 7·07063 | 2·59028 | **96** | ·83333 | 17·51098 | 15·42336 |
| **23** | ·94965 | 15·26930 | 24·36113 | **60** | ·52083 | 25·70966 | 9·63959 | **97** | 1·09201 | 6·61942 | 22·47267 |
| **24** | ·20833 | 4·37774 | 3·85584 | **61** | ·77951 | 14·81810 | 16·68890 | **98** | 1·35069 | 25·25844 | 1·96738 |
| **25** | ·46701 | 23·01677 | 10·90515 | **62** | 1·03819 | 3·92653 | 23·73822 | **99** | 1·60937 | 14·36687 | 9·01668 |
| **26** | ·72569 | 12·12520 | 17·95446 | **63** | 1·29687 | 22·56555 | 3·23292 | **100** | ·86805 | 3·47531 | 16·06600 |
| **27** | ·98437 | 1·23364 | 25·00377 | **64** | ·55555 | 11·67399 | 10·28223 | **200** | ·73611 | 6·95062 | 4·57738 |
| **28** | ·24305 | 19·87267 | 4·49848 | **65** | ·81423 | 0·78243 | 17·33154 | **300** | ·60417 | 10·42593 | 20·64332 |
| **29** | ·50174 | 8·98109 | 11·54779 | **66** | 1·07292 | 19·42144 | 24·38086 | **400** | ·47222 | 13·90124 | 9·15466 |
| **30** | ·76042 | 27·62012 | 18·59710 | **67** | 1·33160 | 8·52989 | 3·87556 | **500** | ·34028 | 17·37655 | 25·22064 |
| **31** | 1·01910 | 16·72856 | 25·64641 | **68** | ·59028 | 27·16891 | 10·92488 | **600** | ·20833 | 20·85186 | 13·73203 |
| **32** | ·27778 | 5·83699 | 5·14112 | **69** | ·84896 | 16·27734 | 17·97419 | **700** | ·07639 | 24·32717 | 2·24342 |
| **33** | ·53646 | 24·47602 | 12·19043 | **70** | 1·10764 | 5·38578 | 25·02350 | **800** | ·94444 | 27·80248 | 18·30941 |
| **34** | ·79514 | 13·58446 | 19·23974 | **71** | 1·36632 | 24·02480 | 4·51820 | **900** | ·81250 | 1·74720 | 6·82080 |
| **35** | 1·05382 | 2·69288 | 26·28905 | **72** | ·62500 | 13·13323 | 11·56751 | **1000** | ·68055 | 5·22251 | 22·88679 |
| **36** | ·31250 | 21·33191 | 5·78376 | **73** | ·88368 | 2·24167 | 18·61682 | **2000** | ·36111 | 10·44502 | 18·21898 |
| **37** | ·57118 | 10·44035 | 12·83307 | **74** | 1·14236 | 20·88070 | 25·66613 | **3000** | ·04166 | 15·66753 | 13·55117 |

## SUBSIDIARY TABLE VII (b)

### Sun's Longitude in days (360° = 365·258756484 days).

| Degrees. | Days. | Degrees. | Days | Minutes of degree. | Fraction of day. | Minutes of degree. | Fraction of day. | Minutes of degree. | Fraction of day. | Minutes of degree. | Fraction of day. | Minutes of degree. | Fraction of day. |
|---|---|---|---|---|---|---|---|---|---|---|---|---|---|
| **1** | 1·0146076 | **40** | 40·5843064 | **1** | ·0169101 | **13** | ·2198316 | **25** | ·4227532 | **37** | ·6256747 | **49** | ·8285962 |
| **2** | 2·0292153 | **50** | 50·7303830 | **2** | ·0338202 | **14** | ·2367418 | **26** | ·4396633 | **38** | ·6425848 | **50** | ·8455063 |
| **3** | 3·0438230 | **40** | 60·8764596 | **3** | ·0507304 | **15** | ·2536519 | **27** | ·4565734 | **39** | ·6594949 | **51** | ·8624165 |
| **4** | 4·0584306 | **70** | 71·0225362 | **4** | ·0676405 | **16** | ·2705620 | **28** | ·4734835 | **40** | ·6764051 | **52** | ·8793266 |
| **5** | 5·0730383 | **80** | 81·1686128 | **5** | ·0845506 | **17** | ·2874722 | **29** | ·4903937 | **41** | ·6933152 | **53** | ·8962367 |
| **6** | 6·0876460 | **90** | 91·3146894 | **6** | ·1014608 | **18** | ·3043823 | **30** | ·5073038 | **42** | ·7102253 | **54** | ·9131468 |
| **7** | 7·1022536 | **100** | 101·4607660 | **7** | ·1183709 | **19** | ·3212924 | **31** | ·5242139 | **43** | ·7271355 | **55** | ·9300570 |
| **8** | 8·1168613 | **200** | 202·9215320 | **8** | ·1352810 | **20** | ·3382025 | **32** | ·5411241 | **44** | ·7440456 | **56** | ·9469671 |
| **9** | 9·1314689 | **300** | 304·3822980 | **9** | ·1521911 | **21** | ·3551127 | **33** | ·5580342 | **45** | ·7609557 | **57** | ·9638772 |
| **10** | 10·1460766 | **360** | 365·2587565 | **10** | ·1691013 | **22** | ·3720228 | **34** | ·5749443 | **46** | ·7778658 | **58** | ·9807874 |
| **20** | 20·2921532 | | | **11** | ·1860114 | **23** | ·3889329 | **35** | ·5918544 | **47** | ·7947760 | **59** | ·9976975 |
| **30** | 30·4382298 | | | **12** | ·2029215 | **24** | ·4058430 | **36** | ·6087646 | **48** | ·8116861 | **60** | 1·0146076 |

# TABLE VIII.—

## CALENDAR FOR ASCERTAINING THE A.D. MONTH AND DAY CORRESPONDING TO ANY SOLAR DATE, LUNAR TITHI, NAKSHATRA, YOGA OR KARANA.

**[N.B.—The Solar days and Fractions of days, Tithis and ending moments of Tithis, and Moon's Anomaly should be added to the elements given in Tables VI and VII for Centuries and odd years. See the Text for instructions on this subject.]**

| Week-day. | Days reckoned from* Jan. 1 | Mar. 1 | Apr. 1 | Day of Solar Month. | Day of Solar Year. | Ending moment of Tithi. | Tithi No. | ☾'s An. for each Tithi. | ☉'s Long. for Nakshatras and Yogas. |
|---|---|---|---|---|---|---|---|---|---|
| | | | | **Chittirai.** | | | **Vaisakha.** | | |
| | | | | (Mal.) Medam. | | | | | |
| | Jan. | Mar. | Apr. | (Beng.) Vaisakha. | | | | | |
| | | | | 0 | **0** | ·9843 | 1 | 0·984 | ... |
| **1** | 1 | 1 | **1** | 1 | **1** | ·9687 | 2 | 1·969 | 29·4150 |
| **2** | 2 | 2 | **2** | 2 | **2** | ·9530 | 3 | 2·953 | 29·4959 |
| **3** | 3 | 3 | **3** | 3 | **3** | ·9374 | 4 | 3·937 | ·0670 |
| **4** | 4 | 4 | **4** | 4 | **4** | ·9217 | 5 | 4·921 | ·1479 |
| **5** | 5 | 5 | **5** | 5 | **5** | ·9061 | 6 | 5·906 | ·2287 |
| **6** | 6 | 6 | **6** | 6 | **6** | ·8904 | 7 | 6·890 | ·3096 |
| **7** | 7 | 7 | **7** | 7 | **7** | ·8748 | 8 | 7·875 | ·3904 |
| **1** | 8 | 8 | **8** | 8 | **8** | ·8592 | 9 | 8·859 | ·4713 |
| **2** | 9 | 9 | **9** | 9 | **9** | ·8435 | 10 | 9·843 | ·5521 |
| **3** | 10 | 10 | **10** | 10 | **10** | ·8279 | 11 | 10·827 | ·6330 |
| **4** | 11 | 11 | **11** | 11 | **11** | ·8122 | 12 | 11·812 | ·7138 |
| **5** | 12 | 12 | **12** | 12 | **12** | ·7966 | 13 | 12·797 | ·7947 |
| **6** | 13 | 13 | **13** | 13 | **13** | ·7809 | 14 | 13·781 | ·8755 |
| **7** | 14 | 14 | **14** | 14 | **14** | ·7652 | 15 | 14·765 | ·9564 |
| **1** | 15 | 15 | **15** | 15 | **15** | ·7496 | **1** | 15·750 | 1·0372 |
| **2** | 16 | 16 | **16** | 16 | **16** | ·7340 | **2** | 16·734 | 1·1181 |
| **3** | 17 | 17 | **17** | 17 | **17** | ·7183 | **3** | 17·718 | 1·1989 |
| **4** | 18 | 18 | **18** | 18 | **18** | ·7027 | **4** | 18·703 | 1·2798 |
| **5** | 19 | 19 | **19** | 19 | **19** | ·6870 | **5** | 19·687 | 1·3606 |
| **6** | 20 | 20 | **20** | 20 | **20** | ·6714 | **6** | 20·671 | 1·4415 |
| **7** | 21 | 21 | **21** | 21 | **21** | ·6557 | **7** | 21·656 | 1·5223 |
| **1** | 22 | 22 | **22** | 22 | **22** | ·6401 | **8** | 22·640 | 1·6032 |
| **2** | 23 | 23 | **23** | 23 | **23** | ·6244 | **9** | 23·624 | 1·6840 |
| **3** | 24 | 24 | **24** | 24 | **24** | ·6088 | **10** | 24·609 | 1·7649 |
| **4** | 25 | 25 | **25** | 25 | **25** | ·5932 | **11** | 25·593 | 1·8457 |
| **5** | 26 | 26 | **26** | 26 | **26** | ·5775 | **12** | 26·577 | 1·9266 |
| **6** | 27 | 27 | **27** | 27 | **27** | ·5619 | **13** | 0·007 | 2·0074 |
| **7** | 28 | 28 | **28** | 28 | **28** | ·5462 | **14** | 0·992 | 2·0882 |
| **1** | 29 | 29 | **29** | 29 | **29** | ·5306 | **15** | 1·976 | 2·1691 |
| | | | | | | | **Jyeshta.** | | |
| **2** | 30 | 30 | **30** | 30 | **30** | ·5149 | 1 | 2·960 | 2·2499 |
| Sankranti. | { S. Sid. 30·9353 days. A. Sid. 30·9250 days. | | | | | | | | |
| | | | | **Vaikasi.** | | | | | |
| | | | | (Mal.) Edavam. | | | | | |
| | Jan. | Mar. | May. | (Beng.) Jyeshta. | | | *Jyeshta (Lunar).* | | |
| **3** | 31 | 31 | **1** | 1 | **31** | ·4993 | 2. | 3·945 | 2·3308 |
| | Feb. | Apr. | | | | | | | |
| **4** | 1 | 1 | **2** | 2 | **32** | ·4836 | 3 | 4·929 | 2·4116 |
| **5** | 2 | 2 | **3** | 3 | **33** | ·4680 | 4 | 5·913 | 2·4925 |
| **6** | 3 | 3 | **4** | 4 | **34** | ·4523 | 5 | 6·898 | 2·5733 |
| **7** | 4 | 4 | **5** | 5 | **35** | ·4367 | 6 | 7·882 | 2·6542 |
| **1** | 5 | 5 | **6** | 6 | **36** | ·4210 | 7 | 8·866 | 2·7350 |
| **2** | 6 | 6 | **7** | 7 | **37** | ·4054 | 8 | 9·851 | 2·8159 |
| **3** | 7 | 7 | **8** | 8 | **38** | ·3898 | 9 | 10·835 | 2·8967 |
| **4** | 8 | 8 | **9** | 9 | **39** | ·3741 | 10 | 11·819 | 2·9776 |
| **5** | 9 | 9 | **10** | 10 | **40** | ·3585 | 11 | 12·804 | 3·0584 |
| **6** | 10 | 10 | **11** | 11 | **41** | ·3428 | 12 | 13·788 | 3·1393 |
| **7** | 11 | 11 | **12** | 12 | **42** | ·3272 | 13 | 14·773 | 3·2201 |

| Week-day. | Days reckoned from* Jan. 1 | Mar. 1 | Apr. 1 | Day of Solar Month. | Day of Solar Year. | Ending moment of Tithi. | Tithi No. | ☾'s An. for each Tithi. | ☉'s Long. for Nakshatras and Yogas. |
|---|---|---|---|---|---|---|---|---|---|
| | Feb. | Apr. | May. | Vai. | | | *Jyeshta (Lunar).* | | |
| **1** | 12 | 12 | **13** | 13 | **43** | ·3115 | 14 | 15·757 | 3·3010 |
| **2** | 13 | 13 | **14** | 14 | **44** | ·2959 | 15 | 16·741 | 3·3818 |
| **3** | 14 | 14 | **15** | 15 | **45** | ·2802 | **1** | 17·726 | 3·4627 |
| **4** | 15 | 15 | **16** | 16 | **46** | ·2646 | **2** | 18·710 | 3·5435 |
| **5** | 16 | 16 | **17** | 17 | **47** | ·2489 | **3** | 19·694 | 3·6244 |
| **6** | 17 | 17 | **18** | 18 | **48** | ·2333 | **4** | 20·679 | 3·7052 |
| **7** | 18 | 18 | **19** | 19 | **49** | ·2176 | **5** | 21·663 | 3·7861 |
| **1** | 19 | 19 | **20** | 20 | **50** | ·2020 | **6** | 22·647 | 3·8669 |
| **2** | 20 | 20 | **21** | 21 | **51** | ·1863 | **7** | 23·632 | 3·9478 |
| **3** | 21 | 21 | **22** | 22 | **52** | ·1707 | **8** | 24·616 | 4·0286 |
| **4** | 22 | 22 | **23** | 23 | **53** | ·1551 | **9** | 25·600 | 4·1095 |
| **5** | 23 | 23 | **24** | 24 | **54** | ·1394 | **10** | 26·585 | 4·1903 |
| **6** | 24 | 24 | **25** | 25 | **55** | ·1238 | **11** | 0·015 | 4·2712 |
| **7** | 25 | 25 | **26** | 26 | **56** | ·1081 | **12** | 0·999 | 4·3520 |
| **1** | 26 | 26 | **27** | 27 | **57** | ·0925 | **13** | 1·983 | 4·4329 |
| **2** | 27 | 27 | **28** | 28 | **58** | ·0768 | **14** | 2·968 | 4·5137 |
| **3** | 28 | 28 | **29** | 29 | **59** | ·0612 | **15** | 3·952 | 4·5945 |
| | | | | | | | **Ashada.** | | |
| **4** | 29, 1 | 29 | **30** | 30 | **60** | ·0455 | 1 | 4·936 | 4·6754 |
| | Mar. | | | | | | | | |
| **5** | 1, 2 | 30 | **31** | 31 | **61** | ·0299 | 2 | 5·921 | 4·7562 |
| | | May. | June. | | | | | | |
| **6** | 2, 3 | 1 | **1** | 31 | **62** | ·0142 | 3 | 6·905 | 4·8371 |
| ... | ... | ... | ... | ... | **62** | ·9986 | 4 | 7·889 | ... |
| Sankranti. | { Day of month. | | | S. Sid. 31·4203 | | | A. Sid. 31·4011 | | |
| | „ year. | | | „ 62·3555 | | | „ 62·3261 | | |
| | | | | **Ani.** | | | | | |
| | | | | (Mal.) Mithunam. | | | | | |
| | Mar. | May. | June. | (Beng.) Ashada. | | | *Ashada (Lunar).* | | |
| **7** | 3, 4 | 2 | **2** | 1 | **63** | ·9829 | 5 | 8·874 | 4·9179 |
| **1** | 4, 5 | 3 | **3** | 2 | **64** | ·9673 | 6 | 9·858 | 4·9988 |
| **2** | 5, 6 | 4 | **4** | 3 | **65** | ·9516 | 7 | 10·842 | 5·0796 |
| **3** | 6, 7 | 5 | **5** | 4 | **66** | ·9360 | 8 | 11·827 | 5·1605 |
| **4** | 7, 8 | 6 | **6** | 5 | **67** | ·9203 | 9 | 12·811 | 5·2413 |
| **5** | 8, 9 | 7 | **7** | 6 | **68** | ·9047 | 10 | 13·795 | 5·3222 |
| **6** | 9,10 | 8 | **8** | 7 | **69** | ·8891 | 11 | 14·780 | 5·4030 |
| **7** | 10,11 | 9 | **9** | 8 | **70** | ·8734 | 12 | 15·764 | 5·4839 |
| **1** | 11,12 | 10 | **10** | 9 | **71** | ·8578 | 13 | 16·749 | 5·5647 |
| **2** | 12,13 | 11 | **11** | 10 | **72** | ·8421 | 14 | 17·733 | 5·6456 |
| **3** | 13,14 | 12 | **12** | 11 | **73** | ·8265 | 15 | 18·717 | 5·7264 |
| **4** | 14,15 | 13 | **13** | 12 | **74** | ·8108 | **1** | 19·702 | 5·8073 |
| **5** | 15,16 | 14 | **14** | 13 | **75** | ·7952 | **2** | 20·686 | 5·8881 |
| **6** | 16,17 | 15 | **15** | 14 | **76** | ·7795 | **3** | 21·670 | 5·9690 |
| **7** | 17,18 | 16 | **16** | 15 | **77** | ·7639 | **4** | 22·655 | 6·0498 |
| **1** | 18,19 | 17 | **17** | 16 | **78** | ·7482 | **5** | 23·639 | 6·1307 |
| **2** | 19,20 | 18 | **18** | 17 | **79** | ·7326 | **6** | 24·623 | 6·2115 |
| **3** | 20,21 | 19 | **19** | 18 | **80** | ·7169 | **7** | 25·608 | 6·2924 |
| **4** | 21,22 | 20 | **20** | 19 | **81** | ·7013 | **8** | 26·592 | 6·3732 |
| **5** | 22,23 | 21 | **21** | 20 | **82** | ·6856 | **9** | 0·022 | 6·4541 |
| **6** | 23,24 | 22 | **22** | 21 | **83** | ·6700 | **10** | 1·006 | 6·5349 |
| **7** | 24,25 | 23 | **23** | 22 | **84** | ·6543 | **11** | 1·991 | 6·6158 |
| **1** | 25,26 | 24 | **24** | 23 | **85** | ·6387 | **12** | 2·975 | 6·6965 |
| **2** | 26,27 | 25 | **25** | 24 | **86** | ·6231 | **13** | 3·959 | 6·7775 |
| **3** | 27,28 | 26 | **26** | 25 | **87** | ;6074 | **14** | 4·944 | 6·8583 |
| **4** | 28,29 | 27 | **27** | 26 | **88** | ·5918 | **15** | 5·928 | 6·9392 |

* **When two dates are given in any of these columns, use the first in a leap year ; otherwise use the second.**

# TABLE VIII.

| Week-day. | Days reckoned from* Jan. 1 | Mar. 1 | Apr. 1 | Day of Solar Month. | Day of Solar Year. | Ending moment of Tithi. | Tithi No. | ☾'s An. for each Tithi. | ☉'s Long. for Nakshatras and Yogas. |
|---|---|---|---|---|---|---|---|---|---|
| | Mar. | May. | June. | Ani. | | **Sravana.** | | | |
| **5** | 29,30 | 28 | **28** | 27 | **89** | ·5761 | 1 | 6·912 | 7·0200 |
| **6** | 30,31 | 29 | **29** | 28 | **90** | ·5605 | 2 | 7·897 | 7·1008 |
| **7** | 31, 1 | 30 | **30** | 29 | **91** | ·5448 | 3 | 8·881 | 7·1817 |
| | Apr. | | July. | | | | | | |
| **1** | 1, 2 | 31 | **1** | 30 | **92** | ·5292 | 4 | 9·865 | 7·2625 |
| | | June. | | | | | | | |
| **2** | 2, 3 | 1 | **2** | 31 | **93** | ·5135 | 5 | 10·850 | 7·3434 |
| **3** | 3, 4 | 2 | **3** | 31 | **94** | ·4979 | 6 | 11·834 | 7·4242 |
| Sankranti. | Day of month. | | | S. Sid. | 31·6447 | | A. Sid. | 31·6072 | |
| | „ year. | | | „ | 94·0003 | | „ | 93·9333 | |
| | | | | **Adi.** | | | | | |
| | | | | (Mal.) Karkatagam. | | | | | |
| | Apr. | June. | Jul. | (Beng.) | Sravana. | | *Sravana (Lunar).* | | |
| **4** | 4, 5 | 3 | **4** | 1 | **95** | ·4822 | 7 | 12·818 | 7·5051 |
| **5** | 5, 6 | 4 | **5** | 2 | **96** | ·4666 | 8 | 13·803 | 7·5859 |
| **6** | 6, 7 | 5 | **6** | 3 | **97** | ·4509 | 9 | 14·787 | 7·6668 |
| **7** | 7, 8 | 6 | **7** | 4 | **98** | ·4353 | 10 | 15·771 | 7·7476 |
| **1** | 8, 9 | 7 | **8** | 5 | **99** | ·4196 | 11 | 16·756 | 7·8285 |
| **2** | 9,10 | 8 | **9** | 6 | **100** | ·4040 | 12 | 17·740 | 7·9093 |
| **3** | 10,11 | 9 | **10** | 7 | **101** | ·3883 | 13 | 18·725 | 7·9902 |
| **4** | 11,12 | 10 | **11** | 8 | **102** | ·3727 | 14 | 19·709 | 8·0710 |
| **5** | 12,13 | 11 | **12** | 9 | **103** | ·3571 | 15 | 20·693 | 8·1518 |
| **6** | 13,14 | 12 | **13** | 10 | **104** | ·3414 | **1** | 21·678 | 8·2327 |
| **7** | 14,15 | 13 | **14** | 11 | **105** | ·3258 | **2** | 22·662 | 8·3136 |
| **1** | 15,16 | 14 | **15** | 12 | **106** | ·3101 | **3** | 23·646 | 8·3944 |
| **2** | 16,17 | 15 | **16** | 13 | **107** | ·2945 | **4** | 24·631 | 8·4753 |
| **3** | 17,18 | 16 | **17** | 14 | **108** | ·2788 | **5** | 25·615 | 8·5561 |
| **4** | 18,19 | 17 | **18** | 15 | **109** | ·2632 | **6** | 26·599 | 8·6370 |
| **5** | 19,20 | 18 | **19** | 16 | **110** | ·2475 | **7** | 0·029 | 8·7178 |
| **6** | 20,21 | 19 | **20** | 17 | **111** | ·2319 | **8** | 1·013 | 8·7987 |
| **7** | 21,22 | 20 | **21** | 18 | **112** | ·2162 | **9** | 1·998 | 8·8795 |
| **1** | 22,23 | 21 | **22** | 19 | **113** | ·2006 | **10** | 2·982 | 8·9604 |
| **2** | 23,24 | 22 | **23** | 20 | **114** | ·1849 | **11** | 3·966 | 9·0412 |
| **3** | 24,25 | 23 | **24** | 21 | **115** | ·1693 | **12** | 4·951 | 9·1221 |
| **4** | 25,26 | 24 | **25** | 22 | **116** | ·1536 | **13** | 5·935 | 9·2029 |
| **5** | 26,27 | 25 | **26** | 23 | **117** | ·1380 | **14** | 6·920 | 9·2838 |
| **6** | 27,28 | 26 | **27** | 24 | **118** | ·1223 | **15** | 7·904 | 9·3646 |
| | | | | | | **Bhadrapada.** | | | |
| **7** | 28,29 | 27 | **28** | 25 | **119** | ·1067 | 1 | 8·888 | 9·4455 |
| **1** | 29,30 | 28 | **29** | 26 | **120** | ·0911 | 2 | 9·873 | 9·5263 |
| **2** | 30,31 | 29 | **30** | 27 | **121** | ·0754 | 3 | 10·857 | 9·6071 |
| | May. | | | | | | | | |
| **3** | 1, 2 | 30 | **31** | 28 | **122** | ·0598 | 4 | 11·841 | 9·6880 |
| | | Jul. | Aug. | | | | | | |
| **4** | 2, 3 | 1 | **1** | 29 | **123** | ·0441 | 5 | 12·826 | 9·7688 |
| **5** | 3, 4 | 2 | **2** | 30 | **124** | ·0285 | 6 | 13·810 | 9·8497 |
| **6** | 4, 5 | 3 | **3** | 31 | **125** | ·0128 | 7 | 14·794 | 9·9305 |
| ... | ... | ... | ... | 31 | **125** | ·9972 | 8 | 15·779 | ... |
| Sankranti. | Day of month. | | | S. Sid. | 31·4753 | | A. Sid. | 31·4677 | |
| | „ year. | | | „ | 125·4755 | | „ | 125·4011 | |
| | | | | **Avani.** | *Bhadrapada (Lunar).* | | | | |
| | | | | (Mal.) Chingam. | | | | | |
| | May. | Jul. | Aug. | (Beng.) | Bhadrapada. | | | | |
| **7** | 5, 6 | 4 | **4** | 1 | **126** | ·9815 | 9 | 16·763 | 10·0114 |
| **1** | 6, 7 | 5 | **5** | 2 | **127** | ·9659 | 10 | 17·747 | 10·0922 |
| **2** | 7, 8 | 6 | **6** | 3 | **128** | ·9502 | 11 | 18·732 | 10·1731 |
| **3** | 8, 9 | 7 | **7** | 4 | **129** | ·9346 | 12 | 19·716 | 10·2539 |
| **4** | 9,10 | 8 | **8** | 5 | **130** | ·9189 | 13 | 20·700 | 10·3348 |
| **5** | 10,11 | 9 | **9** | 6 | **131** | ·9033 | 14 | 21·685 | 10·4156 |
| **6** | 11,12 | 10 | **10** | 7 | **132** | ·8876 | 15 | 22·669 | 10·4965 |
| **7** | 12,13 | 11 | **11** | 8 | **133** | ·8720 | 1 | 23·654 | 10·5773 |

| Week-day. | Days reckoned from* Jan. 1 | Mar. 1 | Apr. 1 | Day of Solar Month. | Day of Solar Year. | Ending moment of Tithi. | Tithi No. | ☾'s An. for each Tithi. | ☉'s Long. for Nakshatras and Yogas. |
|---|---|---|---|---|---|---|---|---|---|
| | May. | Jul. | Aug. | Avani. | | *Bhadrapada (Lunar).* | | | |
| **1** | 13,14 | 12 | **12** | 9 | **134** | ·8563 | **2** | 24·638 | 10·6583 |
| **2** | 14,15 | 13 | **13** | 10 | **135** | ·8407 | **3** | 25·622 | 10·7390 |
| **3** | 15,16 | 14 | **14** | 11 | **136** | ·8251 | **4** | 26·607 | 10·8199 |
| **4** | 16,17 | 15 | **15** | 12 | **137** | ·8094 | **5** | 0·036 | 10·9007 |
| **5** | 17,18 | 16 | **16** | 13 | **138** | ·7938 | **6** | 1·021 | 10·9816 |
| **6** | 18,19 | 17 | **17** | 14 | **139** | ·7781 | **7** | 2·005 | 11·0624 |
| **7** | 19,20 | 18 | **18** | 15 | **140** | ·7625 | **8** | 2·989 | 11·1433 |
| **1** | 20,21 | 19 | **19** | 16 | **141** | ·7468 | **9** | 3·974 | 11·2241 |
| **2** | 21,22 | 20 | **20** | 17 | **142** | ·7312 | **10** | 4·958 | 11·3050 |
| **3** | 22,23 | 21 | **21** | 18 | **143** | ·7155 | **11** | 5·942 | 11·3858 |
| **4** | 23,24 | 22 | **22** | 19 | **144** | ·6999 | **12** | 6·927 | 11·4667 |
| **5** | 24,25 | 23 | **23** | 20 | **145** | ·6842 | **13** | 7·911 | 11·5475 |
| **6** | 25,26 | 24 | **24** | 21 | **146** | ·6686 | **14** | 8·896 | 11·6284 |
| **7** | 26,27 | 25 | **25** | 22 | **147** | ·6529 | **15** | 9·880 | 11·7092 |
| | | | | | | **Asvina.** | | | |
| **1** | 27,28 | 26 | **26** | 23 | **148** | ·6373 | 1 | 10·864 | 11·7908 |
| **2** | 28,29 | 27 | **27** | 24 | **149** | ·6216 | 2 | 11·849 | 11·8706 |
| **3** | 29,30 | 28 | **28** | 25 | **150** | ·6060 | 3 | 12·833 | 11·9515 |
| **4** | 30,31 | 29 | **29** | 26 | **151** | ·5904 | 4 | 13·817 | 12·0321 |
| **5** | 31, 1 | 30 | **30** | 27 | **152** | ·5747 | 5 | 14·802 | 12·1130 |
| | June. | | | | | | | | |
| **6** | 1, 2 | 31 | **31** | 28 | **153** | ·5591 | 6 | 15·786 | 12·1943 |
| | | Aug. | Sep. | | | | | | |
| **7** | 2, 3 | 1 | **1** | 29 | **154** | ·5434 | 7 | 16·770 | 12·2759 |
| **1** | 3, 4 | 2 | **2** | 30 | **155** | ·5278 | 8 | 17·755 | 12·3568 |
| **2** | 4, 5 | 3 | **3** | 31 | **156** | ·5121 | 9 | 18·739 | 12·4367 |
| Sankranti. | Day of month. | | | S. Sid. | 31·0186 | | A. Sid. | 31·0347 | |
| | „ year. | | | „ | 156·4942 | | „ | 156·4358 | |
| | | | | **Purattasi.** | | | | | |
| | | | | (Mal.) Kanni. | | | | | |
| | June. | Aug. | Sep. | (Beng.) | Asvina. | | *Asvina (Lunar).* | | |
| **3** | 5, 6 | 4 | **4** | 1 | **157** | ·4965 | 10 | 19·723 | 12·5177 |
| **4** | 6, 7 | 5 | **5** | 2 | **158** | ·4808 | 11 | 20·708 | 12·5985 |
| **5** | 7, 8 | 6 | **6** | 3 | **159** | ·4652 | 12 | 21·692 | 12·6794 |
| **6** | 8, 9 | 7 | **7** | 4 | **160** | ·4495 | 13 | 22·676 | 12·7602 |
| **7** | 9,10 | 8 | **8** | 5 | **161** | ·4339 | 14 | 23·661 | 12·8411 |
| **1** | 10,11 | 9 | **9** | 6 | **162** | ·4182 | 15 | 24·645 | 12·9219 |
| **2** | 11,12 | 10 | **10** | 7 | **163** | ·4026 | **1** | 25·630 | 13·0028 |
| **3** | 12,13 | 11 | **11** | 8 | **164** | ·3869 | **2** | 26·614 | 13·0836 |
| **4** | 13,14 | 12 | **12** | 9 | **165** | ·3713 | **3** | 0·044 | 13·1645 |
| **5** | 14,15 | 13 | **13** | 10 | **166** | ·3556 | **4** | 1·028 | 13·2453 |
| **6** | 15,16 | 14 | **14** | 11 | **167** | ·3400 | **5** | 2·012 | 13·3262 |
| **7** | 16,17 | 15 | **15** | 12 | **168** | ·3244 | **6** | 2·997 | 13·4070 |
| **1** | 17,18 | 16 | **16** | 13 | **169** | ·3087 | **7** | 3·981 | 13·4879 |
| **2** | 18,19 | 17 | **17** | 14 | **170** | ·2931 | **8** | 4·965 | 13·5687 |
| **3** | 19,20 | 18 | **18** | 15 | **171** | ·2774 | **9** | 5·950 | 13·6496 |
| **4** | 20,21 | 19 | **19** | 16 | **172** | ·2618 | **10** | 6·934 | 13·7304 |
| **5** | 21,22 | 20 | **20** | 17 | **173** | ·2461 | **11** | 7·918 | 13·8113 |
| **6** | 22,23 | 21 | **21** | 18 | **174** | ·2305 | **12** | 8·903 | 13·8921 |
| **7** | 23,24 | 22 | **22** | 19 | **175** | ·2148 | **13** | 9·887 | 13·9730 |
| **1** | 24,25 | 23 | **23** | 20 | **176** | ·1992 | **14** | 10·872 | 14·0531 |
| **2** | 25,26 | 24 | **24** | 21 | **177** | ·1835 | **15** | 11·856 | 14·1347 |
| | | | | | | **Kartika.** | | | |
| **3** | 26,27 | 25 | **25** | 22 | **178** | ·1679 | 1 | 12·840 | 14·2155 |
| **4** | 27,28 | 26 | **26** | 23 | **179** | ·1522 | 2 | 13·825 | 14·2964 |
| **5** | 28,29 | 27 | **27** | 24 | **180** | ·1366 | 3 | 14·809 | 14·3772 |
| **6** | 29,30 | 28 | **28** | 25 | **181** | ·1209 | 4 | 15·793 | 14·4581 |
| **7** | 30, 1 | 29 | **29** | 26 | **182** | ·1053 | 5 | 16·778 | 14·5389 |
| | July. | | | | | | | | |
| **1** | 1, 2 | 30 | **30** | 27 | **183** | ·0896 | 6 | 17·762 | 14·6198 |
| | | | Oct. | | | | | | |
| **2** | 2, 3 | 31 | **1** | 28 | **184** | ·0740 | 7 | 18·746 | 14·7006 |
| | | Sep. | | | | | | | |
| **3** | 3, 4 | 1 | **2** | 29 | **185** | ·0584 | 8 | 19·731 | 14·7814 |
| **4** | 4, 5 | 2 | **3** | 30 | **186** | ·0427 | 9 | 20·715 | 14·8623 |

*** When two dates are given in any of these columns, use the first in a leap year ; otherwise use the second.**

# TABLE VIII.

| Week-day. | * Days reckoned from Jan. 1 | Mar. 1 | Apr. 1 | Day of Solar Month. | Day of Solar Year. | Ending moment of Tithi. | Tithi. | ☾'s An. for each Tithi. | ⊙'s Long. for Nakshatras and Yogas. |
|---|---|---|---|---|---|---|---|---|---|

Sankranti. { Day of month. S. Sid. 30·4414 A. Sid. 30·4567
„ year. „ 186·9355 „ 186·8925

## Aippasi.

(Mal.) Tulam.

| Week-day. | Jul. | Sep. | Oct. | (Beng.) Kartika. | | | Kartika (Lunar). | | |
|---|---|---|---|---|---|---|---|---|---|
| 5 | 5, 6 | 3 | 4 | 1 | 187 | ·0271 | 10 | 21·699 | 14·9431 |
| 6 | 6, 7 | 4 | 5 | 2 | 188 | ·0114 | 11 | 22·684 | 15·0040 |
| ... | ... | ... | ... | ... | 188 | ·9958 | 12 | 23·668 | ... |
| 7 | 7, 8 | 5 | 6 | 3 | 189 | ·9801 | 13 | 24·652 | 15·1248 |
| 1 | 8, 9 | 6 | 7 | 4 | 190 | ·9645 | 14 | 25·637 | 15·1857 |
| 2 | 9,10 | 7 | 8 | 5 | 191 | ·9488 | 15 | 26·621 | 15·2665 |
| 3 | 10,11 | 8 | 9 | 6 | 192 | ·9332 | 1 | 0·051 | 15·3474 |
| 4 | 11,12 | 9 | 10 | 7 | 193 | ·9175 | 2 | 1·035 | 15·4282 |
| 5 | 12,13 | 10 | 11 | 8 | 194 | ·9019 | 3 | 2·020 | 15·5090 |
| 6 | 13,14 | 11 | 12 | 9 | 195 | ·8862 | 4 | 3·004 | 15·5899 |
| 7 | 14,15 | 12 | 13 | 10 | 196 | ·8706 | 5 | 3·988 | 15·6708 |
| 1 | 15,16 | 13 | 14 | 11 | 197 | ·8549 | 6 | 4·973 | 15·7516 |
| 2 | 16,17 | 14 | 15 | 12 | 198 | ·8393 | 7 | 5·957 | 15·8325 |
| 3 | 17,18 | 15 | 16 | 13 | 199 | ·8236 | 8 | 6·941 | 15·9133 |
| 4 | 18,19 | 16 | 17 | 14 | 200 | ·8080 | 9 | 7·926 | 15·9942 |
| 5 | 19,20 | 17 | 18 | 15 | 201 | ·7924 | 10 | 8·910 | 16·0750 |
| 6 | 20,21 | 18 | 19 | 16 | 202 | ·7767 | 11 | 9·894 | 16·1559 |
| 7 | 21,22 | 19 | 20 | 17 | 203 | ·7611 | 12 | 10·879 | 16·2367 |
| 1 | 22,23 | 20 | 21 | 18 | 204 | ·7454 | 13 | 11·863 | 16·3176 |
| 2 | 23,24 | 21 | 22 | 19 | 205 | ·7298 | 14 | 12·848 | 16·3984 |
| 3 | 24,25 | 22 | 23 | 20 | 206 | ·7141 | 15 | 13·832 | 16·4793 |
| | | | | | | | **Margasira.** | | |
| 4 | 25,26 | 23 | 24 | 21 | 207 | ·6985 | 1 | 14·816 | 16·5601 |
| 5 | 26,27 | 24 | 25 | 22 | 208 | ·6828 | 2 | 15·801 | 16·6410 |
| 6 | 27,28 | 25 | 26 | 23 | 209 | ·6672 | 3 | 16·785 | 16·7218 |
| 7 | 28,29 | 26 | 27 | 24 | 210 | ·6515 | 4 | 17·769 | 16·8026 |
| 1 | 29,30 | 27 | 28 | 25 | 211 | ·6359 | 5 | 18·754 | 16·8835 |
| 2 | 30,31 | 28 | 29 | 26 | 212 | ·6202 | 6 | 19·738 | 16·9644 |
| 3 | 31, 1 | 29 | 30 | 27 | 213 | ·6046 | 7 | 20·722 | 17·0452 |
| | Aug. | | | | | | | | |
| 4 | 1, 2 | 30 | 31 | 28 | 214 | ·5889 | 8 | 21·707 | 17·1261 |
| | | Oct. | Nov. | | | | | | |
| 5 | 2, 3 | 1 | 1 | 29 | 215 | ·5733 | 9 | 22·691 | 17·2069 |
| 6 | 3, 4 | 2 | 2 | 29 | 216 | ·5576 | 10 | 23·675 | 17·2877 |

Sankranti. { Day of month. S.Sid. 29·8933 A.Sid. 29·9033
„ year. „ 216·8289 „ 216·7958

## Kartikai.

(Mal.) Vrischikam.

| Week-day. | Aug. | Oct. | Nov. | (Beng.) Margasira. | | | Margasira (Lunar). | | |
|---|---|---|---|---|---|---|---|---|---|
| 7 | 4, 5 | 3 | 3 | 1 | 217 | ·5420 | 11 | 24·660 | 17·3686 |
| 1 | 5, 6 | 4 | 4 | 2 | 218 | ·5264 | 12 | 25·644 | 17·4494 |
| 2 | 6, 7 | 5 | 5 | 3 | 219 | ·5107 | 13 | 26·628 | 17·5303 |
| 3 | 7, 8 | 6 | 6 | 4 | 220 | ·4951 | 14 | 0·058 | 17·6111 |
| 4 | 8, 9 | 7 | 7 | 5 | 221 | ·4794 | 15 | 1·043 | 17·6900 |
| 5 | 9,10 | 8 | 8 | 6 | 222 | ·4638 | 1 | 2·027 | 17·7728 |
| 6 | 10,11 | 9 | 9 | 7 | 223 | ·4481 | 2 | 3·011 | 17·8537 |
| 7 | 11,12 | 10 | 10 | 8 | 224 | ·4325 | 3 | 3·996 | 17·9345 |
| 1 | 12 13 | 11 | 11 | 9 | 225 | ·4168 | 4 | 4·980 | 18·0154 |
| 2 | 13,14 | 12 | 12 | 10 | 226 | ·4012 | 5 | 5·964 | 18·0962 |
| 3 | 14,15 | 13 | 13 | 11 | 227 | ·3855 | 6 | 6·949 | 18·1771 |
| 4 | 15,16 | 14 | 14 | 12 | 228 | ·3699 | 7 | 7·933 | 18·2579 |
| 5 | 16,17 | 15 | 15 | 13 | 229 | ·3542 | 8 | 8·917 | 18·3388 |
| 6 | 17,18 | 16 | 16 | 14 | 230 | ·3386 | 9 | 9·902 | 18·4196 |
| 7 | 18,19 | 17 | 17 | 15 | 231 | ·3229 | 10 | 10·886 | 18·5005 |
| 1 | 19,20 | 18 | 18 | 16 | 232 | ·3073 | 11 | 11·870 | 18·5813 |
| 2 | 20,21 | 19 | 19 | 17 | 233 | ·2916 | 12 | 12·855 | 18·6622 |
| 3 | 21,22 | 20 | 20 | 18 | 234 | ·2760 | 13 | 13·839 | 18·7430 |
| 4 | 22,23 | 21 | 21 | 19 | 235 | ·2604 | 14 | 14·824 | 18·8239 |
| 5 | 23,24 | 22 | 22 | 20 | 236 | ·2447 | 15 | 15·808 | 18·9047 |

| Week-day. | Aug. | Oct. | Nov. | Kartikai. | | | **Pausha.** | | |
|---|---|---|---|---|---|---|---|---|---|
| 6 | 24,25 | 23 | 23 | 21 | 237 | ·2291 | 1 | 16·792 | 18·9856 |
| 7 | 25,26 | 24 | 24 | 22 | 238 | ·2134 | 2 | 17·777 | 19·0664 |
| 1 | 26,27 | 25 | 25 | 23 | 239 | ·1978 | 3 | 18·761 | 19·1473 |
| 2 | 27,28 | 26 | 26 | 24 | 240 | ·1821 | 4 | 19·745 | 19·2281 |
| 3 | 28,29 | 27 | 27 | 25 | 241 | ·1665 | 5 | 20·730 | 19·3090 |
| 4 | 29,30 | 28 | 28 | 26 | 242 | ·1508 | 6 | 21·714 | 19·3898 |
| 5 | 30,31 | 29 | 29 | 27 | 243 | ·1352 | 7 | 22·698 | 19·4707 |
| 6 | 31, 1 | 30 | 30 | 28 | 244 | ·1195 | 8 | 23·683 | 19·5515 |
| | Sep. | | Dec. | | | | | | |
| 7 | 1, 2 | 31 | 1 | 29 | 245 | ·1039 | 9 | 24·667 | 19·6324 |
| | | Nov. | | | | | | | |
| 1 | 2, 3 | 1 | 2 | 30 | 246 | ·0882 | 10 | 25·651 | 19·7132 |

Sankranti. { Day of month. S. Sid. 29·4903 A. Sid. 29·5086
„ year. „ 246·3192 „ 246·3044

## Margali.

(Mal.) Dhanus.

| Week-day. | Sep. | Nov. | Dec. | (Beng.) Pausha. | | | Pausha (Lunar). | | |
|---|---|---|---|---|---|---|---|---|---|
| 2 | 3, 4 | 2 | 3 | 1 | 247 | ·0726 | 11 | 26·636 | 19·7941 |
| 3 | 4, 5 | 3 | 4 | 2 | 248 | ·0569 | 12 | 0·065 | 19·8749 |
| 4 | 5, 6 | 4 | 5 | 3 | 249 | ·0413 | 13 | 1·050 | 19·9557 |
| 5 | 6, 7 | 5 | 6 | 4 | 250 | ·0257 | 14 | 2·034 | 20·0366 |
| 6 | 7, 8 | 6 | 7 | 5 | 251 | ·0108 | 15 | 3·019 | 20·1174 |
| ... | ... | ... | ... | ... | 251 | ·9944 | 1 | 4·003 | ... |
| 7 | 8, 9 | 7 | 8 | 6 | 252 | ·9787 | 2 | 4·987 | 20·1983 |
| 1 | 9,10 | 8 | 9 | 7 | 253 | ·9631 | 3 | 5·972 | 20·2791 |
| 2 | 10,11 | 9 | 10 | 8 | 254 | ·9474 | 4 | 6·956 | 20·3600 |
| 3 | 11,12 | 10 | 11 | 9 | 255 | ·9318 | 5 | 7·940 | 20·4408 |
| 4 | 12,13 | 11 | 12 | 10 | 256 | ·9161 | 6 | 8·925 | 20·5217 |
| 5 | 13,14 | 12 | 13 | 11 | 257 | ·9005 | 7 | 9·909 | 20·6025 |
| 6 | 14,15 | 13 | 14 | 12 | 258 | ·8848 | 8 | 10·893 | 20·6834 |
| 7 | 15,16 | 14 | 15 | 13 | 259 | ·8692 | 9 | 11·878 | 20·7642 |
| 1 | 16,17 | 15 | 16 | 14 | 260 | ·8535 | 10 | 12·862 | 20·8451 |
| 2 | 17,18 | 16 | 17 | 15 | 261 | ·8379 | 11 | 13·846 | 20·9259 |
| 3 | 18,19 | 17 | 18 | 16 | 262 | ·8222 | 12 | 14·831 | 21·0068 |
| 4 | 19,20 | 18 | 19 | 17 | 263 | ·8066 | 13 | 15·815 | 21·0876 |
| 5 | 20,21 | 19 | 20 | 18 | 264 | ·7909 | 14 | 16·799 | 21·1685 |
| 6 | 21,22 | 20 | 21 | 19 | 265 | ·7753 | 15 | 17·784 | 21·2493 |
| | | | | | | | **Magha** | | |
| 7 | 22,23 | 21 | 22 | 20 | 266 | ·7597 | 1 | 18·768 | 21·3302 |
| 1 | 23,24 | 22 | 23 | 21 | 267 | ·7440 | 2 | 19·753 | 21·4110 |
| 2 | 24,25 | 23 | 24 | 22 | 268 | ·7284 | 3 | 20·737 | 21·4919 |
| 3 | 25,26 | 24 | 25 | 23 | 269 | ·7127 | 4 | 21·721 | 21·5727 |
| 4 | 26,27 | 25 | 26 | 24 | 270 | ·6971 | 5 | 22·706 | 21·6536 |
| 5 | 27,28 | 26 | 27 | 25 | 271 | ·6814 | 6 | 23·690 | 21·7344 |
| 6 | 28,29 | 27 | 28 | 26 | 272 | ·6658 | 7 | 24·674 | 21·8153 |
| 7 | 29,30 | 28 | 29 | 27 | 273 | ·6501 | 8 | 25·659 | 21·8961 |
| 1 | 30, 1 | 29 | 30 | 28 | 274 | ·6345 | 9 | 26·643 | 21·9770 |
| | Oct. | | | | | | | | |
| 2 | 1, 2 | 30 | 31 | 29 | 275 | ·6188 | 10 | 0·073 | 22·0578 |

Sankranti. { Day of month. S. Sid. 29·3178 A. Sid. 29·3505
„ year. „ 275·6369 „ 275·6550

## Tai.

(Mal.) Magaram.

| Week-day. | Oct. | Dec. | Jan. | (Beng.) Magha. | | | Magha (Lunar). | | |
|---|---|---|---|---|---|---|---|---|---|
| 3 | 2, 3 | 1 | 1 | 1 | 276 | ·6032 | 11 | 1·057 | 22·1387 |
| 4 | 3, 4 | 2 | 2 | 2 | 277 | ·5875 | 12 | 2·041 | 22·2195 |
| 5 | 4, 5 | 3 | 3 | 3 | 278 | ·5719 | 13 | 3·026 | 22·3004 |
| 6 | 5, 6 | 4 | 4 | 4 | 279 | ·5562 | 14 | 4·010 | 22·3810 |
| 7 | 6, 7 | 5 | 5 | 5 | 280 | ·5406 | 15 | 4·995 | 22·4621 |

* When two dates are given in any of these columns, use the first in a leap year ; otherwise use the second.

# TABLE VIII.

| Week-day. | Days reckoned from* Jan. 1 | Mar. 1 | Apr. 1 | Day of Solar Month. | Day of Solar Year. | Ending moment of Tithi. | Tithi. | ☾'s An. for each Tithi. | ⊙'s Long. for Nakshatras and Yogas. |
|---|---|---|---|---|---|---|---|---|---|
| | Oct. | Dec. | Jan. | Tai. | | | *Magha (Lunar).* | | |
| 1 | 7, 8 | 6 | **6** | 6 | **281** | ·5249 | **1** | 5·979 | 22·5429 |
| 2 | 8, 9 | 7 | **7** | 7 | **282** | ·5093 | **2** | 6·963 | 22·6237 |
| 3 | 9,10 | 8 | **8** | 8 | **283** | ·4937 | **3** | 7·948 | 22·7046 |
| 4 | 10,11 | 9 | **9** | 9 | **284** | ·4780 | **4** | 8·932 | 22·7854 |
| 5 | 11,12 | 10 | **10** | 10 | **285** | ·4624 | **5** | 9·916 | 22·8663 |
| 6 | 12,13 | 11 | **11** | 11 | **286** | ·4467 | **6** | 10·901 | 22·9471 |
| 7 | 13,14 | 12 | **12** | 12 | **287** | ·4311 | **7** | 11·885 | 23·0280 |
| 1 | 14,15 | 13 | **13** | 13 | **288** | ·4154 | **8** | 12·869 | 23·1088 |
| 2 | 15,16 | 14 | **14** | 14 | **289** | ·3998 | **9** | 13·854 | 23·1897 |
| 3 | 16,17 | 15 | **15** | 15 | **290** | ·3841 | **10** | 14·838 | 23·2705 |
| 4 | 17,18 | 16 | **16** | 16 | **291** | ·3685 | **11** | 15·822 | 23·3514 |
| 5 | 18,19 | 17 | **17** | 17 | **292** | ·3528 | **12** | 16·807 | 23·4322 |
| 6 | 19,20 | 18 | **18** | 18 | **293** | ·3372 | **13** | 17·791 | 23·5131 |
| 7 | 20,21 | 19 | **19** | 19 | **294** | ·3215 | **14** | 18·775 | 23·5939 |
| 1 | 21,22 | 20 | **20** | 20 | **295** | ·3059 | **15** | 19·760 | 23·6748 |
| | | | | | | | **Phalguna.** | | |
| 2 | 22,23 | 21 | **21** | 21 | **296** | ·2902 | 1 | 20·744 | 23·7556 |
| 3 | 23,24 | 22 | **22** | 22 | **297** | ·2746 | 2 | 21·729 | 23·8364 |
| 4 | 24,25 | 23 | **23** | 23 | **298** | ·2589 | 3 | 22·713 | 23·9173 |
| 5 | 25,26 | 24 | **24** | 24 | **299** | ·2433 | 4 | 23·697 | 23·9982 |
| 6 | 26,27 | 25 | **25** | 25 | **300** | ·2277 | 5 | 24·682 | 24·0790 |
| 7 | 27,28 | 26 | **26** | 26 | **301** | ·2120 | 6 | 25·666 | 24·1599 |
| 1 | 28,29 | 27 | **27** | 27 | **302** | ·1964 | 7 | 26·650 | 24·2407 |
| 2 | 29,30 | 28 | **28** | 28 | **303** | ·1807 | 8 | 0·080 | 24·3216 |
| 3 | 30,31 | 29 | **29** | 29 | **304** | ·1651 | 9 | 1·064 | 24·4024 |
| 4 | 31, 1 | 30 | **30** | 29 | **305** | ·1494 | 10 | 2·049 | 24·4833 |

Sankranti. { Day of month. S. Sid. 29·4480 A. Sid. 29·4567
{ „ year. „ 305·0850 „ 305·1117

**Masi.**

(Mal.) Kumbam.

| Week-day. | Jan. 1 | Mar. 1 | Apr. 1 | Day of Solar Month. | Day of Solar Year. | Ending moment of Tithi. | Tithi. | ☾'s An. for each Tithi. | ⊙'s Long. for Nakshatras and Yogas. |
|---|---|---|---|---|---|---|---|---|---|
| | Nov. | Dec. | Jan. | (Beng) Phalguna. | | | *Phalguna (Lunar).* | | |
| 5 | 1, 2 | 31 | **31** | 1 | **306** | ·1338 | 11 | 3·033 | 24·5641 |
| | | Jan. | Feb. | | | | | | |
| O | 2, 3 | 1 | **1** | 2 | **307** | ·1181 | 12 | 4·017 | 24·6450 |
| 7 | 3, 4 | 2 | **2** | 3 | **308** | ·1025 | 13 | 5·002 | 24·7258 |
| 1 | 4, 5 | 3 | **3** | 4 | **309** | ·0868 | 14 | 5·986 | 24·8067 |
| 2 | 5, 6 | 4 | **4** | 5 | **310** | ·0712 | 15 | 6·971 | 24·8875 |
| 3 | 6, 7 | 5 | **5** | 6 | **311** | ·0555 | **1** | 7·955 | 24·9683 |
| 4 | 7, 8 | 6 | **6** | 7 | **312** | ·0399 | **2** | 8·939 | 25·0492 |
| 5 | 8, 9 | 7 | **7** | 8 | **313** | ·0242 | **3** | 9·924 | 25·1300 |
| 6 | 9,10 | 8 | **8** | 9 | **314** | ·0086 | **4** | 10·908 | 25·2109 |
| ... | ... | ... | ... | ... | **314** | ·9929 | **5** | 11·892 | ... |
| 7 | 10,11 | 9 | **9** | 10 | **315** | ·9773 | **6** | 12·877 | 25·2917 |
| 1 | 11,12 | 10 | **10** | 11 | **316** | ·9617 | **7** | 13·861 | 25·3726 |
| 2 | 12,13 | 11 | **11** | 12 | **317** | ·9460 | **8** | 14·845 | 25·4534 |
| 3 | 13,14 | 12 | **12** | 13 | **318** | ·9304 | **9** | 15·830 | 25·5343 |
| 4 | 14,15 | 13 | **13** | 14 | **319** | ·9147 | **10** | 16·814 | 25·6151 |
| 5 | 15,16 | 14 | **14** | 15 | **320** | ·8991 | **11** | 17·798 | 25·6960 |
| 6 | 16,17 | 15 | **15** | 16 | **321** | ·8834 | **12** | 18·783 | 25·7768 |
| 7 | 17,18 | 16 | **16** | 17 | **322** | ·8678 | **13** | 19·767 | 25·8577 |
| 1 | 18,19 | 17 | **17** | 18 | **323** | ·8521 | **14** | 20·751 | 25·9385 |
| 2 | 19,20 | 18 | **18** | 19 | **324** | ·8365 | **15** | 21·736 | 26·0194 |
| | | | | | | | **Chaitra** (when there is no Adhika masa). | | |
| 3 | 20,21 | 19 | **19** | 20 | **325** | ·8208 | 1 | 22·720 | 26·1002 |
| 4 | 21,22 | 20 | **20** | 21 | **326** | ·8052 | 2 | 23·705 | 26·1811 |
| 5 | 22,23 | 21 | **21** | 22 | **327** | ·7895 | 3 | 24·689 | 26·2619 |
| 6 | 23,24 | 22 | **22** | 23 | **328** | ·7739 | 4 | 25·673 | 26·3428 |
| 7 | 24,25 | 23 | **23** | 24 | **329** | ·7582 | 5 | 26·658 | 26·4236 |
| 1 | 25,26 | 24 | **24** | 25 | **330** | ·7426 | 6 | 0·087 | 26·5045 |
| 2 | 26,27 | 25 | **25** | 26 | **331** | ·7269 | 7 | 1·072 | 26·5853 |
| 3 | 27,28 | 26 | **26** | 27 | **332** | ·7113 | 8 | 2·056 | 26·6662 |
| 4 | 28,29 | 27 | **27** | 28 | **333** | ·6957 | 9 | 3·040 | 26·7470 |
| 5 | 29,30 | 28 | **28** | 29 | **334** | ·6800 | 10 | 4·025 | 26·8279 |

Sankranti. { Day of month. S. Sid. 29·8203 A. Sid. 29·8083
{ „ year. „ 334·9053 „ 334·9200

**Panguni.**

(Mal.) Meenam.

| Week-day. | Jan. 1 | Mar. 1 | Apr. 1 | Day of Solar Month. | Day of Solar Year. | Ending moment of Tithi. | Tithi. | ☾'s An. for each Tithi. | ⊙'s Long. for Nakshatras and Yogas. |
|---|---|---|---|---|---|---|---|---|---|
| | Nov. | Jan. | Feb. | (Beng.) Chaitra. | | | *Chaitra (Lunar).* | | |
| 6 | 30, 1 | 29 | **29, 1** | 1 | **335** | ·6644 | 11 | 5·009 | 26·9087 |
| | Dec. | | Mar. | | | | | | |
| 7 | 1, 2 | 30 | **1, 2** | 2 | **336** | ·6487 | 12 | 5·993 | 26·9896 |
| 1 | 2, 3 | 31 | **2, 3** | 3 | **337** | ·6331 | 13 | 6·978 | 27·0704 |
| | | Feb. | | | | | | | |
| 2 | 3, 4 | 1 | **3, 4** | 4 | **338** | ·6174 | 14 | 7·962 | 27·1513 |
| 3 | 4, 5 | 2 | **4, 5** | 5 | **339** | ·6018 | 15 | 8·947 | 27·2321 |
| 4 | 5, 6 | 3 | **5, 6** | 6 | **340** | ·5861 | **1** | 9·931 | 27·3130 |
| 5 | 6, 7 | 4 | **6, 7** | 7 | **341** | ·5705 | **2** | 10·915 | 27·3938 |
| 6 | 7, 8 | 5 | **7, 8** | 8 | **342** | ·5548 | **3** | 11·900 | 27·4746 |
| 7 | 8, 9 | 6 | **8, 9** | 9 | **343** | ·5392 | **4** | 12·884 | 27·5555 |
| 1 | 9,10 | 7 | **9,10** | 10 | **344** | ·5235 | **5** | 13·868 | 27·6363 |
| 2 | 10,11 | 8 | **10,11** | 11 | **345** | ·5079 | **6** | 14·853 | 27·7172 |
| 3 | 11,12 | 9 | **11,12** | 12 | **346** | ·4922 | **7** | 15·837 | 27·7980 |
| 4 | 12,13 | 10 | **12,13** | 13 | **347** | ·4766 | **8** | 16·821 | 27·8789 |
| 5 | 13,14 | 11 | **13,14** | 14 | **348** | ·4610 | **9** | 17·806 | 27·9597 |
| 6 | 14,15 | 12 | **14,15** | 15 | **349** | ·4453 | **10** | 18·790 | 28·0406 |
| 7 | 15,16 | 13 | **15,16** | 16 | **350** | ·4297 | **11** | 19·774 | 28·1214 |
| 1 | 16,17 | 14 | **16,17** | 17 | **351** | ·4140 | **12** | 20·759 | 28·2023 |
| 2 | 17,18 | 15 | **17,18** | 18 | **352** | ·3984 | **13** | 21·743 | 28·2831 |
| 3 | 18,19 | 16 | **18,19** | 19 | **353** | ·3827 | **14** | 22·727 | 28·3640 |
| 4 | 19,20 | 17 | **19,20** | 20 | **354** | ·3671 | **15** | 23·712 | 28·4448 |
| | | | | | | | **Chaitra** (when there is Adhika masa). | | |
| 5 | 20,21 | 18 | **20,21** | 21 | **355** | ·3514 | 1 | 24·696 | 28·5257 |
| 6 | 21,22 | 19 | **21,22** | 22 | **356** | ·3358 | 2 | 25·681 | 28·6066 |
| 7 | 22,23 | 20 | **22,23** | 23 | **357** | ·3201 | 3 | 26·665 | 28·6874 |
| 1 | 23,24 | 21 | **23,24** | 24 | **358** | ·3045 | 4 | 0·095 | 28·7683 |
| 2 | 24,25 | 22 | **24,25** | 25 | **359** | ·2888 | 5 | 1·079 | 28·8491 |
| 3 | 25,26 | 23 | **25,26** | 26 | **360** | ·2732 | 6 | 2·063 | 28·9300 |
| 4 | 26,27 | 24 | **26,27** | 27 | **361** | ·2575 | 7 | 3·048 | 29·0108 |
| 5 | 27,28 | 25 | **27,28** | 28 | **362** | ·2419 | 8 | 4·032 | 29·0917 |
| 6 | 28,29 | 26 | **28,29** | 29 | **363** | ·2262 | 9 | 5·016 | 29·1725 |
| 7 | 29,30 | 27 | **29,30** | 30 | **364** | ·2106 | 10 | 6·001 | 29·2533 |
| 1 | 30,31 | 28 | **30,31** | 30 | **365** | ·1950 | 11 | 6·985 | 29·3342 |

Sankranti. { Day of month. S. Sid. 30·3535 A. Sid. 30·3388
{ „ year. „ 365·2587 „ 365·2587

**Chittirai.**

(Mal.) Medam

| Week-day. | Jan. 1 | Mar. 1 | Apr. 1 | Day of Solar Month. | Day of Solar Year. | Ending moment of Tithi. | Tithi. | ☾'s An. for each Tithi. | ⊙'s Long. for Nakshatras and Yogas. |
|---|---|---|---|---|---|---|---|---|---|
| | Dec. | Feb. | Mar. | (Beng.) Vaisakha. | | | *Chaitra (Lunar).* | | |
| 2 | 31, 1 | 29, 1 | **31, 1** | 1 | **366** | ·1793 | 12 | 7·969 | 29·4150 |
| | Jan. | Mar. | Apr. | | | | | | |
| 3 | 1, 2 | 1, 2 | **1, 2** | 2 | **367** | ·1637 | 13 | 8·954 | |
| 4 | 2, 3 | 2, 3 | **2, 3** | 3 | **368** | ·1480 | 14 | 9·938 | |
| 5 | 3, 4 | 3, 4 | **3, 4** | 4 | **369** | ·1324 | 15 | 10·923 | |
| 6 | 4, 5 | 4, 5 | **4, 5** | 5 | **370** | ·1167 | **1** | 11·907 | |
| 7 | 5, 6 | 5, 6 | **5, 6** | 6 | **371** | ·1011 | **2** | 12·891 | |
| 1 | 6, 7 | 6, 7 | **6, 7** | 7 | **372** | ·0854 | **3** | 13·876 | |
| 2 | 7, 8 | 7, 8 | **7, 8** | 8 | **373** | ·0698 | **4** | 14·860 | |
| 3 | 8, 9 | 8, 9 | **8, 9** | 9 | **374** | ·0541 | **5** | 15·844 | |
| 4 | 9,10 | 9,10 | **9,10** | 10 | **375** | ·0385 | **6** | 16·829 | |
| 5 | 10,11 | 10,11 | **10,11** | 11 | **376** | ·0228 | **7** | 17·813 | |
| 6 | 11,12 | 11,12 | **11,12** | 12 | **377** | ·0072 | **8** | 18·797 | |
| ... | ... | ... | ... | ... | **377** | ·9915 | **9** | 19·782 | |
| 7 | 12,13 | 12,13 | **12,13** | 13 | **378** | ·9759 | **10** | 20·766 | |
| 1 | 13,14 | 13,14 | **13,14** | 14 | **379** | ·9602 | **11** | 21·750 | |
| 2 | 14,15 | 14,15 | **14,15** | 15 | **380** | ·9446 | **12** | 22·735 | |
| 3 | 15,16 | 15,16 | **15,16** | 16 | **381** | ·9290 | **13** | 23·719 | |
| 4 | 16,17 | 16,17 | **16,17** | 17 | **382** | ·9133 | **14** | 24·703 | |
| 5 | 17,18 | 17,18 | **17,18** | 18 | **383** | ·8977 | **15** | 25·688 | |

* When two dates are given in any of these columns, use the first in a leap year ; otherwise use the second.

# TABLE IX—(a)—For Tithis.

## Moon's Anomaly and Moon's Equation of the Centre in Days and Fractions of a Day.

### SURYA SIDDHANTA.

| Eqn. | 0 | 1 | 2 | 3 | 4 | 5 | 6 | 7 | 8 | 9 | Eqn. | 9 | 8 | 7 | 6 | 5 | 4 | 3 | 2 | 1 | 0 |
|---|---|---|---|---|---|---|---|---|---|---|---|---|---|---|---|---|---|---|---|---|---|
| — | Moon's Anomaly in days. | | | | | | | | | | — | Moon's Anomaly in days. | | | | | | | | | |
| ·00 | ·000 | ·011 | ·023 | ·034 | ·046 | ·057 | ·069 | ·080 | ·091 | ·102 | ·41 | ... | ... | ... | ... | ... | 7·275 | 7·570 | 7·677 | 7·785 | 7·874 |
| ·01 | ·114 | ·125 | ·137 | ·148 | ·160 | ·171 | ·183 | ·195 | ·206 | ·218 | ·40 | 7·938 | 8·002 | 8·066 | 8·130 | 8·176 | 8·222 | 8·268 | 8·314 | 8·360 | 8·407 |
| ·02 | ·229 | ·241 | ·252 | ·264 | ·275 | ·286 | ·297 | ·309 | ·320 | ·332 | ·39 | 8·443 | 8·479 | 8·515 | 8·552 | 8·588 | 8·623 | 8·659 | 8·694 | 8·724 | 8·754 |
| ·03 | ·343 | ·355 | ·366 | ·378 | ·389 | ·401 | ·413 | ·424 | ·436 | ·447 | ·38 | 8·784 | 8·813 | 8·842 | 8·871 | 8·900 | 8·929 | 8·959 | 8·985 | 9·011 | 9·035 |
| ·04 | ·458 | ·469 | ·481 | ·492 | ·504 | ·516 | ·527 | ·539 | ·550 | ·562 | ·37 | 9·060 | 9·085 | 9·109 | 9·134 | 9·159 | 9·184 | 9·208 | 9·233 | 9·256 | 9·278 |
| ·05 | ·573 | ·585 | ·597 | ·608 | ·619 | ·630 | ·642 | ·654 | ·665 | ·677 | ·36 | 9·299 | 9·321 | 9·343 | 9·365 | 9·387 | 9·409 | 9·430 | 9·452 | 9·474 | 9·495 |
| ·06 | ·689 | ·700 | ·712 | ·724 | ·735 | ·747 | ·759 | ·771 | ·782 | ·793 | ·35 | 9·517 | 9·536 | 9·555 | 9·575 | 9·594 | 9·613 | 9·632 | 9·652 | 9·671 | 9·690 |
| ·07 | ·805 | ·816 | ·828 | ·840 | ·851 | ·863 | ·875 | ·886 | ·898 | ·910 | ·34 | 9·709 | 9·728 | 9·747 | 9·766 | 9·785 | 9·804 | 9·821 | 9·838 | 9·855 | 9·873 |
| ·08 | ·921 | ·933 | ·945 | ·957 | ·968 | ·979 | ·991 | 1·003 | 1·015 | 1·027 | ·33 | 9·890 | 9·907 | 9·925 | 9·943 | 9·960 | 9·977 | 9·994 | 10·012 | 10·029 | 10·046 |
| ·09 | 1·038 | 1·050 | 1·062 | 1·074 | 1·086 | 1·098 | 1·109 | 1·121 | 1·132 | 1·144 | ·32 | 10·063 | 10·079 | 10·095 | 10·111 | 10·127 | 10·143 | 10·159 | 10·175 | 10·191 | 10·206 |
| ·10 | 1·156 | 1·168 | 1·179 | 1·191 | 1·203 | 1·215 | 1·227 | 1·239 | 1·251 | 1·263 | ·31 | 10·222 | 10·238 | 10·254 | 10·269 | 10·285 | 10·301 | 10·317 | 10·332 | 10·347 | 10·362 |
| ·11 | 1·275 | 1·287 | 1·299 | 1·310 | 1·322 | 1·334 | 1·346 | 1·358 | 1·370 | 1·382 | ·30 | 10·376 | 10·391 | 10·406 | 10·421 | 10·435 | 10·450 | 10·464 | 10·479 | 10·493 | 10·508 |
| ·12 | 1·394 | 1·406 | 1·418 | 1·430 | 1·442 | 1·454 | 1·466 | 1·478 | 1·490 | 1·502 | ·29 | 10·523 | 10·537 | 10·552 | 10·566 | 10·581 | 10·595 | 10·610 | 10·624 | 10·638 | 10·651 |
| ·13 | 1·514 | 1·526 | 1·538 | 1·550 | 1·562 | 1·574 | 1·587 | 1·599 | 1·611 | 1·624 | ·28 | 10·665 | 10·679 | 10·692 | 10·706 | 10·719 | 10·733 | 10·747 | 10·760 | 10·774 | 10·787 |
| ·14 | 1·636 | 1·648 | 1·660 | 1·672 | 1·684 | 1·697 | 1·709 | 1·721 | 1·734 | 1·746 | ·27 | 10·801 | 10·815 | 10·829 | 10·843 | 10·857 | 10·870 | 10·883 | 10·896 | 10·909 | 10·922 |
| ·15 | 1·758 | 1·771 | 1·783 | 1·795 | 1·808 | 1·820 | 1·832 | 1·844 | 1·856 | 1·868 | ·26 | 10·934 | 10·947 | 10·960 | 10·973 | 10·985 | 10·998 | 11·011 | 11·024 | 11·037 | 11·050 |
| ·16 | 1·881 | 1·893 | 1·906 | 1·919 | 1·931 | 1·944 | 1·957 | 1·969 | 1·982 | 1·994 | ·25 | 11·063 | 11·076 | 11·088 | 11·101 | 11·114 | 11·127 | 11·139 | 11·152 | 11·164 | 11·176 |
| ·17 | 2·007 | 2·020 | 2·032 | 2·044 | 2·057 | 2·069 | 2·082 | 2·094 | 2·107 | 2·120 | ·24 | 11·188 | 11·200 | 11·213 | 11·225 | 11·237 | 11·249 | 11·261 | 11·273 | 11·285 | 11·297 |
| ·18 | 2·132 | 2·145 | 2·157 | 2·170 | 2·183 | 2·196 | 2·209 | 2·222 | 2·234 | 2·247 | ·23 | 11·309 | 11·321 | 11·334 | 11·346 | 11·358 | 11·370 | 11·382 | 11·395 | 11·407 | 11·419 |
| ·19 | 2·260 | 2·273 | 2·286 | 2·299 | 2·312 | 2·325 | 2·338 | 2·351 | 2·364 | 2·377 | ·22 | 11·430 | 11·442 | 11·453 | 11·465 | 11·476 | 11·488 | 11·499 | 11·511 | 11·522 | 11·534 |
| ·20 | 2·390 | 2·403 | 2·415 | 2·427 | 2·440 | 2·453 | 2·466 | 2·479 | 2·493 | 2·506 | ·21 | 11·545 | 11·557 | 11·569 | 11·581 | 11·592 | 11·604 | 11·615 | 11·627 | 11·638 | 11·649 |
| ·21 | 2·520 | 2·533 | 2·546 | 2·560 | 2·573 | 2·587 | 2·600 | 2·614 | 2·627 | 2·640 | ·20 | 11·661 | 11·672 | 11·684 | 11·695 | 11·706 | 11·717 | 11·728 | 11·739 | 11·750 | 11·761 |
| ·22 | 2·653 | 2·667 | 2·680 | 2·694 | 2·707 | 2·721 | 2·734 | 2·748 | 2·761 | 2·775 | ·19 | 11·772 | 11·783 | 11·794 | 11·805 | 11·816 | 11·827 | 11·838 | 11·849 | 11·860 | 11·871 |
| ·23 | 2·788 | 2·802 | 2·816 | 2·829 | 2·843 | 2·857 | 2·871 | 2·885 | 2·900 | 2·914 | ·18 | 11·882 | 11·893 | 11·905 | 11·916 | 11·927 | 11·938 | 11·949 | 11·960 | 11·970 | 11·981 |
| ·24 | 2·928 | 2·942 | 2·956 | 2·970 | 2·984 | 2·998 | 3·012 | 3·026 | 3·039 | 3·053 | ·17 | 11·991 | 12·002 | 12·013 | 12·023 | 12·034 | 12·044 | 12·055 | 12·066 | 12·077 | 12·088 |
| ·25 | 3·067 | 3·081 | 3·095 | 3·110 | 3·124 | 3·139 | 3·154 | 3·169 | 3·183 | 3·198 | ·16 | 12·099 | 12·109 | 12·120 | 12·130 | 12·141 | 12·152 | 12·162 | 12·173 | 12·184 | 12·194 |
| ·26 | 3·213 | 3·228 | 3·242 | 3·256 | 3·271 | 3·286 | 3·300 | 3·315 | 3·330 | 3·345 | ·15 | 12·205 | 12·215 | 12·226 | 12·237 | 12·247 | 12·257 | 12·268 | 12·278 | 12·288 | 12·299 |
| ·27 | 3·359 | 3·374 | 3·389 | 3·404 | 3·419 | 3·434 | 3·450 | 3·466 | 3·481 | 3·496 | ·14 | 12·309 | 12·319 | 12·330 | 12·340 | 12·350 | 12·361 | 12·371 | 12·382 | 12·393 | 12·403 |
| ·28 | 3·512 | 3·527 | 3·543 | 3·558 | 3·574 | 3·590 | 3·605 | 3·621 | 3·636 | 3·652 | ·13 | 12·413 | 12·424 | 12·434 | 12·444 | 12·455 | 12·465 | 12·475 | 12·485 | 12·495 | 12·505 |
| ·29 | 3·668 | 3·683 | 3·699 | 3·714 | 3·730 | 3·746 | 3·763 | 3·779 | 3·796 | 3·812 | ·12 | 12·515 | 12·525 | 12·536 | 12·546 | 12·557 | 12·567 | 12·577 | 12·587 | 12·597 | 12·607 |
| ·30 | 3·829 | 3·846 | 3·862 | 3·879 | 3·895 | 3·912 | 3·928 | 3·945 | 3·961 | 3·977 | ·11 | 12·617 | 12·627 | 12·637 | 12·647 | 12·657 | 12·667 | 12·677 | 12·688 | 12·698 | 12·708 |
| ·31 | 3·993 | 4·010 | 4·027 | 4·044 | 4·062 | 4·080 | 4·098 | 4·115 | 4·133 | 4·151 | ·10 | 12·718 | 12·729 | 12·738 | 12·748 | 12·758 | 12·768 | 12·778 | 12·788 | 12·797 | 12·807 |
| ·32 | 4·169 | 4·186 | 4·204 | 4·221 | 4·239 | 4·256 | 4·274 | 4·292 | 4·310 | 4·328 | ·09 | 12·817 | 12·827 | 12·838 | 12·848 | 12·857 | 12·867 | 12·877 | 12·887 | 12·897 | 12·907 |
| ·33 | 4·347 | 4·366 | 4·385 | 4·405 | 4·424 | 4·443 | 4·462 | 4·481 | 4·500 | 4·519 | ·08 | 12·916 | 12·926 | 12·936 | 12·946 | 12·956 | 12·966 | 12·976 | 12·986 | 12·996 | 13·006 |
| ·34 | 4·538 | 4·558 | 4·577 | 4·596 | 4·615 | 4·636 | 4·657 | 4·678 | 4·699 | 4·720 | ·07 | 13·015 | 13·025 | 13·035 | 13·044 | 13·054 | 13·064 | 13·073 | 13·083 | 13·093 | 13·102 |
| ·35 | 4·741 | 4·762 | 4·783 | 4·803 | 4·824 | 4·845 | 4·866 | 4·888 | 4·909 | 4·930 | ·06 | 13·112 | 13·122 | 13·132 | 13·142 | 13·152 | 13·162 | 13·171 | 13·181 | 13·191 | 13·200 |
| ·36 | 4·954 | 4·977 | 5·001 | 5·025 | 5·048 | 5·072 | 5·096 | 5·119 | 5·142 | 5·166 | ·05 | 13·210 | 13·220 | 13·229 | 13·239 | 13·249 | 13·258 | 13·268 | 13·278 | 13·288 | 13·298 |
| ·37 | 5·189 | 5·213 | 5·238 | 5·265 | 5·291 | 5·318 | 5·345 | 5·372 | 5·398 | 5·425 | ·04 | 13·307 | 13·317 | 13·326 | 13·336 | 13·345 | 13·355 | 13·365 | 13·374 | 13·384 | 13·393 |
| ·38 | 5·452 | 5·478 | 5·504 | 5·532 | 5·564 | 5·595 | 5·626 | 5·658 | 5·689 | 5·721 | ·03 | 13·403 | 13·413 | 13·423 | 13·432 | 13·442 | 13·452 | 13·461 | 13·471 | 13·480 | 13·490 |
| ·39 | 5·752 | 5·784 | 5·815 | 5·852 | 5·890 | 5·927 | 5·965 | 6·003 | 6·040 | 6·078 | ·02 | 13·499 | 13·509 | 13·518 | 13·528 | 13·537 | 13·547 | 13·557 | 13·567 | 13·576 | 13·586 |
| ·40 | 6·116 | 6·165 | 6·213 | 6·261 | 6·309 | 6·357 | 6·405 | 6·471 | 6·537 | 6·603 | ·01 | 13·595 | 13·605 | 13·614 | 13·624 | 13·634 | 13·643 | 13·653 | 13·662 | 13·672 | 13·681 |
| ·41 | 6·669 | 6·759 | 6·868 | 7·077 | 7·275 | ... | ... | ... | ... | ... | ·00 | 13·691 | 13·701 | 13·710 | 13·720 | 13·730 | 13·739 | 13·749 | 13·758 | 13·768 | 13·777 |

**Maximum equation ·413929** for Anomaly of 7·2752 days.

# TABLE IX—(*b*)—For Tithis.

## SURYA SIDDHANTA.

## Moon's Anomaly and Moon's Equation of the Centre in Days and Fractions of a Day.

| Eqn. + | 0 | 1 | 2 | 3 | 4 | 5 | 6 | 7 | 8 | 9 | Eqn. + | 9 | 8 | 7 | 6 | 5 | 4 | 3 | 2 | 1 | 0 |
|---|---|---|---|---|---|---|---|---|---|---|---|---|---|---|---|---|---|---|---|---|---|
| | Moon's Anomaly in days. | | | | | | | | | | | Moon's Anomaly in days. | | | | | | | | | |
| ·00 | 13·777 | 13·786 | 13·796 | 13·805 | 13·815 | 13·824 | 13·834 | 13·843 | 13·853 | 13·863 | ·41 | ... | ... | ... | ... | ... | 20·280 | 20·577 | 20·686 | 20·795 | 20·885 |
| ·01 | 13·873 | 13·882 | 13·892 | 13·902 | 13·911 | 13·920 | 13·930 | 13·940 | 13·949 | 13·959 | ·40 | 20·951 | 21·017 | 21·083 | 21·149 | 21·197 | 21·245 | 21·293 | 21·341 | 21·389 | 21·438 |
| ·02 | 13·968 | 13·978 | 13·987 | 13·997 | 14·007 | 14·017 | 14·026 | 14·036 | 14·045 | 14·055 | ·39 | 21·476 | 21·514 | 21·552 | 21·589 | 21·627 | 21·664 | 21·702 | 21·739 | 21·770 | 21·802 |
| ·03 | 14·064 | 14·074 | 14·083 | 14·093 | 14·102 | 14·112 | 14·122 | 14·131 | 14·141 | 14·151 | ·38 | 21·833 | 21·865 | 21·896 | 21·928 | 21·959 | 21·990 | 22·022 | 22·050 | 22·076 | 22·102 |
| ·04 | 14·161 | 14·170 | 14·180 | 14·189 | 14·199 | 14·209 | 14·218 | 14·228 | 14·237 | 14·247 | ·37 | 22·129 | 22·156 | 22·182 | 22·209 | 22·236 | 22·263 | 22·289 | 22·316 | 22·341 | 22·365 |
| ·05 | 14·256 | 14·266 | 14·275 | 14·285 | 14·295 | 14·305 | 14·315 | 14·325 | 14·334 | 14·344 | ·36 | 22·388 | 22·412 | 22·436 | 22·458 | 22·482 | 22·506 | 22·529 | 22·553 | 22·577 | 22·600 |
| ·06 | 14·354 | 14·363 | 14·373 | 14·383 | 14·392 | 14·402 | 14·412 | 14·422 | 14·431 | 14·441 | ·35 | 22·624 | 22·645 | 22·666 | 22·688 | 22·709 | 22·730 | 22·751 | 22·771 | 22·792 | 22·813 |
| ·07 | 14·451 | 14·461 | 14·471 | 14·481 | 14·490 | 14·500 | 14·510 | 14·519 | 14·529 | 14·539 | ·34 | 22·834 | 22·855 | 22·876 | 22·897 | 22·918 | 22·939 | 22·958 | 22·977 | 22·996 | 23·016 |
| ·08 | 14·548 | 14·558 | 14·568 | 14·578 | 14·588 | 14·598 | 14·608 | 14·618 | 14·628 | 14·638 | ·33 | 23·035 | 23·054 | 23·073 | 23·092 | 23·111 | 23·130 | 23·149 | 23·169 | 23·188 | 23·207 |
| ·09 | 14·647 | 14·657 | 14·667 | 14·677 | 14·687 | 14·697 | 14·707 | 14·717 | 14·727 | 14·737 | ·32 | 23·226 | 23·244 | 23·262 | 23·280 | 23·298 | 23·315 | 23·333 | 23·350 | 23·368 | 23·385 |
| ·10 | 14·747 | 14·757 | 14·766 | 14·776 | 14·786 | 14·796 | 14·806 | 14·816 | 14·826 | 14·836 | ·31 | 23·403 | 23·421 | 23·439 | 23·456 | 23·474 | 23·492 | 23·510 | 23·527 | 23·544 | 23·561 |
| ·11 | 14·846 | 14·856 | 14·866 | 14·877 | 14·887 | 14·897 | 14·907 | 14·917 | 14·927 | 14·937 | ·30 | 23·577 | 23·594 | 23·610 | 23·626 | 23·642 | 23·659 | 23·675 | 23·692 | 23·708 | 23·725 |
| ·12 | 14·947 | 14·957 | 14·967 | 14·977 | 14·987 | 14·997 | 15·008 | 15·018 | 15·029 | 15·039 | ·29 | 23·742 | 23·758 | 23·775 | 23·791 | 23·808 | 23·824 | 23·841 | 23·856 | 23·871 | 23·886 |
| ·13 | 15·049 | 15·059 | 15·069 | 15·079 | 15·089 | 15·099 | 15·110 | 15·120 | 15·130 | 15·141 | ·28 | 23·902 | 23·918 | 23·933 | 23·949 | 23·964 | 23·980 | 23·996 | 24·011 | 24·027 | 24·042 |
| ·14 | 15·151 | 15·161 | 15·172 | 15·183 | 15·193 | 15·204 | 15·214 | 15·224 | 15·235 | 15·245 | ·27 | 24·058 | 24·074 | 24·089 | 24·104 | 24·120 | 24·135 | 24·150 | 24·165 | 24·180 | 24·195 |
| ·15 | 15·255 | 15·266 | 15·276 | 15·286 | 15·297 | 15·307 | 15·317 | 15·328 | 15·339 | 15·349 | ·26 | 24·209 | 24·224 | 24·239 | 24·254 | 24·268 | 24·283 | 24·298 | 24·312 | 24·326 | 24·341 |
| ·16 | 15·360 | 15·370 | 15·381 | 15·392 | 15·402 | 15·413 | 15·424 | 15·434 | 15·445 | 15·455 | ·25 | 24·356 | 24·371 | 24·385 | 24·400 | 24·415 | 24·430 | 24·444 | 24·459 | 24·473 | 24·487 |
| ·17 | 15·466 | 15·477 | 15·487 | 15·498 | 15·509 | 15·520 | 15·531 | 15·541 | 15·552 | 15·563 | ·24 | 24·501 | 24·515 | 24·528 | 24·542 | 24·556 | 24·570 | 24·584 | 24·598 | 24·612 | 24·626 |
| ·18 | 15·573 | 15·584 | 15·594 | 15·605 | 15·616 | 15·627 | 15·638 | 15·649 | 15·661 | 15·672 | ·23 | 24·640 | 24·654 | 24·669 | 24·683 | 24·697 | 24·711 | 24·725 | 24·738 | 24·752 | 24·766 |
| ·19 | 15·683 | 15·694 | 15·705 | 15·716 | 15·727 | 15·738 | 15·749 | 15·760 | 15·771 | 15·782 | ·22 | 24·779 | 24·793 | 24·806 | 24·820 | 24·833 | 24·847 | 24·860 | 24·874 | 24·887 | 24·901 |
| ·20 | 15·793 | 15·804 | 15·815 | 15·826 | 15·837 | 15·848 | 15·859 | 15·870 | 15·882 | 15·893 | ·21 | 24·914 | 24·928 | 24·941 | 24·954 | 24·967 | 24·981 | 24·994 | 25·008 | 25·021 | 25·034 |
| ·21 | 15·905 | 15·916 | 15·927 | 15·939 | 15·950 | 15·962 | 15·973 | 15·985 | 15·997 | 16·009 | ·20 | 25·048 | 25·061 | 25·075 | 25·088 | 25·101 | 25·114 | 25·127 | 25·139 | 25·151 | 25·164 |
| ·22 | 16·020 | 16·032 | 16·043 | 16·055 | 16·066 | 16·078 | 16·089 | 16·101 | 16·112 | 16·124 | ·19 | 25·177 | 25·190 | 25·203 | 25·216 | 25·229 | 25·242 | 25·255 | 25·268 | 25·281 | 25·294 |
| ·23 | 16·135 | 16·147 | 16·159 | 16·172 | 16·184 | 16·196 | 16·208 | 16·220 | 16·233 | 16·245 | ·18 | 25·307 | 25·320 | 25·332 | 25·345 | 25·358 | 25·371 | 25·384 | 25·397 | 25·409 | 25·422 |
| ·24 | 16·257 | 16·269 | 16·281 | 16·293 | 16·305 | 16·317 | 16·329 | 16·341 | 16·354 | 16·366 | ·17 | 25·434 | 25·447 | 25·460 | 25·472 | 25·485 | 25·497 | 25·510 | 25·522 | 25·534 | 25·547 |
| ·25 | 16·378 | 16·390 | 16·403 | 16·415 | 16·427 | 16·440 | 16·453 | 16·466 | 16·478 | 16·491 | ·16 | 25·560 | 25·572 | 25·585 | 25·597 | 25·610 | 25·623 | 25·635 | 25·648 | 25·661 | 25·673 |
| ·26 | 16·504 | 16·517 | 16·530 | 16·543 | 16·556 | 16·569 | 16·581 | 16·594 | 16·607 | 16·620 | ·15 | 25·686 | 25·698 | 25·710 | 25·722 | 25·734 | 25·746 | 25·759 | 25·771 | 25·783 | 25·796 |
| ·27 | 16·632 | 16·645 | 16·658 | 16·671 | 16·684 | 16·697 | 16·711 | 16·725 | 16·739 | 16·753 | ·14 | 25·808 | 25·820 | 25·833 | 25·845 | 25·857 | 25·870 | 25·882 | 25·893 | 25·906 | 25·918 |
| ·28 | 16·767 | 16·780 | 16·794 | 16·807 | 16·821 | 16·835 | 16·848 | 16·862 | 16·875 | 16·889 | ·13 | 25·930 | 25·943 | 25·955 | 25·967 | 25·980 | 25·992 | 26·004 | 26·016 | 26·028 | 26·040 |
| ·29 | 16·903 | 16·916 | 16·930 | 16·944 | 16·959 | 16·973 | 16·988 | 17·002 | 17·017 | 17·031 | ·12 | 26·052 | 26·064 | 26·076 | 26·088 | 26·100 | 26·112 | 26·124 | 26·136 | 26·148 | 26·160 |
| ·30 | 17·046 | 17·061 | 17·075 | 17·090 | 17·104 | 17·119 | 17·133 | 17·148 | 17·163 | 17·178 | ·11 | 26·172 | 26·184 | 26·196 | 26·208 | 26·220 | 26·232 | 26·244 | 26·255 | 26·267 | 26·279 |
| ·31 | 17·192 | 17·207 | 17·222 | 17·237 | 17·251 | 17·269 | 17·285 | 17·300 | 17·316 | 17·332 | ·10 | 26·291 | 26·303 | 26·315 | 26·327 | 26·339 | 26·351 | 26·363 | 26·375 | 26·386 | 26·398 |
| ·32 | 17·348 | 17·363 | 17·379 | 17·395 | 17·411 | 17·427 | 17·443 | 17·459 | 17·475 | 17·491 | ·09 | 26·410 | 26·422 | 26·433 | 26·445 | 26·456 | 26·468 | 26·480 | 26·492 | 26·504 | 26·516 |
| ·33 | 17·508 | 17·525 | 17·542 | 17·560 | 17·577 | 17·594 | 17·611 | 17·629 | 17·647 | 17·664 | ·08 | 26·527 | 26·539 | 26·551 | 26·563 | 26·575 | 26·587 | 26·598 | 26·609 | 26·621 | 26·633 |
| ·34 | 17·681 | 17·699 | 17·716 | 17·733 | 17·750 | 17·769 | 17·788 | 17·807 | 17·826 | 17·845 | ·07 | 26·644 | 26·656 | 26·668 | 26·679 | 26·691 | 26·703 | 26·714 | 26·726 | 26·738 | 26·749 |
| ·35 | 17·864 | 17·883 | 17·902 | 17·922 | 17·941 | 17·960 | 17·979 | 17·999 | 18·018 | 18·037 | ·06 | 26·761 | 26·772 | 26·783 | 26·795 | 26·807 | 26·819 | 26·830 | 26·842 | 26·854 | 26·865 |
| ·36 | 18·059 | 18·080 | 18·102 | 18·124 | 18·145 | 18·167 | 18·189 | 18·221 | 18·233 | 18·255 | ·05 | 26·877 | 26·889 | 26·900 | 26·912 | 26·924 | 26·935 | 26·946 | 26·957 | 26·969 | 26·981 |
| ·37 | 18·276 | 18·298 | 18·321 | 18·346 | 18·370 | 18·395 | 18·420 | 18·445 | 18·469 | 18·494 | ·04 | 26·992 | 27·004 | 27·015 | 27·027 | 27·038 | 27·050 | 27·062 | 27·073 | 27·085 | 27·096 |
| ·38 | 18·519 | 18·543 | 18·569 | 18·595 | 18·625 | 18·654 | 18·683 | 18·713 | 18·742 | 18·772 | ·03 | 27·108 | 27·119 | 27·130 | 27·141 | 27·153 | 27·165 | 27·176 | 27·188 | 27·199 | 27·211 |
| ·39 | 18·801 | 18·831 | 18·860 | 18·895 | 18·931 | 18·966 | 19·003 | 19·039 | 19·075 | 19·111 | ·02 | 27·222 | 27·234 | 27·245 | 27·257 | 27·268 | 27·280 | 27·291 | 27·302 | 27·313 | 27·325 |
| ·40 | 19·147 | 19·194 | 19·240 | 19·286 | 19·332 | 19·378 | 19·424 | 19·488 | 19·552 | 19·616 | ·01 | 27·336 | 27·348 | 27·359 | 27·371 | 27·383 | 27·394 | 27·406 | 27·417 | 27·429 | 27·440 |
| ·41 | 19·680 | 19·768 | 19·877 | 19·984 | 20·280 | ... | ... | ... | ... | ... | ·00 | 27·452 | 27·463 | 27·474 | 27·485 | 27·497 | 27·508 | 27·520 | 27·531 | 27·543 | 27·555 |

**Maximum equation ·413929** for Anomaly of 20·2797 days.

# TABLE IX—(c)—For Tithis.

## Sun's Anomaly expressed as Days of the Solar Year and Sun's Equation of the Centre in Fractions of a Day.

### (SURYA SIDDHANTA.)

| Eqn. + | 9 | 8 | 7 | 6 | 5 | 4 | 3 | 2 | 1 | 0 | Eqn. — | 0 | 1 | 2 | 3 | 4 | 5 | 6 | 7 | 8 | 9 |
|---|---|---|---|---|---|---|---|---|---|---|---|---|---|---|---|---|---|---|---|---|---|
| | Sun's Anomaly in days of the Solar Year. | | | | | | | | | | | Sun's Anomaly in days of the Solar Year. | | | | | | | | | |
| ·17 | ... | ... | ... | ... | 0·77 | 2·25 | 3·73 | 5·04 | 6·19 | 7·35 | ·00 | 80·55 | 80·87 | 81·19 | 81·51 | 81·83 | 82·15 | 82·47 | 82·79 | 83·11 | 83·43 |
| ·16 | 8·45 | 9·40 | 10·35 | 11·30 | 12·22 | 13·02 | 13·82 | 14·63 | 15·43 | 16·19 | ·01 | 83·74 | 84·05 | 84·37 | 84·70 | 85·02 | 85·34 | 85·66 | 85·99 | 86·31 | 86·63 |
| ·15 | 16·89 | 17·60 | 18·31 | 19·02 | 19·72 | 20·34 | 20·97 | 21·60 | 22·22 | 22·85 | ·02 | 86·94 | 87·26 | 87·59 | 87·91 | 88·23 | 88·56 | 88·88 | 89·21 | 89·53 | 89·86 |
| ·14 | 23·47 | 24·04 | 24·60 | 25·17 | 25·74 | 26·31 | 26·88 | 27·43 | 27·96 | 28·48 | ·03 | 90·18 | 90·50 | 90·83 | 91·15 | 91·48 | 91·81 | 92·13 | 92·46 | 92·80 | 93·13 |
| ·13 | 28·99 | 29·51 | 30·03 | 30·55 | 31·07 | 31·56 | 32·04 | 32·52 | 33·00 | 33·48 | ·04 | 93·45 | 93·78 | 94·11 | 94·44 | 94·77 | 95·10 | 95·43 | 95·76 | 96·10 | 96·44 |
| ·12 | 33·96 | 34·44 | 34·92 | 35·37 | 35·82 | 36·28 | 36·73 | 37·18 | 37·63 | 38·08 | ·05 | 96·77 | 97·11 | 97·45 | 97·79 | 98·13 | 98·47 | 98·81 | 99·14 | 99·48 | 99·83 |
| ·11 | 38·53 | 38·96 | 39·39 | 39·81 | 40·24 | 40·66 | 41·09 | 41·51 | 41·94 | 42·36 | ·06 | 100·17 | 100·51 | 100·86 | 101·21 | 101·56 | 101·90 | 102·25 | 102·60 | 102·95 | 103·29 |
| ·10 | 42·76 | 43·16 | 43·57 | 43·97 | 44·37 | 44·78 | 45·18 | 45·58 | 45·99 | 46·39 | ·07 | 103·64 | 103·99 | 104·35 | 104·71 | 105·07 | 105·42 | 105·78 | 106·14 | 106·49 | 106·85 |
| ·09 | 46·76 | 47·15 | 47·53 | 47·92 | 48·30 | 48·69 | 49·07 | 49·46 | 49·84 | 50·22 | ·08 | 107·20 | 107·57 | 107·94 | 108·30 | 108·67 | 109·04 | 109·41 | 109·78 | 110·15 | 110·52 |
| ·08 | 50·58 | 50·95 | 51·32 | 51·69 | 52·05 | 52·42 | 52·79 | 53·16 | 53·53 | 53·90 | ·09 | 110·88 | 111·26 | 111·64 | 112·03 | 112·41 | 112·80 | 113·18 | 113·57 | 113·95 | 114·33 |
| ·07 | 54·25 | 54·60 | 54·96 | 55·32 | 55·67 | 56·03 | 56·39 | 56·75 | 57·10 | 57·46 | ·10 | 114·71 | 115·11 | 115·51 | 115·92 | 116·32 | 116·72 | 117·13 | 117·53 | 117·93 | 118·34 |
| ·06 | 57·80 | 58·15 | 58·50 | 58·85 | 59·19 | 59·54 | 59·89 | 60·24 | 60·58 | 60·93 | ·11 | 118·74 | 119·16 | 119·59 | 120·01 | 120·44 | 120·86 | 121·29 | 121·71 | 122·14 | 122·57 |
| ·05 | 61·27 | 61·61 | 61·95 | 62·29 | 62·63 | 62·97 | 63·31 | 63·65 | 63·99 | 64·33 | ·12 | 123·01 | 123·46 | 123·92 | 124·37 | 124·82 | 125·27 | 125·73 | 126·18 | 126·66 | 127·14 |
| ·04 | 64·65 | 64·99 | 65·33 | 65·66 | 65·99 | 66·33 | 66·66 | 66·99 | 67·32 | 67·65 | ·13 | 127·61 | 128·09 | 128·58 | 129·06 | 129·54 | 130·02 | 130·55 | 131·07 | 131·59 | 132·11 |
| ·03 | 67·97 | 68·30 | 68·63 | 68·96 | 69·29 | 69·62 | 69·94 | 70·27 | 70·60 | 70·92 | ·14 | 132·62 | 133·14 | 133·66 | 134·22 | 134·79 | 135·35 | 135·92 | 136·49 | 137·06 | 137·63 |
| ·02 | 71·24 | 71·56 | 71·89 | 72·21 | 72·54 | 72·87 | 73·19 | 73·51 | 73·83 | 74·16 | ·15 | 138·25 | 138·87 | 139·50 | 140·13 | 140·75 | 141·38 | 142·08 | 142·79 | 143·50 | 144·21 |
| ·01 | 74·47 | 74·79 | 75·11 | 75·43 | 75·76 | 76·08 | 76·40 | 76·72 | 77·04 | 77·36 | ·16 | 144·91 | 145·67 | 146·47 | 147·27 | 148·08 | 148·88 | 149·80 | 150·75 | 151·70 | 152·65 |
| ·00 | 77·67 | 77·99 | 78·31 | 78·63 | 78·95 | 79·27 | 79·59 | 79·91 | 80·23 | 80·55 | ·17 | 153·75 | 154·90 | 156·06 | 157·37 | 158·85 | 160·33 | 162·31 | 164·41 | 167·82 | ... |

**Maximum equation ·178436** for Anomaly of 171·864 days.

| Eqn. — | 9 | 8 | 7 | 6 | 5 | 4 | 3 | 2 | 1 | 0 | Eqn. + | 0 | 1 | 2 | 3 | 4 | 5 | 6 | 7 | 8 | 9 |
|---|---|---|---|---|---|---|---|---|---|---|---|---|---|---|---|---|---|---|---|---|---|
| | Sun's Anomaly in days of the Solar Year. | | | | | | | | | | | Sun's Anomaly in days of the Solar Year. | | | | | | | | | |
| ·17 | ... | 175·92 | 179·33 | 181·42 | 183·41 | 184·89 | 186·36 | 187·67 | 188·83 | 189·98 | ·00 | 263·18 | 263·50 | 263·81 | 264·13 | 264·45 | 264·77 | 265·08 | 265·40 | 265·72 | 266·04 |
| ·16 | 191·09 | 192·04 | 192·99 | 193·94 | 194·86 | 195·66 | 196·46 | 197·26 | 198·06 | 198·82 | ·01 | 266·37 | 266·68 | 267·00 | 267·32 | 267·64 | 267·96 | 268·28 | 268·60 | 268·92 | 269·23 |
| ·15 | 199·54 | 200·24 | 200·95 | 201·66 | 202·36 | 202·98 | 203·61 | 204·23 | 204·86 | 205·48 | ·02 | 269·57 | 269·89 | 270·21 | 270·53 | 270·85 | 271·18 | 271·50 | 271·82 | 272·15 | 272·47 |
| ·14 | 206·11 | 206·68 | 207·25 | 207·81 | 208·38 | 208·95 | 209·52 | 210·07 | 210·59 | 211·11 | ·03 | 272·81 | 273·13 | 273·45 | 273·78 | 274·10 | 274·43 | 274·75 | 275·08 | 275·41 | 275·74 |
| ·13 | 211·64 | 212·16 | 212·67 | 213·19 | 213·71 | 214·19 | 214·67 | 215·15 | 215·63 | 216·11 | ·04 | 276·08 | 276·41 | 276·74 | 277·06 | 277·39 | 277·72 | 278·05 | 278·38 | 278·72 | 279·05 |
| ·12 | 216·60 | 217·08 | 217·56 | 218·01 | 218·46 | 218·91 | 219·36 | 219·81 | 220·26 | 220·71 | ·05 | 279·40 | 279·74 | 280·08 | 280·41 | 280·75 | 281·09 | 281·42 | 281·76 | 282·10 | 282·44 |
| ·11 | 221·18 | 221·61 | 222·53 | 222·45 | 222·88 | 223·30 | 223·72 | 224·14 | 224·57 | 224·99 | ·06 | 282·80 | 283·14 | 283·49 | 283·83 | 284·18 | 284·52 | 284·87 | 285·21 | 285·56 | 285·91 |
| ·10 | 225·41 | 225·81 | 226·21 | 226·61 | 227·01 | 227·42 | 227·82 | 228·22 | 228·62 | 229·02 | ·07 | 286·27 | 286·62 | 286·98 | 287·33 | 287·69 | 288·04 | 288·40 | 288·75 | 289·11 | 289·46 |
| ·09 | 229·41 | 229·79 | 230·18 | 230·56 | 230·94 | 231·32 | 231·71 | 232·09 | 232·47 | 232·85 | ·08 | 289·83 | 290·20 | 290·56 | 290·93 | 291·30 | 291·66 | 292·03 | 292·40 | 292·76 | 293·13 |
| ·08 | 233·23 | 233·59 | 233·96 | 234·33 | 234·69 | 235·06 | 235·43 | 235·80 | 236·16 | 236·53 | ·09 | 293·51 | 293·88 | 294·27 | 294·65 | 295·03 | 295·42 | 295·80 | 296·18 | 296·56 | 296·95 |
| ·07 | 236·89 | 237·25 | 237·60 | 237·96 | 238·31 | 238·67 | 239·02 | 239·38 | 239·73 | 240·09 | ·10 | 297·34 | 297·74 | 298·14 | 298·54 | 298·94 | 299·34 | 299·74 | 300·15 | 300·55 | 300·95 |
| ·06 | 240·45 | 240·80 | 241·14 | 241·49 | 241·83 | 242·18 | 242·52 | 242·87 | 243·22 | 243·56 | ·11 | 301·37 | 301·79 | 302·21 | 302·64 | 303·06 | 303·48 | 303·90 | 304·33 | 304·75 | 305·18 |
| ·05 | 243·92 | 244·26 | 244·60 | 244·93 | 245·27 | 245·61 | 245·94 | 246·28 | 246·62 | 246·96 | ·12 | 305·64 | 306·09 | 306·54 | 306·99 | 307·44 | 307·89 | 308·34 | 308·79 | 309·27 | 309·75 |
| ·04 | 247·30 | 247·64 | 247·98 | 248·31 | 248·63 | 248·96 | 249·29 | 249·62 | 249·95 | 250·28 | ·13 | 310·24 | 310·72 | 311·20 | 311·68 | 312·16 | 312·64 | 313·16 | 313·68 | 314·20 | 314·72 |
| ·03 | 250·62 | 250·95 | 251·28 | 251·61 | 251·93 | 252·26 | 252·58 | 252·90 | 253·23 | 253·55 | ·14 | 315·25 | 315·77 | 316·29 | 316·84 | 317·41 | 317·98 | 318·54 | 319·11 | 319·68 | 320·24 |
| ·02 | 253·88 | 254·21 | 254·53 | 254·86 | 255·18 | 255·50 | 255·82 | 256·15 | 256·47 | 256·79 | ·15 | 320·88 | 321·50 | 322·13 | 322·75 | 323·37 | 324·00 | 324·70 | 325·41 | 326·11 | 326·82 |
| ·01 | 257·12 | 257·44 | 257·76 | 258·08 | 258·40 | 258·72 | 259·04 | 259·36 | 259·67 | 259·99 | ·16 | 327·54 | 328·29 | 329·10 | 329·90 | 330·70 | 331·50 | 332·42 | 333·37 | 334·31 | 335·26 |
| ·00 | 260·32 | 260·64 | 260·95 | 261·27 | 261·59 | 261·91 | 262·23 | 262·54 | 262·86 | 263·18 | ·17 | 336·38 | 337·53 | 338·68 | 339·99 | 341·47 | 342·95 | 344·93 | 347·03 | 350·44 | ... |
| | | | | | | | | | | | | | | | | | | 363·87 | 361·77 | 358·53 | ... |

**Maximum equation ·178436** for Anomaly of 354·493 days.

# TABLE IX—(*d*)—For Tithis.

## Moon's Anomaly and Equation of the Centre expressed in Days and Fractions of a Day.

### (ARYA SIDDHANTA.)

| Eqn. | 0 | 1 | 2 | 3 | 4 | 5 | 6 | 7 | 8 | 9 | Eqn. | 9 | 8 | 7 | 6 | 5 | 4 | 3 | 2 | 1 | 0 |
|---|---|---|---|---|---|---|---|---|---|---|---|---|---|---|---|---|---|---|---|---|---|
| — | Moon's Anomaly in days. | | | | | | | | | | — | Moon's Anomaly in days. | | | | | | | | | |
| **·00** | ... | ·012 | ·023 | ·035 | ·047 | ·058 | ·070 | ·082 | ·093 | ·104 | **·41** | ... | ... | ... | ... | ... | ... | ... | 7·272 | 7·470 | 7·638 |
| **·01** | ·116 | ·127 | ·139 | ·151 | ·162 | ·174 | ·186 | ·197 | ·209 | ·221 | **·40** | 7·756 | 7·862 | 7·926 | 7·989 | 8·053 | 8·116 | 8·163 | 8·209 | 8·255 | 8·300 |
| **·02** | ·232 | ·244 | ·256 | ·267 | ·278 | ·290 | ·301 | ·313 | ·325 | ·336 | **·39** | 8·346 | 8·392 | 8·431 | 8·468 | 8·504 | 8·539 | 8·575 | 8·611 | 8·647 | 8·683 |
| **·03** | ·348 | ·360 | ·371 | ·383 | ·395 | ·407 | ·418 | ·430 | ·441 | ·452 | **·38** | 8·713 | 8·742 | 8·771 | 8·800 | 8·830 | 8·859 | 8·888 | 8·917 | 8·947 | 8·975 |
| **·04** | ·464 | ·476 | ·488 | ·499 | ·511 | ·523 | ·534 | ·546 | ·558 | ·570 | **·37** | 9·000 | 9·025 | 9·050 | 9·075 | 9·100 | 9·125 | 9·150 | 9·174 | 9·199 | 9·224 |
| **·05** | ·581 | ·593 | ·605 | ·616 | ·627 | ·639 | ·651 | ·663 | ·674 | ·686 | **·36** | 9·248 | 9·269 | 9·291 | 9·313 | 9·335 | 9·356 | 9·377 | 9·399 | 9·420 | 9·441 |
| **·06** | ·698 | ·710 | ·722 | ·733 | ·745 | ·757 | ·769 | ·781 | ·792 | ·803 | **·35** | 9·463 | 9·484 | 9·505 | 9·526 | 9·545 | 9·564 | 9·583 | 9·603 | 9·622 | 9·641 |
| **·07** | ·815 | ·827 | ·839 | ·850 | ·862 | ·874 | ·886 | ·898 | ·909 | ·921 | **·34** | 9·660 | 9·680 | 9·699 | 9·718 | 9·737 | 9·756 | 9·775 | 9·794 | 9·811 | 9·828 |
| **·08** | ·933 | ·945 | ·957 | ·969 | ·980 | ·992 | 1·004 | 1·016 | 1·028 | 1·040 | **·33** | 9·846 | 9·863 | 9·881 | 9·898 | 9·916 | 9·933 | 9·950 | 9·967 | 9·984 | 10·002 |
| **·09** | 1·052 | 1·064 | 1·076 | 1·087 | 1·099 | 1·111 | 1·123 | 1·135 | 1·146 | 1·158 | **·32** | 10·019 | 10·036 | 10·053 | 10·070 | 10·086 | 10·102 | 10·118 | 10·134 | 10·149 | 10·165 |
| **·10** | 1·170 | 1·182 | 1·194 | 1·206 | 1·218 | 1·230 | 1·242 | 1·254 | 1·266 | 1·278 | **·31** | 10·181 | 10·196 | 10·212 | 10·228 | 10·243 | 10·259 | 10·275 | 10·291 | 10·306 | 10·322 |
| **·11** | 1·290 | 1·302 | 1·314 | 1·326 | 1·338 | 1·350 | 1·362 | 1·375 | 1·387 | 1·399 | **·30** | 10·337 | 10·352 | 10·367 | 10·382 | 10·397 | 10·401 | 10·426 | 10·440 | 10·455 | 10·470 |
| **·12** | 1·411 | 1·423 | 1·435 | 1·448 | 1·460 | 1·472 | 1·484 | 1·496 | 1·508 | 1·520 | **·29** | 10·484 | 10·499 | 10·513 | 10·528 | 10·543 | 10·557 | 10·572 | 10·587 | 10·602 | 10·616 |
| **·13** | 1·532 | 1·544 | 1·556 | 1·568 | 1·581 | 1·593 | 1·606 | 1·618 | 1·631 | 1·643 | **·28** | 10·629 | 10·643 | 10·657 | 10·670 | 10·684 | 10·698 | 10·711 | 10·725 | 10·738 | 10·752 |
| **·14** | 1·655 | 1·668 | 1·680 | 1·692 | 1·704 | 1·717 | 1·729 | 1·741 | 1·754 | 1·766 | **·27** | 10·766 | 10·780 | 10·794 | 10·808 | 10·821 | 10·835 | 10·848 | 10·862 | 10·875 | 10·888 |
| **·15** | 1·779 | 1·791 | 1·804 | 1·816 | 1·828 | 1·841 | 1·853 | 1·865 | 1·877 | 1·890 | **·26** | 10·901 | 10·914 | 10·927 | 10·939 | 10·952 | 10·965 | 10·978 | 10·991 | 11·004 | 11·017 |
| **·16** | 1·903 | 1·915 | 1·928 | 1·941 | 1·954 | 1·966 | 1·979 | 1·992 | 2·004 | 2·017 | **·25** | 11·030 | 11·043 | 11·055 | 11·068 | 11·081 | 11·094 | 11·107 | 11·119 | 11·132 | 11·145 |
| **·17** | 2·030 | 2·042 | 2·055 | 2·067 | 2·079 | 2·092 | 2·105 | 2·118 | 2·130 | 2·143 | **·24** | 11·157 | 11·169 | 11·182 | 11·194 | 11·206 | 11·218 | 11·231 | 11·243 | 11·255 | 11·267 |
| **·18** | 2·156 | 2·168 | 2·181 | 2·194 | 2·207 | 2·220 | 2·233 | 2·246 | 2·258 | 2·271 | **·23** | 11·279 | 11·291 | 11·303 | 11·315 | 11·328 | 11·340 | 11·352 | 11·365 | 11·377 | 11·389 |
| **·19** | 2·285 | 2·298 | 2·311 | 2·324 | 2·337 | 2·350 | 2·363 | 2·376 | 2·389 | 2·402 | **·22** | 11·401 | 11·413 | 11·425 | 11·436 | 11·448 | 11·459 | 11·471 | 11·482 | 11·494 | 11·506 |
| **·20** | 2·415 | 2·428 | 2·441 | 2·453 | 2·466 | 2·479 | 2·493 | 2·506 | 2·520 | 2·533 | **·21** | 11·517 | 11·529 | 11·541 | 11·553 | 11·564 | 11·576 | 11·587 | 11·599 | 11·611 | 11·622 |
| **·21** | 2·547 | 2·560 | 2·574 | 2·588 | 2·601 | 2·615 | 2·628 | 2·642 | 2·655 | 2·668 | **·20** | 11·634 | 11·645 | 11·657 | 11·668 | 11·680 | 11·691 | 11·702 | 11·714 | 11·725 | 11·736 |
| **·22** | 2·681 | 2·795 | 2·709 | 2·722 | 2·736 | 2·749 | 2·763 | 2·776 | 2·890 | 2·804 | **·19** | 11·747 | 11·758 | 11·769 | 11·780 | 11·791 | 11·802 | 11·813 | 11·824 | 11·835 | 11·846 |
| **·23** | 2·818 | 2·832 | 2·846 | 2·860 | 2·873 | 2·987 | 2·902 | 2·916 | 2·930 | 2·944 | **·18** | 11·858 | 11·869 | 11·881 | 11·892 | 11·903 | 11·914 | 11·925 | 11·936 | 11·947 | 11·957 |
| **·24** | 2·958 | 2·972 | 2·986 | 3·000 | 3·015 | 3·029 | 3·043 | 3·057 | 3·071 | 3·084 | **·17** | 11·968 | 11·979 | 11·989 | 12·000 | 12·011 | 12·022 | 12·033 | 12·044 | 12·055 | 12·065 |
| **·25** | 3·098 | 3·113 | 3·128 | 3·142 | 3·157 | 3·172 | 3·187 | 3·202 | 3·216 | 3·231 | **·16** | 12·076 | 12·087 | 12·098 | 12·108 | 12·119 | 12·129 | 12·140 | 12·151 | 12·162 | 12·172 |
| **·26** | 3·246 | 3·261 | 3·275 | 3·290 | 3·304 | 3·319 | 3·334 | 3·348 | 3·363 | 3·378 | **·15** | 12·183 | 12·194 | 12·205 | 12·216 | 12·226 | 12·237 | 12·247 | 12·257 | 12·268 | 12·278 |
| **·27** | 3·393 | 3·407 | 3·423 | 3·439 | 3·454 | 3·470 | 3·485 | 3·501 | 3·516 | 3·531 | **·14** | 12·289 | 12·299 | 12·310 | 12·320 | 12·330 | 12·341 | 12·351 | 12·362 | 12·373 | 12·384 |
| **·28** | 3·547 | 3·563 | 3·578 | 3·594 | 3·609 | 3·625 | 3·641 | 3·656 | 3·672 | 3·688 | **·13** | 12·394 | 12·404 | 12·415 | 12·425 | 12·436 | 12·446 | 12·457 | 12·467 | 12·477 | 12·487 |
| **·29** | 3·703 | 3·719 | 3·735 | 3·751 | 3·768 | 3·784 | 3·801 | 3·818 | 3·834 | 3·851 | **·12** | 12·497 | 12·508 | 12·519 | 12·529 | 12·539 | 12·549 | 12·559 | 12·570 | 12·580 | 12·590 |
| **·30** | 3·867 | 3·884 | 3·901 | 3·917 | 3·934 | 3·950 | 3·967 | 3·983 | 3·999 | 4·016 | **·11** | 12·600 | 12·610 | 12·620 | 12·631 | 12·641 | 12·651 | 12·661 | 12·672 | 12·682 | 12·692 |
| **·31** | 4·033 | 4·051 | 4·068 | 4·086 | 4·104 | 4·122 | 4·139 | 4·157 | 4·175 | 4·192 | **·10** | 12·703 | 12·713 | 12·723 | 12·733 | 12·743 | 12·753 | 12·763 | 12·773 | 12·783 | 12·793 |
| **·32** | 4·210 | 4·228 | 4·245 | 4·262 | 4·280 | 4·297 | 4·315 | 4·334 | 4·353 | 4·372 | **·09** | 12·803 | 12·813 | 12·824 | 12·834 | 12·844 | 12·854 | 12·864 | 12·873 | 12·883 | 12·893 |
| **·33** | 4·391 | 4·411 | 4·430 | 4·449 | 4·468 | 4·487 | 4·506 | 4·525 | 4·544 | 4·563 | **·08** | 12·903 | 12·913 | 12·923 | 12·933 | 12·943 | 12·953 | 12·964 | 12·974 | 12·984 | 12·994 |
| **·34** | 4·583 | 4·602 | 4·621 | 4·642 | 4·663 | 4·684 | 4·705 | 4·726 | 4·747 | 4·769 | **·07** | 13·004 | 13·014 | 13·023 | 13·033 | 13·043 | 13·053 | 13·063 | 13·072 | 13·082 | 13·092 |
| **·35** | 4·790 | 4·811 | 4·832 | 4·852 | 4·873 | 4·894 | 4·915 | 4·938 | 4·961 | 4·984 | **·06** | 13·102 | 13·112 | 13·122 | 13·132 | 13·142 | 13·152 | 13·162 | 13·171 | 13·181 | 13·191 |
| **·36** | 5·008 | 5·031 | 5·054 | 5·078 | 5·101 | 5·124 | 5·148 | 5·171 | 5·194 | 5·217 | **·05** | 13·201 | 13·211 | 13·220 | 13·230 | 13·240 | 13·250 | 13·260 | 13·270 | 13·280 | 13·290 |
| **·37** | 5·243 | 5·270 | 5·297 | 5·323 | 5·350 | 5·377 | 5·404 | 5·431 | 5·458 | 5·485 | **·04** | 13·299 | 13·309 | 13·319 | 13·329 | 13·338 | 13·348 | 13·358 | 13·367 | 13·377 | 13·387 |
| **·38** | 5·512 | 5·542 | 5·572 | 5·602 | 5·634 | 5·665 | 5·697 | 5·728 | 5·759 | 5·790 | **·03** | 13·397 | 13·407 | 13·417 | 13·427 | 13·436 | 13·446 | 13·456 | 13·466 | 13·475 | 13·485 |
| **·39** | 5·822 | 5·860 | 5·898 | 5·936 | 5·974 | 6·011 | 6·049 | 6·086 | 6·127 | 6·175 | **·02** | 13·495 | 13·504 | 13·514 | 13·524 | 13·533 | 13·543 | 13·553 | 13·563 | 13·573 | 13·583 |
| **·40** | 6·223 | 6·270 | 6·318 | 6·366 | 6·415 | 6·480 | 6·546 | 6·611 | 6·677 | 6·775 | **·01** | 13·592 | 13·602 | 13·612 | 13·621 | 13·631 | 13·641 | 13·650 | 13·660 | 13·670 | 13·680 |
| **·41** | 6·875 | 6·975 | 7·272 | ... | ... | ... | ... | ... | ... | ... | **·00** | 13·690 | 13·700 | 13·709 | 13·719 | 13·729 | 13·738 | 13·748 | 13·758 | 13·767 | 13·777 |

**Maximum equation ·411719** for Anomaly of 7·2728 days.

# TABLE IX—(*c*)—For Tithis.

## ARYA SIDDHANTA.

### Moon's Anomaly and Equation of the Centre expressed in days and fractions of a day.

| Eqn. + | 0 | 1 | 2 | 3 | 4 | 5 | 6 | 7 | 8 | 9 | Eqn. + | 9 | 8 | 7 | 6 | 5 | 4 | 3 | 2 | 1 | 0 |
|---|---|---|---|---|---|---|---|---|---|---|---|---|---|---|---|---|---|---|---|---|---|
| | Moon's Anomaly in days. | | | | | | | | | | | Moon's Anomaly in days. | | | | | | | | | |
| ·00 | 13·777 | 13·787 | 13·796 | 13·806 | 13·816 | 13·825 | 13·835 | 13·845 | 13·855 | 13·865 | ·41 | ... | ... | ... | ... | ... | ... | ... | 20·282 | 20·520 | 20·650 |
| ·01 | 13·875 | 13·884 | 13·894 | 13·904 | 13·913 | 13·923 | 13·933 | 13·942 | 13·952 | 13·962 | ·40 | 20·770 | 20·878 | 20·944 | 21·009 | 21·075 | 21·140 | 21·189 | 21·237 | 21·285 | 21·338 |
| ·02 | 13·971 | 13·981 | 13·991 | 14·001 | 14·011 | 14·021 | 14·030 | 14·040 | 14·050 | 14·059 | ·39 | 21·380 | 21·428 | 21·469 | 21·506 | 21·544 | 21·581 | 21·620 | 21·657 | 21·695 | 21·733 |
| ·03 | 14·069 | 14·079 | 14·088 | 14·098 | 14·108 | 14·118 | 14·127 | 14·137 | 14·147 | 14·157 | ·38 | 21·765 | 21·796 | 21·827 | 21·858 | 21·890 | 21·921 | 21·952 | 21·983 | 22·013 | 22·043 |
| ·04 | 14·167 | 14·177 | 14·187 | 14·196 | 14·206 | 14·216 | 14·225 | 14·235 | 14·244 | 14·254 | ·37 | 22·070 | 22·097 | 22·124 | 22·151 | 22·178 | 22·205 | 22·232 | 22·258 | 22·285 | 22·312 |
| ·05 | 14·264 | 14·274 | 14·282 | 14·294 | 14·304 | 14·314 | 14·324 | 14·334 | 14·343 | 14·353 | ·36 | 22·338 | 22·361 | 22·385 | 22·408 | 22·431 | 22·454 | 22·477 | 22·501 | 22·524 | 22·547 |
| ·06 | 14·363 | 14·373 | 14·383 | 14·392 | 14·402 | 14·412 | 14·422 | 14·432 | 14·442 | 14·452 | ·35 | 22·571 | 22·594 | 22·617 | 22·640 | 22·661 | 22·682 | 22·703 | 22·723 | 22·744 | 22·765 |
| ·07 | 14·462 | 14·472 | 14·482 | 14·491 | 14·501 | 14·511 | 14·521 | 14·531 | 14·540 | 14·550 | ·34 | 22·786 | 22·808 | 22·829 | 22·850 | 22·871 | 22·892 | 22·913 | 22·934 | 22·953 | 22·972 |
| ·08 | 14·560 | 14·570 | 14·580 | 14·590 | 14·601 | 14·611 | 14·621 | 14·631 | 14·641 | 14·651 | ·33 | 22·992 | 23·011 | 23·030 | 23·049 | 23·068 | 23·087 | 23·106 | 23·125 | 23·144 | 23·164 |
| ·09 | 14·661 | 14·671 | 14·681 | 14·690 | 14·700 | 14·710 | 14·720 | 14·730 | 14·741 | 14·751 | ·32 | 23·183 | 23·202 | 23·221 | 23·240 | 23·258 | 23·275 | 23·293 | 23·310 | 23·327 | 23·345 |
| ·10 | 14·761 | 14·771 | 14·781 | 14·791 | 14·801 | 14·811 | 14·821 | 14·831 | 14·841 | 14·851 | ·31 | 23·363 | 23·380 | 23·398 | 23·416 | 23·433 | 23·451 | 23·469 | 23·487 | 23·504 | 23·522 |
| ·11 | 14·862 | 14·872 | 14·882 | 14·893 | 14·903 | 14·913 | 14·923 | 14·934 | 14·944 | 14·954 | ·30 | 23·539 | 23·556 | 23·572 | 23·588 | 23·605 | 23·621 | 23·638 | 23·654 | 23·671 | 23·688 |
| ·12 | 14·964 | 14·974 | 14·984 | 14·995 | 15·005 | 15·015 | 15·025 | 15·035 | 15·046 | 15·057 | ·29 | 23·704 | 23·721 | 23·737 | 23·754 | 23·771 | 23·787 | 23·804 | 23·820 | 23·836 | 23·852 |
| ·13 | 15·067 | 15·077 | 15·087 | 15·097 | 15·103 | 15·118 | 15·129 | 15·139 | 15·150 | 15·160 | ·28 | 23·867 | 23·883 | 23·899 | 23·914 | 23·930 | 23·946 | 23·961 | 23·977 | 23·992 | 24·008 |
| ·14 | 15·170 | 15·181 | 15·191 | 15·202 | 15·213 | 15·224 | 15·234 | 15·244 | 15·255 | 15·265 | ·27 | 24·024 | 24·039 | 24·054 | 24·070 | 24·085 | 24·101 | 24·116 | 24·132 | 24·147 | 24·162 |
| ·15 | 15·276 | 15·286 | 15·297 | 15·307 | 15·317 | 15·328 | 15·338 | 15·349 | 15·360 | 15·371 | ·23 | 24·177 | 24·192 | 24·207 | 24·221 | 24·236 | 24·251 | 24·266 | 24·280 | 24·294 | 24·309 |
| ·16 | 15·382 | 15·392 | 15·403 | 15·414 | 15·425 | 15·435 | 15·446 | 15·457 | 15·467 | 15·478 | ·25 | 24·324 | 24·339 | 24·353 | 24·368 | 24·383 | 24·398 | 24·413 | 24·427 | 24·442 | 24·457 |
| ·17 | 15·489 | 15·499 | 15·510 | 15·521 | 15·532 | 15·543 | 15·554 | 15·565 | 15·575 | 15·586 | ·24 | 24·471 | 24·485 | 24·498 | 24·512 | 24·526 | 24·540 | 24·555 | 24·569 | 24·583 | 24·597 |
| ·18 | 15·597 | 15·607 | 15·618 | 15·629 | 15·640 | 15·651 | 15·662 | 15·673 | 15·685 | 15·696 | ·23 | 24·611 | 24·625 | 24·639 | 24·653 | 24·668 | 24·682 | 24·696 | 24·709 | 24·723 | 24·737 |
| ·19 | 15·708 | 15·719 | 15·730 | 15·741 | 15·752 | 15·763 | 15·774 | 15·785 | 15·793 | 15·807 | ·22 | 24·751 | 24·765 | 24·779 | 24·792 | 24·806 | 24·820 | 24·833 | 24·846 | 24·860 | 24·874 |
| ·20 | 15·818 | 15·829 | 15·840 | 15·852 | 15·853 | 15·874 | 15·886 | 15·897 | 15·909 | 15·920 | ·21 | 24·887 | 24·900 | 24·913 | 24·927 | 24·940 | 24·954 | 24·967 | 24·981 | 24·995 | 25·008 |
| ·21 | 15·932 | 15·943 | 15·955 | 15·957 | 16·978 | 15·990 | 16·001 | 16·013 | 16·025 | 16·037 | ·20 | 25·022 | 25·035 | 25·049 | 25·062 | 25·075 | 25·088 | 25·101 | 25·114 | 25·127 | 25·140 |
| ·22 | 16·048 | 16·060 | 16·072 | 16·083 | 16·095 | 16·106 | 16·118 | 16·129 | 16·141 | 16·153 | ·19 | 25·153 | 25·166 | 25·179 | 25·192 | 25·205 | 25·218 | 25·231 | 25·244 | 25·257 | 25·270 |
| ·23 | 16·165 | 16·177 | 16·189 | 16·202 | 16·214 | 16·226 | 16·239 | 16·251 | 16·263 | 16·275 | ·18 | 25·233 | 25·296 | 25·309 | 25·322 | 25·335 | 25·348 | 25·361 | 25·374 | 25·387 | 25·399 |
| ·24 | 16·237 | 16·299 | 16·311 | 16·323 | 16·336 | 16·348 | 16·360 | 16·372 | 16·385 | 16·397 | ·17 | 25·412 | 25·425 | 25·437 | 25·450 | 25·463 | 25·476 | 25·488 | 25·500 | 25·513 | 25·525 |
| ·25 | 16·409 | 16·422 | 16·435 | 16·447 | 16·460 | 16·473 | 16·486 | 16·499 | 16·511 | 16·524 | ·16 | 25·533 | 25·551 | 25·563 | 25·576 | 25·589 | 25·601 | 25·614 | 25·627 | 25·640 | 25·652 |
| ·26 | 16·537 | 16·550 | 16·563 | 16·576 | 16·589 | 16·602 | 16·615 | 16·627 | 16·640 | 16·653 | ·15 | 25·655 | 25·677 | 25·689 | 25·702 | 25·714 | 25·727 | 25·739 | 25·751 | 25·764 | 25·776 |
| ·27 | 16·666 | 16·679 | 16·692 | 16·706 | 16·719 | 16·733 | 16·746 | 16·760 | 16·774 | 16·788 | ·14 | 25·789 | 25·801 | 25·814 | 25·826 | 25·838 | 25·851 | 25·863 | 25·875 | 25·887 | 25·900 |
| ·28 | 16·802 | 16·816 | 16·829 | 16·843 | 16·856 | 16·870 | 16·884 | 16·897 | 16·911 | 16·925 | ·13 | 25·912 | 25·924 | 25·936 | 25·949 | 25·962 | 25·974 | 25·987 | 25·999 | 26·011 | 26·023 |
| ·29 | 16·938 | 16·952 | 16·967 | 16·982 | 16·997 | 17·011 | 17·026 | 17·041 | 17·055 | 17·070 | ·12 | 26·035 | 26·047 | 26·059 | 26·071 | 26·083 | 26·095 | 26·107 | 26·120 | 26·132 | 26·144 |
| ·30 | 17·084 | 17·099 | 17·114 | 17·128 | 17·143 | 17·157 | 17·172 | 17·187 | 17·202 | 17·217 | ·11 | 26·156 | 26·168 | 26·180 | 26·193 | 26·205 | 26·217 | 26·229 | 26·241 | 26·253 | 26·265 |
| ·31 | 17·232 | 17·248 | 17·263 | 17·279 | 17·295 | 17·311 | 17·326 | 17·342 | 17·358 | 17·373 | ·10 | 26·277 | 26·289 | 26·301 | 26·313 | 26·325 | 26·337 | 26·349 | 26·361 | 26·373 | 26·385 |
| ·32 | 17·389 | 17·405 | 17·421 | 17·437 | 17·453 | 17·469 | 17·485 | 17·501 | 17·518 | 17·535 | ·09 | 26·397 | 26·409 | 26·420 | 26·432 | 26·444 | 26·456 | 26·468 | 26·479 | 26·491 | 26·503 |
| ·33 | 17·552 | 17·570 | 17·587 | 17·604 | 17·621 | 17·638 | 17·656 | 17·673 | 17·691 | 17·708 | ·08 | 26·515 | 26·527 | 26·539 | 26·551 | 26·563 | 26·575 | 26·586 | 26·598 | 26·610 | 26·622 |
| ·34 | 17·726 | 17·743 | 17·760 | 17·779 | 17·793 | 17·817 | 17·836 | 17·855 | 17·874 | 17·894 | ·07 | 26·634 | 26·646 | 26·657 | 26·669 | 26·681 | 26·693 | 26·705 | 26·716 | 26·728 | 26·740 |
| ·35 | 17·913 | 17·932 | 17·951 | 17·971 | 17·990 | 18·000 | 18·029 | 18·049 | 18·070 | 18·091 | ·06 | 26·752 | 26·763 | 26·774 | 26·786 | 26·798 | 26·810 | 26·822 | 26·833 | 26·845 | 26·857 |
| ·36 | 18·113 | 18·134 | 18·155 | 18·177 | 18·188 | 18·219 | 18·241 | 18·263 | 18·285 | 18·303 | ·05 | 26·869 | 26·881 | 26·892 | 26·904 | 26·916 | 26·928 | 26·939 | 26·950 | 26·962 | 26·974 |
| ·37 | 18·330 | 18·355 | 18·380 | 18·404 | 18·429 | 18·454 | 18·479 | 18·504 | 18·529 | 18·554 | ·04 | 26·985 | 26·997 | 27·009 | 27·021 | 27·032 | 27·044 | 27·056 | 27·067 | 27·079 | 27·091 |
| ·38 | 18·579 | 18·607 | 18·637 | 18·663 | 18·695 | 18·724 | 18·754 | 18·783 | 18·812 | 18·841 | ·03 | 27·102 | 27·113 | 27·125 | 27·137 | 27·148 | 27·160 | 27·172 | 27·184 | 27·195 | 27·207 |
| ·39 | 18·871 | 18·907 | 18·943 | 18·979 | 19·015 | 19·050 | 19·086 | 19·123 | 19·162 | 19·203 | ·02 | 27·212 | 27·230 | 27·242 | 27·254 | 27·265 | 27·277 | 27·288 | 27·299 | 27·311 | 27·323 |
| ·40 | 19·254 | 19·299 | 19·345 | 19·391 | 19·438 | 19·501 | 19·565 | 19·623 | 19·692 | 19·793 | ·01 | 27·334 | 27·346 | 27·358 | 27·369 | 27·381 | 27·393 | 27·404 | 27·416 | 27·428 | 27·439 |
| ·41 | 19·916 | 20·084 | 20·282 | ... | ... | ... | ... | ... | ... | ... | ·00 | 27·451 | 27·462 | 27·473 | 27·485 | 27·497 | 27·508 | 27·520 | 27·532 | 27·543 | 27·555 |

**Maximum equation ·41171** for Anomaly of 20·2817 days.

# TABLE IX—(*h*)—For Tithis.

## Sun's Anomaly and Equation. Surya Siddhanta.

| Ghatikas. | Eqn. + | 50 d. | g. | p. | 40 d. | g. | p. | 30 d. | g. | p. | 20 d. | g. | p. | 10 d. | g. | p. | 0 d. | g. | p. |
|---|---|---|---|---|---|---|---|---|---|---|---|---|---|---|---|---|---|---|---|
| Ghatikas. | 10 | ... | | | 359 | 13 | 0 | 0 | 46 | 30 | 4 | 47 | 0 | 8 | 1 | 10 | 10 | 38 | 50 |
| ,, | 9 | 13 | 6 | 20 | 15 | 21 | 0 | 17 | 23 | 50 | 19 | 22 | 30 | 21 | 6 | 10 | 22 | 51 | 0 |
| ,, | 8 | 24 | 29 | 50 | 26 | 5 | 0 | 27 | 36 | 10 | 29 | 3 | 0 | 30 | 30 | 20 | 31 | 54 | 10 |
| ,, | 7 | 33 | 14 | 30 | 34 | 32 | 50 | 35 | 49 | 30 | 37 | 5 | 30 | 38 | 21 | 30 | 39 | 31 | 0 |
| ,, | 6 | 40 | 42 | 50 | 41 | 54 | 0 | 43 | 2 | 50 | 44 | 10 | 30 | 45 | 15 | 50 | 46 | 23 | 20 |
| ,, | 5 | 47 | 27 | 30 | 48 | 32 | 0 | 49 | 34 | 20 | 50 | 37 | 40 | 51 | 39 | 20 | 52 | 41 | 0 |
| ,, | 4 | 53 | 43 | 10 | 54 | 40 | 40 | 55 | 40 | 30 | 56 | 41 | 0 | 57 | 40 | 10 | 58 | 37 | 0 |
| ,, | 3 | 59 | 35 | 0 | 60 | 33 | 0 | 61 | 31 | 10 | 62 | 28 | 30 | 63 | 23 | 30 | 64 | 19 | 50 |
| ,, | 2 | 65 | 16 | 0 | 66 | 12 | 20 | 67 | 5 | 30 | 68 | 0 | 30 | 68 | 56 | 10 | 69 | 50 | 40 |
| ,, | 1 | 70 | 45 | 40 | 71 | 38 | 20 | 72 | 32 | 40 | 73 | 27 | 10 | 74 | 21 | 0 | 75 | 13 | 0 |
| ,, | 0 | 76 | 6 | 50 | 77 | 0 | 40 | 77 | 53 | 50 | 78 | 48 | 20 | 79 | 39 | 10 | 80 | 33 | 0 |

| Ghatikas. | Eqn. — | 0 d. | g. | p. | 10 d. | g. | p. | 20 d. | g. | p. | 30 d. | g. | p. | 40 d. | g. | p. | 50 d. | g. | p. |
|---|---|---|---|---|---|---|---|---|---|---|---|---|---|---|---|---|---|---|---|
| Ghatikas. | 0 | 80 | 33 | 0 | 81 | 26 | 50 | 82 | 17 | 40 | 83 | 12 | 10 | 84 | 4 | 40 | 84 | 59 | 10 |
| ,, | 1 | 85 | 56 | 40 | 86 | 45 | 0 | 87 | 38 | 50 | 88 | 33 | 30 | 89 | 27 | 40 | 90 | 20 | 20 |
| ,, | 2 | 91 | 14 | 50 | 92 | 9 | 10 | 93 | 5 | 30 | 94 | 0 | 30 | 94 | 53 | 40 | 95 | 49 | 20 |
| ,, | 3 | 96 | 46 | 10 | 97 | 42 | 30 | 98 | 37 | 30 | 99 | 34 | 20 | 100 | 32 | 20 | 101 | 31 | 0 |
| ,, | 4 | 102 | 29 | 0 | 103 | 25 | 10 | 104 | 25 | 0 | 105 | 24 | 50 | 106 | 25 | 20 | 107 | 22 | 50 |
| ,, | 5 | 108 | 24 | 20 | 109 | 26 | 40 | 110 | 28 | 20 | 111 | 31 | 40 | 112 | 34 | 0 | 113 | 38 | 30 |
| ,, | 6 | 114 | 42 | 40 | 115 | 49 | 40 | 116 | 54 | 50 | 118 | 2 | 30 | 119 | 12 | 0 | 120 | 23 | 10 |
| ,, | 7 | 121 | 36 | 0 | 122 | 44 | 30 | 124 | 0 | 30 | 125 | 15 | 50 | 126 | 33 | 10 | 127 | 51 | 0 |
| ,, | 8 | 129 | 11 | 50 | 130 | 35 | 40 | 132 | 3 | 0 | 133 | 29 | 50 | 135 | 1 | 0 | 136 | 35 | 30 |
| ,, | 9 | 138 | 15 | 0 | 139 | 59 | 50 | 141 | 43 | 30 | 143 | 42 | 10 | 145 | 45 | 0 | 147 | 59 | 10 |
| ,, | 10 | 150 | 27 | 10 | 153 | 4 | 50 | 156 | 19 | 0 | 160 | 19 | 30 | 167 | 8 | 0 | ... | | |

**Maximum equation—10 ghatikas 42 palas** for Anomaly of 171 days 51 ghatikas 50 palas.

| Ghatikas. | Eqn. — | 50 d. | g. | p. | 40 d. | g | p. | 30 d. | g. | p. | 20 d. | g. | p | 10 d. | g. | p. | 0 d. | g. | p. |
|---|---|---|---|---|---|---|---|---|---|---|---|---|---|---|---|---|---|---|---|
| Ghatikas. | 10 | ... | | | 176 | 35 | 30 | 183 | 24 | 20 | 187 | 24 | 30 | 190 | 38 | 30 | 193 | 16 | 20 |
| ,, | 9 | 195 | 44 | 10 | 197 | 58 | 40 | 200 | 1 | 20 | 202 | 0 | 20 | 203 | 43 | 40 | 205 | 28 | 50 |
| ,, | 8 | 207 | 8 | 0 | 208 | 43 | 0 | 210 | 13 | 40 | 211 | 40 | 50 | 213 | 8 | 0 | 214 | 31 | 40 |
| ,, | 7 | 215 | 52 | 10 | 217 | 10 | 20 | 218 | 27 | 20 | 219 | 43 | 0 | 220 | 59 | 30 | 222 | 8 | 40 |
| ,, | 6 | 223 | 20 | 10 | 224 | 31 | 40 | 225 | 40 | 50 | 226 | 48 | 20 | 227 | 53 | 50 | 229 | 1 | 10 |
| ,, | 5 | 230 | 5 | 30 | 231 | 9 | 20 | 232 | 11 | 50 | 233 | 15 | 0 | 234 | 17 | 0 | 235 | 19 | 0 |
| ,, | 4 | 236 | 21 | 0 | 237 | 18 | 10 | 238 | 18 | 20 | 239 | 18 | 30 | 240 | 18 | 0 | 241 | 14 | 40 |
| ,, | 3 | 242 | 12 | 20 | 243 | 11 | 20 | 244 | 9 | 10 | 245 | 6 | 20 | 246 | 1 | 0 | 246 | 57 | 40 |
| ,, | 2 | 247 | 54 | 10 | 248 | 49 | 30 | 249 | 43 | 10 | 250 | 38 | 30 | 251 | 34 | 30 | 252 | 28 | 40 |
| ,, | 1 | 253 | 23 | 20 | 254 | 15 | 40 | 255 | 10 | 30 | 256 | 5 | 20 | 256 | 58 | 50 | 257 | 51 | 50 |
| ,, | 0 | 258 | 44 | 50 | 259 | 38 | 20 | 260 | 31 | 50 | 261 | 26 | 0 | 262 | 16 | 40 | 263 | 10 | 40 |

| Ghatikas. | Eqn. + | 0 d. | g. | p. | 10 d. | g. | p. | 20 d. | g. | p. | 30 d. | g. | p. | 40 d. | g. | p. | 50 d. | g. | p. |
|---|---|---|---|---|---|---|---|---|---|---|---|---|---|---|---|---|---|---|---|
| Ghatikas. | 0 | 263 | 10 | 40 | 264 | 4 | 20 | 264 | 55 | 30 | 265 | 49 | 40 | 266 | 42 | 40 | 267 | 36 | 50 |
| ,, | 1 | 268 | 30 | 20 | 269 | 22 | 50 | 270 | 16 | 20 | 271 | 11 | 10 | 272 | 5 | 20 | 272 | 58 | 10 |
| ,, | 2 | 273 | 53 | 0 | 274 | 47 | 10 | 275 | 43 | 10 | 276 | 38 | 30 | 277 | 31 | 30 | 278 | 27 | 30 |
| ,, | 3 | 279 | 24 | 0 | 280 | 20 | 30 | 281 | 15 | 20 | 282 | 12 | 30 | 283 | 10 | 20 | 284 | 8 | 40 |
| ,, | 4 | 285 | 7 | 0 | 286 | 3 | 30 | 287 | 3 | 10 | 288 | 2 | 40 | 289 | 2 | 50 | 290 | 2 | 40 |
| ,, | 5 | 291 | 2 | 40 | 292 | 4 | 30 | 293 | 6 | 0 | 294 | 9 | 10 | 295 | 11 | 40 | 296 | 16 | 0 |
| ,, | 6 | 297 | 20 | 20 | 298 | 26 | 50 | 299 | 32 | 40 | 300 | 40 | 40 | 301 | 49 | 50 | 303 | 1 | 30 |
| ,, | 7 | 304 | 12 | 20 | 305 | 22 | 10 | 306 | 38 | 0 | 307 | 53 | 40 | 309 | 10 | 40 | 310 | 28 | 50 |
| ,, | 8 | 311 | 49 | 20 | 313 | 13 | 0 | 314 | 40 | 40 | 316 | 7 | 50 | 317 | 38 | 40 | 319 | 13 | 40 |
| ,, | 9 | 320 | 52 | 50 | 322 | 38 | 0 | 324 | 21 | 20 | 326 | 19 | 40 | 328 | 22 | 20 | 330 | 37 | 30 |
| ,, | 10 | 333 | 5 | 10 | 335 | 42 | 30 | 338 | 56 | 30 | 342 | 57 | 20 | 349 | 46 | 0 | ... | | |

**Maximum equation + 10 ghatikas 42 palas** for Anomaly of 354 days 29 ghatikas 36 palas.

# TABLE IX—(*i*)—For Yogas.

## Sun's Anomaly and Equation. Surya Siddhanta.

**The Sun's anomaly in this table is expressed in days of the Indian Solar year.**

| | Eqn. — | 50 palas. d. | g. | 40 palas. d. | g. | 30 palas. d. | g. | 20 palas. d. | g. | 10 palas. d. | g. | 0 pala. d. | g. |
|---|---|---|---|---|---|---|---|---|---|---|---|---|---|
| Ghatikas. | 9 | ... | | ... | | ... | | ... | | 359 | 54 | 1 | 59 |
| „ | 8 | 6 | 3 | 9 | 37 | 12 | 23 | 15 | 7 | 17 | 35 | 19 | 43 |
| „ | 7 | 21 | 48 | 23 | 50 | 25 | 36 | 27 | 26 | 29 | 11 | 30 | 47 |
| „ | 6 | 32 | 26 | 33 | 54 | 35 | 25 | 36 | 57 | 38 | 21 | 39 | 45 |
| „ | 5 | 41 | 11 | 42 | 28 | 43 | 51 | 45 | 2 | 46 | 23 | 47 | 33 |
| „ | 4 | 48 | 51 | 50 | 5 | 51 | 13 | 52 | 28 | 53 | 36 | 54 | 46 |
| „ | 3 | 55 | 59 | 57 | 5 | 58 | 14 | 59 | 23 | 60 | 27 | 61 | 37 |
| „ | 2 | 62 | 44 | 63 | 46 | 64 | 54 | 65 | 54 | 67 | 0 | 68 | [illegible] |
| „ | 1 | 69 | 7 | 70 | 12 | 71 | 17 | 72 | 14 | 73 | 22 | 74 | 21 |
| „ | 0 | 75 | 23 | 76 | 23 | 77 | 27 | 78 | 32 | 79 | 29 | 80 | 33 |

| | Eqn. + | 0 pala. d. | g. | 10 palas. d. | g. | 20 palas. d. | g. | 30 palas. d. | g. | 40 palas. d. | g. | 50 palas. d. | g. |
|---|---|---|---|---|---|---|---|---|---|---|---|---|---|
| Ghatikas. | 0 | 80 | 33 | 81 | 37 | 82 | 35 | 83 | 38 | 84 | 44 | 85 | 45 |
| „ | 1 | 86 | 45 | 87 | 44 | 88 | 49 | 89 | 48 | 90 | 54 | 91 | 58 |
| „ | 2 | 92 | 59 | 94 | 5 | 95 | 10 | 96 | 12 | 97 | 20 | 98 | 26 |
| „ | 3 | 99 | 27 | 100 | 38 | 101 | 43 | 102 | 51 | 104 | 1 | 105 | 7 |
| „ | 4 | 106 | 19 | 107 | 29 | 108 | 37 | 109 | 51 | 111 | 0 | 112 | 15 |
| „ | 5 | 113 | 25 | 114 | 43 | 116 | 3 | 117 | 15 | 118 | 35 | 120 | 2 |
| „ | 6 | 121 | 21 | 122 | 44 | 124 | 8 | 125 | 39 | 127 | 12 | 128 | 39 |
| „ | 7 | 130 | 19 | 132 | 3 | 133 | 39 | 135 | 29 | 137 | 15 | 139 | 18 |
| „ | 8 | 141 | 22 | 143 | 30 | 145 | 58 | 148 | 43 | 151 | 30 | 155 | 1 |
| „ | 9 | 159 | 7 | 166 | 27 | ... | | ... | | ... | | ... | |

**Maximum equation + 9 ghatikas 13 palas** for Anomaly of 171 days 52 ghatikas.

| | Eqn. + | 50 palas. d. | g. | 40 palas d. | g | 30 palas. d. | g. | 20 palas. d | g. | 10 palas. d. | g. | 0 pala. d. | g. |
|---|---|---|---|---|---|---|---|---|---|---|---|---|---|
| Ghatikas. | 9 | ... | | ... | | ... | | ... | | 177 | 16 | 184 | 36 |
| „ | 8 | 188 | 42 | 192 | 13 | 195 | 0 | 197 | 45 | 200 | 13 | 202 | 21 |
| „ | 7 | 204 | 25 | 206 | 27 | 208 | 14 | 210 | 5 | 211 | 41 | 213 | 25 |
| „ | 6 | 215 | 3 | 216 | 31 | 218 | 4 | 219 | 35 | 220 | 59 | 222 | 22 |
| „ | 5 | 223 | 42 | 225 | 5 | 226 | 28 | 227 | 41 | 229 | 1 | 230 | 18 |
| „ | 4 | 231 | 28 | 232 | 42 | 233 | 52 | 235 | 7 | 236 | 15 | 237 | 24 |
| „ | 3 | 238 | 36 | 239 | 42 | 240 | 52 | 242 | 0 | 243 | 5 | 244 | 15 |
| „ | 2 | 245 | 17 | 246 | 24 | 247 | 31 | 248 | 32 | 249 | 38 | 250 | 44 |
| „ | 1 | 251 | 45 | 252 | 50 | 253 | 54 | 254 | 53 | 255 | 59 | 256 | 59 |
| „ | 0 | 258 | 2 | 259 | 0 | 260 | 5 | 261 | 9 | 262 | 7 | 263 | 11 |

| | Eqn. — | 0 pala. d. | g. | 10 palas. d. | g. | 20 palas. d. | g. | 30 palas. d. | g. | 40 palas. d. | g. | 50 palas. d. | g. |
|---|---|---|---|---|---|---|---|---|---|---|---|---|---|
| Ghatikas. | 0 | 263 | 11 | 264 | 14 | 265 | 11 | 266 | 15 | 267 | 20 | 268 | 19 |
| „ | 1 | 269 | 23 | 270 | 21 | 271 | 27 | 272 | 26 | 273 | 31 | 274 | 36 |
| „ | 2 | 275 | 37 | 276 | 43 | 277 | 48 | 278 | 50 | 279 | 58 | 281 | 4 |
| „ | 3 | 282 | 7 | 283 | 16 | 284 | 20 | 285 | 30 | 286 | 39 | 287 | 45 |
| „ | 4 | 288 | 57 | 290 | 9 | 291 | 14 | 292 | 29 | 293 | 37 | 294 | 51 |
| „ | 5 | 296 | 2 | 297 | 20 | 298 | 39 | 299 | 52 | 301 | 14 | 302 | 39 |
| „ | 6 | 303 | 59 | 305 | 22 | 306 | 46 | 308 | 16 | 309 | 49 | 311 | 17 |
| „ | 7 | 312 | 56 | 314 | 42 | 316 | 16 | 318 | 6 | 319 | 53 | 321 | 55 |
| „ | 8 | 324 | 1 | 326 | 8 | 328 | 35 | 331 | 21 | 334 | 8 | 337 | 38 |
| „ | 9 | 341 | 44 | 349 | 4 | ... | | ... | | ... | | ... | |

**Maximum equation – 9 ghatikas 13 palas** for Anomaly of 354 days 30 ghatikas.

## TABLE IX—(*i*)—For Yogas.

### Moon's Anomaly and Equation of the Centre in Days, Ghatikas and Palas.

**SURYA SIDDHANTA.**

| Eqn. — Palas. | 0 | 10 | 20 | 30 | 40 | 50 |
|---|---|---|---|---|---|---|
| Ghat. | d. g. p. | d. g. p. | d. g. p. | d. g. p. | d. g. p. | d. g. p. |
| 0 | ... | 0 2 12 | 0 4 25 | 0 6 37 | 0 8 49 | 0 11 2 |
| 1 | 0 13 14 | 0 15 26 | 0 17 38 | 0 19 51 | 0 22 5 | 0 24 19 |
| 2 | 0 26 33 | 0 28 46 | 0 31 0 | 0 33 14 | 0 35 28 | 0 37 42 |
| 3 | 0 39 58 | 0 42 13 | 0 44 29 | 0 46 44 | 0 49 0 | 0 51 16 |
| 4 | 0 53 32 | 0 55 47 | 0 58 3 | 1 0 20 | 1 2 37 | 1 4 54 |
| 5 | 1 7 11 | 1 9 27 | 1 11 44 | 1 14 1 | 1 16 18 | 1 18 37 |
| 6 | 1 20 55 | 1 23 14 | 1 25 33 | 1 27 51 | 1 30 10 | 1 32 28 |
| 7 | 1 34 47 | 1 37 11 | 1 39 34 | 1 42 57 | 1 44 21 | 1 46 44 |
| 8 | 1 49 8 | 1 51 31 | 1 53 55 | 1 56 20 | 1 58 45 | 2 1 10 |
| 9 | 2 3 35 | 2 6 1 | 2 8 26 | 2 10 51 | 2 13 22 | 2 15 52 |
| 10 | 2 18 23 | 2 20 54 | 2 23 25 | 2 25 55 | 2 28 26 | 2 30 57 |
| 11 | 2 33 35 | 2 36 14 | 2 38 52 | 2 41 31 | 2 44 9 | 2 46 47 |
| 12 | 2 49 26 | 2 52 6 | 2 54 47 | 2 57 28 | 3 0 9 | 3 2 49 |
| 13 | 3 5 30 | 3 8 11 | 3 11 3 | 3 13 55 | 3 16 46 | 3 19 38 |
| 14 | 3 22 29 | 3 25 21 | 3 28 21 | 3 31 22 | 3 34 23 | 3 37 23 |
| 15 | 3 40 24 | 3 43 24 | 3 46 37 | 3 49 50 | 3 53 3 | 3 56 16 |
| 16 | 3 59 29 | 4 2 42 | 4 6 10 | 4 9 38 | 4 13 6 | 4 16 34 |
| 17 | 4 20 2 | 4 23 38 | 4 27 14 | 4 30 50 | 4 34 26 | 4 38 2 |
| 18 | 4 42 17 | 4 46 32 | 4 50 47 | 4 55 2 | 4 59 17 | 5 4 1 |
| 19 | 5 8 45 | 5 13 29 | 5 18 12 | 5 23 24 | 5 28 37 | 5 33 49 |
| 20 | 5 40 9 | 5 46 29 | 5 52 43 | 6 0 17 | 6 7 44 | 6 17 19 |
| 21 | 6 26 55 | 6 37 37 | 6 59 13 | ... | ... | ... |

**Maximum equation—21 ghat., 23 palas** for Anomaly of 7 days 16 ghatikas 29 palas.

| Eqn. + | 0 | 10 | 20 | 30 | 40 | 50 |
|---|---|---|---|---|---|---|
| Ghat. | d. g. p. | d. g. p. | d. g. p. | d. g. p. | d. g. p. | d. g. p. |
| 0 | 13 46 38 | 13 48 29 | 13 50 20 | 13 52 11 | 13 54 2 | 13 55 53 |
| 1 | 13 57 44 | 13 59 35 | 14 1 26 | 14 3 15 | 14 5 7 | 14 6 59 |
| 2 | 14 8 51 | 14 10 43 | 14 12 35 | 14 14 27 | 14 16 19 | 14 18 13 |
| 3 | 14 20 6 | 14 22 0 | 14 23 54 | 14 25 48 | 14 27 42 | 14 29 36 |
| 4 | 14 31 30 | 14 33 24 | 14 35 17 | 14 37 12 | 14 39 7 | 14 41 2 |
| 5 | 14 42 57 | 14 44 53 | 14 46 48 | 14 48 43 | 14 50 40 | 14 52 37 |
| 6 | 14 54 34 | 14 56 31 | 14 58 28 | 15 0 25 | 15 2 22 | 15 4 19 |
| 7 | 15 6 15 | 15 8 17 | 15 10 19 | 15 12 21 | 15 14 23 | 15 16 25 |
| 8 | 15 18 27 | 15 20 29 | 15 22 30 | 15 24 33 | 15 26 37 | 15 28 40 |
| 9 | 15 30 43 | 15 32 47 | 15 34 50 | 15 36 53 | 15 39 2 | 15 41 11 |
| 10 | 15 43 21 | 15 45 30 | 15 47 39 | 15 49 49 | 15 51 58 | 15 54 7 |
| 11 | 15 56 24 | 15 58 40 | 16 0 57 | 16 3 14 | 16 5 30 | 16 7 47 |
| 12 | 16 10 4 | 16 12 26 | 16 14 48 | 16 17 10 | 16 19 32 | 16 21 54 |
| 13 | 16 24 16 | 16 26 37 | 16 29 4 | 16 31 31 | 16 33 57 | 16 36 24 |
| 14 | 16 38 51 | 16 41 17 | 16 43 56 | 16 46 35 | 16 49 14 | 16 51 53 |
| 15 | 16 54 32 | 16 57 10 | 17 0 1 | 17 2 52 | 17 5 43 | 17 8 34 |
| 16 | 17 11 25 | 17 14 18 | 17 17 20 | 17 20 23 | 17 23 25 | 17 26 27 |
| 17 | 17 29 30 | 17 32 49 | 17 36 8 | 17 39 27 | 17 42 46 | 17 46 4 |
| 18 | 17 49 57 | 17 53 50 | 17 57 43 | 18 1 36 | 18 5 29 | 18 9 51 |
| 19 | 18 14 12 | 18 18 34 | 18 22 56 | 18 27 47 | 18 32 38 | 18 37 29 |
| 20 | 18 43 27 | 18 49 25 | 18 55 23 | 19 2 30 | 19 9 36 | 19 18 50 |
| 21 | 19 28 3 | 19 38 23 | 19 59 37 | ... | ... | ... |

**Maximum equation+21 ghat., 23 palas** for Anomaly of 20 days 16 ghatikas 47 palas.

| Eqn. — Palas. | 50 | 40 | 30 | 20 | 10 | 0 |
|---|---|---|---|---|---|---|
| Ghat. | d. g. p. | d. g. p. | d. g. p. | d. g. p. | d. g. p. | d. g. p. |
| 21 | ... | ... | ... | 7 33 39 | 7 54 53 | 8 5 13 |
| 20 | 8 14 26 | 8 23 40 | 8 30 46 | 8 37 53 | 8 43 51 | 8 49 49 |
| 19 | 8 55 47 | 9 0 38 | 9 5 29 | 9 10 20 | 9 14 42 | 9 19 3 |
| 18 | 9 23 25 | 9 27 46 | 9 31 39 | 9 35 32 | 9 39 25 | 9 43 18 |
| 17 | 9 47 12 | 9 50 31 | 9 53 50 | 9 57 9 | 10 0 28 | 10 3 46 |
| 16 | 10 6 48 | 10 9 50 | 10 12 53 | 10 15 55 | 10 18 58 | 10 21 49 |
| 15 | 10 24 40 | 10 27 31 | 10 30 22 | 10 33 13 | 10 36 6 | 10 38 45 |
| 14 | 10 41 24 | 10 44 3 | 10 46 42 | 10 49 21 | 10 51 59 | 10 54 26 |
| 13 | 10 56 53 | 10 59 19 | 11 1 46 | 11 4 12 | 11 6 39 | 11 9 1 |
| 12 | 11 11 23 | 11 13 45 | 11 16 7 | 11 18 29 | 11 20 51 | 11 23 12 |
| 11 | 11 25 29 | 11 27 46 | 11 30 2 | 11 32 19 | 11 34 35 | 11 36 52 |
| 10 | 11 39 9 | 11 41 18 | 11 43 27 | 11 45 37 | 11 47 46 | 11 49 55 |
| 9 | 11 52 5 | 11 54 14 | 11 56 23 | 11 58 26 | 12 0 29 | 12 2 33 |
| 8 | 12 4 36 | 12 6 40 | 12 8 43 | 12 10 46 | 12 12 48 | 12 14 50 |
| 7 | 12 16 52 | 12 18 54 | 12 20 56 | 12 22 58 | 12 25 0 | 12 27 1 |
| 6 | 12 28 58 | 12 30 55 | 12 32 52 | 12 34 49 | 12 36 46 | 12 38 43 |
| 5 | 12 40 40 | 12 42 36 | 12 44 31 | 12 46 27 | 12 48 22 | 12 50 18 |
| 4 | 12 52 13 | 12 54 9 | 12 56 4 | 12 57 59 | 12 59 53 | 13 1 47 |
| 3 | 13 3 41 | 13 5 35 | 13 7 29 | 13 9 23 | 13 11 17 | 13 13 11 |
| 2 | 13 15 4 | 13 16 56 | 13 18 48 | 13 20 40 | 13 22 32 | 13 24 24 |
| 1 | 13 26 16 | 13 28 8 | 13 30 1 | 13 31 52 | 13 33 43 | 13 35 34 |
| 0 | 13 37 25 | 13 39 16 | 13 41 7 | 13 42 58 | 13 44 49 | 13 46 38 |

| Eqn. + | 50 | 40 | 30 | 20 | 10 | 0 |
|---|---|---|---|---|---|---|
| Ghat. | d. g. p. | d. g. p. | d. g. p. | d. g. p. | d. g. p. | d. g. p. |
| 21 | ... | ... | ... | 20 34 3 | 20 55 39 | 21 6 18 |
| 20 | 21 16 42 | 21 25 25 | 21 33 23 | 21 40 53 | 21 47 40 | 21 53 27 |
| 19 | 21 59 27 | 22 4 53 | 22 10 17 | 22 15 4 | 22 20 20 | 22 24 45 |
| 18 | 22 29 28 | 22 34 11 | 22 38 35 | 22 42 37 | 22 46 49 | 22 50 40 |
| 17 | 22 54 52 | 22 58 41 | 23 2 20 | 23 6 8 | 23 9 40 | 23 13 30 |
| 16 | 23 17 6 | 23 20 24 | 23 23 45 | 23 27 21 | 23 30 36 | 23 34 0 |
| 15 | 23 37 24 | 23 40 23 | 23 43 30 | 23 46 37 | 23 49 55 | 23 52 44 |
| 14 | 23 55 56 | 23 59 8 | 24 1 55 | 24 5 0 | 24 7 45 | 24 10 45 |
| 13 | 24 13 45 | 24 16 23 | 24 19 20 | 24 22 15 | 24 24 58 | 24 27 50 |
| 12 | 24 30 30 | 24 33 18 | 24 36 6 | 24 38 35 | 24 41 23 | 24 44 15 |
| 11 | 24 46 45 | 24 49 23 | 24 52 11 | 24 54 35 | 24 57 8 | 24 59 42 |
| 10 | 25 2 25 | 25 5 8 | 25 7 26 | 25 10 0 | 25 12 18 | 25 14 54 |
| 9 | 25 17 30 | 25 19 53 | 25 22 37 | 25 24 53 | 25 27 26 | 25 29 49 |
| 8 | 25 32 4 | 25 34 28 | 25 36 51 | 25 39 27 | 25 42 1 | 25 44 9 |
| 7 | 25 46 37 | 25 49 7 | 25 51 19 | 25 53 43 | 25 56 5 | 25 58 39 |
| 6 | 26 0 58 | 26 3 5 | 26 5 31 | 26 7 58 | 26 10 10 | 26 12 33 |
| 5 | 26 14 41 | 26 17 5 | 26 19 29 | 26 21 33 | 26 23 55 | 26 26 21 |
| 4 | 26 28 36 | 26 30 50 | 26 33 2 | 26 35 18 | 26 37 28 | 26 39 51 |
| 3 | 26 42 4 | 26 44 13 | 26 46 31 | 26 48 51 | 26 51 3 | 26 53 26 |
| 2 | 26 55 41 | 26 57 51 | 27 0 6 | 27 2 14 | 27 4 29 | 27 6 51 |
| 1 | 27 8 54 | 27 11 12 | 27 13 29 | 27 15 39 | 27 17 54 | 27 20 5 |
| 0 | 27 22 17 | 27 24 21 | 27 26 40 | 27 28 55 | 27 31 6 | 27 33 17 |

# TABLE IX—(j)—For Nakshatras.

## Moon's Anomaly and Moon's Equation of the Centre in days and fractions of a day.

### SURYA SIDDHANTA.

| Eqn. | 0 | 1 | 2 | 3 | 4 | 5 | 6 | 7 | 8 | 9 | Eqn. | 9 | 8 | 7 | 6 | 5 | 4 | 3 | 2 | 1 | 0 |
|---|---|---|---|---|---|---|---|---|---|---|---|---|---|---|---|---|---|---|---|---|---|
| — | Moon's Anomaly in days. | | | | | | | | | | — | Moon's Anomaly in days. | | | | | | | | | |
| ·00 | ... | ·012 | ·025 | ·037 | ·050 | ·062 | ·074 | ·086 | ·098 | ·110 | ·38 | ... | ... | ... | ... | ... | ... | 7·275 | 7·582 | 7·701 | 7·812 |
| ·01 | ·123 | ·136 | ·148 | ·161 | ·173 | ·186 | ·199 | ·211 | ·223 | ·236 | ·37 | 7·895 | 7·966 | 8·036 | 8·106 | 8·157 | 8·207 | 8·258 | 8·309 | 8·360 | 8·407 |
| ·02 | ·248 | ·261 | ·273 | ·285 | ·297 | ·309 | ·321 | ·324 | ·347 | ·359 | ·36 | 8·447 | 8·487 | 8·527 | 8·566 | 8·604 | 8·643 | 8·680 | 8·716 | 8·749 | 8·788 |
| ·03 | ·371 | ·384 | ·396 | ·409 | ·422 | ·424 | ·445 | ·458 | ·470 | ·482 | ·35 | 8·813 | 8·844 | 8·875 | 8·906 | 8·938 | 8·968 | 8·997 | 9·025 | 9·051 | 9·077 |
| ·04 | ·495 | ·508 | ·520 | ·532 | ·545 | ·557 | ·569 | ·582 | ·595 | ·608 | ·34 | 9·103 | 9·130 | 9·156 | 9·184 | 9·211 | 9·238 | 9·261 | 9·285 | 9·309 | 9·333 |
| ·05 | ·620 | ·632 | ·645 | ·657 | ·669 | ·682 | ·695 | ·707 | ·720 | ·732 | ·33 | 9·356 | 9·380 | 9·403 | 9·426 | 9·450 | 9·474 | 9·497 | 9·519 | 9·540 | 9·562 |
| ·06 | ·745 | ·758 | ·771 | ·783 | ·795 | ·807 | ·820 | ·833 | ·846 | ·858 | ·32 | 9·583 | 9·603 | 9·623 | 9·644 | 9·665 | 9·686 | 9·707 | 9·728 | 9·748 | 9·768 |
| ·07 | ·871 | ·883 | ·896 | ·909 | ·921 | ·934 | ·946 | ·959 | ·971 | ·984 | ·31 | 9·789 | 9·809 | 9·827 | 9·846 | 9·866 | 9·882 | 9·901 | 9·921 | 9·941 | 9·958 |
| ·08 | ·996 | 1·009 | 1·022 | 1·034 | 1·047 | 1·060 | 1·073 | 1·086 | 1·099 | 1·112 | ·30 | 9·977 | 9·996 | 10·015 | 10·033 | 10·052 | 10·070 | 10·087 | 10·104 | 10·122 | 10·139 |
| ·09 | 1·124 | 1·136 | 1·149 | 1·162 | 1·174 | 1·187 | 1·200 | 1·213 | 1·226 | 1·239 | ·29 | 10·156 | 10·173 | 10·191 | 10·208 | 10·225 | 10·242 | 10·259 | 10·275 | 10·292 | 10·309 |
| ·10 | 1·252 | 1·265 | 1·278 | 1·291 | 1·304 | 1·316 | 1·329 | 1·242 | 1·355 | 1·367 | ·28 | 10·326 | 10·342 | 10·359 | 10·375 | 10·391 | 10·407 | 10·423 | 10·438 | 10·454 | 10·469 |
| ·11 | 1·381 | 1·294 | 1·407 | 1·420 | 1·433 | 1·446 | 1·459 | 1·472 | 1·485 | 1·498 | ·27 | 10·485 | 10·501 | 10·517 | 10·533 | 10·549 | 10·564 | 10·579 | 10·595 | 10·611 | 10·626 |
| ·12 | 1·510 | 1·523 | 1·537 | 1·550 | 1·562 | 1·575 | 1·589 | 1·602 | 1·615 | 1·629 | ·26 | 10·641 | 10·656 | 10·671 | 10·686 | 10·701 | 10·715 | 10·730 | 10·744 | 10·758 | 10·773 |
| ·13 | 1·643 | 1·656 | 1·669 | 1·682 | 1·695 | 1·708 | 1·721 | 1·735 | 1·748 | 1·761 | ·25 | 10·787 | 10·803 | 10·818 | 10·833 | 10·848 | 10·863 | 10·877 | 10·891 | 10·905 | 10·919 |
| ·14 | 1·775 | 1·788 | 1·801 | 1·814 | 1·827 | 1·840 | 1·853 | 1·867 | 1·880 | 1·893 | ·24 | 10·933 | 10·946 | 10·960 | 10·974 | 10·988 | 11·002 | 11·016 | 11·030 | 11·044 | 11·058 |
| ·15 | 1·907 | 1·921 | 1·935 | 1·948 | 1·962 | 1·976 | 1·989 | 2·003 | 2·017 | 2·031 | ·23 | 11·072 | 11·085 | 11·099 | 11·113 | 11·127 | 11·140 | 11·153 | 11·167 | 11·180 | 11·193 |
| ·16 | 2·044 | 2·057 | 2·070 | 2·084 | 2·097 | 2·111 | 2·125 | 2·138 | 2·151 | 2·166 | ·22 | 11·206 | 11·220 | 11·233 | 11·245 | 11·258 | 11·272 | 11·285 | 11·298 | 11·311 | 11·324 |
| ·17 | 2·180 | 2·194 | 2·208 | 2·222 | 2·235 | 2·248 | 2·263 | 2·277 | 2·291 | 2·305 | ·21 | 11·337 | 11·350 | 11·363 | 11·377 | 11·390 | 11·403 | 11·416 | 11·429 | 11·441 | 11·453 |
| ·18 | 2·319 | 2·334 | 2·347 | 2·361 | 2·376 | 2·390 | 2·403 | 2·416 | 2·430 | 2·444 | ·20 | 11·466 | 11·478 | 11·490 | 11·503 | 11·516 | 11·528 | 11·540 | 11·553 | 11·566 | 11·579 |
| ·19 | 2·457 | 2·471 | 2·485 | 2·500 | 2·515 | 2·530 | 2·545 | 2·559 | 2·573 | 2·588 | ·19 | 11·591 | 11·604 | 11·616 | 11·628 | 11·640 | 11·653 | 11·665 | 11·677 | 11·690 | 11·702 |
| ·20 | 2·603 | 2·617 | 2·631 | 2·646 | 2·660 | 2·674 | 2·689 | 2·704 | 2·718 | 2·732 | ·18 | 11·714 | 11·726 | 11·738 | 11·750 | 11·761 | 11·773 | 11·785 | 11·797 | 11·809 | 11·821 |
| ·21 | 2·747 | 2·762 | 2·776 | 2·791 | 2·806 | 2·821 | 2·835 | 2·850 | 2·865 | 2·880 | ·17 | 11·824 | 11·844 | 11·856 | 11·869 | 11·881 | 11·894 | 11·905 | 11·919 | 11·929 | 11·940 |
| ·22 | 2·896 | 2·912 | 2·928 | 2·943 | 2·958 | 2·973 | 2·988 | 3·003 | 3·018 | 3·033 | ·16 | 11·953 | 11·964 | 11·975 | 11·986 | 11·998 | 12·010 | 12·021 | 12·033 | 12·044 | 12·056 |
| ·23 | 3·047 | 3·062 | 3·078 | 3·093 | 3·108 | 3·124 | 3·141 | 3·157 | 3·172 | 3·188 | ·15 | 12·068 | 12·079 | 12·091 | 12·103 | 12·114 | 12·125 | 12·137 | 12·149 | 12·160 | 12·172 |
| ·24 | 3·205 | 3·221 | 3·236 | 3·251 | 3·267 | 3·283 | 3·298 | 3·315 | 3·331 | 3·347 | ·14 | 12·183 | 12·194 | 12·206 | 12·217 | 12·228 | 12·240 | 12·251 | 12·263 | 12·274 | 12·285 |
| ·25 | 3·362 | 3·379 | 3·395 | 3·411 | 3·427 | 3·444 | 3·461 | 3·477 | 3·494 | 3·511 | ·13 | 12·296 | 12·307 | 12·318 | 12·330 | 12·341 | 12·352 | 12·363 | 12·375 | 12·386 | 12·397 |
| ·26 | 3·527 | 3·544 | 3·561 | 3·578 | 3·595 | 3·612 | 3·629 | 3·646 | 3·663 | 3·680 | ·12 | 12·408 | 12·420 | 12·431 | 12·442 | 12·453 | 12·464 | 12·475 | 12·486 | 12·497 | 12·508 |
| ·27 | 3·696 | 3·713 | 3·730 | 3·748 | 3·765 | 3·783 | 3·801 | 3·819 | 3·837 | 3·854 | ·11 | 12·519 | 12·530 | 12·542 | 12·553 | 12·564 | 12·575 | 12·586 | 12·596 | 12·607 | 12·618 |
| ·28 | 3·871 | 3·889 | 3·908 | 3·926 | 3·944 | 3·961 | 3·979 | 3·997 | 4·015 | 4·033 | ·10 | 12·628 | 12·639 | 12·650 | 12·661 | 12·672 | 12·684 | 12·695 | 12·706 | 12·717 | 12·728 |
| ·29 | 4·052 | 4·071 | 4·090 | 4·109 | 4·129 | 4·148 | 4·167 | 4·186 | 4·205 | 4·224 | ·09 | 12·738 | 12·749 | 12·759 | 12·770 | 12·781 | 12·792 | 12·802 | 12·813 | 12·824 | 12·835 |
| ·30 | 4·243 | 4·262 | 4·282 | 4·301 | 4·320 | 4·341 | 4·362 | 4·381 | 4·403 | 4·424 | ·08 | 12·846 | 12·856 | 12·867 | 12·877 | 12·888 | 12·899 | 12·910 | 12·920 | 12·931 | 12·942 |
| ·31 | 4·445 | 4·464 | 4·485 | 4·506 | 4·527 | 4·547 | 4·569 | 4·590 | 4·610 | 4·631 | ·07 | 12·952 | 12·963 | 12·974 | 12·985 | 12·995 | 13·006 | 13·016 | 13·027 | 13·037 | 13·047 |
| ·32 | 4·655 | 4·678 | 4·700 | 4·722 | 4·745 | 4·768 | 4·790 | 4·812 | 4·834 | 4·857 | ·06 | 13·058 | 13·069 | 13·079 | 13·090 | 13·100 | 13·111 | 13·122 | 13·132 | 13·143 | 13·153 |
| ·33 | 4·881 | 4·904 | 4·928 | 4·951 | 4·977 | 5·004 | 5·030 | 5·055 | 5·080 | 5·106 | ·05 | 13·164 | 13·174 | 13·185 | 13·195 | 13·206 | 13·217 | 13·227 | 13·237 | 13·248 | 13·258 |
| ·34 | 5·131 | 5·156 | 5·181 | 5·206 | 5·232 | 5·262 | 5·291 | 5·320 | 5·349 | 5·378 | ·04 | 13·269 | 13·279 | 13·290 | 13·301 | 13·311 | 13·321 | 13·332 | 13·342 | 13·352 | 13·363 |
| ·35 | 5·407 | 5·436 | 5·464 | 5·492 | 5·522 | 5·554 | 5·588 | 5·622 | 5·657 | 5·691 | ·03 | 13·373 | 13·383 | 13·393 | 13·404 | 13·415 | 13·425 | 13·435 | 13·446 | 13·457 | 13·467 |
| ·36 | 5·725 | 5·759 | 5·793 | 5·827 | 5·867 | 5·908 | 5·949 | 5·990 | 6·032 | 6·074 | ·02 | 13·477 | 13·487 | 13·497 | 13·508 | 13·518 | 13·528 | 13·538 | 13·549 | 13·560 | 13·570 |
| ·37 | 6·116 | 6·165 | 6·218 | 6·271 | 6·322 | 6·373 | 6·431 | 6·500 | 6·570 | 6·642 | ·01 | 13·580 | 13·591 | 13·601 | 13·611 | 13·622 | 13·633 | 13·643 | 13·653 | 13·663 | 13·674 |
| ·38 | 6·723 | 6·826 | 7·055 | 7·275 | ... | ... | ... | ... | ... | ... | ·00 | 13·684 | 13·694 | 13·705 | 13·716 | 13·726 | 13·736 | 13·747 | 13·757 | 13·767 | 13·777 |

**Maximum equation ·382967** for Anomaly of 7·2752 days.

## TABLE IX—(*k*)—For Nakshatras.

### Moon's Anomaly and Moon's Equation of the Centre in days and fractions of a day.

### SURYA SIDDHANTA.

| Eqn. + | 0 | 1 | 2 | 3 | 4 | 5 | 6 | 7 | 8 | 9 |
|---|---|---|---|---|---|---|---|---|---|---|
| | Moon's Anomaly in days. | | | | | | | | | |
| ·00 | 13·777 | 13·787 | 13·797 | 13·807 | 13·818 | 13·828 | 13·839 | 13·849 | 13·860 | 13·870 |
| ·01 | 13·880 | 13·891 | 13·902 | 13·912 | 13·922 | 13·932 | 13·943 | 13·953 | 13·963 | 13·974 |
| ·02 | 13·984 | 13·994 | 14·005 | 14·016 | 14·026 | 14·036 | 14·046 | 14·057 | 14·067 | 14·077 |
| ·03 | 14·087 | 14·098 | 14·108 | 14·119 | 14·129 | 14·139 | 14·150 | 14·161 | 14·171 | 14·181 |
| ·04 | 14·191 | 14·202 | 14·213 | 14·223 | 14·233 | 14·243 | 14·253 | 14·264 | 14·274 | 14·285 |
| ·05 | 14·296 | 14·306 | 14·317 | 14·328 | 14·338 | 14·348 | 14·359 | 14·369 | 14·380 | 14·390 |
| ·06 | 14·401 | 14·411 | 14·422 | 14·433 | 14·444 | 14·454 | 14·464 | 14·475 | 14·486 | 14·496 |
| ·07 | 14·507 | 14·517 | 14·527 | 14·538 | 14·548 | 14·559 | 14·569 | 14·580 | 14·591 | 14·602 |
| ·08 | 14·613 | 14·624 | 14·634 | 14·644 | 14·655 | 14·666 | 14·677 | 14·687 | 14·698 | 14·708 |
| ·09 | 14·719 | 14·730 | 14·741 | 14·752 | 14·762 | 14·773 | 14·784 | 14·795 | 14·805 | 14·816 |
| ·10 | 14·827 | 14·838 | 14·848 | 14·859 | 14·870 | 14·882 | 14·893 | 14·904 | 14·915 | 14·925 |
| ·11 | 14·936 | 14·947 | 14·958 | 14·968 | 14·979 | 14·990 | 15·001 | 15·012 | 15·024 | 15·035 |
| ·12 | 15·046 | 15·057 | 15·068 | 15·079 | 15·089 | 15·100 | 15·112 | 15·123 | 15·134 | 15·145 |
| ·13 | 15·157 | 15·168 | 15·180 | 15·191 | 15·202 | 15·213 | 15·224 | 15·236 | 15·247 | 15·258 |
| ·14 | 15·269 | 15·280 | 15·292 | 15·303 | 15·314 | 15·326 | 15·337 | 15·348 | 15·359 | 15·370 |
| ·15 | 15·382 | 15·394 | 15·405 | 15·417 | 15·428 | 15·439 | 15·450 | 15·462 | 15·474 | 15·486 |
| ·16 | 15·498 | 15·510 | 15·521 | 15·533 | 15·544 | 15·556 | 15·567 | 15·578 | 15·589 | 15·601 |
| ·17 | 15·613 | 15·625 | 15·637 | 15·649 | 15·661 | 15·673 | 15·685 | 15·697 | 15·709 | 15·721 |
| ·18 | 15·733 | 15·745 | 15·757 | 15·769 | 15·781 | 15·793 | 15·804 | 15·816 | 15·828 | 15·840 |
| ·19 | 15·852 | 15·864 | 15·876 | 15·888 | 15·901 | 15·914 | 15·926 | 15·938 | 15·950 | 15·963 |
| ·20 | 15·975 | 15·988 | 16·001 | 16·014 | 16·026 | 16·038 | 16·051 | 16·064 | 16·076 | 16·088 |
| ·21 | 16·101 | 16·113 | 16·125 | 16·138 | 16·151 | 16·164 | 16·177 | 16·191 | 16·204 | 16·217 |
| ·22 | 16·230 | 16·243 | 16·256 | 16·269 | 16·282 | 16·295 | 16·308 | 16·321 | 16·334 | 16·348 |
| ·23 | 16·361 | 16·374 | 16·387 | 16·401 | 16·414 | 16·427 | 16·441 | 16·455 | 16·469 | 16·483 |
| ·24 | 16·497 | 16·511 | 16·525 | 16·539 | 16·553 | 16·566 | 16·580 | 16·594 | 16·608 | 16·621 |
| ·25 | 16·635 | 16·649 | 16·663 | 16·677 | 16·691 | 16·706 | 16·721 | 16·736 | 16·751 | 16·766 |
| ·26 | 16·780 | 16·795 | 16·810 | 16·824 | 16·839 | 16·854 | 16·869 | 16·834 | 16·899 | 16·914 |
| ·27 | 16·929 | 16·944 | 16·959 | 16·975 | 16·990 | 17·005 | 17·021 | 17·037 | 17·053 | 17·068 |
| ·28 | 17·083 | 17·099 | 17·115 | 17·131 | 17·147 | 17·163 | 17·179 | 17·195 | 17·211 | 17·227 |
| ·29 | 17·244 | 17·261 | 17·278 | 17·295 | 17·312 | 17·329 | 17·346 | 17·363 | 17·381 | 17·398 |
| ·30 | 17·415 | 17·432 | 17·450 | 17·467 | 17·484 | 17·502 | 17·520 | 17·538 | 17·558 | 17·577 |
| ·31 | 17·596 | 17·615 | 17·634 | 17·653 | 17·671 | 17·689 | 17·708 | 17·727 | 17·746 | 17·765 |
| ·32 | 17·786 | 17·807 | 17·827 | 17·847 | 17·868 | 17·889 | 17·909 | 17·930 | 17·951 | 17·972 |
| ·33 | 17·993 | 18·014 | 18·035 | 18·057 | 18·080 | 18·104 | 18·128 | 18·151 | 18·175 | 18·199 |
| ·34 | 18·223 | 18·246 | 18·269 | 18·293 | 18·318 | 18·343 | 18·370 | 18·397 | 18·423 | 18·450 |
| ·35 | 18·477 | 18·504 | 18·530 | 18·557 | 18·586 | 18·615 | 18·648 | 18·680 | 18·711 | 18·742 |
| ·36 | 18·775 | 18·808 | 18·840 | 18·872 | 18·910 | 18·948 | 18·987 | 19·027 | 19·067 | 19·107 |
| ·37 | 19·147 | 19·191 | 19·240 | 19·289 | 19·337 | 19·386 | 19·450 | 19·516 | 19·580 | 19·654 |
| ·38 | 19·732 | 19·841 | 19·951 | 20·280 | ... | ... | ... | ... | ... | ... |

**Maximum equation ·382967** for Anomaly of 20·2797 days.

| Eqn. + | 9 | 8 | 7 | 6 | 5 | 4 | 3 | 2 | 1 | 0 |
|---|---|---|---|---|---|---|---|---|---|---|
| | Moon's Anomaly in days. | | | | | | | | | |
| ·38 | ... | ... | ... | ... | ... | ... | 20·280 | 20·589 | 20·710 | 20·822 |
| ·37 | 20·907 | 20·984 | 21·057 | 21·128 | 21·181 | 21·233 | 21·285 | 21·336 | 21·389 | 21·442 |
| ·36 | 21·480 | 21·522 | 21·564 | 21·604 | 21·643 | 21·682 | 21·720 | 21·759 | 21·795 | 21·829 |
| ·35 | 21·865 | 21·896 | 21·930 | 21·963 | 21·996 | 22·031 | 22·062 | 22·091 | 22·119 | 22·147 |
| ·34 | 22·176 | 22·204 | 22·233 | 22·263 | 22·292 | 22·321 | 22·347 | 22·373 | 22·399 | 22·424 |
| ·33 | 22·449 | 22·474 | 22·501 | 22·526 | 22·550 | 22·575 | 22·600 | 22·625 | 22·650 | 22·673 |
| ·32 | 22·696 | 22·719 | 22·742 | 22·765 | 22·787 | 22·809 | 22·832 | 22·855 | 22·877 | 22·899 |
| ·31 | 22·922 | 22·944 | 22·965 | 22·986 | 23·007 | 23·028 | 23·048 | 23·068 | 23·089 | 23·109 |
| ·30 | 23·130 | 23·151 | 23·172 | 23·192 | 23·213 | 23·234 | 23·253 | 23·272 | 23·292 | 23·311 |
| ·29 | 23·329 | 23·348 | 23·367 | 23·386 | 23·405 | 23·425 | 23·444 | 23·463 | 23·482 | 23·501 |
| ·28 | 23·520 | 23·538 | 23·557 | 23·575 | 23·594 | 23·612 | 23·629 | 23·646 | 23·664 | 23·681 |
| ·27 | 23·699 | 23·717 | 23·735 | 23·753 | 23·771 | 23·789 | 23·806 | 23·824 | 23·841 | 23·858 |
| ·26 | 23·875 | 23·892 | 23·909 | 23·926 | 23·942 | 23·959 | 23·977 | 23·993 | 24·010 | 24·027 |
| ·25 | 24·044 | 24·060 | 24·077 | 24·093 | 24·109 | 24·126 | 24·142 | 24·158 | 24·175 | 24·191 |
| ·24 | 24·207 | 24·223 | 24·239 | 24·255 | 24·271 | 24·287 | 24·303 | 24·318 | 24·333 | 24·349 |
| ·23 | 24·366 | 24·382 | 24·397 | 24·413 | 24·430 | 24·446 | 24·461 | 24·476 | 24·492 | 24·507 |
| ·22 | 24·521 | 24·536 | 24·551 | 24·566 | 24·581 | 24·596 | 24·611 | 24·626 | 24·642 | 24·657 |
| ·21 | 24·673 | 24·688 | 24·703 | 24·718 | 24·732 | 24·747 | 24·763 | 24·778 | 24·792 | 24·806 |
| ·20 | 24·821 | 24·836 | 24·850 | 24·865 | 24·880 | 24·894 | 24·908 | 24·923 | 24·937 | 24·951 |
| ·19 | 24·965 | 24·980 | 24·995 | 25·009 | 25·024 | 25·039 | 25·053 | 25·067 | 25·082 | 25·096 |
| ·18 | 25·110 | 25·124 | 25·138 | 25·151 | 25·164 | 25·178 | 25·193 | 25·207 | 25·221 | 25·235 |
| ·17 | 25·249 | 25·263 | 25·277 | 25·291 | 25·305 | 25·319 | 25·332 | 25·346 | 25·360 | 25·374 |
| ·16 | 25·388 | 25·402 | 25·416 | 25·429 | 25·443 | 25·457 | 25·470 | 25·484 | 25·497 | 25·511 |
| ·15 | 25·524 | 25·537 | 25·551 | 25·565 | 25·578 | 25·591 | 25·605 | 25·619 | 25·633 | 25·647 |
| ·14 | 25·660 | 25·673 | 25·687 | 25·700 | 25·713 | 25·726 | 25·739 | 25·752 | 25·766 | 25·779 |
| ·13 | 25·792 | 25·805 | 25·819 | 25·833 | 25·846 | 25·859 | 25·873 | 25·886 | 25·898 | 25·911 |
| ·12 | 25·924 | 25·938 | 25·951 | 25·965 | 25·979 | 25·992 | 26·005 | 26·018 | 26·031 | 26·044 |
| ·11 | 26·056 | 26·069 | 26·082 | 26·095 | 26·108 | 26·121 | 26·134 | 26·147 | 26·160 | 26·173 |
| ·10 | 26·186 | 26·199 | 26·212 | 26·225 | 26·238 | 26·250 | 26·263 | 26·276 | 26·289 | 26·302 |
| ·09 | 26·315 | 26·328 | 26·341 | 26·354 | 26·367 | 26·379 | 26·392 | 26·405 | 26·418 | 26·430 |
| ·08 | 26·443 | 26·455 | 26·468 | 26·480 | 26·493 | 26·507 | 26·519 | 26·531 | 26·544 | 26·557 |
| ·07 | 26·570 | 26·583 | 26·595 | 26·608 | 26·620 | 26·633 | 26·645 | 26·658 | 26·670 | 26·683 |
| ·06 | 26·696 | 26·709 | 26·721 | 26·734 | 26·747 | 26·759 | 26·771 | 26·783 | 26·796 | 26·808 |
| ·05 | 26·821 | 26·834 | 26·847 | 26·859 | 26·872 | 26·884 | 26·896 | 26·909 | 26·922 | 26·934 |
| ·04 | 26·946 | 26·959 | 26·972 | 26·984 | 26·996 | 27·009 | 27·021 | 27·033 | 27·046 | 27·060 |
| ·03 | 27·072 | 27·084 | 27·096 | 27·108 | 27·120 | 27·132 | 27·145 | 27·158 | 27·171 | 27·183 |
| ·02 | 27·195 | 27·207 | 27·219 | 27·232 | 27·245 | 27·257 | 27·269 | 27·282 | 27·294 | 27·306 |
| ·01 | 27·318 | 27·331 | 27·343 | 27·355 | 27·368 | 27·381 | 27·393 | 27·406 | 27·418 | 27·431 |
| ·00 | 27·443 | 27·455 | 27·467 | 27·480 | 27·492 | 27·504 | 27·517 | 27·529 | 27·542 | 27·555 |

# TABLE IX—(*l*)—For Nakshatras.

## Moon's Anomaly and Equation of the Centre in Days, Ghatikas and Palas.

### SURYA SIDDHANTA.

| Eqn. Palas. — Ghat. | 0 d. g. p. | 10 d. g. p. | 20 d. g. p. | 30 d. g. p. | 40 d. g. p. | 50 d. g. p. |
|---|---|---|---|---|---|---|
| 0 | ... | 2 5 | 4 10 | 6 10 | 8 15 | 10 15 |
| 1 | 12 25 | 14 30 | 16 30 | 18 35 | 20 40 | 22 40 |
| 2 | 24 50 | 26 45 | 28 50 | 31 0 | 33 0 | 35 10 |
| 3 | 37 10 | 39 20 | 41 20 | 43 25 | 45 35 | 47 40 |
| 4 | 49 45 | 51 45 | 53 55 | 56 5 | 58 5 | 1 0 10 |
| 5 | 1 2 20 | 1 4 30 | 1 6 35 | 1 8 45 | 1 10 45 | 1 13 0 |
| 6 | 1 15 5 | 1 17 20 | 1 19 20 | 1 21 35 | 1 23 45 | 1 25 55 |
| 7 | 1 28 5 | 1 30 10 | 1 32 20 | 1 34 30 | 1 36 45 | 1 38 55 |
| 8 | 1 41 5 | 1 43 20 | 1 45 35 | 1 47 55 | 1 50 0 | 1 52 10 |
| 9 | 1 54 25 | 1 56 45 | 1 59 0 | 2 1 15 | 2 3 30 | 2 5 45 |
| 10 | 2 8 5 | 2 10 15 | 2 12 35 | 2 14 55 | 2 17 20 | 2 19 35 |
| 11 | 2 21 55 | 2 24 15 | 2 26 35 | 2 28 55 | 2 31 15 | 2 33 40 |
| 12 | 2 36 10 | 2 38 35 | 2 40 55 | 2 43 20 | 2 45 50 | 2 48 20 |
| 13 | 2 50 45 | 2 53 15 | 2 55 50 | 2 58 25 | 3 0 55 | 3 3 20 |
| 14 | 3 5 55 | 3 8 30 | 3 11 5 | 3 13 50 | 3 15 50 | 3 19 25 |
| 15 | 3 21 45 | 3 24 25 | 3 27 10 | 3 30 0 | 3 32 55 | 3 35 35 |
| 16 | 3 38 25 | 3 41 10 | 3 44 5 | 3 47 0 | 3 50 0 | 3 52 55 |
| 17 | 3 55 55 | 3 58 50 | 4 1 50 | 4 5 5 | 4 8 15 | 4 11 25 |
| 18 | 4 13 30 | 4 17 50 | 4 21 5 | 4 24 35 | 4 28 0 | 4 31 30 |
| 19 | 4 35 0 | 4 38 30 | 4 42 15 | 4 46 5 | 4 49 50 | 4 53 35 |
| 20 | 4 57 35 | 5 1 54 | 5 6 10 | 5 10 25 | 5 14 40 | 5 19 0 |
| 21 | 5 24 25 | 5 29 15 | 5 34 15 | 5 40 0 | 5 45 45 | 5 51 55 |
| 22 | 5 58 40 | 65 20. | 6 13 45 | 6 22 25 | 6 33 35 | 6 47 0 |
| 23 | 7 16 30 | ... | ... | ... | ... | ... |

**Maximum equation 22 ghat 59 palas** for Anomaly of 17 days 16 ghatikas 31 palas.

| Eqn. Palas. + Ghat. | 0 d. g. p. | 10 d. g. p. | 20 d. g. p. | 30 d. g. p. | 40 d. g. p. | 50 d. g. p. |
|---|---|---|---|---|---|---|
| 0 | ... | 13 48 20 | 13 50 5 | 13 51 50 | 13 53 30 | 13 55 15 |
| 1 | 13 57 0 | 13 58 40 | 14 0 25 | 14 2 10 | 14 3 55 | 14 5 35 |
| 2 | 14 7 20 | 14 9 5 | 14 10 50 | 14 12 35 | 14 14 15 | 14 16 0 |
| 3 | 14 17 45 | 14 19 35 | 14 21 15 | 14 23 0 | 14 24 45 | 14 26 35 |
| 4 | 14 28 20 | 14 30 0 | 14 31 45 | 14 33 35 | 14 35 20 | 14 37 5 |
| 5 | 14 38 50 | 14 40 40 | 14 42 25 | 14 44 20 | 14 46 0 | 14 47 50 |
| 6 | 14 49 35 | 14 51 25 | 14 53 15 | 14 55 5 | 14 56 55 | 14 58 40 |
| 7 | 15 0 35 | 15 2 25 | 15 4 15 | 15 6 0 | 15 7 55 | 15 9 45 |
| 8 | 15 11 40 | 15 13 30 | 15 15 25 | 15 17 20 | 15 19 5 | 15 21 0 |
| 9 | 15 22 55 | 15 24 50 | 15 26 45 | 15 28 40 | 15 30 40 | 15 32 30 |
| 10 | 15 34 30 | 15 36 25 | 15 38 20 | 15 40 25 | 15 42 25 | 15 44 20 |
| 11 | 15 46 20 | 15 48 20 | 15 50 20 | 15 52 25 | 15 54 25 | 15 56 25 |
| 12 | 15 58 30 | 16 0 40 | 16 2 40 | 16 4 45 | 16 6 50 | 16 9 0 |
| 13 | 16 11 15 | 16 13 15 | 16 15 35 | 16 17 45 | 16 19 55 | 16 22 0 |
| 14 | 16 24 20 | 16 26 35 | 16 28 55 | 16 31 15 | 16 33 35 | 16 35 50 |
| 15 | 16 38 5 | 16 40 25 | 16 42 50 | 16 45 25 | 16 47 50 | 16 50 20 |
| 16 | 16 52 40 | 16 55 10 | 16 57 45 | 17 0 20 | 17 2 55 | 17 5 35 |
| 17 | 17 8 10 | 17 10 55 | 17 13 30 | 17 16 25 | 17 19 15 | 17 22 0 |
| 18 | 17 24 55 | 17 27 50 | 17 30 45 | 17 33 50 | 17 36 55 | 17 40 10 |
| 19 | 17 43 20 | 17 46 25 | 17 49 55 | 17 53 20 | 17 56 50 | 18 0 10 |
| 20 | 18 3 50 | 18 7 50 | 18 11 45 | 18 15 45 | 18 19 40 | 18 24 0 |
| 21 | 18 28 25 | 18 33 10 | 18 37 55 | 18 43 20 | 18 48 40 | 18 54 25 |
| 22 | 19 0 55 | 19 7 20 | 19 15 20 | 19 23 35 | 19 34 25 | 19 47 30 |
| 23 | 20 16 50 | ... | ... | ... | ... | ... |

**Maximum equation 22 ghat 59 palas** for Anomaly of 20 days 16 ghatikas 47 palas.

| Eqn. Palas. — Ghat. | 50 d. g. p. | 40 d. g. p. | 30 d. g. p. | 20 d. g. p. | 10 d. g. p. | 0 d. g. p. |
|---|---|---|---|---|---|---|
| 23 | ... | ... | ... | ... | ... | 7 16 30 |
| 22 | 7 46 0 | 7 58 55 | 8 9 40 | 8 16 55 | 8 25 50 | 8 32 30 |
| 21 | 8 38 50 | 8 44 30 | 8 49 50 | 8 55 25 | 9 0 10 | 9 4 40 |
| 20 | 9 9 5 | 9 13 40 | 9 17 30 | 9 21 30 | 9 25 25 | 9 29 30 |
| 19 | 9 33 5 | 9 36 25 | 9 39 55 | 9 43 25 | 9 46 55 | 9 49 55 |
| 18 | 9 53 5 | 9 56 25 | 9 59 25 | 10 2 30 | 10 5 25 | 10 8 20 |
| 17 | 10 11 20 | 10 14 5 | 10 16 50 | 10 19 45 | 10 22 25 | 10 25 5 |
| 16 | 10 27 40 | 10 30 20 | 10 32 55 | 10 35 30 | 10 38 5 | 10 40 35 |
| 15 | 10 43 0 | 10 45 25 | 10 47 55 | 10 50 25 | 10 52 50 | 10 55 10 |
| 14 | 10 57 25 | 10 59 45 | 11 2 5 | 11 4 25 | 11 6 40 | 11 8 55 |
| 13 | 11 11 15 | 11 13 20 | 11 15 30 | 11 17 40 | 11 20 0 | 11 22 0 |
| 12 | 11 24 15 | 11 26 25 | 11 28 30 | 11 30 35 | 11 32 30 | 11 34 45 |
| 11 | 11 36 50 | 11 38 55 | 11 40 55 | 11 42 55 | 11 44 55 | 11 46 55 |
| 10 | 11 48 55 | 11 50 50 | 11 52 55 | 11 54 55 | 11 56 55 | 11 58 45 |
| 9 | 12 0 45 | 12 2 35 | 12 4 35 | 12 6 30 | 12 8 25 | 12 10 20 |
| 8 | 12 12 15 | 12 14 10 | 12 16 0 | 12 17 55 | 12 19 45 | 12 21 40 |
| 7 | 12 23 30 | 11 25 25 | 12 27 15 | 12 29 5 | 12 30 50 | 12 32 45 |
| 6 | 12 34 35 | 12 36 25 | 12 38 10 | 12 40 0 | 12 41 50 | 12 43 40 |
| 5 | 12 45 25 | 22 47 20 | 12 49 0 | 12 50 50 | 12 52 35 | 12 54 25 |
| 4 | 12 56 10 | 12 57 55 | 12 59 45 | 13 1 30 | 13 3 15 | 13 4 55 |
| 3 | 13 6 40 | 13 8 30 | 13 10 15 | 13 12 0 | 13 13 40 | 13 15 30 |
| 2 | 13 17 20 | 13 19 0 | 13 20 0 | 13 22 25 | 13 24 10 | 13 25 55 |
| 1 | 13 27 40 | 13 29 20 | 13 31 5 | 13 32 50 | 13 34 35 | 13 36 15 |
| 0 | 13 38 5 | 13 39 45 | 13 41 30 | 13 43 15 | 13 44 55 | ... |

| Eqn. Palas. + Ghat. | 50 d. g. p. | 40 d. g. p. | 30 d. g. p. | 20 d. g. p. | 10 d. g. p. | 0 d. g. p. |
|---|---|---|---|---|---|---|
| 23 | ... | ... | ... | ... | ... | 20 16 50 |
| 22 | 20 46 40 | 20 59 45 | 21 10 50 | 21 19 30 | 21 27 50 | 21 34 40 |
| 21 | 21 41 25 | 21 47 30 | 21 53 5 | 21 59 5 | 22 4 0 | 22 8 50 |
| 20 | 22 13 40 | 22 18 40 | 22 22 50 | 22 27 0 | 22 31 15 | 22 35 45 |
| 19 | 22 39 40 | 22 43 25 | 22 47 5 | 22 51 5 | 22 54 45 | 22 58 15 |
| 18 | 23 1 45 | 23 5 20 | 23 8 45 | 23 12 10 | 23 15 25 | 23 18 40 |
| 17 | 23 21 50 | 23 25 0 | 23 29 20 | 23 31 25 | 23 34 25 | 23 37 15 |
| 16 | 23 40 15 | 23 43 20 | 23 46 15 | 23 49 15 | 23 52 5 | 23 54 55 |
| 15 | 23 57 35 | 24 0 25 | 24 3 15 | 24 6 0 | 24 8 50 | 24 11 30 |
| 14 | 24 14 10 | 24 16 50 | 24 19 25 | 24 22 5 | 24 24 45 | 24 27 25 |
| 13 | 24 30 0 | 24 32 20 | 24 34 55 | 24 37 25 | 24 40 5 | 24 42 30 |
| 12 | 24 44 50 | 24 47 25 | 24 49 55 | 24 52 25 | 24 54 40 | 24 57 5 |
| 11 | 24 59 35 | 25 1 55 | 25 4 20 | 25 6 40 | 25 9 0 | 25 11 20 |
| 10 | 25 13 40 | 25 15 55 | 25 18 25 | 25 20 40 | 25 23 0 | 25 25 10 |
| 9 | 25 27 35 | 25 28 55 | 25 32 0 | 25 34 15 | 25 36 35 | 25 38 50 |
| 8 | 25 41 5 | 25 43 15 | 25 45 25 | 25 47 45 | 25 49 55 | 25 52 10 |
| 7 | 25 54 20 | 25 56 30 | 25 58 40 | 26 0 55 | 26 3 5 | 26 5 10 |
| 6 | 26 7 25 | 26 9 35 | 26 11 45 | 26 13 55 | 26 16 0 | 26 18 10 |
| 5 | 26 20 20 | 26 22 30 | 26 24 35 | 26 26 40 | 26 28 40 | 26 31 0 |
| 4 | 26 33 5 | 26 35 10 | 26 37 15 | 26 39 25 | 26 41 30 | 26 43 30 |
| 3 | 26 45 35 | 26 47 45 | 26 49 50 | 26 51 55 | 26 53 55 | 26 56 5 |
| 2 | 26 58 10 | 27 0 15 | 27 2 20 | 27 4 25 | 27 6 30 | 27 8 30 |
| 1 | 27 10 40 | 27 12 35 | 27 14 45 | 27 16 50 | 27 18 50 | 27 20 50 |
| 0 | 27 23 0 | 27 25 0 | 27 27 10 | 27 29 5 | 27 31 15 | ... |

## TABLE X—

| Kaliyuga. | Vikrama Era. | Saka Era. | Commencement of Solar year. | | First New-Moon in Solar Year. | Anomaly of first New-Moon. | Christian Era. | Week-day of 1st January. | ☉'s Anom col. 5 / ☾'s Anom col. 6 | | | | + 29·53059 / + 1·976 | | | | + 59·06117 / + 3·952 | | | | + 88·59176 / + 5·928 | | | | + 118·12235 / + 7·904 | | | |
|---|---|---|---|---|---|---|---|---|---|---|---|---|---|---|---|---|---|---|---|---|---|---|---|---|---|---|---|---|
| | | | | | | | | | Vaisakha | | | | Jyeshta | | | | Ashada | | | | Sravana | | | | Bhadrapada | | | |
| | | | Month and day A.D. | Fraction of day. | | | | | Week-day | Month | Day | Fraction | Week-day | Month | Day | Fraction | Week-day | Month | Day | Fraction | Week-day | Month | Day | Fraction | Week-day | Month | Day | Fraction |
| 3101 | 57 | | 13M | ·9831 | 9·9981 | 22·736 | B. C. 1 | 5 | 3 | Mr | 23 | ·98 | 5 | Ap | 22 | ·51 | 7 | My | 22 | ·04 | 1 | Je | 20○ | ·57 | 3 | Jl | 20 | ·10 |
| 3102 | 58 | | 14M | ·2419 | 28·6370 | 20·870 | A.D. 1 | 7 | 2 | Ap | 11 | ·88 | 4 | My | 11 | ·41 | 5 | Je | 9●○ | ·94 | 7 | Jl | 9 | ·47 | 2 | Au | 8 | ·00 |
| 3103 | 59 | | 14M | ·5006 | 17·7453 | 17·027 | 2 | 1 | 7 | Ap | 1 | ·25 | 1 | Ap | 30○ | ·78 | 3 | My | 30 | ·31 | 4 | Je | 28 | ·84 | 6 | Jl | 28 | ·37 |
| 3104 | 60 | | 14M | ·7594 | 6·8536 | 13·184 | 3 | 2 | 4 | Mr | 21 | ·61 | 6 | Ap | 20○ | ·14 | 7 | My | 19 | ·67 | 2 | Je | 18 | ·20 | 5 | Au | 16 | ·27 |
| | | | | | | | | | | | | | | | | | | | | | 3 | Jl | 17 | ·73 | | | | |
| 3105 | 61 | | 14M | ·0181 | 25·4925 | 11·317 | 4 | 3 | 3 | Ap | 8●○ | ·51 | 5 | My | 8 | ·04 | 6 | Je | 6 | ·57 | 1 | Jl | 6 | ·10 | 2 | Au | 4 | ·63 |
| 3106 | 62 | | 14M | ·2769 | 14·6008 | 7·475 | 5 | 5 | 7 | Mr | 28● | ·88 | 2 | Ap | 27 | ·41 | 3 | My | 26 | ·94 | 5 | Je | 25 | ·47 | 7 | Jl | 25 | ·00 |
| 3107 | 63 | | 14M | ·5356 | 3·7091 | 3·633 | 6 | 6 | 5 | Mr | 18 | ·24 | 6 | Ap | 16 | ·77 | 1 | My | 16 | ·31 | 4 | Jl | 14 | ·37 | 5 | Au | 12○ | ·90 |
| | | | | | | | | | | | | | | | | | 2 | Je | 14 | ·84 | | | | | | | | |
| 3108 | 64 | | 14M | ·7944 | 22·3479 | 1·765 | 7 | 7 | 4 | Ap | 6 | ·14 | 5 | My | 5 | ·67 | 7 | Je | 4 | ·20 | 1 | Jl | 3 | ·73 | 3 | Au | 2○ | ·26 |
| 3109 | 65 | | 14M | ·0531 | 11·4562 | 25·477 | 8 | 1 | 1 | Mr | 25 | ·51 | 3 | Ap | 24 | ·04 | 4 | My | 23 | ·57 | 6 | Je | 22 | ·10 | 7 | Jl | 21○ | ·63 |
| 3110 | 66 | | 14M | ·3119 | 0·5645 | 21·634 | 9 | 3 | 5 | Mr | 14 | ·88 | 1 | My | 12 | ·94 | 3 | Je | 11 | ·47 | 5 | Jl | 11● | ·00 | 6 | Au | 9 | ·53 |
| | | | | | | | | | 7 | Ap | 13 | ·41 | | | | | | | | | | | | | | | | |
| 3111 | 67 | | 14M | ·5707 | 19·2034 | 19·767 | 10 | 4 | 4 | Ap | 2 | ·77 | 6 | My | 2 | ·30 | 7 | My | 31○ | ·83 | 2 | Je | 30● | ·37 | 3 | Jl | 29 | ·90 |
| 3112 | 68 | | 14M | ·8294 | 8·3117 | 15·925 | 11 | 5 | 2 | Mr | 23 | ·14 | 3 | Ap | 21 | ·67 | 5 | My | 21○ | ·20 | 6 | Je | 19. | ·73 | 1 | Jl | 19 | ·26 |
| | | | | | | | | | | | | | | | | | | | | | | | | | 2 | Au | 17 | ·79 |
| 3113 | 69 | | 14M | ·0882 | 26·9506 | 14·058 | 12 | 6 | 1 | Ap | 10 | ·04 | 2 | My | 9● | ·57 | 4 | Je | 8 | ·10 | 5 | Jl | 7 | ·63 | 7 | Au | 6 | ·16 |
| 3114 | 70 | | 14M | ·3469 | 16·0589 | 10·215 | 13 | 1 | 5 | Mr | 30○ | ·41 | 6 | Ap | 28● | ·94 | 1 | My | 28 | ·47 | 3 | Je | 27 | ·00 | 4 | Jl | 26 | ·53 |
| 3115 | 71 | | 14M | ·6057 | 5·1672 | 6·373 | 14 | 2 | 2 | Mr | 19○ | ·77 | 4 | Ap | 18● | ·30 | 5 | My | 17 | ·83 | 1 | Jl | 15 | ·89 | 3 | Au | 14 | ·43 |
| | | | | | | | | | | | | | | | | | 7 | Je | 16 | ·36 | | | | | | | | |
| 3116 | 72 | | 14M | ·8644 | 23·8061 | 4·506 | 15 | 3 | 1 | Ap | 7 | ·67 | 3 | My | 7 | ·20 | 4 | Je | 5 | ·73 | 6 | Jl | 5 | ·26 | 7 | Au | 3 | ·79 |
| 3117 | 73 | | 14M | ·1232 | 12·9144 | 0·663 | 16 | 4 | 6 | Mr | 27 | ·04 | 7 | Ap | 25 | ·57 | 2 | My | 25 | ·10 | 3 | Je | 23 | ·63 | 5 | Jl | 23 | ·16 |
| 3118 | 74 | | 14M | ·3820 | 2·0227 | 24·375 | 17 | 6 | 3 | Mr | 16 | ·40 | 4 | Ap | 14 | ·93 | 1 | Je | 13 | ·00 | 2 | Jl | 12○ | ·53 | 4 | Au | 11 | ·06 |
| | | | | | | | | | | | | | 6 | My | 14 | ·47 | | | | | | | | | | | | |
| 3119 | 75 | | 14M | ·6407 | 20·6616 | 22·508 | 18 | 7 | 2 | Ap | 4 | ·30 | 3 | My | 3 | ·83 | 5 | Je | 2 | ·36 | 6 | Jl | 1●○ | ·89 | 1 | Jl | 31 | ·42 |
| 3120 | 76 | | 14M | ·8995 | 9·7699 | 18·665 | 19 | 1 | 6 | Mr | 24 | ·67 | 1 | Ap | 23 | ·20 | 2 | My | 22 | ·73 | 4 | Je | 21●○ | ·26 | 5 | Jl | 20 | ·79 |
| 3121 | 77 | | 14M | ·1582 | 28·4088 | 16·799 | 20 | 2 | 5 | Ap | 11 | ·57 | 7 | My | 11○ | ·10 | 1 | Je | 9● | ·63 | 3 | Jl | 9 | ·16 | 4 | Au | 7 | ·69 |
| 3122 | 78 | | 14M | ·4170 | 17·5171 | 12·956 | 21 | 4 | 2 | Mr | 31 | ·93 | 4 | Ap | 30○ | ·46 | 5 | My | 29 | ·99 | 7 | Je | 28 | ·53 | 2 | Jl | 28 | ·06 |
| 3123 | 79 | | 14M | ·6757 | 6·6254 | 9·113 | 22 | 5 | 7 | Mr | 21 | ·30 | 1 | Ap | 19●○ | 83 | 3 | My | 19 | ·36 | 4 | Je | 17 | ·89 | 7 | Au | 15 | ·95 |
| | | | | | | | | | | | | | | | | | | | | | 6 | Jl | 17 | ·42 | | | | |
| 3124 | 80 | | 14M | ·9345 | 25·2647 | 7·246 | 23 | 6 | 6 | Ap | 9 | ·20 | 7 | My | 8 | ·73 | 2 | Je | 7 | ·26 | 3 | Jl | 6 | ·79 | 5 | Au | 5 | ·32 |
| 3125 | 81 | | 14M | ·1932 | 14·3726 | 3·404 | 24 | 7 | 3 | Mr | 28 | ·57 | 5 | Ap | 27 | ·10 | 6 | My | 26 | ·63 | 1 | Je | 25 | ·16 | 2 | Jl | 24 | ·69 |
| 3126 | 82 | | 14M | ·4520 | 3·4809 | 27·115 | 25 | 2 | 7 | Mr | 17 | ·93 | 2 | Ap | 16 | ·46 | 3 | My | 15 | ·99 | 7 | Jl | 14 | ·05 | 1 | Au | 12○ | ·59 |
| | | | | | | | | | | | | | | | | | 5 | Je | 14 | ·52 | | | | | | | | |
| 3127 | 83 | | 14M | ·7108 | 22·1197 | 25·2 | 26 | 3 | 6 | Ap | 5 | ·83 | 1 | My | 5 | ·36 | 2 | Je | 3 | ·89 | 4 | Jl | 3 | ·42 | 5 | Au | 1○ | ·95 |
| 3128 | 84 | | 14M | ·9695 | 11·2280 | 21·[illegible] | 27 | 4 | 4 | Mr | 26 | ·20 | 5 | Ap | 24 | ·73 | 7 | My | 24 | ·26 | 1 | Je | 22 | ·79 | 3 | Jl | 22● | ·32 |
| 3129 | 85 | | 14M | ·2283 | 0·3363 | 17·563 | 28 | 5 | 1 | Mr | 14 | ·56 | 4 | My | 12 | ·63 | 6 | Je | 11○ | ·16 | 7 | Jl | 10● | ·69 | 2 | Au | 9 | ·22 |
| | | | | | | | | | 3 | Ap | 13 | ·09 | | | | | | | | | | | | | | | | |
| 3130 | 86 | | 14M | ·4870 | 18·9752 | 15·696 | 29 | 7 | 7 | Ap | 2 | ·46 | 1 | My | 1 | ·99 | 3 | My | 31○ | ·52 | 5 | Je | 30 | ·05 | 6 | Jl | 29 | ·58 |
| 3131 | 87 | | 14M | ·7458 | 8·0835 | 11·854 | 30 | 1 | 4 | Mr | 22 | ·83 | 6 | Ap | 21 | ·36 | 7 | My | 20●○ | ·89 | 2 | Je | 19 | ·42 | 3 | Jl | 18 | ·95 |
| | | | | | | | | | | | | | | | | | | | | | | | | | 5 | Au | 17 | ·48 |
| 3132 | 88 | | 15M | ·0045 | 26·7224 | 9·987 | 31 | 2 | 3 | Ap | 10○ | ·73 | 5 | My | 10● | ·26 | 6 | Je | 8 | ·79 | 1 | Jl | 8 | ·32 | 2 | Au | 6 | ·85 |
| 3133 | 89 | | 14M | ·2633 | 15·8307 | 6·144 | 32 | 3 | 1 | Mr | 30○ | ·09 | 2 | Ap | 28● | ·62 | 4 | My | 28 | ·15 | 5 | Je | 26 | ·69 | 7 | Jl | 26 | ·22 |
| 3134 | 90 | | 14M | ·5221 | 4·9390 | 2·301 | 33 | 5 | 5 | Mr | 19○ | ·46 | 6 | Ap | 17 | ·99 | 1 | My | 17 | ·52 | 4 | Jl | 15 | ·58 | 6 | Au | 14 | ·11 |
| | | | | | | | | | | | | | | | | | 3 | Je | 16 | ·05 | | | | | | | | |
| 3135 | 91 | | 14M | ·7808 | 23·5779 | 0·435 | 34 | 6 | 4 | Ap | 7 | ·36 | 5 | My | 6 | ·89 | 7 | Je | 5 | ·42 | 1 | Jl | 4 | ·95 | 3 | Au | 3 | ·48 |
| 3136 | 92 | | 15M | ·0396 | 12·6862 | 24·146 | 35 | 7 | 1 | Mr | 27 | ·73 | 3 | Ap | 26 | ·26 | 4 | My | 25 | ·79 | 6 | Je | 24 | ·32 | 7 | Jl | 23○ | ·85 |
| 3137 | 93 | | 14M | ·2983 | 1·7945 | 20·304 | 36 | 1 | 6 | Mr | 16 | ·09 | 7 | Ap | 14 | ·62 | 3 | Je | 12 | ·68 | 5 | Jl | 12●○ | ·21 | 6 | Au | 10 | ·75 |
| | | | | | | | | | | | | | 2 | My | 14 | ·15 | | | | | | | | | | | | |
| 3138 | 94 | | 14M | ·5571 | 20·4334 | 18·437 | 37 | 3 | 4 | Ap | 3 | ·99 | 6 | My | 3 | ·52 | 1 | Je | 2 | ·05 | 2 | Jl | 1●○ | ·58 | 4 | Jl | 31 | ·11 |
| 3139 | 95 | | 14M | ·8158 | 9·5417 | 14·594 | 38 | 4 | 2 | Mr | 24 | ·36 | 3 | Ap | 22 | ·89 | 5 | My | 22 | ·42 | 6 | Je | 20● | ·95 | 1 | Jl | 20 | ·48 |

## Solar years and New moons from B.C. 1 to A.D. 2000 (Surya Siddhanta).*

| Asvina | | | | Kartika | | | | Margasira | | | | Pausha | | | | A.D. | Magha | | | | A.D. | Phalguna | | | | Chaitra | | | |
|---|---|---|---|---|---|---|---|---|---|---|---|---|---|---|---|---|---|---|---|---|---|---|---|---|---|---|---|---|---|
| + 147·65293 | | | | + 177·18353 | | | | + 206·71411 | | | | + 236·24470 | | | | | + 265·77529 | | | | | + 295·30588 | | | | +324·83647 / +354·36705 | | | |
| + 9·880 | | | | + 11·856 | | | | + 13·832 | | | | + 15·808 | | | | | + 17·784 | | | | | + 19·760 | | | | + 21·736 / + 23·712 | | | |
| Week-day | Month | Day | Fraction | Week-day | Month | Day | Fraction | Week-day | Month | Day | Fraction | Week-day | Month | Day | Fraction | | Week-day | Month | Day | Fraction | | Week-day | Month | Day | Fraction | Week-day | Month | Day | Fraction |
| 4 | Au | 18 | ·63 | 7 | O | 16 | ·69 | 2 | N | 15 | ·23 | Pausha. Kshaya. | | | | | 3 | D | 14○ | ·76 | 1 | 5 | Jr | 13 | ·29 | 6 | F | 11 | ·82 |
| 6 | S | 17 | ·16 | | | | | | | | | | | | | | | | | | | | | | | 1 | Mr | 13 | ·35 |
| 3 | S | 6 | ·53 | 5 | O | 6 | ·06 | 6 | N | 4 | ·59 | 1 | D | 4 | ·12 | 2 | 2 | Jr | 2 | ·65 | | 4 | F | 1 | ·18 | 5 | Mr | 2 | ·71 |
| 7 | Au | 26 | ·90 | 2 | S | 25 | ·43 | 3 | O | 24○ | ·96 | 5 | N | 23● | ·49 | | 7 | D | 23 | ·02 | 3 | 1 | Jr | 21 | ·55 | 3 | F | 20 | ·08 |
| 6 | S | 14 | ·80 | 1 | O | 14○ | ·33 | 2 | N | 12 | ·86 | 4 | D | 12 | ·39 | 4 | 5 | Jr | 10 | ·92 | | 7 | F | 9 | ·45 | 1 | Mr | 9 | ·98 |
| 4 | S | 3 | ·16 | 5 | O | 2○ | ·69 | 7 | N | 1 | ·22 | 1 | N | 30 | ·75 | | 3 | D | 30 | ·29 | 5 | 4 | Jr | 28 | ·82 | 6 | F | 27 | ·35 |
| 1 | Au | 23 | ·53 | 3 | S | 22● | ·06 | 4 | O | 21 | ·59 | 6 | N | 20 | ·12 | | 7 | D | 19 | ·65 | 6 | 2 | Jr | 18 | ·18 | 3 | F | 16○ | ·71 |
| 7 | S | 11● | ·43 | 1 | O | 10 | ·96 | 3 | N | 9 | ·49 | 5 | D | 9 | ·02 | 7 | 6 | Jr | 7 | ·55 | | 1 | F | 6●○ | ·08 | 2 | Mr | 7 | ·61 |
| 4 | Au | 31● | ·79 | 6 | S | 30 | ·33 | 7 | O | 29 | ·86 | 2 | N | 28 | ·39 | | 3 | D | 27 | ·92 | 8 | 5 | Jr | 26●○ | ·45 | 6 | F | 24 | ·98 |
| 2 | Au | 20 | ·16 | 3 | S | 18 | ·69 | 5 | O | 18 | ·22 | 6 | N | 16 | ·75 | | 1 | D | 16 | ·28 | 9 | 2 | Jr | 14● | ·81 | 4 | F | 13 | ·35 |
| 1 | S | 8 | ·06 | 2 | O | 7 | ·59 | 4 | N | 6 | ·12 | 5 | D | 5○ | ·65 | 10 | 7 | Jr | 4 | ·18 | | 1 | F | 2 | ·71 | 3 | Mr | 4 | ·24 |
| 5 | Au | 28 | ·43 | 6 | S | 26 | ·96 | 1 | O | 26 | ·49 | 3 | N | 25●○ | ·02 | | 4 | D | 24 | ·55 | 11 | 6 | Jr | 23 | ·08 | 7 | F | 21 | ·61 |
| 4 | S | 16 | ·32 | 5 | O | 15 | ·85 | 7 | N | 14●○ | ·39 | 1 | D | 13 | ·92 | 12 | 3 | Jr | 12 | ·45 | | 4 | F | 10 | ·98 | 6 | Mr | 11 | ·51 |
| 1 | S | 4 | ·69 | 3 | O | 4 | ·22 | 4 | N | 2 | ·75 | 6 | D | 2 | ·28 | | 7 | D | 31 | ·81 | 13 | 2 | Jr | 30 | ·34 | 3 | F | 28 | ·87 |
| 6 | Au | 25 | ·06 | 7 | S | 23○ | ·59 | 2 | O | 23 | ·12 | 3 | N | 21 | ·65 | | 5 | D | 21 | ·18 | 14 | 6 | Jr | 19 | ·71 | 1 | F | 18 | ·24 |
| 4 | S | 12○ | ·96 | 6 | O | 12 | ·49 | 1 | N | 11 | ·02 | 2 | D | 10 | ·55 | 15 | 4 | Jr | 9 | ·08 | | 5 | F | 7 | ·61 | 7 | Mr | 9○ | ·14 |
| 2 | S | 2●○ | ·32 | 3 | O | 1 | ·85 | 5 | O | 31 | ·38 | 6 | N | 29 | ·91 | | 1 | D | 29 | ·45 | 16 | 2 | Jr | 27 | ·98 | 4 | F | 26 | ·51 |
| 6 | Au | 21● | ·69 | 1 | S | 20 | ·22 | 2 | O | 19 | ·75 | 4 | N | 18 | ·28 | | 5 | D | 17 | ·81 | 17 | 7 | Jr | 16○ | ·34 | 1 | F | 14● | ·87 |
| 5 | S | 9 | ·59 | 7 | O | 9 | ·12 | 1 | N | 7 | ·65 | 3 | D | 7 | ·18 | 18 | 4 | Jr | 5○ | ·71 | | 6 | F | 4 | ·24 | 7 | Mr | 5 | ·77 |
| 2 | Au | 29 | ·95 | 4 | S | 28 | ·49 | 6 | O | 28 | ·02 | 7 | N | 26 | ·55 | | 2 | D | 26○ | ·08 | 19 | 3 | Jr | 24 | ·61 | 5 | F | 23 | ·14 |
| 7 | Au | 19 | ·32 | 3 | O | 17 | ·38 | 4 | N | 15 | ·91 | 6 | D | 15● | ·44 | 20 | 7 | Jr | 13 | ·97 | | 2 | F | 12 | ·51 | 4 | Mr | 13 | ·04 |
| 1 | S | 17 | ·85 | | | | | | | | | | | | | | | | | | | | | | | | | | |
| 6 | S | 6 | ·22 | 7 | O | 5 | ·75 | 2 | N | 4○ | ·28 | 3 | D | 3● | ·81 | 21 | 5 | Jr | 2 | ·34 | | 6 | Jr | 31 | ·87 | 1 | Mr | 2 | ·40 |
| 3 | Au | 26 | ·59 | 5 | S | 25 | ·12 | 6 | O | 24○ | ·65 | 1 | N | 23● | ·18 | | 2 | D | 22 | ·71 | 22 | 4 | Jr | 21 | ·24 | 5 | F | 19 | ·77 |
| 2 | S | 14 | ·48 | 4 | O | 14○ | ·01 | 5 | N | 12 | ·55 | 7 | D | 12 | ·08 | 23 | 1 | Jr | 10 | ·61 | | 3 | F | 9 | ·14 | 4 | Mr | 10 | ·67 |
| 6 | S | 3 | ·85 | 1 | O | 3 | ·38 | 2 | N | 1 | ·91 | 4 | D | 1 | ·44 | | 5 | D | 30 | ·97 | 24 | 7 | Jr | 29 | ·50 | 2 | F | 28○ | ·03 |
| 4 | Au | 23○ | ·22 | 5 | S | 21● | ·75 | 7 | O | 21 | ·28 | 1 | N | 19 | ·81 | | 3 | D | 19 | ·34 | 25 | 4 | Jr | 17 | ·87 | 6 | F | 16○ | ·40 |
| 3 | S | 11● | ·12 | 4 | O | 10 | ·65 | 6 | N | 9 | ·18 | 7 | D | 8 | ·71 | 26 | 2 | Jr | 7 | ·24 | | 3 | F | 5●○ | ·77 | 5 | Mr | 7 | ·30 |
| 7 | Au | 31 | ·48 | 2 | S | 30 | ·01 | 3 | O | 29 | ·54 | 5 | N | 28 | ·07 | | 6 | D | 27 | ·61 | 27 | 1 | Jr | 26● | ·14 | 2 | F | 24 | ·67 |
| 4 | Au | 20 | ·85 | 6 | S | 19 | ·38 | 7 | O | 18 | ·91 | 2 | N | 17 | ·44 | | 3 | D | 16○ | ·97 | 28 | 5 | Jr | 15 | ·50 | 7 | F | 14 | ·03 |
| 3 | S | 7 | ·75 | 5 | O | 7 | ·28 | 6 | N | 5 | ·81 | 1 | D | 5○ | ·34 | 29 | 2 | Jr | 3 | ·87 | | 4 | F | 2 | ·40 | 5 | Mr | 3 | ·93 |
| 1 | Au | 28 | ·11 | 2 | S | 26 | ·65 | 4 | O | 26 | ·18 | 5 | N | 24●○ | ·71 | | 7 | D | 24 | ·24 | 30 | 1 | Jr | 22 | ·77 | 3 | F | 21 | ·30 |
| 7 | S | 16 | ·01 | 1 | O | 15 | ·54 | 3 | N | 14● | ·07 | 4 | D | 13 | ·60 | 31 | 6 | Jr | 12 | ·13 | | 7 | F | 10 | ·67 | 2 | Mr | 12 | ·20 |
| 4 | S | 5 | ·38 | 5 | O | 4○ | ·91 | 7 | N | 3 | ·44 | 1 | D | 2 | ·97 | 32 | 3 | Jr | 1 | ·50 | | 5 | Jr | 31 | ·03 | 6 | F | 29 | ·56 |
| 1 | Au | 24 | ·75 | 3 | S | 23○ | ·28 | 4 | O | 22 | ·81 | 6 | N | 21 | ·34 | | 7 | D | 20 | ·87 | 33 | 2 | Jr | 19 | ·40 | 3 | F | 17 | ·93 |
| 7 | S | 12●○ | ·64 | 2 | O | 12 | ·17 | 3 | N | 10 | ·71 | 5 | D | 10 | ·24 | 34 | 6 | Jr | 8 | ·77 | | 1 | F | 7 | ·30 | 2 | Mr | 8● | ·83 |
| 5 | S | 2● | ·01 | 6 | O | 1 | ·54 | 1 | O | 31 | ·07 | 2 | N | 29 | ·60 | | 4 | D | 29 | ·13 | 35 | 5 | Jr | 27○ | ·66 | 7 | F | 26 | ·19 |
| 2 | Au | 22 | ·38 | 3 | S | 20 | ·91 | 5 | O | 20 | ·44 | 6 | N | 18 | ·97 | | 1 | D | 18 | ·30 | 36 | 3 | Jr | 17○ | ·03 | 4 | F | 15● | ·56 |
| 1 | S | 9 | ·28 | 2 | O | 8 | ·81 | 4 | N | 7 | ·34 | 5 | D | 6 | ·87 | 37 | 7 | Jr | 5○ | ·40 | | 1 | F | 3 | ·93 | 3 | Mr | 5 | ·46 |
| 5 | Au | 29 | ·64 | 7 | S | 28 | ·17 | 1 | O | 27 | ·70 | 3 | N | 26 | ·23 | | 4 | D | 25● | ·77 | 38 | 6 | Jr | 24 | ·30 | 7 | F | 22 | ·83 |
| 3 | Au | 19 | ·01 | 6 | O | 17 | ·07 | 7 | N | 15○ | ·60 | 2 | D | 15 | ·13 | 39 | 3 | Jr | 13 | ·66 | | 5 | F | 12 | ·19 | 6 | Mr | 13 | ·72 |
| 4 | S | 17 | ·54 | | | | | | | | | | | | | | | | | | | | | | | | | | |

* From A.D. 500 to A.D. 999 Arya Siddhanta figures have been given in detail, while Surya Siddhanta figures for the same period are given in abstract at pages 76—79 below.

**TABLE X—**

| Kaliyuga. | Vikrama Era. | Saka Era. | Commencement of Solar Year. | | First New-Moon in Solar Year. | Anomaly of first New-Moon. | A.D. | Week-day of 1st January. | ☉'s Anom col. 5<br>☾'s Anom col. 6<br>Vaisakha | | | | + 29·53059<br>+ 1·976<br>Jyeshta | | | | + 59·06117<br>+ 3·952<br>Ashada | | | | + 88·59176<br>+ 5·929<br>Sravana | | | | + 118·12235<br>+ 7·904<br>Bhadrapada | | | |
|---|---|---|---|---|---|---|---|---|---|---|---|---|---|---|---|---|---|---|---|---|---|---|---|---|---|---|---|---|
| | | | Month and day A.D. | Fraction of day. | | | | | Week-day | Month | Day | Fraction | Week-day | Month | Day | Fraction | Week-day | Month | Day | Fraction | Week-day | Month | Day | Fraction | Week-day | Month | Day | Fraction |
| 3140 | 96 | | 15M | ·0746 | 28·1806 | 12·727 | **39** | 5 | 1 | Ap | 12 | ·25 | 2 | My | 11○ | ·79 | 4 | Je | 10 | ·32 | 5 | Jl | 9 | ·85 | 7 | Au | 8 | ·38 |
| 3141 | 97 | | 14M | ·3334 | 17·2889 | 8·885 | **40** | 6 | 5 | Mr | 31 | ·62 | 7 | Ap | 30●○ | ·15 | 1 | My | 29 | ·68 | 3 | Je | 28 | ·21 | 4 | Jl | 27 | ·74 |
| 3142 | 98 | | 14M | ·5921 | 6·3972 | 5·043 | **41** | 1 | 2 | Mr | 20 | ·99 | 4 | Ap | 19● | ·52 | 6 | My | 19 | ·05 | 7 | Je | 17 | ·58 | 3 | Au | 15 | ·64 |
| | | | | | | | | | | | | | | | | | | | | | 2 | Jl | 17 | ·11 | | | | |
| 3143 | 99 | | 14M | ·8509 | 25·0361 | 3·175 | **42** | 2 | 1 | Ap | 8 | ·89 | 3 | My | 8 | ·42 | 4 | Je | 6 | ·95 | 6 | Jl | 6 | ·48 | 1 | Au | 5 | ·01 |
| 3144 | 100 | | 15M | ·1096 | 14·1443 | 26·887 | **43** | 3 | 6 | Mr | 29 | ·25 | 7 | Ap | 27 | ·78 | 2 | My | 27 | ·31 | 3 | Je | 25 | ·85 | 5 | Jl | 25 | ·38 |
| 3145 | 101 | | 14M | ·3684 | 3·2526 | 23·044 | **44** | 4 | 3 | Mr | 17 | ·62 | 5 | Ap | 16 | ·15 | 6 | My | 15 | ·68 | 2 | Jl | 13 | ·74 | 4 | Au | 12○ | ·27 |
| | | | | | | | | | | | | | | | | | 1 | Je | 14 | ·21 | | | | | | | | |
| 3146 | 102 | | 14M | ·6271 | 21·8915 | 21·178 | **45** | 6 | 2 | Ap | 5 | ·52 | 4 | My | 5 | ·05 | 5 | Je | 3 | ·58 | 7 | Jl | 3 | ·11 | 1 | Au | 1● | ·64 |
| 3147 | 103 | | 14M | ·8859 | 10·9998 | 17·335 | **46** | 7 | 6 | Mr | 25 | ·88 | 1 | Ap | 24 | ·42 | 2 | My | 23 | ·95 | 4 | Je | 22○ | ·48 | 6 | Jl | 22● | ·01 |
| 3148 | 104 | | 15M | ·1446 | 0·1081 | 13·493 | **47** | 1 | 4 | Mr | 15 | ·25 | 7 | My | 13 | ·31 | 1 | Je | 11○ | ·84 | 3 | Jl | 11 | ·37 | 4 | Au | 9 | ·91 |
| | | | | | | | | | 5 | Ap | 13 | ·78 | | | | | | | | | | | | | | | | |
| 3149 | 105 | | 14M | ·4034 | 18·7470 | 11·625 | **48** | 2 | 3 | Ap | 2 | ·15 | 4 | My | 1 | ·68 | 6 | My | 31●○ | ·21 | 7 | Je | 29 | ·74 | 2 | Jl | 29 | ·27 |
| 3150 | 106 | | 14M | ·6622 | 7·8553 | 7·783 | **49** | 4 | 7 | Mr | 22 | ·52 | 2 | Ap | 21○ | ·05 | 3 | My | 20● | ·58 | 5 | Je | 19 | ·11 | 6 | Jl | 18 | ·64 |
| | | | | | | | | | | | | | | | | | | | | | | | | | 1 | Au | 17 | ·17 |
| 3151 | 107 | | 14M | ·9209 | 26·4942 | 5·916 | **50** | 5 | 6 | Ap | 10○ | ·41 | 7 | My | 9● | ·95 | 2 | Je | 8 | ·48 | 4 | Jl | 8 | ·01 | 5 | Au | 6 | ·54 |
| 3152 | 108 | | 15M | ·1797 | 15·6025 | 2·073 | **51** | 6 | 3 | Mr | 30○ | ·78 | 5 | Ap | 29 | ·31 | 6 | My | 28 | ·84 | 1 | Je | 27 | ·37 | 2 | Jl | 26 | ·90 |
| 3153 | 109 | | 14M | ·4384 | 4·7108 | 25·785 | **52** | 7 | 1 | Mr | 19● | ·15 | 2 | Ap | 17 | ·68 | 4 | My | 17 | ·21 | 7 | Jl | 15 | ·27 | 1 | Au | 13 | ·80 |
| | | | | | | | | | | | | | | | | | 5 | Je | 15 | ·74 | | | | | | | | |
| 3154 | 110 | | 14M | ·6972 | 23·3497 | 23·918 | **53** | 2 | 7 | Ap | 7 | ·05 | 1 | My | 6 | ·58 | 3 | Je | 5 | ·11 | 4 | Jl | 4 | ·64 | 6 | Au | 3○ | ·17 |
| 3155 | 111 | | 14M | ·9559 | 12·4580 | 20·075 | **54** | 3 | 4 | Mr | 27 | ·41 | 5 | Ap | 25 | ·94 | 7 | My | 25 | ·47 | 2 | Je | 24 | ·01 | 3 | Jl | 23●○ | ·54 |
| 3156 | 112 | | 15M | ·2147 | 1·5663 | 16·233 | **55** | 4 | 1 | Mr | 16 | ·78 | 3 | Ap | 15 | ·31 | 6 | Je | 13 | ·37 | 7 | Jl | 12●○ | ·90 | 2 | Au | 11 | ·43 |
| | | | | | | | | | | | | | 4 | My | 14 | ·84 | | | | | | | | | | | | |
| 3157 | 113 | | 14M | ·4735 | 20·2052 | 14·366 | **56** | 5 | 7 | Ap | 3 | 68 | 2 | My | 3 | ·21 | 3 | Je | 1 | ·74 | 5 | Jl | 1● | ·27 | 6 | Jl | 30 | ·80 |
| 3158 | 114 | | 14M | ·7322 | 9·3136 | 10·523 | **57** | 7 | 5 | Mr | 24 | ·05 | 6 | Ap | 22 | ·58 | 1 | My | 22○ | ·11 | 2 | Je | 20 | ·64 | 4 | Jl | 20 | ·17 |
| 3159 | 115 | | 14M | ·9910 | 27·9524 | 8·656 | **58** | 1 | 3 | Ap | 11 | ·94 | 5 | My | 11●○ | ·47 | 7 | Je | 10 | ·00 | 1 | Jl | 9 | ·53 | 3 | Au | 8 | ·07 |
| 3160 | 116 | | 15M | ·2497 | 17·0607 | 4·814 | **59** | 2 | 1 | Ap | 1 | ·31 | 2 | Ap | 30● | ·84 | 4 | My | 30 | ·37 | 5 | Je | 28 | ·90 | 7 | Jl | 28 | ·43 |
| 3161 | 117 | | 14M | ·5085 | 6·1690 | 0·971 | **60** | 3 | 5 | Mr | 20○ | ·68 | 7 | Ap | 19 | ·21 | 1 | My | 18 | ·74 | 3 | Je | 17 | ·27 | 6 | Au | 15 | ·33 |
| | | | | | | | | | | | | | | | | | | | | | 4 | Jl | 16 | ·80 | | | | |
| 3162 | 118 | | 14M | ·7672 | 24·8078 | 26·659 | **61** | 5 | 4 | Ap | 8 | ·57 | 6 | My | 8 | ·11 | 7 | Je | 6 | ·64 | 2 | Jl | 6 | ·17 | 3 | Au | 4 | ·70 |
| 3163 | 119 | | 15M | ·0260 | 13·9161 | 22·816 | **62** | 6 | 1 | Mr | 28 | ·94 | 3 | Ap | 27 | ·47 | 5 | My | 27 | ·00 | 6 | Je | 25 | ·53 | 1 | Jl | 25 | ·06 |
| 3164 | 120 | | 15M | ·2848 | 3·0244 | 18·973 | **63** | 7 | 6 | Mr | 18 | ·31 | 7 | Ap | 16 | ·84 | 2 | My | 16 | ·37 | 5 | Jl | 14 | ·43 | 6 | Au | 12 | ·96 |
| | | | | | | | | | | | | | | | | | 3 | Je | 14 | ·90 | | | | | | | | |
| 3165 | 121 | | 14M | ·5435 | 21·6633 | 17·107 | **64** | 1 | 5 | Ap | 5 | ·21 | 6 | My | 4 | ·74 | 1 | Je | 3 | ·27 | 2 | Jl | 2○ | ·80 | 4 | Au | 1● | ·33 |
| 3166 | 122 | | 14M | ·8023 | 10·7716 | 13·264 | **65** | 3 | 2 | Mr | 25 | ·57 | 4 | Ap | 24 | ·10 | 5 | My | 23 | ·63 | 7 | Je | 22○ | ·17 | 1 | Jl | 21 | ·70 |
| 3167 | 123 | | 15M | ·0610 | 29·4105 | 11·397 | **66** | 4 | 1 | Ap | 13 | ·47 | 3 | My | 13 | ·00 | 4 | Je | 11○ | ·53 | 6 | Jl | 11 | ·06 | 7 | Au | 9 | ·59 |
| 3168 | 124 | | 15M | ·3198 | 18·5188 | 7·554 | **67** | 5 | 5 | Ap | 2 | ·84 | 7 | My | 2○ | ·37 | 1 | My | 31● | ·90 | 3 | Je | 30 | ·43 | 4 | Jl | 29 | ·96 |
| 3169 | 125 | | 14M | ·5785 | 7·6271 | 3·712 | **68** | 6 | 3 | Mr | 22 | ·21 | 4 | Ap | 20○ | ·74 | 6 | My | 20● | ·27 | 7 | Je | 18 | ·80 | 3 | Au | 16 | ·86 |
| | | | | | | | | | | | | | | | | | | | | | 2 | Jl | 18 | ·33 | | | | |
| 3170 | 126 | | 14M | ·8373 | 26·2660 | 1·845 | **69** | 1 | 2 | Ap | 10○ | ·10 | 3 | My | 9 | ·63 | 5 | Je | 8 | ·16 | 6 | Jl | 7 | ·69 | 1 | Au | 6 | ·22 |
| 3171 | 127 | | 15M | ·0960 | 15·3743 | 25·557 | **70** | 2 | 6 | Mr | 30 | ·47 | 1 | Ap | 29 | ·00 | 2 | My | 28 | ·53 | 4 | Je | 27 | ·06 | 5 | Jl | 26 | ·59 |
| 3172 | 128 | | 15M | ·3548 | 4·4826 | 21·714 | **71** | 3 | 3 | Mr | 19● | ·84 | 5 | Ap | 18 | ·37 | 6 | My | 17 | ·90 | 2 | Jl | 15 | ·96 | 4 | Au | 14○ | ·49 |
| | | | | | | | | | | | | | | | | | 1 | Je | 16 | ·43 | | | | | | | | |
| 3173 | 129 | | 14M | ·6136 | 23·1215 | 19·847 | **72** | 4 | 2 | Ap | 6 | ·73 | 4 | My | 6 | ·27 | 5 | Je | 4 | ·80 | 7 | Jl | 4 | ·33 | 1 | Au | 2●○ | ·86 |
| 3174 | 130 | | 14M | ·8723 | 12·2298 | 16·004 | **73** | 6 | 7 | Mr | 27 | ·10 | 1 | Ap | 25 | ·63 | 3 | My | 25 | ·16 | 4 | Je | 23 | ·69 | 6 | Jl | 23●○ | ·22 |
| 3175 | 131 | | 15M | ·1311 | 1·3381 | 12·163 | **74** | 7 | 4 | Mr | 16 | ·47 | 6 | Ap | 15 | ·00 | 2 | Je | 13 | ·06 | 3 | Jl | 12● | ·59 | 5 | Au | 11 | ·12 |
| | | | | | | | | | | | | | 7 | My | 14 | ·53 | | | | | | | | | | | | |
| 3176 | 132 | | 15M | ·3898 | 19·9770 | 10·295 | **75** | 1 | 3 | Ap | 4 | ·37 | 4 | My | 3 | ·90 | 6 | Je | 2○ | ·43 | 7 | Jl | 1 | ·96 | 2 | Jl | 31 | ·49 |
| 3177 | 133 | | 14M | ·6486 | 9·0853 | 6·452 | **76** | 2 | 7 | Mr | 23 | ·73 | 2 | Ap | 22 | ·26 | 3 | My | 21●○ | ·79 | 5 | Je | 20 | ·33 | 6 | Jl | 19 | ·86 |
| 3178 | 134 | | 14M | ·9073 | 27·7242 | 4·585 | **77** | 4 | 6 | Ap | 11 | ·63 | 1 | My | 11 | ·16 | 2 | Je | 9 | ·69 | 4 | Jl | 9 | ·22 | 5 | Au | 7 | ·72 |
| 3179 | 135 | | 15M | ·1661 | 16·8325 | 0·743 | **78** | 5 | 4 | Ap | 1○ | ·00 | 5 | Ap | 30● | ·53 | 7 | My | 30 | ·06 | 1 | Je | 28 | ·59 | 3 | Jl | 28 | ·12 |

**Surya Siddhanta.**

| Asvina | | | | Kartika | | | | Margasira | | | | Pausha | | | | A.D. | Magha | | | | A.D. | Phalguna | | | | Chaitra | | | |
|---|---|---|---|---|---|---|---|---|---|---|---|---|---|---|---|---|---|---|---|---|---|---|---|---|---|---|---|---|---|
| +147·65293 | | | | + 177·18353 | | | | + 206·71411 | | | | + 236·24470 | | | | | + 265·77529 | | | | | + 295·30588 | | | | + 324·83647<br>+ 354·36705 | | | |
| + 9·880 | | | | + 11·856 | | | | + 13·832 | | | | + 15·808 | | | | | + 17·784 | | | | | + 19·760 | | | | + 21·736<br>+ 23·712 | | | |
| Week-day | Month | Day | Fraction | Week-day | Month | Day | Fraction | Week-day | Month | Day | Fraction | Week-day | Month | Day | Fraction | | Week-day | Month | Day | Fraction | | Week-day | Month | Day | Fraction | Week-day | Month | Day | Fraction |
| 1 | S | 6 | ·91 | 3 | O | 6 | ·44 | 4 | N | 4○ | ·97 | 6 | D | 4● | ·50 | **40** | 1 | Jr | 3 | ·03 | | 2 | F | 1 | ·56 | 4 | Mr | 2 | ·09 |
| 6 | Au | 26 | ·27 | 7 | S | 24 | ·81 | 2 | O | 24○ | ·34 | 3 | N | 22 | ·87 | | 5 | D | 22 | ·40 | **41** | 6 | Jr | 20 | ·93 | 1 | F | 19 | ·46 |
| 5 | S | 14 | ·17 | 6 | O | 13● | ·70 | 1 | N | 12 | ·23 | 2 | D | 11 | ·76 | **42** | 4 | Jr | 10 | ·29 | | 5 | F | 8 | ·83 | 7 | Mr | 10○ | ·36 |
| 2 | S | 3○ | ·54 | 4 | O | 3● | ·07 | 5 | N | 1 | ·60 | 7 | D | 1 | ·13 | | 1 | D | 30 | ·66 | **43** | 3 | Jr | 29 | ·19 | 4 | F | 27●○ | ·72 |
| 6 | Au | 23○ | ·91 | 1 | S | 22 | ·44 | 2 | O | 21 | ·97 | 4 | N | 20 | ·50 | | 6 | D | 20 | ·03 | **44** | 7 | Jr | 18 | ·56 | 2 | F | 17●○ | ·09 |
| 5 | S | 10 | ·80 | 7 | O | 10 | ·33 | 1 | N | 8 | ·87 | 3 | D | 8 | ·40 | **45** | 4 | Jr | 6 | ·93 | | 6 | F | 5 | ·46 | 7 | Mr | 6 | ·99 |
| 3 | Au | 31 | ·17 | 4 | S | 29 | ·70 | 6 | O | 29 | ·23 | 7 | N | 27 | ·76 | | 2 | D | 27○ | ·29 | **46** | 3 | Jr | 25 | ·82 | 5 | F | 24 | ·35 |
| 7 | Au | 20 | ·54 | 2 | S | 19 | ·07 | 3 | O | 18 | ·60 | 5 | N | 17 | ·13 | | 6 | D | 16●○ | ·66 | **47** | 1 | Jr | 15 | ·19 | 2 | F | 13 | ·72 |
| 6 | S | 8 | ·44 | 7 | O | 7 | ·97 | 2 | N | 6 | ·50 | 4 | D | 6○ | ·03 | **48** | 5 | Jr | 4 | ·56 | | 7 | F | 3 | ·09 | 1 | Mr | 3 | ·62 |
| 3 | Au | 27 | ·80 | 5 | S | 26 | ·33 | 6 | O | 25 | ·86 | 1 | N | 24● | ·39 | | 2 | D | 23 | ·93 | **49** | 4 | Jr | 22 | ·46 | 5 | F | 20 | ·99 |
| 2 | S | 15 | ·70 | 4 | O | 15○ | ·23 | 5 | N | 13 | ·76 | 7 | D | 13 | ·29 | **50** | 1 | Jr | 11 | ·82 | | 3 | F | 10 | ·35 | 4 | Mr | 11 | ·88 |
| 7 | S | 5 | ·07 | 1 | O | 4○ | ·60 | 3 | N | 3 | 13 | 4 | D | 2 | ·66 | **51** | 6 | Jr | 1 | ·19 | | 7 | Jr | 30 | ·72 | 2 | Mr | 1 | ·25 |
| 4 | Au | 25 | ·43 | 5 | S | 23●○ | ·97 | 7 | O | 23 | ·50 | 2 | N | 22 | ·03 | | 3 | D | 21 | ·56 | **52** | 5 | Jr | 20 | ·09 | 6 | F | 18 | ·62 |
| 3 | S | 12 | ·33 | 4 | O | 11 | ·86 | 6 | N | 10 | ·39 | 7 | D | 9 | ·92 | **53** | 2 | Jr | 8 | ·45 | | 3 | F | 6○ | ·99 | 5 | Mr | 8● | ·52 |
| 7 | S | 1 | ·70 | 2 | O | 1 | ·23 | 3 | O | 30 | ·76 | 5 | N | 29 | ·29 | | 6 | D | 28 | ·82 | **54** | 1 | Jr | 27○ | ·35 | 2 | F | 25● | ·88 |
| 5 | Au | 22 | ·07 | 6 | S | 20 | ·60 | 1 | O | 20 | ·13 | 2 | N | 18 | ·66 | | 4 | D | 18 | ·19 | **55** | 5 | Jr | 16○ | ·72 | 7 | F | 15 | ·25 |
| 3 | S | 9 | ·96 | 5 | O | 9 | ·49 | 7 | N | 8 | ·03 | 1 | D | 7 | ·56 | **56** | 3 | Jr | 6 | ·09 | | 4 | F | 4 | ·62 | 6 | Mr | 5 | ·15 |
| 1 | Au | 29 | ·33 | 2 | S | 27 | ·86 | 4 | O | 27 | ·39 | 5 | N | 25○ | ·92 | | 7 | D | 25 | ·45 | **57** | 1 | Jr | 23 | ·98 | 3 | F | 22 | ·51 |
| 5 | Au | 18 | ·70 | 1 | O | 16 | ·76 | 3 | N | 15○ | ·29 | 4 | D | 14 | ·82 | **58** | 6 | Jr | 13 | ·35 | | 7 | F | 11 | ·88 | 2 | Mr | 13 | ·41 |
| 7 | S | 17 | ·23 | | | | | | | | | | | | | | | | | | | | | | | | | | |
| 4 | S | 6 | ·60 | 6 | O | 6 | ·13 | 7 | N | 4○ | ·66 | 2 | D | 4 | ·19 | **59** | 3 | Jr | 2 | ·72 | | 5 | F | 1 | ·25 | 6 | Mr | 2 | ·78 |
| 1 | Au | 26 | ·96 | 3 | S | 25 | ·49 | 5 | O | 25● | ·02 | 6 | N | 23 | ·55 | | 1 | D | 23 | ·09 | **60** | 2 | Jr | 21 | ·62 | 4 | F | 20 | ·15 |
| 7 | S | 13○ | ·86 | 2 | O | 13● | ·39 | 3 | N | 11 | ·92 | 5 | D | 11 | ·45 | **61** | 6 | Jr | 9 | ·98 | | 1 | F | 8 | ·51 | 3 | Mr | 10●○ | ·04 |
| 5 | S | 3○ | ·23 | 6 | O | 2● | ·76 | 1 | N | 1 | ·29 | 2 | N | 30 | ·82 | | 4 | D | 30 | ·35 | **62** | 5 | Jr | 28 | ·88 | 7 | F | 27●○ | ·41 |
| 2 | Au | 23○ | ·59 | 4 | S | 22 | ·13 | 5 | O | 21 | ·66 | 7 | N | 20 | ·19 | | 1 | D | 19 | ·72 | **63** | 3 | Jr | 18 | ·25 | 4 | F | 16● | ·78 |
| 1 | S | 11 | ·49 | 3 | O | 11 | ·02 | 4 | N | 9 | ·55 | 6 | D | 9 | ·08 | **64** | 7 | Jr | 7○ | ·61 | | 2 | F | 6 | ·15 | 3 | Mr | 6 | ·68 |
| 5 | Au | 30 | ·86 | 7 | S | 29 | ·39 | 1 | O | 28 | ·92 | 3 | N | 27 | ·45 | | 4 | D | 26○ | ·98 | **65** | 6 | Jr | 25 | ·51 | 1 | F | 24 | ·04 |
| 3 | Au | 20 | ·23 | 4 | S | 18 | ·76 | 6 | O | 18 | ·29 | 7 | N | 16 | ·82 | | 2 | D | 16●○ | ·35 | **66** | 3 | Jr | 14 | ·88 | 5 | F | 13 | ·41 |
| | | | | | | | | | | | | | | | | | | | | | | | | | | 6 | Mr | 14 | ·94 |
| 2 | S | 8 | ·12 | 3 | O | 7 | ·65 | 5 | N | 6 | ·19 | 6 | D | 5 | ·72 | **67** | 1 | Jr | 4 | ·25 | | 2 | F | 2 | ·78 | 4 | Mr | 4 | ·31 |
| 6 | Au | 28 | ·49 | 1 | S | 27 | ·02 | 2 | O | 26○ | ·55 | 4 | N | 25 | ·08 | | 5 | D | 24 | ·61 | **68** | 7 | Jr | 23 | ·14 | 1 | F | 21 | ·67 |
| 5 | S | 15 | ·39 | 6 | O | 14○ | ·92 | 1 | N | 13 | ·45 | 2 | D | 12 | ·98 | **69** | 4 | Jr | 11 | ·51 | | 6 | F | 10 | ·04 | 7 | Mr | 11 | ·57 |
| 2 | S | 4 | ·76 | 4 | O | 4●○ | ·29 | 5 | N | 2 | ·82 | 7 | D | 2 | ·35 | | 1 | D | 31 | ·88 | **70** | 3 | Jr | 30 | ·41 | 4 | F | 28 | ·94 |
| 7 | Au | 25 | ·12 | 1 | S | 23● | ·65 | 3 | O | 23 | ·18 | 4 | N | 21 | ·71 | | 6 | D | 21 | ·25 | **71** | 7 | Jr | 19 | ·78 | 2 | F | 18○ | ·31 |
| 6 | S | 13 | ·02 | 7 | O | 12 | ·55 | 2 | N | 11 | ·08 | 3 | D | 10 | ·61 | **72** | 5 | Jr | 9 | ·14 | | 6 | F | 7○ | ·67 | 1 | Mr | 8 | ·20 |
| 3 | S | 1 | ·39 | 4 | S | 30 | ·92 | 6 | O | 30 | ·45 | 7 | N | 28 | ·98 | | 2 | D | 28 | ·51 | **73** | 4 | Jr | 27○ | ·04 | 5 | F | 25 | ·57 |
| 7 | Au | 21 | ·75 | 2 | S | 20 | ·29 | 3 | O | 19 | ·82 | 5 | N | 18 | ·35 | | 6 | D | 17 | ·88 | **74** | 1 | Jr | 16 | ·41 | 2 | F | 14 | ·94 |
| 6 | S | 9 | ·65 | 1 | O | 9 | ·18 | 2 | N | 7 | ·71 | 4 | D | 7○ | ·24 | **75** | 5 | Jr | 5○ | ·77 | | 7 | F | 4 | ·31 | 1 | Mr | 5 | ·84 |
| 4 | Au | 30 | ·02 | 5 | S | 28 | ·55 | 7 | O | 28 | ·08 | 1 | N | 26○ | ·61 | | 3 | D | 26● | ·14 | **76** | 4 | Jr | 24 | ·67 | 6 | F | 23 | ·20 |
| 1 | Au | 18 | ·39 | 4 | O | 16 | ·45 | 5 | N | 14○ | ·98 | 7 | D | 14 | ·51 | **77** | 2 | Jr | 13 | ·04 | | 3 | F | 11 | ·57 | 5 | Mr | 13 | ·10 |
| 2 | S | 16 | ·92 | | | | | | | | | | | | | | | | | | | | | | | | | | |
| 7 | S | 6 | ·28 | 1 | O | 5 | ·81 | 3 | N | 4 | ·35 | 4 | D | 3 | ·88 | **78** | 6 | Jr | 2 | ·41 | | 7 | Jr | 31 | ·94 | 2 | Mr | 2 | ·47 |
| 4 | Au | 26 | ·65 | 6 | S | 25○ | ·18 | 7 | O | 24● | ·71 | 2 | N | 23 | ·24 | | 3 | D | 22 | ·77 | **79** | 5 | Jr | 21 | ·30 | 6 | F | 19 | ·83 |

## TABLE X–

| Kaliyuga. | Vikrama Era. | Saka Era. | Commencement of Solar Year. Month and day A.D. | Commencement of Solar Year. Fraction of day. | First New-Moon in Solar Year. | Anomaly of first New-Moon. | A.D. | Week-day of 1st January. | Vaisakha Week-day | Vaisakha Month | Vaisakha Day | Vaisakha Fraction | Jyeshta Week-day | Jyeshta Month | Jyeshta Day | Jyeshta Fraction | Ashada Week-day | Ashada Month | Ashada Day | Ashada Fraction | Sravana Week-day | Sravana Month | Sravana Day | Sravana Fraction | Bhadrapada Week-day | Bhadrapada Month | Bhadrapada Day | Bhadrapada Fraction |
|---|---|---|---|---|---|---|---|---|---|---|---|---|---|---|---|---|---|---|---|---|---|---|---|---|---|---|---|---|
| | | | | | | | | | ☉'s Anom col. 6 | | | | + 29·53059 | | | | + 59·06117 | | | | + 88·59176 | | | | + 118·12235 | | | |
| | | | | | | | | | ☾'s Anom col. 7 | | | | + 1·976 | | | | + 3·952 | | | | + 5·928 | | | | + 7·904 | | | |
| 3180 | 136 | 1 | 15M | ·4249 | 5·9408 | 24·455 | 79 | 6 | 1 | Mr | 21○ | ·36 | 2 | Ap | 19 | ·90 | 4 | My | 19 | ·43 | 5 | Je | 17 | ·96 | 2 | Au | 16 | ·02 |
| | | | | | | | | | | | | | | | | | | | | | 7 | Jl | 17 | ·49 | | | | |
| 3181 | 137 | 2 | 14M | ·6836 | 24·5796 | 22·588 | 80 | 7 | 7 | Ap | 8 | ·26 | 1 | My | 7 | ·79 | 3 | Je | 6 | ·32 | 4 | Ji | 5 | ·85 | 6 | Au | 4 | ·39 |
| 3182 | 138 | 3 | 14M | ·9424 | 13·6879 | 18·745 | 81 | 2 | 4 | Mr | 28 | ·63 | 6 | Ap | 27 | ·16 | 7 | My | 26 | ·69 | 2 | Je | 25 | ·22 | 3 | Jl | 24 | ·75 |
| 3183 | 139 | 4 | 15M | ·2011 | 2·7962 | 14·902 | 82 | 3 | 2 | Mr | 18 | ·00 | 3 | Ap | 16 | ·53 | 6 | Je | 14 | ·59 | 1 | Jl | 14○ | ·12 | 2 | Au | 12● | ·65 |
| | | | | | | | | | | | | | 5 | My | 16 | ·06 | | | | | | | | | | | | |
| 3184 | 140 | 5 | 15M | ·4599 | 21·4351 | 13·036 | 83 | 4 | 7 | Ap | 5 | ·89 | 2 | My | 5 | ·43 | 3 | Je | 3 | ·96 | 5 | Jl | 3○ | ·49 | 7 | Au | 2● | ·02 |
| 3185 | 141 | 6 | 14M | ·7186 | 10·5434 | 9·193 | 84 | 5 | 5 | Mr | 25 | ·26 | 6 | Ap | 23 | ·79 | 1 | My | 23 | ·32 | 2 | Je | 21○ | ·85 | 4 | Jl | 21 | ·38 |
| 3186 | 142 | 7 | 14M | ·9774 | 29·1823 | 7·326 | 85 | 7 | 4 | Ap | 13 | ·16 | 5 | My | 12○ | ·69 | 7 | Je | 11● | ·22 | 1 | Jl | 10 | ·75 | 3 | Au | 9 | ·28 |
| 3187 | 143 | 8 | 15M | ·2362 | 18·2906 | 3·483 | 86 | 1 | 1 | Ap | 2 | ·53 | 3 | My | 2○ | ·06 | 4 | My | 31● | ·59 | 6 | Je | 30 | ·12 | 7 | Jl | 29 | ·65 |
| 3188 | 144 | 9 | 15M | ·4949 | 7·3989 | 27·195 | 87 | 2 | 5 | Mr | 22 | ·89 | 7 | Ap | 21○ | ·42 | 1 | My | 20 | ·95 | 3 | Je | 19 | ·48 | 6 | Au | 17 | ·55 |
| | | | | | | | | | | | | | | | | | | | | | 5 | Jl | 19 | ·02 | | | | |
| 3189 | 145 | 10 | 14M | ·7537 | 26·0378 | 25·328 | 88 | 3 | 4 | Ap | 9● | ·79 | 6 | My | 9 | ·32 | 7 | Je | 7 | ·85 | 2 | Jl | 7 | ·38 | 3 | Au | 5 | ·91 |
| 3190 | 146 | 11 | 15M | ·0124 | 15·1461 | 21·486 | 89 | 5 | 2 | Mr | 30● | ·16 | 3 | Ap | 28 | ·69 | 5 | My | 28 | ·22 | 6 | Je | 26 | ·75 | 1 | Jl | 26 | ·28 |
| 3191 | 147 | 12 | 15M | ·2712 | 4·2544 | 17·643 | 90 | 6 | 6 | Mr | 19● | ·52 | 1 | Ap | 18 | ·06 | 2 | My | 17 | ·59 | 5 | Jl | 15 | ·65 | 7 | Au | 14○ | ·18 |
| | | | | | | | | | | | | | | | | | 4 | Je | 16 | ·12 | | | | | | | | |
| 3192 | 148 | 13 | 15M | ·5299 | 22·8933 | 15·776 | 91 | 7 | 5 | Ap | 7 | ·42 | 6 | My | 6 | ·95 | 1 | Je | 5 | ·48 | 3 | Jl | 5 | ·01 | 4 | Au | 3●○ | ·54 |
| 3193 | 149 | 14 | 14M | ·7887 | 12·0016 | 11·933 | 92 | 1 | 2 | Mr | 26 | ·79 | 4 | Ap | 25 | ·32 | 5 | My | 24 | ·85 | 7 | Je | 23 | ·38 | 1 | Jl | 22● | ·91 |
| 3194 | 150 | 15 | 15M | ·0474 | 1·1099 | 8·091 | 93 | 3 | 7 | Mr | 16 | ·16 | 3 | My | 14 | ·22 | 4 | Je | 12○ | ·75 | 6 | Jl | 12 | ·28 | 7 | Au | 10 | ·81 |
| | | | | | | | | | 1 | Ap | 14 | ·69 | | | | | | | | | | | | | | | | |
| 3195 | 151 | 16 | 15M | ·3062 | 19·7488 | 6·224 | 94 | 4 | 6 | Ap | 4 | ·05 | 7 | My | 3 | ·58 | 2 | Je | 2●○ | ·12 | 3 | Jl | 1 | ·65 | 5 | Jl | 31 | ·18 |
| 2196 | 152 | 17 | 15M | ·5650 | 8·8571 | 2·381 | 95 | 5 | 3 | Mr | 24 | ·42 | 4 | Ap | 22 | ·95 | 6 | My | 22●○ | ·48 | 1 | Je | 21 | ·01 | 2 | Jl | 20 | ·54 |
| | | | | | | | | | | | | | | | | | | | | | | | | | 4 | Au | 19 | ·07 |
| 3197 | 153 | 18 | 14M | ·8237 | 27·4959 | 0·514 | 96 | 6 | 2 | Ap | 11 | ·32 | 3 | My | 10● | ·85 | 5 | Je | 9 | ·38 | 6 | Jl | 8 | ·91 | 1 | Au | 7 | ·44 |
| 3198 | 154 | 19 | 15M | ·0825 | 16·6042 | 24·226 | 97 | 1 | 6 | Mr | 31●○ | ·69 | 1 | Ap | 30 | ·22 | 2 | My | 29 | ·75 | 4 | Je | 28 | ·28 | 5 | Jl | 27 | ·81 |
| 3199 | 155 | 20 | 15M | ·3412 | 5·7125 | 20·383 | 98 | 2 | 4 | Mr | 21●○ | ·05 | 5 | Ap | 19 | ·58 | 7 | My | 19 | ·11 | 1 | Je | 17 | ·64 | 4 | Au | 15 | ·71 |
| | | | | | | | | | | | | | | | | | | | | | 3 | Jl | 17 | ·18 | | | | |
| 3200 | 156 | 21 | 15M | ·6000 | 24·3514 | 18·517 | 99 | 3 | 2 | Ap | 8 | ·95 | 4 | My | 8 | ·48 | 6 | Je | 7 | ·01 | 7 | Jl | 6 | ·54 | 2 | Au | 5 | ·07 |
| 3201 | 157 | 22 | 14M | ·8587 | 13·4597 | 14·674 | 100 | 4 | 7 | Mr | 28 | ·32 | 1 | Ap | 26 | ·85 | 3 | My | 26 | ·38 | 4 | Je | 24 | ·91 | 6 | Jl | 24○ | ·44 |
| 3202 | 158 | 23 | 15M | ·1175 | 2·5680 | 10·831 | 101 | 6 | 4 | Mr | 17 | ·68 | 6 | Ap | 16 | ·22 | 2 | Je | 14 | ·28 | 3 | Jl | 13○ | ·81 | 5 | Au | 12● | ·34 |
| | | | | | | | | | | | | | 7 | My | 15 | ·75 | | | | | | | | | | | | |
| 3203 | 159 | 24 | 15M | ·8763 | 21·2069 | 8·964 | 102 | 7 | 3 | Ap | 5 | ·58 | 5 | My | 5 | ·11 | 6 | Je | 3 | ·64 | 1 | Jl | 3○ | ·17 | 2 | Au | 1 | ·70 |
| 3204 | 160 | 25 | 15M | ·6350 | 10·3152 | 5·122 | 103 | 7 | 7 | Mr | 25 | ·95 | 2 | Ap | 24 | ·48 | 4 | My | 24 | ·01 | 5 | Je | 22● | ·54 | 7 | Jl | 22 | ·07 |
| 3205 | 161 | 26 | 14M | ·8938 | 28·9541 | 3·255 | 104 | 2 | 6 | Ap | 12 | ·85 | 1 | My | 12○ | ·38 | 2 | Je | 10● | ·91 | 4 | Jl | 10 | ·44 | 5 | Au | 8 | ·97 |
| 3206 | 162 | 27 | 15M | ·1525 | 18·0624 | 26·967 | 105 | 4 | 4 | Ap | 2 | ·21 | 5 | My | 1○ | ·74 | 7 | My | 31 | ·28 | 1 | Je | 29 | ·81 | 3 | Jl | 29 | ·34 |
| 3207 | 163 | 28 | 15M | ·4113 | 7·1707 | 23·124 | 106 | 5 | 1 | Mr | 22 | ·58 | 3 | Ap | 21● | ·11 | 4 | My | 20 | ·64 | 6 | Je | 19 | ·17 | 2 | Au | 17 | ·23 |
| | | | | | | | | | | | | | | | | | | | | | 7 | Jl | 18 | ·70 | | | | |
| 3208 | 164 | 29 | 15M | ·6700 | 25·8096 | 21·257 | 107 | 6 | 7 | Ap | 10● | ·48 | 2 | My | 10 | ·01 | 3 | Je | 8 | ·54 | 5 | Jl | 8 | ·07 | 6 | Au | 6 | ·60 |
| 3209 | 165 | 30 | 14M | ·9288 | 14·9179 | 17·415 | 108 | 7 | 4 | Mr | 29● | ·85 | 6 | Ap | 28 | ·38 | 7 | My | 27 | ·91 | 2 | Je | 26 | ·44 | 3 | Jl | 25 | ·97 |
| 3210 | 166 | 31 | 15M | ·1876 | 4·0262 | 13·572 | 109 | 2 | 2 | Mr | 19 | ·21 | 3 | Ap | 17 | ·74 | 5 | My | 17 | ·27 | 1 | Jl | 15 | ·34 | 2 | Au | 13●○ | ·87 |
| | | | | | | | | | | | | | | | | | 6 | Je | 15 | ·80 | | | | | | | | |
| 3211 | 167 | 32 | 15M | ·4463 | 22·6651 | 11·705 | 110 | 3 | 1 | Ap | 7 | ·11 | 2 | My | 6 | ·64 | 4 | Je | 5 | ·17 | 5 | Jl | 4 | ·70 | 7 | Au | 3● | ·23 |
| 3212 | 168 | 33 | 15M | ·7051 | 11·7738 | 7·862 | 111 | 4 | 5 | Mr | 27 | ·48 | 7 | Ap | 26 | ·01 | 1 | My | 25 | ·54 | 3 | Je | 24○ | ·07 | 4 | Jl | 23 | ·60 |
| 3213 | 169 | 34 | 14M | ·9638 | 0·8817 | 4·020 | 112 | 5 | 2 | Mr | 15 | ·84 | 5 | My | 13 | ·91 | 7 | Je | 12●○ | ·44 | 1 | Jl | 11 | ·97 | 3 | Au | 10 | ·50 |
| | | | | | | | | | 4 | Ap | 14 | ·38 | | | | | | | | | | | | | | | | |
| 3214 | 170 | 35 | 15M | ·2226 | 19·5206 | 2·153 | 113 | 7 | 1 | Ap | 3 | ·74 | 3 | My | 3 | ·27 | 4 | Je | 1●○ | ·80 | 6 | Jl | 1 | ·33 | 7 | Jl | 30 | ·86 |
| 3215 | 171 | 36 | 15M | ·4813 | 8·6289 | 25·865 | 114 | 1 | 6 | Mr | 24 | ·11 | 7 | Ap | 22 | ·64 | 2 | My | 22● | ·17 | 3 | Je | 20 | ·70 | 5 | Jl | 20 | ·23 |
| | | | | | | | | | | | | | | | | | | | | | | | | | 6 | Au | 18 | ·76 |
| 3216 | 172 | 37 | 15M | ·7401 | 27·2677 | 23·998 | 115 | 2 | 5 | Ap | 12○ | ·01 | 6 | My | 11 | ·54 | 1 | Je | 10 | ·07 | 2 | Jl | 9 | ·60 | 4 | Au | 8 | ·13 |
| 3217 | 173 | 38 | 14M | ·9988 | 16·3760 | 20·155 | 116 | 3 | 2 | Mr | 31●○ | ·37 | 3 | Ap | 29 | ·90 | 5 | My | 29 | ·44 | 6 | Je | 27 | ·97 | 1 | Jl | 27 | ·50 |
| 3218 | 174 | 39 | 15M | ·2576 | 5·4843 | 16·312 | 117 | 5 | 6 | Mr | 20● | ·74 | 1 | Ap | 19 | ·27 | 2 | My | 18 | ·80 | 4 | Je | 17 | ·33 | 7 | Au | 15 | ·39 |
| | | | | | | | | | | | | | | | | | | | | | 5 | Jl | 16 | ·86 | | | | |

**Surya Siddhanta.**

| Asvina | | | | Kartika | | | | Margasira | | | | Pausha | | | | A.D. | Magha | | | | A.D. | Phalguna | | | | Chaitra | | | |
|---|---|---|---|---|---|---|---|---|---|---|---|---|---|---|---|---|---|---|---|---|---|---|---|---|---|---|---|---|---|
| + 147·65293 | | | | + 177·18353 | | | | + 206·71411 | | | | + 236·24470 | | | | | + 265·77529 | | | | | + 295·30558 | | | | + 324·83647<br>+ 354·36705 | | | |
| + 9·880 | | | | + 11·856 | | | | + 13·832 | | | | + 15·808 | | | | | + 17·784 | | | | | + 19·760 | | | | + 21·736<br>+ 23·712 | | | |
| Week-day | Month | Day | Fraction | Week-day | Month | Day | Fraction | Week-day | Month | Day | Fraction | Week-day | Month | Day | Fraction | | Week-day | Month | Day | Fraction | | Week-day | Month | Day | Fraction | Week-day | Month | Day | Fraction |
| 3 | S | 14○ | ·55 | 5 | O | 14● | ·08 | 6 | N | 12 | ·61 | 1 | D | 12 | ·14 | 80 | 2 | Jr | 10 | ·67 | | 4 | F | 9 | ·20 | 5 | Mr | 9●○ | ·78 |
| 7 | S | 2○ | ·92 | 2 | O | 2 | ·45 | 3 | O | 31 | ·98 | 5 | N | 30 | ·51 | | 7 | D | 30 | ·04 | 81 | 1 | Jr | 28 | ·57 | 3 | F | 27● | ·10 |
| 5 | Au | 23● | ·28 | 6 | S | 21 | ·81 | 1 | O | 21 | ·34 | 2 | N | 19 | ·87 | | 4 | D | 19 | ·41 | 82 | 5 | Jr | 17○ | ·94 | 7 | F | 16 | ·47 |
| 4 | S | 11 | ·18 | 5 | O | 10 | ·71 | 7 | N | 9 | ·24 | 1 | D | 8 | ·77 | 83 | 3 | Jr | 7○ | ·30 | | 4 | F | 5 | ·83 | 6 | Mr | 7 | ·36 |
| 1 | Au | 31 | ·55 | 3 | S | 30 | ·08 | 4 | O | 29 | ·61 | 6 | N | 28 | ·14 | | 7 | D | 27●○ | ·67 | 84 | 2 | Jr | 26 | ·20 | 3 | F | 24 | ·73 |
| 5 | Au | 19 | ·91 | 7 | S | 18 | ·45 | 1 | O | 17 | ·98 | 3 | N | 16 | ·51 | | 5 | D | 16● | ·04 | 85 | 6 | Jr | 14 | ·87 | 1 | F | 13 | ·10 |
| | | | | | | | | | | | | | | | | | | | | | | | | | | 2 | Mr | 14 | ·63 |
| 4 | S | 7 | ·81 | 6 | O | 7 | ·34 | 7 | N | 5○ | ·87 | 2 | D | 5 | ·40 | 86 | 3 | Jr | 3 | ·93 | | 5 | F | 2 | ·46 | 7 | Mr | 4 | ·00 |
| 2 | Au | 28 | ·18 | 3 | S | 26 | ·71 | 5 | O | 26○ | ·24 | 6 | N | 24 | ·77 | | 1 | D | 24 | ·30 | 87 | 2 | Jr | 22 | ·83 | 4 | F | 21 | ·36 |
| 1 | S | 16 | ·08 | 2 | O | 15●○ | ·61 | 4 | N | 14 | ·14 | 5 | D | 13 | ·67 | 88 | 7 | Jr | 12 | ·20 | | 1 | F | 10 | ·73 | 3 | Mr | 11 | ·26 |
| 5 | S | 4 | ·44 | 6 | O | 3● | ·97 | 1 | N | 2 | ·50 | 3 | D | 2 | ·04 | | 4 | D | 31 | ·57 | 89 | 6 | Jr | 30 | ·10 | 7 | F | 28○ | ·63 |
| 2 | Au | 24○ | ·81 | 4 | S | 23 | ·34 | 5 | O | 22 | ·87 | 7 | N | 21 | ·40 | | 1 | D | 20 | ·93 | 90 | 3 | Jr | 19 | ·46 | 4 | F | 17○ | ·99 |
| 1 | S | 12 | ·71 | 3 | O | 12 | ·24 | 4 | N | 10 | ·77 | 6 | D | 10 | ·30 | 91 | 7 | Jr | 8 | ·83 | | 2 | F | 7○ | ·36 | 3 | Mr | 8 | ·89 |
| 6 | S | 2 | ·08 | 7 | O | 1 | ·61 | 2 | O | 31 | ·14 | 3 | N | 29 | ·67 | | 5 | D | 29 | ·20 | 92 | 6 | Jr | 27● | ·73 | 1 | F | 26 | ·26 |
| 3 | Au | 21 | ·44 | 4 | S | 19 | ·97 | 6 | O | 19 | ·50 | 1 | N | 18 | ·03 | | 2 | D | 17○ | ·56 | 93 | 4 | Jr | 16 | ·10 | 5 | F | 14 | ·63 |
| 2 | S | 9 | ·34 | 3 | O | 8 | ·87 | 5 | N | 7 | ·40 | 6 | D | 6○ | ·93 | 94 | 1 | Jr | 5● | ·46 | | 2 | F | 3 | ·99 | 4 | Mr | 5 | ·52 |
| 6 | Au | 29 | ·71 | 1 | S | 28 | ·24 | 2 | O | 27 | ·77 | 4 | N | 26○ | ·30 | | 5 | D | 25 | ·83 | 95 | 7 | Jr | 24 | ·36 | 1 | F | 22 | ·89 |
| 5 | S | 17 | ·60 | 7 | O | 17 | ·14 | 1 | N | 15 | ·67 | 3 | D | 15 | ·20 | 96 | 4 | Jr | 13 | ·73 | | 6 | F | 12 | ·26 | 7 | Mr | 12 | ·79 |
| 2 | S | 5 | ·97 | 4 | O | 5○ | ·50 | 6 | N | 4● | ·03 | 7 | D | 3 | ·56 | 97 | 2 | Jr | 2 | ·09 | | 3 | Jr | 31 | ·62 | 5 | Mr | 2 | ·16 |
| 7 | Au | 26 | ·34 | 1 | S | 24○ | ·87 | 3 | O | 24 | ·40 | 4 | N | 22 | ·93 | | 6 | D | 22 | ·46 | 98 | 7 | Jr | 20 | ·99 | 2 | F | 19 | ·52 |
| 6 | S | 14○ | ·24 | 7 | O | 13 | ·77 | 2 | N | 12 | ·30 | 3 | D | 11 | ·83 | 99 | 5 | Jr | 10 | ·36 | | 6 | F | 8 | ·89 | 1 | Mr | 10 | ·42 |
| 3 | S | 3● | ·60 | 5 | O | 3 | ·13 | 6 | N | 1 | ·66 | 1 | D | 1 | ·20 | | 2 | D | 30 | ·73 | 100 | 4 | Jr | 29○ | ·26 | 5 | F | 27 | ·79 |
| 7 | Au | 22● | ·97 | 2 | S | 21 | ·50 | 4 | O | 21 | ·03 | 5 | N | 19 | ·56 | | 7 | D | 19 | ·09 | 01 | 1 | Jr | 17●○ | ·62 | 3 | F | 16 | ·15 |
| 6 | S | 10 | ·87 | 1 | O | 10 | ·40 | 2 | N | 8 | ·93 | 4 | D | 8 | ·46 | 02 | 5 | Jr | 6○ | ·99 | | 7 | F | 5 | 52 | 2 | Mr | 7 | ·05 |
| 4 | Au | 31 | ·24 | 5 | S | 29 | ·77 | 7 | O | 29 | ·30 | 1 | N | 27 | ·83 | | 3 | D | 27● | ·36 | 03 | 4 | Jr | 25 | ·89 | 6 | F | 24 | ·42 |
| 1 | Au | 20 | ·60 | 3 | S | 19 | ·13 | 4 | O | 18 | ·66 | 6 | N | 17○ | ·19 | | 7 | D | 16 | ·72 | 04 | 2 | Jr | 15 | ·26 | 3 | F | 13 | ·79 |
| | | | | | | | | | | | | | | | | | | | | | | | | | | 5 | Mr | 14 | ·32 |
| 7 | S | 7 | ·50 | 2 | O | 7 | ·03 | 3 | N | 5○ | ·56 | 5 | D | 5 | ·09 | 05 | 6 | Jr | 3 | ·62 | | 1 | F | 2 | ·15 | 2 | Mr | 3 | ·68 |
| 4 | Au | 27 | ·87 | 6 | S | 26 | ·40 | 7 | O | 25●○ | ·93 | 2 | N | 24 | ·46 | | 3 | D | 23 | ·99 | 06 | 5 | Jr | 22 | ·52 | 7 | F | 21 | ·05 |
| 3 | S | 15 | ·76 | 5 | O | 15 | ·30 | 6 | N | 13 | ·83 | 1 | D | 13 | ·36 | 07 | 2 | Jr | 11 | ·89 | | 4 | F | 10 | ·42 | 5 | Mr | 11○ | ·95 |
| 1 | S | 5○ | ·13 | 2 | O | 4 | ·66 | 4 | N | 3 | ·19 | 5 | D | 2 | ·72 | 08 | 7 | Jr | 1 | ·25 | | 1 | Jr | 30 | ·78 | 3 | F | 29○ | ·32 |
| 5 | Au | 24●○ | ·50 | 7 | S | 23 | ·03 | 1 | O | 22 | ·56 | 3 | N | 21 | ·09 | | 4 | D | 20 | ·62 | 09 | 6 | Jr | 19 | ·15 | 7 | F | 17○ | ·68 |
| 4 | S | 12 | ·40 | 5 | O | 11 | ·93 | 7 | N | 10 | ·46 | 1 | D | 9 | ·99 | 10 | 3 | Jr | 8 | ·52 | | 5 | F | 7 | ·05 | 6 | Mr | 8 | ·58 |
| 1 | S | 1 | ·76 | 3 | O | 1 | ·29 | 4 | O | 30 | ·82 | 6 | N | 29 | ·36 | | 7 | D | 28○ | ·89 | 11 | 2 | Jr | 27● | ·42 | 3 | F | 25 | ·95 |
| 6 | Au | 22 | ·13 | 7 | S | 20 | ·66 | 2 | O | 20 | ·19 | 3 | N | 18 | ·72 | | 5 | D | 18○ | ·25 | 12 | 6 | Jr | 16 | ·78 | 1 | F | 15 | ·31 |
| 5 | S | 9 | ·03 | 6 | O | 8 | ·56 | 1 | N | 7 | ·09 | 2 | D | 6○ | ·62 | 13 | 4 | Jr | 5 | ·15 | | 5 | F | 3 | ·68 | 7 | Mr | 5 | ·21 |
| 2 | Au | 29 | ·40 | 3 | S | 27 | ·93 | 5 | O | 27 | ·46 | 6 | N | 25● | ·99 | | 1 | D | 25 | ·52 | 14 | 3 | Jr | 24 | ·05 | 4 | F | 22 | ·58 |
| 1 | S | 17 | ·29 | 2 | O | 16○ | ·82 | 4 | N | 15● | ·35 | 5 | D | 14 | ·88 | 15 | 7 | Jr | 13 | ·42 | | 1 | F | 11 | ·95 | 3 | Mr | 13 | 48 |
| 5 | S | 6 | ·66 | 7 | O | 6○ | ·19 | 1 | N | 4● | ·72 | 3 | D | 4 | ·25 | 16 | 4 | Jr | 2 | ·78 | | 6 | F | 1 | ·31 | 7 | Mr | 1 | ·84 |
| 3 | Au | 26 | ·03 | 4 | S | 24○ | ·56 | 6 | O | 24 | ·09 | 7 | N | 22 | ·62 | | 2 | D | 22 | ·15 | 17 | 3 | Jr | 20 | ·68 | 5 | F | 19 | ·21 |
| 1 | S | 13 | ·92 | 3 | O | 13 | ·46 | 4 | N | 11 | ·99 | 6 | D | 11 | ·52 | 18 | 1 | Jr | 10 | ·05 | | 2 | F | 8○ | ·58 | 4 | Mr | 10 | ·10 |

| Kaliyuga | Vikrama Era | Saka Era | Commencement of Solar Year | | First New-Moon in Solar Year | Anomaly of first New-Moon | A.D. | Week-day of 1st January | Vaisakha | | | | Jyeshta | | | | Ashada | | | | Sravana | | | | Bhadrapada | | | |
|---|---|---|---|---|---|---|---|---|---|---|---|---|---|---|---|---|---|---|---|---|---|---|---|---|---|---|---|---|
| | | | | | | | | ☉'s Anom col. 6 | | | | | + 29·53059 | | | | + 59·06117 | | | | + 88·59176 | | | | + 118·12238 | | | |
| | | | | | | | | ☾'s Anom col. 7 | | | | | + 1·976 | | | | + 5·952 | | | | + 5·928 | | | | + 7·904 | | | |
| | | | Month and day A.D. | Fraction of day | | | | | Week-day | Month | Day | Fraction | Week-day | Month | Day | Fraction | Week-day | Month | Day | Fraction | Week-day | Month | Day | Fraction | Week-day | Month | Day | Fraction |
| 3219 | 175 | 40 | 15M | ·5164 | 24·1232 | 14·446 | **118** | 6 | 5 | Ap | 8 | ·64 | 7 | My | 8 | ·17 | 1 | Je | 6 | ·70 | 3 | Jl | 6 | ·23 | 4 | Au | 4○ | ·76 |
| 3220 | 176 | 41 | 15M | ·7751 | 13·2315 | 10·603 | **119** | 7 | 3 | Mr | 29 | ·01 | 4 | Ap | 27 | ·54 | 6 | My | 27 | ·07 | 7 | Je | 25 | ·60 | 2 | Jl | 25○ | ·13 |
| 3221 | 177 | 42 | 15M | ·0339 | 2·3398 | 6·760 | **120** | 1 | 7 | Mr | 17 | ·37 | 1 | Ap | 15 | ·90 | 4 | Je | 13 | ·96 | 6 | Jl | 13○ | ·50 | 1 | Au | 12 | ·03 |
| | | | | | | | | | | | | | 3 | My | 15 | ·43 | | | | | | | | | | | | |
| 3222 | 178 | 43 | 15M | ·2926 | 20·9787 | 4·893 | **121** | 3 | 6 | Ap | 5 | ·27 | 7 | My | 4 | ·80 | 2 | Je | 3 | ·33 | 3 | Jl | 2● | ·86 | 5 | Au | 1 | ·39 |
| 3223 | 179 | 44 | 15M | ·5514 | 10·0370 | 1·051 | **122** | 4 | 3 | Mr | 25 | ·64 | 5 | Ap | 24 | ·17 | 6 | My | 23○ | ·70 | 2 | Jl | 21 | ·76 | 4 | Au | 20 | ·29 |
| | | | | | | | | | | | | | | | | | 1 | Je | 22● | ·23 | | | | | | | | |
| 3224 | 180 | 45 | 15M | ·8101 | 28·7259 | 26·739 | **123** | 5 | 2 | Ap | 13 | ·54 | 4 | My | 13○ | ·07 | 5 | Je | 11 | ·60 | 7 | Jl | 11 | ·13 | 1 | Au | 9 | ·66 |
| 3225 | 181 | 46 | 15M | ·0689 | 17·8342 | 22·896 | **124** | 6 | 6 | Ap | 1 | ·90 | 1 | My | 1● | ·43 | 2 | My | 30 | ·96 | 4 | Je | 29 | ·49 | 6 | Jl | 29 | ·02 |
| 3226 | 182 | 47 | 15M | ·3277 | 6·9425 | 19·053 | **125** | 1 | 4 | Mr | 22○● | ·28 | 5 | Ap | 20● | ·80 | 7 | My | 20 | ·33 | 1 | Je | 18 | ·86 | 4 | Au | 16 | ·92 |
| | | | | | | | | | | | | | | | | | | | | | 3 | Jl | 18 | ·39 | | | | |
| 3227 | 183 | 48 | 15M | ·5864 | 25·5814 | 17·176 | **126** | 2 | 3 | Ap | 10● | ·17 | 4 | My | 9 | ·70 | 6 | Je | 8 | ·23 | 7 | Jl | 7 | ·76 | 2 | Au | 6 | ·29 |
| 3228 | 184 | 49 | 15M | ·8452 | 14·6897 | 13·344 | **127** | 3 | 7 | Mr | 30 | ·53 | 2 | Ap | 29 | ·06 | 3 | My | 28 | ·60 | 5 | Je | 27 | ·13 | 6 | Jl | 26 | ·66 |
| 3229 | 185 | 50 | 15M | ·1039 | 3·7980 | 9·501 | **128** | 4 | 4 | Mr | 18 | ·90 | 6 | Ap | 17 | ·43 | 7 | My | 16 | ·96 | 4 | Jl | 15 | ·02 | 5 | Au | 13 | ·55 |
| | | | | | | | | | | | | | | | | | 2 | Je | 15 | ·49 | | | | | | | | |
| 3230 | 186 | 51 | 15M | ·3627 | 22·4369 | 7·634 | **129** | 6 | 3 | Ap | 6 | ·80 | 5 | My | 6 | ·33 | 6 | Je | 4 | ·86 | 1 | Jl | 4○ | ·39 | 2 | Au | 2 | ·92 |
| 3231 | 187 | 52 | 15M | ·6214 | 11·5452 | 3·791 | **130** | 7 | 1 | Mr | 27 | ·17 | 2 | Ap | 25 | ·70 | 4 | My | 25 | ·23 | 5 | Je | 23●○ | ·76 | 7 | Jl | 23 | ·29 |
| 3232 | 188 | 53 | 15M | ·8802 | 0·6535 | 27·503 | **131** | 1 | 5 | Mr | 16 | ·53 | 1 | My | 14 | ·59 | 3 | Je | 13●○ | ·12 | 4 | Jl | 12 | ·66 | 6 | Au | 11 | ·19 |
| | | | | | | | | | 7 | Ap | 15 | ·06 | | | | | | | | | | | | | | | | |
| 3233 | 189 | 54 | 15M | ·1389 | 19·2923 | 25·636 | **132** | 2 | 4 | Ap | 3 | ·43 | 5 | My | 2 | ·96 | 7 | Je | 1● | ·49 | 2 | Jl | 1 | ·02 | 3 | Jl | 30 | ·55 |
| 3234 | 190 | 55 | 15M | ·3977 | 8·4006 | 21·794 | **133** | 4 | 1 | Mr | 23 | ·80 | 3 | Ap | 22○ | ·33 | 4 | My | 21 | ·86 | 6 | Je | 20 | ·39 | 7 | Jl | 19 | ·92 |
| | | | | | | | | | | | | | | | | | | | | | | | | | 2 | Au | 18 | ·45 |
| 3235 | 191 | 56 | 15M | ·6565 | 27·0395 | 19·927 | **134** | 5 | 7 | Ap | 11●○ | ·70 | 2 | My | 11 | ·23 | 3 | Je | 9 | ·76 | 5 | Jl | 9 | ·29 | 6 | Au | 7 | ·82 |
| 3236 | 192 | 57 | 15M | ·9152 | 16·1478 | 16·084 | **135** | 6 | 5 | Ap | 1●○ | ·06 | 6 | Ap | 30 | ·59 | 1 | My | 30 | ·12 | 2 | Je | 28 | ·65 | 4 | Jl | 28 | ·18 |
| 3237 | 193 | 58 | 15M | ·1740 | 5·2561 | 12·241 | **136** | 7 | 2 | Mr | 20 | ·43 | 3 | Ap | 18 | ·96 | 5 | My | 18 | ·49 | 7 | Je | 17 | ·02 | 3 | Au | 15○ | ·08 |
| | | | | | | | | | | | | | | | | | | | | | 1 | Jl | 16 | ·55 | | | | |
| 3238 | 194 | 59 | 15M | ·4327 | 23·8950 | 10·375 | **137** | 2 | 1 | Ap | 8 | ·33 | 2 | My | 7 | ·86 | 4 | Je | 6 | ·39 | 5 | Jl | 5 | ·92 | 7 | Au | 4○ | ·45 |
| 3239 | 195 | 60 | 15M | ·6915 | 13·0083 | 6·532 | **138** | 3 | 5 | Mr | 28 | ·69 | 7 | Ap | 27 | ·22 | 1 | My | 26 | ·76 | 3 | Je | 25 | ·29 | 4 | Jl | 24○ | ·82 |
| 3240 | 196 | 61 | 15M | ·9502 | 2·1116 | 2·689 | **139** | 4 | 3 | Mr | 18 | ·06 | 4 | Ap | 16 | ·59 | 7 | Je | 14 | ·65 | 2 | Jl | 14 | ·18 | 3 | Au | 12 | ·71 |
| | | | | | | | | | | | | | 6 | My | 16 | ·12 | | | | | | | | | | | | |
| 3241 | 197 | 62 | 15M | ·2090 | 20·7505 | 0·822 | **140** | 5 | 1 | Ap | 4 | ·96 | 3 | My | 4 | ·49 | 5 | Je | 3○ | ·02 | 6 | Jl | 2● | ·55 | 1 | Au | 1 | ·08 |
| 3242 | 198 | 63 | 15M | ·4678 | 9·8589 | 24·534 | **141** | 7 | 6 | Mr | 25 | ·33 | 7 | Ap | 23 | ·86 | 2 | My | 23○ | ·39 | 3 | Je | 21● | ·92 | 5 | Jl | 21 | ·45 |
| 3243 | 199 | 64 | 15M | ·7265 | 28·4977 | 22·667 | **142** | 1 | 5 | Ap | 13 | ·22 | 6 | My | 12●○ | ·75 | 1 | Je | 11 | ·28 | 2 | Jl | 10 | ·82 | 4 | Au | 9 | ·35 |
| 3244 | 200 | 65 | 15M | ·9853 | 17·6060 | 18·825 | **143** | 2 | 2 | Ap | 2○ | ·59 | 4 | My | 2● | ·12 | 5 | My | 31 | ·65 | 7 | Je | 30 | ·18 | 1 | Jl | 29 | ·71 |
| 3245 | 201 | 66 | 15M | ·2440 | 6·7148 | 14·982 | **144** | 3 | 6 | Mr | 21○ | ·96 | 1 | Ap | 20● | ·49 | 3 | My | 20 | ·02 | 4 | Je | 18 | ·55 | 7 | Au | 16 | ·61 |
| | | | | | | | | | | | | | | | | | | | | | 6 | Jl | 18 | ·08 | | | | |
| 3246 | 202 | 67 | 15M | ·5028 | 25·3532 | 13·115 | **145** | 5 | 5 | Ap | 9 | ·86 | 7 | My | 9 | ·39 | 1 | Je | 7 | ·92 | 3 | Jl | 7 | ·45 | 4 | Au | 5 | ·98 |
| 3247 | 203 | 68 | 15M | ·7615 | 14·4615 | 9·272 | **146** | 6 | 2 | Mr | 30 | ·22 | 4 | Ap | 28 | ·75 | 6 | My | 28 | ·28 | 7 | Je | 26 | ·81 | 2 | Jl | 26 | ·34 |
| 3248 | 204 | 69 | 16M | ·0203 | 3·5698 | 5·430 | **147** | 7 | 7 | Mr | 19 | ·59 | 2 | Ap | 18 | ·12 | 5 | Je | 16 | ·18 | 6 | Jl | 15○ | ·71 | 1 | Au | 14 | ·24 |
| | | | | | | | | | | | | | 3 | My | 17 | ·65 | | | | | | | | | | | | |
| 3249 | 205 | 70 | 15M | ·2791 | 22·2087 | 3·563 | **148** | 1 | 6 | Ap | 6 | ·49 | 1 | My | 6 | ·02 | 2 | Je | 4 | ·55 | 4 | Jl | 4●○ | ·08 | 5 | Au | 2 | ·61 |
| 3250 | 206 | 71 | 15M | ·5378 | 11·3170 | 27·275 | **149** | 3 | 3 | Mr | 26 | ·85 | 5 | Ap | 25 | ·38 | 6 | My | 24 | ·92 | 1 | Je | 23●○ | ·45 | 2 | Jl | 22 | ·98 |
| 3251 | 207 | 72 | 15M | ·7966 | 0·4253 | 23·432 | **150** | 4 | 1 | Mr | 16 | ·22 | 4 | My | 14 | ·28 | 5 | Je | 12● | ·81 | 7 | Jl | 12 | ·34 | 1 | Au | 10 | ·87 |
| | | | | | | | | | 2 | Ap | 14 | ·75 | | | | | | | | | | | | | | | | |
| 3252 | 208 | 73 | 16M | ·0553 | 19·0641 | 21·565 | **151** | 5 | 7 | Ap | 4 | ·12 | 1 | My | 3○ | ·65 | 3 | Je | 2 | ·18 | 4 | Jl | 1 | ·71 | 6 | Jl | 31 | ·24 |
| 3253 | 209 | 74 | 15M | ·3141 | 8·1724 | 17·723 | **152** | 6 | 4 | Mr | 23 | ·49 | 6 | Ap | 22●○ | ·02 | 7 | My | 21 | ·55 | 2 | Je | 20 | ·08 | 3 | Jl | 19 | ·61 |
| | | | | | | | | | | | | | | | | | | | | | | | | | 5 | Au | 18 | ·14 |
| 3254 | 210 | 75 | 15M | ·5728 | 26·8113 | 15·856 | **153** | 1 | 3 | Ap | 11●○ | ·38 | 4 | My | 10 | ·91 | 6 | Je | 9 | ·44 | 7 | Jl | 8 | ·98 | 2 | Au | 7 | ·51 |
| 3255 | 211 | 76 | 15M | ·8316 | 15·9196 | 12·013 | **154** | 2 | 7 | Mr | 31● | ·75 | 2 | Ap | 30 | ·28 | 3 | My | 29 | ·81 | 5 | Je | 28 | ·34 | 6 | Jl | 27 | ·87 |
| 3256 | 212 | 77 | 16M | ·0904 | 5·0279 | 8·170 | **155** | 3 | 5 | Mr | 21 | ·12 | 6 | Ap | 19 | ·65 | 1 | My | 19 | ·18 | 4 | Jl | 17 | ·24 | 5 | Au | 15○ | ·77 |
| | | | | | | | | | | | | | | | | | 2 | Je | 17 | ·71 | | | | | | | | |
| 3257 | 213 | 78 | 15M | ·3491 | 23·6668 | 6·304 | **156** | 4 | 4 | Ap | 8 | ·02 | 5 | My | 7 | ·55 | 7 | Je | 6 | ·08 | 1 | Jl | 5 | ·61 | 8 | Au | 4○ | ·14 |
| 3258 | 214 | 79 | 15M | ·6079 | 12·7751 | 2·461 | **157** | 6 | 1 | Mr | 28 | ·38 | 2 | Ap | 26 | ·91 | 4 | My | 26 | ·44 | 5 | Je | 24● | ·97 | 7 | Jl | 24 | ·50 |

**Surya Siddhanta.**

| Asvina | | | Kartika | | | Margasira | | | Pausha | | | A.D. | Magha | | | A.D. | Phalguna | | | Chaitra | | |
|---|---|---|---|---|---|---|---|---|---|---|---|---|---|---|---|---|---|---|---|---|---|---|
| + 147·65293 | | | + 177·18353 | | | + 206·71411 | | | + 236·24470 | | | | + 265·77529 | | | | + 295·34588 | | | { + 324·83647<br>+ 354·36705 | | |
| + 9·880 | | | + 11·856 | | | + 13·832 | | | + 15·808 | | | | + 17·784 | | | | + 19·760 | | | { + 21·736<br>+ 23·712 | | |
| Week-day Month | Day | Fraction | Week-day Month | Day | Fraction | Week-day Month | Day | Fraction | Week-day Month | Day | Fraction | | Week-day Month | Day | Fraction | | Week-day Month | Day | Fraction | Week-day Month | Day | Fraction |
| 6 S | 3● | ·29 | 7 O | 2 | ·82 | 2 N | 1 | ·35 | 3 N | 30 | ·88 | | 5 D | 30 | ·41 | **19** | 6 Jr | 28○ | ·94 | 1 F | 27 | ·48 |
| 3 Au | 23 | ·66 | 5 S | 22 | ·19 | 6 O | 21 | ·72 | 1 N | 20 | ·25 | | 2 D | 19 | ·78 | **20** | 4 Jr | 18●○ | ·31 | 5 F | 16 | ·84 |
| 2 S | 10 | ·56 | 4 O | 10 | ·09 | 5 N | 8 | ·62 | 7 D | 8 | ·15 | **21** | 1 Jr | 6 | ·68 | | 3 F | 5 | ·21 | 4 Mr | 6 | ·74 |
| 6 Au | 30 | ·92 | 1 S | 29 | ·45 | 2 O | 28 | ·98 | 4 N | 27○ | ·52 | | 6 D | 27 | ·05 | **22** | 7 Jr | 25 | ·58 | 2 F | 24 | ·11 |
| 5 S | 18 | ·82 | 7 O | 18 | ·35 | **Margasira Kshaya.** | | | 1 N | 16○ | ·88 | | 3 D | 16 | ·41 | **23** | 4 Jr | 14 | ·94 | { 6 F | 13 | ·47 |
| | | | | | | | | | | | | | | | | | | | | 1 Mr | 15 | ·00 |
| 3 S | 8 | ·19 | 4 O | 7 | ·72 | 6 N | 6●○ | ·25 | 7 D | 5 | ·78 | **24** | 2 Jr | 4 | ·31 | | 3 F | 2 | ·84 | 5 Mr | 3 | ·37 |
| 7 Au | 27 | ·56 | 2 S | 26 | ·09 | 3 O | 25● | ·62 | 5 N | 24 | ·15 | | 6 D | 23 | ·68 | **25** | 1 Jr | 22 | ·21 | 2 F | 20 | ·74 |
| 6 S | 15○ | ·45 | 7 O | 14 | ·98 | 2 N | 13 | ·51 | 4 D | 13 | ·04 | **26** | 5 Jr | 11 | ·58 | | 7 F | 10 | ·11 | 1 Mr | 11○ | ·64 |
| 3 S | 4●○ | ·82 | 5 O | 4 | ·35 | 6 N | 2 | ·88 | 1 D | 2 | ·41 | | 2 D | 31 | ·94 | **27** | 4 Jr | 30 | ·47 | 6 Mr | 1○ | ·00 |
| 1 Au | 25●○ | ·19 | 2 S | 23 | ·72 | 4 O | 23 | ·25 | 5 N | 21 | ·78 | | 7 D | 21 | ·31 | **28** | 1 Jr | 19 | ·84 | 3 F | 18 | ·37 |
| 7 S | 12 | ·08 | 1 O | 11 | ·62 | 3 N | 10 | ·15 | 4 D | 9 | ·68 | **29** | 6 Jr | 8○ | ·21 | | 7 F | 6● | ·74 | 2 Mr | 8 | ·27 |
| 4 S | 1 | ·45 | 5 S | 30 | ·98 | 7 O | 30 | ·51 | 2 N | 29 | ·04 | | 3 D | 28○ | ·57 | **30** | 5 Jr | 27● | ·10 | 6 F | 25 | ·64 |
| 1 Au | 21 | ·82 | 3 S | 20 | ·35 | 4 O | 19 | ·88 | 6 N | 18 | ·41 | | 7 D | 17○ | ·94 | **31** | 2 Jr | 16 | ·47 | 4 F | 15 | ·00 |
| 7 S | 9 | ·72 | 2 O | 9 | ·25 | 3 N | 7 | ·78 | 5 D | 7 | ·31 | **32** | 6 Jr | 5 | ·84 | | 1 F | 4 | ·37 | 2 Mr | 4 | ·90 |
| 5 Au | 29 | ·08 | 6 S | 27 | ·61 | 1 O | 27○ | ·14 | 2 N | 25● | ·68 | | 4 D | 25 | ·21 | **33** | 5 Jr | 23 | ·74 | 7 F | 22 | ·27 |
| 3 S | 16 | ·98 | 5 O | 16○ | ·51 | 7 N | 15● | ·04 | 1 D | 14 | ·57 | **34** | 3 Jr | 13 | ·10 | | 4 F | 11 | ·63 | 6 Mr | 13 | ·16 |
| 1 S | 6 | ·35 | 2 O | 5○ | ·88 | 4 N | 4 | ·41 | 5 D | 3 | ·94 | **35** | 7 Jr | 2 | ·47 | | 2 F | 1 | ·00 | 3 Mr | 2 | ·53 |
| 5 Au | 26 | ·72 | 7 S | 25● | ·25 | 1 O | 24 | ·78 | 3 N | 23 | ·31 | | 4 D | 22 | ·84 | **36** | 6 Jr | 21 | ·37 | 7 F | 19○ | ·90 |
| 4 S | 13● | ·61 | 6 O | 13 | ·14 | 7 N | 11 | ·67 | 2 D | 11 | ·20 | **37** | 3 Jr | 9 | ·74 | | 5 F | 8○ | ·27 | 6 Mr | 9 | ·80 |
| 1 S | 2● | ·98 | 3 O | 2 | ·51 | 5 N | 1 | ·04 | 6 N | 30 | ·57 | | 1 D | 30 | ·10 | **38** | 2 Jr | 28●○ | ·63 | 4 F | 27 | ·16 |
| 6 Au | 23 | ·35 | 7 S | 21 | ·88 | 2 O | 21 | ·41 | 3 N | 19 | ·94 | | 5 D | 19 | ·47 | **39** | 7 Jr | 18● | ·00 | 1 F | 16 | ·53 |
| 5 S | 11 | ·24 | 6 O | 10 | ·78 | 1 N | 9 | ·31 | 2 D | 8○ | ·84 | **40** | 4 Jr | 7 | ·37 | | 5 F | 5 | ·90 | 7 Mr | 6 | ·43 |
| 2 Au | 30 | ·61 | 4 S | 29 | ·14 | 5 O | 28 | ·67 | 7 N | 27○ | ·20 | | 1 D | 26 | ·73 | **41** | 3 Jr | 25 | ·26 | 4 F | 23 | ·80 |
| { 6 Au | 19 | ·98 | 3 O | 18 | ·04 | **Margasira Kshaya.** | | | 4 N | 16●○ | ·57 | | 6 D | 16 | ·10 | **42** | 7 Jr | 14 | ·63 | { 2 F | 13 | ·16 |
| 1 S | 18 | ·51 | | | | | | | | | | | | | | | | | | 3 Mr | 14 | ·69 |
| 5 S | 7 | ·88 | 7 O | 7 | ·41 | 1 N | 5● | ·94 | 3 D | 5 | ·47 | **43** | 5 Jr | 4 | ·00 | | 6 F | 2 | ·53 | 1 Mr | 4 | ·06 |
| 3 Au | 28 | ·24 | 4 S | 26○ | ·77 | 6 O | 26 | ·30 | 7 N | 24 | ·84 | | 2 D | 24 | ·37 | **44** | 3 Jr | 22 | ·90 | 5 F | 21 | ·43 |
| 2 S | 15○ | ·14 | 3 O | 14 | ·67 | 5 N | 13 | ·20 | 6 D | 12 | ·73 | **45** | 1 Jr | 11 | ·26 | | 2 F | 9 | ·79 | 4 Mr | 11○ | ·32 |
| 6 S | 4●○ | ·51 | 1 O | 4 | ·04 | 2 N | 2 | ·57 | 4 D | 2 | ·10 | | 5 D | 31 | ·63 | **46** | 7 Jr | 30 | ·16 | 1 F | 28● | ·69 |
| 3 Au | 24● | ·88 | 5 S | 23 | ·41 | 6 O | 22 | ·94 | 1 N | 21 | ·47 | | 3 D | 21 | ·00 | **47** | 4 Jr | 19○ | ·53 | 6 F | 18● | ·06 |
| 2 S | 12 | ·77 | 4 O | 12 | ·30 | 5 N | 10 | ·83 | 7 D | 10 | ·36 | **48** | 1 Jr | 8○ | ·90 | | 3 F | 7● | ·43 | 4 Mr | 7 | ·96 |
| 7 S | 1 | ·14 | 1 S | 30 | ·67 | 3 O | 30 | ·20 | 4 N | 28 | ·73 | | 6 D | 28○ | ·26 | **49** | 7 Jr | 26 | ·79 | 2 F | 25 | ·32 |
| 4 Au | 21 | ·51 | 7 S | 20 | ·04 | 7 O | 19 | ·57 | 2 N | 18 | ·10 | | 3 D | 17 | ·63 | **50** | 5 Jr | 16 | ·16 | 6 F | 14 | ·69 |
| 3 S | 9 | ·40 | 4 O | 8 | ·94 | 6 N | 7○ | ·47 | 1 D | 7● | ·00 | **51** | 2 Jr | 5 | ·53 | | 4 F | 4 | ·06 | 5 Mr | 5 | ·59 |
| 7 Au | 29 | ·77 | 2 S | 28 | ·30 | 3 O | 27○ | ·83 | 5 N | 26● | ·36 | | 6 D | 25 | ·89 | **52** | 1 Jr | 24 | ·42 | 2 F | 22 | ·96 |
| 6 S | 16 | ·67 | 1 O | 16○ | ·20 | 2 N | 14 | ·73 | 4 D | 14 | ·26 | **53** | 5 Jr | 12 | ·79 | | 7 F | 11 | ·32 | 1 Mr | 12 | ·85 |
| 4 S | 6 | ·04 | 5 O | 5 | ·57 | 7 N | 4 | ·10 | 1 D | 3 | ·63 | **54** | 3 Jr | 2 | ·16 | | 4 Jr | 31 | ·69 | 6 Mr | 2○ | ·22 |
| 1 Au | 26○ | ·40 | 2 S | 24● | ·93 | 4 O | 24 | ·46 | 6 N | 23 | ·00 | | 7 D | 22 | ·53 | **55** | 2 Jr | 21 | ·06 | 3 F | 19○ | ·59 |
| 7 S | [illegible]● | ·30 | 1 O | 13 | ·83 | 3 N | 12 | ·36 | 4 D | 11 | ·89 | **56** | 6 Jr | 10 | ·42 | | 7 F | 8●○ | ·95 | 2 Mr | 9 | ·48 |
| 4 S | 2 | ·67 | 6 O | 2 | ·20 | 7 O | 31 | ·73 | 2 N | 30 | ·26 | | 3 D | 29 | ·79 | **57** | 5 Jr | 28● | ·32 | 6 F | 26 | ·85 |
| 2 Au | 23 | ·04 | 3 S | 21 | ·57 | 5 O | 21 | ·10 | 6 N | 19 | ·63 | | 1 D | 19○ | ·16 | **58** | 2 Jr | 17 | ·69 | 4 F | 16 | ·22 |

## TABLE X—

☉'s Anom col. 6 + 29·53059 + 59·06117 + 88·59176 + 118·12135

☾'s Anom col. 7 + 1·976 + 3·952 + 5·928 + 7·904

| Kaliyuga | Vikrama Era | Saka Era | Commencement of Solar Year: Month and day A.D. | Fraction of day | First New-Moon in Solar Year | Anomaly of first New-Moon | A.D. | Week-day of 1st January | Vaisakha: Week-day | Month | Day | Fraction | Jyeshta: Week-day | Month | Day | Fraction | Ashada: Week-day | Month. | Day | Fraction. | Sravana: Week-day | Month | Day | Fraction | Bhadrapada: Week-day | Month | Day | Fraction |
|---|---|---|---|---|---|---|---|---|---|---|---|---|---|---|---|---|---|---|---|---|---|---|---|---|---|---|---|---|
| 3259 | 215 | 80 | 15M | ·8666 | 1·8834 | 26·173 | **158** | 7 | 5 | Mr | 17 | ·75 | 7 | Ap | 16 | ·28 | 3 | Je | 14○ | ·34 | 4 | Jl | 13● | ·87 | 6 | Au | 12 | ·40 |
| | | | | | | | | | | | | | 1 | My | 15 | ·81 | | | | | | | | | | | | |
| 3260 | 216 | 81 | 16M | ·1254 | 20·5223 | 24·306 | **159** | 1 | 4 | Ap | 5 | ·65 | 6 | My | 5 | ·18 | 7 | Je | 3○ | ·71 | 2 | Jl | 3 | ·24 | 3 | Au | 1 | ·77 |
| 3261 | 217 | 82 | 15M | ·3841 | 9·6306 | 20·463 | **160** | 2 | 2 | Mr | 25 | ·01 | 3 | Ap | 23 | ·54 | 5 | My | 23●○ | ·07 | 6 | Je | 21 | ·61 | 1 | Jl | 21 | ·14 |
| 3262 | 218 | 83 | 15M | ·6429 | 28·2695 | 18·596 | **161** | 4 | 7 | Ap | 12 | ·91 | 2 | My | 12● | ·44 | 3 | Je | 10 | ·97 | 5 | Jl | 10 | ·50 | 7 | Au | 9 | ·03 |
| 3263 | 219 | 84 | 15M | ·9016 | 17·3778 | 14·754 | **162** | 5 | 5 | Ap | 2○ | ·28 | 6 | My | 1● | ·81 | 1 | My | 31 | ·34 | 2 | Je | 29 | ·87 | 4 | Jl | 29 | ·40 |
| 3264 | 220 | 85 | 16M | ·1604 | 6·4861 | 10·911 | **163** | 6 | 2 | Mr | 22○ | ·65 | 4 | Ap | 21 | ·18 | 5 | My | 20 | ·71 | 7 | Je | 19 | ·24 | 3 | Au | 17 | ·30 |
| | | | | | | | | | | | | | | | | | | | | | 1 | Jl | 18 | ·77 | | | | |
| 3265 | 221 | 86 | 15M | ·4192 | 25·1250 | 9·044 | **164** | 7 | 1 | Ap | 9 | ·54 | 3 | My | 9 | ·07 | 4 | Je | 7 | ·60 | 6 | Jl | 7 | ·14 | 7 | Au | 5 | ·67 |
| 3266 | 222 | 87 | 15M | ·6779 | 14·2333 | 5·201 | **165** | 2 | 5 | Mr | 29 | ·91 | 7 | Ap | 28 | ·44 | 1 | My | 27 | ·97 | 3 | Je | 26 | ·50 | 5 | Jl | 26○ | ·03 |
| 3267 | 223 | 88 | 15M | ·9367 | 3·3416 | 1·359 | **166** | 3 | 3 | Mr | 19 | ·28 | 4 | Ap | 17 | ·81 | 7 | Je | 15 | ·87 | 2 | Jl | 15○ | ·40 | 3 | Au | 13 | ·93 |
| | | | | | | | | | | | | | 6 | My | 17 | ·34 | | | | | | | | | | | | |
| 3268 | 224 | 89 | 16M | ·1954 | 21·9805 | 27·047 | **167** | 4 | 2 | Ap | 7 | ·18 | 3 | My | 6 | ·71 | 5 | Je | 5 | ·24 | 6 | Jl | 4●○ | ·77 | 1 | Au | 3 | ·30 |
| 3269 | 225 | 90 | 15M | ·4542 | 11·0888 | 23·204 | **168** | 5 | 6 | Mr | 26 | ·54 | 1 | Ap | 25 | ·07 | 2 | My | 24 | ·60 | 4 | Je | 23● | ·13 | 5 | Jl | 22 | ·66 |
| 3270 | 226 | 91 | 15M | ·7129 | 0·1971 | 19·361 | **169** | 7 | 3 | Mr | 15 | ·91 | 6 | My | 13○ | ·97 | 1 | Je | 12 | ·50 | 3 | Jl | 12 | ·03 | 4 | Au | 10 | ·56 |
| | | | | | | | | | 5 | Ap | 14 | ·44 | | | | | | | | | | | | | | | | |
| 3271 | 227 | 92 | 15M | ·9717 | 18·8359 | 17·494 | **170** | 1 | 2 | Ap | 3 | ·81 | 4 | My | 3●○ | ·34 | 5 | Je | 1 | ·87 | 7 | Jl | 1 | ·40 | 1 | Jl | 30 | ·93 |
| 3272 | 228 | 93 | 16M | ·2305 | 7·9442 | 13·652 | **171** | 2 | 7 | Mr | 24 | ·17 | 1 | Ap | 22●○ | ·70 | 3 | My | 22 | ·24 | 4 | Je | 20 | ·77 | 6 | Jl | 20 | ·30 |
| | | | | | | | | | | | | | | | | | | | | | | | | | 7 | Au | 18 | ·83 |
| 3273 | 229 | 94 | 15M | ·4892 | 26·5831 | 11·785 | **172** | 3 | 6 | Ap | 11 | ·07 | 7 | My | 10 | ·60 | 2 | Je | 9 | ·13 | 3 | Jl | 8 | ·66 | 5 | Au | 7 | ·19 |
| 3274 | 230 | 95 | 15M | ·7480 | 15·6914 | 7·942 | **173** | 5 | 3 | Mr | 31 | ·44 | 4 | Ap | 29 | ·97 | 6 | My | 29 | ·50 | 1 | Je | 28 | ·03 | 2 | Jl | 27 | ·56 |
| 3275 | 231 | 96 | 16M | ·0067 | 4·7997 | 4·099 | **174** | 6 | 7 | Mr | 20 | ·81 | 2 | Ap | 19 | ·34 | 3 | My | 18 | ·87 | 6 | Jl | 16 | ·93 | 1 | Au | 15○ | ·46 |
| | | | | | | | | | | | | | | | | | 5 | Je | 17 | ·40 | | | | | | | | |
| 3276 | 232 | 97 | 16M | ·2655 | 23·4386 | 2·233 | **175** | 7 | 6 | Ap | 8 | ·70 | 1 | My | 8 | ·23 | 2 | Je | 6 | ·76 | 4 | Jl | 6 | ·30 | 5 | Au | 4● | ·83 |
| 3277 | 233 | 98 | 15M | ·5242 | 12·5469 | 25·945 | **176** | 1 | 4 | Mr | 28 | ·07 | 5 | Ap | 26 | ·60 | 7 | My | 26 | ·13 | 1 | Je | 24○ | ·66 | 3 | Jl | 24● | ·19 |
| 3278 | 234 | 99 | 15M | ·7830 | 1·6552 | 22·102 | **177** | 3 | 1 | Mr | 17 | ·44 | 2 | Ap | 15 | ·97 | 6 | Je | 14○ | ·03 | 7 | Jl | 13● | ·56 | 2 | Au | 12 | ·09 |
| | | | | | | | | | | | | | 4 | My | 15 | ·50 | | | | | | | | | | | | |
| 3279 | 235 | 100 | 16M | ·0417 | 20·2941 | 20·235 | **178** | 4 | 7 | Ap | 5 | ·34 | 1 | My | 4 | ·87 | 3 | Je | 3○ | ·40 | 4 | Jl | 2 | ·93 | 6 | Au | 1 | ·46 |
| 3280 | 236 | 101 | 16M | ·3005 | 9·4024 | 16·392 | **179** | 5 | 4 | Mr | 25 | ·70 | 6 | Ap | 24 | ·23 | 7 | My | 23● | ·76 | 2 | Je | 22 | ·29 | 3 | Jl | 21 | ·82 |
| 3281 | 237 | 102 | 15M | ·5593 | 28·0413 | 14·526 | **180** | 6 | 3 | Ap | 12○ | ·60 | 5 | My | 12● | ·13 | 6 | Je | 10 | ·66 | 1 | Jl | 10 | ·19 | 2 | Au | 8 | ·72 |
| 3282 | 238 | 103 | 15M | ·8180 | 17·1496 | 10·683 | **181** | 1 | 7 | Ap | 1○ | ·97 | 2 | My | 1 | ·50 | 4 | My | 31 | ·03 | 5 | Je | 29 | ·56 | 7 | Jl | 29 | ·09 |
| 3283 | 239 | 104 | 16M | ·0768 | 6·2579 | 6·840 | **182** | 2 | 5 | Mr | 22 | ·33 | 6 | Ap | 20 | ·86 | 1 | My | 20 | ·40 | 2 | Je | 18 | ·93 | 5 | Au | 16 | ·99 |
| | | | | | | | | | | | | | | | | | | | | | 4 | Jl | 18 | ·46 | | | | |
| 3284 | 240 | 105 | 16M | ·3355 | 24·8968 | 4·973 | **183** | 3 | 4 | Ap | 10 | ·23 | 5 | My | 9 | ·76 | 7 | Je | 8 | ·29 | 1 | Jl | 7 | ·82 | 3 | Au | 6○ | ·35 |
| 3285 | 241 | 106 | 15M | ·5943 | 14·0051 | 1·130 | **184** | 4 | 1 | Mr | 29 | ·60 | 3 | Ap | 28 | ·13 | 4 | My | 27 | ·66 | 6 | Je | 26 | ·19 | 7 | Jl | 25○ | ·72 |
| 3286 | 242 | 107 | 15M | ·8530 | 3·1134 | 24·842 | **185** | 6 | 5 | Mr | 18 | ·97 | 7 | Ap | 17 | ·50 | 3 | Je | 15 | ·56 | 5 | Jl | 15●○ | ·09 | 6 | Au | 13 | ·62 |
| | | | | | | | | | | | | | 2 | My | 17 | ·03 | | | | | | | | | | | | |
| 3287 | 243 | 108 | 16M | ·1118 | 21·7522 | 22·976 | **186** | 7 | 4 | Ap | 6 | ·86 | 6 | My | 6 | ·39 | 7 | Je | 4 | ·92 | 2 | Jl | 4● | ·46 | 3 | Au | 2 | ·99 |
| 3288 | 244 | 109 | 16M | ·3706 | 10·8605 | 19·133 | **187** | 1 | 2 | Mr | 27 | ·23 | 3 | Ap | 25 | ·76 | 5 | My | 25○ | ·29 | 6 | Je | 23 | ·82 | 1 | Jl | 23 | ·35 |
| 3289 | 245 | 110 | 15M | ·6293 | 29·4994 | 17·266 | **188** | 2 | 6 | Mr | 15 | ·60 | 2 | My | 13●○ | ·66 | 4 | Je | 12 | ·19 | 5 | Jl | 11 | ·72 | 7 | Au | 10 | ·25 |
| | | | | | | | | | 1 | Ap | 14 | ·13 | | | | | | | | | | | | | | | | |
| 3290 | 246 | 111 | 15M | ·8881 | 18·6077 | 13·423 | **189** | 4 | 5 | Ap | 3 | ·50 | 7 | My | 3●○ | ·03 | 1 | Je | 1 | ·56 | 3 | Jl | 1 | ·09 | 4 | Jl | 30 | ·62 |
| 3291 | 247 | 112 | 16M | ·1468 | 7·7160 | 9·581 | **190** | 5 | 2 | Mr | 23○ | ·86 | 4 | Ap | 22● | ·39 | 5 | My | 21 | ·92 | 7 | Je | 20 | ·45 | 1 | Jl | 19 | ·98 |
| | | | | | | | | | | | | | | | | | | | | | | | | | 3 | Au | 18 | ·52 |
| 3292 | 248 | 113 | 16M | ·4056 | 26·3549 | 7·714 | **191** | 6 | 1 | Ap | 11 | ·76 | 3 | My | 11 | ·29 | 4 | Je | 9 | ·82 | 6 | Jl | 9 | ·35 | 7 | Au | 7 | ·88 |
| 3293 | 249 | 114 | 15M | ·6643 | 15·4632 | 3·871 | **192** | 7 | 6 | Mr | 31 | ·13 | 7 | Ap | 29 | ·66 | 2 | My | 29 | ·19 | 3 | Je | 27 | ·72 | 5 | Jl | 27 | ·25 |
| 3294 | 250 | 115 | 15M | ·9231 | 4·5715 | 0·028 | **193** | 2 | 3 | Mr | 20 | ·49 | 5 | Ap | 19 | ·02 | 6 | My | 18 | ·56 | 2 | Jl | 16 | ·62 | 4 | Au | 15 | ·15 |
| | | | | | | | | | | | | | | | | | 1 | Je | 17 | ·09 | | | | | | | | |
| 3295 | 251 | 116 | 16M | ·1819 | 23·2104 | 25·716 | **194** | 3 | 2 | Ap | 8 | ·39 | 3 | My | 7 | ·92 | 5 | Je | 6 | ·45 | 6 | Jl | 5○ | ·98 | 1 | Au | 4● | ·51 |
| 3296 | 252 | 117 | 16M | ·4406 | 12·3187 | 21·873 | **195** | 4 | 6 | Mr | 28 | ·76 | 1 | Ap | 27 | ·29 | 2 | My | 26 | ·82 | 4 | Je | 25○ | ·35 | 5 | Jl | 24● | ·88 |
| 3297 | 253 | 118 | 15M | ·6994 | 1·4270 | 18·031 | **196** | 5 | 4 | Mr | 17 | ·13 | 5 | Ap | 15 | ·66 | 1 | Je | 13○ | ·72 | 3 | Jl | 13 | ·25 | 4 | Au | 11 | ·78 |
| | | | | | | | | | | | | | 7 | My | 15 | ·19 | | | | | | | | | | | | |

**Surya Siddhanta.**

| Asvina | | | | Kartika | | | | Margasira | | | | Pausha | | | | A.D. | Magha | | | | A.D. | Phalguna | | | | Chaitra | | | |
|---|---|---|---|---|---|---|---|---|---|---|---|---|---|---|---|---|---|---|---|---|---|---|---|---|---|---|---|---|---|
| + 147·65293 | | | | + 177·18353 | | | | + 206·71411 | | | | + 236·24470 | | | | | + 265·77529 | | | | | + 295·30588 | | | | +324·83647 +354·36705 | | | |
| + 9·880 | | | | + 11·856 | | | | + 13·832 | | | | + 15·808 | | | | | + 17·784 | | | | | + 19·760 | | | | + 21·736 + 23·712 | | | |
| Week-day | Month | Day | Fraction | Week-day | Month | Day | Fraction | Week-day | Month | Day | Fraction | Week-day | Month | Day | Fraction | | Week-day | Month | Day | Fraction | | Week-day | Month | Day | Fraction | Week-day | Month | Day | Fraction |
| 7 | S | 10 | ·93 | 2 | O | 10 | ·46 | 3 | N | 8 | ·99 | 5 | D | 8○ | ·52 | **59** | 7 | Jr | 7 | ·06 | | 1 | F | 5 | ·59 | 3 | Mr | 7 | ·12 |
| 5 | Au | 31 | ·30 | 6 | S | 29 | ·83 | 1 | O | 29 | ·36 | 2 | N | 27○ | ·89 | | 4 | D | 27 | ·42 | **60** | 5 | Jr | 25 | ·95 | 7 | F | 24 | ·84 |
| {2 | Au | 19 | ·67 | 5 | O | 17 | ·73 | 7 | N | 16 | ·26 | 1 | D | 15 | ·79 | **61** | 3 | Jr | 14 | ·32 | | 4 | F | 12 | ·85 | 6 | Mr | 14 | ·38 |
| {4 | S | 18 | ·20 | | | | | | | | | | | | | | | | | | | | | | | | | | |
| 1 | S | 7 | ·56 | 3 | O | 7○ | ·10 | 4 | N | 5 | ·63 | 6 | D | 5 | ·16 | **62** | 7 | Jr | 3 | ·69 | | 2 | F | 2 | ·22 | 3 | Mr | 3 | ·75 |
| 5 | Au | 27 | ·93 | 7 | S | 26○ | ·46 | 1 | O | 25 | ·99 | 3 | N | 24 | ·52 | | 5 | D | 24 | ·05 | **63** | 6 | Jr | 22 | ·58 | 1 | F | 21 | ·12 |
| 4 | S | 15●○ | ·83 | 6 | O | 15 | ·36 | 7 | N | 13 | ·89 | 2 | D | 13 | ·42 | **64** | 3 | Jr | 11 | ·95 | | 5 | F | 10 | ·48 | 7 | Mr | 11 | ·01 |
| 2 | S | 4● | ·20 | 3 | O | 3 | ·73 | 5 | N | 2 | ·26 | 6 | D | 1 | ·79 | | 1 | D | 31 | ·32 | **65** | 2 | Jr | 29○ | ·85 | 4 | F | 28● | ·38 |
| 6 | Au | 24 | ·56 | 1 | S | 23 | ·09 | 2 | O | 22 | ·62 | 4 | N | 21 | ·15 | | 5 | D | 20 | ·69 | **66** | 4 | Jr | 19○ | ·22 | 1 | F | 17● | ·75 |
| 5 | S | 12 | ·46 | 6 | O | 11 | ·99 | 1 | N | 10 | ·52 | 3 | D | 10 | ·05 | **67** | 4 | Jr | 8○ | ·58 | | 6 | F | 7 | ·11 | 7 | Mr | 8 | ·64 |
| 2 | S | 1 | ·83 | 4 | O | 1 | ·36 | 5 | O | 30 | ·89 | 7 | N | 29 | ·42 | | 1 | D | 28 | ·95 | **68** | 3 | Jr | 27 | ·48 | 5 | F | 26 | ·01 |
| 7 | Au | 21 | ·20 | 1 | S | 19 | ·73 | 3 | O | 19 | ·26 | 4 | N | 17○ | ·79 | | 6 | D | 17● | ·32 | **69** | 7 | Jr | 15 | ·85 | 2 | F | 14 | ·38 |
| 6 | S | 9 | ·09 | 7 | O | 8 | ·62 | 2 | N | 7○ | ·15 | 3 | D | 6● | ·68 | **70** | 5 | Jr | 5 | ·22 | | 6 | F | 3 | ·75 | 1 | Mr | 5 | ·28 |
| 3 | Au | 29 | ·46 | 4 | S | 27 | ·99 | 6 | O | 27○ | ·52 | 1 | N | 26 | ·05 | | 2 | D | 25 | ·58 | **71** | 4 | Jr | 24 | ·11 | 5 | F | 22 | ·64 |
| 2 | S | 17 | ·36 | 3 | O | 16 | ·89 | 5 | N | 15 | ·42 | 6 | D | 14 | ·95 | **72** | 1 | Jr | 13 | ·48 | | 3 | F | 12 | ·01 | 4 | Mr | 12○ | ·54 |
| 6 | S | 5○ | ·72 | 1 | O | 5● | ·26 | 2 | N | 3 | ·79 | 4 | D | 3 | ·32 | **73** | 5 | Jr | 1 | ·85 | | 7 | Jr | 31 | ·38 | 1 | Mr | 1○ | ·91 |
| 4 | Au | 26○ | ·09 | 5 | S | 24 | ·62 | 7 | O | 24 | ·15 | 1 | N | 22 | ·68 | | 3 | D | 22 | ·21 | **74** | 4 | Jr | 20 | ·74 | 6 | F | 19●○ | ·28 |
| 2 | S | 13 | ·99 | 4 | O | 13 | ·52 | 6 | N | 12 | ·05 | 7 | D | 11 | ·58 | **75** | 2 | Jr | 10 | ·11 | | 3 | F | 8● | ·64 | 5 | Mr | 10 | ·17 |
| 7 | S | 3 | ·36 | 1 | O | 2 | ·89 | 5 | N | 1 | ·42 | 4 | N | 30 | ·95 | | 6 | D | 30○ | ·48 | **76** | 1 | Jr | 29 | ·01 | 2 | F | 27 | ·54 |
| 4 | Au | 22 | ·72 | 6 | S | 21 | ·25 | 7 | O | 20 | ·78 | 2 | N | 19 | ·32 | | 3 | D | 18○ | ·85 | **77** | 5 | Jr | 17 | ·38 | 6 | F | 15 | ·91 |
| 3 | S | 10 | ·62 | 5 | O | 10 | ·15 | 6 | N | 8 | ·68 | 1 | D | 8●○ | ·21 | **78** | 2 | Jr | 6 | ·74 | | 4 | F | 5 | ·27 | 5 | Mr | 6 | ·80 |
| 7 | Au | 30 | ·99 | 2 | S | 29 | ·52 | 4 | O | 29 | ·05 | 5 | N | 27● | ·58 | | 7 | D | 27 | ·11 | **79** | 1 | Jr | 25 | ·64 | 3 | F | 24 | ·17 |
| {5 | Au | 20 | ·36 | 1 | O | 18○ | ·42 | 2 | N | 16 | ·95 | 4 | D | 16 | ·48 | **80** | 6 | Jr | 15 | ·01 | | 7 | F | 13 | ·54 | 2 | Mr | 14 | ·07 |
| {6 | S | 18 | ·89 | | | | | | | | | | | | | | | | | | | | | | | | | | |
| 4 | S | 7 | ·25 | 5 | O | 6○ | ·78 | 7 | N | 5 | ·31 | 1 | D | 4 | ·84 | **81** | 3 | Jr | 3 | ·38 | | 4 | F | 1 | ·91 | 6 | Mr | 3 | ·44 |
| 1 | Au | 27 | ·62 | 3 | S | 26●○ | ·15 | 4 | O | 25 | ·68 | 6 | N | 24 | ·21 | | 7 | D | 23 | ·74 | **82** | 2 | Jr | 22 | ·27 | 3 | F | 20 | ·80 |
| 7 | S | 15 | ·52 | 2 | O | 15 | ·05 | 3 | N | 13 | ·58 | 5 | D | 13 | ·11 | **83** | 6 | Jr | 11 | ·64 | | 1 | F | 10○ | ·17 | 2 | Mr | 11● | ·70 |
| 4 | S | 4 | ·88 | 6 | O | 4 | ·42 | 7 | N | 2 | ·95 | 2 | D | 2 | ·48 | **84** | 4 | Jr | 1 | ·01 | | 5 | Jr | 30○ | ·54 | 7 | F | 29● | ·07 |
| 2 | Au | 24 | ·25 | 3 | S | 22 | ·78 | 5 | O | 22 | ·31 | 6 | N | 20 | ·84 | | 1 | D | 20 | ·37 | **85** | 2 | Jr | 18○ | ·90 | 4 | F | 17 | ·44 |
| 1 | S | 12 | ·15 | 2 | O | 11 | ·68 | 4 | N | 10 | ·21 | 5 | D | 9 | ·74 | **86** | 7 | Jr | 8● | ·27 | | 1 | F | 6 | ·80 | 3 | Mr | 8 | ·33 |
| 5 | S | 1 | ·52 | 7 | O | 1 | ·05 | 1 | O | 30 | ·58 | 3 | N | 29○ | ·11 | | 4 | D | 28● | ·64 | **87** | 6 | Jr | 27 | ·17 | 7 | F | 25 | ·70 |
| 2 | Au | 21 | ·88 | 4 | S | 20 | ·41 | 5 | O | 19 | ·94 | 7 | N | 18○ | ·48 | | 2 | D | 18● | ·01 | **88** | 3 | Jr | 16 | ·54 | 5 | F | 15 | ·07 |
| 1 | S | 8 | ·78 | 3 | O | 8 | ·31 | 4 | N | 6○ | ·84 | 6 | D | 6 | ·37 | **89** | 7 | Jr | 4 | ·90 | | 2 | F | 3 | ·43 | 3 | Mr | 4 | ·96 |
| 6 | Au | 29 | ·15 | 7 | S | 27 | ·68 | 2 | O | 27● | ·21 | 3 | N | 25 | ·74 | | 5 | D | 25 | ·27 | **90** | 6 | Jr | 23 | ·80 | 1 | F | 22 | ·33 |
| 5 | S | 17 | ·05 | 6 | O | 16 | ·58 | 1 | N | 15 | ·11 | 2 | D | 14 | ·64 | **91** | 4 | Jr | 13 | ·17 | | 5 | F | 11 | ·70 | 7 | Mr | 13○ | ·23 |
| 2 | S | 6○ | ·41 | 3 | O | 5● | ·94 | 5 | N | 4 | ·47 | 7 | D | 4 | ·00 | **92** | 1 | Jr | 2 | ·54 | | 3 | F | 1 | ·07 | 4 | Mr | 1●○ | ·60 |
| 6 | Au | 25○ | ·78 | 1 | S | 24 | ·31 | 2 | O | 23 | ·84 | 4 | N | 22 | ·37 | | 5 | D | 21 | ·90 | **93** | 7 | Jr | 20 | ·43 | 1 | F | 18● | ·96 |
| 5 | S | 13 | ·68 | 7 | O | 13 | ·21 | 1 | N | 11 | ·74 | 3 | D | 11 | ·27 | **94** | 4 | Jr | 9○ | ·80 | | 6 | F | 8 | ·33 | 7 | Mr | 9 | ·86 |
| 3 | S | 3 | ·04 | 4 | O | 2 | ·58 | 6 | N | 1 | ·11 | 7 | N | 30 | ·64 | | 2 | D | 30○ | ·17 | **95** | 3 | Jr | 28 | ·70 | 5 | F | 27 | ·23 |
| 7 | Au | 23 | ·41 | 1 | S | 21 | ·94 | 3 | O | 21 | ·47 | 5 | N | 20 | ·00 | | 6 | D | 19●○ | ·53 | **96** | 1 | Jr | 18 | ·06 | 2 | F | 16 | ·60 |
| 6 | S | 10 | ·31 | 7 | O | 9 | ·84 | 2 | N | 8 | ·37 | 3 | D | 7● | ·90 | **97** | 5 | Jr | 6 | ·43 | | 6 | F | 4 | ·96 | 1 | Mr | 6 | ·49 |

## TABLE X.

| Kaliyuga. | Vikrama Era. | Saka Era | Commencement of Solar Year: Month and day A.D. | Commencement of Solar Year: Fraction of day. | First New-Moon in Solar Year. | Anomaly of first New-Moon. | A.D. | Week-day of 1st January. | Vaisakha (☉'s Anom col. 6; ☾'s Anom col. 7) Week-day | Month | Day | Fraction | Jyeshta (+ 29·53059; + 1·976) Week-day | Month | Day | Fraction | Ashada (+ 59·06117; + 3·952) Week-day | Month | Day | Fraction | Sravana (+ 18·59176; + 5·928) Week-day | Month | Day | Fraction | Bhadrapada (+ 118·12235; + 7·904) Week-day | Month | Day | Fraction |
|---|---|---|---|---|---|---|---|---|---|---|---|---|---|---|---|---|---|---|---|---|---|---|---|---|---|---|---|---|
| 3298 | 254 | 119 | 15M | ·9581 | 20·0659 | 16·164 | **197** | 7 | 3 | Ap | 5 | ·02 | 7 | My | 4 | ·55 | 6 | Je | 3 | ·08 | 7 | Jl | 2 | ·62 | 2 | Au | 1 | ·15 |
| 3299 | 255 | 120 | 16M | ·2169 | 9·1742 | 12·321 | **198** | 1 | 7 | Mr | 23 | ·39 | 1 | Ap | 23○ | ·92 | 3 | My | 23● | ·45 | 4 | Je | 21 | ·91 | 6 | Jl | 21 | ·51 |
| 3300 | 256 | 121 | 16M | ·4756 | 27·8131 | 10·454 | **199** | 2 | 6 | Ap | 13○ | ·29 | 7 | My | 12 | ·82 | 2 | Je | 11 | ·35 | 3 | Jl | 10 | ·88 | 5 | Au | 9 | ·41 |
| 3301 | 257 | 122 | 15M | ·7344 | 16·9214 | 6·612 | **200** | 3 | 3 | Ap | 1● | ·66 | 5 | My | 1 | ·19 | 6 | My | 30 | ·72 | 1 | Je | 29 | ·25 | 2 | Jl | 28 | ·78 |
| 3302 | 258 | 123 | 15M | ·9932 | 6·0297 | 2·769 | **201** | 5 | 1 | Mr | 22● | ·02 | 2 | Ap | 20 | ·55 | 4 | My | 20 | ·08 | { 5 | Je | 18 | ·61 | 1 | Au | 16○ | ·68 |
| | | | | | | | | | | | | | | | | | | | | | { 7 | Jl | 18 | ·14 | | | | |
| 3303 | 259 | 124 | 16M | ·2519 | 24·6686 | 0·902 | **202** | 6 | 6 | Ap | 9 | ·92 | 1 | My | 9 | ·45 | 2 | Je | 7 | ·98 | 4 | Jl | 7 | ·51 | 6 | Au | 6○ | ·04 |
| 3304 | 260 | 125 | 16M | ·5107 | 13·7769 | 24·614 | **203** | 7 | 4 | Mr | 30 | ·29 | 5 | Ap | 28 | ·82 | 7 | My | 28 | ·35 | 1 | Je | 26 | ·88 | 3 | Jl | 26●○ | ·41 |
| 3305 | 261 | 126 | 15M | ·7694 | 2·8852 | 20·771 | **204** | 1 | 1 | Mr | 18 | ·65 | { 3 | Ap | 17 | ·18 | 6 | Je | 15 | ·25 | 7 | Jl | 14● | ·78 | 2 | Au | 13 | ·31 |
| | | | | | | | | | | | | | { 4 | My | 16 | ·72 | | | | | | | | | | | | |
| 3306 | 262 | 127 | 16M | ·0282 | 21·5240 | 18·904 | **205** | 3 | 7 | Ap | 6 | ·55 | 2 | My | 6 | ·08 | 3 | Je | 4○ | ·61 | 5 | Jl | 4 | ·14 | 6 | Au | 2 | ·67 |
| 3307 | 263 | 128 | 16M | ·2869 | 10·6323 | 15·062 | **206** | 4 | 4 | Mr | 26 | ·92 | 6 | Ap | 25 | ·45 | 7 | My | 24●○ | ·98 | 2 | Je | 23 | ·51 | 4 | Jl | 23 | ·04 |
| 3308 | 264 | 129 | 16M | ·5457 | 29·2712 | 13·195 | **207** | 5 | 3 | Ap | 14 | ·82 | 5 | My | 14●○ | ·35 | 6 | Je | 12 | ·88 | 1 | Jl | 12 | ·41 | 2 | Au | 10 | ·94 |
| 3309 | 265 | 130 | 15M | ·8044 | 18·3795 | 9·352 | **208** | 6 | 1 | Ap | 3○ | ·18 | 2 | My | 2● | ·71 | 4 | Je | 1 | ·24 | 5 | Je | 30 | ·78 | 7 | Jl | 30 | ·31 |
| 3310 | 266 | 131 | 16M | ·0632 | 7·4878 | 5·509 | **209** | 1 | 5 | Mr | 23○ | ·55 | 7 | Ap | 22 | ·08 | 1 | My | 21 | ·61 | { 3 | Je | 20 | ·14 | 6 | Au | 18 | ·20 |
| | | | | | | | | | | | | | | | | | | | | | { 4 | Jl | 19 | ·67 | | | | |
| 3311 | 267 | 132 | 16M | ·8220 | 26·1267 | 3·643 | **210** | 2 | 4 | Ap | 11 | ·45 | 5 | My | 10 | ·98 | 7 | Je | 9 | ·51 | 2 | Jl | 9 | ·04 | 3 | Au | 7 | ·57 |
| 3312 | 268 | 133 | 16M | ·5807 | 15·2350 | 27·355 | **211** | 3 | 1 | Mr | 31 | ·82 | 3 | Ap | 30 | ·35 | 4 | My | 29 | ·88 | 6 | Je | 28 | ·41 | 7 | Jl | 27 | ·94 |
| 3313 | 269 | 134 | 15M | ·8395 | 4·3433 | 23·512 | **212** | 4 | 6 | Mr | 20 | ·18 | 7 | Ap | 18 | ·71 | { 2 | My | 18 | ·24 | 5 | Jl | 16○ | ·30 | 6 | Au | 14● | ·84 |
| | | | | | | | | | | | | | | | | | { 3 | Je | 16 | ·77 | | | | | | | | |
| 3314 | 270 | 135 | 16M | ·0982 | 22·9822 | 21·645 | **213** | 6 | 5 | Ap | 8 | ·08 | 6 | My | 7 | ·61 | 1 | Je | 6 | ·14 | 2 | Jl | 5○ | ·67 | 4 | Au | 4● | ·20 |
| 3315 | 271 | 136 | 16M | ·3570 | 12·0905 | 17·802 | **214** | 7 | 2 | Mr | 28 | ·45 | 3 | Ap | 26 | ·98 | 5 | My | 26 | ·51 | 7 | Je | 25○ | ·04 | 1 | Jl | 24 | ·57 |
| 3316 | 272 | 137 | 16M | ·6157 | 1·1988 | 13·960 | **215** | 1 | 6 | Mr | 17 | ·81 | { 1 | Ap | 16 | ·34 | 4 | Je | 14● | ·41 | 5 | Jl | 13 | ·94 | 7 | Au | 12 | ·47 |
| | | | | | | | | | | | | | { 2 | My | 15 | ·88 | | | | | | | | | | | | |
| 3317 | 273 | 138 | 15M | ·8745 | 19·8377 | 12·093 | **216** | 2 | 5 | Ap | 4 | ·71 | 7 | My | 4○ | ·24 | 1 | Je | 2● | ·77 | 3 | Jl | 2 | ·30 | 4 | Jl | 31 | ·83 |
| 3318 | 274 | 139 | 16M | ·1333 | 8·9460 | 8·250 | **217** | 4 | 3 | Mr | 25 | ·08 | 4 | Ap | 23○ | ·61 | 6 | My | 23 | ·14 | 7 | Je | 21 | ·67 | 2 | Jl | 21 | ·20 |
| 3319 | 275 | 140 | 16M | ·3920 | 27·5849 | 6·383 | **218** | 5 | 1 | Ap | 12● | ·98 | 3 | My | 12 | ·51 | 5 | Je | 11 | ·04 | 6 | Jl | 10 | ·57 | 1 | Au | 9 | ·10 |
| 3320 | 276 | 141 | 16M | ·6508 | 16·6932 | 2·541 | **219** | 6 | 6 | Ap | 2● | ·34 | 7 | My | 1 | ·87 | 2 | My | 31 | ·40 | 3 | Je | 29 | ·94 | 5 | Jl | 29 | ·47 |
| 3321 | 277 | 142 | 15M● | ·9095 | 5·8015 | 26·252 | **220** | 7 | 3 | Mr | 21● | ·71 | 5 | Ap | 20 | ·24 | 6 | My | 19 | ·77 | { 1 | Je | 18 | ·30 | 4 | Au | 16○ | ·36 |
| | | | | | | | | | | | | | | | | | | | | | { 2 | Jl | 17 | ·83 | | | | |
| 3322 | 278 | 143 | 16M | ·1683 | 24·4404 | 24·386 | **221** | 2 | 2 | Ap | 9 | ·61 | 4 | My | 9 | ·14 | 5 | Je | 7 | ·67 | 7 | Jl | 7 | ·20 | 1 | Au | 5●○ | ·73 |
| 3323 | 279 | 144 | 16M | ·4270 | 13·5486 | 20·543 | **222** | 3 | 6 | Mr | 29 | ·98 | 1 | Ap | 28 | ·51 | 3 | My | 28 | ·04 | 4 | Je | 26 | ·57 | 6 | Jl | 26● | ·10 |
| 3324 | 280 | 14[illegible] | 16M | ·6858 | 2·6569 | 16·700 | **223** | 4 | 4 | Mr | 19 | ·34 | { 5 | Ap | 17 | ·87 | 1 | Je | 15○ | ·93 | 3 | Jl | 15 | ·46 | 5 | Au | 14 | ·00 |
| | | | | | | | | | | | | | { 7 | My | 17 | ·40 | | | | | | | | | | | | |
| 3325 | 281 | 146 | 15M | ·9445 | 21·2958 | 14·833 | **224** | 5 | 3 | Ap | 6 | ·24 | 4 | My | 5 | ·77 | 6 | Je | 4●○ | ·30 | 7 | Jl | 3 | ·83 | 2 | Au | 2 | ·36 |
| 3326 | 282 | 147 | 16M | ·2033 | 10·4041 | 10·991 | **225** | 6 | 7 | Mr | 26 | ·61 | 2 | Ap | 25 | ·14 | 3 | My | 24●○ | ·67 | 5 | Je | 23 | ·20 | 6 | Jl | 22 | ·73 |
| 3327 | 283 | 148 | 16M | ·4621 | 29·0430 | 9·124 | **226** | 1 | 6 | Ap | 14 | ·50 | 1 | My | 14 | ·04 | 2 | Je | 12 | ·57 | 4 | Jl | 12 | ·10 | 5 | Au | 10 | ·63 |
| 3328 | 284 | 149 | 16M | ·7208 | 18·1513 | 5·281 | **227** | 2 | 3 | Ap | 3○ | ·87 | 5 | My | 3 | ·40 | 6 | Je | 1 | ·93 | 1 | Jl | 1 | ·46 | 2 | Jl | 30 | ·99 |
| 3329 | 285 | 150 | 15M | ·9796 | 7·2596 | 1·438 | **228** | 3 | 1 | Mr | 23●○ | ·24 | 2 | Ap | 21 | ·77 | 4 | My | 21 | ·30 | { 5 | Je | 19 | ·83 | 1 | Au | 17 | ·89 |
| | | | | | | | | | | | | | | | | | | | | | { 7 | Jl | 19 | ·36 | | | | |
| 3330 | 286 | 151 | 16M | ·2383 | 25·8985 | 27·126 | **229** | 5 | 7 | Ap | 11 | ·14 | 1 | My | 10 | ·67 | 3 | Je | 9 | ·20 | 4 | Jl | 8 | ·73 | 6 | Au | 7 | ·26 |
| 3331 | 287 | 152 | 16M | ·4971 | 15·0068 | 23·284 | **230** | 6 | 4 | Mr | 31 | ·50 | 6 | Ap | 30 | ·03 | 7 | My | 29 | ·56 | 2 | Je | 28 | ·10 | 3 | Jl | 27○ | ·63 |
| 3332 | 288 | 153 | 16M | ·7558 | 4·1151 | 19·441 | **231** | 7 | 1 | Mr | 20 | ·87 | 3 | Ap | 19 | ·40 | { 4 | My | 18 | ·93 | 7 | Jl | 16○ | ·99 | 2 | Au | 15● | ·52 |
| | | | | | | | | | | | | | | | | | { 6 | Je | 17 | ·46 | | | | | | | | |
| 3333 | 289 | 154 | 16M | ·0146 | 22·7540 | 17·574 | **232** | 1 | 7 | Ap | 7 | ·77 | 2 | My | 7 | ·30 | 3 | Je | 5 | ·83 | 5 | Jl | 5○ | ·36 | 6 | Au | 3 | ·89 |
| 3334 | 290 | 155 | 16M | ·2734 | 11·8623 | 13·731 | **233** | 3 | 5 | Mr | 28 | ·14 | 6 | Ap | 26 | ·67 | 1 | My | 26 | ·20 | 2 | Je | 24● | ·73 | 4 | Jl | 24 | ·26 |
| 3335 | 291 | 156 | 16M | ·5321 | 0·9706 | 9·889 | **234** | 4 | { 2 | Mr | 17 | ·50 | 5 | My | 15○ | ·56 | 7 | Je | 14● | ·09 | 1 | Jl | 13 | ·62 | 3 | Au | 12 | ·16 |
| | | | | | | | | | { 4 | Ap | 16 | ·03 | | | | | | | | | | | | | | | | |
| 3336 | 292 | 157 | 16M | ·7909 | 19·6095 | 8·022 | **235** | 5 | 1 | Ap | 5 | ·40 | 2 | My | 4○ | ·93 | 4 | Je | 3● | ·46 | 5 | Jl | 2 | ·99 | 7 | Au | 1 | ·52 |
| 3337 | 293 | 158 | 16M | ·0496 | 8·7178 | 4·179 | **236** | 6 | 5 | Mr | 24 | ·77 | 7 | Ap | 23●○ | ·30 | 1 | My | 22 | ·83 | 3 | Je | 21 | ·36 | { 4 | Jl | 20 | ·89 |
| | | | | | | | | | | | | | | | | | | | | | | | | | { 6 | Au | 19 | ·42 |

**Surya Siddhanta.**

| | | | | | | | | | | | | | | | | | | | | | | |
|---|---|---|---|---|---|---|---|---|---|---|---|---|---|---|---|---|---|---|---|---|---|---|
| + 147·65293 | | | + 177·18353 | | | + 206·71411 | | | + 236·24470 | | | | + 266·77528 | | | | + 295·30588 | | | {+324·83647<br>{+354·36705 | | |
| + 9·880 | | | + 11·856 | | | + 13·832 | | | + 15·808 | | | | + 17·784 | | | | + 19·760 | | | {+ 21·736<br>{+ 23·712 | | |
| **Asvina** | | | **Kartika** | | | **Margasira** | | | **Pausha** | | | **A.D.** | **Magha** | | | **A.D.** | **Phalguna** | | | **Chaitra** | | |
| Week-day Month | Day | Fraction | Week-day Month | Day | Fraction | Week-day Month | Day | Fraction | Week-day Month | Day | Fraction | | Week-day Month | Day | Fraction | | Week-day Month | Day | Fraction | Week-day Month | Day | Fraction |
| 3 Au | 30 | ·68 | 5 S | 29 | ·21 | 6 O | 28○ | ·74 | 1 N | 27○ | ·27 | | 2 D | 26 | ·80 | **98** | 4 Jr | 25 | ·33 | 5 F | 23 | ·8[illegible] |
| {1 Au | 20 | ·04 | 5 O | 18○ | ·10 | 5 N | 16 | ·64 | 7 D | 16 | ·17 | **99** | 1 Jr | 84 | ·70 | | 3 F | 13 | ·23 | 4 Mr | 14 | ·76 |
| {2 S | 18 | ·57 | | | | | | | | | | | | | | | | | | | | |
| 6 S | 7 | ·94 | 1 O | 7●○ | ·47 | 3 N | 6 | ·00 | 4 D | 5 | ·53 | **200** | 6 Jr | 4 | ·06 | | 7 F | 2 | ·59 | 2 Mr | 3 | ·12 |
| 4 Au | 27 | ·31 | 5 S | 25 | ·84 | 7 O | 25 | ·37 | 1 N | 23 | ·90 | | 3 D | 23 | ·43 | **01** | 4 Jr | 21 | ·96 | 6 F | 20○ | ·49 |
| 3 S | 15 | ·21 | 4 O | 14 | ·74 | 6 N | 13 | ·27 | 7 D | 12 | ·80 | **02** | 2 Jr | 11 | ·33 | | 3 F | 9○ | ·86 | 5 Mr | 11● | ·39 |
| 7 S | 4 | ·57 | 2 O | 4 | ·10 | 3 N | 2 | ·63 | 5 D | 2 | ·16 | | 6 D | 31 | ·70 | **03** | 1 Jr | 30○ | ·23 | 2 F | 28 | ·76 |
| 4 Au | 24 | ·94 | 6 S | 23 | ·47 | 1 O | 23 | ·00 | 2 N | 21 | ·53 | | 4 D | 21 | ·06 | **04** | 5 Jr | 19 | ·59 | 7 F | 18 | ·12 |
| 3 S | 11 | ·84 | 5 O | 11 | ·37 | 6 N | 9 | ·90 | 1 D | 9○ | ·43 | **05** | 2 Jr | 7 | ·96 | | 4 F | 6 | ·49 | 6 Mr | 8 | ·02 |
| 1 S | 1 | ·20 | 2 S | 30 | ·74 | 4 O | 30 | ·27 | 5 N | 28○ | ·80 | | 7 D | 28● | ·33 | **06** | 1 Jr | 26 | ·86 | 3 F | 25 | ·39 |
| 5 Au | 21 | ·57 | 7 S | 20 | ·10 | 1 O | 19 | ·63 | 3 N | 18○ | ·16 | | 4 D | 17 | ·69 | **07** | 6 Jr | 16 | ·22 | {7 F | 14 | ·76 |
| | | | | | | | | | | | | | | | | | | | | {2 Mr | 16 | ·29 |
| 4 S | 9 | ·47 | 6 O | 9 | ·00 | 7 N | 7 | ·53 | 2 D | 7 | ·06 | **08** | 3 Jr | 5 | ·59 | | 5 F | 4 | ·12 | 6 Mr | 4 | ·65 |
| 1 Au | 28 | ·84 | 3 S | 27 | ·37 | 4 O | 26● | ·90 | 6 N | 25 | ·43 | | 7 D | 24 | ·96 | **09** | 2 Jr | 23 | ·49 | 4 F | 22 | ·02 |
| 7 S | 16○ | ·73 | 2 O | 16● | ·26 | 3 N | 14 | ·80 | 5 D | 14 | ·33 | **10** | 6 Jr | 12 | ·86 | | 1 F | 11 | ·39 | 2 Mr | 12●○ | ·92 |
| 5 S | 6○ | ·10 | 6 O | 5 | ·63 | 1 N | 4 | ·16 | 2 D | 3 | ·69 | **11** | 4 Jr | 2 | ·22 | | 5 Jr | 31 | ·75 | 7 Mr | 2● | ·28 |
| 2 Au | 26● | ·47 | 4 S | 25 | ·00 | 5 O | 24 | ·53 | 7 N | 23 | ·06 | | 1 D | 22 | ·59 | **12** | 3 Jr | 21○ | ·12 | 4 F | 19 | ·65 |
| 1 S | 13 | ·37 | 2 O | 12 | ·90 | 4 N | 11 | ·43 | 5 D | 10 | ·96 | **13** | 7 Jr | 9○ | ·49 | | 2 F | 8 | ·02 | 3 Mr | 9 | ·55 |
| 5 S | 2 | ·73 | 7 O | 2 | ·26 | 1 O | 31 | ·79 | 3 N | 30 | ·32 | | 4 D | 29○ | ·86 | **14** | 6 Jr | 28 | ·39 | 7 F | 26 | ·92 |
| 3 Au | 23 | ·10 | 4 S | 21 | ·63 | 6 O | 21 | ·16 | 7 N | 19 | ·69 | | 2 D | 19 | ·22 | **15** | 3 Jr | 17 | ·75 | 5 F | 16 | ·28 |
| 2 S | 11 | ·00 | 3 O | 10 | ·53 | 5 N | 9 | ·06 | 6 D | 8 | ·5[illegible] | **16** | 1 Jr | 7 | ·12 | | 2 F | 5 | ·65 | 4 Mr | 6 | ·18 |
| 6 Au | 30 | ·36 | 7 S | 28 | ·90 | 2 O | 28○ | ·43 | 3 N | 26 | ·96 | | 5 D | 26 | ·49 | **17** | 7 Jr | 25 | ·02 | 1 F | 23 | ·55 |
| {3 Au | 19 | ·73 | 6 O | 17●○ | ·79 | 1 N | 16 | ·32 | 2 D | 15 | ·85 | **18** | 4 Jr | 14 | ·38 | | 5 F | 12 | ·92 | 7 Mr | 14 | ·45 |
| {5 S | 18 | ·26 | | | | | | | | | | | | | | | | | | | | |
| 2 S | 7 | ·63 | 4 O | 7●○ | ·16 | 5 N | 5 | ·69 | 7 D | 5 | ·22 | **19** | 1 Jr | 3 | ·75 | | 3 F | 2 | ·28 | 4 Mr | 3○ | ·81 |
| 7 Au | 28○ | ·00 | 1 S | 26 | ·53 | 3 O | 26 | ·06 | 4 N | 24 | ·59 | | 6 D | 24 | ·12 | **20** | 7 Jr | 22 | ·65 | 2 F | 21○ | ·18 |
| 5 S | 14 | ·89 | 7 O | 14 | ·42 | 1 N | 12 | ·96 | 3 D | 12 | ·49 | **21** | 5 Jr | 11 | ·02 | | 6 F | 9○ | ·55 | 1 Mr | 11 | ·08 |
| 3 S | 4 | ·26 | 4 O | 3 | ·79 | 6 N | 2 | ·32 | 7 D | 1 | ·85 | | 2 D | 31 | ·38 | **22** | 3 Jr | 29● | ·91 | 5 F | 28 | ·44 |
| 7 Au | 24 | ·63 | 2 S | 23 | ·16 | 3 O | 22 | ·69 | 5 N | 21 | ·22 | | 6 D | 20○ | ·75 | **23** | 1 Jr | 19● | ·28 | 2 F | 17 | ·81 |
| 6 S | 12 | ·53 | 1 O | 12 | ·06 | 2 N | 10 | ·59 | 4 D | 10○ | ·12 | **24** | 5 Jr | 8● | ·65 | | 7 F | 7 | ·18 | 1 Mr | 7 | ·71 |
| 3 Au | 31 | ·89 | 5 S | 30 | ·42 | 6 O | 29 | ·95 | 1 N | 28○ | ·48 | | 3 D | 28 | ·02 | **25** | 4 Jr | 26 | ·55 | 6 F | 25 | ·08 |
| 1 Au | 21 | ·26 | 2 S | 19 | ·79 | 4 O | 19 | ·32 | 5 N | 17● | ·85 | | 7 D | 17 | ·38 | **26** | 1 Jr | 15 | ·91 | {3 F | 14 | ·44 |
| | | | | | | | | | | | | | | | | | | | | {4 Mr | 15 | ·97 |
| 7 S | 9 | ·16 | 1 O | 8 | ·69 | 3 N | 7● | ·22 | 4 D | 6 | ·75 | **27** | 6 Jr | 5 | ·28 | | 7 F | 3 | ·81 | 2 Mr | 5 | ·34 |
| 4 Au | 29 | ·52 | 6 S | 28○ | ·06 | 7 O | 27 | ·59 | 2 N | 26 | ·12 | | 3 D | 25 | ·65 | **28** | 5 Jr | 24 | ·18 | 6 F | 22 | ·71 |
| 3 S | 16○ | ·42 | 4 O | 15 | ·95 | 6 N | 14 | ·48 | 1 D | 14 | ·01 | **29** | 2 Jr | 12 | ·54 | | 4 F | 11 | ·08 | 5 Mr | 12● | ·61 |
| 7 S | 5 | ·79 | 2 O | 5 | ·32 | 3 N | 3 | ·85 | 5 D | 3 | ·38 | **30** | 6 Jr | 1 | ·91 | | 1 Jr | 31○ | ·44 | 2 Mr | 1 | ·97 |
| 5 Au | 26● | ·16 | 6 S | 24 | ·69 | 1 O | 24 | ·22 | 2 N | 22 | ·75 | | 4 D | 22 | ·28 | **31** | 5 Jr | 20○ | ·81 | 7 F | 19 | ·34 |
| 4 S | 14 | ·05 | 5 O | 13 | ·58 | 7 N | 12 | ·12 | 1 D | 11 | ·65 | **32** | 3 Jr | 10●○ | ·18 | | 4 F | 8 | ·71 | 6 Mr | 9 | ·24 |
| 1 S | 2 | ·42 | 2 O | 1 | ·95 | 4 O | 31 | ·48 | 6 N | 30 | ·01 | | 7 D | 29● | ·54 | **33** | 2 Jr | 28 | ·07 | 3 F | 26 | ·60 |
| 5 Au | 22 | ·79 | 7 S | 21 | ·32 | 1 O | 20 | ·85 | 3 N | 19 | ·38 | | 4 D | 18 | ·91 | **34** | 6 Jr | 17 | ·44 | 7 F | 15 | ·97 |
| 4 S | 10 | ·69 | 6 O | 10 | ·22 | 7 N | 8○ | ·75 | 2 D | 8 | ·28 | **35** | 3 Jr | 6 | ·81 | | 5 F | 5 | ·34 | 6 Mr | 6 | ·87 |
| 2 Au | 31 | ·05 | 3 S | 29 | ·58 | 5 O | 29●○ | ·11 | 6 N | 27 | ·64 | | 1 D | 27 | ·18 | **36** | 2 Jr | 25 | ·71 | 4 F | 24 | ·24 |
| 7 S | 17 | ·95 | 2 O | 17●○ | ·48 | 4 N | 16 | ·01 | 5 D | 15 | ·54 | **37** | 7 Jr | 14 | ·07 | | 1 F | 12 | ·60 | 3 Mr | 14 | ·13 |

| Kaliyuga. | Vikrama Era. | Saka Era. | Commencement of Solar Year. Month and day A.D. | Fraction of day. | First New-Moon in Solar Year. | Anomaly of first New-Moon. | A.D. | Week-day of 1st January. | Vaisakha Week-day | Month | Day | Fraction | Jyeshta Week-day | Month | Day | Fraction | Ashada Week-day | Month | Day | Fraction | Sravana Week-day | Month | Day | Fraction | Bhadrapada Week-day | Month | Day | Fraction |
|---|---|---|---|---|---|---|---|---|---|---|---|---|---|---|---|---|---|---|---|---|---|---|---|---|---|---|---|---|
| | | | | | | | | | ☉'s Anom col. 6 | | | | + 29·53059 | | | | + 59·06117 | | | | + 88·59176 | | | | + 118·12235 | | | |
| | | | | | | | | | ☾'s Anom col. 7 | | | | + 1·976 | | | | + 3·952 | | | | + 5·928 | | | | + 7·904 | | | |
| 3338 | 294 | 159 | 16M | ·3084 | 27·3567 | 2·312 | **237** | 1 | 4 | Ap | 12● | ·66 | 6 | My | 12 | ·20 | 7 | Je | 10 | ·73 | 2 | Jl | 10 | ·26 | 3 | Au | 8 | ·79 |
| 3339 | 295 | 160 | 16M | ·5671 | 16·4650 | 26·024 | **238** | 2 | 2 | Ap | 2● | ·03 | 3 | My | 1 | ·56 | 5 | My | 31 | ·09 | 6 | Je | 29 | ·62 | 1 | Jl | 29 | ·15 |
| 3340 | 296 | 161 | 16M | ·8259 | 5·5733 | 22·181 | **239** | 3 | 6 | Mr | 22 | ·40 | 7 | Ap | 20 | ·93 | 2 | My | 20 | ·46 | 3 | Je | 18 | ·99 | 7 | Au | 17●○ | ·05 |
| | | | | | | | | | | | | | | | | | | | | | 5 | Jl | 18 | ·52 | | | | |
| 3341 | 297 | 162 | 16M | ·0847 | 24·2121 | 20·315 | **240** | 4 | 5 | Ap | 9 | ·30 | 6 | My | 8 | ·83 | 1 | Je | 7 | ·36 | 2 | Jl | 6 | ·89 | 4 | Au | 5● | ·42 |
| 3342 | 298 | 163 | 16M | ·3434 | 13·3204 | 16·472 | **241** | 6 | 2 | Mr | 29 | ·66 | 4 | Ap | 28 | ·19 | 5 | My | 27 | ·72 | 7 | Je | 26○ | ·26 | 1 | Jl | 25 | ·79 |
| 3343 | 299 | 164 | 16M | ·6022 | 2·4287 | 12·629 | **242** | 7 | 7 | Mr | 19 | ·03 | 1 | Ap | 17 | ·56 | 4 | Je | 15●○ | ·62 | 6 | Jl | 15 | ·15 | 7 | Au | 13 | ·68 |
| | | | | | | | | | | | | | 3 | My | 17 | ·09 | | | | | | | | | | | | |
| 3344 | 300 | 165 | 16M | ·8609 | 21·0676 | 10·762 | **243** | 1 | 5 | Ap | 7 | ·93 | 7 | My | 6 | ·46 | 1 | Je | 4●○ | ·99 | 3 | Jl | 4 | ·52 | 5 | Au | 3 | ·05 |
| 3345 | 301 | 166 | 16M | ·1197 | 10·1759 | 6·920 | **244** | 2 | 3 | Mr | 26 | ·30 | 4 | Ap | 24 | ·83 | 6 | My | 24● | 36 | 7 | Je | 22 | ·89 | 2 | Jl | 22 | ·42 |
| 3346 | 302 | 167 | 16M | ·3784 | 28·8148 | 5·053 | **245** | 4 | 2 | Ap | 14○ | ·19 | 3 | My | 13 | ·72 | 5 | Je | 12 | ·25 | 6 | Jl | 11 | ·78 | 1 | Au | 10 | ·32 |
| 3347 | 303 | 168 | 16M | ·6372 | 17·9231 | 1·210 | **246** | 5 | 6 | Ap | 3●○ | ·56 | 1 | My | 3 | ·09 | 2 | Je | 1 | ·62 | 4 | Jl | 1 | ·15 | 5 | Jl | 30 | ·68 |
| 3348 | 304 | 169 | 16M | ·8959 | 7·0314 | 24·922 | **247** | 6 | 3 | Mr | 23● | ·93 | 5 | Ap | 22 | ·46 | 6 | My | 21 | ·99 | 1 | Je | 20 | ·52 | 4 | Au | 18 | ·58 |
| | | | | | | | | | | | | | | | | | | | | | 3 | Jl | 20 | ·05 | | | | |
| 3349 | 305 | 170 | 16M | ·1547 | 25·6703 | 23·055 | **248** | 7 | 2 | Ap | 10 | ·82 | 4 | My | 10 | ·36 | 5 | Je | 8 | ·89 | 7 | Jl | 8 | ·42 | 1 | Au | 6 | ·95 |
| 3350 | 306 | 171 | 16M | ·4135 | 14·7786 | 10·213 | **249** | 2 | 7 | Mr | 31 | ·19 | 1 | Ap | 29 | ·72 | 3 | My | 29 | ·25 | 4 | Je | 27 | ·78 | 6 | Jl | 27○ | ·31 |
| 3351 | 307 | 172 | 16M | ·6722 | 3·8869 | 15·370 | **250** | 3 | 4 | Mr | 20 | ·56 | 6 | Ap | 19 | ·09 | 7 | My | 18 | ·62 | 3 | Jl | 16○ | 68 | 5 | Au | 15 | ·21 |
| | | | | | | | | | | | | | | | | | 2 | Je | 17 | ·15 | | | | | | | | |
| 3352 | 308 | 173 | 16M | ·9310 | 22·5258 | 13·503 | **251** | 4 | 3 | Ap | 8 | ·46 | 4 | My | 7 | ·99 | 6 | Je | 6 | ·52 | 1 | Jl | 6● | ·05 | 2 | Au | 4 | ·58 |
| 3353 | 309 | 174 | 16M | ·1897 | 11·6341 | 9·660 | **252** | 5 | 7 | Mr | 27 | ·82 | 2 | Ap | 26 | ·35 | 3 | My | 25○ | ·88 | 5 | Je | 24● | ·41 | 6 | Jl | 23 | ·95 |
| 3354 | 310 | 175 | 16M | ·4485 | 0·7424 | 5·818 | **253** | 7 | 5 | Mr | 17 | ·19 | 1 | My | 15○ | ·25 | 2 | Je | 13● | ·78 | 4 | Jl | 13 | ·31 | 5 | Au | 11 | ·84 |
| | | | | | | | | | 6 | Ap | 15 | ·72 | | | | | | | | | | | | | | | | |
| 3355 | 311 | 176 | 16M | ·7072 | 19·3813 | 3·951 | **254** | 1 | 4 | Ap | 5 | ·09 | 5 | My | 4●○ | ·62 | 7 | Je | 3 | ·15 | 1 | Jl | 2 | ·68 | 3 | Au | 1 | ·21 |
| 3356 | 312 | 177 | 16M | ·9660 | 8·4896 | 0·108 | **255** | 2 | 1 | Mr | 25 | ·45 | 2 | Ap | 23● | ·99 | 4 | My | 23 | ·52 | 6 | Je | 22 | ·05 | 7 | Jl | 21 | ·58 |
| | | | | | | | | | | | | | | | | | | | | | | | | | 2 | Au | 20 | ·11 |
| 3357 | 313 | 178 | 16M | ·2248 | 27·1285 | 25·796 | **256** | 3 | 7 | Ap | 12● | ·35 | 1 | My | 11 | ·88 | 3 | Je | 10 | ·41 | 4 | Jl | 9 | ·94 | 6 | Au | 8 | ·47 |
| 3358 | 314 | 179 | 16M | ·4835 | 16·2368 | 21·953 | **257** | 5 | 4 | Ap | 1 | ·72 | 6 | My | 1 | ·25 | 7 | My | 30 | ·78 | 2 | Je | 29 | ·31 | 3 | Jl | 28 | ·84 |
| 3359 | 315 | 180 | 16M | ·7423 | 5·3451 | 18·110 | **258** | 6 | 2 | Mr | 22 | ·09 | 3 | Ap | 20 | ·62 | 5 | My | 20 | ·15 | 6 | Je | 18 | ·68 | 2 | Au | 16● | ·74 |
| | | | | | | | | | | | | | | | | | | | | | 1 | Jl | 18 | ·21 | | | | |
| 3360 | 316 | 181 | 17M | ·0010 | 23·9839 | 16·244 | **259** | 7 | 7 | Ap | 9 | ·98 | 2 | My | 9 | ·51 | 4 | Je | 8 | ·05 | 5 | Jl | 7○ | ·58 | 7 | Au | 6● | ·11 |
| 3361 | 317 | 182 | 16M | ·2598 | 13·0922 | 12·401 | **260** | 1 | 5 | Mr | 29 | ·35 | 6 | Ap | 27 | ·88 | 1 | My | 27 | ·41 | 2 | Je | 25○ | ·94 | 4 | Jl | 25 | ·47 |
| 3362 | 318 | 183 | 16M | ·5185 | 2·2005 | 8·558 | **261** | 3 | 2 | Mr | 18 | ·72 | 4 | Ap | 17 | ·25 | 7 | Je | 15● | ·31 | 1 | Jl | 14 | ·84 | 3 | Au | 13 | ·37 |
| | | | | | | | | | | | | | 5 | My | 16 | ·78 | | | | | | | | | | | | |
| 3363 | 319 | 184 | 16M | ·7773 | 20·8394 | 6·691 | **262** | 4 | 1 | Ap | 6 | ·62 | 3 | My | 6 | ·15 | 4 | Je | 4● | ·68 | 6 | Jl | 4 | ·21 | 7 | Au | 2 | ·74 |
| 3364 | 320 | 185 | 17M | ·0361 | 9·9477 | 2·849 | **263** | 5 | 5 | Mr | 26 | ·98 | 7 | Ap | 25○ | ·51 | 2 | My | 25 | ·04 | 3 | Je | 23 | ·57 | 5 | Jl | 23 | ·11 |
| 3365 | 321 | 186 | 16M | ·2948 | 28·5866 | 0·982 | **264** | 6 | 4 | Ap | 13●○ | ·88 | 6 | My | 13 | ·41 | 7 | Je | 11 | ·94 | 2 | Jl | 11 | ·47 | 4 | Au | 10 | ·00 |
| 3366 | 322 | 187 | 16M | ·5536 | 17·6949 | 24·694 | **265** | 1 | 2 | Ap | 3●○ | ·25 | 3 | My | 2 | ·78 | 5 | Je | 1 | ·31 | 6 | Je | 30 | ·84 | 1 | Jl | 30 | ·37 |
| 3367 | 323 | 188 | 16M | ·8123 | 6·8032 | 20·851 | **266** | 2 | 6 | Mr | 23● | ·61 | 1 | Ap | 22 | ·15 | 2 | My | 21 | ·68 | 4 | Je | 20 | ·21 | 7 | Au | 18 | ·27 |
| | | | | | | | | | | | | | | | | | | | | | 5 | Jl | 19 | ·74 | | | | |
| 3368 | 324 | 189 | 17M | ·0711 | 25·4421 | 18·984 | **267** | 3 | 5 | Ap | 11 | ·51 | 7 | My | 11 | ·04 | 1 | Je | 9 | ·57 | 3 | Jl | 9 | ·10 | 4 | Au | 7○ | ·63 |
| 3369 | 325 | 190 | 16M | ·3298 | 14·5504 | 15·141 | **268** | 4 | 2 | Mr | 30 | ·88 | 4 | Ap | 29 | ·41 | 5 | My | 28 | ·94 | 7 | Je | 27 | ·47 | 2 | Jl | 27○ | ·00 |
| 3370 | 326 | 191 | 16M | ·5886 | 3·6587 | 11·299 | **269** | 6 | 7 | Mr | 20 | ·25 | 1 | Ap | 18 | ·78 | 3 | My | 18 | ·31 | 6 | Jl | 16● | ·37 | 7 | Au | 14 | ·90 |
| | | | | | | | | | | | | | | | | | 4 | Je | 16 | ·84 | | | | | | | | |
| 3371 | 327 | 192 | 16M | ·8473 | 22·2975 | 9·432 | **270** | 7 | 6 | Ap | 8 | ·14 | 7 | My | 7 | ·67 | 2 | Je | 6○ | ·21 | 3 | Jl | 5● | ·74 | 5 | Au | 4 | ·27 |
| 3372 | 328 | 193 | 17M | ·1061 | 11·4059 | 5·589 | **271** | 1 | 3 | Mr | 28 | ·51 | 5 | Ap | 27 | ·04 | 6 | My | 26○ | ·57 | 1 | Je | 25● | ·10 | 2 | Jl | 24 | ·63 |
| 3373 | 329 | 194 | 16M | ·3649 | 0·5142 | 1·747 | **272** | 2 | 7 | Mr | 16 | ·88 | 3 | My | 14○ | ·94 | 5 | Je | 13 | ·47 | 7 | Jl | 13 | ·00 | 1 | Au | 11 | ·53 |
| | | | | | | | | | 2 | Ap | 15 | ·41 | | | | | | | | | | | | | | | | |
| 3374 | 330 | 195 | 16M | ·6236 | 19·1531 | 27·434 | **273** | 4 | 6 | Ap | 4 | ·78 | 1 | My | 4● | ·31 | 2 | Je | 2 | ·84 | 4 | Jl | 2 | ·37 | 5 | Jl | 31 | ·90 |
| 3375 | 331 | 196 | 16M | ·8824 | 8·2614 | 23·592 | **274** | 5 | 4 | Mr | 25○ | ·14 | 5 | Ap | 23● | ·67 | 7 | My | 23 | ·20 | 1 | Je | 21 | ·73 | 3 | Jl | 21 | ·27 |
| | | | | | | | | | | | | | | | | | | | | | | | | | 4 | Au | 19 | ·80 |
| 3376 | 332 | 197 | 17M | ·1411 | 26·9002 | 21·725 | **275** | 6 | 3 | Ap | 13 | ·04 | 4 | My | 12 | ·57 | 6 | Je | 11 | ·10 | 7 | Jl | 10 | ·63 | 2 | Au | 9 | ·16 |
| 3377 | 333 | 198 | 16M | ·3999 | 16·0085 | 17·882 | **276** | 7 | 7 | Ap | 1 | ·41 | 1 | Ap | 30 | ·94 | 3 | My | 30 | ·47 | 5 | Je | 29 | ·00 | 6 | Jl | 28 | ·53 |

**Surya Siddhanta.**

| Asvina + 147·65293 + 9·880 | | | | Kartika + 177·18353 + 11·856 | | | | Margasira + 206·71411 + 13·832 | | | | Pausha + 236·24470 + 15·808 | | | | A.D. | Magha + 265·77529 + 17·784 | | | | A.D. | Phalguna + 295·30588 + 19·760 | | | | Chaitra {+324·83647 {+354·36705 {+ 21·736 {+ 23·712 | | | |
|---|---|---|---|---|---|---|---|---|---|---|---|---|---|---|---|---|---|---|---|---|---|---|---|---|---|---|---|---|---|
| Week-day | Month | Day | Fraction | Week-day | Month | Day | Fraction | Week-day | Month | Day | Fraction | Week-day | Month | Day | Fraction | | Week-day | Month | Day | Fraction | | Week-day | Month | Day | Fraction | Week-day | Month | Day | Fraction |
| 5 | S | 7○ | ·32 | 6 | O | 6 | ·85 | 1 | N | 5 | ·38 | 2 | D | 4 | ·91 | 38 | 4 | Jr | 3 | ·44 | | 5 | F | 1 | ·97 | 7 | Mr | 3○ | ·50 |
| 2 | Au | 27○ | ·68 | 4 | S | 26 | ·22 | 5 | O | 25 | ·75 | 7 | N | 24 | ·28 | | 1 | D | 23 | ·81 | 39 | 3 | Jr | 22 | ·34 | 4 | F | 20○ | ·87 |
| 1 | S | 15 | ·58 | 3 | O | 15 | ·11 | 4 | N | 13 | ·64 | 6 | D | 13 | ·17 | 40 | 7 | Jr | 11 | ·70 | | 2 | F | 10○ | ·24 | 3 | Mr | 10 | ·77 |
| 5 | S | 3 | ·95 | 7 | O | 3 | ·48 | 2 | N | 2 | ·01 | 3 | D | 1 | ·54 | | 5 | D | 31○ | ·07 | 41 | 6 | Jr | 29● | ·60 | 1 | F | 28 | ·13 |
| 3 | Au | 24 | ·32 | 4 | S | 22 | ·85 | 6 | O | 22 | ·38 | 7 | N | 20 | ·91 | | 2 | D | 20○ | ·44 | 42 | 3 | Jr | 18 | ·97 | 5 | F | 17 | ·50 |
| 2 | S | 12 | ·21 | 3 | O | 11 | ·74 | 5 | N | 10 | ·28 | 6 | D | 9○ | ·81 | 43 | 1 | Jr | 8 | ·34 | | 2 | F | 6 | ·87 | 4 | Mr | 8 | ·40 |
| 6 | S | 1 | ·58 | 1 | O | 1 | ·11 | 2 | O | 30 | ·64 | 4 | N | 29 | ·17 | | 5 | D | 28 | ·70 | 44 | 7 | Jr | 27 | ·23 | 1 | F | 25 | ·76 |
| 3 | Au | 20 | ·95 | 5 | S | 19 | ·48 | 7 | O | 19 | ·01 | 1 | N | 17 | ·54 | | 3 | D | 17 | ·07 | 45 | 4 | Jr | 15 | ·60 | {6 | F | 14 | ·13 |
| | | | | | | | | | | | | | | | | | | | | | | | | | | {7 | Mr | 15 | ·66 |
| 2 | S | 8 | ·85 | 4 | O | 8○ | ·38 | 5 | N | 6● | ·91 | 7 | D | 6 | ·44 | 46 | 1 | Jr | 4 | ·97 | | 3 | F | 3 | ·50 | 5 | Mr | 5 | ·03 |
| 7 | Au | 29 | ·21 | 1 | S | 27○ | ·74 | 3 | O | 27 | ·27 | 4 | N | 25 | ·80 | | 6 | D | 25 | ·34 | 47 | 7 | Jr | 23 | ·87 | 2 | F | 22 | ·40 |
| 6 | S | 17 | ·11 | 7 | O | 16 | ·64 | 2 | N | 15 | ·17 | 3 | D | 14 | ·70 | 48 | 5 | Jr | 13 | ·23 | | 6 | F | 11○ | ·76 | 1 | Mr | 12 | ·29 |
| 3 | S | 5● | ·48 | 5 | O | 5 | ·01 | 6 | N | 3 | ·54 | 1 | D | 3 | ·07 | 49 | 2 | Jr | 1 | ·60 | | 4 | Jr | 31○ | ·13 | 5 | Mr | 1 | ·66 |
| 7 | Au | 25● | ·84 | 2 | S | 24 | ·38 | 3 | O | 23 | ·91 | 5 | N | 22 | ·44 | | 6 | D | 21 | ·97 | 50 | 1 | Jr | 20●○ | ·50 | 3 | F | 19 | ·03 |
| 6 | S | 13 | ·74 | 1 | O | 13 | ·27 | 2 | N | 11 | ·80 | 4 | D | 11 | ·33 | 51 | 5 | Jr | 9● | ·86 | | 7 | F | 8 | ·40 | 1 | Mr | 9 | ·93 |
| 4 | S | 3 | ·11 | 5 | O | 2 | ·64 | 7 | N | 1 | ·17 | 1 | N | 30 | ·70 | | 3 | D | 30 | ·23 | 52 | 4 | Jr | 28 | ·76 | 6 | F | 27 | ·29 |
| 1 | Au | 22 | ·48 | 3 | S | 21 | ·01 | 4 | O | 20 | ·54 | 6 | N | 19○ | ·07 | | 7 | D | 18 | ·60 | 53 | 2 | Jr | 17 | ·13 | 3 | F | 15 | ·66 |
| 7 | S | 10 | ·37 | 1 | O | 9 | ·90 | 3 | N | 8○ | ·43 | 4 | D | 7 | ·97 | 54 | 6 | Jr | 6 | ·50 | | 1 | F | 5 | ·03 | 2 | Mr | 6 | ·56 |
| 4 | Au | 30 | ·74 | 6 | S | 29 | ·27 | 7 | O | 28●○ | ·80 | 2 | N | 27 | ·33 | | 3 | D | 26 | ·86 | 55 | 5 | Jr | 25 | ·39 | 6 | F | 23 | ·92 |
| 3 | S | 18○ | ·64 | 5 | O | 18 | ·17 | 6 | N | 16 | ·70 | 1 | D | 16 | ·23 | 56 | 2 | Jr | 14 | ·76 | | 4 | F | 13 | ·29 | 5 | Mr | 13○ | ·82 |
| 1 | S | 7○ | ·01 | 2 | O | 6 | ·54 | 4 | N | 5 | ·07 | 5 | D | 4 | ·60 | 57 | 7 | Jr | 3 | ·13 | | 1 | F | 1 | ·66 | 3 | Mr | 3○ | ·19 |
| 5 | Au | 27●○ | ·37 | 6 | S | 25 | ·90 | 1 | O | 25 | ·43 | 2 | N | 23 | ·96 | | 4 | D | 23 | ·49 | 58 | 6 | Jr | 22 | ·03 | 7 | F | 20 | ·56 |
| 4 | S | 15 | ·27 | 5 | O | 14 | ·80 | 7 | N | 13 | ·33 | 1 | D | 12 | ·86 | 59 | 3 | Jr | 11○ | ·39 | | 4 | F | 9 | ·92 | 6 | Mr | 11 | ·45 |
| 1 | S | 4 | ·64 | 3 | O | 4 | ·17 | 4 | N | 2 | ·70 | 6 | D | 2 | ·23 | | 7 | D | 31○ | ·76 | 60 | 2 | Jr | 30● | ·29 | 3 | F | 28 | ·82 |
| 6 | Au | 24 | ·00 | 7 | S | 22 | ·53 | 2 | O | 22 | ·07 | 3 | N | 20 | ·60 | | 5 | D | 20○ | ·13 | 61 | 6 | Jr | 18 | ·66 | 1 | F | 17 | ·19 |
| 4 | S | 11 | ·90 | 6 | O | 11 | ·43 | 7 | N | 9 | ·96 | 2 | D | 9 | ·49 | 62 | 4 | Jr | 8 | ·02 | | 5 | F | 6 | ·55 | 7 | Mr | 8 | ·09 |
| 2 | S | 1 | ·27 | 3 | S | 30 | ·80 | 5 | O | 30 | ·33 | 6 | N | 28● | ·86 | | 1 | D | 28 | ·39 | 63 | 2 | Jr | 26 | ·92 | 4 | F | 25 | ·45 |
| 6 | Au | 21 | ·64 | {1 | S | 20 | ·17 | **Margasira Kshaya.** | | | | 4 | N | 18● | ·23 | | 5 | D | 17 | ·76 | 64 | 7 | Jr | 16 | ·29 | {1 | F | 14 | ·82 |
| | | | | {2 | O | 19○ | ·70 | | | | | | | | | | | | | | | | | | | {3 | Mr | 15 | ·35 |
| 5 | S | 8 | ·53 | 7 | O | 8○ | ·06 | 1 | N | 6 | ·59 | 3 | D | 6 | ·13 | 65 | 4 | Jr | 4 | ·66 | | 6 | F | 3 | 19 | 7 | Mr | 4 | ·72 |
| 2 | Au | 28 | ·90 | 4 | S | 27○ | ·43 | 5 | O | 26 | ·96 | 7 | N | 25 | ·49 | | 2 | D | 25 | ·02 | 66 | 3 | Jr | 23 | ·55 | 5 | F | 22○ | ·08 |
| 1 | S | 16● | ·80 | 3 | O | 16 | ·33 | 4 | N | 14 | ·86 | 6 | D | 14 | ·39 | 67 | 7 | Jr | 12 | ·92 | | 2 | F | 11○ | ·45 | 3 | Mr | 12 | ·98 |
| 6 | S | 6● | ·17 | 7 | O | 5 | ·70 | 2 | N | 4 | ·23 | 3 | D | 3 | ·76 | 68 | 5 | Jr | 2 | ·29 | | 6 | Jr | 31●○ | ·82 | 1 | Mr | 1 | ·35 |
| 3 | Au | 25 | ·53 | 5 | S | 24 | ·06 | 6 | O | 23 | ·59 | 1 | N | 22 | ·12 | | 2 | D | 21 | ·65 | 69 | 4 | Jr | 20 | ·19 | 5 | F | 18 | ·72 |
| 2 | S | 13 | ·43 | 3 | O | 12 | ·96 | 5 | N | 11 | ·49 | 7 | D | 11 | ·02 | 70 | 7 | Jr | 9 | ·55 | | 3 | F | 8 | ·08 | 4 | Mr | 9 | ·61 |
| 6 | S | 2 | ·80 | 1 | O | 2 | ·33 | 2 | O | 31 | ·86 | 4 | N | 30○ | ·39 | | 5 | D | 29 | ·92 | 71 | 7 | Jr | 28 | ·45 | 1 | F | 26 | ·98 |
| 4 | Au | 23 | ·16 | 5 | S | 21 | ·69 | 7 | O | 21 | ·22 | 1 | N | 19●○ | ·76 | | 3 | D | 19 | ·29 | 72 | 4 | Jr | 17 | ·82 | 6 | F | 16 | ·35 |
| 3 | S | 10 | ·06 | 4 | O | 9 | ·59 | 6 | N | 8●○ | ·12 | 7 | D | 7 | ·65 | 73 | 2 | Jr | 6 | ·18 | | 3 | F | 4 | ·71 | 5 | Mr | 6 | ·25 |
| 7 | Au | 30 | ·43 | 1 | S | 28○ | ·96 | 3 | O | 28 | ·49 | 5 | N | 27 | ·02 | | 6 | D | 26 | ·55 | 74 | 1 | Jr | 25 | ·08 | 2 | F | 23 | ·61 |
| 6 | S | 18○ | ·33 | 7 | O | 17 | ·86 | 2 | N | 16 | ·39 | 3 | D | 15 | ·92 | 75 | 5 | Jr | 14 | ·45 | | 6 | F | 12 | ·98 | 1 | Mr | 14○ | ·51 |
| 3 | S | 7●○ | ·69 | 5 | O | 7 | ·22 | 6 | N | 5 | ·75 | 1 | D | 5 | ·29 | 76 | 2 | Jr | 3 | ·82 | | 4 | F | 2 | ·35 | 5 | Mr | 2●○ | 88 |
| 1 | Au | 27● | ·06 | 2 | S | 25 | ·59 | 4 | O | 25 | ·12 | 5 | N | 23 | ·65 | | 7 | D | 23 | ·18 | 77 | 1 | Jr | 21○ | ·71 | 3 | F | 20● | 24 |

**TABLE X—**

| | Vaisakha | Jyeshta | Ashada | Sravana | Bhadrapada |
|---|---|---|---|---|---|
| ☉'s Anom col. 6 | | + 29·33059 | + 59·06117 | + 88·59176 | + 118·12225 |
| ☾'s Anom col. 7 | | + 1·976 | + 3·952 | + 5·928 | + 7·904 |

| Kaliyuga | Vikrama Era | Saka Era | Commencement of Solar Year: Month and day A.D. | Fraction of day | First New-Moon in Solar Year | Anomaly of first New-Moon | A.D. | Week-day of 1st January | Vaisakha Week-day | Month | Day | Fraction | Jyeshta Week-day | Month | Day | Fraction | Ashada Week-day | Month | Day | Fraction | Sravana Week-day | Month | Day | Fraction | Bhadrapada Week-day | Month | Day | Fraction |
|---|---|---|---|---|---|---|---|---|---|---|---|---|---|---|---|---|---|---|---|---|---|---|---|---|---|---|---|---|
| 3378 | 334 | 199 | 16M | ·6586 | 5·1168 | 14·039 | **277** | 2 | 4 | Mr | 21 | ·77 | 6 | Ap | 20 | ·30 | 7 | My | 19 | ·84 | 2 | Je | 18 | ·33 | 5 | Au | 16 | ·43 |
| | | | | | | | | | | | | | | | | | | | | | 3 | Jl | 17 | ·90 | | | | |
| 3379 | 335 | 200 | 16M | ·9174 | 23·7557 | 12·173 | **278** | 3 | 3 | Ap | 9 | ·67 | 5 | My | 9 | ·20 | 6 | Je | 7 | ·73 | 1 | Jl | 7○ | ·26 | 2 | Au | 5 | ·79 |
| 3380 | 336 | 201 | 17M | ·1762 | 12·8640 | 8·330 | **279** | 4 | 1 | Mr | 30 | ·04 | 2 | Ap | 28 | ·57 | 4 | My | 28 | ·10 | 5 | Je | 26●○ | ·63 | 7 | Jl | 26 | ·16 |
| 3381 | 337 | 202 | 16M | ·4349 | 1·9723 | 4·487 | **280** | 5 | 5 | Mr | 18 | ·41 | 6 | Ap | 16 | ·94 | 3 | Je | 15● | ·00 | 4 | Jl | 14 | ·53 | 6 | Au | 13 | ·06 |
| | | | | | | | | | | | | | 1 | My | 16 | ·47 | | | | | | | | | | | | |
| 3382 | 338 | 203 | 16M | ·6937 | 20·6112 | 3·620 (9.620) | **281** | 7 | 4 | Ap | 6 | ·30 | 5 | My | 5○ | ·83 | 7 | Je | 4 | ·37 | 1 | Jl | 3 | ·90 | 3 | Au | 2 | ·43 |
| 3383 | 339 | 204 | 16M | ·9524 | 9·7195 | 26·332 | **282** | 1 | 1 | Mr | 26 | ·67 | 3 | Ap | 25●○ | ·20 | 4 | My | 24 | ·73 | 6 | Je | 23 | ·26 | 7 | Jl | 22 | ·79 |
| 3384 | 340 | 205 | 17M | ·2112 | 28·3584 | 24·465 | **283** | 2 | 7 | Ap | 14●○ | ·57 | 2 | My | 14 | ·10 | 3 | Je | 12 | ·63 | 5 | Jl | 12 | ·16 | 6 | Au | 10 | ·69 |
| 3385 | 341 | 206 | 16M | ·4699 | 17·4667 | 20·623 | **284** | 3 | 4 | Ap | 2● | ·94 | 6 | My | 2 | ·47 | 1 | Je | 1 | ·00 | 2 | Je | 30 | ·53 | 4 | Jl | 30 | ·06 |
| 3386 | 342 | 207 | 16M | ·7287 | 6·5750 | 16·780 | **285** | 5 | 2 | Mr | 23 | ·30 | 3 | Ap | 21 | ·83 | 5 | My | 21 | ·36 | 6 | Je | 19 | ·89 | 2 | Au | 17○ | ·96 |
| | | | | | | | | | | | | | | | | | | | | | 1 | Jl | 19 | ·42 | | | | |
| 3387 | 343 | 208 | 16M | ·9875 | 25·2139 | 14·913 | **286** | 6 | 1 | Ap | 11 | ·20 | 2 | My | 10 | ·73 | 4 | Je | 9 | ·26 | 5 | Jl | 8 | ·79 | 7 | Au | 7○ | ·32 |
| 3388 | 344 | 209 | 17M | ·2462 | 14·3222 | 11·070 | **287** | 7 | 5 | Mr | 31 | ·57 | 7 | Ap | 30 | ·10 | 1 | My | 29 | ·63 | 3 | Je | 28 | ·16 | 4 | Jl | 27●○ | ·69 |
| 3389 | 345 | 210 | 16M | ·5050 | 3·4305 | 7·228 | **288** | 1 | 2 | Mr | 19 | ·93 | 4 | Ap | 18 | ·47 | 7 | Je | 16○ | ·53 | 2 | Jl | 16● | ·06 | 3 | Au | 14 | ·59 |
| | | | | | | | | | | | | | 6 | My | 18 | ·00 | | | | | | | | | | | | |
| 3390 | 346 | 211 | 16M | ·7637 | 22·0694 | 5·361 | **289** | 3 | 1 | Ap | 7 | ·83 | 3 | My | 7 | ·36 | 4 | Je | 5○ | ·89 | 6 | Jl | 5● | ·42 | 7 | Au | 3 | ·95 |
| 3391 | 347 | 212 | 17M | ·0225 | 11·1777 | 1·514 | **290** | 4 | 6 | Mr | 28 | ·20 | 7 | Ap | 26 | ·73 | 2 | My | 26○ | ·26 | 3 | Je | 24 | ·79 | 5 | Jl | 24 | ·32 |
| 3392 | 348 | 213 | 17M | ·2812 | 0·2860 | 25·230 | **291** | 5 | 3 | Mr | 17 | ·57 | 6 | My | 15● | ·63 | 1 | Je | 14 | ·16 | 2 | Jl | 13 | ·69 | 4 | Au | 12 | ·22 |
| | | | | | | | | | 5 | Ap | 16 | ·10 | | | | | | | | | | | | | | | | |
| 3393 | 349 | 214 | 16M | ·5400 | 18·9249 | 23·363 | **292** | 6 | 2 | Ap | 4○ | ·46 | 3 | My | 3● | ·99 | 5 | Je | 2 | ·53 | 7 | Jl | 2 | ·06 | 1 | Jl | 31 | ·59 |
| 3394 | 350 | 215 | 16M | ·7987 | 8·0332 | 19·521 | **293** | 1 | 6 | Mr | 24○ | ·83 | 1 | Ap | 23 | ·36 | 2 | My | 22 | ·89 | 4 | Je | 21 | ·42 | 5 | Jl | 20 | ·95 |
| | | | | | | | | | | | | | | | | | | | | | | | | | 7 | Au | 19 | ·48 |
| 3395 | 351 | 216 | 17M | ·0575 | 26·6720 | 17·654 | **294** | 2 | 5 | Ap | 12 | ·73 | 7 | My | 12 | ·26 | 1 | Je | 10 | ·79 | 3 | Jl | 10 | ·32 | 4 | Au | 8 | ·85 |
| 3396 | 352 | 217 | 17M | ·3163 | 15·7803 | 13·811 | **295** | 3 | 3 | Ap | 2 | ·10 | 4 | My | 1 | ·63 | 6 | My | 31 | ·16 | 7 | Je | 29 | ·69 | 2 | Jl | 29 | ·22 |
| 3397 | 353 | 218 | 16M | ·5750 | 4·8886 | 9·968 | **296** | 4 | 7 | Mr | 21 | ·46 | 1 | Ap | 19 | ·99 | 3 | My | 19 | ·52 | 6 | Jl | 17○ | ·59 | 1 | Au | 16 | ·12 |
| | | | | | | | | | | | | | | | | | 5 | Je | 18 | ·05 | | | | | | | | |
| 3398 | 354 | 219 | 16M | ·8338 | 23·5275 | 8·102 | **297** | 6 | 6 | Ap | 9 | ·36 | 7 | My | 8 | ·89 | 2 | Je | 7 | ·42 | 3 | Jl | 6●○ | ·95 | 5 | Au | 5 | ·48 |
| 3399 | 355 | 220 | 17M | ·0925 | 12·6358 | 4·259 | **298** | 7 | 3 | Mr | 29 | ·73 | 5 | Ap | 28 | ·26 | 6 | My | 27 | ·79 | 1 | Je | 26● | ·32 | 2 | Jl | 25 | ·85 |
| 3400 | 356 | 221 | 17M | ·3513 | 1·7441 | 0·416 | **299** | 1 | 1 | Mr | 19 | ·09 | 2 | Ap | 17 | ·63 | 5 | Je | 15 | ·69 | 7 | Jl | 15 | ·22 | 1 | Au | 13 | ·75 |
| | | | | | | | | | | | | | 4 | My | 17○ | ·16 | | | | | | | | | | | | |
| 3401 | 357 | 222 | 16M | ·6100 | 20·3830 | 26·104 | **300** | 2 | 6 | Ap | 5 | ·99 | 1 | My | 5●○ | ·52 | 3 | Je | 4 | ·05 | 4 | Jl | 3 | ·58 | 6 | Au | 2 | ·11 |
| 3402 | 358 | 223 | 16M | ·8688 | 9·4913 | 22·261 | **301** | 4 | 4 | Mr | 26 | ·36 | 5 | Ap | 24●○ | ·89 | 7 | My | 24 | ·42 | 1 | Je | 22 | ·95 | 3 | Jl | 22 | ·48 |
| 3403 | 359 | 224 | 17M | ·1276 | 28·1302 | 20·394 | **302** | 5 | 3 | Ap | 14 | ·26 | 4 | My | 13 | ·79 | 6 | Je | 12 | 32 | 7 | Jl | 11 | ·85 | 2 | Au | 10 | ·38 |
| 3404 | 360 | 225 | 17M | ·3863 | 17·2385 | 16·552 | **303** | 6 | 7 | Ap | 3 | ·62 | 2 | My | 3 | ·15 | 3 | Je | 1 | ·69 | 5 | Jl | 1 | ·22 | 6 | Jl | 30 | ·75 |
| 3405 | 361 | 226 | 16M | ·6451 | 6·3468 | 12·709 | **304** | 7 | 4 | Mr | 22 | ·99 | 6 | Ap | 21 | ·52 | 1 | My | 21 | ·05 | 2 | Je | 19 | ·58 | 5 | Au | 17○ | ·64 |
| | | | | | | | | | | | | | | | | | | | | | 4 | Jl | 19 | ·11 | | | | |
| 3406 | 362 | 227 | 16M | ·9038 | 24·9860 | 10·842 | **305** | 2 | 3 | Ap | 10 | ·89 | 5 | My | 10 | ·42 | 6 | Je | 8 | ·95 | 1 | Jl | 8 | ·48 | 3 | Au | 7●○ | ·01 |
| 3407 | 363 | 228 | 17M | ·1626 | 14·0940 | 6·999 | **306** | 3 | 1 | Mr | 31 | ·26 | 2 | Ap | 29 | ·79 | 4 | My | 29 | ·32 | 5 | Je | 27○ | ·85 | 7 | Jl | 27● | ·38 |
| 3408 | 364 | 229 | 17M | ·4213 | 3·2023 | 3·157 | **307** | 4 | 5 | Mr | 20 | ·64 | 7 | Ap | 19 | ·15 | 3 | Je | 17○ | ·21 | 4 | Jl | 16● | ·75 | 6 | Au | 15 | ·28 |
| | | | | | | | | | | | | | 1 | My | 18 | ·68 | | | | | | | | | | | | |
| 3409 | 365 | 230 | 16M | ·6801 | 21·8412 | 1·290 | **308** | 5 | 4 | Ap | 7 | ·52 | 6 | My | 7 | ·05 | 7 | Je | 5○ | ·58 | 2 | Jl | 5 | ·11 | 3 | Au | 3 | ·64 |
| 3410 | 366 | 231 | 16M | ·9389 | 10·9495 | 25·002 | **309** | 7 | 1 | Mr | 27 | ·89 | 3 | Ap | 26 | ·42 | 4 | My | 25 | ·95 | 6 | Je | 24 | ·48 | 1 | Jl | 24 | ·01 |
| 3411 | 367 | 232 | 17M | ·1976 | 0·0578 | 21·159 | **310** | 1 | 6 | Mr | 17 | ·25 | 2 | My | 15 | ·32 | 3 | Je | 13 | ·85 | 5 | Jl | 13 | ·38 | 6 | Au | 11 | ·91 |
| | | | | | | | | | 7 | Ap | 15○ | ·79 | | | | | | | | | | | | | | | | |
| 3412 | 368 | 233 | 17M | ·4564 | 18·6966 | 19·292 | **311** | 2 | 5 | Ap | 5○ | ·15 | 6 | My | 4 | ·68 | 1 | Je | 3 | ·21 | 2 | Jl | 2 | ·74 | 4 | Au | 1 | ·27 |
| 3413 | 369 | 234 | 16M | ·7151 | 7·8049 | 15·450 | **312** | 3 | 2 | Mr | 24○ | ·52 | 4 | Ap | 23 | ·05 | 5 | My | 22 | ·58 | 7 | Je | 21 | ·11 | 1 | Jl | 20 | ·64 |
| | | | | | | | | | | | | | | | | | | | | | | | | | 3 | Au | 19 | ·17 |
| 3414 | 370 | 235 | 16M | ·9739 | 26·4438 | 13·583 | **313** | 5 | 1 | Ap | 12 | ·42 | 2 | My | 11 | ·95 | 4 | Je | 10 | ·48 | 6 | Jl | 10 | ·01 | 7 | Au | 8 | ·54 |
| 3415 | 371 | 236 | 17M | ·2326 | 15·5521 | 9·740 | **314** | 6 | 5 | Ap | 1 | ·78 | 7 | My | 1 | ·31 | 1 | My | 30 | ·85 | 3 | Je | 29 | ·38 | 4 | Jl | 28○ | ·91 |
| 3416 | 372 | 237 | 17M | ·4914 | 4·6604 | 5·897 | **315** | 7 | 3 | Mr | 22 | ·15 | 4 | Ap | 20 | ·68 | 6 | My | 20 | ·21 | 2 | Jl | 18○ | ·27 | 3 | Au | 16 | ·80 |
| | | | | | | | | | | | | | | | | | 7 | Je | 18 | ·74 | | | | | | | | |

## Surya Siddhanta.

| Asvina | | | | Kartika | | | | Margasira | | | | Pausha | | | | A.D. | Magha | | | | A.D. | Phalguna | | | | Chaitra | | | |
|---|---|---|---|---|---|---|---|---|---|---|---|---|---|---|---|---|---|---|---|---|---|---|---|---|---|---|---|---|---|
| + 147·65293 | | | | + 177·18353 | | | | + 206·71411 | | | | + 236·24470 | | | | | +, 265·77529 | | | | | + 295·20588 | | | | {+324·88647 +354·36705 | | | |
| + 9·880 | | | | + 11·856 | | | | + 13·832 | | | | + 15·808 | | | | | + 17·784 | | | | | + 19·760 | | | | {+ 21·736 + 23·712 | | | |
| Week-day | Month | Day | Fraction | Week-day | Month | Day | Fraction | Week-day | Month | Day | Fraction | Week-day | Month | Day | Fraction | | Week-day | Month | Day | Fraction | | Week-day | Month | Day | Fraction | Week-day | Month | Day | Fraction |
| 6 | S | 14 | ·96 | 1 | O | 14 | ·49 | 3 | N | 13 | ·02 | 4 | D | 12 | ·55 | **78** | 6 | Jr | 11○ | ·08 | | 7 | F | 9● | ·61 | 2 | Mr | 11 | ·14 |
| 4 | S | 4 | ·33 | 5 | O | 3 | ·86 | 7 | N | 2 | ·39 | 1 | D | 1 | ·92 | | 3 | D | 31○ | ·45 | **79** | 4 | Jr | 29 | ·98 | 6 | F | 28 | ·51 |
| 1 | Au | 24 | ·69 | 3 | S | 23 | ·22 | 4 | O | 22 | ·75 | 6 | N | 21 | ·28 | | 7 | D | 20● | ·81 | **80** | 2 | Jr | 19 | ·35 | 3 | F | 17 | ·88 |
| 7 | S | 11 | ·59 | 2 | O | 11 | ·12 | 3 | N | 9 | ·65 | 5 | D | 9● | ·18 | **81** | 6 | Jr | 7 | ·71 | | 1 | F | 6 | ·24 | 2 | Mr | 7 | ·77 |
| 4 | Au | 31 | ·96 | 6 | S | 30 | ·46 | 1 | O | 30○ | ·02 | 2 | N | 28 | ·55 | | 4 | D | 28 | ·08 | **82** | 5 | Jr | 26 | ·61 | 7 | F | 25 | ·14 |
| {2 | Au | 21 | ·32 | 5 | O | 19○ | ·36 | **Margasira Kshaya.** | | | | 6 | N | 17 | ·92 | | 1 | D | 17 | ·45 | **83** | 2 | Jr | 15 | ·98 | {4 | F | 14 | ·51 |
| {3 | S | 19 | ·85 | | | | | | | | | | | | | | | | | | | | | | | {6 | Mr | 16 | ·04 |
| 1 | S | 9 | ·22 | 2 | O | 8●○ | ·75 | 4 | N | 7 | ·28 | 5 | D | 6 | ·81 | **84** | 7 | Jr | 5 | ·34 | | 1 | F | 3 | ·87 | 3 | Mr | 4 | ·41 |
| 5 | Au | 28 | ·59 | 7 | S | 27● | ·12 | 1 | O | 26 | ·65 | 3 | N | 25 | ·18 | | 4 | D | 24 | ·71 | **85** | 6 | Jr | 23 | ·24 | 7 | F | 21○ | ·77 |
| 4 | S | 16● | ·49 | 6 | O | 16 | ·02 | 7 | N | 14 | ·55 | 2 | D | 14 | ·08 | **86** | 3 | Jr | 12 | ·61 | | 5 | F | 11●○ | ·14 | 6 | Mr | 12 | ·67 |
| 1 | S | 5 | ·85 | 3 | O | 5 | ·38 | 4 | N | 3 | ·91 | 6 | D | 3 | ·45 | **87** | 7 | Jr | 1 | ·98 | | 2 | Jr | 31● | ·51 | 4 | Mr | 2 | ·04 |
| 6 | Au | 26 | ·22 | 7 | S | 24 | ·75 | 2 | O | 24 | ·28 | 3 | N | 22 | ·81 | | 5 | D | 22 | ·34 | **88** | 6 | Jr | 20 | ·87 | 1 | F | 19 | ·40 |
| 5 | S | 13 | ·12 | 6 | O | 12 | ·65 | 1 | N | 11 | ·18 | 2 | D | 10○ | ·71 | **89** | 4 | Jr | 9 | ·24 | | 5 | F | 7 | ·77 | 7 | Mr | 9 | ·30 |
| 2 | S | 2 | ·49 | 4 | O | 2 | ·02 | 5 | O | 31 | ·55 | 7 | N | 30●○ | ·08 | | 1 | D | 29 | ·61 | **90** | 3 | Jr | 28 | ·14 | 4 | F | 26 | ·67 |
| 6 | Au | 22 | ·85 | 1 | S | 21 | ·38 | 2 | O | 20 | ·91 | 4 | N | 19●○ | ·44 | | 5 | D | 18 | ·97 | **91** | 7 | Jr | 17 | ·51 | 2 | F | 16 | ·04 |
| 5 | S | 10 | ·75 | 7 | O | 10○ | ·28 | 1 | N | 8 | ·81 | 3 | D | 8 | ·34 | **92** | 4 | Jr | 6 | ·87 | | 6 | F | 5 | ·40 | 7 | Mr | 5 | ·93 |
| 3 | Au | 30 | ·12 | 4 | S | 28○ | ·65 | 6 | O | 28 | ·18 | 7 | N | 26 | ·71 | | 2 | D | 26 | ·24 | **93** | 3 | Jr | 24 | ·77 | 5 | F | 23 | ·30 |
| 2 | S | 18●○ | ·01 | 3 | O | 17 | ·55 | 5 | N | 16 | ·08 | 6 | D | 15 | ·61 | **94** | 1 | Jr | 14 | ·14 | | 2 | F | 12 | ·67 | 4 | Mr | 14●○ | ·20 |
| 6 | S | 7● | ·38 | 7 | O | 6 | ·91 | 2 | N | 5 | ·44 | 3 | D | 4 | ·97 | **95** | 5 | Jr | 3 | ·50 | | 7 | F | 2○ | ·03 | 1 | Mr | 3● | ·57 |
| 3 | Au | 27 | ·75 | 5 | S | 26 | ·28 | 6 | O | 25. | ·81 | 1 | N | 24 | ·34 | | 2 | D | 23 | ·87 | **96** | 4 | Jr | 22○ | ·40 | 5 | F | 20 | ·93 |
| 2 | S | 14 | ·65 | 4 | O | 14 | ·18 | 5 | N | 12 | ·71 | 7 | D | 12 | ·24 | **97** | 1 | Jr | 10○ | ·77 | | 3 | F | 9 | ·30 | 4 | Mr | 10 | ·83 |
| 7 | S | 4 | ·01 | 1 | O | 3 | ·54 | 3 | N | 2 | ·07 | 4 | D | 1 | ·61 | | 6 | D | 31● | ·14 | **98** | 7 | Jr | 29 | ·67 | 2 | F | 28 | ·20 |
| 4 | Au | 24 | ·38 | 5 | S | 22 | ·91 | 7 | O | 22 | ·44 | 1 | N | 20 | ·97 | | 3 | D | 20● | ·50 | **99** | 5 | Jr | 19 | ·03 | 6 | F | 17 | ·56 |
| 3 | S | 12 | ·28 | 4 | O | 11 | ·81 | 6 | N | 10○ | ·34 | 7 | D | 9● | ·87 | **300** | 3 | Jr | 8 | ·40 | | 4 | F | 6 | ·93 | 6 | Mr | 7 | ·46 |
| 7 | Au | 31 | ·65 | 2 | S | 30 | ·18 | 3 | O | 29○ | ·71 | 5 | N | 28 | ·24 | | 6 | D | 27 | ·77 | **01** | 1 | Jr | 26 | ·30 | 2 | F | 24 | ·83 |
| {5 | Au | 21 | ·01 | 1 | O | 19○ | ·07 | 2 | N | 17 | ·60 | 4 | D | 17 | ·13 | **02** | 5 | Jr | 15 | ·67 | | 7 | F | 14 | ·20 | 1 | Mr | 15 | ·73 |
| {6 | S | 19 | ·54 | | | | | | | | | | | | | | | | | | | | | | | | | | |
| 3 | S | 8 | ·91 | 5 | O | 8● | ·44 | 6 | N | 6 | ·97 | 1 | D | 6 | ·50 | **03** | 3 | Jr | 5 | ·03 | | 4 | F | 3 | ·56 | 6 | Mr | 5○ | ·09 |
| 1 | Au | 29○ | ·28 | 2 | S | 27 | ·81 | 4 | O | 27 | ·34 | 5 | N | 25 | ·87 | | 7 | D | 25 | ·14 | **04** | 1 | Jr | 23 | ·93 | 3 | F | 22●○ | ·46 |
| 7 | S | 16 | ·17 | 1 | O | 15 | ·71 | 3 | N | 14 | ·24 | 4 | D | 13 | ·77 | **05** | 6 | Jr | 12 | ·30 | | 7 | F | 10 | ·83 | 2 | Mr | 12 | ·36 |
| 4 | S | 5 | ·54 | 6 | O | 5 | ·07 | 7 | N | 3 | ·60 | 2 | D | 3 | ·13 | **06** | 3 | Jr | 1 | ·66 | | 5 | Jr | 31 | ·19 | 6 | Mr | 1 | ·73 |
| 1 | Au | 25 | ·91 | 3 | S | 24 | ·44 | 4 | O | 23 | ·97 | 6 | N | 22 | ·50 | | 1 | D | 22○ | ·03 | **07** | 2 | Jr | 20 | ·56 | 4 | F | 19 | ·09 |
| 7 | S | 13 | ·81 | 2 | O | 13 | ·34 | 3 | N | 11 | ·87 | 5 | D | 11○ | ·40 | **08** | 6 | Jr | 9 | ·93 | | 1 | F | 8 | ·46 | 2 | Mr | 8 | ·99 |
| 5 | S | 2 | ·17 | 6 | O | 1 | ·70 | 1 | O | 31 | ·23 | 2 | N | 29●○ | ·77 | | 4 | D | 29 | ·30 | **09** | 5 | Jr | 27 | ·83 | 7 | F | 26 | ·36 |
| 2 | Au | 22 | ·54 | 4 | S | 21 | ·07 | 5 | O | 20 | ·60 | 7 | N | 19 | ·13 | | 1 | D | 18 | ·66 | **10** | 3 | Jr | 17 | ·19 | 4 | F | 15 | ·72 |
| 1 | S | 10 | ·44 | 2 | O | 9○ | ·97 | 4 | N | 8 | ·50 | 6 | D | 8 | ·03 | **11** | 7 | Jr | 6 | ·56 | | 2 | F | 5 | ·09 | 3 | Mr | 6 | ·62 |
| 5 | Au | 30 | ·81 | 7 | S | 29○ | ·34 | 1 | O | 28 | ·87 | 3 | N | 27 | ·40 | | 4 | D | 26 | ·93 | **12** | 6 | Jr | 25 | ·46 | 7 | F | 23 | ·99 |
| 4 | S | 17 | ·70 | 6 | O | 17 | ·23 | 7 | N | 15 | ·76 | 2 | D | 15 | ·29 | **13** | 3 | Jr | 13 | ·83 | | 5 | F | 12○ | ·36 | 6 | Mr | 13 | ·89 |
| 2 | S | 7● | ·07 | 3 | O | 6 | ·60 | 5 | N | 5 | ·13 | 6 | D | 4 | ·66 | **14** | 1 | Jr | 3 | ·19 | | 2 | F | 1○ | ·72 | 4 | Mr | 3● | ·25 |
| 6 | Au | 27 | ·44 | 7 | S | 25 | ·97 | 2 | O | 25 | ·50 | 4 | N | 24 | ·03 | | 5 | D | 23 | ·56 | **15** | 7 | Jr | 22○ | ·09 | 1 | F | 20 | ·62 |
| 5 | S | 15 | ·33 | 6 | O | 14 | ·87 | 1 | N | 13 | ·40 | 2 | D | 12 | ·93 | **16** | 4 | Jr | 11 | ·46 | | 5 | F | 9 | ·99 | 7 | Mr | 10 | ·52 |

**TABLE X—**

| | | | | | | | | | ☉'s Anom col. 6 | | | | + 29·53059 | | | | + 59·06117 | | | | + 88·59176 | | | | + 118·12235 | | | |
|---|---|---|---|---|---|---|---|---|---|---|---|---|---|---|---|---|---|---|---|---|---|---|---|---|---|---|---|---|
| | | | | | | | | | ☾'s Anom col. 7 | | | | + 1·976 | | | | + 3·952 | | | | + 5·928 | | | | + 7·904 | | | |
| | | | Commencement of Solar Year. | | | | | | **Vaisakha** | | | | **Jyeshta** | | | | **Ashada** | | | | **Sravana** | | | | **Bhadrapada** | | | |
| Kaliyuga. | Vikrama Era. | Saka Era. | Month and day A.D. | Fraction of day. | First New-Moon in Solar Year. | Anomaly of first New-Moon | A.D. | Week-day of 1st January. | Week-day | Month | Day | Fraction | Week-day | Month | Day | Fraction | Week-day | Month | Day | Fraction | Week-day | Month | Day | Fraction | Week-day | Month | Day | Fraction |
| 3417 | 373 | 238 | 16M | ·7501 | 23·2993 | 4·031 | **316** | 1 | 2 | Ap | 9 | ·05 | 3 | My | 8 | ·58 | 5 | Je | 7 | ·11 | 6 | Jl | 6● | ·64 | 1 | Au | 5 | ·17 |
| 3418 | 374 | 239 | 17M | ·0089 | 12·4076 | 0·187 | **317** | 3 | 6 | Mr | 29 | ·42 | 7 | Ap | 27 | ·95 | 2 | My | 27○ | ·48 | 4 | Je | 26 | ·01 | 5 | Jl | 25 | ·54 |
| 3419 | 375 | 240 | 17M | ·2677 | 1·5159 | 23·900 | **318** | 4 | 3 | Mr | 18 | ·78 | 5 | Ap | 17 | ·31 | 1 | Je | 15 | ·37 | 2 | Jl | 14 | ·91 | 4 | Au | 13 | ·44 |
| | | | | | | | | | | | | | 6 | My | 16○ | ·84 | | | | | | | | | | | | |
| 3420 | 376 | 241 | 17M | ·5264 | 20·1548 | 22·033 | **319** | 5 | 2 | Ap | 6 | ·68 | 4 | My | 6○ | ·21 | 5 | Je | 4 | ·74 | 7 | Jl | 4 | ·27 | 1 | Au | 2 | ·80 |
| 3421 | 377 | 242 | 16M | ·7852 | 9·2631 | 18·190 | **320** | 6 | 7 | Mr | 26 | ·05 | 1 | Ap | 24● | ·58 | 3 | My | 24 | ·11 | 4 | Je | 22 | ·64 | 6 | Jl | 22 | ·17 |
| 3422 | 378 | 243 | 17M | ·0439 | 27·9020 | 16·323 | **321** | 1 | 5 | Ap | 13 | ·95 | 7 | My | 13 | ·48 | 2 | Je | 12 | ·01 | 3 | Jl | 11 | ·54 | 5 | Au | 10 | ·07 |
| 3423 | 379 | 244 | 17M | ·3027 | 17·0103 | 12·481 | **322** | 2 | 3 | Ap | 3 | ·31 | 4 | My | 2 | ·84 | 6 | Je | 1 | ·37 | 7 | Je | 30 | ·90 | 2 | Jl | 30 | ·43 |
| 3424 | 380 | 245 | 17M | ·5614 | 6·1186 | 8·638 | **323** | 3 | 7 | Mr | 23 | ·68 | 2 | Ap | 22 | ·21 | 3 | My | 21 | ·74 | 5 | Je | 20 | ·27 | 1 | Au | 18○ | ·33 |
| | | | | | | | | | | | | | | | | | | | | | 6 | Jl | 19 | ·80 | | | | |
| 3425 | 381 | 246 | 16M | ·8202 | 24·7575 | 6·771 | **324** | 4 | 6 | Ap | 10 | ·58 | 1 | My | 10 | ·11 | 2 | Je | 8 | ·64 | 4 | Jl | 8○ | ·17 | 5 | Au | 6 | ·70 |
| 3426 | 382 | 247 | 17M | ·0790 | 13·8658 | 2·928 | **325** | 6 | 3 | Mr | 31 | ·94 | 5 | Ap | 29 | ·47 | 7 | My | 29 | ·01 | 1 | Je | 27○ | ·54 | 3 | Jl | 27 | ·07 |
| 3427 | 383 | 248 | 17M | ·3377 | 2·9741 | 26·640 | **326** | 7 | 1 | Mr | 20 | ·34 | 2 | Ap | 18 | ·84 | 5 | Je | 16○ | ·90 | 7 | Jl | 16 | ·43 | 1 | Au | 14 | ·96 |
| | | | | | | | | | | | | | 4 | My | 18 | ·37 | | | | | | | | | | | | |
| 3428 | 384 | 249 | 17M | ·5965 | 21·6130 | 24·773 | **327** | 1 | 7 | Ap | 8 | ·21 | 1 | My | 7 | ·74 | 3 | Je | 6● | ·27 | 4 | Jl | 5 | ·80 | 6 | Au | 4 | ·33 |
| 3429 | 385 | 250 | 16M | ·8552 | 10·7213 | 20·931 | **328** | 2 | 4 | Mr | 27 | ·58 | 6 | Ap | 26○ | ·11 | 7 | My | 25 | ·64 | 2 | Je | 24 | ·17 | 3 | Jl | 23 | ·70 |
| 3430 | 386 | 251 | 17M | ·1140 | 29·3601 | 19·064 | **329** | 4 | 1 | Mr | 16 | ·94 | 5 | My | 15 | ·00 | 6 | Je | 13 | ·53 | 1 | Jl | 13 | ·07 | 2 | Au | 11 | ·60 |
| | | | | | | | | | 3 | Ap | 15○ | ·47 | | | | | | | | | | | | | | | | |
| 3431 | 387 | 252 | 17M | ·3727 | 18·4684 | 15·221 | **330** | 5 | 7 | Ap | 4○ | ·84 | 2 | My | 4 | ·37 | 3 | Je | 2 | ·90 | 5 | Jl | 2 | ·43 | 6 | Jl | 31 | ·96 |
| 3432 | 388 | 253 | 17M | ·6315 | 7·5767 | 11·379 | **331** | 6 | 5 | Mr | 25● | ·21 | 6 | Ap | 23 | ·74 | 1 | My | 23 | ·27 | 2 | Je | 21 | ·80 | 4 | Jl | 21 | ·33 |
| | | | | | | | | | | | | | | | | | | | | | | | | | 5 | Au | 19 | ·86 |
| 3433 | 389 | 254 | 16M | ·8902 | 26·2156 | 9·512 | **332** | 7 | 4 | Ap | 12 | ·11 | 5 | My | 11 | ·64 | 7 | Je | 10 | ·17 | 1 | Jl | 9 | ·70 | 3 | Au | 8○ | ·23 |
| 3434 | 390 | 255 | 17M | ·1490 | 15·3239 | 5·669 | **333** | 2 | 1 | Ap | 1 | ·47 | 3 | My | 1 | ·00 | 4 | My | 30 | ·53 | 6 | Je | 29 | ·06 | 7 | Jl | 28●○ | ·59 |
| 3435 | 391 | 256 | 17M | ·4078 | 4·4322 | 1·826 | **334** | 3 | 5 | Mr | 21 | ·84 | 7 | Ap | 20 | ·37 | 1 | My | 19 | ·90 | 4 | Jl | 17● | ·96 | 6 | Au | 16 | ·49 |
| | | | | | | | | | | | | | | | | | 3 | Je | 18 | ·43 | | | | | | | | |
| 3436 | 392 | 257 | 17M | ·6665 | 23·0711 | 27·514 | **335** | 4 | 4 | Ap | 9 | ·74 | 6 | My | 9 | ·27 | 7 | Je | 7○ | ·80 | 2 | Jl | 7 | ·33 | 3 | Au | 5 | ·86 |
| 3437 | 393 | 258 | 16M | ·9253 | 12·1794 | 23·671 | **336** | 5 | 2 | Mr | 29 | ·10 | 3 | Ap | 27 | ·63 | 5 | My | 27○ | ·17 | 6 | Je | 25 | ·70 | 1 | Jl | 25 | ·23 |
| 3438 | 394 | 259 | 17M | ·1840 | 1·2877 | 19·829 | **337** | 7 | 6 | Mr | 18 | ·47 | 1 | Ap | 17 | ·00 | 4 | Je | 15 | ·06 | 5 | Jl | 14 | ·59 | 7 | Au | 13 | ·12 |
| | | | | | | | | | | | | | 2 | My | 16○ | ·53 | | | | | | | | | | | | |
| 3439 | 395 | 260 | 17M | ·4428 | 19·9266 | 17·962 | **338** | 1 | 5 | Ap | 6 | ·37 | 6 | My | 5● | ·90 | 1 | Je | 4 | ·43 | 2 | Jl | 3 | ·96 | 4 | Au | 2 | ·49 |
| 3440 | 396 | 261 | 17M | ·7015 | 9·0349 | 14·119 | **339** | 2 | 2 | Mr | 26○ | ·74 | 4 | Ap | 25 | ·27 | 5 | My | 24 | ·80 | 7 | Je | 23 | ·33 | 1 | Jl | 22 | ·86 |
| 3441 | 397 | 262 | 16M | ·9603 | 27·6738 | 12·252 | **340** | 3 | 1 | Ap | 13 | ·63 | 3 | My | 13 | ·16 | 4 | Je | 11 | ·69 | 6 | Jl | 11 | ·23 | 7 | Au | 9 | ·76 |
| 3442 | 398 | 263 | 17M | ·2191 | 16·7821 | 8·410 | **341** | 5 | 6 | Ap | 3 | ·00 | 7 | My | 2 | ·53 | 2 | Je | 1 | ·06 | 3 | Je | 30 | ·59 | 5 | Jl | 30 | ·12 |
| 3443 | 399 | 264 | 17M | ·4778 | 5·8904 | 4·567 | **342** | 6 | 3 | Mr | 23 | ·37 | 4 | Ap | 21 | ·90 | 6 | My | 21 | ·43 | 7 | Je | 19 | ·95 | 4 | Au | 18 | ·02 |
| | | | | | | | | | | | | | | | | | | | | | 2 | Jl | 19○ | ·49 | | | | |
| 3444 | 400 | 265 | 17M | ·7366 | 24·5293 | 2·700 | **343** | 7 | 2 | Ap | 11 | ·27 | 3 | My | 10 | ·80 | 5 | Je | 9 | ·33 | 6 | Jl | 8○ | ·86 | 1 | Au | 7 | ·39 |
| 3445 | 401 | 266 | 16M | ·9953 | 13·6376 | 26·412 | **344** | 1 | 6 | Mr | 30 | ·63 | 1 | Ap | 29 | ·16 | 2 | My | 28 | ·69 | 4 | Je | 27○ | ·22 | 5 | Jl | 26 | ·75 |
| 3446 | 402 | 267 | 17M | ·2541 | 2·7459 | 22·569 | **345** | 3 | 4 | Mr | 20 | ·00 | 5 | Ap | 18 | ·53 | 1 | Je | 16 | ·59 | 3 | Jl | 16 | ·12 | 4 | Au | 14 | ·65 |
| | | | | | | | | | | | | | 7 | My | 18 | ·06 | | | | | | | | | | | | |
| 3447 | 403 | 268 | 17M | ·5128 | 21·3848 | 20·702 | **346** | 4 | 2 | Ap | 7 | ·90 | 4 | My | 7○ | ·43 | 5 | Je | 5● | ·96 | 7 | Jl | 5 | ·49 | 2 | Au | 4 | ·02 |
| 3448 | 404 | 269 | 17M | ·7716 | 10·4931 | 16·860 | **347** | 5 | 7 | Mr | 28 | ·26 | 1 | Ap | 26○ | ·65 | 3 | My | 26 | ·33 | 4 | Je | 24 | ·86 | 6 | Jl | 24 | ·39 |
| 3449 | 405 | 270 | 17M | ·0304 | 29·1319 | 14·993 | **348** | 6 | 6 | Ap | 15●○ | ·16 | 7 | My | 14 | ·69 | 2 | Je | 13 | ·22 | 3 | Jl | 12 | ·75 | 5 | Au | 11 | ·28 |
| 3450 | 406 | 271 | 17M | ·2891 | 18·2402 | 11·150 | **349** | 1 | 3 | Ap | 4● | ·53 | 5 | My | 4 | ·06 | 6 | Je | 2 | ·59 | 1 | Jl | 2 | ·12 | 2 | Jl | 31 | ·65 |
| 3451 | 407 | 272 | 17M | ·5479 | 7·3485 | 7·307 | **350** | 9 | 7 | Mr | 24 | ·90 | 2 | Ap | 23 | ·43 | 3 | My | 22 | ·96 | 5 | Je | 21 | ·49 | 7 | Jl | 21 | ·02 |
| | | | | | | | | | | | | | | | | | | | | | | | | | 1 | Au | 19○ | ·55 |
| 3452 | 408 | 273 | 17M | ·8066 | 25·9874 | 5·441 | **351** | 3 | 6 | Ap | 12 | ·79 | 1 | My | 12 | ·32 | 2 | Je | 10 | ·85 | 4 | Jl | 10 | ·39 | 5 | Au | 8○ | ·92 |
| 3453 | 409 | 274 | 17M | ·0654 | 15·0957 | 1·598 | **352** | 4 | 4 | Ap | 1 | ·16 | 5 | Ap | 30 | ·69 | 7 | My | 30 | ·22 | 1 | Je | 28 | ·75 | 3 | Jl | 28○ | ·28 |
| 3454 | 410 | 275 | 17M | ·3241 | 4·2040 | 25·310 | **353** | 6 | 1 | Mr | 21 | ·53 | 3 | Ap | 20 | ·06 | 4 | My | 19 | ·59 | 7 | Jl | 17● | ·65 | 2 | Au | 16 | ·18 |
| | | | | | | | | | | | | | | | | | 6 | Je | 18○ | ·12 | | | | | | | | |
| 3455 | 411 | 276 | 17M | ·5829 | 22·8429 | 23·443 | **354** | 7 | 7 | Ap | 9 | ·43 | 1 | My | 8 | ·96 | 3 | Je | 7○ | ·49 | 5 | Jl | 7 | ·02 | 6 | Au | 5 | ·55 |

**Surya Siddhanta.**

| Asvina | | | | Kartika | | | | Margasira | | | | Pausha | | | | A.D. | Magha | | | | A.D. | Phalguna | | | | Chaitra | | | |
|---|---|---|---|---|---|---|---|---|---|---|---|---|---|---|---|---|---|---|---|---|---|---|---|---|---|---|---|---|---|
| + 147·65293 | | | | + 177·18353 | | | | + 206·71411 | | | | + 236·24470 | | | | | + 265·77529 | | | | | + 295·30588 | | | | {+324·83647 +354·36705 | | | |
| + 9·880 | | | | + 11·856 | | | | + 13·832 | | | | + 15·808 | | | | | + 17·784 | | | | | + 19·760 | | | | {+ 21·736 + 23·712 | | | |
| Week-day | Month | Day | Fraction | Week-day | Month | Day | Fraction | Week-day | Month | Day | Fraction | Week-day | Month | Day | Fraction | | Week-day | Month | Day | Fraction | | Week-day | Month | Day | Fraction | Week-day | Month | Day | Fraction |
| 2 | S | 3 | ·70 | 4 | O | 3 | ·23 | 5 | N | 1 | ·76 | 7 | D | 1 | ·29 | | 1 | D | 30● | ·82 | **17** | 3 | Jr | 29 | ·35 | 4 | F | 27 | ·89 |
| 7 | Au | 24 | ·07 | 1 | S | 22 | ·60 | 3 | O | 22 | ·13 | 4 | N | 20○ | ·66 | | 6 | D | 20 | ·19 | **18** | 7 | Jr | 18 | ·72 | 2 | F | 17 | ·25 |
| 5 | S | 11 | ·97 | 7 | O | 11 | ·50 | 2 | N | 10○ | ·03 | 3 | D | 9 | ·56 | **19** | 5 | Jr | 8 | ·09 | | 6 | F | 6 | ·62 | 1 | Mr | 8 | ·15 |
| 3 | S | 1 | ·33 | 4 | S | 30 | ·86 | 6 | O | 30○ | ·39 | 7 | N | 28 | ·93 | | 2 | D | 28 | ·46 | **20** | 3 | Jr | 26 | ·99 | 5 | F | 25 | ·52 |
| {7 | Au | 20 | ·70 | 3 | O | 18● | ·76 | 5 | N | 17 | ·29 | 6 | D | 16 | ·82 | **21** | 1 | Jr | 15 | ·35 | | 2 | F | 13 | ·88 | 4 | Mr | 15○ | ·41 |
| {2 | S | 19 | ·23 | | | | | | | | | | | | | | | | | | | | | | | | | | |
| 6 | S | 8○ | ·60 | 1 | O | 8 | ·13 | 2 | N | 6 | ·66 | 4 | D | 6 | ·19 | **22** | 5 | Jr | 4 | ·72 | | 7 | F | 3 | ·25 | 1 | Mr | 4○ | ·78 |
| 3 | Au | 28○ | ·97 | 5 | S | 27 | ·50 | 7 | O | 27 | ·03 | 1 | N | 25 | ·56 | | 3 | D | 25 | ·09 | **23** | 4 | Jr | 23 | ·62 | 6 | F | 22 | ·15 |
| 2 | S | 16 | ·86 | 4 | O | 16 | ·39 | 5 | N | 14 | ·92 | 7 | D | 14 | ·45 | **24** | 1 | Jr | 12 | ·99 | | 3 | F | 11● | ·52 | 5 | Mr | 12 | ·05 |
| 7 | S | 5 | ·23 | 1 | O | 4 | ·76 | 3 | N | 3 | ·29 | 4 | D | 2 | ·82 | **25** | 6 | Jr | 1○ | ·35 | | 7 | Jr | 30 | ·88 | 2 | Mr | 1 | ·41 |
| 4 | Au | 25 | ·60 | 6 | S | 24 | ·13 | 7 | O | 23 | ·66 | 2 | N | 22 | ·19 | | 3 | D | 21●○ | ·72 | **26** | 5 | Jr | 20 | ·25 | 6 | F | 18 | ·78 |
| 3 | S | 13 | ·49 | 5 | O | 13 | ·03 | 6 | N | 11 | ·56 | 1 | D | 11●○ | ·09 | **27** | 2 | Jr | 9 | ·62 | | 4 | F | 8 | 15 | 5 | Mr | 9 | ·68 |
| 7 | S | 2 | ·80 | 2 | O | 2 | ·39 | 3 | O | 31 | ·92 | 5 | N | 30 | ·45 | | 6 | D | 29 | ·98 | **28** | 1 | Jr | 28 | ·51 | 3 | F | 27 | ·05 |
| 5 | Au | 22 | ·23 | 6 | S | 20 | ·76 | 1 | O | 20○ | ·29 | 2 | N | 18 | ·82 | | 4 | D | 18 | ·35 | **29** | 5 | Jr | 16 | ·88 | 7 | F | 15 | ·41 |
| 4 | S | 10 | ·13 | 5 | O | 9●○ | ·66 | 7 | N | 8 | ·19 | 1 | D | 7 | ·72 | **30** | 3 | Jr | 6 | ·25 | | 4 | F | 4 | ·78 | 6 | Mr | 6 | ·31 |
| 1 | Au | 30 | ·49 | 3 | S | 29 | ·02 | 4 | O | 28 | ·55 | 6 | N | 27 | ·09 | | 7 | D | 26 | ·62 | **31** | 2 | Jr | 25 | ·15 | 3 | F | 23○ | ·68 |
| 7 | S | 18 | ·39 | 1 | O | 17 | ·92 | 3 | N | 16 | ·45 | 4 | D | 15 | ·98 | **32** | 6 | Jr | 14 | ·51 | | 1 | F | 13○ | ·04 | 2 | Mr | 13● | ·57 |
| 4 | S | 6 | ·76 | 6 | O | 6 | ·29 | 7 | N | 4 | ·82 | 2 | D | 4 | ·35 | **33** | 3 | Jr | 2 | ·88 | | 5 | F | 1●○ | ·41 | 6 | Mr | 2 | ·94 |
| 2 | Au | 27 | ·13 | 3 | S | 25 | ·66 | 5 | O | 25 | ·19 | 6 | N | 23 | ·72 | | 1 | D | 23 | ·25 | **34** | 2 | Jr | 21● | ·78 | 4 | F | 20 | ·31 |
| 1 | S | 15 | ·02 | 2 | O | 14 | ·55 | 4 | N | 13 | ·08 | 5 | D | 12 | ·61 | **35** | 7 | Jr | 11 | ·15 | | 1 | F | 9 | ·68 | 3 | Mr | 11 | ·21 |
| 5 | S | 4 | ·39 | 6 | O | 3 | ·92 | 1 | N | 2 | ·45 | 2 | D | 1○ | ·98 | | 4 | D | 31 | ·51 | **36** | 6 | Jr | 30 | ·04 | 7 | F | 28 | ·57 |
| 2 | Au | 23 | ·76 | 4 | S | 22 | ·29 | 5 | O | 21 | ·82 | 7 | N | 20○ | ·35 | | 1 | D | 19 | ·88 | **37** | 3 | Jr | 18 | ·41 | 4 | F | 16 | ·94 |
| 1 | S | 11 | ·65 | 3 | O | 11 | ·19 | 4 | N | 9○ | ·72 | 6 | D | 9 | ·25 | **38** | 7 | Jr | 7 | ·78 | | 2 | F | 6 | ·31 | 3 | Mr | 7 | ·84 |
| 6 | S | 1 | ·02 | 7 | S | 30 | ·55 | 2 | O | 30 | ·08 | 3 | N | 28 | ·61 | | 5 | D | 28 | ·14 | **39** | 6 | Jr | 26 | ·67 | 1 | F | 25 | ·21 |
| {3 | Au | 21 | ·39 | 6 | O | 19● | ·45 | 7 | N | 17 | ·98 | 2 | D | 17 | ·51 | **40** | 4 | Jr | 16 | ·04 | | 5 | F | 14 | ·57 | 7 | Mr | 15○ | ·10 |
| {4 | S | 19○ | ·92 | | | | | | | | | | | | | | | | | | | | | | | | | | |
| 2 | S | 8○ | ·29 | 3 | O | 7 | ·82 | 5 | N | 6 | ·35 | 6 | D | 5 | ·88 | **41** | 1 | Jr | 4 | ·41 | | 2 | F | 2 | ·94 | 4 | Mr | 4● | ·47 |
| 6 | Au | 28○ | ·65 | 1 | S | 27 | ·18 | 2 | O | 26 | ·71 | 4 | N | 25 | ·25 | | 5 | D | 24 | ·78 | **42** | 7 | Jr | 23 | ·31 | 1 | F | 21 | ·84 |
| 5 | S | 16 | ·55 | 7 | O | 16 | ·08 | 1 | N | 14 | ·61 | 3 | D | 14 | ·14 | **43** | 4 | Jr | 12○ | ·67 | | 6 | F | 11 | ·20 | 7 | Mr | 12 | ·73 |
| 2 | S | 5 | ·92 | 4 | O | 5 | ·45 | 5 | N | 3 | ·98 | 7 | D | 3 | ·51 | **44** | 2 | Jr | 2○ | ·04 | | 3 | Jr | 31 | ·57 | 4 | Mr | 1 | ·10 |
| 7 | Au | 25 | ·29 | 1 | S | 23 | ·82 | 3 | O | 23 | ·35 | 4 | N | 21 | ·88 | | 6 | D | 21○ | ·41 | **45** | 7 | Jr | 19 | ·94 | 2 | F | 18 | ·47 |
| 6 | S | 13 | ·18 | 7 | O | 12 | ·71 | 2 | N | 11 | ·24 | 3 | D | 10 | ·77 | **46** | 5 | Jr | 9 | ·31 | | 6 | F | 7 | ·84 | 1 | Mr | 9 | ·37 |
| 3 | S | 2 | ·55 | 5 | O | 2 | ·08 | 6 | O | 31○ | ·61 | 1 | N | 30 | ·14 | | 2 | D | 29 | ·67 | **47** | 4 | Jr | 28 | ·20 | 5 | F | 26 | ·73 |
| 7 | Au | 22 | ·92 | 2 | S | 21 | ·45 | 3 | O | 20○ | ·98 | 5 | N | 19 | ·51 | | 7 | D | 19 | ·04 | **48** | 1 | Jr | 17 | ·57 | {3 | F | 16 | ·10 |
| | | | | | | | | | | | | | | | | | | | | | | | | | | {4 | Mr | 16 | ·63 |
| 6 | S | 9 | ·81 | 1 | O | 9● | ·35 | 2 | N | 7 | ·88 | 4 | D | 7 | ·41 | **49** | 5 | Jr | 5 | ·94 | | 7 | F | 4 | ·47 | 2 | Mr | 6 | ·00 |
| 4 | Au | 30 | ·18 | 5 | S | 28 | ·71 | 7 | O | 28 | ·24 | 1 | N | 26 | ·77 | | 3 | D | 26 | ·30 | **50** | 4 | Jr | 24 | ·83 | 6 | F | 23○ | ·37 |
| 3 | S | 18 | ·08 | 4 | O | 17 | ·61 | 6 | N | 16 | ·14 | 7 | D | 15 | ·67 | **51** | 2 | Jr | 14 | ·20 | | 3 | F | 12○ | ·73 | 5 | Mr | 14 | ·26 |
| 7 | S | 7 | ·45 | 1 | O | 6 | ·98 | 3 | N | 5 | ·51 | 5 | D | 5 | ·04 | **52** | 6 | Jr | 3 | ·57 | | 1 | F | 2● | ·10 | 2 | Mr | 2 | ·63 |
| 4 | Au | 26 | ·81 | 6 | S | 25 | ·34 | 7 | O | 24 | ·87 | 2 | N | 23 | ·41 | | 3 | D | 22 | ·94 | **53** | 5 | Jr | 21 | ·47 | 7 | F | 20 | ·00 |
| 3 | S | 14 | ·71 | 5 | O | 14 | ·24 | 6 | N | 12 | ·77 | 1 | D | 12○ | ·30 | **54** | 2 | Jr | 10● | ·83 | | 4 | F | 9 | ·36 | 5 | Mr | 10 | ·89 |
| 1 | S | 4 | ·08 | 2 | O | 3 | ·61 | 4 | N | 2 | ·14 | 5 | D | 1○ | ·67 | | 7 | D | 31 | ·20 | **55** | 1 | Jr | 29 | ·73 | 3 | F | 28 | ·26 |

**TABLE X—**

| Kaliyuga. | Vikrama Era. | Saka Era. | Commencement of Solar Year: Month and day A.D. | Fraction of day. | First New-Moon in Solar Year. | Anomaly of first New-Moon. | A.D. | Week-day of 1st January. | Vaisakha (☉'s Anom col. 6; ☾'s Anom col. 7) | | | | Jyeshta (+ 29·53059; + 1·976) | | | | Ashada (+ 59·06117; + 3·952) | | | | Sravana (+ 88·59176; + 5·928) | | | | Bhadrapada (+ 118·12235; + 7·904) | | | |
|---|---|---|---|---|---|---|---|---|---|---|---|---|---|---|---|---|---|---|---|---|---|---|---|---|---|---|---|---|
| | | | | | | | | | Week-day | Month | Day | Fraction | Week-day | Month | Day | Fraction | Week-day | Month | Day | Fraction | Week-day | Month | Day | Fraction | Week-day | Month | Day | Fraction |
| 3456 | 412 | 277 | 17M | ·8416 | 11·9512 | 19·600 | **355** | 1 | 4 | Mr | 29 | ·79 | 6 | Ap | 28 | ·32 | 7 | My | 27●○ | ·85 | 2 | Je | 26 | ·38 | 3 | Jl | 25 | ·91 |
| 3457 | 413 | 278 | 17M | ·1004 | 1·0595 | 15·758 | **356** | 2 | 2 | Mr | 18 | ·16 | 3 | Ap | 16 | ·69 | 6 | Je | 14 | ·75 | 1 | Jl | 14 | ·28 | 2 | Au | 12 | ·81 |
| | | | | | | | | | | | | | 5 | My | 16 | ·22 | | | | | | | | | | | | |
| 3458 | 414 | 279 | 17M | ·3592 | 19·6984 | 13·891 | **357** | 4 | 1 | Ap | 6○ | ·06 | 2 | My | 5 | ·59 | 4 | Je | 4 | ·12 | 5 | Jl | 3 | ·65 | 7 | Au | 2 | ·18 |
| 3459 | 415 | 280 | 17M | ·6179 | 8·8067 | 10·048 | **358** | 5 | 5 | Mr | 26●○ | ·42 | 6 | Ap | 24 | ·95 | 1 | My | 24 | ·49 | 3 | Je | 23 | ·02 | 4 | Jl | 22 | ·55 |
| 3460 | 416 | 281 | 17M | ·8767 | 27·4459 | 8·181 | **359** | 6 | 4 | Ap | 14 | ·32 | 5 | My | 13 | ·85 | 7 | Je | 12 | ·38 | 1 | Jl | 11 | ·91 | 3 | Au | 10 | ·44 |
| 3461 | 417 | 282 | 17M | ·1354 | 16·5539 | 4·339 | **360** | 7 | 1 | Ap | 2 | ·69 | 3 | My | 2 | ·22 | 4 | My | 31 | ·75 | 6 | Je | 30 | ·28 | 7 | Jl | 29 | ·81 |
| 3462 | 418 | 283 | 17M | ·3942 | 5·6622 | 0·496 | **361** | 2 | 6 | Mr | 28 | ·30 | 7 | Ap | 21 | ·59 | 2 | My | 21 | ·12 | 5 | Jl | 19○ | ·18 | 6 | Au | 17● | ·71 |
| | | | | | | | | | | | | | | | | | 3 | Je | 19 | ·65 | | | | | | | | |
| 3463 | 419 | 284 | 17M | ·6529 | 24·3011 | 26·184 | **362** | 3 | 4 | Ap | 10 | ·95 | 6 | My | 10 | ·48 | 1 | Je | 9 | ·01 | 2 | Jl | 8○ | ·55 | 4 | Au | 7 | ·08 |
| 3464 | 420 | 285 | 17M | ·9117 | 13·4094 | 22·341 | **363** | 4 | 2 | Mr | 31 | ·32 | 3 | Ap | 29 | ·85 | 5 | My | 29 | ·38 | 6 | Je | 27 | ·91 | 1 | Jl | 27 | ·44 |
| 3465 | 421 | 286 | 17M | ·1705 | 2·5177 | 18·498 | **364** | 5 | 6 | Mr | 19 | ·69 | 1 | Ap | 18 | ·22 | 4 | Je | 16● | ·28 | 5 | Jl | 15 | ·81 | 7 | Au | 14 | ·34 |
| | | | | | | | | | | | | | 2 | My | 17○ | ·75 | | | | | | | | | | | | |
| 3466 | 422 | 287 | 17M | ·4292 | 21·1565 | 16·631 | **365** | 7 | 5 | Ap | 7 | ·59 | 7 | My | 7○ | ·12 | 1 | Je | 5● | ·65 | 3 | Jl | 5 | ·18 | 4 | Au | 3 | ·71 |
| 3467 | 423 | 288 | 17M | ·6880 | 10·2648 | 12·789 | **366** | 1 | 2 | Mr | 27 | ·95 | 4 | Ap | 26○ | ·48 | 6 | My | 26 | ·01 | 7 | Je | 24 | ·54 | 2 | Jl | 24 | ·07 |
| 3468 | 424 | 289 | 17M | ·9467 | 28·9037 | 10·922 | **367** | 2 | 1 | Ap | 15 | ·85 | 3 | My | 15 | ·38 | 4 | Je | 13 | ·91 | 6 | Jl | 13 | ·44 | 7 | Au | 11 | ·97 |
| 3469 | 425 | 290 | 17M | ·2055 | 18·0120 | 7·079 | **368** | 3 | 6 | Ap | 4● | ·22 | 7 | My | 3 | ·75 | 2 | Je | 2 | ·28 | 3 | Jl | 1 | ·81 | 5 | Jl | 31 | ·34 |
| 3470 | 426 | 291 | 17M | ·4642 | 7·1203 | 3·236 | **369** | 5 | 3 | Mr | 24 | ·58 | 5 | Ap | 23 | ·11 | 6 | My | 22 | ·65 | 1 | Je | 21 | ·18 | 4 | Au | 19○ | ·24 |
| | | | | | | | | | | | | | | | | | | | | | 2 | Jl | 20 | ·71 | | | | |
| 3471 | 427 | 292 | 17M | ·7230 | 25·7592 | 1·370 | **370** | 6 | 2 | Ap | 12 | ·48 | 4 | My | 12 | ·01 | 5 | Je | 10 | ·54 | 7 | Jl | 10 | ·07 | 1 | Au | 8●○ | ·60 |
| 3472 | 428 | 293 | 17M | ·9818 | 14·8675 | 25·082 | **371** | 7 | 6 | Ap | 1 | ·85 | 1 | My | 1 | ·38 | 2 | My | 30 | ·91 | 4 | Je | 29○ | ·44 | 5 | Jl | 28 | ·97 |
| 3473 | 429 | 294 | 17M | ·2405 | 3·9758 | 21·239 | **372** | 1 | 4 | Mr | 21 | ·22 | 5 | Ap | 19 | ·75 | 7 | My | 19 | ·28 | 3 | Jl | 17● | ·34 | 4 | Au | 15 | ·87 |
| | | | | | | | | | | | | | | | | | 1 | Je | 17○ | ·81 | | | | | | | | |
| 3474 | 430 | 295 | 17M | ·4993 | 22·6147 | 19·372 | **373** | 3 | 3 | Ap | 9 | ·11 | 4 | My | 8 | ·64 | 6 | Je | 7●○ | ·17 | 7 | Jl | 6 | ·71 | 2 | Au | 5 | ·24 |
| 3475 | 431 | 296 | 17M | ·7580 | 11·7230 | 15·529 | **374** | 4 | 7 | Mr | 29 | ·48 | 2 | Ap | 28 | ·01 | 3 | My | 27 | ·54 | 5 | Je | 26 | ·07 | 6 | Jl | 25 | ·60 |
| 3476 | 432 | 297 | 18M | ·0168 | 0·8313 | 11·687 | **375** | 5 | 4 | Mr | 18 | ·85 | 7 | My | 16 | ·91 | 2 | Je | 15 | ·44 | 3 | Jl | 14 | ·97 | 5 | Au | 13 | ·50 |
| | | | | | | | | | 6 | Ap | 17○ | ·38 | | | | | | | | | | | | | | | | |
| 3477 | 433 | 298 | 17M | ·2755 | 19·4702 | 9·820 | **376** | 6 | 3 | Ap | 5○ | ·75 | 5 | My | 5 | ·28 | 6 | Je | 3 | ·81 | 1 | Jl | 3 | ·34 | 2 | Au | 1 | ·87 |
| 3478 | 434 | 299 | 17M | ·5343 | 8·5785 | 5·977 | **377** | 1 | 1 | Mr | 26○ | ·11 | 2 | Ap | 24 | ·64 | 4 | My | 24 | ·17 | 5 | Je | 22 | ·70 | 7 | Jl | 22 | ·23 |
| | | | | | | | | | | | | | | | | | | | | | | | | | 1 | Au | 20 | ·77 |
| 3479 | 435 | 300 | 17M | ·7930 | 27·2174 | 4·110 | **378** | 2 | 7 | Ap | 14 | ·01 | 1 | My | 13 | ·54 | 3 | Je | 12 | ·07 | 4 | Jl | 11 | ·60 | 6 | Au | 10 | ·13 |
| 3480 | 436 | 301 | 18M | ·0518 | 16·3259 | 0·268 | **379** | 3 | 4 | Ap | 3 | ·38 | 5 | My | 2 | ·91 | 7 | Je | 1 | ·44 | 1 | Je | 30 | ·97 | 3 | Jl | 30○ | ·50 |
| 3481 | 437 | 302 | 17M | ·3106 | 5·4340 | 23·979 | **380** | 4 | 1 | Mr | 22 | ·74 | 3 | Ap | 21 | ·27 | 4 | My | 20 | ·81 | 7 | Jl | 18○ | ·87 | 2 | Au | 17 | ·40 |
| | | | | | | | | | | | | | | | | | 6 | Je | 19 | ·34 | | | | | | | | |
| 3482 | 438 | 303 | 17M | ·5693 | 24·0729 | 22·113 | **381** | 6 | 7 | Ap | 10 | ·64 | 2 | My | 10 | ·17 | 3 | Je | 8 | ·70 | 5 | Jl | 8● | ·23 | 6 | Au | 6 | ·76 |
| 3483 | 439 | 304 | 17M | ·8281 | 13·1812 | 18·270 | **382** | 7 | 5 | Mr | 31 | ·01 | 6 | Ap | 29 | ·54 | 1 | My | 29○ | ·07 | 2 | Je | 27 | ·60 | 4 | Jl | 27 | ·13 |
| 3484 | 440 | 305 | 18M | ·0868 | 2·2895 | 14·427 | **383** | 1 | 2 | Mr | 20 | ·38 | 3 | Ap | 18 | ·91 | 6 | Je | 16 | ·97 | 1 | Jl | 16 | ·50 | 3 | Au | 15 | ·03 |
| | | | | | | | | | | | | | 5 | My | 18○ | ·44 | | | | | | | | | | | | |
| 3485 | 441 | 306 | 17M | ·3456 | 20·9283 | 12·560 | **384** | 2 | 1 | Ap | 7 | ·27 | 2 | My | 6○ | ·80 | 4 | Je | 5 | ·33 | 5 | Jl | 4 | ·87 | 7 | Au | 3 | ·40 |
| 3486 | 442 | 307 | 17M | ·6043 | 10·0366 | 8·718 | **385** | 4 | 5 | Mr | 27 | ·64 | 7 | Ap | 26● | ·17 | 1 | My | 25 | ·70 | 3 | Je | 24 | ·23 | 4 | Jl | 23 | ·76 |
| 3487 | 443 | 308 | 17M | ·8631 | 28·6755 | 6·851 | **386** | 5 | 4 | Ap | 15● | ·54 | 6 | My | 15 | ·07 | 7 | Je | 13 | ·60 | 2 | Jl | 13 | ·13 | 3 | Au | 11 | ·66 |
| 3488 | 444 | 309 | 18M | ·1219 | 17·7838 | 3·008 | **387** | 6 | 1 | Ap | 4 | ·91 | 3 | My | 4 | ·44 | 4 | Je | 2 | ·97 | 6 | Jl | 2 | ·50 | 1 | Au | 1 | ·03 |
| 3489 | 445 | 310 | 17M | ·3806 | 6·8921 | 26·720 | **388** | 7 | 6 | Mr | 24 | ·27 | 7 | Ap | 22 | ·80 | 2 | My | 22 | ·33 | 3 | Je | 20 | ·86 | 6 | Au | 18●○ | ·93 |
| | | | | | | | | | | | | | | | | | | | | | 5 | Jl | 20 | ·39 | | | | |
| 3490 | 446 | 311 | 17M | ·6394 | 25·5310 | 24·853 | **389** | 2 | 5 | Ap | 12 | ·17 | 6 | My | 11 | ·70 | 1 | Je | 10 | ·23 | 2 | Jl | 9 | ·76 | 4 | Au | 8 | ·29 |
| 3491 | 447 | 312 | 17M | ·8981 | 14·6393 | 21·011 | **390** | 3 | 2 | Ap | 1 | ·54 | 4 | My | 1 | ·07 | 5 | My | 30 | ·60 | 7 | Je | 29○ | ·13 | 1 | Jl | 28 | ·66 |
| 3492 | 448 | 313 | 18M | ·1569 | 3·7476 | 17·168 | **391** | 4 | 6 | Mr | 21 | ·90 | 1 | Ap | 20 | ·43 | 2 | My | 19 | ·97 | 6 | Jl | 18 | ·03 | 7 | Au | 16 | ·56 |
| | | | | | | | | | | | | | | | | | 4 | Je | 18○ | ·50 | | | | | | | | |
| 3493 | 449 | 314 | 17M | ·4156 | 22·3865 | 15·301 | **392** | 5 | 5 | Ap | 8 | ·80 | 7 | My | 8 | ·33 | 1 | Je | 6● | ·86 | 3 | Jl | 6 | ·39 | 4 | Au | 4 | ·92 |
| 3494 | 450 | 315 | 17M | ·6744 | 11·4948 | 11·458 | **393** | 7 | 3 | Mr | 29 | ·17 | 4 | Ap | 27○ | ·70 | 6 | My | 27● | ·23 | 7 | Je | 35 | ·76 | 2 | Jl | 25 | ·29 |

**Surya Siddhanta.**

| Asvina | | | | Kartika | | | | Margasira | | | | Pausha | | | | A.D. | Magha | | | | A.D. | Phalguna | | | | Chaitra | | | |
|---|---|---|---|---|---|---|---|---|---|---|---|---|---|---|---|---|---|---|---|---|---|---|---|---|---|---|---|---|---|
| + 147·65293 | | | | + 177·18353 | | | | + 206·71411 | | | | + 236·24470 | | | | | + 265·77529 | | | | | + 295·30588 | | | | +324·83647 / +354·36705 | | | |
| + 9·880 | | | | + 11·356 | | | | + 13·832 | | | | + 15·808 | | | | | + 17·784 | | | | | + 19·760 | | | | + 21·736 / + 23·712 | | | |
| Week-day | Month | Day | Fraction | Week-day | Month | Day | Fraction | Week-day | Month | Day | Fraction | Week-day | Month | Day | Fraction | | Week-day | Month | Day | Fraction | | Week-day | Month | Day | Fraction | Week-day | Month | Day | Fraction |
| 5 | Au | 24 | ·45 | 6 | S | 22 | ·98 | 1 | O | 22 | ·51 | 3 | N | 21○ | ·04 | | 4 | D | 30 | ·57 | 56 | 6 | Jr | 19 | ·10 | 7 | F | 17 | ·63 |
| 4 | S | 11 | ·34 | 5 | O | 10 | ·87 | 7 | N | 9● | ·40 | 1 | D | 8 | ·93 | 57 | 3 | Jr | 7 | ·47 | | 5 | F | 6 | ·00 | 6 | Mr | 7 | ·53 |
| 1 | Au | 31 | ·71 | 3 | S | 30○ | ·24 | 4 | O | 29 | ·77 | 6 | N | 28 | ·30 | | 7 | D | 27 | ·83 | 58 | 2 | Jr | 26 | ·36 | 3 | F | 24 | ·89 |
| {6 | Au | 21 | ·08 | 2 | O | 19 | ·14 | 3 | N | 17 | ·67 | 5 | D | 17 | ·20 | 59 | 6 | Jr | 15 | ·73 | | 1 | F | 14 | ·26 | 2 | Mr | 15○ | ·79 |
| {7 | S | 19○ | ·61 | | | | | | | | | | | | | | | | | | | | | | | | | | |
| 4 | S | 8●○ | ·97 | 6 | O | 8 | ·31 | 1 | N | 7 | ·04 | 2 | D | 6 | ·57 | 60 | 4 | Jr | 5 | ·10 | | 5 | F | 3 | ·63 | 7 | Mr | 4● | ·16 |
| 2 | Au | 28● | ·34 | 3 | S | 26 | ·87 | 5 | O | 26 | ·40 | 6 | N | 24 | ·93 | | 1 | D | 24 | ·46 | 61 | 2 | Jr | 22○ | ·99 | 4 | F | 21 | ·53 |
| 1 | S | 16 | 24 | 2 | O | 15 | ·77 | 4 | N | 14 | ·30 | 5 | D | 13 | ·83 | 62 | 7 | Jr | 12○ | ·36 | | 1 | F | 10 | ·89 | 3 | Mr | 12 | ·42 |
| 5 | S | 5 | ·61 | 7 | O | 5 | ·14 | 1 | N | 3 | ·67 | 3 | D | 3 | ·20 | 63 | 4 | Jr | 1●○ | ·73 | | 6 | Jr | 31 | ·26 | 7 | Mr | 1 | ·76 |
| 2 | Au | 25 | ·97 | 4 | S | 24 | ·50 | 6 | O | 24 | ·03 | 7 | N | 22 | ·57 | | 2 | D | 22 | ·10 | 64 | 3 | Jr | 20 | ·63 | 5 | F | 19 | ·19 |
| 1 | S | 12 | ·87 | 3 | O | 12 | ·40 | 4 | N | 10○ | ·93 | 6 | D | 10 | ·46 | 65 | 7 | Jr | 8 | ·99 | | 2 | F | 7 | ·52 | 4 | Mr | 9 | ·08 |
| 6 | S | 2 | ·24 | 7 | O | 1 | ·77 | 2 | O | 31○ | ·30 | 3 | N | 29 | ·83 | | 5 | D | 29 | ·36 | 66 | 6 | Jr | 27 | ·89 | 1 | F | 26 | ·42 |
| 3 | Au | 22 | ·61 | 5 | S | 21 | ·14 | 6 | O | 20○ | ·67 | 1 | N | 19 | ·20 | | 2 | D | 18 | ·73 | 67 | {4 | Jr | 17 | ·26 | 7 | Mr | 17 | ·32 |
| | | | | | | | | | | | | | | | | | | | | | | {5 | F | 15 | ·79 | | | | |
| 2 | S | 10 | ·50 | 4 | O | 10● | ·03 | 5 | N | 8 | ·56 | 7 | D | 8 | ·09 | 68 | 1 | Jr | 6 | ·63 | | 3 | F | 5 | ·16 | 4 | Mr | 5○ | ·69 |
| 6 | Au | 29○ | ·87 | 1 | S | 28 | ·40 | 2 | O | 27 | ·93 | 4 | N | 26 | ·45 | | 5 | D | 25 | ·99 | 69 | 7 | Jr | 24 | ·52 | 2 | F | 23○ | ·05 |
| 5 | S | 17 | ·77 | 7 | O | 17 | ·30 | 1 | N | 15 | ·83 | 3 | D | 15 | ·36 | 70 | 4 | Jr | 13 | ·89 | | 6 | F | 12 | ·42 | 7 | Mr | 13 | ·95 |
| 3 | S | 7 | ·13 | 4 | O | 6 | ·67 | 6 | N | 5 | ·20 | 7 | D | 4 | ·73 | 71 | 2 | Jr | 3 | ·26 | | 3 | F | 1● | ·79 | 5 | Mr | 3 | ·32 |
| 7 | Au | 27 | ·50 | 2 | S | 26 | ·03 | 3 | O | 25 | ·56 | 5 | N | 24 | ·09 | | 6 | D | 23○ | ·62 | 72 | 1 | Jr | 22 | ·15 | 2 | F | 20 | ·69 |
| 6 | S | 14 | ·40 | 7 | O | 13 | ·93 | 2 | N | 12 | ·46 | 3 | D | 11○ | ·99 | 73 | 5 | Jr | 10 | ·52 | | 7 | F | 9 | ·05 | 1 | Mr | 10 | ·58 |
| 3 | S | 3 | ·77 | 5 | O | 3 | ·30 | 6 | N | 1 | ·83 | 1 | D | 1○ | ·36 | | 2 | D | 30 | ·89 | 74 | 4 | Jr | 29 | ·42 | 5 | F | 27 | ·95 |
| 1 | Au | 24 | ·13 | 2 | S | 22 | ·66 | 4 | O | 22 | ·19 | 5 | N | 20● | ·73 | | 7 | D | 20 | ·26 | 75 | 1 | Jr | 18 | ·79 | 3 | F | 17 | ·32 |
| 7 | S | 12 | ·03 | 1 | O | 11○ | ·56 | 3 | N | 10● | ·09 | 4 | D | 9 | ·62 | 76 | 6 | Jr | 8 | ·15 | | 7 | F | 6 | ·68 | 2 | Mr | 7 | ·21 |
| 4 | Au | 31 | ·40 | 5 | S | 29○ | ·93 | 7 | O | 29 | ·46 | 1 | N | 27 | ·99 | | 3 | D | 27 | ·52 | 77 | 5 | Jr | 26 | ·05 | 6 | F | 24 | ·58 |
| 3 | S | 19○ | ·30 | 4 | O | 18 | ·83 | 6 | N | 17 | 36 | 7 | D | 16 | ·89 | 78 | 2 | Jr | 15 | ·42 | | 3 | F | 13 | ·95 | 5 | Mr | 15 | ·48 |
| 7 | S | 8● | ·66 | 2 | O | 8 | ·19 | 3 | N | 6 | ·72 | 5 | D | 6 | ·25 | 79 | 6 | Jr | 4 | ·79 | | 1 | F | 3○ | ·32 | 2 | Mr | 4 | ·85 |
| 5 | Au | 29● | ·03 | 6 | S | 27 | ·56 | 1 | O | 27 | ·09 | 2 | N | 25 | ·62 | | 4 | D | 25 | ·15 | 80 | 5 | Jr | 23●○ | ·68 | 7 | F | 22 | ·21 |
| 3 | S | 15 | ·93 | 5 | O | 15 | ·46 | 6 | N | 13 | ·99 | 1 | D | 13 | ·52 | 81 | 3 | Jr | 12●○ | ·05 | | 4 | F | 10 | ·58 | 6 | Mr | 12 | ·11 |
| 1 | S | 5 | ·29 | 2 | O | 4 | ·83 | 4 | N | 3 | ·36 | 5 | D | 2 | ·89 | 82 | 7 | Jr | 1● | ·42 | | 1 | Jr | 30 | ·95 | 3 | Mr | 1 | ·48 |
| 5 | Au | 25 | ·66 | 7 | S | 24 | ·19 | 1 | O | 23 | ·72 | 3 | N | 22○ | ·25 | | 4 | D | 21 | ·78 | 83 | 6 | Jr | 20 | ·31 | 7 | F | 18 | ·85 |
| 4 | S | 13 | ·56 | 6 | O | 13 | ·09 | 7 | N | 11●○ | ·62 | 2 | D | 11 | ·15 | 84 | 3 | Jr | 9 | ·68 | | 5 | F | 8 | ·21 | 6 | Mr | 8 | ·74 |
| 1 | S | 1 | ·93 | 3 | O | 1 | ·46 | 4 | O | 30○ | ·99 | 6 | N | 29 | ·52 | | 1 | D | 29 | ·05 | 85 | 2 | Jr | 27 | ·58 | 4 | F | 26 | ·11 |
| 6 | Au | 22 | ·29 | 7 | S | 20 | ·82 | 2 | O | 20 | ·35 | 3 | N | 18 | ·89 | | 5 | D | 18 | ·42 | 86 | {6 | Jr | 16 | ·95 | 3 | Mr | 17○ | ·01 |
| | | | | | | | | | | | | | | | | | | | | | | {1 | F | 15 | ·48 | | | | |
| 5 | S | 10○ | ·19 | 6 | O | 9 | ·72 | 1 | N | 8 | ·25 | 2 | D | 7 | ·78 | 87 | 4 | Jr | 6 | ·31 | | 5 | F | 4 | ·84 | 7 | Mr | 6●○ | ·37 |
| 2 | Au | 30○ | ·56 | 4 | S | 29 | ·09 | 5 | O | 28 | ·62 | 7 | N | 27 | ·15 | | 1 | D | 26 | ·68 | 88 | 3 | Jr | 25 | ·21 | 4 | F | 23 | ·74 |
| 1 | S | 17 | ·46 | 2 | O | 16 | ·99 | 4 | N | 15 | ·52 | 6 | D | 15 | ·05 | 89 | 7 | Jr | 13 | ·58 | | 2 | F | 12 | ·11 | 3 | Mr | 13 | ·64 |
| 5 | S | 6 | ·82 | 7 | O | 6 | ·35 | 1 | N | 4 | ·88 | 3 | D | 4 | ·41 | 90 | 4 | Jr | 2○ | ·95 | | 6 | F | 1 | ·48 | 1 | Mr | 3 | ·01 |
| 3 | Au | 27 | ·19 | 4 | S | 25 | ·72 | 6 | O | 25 | ·25 | 7 | N | 23 | ·78 | | 2 | D | 23○ | ·31 | 91 | 3 | Jr | 21 | ·84 | 5 | F | 20 | ·37 |
| 2 | S | 15 | ·09 | 3 | O | 14 | ·62 | 5 | N | 13 | ·15 | 6 | D | 12○ | ·68 | 92 | 1 | Jr | 11 | ·21 | | 2 | F | 9 | ·74 | 4 | Mr | 10 | ·27 |
| 6 | S | 3 | ·45 | 7 | O | 2 | ·99 | 2 | N | 1 | ·52 | 4 | D | 1 | ·05 | | 5 | D | 30 | ·58 | 93 | 7 | Jr | 29 | ·11 | 1 | F | 27 | ·64 |
| 3 | Au | 23 | ·82 | 5 | S | 22 | ·35 | 6 | O | 21○ | ·88 | 1 | N | 20● | ·41 | | 2 | D | 19 | ·94 | 94 | 4 | Jr | 18 | ·47 | 6 | F | 17 | ·01 |

☉'s Anom col. 6: + 29·53059 (Jyeshta), + 59·06117 (Ashada), + 88·59176 (Sravana), + 118·12235 (Bhadrapada)

☾'s Anom col. 7: + 1·976 (Jyeshta), + 3·952 (Ashada), + 5·928 (Sravana), + 7·904 (Bhadrapada)

| Kaliyuga. | Vikrama Era. | Saka Era. | Commencement of Solar Year. Month and day A.D. | Fraction of day. | First New-Moon in Solar Year. | Anomaly of first New-Moon. | A.D. | Week-day of 1st January. | Vaisahka. Week-day | Month | Day | Fraction | Jyeshta Week-day | Month | Day | Fraction | Ashada Week-day | Month | Day | Fraction | Sravana Week-day | Month | Day | Fraction | Bhadrapada Week-day | Month | Day | Fraction |
|---|---|---|---|---|---|---|---|---|---|---|---|---|---|---|---|---|---|---|---|---|---|---|---|---|---|---|---|---|
| 3495 | 451 | 316 | 17M | ·9332 | 0·6031 | 7·616 | **394** | 1 | 7 | Mr | 18 | 54 | 3 | My | 16 | ·60 | 5 | Je | 15 | ·13 | 6 | Jl | 14 | ·66 | 1 | Au | 13 | ·19 |
| | | | | | | | | | 2 | Ap | 17●○ | ·67 | | | | | | | | | | | | | | | | |
| 3496 | 452 | 317 | 18M | ·1919 | 19·2420 | 5·749 | **395** | 2 | 6 | Ap | 6●○ | ·43 | 7 | My | 5 | ·96 | 2 | Je | 4 | ·49 | 4 | Jl | 4 | ·03 | 5 | Au | 2 | ·56 |
| 3497 | 453 | 318 | 17M | ·4507 | 8·3503 | 1·906 | **396** | 3 | 3 | Mr | 25 | ·80 | 5 | Ap | 24 | ·33 | 6 | My | 23 | ·86 | 1 | Je | 22 | ·39 | 2 | Jl | 21 | ·92 |
| | | | | | | | | | | | | | | | | | | | | | | | | | 4 | Au | 20 | ·45 |
| 3498 | 454 | 319 | 17M | ·7094 | 26·9892 | 0·039 | **397** | 5 | 2 | Ap | 13 | ·70 | 4 | My | 13 | ·23 | 5 | Je | 11 | ·76 | 7 | Jl | 11 | ·29 | 1 | Au | 9○ | ·82 |
| 3499 | 455 | 320 | 17M | ·9682 | 16·0975 | 23·751 | **398** | 6 | 7 | Ap | 3 | ·07 | 1 | My | 2 | ·60 | 3 | Je | 1 | ·13 | 4 | Je | 30 | ·66 | 6 | Jl | 30○ | ·19 |
| 3500 | 456 | 321 | 18M | ·2269 | 5·2058 | 19·908 | **399** | 7 | 4 | Mr | 23 | ·43 | 5 | Ap | 21 | ·96 | 7 | My | 21 | ·49 | 3 | Jl | 19● | ·55 | 5 | Au | 18 | ·09 |
| | | | | | | | | | | | | | | | | | 2 | Je | 20 | ·02 | | | | | | | | |
| 3501 | 457 | 322 | 17M | ·4857 | 23·8447 | 18·042 | **400** | 1 | 3 | Ap | 10 | ·33 | 4 | My | 9 | ·86 | 6 | Je | 8○ | ·39 | 7 | Jl | 7● | ·92 | 2 | Au | 6 | ·45 |
| 3502 | 458 | 323 | 17M | ·7444 | 12·9530 | 14·199 | **401** | 3 | 7 | Mr | 30 | ·70 | 2 | Ap | 29 | ·23 | 3 | My | 28○ | ·76 | 5 | Je | 27 | ·29 | 6 | Jl | 26 | ·82 |
| 3503 | 459 | 324 | 18M | ·0032 | 2·0613 | 10·356 | **402** | 4 | 5 | Mr | 20 | ·06 | 6 | Ap | 18 | ·59 | 2 | Je | 16 | ·66 | 4 | Jl | 16 | ·19 | 5 | Au | 14 | ·72 |
| | | | | | | | | | | | | | 1 | My | 18●○ | ·13 | | | | | | | | | | | | |
| 3504 | 460 | 325 | 18M | ·2320 | 20·7001 | 8·489 | **403** | 5 | 3 | Ap | 7 | ·96 | 5 | My | 7● | ·49 | 7 | Je | 6 | ·02 | 2 | Jl | 5 | ·55 | 3 | Au | 4 | ·08 |
| 3505 | 461 | 326 | 17M | ·5207 | 9·8084 | 4·647 | **404** | 6 | 1 | Mr | 27○ | ·33 | 2 | Ap | 25 | ·86 | 4 | My | 25 | ·39 | 6 | Je | 24 | ·92 | 7 | Jl | 23 | ·45 |
| 3506 | 462 | 327 | 17M | ·7795 | 28·4473 | 2·780 | **405** | 1 | 7 | Ap | 15 | ·23 | 1 | My | 14 | ·76 | 2 | Je | 13 | ·29 | 4 | Jl | 12 | ·82 | 6 | Au | 11 | ·35 |
| 3507 | 463 | 328 | 18M | ·0382 | 17·5556 | 26·492 | **406** | 2 | 4 | Ap | 4 | ·59 | 6 | My | 4 | ·12 | 7 | Je | 2 | ·65 | 2 | Jl | 2 | ·19 | 3 | Jl | 31 | ·72 |
| 3508 | 464 | 329 | 18M | ·2970 | 6·6639 | 22·649 | **407** | 3 | 1 | Mr | 24 | ·96 | 3 | Ap | 23 | ·49 | 5 | My | 23 | ·02 | 6 | Je | 21 | ·55 | 2 | Au | 19● | ·61 |
| | | | | | | | | | | | | | | | | | | | | | 1 | Jl | 21 | ·08 | | | | |
| 3509 | 465 | 330 | 17M | ·5557 | 25·3028 | 20·782 | **408** | 4 | 7 | Ap | 11 | ·86 | 2 | My | 11 | ·39 | 3 | Je | 9 | ·92 | 5 | Jl | 9○ | ·45 | 6 | Au | 7 | ·98 |
| 3510 | 466 | 331 | 17M | ·8145 | 14·4111 | 16·939 | **409** | 6 | 5 | Ap | 1 | ·23 | 6 | Ap | 30 | ·76 | 1 | My | 30 | ·29 | 2 | Je | 28●○ | ·82 | 4 | Jl | 28 | ·35 |
| 3511 | 467 | 332 | 18M | ·0732 | 3·5194 | 13·097 | **410** | 7 | 2 | Mr | 21 | ·59 | 4 | Ap | 20 | ·12 | 5 | My | 19 | ·65 | 1 | Jl | 17 | ·71 | 3 | Au | 16 | ·25 |
| | | | | | | | | | | | | | | | | | 7 | Je | 18● | ·18 | | | | | | | | |
| 3512 | 468 | 333 | 18M | ·3320 | 22·1583 | 11·230 | **411** | 1 | 1 | Ap | 9 | ·49 | 3 | My | 9○ | ·02 | 4 | Je | 7 | ·55 | 6 | Jl | 7 | ·08 | 1 | Au | 5 | ·61 |
| 3513 | 469 | 334 | 17M | ·5908 | 11·2666 | 7·387 | **412** | 2 | 5 | Mr | 28 | ·86 | 7 | Ap | 27○ | ·39 | 1 | My | 26 | ·92 | 3 | Je | 25 | ·45 | 4 | Jl | 24 | ·98 |
| 3514 | 470 | 335 | 17M | ·8495 | 0·3749 | 3·544 | **413** | 4 | 3 | Mr | 18 | ·22 | 6 | My | 16 | ·28 | 7 | Je | 14 | ·82 | 2 | Jl | 14 | ·35 | 3 | Au | 12 | ·88 |
| | | | | | | | | | 4 | Ap | 16○ | ·75 | | | | | | | | | | | | | | | | |
| 3515 | 471 | 336 | 18M | ·1083 | 19·0138 | 1·678 | **414** | 5 | 2 | Ap | 6● | ·12 | 3 | My | 5 | ·65 | 5 | Je | 4 | ·18 | 6 | Jl | 3 | ·71 | 1 | Au | 2 | ·24 |
| 3516 | 472 | 337 | 18M | ·3670 | 8·1221 | 25·390 | **415** | 6 | 6 | Mr | 26 | ·49 | 1 | Ap | 25 | ·02 | 2 | My | 24 | ·55 | 4 | Je | 23 | ·08 | 5 | Jl | 22 | ·61 |
| | | | | | | | | | | | | | | | | | | | | | | | | | 7 | Au | 21○ | ·14 |
| 3517 | 473 | 338 | 17M | ·6258 | 26·7610 | 23·523 | **416** | 7 | 5 | Ap | 13 | ·39 | 6 | My | 12 | ·92 | 1 | Je | 11 | ·45 | 2 | Jl | 10 | ·98 | 4 | Au | 9○ | ·51 |
| 3518 | 474 | 339 | 17M | ·8845 | 15·8693 | 19·680 | **417** | 2 | 2 | Ap | 2 | ·75 | 4 | My | 2 | ·28 | 5 | My | 31 | ·81 | 7 | Je | 30 | ·34 | 1 | Jl | 29 | ·88 |
| 3519 | 475 | 340 | 18M | ·1433 | 4·9776 | 15·837 | **418** | 3 | 7 | Mr | 23 | ·12 | 1 | Ap | 21 | ·65 | 3 | My | 21 | ·18 | 6 | Jl | 19● | ·24 | 7 | Au | 17 | ·77 |
| | | | | | | | | | | | | | | | | | 4 | Je | 19 | ·71 | | | | | | | | |
| 3520 | 476 | 341 | 18M | ·4021 | 23·6164 | 13·971 | **419** | 4 | 6 | Ap | 11 | ·02 | 7 | My | 10 | ·55 | 2 | Je | 9○ | ·08 | 3 | Jl | 8 | ·61 | 5 | Au | 7 | ·14 |
| 3521 | 477 | 342 | 17M | ·6608 | 12·7247 | 10·128 | **420** | 5 | 3 | Mr | 30 | ·38 | 4 | Ap | 28 | ·92 | 6 | My | 28○ | ·45 | 7 | Je | 26 | ·98 | 2 | Jl | 26 | ·51 |
| 3522 | 478 | 343 | 17M | ·9196 | 1·8330 | 6·285 | **421** | 7 | 7 | Mr | 19 | ·75 | 2 | Ap | 18 | ·28 | 5 | Je | 16 | ·34 | 6 | Jl | 15 | ·87 | 1 | Au | 14 | ·40 |
| | | | | | | | | | | | | | 3 | My | 17 | ·81 | | | | | | | | | | | | |
| 3523 | 479 | 344 | 18M | ·1783 | 20·4719 | 4·418 | **422** | 1 | 6 | Ap | 7○ | ·65 | 1 | My | 7 | ·18 | 2 | Je | 5 | ·71 | 4 | Jl | 5 | ·24 | 5 | Au | 3 | ·77 |
| 3524 | 480 | 345 | 18M | ·4371 | 9·5802 | 0·576 | **423** | 2 | 4 | Mr | 28○ | ·02 | 5 | Ap | 26 | ·55 | 7 | My | 26 | ·08 | 1 | Je | 24 | ·61 | 3 | Jl | 24 | ·14 |
| 3525 | 481 | 346 | 17M | ·6958 | 28·2191 | 26·263 | **424** | 3 | 2 | Ap | 14 | ·91 | 4 | My | 14 | ·44 | 5 | Je | 12 | ·98 | 7 | Jl | 12 | ·51 | 2 | Au | 11 | ·04 |
| 3526 | 482 | 347 | 17M | ·9546 | 17·3274 | 22·421 | **425** | 5 | 7 | Ap | 4 | ·28 | 1 | My | 3 | ·81 | 3 | Je | 2 | ·34 | 4 | Jl | 1 | ·87 | 6 | Jl | 31 | ·40 |
| 3527 | 483 | 348 | 18M | ·2134 | 6·4357 | 18·578 | **426** | 6 | 4 | Mr | 24 | ·65 | 6 | Ap | 23 | ·18 | 7 | My | 22 | ·74 | 2 | Je | 21 | ·24 | 5 | Au | 19● | ·30 |
| | | | | | | | | | | | | | | | | | | | | | 3 | Jl | 20○ | ·77 | | | | |
| 3528 | 484 | 349 | 18M | ·4721 | 25·0746 | 16·711 | **427** | 7 | 3 | Ap | 12 | ·55 | 5 | My | 12 | ·08 | 6 | Je | 10 | 61 | 1 | Jl | 10●○ | ·14 | 2 | Au | 8 | ·67 |
| 3529 | 485 | 350 | 17M | ·7309 | 14·1829 | 12·868 | **428** | 1 | 7 | Mr | 31 | ·91 | 2 | Ap | 30 | ·44 | 3 | My | 29 | ·97 | 5 | Je | 28○ | ·50 | 7 | Jl | 28 | ·04 |
| 3530 | 486 | 351 | 17M | ·9896 | 3·2912 | 9·026 | **429** | 3 | 5 | Mr | 21 | ·28 | 6 | Ap | 19 | ·81 | 2 | Je | 17 | ·87 | 4 | Jl | 17 | ·40 | 5 | Au | 15 | ·93 |
| | | | | | | | | | | | | | 1 | My | 19○ | ·34 | | | | | | | | | | | | |
| 3531 | 487 | 352 | 18M | ·2484 | 21·9301 | 7·159 | **430** | 4 | 4 | Ap | 9 | ·18 | 5 | My | 8○ | ·71 | 7 | Je | 7 | ·24 | 1 | Jl | 6 | ·77 | 3 | Au | 5 | ·30 |
| 3532 | 488 | 353 | 18M | ·5071 | 11·0384 | 3·316 | **431** | 5 | 1 | Mr | 29 | ·54 | 3 | Ap | 28○ | ·08 | 4 | My | 27 | ·61 | 6 | Je | 26 | ·14 | 7 | Jl | 25 | ·67 |
| 3533 | 489 | 354 | 17M | ·7659 | 0·1467 | 27·028 | **432** | 6 | 5 | Mr | 17 | ·91 | 1 | My | 15 | ·97 | 3 | Je | 14 | ·50 | 5 | Jl | 14 | ·03 | 6 | Au | 12 | ·56 |
| | | | | | | | | | 7 | Ap | 16● | ·44 | | | | | | | | | | | | | | | | |

| | | | | | | | | | | | | | | | | | | | | | | | | | | | | | |
|---|---|---|---|---|---|---|---|---|---|---|---|---|---|---|---|---|---|---|---|---|---|---|---|---|---|---|---|---|---|
| + 147·65293 | | | | + 177·18353 | | | | + 206·71411 | | | | + 236·24470 | | | | | + 265·77529 | | | | | + 295·30588 | | | | +324·83647 +354·36705 | | | |
| + 9·880 | | | | + 11·856 | | | | + 13·832 | | | | + 15·808 | | | | | + 17·784 | | | | | + 19·760 | | | | + 21·736 + 23·712 | | | |
| **Asvina** | | | | **Kartika** | | | | **Margasira** | | | | **Pausha** | | | | **A.D.** | **Magha** | | | | **A.D.** | **Phalguna** | | | | **Chaitra** | | | |
| Week-day | Month | Day | Fraction | Week-day | Month | Day | Fraction | Week-day | Month | Day | Fraction | Week-day | Month | Day | Fraction | | Week-day | Month | Day | Fraction | | Week-day | Month | Day | Fraction | Week-day | Month | Day | Fraction |
| 2 | S | 11 | ·72 | 4 | O | 11○ | ·25 | 5 | N | 9 | ·78 | 7 | D | 9 | ·31 | **95** | 1 | Jr | 7 | ·84 | | 3 | F | 6 | ·37 | 4 | Mr | 7 | ·90 |
| 7 | S | 1 | ·09 | 1 | S | 30○ | ·62 | 3 | O | 30 | ·15 | 4 | N | 28 | ·68 | | 6 | D | 28 | ·21 | **96** | 7 | Jr | 26 | ·74 | 2 | F | 25 | ·27 |
| 5 | S | 18 | ·98 | 7 | O | 18 | ·51 | 2 | N | 17 | ·05 | 3 | D | 16 | ·58 | **97** | 5 | Jr | 15 | ·11 | | 6 | F | 13○ | ·64 | 1 | Mr | 15 | ·17 |
| 3 | S | 8 | ·35 | 4 | O | 7 | ·88 | 6 | N | 6 | ·41 | 7 | D | 5 | ·94 | **98** | 2 | Jr | 4 | ·47 | | 4 | F | 3○ | ·00 | 5 | Mr | 4 | ·53 |
| 7 | Au | 28 | ·72 | 2 | S | 27 | ·25 | 3 | O | 26 | ·78 | 5 | N | 25 | ·31 | | 6 | D | 24 | ·84 | **99** | 1 | Jr | 23○ | ·37 | 2 | F | 21 | ·90 |
| 6 | S | 16 | ·62 | 1 | O | 16 | ·15 | 2 | N | 14 | ·68 | 4 | D | 14 | ·21 | **400** | 5 | Jr | 12 | ·74 | | 7 | F | 11 | ·27 | 1 | Mr | 11 | ·80 |
| 3 | S | 4 | ·98 | 5 | O | 4 | ·51 | 7 | N | 3 | ·04 | 1 | D | 2○ | ·57 | **01** | 3 | Jr | 1 | ·11 | | 4 | Jr | 30 | ·64 | 6 | Mr | 1 | ·17 |
| 1 | Au | 25 | ·35 | 2 | S | 23 | ·88 | 4 | O | 23 | ·41 | 5 | N | 21○ | ·94 | | 7 | D | 21 | ·47 | **02** | 2 | Jr | 20 | ·00 | 3 | F | 18 | ·53 |
| 7 | S | 13 | ·25 | 1 | O | 12 | ·78 | 3 | N | 11●○ | ·31 | 4 | D | 10 | ·84 | **03** | 6 | Jr | 9 | ·37 | | 7 | F | 7 | ·90 | 2 | Mr | 9 | ·43 |
| 4 | S | 2 | ·61 | 6 | O | 2 | ·15 | 7 | O | 31 | ·68 | 2 | N | 30 | ·21 | | 3 | D | 29 | ·74 | **04** | 5 | Jr | 28 | ·27 | 6 | F | 26 | ·80 |
| 1 | Au | 21 | ·98 | {3 | S | 20○ | ·51 | **Margasira Kshaya** | | | | 6 | N | 18 | ·57 | | 1 | D | 18 | ·10 | **05** | {2 | Jr | 16 | ·63 | 5 | Mr | 16○ | ·70 |
| | | | | {5 | O | 20 | ·04 | | | | | | | | | | | | | | | {4 | F | 15 | ·17 | | | | |
| 7 | S | 9○ | ·88 | 2 | O | 9 | ·41 | 3 | N | 7 | ·94 | 5 | D | 7 | ·47 | **06** | 7 | Jr | 6 | ·00 | | 1 | F | 4 | ·53 | 3 | Mr | 6○ | ·06 |
| 5 | Au | 30○ | ·25 | 6 | S | 28 | ·73 | 1 | O | 28 | ·31 | 2 | N | 26 | ·84 | | 4 | D | 26 | ·37 | **07** | 5 | Jr | 24 | ·90 | 7 | F | 23● | ·43 |
| 4 | S | 18 | ·14 | 5 | O | 17 | ·67 | 7 | N | 16 | ·21 | 1 | D | 15 | ·74 | **08** | 3 | Jr | 14○ | ·27 | | 4 | F | 12● | ·80 | 6 | Mr | 13 | ·33 |
| 1 | S | 6 | ·51 | 3 | O | 6 | ·04 | 4 | N | 4 | ·57 | 6 | D | 4 | ·10 | **09** | 7 | Jr | 2○ | ·63 | | 2 | F | 1 | ·16 | 3 | Mr | 2 | ·69 |
| 5 | Au | 26 | ·88 | 7 | S | 25 | ·41 | 1 | O | 24 | ·94 | 3 | N | 23 | ·47 | | 5 | D | 23○ | ·00 | **10** | 6 | Jr | 21 | ·53 | 1 | F | 20 | ·06 |
| 4 | S | 14 | ·78 | 6 | O | 14 | ·31 | 7 | N | 12 | ·84 | 2 | D | 12● | ·37 | **11** | 3 | Jr | 10 | ·90 | | 5 | F | 9 | ·43 | 6 | Mr | 10 | ·96 |
| 2 | S | 4 | ·14 | 3 | O | 3 | ·67 | 5 | N | 2○ | ·20 | 6 | D | 1 | ·73 | | 1 | D | 31 | ·27 | **12** | 2 | Jr | 29 | ·80 | 4 | F | 28 | ·33 |
| 6 | Au | 23 | ·51 | 1 | S | 22 | ·04 | 2 | O | 21○ | ·57 | 4 | N | 20 | ·10 | | 5 | D | 19 | ·63 | **13** | 7 | Jr | 18 | ·16 | 1 | E | 16 | ·69 |
| 5 | S | 11 | ·41 | 6 | O | 10●○ | ·94 | 1 | N | 9 | ·47 | 3 | D | 9 | ·00 | **14** | 4 | Jr | 7 | ·53 | | 6 | F | 6 | ·06 | 7 | Mr | 7 | ·59 |
| 2 | Au | 31 | ·77 | 4 | S | 30● | ·30 | 5 | O | 29 | ·84 | 7 | N | 28 | ·37 | | 1 | D | 27 | ·90 | **15** | 3 | Jr | 26 | ·43 | 4 | F | 24○ | ·96 |
| 1 | S | 19● | ·67 | 3 | O | 19 | ·20 | 4 | N | 17 | ·73 | 6 | D | 17 | ·26 | **16** | 7 | Jr | 15 | ·79 | | 2 | F | 14○ | ·32 | 3 | Mr | 14 | ·86 |
| 6 | S | 8 | ·04 | 7 | O | 7 | ·57 | 2 | N | 6 | ·10 | 3 | D | 5 | ·63 | **17** | 5 | Jr | 4 | ·16 | | 6 | F | 2○ | ·69 | 1 | Mr | 4 | ·22 |
| 3 | Au | 28 | ·41 | 4 | S | 26 | ·94 | 6 | O | 26 | ·47 | 1 | N | 25 | ·00 | | 2 | D | 24 | ·53 | **18** | 4 | Jr | 23 | ·06 | 5 | F | 21 | ·59 |
| 2 | S | 16 | ·30 | 3 | O | 15 | ·83 | 5 | N | 14 | ·36 | 6 | D | 13○ | ·90 | **19** | 1 | Jr | 12 | ·43 | | 2 | F | 10 | ·96 | 4 | Mr | 12 | ·49 |
| 6 | S | 5 | ·67 | 1 | O | 5 | ·20 | 2 | N | 3 | ·73 | 4 | D | 3●○ | ·26 | **20** | 5 | Jr | 1 | ·79 | | 7 | Jr | 31 | ·32 | 1 | F | 29 | ·85 |
| 4 | Au | 25 | ·04 | 5 | S | 23 | ·57 | 7 | O | 23 | ·10 | 1 | N | 21○ | ·63 | | 3 | D | 21 | ·16 | **21** | 4 | Jr | 19 | ·69 | 6 | F | 18 | ·22 |
| 2 | S | 12 | ·94 | 4 | O | 12 | ·47 | 6 | N | 11● | ·00 | 7 | D | 10 | ·53 | **22** | 2 | Jr | 9 | ·06 | | 3 | F | 7 | ·59 | 5 | Mr | 9 | ·12 |
| 7 | S | 2 | ·30 | 1 | O | 1○ | ·83 | 2 | O | 31 | ·36 | 4 | N | 29 | ·89 | | 6 | D | 29 | ·42 | **23** | 7 | Jr | 27 | ·96 | 2 | F | 26 | ·49 |
| {4 | Au | 22 | ·67 | 7 | O | 20 | ·73 | 2 | N | 19 | ·26 | **Pausha Kshaya** | | | | | 3 | D | 18 | ·79 | **24** | 5 | Jr | 17 | ·32 | {6 | F | 15 | ·85 |
| {6 | S | 21○ | ·20 | | | | | | | | | | | | | | | | | | | | | | | {1 | Mr | 16○ | ·38 |
| 3 | S | 9○ | ·57 | 5 | O | 9 | ·10 | 6 | N | 7 | ·63 | 1 | D | 7 | ·16 | **25** | 2 | Jr | 5 | ·69 | | 4 | F | 4 | ·22 | 5 | Mr | 5● | ·75 |
| 7 | Au | 29● | ·93 | 2 | S | 28 | ·46 | 4 | O | 28 | ·00 | 5 | N | 26 | ·53 | | 7 | D | 26 | ·06 | **26** | 1 | Jr | 24○ | ·59 | 3 | F | 23 | ·12 |
| 6 | S | 17 | ·83 | 1 | O | 17 | ·36 | 2 | N | 15 | ·89 | 4 | D | 15 | ·42 | **27** | 5 | Jr | 13○ | ·95 | | 7 | F | 12 | ·48 | 2 | Mr | 14 | ·02 |
| 4 | S | 7 | ·20 | 5 | O | 6 | ·73 | 7 | N | 5 | ·26 | 1 | D | 4 | ·79 | **28** | 3 | Jr | 3○ | ·32 | | 4 | F | 1 | ·85 | 6 | Mr | 2 | ·38 |
| 1 | Au | 26 | ·57 | 3 | S | 25 | ·10 | 4 | O | 24 | ·63 | 6 | N | 23 | ·16 | | 7 | D | 22 | ·69 | **29** | 2 | Jr | 21 | ·22 | 3 | F | 19 | ·75 |
| 7 | S | 14 | ·46 | 1 | O | 13 | ·99 | 3 | N | 12○ | ·52 | 5 | D | 12● | ·06 | **30** | 6 | Jr | 10 | ·59 | | 1 | F | 9 | ·12 | 2 | Mr | 10 | ·65 |
| 4 | S | 3 | ·83 | 6 | O | 3 | ·36 | 7 | N | 1○ | ·89 | 2 | D | 1 | ·42 | | 3 | D | 30 | ·95 | **31** | 5 | Jr | 29 | ·48 | 7 | F | 28 | ·01 |
| 2 | Au | 24 | ·20 | 3 | S | 22 | ·73 | 5 | O | 22○ | ·26 | 6 | N | 20 | ·79 | | 1 | D | 20 | ·32 | **32** | 2 | Jr | 18 | ·85 | 4 | F | 17 | ·38 |
| 1 | S | 11 | ·10 | 2 | O | 10● | ·63 | 4 | N | 9 | ·16 | 5 | D | 8 | ·69 | **33** | 7 | Jr | 7 | ·22 | | 1 | F | 5 | ·75 | 3 | Mr | 7○ | ·28 |

| Kaliyuga. | Vikrama Era | Saka Era. | Commencement of Solar Year. Month and day A.D. | Fraction of day. | First New-Moon in Solar Year. | Anomaly of first New-Moon. | A.D. | Week-day of 1st January. | Vaisakha Week-day | Month | Day | Fraction | Jyeshta Week-day | Month | Day | Fraction | Ashada Week-day | Month | Day | Fraction | Sravana Week-day | Month | Day | Fraction | Bhadrapada Week-day | Month | Day | Fraction |
|---|---|---|---|---|---|---|---|---|---|---|---|---|---|---|---|---|---|---|---|---|---|---|---|---|---|---|---|---|
| | | | | | | | | ☉'s Anom col. 6. | | | | | + 29·53059 | | | | + 59·06117 | | | | + 88·59176 | | | | + 118·12235 | | | |
| | | | | | | | | ☾'s Anom col. 7. | | | | | + 1·976 | | | | + 3·952 | | | | + 5·928 | | | | + 7·904 | | | |
| 3534 | 490 | 355 | 18M | ·0246 | 18·7859 | 25·161 | **433** | 1 | 4 | Ap | 5 | ·81 | 6 | My | 5 | ·84 | 7 | Je | 3 | ·87 | 2 | Jl | 3 | ·40 | 3 | Au | 1 | ·93 |
| 3535 | 491 | 356 | 18M | ·2834 | 7·8939 | 21·319 | **434** | 2 | 2 | Mr | 26 | ·18 | 3 | Ap | 24 | ·71 | 5 | My | 24 | ·24 | 6 | Je | 22 | ·77 | 1 | Jl | 22 | ·30 |
| | | | | | | | | | | | | | | | | | | | | | | | | | 2 | Au | 20○ | ·83 |
| 3536 | 492 | 357 | 18M | ·5422 | 26·5328 | 19·452 | **435** | 3 | 1 | Ap | 14 | ·07 | 2 | My | 13 | ·60 | 4 | Je | 12 | ·14 | 5 | Jl | 11 | ·67 | 7 | Au | 10●○ | ·20 |
| 3537 | 493 | 358 | 17M | ·8009 | 15·6411 | 15·609 | **436** | 4 | 5 | Ap | 2 | ·44 | 6 | My | 1 | ·97 | 1 | My | 31 | ·50 | 3 | Je | 30 | ·03 | 4 | Jl | 29 | ·56 |
| 3538 | 494 | 359 | 18M | ·0597 | 4·7494 | 11·766 | **437** | 6 | 2 | Mr | 22 | ·81 | 4 | Ap | 21 | ·34 | 5 | My | 20 | ·87 | 1 | Jl | 18 | ·93 | 3 | Au | 17 | ·42 |
| | | | | | | | | | | | | | | | | | 7 | Je | 19○ | ·40 | | | | | | | | |
| 3539 | 495 | 360 | 18M | ·3184 | 23·3882 | 9·900 | **438** | 7 | 1 | Ap | 10 | ·71 | 3 | My | 10 | ·24 | 4 | Je | 8○ | ·77 | 6 | Jl | 8 | ·30 | 7 | Au | 6 | ·83 |
| 3540 | 496 | 361 | 18M | ·5772 | 12·4965 | 6·057 | **439** | 1 | 6 | Mr | 31 | ·07 | 7 | Ap | 29 | ·60 | 2 | My | 29 | ·13 | 3 | Je | 27 | ·66 | 5 | Jl | 27 | ·20 |
| 3541 | 497 | 362 | 17M | ·8359 | 1·6048 | 2·214 | **440** | 2 | 3 | Mr | 19 | ·44 | 4 | Ap | 17○ | ·97 | 1 | Je | 16 | ·03 | 2 | Jl | 15 | ·56 | 4 | Au | 14 | ·09 |
| | | | | | | | | | | | | | 6 | My | 17● | ·50 | | | | | | | | | | | | |
| 3542 | 498 | 363 | 18M | ·0947 | 20·2437 | 0·347 | **441** | 4 | 2 | Ap | 7○ | ·34 | 3 | My | 6 | ·87 | 5 | Je | 5 | ·40 | 6 | Jl | 4 | ·93 | 1 | Au | 3 | ·46 |
| 3543 | 499 | 364 | 18M | ·3535 | 9·3520 | 24·059 | **442** | 5 | 6 | Mr | 27○ | ·70 | 1 | Ap | 26 | ·24 | 2 | My | 25 | ·77 | 4 | Je | 24 | ·30 | 5 | Jl | 23 | ·83 |
| | | | | | | | | | | | | | | | | | | | | | | | | | 7 | Au | 22 | ·36 |
| 3544 | 500 | 365 | 18M | ·6122 | 27·9909 | 22·192 | **443** | 6 | 5 | Ap | 15 | ·60 | 7 | My | 15 | ·13 | 1 | Je | 13 | ·66 | 3 | Jl | 13 | ·19 | 4 | Au | 11 | ·72 |
| 3545 | 501 | 366 | 17M | ·8710 | 17·0992 | 18·350 | **444** | 7 | 2 | Ap | 3 | ·97 | 4 | My | 3 | ·50 | 6 | Je | 2 | ·03 | 7 | Jl | 1 | ·56 | 2 | Jl | 31○ | ·09 |
| 3546 | 502 | 367 | 18M | ·1297 | 6·2075 | 14·507 | **445** | 2 | 7 | Mr | 24 | ·34 | 1 | Ap | 22 | ·87 | 3 | My | 22 | ·40 | 4 | Je | 20 | ·93 | 7 | Au | 18 | ·99 |
| | | | | | | | | | | | | | | | | | | | | | 6 | Jl | 20○ | ·46 | | | | |
| 3547 | 503 | 368 | 18M | ·3885 | 24·8464 | 12·640 | **446** | 3 | 6 | Ap | 12 | ·23 | 7 | My | 11 | ·76 | 2 | Je | 10 | ·30 | 3 | Jl | 9●○ | ·83 | 5 | Au | 8 | ·36 |
| 3548 | 504 | 369 | 18M | ·6472 | 13·9547 | 8·797 | **447** | 4 | 3 | Ap | 1 | ·60 | 5 | My | 1 | ·13 | 6 | My | 30○ | ·66 | 1 | Je | 29● | ·19 | 2 | Jl | 28 | ·72 |
| 3549 | 505 | 370 | 17M | ·9060 | 3·0630 | 4·955 | **448** | 5 | 7 | Mr | 20 | ·97 | 2 | Ap | 19 | ·50 | 5 | Je | 17 | ·56 | 7 | Jl | 17 | ·09 | 1 | Au | 15 | ·62 |
| | | | | | | | | | | | | | 4 | My | 19○ | ·03 | | | | | | | | | | | | |
| 3550 | 506 | 371 | 18M | ·1648 | 21·7018 | 3·038 | **449** | 7 | 6 | Ap | 8 | ·87 | 1 | My | 8●○ | ·40 | 2 | Je | 6 | ·93 | 4 | Jl | 6 | ·46 | 5 | Au | 4 | ·99 |
| 3551 | 507 | 372 | 18M | ·4235 | 10·8102 | 26·800 | **450** | 1 | 4 | Mr | 29 | ·23 | 5 | Ap | 27 | ·76 | 7 | My | 27 | ·29 | 1 | Je | 25 | ·82 | 3 | Jl | 25 | ·36 |
| 3552 | 508 | 373 | 18M | ·6823 | 29·4491 | 24·933 | **451** | 2 | 1 | Mr | 18○ | ·60 | 4 | My | 16 | ·66 | 6 | Je | 15 | ·19 | 7 | Jl | 14 | ·72 | 2 | Au | 13 | ·25 |
| | | | | | | | | | 3 | Ap | 17 | ·13 | | | | | | | | | | | | | | | | |
| 3553 | 509 | 374 | 17M | ·9410 | 18·5574 | 21·090 | **452** | 3 | 7 | Ap | 5 | ·50 | 2 | My | 5 | ·03 | 3 | Je | 3 | ·56 | 5 | Jl | 3 | ·09 | 6 | Au | 1 | ·62 |
| 3554 | 510 | 375 | 18M | ·1998 | 7·6657 | 17·248 | **453** | 5 | 4 | Mr | 25 | ·86 | 6 | Ap | 24 | ·40 | 7 | My | 23 | ·93 | 2 | Je | 22 | ·46 | 3 | Jl | 21 | ·99 |
| | | | | | | | | | | | | | | | | | | | | | | | | | 5 | Au | 20○ | ·52 |
| 3555 | 511 | 376 | 18M | ·4585 | 26·3046 | 14·381 | **454** | 6 | 3 | Ap | 13 | ·76 | 5 | My | 13 | ·29 | 6 | Je | 11 | ·82 | 1 | Jl | 11 | ·35 | 2 | Au | 9● | ·88 |
| 3556 | 512 | 377 | 18M | ·7173 | 15·4128 | 11·538 | **455** | 7 | 1 | Ap | 3 | ·13 | 2 | My | 2 | ·66 | 4 | Je | 1 | ·19 | 5 | Je | 30○ | ·72 | 7 | Jl | 30● | ·25 |
| 3557 | 513 | 378 | 17M | ·9760 | 4·5211 | 7·695 | **456** | 1 | 5 | Mr | 22 | ·50 | 7 | Ap | 21 | ·03 | 1 | My | 20 | ·56 | 4 | Jl | 18 | ·62 | 6 | Au | 17 | ·15 |
| | | | | | | | | | | | | | | | | | 3 | Je | 19○ | ·09 | | | | | | | | |
| 3558 | 514 | 379 | 18M | ·2348 | 23·1600 | 5·829 | **457** | 3 | 4 | Ap | 10 | ·39 | 5 | My | 9 | ·92 | 7 | Je | 8● | ·46 | 1 | Jl | 7 | ·99 | 3 | Au | 6 | ·52 |
| 3559 | 515 | 380 | 18M | ·4936 | 12·2683 | 1·986 | **458** | 4 | 1 | Mr | 30 | ·76 | 3 | Ap | 29○ | ·29 | 4 | My | 28● | ·82 | 6 | Je | 27 | ·35 | 7 | Jl | 26 | ·88 |
| 3560 | 516 | 381 | 18M | ·7523 | 1·3766 | 25·698 | **459** | 5 | 6 | Mr | 20 | ·13 | 7 | Ap | 18○ | ·66 | 3 | Je | 16 | ·72 | 5 | Jl | 16 | ·25 | 6 | Au | 14 | ·78 |
| | | | | | | | | | | | | | 2 | My | 18● | ·19 | | | | | | | | | | | | |
| 3561 | 517 | 382 | 18M | ·0111 | 20·0155 | 23·831 | **460** | 6 | 5 | Ap | 7●○ | ·03 | 6 | My | 6 | ·56 | 1 | Je | 5 | ·09 | 2 | Jl | 4 | ·62 | 4 | Au | 3 | ·15 |
| 3562 | 518 | 383 | 18M | ·2698 | 9·1238 | 19·988 | **461** | 1 | 2 | Mr | 27● | ·39 | 3 | Ap | 25 | ·92 | 5 | My | 25 | ·45 | 6 | Je | 23 | ·98 | 1 | Jl | 23 | ·52 |
| | | | | | | | | | | | | | | | | | | | | | | | | | 3 | Au | 22 | ·05 |
| 3563 | 519 | 384 | 18M | ·5286 | 27·7627 | 18·121 | **462** | 2 | 1 | Ap | 15 | ·29 | 2 | My | 14 | ·82 | 4 | Je | 13 | ·35 | 5 | Jl | 12 | ·88 | 7 | Au | 11○ | ·41 |
| 3564 | 520 | 385 | 18M | ·7873 | 16·8710 | 14·279 | **463** | 3 | 5 | Ap | 4 | ·66 | 7 | My | 4 | ·19 | 1 | Je | 2 | ·72 | 3 | Jl | 2 | ·25 | 4 | Jl | 31○ | ·78 |
| 3565 | 521 | 386 | 18M | ·0461 | 5·9793 | 10·436 | **464** | 4 | 3 | Mr | 24 | ·02 | 4 | Ap | 22 | ·56 | 6 | My | 22 | ·09 | 7 | Je | 20 | ·62 | 3 | Au | 18 | ·68 |
| | | | | | | | | | | | | | | | | | | | | | 2 | Jl | 20●○ | ·15 | | | | |
| 3566 | 522 | 387 | 18M | ·3049 | 24·6182 | 8·569 | **465** | 6 | 1 | Ap | 11 | ·92 | 3 | My | 11 | ·45 | 4 | Je | 9○ | ·98 | 6 | Jl | 9● | ·51 | 1 | Au | 8 | ·04 |
| 3567 | 523 | 388 | 18M | ·5636 | 13·7242 | 4·726 | **466** | 7 | 6 | Ap | 1 | ·29 | 7 | Ap | 30 | ·82 | 2 | My | 30○ | ·35 | 3 | Je | 28 | ·88 | 5 | Jl | 28 | ·41 |
| 3568 | 524 | 389 | 18M | ·8224 | 2·8347 | 0·884 | **467** | 1 | 3 | Mr | 21 | ·66 | 5 | Ap | 20 | ·19 | 1 | Je | 18 | ·25 | 2 | Jl | 17 | ·78 | 4 | Au | 16 | ·31 |
| | | | | | | | | | | | | | 6 | My | 19●○ | ·72 | | | | | | | | | | | | |
| 3569 | 525 | 390 | 18M | ·0811 | 21·4737 | 26·571 | **468** | 2 | 2 | Ap | 8 | ·55 | 4 | My | 8● | ·08 | 5 | Je | 6 | ·62 | 7 | Jl | 6 | ·15 | 1 | Au | 4 | ·68 |
| 3570 | 526 | 391 | 18M | ·3399 | 10·5826 | 22·729 | **469** | 4 | 6 | Mr | 28○ | ·92 | 1 | Ap | 27 | ·45 | 2 | My | 26 | ·98 | 4 | Je | 25 | ·51 | 6 | Jl | 25 | ·04 |
| 3571 | 527 | 392 | 18M | ·5986 | 29·2209 | 20·862 | **470** | 5 | 4 | Mr | 18○ | ·29 | 7 | My | 16 | ·35 | 1 | Je | 14 | ·88 | 3 | Jl | 14 | ·41 | 4 | Au | 12 | ·94 |
| | | | | | | | | | 5 | Ap | 16 | ·82 | | | | | | | | | | | | | | | | |
| 3572 | 528 | 393 | 18M | ·8574 | 18·3292 | 17·019 | **471** | 6 | 3 | Ap | 6 | ·19 | 4 | My | 5 | ·72 | 6 | Je | 4 | ·25 | 7 | Jl | 3 | ·78 | 2 | Au | 2 | ·31 |

**Surya Siddhanta.**

| Asvina | Kartika | Margasira | Pausha | Magha | Phalguna | Chaitra |
|---|---|---|---|---|---|---|
| + 147·65293 | + 177·18353 | + 206·71411 | + 236·24470 | + 265·77529 | + 295·30588 | +324·83647<br>+354·36705 |
| + 9·880 | + 11·856 | + 13·832 | + 15·808 | + 17·784 | + 19·760 | + 21·736<br>+ 23·712 |

| Asvina Week-day | Month | Day | Fraction | Kartika Week-day | Month | Day | Fraction | Margasira Week-day | Month | Day | Fraction | Pausha Week-day | Month | Day | Fraction | A.D. | Magha Week-day | Month | Day | Fraction | A.D. | Phalguna Week-day | Month | Day | Fraction | Chaitra Week-day | Month | Day | Fraction |
|---|---|---|---|---|---|---|---|---|---|---|---|---|---|---|---|---|---|---|---|---|---|---|---|---|---|---|---|---|---|
| 5 | Au | 31○ | ·46 | 6 | S | 29● | ·99 | 1 | O | 29 | ·52 | 3 | N | 28 | ·05 | | 4 | D | 27 | ·58 | **34** | 6 | Jr | 26 | ·12 | 7 | F | 24●○ | ·65 |
| 4 | S | 19 | ·36 | 5 | O | 18 | ·89 | 7 | N | 17 | ·42 | 1 | D | 16 | ·95 | **35** | 3 | Jr | 15 | ·48 | | 5 | F | 14●○ | ·01 | 6 | Mr | 15 | ·54 |
| 1 | S | 8 | ·73 | 3 | O | 8 | ·26 | 4 | N | 6 | ·79 | 6 | D | 6 | ·32 | **36** | 7 | Jr | 4 | ·85 | | 2 | F | 3● | ·38 | 3 | Mr | 3 | ·91 |
| 6 | Au | 28 | ·09 | 7 | S | 26 | ·62 | 2 | O | 26 | ·16 | 3 | N | 24 | ·69 | | 5 | D | 24○ | ·22 | **37** | 6 | Jr | 22 | ·75 | 1 | F | 21 | ·28 |
| 4 | S | 15 | ·99 | 6 | O | 15 | ·52 | 1 | N | 14 | ·05 | 2 | D | 13○ | ·58 | **38** | 4 | Jr | 12 | ·11 | | 5 | F | 10 | ·64 | 7 | Mr | 12 | ·18 |
| 2 | S | 5 | ·36 | 3 | O | 4 | ·89 | 5 | N | 3 | ·42 | 6 | D | 2●○ | ·95 | **39** | 1 | Jr | 1 | ·48 | | 3 | Jr | 31 | ·01 | 4 | Mr | 1 | ·54 |
| 6 | Au | 25 | ·73 | 1 | S | 24 | ·26 | 2 | O | 23 | ·79 | 4 | N | 22 | ·32 | | 5 | D | 21 | ·85 | **40** | 7 | Jr | 20 | ·38 | 1 | F | 18 | ·91 |
| 5 | S | 12 | ·62 | 7 | O | 12○ | ·15 | 1 | N | 10 | ·68 | 3 | D | 10 | ·22 | **41** | 4 | Jr | 8 | ·75 | | 6 | F | 7 | ·28 | 7 | Mr | 8 | ·81 |
| 2 | S | 1 | ·99 | 4 | O | 1○ | ·52 | 6 | O | 31 | ·05 | 7 | N | 29 | ·58 | | 2 | D | 29 | ·11 | **42** | 3 | Jr | 27 | ·64 | 5 | F | 26 | ·17 |
| 1 | S | 20●○ | ·89 | 3 | O | 20 | ·42 | 4 | N | 18 | ·95 | 6 | D | 18 | ·48 | **43** | 1 | Jr | 17 | ·01 | | 2 | F | 15 | ·54 | 4 | Mr | 17 | ·07 |
| 6 | S | 10 | ·26 | 7 | O | 9 | ·79 | 2 | N | 8 | ·32 | 3 | D | 7 | ·85 | **44** | 5 | Jr | 6 | ·38 | | 6 | F | 4○ | ·91 | 1 | Mr | 5 | ·44 |
| 3 | Au | 29 | ·62 | 5 | S | 28 | ·15 | 6 | O | 27 | ·68 | 1 | N | 26 | ·21 | | 2 | D | 25 | ·74 | **45** | 4 | Jr | 24○ | ·28 | 5 | F | 22 | ·81 |
| 2 | S | 17 | ·52 | 4 | O | 17 | ·05 | 5 | N | 15 | ·58 | 7 | D | 15 | ·11 | **46** | 1 | Jr | 13●○ | ·64 | | 3 | F | 12 | ·17 | 4 | Mr | 13 | ·70 |
| 6 | S | 6 | ·89 | 1 | O | 6 | ·42 | 2 | N | 4 | ·95 | 4 | D | 4 | ·48 | **47** | 6 | Jr | 3 | ·01 | | 7 | F | 1 | ·54 | 2 | Mr | 3 | ·07 |
| 4 | Au | 27 | ·25 | 5 | S | 25 | ·78 | 7 | O | 25 | ·32 | 1 | N | 23○ | ·85 | | 3 | D | 23 | ·38 | **48** | 4 | Jr | 21 | ·91 | 6 | F | 20 | ·44 |
| 3 | S | 14 | ·15 | 4 | O | 13 | ·68 | 6 | N | 12○ | ·21 | 7 | D | 11 | ·74 | **49** | 2 | Jr | 10 | ·27 | | 3 | F | 8 | ·80 | 5 | Mr | 10 | ·34 |
| 7 | S | 3 | ·52 | 2 | O | 3 | ·05 | 3 | N | 1○ | ·58 | 5 | D | 1 | ·11 | | 6 | D | 30 | ·64 | **50** | 1 | Jr | 29 | ·17 | 2 | F | 27 | ·70 |
| 4 | Au | 23 | ·89 | 6 | S | 22 | ·42 | 7 | O | 21 | ·95 | 2 | N | 20 | ·48 | | 4 | D | 20 | ·01 | **51** | 5 | Jr | 18 | ·54 | 7 | F | 17 | ·07 |
| 3 | S | 11○ | ·78 | 5 | O | 11 | ·31 | 6 | N | 9 | ·84 | 1 | D | 9 | ·38 | **52** | 2 | Jr | 7 | ·91 | | 4 | F | 6 | ·44 | 5 | Mr | 6○ | ·97 |
| 1 | Au | 31○ | ·15 | 2 | S | 29 | ·68 | 4 | O | 29 | ·21 | 5 | N | 27 | ·74 | | 7 | D | 27 | ·27 | **53** | 1 | Jr | 25 | ·80 | 3 | F | 24○ | ·33 |
| 7 | S | 19 | ·05 | 1 | O | 18 | ·58 | 3 | N | 17 | ·11 | 4 | D | 16 | ·64 | **54** | 6 | Jr | 15 | ·17 | | 7 | F | 13 | ·70 | 2 | Mr | 15 | ·23 |
| 4 | S | 8 | ·42 | 5 | O | 7 | ·95 | 7 | N | 6 | ·48 | 2 | D | 6 | ·01 | **55** | 3 | Jr | 4○ | ·54 | | 5 | F | 3 | ·07 | 6 | Mr | 4 | ·60 |
| 1 | Au | 28 | ·78 | 3 | S | 27 | ·31 | 4 | O | 26 | ·84 | 6 | N | 25 | ·37 | | 7 | D | 24○ | ·90 | **56** | 2 | Jr | 23 | ·44 | 3 | F | 21 | ·97 |
| 7 | S | 15 | ·68 | 2 | O | 15 | ·21 | 3 | N | 13 | ·74 | 5 | D | 13○ | ·27 | **57** | 6 | Jr | 11 | ·80 | | 1 | F | 10 | ·33 | 2 | Mr | 11 | ·86 |
| 5 | S | 5 | ·05 | 6 | O | 4 | ·58 | 1 | N | 3 | ·11 | 2 | D | 2● | ·64 | **58** | 4 | Jr | 1 | ·17 | | 5 | Jr | 30 | ·70 | 7 | Mr | 1 | ·23 |
| 2 | Au | 25 | ·41 | 3 | S | 23 | ·94 | 5 | O | 23○ | ·48 | 7 | N | 22 | ·01 | | 1 | D | 21 | ·54 | **59** | 3 | Jr | 20 | ·07 | 4 | F | 18 | ·60 |
| 1 | S | 13 | ·31 | 2 | O | 12●○ | ·84 | 4 | N | 11 | ·37 | 5 | D | 10 | ·90 | **60** | 7 | Jr | 9 | ·43 | | 1 | F | 7 | ·96 | 3 | Mr | 8 | ·50 |
| 5 | S | 1 | ·68 | 7 | O | 1○ | ·21 | 1 | O | 30 | ·74 | 3 | N | 29 | ·27 | | 4 | D | 28 | ·80 | **61** | 6 | Jr | 27 | ·33 | 7 | F | 25 | ·86 |
| 4 | S | 20● | ·58 | 6 | O | 20 | ·11 | 7 | N | 18 | ·64 | 2 | D | 18 | ·17 | **62** | 3 | Jr | 16 | ·70 | | 5 | F | 15○ | ·23 | 6 | Mr | 16● | ·76 |
| 1 | S | 9 | ·94 | 3 | O | 9 | ·47 | 5 | N | 8 | ·00 | 6 | D | 7 | ·54 | **63** | 1 | Jr | 6 | ·07 | | 2 | F | 4○ | ·60 | 4 | Mr | 6 | ·13 |
| 6 | Au | 30 | ·31 | 7 | S | 28 | ·84 | 2 | O | 28 | ·37 | 3 | N | 26 | ·90 | | 5 | D | 26 | ·43 | **64** | 6 | Jr | 24○ | ·96 | 1 | F | 23 | ·49 |
| 5 | S | 17 | ·21 | 6 | O | 16 | ·74 | 1 | N | 15 | ·27 | 2 | D | 14 | ·80 | **65** | 4 | Jr | 13● | ·33 | | 5 | F | 11 | ·86 | 7 | Mr | 13 | ·39 |
| 2 | S | 6 | ·58 | 4 | O | 6 | ·11 | 5 | N | 4 | ·64 | 7 | D | 4○ | ·17 | **66** | 1 | Jr | 2 | ·70 | | 3 | F | 1 | ·23 | 4 | Mr | 2 | ·76 |
| 6 | Au | 26 | ·94 | 1 | S | 25 | ·47 | 3 | O | 25 | ·00 | 4 | N | 23○ | ·53 | | 6 | D | 23 | ·06 | **67** | 7 | Jr | 21 | ·60 | 2 | F | 20 | ·13 |
| 5 | S | 14 | ·84 | 7 | O | 14 | ·37 | 1 | N | 12●○ | ·90 | 3 | D | 12 | ·43 | **68** | 4 | Jr | 10 | ·96 | | 6 | F | 9 | ·49 | 1 | Mr | 10 | ·02 |
| 3 | S | 3 | ·21 | 4 | O | 2 | ·74 | 6 | N | 1● | ·27 | 7 | N | 30 | ·80 | | 2 | D | 30 | ·33 | **69** | 3 | Jr | 28 | ·86 | 5 | F | 27 | ·31 |
| 7 | Au | 23 | ·57 | 2 | S | 22○ | ·10 | 3 | O | 21● | ·64 | 5 | N | 20 | ·17 | | 6 | D | 19 | ·70 | **70** | 1 | Jr | 18 | ·23 | 2 | F | 16 | ·76 |
| 6 | S | 11○ | ·47 | 1 | O | 11 | ·00 | 2 | N | 9 | ·53 | 4 | D | 9 | ·06 | **71** | 5 | Jr | 7 | ·59 | | 7 | F | 6 | ·12 | 1 | Mr | 7○ | ·66 |
| 3 | Au | 31○ | ·84 | 5 | S | 30 | ·37 | 6 | O | 29 | ·90 | 1 | N | 28 | ·43 | | 2 | D | 27 | ·96 | **72** | 4 | Jr | 26 | ·49 | 6 | F | 25 | ·02 |

| Kaliyuga | Vikrama Era | Saka Era | Commencement of Solar Year | | First New-Moon in Solar Year | Anomaly of first New-Moon | A.D. | Week-day of 1st January | Vaisakha | | | | Jyeshta | | | | Ashada | | | | Sravana | | | | Bhadrapada | | | |
|---|---|---|---|---|---|---|---|---|---|---|---|---|---|---|---|---|---|---|---|---|---|---|---|---|---|---|---|---|
| | | | | | | | | ⊙'s Anom col. 6 | | | | | + 29·53059 | | | | + 59·06117 | | | | + 88·59176 | | | | + 118·12235 | | | |
| | | | | | | | | ☾'s Anom col. 7 | | | | | + 1·976 | | | | + 3·952 | | | | + 5·928 | | | | + 7·904 | | | |
| | | | Month and day A.D. | Fraction of day. | | | | | Week-day | Month | Day | Fraction | Week-day | Month | Day | Fraction | Week-day | Month | Day | Fraction | Week-day | Month | Day | Fraction | Week-day | Month | Day | Fraction |
| 3573 | 529 | 394 | 18M | ·1162 | 7·4375 | 13·176 | **472** | 7 | 7 | Mr | 25 | ·55 | 2 | Ap | 24 | ·08 | 3 | My | 23 | ·61 | 5 | Je | 22 | ·14 | { 6 | Jl | 21 | ·68 |
| | | | | | | | | | | | | | | | | | | | | | | | | | { 1 | Au | 20● | ·21 |
| 3574 | 530 | 395 | 18M | ·3749 | 26·0763 | 11·310 | **473** | 2 | 6 | Ap | 13 | ·45 | 7 | My | 12 | ·98 | 2 | Je | 11 | ·51 | 4 | Jl | 11○ | ·04 | 5 | Au | 9 | ·57 |
| 3575 | 531 | 396 | 18M | ·6337 | 15·1846 | 7·467 | **474** | 3 | 3 | Ap | 2 | ·82 | 5 | My | 2 | ·35 | 6 | My | 31 | ·88 | 1 | Je | 30○ | ·41 | 2 | Jl | 29 | ·94 |
| 3576 | 532 | 397 | 18M | ·8924 | 4·2929 | 3·624 | **475** | 4 | 1 | Mr | 23 | ·18 | 2 | Ap | 21 | ·72 | { 4 | My | 21 | ·25 | 7 | Jl | 19 | ·31 | 1 | Au | 17 | ·84 |
| | | | | | | | | | | | | | | | | | { 5 | Je | 19●○ | ·78 | | | | | | | | |
| 3577 | 533 | 398 | 18M | ·1512 | 22·9318 | 1·757 | **476** | 5 | 7 | Ap | 10 | ·08 | 1 | My | 9○ | ·61 | 3 | Je | 8 | ·14 | 4 | Jl | 7 | ·67 | 6 | Au | 6 | ·20 |
| 3578 | 534 | 399 | 18M | ·4099 | 12·0401 | 25·469 | **477** | 7 | 4 | Mr | 30 | ·45 | 5 | Ap | 28○ | ·98 | 7 | My | 28 | ·51 | 2 | Je | 27 | ·04 | 3 | Jl | 26 | ·57 |
| 3579 | 535 | 400 | 18M | ·6687 | 1·1484 | 21·627 | **478** | 1 | 1 | Mr | 19 | ·82 | { 3 | Ap | 18○ | ·35 | 6 | Je | 16 | ·41 | 7 | Jl | 15 | ·94 | 2 | Au | 14 | ·47 |
| | | | | | | | | | | | | | { 4 | My | 17 | ·88 | | | | | | | | | | | | |
| 3580 | 536 | 401 | 18M | ·9274 | 19·7873 | 19·760 | **479** | 2 | 7 | Ap | 7● | ·71 | 2 | My | 7 | ·24 | 3 | Je | 5 | ·78 | 5 | Jl | 5 | ·31 | 6 | Au | 3 | ·84 |
| 3581 | 537 | 402 | 18M | ·1862 | 8·8956 | 15·917 | **480** | 3 | 5 | Mr | 27 | ·08 | 6 | Ap | 25 | ·61 | 1 | My | 25 | ·14 | 2 | Je | 23 | ·67 | 4 | Jl | 23 | ·20 |
| 3582 | 538 | 403 | 18M | ·4450 | 27·5345 | 14·050 | **481** | 5 | 3 | Ap | 14 | ·98 | 5 | My | 14 | ·51 | 7 | Je | 13 | ·04 | 1 | Jl | 12 | ·57 | 3 | Au | 11●○ | ·10 |
| 3583 | 539 | 404 | 18M | ·7037 | 16·6428 | 10·208 | **482** | 6 | 1 | Ap | 4 | ·35 | 2 | My | 3 | ·88 | 4 | Je | 2 | ·41 | 5 | Jl | 1 | ·94 | 7 | Jl | 31○ | ·47 |
| 3584 | 540 | 405 | 18M | ·9625 | 5·7511 | 6·365 | **483** | 7 | 5 | Mr | 24 | ·71 | 7 | Ap | 23 | ·24 | 1 | My | 22 | ·77 | { 3 | Je | 21○ | ·20 | 6 | Au | 19 | ·37 |
| | | | | | | | | | | | | | | | | | | | | | { 4 | Jl | 20 | ·84 | | | | |
| 3585 | 541 | 406 | 18M | ·2212 | 24·3900 | 4·498 | **484** | 1 | 4 | Ap | 11 | ·61 | 5 | My | 11 | ·14 | 7 | Je | 9○ | ·67 | 2 | Jl | 9 | ·20 | 3 | Au | 7 | ·73 |
| 3586 | 542 | 407 | 18M | ·4800 | 13·4983 | 0·655 | **485** | 3 | 1 | Mr | 31 | ·98 | 3 | Ap | 30 | ·51 | 5 | My | 30○ | ·04 | 6 | Je | 28 | ·57 | 1 | Jl | 28 | ·10 |
| 3587 | 543 | 408 | 18M | ·7387 | 2·6061 | 24·367 | **486** | 4 | 6 | Mr | 21 | ·34 | { 7 | Ap | 19 | ·88 | 3 | Je | 17 | ·94 | 5 | Jl | 17 | ·47 | 7 | Au | 16 | ·00 |
| | | | | | | | | | | | | | { 2 | My | 19● | ·41 | | | | | | | | | | | | |
| 3588 | 544 | 409 | 18M | ·9975 | 21·2455 | 22·500 | **487** | 5 | 5 | Ap | 9○ | ·24 | 6 | My | 8● | ·77 | 1 | Je | 7 | ·30 | 2 | Jl | 6 | ·83 | 4 | Au | 5 | ·36 |
| 3589 | 545 | 410 | 18M | ·2563 | 10·3538 | 18·658 | **488** | 6 | 2 | Mr | 28●○ | ·61 | 4 | Ap | 27 | ·14 | 5 | My | 26 | ·67 | 7 | Je | 25 | ·20 | 1 | Jl | 24 | ·73 |
| 3590 | 546 | 411 | 18M | ·5150 | 28·9927 | 16·791 | **489** | 1 | 1 | Ap | 1[illegible] | ·51 | 3 | My | 16 | ·04 | 4 | Je | 14 | ·57 | 6 | Jl | 14 | ·10 | 7 | Au | 12 | ·63 |
| 3591 | 547 | 412 | 18M | ·7738 | 18·1010 | 12·948 | **490** | 2 | 5 | Ap | 5 | ·87 | 7 | My | 5 | ·40 | 1 | Je | 3 | ·94 | 3 | Jl | 3 | ·47 | 5 | Au | 2 | ·00 |
| 3592 | 548 | 413 | 19M | ·0325 | 7·2092 | 9·105 | **491** | 3 | 3 | Mr | 26 | ·24 | 4 | Ap | 24 | ·77 | 6 | My | 24 | ·30 | 7 | Je | 22 | ·83 | { 2 | Jl | 22○ | ·36 |
| | | | | | | | | | | | | | | | | | | | | | | | | | { 3 | Au | 20● | ·89 |
| 3593 | 549 | 414 | 18M | ·2913 | 25·8481 | 7·239 | **492** | 4 | 2 | Ap | 13 | ·14 | 3 | My | 12 | ·67 | 5 | Je | 11 | ·20 | 6 | Jl | 10○ | ·73 | 1 | Au | 9 | ·26 |
| 3594 | 550 | 415 | 18M | ·5500 | 14·9564 | 3·396 | **493** | 6 | 6 | Ap | 2 | ·51 | 1 | My | 2 | ·04 | 2 | My | 31 | ·57 | 4 | Je | 30○ | ·10 | 5 | Jl | 29 | ·63 |
| 3595 | 551 | 416 | 18M | ·8088 | 4·0647 | 27·108 | **494** | 7 | 3 | Mr | 22 | ·87 | 5 | Ap | 21 | ·40 | { 6 | My | 20○ | ·93 | 3 | Jl | 19 | ·00 | 4 | Au | 17 | ·53 |
| | | | | | | | | | | | | | | | | | { 1 | Je | 1[illegible] | ·46 | | | | | | | | |
| 3596 | 552 | 417 | 19M | ·0675 | 22·7036 | 25·241 | **495** | 1 | 2 | Ap | 10 | ·77 | 4 | My | 10○ | ·30 | 5 | Je | 8 | ·83 | 7 | Jl | 8 | ·36 | 1 | Au | 6 | ·89 |
| 3597 | 553 | 418 | 18M | ·3263 | 11·8119 | 21·398 | **496** | 2 | 7 | Mr | 30 | ·14 | 1 | Ap | 28○ | ·67 | 3 | My | 28 | ·20 | 4 | Je | 26 | ·73 | 6 | Jl | 26 | ·26 |
| 3598 | 554 | 419 | 18M | ·5851 | 0·9202 | 17·556 | **497** | 4 | 4 | Mr | 19 | ·50 | { 6 | Ap | 18 | ·04 | 2 | Je | 16 | ·10 | 3 | Jl | 15 | ·63 | 5 | Au | 14 | ·11 |
| | | | | | | | | | | | | | { 7 | My | 17 | ·57 | | | | | | | | | | | | |
| 3599 | 555 | 420 | 18M | ·8438 | 19·5591 | 15·689 | **498** | 5 | 3 | Ap | 7 | ·40 | 4 | My | 6 | ·93 | 6 | Je | 5 | ·46 | 7 | Jl | 4 | ·99 | 2 | Au | 3 | ·52 |
| 3600 | 556 | 421 | 19M | ·1026 | 8·6674 | 11·846 | **499** | 6 | 7 | Mr | 27 | ·77 | 2 | Ap | 26 | ·30 | 3 | My | 25 | ·83 | 5 | Je | 24 | ·36 | 6 | Jl | 23 | ·89 |
| **ARYA SIDDHANTA FROM 500 A.D. TO 999 A.D.** | | | | | | | | | | | | | | | | | | | | | | | | | | | | |
| 3601 | 557 | 422 | 18M | ·3611 | 27·28318 | 9·612 | **500** | 7 | 6 | Ap | 14 | ·65 | 1 | My | 14 | ·18 | 2 | Je | 12 | ·71 | 4 | Jl | 12 | ·24 | 5 | Au | 10○ | ·77 |
| 3602 | 558 | 423 | 18M | ·6198 | 16·39162 | 5·770 | **501** | 2 | 4 | Ap | 4 | ·02 | 5 | My | 3 | ·54 | 7 | Je | 2 | ·08 | 1 | Jl | 1 | ·61 | 3 | Jl | 31● | ·40 |
| 3603 | 559 | 424 | 18M | ·8785 | 5·50004 | 1·928 | **502** | 3 | 1 | Mr | 24 | ·38 | 2 | Ap | 22 | ·91 | { 4 | My | 22 | ·44 | 7 | Jl | 20● | ·50 | 2 | Au | 19 | ·08 |
| | | | | | | | | | | | | | | | | | { 5 | Je | 20○ | ·97 | | | | | | | | |
| 3604 | 560 | 425 | 19M | ·1371 | 24·13907 | 0·062 | **503** | 4 | 7 | Ap | 12 | ·28 | 1 | My | 11 | ·81 | 3 | Je | 10●○ | ·34 | 4 | Jl | 9 | ·87 | 6 | Au | 8 | ·40 |
| 3605 | 561 | 426 | 18M | ·3958 | 13·24751 | 23·774 | **504** | 5 | 4 | Mr | 31 | ·65 | 6 | Ap | 30 | ·18 | 7 | My | 29 | ·71 | 2 | Je | 28 | ·24 | 3 | Jl | 27 | ·77 |
| 3606 | 562 | 427 | 18M | ·6545 | 2·35594 | 19·932 | **505** | 7 | 2 | Mr | 21 | ·01 | { 3 | Ap | 19○ | ·54 | 5 | Je | 17 | ·61 | 1 | Jl | 17 | ·14 | 2 | Au | 15 | ·67 |
| | | | | | | | | | | | | | { 5 | My | 19● | ·07 | | | | | | | | | | | | |
| 3607 | 563 | 428 | 18M | ·9132 | 20·99497 | 18·066 | **506** | 1 | 7 | Ap | 8○ | ·91 | 2 | My | 8 | ·44 | 3 | Je | 6 | ·97 | 5 | Jl | 6 | ·50 | 7 | Au | 5 | ·08 |
| 3608 | 564 | 429 | 19M | ·1719 | 10·10341 | 14·224 | **507** | 2 | 5 | Mr | 29○ | ·28 | 6 | Ap | 27 | ·81 | 1 | My | 27 | ·34 | 2 | Je | 25 | ·87 | 4 | Jl | 25 | ·40 |
| 3609 | 565 | 430 | 18M | ·4305 | 28·74242 | 12·357 | **508** | 3 | 4 | Ap | 16 | ·18 | 5 | My | 15 | ·71 | 7 | Je | 14 | ·24 | 1 | Jl | 13 | ·77 | 3 | Au | 12 | ·80 |
| 3610 | 566 | 431 | 18M | ·6892 | 17·85086 | 8·515 | **509** | 5 | 1 | Ap | 5 | ·64 | 3 | My | 5 | ·07 | 4 | Je | 3 | ·60 | 6 | Jl | 3 | ·13 | 7 | Au | 1○ | ·66 |
| 3611 | 567 | 432 | 18M | ·9479 | 6·95930 | 4·673 | **510** | 6 | 5 | Mr | 25 | ·91 | 7 | Ap | 24 | ·44 | 1 | My | 23 | ·97 | { 3 | Je | 22 | ·50 | 6 | Au | 20 | ·56 |
| | | | | | | | | | | | | | | | | | | | | | { 5 | Jl | 22○ | ·03 | | | | |

**N.B.**—For Surya Siddhanta figures for **A.D. 500** to **A.D. 999** see pages **76—79** below.

## Surya Siddhanta.

| + 147·65298 | | | | + 177·18353 | | | | + 206·71411 | | | | + 236·24470 | | | | | + 265·77529 | | | | | + 295·30588 | | | | +324·83647 / +354·36705 | | | |
|---|---|---|---|---|---|---|---|---|---|---|---|---|---|---|---|---|---|---|---|---|---|---|---|---|---|---|---|---|---|
| + 9·880 | | | | + 11·856 | | | | + 13·832 | | | | + 15·806 | | | | | + 17·784 | | | | | + 19·760 | | | | + 21·736 / + 23·712 | | | |
| Asvina | | | | Kartika | | | | Margasira | | | | Pausha | | | | A.D. | Magha | | | | A.D. | Phalguna | | | | Chaitra | | | |
| Week-day | Month | Day | Fraction | Week-day | Month | Day | Fraction | Week-day | Month | Day | Fraction | Week-day | Month | Day | Fraction | | Week-day | Month | Day | Fraction | | Week-day | Month | Day | Fraction | Week-day | Month | Day | Fraction |
| 2 | S | 18 | ·74 | 4 | O | 18 | ·27 | 5 | N | 16 | ·80 | 7 | D | 16 | ·33 | 73 | 1 | Jr | 14○ | ·86 | | 3 | F | 13 | ·39 | 4 | Mr | 14 | ·92 |
| 7 | S | 8 | ·10 | 1 | O | 7 | ·63 | 3 | N | 6 | ·16 | 4 | D | 5 | ·70 | 74 | 6 | Jr | 4●○ | ·23 | | 7 | F | 2 | ·76 | 2 | Mr | 4 | ·29 |
| 4 | Au | 28 | ·47 | 6 | S | 27 | ·00 | 7 | O | 26 | ·53 | 2 | N | 25 | ·06 | | 3 | D | 24○ | ·59 | 75 | 5 | Jr | 23 | ·12 | 6 | F | 21 | ·65 |
| 3 | S | 16 | ·37 | 4 | O | 15 | ·90 | 6 | N | 14 | ·43 | 7 | D | 13● | ·96 | 76 | 2 | Jr | 12 | ·49 | | 4 | F | 11 | ·02 | 5 | Mr | 11 | ·55 |
| 7 | S | 4 | ·74 | 2 | O | 4 | ·27 | 3 | N | 2○ | ·80 | 5 | D | 2 | ·33 | | 6 | D | 31 | ·86 | 77 | 1 | Jr | 30 | ·39 | 2 | F | 28 | ·92 |
| 5 | Au | 25 | ·10 | 6 | S | 23 | ·63 | 1 | O | 23○ | ·16 | 2 | N | 21 | ·69 | | 4 | D | 21 | ·22 | 78 | 5 | Jr | 19 | ·76 | 7 | F | 18 | ·29 |
| 4 | S | 13 | ·00 | 5 | O | 12○ | ·53 | 7 | N | 11 | ·06 | 1 | D | 10 | ·59 | 79 | 3 | Jr | 9 | ·12 | | 4 | F | 7 | ·65 | 6 | Mr | 9 | ·18 |
| 1 | S | 2 | ·37 | 2 | O | 1● | ·90 | 4 | O | 31 | ·43 | 5 | N | 29 | ·96 | | 7 | D | 29 | ·49 | 80 | 2 | Jr | 28 | ·02 | 3 | F | 26○ | ·55 |
| { 5 | Au | 21○ | ·73 | 1 | O | 19 | ·80 | 3 | N | 18 | ·33 | 4 | D | 17 | ·86 | 81 | 6 | Jr | 16 | ·39 | | 7 | F | 14○ | ·92 | 2 | Mr | 16 | ·45 |
| { 7 | S | 20● | ·26 | | | | | | | | | | | | | | | | | | | | | | | | | | |
| 4 | S | 9 | ·63 | 6 | O | 9 | ·16 | 7 | N | 7 | ·69 | 2 | D | 7 | ·22 | 82 | 3 | Jr | 5 | ·75 | | 5 | F | 4○ | ·28 | 6 | Mr | 5 | ·82 |
| 2 | Au | 30 | ·00 | 3 | S | 28 | ·53 | 5 | O | 28 | ·06 | 6 | N | 26 | ·59 | | 1 | D | 26 | ·12 | 83 | 2 | Jr | 24 | ·65 | 4 | F | 23 | ·18 |
| 7 | S | 17 | ·90 | 2 | O | 17 | ·43 | 3 | N | 15 | ·96 | 5 | D | 15○ | ·49 | 84 | 7 | Jr | 14● | ·02 | | 1 | F | 12 | ·55 | 3 | Mr | 13 | ·08 |
| 5 | S | 6 | ·26 | 6 | O | 5 | ·79 | 1 | N | 4 | ·32 | 2 | D | 3○ | ·86 | 85 | 4 | Jr | 2 | ·39 | | 5 | Jr | 31 | ·92 | 7 | Mr | 2 | ·45 |
| 2 | Au | 26 | ·63 | 4 | S | 25 | ·16 | 5 | O | 24 | ·69 | 7 | N | 23●○ | ·22 | | 1 | D | 22 | ·75 | 86 | 3 | Jr | 21 | ·28 | 4 | F | 19 | ·81 |
| 1 | S | 14 | ·53 | 3 | O | 14 | ·06 | 4 | N | 12● | ·59 | 6 | D | 12 | ·12 | 87 | 7 | Jr | 10 | ·65 | | 2 | F | 9 | ·18 | 3 | Mr | 10 | ·71 |
| 5 | S | 3 | ·90 | 7 | O | 3○ | ·43 | 1 | N | 1● | ·96 | 3 | D | 1 | ·49 | | 5 | D | 31 | ·02 | 88 | 6 | Jr | 29 | ·55 | 1 | F | 28 | ·08 |
| 3 | Au | 23 | ·26 | { 4 | S | 21○ | ·79 | 7 | N | 19 | ·85 | **Pausha Kshaya** | | | | | 2 | D | 19 | ·38 | 89 | 3 | Jr | 17 | ·92 | { 5 | F | 16 | ·45 |
| | | | | { 6 | O | 21 | ·32 | | | | | | | | | | | | | | | | | | | { 6 | Mr | 17●○ | ·98 |
| 2 | S | 11●○ | ·16 | 3 | O | 10 | ·69 | 5 | N | 9 | ·22 | 6 | D | 8 | ·75 | 90 | 1 | Jr | 7 | ·28 | | 2 | F | 5 | ·81 | 4 | Mr | 7● | ·34 |
| 6 | Au | 31 | ·53 | 1 | S | 30 | ·06 | 2 | O | 29 | ·59 | 4 | N | 28 | ·12 | | 5 | D | 27 | ·65 | 91 | 7 | Jr | 26○ | ·18 | 1 | F | 24● | ·71 |
| 5 | S | 19 | ·42 | 6 | O | 18 | ·96 | 1 | N | 17 | ·49 | 3 | D | 17 | 02 | 92 | 4 | Jr | 15○ | ·55 | | 6 | F | 14 | ·08 | 7 | Mr | 14 | ·61 |
| 2 | S | 7 | ·79 | 4 | O | 7 | ·32 | 5 | N | 5 | ·85 | 7 | D | 5 | ·38 | 93 | 1 | Jr | 3●○ | ·91 | | 3 | F | 2 | ·44 | 4 | Mr | 3 | ·98 |
| 7 | Au | 28 | ·16 | 1 | S | 26 | ·69 | 3 | O | 26 | ·22 | 4 | N | 24 | ·75 | | 6 | D | 24 | ·28 | 94 | 1 | Jr | 22 | ·81 | 2 | F | 21 | ·34 |
| 6 | S | 16 | ·06 | 7 | O | 15 | ·59 | 2 | N | 14○ | ·12 | 3 | D | 13 | ·65 | 95 | 5 | Jr | 12 | ·18 | | 6 | F | 10 | ·71 | 1 | Mr | 12 | ·24 |
| 3 | S | 5 | ·42 | 4 | O | 4 | ·95 | 6 | N | 3○ | ·48 | 1 | D | 3 | ·02 | 96 | 2 | Jr | 1 | ·55 | | 4 | Jr | 31 | ·08 | 5 | F | 29 | ·61 |
| 7 | Au | 24 | ·79 | 2 | S | 23 | ·32 | 3 | O | 22●○ | ·85 | 5 | N | 21 | ·38 | | 6 | D | 20 | ·91 | 97 | 1 | Jr | 19 | ·44 | 2 | F | 17 | ·97 |
| 6 | S | 12 | ·69 | 1 | O | 12 | ·22 | 2 | N | 10 | ·75 | 4 | D | 10 | ·28 | 98 | 5 | Jr | 8 | ·81 | | 7 | F | 7 | ·34 | 1 | Mr | 8○ | ·87 |
| 4 | S | 2○ | ·06 | 5 | O | 1 | ·59 | 7 | O | 31 | ·12 | 1 | N | 29 | ·65 | | 3 | D | 29 | ·18 | 99 | 4 | Jr | 27 | ·71 | 6 | F | 26○ | ·24 |
| { 1 | Au | 22○ | ·42 | 4 | O | 20 | ·48 | 6 | N | 19 | ·01 | 7 | D | 18 | ·54 | 500 | 2 | Jr | 17 | ·06 | | 3 | F | 15●○ | ·59 | 5 | Mr | 16 | ·12 |
| { 2 | S | 20 | ·95 | | | | | | | | | | | | | | | | | | | | | | | | | | |
| 7 | S | 9 | ·30 | 1 | O | 8 | ·83 | 3 | N | 7 | ·36 | 4 | D | 6 | ·89 | 01 | 6 | Jr | 5 | ·42 | | 7 | F | 3 | ·95 | 2 | Mr | 5 | ·48 |
| 4 | Au | 29 | ·67 | 6 | S | 28 | ·20 | 7 | O | 27 | ·73 | 2 | N | 26 | ·26 | | 3 | D | 25○ | ·79 | 02 | 5 | Jr | 24 | ·32 | 6 | F | 22 | ·85 |
| 3 | S | 17 | ·56 | 5 | O | 17 | ·10 | 6 | N | 15 | ·63 | 1 | D | 15○ | ·16 | 03 | 2 | Jr | 13 | ·69 | | 4 | F | 12 | ·22 | 5 | Mr | 13 | ·75 |
| 7 | S | 6 | ·93 | 2 | O | 6 | ·46 | 3 | N | 4 | ·99 | 5 | D | 4○ | ·52 | 04 | 7 | Jr | 3 | ·05 | | 1 | F | 1 | ·58 | 3 | Mr | 2 | ·12 |
| 5 | Au | 26 | ·30 | 6 | S | 24 | ·83 | 1 | O | 24 | ·36 | 2 | N | 22 | ·89 | | 4 | D | 22 | ·42 | 05 | 5 | Jr | 20 | ·95 | 7 | F | 19 | ·48 |
| 4 | S | 14 | ·20 | 5 | O | 13○ | ·73 | 7 | N | 12 | ·26 | 1 | D | 11 | ·79 | 06 | 3 | Jr | 10 | ·32 | | 4 | F | 8 | ·79 | 6 | Mr | 10 | ·38 |
| 1 | S | 3 | ·56 | 3 | O | 3○ | ·09 | 4 | N | 1● | ·62 | 6 | D | 1 | ·16 | | 7 | D | 30 | ·69 | 07 | 2 | Jr | 29 | ·22 | 3 | F | 27 | ·75 |
| 5 | Au | 23 | ·93 | { 7 | S | 22○ | ·46 | 3 | N | 20 | ·52 | 5 | D | 20 | ·05 | | **Magha Kshaya.** | | | | 08 | { 6 | Jr | 18 | ·58 | 2 | Mr | 17 | ·62 |
| | | | | { 2 | O | 21 | ·99 | | | | | | | | | | | | | | | { 1 | F | 17 | ·10 | | | | |
| 4 | S | 10● | ·83 | 6 | O | 10 | ·36 | 7 | N | 8 | ·99 | 2 | D | 8 | ·42 | 09 | 3 | Jr | 6 | ·95 | | 5 | F | 5○ | ·48 | 7 | Mr | 7 | ·01 |
| 2 | Au | 31● | ·20 | 3 | S | 29 | ·73 | 5 | O | 29 | ·26 | 6 | N | 27 | ·79 | | 1 | D | 27 | ·32 | 10 | 2 | Jr | 25○ | ·85 | 4 | F | 24 | ·38 |
| 1 | S | 19 | ·09 | 2 | O | 18 | ·62 | 4 | N | 17 | ·15 | 5 | D | 16 | ·68 | 11 | 7 | Jr | 15○ | ·22 | | 1 | F | 13 | ·75 | 3 | Mr | 15 | ·28 |

**TABLE X–**

⊙'s Anom col. 6 + 29·53059 + 59·06117 + 88·59176 + 118·12235

☾'s Anom col. 7 + 1·976 + 3·952 + 5·928 + 7·904

| Kaliyuga | Vikrama Era | Saka Era | Commencement of Solar Year: Month and day A.D. | Fraction of day | First New-Moon in Solar Year | Anomaly of first New-Moon | A.D. | Week-day of 1st January | Vaisakha: Week-day | Month | Day | Fraction | Jyeshta: Week-day | Month | Day | Fraction | Ashada: Week-day | Month | Day | Fraction | Sravana: Week-day | Month | Day | Fraction | Bhadrapada: Week-day | Month | Day | Fraction |
|---|---|---|---|---|---|---|---|---|---|---|---|---|---|---|---|---|---|---|---|---|---|---|---|---|---|---|---|---|
| 3612 | 568 | 433 | 19M | ·2066 | 25·59832 | 2·806 | **511** | 7 | 4 | Ap | 13 | ·81 | 6 | My | 13 | ·34 | 7 | Je | 11 | ·87 | 2 | Jl | 11○ | ·40 | 3 | Au | 9 | ·93 |
| 3613 | 569 | 434 | 18M | ·4653 | 14·70676 | 26·518 | **512** | 1 | 2 | Ap | 2 | ·17 | 3 | My | 1 | ·70 | 5 | My | 31○ | ·24 | 6 | Je | 29● | ·77 | 1 | Jl | 29 | ·30 |
| 3614 | 570 | 435 | 18M | ·7240 | 3·81519 | 22·677 | **513** | 3 | 6 | Mr | 22 | ·54 | 1 | Ap | 21 | ·07 | 2 | My | 20○ | ·60 | 5 | Jl | 18 | ·66 | 6 | Au | 17 | ·19 |
| | | | | | | | | | | | | | | | | | 4 | Je | 19● | ·14 | | | | | | | | |
| 3615 | 571 | 436 | 18M | ·9826 | 22·45421 | 20·810 | **514** | 4 | 5 | Ap | 10 | ·44 | 6 | My | 9●○ | ·97 | 1 | Je | 8 | ·50 | 3 | Jl | 8 | ·03 | 4 | Au | 6 | ·56 |
| 3616 | 572 | 437 | 19M | ·2413 | 11·56265 | 16·968 | **515** | 5 | 2 | Mr | 30 | ·81 | 4 | Ap | 29 | ·34 | 5 | My | 28 | ·87 | 7 | Je | 27 | ·40 | 1 | Jl | 26 | ·93 |
| 3617 | 573 | 438 | 18M | ·5000 | 0·67109 | 13·126 | **516** | 6 | 7 | Mr | 19○ | ·17 | 3 | My | 17 | ·23 | 4 | Je | 15 | ·76 | 6 | Jl | 15 | ·30 | 7 | Au | 13 | ·88 |
| | | | | | | | | | 1 | Ap | 17● | ·70 | | | | | | | | | | | | | | | | |
| 3618 | 574 | 439 | 18M | ·7587 | 19·31011 | 11·260 | **517** | 1 | 6 | Ap | 7● | ·07 | 7 | My | 6 | ·60 | 2 | Je | 5 | ·13 | 3 | Jl | 4 | ·66 | 5 | Au | 3 | ·19 |
| 3619 | 575 | 440 | 19M | ·0174 | 8·41855 | 7·417 | **518** | 2 | 3 | Mr | 27 | ·44 | 4 | Ap | 25 | ·97 | 6 | My | 25 | ·50 | 1 | Je | 24 | ·03 | 2 | Jl | 23 | ·56 |
| | | | | | | | | | | | | | | | | | | | | | | | | | 4 | Au | 22●○ | ·09 |
| 3620 | 576 | 441 | 19M | ·2760 | 27·05757 | 5·551 | **519** | 3 | 2 | Ap | 15 | ·34 | 3 | My | 14 | ·87 | 5 | Je | 13 | ·40 | 6 | Jl | 12 | ·93 | 1 | Au | 11● | ·46 |
| 3621 | 577 | 442 | 18M | ·5347 | 16·16600 | 1·709 | **520** | 4 | 6 | Ap | 3 | ·70 | 1 | My | 3 | ·23 | 2 | Je | 1 | ·76 | 4 | Jl | 1○ | ·29 | 5 | Jl | 30 | ·82 |
| 3622 | 578 | 443 | 18M | ·7934 | 5·27444 | 25·421 | **521** | 6 | 4 | Mr | 24 | ·07 | 5 | Ap | 22 | ·60 | 7 | My | 22 | ·13 | 3 | Jl | 20 | ·19 | 4 | Au | 18 | ·72 |
| | | | | | | | | | | | | | | | | | 1 | Je | 20●○ | ·66 | | | | | | | | |
| 3623 | 579 | 444 | 19M | ·0521 | 23·91347 | 23·555 | **522** | 7 | 2 | Ap | 11 | ·97 | 4 | My | 11 | ·50 | 6 | Je | 10● | ·03 | 7 | Jl | 9 | ·56 | 2 | Au | 8 | ·09 |
| 3624 | 580 | 445 | 19M | ·3108 | 13·02189 | 19·713 | **523** | 1 | 7 | Ap | 1 | ·33 | 1 | Ap | 30○ | ·86 | 3 | My | 30 | ·40 | 4 | Je | 28 | ·93 | 6 | Jl | 28 | ·46 |
| 3625 | 581 | 446 | 18M | ·5694 | 2·13033 | 15·870 | **524** | 2 | 4 | Mr | 20 | ·70 | 6 | Ap | 19○ | ·23 | 2 | Je | 17 | ·29 | 4 | Jl | 16 | ·82 | 5 | Au | 15 | ·35 |
| | | | | | | | | | | | | | 7 | My | 18 | ·76 | | | | | | | | | | | | |
| 3626 | 582 | 447 | 18M | ·8281 | 20·76936 | 14·004 | **525** | 4 | 3 | Ap | 8○ | ·60 | 5 | My | 8 | ·13 | 6 | Je | 6 | ·66 | 1 | Jl | 6 | ·19 | 2 | Au | 4 | ·72 |
| 3627 | 583 | 448 | 19M | ·0868 | 9·87779 | 10·162 | **526** | 5 | 7 | Mr | 28 | ·97 | 2 | Ap | 27 | ·50 | 4 | My | 27 | ·03 | 5 | Je | 25 | ·56 | 7 | Jl | 25 | ·09 |
| 3328 | 584 | 449 | 19M | ·3455 | 28·51682 | 8·295 | **527** | 6 | 6 | Ap | 16 | ·86 | 1 | My | 16 | ·39 | 2 | Je | 14 | ·92 | 4 | Jl | 14 | ·46 | 5 | Au | 12○ | ·99 |
| 3629 | 585 | 450 | 18M | ·6042 | 17·62526 | 4·453 | **528** | 7 | 4 | Ap | 5 | ·23 | 5 | My | 4 | ·76 | 7 | Je | 3 | ·29 | 1 | Jl | 2 | ·82 | 3 | Au | 1○ | ·35 |
| 3630 | 586 | 451 | 18M | ·8628 | 6·73368 | 0·611 | **529** | 2 | 1 | Mr | 25 | ·60 | 3 | Ap | 24 | ·13 | 4 | My | 23 | ·66 | 6 | Je | 22 | ·19 | 2 | Au | 20 | ·25 |
| | | | | | | | | | | | | | | | | | | | | | 7 | Jl | 21●○ | ·72 | | | | |
| 3631 | 587 | 452 | 19M | ·1215 | 25·37271 | 26·299 | **530** | 3 | 7 | Ap | 13 | ·50 | 2 | My | 13 | ·03 | 3 | Je | 11 | ·56 | 5 | Jl | 11 | ·09 | 6 | Au | 9 | ·62 |
| 3632 | 588 | 453 | 19M | ·3802 | 14·48115 | 22·457 | **531** | 4 | 4 | Ap | 2 | ·86 | 6 | My | 2 | ·39 | 7 | My | 31○ | ·92 | 2 | Je | 30● | ·45 | 3 | Jl | 29 | ·98 |
| 3633 | 589 | 454 | 18M | ·6389 | 3·58958 | 18·614 | **532** | 5 | 2 | Mr | 22 | ·23 | 3 | Ap | 20 | ·76 | 5 | My | 20○ | ·29 | 1 | Jl | 18 | ·35 | 2 | Au | 16 | ·88 |
| | | | | | | | | | | | | | | | | | 6 | Je | 18 | ·82 | | | | | | | | |
| 3634 | 590 | 455 | 18M | ·8976 | 22·22861 | 16·749 | **533** | 7 | 1 | Ap | 10 | ·13 | 2 | My | 9● | ·66 | 4 | Je | 8 | ·19 | 5 | Jl | 7 | ·72 | 7 | Au | 6 | ·25 |
| 3635 | 591 | 456 | 19M | ·1562 | 11·33705 | 12·906 | **534** | 1 | 5 | Mr | 30○ | ·49 | 7 | Ap | 29● | ·02 | 1 | My | 28 | ·56 | 3 | Je | 27 | ·09 | 4 | Jl | 26 | ·62 |
| 3636 | 592 | 457 | 19M | ·4149 | 0·44547 | 9·064 | **535** | 2 | 2 | Mr | 19○ | ·86 | 5 | My | 17 | ·92 | 7 | Je | 16 | ·45 | 1 | Jl | 15 | ·98 | 3 | Au | 14 | ·51 |
| | | | | | | | | | 4 | Ap | 18 | ·39 | | | | | | | | | | | | | | | | |
| 3637 | 593 | 458 | 18M | ·6736 | 19·08450 | 7·198 | **536** | 3 | 1 | Ap | 6 | ·76 | 3 | My | 6 | ·29 | 4 | Je | 4 | ·82 | 6 | Jl | 4 | ·35 | 7 | Au | 2 | ·88 |
| 3638 | 594 | 459 | 18M | ·9323 | 8·19294 | 3·356 | **537** | 5 | 6 | Mr | 27 | ·13 | 7 | Ap | 25 | ·66 | 2 | My | 25 | ·19 | 3 | Je | 23 | ·72 | 5 | Jl | 23 | ·25 |
| | | | | | | | | | | | | | | | | | | | | | | | | | 6 | Au | 21 | ·78 |
| 3639 | 595 | 460 | 19M | ·1910 | 26·83226 | 1·490 | **538** | 6 | 5 | Ap | 15 | ·02 | 6 | My | 14 | ·55 | 1 | Je | 13 | ·08 | 2 | Jl | 18○ | ·62 | 4 | Au | 11 | ·15 |
| 3640 | 596 | 461 | 19M | ·4496 | 15·94039 | 25·202 | **539** | 7 | 2 | Ap | 4 | ·39 | 3 | My | 3 | ·92 | 5 | Je | 2 | ·45 | 6 | Jl | 1○ | ·98 | 1 | Jl | 31 | ·51 |
| 3641 | 597 | 462 | 18M | ·7083 | 5·04883 | 21·359 | **540** | 1 | 6 | Mr | 23 | ·76 | 1 | Ap | 22 | ·29 | 2 | My | 21 | ·82 | 5 | Jl | 19 | ·88 | 7 | Au | 18 | ·41 |
| | | | | | | | | | | | | | | | | | 4 | Je | 20● | ·35 | | | | | | | | |
| 3642 | 598 | 463 | 18M | ·9670 | 23·68785 | 19·493 | **541** | 3 | 5 | Ap | 11 | ·66 | 7 | My | 11○ | ·19 | 1 | Je | 9● | ·12 | 3 | Jl | 9 | ·25 | 4 | Au | 7 | ·78 |
| 3643 | 599 | 464 | 19M | ·2257 | 12·79629 | 15·651 | **542** | 4 | 3 | Ap | 1 | ·02 | 4 | Ap | 30○ | ·55 | 6 | My | 30 | ·08 | 7 | Je | 28 | ·61 | 2 | Jl | 28 | ·14 |
| 3644 | 600 | 465 | 19M | ·4844 | 1·90473 | 11·809 | **543** | 5 | 7 | Mr | 21 | ·39 | 1 | Ap | 19●○ | ·92 | 4 | Je | 17 | ·95 | 6 | Jl | 17 | ·51 | 1 | Au | 16 | ·04 |
| | | | | | | | | | | | | | 3 | My | 19 | ·45 | | | | | | | | | | | | |
| 3645 | 601 | 466 | 18M | ·7430 | 20·54375 | 9·942 | **544** | 6 | 6 | Ap | 8● | ·29 | 7 | My | 7 | ·82 | 2 | Je | 6 | ·35 | 3 | Jl | 5 | ·88 | 5 | Au | 4 | ·41 |
| 3646 | 602 | 467 | 19M | ·0017 | 9·65218 | 6·100 | **545** | 1 | 3 | Mr | 28● | ·65 | 5 | Ap | 27 | ·28 | 6 | My | 26 | ·72 | 1 | Je | 25 | ·25 | 2 | Jl | 24 | ·78 |
| 3647 | 603 | 468 | 19M | ·2604 | 28·29121 | 4·234 | **546** | 2 | 2 | Ap | 16 | ·55 | 4 | My | 16 | ·08 | 5 | Je | 14 | ·61 | 7 | Jl | 14 | ·14 | 1 | Au | 12○ | ·67 |
| 3648 | 604 | 469 | 19M | ·5191 | 17·39964 | 0·392 | **547** | 3 | 6 | Ap | 5 | ·92 | 1 | My | 5 | ·45 | 2 | Je | 3 | ·98 | 4 | Jl | 3 | ·51 | 6 | Au | 2○ | ·04 |
| 3649 | 605 | 470 | 18M | ·7778 | 6·50808 | 24·104 | **548** | 4 | 4 | Mr | 25 | ·29 | 5 | Ap | 23 | ·82 | 7 | My | 23 | ·35 | 1 | Je | 21 | ·88 | 4 | Au | 19 | ·94 |
| | | | | | | | | | | | | | | | | | | | | | 3 | Jl | 21● | ·41 | | | | |
| 3650 | 606 | 471 | 19M | ·0365 | 25·14711 | 22·238 | **549** | 6 | 3 | Ap | 13 | ·18 | 4 | My | 12 | ·71 | 6 | Je | 11○ | ·24 | 7 | Jl | 10 | ·78 | 2 | Au | 9 | ·31 |

**N.B.**—For Surya Sidddanta-figures for **A.D. 500** to **A.D. 999** see pages **76—79** below.

**Ârya Siddhanta.**

| Asvina | | | Kartika | | | Margasira | | | Pausha | | | A.D. | Magha | | | A.D. | Phalguna | | | Chaitra | | |
|---|---|---|---|---|---|---|---|---|---|---|---|---|---|---|---|---|---|---|---|---|---|---|
| + 147·65293<br>+ 9·880 | | | + 177·18353<br>+ 11·856 | | | + 206·71411<br>+ 13·832 | | | + 236·24470<br>+ 15·808 | | | | + 265·77529<br>+ 17·784 | | | | + 295·30588<br>+ 19·760 | | | +324·83647<br>+354·36705<br>+ 21·736<br>+ 23·712 | | |
| Week-day Month | Day | Fraction | Week-day Month | Day | Fraction | Week-day Month | Day | Fraction | Week-day Month | Day | Fraction | | Week-day Month | Day | Fraction | | Week-day Month | Day | Fraction | Week-day Month | Day | Fraction |
| 5 S | 8 | ·46 | 7 O | 8 | ·09 | 1 N | 6 | ·52 | 3 D | 6 | ·05 | **12** | 4 Jr | 4● | ·58 | | 6 F | 3 | ·11 | 7 Mr | 3 | ·64 |
| 2 Au | 27 | ·83 | 4 S | 26 | ·36 | 5 O | 25 | ·89 | 7 N | 24○ | ·42 | | 1 D | 23 | ·95 | **13** | 3 Jr | 22 | ·48 | 5 F | 21 | ·01 |
| 1 S | 15 | ·72 | 3 O | 15 | ·26 | 4 N | 13○ | ·79 | 6 D | 13 | ·32 | **14** | 7 Jr | 11 | ·85 | | 2 F | 10 | ·38 | 3 Mr | 11 | ·91 |
| 6 S | 5 | ·09 | 7 O | 4 | ·62 | 2 N | 3○ | ·15 | 3 D | 2 | ·68 | **15** | 5 Jr | 1 | ·21 | | 6 Jr | 30 | ·74 | 1 Mr | 1 | ·28 |
| 3 Au | 25 | ·46 | 4 S | 23 | ·99 | 6 O | 23● | ·52 | 1 N | 22 | ·05 | | 2 D | 21 | ·58 | **16** | 4 Jr | 20 | ·11 | 5 F | 18 | ·64 |
| 2 S | 12○ | ·36 | 3 O | 11 | ·89 | 5 N | 10 | ·42 | 6 D | 9 | ·95 | **17** | 1 Jr | 8 | ·48 | | 3 F | 7 | ·01 | 4 Mr | 8○ | ·54 |
| 6 S | 1○ | ·72 | 1 O | 1 | ·25 | 2 O | 30 | ·78 | 4 N | 29 | ·32 | | 5 D | 28 | ·85 | **18** | 7 Jr | 27 | ·38 | 1 F | 25○ | ·91 |
| 5 S | 20 | ·62 | 7 O | 20 | ·15 | 1 N | 18 | ·68 | 3 D | 18 | ·21 | **19** | 4 Jr | 16 | ·74 | | 6 F | 15● | ·27 | 7 Mr | 16 | ·80 |
| 2 S | 9 | ·99 | 4 O | 9 | ·52 | 6 N | 8 | ·05 | 7 D | 7 | ·58 | **20** | 2 Jr | 6○ | ·11 | | 3 F | 4 | ·64 | 5 Mr | 5 | ·17 |
| 7 Au | 29 | ·36 | 1 S | 27 | ·89 | 3 O | 27 | ·42 | 4 N | 25 | ·95 | | 6 D | 25○ | ·48 | **21** | 1 Jr | 24 | ·01 | 2 F | 22 | ·54 |
| 6 S | 17 | ·25 | 7 O | 16 | ·78 | 2 N | 15 | ·31 | 3 D | 14●○ | ·84 | **22** | 5 Jr | 13 | ·38 | | 7 F | 11 | ·91 | 1 Mr | 13 | ·44 |
| 3 S | 6 | ·62 | 5 O | 6 | ·15 | 6 N | 4 | ·68 | 1 D | 4● | ·21 | **23** | 2 Jr | 2 | ·74 | | 4 F | 1 | ·27 | 5 Mr | 2 | ·80 |
| 1 Au | 26 | ·99 | 2 S | 25 | ·52 | 4 O | 25○ | ·05 | 5 N | 23● | ·58 | | 7 D | 23 | ·11 | **24** | 1 Jr | 21 | ·64 | 3 F | 20 | ·17 |
| 6 S | 13 | ·88 | 1 O | 13○ | ·42 | 2 N | 11 | ·95 | 4 D | 11 | ·48 | **25** | 6 Jr | 10 | ·01 | | 7 F | 8 | ·54 | 2 Mr | 10 | ·07 |
| 4 S | 3 | ·25 | 5 O | 2○ | ·78 | 7 N | 1 | ·31 | 1 N | 30 | ·84 | | 3 D | 30 | ·37 | **26** | 4 Jr | 28 | ·90 | 6 F | 27 | ·44 |
| 1 Au | 23 | ·62 | {3 S | 22● | ·15 | 6 N | 20 | ·21 | **Pausha** | | | | 7 D | 19 | ·74 | **27** | {2 Jr | 18 | ·27 | 5 Mr | 18 | ·33 |
| | | | {4 O | 21 | ·68 | | | | **Kshaya** | | | | | | | | {3 F | 16○ | ·80 | | | |
| 7 S | 11 | ·52 | 2 O | 11 | ·05 | 3 N | 9 | ·58 | 5 D | 9 | ·11 | **28** | 6 Jr | 7 | ·64 | | 1 F | 6●○ | ·77 | 2 Mr | 6 | ·70 |
| 4 Au | 30 | ·88 | 6 S | 29 | ·41 | 7 O | 28 | ·94 | 2 N | 27 | ·48 | | 4 D | 27 | ·01 | **29** | 5 Jr | 25○ | ·45 | 7 F | 24 | ·07 |
| 3 S | 18 | ·78 | 5 O | 18 | ·31 | 6 N | 16 | ·84 | 1 D | 16 | ·37 | **30** | 2 Jr | 14● | ·90 | | 4 F | 13 | ·43 | 5 Mr | 14 | ·96 |
| 1 S | 8 | ·15 | 2 O | 7 | ·68 | 4 N | 6 | ·21 | 5 D | 5○ | ·74 | **31** | 7 Jr | 4 | ·27 | | 1 F | 2 | ·80 | 3 Mr | 4 | ·33 |
| 5 Au | 28 | ·52 | 7 S | 27 | ·05 | 1 O | 26 | ·58 | 3 N | 25○ | ·11 | | 4 D | 24 | ·64 | **32** | 6 Jr | 23 | ·17 | 7 F | 21 | ·70 |
| 4 S | 15 | ·41 | 5 O | 14 | ·94 | 7 N | 13●○ | ·47 | 2 D | 13 | ·00 | **33** | 3 Jr | 11 | ·54 | | 5 F | 10 | ·07 | 6 Mr | 11 | ·60 |
| 1 S | 4 | ·78 | 3 O | 4 | ·31 | 4 N | 2 | ·84 | 6 D | 2 | ·37 | | 7 D | 31 | ·90 | **34** | 2 Jr | 30 | ·43 | 3 F | 28 | ·96 |
| 6 Au | 25 | ·15 | 7 S | 23○ | ·68 | 2 O | 23● | ·21 | 3 N | 21 | ·74 | | 5 D | 21 | ·27 | **35** | 6 Jr | 19 | ·80 | 1 F | 18 | ·33 |
| 5 S | 13●○ | ·04 | 6 O | 12 | ·58 | 1 N | 11 | ·11 | 2 D | 10 | ·64 | **36** | 4 Jr | 9 | ·17 | | 5 F | 7 | ·70 | 7 Mr | 8○ | ·23 |
| 2 S | 1○ | ·41 | 3 S | 30 | ·94 | 5 O | 30 | ·47 | 7 N | 29 | ·00 | | 1 D | 28 | ·53 | **37** | 3 Jr | 27 | ·06 | 4 F | 25 | ·60 |
| 1 S | 20 | ·31 | 2 O | 19 | ·84 | 4 N | 18 | ·37 | 5 D | 17 | ·90 | **38** | 7 Jr | 16○ | ·43 | | 1 F | 14● | ·96 | 3 Mr | 16 | ·49 |
| 5 S | 9 | ·68 | 7 O | 9 | ·21 | 1 N | 7 | ·74 | 3 D | 7 | ·27 | **39** | 4 Jr | 5○ | ·80 | | 6 F | 4 | ·38 | 7 Mr | 5 | ·86 |
| 3 Au | 30 | ·04 | 4 S | 28 | ·57 | 6 O | 28 | ·10 | 7 N | 26 | ·64 | | 2 D | 26●○ | ·17 | **40** | 3 Jr | 24 | ·70 | 5 F | 23 | ·23 |
| 1 S | 16 | ·94 | 3 O | 16 | ·47 | 5 N | 15 | ·00 | 6 D | 14● | ·53 | **41** | 1 Jr | 13 | ·06 | | 2 F | 11 | ·59 | 4 Mr | 13 | ·12 |
| 6 S | 6 | ·31 | 7 O | 5 | ·84 | 2 N | 4○ | ·37 | 3 D | 3 | ·90 | **42** | 5 Jr | 2 | ·43 | | 6 Jr | 31 | ·96 | 1 Mr | 2 | ·49 |
| 3 Au | 26 | ·68 | 5 S | 25 | ·21 | 6 O | 24○ | ·74 | 1 N | 23 | ·27 | | 2 D | 22 | ·80 | **43** | 4 Jr | 21 | ·33 | 5 F | 19 | ·86 |
| 2 S | 14 | ·57 | 4 O | 14●○ | ·10 | 5 N | 12 | ·63 | 7 D | 12 | ·16 | **44** | 1 Jr | 10 | ·70 | | 3 F | 9 | ·23 | 4 Mr | 9 | ·76 |
| 6 S | 2 | ·94 | 1 O | 2 | ·47 | 3 N | 1 | ·00 | 4 N | 30 | ·53 | | 6 D | 30 | ·06 | **45** | 7 Jr | 28 | ·59 | 2 F | 27○ | ·12 |
| 4 Au | 23○ | ·31 | {5 S | 21● | ·84 | 1 N | 19 | ·90 | **Pausha** | | | | 3 D | 19 | ·43 | **46** | {4 Jr | 17 | ·96 | 1 Mr | 18 | ·02 |
| | | | {7 O | 21 | ·37 | | | | **Kshaya** | | | | | | | | {6 F | 16○ | ·49 | | | |
| 3 S | 11 | ·20 | 4 O | 10 | ·74 | 6 N | 9 | ·27 | 7 D | 8 | ·80 | **47** | 2 Jr | 7 | ·32 | | 3 F | 5●○ | ·86 | 5 Mr | 7 | ·39 |
| 7 Au | 31 | ·55 | 2 S | 30 | ·10 | 3 O | 29 | ·63 | 5 N | 28 | ·16 | | 6 D | 27 | ·69 | **48** | 1 Jr | 26 | ·22 | 2 F | 24 | ·76 |
| 6 S | 18 | ·47 | 1 O | 18 | ·00 | 2 N | 16 | ·53 | 4 D | 16○ | ·06 | **49** | 5 Jr | 14 | ·59 | | 7 F | 13 | ·12 | 1 Mr | 14 | ·65 |
| 3 S | 7 | ·80 | 5 O | 7 | ·37 | 6 N | 5 | ·90 | 1 D | 5●○ | ·43 | **50** | 2 Jr | 3 | ·96 | | 4 F | 2 | ·49 | 6 Mr | 4 | ·02 |

**TABLE X—**

| Kaliyuga | Vikrama Era | Saka Era | Commencement of Solar Year. Month and day A.D. | Fraction of day. | First New-Moon in Solar Year. | Anomaly of first New-Moon. | A.D. | Week Day of 1st January. | Vaisakha (☉'s Anom col. 6; ☾'s Anom col. 7) | | | | Jyeshta (☉ + 29·53059; ☾ + 1·976) | | | | Ashada (☉ + 59·06117; ☾ + 3·952) | | | | Sravana (☉ + 88·59176; ☾ + 5·928) | | | | Bhadrapada (☉ + 118·12235; ☾ + 7·904) | | | |
|---|---|---|---|---|---|---|---|---|---|---|---|---|---|---|---|---|---|---|---|---|---|---|---|---|---|---|---|---|
| | | | | | | | | | Week-day | Month | Day | Fraction | Week-day | Month | Day | Fraction | Week-day | Month | Day | Fraction | Week-day | Month | Day | Fraction | Week-day | Month | Day | Fraction |
| 3651 | 607 | 472 | 19M | ·2951 | 14·25553 | 18·395 | **550** | 7 | 7 | Ap | 2 | ·55 | 2 | My | 2 | ·08 | 3 | My | 31○ | ·61 | 5 | Je | 30 | ·14 | 6 | Jl | 29 | ·67 |
| 3652 | 608 | 473 | 19M | ·5538 | 3·36397 | 14·553 | **551** | 1 | 4 | Mr | 22 | ·92 | 6 | Ap | 21 | ·45 | 7 | My | 20● | ·98 | 4 | Jl | 19 | ·64 | 5 | Au | 17 | ·57 |
| | | | | | | | | | | | | | | | | | 2 | Je | 19 | ·51 | | | | | | | | |
| 3653 | 609 | 474 | 18M | ·8125 | 22·00300 | 12·688 | **552** | 2 | 3 | Ap | 9○ | ·82 | 5 | My | 9 | ·34 | 6 | Je | 7 | ·88 | 1 | Jl | 7 | ·41 | 2 | Au | 5 | ·94 |
| 3654 | 610 | 475 | 19M | ·0711 | 11·11143 | 8·845 | **553** | 4 | 1 | Mr | 30○ | ·18 | 2 | Ap | 28 | ·71 | 4 | My | 28 | ·24 | 5 | Je | 26 | ·77 | 7 | Jl | 26 | ·30 |
| 3655 | 611 | 476 | 19M | ·3299 | 0·21987 | 5·012 | **554** | 5 | 5 | Mr | 19●○ | ·55 | 1 | My | 17 | ·61 | 3 | Je | 16 | ·14 | 4 | Jl | 15 | ·67 | 6 | Au | 14 | ·20 |
| | | | | | | | | | 7 | Ap | 18 | ·08 | | | | | | | | | | | | | | | | |
| 3656 | 612 | 477 | 19M | ·5885 | 18·85889 | 3·136 | **555** | 6 | 4 | Ap | 7 | ·45 | 5 | My | 6 | ·98 | 7 | Je | 5 | ·51 | 2 | Jl | 5 | ·04 | 3 | Au | 3 | ·57 |
| 3657 | 613 | 478 | 18M | ·8472 | 7·96733 | 26·848 | **556** | 7 | 1 | Mr | 26 | ·81 | 3 | Ap | 25 | ·34 | 4 | My | 24 | ·88 | 6 | Je | 23 | ·41 | 7 | Jl | 22○ | ·94 |
| | | | | | | | | | | | | | | | | | | | | | | | | | 2 | Au | 21 | ·47 |
| 3658 | 614 | 479 | 19M | ·1059 | 26·60636 | 24·982 | **557** | 2 | 7 | Ap | 14 | ·71 | 2 | My | 14 | ·24 | 3 | Je | 12 | ·77 | 5 | Jl | 12○ | ·30 | 6 | Au | 10 | ·83 |
| 3659 | 615 | 480 | 19M | ·3646 | 15·71472 | 21·140 | **558** | 3 | 5 | Ap | 4 | ·08 | 6 | My | 3 | ·61 | 1 | Je | 2 | ·14 | 2 | Jl | 1○ | ·67 | 4 | Jl | 31 | ·20 |
| 3660 | 616 | 481 | 19M | ·6233 | 4·82383 | 17·298 | **559** | 4 | 2 | Mr | 24 | ·45 | 3 | Ap | 22 | ·98 | 5 | My | 22 | ·51 | 1 | Jl | 20 | ·57 | 3 | Au | 19 | ·10 |
| | | | | | | | | | | | | | | | | | 7 | Je | 21● | ·04 | | | | | | | | |
| 3661 | 617 | 482 | 18M | ·8819 | 23·46225 | 15·431 | **560** | 5 | 1 | Ap | 11 | ·34 | 2 | My | 10○ | ·87 | 4 | Je | 9 | ·40 | 5 | Jl | 8 | ·94 | 7 | Au | 7 | ·47 |
| 3662 | 618 | 483 | 19M | ·1406 | 12·57069 | 11·589 | **561** | 7 | 5 | Mr | 31 | ·71 | 7 | Ap | 30●○ | ·24 | 1 | My | 29 | ·77 | 3 | Je | 28 | ·30 | 4 | Jl | 27 | ·83 |
| 3663 | 619 | 484 | 19M | ·3993 | 1·67912 | 7·747 | **562** | 1 | 3 | Mr | 21 | ·08 | 4 | Ap | 19● | ·61 | 7 | Je | 17 | ·67 | 2 | Jl | 17 | ·20 | 3 | Au | 15 | ·73 |
| | | | | | | | | | | | | | 6 | My | 19 | ·14 | | | | | | | | | | | | |
| 3664 | 620 | 485 | 19M | ·6580 | 20·31814 | 5·881 | **563** | 2 | 1 | Ap | 8 | ·98 | 3 | My | 8 | ·51 | 5 | Je | 7 | ·03 | 6 | Jl | 6 | ·56 | 1 | Au | 5 | ·10 |
| 3665 | 621 | 486 | 18M | ·9167 | 9·42658 | 2·038 | **564** | 3 | 6 | Mr | 28 | ·34 | 7 | Ap | 26 | ·87 | 2 | My | 26 | ·40 | 3 | Je | 24 | ·93 | 5 | Jl | 24 | ·46 |
| 3666 | 622 | 487 | 19M | ·1753 | 28·06561 | ~~27·727~~ 0·172 | **565** | 5 | 5 | Ap | 16 | ·24 | 6 | My | 15 | ·77 | 1 | Je | 14 | ·30 | 2 | Jl | 13 | ·83 | 4 | Au | 12○ | ·36 |
| 3367 | 623 | 488 | 19M | ·1440 | 17·17403 | 23·[illegible] | **566** | 6 | 2 | Ap | 5 | ·61 | 4 | My | 5 | ·14 | 5 | Je | 3 | ·67 | 7 | Jl | 3 | ·20 | 1 | Au | 1● | ·73 |
| 3668 | 624 | 489 | 19M | ·6927 | 6·28248 | 20·042 | **567** | 7 | 6 | Mr | 25 | ·97 | 1 | Ap | 24 | ·50 | 3 | My | 24 | ·04 | 4 | Je | 22○ | ·57 | 7 | Au | 20 | ·63 |
| | | | | | | | | | | | | | | | | | | | | | 6 | Jl | 22● | ·10 | | | | |
| 3669 | 625 | 490 | 18M | ·9514 | 24·92150 | 18·1[illegible]6 (7) | **568** | 1 | 5 | Ap | 12 | ·85 | 7 | My | 12 | ·40 | 1 | Je | 10●○ | ·93 | 3 | Jl | 10 | ·46 | 4 | Au | 8 | ·99 |
| 3670 | 626 | 491 | 19M | ·2101 | 14·02993 | 14·333 | **569** | 3 | 3 | Ap | 2 | ·74 | 4 | My | 1 | ·77 | 6 | My | 31○ | ·30 | 7 | Je | 29 | ·83 | 2 | Jl | 29 | ·34 |
| 3671 | 627 | 492 | 19M | ·4687 | 3·13837 | 10·491 | **570** | 4 | 7 | Mr | 22 | ·61 | 2 | Ap | 21○ | ·13 | 3 | My | 20 | ·66 | 6 | Jl | 18 | ·73 | 1 | Au | 17 | ·26 |
| | | | | | | | | | | | | | | | | | 5 | Je | 19 | ·19 | | | | | | | | |
| 3672 | 628 | 493 | 19M | ·7274 | 21·77749 | 8·625 | **571** | 5 | 6 | Ap | 10○ | ·50 | 1 | My | 10 | ·03 | 2 | Je | 8 | ·56 | 4 | Jl | 8 | ·09 | 5 | Au | 6 | ·62 |
| 3673 | 629 | 494 | 18M | ·9861 | 10·88582 | 4·783 | **572** | 6 | 3 | Mr | 29○ | ·87 | 5 | Ap | 28 | ·40 | 6 | My | 27 | ·93 | 1 | Je | 26 | ·46 | 2 | Jl | 25 | ·99 |
| 3674 | 630 | 495 | 19M | ·2448 | 29·52484 | 0·941 | **573** | 1 | 1 | Mr | 19● | ·23 | 4 | My | 17 | ·30 | 5 | Je | 15 | ·83 | 7 | Jl | 15 | ·36 | 1 | Au | 13 | ·89 |
| | | | | | | | | | 2 | Ap | 17 | ·77 | | | | | | | | | | | | | | | | |
| 3675 | 631 | 496 | 19M | ·5035 | 18·63329 | 26·629 | **574** | 2 | 7 | Ap | 7 | ·13 | 1 | My | 6 | ·66 | 3 | Je | 5 | ·19 | 4 | Jl | 4 | ·72 | 6 | Au | 3○ | ·25 |
| 3676 | 632 | 497 | 19M | ·7621 | 7·74172 | 22·787 | **575** | 3 | 4 | Mr | 27 | ·50 | 6 | Ap | 26 | ·03 | 7 | My | 25 | ·56 | 2 | Je | 24 | ·09 | 3 | Jl | 23○ | ·62 |
| | | | | | | | | | | | | | | | | | | | | | | | | | 5 | Au | 22 | ·15 |
| 3677 | 633 | 498 | 19M | ·0208 | 26·38074 | 20·920 | **576** | 4 | 3 | Ap | 14 | ·40 | 4 | My | 13 | ·93 | 6 | Je | 12 | ·46 | 1 | Jl | 11●○ | ·99 | 2 | Au | 10 | ·52 |
| 3678 | 634 | 499 | 19M | ·2795 | 15·48918 | 17·078 | **577** | 6 | 7 | Ap | 3 | ·76 | 2 | My | 3 | ·29 | 3 | Je | 1 | ·83 | 7 | Jl | 1 | ·36 | 7 | Jl | 30 | ·89 |
| 3679 | 635 | 500 | 19M | ·5382 | 4·59761 | 13·236 | **578** | 7 | 5 | Mr | 24 | ·13 | 6 | Ap | 22 | ·66 | 1 | My | 22○ | ·19 | 4 | Jl | 20 | ·25 | 5 | Au | 18 | ·78 |
| | | | | | | | | | | | | | | | | | 3 | Je | 20 | ·72 | | | | | | | | |
| 3680 | 636 | 501 | 19M | ·7969 | 23·23664 | 11·370 | **579** | 1 | 4 | Ap | 12 | ·03 | 5 | My | 11○ | ·56 | 7 | Je | 10 | ·09 | 1 | Jl | 9 | ·62 | 3 | Au | 8 | ·15 |
| 3681 | 637 | 502 | 19M | ·0555 | 12·34508 | 7·527 | **580** | 2 | 1 | Mr | 31 | ·40 | 2 | Ap | 29 | ·93 | 4 | My | 29 | ·46 | 5 | Je | 27 | ·99 | 7 | Jl | 27 | ·52 |
| 3682 | 638 | 503 | 19M | ·3142 | 1·45351 | 3·684 | **581** | 4 | 5 | Mr | 20○ | ·76 | 1 | My | 18 | ·82 | 3 | Je | 17 | ·35 | 4 | Jl | 16 | ·89 | 6 | Au | 15 | ·42 |
| | | | | | | | | | 7 | Ap | 19 | ·29 | | | | | | | | | | | | | | | | |
| 3683 | 639 | 504 | 19M | ·5729 | 20·09253 | 1·819 | **582** | 5 | 4 | Ap | 8 | ·66 | 6 | My | 8 | ·19 | 7 | Je | 6 | ·72 | 2 | Jl | 6 | ·25 | 3 | Au | 4 | ·78 |
| 3684 | 640 | 505 | 19M | ·8316 | 9·20097 | 25·528 | **583** | 6 | 2 | Mr | 29 | ·03 | 3 | Ap | 27 | ·56 | 5 | My | 27 | ·09 | 6 | Je | 25 | ·62 | 1 | Jl | 25 | ·15 |
| | | | | | | | | | | | | | | | | | | | | | | | | | 2 | Au | 23●○ | ·68 |
| 3685 | 641 | 506 | 19M | ·0903 | 27·83999 | 23·665 | **584** | 7 | 7 | Ap | 15 | ·93 | 2 | My | 15 | ·46 | 3 | Je | 13 | ·99 | 5 | Jl | 13 | ·52 | 7 | Au | 12 | ·85 |
| 3686 | 642 | 507 | 19M | ·3490 | 16·94848 | 19·823 | **585** | 2 | 5 | Ap | 5 | ·29 | 6 | My | 4 | ·82 | 1 | Je | 3 | ·35 | 2 | Jl | 2○ | ·88 | 4 | Au | 1● | ·41 |
| 3687 | 643 | 508 | 19M | ·6076 | 6·05687 | 15·980 | **586** | 3 | 2 | Mr | 25 | ·66 | 4 | Ap | 24 | ·19 | 5 | My | 23 | ·72 | 7 | Je | 22○ | ·25 | 3 | Au | 20 | ·31 |
| | | | | | | | | | | | | | | | | | | | | | 1 | Jl | 21 | ·78 | | | | |
| 3688 | 644 | 509 | 19M | ·8663 | 24·69588 | 14·114 | **587** | 4 | 1 | Ap | 13 | ·16 | 3 | My | 13 | ·09 | 4 | Je | 11●○ | ·62 | 6 | Jl | 11 | ·15 | 7 | Au | 9 | ·68 |
| 3689 | 645 | 510 | 19M | ·1250 | 13·80432 | 10.273 | **588** | 5 | 5 | Ap | 1 | ·92 | 7 | My | 1○ | ·45 | 1 | My | 30● | ·99 | 3 | Je | 29 | ·52 | 5 | Jl | 29 | ·05 |

**N.B.**—For Surya Siddhanta figures for **A.D. 500** to **A.D. 999** see pages **76—79** below.

**Arya Siddhanta.**

| Asvina | | | | Kartika | | | | Margasira | | | | Pausha | | | | A.D. | Magha | | | | A.D. | Phalguna | | | | Chaitra | | | |
|---|---|---|---|---|---|---|---|---|---|---|---|---|---|---|---|---|---|---|---|---|---|---|---|---|---|---|---|---|---|
| + 147·65293 | | | | + 177·18353 | | | | + 206·71411 | | | | + 236·24470 | | | | | + 265·77422 | | | | | + 295·30588 | | | | {+324·83647 +354·36765 | | | |
| + 9·880 | | | | + 11·856 | | | | + 13·832 | | | | + 15·808 | | | | | + 17·784 | | | | | + 19·760 | | | | {+ 21·736 + 23·712 | | | |
| Week-day | Month | Day | Fraction | Week-day | Month | Day | Fraction | Week-day | Month | Day | Fraction | Week-day | Month | Day | Fraction | | Week-day | Month | Day | Fraction | | Week-day | Month | Day | Fraction | Week-day | Month | Day | Fraction |
| 1 | Au | 28 | ·20 | 2 | S | 26 | ·73 | 4 | O | 26 | ·26 | 5 | N | 24●○ | ·80 | | 7 | D | 24 | ·33 | **51** | 1 | Jr | 22 | ·86 | 3 | F | 21 | ·89 |
| 7 | S | 16 | ·14 | 1 | O | 15 | ·63 | 3 | N | 14 | ·16 | 4 | D | 13 | ·69 | **52** | 6 | Jr | 12 | ·22 | | 7 | F | 10 | ·75 | 2 | Mr | 11 | ·28 |
| 4 | S | 4 | ·47 | 6 | O | 4○ | ·00 | 7 | N | 2 | ·53 | 2 | D | 2 | ·06 | | 3 | D | 31 | ·59 | **53** | 5 | Jr | 30 | ·12 | 6 | F | 28 | ·65 |
| 1 | Au | 24 | ·84 | 3 | S | 23○ | ·37 | 4 | O | 22 | ·90 | 6 | N | 21 | ·43 | | 7 | D | 20 | ·96 | **54** | 2 | Jr | 19 | ·49 | 4 | F | 18 | ·02 |
| 7 | S | 12○ | ·73 | 2 | O | 12 | ·26 | 3 | N | 10 | ·79 | 5 | D | 10 | ·32 | **55** | 6 | Jr | 8 | ·86 | | 1 | F | 7 | ·39 | 2 | Mr | 8● | ·92 |
| 5 | S | 2 | ·10 | 6 | O | 1 | ·63 | 1 | O | 31 | ·16 | 2 | N | 29 | ·69 | | 4 | D | 29 | ·22 | **56** | 5 | Jr | 27○ | ·75 | 7 | F | 26 | ·28 |
| 4 | S | 20 | ·00 | 5 | O | 19 | ·53 | 7 | N | 18 | ·06 | 1 | D | 17 | ·59 | **57** | 3 | Jr | 16○ | ·12 | | 4 | F | 14 | ·65 | 6 | Mr | 16 | ·18 |
| 1 | S | 9 | ·36 | 2 | O | 8 | ·90 | 4 | N | 7 | ·43 | 5 | D | 6 | ·96 | **58** | 7 | Jr | 5○ | ·49 | | 2 | F | 4 | ·02 | 3 | Mr | 5 | ·55 |
| 5 | Au | 29 | ·73 | 7 | S | 28 | ·25 | 1 | O | 27 | ·79 | 3 | N | 26 | ·32 | | 4 | D | 25 | ·85 | **59** | 6 | Jr | 24 | ·38 | 7 | F | 22 | ·92 |
| 4 | S | 17 | ·63 | 6 | O | 17 | ·15 | 7 | N | 15○ | ·69 | 2 | D | 15 | ·22 | **60** | 3 | Jr | 13 | ·75 | | 5 | F | 12 | ·28 | 6 | Mr | 12 | ·81 |
| 2 | S | 6 | ·00 | 3 | O | 5 | ·53 | 5 | N | 4○ | ·06 | 6 | D | 3● | ·59 | **61** | 1 | Jr | 2 | ·12 | | 2 | Jr | 31 | ·65 | 4 | Mr | 2 | ·18 |
| 6 | Au | 26 | ·36 | 7 | S | 24 | ·89 | 2 | O | 24○ | ·42 | 3 | N | 22 | ·96 | | 5 | D | 22 | ·49 | **62** | 7 | Jr | 21 | ·02 | 4 | F | 19 | ·55 |
| 5 | S | 14 | ·26 | 6 | O | 13● | ·79 | 1 | N | 12 | ·32 | 2 | D | 11 | ·85 | **63** | 4 | Jr | 10 | ·38 | | 5 | F | 8 | ·91 | 7 | Mr | 10○ | ·44 |
| 2 | S | 3○ | ·63 | 4 | O | 3● | ·16 | 5 | N | 1 | ·69 | 7 | D | 1 | ·22 | | 1 | D | 30 | ·75 | **64** | 3 | Jr | 29 | ·28 | 4 | F | 27○ | ·81 |
| {6 | Au | 22○ | ·99 | 3 | O | 21 | ·05 | 5 | N | 19 | ·58 | **Pausha Kshaya** | | | | | 6 | D | 19 | ·12 | **65** | {7 | Jr | 17 | ·65 | 3 | Mr | 17 | ·71 |
| {1 | S | 21 | ·52 | | | | | | | | | | | | | | | | | | | {2 | F | 16○ | ·18 | | | | |
| 5 | S | 10 | ·89 | 7 | O | 10 | ·42 | 1 | N | 8 | ·95 | 3 | D | 8 | ·48 | **66** | 5 | Jr | 7 | ·01 | | 6 | F | 5● | ·54 | 1 | Mr | 7 | ·07 |
| 3 | Au | 31 | ·26 | 4 | S | 29 | ·79 | 6 | O | 29 | ·32 | 7 | N | 27 | ·85 | | 2 | D | 27○ | ·38 | **67** | 4 | Jr | 25 | ·91 | 5 | F | 24 | ·44 |
| 2 | S | 19 | ·16 | 3 | O | 18 | ·69 | 5 | N | 17 | ·22 | 6 | D | 16○ | ·75 | **68** | 1 | Jr | 15 | ·28 | | 2 | F | 13 | ·81 | 4 | Mr | 14 | ·34 |
| 6 | S | 7 | ·20 | 1 | O | 7 | ·05 | 2 | N | 5 | ·59 | 4 | D | 5○ | ·12 | **69** | 5 | Jr | 3 | ·64 | | 7 | F | 2 | ·18 | 1 | Mr | 3 | ·18 |
| 3 | Au | 27 | ·89 | 5 | S | 26 | ·42 | 6 | O | 25 | ·95 | 1 | N | 24● | ·48 | | 3 | D | 24 | ·01 | **70** | 4 | Jr | 22 | ·54 | 6 | F | 21 | ·07 |
| 2 | S | 15 | ·79 | 4 | O | 15○ | ·32 | 5 | N | 13 | ·85 | 7 | D | 13 | ·38 | **71** | 1 | Jr | 11 | ·91 | | 3 | F | 10 | ·44 | 4 | Mr | 11 | ·97 |
| 7 | S | 5 | ·15 | 1 | O | 4○ | ·68 | 3 | N | 3 | ·21 | 4 | D | 2 | ·75 | **72** | 6 | Jr | 1 | ·28 | | 7 | Jr | 30 | ·81 | 2 | F | 29 | ·34 |
| 4 | Au | 24 | ·52 | 6 | S | 23●○ | ·06 | 7 | O | 22 | ·58 | 2 | N | 21 | ·11 | | 3 | D | 20 | ·64 | **73** | 5 | Jr | 19 | ·17 | 6 | F | 17 | ·70 |
| 3 | S | 12● | ·42 | 4 | O | 11 | ·95 | 6 | N | 10 | ·48 | 1 | D | 10 | ·01 | **74** | 2 | Jr | 8 | ·54 | | 4 | F | 7○ | ·07 | 5 | Mr | 8● | ·60 |
| 7 | S | 1● | ·79 | 2 | O | 1 | ·32 | 3 | O | 30 | ·85 | 5 | N | 29 | ·38 | | 6 | D | 28 | ·91 | **75** | 1 | Jr | 27○ | ·44 | 2 | F | 25 | ·97 |
| 6 | S | 20 | ·68 | 1 | O | 20 | ·23 | 2 | N | 18 | ·74 | 4 | D | 18 | ·27 | **76** | 5 | Jr | 16○ | ·81 | | 7 | F | 15 | ·34 | 1 | Mr | 15 | ·87 |
| 4 | S | 9 | ·05 | 5 | O | 8 | ·58 | 7 | N | 7 | ·11 | 1 | D | 6 | ·64 | **77** | 3 | Jr | 5● | ·17 | | 4 | F | 3 | ·70 | 6 | Mr | 5 | ·20 |
| 1 | Au | 29 | ·42 | 2 | S | 27 | ·95 | 4 | O | 27 | ·48 | 6 | N | 26○ | ·01 | | 7 | D | 25● | ·54 | **78** | 2 | Jr | 24 | ·07 | 3 | F | 22 | ·63 |
| 7 | S | 17 | ·31 | 1 | O | 16 | ·85 | 3 | N | 15○ | ·38 | 4 | D | 14 | ·91 | **79** | 6 | Jr | 13 | ·44 | | 7 | F | 11 | ·97 | 2 | Mr | 13 | ·50 |
| 4 | S | 6 | ·63 | 6 | O | 6 | ·21 | 7 | N | 4○ | ·74 | 2 | D | 4 | ·27 | **80** | 3 | Jr | 2 | ·80 | | 5 | F | 1 | ·33 | 6 | Mr | 1 | ·87 |
| 2 | Au | 26 | ·05 | 3 | S | 24 | ·58 | 5 | O | 24● | ·11 | 6 | N | 22 | ·64 | | 1 | D | 22 | ·17 | **81** | 2 | Jr | 20 | ·70 | 4 | F | 19 | ·23 |
| 7 | S | 13○ | ·95 | 2 | O | 13 | ·48 | 4 | N | 12 | ·01 | 5 | D | 11 | ·54 | **82** | 7 | Jr | 10 | ·07 | | 1 | F | 8 | ·60 | 3 | Mr | 10○ | ·13 |
| 5 | S | 3○ | ·31 | 6 | O | 2 | ·84 | 1 | N | 1 | ·37 | 2 | N | 30 | ·91 | | 4 | D | 30 | ·44 | **83** | 6 | Jr | 28 | ·97 | 7 | F | 27○ | ·50 |
| 4 | S | 22 | ·21 | 5 | O | 21 | ·74 | 7 | N | 20 | ·27 | 1 | D | 19 | ·50 | **84** | 3 | Jr | 18 | ·33 | | 4 | F | 16● | ·86 | 6 | Mr | 17 | ·39 |
| 1 | S | 10 | ·58 | 3 | O | 10 | ·11 | 4 | N | 8 | ·64 | 6 | D | 8 | ·17 | **85** | 7 | Jr | 6○ | ·70 | | 2 | F | 5 | ·23 | 3 | Mr | 6 | ·76 |
| 5 | Au | 30 | ·95 | 7 | S | 29 | ·48 | 2 | O | 29 | ·01 | 3 | N | 27 | ·54 | | 5 | D | 27○ | ·07 | **86** | 6 | Jr | 25 | ·60 | 1 | F | 24 | ·13 |
| 4 | S | 18 | ·84 | 6 | O | 18 | ·37 | 7 | N | 16 | ·90 | 2 | D | 16●○ | ·39 | **87** | 3 | Jr | 14 | ·97 | | 5 | F | 13 | ·50 | 7 | Mr | 15 | ·03 |
| 2 | S | 8 | ·21 | 3 | O | 7 | ·74 | 5 | N | 6 | ·27 | 6 | D | 5 | ·80 | **88** | 1 | Jr | 4 | ·33 | | 2 | F | 2 | ·86 | 4 | Mr | 3 | ·39 |
| 6 | Au | 27 | ·58 | 1 | S | 26 | ·11 | 3 | O | 25○ | ·64 | 4 | N | 24 | ·17 | | 5 | D | 23 | ·70 | **89** | 7 | Jr | 22 | ·23 | 1 | F | 20 | ·76 |

**TABLE X—**

| Kaliyuga | Vikrama Era | Saka Era | Commencement of Solar Year: Month and day A.D. | Fraction of day | First New-Moon in Solar Year | Anomaly of first New-Moon | A.D. | Week-day of 1st January | Vaisakha Week-day | Month | Day | Fraction | Jyeshta Week-day | Month | Day | Fraction | Ashada Week-day | Month | Day | Fraction | Sravana Week-day | Month | Day | Fraction | Bhadrapada Week-day | Month | Day | Fraction |
|---|---|---|---|---|---|---|---|---|---|---|---|---|---|---|---|---|---|---|---|---|---|---|---|---|---|---|---|---|
| | | | | | | | | | ☉'s Anom col. 6 | | | | + 29·53059 | | | | + 59·06117 | | | | + 88·59176 | | | | + 118·12233 | | | |
| | | | | | | | | | ☾'s Anom col. 7 | | | | + 1·976 | | | | + 3·952 | | | | + 5·928 | | | | + 7·904 | | | |
| 3690 | 646 | 511 | 19M | ·3837 | 2·91276 | 6·430 | **589** | 7 | 3 | Mr | 22 | ·29 | 4 | Ap | 20○ | ·82 | 7 | Je | 18 | ·88 | 2 | Jl | 18 | ·41 | 3 | Au | 16 | ·94 |
| | | | | | | | | | | | | | 6 | My | 20● | ·35 | | | | | | | | | | | | |
| 3691 | 647 | 512 | 19M | ·6424 | 21·55178 | 4·563 | **590** | 1 | 2 | Ap | 10○ | ·19 | 3 | My | 9 | ·72 | 5 | Je | 8 | ·25 | 6 | Jl | 7 | ·78 | 1 | Au | 6 | ·31 |
| 3692 | 648 | 513 | 19M | ·9010 | 10·66022 | 0·721 | **591** | 2 | 6 | Mr | 30 | ·56 | 1 | Ap | 29 | ·09 | 2 | My | 28 | ·62 | 4 | Je | 27 | ·15 | 5 | Jl | 26 | ·68 |
| 3693 | 649 | 514 | 19M | ·1597 | 29·29925 | 26·409 | **592** | 3 | 4 | Mr | 18● | ·92 | 6 | My | 16 | ·98 | 1 | Je | 15 | ·51 | 3 | Jl | 15 | ·05 | 4 | Au | 13○ | ·58 |
| | | | | | | | | | 5 | Ap | 17 | ·45 | | | | | | | | | | | | | | | | |
| 3694 | 650 | 515 | 19M | ·4184 | 18·40757 | 22·568 | **593** | 5 | 2 | Ap | 6 | ·82 | 4 | My | 6 | ·35 | 5 | Je | 4 | ·88 | 7 | Jl | 4 | ·41 | 1 | Au | 2○ | ·94 |
| 3695 | 651 | 516 | 19M | ·6771 | 7·51611 | 18·725 | **594** | 6 | 7 | Mr | 27 | ·19 | 1 | Ap | 25 | ·75 | 3 | My | 25 | ·25 | 4 | Je | 23 | ·78 | 6 | Jl | 23●○ | ·31 |
| | | | | | | | | | | | | | | | | | | | | | | | | | 7 | Au | 21 | ·84 |
| 3696 | 652 | 517 | 19M | ·9358 | 26·15445 | 16·858 | **595** | 7 | 6 | Ap | 15 | ·09 | 7 | My | 14 | ·62 | 2 | Je | 13 | ·15 | 3 | Jl | 12 | ·68 | 5 | Au | 11 | ·21 |
| 3697 | 653 | 518 | 19M | ·1944 | 15·26357 | 13·016 | **596** | 1 | 3 | Ap | 3 | ·45 | 4 | My | 2 | ·98 | 6 | Je | 1○ | ·51 | 1 | Jl | 1 | ·04 | 2 | Jl | 30 | ·57 |
| 3698 | 654 | 519 | 19M | ·4531 | 4·37201 | 9·174 | **597** | 3 | 7 | Mr | 23 | ·22 | 2 | Ap | 22 | ·35 | 3 | My | 21○ | ·88 | 6 | Jl | 19 | ·94 | 1 | An | 18 | ·47 |
| | | | | | | | | | | | | | | | | | 5 | Je | 20 | .41 | | | | | | | | |
| 3699 | 655 | 520 | 19M | ·7118 | 23·01103 | 7·308 | **598** | 4 | 6 | Ap | 11 | ·72 | 1 | My | 11● | ·25 | 2 | Je | 9 | ·78 | 4 | Jl | 9 | ·31 | 5 | Au | 7 | ·84 |
| 3700 | 656 | 521 | 19M | ·9705 | 12·11946 | 3·466 | **599** | 5 | 4 | Ap | 1○ | ·08 | 5 | Ap | 30● | ·62 | 7 | My | 30 | ·15 | 1 | Je | 28 | ·68 | 3 | Jl | 28 | ·21 |
| 3701 | 657 | 522 | 19M | ·2292 | 1·22790 | 27·178 | **600** | 6 | 1 | Mr | 20○ | ·46 | 4 | My | 18 | ·52 | 6 | Je | 17 | ·05 | 7 | Jl | 16 | ·58 | 2 | Au | 15 | ·11 |
| | | | | | | | | | 2 | Ap | 18 | ·99 | | | | | | | | | | | | | | | | |
| 3702 | 658 | 523 | 19M | ·4878 | 19·86693 | 25·312 | **601** | 1 | 7 | Ap | 8 | ·36 | 1 | My | 7 | ·89 | 3 | Je | 6 | ·42 | 4 | Jl | 5 | ·95 | 6 | Au | 4 | ·48 |
| 3703 | 659 | 524 | 19M | ·7465 | 8·97535 | 21·469 | **602** | 2 | 4 | Mr | 28 | ·72 | 6 | Ap | 27 | ·25 | 7 | My | 26 | ·79 | 2 | Je | 25 | ·32 | 3 | Jl | 24 | ·85 |
| | | | | | | | | | | | | | | | | | | | | | | | | | 5 | Au | 23 | ·38 |
| 3704 | 660 | 525 | 20M | ·0052 | 27·61438 | 19·603 | **603** | 3 | 3 | Ap | 16 | ·62 | 5 | My | 16 | ·15 | 6 | Je | 14 | ·68 | 1 | Jl | 14○ | ·21 | 2 | Au | 12 | ·74 |
| 3705 | 661 | 526 | 19M | ·2639 | 16·72282 | 15·761 | **604** | 4 | 7 | Ap | 4 | ·99 | 2 | My | 4 | ·52 | 4 | Je | 3 | ·05 | 5 | Jl | 2○ | ·58 | 7 | Au | 1 | ·11 |
| 3706 | 662 | 527 | 19M | ·5225 | 5·83125 | 11·919 | **605** | 6 | 5 | Mr | 25 | ·38 | 6 | Ap | 23 | ·89 | 1 | My | 23 | ·42 | 2 | Je | 21●○ | ·95 | 6 | Au | 20 | ·01 |
| | | | | | | | | | | | | | | | | | | | | | 4 | Jl | 21 | ·48 | | | | |
| 3707 | 663 | 528 | 19M | ·7812 | 24·47028 | 10·053 | **606** | 7 | 4 | Ap | 13 | ·25 | 5 | My | 12○ | ·78 | 7 | Je | 11● | ·31 | 1 | Jl | 10 | ·85 | 3 | Au | 9 | ·38 |
| 3708 | 66[illegible] | 529 | 20M | ·0399 | 13·57872 | 6·210 | **607** | 1 | 1 | Ap | 2 | ·62 | 3 | My | 2○ | ·15 | 4 | My | 31 | ·68 | 6 | Je | 30 | ·21 | 7 | Jl | 29 | ·74 |
| 3709 | 665 | 530 | 19M | ·2986 | 2·68714 | 2·368 | **608** | 2 | 5 | Mr | 21 | ·99 | 7 | Ap | 20●○ | ·52 | 3 | Je | 18 | ·58 | 5 | Jl | 18 | ·11 | 6 | Au | 16 | ·64 |
| | | | | | | | | | | | | | 2 | My | 20 | ·05 | | | | | | | | | | | | |
| 3710 | 666 | 531 | 19M | ·5573 | 21·32617 | 0·502 | **609** | 4 | 4 | Ap | 9● | ·89 | 6 | My | 9 | ·42 | 7 | Je | 7 | ·95 | 2 | Jl | 7 | ·48 | 4 | Au | 6 | ·01 |
| 3711 | 667 | 532 | 19M | ·8160 | 10·43461 | 24·215 | **610** | 5 | 2 | Mr | 30 | ·25 | 3 | Ap | 28 | ·78 | 5 | My | 28 | ·31 | 6 | Je | 26 | ·84 | 1 | Jl | 26 | ·37 |
| 3712 | 668 | 533 | 20M | ·0746 | 29·07363 | 22·348 | **611** | 6 | 6 | Mr | 19 | ·62 | 2 | My | 17 | ·68 | 4 | Je | 16 | ·21 | 6 | Jl | 15 | ·74 | 7 | Au | 14○ | ·27 |
| | | | | | | | | | 1 | Ap | 18 | ·15 | | | | | | | | | | | | | | | | |
| 3713 | 669 | 534 | 19M | ·3333 | 18·18207 | 18·500 | **612** | 7 | 5 | Ap | 6 | ·52 | 7 | My | 6 | ·05 | 1 | Je | 4 | ·58 | 3 | Jl | 4 | ·11 | 4 | Au | 2○ | ·64 |
| 3714 | 670 | 535 | 1[illegible]M | ·5920 | 7·29050 | 14·663 | **613** | 2 | 2 | Mr | 26 | ·88 | 4 | Ap | 25 | ·41 | 5 | My | 24 | ·95 | 7 | Je | 23 | ·48 | 2 | Jl | 23● | ·01 |
| | | | | | | | | | | | | | | | | | | | | | | | | | 3 | Au | 21 | ·54 |
| 3715 | 671 | 536 | 19M | ·8507 | 25·92952 | 12·797 | **614** | 3 | 1 | Ap | 14 | ·78 | 3 | My | 14 | ·31 | 4 | Je | 12○ | ·84 | 6 | Jl | 12 | ·37 | 7 | Au | 10 | ·90 |
| 3716 | 672 | 537 | 20M | ·1094 | 15·03796 | 8·955 | **615** | 4 | 6 | Ap | 4 | ·15 | 7 | My | 3 | ·68 | 2 | Je | 2○ | ·21 | 3 | Jl | 1 | ·74 | 5 | Jl | 31 | ·27 |
| 3717 | 673 | 538 | 19M | ·3680 | 4·14640 | 5·112 | **616** | 5 | 3 | Mr | 23 | ·52 | 5 | Ap | 22 | ·05 | 6 | My | 21●○ | ·58 | 2 | Jl | 19 | ·64 | 4 | Au | 18 | ·17 |
| | | | | | | | | | | | | | | | | | 1 | Je | 20 | ·11 | | | | | | | | |
| 3718 | 674 | 539 | 19M | ·6267 | 22·78542 | 3·246 | **617** | 7 | 2 | Ap | 11○ | ·41 | 3 | My | 10 | ·94 | 5 | Je | 9 | ·47 | 7 | Jl | 9 | ·01 | 1 | Au | 7 | ·54 |
| 3719 | 675 | 540 | 19M | ·8854 | 11·89386 | 26·959 | **618** | 1 | 6 | Mr | 31●○ | ·78 | 1 | Ap | 30 | ·31 | 2 | My | 29 | ·84 | 4 | Je | 28 | ·37 | 5 | Jl | 27 | ·90 |
| 3720 | 676 | 541 | 20M | ·1441 | 1·00229 | 23·116 | **619** | 2 | 4 | Mr | 21○ | ·15 | 7 | My | 19 | ·21 | 1 | Je | 17 | ·74 | 3 | Jl | 17 | ·27 | 4 | Au | 15 | ·80 |
| | | | | | | | | | 5 | Ap | 19 | ·68 | | | | | | | | | | | | | | | | |
| 3721 | 677 | 542 | 19M | ·4028 | 19·64131 | 21·250 | **620** | 3 | 3 | Ap | 8 | ·04 | 4 | My | 7 | ·58 | 6 | Je | 6 | ·11 | 7 | Jl | 5 | ·64 | 2 | Au | 4 | ·17 |
| 3722 | 678 | 543 | 19M | ·6614 | 8·74975 | 17·408 | **621** | 5 | 7 | Mr | 28 | ·41 | 1 | Ap | 26 | ·94 | 3 | My | 26 | ·47 | 5 | Je | 25 | ·00 | 6 | Jl | 24○ | ·53 |
| | | | | | | | | | | | | | | | | | | | | | | | | | 1 | Au | 23 | ·07 |
| 3723 | 679 | 544 | 19M | ·9201 | 27·38878 | 15·542 | **622** | 6 | 6 | Ap | 16 | ·31 | 7 | My | 15 | ·84 | 2 | Je | 14 | ·37 | 3 | Jl | 13○ | ·90 | 5 | Au | 12 | ·43 |
| 3724 | 680 | 545 | 20M | ·[illegible]788 | 16·49720 | 11·699 | **623** | 7 | 3 | Ap | 5 | ·68 | 5 | My | 5 | ·21 | 6 | Je | 3 | ·74 | 1 | Jl | 3○ | ·27 | 2 | Au | 1 | ·80 |
| 3725 | 681 | 546 | 19M | ·4375 | 5·60564 | 7·857 | **624** | 1 | 1 | Mr | 25 | ·06 | 2 | Ap | 23 | ·57 | 4 | My | 23○ | ·11 | 5 | Je | 21 | ·64 | 1 | Au | 19 | ·70 |
| | | | | | | | | | | | | | | | | | | | | | 7 | Jl | 21 | ·17 | | | | |
| 3726 | 682 | 547 | 19M | ·6962 | 24·24467 | 5·990 | **625** | 3 | 6 | Ap | 12 | ·94 | 1 | My | 12○ | ·47 | 3 | Je | 11 | ·00 | 4 | Jl | 10 | ·53 | 6 | Au | 9 | ·06 |
| 3727 | 683 | 548 | 19M | ·9548 | 13·35310 | 2·[illegible]49 | **626** | 4 | 4 | Ap | 2 | ·31 | 5 | My | 1○ | ·84 | 7 | My | 31 | ·37 | 1 | Je | 29 | ·90 | 3 | Jl | 29 | ·43 |

**N.B.**— For Surya Siddhanta figures for **A.D. 500** to **A.D. 999** see pages **76—79** below.

**Arya Siddhanta.**

| | | | | | | | | | | | | | | | | | | | | | | | | | | | | | |
|---|---|---|---|---|---|---|---|---|---|---|---|---|---|---|---|---|---|---|---|---|---|---|---|---|---|---|---|---|---|
| + 147·65293 | | | | + 177·18353 | | | | + 206·71411 | | | | + 236·24470 | | | | | + 265·77529 | | | | | + 295·30588 | | | | +324·83647 +354·36705 | | | |
| + 9·880 | | | | + 11·856 | | | | + 13·832 | | | | + 15·808 | | | | | + 17·784 | | | | | + 19·760 | | | | + 21·736 + 23·712 | | | |
| **Asvina** | | | | **Kartika** | | | | **Margasira** | | | | **Pausha** | | | | **A.D.** | **Magha** | | | | **A.D.** | **Phalguna** | | | | **Chaitra** | | | |
| Week-day | Month | Day | Fraction | Week-day | Month | Day | Fraction | Week-day | Month | Day | Fraction | Week-day | Month | Day | Fraction | | Week-day | Month | Day | Fraction | | Week-day | Month | Day | Fraction | Week-day | Month | Day | Fraction |
| 5 | S | 15 | ·47 | 7 | O | 15●○ | ·01 | 1 | N | 13 | ·54 | 3 | D | 13 | ·07 | **90** | 4 | Jr | 11 | ·60 | | 6 | F | 10 | ·13 | 7 | Mr | 11 | ·66 |
| 2 | S | 4 | ·84 | 4 | O | 4●○ | ·37 | 5 | N | 2 | ·90 | 7 | D | 2 | ·43 | | 2 | D | 31 | ·96 | **91** | 3 | Jr | 30 | ·49 | 5 | Mr | 1 | ·03 |
| 7 | Au | 25 | ·21 | 1 | S | 23● | ·74 | 3 | O | 23 | ·27 | 4 | N | 21 | ·80 | | 6 | D | 21 | ·33 | **92** | 7 | Jr | 19 | ·86 | 2 | F | 18○ | ·39 |
| 6 | S | 12 | ·11 | 7 | O | 11 | ·64 | 2 | N | 10 | ·17 | 3 | D | 9 | ·70 | **93** | 5 | Jr | 8 | 23 | | 6 | F | 6○ | ·76 | 1 | Mr | 8 | ·29 |
| 3 | S | 1 | ·47 | 5 | O | 1 | ·00 | 6 | O | 30 | ·53 | 1 | N | 29 | ·07 | | 2 | D | 28 | ·60 | **94** | 4 | Jr | 27●○ | ·13 | 5 | F | 25 | ·66 |
| 2 | S | 20 | ·37 | 3 | O | 19 | ·90 | 5 | N | 18 | ·43 | 6 | D | 17 | ·96 | **95** | 1 | Jr | 16● | ·55 | | 3 | F | 15 | ·02 | 4 | Mr | 16 | ·55 |
| 6 | S | 9 | ·74 | 1 | O | 9 | ·27 | 2 | N | 7 | ·80 | 4 | D | 7○ | ·33 | **96** | 5 | Jr | 5 | ·86 | | 7 | F | 4 | ·39 | 1 | Mr | 4 | ·92 |
| 4 | Au | 29 | ·12 | 5 | S | 27 | ·64 | 7 | O | 27 | ·17 | 1 | N | 25○ | ·70 | | 3 | D | 25● | ·23 | **97** | 4 | Jr | 23 | ·76 | 6 | F | 22 | ·29 |
| 3 | S | 17 | ·00 | 4 | O | 16 | ·53 | 6 | N | 15○ | ·06 | 7 | D | 14 | ·59 | **98** | 2 | Jr | 13 | ·13 | | 3 | F | 11 | ·66 | 5 | Mr | 13 | ·19 |
| 7 | S | 6 | ·37 | 1 | O | 5 | ·90 | 3 | N | 4 | ·43 | 4 | D | 3 | ·96 | **99** | 6 | Jr | 2 | ·49 | | 1 | F | 1 | ·02 | 2 | Mr | 2 | ·55 |
| 4 | Au | 26 | ·74 | 6 | S | 25○ | ·27 | 7 | O | 24 | ·80 | 2 | N | 23 | ·33 | | 6 | D | 22 | ·86 | **600** | 5 | Jr | 21 | ·39 | 6 | F | 19 | ·92 |
| 3 | S | 13○ | ·64 | 5 | O | 13 | ·17 | 6 | N | 11 | ·70 | 1 | D | 11 | ·23 | **01** | 2 | Jr | 9 | ·77 | | 4 | F | 8 | ·30 | 5 | Mr | 9●○ | ·83 |
| 1 | S | 3○ | ·01 | 2 | O | 2 | ·54 | 4 | N | 1 | ·07 | 5 | N | 30 | ·60 | | 7 | D | 30 | ·13 | **02** | 1 | Jr | 28 | ·66 | 3 | F | 27 | ·10 |
| 6 | S | 21 | ·91 | 1 | O | 21 | ·44 | 2 | N | 19 | ·97 | 4 | D | 19 | ·50 | **03** | 6 | Jr | 18○ | ·03 | | 7 | F | 16 | ·56 | 2 | Mr | 18 | ·09 |
| 4 | S | 11 | ·27 | 5 | O | 10 | ·81 | 7 | N | 9 | ·34 | 1 | D | 8 | ·87 | **04** | 3 | Jr | 7●○ | ·40 | | 4 | F | 5 | ·93 | 6 | Mr | 6 | ·46 |
| 1 | Au | 30 | ·64 | 3 | S | 29 | ·17 | 4 | O | 28 | ·70 | 6 | N | 27 | ·23 | | 7 | D | 26●○ | ·16 | **05** | 2 | Jr | 25 | ·29 | 3 | F | 23 | ·83 |
| 7 | S | 18 | ·54 | 2 | O | 18 | ·07 | 3 | N | 16 | ·60 | 5 | D | 16 | ·13 | **06** | 6 | Jr | 14 | ·66 | | 1 | F | 13 | ·19 | 2 | Mr | 14 | ·72 |
| 4 | S | 7 | ·91 | 6 | O | 7 | ·44 | 7 | N | 5○ | ·97 | 2 | D | 5 | ·50 | **07** | 4 | Jr | 4 | ·03 | | 5 | F | 2 | ·56 | 7 | Mr | 4 | ·09 |
| 2 | Au | 28 | ·27 | 3 | S | 26 | ·80 | 5 | O | 26○ | ·33 | 6 | N | 24 | ·87 | | 1 | D | 24 | ·40 | **08** | 2 | Jr | 22 | ·93 | 4 | F | 21 | ·46 |
| 1 | S | 15 | ·17 | 2 | O | 14○ | ·70 | 4 | N | 13 | ·23 | 5 | D | 12 | ·76 | **09** | 7 | Jr | 11 | ·29 | | 7 | F | 9 | ·82 | 3 | Mr | 11 | ·35 |
| 5 | S | 4 | ·54 | 7 | O | 4 | ·07 | 1 | N | 2 | ·60 | 3 | D | 2 | ·13 | | 4 | D | 31 | ·66 | **10** | 5 | Jr | 30 | ·19 | 7 | F | 28○ | ·72 |
| 2 | Au | 24○ | ·91 | {4 | S | 23 | ·44 | **Margasira Kshaya** | | | | 7 | N | 21 | ·50 | | 2 | D | 21 | ·03 | **11** | 3 | Jr | 19 | ·56 | 5 | F | 18○ | ·09 |
| | | | | {5 | O | 22 | ·97 | | | | | | | | | | | | | | | | | | | | | | |
| 1 | S | 12 | ·80 | 3 | O | 12 | ·33 | 4 | N | 10 | ·86 | 6 | D | 10 | ·39 | **12** | 7 | Jr | 8 | ·93 | | 2 | F | 7○ | ·46 | 3 | Mr | 7 | ·99 |
| 6 | S | 1 | ·17 | 7 | S | 30 | ·70 | 2 | O | 30 | ·23 | 3 | N | 28 | ·76 | | 5 | D | 28 | ·29 | **13** | 6 | Jr | 26 | ·82 | 1 | F | 25 | ·35 |
| 5 | S | 20 | ·07 | 6 | O | 19 | ·60 | 1 | N | 18 | ·13 | 2 | D | 17○ | ·66 | **14** | 4 | Jr | 16 | ·19 | | 5 | F | 14 | ·72 | 7 | Mr | 16 | ·25 |
| 2 | S | 9 | ·43 | 3 | O | 8 | ·97 | 5 | N | 7 | ·50 | 7 | D | 7○ | ·03 | **15** | 1 | Jr | 5 | ·56 | | 3 | F | 4 | ·09 | 4 | Mr | 5 | ·62 |
| 6 | Au | 29 | ·80 | 1 | S | 28 | ·33 | 2 | O | 27 | ·86 | 4 | N | 26○ | ·39 | | 5 | D | 25 | ·92 | **16** | 7 | Jr | 24 | ·45 | 2 | F | 23 | ·98 |
| 5 | S | 16 | ·70 | 7 | O | 16 | ·23 | 1 | N | 14● | ·76 | 3 | D | 14 | ·29 | **17** | 4 | Jr | 12 | ·82 | | 6 | F | 11 | ·35 | 7 | Mr | 12 | ·88 |
| 3 | S | 6 | ·07 | 4 | O | 5○ | ·60 | 6 | N | 4● | ·13 | 7 | D | 3 | ·66 | **18** | 2 | Jr | 2 | ·19 | | 3 | Jr | 31 | ·72 | 5 | Mr | 2 | ·25 |
| 7 | Au | 26 | ·43 | 1 | S | 24○ | ·96 | 3 | O | 24● | ·49 | 5 | N | 23 | ·03 | | 6 | D | 22 | ·56 | **19** | 1 | Jr | 21 | ·19 | 2 | F | 19 | ·62 |
| 6 | S | 14○ | ·23 | 7 | O | 13 | ·86 | 2 | N | 12 | ·39 | 3 | D | 11 | ·92 | **20** | 5 | Jr | 10 | ·45 | | 6 | F | 8 | ·98 | 1 | Mr | 9● | ·51 |
| 3 | S | 2● | ·70 | 5 | O | 2 | ·23 | 6 | O | 31 | ·76 | 1 | N | 30 | ·29 | | 2 | D | 29 | ·82 | **21** | 4 | Jr | 28○ | ·35 | 5 | F | 26 | ·88 |
| 2 | S | 21 | ·60 | 4 | O | 21 | ·13 | 5 | N | 19 | ·66 | 7 | D | 19 | ·19 | **22** | 1 | Jr | 17○ | ·72 | | 3 | F | 16 | ·25 | 4 | Mr | 17 | ·78 |
| 6 | S | 10 | ·96 | 1 | O | 10 | ·49 | 3 | N | 9 | ·12 | 4 | D | 8 | ·55 | **23** | 6 | Jr | 7○ | ·09 | | 7 | F | 5 | ·62 | 2 | Mr | 7 | ·15 |
| 4 | Au | 31 | ·33 | 5 | S | 29 | ·86 | 7 | O | 29 | ·39 | 1 | N | 27 | ·92 | | 3 | D | 27● | ·45 | **24** | 4 | Jr | 25 | ·98 | 6 | F | 24 | ·51 |
| 3 | S | 18 | ·23 | 4 | O | 17 | ·73 | 6 | N | 16○ | ·29 | 7 | D | 15● | ·82 | **25** | 2 | Jr | 14 | ·35 | | 3 | F | 12 | ·88 | 5 | Mr | 14 | ·41 |
| 7 | S | 7 | ·60 | 2 | O | 7 | ·13 | 3 | N | 5○ | ·66 | 5 | D | 5 | ·19 | **26** | 6 | Jr | 3 | ·72 | | 1 | F | 2 | ·25 | 2 | Mr | 3 | ·78 |
| 4 | Au | 27 | ·96 | 6 | S | 26 | ·49 | 1 | O | 26●○ | ·02 | 2 | N | 24 | ·55 | | 4 | D | 24 | ·08 | **27** | 5 | Jr | 22 | ·67 | 7 | F | 21 | ·15 |

**TABLE X—**

⊙'s Anom col. 6 + 29·53059 + 59·06117 + 88·59176 + 118·12235

☾'s Anom col. 7 + 1·976 + 3·952 + 5·928 + 7·904

| Kaliyuga | Vikrama Era | Saka Era | Commencement of Solar Year: Month and day A.D. | Commencement of Solar Year: Fraction of day | First New-Moon in Solar Year. | Anomaly of first New-Moon. | A.D. | Week-day of 1st January. | Vaisakha: Week-day | Month | Day | Fraction | Jyeshta: Week-day | Month | Day | Fraction | Ashada: Week-day | Month | Day | Fraction | Sravana: Week-day | Month | Day | Fraction | Bhadrapada: Week-day | Month | Day | Fraction |
|---|---|---|---|---|---|---|---|---|---|---|---|---|---|---|---|---|---|---|---|---|---|---|---|---|---|---|---|---|
| 3728 | 684 | 549 | 20M | ·1213 | 2·46154 | 25·861 | **627** | 5 | 1 | Mr | 22 | ·68 | 3 | Ap | 21● | ·21 | 6 | Je | 19 | ·27 | 7 | Jl | 18 | ·80 | 2 | Au | 17 | ·33 |
| | | | | | | | | | | | | | 4 | My | 20 | ·74 | | | | | | | | | | | | |
| 3729 | 685 | 550 | 19M | ·4722 | 21·10057 | 23·985 | **628** | 6 | 7 | Ap | 9● | ·57 | 2 | My | 9 | ·10 | 3 | Je | 7 | ·63 | 5 | Jl | 7 | ·17 | 6 | Au | 5 | ·70 |
| 3730 | 686 | 551 | 19M | ·7309 | 10·20899 | 20·153 | **629** | 1 | 4 | Mr | 29 | ·94 | 6 | Ap | 28 | ·47 | 1 | My | 28 | ·00 | 2 | Je | 26 | ·53 | 4 | Jl | 26 | ·06 |
| 3731 | 687 | 552 | 19M | ·9896 | 28·84802 | 18·286 | **630** | 2 | 3 | Ap | 17 | ·84 | 5 | My | 17 | ·37 | 6 | Je | 15 | ·90 | 1 | Jl | 15 | ·43 | 2 | Au | 13●○ | ·96 |
| 3732 | 688 | 553 | 20M | ·2483 | 17·95646 | 14·444 | **631** | 3 | 1 | Ap | 7 | ·21 | 2 | My | 6 | ·74 | 4 | Je | 5 | ·27 | 5 | Jl | 4 | ·80 | 7 | Au | 3 | ·33 |
| 3733 | 689 | 554 | 19M | ·5069 | 7·06489 | 10·602 | **632** | 4 | 5 | Mr | 26 | ·56 | 7 | Ap | 25 | ·10 | 1 | My | 24 | ·63 | 3 | Je | 23○ | ·16 | 6 | Au | 21 | ·23 |
| | | | | | | | | | | | | | | | | | | | | | 4 | Jl | 22 | ·69 | | | | |
| 3734 | 690 | 555 | 19M | ·7656 | 25·70392 | 8·735 | **633** | 6 | 4 | Ap | 14 | ·47 | 6 | My | 14 | ·00 | 7 | Je | 12●○ | ·53 | 2 | Jl | 12 | ·06 | 3 | Au | 10 | ·59 |
| 3735 | 691 | 556 | 20M | ·0243 | 14·81236 | 4·893 | **634** | 7 | 1 | Ap | 3 | ·84 | 3 | My | 3 | ·37 | 4 | Je | 1○ | ·90 | 6 | Jl | 1 | ·43 | 7 | Jl | 30 | ·96 |
| 3736 | 692 | 557 | 20M | ·2830 | 3·92078 | 1·051 | **635** | 1 | 6 | Mr | 24 | ·20 | 7 | Ap | 22○ | ·73 | 2 | My | 22 | ·27 | 5 | Jl | 20 | ·33 | 6 | Au | 18 | ·86 |
| | | | | | | | | | | | | | | | | | 3 | Je | 20 | ·88 | | | | | | | | |
| 3737 | 693 | 558 | 19M | ·5417 | 22·55981 | 26·739 | **636** | 2 | 5 | Ap | 11○ | ·10 | 6 | My | 10 | ·63 | 1 | Je | 9 | ·16 | 2 | Jl | 8 | ·69 | 4 | Au | 7 | ·22 |
| 3738 | 694 | 559 | 19M | ·8003 | 11·66825 | 22·897 | **637** | 4 | 2 | Mr | 31●○ | ·47 | 4 | Ap | 30 | ·00 | 5 | My | 29 | ·53 | 7 | Je | 28 | ·07 | 1 | Jl | 27 | ·59 |
| 3739 | 695 | 560 | 20M | ·0590 | 0·77698 | 19·055 | **638** | 5 | 6 | Mr | 20● | ·84 | 2 | My | 18 | ·90 | 4 | Je | 17 | ·43 | 5 | Jl | 16 | ·96 | 7 | Au | 15 | ·49 |
| | | | | | | | | | 1 | Ap | 19 | ·37 | | | | | | | | | | | | | | | | |
| 3740 | 696 | 561 | 20M | ·3177 | 19·41570 | 17·188 | **639** | 6 | 5 | Ap | 8 | ·73 | 7 | My | 8 | ·26 | 1 | Je | 6 | ·79 | 3 | Jl | 6 | ·33 | 4 | Au | 4○ | ·86 |
| 3741 | 697 | 562 | 19M | ·5764 | 8·52414 | 13·346 | **640** | 7 | 3 | Mr | 28 | ·10 | 4 | Ap | 26 | ·63 | 6 | My | 26 | ·16 | 7 | Je | 24 | ·69 | 2 | Jl | 24○ | ·22 |
| | | | | | | | | | | | | | | | | | | | | | | | | | 3 | Au | 22 | ·75 |
| 3742 | 698 | 563 | 19M | ·8351 | 27·16316 | 11·480 | **641** | 2 | 2 | Ap | 16 | ·00 | 3 | My | 15 | ·53 | 5 | Je | 14 | ·06 | 6 | Jl | 13○ | ·59 | 1 | Au | 12 | ·12 |
| 3743 | 699 | 564 | 20M | ·0937 | 16·27160 | 7·638 | **642** | 3 | 6 | Ap | 5 | ·37 | 7 | My | 4 | ·90 | 2 | Je | 3 | ·43 | 3 | Jl | 2 | ·96 | 5 | Au | 1 | ·49 |
| 3744 | 700 | 565 | 20M | ·3524 | 5·38004 | 3·795 | **643** | 4 | 3 | Mr | 25 | ·73 | 5 | Ap | 24 | ·26 | 6 | My | 23○ | ·79 | 2 | Jl | 21 | ·85 | 4 | Au | 20 | ·39 |
| | | | | | | | | | | | | | | | | | 1 | Je | 22● | ·32 | | | | | | | | |
| 3745 | 701 | 566 | 19M | ·6111 | 24·01906 | 1·929 | **644** | 5 | 2 | Ap | 12 | ·63 | 4 | My | 12○ | ·16 | 5 | Je | 10 | ·69 | 7 | Jl | 10 | ·22 | 1 | Au | 8 | ·75 |
| 3746 | 702 | 567 | 19M | ·8698 | 13·12749 | 25·641 | **645** | 7 | 7 | Ap | 2 | ·00 | 1 | My | 1 | ·53 | 3 | My | 31 | ·06 | 4 | Je | 29 | ·59 | 6 | Jl | 29 | ·12 |
| 3747 | 703 | 568 | 20M | ·1285 | 2·23593 | 21·799 | **646** | 1 | 4 | Mr | 22○ | ·36 | 5 | Ap | 20● | ·89 | 1 | Je | 18 | ·96 | 3 | Jl | 18 | ·41 | 5 | Au | 17 | ·02 |
| | | | | | | | | | | | | | 7 | My | 20 | ·43 | | | | | | | | | | | | |
| 3748 | 704 | 569 | 20M | ·3871 | 20·87495 | 19·933 | **647** | 2 | 3 | Ap | 10 | ·26 | 4 | My | 9 | ·79 | 6 | Je | 8 | ·32 | 7 | Jl | 7 | ·85 | 2 | Au | 6 | ·39 |
| 3749 | 705 | 570 | 19M | ·6458 | 9·98339 | 16·091 | **648** | 3 | 7 | Mr | 29 | ·63 | 2 | Ap | 28 | ·16 | 3 | My | 27 | ·69 | 5 | Je | 26 | ·22 | 6 | Jl | 25 | ·75 |
| 3750 | 706 | 571 | 19M | ·9045 | 28·62242 | 14·234 | **649** | 5 | 6 | Ap | 17 | ·53 | 1 | My | 17 | ·06 | 2 | Je | 15 | ·59 | 4 | Jl | 15 | ·12 | 5 | Au | 13 | ·65 |
| 3751 | 707 | 572 | 20M | ·1632 | 17·73084 | 10·383 | **650** | 6 | 3 | Ap | 6 | ·89 | 5 | My | 6 | ·42 | 6 | Je | 4 | ·95 | 1 | Jl | 4○ | ·49 | 3 | Au | 3● | ·02 |
| 3752 | 708 | 572 | 20M | ·4219 | 6·83928 | 6·540 | **651** | 7 | 1 | Mr | 27 | ·26 | 2 | Ap | 25 | ·79 | 4 | My | 25 | ·32 | 5 | Je | 23○ | ·85 | 1 | Au | 21 | ·91 |
| | | | | | | | | | | | | | | | | | | | | | 7 | Jl | 23 | ·38 | | | | |
| 3753 | 709 | 574 | 19M | ·6805 | 25·47831 | 4·694 | **652** | 1 | 7 | Ap | 14 | ·16 | 1 | My | 13 | ·69 | 3 | Je | 12○ | ·22 | 4 | Jl | 11 | ·75 | 6 | Au | 10 | ·28 |
| 3754 | 710 | 575 | 19M | ·9392 | 14·58674 | 0·831 | **653** | 3 | 4 | Ap | 3 | ·53 | 6 | My | 3 | ·06 | 7 | Je | 1● | ·59 | 2 | Jl | 1 | ·12 | 3 | Jl | 30 | ·65 |
| 3755 | 711 | 576 | 20M | ·1979 | 3·69518 | 24·544 | **654** | 4 | 1 | Mr | 23 | ·89 | 3 | Ap | 22○ | ·42 | 4 | My | 21 | ·95 | 1 | Jl | 20 | ·01 | 2 | Au | 18 | ·55 |
| | | | | | | | | | | | | | | | | | 6 | Je | 20 | ·48 | | | | | | | | |
| 3756 | 712 | 577 | 20M | ·4566 | 22·33420 | 22·677 | **655** | 5 | 7 | Ap | 11●○ | ·79 | 2 | My | 11 | ·32 | 3 | Je | 9 | ·85 | 5 | Jl | 9 | ·33 | 6 | Au | 7 | ·91 |
| 3757 | 713 | 578 | 19M | ·7153 | 11·44264 | 18·835 | **656** | 6 | 5 | Mr | 31 | ·16 | 6 | Ap | 29 | ·69 | 1 | My | 29 | ·22 | 2 | Je | 27 | ·75 | 4 | Jl | 27 | ·28 |
| 3758 | 714 | 579 | 19M | ·9738 | 0·55107 | 14·993 | **657** | 1 | 2 | Mr | 20 | ·52 | 5 | My | 18 | ·59 | 7 | Je | 17 | ·12 | 1 | Jl | 16 | ·65 | 3 | Au | 15○ | ·18 |
| | | | | | | | | | 4 | Ap | 19 | ·05 | | | | | | | | | | | | | | | | |
| 3759 | 715 | 580 | 20M | ·2325 | 19·19003 | 13·127 | **658** | 2 | 1 | Ap | 8 | ·42 | 2 | My | 7 | ·95 | 4 | Je | 6 | ·48 | 6 | Jl | 6 | ·01 | 7 | Au | 4○ | ·54 |
| 3760 | 716 | 581 | 20M | ·4913 | 8·29914 | 9·285 | **659** | 3 | 5 | Mr | 28 | ·79 | 7 | Ap | 27 | ·32 | 1 | My | 26 | ·85 | 3 | Je | 25 | ·38 | 4 | Jl | 24●○ | ·91 |
| | | | | | | | | | | | | | | | | | | | | | | | | | 6 | Au | 23 | ·44 |
| 3761 | 717 | 582 | 19M | ·7500 | 26·93756 | 7·418 | **660** | 4 | 4 | Ap | 15 | ·69 | 6 | My | 15 | ·22 | 7 | Je | 13 | ·75 | 2 | Jl | 13● | ·28 | 3 | Au | 11 | ·81 |
| 3762 | 718 | 583 | 20M | ·0087 | 16·04600 | 3·576 | **661** | 6 | 2 | Ap | 5 | ·05 | 3 | My | 4 | ·58 | 5 | Je | 3○ | ·11 | 6 | Jl | 2● | ·65 | 1 | Au | 1 | ·18 |
| 3763 | 719 | 584 | 20M | ·2673 | 5·15443 | 27·288 | **662** | 7 | 6 | Mr | 25 | ·42 | 7 | Ap | 23 | ·95 | 2 | My | 23●○ | ·48 | 5 | Jl | 21 | ·54 | 7 | Au | 20 | ·07 |
| | | | | | | | | | | | | | | | | | 4 | Je | 22 | ·01 | | | | | | | | |
| 3764 | 720 | 585 | 20M | ·5260 | 23·79345 | 25·422 | **663** | 1 | 5 | Ap | 13 | ·32 | 6 | My | 12● | ·85 | 1 | Je | 11 | ·38 | 2 | Jl | 10 | ·91 | 4 | Au | 9 | ·44 |
| 3765 | 721 | 586 | 19M | ·7847 | 12·90189 | 21·580 | **664** | 2 | 2 | Ap | 1○ | ·69 | 4 | My | 1 | ·22 | 5 | My | 30 | ·75 | 7 | Je | 29 | ·28 | 1 | Jl | 28 | ·81 |
| 3766 | 722 | 587 | 20M | ·0434 | 2·01033 | 17·737 | **665** | 4 | 7 | Mr | 22○ | ·05 | 1 | Ap | 20● | ·58 | 4 | Je | 18 | ·64 | 6 | Jl | 18 | ·17 | 7 | Au | 16 | ·71 |
| | | | | | | | | | | | | | 3 | My | 20 | ·11 | | | | | | | | | | | | |

**N.B.**—For Surya Siddhanta figures for **A.D. 500** to **A.D. 999** see pages **76—79** below.

**Arya Siddhanta.**

| | | | | | | | | | | | | | | | | | | | | | | | | | | | | | |
|---|---|---|---|---|---|---|---|---|---|---|---|---|---|---|---|---|---|---|---|---|---|---|---|---|---|---|---|---|---|
| + 147·65298 | | | | + 177·18353 | | | | + 206·71411 | | | | + 236·24470 | | | | | + 265·77529 | | | | | + 295·30588 | | | | +324·88647<br>+354·36705 | | | |
| + 9·880 | | | | + 11·856 | | | | + 13·832 | | | | + 15·808 | | | | | + 17·784 | | | | | + 19·760 | | | | + 21·736<br>+ 23·712 | | | |
| **Asvina** | | | | **Kartika** | | | | **Margasira** | | | | **Pausha** | | | | **A.D.** | **Magha** | | | | **A.D.** | **Phalguna** | | | | **Chaitra** | | | |
| Week-day | Month | Day | Fraction | Week-day | Month | Day | Fraction | Week-day | Month | Day | Fraction | Week-day | Month | Day | Fraction | | Week-day | Month | Day | Fraction | | Week-day | Month | Day | Fraction | Week-day | Month | Day | Fraction |
| 3 | S | 15 | ·86 | 5 | O | 15● | ·39 | 6 | N | 13 | ·92 | 1 | D | 13 | ·45 | **28** | 2 | Jr | 11 | ·98 | | 4 | F | 10 | 51 | 6 | Mr | 11○ | ·04 |
| 1 | S | 4○ | ·23 | 2 | O | 3● | ·76 | 4 | N | 2 | ·29 | 5 | D | 1 | ·82 | | 7 | D | 31 | ·35 | **29** | 1 | Jr | 29 | ·88 | 3 | F | 28○ | ·41 |
| {5 | Au | 24○ | ·59 | 1 | O | 22 | ·65 | 3 | N | 21 | ·19 | | **Pausha** | | | | 4 | D | 20 | ·72 | **30** | 6 | Jr | 19 | ·25 | {7 | F | 17○ | ·78 |
| {7 | S | 23 | ·12 | | | | | | | | | | **Kshaya** | | | | | | | | | | | | | {2 | Mr | 19 | ·31 |
| 4 | S | 12 | ·49 | 6 | O | 12 | ·02 | 7 | N | 10 | ·55 | 2 | D | 10 | ·08 | **31** | 3 | Jr | 8 | ·61 | | 5 | F | 7● | ·14 | 6 | Mr | 8 | ·67 |
| 1 | S | 1 | ·86 | 3 | O | 1 | ·39 | 4 | O | 30 | ·92 | 6 | N | 29 | ·45 | | 7 | D | 28○ | ·98 | **32** | 2 | Jr | 27● | ·51 | 4 | F | 26 | ·05 |
| 7 | S | 19 | ·76 | 2 | O | 19 | ·29 | 3 | N | 17 | ·82 | 5 | D | 17○ | ·35 | **33** | 6 | Jr | 15 | ·88 | | 1 | F | 14 | ·41 | 2 | Mr | 15 | ·94 |
| 5 | S | 9 | ·12 | 6 | O | 8 | ·65 | 1 | N | 7 | ·18 | 2 | D | 6○ | ·71 | **34** | 4 | Jr | 5 | ·25 | | 5 | F | 3 | ·78 | 7 | Mr | 5 | ·31 |
| 2 | Au | 29 | ·49 | 4 | S | 28 | ·02 | 5 | O | 27 | ·55 | 7 | N | 26● | ·08 | | 1 | D | 25 | ·61 | **35** | 3 | Jr | 24 | ·14 | 4 | F | 22 | ·67 |
| 1 | S | 17 | ·39 | 2 | O | 16○ | ·92 | 4 | N | 15 | ·45 | 5 | D | 14 | ·98 | **36** | 7 | Jr | 13 | ·51 | | 2 | F | 12 | ·04 | 3 | Mr | 12 | ·57 |
| 5 | S | 5 | ·75 | 7 | O | 5○ | ·29 | 1 | N | 3 | ·82 | 3 | D | 3 | ·35 | **37** | 4 | Jr | 1 | ·88 | | 6 | Jr | 31 | ·41 | 7 | Mr | 1 | ·94 |
| 3 | Au | 26 | ·12 | 4 | S | 24●○ | ·65 | 6 | O | 24 | ·18 | 7 | N | 22 | ·71 | | 2 | D | 22 | ·24 | **38** | 3 | Jr | 20 | ·77 | 5 | F | 19 | ·30 |
| 2 | S | 14 | ·02 | 3 | O | 13 | ·55 | 5 | N | 12 | ·08 | 6 | D | 11 | ·61 | **39** | 1 | Jr | 10 | ·14 | | 2 | F | 8○ | ·67 | 4 | Mr | 10 | ·20 |
| 6 | S | 3● | ·39 | 7 | O | 2 | ·92 | 2 | N | 1 | ·45 | 4 | N | 30 | ·98 | | 5 | D | 30 | ·51 | **40** | 7 | Jr | 29○ | ·04 | 1 | F | 27 | ·53 |
| 5 | S | 21 | ·28 | 6 | O | 20 | ·81 | 1 | N | 19 | ·35 | 2 | D | 18 | ·88 | **41** | 4 | Jr | 17●○ | ·41 | | 5 | F | 15 | ·94 | 7 | Mr | 17 | ·40 |
| 2 | S | 10 | ·65 | 4 | O | 10 | ·18 | 5 | N | 8 | ·71 | 7 | D | 8 | ·24 | **42** | 1 | Jr | 6 | ·77 | | 3 | F | 5 | ·30 | 4 | Mr | 6 | ·87 |
| 7 | Au | 31 | ·02 | 1 | S | 29 | ·55 | 3 | O | 29 | ·08 | 4 | N | 27○ | ·61 | | 6 | D | 27● | ·14 | **43** | 7 | Jr | 25 | ·67 | 2 | F | 24 | ·72 |
| 5 | S | 18 | ·92 | 7 | O | 18 | ·45 | 1 | N | 16●○ | ·98 | 3 | D | 16 | ·51 | **44** | 5 | Jr | 15 | ·04 | | 6 | F | 13 | ·57 | 1 | Mr | 14 | ·10 |
| 3 | S | 7 | ·28 | 4 | O | 6 | ·81 | 6 | N | 5●○ | ·34 | 7 | D | 4 | ·87 | **45** | 2 | Jr | 3 | ·41 | | 3 | F | 1 | ·94 | 5 | Mr | 3 | ·47 |
| 7 | Au | 27 | ·65 | 2 | S | 26 | ·18 | 3 | O | 25● | ·71 | 5 | N | 24 | ·24 | | 6 | D | 23 | ·77 | **46** | 1 | Jr | 22 | ·30 | 2 | F | 20 | ·83 |
| 6 | S | 15○ | ·55 | 1 | O | 15 | ·08 | 2 | N | 13 | ·61 | 4 | D | 13 | ·14 | **47** | 5 | Jr | 11 | ·67 | | 7 | F | 10 | ·20 | 1 | Mr | 11○ | ·73 |
| 3 | S | 4○ | ·91 | 5 | O | 4 | ·45 | 6 | N | 2 | ·90 | 1 | D | 2 | ·51 | **48** | 3 | Jr | 1 | ·04 | | 4 | Jr | 30 | ·57 | 6 | F | 29●○ | ·10 |
| 1 | Au | 24●○ | ·28 | {2 | S | 22 | ·81 | 5 | N | 20 | ·87 | 7 | D | 20 | ·40 | **49** | 1 | Jr | 18 | ·93 | | 3 | F | 17● | ·47 | 5 | Mr | 19 | ·00 |
| | | | | {4 | O | 22 | ·34 | | | | | | | | | | | | | | | | | | | | | | |
| 7 | S | 12 | ·18 | 1 | O | 11 | ·71 | 3 | N | 10 | ·24 | 4 | D | 9 | ·77 | **50** | 6 | Jr | 8○ | ·30 | | 7 | F | [illegible] | ·83 | 2 | Mr | 8 | ·36 |
| 4 | S | 1 | ·55 | 6 | O | 1 | ·08 | 7 | O | 30 | ·61 | 2 | N | 29 | ·14 | | 3 | D | 28○ | ·67 | **51** | 5 | Jr | 27● | ·20 | 6 | F | 25 | ·73 |
| 3 | S | 20 | ·44 | 4 | O | 19 | ·97 | 6 | N | 18 | ·51 | 1 | D | 18●○ | ·04 | **52** | 2 | Jr | 16 | ·57 | | 4 | F | 15 | ·10 | [illegible] | Mr | 15 | ·63 |
| 7 | S | 8 | ·81 | 2 | O | 8 | ·34 | 3 | N | 6 | ·87 | 5 | D | 6 | ·40 | **53** | 6 | Jr | 4 | ·93 | | 1 | F | 3 | ·46 | 3 | Mr | 4 | ·99 |
| 5 | Au | 29 | ·18 | 6 | S | 27 | ·71 | 1 | O | 27○ | ·24 | 2 | N | 25● | ·77 | | 4 | D | 25 | ·30 | **54** | 5 | Jr | 23 | ·83 | 7 | F | 22 | ·36 |
| 4 | S | 17 | ·08 | 5 | O | 16○ | ·61 | 7 | N | 15 | ·14 | 1 | D | 14 | ·67 | **55** | 3 | Jr | 13 | ·20 | | 4 | F | 11 | ·73 | 6 | Mr | 13 | ·26 |
| 1 | S | 6 | ·44 | 2 | O | 5○ | ·90 | 4 | N | 4 | ·50 | 6 | D | 4 | ·03 | **56** | 7 | Jr | 2 | ·57 | | 2 | F | 1 | ·10 | 3 | Mr | 1 | ·63 |
| 5 | Au | 25 | ·81 | 7 | S | 24 | ·34 | 1 | O | 23 | ·87 | 3 | N | 22 | ·40 | | 4 | D | 21 | ·93 | **57** | 6 | Jr | 20 | ·46 | 7 | F | 18○ | ·99 |
| 4 | S | 13 | ·71 | 6 | O | 13 | ·24 | 7 | N | 11 | ·77 | 2 | D | 11 | ·30 | **58** | 3 | Jr | 9 | ·83 | | 5 | F | 8○ | ·36 | 6 | Mr | 9 | ·89 |
| 2 | S | 3● | ·07 | 3 | O | 2 | ·61 | 5 | N | 1 | ·14 | 6 | N | 30 | ·67 | | 1 | D | 30 | ·20 | **59** | 2 | Jr | 28○ | ·73 | 4 | F | 27 | ·26 |
| 7 | S | 21 | ·97 | 2 | O | 21 | ·50 | 4 | N | 20 | ·03 | 5 | D | 19 | ·56 | **60** | 7 | Jr | 18● | ·09 | | 1 | F | 16 | ·62 | 3 | Mr | 17 | ·16 |
| 5 | S | 10 | ·34 | 6 | O | 9 | ·87 | 1 | N | 8 | ·40 | 2 | D | 7○ | ·93 | **61** | 4 | Jr | 6 | ·46 | | 5 | F | 4 | ·99 | 7 | Mr | 6 | ·59 |
| 2 | Au | 30 | ·71 | 4 | S | 29 | ·24 | 5 | O | 28 | ·77 | 7 | N | 27○ | ·30 | | 1 | D | 26 | ·83 | **62** | 3 | Jr | 25 | ·36 | 4 | F | 23 | ·89 |
| 1 | S | 18 | ·60 | 3 | O | 18 | ·13 | 4 | N | 16○ | ·67 | 6 | D | 16 | ·20 | **63** | 7 | Jr | 14 | ·73 | | 2 | F | 13 | ·26 | 3 | Mr | 14 | ·79 |
| 5 | S | 7 | ·98 | 7 | O | 7 | ·50 | 2 | N | 6 | ·03 | 3 | D | 5 | ·56 | **64** | 5 | Jr | 4 | ·09 | | 6 | F | 2 | ·62 | 1 | Mr | 3 | ·15 |
| 2 | Au | 27 | ·34 | 4 | S | 25○ | ·87 | 6 | O | 25 | ·40 | 7 | N | 23 | ·93 | | 2 | D | 23 | ·46 | **65** | 3 | Jr | 21 | ·99 | 5 | F | 20 | ·52 |
| 2 | S | 15○ | ·24 | 3 | O | 14 | ·77 | 5 | N | 13 | ·30 | 6 | D | 12 | ·83 | **66** | 1 | Jr | 11 | ·36 | | 2 | F | 9 | ·89 | 4 | Mr | 11○ | ·42 |

## TABLE X—

| | ⊙'s Anom col. 6 | ☾'s Anom col. 7 |
|---|---|---|
| Vaisakha | | |
| Jyeshta | + 29·53059 | + 1·976 |
| Ashada | + 59·06117 | + 3·952 |
| Sravana | + 88·59176 | + 5·928 |
| Bhadrapada | + 118·12235 | + 7·904 |

| Kaliyuga. | Vikrama Era. | Saka Era | Commencement of Solar Year. | | First New-Moon in Solar Year. | Anomaly of first New-Moon. | A.D. | Week-day of 1st January. | Vaisakha | | | | Jyeshta | | | | Ashada | | | | Sravana | | | | Bhadrapada | | | |
|---|---|---|---|---|---|---|---|---|---|---|---|---|---|---|---|---|---|---|---|---|---|---|---|---|---|---|---|---|
| | | | Month and day A.D. | Fraction of day | | | | | Week-day | Month | Day | Fraction | Week-day | Month | Day | Fraction | Week-day | Month | Day | Fraction | Week-day | Month | Day | Fraction | Week-day | Month | Day | Fraction |
| 3767 | 723 | 588 | 20M | ·3021 | 20·64934 | 15·871 | **666** | 5 | 5 | Ap | 9 | ·95 | 7 | My | 9 | ·48 | 2 | Je | 8 | ·01 | 3 | Jl | 7 | ·54 | 5 | Au | 6 | ·07 |
| 3768 | 724 | 589 | 20M | ·5608 | 9·75779 | 12·029 | **667** | 6 | 3 | Mr | 30 | ·32 | 4 | Ap | 28 | ·85 | 6 | My | 28 | ·38 | 7 | Je | 26 | ·91 | 2 | Jl | 26 | ·44 |
| 3769 | 725 | 590 | 19M | ·8194 | 28·39681 | 10·163 | **668** | 7 | 2 | Ap | 17 | ·21 | 3 | My | 16 | ·75 | 5 | Je | 15 | ·28 | 6 | Jl | 14○ | ·81 | 1 | Au | 13 | ·34 |
| 3770 | 726 | 591 | 20M | ·0781 | 17·50524 | 6·320 | **669** | 2 | 6 | Ap | 6 | ·58 | 1 | My | 6 | ·11 | 2 | Je | 4 | ·64 | 4 | Jl | 4○ | ·17 | 5 | Au | 2 | ·70 |
| 3771 | 727 | 592 | 20M | ·3368 | 6·61368 | 2·478 | **670** | 3 | 3 | Mr | 26 | ·95 | 5 | Ap | 5 | ·48 | 7 | My | 25 | ·01 | {1 | Je | 23●○ | ·54 | 4 | Au | 21 | ·60 |
| | | | | | | | | | | | | | | | | | | | | | {3 | Jl | 23 | ·07 | | | | |
| 3772 | 728 | 593 | 20M | ·5955 | 25·25280 | 0·612 | **671** | 4 | 2 | Ap | 14 | ·85 | 4 | My | 14 | ·38 | 5 | Je | 12 | ·91 | 7 | Jl | 12 | ·44 | 1 | Au | 10 | ·97 |
| 3773 | 729 | 594 | 19M | ·8542 | 14·36113 | 24·324 | **672** | 5 | 7 | Ap | 3 | ·21 | 1 | My | 2○ | ·74 | 3 | Je | 1● | ·27 | 4 | Je | 30 | ·81 | 6 | Jl | 30 | ·34 |
| 3774 | 730 | 595 | 20M | ·1128 | 3·46957 | 20·483 | **673** | 7 | 4 | Mr | 23 | ·58 | 6 | Ap | 22○ | ·11 | {7 | My | 21 | ·64 | 3 | Jl | 19 | ·70 | 5 | Au | 18 | ·22 |
| | | | | | | | | | | | | | | | | | {2 | Je | 20 | ·17 | | | | | | | | |
| 3775 | 731 | 596 | 20M | ·3715 | 22·10860 | 18·616 | **674** | 1 | 3 | Ap | 11● | ·48 | 5 | My | 11 | ·01 | 6 | Je | 9 | ·54 | 1 | Jl | 9 | ·07 | 2 | Au | 7 | ·60 |
| 3776 | 732 | 597 | 20M | ·6302 | 11·21703 | 14·773 | **675** | 2 | 7 | Mr | 31 | ·85 | 2 | Ap | 30 | ·38 | 3 | My | 29 | ·91 | 5 | Je | 28 | ·44 | 6 | Jl | 27 | ·97 |
| 3777 | 733 | 598 | 19M | ·8889 | 0·32546 | 10·931 | **676** | 3 | {5 | Mr | 20 | ·21 | 1 | My | 18 | ·27 | 2 | Je | 16 | ·80 | 4 | Jl | 16 | ·33 | 6 | Au | 14○ | ·87 |
| | | | | | | | | | {6 | Ap | 18 | ·74 | | | | | | | | | | | | | | | | |
| 3778 | 734 | 599 | 20M | ·1476 | 18·96449 | 9·065 | **677** | 5 | 4 | Ap | 8 | ·11 | 5 | My | 7 | ·64 | 7 | Je | 6 | ·17 | 1 | Jl | 5 | ·70 | 3 | Au | 4○ | ·23 |
| 3779 | 735 | 600 | 20M | ·4062 | 8·07292 | 5·223 | **678** | 6 | 1 | Mr | 28 | ·48 | 3 | Ap | 27 | ·01 | 4 | My | 26 | ·54 | 6 | Je | 25 | ·07 | {7 | Jl | 24● | ·60 |
| | | | | | | | | | | | | | | | | | | | | | | | | | {2 | Au | 23 | ·13 |
| 3780 | 736 | 601 | 20M | ·6649 | 26·71195 | 3·356 | **679** | 7 | 7 | Ap | 16 | ·37 | 1 | My | 15 | ·91 | 3 | Je | 14○ | ·44 | 4 | Jl | 13● | ·97 | 6 | Au | 12 | ·50 |
| 3781 | 737 | 602 | 19M | ·9236 | 15·82039 | 27·069 | **680** | 1 | 4 | Ap | 4 | ·74 | 6 | My | 4 | ·27 | 7 | Je | 2○ | ·80 | 2 | Jl | 2 | ·33 | 3 | Jl | 31 | ·86 |
| 3782 | 738 | 603 | 20M | ·1823 | 4·92882 | 23·226 | **681** | 3 | 2 | Mr | 25 | ·11 | 3 | Ap | 23 | ·64 | {5 | My | 23● | ·17 | 1 | Jl | 21 | ·23 | 2 | Au | 19 | ·76 |
| | | | | | | | | | | | | | | | | | {6 | Je | 21 | ·70 | | | | | | | | |
| 3783 | 739 | 604 | 20M | ·4410 | 23·56784 | 21·360 | **682** | 4 | 1 | Ap | 13 | ·01 | 2 | My | 12● | ·54 | 4 | Je | 11 | ·07 | 5 | Jl | 10 | ·60 | 7 | Au | 9 | ·13 |
| 3784 | 740 | 605 | 20M | ·6996 | 12·67628 | 17·518 | **683** | 5 | 5 | Ap | 2○ | ·37 | 6 | My | 1 | ·90 | 1 | My | 31 | ·43 | 2 | Je | 29 | ·97 | 4 | Jl | 29 | ·50 |
| 3785 | 741 | 606 | 19M | ·9583 | 1·78471 | 13·676 | **684** | 6 | 2 | Mr | 21○ | ·74 | {4 | Ap | 20 | ·27 | 7 | Je | 18 | ·33 | 1 | Jl | 17 | ·86 | 3 | Au | 16 | ·39 |
| | | | | | | | | | | | | | {5 | My | 19 | ·80 | | | | | | | | | | | | |
| 3786 | 742 | 607 | 20M | ·2170 | 20·42374 | 11·809 | **685** | 1 | 1 | Ap | 9 | ·64 | 3 | My | 9 | ·17 | 4 | Je | 7 | ·70 | 6 | Jl | 7 | ·23 | 7 | Au | 5 | ·76 |
| 3787 | 743 | 608 | 20M | ·4757 | 9·53218 | 7·967 | **686** | 2 | 6 | Mr | 30 | ·01 | 7 | Ap | 28 | ·54 | 2 | My | 28 | ·07 | 3 | Je | 26 | ·60 | 5 | Jl | 26○ | ·13 |
| 3788 | 744 | 609 | 20M | ·7344 | 28·17119 | 6·101 | **687** | 3 | 4 | Ap | 17 | ·90 | 6 | My | 17 | ·43 | 7 | Je | 15 | ·96 | 2 | Jl | 15○ | ·49 | 4 | Au | 14 | ·03 |
| 3789 | 745 | 610 | 19M | ·9930 | 17·27963 | 2·259 | **688** | 4 | 2 | Ap | 6 | ·27 | 3 | My | 5 | ·80 | 5 | Je | 4 | ·33 | 6 | Jl | 3●○ | ·86 | 1 | Au | 2 | ·39 |
| 3790 | 746 | 611 | 20M | ·2517 | 6·38807 | 25·971 | **689** | 6 | 6 | Mr | 26 | ·64 | 1 | Ap | 25 | ·17 | 2 | My | 24 | ·70 | {4 | Je | 23 | ·23 | 7 | Au | 21 | ·29 |
| | | | | | | | | | | | | | | | | | | | | | {5 | Jl | 22 | ·76 | | | | |
| 3791 | 747 | 612 | 20M | ·5104 | 25·02709 | 24·105 | **690** | 7 | 5 | Ap | 14 | ·53 | 7 | My | 14○ | ·07 | 1 | Je | 12 | ·60 | 3 | Jl | 12 | ·13 | 4 | Au | 10 | ·66 |
| 3792 | 748 | 613 | 20M | ·7691 | 14·13553 | 20·262 | **691** | 1 | 2 | Ap | 3 | ·90 | 4 | My | 3○ | ·43 | 5 | Je | 1 | ·96 | 7 | Jl | 1 | ·49 | 2 | Jl | 31 | ·02 |
| 3793 | 749 | 614 | 20M | ·0278 | 3·24397 | 16·420 | **692** | 2 | 7 | Mr | 23 | ·27 | 1 | Ap | 21● | ·80 | {3 | My | 21 | ·33 | 6 | Jl | 19 | ·39 | 7 | Au | 17 | ·92 |
| | | | | | | | | | | | | | | | | | {4 | Je | 19 | ·86 | | | | | | | | |
| 3794 | 750 | 615 | 20M | ·2864 | 21·88298 | 14·554 | **693** | 4 | 6 | Ap | 11● | ·17 | 7 | My | 10 | ·70 | 2 | Je | 9 | ·23 | 3 | Jl | 8 | ·76 | 5 | Au | 7 | ·29 |
| 3795 | 751 | 616 | 20M | ·5451 | 10·99142 | 10·712 | **694** | 5 | 3 | Mr | 31 | ·53 | 5 | Ap | 30 | ·06 | 6 | My | 29 | ·59 | 1 | Je | 28 | ·13 | 2 | Jl | 27 | ·66 |
| 3796 | 752 | 617 | 20M | ·8038 | 0·09985 | 6·869 | **695** | 6 | 2 | Ap | 19 | ·43 | 3 | My | 18 | ·96 | 5 | Je | 17 | ·49 | 7 | Jl | 17 | ·02 | 1 | Au | 15○ | ·55 |
| 3797 | 753 | 618 | 20M | ·0625 | 18·73888 | 5·003 | **696** | 7 | 6 | Ap | 7 | ·80 | 1 | My | 7 | ·33 | 2 | Je | 5 | ·86 | 4 | Jl | 5 | ·39 | 5 | Au | 3 | ·92 |
| 3798 | 754 | 619 | 20M | ·3212 | 7·84732 | 1·161 | **697** | 2 | 4 | Mr | 28 | ·17 | 5 | Ap | 26 | ·70 | 7 | My | 26 | ·23 | {1 | Je | 24○ | ·76 | 4 | Au | 22 | ·82 |
| | | | | | | | | | | | | | | | | | | | | | {3 | Jl | 24 | ·29 | | | | |
| 3799 | 755 | 620 | 20M | ·5798 | 26·48634 | 26·849 | **698** | 3 | 3 | Ap | 16 | ·06 | 4 | My | 15 | ·59 | 6 | Je | 14○ | ·12 | 7 | Jl | 13 | ·65 | 2 | Au | 12 | ·19 |
| 3800 | 756 | 621 | 20M | ·8385 | 15·59477 | 23·007 | **699** | 4 | 7 | Ap | 5 | ·43 | 1 | My | 4 | ·96 | 3 | Je | 3○ | ·49 | 5 | Jl | 3 | ·02 | 6 | Au | 1 | ·55 |
| 3801 | 757 | 622 | 20M | ·0972 | 4·70321 | 19·164 | **700** | 5 | 4 | Mr | 24 | ·80 | 6 | Ap | 23 | ·33 | {7 | My | 22● | ·86 | 3 | Jl | 20 | ·92 | 5 | Au | 19 | ·45 |
| | | | | | | | | | | | | | | | | | {2 | Je | 21 | ·39 | | | | | | | | |
| 3802 | 758 | 623 | 20M | ·3559 | 23·34224 | 17·288 | **701** | 1 | 3 | Ap | 12○ | ·70 | 5 | My | 12 | ·23 | 6 | Je | 10 | ·76 | 1 | Jl | 10 | ·29 | 2 | Au | 8 | ·82 |
| 3803 | 759 | 624 | 20M | ·6146 | 12·45066 | 13·456 | **702** | 1 | 1 | Ap | 2●○ | ·07 | 2 | My | 1 | ·60 | 4 | My | 31 | ·13 | 5 | Je | 29 | ·66 | 7 | Jl | 29 | ·19 |
| 3804 | 760 | 625 | 20M | ·8733 | 1·55910 | 9·614 | **703** | 2 | 5 | Mr | 22● | ·43 | {6 | Ap | 20 | ·97 | 3 | Je | 19 | ·03 | 4 | Jl | 18 | ·56 | 6 | Au | 17 | ·09 |
| | | | | | | | | | | | | | {1 | My | 20 | ·50 | | | | | | | | | | | | |

**N.B.**—For Surya Siddhanta figures for **A.D. 500** to **A.D. 999** see pages **76—79** below.

**Arya Siddhanta.**

| Asvina | | | | Kartika | | | | Margasira | | | | Pausha | | | | A.D. | Magha | | | | A.D. | Phalguna | | | | Chaitra | | | |
|---|---|---|---|---|---|---|---|---|---|---|---|---|---|---|---|---|---|---|---|---|---|---|---|---|---|---|---|---|---|
| + 147·65293 | | | | + 177·18353 | | | | + 206·71411 | | | | + 236·24470 | | | | | + 265·77529 | | | | | + 295·30588 | | | | +324·83647 | | | |
| | | | | | | | | | | | | | | | | | | | | | | | | | | +354·36705 | | | |
| + 9·880 | | | | + 11·856 | | | | + 13·832 | | | | + 15·808 | | | | | + 17·784 | | | | | + 19·760 | | | | + 21·736 | | | |
| | | | | | | | | | | | | | | | | | | | | | | | | | | + 23·712 | | | |
| Week-day | Month | Day | Fraction | Week-day | Month | Day | Fraction | Week-day | Month | Day | Fraction | Week-day | Month | Day | Fraction | | Week-day | Month | Day | Fraction | | Week-day | Month | Day | Fraction | Week-day | Month | Day | Fraction |
| 6 | S | 40○ | ·60 | 1 | O | 4 | ·13 | 2 | N | 2 | ·66 | 4 | D | 2 | ·19 | | 5 | D | 31 | ·73 | **67** | 7 | Jr | 30 | ·26 | 1 | F | 28 | ·79 |
| 4 | Au | 24● | ·97 | {5 | S | 23 | ·50 | 1 | N | 21 | ·56 | 3 | D | 21 | ·09 | **68** | 4 | Jr | 49○ | ·62 | | 6 | F | 18 | ·15 | 7 | Mr | 18 | ·68 |
| | | | | {7 | O | 23 | ·03 | | | | | | | | | | | | | | | | | | | | | | |
| 2 | S | 11 | ·87 | 4 | O | 11 | ·40 | 6 | N | 9 | ·93 | 7 | D | 9 | ·46 | **69** | 2 | Jr | 8○ | ·09 | | 3 | F | 6 | ·52 | 5 | Mr | 8 | ·05 |
| 7 | S | 1 | ·23 | 1 | S | 30 | ·77 | 3 | O | 30 | ·30 | 4 | N | 28 | ·83 | | 6 | D | 28○ | ·36 | **70** | 7 | Jr | 26 | ·89 | 2 | F | 25 | ·42 |
| 6 | S | 20 | ·13 | 7 | O | 19 | ·66 | 2 | N | 18 | ·19 | 3 | D | 17● | ·72 | **71** | 5 | Jr | 16 | ·25 | | 6 | F | 14 | ·79 | 1 | Mr | 16 | ·32 |
| 3 | S | 9 | ·50 | 5 | O | 9 | ·03 | 6 | N | 7○ | ·56 | 1 | D | 7● | ·09 | **72** | 2 | Jr | 5 | ·62 | | 4 | F | 4 | ·15 | 5 | Mr | 4 | ·68 |
| 7 | Au | 28 | ·87 | 2 | S | 27 | ·40 | 3 | O | 26○ | ·93 | 5 | N | 25● | ·46 | | 6 | D | 24 | ·99 | **73** | 1 | Jr | 23 | ·52 | 3 | F | 22 | ·05 |
| 6 | S | 16 | ·76 | 1 | O | 16○ | ·29 | 2 | N | 14 | ·83 | 4 | D | 14 | ·36 | **74** | 5 | Jr | 12 | ·89 | | 7 | F | 11 | ·42 | 1 | Mr | 12 | ·95 |
| 4 | S | 6 | ·13 | 5 | O | 5● | ·66 | 7 | N | 4 | ·19 | 1 | D | 3 | ·72 | **75** | 3 | Jr | 2 | ·25 | | 4 | Jr | 31 | ·78 | 6 | Mr | 2○ | ·31 |
| 1 | Au | 26○ | ·50 | 3 | S | 25 | ·03 | 4 | O | 24 | ·56 | 6 | N | 23 | ·09 | | 7 | D | 22 | ·62 | **76** | 2 | Jr | 21 | ·15 | 3 | F | 19○ | ·68 |
| 7 | S | 13 | ·40 | 1 | O | 12 | ·93 | 3 | N | 11 | ·46 | 4 | D | 10 | ·99 | **77** | 6 | Jr | 9 | ·52 | | 1 | F | 8○ | ·05 | 2 | Mr | 9 | ·58 |
| 4 | S | 2 | ·76 | 6 | O | 2 | ·29 | 7 | O | 31 | ·82 | 2 | N | 30 | ·35 | | 3 | D | 29 | ·89 | **78** | 5 | Jr | 28● | ·42 | 6 | F | 26 | ·95 |
| 3 | S | 21 | ·66 | 5 | O | 21 | ·19 | 6 | N | 19 | ·72 | 1 | D | 19○ | ·25 | **79** | 2 | Jr | 17 | ·78 | | 4 | F | 16 | ·31 | 5 | Mr | 17 | ·85 |
| 1 | S | 11 | ·03 | 2 | O | 10 | ·56 | 4 | N | 9 | ·09 | 5 | D | 8○ | ·62 | **80** | 7 | Jr | 7 | ·15 | | 1 | F | 5 | ·68 | 3 | Mr | 6 | ·21 |
| 5 | Au | 30 | ·39 | 6 | S | 28 | ·93 | 1 | O | 28 | ·46 | 2 | N | 26●○ | ·99 | | 4 | D | 26 | ·52 | **81** | 6 | Jr | 25 | ·05 | 7 | F | 23 | ·58 |
| 4 | S | 18 | ·29 | 5 | O | 17 | ·82 | 7 | N | 16● | ·35 | 1 | D | 15 | ·88 | **82** | 3 | Jr | 14 | ·41 | | 4 | F | 12 | ·95 | 6 | Mr | 14 | ·48 |
| 1 | S | 7 | ·66 | 3 | O | 7○ | ·19 | 4 | N | 5● | ·72 | 6 | D | 5 | ·25 | **83** | 7 | Jr | 3 | ·78 | | 2 | F | 2 | ·31 | 3 | Mr | 3 | ·84 |
| 6 | Au | 28 | ·03 | 7 | S | 26○ | ·56 | 2 | O | 26 | ·09 | 3 | N | 24 | ·62 | | 5 | D | 24 | ·15 | **84** | 6 | Jr | 22 | ·68 | 1 | F | 21 | ·21 |
| 4 | S | 14○ | ·92 | 6 | O | 14 | ·45 | 7 | N | 12 | ·99 | 2 | D | 12 | ·52 | **85** | 4 | Jr | 11 | ·05 | | 5 | F | 9 | ·58 | 7 | Mr | 11 | ·11 |
| 2 | S | 4 | ·29 | 3 | O | 3 | ·82 | 5 | N | 2 | ·35 | 6 | D | 1 | ·88 | | 1 | D | 31 | ·41 | **86** | 2 | Jr | 29○ | ·94 | 4 | F | 28● | ·47 |
| 6 | Au | 24 | ·68 | {1 | S | 23 | ·19 | 4 | N | 21 | ·25 | 5 | D | 20 | ·78 | **87** | 7 | Jr | 19○ | ·31 | | 1 | F | 17 | ·84 | 3 | Mr | 19 | ·37 |
| | | | | {2 | O | 22 | ·72 | | | | | | | | | | | | | | | | | | | | | | |
| 5 | S | 12 | ·56 | 7 | O | 12 | ·09 | 1 | N | 10 | .62 | 3 | D | 10 | ·15 | **88** | 4 | Jr | 8○ | ·68 | | 6 | F | 7 | ·21 | 7 | Mr | 7 | ·74 |
| 2 | Au | 31 | ·92 | 4 | S | 30 | ·45 | 5 | O | 29 | ·98 | 7 | N | 28 | ·51 | | 2 | D | 28 | ·05 | **89** | 3 | Jr | 26 | ·52 | 5 | F | 25 | ·11 |
| 1 | S | 19 | ·82 | 3 | O | 19 | ·35 | 4 | N | 17○ | ·88 | 6 | D | 17 | ·41 | **90** | 7 | Jr | 15 | ·94 | | 2 | F | 14 | ·47 | 4 | Mr | 16 | ·00 |
| 6 | S | 9 | ·19 | 7 | O | 8 | ·72 | 2 | N | 7○ | ·75 | 3 | D | 6 | ·78 | **91** | 5 | Jr | 5 | ·31 | | 6 | F | 3 | ·84 | 1 | Mr | 5 | ·37 |
| 3 | Au | 29 | ·55 | 5 | S | 28 | ·09 | 6 | O | 27○ | ·62 | 1 | N | 26 | ·15 | | 2 | D | 25 | ·68 | **92** | 4 | Jr | 24 | ·21 | 5 | F | 22 | ·74 |
| 2 | S | 16 | ·45 | 4 | O | 15 | ·98 | 5 | N | 14 | ·51 | 7 | D | 14 | ·04 | **93** | 1 | Jr | 12 | ·57 | | 3 | F | 11 | ·11 | 4 | Mr | 12○ | ·64 |
| 6 | S | 5○ | ·82 | 1 | O | 5● | ·35 | 2 | N | 3 | ·88 | 4 | D | 3 | ·41 | **94** | 5 | Jr | 1 | ·94 | | 7 | Jr | 31 | ·47 | 2 | Mr | 2○ | ·00 |
| 4 | Au | 26○ | ·19 | 5 | S | 24 | ·72 | 7 | O | 24 | ·25 | 1 | N | 22 | ·78 | | 3 | D | 22 | ·31 | **95** | 4 | Jr | 20 | ·84 | {6 | F | 19●○ | ·37 |
| | | | | | | | | | | | | | | | | | | | | | | | | | | {7 | Mr | 20 | ·90 |
| 3 | S | 14 | ·80 | 4 | O | 13 | ·61 | 6 | N | 12 | ·15 | 7 | D | 11 | ·68 | **96** | 2 | Jr | 10 | ·21 | | 3 | F | 8 | ·74 | 5 | Mr | 9 | ·27 |
| 7 | S | 2 | ·45 | 1 | O | 1 | ·98 | 3 | O | 31 | ·51 | 5 | N | 30 | ·04 | | 6 | D | 29○ | ·57 | **97** | 1 | Jr | 28● | ·10 | 2 | F | 26 | ·63 |
| 6 | S | 21 | ·35 | 7 | O | 20 | ·88 | 2 | N | 19 | ·41 | 3 | D | 18○ | ·94 | **98** | 5 | Jr | 17 | ·47 | | 7 | F | 16 | ·00 | 1 | Mr | 17 | ·53 |
| 3 | S | 10 | ·72 | 5 | O | 10 | ·25 | 6 | N | 8 | ·78 | 1 | D | 8●○ | ·31 | **99** | 2 | Jr | 6 | ·84 | | 4 | F | 5 | ·37 | 5 | Mr | 6 | ·90 |
| 1 | Au | 31 | ·08 | 2 | S | 29 | ·61 | 4 | O | 29 | ·14 | 5 | N | 27● | ·67 | | 7 | D | 27 | ·21 | **700** | 1 | Jr | 25 | ·74 | 3 | F | 24 | ·27 |
| 6 | S | 17 | ·98 | 1 | O | 17○ | ·51 | 3 | N | 16 | ·04 | 4 | D | 15 | ·57 | **01** | 6 | Jr | 14 | ·11 | | 7 | F | 12 | ·64 | 2 | Mr | 14 | ·17 |
| 4 | S | 7 | ·35 | 5 | O | 6○ | ·88 | 7 | N | 5 | ·41 | 1 | D | 4 | ·94 | **02** | 3 | Jr | 3 | ·47 | | 5 | F | 2 | ·01 | 6 | Mr | 3 | ·54 |
| 1 | Au | 27 | ·72 | 3 | S | 26●○ | ·25 | 4 | O | 25 | ·78 | 6 | N | 24 | ·31 | | 7 | D | 23 | ·84 | **03** | 2 | Jr | 22 | ·37 | 3 | F | 20 | ·90 |
| 7 | S | 15 | ·62 | 2 | O | 15 | ·15 | 3 | N | 13 | ·68 | 5 | D | 13 | ·21 | **04** | 6 | Jr | 11 | ·84 | | 1 | F | 10○ | ·27 | 2 | Mr | 10 | ·80 |

| Kaliyuga. | Vikrama Era. | Saka Era. | Commencement of Solar Year. Month and day A.D. | Fraction of day. | First New-Moon in Solar Year | Anomaly of first New-Moon. | A.D. | Week-day of 1st January. | Vaisakha (☉'s Anom col. 6, ☾'s Anom col. 7) Week-day | Month | Day | Fraction | Jyeshta (+ 29·53059, + 1·976) Week-day | Month | Day | Fraction | Ashada (+ 59·06117, + 3·952) Week-day | Month | Day | Fraction | Sravana (+ 88·59176, + 5·928) Week-day | Month | Day | Fraction | Bhadrapada (+ 118·12235, + 7·904) Week-day | Month | Day | Fraction |
|---|---|---|---|---|---|---|---|---|---|---|---|---|---|---|---|---|---|---|---|---|---|---|---|---|---|---|---|---|
| 3805 | 761 | 626 | 20M | ·1319 | 20·19813 | 7·747 | **704** | 3 | 4 | Ap | 9 | ·33 | 5 | My | 8 | ·86 | 7 | Je | 7 | ·89 | 1 | Jl | 6 | ·92 | 3 | Au | 5○ | ·45 |
| 3806 | 762 | 627 | 20M | ·3906 | 9·30656 | 3·905 | **705** | 5 | 1 | Mr | 29 | ·70 | 3 | Ap | 28 | ·23 | 4 | My | 27 | ·76 | 6 | Je | 26 | ·29 | 7 | Jl | 25●○ | ·82 |
| 3807 | 763 | 628 | 20M | ·6493 | 27·94559 | 2·039 | **706** | 6 | 7 | Ap | 17 | ·60 | 2 | My | 17 | ·13 | 3 | Je | 15 | ·66 | 5 | Jl | 15○ | ·19 | 6 | Au | 13 | ·72 |
| 3808 | 764 | 629 | 20M | ·9080 | 17·05403 | 25·751 | **707** | 7 | 4 | Ap | 6 | ·96 | 6 | My | 6 | ·49 | 1 | Je | 5 | ·03 | 3 | Jl | 4● | ·56 | 4 | Au | 3 | ·09 |
| 3809 | 765 | 630 | 20M | ·1667 | 6·16245 | 21·909 | [illegible] | 1 | 2 | Mr | 26 | ·33 | 3 | Ap | 24 | ·86 | 5 | My | 24○ | ·39 | { 6 | Je | 22 | ·92 | 2 | Au | 20 | ·98 |
| | | | | | | | | | | | | | | | | | | | | | { 1 | Jl | 22 | ·45 | | | | |
| 3810 | 766 | 631 | 20M | ·4253 | 24·80148 | 20·04 | [illegible] | 3 | 1 | Ap | 14 | ·23 | 2 | My | 13●○ | ·76 | 4 | Je | 12 | ·29 | 5 | Jl | 11 | ·82 | 7 | Au | 10 | ·35 |
| 3811 | 767 | 632 | 20M | ·6840 | 13·90992 | 16·20[illegible] | [illegible] | 4 | 5 | Ap | 3 | ·60 | 7 | My | 3○ | ·13 | 1 | Je | 1 | ·66 | 3 | Jl | 1 | ·19 | 4 | Jl | 30 | ·74 |
| 3812 | 768 | 633 | 20M | ·9427 | 3·019835 | 12·358 | **711** | 5 | 2 | Mr | 23○ | ·96 | 4 | Ap | 22 | ·49 | { 6 | My | 22 | ·02 | 2 | Jl | 20 | ·09 | 3 | Au | 18 | ·62 |
| | | | | | | | | | | | | | | | | | { 7 | Je | 20 | ·55 | | | | | | | | |
| 3813 | 769 | 634 | 20M | ·2014 | 21·65738 | 10·492 | **712** | 6 | 1 | Ap | 10 | ·86 | 3 | My | 10 | ·39 | 4 | Je | 8 | ·92 | 6 | Jl | 8 | ·45 | 7 | Au | 6 | ·98 |
| 3814 | 770 | 635 | 20M | ·4601 | 10·76581 | 6·650 | **713** | 1 | 6 | Mr | 31 | ·23 | 7 | Ap | 29 | ·76 | 2 | My | 29 | ·29 | 3 | Je | 27 | ·82 | 5 | Jl | 27 | ·35 |
| 3815 | 771 | 636 | 20M | ·7187 | 29·40483 | 4·784 | **714** | 2 | 5 | Ap | 19 | ·13 | 6 | My | 18 | ·66 | 1 | Je | 17 | ·19 | 2 | Jl | 16 | ·72 | 4 | Au | 15● | ·25 |
| 3816 | 772 | 637 | 20M | ·9774 | 18·51327 | 0·941 | **715** | 3 | 2 | Ap | 8 | ·49 | 4 | My | 8 | ·02 | 5 | Je | 6 | ·55 | 7 | Jl | 6○ | ·08 | 1 | Au | 4● | ·61 |
| 3817 | 773 | 638 | 20M | ·2361 | 7·62171 | 24·653 | **716** | 4 | 6 | Mr | 27 | ·86 | 1 | Ap | 26 | ·39 | 2 | My | 25 | ·92 | { 4 | Je | 24○ | ·45 | 7 | Au | 22 | ·51 |
| | | | | | | | | | | | | | | | | | | | | | { 5 | Jl | 23● | ·98 | | | | |
| 3818 | 774 | 639 | 20M | ·4948 | 26·26073 | 22·788 | **717** | 6 | 5 | Ap | 15 | ·76 | 7 | My | 15 | ·28 | 1 | Je | 13○ | ·82 | 3 | Jl | 13 | ·35 | 4 | Au | 11 | ·88 |
| 3819 | 775 | 640 | 20M | ·7535 | 15·36907 | 18·945 | **718** | 7 | 3 | Ap | 5 | ·12 | 4 | My | 4 | ·65 | 6 | Je | 3 | ·19 | 7 | Jl | 2 | ·72 | 2 | Au | 1 | ·25 |
| 3820 | 776 | 641 | 21M | ·0121 | 4·47760 | 15·103 | **719** | 1 | 7 | Mr | 25 | ·49 | 2 | Ap | 24○ | ·02 | { 3 | My | 23● | ·55 | 6 | Jl | 21 | ·61 | 1 | Au | 20 | ·14 |
| | | | | | | | | | | | | | | | | | { 5 | Je | 22 | ·08 | | | | | | | | |
| 3821 | 777 | 642 | 20M | ·2708 | 23·11662 | 13·237 | **720** | 2 | 6 | Ap | 12○ | ·39 | 7 | My | 11 | ·92 | 2 | Je | 10 | ·45 | 3 | Jl | 9 | ·98 | 5 | Au | 8 | ·51 |
| 3822 | 778 | 643 | 20M | ·5295 | 12·22506 | 9·394 | **721** | 4 | 3 | Ap | 1 | ·76 | 5 | My | 1 | ·29 | 6 | My | 30 | ·82 | 1 | Je | 29 | ·35 | 2 | Jl | 28 | ·88 |
| 3823 | 779 | 644 | 20M | ·7882 | 1·33350 | 5·552 | **722** | 5 | { 1 | Mr | 22 | ·12 | 4 | My | 20 | ·18 | 5 | Je | 18 | ·71 | 7 | Jl | 18 | ·25 | 1 | Au | 16○ | ·78 |
| | | | | | | | | | { 2 | Ap | 20 | ·65 | | | | | | | | | | | | | | | | |
| 3824 | 780 | 645 | 21M | ·0469 | 19·97251 | 3·686 | **723** | 6 | 7 | Ap | 10 | ·02 | 1 | My | 9 | ·55 | 3 | Je | 8 | ·08 | 4 | Jl | 7 | ·61 | 6 | Au | 6○ | ·14 |
| 3825 | 781 | 646 | 20M | ·3055 | 9·08095 | 27·398 | **724** | 7 | 4 | Mr | 29 | ·39 | 5 | Ap | 27 | ·92 | 7 | My | 27 | ·45 | 1 | Je | 25 | ·98 | { 3 | Jl | 25●○ | ·51 |
| | | | | | | | | | | | | | | | | | | | | | | | | | { 5 | Au | 24 | ·04 |
| 3826 | 782 | 647 | 20M | ·5642 | 27·71998 | 25·532 | **725** | 2 | 3 | Ap | 17 | ·29 | 4 | My | 16 | ·82 | 6 | Je | 15 | ·35 | 7 | Jl | 14● | ·88 | 2 | Au | 13 | ·41 |
| 3827 | 783 | 648 | 20M | ·8229 | 16·82841 | 21·690 | **726** | 3 | 7 | Ap | 6 | ·65 | 2 | My | 6 | ·18 | 3 | Je | 4○ | ·71 | 5 | Jl | 4● | ·24 | 6 | Au | 2 | ·77 |
| 3828 | 784 | 649 | 21M | ·0816 | 5·93685 | 17·847 | **727** | 4 | 5 | Mr | 27 | ·02 | 6 | Ap | 25 | ·55 | 1 | My | 25●○ | ·08 | { 2 | Je | 23 | ·61 | 5 | Au | 21 | ·67 |
| | | | | | | | | | | | | | | | | | | | | | { 4 | Jl | 23 | ·14 | | | | |
| 3829 | 785 | 650 | 20M | ·3493 | 24·57588 | 15·981 | **728** | 5 | 8 | Ap | 13 | ·92 | 5 | My | 13○ | ·45 | 6 | Je | 11 | ·98 | 1 | Jl | 11 | ·51 | 3 | Au | 10 | ·04 |
| 3830 | 786 | 651 | 20M | ·5220 | 13·68430 | 12·1[illegible]9 | **729** | 7 | 1 | Ap | 3○ | ·28 | 2 | My | 2 | ·81 | 4 | Je | 1 | ·35 | 5 | Je | 30 | ·88 | 7 | Jl | 30 | ·41 |
| 3831 | 787 | 652 | 20M | ·8576 | 2·79274 | 8·296 | **730** | 1 | 5 | Mr | 23○ | ·65 | { 7 | Ap | 22 | ·18 | 3 | Je | 20 | ·24 | 4 | Jl | 19 | ·77 | 6 | Au | 18 | ·30 |
| | | | | | | | | | | | | | { 1 | My | 21 | ·71 | | | | | | | | | | | | |
| 3832 | 788 | 653 | 21M | ·1163 | 21·48177 | 6·430 | **731** | 2 | 4 | Ap | 11 | ·55 | 6 | My | 11 | ·08 | 7 | Je | 9 | ·61 | 2 | Jl | 9 | ·14 | 3 | Au | 7 | ·67 |
| 3833 | 789 | 654 | 20M | ·3749 | 10·54020 | 2·588 | **732** | 3 | 1 | Mr | 30 | ·92 | 3 | Ap | 29 | ·45 | 4 | My | 28 | ·98 | 6 | Je | 27 | ·51 | 1 | Jl | 27 | ·04 |
| 3834 | 790 | 655 | 20M | ·6337 | 29·17923 | 0·722 | **733** | 5 | 7 | Ap | 18 | ·81 | 2 | My | 18 | ·34 | 3 | Je | 16 | ·87 | 5 | Jl | 16○ | ·41 | 6 | Au | 14● | ·94 |
| 3835 | 791 | 656 | 20M | ·8924 | 18·28767 | 24·434 | **734** | 6 | 5 | Ap | 8 | ·18 | 6 | My | 7 | ·71 | 1 | Je | 6 | ·24 | 2 | Jl | 5○ | ·77 | 4 | Au | 4 | ·30 |
| 3836 | 792 | 657 | 21M | ·1510 | 7·39609 | 20·592 | **735** | 7 | 2 | Mr | 28 | ·55 | 4 | Ap | 27 | ·08 | 5 | My | 26 | ·61 | { 7 | Je | 25●○ | ·14 | 3 | Au | 23 | ·20 |
| | | | | | | | | | | | | | | | | | | | | | { 1 | Jl | 24 | ·67 | | | | |
| 3837 | 793 | 658 | 20M | ·4097 | 26·03512 | 18·725 | **736** | 1 | 1 | Ap | 15 | ·45 | 2 | My | 14 | ·98 | 4 | Je | 13 | ·51 | 6 | Jl | 13 | ·04 | 7 | Au | 11 | ·57 |
| 3838 | 794 | 659 | 20M | ·6684 | 15·14356 | 14·883 | **737** | 3 | 5 | Ap | 4 | ·81 | 7 | My | 4○ | ·34 | 1 | Je | 2 | ·87 | 3 | Jl | 2 | ·40 | 4 | Jl | 31 | ·94 |
| 3839 | 795 | 660 | 20M | ·9271 | 4·25229 | 11·042 | **738** | 4 | 3 | Mr | 25 | ·18 | 4 | Ap | 23○ | ·71 | { 6 | My | 23 | ·24 | 2 | Jl | 21 | ·30 | 3 | Au | 19 | ·83 |
| | | | | | | | | | | | | | | | | | { 7 | Je | 21 | ·77 | | | | | | | | |
| 3840 | 796 | 661 | 21M | ·1858 | 22·89101 | 9·175 | **739** | 5 | 2 | Ap | 13 | ·08 | 3 | My | 12 | ·61 | 5 | Je | 11 | ·14 | 7 | Jl | 10 | ·67 | 1 | Au | 9 | ·20 |
| 3841 | 797 | 662 | 20M | ·4444 | 11·99945 | 5·333 | **740** | 6 | 6 | Ap | 1● | ·44 | 7 | Ap | 30 | ·97 | 2 | My | 30 | ·51 | 4 | Je | 29 | ·04 | 5 | Jl | 28 | ·57 |
| 3842 | 798 | 663 | 20M | ·7031 | 1·10788 | 1·490 | **741** | 1 | { 3 | Mr | 21 | ·81 | 6 | My | 19 | ·87 | 1 | Je | 18 | ·41 | 2 | Jl | 17 | ·93 | 4 | Au | 16○ | ·46 |
| | | | | | | | | | { 5 | Ap | 20 | ·34 | | | | | | | | | | | | | | | | |
| 3843 | 799 | 664 | 20M | ·9618 | 19·74691 | 27·179 | **742** | 2 | 2 | Ap | 9 | ·71 | 4 | My | 9 | ·24 | 5 | Je | 7 | ·77 | 7 | Jl | 7 | ·30 | 1 | Au | 5●○ | ·83 |

**N.B.**—For Surya Siddhanta figures for **A.D. 500** to **A.D. 999** see pages **76—79** below.

## Ārya Siddhanta.

| Asvina | | | | Kartika | | | | Margasira | | | | Pausha | | | | A.D. | Magha | | | | A.D. | Phalguna | | | | Chaitra | | | |
|---|---|---|---|---|---|---|---|---|---|---|---|---|---|---|---|---|---|---|---|---|---|---|---|---|---|---|---|---|---|
| + 147·65298 | | | | + 177·18353 | | | | + 206·71411 | | | | + 236·24470 | | | | | + 265·77529 | | | | | + 295·30588 | | | | { +324·83647<br>+354·36706 | | | |
| + 9·880 | | | | + 11·856 | | | | + 13·832 | | | | + 15·808 | | | | | + 17·784 | | | | | + 19·760 | | | | { + 21·736<br>+ 23·712 | | | |
| Week-day | Month | Day | Fraction | Week-day | Month | Day | Fraction | Week-day | Month | Day | Fraction | Week-day | Month | Day | Fraction | | Week-day | Month | Day | Fraction | | Week-day | Month | Day | Fraction | Week-day | Month | Day | Fraction |
| 5 | S | 3● | ·99 | 6 | O | 8 | ·52 | 1 | N | 2 | ·05 | 2 | D | 1 | ·58 | | 4 | D | 31 | ·11 | 05 | 5 | Jr | 29ↄ | ·64 | 7 | F | 28● | ·17 |
| {2 | Au | 24 | ·35 | 5 | O | 22 | ·42 | 6 | N | 20 | ·94 | 1 | D | 20 | ·47 | 06 | 3 | Jr | 19●○ | ·01 | | 4 | F | 17 | ·54 | 6 | Mr | 19 | ·07 |
| {3 | S | 22 | ·88 | | | | | | | | | | | | | | | | | | | | | | | | | | |
| 1 | S | 12 | ·25 | 2 | O | 11 | ·78 | 4 | N | 10 | ·31 | 5 | D | 9 | ·84 | 07 | 7 | Jr | 8 | ·37 | | 1 | F | 6 | ·90 | 3 | Mr | 8 | ·43 |
| 5 | S | 1 | ·62 | 7 | O | 1 | ·15 | 1 | O | 30 | ·68 | 3 | N | 29ↄ | ·21 | | 4 | D | 28● | ·74 | 08 | 6 | Jr | 27 | ·27 | 7 | F | 25 | ·80 |
| 4 | S | 19 | ·51 | 6 | O | 19 | ·05 | 7 | N | 17○ | ·58 | 2 | D | 17 | ·11 | 09 | 3 | Jr | 15 | ·64 | | 5 | F | 14 | ·17 | 6 | Mr | 15 | ·70 |
| 1 | S | 8 | ·88 | 3 | O | 8 | ·41 | 4 | N | 6ↄ | ·94 | 6 | D | 6 | ·47 | 10 | 1 | Jr | 5 | ·00 | | 2 | F | 3 | ·53 | 4 | Mr | 5 | ·07 |
| 6 | Au | 29 | ·25 | 7 | S | 27 | ·78 | 2 | O | 27● | ·31 | 3 | N | 25 | ·84 | | 5 | D | 25 | ·37 | 11 | 6 | Jr | 23 | ·90 | 1 | F | 22 | ·43 |
| 5 | S | 17 | ·15 | 6 | O | 16 | ·68 | 1 | N | 15 | ·21 | 2 | D | 14 | ·74 | 12 | 4 | Jr | 13 | ·27 | | 5 | F | 11 | ·80 | 7 | Mr | 12○ | ·33 |
| 2 | S | 5○ | ·52 | 4 | O | 5● | ·04 | 5 | N | 3 | ·57 | 7 | D | 3 | ·11 | 13 | 1 | Jr | 1 | ·64 | | 3 | Jr | 31 | ·17 | 4 | Mr | 1○ | ·70 |
| 6 | Au | 25○ | ·88 | 1 | S | 24 | ·41 | 2 | O | 23 | ·94 | 4 | N | 22 | ·47 | | 6 | D | 22 | ·00 | 14 | 7 | Jr | 20 | ·53 | {2 | F | 19● | ·96 |
| | | | | | | | | | | | | | | | | | | | | | | | | | | {3 | Mr | 20 | ·59 |
| 5 | S | 13 | ·78 | 7 | O | 13 | ·31 | 1 | N | 11 | ·84 | 3 | D | 11 | ·37 | 15 | 4 | Jr | 9ↄ | ·90 | | 6 | F | 8 | ·43 | 7 | Mr | 9 | ·96 |
| 3 | S | 3 | ·15 | 4 | O | 2 | ·68 | 6 | N | 1 | ·21 | 7 | N | 30 | ·74 | | 2 | D | 30ↄ | ·27 | 16 | 3 | Jr | 28 | ·80 | 5 | F | 27 | ·33 |
| 2 | S | 21 | ·04 | 3 | O | 20 | ·57 | 5 | N | 19 | ·10 | 6 | D | 18○ | ·63 | 17 | 1 | Jr | 17 | ·17 | | 2 | F | 15 | ·70 | 4 | Mr | 17 | ·23 |
| 6 | S | 10 | ·41 | 7 | O | 9 | ·94 | 2 | N | 8 | ·47 | 4 | D | 8 | ·00 | 18 | 5 | Jr | 6 | ·53 | | 7 | F | 5 | ·06 | 1 | Mr | 6 | ·59 |
| 3 | Au | 30 | ·78 | 5 | S | 29 | ·31 | 6 | O | 28ↄ | ·84 | 1 | N | 27 | ·37 | | 2 | D | 26 | ·90 | 19 | 4 | Jr | 25 | ·43 | 5 | F | 23 | ·96 |
| 3 | S | 18 | ·67 | 4 | O | 18○ | ·21 | 5 | N | 16 | ·74 | 7 | D | 16 | ·27 | 20 | 1 | Jr | 14 | ·80 | | 3 | F | 13 | ·33 | 4 | Mr | 13 | ·86 |
| 7 | S | 7 | ·04 | 1 | O | 6ↄ | ·57 | 3 | N | 5 | ·10 | 4 | D | 4 | ·63 | 21 | 6 | Jr | 3 | ·16 | | 7 | F | 1 | ·69 | 2 | Mr | 3 | ·23 |
| 4 | Au | 27 | ·41 | 5 | S | 25● | ·94 | 7 | O | 25 | ·47 | 2 | N | 24 | ·00 | | 3 | D | 23 | ·53 | 22 | 5 | Jr | 22 | ·06 | 6 | F | 20○ | ·59 |
| 3 | S | 15 | ·31 | 4 | O | 14 | ·84 | 6 | N | 13 | ·37 | 7 | D | 12 | ·90 | 23 | 2 | Jr | 11 | ·43 | | 3 | F | 9ↄ | ·96 | 5 | Mr | 11 | ·49 |
| 7 | S | 4 | ·67 | 2 | O | 4 | ·20 | 3 | N | 2 | ·73 | 5 | D | 2 | ·27 | | 6 | D | 31 | ·80 | 24 | 1 | Jr | 30ↄ | ·33 | 2 | F | 28 | ·86 |
| 6 | S | 22 | ·57 | 1 | O | 22 | ·10 | 2 | N | 20 | ·63 | 4 | D | 20 | ·16 | 25 | 5 | Jr | 18● | ·69 | | 7 | F | 17 | ·22 | 1 | Mr | 18 | ·75 |
| 3 | S | 11 | ·94 | 5 | O | 11 | ·47 | 7 | N | 10 | ·00 | 1 | D | 9ↄ | ·53 | 26 | 3 | Jr | 8● | ·06 | | 4 | F | 6 | ·59 | 6 | Mr | 8 | ·12 |
| 1 | S | 1 | ·31 | 2 | S | 30 | ·84 | 4 | O | 30 | ·37 | 5 | N | 28ↄ | ·90 | | 7 | D | 28● | ·43 | 27 | 1 | Jr | 26 | ·96 | 3 | F | 25 | ·49 |
| 7 | S | 20 | ·20 | 1 | O | 19 | ·73 | 3 | N | 18○ | ·26 | 4 | D | 17 | ·79 | 28 | 6 | Jr | 16 | ·33 | | 7 | F | 14 | ·86 | 2 | Mr | 15 | ·39 |
| 4 | S | 8 | ·57 | 6 | O | 8 | ·10 | 7 | N | 6● | ·63 | 2 | D | 6 | ·16 | 29 | 3 | Jr | 4 | ·69 | | 5 | F | 3 | ·22 | 6 | Mr | 4 | ·75 |
| 1 | Au | 28 | ·94 | 3 | S | 27 | ·47 | 5 | O | 27● | ·00 | 6 | N | 25 | ·53 | | 1 | D | 25 | ·06 | 30 | 2 | Jr | 23 | ·59 | 4 | F | 22 | ·12 |
| 7 | S | 16ↄ | ·83 | 2 | O | 16 | ·37 | 3 | N | 14 | ·90 | 5 | D | 14 | ·43 | 31 | 6 | Jr | 12 | ·96 | | 1 | F | 11 | ·49 | 3 | Mr | 13○ | ·02 |
| 5 | S | 6ↄ | ·20 | 6 | O | 5 | ·73 | 1 | N | 4 | ·26 | 2 | D | 3 | ·79 | 32 | 4 | Jr | 2 | ·32 | | 5 | Jr | 31 | ·85 | 7 | Mr | 1 | ·39 |
| 2 | Au | 25● | ·57 | 4 | S | 24 | ·10 | 5 | O | 23 | ·63 | 7 | N | 22 | ·16 | | 1 | D | 21 | ·69 | 33 | 3 | Jr | 20ↄ | ·22 | {4 | F | 18 | ·75 |
| | | | | | | | | | | | | | | | | | | | | | | | | | | {6 | Mr | 20 | ·28 |
| 1 | S | 13 | ·47 | 3 | O | 13 | ·00 | 4 | N | 11 | ·53 | 6 | D | 11 | ·06 | 34 | 7 | Jr | 9ↄ | ·59 | | 2 | F | 8 | ·12 | 3 | Mr | 9 | ·65 |
| 5 | S | 2 | ·33 | 7 | O | 2 | ·36 | 1 | O | 31 | ·89 | 3 | N | 30 | ·43 | | 4 | D | 29●ↄ | ·96 | 35 | 6 | Jr | 28 | ·49 | 1 | F | 27 | ·02 |
| 4 | S | 21 | ·73 | 6 | O | 21 | ·26 | 7 | N | 19 | ·79 | 2 | D | 19● | ·33 | 36 | 3 | Jr | 17 | ·85 | | 5 | F | 16 | ·38 | 6 | Mr | 16 | ·91 |
| 2 | S | 10 | ·10 | 3 | O | 9 | ·63 | 5 | N | 8○ | ·15 | 6 | D | 7 | ·69 | 37 | 1 | Jr | 6 | ·22 | | 2 | F | 4 | ·75 | 4 | Mr | 6 | ·28 |
| 6 | Au | 30 | ·47 | 1 | S | 29 | ·00 | 2 | O | 28●ↄ | ·53 | 4 | N | 27 | ·06 | | 5 | D | 26 | ·59 | 38 | 7 | Jr | 25 | ·12 | 1 | F | 23 | ·65 |
| 5 | S | 18 | ·36 | 6 | O | 17○ | ·89 | 1 | N | 16 | ·42 | 2 | D | 15 | ·95 | 39 | 4 | Jr | 14 | ·49 | | 6 | F | 13 | ·02 | 7 | Mr | 14 | ·55 |
| 2 | S | 7 | ·73 | 4 | O | 7 | ·26 | 5 | N | 5 | ·79 | 7 | D | 5 | ·32 | 40 | 1 | Jr | 3 | ·85 | | 3 | F | 2 | ·38 | 4 | Mr | 2ↄ | ·91 |
| 7 | Au | 27○ | ·10 | 1 | S | 25 | ·63 | 3 | O | 25 | ·16 | 4 | N | 23 | ·69 | | 6 | D | 23 | ·22 | 41 | 7 | Jr | 21 | ·75 | 2 | F | 20ↄ | ·28 |
| 5 | S | 14 | ·99 | 7 | O | 14 | ·53 | 2 | N | 13 | ·06 | 3 | D | 12 | ·59 | 42 | 5 | Jr | 11 | ·12 | | 6 | F | 9○ | ·65 | 1 | Mr | 11 | ·18 |
| 3 | S | 4 | ·86 | 4 | O | 3 | ·89 | 6 | N | 2 | ·42 | 7 | D | 1 | ·95 | | 2 | D | 31 | ·48 | 43 | 4 | Jr | 30 | ·01 | 5 | F | 28 | ·55 |

**TABLE X–**

☉'s Anom col. 6 + 29·53059 + 59·06117 + 88·59176 + 118·12235

☾'s Anom col. 7 + 1·976 + 3·952 + 5·928 + 7·904

| Kaliyuga. | Vikrama Era. | Saka Era. | Commencement of Solar Year. Month and day A.D. | Fraction of day. | First New-Moon in Solar Year. | Anomaly of first New-Moon. | A.D. | Week-day of 1st January. | Vaisakha Week-day | Month | Day | Fraction | Jyeshta Week-day | Month | Day | Fraction | Ashada Week-day | Month | Day | Fraction | Sravana Week-day | Month | Day | Fraction | Bhadrapada Week-day | Month | Day | Fraction |
|---|---|---|---|---|---|---|---|---|---|---|---|---|---|---|---|---|---|---|---|---|---|---|---|---|---|---|---|---|
| 3844 | 800 | 665 | 21M | ·2205 | 8·85525 | 23·337 | **743** | 3 | 7 | Mr | 30 | ·08 | 1 | Ap | 28 | ·61 | 3 | My | 28 | ·14 | 4 | Je | 26 | ·67 | 6 | Jl | 26 | ·20 |
| | | | | | | | | | | | | | | | | | | | | | | | | | 7 | Au | 24 | ·73 |
| 3845 | 801 | 666 | 20M | ·4792 | 27·49437 | 21·470 | **744** | 4 | 5 | Ap | 16 | ·97 | 7 | My | 16 | ·50 | 2 | Je | 15○ | ·03 | 3 | Jl | 14 | ·57 | 5 | Au | 13 | ·10 |
| 3846 | 802 | 667 | 20M | ·7378 | 16·60280 | 17·629 | **745** | 6 | 3 | Ap | 6 | ·84 | 3 | My | 5 | ·87 | 6 | Je | 4○ | ·40 | 7 | Jl | 3 | ·93 | 2 | Au | 2 | ·46 |
| 3847 | 803 | 668 | 20M | ·9995 | 5·71124 | 13·786 | **746** | 7 | 7 | Mr | 26 | ·71 | 2 | Ap | 25 | ·24 | 3 | My | 24●○ | ·77 | 5 | Je | 23 | ·30 | [illegible] | Au | 21 | ·36 |
| | | | | | | | | | | | | | | | | | | | | | 6 | Jl | 22 | ·83 | | | | |
| 3848 | 804 | 669 | 21M | ·2552 | 24·35026 | 11·920 | **747** | 1 | 6 | Ap | 14○ | ·61 | 1 | My | 14● | ·14 | 2 | Je | 12 | ·67 | 4 | Jl | 12 | ·20 | 5 | Au | 10 | ·73 |
| 3849 | 805 | 670 | 20M | ·5132 | 13·45870 | 8·077 | **748** | 2 | 3 | Ap | 2○ | ·97 | 5 | My | 2 | ·50 | 7 | Je | 1 | ·03 | 1 | Je | 30 | ·56 | 3 | Jl | 30 | ·09 |
| 3850 | 806 | 671 | 20M | ·7726 | 2·56714 | 4·235 | **749** | 4 | 1 | Mr | 23●○ | ·21 | 2 | Ap | 21 | ·17 | 5 | Je | 19 | ·93 | 7 | Jl | 19 | ·46 | 1 | Au | 17 | ·99 |
| | | | | | | | | | | | | | 4 | My | 21 | ·46 | | | | | | | | | | | | |
| 3851 | 807 | 672 | 21M | ·0312 | 21·20615 | 2·369 | **750** | 5 | 7 | Ap | 11 | ·24 | 1 | My | 10 | ·77 | 3 | Je | 9 | ·30 | 4 | Jl | 8 | ·83 | 6 | Au | 7 | ·36 |
| 3852 | 808 | 673 | 21M | ·2899 | 10·31459 | 26·081 | **751** | 6 | 4 | Mr | 31 | ·60 | 6 | Ap | 30 | ·13 | 7 | My | 29 | ·67 | 2 | Je | 28 | ·20 | 3 | Jl | 27○ | ·73 |
| 3853 | 809 | 674 | 20M | ·5486 | 28·95362 | 24·215 | **752** | 7 | 3 | Ap | 18 | ·50 | 5 | My | 18 | ·03 | 6 | Je | 16 | ·56 | 1 | Jl | 16○ | ·09 | 2 | Au | 14 | ·62 |
| 3854 | 810 | 675 | 20M | ·8073 | 18·06205 | 20·373 | **753** | 2 | 7 | Ap | 7 | ·87 | 2 | My | 7 | ·40 | 3 | Je | 5 | ·93 | 5 | Jl | 5○ | ·46 | 6 | Au | 3 | ·99 |
| 3855 | 811 | 676 | 21M | ·0660 | 7·17049 | 16·530 | **754** | 3 | 5 | Mr | 28 | ·24 | 6 | Ap | 26 | ·77 | 1 | My | 26 | ·30 | 2 | Je | 24● | ·83 | 5 | Au | 22 | ·89 |
| | | | | | | | | | | | | | | | | | | | | | 4 | Jl | 24 | ·36 | | | | |
| 3856 | 812 | 677 | 21M | ·3246 | 25·80951 | 14·664 | **755** | 4 | 4 | Ap | 16 | ·13 | 5 | My | 15○ | ·66 | 7 | Je | 14 | ·19 | 1 | Jl | 13 | ·73 | 3 | Au | 12 | ·26 |
| 3857 | 813 | 678 | 20M | ·5833 | 14·91795 | 10·822 | **756** | 5 | 1 | Ap | 4 | ·50 | 3 | My | 4○ | ·03 | 4 | Je | 2 | ·56 | 6 | Jl | 2 | ·09 | 7 | Jl | 31 | ·62 |
| 3858 | 814 | 679 | 20M | ·3420 | 4·02638 | 6·980 | **757** | 7 | 5 | Mr | 24 | ·87 | 7 | Ap | 23●○ | ·40 | 1 | My | 22 | ·93 | 4 | Jl | 20 | ·99 | 6 | Au | 19 | ·52 |
| | | | | | | | | | | | | | | | | | 3 | Je | 21 | ·44 | | | | | | | | |
| 3859 | 815 | 680 | 21M | ·1007 | 22·66534 | 5·113 | **758** | 1 | 4 | Ap | 12 | ·77 | 6 | My | 12 | ·30 | 7 | Je | 10 | ·83 | 2 | Jl | 10 | ·36 | 3 | Au | 8 | ·89 |
| 3860 | 816 | 681 | 21M | ·3594 | 11·77445 | 1·272 | **759** | 2 | 2 | Ap | 2● | ·13 | 3 | My | 1 | ·66 | 5 | My | 31 | ·19 | 6 | Je | 29 | ·72 | 1 | Jl | 29 | ·25 |
| 3861 | 817 | 682 | 20M | ·6180 | 0·88228 | 24·983 | **760** | 3 | 6 | Mr | 21 | ·50 | 2 | My | 19 | ·56 | 4 | Je | 18 | ·09 | 5 | Jl | 17 | ·62 | 7 | Au | 16○ | ·15 |
| | | | | | | | | | 1 | Ap | 20 | ·03 | | | | | | | | | | | | | | | | |
| 3862 | 818 | 683 | 20M | ·8767 | 19·52131 | 23·117 | **761** | 5 | 5 | Ap | 9 | ·40 | 6 | My | 8 | ·93 | 1 | Je | 7 | ·46 | 2 | Jl | 6 | ·99 | 4 | Au | 5● | ·52 |
| 3863 | 819 | 684 | 21M | ·1354 | 8·62974 | 19·275 | **762** | 6 | 2 | Mr | 29 | ·77 | 4 | Ap | 28 | ·30 | 5 | My | 27 | ·83 | 7 | Je | 26○ | ·36 | 1 | Jl | 25 | ·89 |
| | | | | | | | | | | | | | | | | | | | | | | | | | 3 | Au | 24 | ·42 |
| 3864 | 820 | 685 | 21M | ·3941 | 27·26876 | 17·409 | **763** | 7 | 1 | Ap | 17 | ·67 | 3 | My | 17 | ·20 | 4 | Je | 15○ | ·73 | 6 | Jl | 15 | ·26 | 7 | Au | 13 | ·79 |
| 3865 | 821 | 686 | 20M | ·6528 | 16·37720 | 13·566 | **764** | 1 | 6 | Ap | 6 | ·03 | 7 | My | 5 | ·56 | 2 | Je | 4●○ | ·09 | 3 | Jl | 3 | ·62 | 5 | Au | 2 | ·15 |
| 3866 | 822 | 687 | 20M | ·9114 | 5·48564 | 9·724 | **765** | 3 | 3 | Mr | 26 | ·40 | 4 | Ap | 24○ | ·93 | 6 | My | 24 | ·46 | 7 | Je | 22 | ·99 | 4 | Au | 21 | ·05 |
| | | | | | | | | | | | | | | | | | | | | | 2 | Jl | 22 | ·52 | | | | |
| 3867 | 823 | 688 | 21M | ·1701 | 24·12465 | 7·858 | **766** | 4 | 2 | Ap | 14○ | ·90 | 3 | My | 13 | ·83 | 5 | Je | 12 | ·36 | 6 | Jl | 11 | ·89 | 1 | Au | 10 | ·42 |
| 3868 | 824 | 689 | 21M | ·3288 | 13·23310 | 4·016 | **767** | 5 | 6 | Ap | 3●○ | ·66 | 1 | My | 3 | ·19 | 2 | Je | 1 | ·73 | 4 | Jl | 1 | ·26 | 5 | Jl | 30 | ·79 |
| 3869 | 825 | 690 | 20M | ·6875 | 2·34153 | 0·173 | **768** | 6 | 4 | Mr | 23● | ·03 | 6 | Ap | 21 | ·56 | 1 | Je | 19 | ·62 | 3 | Jl | 19 | ·15 | 4 | Au | 17 | ·68 |
| | | | | | | | | | | | | | 7 | My | 21 | ·09 | | | | | | | | | | | | |
| 3870 | 826 | 691 | 20M | ·9362 | 20·98055 | 25·862 | **769** | 1 | 2 | Ap | 10 | ·93 | 4 | My | 10 | ·46 | 5 | Je | 8 | ·99 | 7 | Jl | 8 | ·52 | 2 | Au | 7○ | ·05 |
| 3871 | 827 | 692 | 21M | ·2049 | 10·08899 | 22·019 | **770** | 2 | 7 | Mr | 31 | ·30 | 1 | Ap | 29 | ·83 | 3 | My | 29 | ·36 | 4 | Je | 27 | ·89 | 2 | Jl | 27○ | ·42 |
| 3872 | 828 | 693 | 21M | ·4635 | 28·72811 | 20·153 | **771** | 3 | 6 | Ap | 19 | ·19 | 7 | My | 18 | ·72 | 2 | Je | 17 | ·25 | 3 | Jl | 16○ | ·79 | 5 | Au | 15 | ·32 |
| 3873 | 829 | 694 | 20M | ·7222 | 17·83644 | 16·311 | **772** | 4 | 3 | Ap | 7 | ·56 | 5 | My | 7 | ·09 | 6 | Je | 5 | ·62 | 1 | Jl | 5● | ·15 | 2 | Au | 3 | ·68 |
| 3874 | 830 | 695 | 20M | ·9809 | 6·94488 | 12·489 | **773** | 6 | 7 | Mr | 27 | ·93 | 2 | Ap | 26 | ·46 | 3 | My | 25○ | ·99 | 5 | Je | 24 | ·52 | 1 | Au | 22 | ·58 |
| | | | | | | | | | | | | | | | | | | | | | 7 | Jl | 24 | ·05 | | | | |
| 3875 | 831 | 696 | 21M | ·2396 | 25·58391 | 10·602 | **774** | 7 | 6 | Ap | 15 | ·83 | 1 | My | 15○ | ·36 | 2 | Je | 13 | ·89 | 4 | Jl | 13 | ·42 | 5 | Au | 11 | ·95 |
| 3876 | 832 | 697 | 21M | ·4983 | 14·69234 | 6·760 | **775** | 1 | 4 | Ap | 5 | ·19 | 5 | My | 4●○ | ·72 | 7 | Je | 3 | ·25 | 1 | Jl | 2 | ·78 | 3 | Au | 1 | ·31 |
| 3877 | 833 | 698 | 20M | ·7569 | 3·80077 | 2·912 | **776** | 2 | 1 | Mr | 24○ | ·56 | 3 | Ap | 23 | ·09 | 4 | My | 22 | ·62 | 7 | Jl | 20 | ·68 | 2 | Au | 19 | ·21 |
| | | | | | | | | | | | | | | | | | 6 | Je | 21 | ·15 | | | | | | | | |
| 3878 | 834 | 699 | 21M | ·0156 | 22·43980 | 1·052 | **777** | 4 | 7 | Ap | 12 | ·46 | 1 | My | 11 | ·99 | 3 | Je | 10 | ·52 | 5 | Jl | 10 | ·05 | 6 | Au | 8 | ·58 |
| 3879 | 835 | 700 | 21M | ·2743 | 11·54823 | 24·764 | **778** | 5 | 4 | Ap | 1 | ·82 | 6 | My | 1 | ·35 | 7 | My | 30 | ·88 | 2 | Je | 29 | ·42 | 3 | Jl | 28 | ·95 |
| 3880 | 836 | 701 | 21M | ·5330 | 0·65667 | 20·922 | **779** | 6 | 2 | Mr | 22 | ·19 | 5 | My | 20 | ·25 | 6 | Je | 18 | ·78 | 1 | Jl | 18 | ·31 | 2 | Au | 16● | ·84 |
| | | | | | | | | | 3 | Ap | 20 | ·72 | | | | | | | | | | | | | | | | |
| 3881 | 837 | 702 | 20M | ·7917 | 19·29570 | 19·055 | **780** | 1 | 1 | Ap | 9 | ·09 | 2 | My | 8 | ·62 | 4 | Je | 7 | ·15 | 5 | Jl | 6○ | ·68 | 7 | Au | 5● | ·21 |
| 3862 | 838 | 703 | 21M | ·0503 | 8·40413 | 15·213 | **781** | 2 | 5 | Mr | 29 | ·46 | 6 | Ap | 27 | ·99 | 1 | My | 27 | ·52 | 3 | Je | 26●○ | ·05 | 4 | Jl | 25 | ·58 |
| | | | | | | | | | | | | | | | | | | | | | | | | | 6 | Au | 24 | ·11 |
| 3888 | 839 | 704 | 21M | ·3090 | 27·04315 | 13·347 | **782** | 3 | 4 | Ap | 17 | ·35 | 5 | My | 16 | ·88 | 7 | Je | 15○ | ·41 | 1 | Jl | 14 | ·95 | 3 | Au | 13 | ·48 |

**N.B.**—For Surya Siddhanta figures for **A.D. 500** to **A.D. 999** see pages **76—79** below.

**Arya Siddhanta.**

| + 147·65293 | | | | + 177·18353 | | | | + 206·71411 | | | | + 236·24470 | | | | | + 265·77529 | | | | | + 295·30588 | | | | {+324·82647 +354·36705} | | | |
|---|---|---|---|---|---|---|---|---|---|---|---|---|---|---|---|---|---|---|---|---|---|---|---|---|---|---|---|---|---|
| + 9·880 | | | | + 11·856 | | | | + 13·832 | | | | + 15·808 | | | | | + 17·784 | | | | | + 19·760 | | | | {+ 21·736 + 23·712} | | | |
| **Asvina** | | | | **Kartika** | | | | **Margasira** | | | | **Pausha** | | | | **A.D.** | **Magha** | | | | **A.D.** | **Phalguna** | | | | **Chaitra** | | | |
| Week-day | Month | Day | Fraction | Week-day | Month | Day | Fraction | Week-day | Month | Day | Fraction | Week-day | Month | Day | Fraction | | Week-day | Month | Day | Fraction | | Week-day | Month | Day | Fraction | Week-day | Month | Day | Fraction |
| 2 | S | 23 | ·26 | 3 | O | 22 | ·79 | 5 | N | 21 | ·32 | 6 | D | 20○ | ·85 | **44** | 1 | Jr | 19 | ·38 | | 2 | F | 17 | ·91 | 4 | Mr | 18 | ·44 |
| 6 | S | 11 | ·63 | 1 | O | 11 | ·16 | 2 | N | 9 | ·69 | 4 | D | 9○ | ·22 | **45** | 5 | Jr | 7 | ·75 | | 7 | F | 6 | ·28 | 1 | Mr | 7 | ·81 |
| 3 | Au | 31 | ·99 | 5 | S | 30 | ·52 | 7 | O | 30 | ·05 | 1 | N | 28○ | ·59 | | 3 | D | 28 | ·12 | **46** | 4 | Jr | 26 | ·65 | 6 | F | 25 | ·18 |
| 2 | S | 19 | ·89 | 4 | O | 19 | ·42 | 5 | N | 17 | ·95 | 7 | D | 17 | ·48 | **47** | 2 | Jr | 16 | ·01 | | 3 | F | 14 | ·11 | 5 | Mr | 16 | ·07 |
| 7 | S | 9 | ·26 | 1 | O | 8 | ·79 | 3 | N | 7● | ·32 | 4 | D | 6 | ·85 | **48** | 6 | Jr | 5 | ·38 | | 7 | F | 3 | ·93 | 2 | Mr | 4 | ·44 |
| 4 | Au | 28 | ·63 | 6 | S | 27○ | ·16 | 7 | O | 26 | ·69 | 2 | N | 25 | ·22 | | 3 | D | 24 | ·75 | **49** | 5 | Jr | 23 | ·28 | 6 | F | 21 | ·81 |
| 3 | S | 16○ | ·52 | 5 | O | 16 | ·05 | 6 | N | 14 | ·58 | 1 | D | 14 | ·11 | **50** | 2 | Jr | 12 | ·62 | | 4 | F | 11 | ·18 | 5 | Mr | 12 | ·71 |
| 7 | S | 5 | ·89 | 2 | O | 5 | ·42 | 3 | N | 3 | ·96 | 5 | D | 3 | ·48 | **51** | 7 | Jr | 2 | ·01 | | 1 | Jr | 31○ | ·54 | 3 | Mr | 2 | ·07 |
| 5 | Au | 26 | ·26 | {6 | S | 24 | ·79 | **Margasira** | | | | 2 | N | 22 | ·85 | | 4 | D | 22 | ·38 | **52** | 5 | Jr | 20○ | ·91 | {7 | F | 19 | ·44 |
| | | | | {1 | O | 24 | ·32 | **Kshaya** | | | | | | | | | | | | | | | | | | {1 | Mr | 19 | ·97 |
| 4 | S | 13 | ·15 | 5 | O | 12 | ·69 | 7 | N | 11 | ·22 | 1 | D | 10 | ·75 | **53** | 3 | Jr | 9●○ | ·28 | | 4 | F | 7 | ·81 | 6 | Mr | 9 | ·34 |
| 1 | S | 2 | ·52 | 3 | O | 2 | ·05 | 4 | O | 31 | ·58 | 6 | N | 30 | ·11 | | 7 | D | 29● | ·64 | **54** | 2 | Jr | 28 | ·17 | 3 | F | 26 | ·71 |
| 7 | S | 21 | ·42 | 1 | O | 20 | ·95 | 3 | N | 19○ | ·48 | 5 | D | 19 | 01 | **55** | 6 | Jr | 17 | ·54 | | 1 | F | 16 | ·07 | 2 | Mr | 17 | ·60 |
| 4 | S | 10 | ·79 | 6 | O | 10 | ·32 | 7 | N | 8○ | ·85 | 2 | D | 8 | ·38 | **56** | 3 | Jr | 6● | ·91 | | 5 | F | 5 | ·44 | 6 | Mr | 5 | ·97 |
| 2 | Au | 30 | ·15 | 3 | S | 28 | ·68 | 5 | O | 28●○ | ·21 | 6 | N | 26 | ·75 | | 1 | D | 26 | ·28 | **57** | 2 | Jr | 24 | ·81 | 4 | F | 23 | ·34 |
| 1 | S | 18 | ·05 | 2 | O | 17 | ·58 | 4 | N | 16 | ·11 | 5 | D | 15 | ·64 | **58** | 7 | Jr | 14 | ·17 | | 1 | F | 12 | ·70 | 3 | Mr | 14○ | ·73 |
| 5 | S | 7○ | ·42 | 6 | O | 6● | ·95 | 1 | N | 5 | ·48 | 3 | D | 5 | ·01 | **59** | 4 | Jr | 3 | ·54 | | 6 | F | 2 | ·07 | 7 | Mr | 3○ | ·60 |
| 2 | Au | 27○ | ·79 | 4 | S | 26 | ·32 | 5 | O | 25 | ·85 | 7 | N | 24 | ·38 | | 1 | D | 23 | ·91 | **60** | 3 | Jr | 22 | ·44 | 4 | F | 20●○ | ·97 |
| 1 | S | 14 | ·68 | 3 | O | 14 | ·21 | 4 | N | 12 | ·74 | 6 | D | 12 | ·27 | **61** | 7 | Jr | 10 | ·81 | | 2 | F | 9 | ·34 | 3 | Mr | 10 | ·87 |
| 6 | S | 4 | ·05 | 7 | O | 3 | ·58 | 2 | N | 2 | ·11 | 3 | D | 1 | ·64 | | 5 | D | 31○ | ·18 | **62** | 6 | Jr | 29● | ·71 | 1 | F | 28 | ·24 |
| 4 | S | 22 | ·95 | 6 | O | 22 | ·48 | 1 | N | 21 | ·01 | 2 | D | 20○ | ·54 | **63** | 4 | Jr | 19● | ·07 | | 5 | F | 17 | ·60 | 7 | Mr | 19 | ·13 |
| 2 | S | 12 | ·32 | 3 | O | 11 | ·85 | 5 | N | 10 | ·38 | 6 | D | 9○ | ·91 | **64** | 1 | Jr | 8 | ·44 | | 2 | F | 6 | ·97 | 4 | Mr | 7 | ·50 |
| 6 | Au | 31 | ·69 | 1 | S | 30 | ·22 | 2 | O | 29 | ·75 | 4 | N | 28● | ·28 | | 5 | D | 27 | ·81 | **65** | 7 | Jr | 26 | ·34 | 1 | F | 24 | ·87 |
| 5 | S | 19 | ·58 | 7 | O | 19 | ·11 | 1 | N | 17 | ·64 | 3 | D | 17 | ·17 | **66** | 4 | Jr | 15 | ·71 | | 6 | F | 14 | ·24 | 7 | Mr | 15 | ·77 |
| 2 | S | 8 | ·95 | 4 | O | 8○ | ·48 | 6 | N | 7● | ·01 | 7 | D | 6 | ·54 | **67** | 2 | Jr | 5 | ·07 | | 3 | F | 3 | ·60 | 5 | Mr | 5 | ·13 |
| 7 | Au | 29 | ·32 | 1 | S | 27○ | ·85 | 3 | O | 27 | ·38 | 4 | N | 25 | ·91 | | 6 | D | 25 | ·44 | **68** | 7 | Jr | 23 | ·97 | 2 | F | 22 | ·50 |
| 6 | S | 16 | ·25 | 7 | O | 15 | ·75 | 2 | N | 14 | ·28 | 3 | D | 13 | ·81 | **69** | 5 | Jr | 12 | ·34 | | 6 | F | 10○ | ·87 | 1 | Mr | 12 | ·40● |
| 3 | S | 5● | ·58 | 5 | O | 5 | ·11 | 6 | N | 3 | ·64 | 1 | D | 3 | ·17 | **70** | 2 | Jr | 1 | ·70 | | 4 | Jr | 31○ | ·23 | 5 | Mr | 1 | ·77 |
| {7 | Au | 25● | ·95 | 4 | O | 24 | ·01 | 5 | N | 22 | ·54 | **Pausha.** | | | | | 7 | D | 22 | ·07 | **71** | 1 | Jr | 20○ | ·60 | {3 | F | 19 | ·13 |
| {2 | S | 24 | ·48 | | | | | | | | | **Kshaya** | | | | | | | | | | | | | | {4 | Mr | 20 | ·66 |
| 6 | S | 13 | ·85 | 1 | O | 13 | ·38 | 2 | N | 11 | ·91 | 4 | D | 11 | ·44 | **72** | 5 | Jr | 9 | ·97 | | 7 | [illegible] | 8 | ·50 | 2 | Mr | 9 | ·03 |
| 4 | S | 2 | ·21 | 5 | O | 1 | ·74 | 6 | O | 31 | ·27 | 1 | N | 29○ | ·81 | | 3 | D | 29● | ·34 | **73** | 4 | Jr | 27 | ·87 | 6 | F | 26 | ·40 |
| 3 | S | 21 | ·11 | 5 | O | 20 | ·64 | 9 | N | 19○ | ·17 | 7 | D | 18 | ·70 | **74** | 2 | Jr | 17 | ·23 | | 3 | F | 15 | ·76 | 5 | Mr | 17 | ·29 |
| 7 | S | 10 | ·48 | 2 | O | 10 | ·01 | 3 | N | 8○ | ·54 | 5 | D | 8 | ·07 | **75** | 6 | Jr | 6 | ·60 | | 1 | F | 5 | ·13 | 2 | Mr | 6 | ·66 |
| 4 | Au | 30 | ·85 | 6 | S | 29 | ·38 | 7 | O | 28● | ·91 | 2 | N | 27 | ·44 | | 3 | D | 26 | ·97 | **76** | 5 | Jr | 25 | ·50 | 7 | F | 24 | ·03 |
| 3 | S | 17○ | ·74 | 5 | O | 17 | ·27 | 6 | N | 15 | ·80 | 1 | D | 15 | ·33 | **77** | 2 | Jr | 13 | ·87 | | 4 | F | 12 | ·40 | 5 | Mr | 13○ | ·93 |
| 1 | S | 7○ | ·11 | 2 | O | 6 | ·64 | 4 | N | 5 | ·17 | 5 | D | 4 | ·70 | **78** | 7 | Jr | 3 | ·23 | | 1 | F | 1 | ·76 | 3 | Mr | 3○ | ·29 |
| 5 | Au | 27○ | ·48 | 7 | S | 26 | ·01 | 1 | O | 25 | ·54 | 3 | N | 24 | ·07 | | 4 | D | 23 | ·60 | **79** | 6 | Jr | 22 | ·13 | 7 | F | 20● | ·66 |
| 4 | S | 15 | ·37 | 5 | O | 14 | ·91 | 7 | N | 13 | ·44 | 1 | D | 12 | ·97 | **80** | 3 | Jr | 11○ | ·50 | | 5 | F | 10● | ·03 | 6 | Mr | 10 | ·56 |
| 1 | S | 3 | ·74 | 3 | O | 3 | ·27 | 4 | N | 1 | ·80 | 6 | D | 1 | ·33 | | 7 | D | 30○ | ·86 | **81** | 2 | Jr | 29 | ·39 | 3 | F | 27 | ·92 |
| 7 | S | 22 | ·64 | 2 | O | 22 | ·17 | 3 | N | 20 | ·70 | 5 | D | 20○ | ·23 | **82** | 6 | Jr | 18 | ·76 | | 1 | F | 17 | ·29 | 2 | Mr | 18 | ·82 |
| 5 | S | 12 | ·01 | 6 | O | 11 | ·54 | 1 | N | 10 | ·07 | 2 | D | 9● | ·60 | **83** | 4 | Jr | 8 | ·13 | | 5 | F | 6 | ·66 | 7 | Mr | 8 | ·19 |

**TABLE X—**

⊙'s Anom col. 6 + 29·53059 + 59·06117 + 88·59176 + 118·12235

☾'s Anom col. 7 + 1·976 + 3·952 + 5·928 + 7·904

| Kaliyuga | Vikrama Era | Saka Era | Commencement of Solar Year: Month and day A.D. | Fraction of day | First New-Moon in Solar Year | Anomaly of first New-Moon | A.D. | Week-day of 1st January | Vaisakha: Week-day | Month | Day | Fraction | Jyeshta: Week-day | Month | Day | Fraction | Ashada: Week-day | Month | Day | Fraction | Sravana: Week-day | Month | Day | Fraction | Bhadrapada: Week-day | Month | Day | Fraction |
|---|---|---|---|---|---|---|---|---|---|---|---|---|---|---|---|---|---|---|---|---|---|---|---|---|---|---|---|---|
| 3884 | 840 | 705 | 21M | ·5677 | 16·15159 | 9·505 | **783** | 4 | 1 | Ap | 6 | ·72 | 3 | My | 6 | ·25 | 4 | Je | 4 | ·78 | 6 | Jl | 4 | ·31 | 7 | Au | 2 | ·84 |
| 3885 | 841 | 706 | 20M | ·8264 | 5·26002 | 5·662 | **784** | 5 | 6 | Mr | 26 | ·09 | 7 | Ap | 24○ | ·62 | 2 | My | 24 | ·15 | 5 | Jl | 22 | ·21 | 6 | Au | 20 | ·74 |
| | | | | | | | | | | | | | | | | | 3 | Je | 22 | ·68 | | | | | | | | |
| 3886 | 842 | 707 | 21M | ·0851 | 23·89905 | 3·796 | **785** | 7 | 4 | Ap | 13○ | ·99 | 6 | My | 13 | ·52 | 1 | Je | 12 | ·05 | 2 | Jl | 11 | ·58 | 4 | Au | 10 | ·11 |
| 3887 | 843 | 708 | 21M | ·3437 | 13·00749 | 27·508 | **786** | 1 | 2 | Ap | 3● | ·35 | 3 | My | 2 | ·88 | 5 | Je | 1 | ·41 | 6 | Je | 30 | ·94 | 1 | Jl | 30 | ·47 |
| 3888 | 844 | 709 | 21M | ·6024 | 2·11591 | 23·666 | **787** | 2 | 6 | Mr | 23● | ·72 | 1 | Ap | 22 | ·25 | 4 | Je | 20 | ·31 | 5 | Jl | 19 | ·84 | 7 | Au | 18○ | ·37 |
| | | | | | | | | | | | | | 2 | My | 21 | ·78 | | | | | | | | | | | | |
| 3889 | 845 | 710 | 20M | ·8611 | 20·75494 | 21·800 | **788** | 3 | 5 | Ap | 10 | ·62 | 7 | My | 10 | ·15 | 1 | Je | 8 | ·68 | 3 | Jl | 8 | ·21 | 4 | Au | 6○ | ·74 |
| 3890 | 846 | 711 | 21M | ·1198 | 9·86338 | 17·958 | **789** | 5 | 2 | Mr | 30 | ·98 | 4 | Ap | 29 | ·51 | 6 | My | 29 | ·05 | 7 | Je | 27 | ·58 | 2 | Jl | 27●○ | ·11 |
| 3891 | 847 | 712 | 21M | ·3785 | 28·50240 | 16·091 | **790** | 6 | 1 | Ap | 18 | ·88 | 3 | My | 18 | ·41 | 4 | Je | 16 | ·94 | 6 | Jl | 16 | ·47 | 1 | Au | 15 | ·00 |
| 3892 | 848 | 713 | 21M | ·6371 | 17·61084 | 12·249 | **791** | 7 | 6 | Ap | 8 | ·25 | 7 | My | 7 | ·78 | 2 | Je | 6○ | ·31 | 3 | Jl | 5● | ·84 | 5 | Au | 4 | ·37 |
| 3893 | 849 | 714 | 20M | ·8958 | 6·71928 | 8·407 | **792** | 1 | 3 | Mr | 27 | ·62 | 5 | Ap | 26 | ·15 | 6 | My | 25○ | ·68 | 1 | Je | 24 | ·21 | 4 | Au | 22 | ·27 |
| | | | | | | | | | | | | | | | | | | | | | 2 | Jl | 23 | ·74 | | | | |
| 3894 | 850 | 715 | 21M | ·1545 | 25·35829 | 6·541 | **793** | 3 | 2 | Ap | 15 | ·51 | 4 | My | 15○ | ·04 | 5 | Je | 13 | ·57 | 7 | Jl | 13 | ·11 | 1 | Au | 11 | ·64 |
| 3895 | 851 | 716 | 21M | ·4132 | 14·46673 | 2·698 | **794** | 4 | 6 | Ap | 4 | ·88 | 1 | My | 4● | ·41 | 2 | Je | 2 | ·94 | 4 | Jl | 2 | ·47 | 6 | Au | 1 | ·00 |
| 3896 | 852 | 717 | 21M | ·6719 | 3·57448 | 26·410 | **795** | 5 | 4 | Mr | 25○ | ·25 | 5 | Ap | 23 | ·78 | 7 | My | 23 | ·31 | 3 | Jl | 21 | ·37 | 4 | Au | 19 | ·90 |
| | | | | | | | | | | | | | | | | | 1 | Je | 21 | ·84 | | | | | | | | |
| 3897 | 853 | 718 | 20M | ·9305 | 22·21419 | 24·544 | **796** | 6 | 3 | Ap | 12 | ·15 | 4 | My | 11 | ·68 | 6 | Je | 10 | ·21 | 7 | Jl | 9 | ·74 | 2 | Au | 8 | ·27 |
| 3898 | 854 | 719 | 21M | ·1892 | 11·32263 | 20·702 | **797** | 1 | 7 | Ap | 1 | ·51 | 2 | My | 1 | ·04 | 3 | My | 30 | ·57 | 5 | Je | 29 | ·10 | 6 | Jl | 28 | ·63 |
| 3899 | 855 | 720 | 21M | ·4479 | 0·43106 | 16·860 | **798** | 2 | 4 | Mr | 21 | ·88 | 7 | My | 19 | ·94 | 2 | Je | 18 | ·47 | 4 | Jl | 18○ | ·00 | 5 | Au | 16 | ·53 |
| | | | | | | | | | 6 | Ap | 20 | ·41 | | | | | | | | | | | | | | | | |
| 3900 | 856 | 721 | 21M | ·7066 | 19·07008 | 14·993 | **799** | 3 | 3 | Ap | 9 | ·78 | 5 | My | 9 | ·31 | 6 | Je | 7 | ·84 | 1 | Jl | 7○ | ·37 | 2 | Au | 5 | ·90 |
| 3901 | 857 | 722 | 20M | ·9653 | 8·17852 | 11·151 | **800** | 4 | 1 | Mr | 29 | ·15 | 2 | Ap | 27 | ·68 | 4 | My | 27 | ·21 | 5 | Je | 25●○ | ·74 | 7 | Jl | 25 | ·27 |
| | | | | | | | | | | | | | | | | | | | | | | | | | 1 | Au | 23 | ·80 |
| 3902 | 858 | 723 | 21M | ·2240 | 26·81755 | 9·285 | **801** | 6 | 7 | Ap | 17 | ·04 | 1 | My | 16 | ·57 | 3 | Je | 15● | ·11 | 4 | Jl | 14 | ·64 | 6 | Au | 13 | ·17 |
| 3903 | 859 | 724 | 21M | ·4826 | 15·92587 | 5·443 | **802** | 7 | 4 | Ap | 6 | ·41 | 5 | My | 5○ | ·94 | 7 | Je | 4● | ·47 | 2 | Jl | 4 | ·00 | 3 | Au | 2 | ·53 |
| 3904 | 860 | 725 | 21M | ·7413 | 5·03441 | 1·601 | **803** | 1 | 1 | Mr | 26 | ·78 | 3 | Ap | 25●○ | ·31 | 4 | My | 24 | ·84 | 7 | Jl | 22 | ·90 | 2 | Au | 21 | ·43 |
| | | | | | | | | | | | | | | | | | 6 | Je | 23 | ·37 | | | | | | | | |
| 3905 | 861 | 726 | 21M | ·0000 | 23·67344 | 27·289 | **804** | 2 | 7 | Ap | 13 | ·68 | 2 | My | 13 | ·21 | 3 | Je | 11 | ·74 | 5 | Jl | 11 | ·27 | 6 | Au | 9 | ·80 |
| 3906 | 862 | 727 | 21M | ·2587 | 12·78187 | 23·447 | **805** | 4 | 5 | Ap | 3 | ·04 | 6 | My | 2 | ·57 | 1 | Je | 1 | ·10 | 2 | Je | 30 | ·63 | 4 | Jl | 30 | ·17 |
| 3907 | 863 | 728 | 21M | ·5174 | 1·89031 | 19·604 | **806** | 5 | 2 | Mr | 23 | ·41 | 3 | Ap | 21 | ·94 | 7 | Je | 20 | ·00 | 1 | Jl | 19 | ·53 | 3 | Au | 18○ | ·06 |
| | | | | | | | | | | | | | 5 | My | 21 | ·47 | | | | | | | | | | | | |
| 3908 | 864 | 729 | 21M | ·7760 | 20·52934 | 17·738 | **807** | 6 | 1 | Ap | 11 | ·31 | 2 | My | 10 | ·84 | 4 | Je | 9 | ·37 | 5 | Jl | 8 | ·90 | 7 | Au | 7○ | ·43 |
| 3909 | 865 | 730 | 21M | ·0347 | 9·63776 | 13·896 | **808** | 7 | 5 | Mr | 30 | ·67 | 7 | Ap | 29 | ·21 | 1 | My | 28 | ·74 | 3 | Je | 27 | ·27 | 4 | Jl | 26● | ·80 |
| 3910 | 866 | 731 | 21M | ·2934 | 28·27679 | 12·080 | **809** | 2 | 4 | Ap | 18 | ·57 | 6 | My | 18 | ·10 | 7 | Je | 16○ | ·63 | 2 | Jl | 16● | ·16 | 3 | Au | 14 | ·69 |
| 3911 | 867 | 732 | 21M | ·5521 | 17·38523 | 8·187 | **810** | 3 | 1 | Ap | 7 | ·94 | 3 | My | 7 | ·47 | 5 | Je | 6○ | ·00 | 6 | Jl | 5 | ·53 | 1 | Au | 4 | ·06 |
| 3912 | 868 | 733 | 21M | ·8108 | 6·49366 | 4·345 | **811** | 4 | 6 | Mr | 28 | ·31 | 7 | Ap | 26 | ·84 | 2 | My | 26○ | ·37 | 3 | Je | 24 | ·90 | 6 | Au | 22 | ·96 |
| | | | | | | | | | | | | | | | | | | | | | 5 | Jl | 24 | ·43 | | | | |
| 3913 | 869 | 734 | 21M | ·0694 | 25·13269 | 2·479 | **812** | 5 | 5 | Ap | 15 | ·20 | 6 | My | 14● | ·73 | 1 | Je | 13 | ·27 | 2 | Jl | 12 | ·80 | 4 | Au | 11 | ·33 |
| 3914 | 870 | 735 | 21M | ·3281 | 14·24112 | 26·191 | **813** | 7 | 2 | Ap | 4○ | ·57 | 4 | My | 4● | ·10 | 5 | Je | 2 | ·63 | 7 | Jl | 2 | ·16 | 1 | Jl | 31 | ·69 |
| 3915 | 871 | 736 | 21M | ·5868 | 3·34955 | 22·349 | **814** | 1 | 6 | Mr | 24●○ | ·94 | 1 | Ap | 23 | ·47 | 3 | My | 23 | ·00 | 6 | Jl | 21 | ·06 | 7 | Au | 19 | ·59 |
| | | | | | | | | | | | | | | | | | 4 | Je | 21 | ·53 | | | | | | | | |
| 3916 | 872 | 737 | 21M | ·8455 | 21·98858 | 20·483 | **815** | 2 | 5 | Ap | 12 | ·84 | 7 | My | 12 | ·37 | 1 | Je | 10 | ·90 | 3 | Jl | 10 | ·43 | 4 | Au | 8 | ·96 |
| 3917 | 873 | 738 | 21M | ·1042 | 11·09702 | 16·640 | **816** | 3 | 3 | Ap | 1 | ·20 | 4 | Ap | 30 | ·73 | 6 | My | 30 | ·26 | 7 | Je | 28 | ·79 | 2 | Jl | 28 | ·33 |
| 3918 | 874 | 739 | 21M | ·3628 | 0·20545 | 12·799 | **817** | 5 | 7 | Mr | 21 | ·57 | 3 | My | 19 | ·63 | 5 | Je | 18 | ·16 | 6 | Jl | 17○ | ·69 | 1 | Au | 16 | ·22 |
| | | | | | | | | | 2 | Ap | 20 | ·10 | | | | | | | | | | | | | | | | |
| 3919 | 875 | 740 | 21M | ·6215 | 18·84448 | 10·982 | **818** | 6 | 6 | Ap | 9 | ·47 | 1 | My | 9 | ·00 | 2 | Je | 7 | ·53 | 4 | Jl | 7●○ | ·06 | 5 | Au | 5 | ·59 |
| 3920 | 876 | 741 | 21M | ·8802 | 7·95291 | 7·089 | **819** | 7 | 3 | Mr | 29 | ·83 | 5 | Ap | 28 | ·37 | 6 | My | 27 | ·90 | 1 | Je | 26● | ·43 | 2 | Jl | 25 | ·96 |
| | | | | | | | | | | | | | | | | | | | | | | | | | 4 | Au | 24 | ·49 |
| 3921 | 877 | 742 | 21M | ·1389 | 26·59193 | 5·223 | **820** | 1 | 2 | Ap | 16 | ·73 | 4 | My | 16○ | ·26 | 5 | Je | 14 | ·79 | 7 | Jl | 14 | 32 | 1 | Au | 12 | ·85 |
| [illegible] | 878 | 743 | 21M | ·3976 | 15·70037 | 1·381 | **821** | 3 | 7 | Ap | 6 | ·10 | 1 | My | 5●○ | ·63 | 3 | Je | 4 | ·16 | 4 | Jl | 3 | 69 | 6 | Au | 2 | ·22 |

**N.B.**—For Surya Siddhanta figures for **A.D. 500** to **A.D. 999** see pages **76—79** below.

**Arya Siddhanta.**

| Asvina | | | | Kartika | | | | Margasira | | | | Pausha | | | | A.D. | Magha | | | | A.D. | Phalguna | | | | Chaitra | | | |
|---|---|---|---|---|---|---|---|---|---|---|---|---|---|---|---|---|---|---|---|---|---|---|---|---|---|---|---|---|---|
| + 147·65293 | | | | + 177·18353 | | | | + 206·71411 | | | | + 236·24470 | | | | | + 265·77529 | | | | | + 295·30588 | | | | +324·83647<br>+354·36706 | | | |
| + 9·880 | | | | + 11·856 | | | | + 13·832 | | | | + 15·808 | | | | | + 17·784 | | | | | + 19·760 | | | | + 21·736<br>+ 23·712 | | | |
| Week-day | Month | Day | Fraction | Week-day | Month | Day | Fraction | Week-day | Month | Day | Fraction | Week-day | Month | Day | Fraction | | Week-day | Month | Day | Fraction | | Week-day | Month | Day | Fraction | Week-day | Month | Day | Fraction |
| 2 | S | 1 | ·37 | 3 | S | 30 | ·90 | 5 | O | 30 | ·48 | 6 | N | 28● | ·97 | | 1 | D | 28 | ·50 | 84 | 3 | Jr | 27 | ·03 | 4 | F | 25 | ·56 |
| 1 | S | 19 | ·27 | 2 | O | 18○ | ·80 | 4 | N | 17 | ·33 | 5 | D | 16 | ·86 | 85 | 7 | Jr | 15 | ·39 | | 1 | F | 13 | ·92 | 3 | Mr | 15 | ·45 |
| 5 | S | 8 | ·64 | 7 | O | 8○ | ·17 | 1 | N | 6 | ·70 | 3 | D | 6 | ·23 | 86 | 4 | Jr | 4 | ·76 | | 6 | F | 3 | ·29 | 7 | Mr | 4 | ·82 |
| 3 | Au | 29 | ·01 | 4 | S | 27● | ·54 | 6 | O | 27 | ·07 | 7 | N | 25 | ·60 | | 2 | D | 25 | ·13 | 87 | 3 | Jr | 23 | ·66 | 5 | F | 22○ | ·19 |
| 1 | S | 16● | ·90 | 3 | O | 16 | ·43 | 4 | N | 14 | ·96 | 6 | D | 14 | ·49 | 88 | 1 | Jr | 13 | ·03 | | 2 | F | 11○ | ·56 | 4 | Mr | 12 | ·09 |
| 6 | S | 5 | ·27 | 7 | O | 4 | ·80 | 2 | N | 3 | ·33 | 3 | D | 2 | ·86 | 89 | 5 | Jr | 1 | ·39 | | 6 | Jr | 30●○ | ·92 | 1 | Mr | 1 | ·45 |
| 3 | Au | 25 | ·64 | 6 | O | 23 | ·70 | 1 | N | 22 | ·23 | 2 | D | 21 | ·76 | 90 | 4 | Jr | 20● | ·29 | | 5 | F | 18 | ·82 | 7 | Mr | 20 | ·35 |
| 5 | S | 24 | ·17 | | | | | | | | | | | | | | | | | | | | | | | | | | |
| 2 | S | 13 | ·53 | 4 | O | 13 | ·07 | 5 | N | 11 | ·60 | 7 | D | 11○ | ·13 | 91 | 1 | Jr | 9● | ·66 | | 3 | F | 8 | ·19 | 4 | Mr | 9 | ·72 |
| 6 | S | 2 | ·90 | 1 | O | 2 | ·43 | 2 | O | 31 | ·96 | 4 | N | 30○ | ·49 | | 6 | D | 30 | ·02 | 92 | 7 | Jr | 28 | ·55 | 2 | F | 27 | ·09 |
| 5 | S | 20 | ·80 | 7 | O | 20 | ·33 | 1 | N | 18●○ | ·86 | 3 | D | 18 | ·39 | 93 | 4 | Jr | 16 | ·92 | | 6 | F | 15 | ·45 | 7 | Mr | 16 | ·98 |
| 3 | S | 10 | ·17 | 4 | O | 9 | ·70 | 6 | N | 8 | ·23 | 7 | D | 7 | ·76 | 94 | 2 | Jr | 6 | ·29 | | 3 | F | 4 | ·82 | 5 | Mr | 6 | ·35 |
| 7 | Au | 30 | ·53 | 2 | S | 29○ | ·06 | 3 | O | 28 | ·59 | 5 | N | 27 | ·13 | | 6 | D | 26 | ·66 | 95 | 1 | Jr | 25 | ·19 | 2 | F | 23 | ·72 |
| 6 | S | 18○ | ·43 | 7 | O | 17 | ·96 | 2 | N | 16 | ·50 | 4 | D | 16 | ·02 | 96 | 5 | Jr | 14 | ·55 | | 7 | F | 13 | ·08 | 1 | Mr | 13○ | ·6 |
| 3 | S | 6●○ | ·80 | 5 | O | 6 | ·33 | 6 | N | 4 | ·86 | 1 | D | 4 | ·39 | 97 | 2 | Jr | 2 | ·92 | | 4 | F | 1 | ·45 | 5 | Mr | 2 | ·98 |
| 1 | Au | 27 | ·17 | 2 | S | 25 | ·70 | 4 | O | 25 | ·23 | 5 | N | 23 | ·76 | | 7 | D | 23 | ·29 | 98 | 1 | Jr | 21○ | ·82 | 3 | F | 20 | ·35 |
| 7 | S | 15 | ·06 | 1 | O | 14 | ·59 | 3 | N | 13 | ·12 | 4 | D | 12 | ·65 | 99 | 6 | Jr | 11○ | ·19 | | 7 | F | 9 | ·72 | 2 | Mr | 11 | ·25 |
| 4 | S | 4 | ·43 | 5 | O | 3 | ·96 | 7 | N | 2 | ·49 | 2 | D | 2 | ·02 | | 3 | D | 31○ | ·55 | 800 | 5 | Jr | 30 | ·08 | 6 | F | 28 | ·61 |
| 3 | S | 22 | ·32 | 4 | O | 21 | ·86 | 6 | N | 20 | ·39 | 7 | D | 19 | ·92 | 01 | 2 | Jr | 18 | ·45 | | 3 | F | 16 | ·98 | 5 | Mr | 18 | ·51 |
| 7 | S | 11 | ·70 | 2 | O | 11 | ·23 | 3 | N | 9 | ·76 | 5 | D | 9 | ·29 | 02 | 6 | Jr | 7 | ·82 | | 1 | F | 6 | ·35 | 2 | Mr | 7 | ·88 |
| 5 | S | 1 | ·06 | 6 | S | 30 | ·59 | 1 | O | 30○ | ·13 | 2 | N | 28● | ·66 | | 4 | D | 28 | ·19 | 03 | 5 | Jr | 26 | ·72 | 7 | F | 25 | ·25 |
| 3 | S | 19 | ·96 | 5 | O | 19○ | ·49 | 7 | N | 18 | ·02 | 1 | D | 17 | ·55 | 04 | 3 | Jr | 16 | ·09 | | 4 | F | 14 | ·61 | 6 | Mr | 15 | ·15 |
| 1 | S | 8 | ·33 | 2 | O | 7 | ·86 | 4 | N | 6 | ·39 | 5 | D | 5 | ·92 | 05 | 7 | Jr | 4 | ·45 | | 1 | F | 2 | ·98 | 3 | Mr | 4 | ·51 |
| 5 | Au | 28○ | ·70 | 7 | S | 27 | ·23 | 1 | O | 26 | ·76 | 3 | N | 25 | ·29 | | 4 | D | 24 | ·82 | 06 | 6 | Jr | 23 | ·35 | 1 | F | 21○ | ·96 |
| 4 | S | 16● | ·59 | 6 | O | 16 | ·12 | 7 | N | 14 | ·65 | 2 | D | 14 | ·19 | 07 | 3 | Jr | 12 | ·72 | | 5 | F | 11●○ | ·25 | 6 | Mr | 12 | ·78 |
| 1 | S | 5 | ·96 | 3 | O | 5 | ·49 | 5 | N | 4 | ·02 | 6 | D | 3 | ·55 | 08 | 1 | Jr | 2 | ·08 | | 2 | Jr | 31● | ·61 | 4 | Mr | 1 | ·14 |
| 6 | Au | 25 | ·32 | 2 | O | 23 | ·39 | 3 | N | 21 | ·92 | 5 | D | 21○ | ·45 | 09 | 6 | Jr | 19 | ·98 | | 1 | F | 18 | ·51 | 3 | Mr | 20 | ·04 |
| 7 | S | 23 | ·86 | | | | | | | | | | | | | | | | | | | | | | | | | | |
| 5 | S | 13 | ·23 | 6 | O | 12 | ·76 | 1 | N | 11 | ·29 | 2 | D | 10○ | ·82 | 10 | 4 | Jr | 9 | ·35 | | 5 | F | 7 | ·88 | 7 | Mr | 9 | ·41 |
| 2 | S | 2 | ·59 | 4 | O | 2 | ·12 | 5 | O | 31 | ·65 | 7 | N | 30●○ | ·18 | | 1 | D | 29 | ·71 | 11 | 3 | Jr | 28 | ·25 | 4 | F | 26 | ·78 |
| 1 | S | 21 | ·49 | 3 | O | 21 | ·02 | 4 | N | 19 | ·55 | 6 | D | 19 | ·08 | 12 | 7 | Jr | 17 | ·61 | | 2 | F | 16 | ·14 | 3 | Mr | 16 | ·67 |
| 5 | S | 9 | ·86 | 7 | O | 9○ | ·39 | 1 | N | 7● | ·92 | 3 | D | 7 | ·45 | 13 | 4 | Jr | 5 | ·98 | | 6 | F | 4 | ·51 | 1 | Mr | 6 | ·04 |
| 3 | Au | 30 | ·22 | 4 | S | 28○ | ·75 | 6 | O | 28 | ·29 | 7 | N | 26 | ·82 | | 2 | D | 26 | ·35 | 14 | 3 | Jr | 24 | ·88 | 5 | F | 23 | ·41 |
| 2 | S | 18○ | ·12 | 3 | O | 17 | ·65 | 5 | N | 16 | ·18 | 6 | D | 15 | ·71 | 15 | 1 | Jr | 14 | ·24 | | 2 | F | 12 | ·77 | 4 | Mr | 14 | ·31 |
| 6 | S | 7● | ·49 | 1 | O | 7 | ·02 | 2 | N | 5 | ·55 | 4 | D | 5 | ·08 | 16 | 5 | Jr | 3 | ·61 | | 7 | F | 2○ | ·14 | 1 | Mr | 2● | ·67 |
| 3 | Au | 26 | ·86 | 5 | S | 25 | ·39 | 6 | O | 24 | ·92 | 1 | N | 23 | ·45 | | 2 | D | 22 | ·98 | 17 | 4 | Jr | 21○ | ·51 | 6 | F | 20● | ·04 |
| 2 | S | 14 | ·75 | 4 | O | 14 | ·28 | 5 | N | 12 | ·81 | 7 | D | 12 | ·35 | 18 | 1 | Jr | 10○ | ·88 | | 3 | F | 9 | ·41 | 4 | Mr | 10 | ·94 |
| 7 | S | 4 | ·12 | 1 | O | 3 | 65 | 3 | N | 2 | ·18 | 4 | D | 1 | ·71 | | 6 | D | 31● | ·24 | 19 | 7 | Jr | 29 | ·77 | 2 | F | 28 | ·30 |
| 6 | S | 23 | ·02 | 7 | O | 22 | ·55 | 2 | N | 21 | ·08 | 3 | D | 20 | ·61 | 20 | 5 | Jr | 19 | ·14 | | 6 | F | 17 | ·67 | 1 | Mr | 18 | ·20 |
| 3 | S | 11 | ·39 | 4 | O | 10 | ·92 | 6 | N | 9○ | ·45 | 7 | D | 8● | ·98 | 21 | 2 | Jr | 7 | ·51 | | 4 | F | 6 | ·04 | [illegible] | Mr | 7 | ·57 |
| 7 | Au | 31 | ·75 | 2 | S | 30 | ·28 | 3 | O | 29○ | ·81 | 5 | N | 28 | ·34 | | 6 | D | 27 | ·87 | 22 | 1 | Jr | 26 | ·41 | 2 | F | 24 | ·94 |

## TABLE X—

| Kaliyuga | Vikrama Era | Saka Era | Commencement of Solar Year: Month and day A.D. | Fraction of day | First New-Moon in Solar Year | Anomaly of first New-Moon | A.D. | Week-day of 1st January | Vaisakha: Week-day | Month | Day | Fraction | Jyeshta: Week-day | Month | Day | Fraction | Ashada: Week-day | Month | Day | Fraction | Sravana: Week-day | Month | Day | Fraction | Bhadrapada: Week-day | Month | Day | Fraction |
|---|---|---|---|---|---|---|---|---|---|---|---|---|---|---|---|---|---|---|---|---|---|---|---|---|---|---|---|---|
| | | | | | | | | | ☉'s Anom col. 6 | | | | + 29·53059 | | | | + 59·06117 | | | | + 88·59176 | | | | + 118·12235 | | | |
| | | | | | | | | | ☾'s Anom col. 7 | | | | + 1·976 | | | | + 3·952 | | | | + 5·928 | | | | + 7·904 | | | |
| 3923 | 879 | 744 | 21M | ·6562 | 4·81881 | 25·103 | **822** | 4 | 4 | Mr | 26 | ·47 | 6 | Ap | 25● | ·00 | 7 | My | 24 | ·53 | 3 | Jl | 22 | ·59 | 5 | Au | 21 | ·12 |
| | | | | | | | | | | | | | | | | | 2 | Je | 23 | ·06 | | | | | | | | |
| 3924 | 880 | 745 | 21M | ·9149 | 23·44782 | 23·227 | **823** | 5 | 3 | Ap | 14 | ·36 | 4 | My | 13 | ·89 | 6 | Je | 12 | ·43 | 7 | Jl | 11 | ·96 | 2 | Au | 10 | ·49 |
| 3925 | 881 | 746 | 21M | ·1736 | 12·55626 | 19·385 | **824** | 6 | 7 | Ap | 2 | ·73 | 2 | My | 2 | ·26 | 3 | My | 31 | ·79 | 5 | Je | 30 | ·32 | 6 | Jl | 29 | ·85 |
| 3926 | 882 | 747 | 21M | ·4823 | 1·66470 | 15·543 | **825** | 1 | 5 | Mr | 23 | ·10 | 6 | Ap | 21 | ·63 | 2 | Je | 19 | ·69 | 4 | Jl | 19 | ·22 | 5 | Au | 17○ | ·75 |
| | | | | | | | | | | | | | 1 | My | 21 | ·16 | | | | | | | | | | | | |
| 3927 | 883 | 748 | 21M | ·6910 | 20·30372 | 13·676 | **826** | 2 | 4 | Ap | 11 | ·00 | 5 | My | 10 | ·53 | 7 | Je | 9 | ·06 | 1 | Jl | 8 | ·59 | 3 | Au | 7● | ·12 |
| 3928 | 884 | 749 | 21M | ·9496 | 9·41246 | 9·834 | **827** | 3 | 1 | Mr | 31 | ·36 | 2 | Ap | 29 | ·89 | 4 | My | 29 | ·42 | 5 | Je | 27○ | ·95 | 7 | Jl | 27 | ·49 |
| 3929 | 885 | 750 | 21M | ·2083 | 28·05119 | 7·968 | **828** | 4 | 7 | Ap | 18 | ·26 | 1 | My | 17 | ·79 | 3 | Je | 16○ | ·32 | 4 | Jl | 15 | ·85 | 6 | Au | 14 | ·38 |
| 3930 | 886 | 751 | 21M | ·4670 | 17·15961 | 4·126 | **829** | 6 | 4 | Ap | 7 | ·63 | 6 | My | 7 | ·16 | 7 | Je | 5●○ | ·69 | 2 | Jl | 5 | ·22 | 3 | Au | 3 | ·75 |
| 3931 | 887 | 752 | 21M | ·7257 | 6·26805 | 0·283 | **830** | 7 | 1 | Mr | 27 | ·99 | 3 | Ap | 26 | ·53 | 5 | My | 26 | ·06 | 6 | Je | 24 | ·59 | 2 | Au | 22 | ·65 |
| | | | | | | | | | | | | | | | | | | | | | 1 | Jl | 24 | ·12 | | | | |
| 3932 | 888 | 753 | 21M | ·9844 | 24·90708 | 25·971 | **831** | 1 | 7 | Ap | 15○ | ·89 | 2 | My | 15● | ·42 | 3 | Je | 13 | ·95 | 5 | Jl | 13 | ·48 | 7 | Au | 12 | ·01 |
| 3933 | 889 | 754 | 21M | ·2431 | 14·01551 | 22·129 | **832** | 2 | 5 | Ap | 4○ | ·26 | 6 | My | 3 | ·79 | 1 | Je | 2 | ·32 | 2 | Jl | 1 | ·85 | 4 | Jl | 31 | ·38 |
| 3934 | 890 | 755 | 21M | ·5017 | 3·12395 | 18·287 | **833** | 4 | 2 | Mr | 24●○ | ·63 | 4 | Ap | 23 | ·16 | 5 | My | 22 | ·69 | 1 | Jl | 20 | ·75 | 3 | Au | 19 | ·28 |
| | | | | | | | | | | | | | | | | | 7 | Je | 21 | ·22 | | | | | | | | |
| 3935 | 891 | 756 | 21M | ·7604 | 21·76298 | 16·421 | **834** | 5 | 1 | Ap | 12 | ·52 | 3 | My | 12 | ·05 | 4 | Je | 10 | ·59 | 6 | Jl | 10 | ·12 | 7 | Au | 8 | ·65 |
| 3936 | 892 | 757 | 22M | ·0191 | 10·87140 | 12·579 | **835** | 6 | 5 | Ap | 1 | ·89 | 7 | My | 1 | ·42 | 2 | My | 30 | ·95 | 3 | Je | 29 | ·48 | 5 | Jl | 29○ | ·01 |
| 3937 | 893 | 758 | 21M | ·2778 | 29·51043 | 8·736 | **836** | 7 | 4 | Ap | 19 | ·79 | 6 | My | 19 | ·32 | 7 | Je | 17 | ·85 | 2 | Jl | 17●○ | ·38 | 3 | Au | 15 | ·91 |
| 3938 | 894 | 759 | 21M | ·5365 | 18·61887 | 6·870 | **837** | 2 | 2 | Ap | 9 | ·16 | 3 | My | 8 | ·69 | 5 | Je | 7 | ·22 | 6 | Jl | 6 | ·75 | 1 | Au | 5 | ·28 |
| 3939 | 895 | 760 | 21M | ·7951 | 7·72760 | 3·028 | **838** | 3 | 6 | Mr | 29 | ·52 | 1 | Ap | 28 | ·05 | 2 | My | 27○ | ·58 | 4 | Je | 26 | ·11 | 7 | Au | 24 | ·18 |
| | | | | | | | | | | | | | | | | | | | | | 5 | Jl | 25 | ·65 | | | | |
| 3940 | 896 | 761 | 22M | ·0538 | 26·36632 | 1·161 | **839** | 4 | 5 | Ap | 17 | ·42 | 6 | My | 16○ | ·95 | 1 | Je | 15 | ·48 | 3 | Jl | 15 | ·01 | 4 | Au | 13 | ·54 |
| 3941 | 897 | 732 | 21M | ·3125 | 15·47476 | 24·874 | **840** | 5 | 2 | Ap | 5 | ·79 | 4 | My | 5●○ | ·32 | 5 | Je | 3 | ·85 | 7 | Jl | 3 | ·38 | 1 | Au | 1 | ·91 |
| 3942 | 898 | 763 | 21M | ·5712 | 4·58319 | 21·032 | **841** | 7 | 7 | Mr | 26 | ·15 | 1 | Ap | 24● | ·69 | 3 | My | 24 | ·22 | 6 | Jl | 22 | ·28 | 7 | Au | 20 | ·81 |
| | | | | | | | | | | | | | | | | | 4 | Je | 22 | ·75 | | | | | | | | |
| 3943 | 899 | 764 | 21M | ·8299 | 23·22222 | 19·176 | **842** | 1 | 6 | Ap | 14 | ·05 | 7 | My | 13 | ·58 | 2 | Je | 12 | ·11 | 3 | Jl | 11 | ·64 | 5 | Au | 10 | ·17 |
| 3944 | 900 | 765 | 22M | ·0885 | 12·33066 | 15·323 | **843** | 2 | 3 | Ap | 3 | ·42 | 4 | My | 2 | ·95 | 6 | Je | 1 | ·48 | 1 | Jl | 1 | ·01 | 2 | Jl | 30 | ·54 |
| 3945 | 901 | 766 | 21M | ·3472 | 1·43909 | 11·481 | **844** | 3 | 7 | Mr | 22 | ·79 | 2 | Ap | 21 | ·32 | 5 | Je | 19 | ·38 | 6 | Jl | 18 | ·91 | 1 | Au | 17 | .44 |
| | | | | | | | | | | | | | 3 | My | 20 | ·85 | | | | | | | | | | | | |
| 3946 | 902 | 767 | 21M | ·6059 | 20·07811 | 9·615 | **845** | 5 | 6 | Ap | 10 | ·68 | 1 | My | 10 | ·21 | 2 | Je | 8 | ·75 | 4 | Jl | 8○ | ·28 | 5 | Au | 6● | ·81 |
| 3947 | 903 | 768 | 21M | ·8646 | 9·18655 | 5·773 | **846** | 6 | 4 | Mr | 31 | ·05 | 6 | Ap | 29 | ·58 | 7 | My | 29 | ·11 | 1 | Je | 27○ | 64 | 3 | Jl | 27 | ·17 |
| 3948 | 904 | 769 | 22M | ·1233 | 27·82557 | 3·906 | **847** | 7 | 2 | Ap | 18 | ·95 | 4 | My | 18 | ·48 | 6 | Je | 17○ | ·01 | 7 | Jl | 16 | ·54 | 2 | Au | 15 | ·07 |
| 3949 | 905 | 770 | 21M | ·3819 | 16·93401 | 0·064 | **848** | 1 | 7 | Ap | 7 | ·32 | 1 | My | 6 | ·85 | 3 | Je | 5● | ·38 | 4 | Jl | 4 | ·91 | 6 | Au | 3 | ·44 |
| 3950 | 906 | 771 | 21M | ·6406 | 6·04245 | 23·776 | **849** | 3 | 4 | Mr | 27 | ·68 | 6 | Ap | 26○ | ·21 | 7 | My | 25 | ·74 | 2 | Je | 24 | ·27 | 5 | Au | 22 | ·34 |
| | | | | | | | | | | | | | | | | | | | | | 3 | Jl | 23 | ·81 | | | | |
| 3951 | 907 | 772 | 21M | ·8993 | 24·68146 | 21·910 | **850** | 4 | 3 | Ap | 15○ | ·58 | 5 | My | 15 | ·11 | 6 | Je | 13 | ·64 | 1 | Jl | 13 | ·17 | 2 | Au | 11 | ·70 |
| 3952 | 908 | 773 | 22M | ·1580 | 13·78990 | 18·068 | **851** | 5 | 7 | Ap | 4●○ | ·95 | 2 | My | 4 | ·48 | 4 | Je | 3 | ·01 | 5 | Jl | 2 | ·54 | 7 | Au | 1 | ·07 |
| 3953 | 909 | 774 | 21M | ·4167 | 2·89834 | 14·226 | **852** | 6 | 5 | Mr | 24 | ·31 | 6 | Ap | 22 | ·85 | 1 | My | 22 | ·38 | 4 | Jl | 20 | ·44 | 5 | Au | 18 | ·97 |
| | | | | | | | | | | | | | | | | | 2 | Je | 20 | ·91 | | | | | | | | |
| 3954 | 910 | 775 | 21M | ·6753 | 21·53736 | 12·359 | **853** | 1 | 4 | Ap | 12 | ·21 | 5 | My | 11 | ·74 | 7 | Je | 10 | ·27 | 1 | Jl | 9 | ·80 | 3 | Au | 8○ | ·33 |
| 3955 | 911 | 776 | 21M | ·9340 | 10·64580 | 8·517 | **854** | 2 | 1 | Ap | 1 | ·58 | 3 | My | 1 | ·11 | 4 | My | 30 | ·64 | 6 | Je | 29 | ·17 | 7 | Jl | 28○ | ·70 |
| 3956 | 912 | 777 | 22M | ·1927 | 29·28482 | 6·651 | **855** | 3 | 7 | Ap | 20 | ·48 | 2 | My | 20 | ·01 | 3 | Je | 18 | ·54 | 5 | Jl | 18 | ·07 | 6 | Au | 16 | ·60 |
| 3957 | 913 | 778 | 21M | ·4514 | 18·39326 | 2·809 | **856** | 4 | 4 | Ap | 8 | ·84 | 6 | My | 8 | ·37 | 7 | Je | 6○ | ·91 | 2 | Jl | 6● | ·44 | 3 | Au | 4 | ·97 |
| 3958 | 914 | 779 | 21M | ·7101 | 7·50169 | 26·521 | **857** | 6 | 2 | Mr | 29 | ·21 | 3 | Ap | 27 | .74 | 5 | My | 27○ | ·27 | 6 | Je | 25 | ·80 | 2 | Au | 23 | ·86 |
| | | | | | | | | | | | | | | | | | | | | | 1 | Jl | 25 | ·33 | | | | |
| 3959 | 915 | 780 | 21M | ·9687 | 26·14065 | 24·655 | **858** | 7 | 1 | Ap | 17 | ·11 | 2 | My | 16○ | ·64 | 4 | Je | 15 | ·17 | 5 | Jl | 14 | ·70 | 7 | Au | 13 | ·23 |
| 3960 | 916 | 781 | 22M | ·2274 | 15·24976 | 20·813 | **859** | 1 | 5 | Ap | 6 | ·48 | 7 | My | 6● | ·01 | 1 | Je | 4 | ·54 | 3 | Jl | 4 | ·07 | 4 | Au | 2 | ·60 |
| 3961 | 917 | 782 | 21M | ·4861 | 4·35759 | 16·970 | **860** | 2 | 2 | Mr | 25○ | ·84 | 4 | Ap | 24 | ·37 | 5 | My | 23 | ·90 | 1 | Jl | 21 | ·97 | 3 | Au | 20 | ·50 |
| | | | | | | | | | | | | | | | | | 7 | Je | 22 | ·43 | | | | | | | | |

**N.B.**—For Surya Siddhanta figures for **A.D. 500** to **A.D. 999** see pages **76—79** below.

**Ârya Siddhanta.**

| Asvina | | | | Kartika | | | | Margasira | | | | Pausha | | | | A.D. | Magha | | | | A.D. | Phalguna | | | | Chaitra | | | |
|---|---|---|---|---|---|---|---|---|---|---|---|---|---|---|---|---|---|---|---|---|---|---|---|---|---|---|---|---|---|
| + 147·65293 | | | | + 177·18353 | | | | + 206·71411 | | | | + 236·24470 | | | | | + 265·77529 | | | | | + 295·30588 | | | | +324·83647<br>+354·36705 | | | |
| + 9·880 | | | | + 11·856 | | | | + 13·832 | | | | + 15·808 | | | | | + 17·784 | | | | | + 19·760 | | | | + 21·736<br>+ 23·712 | | | |
| Week-day | Month | Day | Fraction | Week-day | Month | Day | Fraction | Week-day | Month | Day | Fraction | Week-day | Month | Day | Fraction | | Week-day | Month | Day | Fraction | | Week-day | Month | Day | Fraction | Week-day | Month | Day | Fraction |
| 6 | S | 19 | ·65 | 1 | O | 19○ | ·18 | 2 | N | 17 | ·71 | 4 | D | 17 | ·24 | **23** | 5 | Jr | 15 | ·27 | | 7 | F | 14 | ·30 | 1 | Mr | 15 | ·83 |
| 4 | S | 9○ | ·02 | 5 | O | 8● | ·55 | 7 | N | 7 | ·08 | 1 | D | 6 | ·61 | **24** | 3 | Jr | 5 | ·14 | | 4 | F | 3 | ·67 | 6 | Mr | 4○ | ·20 |
| 1 | Au | 28○ | ·38 | 2 | S | 26● | ·91 | 4 | O | 26 | ·45 | 5 | N | 24 | ·98 | | 7 | D | 24 | ·51 | **25** | 2 | Jr | 23 | ·04 | 3 | F | 21○ | ·57 |
| 7 | S | 16 | ·28 | 1 | O | 15 | ·81 | 3 | N | 14 | ·34 | 4 | D | 13 | ·87 | **26** | 6 | Jr | 12 | ·40 | | 7 | F | 10 | ·93 | 2 | Mr | 12 | ·47 |
| 4 | S | 5 | ·65 | 6 | O | 5 | ·18 | 7 | N | 3 | ·71 | 2 | D | 3 | ·24 | **27** | 3 | Jr | 1○ | ·77 | | 5 | Jr | 31 | ·30 | 6 | Mr | 1 | ·83 |
| {2 | Au | 26 | ·02 | 5 | O | 24 | ·08 | 6 | N | 22 | ·61 | 1 | D | 22○ | ·14 | **28** | 2 | Jr | 20 | ·67 | | 4 | F | 19 | ·20 | 5 | Mr | 19 | ·73 |
| {3 | S | 24 | ·55 | | | | | | | | | | | | | | | | | | | | | | | | | | |
| 1 | S | 12 | ·91 | 2 | O | 12 | ·44 | 4 | N | 10 | ·97 | 5 | D | 10○ | ·51 | **29** | 7 | Jr | 9 | ·04 | | 1 | F | 7 | ·57 | 3 | Mr | 9 | ·10 |
| 5 | S | 2 | ·28 | 6 | O | 1 | ·81 | 1 | O | 31 | ·34 | 2 | N | 29● | ·87 | | 4 | D | 29 | ·40 | **30** | 5 | Jr | 27 | ·93 | 7 | F | 26 | ·46 |
| 4 | S | 21 | ·18 | 5 | O | 20○ | ·71 | 7 | N | 19 | ·24 | 1 | D | 18 | ·77 | **31** | 3 | Jr | 17 | ·30 | | 5 | F | 15 | ·83 | 6 | Mr | 17 | ·36 |
| 1 | S | 10 | ·55 | 3 | O | 10○ | ·08 | 4 | N | 8 | ·61 | 6 | D | 8 | ·14 | **32** | 7 | Jr | 6 | ·67 | | 2 | F | 5 | ·20 | 3 | Mr | 5 | ·73 |
| 5 | Au | 29 | ·91 | 7 | S | 28○ | ·44 | 1 | O | 27 | ·97 | 3 | N | 26 | ·50 | | 5 | D | 26 | ·03 | **33** | 6 | Jr | 24 | ·57 | 1 | F | 23 | ·10 |
| 4 | S | 17● | ·81 | 6 | O | 17 | ·34 | 7 | N | 15 | ·87 | 2 | D | 15 | ·40 | **34** | 3 | Jr | 13 | ·93 | | 5 | F | 12○ | ·46 | 6 | Mr | 13● | ·99 |
| 2 | S | 7● | ·18 | 3 | O | 6 | ·71 | 5 | N | 5 | ·24 | 6 | D | 4 | ·77 | **35** | 1 | Jr | 3 | ·30 | | 2 | F | 1○ | ·83 | 4 | Mr | 3● | [illegible] |
| 6 | Au | 27 | ·54 | 1 | S | 26 | ·07 | 2 | O | 25 | ·61 | 4 | N | 24 | ·14 | | 5 | D | 23 | ·67 | **36** | 7 | Jr | 22○ | ·20 | {1 | F | 20 | ·73 |
| | | | | | | | | | | | | | | | | | | | | | | | | | | {3 | Mr | 21 | ·26 |
| 5 | S | 14 | ·44 | 6 | O | 13 | ·97 | 1 | N | 12 | ·50 | 3 | D | 12 | ·03 | **37** | 4 | Jr | 10 | ·56 | | 6 | F | 9 | ·09 | 7 | Mr | 10 | ·63 |
| 2 | S | 3 | ·83 | 4 | O | 3 | ·34 | 5 | N | 1 | ·87 | 7 | D | 1 | ·40 | | 1 | D | 30● | ·93 | **38** | 3 | Jr | 29 | ·46 | 4 | F | 27 | ·99 |
| 1 | S | 22 | ·71 | 3 | O | 22 | ·24 | 4 | N | 20○ | ·77 | 6 | D | 20 | ·30 | **39** | 7 | Jr | 18 | ·83 | | 2 | F | 17 | ·36 | 3 | Mr | 18 | ·89 |
| 6 | S | 12 | ·07 | 7 | O | 11 | ·60 | 2 | N | 10○ | ·13 | 3 | D | 9 | ·67 | **40** | 5 | Jr | 8 | ·20 | | 6 | F | 6 | ·73 | 1 | Mr | 7 | ·26 |
| 3 | Au | 31 | ·44 | 4 | S | 29 | ·97 | 6 | O | 29●○ | ·50 | 1 | N | 28 | ·03 | | 2 | D | 27 | ·56 | **41** | 4 | Jr | 26 | ·09 | 5 | F | 24 | ·62 |
| 2 | S | 19 | ·34 | 3 | O | 18● | ·87 | 5 | N | 17 | ·40 | 6 | D | 16 | ·93 | **42** | 1 | Jr | 15 | ·46 | | 2 | F | 13 | ·99 | 4 | Mr | 15○ | ·52 |
| 6 | S | 8○ | ·71 | 1 | O | 8 | ·24 | 2 | N | 6 | ·77 | 4 | D | 6 | ·30 | **43** | 5 | Jr | 4 | ·83 | | 7 | F | 3 | ·36 | 1 | Mr | 4●○ | ·89 |
| 4 | Au | 29●○ | ·07 | 5 | S | 27 | ·60 | 7 | O | 27 | ·13 | 1 | N | 25 | ·66 | | 3 | D | 25 | ·19 | **44** | 4 | Jr | 23 | ·73 | 6 | F | 22● | ·26 |
| 2 | S | 15 | ·97 | 4 | O | 15 | ·50 | 6 | N | 14 | ·03 | 7 | D | 13 | ·56 | **45** | 2 | Jr | 12○ | ·09 | | 3 | F | 10● | ·62 | 5 | Mr | 12 | ·15 |
| 7 | S | 5 | ·34 | 1 | O | 4 | ·87 | 3 | N | 3 | ·40 | 4 | D | 2 | ·93 | **46** | 6 | Jr | 1○ | ·46 | | 7 | Jr | 30 | ·99 | 2 | Mr | 1 | ·52 |
| 4 | Au | 25 | ·70 | 7 | O | 23 | ·77 | 2 | N | 22 | ·30 | 3 | D | 21●○ | ·83 | **47** | 5 | Jr | 20 | ·36 | | 6 | F | 18 | ·89 | 1 | Mr | 20 | ·42 |
| 6 | S | 24 | ·23 | | | | | | | | | | | | | | | | | | | | | | | | | | |
| 3 | S | 13 | ·60 | 5 | O | 13 | ·13 | 6 | N | 11 | ·66 | 1 | D | 11 | ·19 | **48** | 2 | Jr | 9 | ·72 | | 4 | F | 8 | ·25 | 5 | Mr | 8 | ·79 |
| 7 | S | 1 | ·97 | 2 | O | 1 | ·50 | 4 | O | 31○ | ·03 | 5 | N | 29 | ·56 | | 7 | D | 29 | ·09 | **49** | 1 | Jr | 27 | ·62 | 3 | F | 26 | ·15 |
| 6 | S | 20 | ·87 | 1 | O | 20○ | ·40 | 2 | N | 18 | ·93 | 4 | D | 18 | ·46 | **50** | 5 | Jr | 16 | ·99 | | 7 | F | 15 | ·52 | 2 | Mr | 17 | ·05 |
| 4 | S | 10 | ·23 | 5 | O | 9●○ | ·76 | 7 | N | 8 | ·29 | 1 | D | 7 | ·83 | **51** | 3 | Jr | 6 | ·36 | | 4 | F | 4 | ·89 | 6 | Mr | 6 | ·42 |
| 1 | Au | 30 | ·60 | 3 | S | 29 | ·13 | 4 | O | 28 | ·66 | 6 | N | 27 | ·19 | | 7 | D | 26 | ·72 | **52** | 2 | Jr | 25 | ·25 | 3 | F | 23○ | ·78 |
| 7 | S | 17 | ·50 | 2 | O | 17 | ·03 | 3 | N | 15 | ·56 | 5 | D | 15 | ·09 | **53** | 6 | Jr | 13 | ·62 | | 1 | F | 12○ | ·15 | 2 | Mr | 13 | ·68 |
| 4 | S | 6● | ·87 | 6 | O | 6 | ·40 | 7 | N | 4 | ·93 | 2 | D | 4 | ·46 | **54** | 3 | Jr | 2 | ·99 | | 5 | F | 1●○ | ·52 | 7 | Mr | 3 | ·05 |
| 2 | Au | 27 | ·23 | 3 | S | 25 | ·76 | 5 | O | 25 | ·29 | 6 | N | 23 | ·82 | | 1 | D | 23 | ·35 | **55** | 2 | Jr | 21 | ·89 | {4 | F | 20 | ·42 |
| | | | | | | | | | | | | | | | | | | | | | | | | | | {5 | Mr | 21 | ·95 |
| 1 | S | 15 | ·13 | 2 | O | 14 | ·66 | 4 | N | 13 | ·19 | 5 | D | 12 | ·72 | **56** | 7 | Jr | 11 | ·25 | | 1 | F | 9 | ·78 | 3 | Mr | 10 | ·31 |
| 5 | S | 3 | ·50 | 7 | O | 3 | ·03 | 1 | N | 1 | ·56 | 3 | D | 1○ | ·09 | | 4 | D | 30● | ·62 | **57** | 6 | Jr | 29 | ·18 | 7 | F | 27 | ·68 |
| 4 | S | 22 | ·39 | 5 | O | 21 | ·93 | 7 | N | 20○ | ·46 | 1 | D | 19 | ·99 | **58** | 3 | Jr | 18 | ·52 | | 5 | F | 17 | ·05 | 6 | Mr | 18 | ·58 |
| 1 | S | 11 | ·76 | 3 | O | 11 | ·29 | 4 | N | 9○ | ·82 | 6 | D | 9 | ·35 | **59** | 7 | Jr | 7 | ·88 | | 2 | F | 6 | ·41 | 3 | Mr | 7 | ·95 |
| 6 | S | 1 | ·13 | 7 | S | 30 | ·66 | 2 | O | 30 | ·19 | 3 | N | 28 | ·72 | | 5 | D | 28 | ·25 | **60** | 6 | Jr | 26 | ·78 | 1 | F | 25 | ·31 |
| 5 | S | 19○ | ·03 | 6 | O | 18● | ·56 | 1 | N | 17 | ·09 | 2 | D | 16 | ·62 | **61** | 4 | Jr | 15 | ·15 | | 5 | F | 13 | ·68 | 7 | Mr | 15●○ | ·21 |

**TABLE X-**

| Kaliyuga. | Vikrama Era. | Saka Era. | Commencement of Solar Year. Month and day A.D. | Fraction of day. | First New-Moon in Solar Year. | Anomaly of first New-Moon. | A.D. | Week-day of 1st January. | Vaisakha ☉'s Anom col. 6 ☾'s Anom col. 7 Week-day | Month | Day | Fraction | Jyeshta + 29·53059 + 1·976. Week-day | Month | Day | Fraction | Ashada + 59·06117 + 3·952 Week-day | Month | Day | Fraction | Sravana + 88·59176 + 5·928 Week-day | Month | Day | Fraction | Bhadrapada + 118·12235 + 7·904 Week-day | Month | Day | Fraction |
|---|---|---|---|---|---|---|---|---|---|---|---|---|---|---|---|---|---|---|---|---|---|---|---|---|---|---|---|---|
| 3962 | 918 | 783 | 21M | ·7448 | 22·99662 | 15·104 | **861** | 4 | 1 | Ap | 13 | ·74 | 3 | My | 13 | ·27 | 4 | Je | 11 | 80 | 6 | Jl | 11 | ·33 | 7 | Au | 9 | ·86 |
| 3963 | 919 | 784 | 22M | ·0035 | 12·10505 | 11·262 | **862** | 5 | 6 | Ap | 3 | ·11 | 7 | My | 2 | ·64 | 2 | Je | 1 | 17 | 3 | Je | 30 | ·70 | 5 | Jl | 30 | ·23 |
| 3964 | 920 | 785 | 22M | ·2621 | 1·21348 | 7·419 | **863** | 6 | 3 | Mr | 23 | ·47 | 6 | My | 21 | ·54 | 1 | Je | 20 | 65 | 2 | Jl | 19○ | ·60 | 4 | Au | 18● | ·13 |
| | | | | | | | | | 5 | Ap | 22 | ·01 | | | | | | | | | | | | | | | | |
| 3965 | 921 | 786 | 21M | ·5208 | 19·85251 | 5·553 | **864** | 7 | 2 | Ap | 10 | ·37 | 3 | My | 9 | ·90 | 5 | Je | 8 | ·43 | 6 | Jl | 7○ | ·96 | 1 | Au | 6● | ·49 |
| 3966 | 922 | 787 | 21M | ·7795 | 8·96095 | 1·711 | **865** | 2 | 6 | Mr | 30 | ·74 | 1 | Ap | 29 | ·27 | 2 | My | 28 | ·80 | 4 | Je | 27○ | ·33 | 5 | Jl | 26 | ·86 |
| | | | | | | | | | | | | | | | | | | | | | | | | | 7 | Au | 25 | ·39 |
| 3967 | 923 | 788 | 22M | ·0382 | 27·59996 | 27·399 | **866** | 3 | 5 | Ap | 18 | ·64 | 7 | My | 18 | ·17 | 1 | Je | 16● | ·70 | 3 | Jl | 16 | ·23 | 4 | Au | 14 | ·76 |
| 3968 | 924 | 789 | 22M | ·2969 | 16·70841 | 23·557 | **867** | 4 | 3 | Ap | 8 | ·00 | 4 | My | 7○ | ·53 | 6 | Je | 6● | ·07 | 7 | Jl | 5 | ·60 | 2 | Au | 4 | ·13 |
| 3969 | 925 | 790 | 21M | ·5556 | 5·81684 | 19·715 | **868** | 5 | 7 | Mr | 27 | ·37 | 1 | Ap | 25○ | ·90 | 3 | My | 25 | ·43 | 4 | Je | 23 | ·96 | 1 | Au | 22 | ·02 |
| | | | | | | | | | | | | | | | | | | | | | 6 | Jl | 23 | ·49 | | | | |
| 3970 | 926 | 791 | 21M | ·8142 | 24·45586 | 17·848 | **869** | 7 | 6 | Ap | 15○ | ·27 | 7 | My | 14 | ·80 | 2 | Je | 13 | ·33 | 3 | Jl | 12 | ·86 | 5 | Au | 11 | ·39 |
| 3971 | 927 | 792 | 22M | ·0729 | 13·56430 | 14·006 | **870** | 1 | 3 | Ap | 4 | ·64 | 5 | My | 4 | ·17 | 6 | Je | 2 | ·70 | 1 | Jl | 2 | ·23 | 2 | Jl | 31 | ·76 |
| 3972 | 928 | 793 | 22M | ·3316 | 2·67283 | 10·164 | **871** | 2 | 1 | Mr | 25 | ·00 | 2 | Ap | 23 | ·53 | 5 | Je | 21 | ·59 | 7 | Jl | 21 | ·12 | 1 | Au | 19○ | ·66 |
| | | | | | | | | | | | | | 4 | My | 23 | ·06 | | | | | | | | | | | | |
| 3973 | 929 | 794 | 21M | ·5903 | 21·31175 | 8·298 | **872** | 3 | 6 | Ap | 11 | ·90 | 1 | My | 11 | ·43 | 2 | Je | 9 | ·96 | 4 | Jl | 9 | ·49 | 6 | Au | 8○ | ·09 |
| 3974 | 930 | 795 | 21M | ·8490 | 10·42019 | 4·455 | **873** | 5 | 4 | Ap | 1 | ·27 | 5 | Ap | 30 | ·80 | 7 | My | 30 | ·33 | 1 | Je | 28 | ·86 | 3 | Jl | 28● | ·32 |
| 3975 | 931 | 796 | 22M | ·1076 | 29·05922 | 2·589 | **874** | 6 | 3 | Ap | 20 | ·16 | 4 | My | 19 | ·70 | 6 | Je | 18○ | ·23 | 7 | Jl | 17● | ·76 | 2 | Au | 16 | ·29 |
| 3976 | 932 | 797 | 22M | ·3663 | 18·16765 | 26·301 | **875** | 7 | 7 | Ap | 9 | ·53 | 2 | My | 9 | ·06 | 3 | Je | 7○ | ·59 | 5 | Jl | 7 | ·12 | 6 | Au | 5 | ·65 |
| 3977 | 933 | 798 | 21M | ·6250 | 7·27608 | 22·459 | **876** | 1 | 4 | Mr | 28 | ·90 | 6 | Ap | 27 | ·43 | 7 | My | 26●○ | ·96 | 2 | Je | 25 | ·49 | 5 | Au | 23 | ·55 |
| | | | | | | | | | | | | | | | | | | | | | 4 | Jl | 25 | ·02 | | | | |
| 3978 | 934 | 799 | 21M | ·8837 | 25·91511 | 20·593 | **877** | 3 | 3 | Ap | 16 | ·80 | 5 | My | 16 | ·32 | 6 | Je | 14 | ·86 | 1 | Jl | 14 | ·39 | 2 | Au | 12 | ·92 |
| 3979 | 935 | 800 | 22M | ·1424 | 15·02354 | 16·751 | **878** | 4 | 1 | Ap | 6○ | ·16 | 2 | My | 5● | ·69 | 4 | Je | 4 | ·22 | 5 | Jl | 3 | ·76 | 7 | Au | 2 | ·29 |
| 3980 | 936 | 801 | 22M | ·4009 | 4·13198 | 12·908 | **879** | 5 | 5 | Mr | 26○ | ·53 | 7 | Ap | 25 | ·06 | 1 | My | 24 | ·59 | 4 | Jl | 22 | ·65 | 6 | Au | 21 | ·18 |
| | | | | | | | | | | | | | | | | | 3 | Je | 23 | ·12 | | | | | | | | |
| 3981 | 937 | 802 | 21M | ·6597 | 22·77101 | 11·042 | **880** | 6 | 4 | Ap | 13 | ·43 | 5 | My | 12 | ·96 | 7 | Je | 11 | ·49 | 2 | Jl | 11 | ·02 | 3 | Au | 9 | ·55 |
| 3982 | 938 | 803 | 21M | ·9184 | 11·87944 | 7·200 | **881** | 1 | 1 | Ap | 2 | ·80 | 3 | My | 2 | ·33 | 4 | My | 31 | ·86 | 6 | Je | 30 | ·39 | 7 | Jl | 29○ | ·92 |
| 3983 | 939 | 804 | 22M | ·1771 | 0·98787 | 3·358 | **882** | 2 | 6 | Mr | 23 | ·16 | 2 | My | 21 | ·22 | 3 | Je | 19 | ·75 | 5 | Jl | 19○ | ·28 | 6 | Au | 17 | ·82 |
| | | | | | | | | | 7 | Ap | 21 | ·69 | | | | | | | | | | | | | | | | |
| 3984 | 940 | 805 | 22M | ·4358 | 19·62690 | 1·491 | **883** | 3 | 5 | Ap | 11 | ·06 | 6 | My | 10 | ·59 | 1 | Je | 9 | ·12 | 2 | Jl | 8●○ | ·65 | 4 | Au | 7 | ·18 |
| 3985 | 941 | 806 | 21M | ·6944 | 8·73533 | 25·204 | **884** | 4 | 2 | Mr | 30 | ·43 | 3 | Ap | 28 | ·96 | 5 | My | 28 | ·49 | 7 | Je | 27 | ·05 | 1 | Jl | 26 | ·55 |
| | | | | | | | | | | | | | | | | | | | | | | | | | 3 | Au | 25 | ·08 |
| 3986 | 942 | 807 | 21M | ·9531 | 27·37436 | 23·337 | **885** | 6 | 1 | Ap | 18 | ·32 | 2 | My | 17○ | ·86 | 4 | Je | 16● | ·39 | 5 | Jl | 15 | ·92 | 7 | Au | 14 | ·45 |
| 3987 | 943 | 808 | 22M | ·2118 | 16·48280 | 19·495 | **886** | 7 | 5 | Ap | 7 | ·69 | 7 | My | 7○ | ·22 | 1 | Je | 5 | ·75 | 3 | Jl | 5 | ·28 | 4 | Au | 3 | ·81 |
| 3988 | 944 | 809 | 22M | ·4705 | 5·59122 | 15·653 | **887** | 1 | 3 | Mr | 28 | ·06 | 4 | Ap | 26○ | ·59 | 6 | My | 26 | ·12 | 7 | Je | 24 | ·65 | 3 | Au | 22 | .71 |
| | | | | | | | | | | | | | | | | | | | | | 2 | Jl | 24 | ·18 | | | | |
| 3989 | 945 | 810 | 21M. | ·7292 | 24·23025 | 13·787 | **888** | 2 | 1 | Ap | 14● | ·96 | 3 | My | 14 | ·49 | 5 | Je | 13 | ·02 | 6 | Jl | 12 | ·55 | 1 | Au | 11 | ·08 |
| 3990 | 946 | 811 | 21M | ·9878 | 13·33869 | 9·944 | **889** | 4 | 6 | Ap | 4● | ·32 | 7 | My | 3 | ·85 | 2 | Je | 2 | ·38 | 3 | Jl | 1 | ·92 | 5 | Jl | 31 | ·45 |
| 3991 | 947 | 812 | 22M | ·2465 | 2·44712 | 6·102 | **890** | 5 | 3 | Mr | 24 | ·69 | 5 | Ap | 23 | ·22 | 1 | Je | 21 | ·28 | 2 | Jl | 20 | ·81 | 4 | Au | 19●○ | ·34 |
| | | | | | | | | | | | | | 6 | My | 22 | ·75 | | | | | | | | | | | | |
| 3992 | 948 | 813 | 22M | ·5052 | 21·08615 | 4·236 | **891** | 6 | 2 | Ap | 12 | ·59 | 4 | My | 12 | ·12 | 5 | Je | 10 | ·65 | 7 | Jl | 10 | ·18 | 1 | Au | 8●○ | ·71 |
| 3993 | 949 | 814 | 21M | ·7639 | 10·19459 | 0·394 | **892** | 7 | 6 | Mr | 31 | ·96 | 1 | Ap | 30 | ·49 | 3 | My | 30 | ·02 | 4 | Je | 28○ | ·55 | 6 | Jl | 28 | ·08 |
| 3994 | 950 | 815 | 22M | ·0226 | 28·83360 | 26·082 | **893** | 2 | 5 | Ap | 19 | ·85 | 7 | My | 19 | ·38 | 1 | Je | 17○ | ·91 | 3 | Jl | 17 | ·44 | 4 | Au | 15 | ·98 |
| 3995 | 951 | 816 | 22M | ·2812 | 17·94204 | 22·240 | **894** | 3 | 3 | Ap | 9 | ·22 | 4 | My | 8 | ·75 | 6 | Je | 7●○ | ·28 | 7 | Jl | 6 | ·81 | 2 | Au | 5 | ·34 |
| 3996 | 952 | 817 | 22M | ·5399 | 7·04979 | 18·397 | **895** | 4 | 7 | Mr | 29 | ·59 | 2 | Ap | 28 | ·12 | 3 | My | 27● | ·65 | 5 | Je | 26 | ·18 | 1 | Au | 24 | ·24 |
| | | | | | | | | | | | | | | | | | | | | | 6 | Jl | 25 | ·71 | | | | |
| 3997 | 953 | 818 | 21M | ·7986 | 25·68950 | 16·531 | **896** | 5 | 6 | Ap | 16○ | ·48 | 1 | My | 16 | ·02 | 2 | Je | 14 | ·55 | 4 | Jl | 14 | ·08 | 5 | Au | 12 | ·61 |
| 3998 | 954 | 819 | 22M | ·0573 | 14·79794 | 12·689 | **897** | 8 | 3 | Ap | 5●○ | ·85 | 5 | My | 5 | ·38 | 6 | Je | 3 | ·91 | 1 | Jl | 3 | ·44 | 2 | Au | 1 | ·97 |
| 3999 | 955 | 820 | 22M | ·3160 | 3·90637 | 8·846 | **898** | 1 | 1 | Mr | 26●○ | ·22 | 2 | Ap | 24 | ·75 | 4 | My | 24 | ·28 | 7 | Jl | 22 | ·34 | 1 | Au | 20 | ·87 |
| | | | | | | | | | | | | | | | | | 5 | Je | 22 | ·81 | | | | | | | | |
| 4000 | 956 | 821 | 22M | ·5746 | 22·54539 | 6·980 | **899** | 2 | 7 | Ap | 14 | ·12 | 1 | My | 13 | ·65 | 3 | Je | 12 | ·18 | 4 | Jl | 11 | ·71 | 6 | Au | 10○ | ·24 |

**N.B.**—For Surya Siddhanta figures for **A.D. 500** to **A.D. 999** see pages **76—79** below.

**Ārya Siddhānta.**

| Asvina | | | | Kartika | | | | Margasira | | | | Pausha | | | | A.D. | Magha | | | | A.D. | Phalguna | | | | Chaitra | | | |
|---|---|---|---|---|---|---|---|---|---|---|---|---|---|---|---|---|---|---|---|---|---|---|---|---|---|---|---|---|---|
| + 147·65293 | | | | + 177·18353 | | | | + 206·71411 | | | | + 236·24470 | | | | | + 265·77529 | | | | | + 295·30588 | | | | { +324·83647 { +354·36705 | | | |
| + 9·880 | | | | + 11·856 | | | | + 13·832 | | | | + 15·808 | | | | | + 17·784 | | | | | + 19·760 | | | | { + 21·736 { + 23·712 | | | |
| Week-day | Month | Day | Fraction | Week-day | Month | Day | Fraction | Week-day | Month | Day | Fraction | Week-day | Month | Day | Fraction | | Week-day | Month | Day | Fraction | | Week-day | Month | Day | Fraction | Week-day | Month | Day | Fraction |
| 2 | S | 8○ | ·39 | 3 | O | 7 | ·92 | 5 | N | 6 | ·45 | 7 | D | 6 | ·09 | 62 | 1 | Jr | 4 | ·52 | | 3 | F | 3 | ·05 | 4 | Mr | 4● | ·58 |
| 6 | Au | 28● | ·76 | 1 | S | 27 | ·29 | 2 | O | 26 | ·82 | 4 | N | 25 | ·35 | | 5 | D | 24 | ·88 | 63 | 7 | Jr | 23○ | ·41 | 1 | F | 21 | ·94 |
| 5 | S | 16 | ·66 | 7 | O | 16 | ·19 | 1 | N | 14 | ·72 | 3 | D | 14 | ·25 | 64 | 4 | Jr | 12○ | ·78 | | 6 | F | 11 | ·31 | 7 | Mr | 11 | ·84 |
| 3 | S | 5 | ·03 | 4 | O | 4 | ·56 | 6 | N | 3 | ·09 | 7 | D | 2 | ·62 | 65 | 2 | Jr | 1○ | ·15 | | 3 | Jr | 30 | ·68 | 5 | Mr | 1 | ·21 |
| 1 | S | 23 | ·92 | 3 | O | 23 | ·45 | 4 | N | 21 | ·98 | 6 | D | 21 | ·51 | 66 | 1 | Jr | 20 | ·05 | | 2 | F | 18 | ·58 | 4 | Mr | 20 | ·11 |
| 6 | S | 13 | ·29 | 7 | O | 12 | ·82 | 2 | N | 11○ | ·35 | 3 | D | 10● | ·88 | 67 | 5 | Jr | 9 | ·41 | | 6 | F | 7 | ·94 | 1 | Mr | 9 | ·47 |
| 3 | S | 2 | ·66 | 5 | O | 2 | ·19 | 6 | O | 31○ | ·72 | 1 | N | 30 | ·25 | | 2 | D | 29 | ·78 | 68 | 4 | Jr | 28 | ·31 | 5 | F | 26 | ·84 |
| 2 | S | 20 | ·55 | 4 | O | 20○ | ·09 | 5 | N | 18 | ·62 | 7 | D | 18 | ·15 | 69 | 1 | Jr | 16 | ·68 | | 3 | F | 15 | ·21 | 4 | Mr | 16 | ·74 |
| 6 | S | 9 | ·92 | 1 | O | 9● | ·45 | 2 | N | 7 | ·98 | 4 | D | 7 | ·51 | 70 | 6 | Jr | 6 | ·04 | | 7 | F | 4 | ·57 | 2 | Mr | 6○ | ·11 |
| 4 | Au | 30 | ·29 | 5 | S | 28 | ·82 | 7 | O | 28 | ·35 | 1 | N | 26 | ·88 | | 3 | D | 26 | ·41 | 71 | 4 | Jr | 24 | ·94 | 6 | F | 23○ | ·47 |
| 3 | S | 18 | ·19 | 4 | O | 17 | ·72 | 6 | N | 16 | ·25 | 7 | D | 15 | ·78 | 72 | 2 | Jr | 14 | ·31 | | 3 | F | 12○ | ·84 | 5 | Mr | 13 | ·37 |
| 7 | S | 6 | ·55 | 2 | O | 6 | ·08 | 3 | N | 4 | ·61 | 5 | D | 4 | ·14 | 73 | 6 | Jr | 2 | ·68 | | 1 | F | 1● | ·21 | 2 | Mr | 2 | ·74 |
| 4 | Au | 26 | ·92 | 6 | S | 25 | ·45 | 7 | O | 24 | ·98 | 2 | N | 23 | ·51 | | 4 | D | 23 | ·04 | 74 | 5 | Jr | 21 | ·57 | { 7 | F | 20 | ·10 |
| | | | | | | | | | | | | | | | | | | | | | | | | | | { 1 | Mr | 21 | ·63 |
| 3 | S | 14 | ·82 | 5 | O | 14 | ·35 | 6 | N | 12 | ·89 | 1 | D | 12○ | ·41 | 75 | 2 | Jr | 10 | ·94 | | 4 | F | 9 | ·47 | 6 | Mr | 11 | ·00 |
| 1 | S | 4 | ·19 | 2 | O | 3 | ·72 | 4 | N | 2 | ·25 | 5 | D | 1○ | ·78 | | 7 | D | 31 | ·31 | 76 | 1 | Jr | 29 | ·84 | 3 | F | 28 | ·37 |
| 7 | S | 22 | ·08 | 1 | O | 21 | ·61 | 3 | N | 20○ | ·14 | 4 | D | 19 | ·67 | 77 | 6 | Jr | 18 | ·20 | | 7 | F | 16 | ·74 | 2 | Mr | 18 | ·27 |
| 4 | S | 11 | ·45 | 5 | O | 10 | ·98 | 7 | N | 9● | ·51 | 2 | D | 9 | ·04 | 78 | 3 | Jr | 7 | ·57 | | 5 | F | 6 | ·10 | 6 | Mr | 7 | ·63 |
| 1 | Au | 31 | ·82 | 3 | S | 30○ | ·35 | 4 | O | 29 | ·88 | 6 | N | 28 | ·41 | | 7 | D | 27 | ·94 | 79 | 2 | Jr | 26 | ·47 | 4 | F | 25 | ·00 |
| 7 | S | 19○ | ·71 | 2 | O | 19 | ·24 | 3 | N | 17 | ·78 | 5 | D | 17 | ·31 | 80 | 6 | Jr | 15 | ·84 | | 1 | F | 14 | ·37 | 2 | Mr | 14○ | ·90 |
| 5 | S | 8●○ | ·08 | 6 | O | 7 | ·61 | 1 | N | 6 | ·14 | 2 | D | 5 | ·67 | 81 | 4 | Jr | 4 | ·20 | | 5 | F | 2○ | ·73 | 7 | Mr | 4 | ·26 |
| 2 | Au | 28 | ·45 | 3 | S | 26 | ·98 | 5 | O | 26 | ·51 | 7 | N | 25 | ·04 | | 1 | D | 24 | ·57 | 82 | 3 | Jr | 23○ | ·10 | 4 | F | 21 | ·63 |
| 1 | S | 16 | ·35 | 2 | O | 15 | ·88 | 4 | N | 14 | ·41 | 5 | D | 13 | ·94 | 83 | 7 | Jr | 12○ | ·47 | | 2 | F | 11 | ·00 | 3 | Mr | 12 | ·53 |
| 5 | S | 5 | ·71 | 7 | O | 5 | ·24 | 1 | N | 3 | ·77 | 3 | D | 3 | ·30 | 84 | 4 | Jr | 1● | ·84 | | 6 | Jr | 31 | ·37 | 7 | F | 29 | ·90 |
| 4 | S | 23 | ·61 | 6 | O | 23 | ·14 | 7 | N | 21○ | ·67 | 2 | D | 21● | ·20 | 85 | 3 | Jr | 19 | ·73 | | 5 | F | 18 | ·26 | 6 | Mr | 19 | ·79 |
| 1 | S | 12 | ·98 | 3 | O | 12 | ·51 | 5 | N | 11○ | ·04 | 6 | D | 10 | ·57 | 86 | 1 | Jr | 9 | ·10 | | 2 | F | 7 | ·63 | 4 | Mr | 9 | ·16 |
| 6 | S | 2 | ·34 | 7 | O | 1 | ·88 | 2 | O | 31○ | ·41 | 3 | N | 29 | ·94 | | 5 | D | 29 | ·47 | 87 | 7 | Jr | 28 | ·00 | 1 | F | 26 | ·53 |
| 5 | S | 21 | ·24 | 6 | O | 20 | ·77 | 1 | N | 19 | ·30 | 2 | D | 18 | ·83 | 88 | 4 | Jr | 17 | ·36 | | 5 | F | 15 | ·90 | 7 | Mr | 16○ | ·43 |
| 2 | S | 9 | ·61 | 4 | O | 9● | ·14 | 5 | N | 7 | ·67 | 7 | D | 7 | ·20 | 89 | 1 | Jr | 5 | ·73 | | 3 | F | 4 | ·26 | 4 | Mr | 5○ | ·79 |
| 6 | Au | 29○ | ·98 | 1 | S | 28 | ·51 | 3 | O | 28 | ·04 | 4 | N | 26 | ·57 | | 6 | D | 26 | ·10 | 90 | 7 | Jr | 24 | ·63 | 2 | F | 23○ | ·16 |
| 5 | S | 17 | ·87 | 7 | O | 17 | ·40 | 1 | N | 15 | ·94 | 3 | D | 15 | ·47 | 91 | 5 | Jr | 14 | ·00 | | 6 | F | 12 | ·53 | 1 | Mr | 14 | ·06 |
| 3 | S | 7 | ·24 | 4 | O | 6 | ·77 | 6 | N | 5 | ·30 | 7 | D | 4 | ·83 | 92 | 2 | Jr | 3 | ·36 | | 3 | F | 1● | ·89 | 5 | Mr | 2 | ·42 |
| 7 | Au | 26 | ·61 | { 2 | S | 25 | ·14 | **Margasira Kshaya** | | | | 5 | N | 23 | ·20 | | 6 | D | 22○ | ·73 | 93 | 1 | Jr | 21 | ·26 | { 2 | F | 19 | ·79 |
| | | | | { 3 | O | 24 | ·67 | | | | | | | | | | | | | | | | | | | { 4 | Mr | 21 | ·32 |
| 6 | S | 14 | ·51 | 1 | O | 14 | ·04 | 2 | N | 12 | ·57 | 4 | D | 12○ | ·10 | 94 | 5 | Jr | 10 | ·63 | | 7 | F | 9 | ·16 | 1 | Mr | 10 | ·69 |
| 3 | S | 3 | ·87 | 5 | O | 3 | ·40 | 6 | N | 1 | ·93 | 1 | D | 1●○ | ·46 | | 3 | D | 31 | ·00 | 95 | 4 | Jr | 29 | ·53 | 6 | F | 28 | ·06 |
| 2 | S | 22 | ·77 | 4 | O | 22 | ·30 | 5 | N | 20● | ·83 | 7 | D | 20 | 36 | 96 | 1 | Jr | 18 | ·89 | | 3 | F | 17 | ·42 | 4 | Mr | 17 | ·95 |
| 7 | S | 11 | ·14 | 1 | O | 10○ | ·67 | 3 | N | 9 | ·20 | 4 | D | 8 | ·73 | 97 | 6 | Jr | 7 | ·26 | | 7 | F | 5 | ·79 | 2 | Mr | 7 | ·32 |
| 4 | Au | 31 | ·50 | 6 | S | 30○ | ·04 | 7 | O | 29 | ·57 | 2 | N | 28 | ·10 | | 3 | D | 27 | ·63 | 98 | 5 | Jr | 26 | ·16 | 6 | F | 24 | ·69 |
| 3 | S | 19○ | ·40 | 4 | O | 18 | ·93 | 6 | N | 17 | ·46 | 7 | D | 16 | ·99 | 99 | 2 | Jr | 15 | ·52 | | 4 | F | 14 | ·06 | 5 | Mr | 15● | ·59 |
| 7 | S | 8 | ·77 | 2 | O | 8 | ·30 | 3 | N | 6 | ·83 | 5 | D | 6 | ·36 | 900 | 6 | Jr | 4 | ·89 | | 1 | F | 3○ | ·42 | 2 | Mr | 3 | ·95 |

**TABLE X—**

| Kaliyuga. | Vikrama Era. | Saka Era. | Commencement of Solar Year. Month and day A.D. | Fraction of day. | First New-Moon in Solar Year. | Anomaly of first New-Moon. | A.D. | Week-day of 1st January. | Vaisakha Week-day | Month | Day | Fraction | Jyeshta Week-day | Month | Day | Fraction | Ashada Week-day | Month | Day | Fraction | Sravana Week-day | Month | Day | Fraction | Bhadrapada Week-day | Month | Day | Fraction |
|---|---|---|---|---|---|---|---|---|---|---|---|---|---|---|---|---|---|---|---|---|---|---|---|---|---|---|---|---|
| | | | | | | | | | ⊙'s Anom col. 6 | | | | + 29·53059 | | | | + 59·06117 | | | | + 88·59176 | | | | + 118·12235 | | | |
| | | | | | | | | | ☾'s Anom col. 7 | | | | + 1·976 | | | | + 3·952 | | | | + 5·928 | | | | + 7·904 | | | |
| 4001 | 957 | 822 | 21M | ·8333 | 11·65384 | 3·138 | **900** | 3 | 4 | Ap | 2 | ·49 | 6 | My | 2 | ·02 | 7 | My | 31 | ·55 | 2 | Je | 30 | ·08 | 3 | Jl | 29○ | ·61 |
| 4002 | 958 | 823 | 22M | ·0920 | 0·76328 | 26·850 | **901** | 5 | 1 | Mr | 22 | ·86 | 4 | My | 20 | ·92 | 6 | Je | 19 | ·45 | 7 | Jl | 18○ | ·98 | 2 | Au | 17 | ·51 |
| | | | | | | | | | 3 | Ap | 21 | ·39 | | | | | | | | | | | | | | | | |
| 4003 | 959 | 824 | 22M | ·3507 | 19·40129 | 24·984 | **902** | 6 | 7 | Ap | 10 | ·75 | 2 | My | 10 | ·28 | 3 | Je | 8 | ·82 | 5 | Jl | 8● | ·35 | 6 | Au | 6 | ·88 |
| 4004 | 960 | 825 | 22M | ·6094 | 8·50973 | 21·142 | **903** | 7 | 5 | Mr | 31 | ·12 | 6 | Ap | 29 | ·65 | 1 | My | 29○ | ·18 | 2 | Je | 27 | ·71 | 4 | Jl | 27 | ·24 |
| | | | | | | | | | | | | | | | | | | | | | | | | | 5 | Au | 25 | ·77 |
| 4005 | 961 | 826 | 21M | ·8680 | 27·14876 | 19·275 | **904** | 1 | 4 | Ap | 18 | ·02 | 5 | My | 17○ | ·55 | 7 | Je | 16 | ·08 | 1 | Jl | 15 | ·61 | 3 | Au | 14 | ·14 |
| 4006 | 962 | 827 | 22M | ·1267 | 16·25719 | 15·433 | **905** | 3 | 1 | Ap | 7 | ·39 | 2 | My | 6●○ | ·92 | 4 | Je | 5 | ·45 | 5 | Jl | 4 | ·98 | 7 | Au | 3 | ·51 |
| 4007 | 963 | 828 | 22M | ·3854 | 5·36563 | 11·591 | **906** | 4 | 5 | Mr | 27 | ·75 | 7 | Ap | 26● | ·28 | 1 | My | 25 | ·81 | 3 | Je | 24 | ·34 | 6 | Au | 22 | ·41 |
| | | | | | | | | | | | | | | | | | | | | | 4 | Jl | 23 | ·88 | | | | |
| 4008 | 964 | 829 | 22M | ·6441 | 24·00466 | 9·725 | **907** | 5 | 4 | Ap | 15 | ·65 | 6 | My | 15 | ·18 | 7 | Je | 13 | ·71 | 2 | Jl | 13 | ·24 | 3 | Au | 11 | ·77 |
| 4009 | 965 | 830 | 21M | ·9028 | 13·11308 | 5·883 | **908** | 6 | 2 | Ap | 4 | ·02 | 3 | My | 3 | ·55 | 5 | Je | 2 | ·08 | 6 | Jl | 1 | ·61 | 1 | Jl | 31 | ·14 |
| 4010 | 966 | 831 | 22M | ·1614 | 2·22152 | 2·040 | **909** | 1 | 6 | Mr | 24 | ·39 | 7 | Ap | 22 | ·92 | 3 | Je | 20 | ·98 | 5 | Jl | 20 | ·51 | 7 | Au | 19○ | ·04 |
| | | | | | | | | | | | | | 2 | My | 22 | ·45 | | | | | | | | | | | | |
| 4011 | 967 | 832 | 22M | ·4201 | 20·86055 | 0·174 | **910** | 2 | 5 | Ap | 12 | ·28 | 6 | My | 11 | ·81 | 1 | Je | 10 | ·34 | 2 | Jl | 9○ | ·87 | 4 | Au | 8 | ·40 |
| 4012 | 968 | 833 | 22M | ·6788 | 9·96898 | 23·886 | **911** | 3 | 2 | Ap | 1 | ·65 | 4 | My | 1 | ·18 | 5 | My | 30 | ·71 | 7 | Je | 29○ | ·24 | 1 | Jl | 28 | ·77 |
| 4013 | 969 | 834 | 21M | ·9375 | 28·60801 | 22·020 | **912** | 4 | 1 | Ap | 19 | ·55 | 3 | My | 19 | ·08 | 4 | Je | 17○ | ·61 | 6 | Jl | 17 | ·14 | 7 | Au | 15 | ·67 |
| 4014 | 970 | 835 | 22M | ·1962 | 17·71644 | 18·178 | **913** | 6 | 5 | Ap | 8 | ·92 | 7 | My | 8 | ·44 | 1 | Je | 6● | ·98 | 3 | Jl | 6 | ·51 | 5 | Au | 5 | ·04 |
| 4015 | 971 | 836 | 22M | ·4549 | 6·82487 | 14·335 | **914** | 7 | 3 | Mr | 29 | ·28 | 4 | Ap | 27○ | ·81 | 6 | My | 27 | ·34 | 7 | Je | 25 | ·88 | 4 | Au | 24 | ·94 |
| | | | | | | | | | | | | | | | | | | | | | 2 | Jl | 25 | ·40 | | | | |
| 4016 | 972 | 837 | 22M | ·7135 | 25·46390 | 12·469 | **915** | 1 | 2 | Ap | 17○ | ·18 | 3 | My | 16 | ·71 | 5 | Je | 15 | ·24 | 6 | Jl | 14 | ·77 | 1 | Au | 13 | ·30 |
| 4017 | 973 | 838 | 21M | ·9722 | 14·57234 | 8·627 | **916** | 2 | 6 | Ap | 5●○ | ·55 | 1 | My | 5 | ·08 | 2 | Je | 3 | ·61 | 4 | Jl | 3 | ·14 | 5 | Au | 1 | ·67 |
| 4018 | 974 | 839 | 22M | ·2309 | 3·68077 | 4·785 | **917** | 4 | 3 | Mr | 25 | ·91 | 5 | Ap | 24 | ·44 | 6 | My | 23 | ·97 | 3 | Jl | 22 | ·04 | 4 | Au | 20 | ·57 |
| | | | | | | | | | | | | | | | | | 1 | Je | 22 | ·50 | | | | | | | | |
| 4019 | 975 | 840 | 22M | ·4896 | 22·31980 | 2·910 | **918** | 5 | 2 | Ap | 13 | ·81 | 4 | My | 13 | ·34 | 5 | Je | 11 | ·87 | 7 | Jl | 11 | ·40 | 2 | Au | 10○ | ·93 |
| 4020 | 976 | 841 | 22M | ·7483 | 11·42823 | 26·631 | **919** | 6 | 7 | Ap | 3 | ·18 | 1 | My | 2 | ·71 | 3 | Je | 1 | ·24 | 4 | Je | 30 | ·77 | 6 | Jl | 30○ | ·30 |
| 4021 | 977 | 842 | 22M | ·0069 | 0·53666 | 22·789 | **920** | 7 | 4 | Mr | 22 | ·54 | 7 | My | 20 | ·61 | 2 | Je | 19 | ·14 | 3 | Jl | 18● | ·67 | 5 | Au | 17 | ·20 |
| | | | | | | | | | 6 | Ap | 21 | ·08 | | | | | | | | | | | | | | | | |
| 4022 | 978 | 843 | 22M | ·2656 | 19·17569 | 20·922 | **921** | 2 | 3 | Ap | 10 | ·44 | 4 | My | 9 | ·97 | 6 | Je | 8○ | ·50 | 1 | Jl | 8● | ·03 | 2 | Au | 6 | ·56 |
| 4023 | 979 | 844 | 22M | ·5243 | 8·28413 | 17·080 | **922** | 3 | 7 | Mr | 30 | ·81 | 2 | Ap | 29 | ·34 | 3 | My | 28○ | ·87 | 5 | Je | 27 | ·40 | 6 | Jl | 26 | ·93 |
| | | | | | | | | | | | | | | | | | | | | | | | | | 1 | Au | 25 | ·46 |
| 4024 | 980 | 845 | 22M | ·7830 | 26·92314 | 15·214 | **923** | 4 | 6 | Ap | 18 | ·71 | 1 | My | 18○ | ·24 | 2 | Je | 16 | ·77 | 4 | Jl | 16 | ·30 | 5 | Au | 14 | ·83 |
| 4025 | 981 | 846 | 22M | ·0417 | 16·03158 | 11·372 | **924** | 5 | 4 | Ap | 7 | ·07 | 5 | My | 6 | ·60 | 7 | Je | 5 | ·14 | 1 | Jl | 4 | ·67 | 3 | Au | 3 | ·20 |
| 4026 | 982 | 847 | 22M | ·3003 | 5·14002 | 7·529 | **925** | 7 | 1 | Mr | 27○ | ·44 | 2 | Ap | 25 | ·97 | 4 | My | 25 | ·50 | 7 | Jl | 23 | ·56 | 2 | Au | 22 | ·09 |
| | | | | | | | | | | | | | | | | | 6 | Je | 24 | ·03 | | | | | | | | |
| 4027 | 983 | 848 | 22M | ·5590 | 23·77904 | 5·663 | **926** | 1 | 7 | Ap | 15 | ·34 | 1 | My | 14 | ·87 | 3 | Je | 13 | ·40 | 4 | Jl | 12 | ·93 | 6 | Au | 11 | ·46 |
| 4028 | 984 | 849 | 22M | ·8177 | 12·88748 | 1·821 | **927** | 2 | 4 | Ap | 4 | ·71 | 6 | My | 4 | ·24 | 7 | Je | 2 | ·77 | 2 | Jl | 2 | ·30 | 3 | Jl | 31 | ·83 |
| 4029 | 985 | 850 | 22M | ·0764 | 1·99592 | 25·533 | **928** | 3 | 2 | Mr | 24 | ·07 | 3 | Ap | 22 | ·60 | 6 | Je | 20 | ·66 | 1 | Jl | 20○ | ·20 | 2 | Au | 18● | ·73 |
| | | | | | | | | | | | | | 5 | My | 22 | ·13 | | | | | | | | | | | | |
| 4030 | 986 | 851 | 22M | ·3351 | 20·63493 | 23·667 | **929** | 5 | 7 | Ap | 11 | ·97 | 2 | My | 11 | ·50 | 4 | Je | 10 | ·03 | 5 | Jl | 9○ | ·56 | 7 | Au | 8 | ·09 |
| 4031 | 987 | 852 | 22M | ·5937 | 9·74337 | 19·822 | **930** | 6 | 5 | Ap | 1 | ·34 | 6 | Ap | 30 | ·87 | 1 | My | 30 | ·40 | 2 | Je | 28●○ | ·93 | 4 | Jl | 28 | ·46 |
| 4032 | 988 | 853 | 22M | ·8524 | 28·38240 | 17·959 | **931** | 7 | 4 | Ap | 20 | ·24 | 5 | My | 19 | ·77 | 7 | Je | 18 | ·30 | 1 | Jl | 17 | ·83 | 3 | Au | 16 | ·36 |
| 4033 | 989 | 854 | 22M | ·1111 | 17·49083 | 14·116 | **932** | 1 | 1 | Ap | 8 | ·60 | 3 | My | 8○ | ·13 | 4 | Je | 6 | ·66 | 6 | Jl | 6 | ·19 | 7 | Au | 4 | ·72 |
| 4034 | 990 | 855 | 22M | ·3698 | 6·59927 | 10·274 | **933** | 3 | 5 | Mr | 28 | ·97 | 7 | Ap | 27○ | ·50 | 2 | My | 27 | ·03 | 3 | Je | 25 | ·56 | 6 | Au | 23 | ·62 |
| | | | | | | | | | | | | | | | | | | | | | 5 | Jl | 25 | ·09 | | | | |
| 4035 | 991 | 856 | 22M | ·6285 | 25·23830 | 8·408 | **934** | 4 | 4 | Ap | 16○ | ·87 | 6 | My | 16 | ·40 | 7 | Je | 14 | ·93 | 2 | Jl | 14 | ·46 | 3 | Au | 12 | ·99 |
| 4036 | 992 | 857 | 22M | ·8871 | 14·34672 | 4·565 | **935** | 5 | 2 | Ap | 6● | ·23 | 3 | My | 5 | ·76 | 5 | Je | 4 | ·30 | 6 | Jl | 3 | ·83 | 1 | Au | 2 | ·36 |
| 4037 | 993 | 858 | 22M | ·1458 | 3·45516 | 0·723 | **936** | 6 | 6 | Mr | 25 | ·60 | 1 | Ap | 24 | ·13 | 4 | Je | 22 | ·19 | 5 | Jl | 21 | ·72 | 7 | Au | 20○ | ·25 |
| | | | | | | | | | | | | | 2 | My | 23 | ·66 | | | | | | | | | | | | |
| 4038 | 994 | 859 | 22M | ·4045 | 22·09419 | 26·412 | **937** | 1 | 5 | Ap | 13 | ·50 | 1 | My | 13 | ·03 | 1 | Je | 11 | ·56 | 3 | Jl | 11 | ·09 | 4 | Au | 9○ | ·62 |
| 4039 | 995 | 860 | 22M | ·6632 | 11·20292 | 22·570 | **938** | 2 | 2 | Ap | 2 | ·87 | 4 | My | 2 | ·40 | 5 | My | 31 | ·93 | 7 | Je | 30 | ·46 | 1 | Jl | 29 | ·99 |

**N.B.**—For Surya Siddhanta figures for **A.D. 500** to **A.D. 999** see pages **76—78** below.

**Arya Siddhanta.**

| Asvina | | | | Kartika | | | | Margasira | | | | Pausha | | | | A.D. | Magha | | | | A.D. | Phalguna | | | | Chaitra | | | |
|---|---|---|---|---|---|---|---|---|---|---|---|---|---|---|---|---|---|---|---|---|---|---|---|---|---|---|---|---|---|
| + 147·65293 | | | | + 177·18353 | | | | + 206·71411 | | | | + 236·24470 | | | | | + 265·77529 | | | | | + 295·30588 | | | | +324·83647 +354·36705 | | | |
| + 9·880 | | | | + 11·856 | | | | + 13·832 | | | | + 15·808 | | | | | + 17·784 | | | | | + 19·760 | | | | + 21·736 + 23·712 | | | |
| Week-day | Month | Day | Fraction | Week-day | Month | Day | Fraction | Week-day | Month | Day | Fraction | Week-day | Month | Day | Fraction | | Week-day | Month | Day | Fraction | | Week-day | Month | Day | Fraction | Week-day | Month | Day | Fraction |
| 5 | Au | 28 | ·14 | 6 | S | 26 | ·67 | 1 | O | 26 | ·20 | 2 | N | 24 | ·73 | | 4 | D | 24 | ·26 | **01** | 5 | Jr | 22●○ | ·80 | 7 | F | 21 | ·33 |
| 4 | S | 16 | ·04 | 5 | O | 15 | ·57 | 7 | N | 14 | ·10 | 1 | D | 13 | ·63 | **02** | 3 | Jr | 12 | ·16 | | 4 | F | 10 | ·69 | 6 | Mr | 12 | ·22 |
| 1 | S | 5 | ·41 | 2 | O | 4 | ·94 | 4 | N | 3 | ·47 | 6 | D | 3ↄ | ·00 | **03** | 7 | Jr | 1 | ·53 | | 2 | Jr | 31 | ·06 | 3 | Mr | 1 | ·59 |
| 7 | S | 24 | ·30 | 1 | O | 23 | ·81 | 3 | N | 22ↄ | ·37 | 4 | D | 21 | ·90 | **04** | 6 | Jr | 20 | ·43 | | 7 | F | 18 | ·96 | 2 | Mr | 19 | ·49 |
| 4 | S | 12 | ·67 | 6 | O | 12 | ·20 | 7 | N | 10●○ | ·73 | 2 | D | 10 | ·26 | **05** | 3 | Jr | 8 | ·79 | | 5 | F | 7 | ·32 | 6 | Mr | 8 | ·86 |
| 2 | S | 2 | ·04 | 3 | O | 1 | ·57 | 5 | O | 31 | ·10 | 6 | N | 29 | ·63 | | 1 | D | 29 | ·16 | **06** | 2 | Jr | 27 | ·69 | 4 | F | 26 | ·22 |
| 7 | S | 20 | ·94 | 2 | O | 20 | ·47 | 4 | N | 19 | ·00 | 5 | D | 18 | ·53 | **07** | 7 | Jr | 17 | ·06 | | 1 | F | 15 | ·59 | 3 | Mr | 17○ | ·12 |
| 5 | S | 10○ | ·30 | 6 | O | 9● | ·83 | 1 | N | 8 | ·36 | 1 | D | 7 | ·90 | **08** | 4 | Jr | 6 | ·43 | | 5 | F | 4 | ·96 | 7 | Mr | 5●○ | ·49 |
| 2 | Au | 29○ | ·67 | 4 | S | 28 | ·20 | 5 | O | 27 | ·73 | 7 | N | 26 | ·26 | | 1 | D | 25 | ·79 | **09** | 3 | Jr | 24 | ·32 | 4 | F | 22 | ·85 |
| 1 | S | 17 | ·57 | 3 | O | 17 | ·10 | 4 | N | 15 | ·63 | 6 | D | 15 | ·16 | **10** | 7 | Jr | 13 | ·69 | | 2 | F | 12 | ·22 | 3 | Mr | 13 | ·75 |
| 5 | S | 6 | ·94 | 6 | O | 6 | ·47 | 2 | N | 5 | ·00 | 3 | D | 4 | ·53 | **11** | 5 | Jr | 3○ | ·06 | | 6 | F | 1● | ·59 | 1 | Mr | 3 | ·12 |
| {3 | Au | 27 | ·30 | 6 | O | 25 | ·36 | 7 | N | 23 | ·90 | **Pausha Kshaya** | | | | | 2 | D | 23○ | ·42 | **12** | 3 | Jr | 21 | ·96 | {5 | F | 20 | ·49 |
| {4 | S | 25 | ·84 | | | | | | | | | | | | | | | | | | | | | | | {7 | Mr | 21 | ·02 |
| 2 | S | 14 | ·20 | 3 | O | 13 | ·73 | 5 | N | 12 | ·26 | 6 | D | 11○ | ·79 | **13** | 1 | Jr | 10 | ·32 | | 2 | F | 8 | ·86 | 4 | Mr | 10 | [illegible] |
| 6 | S | 3 | ·57 | 1 | O | 3 | ·10 | 2 | N | 1 | ·63 | 4 | D | 1 | ·16 | | 5 | D | 30 | ·69 | **14** | 7 | Jr | 29 | ·22 | 1 | F | 27 | ·75 |
| 5 | S | 22 | ·46 | 7 | O | 22○ | ·00 | 1 | N | 20● | ·53 | 3 | D | 20 | ·06 | **15** | 4 | Jr | 18 | ·62 | | 6 | F | 17 | ·12 | 7 | Mr | 18 | ·65 |
| 2 | S | 11 | ·83 | 4 | O | 11○ | ·36 | 5 | N | 9 | ·90 | 7 | D | 9 | ·42 | **16** | 1 | Jr | 7 | ·96 | | 3 | F | 6 | ·48 | 5 | Mr | 7 | ·02 |
| 7 | Au | 31 | ·20 | 1 | S | 29○● | ·73 | 3 | O | 29 | ·26 | 4 | N | 27 | ·79 | | 6 | D | 27 | ·32 | **17** | 7 | Jr | 25 | ·85 | 2 | F | 24 | ·38 |
| 6 | S | 19● | ·10 | 7 | O | 18 | ·63 | 2 | N | 17 | ·16 | 3 | D | 16 | ·69 | **18** | 5 | Jr | 15 | ·22 | | 6 | F | 13○ | ·75 | 1 | Mr | 15 | ·28 |
| 3 | S | 8● | ·46 | 4 | O | 7 | ·99 | 6 | N | 6 | ·52 | 1 | D | 6 | ·06 | **19** | 1 | Jr | 4 | ·59 | | 4 | F | 3○ | ·12 | 5 | Mr | 4 | ·65 |
| 7 | Au | 28 | ·83 | 2 | S | 27 | ·36 | 3 | O | 26 | ·89 | 5 | N | 25 | ·42 | | 6 | D | 24 | ·95 | **20** | 1 | Jr | 23● | ·48 | 3 | F | 22 | ·01 |
| 6 | S | 15 | ·73 | 1 | O | 15 | ·26 | 2 | N | 13 | ·79 | 4 | D | 13○ | ·32 | **21** | 5 | Jr | 11● | ·85 | | 7 | F | 10 | ·38 | 1 | Mr | 11 | ·91 |
| 4 | S | 5 | ·10 | 5 | O | 4 | ·63 | 7 | N | 3 | ·16 | 1 | D | 2○ | ·69 | **22** | 3 | Jr | 1 | ·22 | | 4 | Jr | 30 | ·75 | 6 | Mr | 1 | ·28 |
| 3 | S | 24 | ·00 | 4 | O | 23 | ·53 | 6 | N | 22ↄ | ·05 | 7 | D | 21 | ·58 | **23** | 2 | Jr | 20 | ·12 | | 3 | F | 18 | ·65 | 5 | Mr | 20 | ·18 |
| 7 | S | 13 | ·36 | 1 | O | 12 | ·89 | 3 | N | 11● | ·42 | 4 | D | 10 | ·95 | **24** | 6 | Jr | 9 | ·48 | | 1 | F | 8 | ·01 | 2 | Mr | 8 | ·54 |
| 4 | S | 1 | ·73 | 6 | O | 1 | ·26 | 7 | O | 30 | ·79 | 2 | N | 29 | ·32 | | 3 | D | 28 | ·85 | **25** | 5 | Jr | 27 | ·38 | 6 | F | 25 | ·91 |
| 3 | S | 20○ | ·62 | 5 | O | 20 | ·16 | 6 | N | 18 | ·69 | 1 | D | 18 | ·22 | **26** | 2 | Jr | 16 | ·75 | | 4 | F | 15 | ·28 | 5 | Mr | 16ↄ | ·81 |
| 7 | S | 9○ | ·99 | 2 | O | 9 | ·52 | 4 | N | 8 | ·05 | 5 | D | 7 | ·58 | **27** | 7 | Jr | 6 | ·11 | | 1 | F | 4 | ·64 | 3 | Mr | 6● | ·18 |
| 5 | Au | 30●○ | ·36 | 6 | S | 28 | ·89 | 1 | O | 28 | ·42 | 2 | N | 26 | ·95 | | 4 | D | 26 | ·48 | **28** | 6 | Jr | 25 | ·01 | 7 | F | 23● | ·54 |
| 4 | S | 17 | ·26 | 5 | O | 16 | ·79 | 7 | N | 15 | ·32 | 1 | D | 14 | ·85 | **29** | 3 | Jr | 13○ | ·38 | | 4 | F | 11 | ·91 | 6 | Mr | 13 | ·44 |
| 1 | S | 6 | ·62 | 3 | O | 6 | ·15 | 4 | N | 4 | ·68 | 6 | D | 4 | ·22 | **30** | 7 | Jr | 2○ | ·75 | | 2 | F | 1 | ·28 | 3 | Mr | 2 | ·81 |
| {5 | Au | 26 | ·99 | 2 | O | 25 | ·05 | 3 | N | 23 | ·58 | 5 | D | 23○ | ·11 | **31** | 6 | Jr | 21 | ·64 | | 1 | F | 20 | ·17 | 2 | Mr | 21 | ·70 |
| {7 | S | 25 | ·52 | | | | | | | | | | | | | | | | | | | | | | | | | | |
| 4 | S | 14 | ·89 | 6 | O | 14 | ·42 | 7 | N | 12 | ·95 | 2 | D | 12● | ·57 | **32** | 4 | Jr | 11 | ·01 | | 5 | F | 9 | ·54 | 7 | Mr | 10 | ·07 |
| 2 | S | 3 | ·26 | 3 | O | 2 | ·79 | 5 | N | 1○ | ·32 | 6 | N | 30 | ·85 | | 1 | D | 30 | ·38 | **33** | 2 | Jr | 28 | ·91 | 4 | F | 27 | ·44 |
| 1 | S | 22 | ·15 | 2 | O | 21○ | ·69 | 4 | N | 20 | ·21 | 5 | D | 19 | ·74 | **34** | 7 | Jr | 18 | ·28 | | 1 | F | 16 | ·81 | 3 | Mr | 18 | ·34 |
| 5 | S | 11 | ·52 | 7 | O | 11○ | ·05 | 1 | N | 9 | ·58 | 3 | D | 9 | ·11 | **35** | 4 | Jr | 7 | ·64 | | 6 | F | 6 | ·17 | 7 | Mr | 7 | ·70 |
| 2 | Au | 31 | ·89 | 4 | S | 30● | ·42 | 5 | O | 29 | ·95 | 7 | N | 28 | ·48 | | 2 | D | 28 | ·01 | **36** | 3 | Jr | 26 | ·54 | 5 | F | 25○ | ·07 |
| 1 | S | 18● | ·78 | 3 | O | 18 | ·32 | 4 | N | 16 | ·85 | 6 | D | 16 | ·38 | **37** | 7 | Jr | 14 | ·91 | | 2 | F | 13●○ | ·44 | 3 | Mr | 14 | ·97 |
| 6 | S | 8 | ·15 | 7 | O | 7 | ·68 | 2 | N | 6 | ·21 | 3 | D | 5 | ·74 | **38** | 5 | Jr | 4 | ·27 | | 6 | F | 2●ↄ | ·80 | 1 | Mr | 4 | ·34 |
| 3 | Au | 28 | ·52 | 5 | S | 27 | ·05 | 6 | O | 26 | ·58 | 1 | N | 25 | ·11 | | 2 | D | 24○ | ·64 | **39** | 4 | Jr | 23● | ·17 | 5 | F | 21 | ·70 |

**TABLE X–**

| Kaliyuga. | Vikrama Era. | Saka Era. | Commencement of Solar Year. | | First New-Moon in Solar Year. | Anomaly of first New-Moon. | A.D. | Week-day of 1st January. | Vaisakha | | | | Jyeshta | | | | Ashada | | | | Sravana | | | | Bhadrapada | | | |
|---|---|---|---|---|---|---|---|---|---|---|---|---|---|---|---|---|---|---|---|---|---|---|---|---|---|---|---|---|
| | | | | | | | | | ☉'s Anom col. 6 | | | | + 29·53059 | | | | + 59·06117 | | | | + 88·59176 | | | | + 118·12235 | | | |
| | | | | | | | | | ☾'s Anom col. 7 | | | | + 1·976 | | | | + 3·952 | | | | + 5·928 | | | | + 7·904 | | | |
| | | | Month and day A.D. | Fraction of day. | | | | | Week-day | Month | Day | Fraction | Week-day | Month | Day | Fraction | Week-day | Month | Day | Fraction | Week-day | Month | Day | Fraction | Week-day | Month | Day | Fraction |
| 4040 | 996 | 861 | 22M | ·9219 | 0·31105 | 18·727 | **939** | 3 | 7 | Mr | 23 | ·23 | 3 | My | 21 | ·29 | 4 | Je | 19○ | ·82 | 6 | Jl | 19● | ·36 | 7 | Au | 17 | ·89 |
| | | | | | | | | | 1 | Ap | 21 | ·76 | | | | | | | | | | | | | | | | |
| 4041 | 997 | 862 | 22M | ·1805 | 18·95008 | 16·861 | **940** | 4 | 6 | Ap | 10 | ·13 | 7 | My | 9 | ·66 | 2 | Je | 8○ | ·19 | 3 | Jl | 7● | ·72 | 5 | Au | 6 | ·25 |
| 4042 | 998 | 863 | 22M | ·4392 | 8·05851 | 13·019 | **941** | 6 | 3 | Mr | 30 | ·50 | 5 | Ap | 29 | ·03 | 6 | My | 28○ | ·56 | 1 | Je | 27 | ·09 | 2 | Jl | 26 | ·62 |
| | | | | | | | | | | | | | | | | | | | | | | | | | 4 | Au | 25 | ·15 |
| 4043 | 999 | 864 | 22M | ·6979 | 26·69754 | 11·152 | **942** | 7 | 2 | Ap | 18 | ·40 | 3 | My | 17● | ·93 | 5 | Je | 16 | ·46 | 6 | Jl | 15 | ·99 | 1 | Au | 14 | ·52 |
| 4044 | 1000 | 865 | 22M | ·9566 | 15·80598 | 7·310 | **943** | 1 | 6 | Ap | 7○ | ·76 | 1 | My | 7● | ·29 | 2 | Je | 5 | ·82 | 4 | Jl | 5 | ·35 | 5 | Au | 3 | ·88 |
| 4045 | 1001 | 866 | 22M | ·2153 | 4·91441 | 3·468 | **944** | 2 | 4 | Mr | 27○ | ·13 | 5 | Ap | 25 | ·67 | 7 | My | 25 | ·19 | 3 | Jl | 23 | ·25 | 4 | Au | 21 | ·78 |
| | | | | | | | | | | | | | | | | | 1 | Je | 23 | ·72 | | | | | | | | |
| 4046 | 1002 | 867 | 22M | ·4739 | 23·55343 | 1·601 | **945** | 4 | 3 | Ap | 15 | ·83 | 4 | My | 14 | ·56 | 6 | Je | 13 | ·09 | 7 | Jl | 12 | ·62 | 2 | Au | 11 | ·15 |
| 4047 | 1003 | 868 | 22M | ·7326 | 12·66187 | 25·314 | **946** | 5 | 7 | Ap | 4 | ·39 | 1 | My | 3 | ·92 | 3 | Je | 2 | ·46 | 4 | Jl | 1 | ·99 | 6 | Jl | 31 | ·52 |
| 4048 | 1004 | 869 | 22M | ·9913 | 1·77030 | 21·472 | **947** | 6 | 4 | Mr | 24 | ·76 | 6 | Ap | 23 | ·29 | 2 | Je | 21 | ·35 | 3 | Jl | 20○ | ·88 | 5 | Au | 19 | ·41 |
| | | | | | | | | | | | | | 7 | My | 22 | ·82 | | | | | | | | | | | | |
| 4049 | 1005 | 870 | 22M | ·2500 | 20·40933 | 19·605 | **948** | 7 | 3 | Ap | 11 | ·66 | 5 | My | 11 | ·19 | 6 | Je | 9 | ·72 | 1 | Jl | 9●○ | ·25 | 7 | Au | 7 | ·78 |
| 4050 | 1006 | 871 | 22M | ·5087 | 9·51777 | 15·763 | **949** | 2 | 1 | Ap | 1 | ·03 | 2 | Ap | 30 | ·56 | 4 | My | 30 | ·09 | 5 | Je | 28● | ·62 | 7 | Jl | 28 | ·15 |
| 4051 | 1007 | 872 | 22M | ·7674 | 28·15678 | 13·896 | **950** | 3 | 6 | Ap | 19 | ·92 | 1 | My | 19○ | ·45 | 2 | Je | 17● | ·98 | 4 | Ji | 17 | ·52 | 6 | Au | 16 | ·05 |
| 4052 | 1008 | 873 | 23M | ·0260 | 17·26522 | 10·055 | **951** | 4 | 4 | Ap | 9 | ·29 | 5 | My | 8○ | ·82 | 6 | Je | 7 | ·35 | 1 | Jl | 6 | ·88 | 3 | Au | 5 | ·41 |
| 4053 | 1009 | 874 | 22M | ·2847 | 6·37366 | 6·212 | **952** | 5 | 1 | Mr | 28 | ·66 | 3 | Ap | 27●○ | ·19 | 4 | My | 26 | ·72 | 6 | Je | 25 | ·25 | 2 | Au | 23 | ·31 |
| | | | | | | | | | | | | | | | | | | | | | 7 | Jl | 24 | ·78 | | | | |
| 4054 | 1010 | 875 | 22M | ·5434 | 25·01368 | 4·346 | **953** | 7 | 7 | Ap | 16● | ·56 | 2 | My | 16 | ·09 | 3 | Je | 14 | ·62 | 5 | Jl | 14 | ·15 | 6 | Au | 12 | ·68 |
| 4055 | 1011 | 876 | 22M | ·8021 | 14·12112 | 0·504 | **954** | 1 | 4 | Ap | 5 | ·92 | 6 | My | 5 | ·45 | 7 | Je | 3 | ·98 | 2 | Jl | 3 | ·51 | 4 | Au | 2 | ·04 |
| 4056 | 1012 | 877 | 23M | ·0608 | 3·22955 | 24·216 | **955** | 2 | 2 | Mr | 26 | ·29 | 3 | Ap | 24 | ·82 | 6 | Je | 22 | ·88 | 1 | Jl | 22 | ·41 | 2 | Au | 20○ | ·94 |
| | | | | | | | | | | | | | 5 | My | 24 | ·35 | | | | | | | | | | | | |
| 4057 | 1013 | 878 | 22M | ·3194 | 21·86858 | 22·350 | **956** | 3 | 1 | Ap | 13 | ·19 | 2 | My | 12 | ·72 | 4 | Je | 11 | ·25 | 5 | Jl | 10 | ·78 | 7 | Au | 9 | ·31 |
| 4058 | 1014 | 879 | 22M | ·5781 | 10·97701 | 18·508 | **957** | 5 | 5 | Ap | 2 | ·55 | 7 | My | 2 | ·08 | 1 | My | 31 | ·62 | 3 | Je | 30 | ·15 | 4 | Jl | 29 | ·68 |
| 4059 | 1015 | 880 | 22M | ·8368 | 0·08538 | 14·665 | **958** | 6 | 2 | Mr | 22 | ·92 | 5 | My | 20 | ·98 | 7 | Je | 19○ | ·51 | 2 | Jl | 19● | ·04 | 3 | Au | 17 | ·57 |
| | | | | | | | | | 4 | Ap | 21 | ·45 | | | | | | | | | | | | | | | | |
| 4060 | 1016 | 881 | 23M | ·0955 | 18·72508 | 12·800 | **959** | 7 | 1 | Ap | 10 | ·82 | 3 | My | 10 | ·35 | 4 | Je | 8●○ | ·88 | 6 | Jl | 8 | ·41 | 7 | Au | 6 | ·94 |
| 4061 | 1017 | 882 | 22M | ·3542 | 7·83291 | 8·957 | **960** | 1 | 6 | Mr | 30 | ·19 | 7 | Ap | 28 | ·72 | 2 | My | 28● | ·25 | 3 | Je | 26 | ·78 | 5 | Jl | 26 | ·31 |
| | | | | | | | | | | | | | | | | | | | | | | | | | 6 | Au | 24 | ·84 |
| 4062 | 1018 | 883 | 22M | ·6128 | 26·47194 | 7·161 | **961** | 3 | 5 | Ap | 18○ | ·08 | 6 | My | 17● | ·61 | 1 | Je | 16 | ·14 | 2 | Jl | 15 | ·68 | 4 | Au | 14 | ·21 |
| 4063 | 1019 | 884 | 22M | ·8715 | 15·58037 | 3·248 | **962** | 4 | 2 | Ap | 7○ | ·45 | 3 | My | 6 | ·98 | 5 | Je | 5 | ·51 | 7 | Jl | 5 | ·04 | 1 | Au | 3 | ·57 |
| 4064 | 1020 | 885 | 23M | ·1302 | 4·68880 | 26·961 | **963** | 5 | 6 | Mr | 27○ | ·82 | 1 | Ap | 26 | ·35 | 2 | My | 25 | ·88 | 5 | Jl | 23 | ·94 | 7 | Au | 22 | ·47 |
| | | | | | | | | | | | | | | | | | 4 | Je | 24 | ·41 | | | | | | | | |
| 4065 | 1021 | 886 | 22M | ·3889 | 23·32783 | 25·094 | **964** | 7 | 5 | Ap | 14 | ·72 | 7 | My | 14 | ·25 | 1 | Je | 12 | ·78 | 3 | Jl | 12 | ·31 | 4 | Au | 10 | ·84 |
| 4066 | 1022 | 887 | 22M | ·6476 | 12·43627 | 21·252 | **965** | 1 | 3 | Ap | 4 | ·08 | 4 | My | 3 | ·61 | 6 | Je | 2 | ·14 | 7 | Jl | 1 | ·67 | 2 | Jl | 31○ | ·20 |
| 4067 | 1023 | 888 | 22M | ·9062 | 1·54469 | 17·410 | **966** | 2 | 7 | Mr | 24 | ·45 | 1 | Ap | 22 | ·98 | 5 | Je | 21 | ·04 | 6 | Jl | 20○ | ·57 | 1 | Au | 19 | ·10 |
| | | | | | | | | | | | | | 3 | My | 22 | ·51 | | | | | | | | | | | | |
| 4068 | 1024 | 889 | 23M | ·1649 | 20·18373 | 15·544 | **967** | 3 | 6 | Ap | 12 | ·35 | 7 | My | 11 | ·88 | 2 | Je | 10 | ·41 | 3 | Jl | 9● | ·94 | 5 | Au | 8 | ·47 |
| 4069 | 1025 | 890 | 22M | ·4236 | 9·29216 | 11·701 | **968** | 4 | 3 | Mr | 31 | ·71 | 5 | Ap | 30 | ·24 | 6 | My | 29○ | ·78 | 1 | Je | 28 | ·31 | 2 | Jl | 27 | ·84 |
| 4070 | 1026 | 891 | 22M | ·6823 | 27·93118 | 9·835 | **969** | 6 | 2 | Ap | 19 | ·61 | 4 | My | 19○ | ·14 | 5 | Je | 17 | ·67 | 7 | Jl | 17 | ·20 | 1 | Au | 15 | ·78 |
| 4071 | 1027 | 892 | 22M | ·9410 | 17·03962 | 5·993 | **970** | 7 | 6 | Ap | 8 | ·98 | 1 | My | 8●○ | ·51 | 3 | Je | 7 | ·04 | 4 | Jl | 6 | ·57 | 6 | Au | 5 | ·10 |
| 4072 | 1028 | 893 | 23M | ·1996 | 6·14815 | 2·151 | **971** | 1 | 4 | Mr | 29 | ·35 | 5 | Ap | 27 | ·88 | 7 | My | 27 | ·41 | 1 | Je | 25 | ·94 | 5 | Au | 24 | ·00 |
| | | | | | | | | | | | | | | | | | | | | | 3 | Jl | 25 | ·47 | | | | |
| 4073 | 1029 | 894 | 22M | ·4583 | 24·78707 | 0·284 | **972** | 2 | 3 | Ap | 16● | ·24 | 4 | My | 15 | ·77 | 6 | Je | 14 | ·30 | 7 | Jl | 13 | ·84 | 2 | Au | 12 | ·37 |
| 4074 | 1030 | 895 | 22M | ·7170 | 13·89551 | 23·997 | **973** | 4 | 7 | Ap | 5 | ·61 | 2 | My | 5 | ·14 | 3 | Je | 3 | ·67 | 5 | Jl | 3 | ·20 | 6 | Au | 1 | ·73 |
| 4075 | 1031 | 896 | 22M | ·9757 | 3·00395 | 20·154 | **974** | 5 | 4 | Mr | 25 | ·98 | 6 | Ap | 24 | ·51 | 2 | Je | 22 | ·57 | 4 | Jl | 22 | ·10 | 5 | Au | 20● | ·63 |
| | | | | | | | | | | | | | 1 | My | 24 | ·04 | | | | | | | | | | | | |
| 4076 | 1032 | 897 | 23M | ·2344 | 21·64287 | 18·288 | **975** | 6 | 3 | Ap | 13 | ·88 | 5 | My | 13 | ·41 | 6 | Je | 11 | ·94 | 1 | Jl | 11 | ·47 | 3 | Au | 10● | ·00 |
| 4077 | 1033 | 898 | 22M | ·4930 | 10·75140 | 14·446 | **976** | 7 | 1 | Ap | 2 | ·24 | 2 | My | 1 | ·77 | 4 | My | 31 | ·30 | 5 | Je | 29○ | ·83 | 7 | Jl | 29 | ·36 |

**N.B.**—For Surya Siddhanta figures for **A.D. 500** to **A.D. 999** see pages **76—79** below.

**Arya Siddhanta.**

| Week-day | Month | Day | Fraction | Week-day | Month | Day | Fraction | Week-day | Month | Day | Fraction | Week-day | Month | Day | Fraction | A.D. | Week-day | Month | Day | Fraction | A.D. | Week-day | Month | Day | Fraction | Week-day | Month | Day | Fraction |
|---|---|---|---|---|---|---|---|---|---|---|---|---|---|---|---|---|---|---|---|---|---|---|---|---|---|---|---|---|---|
| + 147·65293 | | | | + 177·18353 | | | | + 206·71411 | | | | + 236·24470 | | | | | + 265·77529 | | | | | + 295·30588 | | | | +324·83647<br>+354·36705 | | | |
| + 9·880 | | | | + 11·856 | | | | + 13·832 | | | | + 15·808 | | | | | + 17·784 | | | | | + 19·760 | | | | + 21·736<br>+ 23·712 | | | |
| **Asvina** | | | | **Kartika** | | | | **Margasira** | | | | **Pausha** | | | | **A.D.** | **Magha** | | | | **A.D.** | **Phalguna** | | | | **Chaitra** | | | |
| 2 | S | 16 | ·42 | 3 | O | 15 | ·95 | 5 | N | 14 | ·48 | 7 | D | 14○ | ·01 | **40** | 1 | Jr | 12 | ·54 | | 3 | F | 11 | ·07 | 4 | Mr | 11 | ·60 |
| 6 | S | 4 | ·78 | 1 | O | 4 | ·31 | 2 | N | 2 | ·84 | 4 | D | 2○ | ·38 | | 5 | D | 31 | ·91 | **41** | 7 | Jr | 30 | ·44 | 1 | F | 28 | ·97 |
| 5 | S | 23 | ·68 | 7 | O | 23 | ·21 | 1 | N | 21 | ·74 | 3 | D | 21 | ·27 | **42** | 4 | Jr | 19 | ·80 | | 6 | F | 18 | ·33 | 7 | Mr | 19 | ·86 |
| 3 | S | 13 | ·05 | 4 | O | 12 | ·58 | 6 | N | 11● | ·11 | 7 | D | 10 | ·64 | **43** | 2 | Jr | 9 | ·17 | | 3 | F | 7 | ·70 | 5 | Mr | 9 | ·23 |
| 7 | S | 2 | ·42 | 1 | O | 1○ | ·95 | 3 | O | 31 | ·48 | 5 | N | 30 | ·01 | | 6 | D | 29 | ·54 | **44** | 1 | Jr | 28 | ·07 | 2 | F | 26 | ·60 |
| 6 | S | 20●○ | ·31 | 7 | O | 19 | ·84 | 2 | N | 18 | ·37 | 3 | D | 17 | ·90 | **45** | 5 | Jr | 16 | ·44 | | 6 | F | 14 | ·96 | 1 | Mr | 16 | ·50 |
| 3 | S | 9●○ | ·68 | 5 | O | 9 | ·21 | 6 | N | 7 | ·74 | 1 | D | 7 | ·27 | **46** | 2 | Jr | 5 | ·80 | | 4 | F | 4 | ·33 | 5 | Mr | 5● | ·86 |
| 1 | Au | 30 | ·05 | 2 | S | 28 | ·58 | 4 | O | 28 | ·11 | 5 | N | 26 | ·64 | | 7 | D | 26 | ·17 | **47** | 1 | Jr | 24○ | ·70 | 3 | F | 23 | ·23 |
| 6 | S | 17 | ·94 | 1 | O | 17 | ·48 | 3 | N | 16 | ·01 | 4 | D | 15 | ·54 | **48** | 6 | Jr | 14○ | ·07 | | 7 | F | 12 | ·60 | 2 | Mr | 13 | ·13 |
| 4 | S | 6 | ·31 | 5 | O | 5 | ·84 | 7 | N | 4 | ·37 | 1 | D | 3 | ·90 | **49** | 3 | Jr | 2○ | ·43 | | 4 | Jr | 31 | ·96 | 6 | Mr | 2 | ·50 |
| {1 | Au | 26 | ·68 | 4 | O | 24 | ·74 | 6 | N | 23 | ·27 | 7 | D | 22● | ·80 | **50** | 2 | Jr | 21 | ·33 | | 3 | F | 19 | ·86 | 5 | Mr | 21 | ·39 |
| {3 | S | 25 | ·21 | | | | | | | | | | | | | | | | | | | | | | | | | | |
| 7 | S | 14 | ·58 | 2 | O | 14 | ·11 | 3 | N | 12○ | ·64 | 8 | D | 12 | ·17 | **51** | 6 | Jr | 10 | ·70 | | 1 | F | 9 | ·23 | 2 | Mr | 10 | ·76 |
| 4 | S | 3 | ·94 | 6 | O | 3 | ·47 | 1 | N | 2○ | ·00 | 2 | D | 1 | ·54 | | 4 | D | 31 | ·07 | **52** | 5 | Jr | 29 | ·60 | 7 | F | 28 | ·13 |
| 3 | S | 21 | ·84 | 5 | O | 21○ | ·37 | 6 | N | 19 | ·90 | 1 | D | 19 | ·43 | **53** | 2 | Jr | 17 | ·96 | | 4 | F | 16 | ·49 | 6 | Mr | 18 | ·02 |
| 1 | S | 11 | ·21 | 2 | O | 10 | ·74 | 4 | N | 9 | ·27 | 5 | D | 8 | ·80 | **54** | 7 | Jr | 7 | ·33 | | 1 | F | 5 | ·86 | 3 | Mr | 7○ | ·39 |
| 5 | Au | 31○ | ·58 | 7 | S | 30 | ·11 | 1 | O | 29 | ·64 | 3 | N | 28 | ·17 | | 4 | D | 27 | ·70 | **55** | 6 | Jr | 26 | ·23 | 7 | F | 24●○ | ·76 |
| 4 | S | 19 | ·47 | 6 | O | 19 | ·00 | 7 | N | 17 | ·53 | 2 | D | 17 | ·06 | **56** | 3 | Jr | 15 | ·60 | | 5 | F | 14○ | ·13 | 6 | Mr | 14 | ·06 |
| 1 | S | 7 | ·84 | 3 | O | 7 | ·37 | 4 | N | 5 | ·90 | 6 | D | 5 | ·43 | **57** | 7 | Jr | 3○ | ·96 | | 2 | F | 2 | ·49 | 4 | Mr | 4 | ·02 |
| 6 | Au | 28 | ·21 | 7 | S | 26 | ·74 | 2 | O | 26 | ·27 | 3 | N | 24 | ·80 | | 5 | D | 24○ | ·33 | **58** | 6 | Jr | 22 | ·86 | 1 | F | 21 | ·39 |
| 5 | S | 16 | ·10 | 6 | O | 15 | ·64 | 1 | N | 14 | ·17 | 2 | D | 13●○ | ·70 | **59** | 4 | Jr | 12 | ·23 | | 5 | F | 10 | ·76 | 7 | Mr | 12 | ·29 |
| 2 | S | 5 | ·47 | 4 | O | 5 | ·00 | 5 | N | 3 | ·53 | 7 | D | 3 | ·06 | **60** | 1 | Jr | 1 | ·59 | | 3 | Jr | 31 | ·12 | 4 | F | 29 | ·66 |
| 1 | S | 23 | ·37 | 2 | O | 22 | ·90 | 4 | N | 21 | ·43 | 5 | D | 20 | ·96 | **61** | 7 | Jr | 19 | ·49 | | 2 | F | 18 | ·02 | 3 | Mr | 19 | ·55 |
| 5 | S | 12 | ·74 | 7 | O | 12○ | ·27 | 1 | N | 10 | ·80 | 3 | D | 10 | ·33 | **62** | 4 | Jr | 8 | ·86 | | 6 | F | 7 | ·39 | 7 | Mr | 8 | ·92 |
| 3 | S | 2 | ·10 | 4 | O | 1○ | ·63 | 6 | O | 31 | ·16 | 7 | N | 29 | ·70 | | 2 | D | 29 | ·23 | **63** | 3 | Jr | 27 | ·76 | 5 | F | 26 | ·29 |
| 2 | S | 21○ | ·00 | 3 | O | 20 | ·53 | 5 | N | 19 | ·06 | 6 | D | 18 | ·59 | **64** | 1 | Jr | 17 | ·12 | | 2 | F | 15 | ·65 | 4 | Mr | 16 | ·18 |
| 6 | S | 9 | ·37 | 7 | O | 8 | ·90 | 2 | N | 7 | ·43 | 3 | D | 6 | ·96 | **65** | 5 | Jr | 5 | ·49 | | 7 | F | 4○ | ·02 | 1 | Mr | 5● | ·55 |
| 3 | Au | 29 | ·74 | 5 | S | 28 | ·27 | 6 | O | 27 | ·80 | 1 | N | 26 | ·33 | | 2 | D | 25 | ·86 | **66** | 4 | Jr | 24○ | ·39 | 6 | F | 22 | ·92 |
| 2 | S | 17 | ·83 | 4 | O | 17 | ·16 | 5 | N | 15 | ·69 | 7 | D | 15 | ·22 | **67** | 1 | Jr | 13○ | ·76 | | 3 | F | 12 | ·29 | 4 | Mr | 13 | ·82 |
| 7 | S | 7 | ·00 | 1 | O | 6 | ·53 | 3 | N | 5 | ·06 | 4 | D | 4 | ·59 | **68** | 6 | Jr | 3 | ·12 | | 7 | F | 1 | ·65 | 2 | Mr | 2 | ·18 |
| {4 | Au | 26 | ·37 | 7 | O | 24 | ·43 | 1 | N | 22○ | ·96 | 3 | D | 22● | ·49 | **69** | 5 | Jr | 21 | ·02 | | 6 | F | 19 | ·55 | 1 | Mr | 21 | ·08 |
| {5 | S | 24 | ·90 | | | | | | | | | | | | | | | | | | | | | | | | | | |
| 3 | S | 14 | ·26 | 4 | O | 13 | ·80 | 6 | N | 12○ | ·33 | 7 | D | 11 | ·87 | **70** | 2 | Jr | 10 | ·39 | | 3 | F | 8 | ·92 | 5 | Mr | 10 | ·45 |
| 7 | S | 3 | ·63 | 2 | O | 3 | ·16 | 3 | N | 1●○ | ·69 | 5 | D | 1 | ·22 | | 6 | D | 30 | ·75 | **71** | 1 | Jr | 29 | ·28 | 2 | F | 27 | ·82 |
| 6 | S | 22 | ·53 | 1 | O | 22● | ·06 | 2 | N | 20 | ·56 | 4 | D | 20 | ·12 | **72** | 5 | Jr | 18 | ·65 | | 7 | F | 17 | ·18 | 1 | Mr | 17○ | ·71 |
| 3 | S | 10○ | ·90 | 5 | O | 10● | ·43 | 7 | N | 8 | ·96 | 1 | D | 8 | ·49 | **73** | 3 | Jr | 7 | ·02 | | 4 | F | 5 | ·55 | 6 | Mr | 7○ | ·08 |
| 1 | Au | 31○ | ·26 | 2 | S | 29 | ·79 | 4 | O | 29 | ·32 | 5 | N | 27 | ·86 | | 7 | D | 27 | ·39 | **74** | 1 | Jr | 25 | ·92 | 3 | F | 24●○ | ·45 |
| 7 | S | 19 | ·16 | 1 | O | 18 | ·69 | 3 | N | 17 | ·22 | 4 | D | 16 | ·75 | **75** | 6 | Jr | 15 | ·28 | | 7 | F | 13● | ·81 | 2 | Mr | 15 | ·34 |
| 4 | S | 8 | ·53 | 6 | O | 8 | ·06 | 7 | N | 6 | ·59 | 2 | D | 6 | ·12 | **76** | 3 | Jr | 4○ | ·65 | | 5 | F | 3 | ·18 | 6 | Mr | 3 | ·71 |
| 1 | Au | 27 | ·90 | 3 | S | 26 | ·43 | 4 | O | 25 | ·96 | 6 | N | 24 | ·49 | | 1 | D | 24○ | ·02 | **77** | 2 | Jr | 22 | ·55 | {4 | F | 21 | ·08 |
| | | | | | | | | | | | | | | | | | | | | | | | | | | {5 | Mr | 22 | ·61 |

## TABLE X—

| | Vaisakha | Jyeshta | Ashada | Sravana | Bhadrapada |
|---|---|---|---|---|---|
| ⊙'s Anom col. 6 | | + 29·53059 | + 59·06117 | + 88·59176 | + 118·12235 |
| ☾'s Anom col. 7 | | + 1·976 | + 3·952 | + 5·928 | + 7·904 |

| Kaliyuga. | Vikrama Era. | Saka Era. | Commencement of Solar Year. Month and day A. D. | Fraction of day. | First New-Moon in Solar Year. | Anomaly of first New-Moon. | A. D. | Week-day of 1st January. | Vaisakha Week-day | Month | Day | Fraction | Jyeshta Week-day | Month | Day | Fraction | Ashada Week-day | Month | Day | Fraction | Sravana Week-day | Month | Day | Fraction | Bhadrapada Week-day | Month | Day | Fraction |
|---|---|---|---|---|---|---|---|---|---|---|---|---|---|---|---|---|---|---|---|---|---|---|---|---|---|---|---|---|
| 4078 | 1034 | 899 | 22M | ·7517 | 29·39043 | 12·580 | **977** | 2 | 7 | Ap | 21 | ·14 | 1 | My | 20 | ·67 | 3 | Je | 19○ | ·20 | 4 | Jl | 18 | ·78 | 6 | Au | 17 | ·26 |
| 4079 | 1035 | 900 | 23M | ·0104 | 18·49886 | 8·737 | **978** | 3 | 4 | Ap | 10 | ·51 | 6 | My | 10 | ·04 | 7 | Je | 8● | ·57 | 2 | Jl | 8 | ·10 | 3 | Au | 6 | ·63 |
| 4080 | 1036 | 901 | 23M | ·2691 | 7·60730 | 4·895 | **979** | 4 | 1 | Mr | 30 | ·87 | 3 | Ap | 29○ | ·40 | 4 | My | 28 | ·94 | { 6 | Je | 27 | ·47 | 2 | Au | 25 | ·53 |
| | | | | | | | | | | | | | | | | | | | | | { 1 | Jl | 27 | ·00 | | | | |
| 4081 | 1037 | 902 | 22M | ·5278 | 26·24633 | 3·029 | **980** | 5 | 7 | Ap | 17○ | ·77 | 2 | My | 17● | ·30 | 3 | Je | 15 | ·83 | 5 | Jl | 15 | ·36 | 6 | Au | 13 | ·89 |
| 4082 | 1038 | 903 | 22M | ·7864 | 15·35476 | 26·741 | **981** | 7 | 5 | Ap | 7●○ | ·14 | 6 | My | 6 | ·67 | 1 | Je | 5 | ·20 | 2 | Jl | 4 | ·73 | 4 | Au | 3 | ·26 |
| 4083 | 1039 | 904 | 23M | ·0451 | 4·46319 | 22·899 | **982** | 1 | 2 | Mr | 27● | ·51 | 4 | Ap | 26 | ·04 | { 5 | My | 25 | ·57 | 1 | Jl | 23 | ·63 | 3 | Au | 22 | ·16 |
| | | | | | | | | | | | | | | | | | { 7 | Je | 24 | ·10 | | | | | | | | |
| 4084 | 1040 | 905 | 23M | ·3038 | 23·10222 | 21·033 | **983** | 2 | 1 | Ap | 15 | ·40 | 2 | My | 14 | ·93 | 4 | Je | 13 | ·46 | 6 | Jl | 13 | ·00 | 7 | Au | 11○ | ·53 |
| 4085 | 1041 | 906 | 22M | ·5625 | 12·21065 | 17·190 | **984** | 3 | 5 | Ap | 3 | ·77 | 7 | My | 3 | ·30 | 1 | Je | 1 | ·83 | 3 | Jl | 1 | ·36 | 4 | Jl | 30○● | ·89 |
| 4086 | 1042 | 907 | 22M | ·8212 | 1·31909 | 13·348 | **985** | 5 | 3 | Mr | 24 | ·14 | { 4 | Ap | 22 | ·69 | 7 | Je | 20 | ·73 | 2 | Jl | 20○ | ·26 | 3 | Au | 18 | ·79 |
| | | | | | | | | | | | | | { 6 | My | 22 | ·20 | | | | | | | | | | | | |
| 4087 | 1043 | 908 | 23M | ·0799 | 19·95812 | 11·482 | **986** | 6 | 2 | Ap | 12 | ·03 | 3 | My | 11 | ·58 | 5 | Je | 10○ | ·10 | 6 | Jl | 9 | ·63 | 1 | Au | 8 | ·16 |
| 4088 | 1044 | 909 | 23M | ·3385 | 9·06654 | 7·640 | **987** | 7 | 6 | Ap | 1 | ·40 | 7 | Ap | 30 | ·93 | 2 | My | 30○ | ·46 | 3 | Je | 28 | ·99 | 5 | Jl | 28 | ·52 |
| 4089 | 1045 | 910 | 22M | ·5972 | 27·70557 | 5·773 | **988** | 1 | 5 | Ap | 19 | ·30 | 6 | My | 18●○ | ·83 | 1 | Je | 17 | ·36 | 2 | Jl | 16 | ·89 | 4 | Au | 15 | ·42 |
| 4090 | 1046 | 911 | 22M | ·8559 | 16·81401 | 1·931 | **989** | 3 | 2 | Ap | 8 | ·67 | 4 | My | 8● | ·20 | 5 | Je | 6 | ·73 | 7 | Jl | 6 | ·26 | 1 | Au | 4 | ·79 |
| 4091 | 1047 | 912 | 23M | ·1146 | 5·92244 | 25·643 | **990** | 4 | 7 | Mr | 29○ | ·03 | 1 | Ap | 27 | ·56 | 3 | My | 27 | ·09 | { 4 | Je | 25 | ·62 | 7 | Au | 23 | ·69 |
| | | | | | | | | | | | | | | | | | | | | | { 6 | Jl | 25 | ·16 | | | | |
| 4092 | 1048 | 913 | 23M | ·3733 | 24·56147 | 23·777 | **991** | 5 | 5 | Ap | 16 | ·93 | 7 | My | 16 | ·46 | 1 | Je | 14 | ·99 | 3 | Jl | 14 | ·52 | 5 | Au | 13 | ·05 |
| 4093 | 1049 | 914 | 22M | ·6319 | 13·66991 | 19·935 | **992** | 6 | 3 | Ap | 5 | ·30 | 4 | My | 4 | ·83 | 6 | Je | 3 | ·36 | 7 | Jl | 2 | ·89 | 2 | Au | 1 | ·42 |
| 4094 | 1050 | 915 | 22M | ·8906 | 2·77833 | 16·093 | **993** | 1 | 7 | Mr | 25 | ·66 | { 2 | Ap | 24 | ·20 | 5 | Je | 22 | ·26 | 6 | Jl | 21 | ·79 | 1 | Au | 20● | ·32 |
| | | | | | | | | | | | | | { 3 | My | 23 | ·73 | | | | | | | | | | | | |
| 4095 | 1051 | 916 | 23M | ·1493 | 21·41764 | 14·227 | **994** | 2 | 6 | Ap | 13 | ·56 | 1 | My | 13 | ·10 | 1 | Je | 11 | ·62 | 4 | Jl | 11○ | ·15 | 5 | Au | 9 | ·68 |
| 4096 | 1052 | 917 | 23M | ·4080 | 10·52511 | 10·385 | **995** | 3 | 3 | Ap | 2 | ·93 | 5 | My | 2 | ·46 | 6 | My | 31 | ·99 | 1 | Je | 30○ | ·52 | 3 | Jl | 30 | ·05 |
| 4097 | 1053 | 918 | 22M | ·6667 | 29·16482 | 8·518 | **996** | 4 | 2 | Ap | 20 | ·83 | 4 | My | 20 | ·36 | 5 | Je | 18 | ·89 | 7 | Jl | 18 | ·42 | 1 | Au | 16 | ·95 |
| 4098 | 1054 | 919 | 22M | ·9253 | 18·27326 | 4·676 | **997** | 6 | 7 | Ap | 10 | ·19 | 1 | My | 9○ | ·72 | 3 | Je | 8 | ·26 | 4 | Jl | 7 | ·79 | 6 | Au | 6 | ·32 |
| 4099 | 1055 | 920 | 23M | ·1840 | 7·38169 | 0·833 | **998** | 7 | 4 | Mr | 30 | ·56 | 6 | Ap | 29○ | ·09 | 7 | My | 28 | ·62 | { 2 | Je | 27 | ·15 | 5 | Au | 25 | ·21 |
| | | | | | | | | | | | | | | | | | | | | | { 3 | Jl | 26 | ·68 | | | | |
| 4100 | 1056 | 921 | 23M | ·4427 | 26·02072 | 26·522 | **999** | 1 | 3 | Ap | 18○ | ·46 | 4 | My | 17 | ·99 | 6 | Je | 16 | ·52 | 1 | Jl | 16 | ·05 | 2 | Au | 14 | ·58 |

**N.B.**—*For Surya Siddhanta figures for* **A.D. 500** to **A.D. 999** see pages **76—79** below.

**Ārya Siddhānta.**

| Asvina | | | | Kartika | | | | Margasira | | | | Pausha | | | | A.D. | Magha | | | | A.D. | Phalguna | | | | Chaitra | | | |
|---|---|---|---|---|---|---|---|---|---|---|---|---|---|---|---|---|---|---|---|---|---|---|---|---|---|---|---|---|---|
| + 147·65293 | | | | + 177·18353 | | | | + 206·71411 | | | | + 236·24470 | | | | | + 265·77529 | | | | | + 295·30588 | | | | +324·83647 / +354·36705 | | | |
| + 9·880 | | | | + 11·856 | | | | + 13·832 | | | | + 15·808 | | | | | + 17·784 | | | | | + 19·760 | | | | + 21·736 / + 23·712 | | | |
| Week-day | Month | Day | Fraction | Week-day | Month | Day | Fraction | Week-day | Month | Day | Fraction | Week-day | Month | Day | Fraction | | Week-day | Month | Day | Fraction | | Week-day | Month | Day | Fraction | Week-day | Month | Day | Fraction |
| 7 | S | 15 | ·79 | 2 | O | 15 | ·32 | 3 | N | 13 | ·85 | 5 | D | 13● | ·39 | **78** | 6 | Jr | 11 | ·91 | | 1 | F | 10 | ·45 | 2 | Mr | 11 | ·98 |
| 5 | S | 5 | ·16 | 6 | O | 4 | ·69 | 1 | N | 3 | ·22 | 2 | D | 2● | ·75 | **79** | 4 | Jr | 1 | ·28 | | 5 | Jr | 30 | ·81 | 7 | Mr | 1 | ·34 |
| 4 | S | 24 | ·06 | 5 | O | 23○ | ·59 | 7 | N | 22 | ·12 | 1 | D | 21 | ·65 | **80** | 3 | Jr | 20 | ·18 | | 4 | F | 18 | ·71 | 6 | Mr | 19 | ·24 |
| 1 | S | 12 | ·42 | 2 | O | 11○ | ·96 | 4 | N | 10 | ·49 | 6 | D | 10 | ·02 | **81** | 7 | Jr | 8 | ·55 | | 2 | F | 7 | ·08 | 3 | Mr | 8 | ·61 |
| 5 | S | 1 | ·79 | 7 | O | 1○ | ·32 | 1 | O | 30 | ·85 | 3 | N | 29 | ·38 | | 4 | D | 28 | ·91 | **82** | 6 | Jr | 27 | ·44 | 7 | F | 25 | ·98 |
| 4 | S | 20● | ·69 | 6 | O | 20 | ·22 | 7 | N | 18 | ·75 | 2 | D | 18 | ·28 | **83** | 3 | Jr | 16 | ·81 | | 5 | F | 15○ | ·34 | 6 | Mr | 16 | ·87 |
| 2 | S | 10 | ·06 | 3 | O | 9 | ·59 | 5 | N | 8 | ·12 | 6 | D | 7 | ·65 | **84** | 1 | Jr | 6 | ·18 | | 2 | F | 4○ | ·71 | 4 | Mr | 5 | ·24 |
| 6 | Au | 29 | ·42 | 7 | S | 27 | ·95 | 2 | O | 27 | ·48 | 4 | N | 26 | ·02 | | 5 | D | 25 | ·55 | **85** | 7 | Jr | 24○ | ·08 | 1 | F | 22 | ·61 |
| 5 | S | 17 | ·32 | 6 | O | 16 | ·85 | 1 | N | 15 | ·38 | 2 | D | 14 | ·91 | **86** | 4 | Jr | 13● | ·44 | | 5 | F | 11 | ·97 | 7 | Mr | 13 | ·50 |
| 2 | S | 6 | ·69 | 4 | O | 6 | ·22 | 5 | N | 4 | ·75 | 7 | D | 4○ | ·28 | **87** | 1 | Jr | 2 | ·81 | | 3 | F | 1 | ·34 | 4 | Mr | 2 | ·87 |
| {7 | Au | 27 | ·05 | 3 | O | 25 | ·12 | 4 | N | 23○ | ·65 | 6 | D | 23 | ·18 | **88** | 7 | Jr | 21 | ·71 | | 2 | F | 20 | ·24 | 3 | Mr | 20 | ·77 |
| {1 | S | 25 | ·58 | | | | | | | | | | | | | | | | | | | | | | | | | | |
| 5 | S | 13 | ·96 | 7 | O | 13 | ·48 | 2 | N | 12●○ | ·01 | 3 | D | 11 | ·54 | **89** | 5 | Jr | 10 | ·07 | | 6 | F | 8 | ·60 | 1 | Mr | 10 | ·14 |
| 3 | S | 3 | ·32 | 4 | O | 2 | ·84 | 6 | N | 1● | ·31 | 7 | N | 30 | ·91 | | 2 | D | 30 | ·44 | **90** | 3 | Jr | 28 | ·97 | 5 | F | 27 | ·50 |
| 2 | S | 22○ | ·22 | 3 | O | 21● | ·75 | 5 | N | 20 | ·28 | 6 | D | 19 | ·81 | **91** | 1 | Jr | 18 | ·34 | | 2 | F | 16 | ·87 | 4 | Mr | 18●○ | ·40 |
| 6 | S | 11○ | ·58 | 1 | O | 11 | ·11 | 2 | N | 9 | ·64 | 4 | D | 9 | ·18 | **92** | 5 | Jr | 7 | ·71 | | 7 | F | 6 | ·24 | 1 | Mr | 6●○ | ·77 |
| 3 | Au | 30○ | ·95 | 5 | S | 29 | ·48 | 7 | O | 29 | ·01 | 1 | N | 27 | ·54 | | 3 | D | 27 | ·07 | **93** | 4 | Jr | 25 | ·60 | 6 | F | 24● | ·13 |
| 2 | S | 18 | ·85 | 4 | O | 18 | ·38 | 5 | N | 16 | ·91 | 7 | D | 16 | ·44 | **94** | 1 | Jr | 14○ | ·97 | | 3 | F | 13 | ·50 | 5 | Mr | 15 | ·03 |
| 7 | S | 8 | ·22 | 1 | O | 7 | ·75 | 3 | N | 6 | ·28 | 4 | D | 5 | ·81 | **95** | 6 | Jr | 4●○ | ·34 | | 5 | F | 2 | .87 | 2 | Mr | 4 | 40 |
| 4 | Au | 28 | ·58 | 6 | S | 27 | ·11 | 7 | O | 26 | ·64 | 2 | N | 25 | ·17 | | 3 | D | 24 | ·70 | **96** | 5 | Jr | 23 | ·24 | {6 | F | 21 | ·77 |
| | | | | | | | | | | | | | | | | | | | | | | | | | | {1 | Mr | 23 | ·30 |
| 3 | S | 15 | ·48 | 5 | O | 15 | ·01 | 6 | N | 13 | ·54 | 1 | D | 13● | ·07 | **97** | 2 | Jr | 11 | ·60 | | 4 | F | 10 | ·13 | 5 | Mr | 11 | ·66 |
| 7 | S | 4 | ·85 | 2 | O | 4 | ·38 | 3 | N | 2○ | ·91 | 5 | D | 2 | ·44 | | 6 | D | 31 | ·97 | **98** | 1 | Jr | 30 | ·50 | 3 | Mr | 1 | ·03 |
| 6 | S | 23 | ·74 | 1 | O | 23●○ | ·28 | 2 | N | 21 | ·81 | 4 | D | 21 | ·34 | **99** | 5 | Jr | 19 | ·87 | | 7 | F | 18 | ·40 | 1 | Mr | 19 | ·93 |
| 4 | S | 13 | ·11 | 5 | O | 12●○ | ·64 | 7 | N | 11 | ·17 | 1 | D | 10 | ·70 | **10** | 3 | Jr | 9 | ·23 | **00** | 4 | F | 7 | ·76 | 6 | Mr | 8 | ·30 |

| A.D. | Kaliyuga | Week day of 1st January | Commencement of Indian Solar Year. Day | Fraction of day | First New-Moon in Solar Year | Anomaly of first New-Moon | A.D. | Kaliyuga | Week-day of 1st January | Commencement of Indian Solar Year. Day | Fraction of day | First New-Moon in Solar Year | Anomaly of first New-Moon | A.D. | Kaliyuga | Week-day of 1st January | Commencement of Indian Solar Year. Day | Fraction of day |
|---|---|---|---|---|---|---|---|---|---|---|---|---|---|---|---|---|---|---|
| **500** | 3601 | 7 | 18M | ·3613 | 27·3063 | 9·979 | **550** | 3651 | 7 | 19M | ·2991 | 14·2718 | 18·737 | **600** | 3701 | 7 | 19M | ·2369 |
| **501** | 3602 | 1 | 18M | ·6200 | 16·4146 | 6·137 | **551** | 3652 | 1 | 19M | ·5579 | 3·3801 | 14·894 | **601** | 3702 | 1 | 19M | 4957 |
| **502** | 3603 | 2 | 18M | ·8788 | 5·5229 | 2·294 | **552** | 3653 | 2 | 18M | ·8166 | 22·0190 | 13·028 | **602** | 3703 | 2 | 19M | ·7545 |
| **503** | 3604 | 4 | 19M | ·1375 | 24·1618 | 0·427 | **553** | 3654 | 4 | 19M | ·0754 | 11·1273 | 9·185 | **603** | 3704 | 4 | 20M | ·0132 |
| **504** | 3605 | 5 | 18M | ·3963 | 13·2701 | 24·139 | **554** | 3655 | 5 | 19M | ·3341 | 0·2356 | 5·343 | **604** | 3705 | 5 | 19M | ·2720 |
| **505** | 3606 | 6 | 18M | ·6550 | 2·3784 | 20·296 | **555** | 3656 | 6 | 19M | ·5929 | 18·8745 | 3·476 | **605** | 3706 | 6 | 19M | ·5307 |
| **506** | 3607 | 7 | 18M | ·9138 | 21·0173 | 18·429 | **556** | 3657 | 7 | 18M | ·8516 | 7·9828 | 27·188 | **606** | 3707 | 7 | 19M | ·7895 |
| **507** | 3608 | 2 | 19M | ·1726 | 10·1256 | 14·587 | **557** | 3658 | 2 | 19M | ·1104 | 26·6217 | 25·321 | **607** | 3708 | 2 | 20M | ·0482 |
| **508** | 3609 | 3 | 18M | ·4313 | 28·7644 | 12·720 | **558** | 3659 | 3 | 19M | ·3692 | 15·7300 | 21·478 | **608** | 3709 | 3 | 19M | ·3070 |
| **509** | 3610 | 4 | 18M | ·6901 | 17·8727 | 8·877 | **559** | 3660 | 4 | 19M | ·6279 | 4·8383 | 17·635 | **609** | 3710 | 4 | 19M | ·5658 |
| **510** | 3611 | 5 | 18M | ·9488 | 6·9810 | 5·034 | **560** | 3661 | 5 | 18M | ·8867 | 23·4772 | 15·769 | **610** | 3711 | 5 | 19M | ·8245 |
| **511** | 3612 | 7 | 19M | ·2076 | 25·6199 | 3·168 | **561** | 3662 | 7 | 19M | ·1454 | 12·5854 | 11·926 | **611** | 3712 | 7 | 20M | ·0833 |
| **512** | 3613 | 1 | 18M | ·4663 | 14·7282 | 26·880 | **562** | 3663 | 1 | 19M | ·4042 | 1·6937 | 8·083 | **612** | 3713 | 1 | 19M | ·3420 |
| **513** | 3614 | 2 | 18M | ·7251 | 3·8365 | 23·037 | **563** | 3664 | 2 | 19M | ·6629 | 20·3326 | 6·216 | **613** | 3714 | 2 | 19M | ·6008 |
| **514** | 3615 | 3 | 18M | ·9869 | 22·4754 | 21·170 | **564** | 3665 | 3 | 18M | ·9217 | 9·4409 | 2·374 | **614** | 3715 | 3 | 19M | ·8595 |
| **515** | 3616 | 5 | 19M | ·2426 | 11·5837 | 17·327 | **565** | 3666 | 5 | 19M | ·1805 | 28·0798 | 0·507 | **615** | 3716 | 5 | 20M | ·1183 |
| **516** | 3617 | 6 | 18M | ·5014 | 0·6920 | 13·485 | **566** | 3667 | 6 | 19M | ·4392 | 17·1881 | 24·219 | **616** | 3717 | 6 | 19M | ·3771 |
| **517** | 3618 | 7 | 18M | ·7601 | 19·3309 | 11·618 | **567** | 3668 | 7 | 19M | ·6980 | 6·2964 | 20·376 | **617** | 3718 | 7 | 19M | ·6358 |
| **518** | 3619 | 2 | 19M | ·0189 | 8·4392 | 7·775 | **568** | 3669 | 1 | 18M | ·9567 | 24·9353 | 18·509 | **618** | 3719 | 1 | 19M | ·8946 |
| **519** | 3620 | 3 | 19M | ·2776 | 27·0781 | 5·908 | **569** | 3670 | 3 | 19M | ·2155 | 14·0436 | 14·666 | **619** | 3720 | 3 | 20M | ·1533 |
| **520** | 3621 | 4 | 18M | ·5364 | 16·1864 | 2·066 | **570** | 3671 | 4 | 19M | ·4742 | 3·1519 | 10·824 | **620** | 3721 | 4 | 19M | 4121 |
| **521** | 3622 | 5 | 18M | ·7952 | 5·2947 | 25·777 | **571** | 3672 | 5 | 19M | ·7330 | 21·7908 | 8·957 | **621** | 3722 | 5 | 19M | ·6708 |
| **522** | 3623 | 7 | 19M | ·0569 | 23·9336 | 23·911 | **572** | 3673 | 6 | 18M | ·9918 | 10·8991 | 5·114 | **622** | 3723 | 6 | 19M | ·9296 |
| **523** | 3624 | 1 | 19M | ·3127 | 13·0419 | 20·068 | **573** | 3674 | 1 | 19M | ·2505 | 0·0074 | 1·271 | **623** | 3724 | 1 | 20M | ·1884 |
| **524** | 3625 | 2 | 18M | ·5714 | 2·1502 | 16·225 | **574** | 3675 | 2 | 19M | ·5093 | 18·6463 | 26·959 | **624** | 3725 | 2 | 19M | ·4471 |
| **525** | 3626 | 3 | 18M | ·8302 | 20·7891 | 14·358 | **575** | 3676 | 3 | 19M | ·7680 | 7·7546 | 23·117 | **625** | 3726 | 3 | 19M | ·7059 |
| **526** | 3627 | 5 | 19M | ·0889 | 9·8974 | 10·516 | **576** | 3677 | 5 | 19M | ·0268 | 26·3935 | 21·250 | **626** | 3727 | 4 | 19M | ·9646 |
| **527** | 3628 | 6 | 19M | ·3477 | 28·5362 | 8·649 | **577** | 3678 | 6 | 19M | ·2855 | 15·5018 | 17·407 | **627** | 3728 | 6 | 20M | ·2234 |
| **528** | 3629 | 7 | 18M | ·6065 | 17·6445 | 4·806 | **578** | 3679 | 7 | 19M | ·5443 | 4·6101 | 13·564 | **628** | 3729 | 7 | 19M | ·4821 |
| **529** | 3630 | 1 | 18M | ·8652 | 6·7528 | 0·963 | **579** | 3680 | 1 | 19M | ·8031 | 23·2489 | 11·698 | **629** | 3730 | 1 | 19M | ·7409 |
| **530** | 3631 | 3 | 19M | ·1240 | 25·3917 | 26·651 | **580** | 3681 | 3 | 19M | ·0618 | 12·3572 | 7·855 | **630** | 3731 | 2 | 19M | ·9996 |
| **531** | 3632 | 4 | 19M | ·3827 | 14·5000 | 22·809 | **581** | 3682 | 4 | 19M | ·3206 | 1·4655 | 4·013 | **631** | 3732 | 4 | 20M | ·2584 |
| **532** | 3633 | 5 | 18M | ·6415 | 3·6083 | 18·966 | **582** | 3683 | 5 | 19M | ·5793 | 20·1044 | 2·145 | **632** | 3733 | 5 | 19M | ·5172 |
| **533** | 3634 | 6 | 18M | ·9002 | 22·2472 | 17·099 | **583** | 3684 | 6 | 19M | ·8381 | 9·2127 | 25·857 | **633** | 3734 | 6 | 19M | ·7759 |
| **534** | 3635 | 1 | 19M | ·1590 | 11·3555 | 13·256 | **584** | 3685 | 1 | 19M | ·0958 | 27·8516 | 23·990 | **634** | 3735 | 1 | 20M | ·0347 |
| **535** | 3636 | 2 | 19M | ·4177 | 0·4638 | 9·414 | **585** | 3686 | 2 | 19M | ·3556 | 16·9599 | 20·148 | **635** | 3736 | 2 | 20M | ·2934 |
| **536** | 3637 | 3 | 18M | ·6765 | 19·1027 | 7·547 | **586** | 3687 | 3 | 19M | ·143 | 6·0682 | 16·305 | **636** | 3737 | 3 | 19M | ·5522 |
| **537** | 3638 | 4 | 18M | ·9353 | 8·2110 | 3·704 | **587** | 3688 | 4 | 19M | ·731 | 24·7071 | 14·438 | **637** | 3738 | 4 | 19M | ·8109 |
| **538** | 3639 | 6 | 19M | ·1940 | 26·8499 | 1·837 | **588** | 3689 | 6 | 19M | ·1319 | 13·8154 | 10·595 | **638** | 3739 | 6 | 20M | ·0697 |
| **539** | 3640 | 7 | 19M | ·4518 | 15·9582 | 25·549 | **589** | 3690 | 7 | 19M | ·3906 | 2·9237 | 6·753 | **639** | 3740 | 7 | 20M | ·3285 |
| **540** | 3641 | 1 | 18M | ·7115 | 5·0665 | 21·706 | **590** | 3691 | 1 | 19M | ·6494 | 21·5626 | 4·886 | **640** | 3741 | 1 | 19M | ·5872 |
| **541** | 3642 | 2 | 18M | ·9703 | 23·7054 | 19·840 | **591** | 3692 | 2 | 19M | ·9081 | 10·6709 | 1·043 | **641** | 3742 | 2 | 19M | ·8460 |
| **542** | 3643 | 4 | 19M | ·2290 | 12·8137 | 15·997 | **592** | 3693 | 4 | 19M | ·1669 | 29·3098 | 26·731 | **642** | 3743 | 4 | 20M | ·1047 |
| **543** | 3644 | 5 | 19M | ·4868 | 1·9220 | 12·154 | **593** | 3694 | 5 | 19M | ·4256 | 18·4181 | 22·888 | **643** | 3744 | 5 | 20M | ·3635 |
| **544** | 3645 | 6 | 18M | ·7466 | 20·5608 | 10·277 | **594** | 3695 | 6 | 19M | ·6844 | 7·5264 | 19·045 | **644** | 3745 | 6 | 19M | ·6222 |
| **545** | 3646 | 1 | 19M | ·0053 | 9·6691 | 6·445 | **595** | 3696 | 7 | 19M | ·9432 | 26·1653 | 17·179 | **645** | 3746 | 7 | 19M | ·8810 |
| **546** | 3647 | 2 | 19M | ·2641 | 28·3080 | 4·578 | **596** | 3697 | 2 | 19M | ·2019 | 15·2736 | 13·336 | **646** | 3747 | 2 | 20M | ·1298 |
| **547** | 3648 | 3 | 19M | ·5228 | 17·4163 | 0·735 | **597** | 3698 | 3 | 19M | ·4607 | 4·3819 | 9·493 | **647** | 3748 | 3 | 20M | ·3985 |
| **548** | 3649 | 4 | 18M | ·7816 | 6·5246 | 24·447 | **598** | 3699 | 4 | 19M | ·7194 | 23·0207 | 7·627 | **648** | 3749 | 4 | 19M | ·6573 |
| **549** | 3650 | 6 | 19M | ·0403 | 25·1635 | 22·580 | **599** | 3700 | 5 | 19M | ·9782 | 12·1290 | 3·784 | **649** | 3750 | 5 | 19M | ·9163 |

## for A.D. 500 to A.D. 999.

| First New-Moon in Solar Year | Anomaly of first New-Moon | A.D. | Kaliyuga | Week-day of 1st January | Commencement of Indian Solar Year. Day | Fraction of day | First New-Moon in Solar Year | Anomaly of first New-Moon |
|---|---|---|---|---|---|---|---|---|
| 1·2373 | 27·496 | **650** | 3751 | 7 | 20M | ·1748 | 17·7334 | 10·675 |
| 19·8762 | 25·629 | **651** | 3752 | 1 | 20M | ·4335 | 6·8417 | 6·838 |
| 8·9845 | 21·786 | **652** | 3753 | 2 | 19M | ·6923 | 25·4806 | 4·966 |
| 27·6234 | 19·919 | **653** | 3754 | 3 | 19M | ·9511 | 14·5889 | 1·123 |
| 16·7317 | 16·077 | **654** | 3755 | 5 | 20M | ·2098 | 3·6972 | 24·835 |
| 5·8400 | 12·234 | **655** | 3756 | 6 | 20M | ·4686 | 22·3361 | 22·968 |
| 24·4789 | 10·367 | **656** | 3757 | 7 | 19M | ·7273 | 11·4444 | 19·125 |
| 13·5872 | 6·524 | **657** | 3758 | 1 | 19M | ·9861 | 0·5527 | 15·283 |
| 2·6955 | 2·681 | **658** | 3759 | 3 | 20M | ·2448 | 19·1916 | 13·416 |
| 21·3344 | 0·815 | **659** | 3760 | 4 | 20M | ·5036 | 8·2999 | 9·573 |
| 10·4427 | 24·527 | **660** | 3761 | 5 | 19M | ·7623 | 26·9388 | 7·706 |
| 29·0816 | 22·660 | **661** | 3762 | 7 | 20M | ·0211 | 16·0471 | 3·864 |
| 18·1899 | 18·817 | **662** | 3763 | 1 | 20M | ·2799 | 5·1554 | 0·021 |
| 7·2982 | 14·974 | **663** | 3764 | 2 | 20M | ·5386 | 23·7943 | 25·709 |
| 25·9370 | 13·107 | **664** | 3765 | 3 | 19M | ·7974 | 12·9026 | 21·866 |
| 15·0453 | 9·265 | **665** | 3766 | 5 | 20M | ·0561 | 2·0109 | 18·023 |
| 4·1536 | 5·423 | **666** | 3767 | 6 | 20M | ·3149 | 20·6498 | 16·156 |
| 22·7925 | 3·555 | **667** | 3768 | 7 | 20M | ·5737 | 9·7581 | 12·314 |
| 11·9008 | 27·267 | **668** | 3769 | 1 | 19M | ·8324 | 28·3969 | 10·447 |
| 1·0091 | 23·425 | **669** | 3770 | 3 | 20M | ·0912 | 17·5052 | 6·604 |
| 19·6480 | 21·558 | **670** | 3771 | 4 | 20M | ·3499 | 6·6135 | 2·761 |
| 8·7563 | 17·715 | **671** | 3772 | 5 | 20M | ·6087 | 25·2524 | 0·895 |
| 27·3952 | 15·848 | **672** | 3773 | 6 | 19M | ·8674 | 14·3607 | 24·606 |
| 16·5035 | 12·006 | **673** | 3774 | 1 | 20M | ·1262 | 3·4690 | 20·764 |
| 5·6118 | 8·163 | **674** | 3775 | 2 | 20M | ·3850 | 22·1079 | 18·897 |
| 24·2507 | 6·296 | **675** | 3776 | 3 | 20M | ·6437 | 11·2162 | 15·054 |
| 13·3590 | 2·453 | **676** | 3777 | 4 | 19M | ·9025 | 0·3245 | 11·212 |
| 2·4673 | 26·165 | **677** | 3778 | 6 | 20M | ·1612 | 18·9634 | 9·345 |
| 21·1062 | 24·298 | **678** | 3779 | 7 | 20M | ·4200 | 8·0717 | 5·502 |
| 10·2145 | 20·456 | **679** | 3780 | 1 | 20M | ·6787 | 26·7106 | 3·635 |
| 28·8534 | 18·589 | **680** | 3781 | 2 | 19M | ·9375 | 15·8189 | 27·347 |
| 17·9617 | 14·746 | **681** | 3782 | 4 | 20M | ·1962 | 4·9272 | 23·504 |
| 7·0700 | 10·903 | **682** | 3783 | 5 | 20M | ·4550 | 23·5661 | 21·638 |
| 25·7088 | 9·037 | **683** | 3784 | 6 | 20M | ·7138 | 12·6744 | 17·795 |
| 14·8171 | 5·194 | **684** | 3785 | 7 | 19M | ·9725 | 1·7827 | 13·952 |
| 3·9254 | 1·351 | **685** | 3786 | 2 | 20M | ·2313 | 20·4216 | 12·085 |
| 22·5643 | 27·039 | **686** | 3787 | 3 | 20M | ·4900 | 9·5299 | 8·242 |
| 11·6726 | 23·196 | **687** | 3788 | 4 | 20M | ·7488 | 28·1687 | 6·376 |
| 0·7809 | 19·354 | **688** | 3789 | 5 | 20M | ·0075 | 17·2770 | 2·533 |
| 19·4198 | 17·487 | **689** | 3790 | 7 | 20M | ·2663 | 6·3853 | 26·245 |
| 8·5281 | 13·644 | **690** | 3791 | 1 | 20M | ·5251 | 25·0242 | 24·378 |
| 27·1670 | 11·777 | **691** | 3792 | 2 | 20M | ·7838 | 14·1325 | 20·535 |
| 16·2753 | 7·935 | **692** | 3793 | 4 | 20M | ·0426 | 3·2408 | 16·693 |
| 5·3836 | 4·093 | **693** | 3794 | 5 | 20M | ·3013 | 21·8797 | 14·826 |
| 24·0225 | 2·225 | **694** | 3795 | 6 | 20M | ·5601 | 10·9808 | 10·983 |
| 13·1308 | 25·937 | **695** | 3796 | 7 | 20M | ·8188 | 0·0963 | 7·140 |
| 2·2391 | 22·094 | **696** | 3797 | 2 | 20M | ·0776 | 18·7352 | 5·274 |
| 20·8780 | 20·227 | **697** | 3798 | 3 | 20M | ·3364 | 7·8435 | 1·431 |
| 9·9863 | 16·384 | **698** | 3799 | 4 | 20M | ·5951 | 26·4824 | 27·119 |
| 28·6251 | 14·518 | **699** | 3800 | 5 | 20M | ·8539 | 15·5907 | 23·276 |

| A.D. | Kaliyuga | Week-day of 1st January | Commencement of Indian Solar Year. Day | Fraction of day | First New-Moon in Solar Year | Anomaly of first New-Moon |
|---|---|---|---|---|---|---|
| **700** | 3801 | 7 | 20M | ·1126 | 4·6990 | 19·433 |
| **701** | 3802 | 1 | 20M | ·3714 | 23·3379 | 17·567 |
| **702** | 3803 | 2 | 20M | ·6301 | 12·4462 | 13·724 |
| **703** | 3804 | 3 | 20M | ·8889 | 1·5545 | 9·881 |
| **704** | 3805 | 5 | 20M | ·1477 | 20·1933 | 8·014 |
| **705** | 3806 | 6 | 20M | ·4064 | 9·3016 | 4·172 |
| **706** | 3807 | 7 | 20M | ·6652 | 27·9405 | 2·305 |
| **707** | 3808 | 1 | 20M | ·9209 | 17·0488 | 26·017 |
| **708** | 3809 | 3 | 20M | ·1827 | 6·1571 | 22·174 |
| **709** | 3810 | 4 | 20M | ·4414 | 24·7960 | 20·307 |
| **710** | 3811 | 5 | 20M | ·7002 | 13·9043 | 16·464 |
| **711** | 3812 | 6 | 20M | ·9590 | 3·0126 | 12·622 |
| **712** | 3813 | 1 | 20M | ·2177 | 21·6515 | 10·755 |
| **713** | 3814 | 2 | 20M | ·4765 | 10·7598 | 6·912 |
| **714** | 3815 | 3 | 20M | ·7352 | 29·3987 | 5·045 |
| **715** | 3816 | 4 | 20M | ·9940 | 18·5070 | 1·203 |
| **716** | 3817 | 6 | 20M | ·2527 | 7·6153 | 24·914 |
| **717** | 3818 | 7 | 20M | ·5115 | 26·2542 | 23·048 |
| **718** | 3819 | 1 | 20M | ·7703 | 15·3625 | 19·205 |
| **719** | 3820 | 3 | 21M | ·0290 | 4·4708 | 15·362 |
| **720** | 3821 | 4 | 20M | ·2878 | 23·1097 | 13·495 |
| **721** | 3822 | 5 | 20M | ·5465 | 12·2180 | 9·653 |
| **722** | 3823 | 6 | 20M | ·8053 | 1·3263 | 5·810 |
| **723** | 3824 | 1 | 21M | ·0640 | 19·9651 | 3·9[illegible]3 |
| **724** | 3825 | 2 | 20M | ·3228 | 9·0734 | 0·100 |
| **725** | 3826 | 3 | 20M | ·5815 | 27·7123 | 25·788 |
| **726** | 3827 | 4 | 20M | ·8403 | 16·8206 | 21·946 |
| **727** | 3828 | 6 | 21M | ·0991 | 5·9289 | 18·103 |
| **728** | 3829 | 7 | 20M | ·3578 | 24·5678 | 16·236 |
| **729** | 3830 | 1 | 20M | ·6166 | 13·6761 | 12·393 |
| **730** | 3831 | 2 | 20M | ·8753 | 2·7844 | 8·551 |
| **731** | 3832 | 4 | 21M | ·1341 | 21·4233 | 6·6[illegible]4 |
| **732** | 3833 | 5 | 20M | ·3928 | 10·5316 | 2·841 |
| **733** | 3834 | 6 | 20M | ·6516 | 29·1705 | 0·974 |
| **734** | 3835 | 7 | 20M | ·9104 | 18·2788 | 24·686 |
| **735** | 3836 | 2 | 21M | ·1691 | 7·3871 | 20·843 |
| **736** | 3837 | 3 | 20M | ·4279 | 26·0260 | 18·976 |
| **737** | 3838 | 4 | 20M | ·6866 | 15·1343 | 15·134 |
| **738** | 3839 | 5 | 20M | ·9454 | 4·2426 | 11·291 |
| **739** | 3840 | 7 | 21M | ·2041 | 22·8815 | 9·424 |
| **740** | 3841 | 1 | 20M | ·4629 | 11·9897 | 5·582 |
| **741** | 3842 | 2 | 20M | ·7217 | 1·0980 | 1·739 |
| **742** | 3843 | 3 | 20M | ·9804 | 19·7369 | 27·427 |
| **743** | 3844 | 5 | 21M | ·2392 | 8·8452 | 23·584 |
| **744** | 3845 | 6 | 20M | ·4979 | 27·4841 | 21·717 |
| **745** | 3846 | 7 | 20M | ·7567 | 16·5924 | 17·875 |
| **746** | 3847 | 1 | 21M | ·0154 | 5·7007 | 14·032 |
| **747** | 3848 | 3 | 21M | ·2742 | 24·3396 | 12·165 |
| **748** | 3849 | 4 | 20M | ·5330 | 13·4479 | 8·322 |
| **749** | 3850 | 5 | 20M | ·7917 | 2·5562 | 4·480 |

## SURYA SIDDHANTA figures

| A.D. | Kaliyuga. | Week-day of 1st January | Commencement of Indian Solar Year. Day | Commencement of Indian Solar Year. Fraction of day | First New-Moon in Solar Year | Anomaly of first New-Moon |
|---|---|---|---|---|---|---|
| 750 | 3851 | 7 | 21M | ·0505 | 21·1951 | 2·613 |
| 751 | 3852 | 1 | 21M | ·3092 | 10·3034 | 26·325 |
| 752 | 3853 | 2 | 20M | ·5680 | 28·9423 | 24·458 |
| 753 | 3854 | 3 | 20M | ·8267 | 18·0506 | 20·615 |
| 754 | 3855 | 5 | 21M | ·0855 | 7·1589 | 16·772 |
| 755 | 3856 | 6 | 21M | ·3443 | 25·7978 | 14·906 |
| 756 | 3857 | 7 | 20M | ·6030 | 14·9061 | 11·063 |
| 757 | 3858 | 1 | 20M | ·8618 | 4·0144 | 7·220 |
| 758 | 3859 | 3 | 21M | ·1205 | 22·6532 | 5·353 |
| 759 | 3860 | 4 | 21M | ·3793 | 11·7615 | 1·511 |
| 760 | 3861 | 5 | 20M | ·6380 | 0·8698 | 25·223 |
| 761 | 3862 | 6 | 20M | ·8968 | 19·5087 | 23·356 |
| 762 | 3863 | 1 | 21M | ·1556 | 8·6170 | 19·513 |
| 763 | 3864 | 2 | 21M | ·4143 | 27·2559 | 17·646 |
| 764 | 3865 | 3 | 20M | ·6731 | 16·3642 | 13·804 |
| 765 | 3866 | 4 | 20M | ·9318 | 5·4725 | 9·961 |
| 766 | 3867 | 6 | 21M | ·1906 | 24·1114 | 8·094 |
| 767 | 3868 | 7 | 21M | ·4493 | 13·2197 | 4·251 |
| 768 | 3869 | 1 | 20M | ·7081 | 2·3280 | 0·409 |
| 769 | 3870 | 2 | 20M | ·9668 | 20·9669 | 26·096 |
| 770 | 3871 | 4 | 21M | ·2256 | 10·0752 | 22·254 |
| 771 | 3872 | 5 | 21M | ·4844 | 28·7141 | 20·387 |
| 772 | 3873 | 6 | 20M | ·7431 | 17·8224 | 16·544 |
| 773 | 3874 | 7 | 21M | ·0019 | 6·9307 | 12·701 |
| 774 | 3875 | 2 | 21M | ·2606 | 25·5696 | 10·835 |
| 775 | 3876 | 3 | 21M | ·5194 | 14·6779 | 6·992 |
| 776 | 3877 | 4 | 20M | ·7781 | 3·7862 | 3·149 |
| 777 | 3878 | 6 | 21M | ·0369 | 22·4250 | 1·282 |
| 778 | 3879 | 7 | 21M | ·2957 | 11·5333 | 24·994 |
| 779 | 3880 | 1 | 21M | ·5544 | 0·6416 | 21·152 |
| 780 | 3881 | 2 | 20M | ·8132 | 19·2805 | 19·284 |
| 781 | 3882 | 4 | 21M | ·0719 | 8·3888 | 15·442 |
| 782 | 3883 | 5 | 21M | ·3307 | 27·0277 | 13·575 |
| 783 | 3884 | 6 | 21M | ·5894 | 16·1360 | 9·738 |
| 784 | 3885 | 7 | 20M | ·8482 | 5·2443 | 5·890 |
| 785 | 3886 | 2 | 21M | ·1070 | 23·8832 | 4·023 |
| 786 | 3887 | 3 | 21M | ·3657 | 12·9915 | 0·180 |
| 787 | 3888 | 4 | 21M | ·6245 | 2·0998 | 23·892 |
| 788 | 3889 | 5 | 20M | ·8832 | 20·7387 | 22·025 |
| 789 | 3890 | 7 | 21M | ·1420 | 9·8470 | 18·183 |
| 790 | 3891 | 1 | 21M | ·4007 | 28·4859 | 16·316 |
| 791 | 3892 | 2 | 21M | ·6595 | 17·5942 | 12·473 |
| 792 | 3893 | 3 | 20M | ·9183 | 6·7025 | 8·630 |
| 793 | 3894 | 5 | 21M | ·1770 | 25·3413 | 6·764 |
| 794 | 3895 | 6 | 21M | ·4358 | 14·4496 | 2·921 |
| 795 | 3896 | 7 | 21M | ·6945 | 3·5579 | 26·633 |
| 796 | 3897 | 1 | 20M | ·9533 | 22·1968 | 24·766 |
| 797 | 3898 | 3 | 21M | ·2120 | 11·3051 | 20·923 |
| 798 | 3899 | 4 | 21M | ·4708 | 0·4134 | 17·080 |
| 799 | 3900 | 5 | 21M | ·7296 | 19·0623 | 15·214 |

| A.D. | Kaliyuga | Week-day of 1st January | Commencement of Indian Solar Year. Day | Commencement of Indian Solar Year. Fraction of day | First New-Moon in Solar Year | Anomaly of first New-Moon |
|---|---|---|---|---|---|---|
| 800 | 3901 | 6 | 20M | ·9883 | 8·1606 | 11·371 |
| 801 | 3902 | 1 | 21M | ·2471 | 26·7995 | 9·504 |
| 802 | 3903 | 2 | 21M | ·5058 | 15·9078 | 5·661 |
| 803 | 3904 | 3 | 21M | ·7646 | 5·0161 | 1·819 |
| 804 | 3905 | 5 | 21M | ·0233 | 23·6550 | 27·507 |
| 805 | 3906 | 6 | 21M | ·2821 | 12·7633 | 23·664 |
| 806 | 3907 | 7 | 21M | ·5409 | 1·8716 | 19·821 |
| 807 | 3908 | 1 | 21M | ·7996 | 20·5105 | 17·954 |
| 808 | 3909 | 3 | 21M | ·0584 | 9·6188 | 14·112 |
| 809 | 3910 | 4 | 21M | ·3171 | 28·2577 | 12·245 |
| 810 | 3911 | 5 | 21M | ·5759 | 17·3660 | 8·402 |
| 811 | 3912 | 6 | 21M | ·8346 | 6·4743 | 4·559 |
| 812 | 3913 | 1 | 21M | ·0934 | 25·1131 | 2·693 |
| 813 | 3914 | 2 | 21M | ·3522 | 14·2214 | 26·404 |
| 814 | 3915 | 3 | 21M | ·6109 | 3·3297 | 22·562 |
| 815 | 3916 | 4 | 21M | ·8697 | 21·9686 | 20·695 |
| 816 | 3917 | 6 | 21M | ·1284 | 11·0769 | 16·852 |
| 817 | 3918 | 7 | 21M | ·3872 | 0·1852 | 13·009 |
| 818 | 3919 | 1 | 21M | ·6459 | 18·8241 | 11·143 |
| 819 | 3920 | 2 | 21M | ·9047 | 7·9324 | 7·300 |
| 820 | 3921 | 4 | 21M | ·1634 | 26·5713 | 5·433 |
| 821 | 3922 | 5 | 21M | ·4222 | 15·6796 | 1·590 |
| 822 | 3923 | 6 | 21M | ·6810 | 4·7879 | 25·302 |
| 823 | 3924 | 7 | 21M | ·9397 | 23·4268 | 23·435 |
| 824 | 3925 | 2 | 21M | ·1985 | 12·5351 | 19·593 |
| 825 | 3926 | 3 | 21M | ·4572 | 1·6434 | 15·750 |
| 826 | 3927 | 4 | 21M | ·7160 | 20·2823 | 13·883 |
| 827 | 3928 | 5 | 21M | ·9747 | 9·3906 | 10·040 |
| 828 | 3929 | 7 | 21M | ·2335 | 28·0295 | 8·174 |
| 829 | 3930 | 1 | 21M | ·4923 | 17·1378 | 4·331 |
| 830 | 3931 | 2 | 21M | ·7510 | 6·2461 | 0·488 |
| 831 | 3932 | 3 | 22M | ·0098 | 24·8849 | 26·176 |
| 832 | 3933 | 5 | 21M | ·2685 | 13·9932 | 22·333 |
| 833 | 3934 | 6 | 21M | ·5273 | 3·1015 | 18·491 |
| 834 | 3935 | 7 | 21M | ·7860 | 21·7404 | 16·624 |
| 835 | 3936 | 2 | 22M | ·0448 | 10·8487 | 12·781 |
| 836 | 3937 | 3 | 21M | ·3036 | 29·4876 | 10·914 |
| 837 | 3938 | 4 | 21M | ·5623 | 18·5959 | 7·072 |
| 838 | 3939 | 5 | 21M | ·8211 | 7·7042 | 3·229 |
| 839 | 3940 | 7 | 22M | ·0798 | 26·3431 | 1·362 |
| 840 | 3941 | 1 | 21M | ·3386 | 15·4514 | 25·074 |
| 841 | 3942 | 2 | 21M | ·5973 | 4·5597 | 21·231 |
| 842 | 3943 | 3 | 21M | ·8561 | 23·1986 | 19·364 |
| 843 | 3944 | 5 | 22M | ·1149 | 12·3069 | 15·522 |
| 844 | 3945 | 6 | 21M | ·3736 | 1·4152 | 11·679 |
| 845 | 3946 | 7 | 21M | ·6324 | 20·0541 | 9·812 |
| 846 | 3947 | 1 | 21M | ·8911 | 9·1624 | 5·969 |
| 847 | 3948 | 3 | 22M | ·1499 | 27·8012 | 4·103 |
| 848 | 3949 | 4 | 21M | ·4086 | 16·9095 | 0·260 |
| 849 | 3950 | 5 | 21M | ·6674 | 6·0178 | 23·972 |

| A.D. | Kaliyuga | Week-day of 1st January | Commencement of Indian Solar Year. Day | Commencement of Indian Solar Year. Fraction of day |
|---|---|---|---|---|
| 850 | 3951 | 6 | 21M | ·9262 |
| 851 | 3952 | 1 | 22M | ·1849 |
| 852 | 3953 | 2 | 21M | ·4437 |
| 853 | 3954 | 3 | 21M | ·7024 |
| 854 | 3955 | 4 | 21M | ·9612 |
| 855 | 3956 | 6 | 22M | ·2199 |
| 856 | 3957 | 7 | 21M | ·4787 |
| 857 | 3958 | 1 | 21M | ·7375 |
| 858 | 3959 | 2 | 21M | ·9962 |
| 859 | 3960 | 4 | 22M | ·2550 |
| 860 | 3961 | 5 | 21M | ·5137 |
| 861 | 3962 | 6 | 21M | ·7725 |
| 862 | 3963 | 1 | 22M | ·0312 |
| 863 | 3964 | 2 | 22M | ·2900 |
| 864 | 3965 | 3 | 21M | ·5487 |
| 865 | 3966 | 4 | 21M | ·8075 |
| 866 | 3967 | 6 | 22M | ·0663 |
| 867 | 3968 | 7 | 22M | ·3250 |
| 868 | 3969 | 1 | 21M | ·5838 |
| 869 | 3970 | 2 | 21M | ·8425 |
| 870 | 3971 | 4 | 22M | ·1013 |
| 871 | 3972 | 5 | 22M | ·3600 |
| 872 | 3973 | 6 | 21M | ·6188 |
| 873 | 3974 | 7 | 21M | ·8776 |
| 874 | 3975 | 2 | 22M | ·1363 |
| 875 | 3976 | 3 | 22M | ·3951 |
| 876 | 3977 | 4 | 21M | ·6538 |
| 877 | 3978 | 5 | 21M | ·9126 |
| 878 | 3979 | 7 | 22M | ·1713 |
| 879 | 3980 | 1 | 22M | ·4301 |
| 880 | 3981 | 2 | 21M | ·6889 |
| 881 | 3982 | 3 | 21M | ·9476 |
| 882 | 3983 | 5 | 22M | ·2064 |
| 883 | 3984 | 6 | 22M | ·4651 |
| 884 | 3985 | 7 | 21M | ·7239 |
| 885 | 3986 | 1 | 21M | ·9826 |
| 886 | 3987 | 3 | 22M | ·2414 |
| 887 | 3988 | 4 | 22M | ·5002 |
| 888 | 3989 | 5 | 21M | ·7589 |
| 889 | 3990 | 6 | 22M | ·0177 |
| 890 | 3991 | 1 | 22M | ·2764 |
| 891 | 3992 | 2 | 22M | ·5352 |
| 892 | 3993 | 3 | 21M | ·7939 |
| 893 | 3994 | 5 | 22M | ·0527 |
| 894 | 3995 | 6 | 22M | ·3115 |
| 895 | 3996 | 7 | 22M | ·5702 |
| 896 | 3997 | 1 | 21M | ·8290 |
| 897 | 3998 | 3 | 22M | ·0877 |
| 898 | 3999 | 4 | 22M | ·3465 |
| 899 | 4000 | 5 | 22M | ·6052 |

| First New-Moon in Solar Year (continued) | Anomaly of first New-Moon (continued) | A.D. | Kaliyuga | Week-day of 1st January | Commencement of Indian Solar Year: Day | Commencement of Indian Solar Year: Fraction of day | First New-Moon in Solar Year | Anomaly of first New-Moon |
|---|---|---|---|---|---|---|---|---|
| 24·6567 | 22·105 | **900** | 4001 | 6 | 21M | ·8640 | 11·6220 | 3·309 |
| 13·7650 | 18·262 | **901** | 4002 | 1 | 22M | ·1228 | 0·7306 | 27·021 |
| 2·8733 | 14·419 | **902** | 4003 | 2 | 22M | ·3815 | 19·3694 | 25·154 |
| 21·5122 | 12·552 | **903** | 4004 | 3 | 22M | ·6403 | 8·4777 | 21·311 |
| 10·6205 | 8·1[illegible]0 | **904** | 4005 | 4 | 21M | ·8090 | 27·1166 | 19·444 |
| 29·2594 | 6·843 | **905** | 4006 | 6 | 22M | ·1578 | 16·2249 | 15·602 |
| 18·3677 | 3·000 | **906** | 4007 | 7 | 22M | ·4165 | 5·3332 | 11·752 |
| 7·4760 | 26·712 | **907** | 4008 | 1 | 22M | ·6753 | 23·9721 | 9·892 |
| 26·1149 | 24·845 | **908** | 4009 | 2 | 21M | ·9341 | 13·0804 | 6·049 |
| 15·2232 | 21·003 | **909** | 4010 | 4 | 22M | ·1928 | 2·1887 | 2·207 |
| 4·3315 | 17·160 | **910** | 4011 | 5 | 22M | ·4516 | 20·8276 | 0·340 |
| 22·9704 | 15·293 | **911** | 4012 | 6 | 22M | ·7103 | 9·9359 | 24·052 |
| 12·0787 | 11·450 | **912** | 4013 | 7 | 21M | ·9691 | 28·5748 | 22·185 |
| 1·1870 | 7·608 | **913** | 4014 | 2 | 22M | ·2278 | 17·6831 | 18·342 |
| 19·8259 | 5·741 | **914** | 4015 | 3 | 22M | ·4866 | 6·7914 | 14·499 |
| 8·9342 | 1·898 | **915** | 4016 | 4 | 22M | ·7453 | 25·4303 | 12·633 |
| 27·5730 | 0·031 | **916** | 4017 | 5 | 21M | ·0041 | 14·5386 | 8·790 |
| 16·6813 | 23·744 | **917** | 4018 | 7 | 22M | ·2629 | 3·6469 | 4·947 |
| 5·7896 | 19·901 | **918** | 4019 | 1 | 22M | ·5216 | 22·2858 | 3·080 |
| 24·4285 | 18·034 | **919** | 4020 | 2 | 22M | ·7804 | 11·3941 | 26·792 |
| 13·5368 | 14·191 | **920** | 4021 | 4 | 22M | ·0391 | 0·5024 | 22·949 |
| 2·6451 | 10·348 | **921** | 4022 | 5 | 22M | ·2979 | 19·1412 | 21·083 |
| 21·2840 | 8·481 | **922** | 4023 | 6 | 22M | ·5566 | 8·2495 | 17·240 |
| 10·3923 | 4·639 | **923** | 4024 | 7 | 22M | ·8154 | 26·8884 | 15·373 |
| 29·0312 | 2·772 | **924** | 4025 | 2 | 22M | ·0742 | 15·9967 | 11·530 |
| 18·1395 | 26·484 | **925** | 4026 | 3 | 22M | ·3329 | 5·1050 | 7·688 |
| 7·2478 | 22·641 | **926** | 4027 | 4 | 22M | ·5917 | 23·7439 | 5·828 |
| 25·8867 | 20·774 | **927** | 4028 | 5 | 22M | ·8504 | 12·8522 | 1·978 |
| 14·9950 | 16·932 | **928** | 4029 | 7 | 22M | ·1092 | 1·9605 | 25·690 |
| 4·1033 | 13·089 | **929** | 4030 | 1 | 22M | ·3679 | 20·5994 | 23·823 |
| 22·7422 | 11·222 | **930** | 4031 | 2 | 22M | ·6267 | 9·7077 | 19·980 |
| 11·8505 | 7·379 | **931** | 4032 | 3 | 22M | ·8855 | 28·3466 | 18·114 |
| 0·9588 | 3·537 | **932** | 4033 | 5 | 22M | ·1442 | 17·4549 | 14·271 |
| 19·5977 | 1·670 | **933** | 4034 | 6 | 22M | ·4020 | 6·5632 | 10·428 |
| 8·7060 | 25·382 | **934** | 4035 | 7 | 22M | ·6617 | 25·2021 | 8·561 |
| 27·3448 | 23·515 | **935** | 4036 | 1 | 22M | ·9205 | 14·3104 | 4·719 |
| 16·4531 | 19·672 | **936** | 4037 | 3 | 22M | ·1792 | 3·4187 | 0·876 |
| 5·5614 | 15·829 | **937** | 4038 | 4 | 22M | ·4380 | 22·0575 | 26·564 |
| 24·2003 | 13·963 | **938** | 4039 | 5 | 22M | ·6968 | 11·1658 | 22·721 |
| 13·3086 | 10·120 | **939** | 4040 | 6 | 22M | ·9555 | 0·2741 | 18·878 |
| 2·4169 | 6·277 | **940** | 4041 | 1 | 22M | ·2143 | 18·9130 | 17·011 |
| 21·0558 | 4·410 | **941** | 4042 | 2 | 22M | ·4730 | 8·0213 | 13·169 |
| 10·1641 | 0·568 | **942** | 4043 | 3 | 22M | ·7318 | 26·6602 | 11·299 |
| 28·8030 | 26·256 | **943** | 4044 | 4 | 22M | ·9904 | 15·7685 | 7·459 |
| 17·9113 | 22·413 | **944** | 4045 | 5 | 22M | ·2493 | 4·8768 | 3·617 |
| 7·0196 | 18·570 | **945** | 4046 | 7 | 22M | ·5081 | 23·5157 | 1·750 |
| 25·6585 | 16·703 | **946** | 4047 | 1 | 22M | ·7668 | 12·6240 | 25·463 |
| 14·7668 | 12·861 | **947** | 4048 | 2 | 22M | ·0256 | 1·7323 | 21·612 |
| 3·8751 | 9·018 | **948** | 4049 | 4 | 22M | ·2843 | 20·3712 | 19·752 |
| 22·5140 | 7·151 | **949** | 4050 | 5 | 22M | ·5431 | 9·4795 | 15·909 |

| A.D. | Kaliyuga | Week-day of 1st January | Commencement of Indian Solar Year: Day | Commencement of Indian Solar Year: Fraction of day | First New-Moon in Solar Year | Anomaly of first New-Moon |
|---|---|---|---|---|---|---|
| **950** | 4051 | 6 | 22M | ·8018 | 28·1184 | 14·043 |
| **951** | 4052 | 1 | 23M | ·0606 | 17·2267 | 10·200 |
| **952** | 4053 | 2 | 22M | ·3194 | 6·3350 | 6·357 |
| **953** | 4054 | 3 | 22M | ·5781 | 24·9739 | 4·490 |
| **954** | 4055 | 4 | 22M | ·8369 | 14·0822 | 0·648 |
| **955** | 4056 | 6 | 23M | ·0956 | 3·1905 | 24·359 |
| **956** | 4057 | 7 | 22M | ·3544 | 21·8293 | 22·493 |
| **957** | 4058 | 1 | 22M | ·6131 | 10·9376 | 18·650 |
| **958** | 4059 | 2 | 22M | ·8719 | 0·0459 | 14·807 |
| **959** | 4060 | 4 | 23M | ·1306 | 18·6848 | 12·940 |
| **960** | 4061 | 5 | 22M | ·3894 | 7·7931 | 9·098 |
| **961** | 4062 | 6 | 22M | ·6482 | 26·4320 | 7·231 |
| **962** | 4063 | 7 | 22M | ·9069 | 15·5403 | 3·388 |
| **963** | 4064 | 2 | 23M | ·1657 | 4·6486 | 27·100 |
| **964** | 4065 | 3 | 22M | ·4244 | 23·2875 | 25·233 |
| **965** | 4066 | 4 | 22M | ·6832 | 12·3958 | 21·391 |
| **966** | 4067 | 5 | 22M | ·9419 | 1·5041 | 17·548 |
| **967** | 4068 | 7 | 23M | ·2007 | 20·1430 | 15·681 |
| **968** | 4069 | 1 | 22M | ·4595 | 9·2513 | 11·838 |
| **869** | 4070 | 2 | 22M | ·7182 | 27·8902 | 9·972 |
| **970** | 4071 | 3 | 22M | ·9770 | 16·9985 | 6·129 |
| **971** | 4072 | 5 | 23M | ·2357 | 6·1068 | 2·286 |
| **972** | 4073 | 6 | 22M | ·4945 | 24·7457 | 0·419 |
| **973** | 4074 | 7 | 22M | ·7532 | 13·8540 | 24·131 |
| **974** | 4075 | 1 | 23M | ·0120 | 2·9623 | 20·288 |
| **975** | 4076 | 3 | 23M | ·2708 | 21·6011 | 18·422 |
| **976** | 4077 | 4 | 22M | ·6295 | 10·7094 | 14·579 |
| **977** | 4078 | 5 | 22M | ·7883 | 29·3483 | 12·712 |
| **978** | 4079 | 7 | 23M | ·0470 | 18·4566 | 8·869 |
| **979** | 4080 | 1 | 23M | ·3058 | 7·5649 | 5·080 |
| **980** | 4081 | 2 | 22M | ·5645 | 26·2038 | 3·160 |
| **981** | 4082 | 3 | 22M | ·8233 | 15·3121 | 26·872 |
| **982** | 4083 | 5 | 23M | ·0821 | 4·4204 | 23·029 |
| **983** | 4084 | 5 | 23M | ·3408 | 23·0593 | 21·162 |
| **984** | 4085 | 7 | 22M | ·5996 | 12·1676 | 17·319 |
| **985** | 4086 | 1 | 22M | ·8583 | 1·2759 | 13·477 |
| **986** | 4087 | 3 | 23M | ·1171 | 19·9148 | 11·610 |
| **987** | 4088 | 4 | 23M | ·3758 | 9·0231 | 7·767 |
| **988** | 4089 | 5 | 22M | ·6346 | 27·6620 | 5·900 |
| **989** | 4090 | 6 | 22M | ·8994 | 16·7703 | 2·058 |
| **990** | 4091 | 1 | 23M | ·1521 | 5·8786 | 25·769 |
| **991** | 4092 | 2 | 23M | ·4109 | 24·5174 | 23·903 |
| **992** | 4093 | 3 | 22M | ·6696 | 13·6257 | 20·060 |
| **993** | 4094 | 4 | 22M | ·9284 | 2·7340 | 16·217 |
| **994** | 4095 | 6 | 23M | ·1871 | 21·3729 | 14·350 |
| **995** | 4096 | 7 | 23M | ·4459 | 10·4812 | 10·508 |
| **996** | 4097 | 1 | 22M | ·7047 | 29·1201 | 8·641 |
| **997** | 4098 | 2 | 22M | ·9634 | 18·2284 | 4·798 |
| **998** | 4099 | 4 | 23M | ·2222 | 7·3367 | 0·955 |
| **999** | 4100 | 5 | 23M | ·4809 | 25·9756 | 26·643 |

## TABLE X.

☉'s Anom col. 6: + 29·53059, + 59·06117, + 88·59176, + 118·12235

☾'s Anom col. 7: + 1·976, + 3·952, + 5·928, + 7·904

| Kaliyuga. | Vikrama Era. | Saka Era. | Commencement of Solar Year. Month and day A.D. | Fraction of day. | First New-Moon in Solar Year. | Anomaly of first New-Moon. | A.D. | Week-day of 1st January. | Vaisakha Week-day | Month | Day | Fraction | Jyeshta Week-day | Month | Day | Fraction | Ashada Week-day | Month | Day | Fraction | Sravana Week-day | Month | Day | Fraction | Bhadrapada Week-day | Month | Day | Fraction |
|---|---|---|---|---|---|---|---|---|---|---|---|---|---|---|---|---|---|---|---|---|---|---|---|---|---|---|---|---|
| 4101 | 1057 | 922 | 22M | ·7396 | 15·0839 | 22·801 | **1000** | 2 | 7 | Ap | 6● | ·82 | 2 | My | 6 | ·35 | 3 | Je | 4 | ·88 | 5 | Jl | 4 | ·41 | 6 | Au | 2 | ·9 |
| 4102 | 1058 | 923 | 22M | ·9983 | 4·1922 | 18·959 | **1001** | 4 | 5 | Mr | 27 | ·19 | 6 | Ap | 25 | ·72 | 1 | My | 25 | ·25 | 4 | Jl | 23 | ·31 | 5 | Au | 21○ | ·8 |
| | | | | | | | | | | | | | | | | | 2 | Je | 23 | ·78 | | | | | | | | |
| 4103 | 1059 | 924 | 23M | ·2571 | 22·8311 | 17·091 | **1002** | 5 | 4 | Ap | 15 | ·09 | 5 | My | 14 | ·62 | 7 | Je | 13 | ·15 | 1 | Jl | 12 | ·68 | 3 | Au | 11●○ | ·2 |
| 4104 | 1060 | 925 | 23M | ·5158 | 11·9394 | 13·249 | **1003** | 6 | 1 | Ap | 4 | ·45 | 2 | My | 3 | ·99 | 4 | Je | 2 | ·52 | 6 | Jl | 2 | ·05 | 7 | Jl | 31○ | ·5 |
| 4105 | 1061 | 926 | 22M | ·7746 | 1·0477 | 9·406 | **1004** | 7 | 5 | Mr | 23 | ·82 | 1 | My | 21 | ·88 | 3 | Je | 20○ | ·41 | 4 | Jl | 19● | ·94 | 6 | Au | 18 | ·4 |
| | | | | | | | | | 7 | Ap | 22 | ·35 | | | | | | | | | | | | | | | | |
| 4106 | 1062 | 927 | 23M | ·0334 | 19·6866 | 7·539 | **1005** | 2 | 4 | Ap | 11 | ·72 | 6 | My | 11 | ·25 | 7 | Je | 9○ | ·78 | 2 | Jl | 9 | ·31 | 3 | Au | 7 | ·8 |
| 4107 | 1063 | 928 | 23M | ·2921 | 8·7949 | 3·696 | **1006** | 3 | 2 | Ap | 1 | ·09 | 3 | Ap | 30 | ·62 | 5 | My | 30○ | ·15 | 6 | Je | 28 | ·68 | 1 | Jl | 28 | ·2 |
| | | | | | | | | | | | | | | | | | | | | | | | | | 2 | Au | 26 | ·7 |
| 4108 | 1064 | 929 | 23M | ·5509 | 27·4338 | 1·830 | **1007** | 4 | 7 | Ap | 19 | ·98 | 2 | My | 19● | ·51 | 4 | Je | 18 | ·05 | 5 | Jl | 17 | ·58 | 7 | Au | 16 | ·1 |
| 4109 | 1065 | 930 | 22M | ·8096 | 16·5421 | 25·542 | **1008** | 5 | 5 | Ap | 8○ | ·35 | 6 | My | 7 | ·88 | 1 | Je | 6 | ·41 | 2 | Jl | 5 | ·94 | 4 | Au | 4 | ·4 |
| 4110 | 1066 | 931 | 23M | ·0684 | 5·6504 | 21·699 | **1009** | 7 | 2 | Mr | 28○ | ·72 | 4 | Ap | 27 | ·25 | 5 | My | 26 | ·78 | 7 | Je | 25 | ·31 | 3 | Au | 23 | ·3 |
| | | | | | | | | | | | | | | | | | | | | | 1 | Jl | 24 | ·84 | | | | |
| 4111 | 1067 | 932 | 23M | ·3271 | 24·2892 | 19·832 | **1010** | 1 | 1 | Ap | [illegible] | ·62 | 3 | My | 16 | ·15 | 4 | Je | 14 | ·68 | 6 | Jl | 14 | ·21 | 7 | Au | 12 | ·7 |
| 4112 | 1068 | 933 | 23M | ·5859 | 13·3975 | 15·389 | **1011** | 2 | 5 | Ap | 5 | ·98 | 7 | My | 5 | ·51 | 2 | Je | 4 | ·04 | 3 | Jl | 3 | ·57 | 5 | Au | 2 | ·1 |
| 4113 | 1069 | 934 | 22M | ·8446 | 2·5058 | 12·147 | **1012** | 3 | 3 | Mr | 25 | ·35 | 4 | Ap | 23 | ·88 | 7 | Je | 21 | ·94 | 2 | Jl | 21○ | ·47 | 4 | Au | 20● | ·0 |
| | | | | | | | | | | | | | 6 | My | 23 | ·41 | | | | | | | | | | | | |
| 4114 | 1070 | 935 | 23M | ·1084 | 21·1447 | 10·280 | **1013** | 5 | 2 | Ap | 13 | ·25 | 3 | My | 12 | ·78 | 5 | Je | 11 | ·31 | 6 | Jl | 10○ | ·84 | 1 | Au | 9 | ·3 |
| 4115 | 1071 | 936 | 23M | ·3622 | 10·2530 | 6·437 | **1014** | 6 | 6 | Ap | 2 | ·61 | 1 | My | 2 | ·15 | 2 | My | 31 | ·68 | 4 | Je | 30●○ | ·21 | 5 | Jl | 29 | ·7 |
| 4116 | 1072 | 937 | 23M | ·6209 | 28·8919 | 4·570 | **1015** | 7 | 5 | Ap | 21 | ·51 | 7 | My | 21○ | ·04 | 1 | Je | 19● | ·57 | 3 | Jl | 19 | ·10 | 4 | Au | 17 | ·6 |
| 4117 | 1073 | 938 | 22M | ·8797 | 18·0002 | 0·727 | **1016** | 1 | 2 | Ap | 9 | ·88 | 4 | My | 9○ | ·41 | 5 | Je | 7 | ·94 | 7 | Jl | 7 | ·47 | 2 | Au | 6 | ·0 |
| 4118 | 1074 | 939 | 23M | ·1384 | 7·1085 | 24·439 | **1017** | 3 | 7 | Mr | 30 | ·25 | 1 | Ap | 28○ | ·78 | 3 | My | 28 | ·31 | 4 | Je | 26 | ·84 | 7 | Au | 24 | ·9 |
| | | | | | | | | | | | | | | | | | | | | | 6 | Jl | 26 | ·37 | | | | |
| 4119 | 1075 | 940 | 23M | ·3972 | 25·7474 | 22·573 | **1018** | 4 | 6 | Ap | 18 | ·14 | 7 | My | 17 | ·67 | 2 | Je | 16 | ·21 | 3 | Jl | 15 | ·74 | 5 | Au | 14 | ·2 |
| 4120 | 1076 | 941 | 23M | ·6559 | 14·8557 | 18·730 | **1019** | 5 | 3 | Ap | 7● | ·51 | 5 | My | 7 | ·04 | 6 | Je | 5 | ·57 | 1 | Jl | 5 | ·10 | 2 | Au | 3 | ·6 |
| 4121 | 1077 | 942 | 22M | ·9147 | 3·9640 | 14·887 | **1020** | 6 | 7 | Mr | 26 | ·88 | 2 | Ap | 25 | ·41 | 3 | My | 24 | ·94 | 7 | Jl | 23 | ·00 | 1 | Au | 21○ | ·5 |
| | | | | | | | | | | | | | | | | | 5 | Je | 23 | ·47 | | | | | | | | |
| 4122 | 1078 | 943 | 23M | ·1735 | 22·6029 | 13·020 | **1021** | 1 | 6 | Ap | 14 | ·78 | 1 | My | 14 | ·31 | 2 | Je | 12 | ·84 | 4 | Jl | 12 | ·37 | 5 | Au | 10●○ | ·9 |
| 4123 | 1079 | 944 | 23M | ·4322 | 11·7112 | 9·178 | **1022** | 2 | 4 | Ap | 4 | ·14 | 5 | My | 3 | ·67 | 7 | Je | 2 | ·20 | 1 | Jl | 1○ | ·73 | 3 | Jl | 31 | ·2 |
| 4124 | 1080 | 945 | 23M | ·6910 | 0·8195 | 5·334 | **1023** | 3 | 1 | Mr | 24 | ·51 | 4 | My | 22 | ·57 | 6 | Je | 21○ | ·10 | 7 | Jl | 20 | ·63 | 2 | Au | 19 | ·1 |
| | | | | | | | | | 3 | Ap | 23 | ·04 | | | | | | | | | | | | | | | | |
| 4125 | 1081 | 946 | 22M | ·9497 | 19·4584 | 3·468 | **1024** | 4 | 7 | Ap | 11 | ·41 | 1 | My | 10 | ·94 | 3 | Je | 9●○ | ·47 | 5 | Jl | 9 | ·00 | 6 | Au | 7 | ·5 |
| 4126 | 1082 | 947 | 23M | ·2085 | 8·5667 | 27·180 | **1025** | 6 | 4 | Mr | 31 | ·77 | 6 | Ap | 30 | ·31 | 7 | My | 29 | ·84 | 2 | Je | 28 | ·37 | 3 | Jl | 27 | ·9 |
| | | | | | | | | | | | | | | | | | | | | | | | | | 5 | Au | 26 | ·4 |
| 4127 | 1083 | 948 | 23M | ·4672 | 27·2056 | 25·313 | **1026** | 7 | 3 | Ap | 19○ | ·67 | 5 | My | 19● | ·20 | 6 | Je | 17 | ·73 | 1 | Jl | 17 | ·26 | 2 | Au | 15 | ·7 |
| 4128 | 1084 | 949 | 23M | ·7260 | 16·3139 | 21·470 | **1027** | 1 | 1 | Ap | 9○ | ·04 | 2 | My | 8 | ·57 | 4 | Je | 7 | ·10 | 5 | Jl | 6 | ·63 | 7 | Au | 5 | ·1 |
| 4129 | 1085 | 950 | 22M | ·9848 | 5·4222 | 17·628 | **1028** | 2 | 5 | Mr | 28○ | ·41 | 6 | Ap | 26 | ·94 | 1 | My | 26 | ·47 | 3 | Je | 25 | ·00 | 6 | Au | 23 | ·0 |
| | | | | | | | | | | | | | | | | | | | | | 4 | Jl | 24 | ·53 | | | | |
| 4130 | 1086 | 951 | 23M | ·2435 | 24·0610 | 15·761 | **1029** | 4 | 4 | Ap | 16 | ·30 | 5 | My | 15 | ·83 | 7 | Je | 14 | ·37 | 1 | Jl | 13 | ·90 | 3 | Au | 12 | ·4 |
| 4131 | 1087 | 952 | 23M | ·5023 | 13·1693 | 11·918 | **1030** | 5 | 1 | Ap | 5 | ·67 | 3 | My | 5 | ·20 | 4 | Je | 3 | ·73 | 6 | Jl | 3 | ·26 | 7 | Au | 1○ | ·7 |
| 4132 | 1088 | 953 | 23M | ·7610 | 2·2776 | 8·075 | **1031** | 6 | 6 | Mr | 26 | ·04 | 7 | Ap | 24 | ·57 | 3 | Je | 22 | ·63 | 5 | Jl | 22○ | ·16 | 6 | Au | 20 | ·6 |
| | | | | | | | | | | | | | 2 | My | 24 | ·10 | | | | | | | | | | | | |
| 4133 | 1089 | 954 | 22M | ·0198 | 20·9165 | 6·209 | **1032** | 7 | 4 | Ap | 12 | ·94 | 6 | My | 12 | ·47 | 1 | Je | 11 | ·00 | 2 | Jl | 10●○ | ·53 | 4 | Au | 9 | ·0 |
| 4134 | 1090 | 955 | 23M | ·2785 | 10·0218 | 2·366 | **1033** | 2 | 2 | Ap | 2 | ·30 | 3 | My | 1 | ·83 | 5 | My | 31○ | ·36 | 6 | Je | 29● | ·89 | 1 | Jl | 29 | ·4 |
| 4135 | 1091 | 956 | 23M | ·5373 | 28·6637 | 0·499 | **1034** | 3 | 1 | Ap | 21 | ·20 | 2 | My | 20○ | ·73 | 4 | Je | 19● | ·26 | 5 | Jl | 18 | ·79 | 7 | Au | 17 | ·32 |
| 4136 | 1092 | 957 | 23M | ·7960 | 17·7720 | 24·211 | **1035** | 4 | 5 | Ap | 10 | ·57 | 7 | My | 10●○ | ·10 | 1 | Je | 8 | ·63 | 3 | Jl | 8 | ·16 | 4 | Au | 6 | ·69 |
| 4137 | 1093 | 958 | 23M | ·0548 | 6·8803 | 20·368 | **1036** | 5 | 2 | Mr | 29 | ·93 | 4 | Ap | 28● | ·47 | 6 | My | 28 | ·00 | 7 | Je | 26 | ·53 | 3 | Au | 24 | ·5 |
| | | | | | | | | | | | | | | | | | | | | | 2 | Jl | 26 | ·06 | | | | |
| 4138 | 1094 | 959 | 23M | ·3136 | 25·5192 | 18·501 | **1037** | 7 | 1 | Ap | 17 | ·83 | 3 | My | 17 | ·36 | 4 | Je | 15 | ·89 | 6 | Jl | 15 | ·42 | 7 | Au | 13 | ·9 |
| 4139 | 1095 | 960 | 23M | ·5723 | 14·6275 | 14·659 | **1038** | 1 | 6 | Ap | 7 | ·20 | 7 | My | 6 | ·73 | 2 | Je | 5 | ·26 | 3 | Jl | 4 | ·79 | 5 | Au | 3 | ·3 |

**Surya Siddhanta.**

| Asvina | | | | Kartika | | | | Margasira | | | | Pausha | | | | A.D. | Magha | | | | A.D. | Phalguna | | | | Chaitra | | | |
|---|---|---|---|---|---|---|---|---|---|---|---|---|---|---|---|---|---|---|---|---|---|---|---|---|---|---|---|---|---|
| + 147·65293<br>+ 9·880 | | | | + 177·18353<br>+ 11·856 | | | | + 206·71411<br>+ 13·832 | | | | + 236·24470<br>+ 15·808 | | | | | + 265·77529<br>+ 17·784 | | | | | + 295·30588<br>+ 19·760 | | | | {+324·83647<br>+354·36705}<br>{+ 21·736<br>+ 23·712} | | | |
| Week-day | Month | Day | Fraction | Week-day | Month | Day | Fraction | Week-day | Month | Day | Fraction | Week-day | Month | Day | Fraction | | Week-day | Month | Day | Fraction | | Week-day | Month | Day | Fraction | Week-day | Month | Day | Fraction |
| 1 | S | 1 | ·48 | 3 | O | 1● | ·01 | 4 | O | 30 | ·54 | 6 | N | 29 | ·07 | | 7 | D | 28 | ·60 | 01 | 2 | Jr | 27 | ·13 | 3 | F | 25○ | ·66 |
| 7 | S | 20● | ·37 | 1 | O | 19 | ·90 | 3 | N | 18 | ·43 | 4 | D | 17 | ·97 | 02 | 6 | Jr | 16 | ·50 | | 1 | F | 15○ | ·03 | 2 | Mr | 16 | ·56 |
| 4 | S | 9 | ·74 | 6 | O | 9 | ·27 | 7 | N | 7 | ·80 | 2 | D | 7 | ·33 | 03 | 3 | Jr | 5 | ·86 | | 5 | F | 4○ | ·39 | 6 | Mr | 5 | ·92 |
| 2 | Au | 30 | ·11 | 3 | S | 28 | ·64 | 5 | O | 28 | ·17 | 6 | N | 26 | ·70 | | 1 | D | 26 | ·23 | 04 | 2 | Jr | 24 | ·76 | 4 | F | 23 | ·29 |
| 1 | S | 17 | ·01 | 2 | O | 16 | ·54 | 4 | N | 15 | ·07 | 5 | D | 14○ | ·60 | 05 | 7 | Jr | 13● | ·13 | | 1 | F | 11 | ·66 | 3 | Mr | 13 | ·19 |
| 5 | S | 6 | ·37 | 6 | O | 5 | ·90 | 1 | N | 4 | ·43 | 2 | D | 3○ | ·96 | 06 | 4 | Jr | 2 | ·49 | | 6 | F | 1 | ·03 | 7 | Mr | 2 | ·56 |
| 4 | S | 25 | ·27 | 5 | O | 24 | ·80 | 7 | N | 23○ | ·33 | 1 | D | 22 | ·86 | 07 | 3 | Jr | 21 | ·39 | | 4 | F | 19 | ·92 | 6 | Mr | 21 | ·45 |
| 1 | S | 14 | ·64 | 3 | O | 14 | ·17 | 4 | N | 12 | ·70 | 6 | D | 12 | ·23 | 08 | 7 | Jr | 10 | ·76 | | 2 | F | 9 | 29 | 3 | Mr | 9 | ·82 |
| 6 | S | 3 | ·00 | 7 | O | 2○ | ·53 | 2 | N | 1 | ·07 | 3 | N | 30 | ·60 | | 5 | D | 30 | ·13 | 09 | 6 | Jr | 28 | ·66 | 1 | F | 27 | ·19 |
| 4 | S | 21○ | ·90 | 6 | O | 21 | ·43 | 7 | N | 19 | ·96 | 2 | D | 19 | ·49 | 10 | 4 | Jr | 18 | ·02 | | 5 | F | 16 | ·55 | 7 | Mr | 18○ | ·09 |
| 2 | S | 11○ | ·27 | 3 | O | 10 | ·80 | 5 | N | 9 | ·33 | 6 | D | 8 | ·86 | 11 | 1 | Jr | 7 | ·39 | | 2 | F | 5 | ·92 | 4 | Mr | 7 | ·45 |
| 6 | Au | 31 | ·64 | 1 | S | 30 | ·17 | 2 | O | 29 | ·70 | 4 | N | 28 | ·23 | | 5 | D | 27 | ·76 | 12 | 7 | Jr | 26○ | ·29 | 1 | F | 24 | ·82 |
| 5 | S | 18 | ·53 | 7 | O | 18 | ·06 | 1 | N | 16 | ·59 | 3 | D | 16 | ·13 | 13 | 4 | Jr | 14○ | ·66 | | 6 | F | 13 | ·19 | 7 | Mr | 14 | ·72 |
| 2 | S | 7 | ·90 | 4 | C | 7 | ·43 | 5 | N | 5 | ·96 | 7 | D | 5 | ·49 | 14 | 2 | Jr | 4● | ·02 | | 3 | F | 2 | ·55 | 5 | Mr | 4 | ·08 |
| 7 | Au | 28 | ·27 | 1 | S | 26 | ·90 | 3 | O | 26 | ·33 | 4 | N | 24 | ·86 | | 6 | D | 24 | ·39 | 15 | 7 | Jr | 22 | ·92 | {2 | F | 21 | ·45 |
| | | | | | | | | | | | | | | | | | | | | | | | | | | {3 | Mr | 22 | ·98 |
| 6 | S | 16 | ·17 | 7 | O | 15 | ·70 | 2 | N | 14○ | ·23 | 3 | D | 13 | ·76 | 16 | 5 | Jr | 12 | ·29 | | 6 | F | 10 | ·82 | 1 | Mr | 11 | ·35 |
| 3 | S | 4 | ·53 | 5 | O | 4 | ·06 | 6 | N | 2○ | ·59 | 1 | D | 2 | ·12 | | 2 | D | 31 | ·65 | 17 | 4 | Jr | 30 | ·19 | 5 | F | 28 | ·72 |
| 2 | S | 23 | ·43 | 3 | O | 22○ | ·96 | 5 | N | 21 | ·49 | 7 | D | 21 | ·02 | 18 | 1 | Jr | 19 | ·55 | | 3 | F | 18 | ·08 | 4 | Mr | 19 | ·61 |
| 6 | S | 12 | ·83 | 1 | O | 12 | ·33 | 2 | N | 10 | ·86 | 4 | D | 10 | ·39 | 19 | 5 | Jr | 8 | ·92 | | 7 | F | 7 | ·45 | 1 | Mr | 8○ | ·98 |
| 4 | S | 2○ | ·16 | 5 | O | 1 | ·69 | 7 | O | 31 | ·23 | 1 | N | 29 | ·76 | | 3 | D | 29 | ·29 | 20 | 4 | Jr | 27 | ·82 | 6 | F | 26○ | ·35 |
| 3 | S | 20 | ·06 | 4 | O | 19 | ·59 | 6 | N | 18 | ·12 | 7 | D | 17 | ·65 | 21 | 2 | Jr | 16 | ·18 | | 3 | F | 14○ | ·71 | 5 | Mr | 16 | ·25 |
| 7 | S | 9 | ·43 | 1 | O | 8 | ·96 | 3 | N | 7 | ·49 | 5 | D | 7 | ·02 | 22 | 6 | Jr | 5 | ·55 | | 1 | F | 4 | ·08 | 2 | Mr | 5 | ·61 |
| 4 | Au | 29 | ·80 | 6 | S | 28 | ·33 | 7 | O | 27 | ·86 | 2 | N | 26 | ·39 | | 3 | D | 25○ | ·92 | 23 | 5 | Jr | 24 | ·45 | 6 | F | 22 | ·98 |
| 3 | S | 17 | ·69 | 5 | O | 17 | ·22 | 6 | N | 15 | ·75 | 1 | D | 15○ | ·29 | 24 | 2 | Jr | 13 | ·82 | | 4 | F | 12 | ·35 | 5 | Mr | 12 | ·88 |
| 1 | S | 6 | ·06 | 2 | O | 5 | ·59 | 4 | N | 4 | ·12 | 5 | D | 3●○ | ·65 | 25 | 7 | Jr | 2 | ·18 | | 1 | Jr | 31 | ·71 | 3 | Mr | 2 | ·24 |
| 6 | S | 24 | ·96 | 1 | O | 24 | ·49 | 3 | N | 23● | ·02 | 4 | D | 22 | ·55 | 26 | 6 | Jr | 21 | ·08 | | 7 | F | 19 | ·61 | 2 | Mr | 21 | ·14 |
| 4 | S | 14 | ·33 | 5 | O | 13○ | ·86 | 7 | N | 12● | ·39 | 1 | D | 11 | ·92 | 27 | 3 | Jr | 10 | ·45 | | 4 | F | 8 | ·98 | 6 | Mr | 10 | ·51 |
| 1 | S | 3 | ·69 | 3 | O | 3○ | ·22 | 4 | N | 1● | ·75 | 6 | D | 1 | ·28 | | 7 | D | 30 | ·81 | 28 | 2 | Jr | 29 | ·35 | 3 | F | 27 | ·88 |
| 7 | S | 21●○ | ·59 | 2 | O | 21 | ·12 | 3 | N | 19 | ·65 | 5 | D | 19 | ·18 | 29 | 6 | Jr | 17 | ·71 | | 1 | F | 16 | ·24 | 2 | Mr | 17 | ·77 |
| 4 | S | 10● | ·96 | 6 | O | 10 | ·49 | 1 | N | 9 | ·02 | 2 | D | 8 | ·55 | 30 | 4 | Jr | 7 | ·08 | | 5 | F | 5○ | ·61 | 7 | Mr | 7 | ·14 |
| 2 | Au | 31 | ·32 | 3 | S | 29 | ·85 | 5 | O | 29 | ·39 | 6 | N | 27 | ·92 | | 1 | D | 27 | ·45 | 31 | 2 | Jr | 25○ | ·98 | 4 | F | 24 | ·51 |
| 1 | S | 19 | ·22 | 2 | O | 18 | ·75 | 4 | N | 17 | ·28 | 5 | D | 16 | ·81 | 32 | 7 | Jr | 15● | ·34 | | 1 | F | 13 | ·87 | 3 | Mr | 14 | ·41 |
| 5 | S | 7 | ·59 | 7 | O | 7 | ·12 | 1 | N | 5 | ·65 | 3 | D | 5 | ·18 | 33 | 4 | Jr | 3● | ·71 | | 6 | F | 2 | ·24 | 7 | Mr | 3 | ·77 |
| 2 | Au | 27 | ·96 | 4 | S | 26 | ·49 | 6 | O | 26 | ·02 | 7 | N | 24○ | ·55 | | 2 | D | 24 | ·08 | 34 | 3 | Jr | 22 | ·61 | {5 | F | 21 | ·14 |
| | | | | | | | | | | | | | | | | | | | | | | | | | | {6 | Mr | 22 | ·67 |
| 1 | S | 15 | ·85 | 3 | O | 15 | ·38 | 4 | N | 13○ | ·91 | 6 | D | 13 | ·45 | 35 | 7 | Jr | 11 | ·98 | | 2 | F | 10 | ·51 | 4 | Mr | 12 | ·04 |
| 6 | S | 5 | ·22 | 7 | O | 4 | ·75 | 2 | N | 3○ | ·28 | 3 | D | 2 | ·81 | 36 | 5 | Jr | 1 | ·34 | | 6 | Jr | 30 | ·87 | 1 | F | 29 | ·40 |
| 5 | S | 23 | ·12 | 6 | O | 22● | ·65 | 1 | N | 21 | ·18 | 2 | D | 20 | ·71 | 37 | 4 | Jr | 19 | ·24 | | 5 | F | 17 | ·77 | 7 | Mr | 19○ | ·30 |
| 2 | S | 12○ | ·49 | 4 | O | 12 | ·02 | 5 | N | 10 | ·55 | 7 | D | 10 | ·08 | 38 | 1 | Jr | 8 | ·61 | | 3 | F | 7 | ·14 | 4 | Mr | 8○ | ·67 |
| 6 | S | 1○ | ·85 | 1 | O | 1 | ·38 | 2 | O | 30 | ·91 | 4 | N | 29 | ·44 | | 5 | D | 28 | ·97 | 39 | 7 | Jr | 27 | ·51 | 2 | F | 26○ | ·04 |

☉'s Anom col. 6 + 29·53059 + 59·06117 + 88·59176 + 118·12235

☾'s Anom col. 7 + 1·976 + 3·952 + 5·928 + 7·904

| Kaliyuga. | Vikrama Era. | Saka Era. | Commencement of Solar Year. Month and day A.D. | Commencement of Solar Year. Fraction of day. | First New-Moon in Solar Year. | Anomaly of first New-Moon. | A.D. | Week-day of 1st January. | Vaisakha Week-day | Vaisakha Month | Vaisakha Day | Vaisakha Fraction | Jyeshta Week-day | Jyeshta Month | Jyeshta Day | Jyeshta Fraction | Ashada Week-day | Ashada Month | Ashada Day | Ashada Fraction | Sravana Week-day | Sravana Month | Sravana Day | Sravana Fraction | Bhadrapada Week-day | Bhadrapada Month | Bhadrapada Day | Bhadrapada Fraction |
|---|---|---|---|---|---|---|---|---|---|---|---|---|---|---|---|---|---|---|---|---|---|---|---|---|---|---|---|---|
| 4140 | 1096 | 961 | 23M | ·8811 | 3·7358 | 10·816 | **1039** | 2 | 3 | Mr | 27 | ·57 | 5 | Ap | 26 | ·10 | 6 | My | 25 | ·63 | 2 | Jl | 23 | ·69 | 4 | Au | 22●○ | ·22 |
| | | | | | | | | | | | | | | | | | 1 | Je | 24 | ·16 | | | | | | | | |
| 4141 | 1097 | 962 | 23M | ·0898 | 22·3747 | 9·949 | **1040** | 3 | 2 | Ap | 14 | ·46 | 3 | My | 13 | ·99 | 5 | Je | 12 | ·53 | 7 | Jl | 12 | ·06 | 1 | Au | 10 | ·59 |
| 4142 | 1098 | 963 | 23M | ·3486 | 11·4830 | 5·106 | **1041** | 5 | 6 | Ap | 3 | ·83 | 1 | My | 3 | ·36 | 2 | Je | 1 | ·89 | 4 | Jl | 1○ | ·42 | 5 | Jl | 30 | ·95 |
| 4143 | 1099 | 964 | 23M | ·6073 | 0·5913 | 1·264 | **1042** | 6 | 4 | Mr | 24 | ·20 | 7 | My | 22 | ·26 | 1 | Je | 20●○ | ·79 | 3 | Jl | 20 | ·32 | 4 | Au | 18 | ·85 |
| | | | | | | | | | 5 | Ap | 22 | ·73 | | | | | | | | | | | | | | | | |
| 4144 | 1100 | 965 | 23M | ·8661 | 19·2302 | 26·952 | **1043** | 7 | 3 | Ap | 12 | ·10 | 4 | My | 11 | ·63 | 6 | Je | 10● | ·16 | 7 | Jl | 9 | ·69 | 2 | Au | 8 | ·22 |
| 4145 | 1101 | 966 | 23M | ·1249 | 8·3385 | 23·109 | **1044** | 1 | 7 | Mr | 31 | ·46 | 1 | Ap | 29○ | ·99 | 3 | My | 29 | ·52 | 5 | Je | 28 | ·05 | 6 | Jl | 27 | ·59 |
| | | | | | | | | | | | | | | | | | | | | | | | | | 1 | Au | 26 | ·12 |
| 4146 | 1102 | 967 | 23M | ·3836 | 26·9773 | 21·242 | **1045** | 3 | 6 | Ap | 19●○ | ·36 | 7 | My | 18 | ·89 | 2 | Je | 17 | ·42 | 3 | Jl | 16 | ·95 | 5 | Au | 15 | ·48 |
| 4147 | 1103 | 968 | 23M | ·6424 | 16·0856 | 17·399 | **1046** | 4 | 3 | Ap | 8●○ | ·73 | 5 | My | 8 | ·26 | 6 | Je | 6 | ·79 | 1 | Jl | 6 | ·32 | 2 | Au | 4 | ·85 |
| 4148 | 1104 | 969 | 23M | ·9011 | 5·1939 | 13·557 | **1047** | 5 | 1 | Mr | 29● | ·09 | 2 | Ap | 27 | ·63 | 4 | My | 27 | ·16 | 5 | Je | 25 | ·69 | 1 | Au | 23 | ·75 |
| | | | | | | | | | | | | | | | | | | | | | 7 | Jl | 25 | ·22 | | | | |
| 4149 | 1105 | 970 | 23M | ·1599 | 23·8328 | 11·690 | **1048** | 6 | 6 | Ap | 15 | ·99 | 1 | My | 15 | ·52 | 3 | Je | 14 | ·05 | 4 | Jl | 13 | ·58 | 6 | Au | 12○ | ·11 |
| 4150 | 1106 | 971 | 23M | ·4186 | 12·9411 | 7·847 | **1049** | 1 | 4 | Ap | 5 | ·36 | 5 | My | 4 | ·89 | 7 | Je | 3 | ·42 | 1 | Jl | 2 | ·95 | 3 | Au | 1○ | ·48 |
| 4151 | 1107 | 972 | 23M | ·6774 | 2·0494 | 4·004 | **1050** | 2 | 1 | Mr | 25 | ·73 | 3 | Ap | 24 | ·26 | 6 | Je | 22 | ·32 | 7 | Jl | 21○ | ·85 | 2 | Au | 20 | ·38 |
| | | | | | | | | | | | | | 4 | My | 23 | ·79 | | | | | | | | | | | | |
| 4152 | 1108 | 973 | 23M | ·9362 | 20·6883 | 2·138 | **1051** | 3 | 7 | Ap | 13 | ·62 | 2 | My | 13 | ·15 | 3 | Je | 11○ | ·69 | 5 | Jl | 11 | ·22 | 6 | Au | 9 | ·75 |
| 4153 | 1109 | 974 | 23M | ·1949 | 9·7966 | 25·849 | **1052** | 4 | 4 | Ap | 1 | ·99 | 6 | My | 1 | ·52 | 1 | My | 31○ | ·05 | 2 | Je | 29 | ·58 | 4 | Jl | 29 | ·11 |
| 4154 | 1110 | 975 | 23M | ·4587 | 28·4355 | 23·983 | **1053** | 6 | 3 | Ap | 20 | ·89 | 5 | My | 20○ | ·42 | 6 | Je | 18 | ·95 | 1 | Jl | 18 | ·48 | 3 | Au | 17 | ·01 |
| 4155 | 1111 | 976 | 23M | ·7124 | 17·5438 | 20·140 | **1054** | 7 | 1 | Ap | 10 | ·26 | 2 | My | 9● | ·79 | 4 | Je | 8 | ·32 | 5 | Jl | 7 | ·85 | 7 | Au | 6 | ·38 |
| 4156 | 1112 | 977 | 23M | ·9711 | 6·6521 | 16·297 | **1055** | 1 | 5 | Mr | 30○ | ·62 | 7 | Ap | 29 | ·15 | 1 | My | 28 | ·68 | 3 | Je | 27 | ·21 | 6 | Au | 25 | ·28 |
| | | | | | | | | | | | | | | | | | | | | | 4 | Jl | 26 | ·75 | | | | |
| 4157 | 1113 | 978 | 23M | ·2299 | 25·2910 | 14·430 | **1056** | 2 | 4 | Ap | 17 | ·52 | 6 | My | 17 | ·05 | 7 | Je | 15 | ·58 | 2 | Jl | 15 | ·11 | 3 | Au | 13 | ·64 |
| 4158 | 1114 | 979 | 23M | ·4887 | 14·3993 | 10·588 | **1057** | 4 | 1 | Ap | 6 | ·89 | 3 | My | 6 | ·42 | 4 | Je | 4 | ·95 | 6 | Jl | 4 | ·48 | 1 | Au | 3 | ·01 |
| 4159 | 1115 | 980 | 23M | ·7474 | 3·5076 | 6·745 | **1058** | 5 | 6 | Mr | 27 | ·25 | 7 | Ap | 25 | ·79 | 3 | Je | 23 | ·85 | 5 | Jl | 23 | ·38 | 6 | Au | 21● | ·91 |
| | | | | | | | | | | | | | 2 | My | 25 | ·32 | | | | | | | | | | | | |
| 4160 | 1116 | 981 | 24M | ·0062 | 22·1465 | 4·878 | **1059** | 6 | 5 | Ap | 15 | ·15 | 6 | My | 14 | ·68 | 1 | Je | 13 | ·21 | 2 | Jl | 12○ | ·74 | 4 | Au | 11● | ·27 |
| 4161 | 1117 | 982 | 23M | ·2650 | 11·2548 | 1·035 | **1060** | 7 | 2 | Ap | 3 | ·52 | 4 | My | 3 | ·05 | 5 | Je | 1 | ·58 | 7 | Jl | 1○ | ·11 | 1 | Jl | 30 | ·64 |
| 4162 | 1118 | 983 | 23M | ·5237 | 0·3631 | 24·747 | **1061** | 2 | 6 | Mr | 23 | ·89 | 2 | My | 21 | ·95 | 4 | Je | 20● | ·48 | 6 | Jl | 20 | ·01 | 7 | Au | 18 | ·54 |
| | | | | | | | | | 1 | Ap | 22 | ·42 | | | | | | | | | | | | | | | | |
| 4163 | 1119 | 984 | 23M | ·7825 | 19·0020 | 22·880 | **1062** | 3 | 5 | Ap | 11 | ·78 | 7 | My | 11○ | ·31 | 1 | Je | 9 | ·85 | 3 | Jl | 9 | ·38 | 4 | Au | 7 | ·91 |
| 4164 | 1120 | 985 | 24M | ·0412 | 8·1103 | 19·038 | **1063** | 4 | 3 | Ap | 1 | ·15 | 4 | Ap | 30○ | ·68 | 6 | My | 30 | ·21 | 7 | Je | 28 | ·74 | 2 | Jl | 28 | ·27 |
| | | | | | | | | | | | | | | | | | | | | | | | | | 3 | Au | 26 | ·80 |
| 4165 | 1121 | 986 | 23M | ·3000 | 26·7491 | 17·171 | **1064** | 5 | 2 | Ap | 19●○ | ·05 | 3 | My | 18 | ·58 | 5 | Je | 17 | ·11 | 6 | Jl | 16 | ·64 | 1 | Au | 15 | ·17 |
| 4166 | 1122 | 987 | 23M | ·5587 | 15·8574 | 13·328 | **1065** | 7 | 6 | Ap | 8 | ·42 | 7 | My | 7 | ·95 | 2 | Je | 6 | ·48 | 4 | Jl | 6 | ·01 | 5 | Au | 4 | ·54 |
| 4167 | 1123 | 988 | 23M | ·8175 | 4·9657 | 9·485 | **1066** | 1 | 3 | Mr | 28 | ·78 | 5 | Ap | 27 | ·31 | 6 | My | 26 | ·84 | 2 | Jl | 24 | ·91 | 4 | Au | 23○ | ·44 |
| | | | | | | | | | | | | | | | | | 1 | Je | 25 | ·37 | | | | | | | | |
| 4168 | 1124 | 989 | 24M | ·0763 | 23·6046 | 7·619 | **1067** | 2 | 2 | Ap | 16 | ·68 | 4 | My | 16 | ·21 | 5 | Je | 14 | ·74 | 7 | Jl | 14 | ·27 | 1 | Au | 12○ | ·80 |
| 4169 | 1125 | 990 | 23M | ·3350 | 12·7129 | 3·776 | **1068** | 3 | 7 | Ap | 5 | ·05 | 1 | My | 4 | ·58 | 3 | Je | 3 | ·11 | 4 | Jl | 2 | ·64 | 6 | Au | 1○ | ·17 |
| 4170 | 1126 | 991 | 23M | ·5938 | 1·8212 | 27·488 | **1069** | 5 | 4 | Mr | 25 | ·41 | 5 | Ap | 23 | ·95 | 2 | Je | 22 | ·01 | 3 | Jl | 21● | ·54 | 5 | Au | 20 | ·07 |
| | | | | | | | | | | | | | 7 | My | 23 | ·48 | | | | | | | | | | | | |
| 4171 | 1127 | 992 | 23M | ·8525 | 20·4601 | 25·621 | **1070** | 6 | 3 | Ap | 13 | ·31 | 4 | My | 12 | ·84 | 6 | Je | 11○ | ·37 | 7 | Jl | 10● | ·90 | 2 | Au | 9 | ·43 |
| 4172 | 1128 | 993 | 24M | ·1113 | 9·5684 | 21·778 | **1071** | 7 | 7 | Ap | 2 | ·68 | 2 | My | 2 | ·21 | 3 | My | 31○ | ·74 | 5 | Je | 30 | ·27 | 6 | Jl | 29 | ·80 |
| 4173 | 1129 | 994 | 23M | ·3700 | 2·82073 | 19·912 | **1072** | 1 | 6 | Ap | 20 | ·58 | 1 | My | 20 | ·11 | 2 | Je | 18 | ·64 | 4 | Jl | 18 | ·17 | 5 | Au | 16 | ·70 |
| 4174 | 1130 | 995 | 23M | ·6288 | 17·3156 | 16·069 | **1073** | 3 | 3 | Ap | 9○ | ·94 | 5 | My | 9● | ·47 | 7 | Je | 8 | ·00 | 1 | Jl | 7 | ·54 | 3 | Au | 6 | ·07 |
| 4175 | 1131 | 996 | 23M | ·8876 | 6·4239 | 12·226 | **1074** | 4 | 1 | Mr | 30○ | ·31 | 2 | Ap | 28● | ·84 | 4 | My | 28 | ·37 | 5 | Je | 26 | ·90 | 1 | Au | 24 | ·96 |
| | | | | | | | | | | | | | | | | | | | | | 7 | Jl | 26 | ·43 | | | | |
| 4176 | 1132 | 997 | 24M | ·1463 | 25·0628 | 10·359 | **1075** | 5 | 7 | Ap | 18 | ·21 | 1 | My | 17 | ·74 | 3 | Je | 16 | ·27 | 4 | Jl | 15 | ·80 | 6 | Au | 14 | ·33 |
| 4177 | 1133 | 998 | 23M | ·4051 | 14·1711 | 6·517 | **1076** | 6 | 4 | Ap | 6 | ·58 | 6 | My | 6 | ·11 | 7 | Je | 4 | ·64 | 2 | Jl | 4 | ·17 | 3 | Au | 2 | ·70 |

## Surya Siddhanta.

| Asvina | | | | Kartika | | | | Margasira | | | | Pausha | | | | A.D. | Magha | | | | A.D. | Phalguna | | | | Chaitra | | | |
|---|---|---|---|---|---|---|---|---|---|---|---|---|---|---|---|---|---|---|---|---|---|---|---|---|---|---|---|---|---|
| + 147·65293 | | | | + 177·18353 | | | | + 206·71411 | | | | + 236·24470 | | | | | + 265·77529 | | | | | + 295·30588 | | | | +324·83647 / +354·36705 | | | |
| + 9·880 | | | | + 11·856 | | | | + 13·832 | | | | + 15·808 | | | | | + 17·784 | | | | | + 19·760 | | | | + 21·736 / + 23·712 | | | |
| Week-day | Month | Day | Fraction | Week-day | Month | Day | Fraction | Week-day | Month | Day | Fraction | Week-day | Month | Day | Fraction | | Week-day | Month | Day | Fraction | | Week-day | Month | Day | Fraction | Week-day | Month | Day | Fraction |
| 5 | S | 20 | ·75 | 7 | O | 20 | ·28 | 1 | N | 18 | ·81 | 3 | D | 18 | ·34 | 40 | 4 | Jr | 16 | ·87 | | 6 | F | 15● | ·40 | 7 | Mr | 15 | ·98 |
| 3 | S | 9 | ·12 | 4 | O | 8 | ·65 | 6 | N | 7 | ·18 | 7 | D | 6 | ·71 | 41 | 2 | Jr | 5○ | ·24 | | 3 | F | 3 | ·77 | 5 | Mr | 5 | ·80 |
| 7 | Au | 29 | ·48 | 2 | S | 28 | ·01 | 3 | O | 27 | ·55 | 5 | N | 26 | ·08 | | 6 | D | 25○ | ·61 | 42 | 1 | Jr | 24 | ·14 | 2 | F | 22 | ·67 |
| 6 | S | 17 | ·38 | 7 | O | 16 | ·91 | 2 | N | 15 | ·44 | 3 | D | 14●○ | ·97 | 43 | 5 | Jr | 13 | ·50 | | 7 | F | 12 | ·03 | 1 | Mr | 13 | ·57 |
| 3 | S | 6 | ·75 | 5 | O | 6 | ·28 | 6 | N | 4 | ·81 | 1 | D | 4● | ·34 | 44 | 2 | Jr | 2 | ·87 | | 4 | F | 1 | ·40 | 5 | Mr | 1 | ·93 |
| 2 | S | 24 | ·65 | 4 | O | 24○ | ·18 | 5 | N | 22● | ·71 | 7 | D | 22 | ·24 | 45 | 1 | Jr | 20 | ·77 | | 3 | F | 19 | ·30 | 4 | Mr | 20 | ·88 |
| 7 | S | 14 | ·01 | 1 | O | 13○ | ·54 | 3 | N | 12 | ·07 | 4 | D | 11 | ·61 | 46 | 6 | Jr | 10 | ·14 | | 7 | F | 8 | ·67 | 2 | Mr | 10 | ·20 |
| 4 | S | 3 | ·38 | 5 | O | 2○ | ·91 | 7 | N | 1 | ·44 | 1 | N | 30 | ·97 | | 3 | D | 30 | ·50 | 47 | 5 | Jr | 29 | ·03 | 6 | F | 27 | ·56 |
| 3 | S | 22● | ·28 | 4 | O | 21 | ·81 | 6 | N | 20 | ·34 | 7 | D | 19 | ·87 | 48 | 2 | Jr | 18 | ·40 | | 3 | F | 16○ | ·98 | 5 | Mr | 17● | ·46 |
| 7 | S | 10 | ·65 | 2 | O | 10 | ·18 | 3 | N | 8 | ·79 | 5 | D | 8 | ·24 | 49 | 6 | Jr | 6 | ·77 | | 1 | F | 5●○ | ·30 | 2 | Mr | 6 | ·83 |
| 5 | Au | 31 | ·01 | 6 | S | 29 | ·54 | 1 | O | 29 | ·07 | 2 | N | 27 | ·60 | | 4 | D | 27 | ·13 | 50 | 5 | Jr | 25 | ·67 | 7 | F | 24 | ·20 |
| 3 | S | 18 | ·91 | 5 | O | 18 | ·44 | 6 | N | 16 | ·97 | 1 | D | 16 | ·50 | 51 | 3 | Jr | 15● | ·08 | | 4 | F | 13 | ·56 | 6 | Mr | 15 | ·09 |
| 1 | S | 8 | ·28 | 2 | O | 7 | ·81 | 4 | N | 6 | ·34 | 5 | D | 5○ | ·87 | 52 | 7 | Jr | 4 | ·40 | | 1 | F | 2 | ·98 | 3 | Mr | 8 | ·46 |
| {5 | Au | 27 | ·64 | 1 | O | 25 | ·71 | 3 | N | 24●○ | ·24 | Pausha Kshaya | | | | | 4 | D | 23 | ·77 | 53 | 6 | Jr | 22 | ·30 | {7 | F | 20 | ·83 |
| {7 | S | 26 | ·17 | | | | | | | | | | | | | | | | | | | | | | | {2 | Mr | 22 | ·36 |
| 4 | S | 15 | ·54 | 6 | O | 15 | ·07 | 7 | N | 13●○ | ·60 | 2 | D | 13 | ·13 | 54 | 3 | Jr | 11 | ·66 | | 5 | F | 10 | ·19 | 6 | Mr | 11 | ·73 |
| 1 | S | 4 | ·91 | 3 | O | 4 | ·44 | 4 | N | 2● | ·97 | 6 | D | 2 | ·50 | 55 | 1 | Jr | 1 | ·03 | | 2 | Jr | 30 | ·56 | 4 | Mr | 1 | ·09 |
| 7 | S | 23○ | ·81 | 2 | O | 23● | ·34 | 3 | N | 21 | ·87 | 5 | D | 21 | ·40 | 56 | 6 | Jr | 19 | ·93 | | 1 | F | 18 | ·46 | 2 | Mr | 18○ | ·99 |
| 5 | S | 12●○ | ·17 | 6 | O | 11 | ·70 | 1 | N | 10 | ·23 | 2 | D | 9 | ·77 | 57 | 4 | Jr | 8 | ·30 | | 5 | F | 6 | ·83 | 7 | Mr | 8○ | ·86 |
| 2 | S | 1○ | ·54 | 4 | O | 1 | ·07 | 5 | O | 30 | ·60 | 7 | N | 29 | ·13 | | 1 | D | 28 | ·66 | 58 | 3 | Jr | 27 | ·19 | 4 | F | 25 | ·72 |
| 1 | S | 20 | ·44 | 2 | O | 19 | ·97 | 4 | N | 18 | ·50 | 6 | D | 18 | ·03 | 59 | 7 | Jr | 16○ | ·56 | | 2 | F | 15● | ·09 | 3 | Mr | 16 | ·62 |
| 5 | S | 9 | ·81 | 7 | O | 9 | ·34 | 1 | N | 7 | ·87 | 3 | D | 7 | ·40 | 60 | 4 | Jr | 5○ | ·93 | | 6 | F | 4 | ·46 | 7 | Mr | 4 | ·99 |
| 3 | Au | 29 | ·17 | 4 | S | 27 | ·70 | 6 | O | 2[illegible] | ·23 | 7 | N | 25 | ·76 | | 2 | D | 25○ | ·29 | 61 | 3 | Jr | 23 | ·83 | 5 | F | 22 | ·36 |
| 2 | S | 17 | ·07 | 3 | O | 16 | ·60 | 5 | N | 1[illegible] | ·13 | 6 | D | 14 | ·66 | 62 | 1 | Jr | 13 | ·19 | | 2 | F | 11 | ·72 | 4 | Mr | 13 | ·25 |
| 6 | S | 6 | ·44 | 7 | O | 5 | ·97 | 2 | N | 4○ | ·50 | 4 | D | 4 | ·03 | 63 | 5 | Jr | 2 | ·56 | | 7 | F | 1 | ·09 | 1 | Mr | 2 | ·62 |
| 5 | S | 25 | ·33 | 6 | O | 24○ | ·87 | 1 | N | 23 | ·40 | 2 | D | 22 | ·93 | 64 | 4 | Jr | 21 | ·46 | | 5 | F | 19 | ·99 | 7 | Mr | 20 | ·52 |
| 2 | S | 13 | ·70 | 4 | O | 13●○ | ·23 | 5 | N | 11 | ·76 | 7 | D | 11 | ·29 | 65 | 1 | Jr | 9 | ·82 | | 3 | F | 8 | ·35 | 4 | Mr | 9 | ·89 |
| 7 | S | 3 | ·07 | 1 | O | 2 | ·60 | 3 | N | 1 | ·13 | 4 | N | 30 | ·66 | | 5 | D | 30 | ·19 | 66 | 7 | Jr | 28 | ·72 | 2 | F | 27○ | ·25 |
| 5 | S | 21● | ·97 | 7 | O | 21 | ·50 | 2 | N | 20 | ·03 | 3 | D | 19 | ·56 | 67 | 5 | Jr | 18 | ·09 | | 6 | F | 16○ | ·62 | 1 | Mr | 18 | ·15 |
| 3 | S | 11 | ·33 | 1 | O | 10 | ·86 | 6 | N | 9 | ·39 | 7 | D | 8 | ·93 | 68 | 2 | Jr | 7 | ·46 | | 3 | F | 5●○ | ·99 | 5 | Mr | 6 | ·52 |
| 7 | Au | 30 | ·70 | 2 | S | 29 | ·23 | 3 | O | 28 | ·76 | 5 | N | 27 | ·29 | | 6 | D | 26 | ·82 | 69 | 1 | Jr | 25 | ·35 | 2 | F | 23 | ·88 |
| 6 | S | 18 | ·60 | 1 | O | 18 | ·13 | 2 | N | 16 | ·66 | 4 | D | 16○ | ·19 | 70 | 5 | Jr | 14 | ·72 | | 7 | F | 13 | ·25 | 1 | Mr | 14 | ·78 |
| 3 | S | 7 | ·79 | 5 | O | 7 | ·50 | 7 | N | 6 | ·03 | 1 | D | 5○ | ·56 | 71 | 3 | Jr | 4 | ·09 | | 4 | F | 2 | ·62 | 6 | Mr | 4 | ·15 |
| {1 | Au | 28 | ·33 | 4 | O | 26 | ·39 | 5 | N | 24○ | ·92 | 7 | D | 24 | ·45 | 72 | 1 | Jr | 22 | ·98 | | 3 | F | 21 | ·52 | 5 | Mr | 22 | ·05 |
| {2 | S | 26 | ·86 | | | | | | | | | | | | | | | | | | | | | | | | | | |
| 7 | S | 15 | ·23 | 1 | O | 14 | ·76 | 3 | N | 13 | ·29 | 4 | D | 12 | ·82 | 73 | 6 | Jr | 11 | ·35 | | 7 | F | 9 | ·88 | 2 | Mr | 11 | ·41 |
| 4 | S | 4 | ·60 | 6 | O | 4○ | ·13 | 7 | N | 2 | ·66 | 2 | D | 2 | ·19 | | 3 | D | 31 | ·72 | 74 | 5 | Jr | 30 | ·25 | 6 | F | 28 | ·78 |
| 3 | S | 23○ | ·49 | 5 | O | 23 | ·02 | 6 | N | 21 | ·56 | 1 | D | 21 | ·09 | 75 | 2 | Jr | 19 | ·62 | | 4 | F | 18 | ·15 | 5 | Mr | 19●○ | ·68 |
| 7 | S | 12●○ | ·86 | 2 | O | 12 | ·39 | 3 | N | 10 | ·92 | 5 | D | 10 | ·45 | 76 | 6 | Jr | 8 | ·98 | | 1 | F | 7 | ·51 | 3 | Mr | 8 | ·04 |
| 5 | S | 1● | ·23 | 6 | S | 30 | ·76 | 1 | O | 30 | ·29 | 2 | N | 28 | ·82 | | 4 | D | 28 | ·35 | 77 | 5 | Jr | 26○ | ·88 | 7 | F | 25 | ·41 |

☉'s Anom col. 6 + 29·53059 + 59·06117 + 88·59176 + 118·12235

☾'s Anom col. 7 + 1·976 + 3·958 + 5·928 + 7·904

| Kaliyuga | Vikrama Era | Saka Era | Commencement of Solar Year: Month and day A.D. | Commencement of Solar Year: Fraction of day | First New-Moon in Solar Year | Anomaly of first New-Moon | A.D. | Week-day of 1st January | Vaisakha: Week-day | Month | Day | Fraction | Jyeshta: Week-day | Month | Day | Fraction | Ashada: Week-day | Month | Day | Fraction | Sravana: Week-day | Month | Day | Fraction | Bhadrapada: Week-day | Month | Day | Fraction |
|---|---|---|---|---|---|---|---|---|---|---|---|---|---|---|---|---|---|---|---|---|---|---|---|---|---|---|---|---|
| 4178 | 1134 | 999 | 23M | ·6638 | 3·2794 | 2·674 | **1077** | 1 | 1 | Mr | 26 | ·94 | 3 | Ap | 25 | ·47 | 6 | Je | 23 | ·53 | 1 | Jl | 23○ | ·06 | 2 | Au | 21 | ·60 |
| | | | | | | | | | | | | | 5 | My | 25 | ·00 | | | | | | | | | | | | |
| 4179 | 1135 | 1000 | 23M | ·9226 | 21·9183 | 0·807 | **1078** | 2 | 7 | Ap | 14 | ·84 | 2 | My | 14 | ·37 | 3 | Je | 12 | ·90 | 5 | Jl | 12○ | ·43 | 6 | Au | 10 | ·96 |
| 4180 | 1136 | 1001 | 24M | ·1813 | 11·0266 | 24·519 | **1079** | 3 | 5 | Ap | 4 | ·21 | 6 | My | 3 | ·74 | 1 | Je | 2 | ·27 | 2 | Jl | 1● | ·80 | 4 | Jl | 31 | ·33 |
| 4181 | 1137 | 1002 | 23M | ·4401 | 0·1349 | 20·676 | **1080** | 4 | 2 | Mr | 23 | ·57 | 5 | My | 21○ | ·64 | 7 | Je | 20● | ·17 | 1 | Jl | 19 | ·70 | 3 | Au | 18 | ·23 |
| | | | | | | | | | 4 | Ap | 22 | ·10 | | | | | | | | | | | | | | | | |
| 4182 | 1138 | 1003 | 23M | ·6988 | 18·7738 | 18·809 | **1081** | 6 | 1 | Ap | 11 | ·47 | 3 | My | 11○ | ·00 | 4 | Je | 9 | ·53 | 6 | Jl | 9 | ·06 | 7 | Au | 7 | ·59 |
| 4183 | 1139 | 1004 | 23M | ·9576 | 7·8820 | 14·967 | **1082** | 7 | 5 | Mr | 31 | ·84 | [illegible] | Ap | 30○ | ·37 | 1 | My | 29 | ·90 | 3 | Je | 28 | ·43 | 4 | Jl | 27 | ·96 |
| | | | | | | | | | | | | | | | | | | | | | | | | | 6 | Au | 26 | ·49 |
| 4184 | 1140 | 1005 | 24M | ·2164 | 26·5209 | 13·100 | **1083** | 1 | 4 | Ap | 19 | ·74 | 6 | My | 19 | ·27 | 7 | Je | 17 | ·80 | 2 | Jl | 17 | ·33 | 3 | Au | 15 | ·86 |
| 4185 | 1141 | 1006 | 23M | ·4751 | 15·6292 | 9·257 | **1084** | 2 | 2 | Ap | 8 | ·10 | 3 | My | 7 | ·63 | 5 | Je | 6 | ·16 | 6 | Jl | 5 | ·70 | 1 | Au | 24 | ·23 |
| 4186 | 1142 | 1007 | 23M | ·7339 | 4·7375 | 5·414 | **1085** | 4 | 6 | Mr | 28 | ·47 | 1 | Ap | 27 | ·00 | 2 | My | 26 | ·53 | 5 | Jl | 24 | ·59 | 7 | Au | 23○ | ·12 |
| | | | | | | | | | | | | | | | | | 4 | Je | 25 | ·06 | | | | | | | | |
| 4187 | 1143 | 1008 | 23M | ·9926 | 23·3764 | 3·548 | **1086** | 5 | 5 | Ap | 16 | ·37 | 6 | My | 15 | ·90 | 1 | Je | 14 | ·43 | 2 | Jl | 13 | ·96 | 4 | Au | 12●○ | ·49 |
| 4188 | 1144 | 1009 | 24M | ·2514 | 12·4847 | 27·259 | **1087** | 6 | 2 | Ap | 5 | ·74 | 4 | My | 5 | ·27 | 5 | Je | 3 | ·80 | 7 | Jl | 3 | ·33 | 1 | Au | 1● | ·86 |
| 4189 | 1145 | 1010 | 23M | ·5101 | 1·5930 | 23·417 | **1088** | 7 | 7 | Mr | 25 | ·10 | 1 | Ap | 23 | ·63 | 4 | Je | 21○ | ·69 | 6 | Jl | 21 | ·22 | 7 | Au | 19 | ·76 |
| | | | | | | | | | | | | | 3 | My | 23 | ·16 | | | | | | | | | | | | |
| 4190 | 1146 | 1011 | 23M | ·7689 | 20·2319 | 21·550 | **1089** | 2 | 6 | Ap | 13 | ·00 | 7 | My | 12 | ·53 | 2 | Je | 11●○ | ·06 | 3 | Jl | 10 | ·59 | 5 | Au | 9 | ·12 |
| 4191 | 1147 | 1012 | 24M | ·0277 | 9·3402 | 17·707 | **1090** | 3 | 3 | Ap | 2 | ·37 | 4 | My | 1 | ·90 | 6 | My | 31 | ·43 | 7 | Je | 29 | ·96 | 2 | Jl | 29 | ·49 |
| 4192 | 1148 | 1013 | 24M | ·2864 | 27·9791 | 15·840 | **1091** | 4 | 2 | Ap | 21○ | ·26 | 3 | My | 20● | ·80 | 5 | Je | 19 | ·33 | 6 | Jl | 18 | ·86 | 1 | Au | 17 | ·39 |
| 4193 | 1149 | 1014 | 23M | ·5452 | 17·0874 | 11·998 | **1092** | 5 | 6 | Ap | 9○ | ·63 | 1 | My | 9 | ·16 | 2 | Je | 7 | ·69 | 4 | Jl | 7 | ·22 | 5 | Au | 5 | ·75 |
| 4194 | 1150 | 1015 | 23M | ·8039 | 6·1957 | 8·155 | **1093** | 7 | 4 | Mr | 30○ | ·00 | 5 | Ap | 28 | ·53 | 7 | My | 28 | ·06 | 1 | Je | 26 | ·59 | 4 | Au | 24 | ·65 |
| | | | | | | | | | | | | | | | | | | | | | 3 | Jl | 26 | .12 | | | | |
| 4195 | 1151 | 1016 | 24M | ·0627 | 24·8346 | 6·288 | **1094** | 1 | 2 | Ap | 17 | ·90 | 4 | My | 17 | ·43 | 5 | Je | 15 | ·96 | 7 | Jl | 15 | ·49 | 2 | Au | 14 | ·02 |
| 4196 | 1152 | 1017 | 24M | ·3214 | 13·9429 | 2·445 | **1095** | 2 | 7 | Ap | 7 | ·26 | 1 | My | 6 | ·79 | 3 | Je | 5 | ·32 | 4 | Jl | 4 | ·86 | 6 | Au | 3○ | ·39 |
| 4197 | 1153 | 1018 | 23M | ·5802 | 3·0512 | 26·157 | **1096** | 3 | 4 | Mr | 26 | ·63 | 6 | Ap | 25 | ·16 | 2 | Je | 23 | ·22 | 3 | Jl | 22○ | ·75 | 5 | Au | 21 | ·28 |
| | | | | | | | | | | | | | 7 | My | 24 | ·69 | | | | | | | | | | | | |
| 4198 | 1154 | 1019 | 23M | ·8390 | 21·6901 | 24·291 | **1097** | 5 | 3 | Ap | 14 | ·53 | 5 | My | 14 | ·06 | 6 | Je | 12 | ·59 | 1 | Jl | 12 | ·12 | 2 | Au | 10 | ·65 |
| 4199 | 1155 | 1020 | 24M | ·0977 | 10·7984 | 20·448 | **1098** | 6 | 7 | Ap | 3 | ·90 | 2 | My | 3 | ·43 | 3 | Je | 1 | ·96 | 5 | Jl | 1 | ·49 | 7 | Jl | 31 | ·02 |
| 4200 | 1156 | 1021 | 24M | ·3565 | 29·4372 | 18·581 | **1099** | 7 | 5 | Mr | 24 | ·26 | 1 | My | 22○ | ·32 | 2 | Je | 20 | ·85 | 4 | Jl | 20 | ·38 | 5 | Au | 18 | ·92 |
| | | | | | | | | | 6 | Ap | 22 | ·79 | | | | | | | | | | | | | | | | |
| 4201 | 1157 | 1022 | 23M | ·6152 | 18·5456 | 14·739 | **1100** | 1 | 4 | Ap | 11 | ·16 | 5 | My | 10●○ | ·69 | 7 | Je | 9 | ·22 | 1 | Jl | 8 | ·75 | 3 | Au | 7 | ·28 |
| 4202 | 1158 | 1023 | 23M | ·8740 | 7·6538 | 10·896 | **1101** | 3 | 1 | Mr | 31 | ·53 | 3 | Ap | 30● | ·06 | 4 | My | 29 | ·59 | 6 | Je | 28 | ·12 | 7 | Jl | 27 | ·65 |
| | | | | | | | | | | | | | | | | | | | | | | | | | 2 | Au | 26 | ·18 |
| 4203 | 1159 | 1024 | 24M | ·1327 | 26·2927 | 9·029 | **1102** | 4 | 7 | Ap | 19● | ·42 | 1 | My | 18 | ·96 | 3 | Je | 17 | ·49 | 5 | Jl | 17 | ·02 | 6 | Au | 15 | ·55 |
| 4204 | 1160 | 1025 | 24M | ·3915 | 15·4010 | 5·186 | **1103** | 5 | 4 | Ap | 8 | ·79 | 6 | My | 8 | ·32 | 7 | Je | 6 | ·85 | 2 | Jl | 6 | ·38 | 3 | Au | 4 | ·91 |
| 4205 | 1161 | 1026 | 23M | ·6502 | 4·5093 | 1·344 | **1104** | 6 | 2 | Mr | 28 | ·16 | 3 | Ap | 26 | ·69 | 5 | My | 26 | ·22 | 1 | Jl | 24 | ·28 | 2 | Au | 22○ | ·81 |
| | | | | | | | | | | | | | | | | | 6 | Je | 24 | ·75 | | | | | | | | |
| 4206 | 1162 | 1027 | 23M | ·9090 | 23·1482 | 27·031 | **1105** | 1 | 1 | Ap | 16 | ·06 | 2 | My | 15 | ·59 | 4 | Je | 14 | ·12 | 5 | Jl | 13 | ·65 | 7 | Au | 12 | ·18 |
| 4207 | 1163 | 1028 | 24M | ·1678 | 12·2565 | 23·189 | **1106** | 2 | 5 | Ap | 5 | ·42 | 6 | My | 4 | ·95 | 1 | Je | 3 | ·48 | 3 | Jl | 3○ | ·02 | 4 | Au | 1● | ·55 |
| 4208 | 1164 | 1029 | 24M | ·4265 | 1·3648 | 19·346 | **1107** | 3 | 2 | Mr | 25 | ·79 | 4 | Ap | 24 | ·32 | 7 | Je | 22○ | ·38 | 1 | Jl | 21 | ·91 | 3 | Au | 20 | ·44 |
| | | | | | | | | | | | | | 5 | My | 23 | ·85 | | | | | | | | | | | | |
| 4209 | 1165 | 1030 | 23M | ·6853 | 20·0037 | 17·479 | **1108** | 4 | 1 | Ap | 12 | ·69 | 3 | My | 12 | ·22 | 4 | Je | 10● | ·75 | 6 | Jl | 10 | ·28 | 7 | Au | 8 | ·81 |
| 4210 | 1166 | 1031 | 23M | ·9440 | 9·1120 | 13·636 | **1109** | 6 | 6 | Ap | 2 | ·06 | 7 | My | 1○ | ·59 | 2 | My | 31● | ·12 | 3 | Je | 22 | ·65 | 5 | Jl | 29 | ·18 |
| 4211 | 1167 | 1032 | 24M | ·2028 | 27·7509 | 11·770 | **1110** | 7 | 4 | Ap | 20○ | ·95 | 6 | My | 20 | ·48 | 1 | Je | 19 | ·01 | 2 | Jl | 18 | ·54 | 4 | Au | 17 | ·08 |
| 4212 | 1168 | 1033 | 24M | ·4615 | 16·8592 | 7·927 | **1111** | 1 | 2 | Ap | 10○ | ·32 | 3 | My | 9 | ·85 | 5 | Je | 8 | ·38 | 6 | Jl | 7 | ·91 | 1 | Au | 6 | ·44 |
| 4213 | 1169 | 1034 | 23M | ·7203 | 5·9675 | 4·084 | **1112** | 2 | 6 | Mr | 29 | ·69 | 1 | Ap | 28 | ·22 | 2 | My | 27 | ·75 | 4 | Je | 26 | ·28 | 7 | Au | 24 | ·34 |
| | | | | | | | | | | | | | | | | | | | | | 5 | Jl | 25 | ·81 | | | | |
| 4214 | 1170 | 1035 | 23M | ·9791 | 24·6064 | 2·217 | **1113** | 4 | 5 | Ap | 17 | ·58 | 7 | My | 17 | ·12 | 1 | Je | 15 | ·65 | 3 | Jl | 15 | ·18 | 4 | Au | 13○ | ·71 |
| 4215 | 1171 | 1036 | 24M | ·2378 | 13·7147 | 25·929 | **1114** | 5 | 2 | Ap | 6 | ·95 | 4 | My | 6 | ·48 | 6 | Je | 5 | ·01 | 7 | Jl | 4 | ·54 | 2 | Au | 3○ | ·07 |
| 4216 | 1172 | 1037 | 24M | ·4966 | 2·8230 | 22·087 | **1115** | 6 | 7 | Mr | 27 | ·32 | 1 | Ap | 25 | ·85 | 4 | Je | 23 | ·91 | 6 | Jl | 23●○ | ·44 | 7 | Au | 21 | ·97 |
| | | | | | | | | | | | | | 3 | My | 25 | ·38 | | | | | | | | | | | | |

**Surya Siddhanta.**

| Asvina | | | | Kartika | | | | Margasira | | | | Pausha | | | | A.D. | Magha | | | | A.D. | Phalguna | | | | Chaitra | | | |
|---|---|---|---|---|---|---|---|---|---|---|---|---|---|---|---|---|---|---|---|---|---|---|---|---|---|---|---|---|---|
| + 147·65298 | | | | + 177·18353 | | | | + 206·71411 | | | | + 236·24470 | | | | | + 265·77529 | | | | | + 295·30588 | | | | { +324·83647 +354·36705 | | | |
| + 9·880 | | | | + 11·856 | | | | + 13·832 | | | | + 15·808 | | | | | + 17·784 | | | | | + 19·760 | | | | { + 21·736 + 23·712 | | | |
| Week-day | Month | Day | Fraction | Week-day | Month | Day | Fraction | Week-day | Month | Day | Fraction | Week-day | Month | Day | Fraction | | Week-day | Month | Day | Fraction | | Week-day | Month | Day | Fraction | Week-day | Month | Day | Fraction |
| 4 | S | 20 | ·13 | 5 | O | 19 | ·66 | 7 | N | 18 | ·19 | 1 | D | 17 | ·72 | 78 | 3 | Jr | 16○ | ·25 | | 4 | F | 14 | ·78 | 6 | Mr | 16 | ·31 |
| 1 | S | 9 | ·49 | 3 | O | 9 | ·02 | 4 | N | 7 | ·55 | 6 | D | 7 | ·08 | 79 | 7 | Jr | 5○ | ·61 | | 2 | F | 4 | ·15 | 3 | Mr. | 5 | ·68 |
| 5 | Au | 29 | ·86 | 7 | S | 28 | ·39 | 1 | O | 27 | ·92 | 3 | N | 26 | ·45 | | 4 | D | 25● | ·98 | 80 | 6 | Jr | 24 | ·51 | 1 | F | 23 | ·04 |
| 4 | S | 16 | ·76 | 6 | O | 16 | ·29 | 7 | N | 14○ | ·82 | 2 | D | 14● | ·35 | 81 | 3 | Jr | 12 | ·88 | | 5 | F | 11 | ·41 | 6 | Mr | 12 | ·94 |
| 2 | S | 6 | ·12 | 3 | O | 5 | ·66 | 5 | N | 4○ | ·19 | 6 | D | 3● | ·72 | 82 | 1 | Jr | 2 | ·25 | | 2 | Jr | 31 | ·78 | 4 | Mr | 2 | ·31 |
| 1 | S | 25 | ·02 | 2 | O | 24○ | ·55 | 4 | N | 23 | ·08 | 5 | D | 22 | ·61 | 83 | 7 | Jr | 21 | ·14 | | 1 | F | 19 | ·68 | 3 | Mr | 21 | ·21 |
| 5 | S | 14 | ·39 | 6 | O | 13● | ·92 | 1 | N | 12 | ·45 | 2 | D | 11 | ·98 | 84 | 4 | Jr | 10 | ·51 | | 6 | F | 9 | ·04 | 7 | Mr | 9○ | ·57 |
| 2 | S | 2○ | ·76 | 4 | O | 2 | ·29 | 5 | O | 31 | ·82 | 7 | N | 30 | ·35 | | 1 | D | 29 | ·88 | 85 | 3 | Jr | 32 | ·41 | 4 | F | 26○ | ·94 |
| 1 | S | 21 | ·65 | 3 | O | 21 | ·18 | 4 | N | 19 | ·72 | 6 | D | 19 | ·25 | 86 | 7 | Jr | 17 | ·78 | | 2 | F | 16○ | ·31 | 3 | Mr | 17 | ·84 |
| 6 | S | 11 | ·92 | 7 | O | 10 | ·55 | 2 | N | 9 | ·08 | 3 | D | 8 | ·61 | 87 | 5 | Jr | 7 | ·14 | | 6 | F | 5● | ·67 | 1 | Mr | 7 | ·20 |
| 3 | Au | 31 | ·39 | 4 | S | 29 | ·92 | 6 | O | 29 | ·45 | 7 | N | 27 | ·98 | | 2 | D | 27○ | ·51 | 88 | 4 | Jr | 26 | ·04 | 5 | F | 24 | ·57 |
| 2 | S | 18 | ·29 | 3 | O | 17 | ·82 | 5 | N | 16 | ·35 | 6 | D | 15○ | ·88 | 89 | 1 | Jr | 14 | ·41 | | 2 | F | 12 | ·94 | 4 | Mr | 14 | ·47 |
| 6 | S | 7 | ·65 | 1 | O | 7 | ·18 | 2 | N | 5 | ·71 | 4 | D | 5○ | ·24 | 90 | 5 | Jr | 3 | ·78 | | 7 | F | 2 | ·31 | 1 | Mr | 3 | ·84 |
| {4 | Au | 28 | ·02 | 7 | O | 26 | ·08 | 1 | N | 24● | ·61 | 3 | D | 24 | ·14 | 91 | 4 | Jr | 22 | ·67 | | 6 | F | 21 | ·20 | 7 | Mr | 22 | ·73 |
| {5 | S | 26 | ·55 | | | | | | | | | | | | | | | | | | | | | | | | | | |
| 2 | S | 15 | ·92 | 4 | O | 15○ | ·45 | 5 | N | 13 | ·98 | 7 | D | 13 | ·51 | 92 | 2 | Jr | 12 | ·04 | | 3 | F | 10 | ·57 | 5 | Mr | 11 | ·10 |
| 7 | S | 4 | ·28 | 1 | O | 3○ | ·82 | 3 | N | 2 | ·35 | 4 | D | 1 | ·88 | | 6 | D | 31 | ·41 | 93 | 7 | Jr | 29 | ·94 | 2 | F | 28 | ·47 |
| 6 | S | 23●○ | ·18 | 7 | O | 22 | ·71 | 2 | N | 21 | ·24 | 3 | D | 20 | ·77 | 94 | 5 | Jr | 19 | ·80 | | 6 | F | 17 | ·84 | 1 | Mr | 19● | ·37 |
| 3 | S | 12 | ·55 | 5 | O | 12 | ·08 | 6 | N | 10 | ·61 | 1 | D | 10 | ·14 | 95 | 2 | Jr | 8 | ·67 | | 4 | F | 7○ | ·20 | 5 | Mr | 8 | ·78 |
| 7 | S | 1 | ·92 | 2 | O | 1 | ·45 | 3 | O | 30 | ·98 | 5 | N | 29 | ·51 | | 7 | D | 29 | ·04 | 96 | 1 | Jr | 27○ | ·57 | 3 | F | 26 | ·10 |
| 6 | S | 19 | ·81 | 1 | O | 19 | ·34 | 2 | N | 17 | ·88 | 4 | D | 17 | ·41 | 97 | 5 | Jr | 15●○ | ·94 | | 7 | F | 14 | ·47 | 2 | Mr | 16 | ·00 |
| 4 | S | 9 | ·18 | 5 | O | 8 | ·71 | 7 | N | 7 | ·24 | 1 | D | 6 | ·77 | 98 | 3 | Jr | 5● | ·30 | | 4 | F | 3 | ·83 | 6 | Mr | 5 | ·36 |
| 1 | Au | 29 | ·55 | 3 | S | 28 | ·08 | 4 | O | 27 | ·61 | 6 | N | 26○ | ·14 | | 7 | D | 25 | ·67 | 99 | 2 | Jr | 24 | ·20 | 3 | F | 22 | ·73 |
| 7 | S | 17 | ·45 | 1 | O | 16 | ·98 | 3 | N | 15○ | ·51 | 5 | D | 15 | ·04 | 11 | 6 | Jr | 13 | ·57 | 00 | 1 | F | 12 | ·10 | 2 | Mr | 12 | ·63 |
| 4 | S | 5 | ·81 | 6 | O | 5 | ·34 | 7 | N | 3○ | ·87 | 2 | D | 3 | ·40 | 01 | 3 | Jr | 1 | ·94 | | 5 | Jr | 31 | ·47 | 7 | Mr | 2 | ·00 |
| 3 | S | 24 | ·71 | 5 | O | 24● | ·24 | 6 | N | 22 | ·77 | 1 | D | 22 | ·30 | 02 | 2 | Jr | 20 | ·83 | | 4 | F | 19 | ·36 | 5 | Mr | 20○ | ·89 |
| 1 | S | 14○ | ·08 | 2 | O | 13 | ·61 | 4 | N | 12 | ·14 | 5 | D | 11 | ·67 | 03 | 7 | Jr | 10 | ·20 | | 1 | F | 8 | ·73 | 3 | Mr | 10●○ | ·26 |
| 5 | S | 3○ | ·44 | 6 | O | 2 | ·98 | 1 | N | 1 | ·51 | 3 | D | 1 | ·04 | | 4 | D | 30 | ·57 | 04 | 6 | Jr | 29 | ·10 | 7 | F | 27○ | ·63 |
| 4 | S | 21 | ·34 | 6 | O | 20 | ·87 | 7 | N | 19 | ·40 | 1 | D | 18 | ·93 | 05 | 3 | Jr | 17 | ·46 | | 5 | F | 16 | ·00 | 6 | Mr | 17 | ·56 |
| 1 | S | 10 | ·71 | 3 | O | 10 | ·24 | 4 | N | 8 | ·77 | 6 | D | 8 | ·30 | 06 | 7 | Jr | 6○ | ·83 | | 2 | F | 5 | ·36 | 3 | Mr | 6 | ·89 |
| 6 | Au | 31 | ·08 | 7 | S | 29 | ·61 | 2 | O | 29 | ·14 | 3 | N | 27 | ·67 | | 5 | D | 27●○ | ·20 | 07 | 6 | Jr | 25 | ·73 | 1 | F | 24 | ·26 |
| 4 | S | 18 | ·97 | 6 | O | 18 | ·50 | 1 | N | 17 | ·04 | 2 | D | 16●○ | ·56 | 08 | 4 | Jr | 15 | ·10 | | 5 | F | 13 | ·63 | 7 | Mr | 14 | ·16 |
| 2 | S | 7 | ·34 | 3 | O | 6 | ·87 | 5 | N | 5 | ·40 | 6 | D | 4 | ·93 | 09 | 1 | Jr | 3 | ·46 | | 2 | F | 1 | ·99 | 4 | Mr | 3 | ·52 |
| {6 | Au | 27 | ·71 | 2 | O | 25○ | ·77 | 4 | N | 24● | ·30 | 5 | D | 23 | ·83 | 10 | 7 | Jr | 22 | ·36 | | 1 | F | 20 | ·89 | 3 | Mr | 22 | ·42 |
| {1 | S | 26 | ·24 | | | | | | | | | | | | | | | | | | | | | | | | | | |
| 5 | S | 15 | ·61 | 7 | O | 15●○ | ·14 | 1 | N | 13 | ·67 | 3 | D | 13 | ·20 | 11 | 4 | Jr | 11 | ·73 | | 6 | F | 10 | ·26 | 7 | Mr | 11 | ·79 |
| 2 | S | 4 | ·97 | 4 | O | 4○ | ·50 | 6 | N | 3 | ·03 | 7 | D | 2 | ·56 | 12 | 2 | Jr | 1 | ·10 | | 3 | Jr | 30 | ·63 | 5 | F | 29 | ·16 |
| 1 | S | 22 | ·87 | 3 | O | 22 | ·40 | 4 | N | 20 | ·93 | 6 | D | 20 | ·46 | 13 | 7 | Jr | 18 | ·99 | | 2 | F | 17○ | ·52 | 4 | Mr | 19● | ·05 |
| 6 | S | 12 | ·24 | 7 | O | 11 | ·77 | 2 | N | 10 | ·30 | 3 | D | 9 | ·83 | 14 | 5 | Jr | 8 | ·36 | | 6 | F | 6○ | ·89 | 1 | Mr | 8 | ·42 |
| 3 | S | 1 | ·60 | 5 | O | 1 | ·14 | 6 | O | 30 | ·67 | 1 | N | 29 | ·20 | | 2 | D | 28 | ·73 | 15 | 4 | Jr | 27○ | ·26 | 5 | F | 25 | ·79 |
| 2 | S | 20 | ·50 | 4 | O | 20 | ·03 | 5 | N | 18 | ·56 | 7 | D | 18 | ·09 | 16 | 1 | Jr | 16 | ·62 | | 3 | F | 15 | ·16 | 4 | Mr | 15 | ·69 |

## TABLE X—

☉'s Anom col. 6 + 29·53059 + 59·06117 + 88·59176 + 118·12235

☾'s Anom col. 7 + 1·976 + 3·952 + 5·928 + 7·904

| Kaliyuga | Vikrama Era | Saka Era | Commencement of Solar Year: Month and day A.D. | Commencement of Solar Year: Fraction of day | First New-Moon in Solar Year | Anomaly of first New-Moon | A.D. | Week-day of 1st January | Vaisakha: Week-day | Month | Day | Fraction | Jyeshta: Week-day | Month | Day | Fraction | Ashada: Week-day | Month | Day | Fraction | Sravana: Week-day | Month | Day | Fraction | Bhadrapada: Week-day | Month | Day | Fraction |
|---|---|---|---|---|---|---|---|---|---|---|---|---|---|---|---|---|---|---|---|---|---|---|---|---|---|---|---|---|
| 4217 | 1173 | 1038 | 23M | ·7553 | 21·4619 | 20·220 | **1116** | 7 | 6 | Ap | 14 | ·22 | 7 | My | 13 | ·75 | 2 | Je | 12 | ·28 | 3 | Jl | 11 | ·81 | 5 | Au | 10 | ·34 |
| 4218 | 1174 | 1039 | 23M | ·0141 | 10·5702 | 16·377 | **1117** | 2 | 3 | Ap | 3 | ·58 | 5 | My | 3 | ·11 | 6 | Je | 1○ | ·64 | 1 | Jl | 1 | ·18 | 2 | Jl | 30 | ·71 |
| 4219 | 1175 | 1040 | 24M | ·2728 | 29·2090 | 14·510 | **1118** | 3 | 2 | Ap | 22 | ·48 | 4 | My | 22●○ | ·01 | 5 | Je | 20 | ·54 | 7 | Jl | 20 | ·07 | 1 | Au | 18 | ·60 |
| 4220 | 1176 | 1041 | 24M | ·5316 | 18·3173 | 10·668 | **1119** | 4 | 6 | Ap | 11 | ·85 | 1 | My | 11● | ·38 | 2 | Je | 9 | ·91 | 4 | Jl | 9 | ·44 | 5 | Au | 7 | ·97 |
| 4221 | 1177 | 1042 | 23M | ·7904 | 7·4256 | 6·825 | **1120** | 5 | 4 | Mr | 31○ | ·22 | 15 | Ap | 29 | ·75 | 7 | My | 29 | ·28 | 1 | Je | 27 | ·81 | { 3 | Jl | 27 | ·34 |
| | | | | | | | | | | | | | | | | | | | | | | | | | { 4 | Au | 26 | ·87 |
| 4222 | 1178 | 1043 | 24M | ·0491 | 26·0645 | 4·958 | **1121** | 7 | 3 | Ap | 19 | ·11 | 4 | My | 18 | ·64 | 6 | Je | 17 | ·17 | 7 | Jl | 16 | ·70 | 2 | Au | 15 | ·24 |
| 4223 | 1179 | 1044 | 24M | ·3079 | 15·1728 | 1·115 | **1122** | 1 | 7 | Ap | 8 | 48 | 2 | My | 8 | ·01 | 3 | Je | 6 | ·54 | 5 | Jl | 6 | ·07 | 6 | Au | 4 | ·60 |
| 4224 | 1180 | 1045 | 24M | ·5666 | 4·2811 | 24·827 | **1123** | 2 | 4 | Mr | 28 | ·85 | 6 | Ap | 27 | ·38 | { 7 | My | 26 | ·91 | 3 | Jl | 24 | ·97 | 5 | Au | 23● | ·50 |
| | | | | | | | | | | | | | | | | | { 2 | Je | 25 | ·44 | | | | | | | | |
| 4225 | 1181 | 1046 | 23M | ·8254 | 22·9200 | 22·960 | **1124** | 3 | 3 | Ap | 15 | ·74 | 5 | My | 15 | ·27 | 6 | Jo | 13 | ·80 | 1 | Jl | 13○ | ·34 | 2 | Au | 11● | ·87 |
| 4226 | 1182 | 1047 | 24M | ·0841 | 12·0283 | 19·118 | **1125** | 5 | 1 | Ap | 5 | ·11 | 2 | My | 4 | ·64 | 4 | Je | 3 | ·17 | 5 | Jl | 2○ | ·70 | 7 | Au | 1 | ·23 |
| 4227 | 1183 | 1048 | 24M | ·3429 | 1·1366 | 15·275 | **1126** | 6 | 5 | Mr | 25 | ·48 | { 7 | Ap | 24 | ·01 | 3 | Je | 22●○ | ·07 | 4 | Jl | 21 | ·60 | 6 | Au | 20 | ·13 |
| | | | | | | | | | | | | | { 1 | My | 23 | ·54 | | | | | | | | | | | | |
| 4228 | 1184 | 1049 | 24M | ·6016 | 19·7755 | 13·408 | **1127** | 7 | 4 | Ap | 13 | ·38 | 5 | My | 12○ | ·91 | 7 | Je | 11 | ·44 | 1 | Jl | 10 | ·97 | 3 | Au | 9 | ·50 |
| 4229 | 1185 | 1050 | 23M | ·8604 | 8·8838 | 9·565 | **1128** | 1 | 1 | Ap | 1 | ·74 | 3 | My | 1○ | ·27 | 4 | My | 30 | ·80 | 6 | Je | 29 | ·34 | 7 | Jl | 28 | ·87 |
| 4230 | 1186 | 1051 | 24M | ·1192 | 27·5227 | 7·699 | **1129** | 3 | 7 | Ap | 20●○ | ·64 | 2 | My | 20 | ·17 | 3 | Je | 18 | ·70 | 5 | Jl | 18 | ·23 | 6 | Au | 16 | ·76 |
| 4231 | 1187 | 1052 | 24M | ·3779 | 16·6310 | 3·856 | **1130** | 4 | 5 | Ap | 10 | ·01 | 6 | My | 9 | ·54 | 1 | Je | 8 | ·07 | 2 | Jl | 7 | ·60 | 4 | Au | 6 | ·13 |
| 4232 | 1188 | 1053 | 24M | ·6367 | 5·7393 | 0·013 | **1131** | 5 | 2 | Mr | 30 | ·38 | 3 | Ap | 28 | ·91 | { 5 | My | 28 | ·44 | 1 | Jl | 26 | ·50 | 3 | Au | 25○ | ·03 |
| | | | | | | | | | | | | | | | | | { 6 | Je | 26 | ·97 | | | | | | | | |
| 4233 | 1189 | 1054 | 23M | ·8954 | 24·3782 | 25·701 | **1132** | 6 | 1 | Ap | 17 | ·27 | 2 | My | 16 | ·80 | 4 | Je | 15 | ·33 | 5 | Jl | 14 | ·86 | 7 | Au | 13○ | ·40 |
| 4234 | 1190 | 1055 | 24M | ·1542 | 13·4865 | 21·858 | **1133** | 1 | 5 | Ap | 6 | ·64 | 7 | My | 6 | ·17 | 1 | Je | 4 | ·70 | 3 | Jl | 4 | ·23 | 4 | Au | 2●○ | ·76 |
| 4235 | 1191 | 1056 | 24M | ·4129 | 2·5948 | 18·015 | **1134** | 2 | 3 | Mr | 27 | ·01 | { 4 | Ap | 25 | ·54 | 7 | Je | 23 | ·60 | 2 | Jl | 23● | ·13 | 3 | Au | 21 | ·66 |
| | | | | | | | | | | | | | { 6 | My | 25 | ·07 | | | | | | | | | | | | |
| 4236 | 1992 | 1057 | 24M | ·6717 | 21·2336 | 16·149 | **1135** | 3 | 1 | Ap | 14 | ·90 | 3 | My | 14 | ·44 | 4 | Je | 12○ | ·97 | 6 | Jl | 12 | ·50 | 1 | Au | 11 | ·03 |
| 4237 | 1193 | 1058 | 23M | ·9305 | 10·3419 | 12·306 | **1136** | 4 | 6 | Ap | 3 | ·27 | 7 | My | 2 | ·80 | 2 | Je | 1○ | ·33 | 3 | Je | 30 | ·86 | 5 | Jl | 30 | ·39 |
| 4238 | 1194 | 1059 | 24M | ·1892 | 28·9808 | 10·439 | **1137** | 6 | 5 | Ap | 22 | ·17 | 6 | My | 21 | ·70 | 1 | Je | 20 | ·23 | 2 | Jl | 19 | ·76 | 4 | Au | 18 | ·29 |
| 4239 | 1195 | 1060 | 24M | ·4480 | 18·0891 | 6·596 | **1138** | 7 | 2 | Ap | 11○ | ·54 | 4 | My | 11 | ·07 | 5 | Je | 9 | ·60 | 7 | Jl | 9 | ·13 | 1 | Au | 7 | ·66 |
| 4240 | 1196 | 1061 | 24M | ·7067 | 7·1974 | 2·754 | **1139** | 1 | 6 | Mr | 31○ | ·90 | 1 | Ap | 30 | ·43 | 2 | My | 29 | ·96 | { 4 | Je | 28 | ·50 | 7 | Au | 26 | ·56 |
| | | | | | | | | | | | | | | | | | | | | | { 6 | Jl | 28 | ·03 | | | | |
| 4241 | 1197 | .62 | 23M | ·9655 | 25·8363 | 0·887 | **1140** | 2 | 5 | Ap | 18 | ·80 | 7 | My | 18 | ·33 | 1 | Je | 16 | ·86 | 3 | Jl | 16 | ·39 | 4 | Au | 14 | ·92 |
| 4242 | 1198 | 1063 | 24M | ·2242 | 14·9446 | 24·599 | **1141** | 4 | 3 | Ap | 8 | ·17 | 4 | My | 7 | ·70 | 6 | Je | 6 | ·23 | 7 | Jl | 5 | ·76 | 2 | Au | 4 | ·29 |
| 4243 | 1199 | 1064 | 24M | ·4830 | 4·0529 | 20·756 | **1142** | 5 | 7 | Mr | 28 | ·54 | 2 | Ap | 27 | ·07 | { 3 | My | 26 | ·60 | 6 | Jl | 24○ | ·66 | 1 | Au | 23 | ·19 |
| | | | | | | | | | | | | | | | | | { 5 | Je | 25 | ·13 | | | | | | | | |
| 4244 | 1200 | 1065 | 24M | ·7418 | 22·6918 | 18·889 | **1143** | 6 | 6 | Ap | 16 | ·43 | 7 | My | 15 | ·96 | 2 | Je | 14 | ·49 | 4 | Jl | 14○ | ·02 | 5 | Au | 12● | ·56 |
| 4245 | 1201 | 1066 | 24M | ·0005 | 11·8001 | 15·047 | **1144** | 7 | 3 | Ap | 4 | ·80 | 5 | My | 4 | ·33 | 6 | Je | 2 | ·86 | 1 | Jl | 2○ | ·39 | 2 | Jl | 31 | ·92 |
| 4246 | 1202 | 1067 | 24M | 2593 | 0·9084 | 11·204 | **1145** | 2 | { 1 | Mr | 25 | ·17 | 4 | My | 23○ | ·23 | 5 | Je | 21● | ·76 | 7 | Jl | 21 | ·29 | 1 | Au | 19 | ·82 |
| | | | | | | | | | { 2 | Ap | 23 | ·70 | | | | | | | | | | | | | | | | |
| 4247 | 1203 | 1068 | 24M | ·5180 | 19·5473 | 9·337 | **1146** | 3 | 7 | Ap | 13 | ·06 | 1 | My | 12○ | ·60 | 3 | Je | 11● | ·13 | 4 | Jl | 10 | ·66 | 6 | Au | 9 | ·19 |
| 4248 | 1204 | 1069 | 24M | ·7768 | 8·6555 | 5·494 | **1147** | 4 | 4 | Ap | 2 | ·43 | 5 | My | 1○ | ·96 | 7 | My | 31 | ·49 | 2 | Je | 30 | ·02 | { 3 | Jl | 29 | ·55 |
| | | | | | | | | | | | | | | | | | | | | | | | | | { 5 | Au | 28 | ·08 |
| 4249 | 1205 | 1070 | 24M | ·0355 | 27·2945 | 3·628 | **1148** | 5 | 3 | Ap | 20● | ·33 | 4 | My | 19 | ·86 | 6 | Je | 18 | ·39 | 7 | Jl | 17 | ·92 | 2 | Au | 16 | ·45 |
| 4250 | 1206 | 1071 | 24M | ·2943 | 16·4028 | 27·339 | **1149** | 7 | 7 | Ap | 9 | ·70 | 2 | My | 9 | ·23 | 3 | Je | 7 | ·76 | 5 | Jl | 7 | ·29 | 6 | Au | 5 | ·82 |
| 4251 | 1207 | 1072 | 24M | ·5530 | 5·5111 | 23·497 | **1150** | 1 | 5 | Mr | 30 | ·06 | 6 | Ap | 28 | ·59 | { 1 | My | 28 | ·12 | 4 | Jl | 26 | ·18 | 5 | Au | 24○ | ·72 |
| | | | | | | | | | | | | | | | | | { 2 | Je | 26 | ·66 | | | | | | | | |
| 4252 | 1208 | 1073 | 24M | ·8118 | 24·1500 | 21·630 | **1151** | 2 | 3 | Ap | 17 | ·96 | 5 | My | 17 | ·49 | 7 | Je | 16 | ·02 | 1 | Jl | 15 | ·55 | 3 | Au | 14○ | ·08 |
| 4253 | 1209 | 1074 | 24M | ·0706 | 13·2583 | 17·787 | **1152** | 3 | 1 | Ap | 6 | ·33 | 2 | My | 5 | ·86 | 4 | Je | 4 | ·39 | 5 | Jl | 3 | ·92 | 7 | Au | 2 | ·45 |
| 4254 | 1210 | 1075 | 24M | ·3293 | 2·3666 | 13·944 | **1153** | 5 | 5 | Mr | 26 | ·70 | { 7 | Ap | 25 | ·23 | 3 | Je | 23○ | ·29 | 4 | Jl | 22● | ·82 | 6 | Au | 21 | ·35 |
| | | | | | | | | | | | | | { 1 | My | 24 | ·76 | | | | | | | | | | | | |
| 4255 | 1211 | 1076 | 24M | ·5881 | 21·0054 | 12·078 | **1154** | 6 | 4 | Ap | 14 | ·59 | 6 | My | 14 | ·12 | 7 | Je | 12○ | ·65 | 2 | Jl | 12 | ·18 | 3 | Au | 10 | ·72 |

**Surya Siddhanta**

| Asvina | | | | Kartika | | | | Margasira | | | | Pausha | | | | A.D. | Magha | | | | A.D. | Phalguna | | | | Chaitra | | | |
|---|---|---|---|---|---|---|---|---|---|---|---|---|---|---|---|---|---|---|---|---|---|---|---|---|---|---|---|---|---|
| + 147·65293 | | | | + 177·18353 | | | | + 206·71411 | | | | + 236·24470 | | | | | + 265·77429 | | | | | + 295·30588 | | | | +324·83647<br>+354·36705 | | | |
| + 9·880 | | | | + 11·856 | | | | + 13·832 | | | | + 15·808 | | | | | + 17·784 | | | | | + 19·760 | | | | + 21·736<br>+ 23·712 | | | |
| Week-day | Month | Day | Fraction | Week-day | Month | Day | Fraction | Week-day | Month | Day | Fraction | Week-day | Month | Day | Fraction | | Week-day | Month | Day | Fraction | | Week-day | Month | Day | Fraction | Week-day | Month | Day | Fraction |
| 6 | S | 8 | ·87 | 1 | O | 8 | ·40 | 2 | N | 6 | ·93 | 4 | D | 6○ | ·46 | 17 | 5 | Jr | 4 | ·99 | | 7 | F | 3 | ·52 | 2 | Mr | 5 | ·05 |
| 4 | Au | 29 | ·24 | 5 | S | 27 | ·77 | 7 | O | 27 | ·30 | 1 | N | 25○ | ·83 | | 3 | D | 25 | ·36 | 18 | 4 | Jr | 23 | ·89 | 6 | F | 22 | ·42 |
| | | | | | | | | | | | | | | | | | | | | | | | | | | 7 | Mr | 23 | ·95 |
| 3 | S | 17 | ·13 | 4 | O | 16 | ·66 | 6 | N | 15●○ | ·20 | 7 | D | 14 | ·73 | 19 | 2 | Jr | 13 | ·26 | | 3 | F | 11 | ·79 | 5 | Mr | 13 | ·32 |
| 7 | S | 6 | ·50 | 2 | O | 6 | ·03 | 3 | N | 4 | ·56 | 5 | D | 4 | ·09 | 20 | 6 | Jr | 2 | ·62 | | 1 | F | 1 | ·15 | 2 | Mr | 1 | ·68 |
| 6 | S | 24○ | ·40 | 7 | O | 23● | ·93 | 2 | N | 22 | ·46 | 3 | D | 21 | ·99 | 21 | 5 | Jr | 20 | ·52 | | 7 | F | 19 | ·05 | 1 | Mr | 20○ | ·58 |
| 3 | S | 13○ | ·77 | 5 | O | 13 | ·30 | 6 | N | 11 | ·83 | 1 | D | 11 | ·36 | 22 | 2 | Jr | 9 | ·89 | | 4 | F | 8 | ·42 | 5 | Mr | 9●○ | ·95 |
| 1 | S | 3○ | ·13 | 2 | O | 2 | ·66 | 4 | N | 1 | ·19 | 5 | N | 30 | ·72 | | 7 | D | 30 | ·26 | 23 | 1 | Jr | 28 | ·79 | 3 | F | 27 | ·32 |
| 7 | S | 22 | ·03 | 1 | O | 21 | ·56 | 3 | N | 20 | ·09 | 4 | D | 19 | ·62 | 24 | 6 | Jr | 18○ | ·15 | | 7 | F | 16 | ·68 | 2 | Mr | 17 | ·21 |
| 4 | S | 10 | ·40 | 5 | O | 9 | ·93 | 7 | N | 8 | ·46 | 1 | D | 7 | ·99 | 25 | 3 | Jr | 6○ | ·52 | | 5 | F | 5 | ·08 | 6 | Mr | 6 | ·58 |
| 1 | Au | 30 | ·76 | 3 | S | 29 | ·30 | 4 | O | 28 | ·83 | 6 | N | 27 | ·36 | | 7 | D | 26○ | ·89 | 26 | 2 | Jr | 25 | ·42 | 3 | F | 23 | ·95 |
| 7 | S | 18 | ·66 | 2 | O | 18 | ·19 | 3 | N | 16 | ·72 | 5 | D | 16 | ·25 | 27 | 6 | Jr | 14 | ·78 | | 1 | F | 13 | ·32 | 2 | Mr | 14 | ·85 |
| 5 | S | 8 | ·03 | 6 | O | 7 | ·56 | 1 | N | 6○ | ·09 | 2 | D | 5 | ·62 | 28 | 4 | Jr | 4 | ·15 | | 5 | F | 2 | ·68 | 7 | Mr | 3 | ·21 |
| 2 | Au | 27 | ·40 | 5 | O | 25○ | ·46 | 6 | N | 23 | ·99 | 1 | D | 23 | ·52 | 29 | 3 | Jr | 22 | ·05 | | 4 | F | 20 | ·58 | 6 | Mr | 22 | ·11 |
| 3 | S | 25 | ·93 | | | | | | | | | | | | | | | | | | | | | | | | | | |
| 1 | S | 15 | ·29 | 2 | O | 14●○ | ·82 | 4 | N | 13 | ·36 | 5 | D | 12 | ·89 | 30 | 7 | Jr | 11 | ·42 | | 1 | F | 9 | ·95 | 3 | Mr | 11 | ·48 |
| 5 | S | 4 | ·66 | 7 | O | 4● | ·19 | 1 | N | 2 | ·72 | 3 | D | 2 | ·25 | | 4 | D | 31 | ·78 | 31 | 6 | Jr | 30 | ·31 | 7 | F | 28○ | ·84 |
| 4 | S | 23● | ·56 | 6 | O | 23 | ·09 | 7 | N | 21 | ·62 | 2 | D | 21 | ·15 | 32 | 3 | Jr | 19 | ·68 | | 5 | F | 18○ | ·21 | 6 | Mr | 18 | ·74 |
| 1 | S | 11 | ·93 | 3 | O | 11 | ·46 | 4 | N | 9 | ·99 | 6 | D | 9 | ·52 | 33 | 1 | Jr | 8 | ·05 | | 2 | F | 6○ | ·58 | 4 | Mr | 8 | ·11 |
| 6 | S | 1 | ·29 | 7 | S | 30 | ·82 | 2 | O | 30 | ·35 | 3 | N | 28 | ·88 | | 5 | D | 28 | ·42 | 34 | 6 | Jr | 26● | ·95 | 1 | F | 25 | ·48 |
| 5 | S | 20 | ·19 | 6 | O | 19 | ·72 | 1 | N | 18 | ·25 | 2 | D | 17○ | ·78 | 35 | 4 | Jr | 16● | ·31 | | 5 | F | 14 | ·84 | 7 | Mr | 16 | ·37 |
| 2 | S | 9 | ·56 | 4 | O | 9 | ·09 | 5 | N | 7 | ·62 | 7 | D | 7○ | ·15 | 36 | 1 | Jr | 5 | ·68 | | 3 | F | 4 | ·21 | 4 | Mr | 4 | ·74 |
| 6 | Au | 28 | ·92 | 1 | S | 27 | ·46 | 2 | O | 26 | ·99 | 4 | N | 25○ | ·52 | | 6 | D | 25 | ·05 | 37 | 7 | Jr | 23 | ·58 | 2 | F | 22 | ·11 |
| | | | | | | | | | | | | | | | | | | | | | | | | | | 3 | Mr | 23 | ·64 |
| 5 | S | 16 | ·82 | 7 | O | 16 | ·35 | 1 | N | 14● | ·88 | 3 | D | 14 | ·41 | 38 | 4 | Jr | 12 | ·94 | | 6 | F | 11 | ·48 | 1 | Mr | 13 | ·01 |
| 3 | S | 6 | ·19 | 4 | O | 5○ | ·72 | 6 | N | 4 | ·25 | 7 | D | 3 | ·78 | 39 | 2 | Jr | 2 | ·31 | | 3 | Jr | 31 | ·84 | 5 | Mr | 2 | ·37 |
| 2 | S | 25○ | ·09 | 3 | O | 24 | ·62 | 5 | N | 23 | ·15 | 6 | D | 22 | ·68 | 40 | 1 | Jr | 21 | ·21 | | 2 | F | 19 | ·74 | 4 | Mr | 20○ | ·27 |
| 6 | S | 13●○ | ·45 | 7 | O | 12 | ·98 | 2 | N | 11 | ·52 | 4 | D | 11 | ·05 | 41 | 5 | Jr | 9 | ·58 | | 7 | F | 8 | ·11 | 1 | Mr | 9● | ·64 |
| 3 | S | 2● | ·82 | 5 | O | 2 | ·35 | 6 | O | 31 | ·88 | 1 | N | 30 | ·41 | | 2 | D | 29 | ·94 | 42 | 4 | Jr | 28○ | ·47 | 6 | F | 27 | ·00 |
| 2 | S | 21 | ·72 | 4 | O | 21 | ·25 | 5 | N | 19 | ·78 | 7 | D | 19 | ·31 | 43 | 1 | Jr | 17○ | ·84 | | 3 | F | 16 | ·37 | 4 | Mr | 17 | ·90 |
| 7 | S | 11 | ·09 | 1 | O | 10 | ·62 | 3 | N | 9 | ·15 | 4 | D | 8 | ·68 | 44 | 6 | Jr | 7○ | ·21 | | 7 | F | 5 | ·74 | 2 | Mr | 6 | ·27 |
| 4 | Au | 30 | ·45 | 5 | S | 28 | ·98 | 7 | O | 28 | ·51 | 2 | N | 27 | ·04 | | 3 | D | 26● | ·58 | 45 | 5 | Jr | 25 | ·11 | 6 | F | 23 | ·64 |
| 3 | S | 18 | ·35 | 4 | O | 17 | ·88 | 6 | N | 16○ | ·41 | 7 | D | 15 | ·94 | 46 | 2 | Jr | 14 | ·47 | | 4 | F | 13 | ·00 | 5 | Mr | 14 | ·53 |
| 7 | S | 7 | ·72 | 2 | O | 7 | ·25 | 3 | N | 5○ | ·78 | 5 | D | 5 | ·31 | 47 | 6 | Jr | 3 | ·84 | | 1 | F | 2 | ·37 | 2 | Mr | 3 | ·90 |
| 6 | S | 26 | ·62 | 1 | O | 26●○ | ·15 | 2 | N | 24 | ·68 | 4 | D | 24 | ·21 | 48 | 5 | Jr | 22 | ·74 | | 7 | F | 21 | ·27 | 1 | Mr | 21 | ·80 |
| 3 | S | 14 | ·98 | 5 | O | 14 | ·51 | 7 | N | 13 | ·04 | 1 | D | 12 | ·57 | 49 | 3 | Jr | 11 | ·10 | | 4 | F | 9 | ·64 | 6 | Mr | 11○ | ·17 |
| 1 | S | 4○ | ·35 | 2 | O | 3 | ·88 | 4 | N | 2 | ·41 | 5 | D | 1 | ·94 | | 7 | D | 31 | ·47 | 50 | 2 | Jr | 30 | ·00 | 3 | F | 28○ | ·53 |
| 7 | S | 23 | ·25 | 1 | O | 22 | ·78 | 3 | N | 21 | ·31 | 4 | D | 20 | ·84 | 51 | 6 | Jr | 19 | ·37 | | 7 | F | 17●○ | ·90 | 2 | Mr | 19 | ·43 |
| 4 | S | 12 | ·61 | 6 | O | 12 | ·14 | 7 | N | 10 | ·68 | 2 | D | 10 | ·21 | 52 | 3 | Jr | 8 | ·74 | | 5 | F | 7● | ·27 | 6 | Mr | 7 | ·80 |
| 1 | Au | 31 | ·98 | 3 | S | 30 | ·51 | 5 | O | 30 | ·04 | 6 | N | 28 | ·57 | | 1 | D | 28○ | ·10 | 53 | 2 | Jr | 26● | ·63 | 4 | F | 25 | ·16 |
| 7 | S | 19 | ·88 | 2 | O | 19 | ·41 | 3 | N | 17 | ·94 | 5 | D | 17○ | ·47 | 54 | 7 | Jr | 16 | ·00 | | 1 | F | 14 | ·53 | 3 | Mr | 16 | ·06 |
| 5 | S | 9 | ·25 | 6 | O | 8 | ·78 | 1 | N | 7 | ·31 | 2 | D | 6○ | ·84 | 55 | 4 | Jr | 5 | ·37 | | 5 | F | 3 | ·90 | 7 | Mr | 5 | ·43 |

## TABLE X—

| | | | | | | | | | ⊙'s Anom col. 6 | | | | + 29·53059 | | | | + 59·06117 | | | | + 88·59176 | | | | + 118·12235 | | | |
|---|---|---|---|---|---|---|---|---|---|---|---|---|---|---|---|---|---|---|---|---|---|---|---|---|---|---|---|---|
| | | | | | | | | | ☾'s Anom col. 7 | | | | + 1·976 | | | | + 3·952 | | | | + 5·928 | | | | + 7·904 | | | |
| | | | Commencement of Solar Year. | | | | | | Vaisakha | | | | Jyeshta | | | | Ashada | | | | Sravana | | | | Bhadrapada | | | |
| Kaliyuga | Vikrama Era | Saka Era | Month and day A.D. | Fraction of day | First New-Moon in Solar Year. | Anomaly of first New-Moon. | A.D. | Week-day of 1st January | Week-day | Month | Day | Fraction | Week-day | Month | Day | Fraction | Week-day | Month | Day | Fraction | Week-day | Month | Day | Fraction | Week-day | Month | Day | Fraction |
| 4256 | 1212 | 1077 | 24M | ·8468 | 10·1137 | 8·235 | **1155** | 7 | 1 | Ap | 3 | ·96 | 3 | My | 3 | ·49 | 5 | Je | 2●○ | ·02 | 6 | Jl | 1 | ·55 | 1 | Jl | 31 | ·08 |
| 4257 | 1213 | 1078 | 24M | ·1056 | 28·7526 | 6·368 | **1156** | 1 | 7 | Ap | 21○ | ·86 | 2 | My | 21● | ·39 | 3 | Je | 19 | ·92 | 5 | Jl | 19 | ·45 | 6 | Au | 17 | ·98 |
| 4258 | 1214 | 1079 | 24M | ·3643 | 17·8609 | 2·525 | **1157** | 3 | 5 | Ap | 11○ | ·22 | 6 | My | 10 | ·76 | 1 | Je | 9 | ·29 | 2 | Jl | 8 | ·82 | 4 | Au | 7 | ·35 |
| 4259 | 1215 | 1080 | 24M | ·6231 | 6·9692 | 26·237 | **1158** | 4 | 2 | Mr | 31○ | ·59 | 4 | Ap | 30 | ·12 | 5 | My | 29 | ·65 | 7 | Je | 28 | ·18 | 3 | Au | 26 | ·24 |
| | | | | | | | | | | | | | | | | | | | | | 1 | Jl | 27 | ·71 | | | | |
| 4260 | 1216 | 1081 | 24M | ·8819 | 25·6081 | 24·370 | **1159** | 5 | 1 | Ap | 19 | ·49 | 3 | My | 19 | ·02 | 4 | Je | 17 | ·55 | 6 | Jl | 17 | ·08 | 7 | Au | 15 | ·61 |
| 4261 | 1217 | 1082 | 24M | ·1406 | 14·7164 | 20·527 | **1160** | 6 | 5 | Ap | 7 | ·86 | 7 | My | 7 | ·39 | 1 | Je | 5 | ·92 | 3 | Jl | 5 | ·45 | 4 | Au | 3○ | ·98 |
| 4262 | 1218 | 1083 | 24M | ·3994 | 3·8247 | 16·685 | **1161** | 1 | 3 | Mr | 28 | ·22 | 4 | Ap | 26 | ·75 | 6 | My | 26 | ·28 | 2 | Jl | 24○ | ·35 | 3 | Au | 22 | ·88 |
| | | | | | | | | | | | | | | | | | 7 | Je | 24 | ·82 | | | | | | | | |
| 4263 | 1219 | 1084 | 24M | ·6581 | 22·4636 | 14·818 | **1162** | 2 | 2 | Ap | 16 | ·12 | 3 | My | 15 | ·65 | 5 | Je | 14 | ·18 | 6 | Jl | 13●○ | ·71 | 1 | Au | 12 | ·24 |
| 4264 | 1220 | 1085 | 24M | ·9169 | 11·5719 | 10·975 | **1163** | 3 | 6 | Ap | 5 | ·49 | 1 | My | 5 | ·02 | 2 | Je | 3○ | ·55 | 4 | Jl | 3● | ·08 | 5 | Au | 1 | ·61 |
| 4265 | 1221 | 1086 | 24M | ·1756 | 0·6802 | 7·183 | **1164** | 4 | 3 | Mr | 24 | ·86 | 6 | My | 22○ | ·92 | 1 | Je | 21● | ·45 | 2 | Jl | 20 | ·98 | 4 | Au | 19 | ·51 |
| | | | | | | | | | 5 | Ap | 23 | ·39 | | | | | | | | | | | | | | | | |
| 4266 | 1222 | 1087 | 24M | ·4344 | 19·3191 | 5·266 | **1165** | 6 | 2 | Ap | 12 | ·75 | 4 | My | 12○ | ·28 | 5 | Je | 10 | ·81 | 7 | Jl | 10 | ·34 | 1 | Au | 8 | ·88 |
| 4267 | 1223 | 1088 | 24M | ·6932 | 8·4273 | 1·423 | **1166** | 7 | 7 | Ap | 2 | ·12 | 1 | My | 1● | ·65 | 3 | My | 31 | ·18 | 4 | Je | 29 | ·71 | 6 | Jl | 29 | ·24 |
| | | | | | | | | | | | | | | | | | | | | | | | | | 7 | Au | 27 | ·77 |
| 4268 | 1224 | 1089 | 24M | ·9519 | 27·0663 | 27·111 | **1167** | 1 | 6 | Ap | 21● | ·02 | 7 | My | 20 | ·55 | 2 | Je | 19 | ·08 | 3 | Jl | 18 | ·61 | 5 | Au | 17 | ·14 |
| 4269 | 1225 | 1090 | 24M | ·2107 | 16·1746 | 23·268 | **1168** | 2 | 3 | Ap | 9 | ·38 | 4 | My | 8 | ·92 | 6 | Je | 7 | ·45 | 7 | Jl | 6 | ·98 | 2 | Au | 5 | ·51 |
| 4270 | 1226 | 1091 | 24M | ·4694 | 5·2829 | 19·426 | **1169** | 4 | 7 | Mr | 29 | ·75 | 2 | Ap | 28 | ·28 | 3 | My | 27 | ·81 | 5 | Je | 26 | ·34 | 1 | Au | 24●○ | ·40 |
| | | | | | | | | | | | | | | | | | | | | | 6 | Jl | 25 | ·87 | | | | |
| 4271 | 1227 | 1092 | 24M | ·7282 | 23·9218 | 17·559 | **1170** | 5 | 6 | Ap | 17 | ·65 | 1 | My | 17 | ·18 | 2 | Je | 15 | ·71 | 4 | Jl | 15 | ·24 | 5 | Au | 13 | ·77 |
| 4272 | 1228 | 1093 | 24M | ·9869 | 13·0301 | 13·716 | **1171** | 6 | 4 | Ap | 7 | ·02 | 5 | My | 6 | ·55 | 7 | Je | 5 | ·08 | 1 | Jl | 4○ | ·61 | 3 | Au | 3 | ·14 |
| 4273 | 1229 | 1094 | 24M | ·2457 | 2·1384 | 9·873 | **1172** | 7 | 1 | Mr | 26 | ·38 | 2 | Ap | 24 | ·91 | 5 | Je | 22○ | ·98 | 7 | Jl | 22 | ·51 | 2 | Au | 21 | ·04 |
| | | | | | | | | | | | | | 4 | My | 24 | ·44 | | | | | | | | | | | | |
| 4274 | 1230 | 1095 | 24M | ·5044 | 20·7772 | 8·006 | **1173** | 2 | 7 | Ap | 14 | ·28 | 1 | My | 13 | ·81 | 3 | Je | 12●○ | ·34 | 4 | Jl | 11 | ·87 | 6 | Au | 10 | ·40 |
| 4275 | 1231 | 1096 | 24M | ·7632 | 9·8855 | 4·164 | **1174** | 3 | 4 | Ap | 3 | ·65 | 6 | My | 3○ | ·18 | 7 | Je | 1● | ·71 | 2 | Jl | 1 | ·24 | 3 | Jl | 30 | ·77 |
| 4276 | 1232 | 1097 | 24M | ·0220 | 28·5244 | 2·297 | **1175** | 4 | 3 | Ap | 22○ | ·55 | 5 | My | 22 | ·08 | 6 | Je | 20 | ·61 | 1 | Jl | 20 | ·14 | 2 | Au | 18 | ·67 |
| 4277 | 1233 | 1098 | 24M | ·2807 | 17·6327 | 26·008 | **1176** | 5 | 7 | Ap | 10●○ | ·91 | 2 | My | 10 | ·44 | 3 | Je | 8 | ·97 | 5 | Jl | 8 | ·50 | 7 | Au | 7 | ·04 |
| 4278 | 1234 | 1099 | 24M | ·5395 | 6·7410 | 22·171 | **1177** | 7 | 5 | Mr | 31 | ·28 | 6 | Ap | 29 | ·81 | 1 | My | 29 | ·34 | 2 | Je | 27 | ·87 | 5 | Au | 25 | ·93 |
| | | | | | | | | | | | | | | | | | | | | | 4 | Jl | 27 | ·40 | | | | |
| 4279 | 1235 | 1100 | 24M | ·7982 | 25·3799 | 20·299 | **1178** | 1 | 4 | Ap | 19 | ·18 | 5 | My | 18 | ·71 | 7 | Je | 17 | ·24 | 1 | Jl | 16 | ·77 | 3 | Au | 15○ | ·30 |
| 4280 | 1236 | 1101 | 25M | ·0570 | 14·4882 | 16·457 | **1179** | 2 | 1 | Ap | 8 | ·54 | 3 | My | 8 | ·08 | 4 | Je | 6 | ·61 | 6 | Jl | 6 | ·14 | 7 | Au | 4○ | ·67 |
| 4281 | 1237 | 1102 | 24M | ·3157 | 3·5965 | 12·614 | **1180** | 3 | 5 | Mr | 27 | ·91 | 7 | Ap | 26 | ·44 | 1 | My | 25 | ·97 | 5 | Jl | 24●○ | ·03 | 6 | Au | 22 | ·56 |
| | | | | | | | | | | | | | | | | | 3 | Je | 24 | ·50 | | | | | | | | |
| 4282 | 1238 | 1103 | 24M | ·5745 | 22·2354 | 10·747 | **1181** | 5 | 4 | Ap | 15 | ·81 | 6 | My | 15 | ·34 | 7 | Je | 13 | ·87 | 2 | Jl | 13 | ·40 | 3 | Au | 11 | ·93 |
| 4283 | 1239 | 1104 | 24M | ·8333 | 11·3437 | 6·904 | **1182** | 6 | 2 | Ap | 5 | ·18 | 3 | My | 4 | ·71 | 5 | Je | 3○ | ·24 | 6 | Jl | 2 | ·77 | 1 | Au | 1 | ·30 |
| 4284 | 1240 | 1105 | 25M | ·0920 | 0·4520 | 3·062 | **1183** | 7 | 6 | Mr | 25 | ·54 | 2 | My | 23●○ | ·60 | 4 | Je | 22 | ·14 | 5 | Jl | 21 | ·67 | 7 | Au | 20 | ·20 |
| | | | | | | | | | 1 | Ap | 24 | ·07 | | | | | | | | | | | | | | | | |
| 4285 | 1241 | 1106 | 24M | ·3508 | 19·0909 | 1·195 | **1184** | 1 | 5 | Ap | 12 | ·44 | 6 | My | 11 | ·97 | 1 | Je | 10 | ·50 | 3 | Jl | 10 | ·03 | 4 | Au | 8 | ·56 |
| 4286 | 1242 | 1107 | 24M | ·6095 | 8·1992 | 24·907 | **1185** | 3 | 2 | Ap | 1○ | ·81 | 4 | My | 1● | ·34 | 5 | My | 30 | ·87 | 7 | Je | 29 | ·40 | 1 | Jl | 28 | ·93 |
| | | | | | | | | | | | | | | | | | | | | | | | | | 3 | Au | 27 | ·46 |
| 4287 | 1243 | 1108 | 24M | ·8683 | 26·8381 | 23·040 | **1186** | 4 | 1 | Ap | 20 | ·71 | 3 | My | 20 | ·24 | 4 | Je | 18 | ·77 | 6 | Jl | 18 | ·30 | 7 | Au | 16 | ·83 |
| 4288 | 1244 | 1109 | 25M | ·1270 | 15·9464 | 19·197 | **1187** | 5 | 6 | Ap | 10 | ·07 | 7 | My | 9 | ·60 | 2 | Je | 8 | ·13 | 3 | Jl | 7 | ·66 | 5 | Au | 6 | ·20 |
| 4289 | 1245 | 1110 | 24M | ·3858 | 5·0547 | 15·354 | **1188** | 6 | 3 | Mr | 29 | ·44 | 4 | Ap | 27 | ·97 | 6 | My | 27 | ·50 | 1 | Je | 26 | ·03 | 4 | Au | 24● | ·10 |
| | | | | | | | | | | | | | | | | | | | | | 2 | Jl | 25 | ·56 | | | | |
| 4290 | 1246 | 1111 | 24M | ·6445 | 23·6936 | 13·488 | **1189** | 1 | 2 | Ap | 17 | ·34 | 3 | My | 16 | ·87 | 5 | Je | 15 | ·40 | 6 | Jl | 14○ | ·93 | 1 | Au | 13 | ·46 |
| 4291 | 1247 | 1112 | 24M | ·9033 | 12·8019 | 9·645 | **1190** | 2 | 6 | Ap | 6 | ·70 | 1 | My | 6 | ·24 | 2 | Je | 4 | ·77 | 4 | Jl | 4●○ | ·30 | 5 | Au | 2 | ·83 |
| 4292 | 1248 | 1113 | 25M | ·1621 | 1·9101 | 5·802 | **1191** | 3 | 4 | Mr | 27 | ·07 | 5 | Ap | 25 | ·60 | 1 | Je | 23●○ | ·66 | 3 | Jl | 23 | ·19 | 4 | Au | 21 | ·72 |
| | | | | | | | | | | | | | 7 | My | 25 | ·13 | | | | | | | | | | | | |
| 4293 | 1249 | 1114 | 24M | ·4208 | 20·5490 | 3·935 | **1192** | 4 | 2 | Ap | 13 | ·97 | 4 | My | 13○ | ·50 | 6 | Je | 12 | ·03 | 7 | Jl | 11 | ·56 | 2 | Au | 10 | ·09 |
| 4294 | 1250 | 1115 | 24M | ·6796 | 9·6573 | 0·093 | **1193** | 6 | 7 | Ap | 3 | ·34 | 1 | My | 2○ | ·87 | 3 | Je | 1● | ·40 | 4 | Je | 30 | ·93 | 6 | Jl | 30 | ·46 |

**Surya Siddhanta.**

| Asvina | | | | Kartika | | | | Margasira | | | | Pausha | | | | A.D. | Magha | | | | A.D. | Phalguna | | | | Chaitra | | | |
|---|---|---|---|---|---|---|---|---|---|---|---|---|---|---|---|---|---|---|---|---|---|---|---|---|---|---|---|---|---|
| + 147·65293 | | | | +177·18353 | | | | + 206·71411 | | | | + 236·24470 | | | | | + 265·77529 | | | | | + 295·30588 | | | | +324·83647<br>+354·36705 | | | |
| + 9·880 | | | | + 11·856 | | | | + 13·832 | | | | + 15·808 | | | | | + 17·784 | | | | | + 19·760 | | | | + 21·736<br>+ 23·712 | | | |
| Week-day | Month | Day | Fraction | Week-day | Month | Day | Fraction | Week-day | Month | Day | Fraction | Week-day | Month | Day | Fraction | | Week-day | Month | Day | Fraction | | Week-day | Month | Day | Fraction | Week-day | Month | Day | Fraction |
| 2 | Au | 29 | ·61 | 4 | S | 28 | ·14 | 5 | O | 27 | ·67 | 7 | N | 26● | ·20 | | 1 | D | 25 | ·74 | 56 | 3 | Jr | 24 | ·27 | 6 | Mr | 23 | ·33 |
| | | | | | | | | | | | | | | | | | | | | | | 4 | F | 22 | ·80 | | | | |
| 1 | S | 16 | ·51 | 3 | O | 16○ | ·04 | 4 | N | 14 | ·57 | 6 | D | 14 | ·10 | 57 | 7 | Jr | 12 | ·63 | | 2 | F | 11 | ·16 | 3 | Mr | 12 | ·65 |
| 5 | S | 5 | ·88 | 7 | O | 5○ | ·41 | 1 | N | 3 | ·94 | 3 | D | 3 | ·47 | 58 | 5 | Jr | 2 | ·00 | | 6 | Jr | 31 | ·53 | 1 | Mr | 2 | ·06 |
| 4 | S | 24○ | ·78 | 6 | O | 24 | ·31 | 7 | N | 22 | ·84 | 2 | D | 22 | ·37 | 59 | 3 | Jr | 20 | ·90 | | 5 | F | 19 | ·43 | 6 | Mr | 20 | ·96 |
| 2 | S | 14 | ·14 | 3 | O | 13 | ·67 | 5 | N | 12 | ·20 | 6 | D | 11 | ·73 | 60 | 1 | Jr | 10 | ·26 | | 2 | F | 8○ | ·80 | 4 | Mr | 9 | ·33 |
| 6 | S | 2● | ·51 | 1 | O | 2 | ·04 | 2 | O | 31 | ·57 | 4 | N | 30 | ·10 | | 5 | D | 29 | ·63 | 61 | 7 | Jr | 28●○ | ·16 | 1 | F | 26 | ·69 |
| 5 | S | 21 | ·41 | 6 | O | 20 | ·94 | 1 | N | 19 | ·47 | 3 | D | 19 | ·00 | 62 | 4 | Jr | 17●○ | ·53 | | 6 | F | 16 | ·06 | 7 | Mr | 17 | ·59 |
| 2 | S | 10 | ·77 | 4 | O | 10 | ·30 | 5 | N | 8 | ·84 | 7 | D | 8 | ·37 | 63 | 1 | Jr | 6 | ·90 | | 3 | F | 5 | ·43 | 4 | Mr | 6 | ·96 |
| 7 | Au | 31 | ·14 | 1 | S | 29 | ·67 | 3 | O | 29 | ·20 | 4 | N | 27○ | ·73 | | 6 | D | 27 | ·26 | 64 | 7 | Jr | 25 | ·79 | 2 | F | 24 | ·32 |
| 6 | S | 18 | ·04 | 7 | O | 17 | ·57 | 2 | N | 16●○ | ·10 | 3 | D | 15 | ·63 | 65 | 5 | Jr | 14 | ·16 | | 6 | F | 12 | ·69 | 1 | Mr | 14 | ·22 |
| 3 | S | 7 | ·41 | 4 | O | 6 | ·94 | 6 | N | 5○ | ·47 | 1 | D | 5 | ·00 | 66 | 2 | Jr | 3 | ·53 | | 4 | F | 2 | ·06 | 5 | Mr | 3 | ·59 |
| 2 | S | 26 | ·30 | 3 | O | 25 | ·83 | 5 | N | 24 | ·36 | 6 | D | 23 | ·90 | 67 | 1 | Jr | 22 | ·43 | | 2 | F | 20 | ·96 | 4 | Mr | 22○ | ·49 |
| 6 | S | 15○ | ·67 | 1 | O | 15 | ·20 | 2 | N | 13 | ·73 | 4 | D | 13 | ·26 | 68 | 5 | Jr | 11 | ·79 | | 7 | F | 10 | ·32 | 1 | Mr | 10○ | ·85 |
| 4 | S | 4●○ | ·04 | 5 | O | 3 | ·57 | 7 | N | 2 | ·10 | 1 | D | 1 | ·63 | | 3 | D | 31 | ·16 | 69 | 4 | Jr | 29 | ·69 | 6 | F | 28○ | ·22 |
| 2 | S | 22 | ·93 | 4 | O | 22 | ·47 | 6 | N | 21 | ·00 | 7 | D | 20 | ·53 | 70 | 2 | Jr | 19 | ·06 | | 3 | F | 17 | ·59 | 5 | Mr | 19 | ·12 |
| 7 | S | 12 | ·30 | 1 | O | 11 | ·83 | 3 | N | 10 | ·36 | 4 | D | 9 | ·89 | 71 | 6 | Jr | 8○ | ·42 | | 7 | F | 6 | ·96 | 2 | Mr | 8 | ·49 |
| 4 | S | 1 | ·67 | 6 | O | 1 | ·20 | 7 | O | 30 | ·73 | 2 | N | 29 | ·26 | | 3 | D | 28○ | ·79 | 72 | 5 | Jr | 27● | ·32 | 6 | F | 25 | ·85 |
| 3 | S | 19 | ·57 | 5 | O | 19 | ·10 | 6 | N | 17 | ·63 | 1 | D | 17○ | ·16 | 73 | 2 | Jr | 15 | ·69 | | 4 | F | 14 | ·22 | 5 | Mr | 15 | ·75 |
| 7 | S | 8 | ·93 | 2 | O | 8 | ·46 | 4 | N | 7 | ·00 | 5 | D | 6 | ·53 | 74 | 7 | Jr | 5 | ·06 | | 1 | F | 3 | ·59 | 3 | Mr | 5 | ·12 |
| 5 | Au | 29 | ·30 | 6 | S | 27 | ·83 | 1 | O | 27○ | ·36 | 2 | N | 25● | ·89 | | 4 | D | 25 | ·42 | 75 | 5 | Jr | 23 | ·95 | 7 | F | 22 | ·48 |
| | | | | | | | | | | | | | | | | | | | | | | | | | | 2 | Mr | 24 | ·01 |
| 4 | S | 17 | ·20 | 5 | O | 16○ | ·73 | 7 | N | 15 | ·26 | 1 | D | 14 | ·79 | 76 | 3 | Jr | 13 | ·32 | | 4 | F | 11 | ·85 | 6 | Mr | 12 | ·38 |
| 1 | S | 5 | ·57 | 3 | O | 5○ | ·10 | 4 | N | 3 | ·63 | 6 | D | 3 | ·16 | 77 | 7 | Jr | 1 | ·69 | | 2 | Jr | 31 | ·22 | 3 | Mr | 1 | ·75 |
| 7 | S | 24 | ·46 | 1 | O | 23 | ·99 | 3 | N | 22 | ·52 | 5 | D | 22 | ·06 | 78 | 6 | Jr | 20 | ·59 | | 1 | F | 19○ | ·12 | 2 | Mr | 20● | ·65 |
| 4 | S | 13● | ·83 | 6 | O | 13 | ·36 | 7 | N | 11 | ·89 | 2 | D | 11 | ·42 | 79 | 3 | Jr | 9 | ·95 | | 5 | F | 8○ | ·48 | 7 | Mr | 10 | ·01 |
| 2 | S | 3 | ·20 | 3 | O | 2 | ·73 | 5 | N | 1 | ·26 | 6 | N | 30 | ·79 | | 1 | D | 30 | ·32 | 80 | 2 | Jr | 28○ | ·85 | 4 | F | 27 | ·38 |
| 1 | S | 21 | ·10 | 2 | O | 20 | ·63 | 4 | N | 19 | ·16 | 5 | D | 18 | ·69 | 81 | 7 | Jr | 17● | ·22 | | 1 | F | 15 | ·75 | 3 | Mr | 17 | ·28 |
| 5 | S | 10 | ·46 | 6 | O | 9 | ·99 | 1 | N | 8 | ·52 | 3 | D | 8○ | ·05 | 82 | 4 | Jr | 6 | ·58 | | 6 | F | 5 | ·12 | 7 | Mr | 6 | ·65 |
| 2 | Au | 30 | ·83 | 4 | S | 29 | ·36 | 5 | O | 28 | ·89 | 7 | N | 27○ | ·42 | | 1 | D | 26 | ·95 | 83 | 3 | Jr | 25 | ·48 | 5 | F | 24 | ·01 |
| 1 | S | 18 | ·73 | 3 | O | 18 | ·26 | 4 | N | 16●○ | ·79 | 6 | D | 16 | ·32 | 84 | 7 | Jr | 14 | ·85 | | 2 | F | 13 | ·38 | 3 | Mr | 13 | ·91 |
| 6 | S | 7 | ·09 | 7 | O | 6 | ·62 | 2 | N | 5 | ·16 | 3 | D | 4 | ·69 | 85 | 5 | Jr | 3 | ·22 | | 6 | F | 1 | ·75 | 1 | Mr | 3 | ·28 |
| 4 | S | 25○ | ·99 | 6 | O | 25● | ·52 | 1 | N | 24 | ·05 | 2 | D | 23 | ·58 | 86 | 4 | Jr | 22 | ·11 | | 5 | F | 20 | ·64 | 7 | Mr | 22○ | ·18 |
| 3 | S | 15○ | ·36 | 4 | O | 14 | ·89 | 5 | N | 13 | ·42 | 6 | D | 12 | ·95 | 87 | 1 | Jr | 11 | ·48 | | 3 | F | 10 | ·01 | 4 | Mr | 11○ | ·54 |
| 6 | S | 4●○ | ·73 | 1 | O | 4 | ·26 | 2 | N | 2 | ·79 | 4 | D | 2 | ·32 | | 5 | D | 31 | ·85 | 88 | 7 | Jr | 30 | ·38 | 1 | F | 28● | ·91 |
| 5 | S | 22 | ·62 | 7 | O | 22 | ·15 | 1 | N | 20 | ·68 | 3 | D | 20 | ·22 | 89 | 4 | Jr | 18○ | ·75 | | 6 | F | 17● | ·28 | 7 | Mr | 18 | ·81 |
| 2 | S | 11 | ·99 | 4 | O | 11 | ·52 | 6 | N | 10 | ·05 | 7 | D | 9 | ·58 | 90 | 2 | Jr | 8○ | ·11 | | 3 | F | 6 | ·64 | 5 | Mr | 8 | ·17 |
| 7 | S | 1 | ·36 | 1 | S | 30 | ·89 | 3 | O | 30 | ·42 | 4 | N | 28 | ·95 | | 6 | D | 28○ | ·48 | 91 | 1 | Jr | 27 | ·01 | 2 | F | 25 | ·54 |
| 6 | S | 20 | ·26 | 7 | O | 19 | ·79 | 2 | N | 18 | ·32 | 3 | D | 17● | ·85 | 92 | 5 | Jr | 16 | ·38 | | 6 | F | 14 | ·91 | 1 | Mr | 15 | ·44 |
| 3 | S | 8 | ·62 | 5 | O | 8 | ·15 | 6 | N | 6○ | ·68 | 1 | D | 6 | ·21 | 93 | 2 | Jr | 4 | ·74 | | 4 | F | 3 | ·28 | 5 | Mr | 4 | ·81 |
| 7 | Au | 28 | ·99 | 4 | O | 27○ | ·05 | 5 | N | 25 | ·58 | **Pausha Kshaya** | | | | | 7 | D | 25 | ·11 | 94 | 1 | Jr | 23 | ·64 | 3 | F | 22 | ·17 |
| 2 | S | 27 | ·52 | | | | | | | | | | | | | | | | | | | | | | | 4 | Mr | 23 | ·70 |

⊙'s Anom col. 6 + 29·53059 + 59·06117 + 88·59176 + 118·12235

☾'s Anom col. 7 + 1·976 + 3·952 + 5·928 + 7·904

| Kaliyuga. | Vikrama Era. | Saka Era. | Commencement of Solar Year. Month and Day A.D. | Fraction of day. | First New-Moon in Solar Year. | Anomaly of first New-Moon. | A.D. | Week-day of 1st January. | Vaisakha Week-day | Month | Day | Fraction | Jyeshta Week-day | Month | Day | Fraction | Ashada Week-day | Month | Day | Fraction | Sravana Week-day | Month | Day | Fraction | Bhadrapada Week-day | Month | Day | Fraction |
|---|---|---|---|---|---|---|---|---|---|---|---|---|---|---|---|---|---|---|---|---|---|---|---|---|---|---|---|---|
| 4295 | 1251 | 1116 | 24M | ·9383 | 28·2962 | 25·780 | **1194** | 7 | 6 | Ap | 22○ | ·23 | 7 | My | 21 | ·76 | 2 | Je | 20 | ·30 | 3 | Jl | 19 | ·83 | 5 | Au | 18 | ·36 |
| 4296 | 1252 | 1117 | 25M | ·1971 | 17·4045 | 21·938 | **1195** | 1 | 3 | Ap | 11● | ·60 | 5 | My | 11 | ·13 | 6 | Je | 9 | ·66 | 1 | Jl | 9 | ·19 | 2 | Au | 7 | ·72 |
| 4297 | 1253 | 1118 | 24M | ·4558 | 6·5128 | 18·095 | **1196** | 2 | 7 | Mr | 30 | ·97 | 2 | Ap | 29 | ·50 | 4 | My | 29 | ·03 | 5 | Je | 27 | ·56 | 1 | Au | 25○ | ·62 |
| | | | | | | | | | | | | | | | | | | | | | 7 | Jl | 27 | ·09 | | | | |
| 4298 | 1254 | 1119 | 24M | ·7146 | 25·1517 | 16·228 | **1197** | 4 | 6 | Ap | 18 | ·87 | 1 | My | 18 | ·40 | 2 | Je | 16 | ·93 | 4 | Jl | 16 | ·46 | 5 | Au | 14○ | ·99 |
| 4299 | 1255 | 1120 | 24M | ·9734 | 14·2600 | 12·385 | **1198** | 5 | 4 | Ap | 8 | ·23 | 5 | My | 7 | ·76 | 7 | Je | 6 | ·29 | 1 | Jl | 5 | ·82 | 3 | Au | 4○ | ·36 |
| 4300 | 1256 | 1121 | 25M | ·2321 | 3·3683 | 8·543 | **1199** | 6 | 1 | Mr | 28 | ·60 | 3 | Ap | 27 | ·13 | 4 | My | 26 | ·66 | 7 | Jl | 24 | ·72 | 2 | Au | 23 | ·25 |
| | | | | | | | | | | | | | | | | | 6 | Je | 25 | ·19 | | | | | | | | |
| 4301 | 1257 | 1122 | 24M | ·4909 | 22·0072 | 6·676 | **1200** | 7 | 7 | Ap | 15 | ·50 | 2 | My | 15 | ·03 | 3 | Je | 13ↄ | ·56 | 5 | Jl | 13 | ·09 | 6 | Au | 11 | ·62 |
| 4302 | 1258 | 1123 | 24M | ·7496 | 11·1155 | 2·834 | **1201** | 2 | 4 | Ap | 4 | ·86 | 6 | My | 4 | ·40 | 7 | Je | 2ↄ | ·93 | 2 | Jl | 2 | ·46 | 3 | Jl | 31 | ·99 |
| 4303 | 1259 | 1124 | 25M | ·0084 | 0·2238 | 26·545 | **1202** | 3 | 2 | Mr | 25 | ·23 | 5 | My | 23● | ·29 | 6 | Je | 21 | ·82 | 1 | Jl | 21 | ·35 | 2 | Au | 19 | ·88 |
| | | | | | | | | | 3 | Ap | 23 | ·76 | | | | | | | | | | | | | | | | |
| 4304 | 1260 | 1125 | 25M | ·2671 | 18·8627 | 24·679 | **1203** | 4 | 1 | Ap | 13○ | ·13 | 2 | My | 12 | ·66 | 4 | Je | 11 | ·19 | 5 | Jl | 10 | ·72 | 7 | Au | 9 | ·25 |
| 4305 | 1261 | 1126 | 24M | ·5259 | 7·9710 | 20·836 | **1204** | 5 | 5 | Ap | 1○ | ·50 | 7 | My | 1 | ·03 | 1 | My | 30 | ·56 | 3 | Je | 29 | ·09 | 4 | Jl | 28 | ·62 |
| | | | | | | | | | | | | | | | | | | | | | | | | | 6 | Au | 27 | ·15 |
| 4306 | 1262 | 1127 | 24M | ·7847 | 26·6099 | 18·969 | **1205** | 7 | 4 | Ap | 20 | ·39 | 5 | My | 19 | ·92 | 7 | Je | 18 | ·46 | 1 | Jl | 17 | ·99 | 3 | Au | 16 | ·52 |
| 4307 | 1263 | 1128 | 25M | ·0434 | 15·7182 | 15·126 | **1206** | 1 | 1 | Ap | 9 | ·76 | 3 | My | 9 | ·29 | 4 | Je | 7 | ·82 | 6 | Jl | 7 | ·35 | 7 | Au | 5 | ·88 |
| 4308 | 1264 | 1129 | 25M | ·3022 | 4·8265 | 11·284 | **1207** | 2 | 6 | Mr | 30 | ·13 | 7 | Ap | 28 | ·66 | 2 | My | 28 | ·19 | 5 | Jl | 26ↄ | ·25 | 6 | Au | 24● | ·78 |
| | | | | | | | | | | | | | | | | | 3 | Je | 26 | ·72 | | | | | | | | |
| 4309 | 1265 | 1130 | 24M | ·5609 | 23·4653 | 9·417 | **1208** | 3 | 5 | Ap | 17 | ·03 | 6 | My | 16 | ·56 | 1 | Je | 15 | ·09 | 2 | Jl | 14ↄ | ·62 | 4 | Au | 13 | ·15 |
| 4310 | 1266 | 1131 | 24M | ·8197 | 12·5736 | 5·574 | **1209** | 5 | 2 | Ap | 6 | ·39 | 3 | My | 5 | ·92 | 5 | Je | 4 | ·45 | 6 | Jl | 3ↄ | ·98 | 1 | Au | 2 | ·52 |
| 4311 | 1267 | 1132 | 25M | ·0784 | 1·6819 | 1·731 | **1210** | 6 | 6 | Mr | 26 | ·76 | 1 | Ap | 25 | ·29 | 4 | Je | 23 | ·35 | 5 | Jl | 22 | ·88 | 7 | Au | 21 | ·41 |
| | | | | | | | | | | | | | 2 | My | 24 | ·82 | | | | | | | | | | | | |
| 4312 | 1268 | 1133 | 25M | ·3372 | 20·3208 | 27·419 | **1211** | 7 | 5 | Ap | 14 | ·66 | 7 | My | 14○ | ·19 | 1 | Je | 12 | ·72 | 3 | Jl | 12 | ·25 | 4 | Au | 10 | ·78 |
| 4313 | 1269 | 1134 | 24M | ·5959 | 9·4291 | 23·577 | **1212** | 1 | 3 | Ap | 3 | ·02 | 4 | My | 2○ | ·56 | 6 | Je | 1 | ·09 | 7 | Je | 30 | ·62 | 2 | Jl | 30 | ·15 |
| 4314 | 1270 | 1135 | 24M | ·8547 | 28·0680 | 21·710 | **1213** | 3 | 1 | Ap | 21● | ·92 | 3 | My | 21 | ·45 | 4 | Je | 19 | ·98 | 6 | Jl | 19 | ·51 | 1 | Au | 18 | ·04 |
| 4315 | 1271 | 1136 | 25M | ·1135 | 17·1763 | 17·867 | **1214** | 4 | 6 | Ap | 11 | ·29 | 7 | My | 10 | ·82 | 2 | Je | 9 | ·35 | 3 | Jl | 8 | ·88 | 5 | Au | 7 | ·41 |
| 4316 | 1272 | 1137 | 25M | ·3722 | 6·2846 | 14·024 | **1215** | 5 | 3 | Mr | 31 | ·66 | 5 | Ap | 30 | ·19 | 6 | My | 29 | ·72 | 1 | Je | 28 | ·25 | 4 | Au | 26ↄ | ·31 |
| | | | | | | | | | | | | | | | | | | | | | 2 | Jl | 27 | ·78 | | | | |
| 4317 | 1273 | 1138 | 24M | ·6810 | 24·9235 | 12·158 | **1216** | 6 | 2 | Ap | 18 | ·55 | 4 | My | 18 | ·08 | 5 | Je | 16 | ·62 | 7 | Jl | 16 | ·15 | 1 | Au | 14○ | ·68 |
| 4318 | 1274 | 1139 | 24M | ·8897 | 14·0318 | 8·315 | **1217** | 1 | 6 | Ap | 7 | ·92 | 1 | My | 7 | ·45 | 2 | Je | 5 | ·98 | 4 | Jl | 5 | ·51 | 6 | Au | 4● | ·04 |
| 4319 | 1275 | 1140 | 25M | ·1485 | 3·1401 | 4·472 | **1218** | 2 | 4 | Mr | 28 | ·29 | 5 | Ap | 26 | ·82 | 1 | Je | 24○ | ·88 | 3 | Jl | 24● | ·41 | 4 | Au | 22 | ·94 |
| | | | | | | | | | | | | | 7 | My | 26 | ·35 | | | | | | | | | | | | |
| 4320 | 1276 | 1141 | 25M | ·4072 | 21·7790 | 2·605 | **1219** | 3 | 3 | Ap | 16 | ·19 | 4 | My | 15 | ·72 | 6 | Je | 14○ | ·25 | 7 | Jl | 13 | ·78 | 2 | Au | 12 | ·31 |
| 4321 | 1277 | 1142 | 24M | ·6660 | 10·8873 | 26·317 | **1220** | 4 | 7 | Ap | 4 | ·55 | 2 | My | 4 | ·08 | 3 | Je | 2● | ·61 | 5 | Jl | 2 | ·14 | 6 | Jl | 31 | ·68 |
| 4322 | 1278 | 1143 | 24M | ·9248 | 29·5262 | 24·450 | **1221** | 6 | 4 | Mr | 24 | ·92 | 7 | My | 22● | ·98 | 2 | Je | 21 | ·51 | 4 | Jl | 21 | ·04 | 5 | Au | 19 | ·57 |
| | | | | | | | | | 6 | Ap | 23 | ·45 | | | | | | | | | | | | | | | | |
| 4323 | 1279 | 1144 | 25M | ·1835 | 18·6345 | 20·608 | **1222** | 7 | 3 | Ap | 12ↄ | ·82 | 5 | My | 12 | ·35 | 6 | Je | 10 | ·88 | 1 | Jl | 10 | ·41 | 2 | Au | 8 | ·94 |
| 4324 | 1280 | 1145 | 25M | ·4423 | 7·7428 | 16·765 | **1223** | 1 | 1 | Ap | 2ↄ | ·18 | 2 | My | 1 | ·72 | 4 | My | 31 | ·25 | 5 | Je | 29 | ·78 | 7 | Jl | 29 | ·31 |
| | | | | | | | | | | | | | | | | | | | | | | | | | 1 | Au | 27 | ·84 |
| 4325 | 1281 | 1146 | 24M | ·7010 | 26·3817 | 14·898 | **1224** | 2 | 7 | Ap | 20 | ·08 | 1 | My | 19 | ·61 | 3 | Je | 18 | ·14 | 4 | Jl | 17 | ·67 | 6 | Au | 16 | ·20 |
| 4326 | 1282 | 1147 | 24M | ·9598 | 15·4900 | 11·055 | **1225** | 4 | 4 | Ap | 9 | ·45 | 5 | My | 8 | ·98 | 7 | Je | 7 | ·51 | 2 | Jl | 7 | ·04 | 3 | Au | 5○ | ·57 |
| 4327 | 1283 | 1148 | 25M | ·2185 | 4·5982 | 7·213 | **1226** | 5 | 1 | Mr | 29 | ·82 | 3 | Ap | 28 | ·35 | 4 | My | 27 | ·88 | 7 | Jl | 25ↄ | ·94 | 2 | Au | 24 | ·47 |
| | | | | | | | | | | | | | | | | | 6 | Je | 26 | ·41 | | | | | | | | |
| 4328 | 1284 | 1149 | 25M | ·4773 | 23·2371 | 5·346 | **1227** | 6 | 7 | Ap | 17 | ·71 | 2 | My | 17 | ·24 | 3 | Je | 15 | ·78 | 5 | Jl | 15●○ | ·31 | 6 | Au | 13 | ·84 |
| 4329 | 1285 | 1150 | 24M | ·7361 | 12·3454 | 1·503 | **1228** | 7 | 5 | Ap | 6 | ·08 | 6 | My | 5 | ·61 | 1 | Je | 4 | ·14 | 2 | Jl | 3● | ·67 | 4 | Au | 2 | ·20 |
| 4330 | 1286 | 1151 | 24M | ·9948 | 1·4537 | 25·215 | **1229** | 2 | 2 | Mr | 26 | ·45 | 3 | Ap | 24 | ·98 | 7 | Je | 23 | ·04 | 1 | Jl | 22 | ·57 | 3 | Au | 21 | ·10 |
| | | | | | | | | | | | | | 5 | My | 24ↄ | ·51 | | | | | | | | | | | | |
| 4331 | 1287 | 1152 | 25M | ·2536 | 20·0926 | 23·348 | **1230** | 3 | 1 | Ap | 14 | ·35 | 2 | My | 13●○ | ·88 | 4 | Je | 12 | ·41 | 5 | Jl | 11 | ·94 | 7 | Au | 10 | ·47 |
| 4332 | 1288 | 1153 | 25M | ·5123 | 9·2009 | 19·506 | **1231** | 4 | 5 | Ap | 3 | ·71 | 7 | My | 3 | ·24 | 1 | Je | 1 | ·77 | 3 | Jl | 1 | ·30 | 4 | Jl | 30 | ·84 |
| 4333 | 1289 | 1154 | 24M | ·7711 | 27·8398 | 17·639 | **1232** | 5 | 4 | Ap | 21● | ·61 | 6 | My | 21 | ·14 | 7 | Je | 19 | ·67 | 2 | Jl | 19 | ·20 | 3 | Au | 17 | ·73 |

## Surya Siddhanta.

| Asvina | | | | Kartika | | | | Margasira | | | | Pausha | | | | A.D. | Magha | | | | A.D. | Phalguna | | | | Chaitra | | | |
|---|---|---|---|---|---|---|---|---|---|---|---|---|---|---|---|---|---|---|---|---|---|---|---|---|---|---|---|---|---|
| + 147·65293 | | | | + 177·18353 | | | | + 206·71411 | | | | + 236·24470 | | | | | + 265·77529 | | | | | + 295·30588 | | | | +324·83647 / +354·36705 | | | |
| + 9·880 | | | | + 11·856 | | | | + 13·832 | | | | + 15·808 | | | | | + 17·784 | | | | | + 19·760 | | | | + 21·736 / + 23·712 | | | |
| Week-day | Month | Day | Fraction | Week-day | Month | Day | Fraction | Week-day | Month | Day | Fraction | Week-day | Month | Day | Fraction | | Week-day | Month | Day | Fraction | | Week-day | Month | Day | Fraction | Week-day | Month | Day | Fraction |
| 6 | S | 16 | ·89 | 1 | O | 16ↄ | ·42 | 2 | N | 14 | ·95 | 4 | D | 14 | ·48 | 95 | 6 | Jr | 13 | ·00 | | 7 | F | 11 | ·54 | 2 | Mr | 13 | ·07 |
| 4 | S | 6 | ·25 | 5 | O | 5● | ·78 | 7 | N | 4 | ·32 | 1 | D | 3 | ·85 | 96 | 3 | Jr | 2 | ·38 | | 4 | Jr | 31 | ·91 | 6 | Mr | 1○ | ·44 |
| 3 | S | 24 | ·15 | 4 | O | 23 | ·68 | 6 | N | 22 | ·21 | 7 | D | 21 | ·74 | 97 | 2 | Jr | 20 | ·27 | | 3 | F | 18○ | ·80 | 5 | Mr | 20 | ·34 |
| 7 | S | 13● | ·52 | 2 | O | 13 | ·05 | 3 | N | 11 | ·58 | 5 | D | 11 | ·11 | 98 | 6 | Jr | 9 | ·64 | | 1 | F | 8●○ | ·17 | 2 | Mr | 9 | ·70 |
| 4 | S | 2 | ·89 | 6 | O | 2 | ·42 | 7 | O | 31 | ·95 | 2 | N | 30 | ·48 | | 4 | D | 30 | ·01 | 99 | 5 | Jr | 28● | ·54 | 7 | F | 27 | ·07 |
| 3 | S | 21 | ·78 | 5 | O | 21 | ·31 | 6 | N | 19 | ·84 | 1 | D | 19ↄ | ·38 | 12 | 2 | Jr | 17 | ·91 | 00 | 4 | F | 16 | ·44 | 5 | Mr | 16 | ·97 |
| 1 | S | 10 | ·15 | 2 | O | 9 | ·68 | 4 | N | 8 | ·21 | 5 | D | 7○ | ·74 | 01 | 7 | Jr | 6 | ·27 | | 1 | F | 4 | ·80 | 3 | Mr | 6 | ·38 |
| 5 | Au | 30 | ·52 | 7 | S | 29 | ·05 | 1 | O | 28 | ·58 | 3 | N | 27●○ | ·11 | | 4 | D | 26 | ·64 | 02 | 6 | Jr | 25 | ·17 | 7 | F | 23 | ·70 |
| 4 | S | 18 | ·42 | 5 | O | 17 | ·95 | 7 | N | 16● | ·48 | 2 | D | 16 | ·01 | 03 | 3 | Jr | 14 | ·54 | | 5 | F | 13 | ·07 | 6 | Mr | 14 | ·60 |
| 1 | S | 7 | ·78 | 3 | O | 7○ | ·31 | 4 | N | 5 | ·84 | 6 | D | 5 | ·37 | 04 | 7 | Jr | 3 | ·90 | | 2 | F | 2 | ·44 | 3 | Mr | 2 | ·97 |
| 7 | S | 25ↄ | ·68 | 2 | O | 25 | ·21 | 3 | N | 23 | ·74 | 5 | D | 23 | ·27 | 05 | 6 | Jr | 21 | ·80 | | 1 | F | 20 | ·33 | 2 | Mr | 21●○ | ·86 |
| 5 | S | 15ↄ | ·05 | 6 | O | 14 | ·58 | 1 | N | 13 | ·11 | 2 | D | 12 | ·64 | 06 | 4 | Jr | 11 | ·17 | | 5 | F | 9 | ·70 | 7 | Mr | 11● | ·23 |
| 2 | S | 4● | ·41 | 3 | O | 3 | ·94 | 5 | N | 2 | ·48 | 7 | D | 2 | ·01 | | 1 | D | 31 | ·54 | 07 | 3 | Jr | 30ↄ | ·07 | 4 | F | 28● | ·60 |
| 1 | S | 23 | ·31 | 2 | O | 22 | ·84 | 4 | N | 21 | ·37 | 5 | D | 20 | ·90 | 08 | 7 | Jr | 19ↄ | ·43 | | 1 | F | 17 | 96 | 3 | Mr | 18 | ·50 |
| 5 | S | 11 | ·68 | 6 | O | 11 | ·21 | 1 | N | 9 | ·74 | 3 | D | 9 | ·27 | 09 | 4 | Jr | 7○ | ·80 | | 6 | F | 6 | ·33 | 7 | Mr | 7 | ·86 |
| 3 | S | 1 | ·05 | 4 | S | 30 | ·58 | 6 | O | 30 | ·11 | 7 | N | 28 | ·64 | | 2 | D | 28 | ·17 | 10 | 3 | Jr | 26 | ·70 | 5 | F | 25 | ·23 |
| 1 | S | 19 | ·94 | 3 | O | 19 | ·47 | 5 | N | 18ↄ | ·00 | 6 | D | 17 | ·54 | 11 | 1 | Jr | 16 | ·07 | | 2 | F | 14 | ·60 | 4 | Mr | 16 | ·13 |
| 6 | S | 9 | ·31 | 7 | O | 8 | ·84 | 2 | N | 7○ | ·37 | 3 | D | 6● | ·90 | 12 | 5 | Jr | 5 | ·43 | | 6 | F | 3 | ·96 | 1 | Mr | 4 | ·49 |
| {3 | Au | 28 | ·68 | 6 | O | 26ↄ | ·74 | 1 | N | 25 | ·27 | 2 | D | 24 | ·80 | 13 | 4 | Jr | 23 | ·33 | | 5 | F | 21 | ·86 | 7 | Mr | 23 | ·39 |
| {5 | S | 27 | ·21 | | | | | | | | | | | | | | | | | | | | | | | | | | |
| 2 | S | 16 | ·58 | 4 | O | 16 | ·11 | 5 | N | 14 | ·64 | 7 | D | 14 | ·17 | 14 | 1 | Jr | 12 | ·70 | | 3 | F | 11 | ·23 | 4 | Mr | 12○ | ·76 |
| 6 | S | 5ↄ | ·94 | 1 | O | 5● | ·47 | 3 | N | 4 | ·00 | 4 | D | 3 | ·53 | 15 | 6 | Jr | 2 | ·06 | | 7 | Jr | 31 | ·60 | 2 | Mr | 2○ | ·13 |
| 5 | S | 24 | ·84 | 7 | O | 24 | ·37 | 1 | N | 22 | ·90 | 3 | D | 22 | ·43 | 16 | 4 | Jr | 20 | ·96 | | 6 | F | 19●○ | ·49 | 1 | Mr | 20 | ·02 |
| 3 | S | 13 | ·21 | 4 | O | 12 | ·74 | 6 | N | 11 | ·27 | 7 | D | 10 | ·80 | 17 | 2 | Jr | 9 | ·33 | | 3 | F | 7 | ·86 | 5 | Mr | 9 | ·39 |
| 7 | S | 2 | ·57 | 2 | O | 2 | ·10 | 3 | O | 31 | ·64 | 5 | N | 30 | ·17 | | 6 | D | 29ↄ | ·70 | 18 | 1 | Jr | 28● | ·23 | 2 | F | 26 | ·76 |
| 6 | S | 21 | ·47 | 1 | O | 21 | ·00 | 2 | N | 19 | ·53 | 4 | D | 19ↄ | ·06 | 19 | 5 | Jr | 17 | ·59 | | 7 | F | 16 | ·12 | 1 | Mr | 17 | ·66 |
| 3 | S | 10 | ·84 | 5 | O | 10 | ·37 | 6 | N | 8 | ·90 | 1 | D | 8ↄ | ·43 | 20 | 2 | Jr | 6 | ·96 | | 4 | F | 5 | ·49 | 6 | Mr | 6 | ·02 |
| 1 | Au | 30 | ·21 | 2 | S | 28 | ·74 | 4 | O | 28 | ·27 | 5 | N | 26 | ·80 | | 7 | D | 26 | ·33 | 21 | 1 | Jr | 24 | ·86 | 3 | F | 23 | ·39 |
| 7 | S | 18 | ·10 | 1 | O | 17○ | ·63 | 3 | N | 16 | ·16 | 4 | D | 15 | ·70 | 22 | 6 | Jr | 14 | ·23 | | 7 | F | 12 | ·76 | 2 | Mr | 14 | ·29 |
| 4 | S | 7 | ·47 | 6 | O | 7○ | ·00 | 7 | N | 5 | ·53 | 2 | D | 5 | ·06 | 23 | 3 | Jr | 3 | ·59 | | 5 | F | 2 | ·12 | 6 | Mr | 3 | ·65 |
| 3 | S | 26●○ | ·37 | 4 | O | 25 | ·90 | 6 | N | 24 | ·43 | 7 | D | 23 | ·96 | 24 | 2 | Jr | 22 | ·49 | | 4 | F | 21 | ·02 | 5 | Mr | 21 | ·55 |
| 7 | S | 14 | ·74 | 2 | O | 14 | ·27 | 3 | N | 12 | ·80 | 5 | D | 12 | ·33 | 25 | 6 | Jr | 10 | ·86 | | 1 | F | 9ↄ | ·39 | 2 | Mr | 10 | ·92 |
| 5 | S | 4 | ·10 | 6 | O | 3 | ·63 | 1 | N | 2 | ·16 | 2 | D | 1 | ·69 | | 4 | D | 31 | ·22 | 26 | 5 | Jr | 29ↄ | ·76 | 7 | F | 28● | ·29 |
| 4 | S | 23 | ·00 | 5 | O | 22 | ·53 | 7 | N | 21 | ·06 | 1 | D | 20 | ·59 | 27 | 3 | Jr | 19●○ | ·12 | | 4 | F | 17 | ·65 | 6 | Mr | 19 | ·18 |
| 1 | S | 12 | ·37 | 2 | O | 11 | ·90 | 4 | N | 10 | ·43 | 5 | D | 9 | ·96 | 28 | 7 | Jr | 8 | ·49 | | 2 | F | 7 | ·02 | 3 | Mr | 7 | ·55 |
| 5 | Au | 31 | ·73 | 7 | S | 30 | ·26 | 1 | O | 29 | ·80 | 3 | N | 28ↄ | ·33 | | 4 | D | 27● | ·86 | 29 | 6 | Jr | 26 | ·39 | 7 | F | 24 | ·92 |
| 4 | S | 19 | ·63 | 6 | O | 19 | ·16 | 7 | N | 17○ | ·69 | 2 | D | 17 | ·22 | 30 | 3 | Jr | 15 | ·75 | | 5 | F | 14 | ·28 | 6 | Mr | 15 | ·81 |
| 2 | S | 9 | ·00 | 3 | O | 8 | ·53 | 5 | N | 7ↄ | ·06 | 6 | D | 6 | ·59 | 31 | 1 | Jr | 5 | ·12 | | 2 | F | 3 | ·65 | 4 | Mr | 5 | ·18 |
| {6 | Au | 29 | ·37 | 2 | O | 27 | ·43 | 3 | N | 25 | ·96 | 5 | D | 25 | ·49 | 32 | 7 | Jr | 24 | ·02 | | 1 | F | 22 | ·55 | 3 | Mr | 23○ | ·08 |
| {7 | S | 27 | ·90 | | | | | | | | | | | | | | | | | | | | | | | | | | |
| 5 | S | 16ↄ | ·26 | 6 | O | 15 | ·79 | 1 | N | 14 | ·32 | 2 | D | 13 | ·86 | 33 | 4 | Jr | 12 | ·39 | | 5 | F | 10 | ·92 | 7 | Mr | 12○ | ·45 |

| Kaliyuga. | Vikrama Era. | Saka Era. | Commencement of Solar Year. Month and day A.D. | Commencement of Solar Year. Fraction of day. | First New-Moon in Solar Year. | Anomaly of first New-Moon. | A.D. | Week-day of 1st January. | Vaisakha | | | | Jyeshta | | | | Ashada | | | | Sravana | | | | Bhadrapada | | | |
|---|---|---|---|---|---|---|---|---|---|---|---|---|---|---|---|---|---|---|---|---|---|---|---|---|---|---|---|---|
| | | | | | | | | ☉'s Anom col. 6 | | | | | + 29·53059 | | | | + 59·06117 | | | | + 88·59176 | | | | + 118·12235 | | | |
| | | | | | | | | ☾'s Anom col. 7 | | | | | + 1·976 | | | | + 3·952 | | | | + 5·928 | | | | + 7·904 | | | |
| | | | | | | | | | Week-day | Month | Day | Fraction | Week-day | Month | Day | Fraction | Week-day | Month | Day | Fraction | Week-day | Month | Day | Fraction | Week-day | Month | Day | Fraction |
| 4334 | 1290 | 1155 | 25M | ·0298 | 16·9481 | 13·796 | **1233** | 7 | 1 | Ap | 10 | ·98 | 3 | My | 10 | ·51 | 5 | Je | 9 | ·04 | 6 | Jl | 8 | ·57 | 1 | Au | 7 | ·10 |
| 4335 | 1291 | 1156 | 25M | ·2886 | 6·0564 | 9·953 | **1234** | 1 | 6 | Mr | 31 | ·34 | 7 | Ap | 29 | ·88 | 2 | My | 29 | ·41 | 3 | Je | 27 | ·94 | 7 | Au | 26●○ | ·00 |
| | | | | | | | | | | | | | | | | | | | | | 5 | Jl | 27 | ·47 | | | | |
| 4336 | 1292 | 1157 | 25M | ·5473 | 24·6953 | 8·087 | **1235** | 2 | 5 | Ap | 19 | ·24 | 6 | My | 18 | ·77 | 1 | Je | 17 | ·30 | 2 | Jl | 16 | ·83 | 4 | Au | 15● | ·36 |
| 4337 | 1293 | 1158 | 24M | ·8061 | 13·8036 | 4·244 | **1236** | 3 | 2 | Ap | 7 | ·61 | 4 | My | 7 | ·14 | 5 | Je | 5 | ·67 | 7 | Jl | 5○ | ·20 | 1 | Au | 3● | ·73 |
| 4338 | 1294 | 1159 | 25M | ·0649 | 2·9119 | 0·401 | **1237** | 5 | 6 | Mr | 27 | ·98 | 1 | Ap | 26 | ·51 | 4 | Je | 24○ | ·57 | 6 | Jl | 24 | ·10 | 7 | Au | 22 | ·63 |
| | | | | | | | | | | | | | 3 | My | 26 | ·04 | | | | | | | | | | | | |
| 4339 | 1295 | 1160 | 25M | ·3236 | 21·5508 | 26·089 | **1238** | 6 | 5 | Ap | 15 | ·87 | 7 | My | 15 | ·40 | 1 | Je | 13○ | ·94 | 3 | Jl | 13 | ·47 | 5 | Au | 12 | ·00 |
| 4340 | 1296 | 1161 | 25M | ·5824 | 10·6591 | 22·246 | **1239** | 7 | 3 | Ap | 5 | ·24 | 4 | My | 4 | ·77 | 6 | Je | 3● | ·30 | 7 | Jl | 2 | ·83 | 2 | Au | 1 | ·36 |
| 4341 | 1297 | 1162 | 24M | ·8411 | 29·2980 | 20·379 | **1240** | 1 | 7 | Mr | 24 | ·61 | 3 | My | 22● | ·67 | 5 | Je | 21 | ·20 | 6 | Jl | 20 | ·73 | 1 | Au | 19 | ·26 |
| | | | | | | | | | 2 | Ap | 23○ | ·14 | | | | | | | | | | | | | | | | |
| 4342 | 1298 | 1163 | 25M | ·0999 | 18·4063 | 16·537 | **1241** | 3 | 6 | Ap | 12○ | ·51 | 1 | My | 12 | ·04 | 2 | Je | 10 | ·57 | 4 | Jl | 10 | ·10 | 5 | Au | 8 | ·63 |
| 4343 | 1299 | 1164 | 25M | ·3586 | 7·5146 | 12·694 | **1242** | 4 | 3 | Ap | 1 | ·87 | 5 | My | 1 | ·40 | 6 | My | 30 | ·93 | 1 | Je | 29 | ·46 | 3 | Jl | 29 | ·00 |
| | | | | | | | | | | | | | | | | | | | | | | | | | 4 | Au | 27 | ·53 |
| 4344 | 1300 | 1165 | 25M | ·6174 | 26·1534 | 10·827 | **1243** | 5 | 2 | Ap | 20 | ·77 | 4 | My | 20 | ·30 | 5 | Je | 18 | ·83 | 7 | Jl | 18 | ·36 | 1 | Au | 16○ | ·89 |
| 4345 | 1301 | 1166 | 24M | ·8762 | 15·2617 | 6·984 | **1244** | 6 | 7 | Ap | 9 | ·14 | 1 | My | 8 | ·67 | 3 | Je | 7 | ·20 | 4 | Jl | 6 | ·73 | 6 | Au | 5○ | ·26 |
| 4346 | 1302 | 1167 | 25M | ·1349 | 4·3700 | 3·143 | **1245** | 1 | 4 | Mr | 29 | ·50 | 6 | Ap | 28 | ·03 | 7 | My | 27 | ·57 | 3 | Jl | 25●○ | ·63 | 5 | Au | 24 | ·16 |
| | | | | | | | | | | | | | | | | | 2 | Je | 26 | ·10 | | | | | | | | |
| 4347 | 1303 | 1168 | 25M | ·3937 | 23·0089 | 1·275 | **1246** | 2 | 3 | Ap | 17 | ·40 | 4 | My | 16 | ·93 | 6 | Je | 15 | ·46 | 7 | Jl | 14 | ·99 | 2 | Au | 13 | ·52 |
| 4348 | 1304 | 1169 | 25M | ·6524 | 12·1172 | 24·987 | **1247** | 3 | 7 | Ap | 6 | ·77 | 2 | My | 6 | ·30 | 3 | Je | 4○ | ·83 | 5 | Jl | 4● | ·36 | 6 | Au | 2 | ·89 |
| 4349 | 1305 | 1170 | 24M | ·9112 | 1·2255 | 21·144 | **1248** | 4 | 5 | Mr | 26 | ·14 | 6 | Ap | 24 | ·67 | 2 | Je | 22 | ·73 | 4 | Jl | 22 | ·26 | 5 | Au | 20 | ·79 |
| | | | | | | | | | | | | | 1 | My | 24●○ | ·20 | | | | | | | | | | | | |
| 4350 | 1306 | 1171 | 25M | ·1699 | 19·8644 | 19·277 | **1249** | 6 | 4 | Ap | 14 | ·03 | 5 | My | 13● | ·56 | 7 | Je | 12 | ·09 | 1 | Jl | 11 | ·63 | 3 | Au | 10 | ·16 |
| 4351 | 1307 | 1172 | 25M | ·4287 | 8·9727 | 15·434 | **1250** | 7 | 1 | Ap | 3○ | ·40 | 2 | My | 2● | ·93 | 4 | Je | 1 | ·46 | 5 | Je | 30 | ·99 | 7 | Jl | 30 | ·52 |
| 4352 | 1308 | 1173 | 25M | ·6875 | 27·6117 | 13·568 | **1251** | 1 | 7 | Ap | 22 | ·30 | 1 | My | 21 | ·83 | 3 | Je | 20 | ·36 | 4 | Jl | 19 | ·89 | 6 | Au | 18 | ·42 |
| 4353 | 1309 | 1174 | 24M | ·9462 | 16·7199 | 9·725 | **1252** | 2 | 4 | Ap | 10 | ·67 | 6 | My | 10 | ·20 | 7 | Je | 8 | ·73 | 2 | Jl | 8 | ·26 | 3 | Au | 6 | ·79 |
| 4354 | 1310 | 1175 | 25M | ·2050 | 5·8282 | 5·883 | **1253** | 4 | 2 | Mr | 31 | ·03 | 3 | Ap | 29 | ·56 | 5 | My | 29 | ·09 | 6 | Je | 27 | ·62 | 2 | Au | 25 | ·69 |
| | | | | | | | | | | | | | | | | | | | | | 1 | Jl | 27 | ·15 | | | | |
| 4355 | 1311 | 1176 | 25M | ·4637 | 24·4671 | 4·015 | **1254** | 5 | 7 | Ap | 18 | ·93 | 2 | My | 18 | ·46 | 3 | Je | 16 | ·99 | 5 | Jl | 16○ | ·52 | 6 | Au | 14 | ·05 |
| 4356 | 1312 | 1177 | 25M | ·7225 | 13·5754 | 0·173 | **1255** | 6 | 5 | Ap | 8 | ·30 | 6 | My | 7 | ·83 | 1 | Je | 6 | ·36 | 2 | Jl | 5○ | ·89 | 4 | Au | 4 | ·42 |
| 4357 | 1313 | 1178 | 24M | ·9812 | 2·6887 | 23·885 | **1256** | 7 | 2 | Mr | 27 | ·66 | 4 | Ap | 26 | ·19 | 7 | Je | 24●○ | ·26 | 1 | Jl | 23 | ·79 | 3 | Au | 22 | ·32 |
| | | | | | | | | | | | | | 5 | My | 25 | ·73 | | | | | | | | | | | | |
| 4358 | 1314 | 1179 | 25M | ·2400 | 21·3226 | 22·018 | **1257** | 2 | 1 | Ap | 15 | ·56 | 3 | My | 15 | ·09 | 4 | Je | 13 | ·62 | 6 | Jl | 13 | ·15 | 7 | Au | 11 | ·68 |
| 4359 | 1315 | 1180 | 25M | ·4987 | 10·4309 | 18·175 | **1258** | 3 | 5 | Ap | 4 | ·93 | 7 | My | 4○ | ·46 | 1 | Je | 2● | ·99 | 3 | Jl | 2 | ·52 | 5 | Au | 1 | ·05 |
| 4360 | 1316 | 1181 | 25M | ·7575 | 29·0698 | 16·308 | **1259** | 4 | 4 | Ap | 23○ | ·83 | 6 | My | 23 | ·36 | 7 | Je | 21 | ·89 | 2 | Jl | 21 | ·42 | 3 | Au | 19 | ·95 |
| 4361 | 1317 | 1182 | 25M | ·0163 | 18·1781 | 12·466 | **1260** | 5 | 2 | Ap | 12● | ·19 | 3 | My | 11 | ·72 | 5 | Je | 10 | ·25 | 6 | Jl | 9 | ·79 | 1 | Au | 8 | ·32 |
| 4362 | 1318 | 1183 | 25M | ·2750 | 7·2864 | 8·623 | **1261** | 7 | 6 | Ap | 1● | ·56 | 1 | My | 1 | ·09 | 2 | My | 30 | ·62 | 4 | Je | 29 | ·15 | 5 | Jl | 28 | ·68 |
| | | | | | | | | | | | | | | | | | | | | | | | | | 7 | Au | 27○ | ·21 |
| 4363 | 1319 | 1184 | 25M | ·5338 | 25·9252 | 6·756 | **1262** | 1 | 5 | Ap | 20 | ·46 | 6 | My | 19 | ·99 | 1 | Je | 18 | ·52 | 3 | Jl | 18 | ·05 | 4 | Au | 16●○ | ·58 |
| 4364 | 1320 | 1185 | 25M | ·7925 | 15·0835 | 2·913 | **1263** | 2 | 2 | Ap | 9 | ·83 | 4 | My | 9 | ·36 | 5 | Je | 7 | ·89 | 7 | Jl | 7 | ·42 | 1 | Au | 5○ | ·95 |
| 4365 | 1321 | 1186 | 25M | ·0513 | 4·1418 | 26·625 | **1264** | 3 | 7 | Mr | 29 | ·19 | 1 | Ap | 27 | ·72 | 3 | My | 27 | ·25 | 6 | Jl | 25 | ·31 | 7 | Au | 23 | ·85 |
| | | | | | | | | | | | | | | | | | 4 | Je | 25 | ·78 | | | | | | | | |
| 4366 | 1322 | 1187 | 25M | ·3100 | 22·7807 | 24·758 | **1265** | 5 | 6 | Ap | 17 | ·09 | 7 | My | 16 | ·62 | 2 | Je | 15○ | ·15 | 3 | Jl | 14 | ·68 | 5 | Au | 13 | ·21 |
| 4367 | 1323 | 1188 | 25M | ·5688 | 11·8890 | 20·916 | **1266** | 6 | 3 | Ap | 6 | ·46 | 4 | My | 5 | ·99 | 6 | Je | 4○ | ·52 | 1 | Jl | 4 | ·05 | 2 | Au | 2 | ·58 |
| 4368 | 1324 | 1189 | 25M | ·8276 | 0·9973 | 17·073 | **1267** | 7 | 7 | Mr | 26 | ·82 | 2 | Ap | 25 | ·35 | 5 | Je | 23 | ·42 | 6 | Jl | 22 | ·95 | 1 | Au | 21 | ·48 |
| | | | | | | | | | | | | | 3 | My | 24●○ | ·89 | | | | | | | | | | | | |
| 4369 | 1325 | 1190 | 25M | ·0863 | 19·6362 | 15·206 | **1268** | 1 | 6 | Ap | 13○ | ·72 | 1 | My | 13 | ·25 | 2 | Je | 11 | ·78 | 4 | Jl | 11 | ·31 | 5 | Au | 9 | ·84 |
| 4370 | 1326 | 1191 | 25M | ·3451 | 8·7445 | 11·363 | **1269** | 3 | 4 | Ap | 3○ | ·08 | 5 | My | 2 | ·62 | 7 | Je | 1 | ·15 | 1 | Je | 30 | ·68 | 3 | Jl | 30 | ·21 |
| 4371 | 1327 | 1192 | 25M | ·6038 | 27·3834 | 9·497 | **1270** | 4 | 2 | Ap | 21 | ·99 | 4 | My | 21 | ·52 | 6 | Je | 20 | ·05 | 7 | Jl | 19 | ·58 | 2 | Au | 18 | ·11 |
| 4372 | 1328 | 1193 | 25M | ·8626 | 16·4917 | 5·654 | **1271** | 5 | 7 | Ap | 11 | ·35 | 1 | My | 10 | ·88 | 3 | Je | 9 | ·41 | 4 | Jl | 8 | ·95 | 6 | Au | 7 | ·48 |

## urya Siddhanta.

| Week-day | Month | Day | Fraction | Week-day | Month | Day | Fraction | Week-day | Month | Day | Fraction | Week-day | Month | Day | Fraction | A.D. | Week-day | Month | Day | Fraction | A.D. | Week-day | Month | Day | Fraction | Week-day | Month | Day | Fraction |
|---|---|---|---|---|---|---|---|---|---|---|---|---|---|---|---|---|---|---|---|---|---|---|---|---|---|---|---|---|---|
| + 147·65293 | | | | + 177·18353 | | | | + 206·71411 | | | | + 236·24470 | | | | | + 265·77529 | | | | | + 295·30588 | | | | +324·83647 / +354·36705 | | | |
| + 9·880 | | | | + 11·856 | | | | + 13·832 | | | | + 15·808 | | | | | + 17·784 | | | | | + 19·760 | | | | + 21·736 / + 23·712 | | | |
| Asvina | | | | Kartika | | | | Margasira | | | | Pausha | | | | A.D. | Magha | | | | A.D. | Phalguna | | | | Chaitra | | | |
| 2 | S | 5○ | ·63 | 4 | O | 5● | ·16 | 5 | N | 3 | ·69 | 7 | D | 3 | ·22 | 34 | 1 | Jr | 1 | ·75 | | 3 | Jr | 31 | ·28 | 4 | Mr | 1○ | ·81 |
| 1 | S | 24 | ·53 | 3 | O | 24 | ·06 | 4 | N | 22 | ·59 | 6 | D | 22 | ·12 | 35 | 7 | Jr | 20 | ·65 | | 2 | F | 19● | ·18 | 3 | Mr | 20 | ·71 |
| 5 | S | 13 | ·90 | 7 | O | 13 | ·43 | 1 | N | 11 | ·96 | 3 | D | 11 | ·49 | 36 | 5 | Jr | 10ↄ | ·02 | | 6 | F | 8 | ·55 | 1 | Mr | 9 | ·08 |
| 3 | S | 2 | ·26 | 4 | O | 1 | ·79 | 6 | O | 31 | ·32 | 7 | N | 29 | ·85 | | 2 | D | 29○ | ·38 | 37 | 3 | Jr | 27 | ·92 | 5 | F | 26 | ·45 |
| 2 | S | 21 | ·16 | 3 | O | 20 | ·69 | 5 | N | 19 | ·22 | 6 | D | 18●○ | ·75 | 38 | 1 | Jr | 17 | ·28 | | 2 | F | 15 | ·81 | 4 | Mr | 17 | ·34 |
| 6 | S | 10 | ·53 | 1 | O | 10 | ·06 | 2 | N | 8 | ·59 | 4 | D | 8● | ·12 | 39 | 5 | Jr | 6 | ·65 | | 7 | F | 5 | ·18 | 1 | Mr | 6 | ·71 |
| 3 | Au | 30 | ·89 | 5 | S | 29 | ·42 | 6 | O | 28ↄ | ·95 | 1 | N | 27● | ·49 | | 3 | D | 27 | ·02 | 40 | 4 | Jr | 25 | ·55 | 6 | F | 24 | ·08 |
| 2 | S | 17 | ·79 | 4 | O | 17○ | ·32 | 5 | N | 15 | ·85 | 7 | D | 15 | ·38 | 41 | 1 | Jr | 13 | ·91 | | 3 | F | 12 | ·44 | 4 | Mr | 13 | ·97 |
| 7 | S | 7 | ·16 | 1 | O | 6●ↄ | ·69 | 3 | O | 5 | ·22 | 4 | D | 4 | ·75 | 42 | 6 | Jr | 3 | ·28 | | 7 | F | 1 | ·81 | 2 | Mr | 3 | ·34 |
| 6 | S | 26● | ·06 | 7 | O | 25 | ·59 | 2 | N | 24 | ·12 | 3 | D | 23 | ·65 | 43 | 5 | Jr | 22 | ·18 | | 6 | F | 20ↄ | ·71 | 1 | Mr | 22● | ·24 |
| 3 | S | 15 | ·42 | 4 | O | 14 | ·95 | 6 | N | 13 | ·48 | 1 | D | 13 | ·01 | 44 | 2 | Jr | 11 | ·55 | | 4 | F | 10ↄ | ·08 | 5 | Mr | 10 | ·61 |
| 7 | S | 3 | ·79 | 2 | O | 3 | ·32 | 3 | N | 1 | ·85 | 5 | D | 1 | ·38 | | 6 | D | 30 | ·91 | 45 | 1 | Jr | 29ↄ | ·44 | 2 | F | 27 | ·97 |
| 6 | S | 22 | 69 | 1 | O | 22 | ·22 | 2 | N | 20 | ·75 | 4 | D | 20 | ·28 | 46 | 5 | Jr | 18● | ·81 | | 7 | F | 17 | ·34 | 1 | Mr | 18 | ·87 |
| 4 | S | 12 | ·05 | 5 | O | 11 | ·59 | 7 | N | 10 | ·12 | 1 | D | 9ↄ | ·65 | 47 | 3 | Jr | 8 | ·18 | | 4 | F | 6 | ·71 | 6 | Mr | 8 | ·24 |
| 1 | S | 1 | ·42 | 2 | S | 30 | ·95 | 4 | O | 30 | ·48 | 6 | N | 29ↄ | ·01 | | 7 | D | 28 | ·54 | 48 | 2 | Jr | 27 | ·07 | 3 | F | 25 | ·61 |
| 7 | S | 19 | ·32 | 1 | O | 18 | ·85 | 3 | N | 17ↄ | ·38 | 4 | D | 16 | ·91 | 49 | 6 | Jr | 15 | ·44 | | 7 | F | 13 | ·97 | 2 | Mr | 15 | ·50 |
| 4 | S | 8 | ·69 | 6 | O | 8 | ·22 | 7 | N | 6● | ·75 | 2 | D | 6 | ·28 | 50 | 3 | Jr | 4 | ·81 | | 5 | F | 3 | ·34 | 6 | Mr | 4 | ·87 |
| 2 | Au | 29 | ·05 | 5 | O | 27 | ·11 | 6 | N | 25 | ·65 | 1 | D | 25 | ·18 | 51 | 2 | Jr | 23 | ·71 | | 4 | F | 22 | ·24 | 5 | Mr | 23ↄ | ·77 |
| 3 | S | 27○ | ·58 | | | | | | | | | | | | | | | | | | | | | | | | | | |
| 7 | S | 16○ | ·95 | 2 | O | 16 | ·48 | 4 | N | 15 | ·01 | 5 | D | 14 | ·54 | 52 | 7 | Jr | 13 | ·07 | | 1 | F | 11 | ·60 | 3 | Mr | 12ↄ | ·13 |
| 5 | S | 5○ | ·32 | 6 | O | 4 | ·85 | 1 | N | 3 | ·38 | 2 | D | 2 | ·91 | 53 | 4 | Jr | 1 | ·44 | | 6 | Jr | 31 | ·97 | 7 | Mr | 1● | ·50 |
| 4 | S | 24 | ·22 | 5 | O | 23 | ·75 | 7 | N | 22 | ·28 | 1 | D | 21 | ·81 | 54 | 3 | Jr | 20ↄ | ·34 | | 4 | F | 18 | ·87 | 6 | Mr | 20 | ·40 |
| 1 | S | 13 | ·58 | 3 | O | 13 | ·11 | 4 | N | 11 | ·64 | 6 | D | 11 | ·17 | 55 | 7 | Jr | 9●ↄ | ·71 | | 2 | F | 8 | ·24 | 3 | Mr | 9 | ·77 |
| 5 | S | 2 | ·95 | 7 | O | 2 | ·48 | 2 | N | 1 | ·01 | 3 | N | 30 | ·54 | | 5 | D | 30ↄ | ·07 | 56 | 6 | Jr | 28 | ·60 | 1 | F | 27 | ·13 |
| 4 | S | 20 | ·85 | 6 | O | 20 | ·38 | 7 | N | 18 | ·91 | 2 | D | 18 | ·44 | 57 | 3 | Jr | 16 | ·97 | | 5 | F | 15 | ·50 | 7 | Mr | 17 | ·03 |
| 2 | S | 10 | ·21 | 3 | O | 9 | ·75 | 5 | N | 8ↄ | ·28 | 6 | D | 7 | ·81 | 58 | 1 | Jr | 6 | ·34 | | 2 | F | 4 | ·87 | 4 | Mr | 6 | ·40 |
| 6 | Au | 30 | ·58 | 1 | S | 29 | ·11 | 4 | N | 27 | ·17 | Pausha | | | | | 5 | D | 26 | ·70 | 59 | 7 | Jr | 25 | ·23 | 1 | F | 23 | ·77 |
| | | | | 2 | O | 28ↄ | ·64 | | | | | Kshaya. | | | | | | | | | | | | | | 3 | Mr | 25 | ·30 |
| 5 | S | 18 | ·48 | 7 | O | 18○ | ·01 | 1 | N | 16 | ·54 | 3 | D | 16 | ·07 | 60 | 4 | Jr | 14 | ·60 | | 6 | F | 13 | ·13 | 7 | Mr | 13 | ·66 |
| 2 | S | 6 | ·85 | 4 | O | 6● | ·38 | 5 | N | 4 | ·91 | 7 | D | 4 | ·44 | 61 | 2 | Jr | 3 | ·97 | | 3 | F | 1 | ·50 | 5 | Mr | 3ↄ | ·03 |
| 1 | S | 25● | ·74 | 3 | O | 25 | ·27 | 4 | N | 23 | ·81 | 6 | D | 23 | ·34 | 62 | 7 | Jr | 21 | ·87 | | 2 | F | 20ↄ | ·40 | 3 | Mr | 21 | ·93 |
| 6 | S | 15 | ·11 | 7 | O | 14 | ·64 | 2 | N | 13 | ·17 | 3 | D | 12 | ·70 | 63 | 5 | Jr | 11 | ·23 | | 6 | F | 9ↄ | ·76 | 1 | Mr | 11 | ·29 |
| 3 | S | 4 | ·48 | 5 | O | 4 | ·01 | 6 | N | 2 | ·54 | 1 | D | 2 | ·07 | | 2 | D | 31 | ·60 | 64 | 4 | Jr | 30 | ·13 | 5 | F | 28 | ·66 |
| 2 | S | 22 | ·38 | 3 | O | 21 | ·91 | 5 | N | 20 | ·44 | 6 | D | 19ↄ | ·97 | 65 | 1 | Jr | 18● | ·50 | | 3 | F | 17 | ·03 | 4 | Mr | 18 | ·56 |
| 6 | S | 11 | ·74 | 1 | O | 11 | ·27 | 2 | N | 9 | ·80 | 4 | D | 9ↄ | ·33 | 66 | 5 | Jr | 7● | ·87 | | 7 | F | 6 | ·40 | 1 | Mr | 7 | ·93 |
| 4 | S | 1 | ·11 | 5 | S | 30 | ·64 | 7 | O | 30 | ·17 | 1 | N | 28○ | ·70 | | 3 | D | 28 | ·23 | 67 | 4 | Jr | 26 | ·76 | 6 | F | 25 | ·29 |
| 3 | S | 20 | ·01 | 4 | O | 19 | ·54 | 6 | N | 18 | ·07 | 7 | D | 17 | ·60 | 68 | 2 | Jr | 16 | ·13 | | 3 | F | 14 | ·66 | 5 | Mr | 15 | ·19 |
| 7 | S | 8 | ·37 | 1 | O | 7○ | ·91 | 3 | N | 6● | ·44 | 4 | D | 5 | ·97 | 69 | 6 | Jr | 4 | ·50 | | 1 | F | 3 | ·03 | 2 | Mr | 4 | ·56 |
| 4 | Au | 28 | ·74 | 7 | O | 26 | ·80 | 2 | N | 25 | ·33 | 3 | D | 24 | ·86 | 70 | 5 | Jr | 23 | ·39 | | 6 | F | 21 | ·93 | 1 | Mr | 23●ↄ | ·46 |
| 6 | S | 27○ | ·27 | | | | | | | | | | | | | | | | | | | | | | | | | | |
| 3 | S | 16○ | ·64 | 5 | O | 16 | ·17 | 6 | N | 14 | ·70 | 1 | D | 14 | ·23 | 71 | 2 | Jr | 12 | ·76 | | 4 | F | 11 | ·29 | 5 | Mr | 12 | ·82 |
| 1 | S | 6● | ·01 | 2 | O | 5 | ·54 | 4 | N | 4 | ·07 | 5 | D | 3 | ·60 | 72 | 7 | Jr | 2 | ·13 | | 1 | Jr | 31○ | ·66 | 3 | Mr | 1● | ·19 |

☉'s Anom col. 6 — Vaisakha: + 29·53059; Jyeshta: + 59·06117; Ashada: + 88·59176; Sravana: + 118·12235

☾'s Anom col. 7 — Vaisakha: + 1·976; Jyeshta: + 3·952; Ashada: + 5·928; Sravana: + 7·904

| Kaliyuga. | Vikrama Era. | Saka Era. | Commencement of Solar Year. Month and day A.D. | Commencement of Solar Year. Fraction of day. | First New-Moon in Solar Year. | Anomaly of first New-Moon. | A.D. | Week-day of 1st January. | Vaisakha Week-day | Vaisakha Month | Vaisakha Day | Vaisakha Fraction | Jyeshta Week-day | Jyeshta Month | Jyeshta Day | Jyeshta Fraction | Ashada Week-day | Ashada Month | Ashada Day | Ashada Fraction | Sravana Week-day | Sravana Month | Sravana Day | Sravana Fraction | Bhadrapada Week-Day | Bhadrapada Month | Bhadrapada Day | Bhadrapada Fraction |
|---|---|---|---|---|---|---|---|---|---|---|---|---|---|---|---|---|---|---|---|---|---|---|---|---|---|---|---|---|
| 4373 | 1329 | 1194 | 25M | ·1213 | 5·6000 | 1·811 | **1272** | 6 | 4 | Mr | 30 | ·72 | 6 | Ap | 29 | ·25 | 7 | My | 28 | ·78 | 3 | Jl | 26☽ | ·84 | 5 | Au | 25● | ·37 |
| | | | | | | | | | | | | | | | | | 2 | Je | 27 | ·31 | | | | | | | | |
| 4374 | 1330 | 1195 | 25M | ·3801 | 24·2389 | 27·499 | **1273** | 1 | 3 | Ap | 18 | ·62 | 5 | My | 18 | ·15 | 6 | Je | 16 | ·68 | 1 | Jl | 16☽ | ·21 | 2 | Au | 14 | ·74 |
| 4375 | 1331 | 1196 | 25M | ·6389 | 13·3472 | 23·656 | **1274** | 2 | 7 | Ap | 7 | ·99 | 2 | My | 7 | ·52 | 4 | Je | 6 | ·05 | 5 | Jl | 5●○ | ·58 | 7 | Au | 4 | ·11 |
| 4376 | 1332 | 1197 | 25M | ·8976 | 2·4555 | 19·814 | **1275** | 3 | 5 | Mr | 28 | ·35 | 6 | Ap | 26 | ·88 | 2 | Je | 24● | ·94 | 4 | Jl | 24 | ·47 | 6 | Au | 23 | ·01 |
| | | | | | | | | | | | | | 1 | My | 26 | ·41 | | | | | | | | | | | | |
| 4377 | 1333 | 1198 | 25M | ·1564 | 21·0944 | 17·947 | **1276** | 4 | 4 | Ap | 15 | ·25 | 5 | My | 14○ | ·78 | 7 | Je | 13 | ·31 | 1 | Jl | 12 | ·84 | 3 | Au | 11 | ·37 |
| 4378 | 1334 | 1199 | 25M | ·4151 | 10·2027 | 14·104 | **1277** | 6 | 1 | Ap | 4 | ·62 | 3 | My | 4☽ | ·15 | 4 | Je | 2 | ·68 | 6 | Jl | 2 | ·21 | 7 | Jl | 31 | ·74 |
| 4379 | 1335 | 1200 | 25M | ·6739 | 28·8416 | 12·237 | **1278** | 7 | 7 | Ap | 23 | ·51 | 2 | My | 23 | ·05 | 3 | Je | 21 | ·58 | 5 | Jl | 21 | ·11 | 6 | Au | 19 | ·64 |
| 4380 | 1336 | 1201 | 25M | ·9336 | 17·9498 | 8·395 | **1279** | 1 | 4 | Ap | 12 | ·88 | 6 | My | 12 | ·41 | 7 | Je | 10 | ·94 | 2 | Jl | 10 | ·47 | 4 | Au | 9 | ·00 |
| 4381 | 1337 | 1202 | 25M | ·1914 | 7·0581 | 4·552 | **1280** | 2 | 2 | Ap | 1● | ·25 | 3 | Ap | 30 | ·78 | 5 | My | 30 | ·31 | 6 | Je | 28 | ·84 | 2 | Au | 26☽ | ·90 |
| | | | | | | | | | | | | | | | | | | | | | 1 | Jl | 28 | ·37 | | | | |
| 4382 | 1338 | 1203 | 25M | ·4501 | 25·6970 | 2·685 | **1281** | 4 | 1 | Ap | 20 | ·15 | 2 | My | 19 | ·68 | 4 | Je | 18 | ·21 | 5 | Jl | 17 | ·74 | 7 | Au | 16☽ | ·27 |
| 4383 | 1339 | 1204 | 25M | ·7089 | 14·8053 | 26·397 | **1282** | 5 | 5 | Ap | 9 | ·51 | 7 | My | 9 | ·04 | 1 | Je | 7 | ·57 | 3 | Jl | 7 | ·11 | 4 | Au | 5● | ·64 |
| 4384 | 1340 | 1205 | 25M | ·9677 | 3·9136 | 22·554 | **1283** | 6 | 2 | Mr | 29 | ·88 | 4 | Ap | 28 | ·41 | 5 | My | 27 | ·94 | 2 | Jl | 26 | ·00 | 3 | Au | 24 | ·53 |
| | | | | | | | | | | | | | | | | | 7 | Je | 26☽ | ·47 | | | | | | | | |
| 4385 | 1341 | 1206 | 25M | ·2264 | 22·5525 | 20·688 | **1284** | 7 | 1 | Ap | 16 | ·78 | 3 | My | 16 | ·31 | 4 | Je | 14●☽ | ·84 | 6 | Jl | 14 | ·37 | 7 | Au | 12 | ·90 |
| 4386 | 1342 | 1207 | 25M | ·4852 | 11·6608 | 16·845 | **1285** | 2 | 6 | Ap | 6 | ·15 | 7 | My | 5 | ·68 | 2 | Je | 4☽ | ·21 | 3 | Jl | 3 | ·74 | 5 | Au | 2 | ·27 |
| 4387 | 1343 | 1208 | 25M | ·7439 | 0·7691 | 13·002 | **1286** | 3 | 3 | Mr | 26 | ·51 | 6 | My | 24 | ·57 | 1 | Je | 23 | ·10 | 2 | Jl | 22 | ·63 | 4 | Au | 21 | ·17 |
| | | | | | | | | | 5 | Ap | 25☽ | ·04 | | | | | | | | | | | | | | | | |
| 4388 | 1344 | 1209 | 26M | ·0027 | 19·4080 | 11·135 | **1287** | 4 | 2 | Ap | 14○ | ·41 | 3 | My | 13 | ·94 | 5 | Je | 12 | ·47 | 7 | Jl | 12 | ·00 | 1 | Au | 10 | ·53 |
| 4389 | 1345 | 1210 | 25M | ·2614 | 8·5163 | 7·292 | **1288** | 5 | 6 | Ap | 2☽ | ·78 | 1 | My | 2 | ·31 | 2 | My | 31 | ·84 | 4 | Je | 30 | ·37 | 5 | Jl | 29 | ·90 |
| | | | | | | | | | | | | | | | | | | | | | | | | | 7 | Au | 28 | ·43 |
| 4390 | 1346 | 1211 | 25M | ·5202 | 27·1552 | 5·426 | **1289** | 7 | 5 | Ap | 21 | ·67 | 7 | My | 21 | ·21 | 1 | Je | 19 | ·74 | 3 | Jl | 19 | ·27 | 4 | Au | 17 | ·80 |
| 4391 | 1347 | 1212 | 25M | ·7790 | 16·2635 | 1·583 | **1290** | 1 | 3 | Ap | 11 | ·04 | 4 | My | 10 | ·57 | 6 | Je | 9 | ·10 | 7 | Jl | 8 | ·63 | 2 | Au | 7☽ | ·16 |
| 4392 | 1348 | 1213 | 26M | ·0377 | 5·3718 | 25·295 | **1291** | 2 | 7 | Mr | 31 | ·41 | 2 | Ap | 30 | ·94 | 3 | My | 29 | ·47 | 6 | Jl | 27☽ | ·53 | 1 | Au | 26● | ·06 |
| | | | | | | | | | | | | | | | | | 5 | Je | 28 | ·00 | | | | | | | | |
| 4393 | 1349 | 1214 | 25M | ·2965 | 24·0107 | 23·428 | **1292** | 3 | 6 | Ap | 18 | ·31 | 7 | My | 17 | ·84 | 2 | Je | 16 | ·37 | 3 | Jl | 15☽ | ·90 | 5 | Au | 14 | ·43 |
| 4394 | 1350 | 1215 | 25M | ·5552 | 13·1190 | 19·585 | **1293** | 5 | 3 | Ap | 7 | ·67 | 5 | My | 7 | ·20 | 6 | Je | 5 | ·73 | 1 | Jl | 5● | ·27 | 2 | Au | 3 | ·80 |
| 4395 | 1351 | 1216 | 25M | ·8140 | 2·2273 | 15·744 | **1294** | 6 | 1 | Mr | 28 | ·04 | 2 | Ap | 26 | ·57 | 5 | Je | 24● | ·63 | 7 | Jl | 24 | ·16 | 1 | Au | 22 | ·69 |
| | | | | | | | | | | | | | 4 | My | 26☽ | ·10 | | | | | | | | | | | | |
| 4396 | 1352 | 1217 | 26M | ·0727 | 20·8662 | 13·876 | **1295** | 7 | 6 | Ap | 15 | ·94 | 1 | My | 15○ | ·47 | 3 | Je | 14 | ·00 | 4 | Jl | 13 | ·53 | 6 | Au | 12 | ·06 |
| 4397 | 1353 | 1218 | 25M | ·3315 | 9·9745 | 10·033 | **1296** | 1 | 4 | Ap | 4 | ·31 | 5 | My | 3○ | ·84 | 7 | Je | 2 | ·37 | 1 | Jl | 1 | ·90 | 3 | Jl | 31 | ·43 |
| 4398 | 1354 | 1219 | 25M | ·5902 | 28·6133 | 8·166 | **1297** | 3 | 3 | Ap | 23● | ·20 | 4 | My | 22 | ·73 | 6 | Je | 21 | ·26 | 7 | Jl | 20 | ·79 | 2 | Au | 19 | ·33 |
| 4399 | 1355 | 1220 | 25M | ·8490 | 17·7216 | 4·324 | **1298** | 4 | 7 | Ap | 12 | ·57 | 2 | My | 12 | ·10 | 3 | Je | 10 | ·63 | 5 | Jl | 10 | ·16 | 6 | Au | 8 | ·69 |
| 4400 | 1356 | 1221 | 26M | ·1078 | 6·8299 | 0·481 | **1299** | 5 | 4 | Ap | 1 | ·94 | 6 | My | 1 | ·47 | 1 | My | 31 | ·00 | 2 | Je | 29 | ·53 | 5 | Au | 27●○ | ·59 |
| | | | | | | | | | | | | | | | | | | | | | 4 | Jl | 29 | ·06 | | | | |
| 4401 | 1357 | 1222 | 25M | ·3665 | 25·4688 | 26·169 | **1300** | 6 | 3 | Ap | 19 | ·83 | 5 | My | 19 | ·37 | 6 | Je | 17 | ·90 | 1 | Jl | 17 | ·43 | 2 | Au | 15● | ·96 |
| 4402 | 1358 | 1223 | 25M | ·6253 | 14·5771 | 22·326 | **1301** | 1 | 1 | Ap | 9 | ·20 | 2 | My | 8 | ·73 | 4 | Je | 7 | ·26 | 5 | Jl | 6○ | ·79 | 7 | Au | 5● | ·32 |
| 4403 | 1359 | 1224 | 25M | ·8840 | 3·6854 | 18·483 | **1302** | 2 | 5 | Mr | 29 | ·57 | 7 | Ap | 28 | ·10 | 1 | My | 27 | ·63 | 4 | Jl | 25 | ·69 | 6 | Au | 24 | ·22 |
| | | | | | | | | | | | | | | | | | 3 | Je | 26●○ | ·16 | | | | | | | | |
| 4404 | 1360 | 1225 | 26M | ·1428 | 22·3243 | 16·616 | **1303** | 3 | 4 | Ap | 17 | ·47 | 6 | My | 17 | ·00 | 7 | Je | 15●○ | ·53 | 2 | Jl | 15 | ·06 | 3 | Au | 13 | ·59 |
| 4405 | 1361 | 1226 | 25M | ·4015 | 11·4326 | 12·774 | **1304** | 4 | 1 | Ap | 5 | ·83 | 3 | My | 5○ | ·36 | 4 | Je | 3● | ·90 | 6 | Jl | 3 | ·43 | 7 | Au | 1 | ·96 |
| 4406 | 1362 | 1227 | 25M | ·6603 | 0·5409 | 8·931 | **1305** | 6 | 6 | Mr | 26 | ·20 | 2 | My | 24 | ·26 | 3 | Je | 22 | ·79 | 5 | Jl | 23 | ·32 | 6 | Au | 20 | ·85 |
| | | | | | | | | | 7 | Ap | 24○ | ·73 | | | | | | | | | | | | | | | | |
| 4407 | 1363 | 1228 | 25M | ·9191 | 19·1798 | 7·064 | **1306** | 7 | 5 | Ap | 14○ | ·10 | 6 | My | 13 | ·63 | 1 | Je | 12 | ·16 | 2 | Jl | 11 | ·69 | 4 | Au | 10 | ·22 |
| 4408 | 1364 | 1229 | 26M | ·1778 | 8·2881 | 3·221 | **1307** | 1 | 2 | Ap | 3● | ·47 | 4 | My | 3 | ·00 | 5 | Je | 1 | ·53 | 7 | Jl | 1 | ·06 | 1 | Jl | 30 | ·59 |
| | | | | | | | | | | | | | | | | | | | | | | | | | 3 | Au | 29 | ·12 |
| 4409 | 1365 | 1230 | 25M | ·4366 | 26·9270 | 1·355 | **1308** | 2 | 1 | Ap | 21 | ·36 | 2 | My | 20 | ·89 | 4 | Je | 19 | ·43 | 5 | Jl | 18 | ·96 | 7 | Au | 17☽ | 49 |
| 4410 | 1366 | 1231 | 25M | ·6953 | 16·0353 | 25·067 | **1309** | 4 | 5 | Ap | 10 | ·73 | 7 | My | 10 | ·26 | 1 | Je | 8 | ·79 | 3 | Jl | 8 | ·32 | 4 | Au | 6☽ | ·85 |
| 4411 | 1367 | 1232 | 25M | ·9541 | 5·1436 | 21·224 | **1310** | 5 | 3 | Mr | 31 | ·10 | 4 | Ap | 29 | ·63 | 6 | My | 29 | ·16 | 2 | Jl | 27●○ | ·22 | 3 | Au | 25 | ·75 |
| | | | | | | | | | | | | | | | | | 7 | Je | 27 | ·69 | | | | | | | | |

| | | | | | | | | | | | | | | | | | | | | | | | | | | | | | |
|---|---|---|---|---|---|---|---|---|---|---|---|---|---|---|---|---|---|---|---|---|---|---|---|---|---|---|---|---|---|
| + 147·65293 | | | | + 177·18353 | | | | + 206·71411 | | | | + 236·24470 | | | | | + 265·77529 | | | | | + 295·30588 | | | | {+324·83647 +354·36705 | | | |
| + 9·880 | | | | + 11·856 | | | | + 13·832 | | | | + 15·808 | | | | | + 17·784 | | | | | + 19·760 | | | | {+ 21·736 + 23·712 | | | |
| **Asvina** | | | | **Kartika** | | | | **Margasira** | | | | **Pausha** | | | | **A.D.** | **Magha** | | | | **A.D.** | **Phalguna** | | | | **Chaitra** | | | |
| Week-day | Month | Day | Fraction | Week-day | Month | Day | Fraction | Week-day | Month | Day | Fraction | Week-day | Month | Day | Fraction | | Week-day | Month | Day | Fraction | | Week-day | Month | Day | Fraction | Week-day | Month | Day | Fraction |
| 6 | S | 23 | ·90 | 1 | O | 23 | ·43 | 2 | N | 21 | ·97 | 4 | D | 21 | ·50 | **73** | 6 | Jr | 20ↄ | ·03 | | 7 | F | 18 | ·56 | 2 | Mr | 20 | ·09 |
| 4 | S | 13 | ·27 | 5 | O | 12 | ·80 | 7 | N | 11 | ·33 | 1 | D | 10 | ·86 | **74** | 3 | Jr | 9ↄ | ·39 | | 4 | F | 7 | ·92 | 6 | Mr | 9 | ·45 |
| 1 | S | 2 | ·64 | 3 | O | 2 | ·17 | 4 | O | 31 | ·70 | 6 | N | 30 | ·23 | | 7 | D | 29 | ·76 | **75** | 2 | Jr | 28 | ·29 | 3 | F | 26 | ·82 |
| 7 | S | 21 | ·54 | 2 | O | 21 | ·07 | 3 | N | 19ↄ | ·60 | 5 | D | 19 | ·13 | **76** | 6 | Jr | 17 | ·66 | | 1 | F | 16 | ·19 | 2 | Mr | 16 | ·72 |
| 4 | S | 9 | ·90 | 6 | O | 9 | ·43 | 7 | N | 7ↄ | ·96 | 2 | D | 7 | ·49 | **77** | 4 | Jr | 6 | ·03 | | 5 | F | 4 | ·56 | 7 | Mr | 6 | ·09 |
| 2 | Au | 30 | ·27 | {3 | S | 28 | ·80 | 6 | N | 26 | ·86 | **Pausha Kshaya.** | | | | | 1 | D | 26 | 39 | **78** | {2 | Jr | 24 | ·92 | 5 | Mr | 24 | ·98 |
| | | | | {5 | O | 28●ↄ | ·33 | | | | | | | | | | | | | | | {4 | F | 23 | ·45 | | | | |
| 1 | S | 18 | ·17 | 2 | O | 17 | ·70 | 4 | N | 16 | ·23 | 5 | D | 15 | ·76 | **79** | 7 | Jr | 14 | ·29 | | 1 | F | 12 | ·82 | 3 | Mr | 14ↄ | ·35 |
| 5 | S | 7ↄ | ·53 | 6 | O | 7 | ·07 | 1 | N | 5 | ·60 | 3 | D | 5 | ·13 | **80** | 4 | Jr | 3 | ·66 | | 6 | F | 2 | ·19 | 7 | Mr | 2ↄ | ·72 |
| 4 | S | 25 | ·43 | 5 | O | 24 | ·96 | 7 | N | 23 | ·49 | 2 | D | 23 | ·02 | **81** | 3 | Jr | 21 | ·55 | | 5 | F | 20●ↄ | ·09 | 6 | Mr | 21 | ·62 |
| 1 | S | 14 | ·80 | 3 | O | 14 | ·33 | 4 | N | 12 | ·86 | 3 | D | 12 | ·39 | **82** | 7 | Jr | 10 | ·92 | | 2 | F | 9● | ·45 | 3 | Mr | 10 | ·98 |
| 6 | S | 4 | ·17 | 7 | O | 3 | ·70 | 2 | N | 2 | ·23 | · | D | 1 | ·76 | | 5 | D | 31○ | ·29 | **83** | 6 | Jr | 29● | ·82 | 1 | F | 28 | ·35 |
| 5 | S | 23 | ·06 | 6 | O | 22 | ·59 | 1 | N | 21 | ·13 | 2 | D | 20ↄ | ·66 | **84** | 4 | Jr | 19 | ·19 | | 5 | F | 17 | ·72 | 7 | Mr | 18 | ·25 |
| 2 | S | 11 | ·43 | 3 | O | 10 | ·96 | 5 | N | 9 | ·49 | 7 | D | 9ↄ | ·02 | **85** | 1 | Jr | 7 | ·55 | | 3 | F | 6 | ·08 | 4 | Mr | 7 | ·61 |
| 6 | Au | 31 | ·80 | 1 | S | 30 | ·33 | 2 | O | 29 | ·86 | 4 | N | 28● | ·39 | | 5 | D | 27 | ·92 | **86** | 7 | Jr | 26 | ·45 | 1 | F | 24 | ·98 |
| 5 | S | 19 | ·70 | 7 | O | 19ↄ | ·23 | 1 | N | 17 | ·76 | 3 | D | 17 | ·29 | **87** | 4 | Jr | 15 | ·82 | | 6 | F | 14 | ·35 | 7 | Mr | 15 | ·88 |
| 3 | S | 9 | ·06 | 5 | O | 8ↄ | ·59 | 6 | N | 7● | ·12 | 7 | D | 6 | ·65 | **88** | 2 | Jr | 5 | ·19 | | 3 | F | 3 | ·72 | 5 | Mr | 4 | ·25 |
| 1 | S | 26ↄ | ·96 | 3 | O | 26 | ·49 | 5 | N | 25 | ·02 | 6 | D | 24 | ·55 | **89** | 1 | Jr | 23 | ·08 | | 2 | F | 21 | ·61 | 4 | Mr | 23● | ·14 |
| 6 | S | 16● | ·33 | 7 | O | 15 | ·86 | 2 | N | 14 | ·39 | 3 | D | 13 | ·92 | **90** | 5 | Jr | 12 | ·45 | | 6 | F | 10ↄ | ·98 | 1 | M | 12 | ·51 |
| 3 | S | 5● | ·69 | 5 | O | 5 | ·23 | 6 | N | 3 | ·76 | 1 | D | 3 | ·29 | **91** | 2 | Jr | 1 | ·82 | | 4 | Jr | 31ↄ | ·35 | 5 | Mr | 1 | ·88 |
| 2 | S | 24 | ·59 | 4 | O | 24 | ·12 | 5 | N | 22 | ·65 | 7 | D | 22 | ·18 | **92** | 1 | Jr | 20●ↄ | ·71 | | 3 | F | 19 | ·25 | 4 | Mr | 19 | ·78 |
| 6 | S | 12 | ·96 | 1 | O | 12 | ·49 | 3 | N | 11 | ·02 | 4 | D | 10 | ·55 | **93** | 6 | Jr | 9● | ·08 | | 7 | F | 7 | ·61 | 2 | Mr | 9 | ·14 |
| 4 | S | 2 | ·33 | 5 | O | 1 | ·86 | 7 | O | 31 | ·39 | 1 | N | 29ↄ | ·92 | | 3 | D | 29● | ·45 | **94** | 4 | Jr | 27 | ·98 | 6 | F | 26 | ·51 |
| 3 | S | 21 | ·22 | 4 | O | 20 | ·75 | 6 | N | 19ↄ | ·29 | 7 | D | 18 | ·82 | **95** | 2 | Jr | 17 | ·35 | | 3 | F | 15 | ·88 | 5 | Mr | 17 | ·41 |
| 7 | S | 10 | ·59 | 2 | O | 10 | ·12 | 3 | N | 8ↄ | ·65 | 5 | D | 8 | ·18 | **96** | 6 | Jr | 6 | ·71 | | 1 | F | 5 | ·24 | 2 | Mr | 5 | ·77 |
| 4 | Au | 29 | ·96 | 6 | S | 28 | ·49 | {1 | O | 28● | ·02 | **Pausha Kshaya.** | | | | | 4 | D | 26 | ·08 | **97** | {5 | Jr | 24 | ·61 | 1 | M. | 24○ | ·67 |
| | | | | | | | | {2 | N | 26 | ·55 | | | | | | | | | | | {7 | F | 23 | ·14 | | | | |
| 3 | S | 17○ | ·86 | 5 | O | 17 | ·39 | 6 | N | 15 | ·92 | 1 | D | 15 | ·45 | **98** | 2 | Jr | 13 | ·98 | | 4 | F | 12 | ·51 | 6 | Mr | 14○ | ·04 |
| 1 | S | 7○ | ·22 | 2 | O | 6 | ·75 | 4 | N | 5 | ·28 | 5 | D | 5 | ·81 | **99** | 7 | Jr | 3 | ·35 | | 1 | F | 1 | ·88 | 3 | Mr | 3○ | ·41 |
| 7 | S | 26 | ·12 | 1 | O | 25 | ·65 | 3 | N | 24 | ·18 | 4 | D | 23 | ·71 | **13** | 6 | Jr | 22 | ·24 | **00** | 7 | F | 20● | ·77 | 2 | Mr | 21 | ·30 |
| 4 | S | 14 | ·49 | 6 | O | 14 | ·02 | 7 | N | 12 | ·55 | 2 | D | 12 | ·08 | **01** | 3 | Jr | 10○ | ·61 | | 5 | F | 9 | ·14 | 6 | Mr | 10 | ·67 |
| 1 | S | 3 | ·85 | 3 | O | 3 | ·39 | 4 | N | 1 | ·92 | 6 | D | 1 | ·45 | | 7 | D | 30ↄ | ·98 | **02** | 2 | Jr | 29 | ·51 | 4 | F | 28 | ·04 |
| 7 | S | 22 | ·75 | 2 | O | 22 | ·28 | 3 | N | 20 | ·81 | 5 | D | 20ↄ | ·34 | **03** | 6 | Jr | 18 | ·87 | | 1 | F | 17 | ·41 | 2 | Mr | 18 | ·94 |
| 5 | S | 12 | ·12 | 6 | O | 11 | ·65 | 1 | N | 10 | ·18 | 2 | D | 9● | ·71 | **04** | 4 | Jr | 8 | ·24 | | 5 | F | 6 | ·77 | 7 | Mr | 7 | ·30 |
| 2 | Au | 31 | ·49 | 4 | S | 30 | ·02 | 5 | O | 29ↄ | ·55 | 7 | N | 28● | ·08 | | 1 | D | 27 | ·61 | **05** | 3 | Jr | 26 | ·14 | 4 | F | 24 | ·67 |
| 1 | S | 19 | ·38 | 2 | O | 18ↄ | ·92 | 4 | N | 17 | ·45 | 5 | D | 16 | ·98 | **06** | 7 | Jr | 15 | ·51 | | 2 | F | 14 | ·04 | 3 | Mr | 15 | ·57 |
| 5 | S | 8 | ·75 | 7 | O | 8ↄ | ·28 | 1 | N | 6 | ·81 | 3 | D | 6 | ·34 | **07** | 4 | Jr | 4 | ·87 | | 6 | F | 3 | ·40 | 7 | Mr | 4 | ·94 |
| 4 | S | 27 | ·65 | 6 | O | 27 | ·18 | 7 | N | 25 | ·71 | 2 | D | 25 | ·24 | **08** | 3 | Jr | 23 | ·77 | | 5 | F | 22ↄ | ·30 | 6 | Mr | 22 | ·83 |
| 2 | S | 16 | ·02 | 3 | O | 15 | ·55 | 5 | N | 14 | ·08 | 6 | D | 13 | ·61 | **09** | 1 | Jr | 12 | ·14 | | 2 | F | 10ↄ | ·67 | 4 | Mr | 12 | ·20 |
| 6 | S | 5 | ·38 | 7 | O | 4 | ·91 | 2 | N | 3 | ·44 | 3 | D | 2 | ·98 | **10** | 5 | Jr | 1 | ·51 | | 7 | Jr | 31ↄ | ·04 | 1 | Mr | 1 | ·57 |
| 5 | S | 24 | ·28 | 6 | O | 23 | ·81 | 1 | N | 22 | ·34 | 2 | D | 21 | ·87 | **11** | 4 | Jr | 20 | ·40 | | 5 | F | 18 | ·93 | 7 | Mr | 20 | ·47 |

# TABLE X–

☉'s Anom col. 6 + 29·53059 + 59·06117 + 88·59176 + 118·12235

☾'s Anom col. 7 + 1·976 + 3·952 + 5·928 + 7·904

| Kaliyuga. | Vikrama Era. | Saka Era. | Commencement of Solar Year. Month and day A.D. | Fraction of day. | First New-Moon in Solar Year. | Anomaly of first New-Moon. | A.D. | Week-day of 1st January. | Vaisakha Week-day | Month | Day | Fraction | Jyeshta Week-day | Month | Day | Fraction | Ashada Week-day | Month | Day | Fraction | Sravana Week-day | Month | Day | Fraction | Bhadrapada Week-day | Month | Day | Fraction |
|---|---|---|---|---|---|---|---|---|---|---|---|---|---|---|---|---|---|---|---|---|---|---|---|---|---|---|---|---|
| 4412 | 1368 | 1233 | 26M | ·2128 | 23·7825 | 19·357 | **1311** | 6 | 2 | Ap | 19 | ·00 | 3 | My | 18 | ·53 | 5 | Je | 17 | ·06 | 6 | Jl | 16 | ·59 | 1 | Au | 15 | ·12 |
| 4413 | 1369 | 1234 | 25M | ·4716 | 12·8908 | 15·514 | **1312** | 7 | 6 | Ap | 7 | ·36 | 7 | My | 6 | ·89 | 2 | Je | 5○ | ·42 | 3 | Jl | 4● | ·96 | 5 | Au | 3 | ·49 |
| 4414 | 1370 | 1235 | 25M | ·7304 | 1·9991 | 11·672 | **1313** | 2 | 3 | Mr | 27 | ·73 | 5 | Ap | 26 | ·26 | 1 | Je | 24 | ·32 | 2 | Jl | 23 | ·85 | 4 | Au | 22 | ·38 |
| | | | | | | | | | | | | | 6 | My | 25○ | ·79 | | | | | | | | | | | | |
| 4415 | 1371 | 1236 | 25M | ·9892 | 20·6380 | 9·805 | **1314** | 3 | 2 | Ap | 15 | ·63 | 4 | My | 15●○ | ·16 | 5 | Je | 13 | ·69 | 7 | Jl | 13 | ·22 | 1 | Au | 11 | ·75 |
| 4416 | 1372 | 1237 | 26M | ·2479 | 9·7463 | 5·962 | **1315** | 4 | 7 | Ap | 5○ | ·00 | 1 | My | 4● | ·53 | 3 | Je | 3 | ·06 | 4 | Jl | 2 | ·59 | 6 | Au | 1 | ·12 |
| 4417 | 1373 | 1238 | 25M | ·5066 | 28·3851 | 4·095 | **1316** | 5 | 5 | Ap | 22 | ·89 | 7 | My | 22 | ·42 | 1 | Je | 20 | ·95 | 3 | Jl | 20 | ·48 | 5 | Au | 19 | ·02 |
| 4418 | 1374 | 1239 | 25M | ·7654 | 17·4934 | 0·254 | **1317** | 7 | 3 | Ap | 12 | ·26 | 4 | My | 11 | ·79 | 6 | Je | 10 | ·32 | 7 | Jl | 9 | ·85 | 2 | Au | 8 | ·38 |
| 4419 | 1375 | 1240 | 26M | ·0241 | 6·6017 | 23·964 | **1318** | 1 | 7 | Ap | 1 | ·63 | 2 | My | 1 | ·16 | 3 | My | 30 | ·69 | 5 | Je | 29 | ·22 | 1 | Au | 27 | ·28 |
| | | | | | | | | | | | | | | | | | | | | | 6 | Jl | 28 | ·75 | | | | |
| 4420 | 1376 | 1241 | 26M | ·2829 | 25·2406 | 22·098 | **1319** | 2 | 6 | Ap | 20 | ·53 | 1 | My | 20 | ·06 | 2 | Je | 18 | ·59 | 4 | Jl | 18○ | ·12 | 5 | Au | 16● | ·65 |
| 4421 | 1377 | 1242 | 25M | ·5416 | 14·3489 | 18·255 | **1320** | 3 | 3 | Ap | 8 | ·89 | 5 | My | 8 | ·42 | 6 | Je | 6 | ·95 | 1 | Jl | 6○ | ·48 | 3 | Au | 5 | ·01 |
| 4422 | 1378 | 1243 | 25M | ·8004 | 3·4572 | 14·412 | **1321** | 5 | 1 | Mr | 29 | ·26 | 2 | Ap | 27 | ·79 | 4 | My | 27 | ·32 | 7 | Jl | 25 | ·38 | 1 | Au | 23 | ·91 |
| | | | | | | | | | | | | | | | | | 5 | Je | 25●○ | ·85 | | | | | | | | |
| 4423 | 1379 | 1244 | 26M | ·0591 | 22·0961 | 12·545 | **1322** | 6 | 7 | Ap | 17 | ·16 | 1 | My | 16 | ·69 | 3 | Je | 15 | ·22 | 4 | Jl | 14 | ·75 | 6 | Au | 13 | ·28 |
| 4424 | 1380 | 1245 | 26M | ·3179 | 11·2044 | 8·702 | **1323** | 7 | 4 | Ap | 6 | ·52 | 6 | My | 6○ | ·06 | 7 | Je | 4 | ·59 | 2 | Jl | 4 | ·12 | 3 | Au | 2 | ·65 |
| 4425 | 1381 | 1246 | 25M | ·5767 | 0·3127 | 4·860 | **1324** | 1 | 1 | Mr | 25 | ·89 | 4 | My | 23 | ·95 | 6 | Je | 22 | ·48 | 1 | Jl | 22 | ·01 | 2 | Au | 20 | ·54 |
| | | | | | | | | | 3 | Ap | 24●○ | ·42 | | | | | | | | | | | | | | | | |
| 4426 | 1382 | 1247 | 25M | ·8354 | 18·9516 | 2·994 | **1325** | 3 | 7 | Ap | 13 | ·79 | 2 | My | 13 | ·32 | 3 | Je | 11 | ·85 | 5 | Jl | 11 | ·38 | 6 | Au | 9 | ·91 |
| 4427 | 1383 | 1248 | 26M | ·0942 | 8·0599 | 26·705 | **1326** | 4 | 5 | Ap | 3● | ·16 | 6 | My | 2 | ·69 | 1 | Je | 1 | ·22 | 2 | Je | 30 | ·75 | 4 | Jl | 30 | ·28 |
| | | | | | | | | | | | | | | | | | | | | | | | | | 5 | Au | 28○ | ·81 |
| 4428 | 1384 | 1249 | 26M | ·3528 | 26·6988 | 24·838 | **1327** | 5 | 4 | Ap | 22 | ·05 | 5 | My | 21 | ·59 | 7 | Je | 20 | ·12 | 1 | Jl | 19 | ·65 | 3 | Au | 18○ | ·18 |
| 4429 | 1385 | 1250 | 25M | ·6117 | 15·8071 | 20·995 | **1328** | 6 | 1 | Ap | 10 | ·42 | 2 | My | 9 | ·95 | 4 | Je | 8 | ·48 | 6 | Jl | 8 | ·01 | 7 | Au | 6●○ | ·54 |
| 4430 | 1386 | 1251 | 25M | ·8705 | 4·9154 | 17·153 | **1329** | 1 | 5 | Mr | 30 | ·79 | 7 | Ap | 29 | ·32 | 1 | My | 28 | ·85 | 4 | Jl | 26● | ·91 | 6 | Au | 25 | ·44 |
| | | | | | | | | | | | | | | | | | 3 | Je | 27 | ·38 | | | | | | | | |
| 4431 | 1387 | 1252 | 26M | ·1292 | 23·5543 | 15·286 | **1330** | 2 | 4 | Ap | 18 | ·69 | 6 | My | 18 | ·22 | 7 | Je | 16○ | ·75 | 2 | Jl | 16 | ·28 | 3 | Au | 14 | ·81 |
| 4432 | 1388 | 1253 | 26M | ·3880 | 12·6626 | 11·443 | **1331** | 3 | 2 | Ap | 8 | ·05 | 3 | My | 7 | ·58 | 5 | Je | 6○ | ·11 | 6 | Jl | 5 | ·65 | 1 | Au | 4 | ·18 |
| 4433 | 1389 | 1254 | 25M | ·6467 | 1·7709 | 7·600 | **1332** | 4 | 6 | Mr | 27 | ·42 | 7 | Ap | 25 | ·95 | 4 | Je | 24 | ·01 | 5 | Jl | 23 | ·54 | 7 | Au | 22 | ·07 |
| | | | | | | | | | | | | | 2 | My | 25●○ | ·48 | | | | | | | | | | | | |
| 4434 | 1390 | 1255 | 25M | ·9055 | 20·4097 | 5·734 | **1333** | 6 | 5 | Ap | 15○ | ·32 | 6 | My | 14 | ·85 | 1 | Je | 13 | ·38 | 2 | Jl | 12 | ·91 | 4 | Au | 11 | ·44 |
| 4435 | 1391 | 1256 | 26M | ·1642 | 9·5180 | 1·891 | **1334** | 7 | 2 | Ap | 4○ | ·69 | 4 | My | 4● | ·22 | 5 | Je | 2 | ·75 | 7 | Jl | 2 | ·28 | 1 | Jl | 31 | ·81 |
| 4436 | 1392 | 1257 | 26M | ·4230 | 28·1569 | 0·024 | **1335** | 1 | 1 | Ap | 23 | ·58 | 3 | My | 23 | ·11 | 4 | Je | 21 | ·64 | 6 | Jl | 21 | ·18 | 7 | Au | 19 | ·71 |
| 4437 | 1393 | 1258 | 25M | ·6818 | 17·2652 | 23·736 | **1336** | 2 | 5 | Ap | 11 | ·95 | 7 | My | 11 | ·48 | 2 | Je | 10 | ·01 | 3 | Jl | 9 | ·54 | 5 | Au | 8 | ·07 |
| 4438 | 1394 | 1259 | 25M | ·9405 | 6·3735 | 19·893 | **1337** | 4 | 3 | Ap | 1 | ·32 | 4 | Ap | 30 | ·85 | 6 | My | 30 | ·38 | 7 | Je | 28 | ·91 | 3 | Au | 26 | ·97 |
| | | | | | | | | | | | | | | | | | | | | | 2 | Jl | 28○ | ·44 | | | | |
| 4439 | 1395 | 1260 | 26M | ·1993 | 25·0124 | 18·027 | **1338** | 5 | 2 | Ap | 20 | ·22 | 3 | My | 19 | ·75 | 5 | Je | 18 | ·28 | 6 | Jl | 17○ | ·81 | 1 | Au | 16 | ·34 |
| 4440 | 1396 | 1261 | 26M | ·4580 | 14·1207 | 14·184 | **1339** | 6 | 6 | Ap | 9 | ·58 | 1 | My | 9 | ·11 | 2 | Je | 7 | ·64 | 4 | Jl | 7●○ | ·17 | 5 | Au | 5 | ·71 |
| 4441 | 1397 | 1262 | 25M | ·7168 | 3·2290 | 10·341 | **1340** | 7 | 3 | Mr | 28 | ·95 | 5 | Ap | 27 | ·48 | 7 | My | 27 | ·01 | 3 | Jl | 25 | ·07 | 4 | Au | 23 | ·60 |
| | | | | | | | | | | | | | | | | | 1 | Je | 25 | ·54 | | | | | | | | |
| 4442 | 1398 | 1263 | 25M | ·9755 | 21·8679 | 8·474 | **1341** | 2 | 2 | Ap | 16 | ·85 | 4 | My | 16○ | ·38 | 5 | Je | 14 | ·91 | 7 | Jl | 14 | ·44 | 1 | Au | 12 | ·97 |
| 4443 | 1399 | 1264 | 26M | ·2343 | 10·9762 | 4·632 | **1342** | 3 | 7 | Ap | 6 | ·21 | 1 | My | 5●○ | ·75 | 3 | Je | 4 | ·28 | 4 | Jl | 3 | ·81 | 6 | Au | 2 | ·34 |
| 4444 | 1400 | 1265 | 26M | ·4930 | 0·0845 | 0·789 | **1343** | 4 | 4 | Mr | 26 | ·58 | 7 | My | 24 | ·64 | 2 | Je | 23 | ·17 | 3 | Jl | 22 | ·70 | 5 | Au | 21 | ·23 |
| | | | | | | | | | 6 | Ap | 25● | ·11 | | | | | | | | | | | | | | | | |
| 4445 | 1401 | 1266 | 25M | ·7518 | 18·7234 | 26·477 | **1344** | 5 | 3 | Ap | 13 | ·48 | 5 | My | 13 | ·01 | 6 | Je | 11 | ·54 | 1 | Jl | 11 | ·07 | 2 | Au | 9 | ·60 |
| 4446 | 1402 | 1267 | 26M | ·0106 | 7·8317 | 22·634 | **1345** | 7 | 7 | Ap | 2 | ·84 | 2 | My | 2 | ·38 | 3 | My | 31 | ·91 | 5 | Je | 30 | ·44 | 6 | Jl | 29 | ·97 |
| | | | | | | | | | | | | | | | | | | | | | | | | | 1 | Au | 28○ | ·50 |
| 4447 | 1403 | 1268 | 26M | ·2693 | 26·4706 | 20·767 | **1346** | 1 | 6 | Ap | 21 | ·74 | 1 | My | 21 | ·28 | 2 | Je | 19 | ·81 | 4 | Jl | 19 | ·34 | 5 | Au | 17○ | ·87 |
| 4448 | 1404 | 1269 | 26M | ·5281 | 15·5789 | 16·924 | **1347** | 2 | 4 | Ap | 11 | ·11 | 5 | My | 10 | ·64 | 7 | Je | 9 | ·17 | 1 | Jl | 8 | ·70 | 3 | Au | 7● | ·23 |
| 4449 | 1405 | 1270 | 25M | ·7868 | 4·6872 | 13·082 | **1348** | 2 | 1 | Mr | 30 | ·48 | 3 | Ap | 29 | ·01 | 4 | My | 28 | ·54 | 7 | Jl | 26● | ·60 | 2 | Au | 25 | ·13 |
| | | | | | | | | | | | | | | | | | 6 | Je | 27○ | ·07 | | | | | | | | |
| 4450 | 1406 | 1271 | 26M | ·0456 | 23·3261 | 11·215 | **1349** | 5 | 7 | Ap | 18 | ·38 | 1 | My | 17 | ·91 | 3 | Je | 16○ | ·44 | 4 | Jl | 15 | ·97 | 6 | Au | 14 | ·50 |

## Surya Siddhanta.

| Asvina | | | | Kartika | | | | Margasira | | | | Pausha | | | | A.D. | Magha | | | | A.D. | Phalguna | | | | Chaitra | | | |
|---|---|---|---|---|---|---|---|---|---|---|---|---|---|---|---|---|---|---|---|---|---|---|---|---|---|---|---|---|---|
| + 147·65293 | | | | + 177·18353 | | | | + 206·71411 | | | | + 236·24470 | | | | | + 265·77529 | | | | | + 295·30588 | | | | +324·83647 / +354·36705 | | | |
| + 9·880 | | | | + 11·856 | | | | + 13·832 | | | | + 15·808 | | | | | + 17·784 | | | | | + 19·760 | | | | + 21·736 / + 23·712 | | | |
| Week-day | Month | Day | Fraction | Week-day | Month | Day | Fraction | Week-day | Month | Day | Fraction | Week-day | Month | Day | Fraction | | Week-day | Month | Day | Fraction | | Week-day | Month | Day | Fraction | Week-day | Month | Day | Fraction |
| 2 | S | 13 | ·65 | 4 | O | 13 | ·18 | 5 | N | 11 | ·71 | 7 | D | 11○ | ·24 | **12** | 1 | Jr | 9 | ·77 | | 3 | F | 8 | ·30 | 4 | Mr | 8 | ·83 |
| 7 | S | 2 | ·02 | 1 | O | 1 | ·55 | 3 | O | 31 | ·08 | 4 | N | 29○ | ·61 | | 6 | D | 29 | ·14 | **13** | 7 | Jr | 27 | ·67 | 2 | F | 26 | ·20 |
| 5 | S | 20 | ·91 | 7 | O | 20 | ·44 | 1 | N | 18○ | ·98 | 3 | D | 18 | ·51 | **14** | 5 | Jr | 17 | ·04 | | 6 | F | 15 | ·57 | 1 | Mr | 17 | ·10 |
| 3 | S | 10 | ·28 | 4 | O | 9 | ·81 | 6 | N | 8 | ·34 | 7 | D | 7 | ·87 | **15** | 2 | Jr | 6 | ·40 | | 3 | F | 4 | ·93 | 5 | Mr | 6 | ·46 |
| 7 | Au | 30 | ·65 | {2 | S | 29○ | ·18 | **Margasira** | | | | 5 | N | 27 | ·24 | | 6 | D | 26 | ·77 | **16** | {1 | Jr | 25 | ·30 | 4 | Mr | 24○ | ·36 |
| | | | | {3 | O | 28● | ·71 | **Kshaya** | | | | | | | | | | | | | | {2 | F | 23 | ·83 | | | | |
| 6 | S | 17○ | ·55 | 1 | O | 17 | ·08 | 2 | N | 15 | ·61 | 4 | D | 15 | ·14 | **17** | 5 | Jr | 13 | ·67 | | 7 | F | 12 | ·20 | 1 | Mr | 13○ | ·73 |
| 3 | S | 6●○ | ·91 | 5 | O | 6 | ·44 | 6 | N | 4 | ·97 | 1 | D | 4 | ·50 | **18** | 3 | Jr | 3 | ·04 | | 4 | F | 1 | ·57 | 6 | Mr | 3 | ·10 |
| 2 | S | 25● | ·81 | 4 | O | 25 | ·34 | 5 | N | 23 | ·87 | 7 | D | 23 | ·40 | **19** | 1 | Jr | 21○ | ·93 | | 3 | F | 20● | ·46 | 5 | Mr | 22 | ·63 |
| 7 | S | 15 | ·18 | 1 | O | 14 | ·71 | 3 | N | 13 | ·24 | 4 | D | 12 | ·77 | **20** | 6 | Jr | 11○ | ·30 | | 7 | F | 9● | ·83 | 2 | Mr | 10 | ·36 |
| 4 | S | 3 | ·55 | 6 | O | 3 | ·08 | 7 | N | 1 | ·61 | 2 | D | 1 | ·14 | | 3 | D | 30○ | ·67 | **21** | 5 | Jr | 29 | ·20 | 6 | F | 27 | ·73 |
| 3 | S | 22 | ·44 | 4 | O | 21 | ·97 | 6 | N | 20 | ·50 | 1 | D | 20 | ·04 | **22** | 2 | Jr | 18 | ·57 | | 4 | F | 17 | ·10 | 5 | Mr | 18 | ·63 |
| 7 | S | 11 | ·81 | 2 | O | 11 | ·34 | 3 | N | 9 | ·87 | 5 | D | 9● | ·40 | **23** | 6 | Jr | 7 | ·93 | | 1 | F | 6 | ·46 | 2 | Mr | 7 | ·99 |
| 5 | S | 1 | ·18 | 6 | S | 30 | ·71 | 1 | O | 30○ | ·24 | 2 | N | 28 | ·77 | | 4 | D | 28 | ·30 | **24** | 5 | Jr | 26 | ·83 | 7 | F | 25 | ·36 |
| 4 | S | 19 | ·08 | 5 | O | 18○ | ·61 | 7 | N | 17 | ·14 | 1 | D | 16 | ·67 | **25** | 3 | Jr | 15 | ·20 | | 4 | F | 13 | ·73 | 6 | Mr | 15 | ·26 |
| 1 | S | 8 | ·44 | 2 | O | 7● | ·97 | 4 | N | 6 | ·50 | 6 | D | 6 | ·03 | **26** | 7 | Jr | 4 | ·56 | | 2 | F | 3 | ·10 | 3 | Mr | 4○ | ·63 |
| 7 | S | 27 | ·34 | 1 | O | 26 | ·87 | 3 | N | 25 | ·40 | 4 | D | 24 | ·93 | **27** | 6 | Jr | 23 | ·46 | | 1 | F | 22○ | ·99 | 2 | Mr | 23 | ·52 |
| 4 | S | 16 | ·71 | 6 | O | 16 | ·24 | 7 | N | 14 | ·77 | 2 | D | 14 | ·30 | **28** | 3 | Jr | 12 | ·83 | | 5 | F | 11○ | ·36 | 6 | Mr | 11 | ·89 |
| 2 | S | 5 | ·07 | 3 | O | 4 | ·61 | 5 | N | 3 | ·14 | 6 | D | 2 | ·67 | **29** | 1 | Jr | 1 | ·20 | | 2 | Jr | 30 | ·73 | 4 | Mr | 1 | ·26 |
| 7 | S | 23 | ·97 | 2 | O | 23 | ·50 | 4 | N | 22 | ·03 | 5 | D | 21○ | ·56 | **30** | 7 | Jr | 20 | ·09 | | 1 | F | 18 | ·63 | 3 | Mr | 20 | ·16 |
| 5 | S | 13 | ·34 | 6 | O | 12 | ·87 | 1 | N | 11 | ·40 | 2 | D | 10○ | ·93 | **31** | 4 | Jr | 9 | ·46 | | 5 | F | 7 | ·99 | 7 | Mr | 9 | ·52 |
| 2 | S | 2 | ·71 | 4 | O | 2 | ·24 | 5 | O | 31 | ·77 | 7 | N | 30●○ | ·30 | | 1 | D | 29 | ·83 | **32** | 3 | Jr | 28 | ·36 | 4 | F | 26 | ·89 |
| 1 | S | 20 | ·60 | 3 | O | 20 | ·14 | 4 | N | 18 | ·67 | 6 | D | 18 | ·20 | **33** | 7 | Jr | 16 | ·73 | | 2 | F | 15 | ·26 | 3 | Mr | 16 | ·79 |
| 5 | S | 9 | ·97 | 7 | O | 9○ | ·50 | 2 | N | 8 | ·03 | 3 | D | 7 | ·56 | **34** | 5 | Jr | 6 | ·09 | | 6 | F | 4 | ·62 | 1 | Mr | 6 | ·16 |
| {3 | Au | 30 | ·34 | 6 | O | 28 | ·40 | 7 | N | 26 | ·93 | **Pausha** | | | | | 2 | D | 26 | ·46 | **35** | {3 | Jr | 24 | ·99 | 7 | Mr | 25●○ | ·05 |
| {4 | S | 28○ | ·87 | | | | | | | | | **Kshaya** | | | | | | | | | | {5 | F | 23 | ·52 | | | | |
| 2 | S | 18○ | ·24 | 3 | O | 17 | ·77 | 5 | N | 16 | ·30 | 6 | D | 15 | ·83 | **36** | 1 | Jr | 14 | ·36 | | 2 | F | 12 | ·89 | 4 | Mr | 13 | ·42 |
| 6 | S | 6● | ·60 | 1 | O | 6 | ·13 | 2 | N | 4 | ·66 | 4 | D | 4 | ·20 | **37** | 5 | Jr | 2 | ·73 | | 7 | F | 1○ | ·26 | 1 | Mr | 2● | ·79 |
| 5 | S | 25 | ·50 | 7 | O | 25 | ·03 | 1 | N | 23 | ·56 | 3 | D | 23 | ·09 | **38** | 4 | Jr | 21○ | ·62 | | 6 | F | 20 | ·15 | 7 | Mr | 21 | ·68 |
| 2 | S | 14 | ·87 | 4 | O | 14 | ·40 | 5 | N | 12 | ·93 | 7 | D | 12 | ·46 | **39** | 1 | Jr | 10○ | ·99 | | 3 | F | 9 | ·52 | 5 | Mr | 11 | ·05 |
| 7 | S | 4 | ·24 | 1 | O | 3 | ·77 | 3 | N | 2 | ·30 | 4 | D | 1 | ·83 | | 6 | D | 31● | ·36 | **40** | 7 | Jr | 29 | ·89 | 2 | F | 28 | ·42 |
| 6 | S | 22 | ·13 | 7 | O | 21 | ·66 | 2 | N | 20 | ·19 | 3 | D | 19 | ·73 | **41** | 5 | Jr | 18 | ·26 | | 6 | F | 16 | ·79 | 1 | Mr | 18 | ·32 |
| 3 | S | 11 | ·50 | 5 | O | 11 | ·03 | 6 | N | 9○ | ·56 | 7 | D | 9● | ·09 | **42** | 2 | Jr | 7 | ·62 | | 4 | F | 6 | ·15 | 5 | Mr | 7 | ·68 |
| 7 | Au | 31 | ·87 | 2 | S | 30 | ·40 | 3 | O | 29○ | ·93 | 5 | N | 28 | ·46 | | 6 | D | 27 | ·99 | **43** | 1 | Jr | 26 | ·52 | 3 | F | 25 | ·05 |
| 6 | S | 19 | ·77 | 1 | O | 19● | ·30 | 2 | N | 17 | ·83 | 4 | D | 17 | ·36 | **44** | 5 | Jr | 15 | ·89 | | 7 | F | 14 | ·42 | 1 | Mr | 14 | ·95 |
| 4 | S | 8○ | ·13 | 5 | O | 7● | ·66 | 7 | N | 6 | ·19 | 1 | D | 5 | ·72 | **45** | 3 | Jr | 4 | ·26 | | 4 | F | 2 | ·79 | 6 | Mr | 4○ | ·32 |
| 3 | S | 27● | ·03 | 4 | O | 26 | ·56 | 6 | N | 25 | ·09 | 7 | D | 24 | ·62 | **46** | 2 | Jr | 23 | ·15 | | 3 | F | 21●○ | ·68 | 5 | Mr | 23 | ·21 |
| 7 | S | 16 | ·40 | 1 | O | 15 | ·93 | 3 | N | 14 | ·46 | 4 | D | 13 | ·99 | **47** | 6 | Jr | 12 | ·52 | | 1 | F | 11● | ·05 | 2 | Mr | 12 | ·58 |
| 4 | S | 5 | ·76 | 6 | O | 5 | ·30 | 7 | N | 3 | ·83 | 2 | D | 3 | ·36 | **48** | 3 | Jr | 1○ | ·89 | | 5 | Jr | 31 | ·42 | 6 | F | 29 | ·95 |
| 3 | S | 23 | ·66 | 5 | O | 23 | ·19 | 6 | N | 21 | ·72 | 1 | D | 21○ | ·25 | **49** | 2 | Jr | 19 | ·79 | | 4 | F | 18 | ·32 | 5 | Mr | 19 | ·85 |
| 1 | S | 13 | ·03 | 2 | O | 12 | ·56 | 4 | N | 11 | ·09 | 5 | D | 10○ | ·62 | **50** | 7 | Jr | 9 | ·15 | | 1 | F | 7 | ·68 | 3 | Mr | 9 | ·21 |

**TABLE X-**

| Kaliyuga. | Vikrama Era. | Saka Era. | Commencement of Solar Year. Month and day A.D. | Fraction of day. | First New-Moon in Solar Year. | Anomaly of first New-Moon. | A.D. | Week-day of 1st January. | Vaisakha Week-day | Month | Day | Fraction | Jyeshta Week-day | Month | Day | Fraction | Ashada Week-day | Month | Day | Fraction | Sravana Week-day | Month | Day | Fraction | Bhadrapada Week-day | Month | Day | Fraction |
|---|---|---|---|---|---|---|---|---|---|---|---|---|---|---|---|---|---|---|---|---|---|---|---|---|---|---|---|---|
| | | | | | | | | | ☉'s Anom col. 6 | | | | + 29·53059 | | | | + 59·06117 | | | | + 88·59176 | | | | + 118·12235 | | | |
| | | | | | | | | | ☾'s Anom col. 7 | | | | + 1·976 | | | | + 3·952 | | | | + 5·928 | | | | + 7·904 | | | |
| 4451 | 1407 | 1272 | 26M | ·3043 | 12·4344 | 7·372 | **1350** | 6 | 4 | Ap | 7 | ·74 | 6 | My | 7 | ·27 | 7 | Je | 5○ | ·81 | 2 | Jl | 5 | ·34 | 3 | Au | 3 | ·87 |
| 4452 | 1408 | 1273 | 26M | ·5731 | 1·5427 | 3·529 | **1351** | 7 | 2 | Mr | 28 | ·11 | 5 | My | 26 | ·17 | 6 | Je | 24 | ·70 | 1 | Jl | 24 | ·23 | 2 | Au | 22 | ·76 |
| | | | | | | | | | 3 | Ap | 26 | ·64 | | | | | | | | | | | | | | | | |
| 4453 | 1409 | 1274 | 25M | ·8219 | 20·1815 | 1·663 | **1352** | 1 | 1 | Ap | 15○ | ·01 | 2 | My | 14 | ·54 | 4 | Je | 13 | ·07 | 5 | Jl | 12 | ·60 | 7 | Au | 11 | ·13 |
| 4454 | 1410 | 1275 | 26M | ·0806 | 9·2898 | 25·375 | **1353** | 3 | 5 | Ap | 4○ | ·38 | 6 | My | 3 | ·91 | 1 | Je | 2 | ·44 | 2 | Jl | 1 | ·97 | 4 | Jl | 31 | ·50 |
| | | | | | | | | | | | | | | | | | | | | | | | | | 6 | Au | 30 | ·03 |
| 4455 | 1411 | 1276 | 26M | ·3394 | 27·9287 | 23·508 | **1354** | 4 | 4 | Ap | 23 | ·27 | 5 | My | 22 | ·80 | 7 | Je | 21 | ·33 | 1 | Jl | 20 | ·87 | 3 | Au | 19 | ·40 |
| 4456 | 1412 | 1277 | 26M | ·5981 | 17·0370 | 19·665 | **1355** | 5 | 1 | Ap | 12 | ·64 | 3 | My | 12 | ·17 | 4 | Je | 10 | ·70 | 6 | Jl | 10 | ·23 | 7 | Au | 8○ | ·76 |
| 4457 | 1413 | 1278 | 25M | ·8565 | 6·1453 | 15·823 | **1356** | 6 | 6 | Ap | 1 | ·01 | 7 | Ap | 30 | ·54 | 2 | My | 30 | ·07 | 3 | Je | 28 | ·60 | 6 | Au | 26 | ·66 |
| | | | | | | | | | | | | | | | | | | | | | 5 | Jl | 28○ | ·13 | | | | |
| 4458 | 1414 | 1279 | 26M | ·1156 | 24·7842 | 13·956 | **1357** | 1 | 4 | Ap | 19 | ·91 | 6 | My | 19 | ·44 | 7 | Je | 17 | ·97 | 2 | Jl | 17○ | ·50 | 4 | Au | 16 | ·03 |
| 4459 | 1415 | 1280 | 26M | ·3744 | 13·8925 | 10·113 | **1358** | 2 | 2 | Ap | 9 | ·27 | 3 | My | 8 | ·80 | 5 | Je | 7 | ·33 | 6 | Jl | 6● | ·86 | 1 | Au | 5 | ·39 |
| 4460 | 1416 | 1281 | 26M | ·6331 | 3·0008 | 6·270 | **1359** | 3 | 6 | Mr | 29 | ·64 | 1 | Ap | 28 | ·17 | 4 | Je | 26● | ·23 | 5 | Jl | 25 | ·76 | 7 | Au | 24 | ·29 |
| | | | | | | | | | | | | | 2 | My | 27○ | ·70 | | | | | | | | | | | | |
| 4461 | 1417 | 1282 | 25M | ·8919 | 21·6397 | 4·403 | **1360** | 4 | 5 | Ap | 16 | ·54 | 7 | My | 16○ | ·07 | 1 | Je | 14 | ·60 | 3 | Jl | 14 | ·13 | 4 | Au | 12 | ·66 |
| 4462 | 1418 | 1283 | 26M | ·1507 | 10·7480 | 0·561 | **1361** | 6 | 2 | Ap | 5 | ·90 | 4 | My | 5● | ·43 | 5 | Je | 3 | ·97 | 7 | Jl | 3 | ·50 | 2 | Au | 2 | ·03 |
| 4463 | 1419 | 1284 | 26M | ·4094 | 29·3869 | 26·248 | **1362** | 7 | 7 | Mr | 26 | ·27 | 3 | My | 24 | ·33 | 4 | Je | 22 | ·86 | 6 | Jl | 22 | ·39 | 7 | Au | 20 | ·92 |
| | | | | | | | | | 1 | Ap | 24● | ·90 | | | | | | | | | | | | | | | | |
| 4464 | 1420 | 1285 | 26M | ·6682 | 18·4952 | 22·406 | **1363** | 1 | 6 | Ap | 14 | ·17 | 7 | My | 13 | ·70 | 2 | Je | 12 | ·23 | 3 | Jl | 11 | ·76 | 5 | Au | 10 | ·29 |
| 4465 | 1421 | 1286 | 25M | ·9269 | 7·6035 | 18·563 | **1364** | 2 | 3 | Ap | 2 | ·54 | 5 | My | 2 | ·07 | 6 | My | 31 | ·60 | 1 | Jl | 30 | ·13 | 2 | Jl | 29 | ·66 |
| | | | | | | | | | | | | | | | | | | | | | | | | | 4 | Au | 28○ | ·19 |
| 4466 | 1422 | 1287 | 26M | ·1857 | 26·2424 | 16·696 | **1365** | 4 | 2 | Ap | 21 | ·43 | 3 | My | 20 | ·96 | 5 | Je | 19 | ·49 | 7 | Jl | 19 | ·03 | 1 | Au | 17 | ·56 |
| 4467 | 1423 | 1288 | 26M | ·4444 | 15·3507 | 12·853 | **1366** | 5 | 6 | Ap | 10 | ·80 | 1 | My | 10 | ·33 | 2 | Je | 8 | ·86 | 4 | Jl | 8○ | ·39 | 5 | Au | 6● | ·92 |
| 4468 | 1424 | 1289 | 26M | ·7032 | 4·4590 | 9·011 | **1367** | 6 | 4 | Mr | 31 | ·17 | 5 | Ap | 29 | ·70 | 7 | My | 29 | ·23 | 3 | Jl | 27● | ·29 | 4 | Au | 25 | ·82 |
| | | | | | | | | | | | | | | | | | 1 | Je | 27○ | ·76 | | | | | | | | |
| 4469 | 1425 | 1290 | 25M | ·9620 | 28·0979 | 7·144 | **1368** | 7 | 3 | Ap | 18 | ·07 | 4 | My | 17 | ·60 | 6 | Je | 16○ | ·13 | 7 | Jl | 15 | ·66 | 2 | Au | 14 | ·19 |
| 4470 | 1426 | 1291 | 26M | ·2207 | 12·2062 | 3·301 | **1369** | 2 | 7 | Ap | 7 | ·43 | 1 | My | 6 | ·96 | 3 | Je | 5● | ·49 | 5 | Jl | 5 | ·02 | 6 | Au | 3 | ·55 |
| 4471 | 1427 | 1292 | 26M | ·4795 | 1·3144 | 27·013 | **1370** | 3 | 4 | Mr | 27 | ·80 | 7 | My | 25 | ·86 | 2 | Je | 24 | ·39 | 3 | Jl | 23 | ·92 | 5 | Au | 22 | ·45 |
| | | | | | | | | | 6 | Ap | 26○ | ·33 | | | | | | | | | | | | | | | | |
| 4472 | 1428 | 1293 | 26M | ·7382 | 19·9533 | 25·146 | **1371** | 4 | 3 | Ap | 15○ | ·70 | 5 | My | 15 | ·23 | 6 | Je | 13 | ·76 | 1 | Jl | 13 | ·29 | 2 | Au | 11 | ·82 |
| 4473 | 1429 | 1294 | 25M | ·9970 | 9·0616 | 21·304 | **1372** | 5 | 1 | Ap | 4 | ·06 | 2 | My | 3 | ·59 | 4 | Je | 2 | ·13 | 5 | Jl | 1 | ·66 | 7 | Jl | 31 | ·19 |
| | | | | | | | | | | | | | | | | | | | | | | | | | 1 | Au | 29 | ·72 |
| 4474 | 1430 | 1295 | 26M | ·2557 | 27·7005 | 19·437 | **1373** | 7 | 6 | Ap | 22 | ·96 | 1 | My | 22 | ·49 | 3 | Je | 21 | ·02 | 4 | Jl | 20 | ·55 | 6 | Au | 19 | ·08 |
| 4475 | 1431 | 1296 | 26M | ·5145 | 16·8088 | 15·594 | **1374** | 1 | 4 | Ap | 12 | ·33 | 5 | My | 11 | ·86 | 7 | Je | 10 | ·39 | 1 | Jl | 9 | ·92 | 3 | Au | 8○ | ·45 |
| 4476 | 1432 | 1297 | 26M | ·7733 | 5·9171 | 11·751 | **1375** | 2 | 1 | Ap | 1 | ·70 | 3 | My | 1 | ·23 | 4 | My | 30 | ·76 | 6 | Je | 29 | ·29 | 2 | Au | 27 | ·35 |
| | | | | | | | | | | | | | | | | | | | | | 7 | Jl | 28●○ | ·82 | | | | |
| 4477 | 1433 | 1298 | 26M | ·0323 | 24·5560 | 9·885 | **1376** | 3 | 7 | Ap | 19 | ·59 | 2 | My | 19 | ·12 | 3 | Je | 17 | ·65 | 5 | Jl | 17● | ·19 | 6 | Au | 15 | ·72 |
| 4478 | 1434 | 1299 | 26M | ·2908 | 13·6643 | 6·042 | **1377** | 5 | 4 | Ap | 8 | ·96 | 6 | My | 8 | ·49 | 1 | Je | 7○ | ·02 | 2 | Jl | 6● | ·55 | 4 | Au | 5 | ·08 |
| 4479 | 1235 | 1300 | 26M | ·5495 | 2·7726 | 2·199 | **1378** | 6 | 2 | Mr | 29 | ·33 | 3 | Ap | 27 | ·86 | 6 | Je | 25 | ·92 | 1 | Jl | 25 | ·45 | 2 | Au | 23 | ·98 |
| | | | | | | | | | | | | | 5 | My | 27●○ | ·39 | | | | | | | | | | | | |
| 4480 | 1436 | 1301 | 26M | ·8083 | 21·4115 | 0·332 | **1379** | 7 | 1 | Ap | 17 | ·23 | 2 | My | 16○ | ·76 | 4 | Je | 15 | ·29 | 5 | Jl | 14 | ·82 | 7 | Au | 13 | ·35 |
| 4481 | 1437 | 1302 | 26M | ·0670 | 10·5198 | 24·044 | **1380** | 1 | 5 | Ap | 5 | ·59 | 7 | My | 5● | ·12 | 1 | Je | 3 | ·65 | 3 | Jl | 3 | ·18 | 4 | Au | 1 | ·71 |
| 4482 | 1438 | 1303 | 26M | ·3258 | 29·1587 | 22·177 | **1381** | 3 | 2 | Mr | 25○ | ·96 | 6 | My | 24 | ·02 | 7 | Je | 22 | ·55 | 2 | Jl | 22 | ·08 | 3 | Au | 20 | ·61 |
| | | | | | | | | | 4 | Ap | 24 | ·49 | | | | | | | | | | | | | | | | |
| 4483 | 1439 | 1304 | 26M | ·5845 | 18·2670 | 18·335 | **1382** | 4 | 1 | Ap | 13 | ·86 | 3 | My | 13 | ·39 | 4 | Je | 11 | ·92 | 6 | Jl | 11 | ·45 | 7 | Au | 9 | ·98 |
| 4484 | 1440 | 1305 | 26M | ·8433 | 7·3753 | 14·492 | **1383** | 5 | 6 | Ap | 3 | ·22 | 7 | My | 2 | ·75 | 2 | Je | 1 | ·29 | 3 | Je | 30 | ·82 | 5 | Jl | 30 | ·35 |
| | | | | | | | | | | | | | | | | | | | | | | | | | 6 | Au | 28● | ·88 |
| 4485 | 1441 | 1306 | 26M | ·1021 | 26·0142 | 12·625 | **1384** | 6 | 5 | Ap | 21 | ·12 | 6 | My | 20 | ·65 | 1 | Je | 19 | ·18 | 2 | Jl | 18○ | ·71 | 4 | Au | 17● | ·24 |
| 4486 | 1442 | 1307 | 26M | ·3608 | 15·1225 | 8·782 | **1385** | 1 | 2 | Ap | 10 | ·49 | 4 | My | 10 | ·02 | 5 | Je | 8 | ·55 | 7 | Jl | 8○ | ·08 | 1 | Au | 6 | ·61 |
| 4487 | 1443 | 1308 | 26M | ·6196 | 4·2308 | 4·940 | **1386** | 2 | 6 | Mr | 34 | ·86 | 1 | Ap | 20 | ·39 | 2 | My | 28 | ·92 | 5 | Jl | 26 | ·98 | 7 | Au | 25 | ·51 |
| | | | | | | | | | | | | | | | | | 4 | Je | 27●○ | ·45 | | | | | | | | |
| 4488 | 1444 | 1309 | 26M | ·8783 | 22·8696 | 3·073 | **1387** | 3 | 5 | Ap | 18 | ·75 | 7 | My | 18 | ·28 | 1 | Je | 16● | ·81 | 3 | Jl | 16 | ·35 | 4 | Au | 14 | ·88 |
| 4489 | 1445 | 1310 | 26M | ·1371 | 11·9779 | 26·785 | **1388** | 4 | 3 | Ap | 7 | ·12 | 4 | My | 6○ | ·65 | 6 | Je | 5● | ·18 | 7 | Jl | 4 | ·71 | 2 | Au | 3 | ·24 |

**Surya Siddhanta.**

| | Asvina | | | | Kartika | | | | Margasira | | | | Pausha | | | A.D. | Magha | | | | A.D. | Phalguna | | | | Chaitra | | | |
|---|---|---|---|---|---|---|---|---|---|---|---|---|---|---|---|---|---|---|---|---|---|---|---|---|---|---|---|---|---|
| | + 147·65293 | | | | + 177·18353 | | | | + 206·71411 | | | | + 236·24470 | | | | + 265·77529 | | | | | + 295·30588 | | | | +324·83647 +354·36705 | | | |
| | + 9·880 | | | | + 11·856 | | | | + 13·832 | | | | + 15·808 | | | | + 17·784 | | | | | + 19·760 | | | | + 21·736 + 23·712 | | | |
| Week-day | Month | Day | Fraction | Week-day | Month | Day | Fraction | Week-day | Month | Day | Fraction | Week-day | Month | Day | Fraction | | Week-day | Month | Day | Fraction | | Week-day | Month | Day | Fraction | Week-day | Month | Day | Fraction |
| 5 | S | 2 | ·40 | 6 | O | 1 | ·93 | 1 | O | 31 | ·46 | 2 | N | 29● | ·99 | | 4 | D | 29 | ·52 | 51 | 6 | Jr | 28 | ·05 | 7 | F | 26 | ·58 |
| 4 | S | 21 | ·29 | 5 | O | 20ↄ | ·83 | 7 | N | 19 | ·36 | 1 | D | 18 | ·89 | 52 | 3 | Jr | 17 | ·42 | | 4 | F | 15 | ·95 | 6 | Mr | 16 | ·48 |
| 1 | S | 9 | ·66 | 3 | O | 9ↄ | ·19 | 4 | N | 7 | ·72 | 6 | D | 7 | ·25 | 53 | 7 | Jr | 5 | ·78 | | 2 | F | 4 | ·31 | 3 | Mr | 5 | ·85 |
| 7 | S | 28ↄ | ·56 | 2 | O | 28 | ·09 | 3 | N | 26 | ·62 | 5 | D | 26 | ·15 | 54 | 6 | Jr | 24 | ·68 | | 1 | F | 23 | ·21 | 2 | Mr | 24● | ·74 |
| 4 | S | 17● | ·93 | 6 | O | 17 | ·46 | 7 | N | 15 | ·99 | 2 | D | 15 | ·52 | 55 | 4 | Jr | 14 | ·05 | | 5 | F | 12○ | ·58 | 7 | Mr | 14 | ·11 |
| 2 | S | 7● | ·29 | 3 | O | 6 | ·82 | 5 | N | 5 | ·35 | 6 | D | 4 | ·89 | 56 | 1 | Jr | 3 | ·42 | | 2 | F | 1○ | ·95 | 4 | Mr | 2 | ·48 |
| 1 | S | 25 | ·19 | 2 | O | 24 | ·72 | 4 | N | 23 | ·25 | 5 | D | 22 | ·78 | 57 | 7 | Jr | 21○ | ·31 | | 1 | F | 19 | ·84 | 3 | Mr | 21 | ·37 |
| 5 | S | 14 | ·56 | 7 | O | 44 | ·09 | 1 | N | 12 | ·62 | 3 | D | 12 | ·15 | 58 | 4 | Jr | 10● | ·68 | | 6 | F | 9 | ·21 | 7 | Mr | 10 | ·74 |
| 2 | S | 3 | ·93 | 4 | O | 3 | ·46 | 5 | N | 1 | ·99 | 7 | D | 1 | ·52 | | 2 | D | 31● | ·05 | 59 | 3 | Jr | 29 | ·58 | 5 | F | 28 | ·11 |
| 1 | S | 22 | ·83 | 3 | O | 22 | ·35 | 4 | N | 20○ | ·88 | 6 | D | 20 | ·41 | 60 | 7 | Jr | 18 | ·95 | | 2 | F | 17 | ·48 | 4 | Mr | 18 | ·01 |
| 6 | S | 11 | ·19 | 7 | O | 10 | ·72 | 2 | N | 9ↄ | ·25 | 3 | D | 8 | ·78 | 61 | 5 | Jr | 7 | ·31 | | 6 | F | 5 | ·84 | 1 | Mr | 7 | ·37 |
| 3 | Au | 31 | ·56 | 5 | S | 30 | ·09 | 6 | O | 29 | ·62 | 1 | N | 28 | ·15 | | 2 | D | 27 | ·68 | 62 | 4 | Jr | 26 | ·21 | 5 | F | 24 | ·74 |
| 2 | S | 19ↄ | ·45 | 3 | O | 18 | ·99 | 5 | N | 17 | ·52 | 7 | D | 17 | ·05 | 63 | 1 | Jr | 15 | ·58 | | 3 | F | 14 | ·11 | 4 | Mr | 15○ | ·64 |
| 6 | S | 8ↄ | ·82 | 1 | O | 8 | ·35 | 2 | N | 6 | ·88 | 4 | D | 6 | ·41 | 64 | 5 | Jr | 4 | ·94 | | 7 | F | 3 | ·47 | 2 | Mr | 4●○ | ·01 |
| 5 | S | 26 | ·72 | 7 | O | 26 | ·25 | 1 | N | 24 | ·78 | 3 | D | 24 | ·31 | 65 | 4 | Jr | 22 | ·84 | | 6 | F | 21● | ·37 | 7 | Mr | 22 | ·90 |
| 3 | S | 16 | ·09 | 4 | O | 15 | ·62 | 6 | N | 14 | ·15 | 7 | D | 13 | ·68 | 66 | 2 | Jr | 12ↄ | ·21 | | 3 | F | 10 | ·74 | 5 | Mr | 12 | ·27 |
| 7 | S | 5 | ·45 | 1 | O | 4 | ·98 | 3 | N | 3 | ·51 | 5 | D | 3 | ·05 | 67 | 6 | Jr | 1ↄ | ·58 | | 1 | Jr | 31 | ·11 | 2 | Mr | 1 | ·64 |
| 6 | S | 24 | ·35 | 7 | O | 23 | ·88 | 2 | N | 22 | ·41 | 3 | D | 21●ↄ | ·94 | 68 | 5 | Jr | 20 | ·47 | | 7 | F | 19 | ·00 | 1 | Mr | 19 | ·53 |
| 3 | S | 12 | ·72 | 5 | O | 12 | ·25 | 6 | N | 10 | ·78 | 1 | D | 10 | ·31 | 69 | 2 | Jr | 8 | ·84 | | 4 | F | 7 | ·37 | 5 | Mr | 8 | ·90 |
| 1 | S | 2 | ·09 | 2 | O | 1 | ·62 | 4 | O | 31○ | ·15 | 5 | N | 29● | ·68 | | 7 | D | 29 | ·21 | 70 | 1 | Jr | 27 | ·74 | 3 | F | 26 | ·27 |
| 6 | S | 20 | ·98 | 1 | O | 20ↄ | ·51 | 3 | N | 19 | ·04 | 4 | D | 18 | ·57 | 71 | 6 | Jr | 17 | ·11 | | 7 | F | 15 | ·64 | 2 | Mr | 17 | ·17 |
| 4 | S | 10 | ·35 | 6 | O | 10●ↄ | ·88 | 7 | N | 8 | ·41 | 1 | D | 7 | ·94 | 72 | 3 | Jr | 6 | ·47 | | 5 | F | 5 | ·00 | 6 | Mr | 5 | ·53 |
| 3 | S | 28 | ·25 | 4 | O | 27 | ·78 | 6 | N | 26 | ·31 | 7 | D | 25 | ·84 | 73 | 2 | Jr | 24 | ·37 | | 3 | F | 22ↄ | ·90 | 5 | Mr | 24● | ·43 |
| 7 | S | 17● | ·61 | 2 | O | 17 | ·15 | 3 | N | 15 | ·68 | 5 | D | 15 | ·21 | 74 | 6 | Jr | 13 | ·74 | | 1 | F | 12ↄ | ·27 | 2 | Mr | 13● | ·80 |
| 4 | S | 6 | ·98 | 6 | O | 6 | ·51 | 1 | N | 5 | ·04 | 2 | D | 3 | ·57 | 75 | 4 | Jr | 3 | ·10 | | 5 | F | 1●○ | ·63 | 7 | Mr | 3 | ·17 |
| 3 | S | 25 | ·88 | 5 | O | 25 | ·41 | 6 | N | 23 | ·94 | 1 | D | 23 | ·47 | 76 | 3 | Jr | 22 | ·00 | | 4 | F | 20 | ·53 | 6 | Mr | 21 | ·06 |
| 1 | S | 14 | ·25 | 2 | O | 13 | ·78 | 4 | N | 12 | ·31 | 5 | D | 11 | ·84 | 77 | 7 | Jr | 10● | ·37 | | 1 | F | 8 | ·90 | 3 | Mr | 10 | ·43 |
| 5 | S | 3 | ·61 | 7 | O | 3 | ·14 | 1 | N | 1 | ·67 | 3 | D | 1ↄ | ·21 | | 4 | D | 30● | ·74 | 78 | 6 | Jr | 29 | ·27 | 7 | F | 27 | ·80 |
| 4 | S | 22 | ·51 | 6 | O | 22 | ·04 | 7 | N | 20ↄ | ·57 | 2 | D | 20 | ·10 | 79 | 3 | Jr | 18 | ·63 | | 5 | F | 17 | ·16 | 6 | Mr | 18 | ·69 |
| 1 | S | 11ↄ | ·88 | 3 | O | 11 | ·41 | 4 | N | 9 | ·94 | 6 | D | 9 | ·47 | 80 | 1 | Jr | 8 | ·00 | | 2 | F | 6 | ·53 | 4 | Mr | 7 | ·06 |
| 6 | Au | 31 | ·25 | 7 | S | 29ↄ | ·78 | Margasira Kshaya | | | | 3 | N | 27 | ·84 | | 5 | D | 27 | ·37 | 81 | 6 | Jr | 25 | ·90 | 1 | F | 24 | ·43 |
| | | | | 2 | O | 29 | ·31 | | | | | | | | | | | | | | | | | | | | | | |
| 5 | S | 19ↄ | ·14 | 6 | O | 18● | ·67 | 1 | N | 17 | ·20 | 2 | D | 16 | ·73 | 82 | 4 | Jr | 15 | ·27 | | 5 | F | 13 | ·80 | 7 | Mr | 15○ | ·33 |
| 2 | S | 8○ | ·51 | 4 | O | 8 | ·04 | 5 | N | 6 | ·57 | 7 | D | 6 | ·10 | 83 | 1 | Jr | 4 | ·63 | | 3 | F | 3 | ·16 | 4 | Mr | 4 | ·69 |
| 1 | S | 27 | ·41 | 2 | O | 26 | ·94 | 4 | N | 25 | ·47 | 6 | D | 25 | ·00 | 84 | 7 | Jr | 23○ | ·53 | | 2 | F | 22 | ·06 | 3 | Mr | 22 | ·59 |
| 5 | S | 15 | ·77 | 7 | O | 15 | ·31 | 1 | N | 13 | ·84 | 3 | D | 13 | ·37 | 85 | 4 | Jr | 11○ | ·90 | | 6 | F | 10 | ·43 | 7 | Mr | 11 | ·96 |
| 3 | S | 5 | ·14 | 4 | O | 4 | ·67 | 6 | N | 3 | ·20 | 7 | D | 2 | ·73 | 86 | 2 | Jr | 1●ↄ | ·26 | | 3 | Jr | 30 | ·79 | 5 | Mr | 1 | ·33 |
| 2 | S | 24 | ·04 | 3 | O | 23 | ·57 | 5 | N | 22 | ·10 | 6 | D | 21● | ·63 | 87 | 1 | Jr | 20 | ·16 | | 2 | F | 18 | ·69 | 4 | Mr | 20 | ·22 |
| 6 | S | 13 | ·41 | 7 | O | 12 | ·94 | 2 | N | 11ↄ | ·47 | 4 | D | 11● | ·00 | 88 | 5 | Jr | 9 | ·53 | | 7 | F | 8 | ·06 | 1 | Mr | 8 | ·59 |
| 3 | S | 1 | ·77 | 5 | O | 1 | ·30 | 6 | O | 30ↄ | ·83 | 1 | N | 29 | ·37 | | 2 | D | 28 | ·90 | 89 | 4 | Jr | 27 | ·43 | 5 | F | 25 | ·96 |

**TABLE X—**

| Kaliyuga | Vikrama Era | Saka Era | Commencement of Solar Year: Month and Day A.D. | Commencement of Solar Year: Fraction of Day | First New-Moon in Solar Year | Anomaly of first New-Moon | A.D. | Week-day of 1st January | Vaisakha Week-day | Vaisakha Month | Vaisakha Day | Vaisakha Fraction | Jyeshta Week-day | Jyeshta Month | Jyeshta Day | Jyeshta Fraction | Ashada Week-day | Ashada Month | Ashada Day | Ashada Fraction | Sravana Week-day | Sravana Month | Sravana Day | Sravana Fraction | Bhadrapada Week-day | Bhadrapada Month | Bhadrapada Day | Bhadrapada Fraction |
|---|---|---|---|---|---|---|---|---|---|---|---|---|---|---|---|---|---|---|---|---|---|---|---|---|---|---|---|---|
| | | | | | | | | | ☉'s Anom col. 6 | | | | + 29·53059 | | | | + 59·06117 | | | | + 88·59176 | | | | + 118·12235 | | | |
| | | | | | | | | | ☾'s Anom col. 7 | | | | + 1·976 | | | | + 3·952 | | | | + 5·928 | | | | + 7·904 | | | |
| 4490 | 1446 | 1311 | 26M | ·3958 | 1·0862 | 22·942 | **1389** | 6 | 7 | Mr | 27 | ·49 | 2 | Ap | 26●○ | ·02 | 5 | Je | 24 | ·08 | 6 | Jl | 23 | ·61 | 1 | Au | 22 | ·14 |
| | | | | | | | | | | | | | 3 | My | 25 | ·55 | | | | | | | | | | | | |
| 4491 | 1447 | 1312 | 26M | ·6546 | 19·7251 | 21·075 | **1390** | 7 | 6 | Ap | 15○ | ·39 | 7 | My | 14 | ·92 | 2 | Je | 13 | ·45 | 3 | Jl | 12 | ·98 | 5 | Au | 11 | ·51 |
| 4492 | 1448 | 1313 | 26M | ·9134 | 8 8334 | 17·232 | **1391** | 1 | 3 | Ap | 4● | ·75 | 5 | My | 4 | ·28 | 6 | Je | 2 | ·81 | 1 | Jl | 2 | ·34 | 2 | Jl | 31 | ·87 |
| | | | | | | | | | | | | | | | | | | | | | | | | | 4 | Au | 30 | ·41 |
| 4493 | 1449 | 1314 | 26M | ·1721 | 27·4723 | 15·366 | **1392** | 2 | 2 | Ap | 22 | ·65 | 4 | My | 22 | ·18 | 5 | Je | 20 | ·71 | 7 | Jl | 20 | ·24 | 1 | Au | 18○ | ·77 |
| 4494 | 1450 | 1315 | 26M | ·4309 | 16·5806 | 11·523 | **1393** | 4 | 7 | Ap | 12 | ·02 | 1 | My | 11 | ·55 | 3 | Je | 10 | ·08 | 4 | Jl | 9 | ·61 | 6 | Au | 8●○ | ·14 |
| 4495 | 1451 | 1316 | 26M | ·6896 | 5·6889 | 7·680 | **1394** | 5 | 4 | Ap | 1 | ·38 | 6 | My | 1 | ·91 | 7 | My | 30 | ·45 | 1 | Je | 28 | ·98 | 5 | Au | 27 | ·04 |
| | | | | | | | | | | | | | | | | | | | | | 3 | Jl | 28 | ·51 | | | | |
| 4496 | 1452 | 1317 | 26M | ·9484 | 24·3278 | 5·813 | **1395** | 6 | 3 | Ap | 20 | ·28 | 4 | My | 19 | ·81 | 6 | Je | 18○ | ·34 | 7 | Jl | 17 | ·87 | 2 | Au | 16 | ·40 |
| 4497 | 1453 | 1318 | 26M | ·2071 | 13·4361 | 1·970 | **1396** | 7 | 7 | Ap | 8 | ·65 | 2 | My | 8 | ·18 | 3 | Je | 6○ | ·71 | 5 | Jl | 6 | ·24 | 6 | Au | 4 | ·77 |
| 4498 | 1454 | 1319 | 26M | ·4659 | 2·5444 | 25·683 | **1397** | 2 | 5 | Mr | 29 | ·02 | 6 | Ap | 27 | ·55 | 2 | Je | 25 | ·61 | 4 | Jl | 25 | ·14 | 5 | Au | 23 | ·67 |
| | | | | | | | | | | | | | 1 | My | 27●○ | ·08 | | | | | | | | | | | | |
| 4499 | 1455 | 1320 | 26M | ·7247 | 21·1833 | 23·816 | **1398** | 3 | 3 | Ap | 16 | .91 | 5 | My | 16 | ·44 | 6 | Je | 14 | ·97 | 1 | Jl | 14 | ·51 | 3 | Au | 13 | ·04 |
| 4500 | 1456 | 1321 | 26M | ·9834 | 10·2916 | 19·973 | **1399** | 4 | 1 | Ap | 6○ | ·28 | 2 | My | 5 | ·81 | 4 | Je | 4 | ·34 | 5 | Jl | 3 | ·87 | 7 | Au | 2 | ·40 |
| 4501 | 1457 | 1322 | 26M | ·2422 | 28·9305 | 18·106 | **1400** | 5 | 7 | Ap | 24 | ·18 | 1 | My | 23 | ·71 | 3 | Je | 22 | ·24 | 4 | Jl | 21 | ·77 | 6 | Au | 20 | ·30 |
| 4502 | 1458 | 1323 | 26M | ·5009 | 18·0388 | 14·264 | **1401** | 7 | 4 | Ap | 13 | ·55 | 6 | My | 13 | ·08 | 7 | Je | 11 | ·61 | 2 | Jl | 11 | ·14 | 3 | Au | 9 | ·67 |
| 4503 | 1459 | 1324 | 26M | ·7597 | 7·1471 | 10·421 | **1402** | 1 | 1 | Ap | 2 | ·91 | 3 | My | 2 | ·44 | 4 | My | 31 | ·97 | 6 | Je | 30 | ·50 | 1 | Jl | 30○ | ·03 |
| | | | | | | | | | | | | | | | | | | | | | | | | | 2 | Au | 28 | ·57 |
| 4504 | 1460 | 1325 | 27M | ·0184 | 25·7860 | 8·554 | **1403** | 2 | 7 | Ap | 21 | ·81 | 2 | My | 21 | ·34 | 3 | Je | 19 | ·87 | 5 | Jl | 19○ | ·40 | 6 | Au | 17 | ·93 |
| 4505 | 1461 | 1326 | 26M | ·2772 | 14·8943 | 4·711 | **1404** | 3 | 5 | Ap | 10 | ·18 | 6 | My | 9 | ·71 | 1 | Je | 8 | ·24 | 2 | Jl | 7○ | ·77 | 4 | Au | 6 | ·30 |
| 4506 | 1462 | 1327 | 26M | ·5360 | 4·0026 | 0·869 | **1405** | 5 | 2 | Mr | 30 | ·54 | 4 | Ap | 29 | ·07 | 5 | My | 28 | ·61 | 1 | Jl | 26 | ·67 | 3 | Au | 25 | ·20 |
| | | | | | | | | | | | | | | | | | 7 | Je | 27 | ·14 | | | | | | | | |
| 4507 | 1463 | 1328 | 26M | ·7947 | 22·6414 | 26·556 | **1406** | 6 | 1 | Ap | 18 | ·44 | 2 | My | 17○ | ·97 | 4 | Je | 16● | ·50 | 6 | Jl | 16 | ·03 | 7 | Au | 14 | ·56 |
| 4508 | 1464 | 1329 | 27M | ·0535 | 11·7497 | 22·714 | **1407** | 7 | 5 | Ap | 7 | ·81 | 7 | My | 7○ | ·34 | 1 | Je | 5● | ·87 | 3 | Jl | 5 | ·40 | 4 | Au | 3 | ·93 |
| 4500 | 1465 | 1330 | 26M | ·3122 | 0·8580 | 18·871 | **1408** | 1 | 3 | Mr | 27 | ·18 | 6 | My | 25 | ·24 | 7 | Je | 23 | ·77 | 2 | Jl | 23 | ·30 | 3 | Au | 21 | ·83 |
| | | | | | | | | | 4 | Ap | 25●○ | ·71 | | | | | | | | | | | | | | | | |
| 4510 | 1466 | 1331 | 26M | ·5710 | 19·4969 | 17·004 | **1409** | 3 | 2 | Ap | 15 | ·07 | 3 | My | 14 | ·60 | 5 | Je | 13 | ·13 | 6 | Jl | 12 | ·67 | 1 | Au | 11 | ·20 |
| 4511 | 1467 | 1332 | 26M | ·8297 | 8·6052 | 13·161 | **1410** | 4 | 6 | Ap | 4 | ·44 | 7 | My | 3 | ·97 | 2 | Je | 2 | ·50 | 4 | Jl | 2 | ·03 | 5 | Jl | 31 | ·56 |
| | | | | | | | | | | | | | | | | | | | | | | | | | 7 | Au | 30○ | ·09 |
| 4512 | 1468 | 1333 | 27M | ·0885 | 27·2441 | 11·295 | **1411** | 5 | 5 | Ap | 23 | ·34 | 6 | My | 22 | ·87 | 1 | Je | 21 | ·40 | 2 | Jl | 20 | ·93 | 4 | Au | 19○ | ·46 |
| 4513 | 1469 | 1334 | 26M | ·3472 | 16·3524 | 7·453 | **1412** | 6 | 2 | Ap | 11 | ·71 | 4 | My | 11 | ·24 | 5 | Je | 9 | ·77 | 7 | Jl | 9 | ·30 | 1 | Au | 7 | ·83 |
| 4514 | 1470 | 1335 | 26M | ·6060 | 5·4607 | 3·609 | **1413** | 1 | 7 | Ap | 1 | ·07 | 1 | Ap | 30 | ·60 | 3 | My | 30 | ·13 | 6 | Jl | 28 | ·19 | 7 | Au | 26 | ·72 |
| | | | | | | | | | | | | | | | | | 4 | Je | 28○ | ·66 | | | | | | | | |
| 4515 | 1471 | 1336 | 26M | ·8648 | 24·0996 | 1·742 | **1414** | 2 | 5 | Ap | 19 | ·97 | 7 | My | 19 | ·50 | 2 | Je | 18○ | ·03 | 3 | Jl | 17 | ·56 | 5 | Au | 16 | ·09 |
| 4516 | 1472 | 1337 | 27M | ·1235 | 13·2079 | 25·454 | **1415** | 3 | 3 | Ap | 9 | ·34 | 4 | My | 8 | ·87 | 6 | Je | 7●○ | ·40 | 7 | Jl | 6 | ·93 | 2 | Au | 5 | ·46 |
| 4517 | 1473 | 1338 | 26M | ·3823 | 2·3162 | 21·612 | **1416** | 4 | 7 | Mr | 28 | ·70 | 2 | Ap | 27 | ·23 | 5 | Je | 25 | ·30 | 6 | Jl | 24 | ·83 | 1 | Au | 23 | ·36 |
| | | | | | | | | | | | | | 3 | My | 26● | ·77 | | | | | | | | | | | | |
| 4518 | 1474 | 1339 | 26M | ·6410 | 20·9551 | 19·745 | **1417** | 6 | 6 | Ap | 16○ | ·60 | 1 | My | 16 | ·13 | 2 | Je | 14 | ·66 | 4 | Jl | 14 | ·19 | 5 | Au | 12 | ·72 |
| 4519 | 1475 | 1340 | 26M | ·8998 | 10·0634 | 15·902 | **1418** | 7 | 3 | Ap | 5○ | ·97 | 5 | My | 5 | ·50 | 7 | Je | 4 | ·03 | 1 | Jl | 3 | ·56 | 3 | Au | 2 | ·09 |
| 4520 | 1476 | 1341 | 27M | ·1585 | 28·7023 | 14·035 | **1419** | 1 | 2 | Ap | 24 | ·87 | 4 | My | 24 | ·40 | 5 | Je | 22 | ·93 | 7 | Jl | 22 | ·46 | 1 | Au | 20 | ·99 |
| 4521 | 1477 | 1342 | 26M | ·4173 | 17·8106 | 10·193 | **1420** | 2 | 7 | Ap | 13 | ·23 | 1 | My | 12 | ·76 | 3 | Je | 11 | ·29 | 4 | Jl | 10 | ·83 | 6 | Au | 9○ | ·36 |
| 4522 | 1478 | 1343 | 26M | ·6761 | 6·9189 | 6·350 | **1421** | 4 | 4 | Ap | 2 | ·60 | 6 | My | 2 | ·13 | 7 | My | 31 | ·66 | 2 | Je | 30 | ·19 | 5 | Au | 28● | ·25 |
| | | | | | | | | | | | | | | | | | | | | | 3 | Jl | 29○ | ·72 | | | | |
| 4523 | 1479 | 1344 | 26M | ·9348 | 23·5578 | 4·483 | **1422** | 5 | 3 | Ap | 21 | ·50 | 5 | My | 21 | ·03 | 6 | Je | 19 | ·56 | 1 | Jl | 19○ | ·09 | 2 | Au | 17 | ·62 |
| 4524 | 1480 | 1345 | 27M | ·1936 | 14·6660 | 0·640 | **1423** | 6 | 7 | Ap | 10 | ·87 | 2 | My | 10 | ·40 | 3 | Je | 8 | ·93 | 5 | Jl | 8● | ·46 | 6 | Au | 6 | ·99 |
| 4525 | 1481 | 1346 | 26M | ·4523 | 3·7743 | 24·352 | **1424** | 7 | 5 | Mr | 30 | ·23 | 6 | Ap | 28 | ·76 | 1 | My | 28○ | ·29 | 4 | Jl | 26 | ·35 | 5 | Au | 24 | ·89 |
| | | | | | | | | | | | | | | | | | 2 | Je | 26 | ·82 | | | | | | | | |
| 4526 | 1482 | 1347 | 26M | ·7111 | 22·4132 | 22·485 | **1425** | 2 | 4 | Ap | 18 | ·13 | 5 | My | 17○ | ·66 | 7 | Je | 16 | ·19 | 1 | Jl | 15 | ·72 | 3 | Au | 14 | ·25 |
| 4527 | 1483 | 1348 | 26M | ·9698 | 11·5215 | 18·644 | **1426** | 3 | 1 | Ap | 7 | ·50 | 3 | My | 7○ | ·03 | 4 | Je | 5 | ·56 | 6 | Jl | 5 | ·09 | 7 | Au | 3 | ·62 |
| 4528 | 1484 | 1349 | 27M | ·2286 | 0·6298 | 14·800 | **1427** | 4 | 5 | Mr | 27○ | ·86 | 1 | My | 25 | ·93 | 3 | Je | 24 | ·46 | 4 | Jl | 23 | ·99 | 6 | Au | 22 | ·52 |
| | | | | | | | | | 7 | Ap | 26 | ·39 | | | | | | | | | | | | | | | | |

**Śurya Siddhanta.**

| Asvina | | | | Kartika | | | | Margasira | | | | Pausha | | | | A.D. | Magha | | | | A.D. | Phalguna | | | | Chaitra | | | |
|---|---|---|---|---|---|---|---|---|---|---|---|---|---|---|---|---|---|---|---|---|---|---|---|---|---|---|---|---|---|
| + 147·65293 | | | | + 177·18353 | | | | + 206·71411 | | | | + 236·24470 | | | | | + 265·77529 | | | | | + 295·80588 | | | | +324·83647<br>+354·36705 | | | |
| + 9·880 | | | | + 11·856 | | | | + 13·832 | | | | + 15·808 | | | | | + 17·784 | | | | | + 19·760 | | | | +21·736<br>+23·712 | | | |
| Week-day | Month | Day | Fraction | Week-day | Month | Day | Fraction | Week-day | Month | Day | Fraction | Week-day | Month | Day | Fraction | | Week-day | Month | Day | Fraction | | Week-day | Month | Day | Fraction | Week-day | Month | Day | Fraction |
| 2 | S | 20 | ·67 | 4 | O | 20ↄ | ·20 | 5 | N | 18 | ·73 | 7 | D | 18 | ·26 | **90** | 1 | Jr | 16 | ·79 | | 3 | F | 15 | ·32 | 4 | Mr | 16 | ·85 |
| 7 | S | 10 | ·04 | 1 | O | 9● | ·57 | 3 | N | 8 | ·10 | 4 | D | 7 | ·63 | **91** | 6 | Jr | 6 | ·16 | | 7 | F | 4 | ·69 | 2 | Mr | 6ↄ | ·22 |
| 5 | S | 28 | ·94 | 7 | O | 28 | ·47 | 2 | N | 27 | ·00 | 3 | D | 26 | ·53 | **92** | 5 | Jr | 25 | ·06 | | 6 | F | 23ↄ | ·59 | 1 | Mr | 24 | ·12 |
| 3 | S | 17 | ·30 | 4 | O | 16 | ·83 | 6 | N | 15 | ·36 | 7 | D | 14 | ·89 | **93** | 2 | Jr | 13 | ·43 | | 3 | F | 11○ | ·96 | 5 | Mr | 13 | ·49 |
| 7 | S | 6 | ·67 | 2 | O | 6 | ·20 | 3 | N | 4 | ·73 | 5 | D | 4 | ·26 | **94** | 6 | Jr | 2 | ·79 | | 1 | F | 1● | ·32 | 2 | Mr | 2 | ·85 |
| 6 | S | 25 | ·57 | 1 | O | 25 | ·10 | 2 | N | 23 | ·63 | 4 | D | 23 | ·16 | **95** | 5 | Jr | 21 | ·69 | | 7 | F | 20 | ·22 | 1 | Mr | 21 | ·75 |
| 3 | S | 14 | ·93 | 5 | O | 14 | ·47 | 7 | N | 13 | ·00 | 1 | D | 12○ | ·53 | **96** | 3 | Jr | 11 | ·06 | | 4 | F | 9 | ·59 | 6 | Mr | 10 | ·12 |
| 1 | S | 3 | ·30 | 2 | O | 2 | ·83 | 4 | N | 1 | ·36 | 5 | N | 30ↄ | ·89 | | 7 | D | 30 | ·42 | **97** | 1 | Jr | 28 | ·95 | 3 | F | 27 | ·49 |
| 7 | S | 22 | ·20 | 1 | O | 21 | ·73 | 3 | N | 20 | ·26 | 4 | D | 18 | ·79 | **98** | 6 | Jr | 18 | ·32 | | 7 | F | 16 | ·85 | 2 | Mr | 18 | ·38 |
| 4 | S | 11 | ·57 | 6 | O | 11ↄ | ·10 | 7 | N | 9● | ·63 | 2 | D | 9 | ·16 | **99** | 3 | Jr | 7 | ·69 | | 5 | F | 6 | ·22 | 6 | Mr | 7 | ·75 |
| 1 | Au | 31 | ·93 | 3 | S | 30ↄ | ·46 | 6 | N | 28 | ·53 | **Pausha.** | | | | **14** | 1 | D | 28 | ·06 | **00** | 2 | Jr | 26 | ·59 | 4 | Fr | 25 | ·12 |
| | | | | 4 | O | 29 | ·99 | | | | | **Kshaya.** | | | | | | | | | | | | | | 5 | Mr | 25●○ | ·65 |
| 7 | S | 18ↄ | ·83 | 2 | O | 18 | ·36 | 3 | N | 16 | ·89 | 5 | D | 16 | ·42 | **01** | 6 | Jr | 14 | ·95 | | 1 | F | 13 | ·48 | 3 | Mr | 15● | ·01 |
| 5 | S | 8● | ·20 | 6 | O | 7 | ·73 | 1 | N | 6 | ·26 | 2 | D | 5 | ·79 | **02** | 4 | Jr | 4 | ·32 | | 5 | F | 2○ | ·85 | 7 | Mr | 4● | ·38 |
| 4 | S | 27 | ·10 | 5 | O | 26 | ·63 | 7 | N | 25 | ·16 | 1 | D | 24 | ·69 | **03** | 3 | Jr | 23ↄ | ·22 | | 4 | F | 21 | ·75 | 6 | Mr | 23 | ·28 |
| 1 | S | 16 | ·46 | 2 | O | 15 | ·99 | 4 | N | 14 | ·52 | 6 | D | 14 | ·05 | **04** | 7 | Jr | 12ↄ | ·59 | | 2 | F | 11 | ·12 | 3 | Mr | 11 | ·65 |
| 5 | S | 4 | ·83 | 7 | O | 4 | ·36 | 1 | N | 2 | ·89 | 3 | D | 2 | ·42 | | 4 | D | 31● | ·95 | **05** | 6 | Jr | 30 | ·48 | 1 | Mr | 1 | ·01 |
| 4 | S | 23 | ·73 | 6 | O | 23 | ·26 | 7 | N | 21○ | ·79 | 2 | D | 21 | ·32 | **06** | 3 | Jr | 19 | ·85 | | 5 | F | 18 | ·38 | 6 | Mr | 19 | ·91 |
| 2 | S | 13 | ·09 | 3 | O | 12 | ·63 | 5 | N | 11ↄ | ·16 | 6 | D | 10 | ·69 | **07** | 1 | Jr | 9 | ·22 | | 2 | F | 7 | ·75 | 4 | Mr | 9 | ·28 |
| 6 | S | 2 | ·46 | 7 | O | 1 | ·99 | 2 | O | 31ↄ | ·52 | 4 | N | 30 | ·05 | | 5 | D | 29 | ·58 | **08** | 7 | Jr | 28 | ·11 | 1 | F | 26 | ·64 |
| 5 | S | 20 | ·36 | 6 | O | 19● | ·89 | 1 | N | 18 | ·42 | 2 | D | 17 | ·95 | **09** | 4 | Jr | 16 | ·48 | | 6 | F | 15 | ·01 | 7 | Mr | 16ↄ | ·54 |
| 2 | S | 9 | ·73 | 4 | O | 9● | ·26 | 5 | N | 7 | ·79 | 7 | D | 7 | ·32 | **10** | 1 | Jr | 5 | ·85 | | 3 | F | 4 | ·38 | 4 | Mr | 5○ | ·91 |
| 1 | S | 28 | ·62 | 3 | O | 28 | ·15 | 4 | N | 26 | ·69 | 6 | D | 26 | ·22 | **11** | 7 | Jr | 24 | ·75 | | 2 | F | 23ↄ | ·28 | 3 | Mr | 24 | ·81 |
| 5 | S | 17 | ·99 | 7 | O | 17 | ·52 | 2 | N | 16 | ·05 | 3 | D | 15 | ·58 | **12** | 5 | Jr | 14 | ·11 | | 6 | F | 12● | ·34 | 1 | Mr | 13 | ·17 |
| 3 | S | 6 | ·36 | 4 | O | 5 | ·89 | 6 | N | 4 | ·42 | 7 | D | 3 | ·95 | **13** | 2 | Jr | 2 | ·48 | | 4 | F | 1● | ·01 | 5 | Mr | 2 | ·54 |
| 2 | S | 25 | ·26 | 3 | O | 24 | ·79 | 5 | N | 23 | ·32 | 6 | D | 22ↄ | ·85 | **14** | 1 | Jr | 21 | ·38 | | 2 | F | 19 | ·91 | 4 | Mr | 21 | ·44 |
| 6 | S | 14 | ·62 | 1 | O | 14 | ·15 | 2 | N | 12 | ·68 | 4 | D | 12ↄ | ·21 | **15** | 5 | Jr | 18 | ·75 | | 7 | F | 9 | ·28 | 1 | Mr | 10 | ·81 |
| 3 | S | 3 | ·99 | 5 | O | 3 | ·52 | 7 | N | 2 | ·05 | 1 | D | 1 | ·58 | | 3 | D | 31 | ·11 | **16** | 4 | Jr | 29 | ·64 | 6 | Fr | 28 | ·17 |
| 2 | S | 21 | ·89 | 4 | O | 21○ | ·42 | 5 | N | 19 | ·95 | 7 | D | 19 | ·48 | **17** | 2 | Jr | 13 | ·01 | | 3 | F | 16 | ·54 | 5 | Mr | 18 | ·07 |
| 7 | S | 11 | ·25 | 1 | O | 10ↄ | ·79 | 3 | N | 9 | ·32 | 4 | D | 8 | ·85 | **18** | 6 | Jr | 7 | ·38 | | 7 | F | 5 | ·91 | 2 | Mr | 7 | ·44 |
| 4 | Au | 31 | ·62 | 6 | S | 30○ | ·15 | 2 | N | 28 | ·21 | 3 | D | 27 | ·74 | **19** | 5 | Jr | 26 | ·27 | | 6 | F | 24 | ·81 | 1 | Mr | 26●○ | ·34 |
| | | | | 7 | O | 29 | ·68 | | | | | | | | | | | | | | | | | | | | | | |
| 3 | S | 19 | ·52 | 5 | O | 19 | ·05 | 6 | N | 17 | ·58 | 1 | D | 17 | ·11 | **20** | 2 | Jr | 15 | ·64 | | 4 | F | 14ↄ | ·17 | 5 | Mr | 14 | ·70 |
| 7 | S | 7● | ·89 | 2 | O | 7 | ·42 | 3 | N | 5 | ·95 | 5 | D | 5 | ·48 | **21** | 7 | Jr | 4 | ·01 | | 1 | F | 2○ | ·54 | 3 | Mr | 4 | ·07 |
| 6 | S | 26 | ·78 | 1 | O | 26 | ·31 | 2 | N | 24 | ·85 | 4 | D | 24 | ·38 | **22** | 5 | Jr | 22●○ | ·91 | | 7 | F | 21 | ·44 | 1 | Mr | 22 | ·97 |
| 4 | S | 16 | ·15 | 5 | O | 15 | ·68 | 7 | N | 14 | ·21 | 1 | D | 13 | ·74 | **23** | 3 | Jr | 12 | ·27 | | 4 | F | 10 | ·80 | 6 | Mr | 12 | ·33 |
| 1 | S | 5 | ·52 | 3 | O | 5 | ·05 | 4 | N | 3 | ·58 | 6 | D | 3ↄ | ·11 | **24** | 7 | Jr | 1● | ·64 | | 2 | Jr | 31 | ·17 | 3 | F | 29 | ·70 |
| 7 | S | 23 | ·42 | 1 | O | 22 | ·95 | 3 | N | 21ↄ | ·48 | 5 | D | 21 | ·01 | **25** | 6 | Jr | 19 | ·54 | | 1 | F | 18 | ·07 | 2 | Mr | 19 | ·60 |
| 4 | S | 12 | ·78 | 6 | O | 12 | ·31 | 7 | N | 10●○ | ·84 | 2 | D | 10 | ·37 | **26** | 3 | Jr | 8 | ·91 | | 5 | F | 7 | ·44 | 6 | Mr | 8 | ·97 |
| 2 | S | 2 | ·15 | 3 | O | 1 | ·68 | 5 | O | 31 | ·21 | 6 | N | 29 | ·74 | | 1 | D | 29 | ·27 | **27** | 2 | Jr | 27 | ·80 | 4 | F | 26 | ·38 |
| 1 | S | 21 | ·05 | 2 | O | 20 | ·58 | 4 | N | 19 | ·11 | 5 | D | 18 | ·64 | **28** | 7 | Jr | 17 | ·17 | | 1 | F | 15 | ·70 | 3 | Mr | 16○ | ·23 |

⊙'s Anom col. 6 + 29·53059 + 59·06117 + 88·59176 + 118·12235

☾'s Anom col. 7 + 1·976 + 3·952 + 5·928 + 7·904

| Kaliyuga. | Vikrama Era. | Saka Era. | Commencement of Solar Year. Month and day A. D. | Fraction of day. | First New-Moon in Solar Year. | Anomaly of first New-Moon. | A.D. | Week-day of 1st January. | Vaisakha Week-day | Month | Day | Fraction | Jyeshta Week-day | Month | Day | Fraction | Ashada Week-day | Month | Day | Fraction | Sravana Week-day | Month | Day | Fraction | Bhadrapada Week-day | Month | Day | Fraction |
|---|---|---|---|---|---|---|---|---|---|---|---|---|---|---|---|---|---|---|---|---|---|---|---|---|---|---|---|---|
| 4529 | 1485 | 1350 | 26M | ·4873 | 19·2687 | 12·933 | **1428** | 5 | 4 | Ap | 14 | ·76 | 6 | My | 14 | ·29 | 7 | Je | 12 | ·82 | 2 | Jl | 12 | ·35 | 3 | Au | 10 | ·88 |
| 4530 | 1486 | 1351 | 26M | ·7461 | 8·3770 | 9·090 | **1429** | 7 | 2 | Ap | 4 | ·13 | 3 | My | 3 | ·66 | 5 | Je | 2 | ·19 | 6 | Jl | 1 | ·72 | { 1 | Jl | 31 | ·25 |
| | | | | | | | | | | | | | | | | | | | | | | | | | { 2 | Au | 29○ | ·78 |
| 4531 | 1487 | 1352 | 27M | ·0049 | 27·0159 | 7·224 | **1430** | 1 | 1 | Ap | 23 | ·03 | 2 | My | 22 | ·56 | 4 | Je | 21 | ·09 | 5 | Jl | 20 | ·62 | 7 | Au | 19●○ | ·15 |
| 4532 | 1488 | 1353 | 27M | ·2636 | 16·1242 | 8·381 | **1431** | 2 | 5 | Ap | 12 | ·39 | 6 | My | 11 | ·92 | 1 | Je | 10 | ·45 | 2 | Jl | 9○ | ·99 | 4 | Au | 8● | ·52 |
| 4533 | 1489 | 1354 | 26M | ·5224 | 5·2325 | 27·093 | **1432** | 3 | 2 | Mr | 31 | ·76 | 4 | Ap | 30 | ·29 | { 5 | My | 29 | ·82 | 1 | Jl | 27 | ·88 | 3 | Au | 26 | ·41 |
| | | | | | | | | | | | | | | | | | { 7 | Je | 28○ | ·35 | | | | | | | | |
| 4534 | 1490 | 1355 | 26M | ·7811 | 23·8714 | 25·226 | **1433** | 5 | 1 | Ap | 19 | ·66 | 3 | My | 19 | ·19 | 4 | Je | 17○ | ·72 | 6 | Jl | 17 | ·25 | 7 | Au | 15 | ·78 |
| 4535 | 1491 | 1356 | 27M | ·0399 | 12·9797 | 21·383 | **1434** | 6 | 6 | Ap | 9 | ·03 | 7 | My | 8 | ·56 | 2 | Je | 7● | ·09 | 3 | Jl | 6 | ·62 | 5 | Au | 5 | ·15 |
| 4536 | 1492 | 1357 | 27M | ·2986 | 2·0880 | 17·541 | **1435** | 7 | 3 | Mr | 29 | ·39 | { 4 | Ap | 27○ | ·92 | 7 | Je | 25 | ·98 | 2 | Jl | 25 | ·51 | 4 | Au | 24 | ·05 |
| | | | | | | | | | | | | | { 6 | My | 27 | ·45 | | | | | | | | | | | | |
| 4537 | 1493 | 1358 | 26M | ·5574 | 20·7269 | 15·674 | **1436** | 1 | 2 | Ap | 16○ | ·29 | 3 | My | 15 | ·82 | 5 | Je | 14 | ·35 | 6 | Jl | 13 | ·88 | 1 | Au | 12 | ·41 |
| 4538 | 1494 | 1359 | 26M | ·8162 | 9·8352 | 11·831 | **1437** | 3 | 6 | Ap | 5○ | ·66 | 1 | My | 5 | ·19 | 2 | Je | 8 | ·72 | 4 | Jl | 3 | ·25 | 5 | Au | 1 | ·78 |
| 4539 | 1495 | 1360 | 27M | ·0749 | 28·4741 | 8·964 | **1438** | 4 | 5 | Ap | 24 | ·55 | 7 | My | 24 | ·09 | 1 | Je | 22 | ·62 | 3 | Jl | 22 | ·15 | 4 | Au | 20 | ·68 |
| 4540 | 1496 | 1361 | 27M | ·3337 | 17·5824 | 6·122 | **1439** | 5 | 2 | Ap | 13 | ·92 | 4 | My | 13 | ·45 | 5 | Je | 11 | ·98 | 7 | Jl | 11 | ·51 | 2 | Au | 10○ | .04 |
| 4541 | 1497 | 1362 | 26M | ·5924 | 6·6907 | 2·279 | **1440** | 6 | 7 | Ap | 2 | ·29 | 1 | My | 1 | ·81 | 3 | My | 31 | ·35 | { 4 | Je | 29 | ·88 | 7 | Au | 27 | ·94 |
| | | | | | | | | | | | | | | | | | | | | | { 6 | Jl | 29○ | ·41 | | | | |
| 4542 | 1498 | 1363 | 26M | ·8512 | 25·3295 | 0·412 | **1441** | 1 | 6 | Ap | 21 | ·19 | 7 | My | 20 | ·72 | 2 | Je | 19 | ·25 | 3 | Jl | 18● | ·78 | 5 | Au | 17 | ·31 |
| 4543 | 1499 | 1364 | 27M | ·1099 | 14·4378 | 24·124 | **1442** | 2 | 3 | Ap | 10 | ·55 | 5 | My | 10 | ·08 | 6 | Je | 8○ | ·61 | 1 | Jl | 8 | ·15 | 2 | Au | 6 | ·68 |
| 4544 | 1500 | 1365 | 27M | ·3687 | 3·5461 | 20·281 | **1443** | 3 | 7 | Mr | 30 | ·92 | 2 | Ap | 29 | ·45 | { 3 | My | 28○ | ·98 | 7 | Jl | 27 | ·04 | 1 | Au | 25 | ·57 |
| | | | | | | | | | | | | | | | | | { 5 | Je | 27 | ·51 | | | | | | | | |
| 4545 | 1501 | 1366 | 26M | ·6275 | 22·1850 | 18·414 | **1444** | 4 | 6 | Ap | 17 | ·82 | 1 | My | 17○ | ·35 | 2 | Je | 15 | ·88 | 4 | Jl | 15 | ·41 | 5 | Au | 13 | ·94 |
| 4546 | 1502 | 1367 | 26M | ·8862 | 11·2933 | 14·572 | **1445** | 6 | 4 | Ap | 7 | ·18 | 5 | My | 6● | ·71 | 7 | Je | 5 | ·25 | 1 | Jl | 4 | ·78 | 3 | Au | 3 | ·31 |
| 4547 | 1503 | 1368 | 27M | ·1450 | 0·4016 | 10·729 | **1446** | 7 | { 1 | Mr | 27○ | ·55 | 4 | My | 25 | ·61 | 6 | Je | 24 | ·14 | 7 | Jl | 23 | ·67 | 2 | Au | 22 | ·20 |
| | | | | | | | | | { 3 | Ap | 26● | ·08 | | | | | | | | | | | | | | | | |
| 4548 | 1504 | 1369 | 27M | ·4087 | 19·0405 | 8·862 | **1447** | 1 | 7 | Ap | 15 | ·45 | 1 | My | 14 | ·98 | 3 | Je | 13 | ·51 | 5 | Jl | 13 | ·04 | 6 | Au | 11 | ·57 |
| 4549 | 1505 | 1370 | 26M | ·6625 | 8·1488 | 5·019 | **1448** | 2 | 4 | Ap | 3 | ·82 | 6 | My | 3 | ·35 | 7 | Je | 1 | ·88 | 2 | Jl | 1 | ·41 | { 3 | Jl | 30 | ·94 |
| | | | | | | | | | | | | | | | | | | | | | | | | | { 5 | Au | 29●○ | ·47 |
| 4550 | 1506 | 1371 | 26M | ·9212 | 26·7877 | 3·153 | **1449** | 4 | 3 | Ap | 22 | ·71 | 5 | My | 22 | ·24 | 6 | Je | 20 | ·77 | 1 | Jl | 20○ | ·31 | 2 | Au | 18 | ·84 |
| 4551 | 1507 | 1372 | 27M | ·1800 | 15·8960 | 26·864 | **1450** | 5 | 1 | Ap | 12 | ·08 | 2 | My | 11 | ·61 | 4 | Je | 10 | ·14 | 5 | Jl | 9○ | ·67 | 7 | Au | 8 | ·20 |
| 4552 | 1508 | 1373 | 27M | ·4387 | 5·0043 | 23·022 | **1451** | 6 | 5 | Ap | 1 | ·45 | 6 | Ap | 30 | ·98 | { 1 | My | 30 | ·51 | 4 | Jl | 28 | ·57 | 6 | Au | 27 | ·10 |
| | | | | | | | | | | | | | | | | | { 3 | Je | 29○ | ·04 | | | | | | | | |
| 4553 | 1509 | 1374 | 26M | ·6975 | 23·6432 | 21·155 | **1452** | 7 | 4 | Ap | 19 | ·35 | 5 | My | 18 | ·88 | 7 | Je | 17 | ·41 | 1 | Jl | 16 | ·94 | 3 | Au | 15 | ·47 |
| 4554 | 1510 | 1375 | 26M | ·9563 | 12·7515 | 17·312 | **1453** | 2 | 1 | Ap | 8 | ·71 | 2 | My | 8○ | ·24 | 4 | Je | 6● | ·77 | 6 | Jl | 6 | ·30 | 7 | Au | 4 | ·83 |
| 4555 | 1511 | 1376 | 27M | ·2150 | 1·8598 | 13·469 | **1454** | 3 | 6 | Mr | 29 | ·08 | { 7 | Ap | 27●○ | ·61 | 3 | Je | 25 | ·67 | 5 | Jl | 25 | ·20 | 6 | Au | 23 | ·73 |
| | | | | | | | | | | | | | { 2 | My | 27 | ·14 | | | | | | | | | | | | |
| 4556 | 1512 | 1377 | 27M | ·4738 | 20·4987 | 11·603 | **1455** | 4 | 4 | Ap | 16●○ | ·98 | 6 | My | 16 | ·51 | 1 | Je | 15 | ·04 | 2 | Jl | 14 | ·57 | 4 | Au | 13 | ·10 |
| 4557 | 1513 | 1378 | 26M | ·7325 | 9·6070 | 7·760 | **1456** | 5 | 2 | Ap | 5● | ·34 | 3 | My | 4 | ·87 | 5 | Je | 3 | ·41 | 6 | Jl | 2 | ·94 | 1 | Au | 1 | ·47 |
| 4558 | 1514 | 1379 | 26M | ·9913 | 28·2459 | 5·893 | **1457** | 7 | 1 | Ap | 24 | ·24 | 2 | My | 23 | ·77 | 4 | Je | 22 | ·30 | 5 | Jl | 21 | ·83 | 7 | Au | 20○ | ·36 |
| 4559 | 1515 | 1380 | 27M | ·2500 | 17·3542 | 2·050 | **1458** | 1 | 5 | Ap | 13 | ·61 | 7 | My | 13 | ·14 | 1 | Je | 11 | ·67 | 3 | Jl | 11 | ·20 | 4 | Au | 9○ | ·73 |
| 4560 | 1516 | 1381 | 27M | ·5088 | 6·4625 | 25·762 | **1459** | 2 | 2 | Ap | 2 | ·98 | 4 | My | 2 | ·51 | 6 | Je | 1 | ·04 | { 7 | Je | 30 | ·57 | 3 | Au | 28 | ·63 |
| | | | | | | | | | | | | | | | | | | | | | { 2 | Jl | 30 | ·10 | | | | |
| 4561 | 1517 | 1382 | 26M | ·7676 | 25·1013 | 23·896 | **1460** | 3 | 1 | Ap | 20 | ·87 | 3 | My | 20 | ·40 | 4 | Je | 18○ | ·93 | 6 | Jl | 18● | ·47 | 1 | Au | 17 | ·00 |
| 4562 | 1518 | 1383 | 27M | ·0253 | 14·2096 | 20·053 | **1461** | 5 | 6 | Ap | 10 | ·24 | 7 | My | 9 | ·77 | 2 | Je | 8○ | ·30 | 3 | Jl | 7● | ·83 | 5 | Au | 6 | ·36 |
| 4563 | 1519 | 1384 | 27M | ·2851 | 3·3179 | 16·210 | **1462** | 6 | 3 | Mr | 30 | ·61 | 5 | Ap | 29 | ·14 | { 6 | My | 28●○ | ·67 | 2 | Jl | 26 | ·73 | 4 | Au | 25 | ·26 |
| | | | | | | | | | | | | | | | | | { 1 | Je | 27 | ·20 | | | | | | | | |
| 4564 | 1520 | 1385 | 27M | ·5438 | 21·9568 | 14·343 | **1463** | 7 | 2 | Ap | 18 | ·51 | 4 | My | 18● | ·04 | 5 | Je | 16 | ·57 | 7 | Jl | 16 | ·10 | 1 | Au | 14 | ·63 |
| 4565 | 1521 | 1386 | 26M | ·8026 | 11·0651 | 10·501 | **1464** | 1 | 6 | Ap | 6○ | ·87 | 1 | My | 6● | ·40 | 2 | Je | 4 | ·93 | 4 | Jl | 4 | ·46 | 5 | Au | 2 | ·99 |
| 4566 | 1522 | 1387 | 27M | ·0613 | 0·1734 | 6·658 | **1465** | 3 | 5 | Ap | 25 | ·77 | 7 | My | 25 | ·30 | 1 | Je | 23 | ·83 | 3 | Jl | 23 | ·36 | 4 | Au | 21 | ·89 |
| 4567 | 1523 | 1388 | 27M | ·3201 | 18·8123 | 4·791 | **1466** | 4 | 3 | Ap | 15 | ·15 | 4 | My | 14 | ·67 | 6 | Je | 13 | ·20 | 7 | Jl | 12 | ·73 | 2 | Au | 11 | ·26 |

**Surya Siddhanta.**

| Asvina | | | | Kartika | | | | Margasira | | | | Pausha | | | | A.D. | Magha | | | | A.D. | Phalguna | | | | Chaitra | | | |
|---|---|---|---|---|---|---|---|---|---|---|---|---|---|---|---|---|---|---|---|---|---|---|---|---|---|---|---|---|---|
| + 147·652?3 | | | | + 177·18353 | | | | +206·71411 | | | | + 236·24470 | | | | | + 265·77529 | | | | | + 295·30588 | | | | +324·83647<br>+354·??705 | | | |
| + 9·880 | | | | + 11·856 | | | | + 13·832 | | | | + 15·808 | | | | | + 17·784 | | | | | + 19·760 | | | | +21·736<br>+23·712 | | | |
| Week-day | Month | Day | Fraction | Week-day | Month | Day | Fraction | Week-day | Month | Day | Fraction | Week-day | Month | Day | Fraction | | Week-day | Month | Day | Fraction | | Week-day | Month | Day | Fraction | Week-day | Month | Day | Fraction |
| 5 | S | 9○ | ·41 | 6 | O | 8● | ·95 | 1 | N | 7 | ·48 | 3 | D | 7 | ·01 | **29** | 4 | Jr | 5 | ·54 | | 6 | F | 4 | ·07 | 7 | Mr | 5●○ | ·60 |
| 4 | S | 28 | ·31 | 5 | O | 27 | ·84 | 7 | N | 26 | ·37 | 1 | D | 25 | ·90 | **30** | 3 | Jr | 24 | ·43 | | 4 | F | 22 | ·96 | 6 | Mr | 24 | ·50 |
| 1 | S | 17 | ·68 | 3 | O | 17 | ·21 | 4 | N | 15 | ·74 | 6 | D | 15 | ·27 | **31** | 7 | Jr | 13 | ·80 | | 2 | F | 12 | ·33 | 3 | Mr | 13 | ·86 |
| 6 | S | 7 | ·05 | 7 | O | 6 | ·58 | 2 | N | 5 | ·11 | 3 | D | 4 | ·64 | **32** | 5 | Jr | 3○ | ·17 | | 6 | F | 1● | ·70 | 1 | Mr | 2 | ·23 |
| 4 | S | 24 | ·94 | 6 | O | 24 | ·47 | 1 | N | 23 | ·01 | 2 | D | 22○ | ·54 | **33** | 4 | Jr | 21 | ·07 | | 5 | F | 19 | ·60 | 7 | Mr | 21 | ·13 |
| 2 | S | 14 | ·31 | 3 | O | 13 | ·84 | 5 | N | 12 | ·37 | 6 | D | 11 | ·90 | **34** | 1 | Jr | 10 | ·43 | | 2 | F | 8 | ·96 | 4 | Mr | 10 | ·49 |
| 6 | S | 3 | ·68 | 1 | O | 3 | ·21 | 2 | N | 1 | ·74 | 4 | D | 1 | ·27 | | 5 | D | 30 | ·80 | **35** | 7 | Jr | 29 | ·33 | 1 | F | 27 | ·86 |
| 5 | S | 22 | ·58 | 7 | O | 22○ | ·11 | 1 | N | 20● | ·64 | 3 | D | 20 | ·17 | **36** | 4 | Jr | 18 | ·70 | | 6 | F | 17 | ·23 | 7 | Mr | 17 | ·76 |
| 2 | S | 10 | ·94 | 4 | O | 10○ | ·47 | 6 | N | 9 | ·00 | 7 | D | 8 | ·53 | **37** | 2 | Jr | 7 | ·07 | | 3 | F | 5 | ·60 | 5 | Mr | 7 | ·13 |
| 7 | Au | 31 | ·31 | 1 | S | 29● | ·84 | 4 | N | 27 | ·90 | 6 | D | 27 | ·43 | **38** | 7 | Jr | 25 | ·96 | | 2 | F | 24○ | ·49 | 4 | Mr | 26 | ·02 |
| | | | | 3 | O | 29 | ·37 | | | | | | | | | | | | | | | | | | | | | | |
| 6 | S | 19● | ·21 | 7 | O | 18 | ·74 | 2 | N | 17 | ·27 | 3 | D | 16 | ·80 | **39** | 5 | Jr | 15 | ·33 | | 6 | F | 13○ | ·86 | 1 | Mr | 15 | ·39 |
| 3 | S | 8 | ·57 | 5 | O | 8 | ·11 | 6 | N | 6 | ·64 | 1 | D | 6 | ·17 | **40** | 2 | Jr | 4 | ·70 | | 4 | F | 3○ | ·23 | 5 | Mr | 3 | ·76 |
| 2 | S | 26 | ·47 | 4 | O | 26 | ·00 | 5 | N | 24 | ·53 | 7 | D | 24 | ·06 | **41** | 1 | Jr | 22● | ·59 | | 3 | F | 21 | ·13 | 4 | Mr | 22 | ·66 |
| 6 | S | 15 | ·84 | 1 | O | 15 | ·37 | 2 | N | 13 | ·90 | 4 | D | 13○ | ·43 | **42** | 5 | Jr | 11● | ·96 | | 7 | F | 10 | ·49 | 2 | Mr | 12 | ·02 |
| 4 | S | 5 | ·21 | 5 | O | 4 | ·74 | 7 | N | 3 | ·27 | 1 | D | 2○ | ·80 | **43** | 3 | Jr | 1 | ·33 | | 4 | Jr | 30 | ·86 | 6 | Mr | 1 | ·39 |
| 3 | S | 24 | ·10 | 4 | O | 23 | ·63 | 6 | N | 22○ | ·17 | 7 | D | 21 | ·70 | **44** | 2 | Jr | 20 | ·23 | | 3 | F | 18 | ·76 | 5 | Mr | 19 | ·29 |
| 7 | S | 12 | ·47 | 2 | O | 12 | ·00 | 3 | N | 10● | ·53 | 5 | D | 10 | ·06 | **45** | 6 | Jr | 8 | ·59 | | 1 | F | 7 | ·12 | 2 | Mr | 8 | ·65 |
| 4 | S | 1 | ·84 | 6 | O | 1 | ·37 | 7 | O | 30 | ·90 | 2 | N | 29 | ·43 | | 3 | D | 28 | ·96 | **46** | 5 | Jr | 27 | ·49 | 7 | F | 26 | ·02 |
| 3 | S | 20○ | ·73 | 5 | O | 20 | ·27 | 6 | N | 18 | ·80 | 1 | D | 18 | ·33 | **47** | 2 | Jr | 16 | ·86 | | 4 | F | 15 | ·39 | 5 | Mr | 16○ | ·92 |
| 1 | S | 10●○ | ·10 | 2 | O | 9 | ·63 | 4 | N | 8 | ·16 | 5 | D | 7 | ·69 | **48** | 7 | Jr | 6 | ·22 | | 1 | F | 4 | ·75 | 3 | Mr | 5● | ·29 |
| 7 | S | 28 | ·00 | 1 | O | 27 | ·53 | 3 | N | 26 | ·06 | 4 | D | 25 | ·59 | **49** | 6 | Jr | 24 | ·12 | | 7 | F | 22 | ·65 | 2 | Mr | 24 | ·18 |
| 4 | S | 17 | ·37 | 5 | O | 16 | ·90 | 7 | N | 15 | ·43 | 1 | D | 14 | ·96 | **50** | 3 | Jr | 13○ | ·49 | | 5 | F | 12 | ·02 | 6 | Mr | 13 | ·55 |
| 1 | S | 6 | ·73 | 3 | O | 6 | ·26 | 4 | N | 4 | ·79 | 6 | D | 4 | ·33 | **51** | 7 | Jr | 2○ | ·86 | | 2 | F | 1 | ·39 | 3 | Mr | 2 | ·92 |
| 7 | S | 25 | ·63 | 2 | O | 25 | ·16 | 3 | N | 23 | ·62 | 5 | D | 23● | ·22 | **52** | 6 | Jr | 21 | ·75 | | 1 | F | 20 | ·28 | 2 | Mr | 20 | ·81 |
| 5 | S | 14 | ·00 | 6 | O | 13 | ·53 | 1 | N | 12 | ·06 | 2 | D | 11● | ·59 | **53** | 4 | Jr | 19 | ·12 | | 5 | F | 8 | ·65 | 7 | Mr | 10 | ·18 |
| 2 | S | 3 | ·37 | 3 | O | 2 | ·90 | 5 | N | 1○ | ·43 | 6 | N | 30 | ·96 | | 1 | D | 30 | ·49 | **54** | 3 | Jr | 29 | ·02 | 4 | F | 27 | ·55 |
| 1 | S | 22 | ·26 | 2 | O | 21○ | ·79 | 4 | N | 20 | ·32 | 5 | D | 19 | ·85 | **55** | 7 | Jr | 18 | ·39 | | 1 | F | 16 | ·92 | 3 | Mr | 18 | ·45 |
| 5 | S | 11 | ·63 | 7 | O | 11 | ·16 | 1 | N | 9 | ·69 | 3 | D | 9 | ·22 | **56** | 4 | Jr | 7 | ·75 | | 6 | F | 6 | ·28 | 7 | Mr | 6○ | ·81 |
| 3 | Au | 31 | ·00 | 4 | S | 29 | ·53 | 7 | N | 27 | ·59 | 2 | D | 27 | ·12 | **57** | 3 | Jr | 25 | ·65 | | 5 | F | 24○ | ·18 | 6 | Mr | 25 | ·71 |
| | | | | 6 | O | 29 | ·06 | | | | | | | | | | | | | | | | | | | | | | |
| 1 | S | 18 | ·89 | 3 | O | 18 | ·43 | 4 | N | 16 | ·96 | 6 | D | 16 | ·49 | **58** | 1 | Jr | 15 | ·02 | | 2 | F | 13○ | ·55 | 4 | Mr | 15 | ·08 |
| 6 | S | 8 | ·26 | 7 | O | 7 | ·79 | 2 | N | 6 | ·32 | 3 | D | 5 | ·85 | **59** | 5 | Jr | 4 | ·38 | | 6 | F | 2● | ·91 | 1 | Mr | 4 | ·45 |
| 5 | S | 27 | ·16 | 6 | O | 26 | ·69 | 1 | N | 25 | ·22 | 2 | D | 24○ | ·75 | **60** | 4 | Jr | 13 | ·28 | | 5 | F | 21 | ·81 | 7 | Mr | 22 | ·34 |
| 2 | S | 15 | ·53 | 4 | O | 15 | ·06 | 5 | N | 13 | ·59 | 7 | D | 13○ | ·12 | **61** | 1 | Jr | 11 | ·65 | | 3 | F | 10 | ·18 | 4 | Mr | 11 | ·71 |
| 6 | S | 4 | ·89 | 1 | O | 4 | ·42 | 2 | N | 2 | ·95 | 4 | D | 2●○ | ·49 | **62** | 6 | Jr | 1 | ·02 | | 7 | Jr | 30 | ·55 | 2 | Mr | 1 | ·08 |
| 5 | S | 23 | ·79 | 7 | O | 23 | ·32 | 1 | N | 21● | ·85 | 3 | D | 21 | ·38 | **63** | 4 | Jr | 19 | ·91 | | 6 | F | 18 | ·44 | 7 | Mr | 19 | ·97 |
| 3 | S | 13 | ·16 | 4 | O | 12 | ·69 | 6 | N | 11● | ·22 | 7 | D | 10 | ·75 | **64** | 2 | Jr | 9 | ·28 | | 3 | F | 7 | ·81 | 5 | Mr | 8 | ·34 |
| 7 | S | 1 | ·53 | 2 | O | 1○ | ·06 | 3 | O | 30 | ·59 | 5 | N | 29 | ·12 | | 6 | D | 28 | ·65 | **65** | 1 | Jr | 27 | ·18 | 2 | F | 25 | ·71 |
| | | | | | | | | | | | | | | | | | | | | | | | | | | 4 | Mr | 27○ | ·24 |
| 6 | S | 20○ | ·42 | 7 | O | 19 | ·95 | 2 | N | 16 | ·48 | 4 | D | 18 | ·01 | **66** | 5 | Jr | 16 | ·55 | | 7 | F | 15 | ·08 | 1 | Mr | 16 | ·61 |
| 8 | S | 9○ | ·79 | 5 | O | 9 | ·32 | 6 | N | 7 | ·85 | 1 | D | 7 | ·38 | **67** | 2 | Jr | 5 | ·91 | | 4 | F | 4 | ·44 | 5 | Mr | 5● | ·97 |

## TABLE X—

☉'s Anom col. 6 + 29·53059 + 52·06117 + 88·59176 + 118·12235

☾'s Anom col. 7 + 1·976 + 3·952 + 5·928 + 7·904

| Kaliyuga | Vikrama Era | Saka Era | Commencement of Solar Year: Month and day A. D. | Fraction of day | First New-Moon in Solar Year | Anomaly of first New-Moon | A. D. | Week-day of 1st January | Vaisakha: Week-day | Month | Day | Fraction | Jyeshta: Week-day | Month | Day | Fraction | Ashada: Week-day | Month | Day | Fraction | Sravana: Week-day | Month | Day | Fraction | Bhadrapada: Week-day | Month | Day | Fraction |
|---|---|---|---|---|---|---|---|---|---|---|---|---|---|---|---|---|---|---|---|---|---|---|---|---|---|---|---|---|
| 4568 | 1524 | 1389 | 27M | ·5789 | 7·9206 | 0·948 | **1467** | 5 | 7 | Ap | 4 | ·50 | 2 | My | 4 | ·08 | 3 | Je | 2 | ·56 | 5 | Jl | 2 | ·09 | 6 | Jl | 31○ | ·63 |
| | | | | | | | | | | | | | | | | | | | | | | | | | 1 | Au | 30 | ·16 |
| 4569 | 1525 | 1390 | 26M | ·8376 | 26·5595 | 26·636 | **1468** | 6 | 6 | Ap | 22 | ·40 | 7 | My | 21 | ·93 | 2 | Je | 20 | ·46 | 3 | Jl | 19○ | ·99 | 5 | Au | 18 | ·52 |
| 4570 | 1526 | 1391 | 27M | ·0964 | 15·6578 | 22·793 | **1469** | 1 | 3 | Ap | 11 | ·77 | 5 | My | 11 | ·30 | 6 | Je | 9 | ·83 | 1 | Jl | 9●○ | ·36 | 2 | Au | 7 | ·89 |
| 4571 | 1527 | 1392 | 27M | ·3551 | 4·7761 | 18·951 | **1470** | 2 | 1 | Ap | 1 | ·14 | 2 | Ap | 30 | ·67 | 4 | My | 30 | ·20 | 7 | Jl | 28 | ·26 | 1 | Au | 26 | ·79 |
| | | | | | | | | | | | | | | | | | 5 | Je | 28● | ·73 | | | | | | | | |
| 4572 | 1528 | 1393 | 27M | ·6139 | 23·4150 | 17·084 | **1471** | 3 | 7 | Ap | 20 | ·03 | 1 | My | 19○ | ·56 | 3 | Je | 18 | ·09 | 4 | Jl | 17 | ·62 | 6 | Au | 16 | ·16 |
| 4573 | 1529 | 1394 | 26M | ·8726 | 12·5233 | 13·241 | **1472** | 4 | 4 | Ap | 8 | ·40 | 5 | My | 7○ | ·93 | 7 | Je | 6 | ·46 | 1 | Jl | 5 | ·99 | 3 | Au | 4 | ·52 |
| 4574 | 1530 | 1395 | 27M | ·1314 | 1·6316 | 9·398 | **1473** | 6 | 1 | Mr | 28 | ·77 | 3 | Ap | 27●○ | ·30 | 6 | Je | 25 | ·36 | 7 | Jl | 24 | ·89 | 2 | Au | 23 | ·42 |
| | | | | | | | | | | | | | 4 | My | 26 | ·83 | | | | | | | | | | | | |
| 4575 | 1531 | 1396 | 27M | ·3901 | 20·2705 | 7·532 | **1474** | 7 | 7 | Ap | 16● | ·67 | 2 | My | 16 | ·20 | 3 | Je | 14 | ·73 | 5 | Jl | 14 | ·26 | 6 | Au | 12 | ·79 |
| 4576 | 1532 | 1397 | 27M | ·6489 | 9·3788 | 3·689 | **1475** | 1 | 5 | Ap | 6 | ·03 | 6 | My | 5 | ·56 | 1 | Je | 4 | ·09 | 2 | Jl | 3 | ·62 | 4 | Au | 2 | ·15 |
| 4577 | 1533 | 1398 | 26M | ·9077 | 28·0176 | 1·822 | **1476** | 2 | 3 | Ap | 23 | ·93 | 5 | My | 23 | ·46 | 6 | Je | 21 | ·99 | 1 | Jl | 21 | ·52 | 3 | Au | 20○ | ·05 |
| 4578 | 1534 | 1399 | 27M | ·1654 | 17·1259 | 25·534 | **1477** | 4 | 1 | Ap | 13 | ·30 | 2 | My | 12 | ·83 | 4 | Je | 11 | ·36 | 5 | Jl | 10 | ·89 | 7 | Au | 9 | ·42 |
| 4579 | 1535 | 1400 | 27M | ·4252 | 6·2342 | 21·691 | **1478** | 5 | 5 | Ap | 2 | ·66 | 7 | My | 2 | ·19 | 1 | My | 31 | ·72 | 3 | Je | 30○ | ·25 | 6 | Au | 28 | ·31 |
| | | | | | | | | | | | | | | | | | | | | | 4 | Jl | 29● | ·78 | | | | |
| 4580 | 1536 | 1401 | 27M | ·6839 | 24·8731 | 19·825 | **1479** | 6 | 4 | Ap | 21 | ·56 | 6 | My | 21 | ·09 | 7 | Je | 19○ | ·62 | 2 | Jl | 19 | ·15 | 3 | Au | 17 | ·68 |
| 4581 | 1537 | 1402 | 26M | ·9427 | 13·9814 | 15·982 | **1480** | 7 | 1 | Ap | 9 | ·93 | 3 | My | 9 | ·46 | 4 | Je | 7●○ | ·99 | 6 | Jl | 7 | ·52 | 1 | Au | 6 | ·05 |
| 4582 | 1538 | 1403 | 27M | ·2014 | 3·0897 | 12·139 | **1481** | 2 | 6 | Mr | 30 | ·30 | 7 | Ap | 28 | ·83 | 2 | My | 28 | ·36 | 5 | Jl | 26 | ·42 | 6 | Au | 24 | ·95 |
| | | | | | | | | | | | | | | | | | 3 | Je | 26 | ·89 | | | | | | | | |
| 4583 | 1539 | 1404 | 27M | ·4602 | 21·7286 | 10·272 | **1482** | 3 | 5 | Ap | 18○ | ·19 | 6 | My | 17 | ·72 | 1 | Je | 16 | ·25 | 2 | Jl | 15 | ·78 | 4 | Au | 14 | ·31 |
| 4584 | 1540 | 1405 | 27M | ·7190 | 10·8369 | 6·430 | **1483** | 4 | 2 | Ap | 7○ | ·56 | 4 | My | 7 | ·09 | 5 | Je | 5 | ·62 | 7 | Jl | 5 | ·15 | 1 | Au | 3 | ·68 |
| 4585 | 1541 | 1406 | 26M | ·9777 | 29·4758 | 4·563 | **1484** | 5 | 1 | Ap | 25 | ·46 | 2 | My | 24 | ·99 | 4 | Je | 23 | ·52 | 6 | Jl | 23 | ·05 | 7 | Au | 21 | ·58 |
| 4586 | 1542 | 1407 | 27M | ·2365 | 18·5841 | 0·720 | **1485** | 7 | 5 | Ap | 14 | ·83 | 7 | My | 14 | ·36 | 1 | Je | 12 | ·89 | 3 | Jl | 12 | ·42 | 4 | Au | 10○ | ·95 |
| 4587 | 1543 | 1408 | 27M | ·4952 | 7·6924 | 24·432 | **1486** | 1 | 3 | Ap | 4 | ·19 | 4 | My | 3 | ·72 | 6 | Je | 2 | ·25 | 7 | Jl | 1 | ·78 | 3 | Au | 29 | ·84 |
| | | | | | | | | | | | | | | | | | | | | | 2 | Jl | 31○ | ·31 | | | | |
| 4588 | 1544 | 1409 | 27M | ·7540 | 26·3313 | 22·565 | **1487** | 2 | 2 | Ap | 23 | ·09 | 3 | My | 22 | ·62 | 5 | Je | 21 | ·15 | 6 | Jl | 20●○ | ·68 | 1 | Au | 19 | ·21 |
| 4589 | 1545 | 1410 | 27M | ·0127 | 15·4396 | 18·722 | **1488** | 3 | 6 | Ap | 11 | ·46 | 7 | My | 10 | ·99 | 2 | Je | 9 | ·52 | 4 | Jl | 9● | ·05 | 5 | Au | 7 | ·58 |
| 4590 | 1546 | 1411 | 27M | ·2715 | 4·5479 | 14·880 | **1489** | 5 | 3 | Mr | 31 | ·82 | 5 | Ap | 30 | ·35 | 6 | My | 29○ | ·88 | 2 | Jl | 27 | ·95 | 4 | Au | 26 | ·47 |
| | | | | | | | | | | | | | | | | | 1 | Je | 28 | ·41 | | | | | | | | |
| 4591 | 1547 | 1412 | 27M | ·5303 | 23·1868 | 13·013 | **1490** | 6 | 2 | Ap | 19 | ·72 | 4 | My | 19○ | ·25 | 5 | Je | 17 | ·78 | 7 | Jl | 17 | ·31 | 1 | Au | 15 | ·84 |
| 4592 | 1548 | 1413 | 27M | ·7890 | 12·2951 | 9·173 | **1491** | 7 | 7 | Ap | 9 | ·09 | 1 | My | 8●○ | ·62 | 3 | Je | 7 | ·15 | 4 | Jl | 6 | ·68 | 6 | Au | 5 | ·21 |
| 4593 | 1549 | 1414 | 27M | ·0478 | 1·4034 | 5·327 | **1492** | 1 | 4 | Mr | 28 | ·46 | 7 | My | 26 | ·52 | 2 | Je | 25 | ·05 | 3 | Jl | 24 | ·58 | 5 | Au | 23 | ·11 |
| | | | | | | | | | 5 | Ap | 26 | ·99 | | | | | | | | | | | | | | | | |
| 4594 | 1550 | 1415 | 27M | ·3065 | 20·0423 | 3·461 | **1493** | 3 | 3 | Ap | 16● | ·35 | 4 | My | 15 | ·88 | 6 | Je | 14 | ·41 | 7 | Jl | 13 | ·94 | 2 | Au | 12 | ·47 |
| 4595 | 1551 | 1416 | 27M | ·5653 | 9·1506 | 27·173 | **1494** | 4 | 7 | Ap | 5 | ·72 | 2 | My | 5 | ·25 | 3 | Je | 3 | ·78 | 5 | Jl | 3 | ·31 | 6 | Au | 1 | ·84 |
| | | | | | | | | | | | | | | | | | | | | | | | | | 1 | Au | 31○ | ·37 |
| 4596 | 1552 | 1417 | 27M | ·8240 | 27·7894 | 25·306 | **1495** | 5 | 6 | Ap | 24 | ·62 | 1 | My | 24 | ·15 | 2 | Je | 22 | ·68 | 4 | Jl | 22 | ·21 | 5 | Au | 20● | ·74 |
| 4597 | 1553 | 1418 | 27M | ·0828 | 16·8977 | 21·463 | **1496** | 6 | 3 | Ap | 12 | ·98 | 5 | My | 12 | ·52 | 7 | Je | 11 | ·05 | 1 | Jl | 10 | ·58 | 3 | Au | 9 | ·11 |
| 4598 | 1554 | 1419 | 27M | ·3415 | 6·0060 | 17·620 | **1497** | 1 | 1 | Ap | 2 | ·35 | 2 | My | 1 | ·88 | 4 | My | 31 | ·41 | 5 | Je | 29○ | ·94 | 2 | Au | 28 | ·00 |
| | | | | | | | | | | | | | | | | | | | | | 7 | Jl | 29● | ·49 | | | | |
| 4599 | 1555 | 1420 | 27M | ·6003 | 24·6449 | 15·754 | **1498** | 2 | 7 | Ap | 21 | ·25 | 1 | My | 20 | ·78 | 3 | Je | 19○ | ·31 | 4 | Jl | 18 | ·84 | 6 | Au | 17 | ·37 |
| 4600 | 1556 | 1421 | 27M | ·8591 | 13·7532 | 11·911 | **1499** | 3 | 4 | Ap | 10 | ·62 | 6 | My | 10 | ·15 | 7 | Je | 8● | ·68 | 2 | Jl | 8 | ·21 | 3 | Au | 6 | ·74 |
| 4601 | 1557 | 1422 | 27M | ·1178 | 2·8615 | 8·068 | **1500** | 4 | 1 | Mr | 29 | ·98 | 3 | Ap | 28○ | ·51 | 6 | Je | 26 | ·57 | 1 | Jl | 26 | ·10 | 2 | Au | 24 | ·63 |
| | | | | | | | | | | | | | 5 | My | 28● | ·04 | | | | | | | | | | | | |
| 4602 | 1558 | 1423 | 27M | ·3766 | 21·5004 | 6·201 | **1501** | 6 | 7 | Ap | 17○ | ·88 | 2 | My | 17 | ·41 | 3 | Je | 15 | ·94 | 5 | Jl | 15 | ·47 | 7 | Au | 14 | ·00 |
| 4603 | 1559 | 1424 | 27M | ·6353 | 10·6087 | 2·359 | **1502** | 7 | 5 | Ap | 7● | ·24 | 6 | My | 6 | ·77 | 1 | Je | 5 | ·30 | 2 | Jl | 4 | ·84 | 4 | Au | 3 | ·37 |
| 4604 | 1560 | 1425 | 27M | ·8941 | 29·2476 | 0·492 | **1503** | 1 | 4 | Ap | 26 | ·14 | 5 | My | 25 | ·67 | 7 | Je | 24 | ·20 | 1 | Jl | 23 | ·73 | 3 | Au | 22 | ·26 |
| 4605 | 1561 | 1426 | 27M | ·1528 | 18·3559 | 24·204 | **1504** | 2 | 1 | Ap | 14 | ·51 | 3 | My | 14 | ·04 | 4 | Je | 12 | ·57 | 6 | Jl | 12 | ·10 | 7 | Au | 10○ | ·63 |
| 4606 | 1562 | 1427 | 27M | ·4116 | 7·4642 | 20·361 | **1505** | 4 | 5 | Ap | 3 | ·88 | 7 | My | 3 | ·41 | 1 | Je | 1 | ·94 | 3 | Jl | 1 | ·47 | 6 | Au | 29 | ·53 |
| | | | | | | | | | | | | | | | | | | | | | 5 | Jl | 31○ | ·00 | | | | |

**Surya Siddhanta.**

| | | | | | | | | | | | | | | | | | | | | | | | | | | | | | |
|---|---|---|---|---|---|---|---|---|---|---|---|---|---|---|---|---|---|---|---|---|---|---|---|---|---|---|---|---|---|
| + 147·65293 | | | | + 177·18353 | | | | + 206·71411 | | | | + 236·24470 | | | | | + 265·77529 | | | | | + 295·30588 | | | | {324·83647 354·36705 | | | |
| + 9·880 | | | | + 11·856 | | | | + 13·832 | | | | + 15·808 | | | | | + 17·784 | | | | | + 19·760 | | | | {21·736 23·712 | | | |
| **Asvina** | | | | **Kartika** | | | | **Margasira** | | | | **Pausha** | | | | **A.D.** | **Magha** | | | | **A.D.** | **Phalguna** | | | | **Chaitra** | | | |
| Week-day | Month | Day | Fraction | Week-day | Month | Day | Fraction | Week-day | Month | Day | Fraction | Week-day | Month | Day | Fraction | | Week-day | Month | Day | Fraction | | Week-day | Month | Day | Fraction | Week-day | Month | Day | Fraction |
| 2 | S | 28 | ·69 | 4 | O | 28 | ·22 | 5 | N | 26 | ·75 | 7 | D | 26 | ·28 | **68** | 1 | Jr | 24○ | ·81 | | 3 | F | 23 | ·34 | 4 | Mr | 23 | ·87 |
| 7 | S | 17 | ·05 | 1 | O | 16 | ·58 | 3 | N | 15 | ·11 | 4 | D | 14 | ·64 | **69** | 6 | Jr | 13○ | ·17 | | 7 | F | 11 | ·70 | 2 | Mr | 13 | ·23 |
| 4 | S | 6 | ·42 | 5 | O | 5 | ·95 | 7 | N | 4 | ·48 | 2 | D | 4 | ·01 | **70** | 3 | Jr | 2○ | ·54 | | 5 | F | 1 | ·07 | 6 | Mr | 2 | ·60 |
| 3 | S | 25 | ·32 | 4 | O | 24 | ·85 | 6 | N | 23 | ·38 | 7 | D | 22 | ·91 | **71** | 2 | Jr | 21 | ·44 | | 3 | F | 19 | ·97 | 5 | Mr | 21 | ·50 |
| 7 | S | 14 | ·69 | 2 | O | 14 | ·22 | 3 | N | 12○ | ·75 | 5 | D | 12 | ·28 | **72** | 6 | Jr | 10 | ·81 | | 1 | F | 9 | ·34 | 2 | Mr | 9 | ·87 |
| 5 | S | 3 | ·05 | 6 | O | 2 | ·58 | 1 | N | 1○ | ·11 | 2 | N | 30 | ·64 | | 4 | D | 30 | ·17 | **73** | 5 | Jr | 28 | ·70 | 7 | F | 27 | ·23 |
| 3 | S | 21 | ·95 | 5 | O | 21○ | ·48 | 7 | N | 20 | ·01 | 1 | D | 19 | ·54 | **74** | 3 | Jr | 18 | ·07 | | 4 | F | 16 | ·60 | 6 | Mr | 18 | ·13 |
| 1 | S | 11 | ·32 | 2 | O | 10● | ·85 | 4 | N | 9 | ·38 | 5 | D | 8 | ·91 | **75** | 7 | Jr | 7 | ·44 | | 1 | F | 5 | ·97 | 3 | Mr | 7○ | ·50 |
| {5 | Au | 31○ | ·68 | 1 | O | 29 | ·74 | 3 | N | 28 | ·27 | 4 | D | 27 | ·81 | | **Magha** | | | | **76** | {6 | Jr• | 26 | ·33 | 2 | Mr | 25 | ·39 |
| {7 | S | 30● | ·21 | | | | | | | | | | | | | | **Kshaya** | | | | | {7 | F | 24●○ | ·86 | | | | |
| 4 | S | 18 | ·58 | 6 | O | 18 | ·11 | 7 | N | 16 | ·64 | 2 | D | 16 | ·17 | **77** | 3 | Jr | 15 | ·70 | | 5 | F | 13 | ·23 | 6 | Mr | 14 | ·76 |
| 1 | S | 7 | ·95 | 3 | O | 7 | ·48 | 5 | N | 6 | ·01 | 6 | D | 5 | ·54 | **78** | 1 | Jr | 4 | ·07 | | 2 | F | 2 | ·60 | 4 | Mr | 4 | ·13 |
| 7 | S | 26 | ·84 | 2 | O | 26 | ·37 | 3 | N | 24 | ·90 | 5 | D | 24○ | ·43 | **79** | 6 | Jr | 22 | ·96 | | 1 | F | 21 | ·49 | 3 | Mr | 23 | ·02 |
| 5 | S | 16 | ·21 | 6 | O | 15 | ·74 | 1 | N | 14 | ·27 | 2 | D | 13●○ | ·80 | **80** | 4 | Jr | 12 | ·33 | | 5 | F | 10 | ·86 | 7 | Mr | 11 | ·39 |
| 2 | S | 4 | ·58 | 4 | O | 4 | ·11 | 5 | N | 2 | ·64 | 7 | D | 2 | ·17 | | 1 | D | 31 | ·70 | **81** | 3 | Jr | 30 | ·23 | 4 | F | 28 | ·76 |
| 1 | S | 23 | ·48 | 3 | O | 23 | ·01 | 4 | N | 21● | ·54 | 6 | D | 21 | ·07 | **82** | 7 | Jr | 19 | ·60 | | 2 | F | 18 | ·13 | 3 | Mr | 19 | ·66 |
| 5 | S | 12 | ·85 | 7 | O | 12○ | ·38 | 1 | N | 10● | ·91 | 3 | D | 10 | ·44 | **83** | 4 | Jr | 8 | ·97 | | 6 | F | 7 | ·50 | 1 | Mr | 6 | ·03 |
| 3 | S | 2 | ·21 | 4 | O | 1○ | ·74 | 6 | O | 31 | ·27 | 7 | N | 29 | ·80 | | 2 | D | 29 | ·33 | **84** | 3 | Jr | 27 | ·86 | {5 | F | 26 | ·39 |
| | | | | | | | | | | | | | | | | | | | | | | | | | | {6 | Mr | 26 | ·92 |
| 2 | S | 20●○ | ·11 | 3 | O | 19 | ·64 | 5 | N | 18 | ·17 | 6 | D | 17 | ·70 | **85** | 1 | Jr | 16 | ·23 | | 2 | F | 14 | ·76 | 4 | Mr | 16 | ·29 |
| 6 | S | 9● | ·48 | 1 | O | 9 | ·01 | 2 | N | 7 | ·54 | 4 | D | 7 | ·07 | **86** | 5 | Jr | 5 | ·60 | | 7 | F | 4○ | ·13 | 1 | Mr | 5● | ·66 |
| 5 | S | 28 | ·37 | 6 | O | 27 | ·90 | 1 | N | 26 | ·43 | 2 | D | 25 | ·96 | **87** | 4 | Jr | 24○ | ·49 | | 6 | F | 23 | ·02 | 7 | Mr | 24 | ·55 |
| 2 | S | 17 | ·74 | 4 | O | 17 | ·27 | 5 | N | 15 | ·80 | 7 | D | 15 | ·33 | **88** | 1 | Jr | 13○ | ·86 | | 3 | F | 12 | ·39 | 4 | Mr | 12 | ·92 |
| 7 | S | 6 | ·11 | 1 | O | 5 | ·64 | 3 | N | 4 | ·17 | 4 | D | 3 | ·70 | **89** | 6 | Jr | 2 | ·23 | | 7 | Jr | 31 | ·76 | 2 | Mr | 2 | ·29 |
| 6 | S | 25 | ·00 | 7 | O | 24 | ·53 | 2 | N | 23○ | ·06 | 3 | D | 22● | ·59 | **90** | 5 | Jr | 21 | ·12 | | 6 | F | 19 | ·65 | 1 | Mr | 21 | ·19 |
| 3 | S | 14 | ·37 | 4 | O | 13 | ·90 | 6 | N | 12○ | ·43 | 7 | D | 11 | ·96 | **91** | 2 | Jr | 10 | ·49 | | 4 | F | 9 | ·02 | 5 | Mr | 10 | ·55 |
| 7 | S | 3 | ·74 | 2 | O | 3 | ·27 | 3 | N | 1●○ | ·80 | 5 | D | 1 | ·33 | | 6 | D | 30 | ·86 | **92** | 1 | Jr | 29 | ·39 | 2 | F | 27 | ·92 |
| 6 | S | 21 | ·64 | 1 | O | 21● | ·17 | 2 | N | 19 | ·70 | 4 | D | 19 | ·23 | **93** | 5 | Jr | 17 | ·76 | | 7 | F | 16 | ·29 | 1 | Mr | 17○ | ·82 |
| 4 | S | 11○ | ·00 | 5 | O | 10 | ·53 | 7 | N | 9 | ·06 | 1 | D | 8 | ·59 | **94** | 3 | Jr | 7 | ·12 | | 4 | F | 5 | ·65 | 6 | Mr | 7○ | ·18 |
| 2 | S | 29 | ·90 | 4 | O | 29 | ·43 | 5 | N | 27 | ·96 | 7 | D | 27 | ·49 | **95** | 2 | Jr | 26 | ·03 | | 3 | F | 24● | ·56 | 5 | Mr | 26 | ·09 |
| 7 | S | 19 | ·27 | 1 | O | 18 | ·80 | 3 | N | 17 | ·33 | 4 | D | 16 | ·86 | **96** | 6 | Jr | 15○ | ·39 | | 7 | F | 13● | ·92 | 2 | Mr | 14 | ·45 |
| 4 | S | 7 | ·64 | 6 | O | 7 | ·17 | 7 | N | 5 | ·70 | 2 | D | 5 | ·23 | **97** | 3 | Jr | 3○ | ·76 | | 5 | F | 2 | ·29 | 6 | Mr | 3 | ·82 |
| 3 | S | 26 | ·54 | 5 | O | 26 | ·07 | 6 | N | 24 | ·60 | 1 | D | 24○ | ·13 | **98** | 2 | Jr | 22 | ·66 | | 4 | F | 21 | ·19 | 5 | Mr | 22 | ·72 |
| 7 | S | 15 | ·90 | 2 | O | 15 | ·43 | 3 | N | 13 | ·96 | 5 | D | 13● | ·49 | **99** | 7 | Jr | 12 | ·02 | | 1 | F | 10 | ·55 | 3 | Mr | 12 | ·09 |
| 5 | S | 5 | ·27 | 6 | O | 4 | ·80 | 1 | N | 3 | ·33 | 2 | D | 2 | ·86 | **15** | 4 | Jr | 1 | ·39 | **00** | 5 | Jr | 30 | ·92 | 7 | F | 29 | ·45 |
| 4 | S | 23 | ·16 | 5 | O | 22○ | ·69 | 7 | N | 21 | ·22 | 1 | D | 20 | ·75 | **01** | 3 | Jr | 19 | ·28 | | 4 | F | 17 | ·81 | 6 | Mr | 19 | ·34 |
| 1 | S | 12 | ·53 | 3 | O | 12●○ | ·06 | 4 | N | 10 | ·59 | 6 | D | 10 | ·12 | **02** | 7 | Jr | 8 | ·65 | | 2 | F | 7 | ·18 | 3 | Mr | 8 | ·71 |
| 5 | S | 1 | ·89 | 7 | O | 1●○ | ·42 | 1 | O | 30 | ·95 | 3 | N | 29 | ·48 | | 5 | D | 29 | ·01 | **03** | 6 | Jr | 27 | ·54 | {1 | F | 26 | ·07 |
| | | | | | | | | | | | | | | | | | | | | | | | | | | {2 | Mr | 27● | ·61 |
| 4 | S | 20● | ·79 | 6 | O | 20 | ·32 | 7 | N | 18 | ·85 | 2 | D | 18 | ·38 | **04** | 3 | Jr | 16 | ·91 | | 5 | F | 15○ | ·44 | 6 | Mr | 15 | ·97 |
| 2 | S | 9 | ·16 | 3 | O | 8 | ·69 | 5 | N | 7 | ·22 | 6 | D | 6 | ·75 | **05** | 1 | Jr | 5 | ·28 | | 2 | F | 3○ | ·81 | 4 | Mr | 5 | ·34 |
| 1 | S | 28 | ·06 | 2 | O | 27 | ·59 | 4 | N | 26 | ·12 | 5 | D | 25 | ·65 | **06** | 7 | Jr | 24●○ | ·18 | | 1 | F | 22 | ·71 | 3 | Mr | 24 | ·24 |

| | | | | | | | | | ⊙'s Anom. col. 6 | | | | + 29·53059 | | | | + 59·06117 | | | | + 88·59176 | | | | + 118·12235 | | | |
|---|---|---|---|---|---|---|---|---|---|---|---|---|---|---|---|---|---|---|---|---|---|---|---|---|---|---|---|---|
| | | | | | | | | | ☾'s Anom. col. 7 | | | | + 1·976 | | | | + 3·952 | | | | + 5·928 | | | | + 7·904 | | | |
| Kaliyuga. | Vikrama Era. | Saka Era. | Commencement of Solar Year. Month and day A. D. | Fraction of day. | First New-Moon in Solar Year. | Anomaly of first New-Moon. | A.D. | Week-day of 1st January. | Vaisakha | | | | Jyeshta | | | | Ashada | | | | Sravana | | | | Bhadrapada | | | |
| | | | | | | | | | Week-day | Month | Day | Fraction | Week-day | Month | Day | Fraction | Week-day | Month | Day | Fraction | Week-day | Month | Day | Fraction | Week-day | Month | Day | Fraction |
| 4607 | 1563 | 1428 | 27M | ·6704 | 26·1031 | 18·494 | **1506** | 5 | 4 | Ap | 22 | ·77 | 6 | My | 22 | ·30 | 7 | Je | 20 | ·83 | 2 | Jl | 20● | ·36 | 3 | Au | 18 | ·90 |
| 4608 | 1564 | 1429 | 27M | ·9291 | 15·2114 | 14·651 | **1507** | 6 | 2 | Ap | 12 | ·14 | 3 | My | 11 | ·67 | 5 | Je | 10ↄ | ·20 | 6 | Jl | 9● | ·73 | 1 | Au | 8 | ·26 |
| 4609 | 1565 | 1430 | 27M | ·1879 | 4·3197 | 10·809 | **1508** | 7 | 6 | Mr | 31 | ·51 | 1 | Ap | 30 | ·04 | 2 | My | 29ↄ | ·57 | 5 | Jl | 27 | ·63 | 7 | Au | 26 | ·16 |
| | | | | | | | | | | | | | | | | | 4 | Je | 28 | ·10 | | | | | | | | |
| 4610 | 1566 | 1431 | 27M | ·4466 | 22·9586 | 8·942 | **1509** | 2 | 5 | Ap | 19 | ·40 | 6 | My | 18ↄ | ·94 | 1 | Je | 17 | ·47 | 3 | Jl | 17 | ·00 | 4 | Au | 15 | ·53 |
| 4611 | 1567 | 1432 | 27M | ·7054 | 12·0669 | 5·099 | **1510** | 3 | 2 | Ap | 8 | ·77 | 4 | My | 8● | ·30 | 5 | Je | 6 | ·83 | 7 | Jl | 6 | ·36 | 1 | Au | 4 | ·89 |
| 4612 | 1568 | 1433 | 27M | ·9641 | 1·1752 | 1·256 | **1511** | 4 | 7 | Mr | 29ↄ | ·14 | 3 | My | 27 | ·20 | 4 | Je | 25 | ·73 | 6 | Jl | 25 | ·26 | 7 | Au | 23 | ·79 |
| | | | | | | | | | 1 | Ap | 27 | ·67 | | | | | | | | | | | | | | | | |
| 4613 | 1569 | 1434 | 27M | ·2229 | 19·8141 | 26·944 | **1512** | 5 | 6 | Ap | 16 | ·04 | 7 | My | 15 | ·57 | 2 | Je | 14 | ·10 | 3 | Jl | 13 | ·63 | 5 | Au | 12 | ·16 |
| 4614 | 1570 | 1435 | 27M | ·4816 | 8·9224 | 23·101 | **1513** | 7 | 3 | Ap | 5 | ·40 | 4 | My | 4 | ·93 | 6 | Je | 3 | ·46 | 1 | Jl | 3 | ·00 | 2 | Au | 1 | ·53 |
| | | | | | | | | | | | | | | | | | | | | | | | | | 4 | Au | 31ↄ | ·06 |
| 4615 | 1571 | 1436 | 27M | ·7404 | 27·5612 | 21·235 | **1514** | 1 | 2 | Ap | 24 | ·30 | 3 | My | 23 | ·83 | 5 | Je | 22 | ·36 | 6 | Jl | 21 | ·89 | 1 | Au | 20● | ·42 |
| 4616 | 1572 | 1437 | 27M | ·9992 | 16·6695 | 17·392 | **1515** | 2 | 6 | Ap | 13 | ·67 | 1 | My | 13 | ·20 | 2 | Je | 11 | ·73 | 4 | Jl | 11ↄ | ·26 | 5 | Au | 9 | ·79 |
| 4617 | 1573 | 1438 | 27M | ·2579 | 5·7778 | 13·549 | **1516** | 3 | 4 | Ap | 2 | ·04 | 5 | My | 1 | ·57 | 7 | My | 31 | ·10 | 1 | Je | 29ↄ | ·63 | 4 | Au | 27 | ·69 |
| | | | | | | | | | | | | | | | | | | | | | 3 | Jl | 29 | ·16 | | | | |
| 4618 | 1574 | 1439 | 27M | ·5167 | 24·4167 | 11·682 | **1517** | 5 | 2 | Ap | 20 | ·93 | 4 | My | 20 | ·46 | 5 | Je | 18● | ·99 | 7 | Jl | 18 | ·52 | 2 | Au | 17 | ·06 |
| 4619 | 1575 | 1440 | 27M | ·7754 | 13·5250 | 7·840 | **1518** | 6 | 7 | Ap | 10 | ·30 | 1 | My | 9ↄ | ·82 | 3 | Je | 8● | ·36 | 4 | Jl | 7 | ·89 | 6 | Au | 6 | ·42 |
| 4620 | 1576 | 1441 | 27M | ·0342 | 2·6333 | 3·997 | **1519** | 7 | 4 | Mr | 30 | ·67 | 6 | Ap | 29○ | ·20 | 2 | Je | 27 | ·26 | 3 | Jl | 26 | ·79 | 5 | Au | 25 | ·32 |
| | | | | | | | | | | | | | 7 | My | 28 | ·73 | | | | | | | | | | | | |
| 4621 | 1577 | 1442 | 27M | ·2929 | 21·2722 | 2·130 | **1520** | 1 | 3 | Ap | 17○ | ·56 | 5 | My | 17 | ·10 | 6 | Je | 15 | ·63 | 1 | Jl | 15 | ·16 | 2 | Au | 13 | ·69 |
| 4622 | 1578 | 1443 | 27M | ·5517 | 10·3805 | 25·842 | **1521** | 3 | 7 | Ap | 6● | ·93 | 2 | My | 6 | ·46 | 3 | Je | 4 | ·99 | 5 | Jl | 4 | ·52 | 7 | Au | 3 | ·05 |
| 4623 | 1579 | 1444 | 27M | ·8105 | 29·0194 | 23·975 | **1522** | 4 | 5 | Mr | 27 | ·30 | 1 | My | 25 | ·36 | 2 | Je | 20 | ·89 | 4 | Jl | 23 | ·42 | 5 | Au | 21ↄ | ·95 |
| | | | | | | | | | 6 | Ap | 25 | ·83 | | | | | | | | | | | | | | | | |
| 4624 | 1580 | 1·45 | 27M | ·0692 | 18·1277 | 20·133 | **1523** | 5 | 4 | Ap | 15 | ·20 | 5 | My | 14 | ·73 | 7 | Je | 13 | ·26 | 1 | Jl | 12 | ·79 | 3 | Au | 11●○ | ·32 |
| 4625 | 1581 | 1446 | 27M | ·3280 | 7·2360 | 16·290 | **1524** | 6 | 1 | Ap | 3 | ·56 | 3 | My | 3 | ·09 | 4 | Je | 1 | ·62 | 6 | Jl | 1 | ·16 | 7 | Jl | 30 | 69 |
| | | | | | | | | | | | | | | | | | | | | | | | | | 2 | Au | 29 | ·22 |
| 4626 | 1582 | 1447 | 27M | ·5867 | 25·8749 | 14·423 | **1525** | 1 | 7 | Ap | 22 | ·46 | 2 | My | 22 | ·00 | 3 | Je | 20ↄ | ·52 | 5 | Jl | 20 | ·05 | 6 | Au | 18 | ·58 |
| 4627 | 1583 | 1448 | 27M | ·8455 | 14·9832 | 10·580 | **1526** | 2 | 4 | Ap | 11 | ·83 | 6 | My | 11 | ·36 | 7 | Je | 9ↄ | ·89 | 2 | Jl | 9 | ·42 | 3 | Au | 7 | ·95 |
| 4628 | 1584 | 1449 | 28M | ·1042 | 4·0915 | 6·738 | **1527** | 3 | 2 | Ap | 1 | ·20 | 3 | Ap | 30 | ·73 | 5 | My | 30●○ | ·26 | 1 | Jl | 28 | ·32 | 2 | Au | 26 | ·85 |
| | | | | | | | | | | | | | | | | | 6 | Je | 28 | ·79 | | | | | | | | |
| 4629 | 1585 | 1450 | 27M | ·3630 | 22·7304 | 4·871 | **1528** | 4 | 1 | Ap | 19 | ·09 | 2 | My | 18● | ·62 | 4 | Je | 17 | ·15 | 5 | Jl | 16 | ·68 | 7 | Au | 15 | ·22 |
| 4630 | 1586 | 1451 | 27M | ·6218 | 11·8387 | 1·028 | **1529** | 6 | 5 | Ap | 8○ | ·46 | 6 | My | 7 | ·99 | 1 | Je | 6 | ·52 | 3 | Jl | 6 | ·05 | 4 | Au | 4 | ·58 |
| 4631 | 1587 | 1·452 | 27M | ·8805 | 0·9470 | 24·740 | **1530** | 7 | 2 | Mr | 28●○ | ·83 | 5 | My | 26 | ·89 | 7 | Je | 25 | ·42 | 1 | Jl | 24 | 95 | 3 | Au | 23 | ·48 |
| | | | | | | | | | 4 | Ap | 27 | ·36 | | | | | | | | | | | | | | | | |
| 4632 | 1588 | 1453 | 28M | ·1393 | 19·5858 | 22·873 | **1531** | 1 | 1 | Ap | 16 | ·72 | 3 | My | 16 | ·26 | 4 | Je | 14 | ·79 | 6 | Jl | 14 | ·32 | 7 | Au | 12 | ·85 |
| 4632 | 1589 | 1454 | 27M | ·3980 | 8·6941 | 19·030 | **1532** | 2 | 6 | Ap | 5 | ·09 | 7 | My | 4 | ·62 | 2 | Je | 3 | ·15 | 3 | Jl | 2 | ·68 | 5 | Au | 1 | ·21 |
| | | | | | | | | | | | | | | | | | | | | | | | | | 6 | Au | 30● | ·74 |
| 4634 | 1590 | 1455 | 27M | ·6568 | 27·3330 | 17·165 | **1533** | 4 | 4 | Ap | 22 | ·99 | 6 | My | 23 | ·52 | 1 | Je | 22 | ·05 | 2 | Jl | 21○ | ·58 | 4 | Au | 20● | ·11 |
| 4635 | 1591 | 1456 | 27M | ·9155 | 16·4413 | 13·321 | **1534** | 5 | 2 | Ap | 13 | ·36 | 3 | My | 12 | ·89 | 5 | Je | 11 | ·42 | 6 | Jl | 10ↄ | ·95 | 1 | Au | 9 | ·48 |
| 4636 | 1592 | 1457 | 28M | ·1743 | 5·5496 | 9·478 | **1535** | 6 | 6 | Ap | 2 | ·72 | 1 | My | 2 | ·25 | 2 | My | 31 | ·78 | 4 | Jl | 30● | ·32 | 7 | Au | 28 | ·38 |
| | | | | | | | | | | | | | | | | | | | | | 5 | Jl | 29 | ·85 | | | | |
| 4637 | 1593 | 1458 | 27M | ·4330 | 24·1885 | 7·611 | **1536** | 7 | 5 | Ap | 20 | ·62 | 7 | My | 20ↄ | ·15 | 1 | Je | 18 | ·68 | 3 | Jl | 18● | ·21 | 4 | Au | 16 | ·74 |
| 4638 | 1594 | 1459 | 27M | ·6918 | 13·2968 | 3·769 | **1537** | 2 | 2 | Ap | 9 | ·99 | 4 | My | 9ↄ | ·52 | 6 | Je | 8 | ·05 | 7 | Jl | 7 | ·58 | 2 | Au | 6 | ·11 |
| 4639 | 1595 | 1460 | 27M | ·9506 | 2·4057 | 27·481 | **1538** | 3 | 7 | Mr | 30 | ·36 | 1 | Ap | 28ↄ | ·89 | 4 | Je | 26 | ·95 | 6 | Jl | 26 | ·48 | 1 | Au | 25 | ·01 |
| | | | | | | | | | | | | | 3 | My | 28 | ·42 | | | | | | | | | | | | |
| 4640 | 1596 | 1461 | 28M | ·2093 | 21·0440 | 25·614 | **1539** | 4 | 6 | Ap | 18 | ·25 | 7 | My | 17 | ·78 | 2 | Je | 16 | ·31 | 3 | Jl | 15 | ·84 | 5 | Au | 14 | ·38 |
| 4641 | 1597 | 1462 | 27M | ·4681 | 10·1523 | 21·771 | **1540** | 5 | 3 | Ap | 6● | ·62 | 5 | My | 6 | ·15 | 6 | Je | 4 | ·68 | 1 | Jl | 4 | ·21 | 2 | Au | 2 | ·74 |
| 4642 | 1598 | 1463 | 27M | ·7268 | 28·7912 | 19·904 | **1541** | 7 | 2 | Ap | 25 | ·52 | 4 | My | 25 | ·05 | 5 | Je | 23 | ·58 | 7 | Jl | 23 | ·11 | 1 | Au | 21●ↄ | ·64 |
| 4643 | 1599 | 1464 | 27M | ·9856 | 17·8995 | 16·062 | **1542** | 1 | 6 | Ap | 14 | ·88 | 1 | My | 14 | ·42 | 2 | Je | 12 | ·95 | 4 | Jl | 12 | ·48 | 6 | Au | 11● | ·01 |
| 4644 | 1600 | 1465 | 28M | ·2443 | 7·0078 | 12·219 | **1543** | 2 | 4 | Ap | 4 | ·25 | 5 | My | 3 | ·78 | 7 | Je | 2 | ·31 | 1 | Jl | 1○ | ·84 | 4 | Au | 29 | ·90 |
| | | | | | | | | | | | | | | | | | | | | | 3 | Jl | 31 | ·37 | | | | |

## Surya Siddhanta.

+ 147·65293 + 177·18353 + 206·71411 + 236·24470 + 265·77529 + 295·30588 {+324·83647 {+354·36705

+ 9·880 + 11·856 + 13·832 + 15·808 + 17·784 + 19·760 {+ 21·736 {+ 23·712

| Asvina | | | | Kartika | | | | Margasira | | | | Pausha | | | | A.D. | Magha | | | | A.D. | Phalguna | | | | Chaitra | | | |
|---|---|---|---|---|---|---|---|---|---|---|---|---|---|---|---|---|---|---|---|---|---|---|---|---|---|---|---|---|---|
| Week-day | Month | Day | Fraction | Week-day | Month | Day | Fraction | Week-day | Month | Day | Fraction | Week-day | Month | Day | Fraction | | Week-day | Month | Day | Fraction | | Week-day | Month | Day | Fraction | Week-day | Month | Day | Fraction |
| 5 | S | 17 | ·43 | 6 | O | 16 | ·96 | 1 | N | 15 | ·49 | 3 | D | 15 | ·02 | **07** | 4 | Jr | 13● | ·55 | | 6 | F | 12 | ·08 | 7 | Mr | 13 | ·61 |
| 2 | S | 6 | ·79 | 4 | O | 6 | ·32 | 5 | N | 4 | ·85 | 7 | D | 4ɔ | ·38 | **08** | 1 | Jr | 2 | ·92 | | 3 | F | 1 | ·45 | 4 | Mr | 1 | ·98 |
| 1 | S | 24 | ·69 | 3 | O | 24 | ·22 | 4 | N | 22ɔ | ·75 | 6 | D | 22 | ·28 | **09** | 7 | Jr | 20 | ·81 | | 2 | F | 19 | ·34 | 3 | Mr | 20 | ·87 |
| 6 | S | 14 | ·06 | 7 | O | 13 | ·59 | 2 | N | 12●ɔ | ·12 | 3 | D | 11 | ·65 | **10** | 5 | Jr | 10 | ·18 | | 6 | F | 8 | ·71 | 1 | Mr | 10 | ·24 |
| 3 | S | 3 | ·42 | 4 | O | 2 | ·96 | 6 | N | 1 | ·49 | 1 | D | 1 | ·02 | | 2 | D | 30 | ·55 | **11** | 4 | Jr | 29 | ·08 | 5 | F | 27 | ·61 |
| 2 | S | 22ɔ | ·32 | 3 | O | 21 | ·85 | 5 | N | 20 | ·38 | 6 | D | 19 | ·91 | **12** | 1 | Jr | 18 | ·44 | | 2 | F | 16 | ·98 | 4 | Mr | 17○ | ·51 |
| 6 | S | 10ɔ | ·69 | 1 | O | 10 | ·22 | 2 | N | 8 | ·75 | 4 | D | 8 | ·28 | **13** | 5 | Jr | 6 | ·81 | | 7 | F | 5 | ·34 | 1 | Mr | 6●○ | ·87 |
| 5 | S | 29 | ·59 | 7 | O | 29 | ·12 | 1 | N | 27 | ·65 | 3 | D | 27 | ·18 | **14** | 4 | Jr | 25ɔ | ·71 | | 6 | F | 24 | ·24 | 7 | Mr | 25 | ·77 |
| 2 | S | 18 | ·95 | 4 | O | 18 | ·48 | 6 | N | 17 | ·02 | 7 | D | 16 | ·55 | **15** | 2 | Jr | 15ɔ | ·08 | | 3 | F | 13 | ·61 | 5 | Mr | 15 | ·14 |
| 7 | S | 8 | ·32 | 1 | O | 7 | ·85 | 3 | N | 6 | ·38 | 4 | D | 5 | ·91 | **16** | 6 | Jr | 4●ɔ | ·44 | | 7 | F | 2 | ·97 | 2 | Mr | 3 | ·50 |
| 6 | S | 26 | ·22 | 7 | O | 25 | ·75 | 2 | N | 24 | ·28 | 3 | D | 23 | ·81 | **17** | 5 | Jr | 22 | ·34 | | 6 | F | 20 | ·87 | 1 | Mr | 22 | ·40 |
| 3 | S | 15 | ·57 | 5 | O | 15 | ·12 | 6 | N | 13 | ·65 | 1 | D | 13● | ·18 | **18** | 2 | Jr | 11 | ·71 | | 4 | F | 10 | ·24 | 5 | Mr | 11 | ·77 |
| 7 | S | 4 | ·95 | 2 | O | 4 | ·48 | 4 | N | 3ɔ | ·01 | 5 | D | 2 | ·54 | **19** | 7 | Jr | 1 | ·08 | | 1 | Jr | 30 | ·61 | 3 | Mr | 1 | ·14 |
| 6 | S | 23 | ·85 | 1 | O | 23ɔ | ·38 | 2 | N | 21 | ·91 | 4 | D | 21 | ·44 | **20** | 5 | Jr | 19 | ·97 | | 7 | F | 18 | ·50 | 2 | Mr | 19 | ·08 |
| 4 | S | 12 | ·22 | 5 | O | 11ɔ | ·75 | 7 | N | 10 | ·28 | 1 | D | 9 | ·81 | **21** | 3 | Jr | 8 | ·34 | | 4 | F | 6 | ·87 | 6 | Mr | 8 | ·40 |
| 1 | S | 1 | ·58 | {3 | O | 1 | ·12 | **Margasira Kshaya** | | | | 6 | N | 29 | ·18 | | 7 | D | 28 | ·71 | **22** | 2 | Jr | 27 | ·24 | 3 | F | 25ɔ | ·77 |
| | | | | {4 | O | 30 | ·65 | | | | | | | | | | | | | | | | | | | | | | |
| 7 | S | 20 | ·48 | 2 | O | 20 | ·01 | 3 | N | 18 | ·54 | 5 | D | 18 | ·07 | **23** | 6 | Jr | 16 | ·60 | | 1 | F | 15ɔ | ·14 | 2 | Mr | 16 | ·67 |
| 4 | S | 9 | ·85 | 6 | O | 9 | ·38 | 7 | N | 7 | ·91 | 2 | D | 7 | ·44 | **24** | 3 | Jr | 5 | ·97 | | 5 | F | 4ɔ | ·50 | 7 | Mr | 5 | ·03 |
| 3 | S | 27 | ·75 | 5 | O | 27 | ·28 | 6 | N | 25 | ·81 | 1 | D | 25 | ·34 | **25** | 2 | Jr | 23 | ·87 | | 4 | F | 22 | ·40 | 5 | Mr | 23 | ·93 |
| 1 | S | 17 | ·11 | 2 | O | 16 | ·64 | 4 | N | 15 | ·18 | 5 | D | 14ɔ | ·71 | **26** | 7 | Jr | 13● | ·24 | | 1 | F | 11 | ·77 | 3 | Mr | 13 | ·30 |
| 5 | S | 6 | ·48 | 7 | O | 6 | ·01 | 1 | N | 4 | ·54 | 3 | D | 4ɔ | ·07 | **27** | 4 | Jr | 2 | ·60 | | 6 | F | 1 | ·13 | 7 | Mr | 2 | ·66 |
| 4 | S | 25 | ·38 | 5 | O | 24 | ·91 | 7 | N | 23○ | ·44 | 1 | D | 22 | ·97 | **28** | 3 | Jr | 21 | ·50 | | 5 | F | 20 | ·03 | 6 | Mr | 20 | ·56 |
| 1 | S | 13 | ·75 | 3 | O | 13 | ·28 | 4 | N | 11● | ·31 | 6 | D | 11 | ·34 | **29** | 7 | Jr | 9 | ·87 | | 2 | F | 8 | ·40 | 3 | Mr | 9 | ·93 |
| 6 | S | 3 | ·11 | 7 | O | 2ɔ | ·64 | 2 | N | 1● | ·17 | 3 | N | 30 | ·70 | | 5 | D | 30 | ·24 | **30** | 6 | Jr | 28 | ·77 | 1 | F | 27 | ·30 |
| 5 | S | 22ɔ | ·01 | 6 | O | 21 | ·54 | 1 | N | 20 | ·07 | 2 | D | 19 | ·60 | **31** | 4 | Jr | 18 | ·13 | | 5 | F | 16 | ·66 | 7 | Mr | 18○ | ·19 |
| 2 | S | 11ɔ | ·38 | 3 | O | 10 | ·91 | 5 | N | 9 | ·44 | 6 | D | 8 | ·97 | **32** | 1 | Jr | 7 | ·50 | | 3 | F | 6 | ·03 | 4 | Mr | 6 | ·56 |
| 1 | S | 29 | ·28 | 2 | O | 28 | ·81 | 4 | N | 27 | ·34 | 5 | D | 26 | ·87 | **33** | 7 | Jr | 25ɔ | ·40 | | 1 | F | 23 | ·93 | 3 | Mr | 25 | ·46 |
| 5 | S | 18 | ·64 | 7 | O | 18 | ·17 | 1 | N | 16 | ·70 | 3 | D | 16 | ·23 | **34** | 4 | Jr | 14ɔ | ·76 | | 6 | F | 13 | ·30 | 7 | Mr | 14 | ·83 |
| 3 | S | 8 | ·01 | 4 | O | 7 | ·54 | 6 | N | 6 | ·07 | 7 | D | 5 | ·60 | **35** | 2 | Jr | 4 | ·13 | | 3 | F | 2 | ·66 | 5 | Mr | 4 | ·19 |
| 1 | S | 26 | ·91 | 3 | O | 26 | ·44 | 4 | N | 24 | ·97 | 6 | D | 24 | ·50 | **36** | 1 | Jr | 23 | ·03 | | 2 | F | 21 | ·56 | 4 | Mr | 22 | ·09 |
| 6 | S | 15 | ·27 | 7 | O | 14 | ·80 | 2 | N | 13ɔ | ·34 | 3 | D | 12 | ·87 | **37** | 5 | Jr | 11 | ·40 | | 6 | F | 9 | ·93 | 1 | Mr | 11 | ·46 |
| 3 | S | 4 | ·64 | 5 | O | 4 | ·17 | 6 | N | 2ɔ | ·70 | 1 | D | 2 | ·23 | | 2 | D | 31 | ·76 | **38** | 4 | Jr | 30 | ·29 | 5 | F | 28 | ·82 |
| 2 | S | 23 | ·54 | 4 | O | 23ɔ | ·07 | 5 | N | 21 | ·60 | 7 | D | 21 | ·13 | **39** | 1 | Jr | 19 | ·66 | | 3 | F | 18 | ·19 | 4 | Mr | 19 | ·72 |
| 6 | S | 12 | ·91 | 1 | O | 12● | ·44 | 2 | N | 10 | ·97 | 4 | D | 10 | ·50 | **40** | 6 | Jr | 9 | ·03 | | 7 | F | 7 | ·56 | 2 | Mr | 8○ | ·09 |
| {4 | S | 1○ | ·27 | 7 | O | 30 | ·33 | 1 | N | 28 | ·86 | **Pausha Kshaya** | | | | | 3 | D | 28 | ·40 | **41** | 4 | Jr | 26 | ·93 | {6 | F | 25○ | ·46 |
| {5 | S | 30 | ·80 | | | | | | | | | | | | | | | | | | | | | | | {7 | Mr | 26 | ·99 |
| 3 | S | 20 | ·17 | 4 | O | 19 | ·70 | 6 | N | 18 | ·23 | 7 | D | 17 | ·76 | **42** | 2 | Jr | 16 | ·29 | | 3 | F | 14○ | ·82 | 5 | Mr | 16 | ·35 |
| 7 | S | 9 | ·54 | 2 | O | 9 | ·07 | 3 | N | 7 | ·60 | 5 | D | 7 | ·13 | **43** | 6 | Jr | 5 | ·66 | | 1 | F | 4 | ·19 | 2 | Mr | 5 | ·72 |
| 6 | S | 28 | ·44 | 7 | O | 27 | ·97 | 2 | N | 26 | ·50 | 4 | D | 26○ | ·03 | **44** | 5 | Jr | 24● | ·56 | | 7 | F | 23 | ·09 | 1 | Mr | 23 | ·62 |

☉'s Anom col. 6 + 29·53059 + 59·06117 + 88·59176 + 118·12235

☾'s Anom col. 7 + 1·976 + 3·952 + 5·928 + 7·904

| Kaliyuga. | Vikrama Era. | Saka Era. | Commencement of Solar Year. Month and day A. D. | Fraction of day. | First New-Moon in Solar Year. | Anomaly of first New-Moon. | A.D. | Week-day of 1st January. | Vaisakha Week-day | Month | Day | Fraction | Jyeshta Week-day | Month | Day | Fraction | Ashada Week-day | Month | Day | Fraction | Sravana Week-day | Month | Day | Fraction | Bhadrapada Week-day | Month | Day | Fraction |
|---|---|---|---|---|---|---|---|---|---|---|---|---|---|---|---|---|---|---|---|---|---|---|---|---|---|---|---|---|
| 4645 | 1601 | 1466 | 27M | ·5031 | 25·6467 | 10·352 | **1544** | 3 | 3 | Ap | 22 | ·15 | 4 | My | 21 | ·68 | 6 | Je | 20ↄ | ·21 | 7 | Jl | 19 | ·74 | 2 | Au | 18 | ·27 |
| 4646 | 1602 | 1467 | 27M | ·7619 | 14·7550 | 6·509 | **1545** | 5 | 7 | Ap | 11 | ·52 | 2 | My | 11 | ·05 | 3 | Je | 9●ↄ | ·58 | 5 | Jl | 9 | ·11 | 6 | Au | 7 | ·64 |
| 4647 | 1603 | 1468 | 28M | ·0206 | 3·8633 | 2·667 | **1546** | 6 | 4 | Mr | 31 | ·88 | 6 | Ap | 30 | ·41 | { 7 | My | 29 | ·94 | 4 | Jl | 28 | ·01 | 5 | Au | 26 | ·54 |
| | | | | | | | | | | | | | | | | | { 2 | Je | 28 | ·47 | | | | | | | | |
| 4648 | 1604 | 1469 | 28M | ·2794 | 22·5022 | 0·800 | **1547** | 7 | 3 | Ap | 19ↄ | ·78 | 5 | My | 19● | ·31 | 6 | Je | 17 | ·84 | 1 | Jl | 17 | ·37 | 2 | Au | 15 | ·90 |
| 4649 | 1605 | 1470 | 27M | ·5381 | 11·6105 | 24·512 | **1548** | 1 | 1 | Ap | 8ↄ | ·15 | 2 | My | 7 | ·68 | 4 | Je | 6 | ·21 | 5 | Jl | 5 | ·74 | 7 | Au | 4 | ·27 |
| 4650 | 1606 | 1471 | 27M | ·7969 | 0·7188 | 20·669 | **1549** | 3 | { 5 | Mr | 23●ↄ | ·52 | 1 | My | 26 | ·58 | 3 | Je | 26 | ·11 | 4 | Jl | 24 | ·64 | 6 | Au | 23 | ·17 |
| | | | | | | | | | { 7 | Ap | 27 | ·05 | | | | | | | | | | | | | | | | |
| 4651 | 1607 | 1472 | 28M | ·0556 | 19·3576 | 18·802 | **1550** | 4 | 4 | Ap | 16 | ·41 | 5 | My | 15 | ·94 | 7 | Je | 14 | ·47 | 2 | Jl | 14 | ·00 | 3 | Au | 12 | ·54 |
| 4652 | 1608 | 1473 | 28M | ·3144 | 8·4659 | 14·959 | **1551** | 5 | 1 | Ap | 5 | ·78 | 3 | My | 5 | ·31 | 4 | Je | 3 | ·84 | 6 | Jl | 3 | ·37 | { 7 | Au | 1ↄ | ·90 |
| | | | | | | | | | | | | | | | | | | | | | | | | | { 2 | Au | 31● | ·43 |
| 4653 | 1609 | 1474 | 27M | ·5732 | 27·1048 | 13·093 | **1552** | 6 | 7 | Ap | 23 | ·68 | 2 | My | 23 | ·21 | 3 | Je | 21 | ·74 | 5 | Jl | 21ↄ | ·27 | 6 | Au | 19 | ·80 |
| 4654 | 1610 | 1475 | 27M | ·8319 | 16·2131 | 9·250 | **1553** | 1 | 5 | Ap | 13 | ·04 | 6 | My | 12 | ·58 | 1 | Je | 11 | ·11 | 2 | Jl | 10 | ·64 | 4 | Au | 9 | ·17 |
| 4655 | 1611 | 1476 | 28M | ·0907 | 5·3214 | 5·407 | **1554** | 2 | 1 | Ap | 2 | ·41 | 3 | My | 1 | ·94 | { 5 | My | 31ↄ | ·47 | 1 | Jl | 29 | ·54 | 3 | Au | 28 | ·07 |
| | | | | | | | | | | | | | | | | | { 7 | Je | 30 | ·00 | | | | | | | | |
| 4656 | 1612 | 1477 | 28M | ·3494 | 23·9603 | 3·540 | **1555** | 3 | 1 | Ap | 21 | ·31 | 2 | My | 20ↄ | ·84 | 4 | Je | 19● | ·37 | 5 | Jl | 18 | ·90 | 7 | Au | 17 | ·43 |
| 4657 | 1613 | 1478 | 27M | ·6082 | 13·0686 | 27·253 | **1556** | 4 | 5 | Ap | 9 | ·68 | 7 | My | 9●ↄ | ·21 | 1 | Je | 7 | ·74 | 3 | Jl | 7 | ·27 | 4 | Au | 5 | ·80 |
| 4658 | 1614 | 1479 | 27M | ·8669 | 2·1769 | 23·410 | **1557** | 6 | 3 | Mr | 30 | ·04 | { 4 | Ap | 28 | ·57 | 7 | Je | 26 | ·64 | 2 | Jl | 26 | ·17 | 3 | Au | 28 | ·70 |
| | | | | | | | | | | | | | { 6 | My | 28 | ·10 | | | | | | | | | | | | |
| 4659 | 1615 | 1480 | 28M | ·1257 | 20·8158 | 21·543 | **1558** | 7 | 1 | Ap | 17● | ·94 | 3 | My | 17 | ·47 | 5 | Je | 16 | ·00 | 6 | Jl | 15 | ·53 | 1 | Au | 14 | ·06 |
| 4660 | 1616 | 1481 | 28M | ·3844 | 9·9241 | 17·700 | **1559** | 1 | 6 | Ap | 7 | ·31 | 7 | My | 6 | ·84 | 2 | Je | 5 | ·37 | 3 | Jl | 4 | ·90 | 5 | Au | 3 | ·43 |
| 4661 | 1617 | 1482 | 27M | ·6432 | 28·5630 | 15·833 | **1560** | 2 | 5 | Ap | 25 | ·21 | 6 | My | 24 | ·74 | 1 | Je | 23 | ·27 | 2 | Jl | 22 | ·80 | 4 | Au | 21●ↄ | ·33 |
| 4662 | 1618 | 1483 | 27M | ·9020 | 17·6713 | 11·991 | **1561** | 4 | 2 | Ap | 14 | ·57 | 4 | My | 14 | ·10 | 5 | Je | 12 | ·63 | 7 | Jl | 12 | ·16 | 1 | Au | 10● | ·70 |
| 4663 | 1619 | 1484 | 28M | ·1607 | 6·7796 | 8·148 | **1562** | 5 | 6 | Ap | 3 | ·94 | 1 | My | 3 | ·47 | 3 | Je | 2 | ·00 | { 4 | Jl | 1ↄ | ·53 | 7 | Au | 29 | ·59 |
| | | | | | | | | | | | | | | | | | | | | | { 6 | Jl | 31 | ·06 | | | | |
| 4664 | 1620 | 1485 | 28M | ·4195 | 25·4185 | 6·281 | **1563** | 6 | 5 | Ap | 22 | ·84 | 7 | My | 22 | ·37 | 1 | Je | 20ↄ | ·90 | 3 | Jl | 20 | ·43 | 4 | Au | 18 | ·96 |
| 4665 | 1621 | 1486 | 27M | ·6782 | 14·5268 | 2·438 | **1564** | 7 | 3 | Ap | 11 | ·20 | 4 | My | 10 | ·73 | 6 | Je | 9● | ·27 | 7 | Jl | 8 | ·80 | 2 | Au | 7 | ·33 |
| 4666 | 1622 | 1487 | 27M | ·9370 | 3·6351 | 26·150 | **1565** | 2 | 7 | Mr | 31 | ·57 | 2 | Ap | 30ↄ | ·10 | { 3 | My | 29 | ·63 | 6 | Jl | 27 | ·69 | 1 | Au | 26 | ·22 |
| | | | | | | | | | | | | | | | | | { 5 | Je | 28 | ·16 | | | | | | | | |
| 4667 | 1623 | 1488 | 28M | ·1957 | 22·2740 | 24·283 | **1566** | 3 | 6 | Ap | 19ↄ | ·47 | 1 | My | 19 | ·00 | 2 | Je | 17 | ·53 | 4 | Jl | 17 | ·06 | 5 | Au | 15 | ·59 |
| 4668 | 1624 | 1489 | 28M | ·4545 | 11·3822 | 20·440 | **1567** | 4 | 3 | Ap | 8●ↄ | ·84 | 5 | My | 8 | ·37 | 6 | Je | 6 | ·90 | 1 | Jl | 6 | ·43 | 2 | Au | 4 | ·96 |
| 4669 | 1625 | 1490 | 27M | ·7133 | 0·4905 | 16·598 | **1568** | 5 | { 1 | Mr | 28 | ·20 | 4 | My | 26 | ·26 | 5 | Je | 24 | ·79 | 7 | Jl | 24 | ·33 | 1 | Au | 22 | ·86 |
| | | | | | | | | | { 2 | Ap | 26 | ·73 | | | | | | | | | | | | | | | | |
| 4670 | 1626 | 1491 | 27M | ·9720 | 19·1294 | 14·731 | **1569** | 7 | 7 | Ap | 16 | ·10 | 1 | My | 15 | ·63 | 3 | Je | 14 | ·16 | 4 | Jl | 13 | ·69 | 6 | Au | 12ↄ | ·22 |
| 4671 | 1627 | 1492 | 28M | ·2308 | 8·2377 | 10·888 | **1570** | 1 | 4 | Ap | 5 | ·47 | 6 | My | 5 | ·00 | 7 | Je | 3 | ·53 | 2 | Jl | 3 | ·06 | { 3 | Au | 1ↄ | ·59 |
| | | | | | | | | | | | | | | | | | | | | | | | | | { 5 | Au | 31 | ·12 |
| 4672 | 1628 | 1493 | 28M | ·4895 | 26·8766 | 9·022 | **1571** | 2 | 3 | Ap | 24 | ·37 | 4 | My | 23 | ·90 | 6 | Je | 22 | ·43 | 7 | Jl | 21●ↄ | ·96 | 2 | Au | 20 | ·49 |
| 4673 | 1629 | 1494 | 27M | ·7483 | 15·9849 | 5·179 | **1572** | 3 | 7 | Ap | 12 | ·73 | 2 | My | 12 | ·26 | 3 | Je | 10ↄ | ·79 | 5 | Jl | 10● | ·32 | 6 | Au | 8 | ·85 |
| 4674 | 1630 | 1495 | 28M | ·0070 | 5·0932 | 1·336 | **1573** | 5 | 5 | Ap | 2 | ·10 | 6 | My | 1 | ·63 | { 1 | My | 31ↄ | ·16 | 4 | Jl | 29 | ·22 | 5 | Au | 27 | ·75 |
| | | | | | | | | | | | | | | | | | { 2 | Je | 29 | ·69 | | | | | | | | |
| 4675 | 1631 | 1496 | 28M | ·2658 | 23·7321 | 27·024 | **1574** | 6 | 4 | Ap | 21 | ·00 | 5 | My | 20ↄ | ·53 | 7 | Je | 19 | ·06 | 1 | Jl | 18 | ·59 | 3 | Au | 17 | ·12 |
| 4676 | 1632 | 1497 | 28M | ·5246 | 12·8404 | 23·181 | **1575** | 7 | 1 | Ap | 10 | ·36 | 2 | My | 9● | ·89 | 4 | Je | 8 | ·42 | 5 | Jl | 7 | ·95 | 7 | Au | 6 | ·48 |
| 4677 | 1633 | 1498 | 27M | ·7833 | 1·9487 | 19·339 | **1576** | 1 | 5 | Mr | 29ↄ | ·73 | { 7 | Ap | 28 | ·26 | 3 | Je | 26 | ·32 | 4 | Jl | 25 | ·85 | 6 | Au | 24 | ·38 |
| | | | | | | | | | | | | | { 1 | My | 27 | ·79 | | | | | | | | | | | | |
| 4678 | 1634 | 1499 | 28M | ·0421 | 20·5876 | 17·472 | **1577** | 3 | 4 | Ap | 17 | ·63 | 6 | My | 17 | ·16 | 7 | Je | 15 | ·69 | 2 | Jl | 15 | ·22 | 3 | Au | 13 | ·75 |
| 4679 | 1635 | 1500 | 28M | ·3008 | 9·6959 | 13·629 | **1578** | 4 | 2 | Ap | 7 | ·00 | 3 | My | 6 | ·53 | 5 | Je | 5 | ·06 | 6 | Jl | 4 | ·59 | 1 | Au | 3 | ·12 |
| 4680 | 1636 | 1501 | 28M | ·5596 | 28·3348 | 11·762 | **1579** | 5 | 7 | Ap | 25 | ·89 | 2 | My | 25 | ·42 | 3 | Je | 23 | ·95 | 5 | Jl | 23 | ·49 | 7 | Au | 22● | ·02 |
| 4681 | 1637 | 1502 | 27M | ·8183 | 17·4431 | 7·919 | **1580** | 6 | 5 | Ap | 14 | ·26 | 6 | My | 13 | ·79 | 1 | Je | 12 | ·32 | 2 | Jl | 11○ | ·85 | 4 | Au | 10 | ·38 |
| 4682 | 1638 | 1503 | 28M | ·0771 | 6·5514 | 4·077 | **1581** | 1 | 2 | Ap | 3 | ·63 | 4 | My | 3 | ·16 | 5 | Je | 1 | ·69 | { 7 | Jl | 1○ | ·22 | 3 | Au | 29 | ·28 |
| | | | | | | | | | | | | | | | | | | | | | { 1 | Jl | 30 | ·75 | | | | |
| 4683 | 1639 | 1504 | 28M | ·3368 | 25·1903 | 2·210 | **1582** | 2 | 1 | Ap | 22 | ·58 | 3 | My | 22 | ·06 | 4 | Je | 20● | ·59 | 6 | Jl | 20 | ·12 | 7 | Au | 18 | ·65 |

**Surya Siddhanta.**

| | | | | | | | | | | | | | | | | | | | | | | | | | | | | | |
|---|---|---|---|---|---|---|---|---|---|---|---|---|---|---|---|---|---|---|---|---|---|---|---|---|---|---|---|---|---|
| + 147·65293 | | | | + 177·18353 | | | | + 206·71411 | | | | + 236·24470 | | | | | + 265·77529 | | | | | + 295·30588 | | | | {+324·83647 +354·36705 | | | |
| + 9·880 | | | | + 11·856 | | | | + 13·832 | | | | + 15·808 | | | | | + 17·784 | | | | | + 19·760 | | | | {+ 21·736 + 23·712 | | | |
| **Asvina** | | | | **Kartika** | | | | **Margasira** | | | | **Pausha** | | | | **A.D.** | **Magha** | | | | **A.D.** | **Phalguna** | | | | **Chaitra** | | | |
| Week-day | Month | Day | Fraction | Week-day | Month | Day | Fraction | Week-day | Month | Day | Fraction | Week-day | Month | Day | Fraction | | Week-day | Month | Day | Fraction | | Week-day | Month | Day | Fraction | Week-day | Month | Day | Fraction |
| 3 | S | 16 | ·80 | 5 | O | 16 | ·33 | 6 | N | 14 | ·86 | 1 | D | 14ↄ | ·39 | **45** | 2 | Jr | 12 | ·92 | | 4 | F | 11 | ·45 | 5 | Mr | 12 | ·98 |
| 1 | S | 6 | ·17 | 2 | O | 5 | ·70 | 4 | N | 4 | ·23 | 5 | D | 3●ↄ | ·76 | **46** | 7 | Jr | 2 | ·29 | | 1 | Jr | 31 | ·82 | 3 | Mr | 2 | ·35 |
| 7 | S | 25 | ·07 | 1 | O | 24 | ·60 | 3 | N | 23● | ·13 | 4 | D | 22 | ·66 | **47** | 6 | Jr | 21 | ·19 | | 7 | F | 19 | ·72 | 2 | Mr | 21 | ·25 |
| 4 | S | 14 | ·43 | 5 | O | 13ↄ | ·96 | 7 | N | 12 | ·49 | 2 | D | 12 | ·02 | **48** | 3 | Jr | 10 | ·55 | | 5 | F | 9 | ·08 | 6 | Mr | 9 | ·62 |
| 1 | S | 2 | ·80 | 3 | O | 2ↄ | ·33 | 4 | O | 31 | ·86 | 6 | N | 30 | ·39 | | 7 | D | 29 | ·92 | **49** | 2 | Jr | 28 | ·45 | 3 | F | 26 | ·98 |
| 7 | S | 21●ↄ | ·70 | 2 | O | 31 | ·23 | 3 | N | 19 | ·76 | 5 | D | 19 | ·29 | **50** | 6 | Jr | 17 | ·82 | | 1 | F | 16 | ·35 | 2 | Mr | 17● | ·88 |
| 5 | S | 11 | ·07 | 6 | O | 10 | ·60 | 1 | N | 9 | ·13 | 2 | D | 8 | ·66 | **51** | 4 | Jr | 7 | ·19 | | 5 | F | 5ↄ | ·72 | 7 | Mr | 7 | ·25 |
| 3 | S | 29 | ·96 | 5 | O | 29 | ·49 | 7 | N | 28 | ·02 | 1 | D | 27 | ·55 | **52** | 3 | Jr | 26ↄ | ·08 | | 4 | F | 24 | ·61 | 6 | Mr | 25 | ·14 |
| 1 | S | 18 | ·33 | 2 | O | 17 | ·86 | 4 | N | 16 | ·39 | 5 | D | 15 | ·92 | **53** | 7 | Jr | 14● | ·45 | | 1 | F | 12 | ·98 | 3 | Mr | 14 | ·51 |
| 5 | S | 7 | ·70 | 7 | O | 7 | ·23 | 1 | N | 5 | ·76 | 3 | D | 5 | ·29 | **54** | 4 | Jr | 3 | ·82 | | 6 | F | 2 | ·35 | 7 | Mr | 3 | ·88 |
| 4 | S | 26 | ·60 | 6 | O | 26 | ·13 | 7 | N | 24ↄ | ·66 | 2 | D | 24 | ·19 | **55** | 3 | Jr | 22 | ·72 | | 5 | F | 21 | ·25 | 6 | Mr | 22 | ·78 |
| 1 | S | 15 | ·96 | 3 | O | 15 | ·49 | 5 | N | 14●ↄ | ·02 | 6 | D | 13 | ·55 | **56** | 1 | Jr | 12 | ·08 | | 2 | F | 10 | ·61 | 4 | Mr | 11 | ·14 |
| 6 | S | 4 | ·33 | 7 | O | 3 | ·86 | 2 | N | 2●ↄ | ·39 | 3 | D | 1 | ·92 | | 5 | D | 31 | ·45 | **57** | 6 | Jr | 29 | ·98 | 1 | F | 28 | ·51 |
| 5 | S | 23 | ·22 | 6 | O | 22● | ·76 | 1 | N | 21 | ·29 | 2 | D | 20 | ·82 | **58** | 4 | Jr | 19 | ·35 | | 5 | F | 17 | ·88 | 7 | Mr | 19ↄ | ·41 |
| 2 | S | 12ↄ | ·59 | 4 | O | 12 | ·12 | 5 | N | 10 | ·65 | 7 | D | 10 | ·18 | **59** | 1 | Jr | 8 | ·71 | | 3 | F | 7 | ·24 | 4 | Mr | 8ↄ | ·77 |
| {6 | S | 1ↄ | ·96 | 3 | O | 31 | ·02 | 4 | N | 29 | ·55 | 6 | D | 29 | ·08 | **60** | 7 | Jr | 27 | ·61 | | 2 | F | 26●ↄ | ·14 | 3 | Mr | 26 | ·67 |
| {1 | O | 1 | ·49 | | | | | | | | | | | | | | | | | | | | | | | | | | |
| 5 | S | 19 | ·86 | 7 | O | 19 | ·39 | 1 | N | 17 | ·92 | 3 | D | 17 | ·45 | **61** | 4 | Jr | 15 | ·98 | | 6 | F | 14● | ·51 | 1 | Mr | 16 | ·04 |
| 3 | S | 9 | ·22 | 4 | O | 8 | ·75 | 6 | N | 7 | ·28 | 7 | D | 6 | ·81 | **62** | 2 | Jr | 5ↄ | ·34 | | 3 | F | 3 | ·87 | 5 | Mr | 5 | ·40 |
| 2 | S | 28 | ·12 | 3 | O | 27 | ·65 | 5 | N | 26 | ·18 | 6 | D | 25ↄ | ·71 | **63** | 1 | Jr | 24 | ·24 | | 2 | F | 22 | ·77 | 4 | Mr | 24 | ·30 |
| 6 | S | 17 | ·49 | 1 | O | 17 | ·02 | 2 | N | 15 | ·55 | 4 | D | 15●ↄ | ·08 | **64** | 5 | Jr | 13 | ·61 | | 7 | F | 12 | ·14 | 1 | Mr | 12 | ·67 |
| 3 | S | 5 | ·86 | 5 | O | 5 | ·39 | 6 | N | 3 | ·92 | 1 | D | 3 | ·45 | **65** | 2 | Jr | 1 | ·98 | | 4 | Jr | 31 | ·51 | 6 | Mr | 2 | ·04 |
| 2 | S | 24 | ·75 | 4 | O | 24ↄ | ·28 | 5 | N | 22 | ·81 | 7 | D | 22 | ·34 | **66** | 1 | Jr | 20 | ·87 | | 3 | F | 19 | ·40 | 4 | Mr | 20 | ·93 |
| 7 | S | 14 | ·12 | 1 | O | 13ↄ | ·65 | 3 | N | 12 | ·18 | 4 | D | 11 | ·71 | **67** | 6 | Jr | 10 | ·24 | | 7 | F | 8 | ·77 | 2 | Mr | 10 | ·30 |
| 4 | S | 3 | ·49 | 6 | O | 3ↄ | ·02 | 7 | N | 1 | ·55 | 2 | D | 1 | ·08 | | 3 | D | 30 | ·61 | **68** | 5 | Jr | 29 | ·14 | 6 | F | 27 | ·67 |
| 3 | S | 21● | ·38 | 4 | O | 20 | ·91 | 6 | N | 19 | ·44 | 7 | D | 18 | ·97 | **69** | 2 | Jr | 17 | ·50 | | 4 | F | 16ↄ | ·08 | 5 | Mr | 17 | ·56 |
| 7 | S | 10 | ·75 | 2 | O | 10 | ·28 | 3 | N | 8 | ·81 | 5 | D | 8 | ·34 | **70** | 6 | Jr | 6 | ·87 | | 1 | F | 5●ↄ | ·40 | 2 | Mr | 6 | ·93 |
| 6 | S | 29 | ·65 | 1 | O | 29 | ·18 | 2 | N | 27 | ·71 | 4 | D | 27 | ·24 | **71** | 5 | Jr | 25 | ·77 | | 7 | F | 24 | ·30 | 1 | Mr | 25 | ·83 |
| 4 | S | 19 | ·02 | 5 | O | 18 | ·55 | 7 | N | 17 | ·08 | 1 | D | 16 | ·61 | **72** | 3 | Jr | 15● | ·14 | | 4 | F | 18 | ·67 | 6 | Mr | 14 | ·20 |
| 1 | S | 7 | ·38 | 2 | O | 6 | ·91 | 4 | N | 5 | ·44 | 5 | D | 4ↄ | ·97 | **73** | 7 | Jr | 3 | ·50 | | 2 | F | 2 | ·03 | 3 | Mr | 3 | ·56 |
| 7 | S | 26 | ·28 | 1 | O | 25 | ·81 | 3 | N | 24ↄ | ·34 | 4 | D | 23 | ·87 | **74** | 6 | Jr | 22 | ·40 | | 7 | F | 20 | ·93 | 2 | Mr | 22 | ·46 |
| 4 | S | 15 | ·65 | 6 | O | 15 | ·18 | 7 | N | 13ↄ | ·71 | 2 | D | 13 | ·24 | **75** | 3 | Jr | 11 | ·77 | | 5 | F | 10 | ·30 | 6 | Mr | 11 | ·88 |
| 2 | S | 5 | ·01 | 3 | O | 4 | ·54 | 5 | N | 3 | ·07 | 6 | D | 2 | ·60 | **76** | 1 | Jr | 1 | ·13 | | 2 | Jr | 30 | ·66 | 4 | F | 29 | ·19 |
| 7 | S | 22ↄ | ·91 | 2 | O | 22 | ·44 | 3 | N | 20 | ·97 | 5 | D | 20 | ·50 | **77** | 7 | Jr | 19 | ·03 | | 1 | F | 17 | ·56 | 3 | Mr | 19ↄ | ·09 |
| 5 | S | 12ↄ | ·28 | 6 | O | 11 | ·81 | 1 | N | 10 | ·34 | 2 | D | 9 | ·87 | **78** | 4 | Jr | 8 | ·40 | | 5 | F | 5 | ·93 | 7 | Mr | 8●ↄ | ·46 |
| {2 | S | 1ↄ | ·65 | 5 | O | 30 | ·71 | 7 | N | 29 | ·24 | 1 | D | 28 | ·77 | **79** | 3 | Jr | 27 | ·30 | | 4 | F | 25 | ·83 | 6 | Mr | 27 | ·36 |
| {4 | O | 1 | ·18 | | | | | | | | | | | | | | | | | | | | | | | | | | |
| 1 | S | 20 | ·54 | 3 | O | 20 | ·07 | 4 | N | 18 | ·61 | 6 | D | 18 | ·13 | **80** | 7 | Jr | 16ↄ | ·66 | | 2 | F | 15● | ·19 | 3 | Mr | 15 | ·72 |
| 5 | S | 8 | ·91 | 7 | O | 8 | ·44 | 1 | N | 6 | ·97 | 3 | D | 6 | ·50 | **81** | 5 | Jr | 5ↄ | ·03 | | 6 | F | 3 | ·56 | 1 | Mr | 5 | ·09 |
| 4 | S | 27 | ·81 | 6 | O | 27 | ·34 | 7 | N | 25 | ·87 | 2 | D | 25ↄ | ·40 | **82** | 3 | Jr | 28 | ·93 | | 5 | F | 22 | ·46 | 6 | Mr | 23 | ·99 |
| 2 | S | 17 | ·18 | 3 | O | 16 | ·71 | 5 | N | 15 | ·24 | 6 | D | 14● | ·77 | **83** | 1 | Jr | 13 | ·80 | | 2 | F | 11 | ·83 | 4 | Mr | 13 | ·36 |

**TABLE X–**

| Kaliyuga. | Vikrama Era. | Saka Era. | Commencement of Solar Year. Month and day A. D. | Fraction of day. | First New-Moon in Solar Year. | Anomaly of first New-Moon. | A. D. | Week-day of 1st January. | Vaisakha Week-day | Month | Day | Fraction | Jyeshta Week-day | Month | Day | Fraction | Ashada Week-day | Month | Day | Fraction | Sravana Week-day | Month | Day | Fraction | Bhadrapada Week-day | Month | Day | Fraction |
|---|---|---|---|---|---|---|---|---|---|---|---|---|---|---|---|---|---|---|---|---|---|---|---|---|---|---|---|---|
| | | | | | | | | ⊙'s Anom col. 6 | | | | | + 29·53059 | | | | + 59·06117 | | | | + 88·59176 | | | | + 118·12235 | | | |
| | | | | | | | | ☾'s Anom col. 7 | | | | | + 1·976 | | | | + 3·952 | | | | + 5·928 | | | | + 7·904 | | | |
| 4684 | 1640 | 1505 | 28M | ·5946 | 14·2986 | 25·922 | **1583** | 3 | 5 | Ap | 11 | ·89 | 7 | My | 11○ | ·42 | 1 | Je | 9 | ·95 | 3 | Jl | 9 | ·48 | 5 | Au | 8 | ·02 |
| 4685 | 1641 | 1506 | 27M | ·8534 | 3·4069 | 22·079 | **1584** | 4 | 3 | Mr | 31 | ·26 | 4 | Ap | 29ↄ | ·79 | 6 | My | 29 | ·32 | 2 | Jl | 27 | ·38 | 3 | Au | 25 | ·91 |
| | | | | | | | | | | | | | | | | | 7 | Je | 27 | ·85 | | | | | | | | |
| 4686 | 1642 | 1507 | 28M | ·1121 | 22·0457 | 20·212 | **1585** | 6 | 2 | Ap | 19ↄ | ·16 | 3 | My | 18 | ·69 | 5 | Je | 17 | ·22 | 6 | Jl | 16 | ·75 | 1 | Au | 15 | ·28 |
| 4687 | 1643 | 1508 | 28M | ·3709 | 11·1540 | 16·370 | **1586** | 7 | 6 | Ap | 8 | ·52 | 1 | My | 8 | ·05 | 2 | Je | 6 | ·59 | 4 | Jl | 6 | ·12 | 5 | Au | 4 | ·65 |
| 4688 | 1644 | 1509 | 28M | ·6296 | 0·2623 | 12·527 | **1587** | 1 | 3 | Mr | 28 | ·89 | 6 | My | 26 | ·95 | 1 | Je | 25 | ·48 | 3 | Jl | 25 | ·01 | 4 | Au | 23ↄ | ·54 |
| | | | | | | | | | 5 | Ap | 27 | ·42 | | | | | | | | | | | | | | | | |
| 4689 | 1645 | 1510 | 27M | ·8884 | 18·9012 | 10·660 | **1588** | 2 | 2 | Ap | 15 | ·79 | 4 | My | 15 | ·32 | 5 | Je | 13 | ·85 | 7 | Jl | 13 | ·38 | 1 | Au | 11ↄ | ·91 |
| 4690 | 1646 | 1511 | 28M | ·1471 | 8·0095 | 6·817 | **1589** | 4 | 7 | Ap | 5 | ·16 | 1 | My | 4 | ·69 | 3 | Je | 3 | ·22 | 4 | Jl | 2 | ·75 | 6 | Au | 1●ↄ | ·28 |
| | | | | | | | | | | | | | | | | | | | | | | | | | 7 | Au | 30 | ·81 |
| 4691 | 1647 | 1512 | 28M | ·4059 | 26·6484 | 4·951 | **1590** | 5 | 6 | Ap | 24 | ·05 | 7 | My | 23 | ·58 | 2 | Je | 22ↄ | ·11 | 3 | Jl | 21● | ·65 | 5 | Au | 20 | ·13 |
| 4692 | 1648 | 1513 | 28M | ·6647 | 15·7567 | 1·108 | **1591** | 6 | 3 | Ap | 13 | ·42 | 4 | My | 12 | ·95 | 6 | Je | 11ↄ | ·48 | 1 | Jl | 11 | ·01 | 2 | Au | 9 | ·54 |
| 4693 | 1649 | 1514 | 27M | ·9234 | 4·8650 | 24·820 | **1592** | 7 | 7 | Ap | 1 | ·79 | 2 | My | 1 | ·32 | 3 | My | 30ↄ | ·85 | 6 | Jl | 28 | ·91 | 1 | Au | 27 | ·44 |
| | | | | | | | | | | | | | | | | | 5 | Je | 29 | ·38 | | | | | | | | |
| 4694 | 1650 | 1515 | 28M | ·1822 | 23·5039 | 22·953 | **1593** | 2 | 6 | Ap | 20 | ·69 | 1 | My | 20● | ·22 | 2 | Je | 18 | ·75 | 4 | Jl | 18 | ·28 | 5 | Au | 16 | ·81 |
| 4695 | 1651 | 1516 | 28M | ·4409 | 12·6122 | 19·110 | **1594** | 3 | 4 | Ap | 10○ | ·05 | 5 | My | 9● | ·58 | 7 | Je | 8 | ·11 | 1 | Jl | 7 | ·64 | 3 | Au | 6 | ·17 |
| 4696 | 1652 | 1517 | 28M | ·6997 | 1·7205 | 15·268 | **1595** | 4 | 1 | Mr | 30ↄ | ·42 | 2 | Ap | 28 | ·95 | 6 | Je | 27 | ·01 | 7 | Jl | 26 | ·54 | 2 | Au | 25 | ·07 |
| | | | | | | | | | | | | | 4 | My | 28 | ·48 | | | | | | | | | | | | |
| 4697 | 1653 | 1518 | 27M | ·9584 | 20·3594 | 13·401 | **1596** | 5 | 7 | Ap | 17 | ·32 | 1 | My | 16 | ·85 | 3 | Je | 15 | ·38 | 4 | Jl | 14 | ·91 | 6 | Au | 13 | ·44 |
| 4698 | 1654 | 1519 | 28M | ·2172 | 9·4677 | 9·558 | **1597** | 7 | 4 | Ap | 6 | ·68 | 6 | My | 6 | ·21 | 7 | Je | 4 | ·75 | 2 | Jl | 4 | ·28 | 3 | Au | 2 | ·81 |
| 4699 | 1655 | 1520 | 28M | ·4759 | 28·1066 | 7·691 | **1598** | 1 | 3 | Ap | 25 | ·58 | 5 | My | 25 | ·11 | 6 | Je | 23 | ·64 | 1 | Jl | 23ↄ | ·17 | 2 | Au | 21 | ·70 |
| 4700 | 1656 | 1521 | 28M | ·7347 | 17·2149 | 3·848 | **1599** | 2 | 7 | Ap | 14 | ·95 | 2 | My | 14 | ·48 | 4 | Je | 13 | ·01 | 5 | Jl | 12○ | ·54 | 7 | Au | 11 | ·07 |
| 4701 | 1657 | 1522 | 27M | ·9935 | 6·3233 | 0·128 | **1600** | 3 | 5 | Ap | 13 | ·32 | 6 | My | 2 | ·85 | 1 | Je | 1 | ·38 | 2 | Je | 30◗ | .91 | 5 | Au | 28 | ·97 |
| | | | | | | | | | | | | | | | | | | | | | 4 | Jl | 30 | ·44 | | | | |
| 4702 | 1658 | 1523 | 28M | ·2522 | 24·9621 | 25·816 | **1601** | 5 | 4 | Ap | 22 | ·21 | 5 | My | 21○ | ·74 | 7 | Je | 20● | ·27 | 1 | Je | 19 | ·81 | 3 | Au | 18 | ·34 |
| 4703 | 1659 | 1524 | 28M | ·5110 | 14·0704 | 21·973 | **1602** | 6 | 1 | Ap | 11 | ·58 | 3 | My | 11○ | ·11 | 4 | Je | 9 | ·64 | 6 | Jl | 9 | ·17 | 7 | Au | 7 | ·70 |
| 4704 | 1660 | 1525 | 28M | ·7697 | 3·1787 | 18·130 | **1603** | 7 | 5 | Mr | 31 | ·95 | 7 | Ap | 30●○ | ·48 | 2 | My | 30 | ·01 | 5 | Jl | 28 | ·07 | 6 | Au | 26 | ·60 |
| | | | | | | | | | | | | | | | | | 3 | Je | 28 | ·54 | | | | | | | | |
| 4705 | 1661 | 1526 | 28M | ·0285 | 21·8175 | 16·264 | **1604** | 1 | 4 | Ap | 18● | ·85 | 6 | My | 18 | ·38 | 7 | Je | 16 | ·91 | 2 | Jl | 16 | ·44 | 3 | Au | 14 | ·97 |
| 4706 | 1662 | 1527 | 28M | ·2872 | 10·9258 | 14·424 | **1605** | 3 | 2 | Ap | 8● | ·21 | 3 | My | 7 | ·74 | 5 | Je | 6 | ·27 | 6 | Jl | 5 | ·80 | 1 | Au | 4 | ·33 |
| 4707 | 1663 | 1528 | 28M | ·5460 | 0·0341 | 8·578 | **1606** | 4 | 1 | Ap | 27 | ·11 | 2 | My | 26 | ·64 | 4 | Je | 25 | ·17 | 5 | Jl | 24 | ·70 | 7 | Au | 23ↄ | ·23 |
| 4708 | 1664 | 1529 | 28M | ·8048 | 18·6730 | 6·711 | **1607** | 5 | 5 | Ap | 16 | ·48 | 7 | My | 16 | ·01 | 1 | Je | 14 | ·54 | 3 | Jl | 14 | ·07 | 4 | Au | 12ↄ | ·60 |
| 4709 | 1665 | 1530 | 28M | ·0635 | 7·7818 | 2·868 | **1608** | 6 | 2 | Ap | 4 | ·84 | 4 | My | 4 | ·37 | 5 | Je | 2 | ·91 | 7 | Jl | 2ↄ | ·44 | 1 | Jl | 31 | ·97 |
| | | | | | | | | | | | | | | | | | | | | | | | | | 3 | Au | 30 | ·50 |
| 4710 | 1666 | 1531 | 28M | ·3223 | 26·4202 | 1·002 | **1609** | 1 | 1 | Ap | 23 | ·74 | 3 | My | 23 | ·27 | 4 | Je | 21○ | ·80 | 6 | Jl | 21 | ·33 | 7 | Au | 9 | ·86 |
| 4711 | 1667 | 1532 | 28M | ·5810 | 15·5285 | 24·714 | **1610** | 2 | 6 | Ap | 13 | ·11 | 7 | My | 12 | ·64 | 2 | Je | 11●ↄ | ·17 | 3 | Jl | 10 | ·70 | 5 | Au | 9 | ·23 |
| 4712 | 1668 | 1533 | 28M | ·8398 | 4·6368 | 20·871 | **1611** | 3 | 3 | Ap | 2 | ·48 | 5 | My | 2 | ·01 | 6 | My | 31 | ·54 | 2 | Jl | 29 | ·60 | 4 | Au | 28 | ·13 |
| | | | | | | | | | | | | | | | | | 1 | Je | 30 | ·07 | | | | | | | | |
| 4713 | 1669 | 1534 | 28M | ·0985 | 23·2757 | 19·004 | **1612** | 4 | 2 | Ap | 30ↄ | ·37 | 3 | My | 19● | ·90 | 5 | Je | 18 | ·43 | 6 | Jl | 17 | ·97 | 1 | Au | 16 | ·50 |
| 4714 | 1670 | 1535 | 28M | ·3573 | 12·3840 | 15·161 | **1613** | 6 | 6 | Ap | 9ↄ | ·74 | 1 | My | 9 | ·27 | 2 | Je | 7 | ·80 | 4 | Jl | 7 | ·33 | 5 | Au | 5 | ·86 |
| 4715 | 1671 | 1536 | 28M | ·6161 | 1·4923 | 11·319 | **1614** | 7 | 4 | Mr | 30ↄ | ·11 | 5 | Ap | 28 | ·64 | 1 | Je | 26 | ·70 | 3 | Jl | 26 | ·23 | 4 | Au | 24 | ·76 |
| | | | | | | | | | | | | | 7 | My | 28 | ·17 | | | | | | | | | | | | |
| 4716 | 1672 | 1537 | 28M | ·8748 | 20·1312 | 9·452 | **1615** | 1 | 3 | Ap | 18 | ·01 | 4 | My | 17 | ·54 | 6 | Je | 16 | ·07 | 7 | Jl | 15 | ·60 | 2 | Au | 14 | ·13 |
| 4717 | 1673 | 1538 | 28M | ·1336 | 9·2395 | 5·609 | **1616** | 2 | 7 | Ap | 6 | ·37 | 1 | My | 5 | ·90 | 3 | Je | 4 | ·43 | 4 | Jl | 3 | ·96 | 6 | Au | 2○ | ·49 |
| 4718 | 1674 | 1539 | 28M | ·3923 | 27·8784 | 3·742 | **1617** | 4 | 6 | Ap | 25 | ·27 | 7 | My | 24 | ·80 | 2 | Je | 23 | ·33 | 3 | Jl | 22●ↄ | ·86 | 5 | Au | 31 | ·39 |
| 4719 | 1675 | 1540 | 28M | ·6511 | 16·9867 | 27·454 | **1618** | 5 | 3 | Ap | 14 | ·64 | 5 | My | 14 | ·17 | 6 | Je | 12 | ·70 | 1 | Jl | 12 | ·23 | 2 | Au | 10 | ·76 |
| 4720 | 1676 | 1541 | 28M | ·9098 | 6·0950 | 23·611 | **1619** | 6 | 1 | Ap | 4 | ·00 | 2 | My | 3 | ·53 | 4 | Je | 2ↄ | ·07 | 5 | Jl | 1● | ·60 | 1 | Au | 29 | ·66 |
| | | | | | | | | | | | | | | | | | | | | | 7 | Jl | 31 | ·13 | | | | |
| 4721 | 1677 | 1542 | 28M | ·1686 | 24·7338 | 21·745 | **1620** | 7 | 6 | Ap | 21 | ·90 | 1 | My | 21ↄ | ·43 | 2 | Je | 19 | ·96 | 4 | Jl | 19 | ·49 | 6 | Au | 18 | ·02 |
| 4722 | 1678 | 1543 | 28M | ·4274 | 13·8421 | 17·902 | **1621** | 2 | 4 | Ap | 11 | ·27 | 5 | My | 10●ↄ | ·80 | 7 | Je | 9 | ·33 | 1 | Jl | 8 | ·86 | 3 | Au | 7 | ·39 |

## Surya Siddhanta.

| Asvina | | | | Kartika | | | | Margasira | | | | Pausha | | | | A.D. | Magha | | | | A.D. | Phalguna | | | | Chaitra | | | |
|---|---|---|---|---|---|---|---|---|---|---|---|---|---|---|---|---|---|---|---|---|---|---|---|---|---|---|---|---|---|
| + 147·65298 | | | | + 177·18353 | | | | + 206·71411 | | | | + 236·24470 | | | | | + 265·77529 | | | | | + 295·30588 | | | | +324·83647 / +354·36705 | | | |
| + 9·880 | | | | + 11·856 | | | | + 13·832 | | | | + 15·808 | | | | | + 17·784 | | | | | + 19·760 | | | | +21·736 / +23·712 | | | |
| Week-day | Month | Day | Fraction | Week-day | Month | Day | Fraction | Week-day | Month | Day | Fraction | Week-day | Month | Day | Fraction | | Week-day | Month | Day | Fraction | | Week-day | Month | Day | Fraction | Week-day | Month | Day | Fraction |
| 6 | S | 6 | ·55 | 1 | O | 6 | ·08 | 2 | N | 4ↄ | ·61 | 4 | D | 4● | ·14 | 84 | 5 | Jr | 2 | ·67 | | 7 | F | 1 | ·20 | 1 | Mr | 1 | ·73 |
| 5 | S | 24 | ·44 | 6 | O | 23ↄ | ·97 | 1 | N | 22 | ·50 | 3 | D | 22 | ·03 | 85 | 4 | Jr | 20 | ·57 | | 6 | F | 19 | ·10 | 7 | Mr | 20 | ·63 |
| 2 | S | 13 | ·81 | 4 | O | 13ↄ | ·34 | 5 | N | 11 | ·87 | 7 | D | 11 | ·40 | 86 | 1 | Jr | 9 | ·93 | | 3 | F | 8 | ·46 | 4 | Mr | 9 | ·99 |
| 7 | S | 3 | ·18 | 1 | O | 2 | ·71 | 3 | N | 1 | ·24 | 4 | N | 30 | ·77 | | 6 | D | 30 | ·30 | 87 | 7 | Jr | 28 | ·83 | 2 | F | 27○ | ·36 |
| 6 | S | 22● | ·07 | 7 | O | 21 | ·61 | 2 | N | 20 | ·14 | 3 | D | 19 | ·67 | 88 | 5 | Jr | 18 | ·20 | | 6 | F | 16ↄ | ·73 | 1 | Mr | 17 | ·26 |
| 3 | S | 10 | ·44 | 4 | O | 9 | ·97 | 6 | N | 8 | ·50 | 1 | D | 8 | ·03 | 89 | 2 | Jr | 6 | ·56 | | 4 | F | 5● | ·09 | 5 | Mr | 6 | ·63 |
| 2 | S | 29 | ·34 | 3 | O | 23 | ·87 | 6 | N | 27 | ·40 | 6 | D | 26 | ·93 | 90 | 1 | Jr | 25 | ·46 | | 2 | F | 23 | ·99 | 4 | Mr | 25 | ·52 |
| 6 | S | 18 | ·71 | 1 | O | 18 | ·24 | 2 | N | 16 | ·77 | 4 | D | 16ↄ | ·30 | 91 | 5 | Jr | 14 | ·83 | | 7 | F | 13 | ·36 | 1 | Mr | 14 | ·89 |
| 4 | S | 8 | ·07 | 5 | O | 7 | ·60 | 7 | N | 6 | ·13 | 1 | D | 5ↄ | ·67 | 92 | 3 | Jr | 4 | ·20 | | 4 | F | 2 | ·73 | 6 | Mr | 3 | ·26 |
| 2 | S | 25 | ·97 | 4 | O | 25 | ·50 | 6 | N | 24ↄ | ·03 | 7 | D | 23 | ·56 | 93 | 2 | Jr | 28 | ·09 | | 3 | F | 20 | ·62 | 5 | Mr | 22 | ·15 |
| 7 | S | 15 | ·34 | 1 | O | 14 | ·87 | 3 | N | 13● | ·40 | 4 | D | 12 | ·93 | 94 | 6 | Jr | 11 | ·46 | | 7 | F | 9 | ·99 | 2 | Mr | 11 | ·52 |
| 4 | S | 4 | ·71 | 6 | O | 4ↄ | ·24 | 7 | N | 2 | ·77 | 2 | D | 2 | ·30 | | 3 | D | 31 | ·83 | 95 | 5 | Jr | 30 | ·36 | 6 | F | 28 | ·89 |
| 3 | S | 23●ↄ | ·60 | 5 | O | 23 | ·13 | 6 | N | 21 | ·66 | 1 | D | 21 | ·19 | 96 | 2 | Jr | 19 | ·73 | | 4 | F | 18 | ·26 | 5 | Mr | 18ↄ | ·79 |
| 7 | S | 11●ↄ | ·97 | 2 | O | 11 | ·50 | 4 | N | 10 | ·03 | 5 | D | 9 | ·56 | 97 | 7 | Jr | 8 | ·09 | | 1 | F | 6 | ·62 | 3 | Mr | 8● | ·15 |
| {5 | S | 1 | ·34 | 1 | O | 30 | ·40 | 2 | N | 28 | ·93 | 4 | D | 28 | ·46 | 98 | 5 | Jr | 26ↄ | ·99 | | 7 | F | 25 | ·52 | 2 | Mr | 27 | ·05 |
| {6 | S | 30 | ·87 | | | | | | | | | | | | | | | | | | | | | | | | | | |
| 4 | S | 20 | ·28 | 5 | O | 19 | ·77 | 7 | N | 18 | ·30 | 1 | D | 17 | ·83 | 99 | 3 | Jr | 16ↄ | ·36 | | 4 | F | 14● | ·89 | 6 | Mr | 16 | ·42 |
| 1 | S | 9 | ·60 | 3 | O | 9 | ·13 | 4 | N | 7 | ·66 | 6 | D | 7 | ·19 | 16 | 7 | Jr | 5ↄ | ·72 | 00 | 2 | F | 4 | ·25 | 3 | Mr | 4 | ·79 |
| 7 | S | 27 | ·50 | 2 | O | 27 | ·03 | 3 | N | 25 | ·56 | 5 | D | 25● | ·09 | 01 | 6 | Jr | 23 | ·62 | | 1 | F | 22 | ·15 | 2 | Mr | 23 | ·68 |
| 4 | S | 16 | ·87 | 6 | O | 16 | ·40 | 7 | N | 14ↄ | ·93 | 2 | D | 14 | ·46 | 02 | 3 | Jr | 12 | ·99 | | 5 | F | 11 | ·52 | 7 | Mr | 13 | ·05 |
| 2 | S | 6 | ·23 | 3 | O | 5 | ·76 | 5 | N | 4ↄ | ·29 | 6 | D | 3 | ·83 | 03 | 1 | Jr | 2 | ·36 | | 2 | Jr | 31 | ·89 | 4 | Mr | 2 | ·42 |
| 1 | S | 25 | ·13 | 2 | O | 24ↄ | ·66 | 4 | N | 23 | ·19 | 5 | D | 22 | ·72 | 04 | 7 | Jr | 21 | ·25 | | 1 | F | 19 | ·78 | 3 | Mr | 20 | ·31 |
| 5 | S | 13 | ·50 | 7 | O | 13 | ·03 | 1 | N | 11 | ·56 | 3 | D | 11 | ·09 | 05 | 4 | Jr | 9 | ·62 | | 6 | F | 8 | ·15 | 7 | Mr | 9ↄ | ·68 |
| 2 | S | 2ↄ | ·87 | 4 | O | 2 | ·40 | 5 | O | 31 | ·93 | 7 | N | 30 | ·46 | | 1 | D | 29 | ·99 | 06 | 3 | Jr | 28 | ·52 | {5 | F | 27○ | ·05 |
| | | | | | | | | | | | | | | | | | | | | | | | | | | {6 | Mr | 28 | ·58 |
| 1 | S | 21 | ·76 | 3 | O | 21 | ·29 | 4 | N | 19 | ·82 | 6 | D | 19 | ·35 | 07 | 7 | Jr | 17 | ·89 | | 2 | F | 16● | ·42 | 3 | Mr | 17 | ·95 |
| 6 | S | 11 | ·13 | 7 | O | 10 | ·66 | 2 | N | 9 | ·19 | 3 | D | 8 | ·72 | 08 | 5 | Jr | 7 | ·25 | | 6 | F | 5● | ·78 | 1 | Mr | 6 | ·31 |
| 5 | S | 29 | ·03 | 6 | O | 28 | ·56 | 1 | N | 27 | ·09 | 2 | D | 26ↄ | ·62 | 09 | 4 | Jr | 24 | ·15 | | 5 | F | 23 | ·68 | 7 | Mr | 25 | ·21 |
| 2 | S | 18 | ·39 | 3 | O | 17 | ·93 | 5 | N | 16 | ·46 | 6 | D | 15●ↄ | ·99 | 10 | 1 | Jr | 14 | ·52 | | 3 | F | 13 | ·05 | 4 | Mr | 14 | ·58 |
| 6 | S | 7 | ·76 | 1 | O | 7 | ·29 | 2 | N | 5 | ·82 | 4 | D | 5●ↄ | ·35 | 11 | 5 | Jr | 3 | ·88 | | 7 | F | 2 | ·41 | 1 | Mr | 4 | ·95 |
| 5 | S | 26 | ·66 | 7 | O | 26 | ·19 | 1 | N | 24● | ·72 | 3 | D | 24 | ·25 | 12 | 4 | Jr | 22 | ·73 | | 6 | F | 21 | ·31 | 7 | Mr | 21 | ·84 |
| 3 | S | 15 | ·03 | 4 | O | 14ↄ | ·56 | 6 | N | 13 | ·09 | 7 | D | 12 | ·62 | 13 | 2 | Jr | 11 | ·15 | | 3 | F | 9 | ·68 | 5 | Mr | 11 | ·21 |
| 7 | S | 4 | ·39 | 1 | O | 3ↄ | ·92 | 3 | N | 2 | ·45 | 4 | D | 1 | ·99 | | 6 | D | 31 | ·52 | 14 | 1 | Jr | 30 | ·05 | 2 | F | 28 | ·58 |
| 6 | S | 23●ↄ | ·29 | 7 | O | 22 | ·82 | 2 | N | 21 | ·35 | 3 | D | 28 | ·88 | 15 | 5 | Jr | 19 | ·41 | | 6 | F | 17 | ·94 | 1 | Mr | 19● | ·47 |
| 3 | S | 12 | ·66 | 5 | O | 12 | ·19 | 6 | N | 10 | ·72 | 1 | D | 10 | ·25 | 16 | 2 | Jr | 8 | ·78 | | 4 | F | 7○ | ·31 | 5 | Mr | 7 | ·84 |
| {1 | S | 1● | ·08 | 4 | O | 30 | ·09 | 5 | N | 28 | ·62 | 7 | D | 28 | ·15 | 17 | 1 | Jr | 26ↄ | ·68 | | 3 | F | 25 | ·21 | 4 | Mr | 26 | ·74 |
| {2 | S | 30 | ·56 | | | | | | | | | | | | | | | | | | | | | | | | | | |
| 6 | S | 19 | ·92 | 1 | O | 19 | ·45 | 2 | N | 17 | ·98 | 4 | D | 17 | ·51 | 18 | 6 | Jr | 16ↄ | ·05 | | 7 | F | 14 | ·58 | 2 | Mr | 14 | ·11 |
| 4 | S | 9 | ·29 | 5 | O | 8 | ·82 | 7 | N | 7 | ·35 | 1 | D | 6 | ·88 | 19 | 3 | Jr | 5 | ·41 | | 4 | F | 3 | ·94 | 6 | Mr | 5 | ·47 |
| 3 | S | 28 | ·19 | 4 | O | 27 | ·72 | 6 | N | 26ↄ | ·25 | 7 | D | 25 | ·78 | 20 | 2 | Jr | 24 | ·31 | | 3 | F | 22 | ·84 | 5 | Mr | 23 | ·37 |
| 7 | S | 16 | ·55 | 2 | O | 16 | ·09 | 3 | N | 14ↄ | ·62 | 5 | D | 14 | ·15 | 21 | 6 | Jr | 12 | ·68 | | 1 | F | 11 | ·21 | 2 | Mr | 12 | ·74 |
| 4 | S | 5 | ·92 | 6 | O | 5 | ·45 | 7 | N | 3ↄ | ·98 | 2 | D | 3 | ·51 | 22 | 4 | Jr | 2 | ·04 | | 5 | Jr | 31 | ·57 | 7 | Mr | 2 | ·11 |

| Kaliyuga. | Vikrama Era. | Saka Era. | Commencement of Solar Year. Month and day A.D. | Fraction of day. | First New-Moon in Solar Year. | Anomaly of first New-Moon. | A.D. | Week-day of 1st January. | Vaisakha. Week-day | Month | Day | Fraction | Jyeshta. Week-day | Month | Day | Fraction | Ashada. Week-day | Month | Day | Fraction | Sravana. Week-day | Month | Day | Fraction | Bhadrapada. Week-day | Month | Day | Fraction |
|---|---|---|---|---|---|---|---|---|---|---|---|---|---|---|---|---|---|---|---|---|---|---|---|---|---|---|---|---|
| | | | | | | | | ☉'s Anom col. 6 | + 29·53059 | | | | + 59·06117 | | | | + 88·59176 | | | | + 118·12235 | | | | | | | |
| | | | | | | | | ☾'s Anom col. 7 | + 1·976 | | | | + 3·952 | | | | + 5·928 | | | | + 7·904 | | | | | | | |
|4723|1679|1544|28M|·6861|2·9504|14·059|**1622**|3|1|Mr|31|·64|3|Ap|30|·17|4|My|29|·70|7|Jl|27|·76|2|Au|26|·29|
| | | | | | | | | | | | | | | | | |6|Je|28|·23| | | | | | | | |
|4724|1680|1545|28M|·9449|21·5893|12·192|**1623**|4|7|Ap|19|·53|2|My|19|·06|3|Je|17|·59|5|Jl|17|·13|6|Au|15|66|
|4725|1681|1546|28M|·2036|10·6976|8·350|**1624**|5|4|Ap|7|·90|6|My|7|·43|7|Je|5|·96|2|Jl|5|·49|4|Au|4|·02|
|4726|1682|1547|28M|·4624|29·3365|6·483|**1625**|7|3|Ap|26|·80|5|My|26|·33|6|Je|24|·86|1|Jl|24|·39|2|Au|22○|·92|
|4727|1683|1548|28M|·7211|18·4448|2·640|**1626**|1|1|Ap|16|·17|2|My|15|·70|4|Je|14|·23|5|Jl|13|·76|7|Au|12|·29|
|4728|1684|1549|28M|·9799|7·5531|26·352|**1627**|2|5|Ap|5|·53|7|My|5|·06|1|Je|3|·59|3|Jl|3○|·12|6|Au|31|·19|
| | | | | | | | | | | | | | | | | | | | | |4|Au|1●|·65| | | | |
|4729|1685|1550|28M|·2386|26·1920|24·485|**1628**|3|4|Ap|23|·43|5|My|22|·96|7|Je|21○|·49|2|Jl|21|·02|3|Au|19|·55|
|4730|1686|1551|28M|·4974|15·3008|20·643|**1629**|5|1|Ap|12|·80|3|My|12|·33|4|Je|10●|·86|6|Jl|10|·39|7|Au|8|·92|
|4731|1687|1552|28M|·7562|4·4086|16·800|**1630**|6|6|Ap|2|·16|7|My|1○|·69|2|My|31|·23|5|Jl|29|·29|6|Au|27|·82|
| | | | | | | | | | | | | | | | | |3|Je|29|·76| | | | | | | | |
|4732|1688|1553|29M|·0149|23·0475|14·933|**1631**|7|5|Ap|21○|·06|6|My|20●|·59|1|Je|19|·12|2|Jl|18|·65|4|Au|17|·18|
|4733|1689|1554|28M|·2737|12·1558|11·090|**1632**|1|2|Ap|9●○|·43|3|My|8|·96|5|Je|7|·49|7|Jl|7|·02|1|Au|5|·55|
|4734|1690|1555|28M|·5324|1·2641|7·248|**1633**|3|6|Mr|29|·80|2|My|27|·86|4|Je|26|·39|5|Jl|25|·92|7|Au|24|·45|
| | | | | | | | | |1|Ap|28|·33| | | | | | | | | | | | | | | | |
|4735|1691|1556|28M|·7912|19·9030|5·381|**1634**|4|5|Ap|17|·69|7|My|17|·22|1|Je|15|·75|3|Jl|15|·29|4|Au|13○|·82|
|4736|1692|1557|29M|·0499|9·0113|1·538|**1635**|5|3|Ap|7|·06|4|My|6|·59|6|Je|5|·12|7|Jl|4|·65|2|Au|3○|·18|
| | | | | | | | | | | | | | | | | | | | | | | | | |3|S|1|·71|
|4737|1693|1558|28M|·3087|27·6502|27·226|**1636**|6|1|Ap|24|·96|3|My|24|·49|5|Je|23|·02|6|Jl|22●○|·55|1|Au|21|·08|
|4738|1694|1559|28M|·5675|16·7585|23·383|**1637**|1|6|Ap|14|·33|7|My|13|·86|2|Je|12|·39|3|Jl|11|.92|5|Au|10|·45|
|4739|1695|1560|28M|·8262|5·8668|19·540|**1638**|2|3|Ap|3|·69|5|My|3|·22|6|Je|1○|·75|1|Jl|1|·28|4|Au|29|·35|
| | | | | | | | | | | | | | | | | | | | | |2|Jl|30|·81| | | | |
|4740|1696|1561|29M|·0850|24·5056|17·674|**1639**|3|2|Ap|22|·59|4|My|22○|·12|5|Je|20|·65|7|Jl|20|·18|1|Au|18|·71|
|4741|1697|1562|28M|·3437|13·6139|13·831|**1640**|4|6|Ap|10|·96|1|My|10|·49|3|Je|9|·02|4|Jl|8|·55|6|Au|7|·08|
|4742|1698|1563|28M|·6025|2·7222|9·988|**1641**|6|4|Mr|31○|·32|5|Ap|29|·85|1|Je|27|·92|3|Jl|27|·45|4|Au|25|·98|
| | | | | | | | | | | | | |7|My|27|·39| | | | | | | | | | | | |
|4743|1699|1564|28M|·8612|21·3611|8·121|**1642**|7|3|Ap|19|·22|4|My|18|·75|6|Je|17|·28|7|Jl|16|·81|2|Au|15|·34|
|4744|1700|1565|29M|·1200|10·4694|4·279|**1643**|1|7|Ap|8|·59|2|My|8|·12|3|Je|6|·65|5|Jl|6|·18|6|Au|4|·71|
|4745|1701|1566|28M|·3787|29·1083|2·412|**1644**|2|6|Ap|26|·49|1|My|26|·02|2|Je|24|·55|4|Jl|24|·08|5|Au|22●|·61|
|4746|1702|1567|28M|·6375|18·2166|26·124|**1645**|4|3|Ap|15|·85|5|My|15|·38|6|Je|13|·91|1|Jl|13○|·45|2|Au|11●|·98|
|4747|1703|1568|28M|·8963|7·3249|22·281|**1646**|5|1|Ap|5|·22|2|My|4|·75|4|Je|3|·28|5|Jl|2○|·81|1|Au|30|·87|
| | | | | | | | | | | | | | | | | | | | | |7|Au|1|·34| | | | |
|4748|1704|1569|29M|·1550|25·9638|20·414|**1647**|6|7|Ap|24|·12|1|My|23|·65|3|Je|22●|·18|4|Jl|21|·71|6|Au|20|·24|
|4749|1705|1570|28M|·4138|15·0721|16·571|**1648**|7|4|Ap|12|·49|6|My|12○|·02|7|Je|10●|·55|2|Jl|10|·08|3|Au|8|·61|
|4750|1706|1571|28M|·6725|4·1804|12·729|**1649**|2|1|Ap|1|·85|3|My|1○|·38|4|My|30|·91|7|Jl|28|·97|2|Au|30|·51|
| | | | | | | | | | | | | | | | | |6|Je|29|·44| | | | | | | | |
|4751|1707|1572|28M|·9313|22·8193|10·862|**1650**|3|7|Ap|20○|·75|2|My|20|·28|3|Je|18|·81|5|Jl|18|·34|6|Au|16|·87|
|4752|1708|1573|29M|·1900|11·9276|7·019|**1651**|4|5|Ap|10|·12|6|My|9|·65|1|Je|8|·18|2|Jl|7|·71|4|Au|6|·24|
|4753|1709|1574|28M|·4488|1·0359|3·176|**1652**|5|2|Mr|29●|·48|5|My|27|·55|7|Je|26|·08|1|Jl|25|·61|3|Au|24○|·14|
| | | | | | | | | |4|Ap|28|·01| | | | | | | | | | | | | | | | |
|4754|1710|1575|28M|·7076|19·6748|1·310|**1653**|7|1|Ap|17|·38|2|My|16|·91|4|Je|15|·44|5|Jl|14|·97|7|Au|13○|·50|
|4755|1711|1576|28M|·9663|8·7831|25·022|**1654**|1|5|Ap|6|·75|7|My|6|·28|1|Je|4|·81|3|Jl|4|·34|4|Au|2●○|·87|
| | | | | | | | | | | | | | | | | | | | | | | | | |6|S|1|·40|
|4756|1712|1577|29M|·2251|27·4220|23·155|**1655**|2|4|Ap|25|·65|6|My|25|·18|7|Je|23|·71|2|Jl|23●|·24|3|Au|21|·77|
|4757|1713|1578|28M|·4838|16·5303|19·312|**1656**|3|2|Ap|14|·01|3|My|13|·54|5|Je|12○|·07|6|Jl|11|·61|1|Au|10|·14|
|4758|1714|1579|28M|·7426|5·6386|15·469|**1657**|5|6|Ap|3|·38|7|My|2|·91|2|Je|1●○|·44|3|Jl|1|·97|7|Au|27|·03|
| | | | | | | | | | | | | | | | | | | | | |5|Jl|30|·50| | | | |
|4759|1715|1580|29M|·0013|24·2774|13·603|**1658**|6|5|Ap|22|·28|6|My|21●|·81|1|Je|20|·34|2|Jl|19|·87|4|Au|18|·40|
|4760|1716|1581|29M|·2601|13·3857|9·760|**1659**|7|2|Ap|11○|·65|4|My|11●|·18|5|Je|9|·71|7|Jl|9|·24|1|Au|7|·77|
|4761|1717|1582|28M|·5189|2·4940|5·917|**1660**|1|7|Mr|31○|·01|1|Ap|29|·54|4|Je|27|·60|6|Jl|27|·13|7|Au|25|·67|
| | | | | | | | | | | | | |3|My|29|·07| | | | | | | | | | | | |

# Surya Siddhanta.

| Asvina | | | | Kartika | | | | Margasira | | | | Pausha | | | | A.D. | Magha | | | | A.D. | Phalguna | | | | Chaitra | | | |
|---|---|---|---|---|---|---|---|---|---|---|---|---|---|---|---|---|---|---|---|---|---|---|---|---|---|---|---|---|---|
| + 147·65293 | | | | + 177·18353 | | | | + 206·71411 | | | | + 236·24470 | | | | | + 265·77529 | | | | | + 295·30588 | | | | +324·83647 / +354·36705 | | | |
| + 9·880 | | | | + 11·856 | | | | + 13·832 | | | | + 15·808 | | | | | + 17·784 | | | | | + 19·760 | | | | + 21·736 / + 23·712 | | | |
| Week-day | Month | Day | Fraction | Week-day | Month | Day | Fraction | Week-day | Month | Day | Fraction | Week-day | Month | Day | Fraction | | Week-day | Month | Day | Fraction | | Week-day | Month | Day | Fraction | Week-day | Month | Day | Fraction |
| 3 | S | 24 | ·82 ● | 5 | O | 24● | ·35 | 6 | N | 22 | ·88 | 1 | D | 22 | ·41 | **23** | 2 | Jr | 20 | ·94 | | 4 | F | 19 | ·47 | 6 | Mr | 21○ | ·00 |
| 1 | S | 14ↄ | ·19 | 2 | O | 13 | ·72 | 4 | N | 12 | ·25 | 5 | D | 11 | ·78 | **24** | 7 | Jr | 10 | ·31 | | 1 | F | 8 | ·84 | 3 | Mr | 9●○ | ·37 |
| 5 | S | 2○ | ·55 | 7 | O | 2 | ·08 | 1 | O | 31 | ·61 | 3 | N | 30 | ·15 | | 4 | D | 29 | ·68 | **25** | 6 | Jr | 28 | ·21 | 7 | F | 26○ | ·74 |
| | | | | | | | | | | | | | | | | | | | | | | | | | | 2 | Mr | 28 | ·27 |
| 4 | S | 21 | ·45 | 5 | O | 21 | ·98 | 7 | N | 19 | ·51 | 2 | D | 19 | ·04 | **26** | 3 | Jr | 17 | ·57 | | 5 | F | 16● | ·10 | 6 | Mr | 17 | ·63 |
| 1 | S | 10 | ·82 | 3 | O | 10 | ·35 | 4 | N | 8 | ·88 | 6 | D | 8 | ·41 | **27** | 7 | Jr | 6○ | ·94 | | 2 | F | 5 | ·47 | 4 | Mr | 7 | ·00 |
| 7 | S | 29 | ·72 | 2 | O | 29 | ·25 | 3 | N | 27 | ·78 | 5 | D | 27○ | ·31 | **28** | 6 | Jr | 25 | ·84 | | 1 | F | 24 | ·37 | 2 | Mr | 24 | ·90 |
| 5 | S | 18 | ·08 | 6 | O | 17 | ·61 | 1 | N | 16 | ·14 | 2 | D | 15ↄ | ·67 | **29** | 4 | Jr | 14 | ·21 | | 5 | F | 12 | ·74 | 7 | Mr | 14 | ·27 |
| 2 | S | 7 | ·45 | 3 | O | 6 | ·98 | 5 | N | 5 | ·51 | 7 | D | 5 | ·04 | **30** | 1 | Jr | 3 | ·57 | | 3 | F | 2 | ·10 | 4 | Mr | 3 | ·63 |
| 1 | S | 26 | ·35 | 2 | O | 25○ | ·88 | 4 | N | 24● | ·41 | 5 | D | 23 | ·94 | **31** | 7 | Jr | 22 | ·47 | | 2 | F | 21 | ·00 | 3 | Mr | 22 | ·53 |
| 5 | S | 15 | ·71 | 7 | O | 15●○ | ·25 | 1 | N | 13 | ·78 | 3 | D | 13 | ·31 | **32** | 4 | Jr | 11 | ·84 | | 6 | F | 10 | ·37 | 7 | Mr | 10 | ·90 |
| 3 | S | 4 | ·08 | 4 | O | 3ↄ | ·61 | 6 | N | 2 | ·14 | 7 | D | 1 | ·67 | | 2 | D | 31 | ·20 | **33** | 3 | Jr | 29 | ·73 | 5 | F | 28 | ·27 |
| 1 | S | 22● | ·98 | 3 | O | 22 | ·51 | 5 | N | 21 | ·04 | 6 | D | 20 | ·57 | **34** | 1 | Jr | 19 | ·10 | | 2 | F | 17○ | ·63 | 4 | Mr | 19● | ·16 |
| 6 | S | 12 | ·35 | 7 | O | 11 | ·88 | 2 | N | 10 | ·41 | 3 | D | 9 | ·94 | **35** | 5 | Jr | 8 | ·47 | | 7 | F | 7○ | ·00 | 1 | Mr | 8 | ·53 |
| 5 | O | 1 | ·24 | 6 | O | 30 | ·77 | 1 | N | 29 | ·31 | 2 | D | 28 | ·84 | **36** | 4 | Jr | 27○ | ·37 | | 5 | F | 25 | ·90 | 7 | Mr | 26 | ·43 |
| 2 | S | 19 | ·61 | 4 | O | 19 | ·14 | 5 | N | 17 | ·67 | 7 | D | 17 | ·20 | **37** | 1 | Jr | 15● | ·73 | | 3 | F | 14 | ·26 | 4 | Mr | 15 | ·79 |
| 6 | S | 8 | ·98 | 1 | O | 8 | ·51 | 3 | N | 7 | ·04 | 4 | D | 6ↄ | ·57 | **38** | 6 | Jr | 5● | ·10 | | 7 | F | 3 | ·63 | 2 | Mr | 5 | ·16 |
| 5 | S | 27 | ·88 | 7 | O | 27 | ·41 | 1 | N | 25○ | ·94 | 3 | D | 25 | ·47 | **39** | 5 | Jr | 24 | ·00 | | 6 | F | 22 | ·53 | 1 | Mr | 24 | ·06 |
| 3 | S | 17 | ·24 | 4 | O | 16 | ·77 | 6 | N | 15ↄ | ·30 | 7 | D | 14 | ·83 | **40** | 2 | Jr | 13 | ·37 | | 3 | F | 11 | ·90 | 5 | Mr | 12 | ·43 |
| 7 | S | 5 | ·61 | 2 | O | 5 | ·14 | 3 | N | 3 | ·67 | 5 | D | 3 | ·20 | **41** | 6 | Jr | 1 | ·73 | | 1 | Jr | 30 | ·26 | 2 | Mr | 1 | ·79 |
| 6 | S | 24ↄ | ·51 | 1 | O | 24● | ·04 | 2 | N | 22 | ·57 | 4 | D | 22 | ·10 | **42** | 5 | Jr | 20 | ·63 | | 7 | F | 19 | ·16 | 1 | Mr | 20○ | ·69 |
| 3 | S | 13ↄ | ·87 | 5 | O | 3 | ·41 | 6 | N | 11 | ·94 | 1 | D | 11 | ·47 | **43** | 3 | Jr | 9 | ·00 | | 4 | F | 8 | ·53 | 6 | Mr | 10●○ | ·06 |
| 1 | S | 3●○ | ·24 | 2 | O | 2 | ·77 | 4 | N | 1 | ·30 | 5 | N | 30 | ·83 | | 7 | D | 30 | ·36 | **44** | 1 | Jr | 28 | ·89 | 3 | F | 27 | ·43 |
| | | | | | | | | | | | | | | | | | | | | | | | | | | 4 | Mr | 27 | ·96 |
| 7 | S | 21 | ·14 | 1 | O | 20 | ·67 | 3 | N | 19 | ·20 | 4 | D | 18 | ·73 | **45** | 6 | Jr | 17○ | ·26 | | 7 | F | 15 | ·79 | 2 | Mr | 17 | ·32 |
| 4 | S | 10 | ·51 | 6 | O | 10 | ·04 | 7 | N | 8 | ·57 | 2 | D | 8 | ·10 | **46** | 3 | Jr | 6○ | ·63 | | 5 | F | 5 | ·16 | 6 | Mr | 6 | ·69 |
| 3 | S | 29 | ·40 | 4 | O | 28 | ·93 | 6 | N | 27 | ·47 | 1 | D | 27○ | ·00 | **47** | 2 | Jr | 25 | ·53 | | 4 | F | 24 | ·06 | 5 | Mr | 25 | ·59 |
| 7 | S | 18 | ·77 | 2 | O | 18 | ·30 | 3 | N | 16 | ·83 | 5 | D | 16● | ·36 | **48** | 6 | Jr | 14 | ·89 | | 1 | F | 13 | ·42 | 2 | Mr | 13 | ·95 |
| 5 | S | 7 | ·14 | 6 | O | 6 | ·67 | 1 | N | 5ↄ | ·20 | 2 | D | 4 | ·73 | **49** | 4 | Jr | 3 | ·26 | | 5 | F | 1 | ·79 | 7 | Mr | 3 | ·32 |
| 4 | S | 26 | ·04 | 5 | O | 25○ | ·57 | 7 | N | 24 | ·10 | 1 | D | 23 | ·63 | **50** | 3 | Jr | 22 | ·16 | | 4 | F | 20 | ·69 | 6 | Mr | 22 | ·22 |
| 1 | S | 15 | ·40 | 2 | O | 14●○ | ·93 | 4 | N | 13 | ·46 | 5 | D | 12 | ·99 | **51** | 7 | Jr | 11 | ·53 | | 2 | F | 10 | ·06 | 3 | Mr | 11 | ·59 |
| 5 | S | 4 | ·77 | 7 | O | 4 | ·30 | 1 | N | 2 | ·83 | 3 | D | 2 | ·36 | | 4 | D | 31 | ·89 | **52** | 6 | Jr | 30 | ·42 | 7 | F | 28○ | ·95 |
| 4 | S | 22 | ·67 | 6 | O | 22 | ·20 | 7 | N | 20 | ·73 | 2 | D | 20 | ·26 | **53** | 3 | Jr | 18 | ·79 | | 5 | F | 17○ | ·32 | 6 | Mr | 18● | ·85 |
| 2 | S | 12 | ·03 | 3 | O | 11 | ·57 | 5 | N | 10 | ·10 | 6 | D | 9 | ·63 | **54** | 1 | Jr | 8 | ·16 | | 2 | F | 6●○ | ·69 | 4 | Mr | 8 | ·22 |
| 7 | S | 30 | ·93 | 2 | O | 30 | ·46 | 3 | N | 28 | ·99 | 5 | D | 28 | ·52 | **55** | 7 | Jr | 27● | ·05 | | 1 | F | 25 | ·59 | 3 | Mr | 27 | ·12 |
| 5 | S | 20 | ·30 | 6 | O | 19 | ·83 | 1 | N | 18 | ·36 | 2 | D | 17○ | ·89 | **56** | 4 | Jr | 16 | ·42 | | 5 | F | 14 | ·95 | 7 | Mr | 15 | ·48 |
| 2 | S | 8 | ·67 | 4 | O | 8 | ·20 | 5 | N | 6 | ·73 | 7 | D | 6ↄ | ·26 | **57** | 1 | Jr | 4 | ·79 | | 3 | F | 3 | ·32 | 4 | Mr | 4 | ·85 |
| 1 | S | 27 | ·56 | 3 | O | 27 | ·09 | 4 | N | 25ↄ | ·63 | 6 | D | 25 | ·16 | **58** | 7 | Jr | 23 | ·69 | | 2 | F | 22 | ·22 | 3 | Mr | 23 | ·75 |
| 5 | S | 16 | ·93 | 7 | O | 16 | ·46 | 1 | N | 14 | ·99 | 3 | D | 14 | ·52 | **59** | 5 | Jr | 13 | ·05 | | 6 | F | 11 | ·58 | 1 | Mr | 13 | ·11 |
| 3 | S | 6 | ·30 | 4 | O | 5ↄ | ·83 | 6 | N | 4 | ·36 | 7 | D | 3 | ·89 | **60** | 2 | Jr | 2 | ·42 | | 3 | Jr | 31 | ·95 | 5 | Mr | 1 | ·48 |
| 2 | S | 24○ | ·20 | 3 | O | 23 | ·73 | 5 | N | 22 | ·26 | 6 | D | 21 | ·79 | **61** | 1 | Jr | 20 | ·32 | | 2 | F | 18 | ·85 | 4 | Mr | 20●○ | ·38 |

**TABLE X–**

| Kaliyuga. | Vikrama Era. | Saka Era. | Commencement of Solar Year. Month and day A.D. | Fraction of day. | First New-Moon in Solar Year. | Anomaly of first New-Moon. | A.D. | Week-day of 1st January. | ☉'s Anom col. 6 / ☾'s Anom col. 7 | | | | + 29·53059 / + 1·976 | | | | + 59·06117 / + 3·952 | | | | + 88·59176 / + 5·928 | | | | + 118·12235 / + 7·904 | | | |
|---|---|---|---|---|---|---|---|---|---|---|---|---|---|---|---|---|---|---|---|---|---|---|---|---|---|---|---|---|
| | | | | | | | | | **Vaisakha** | | | | **Jyeshta** | | | | **Ashada** | | | | **Sravana** | | | | **Bhadrapada** | | | |
| | | | | | | | | | Week-day | Month | Day | Fraction | Week-day | Month | Day | Fraction | Week-day | Month | Day | Fraction | Week-day | Month | Day | Fraction | Week-day | Month | Day | Fraction |
| 4762 | 1718 | 1583 | 28M | ·7776 | 21·1329 | 4·050 | **1661** | 3 | 5 | Ap | 18 | ·91 | 7 | My | 18 | ·44 | 1 | Je | 16 | ·97 | 3 | Jl | 16 | ·50 | 5 | Au | 15 | ·03 |
| 4763 | 1719 | 1584 | 29M | ·0364 | 10·2412 | 0·208 | **1662** | 4 | 3 | Ap | 8 | ·28 | 4 | My | 7 | ·81 | 6 | Je | 6 | ·34 | 7 | Jl | 5 | ·87 | 2 | Au | 4 | ·40 |
| 4764 | 1720 | 1585 | 29M | ·2951 | 28·8801 | 25·895 | **1663** | 5 | 2 | Ap | 27 | ·17 | 3 | My | 26 | ·71 | 5 | Je | 25 | ·24 | 6 | Jl | 24ↄ | ·77 | 1 | Au | 23 | ·30 |
| 4765 | 1721 | 1586 | 28M | ·5539 | 17·9884 | 22·053 | **1664** | 6 | 6 | Ap | 15 | ·54 | 1 | My | 15 | ·07 | 2 | Je | 13 | ·60 | 4 | Jl | 13ↄ | ·13 | 5 | Au | 11 | ·66 |
| 4766 | 1722 | 1587 | 28M | ·8126 | 7·0967 | 18·210 | **1665** | 1 | 3 | Ap | 4 | ·91 | 5 | My | 4 | ·44 | 6 | Je | 2 | ·97 | { 1 | Jl | 2 | ·50 | 4 | Au | 30 | ·56 |
| | | | | | | | | | | | | | | | | | | | | | { 3 | Au | 1 | ·03 | | | | |
| 4767 | 1723 | 1588 | 29M | ·0714 | 25·7356 | 16·343 | **1666** | 2 | 2 | Ap | 23 | ·81 | 4 | My | 23○ | ·34 | 5 | Je | 21● | ·87 | 7 | Jl | 21 | ·40 | 1 | Au | 19 | ·93 |
| 4768 | 1724 | 1589 | 29M | ·3301 | 14·8439 | 12·500 | **1667** | 3 | 7 | Ap | 13 | ·17 | 1 | My | 12ↄ | ·70 | 3 | Je | 11● | ·23 | 4 | Jl | 10 | ·77 | 6 | Au | 9 | ·30 |
| 4769 | 1725 | 1590 | 28M | ·5889 | 3·9522 | 8·658 | **1668** | 4 | 4 | Ap | 1 | ·54 | 6 | My | 1○ | ·07 | { 7 | My | 30 | ·60 | 3 | Jl | 28 | ·66 | 5 | Au | 27 | ·19 |
| | | | | | | | | | | | | | | | | | { 2 | Je | 29 | ·13 | | | | | | | | |
| 4770 | 1726 | 1591 | 28M | ·8477 | 22·5911 | 6·791 | **1669** | 6 | 3 | Ap | 20● | ·44 | 4 | My | 19 | ·97 | 6 | Je | 18 | ·50 | 1 | Jl | 18 | ·03 | 2 | Au | 16 | ·56 |
| 4771 | 1727 | 1592 | 29M | ·1064 | 11·6994 | 2·948 | **1670** | 7 | 7 | Ap | 9 | ·81 | 2 | My | 9 | ·34 | 3 | Je | 7 | ·87 | 5 | Jl | 7 | ·40 | 6 | Au | 5 | ·93 |
| 4772 | 1728 | 1593 | 29M | ·3652 | 0·8077 | 26·660 | **1671** | 1 | { 5 | Mr | 30 | ·17 | 1 | My | 28 | ·23 | 2 | Je | 26 | ·76 | 4 | Jl | 26 | ·29 | 5 | Au | 24●○ | ·83 |
| | | | | | | | | | { 6 | Ap | 28 | ·70 | | | | | | | | | | | | | | | | |
| 4773 | 1729 | 1594 | 28M | ·6289 | 19·4466 | 24·793 | **1672** | 2 | 4 | Ap | 17 | ·07 | 5 | My | 16 | ·60 | 7 | Je | 15 | ·13 | 1 | Jl | 14 | ·66 | 3 | Au | 13○ | ·19 |
| 4774 | 1730 | 1595 | 28M | ·8827 | 8·5549 | 20·951 | **1673** | 4 | 1 | Ap | 6 | ·44 | 2 | My | 5 | ·97 | 4 | Je | 4 | ·50 | 6 | Jl | 4 | ·03 | { 7 | Au | 2● | ·56 |
| | | | | | | | | | | | | | | | | | | | | | | | | | { 2 | S | 1 | ·09 |
| 4775 | 1731 | 1596 | 29M | ·1414 | 27·1937 | 19·084 | **1674** | 5 | 7 | Ap | 25 | ·33 | 1 | My | 24 | ·87 | 3 | Je | 23ↄ | ·40 | 4 | Jl | 22● | ·93 | 6 | Au | 21 | ·46 |
| 4776 | 1732 | 1597 | 29M | ·4002 | 16·3020 | 15·241 | **1675** | 6 | 4 | Ap | 14 | ·70 | 6 | My | 14 | ·23 | 7 | Je | 12●ↄ | ·76 | 2 | Jl | 12 | ·29 | 3 | Au | 10 | ·82 |
| 4777 | 1733 | 1598 | 28M | ·6590 | 5·4103 | 11·398 | **1676** | 7 | 2 | Ap | 3 | ·07 | 3 | My | 2 | ·60 | 5 | Je | 1● | ·13 | { 6 | Je | 30 | ·66 | 2 | Au | 28 | ·72 |
| | | | | | | | | | | | | | | | | | | | | | { 1 | Jl | 30 | ·19 | | | | |
| 4778 | 1734 | 1599 | 28M | ·9177 | 24·0492 | 9·531 | **1677** | 2 | 7 | Ap | 21○ | ·97 | 2 | My | 21● | ·50 | 4 | Je | 20 | ·03 | 5 | Jl | 19 | ·56 | 7 | Au | 18 | ·09 |
| 4779 | 1735 | 1600 | 29M | ·1765 | 13·1575 | 5·689 | **1678** | 3 | 5 | Ap | 11ↄ | ·33 | 6 | My | 10 | ·86 | 1 | Je | 9 | ·39 | 2 | Jl | 8 | ·93 | 4 | Au | 7 | ·46 |
| 4780 | 1736 | 1601 | 29M | ·4352 | 2·2658 | 1·846 | **1679** | 4 | 2 | Mr | 31○ | ·70 | { 4 | Ap | 30 | ·23 | 7 | Je | 28 | ·29 | 1 | Jl | 27 | ·82 | 3 | Au | 26 | ·35 |
| | | | | | | | | | | | | | { 5 | My | 29 | ·76 | | | | | | | | | | | | |
| 4781 | 1737 | 1602 | 28M | ·6940 | 20·9047 | 27·534 | **1680** | 5 | 1 | Ap | 18 | ·60 | 3 | My | 18 | ·13 | 4 | Je | 16 | ·66 | 6 | Jl | 16 | ·19 | 7 | Au | 14 | ·72 |
| 4782 | 1738 | 1603 | 28M | ·9527 | 10·0130 | 23·691 | **1681** | 7 | 5 | Ap | 7 | ·97 | 7 | My | 7 | ·50 | 2 | Je | 6 | ·03 | 3 | Jl | 5 | ·56 | 5 | Au | 4ↄ | ·09 |
| 4783 | 1739 | 1604 | 29M | ·2115 | 28·6519 | 21·824 | **1682** | 1 | 4 | Ap | 26 | ·86 | 6 | My | 26 | ·39 | 7 | Je | 24 | ·92 | 2 | Jl | 24ↄ | ·45 | 3 | Au | 22 | ·99 |
| 4784 | 1740 | 1605 | 29M | ·4703 | 17·7602 | 17·982 | **1683** | 2 | 2 | Ap | 16 | ·23 | 3 | My | 15 | ·76 | 5 | Je | 14 | ·29 | 6 | Jl | 13●○ | ·82 | 1 | Au | 12 | ·35 |
| 4785 | 1741 | 1606 | 28M | ·7290 | 6·8685 | 14·139 | **1684** | 3 | 6 | Ap | 4 | ·60 | 1 | My | 4 | ·13 | 2 | Je | 2ↄ | ·66 | { 4 | Jl | 2 | ·19 | 7 | Au | 30 | ·25 |
| | | | | | | | | | | | | | | | | | | | | | { 5 | Jl | 31 | ·72 | | | | |
| 4786 | 1742 | 1607 | 28M | ·9878 | 25·5074 | 12·272 | **1685** | 5 | 5 | Ap | 23 | ·49 | 7 | My | 23ↄ | ·03 | 1 | Je | 21 | ·56 | 3 | Jl | 21 | ·09 | 4 | Au | 19 | ·62 |
| 4787 | 1743 | 1608 | 29M | ·2465 | 14·6157 | 8·429 | **1686** | 6 | 2 | Ap | 12 | ·86 | 4 | My | 12●ↄ | ·39 | 5 | Je | 10 | ·92 | 7 | Jl | 10 | ·45 | 1 | Au | 8 | ·98 |
| 4788 | 1744 | 1609 | 29M | ·5053 | 3·7240 | 4·587 | **1687** | 7 | 7 | Ap | 2 | ·23 | 1 | My | 1● | ·76 | { 3 | My | 31 | ·29 | 6 | Jl | 29 | ·35 | 7 | Au | 27 | ·88 |
| | | | | | | | | | | | | | | | | | { 4 | Je | 29 | ·82 | | | | | | | | |
| 4789 | 1745 | 1610 | 28M | ·7640 | 22·3629 | 2·720 | **1688** | 1 | 6 | Ap | 20● | ·13 | 7 | My | 19 | ·66 | 2 | Je | 18 | ·19 | 3 | Jl | 17 | ·72 | 5 | Au | 16 | ·25 |
| 4790 | 1746 | 1611 | 29M | ·0228 | 11·4712 | 26·432 | **1689** | 3 | 3 | Ap | 9 | ·49 | 5 | My | 9 | ·02 | 6 | Je | 7 | ·55 | 1 | Jl | 7 | ·09 | 2 | Au | 5 | ·62 |
| 4791 | 1747 | 1612 | 29M | ·2815 | 0·5795 | 22·589 | **1690** | 4 | { 7 | Mr | 29 | ·86 | 3 | My | 27 | ·92 | 5 | Je | 26 | ·45 | 6 | Jl | 25 | ·98 | 1 | Au | 24●○ | ·51 |
| | | | | | | | | | { 2 | Ap | 28 | ·39 | | | | | | | | | | | | | | | | |
| 4792 | 1748 | 1613 | 29M | ·5403 | 19·2184 | 20·722 | **1691** | 5 | 6 | Ap | 17 | ·76 | 1 | My | 17 | ·29 | 2 | Je | 15 | ·82 | 4 | Jl | 15 | ·35 | 5 | Au | 13 | ·88 |
| 4793 | 1749 | 1614 | 28M | ·7991 | 8·3267 | 16·880 | **1692** | 6 | 4 | Ap | 6 | ·13 | 5 | My | 5 | ·66 | 7 | Je | 4 | ·19 | 1 | Jl | 3ↄ | ·72 | { 3 | Au | 2 | ·25 |
| | | | | | | | | | | | | | | | | | | | | | | | | | { 4 | Au | 31 | ·78 |
| 4794 | 1750 | 1615 | 29M | ·0578 | 26·9655 | 15·013 | **1693** | 1 | 3 | Ap | 25 | ·02 | 4 | My | 24 | ·55 | 6 | Je | 23●○ | ·08 | 7 | Jl | 22 | ·61 | 2 | Au | 21 | ·15 |
| 4795 | 1751 | 1616 | 29M | ·3166 | 16·0738 | 11·170 | **1694** | 2 | 7 | Ap | 14 | ·39 | 1 | My | 13 | ·92 | 3 | Je | 12ↄ | ·45 | 4 | Jl | 11 | ·98 | 6 | Au | 10 | ·51 |
| 4796 | 1752 | 1617 | 29M | ·5753 | 5·1821 | 7·327 | **1695** | 3 | 4 | Ap | 3 | ·76 | 6 | My | 3○ | ·29 | { 7 | Je | 1 | ·82 | 3 | Jl | 30 | ·88 | 5 | Au | 29 | ·41 |
| | | | | | | | | | | | | | | | | | { 2 | Jl | 1 | ·35 | | | | | | | | |
| 4797 | 1753 | 1618 | 28M | ·8341 | 23·8210 | 5·461 | **1696** | 4 | 3 | Ap | 21○ | ·65 | 5 | My | 21 | ·19 | 6 | Je | 19 | ·72 | 1 | Jl | 19 | ·25 | 2 | Au | 17 | ·78 |
| 4798 | 1754 | 1619 | 29M | ·0928 | 12·9293 | 1·618 | **1697** | 6 | 1 | Ap | 11●○ | ·02 | 2 | My | 10 | ·55 | 4 | Je | 9 | ·08 | 5 | Jl | 8 | ·61 | 7 | Au | 7 | ·14 |
| 4799 | 1755 | 1620 | 29M | ·3516 | 2·0870 | 25·330 | **1698** | 7 | 5 | Mr | 31 | ·39 | { 6 | Ap | 29 | ·92 | 2 | Je | 27 | ·98 | 4 | Jl | 27 | ·51 | 6 | Au | 26 | ·04 |
| | | | | | | | | | | | | | { 1 | My | 29 | ·45 | | | | | | | | | | | | |
| 4800 | 1756 | 1621 | 29M | ·6104 | 20·6765 | 25·463 | **1699** | 1 | 4 | Ap | 19 | ·29 | 5 | My | 18 | ·82 | 7 | Je | 17 | ·35 | 1 | Jl | 16 | ·88 | 3 | Au | 15ↄ | ·41 |

## Surya Siddhanta.

+ 147·65293   + 177·18353   + 206·71411   + 236·24470   + 265·77529   + 295·30588   {+324·83647 / +354·36705}

+ 9·880   + 11·856   + 13·832   + 15·808   + 17·784   + 19·760   {+21·736 / +23·712}

| Asvina | | | | Kartika | | | | Margasira | | | | Pausha | | | | A.D. | Magha | | | | A.D. | Phalguna | | | | Chaitra | | | |
|---|---|---|---|---|---|---|---|---|---|---|---|---|---|---|---|---|---|---|---|---|---|---|---|---|---|---|---|---|---|---|
| Week-day | Month | Day | Fraction | Week-day | Month | Day | Fraction | Week-day | Month | Day | Fraction | Week-day | Month | Day | Fraction | | Week-day | Month | Day | Fraction | | Week-day | Month | Day | Fraction | Week-day | Month | Day | Fraction |
| 6 | S | 13ↄ | ·56 | 1 | O | 13 | ·09 | 2 | N | 11 | ·62 | 4 | D | 11 | ·15 | 62 | 5 | Jr | 9 | ·69 | | 7 | F | 8 | ·22 | 1 | Mr | 9● | ·75 |
| 3 | S | 2● | ·93 | 5 | O | 2 | ·46 | 6 | O | 31 | ·99 | 1 | N | 30 | ·52 | | 3 | D | 30 | ·05 | 63 | 4 | Jr | 28○ | ·58 | {6 | F | 27 | ·11 |
| | | | | | | | | | | | | | | | | | | | | | | | | | | {7 | Mr | 28 | ·64 |
| 2 | S | 21 | ·83 | 4 | O | 21 | ·36 | 5 | N | 19 | ·89 | 7 | D | 19 | ·42 | 64 | 1 | Jr | 17●○ | ·95 | | 3 | F | 16 | ·48 | 5 | Mr | 17 | ·01 |
| 7 | S | 10 | ·19 | 1 | O | 9 | ·73 | 3 | N | 8 | ·26 | 4 | D | 7 | ·79 | 65 | 6 | Jr | 6●○ | ·32 | | 7 | F | 4 | ·85 | 2 | Mr | 6 | ·38 |
| 6 | S | 29 | ·09 | 7 | O | 28 | ·62 | 2 | N | 27 | ·15 | 3 | D | 26● | ·68 | 66 | 5 | Jr | 25 | ·21 | | 6 | F | 23 | ·75 | 1 | Mr | 25 | ·28 |
| 3 | S | 18 | ·46 | 4 | O | 17 | ·99 | 6 | N | 16ↄ | ·52 | 1 | D | 16 | ·05 | 67 | 2 | Jr | 14 | ·58 | | 4 | F | 13 | ·11 | 5 | Mr | 14 | ·64 |
| 7 | S | 7 | ·83 | 2 | O | 7 | ·36 | 3 | N | 5ↄ | ·89 | 5 | D | 5 | ·42 | 68 | 6 | Jr | 3 | ·95 | | 1 | F | 2 | ·48 | 3 | Mr | 3 | ·01 |
| 6 | S | 25 | ·72 | 1 | O | 25○ | ·25 | 2 | N | 23 | ·79 | 4 | D | 23 | ·32 | 69 | 5 | Jr | 21 | ·85 | | 7 | F | 20 | ·38 | 1 | Mr | 21 | ·91 |
| 4 | S | 15 | ·09 | 5 | O | 14 | ·62 | 7 | N | 13 | ·15 | 1 | D | 12 | ·68 | 70 | 3 | Jr | 11 | ·21 | | 4 | F | 9 | ·74 | 6 | Mr | 11○ | ·27 |
| 1 | S | 4○ | ·46 | 2 | O | 3 | ·99 | 4 | N | 2 | ·52 | 6 | D | 2 | ·05 | | 7 | D | 31 | ·58 | 71 | 2 | Jr | 30 | ·11 | 3 | F | 28○ | ·64 |
| 7 | S | 23 | ·36 | 1 | O | 22 | ·89 | 3 | N | 21 | ·42 | 4 | D | 20 | ·95 | 72 | 6 | Jr | 19 | ·48 | | 1 | F | 18○ | ·01 | 2 | Mr | 18 | ·54 |
| 4 | S | 11 | ·72 | 6 | O | 11 | ·25 | 7 | N | 9 | ·78 | 2 | D | 9 | ·31 | 73 | 3 | Jr | 7 | ·85 | | 5 | F | 6 | ·38 | 6 | Mr | 7 | ·91 |
| 3 | S | 30 | ·62 | 5 | O | 30 | ·15 | 6 | N | 28 | ·68 | 1 | D | 28ↄ | ·21 | 74 | 2 | Jr | 26 | ·74 | | 4 | F | 25 | ·27 | 5 | Mr | 26 | ·80 |
| 7 | S | 19 | ·99 | 2 | O | 19 | ·52 | 4 | N | 18 | ·05 | 5 | D | 17○ | ·58 | 75 | 7 | Jr | 16 | ·11 | | 1 | F | 14 | ·64 | 3 | Mr | 16 | ·17 |
| 5 | S | 9 | ·35 | 6 | O | 8 | ·89 | 1 | N | 7 | ·42 | 2 | D | 6ↄ | ·95 | 76 | 4 | Jr | 5 | ·48 | | 6 | F | 4 | ·01 | 7 | Mr | 4 | ·54 |
| 4 | S | 27 | ·25 | 5 | O | 26 | ·78 | 7 | N | 25● | ·31 | 1 | D | 24 | ·84 | 77 | 3 | Jr | 23 | ·37 | | 4 | F | 21 | ·91 | 6 | Mr | 23 | ·44 |
| 1 | S | 16 | ·62 | 3 | O | 16ↄ | ·15 | 4 | N | 14 | ·68 | 6 | D | 14 | ·21 | 78 | 7 | Jr | 12 | ·74 | | 2 | F | 11 | ·27 | 3 | Mr | 12 | ·80 |
| 5 | S | 5 | ·99 | 7 | O | 5ↄ | ·52 | 2 | N | 4 | ·05 | 3 | D | 3 | ·58 | 79 | 5 | Jr | 2 | ·11 | | 6 | Jr | 31 | ·64 | 1 | Mr | 2 | ·17 |
| 4 | S | 24○ | ·88 | 6 | O | 24 | ·41 | 7 | N | 22 | ·95 | 2 | D | 22 | ·48 | 80 | 4 | Jr | 21 | ·01 | | 5 | F | 19 | ·54 | 7 | Mr | 20● | ·07 |
| 2 | S | 13 | ·25 | 3 | O | 12 | ·78 | 5 | N | 11 | ·31 | 6 | D | 10 | ·84 | 81 | 1 | Jr | 9 | ·37 | | 2 | F | 7○ | ·90 | 4 | Mr | 9 | ·43 |
| {6 | S | 2● | ·62 | 2 | O | 31 | ·68 | 4 | N | 30 | ·21 | | Pausha | | | | 5 | D | 29 | ·74 | 82 | 7 | Jr | 28○ | ·27 | {1 | F | 26 | ·80 |
| {1 | O | 2 | ·15 | | | | | | | | | | Kshaya | | | | | | | | | | | | | {3 | Mr | 28 | .33 |
| 5 | S | 21 | ·52 | 7 | O | 21 | ·05 | 1 | N | 19 | ·58 | 3 | D | 19 | ·11 | 83 | 4 | Jr | 17○ | ·64 | | 6 | F | 16 | ·17 | 7 | Mr | 17 | ·70 |
| 2 | S | 10 | ·88 | 4 | O | 10 | ·41 | 5 | N | 8 | ·94 | 7 | D | 8 | ·47 | 84 | 2 | Jr | 7 | ·01 | | 3 | F | 5 | ·54 | 5 | Mr | 6 | .07 |
| 1 | S | 28 | ·78 | 3 | O | 28 | ·31 | 4 | N | 26ↄ | ·84 | 6 | D | 26 | ·37 | 85 | 7 | Jr | 24 | ·90 | | 2 | F | 23 | ·43 | 3 | Mr | 24 | ·96 |
| 6 | S | 18 | ·15 | 7 | O | 17 | ·68 | 2 | N | 16●○ | ·21 | 3 | D | 15 | ·74 | 86 | 5 | Jr | 14 | ·27 | | 6 | F | 12 | ·80 | 1 | Mr | 14 | ·33 |
| 3 | S | 7 | ·51 | 5 | O | 7 | ·05 | 6 | N | 5ↄ | ·58 | 1 | D | 5 | ·11 | 87 | 2 | Jr | 3 | ·64 | | 4 | F | 2 | ·17 | 5 | Mr | 3 | ·70 |
| 2 | S | 26 | ·41 | 3 | O | 25● | ·94 | 5 | N | 24 | ·47 | 7 | D | 24 | ·00 | 88 | 1 | Jr | 22 | ·53 | | 3 | F | 21 | ·07 | 4 | Mr | 21○ | ·60 |
| 6 | S | 14○ | ·78 | 1 | O | 14 | ·31 | 2 | N | 12 | ·84 | 4 | D | 12 | ·37 | 89 | 5 | Jr | 10 | ·90 | | 7 | F | 9 | ·43 | 1 | Mr | 10○ | ·96 |
| 4 | S | 4○ | ·15 | 5 | O | 3 | ·68 | 7 | N | 2 | ·21 | 1 | D | 1 | ·74 | | 3 | D | 31 | ·27 | 90 | 4 | Jr | 29 | ·80 | 6 | F | 28○ | ·33 |
| 3 | S | 23 | ·04 | 4 | O | 22 | ·57 | 6 | N | 21 | ·11 | 7 | D | 20 | ·64 | 91 | 2 | Jr | 19 | ·17 | | 3 | F | 17● | ·70 | 5 | Mr | 19 | ·23 |
| 7 | S | 12 | ·41 | 1 | O | 11 | ·94 | 3 | N | 10 | ·47 | 5 | D | 10 | ·00 | 92 | 6 | Jr | 8○ | ·53 | | 1 | F | 7● | ·06 | 2 | Mr | 7 | ·59 |
| 6 | S | 30 | ·31 | 7 | O | 29 | ·84 | 2 | N | 28 | ·37 | 3 | D | 27○ | ·90 | 93 | 5 | Jr | 26 | ·43 | | 6 | F | 24 | ·96 | 1 | Mr | 26 | ·49 |
| 3 | S | 19 | ·68 | 5 | O | 19 | ·21 | 6 | N | 17 | ·74 | 1 | D | 17○ | ·27 | 94 | 2 | Jr | 15 | ·80 | | 4 | F | 14 | ·33 | 5 | Mr | 15 | ·86 |
| 1 | S | 9 | ·04 | 2 | O | 8 | ·57 | 4 | N | 7 | ·10 | 5 | D | 6 | ·63 | 95 | 7 | Jr | 5 | ·17 | | 1 | F | 3 | ·70 | 3 | Mr | 5 | ·23 |
| 6 | S | 27 | ·94 | 1 | O | 27○ | ·47 | 3 | N | 26● | ·00 | 4 | D | 25 | ·53 | 96 | 6 | Jr | 24 | ·06 | | 7 | F | 22 | ·59 | 2 | Mr | 23 | ·12 |
| 4 | S | 16 | ·31 | 5 | O | 15ↄ | ·84 | 7 | N | 14 | ·37 | 1 | D | 13 | ·90 | 97 | 3 | Jr | 12 | ·43 | | 4 | F | 10 | ·96 | 6 | Mr | 12 | ·49 |
| 1 | S | 5 | ·67 | 3 | O | 5●ↄ | ·21 | 4 | N | 3 | ·74 | 6 | D | 3 | ·27 | 98 | 7 | Jr | 1 | ·80 | | 2 | Jr | 31 | ·33 | 3 | Mr | 1 | ·86 |
| 7 | S | 24● | ·57 | 2 | O | 24 | ·10 | 3 | N | 22 | ·63 | 5 | D | 22 | ·16 | 99 | 6 | Jr | 20 | ·69 | | 1 | F | 19○ | ·23 | 2 | Mr | 20● | ·76 |
| 4 | S | 13● | ·94 | 6 | O | 13 | ·47 | 1 | N | 12 | ·00 | 2 | D | 11 | ·53 | 17 | 4 | Jr | 10 | ·06 | 00 | 5 | F | 8○ | ·59 | 7 | Mr | 9 | ·12 |

| Kaliyuga. | Vikrama Era. | Saka Era. | Commencement of Solar Year. Month and day A.D. | Fraction of day. | First New-Moon in Solar Year. | Anomaly of first New-Moon. | A.D. | Week day of 1st January. | Vaisakha | | | | Jyeshta | | | | Ashada | | | | Sravana | | | | Bhadrapada | | | |
|---|---|---|---|---|---|---|---|---|---|---|---|---|---|---|---|---|---|---|---|---|---|---|---|---|---|---|---|---|
| | | | | | | | | ⊙'s Anom col. 6 | | | | | + 29·53059 | | | | + 59·06117 | | | | + 88·59176 | | | | + 118·12235 | | | |
| | | | | | | | | ☾'s Anom col. 7 | | | | | + 1·976 | | | | + 3·952 | | | | + 5·928 | | | | + 7·904 | | | |
| | | | | | | | | | Week-day | Month | Day | Fraction | Week-day | Month | Day | Fraction | Week-day | Month | Day | Fraction | Week-day | Month | Day | Fraction | Week-day | Month | Day | Fraction |
| 4801 | 1757 | 1622 | 28M | ·8691 | 9·7848 | 19·623 | **1700** | 2 | 1 | Ap | 7 | ·65 | 3 | My | 7 | ·18 | 4 | Je | 5 | ·71 | 6 | Jl | 5 | ·24 | 7 | Au | 30○ | ·78 |
| 4802 | 1758 | 1623 | 29M | ·1279 | 28·4237 | 17·756 | **1701** | 4 | 7 | Ap | 26 | ·55 | 2 | My | 26 | ·08 | 3 | Je | 24 | ·61 | 5 | Jl | 24●○ | ·14 | 6 | Au | 22 | ·67 |
| 4803 | 1759 | 1624 | 29M | ·3866 | 17·5320 | 13·913 | **1702** | 5 | 4 | Ap | 15 | ·92 | 6 | My | 15 | ·45 | 7 | Je | 13 | ·98 | 2 | Jl | 13 | ·51 | 4 | Au | 12 | ·04 |
| 4804 | 1760 | 1625 | 29M | ·6454 | 6·6403 | 10·071 | **1703** | 6 | 2 | Ap | 5 | ·28 | 3 | My | 4 | ·82 | 5 | Je | 30○ | ·85 | 6 | Jl | 2 | ·88 | 2 | Au | 30 | ·94 |
| | | | | | | | | | | | | | | | | | | | | | 1 | Au | 1 | ·41 | | | | |
| 4805 | 1761 | 1626 | 28M | ·9041 | 25·2792 | 8·204 | **1704** | 1 | 1 | Ap | 23 | ·18 | 2 | My | 22○ | ·71 | 4 | Je | 21 | ·24 | 5 | Jl | 20 | ·77 | 7 | Au | 19 | ·31 |
| 4806 | 1762 | 1627 | 29M | ·1629 | 14·3875 | 4·361 | **1705** | 2 | 5 | Ap | 12 | ·55 | 7 | My | 12 | ·08 | 1 | Je | 10 | ·61 | 3 | Jl | 10 | ·14 | 4 | Au | 8 | ·67 |
| 4807 | 1763 | 1628 | 29M | ·4217 | 3·4958 | 0·518 | **1706** | 3 | 2 | Ap | 1○ | ·92 | 4 | My | 1● | ·45 | 7 | Je | 29 | ·51 | 2 | Jl | 29 | ·04 | 3 | Au | 27 | ·57 |
| | | | | | | | | | | | | | 5 | My | 30 | ·98 | | | | | | | | | | | | |
| 4808 | 1764 | 1629 | 29M | ·6804 | 22·1347 | 26·206 | **1707** | 4 | 1 | Ap | 20● | ·81 | 3 | My | 20 | ·35 | 4 | Je | 18 | ·88 | 6 | Jl | 18 | ·41 | 7 | Au | 16 | ·94 |
| 4809 | 1765 | 1630 | 28M | ·9392 | 11·2430 | 22·363 | **1708** | 5 | 6 | Ap | 9 | ·18 | 7 | My | 8 | ·71 | 2 | Je | 7 | ·24 | 3 | Jl | 6 | ·77 | 5 | Au | 5 | ·30 |
| 4810 | 1766 | 1631 | 29M | ·1979 | 0·3513 | 18·521 | **1709** | 7 | 3 | Mr | 29 | ·55 | 6 | My | 27 | ·61 | 1 | Je | 26 | ·14 | 2 | Jl | 25 | ·67 | 4 | Au | 24● | ·20 |
| | | | | | | | | | 5 | Ap | 28 | ·08 | | | | | | | | | | | | | | | | |
| 4811 | 1767 | 1632 | 29M | ·4567 | 18·9902 | 16·654 | **1710** | 1 | 2 | Ap | 17 | ·45 | 3 | My | 16 | ·98 | 5 | Je | 15 | ·51 | 7 | Jl | 15○ | ·04 | 1 | Au | 13 | ·57 |
| 4812 | 1768 | 1633 | 29M | ·7154 | 8·0984 | 12·811 | **1711** | 2 | 6 | Ap | 6 | ·81 | 1 | My | 6 | ·34 | 2 | Je | 4 | ·87 | 4 | Jl | 4○ | ·40 | 5 | Au | 2 | ·94 |
| | | | | | | | | | | | | | | | | | | | | | | | | | 7 | S | 1 | ·47 |
| 4813 | 1769 | 1634 | 28M | ·9742 | 26·7373 | 10·944 | **1712** | 3 | 5 | Ap | 24 | ·71 | 7 | My | 24 | ·24 | 1 | Je | 22●○ | ·77 | 3 | Jl | 22 | ·30 | 4 | Au | 20 | ·83 |
| 4814 | 1770 | 1635 | 29M | ·2329 | 15·8456 | 7·102 | **1713** | 5 | 3 | Ap | 14 | ·08 | 4 | My | 13○ | ·61 | 6 | Je | 12 | ·14 | 7 | Jl | 11 | ·67 | 2 | Au | 10 | ·20 |
| 4815 | 1771 | 1636 | 29M | ·4917 | 4·9539 | 3·259 | **1714** | 6 | 7 | Ap | 3 | ·45 | 1 | My | 2○ | ·98 | 3 | Je | 1 | ·51 | 6 | Jl | 30 | ·57 | 1 | Au | 29 | ·10 |
| | | | | | | | | | | | | | | | | | 5 | Jl | 1 | ·04 | | | | | | | | |
| 4816 | 1772 | 1637 | 29M | ·7505 | 23·5928 | 1·392 | **1715** | 7 | 6 | Ap | 22●○ | ·34 | 7 | My | 21 | ·87 | 2 | Je | 20 | ·40 | 3 | Jl | 19 | ·93 | 5 | Au | 18 | ·47 |
| 4817 | 1773 | 1638 | 29M | ·0092 | 12·7011 | 25·104 | **1716** | 1 | 3 | Ap | 10● | ·71 | 5 | My | 10 | ·24 | 6 | Je | 8 | ·77 | 1 | Jl | 8 | ·30 | 2 | Au | 6 | ·83 |
| 4818 | 1774 | 1639 | 29M | ·2680 | 1·8094 | 21·261 | **1717** | 3 | 1 | Mr | 31 | ·08 | 2 | Ap | 29 | ·61 | 5 | Je | 27 | ·67 | 7 | Jl | 27 | ·20 | 1 | Au | 25○ | ·73 |
| | | | | | | | | | | | | | 4 | My | 29 | ·14 | | | | | | | | | | | | |
| 4819 | 1775 | 1640 | 29M | ·5267 | 20·4483 | 19·395 | **1718** | 4 | 6 | Ap | 18 | ·97 | 1 | My | 18 | ·51 | 3 | Je | 17 | ·04 | 4 | Jl | 16 | ·57 | 6 | Au | 15○ | ·10 |
| 4820 | 1776 | 1641 | 29M | ·7855 | 9·5566 | 15·552 | **1719** | 5 | 4 | Ap | 8 | ·34 | 5 | My | 7 | ·87 | 7 | Je | 6 | ·40 | 1 | Jl | 5 | ·93 | 3 | Au | 4○ | ·46 |
| 4821 | 1777 | 1642 | 29M | ·0442 | 28·1955 | 13·685 | **1720** | 6 | 3 | Ap | 26 | ·24 | 4 | My | 25 | ·77 | 6 | Je | 24 | ·30 | 7 | Jl | 23● | ·83 | 2 | Au | 22 | ·36 |
| 4822 | 1778 | 1643 | 29M | ·3030 | 17·3038 | 9·842 | **1721** | 1 | 7 | Ap | 15 | ·61 | 2 | My | 15 | ·14 | 3 | Je | 13○ | ·67 | 5 | Jl | 13● | ·20 | 6 | Au | 11 | ·73 |
| 4823 | 1779 | 1644 | 29M | ·5618 | 6·4121 | 6·000 | **1722** | 2 | 4 | Ap | 4 | ·97 | 6 | My | 4 | ·50 | 1 | Je | 3○ | ·03 | 2 | Jl | 2 | ·57 | 5 | Au | 30 | ·63 |
| | | | | | | | | | | | | | | | | | | | | | 4 | Au | 1 | ·10 | | | | |
| 4824 | 1780 | 1645 | 29M | ·8205 | 25·0510 | 4·133 | **1723** | 3 | 3 | Ap | 23 | ·87 | 5 | My | 23● | ·40 | 6 | Je | 21 | ·93 | 1 | Jl | 21 | ·46 | 2 | Au | 19 | ·99 |
| 4825 | 1781 | 1646 | 29M | ·0793 | 14·1593 | 0·290 | **1724** | 4 | 1 | Ap | 12○ | ·24 | 2 | My | 11 | ·77 | 4 | Je | 10 | ·30 | 5 | Jl | 9 | ·83 | 7 | Au | 8 | ·36 |
| 4826 | 1782 | 1647 | 29M | ·3380 | 3·2676 | 24·002 | **1725** | 6 | 5 | Ap | 1○ | ·61 | 7 | My | 1 | ·14 | 3 | Je | 29 | ·20 | 4 | Jl | 28 | ·73 | 6 | Au | 27 | ·26 |
| | | | | | | | | | | | | | 1 | My | 30 | ·67 | | | | | | | | | | | | |
| 4827 | 1783 | 1648 | 29M | ·5968 | 21·9065 | 22·135 | **1726** | 7 | 4 | Ap | 20 | ·50 | 6 | My | 20 | ·03 | 7 | Je | 18 | ·56 | 2 | Jl | 18 | ·09 | 3 | Au | 16 | ·63 |
| 4828 | 1784 | 1649 | 29M | ·8555 | 11·0148 | 18·292 | **1727** | 1 | 1 | Ap | 9 | ·87 | 3 | My | 9 | ·40 | 4 | Je | 7 | ·93 | 6 | Jl | 7 | ·46 | 7 | Au | 5 | ·99 |
| 4829 | 1785 | 1650 | 29M | ·1143 | 0·1231 | 14·450 | **1728** | 2 | 6 | Mr | 29 | ·24 | 2 | My | 27 | ·30 | 3 | Je | 25 | ·83 | 5 | Jl | 25○ | ·36 | 6 | Au | 23● | ·89 |
| | | | | | | | | | 7 | Ap | 27 | ·77 | | | | | | | | | | | | | | | | |
| 4830 | 1786 | 1651 | 29M | ·3731 | 18·7619 | 12·583 | **1729** | 4 | 5 | Ap | 17 | ·13 | 6 | My | 16 | ·66 | 1 | Je | 15 | ·20 | 2 | Jl | 14○ | ·73 | 4 | Au | 13 | ·26 |
| 4831 | 1787 | 1652 | 29M | ·6318 | 7·8702 | 8·743 | **1730** | 5 | 2 | Ap | 6 | ·50 | 4 | My | 6 | ·03 | 5 | Je | 4 | ·56 | 7 | Jl | 4●○ | ·09 | 1 | Au | 2 | ·62 |
| | | | | | | | | | | | | | | | | | | | | | | | | | 3 | S | 1 | ·15 |
| 4832 | 1788 | 1653 | 29M | ·8906 | 26·5091 | 6·873 | **1731** | 6 | 1 | Ap | 25 | ·40 | 2 | My | 24○ | ·93 | 4 | Je | 23● | ·46 | 5 | Jl | 22 | ·99 | 7 | Au | 21 | ·52 |
| 4833 | 1789 | 1654 | 29M | ·1493 | 15·6174 | 3·031 | **1732** | 7 | 5 | Ap | 13 | ·77 | 7 | My | 13○ | ·30 | 1 | Je | 11 | ·83 | 3 | Jl | 11 | ·36 | 4 | Au | 9 | ·89 |
| 4834 | 1790 | 1655 | 29M | ·4081 | 4·7257 | 26·742 | **1733** | 2 | 3 | Ap | 3 | ·13 | 4 | My | 2○ | ·66 | 6 | Je | 1 | ·19 | 2 | Jl | 30 | ·26 | 3 | Au | 28 | ·72 |
| | | | | | | | | | | | | | | | | | 7 | Je | 30 | ·72 | | | | | | | | |
| 4835 | 1791 | 1656 | 29M | ·6668 | 23·3646 | 24·876 | **1734** | 3 | 2 | Ap | 22● | ·03 | 3 | My | 21 | ·56 | 5 | Je | 20 | ·09 | 6 | Jl | 19 | ·62 | 1 | Au | 18 | ·15 |
| 4836 | 1792 | 1657 | 29M | ·9256 | 12·4729 | 21·033 | **1735** | 4 | 6 | Ap | 11 | ·40 | 7 | My | 10 | ·93 | 2 | Je | 9 | ·46 | 3 | Jl | 8 | ·99 | 5 | Au | 7 | ·52 |
| 4837 | 1793 | 1658 | 29M | ·1843 | 1·5812 | 17·190 | **1736** | 5 | 3 | Mr | 30 | ·76 | 5 | Ap | 29 | ·30 | 1 | Je | 27 | ·36 | 2 | Jl | 26 | ·89 | 4 | Au | 25○ | ·42 |
| | | | | | | | | | | | | | 6 | My | 28 | ·83 | | | | | | | | | | | | |
| 4838 | 1794 | 1659 | 29M | ·4431 | 20·2201 | 15·323 | **1737** | 7 | 2 | Ap | 18 | ·66 | 4 | My | 18 | ·19 | 5 | Je | 16 | ·72 | 7 | Jl | 16 | ·25 | 1 | Au | 14●○ | ·78 |
| 4839 | 1795 | 1660 | 29M | ·7019 | 9·3284 | 11·481 | **1738** | 1 | 7 | Ap | 8 | ·03 | 1 | My | 7 | ·56 | 3 | Je | 6 | ·09 | 4 | Jl | 5 | ·62 | 6 | Au | 4● | ·15 |

## Surya Siddhanta.

| Asvina | | | | Kartika | | | | Margasira | | | | Pausha | | | | A.D. | Magha | | | | A.D. | Phalguna | | | | Chaitra | | | |
|---|---|---|---|---|---|---|---|---|---|---|---|---|---|---|---|---|---|---|---|---|---|---|---|---|---|---|---|---|---|
| + 147·65293 | | | | + 177·18353 | | | | + 206·71411 | | | | + 236·24470 | | | | | + 265·77529 | | | | | + 295·30588 | | | | +324·83647<br>+354·36705 | | | |
| + 9·880 | | | | + 11·856 | | | | + 13·832 | | | | + 15·808 | | | | | + 17·784 | | | | | + 19·760 | | | | + 21·736<br>+ 23·712 | | | |
| Week-day | Month | Day | Fraction | Week-day | Month | Day | Fraction | Week-day | Month | Day | Fraction | Week-day | Month | Day | Fraction | | Week-day | Month | Day | Fraction | | Week-day | Month | Day | Fraction | Week-day | Month | Day | Fraction |
| 2 | S | 2 | ·31 | 5 | O | 31 | ·37 | 6 | N | 29 | ·90 | 1 | D | 29 | ·43 | 01 | 2 | Jr | 27○ | ·96 | | 4 | F | 26 | ·49 | 6 | Mr | 28 | ·02 |
| 3 | O | 1 | ·84 | | | | | | | | | | | | | | | | | | | | | | | | | | |
| 1 | S | 21 | ·20 | 2 | O | 20 | ·73 | 4 | N | 19 | ·26 | 5 | D | 18 | ·80 | 02 | 7 | Jr | 17● | ·33 | | 1 | F | 15 | ·86 | 3 | Mr | 17 | ·39 |
| 5 | S | 10 | ·57 | 7 | O | 10 | ·10 | 1 | N | 8 | ·63 | 3 | D | 8ↄ | ·16 | 03 | 4 | Jr | 6● | ·69 | | 6 | F | 5 | ·22 | 7 | Mr | 6 | ·75 |
| 4 | S | 29 | ·47 | 6 | O | 29 | ·00 | 7 | N | 27○ | ·53 | 2 | D | 27 | ·06 | 04 | 3 | Jr | 25 | ·59 | | 5 | F | 24 | ·12 | 6 | Mr | 24 | ·65 |
| 1 | S | 17 | ·84 | 3 | O | 17 | ·37 | 4 | N | 15●ↄ | ·90 | 6 | D | 15 | ·43 | 05 | 7 | Jr | 13 | ·96 | | 2 | F | 12 | ·49 | 4 | Mr | 14 | ·02 |
| 6 | S | 7 | ·20 | 7 | O | 6 | ·73 | 2 | N | 5 | ·26 | 3 | D | 4 | ·79 | 06 | 5 | Jr | 3 | ·32 | | 6 | F | 1 | ·86 | 1 | Mr | 3 | ·39 |
| 5 | S | 26ↄ | ·10 | 6 | O | 25 | ·63 | 1 | N | 24 | ·16 | 2 | D | 23 | ·69 | 07 | 4 | Jr | 22 | ·22 | | 5 | F | 20 | ·75 | 7 | Mr | 22ↄ | ·28 |
| 2 | S | 15○ | ·47 | 4 | O | 15 | ·00 | 5 | N | 13 | ·53 | 7 | D | 13 | ·06 | 08 | 1 | Jr | 11 | ·59 | | 3 | F | 10 | ·12 | 4 | Mr | 10●○ | ·65 |
| 6 | S | 3●○ | ·83 | 1 | O | 3 | ·36 | 2 | N | 1 | ·90 | 4 | D | 1 | ·43 | | 5 | D | 30 | ·96 | 09 | 7 | Jr | 29 | ·49 | 2 | F | 28● | ·02 |
| 5 | S | 22 | ·73 | 7 | O | 22 | ·26 | 1 | N | 20 | ·79 | 3 | D | 20 | ·32 | 10 | 4 | Jr | 18ↄ | ·85 | | 6 | F | 17 | ·39 | 7 | Mr | 18 | ·92 |
| 3 | S | 12 | ·10 | 4 | O | 11 | ·63 | 6 | N | 10 | ·16 | 7 | D | 9 | ·69 | 11 | 2 | Jr | 8○ | ·22 | | 3 | F | 6 | ·75 | 5 | Mr | 8 | ·28 |
| 2 | O | 1 | ·00 | 3 | O | 30 | ·53 | 5 | N | 29 | ·06 | 6 | D | 28●○ | ·59 | 12 | 1 | Jr | 27 | ·12 | | 2 | F | 25 | ·65 | 4 | Mr | 26 | ·18 |
| 6 | S | 19 | ·36 | 7 | O | 18 | ·89 | 2 | N | 17 | ·43 | 3 | D | 16● | ·96 | 13 | 5 | Jr | 15 | ·49 | | 7 | F | 14 | ·02 | 1 | Mr | 15 | ·55 |
| 3 | S | 8 | ·73 | 5 | O | 8 | ·26 | 6 | N | 6ↄ | ·79 | 1 | D | 6 | ·32 | 14 | 2 | Jr | 4 | ·85 | | 4 | F | 3 | ·38 | 5 | Mr | 4 | ·91 |
| 2 | S | 27 | ·63 | 4 | O | 27ↄ | ·16 | 5 | N | 25 | ·69 | 7 | D | 25 | ·22 | 15 | 1 | Jr | 23 | ·75 | | 3 | F | 22 | ·28 | 4 | Mr | 23 | ·81 |
| 7 | S | 17 | ·00 | 1 | O | 16ↄ | ·53 | 3 | N | 15 | ·06 | 4 | D | 14 | ·59 | 16 | 6 | Jr | 13 | ·12 | | 7 | F | 11 | ·65 | 2 | Mr | 12 | ·18 |
| 4 | S | 5 | ·36 | 5 | O | 4● | ·89 | 7 | N | 3 | ·42 | 1 | D | 2 | ·95 | 17 | 3 | Jr | 1 | ·49 | | 5 | Jr | 31 | ·02 | 6 | Mr | 1○ | ·55 |
| 3 | S | 24 | ·26 | 4 | O | 23 | ·79 | 6 | N | 22 | ·32 | 7 | D | 21 | ·85 | 18 | 2 | Jr | 20 | ·38 | | 3 | F | 18○ | ·91 | 5 | Mr | 20 | ·44 |
| 7 | S | 13● | ·63 | 2 | O | 13 | ·16 | 3 | N | 11 | ·69 | 5 | D | 11 | ·22 | 19 | 6 | Jr | 9 | ·75 | | 1 | F | 8●○ | ·28 | 2 | Mr | 9 | ·81 |
| 4 | S | 2 | ·99 | 1 | N | 1 | ·06 | 2 | N | 30 | ·59 | 4 | D | 30 | ·12 | 20 | 5 | Jr | 28● | ·65 | | 7 | F | 27 | ·18 | 1 | Mr | 27 | ·71 |
| 6 | O | 2 | ·53 | | | | | | | | | | | | | | | | | | | | | | | | | | |
| 3 | S | 20 | ·89 | 5 | O | 20 | ·42 | 6 | N | 18 | ·95 | 1 | D | 18ↄ | ·48 | 21 | 3 | Jr | 17 | ·01 | | 4 | F | 15 | ·55 | 6 | Mr | 17 | ·08 |
| 1 | S | 10 | ·26 | 2 | O | 9 | ·79 | 4 | N | 8 | ·32 | 5 | D | 7ↄ | ·85 | 22 | 7 | Jr | 6 | ·38 | | 1 | F | 4 | ·91 | 3 | Mr | 6 | ·44 |
| 7 | S | 29 | ·16 | 1 | O | 28 | ·69 | 3 | N | 27ↄ | ·22 | 4 | D | 26 | ·75 | 23 | 6 | Jr | 25 | ·28 | | 7 | F | 23 | ·01 | 2 | Mr | 25 | ·34 |
| 4 | S | 18 | ·52 | 6 | O | 18 | ·05 | 7 | N | 16 | ·59 | 2 | D | 16 | ·12 | 24 | 3 | Jr | 14 | ·65 | | 5 | F | 13 | ·18 | 6 | Mr | 13 | ·71 |
| 1 | S | 6 | ·89 | 3 | O | 6○ | ·42 | 4 | N | 4 | ·95 | 6 | D | 4 | ·48 | 25 | 1 | Jr | 3 | ·01 | | 2 | F | 1 | ·54 | 4 | Mr | 3 | ·07 |
| 7 | S | 25ↄ | ·79 | 2 | O | 25 | ·32 | 3 | N | 23 | ·85 | 5 | D | 23 | ·38 | 26 | 6 | Jr | 21 | ·91 | | 1 | F | 20 | ·44 | 2 | Mr | 21○ | ·97 |
| 5 | S | 15○ | ·16 | 6 | O | 14 | ·69 | 1 | N | 13 | ·22 | 2 | D | 12 | ·75 | 27 | 4 | Jr | 11 | ·28 | | 5 | F | 9 | ·81 | 7 | Mr | 11 | ·34 |
| 2 | S | 4● | ·52 | 4 | O | 4 | ·05 | 5 | N | 2 | ·58 | 7 | D | 2 | ·11 | | 1 | D | 31 | ·64 | 28 | 3 | Jr | 30ↄ | ·18 | 4 | F | 28 | ·71 |
| 1 | S | 22 | ·42 | 2 | O | 21 | ·95 | 4 | N | 20 | ·48 | 6 | D | 20 | ·01 | 29 | 7 | Jr | 18ↄ | ·54 | | 2 | F | 17 | ·07 | 3 | Mr | 18 | ·60 |
| 5 | S | 11 | ·79 | 7 | O | 11 | ·32 | 1 | N | 9 | ·85 | 3 | D | 9 | ·38 | 30 | 4 | Jr | 7○ | ·91 | | 6 | F | 6 | ·44 | 7 | Mr | 7 | ·97 |
| 4 | S | 30 | ·68 | 6 | O | 30 | ·22 | 7 | N | 28 | ·75 | 2 | D | 28● | ·28 | 31 | 3 | Jr | 26 | ·81 | | 5 | F | 25 | ·34 | 6 | Mr | 26 | ·87 |
| 2 | S | 20 | ·05 | 3 | O | 19 | ·58 | 5 | N | 18○ | ·11 | 6 | D | 17● | ·64 | 32 | 1 | Jr | 16 | ·17 | | 2 | F | 14 | ·70 | 4 | Mr | 15 | ·24 |
| 6 | S | 8 | ·42 | 7 | O | 7 | ·95 | 2 | N | 6ↄ | ·48 | 4 | D | 6 | ·01 | 33 | 5 | Jr | 4 | ·54 | | 7 | F | 3 | ·07 | 1 | Mr | 4 | ·60 |
| 5 | S | 27 | ·32 | 6 | O | 26ↄ | ·85 | 1 | N | 25 | ·38 | 2 | D | 24 | ·91 | 34 | 4 | Jr | 23 | ·44 | | 5 | F | 21 | ·97 | 7 | Mr | 23 | ·50 |
| 2 | S | 16 | ·68 | 4 | O | 16 | ·21 | 5 | N | 14 | ·74 | 7 | D | 14 | ·28 | 35 | 1 | Jr | 12 | ·81 | | 3 | F | 11 | ·34 | 4 | Mr | 12ↄ | ·87 |
| 7 | S | 6ↄ | ·05 | 1 | O | 5● | ·58 | 3 | N | 4 | ·11 | 4 | D | 3 | ·64 | 36 | 6 | Jr | 2 | ·17 | | 7 | Jr | 31 | ·70 | 2 | Mr | 1○ | ·23 |
| 5 | S | 23 | ·95 | 7 | O | 23 | ·48 | 2 | N | 22 | ·01 | 3 | D | 21 | ·54 | 37 | 5 | Jr | 20 | ·07 | | 6 | F | 18ↄ | ·60 | 1 | Mr | 20 | ·13 |
| 3 | S | 13 | ·32 | 4 | O | 12 | ·85 | 6 | N | 11 | ·38 | 7 | D | 10 | ·91 | 38 | 2 | Jr | 9 | ·44 | | 3 | F | 7 | ·97 | 5 | Mr | 9 | ·50 |
| 7 | S | 2 | ·68 | 3 | O | 31 | ·74 | 5 | N | 30 | ·27 | 6 | D | 29ↄ | ·80 | 39 | 1 | Jr | 28 | ·34 | | 2 | F | 26 | ·87 | 4 | Mr | 28 | ·40 |
| 2 | O | 2 | ·21 | | | | | | | | | | | | | | | | | | | | | | | | | | |

**TABLE X—**

| | | | | | | | | | ☉'s Anom col. 6 | | | | + 29·53059 | | | | + 59·06117 | | | | + 88·59176 | | | | + 118·12235 | | | |
|---|---|---|---|---|---|---|---|---|---|---|---|---|---|---|---|---|---|---|---|---|---|---|---|---|---|---|---|---|
| | | | | | | | | | ☾'s Anom col. 7 | | | | + 1·976 | | | | + 3·952 | | | | + 5·928 | | | | + 7·904 | | | |
| | | | Commencement of Solar Year. | | | | | | **Vaisakha** | | | | **Jyeshta** | | | | **Ashada** | | | | **Sravana** | | | | **Bhadrapada** | | | |
| Kaliyuga. | Vikrama Era. | Saka Era. | Month and day A.D. | Fraction of day. | First New-Moon in Solar Year. | Anomaly of first New-Moon. | A.D. | Week-day of 1st January. | Week-day | Month | Day | Fraction | Week-day | Month | Day | Fraction | Week-day | Month | Day | Fraction | Week-day | Month | Day | Fraction | Week-day | Month | Day | Fraction |
| 4840 | 1796 | 1661 | 29M | ·9606 | 27·9673 | 9·614 | **1739** | 2 | 5 | Ap | 26 | ·93 | 7 | My | 26 | ·46 | 1 | Je | 24○ | ·99 | 3 | Jl | 24 | ·52 | 5 | Au | 23 | ·05 |
| 4841 | 1797 | 1662 | 29M | ·2194 | 17·0756 | 5·771 | **1740** | 3 | 3 | Ap | 15 | ·29 | 4 | My | 14 | ·82 | 6 | Je | 13○ | ·36 | 7 | Jl | 12 | ·89 | 2 | Au | 11 | ·42 |
| 4842 | 1798 | 1663 | 29M | ·4781 | 6·1839 | 1·929 | **1741** | 5 | 7 | Ap | 4 | ·66 | 2 | My | 4 | ·19 | 3 | Je | 2● | ·72 | 5 | Jl | 2 | ·25 | 1 | Au | 30 | ·31 |
| | | | | | | | | | | | | | | | | | | | | | 6 | Jl | 31 | ·78 | | | | |
| 4843 | 1799 | 1664 | 29M | ·7369 | 24·8228 | 0·062 | **1742** | 6 | 6 | Ap | 23○ | ·56 | 1 | My | 23● | ·09 | 2 | Je | 21 | ·62 | 4 | Jl | 21 | ·15 | 5 | Au | 19 | ·68 |
| 4844 | 1800 | 1665 | 29M | ·9956 | 13·9311 | 23·774 | **1743** | 7 | 3 | Ap | 12○ | ·93 | 5 | My | 12 | ·46 | 6 | Je | 10 | ·99 | 1 | Jl | 10 | ·52 | 3 | Au | 9 | ·05 |
| 4845 | 1801 | 1666 | 29M | ·2544 | 3·0394 | 19·931 | **1744** | 1 | 1 | Ap | 1○ | ·29 | 2 | Ap | 30 | ·82 | 4 | My | 30 | ·35 | 7 | Jl | 28 | ·42 | 1 | Au | 26 | ·95 |
| | | | | | | | | | | | | | | | | | 5 | Je | 28 | ·88 | | | | | | | | |
| 4846 | 1802 | 1667 | 29M | ·5132 | 21·6783 | 18·064 | **1745** | 3 | 7 | Ap | 20 | ·19 | 1 | My | 19 | ·72 | 3 | Je | 18 | ·25 | 4 | Jl | 17 | ·78 | 6 | Au | 16 | ·31 |
| 4847 | 1803 | 1668 | 29M | ·7719 | 10·7866 | 14·221 | **1746** | 4 | 4 | Ap | 9 | ·56 | 6 | My | 9 | ·09 | 7 | Je | 7 | ·62 | 2 | Jl | 7 | ·15 | 3 | Au | 5○ | ·68 |
| 4848 | 1804 | 1669 | 30M | ·0307 | 29·4254 | 12·355 | **1747** | 5 | 3 | Ap | 28 | ·46 | 4 | My | 27 | ·99 | 6 | Je | 26 | ·52 | 1 | Jl | 26○ | ·05 | 2 | Au | 24● | ·58 |
| 4849 | 1805 | 1670 | 29M | ·2894 | 18·5337 | 8·512 | **1748** | 6 | 7 | Ap | 16 | ·82 | 2 | My | 16 | ·35 | 3 | Je | 14 | ·88 | 5 | Jl | 14●○ | ·41 | 6 | Au | 12 | ·94 |
| 4850 | 1806 | 1671 | 29M | ·5482 | 7·6420 | 4·669 | **1749** | 1 | 5 | Ap | 6 | ·19 | 6 | My | 5 | ·72 | 1 | Je | 4○ | ·25 | 2 | Jl | 3 | ·78 | 4 | Au | 2 | ·31 |
| | | | | | | | | | | | | | | | | | | | | | | | | | 5 | Au | 31 | ·84 |
| 4851 | 1807 | 1672 | 29M | ·8069 | 26·2809 | 2·802 | **1750** | 2 | 4 | Ap | 25 | ·09 | 5 | My | 24○ | ·62 | 7 | Je | 23 | ·15 | 1 | Jl | 22 | ·68 | 3 | Au | 21 | ·21 |
| 4852 | 1808 | 1673 | 30M | ·0657 | 15·3892 | 26·514 | **1751** | 3 | 1 | Ap | 14 | ·45 | 2 | My | 13●○ | ·98 | 4 | Je | 12 | ·52 | 6 | Jl | 12 | ·05 | 7 | Au | 10 | ·58 |
| 4853 | 1809 | 1674 | 29M | ·3244 | 4·4975 | 22·671 | **1752** | 4 | 5 | Ap | 2 | ·82 | 7 | My | 2 | ·35 | 1 | My | 31 | ·88 | 4 | Jl | 29 | ·94 | 6 | Au | 28 | ·47 |
| | | | | | | | | | | | | | | | | | 3 | Je | 30 | ·41 | | | | | | | | |
| 4854 | 1810 | 1675 | 9A | ·5832 | 23·1364 | 20·805 | **1753** | 6 | 4 | My | 2● | ·72 | 6 | Je | 1 | ·25 | 7 | Je | 30 | ·78 | 2 | Jl | 30 | ·31 | 3 | Au | 28 | ·84 |
| 4855 | 1811 | 1676 | 9A | ·8420 | 12·2447 | 16·962 | **1754** | 6 | 2 | Ap | 22 | ·09 | 3 | My | 21 | ·62 | 5 | Je | 20 | ·15 | 6 | Jl | 19 | ·68 | 1 | Au | 18 | ·21 |
| 4856 | 1812 | 1677 | 10A | ·1007 | 1·3530 | 13·119 | **1755** | 4 | 6 | Ap | 11 | ·45 | 7 | My | 10 | ·98 | 4 | Jl | 9 | ·04 | 5 | Au | 7 | ·58 | 7 | S | 6●○ | ·11 |
| | | | | | | | | | | | | | 2 | Je | 9 | ·51 | | | | | | | | | | | | |
| 4857 | 1813 | 1678 | 9A | ·3595 | 19·9919 | 11·252 | **1756** | 5 | 5 | Ap | 29 | ·35 | 6 | My | 28 | ·88 | 1 | Je | 27 | ·41 | 2 | Jl | 26 | ·94 | 4 | Au | 25 | ·47 |
| 4858 | 1814 | 1679 | 9A | ·6182 | 9·1002 | 7·410 | **1757** | 7 | 2 | Ap | 18 | ·72 | 4 | My | 18 | ·25 | 5 | Je | 16 | ·78 | 7 | Jl | 16○ | ·31 | 1 | Au | 14 | ·84 |
| 4859 | 1815 | 1680 | 9A | ·8770 | 27·7391 | 5·543 | **1758** | 1 | 1 | My | 7 | ·62 | 3 | Je | 6 | ·15 | 4 | Jl | 5○ | ·68 | 6 | Au | 4 | ·21 | 7 | S | 2 | ·74 |
| 4860 | 1816 | 1681 | 10A | ·1357 | 16·8474 | 1·700 | **1759** | 2 | 5 | Ap | 26 | ·98 | 7 | My | 26 | ·51 | 2 | Je | 25 | ·04 | 3 | Jl | 24 | ·57 | 5 | Au | 23 | ·10 |
| 4861 | 1817 | 1682 | 9A | ·3945 | 5·9557 | 25·412 | **1760** | 3 | 3 | Ap | 15 | ·35 | 4 | My | 14○ | ·88 | 6 | Je | 13● | ·41 | 7 | Jl | 12 | ·94 | 4 | S | 10 | ·00 |
| | | | | | | | | | | | | | | | | | | | | | 2 | Au | 11 | ·47 | | | | |
| 4862 | 1818 | 1683 | 9A | ·6533 | 24·5946 | 23·545 | **1761** | 5 | 2 | My | 4○ | ·25 | 3 | Je | 2● | ·78 | 5 | Jl | 2 | ·31 | 6 | Jl | 31 | ·84 | 1 | Au | 30 | ·37 |
| 4863 | 1819 | 1684 | 9A | ·9120 | 13·7029 | 19·703 | **1762** | 6 | 6 | Ap | 23●○ | ·61 | 1 | My | 23 | ·14 | 2 | Je | 21 | ·68 | 4 | Jl | 21 | ·21 | 5 | Au | 19 | ·74 |
| 4864 | 1820 | 1685 | 10A | ·1708 | 2·8112 | 15·860 | **1763** | 7 | 3 | Ap | 12● | ·98 | 5 | My | 12 | ·51 | 1 | Jl | 10 | ·57 | 3 | Au | 9 | ·10 | 4 | S | 7 | ·63 |
| | | | | | | | | | | | | | 7 | Je | 11 | ·04 | | | | | | | | | | | | |
| 4865 | 1821 | 1686 | 9A | ·4295 | 21·4500 | 13·993 | **1764** | 1 | 2 | Ap | 30 | ·88 | 4 | My | 30 | ·41 | 5 | Je | 28 | ·94 | 7 | Jl | 28 | ·47 | 2 | Au | 27○ | ·00 |
| 4866 | 1822 | 1687 | 9A | ·6883 | 10·5583 | 10·150 | **1765** | 3 | 7 | Ap | 20 | ·25 | 1 | My | 19 | ·78 | 3 | Je | 18 | ·31 | 4 | Jl | 17 | ·84 | 6 | Au | 16○ | ·37 |
| 4867 | 1823 | 1688 | 9A | ·9470 | 29·1972 | 8·284 | **1766** | 4 | 6 | My | 9 | ·14 | 7 | Je | 7 | ·67 | 2 | Jl | 7 | ·20 | 3 | Au | 5○ | ·74 | 5 | S | 4 | ·27 |
| 4868 | 1824 | 1689 | 10A | ·2058 | 18·3055 | 4·441 | **1767** | 5 | 3 | Ap | 28 | ·51 | 5 | My | 28 | ·04 | 6 | Je | 26 | ·57 | 1 | Jl | 26 | ·10 | 2 | Au | 24 | ·63 |
| 4869 | 1825 | 1690 | 9A | ·4646 | 7·4138 | 0·598 | **1768** | 6 | 7 | Ap | 16 | ·88 | 2 | My | 16 | ·41 | 3 | Je | 14○ | ·94 | 5 | Jl | 14● | ·47 | 1 | S | 11 | ·53 |
| | | | | | | | | | | | | | | | | | | | | | 7 | Au | 13 | ·00 | | | | |
| 4870 | 1826 | 1691 | 9A | ·7233 | 26·0527 | 26·286 | **1769** | 1 | 6 | My | 5 | ·78 | 1 | Je | 4●○ | ·31 | 2 | Jl | 3 | ·84 | 4 | Au | 2 | ·37 | 5 | Au | 31 | ·90 |
| 4871 | 1827 | 1692 | 9A | ·9821 | 15·1610 | 22·443 | **1770** | 2 | 4 | Ap | 25 | ·14 | 5 | My | 24● | ·67 | 7 | Je | 23 | ·20 | 1 | Jl | 22 | ·73 | 3 | Au | 21 | ·26 |
| 4872 | 1828 | 1693 | 10A | ·2408 | 4·2693 | 18·600 | **1771** | 3 | 1 | Ap | 14○ | ·51 | 3 | My | 14 | ·04 | 4 | Je | 12 | ·57 | 7 | Au | 10 | ·63 | 2 | S | 9 | ·16 |
| | | | | | | | | | | | | | | | | | 6 | Jl | 12 | ·10 | | | | | | | | |
| 4873 | 1829 | 1694 | 9A | ·4996 | 22·9082 | 16·734 | **1772** | 4 | 7 | My | 2 | ·41 | 1 | My | 31 | ·94 | 3 | Je | 30 | ·47 | 5 | Jl | 30 | ·00 | 6 | Au | 28 | ·53 |
| 4874 | 1830 | 1695 | 9A | ·7583 | 12·0165 | 12·891 | **1773** | 6 | 4 | Ap | 21 | ·77 | 6 | My | 21 | ·30 | 7 | Je | 19 | ·84 | 2 | Jl | 19 | ·37 | 3 | Au | 17 | ·90 |
| 4875 | 1831 | 1696 | 10A | ·0171 | 1·1248 | 9·048 | **1774** | 7 | 2 | Ap | 11 | ·14 | 5 | Je | 9 | ·20 | 6 | Jl | 8 | ·73 | 1 | Au | 7 | ·26 | 2 | S | 5● | ·79 |
| | | | | | | | | | 3 | My | 10 | ·67 | | | | | | | | | | | | | | | | |
| 4876 | 1832 | 1697 | 10A | ·2758 | 19·7637 | 7·181 | **1775** | 1 | 1 | Ap | 30 | ·04 | 2 | My | 29 | ·57 | 4 | Je | 28 | ·10 | 5 | Jl | 27○ | ·63 | 7 | Au | 26● | ·16 |
| 4877 | 1833 | 1698 | 9A | ·5346 | 8·8720 | 3·339 | **1776** | 2 | 5 | Ap | 18 | ·41 | 6 | My | 17 | ·94 | 1 | Je | 16 | ·47 | 3 | Jl | 16○ | ·00 | 4 | Au | 14 | ·53 |
| | | | | | | | | | | | | | | | | | | | | | | | | | 6 | S | 13 | ·06 |
| 4878 | 1834 | 1699 | 9A | ·7934 | 27·5109 | 1·472 | **1777** | 4 | 4 | My | 7 | ·30 | 5 | Je | 5 | ·83 | 7 | Jl | 5●○ | ·36 | 1 | Au | 3 | ·90 | 3 | S | 2 | ·43 |
| 4879 | 1835 | 1700 | 10A | ·0521 | 16·6192 | 25·184 | **1778** | 5 | 1 | Ap | 26 | ·67 | 3 | My | 26 | ·20 | 4 | Je | 24 | ·73 | 6 | Jl | 24 | ·26 | 7 | Au | 22 | ·79 |
| 4880 | 1836 | 1701 | 10A | ·3109 | 5·7275 | 21·341 | **1779** | 6 | 6 | Ap | 16 | ·04 | 7 | My | 15○ | ·57 | 2 | Je | 14 | ·10 | 3 | Jl | 13 | ·63 | 6 | S | 10 | ·69 |
| | | | | | | | | | | | | | | | | | | | | | 5 | Au | 12 | ·16 | | | | |

**Surya Siddhanta.**

| | | | | | | | | | | | | | | | | | | | | | | | | | | | | | |
|---|---|---|---|---|---|---|---|---|---|---|---|---|---|---|---|---|---|---|---|---|---|---|---|---|---|---|---|---|---|
| + 147·65293 | | | | + 177·18353 | | | | + 206·71411 | | | | | + 236·24470 | | | | | + 265·77529 | | | | + 295·30588 | | | | { +324·83647 +354·36705 | | | |
| + 9·880 | | | | + 11·856 | | | | + 13·832 | | | | | + 15·808 | | | | | + 17·784 | | | | + 19·760 | | | | { + 21·736 + 23·712 | | | |
| **Asvina** | | | | **Kartika** | | | | **Margasira** | | | | **A.D.** | **Pausha** | | | | **A.D.** | **Magha** | | | | **Phalguna** | | | | **Chaitra** | | | |
| Week-day | Month | Day | Fraction | Week-day | Month | Day | Fraction | Week-day | Month | Day | Fraction | | Week-day | Month | Day | Fraction | | Week-day | Month | Day | Fraction | Week-day | Month | Day | Fraction | Week-day | Month | Day | Fraction |
| 6 | S | 21 | ·58 | 1 | O | 21 | ·11 | 2 | N | 19 | ·64 | | 4 | D | 19●○ | ·17 | **40** | 5 | Jr | 17 | ·70 | 7 | F | 16 | .28 | 1 | Mr | 16 | ·76 |
| 3 | S | 9 | ·95 | 5 | O | 9 | ·48 | 7 | N | 8 | ·01 | | 1 | D | 7○ | ·54 | **41** | 3 | Jr | 6 | ·07 | 4 | F | 4 | ·60 | 6 | Mr | 6 | ·13 |
| 2 | S | 28 | ·84 | 4 | O | 28 | ·88 | 5 | N | 26● | ·91 | | 7 | D | 26 | ·44 | **42** | 1 | Jr | 24 | ·97 | 3 | F | 23 | ·50 | 5 | Mr | 25 | ·03 |
| 7 | S | 18 | ·21 | 1 | O | 17○ | ·74 | 3 | N | 16 | ·27 | | 4 | D | 15 | ·80 | **43** | 6 | Jr | 14 | ·33 | 7 | F | 12 | ·86 | 2 | Mr | 14 | ·40 |
| 4 | S | 7 | ·58 | 6 | O | 7○ | ·11 | 7 | N | 5 | ·64 | | 2 | D | 5 | ·17 | **44** | 3 | Jr | 3 | ·70 | 5 | F | 2 | ·23 | 6 | Mr | 2 | ·76 |
| 3 | S | 25●○ | ·48 | 5 | O | 25 | ·01 | 6 | N | 23 | ·54 | | 1 | D | 23 | ·07 | **45** | 2 | Jr | 21 | ·60 | 4 | F | 20 | ·13 | 5 | Mr | 21● | ·66 |
| 7 | S | 14 | ·84 | 2 | O | 14 | ·37 | 3 | N | 12 | ·91 | | 5 | D | 12 | ·44 | **46** | 6 | Jr | 10 | ·97 | 1 | F | 9○ | ·50 | 3 | Mr | 11● | ·03 |
| 5 | S | 4 | ·21 | 6 | O | 3 | ·74 | 1 | N | 2 | ·27 | | 2 | D | 1 | ·80 | | 4 | D | 31 | ·33 | **47** 5 | Jr | 29Ↄ | ·86 | { 7 | F | 28 | ·39 |
| | | | | | | | | | | | | | | | | | | | | | | | | | | { 1 | Mr | 29 | ·92 |
| 4 | S | 23 | ·11 | 5 | O | 22 | ·64 | 7 | N | 21 | ·17 | | 1 | D | 20 | ·70 | **48** | 3 | Jr | 19Ↄ | ·23 | 4 | F | 17 | ·76 | 6 | Mr | 18 | ·29 |
| 1 | S | 11 | ·48 | 3 | O | 11 | ·01 | 4 | N | 9 | ·54 | | 6 | D | 9 | ·07 | **49** | 7 | Jr | 7 | ·60 | 2 | F | 6 | ·13 | 3 | Mr | 7 | ·66 |
| 7 | S | 30 | ·37 | 1 | O | 29 | ·90 | 3 | N | 28Ↄ | ·43 | | 4 | D | 27● | ·96 | **50** | 6 | Jr | 26 | ·50 | 1 | F | 25 | ·03 | 2 | Mr | 26 | ·56 |
| 4 | S | 19 | ·74 | 6 | O | 19 | ·27 | 7 | N | 17○ | ·80 | | 2 | D | 17 | ·33 | **51** | 3 | Jr | 15 | ·86 | 5 | F | 14 | ·39 | 6 | Mr | 15 | ·92 |
| 2 | S | 9 | ·11 | 3 | O | 8 | ·64 | 5 | N | 7○ | ·17 | | 6 | D | 6 | ·70 | **52** | 1 | Jr | 5 | ·28 | 2 | F | 3 | ·76 | 4 | Mr | 4 | ·29 |
| 1 | O* | 8 | ·00 | 2 | N | 6● | ·54 | 4 | D | 6 | ·07 | **53** | 5 | Jr | 4 | ·60 | | 7 | F | 3 | ·13 | 1 | Mr | 4 | ·66 | 3 | Ap | 30○ | ·19 |
| 5 | S | 27Ↄ | ·37 | 6 | O | 26● | ·90 | 1 | N | 25 | ·43 | | 2 | D | 24 | ·96 | **54** | 4 | Jr | 23 | ·49 | 6 | F | 22 | ·02 | 7 | Mr | 23Ↄ | ·56 |
| 2 | S | 16Ↄ | ·74 | 4 | O | 16 | ·27 | 5 | N | 14 | ·80 | | 7 | D | 14 | ·33 | **55** | 1 | Jr | 12 | ·86 | 3 | F | 11 | ·39 | 4 | Mr | 12Ↄ | ·92 |
| 1 | O | 5 | ·64 | 3 | N | 4 | ·17 | 4 | D | 3 | ·70 | **56** | 6 | Jr | 2 | ·23 | | 7 | Jr | 31 | ·76 | 2 | Mr | 1● | ·29 | 3 | Mr | 30 | ·82 |
| 6 | S | 24 | ·00 | 7 | O | 23 | ·53 | 2 | N | 22 | ·06 | | 3 | D | 21 | ·60 | **57** | 5 | Jr | 20Ↄ | ·13 | 6 | F | 18 | ·66 | 1 | Mr | 20 | ·19 |
| { 3 | S | 13 | ·37 | 6 | N | 11 | ·43 | 7 | D | 10 | ·96 | **58** | 2 | Jr | 9Ↄ | ·49 | | 4 | F | 8 | ·02 | 5 | Mr | 9 | ·55 | 7 | Ap | 8 | ·08 |
| { 4 | O | 12 | ·90 | | | | | | | | | | | | | | | | | | | | | | | | | | |
| 2 | O | 2 | ·27 | 3 | O | 31 | ·80 | 5 | N | 30 | ·33 | | 6 | D | 29●Ↄ | ·86 | **59** | 1 | Jr | 28 | ·39 | 2 | F | 26 | ·92 | 4 | Mr | 28 | ·45 |
| 6 | S | 21 | ·64 | 1 | O | 21 | ·17 | 2 | N | 19 | ·70 | | 4 | D | 19 | ·23 | **60** | 5 | Jr | 17 | ·76 | 7 | F | 16 | ·29 | 1 | Mr | 16 | ·82 |
| 5 | O | 9 | ·53 | 7 | N | 8Ↄ | ·06 | 1 | D | 7 | ·59 | **61** | 3 | Jr | 6 | ·12 | | 4 | F | 4 | ·66 | 6 | Mr | 6 | ·19 | 7 | Ap | 4 | ·72 |
| 2 | S | 28 | ·90 | 4 | O | 28Ↄ | ·43 | 5 | N | 26 | ·96 | | 7 | D | 26 | ·49 | **62** | 2 | Jr | 25 | ·02 | 3 | F | 23 | ·55 | 5 | Mr | 25 | ·08 |
| 7 | S | 18 | ·27 | 1 | O | 17●Ↄ | ·80 | 3 | N | 16 | ·33 | | 4 | D | 15 | ·86 | **63** | 6 | Jr | 14 | ·39 | 7 | F | 12 | ·92 | 2 | Mr | 14 | ·45 |
| 6 | O | 7● | ·16 | 7 | N | 5 | ·70 | 2 | D | 5 | ·23 | **64** | 3 | Jr | 3 | ·76 | | 5 | F | 2 | ·29 | 6 | Mr | 2○ | ·82 | 1 | Ap | 1● | ·35 |
| 3 | S | 25 | ·53 | 5 | O | 25 | ·06 | 6 | N | 23 | ·59 | | 1 | D | 23 | ·12 | **65** | 2 | Jr | 21 | ·65 | 4 | F | 20Ↄ | ·18 | 5 | Mr | 21 | ·72 |
| 7 | S | 14 | ·90 | 2 | O | 14 | ·43 | 3 | N | 12 | ·96 | | 5 | D | 12 | ·49 | **66** | 7 | Jr | 11 | ·02 | 1 | F | 9●○ | ·55 | { 3 | Mr | 11 | ·08 |
| | | | | | | | | | | | | | | | | | | | | | | | | | | { 4 | Ap | 9 | ·61 |
| 6 | O | 3 | ·80 | 1 | N | 2 | ·33 | 2 | D | 1 | ·86 | | 4 | D | 31 | ·39 | **67** | 5 | Jr | 29● | ·92 | 7 | F | 28 | ·45 | 1 | Mr | 29 | ·98 |
| 4 | S | 23 | ·16 | 5 | O | 22 | ·69 | [illegible] | N | 21 | ·22 | | 1 | D | 20Ↄ | ·76 | **68** | 3 | Jr | 19 | ·29 | 4 | F | 17 | ·82 | 6 | Mr | 18 | ·35 |
| 3 | O | 11 | ·06 | 4 | N | 9 | ·59 | 6 | D | 9Ↄ | ·12 | **69** | 7 | Jr | 7● | ·65 | | 2 | F | 6 | ·18 | 3 | Mr | 7 | ·71 | 5 | Ap | 6 | ·24 |
| 7 | S | 30 | ·43 | 1 | O | 29 | ·96 | 3 | N | 28Ↄ | ·49 | | 5 | D | 28 | ·02 | **70** | 6 | Jr | 26 | ·55 | 1 | F | 25 | ·08 | 2 | Mr | 26 | ·61 |
| 4 | S | 19 | ·80 | 6 | O | 19 | ·33 | 7 | N | 17● | ·86 | | 2 | D | 17 | ·39 | **71** | 3 | Jr | 15 | ·92 | 5 | F | 14 | ·45 | 6 | Mr | 15 | ·98 |
| 3 | O | 8Ↄ | ·69 | 5 | N | 7 | ·22 | 6 | D | 6 | ·75 | **72** | 1 | Jr | 5 | ·28 | | 2 | F | 3 | ·82 | 4 | Mr | 4 | ·35 | 5 | Ap | 2Ↄ | ·88 |
| 1 | S | 27Ↄ | ·06 | 2 | O | 26● | ·59 | 4 | N | 25 | ·12 | | 5 | D | 24 | ·65 | **73** | 7 | Jr | 23 | ·18 | 1 | F | 21 | ·71 | 3 | Mr | 23●○ | ·24 |
| 5 | S | 16Ↄ | ·43 | 6 | O | 15 | ·96 | 1 | N | 14 | ·49 | | 3 | D | 14 | ·02 | **74** | 4 | Jr | 12 | ·55 | 6 | F | 11 | ·08 | 7 | Mr | 12● | ·61 |
| 4 | O | 5 | ·32 | 5 | N | 3 | ·86 | 7 | D | 3 | ·39 | **75** | 1 | Jr | 1 | ·92 | | 3 | Jr | 31○ | ·45 | 4 | Mr | 1 | ·98 | 6 | Mr | 31 | ·51 |
| 1 | S | 24 | .69 | 3 | O | 24 | ·22 | 4 | N | 22 | ·75 | | 6 | D | 22 | ·28 | **76** | 7 | Jr | 20●○ | ·81 | 2 | F | 19 | ·34 | 3 | Mr | 19 | ·88 |
| 7 | O | 12 | ·59 | 2 | N | 11 | ·12 | 3 | D | 10 | ·65 | **77** | 5 | Jr | 9Ↄ | ·18 | | 6 | F | 7 | ·71 | 1 | Mr | 9 | ·24 | 2 | Ap | 7 | ·77 |
| 4 | O | 1 | ·96 | 6 | O | 31 | ·49 | 1 | N | 30 | ·02 | | 2 | D | 29 | ·55 | **78** | 4 | Jr | 28 | ·08 | 5 | F | 26 | ·61 | 7 | Mr | 28 | ·14 |
| 2 | S | 21 | ·32 | 3 | O | 20 | ·85 | 5 | N | 19Ↄ | ·38 | | 6 | D | 18 | ·92 | **79** | 1 | Jr | 17 | ·45 | 2 | F | 15 | ·98 | 4 | Mr | 17 | ·51 |
| 1 | O | 10 | ·22 | 2 | N | 8Ↄ | ·75 | 4 | D | 8 | ·28 | **80** | 5 | Jr | 6 | ·81 | | 7 | F | 5 | ·34 | 1 | Mr | 5 | ·87 | 3 | Ap | 4 | ·40 |

* In this year **A.D. 1752**, the **New Style** was introduced by an Act of the British Parliament and **11** days were dropped out, that is, the next day after Wednesday, **2** September **1752** was declared to be Thursday, **14** September **1752**. Hence the New-Moon after **28** August **1752** which would ordinarily have occurred on **26** September **1752** actually occurred on **8** October **1752**.

**TABLE X—**

| | Vaisakha | Jyeshta | Ashada | Sravana | Bhadrapada |
|---|---|---|---|---|---|
| ☉'s Anom col. 6 | | + 29·53029 | + 59·06117 | + 88·59176 | + 118·12235 |
| ☾'s Anom col. 7 | | + 1·976 | + 3·952 | + 5·928 | + 7·904 |

| Kaliyuga. | Vikrama Era. | Saka Era. | Commencement of Solar Year. Month and day A.D. | Fraction of day. | First New-Moon in Solar Year. | Anomaly of first New-Moon. | A.D. | Week-day of 1st January. | Vaisakha Week-day | Month | Day | Fraction | Jyeshta Week-day | Month | Day | Fraction | Ashada Week-day | Month | Day | Fraction | Sravana Week-day | Month | Day | Fraction | Bhadrapada Week-day | Month | Day | Fraction |
|---|---|---|---|---|---|---|---|---|---|---|---|---|---|---|---|---|---|---|---|---|---|---|---|---|---|---|---|---|
| 4881 | 1837 | 1702 | 9A | ·5696 | 24·3664 | 19·474 | **1780** | 7 | 4 | My | 3○ | ·94 | 6 | Je | 2 | ·47 | 1 | Jl | 2 | ·00 | 2 | Jl | 31 | ·53 | 4 | Au | 30 | ·06 |
| 4882 | 1838 | 1703 | 9A | ·8284 | 13·4747 | 15·632 | **1781** | 2 | 2 | Ap | 23 | ·30 | 3 | My | 22 | ·83 | 5 | Je | 21 | ·36 | 6 | Jl | 20 | ·89 | 1 | Au | 19 | ·42 |
| 4883 | 1839 | 1704 | 10A | ·0871 | 2·5880 | 11·789 | **1782** | 3 | 6 | Ap | 12 | ·67 | 1 | My | 12 | ·20 | 4 | Jl | 10 | ·26 | 5 | Au | 8 | ·79 | 7 | S | 7◑ | ·32 |
| | | | | | | | | | | | | | 2 | Je | 10 | ·73 | | | | | | | | | | | | |
| 4884 | 1840 | 1705 | 10A | ·3459 | 21·2218 | 9·922 | **1783** | 4 | 5 | My | 1 | ·57 | 7 | My | 31 | ·10 | 1 | Je | 29 | ·63 | 3 | Jl | 29 | ·16 | 4 | Au | 27○ | ·69 |
| 4885 | 1841 | 1706 | 9A | ·6047 | 10·3301 | 6·079 | **1784** | 5 | 2 | Ap | 19 | ·93 | 4 | My | 19 | ·46 | 6 | Je | 18 | ·00 | 7 | Jl | 17 | ·53 | 2 | Au | 16●○ | ·06 |
| 4886 | 1842 | 1707 | 9A | ·8634 | 28·9690 | 4·213 | **1785** | 7 | 1 | My | 8 | ·83 | 3 | Je | 7 | ·36 | 4 | Jl | 6 | ·89 | 6 | Au | 5● | ·42 | 7 | S | 3 | ·95 |
| 4887 | 1843 | 1708 | 10A | ·1222 | 18·0773 | 0·370 | **1786** | 1 | 6 | Ap | 28 | ·20 | 7 | My | 27 | ·73 | 2 | Je | 26◑ | ·26 | 3 | Jl | 25 | ·79 | 5 | Au | 24 | ·32 |
| 4888 | 1844 | 1709 | 10A | ·3809 | 7·1856 | 24·082 | **1787** | 2 | 3 | Ap | 17 | ·57 | 5 | My | 17 | ·10 | 6 | Je | 15◑ | ·63 | 1 | Jl | 15 | ·16 | 4 | S | 12 | ·22 |
| | | | | | | | | | | | | | | | | | | | | | 2 | Au | 13 | ·69 | | | | |
| 4889 | 1845 | 1710 | 9A | ·6397 | 25·8245 | 22·215 | **1788** | 3 | 2 | My | 5 | ·46 | 3 | Je | 3● | ·99 | 5 | Jl | 3 | ·52 | 7 | Au | 2 | ·06 | 1 | Au | 31 | ·59 |
| 4890 | 1846 | 1711 | 9A | ·8984 | 14·9328 | 18·372 | **1789** | 5 | 6 | Ap | 24◑ | ·83 | 1 | My | 24 | ·36 | 2 | Je | 22 | ·89 | 4 | Jl | 22 | ·42 | 5 | Au | 20 | ·95 |
| 4891 | 1847 | 1712 | 10A | ·1572 | 4·0411 | 14·529 | **1790** | 6 | 4 | Ap | 14○ | ·20 | 5 | My | 13 | ·73 | 7 | Je | 12 | ·26 | 3 | Au | 10 | ·32 | 4 | S | 8 | ·85 |
| | | | | | | | | | | | | | | | | | 1 | Jl | 11 | ·79 | | | | | | | | |
| 4892 | 1848 | 1713 | 10A | ·4160 | 12·6800 | 2·663 | **1791** | 7 | 3 | My | 3 | ·10 | 4 | Je | 1 | ·63 | 6 | Jl | 1 | ·16 | 7 | Jl | 30 | ·69 | 2 | Au | 29 | ·22 |
| 4893 | 1849 | 1714 | 9A | ·6747 | 11·7883 | 8·820 | **1792** | 1 | 7 | Ap | 21 | ·46 | 1 | My | 20 | ·99 | 3 | Je | 19 | ·52 | 5 | Jl | 19 | ·05 | 6 | Au | 17 | ·58 |
| 4894 | 1850 | 1715 | 9A | ·9335 | 0·8966 | 4·977 | **1793** | 3 | 4 | Ap | 10 | ·83 | 7 | Je | 8 | ·89 | 2 | Jl | 8 | ·42 | 3 | Au | 6◑ | ·95 | 5 | S | 5● | ·48 |
| | | | | | | | | | 6 | My | 10 | ·36 | | | | | | | | | | | | | | | | |
| 4895 | 1851 | 1716 | 10A | ·1922 | 19·5355 | 3·110 | **1794** | 4 | 3 | Ap | 29 | ·73 | 5 | My | 29 | ·26 | 6 | Je | 27 | ·79 | 1 | Jl | 27◑ | ·32 | 2 | Au | 25● | ·85 |
| 4896 | 1852 | 1717 | 10A | ·4510 | 8·6438 | 26·822 | **1795** | 5 | 1 | Ap | 19 | ·09 | 2 | My | 18 | ·62 | 4 | Je | 17 | ·16 | 5 | Jl | 16●◑ | ·69 | 7 | Au | 15 | ·22 |
| | | | | | | | | | | | | | | | | | | | | | | | | | 1 | S | 13 | ·75 |
| 4897 | 1853 | 1718 | 9A | ·7097 | 27·2827 | 24·955 | **1796** | 6 | 6 | My | 6 | ·99 | 1 | Je | 5 | ·52 | 3 | Jl | 5● | ·05 | 4 | Au | 3 | ·58 | 6 | S | 2 | ·11 |
| 4898 | 1854 | 1719 | 9A | ·9685 | 16·3910 | 21·113 | **1797** | 1 | 4 | Ap | 26 | ·36 | 5 | My | 25◑ | ·89 | 7 | Je | 24 | .42 | 1 | Jl | 23 | ·95 | 3 | Au | 22 | ·48 |
| 4899 | 1855 | 1720 | 10A | ·2272 | 5·4993 | 17·270 | **1798** | 2 | 1 | Ap | 15 | ·73 | 3 | My | 15◑ | ·26 | 4 | Je | 13 | ·79 | 6 | Jl | 13 | ·32 | 2 | S | 10 | ·38 |
| | | | | | | | | | | | | | | | | | | | | | 7 | Au | 11 | ·85 | | | | |
| 4900 | 1856 | 1721 | 10A | ·4860 | 24·1382 | 15·403 | **1799** | 3 | 7 | My | 4● | ·62 | 2 | Je | 3 | ·15 | 3 | Jl | 2 | ·68 | 5 | Au | 1 | ·22 | 6 | Au | 30 | ·75 |
| 4901 | 1857 | 1722 | 10A | ·7448 | 13·2464 | 11·563 | **1800** | 4 | 4 | Ap | 23● | ·99 | 6 | My | 23 | ·52 | 1 | Je | 22 | ·05 | 2 | Jl | 21 | ·58 | 4 | Au | 20 | ·11 |
| 4902 | 1858 | 1723 | 11A | ·0035 | 2·3548 | 7·720 | **1801** | 5 | 2 | Ap | 13● | ·36 | 3 | My | 12 | ·89 | 6 | Jl | 10 | ·95 | 1 | Au | 9 | ·48 | 3 | S | 8◑ | ·01 |
| | | | | | | | | | | | | | 5 | Je | 11 | ·42 | | | | | | | | | | | | |
| 4903 | 1859 | 1724 | 11A | ·2623 | 20·9936 | 5·854 | **1802** | 6 | 1 | My | 2 | ·26 | 2 | My | 31 | ·79 | 4 | Je | 30 | ·32 | 5 | Jl | 29 | ·85 | 7 | Au | 28●○ | ·38 |
| 4904 | 1860 | 1725 | 11A | ·5210 | 10·1019 | 2·011 | **1803** | 7 | 5 | Ap | 21 | ·62 | 7 | My | 21 | ·15 | 1 | Je | 19 | ·68 | 3 | Jl | 19 | ·21 | 4 | Au | 17● | ·74 |
| 4905 | 1861 | 1726 | 10A | ·7798 | 28·7408 | 0·144 | **1804** | 1 | 4 | My | 9 | ·52 | 6 | Je | 8 | ·05 | 7 | Jl | 7○ | ·58 | 2 | Au | 6 | ·11 | 3 | S | 4 | ·64 |
| 4906 | 1862 | 1727 | 11A | ·0385 | 17·8491 | 23·856 | **1805** | 3 | 1 | Ap | 28 | ·89 | 3 | My | 28 | ·42 | 4 | Je | 26●○ | ·95 | 6 | Jl | 26 | ·48 | 1 | Au | 25 | ·01 |
| 4907 | 1863 | 1728 | 11A | ·2973 | 6·9574 | 20·013 | **1806** | 4 | 6 | Ap | 18 | ·25 | 7 | My | 17 | ·78 | 2 | Je | 16 | ·32 | 3 | Jl | 15 | ·85 | 6 | S | 12 | ·91 |
| | | | | | | | | | | | | | | | | | | | | | 5 | Au | 14 | ·38 | | | | |
| 4908 | 1864 | 1729 | 11A | ·5561 | 25·5963 | 18·147 | **1807** | 5 | 5 | My | 7○ | ·15 | 6 | Je | 5● | ·68 | 1 | Jl | 5 | ·21 | 2 | Au | 3 | ·74 | 4 | S | 2 | ·27 |
| 4909 | 1865 | 1730 | 10A | ·8148 | 14·7046 | 14·304 | **1808** | 6 | 2 | Ap | 25○ | ·52 | 4 | My | 25 | ·05 | 5 | Je | 23 | ·58 | 7 | Jl | 23 | ·11 | 1 | Au | 21 | ·64 |
| 4910 | 1866 | 1731 | 11A | ·0736 | 3·8129 | 10·461 | **1809** | 1 | 6 | Ap | 14○ | ·89 | 1 | My | 14 | ·42 | 2 | Je | 12 | ·95 | 6 | Au | 11 | ·01 | 7 | S | 9 | ·54 |
| | | | | | | | | | | | | | | | | | 4 | Jl | 12 | ·48 | | | | | | | | |
| 4911 | 1867 | 1732 | 11A | ·3323 | 22·4518 | 8·594 | **1810** | 2 | 5 | My | 3 | ·78 | 7 | Je | 2 | ·31 | 1 | Jl | 1 | ·84 | 3 | Jl | 31 | ·37 | 4 | Au | 29 | ·91 |
| 4912 | 1868 | 1733 | 11A | ·5911 | 11·5601 | 4·752 | **1811** | 3 | 3 | Ap | 23 | ·15 | 4 | My | 22 | ·68 | 6 | Je | 21 | ·21 | 7 | Jl | 20 | ·74 | 2 | Au | 19◑ | ·27 |
| 4913 | 1869 | 1734 | 10A | ·8498 | 0·6684 | 0·909 | **1812** | 4 | 7 | Ap | 11 | ·52 | 3 | Je | 9 | ·58 | 5 | Jl | 9 | ·11 | 6 | Au | 7◑ | ·64 | 1 | S | 6 | ·17 |
| | | | | | | | | | 2 | My | 11 | ·05 | | | | | | | | | | | | | | | | |
| 4914 | 1870 | 1735 | 11A | ·1086 | 19·3073 | 26·597 | **1813** | 6 | 6 | Ap | 30 | ·42 | 7 | My | 29 | ·95 | 2 | Je | 28 | ·48 | 4 | Jl | 28○ | ·00 | 5 | Au | 26 | ·54 |
| 4915 | 1871 | 1736 | 11A | ·3674 | 8·4156 | 22·754 | **1814** | 7 | 3 | Ap | 19 | ·78 | 5 | My | 19 | ·31 | 6 | Je | 17 | ·84 | 1 | Jl | 17● | ·37 | 2 | Au | 15 | ·90 |
| | | | | | | | | | | | | | | | | | | | | | | | | | 4 | S | 14 | ·44 |
| 4916 | 1872 | 1737 | 11A | ·6261 | 27·0545 | 20·887 | **1815** | 1 | 2 | My | 8 | ·68 | 4 | Je | 7○ | ·21 | 5 | Jl | 6● | ·74 | 7 | Au | 5 | ·27 | 1 | S | 3 | ·80 |
| 4917 | 1873 | 1738 | 10A | ·8849 | 16·1628 | 17·044 | **1816** | 2 | 7 | Ap | 27 | ·05 | 1 | My | 26○ | ·58 | 3 | Je | 25 | ·11 | 4 | Jl | 24 | ·64 | 6 | Au | 23 | ·17 |
| 4918 | 1874 | 1739 | 11A | ·1336 | 5·2711 | 13·202 | **1817** | 4 | 4 | Ap | 16 | ·41 | 5 | My | 15● | ·94 | 7 | Je | 14 | ·48 | 2 | Jl | 14 | ·01 | 5 | S | 11 | ·07 |
| | | | | | | | | | | | | | | | | | | | | | 3 | Au | 12 | ·54 | | | | |
| 4919 | 1875 | 1740 | 11A | ·4024 | 23·9099 | 11·335 | **1818** | 5 | 3 | My | 5● | ·31 | 4 | Je | 3 | ·84 | 6 | Jl | 3 | ·37 | 7 | Au | 1 | ·90 | 2 | Au | 31 | ·43 |

**Surya Siddhanta.**

| Asvina | | | | Kartika | | | | Margasira | | | | A.D. | Pausha | | | | A.D. | Magha | | | | Phalguna | | | | Chaitra | | | |
|---|---|---|---|---|---|---|---|---|---|---|---|---|---|---|---|---|---|---|---|---|---|---|---|---|---|---|---|---|---|
| + 147·65293 | | | | + 177·18353 | | | | + 206·71411 | | | | | + 236·24470 | | | | | + 265·77529 | | | | + 295·30588 | | | | +324·83647 +354·36705 | | | |
| + 9·880 | | | | + 11·856 | | | | + 13·832 | | | | | + 15·808 | | | | | + 17·784 | | | | + 19·760 | | | | +21·736 +23·712 | | | |
| Week-day | Month | Day | Fraction | Week-day | Month | Day | Fraction | Week-day | Month | Day | Fraction | | Week-day | Month | Day | Fraction | | Week-day | Month | Day | Fraction | Week-day | Month | Day | Fraction | Week-day | Month | Day | Fraction |
| 5 | S | 28 | ·59 | 7 | O | 28○ | ·12 | 1 | N | 26 | ·65 | | 3 | D | 26 | ·18 | 81 | 4 | Jr | 24 | ·71 | 6 | F | 23 | ·24 | 7 | Mr | 24 | ·77 |
| 2 | S | 17 | ·96 | 4 | O | 17● | ·49 | 6 | N | 16 | ·02 | | 7 | D | 15 | ·55 | 82 | 2 | Jr | 14 | ·08 | 3 | F | 12 | ·61 | 5 | Mr | 14○ | ·14 |
| 1 | O | 6● | ·85 | 3 | N | 5 | ·38 | 4 | D | 4 | ·91 | 83 | 6 | Jr | 3 | ·44 | | 7 | F | 1 | ·98 | 2 | Mr | 3○ | ·51 | 4 | Ap | 2 | ·04 |
| 6 | S | 26 | ·22 | 7 | O | 25 | ·75 | 2 | N | 24 | ·28 | | 3 | D | 23 | ·81 | 84 | 5 | Jr | 22 | ·34 | 6 | F | 20○ | ·87 | 1 | Mr | 21 | ·40 |
| 3 | S | 14 | ·59 | 5 | O | 14 | ·12 | 6 | N | 12 | ·65 | | 1 | D | 12 | ·18 | 85 | 2 | Jr | 10 | ·71 | 4 | F | 9● | ·24 | {5 | Mr | 10 | ·77 |
| | | | | | | | | | | | | | | | | | | | | | | | | | | {7 | Ap | 9 | ·30 |
| 2 | O | 3 | ·48 | 4 | N | 2 | ·02 | 5 | D | 1 | ·55 | | 7 | D | 31○ | ·08 | 86 | 1 | Jr | 29● | ·61 | 3 | F | 28 | ·14 | 4 | Mr | 29 | ·67 |
| 6 | S | 22 | ·85 | 1 | O | 22 | ·38 | 2 | N | 20 | ·91 | | 4 | D | 20○ | ·44 | 87 | 5 | Jr | 18 | ·97 | 7 | F | 17 | ·50 | 2 | Mr | 19 | ·04 |
| 5 | O | 11 | ·75 | 7 | N | 10 | ·28 | 1 | D | 9○ | ·81 | 88 | 3 | Jr | 8 | ·34 | | 4 | F | 6 | ·87 | 6 | Mr | 7 | ·40 | 7 | Ap | 5 | ·93 |
| 3 | S | 30 | ·12 | 4 | O | 29 | ·65 | 6 | N | 28 | ·18 | | 7 | D | 27 | ·71 | 89 | 2 | Jr | 26 | ·24 | 3 | F | 24 | ·77 | 5 | Mr | 26 | ·30 |
| 7 | S | 19 | ·48 | 2 | O | 19○ | ·01 | 3 | N | 17● | ·54 | | 5 | D | 17 | ·08 | 90 | 6 | Jr | 15 | ·61 | 1 | F | 14 | ·14 | 2 | Mr | 15 | ·67 |
| 6 | O | 8○ | ·38 | 7 | N | 6 | ·91 | 2 | D | 6 | ·44 | 91 | 3 | Jr | 4 | ·97 | | 5 | F | 3 | ·50 | 7 | Mr | 5 | ·03 | 1 | Ap | 3●○ | ·56 |
| 3 | S | 27●○ | ·75 | 5 | O | 27 | ·28 | 6 | N | 25 | ·81 | | 1 | D | 25 | ·34 | 92 | 2 | Jr | 23 | ·87 | 4 | F | 22 | ·40 | 5 | Mr | 22 | ·93 |
| 1 | S | 16● | ·12 | 2 | O | 15 | ·65 | 4 | N | 14 | ·18 | | 5 | D | 13 | ·71 | 93 | 7 | Jr | 12 | ·24 | 1 | F | 10○ | ·77 | 3 | Mr | 12● | ·30 |
| 7 | O | 5 | ·01 | 1 | N | 3 | ·54 | 3 | D | 3 | ·07 | 94 | 4 | Jr | 1 | ·60 | | 6 | Jr | 31○ | ·14 | 7 | Mr | 1 | ·67 | 2 | Mr | 31 | ·20 |
| 4 | S | 24 | ·38 | 5 | O | 23 | ·91 | 7 | N | 22 | ·44 | | 1 | D | 21 | ·97 | 95 | 3 | Jr | 20●○ | ·50 | 5 | F | 19 | ·03 | 6 | Mr | 20 | ·56 |
| 3 | O | 13 | ·28 | 4 | N | 11 | ·81 | 6 | D | 11 | ·34 | 96 | 7 | Jr | 9● | ·87 | | 2 | F | 8 | ·40 | 3 | Mr | 8 | ·93 | 5 | Ap | 7 | ·46 |
| 7 | O | 1 | ·64 | 2 | O | 31 | ·18 | 3 | N | 29○ | ·71 | | 5 | D | 29 | ·24 | 97 | 6 | Jr | 27 | ·77 | 1 | F | 26 | ·30 | 2 | Mr | 27 | ·33 |
| 5 | S | 21 | ·01 | 6 | O | 20 | ·54 | 1 | N | 19○ | ·07 | | 2 | D | 18 | ·60 | 98 | 4 | Jr | 17 | ·13 | 5 | F | 15 | ·66 | 7 | Mr | 17 | ·20 |
| 3 | O | 9 | ·91 | 5 | N | 8●○ | ·44 | 6 | D | 7 | ·97 | 99 | 1 | Jr | 6 | ·50 | | 3 | F | 5 | ·03 | 4 | Mr | 6 | ·56 | 6 | Ap | 5 | ·09 |
| 1 | S | 29 | ·28 | 2 | O | 28 | ·81 | 4 | N | 27 | ·34 | 18 | 5 | D | 26 | ·87 | 00 | 7 | Jr | 25 | ·40 | 1 | F | 23 | ·93 | 3 | Mr | 25○ | ·46 |
| 5 | S | 18○ | ·64 | 7 | O | 18 | ·17 | 1 | N | 16 | ·70 | | 3 | D | 16 | ·24 | 01 | 4 | Jr | 14 | ·77 | 6 | F | 13 | ·30 | 7 | Mr | 14○ | ·83 |
| 4 | O | 7 | ·54 | 6 | N | 6 | ·07 | 7 | D | 5 | ·60 | 02 | 2 | Jr | 4 | ·13 | | 3 | F | 2 | ·66 | 5 | Mr | 4○ | ·19 | 6 | Ap | 2 | ·72 |
| 1 | S | 26 | ·91 | 3 | O | 26 | ·44 | 4 | N | 24 | ·97 | | 6 | D | 24 | ·50 | 03 | 1 | Jr | 23 | ·03 | 2 | F | 21 | ·56 | 4 | Mr | 23 | ·09 |
| 6 | S | 16 | ·28 | 7 | O | 15 | ·81 | 2 | N | 14 | ·34 | | 3 | D | 13 | ·87 | 04 | 5 | Jr | 12○ | ·40 | 6 | F | 10● | ·93 | {1 | Mr | 11 | ·46 |
| | | | | | | | | | | | | | | | | | | | | | | | | | | {2 | Ap | 9 | ·99 |
| 5 | O | 4 | ·17 | 6 | N | 2 | ·70 | 1 | D | 2 | ·23 | | 2 | D | 31○ | ·76 | 05 | 4 | Jr | 30 | ·30 | 5 | F | 28 | ·83 | 7 | Mr | 30 | ·36 |
| 2 | S | 23 | ·54 | 4 | O | 23 | ·07 | 5 | N | 21 | ·60 | | 7 | D | 21○ | ·18 | 06 | 1 | Jr | 19 | ·66 | 3 | F | 18 | ·19 | 4 | Mr | 19 | ·72 |
| 1 | O | 12 | ·44 | 2 | N | 10 | ·97 | 4 | D | 10● | ·50 | 07 | 6 | Jr | 9 | ·03 | | 7 | F | 7 | ·56 | 2 | Mr | 9 | ·09 | 3 | Ap | 7 | ·62 |
| 5 | O | 1 | ·80 | 7 | O | 31○ | ·34 | 1 | N | 29● | ·87 | | 3 | D | 29 | ·40 | 08 | 4 | Jr | 27 | ·93 | 6 | F | 26 | ·46 | 7 | Mr | 26 | ·99 |
| 3 | S | 20 | ·17 | 4 | O | 19○ | ·70 | 6 | N | 18● | ·23 | | 7 | D | 17 | ·76 | 09 | 2 | Jr | 16 | ·29 | 3 | F | 14 | ·82 | 5 | Mr | 16 | ·36 |
| 2 | O | 9○ | ·07 | 3 | N | 7 | ·60 | 5 | D | 7 | ·13 | 10 | 6 | Jr | 5 | ·66 | | 1 | F | 4 | ·19 | 2 | Mr | 5 | ·72 | 4 | Ap | 4● | ·25 |
| 6 | S | 28 | ·44 | 7 | O | 27 | ·97 | 2 | N | 26 | ·50 | | 4 | D | 26 | ·03 | 11 | 5 | Jr | 24 | ·56 | 7 | F | 23○ | ·09 | 1 | Mr | 24 | ·62 |
| 3 | S | 17 | ·80 | 5 | O | 17 | ·33 | 6 | N | 15 | ·86 | | 1 | D | 15 | ·40 | 12 | 2 | Jr | 13 | ·93 | 4 | F | 12○ | ·46 | 5 | Mr | 12 | ·99 |
| 2 | O | 5 | ·70 | 4 | N | 4 | ·23 | 5 | D | 3 | ·76 | 13 | 7 | Jr | 2 | ·29 | | 1 | Jr | 31●○ | ·82 | 3 | Mr | 2 | ·35 | 4 | Mr | 31 | ·88 |
| 7 | S | 25 | ·07 | 1 | O | 24 | ·60 | 3 | N | 23 | ·13 | | 4 | D | 22 | ·66 | 14 | 6 | Jr | 21 | ·19 | 7 | F | 19 | ·72 | 2 | Mr | 21 | ·25 |
| 5 | O | 13 | ·97 | 7 | N | 12 | ·50 | 2 | D | 12○ | ·03 | 15 | 3 | Jr | 10 | ·56 | | 5 | F | 9 | ·09 | 6 | Mr | 10 | ·62 | 1 | Ap | 9 | ·15 |
| 3 | O | 3 | ·33 | 4 | N | 1 | ·86 | 6 | D | 1○ | ·39 | | 7 | D | 30 | ·92 | 16 | 2 | Jr | 29 | ·46 | 3 | F | 27 | ·99 | 5 | Mr | 28 | ·52 |
| 7 | S | 21 | ·70 | 2 | O | 21 | ·23 | 3 | N | 19●○ | ·76 | | 5 | D | 19 | ·29 | 17 | 6 | Jr | 17 | ·82 | 1 | F | 16 | ·35 | 2 | Mr | 17 | ·88 |
| 6 | O | 10 | ·60 | 1 | N | 9● | ·13 | 2 | D | 8 | ·66 | 18 | 4 | Jr | 7 | ·19 | | 5 | F | 5 | ·72 | 7 | Mr | 7 | ·25 | 1 | Ap | 5○ | ·78 |
| 3 | S | 29○ | ·96 | 5 | O | 29 | ·50 | 7 | N | 28 | ·03 | | 1 | D | 27 | ·56 | 19 | 3 | Jr | 26 | ·09 | 4 | F | 24 | ·62 | 6 | Mr | 26○ | 15 |

☉'s Anom col. 6 + 29·53059 + 59·06117 + 88·59176 + 118·12235

☾'s Anom col. 7 + 1·976 + 3·952 + 5·928 + 7·904

| Kaliyuga. | Vikrama Era. | Saka Era. | Commencement of Solar Year. Month and day A.D. | Fraction of day. | First New-Moon in Solar Year. | Anomaly of first New-Moon. | A.D. | Week-day of 1st January. | Vaisakha | | | | Jyeshta | | | | Ashada | | | | Sravana | | | | Bhadrapada | | | |
|---|---|---|---|---|---|---|---|---|---|---|---|---|---|---|---|---|---|---|---|---|---|---|---|---|---|---|---|---|
| | | | | | | | | | Week-day | Month | Day | Fraction | Week-day | Month | Day | Fraction | Week-day | Month | Day | Fraction | Week-day | Month | Day | Fraction | Week-day | Month | Day | Fraction |
| 4920 | 1876 | 1741 | 11A | ·6611 | 13·0182 | 7·492 | **1819** | 6 | 7 | Ap | 24 | ·68 | 2 | My | 24 | ·21 | 3 | Je | 22 | ·74 | 5 | Jl | 22 | ·27 | 6 | Au | 20 | ·80 |
| 4921 | 1877 | 1742 | 10A | ·9199 | 2·1265 | 3·649 | **1820** | 7 | 5 | Ap | 13 | ·05 | 6 | My | 12 | ·58 | 2 | Jl | 10 | ·64 | 4 | Au | 9 | ·17 | 5 | S | 7○ | ·70 |
| | | | | | | | | | | | | | 1 | Je | 11 | ·11 | | | | | | | | | | | | |
| 4922 | 1878 | 1743 | 11A | ·1787 | 20·7654 | 1·783 | **1821** | 2 | 3 | My | 1 | ·94 | 5 | My | 31 | ·47 | 7 | Je | 30 | ·00 | 1 | Jl | 29 | ·54 | 3 | Au | 28 | ·07 |
| 4923 | 1879 | 1744 | 11A | ·4374 | 9·8737 | 25·495 | **1822** | 3 | 1 | Ap | 21 | ·31 | 2 | My | 20 | ·84 | 4 | Je | 19 | ·37 | 5 | Jl | 18○ | ·90 | 7 | Au | 17 | ·48 |
| 4924 | 1880 | 1745 | 11A | ·6962 | 28·5126 | 23·628 | **1823** | 4 | 7 | My | 10 | ·21 | 1 | Je | 8 | ·74 | 3 | Jl | 8○ | ·27 | 4 | Au | 6 | ·80 | 6 | S | 5 | ·33 |
| 4925 | 1881 | 1746 | 10A | ·9549 | 17·6209 | 19·785 | **1824** | 5 | 4 | Ap | 28 | ·58 | 6 | My | 28 | ·11 | 7 | Je | 26●○ | ·64 | 2 | Jl | 26 | ·17 | 3 | Au | 24 | ·70 |
| 4926 | 1882 | 1747 | 11A | ·2137 | 6·7292 | 15·942 | **1825** | 7 | 1 | Ap | 17 | ·94 | 3 | My | 17○ | ·47 | 5 | Je | 16● | ·00 | 6 | Jl | 15 | ·53 | 2 | S | 12 | ·60 |
| | | | | | | | | | | | | | | | | | | | | | 1 | Au | 14 | ·06 | | | | |
| 4927 | 1883 | 1748 | 11A | ·4724 | 25·3681 | 14·076 | **1826** | 1 | 7 | My | 6○ | ·84 | 2 | Je | 5 | ·37 | 3 | Jl | 4 | ·90 | 5 | Au | 3 | ·43 | 6 | S | 1 | ·96 |
| 4928 | 1884 | 1749 | 11A | ·7312 | 14·4764 | 10·233 | **1827** | 2 | 5 | Ap | 26●○ | ·21 | 6 | My | 25 | ·74 | 1 | Je | 24 | ·27 | 2 | Jl | 23 | ·80 | 4 | Au | 22 | ·33 |
| 4929 | 1885 | 1750 | 10A | ·9899 | 3·5847 | 6·390 | **1828** | 3 | 2 | Ap | 14● | ·57 | 4 | My | 14 | ·10 | 5 | Je | 12 | ·64 | 1 | Au | 10 | ·70 | 3 | S | 9 | ·23 |
| | | | | | | | | | | | | | | | | | 7 | Jl | 12 | ·17 | | | | | | | | |
| 4930 | 1886 | 1751 | 11A | ·2487 | 22·2236 | 4·523 | **1829** | 5 | 1 | My | 3 | ·47 | 3 | Je | 2 | ·00 | 4 | Jl | 1 | ·53 | 6 | Jl | 31 | ·06 | 7 | Au | 29○ | ·59 |
| 4931 | 1887 | 1752 | 11A | ·5075 | 11·3319 | 0·681 | **1830** | 6 | 5 | Ap | 22 | ·84 | 7 | My | 22 | ·37 | 1 | Je | 20 | ·90 | 3 | Jl | 20 | ·43 | 4 | Au | 18○ | ·96 |
| 4932 | 1888 | 1753 | 11A | ·7662 | 0·4402 | 24·392 | **1831** | 7 | 3 | Ap | 12 | ·21 | 6 | Je | 10 | ·27 | 7 | Jl | 9 | ·80 | 2 | Au | 8○ | ·33 | 3 | S | 6 | ·86 |
| | | | | | | | | | 4 | My | 11 | ·74 | | | | | | | | | | | | | | | | |
| 4933 | 1889 | 1754 | 11A | ·0250 | 19·0791 | 22·526 | **1832** | 1 | 2 | Ap | 30 | ·10 | 3 | My | 29 | ·63 | 5 | Je | 28 | ·16 | 6 | Jl | 27● | ·70 | 1 | Au | 26 | ·23 |
| 4934 | 1890 | 1755 | 11A | ·2837 | 8·1874 | 18·683 | **1833** | 3 | 6 | Ap | 19 | ·47 | 1 | My | 19 | ·00 | 2 | Je | 17○ | ·53 | 4 | Jl | 17● | ·06 | 5 | Au | 15 | ·59 |
| | | | | | | | | | | | | | | | | | | | | | | | | | 7 | S | 14 | ·12 |
| 4935 | 1891 | 1756 | 11A | ·5425 | 26·8263 | 16·816 | **1834** | 4 | 5 | My | 8 | ·37 | 6 | Je | 6○ | ·90 | 1 | Jl | 6 | ·43 | 2 | Au | 4 | ·96 | 4 | S | 3 | ·49 |
| 4936 | 1892 | 1757 | 11A | ·8012 | 15·9346 | 12·973 | **1835** | 5 | 2 | Ap | 27 | ·74 | 4 | My | 27○ | ·27 | 5 | Je | 25 | ·80 | 7 | Jl | 25 | ·33 | 1 | Au | 23 | ·86 |
| 4937 | 1893 | 1758 | 11A | ·0600 | 5·0429 | 9·131 | **1836** | 6 | 7 | Ap | 16○ | ·10 | 1 | My | 15 | ·63 | 3 | Je | 1[illegible] | ·16 | 6 | Au | 12 | ·22 | 7 | S | 10 | ·76 |
| | | | | | | | | | | | | | | | | | 4 | Jl | 13 | ·69 | | | | | | | | |
| 4938 | 1894 | 1759 | 11A | ·3188 | 23·6819 | 7·264 | **1837** | 1 | 6 | My | 5 | ·00 | 7 | Je | 3 | ·53 | 2 | Jl | 3 | ·06 | 3 | Au | 1 | ·59 | 5 | Au | 31 | ·12 |
| 4939 | 1895 | 1760 | 11A | ·5775 | 12·7900 | 3·421 | **1838** | 2 | 3 | Ap | 24 | ·37 | 4 | My | 23 | ·90 | 6 | Je | 22 | ·43 | 7 | Jl | 21 | ·96 | 2 | Au | 20 | ·49 |
| 4940 | 1896 | 1761 | 11A | ·8363 | 1·8983 | 27·133 | **1839** | 3 | 7 | Ap | 13 | ·73 | 2 | My | 13 | ·26 | 5 | Jl | 11 | ·33 | 6 | Au | 9 | ·86 | 1 | S | 8 | ·39 |
| | | | | | | | | | | | | | 3 | Je | 11 | ·80 | | | | | | | | | | | | |
| 4941 | 1897 | 1762 | 11A | ·0950 | 20·5372 | 25·266 | **1840** | 4 | 6 | My | 1 | ·63 | 1 | My | 31 | ·16 | 2 | Je | 29 | ·69 | 4 | Jl | 29○ | ·22 | 5 | Au | 27● | ·75 |
| 4942 | 1898 | 1763 | 11A | ·3538 | 9·6455 | 21·424 | **1841** | 6 | 4 | Ap | 21 | ·00 | 5 | My | 20 | ·53 | 7 | Je | 19 | ·06 | 1 | Jl | 18○ | ·59 | 3 | Au | 17 | ·12 |
| 4943 | 1899 | 1764 | 11A | ·6125 | 28·2844 | 19·557 | **1842** | 7 | 2 | My | 9 | ·90 | 4 | Je | 8 | ·43 | 5 | Jl | 7●○ | ·96 | 7 | Au | 6 | ·49 | 2 | S | 5 | ·02 |
| 4944 | 1900 | 1765 | 11A | ·8713 | 17·3927 | 15·714 | **1843** | 1 | 7 | Ap | 29 | ·26 | 1 | My | 28 | ·79 | 3 | Je | 27 | ·32 | 4 | Jl | 26 | ·86 | 6 | Au | 25 | ·39 |
| 4945 | 1901 | 1766 | 11A | ·1300 | 6·5010 | 11·871 | **1844** | 2 | 4 | Ap | 17 | ·63 | 6 | My | 17○ | ·16 | 7 | Je | 15 | ·69 | 2 | Jl | 15 | ·22 | 5 | S | 12 | ·28 |
| | | | | | | | | | | | | | | | | | | | | | 3 | Au | 13 | ·75 | | | | |
| 4946 | 1902 | 1767 | 11A | ·3888 | 25·1399 | 10·004 | **1845** | 4 | 3 | My | 6●○ | ·53 | 5 | Je | 5 | ·06 | 6 | Jl | 4 | ·59 | 1 | Au | 3 | ·12 | 2 | S | 1 | ·63 |
| 4947 | 1903 | 1768 | 11A | ·6476 | 14·2482 | 6·162 | **1846** | 5 | 7 | Ap | 25 | ·90 | 2 | My | 25 | ·43 | 3 | Je | 23 | ·96 | 5 | Jl | 23 | ·49 | 7 | Au | 22 | ·02 |
| 4948 | 1904 | 1769 | 11A | ·9063 | 3·3565 | 2·319 | **1847** | 6 | 5 | Ap | 15● | ·26 | 6 | My | 14 | ·79 | 2 | Jl | 12 | ·85 | 4 | Au | 11 | ·38 | 5 | S | 9○ | ·92 |
| | | | | | | | | | | | | | 1 | Je | 13 | ·32 | | | | | | | | | | | | |
| 4949 | 1905 | 1770 | 11A | ·1651 | 21·9954 | 0·452 | **1848** | 7 | 4 | My | 3 | ·16 | 5 | Je | 1 | ·69 | 7 | Jl | 1 | ·22 | 1 | Jl | 30 | ·75 | 3 | Au | 29○ | ·28 |
| 4950 | 1906 | 1771 | 11A | ·4238 | 11·1037 | 24·164 | **1849** | 2 | 1 | Ap | 22 | ·53 | 3 | My | 22 | ·06 | 4 | Je | 20 | ·59 | 6 | Jl | 20 | ·12 | 7 | Au | 18●○ | ·65 |
| 4951 | 1907 | 1772 | 11A | ·6826 | 0·2120 | 20·321 | **1850** | 3 | 5 | Ap | 11 | ·89 | 1 | Je | 9 | ·96 | 3 | Jl | 9 | ·49 | 5 | Au | 8 | ·02 | 6 | S | 6 | ·55 |
| | | | | | | | | | 7 | My | 11 | ·42 | | | | | | | | | | | | | | | | |
| 4952 | 1908 | 1773 | 11A | ·9413 | 18·8509 | 18·454 | **1851** | 4 | 4 | Ap | 30 | ·79 | 6 | My | 30 | ·32 | 7 | Je | 28○ | ·85 | 2 | Jl | 28 | ·38 | 3 | Au | 26 | ·91 |
| 4953 | 1909 | 1774 | 11A | ·2001 | 7·9592 | 14·612 | **1852** | 5 | 2 | Ap | 19 | ·16 | 3 | My | 18 | ·69 | 5 | Je | 17○ | ·22 | 6 | Jl | 16 | ·75 | 1 | Au | 15 | ·28 |
| | | | | | | | | | | | | | | | | | | | | | | | | | 2 | S | 13 | ·81 |
| 4954 | 1910 | 1775 | 11A | ·4589 | 26·5981 | 12·745 | **1853** | 7 | 1 | My | 8 | ·06 | 2 | Je | 6○ | ·59 | 4 | Jl | 6 | ·12 | 5 | Au | 4 | ·65 | 7 | S | 3 | ·18 |
| 4955 | 1911 | 1776 | 11A | ·7176 | 15·7064 | 8·902 | **1854** | 1 | 5 | Ap | 27○ | ·42 | 6 | My | 26 | ·95 | 1 | Je | 25 | ·48 | 3 | Jl | 25 | ·01 | 4 | Au | 23 | ·55 |
| 4956 | 1912 | 1777 | 11A | ·9764 | 4·8147 | 5·060 | **1855** | 2 | 2 | Ap | 16○ | ·79 | 4 | My | 16● | ·32 | 5 | Je | 14 | ·85 | 1 | Au | 12 | ·91 | 3 | S | 11 | ·44 |
| | | | | | | | | | | | | | | | | | 7 | Jl | 14 | ·38 | | | | | | | | |
| 4957 | 1913 | 1778 | 11A | ·2351 | 23·4535 | 3·193 | **1856** | 3 | 1 | My | 4 | ·69 | 3 | Je | 3 | ·22 | 4 | Jl | 2 | ·75 | 6 | Au | 1 | ·28 | 7 | Au | 30 | ·81 |
| 4958 | 1914 | 1779 | 11A | ·4939 | 12·5618 | 26·905 | **1857** | 5 | 6 | Ap | 24 | ·06 | 7 | My | 23 | ·59 | 2 | Je | 22 | ·12 | 3 | Jl | 21 | ·65 | 5 | Au | 20 | ·18 |
| 4959 | 1915 | 1780 | 11A | ·7526 | 1·6701 | 23·062 | **1858** | 6 | 3 | Ap | 13 | ·42 | 4 | My | 12 | ·95 | 1 | Jl | 11 | ·01 | 2 | Au | 9○ | ·54 | 4 | S | 8 | ·08 |
| | | | | | | | | | | | | | 6 | Je | 11 | ·48 | | | | | | | | | | | | |

**Surya Siddhanta.**

| Asvina | | | | Kartika | | | | Margasira | | | | A.D. | Pausha | | | | A.D. | Magha | | | | Phalguna | | | | Chaitra | | | |
|---|---|---|---|---|---|---|---|---|---|---|---|---|---|---|---|---|---|---|---|---|---|---|---|---|---|---|---|---|---|
| + 147·65293 | | | | + 177·18353 | | | | + 206·71411 | | | | | + 236·24470 | | | | | + 265·77529 | | | | + 295·30588 | | | | {+324·83647<br>{+354·36705 | | | |
| + 9·880 | | | | + 11·856 | | | | + 13·832 | | | | | + 15·808 | | | | | + 17·784 | | | | + 19·760 | | | | {+21·736<br>{+23·712 | | | |
| Week-day | Month | Day | Fraction | Week-day | Month | Day | Fraction | Week-day | Month | Day | Fraction | | Week-day | Month | Day | Fraction | | Week-day | Month | Day | Fraction | Week-day | Month | Day | Fraction | Week-day | Month | Day | Fraction |
| 1 | S | 19●○ | ·33 | 2 | O | 18 | ·86 | 4 | N | 17 | ·39 | | 5 | D | 16 | ·92 | 20 | 7 | Jr | 15 | ·45 | 1 | F | 13 | ·98 | 3 | Mr | 14○ | ·52 |
| 7 | O | 7 | ·23 | 1 | N | 5 | ·76 | 3 | D | 5 | ·29 | 21 | 4 | Jr | 3 | ·82 | | 6 | F | 2 | ·35 | 7 | Mr | 3● | ·88 | 2 | Ap | 2 | ·41 |
| 4 | S | 26 | ·60 | 6 | O | 26 | ·13 | 7 | N | 24 | ·66 | | 2 | D | 24 | ·19 | 22 | 3 | Jr | 22○ | ·72 | 5 | F | 21 | ·25 | 6 | Mr | 22 | ·78 |
| {1 | S | 15 | .96 | 5 | N | 14 | ·02 | 6 | D | 13 | ·56 | | Pausha | | | | 23 | 1 | Jr | 12○ | ·09 | 2 | F | 10● | ·62 | {4 | Mr | 12 | ·15 |
| {3 | O | 15 | ·49 | | | | | | | | | | Kshaya | | | | | | | | | | | | | {5 | Ap | 10 | ·68 |
| 7 | O | 4 | ·86 | 2 | N | 3 | ·39 | 3 | D | 2 | ·92 | 24 | 5 | Jr | 1○ | ·45 | | 6 | Jr | 30 | ·98 | 1 | F | 29 | ·51 | 3 | Mr | 30 | ·04 |
| 5 | S | 23 | ·23 | 6 | O | 22 | ·76 | 1 | N | 21 | ·29 | | 2 | D | 20● | ·83 | 25 | 4 | Jr | 19 | ·35 | 5 | F | 17 | ·88 | 7 | Mr | 19 | ·41 |
| 4 | O | 12 | ·13 | 5 | N | 10○ | ·66 | 7 | D | 10 | ·19 | 26 | 1 | Jr | 8 | ·72 | | 3 | F | 7 | ·25 | 4 | Mr | 8 | ·78 | 6 | Ap | 7 | ·31 |
| 1 | O | 1 | ·49 | 3 | O | 31○ | ·02 | 4 | N | 29 | ·55 | | 6 | D | 29 | ·08 | 27 | 7 | Jr | 27 | ·62 | 2 | F | 26 | ·15 | 3 | Mr | 27 | ·68 |
| 5 | S | 20 | ·86 | 7 | O | 20○ | ·39 | 1 | N | 18 | ·92 | | 3 | D | 18 | ·45 | 28 | 4 | Jr | 16 | ·98 | 6 | F | 15 | ·51 | 1 | Mr | 16 | ·04 |
| 4 | O | 8● | ·76 | 6 | N | 7 | ·29 | 7 | D | 6 | ·82 | 29 | 2 | Jr | 5 | ·35 | | 3 | F | 3 | ·88 | 5 | Mr | 5○ | ·41 | 6 | Ap | 3 | ·94 |
| 2 | S | 28● | ·12 | 3 | O | 27 | ·66 | 5 | N | 26 | ·19 | | 6 | D | 25 | ·72 | 30 | 1 | Jr | 24 | ·25 | 2 | F | 22●○ | ·78 | 4 | Mr | 24 | ·31 |
| 6 | S | 17 | ·49 | 1 | O | 17 | ·02 | 2 | N | 15 | ·55 | | 4 | D | 15 | ·08 | 31 | 5 | Jr | 13 | ·61 | 7 | F | 12○ | ·14 | 1 | Mr | 13 | ·68 |
| 5 | O | 6 | ·39 | 6 | N | 4 | ·92 | 1 | D | 4 | ·45 | 32 | 2 | Jr | 2 | ·98 | | 4 | F | 1 | ·51 | 6 | Mr | 2 | ·04 | 7 | Mr | 31 | ·57 |
| 2 | S | 24 | ·76 | 4 | O | 24 | ·29 | 5 | N | 22 | ·82 | | 7 | D | 22○ | ·35 | 33 | 1 | Jr | 20 | ·88 | 3 | F | 19 | ·41 | 4 | Mr | 20 | ·94 |
| 1 | O | 13 | ·65 | 3 | N | 12 | ·18 | 4 | D | 11○ | ·72 | 34 | 6 | Jr | 10 | .25 | | 7 | F | 8 | ·78 | 2 | Mr | 10 | ·31 | 3 | Ap | 8 | ·84 |
| 6 | O | 3 | ·02 | 7 | N | 1 | ·55 | 2 | D | 1○ | ·08 | | 3 | D | 30 | ·61 | 35 | 5 | Jr | 29 | ·14 | 6 | F | 27 | ·67 | 1 | Mr | 29 | ·20 |
| 3 | S | 22 | ·39 | 4 | O | 21 | ·92 | 6 | N | 20● | ·45 | | 7 | D | 19 | ·98 | 36 | 2 | Jr | 18 | ·51 | 4 | F | 17 | ·04 | 5 | Mr | 17 | ·57 |
| 2 | O | 10○ | ·29 | 3 | N | 8● | ·82 | 5 | D | 8 | ·35 | 37 | 6 | Jr | 6 | ·88 | | 1 | F | 5 | ·41 | 2 | Mr | 6 | ·94 | 4 | Ap | 5○ | ·47 |
| 6 | S | 29○ | ·65 | 1 | O | 29 | ·18 | 2 | N | 27 | ·71 | | 4 | D | 27 | ·24 | 38 | 5 | Jr | 25 | ·78 | 7 | F | 24 | ·31 | 1 | Mr | 25○ | ·84 |
| 4 | S | 19○ | ·02 | 5 | O | 18 | ·55 | 7 | N | 17 | 08 | | 1 | D | 16 | ·61 | 39 | 3 | Jr | 15 | ·14 | 4 | F | 13 | ·67 | 6 | Mr | 15 | ·20 |
| 2 | O | 7 | ·92 | 4 | N | 6 | ·45 | 5 | D | 5 | ·98 | 40 | 7 | Jr | 4 | ·51 | | 2 | F | 3○ | ·04 | 3 | Mr | 3● | ·57 | 5 | Ap | 2 | ·10 |
| 7 | S | 26 | ·28 | 1 | O | 25 | ·82 | 3 | N | 24 | ·35 | | 4 | D | 23 | ·88 | 41 | 6 | Jr | 22○ | ·41 | 7 | F | 20 | ·94 | 2 | Mr | 22 | ·47 |
| {4 | S | 15 | ·65 | 7 | N | 13 | ·71 | 2 | D | 13 | ·24 | 42 | 3 | Jr | 11○ | ·77 | | 5 | F | 10 | ·30 | 6 | Mr | 11 | [illegible] | 1 | Ap | 10 | ·37 |
| {6 | O | 15 | ·18 | | | | | | | | | | | | | | | | | | | | | | | | | | |
| 3 | O | 4 | ·55 | 5 | N | 3 | ·08 | 6 | D | 2 | ·61 | 43 | 1 | Jr | 1 | ·14 | | 2 | Jr | 30 | ·67 | 4 | Mr | 1 | ·20 | 5 | Mr | 30 | ·73 |
| 7 | S | 23 | ·92 | 2 | O | 23 | ·45 | 3 | N | 21○ | ·98 | | 5 | D | 21● | ·51 | 44 | 7 | Jr | 20 | ·04 | 1 | F | 18 | ·57 | 3 | Mr | 19 | ·10 |
| 6 | O | 11 | ·81 | 1 | N | 10○ | ·34 | 2 | D | 9 | ·88 | 45 | 4 | Jr | 8 | ·41 | | 5 | F | 6 | ·94 | 7 | Mr | 8 | ·47 | 2 | Ap | 7 | ·00 |
| 4 | O | 1 | ·18 | 5 | O | 30○ | ·71 | 7 | N | 29 | ·24 | | 1 | D | 28 | ·77 | 46 | 3 | Jr | 27 | ·30 | 4 | F | 25 | ·83 | 6 | Mr | 27 | ·36 |
| 1 | S | 20 | ·55 | 3 | O | 20● | ·08 | 4 | N | 18 | ·61 | | 6 | D | 18 | ·14 | 47 | 7 | Jr | 16 | ·67 | 2 | F | 15 | ·20 | 3 | Mr | 16○ | ·73 |
| 7 | O | 9● | ·45 | 1 | N | 7 | ·98 | 3 | D | 7 | ·51 | 48 | 5 | Jr | 6 | ·04 | | 6 | F | 4 | ·57 | 1 | Mr | 5○ | ·10 | 2 | Ap | 3 | ·63 |
| 4 | S | 27● | ·81 | 6 | O | 27 | ·34 | 7 | N | 25 | ·87 | | 2 | D | 25 | ·40 | 49 | 3 | Jr | 23 | ·94 | 5 | F | 22●○ | ·47 | 7 | Mr | 24 | ·00 |
| 2 | S | 17 | ·18 | 3 | O | 16 | ·71 | 5 | N | 15 | ·24 | | 6 | D | 14 | ·77 | 50 | 1 | Jr | 13 | ·30 | 2 | F | 11● | ·83 | 4 | Mr | 13 | ·36 |
| 1 | O | 6 | ·08 | 2 | N | 4 | ·61 | 4 | D | 4 | ·14 | 51 | 5 | Jr | 2○ | ·67 | | 7 | F | 1 | ·20 | 1 | Mr | 2 | ·73 | 3 | Ap | 1 | ·26 |
| 5 | S | 25 | ·44 | 6 | O | 24 | ·98 | 1 | N | 23 | ·51 | | 3 | D | 23○ | ·04 | 52 | 4 | Jr | 21 | ·57 | 6 | F | 20 | ·10 | 7 | Mr | 20 | ·63 |
| 4 | O | 13 | ·34 | 5 | N | 11 | ·87 | 7 | D | 11●○ | ·40 | 53 | 1 | Jr | 9 | ·93 | | 3 | F | 8 | ·46 | 5 | Mr | 10 | ·00 | 6 | Ap | 8 | ·53 |
| 1 | O | 2 | ·71 | 3 | N | 1 | ·24 | 4 | N | 30 | ·77 | | 6 | D | 30 | ·30 | 54 | 7 | Jr | 28 | ·83 | 2 | F | 27 | ·36 | 3 | Mr | 28 | ·89 |
| 6 | S | 22 | ·08 | 7 | O | 21○ | ·61 | 2 | N | 20 | ·14 | | 3 | D | 19 | ·67 | 55 | 5 | Jr | 18 | ·20 | 6 | F | 16 | ·73 | 1 | Mr | 18 | ·26 |
| 4 | O | 10○ | ·97 | 6 | N | 9 | ·50 | 1 | D | 9 | ·04 | 56 | 2 | Jr | 7 | ·57 | | 4 | F | 6 | ·10 | 5 | Mr | 6 | ·63 | 7 | Ap | 5●○ | ·16 |
| 2 | S | 29●○ | ·34 | 3 | O | 28 | ·87 | 5 | N | 27 | ·40 | | 6 | D | 26 | ·93 | 57 | 1 | Jr | 25 | ·46 | 2 | F | 23 | ·99 | 4 | Mr | 25 | ·52 |
| 6 | S | 18● | ·71 | 1 | O | 18 | ·24 | 2 | N | 16 | ·77 | | 4 | D | 16 | ·30 | 58 | 5 | Jr | 14 | ·83 | 7 | F | 13○ | ·36 | 1 | Mr | 14● | 89 |
| 5 | O | 7 | ·61 | 7 | N | 6 | ·14 | 1 | D | 5 | ·67 | 59 | 3 | Jr | 4 | ·20 | | 4 | F | 2○ | ·73 | 6 | Mr | 4 | ·26 | 7 | Ap | 2 | 79 |

**TABLE X—**

| Kaliyuga. | Vikrama Era. | Saka Era. | Commencement of Solar Year. Month and day A. D. | Fraction of day. | First New-Moon in Solar Year. | Anomaly of first New-Moon. | A.D. | Week-day of 1st January. | Vaisakha ⊙'s Anom col. 6 / ☾'s Anom col. 7 | | | | Jyeshta + 29·53059 / + 1·976 | | | | Ashada + 59·06117 / + 3·952 | | | | Sravana + 88·59176 / + 5·928 | | | | Bhadrapada + 118·12235 / + 7·904 | | | |
|---|---|---|---|---|---|---|---|---|---|---|---|---|---|---|---|---|---|---|---|---|---|---|---|---|---|---|---|---|
| | | | | | | | | | Week-day | Month | Day | Fraction | Week-day | Month | Day | Fraction | Week-day | Month | Day | Fraction | Week-day | Month | Day | Fraction | Week-day | Month | Day | Fraction |
| 4960 | 1916 | 1781 | 12A | ·0114 | 20·3090 | 21·195 | **1859** | 7 | 2 | My | 2 | ·32 | 3 | My | 31 | ·85 | 5 | Je | 30 | ·38 | 6 | Jl | 29○ | ·91 | 1 | Au | 28 | ·44 |
| 4961 | 1917 | 1782 | 11A | ·2702 | 9·4173 | 17·352 | **1860** | 1 | 6 | Ap | 20 | ·69 | 1 | My | 20 | ·22 | 2 | Je | 18 | ·75 | 4 | Jl | 18○ | ·28 | 5 | Au | 16 | ·81 |
| 4962 | 1918 | 1783 | 11A | ·5289 | 28·0562 | 15·487 | **1861** | 3 | 5 | My | 9 | ·58 | 7 | Je | 8 | ·12 | 1 | Jl | 7● | ·65 | 3 | Au | 6 | ·18 | 4 | S | 4 | ·71 |
| 4963 | 1919 | 1784 | 11A | ·7877 | 17·1645 | 11·643 | **1862** | 4 | 2 | Ap | 28 | ·95 | 4 | My | 28○ | ·48 | 6 | Je | 27 | ·01 | 7 | Jl | 26 | ·54 | 2 | Au | 25 | ·07 |
| 4964 | 1920 | 1785 | 12A | ·0464 | 6·2728 | 7·800 | **1863** | 5 | 7 | Ap | 18 | ·32 | 1 | My | 17○ | ·85 | 3 | Je | 16 | ·38 | { 4 | Jl | 15 | ·91 | 7 | S | 12 | ·97 |
| | | | | | | | | | | | | | | | | | | | | | { 6 | Au | 14 | ·44 | | | | |
| 4965 | 1921 | 1786 | 11A | ·3052 | 24·9117 | 5·933 | **1864** | 6 | 6 | My | 6● | ·22 | 7 | Je | 4 | ·75 | 2 | Jl | 4 | ·28 | 3 | Au | 2 | ·81 | 5 | S | 1 | ·34 |
| 4966 | 1922 | 1787 | 11A | ·5639 | 14·0200 | 2·091 | **1865** | 1 | 3 | Ap | 25 | ·58 | 5 | My | 25 | ·11 | 6 | Je | 23 | ·64 | 1 | Jl | 23 | ·18 | 2 | Au | 21 | ·71 |
| 4967 | 1923 | 1788 | 11A | ·8227 | 3·1283 | 25·803 | **1866** | 2 | 7 | Ap | 14 | ·95 | { 2 | My | 14 | ·48 | 5 | Jl | 12 | ·54 | 7 | Au | 11 | ·07 | 1 | S | 9○ | ·60 |
| | | | | | | | | | | | | | { 4 | Je | 13 | ·01 | | | | | | | | | | | | |
| 4968 | 1924 | 1789 | 12A | ·0814 | 21·7672 | 23·936 | **1867** | 3 | 6 | My | 3 | ·85 | 1 | Je | 2 | ·38 | 2 | Jl | 1 | ·91 | 4 | Jl | 31 | ·44 | 5 | Au | 29○ | ·97 |
| 4969 | 1925 | 1790 | 11A | ·3402 | 10·8755 | 20·098 | **1868** | 4 | 4 | Ap | 22 | ·22 | 5 | My | 21 | ·75 | 7 | Je | 20 | ·28 | 1 | Jl | 19 | ·81 | 3 | Au | 18● | ·34 |
| 4970 | 1926 | 1791 | 11A | ·5990 | 29·5144 | 16·250 | **1869** | 6 | { 1 | Ap | 11 | ·58 | 4 | Je | 9 | ·64 | 6 | Jl | 9○ | ·17 | 7 | Au | 7 | ·70 | 2 | S | 6 | ·24 |
| | | | | | | | | | { 3 | My | 11 | ·11 | | | | | | | | | | | | | | | | |
| 4971 | 1927 | 1792 | 11A | ·8577 | 18·6227 | 14·384 | **1870** | 7 | 7 | Ap | 30 | ·48 | 2 | My | 30 | ·01 | 3 | Je | 28○ | ·54 | 5 | Jl | 28 | ·07 | 6 | Au | 26 | ·60 |
| 4972 | 1928 | 1793 | 12A | ·1165 | 7·7310 | 10·541 | **1871** | 1 | 4 | Ap | 19 | ·85 | 6 | My | 19 | ·38 | 7 | Je | 17●○ | ·91 | 2 | Jl | 17 | ·44 | { 3 | Au | 15 | ·97 |
| | | | | | | | | | | | | | | | | | | | | | | | | | { 5 | S | 14 | .50 |
| 4973 | 1929 | 1794 | 11A | ·3752 | 26·3698 | 8·674 | **1872** | 2 | 3 | My | 7○ | ·74 | 5 | Je | 6● | ·28 | 6 | Jl | 5 | ·81 | 1 | Au | 4 | ·34 | 2 | S | 2 | .87 |
| 4974 | 1930 | 1795 | 11A | ·6340 | 15·4781 | 4·831 | **1873** | 4 | 1 | Ap | 27○ | ·11 | 2 | My | 26 | ·64 | 4 | Je | 25 | ·17 | 5 | Jl | 24 | ·70 | 7 | Au | 23 | ·23 |
| 4975 | 1931 | 1796 | 11A | ·8927 | 4·5864 | 0·989 | **1874** | 5 | 5 | Ap | 16○ | ·48 | 7 | My | 16 | ·01 | { 1 | Je | 14 | ·54 | 4 | Au | 12 | ·60 | 6 | S | 11 | ·13 |
| | | | | | | | | | | | | | | | | | { 3 | Jl | 14 | ·07 | | | | | | | | |
| 4976 | 1932 | 1797 | 12A | ·1515 | 23·2254 | 26·676 | **1875** | 6 | 4 | My | 5 | ·38 | 5 | Je | 3 | ·91 | 7 | Jl | 3 | ·44 | 1 | Au | 1 | ·97 | 3 | Au | 31 | ·50 |
| 4977 | 1933 | 1798 | 11A | ·4103 | 12·3336 | 22·834 | **1876** | 7 | 1 | Ap | 23 | ·74 | 3 | My | 23 | ·27 | 4 | Je | 21 | ·80 | 6 | Jl | 21 | ·34 | 7 | Au | 19○ | ·87 |
| 4978 | 1934 | 1799 | 11A | ·6690 | 1·4419 | 18·991 | **1877** | 2 | 6 | Ap | 13 | ·11 | { 7 | My | 12 | ·64 | 3 | Jl | 10 | ·70 | 5 | Au | 9○ | ·23 | 6 | S | 7 | ·76 |
| | | | | | | | | | | | | | { 2 | Je | 11 | ·17 | | | | | | | | | | | | |
| 4979 | 1935 | 1800 | 11A | ·9278 | 20·0808 | 17·124 | **1878** | 3 | 5 | My | 2 | ·01 | 6 | My | 31 | ·54 | 1 | Je | 30 | ·07 | 2 | Jl | 29○ | ·60 | 4 | Au | 28 | ·13 |
| 4980 | 1936 | 1801 | 12A | ·1865 | 9·1891 | 13·281 | **1879** | 4 | 2 | Ap | 21 | ·38 | 3 | My | 20 | ·91 | 5 | Je | 19 | ·44 | 6 | Jl | 18● | ·97 | 1 | Au | 17 | ·50 |
| 4981 | 1937 | 1802 | 11A | ·4453 | 27·8280 | 11·415 | **1880** | 5 | 1 | My | 9 | ·27 | 2 | Je | 7○ | ·80 | 4 | Jl | 7 | ·33 | 5 | Au | 5 | ·86 | 7 | S | 4 | ·40 |
| 4982 | 1938 | 1803 | 11A | ·7040 | 16·9363 | 7·572 | **1881** | 7 | 5 | Ap | 28 | ·64 | 7 | My | 28●○ | ·17 | 1 | Je | 26 | ·70 | 3 | Jl | 26 | ·23 | 4 | Au | 24 | ·76 |
| 4983 | 1939 | 1804 | 11A | ·9628 | 6·0446 | 3·729 | **1882** | 1 | 3 | Ap | 18 | ·01 | 4 | My | 17● | ·54 | 6 | Je | 16 | ·07 | { 7 | Jl | 15 | ·60 | 3 | S | 12 | ·66 |
| | | | | | | | | | | | | | | | | | | | | | { 2 | Au | 14 | ·13 | | | | |
| 4984 | 1940 | 1805 | 12A | ·2216 | 24·6835 | 1·862 | **1883** | 2 | 1 | My | 6 | ·90 | 3 | Je | 5 | ·44 | 4 | Jl | 4 | ·97 | 6 | Au | 3 | ·50 | 1 | S | 2 | ·03 |
| 4985 | 1941 | 1806 | 11A | ·4803 | 13·7918 | 25·574 | **1884** | 3 | 6 | Ap | 25 | ·27 | 7 | My | 24 | ·80 | 2 | Je | 23 | ·33 | 3 | Jl | 22 | ·86 | 5 | Au | 21 | ·39 |
| 4986 | 1942 | 1807 | 11A | ·7391 | 2·9008 | 21·732 | **1885** | 5 | 3 | Ap | 14 | ·64 | { 5 | My | 14 | ·17 | 1 | Jl | 12 | ·23 | 2 | Au | 10 | ·76 | 4 | S | 9○ | ·29 |
| | | | | | | | | | | | | | { 6 | Je | 12 | ·70 | | | | | | | | | | | | |
| 4987 | 1943 | 1808 | 11A | ·9978 | 21·5390 | 19·865 | **1886** | 6 | 2 | My | 3 | ·54 | 4 | Je | 2 | ·07 | 5 | Jl | 1 | ·60 | 7 | Jl | 31 | ·13 | 1 | Au | 29 | ·66 |
| 4988 | 1944 | 1809 | 12A | ·2566 | 10·6473 | 16·022 | **1887** | 7 | 6 | Ap | 22 | ·90 | 1 | My | 22 | ·43 | 2 | Je | 20 | ·96 | 4 | Jl | 20○ | ·50 | 6 | Au | 19● | ·03 |
| 4989 | 1945 | 1810 | 11A | ·5153 | 29·2862 | 14·155 | **1888** | 1 | 5 | My | 10 | ·80 | 7 | Je | 9 | ·33 | 1 | Jl | 8○ | ·86 | 3 | Au | 7 | ·39 | 4 | S | 5 | ·92 |
| 4990 | 1946 | 1811 | 11A | ·7741 | 18·3945 | 10·313 | **1889** | 3 | 3 | Ap | 30 | ·17 | 4 | My | 29 | ·70 | 6 | Je | 28●○ | ·23 | 7 | Jl | 27 | ·76 | 2 | Au | 26 | ·29 |
| 4991 | 1947 | 1812 | 12A | ·0328 | 7·5028 | 6·470 | **1890** | 4 | 7 | Ap | 19 | ·54 | 2 | My | 19○ | ·07 | 3 | Je | 17● | ·60 | 5 | Jl | 17 | ·13 | { 6 | Au | 15 | ·66 |
| | | | | | | | | | | | | | | | | | | | | | | | | | { 1 | S | 14 | ·19 |
| 4992 | 1948 | 1813 | 12A | ·2916 | 26·1416 | 4·603 | **1891** | 5 | 6 | My | 8○ | ·43 | 7 | Je | 6 | ·96 | 2 | Jl | 6 | ·49 | 4 | Au | 5 | ·02 | 5 | S | 3 | ·55 |
| 4993 | 1949 | 1814 | 11A | ·5504 | 15·2499 | 0·760 | **1892** | 6 | 3 | Ap | 26○ | ·80 | 5 | My | 26 | ·33 | 6 | Je | 24 | ·86 | 1 | Jl | 24 | ·39 | 2 | Au | 22 | ·92 |
| 4994 | 1950 | 1815 | 11A | ·8091 | 4·3582 | 24·472 | **1893** | 1 | 1 | Ap | 16 | ·17 | 2 | My | 15 | ·70 | { 4 | Je | 14 | ·23 | 7 | Au | 12 | ·29 | 1 | S | 10 | ·82 |
| | | | | | | | | | | | | | | | | | { 5 | Jl | 13 | ·76 | | | | | | | | |
| 4995 | 1951 | 1816 | 12A | ·0679 | 22·9971 | 22·605 | **1894** | 2 | 7 | My | 5 | ·06 | 1 | Je | 3 | ·59 | 3 | Jl | 3 | ·13 | 4 | Au | 1 | ·66 | 6 | Au | 31○ | ·19 |
| 4996 | 1952 | 1817 | 12A | ·3266 | 12·1054 | 18·763 | **1895** | 3 | 4 | Ap | 24 | ·43 | 5 | My | 23 | ·96 | 7 | Je | 22 | ·49 | 2 | Jl | 22 | ·02 | 3 | Au | 20●○ | ·55 |
| 4997 | 1953 | 1818 | 11A | ·5854 | 1·2137 | 14·920 | **1896** | 4 | 1 | Ap | 12 | ·80 | { 3 | My | 12 | ·33 | 6 | Jl | 10 | ·39 | 7 | Au | 8●○ | ·92 | 2 | S | 7 | ·45 |
| | | | | | | | | | | | | | { 4 | Je | 10 | ·86 | | | | | | | | | | | | |
| 4998 | 1954 | 1819 | 11A | ·8441 | 19·8526 | 13·053 | **1897** | 6 | 7 | My | 1 | ·70 | 2 | My | 31 | ·23 | 3 | Je | 29 | ·76 | 5 | Jl | 29 | ·29 | 6 | Au | 27 | ·82 |

**Surya Siddhanta.**

| Asvina | | | | Kartika | | | | Margasira | | | | A.D. | Pausha | | | | A.D. | Magha | | | | Phalguna | | | | Chaitra | | | |
|---|---|---|---|---|---|---|---|---|---|---|---|---|---|---|---|---|---|---|---|---|---|---|---|---|---|---|---|---|---|
| + 147·65293 | | | | + 177·18353 | | | | + 206·71411 | | | | | + 236·24470 | | | | | + 265·77529 | | | | + 295·30588 | | | | +324·83647<br>+354·36705 | | | |
| + 9·880 | | | | + 11·856 | | | | + 13·832 | | | | | + 15·808 | | | | | + 17·784 | | | | + 19·760 | | | | + 21·736<br>+ 23·712 | | | |
| Week-day | Month | Day | Fraction | Week-day | Month | Day | Fraction | Week-day | Month | Day | Fraction | | Week-day | Month | Day | Fraction | | Week-day | Month | Day | Fraction | Week-day | Month | Day | Fraction | Week-day | Month | Day | Fraction |
| 2 | S | 26 | ·97 | 4 | O | 26 | ·50 | 6 | N | 25 | ·03 | | 7 | D | 24 | ·56 | **60** | 2 | Jr | 23○ | ·10 | 3 | F | 21 | ·63 | 5 | Mr | 22 | ·16 |
| {7 | S | 15 | ·34 | 3 | N | 13 | ·40 | 4 | D | 12 | ·93 | **61** | 6 | Jr | 11● | ·46 | | 7 | F | 9 | ·99 | 2 | Mr | 11 | ·52 | 4 | Ap | 10 | ·05 |
| {1 | O | 14 | ·87 | | | | | | | | | | | | | | | | | | | | | | | | | | |
| 6 | O | 4 | ·24 | 7 | N | 2 | ·77 | 2 | D | 2○ | ·30 | | 3 | D | 31 | ·83 | **62** | 5 | Jr | 30 | ·36 | 6 | F | 28 | ·89 | 1 | Mr | 30 | ·42 |
| 3 | S | 23 | ·60 | 5 | O | 23 | ·14 | 6 | N | 21○ | ·67 | | 1 | D | 21● | ·20 | **63** | 2 | Jr | 19 | ·73 | 4 | F | 18 | ·26 | 5 | Mr | 19 | 79 |
| 2 | O | 12 | ·50 | 4 | N | 11○ | ·03 | 5 | D | 10 | ·56 | **64** | 7 | Jr | 9 | ·09 | | 1 | F | 7 | ·62 | 3 | Mr | 8 | ·16 | 4 | Ap | 6 | ·69 |
| 6 | S | 30 | ·87 | 1 | O | 30 | ·40 | 2 | N | 28 | ·93 | | 4 | D | 28 | ·46 | **65** | 5 | Jr | 26 | ·99 | 7 | F | 25 | ·52 | 2 | Mr | 27○ | ·05 |
| 4 | S | 20○ | ·24 | 5 | O | 19 | ·77 | 7 | N | 18 | ·30 | | 1 | D | 17 | ·83 | **66** | 3 | Jr | 16 | ·36 | 4 | F | 14 | ·89 | 6 | Mr | 16○ | ·42 |
| 3 | O | 9 | ·13 | 4 | N | 7 | ·66 | 6 | D | 7 | ·20 | **67** | 7 | Jr | 5 | ·73 | | 2 | F | 4 | ·26 | 3 | Mr | 5●○ | ·79 | 5 | Ap | 4 | ·32 |
| 7 | S | 28 | ·50 | 2 | O | 28 | ·03 | 3 | N | 26 | ·56 | | 5 | D | 26 | ·09 | **68** | 6 | Jr | 24 | ·62 | 1 | F | 23 | ·15 | 2 | Mr | 23 | ·69 |
| 4 | S | 16 | ·87 | 6 | O | 16 | ·40 | 7 | N | 14 | ·93 | | 2 | D | 14 | ·46 | **69** | 3 | Jr | 12○ | ·99 | 5 | F | 11 | ·52 | 7 | Mr | 13 | ·05 |
| 3 | O | 5 | ·77 | 5 | N | 4 | ·30 | 6 | D | 3 | ·83 | **70** | 1 | Jr | 2○ | ·36 | | 2 | Jr | 31 | ·89 | 4 | Mr. | 2 | ·42 | 5 | Mr | 31 | ·95 |
| 1 | S | 25 | ·13 | 2 | O | 24 | ·66 | 4 | N | 23 | ·19 | | 5 | D | 22○ | ·72 | **71** | 7 | Jr | 21 | ·26 | 1 | F | 19 | ·79 | 3 | Mr | 21 | ·32 |
| 7 | O | 14 | ·03 | 1 | N | 12 | ·56 | 3 | D | 12● | ·09 | **72** | 4 | Jr | 10 | ·62 | | 6 | F | 9 | ·15 | 7 | Mr | 9 | ·68 | 2 | Ap | 8 | ·21 |
| 4 | O | 2 | ·40 | 5 | O | 31○ | ·93 | 7 | N | 30 | ·46 | | 1 | D | 29 | ·99 | **73** | 3 | Jr | 28 | ·52 | 5 | F | 27 | ·05 | 6 | Mr | 28 | ·58 |
| 1 | S | 21 | ·76 | 3 | O | 21○ | ·30 | 4 | N | 19 | ·83 | | 6 | D | 19 | ·36 | **74** | 7 | Jr | 17 | ·89 | 2 | F | 16 | ·42 | 3 | Mr | 17 | ·95 |
| 7 | O | 10●○ | ·66 | 2 | N | 9 | ·19 | 3 | D | 8 | ·72 | **75** | 5 | Jr | 7 | ·25 | | 6 | F | 5 | ·78 | 1 | Mr | 7 | ·31 | 2 | Ap | 5● | ·85 |
| 5 | S | 30● | ·03 | 6 | O | 29 | ·56 | 1 | N | 28 | ·09 | | 2 | D | 27 | ·62 | **76** | 4 | Jr | 26 | ·15 | 5 | F | 24○ | ·68 | 7 | Mr | 25 | ·21 |
| 2 | S | 18 | ·40 | 3 | O | 17 | ·93 | 5 | N | 16 | ·46 | | 6 | D | 15 | ·99 | **77** | 1 | Jr | 14 | ·52 | 3 | F | 13○ | ·05 | 4 | Mr | 14● | ·58 |
| 1 | O | 7 | ·29 | 2 | N | 5 | ·82 | 4 | D | 5 | ·36 | **78** | 5 | Jr | 3 | ·89 | | 7 | F | 2○ | ·42 | 1 | Mr | 3 | ·95 | 3 | Ap | 3 | ·48 |
| 5 | S | 26 | ·66 | 7 | O | 26 | ·19 | 1 | N | 24 | ·72 | | 3 | D | 24 | ·25 | **79** | 4 | Jr | 22● | ·78 | 6 | F | 21 | ·31 | 7 | Mr | 22 | ·84 |
| {3 | S | 16 | ·03 | 6 | N | 14 | ·09 | 7 | D. | 13○ | ·62 | **80** | 2 | Jr | 12 | ·15 | | 3 | F | 10 | ·68 | 5 | Mr | 11 | ·21 | 6 | Ap | 9 | ·74 |
| {4 | O | 15 | ·56 | | | | | | | | | | | | | | | | | | | | | | | | | | |
| 1 | O | 3 | ·93 | 3 | N | 2 | ·46 | 4 | D | 1○ | ·99 | | 6 | D | 31 | ·52 | **81** | 1 | Jr | 30 | ·05 | 2 | F | 28 | ·58 | 4 | Mr | 30 | ·11 |
| 6 | S | 23 | ·29 | 7 | O | 22 | ·82 | 2 | N | 21○ | ·35 | | 3 | D | 20 | ·88 | **82** | 5 | Jr | 19 | ·42 | 6 | F | 17 | ·95 | 1 | Mr | 19 | ·48 |
| 5 | O | 12 | ·19 | 6 | N | 10 | ·72 | 1 | D | 10 | ·25 | **83** | 2 | Jr | 8 | ·78 | | 4 | F | 7 | ·31 | 5 | Mr | 8 | ·84 | 7 | Ap | 7○ | ·37 |
| 2 | O | 1○ | ·56 | 4 | O | 31 | ·09 | 5 | N | 29 | ·62 | | 7 | D | 29 | ·15 | **84** | 1 | Jr | 27 | ·68 | 3 | F | 26 | ·21 | 4 | Mr | 26○ | ·74 |
| 6 | S | 19○ | ·92 | 1 | O | 19 | ·46 | 2 | N | 17 | ·99 | | 4 | D | 17 | ·52 | **85** | 6 | Jr | 16 | ·05 | 7 | F | 14 | ·58 | 2 | Mr | 16○ | ·11 |
| 5 | O | 8 | ·82 | 7 | N | 7 | ·35 | 1 | D | 6 | ·88 | **86** | 3 | Jr | 5 | ·41 | | 4 | F | 3 | ·94 | 6 | Mr | 5 | ·48 | 1 | Ap | 4 | ·01 |
| 3 | S | 28 | ·19 | 4 | O | 27 | ·72 | 6 | N | 26 | ·25 | | 7 | D | 25 | ·78 | **87** | 2 | Jr | 24○ | ·31 | 3 | F | 22 | ·84 | 5 | Mr | 24 | ·37 |
| 7 | S | 17 | ·56 | 2 | O | 17 | ·09 | 3 | N | 15 | ·62 | | 5 | D | 15 | ·15 | **88** | 6 | Jr | 13○ | ·68 | 1 | F | 12 | ·21 | {2 | Mr | 12 | ·74 |
| | | | | | | | | | | | | | | | | | | | | | | | | | | {4 | Ap | 11 | ·27 |
| 6 | O | 5 | ·45 | 7 | N | 3 | ·98 | 2 | D | 3 | ·52 | **89** | 4 | Jr | 2○ | ·05 | | 5 | Jr | 31 | ·58 | 7 | Mr | 2 | ·11 | 1 | Mr | 31 | ·64 |
| 3 | S | 24 | ·82 | 5 | O | 24 | ·35 | 6 | N | 22 | ·88 | | 1 | D | 22 | ·41 | **90** | 2 | Jr | 20 | ·94 | 4 | F | 19 | ·47 | 6 | Mr | 21 | ·00 |
| 2 | O | 13 | ·72 | 4 | N | 12○ | ·25 | 5 | D | 11● | ·78 | **91** | 7 | Jr | 10 | ·31 | | 1 | F | 8 | ·84 | 3 | Mr | 10 | ·37 | 4 | Ap | 8 | ·90 |
| 7 | O | 3 | ·09 | 1 | N | 1○ | ·62 | 3 | D | 1 | ·15 | | 4 | D | 30 | ·68 | **92** | 6 | Jr | 29 | ·21 | 7 | F | 27 | ·74 | 2 | Mr | 28 | ·27 |
| 4 | S | 21 | ·45 | 5 | O | 20○ | ·98 | 7 | N | 19 | ·51 | | 2 | D | 19 | ·04 | **93** | 3 | Jr | 17 | ·57 | 5 | F | 16 | ·11 | 6 | Mr | 17 | ·64 |
| 3 | O | 10 | ·35 | 4 | N | 8 | ·88 | 6 | D | 8 | ·41 | **94** | 7 | Jr | 6 | ·94 | | 2 | F | 5 | ·47 | 4 | Mr | 7○ | ·00 | 5 | Ap | 5● | ·53 |
| 7 | S | 29● | ·72 | 2 | O | 29 | ·25 | 3 | N | 27 | ·78 | | 5 | D | 27 | ·31 | **95** | 6 | Jr | 25 | ·84 | 1 | F | 24○ | ·37 | 2 | Mr | 25 | ·90 |
| 5 | S | 19 | ·08 | 6 | O | 18 | ·61 | 1 | N | 17 | ·15 | | 2 | D | 16 | ·68 | **96** | 4 | Jr | 15 | ·21 | 5 | F | 13○ | ·74 | 7 | Mr | 14 | ·27 |
| 3 | O | 6 | ·98 | 5 | N | 5 | ·51 | 7 | D | 5 | ·04 | **97** | 1 | Jr | 3 | ·57 | | 3 | F | 2 | ·10 | 4 | Mr | 3 | ·63 | 6 | Ap | 2 | ·17 |
| 1 | S | 26 | ·35 | 2 | O | 25 | ·88 | 4 | N | 24 | ·41 | | 5 | D | 23○ | ·94 | **98** | 7 | Jr | 22● | ·47 | 2 | F | 21 | ·00 | 3 | Mr | 22 | ·58 |

| Kaliyuga. | Vikrama Era. | Saka Era. | Commencement of Solar Year. Month and day A.D. | Fraction of day. | First New-Moon in Solar Year. | Anomaly of first New-Moon. | A.D. | Week-day of 1st January. | Vaisakha ☉'s Anom col. 6 ☾'s Anom col. 7 | | | | Jyeshta + 29·53059 + 1·976 | | | | Ashada + 59·06117 + 3·952 | | | | Sravana + 88·59176 + 5·928 | | | | Bhadrapada + 118·12235 + 7·904 | | | |
|---|---|---|---|---|---|---|---|---|---|---|---|---|---|---|---|---|---|---|---|---|---|---|---|---|---|---|---|---|
| | | | | | | | | | Week-day | Month | Day | Fraction | Week-day | Month | Day | Fraction | Week-day | Month | Day | Fraction | Week-day | Month | Day | Fraction | Week-day | Month | Day | Fraction |
| 4999 | 1955 | 1820 | 12A | ·1029 | 8·9609 | 9·210 | **1898** | 7 | 5 | Ap | 21 | ·06 | 6 | My | 20 | ·59 | 1 | Je | 19○ | ·12 | 2 | Jl | 18 | ·65 | 4 | Au | 17 | ·19 |
| 5000 | 1956 | 1821 | 12A | ·3617 | 27·5998 | 7·344 | **1899** | 1 | 3 | My | 9 | ·96 | 5 | Je | 8○ | ·49 | 7 | Jl | 8 | ·02 | 1 | Au | 6 | ·55 | 3 | S | 5 | ·08 |
| 5001 | 1957 | 1822 | 12A | ·6204 | 16·7081 | 3·504 | **1900** | 2 | 1 | Ap | 29 | ·33 | 2 | My | 28●○ | ·86 | 4 | Je | 27 | ·39 | 5 | Jl | 26 | ·92 | 7 | Au | 25 | ·45 |
| 5002 | 1958 | 1823 | 12A | ·8792 | 5·8164 | 27·215 | **1901** | 3 | 5 | Ap | 18 | ·69 | 7 | My | 18● | ·23 | 1 | Je | 16 | ·76 | 3 | Jl | 16 | ·29 | 6 | S | 13 | ·35 |
| | | | | | | | | | | | | | | | | | | | | | 4 | Au | 14 | 82 | | | | |
| 5003 | 1959 | 1824 | 13A | ·1379 | 24·4553 | 25·349 | **1902** | 4 | 4 | My | 7 | ·59 | 6 | Je | 6 | ·12 | 7 | Jl | 5 | ·65 | 2 | Au | 4 | ·18 | 3 | S | 2 | ·71 |
| 5004 | 1960 | 1825 | 13A | ·3967 | 13·5636 | 21·506 | **1903** | 5 | 1 | Ap | 26 | ·96 | 3 | My | 26 | ·49 | 5 | Je | 25 | ·02 | 6 | Jl | 24 | ·55 | 1 | Au | 23 | ·08 |
| 5005 | 1961 | 1826 | 12A | ·6554 | 2·6719 | 17·663 | **1904** | 6 | 6 | Ap | 15 | ·33 | 7 | My | 14 | ·86 | 3 | Jl | 12 | ·92 | 5 | Au | 11 | ·45 | 6 | S | 9 | ·98 |
| | | | | | | | | | | | | | 2 | Je | 13 | ·39 | | | | | | | | | | | | |
| 5006 | 1962 | 1827 | 12A | ·9142 | 21·3107 | 15·796 | **1905** | 1 | 5 | My | 4 | ·22 | 6 | Je | 2 | ·75 | 1 | Jl | 2 | ·29 | 2 | Jl | 31○ | ·82 | 4 | Au | 30● | ·35 |
| 5007 | 1963 | 1828 | 13A | ·1729 | 10·4190 | 11·954 | **1906** | 2 | 2 | Ap | 23 | ·59 | 4 | My | 23 | ·12 | 5 | Je | 21 | ·65 | 7 | Jl | 21○ | ·18 | 1 | Au | 19● | ·71 |
| 5008 | 1964 | 1829 | 13A | ·4317 | 29·0579 | 10·087 | **1907** | 3 | 1 | My | 12 | ·49 | 3 | Je | 11 | ·02 | 4 | Jl | 10○ | ·55 | 6 | Au | 9 | ·08 | 7 | S | 7 | ·61 |
| 5009 | 1965 | 1830 | 12A | ·6905 | 18·1662 | 6·244 | **1908** | 4 | 5 | Ap | 30 | ·86 | 7 | My | 30 | ·39 | 1 | Je | 28● | ·92 | 3 | Jl | 28 | ·45 | 4 | Au | 26 | ·98 |
| 5010 | 1966 | 1831 | 12A | ·9492 | 7·2745 | 2·401 | **1909** | 6 | 3 | Ap | 20 | ·22 | 4 | My | 19○ | ·75 | 6 | Je | 18● | ·28 | 7 | Jl | 17 | ·81 | 3 | S | 14 | ·88 |
| | | | | | | | | | | | | | | | | | | | | | 2 | Au | 16 | ·35 | | | | |
| 5011 | 1967 | 1832 | 13A | ·2080 | 25·9134 | 0·535 | **1910** | 7 | 2 | My | 9○ | ·12 | 3 | Je | 7 | ·65 | 5 | Jl | 7 | ·18 | 6 | Au | 5 | ·71 | 1 | S | 4 | ·24 |
| 5012 | 1968 | 1833 | 13A | ·4667 | 15·0217 | 24·246 | **1911** | 1 | 6 | Ap | 28 | ·49 | 1 | My | 28 | ·02 | 2 | Je | 26 | ·55 | 4 | Jl | 26 | ·08 | 5 | Au | 24 | ·61 |
| 5013 | 1969 | 1834 | 12A | ·7255 | 4·1300 | 20·404 | **1912** | 2 | 3 | Ap | 16● | ·85 | 5 | My | 16 | ·39 | 6 | Je | 14 | ·92 | 2 | Au | 12 | ·98 | 4 | S | 11○ | ·51 |
| | | | | | | | | | | | | | | | | | 1 | Jl | 14 | ·45 | | | | | | | | |
| 5014 | 1970 | 1835 | 12A | ·9842 | 22·7689 | 18·537 | **1913** | 4 | 2 | My | 5 | ·75 | 4 | Je | 4 | ·28 | 5 | Jl | 3 | ·81 | 7 | Au | 2 | ·34 | 1 | Au | 31○ | ·87 |
| 5015 | 1971 | 1836 | 13A | ·2430 | 11·8772 | 14·694 | **1914** | 5 | 7 | Ap | 25 | ·12 | 1 | My | 24 | ·65 | 3 | Je | 23 | ·18 | 4 | Jl | 22 | ·71 | 6 | Au | 21●○ | ·24 |
| 5016 | 1972 | 1837 | 13A | ·5018 | 0·9855 | 10·851 | **1915** | 6 | 4 | Ap | 14 | ·49 | 7 | Je | 12 | ·55 | 2 | Jl | 12 | ·08 | 3 | Au | 10● | ·61 | 5 | S | 9 | ·14 |
| | | | | | | | | | 6 | My | 14 | ·02 | | | | | | | | | | | | | | | | |
| 5017 | 1973 | 1838 | 12A | ·7605 | 19·6244 | 8·985 | **1916** | 7 | 3 | My | 2 | ·38 | 4 | My | 31 | ·91 | 6 | Je | 30○ | ·46 | 7 | Jl | 29 | ·98 | 2 | Au | 28 | ·51 |
| 5018 | 1974 | 1839 | 13A | ·0193 | 8·7327 | 5·142 | **1917** | 2 | 7 | Ap | 21 | ·75 | 2 | My | 21 | ·28 | 3 | Je | 19●○ | ·81 | 5 | Jl | 19 | ·34 | 6 | Au | 17 | ·87 |
| | | | | | | | | | | | | | | | | | | | | | | | | | 1 | S | 16 | ·40 |
| 5019 | 1975 | 1840 | 13A | ·2780 | 27·3715 | 3·275 | **1918** | 3 | 6 | My | 10 | ·65 | 1 | Je | 9●○ | ·18 | 2 | Jl | 8 | ·71 | 4 | Au | 7 | ·24 | 5 | S | 5 | ·77 |
| 5020 | 1976 | 1841 | 13A | ·5368 | 16·4798 | 26·987 | **1919** | 4 | 4 | Ap | 30 | ·02 | 5 | My | 29● | ·55 | 7 | Je | 28 | ·08 | 1 | Jl | 27 | ·61 | 3 | Au | 26 | ·14 |
| 5021 | 1977 | 1842 | 12A | ·7955 | 5·5881 | 23·144 | **1920** | 5 | 1 | Ap | 18○ | ·38 | 2 | My | 17 | ·91 | 4 | Je | 16 | ·44 | 5 | Jl | 15 | ·97 | 2 | S | 13 | ·04 |
| | | | | | | | | | | | | | | | | | | | | | 7 | Au | 14 | ·51 | | | | |
| 5022 | 1978 | 1843 | 13A | .0543 | 24·2270 | 21·278 | **1921** | 7 | 7 | My | 7 | ·28 | 1 | Je | 5 | ·81 | 3 | Jl | 5 | ·34 | 4 | Au | 3 | ·87 | 6 | S | 2 | ·40 |
| 5023 | 1979 | 1844 | 13A | ·3130 | 13·3353 | 17·435 | **1922** | 1 | 4 | Ap | 26 | ·65 | 6 | My | 26 | ·18 | 7 | Je | 24 | ·71 | 2 | Jl | 24 | ·24 | 3 | Au | 22 | ·77 |
| 5024 | 1980 | 1845 | 13A | ·5718 | 2·4436 | 13·592 | **1923** | 2 | 2 | Ap | 16 | ·01 | 3 | My | 15 | ·55 | 6 | Jl | 13 | ·61 | 1 | Au | 12○ | ·14 | 2 | S | 10● | ·67 |
| | | | | | | | | | | | | | 5 | Je | 14 | ·08 | | | | | | | | | | | | |
| 5025 | 1981 | 1846 | 12A | ·8306 | 21·0825 | 11·725 | **1924** | 3 | 7 | My | 3 | ·91 | 2 | Je | 2 | ·44 | 3 | Jl | 1 | ·97 | 5 | Jl | 31○ | ·50 | 7 | Au | 30● | ·03 |
| 5026 | 1982 | 1847 | 13A | ·0893 | 10·1908 | 7·883 | **1925** | 5 | 5 | Ap | 23 | ·28 | 6 | My | 22 | ·81 | 1 | Je | 21 | ·34 | 2 | Jl | 20○ | ·87 | 4 | Au | 19 | ·40 |
| 5027 | 1983 | 1848 | 13A | ·3481 | 28·8297 | 6·016 | **1926** | 6 | 4 | My | 12 | ·18 | 5 | Je | 10 | ·71 | 7 | Jl | 10● | ·24 | 1 | Au | 8 | ·77 | 3 | S | 7 | ·30 |
| 5028 | 1984 | 1849 | 13A | ·6068 | 17·9380 | 2·173 | **1927** | 7 | 1 | My | 1 | ·54 | 3 | My | 31○ | ·07 | 4 | Je | 29● | ·61 | 6 | Jl | 29 | ·14 | 7 | Au | 27 | ·67 |
| 5029 | 1985 | 1850 | 12A | ·8656 | 7·0463 | 25·885 | **1928** | 1 | 5 | Ap | 19 | ·91 | 7 | My | 19●○ | ·44 | 1 | Je | 17 | ·97 | 3 | Jl | 17 | ·50 | 6 | S | 14 | ·56 |
| | | | | | | | | | | | | | | | | | | | | | 5 | Au | 16 | ·03 | | | | |
| 5030 | 1986 | 1851 | 13A | ·1243 | 25·9852 | 24·018 | **1929** | 3 | 4 | My | 8●○ | ·81 | 6 | Je | 7 | ·34 | 7 | Jl | 6 | ·87 | 2 | Au | 5 | ·40 | 3 | S | 3 | ·93 |
| 5031 | 1987 | 1852 | 13A | ·3831 | 14·7935 | 20·175 | **1930** | 4 | 2 | Ap | 28 | ·18 | 3 | My | 27 | ·71 | 5 | Je | 26 | ·24 | 6 | Jl | 25 | ·77 | 1 | Au | 24 | ·30 |
| 5032 | 1988 | 1853 | 13A | ·6419 | 3·9018 | 16·333 | **1931** | 5 | 6 | Ap | 17● | ·54 | 1 | My | 17 | ·07 | 2 | Je | 15 | ·60 | 5 | Au | 13 | ·67 | 7 | S | 12○ | ·20 |
| | | | | | | | | | | | | | | | | | 4 | Jl | 15 | ·13 | | | | | | | | |
| 5033 | 1989 | 1854 | 12A | ·9006 | 22·5406 | 14·467 | **1932** | 6 | 5 | My | 5 | ·44 | 6 | Je | 3 | ·97 | 1 | Jl | 3 | ·50 | 3 | Au | 2 | ·03 | 4 | Au | 31○ | ·56 |
| 5034 | 1990 | 1855 | 13A | ·1594 | 11·6489 | 10·623 | **1933** | 1 | 2 | Ap | 24 | ·81 | 4 | My | 24 | ·34 | 5 | Je | 22 | ·87 | 7 | Jl | 22 | ·40 | 1 | Au | 20● | ·93 |
| 5035 | 1991 | 1856 | 13A | ·4181 | 0·7572 | 6·730 | **1934** | 2 | 7 | Ap | 14 | ·17 | 3 | Je | 12 | ·24 | 4 | Jl | 11○ | ·77 | 6 | Au | 10● | ·30 | 7 | S | 8 | ·83 |
| | | | | | | | | | 1 | My | 13 | ·71 | | | | | | | | | | | | | | | | |
| 5036 | 1992 | 1857 | 13A | ·6769 | 19·3962 | 4·914 | **1935** | 3 | 6 | My | 3 | ·07 | 7 | Je | 1 | ·60 | 2 | Jl | 1○ | ·13 | 3 | Jl | 30 | ·66 | 5 | Au | 29 | ·19 |
| 5037 | 1993 | 1858 | 12A | ·9356 | 8·5044 | 1·071 | **1936** | 4 | 3 | Ap | 21 | ·44 | 4 | My | 20 | ·97 | 6 | Je | 19●○ | ·50 | 1 | Jl | 19 | ·03 | 2 | Au | 17 | ·56 |
| | | | | | | | | | | | | | | | | | | | | | | | | | 4 | S | 16 | ·09 |

**Surya Siddhanta.**

| | | | | | | | | | | | | | | | | | | | | | | | | | | | | | |
|---|---|---|---|---|---|---|---|---|---|---|---|---|---|---|---|---|---|---|---|---|---|---|---|---|---|---|---|---|---|
| + 147·65293 | | | | + 177·18353 | | | | + 206·71411 | | | | | + 236·24470 | | | | | + 265·77529 | | | | + 295·30588 | | | | +324·88647<br>+354·36705 | | | |
| + 9·880 | | | | + 11·856 | | | | + 13·832 | | | | | + 15·808 | | | | | + 17·784 | | | | + 19·760 | | | | + 21·736<br>+ 23·712 | | | |
| **Asvina** | | | | **Kartika** | | | | **Margasira** | | | | **A.D.** | **Pausha** | | | | **A.D.** | **Magha** | | | | **Phalguna** | | | | **Chaitra** | | | |
| Week-day | Month | Day | Fraction | Week-day | Month | Day | Fraction | Week-day | Month | Day | Fraction | | Week-day | Month | Day | Fraction | | Week-day | Month | Day | Fraction | Week-day | Month | Day | Fraction | Week-day | Month | Day | Fraction |
| 5 | S | 15 | ·72 | 1 | N | 13 | ·78 | 3 | D | 13○ | ·31 | **99** | 4 | Jr | 11 | ·84 | | 6 | F | 10 | ·37 | 7 | Mr | 11 | ·90 | 2 | Ap | 10 | ·43 |
| 7 | O | 15 | ·25 | | | | | | | | | | | | | | | | | | | | | | | | | | |
| 4 | O | 4 | ·61 | 6 | N | 3 | ·14 | 7 | D | 2○ | ·67 | **19** | 2 | Jr | 1 | ·21 | **00** | 3 | Jr | 30 | ·74 | 5 | Mr | 1 | ·27 | 6 | Mr | 30 | ·80 |
| 1 | S | 23 | ·98 | 3 | O | 23 | ·51 | 5 | N | 22● | ·04 | | 6 | D | 21 | ·57 | **01** | 1 | Jr | 20 | ·10 | 2 | F | 18 | ·63 | 4 | Mr | 20 | ·16 |
| 7 | O | 12○ | ·88 | 2 | N | 11● | ·41 | 3 | D | 10 | ·94 | **02** | 5 | Jr | 9 | ·47 | | 7 | F | 8 | ·00 | 1 | Mr | 9 | ·53 | 3 | Ap | 8○ | ·06 |
| 5 | O | 2○ | ·25 | 6 | O | 31● | ·78 | 1 | N | 30 | ·31 | | 2 | D | 29 | ·84 | **03** | 4 | Jr | 28 | ·37 | 5 | F | 26 | ·90 | 7 | Mr | 28●○ | ·43 |
| 2 | S | 21●○ | ·61 | 4 | O | 21 | ·14 | 5 | N | 19 | ·67 | | 7 | D | 19 | ·20 | **04** | 1 | Jr | 17 | ·73 | 3 | F | 16 | ·27 | 4 | Mr | 16● | ·80 |
| 1 | O | 9 | ·51 | 3 | N | 8 | ·04 | 4 | D | 7 | ·57 | **05** | 6 | Jr | 6 | ·10 | | 7 | F | 4○ | ·63 | 2 | Mr | 6 | ·16 | 3 | Ap | 4 | ·69 |
| 5 | S | 28 | ·88 | 7 | O | 28 | ·41 | 1 | N | 26 | ·94 | | 3 | D | 26 | ·47 | **06** | 5 | Jr | 25○ | ·00 | 6 | F | 23 | ·53 | 1 | Mr | 25 | ·06 |
| 3 | S | 18 | ·24 | 4 | O | 17 | ·77 | 6 | N | 16 | ·31 | | 7 | D | 15 | ·84 | **07** | 2 | Jr | 14●○ | ·37 | 3 | F | 12 | ·90 | 5 | Mr | 14 | ·43 |
| | | | | | | | | | | | | | | | | | | | | | | | | | | 6 | Ap | 12 | ·96 |
| 2 | O | 7 | ·14 | 3 | N | 5 | ·67 | 5 | D | 5 | ·20 | **08** | 6 | Jr | 3 | ·73 | | 1 | F | 2 | ·26 | 2 | Mr | 2 | ·79 | 4 | Ap | 1 | ·33 |
| 6 | S | 25 | ·51 | 1 | O | 25 | ·04 | 2 | N | 23 | ·57 | | 4 | D | 23● | ·10 | **09** | 5 | Jr | 21 | ·63 | 7 | F | 20 | ·16 | 1 | Mr | 21 | ·69 |
| 5 | O | 14 | ·41 | 6 | N | 12○ | ·94 | 1 | D | 12 | ·47 | **10** | 3 | Jr | 11 | ·00 | | 4 | F | 9 | ·53 | 6 | Mr | 11 | ·06 | 7 | Ap | 9 | ·59 |
| 2 | O | 3 | ·77 | 4 | N | 2●○ | ·30 | 5 | D | 1 | ·83 | | 7 | D | 31 | ·37 | **11** | 1 | Jr | 29 | ·90 | 3 | F | 28 | ·43 | 4 | Mr | 29 | ·96 |
| 7 | S | 23 | ·14 | 1 | O | 22● | ·67 | 3 | N | 21 | ·20 | | 4 | D | 20 | ·73 | **12** | 6 | Jr | 19 | ·26 | 7 | F | 17 | ·79 | 2 | Mr | 18○ | ·32 |
| 6 | O | 11● | ·04 | 7 | N | 9 | ·57 | 2 | D | 9 | ·10 | **13** | 3 | Jr | 7 | ·63 | | 5 | F | 6 | ·16 | 6 | Mr | 7○ | ·69 | 1 | Ap | 6 | ·22 |
| 3 | S | 30 | ·41 | 4 | O | 29 | ·94 | 6 | N | 28 | ·47 | | 1 | D | 28 | ·00 | **14** | 2 | Jr | 26 | ·53 | 4 | F | 25○ | ·06 | 5 | Mr | 26 | ·59 |
| 7 | S | 19 | ·77 | 2 | O | 19 | ·30 | 3 | N | 17 | ·83 | | 5 | D | 17 | ·36 | **15** | 6 | Jr | 15 | ·89 | 1 | F | 14● | ·43 | 2 | Mr | 15 | ·96 |
| 6 | O | 8 | ·67 | 1 | N | 7 | ·20 | 2 | D | 6 | ·73 | **16** | 4 | Jr | 5○ | ·26 | | 5 | F | 3● | ·79 | 7 | Mr | 4 | ·32 | 1 | Ap | 2 | ·85 |
| 4 | S | 27 | ·04 | 5 | O | 26 | ·57 | 7 | N | 25 | ·10 | | 1 | D | 24○ | ·63 | **17** | 3 | Jr | 23● | ·16 | 4 | F | 21 | ·69 | 6 | Mr | 23 | ·22 |
| 2 | O | 15 | ·93 | 4 | N | 14 | ·47 | 6 | D | 14○ | ·00 | **18** | 7 | Jr | 12 | ·53 | | 2 | F | 11 | ·06 | 3 | Mr | 12 | ·59 | 5 | Ap | 11 | ·12 |
| 7 | O | 5 | ·30 | 1 | N | 3 | ·83 | 3 | D | 3● | ·36 | **19** | 4 | Jr | 1 | ·89 | | 6 | Jr | 31 | ·42 | 7 | Mr | 1 | ·95 | 2 | Mr | 31 | ·49 |
| 4 | S | 24 | ·67 | 6 | O | 24○ | ·20 | 7 | N | 22● | ·73 | | 2 | D | 22 | ·26 | **20** | 3 | Jr | 20 | ·79 | 5 | F | 19 | ·32 | 6 | Mr | 19 | ·85 |
| 3 | O | 12○ | ·57 | 5 | N | 11● | ·10 | 6 | D | 10 | ·63 | **21** | 1 | Jr | 9 | ·16 | | 2 | F | 7 | ·69 | 4 | Mr | 9 | ·22 | 5 | Ap | 7●○ | ·75 |
| 7 | O | 1●○ | ·93 | 2 | O | 31 | ·46 | 3 | N | 29 | ·99 | | 5 | D | 29 | ·53 | **22** | 7 | Jr | 28 | ·06 | 1 | F | 26 | ·59 | 3 | Mr | 28● | ·12 |
| 5 | S | 21 | ·30 | 6 | O | 20 | ·83 | 1 | N | 19 | ·36 | | 2 | D | 18 | ·89 | **23** | 4 | Jr | 17 | ·42 | 5 | F | 15○ | ·95 | 7 | Mr | 17● | ·48 |
| 4 | O | 10 | ·20 | 5 | N | 8 | ·73 | 7 | D | 8 | ·26 | **24** | 1 | Jr | 6 | ·79 | | 3 | F | 5○ | ·32 | 4 | Mr | 5 | ·85 | 6 | Ap | 4 | ·38 |
| 1 | S | 28 | ·57 | 3 | O | 28 | ·10 | 4 | N | 26 | ·63 | | 6 | D | 26 | ·16 | **25** | 7 | Jr | 24●○ | ·69 | 2 | F | 23 | ·22 | 3 | Mr | 24 | ·75 |
| 5 | S | 17 | ·93 | 7 | O | 17 | ·46 | 1 | N | 15 | ·99 | | 3 | D | 15 | ·52 | **26** | 5 | Jr | 14● | ·05 | 6 | F | 12 | ·59 | 1 | Mr | 14 | ·12 |
| | | | | | | | | | | | | | | | | | | | | | | | | | | 2 | Ap | 12 | ·65 |
| 4 | O | 6 | ·83 | 6 | N | 5 | ·36 | 7 | D | 4○ | ·89 | **27** | 2 | Jr | 3 | ·42 | | 3 | F | 1 | ·95 | 5 | Mr | 3 | ·48 | 7 | Ap | 2 | ·01 |
| 2 | S | 26 | ·20 | 3 | O | 25 | ·73 | 5 | N | 24○ | ·26 | | 6 | D | 23 | ·79 | **28** | 1 | Jr | 22 | ·32 | 2 | F | 20 | ·85 | 4 | Mr | 21 | ·38 |
| 1 | O | 14 | ·09 | 2 | N | 12●○ | ·63 | 4 | D | 12 | ·16 | **29** | 5 | Jr | 10 | ·69 | | 7 | F | 9 | ·22 | 1 | Mr | 10 | ·75 | 3 | Ap | 9 | ·28 |
| 5 | O | 3 | ·46 | 6 | N | 1● | ·99 | 1 | D | 1 | ·52 | | 3 | D | 31 | ·05 | **30** | 4 | Jr | 29 | ·58 | 6 | F | 28 | ·11 | 7 | Mr | 29○ | ·65 |
| 2 | S | 22○ | ·83 | 4 | O | 22 | ·36 | 5 | N | 20 | ·89 | | 7 | D | 20 | ·42 | **31** | 1 | Jr | 18 | ·95 | 3 | F | 17 | ·48 | 5 | Mr | 19○ | ·01 |
| 1 | O | 11 | ·73 | 3 | N | 10 | ·26 | 4 | D | 9 | ·79 | **32** | 6 | Jr | 8 | ·32 | | 7 | F | 6 | ·85 | 2 | Mr | 7○ | ·38 | 3 | Ap | 5 | ·91 |
| 6 | S | 30 | ·09 | 7 | O | 29 | ·62 | 2 | N | 28 | ·15 | | 3 | D | 27 | ·69 | **33** | 5 | Jr | 26 | ·22 | 6 | F | 24● | ·75 | 1 | Mr | 26 | ·28 |
| 3 | S | 19 | ·46 | 4 | O | 18 | ·99 | 6 | N | 17 | ·52 | | 1 | D | 17 | ·05 | **34** | 2 | Jr | 15○ | ·58 | 4 | F | 14● | ·11 | 5 | Mr | 15 | ·64 |
| 2 | O | 8 | ·36 | 3 | N | 6 | ·89 | 5 | D | 6 | ·42 | **35** | 6 | Jr | 4○ | ·95 | | 1 | F | 3 | ·48 | 3 | Mr | 5 | ·01 | 4 | Ap | 3 | ·54 |
| 6 | S | 27 | ·73 | 1 | O | 27 | ·26 | 2 | N | 25 | ·79 | | 4 | D | 25○ | ·32 | **36** | 5 | Jr | 23 | ·85 | 7 | F | 22 | ·38 | 1 | Mr | 22 | ·91 |
| 5 | O | 15 | ·62 | 7 | N | 14 | ·15 | 1 | D | 13 | ·68 | **37** | 3 | Jr | 12 | ·21 | | 4 | F | 10 | ·75 | 6 | Mr | 12 | ·28 | 7 | Ap | 10 | ·81 |

*NOTE.*—From **1900 A.D.** to **1999 A.D.** all eclipses of the Sun, visible anywhere in the world, are shown. From **1 B.C.** to **300 A.D.** the same observation applies. Between **300 A.D.** and **1900 A.D.** only those eclipses of the Sun which are visible in India are entered in this table.

| Kaliyuga. | Vikrama Era. | Saka Era. | Commencement of Solar Year. Month and day A.D. | Commencement of Solar Year. Fraction of day. | First New-Moon in Solar Year. | Anomaly of first New-Moon. | A.D. | Week-day of 1st January. | Vaisakha Week-day | Vaisakha Month | Vaisakha Day | Vaisakha Fraction | Jyeshta Week-day | Jyeshta Month | Jyeshta Day | Jyeshta Fraction | Ashada Week-day | Ashada Month | Ashada Day | Ashada Fraction | Sravana Week-day | Sravana Month | Sravana Day | Sravana Fraction | Bhadrapada Week-day | Bhadrapada Month | Bhadrapada Day | Bhadrapada Fraction |
|---|---|---|---|---|---|---|---|---|---|---|---|---|---|---|---|---|---|---|---|---|---|---|---|---|---|---|---|---|
| | | | | | | | | | ⊙'s Anom col. 6 | | | | + 29·53059 | | | | + 59·06117 | | | | + 88·59176 | | | | + 118·12235 | | | |
| | | | | | | | | | ☾'s Anom col. 7 | | | | + 1·976 | | | | + 3·952 | | | | + 5·928 | | | | + 7·904 | | | |
| 5038 | 1994 | 1659 | 13A | ·1944 | 27·1433 | 26·759 | **1937** | 6 | 2 | My | 10 | ·34 | 3 | Je | 8 | ·87 | 5 | Jl | 8 | ·40 | 6 | Au | 6 | ·93 | 1 | S | 5 | ·46 |
| 5039 | 1995 | 1860 | 13A | ·4531 | 16·2516 | 22·916 | **1938** | 7 | 6 | Ap | 29○ | ·70 | 1 | My | 29 | ·23 | 2 | Je | 27 | ·77 | 4 | Jl | 27 | ·30 | 5 | Au | 25 | ·83 |
| 5040 | 1996 | 1861 | 13A | ·7119 | 5·3599 | 19·073 | **1939** | 1 | 4 | Ap | 19●○ | ·07 | 5 | My | 18 | ·60 | 7 | Je | 17 | ·13 | 1 | Jl | 16 | ·66 | 4 | S | 13 | ·72 |
| | | | | | | | | | | | | | | | | | | | | | 3 | Au | 15 | ·19 | | | | |
| 5041 | 1997 | 1862 | 12A | ·9707 | 23·9988 | 17·207 | **1940** | 2 | 2 | My | 6 | ·97 | 4 | Je | 5 | ·50 | 6 | Jl | 5 | ·03 | 7 | Au | 3 | ·56 | 2 | S | 2 | ·09 |
| 5042 | 1998 | 1863 | 13A | ·2294 | 13·1071 | 13·364 | **1941** | 4 | 7 | Ap | 26 | ·34 | 1 | My | 25 | ·87 | 3 | Je | 24 | ·40 | 4 | Jl | 23 | ·93 | 6 | Au | 22○ | ·46 |
| 5043 | 1999 | 1864 | 13A | ·4882 | 2·2154 | 9·521 | **1942** | 5 | 4 | Ap | 15 | ·70 | 6 | My | 15 | ·23 | 2 | Jl | 13 | ·29 | 3 | Au | 11○ | ·83 | 5 | S | 10● | ·36 |
| | | | | | | | | | | | | | 7 | Je | 13 | ·76 | | | | | | | | | | | | |
| 5044 | 2000 | 1865 | 13A | ·7469 | 20·8543 | 7·654 | **1943** | 6 | 3 | My | 4 | ·60 | 5 | Je | 3 | ·13 | 6 | Jl | 2 | ·66 | 1 | Au | 1○ | ·19 | 2 | Au | 30 | ·72 |
| 5045 | 2001 | 1866 | 13A | ·0057 | 9·9626 | 3·812 | **1944** | 7 | 7 | Ap | 22 | ·97 | 2 | My | 22 | ·50 | 4 | Je | 21 | ·03 | 5 | Jl | 20● | ·56 | 7 | Au | 19 | ·09 |
| 5046 | 2002 | 1867 | 13A | ·2644 | 28·6614 | 1·845 | **1945** | 2 | 6 | My | 11 | ·86 | 1 | Je | 10○ | ·39 | 2 | Jl | 9● | ·93 | 4 | Au | 8 | ·46 | 5 | S | 6 | ·99 |
| 5047 | 2003 | 1868 | 13A | ·5232 | 17·7097 | 25·657 | **1946** | 3 | 4 | My | 1 | ·23 | 5 | My | 30○ | ·76 | 7 | Je | 29● | ·29 | 1 | Jl | 28 | ·82 | 3 | Au | 27 | ·35 |
| 5048 | 2004 | 1869 | 13A | ·7820 | 6·8180 | 21·814 | **1947** | 4 | 1 | Ap | 20 | ·60 | 3 | My | 20●○ | ·13 | 4 | Je | 18 | ·66 | 6 | Jl | 18 | ·19 | 2 | S | 15 | ·25 |
| | | | | | | | | | | | | | | | | | | | | | 7 | Au | 16 | ·72 | | | | |
| 5049 | 2005 | 1870 | 13A | ·0407 | 25·4569 | 19·947 | **1948** | 5 | 7 | My | 8● | ·50 | 2 | Je | 7 | ·08 | 3 | Jl | 6 | ·56 | 5 | Au | 5 | ·09 | 6 | S | 3 | ·62 |
| 5050 | 2006 | 1871 | 13A | ·2995 | 14·5652 | 16·104 | **1949** | 7 | 4 | Ap | 27● | ·86 | 6 | My | 27 | ·39 | 7 | Je | 25 | ·92 | 2 | Jl | 25 | ·45 | 3 | Au | 23 | ·99 |
| 5051 | 2007 | 1872 | 13A | ·5582 | 3·6735 | 12·262 | **1950** | 1 | 2 | Ap | 17 | ·23 | 3 | My | 16 | ·76 | 5 | Je | 15 | ·29 | 1 | Au | 13 | ·35 | 2 | S | 11●○ | ·88 |
| | | | | | | | | | | | | | | | | | 6 | Jl | 14 | ·82 | | | | | | | | |
| 5052 | 2008 | 1873 | 13A | ·8170 | 22·3124 | 10·395 | **1951** | 2 | 1 | My | 6 | ·13 | 2 | Je | 4 | ·66 | 4 | Jl | 4 | ·19 | 5 | Au | 2 | ·72 | 7 | S | 1● | ·25 |
| 5053 | 2009 | 1874 | 13A | ·0757 | 11·4207 | 6·552 | **1952** | 3 | 5 | Ap | 24 | ·49 | 7 | My | 24 | ·03 | 1 | Je | 22 | ·56 | 3 | Jl | 22○ | ·09 | 4 | Au | 20● | ·62 |
| 5054 | 2010 | 1875 | 13A | ·3345 | 0·5290 | 2·709 | **1953** | 5 | 2 | Ap | 13 | ·86 | 5 | Je | 11 | ·92 | 7 | Jl | 11●○ | ·45 | 1 | Au | 9 | ·98 | 3 | S | 8 | ·51 |
| | | | | | | | | | 4 | My | 13 | ·39 | | | | | | | | | | | | | | | | |
| 5055 | 2011 | 1876 | 13A | ·5933 | 19·1679 | 0·843 | **1954** | 6 | 1 | My | 2 | ·76 | 3 | Je | 1 | ·29 | 4 | Je | 30●○ | ·82 | 6 | Jl | 30 | ·85 | 7 | Au | 28 | ·88 |
| 5056 | 2012 | 1877 | 13A | ·8520 | 8·2762 | 24·554 | **1955** | 7 | 6 | Ap | 22 | ·13 | 7 | My | 21 | ·66 | 2 | Je | 20● | ·19 | 3 | Jl | 19 | ·72 | 5 | Au | 18 | ·25 |
| | | | | | | | | | | | | | | | | | | | | | | | | | 6 | S | 16 | ·78 |
| 5057 | 2013 | 1878 | 13A | ·1108 | 26·9151 | 22·688 | **1956** | 1 | 5 | My | 10○ | ·02 | 6 | Je | 8 | ·55 | 1 | Jl | 8 | ·09 | 2 | Au | 6 | ·62 | 4 | S | 5 | ·15 |
| 5058 | 2014 | 1879 | 13A | ·3695 | 16·0234 | 18·845 | **1957** | 3 | 2 | Ap | 29○ | ·39 | 3 | My | 28 | ·92 | 5 | Je | 27 | ·45 | 6 | Jl | 26 | ·98 | 1 | Au | 25 | ·51 |
| 5059 | 2015 | 1880 | 13A | ·6283 | 5·1817 | 15·002 | **1958** | 4 | 6 | Ap | 18●○ | ·76 | 1 | My | 18 | ·29 | 2 | Je | 16 | ·82 | 4 | Jl | 16 | ·35 | 7 | S | 13 | ·41 |
| | | | | | | | | | | | | | | | | | | | | | 5 | Au | 14 | ·88 | | | | |
| 5060 | 2016 | 1881 | 13A | ·8870 | 23·7705 | 13·135 | **1959** | 5 | 5 | My | 7 | ·66 | 7 | Je | 6 | ·19 | 1 | Jl | 5 | ·72 | 3 | Au | 4 | ·25 | 4 | S | 2○ | ·78 |
| 5061 | 2017 | 1882 | 13A | ·1458 | 12·8788 | 9·293 | **1960** | 6 | 3 | Ap | 26 | ·02 | 4 | My | 25 | ·55 | 6 | Je | 24 | ·08 | 7 | Jl | 23 | ·61 | 2 | Au | 22○ | ·15 |
| 5062 | 2018 | 1883 | 13A | ·4045 | 1·9871 | 5·450 | **1961** | 1 | 7 | Ap | 15 | ·39 | 1 | My | 14 | ·92 | 4 | Jl | 12 | ·98 | 6 | Au | 11●○ | ·51 | 1 | S | 10 | ·04 |
| | | | | | | | | | | | | | 3 | Je | 13 | ·45 | | | | | | | | | | | | |
| 5063 | 2019 | 1884 | 13A | ·6633 | 20·6260 | 3·583 | **1962** | 2 | 6 | My | 4 | ·29 | 7 | Je | 2 | ·82 | 2 | Jl | 2 | ·35 | 3 | Jl | 31● | ·88 | 5 | Au | 30 | ·41 |
| 5064 | 2020 | 1885 | 13A | ·9221 | 9·7343 | 27·295 | **1963** | 3 | 3 | Ap | 23 | ·65 | 5 | My | 23 | ·19 | 6 | Je | 21○ | ·72 | 1 | Jl | 21 | ·25 | 2 | Au | 19 | ·78 |
| 5065 | 2021 | 1886 | 13A | ·1808 | 28·3732 | 25·429 | **1964** | 4 | 2 | My | 11 | ·55 | 4 | Je | 10○ | ·08 | 5 | Jl | 9● | ·61 | 7 | Au | 8 | ·14 | 1 | S | 6 | ·67 |
| 5066 | 2022 | 1887 | 13A | ·4396 | 17·4815 | 21·586 | **1965** | 6 | 6 | Ap | 30 | ·92 | 1 | My | 30○ | ·45 | 2 | Je | 28 | ·98 | 4 | Jl | 28 | ·51 | 6 | Au | 27 | ·04 |
| 5067 | 2023 | 1888 | 13A | ·6983 | 6·5898 | 17·743 | **1966** | 7 | 4 | Ap | 20○ | ·29 | 5 | My | 19● | ·82 | 7 | Je | 18 | ·35 | 1 | Jl | 17 | ·88 | 4 | S | 14 | ·94 |
| | | | | | | | | | | | | | | | | | | | | | 3 | Au | 16 | ·41 | | | | |
| 5068 | 2024 | 1889 | 13A | ·9571 | 25·2287 | 15·876 | **1967** | 1 | 3 | My | 9● | ·18 | 4 | Je | 7 | ·71 | 6 | Jl | 7 | ·25 | 7 | Au | 5 | ·78 | 2 | S | 4 | ·31 |
| 5069 | 2025 | 1890 | 13A | ·2158 | 14·3370 | 12·031 | **1968** | 2 | 7 | Ap | 27 | ·55 | 2 | My | 27 | ·08 | 3 | Je | 25 | ·61 | 5 | Jl | 25 | ·14 | 6 | Au | 23 | ·67 |
| 5070 | 2026 | 1891 | 13A | ·4746 | 3·4453 | 8·191 | **1969** | 4 | 4 | Ap | 16 | ·92 | 6 | My | 16 | ·45 | 7 | Je | 14 | ·98 | 4 | Au | 13 | ·04 | 5 | S | 11 | ·57 |
| | | | | | | | | | | | | | | | | | 2 | Jl | 14 | ·51 | | | | | | | | |
| 5071 | 2027 | 1892 | 13A | ·7334 | 22·0842 | 6·324 | **1970** | 5 | 3 | My | 5 | ·82 | 5 | Je | 4 | ·35 | 6 | Jl | 3 | ·88 | 1 | Au | 2○ | ·41 | 2 | Au | 31 | ·94 |
| 5072 | 2028 | 1893 | 13A | ·9921 | 11·1925 | 2·481 | **1971** | 6 | 1 | Ap | 25 | ·18 | 2 | My | 24 | ·71 | 4 | Je | 23 | ·24 | 5 | Jl | 22●○ | ·77 | 7 | Au | 21 | ·31 |
| 5073 | 2029 | 1894 | 13A | ·2509 | 0·3008 | 26·193 | **1972** | 7 | 5 | Ap | 13 | ·55 | 1 | Je | 11 | ·61 | 3 | Jl | 11○ | ·14 | 4 | Au | 9 | ·67 | 6 | S | 8 | ·20 |
| | | | | | | | | | 7 | My | 13 | ·08 | | | | | | | | | | | | | | | | |
| 5074 | 2030 | 1895 | 13A | ·5096 | 18·9396 | 24·326 | **1973** | 2 | 4 | My | 2 | ·45 | 5 | My | 31 | ·98 | 7 | Je | 30● | ·51 | 2 | Jl | 30 | ·04 | 3 | Au | 28 | ·57 |
| 5075 | 2031 | 1896 | 13A | ·7684 | 8·0479 | 20·483 | **1974** | 3 | 1 | Ap | 21 | ·81 | 3 | My | 21○ | ·35 | 4 | Je | 19 | ·88 | 6 | Jl | 19 | ·41 | 7 | Au | 17 | ·94 |
| | | | | | | | | | | | | | | | | | | | | | | | | | 2 | S | 16 | ·47 |

## Surya Siddhanta.

| Asvina | | | | Kartika | | | | Margasira | | | | A.D. | Pausha | | | | A.D. | Magha | | | | Phalguna | | | | Chaitra | | | |
|---|---|---|---|---|---|---|---|---|---|---|---|---|---|---|---|---|---|---|---|---|---|---|---|---|---|---|---|---|---|
| + 147·65293 | | | | + 177·18353 | | | | + 206·71411 | | | | | + 236·24470 | | | | | + 265·77529 | | | | + 295·30588 | | | | +324·83647<br>+354·86705 | | | |
| + 9·880 | | | | + 11·856 | | | | + 13·832 | | | | | + 15·808 | | | | | + 17·784 | | | | + 19·760 | | | | +21·736<br>+23·712 | | | |
| Week-day | Month | Day | Fraction | Week-day | Month | Day | Fraction | Week-day | Month | Day | Fraction | | Week-day | Month | Day | Fraction | | Week-day | Month | Day | Fraction | Week-day | Month | Day | Fraction | Week-day | Month | Day | Fraction |
| 2 | O | 4 | ·99 | 4 | N | 3○ | ·52 | 6 | D | 3● | ·05 | 38 | 7 | Jr | 1 | ·58 | | 2 | Jr | 31 | ·11 | 3 | Mr | 1 | ·64 | 5 | Mr | 31 | ·17 |
| 7 | S | 24 | ·36 | 1 | O | 23○ | ·89 | 3 | N | 22● | ·42 | | 4 | D | 21 | ·95 | 39 | 6 | Jr | 20 | ·48 | 1 | F | 19 | ·01 | 2 | Mr | 20 | ·54 |
| 6 | O | 13○ | ·25 | 7 | N | 11 | ·79 | 2 | D | 11 | ·32 | 40 | 3 | Jr | 9 | ·85 | | 5 | F | 8 | ·38 | 6 | Mr | 8 | ·91 | 1 | Ap | 7○ | ·44 |
| 3 | O | 1● | ·62 | 5 | O | 31 | ·15 | 6 | N | 29 | ·68 | | 1 | D | 29 | ·21 | 41 | 2 | Jr | 27 | ·74 | 4 | F | 26○ | ·27 | 5 | Mr | 27 | ·81 |
| 7 | S | 20● | ·99 | 2 | O | 20 | ·52 | 4 | N | 19 | ·05 | | 5 | D | 18 | ·58 | 42 | 7 | Jr | 17 | ·11 | 1 | F | 15○ | ·64 | 3 | Mr | 17 | ·17 |
| 6 | O | 9 | ·89 | 1 | N | 8 | ·42 | 2 | D | 7 | ·95 | 43 | 4 | Jr | 6 | ·48 | | 6 | F | 5●○ | ·01 | 7 | Mr | 6 | ·54 | 2 | Ap | 5 | ·07 |
| 4 | S | 29 | ·25 | 5 | O | 28 | ·78 | 7 | N | 27 | ·31 | | 1 | D | 26 | ·85 | 44 | 3 | Jr | 25● | ·38 | 4 | F | 23 | ·91 | 6 | Mr | 24 | ·44 |
| 1 | S | 17 | ·62 | 3 | O | 17 | ·15 | 4 | N | 15 | ·68 | | 6 | D | 15○ | ·21 | 45 | 7 | Jr | 13● | ·74 | 2 | F | 12 | ·27 | {3 | Mr | 13 | ·80 |
| | | | | | | | | | | | | | | | | | | | | | | | | | | {5 | Ap | 12 | ·33 |
| 7 | O | 6 | ·52 | 2 | N | 5 | ·05 | 3 | D | 4○ | ·58 | 46 | 5 | Jr | 3 | ·11 | | 6 | F | 1 | ·64 | 1 | Mr | 3 | ·17 | 2 | Ap | 1 | ·70 |
| 4 | S | 25 | ·88 | 6 | O | 25 | ·41 | 7 | N | 23○ | ·95 | | 2 | D | 23 | ·48 | 47 | 4 | Jr | 22 | ·01 | 5 | F | 20 | ·54 | 7 | Mr | 22 | ·07 |
| 3 | O | 14 | ·78 | 5 | N | 13 | ·31 | 6 | D | 12 | ·84 | 48 | 1 | Jr | 11 | ·37 | | 2 | F | 9 | ·90 | 4 | Mr | 10 | ·43 | 5 | Ap | 8○ | ·97 |
| 1 | O | 3○ | ·15 | 2 | N | 1● | ·68 | 4 | D | 1 | ·21 | | 5 | D | 30 | ·74 | 49 | 7 | Jr | 29 | ·27 | 1 | F | 27 | ·80 | 3 | Mr | 29○ | ·33 |
| 5 | S | 22○ | ·52 | 7 | O | 22 | ·05 | 1 | N | 20 | ·58 | | 3 | D | 20 | ·11 | 50 | 4 | Jr | 18 | ·64 | 6 | F | 17 | ·17 | 7 | Mr | 18○ | ·70 |
| 4 | O | 11 | ·41 | 5 | N | 9 | ·94 | 7 | D | 9 | ·47 | 51 | 2 | Jr | 8 | ·01 | | 3 | F | 6 | ·54 | 5 | Mr | 8 | ·07 | 6 | Ap | 6 | ·60 |
| 1 | S | 30 | ·78 | 3 | O | 30 | ·31 | 4 | N | 28 | ·84 | | 6 | D | 28 | ·37 | 52 | 7 | Jr | 26○ | ·90 | 2 | F | 25● | ·43 | 3 | Mr | 25 | ·96 |
| 6 | S | 19 | ·15 | 7 | O | 18 | ·68 | 2 | N | 17 | ·21 | | 3 | D | 16 | ·74 | 53 | 5 | Jr | 15○ | ·27 | 6 | F | 13● | ·80 | 1 | Mr | 15 | ·33 |
| 5 | O | 8 | ·05 | 6 | N | 6 | ·58 | 1 | D | 6 | ·11 | 54 | 2 | Jr | 4○ | ·64 | | 4 | F | 3 | ·17 | 5 | Mr | 4 | ·70 | 7 | Ap | 3 | ·23 |
| 2 | S | 27 | ·41 | 3 | O | 26 | ·94 | 5 | N | 25 | ·47 | | 7 | D | 25● | ·00 | 55 | 1 | Jr | 23 | ·53 | 3 | F | 22 | ·07 | 4 | Mr | 23 | ·60 |
| 1 | O | 16 | ·31 | 2 | N | 14○ | ·84 | 4 | D | 14● | ·37 | 56 | 5 | Jr | 12 | ·90 | | 7 | F | 11 | ·43 | 1 | Mr | 11 | ·96 | 3 | Ap | 10 | ·49 |
| 5 | O | 4 | ·68 | 7 | N | 3○ | ·21 | 1 | D | 2● | ·74 | 57 | 3 | Jr | 1 | ·27 | | 4 | Jr | 30 | ·80 | 6 | Mr | 1 | ·33 | 7 | Mr | 30 | ·86 |
| 3 | S | 24 | ·04 | 4 | O | 23●○ | ·57 | 6 | N | 22 | ·10 | | 7 | D | 21 | ·64 | 58 | 2 | Jr | 20 | ·17 | 3 | F | 18 | ·70 | 5 | Mr | 20 | ·23 |
| 1 | O | 12 | ·94 | 3 | N | 11 | ·47 | 5 | D | 11 | ·00 | 59 | 6 | Jr | 9 | ·53 | | 1 | F | 8 | ·06 | 2 | Mr | 9○ | ·59 | 4 | Ap | 8 | ·12 |
| 6 | O | 2● | ·31 | 7 | O | 31 | ·84 | 2 | N | 30 | ·37 | | 3 | D | 29 | ·90 | 60 | 5 | Jr | 28 | ·43 | 6 | F | 26○ | ·96 | 1 | Mr | 27 | ·49 |
| 3 | S | 20● | ·68 | 5 | O | 20 | ·21 | 6 | N | 18 | ·74 | | 1 | D | 18 | ·27 | 61 | 2 | Jr | 16 | ·80 | 4 | F | 15○ | 33 | 5 | Mr | 16 | ·86 |
| 2 | O | 9 | ·57 | 4 | N | 8 | ·10 | 5 | D | 7 | ·63 | 62 | 7 | Jr | 6 | ·17 | | 1 | F | 4● | ·70 | 3 | Mr | 6 | ·23 | 4 | Ap | 4 | ·76 |
| 6 | S | 28 | ·94 | 1 | O | 28 | ·47 | 3 | N | 27 | ·00 | | 4 | D | 26○ | ·53 | 63 | 6 | Jr | 25● | ·06 | 7 | F | 23 | ·59 | 2 | Mr | 25 | ·12 |
| {4 | S | 18 | ·31 | 7 | N | 16 | ·37 | 1 | D | 15○ | ·90 | | | Pausha | | | 64 | 3 | Jr | 14 | ·43 | 4 | F | 12 | ·96 | {6 | Mr | 13 | ·49 |
| {5 | O | 17 | ·84 | | | | | | | | | | | Kshaya | | | | | | | | | | | | {1 | Ap | 12 | ·02 |
| 3 | O | 6 | ·21 | 4 | N | 4 | ·74 | 6 | D | 4●○ | ·27 | 65 | 7 | Jr | 2 | ·80 | | 2 | F | 1 | ·33 | 3 | Mr | 2 | ·86 | 5 | Ap | 1 | ·39 |
| 7 | S | 25 | ·57 | 2 | O | 25 | ·10 | 3 | N | 23● | ·63 | | 5 | D | 23 | ·16 | 66 | 6 | Jr | 21 | ·69 | 1 | F | 20 | ·23 | 2 | Mr | 21 | ·76 |
| 6 | O | 14○ | ·47 | 1 | N | 13● | ·00 | 2 | D | 12 | ·53 | 67 | 4 | Jr | 11 | ·06 | | 5 | F | 9 | ·59 | 7 | Mr | 11 | ·12 | 1 | Ap | 9○ | ·65 |
| 3 | O | 3○ | ·84 | 5 | N | 2 | ·37 | 6 | D | 1 | ·90 | | 1 | D | 31 | ·43 | 68 | 2 | Jr | 29 | ·96 | 4 | F | 28 | ·49 | 6 | Mr | 29○ | ·02 |
| 1 | S | 22○ | ·20 | 2 | O | 21 | ·73 | 4 | N | 20 | ·27 | | 5 | D | 19 | ·80 | 69 | 7 | Jr | 18 | ·33 | 1 | F | 16 | ·86 | 3 | Mr | 18● | ·39 |
| 7 | O | 11 | ·10 | 1 | N | 9 | ·63 | 3 | D | 9 | ·16 | 70 | 4 | Jr | 7 | ·69 | | 6 | F | 6○ | ·22 | 7 | Mr | 7● | ·75 | 2 | Ap | 6 | ·29 |
| 4 | S | 30 | ·47 | 6 | O | 30 | ·00 | 7 | N | 28 | ·53 | | 2 | D | 28 | ·06 | 71 | 3 | Jr | 26○ | ·59 | 5 | F | 25● | ·12 | 6 | Mr | 26 | ·65 |
| 1 | S | 19 | ·84 | 3 | O | 19 | ·37 | 4 | N | 17 | ·90 | | 6 | D | 17 | ·43 | 72 | 7 | Jr | 15○ | ·96 | 2 | F | 14 | ·49 | 4 | Mr | 15 | ·02 |
| 7 | O | 7 | ·73 | 2 | N | 6 | ·26 | 3 | D | 5 | ·79 | 73 | 5 | Jr | 4● | ·33 | | 6 | F | 2 | ·86 | 1 | Mr | 4 | ·39 | 2 | Ap | 2 | ·92 |
| 5 | S | 27 | ·10 | 6 | O | 26 | ·63 | 1 | N | 25○ | ·16 | | 2 | D | 24● | ·69 | 74 | 4 | Jr | 23 | ·23 | 5 | F | 21 | ·75 | 7 | Mr | 23 | ·28 |
| 4 | O | 16 | ·00 | 5 | N | 14○ | ·53 | 7 | D | 14● | ·06 | 75 | 1 | Jr | 12 | ·59 | | 3 | F | 11 | ·12 | 4 | Mr | 12 | ·65 | 6 | Ap | 11 | ·18 |

## TABLE X.–

| Kaliyuga. | Vikrama Era. | Saka Era. | Commencement of Solar Year. Month and day A.D. | Commencement of Solar Year. Fraction of day. | First New-Moon in Solar Year. | Anomaly of first New-Moon. | A.D. | Week-day of 1st January. | Vaisakha |  |  |  | Jyeshta |  |  |  | Ashada |  |  |  | Sravana |  |  |  | Bhadrapada |  |  |  |
|---|---|---|---|---|---|---|---|---|---|---|---|---|---|---|---|---|---|---|---|---|---|---|---|---|---|---|---|---|
|  |  |  |  |  |  |  |  | ⊙'s Anom col. 6 |  |  |  |  | + 29·53059 |  |  |  | + 59·06117 |  |  |  | + 88·59176 |  |  |  | + 118·12236 |  |  |  |
|  |  |  |  |  |  |  |  | ☾'s Anom col. 7 |  |  |  |  | + 1·976 |  |  |  | + 3·952 |  |  |  | + 5·928 |  |  |  | + 7·904 |  |  |  |
|  |  |  |  |  |  |  |  |  | Week-day | Month | Day | Fraction | Week-day | Month | Day | Fraction | Week-day | Month | Day | Fraction | Week-day | Month | Day | Fraction | Week-day | Month | Day | Fraction |
| 5076 | 2032 | 1897 | 14A | ·0271 | 26·6868 | 18·617 | **1975** | 4 | 7 | My | 10●○ | ·71 | 2 | Je | 9 | ·24 | 3 | Jl | 8 | ·77 | 5 | Au | 7 | ·30 | 6 | S | 5 | ·83 |
| 5077 | 2033 | 1898 | 13A | ·2859 | 15·7951 | 14·774 | **1976** | 5 | 5 | Ap | 29●○ | ·03 | 6 | My | 28 | ·61 | 1 | Je | 27 | ·14 | 2 | Jl | 26 | ·67 | 4 | Au | 25 | ·20 |
| 5078 | 2034 | 1899 | 13A | ·5447 | 4·9034 | 10·931 | **1977** | 7 | 2 | Ap | 18● | ·45 | 3 | My | 17 | ·98 | 5 | Je | 16 | ·51 | 1 | Au | 14 | ·57 | 3 | S | 13○ | ·10 |
|  |  |  |  |  |  |  |  |  |  |  |  |  |  |  |  |  | 7 | Jl | 16 | ·04 |  |  |  |  |  |  |  |  |
| 5079 | 2035 | 1900 | 13A | ·8034 | 23·5423 | 9·064 | **1978** | 1 | 1 | My | 7 | ·34 | 2 | Je | 5 | ·87 | 4 | Jl | 5 | ·41 | 5 | Au | 3 | ·94 | 7 | S | 2○ | ·47 |
| 5080 | 2036 | 1901 | 14A | ·0622 | 12·6506 | 5·222 | **1979** | 2 | 5 | Ap | 26 | ·71 | 7 | My | 26 | ·24 | 1 | Je | 24 | ·77 | 3 | Jl | 24 | ·30 | 4 | Au | 22○ | ·83 |
| 5081 | 2037 | 1902 | 13A | ·3209 | 1·7589 | 1·379 | **1980** | 3 | 3 | Ap | 15 | ·08 | 4 | My | 14 | ·61 | 7 | Jl | 12 | ·67 | 2 | Au | 11 | ·20 | 3 | S | 9 | ·73 |
|  |  |  |  |  |  |  |  |  |  |  |  |  | 6 | Je | 13 | ·14 |  |  |  |  |  |  |  |  |  |  |  |  |
| 5082 | 2038 | 1903 | 13A | ·5797 | 20·3978 | 27·067 | **1981** | 5 | 1 | My | 3 | ·98 | 3 | Je | 2 | ·51 | 5 | Jl | 2○ | ·04 | 6 | Jl | 31● | ·57 | 1 | Au | 30 | ·10 |
| 5083 | 2039 | 1904 | 13A | ·8384 | 9·5061 | 23·224 | **1982** | 6 | 6 | Ap | 23 | ·34 | 7 | My | 22 | ·87 | 2 | Je | 21○ | ·40 | 3 | Jl | 20● | ·93 | 5 | Au | 19 | ·47 |
| 5084 | 2040 | 1905 | 14A | ·0972 | 28·1450 | 21·357 | **1983** | 7 | 5 | My | 12 | ·24 | 6 | Je | 10●○ | ·77 | 1 | Jl | 10 | ·30 | 2 | Au | 8 | ·83 | 4 | S | 7 | ·36 |
| 5085 | 2041 | 1906 | 13A | ·3559 | 17·2533 | 17·514 | **1984** | 1 | 2 | Ap | 30 | ·61 | 4 | My | 30● | ·14 | 5 | Je | 23 | ·67 | 7 | Jl | 23 | ·20 | 1 | Au | 26 | ·73 |
| 5086 | 2042 | 1907 | 13A | ·6147 | 6·3616 | 13·672 | **1985** | 3 | 6 | Ap | 19○ | ·97 | 1 | My | 19 | ·51 | 3 | Je | 18 | ·04 | 4 | Jl | 17 | ·57 | 7 | S | 14 | ·63 |
|  |  |  |  |  |  |  |  |  |  |  |  |  |  |  |  |  |  |  |  |  | 6 | Au | 16 | ·10 |  |  |  |  |
| 5087 | 2043 | 1908 | 13A | ·8735 | 25·0004 | 11·805 | **1986** | 4 | 5 | My | 8 | ·87 | 7 | Je | 7 | ·40 | 1 | Jl | 6 | ·93 | 3 | Au | 5 | ·46 | 4 | S | 3 | ·99 |
| 5088 | 2044 | 1909 | 14A | ·1322 | 14·1087 | 7·962 | **1987** | 5 | 3 | Ap | 28 | ·24 | 4 | My | 27 | ·77 | 6 | Je | 26 | ·30 | 7 | Jl | 25 | ·83 | 2 | Au | 24 | ·36 |
| 5089 | 2045 | 1910 | 13A | ·3910 | 3·2170 | 4·119 | **1988** | 6 | 7 | Ap | 16 | ·61 | 2 | My | 16 | ·14 | 5 | Jl | 14 | ·20 | 6 | Au | 12○ | ·73 | 1 | S | 11● | ·26 |
|  |  |  |  |  |  |  |  |  |  |  |  |  | 3 | Je | 14 | ·67 |  |  |  |  |  |  |  |  |  |  |  |  |
| 5090 | 2046 | 1911 | 13A | ·6497 | 21·8559 | 2·253 | **1989** | 1 | 6 | My | 5 | ·50 | 1 | Je | 4 | ·03 | 2 | Jl | 3 | ·57 | 4 | Au | 2○ | ·10 | 5 | Au | 31 | ·63 |
| 5091 | 2047 | 1912 | 13A | ·9085 | 10·9642 | 25·965 | **1990** | 2 | 3 | Ap | 24 | ·87 | 5 | My | 24 | ·40 | 6 | Je | 22 | ·93 | 1 | Jl | 22●○ | ·46 | 2 | Au | 20 | ·99 |
| 5092 | 2048 | 1913 | 14A | ·1672 | 0·0725 | 22·122 | **1991** | 3 | 1 | Ap | 14 | ·24 | 4 | Je | 12 | ·30 | 5 | Jl | 11 | ·83 | 7 | Au | 10 | ·36 | 1 | S | 8 | ·89 |
|  |  |  |  |  |  |  |  |  | 2 | My | 13 | ·77 |  |  |  |  |  |  |  |  |  |  |  |  |  |  |  |  |
| 5093 | 2049 | 1914 | 13A | ·4260 | 10·7114 | 20·255 | **1992** | 4 | 7 | My | 2 | ·14 | 1 | My | 31○ | ·67 | 3 | Je | 30 | ·20 | 4 | Jl | 29 | ·73 | 6 | Au | 28 | ·26 |
| 5094 | 2050 | 1915 | 13A | ·6848 | 7·8197 | 16·412 | **1993** | 6 | 4 | Ap | 21 | ·50 | 6 | My | 21●○ | ·03 | 7 | Je | 19 | ·56 | 2 | Jl | 19 | ·09 | 3 | Au | 17 | ·63 |
|  |  |  |  |  |  |  |  |  |  |  |  |  |  |  |  |  |  |  |  |  |  |  |  |  | 5 | S | 16 | ·16 |
| 5095 | 2051 | 1916 | 13A | ·9435 | 26·4586 | 14·546 | **1994** | 7 | 3 | My | 10●○ | ·40 | 4 | Je | 8 | ·93 | 6 | Jl | 8 | ·46 | 7 | Au | 6 | ·99 | 2 | S | 5 | ·52 |
| 5096 | 2052 | 1917 | 14A | ·2023 | 15·5669 | 10·703 | **1995** | 1 | 7 | Ap | 29● | ·77 | 2 | My | 29 | ·30 | 3 | Je | 27 | ·83 | 5 | Jl | 27 | ·36 | 6 | Au | 25 | ·89 |
| 5097 | 2053 | 1918 | 13A | ·4610 | 4·6752 | 6·860 | **1996** | 2 | 5 | Ap | 18 | ·13 | 6 | My | 17 | ·67 | 1 | Je | 16 | ·20 | 4 | Au | 14 | ·26 | 5 | S | 12○ | ·79 |
|  |  |  |  |  |  |  |  |  |  |  |  |  |  |  |  |  | 2 | Jl | 15 | ·73 |  |  |  |  |  |  |  |  |
| 5098 | 2054 | 1919 | 13A | ·7198 | 23·3141 | 4·993 | **1997** | 4 | 4 | My | 7 | ·03 | 5 | Je | 5 | ·56 | 7 | Jl | 5 | ·09 | 1 | Au | 3 | ·62 | 3 | S | 2○ | ·15 |
| 5099 | 2055 | 1920 | 13A | ·9785 | 12·4224 | 1·151 | **1998** | 5 | 1 | Ap | 26 | ·40 | 2 | My | 25 | ·93 | 4 | Je | 24 | ·46 | 5 | Jl | 23 | ·99 | 7 | Au | 22● | ·52 |
| 5100 | 2056 | 1921 | 14A | ·2373 | 1·5307 | 24·862 | **1999** | 6 | 5 | Ap | 15 | ·77 | 7 | My | 15 | ·30 | 3 | Jl | 13○ | ·36 | 4 | Au | 11● | ·89 | 6 | S | 10 | ·42 |
|  |  |  |  |  |  |  |  |  |  |  |  |  | 1 | Je | 13 | ·83 |  |  |  |  |  |  |  |  |  |  |  |  |

## TABLE X—A.

### Distance of Sun from the Moon's Node at the Commencement of each Solar Year.

INCREASE FOR CENTURIES.

| Christian Era. | Kaliyuga. | Days. |  | Christian Era. | Kaliyuga. | Days. |  | INCREASE Yr. | Incr. Days. |
|---|---|---|---|---|---|---|---|---|---|
| B.C. **1** | 3101 | 62·91 | Modern figures (less the Sodhya of Surya Siddhanta). | **1000** | 4101 | 159·27 | Surya Siddhanta without bija but less Sodhya. | **1** | 18·63 |
| A.D. **100** | 3201 | 20·45 |  | **1100** | 4201 | 116·46 |  | **2** | 37·27 |
| **200** | 3301 | 151·30 |  | **1200** | 4301 | 72·44 |  | **3** | 55·90 |
| **300** | 3401 | 109·18 |  | **1300** | 4401 | 29·63 |  | **4** | 74·54 |
| **400** | 3501 | 65·68 |  | **1400** | 4501 | 159·97 |  | **5** | 93·17 |
| **500** | 3601 | 23·76 | Arya Siddhanta (less the Sodhya of Arya Siddhanta). | **1500** | 4601 | 118·20 | Surya Siddhanta with bija but less Sodhya. | **6** | 111·80 |
| **600** | 3701 | 154·79 |  | **1600** | 4701 | 75·91 |  | **7** | 130·44 |
| **700** | 3801 | 112·38 |  | **1700** | 4801 | 33·45 |  | **8** | 149·07 |
| **800** | 3901 | 70·04 |  | **1800** | 4901 | 164·12 |  | **9** | 167·71 |
| **900** | 4001 | 27·59 |  | **1900** | 5001 | 120·62 |  | **10** | 18·03 |

## The Canons of Eclipses.

A Solar Eclipse is certain if ⊙ from Node is between 0 and 15·598 or between 157·714 and 173·312 days.
" " doubtful " " " 15·771 " 18·198 " 157·541 " 165·114 "
" " impossible " " " 18·371 " 154·939 days.

**Surya Siddhanta.**

| Asvina | | | | Kartika | | | | Margasira | | | | | Pausha | | | | | Magha | | | | Phalguna | | | | Chaitra | | | |
|---|---|---|---|---|---|---|---|---|---|---|---|---|---|---|---|---|---|---|---|---|---|---|---|---|---|---|---|---|---|
| + 147·65293 | | | | + 177·18353 | | | | + 206·71411 | | | | | + 236·24470 | | | | | + 265·77529 | | | | + 295·30588 | | | | { +324·83647 +354·36705 | | | |
| + 9·880 | | | | + 11·856 | | | | + 13·832 | | | | | + 15·808 | | | | | + 17·784 | | | | + 19·760 | | | | { + 21·736 + 23·712 | | | |
| Week-day | Month | Day | Fraction | Week-day | Month | Day | Fraction | Week-day | Month | Day | Fraction | A.D. | Week-day | Month | Day | Fraction | A.D. | Week-day | Month | Day | Fraction | Week-day | Month | Day | Fraction | Week-day | Month | Day | Fraction |
| 1 | O | 5 | ·37 | 2 | N | 30○ | ·90 | 4 | D | 3 | ·43 | **76** | 5 | Jr | 1 | ·96 | | 7 | Jr | 31 | ·49 | 2 | Mr | 1 | ·02 | 3 | Mr | 30 | ·55 |
| 5 | S | 23 | ·73 | 7 | O | 23● | ·26 | 1 | N | 21 | ·79 | | 3 | D | 21 | ·32 | **77** | 4 | Jr | 19 | ·85 | 6 | F | 18 | ·39 | 7 | Mr | 19○ | ·92 |
| 4 | O | 12 | ·63 | 6 | N | 11 | ·16 | 7 | D | 10 | ·69 | **78** | 2 | Jr | 9 | ·22 | | 3 | F | 7 | ·75 | 5 | Mr | 9○ | ·28 | 6 | Ap | 7 | ·81 |
| 2 | O | 2● | ·00 | 3 | O | 31 | ·53 | 5 | N | 30 | ·06 | | 6 | D | 29 | ·59 | **79** | 1 | Jr | 28 | ·12 | 2 | F | 26●○ | ·65 | 4 | Mr | 28 | ·18 |
| 6 | S | 21 | ·36 | 7 | O | 20 | ·89 | 2 | N | 19 | ·43 | | 3 | D | 18 | ·96 | **80** | 5 | Jr | 17 | ·49 | 7 | F | 16● | ·02 | 1 | Mr | 16 | ·55 |
| 5 | O | 9 | ·26 | 6 | N | 7 | ·79 | 1 | D | 7 | ·32 | **81** | 2 | Jr | 5 | ·85 | | 4 | F | 4 | ·38 | 5 | Mr | 5 | ·91 | 7 | Ap | 4 | ·45 |
| 2 | S | 28 | ·63 | 4 | O | 28 | ·16 | 5 | N | 26 | ·69 | | 7 | D | 26○ | ·22 | **82** | 1 | Jr | 24 | ·75 | 3 | F | 23 | ·28 | 4 | Mr | 24 | ·81 |
| { 7 | S | 18 | ·00 | 3 | N | 16 | ·06 | 4 | D | 15●○ | ·59 | **83** | 6 | Jr | 14 | ·12 | | **Magha** | | | | { 7 | F | 12 | ·65 | 3 | Ap | 12 | ·71 |
| { 1 | O | 17 | ·53 | | | | | | | | | | | | | | | **Kshaya** | | | | { 2 | Mr | 14 | ·18 | | | | |
| 5 | O | 6 | ·89 | 7 | N | 5 | .42 | 1 | D | 4● | ·95 | **84** | 3 | Jr | 3 | ·49 | | 5 | F | 2 | ·02 | 6 | Mr | 2 | ·55 | 1 | Ap | 1 | ·08 |
| 3 | S | 25 | ·26 | 4 | O | 24 | .79 | 6 | N | 23 | ·32 | | 7 | D | 22 | ·85 | **85** | 2 | Jr | 21 | ·38 | 3 | F | 19 | ·91 | 5 | Mr | 21 | ·44 |
| 2 | O | 14○ | ·16 | 3 | N | 12● | ·69 | 5 | D | 12 | ·22 | **86** | 6 | Jr | 10 | ·75 | | 1 | F | 9 | ·28 | 2 | Mr | 10 | ·81 | 4 | Ap | 9○ | ·34 |
| 6 | O | 3○ | ·53 | 1 | N | 2 | .06 | 2 | D | 1 | ·59 | | 4 | D | 31 | ·12 | **87** | 5 | Jr | 29 | ·65 | 7 | F | 28 | ·18 | 1 | Mr | 29● | 71 |
| 3 | S | 22● | ·89 | 5 | O | 22 | ·42 | 6 | N | 20 | ·95 | | 1 | D | 20 | ·48 | **88** | 3 | Jr | 19 | ·01 | 4 | F | 17 | ·55 | 6 | Mr | 18● | ·08 |
| 2 | O | 10 | ·79 | 4 | N | 9 | ·32 | 5 | D | 8 | ·85 | **89** | 7 | Jr | 7 | ·38 | | 1 | F | 5○ | ·91 | 3 | Mr | 7 | ·44 | 4 | Ap | 5 | ·97 |
| 7 | S | 30 | ·16 | 1 | O | 29 | ·69 | 3 | N | 28 | ·22 | | 4 | D | 27 | ·75 | **90** | 6 | Jr | 26○ | ·28 | 7 | F | 24 | ·81 | 2 | Mr | 26 | 34 |
| 4 | S | 19 | ·52 | 6 | O | 19 | ·05 | 7 | N | 17 | ·59 | | 2 | D | 17 | ·12 | **91** | 3 | Jr | 15○ | ·65 | 5 | F | 14 | ·18 | 6 | Mr | 15 | ·71 |
| 3 | O | 8 | ·42 | 4 | N | 6 | ·95 | 6 | D | 6○ | ·48 | **92** | 1 | Jr | 5 | ·01 | | 2 | F | 3 | ·54 | 4 | Mr | 4 | ·07 | 5 | Ap | 2 | ·61 |
| 7 | S | 26 | ·79 | 2 | O | 26 | ·32 | 3 | N | 24○ | ·85 | | 5 | D | 24● | ·38 | **93** | 6 | Jr | 22 | ·91 | 1 | F | 21 | ·44 | 2 | Mr | 22 | ·97 |
| 6 | O | 15 | ·69 | 1 | N | 14○ | ·22 | 2 | D | 13 | ·75 | **94** | 4 | Jr | 12 | ·28 | | 5 | F | 10 | ·81 | 7 | Mr | 12 | ·34 | 1 | Ap | 10 | ·87 |
| 4 | O | 5 | ·05 | 5 | N | 3● | ·58 | 7 | D | 3 | ·11 | **95** | 1 | Jr | 1 | ·65 | | 3 | Jr | 31 | ·18 | 4 | Mr | 1 | ·71 | 6 | Mr | 31○ | ·24 |
| 1 | S | 24 | ·42 | 2 | O | 23● | ·95 | 4 | N | 22 | ·48 | | 6 | D | 22 | ·01 | **96** | 7 | Jr | 20 | ·54 | 2 | F | 19 | ·07 | 3 | Mr | 19○ | ·60 |
| 7 | O | 12● | ·32 | 1 | N | 10 | ·85 | 3 | D | 10 | ·38 | **97** | 4 | Jr | 8 | ·91 | | 6 | F | 7 | ·44 | 7 | Mr | 8● | ·97 | 2 | Ap | 7 | ·50 |
| 4 | O | 1 | ·69 | 6 | O | 31 | ·22 | 7 | N | 29 | ·75 | | 2 | D | 29 | ·28 | **98** | 3 | Jr | 27 | ·81 | 5 | F | 26● | ·34 | 6 | Mr | 27 | ·87 |
| 2 | S | 21 | ·05 | 3 | O | 20 | ·58 | 5 | N | 19 | ·11 | | 6 | D | 18 | ·64 | **99** | 1 | Jr | 17 | ·17 | 2 | F | 15● | ·71 | 4 | Mr | 17 | ·24 |
| 7 | O | 9 | ·95 | 2 | N | 8 | ·48 | 4 | D | 8 | ·01 | **20** | 5 | Jr | 6○ | ·54 | **00** | 7 | F | 5 | ·07 | 1 | Mr | 5 | ·60 | 3 | Ap | 4 | ·13 |

## TABLE X—A.

### Distance of Sun from the Moon's Node at the Commencement of each Solar Year.

FOR ODD YEARS. INCREASE FOR ODD YEARS.

| Yr. | Increase. days. | Yr. | Increase. days. | Yr. | Increase. days. | Yr. | Increase. days. | Yr. | Increase. days. | Yr. | Increase. days. | Yr. | Increase. days. | Yr. | Increase. days. | Yr. | Incr. days. |
|---|---|---|---|---|---|---|---|---|---|---|---|---|---|---|---|---|---|
| **11** | 31·66 | **21** | 44·69 | **31** | 57·72 | **41** | 70·75 | **51** | 83·78 | **61** | 96·81 | **71** | 109·84 | **81** | 122·86 | **91** | 135·89 |
| **12** | 50·30 | **22** | 63·33 | **32** | 76·35 | **42** | 89·38 | **52** | 102·41 | **62** | 115·44 | **72** | 128·47 | **82** | 141·50 | **92** | 154·53 |
| **13** | 68·93 | **23** | 81·96 | **33** | 94·99 | **43** | 108·01 | **53** | 121·05 | **63** | 134·07 | **73** | 147·10 | **83** | 160·13 | **93** | 173·16 |
| **14** | 87·56 | **24** | 100·59 | **34** | 113·62 | **44** | 126·65 | **54** | 139·68 | **64** | 152·71 | **74** | 165·74 | **84** | 5·45 | **94** | 18·48 |
| **15** | 106·20 | **25** | 119·23 | **35** | 132·66 | **45** | 145·29 | **55** | 158·31 | **65** | 171·34 | **75** | 11·06 | **85** | 24.08 | **95** | 37·12 |
| **16** | 124·83 | **26** | 137·86 | **36** | 150·89 | **46** | 163·92 | **56** | 3·64 | **66** | 16·66 | **76** | 29·69 | **86** | 42·72 | **96** | 55·75 |
| **17** | 143·47 | **27** | 156·50 | **37** | 169·52 | **47** | 9·24 | **57** | 22·27 | **67** | 35·30 | **77** | 48·33 | **87** | 61·36 | **97** | 74·39 |
| **18** | 162·10 | **28** | 1·82 | **38** | 14·85 | **48** | 27·88 | **58** | 40·90 | **68** | 53·93 | **78** | 66·96 | **88** | 79·99 | **98** | 93·02 |
| **19** | 7·42 | **29** | 20·45 | **39** | 33·48 | **49** | 46·51 | **59** | 59·54 | **69** | 72·57 | **79** | 85·60 | **89** | 98·63 | **99** | 111·65 |
| **20** | 26·06 | **30** | 39·09 | **40** | 52·11 | **50** | 65·14 | **60** | 78·17 | **70** | 91·20 | **80** | 104·23 | **90** | 117·26 | | |

## The Canons of Eclipses.

A **Lunar** Eclipse is certain if ⊙ from Node is between 0 and 10·052 or between 163·260 and 173·312 days.
,, ,, doubtful ,, ,, ,, 10·225 ,, 13·345 ,, 163·087 ,, 159·967 ,,
,, ,, impossible ,, ,, ,, 13·518 ,, 159·794 days.

# TABLE XI.— LONGEST INTERVAL IN DAYS FROM NEW-MOON TO ENDING MOMENT

[The largest correction is—2·20890 for Nakshatras

## NAKSHATRAS.

| Names of Nakshatras. | Order | I Ordinarily Vaisakha | Order | II Ordinarily Jyeshta | Order | III Ordinarily Ashada | Order | IV Ordinarily Sravana | Order | V Ordinarily Bhadrapada | Order | VI Ordinarily Asvina | Order | VII Ordinarily Kartika | Order | VIII Ordinarily Margasira |
|---|---|---|---|---|---|---|---|---|---|---|---|---|---|---|---|---|
| Revati ... ... | 27 | 0·1624 | 3 | 0·9892 | 5 | 0·8041 | 7 | 0·6190 | 9 | 0·4339 | 11 | 0·2489 | 13 | 0·0638 | 16 | 0·8906 |
| Asvini ... ... | 1 | 1·1743 | 4 | 2·0011 | 6 | 1·8160 | 8 | 1·6309 | 10 | 1·4459 | 12 | 1·2608 | 14 | 1·0757 | 17 | 1·9025 |
| Bharani ... ... | 2 | 2·1862 | 5 | 3·0130 | 7 | 2·8279 | 9 | 2·6428 | 11 | 2·4578 | 13 | 2·2727 | 15 | 2·0876 | 18 | 2·9144 |
| Krittika ... ... | 3 | 3·1981 | 6 | 4·0249 | 8 | 3·8398 | 10 | 3·6548 | 12 | 3·4697 | 14 | 3·2846 | 16 | 3·0995 | 19 | 3·9263 |
| Rohini ... ... | 4 | 4·2100 | 7 | 5·0368 | 9 | 4·8518 | 11 | 4·6667 | 13 | 4·4816 | 15 | 4·2965 | 17 | 4·1114 | 20 | 4·9383 |
| Mrigasira ... ... | 5 | 5·2219 | 8 | 6·0488 | 10 | 5·8637 | 12 | 5·6786 | 14 | 5·4935 | 16 | 5·3084 | 18 | 5·1233 | 21 | 5·9502 |
| Ardhra ... ... | 6 | 6·2338 | 9 | 7·0607 | 11 | 6·8756 | 13 | 6·6905 | 15 | 6·5054 | 17 | 6·3203 | 19 | 6·1353 | 22 | 6·9621 |
| Punarvasu ... ... | 7 | 7·2458 | 10 | 8·0726 | 12 | 7·8875 | 14 | 7·7024 | 16 | 7·5173 | 18 | 7·3323 | 20 | 7·1472 | 23 | 7·9740 |
| Pushya ... ... | 8 | 8·2577 | 11 | 9·0845 | 13 | 8·8994 | 15 | 8.7143 | 17 | 8·5292 | 19 | 8·3442 | 21 | 8·1591 | 24 | 8·9859 |
| Aslesha ... ... | 9 | 9·2696 | 12 | 10·0964 | 14 | 9·9113 | 16 | 9·7262 | 18 | 9·5412 | 20 | 9·3561 | 22 | 9·1710 | 25 | 9·9978 |
| Magha ... ... | 10 | 10·2815 | 13 | 11·1083 | 15 | 10·9232 | 17 | 10·7382 | 19 | 10·5531 | 21 | 10·3680 | 23 | 10·1829 | 26 | 11·0097 |
| Purna Phalguni ... | 11 | 11·2934 | 14 | 12·1202 | 16 | 11·9352 | 18 | 11·7501 | 20 | 11·5650 | 22 | 11·3799 | 24 | 11·1948 | 27 | 12·0217 |
| Uttara Phalguni ... | 12 | 12·3053 | 15 | 13·1322 | 17 | 12·9471 | 19 | 12·7620 | 21 | 12·5769 | 23 | 12·3918 | 25 | 12·2067 | 1 | 13·0336 |
| Hasta ... ... | 13 | 13·3172 | 16 | 14·1441 | 18 | 13·9590 | 20 | 13·7739 | 22 | 13·5888 | 24 | 13·4037 | 26 | 13·2187 | 2 | 14·0455 |
| Chittra ... ... | 14 | 14·3292 | 17 | 15·1567 | 19 | 14·9709 | 21 | 14·7858 | 23 | 14·6007 | 25 | 14·4156 | 27 | 14·2306 | 3 | 15·0574 |
| Svati ... ... ... | 15 | 15·3411 | 18 | 16·1679 | 20 | 15·9828 | 22 | 15·7977 | 24 | 15·6126 | 26 | 15·4276 | 1 | 15·2425 | 4 | 16·0693 |
| Visakha ... ... | 16 | 16·3530 | 19 | 17·1798 | 21 | 16·9947 | 23 | 16·8096 | 25 | 16·6246 | 27 | 16·4395 | 2 | 16·2544 | 5 | 17·0812 |
| Anuradha ... ... | 17 | 17·3649 | 20 | 18·1917 | 22 | 18·0066 | 24 | 17·8216 | 26 | 17·6365 | 1 | 17·4514 | 3 | 17·2663 | 6 | 18·0931 |
| Jyeshtha ... ... | 18 | 18·3768 | 21 | 19·2036 | 23 | 19·0186 | 25 | 18·8335 | 27 | 18·6484 | 2 | 18·4633 | 4 | 18·2782 | 7 | 19·1051 |
| Mula ... ... ... | 19 | 19·3887 | 22 | 20·2156 | 24 | 20·0305 | 26 | 19·8454 | 1 | 19·6603 | 3 | 19·4752 | 5 | 19·2901 | 8 | 20·1170 |
| Purva Ashada ... ... | 20 | 20·4006 | 23 | 21·2275 | 25 | 21·0424 | 27 | 20·8573 | 2 | 20·6722 | 4 | 20·4871 | 6 | 20·3020 | 9 | 21·1289 |
| Uttara Ashada ... ... | 21 | 21·4126 | 24 | 22·2394 | 26 | 22·0543 | 1 | 21·8692 | 3 | 21·6841 | 5 | 21·4990 | 7 | 21·3140 | 10 | 22·1408 |
| Sravana ... ... | 22 | 22 4245 | 25 | 23·2513 | 27 | 23·0662 | 2 | 22·8811 | 4 | 22·6960 | 6 | 22·5110 | 8 | 22·3259 | 11 | 23·1527 |
| Sravishta or Danishta ... | 23 | 23·4364 | 26 | 24·2632 | 1 | 24·0781 | 3 | 23·8930 | 5 | 23·7080 | 7 | 23·5229 | 9 | 23·3378 | 12 | 24·1646 |
| Satabhisaj or Sataraka ... | 24 | 24·4483 | 27 | 25·2751 | 2 | 25·0900 | 4 | 24·9050 | 6 | 24·7199 | 8 | 24·5348 | 10 | 24·3497 | 13 | 25·1765 |
| Purva Bhadrapada ... | 25 | 25·4602 | 1 | 26·2870 | 3 | 26·1020 | 5 | 25·9169 | 7 | 25·7318 | 9 | 25·5467 | 11 | 25·3616 | 14 | 26·1884 |
| Uttara Bhadrapada ... | 26 | 26·4721 | 2 | 27·2990 | 4 | 27·1139 | 6 | 26·9288 | 8 | 26·7437 | 10 | 26·5586 | 12 | 26·3735 | 15 | 27·2004 |
| Revati ... ... | 27 | 27·4840 | 3 | 28·3109 | 5 | 28·1258 | 7 | 27·9407 | 9 | 27·7556 | 11 | 27·5705 | 13 | 27·3854 | 16 | 28·2123 |
| | 1 | 28·4960 | 4 | 29·3228 | 6 | 29·1377 | 8 | 28·9526 | 10 | 28·7675 | 12 | 28·5824 | 14 | 28·3974 | 17 | 29·2242 |
| | 2 | 29·5079 | | | | | | | | | | | 15 | 29·4093 | | |

## YOGAS.

| Names of Yogas. | Order | I | Order | II | Order | III | Order | IV | Order | V | Order | VI | Order | VII | Order | VIII |
|---|---|---|---|---|---|---|---|---|---|---|---|---|---|---|---|---|
| Vaidhriti ... ... | 27 | 0·3021 | 5 | 0·8992 | 9 | 0·5548 | 13 | 0·2104 | 18 | 0·8075 | 22 | 0·4631 | 26 | 0·1187 | 4 | 0·7157 |
| Vishkamba ... ... | 1 | 1·2436 | 6 | 1·8407 | 10 | 1·4963 | 14 | 1·1519 | 19 | 1·7490 | 23 | 1·4046 | 27 | 1·0601 | 5 | 1·6572 |
| Priti ... ... ... | 2 | 2·1851 | 7 | 2·7822 | 11 | 2·4378 | 15 | 2·0934 | 20 | 2·6905 | 24 | 2·3460 | 1 | 2·0016 | 6 | 2·5987 |
| Ayushmat ... ... | 3 | 3·1266 | 8 | 3·7237 | 12 | 3·3793 | 16 | 3·0349 | 21 | 3·6319 | 25 | 3·2875 | 2 | 2·9431 | 7 | 3·5402 |
| Saubhagya ... ... | 4 | 4·0681 | 9 | 4·6652 | 13 | 4·3208 | 17 | 3·9764 | 22 | 4·5734 | 26 | 4·2290 | 3 | 3·8846 | 8 | 4·4817 |
| Sobhana ... ... | 5 | 5.0096 | 10 | 5·6067 | 14 | 5·2623 | 18 | 4·9178 | 23 | 5·5149 | 27 | 5·1705 | 4 | 4·8261 | 9 | 5·4232 |
| Atiganda ... ... | 6 | 5·9511 | 11 | 6·5482 | 15 | 6·2037 | 19 | 5·8593 | 24 | 6·4564 | 1 | 6·1120 | 5 | 5·7676 | 10 | 6·3647 |
| Sukarman ... ... | 7 | 6·8926 | 12 | 7·4896 | 16 | 7·1452 | 20 | 6·8008 | 25 | 7·3979 | 2 | 7·0535 | 6 | 6·7091 | 11 | 7·3062 |
| Dhriti ... ... | 8 | 7·8341 | 13 | 8·4311 | 17 | 8·0867 | 21 | 7·7423 | 26 | 8·3394 | 3 | 7·9950 | 7 | 7·6506 | 12 | 8·2477 |
| Sula ... ... ... | 9 | 8·7755 | 14 | 9·3726 | 18 | 9·0282 | 22 | 8·6838 | 27 | 9·2809 | 4 | 8·9365 | 8 | 8·5921 | 13 | 9·1891 |
| Ganda ... ... | 10 | 9·7170 | 15 | 10·3141 | 19 | 9·9697 | 23 | 9·6253 | 1 | 10·2224 | 5 | 9·8780 | 9 | 9·5335 | 14 | 10·1306 |
| Vriddhi ... ... | 11 | 10·6585 | 16 | 11·2556 | 20 | 10·9112 | 24 | 10·5668 | 2 | 11·1639 | 6 | 10·8194 | 10 | 10·4750 | 15 | 11·0721 |
| Dhruva ... ... | 12 | 11·6000 | 17 | 12·1971 | 21 | 11·8527 | 25 | 11·5083 | 3 | 12·1053 | 7 | 11·7609 | 11 | 11·4165 | 16 | 12·0136 |
| Vyaghata ... ... | 13 | 12·5415 | 18 | 13·1386 | 22 | 12·7942 | 26 | 12·4498 | 4 | 13·0468 | 8 | 12·7024 | 12 | 12·3580 | 17 | 12·9551 |
| Harshana ... ... | 14 | 13·4830 | 19 | 14·0800 | 23 | 13·7357 | 27 | 13·3912 | 5 | 13 9813 | 9 | 13·6439 | 13 | 13·2995 | 18 | 13·8966 |
| Vajra ... ... | 15 | 14·4245 | 20 | 15·0216 | 24 | 14·6771 | 1 | 14·3327 | 6 | 14·9298 | 10 | 14·5854 | 14 | 14·2410 | 19 | 14·8381 |
| Siddhi ... ... | 16 | 15·3660 | 21 | 15·9630 | 25 | 15·6186 | 2 | 15·2742 | 7 | 15·8713 | 11 | 15·5269 | 15 | 15·1825 | 20 | 15·7796 |
| Vyatipata ... ... | 17 | 16·3075 | 22 | 16·9045 | 26 | 16·5601 | 3 | 16·2157 | 8 | 16·8128 | 12 | 16·4684 | 16 | 16·1240 | 21 | 16·7211 |
| Variyas ... ... | 18 | 17·2489 | 23 | 17·8460 | 27 | 17·5016 | 4 | 17·1572 | 9 | 17·7543 | 13 | 17·4099 | 17 | 17·0655 | 22 | 17·6625 |
| Parigha ... ... | 19 | 18·1904 | 24 | 18·7875 | 1 | 18·4431 | 5 | 18·0987 | 10 | 18·6958 | 14 | 18·3514 | 18 | 18·0070 | 23 | 18·6040 |
| Siva ... ... ... | 20 | 19·1319 | 25 | 19·7290 | 2 | 19·3846 | 6 | 19·0402 | 11 | 19·6373 | 15 | 19·2929 | 19 | 18·9484 | 24 | 19·5455 |
| Siddha ... ... | 21 | 20·0734 | 26 | 20·6705 | 3 | 20·3261 | 7 | 19·9817 | 12 | 20·5788 | 16 | 20·2343 | 20 | 19·8899 | 25 | 20·4870 |
| Sadhya ... ... | 22 | 21·0149 | 27 | 21·6120 | 4 | 21·2676 | 8 | 20·9232 | 13 | 21·5202 | 17 | 21·1758 | 21 | 20·8314 | 26 | 21·4285 |
| Subha ... ... | 23 | 21·9564 | 1 | 22·5535 | 5 | 22·2091 | 9 | 21·8647 | 14 | 22·4617 | 18 | 22·1173 | 22 | 21·7729 | 27 | 22·3700 |
| Sukla ... ... | 24 | 22·8979 | 2 | 23·4950 | 6 | 23·1506 | 10 | 22·8061 | 15 | 23·4032 | 19 | 23·0588 | 23 | 22.7144 | 1 | 23·3115 |
| Brahman ... ... | 25 | 23·8394 | 3 | 24·4665 | 7 | 24·0920 | 11 | 23·7476 | 16 | 24·3447 | 20 | 24·0003 | 24 | 23·6559 | 2 | 24·2530 |
| Indra ... ... ... | 26 | 24·7809 | 4 | 25·3780 | 8 | 25·0335 | 12 | 24·6891 | 17 | 25·2862 | 21 | 24·9418 | 25 | 24·5974 | 3 | 25·1945 |
| Vaidhriti ... ... | 27 | 25·7224 | 5 | 26·3194 | 9 | 25·8750 | 13 | 25·6306 | 18 | 26·2277 | 22 | 25·8833 | 26 | 25·5389 | 4 | 26·1360 |
| | 1 | 26·6638 | 6 | 27·2609 | 10 | 26·9165 | 14 | 26·5721 | 19 | 27·1692 | 23 | 26·8248 | 27 | 26·4804 | 5 | 27·0774 |
| | 2 | 27·6053 | 7 | 28·2024 | 11 | 27·8580 | 15 | 27·5136 | 20 | 28·1107 | 24 | 27·7663 | 1 | 27·4219 | 6 | 28·0189 |
| | 3 | 28·5468 | 8 | 29·1439 | 12 | 28·7995 | 16 | 28·4551 | 21 | 29·0522 | 25 | 28·7078 | 2 | 28·3633 | 7 | 28·9604 |
| | 4 | 29·4883 | | | | | 17 | 29·3966 | | | | | 3 | 29·3048 | | |

## OF EACH NAKSHATRA AND YOGA

and—4·11036 for Yogas.]

### NAKSHATRAS.

| Order | IX Ordinarily Pausha | Order | X Ordinarily Magha | Order | XI Ordinarily Phalguna | Order | XII Ordinarily Chaitra when no Adhika Masa | Order | XIII Chaitra when there is Adhika Masa |
|---|---|---|---|---|---|---|---|---|---|
| **18** | 0·7055 | **20** | 0·5204 | **22** | 0·3353 | **24** | 0·1503 | **27** | 0·9771 |
| **19** | 1·7174 | **21** | 1·5323 | **23** | 1·3473 | **25** | 1·1622 | **1** | 1·9890 |
| **20** | 2·7293 | **22** | 2·5443 | **24** | 2·3592 | **26** | 2·1741 | **2** | 3·0009 |
| **21** | 3·7413 | **23** | 3·5562 | **25** | 3·3711 | **27** | 3·1860 | **3** | 4·0128 |
| **22** | 4·7532 | **24** | 4·5682 | **26** | 4·3830 | **1** | 4·1979 | **4** | 5·0247 |
| **23** | 5·7651 | **25** | 5·5800 | **27** | 5·3949 | **2** | 5·2098 | **5** | 6·0367 |
| **24** | 6·7770 | **26** | 6·5919 | **1** | 6·4068 | **3** | 6·2217 | **6** | 7·0486 |
| **25** | 7·7889 | **27** | 7·6038 | **2** | 7·4187 | **4** | 7·2337 | **7** | 8·0605 |
| **26** | 8·8008 | **1** | 8·6157 | **3** | 8·4307 | **5** | 8·2456 | **8** | 9·0724 |
| **27** | 9·8127 | **2** | 9·6277 | **4** | 9·4426 | **6** | 9·2575 | **9** | 10·0843 |
| **1** | 10·8247 | **3** | 10·6396 | **5** | 10·4545 | **7** | 10·2694 | **10** | 11·0962 |
| **2** | 11·8366 | **4** | 11·6515 | **6** | 11·4664 | **8** | 11·2813 | **11** | 12·1081 |
| **3** | 12·8485 | **5** | 12·6634 | **7** | 12·4783 | **9** | 12·2932 | **12** | 13·1201 |
| **4** | 13·8604 | **6** | 13·6753 | **8** | 13·4902 | **10** | 13·3051 | **13** | 14·1320 |
| **5** | 14·8723 | **7** | 14·6872 | **9** | 14·5021 | **11** | 14·3171 | **14** | 15·1439 |
| **6** | 15·8842 | **8** | 15·6991 | **10** | 15·5141 | **12** | 15·3290 | **15** | 16·1558 |
| **7** | 16·8961 | **9** | 16·7111 | **11** | 16·5260 | **13** | 16·3409 | **16** | 17·1677 |
| **8** | 17·9081 | **10** | 17·7230 | **12** | 17·5379 | **14** | 17·3528 | **17** | 18·1796 |
| **9** | 18·9200 | **11** | 18·7349 | **13** | 18·5498 | **15** | 18·3647 | **18** | 19·1915 |
| **10** | 19·9319 | **12** | 19·7468 | **14** | 19·5617 | **16** | 19·3766 | **19** | 20·2035 |
| **11** | 20·9438 | **13** | 20·7587 | **15** | 20·5736 | **17** | 20·3885 | **20** | 21·2154 |
| **12** | 21·9557 | **14** | 21·7706 | **16** | 21·5855 | **18** | 21·4005 | **21** | 22·2273 |
| **13** | 22·9676 | **15** | 22·7825 | **17** | 22·5975 | **19** | 22·4124 | **22** | 23·2392 |
| **14** | 23·9795 | **16** | 23·7945 | **18** | 23·6094 | **20** | 23·4243 | **23** | 24·2511 |
| **15** | 24·9915 | **17** | 24·8064 | **19** | 24·6213 | **21** | 24·4362 | **24** | 25·2630 |
| **16** | 26·0034 | **18** | 25·8183 | **20** | 25·6332 | **22** | 25·4481 | **25** | 26·2749 |
| **17** | 27·0153 | **19** | 26·8302 | **21** | 26·6451 | **23** | 26·4600 | **26** | 27·2869 |
| **18** | 28·0272 | **20** | 27·8421 | **22** | 27·6570 | **24** | 27·4719 | **27** | 28·2988 |
| **19** | 29·0391 | **21** | 28·8540 | **23** | 28·6689 | **25** | 28·4839 | **1** | 29·3107 |
|  |  |  |  |  |  | **26** | 29·4958 |  |  |

### YOGAS.

| Order | IX | Order | X | Order | XI | Order | XII | Order | XIII |
|---|---|---|---|---|---|---|---|---|---|
| **8** | 0·3713 | **12** | 0·0269 | **17** | 0·6240 | **21** | 0·2796 | **26** | 0·8767 |
| **9** | 1·3128 | **13** | 0·9684 | **18** | 1·5655 | **22** | 1·2211 | **27** | 1·8182 |
| **10** | 2·2543 | **14** | 1·9099 | **19** | 2·5070 | **23** | 2·1626 | **1** | 2·7596 |
| **11** | 3·1958 | **15** | 2·8514 | **20** | 3·4485 | **24** | 3·1041 | **2** | 3·7011 |
| **12** | 4·1373 | **16** | 3·7929 | **21** | 4·3900 | **25** | 4·0455 | **3** | 4·6426 |
| **13** | 5·0788 | **17** | 4·7344 | **22** | 5·3314 | **26** | 4·9870 | **4** | 5·5841 |
| **14** | 6·0203 | **18** | 5·6759 | **23** | 6·2729 | **27** | 5·9285 | **5** | 6·5256 |
| **15** | 6·9618 | **19** | 6·6173 | **24** | 7·2144 | **1** | 6·8700 | **6** | 7·4671 |
| **16** | 7·9032 | **20** | 7·5588 | **25** | 8·1559 | **2** | 7·8115 | **7** | 8·4086 |
| **17** | 8·8447 | **21** | 8·5003 | **26** | 9·0974 | **3** | 8·7530 | **8** | 9·3501 |
| **18** | 9·7862 | **22** | 9·4418 | **27** | 10·0389 | **4** | 9·6945 | **9** | 10·2916 |
| **19** | 10·7277 | **23** | 10·3833 | **1** | 10·9804 | **5** | 10·6360 | **10** | 11·2330 |
| **20** | 11·6692 | **24** | 11·3248 | **2** | 11·9219 | **6** | 11·5775 | **11** | 12·1745 |
| **21** | 12·6107 | **25** | 12·2663 | **3** | 12·8634 | **7** | 12·5189 | **12** | 13·1160 |
| **22** | 13·5522 | **26** | 13·2078 | **4** | 13·8048 | **8** | 13·4604 | **13** | 14·0575 |
| **23** | 14·4937 | **27** | 14·1493 | **5** | 14·7463 | **9** | 14·4019 | **14** | 14·9990 |
| **24** | 15·4352 | **1** | 15·0907 | **6** | 15·6878 | **10** | 15·3434 | **15** | 15·9405 |
| **25** | 16·3766 | **2** | 16·0322 | **7** | 16·6293 | **11** | 16·2849 | **16** | 16·8820 |
| **26** | 17·3181 | **3** | 16·9737 | **8** | 17·5708 | **12** | 17·2264 | **17** | 17·8235 |
| **27** | 18·2596 | **4** | 17·9152 | **9** | 18·5123 | **13** | 18·1679 | **18** | 18·7650 |
| **1** | 19·2011 | **5** | 18·8567 | **10** | 19·4538 | **14** | 19·1094 | **19** | 19·7065 |
| **2** | 20·1426 | **6** | 19·7982 | **11** | 20·3953 | **15** | 20·0509 | **20** | 20·6479 |
| **3** | 21·0841 | **7** | 20·7397 | **12** | 21·3868 | **16** | 20·9924 | **21** | 21·5894 |
| **4** | 22·0256 | **8** | 21·6812 | **13** | 22·2783 | **17** | 21·9338 | **22** | 22·5309 |
| **5** | 22·9671 | **9** | 22·6227 | **14** | 23·2197 | **18** | 22·8753 | **23** | 23·4724 |
| **6** | 23·9086 | **10** | 23·5646 | **15** | 24·1612 | **19** | 23·8168 | **24** | 24·4139 |
| **7** | 24·8501 | **11** | 24·5056 | **16** | 25·1027 | **20** | 24·7583 | **25** | 25·3554 |
| **8** | 25·7915 | **12** | 25·4471 | **17** | 26·0442 | **21** | 25·6999 | **26** | 26·2969 |
| **9** | 26·7330 | **13** | 26·3886 | **18** | 26·9857 | **22** | 26·6413 | **27** | 27·2384 |
| **10** | 27·6745 | **14** | 27·3301 | **19** | 27·9272 | **23** | 27·5828 | **1** | 28·1799 |
| **11** | 28·6160 | **15** | 28·2716 | **20** | 28·8687 | **24** | 28·5243 | **2** | 29·1214 |
|  |  | **16** | 29·2131 |  |  | **25** | 29·4658 |  |  |

## ANNUAL CORRECTION

**Argument** :—Date of appearance of 1st New-Moon in each Solar Year according to Table X.

### NAKSHATRAS.

| Arg. | Corrn. | Arg. | Corrn. | Arg. | Corrn. | Arg. | Corrn. | Arg. | Corrn. |
|---|---|---|---|---|---|---|---|---|---|
| **1** | ·07480 | **·01** | ·00075 | **·30** | ·02244 | **·59** | ·04413 | **·88** | ·06582 |
| **2** | ·14960 | **.02** | ·00150 | **·31** | ·02319 | **·60** | ·04488 | **·89** | ·06657 |
| **3** | ·22440 | **·03** | ·00224 | **·32** | ·02394 | **·61** | ·04563 | **·90** | ·06732 |
| **4** | ·29920 | **·04** | ·00299. | **·33** | ·02468 | **·62** | ·04638 | **·91** | ·06807 |
| **5** | ·37400 | **·05** | ·00374 | **·34** | ·02543 | **·63** | ·04712 | **·92** | ·06882 |
| **6** | ·44880 | **·06** | ·00449 | **·35** | ·02618 | **·64** | ·04787 | **·93** | ·06956 |
| **7** | ·52361 | **·07** | ·00524 | **·36** | ·02693 | **·65** | ·04862 | **·94** | ·07031 |
| **8** | ·59841 | **·08** | .00598 | **·37** | ·02768 | **·66** | ·04937 | **·95** | ·07106 |
| **9** | ·67321 | **·09** | ·00673 | **·38** | ·02342 | **·67** | ·05012 | **·96** | ·07181 |
| **10** | ·74801 | **·10** | ·00748 | **·39** | ·02917 | **·68** | ·05086 | **·97** | ·07256 |
| **11** | ·82281 | **·11** | ·00823 | **·40** | ·02992 | **·69** | ·05161 | **·98** | ·07330 |
| **12** | ·89761 | **·12** | ·00898 | **·41** | ·03067 | **·70** | ·05236 | **·99** | ·07405 |
| **13** | ·97241 | **·13** | ·00972 | **·42** | ·03142 | **·71** | ·05311 |  |  |
| **14** | 1·04721 | **·14** | ·01047 | **·43** | ·03216 | **·72** | ·05386 |  |  |
| **15** | 1·12201 | **·15** | ·01122 | **·44** | ·03291 | **·73** | ·05460 |  |  |
| **16** | 1·19681 | **·16** | ·01197 | **·45** | ·03366 | **·74** | ·05535 |  |  |
| **17** | 1·27161 | **·17** | ·01272 | **·46** | ·03441 | **·75** | ·05610 |  |  |
| **18** | 1·34641 | **·18** | ·01346 | **·47** | ·03516 | **·76** | ·05685 |  |  |
| **19** | 1·42121 | **·19** | ·01421 | **·48** | ·03590 | **·77** | ·05760 |  |  |
| **20** | 1·49602 | **·20** | ·01496 | **·49** | ·03665 | **·78** | ·05834 |  |  |
| **21** | 1·57082 | **·21** | ·01571 | **·50** | ·03740 | **·79** | ·05909 |  |  |
| **22** | 1·64562 | **·22** | ·01646 | **·51** | ·03815 | **·80** | ·05984 |  |  |
| **23** | 1·72042 | **·23** | ·01720 | **·52** | ·03890 | **·81** | ·06059 |  |  |
| **24** | 1·79522 | **·24** | ·01795 | **·53** | ·03964 | **·82** | ·06134 |  |  |
| **25** | 1·87002 | **·25** | ·01870 | **·54** | ·04039 | **·83** | ·06208 |  |  |
| **26** | 1·94482 | **·26** | ·01945 | **.55** | ·04114 | **·84** | ·06283 |  |  |
| **27** | 2·01962 | **·27** | ·02020 | **.56** | ·04189 | **·85** | ·06358 |  |  |
| **28** | 2·09442 | **·28** | ·02094 | **.57** | ·04264 | **·86** | ·06433 |  |  |
| **29** | 2·16922 | **·29** | ·02169 | **.58** | ·04338 | **·87** | ·06508 |  |  |

### YOGAS.

| Arg. | Corrn. | Arg. | Corrn. | Arg. | Corrn. | Arg. | Corrn. | Arg. | Corrn. |
|---|---|---|---|---|---|---|---|---|---|
| **1** | 0·13919 | **·01** | ·00139 | **·30** | ·04176 | **·59** | ·08212 | **·88** | ·12249 |
| **2** | 0·27838 | **·02** | ·00278 | **·31** | ·04315 | **·60** | ·08351 | **·89** | ·12388 |
| **3** | 0·41757 | **·03** | ·00417 | **·32** | ·04454 | **·61** | ·08491 | **·90** | ·12527 |
| **4** | 0·55676 | **·04** | ·00557 | **·33** | ·04593 | **·62** | ·08630 | **·91** | ·12666 |
| **5** | 0·69595 | **·05** | ·00696 | **·34** | ·04732 | **·63** | ·08769 | **·92** | ·12805 |
| **6** | 0·83514 | **·06** | ·00835 | **·35** | ·04872 | **·64** | ·08908 | **·93** | ·12945 |
| **7** | 0·97433 | **·07** | ·00974 | **·36** | ·05011 | **·65** | ·09047 | **·94** | ·13084 |
| **8** | 1·11352 | **·08** | ·01113 | **·37** | ·05150 | **·66** | ·09186 | **·95** | ·13223 |
| **9** | 1·25271 | **·09** | ·01253 | **·38** | ·05289 | **·67** | ·09326 | **·96** | ·13362 |
| **10** | 1·39190 | **·10** | ·01392 | **·39** | ·05428 | **·68** | ·09465 | **·97** | ·13601 |
| **11** | 1·53109 | **·11** | ·01530 | **·40** | ·05568 | **·69** | ·09604 | **·98** | ·13641 |
| **12** | 1·67028 | **·12** | ·01670 | **·41** | ·05707 | **·70** | ·09743 | **·99** | ·13780 |
| **13** | 1·80947 | **·13** | ·01809 | **·42** | ·05846 | **·71** | ·09882 |  |  |
| **14** | 1·94866 | **·14** | ·01949 | **·43** | ·05985 | **·72** | ·10022 |  |  |
| **15** | 2·08785 | **·15** | ·02088 | **·44** | ·06124 | **·73** | ·10161 |  |  |
| **16** | 2·22704 | **·16** | ·02227 | **·45** | ·06263 | **·74** | ·10300 |  |  |
| **17** | 2·36623 | **·17** | ·02366 | **·46** | ·06403 | **·75** | ·10439 |  |  |
| **18** | 2·50542 | **·18** | ·02505 | **·47** | ·06542 | **·76** | ·10578 |  |  |
| **19** | 2·64461 | **·19** | ·02645 | **·48** | ·06681 | **·77** | ·10798 |  |  |
| **20** | 2·78380 | **·20** | .02784 | **·49** | ·06820 | **·78** | ·10851 |  |  |
| **21** | 2·92299 | **·21** | .02923 | **·50** | ·06959 | **·79** | ·10996 |  |  |
| **22** | 3·06218 | **·22** | ·03063 | **·51** | ·07098 | **·80** | ·11135 |  |  |
| **23** | 3·20137 | **·23** | ·03201 | **·52** | ·07238 | **·81** | ·11274 |  |  |
| **24** | 3·34056 | **·24** | ·03340 | **·53** | ·07377 | **·82** | ·11413 |  |  |
| **25** | 3·47975 | **·25** | ·03480 | **·54** | ·07516 | **·83** | ·11553 |  |  |
| **26** | 3·61894 | **·26** | ·03619 | **·55** | ·07655 | **·84** | ·11692 |  |  |
| **27** | 3·75813 | **·27** | ·03758 | **·56** | ·07795 | **·85** | ·11831 |  |  |
| **28** | 3·89732 | **·28** | ·03897 | **·57** | ·07934 | **·86** | ·11970 |  |  |
| **29** | 4·03651 | **·29** | ·04036 | **·58** | ·08073 | **·87** | ·12109 |  |  |

## TABLE XI—A. LONGEST INTERVAL IN DAYS, GHATIKAS AND PALAS

**The largest correction for**

**NAKSHATRAS.**

| Names of Nakshatras. | Order | Ordinarily Vaisakha I (d. g. p.) | Order | Ordinarily Jyeshta II (d. g. p.) | Order | Ordinarily Ashada III (d. g. p.) | Order | Ordinarily Sravana IV (d. g. p.) | Order | Ordinarily Bhadrapada V (d. g. p.) |
|---|---|---|---|---|---|---|---|---|---|---|
| Revati ... ... ... ... | **27** | 0 9 45 | **3** | 0 59 21 | **5** | 0 45 15 | **7** | 0 37 8 | **9** | 0 26 2 |
| Asvini ... ... ... ... | **1** | 1 10 28 | **4** | 2 0 4 | **6** | 1 48 53 | **8** | 1 37 51 | **10** | 1 26 45 |
| Bharani ... ... ... ... | **2** | 2 11 10 | **5** | 3 0 47 | **7** | 2 49 41 | **9** | 2 38 34 | **11** | 2 27 28 |
| Krittika (*Tamil* Kiruttigai) ... ... ... | **3** | 3 11 53 | **6** | 4 1 30 | **8** | 3 50 23 | **10** | 3 39 17 | **12** | 3 28 11 |
| Rohini ... ... ... ... | **4** | 4 12 36 | **7** | 5 2 13 | **9** | 4 51 7 | **11** | 4 40 0 | **13** | 4 28 54 |
| Mirgasira (*Tamil* Mirugasiram) ... ... | **5** | 5 13 19 | **8** | 6 2 56 | **10** | 5 51 49 | **12** | 5 40 43 | **14** | 5 29 37 |
| Ardhra (*Tamil* arudra or Tiruvadirai) ... | **6** | 6 14 2 | **9** | 7 3 39 | **11** | 6 52 32 | **13** | 6 41 26 | **15** | 6 30 20 |
| Punarvasu ... ... ... ... | **7** | 7 14 45 | **10** | 8 4 21 | **12** | 7 53 15 | **14** | 7 42 9 | **16** | 7 31 2 |
| Pushya (*Tamil* Pusam) ... ... | **8** | 8 15 28 | **11** | 9 5 4 | **13** | 8 53 58 | **15** | 8 42 52 | **17** | 8 31 45 |
| Aslesha (*Tamil* Ayilyam) ... ... | **9** | 9 16 11 | **12** | 10 5 47 | **14** | 9 54 41 | **16** | 9 43 34 | **18** | 9 32 28 |
| Magha (*Tamil* Magham) ... ... | **10** | 10 16 53 | **13** | 11 6 30 | **15** | 10 55 24 | **17** | 10 44 18 | **19** | 10 33 11 |
| Purva Phalguni (*Tamil* Puram) ... ... | **11** | 11 17 36 | **14** | 12 7 13 | **16** | 11 56 7 | **18** | 11 45 0 | **20** | 11 33 54 |
| Uttara Phalguni (*Tamil* Uttiram) ... ... | **12** | 12 18 19 | **15** | 13 7 56 | **17** | 12 56 50 | **19** | 12 45 43 | **21** | 12 34 37 |
| Hasta (*Tamil* Hastam) ... ... | **13** | 13 19 2 | **16** | 14 8 39 | **18** | 13 57 32 | **20** | 13 46 26 | **22** | 13 35 20 |
| Chittra (*Tamil* Chittirai) ... ... | **14** | 14 19 45 | **17** | 15 9 24 | **19** | 14 58 15 | **21** | 14 47 9 | **23** | 14 36 3 |
| Svati ... ... ... ... | **15** | 15 20 28 | **18** | 16 10 5 | **20** | 15 58 58 | **22** | 15 47 52 | **24** | 15 36 45 |
| Visakha (*Tamil* Visakam) ... ... | **16** | 16 21 11 | **19** | 17 10 47 | **21** | 16 59 41 | **23** | 16 48 35 | **25** | 16 37 20 |
| Anuradha (*Tamil* Anusham) ... ... | **17** | 17 21 54 | **20** | 18 11 30 | **22** | 18 0 24 | **24** | 17 49 18 | **26** | 17 38 11 |
| Jyeshta (*Tamil* Kettai) ... ... | **18** | 18 22 37 | **21** | 19 12 13 | **23** | 19 1 7 | **25** | 18 50 1 | **27** | 18 38 54 |
| Mula (*Tamil* Mulam) ... ... | **19** | 19 23 19 | **22** | 20 12 56 | **24** | 20 1 50 | **26** | 19 50 44 | **1** | 19 39 37 |
| Purva Ashada (*Tamil* Puradam) ... ... | **20** | 20 24 2 | **23** | 21 13 39 | **25** | 21 2 33 | **27** | 20 51 26 | **2** | 20 40 20 |
| Uttara Ashada (*Tamil* Uttiradam) ... ... | **21** | 21 24 45 | **24** | 22 14 22 | **26** | 22 3 16 | **1** | 21 52 9 | **3** | 21 41 3 |
| Sravana (*Tamil* Tiruvonam) ... ... | **22** | 22 25 28 | **25** | 23 15 5 | **27** | 23 3 58 | **2** | 22 52 52 | **4** | 22 41 46 |
| Sravishta or Danishta (*Tamil* Avittam) ... | **23** | 23 26 11 | **26** | 24 15 48 | **1** | 24 4 41 | **3** | 23 53 35 | **5** | 23 42 29 |
| Satabhisaj or Sataraka (*Tamil* Sadayam) ... | **24** | 24 26 54 | **27** | 25 16 30 | **2** | 25 5 24 | **4** | 24 54 18 | **6** | 24 43 2 |
| Purva Bhadrapada (*Tamil* Purattadi) | **25** | 25 27 37 | **1** | 26 17 13 | **3** | 26 6 7 | **5** | 25 55 1 | **7** | 25 43 55 |
| Uttara Bhadrapada (*Tamil* Uttirattadi) ... | **26** | 26 28 20 | **2** | 27 17 56 | **4** | 27 6 50 | **6** | 26 55 44 | **8** | 26 44 37 |
| Revati ... ... ... ... | **27** | 27 29 2 | **3** | 28 18 39 | **5** | 28 7 33 | **7** | 27 56 27 | **9** | 27 45 20 |
| | **1** | 28 29 46 | **4** | 29 19 22 | **6** | 29 8 16 | **8** | 28 57 9 | **10** | 28 46 3 |
| | **2** | 29 30 29 | | | | | | | | |

**YOGAS.**

| Names of Yogas. | Order | I (d. g. p.) | Order | II (d. g. p.) | Order | III (d. g. p.) | Order | IV (d. g. p.) | Order | V (d. g. p.) |
|---|---|---|---|---|---|---|---|---|---|---|
| Vaidhriti ... ... ... ... ... | **27** | 0 18 8 | **5** | 0 53 57 | **9** | 0 33 17 | **13** | 0 12 38 | **18** | 0 48 27 |
| Vishkamba ... ... ... ... | **1** | 1 14 37 | **6** | 1 50 27 | **10** | 1 29 47 | **14** | 1 9 7 | **19** | 1 44 56 |
| Priti ... ... ... ... ... | **2** | 2 11 6 | **7** | 2 46 56 | **11** | 2 26 16 | **15** | 2 5 36 | **20** | 2 41 26 |
| Ayushmat ... ... ... ... | **3** | 3 7 36 | **8** | 3 43 25 | **12** | 3 22 46 | **16** | 3 2 6 | **21** | 3 37 55 |
| Saubhagya ... ... ... ... | **4** | 4 4 5 | **9** | 4 39 55 | **13** | 4 19 15 | **17** | 3 58 35 | **22** | 4 34 24 |
| Sobhana ... ... ... ... ... | **5** | 5 0 35 | **10** | 5 36 24 | **14** | 5 15 44 | **18** | 4 55 4 | **23** | 5 30 54 |
| Atiganda ... ... ... ... ... | **6** | 5 57 4 | **11** | 6 32 54 | **15** | 6 12 13 | **19** | 5 51 34 | **24** | 6 27 23 |
| Sukarman ... ... ... ... | **7** | 6 53 33 | **12** | 7 29 23 | **16** | 7 8 43 | **20** | 6 48 3 | **25** | 7 23 53 |
| Dhriti ... ... ... ... ... | **8** | 7 50 3 | **13** | 8 25 52 | **17** | 8 5 12 | **21** | 7 44 32 | **26** | 8 20 22 |
| Sula ... ... ... ... ... | **9** | 8 46 32 | **14** | 9 22 21 | **18** | 9 1 42 | **22** | 8 41 2 | **27** | 9 16 51 |
| Ganda ... ... ... ... ... | **10** | 9 43 1 | **15** | 10 18 51 | **19** | 9 58 11 | **23** | 9 37 31 | **1** | 10 13 21 |
| Vriddhi ... ... ... ... ... | **11** | 10 39 31 | **16** | 11 15 20 | **20** | 10 54 40 | **24** | 10 34 1 | **2** | 11 9 50 |
| Dhruva ... ... ... ... ... | **12** | 11 36 0 | **17** | 12 11 50 | **21** | 11 51 10 | **25** | 11 30 30 | **3** | 12 6 19 |
| Vyaghata ... ... ... ... ... | **13** | 12 32 29 | **18** | 13 8 19 | **22** | 12 47 39 | **26** | 12 26 59 | **4** | 13 2 49 |
| Harshana ... ... ... ... | **14** | 13 28 59 | **19** | 14 4 48 | **23** | 13 44 9 | **27** | 13 23 28 | **5** | 13 59 56 |
| Vajra ... ... ... ... ... | **15** | 14 25 28 | **20** | 15 1 18 | **24** | 14 40 38 | **1** | 14 19 58 | **6** | 14 55 47 |
| Siddhi ... ... ... ... ... | **16** | 15 21 58 | **21** | 15 57 47 | **25** | 15 37 7 | **2** | 15 16 27 | **7** | 15 52 17 |
| Vyatipata ... ... ... ... | **17** | 16 18 27 | **22** | 16 54 16 | **26** | 16 33 36 | **3** | 16 12 57 | **8** | 16 48 46 |
| Variyas ... ... ... ... ... | **18** | 17 14 56 | **23** | 17 50 46 | **27** | 17 30 6 | **4** | 17 9 26 | **9** | 17 45 16 |
| Parigha ... ... ... ... ... | **19** | 18 11 26 | **24** | 18 47 15 | **1** | 18 26 35 | **5** | 18 5 55 | **10** | 18 41 45 |
| Siva ... ... ... ... ... | **20** | 19 7 55 | **25** | 19 43 44 | **2** | 19 23 5 | **6** | 19 2 25 | **11** | 19 38 14 |
| Siddha ... ... ... ... ... | **21** | 20 4 24 | **26** | 20 40 14 | **3** | 20 19 34 | **7** | 19 58 54 | **12** | 20 34 44 |
| Sadhya ... ... ... ... ... | **22** | 21 0 54 | **27** | 21 36 43 | **4** | 21 16 3 | **8** | 20 55 24 | **13** | 21 31 13 |
| Subha ... ... ... ... ... | **23** | 21 57 23 | **1** | 22 33 13 | **5** | 22 12 33 | **9** | 21 51 53 | **14** | 22 27 42 |
| Sukla ... ... ... ... ... | **24** | 22 53 53 | **2** | 23 29 42 | **6** | 23 9 2 | **10** | 22 48 22 | **15** | 23 24 12 |
| Brahman ... ... ... ... ... | **25** | 23 50 22 | **3** | 24 27 59 | **7** | 24 5 31 | **11** | 23 44 51 | **16** | 24 20 41 |
| Indra ... ... ... ... ... | **26** | 24 46 51 | **4** | 25 22 41 | **8** | 25 2 1 | **12** | 24 41 21 | **17** | 25 17 10 |
| Vaidhriti ... ... ... ... ... | **27** | 25 43 21 | **5** | 26 19 10 | **9** | 25 58 30 | **13** | 25 37 50 | **18** | 26 13 40 |
| | **1** | 26 39 50 | **6** | 27 15 39 | **10** | 26 54 59 | **14** | 26 31 20 | **19** | 27 10 9 |
| | **2** | 27 36 19 | **7** | 28 12 9 | **11** | 27 51 29 | **15** | 27 30 49 | **20** | 28 6 39 |
| | **3** | 28 32 49 | **8** | 29 8 38 | **12** | 28 47 58 | **16** | 28 27 18 | **21** | 29 3 8 |
| | **4** | 29 29 18 | | | | | **17** | 29 23 48 | | |

# FROM NEW-MOON TO ENDING MOMENT OF EACH NAKSHATRA AND YOGA.

**Nakshatras is—2d., 12gh., 32p, and—4d, 6gh, 38p, for yogas.**

## NAKSHATRAS.

| Order | Ordinarily Asvina VI | Order | Ordinarily Kartika VII | Order | Ordinarily Margasira VIII | Order | Ordinarily Pausha IX | Order | Ordinarily Magha X | Order | Ordinarily Phalguna XI | Order | Chaitra, when no Adhika Masam. XII | Order | Chaitra whenthere is Adhika Masam. XIII |
|---|---|---|---|---|---|---|---|---|---|---|---|---|---|---|---|
| | d. g. p. | | d. g. p. | | d. g. p. | | d. g. p. | | d. g. p. | | d. g. p. | | d. g. p. | | d. g. p. |
| 11 | 0 14 56 | 13 | 0 3 50 | 16 | 0 53 26 | 18 | 0 42 20 | 20 | 0 31 14 | 22 | 0 20 7 | 24 | 0 9 1 | 27 | 0 58 38 |
| 12 | 1 15 39 | 14 | 1 4 33 | 17 | 1 54 9 | 19 | 1 43 3 | 21 | 1 31 56 | 23 | 1 20 50 | 25 | 1 9 44 | 1 | 1 59 20 |
| 13 | 2 16 22 | 15 | 2 5 15 | 18 | 2 54 52 | 20 | 2 43 46 | 22 | 2 32 40 | 24 | 2 21 33 | 26 | 2 10 27 | 2 | 3 0 3 |
| 14 | 3 17 5 | 16 | 3 5 58 | 19 | 3 55 35 | 21 | 3 44 29 | 23 | 3 33 22 | 25 | 3 22 16 | 27 | 3 11 10 | 3 | 4 0 46 |
| 15 | 4 17 47 | 17 | 4 6 41 | 20 | 4 56 18 | 22 | 4 45 12 | 24 | 4 34 6 | 26 | 4 22 59 | 1 | 4 11 53 | 4 | 5 1 29 |
| 16 | 5 18 30 | 18 | 5 7 24 | 21 | 5 57 1 | 23 | 5 45 54 | 25 | 5 34 48 | 27 | 5 23 42 | 2 | 5 12 35 | 5 | 6 2 12 |
| 17 | 6 19 13 | 19 | 6 8 7 | 22 | 6 57 44 | 24 | 6 46 37 | 26 | 6 35 31 | 1 | 6 24 25 | 3 | 6 13 18 | 6 | 7 2 55 |
| 18 | 7 19 56 | 20 | 7 8 50 | 23 | 7 58 26 | 25 | 7 47 20 | 27 | 7 36 14 | 2 | 7 25 7 | 4 | 7 14 1 | 7 | 8 3 38 |
| 19 | 8 20 3 | 21 | 8 9 33 | 24 | 8 59 9 | 26 | 8 48 3 | 1 | 8 36 57 | 3 | 8 25 51 | 5 | 8 14 44 | 8 | 9 4 21 |
| 20 | 9 21 22 | 22 | 9 10 16 | 25 | 9 59 52 | 27 | 9 48 46 | 2 | 9 37 40 | 4 | 9 26 33 | 6 | 9 15 27 | 9 | 10 5 4 |
| 21 | 10 22 5 | 23 | 10 10 59 | 26 | 11 0 35 | 1 | 10 49 29 | 3 | 10 38 23 | 5 | 10 27 16 | 7 | 10 16 10 | 10 | 11 5 46 |
| 22 | 11 22 48 | 24 | 11 11 41 | 27 | 12 1 18 | 2 | 11 50 12 | 4 | 11 39 5 | 6 | 11 27 59 | 8 | 11 16 53 | 11 | 12 6 29 |
| 23 | 12 23 31 | 25 | 12 12 24 | 1 | 13 2 1 | 3 | 12 50 55 | 5 | 12 39 48 | 7 | 12 28 42 | 9 | 12 17 36 | 12 | 13 7 12 |
| 24 | 13 24 13 | 26 | 13 13 7 | 2 | 14 2 44 | 4 | 13 51 38 | 6 | 13 40 31 | 8 | 13 29 25 | 10 | 13 18 18 | 13 | 14 7 55 |
| 25 | 14 24 56 | 27 | 14 13 50 | 3 | 15 3 27 | 5 | 14 52 20 | 7 | 14 41 14 | 9 | 14 30 8 | 11 | 14 19 2 | 14 | 15 8 38 |
| 26 | 15 25 39 | 1 | 15 14 33 | 4 | 16 4 9 | 6 | 15 53 3 | 8 | 15 41 57 | 10 | 15 30 51 | 12 | 15 19 44 | 15 | 16 9 21 |
| 27 | 16 26 22 | 2 | 16 15 16 | 5 | 17 4 52 | 7 | 16 53 46 | 9 | 16 42 40 | 11 | 16 31 34 | 13 | 16 20 27 | 16 | 17 10 4 |
| 1 | 17 27 5 | 3 | 17 15 59 | 6 | 18 5 35 | 8 | 17 54 29 | 10 | 17 43 23 | 12 | 17 32 17 | 14 | 17 21 10 | 17 | 18 10 47 |
| 2 | 18 27 48 | 4 | 18 16 42 | 7 | 19 6 18 | 9 | 18 55 12 | 11 | 18 44 6 | 13 | 18 32 59 | 15 | 18 21 53 | 18 | 19 11 29 |
| 3 | 19 28 31 | 5 | 19 17 24 | 8 | 20 7 1 | 10 | 19 55 55 | 12 | 19 44 49 | 14 | 19 33 42 | 16 | 19 22 36 | 19 | 20 12 13 |
| 4 | 20 29 14 | 6 | 20 18 7 | 9 | 21 7 44 | 11 | 20 56 38 | 13 | 20 45 31 | 15 | 20 34 25 | 17 | 20 23 19 | 20 | 21 12 56 |
| 5 | 21 29 56 | 7 | 21 18 50 | 10 | 22 8 27 | 12 | 21 57 21 | 14 | 21 46 14 | 16 | 21 35 8 | 18 | 21 24 2 | 21 | 22 13 38 |
| 6 | 22 30 40 | 8 | 22 19 33 | 11 | 23 9 10 | 13 | 22 58 3 | 15 | 22 46 57 | 17 | 22 35 51 | 19 | 22 24 45 | 22 | 23 14 21 |
| 7 | 23 31 23 | 9 | 23 20 16 | 12 | 24 9 53 | 14 | 23 58 46 | 16 | 23 47 40 | 18 | 23 36 34 | 20 | 23 25 28 | 23 | 24 15 4 |
| 8 | 24 32 5 | 10 | 24 20 59 | 13 | 25 10 35 | 15 | 24 59 29 | 17 | 24 48 23 | 19 | 24 37 17 | 21 | 24 26 10 | 24 | 25 15 47 |
| 9 | 25 32 48 | 11 | 25 21 42 | 14 | 26 11 18 | 16 | 25 0 12 | 18 | 25 49 6 | 20 | 25 38 0 | 22 | 25 26 53 | 25 | 26 16 30 |
| 10 | 26 33 31 | 12 | 26 22 25 | 15 | 27 12 1 | 17 | 26 0 55 | 19 | 26 49 49 | 21 | 26 38 42 | 23 | 26 27 36 | 26 | 27 17 13 |
| 11 | 27 34 14 | 13 | 27 23 8 | 16 | 28 12 44 | 18 | 27 1 38 | 20 | 27 50 32 | 22 | 27 39 25 | 24 | 27 28 19 | 27 | 28 17 56 |
| 12 | 28 34 57 | 14 | 28 23 51 | 17 | 29 13 27 | 19 | 28 2 21 | 21 | 28 51 14 | 23 | 28 40 8 | 25 | 28 29 2 | 1 | 29 18 39 |
| | | 15 | 29 24 34 | | | | | | | | | 26 | 29 29 45 | | |

## YOGAS.

| Order | VI | Order | VII | Order | VIII | Order | IX | Order | X | Order | XI | Order | XII | Order | XIII |
|---|---|---|---|---|---|---|---|---|---|---|---|---|---|---|---|
| 22 | 0 27 47 | 26 | 0 7 7 | 4 | 0 42 57 | 8 | 0 22 17 | 12 | 0 1 37 | 17 | 0 37 26 | 21 | 0 16 47 | 26 | 0 52 36 |
| 23 | 1 24 17 | 27 | 1 3 36 | 5 | 1 39 26 | 9 | 1 18 46 | 13 | 0 58 6 | 18 | 1 33 56 | 22 | 1 13 16 | 27 | 1 49 6 |
| 24 | 2 20 46 | 1 | 2 0 6 | 6 | 2 35 55 | 10 | 2 15 16 | 14 | 1 54 36 | 19 | 2 30 25 | 23 | 2 9 45 | 1 | 2 45 35 |
| 25 | 3 17 15 | 2 | 2 56 35 | 7 | 3 32 25 | 11 | 3 11 45 | 15 | 2 51 5 | 20 | 3 26 55 | 24 | 3 6 15 | 2 | 3 42 4 |
| 26 | 4 13 44 | 3 | 3 53 5 | 8 | 4 28 54 | 12 | 4 8 14 | 16 | 3 47 35 | 21 | 4 23 24 | 25 | 4 2 44 | 3 | 4 38 33 |
| 27 | 5 10 14 | 4 | 4 49 34 | 9 | 5 25 24 | 13 | 5 4 44 | 17 | 4 44 4 | 22 | 5 19 53 | 26 | 4 59 13 | 4 | 5 35 3 |
| 1 | 6 6 43 | 5 | 5 46 8 | 10 | 6 21 53 | 14 | 6 1 13 | 18 | 5 40 33 | 23 | 6 16 23 | 27 | 5 55 43 | 5 | 6 31 32 |
| 2 | 7 3 13 | 6 | 6 42 33 | 11 | 7 18 22 | 15 | 6 57 48 | 19 | 6 37 2 | 24 | 7 12 52 | 1 | 6 52 12 | 6 | 7 28 2 |
| 3 | 7 59 42 | 7 | 7 39 2 | 12 | 8 14 52 | 16 | 7 54 12 | 20 | 7 33 32 | 25 | 8 9 21 | 2 | 7 48 41 | 7 | 8 24 31 |
| 4 | 8 56 11 | 8 | 8 35 32 | 13 | 9 11 21 | 17 | 8 50 41 | 21 | 8 30 1 | 26 | 9 5 51 | 3 | 8 45 11 | 8 | 9 21 0 |
| 5 | 9 52 41 | 9 | 9 32 1 | 14 | 10 7 50 | 18 | 9 47 10 | 22 | 9 26 31 | 27 | 10 2 20 | 4 | 9 41 40 | 9 | 10 17 30 |
| 6 | 10 49 10 | 10 | 10 28 30 | 15 | 11 4 20 | 19 | 10 48 40 | 23 | 10 23 0 | 1 | 10 58 50 | 5 | 10 38 10 | 10 | 11 13 23 |
| 7 | 11 45 39 | 11 | 11 24 59 | 16 | 12 0 49 | 20 | 11 40 9 | 24 | 11 19 29 | 2 | 11 55 19 | 6 | 11 34 39 | 11 | 12 10 28 |
| 8 | 12 42 9 | 12 | 12 21 29 | 17 | 12 57 18 | 21 | 12 36 39 | 25 | 12 15 59 | 3 | 12 51 48 | 7 | 12 31 8 | 12 | 13 6 58 |
| 9 | 13 38 38 | 13 | 13 17 58 | 18 | 13 53 48 | 22 | 13 33 8 | 26 | 13 12 28 | 4 | 13 48 17 | 8 | 13 27 38 | 13 | 14 3 27 |
| 10 | 14 35 8 | 14 | 14 14 28 | 19 | 14 50 17 | 23 | 14 29 37 | 27 | 14 8 58 | 5 | 14 44 47 | 9 | 14 24 7 | 14 | 14 59 56 |
| 11 | 15 31 37 | 15 | 15 10 57 | 20 | 15 46 47 | 24 | 15 26 7 | 1 | 15 5 27 | 6 | 15 41 16 | 10 | 15 20 36 | 15 | 15 56 26 |
| 12 | 16 28 6 | 16 | 16 7 26 | 21 | 16 43 16 | 25 | 16 22 36 | 2 | 16 1 56 | 7 | 16 37 46 | 11 | 16 17 6 | 16 | 16 52 55 |
| 13 | 17 24 36 | 17 | 17 3 56 | 22 | 17 39 45 | 26 | 17 19 5 | 3 | 16 58 25 | 8 | 17 34 15 | 12 | 17 13 35 | 17 | 17 49 25 |
| 14 | 18 21 5 | 18 | 18 0 25 | 23 | 18 36 14 | 27 | 18 15 35 | 4 | 17 54 55 | 9 | 18 30 44 | 13 | 18 10 5 | 18 | 18 45 54 |
| 15 | 19 17 35 | 19 | 18 56 54 | 24 | 19 32 44 | 1 | 19 12 4 | 5 | 18 51 24 | 10 | 19 27 14 | 14 | 19 6 34 | 19 | 19 42 23 |
| 16 | 20 18 28 | 20 | 19 53 24 | 25 | 20 29 13 | 2 | 20 8 33 | 6 | 19 47 54 | 11 | 20 23 43 | 15 | 20 3 3 | 20 | 20 38 53 |
| 17 | 21 10 33 | 21 | 20 49 53 | 26 | 21 25 43 | 3 | 21 5 3 | 7 | 20 44 23 | 12 | 21 20 13 | 16 | 20 59 33 | 21 | 21 35 22 |
| 18 | 22 7 2 | 22 | 21 46 23 | 27 | 22 22 12 | 4 | 22 1 32 | 8 | 21 40 52 | 13 | 22 16 42 | 17 | 21 56 2 | 22 | 22 31 51 |
| 19 | 23 3 32 | 23 | 22 42 52 | 1 | 23 18 41 | 5 | 22 58 2 | 9 | 22 37 22 | 14 | 23 13 11 | 18 | 22 52 31 | 23 | 23 28 21 |
| 20 | 24 0 1 | 24 | 23 39 21 | 2 | 24 15 11 | 6 | 23 54 31 | 10 | 23 33 53 | 15 | 24 9 40 | 19 | 23 49 1 | 24 | 24 24 50 |
| 21 | 24 56 31 | 25 | 24 35 51 | 3 | 25 11 40 | 7 | 24 51 0 | 11 | 24 30 20 | 16 | 25 6 10 | 20 | 24 45 30 | 25 | 25 21 20 |
| 22 | 25 53 0 | 26 | 25 32 20 | 4 | 26 8 10 | 8 | 25 47 29 | 12 | 25 26 50 | 17 | 26 2 39 | 21 | 25 42 0 | 26 | 26 17 49 |
| 23 | 26 49 29 | 27 | 26 28 50 | 5 | 27 4 39 | 9 | 26 43 59 | 13 | 26 23 19 | 18 | 26 59 9 | 22 | 26 38 29 | 27 | 27 14 18 |
| 24 | 27 45 59 | 1 | 27 25 19 | 6 | 28 1 8 | 10 | 27 40 28 | 14 | 27 19 48 | 19 | 27 55 38 | 23 | 27 34 58 | 1 | 28 10 48 |
| 25 | 28 42 28 | 2 | 28 21 48 | 7 | 28 57 38 | 11 | 28 36 58 | 15 | 28 16 18 | 20 | 28 52 7 | 24 | 28 31 28 | 2 | 29 7 17 |
| | | 3 | 29 18 17 | | | | | 16 | 29 12 47 | | | 25 | 29 27 57 | | |

# TABLE XII.—EPHEMERIS FOR BENGAL.

**Dates of Commencement of Bengal Solar Months, A. D. 1840 to A. D. 1920.**

| A.D. | Vaisakha (April) | Jyeshta (May) | Ashadha (June) | Sravana (July) | Bhadrapada (August) | Asvina (September) | Karttika (October) | Margasira (November) | Pausha (December) | A.D. | Magha (January) | Phalguna (February) | Chaitra (March) |
|---|---|---|---|---|---|---|---|---|---|---|---|---|---|
| 1840 | 12 | 13 | 13 | 15 | 15 | 15 | 16 | 15 | 14 | 41 | 13 | 11 | 13 |
| 1841 | 12 | 13 | 14 | 15 | 16 | 16 | 16 | 15 | 14 | 42 | 13 | 11 | 13 |
| 1842 | 12 | 13 | 14 | 15 | 16 | 16 | 16 | 15 | 15 | 43 | 13 | 11 | 13 |
| 1843 | 13 | 14 | 14 | 16 | 16 | 16 | 17 | 15 | 15 | 44 | 13 | 12 | 13 |
| 1844 | 12 | 13 | 13 | 15 | 15 | 15 | 16 | 15 | 14 | 45 | 13 | 11 | 13 |
| 1845 | 12 | 13 | 14 | 15 | 16 | 16 | 16 | 15 | 15 | 46 | 13 | 11 | 13 |
| 1846 | 12 | 13 | 14 | 15 | 16 | 16 | 16 | 15 | 15 | 47 | 13 | 12 | 13 |
| 1847 | 13 | 14 | 14 | 16 | 16 | 16 | 17 | 16 | 15 | 48 | 13 | 12 | 13 |
| 1848 | 12 | 13 | 13 | 15 | 15 | 15 | 16 | 15 | 14 | 49 | 13 | 11 | 13 |
| 1849 | 12 | 13 | 14 | 15 | 16 | 16 | 16 | 15 | 15 | 50 | 13 | 11 | 13 |
| 1850 | 12 | 13 | 14 | 15 | 16 | 16 | 16 | 15 | 15 | 51 | 13 | 12 | 13 |
| 1851 | 13 | 14 | 14 | 16 | 16 | 16 | 17 | 16 | 15 | 52 | 13 | 12 | 13 |
| 1852 | 12 | 13 | 13 | 15 | 15 | 15 | 16 | 15 | 14 | 53 | 13 | 11 | 13 |
| 1853 | 12 | 13 | 14 | 15 | 16 | 16 | 16 | 15 | 15 | 54 | 13 | 11 | 13 |
| 1854 | 13 | 13 | 14 | 16 | 16 | 16 | 16 | 15 | 15 | 55 | 13 | 12 | 13 |
| 1855 | 13 | 14 | 14 | 16 | 16 | 16 | 17 | 16 | 15 | 56 | 13 | 12 | 13 |
| 1856 | 12 | 13 | 13 | 15 | 16 | 16 | 16 | 15 | 14 | 57 | 13 | 11 | 13 |
| 1857 | 12 | 13 | 14 | 15 | 16 | 16 | 16 | 15 | 15 | 58 | 13 | 11 | 13 |
| 1858 | 13 | 13 | 14 | 16 | 16 | 16 | 16 | 15 | 15 | 59 | 13 | 12 | 13 |
| 1859 | 13 | 14 | 14 | 16 | 16 | 16 | 17 | 16 | 15 | 60 | 13 | 12 | 13 |
| 1860 | 12 | 13 | 13 | 15 | 16 | 16 | 16 | 15 | 14 | 61 | 13 | 11 | 13 |
| 1861 | 12 | 13 | 14 | 15 | 16 | 16 | 16 | 15 | 15 | 62 | 13 | 11 | 13 |
| 1862 | 13 | 14 | 14 | 16 | 16 | 16 | 17 | 15 | 15 | 63 | 13 | 12 | 13 |
| 1863 | 13 | 14 | 14 | 16 | 16 | 16 | 17 | 16 | 15 | 64 | 13 | 12 | 13 |
| 1864 | 12 | 13 | 13 | 15 | 16 | 16 | 16 | 15 | 14 | 65 | 13 | 11 | 13 |
| 1865 | 12 | 13 | 14 | 15 | 16 | 16 | 16 | 15 | 15 | 66 | 13 | 11 | 13 |
| 1866 | 13 | 14 | 14 | 16 | 16 | 16 | 17 | 15 | 15 | 67 | 13 | 12 | 14 |
| 1867 | 13 | 14 | 14 | 16 | 16 | 16 | 17 | 16 | 15 | 68 | 13 | 12 | 13 |
| 1868 | 12 | 13 | 13 | 15 | 16 | 16 | 16 | 15 | 14 | 69 | 13 | 11 | 13 |
| 1869 | 12 | 13 | 14 | 15 | 16 | 16 | 16 | 15 | 15 | 70 | 13 | 11 | 13 |
| 1870 | 13 | 14 | 14 | 16 | 16 | 16 | 17 | 15 | 15 | 71 | 13 | 12 | 14 |
| 1871 | 13 | 14 | 14 | 16 | 16 | 16 | 17 | 16 | 15 | 72 | 14 | 12 | 13 |
| 1872 | 12 | 13 | 14 | 15 | 16 | 16 | 16 | 15 | 14 | 73 | 13 | 11 | 13 |
| 1873 | 12 | 13 | 14 | 15 | 16 | 16 | 16 | 15 | 15 | 74 | 13 | 12 | 13 |
| 1874 | 13 | 14 | 14 | 16 | 16 | 16 | 17 | 16 | 15 | 75 | 13 | 12 | 14 |
| 1875 | 13 | 14 | 14 | 15 | 16 | 16 | 17 | 16 | 15 | 76 | 14 | 12 | 13 |
| 1876 | 12 | 13 | 14 | 15 | 16 | 16 | 16 | 15 | 15 | 77 | 13 | 11 | 13 |
| 1877 | 12 | 13 | 14 | 15 | 16 | 16 | 16 | 15 | 15 | 78 | 13 | 12 | 13 |
| 1878 | 13 | 13 | 14 | 16 | 16 | 16 | 17 | 16 | 15 | 79 | 13 | 12 | 14 |
| 1879 | 13 | 14 | 14 | 16 | 16 | 16 | 17 | 16 | 15 | 80 | 14 | 12 | 13 |

| A.D. | Vaisakha (April) | Jyeshta (May) | Ashadha (June) | Sravana (July) | Bhadrapada (August) | Asvina (September) | Karttika (October) | Margasira (November) | Pausha (December) | A.D. | Magha (January) | Phalguna (February) | Chaitra (March) |
|---|---|---|---|---|---|---|---|---|---|---|---|---|---|
| 1880 | 12 | 13 | 14 | 15 | 16 | 16 | 16 | 15 | 15 | 81 | 13 | 11 | 13 |
| 1881 | 12 | 13 | 14 | 15 | 16 | 16 | 16 | 15 | 15 | 82 | 13 | 12 | 13 |
| 1882 | 13 | 14 | 14 | 16 | 16 | 16 | 17 | 16 | 15 | 83 | 13 | 12 | 14 |
| 1883 | 13 | 14 | 14 | 16 | 16 | 17 | 17 | 16 | 15 | 84 | 14 | 12 | 13 |
| 1884 | 12 | 13 | 14 | 15 | 16 | 16 | 16 | 15 | 15 | 85 | 13 | 11 | 13 |
| 1885 | 13 | 13 | 14 | 16 | 16 | 16 | 16 | 15 | 15 | 86 | 13 | 12 | 13 |
| 1886 | 13 | 14 | 14 | 16 | 16 | 16 | 17 | 16 | 15 | 87 | 13 | 12 | 14 |
| 1887 | 13 | 14 | 14 | 16 | 17 | 17 | 17 | 16 | 15 | 88 | 14 | 12 | 13 |
| 1888 | 12 | 13 | 14 | 15 | 16 | 16 | 16 | 15 | 15 | 89 | 13 | 11 | 13 |
| 1889 | 13 | 14 | 14 | 16 | 16 | 16 | 17 | 15 | 15 | 90 | 13 | 12 | 13 |
| 1890 | 13 | 14 | 14 | 16 | 16 | 16 | 17 | 16 | 15 | 91 | 13 | 12 | 14 |
| 1891 | 13 | 14 | 14 | 16 | 17 | 17 | 17 | 16 | 15 | 92 | 14 | 12 | 13 |
| 1892 | 12 | 13 | 14 | 15 | 16 | 16 | 16 | 15 | 15 | 93 | 13 | 11 | 13 |
| 1893 | 13 | 14 | 14 | 16 | 16 | 16 | 17 | 15 | 15 | 94 | 13 | 12 | 13 |
| 1894 | 13 | 14 | 14 | 16 | 16 | 16 | 17 | 16 | 15 | 95 | 14 | 12 | 14 |
| 1895 | 13 | 14 | 14 | 16 | 17 | 17 | 17 | 16 | 15 | 96 | 14 | 12 | 13 |
| 1896 | 12 | 13 | 14 | 15 | 16 | 16 | 16 | 15 | 15 | 97 | 13 | 11 | 13 |
| 1897 | 13 | 14 | 14 | 16 | 16 | 16 | 17 | 15 | 15 | 98 | 13 | 12 | 14 |
| 1898 | 13 | 14 | 14 | 16 | 16 | 16 | 17 | 16 | 15 | 99 | 14 | 12 | 14 |
| 1899 | 13 | 14 | 15 | 16 | 17 | 17 | 17 | 16 | 15 | 1900 | 14 | 12 | 14 |
| 1900 | 13 | 14 | 15 | 16 | 17 | 17 | 17 | 16 | 16 | 01 | 14 | 12 | 14 |
| 1901 | 14 | 15 | 15 | 17 | 17 | 17 | 18 | 17 | 16 | 02 | 14 | 13 | 15 |
| 1902 | 14 | 15 | 15 | 17 | 17 | 17 | 18 | 17 | 16 | 03 | 15 | 13 | 15 |
| 1903 | 14 | 15 | 16 | 17 | 18 | 18 | 18 | 17 | 17 | 04 | 15 | 13 | 14 |
| 1904 | 13 | 14 | 15 | 16 | 17 | 17 | 17 | 16 | 16 | 05 | 14 | 13 | 14 |
| 1905 | 14 | 15 | 15 | 17 | 17 | 17 | 18 | 17 | 16 | 06 | 14 | 13 | 15 |
| 1906 | 14 | 15 | 15 | 17 | 17 | 17 | 18 | 17 | 16 | 07 | 15 | 13 | 15 |
| 1907 | 14 | 15 | 16 | 17 | 18 | 18 | 18 | 17 | 17 | 08 | 15 | 13 | 14 |
| 1908 | 13 | 14 | 15 | 16 | 17 | 17 | 17 | 16 | 16 | 09 | 14 | 13 | 14 |
| 1909 | 14 | 15 | 15 | 17 | 17 | 17 | 18 | 17 | 16 | 10 | 14 | 13 | 15 |
| 1910 | 14 | 15 | 15 | 17 | 17 | 17 | 18 | 17 | 16 | 11 | 15 | 13 | 15 |
| 1911 | 14 | 15 | 16 | 17 | 18 | 18 | 18 | 17 | 17 | 12 | 15 | 13 | 14 |
| 1912 | 14 | 14 | 15 | 17 | 17 | 17 | 17 | 16 | 16 | 13 | 14 | 13 | 14 |
| 1913 | 14 | 15 | 15 | 17 | 17 | 17 | 18 | 17 | 16 | 14 | 14 | 13 | 15 |
| 1914 | 14 | 15 | 15 | 17 | 18 | 18 | 18 | 17 | 16 | 15 | 15 | 13 | 15 |
| 1915 | 14 | 15 | 16 | 17 | 18 | 18 | 18 | 17 | 17 | 16 | 15 | 13 | 14 |
| 1916 | 14 | 14 | 15 | 17 | 17 | 17 | 17 | 16 | 16 | 17 | 14 | 13 | 14 |
| 1917 | 14 | 15 | 15 | 17 | 17 | 17 | 18 | 17 | 16 | 18 | 14 | 13 | 15 |
| 1918 | 14 | 15 | 15 | 17 | 18 | 18 | 18 | 17 | 16 | 19 | 15 | 13 | 15 |
| 1919 | 14 | 15 | 16 | 17 | 18 | 18 | 18 | 17 | 17 | 20 | 15 | 13 | 14 |

NOTE.—It will be seen from a comparison of the above table with the General Ephemeris given below that solar months in Bengal seldom commence on the same day as in the Tamil and Malayalam countries. In the first place the Sankrantis which determine the commencement of solar months in Bengal are those fixed by the Surya Siddhanta, whereas in Southern India the solar months and their Sankrantis are reckoned according to the Arya Siddhanta.

Secondly, the rule in Bengal is that when the Sankranti occurs less than 45 ghatikas after sunrise, the civil solar month commences on the next day, and when the Sankranti occurs 45 ghatikas after Sunrise or later, the civil solar month commences two days after the date of the Sankranti. It follows from this rule that in Bengal the civil month never begins on the date of the Sankranti which in that province invariably belongs to the previous month whereas in Southern India, the civil month very often begins on the day of the Sankranti. In practice, the rule in Bengal appears to be to begin the Solar month on the third day after Sankranti when the fraction of day at which Sankranti occurs is ·70 of a day or more. The above table has been constructed on this principle.

## TABLE XII.—General Ephemeris A.D. 1840—A.D. 1845.

| Commencement of years in different eras. | Tamil month. | Tamil Month Beginning Week-day | Month | Day | Last Day | End Week-day | Month | Day | Lunar month. | New Moon Ending moment of New Moon Tithi. Mean Week-day | Month | Day (mean) | Ghatikas | Palas | Actual Day (actual) | Ghatikas | Palas | ☾'s Anom. Day | Ghatikas | Palas | ☉'s Anom. Day | Ghatikas | Palas | Deduct for Nakshatras | Deduct for Yogas. |
|---|---|---|---|---|---|---|---|---|---|---|---|---|---|---|---|---|---|---|---|---|---|---|---|---|---|
| **K. Y. 4941** | Chittirai . | 7 | Ap | 11 | 31 | 2 | My | 11 | Vaisakha ... | 6 | My | 1 | 37 | 56 | ... | 57 | 58 | 25 | 15 | 58 | 20 | 32 | 14 | 1 day, 32 ghat., 10 palas. | 2 „ 51 „ 31 „ |
| Sak. 1762 | Vaikasi ... | 3 | My | 12 | 31 | 5 | Je | 11 | Jyeshta ... | 1 | My | 31 | 9 | 46 | ... | 16 | 22 | 27 | 14 | 31 | 50 | 4 | 4 | | |
| Vik. 1897 | Ani ... | 6 | Je | 12 | 32 | 2 | Jl | 13 | Ashada ... | 2 | Je | 29 | 41 | 36 | ... | 33 | 14 | 1 | 39 | 48 | 79 | 35 | 54 | | |
| (S.) Sarvari | Adi ... | 3 | Jl | 14 | 31 | 5 | Au | 13 | Sravana ... | 4 | Jl | 29 | 13 | 26 | 28○ | 51 | 20 | 3 | 38 | 22 | 109 | 7 | 44 | | |
| K. 1015 (S.M.) | Avani ... | 6 | Au | 14 | 31 | 1 | S | 13 | Bhadrapada . | 5 | Au | 27 | 45 | 17 | ...● | 13 | 25 | 5 | 36 | 56 | 138 | 39 | 34 | | |
| K. 1016 (N.M.) | Purattasi . | 2 | S | 14 | 31 | 4 | O | 14 | Asvina .. | 7 | S | 26 | 17 | 7 | 25 | 41 | 40 | 7 | 35 | 29 | 168 | 11 | 24 | | |
| 59 g, 35 p. | Aippasi ... | 5 | O | 15 | 30 | 6 | N | 13 | Kartika .. | 1 | O | 25 | 48 | 57 | ... | 17 | 28 | 9 | 34 | 2 | 197 | 43 | 15 | | |
| B. S. 1246 | Kartigai . | 7 | N | 14 | 29 | 7 | D | 12 | Margasira ... | 3 | N | 24 | 20 | 47 | ... | 0 | 50 | 11 | 32 | 36 | 227 | 15 | 5 | | |
| (N.)Virodhakrit | Margali ... | 1 | D | 13 | 30 | 2 | Ja | 11 | Pausha ... | 4 | D | 23 | 52 | 37 | ... | 49 | 40 | 13 | 31 | 10 | 256 | 46 | 55 | | |
| **A.D. 1841** | Tai ... | 3 | Ja | 12 | 29 | 3 | F | 9 | Magha ... | 6 | Ja | 22 | 24 | 27 | ...○ | 39 | 23 | 15 | 29 | 43 | 286 | 18 | 45 | | |
| H. 1257 | Masi ... | 4 | F | 10 | 30 | 5 | Mr | 11 | Phalguna ... | 7 | F | 20 | 56 | 17 | 21 | 24 | 55 | 17 | 28 | 17 | 315 | 50 | 35 | | |
| Feb. 22, 23 | Panguni . | 6 | Mr | 12 | 30 | 7 | Ap | 10 | Chaitra ... | 2 | Mr | 22 | 28 | 7 | 23 | 33 | 13 | 19 | 26 | 50 | 345 | 22 | 25 | | |
| **K. Y. 4942** | Chittirai . | 1 | Ap | 11 | 31 | 3 | My | 11 | Vaisakha ... | 4 | Ap | 21 | 0 | 0 | ... | 3 | 48 | 21 | 25 | 24 | 9 | 38 | 45 | 0 day, 43 ghat., 17 palas. | 1 „ 20 „ 33 „ |
| Sak. 1763 | Vaikasi ... | 4 | My | 12 | 32 | 7 | Je | 12 | Jyeshta ... | 5 | My | 20 | 31 | 49 | ... | 58 | 37 | 23 | 23 | 58 | 39 | 10 | 34 | | |
| Vik. 1898 | Ani ... | 1 | Je | 13 | 31 | 3 | Jl | 13 | Ashada ... | 7 | Je | 19 | 3 | 39 | ... | 16 | 44 | 25 | 22 | 31 | 68 | 42 | 24 | | |
| (S.) Plava | Adi ... | 4 | Jl | 14 | 32 | 7 | Au | 14 | Sravana ... | 1 | Jl | 18 | 35 | 28 | ...○ | 33 | 33 | 27 | 21 | 5 | 98 | 14 | 14 | | |
| K. 1016 (S. M.) | Avani ... | 1 | Au | 15 | 31 | 3 | S | 14 | Bhadrapada . | 3 | Au | 17 | 7 | 18 | 16 | 51 | 12 | 1 | 46 | 22 | 127 | 46 | 4 | | |
| K. 1016 (N. M.) | Purattasi . | 4 | S | 15 | 30 | 5 | O | 14 | Asvina ... | 4 | S | 15 | 39 | 8 | ... | 11 | 43 | 3 | 44 | 55 | 157 | 17 | 54 | | |
| | | | | | | | | | | 6 | O | 15 | 10 | 58 | 14 | 37 | 36 | 5 | 43 | 29 | 186 | 49 | 44 | | |
| 15 g, 6 p | Aippasi ... | 6 | O | 15 | 30 | 7 | N | 13 | Kartika ... | 7 | N | 13 | 42 | 48 | ... | 10 | 18 | 7 | 42 | 3 | 216 | 21 | 34 | | |
| B. S. 1247 | Kartigai . | 1 | N | 14 | 30 | 2 | D | 13 | Margasira ... | 2 | D | 13 | 14 | 38 | 12 | 50 | 13 | 9 | 40 | 36 | 245 | 53 | 25 | | |
| (N.)Paridhavin. | Margali ... | 3 | D | 14 | 29 | 3 | Jr | 11 | Pausha ... | 3 | Ja | 11 | 46 | 29 | ...○ | 36 | 22 | 11 | 39 | 10 | 275 | 25 | 15 | | |
| **A. D. 1842** | Tai ... | 4 | Ja | 12 | 29 | 4 | F | 9 | Magha ... | 5 | F | 10 | 18 | 19 | ... | 24 | 54 | 13 | 37 | 43 | 304 | 57 | 5 | | |
| H. 1258 | Masi ... | 5 | F | 10 | 30 | 6 | Mr | 11 | Phalguna ... | 6 | Mr | 11 | 50 | 9 | 12 | 12 | 6 | 15 | 36 | 17 | 334 | 28 | 55 | | |
| Feb. 12 | Panguni . | 7 | Mr | 12 | 31 | 2 | Ap | 11 | Chaitra ... | 1 | Ap | 10 | 21 | 59 | ... | 53 | 11 | 17 | 34 | 51 | 364 | 0 | 45 | | |
| **K. Y. 4943** | Chittirai . | 3 | Ap | 12 | 30 | 4 | My | 11 | Vaisakha ... | 2 | My | 9 | 53 | 49 | 10 | 26 | 50 | 19 | 33 | 24 | 28 | 17 | 4 | 2 days, 0 ghat., 57 palas. | 3 „ 56 „ 13 „ |
| Sak. 1764 | Vaikasi ... | 5 | My | 12 | 32 | 1 | Je | 12 | Jyeshta ... | 4 | Je | 8 | 25 | 39 | ... | 53 | 33 | 21 | 31 | 58 | 57 | 48 | 54 | | |
| Vik. 1899 | Ani ... | 2 | Je | 13 | 31 | 4 | Jl | 13 | Ashada ... | 5 | Jl | 7 | 57 | 29 | 8●○ | 15 | 5 | 23 | 30 | 31 | 87 | 20 | 44 | | |
| (S.) Subhakrit. | Adi ... | 5 | Jl | 14 | 32 | 1 | Au | 14 | Sravana ... | 7 | Au | 6 | 29 | 19 | ... | 34 | 0 | 25 | 29 | 5 | 116 | 52 | 34 | | |
| K. 1017 (S. M.) | Avani ... | 2 | Au | 15 | 31 | 4 | S | 14 | Bhadrapada . | 2 | S | 5 | 1 | 10 | 4 | 52 | 48 | 27 | 27 | 39 | 146 | 24 | 24 | | |
| K. 1017 (N. M.) | Purattasi . | 5 | S | 15 | 30 | 6 | O | 14 | Asvina ... | 3 | O | 4 | 33 | 0 | ... | 13 | 36 | 1 | 52 | 56 | 175 | 56 | 14 | | |
| 30 g, 37 p. | Aippasi ... | 7 | O | 15 | 30 | 1 | N | 13 | Kartika ... | 5 | N | 3 | 4 | 49 | 2 | 38 | 16 | 3 | 51 | 29 | 205 | 28 | 4 | | |
| B. S. 1248 | Kartigai . | 2 | N | 14 | 30 | 3 | D | 13 | Margasira ... | 6 | D | 2 | 36 | 40 | ... | 8 | 13 | 5 | 50 | 3 | 234 | 59 | 55 | | |
| (N.) Parmadin | Margali ... | 4 | D | 14 | 29 | 4 | Jr | 11 | Pausha ... | 1 | Ja | 1 | 8 | 30 | D.31 | 44 | 7 | 7 | 48 | 36 | 264 | 31 | 45 | | |
| **A. D. 1843** | Tai ... | 5 | Ja | 12 | 30 | 6 | F | 10 | Magha ... | 2 | Ja | 30 | 40 | 20 | ... | 25 | 26 | 9 | 47 | 10 | 294 | 3 | 35 | | |
| H. 1259 | Masi ... | 7 | F | 11 | 29 | 7 | Mr | 11 | Phalguna ... | 4 | Mr | 1 | 12 | 10 | ... | 10 | 15 | 11 | 45 | 44 | 323 | 35 | 25 | | |
| Feb. 1 | Panguni . | 1 | Mr | 12 | 31 | 3 | Ap | 11 | Chaitra ... | 5 | Mr | 30 | 44 | 0 | ... | 55 | 32 | 13 | 44 | 17 | 353 | 7 | 15 | | |
| **K. Y. 4944** | Chittirai . | 4 | Ap | 12 | 31 | 6 | My | 12 | Vaisakha ... | 7 | Ap | 29 | 15 | 50 | ... | 37 | 46 | 15 | 42 | 51 | 17 | 23 | 34 | 1 day, 18 ghat., 4 palas. | 2 „ 25 „ 15 „ |
| Sak. 1765 | Vaikasi ... | 7 | My | 13 | 31 | 2 | Je | 12 | Jyeshta ... | 1 | My | 28 | 47 | 41 | 29 | 15 | 17 | 17 | 41 | 24 | 46 | 55 | 24 | | |
| Vik. 1900 | Ani ... | 3 | Je | 13 | 32 | 6 | Jl | 14 | Ashada ... | 3 | Je | 27 | 19 | 31 | ... | 44 | 52 | 19 | 39 | 58 | 76 | 27 | 14 | | |
| (S.) Sobhakrit | Adi ... | 7 | Jl | 15 | 31 | 2 | Au | 14 | Sravana ... | 4 | Jl | 26 | 51 | 21 | 27 | 10 | 40 | 21 | 38 | 32 | 105 | 59 | 4 | | |
| K. 1018 (S. M.) | Avani ... | 3 | Au | 15 | 31 | 5 | S | 14 | Bhadrapada . | 6 | Au | 25 | 23 | 11 | ... | 33 | 23 | 23 | 37 | 5 | 135 | 30 | 54 | | |
| K. 1018 (N. M.) | Purattasi . | 6 | S | 15 | 31 | 1 | O | 15 | Asvina ... | 7 | S | 23 | 55 | 1 | ... | 55 | 14 | 25 | 35 | 39 | 165 | 2 | 44 | | |
| 46 g, 9 p | Aippasi ... | 2 | O | 16 | 30 | 3 | N | 14 | Kartika ... | 2 | O | 23 | 26 | 51 | ... | 17 | 46 | 0 | 0 | 20 | 194 | 34 | 34 | | |
| B. S. 1249 | Kartigai . | 4 | N | 15 | 29 | 4 | D | 13 | Margasira ... | 3 | N | 21 | 58 | 41 | ...○ | 42 | 21 | 1 | 59 | 29 | 224 | 6 | 24 | | |
| (N.) Ananda | Margali ... | 5 | D | 14 | 29 | 5 | Jr | 11 | Pausha ... | 5 | D | 21 | 30 | 31 | ...● | 10 | 17 | 3 | 58 | 3 | 253 | 38 | 15 | | |
| **A. D. 1844** | Tai ... | 6 | Ja | 12 | 30 | 7 | F | 10 | Magha ... | 7 | Ja | 20 | 2 | 21 | 19 | 42 | 13 | 5 | 56 | 37 | 283 | 10 | 5 | | |
| H. 1260 | Masi ... | 1 | F | 11 | 30 | 2 | Mr | 11 | Phalguna ... | 1 | F | 18 | 34 | 12 | ... | 17 | 52 | 7 | 55 | 10 | 312 | 41 | 55 | | |
| Ja. 22 | Panguni . | 3 | Mr | 12 | 30 | 4 | Ap | 10 | Chaitra ... | 3 | Mr | 19 | 6 | 2 | 18 | 56 | 47 | 9 | 53 | 44 | 342 | 13 | 45 | | |
| **K. Y. 4945** | Chittirai . | 5 | Ap | 11 | 31 | 7 | My | 11 | Vaisakha ... | 4 | Ap | 17 | 37 | 52 | ... | 37 | 39 | 11 | 52 | 17 | 6 | 30 | 4 | 0 day, 29 ghat., 11 palas. | 0 „ 54 „ 18 „ |
| Sak. 1766 | Vaikasi ... | 1 | My | 12 | 31 | 3 | Je | 11 | Jyeshta ... | 6 | My | 17 | 9 | 42 | ...○ | 18 | 23 | 13 | 50 | 51 | 36 | 1 | 54 | | |
| Vik. 1901 | Ani ... | 4 | Je | 12 | 32 | 7 | Jl | 13 | Ashada ... | 7 | Je | 15 | 41 | 32 | ... | 56 | 44 | 15 | 49 | 25 | 65 | 33 | 44 | | |
| (S.) Krodhi | Adi ... | 1 | Jl | 14 | 31 | 3 | Au | 13 | Sravana ... | 2 | Jl | 15 | 13 | 22 | ... | 31 | 19 | 17 | 47 | 58 | 95 | 5 | 34 | | |
| | | | | | | | | | | 3 | Au | 13 | 45 | 12 | 14 | 2 | 21 | 19 | 46 | 32 | 124 | 37 | 24 | | |
| K. 1019 (S. M.) | Avani ... | 4 | Au | 14 | 31 | 6 | S | 13 | Bhadrapada . | 5 | S | 12 | 17 | 2 | ... | 30 | 36 | 21 | 45 | 5 | 154 | 9 | 14 | | |
| K. 1019 (N. M.) | Purattasi . | 7 | S | 14 | 31 | 2 | O | 14 | Asvina ... | 6 | O | 11 | 48 | 53 | ... | 57 | 2 | 23 | 43 | 39 | 188 | 41 | 4 | | |
| 1 g, 40 p. | Aippasi ... | 3 | O | 15 | 30 | 4 | N | 13 | Kartika ... | 1 | N | 10 | 20 | 43 | ...○ | 22 | 40 | 25 | 42 | 13 | 213 | 12 | 54 | | |
| B. S. 1250 | Kartigai . | 5 | N | 14 | 29 | 5 | D | 12 | Margasira ... | 2 | D | 9 | 52 | 33 | ... | 48 | 29 | 0 | 7 | 30 | 242 | 44 | 45 | | |
| (N.) Rakshasa | Margali ... | 6 | D | 13 | 30 | 7 | Jr | 11 | Pausha ... | 4 | Ja | 8 | 24 | 23 | ... | 15 | 17 | 2 | 6 | 3 | 272 | 16 | 35 | | |
| **A. D. 1845** | Tai ... | 1 | Ja | 12 | 29 | 1 | F | 9 | Magha ... | 5 | F | 6 | 56 | 13 | ... | 43 | 37 | 4 | 4 | 37 | 301 | 48 | 25 | | |
| H. 1261 Ja. 10 | Masi ... | 2 | F | 10 | 30 | 3 | Mr | 11 | Phalguna ... | 7 | Mr | 8 | 28 | 3 | ... | 13 | 48 | 6 | 3 | 11 | 331 | 20 | 15 | | |
| H. 1262 Dec. 30 | Panguni . | 4 | Mr | 12 | 30 | 5 | Ap | 10 | Chaitra ... | 2 | Ap | 7 | 0 | 0 | 6 | 46 | 15 | 8 | 1 | 44 | 360 | 52 | 5 | | |

# TABLE XII.—General Ephemeris A.D. 1845—A.D. 1850.

| Commencement of years in different eras. | Tamil month. | Tamil Month Beginning Week-day | Month | Day | Last Day | Tamil Month End Week-day | Month | Day | Lunar month. | New Moon Week-day | Month | Day (mean) | Ending moment of New Moon Tithi Mean Ghatikas | Palas | Day (actual) | Actual Ghatikas | Palas | ☾'s Anom. Day | Ghatikas | Palas | ☉'s Anom. Day | Ghatikas | Palas | Deduct for Nakshatras | Deduct for Yogas. |
|---|---|---|---|---|---|---|---|---|---|---|---|---|---|---|---|---|---|---|---|---|---|---|---|---|---|
| **K. Y. 4946** | Chittirai | 6 | Ap | 11 | 31 | 1 | My | 11 | Vaisakha | 3 | My | 6 | 31 | 43 | ...●○ | 21 | 4 | 10 | 0 | 18 | 25 | 8 | 24 | 1 day, 52 ghat., 50 palas. | 3 ,, 29 ,, 57 ,, |
| Sak. 1767 | Vaikasi | 2 | My | 12 | 32 | 5 | Je | 12 | Jyeshta | 5 | Je | 5 | 3 | 33 | 4 | 57 | 41 | 11 | 58 | 51 | 54 | 40 | 14 | | |
| Vik. 1902 | Ani | 6 | Je | 13 | 31 | 1 | Jl | 13 | Ashada | 6 | Jl | 4 | 35 | 24 | ... | 35 | 46 | 13 | 57 | 25 | 84 | 12 | 4 | | |
| (S.) Visvavasu | Adi | 2 | Jl | 14 | 32 | 5 | Au | 14 | Sravana | 1 | Au | 3 | 7 | 14 | ... | 13 | 37 | 15 | 55 | 59 | 113 | 43 | 54 | | |
| K. 1020 (S. M.) | Avani | 6 | Au | 15 | 31 | 1 | S | 14 | Bhadrapada | 2 | S | 1 | 39 | 4 | ... | 49 | 14 | 17 | 54 | 32 | 143 | 15 | 44 | | |
| K. 1020 (N. M.) | Purattasi | 2 | S | 15 | 30 | 3 | O | 14 | Asvina | 4 | O | 1 | 10 | 54 | ... | 24 | 49 | 19 | 53 | 6 | 172 | 47 | 34 | | |
| 17 g, 11 p. | Aippasi | 4 | O | 15 | 30 | 5 | N | 13 | Kartika | 5 | O | 30 | 42 | 44 | ...○ | 57 | 1 | 21 | 51 | 39 | 202 | 19 | 24 | | |
| B. S. 1251 | Kartigai | 6 | N | 14 | 30 | 7 | D | 13 | Margasira | 7 | N | 29 | 14 | 34 | ... | 26 | 58 | 23 | 50 | 13 | 231 | 51 | 14 | | |
| (N.) Anala | Margali | 1 | D | 14 | 29 | 1 | Ja | 11 | Pausha | 1 | D | 28 | 46 | 24 | ... | 55 | 1 | 25 | 48 | 47 | 261 | 23 | 5 | | |
| **A. D. 1846** | Tai | 2 | Ja | 12 | 29 | 2 | F | 9 | Magha | 3 | Ja | 27 | 18 | 14 | ... | 21 | 33 | 0 | 14 | 4 | 290 | 54 | 55 | | |
| H. 1263 | Masi | 3 | F | 10 | 30 | 4 | Mr | 11 | Phalguna | 4 | F | 25 | 50 | 5 | ... | 48 | 1 | 2 | 12 | 37 | 320 | 26 | 45 | | |
| Dec. 20 | Panguni | 5 | Mr | 12 | 31 | 7 | Ap | 11 | Chaitra | 6 | Mr | 27 | 21 | 55 | ... | 12 | 43 | 4 | 11 | 11 | 349 | 58 | 35 | | |
| **K. Y. 4947** | Chittirai | 1 | Ap | 12 | 30 | 2 | My | 11 | Vaisakha | 7 | Ap | 25 | 53 | 45 | ... | 39 | 14 | 6 | 9 | 44 | 14 | 14 | 54 | 1 day, 3 ghat., 57 palas. | 1 ,, 59 ,, 0 ,, |
| Sak. 1768 | Vaikasi | 3 | My | 12 | 32 | 6 | Je | 12 | Jyeshta | 2 | My | 25 | 25 | 35 | ... | 7 | 46 | 8 | 8 | 18 | 43 | 46 | 44 | | |
| Vik. 1903 | Ani | 7 | Je | 13 | 31 | 2 | Jl | 13 | Ashada | 3 | Je | 23 | 57 | 25 | ... | 39 | 19 | 10 | 6 | 52 | 73 | 18 | 34 | | |
| (S.) Parabhava. | Adi | 3 | Jl | 14 | 32 | 6 | Au | 14 | Sravana | 5 | Jl | 23 | 29 | 15 | ... | 14 | 36 | 12 | 5 | 25 | 102 | 50 | 24 | | |
| K. 1021 (S. M.) | Avani | 7 | Au | 15 | 31 | 2 | S | 14 | Bhadrapada | 7 | Au | 22 | 1 | 6 | 21 | 53 | 39 | 14 | 3 | 59 | 132 | 22 | 14 | | |
| K. 1021 (N. M.) | Purattasi | 3 | S | 15 | 30 | 4 | O | 14 | Asvina | 1 | S | 20 | 32 | 56 | ... | 34 | 48 | 16 | 2 | 32 | 161 | 54 | 4 | | |
| 32 g, 42 p. | Aippasi | 5 | O | 15 | 30 | 6 | N | 13 | Kartika | 3 | O | 20 | 4 | 46 | ...● | 15 | 36 | 18 | 1 | 6 | 191 | 25 | 54 | | |
| B. S. 1252 | Kartigai | 7 | N | 14 | 30 | 1 | D | 13 | Margasira | 4 | N | 18 | 36 | 36 | ... | 54 | 17 | 19 | 59 | 40 | 220 | 57 | 44 | | |
| (N.) Pingala | Margali | 2 | D | 14 | 29 | 2 | Ja | 11 | Pausha | 6 | D | 18 | 8 | 26 | ... | 29 | 14 | 21 | 58 | 13 | 250 | 29 | 34 | | |
| **A. D. 1847** | Tai | 3 | Ja | 12 | 30 | 4 | F | 10 | Magha | 7 | Ja | 16 | 40 | 16 | 17 | 0 | 22 | 23 | 56 | 47 | 280 | 1 | 25 | | |
| H. 1264 | Masi | 5 | F | 11 | 29 | 5 | Mr | 11 | Phalguna | 2 | F | 15 | 12 | 6 | ... | 27 | 37 | 25 | 55 | 20 | 309 | 33 | 15 | | |
| Dec. 9 | Panguni | 6 | Mr | 12 | 31 | 1 | Ap | 11 | Chaitra | 3 | Mr | 16 | 43 | 56 | ...○ | 51 | 35 | 0 | 20 | 37 | 339 | 5 | 5 | | |
| **K. Y. 4948** | Chittirai | 2 | Ap | 12 | 31 | 4 | My | 12 | Vaisakha | 5 | Ap | 15 | 15 | 46 | ...● | 13 | 41 | 2 | 19 | 11 | 3 | 21 | 24 | 0 day, 15 ghat., 4 palas. | 0 ,, 28 ,, 2 ,, |
| Sak. 1769 | Vaikasi | 5 | My | 13 | 31 | 7 | Je | 12 | Jyeshta | 6 | My | 14 | 47 | 36 | ... | 35 | 27 | 4 | 17 | 31 | 32 | 53 | 14 | | |
| | | | | | | | | | | 1 | Je | 13 | 19 | 26 | 12 | 58 | 30 | 6 | 16 | 8 | 62 | 25 | 4 | | |
| Vik. 1904 | Ani | 1 | Je | 13 | 32 | 4 | Jl | 14 | Ashada | 2 | Jl | 12 | 51 | 17 | ... | 24 | 54 | 8 | 14 | 52 | 91 | 56 | 54 | | |
| (S.) Plavanga | Adi | 5 | Jl | 15 | 31 | 7 | Au | 14 | Sravana | 4 | Au | 11 | 23 | 7 | 10 | 56 | 33 | 10 | 13 | 25 | 121 | 28 | 44 | | |
| K. 1022 (S.M.) | Avani | 1 | Au | 15 | 31 | 3 | S | 14 | Bhadrapada | 5 | S | 9 | 54 | 57 | ...○ | 34 | 23 | 12 | 11 | 59 | 151 | 0 | 34 | | |
| K. 1022 (N.M.) | Purattasi | 4 | S | 15 | 31 | 6 | O | 15 | Asvina | 7 | O | 9 | 26 | 47 | ...● | 17 | 35 | 14 | 10 | 33 | 180 | 32 | 24 | | |
| 48 g, 14 p. | Aippasi | 7 | O | 16 | 30 | 1 | N | 14 | Kartika | 1 | N | 7 | 58 | 37 | 8 | 3 | 14 | 16 | 9 | 6 | 210 | 4 | 14 | | |
| B. S. 1253 | Kartigai | 2 | N | 15 | 29 | 2 | D | 13 | Margasira | 3 | D | 7 | 30 | 27 | ... | 47 | 46 | 18 | 7 | 40 | 239 | 36 | 4 | | |
| (N.) Kalayukta. | Margali | 3 | D | 14 | 29 | 3 | Ja | 11 | Pausha | 5 | Ja | 6 | 2 | 18 | ... | 28 | 12 | 20 | 6 | 13 | 269 | 7 | 54 | | |
| **A. D. 1848** | Tai | 4 | Ja | 12 | 30 | 5 | F | 10 | Magha | 6 | F | 4 | 34 | 8 | 5 | 1 | 5 | 22 | 4 | 47 | 298 | 39 | 45 | | |
| H. 1265 | Masi | 6 | F | 11 | 30 | 7 | Mr | 11 | Phalguna | 1 | Mr | 5 | 5 | 58 | ...○ | 29 | 44 | 24 | 3 | 19 | 328 | 11 | 35 | | |
| Nov. 27 | Panguni | 1 | Mr | 12 | 30 | 2 | Ap | 10 | Chaitra | 2 | Ap | 3 | 37 | 48 | ... | 55 | 27 | 26 | 1 | 54 | 357 | 43 | 25 | | |
| **K. Y. 4949** | Chittirai | 3 | Ap | 11 | 31 | 5 | My | 11 | Vaisakha | 4 | My | 3 | 9 | 38 | ... | 15 | 32 | 0 | 27 | 11 | 21 | 59 | 45 | 1 day, 36 ghat., 43 palas. | 3 ,, 3 ,, 41 ,, |
| Sak. 1770 | Vaikasi | 6 | My | 12 | 31 | 1 | Je | 11 | Jyeshta | 5 | Je | 1 | 41 | 28 | ... | 34 | 4 | 2 | 25 | 45 | 51 | 31 | 34 | | |
| Vik. 1905 | Ani | 2 | Je | 12 | 32 | 5 | Jl | 13 | Ashada | 7 | Jl | 1 | 13 | 18 | Je. 30 | 53 | 32 | 4 | 24 | 18 | 81 | 3 | 24 | | |
| (S.) Kilaka | Adi | 6 | Jl | 14 | 31 | 1 | Au | 13 | Sravana | 1 | Jl | 30 | 45 | 8 | ... | 15 | 34 | 6 | 22 | 52 | 110 | 35 | 14 | | |
| K. 1023 (S.M.) | Avani | 2 | Au | 14 | 31 | 4 | S | 13 | Bhadrapada | 3 | Au | 29 | 16 | 58 | 28○ | 43 | 31 | 8 | 21 | 26 | 140 | 7 | 4 | | |
| K. 1023 (N.M.) | Purattasi | 5 | S | 14 | 31 | 7 | O | 14 | Asvina | 4 | S | 27 | 48 | 48 | ...● | 13 | 43 | 10 | 19 | 59 | 169 | 38 | 54 | | |
| 3 g, 45 p. | Aippasi | 1 | O | 15 | 30 | 2 | N | 13 | Kartika | 6 | O | 27 | 20 | 38 | ... | 1 | 13 | 12 | 18 | 33 | 199 | 10 | 44 | | |
| B. S. 1254 | Kartigai | 3 | N | 14 | 29 | 3 | D | 12 | Margasira | 7 | N | 25 | 52 | 29 | ... | 48 | 59 | 14 | 17 | 6 | 228 | 42 | 35 | | |
| (N) Siddharthin. | Margali | 4 | D | 13 | 30 | 5 | Ja | 11 | Pausha | 2 | D | 25 | 24 | 19 | ... | 37 | 43 | 16 | 15 | 40 | 258 | 14 | 25 | | |
| **A. D. 1849** | Tai | 6 | Ja | 12 | 29 | 6 | F | 9 | Magha | 3 | Ja | 23 | 56 | 9 | 24 | 22 | 53 | 18 | 14 | 14 | 287 | 46 | 15 | | |
| H. 1266 | Masi | 7 | F | 10 | 30 | 1 | Mr | 11 | Phalguna | 5 | F | 22 | 27 | 59 | 23●○ | 1 | 26 | 20 | 12 | 47 | 317 | 18 | 5 | | |
| Nov. 17 | Panguni | 2 | Mr | 12 | 30 | 3 | Ap | 10 | Chaitra | 7 | Mr | 24 | 0 | 0 | ... | 32 | 36 | 22 | 11 | 21 | 346 | 49 | 55 | | |
| **K. Y. 4950** | Chittirai | 4 | Ap | 11 | 31 | 6 | My | 11 | Vaisakha | 1 | Ap | 22 | 31 | 39 | ... | 56 | 16 | 24 | 9 | 54 | 11 | 6 | 14 | 0 day, 49 ghat., 50 palas. | 1 ,, 32 ,, 44 ,, |
| Sak. 1771 | Vaikasi | 7 | My | 12 | 32 | 3 | Je | 12 | Jyeshta | 3 | My | 22 | 3 | 30 | ... | 17 | 5 | 26 | 8 | 28 | 40 | 38 | 4 | | |
| Vik. 1906 | Ani | 4 | Je | 13 | 31 | 6 | Jl | 13 | Ashada | 4 | Je | 20 | 35 | 19 | ... | 34 | 8 | 0 | 33 | 45 | 70 | 9 | 54 | | |
| (S.) Saumya. | Adi | 7 | Jl | 14 | 32 | 3 | Au | 14 | Sravana | 6 | Jl | 20 | 7 | 10 | 19 | 51 | 0 | 2 | 32 | 19 | 99 | 41 | 44 | | |
| K. 1024 (S. M.) | Avani | 4 | Au | 15 | 31 | 6 | S | 14 | Bhadrapada | 7 | Au | 18 | 39 | 0 | ...●○ | 10 | 15 | 4 | 30 | 52 | 129 | 13 | 34 | | |
| K. 1024 (N. M.) | Purattasi | 7 | S | 15 | 30 | 1 | O | 14 | Asvina | 2 | S | 17 | 10 | 50 | 16 | 36 | 9 | 6 | 29 | 26 | 158 | 45 | 24 | | |
| 19 g, 16 p. | Aippasi | 2 | O | 15 | 30 | 3 | N | 13 | Kartika | 3 | O | 16 | 42 | 40 | ... | 8 | 14 | 8 | 27 | 59 | 188 | 17 | 14 | | |
| B. S. 1255 | Kartigai | 4 | N | 14 | 30 | 5 | D | 13 | Margasira | 5 | N | 15 | 14 | 30 | 14 | 48 | 9 | 10 | 26 | 33 | 217 | 49 | 4 | | |
| (N.) Raudra. | Margali | 6 | D | 14 | 29 | 6 | Ja | 11 | Pausha | 6 | D | 14 | 46 | 20 | ... | 34 | 49 | 12 | 25 | 7 | 247 | 20 | 54 | | |
| **A. D. 1850** | Tai | 7 | Ja | 12 | 29 | 7 | F | 9 | Magha | 1 | Ja | 13 | 18 | 10 | ... | 24 | 45 | 14 | 23 | 40 | 276 | 52 | 44 | | |
| H. 1267 | Masi | 1 | F | 10 | 30 | 2 | Mr | 11 | Phalguna | 2 | F | 11 | 50 | 0 | 12● | 12 | 50 | 16 | 22 | 14 | 306 | 24 | 35 | | |
| Nov. 6 | Panguni | 3 | Mr | 12 | 31 | 5 | Ap | 11 | Chaitra | 4 | Mr | 13 | 21 | 50 | ... | 54 | 47 | 18 | 20 | 47 | 335 | 56 | 25 | | |

# TABLE XII.—General Ephemeris A.D. 1850—A.D. 1855.

| Commencement of years in different eras. | Tamil month. | Tamil Month Beginning Week-day | Month | Day | Last Day | End Week-day | Month | Day | Lunar month. | New Moon Ending moment of New Moon Tithi. Mean Week-day | Month | Day (mean) | Ghatikas | Palas | Actual Day (actual) | Ghatikas | Palas | ☾'s Anom. Day | Ghatikas | Palas | ⊙'s Anom. Day | Ghatikas | Palas | Deduct for Nakshatras | Deduct for Yogas. |
|---|---|---|---|---|---|---|---|---|---|---|---|---|---|---|---|---|---|---|---|---|---|---|---|---|---|
| **K. Y. 4951** | Chittirai | 6 | Ap | 12 | 31 | 1 | My | 12 | Vaisakha | 5 | Ap | 11 | 53 | 41 | 12 | 28 | 57 | 20 | 19 | 17 | 0 | 12 | 43 | 0 day, 0 ghat., 57 palas. | 0 „ 1 „ 46 „ |
| | | | | | | | | | | 7 | My | 11 | 25 | 31 | ... | 55 | 46 | 22 | 17 | 51 | 29 | 44 | 33 | | |
| Sak. 1772 | Vaikasi | 2 | My | 13 | 31 | 4 | Je | 12 | Jyeshta | 1 | Je | 9 | 57 | 21 | 10 | 16 | 54 | 24 | 16 | 24 | 59 | 16 | 24 | | |
| Vik. 1907 | Ani | 5 | Je | 13 | 32 | 1 | Jl | 14 | Ashada | 3 | Jl | 9 | 29 | 11 | ... | 34 | 31 | 26 | 14 | 58 | 88 | 48 | 14 | | |
| (S.) Sadharana. | Adi | 2 | Jl | 15 | 31 | 4 | Au | 14 | Sravana | 5 | Au | 8 | 1 | 1 | 7 | 51 | 31 | 0 | 40 | 15 | 118 | 20 | 4 | | |
| K. 1025 (S. M.) | Avani | 5 | Au | 15 | 31 | 7 | S | 14 | Bhadrapada | 6 | S | 6 | 32 | 51 | ... | 10 | 39 | 2 | 38 | 49 | 147 | 51 | 54 | | |
| K. 1025 (N. M.) | Purattasi | 1 | S | 15 | 30 | 2 | O | 14 | Asvina | 1 | O | 6 | 4 | 41 | 5 | 33 | 56 | 4 | 37 | 22 | 177 | 23 | 44 | | |
| 84 g., 47 p. | Aippasi | 3 | O | 15 | 30 | 4 | N | 13 | Kartika | 2 | N | 4 | 36 | 31 | ... | 3 | 16 | 6 | 35 | 56 | 206 | 55 | 34 | | |
| B. S. 1256 | Kartigai | 5 | N | 14 | 30 | 6 | D | 13 | Margasira | 4 | D | 4 | 8 | 22 | 3 | 39 | 41 | 8 | 34 | 29 | 236 | 27 | 24 | | |
| N. Durmati | Margali | 7 | D | 14 | 29 | 7 | Ja | 11 | Pausha | 5 | Ja | 2 | 40 | 12 | ...○ | 22 | 55 | 10 | 33 | 3 | 265 | 59 | 14 | | |
| **A. D. 1851** | Tai | 1 | Ja | 12 | 30 | 2 | F | 10 | Magha | 7 | F | 1 | 12 | 2 | ... | 10 | 38 | 12 | 31 | 37 | 295 | 31 | 4 | | |
| H. 1268 | Masi | 3 | F | 11 | 29 | 3 | Mr | 11 | Phalguna | 1 | Mr | 2 | 43 | 52 | ... | 56 | 35 | 14 | 30 | 10 | 325 | 2 | 54 | | |
| Oct. 26, 27 | Panguni | 4 | Mr | 12 | 31 | 6 | Ap | 11 | Chaitra | 3 | Ap | 1 | 15 | 42 | ... | 42 | 34 | 16 | 28 | 43 | 354 | 34 | 44 | | |
| **K. Y. 4952** | Chittirai | 7 | Ap | 12 | 31 | 2 | My | 12 | Vaisakha | 4 | Ap | 30 | 47 | 32 | My.1. | 19 | 56 | 18 | 27 | 17 | 18 | 51 | 4 | 1 day, 24 ghat., 36 palas. | 2 „ 37 „ 26 „ |
| Sak. 1773 | Vaikasi | 3 | My | 13 | 31 | 5 | Je | 12 | Jyeshta | 6 | My | 30 | 19 | 22 | ... | 49 | 8 | 20 | 25 | 51 | 48 | 22 | 54 | | |
| Vik. 1908 | Ani | 6 | Je | 13 | 32 | 2 | Jl | 14 | Ashada | 7 | Je | 28 | 51 | 12 | 29○ | 14 | 44 | 22 | 24 | 25 | 77 | 54 | 44 | | |
| (S.) Virodhakrit. | Adi | 3 | Jl | 15 | 31 | 5 | Au | 14 | Sravana | 2 | Jl | 28 | 23 | 2 | ... | 33 | 57 | 24 | 22 | 58 | 107 | 26 | 34 | | |
| K. 1026 (S. M.) | Avani | 6 | Au | 15 | 31 | 1 | S | 14 | Bhadrapada | 3 | Au | 26 | 54 | 53 | ... | 52 | 58 | 26 | 21 | 32 | 136 | 58 | 24 | | |
| K. 1026 (N. M.) | Purattasi | 2 | S | 15 | 31 | 4 | O | 15 | Asvina | 5 | S | 25 | 26 | 43 | ... | 12 | 53 | 0 | 46 | 49 | 166 | 30 | 14 | | |
| 50 g., 19 p. | Aippasi | 5 | O | 16 | 30 | 6 | N | 14 | Kartika | 6 | O | 24 | 58 | 33 | ... | 35 | 50 | 2 | 45 | 22 | 196 | 2 | 4 | | |
| B. S. 1257 | Kartigai | 7 | N | 15 | 29 | 7 | D | 13 | Margasira | 1 | N | 23 | 30 | 23 | ... | 3 | 12 | 4 | 43 | 56 | 225 | 33 | 54 | | |
| (N) Dundubhi | Margali | 1 | D | 14 | 29 | 1 | Ja | 11 | Pausha | 3 | D | 23 | 2 | 13 | 22○ | 36 | 7 | 6 | 42 | 30 | 255 | 5 | 44 | | |
| **A. D. 1852** | Tai | 2 | Ja | 12 | 30 | 3 | F | 10 | Magha | 4 | Ja | 21 | 34 | 3 | ... | 14 | 34 | 8 | 41 | 3 | 284 | 37 | 33 | | |
| H. 1269 | Masi | 4 | F | 11 | 30 | 5 | Mr | 11 | Phalguna | 6 | F | 20 | 5 | 53 | 19 | 57 | 23 | 10 | 39 | 37 | 314 | 9 | 23 | | |
| Oct. 15 | Panguni | 6 | Mr | 12 | 30 | 7 | Ap | 10 | Chaitra | 7 | Mr | 20 | 37 | 43 | ... | 42 | 16 | 12 | 38 | 10 | 343 | 41 | 13 | | |
| **K. Y. 4953** | Chittirai | 1 | Ap | 11 | 31 | 3 | My | 11 | Vaisakha | 2 | Ap | 19 | 9 | 34 | ... | 25 | 56 | 14 | 36 | 44 | 7 | 57 | 33 | 0 day, 35 ghat., 43 palas. | 1 „ 6 „ 28 „ |
| Sak. 1774 | Vaikasi | 4 | My | 12 | 31 | 6 | Je | 14 | Jyeshta | 3 | My | 18 | 41 | 24 | 19 | 5 | 49 | 16 | 35 | 18 | 37 | 29 | 23 | | |
| Vik. 1909 | Ani | 7 | Je | 12 | 32 | 3 | Jl | 13 | Ashada | 5 | Je | 17 | 13 | 14 | ...○ | 38 | 45 | 18 | 33 | 51 | 67 | 1 | 13 | | |
| (S.) Paridhavi. | Adi | 4 | Jl | 14 | 31 | 6 | Au | 13 | Sravana | 6 | Jl | 16 | 45 | 4 | 17 | 6 | 52 | 20 | 32 | 25 | 96 | 33 | 4 | | |
| K. 1027 (S. M.) | Avani | 7 | Au | 14 | 32 | 3 | S | 14 | Bhadrapada | 1 | Au | 15 | 16 | 54 | ... | 31 | 27 | 22 | 30 | 58 | 126 | 4 | 54 | | |
| | | | | | | | | | | 2 | S | 13 | 48 | 44 | ... | 54 | 7 | 24 | 29 | 32 | 155 | 36 | 44 | | |
| K. 1027 (N. M.) | Purattasi | 4 | S | 15 | 30 | 5 | O | 14 | Asvina | 4 | O | 13 | 20 | 34 | ... | 16 | 40 | 26 | 28 | 6 | 185 | 8 | 34 | | |
| 5 g., 50 p. | Aippasi | 6 | O | 15 | 30 | 7 | N | 13 | Kartika | 5 | N | 11 | 52 | 24 | ... | 40 | 28 | 0 | 53 | 23 | 214 | 40 | 24 | | |
| B. S. 1258 | Kartigai | 1 | N | 14 | 29 | 1 | D | 12 | Margasira | 7 | D | 11 | 24 | 14 | ...●○ | 6 | 50 | 2 | 51 | 59 | 244 | 12 | 14 | | |
| (N) Rudhirodgarin. | Margali | 2 | D | 13 | 30 | 3 | Ja | 11 | Pausha | 1 | Ja | 9 | 56 | 5 | ... | 36 | 44 | 4 | 50 | 30 | 273 | 44 | 4 | | |
| **A. D. 1853** | Tai | 4 | Ja | 12 | 29 | 4 | F | 9 | Magha | 3 | F | 8 | 27 | 55 | ... | 10 | 3 | 6 | 49 | 3 | 303 | 15 | 54 | | |
| H. 1270 | Masi | 5 | F | 10 | 30 | 6 | Mr | 11 | Phalguna | 5 | Mr | 10 | 0 | 0 | 9 | 46 | 45 | 8 | 47 | 37 | 332 | 47 | 45 | | |
| Oct. 4 | Panguni | 7 | Mr | 12 | 30 | 1 | Ap | 10 | Chaitra | 6 | Ap | 8 | 31 | 35 | ... | 26 | 5 | 10 | 46 | 11 | 36[illegible] | 19 | 35 | | |
| **K. Y. 4954** | Chittirai | 2 | Ap | 11 | 31 | 4 | My | 11 | Vaisakha | 1 | My | 8 | 3 | 25 | ... | 6 | 32 | 12 | 44 | 44 | 26 | 35 | 55 | 1 day, 59 ghat., 23 palas. | 3 „ 42 „ 8 „ |
| Sak. 1775 | Vaikasi | 5 | My | 12 | 32 | 1 | Je | 12 | Jyeshta | 2 | Je | 6 | 35 | 15 | ...○ | 45 | 58 | 14 | 43 | 18 | 56 | 7 | 45 | | |
| Vik. 1910 | Ani | 2 | Je | 13 | 31 | 4 | Jl | 13 | Ashada | 4 | Jl | 6 | 7 | 5 | ... | 22 | 33 | 16 | 41 | 51 | 85 | 39 | 35. | | |
| (S.) Pramadi | Adi | 5 | Jl | 14 | 32 | 1 | Au | 14 | Sravana | 5 | Au | 4 | 38 | 55 | ... | 55 | 47 | 18 | 40 | 25 | 115 | 11 | 24 | | |
| K. 1028 (S. M.) | Avani | 2 | Au | 15 | 31 | 4 | S | 14 | Bhadrapada | 7 | S | 3 | 10 | 46 | ... | 25 | 57 | 20 | 38 | 59 | 144 | 43 | 14 | | |
| K. 1028 (N. M.) | Purattasi | 5 | S | 15 | 30 | 6 | O | 14 | Asvina | 1 | O | 2 | 42 | 36 | ... | 53 | 54 | 22 | 37 | 32 | 174 | 15 | 4 | | |
| 21 g., 21 p. | Aippasi | 7 | O | 15 | 30 | 1 | N | 13 | Kartika | 3 | N | 1 | 14 | 26 | ... | 20 | 26 | 24 | 36 | 6 | 203 | 46 | 54 | | |
| B. S. 1259 | Kartigai | 2 | N | 14 | 30 | 3 | D | 13 | Margasira | 4 | N | 30 | 46 | 6 | ... | 46 | 20 | 26 | 34 | 39 | 233 | 18 | 44 | | |
| (N.) Raktaksha. | Margali | 4 | D | 14 | 29 | 4 | Ja | 11 | Pausha | 6 | D | 30 | 18 | 6 | ... | 12 | 52 | 0 | 59 | 56 | 262 | 50 | 34 | | |
| **A. D. 1854** | Tai | 5 | Ja | 12 | 29 | 5 | F | 9 | Magha | 7 | Ja | 28 | 49 | 56 | ... | 40 | 55 | 2 | 58 | 30 | 292 | 22 | 24 | | |
| H. 1271. | Masi | 6 | F | 10 | 30 | 7 | Mr | 11 | Phalguna | 2 | F | 27 | 21 | 46 | ... | 9 | 42 | 4 | 57 | 4 | 321 | 54 | 14 | | |
| Sep. 24 | Panguni | 1 | Mr | 12 | 31 | 3 | Ap | 11 | Chaitra | 3 | Mr | 28 | 53 | 36 | ... | 39 | 38 | 6 | 55 | 37 | 351 | 26 | 4 | | |
| **K. Y. 4955** | Chittirai | 4 | Ap | 12 | 31 | 6 | My | 12 | Vaisakha | 5 | Ap | 27 | 25 | 26 | ...○ | 12 | 20 | 8 | 54 | 11 | 15 | 42 | 23 | 1 day, 10 ghat., 29 palas. | 2 „ 11 „ 10 „ |
| Sak. 1776 | Vaikasi | 7 | My | 13 | 31 | 2 | Je | 12 | Jyeshta | 6 | My | 26 | 57 | 16 | ... | 47 | 31 | 10 | 52 | 44 | 45 | 14 | 13 | | |
| Vik. 1911 | Ani | 3 | Je | 13 | 32 | 6 | Jl | 14 | Ashada | 1 | Je | 25 | 29 | 6 | ... | 24 | 34 | 12 | 51 | 18 | 74 | 46 | 3 | | |
| (S.) Ananda | Adi | 7 | Jl | 15 | 31 | 2 | Au | 14 | Sravana | 3 | Jl | 25 | 0 | 57 | ... | 2 | 42 | 14 | 49 | 52 | 104 | 17 | 53 | | |
| K. 1029 (S. M.) | Avani | 3 | Au | 15 | 31 | 5 | S | 14 | Bhadrapada | 4 | Au | 23 | 32 | 47 | ... | 40 | 36 | 16 | 48 | 25 | 133 | 49 | 44 | | |
| K. 1029 (N. M.) | Purattasi | 6 | S | 15 | 31 | 1 | O | 15 | Asvina | 6 | S | 22 | 4 | 37 | ... | 17 | 4 | 18 | 46 | 59 | 163 | 21 | 34 | | |
| 36 g., 52 p. | Aippasi | 2 | O | 16 | 29 | 2 | N | 13 | Kartika | 7 | O | 21 | 36 | 27 | ...○ | 52 | 8 | 20 | 45 | 32 | 192 | 53 | 24 | | |
| B. S. 1260 | Kartigai | 3 | N | 14 | 30 | 4 | D | 13 | Margasira | 2 | N | 20 | 8 | 17 | ... | 23 | 58 | 22 | 44 | 6 | 222 | 25 | 14 | | |
| (N.) Krodhana. | Margali | 5 | D | 14 | 29 | 5 | Ja | 11 | Pausha | 3 | D | 19 | 40 | 7 | ... | 52 | 13 | 24 | 42 | 40 | 251 | 57 | 4 | | |
| **A. D. 1855** | Tai | 6 | Ja | 12 | 30 | 7 | F | 10 | Magha | 5 | Ja | 18 | 11 | 58 | ... | 19 | 35 | 26 | 41 | 13 | 281 | 28 | 54 | | |
| H. 1272 | Masi | 1 | F | 11 | 30 | 2 | Mr | 12 | Phalguna | 6 | F | 16 | 43 | 48 | ... | 45 | 15 | 1 | 6 | 30 | 311 | 0 | 44 | | |
| Sep. 13 | Panguni | 3 | Mr | 13 | 30 | 4 | Ap | 11 | Chaitra | 1 | Mr | 18 | 15 | 38 | ... | 10 | 15 | 3 | 5 | 4 | 340 | 32 | 34 | | |

## TABLE XII.—General Ephemeris A.D. 1855—A.D. 1860.

| Commencement of years in different eras. | Tamil month. | Tamil Month Beginning Week-day | Month | Day | Last Day | End Week-day | Month | Day | Lunar month. | New Moon Ending moment of New Moon Tithi Week-day | Month | Mean Day (mean) | Ghatikas | Palas | Actual Day (actual) | Ghatikas | Palas | ☾'s Anom. Day | Ghatikas | Palas | ☉'s Anom. Day | Ghatikas | Palas | Deduct for Nakshatras | Deduct for Yogas. |
|---|---|---|---|---|---|---|---|---|---|---|---|---|---|---|---|---|---|---|---|---|---|---|---|---|---|
| **K. Y. 4956** | Chittirai... | 5 | Ap | 12 | 31 | 7 | My | 12 | Vaisakha ... | 2 | Ap | 16 | 47 | 28 | ... ○ | 35 | 30 | 5 | 3 | 38 | 4 | 48 | 53 | 0 day, 21 ghat., 36 palas. | 0 „ 40 „ 12 „ |
| Sak. 1777 | Vaikasi ... | 1 | My | 13 | 31 | 3 | Je | 12 | Jyeshta ... | 4 | My | 16 | 19 | 18 | ... ● | 2 | 12 | 7 | 2 | 11 | 34 | 20 | 43 | | |
| Vik. 1912 | Ani ... | 4 | Je | 13 | 32 | 7 | Jl | 14 | Ashada ... | 5 | Jo | 14 | 51 | 8 | ... | 31 | 29 | 9 | 0 | 45 | 63 | 52 | 33 | | |
| | | | | | | | | | | 7 | Jl | 14 | 22 | 58 | ... | 4 | 30 | 10 | 59 | 18 | 93 | 24 | 23 | | |
| (S.) Rakshasa | Adi ... | 1 | Jl | 15 | 31 | 3 | Au | 14 | Sravana ... | 1 | Au | 12 | 54 | 48 | ... | 41 | 49 | 12 | 57 | 52 | 122 | 56 | 18 | | |
| K. 1030 (S. M.) | Avani ... | 4 | Au | 15 | 31 | 6 | S | 14 | Bhadrapada . | 3 | S | 11 | 26 | 38 | ... | 22 | 38 | 14 | 56 | 25 | 152 | 28 | 3 | | |
| K. 1030 (N. M.) | Purattasi . | 7 | S | 15 | 31 | 2 | O | 15 | Asvina ... | 4 | O | 10 | 58 | 28 | 11 ○ | 4 | 38 | 16 | 54 | 59 | 181 | 59 | 54 | | |
| 52 g., 24 p. | Aippasi ... | 3 | O | 16 | 30 | 4 | N | 14 | Kartika ... | 6 | N | 9 | 30 | 18 | ... | 45 | 13 | 18 | 53 | 33 | 211 | 31 | 44 | | |
| B. S. 1261 | Kartigai... | 5 | N | 15 | 29 | 5 | D | 13 | Margasira ... | 1 | D | 9 | 2 | 10 | ... | 22 | 47 | 20 | 52 | 7 | 241 | 3 | 34 | | |
| (N.) Kshaya | Margali ... | 6 | D | 14 | 30 | 7 | Ja | 12 | Pausha ... | 2 | Ja | 7 | 33 | 59 | ... | 56 | 13 | 22 | 50 | 40 | 270 | 35 | 24 | | |
| **A. D. 1856** | Tai ... | 1 | Ja | 13 | 29 | 1 | F | 10 | Magha ... | 4 | F | 6 | 5 | 49 | ... | 25 | 21 | 24 | 49 | 14 | 300 | 7 | 14 | | |
| H. 1273 | Masi ... | 2 | F | 11 | 30 | 3 | Mr | 11 | Phalguna ... | 5 | Mr | 6 | 37 | 39 | ... | 50 | 33 | 26 | 47 | 47 | 329 | 39 | 4 | | |
| Sep. 1 | Panguni ... | 4 | Mr | 12 | 30 | 5 | Ap | 10 | Chaitra ... | 7 | Ap | 5 | 9 | 29 | ...●○ | 13 | 47 | 1 | 13 | 4 | 359 | 10 | 54 | | |
| **K. Y. 4957** | Chittirai... | 6 | Ap | 11 | 31 | 1 | My | 11 | Vaisakha ... | 1 | My | 4 | 41 | 19 | ... | 34 | 7 | 3 | 11 | 38 | 23 | 27 | 13 | 1 day, 45 ghat., 16 palas. | 3 „ 15 „ 52 „ |
| Sak. 1778 | Vaikasi ... | 2 | My | 12 | 31 | 4 | Je | 11 | Jyeshta ... | 3 | Je | 3 | 13 | 9 | 2 | 55 | 45 | 5 | 10 | 11 | 52 | 59 | 3 | | |
| Vik. 1913 | Ani ... | 5 | Je | 12 | 32 | 1 | Jl | 13 | Ashada ... | 4 | Jl | 2 | 44 | 59 | ... | 19 | 49 | 7 | 8 | 45 | 82 | 30 | 53 | | |
| (S.) Anala | Adi ... | 2 | Jl | 14 | 32 | 5 | Au | 14 | Sravana ... | 6 | Au | 1 | 16 | 50 | Jl. 31. | 48 | 26 | 9 | 7 | 19 | 112 | 2 | 43 | | |
| K. 1031 (S. M.) | Avani ... | 6 | Au | 15 | 31 | 1 | S | 14 | Bhadrapada . | 7 | Au | 30 | 48 | 40 | ... | 23 | 15 | 11 | 5 | 52 | 141 | 34 | 33 | | |
| K. 1031 (N. M.) | Purattasi... | 2 | S | 15 | 30 | 3 | O | 14 | Asvina ... | 2 | S | 29 | 20 | 30 | ...●○ | 4 | 19 | 13 | 4 | 26 | 171 | 6 | 23 | | |
| 7 g., 55 p. | Aippasi ... | 4 | O | 15 | 30 | 5 | N | 13 | Kartika ... | 3 | O | 28 | 52 | 20 | ... | 49 | 46 | 15 | 2 | 59 | 200 | 38 | 13 | | |
| B. S. 1262 | Kartigai ... | 6 | N | 14 | 29 | 6 | D | 12 | Margasira ... | 5 | N | 27 | 24 | 10 | ... | 35 | 51 | 17 | 1 | 33 | 230 | 10 | 3 | | |
| (N.) Prabhava | Margali ... | 7 | D | 13 | 30 | 1 | Ja | 11 | Pausha ... | 6 | D | 26 | 56 | 0 | 27 | 19 | 6 | 19 | 0 | 6 | 259 | 41 | 54 | | |
| **A. D. 1857** | Tai ... | 2 | Ja | 12 | 29 | 2 | F | 9 | Magha ... | 1 | Ja | 25 | 27 | 50 | ... | 56 | 59 | 20 | 58 | 40 | 289 | 13 | 44 | | |
| H. 1274 | Masi ... | 3 | F | 10 | 30 | 4 | Mr | 11 | Phalguna ... | 2 | F | 23 | 59 | 40 | 24 | 28 | 35 | 22 | 57 | 14 | 318 | 45 | 34 | | |
| Aug. 22 | Panguni ... | 5 | Mr | 12 | 30 | 6 | Ap | 10 | Chaitra ... | 4 | Mr | 25 | 31 | 30 | ... | 54 | 25 | 24 | 55 | 47 | 348 | 17 | 24 | | |
| **K. Y. 4958** | Chittirai... | 7 | Ap | 11 | 31 | 2 | My | 11 | Vaisakha ... | 6 | Ap | 24 | 3 | 22 | ... | 15 | 47 | 26 | 54 | 21 | 12 | 33 | 43 | 0 day, 56 ghat., 23 palas. | 1 „ 44 „ 54 „ |
| Sak. 1779 | Vaikasi ... | 3 | My | 12 | 32 | 6 | Je | 12 | Jyeshta ... | 7 | My | 23 | 35 | 11 | ... | 34 | 23 | 1 | 19 | 38 | 42 | 5 | 33 | | |
| Vik. 1914 | Ani ... | 7 | Je | 13 | 31 | 2 | Jl | 13 | Ashada ... | 2 | Je | 22 | 7 | 1 | 21 | 52 | 36 | 3 | 18 | 12 | 71 | 37 | 23 | | |
| (S.) Pingala | Adi ... | 3 | Jl | 14 | 32 | 6 | Au | 14 | Sravana ... | 3 | Jl | 21 | 38 | 51 | ... | 12 | 19 | 5 | 16 | 45 | 101 | 9 | 13 | | |
| K. 1032. (S. M.) | Avani ... | 7 | Au | 15 | 31 | 2 | S | 14 | Bhadrapada . | 5 | Au | 20 | 10 | 41 | 19 | 37 | 43 | 7 | 15 | 19 | 130 | 41 | 3 | | |
| K. 1032 (N. M.) | Purattasi... | 3 | S | 15 | 30 | 4 | O | 14 | Asvina ... | 6 | S | 18 | 42 | 31 | ... ● | 9 | 16 | 9 | 13 | 52 | 160 | 12 | 53 | | |
| 23 g., 26 p. | Aippasi ... | 5 | O | 15 | 30 | 6 | N | 13 | Kartika ... | 1 | O | 18 | 14 | 21 | 17 | 48 | 29 | 11 | 12 | 26 | 189 | 44 | 43 | | |
| B. S. 1263 | Kartigai ... | 7 | N | 14 | 30 | 1 | D | 13 | Margasira ... | 2 | N | 16 | 46 | 11 | ... | 34 | 21 | 13 | 11 | 0 | 219 | 16 | 33 | | |
| (N.) Sukla | Margali ... | 2 | D | 14 | 29 | 2 | Ja | 11 | Pausha ... | 4 | D | 16 | 18 | 2 | ... | 23 | 33 | 15 | 9 | 33 | 248 | 48 | 23 | | |
| **A. D. 1858** | Tai ... | 3 | Ja | 12 | 30 | 4 | F | 10 | Magha ... | 5 | Ja | 14 | 49 | 52 | 15 | 11 | 13 | 17 | 8 | 7 | 278 | 20 | 14 | | |
| H. 1275 | Masi ... | 5 | F | 11 | 29 | 5 | Mr | 11 | Phalguna ... | 7 | F | 13 | 21 | 42 | ... ○ | 53 | 19 | 19 | 6 | 40 | 307 | 52 | 4 | | |
| Aug. 11 | Panguni ... | 6 | Mr | 12 | 31 | 1 | Ap | 11 | Chaitra ... | 1 | Mr | 14 | 53 | 32 | 15 ● | 28 | 1 | 21 | 5 | 14 | 337 | 23 | 54 | | |
| **K. Y. 4959** | Chittirai ... | 2 | Ap | 12 | 31 | 4 | My | 12 | Vaisakha ... | 3 | Ap | 13 | 25 | 22 | ... | 55 | 32 | 23 | 3 | 48 | 1 | 40 | 13 | 0 day, 7 ghat., 30 palas. | 0 „ 13 „ 57 „ |
| Sak. 1780 | Vaikasi ... | 5 | My | 13 | 31 | 7 | Je | 12 | Jyeshta ... | 4 | My | 12 | 57 | 12 | 13 | 17 | 15 | 25 | 2 | 21 | 31 | 12 | 3 | | |
| | | | | | | | | | | 6 | Je | 11 | 29 | 2 | ... | 35 | 26 | 27 | 0 | 55 | 60 | 43 | 53 | | |
| Vik. 1915 | Ani ... | 1 | Je | 13 | 32 | 4 | Jl | 14 | Ashada ... | 1 | Jl | 11 | 0 | 52 | 10 | 51 | 47 | 1 | 26 | 12 | 90 | 15 | 43 | | |
| (S.) Kalayukta. | Adi ... | 5 | Jl | 15 | 31 | 7 | Au | 14 | Sravana ... | 2 | Au | 9 | 32 | 42 | ... ○ | 9 | 59 | 3 | 24 | 45 | 119 | 47 | 33 | | |
| K. 1033 (S. M.) | Avani ... | 1 | Au | 15 | 31 | 3 | S | 14 | Bhadrapada . | 4 | S | 8 | 4 | 33 | 7 | 32 | 22 | 5 | 23 | 19 | 149 | 19 | 23 | | |
| K. 1033 (N. M.) | Purattasi... | 4 | S | 15 | 31 | 6 | O | 15 | Asvina ... | 5 | O | 7 | 36 | 23 | ... | 0 | 58 | 7 | 21 | 53 | 178 | 51 | 13 | | |
| 38 g., 57 p. | Aippasi ... | 7 | O | 16 | 29 | 7 | N | 13 | Kartika ... | 7 | N | 6 | 8 | 13 | 5 | 37 | 6 | 9 | 20 | 26 | 208 | 23 | 3 | | |
| B. S. 1264 | Kartigai ... | 1 | N | 14 | 30 | 2 | D | 13 | Margasira ... | 1 | D | 5 | 40 | 3 | ... | 20 | 48 | 11 | 19 | 0 | 237 | 54 | 53 | | |
| (N.) Pramoda | Margali ... | 3 | D | 14 | 29 | 3 | Ja | 11 | Pausha ... | 3 | Ja | 4 | 11 | 53 | ... | 9 | 43 | 13 | 17 | 33 | 267 | 26 | 43 | | |
| **A. D. 1859** | Tai ... | 4 | Ja | 12 | 30 | 5 | F | 10 | Magha ... | 4 | F | 2 | 43 | 43 | ... ○ | 59 | 18 | 15 | 16 | 7 | 296 | 58 | 33 | | |
| H. 1276 | Masi ... | 6 | F | 11 | 30 | 7 | Mr | 12 | Phalguna ... | 6 | Mr | 4 | 15 | 33 | ... | 44 | 29 | 17 | 14 | 41 | 326 | 30 | 24 | | |
| July 31 | Panguni ... | 1 | Mr | 13 | 30 | 2 | Ap | 11 | Chaitra ... | 7 | Ap | 2 | 47 | 23 | 3 | 22 | 23 | 19 | 13 | 14 | 356 | 2 | 14 | | |
| **K. Y. 4960** | Chittirai... | 3 | Ap | 12 | 31 | 5 | My | 12 | Vaisakha ... | 2 | My | 2 | 19 | 14 | ... | 52 | 36 | 21 | 11 | 48 | 20 | 18 | 33 | 1 day, 31 ghat., 9 palas. | 2 „ 49 „ 51 „ |
| Sak. 1781 | Vaikasi ... | 6 | My | 13 | 31 | 1 | Je | 12 | Jyeshta ... | 3 | My | 31 | 51 | 4 | Je.1 | 16 | 3 | 23 | 10 | 21 | 49 | 50 | 23 | | |
| Vik. 1916 | Ani ... | 2 | Je | 13 | 32 | 5 | Jl | 14 | Ashada ... | 5 | Je | 30 | 22 | 54 | ... | 35 | 11 | 25 | 8 | 55 | 79 | 22 | 13 | | |
| (S.) Siddharthi. | Adi ... | 6 | Jl | 15 | 31 | 1 | Au | 14 | Sravana ... | 6 | Jl | 29 | 54 | 44 | ... ○ | 52 | 21 | 27 | 7 | 29 | 108 | 54 | 3 | | |
| K. 1034 (S. M.) | Avani ... | 2 | Au | 15 | 31 | 4 | S | 14 | Bhadrapada . | 1 | Au | 28 | 26 | 34 | ... | 10 | 17 | 1 | 32 | 46 | 138 | 25 | 53 | | |
| K. 1034 (N. M.) | Purattasi... | 5 | S | 15 | 31 | 7 | O | 15 | Asvina ... | 2 | S | 26 | 58 | 24 | ... | 31 | 36 | 3 | 31 | 19 | 167 | 57 | 43 | | |
| 54 g., 29 p. | Aippasi ... | 1 | O | 16 | 30 | 2 | N | 14 | Kartika ... | 4 | O | 26 | 30 | 14 | 25 | 58 | 0 | 5 | 29 | 52 | 197 | 29 | 33 | | |
| B. S. 1265. | Kartigai ... | 3 | N | 15 | 29 | 3 | D | 13 | Margasira ... | 6 | N | 25 | 2 | 4 | 24 | 30 | 58 | 7 | 28 | 26 | 227 | 1 | 23 | | |
| (N.) Prajapati. | Margali ... | 4 | D | 14 | 30 | 4 | Ja | 12 | Pausha ... | 7 | D | 24 | 33 | 54 | ... | 10 | 52 | 9 | 27 | 0 | 256 | 33 | 13 | | |
| **A. D. 1860** | Tai ... | 6 | Ja | 13 | 29 | 6 | F | 10 | Magha ... | 2 | Ja | 23 | 5 | 45 | 22 ○ | 57 | 7 | 11 | 25 | 34 | 286 | 5 | 3 | | |
| H. 1277. | Masi ... | 7 | F | 11 | 30 | 1 | Mr | 11 | Phalguna ... | 3 | F | 21 | 37 | 35 | ... | 44 | 32 | 13 | 24 | 7 | 315 | 36 | 54 | | |
| July 20 | Panguni ... | 2 | Mr | 12 | 30 | 3 | Ap | 10 | Chaitra ... | 5 | Mr | 22 | 9 | 25 | ... | 30 | 40 | 15 | 22 | 41 | 345 | 8 | 44 | | |

## TABLE XII.—General Ephemeris A.D. 1860—A.D. 1865.

| Commencement of years in different eras. | Tamil month. | Tamil Month Beginning Week-day | Month | Day | Last Day | End Week-day | Month | Day | Lunar month. | New Moon Ending moment of New Moon Tithi. Week-day | Month | Mean Day (mean) | Ghatikas | Palas | Actual Day (actual) | Ghatikas | Palas | ☾'s Anom. Day | Ghatikas | Palas | ☉'s Anom. Day | Ghatikas | Palas | Deduct for Nakshatras | Deduct for Yogas. |
|---|---|---|---|---|---|---|---|---|---|---|---|---|---|---|---|---|---|---|---|---|---|---|---|---|---|
| **K. Y. 4961** | Chittirai . | 4 | Ap | 11 | 31 | 6 | My | 11 | Vaisakha ... | 6 | Ap | 20 | 41 | 15 | 21 | 11 | 11 | 17 | 21 | 5 | 9 | 25 | 3 | 0 day, 42 ghat., 16 palas. | 1 " 18 " 39 " |
| Sak. 1782 | Vaikasi ... | 7 | My | 12 | 31 | 2 | Je | 11 | Jyeshta ... | 1 | My | 20 | 13 | 5 | ... | 44 | 37 | 19 | 19 | 40 | 38 | 56 | 53 | | |
| Vik. 1917 | Ani ... | 3 | Je | 12 | 32 | 6 | Jl | 13 | Ashada ... | 2 | Je | 18 | 44 | 55 | 19 | 11 | 17 | 21 | 18 | 14 | 68 | 28 | 48 | | |
| (S.) Raudra | Adi ... | 7 | Jl | 14 | 32 | 3 | Au | 14 | Sravana ... | 4 | Jl | 18 | 16 | 45 | ...○ | 33 | 16 | 23 | 16 | 48 | 98 | 0 | 38 | | |
| K. 1035 (S. M.) | Avani ... | 4 | Au | 15 | 31 | 6 | S | 14 | Bhadrapada . | 5 | Au | 16 | 48 | 35 | ... | 52 | 58 | 25 | 15 | 22 | 127 | 32 | 23 | | |
| K. 1035 (N. M.) | Purattasi . | 7 | S | 15 | 30 | 1 | O | 14 | Asvina ... | 7 | S | 15 | 20 | 25 | ... | 12 | 39 | 27 | 13 | 56 | 157 | 4 | 13 | | |
| | | | | | | | | | | 1 | O | 14 | 52 | 16 | ... | 34 | 16 | 1 | 39 | 12 | 186 | 36 | 3 | | |
| 10 g., 0 p. | Aippasi ... | 2 | O | 15 | 30 | 3 | N | 13 | Kartika ... | 3 | N | 13 | 24 | 6 | 12 | 59 | 33 | 3 | 37 | 46 | 216 | 7 | 53 | | |
| B. S. 1266 | Kartigai ... | 4 | N | 14 | 29 | 4 | D | 12 | Margasira ... | 4 | D | 12 | 55 | 56 | ... | 29 | 41 | 5 | 36 | 19 | 245 | 39 | 43 | | |
| (N.) Angiras | Margali ... | 5 | D | 13 | 30 | 6 | Ja | 11 | Pausha ... | 6 | Ja | 11 | 27 | 46 | ...● | 5 | 17 | 7 | 34 | 53 | 275 | 11 | 33 | | |
| **A. D. 1861** | Tai ... | 7 | Ja | 12 | 29 | 7 | F | 9 | Magha ... | 7 | F | 9 | 59 | 36 | ... | 45 | 37 | 9 | 33 | 27 | 304 | 43 | 24 | | |
| H. 1278 | Masi ... | 1 | F | 10 | 30 | 2 | Mr | 11 | Phalguna ... | 2 | Mr | 11 | 31 | 26 | ... | 29 | 10 | 11 | 32 | 0 | 334 | 15 | 14 | | |
| July 9 | Panguni ... | 3 | Mr | 12 | 30 | 4 | Ap | 10 | Chaitra ... | 4 | Ap | 10 | 3 | 16 | ... | 13 | 18 | 13 | 30 | 36 | 363 | 47 | 4 | | |
| **K. Y. 4962** | Chittirai... | 5 | Ap | 11 | 31 | 7 | My | 11 | Vaisakha ... | 5 | My | 9 | 35 | 6 | ... | 54 | 39 | 15 | 29 | 7 | 28 | 8 | 23 | 2 days, 5 ghat., 55 palas. | 3 " 54 " 19 " |
| Sak. 1783 | Vaikasi ... | 1 | My | 12 | 32 | 4 | Je | 12 | Jyeshta ... | 7 | Je | 8 | 6 | 57 | ... | 31 | 1 | 17 | 27 | 41 | 57 | 35 | 13 | | |
| Vik. 1918 | Ani ... | 5 | Je | 13 | 31 | 7 | Jl | 13 | Ashada ... | 1 | Jl | 7 | 38 | 47 | 5● | 1 | 54 | 19 | 26 | 15 | 87 | 7 | 8 | | |
| (S.) Durmati | Adi ... | 1 | Jl | 14 | 32 | 4 | Au | 14 | Sravana ... | 3 | Au | 6 | 10 | 37 | ... | 28 | 29 | 21 | 24 | 48 | 116 | 38 | 53 | | |
| K. 1036 (S. M.) | Avani ... | 5 | Au | 15 | 31 | 7 | S | 14 | Bhadrapada . | 4 | S | 4 | 42 | 27 | ... | 52 | 29 | 23 | 23 | 22 | 146 | 10 | 43 | | |
| K. 1036 (N. M.) | Purattasi . | 1 | S | 15 | 30 | 2 | O | 14 | Asvina ... | 6 | O | 4 | 14 | 17 | ... | 15 | 30 | 25 | 21 | 55 | 175 | 42 | 33 | | |
| 25 g., 31 p. | Aippasi ... | 3 | O | 15 | 30 | 4 | N | 13 | Kartika ... | 7 | N | 2 | 46 | 7 | ... | 39 | 0 | 27 | 20 | 29 | 205 | 14 | 23 | | |
| B. S. 1267 | Kartigai ... | 5 | N | 14 | 30 | 6 | D | 13 | Margasira ... | 2 | D | 2 | 17 | 57 | ...○ | 4 | 16 | 1 | 45 | 46 | 234 | 46 | 13 | | |
| (N.) Srimukha | Margali ... | 7 | D | 14 | 30 | 7 | Ja | 11 | Pausha ... | 3 | D | 31 | 49 | 47 | ... | 32 | 19 | 3 | 44 | 20 | 264 | 18 | 3 | | |
| **A. D. 1862** | Tai ... | 1 | Ja | 12 | 30 | 2 | F | 10 | Magha ... | 5 | Ja | 30 | 21 | 38 | ... | 3 | 35 | 5 | 42 | 53 | 293 | 49 | 53 | | |
| H. 1279 | Masi ... | 3 | F | 11 | 29 | 3 | Mr | 11 | Phalguna ... | 6 | F | 28 | 53 | 28 | ... | 38 | 5 | 7 | 41 | 27 | 323 | 21 | 44 | | |
| June 28, 29 | Panguni ... | 4 | Mr | 12 | 31 | 6 | Ap | 11 | Chaitra ... | 1 | Mr | 30 | 25 | 18 | ... | 15 | 29 | 9 | 40 | 0 | 352 | 53 | 34 | | |
| **K. Y. 4963** | Chittirai . | 7 | Ap | 12 | 31 | 2 | My | 12 | Vaisakha ... | 2 | Ap | 28 | 57 | 8 | ... | 54 | 51 | 11 | 38 | 34 | 17 | 9 | 53 | 1 day, 17 ghat., 2 palas. | 2 " 23 " 21 " |
| Sak. 1784 | Vaikasi ... | 3 | My | 13 | 31 | 5 | Je | 12 | Jyeshta ... | 4 | My | 28 | 28 | 58 | ...○ | 34 | 33 | 13 | 37 | 8 | 46 | 41 | 43 | | |
| Vik. 1919 | Ani ... | 6 | Je | 13 | 32 | 2 | Jl | 14 | Ashada ... | 6 | Je | 27 | 0 | 48 | ... | 12 | 37 | 15 | 35 | 41 | 76 | 13 | 33 | | |
| (S.) Dundubhi | Adi ... | 3 | Jl | 15 | 31 | 5 | Au | 14 | Sravana ... | 7 | Jl | 26 | 32 | 38 | ... | 47 | 57 | 17 | 34 | 15 | 105 | 45 | 23 | | |
| K. 1037 (S. M.) | Avani ... | 6 | Au | 15 | 31 | 1 | S | 14 | Bhadrapada . | 2 | Au | 25 | 4 | 29 | ... | 20 | 7 | 19 | 32 | 48 | 135 | 17 | 13 | | |
| K. 1037 (N. M.) | Purattasi . | 2 | S | 15 | 31 | 4 | O | 15 | Asvina ... | 3 | S | 23 | 36 | 19 | ... | 49 | 49 | 21 | 31 | 22 | 164 | 49 | 3 | | |
| 41 g., 2 p. | Aippasi ... | 5 | O | 16 | 29 | 5 | N | 13 | Kartika ... | 5 | O | 23 | 8 | 9 | ... | 17 | 36 | 23 | 29 | 56 | 194 | 20 | 53 | | |
| B. S. 1268 | Kartigai ... | 6 | N | 14 | 30 | 7 | D | 13 | Margasira ... | 6 | N | 21 | 39 | 59 | ...○ | 44 | 19 | 25 | 28 | 29 | 223 | 52 | 43 | | |
| (N.) Bhava | Margali ... | 1 | D | 14 | 29 | 1 | Ja | 11 | Pausha ... | 1 | D | 21 | 11 | 49 | ...● | 10 | 42 | 27 | 27 | 3 | 253 | 24 | 33 | | |
| **A. D. 1863** | Tai ... | 2 | Ja | 12 | 30 | 3 | F | 10 | Magha ... | 2 | Ja | 19 | 43 | 39 | ... | 37 | 26 | 1 | 52 | 20 | 282 | 56 | 23 | | |
| H. 1280 | Masi ... | 4 | F | 11 | 30 | 5 | Mr | 12 | Phalguna ... | 4 | F | 18 | 15 | 29 | ... | 5 | 0 | 3 | 50 | 54 | 312 | 28 | 13 | | |
| June 18 | Panguni ... | 6 | Mr | 13 | 30 | 7 | Ap | 11 | Chaitra ... | 5 | Mr | 19 | 47 | 19 | ... | 33 | 58 | 5 | 49 | 27 | 342 | 0 | 3 | | |
| **K. Y. 4964** | Chittirai . | 1 | Ap | 12 | 31 | 3 | My | 12 | Vaisakha ... | 7 | Ap | 18 | 19 | 9 | ... | 4 | 46 | 7 | 48 | 1 | 6 | 16 | 22 | 0 day, 28 ghat., 9 palas. | 0 " 52 " 23 " |
| Sak. 1785 | Vaikasi ... | 4 | My | 13 | 31 | 6 | Je | 12 | Jyeshta ... | 1 | My | 17 | 50 | 59 | ...○ | 38 | 10 | 9 | 46 | 34 | 35 | 48 | 12 | | |
| Vik. 1920 | Ani ... | 7 | Je | 13 | 32 | 3 | Jl | 14 | Ashada ... | 3 | Je | 16 | 22 | 50 | ... | 13 | 51 | 11 | 45 | 8 | 65 | 20 | 2 | | |
| (S.) Rudhirogari. | Adi ... | 4 | Jl | 15 | 31 | 6 | Au | 14 | Sravana ... | 4 | Jl | 15 | 54 | 40 | ... | 51 | 25 | 13 | 43 | 42 | 94 | 51 | 53 | | |
| | | | | | | | | | | 6 | Au | 14 | 26 | 30 | ... | 29 | 58 | 15 | 42 | 15 | 124 | 23 | 48 | | |
| K. 1038 (S. M.) | Avani ... | 7 | Au | 15 | 31 | 2 | S | 14 | Bhadrapada . | 7 | S | 12 | 58 | 20 | 13 | 9 | 56 | 17 | 40 | 49 | 153 | 55 | 38 | | |
| K. 1038 (N. M.) | Purattasi . | 3 | S | 15 | 31 | 5 | O | 15 | Asvina ... | 2 | O | 12 | 30 | 10 | ... | 44 | 4 | 19 | 39 | 22 | 183 | 27 | 23 | | |
| 56 g., 34 p. | Aippasi ... | 6 | O | 16 | 30 | 7 | N | 14 | Kartika ... | 4 | N | 11 | 2 | 0 | ...○ | 17 | 40 | 21 | 37 | 56 | 212 | 59 | 13 | | |
| B. S. 1269. | Kartigai ... | 1 | N | 15 | 29 | 1 | D | 13 | Margasira ... | 5 | D | 10 | 33 | 50 | ... | 48 | 43 | 23 | 36 | 30 | 242 | 31 | 3 | | |
| (N.) Yuvan | Margali ... | 2 | D | 14 | 30 | 3 | Ja | 12 | Pausha ... | 7 | Ja | 9 | 5 | 40 | ... | 17 | 14 | 25 | 35 | 3 | 272 | 2 | 58 | | |
| **A. D. 1864** | Tai ... | 4 | Ja | 13 | 29 | 4 | F | 10 | Magha ... | 1 | F | 7 | 37 | 30 | ... | 43 | 32 | 0 | 0 | 20 | 301 | 34 | 43 | | |
| H. 1281 | Masi ... | 5 | F | 11 | 30 | 6 | Mr | 11 | Phalguna ... | 3 | Mr | 8 | 9 | 21 | ... | 8 | 20 | 1 | 58 | 54 | 331 | 6 | 33 | | |
| June 6 | Panguni ... | 7 | Mr | 12 | 30 | 1 | Ap | 10 | Chaitra ... | 4 | Ap | 6 | 41 | 11 | ... | 32 | 44 | 3 | 57 | 27 | 360 | 38 | 23 | | |
| **K. Y. 4965** | Chittirai . | 2 | Ap | 11 | 31 | 4 | My | 11 | Vaisakha ... | 6 | My | 6 | 13 | 1 | 5● | 57 | 51 | 5 | 56 | 1 | 24 | 54 | 42 | 1 day, 51 ghat., 48 palas. | 3 " 28 " 3 " |
| Sak. 1786 | Vaikasi ... | 5 | My | 12 | 32 | 1 | Je | 12 | Jyeshta ... | 7 | Je | 4 | 44 | 51 | ... | 25 | 3 | 7 | 54 | 35 | 54 | 26 | 32 | | |
| Vik. 1921 | Ani ... | 2 | Je | 13 | 31 | 4 | Jl | 13 | Ashada ... | 2 | Jl | 4 | 16 | 41 | ... | 55 | 38 | 9 | 53 | 8 | 83 | 58 | 22 | | |
| (S.) Raktaksha. | Adi ... | 5 | Jl | 14 | 32 | 1 | Au | 14 | Sravana ... | 3 | Au | 2 | 48 | 31 | ... | 30 | 44 | 11 | 51 | 42 | 113 | 30 | 13 | | |
| K. 1039 (S. M.) | Avani ... | 2 | Au | 15 | 31 | 4 | S | 14 | Bhadrapada . | 5 | S | 1 | 20 | 21 | ... | 10 | 53 | 13 | 50 | 15 | 143 | 2 | 3 | | |
| K. 1039 (N. M.) | Purattasi . | 5 | S | 15 | 30 | 6 | O | 14 | Asvina ... | 6 | S | 30 | 52 | 11 | ... | 53 | 12 | 15 | 48 | 49 | 172 | 33 | 53 | | |
| 12 g., 5 p. | Aippasi ... | 7 | O | 15 | 30 | 1 | N | 13 | Kartika ... | 1 | O | 30 | 24 | 1 | ... | 34 | 57 | 17 | 47 | 23 | 202 | 5 | 43 | | |
| B. S. 1270 | Kartigai ... | 2 | N | 14 | 30 | 3 | D | 13 | Margasira ... | 2 | N | 28 | 55 | 52 | 29 | 14 | 52 | 19 | 45 | 56 | 231 | 37 | 33 | | |
| (N.) Dhatri | Margali ... | 4 | D | 14 | 29 | 4 | Ja | 11 | Pausha ... | 4 | D | 28 | 27 | 42 | ... | 50 | 51 | 21 | 44 | 30 | 261 | 9 | 23 | | |
| **A. D. 1865** | Tai ... | 5 | Ja | 12 | 29 | 5 | F | 9 | Magha ... | 5 | Ja | 26 | 59 | 32 | 27 | 21 | 12 | 23 | 43 | 3 | 290 | 41 | 13 | | |
| H. 1282 | Masi ... | 6 | F | 10 | 30 | 7 | Mr | 11 | Phalguna ... | 7 | F | 25 | 31 | 22 | ... | 49 | 5 | 25 | 41 | 37 | 320 | 13 | 3 | | |
| May 27 | Panguni ... | 1 | Mr | 12 | 30 | 2 | Ap | 10 | Chaitra ... | 2 | Mr | 27 | 3 | 12 | ...○ | 12 | 20 | 0 | 6 | 54 | 349 | 44 | 53 | | |

# TABLE XII.—General Ephemeris A.D. 1865—A.D. 1870.

| Commencement of years in different eras. | Tamil month. | Tamil Month Beginning Week-day | Month | Day | Last day | Tamil Month End Week-day | Month | Day | Lunar month. | New Moon Ending moment of New Moon Tithi. Mean Week-day | Month | Day (mean) | Ghatikas | Palas | Actual Day (actual) | Ghatikas | Palas | ☾'s Anom. Day | Ghatikas | Palas | ⊙'s Anom. Day | Ghatikas | Palas | Deduct for Nakshatras | Deduct for Yogas |
|---|---|---|---|---|---|---|---|---|---|---|---|---|---|---|---|---|---|---|---|---|---|---|---|---|---|
| **K. Y. 4966** | Chittirai . | 3 | Ap | 11 | 31 | 5 | M | 11 | Vaisakha ... | 3 | Ap | 25 | 35 | 2 | ... | 34 | 12 | 2 | 5 | 28 | 14 | 1 | 12 | 1 day, 2 ghat., 55 palas. | 1 " 57 " 7 " |
| Sak. 1787 | Vaikasi ... | 6 | My | 12 | 32 | 2 | Je | 12 | Jyeshta ... | 5 | My | 25 | 6 | 52 | 24 | 54 | 5 | 4 | 4 | 1 | 43 | 33 | 2 | | |
| Vik. 1922 | Ani ... | 3 | Je | 13 | 31 | 5 | Jl | 13 | Ashada ... | 6 | Je | 23 | 38 | 42 | ... | 16 | 11 | 6 | 2 | 35 | 73 | 4 | 52 | | |
| (S.) Krodhana. | Adi ... | 6 | Jl | 14 | 32 | 2 | Au | 14 | Sravana ... | 1 | Jl | 23 | 10 | 33 | 22 | 42 | 0 | 8 | 1 | 8 | 102 | 36 | 42 | | |
| K. 1040 (S. M.) | Avani ... | 3 | Au | 15 | 31 | 5 | S | 14 | Bhadrapada . | 2 | Au | 21 | 42 | 23 | ... | 13 | 32 | 9 | 59 | 41 | 132 | 8 | 32 | | |
| K. 1040 (N. M.) | Purattasi . | 6 | S | 15 | 30 | 7 | O | 14 | Asvina ... | 4 | S | 20 | 14 | 13 | 19 ○ | 52 | 28 | 11 | 58 | 16 | 161 | 40 | 22 | | |
| 27 g., 36 p. | Aippasi ... | 1 | O | 15 | 30 | 2 | N | 13 | Kartika ... | 5 | O | 19 | 46 | 3 | ... | 35 | 55 | 13 | 56 | 49 | 191 | 12 | 13 | | |
| B. S. 1271 | Kartigai ... | 3 | N | 14 | 30 | 4 | D | 13 | Margasira ... | 7 | N | 18 | 17 | 53 | ... | 22 | 44 | 15 | 55 | 23 | 220 | 44 | 3 | | |
| (N.) Isvara | Margali ... | 5 | D | 14 | 29 | 5 | Ja | 11 | Pausha ... | 1 | D | 17 | 49 | 43 | 18 | 8 | 18 | 17 | 53 | 56 | 250 | 15 | 53 | | |
| **A. D. 1866** | Tai ... | 6 | Ja | 12 | 30 | 7 | F | 10 | Magha ... | 3 | Ja | 16 | 21 | 33 | ... | 49 | 20 | 19 | 52 | 30 | 279 | 47 | 43 | | |
| H. 1283 | Masi ... | 1 | F | 11 | 29 | 1 | Mr | 11 | Phalguna ... | 4 | F | 14 | 53 | 23 | 15 | 24 | 8 | 21 | 51 | 4 | 309 | 19 | 33 | | |
| May 16 | Panguni ... | 2 | Mr | 12 | 31 | 4 | Ap | 11 | Chaitra ... | 6 | Mr | 16 | 25 | 14 | ... ○ | 54 | 14 | 23 | 49 | 37 | 338 | 51 | 23 | | |
| **K. Y. 4967** | Chittirai . | 5 | Ap | 12 | 31 | 7 | My | 12 | Vaisakha ... | 7 | Ap | 14 | 57 | 4 | 15 | 15 | 36 | 25 | 48 | 11 | 3 | 7 | 42 | 0 day, 14 ghat., 2 palas. | 0 " 26 " 8 " |
| Sak. 1788 | Vaikasi ... | 1 | My | 13 | 31 | 3 | Je | 12 | Jyeshta ... | 2 | My | 14 | 28 | 54 | ... | 34 | 56 | 0 | 13 | 28 | 32 | 39 | 32 | | |
| | | | | | | | | | | 4 | Je | 13 | 0 | 44 | 12 | 52 | 42 | 2 | 12 | 2 | 62 | 11 | 22 | | |
| Vik. 1923 | Ani ... | 4 | Je | 13 | 32 | 7 | Jl | 14 | Ashada ... | 5 | Jl | 12 | 32 | 34 | ... | 11 | 23 | 4 | 10 | 35 | 91 | 43 | 12 | | |
| (S.) Kshaya | Adi ... | 1 | Jl | 15 | 31 | 3 | Au | 14 | Sravana ... | 7 | Au | 11 | 4 | 24 | 10 | 33 | 33 | 6 | 9 | 9 | 121 | 15 | 2 | | |
| K. 1041 (S. M.) | Avani ... | 4 | Au | 15 | 31 | 6 | S | 14 | Bhadrapada . | 1 | S | 9 | 36 | 14 | ... ○ | 1 | 42 | 8 | 7 | 42 | 150 | 46 | 52 | | |
| K. 1041 (N. M.) | Purattasi . | 7 | S | 15 | 31 | 2 | O | 15 | Asvina ... | 3 | O | 9 | 8 | 4 | 8 | 37 | 52 | 10 | 6 | 16 | 180 | 18 | 42 | | |
| 43 g., 7 p. | Aippasi ... | 3 | O | 16 | 30 | 4 | N | 14 | Kartika ... | 4 | N | 7 | 39 | 54 | ... | 20 | 18 | 12 | 4 | 49 | 209 | 50 | 33 | | |
| B. S. 1272 | Kartigai ... | 5 | N | 15 | 29 | 5 | D | 13 | Margasira ... | 6 | D | 7 | 11 | 45 | ... | 8 | 44 | 14 | 3 | 23 | 239 | 22 | 23 | | |
| (N) Bahudhanya. | Margali ... | 6 | D | 14 | 29 | 6 | Ja | 11 | Pausha ... | 7 | Ja | 5 | 43 | 35 | ... | 58 | 2 | 16 | 1 | 57 | 268 | 54 | 13 | | |
| **A. D. 1867** | Tai ... | 7 | Ja | 12 | 30 | 1 | F | 10 | Magha ... | 2 | F | 4 | 15 | 25 | ... | 43 | 20 | 18 | 0 | 30 | 298 | 26 | 3 | | |
| H. 1284 | Masi ... | 2 | F | 11 | 30 | 3 | Mr | 12 | Phalguna ... | 3 | Mr | 5 | 47 | 15 | 6●○ | 21 | 40 | 19 | 59 | 4 | 327 | 57 | 58 | | |
| May 5 | Panguni ... | 4 | Mr | 13 | 30 | 5 | Ap | 11 | Chaitra ... | 5 | Ap | 4 | 19 | 5 | ... | 53 | 5 | 21 | 57 | 38 | 357 | 29 | 43 | | |
| **K. Y. 4968** | Chittirai . | 6 | Ap | 12 | 31 | 1 | My | 12 | Vaisakha ... | 6 | My | 3 | 50 | 55 | 4 | 16 | 39 | 23 | 56 | 11 | 21 | 46 | 2 | 1 day, 37 ghat., 42 palas. | 3 " 1 " 47 " |
| Sak. 1789 | Vaikasi ... | 2 | My | 13 | 31 | 4 | Je | 12 | Jyeshta ... | 1 | Je | 2 | 22 | 45 | ... | 35 | 59 | 25 | 54 | 44 | 51 | 17 | 52 | | |
| Vik. 1924 | Ani ... | 5 | Je | 13 | 32 | 1 | Jl | 14 | Ashada ... | 2 | Jl | 1 | 54 | 35 | ... | 52 | 48 | 0 | 20 | 2 | 80 | 49 | 42 | | |
| (S.) Prabhava | Adi ... | 2 | Jl | 15 | 31 | 4 | Au | 14 | Sravana ... | 4 | Jl | 31 | 26 | 25 | ... | 9 | 53 | 2 | 18 | 35 | 110 | 21 | 32 | | |
| K. 1042 (S. M.) | Avani ... | 5 | Au | 15 | 31 | 7 | S | 14 | Bhadrapada . | 5 | Au | 29 | 58 | 16 | ... ○ | 30 | 2 | 4 | 17 | 19 | 139 | 53 | 23 | | |
| K. 1042 (N. M.) | Purattasi . | 1 | S | 15 | 31 | 3 | O | 15 | Asvina ... | 7 | S | 28 | 30 | 6 | 27 | 55 | 28 | 6 | 15 | 42 | 169 | 25 | 13 | | |
| 58 g., 39 p. | Aippasi ... | 4 | O | 16 | 30 | 5 | N | 14 | Kartika ... | 2 | O | 28 | 1 | 56 | 27 | 27 | 57 | 8 | 14 | 16 | 198 | 57 | 3 | | |
| B. S. 1273 | Kartigai ... | 6 | N | 15 | 29 | 6 | D | 13 | Margasira ... | 3 | N | 26 | 33 | 46 | ... | 8 | 8 | 10 | 12 | 50 | 228 | 28 | 53 | | |
| (N.) Pramathin. | Margali ... | 7 | D | 14 | 30 | 1 | Ja | 12 | Pausha ... | 5 | D | 26 | 5 | 36 | 25 | 58 | 33 | 12 | 11 | 23 | 258 | 0 | 43 | | |
| **A. D. 1868** | Tai ... | 2 | Ja | 13 | 29 | 2 | F | 10 | Magha ... | 6 | Ja | 24 | 37 | 26 | ... | 44 | 45 | 14 | 9 | 57 | 287 | 32 | 33 | | |
| H. 1285 | Masi ... | 3 | F | 11 | 30 | 4 | Mr | 11 | Phalguna ... | 1 | F | 23 | 9 | 16 | ... | 32 | 24 | 16 | 8 | 30 | 317 | 4 | 23 | | |
| April 24 | Panguni ... | 5 | Mr | 12 | 30 | 6 | Ap | 10 | Chaitra ... | 2 | Mr | 23 | 41 | 6 | 24 | 13 | 56 | 18 | 7 | 4 | 346 | 36 | 13 | | |
| **K. Y. 4969** | Chittirai ... | 7 | Ap | 11 | 31 | 2 | My | 11 | Vaisakha ... | 4 | Ap | 22 | 12 | 57 | ... | 47 | 46 | 20 | 5 | 38 | 10 | 52 | 32 | 0 day, 48 ghat., 49 palas. | 1 " 30 " 50 " |
| Sak. 1790 | Vaikasi ... | 3 | My | 12 | 32 | 6 | Je | 12 | Jyeshta ... | 5 | My | 21 | 44 | 47 | 22 | 14 | 8 | 22 | 4 | 12 | 40 | 24 | 22 | | |
| Vik. 1925 | Ani ... | 7 | Je | 13 | 32 | 2 | Jl | 13 | Ashada ... | 7 | Je | 20 | 16 | 37 | ... | 35 | 39 | 24 | 2 | 45 | 69 | 56 | 12 | | |
| (S.) Vibhava | Adi ... | 3 | Jl | 14 | 32 | 6 | Au | 14 | Sravana ... | 1 | Jl | 19 | 48 | 27 | ... | 53 | 9 | 26 | 1 | 18 | 99 | 28 | 2 | | |
| K. 1043 (S. M.) | Avani ... | 7 | Au | 15 | 31 | 2 | S | 14 | Bhadrapada . | 3 | Au | 18 | 20 | 17 | ... ● | 10 | 41 | 0 | 26 | 35 | 128 | 59 | 52 | | |
| K. 1043 (N. M.) | Purattasi... | 3 | S | 15 | 30 | 4 | O | 14 | Asvina ... | 4 | S | 16 | 52 | 7 | ... | 31 | 33 | 2 | 25 | 9 | 158 | 31 | 42 | | |
| 14 g., 10 p. | Aippasi ... | 5 | O | 15 | 30 | 6 | N | 13 | Kartika ... | 6 | O | 16 | 23 | 57 | 15 | 54 | 19 | 4 | 23 | 43 | 188 | 3 | 32 | | |
| B. S. 1274 | Kartigai ... | 7 | N | 14 | 30 | 1 | D | 13 | Margasira ... | 7 | N | 14 | 55 | 48 | ... | 24 | 14 | 6 | 22 | 16 | 217 | 35 | 23 | | |
| (N.) Vikrama | Margali ... | 2 | D | 14 | 29 | 2 | Ja | 11 | Pausha ... | 2 | D | 14 | 27 | 38 | ... | 0 | 31 | 8 | 20 | 50 | 247 | 7 | 13 | | |
| **A. D. 1869** | Tai ... | 3 | Ja | 12 | 29 | 3 | F | 9 | Magha ... | 3 | Ja | 12 | 59 | 28 | ... ○ | 43 | 22 | 10 | 19 | 24 | 276 | 39 | 3 | | |
| H. 1286 | Masi ... | 4 | F | 10 | 30 | 5 | Mr | 11 | Phalguna ... | 5 | F | 11 | 31 | 18 | ... | 30 | 16 | 12 | 17 | 59 | 306 | 10 | 53 | | |
| April 13 | Panguni ... | 6 | Mr | 12 | 30 | 7 | Ap | 10 | Chaitra ... | 7 | Mr | 13 | 3 | 8 | ... | 17 | 27 | 14 | 16 | 31 | 335 | 42 | 43 | | |
| **K. Y. 4970** | Chittirai . | 1 | Ap | 11 | 31 | 3 | My | 11 | Vaisakha ... | 1 | Ap | 11 | 34 | 58 | 12 | 0 | 43 | 16 | 15 | 4 | 365 | 14 | 33 | 2 days, 12 ghat., 28 palas. | 4 " 6 " 29 " |
| | | | | | | | | | | 3 | My | 11 | 6 | 48 | ... | 37 | 31 | 18 | 13 | 38 | 29 | 30 | 52 | | |
| Sak. 1791 | Vaikasi ... | 4 | My | 12 | 32 | 7 | Je | 12 | Jyeshta ... | 4 | Je | 9 | 38 | 39 | 10 | 7 | 16 | 20 | 12 | 12 | 59 | 2 | 42 | | |
| Vik. 1926 | Ani ... | 1 | Je | 13 | 31 | 3 | Jl | 13 | Ashada ... | 6 | Jl | 9 | 10 | 29 | ... ○ | 31 | 40 | 22 | 10 | 45 | 88 | 34 | 32 | | |
| (S.) Sukla | Adi ... | 4 | Jl | 14 | 32 | 7 | Au | 14 | Sravana ... | 7 | Au | 7 | 42 | 19 | ... | 53 | 11 | 24 | 9 | 19 | 118 | 6 | 22 | | |
| K. 1044 (S. M.) | Avani ... | 1 | Au | 15 | 31 | 3 | S | 14 | Bhadrapada . | 2 | S | 6 | 14 | 9 | ... | 12 | 31 | 26 | 7 | 52 | 147 | 38 | 12 | | |
| K. 1044 (N. M.) | Purattasi . | 4 | S | 15 | 30 | 5 | O | 14 | Asvina ... | 3 | O | 5 | 45 | 59 | ... | 33 | 22 | 0 | 33 | 9 | 177 | 10 | 2 | | |
| 29 g., 41 p. | Aippasi ... | 6 | O | 15 | 30 | 7 | N | 13 | Kartika ... | 5 | N | 4 | 17 | 49 | 3 | 56 | 57 | 2 | 31 | 43 | 206 | 41 | 52 | | |
| B. S. 1275 | Kartigai ... | 1 | N | 14 | 30 | 2 | D | 13 | Margasira ... | 6 | D | 3 | 49 | 39 | ... | 24 | 46 | 4 | 30 | 17 | 236 | 13 | 43 | | |
| (N.) Vrisha | Margali ... | [illegible] | D | 14 | 29 | 3 | Ja | 11 | Pausha ... | 1 | Ja | 2 | 21 | 29 | 1 ○ | 57 | 38 | 6 | 28 | 50 | 265 | 45 | 33 | | |
| **A. D. 1870** | Tai ... | 4 | Ja | 12 | 30 | 5 | F | 10 | Magha ... | 2 | Ja | 31 | 53 | 19 | ... | 35 | 14 | 8 | 27 | 24 | 295 | 17 | 23 | | |
| H. 1287 | Masi ... | 6 | F | 11 | 29 | 6 | Mr | 11 | Phalguna ... | 4 | Mr | 2 | 25 | 9 | ... | 16 | 50 | 10 | 25 | 57 | 324 | 49 | 13 | | |
| April 2, 3 | Panguni . | 7 | Mr | 12 | 31 | 2 | Ap | 11 | Chaitra ... | 5 | Mr | 31 | 56 | 59 | Ap. 1. | 0 | 21 | 12 | 24 | 31 | 354 | 21 | 3 | | |

## TABLE XII.—General Ephemeris A.D. 1870—A.D. 1875.

| Commencement of years in different eras. | Tamil month. | Tamil Month Beginning Week-day | Month | Day | Last Day | Tamil Month End Week-day | Month | Day | Lunar month. | New Moon Week-day | Month | Ending moment of New Moon Tithi Mean Day (mean) | Ghatikas | Palas | Actual Day (actual) | Ghatikas | Palas | ☾'s Anom. Day | Ghatikas | Palas | ☉'s Anom. Day | Ghatikas | Palas | Deduct for Nakshatras | Deduct for Yogas. |
|---|---|---|---|---|---|---|---|---|---|---|---|---|---|---|---|---|---|---|---|---|---|---|---|---|---|
| **K. Y. 4971** | Chittirai ... | 3 | Ap | 12 | 31 | 5 | My | 12 | Vaisakha ... | 7 | Ap | 30 | 28 | 49 | ... | 42 | 58 | 14 | 23 | 5 | 18 | 37 | 22 | 1 day, 23 ghat., 35 palas. | 2 „ 35 „ 32 „ |
| Sak. 1792 | Vaikasi ... | 6 | My | 13 | 31 | 1 | Je | 12 | Jyeshta ... | 2 | My | 30 | 0 | 40 | ... | 21 | 46 | 16 | 21 | 38 | 48 | 9 | 12 | | |
| Vik. 1927 | Ani ... | 2 | Je | 13 | 32 | 5 | Jl | 14 | Ashada ... | 3 | Je | 28 | 32 | 30 | ... ○ | 55 | 24 | 18 | 20 | 12 | 77 | 41 | 2 | | |
| (S.) Pramoda | Adi ... | 6 | Jl | 15 | 31 | 1 | Au | 14 | Sravana ... | 5 | Jl | 28 | 4 | 20 | ... | 24 | 21 | 20 | 18 | 45 | 107 | 12 | 52 | | |
| K. 1045 (S.M.) | Avani ... | 2 | Au | 15 | 31 | 4 | S | 14 | Bhadrapada . | 6 | Au | 26 | 36 | 10 | ... | 50 | 2 | 22 | 17 | 19 | 136 | 44 | 42 | | |
| K. 1045 (N.M.) | Purattasi... | 5 | S | 15 | 31 | 7 | O | 15 | Asvina ... | 1 | S | 25 | 8 | 0 | ... | 13 | 59 | 24 | 15 | 53 | 166 | 16 | 32 | | |
| 45 g, 12 p. | Aippasi ... | 1 | O | 16 | 30 | 2 | N | 14 | Kartika ... | 2 | O | 24 | 39 | 50 | ... | 37 | 37 | 26 | 14 | 27 | 195 | 48 | 22 | | |
| B.S. 1276 | Kartigai ... | 3 | N | 15 | 29 | 3 | D | 13 | Margasira ... | 4 | N | 23 | 11 | 40 | ... | 2 | 13 | 0 | 39 | 43 | 225 | 20 | 12 | | |
| (N.) Chitrabhanu. | Margali ... | 4 | D | 14 | 29 | 4 | Ja | 11 | Pausha ... | 5 | D | 22 | 43 | 31 | ... ○ | 28 | 58 | 2 | 38 | 17 | 254 | 52 | 2 | | |
| **A. D. 1871** | Tai ... | 5 | Ja | 12 | 30 | 6 | F | 10 | Magha ... | 7 | Ja | 21 | 15 | 21 | 20 | 58 | 25 | 4 | 36 | 50 | 284 | 23 | 53 | | |
| H. 1288 | Masi ... | 7 | F | 11 | 30 | 1 | Mr | 12 | Phalguna ... | 1 | F | 19 | 47 | 11 | ... | 30 | 48 | 6 | 35 | 24 | 313 | 55 | 43 | | |
| Mr. 23 | Panguni ... | 2 | Mr | 13 | 30 | 3 | Ap | 11 | Chaitra ... | 3 | Mr | 21 | 19 | 1 | ... | 6 | 7 | 8 | 33 | 58 | 343 | 27 | 33 | | |
| **K. Y. 4972** | Chittirai... | 4 | Ap | 12 | 31 | 6 | My | 12 | Vaisakha ... | 4 | Ap | 19 | 50 | 51 | ... | 43 | 54 | 10 | 32 | 31 | 7 | 43 | 52 | 0 day, 34 ghat., 42 palas. | 1 „ 4 „ 34 „ |
| Sak. 1793 | Vaikasi ... | 7 | My | 13 | 31 | 2 | Je | 12 | Jyeshta ... | 6 | My | 19 | 22 | 41 | ... | 22 | 59 | 12 | 31 | 5 | 37 | 15 | 42 | | |
| Vik. 1928 | Ani ... | 3 | Je | 13 | 32 | 6 | Jl | 14 | Ashada ... | 7 | Je | 17 | 54 | 31 | 18 ●○ | 1 | 47 | 14 | 29 | 38 | 66 | 47 | 32 | | |
| (S.) Prajapati | Adi ... | 7 | Jl | 15 | 31 | 2 | Au | 14 | Sravana ... | 2 | Jl | 17 | 26 | 21 | ... | 38 | 38 | 16 | 28 | 12 | 96 | 19 | 22 | | |
| K. 1046 (S.M.) | Avani ... | 3 | Au | 15 | 31 | 5 | S | 14 | Bhadrapada.. | 3 | Au | 15 | 58 | 11 | 16 | 12 | 55 | 18 | 26 | 46 | 125 | 51 | 12 | | |
| | | | | | | | | | | 5 | S | 14 | 30 | 1 | ... | 44 | 33 | 20 | 25 | 19 | 155 | 23 | 2 | | |
| K. 1046 (N.M.) | Purattasi... | 6 | S | 15 | 31 | 1 | O | 15 | Asvina ... | 7 | O | 14 | 1 | 52 | ... | 13 | 33 | 22 | 23 | 53 | 184 | 54 | 52 | | |
| 0 g, 44 p. | Aippasi ... | 2 | O | 16 | 30 | 3 | N | 14 | Kartika ... | 1 | N | 12 | 33 | 42 | ... | 41 | 41 | 24 | 22 | 26 | 214 | 26 | 42 | | |
| B. S. 1277 | Kartigai ... | 4 | N | 15 | 29 | 4 | D | 13 | Margasira ... | 3 | D | 12 | 5 | 32 | ... ● | 8 | 33 | 26 | 21 | 0 | 243 | 58 | 33 | | |
| (N.) Subhanu. | Margali ... | 5 | D | 14 | 30 | 6 | Ja | 12 | Pausha ... | 4 | Ja | 10 | 37 | 22 | ... | 35 | 6 | 0 | 46 | 17 | 273 | 30 | 23 | | |
| **A. D. 1872** | Tai ... | 7 | Ja | 13 | 29 | 7 | F | 10 | Magha ... | 6 | F | 9 | 9 | 12 | ... | 1 | 55 | 2 | 44 | 51 | 303 | 2 | 13 | | |
| H. 1289 | Masi ... | 1 | F | 11 | 30 | 2 | Mr | 11 | Phalguna ... | 7 | Mr | 9 | 41 | 2 | ... | 29 | 34 | 4 | 43 | 24 | 332 | 34 | 3 | | |
| Mr. 11 | Panguni ... | 3 | Mr | 12 | 30 | 4 | Ap | 10 | Chaitra ... | 2 | Ap | 8 | 12 | 52 | 7 | 58 | 45 | 6 | 41 | 58 | 362 | 5 | 53 | | |
| **K. Y. 4973** | Chittirai... | 5 | Ap | 11 | 31 | 7 | My | 11 | Vaisakha ... | 3 | My | 7 | 44 | 42 | ... ○ | 30 | 3 | 8 | 40 | 31 | 26 | 22 | 12 | 1 day, 56 ghat., 21 palas. | 3 „ 40 „ 14 „ |
| Sak. 1794 | Vaikasi ... | 1 | My | 12 | 32 | 4 | Je | 12 | Jyeshta ... | 5 | Je | 6 | 16 | 33 | ... ● | 3 | 55 | 10 | 39 | 5 | 55 | 54 | 2 | | |
| Vik. 1929 | Ani ... | 5 | Je | 13 | 31 | 7 | Jl | 13 | Ashada ... | 6 | Jl | 5 | 48 | 23 | ... | 40 | 17 | 12 | 37 | 39 | 85 | 25 | 52 | | |
| (S.) Angirasa | Adi ... | 1 | Jl | 14 | 32 | 4 | Au | 14 | Sravana ... | 1 | Au | 4 | 20 | 13 | ... | 19 | 8 | 14 | 36 | 12 | 114 | 57 | 42 | | |
| K. 1047 (S.M.) | Avani ... | 5 | Au | 15 | 31 | 7 | S | 14 | Bhadrapada... | 2 | S | 2 | 52 | 3 | ... | 57 | 40 | 16 | 34 | 40 | 144 | 29 | 32 | | |
| K. 1047 (N.M.) | Purattasi... | 1 | S | 15 | 30 | 2 | O | 14 | Asvina ... | 4 | O | 2 | 23 | 53 | ... | 35 | 39 | 18 | 33 | 19 | 174 | 1 | 22 | | |
| 16 g, 15 p. | Aippasi ... | 3 | O | 15 | 30 | 4 | N | 13 | Kartika ... | 5 | O | 31 | 55 | 43 | N.10 | 11 | 20 | 20 | 31 | 53 | 203 | 33 | 12 | | |
| B. S. 1278 | Kartigai ... | 5 | N | 14 | 30 | 6 | D | 13 | Margasira ... | 7 | N | 30 | 27 | 33 | ... | 44 | 16 | 22 | 30 | 27 | 233 | 5 | 2 | | |
| (N.) Tarana | Margali ... | 7 | D | 14 | 29 | 7 | Ja | 11 | Pausha ... | 1 | D | 29 | 59 | 23 | 30 | 14 | 55 | 24 | 29 | 0 | 262 | 36 | 52 | | |
| **A. D. 1873** | Tai ... | 1 | Ja | 12 | 29 | 1 | F | 9 | Magha ... | 3 | Ja | 28 | 31 | 14 | ... | 41 | 40 | 26 | 27 | 34 | 292 | 8 | 42 | | |
| H. 1290 | Masi ... | 2 | F | 10 | 30 | 3 | Mr | 11 | Phalguna ... | 5 | F | 27 | 3 | 4 | ... | 6 | 47 | 0 | 52 | 51 | 321 | 40 | 32 | | |
| Mr. 1 | Panguni ... | 4 | Mr | 12 | 31 | 6 | Ap | 11 | Chaitra ... | 6 | Mr | 28 | 34 | 54 | ... | 30 | 42 | 2 | 51 | 24 | 351 | 12 | 22 | | |
| **K. Y. 4974** | Chittirai... | 7 | Ap | 12 | 30 | 1 | My | 11 | Vaisakha ... | 1 | Ap | 27 | 6 | 44 | 26 ○ | 54 | 41 | 4 | 49 | 58 | 15 | 28 | 41 | 1 day, 9 ghat., 28 palas. | 2 „ 9 „ 16 „ |
| Sak. 1795 | Vaikasi ... | 2 | My | 12 | 32 | 5 | Je | 12 | Jyeshta ... | 2 | My | 26 | 38 | 34 | ... | 20 | 3 | 6 | 48 | 32 | 45 | 0 | 31 | | |
| Vik. 1930 | Ani ... | 6 | Je | 13 | 31 | 1 | Jl | 13 | Ashada ... | 4 | Je | 25 | 10 | 24 | 24 | 48 | 14 | 8 | 47 | 5 | 74 | 32 | 21 | | |
| (S.) Srimukha | Adi ... | 2 | Jl | 14 | 32 | 5 | Au | 14 | Sravana ... | 5 | Jl | 24 | 42 | 14 | ... | 20 | 43 | 10 | 45 | 39 | 104 | 4 | 12 | | |
| K. 1048 (S.M.) | Avani ... | 6 | Au | 15 | 31 | 1 | S | 14 | Bhadrapada . | 7 | Au | 23 | 14 | 4 | 22 | 58 | 15 | 12 | 44 | 13 | 133 | 36 | 2 | | |
| K. 1048 (N.M.) | Purattasi... | 2 | S | 15 | 30 | 3 | O | 14 | Asvina ... | 1 | S | 21 | 45 | 54 | ... | 40 | 3 | 14 | 42 | 46 | 163 | 7 | 52 | | |
| 31 g, 46 p. | Aippasi ... | 4 | O | 15 | 30 | 5 | N | 13 | Kartika ... | 3 | O | 21 | 17 | 45 | ... ○ | 23 | 23 | 16 | 41 | 20 | 192 | 39 | 42 | | |
| B. S. 1279 | Kartigai ... | 6 | N | 14 | 30 | 7 | D | 13 | Margasira ... | 4 | N | 19 | 49 | 35 | 20 | 5 | 26 | 18 | 39 | 53 | 222 | 11 | 32 | | |
| (N.) Parthiva | Margali ... | 1 | D | 14 | 29 | 1 | Ja | 11 | Pausha ... | 6 | D | 19 | 21 | 25 | ... | 44 | 1 | 20 | 38 | 26 | 251 | 43 | 22 | | |
| **A. D. 1874** | Tai ... | 2 | Ja | 12 | 30 | 3 | F | 10 | Magha ... | 7 | Ja | 17 | 53 | 15 | 18 | 18 | 0 | 22 | 37 | 0 | 281 | 15 | 12 | | |
| H. 1291 | Masi ... | 4 | F | 11 | 29 | 4 | Mr | 11 | Phalguna .. | 2 | F | 16 | 25 | 5 | ... | 46 | 56 | 24 | 35 | 34 | 310 | 47 | 2 | | |
| Feb. 18 | Panguni ... | 5 | Mr | 12 | 31 | 7 | Ap | 11 | Chaitra ... | 3 | Mr | 17 | 56 | 55 | 18 | 12 | 32 | 26 | 34 | 8 | 340 | 18 | 52 | | |
| **K. Y. 4975** | Chittirai... | 1 | Ap | 12 | 31 | 3 | My | 12 | Vaisakha ... | 5 | Ap | 16 | 28 | 45 | ... ○ | 33 | 4 | 0 | 59 | 25 | 4 | 35 | 11 | 0 day, 20 ghat., 35 palas. | 0 „ 38 „ 18 „ |
| Sak. 1796 | Vaikasi ... | 4 | My | 13 | 31 | 6 | Je | 12 | Jyeshta ... | 7 | My | 16 | 0 | 35 | 15 | 53 | 11 | 2 | 57 | 58 | 34 | 7 | 1 | | |
| Vik. 1931 | Ani ... | 7 | Je | 13 | 32 | 3 | Jl | 14 | Ashada ... | 1 | Je | 14 | 32 | 26 | ... | 13 | 51 | 4 | 56 | 32 | 63 | 38 | 51 | | |
| | | | | | | | | | | 3 | Jl | 14 | 4 | 16 | 13 | 37 | 13 | 6 | 55 | 6 | 93 | 10 | 42 | | |
| (S.) Bhava | Adi ... | 4 | Jl | 15 | 31 | 6 | Au | 14 | Sravana ... | 4 | Au | 12 | 36 | 6 | ... | 5 | 34 | 8 | 53 | 39 | 122 | 42 | 32 | | |
| K. 1049 (S.M.) | Avani ... | 7 | Au | 15 | 31 | 2 | S | 14 | Bhadrapada . | 6 | S | 11 | 7 | 56 | 10 | 40 | 38 | 10 | 52 | 13 | 152 | 14 | 22 | | |
| K. 1049 (N.M.) | Purattasi... | 3 | S | 15 | 31 | 5 | O | 15 | Asvina ... | 7 | O | 10 | 39 | 46 | ●○ | 22 | 23 | 12 | 50 | 46 | 181 | 46 | 12 | | |
| 47 g, 17 p. | Aippasi ... | 6 | O | 16 | 30 | 7 | N | 14 | Kartika ... | 2 | N | 9 | 11 | 36 | ... | 8 | 56 | 14 | 49 | 20 | 211 | 18 | 2 | | |
| B. S. 1280 | Kartigai ... | 1 | N | 15 | 29 | 1 | D | 13 | Margasira ... | 3 | D | 8 | 43 | 26 | ... | 56 | 0 | 16 | 47 | 54 | 240 | 49 | 52 | | |
| (N.) Vyaya | Margali ... | 2 | D | 14 | 29 | 2 | Ja | 11 | Pausha ... | 5 | Ja | 7 | 15 | 16 | ... | 40 | 0 | 18 | 46 | 27 | 270 | 21 | 42 | | |
| **A. D. 1875** | Tai ... | 3 | Ja | 12 | 30 | 4 | F | 10 | Magha ... | 6 | F | 5 | 47 | 6 | 6 | 18 | 6 | 20 | 45 | 1 | 299 | 53 | 32 | | |
| H. 1292 | Masi ... | 5 | F | 11 | 30 | 6 | Mr | 12 | Phalguna ... | 1 | Mr | 7 | 18 | 56 | ... | 49 | 28 | 22 | 43 | 34 | 329 | 25 | 22 | | |
| Feb. 7 | Panguni ... | 7 | Mr | 13 | 30 | 1 | Ap | 11 | Chaitra ... | 2 | Ap | 5 | 50 | 46 | 6 ● | 14 | 44 | 24 | 42 | 8 | 358 | 57 | 11 | | |

# TABLE XII.—General Ephemeris A.D. 1875—A.D. 1880.

| Commencement of years in different eras. | Tamil month. | Tamil Month Beginning Week-day | Month | Day | Last Day | Tamil Month End Week-day | Month | Day | Lunar Month. | New Moon Ending moment of New-Moon Tithi. Mean Week-day | Month | Day (mean) | Ghatikas | Palas | Actual Day (actual) | Ghatikas | Palas | ☾'s Anom. Day | Ghatikas | Palas | ⊙'s Anom. Day | Ghatikas | Palas | Deduct for Nakshatras | Deduct for Yogas. |
|---|---|---|---|---|---|---|---|---|---|---|---|---|---|---|---|---|---|---|---|---|---|---|---|---|---|
| **K. Y. 4976** | Chittirai ... | 2 | Ap | 12 | 31 | 4 | My | 12 | Vaisakha ... | 4 | My | 5 | 22 | 36 | ... | 35 | 22 | 26 | 40 | 42 | 23 | 13 | 31 | 1 day, 44 ghat., 14 palas. | 3 „ 13 „ 58 „ |
| Sak. 1797 | Vaikasi ... | 5 | My | 13 | 31 | 7 | Je | 12 | Jyesta ... | 5 | Je | 3 | 54 | 27 | ... | 53 | 18 | 1 | 5 | 59 | 52 | 45 | 21 | | |
| Vik. 1932 | Ani ... | 1 | Je | 13 | 32 | 4 | Jl | 14 | Ashada ... | 7 | Jl | 3 | 26 | 17 | ... | 10 | 56 | 3 | 4 | 32 | 82 | 17 | 11 | | |
| (S.) Yuva | Adi ... | 5 | Jl | 15 | 31 | 7 | Au | 14 | Sravana ... | 1 | Au | 1 | 58 | 7 | ... | 30 | 57 | 5 | 3 | 6 | 111 | 49 | 1 | | |
| K. 1050 (S. M.) | Avani ... | 1 | Au | 15 | 31 | 3 | S | 14 | Bhadrapada . | 3 | Au | 31 | 29 | 57 | 30 | 55 | 57 | 7 | 1 | 39 | 141 | 20 | 52 | | |
| K. 1050 (N.M.) | Purattasi . | 4 | S | 15 | 31 | 6 | O | 15 | Asvina ... | 5 | S | 30 | 1 | 47 | 29● | 27 | 50 | 9 | 0 | 13 | 170 | 52 | 42 | | |
| 2 g, 49 p. | Aippasi ... | 7 | O | 16 | 30 | 1 | N | 14 | Kartika ... | 6 | O | 29 | 33 | 37 | ... | 7 | 30 | 10 | 58 | 47 | 200 | 24 | 32 | | |
| B. S. 1281 | Kartigai ... | 2 | N | 15 | 29 | 2 | D | 13 | Margasira ... | 1 | N | 28 | 5 | 27 | 27 | 53 | 54 | 12 | 57 | 20 | 229 | 56 | 22 | | |
| (N.) Sarvajit | Margali ... | 3 | D | 14 | 30 | 4 | Ja | 12 | Pausha ... | 2 | D | 27 | 37 | 17 | ... | 43 | 37 | 14 | 55 | 54 | 259 | 28 | 12 | | |
| **A. D. 1876** | Tai ... | 5 | Ja | 13 | 29 | 5 | F | 10 | Magha ... | 4 | Ja | 26 | 9 | 8 | ... | 31 | 33 | 16 | 54 | 27 | 289 | 0 | 2 | | |
| H. 1293 | Masi ... | 6 | F | 11 | 30 | 7 | Mr | 11 | Phalguna ... | 5 | F | 24 | 40 | 58 | 25○ | 13 | 24 | 18 | 53 | 1 | 318 | 31 | 52 | | |
| Ja. 28 | Panguni ... | 1 | Mr | 12 | 30 | 2 | Ap | 10 | Chaitra ... | 7 | Mr | 25 | 12 | 48 | ... | 47 | 54 | 20 | 51 | 35 | 348 | 3 | 42 | | |
| **K. Y. 4977** | Chittirai . | 3 | Ap | 11 | 31 | 5 | My | 11 | Vaisaka ... | 1 | Ap | 23 | 44 | 38 | 24 | 14 | 58 | 22 | 50 | 8 | 12 | 20 | 1 | 0 day, 55 ghat., 21 palas. | 1 „ 43 „ 0 „ |
| Sak. 1798 | Vaikasi ... | 6 | My | 12 | 32 | 2 | Je | 12 | Jyeshta ... | 3 | My | 23 | 16 | 28 | ... | 36 | 14 | 24 | 48 | 42 | 41 | 51 | 51 | | |
| Vik. 1933 | Ani ... | 3 | Je | 13 | 31 | 5 | Jl | 13 | Ashada ... | 4 | Je | 21 | 48 | 18 | ... | 53 | 52 | 26 | 47 | 15 | 71 | 28 | 41 | | |
| (S.) Dhatri | Adi ... | 6 | Jl | 14 | 32 | 2 | Au | 14 | Sravana ... | 6 | Jl | 21 | 20 | 8 | ... | 10 | 27 | 1 | 12 | 32 | 100 | 55 | 31 | | |
| K. 1051 (S. M.) | Avani ... | 3 | Au | 15 | 31 | 5 | S | 14 | Bhadrapada . | 7 | Au | 19 | 51 | 58 | ...○ | 28 | 54 | 3 | 11 | 5 | 130 | 27 | 21 | | |
| K. 1051 (N. M.) | Purattasi . | 6 | S | 15 | 30 | 7 | O | 14 | Asvina ... | 2 | S | 18 | 23 | 49 | 17 | 51 | 38 | 5 | 9 | 38 | 159 | 59 | 11 | | |
| 18 g, 20 p. | Aippasi ... | 1 | O | 15 | 30 | 2 | N | 13 | Kartika ... | 3 | O | 17 | 55 | 39 | ... | 20 | 42 | 7 | 8 | 11 | 189 | 31 | 2 | | |
| B. S. 1282 | Kartigai . | 3 | N | 14 | 30 | 4 | D | 13 | Margasira ... | 5 | N | 16 | 27 | 29 | 15 | 57 | 8 | 9 | 6 | 44 | 219 | 2 | 52 | | |
| (N.) Sarvadharin. | Margali ... | 5 | D | 14 | 29 | 5 | Ja | 11 | Pausha ... | 6 | D | 15 | 59 | 19 | ... | 40 | 57 | 11 | 5 | 17 | 248 | 34 | 42 | | |
| **A. D. 1877** | Tai ... | 6 | Ja | 12 | 29 | 6 | F | 9 | Magha ... | 1 | Ja | 14 | 31 | 9 | ... | 29 | 42 | 13 | 3 | 50 | 278 | 6 | 32 | | |
| H. 1294 | Masi ... | 7 | F | 10 | 30 | 1 | Mr | 11 | Phalguna ... | 3 | F | 13 | 2 | 59 | ...○ | 17 | 25 | 15 | 2 | 23 | 307 | 38 | 22 | | |
| Ja. 16 | Panguni ... | 2 | Mr | 12 | 31 | 4 | Ap | 11 | Chaitra ... | 4 | Mr | 14 | 34 | 49 | 15● | 3 | 35 | 17 | 0 | 56 | 337 | 10 | 12 | | |
| **K. Y. 4978** | Chittirai . | 5 | Ap | 12 | 30 | 6 | My | 11 | Vaisaka .. | 6 | Ap | 13 | 6 | 39 | ... | 41 | 8 | 18 | 59 | 29 | 1 | 26 | 31 | 0 day, 6 ghat., 28 palas. | 0 „ 12 „ 3 „ |
| Sak. 1799 | Vaikasi ... | 7 | My | 12 | 32 | 3 | Je | 12 | Jyeshta ... | 7 | My | 12 | 38 | 29 | 13 | 11 | 1 | 20 | 58 | 2 | 30 | 58 | 21 | | |
| | | | | | | | | | | 2 | Je | 11 | 10 | 19 | ... | 34 | 29 | 22 | 56 | 35 | 60 | 30 | 11 | | |
| Vik. 1934 | Ani ... | 4 | Je | 13 | 31 | 6 | Jl | 13 | Ashada ... | 3 | Jl | 10 | 42 | 10 | ... | 53 | 43 | 24 | 55 | 8 | 90 | 2 | 1 | | |
| (S.) Isvara. | Adi ... | 7 | Jl | 14 | 32 | 3 | Au | 14 | Sravana ... | 5 | Au | 9 | 14 | 0 | ...○ | 11 | 24 | 26 | 53 | 42 | 119 | 33 | 51 | | |
| K. 1052 (S. M.) | Avani ... | 4 | Au | 15 | 31 | 6 | S | 14 | Bhadrapada . | 6 | S | 7 | 45 | 50 | ... | 30 | 3 | 1 | 18 | 59 | 149 | 5 | 41 | | |
| K. 1052(N. M.) | Purattasi | 7 | S | 15 | 30 | 1 | O | 14 | Asvina ... | 1 | O | 7 | 17 | 40 | 6 | 51 | 56 | 3 | 17 | 29 | 178 | 37 | 31 | | |
| 33 g, 51 p. | Aippasi ... | 2 | O | 15 | 30 | 3 | N | 13 | Kartika ... | 2 | N | 5 | 49 | 30 | ... | 18 | 47 | 5 | 16 | 3 | 208 | 9 | 21 | | |
| B. S. 1283 | Kartigai ... | 4 | N | 14 | 30 | 5 | D | 13 | Margasira ... | 4 | D | 5 | 21 | 20 | 4 | 51 | 56 | 7 | 14 | 36 | 237 | 41 | 11 | | |
| (N.) Virodhin | Margali ... | 6 | D | 14 | 29 | 6 | Ja | 11 | Pausha ... | 5 | Ja | 3 | 53 | 10 | ... | 31 | 34 | 9 | 13 | 10 | 267 | 13 | 1 | | |
| **A. D. 1878** | Tai ... | 7 | Ja | 12 | 30 | 1 | F | 10 | Magha ... | 7 | F | 2 | 25 | 0 | ...○ | 16 | 24 | 11 | 11 | 44 | 296 | 44 | 51 | | |
| H. 1295 | Masi ... | 2 | F | 11 | 29 | 2 | Mr | 11 | Phalguna ... | 1 | Mr | 3 | 56 | 50 | 4 | 3 | 25 | 13 | 10 | 17 | 326 | 16 | 41 | | |
| Ja. 5 | Panguni ... | 3 | Mr | 12 | 31 | 5 | Ap | 11 | Chaitra ... | 3 | Ap | 2 | 28 | 40 | ... | 49 | 38 | 15 | 8 | 50 | 355 | 48 | 31 | | |
| **K. Y. 4979** | Chittirai ... | 6 | Ap | 12 | 31 | 1 | My | 12 | Vaisaka ... | 5 | My | 2 | 0 | 31 | ... | 28 | 38 | 17 | 7 | 24 | 20 | 4 | 51 | 1 day, 30 ghat., 8 palas. | 2 „ 47 „ 42 „ |
| Sak. 1800 | Vaikasi ... | 2 | My | 13 | 31 | 4 | Je | 12 | Jyesta ... | 6 | My | 31 | 32 | 21 | Je. 1. | 1 | 53 | 19 | 5 | 58 | 49 | 36 | 41 | | |
| Vik. 1935 | Ani ... | 5 | Je | 13 | 32 | 1 | Jl | 14 | Ashada ... | 1 | Je | 30 | 4 | 11 | ... | 28 | 53 | 21 | 4 | 32 | 79 | 8 | 31 | | |
| (S.) Bahudhanya. | Adi ... | 2 | Jl | 15 | 31 | 4 | Au | 14 | Sravana ... | 2 | Jl | 29 | 36 | 1 | ...○ | 51 | 34 | 23 | 3 | 5 | 108 | 40 | 21 | | |
| K. 1053 (S. M.) | Avani ... | 5 | Au | 15 | 31 | 7 | S | 14 | Bhadrapada . | 4 | Au | 28 | 7 | 51 | ... | 12 | 14 | 25 | 1 | 39 | 138 | 12 | 11 | | |
| K. 1053 (N. M.) | Purattasi . | 1 | S | 15 | 31 | 3 | O | 15 | Asvina ... | 5 | S | 26 | 39 | 41 | ... | 32 | 55 | 27 | 0 | 12 | 167 | 44 | 1 | | |
| 49 g, 22 p. | Aippasi .. | 4 | O | 16 | 30 | 5 | N | 14 | Kartika ... | 7 | O | 26 | 11 | 32 | 25 | 55 | 21 | 1 | 25 | 30 | 197 | 15 | 51 | | |
| B. S. 1284 | Kartigai ... | 6 | N | 15 | 29 | 6 | D | 13 | Margasira ... | 1 | N | 24 | 43 | 22 | ... | 21 | 10 | 3 | 24 | 3 | 226 | 47 | 42 | | |
| (N.) Vikrita | Margali ... | 7 | D | 14 | 29 | 7 | Ja | 11 | Pausha ... | 3 | D | 24 | 15 | 12 | 23 | 51 | 22 | 5 | 22 | 37 | 256 | 19 | 32 | | |
| **A. D. 1879** | Tai ... | 1 | Ja | 12 | 30 | 2 | F | 10 | Magha ... | 4 | Ja | 22 | 47 | 2 | ...● | 27 | 23 | 7 | 21 | 10 | 285 | 51 | 22 | | |
| H. 1297 | Masi ... | 3 | F | 11 | 30 | 4 | Mr | 12 | Phalguna ... | 6 | F | 21 | 18 | 52 | ... | 5 | 34 | 9 | 19 | 44 | 315 | 23 | 12 | | |
| Dec. 15 | Panguni ... | 5 | Mr | 13 | 30 | 6 | Ap | 11 | Chaitra ... | 7 | Mr | 22 | 50 | 42 | .. | 47 | 28 | 11 | 18 | 18 | 344 | 55 | 2 | | |
| **K. Y. 4980** | Chittirai ... | 7 | Ap | 12 | 31 | 2 | My | 12 | Vaisakha ... | 2 | Ap | 21 | 22 | 32 | ... | 30 | 25 | 13 | 16 | 51 | 9 | 11 | 21 | 0 day, 41 ghat., 14 palas. | 1 „ 16 „ 44 „ |
| Sak. 1801 | Vaikasi ... | 3 | My | 13 | 31 | 5 | Je | 12 | Jyesta ... | 3 | My | 20 | 54 | 22 | 21 | 11 | 5 | 15 | 15 | 25 | 38 | 43 | 11 | | |
| Vik. 1936 | Ani ... | 6 | Je | 13 | 32 | 2 | Jl | 14 | Ashada ... | 5 | Je | 19 | 26 | 13 | ... | 47 | 19 | 17 | 13 | 58 | 68 | 15 | 1 | | |
| (S.) Pramathi | Adi ... | 3 | Jl | 15 | 31 | 5 | Au | 14 | Sravana ... | 6 | Jl | 18 | 58 | 3 | 19 | 18 | 54 | 19 | 12 | 32 | 97 | 46 | 51 | | |
| K. 1054 (S. M.) | Avani ... | 6 | Au | 15 | 32 | 2 | S | 16 | Bhadrapada . | 1 | Au | 17 | 29 | 53 | ... | 46 | 35 | 21 | 11 | 6 | 127 | 18 | 41 | | |
| K. 1054 (N. M.) | Purattasi... | 3 | S | 16 | 30 | 4 | O | 15 | Asvina ... | 3 | S | 16 | 1 | 43 | ... | 11 | 55 | 23 | 9 | 39 | 156 | 50 | 31 | | |
| | | | | | | | | | | 4 | O | 15 | 33 | 33 | ... | 36 | 9 | 25 | 8 | 13 | 186 | 22 | 21 | | |
| 4 g, 54 p. | Aippasi ... | 5 | O | 16 | 30 | 6 | N | 14 | Kartika ... | 6 | N | 14 | 5 | 23 | ...○ | 0 | 39 | 27 | 6 | 46 | 215 | 54 | 12 | | |
| B. S. 1285 | Kartigai ... | 7 | N | 15 | 29 | 7 | D | 13 | Margasira ... | 7 | D | 13 | 37 | 13 | ... | 26 | 24 | 1 | 32 | 3 | 245 | 26 | 2 | | |
| (N.) Khara | Margali ... | 1 | D | 14 | 30 | 2 | Ja | 12 | Pausha ... | 2 | Ja | 12 | 9 | 3 | 11 | 54 | 22 | 3 | 30 | 37 | 274 | 57 | 52 | | |
| **A. D. 1880** | Tai ... | 3 | Ja | 13 | 29 | 3 | F | 10 | Magha ... | 3 | F | 10 | 40 | 53 | ... | 24 | 50 | 5 | 29 | 11 | 304 | 29 | 42 | | |
| H. 1298 | Masi ... | 4 | F | 11 | 30 | 5 | Mr | 11 | Phalguna ... | 5 | Mr | 11 | 12 | 44 | 10 | 58 | 5 | 7 | 27 | 44 | 334 | 1 | 32 | | |
| Dec. 4. | Panguni ... | 6 | Mr | 12 | 30 | 7 | Ap | 10 | Chaitra - ... | 6 | Ap | 9 | 44 | 34 | ... | 34 | 9 | 9 | 26 | 18 | 363 | 33 | 22 | | |

# TABLE XII.—General Ephemeris A.D. 1880—A.D. 1885.

| Commencement of years in different eras. | Tamil month. | Tamil Month Beginning: Week-day | Beginning: Month | Beginning: Day | Last Day | End: Week-day | End: Month | End: Day | Lunar month. | New Moon: Week-day | New Moon: Month | Ending moment of New Moon Tithi, Mean: Day (mean) | Mean: Ghatikas | Mean: Palas | Actual: Day (actual) | Actual: Ghatikas | Actual: Palas | ☾'s Anom.: Day | ☾'s Anom.: Ghatikas | ☾'s Anom.: Palas | ⊙'s Anom.: Day | ⊙'s Anom.: Ghatikas | ⊙'s Anom.: Palas | Deduct for Nakshatras | Deduct for Yogas. |
|---|---|---|---|---|---|---|---|---|---|---|---|---|---|---|---|---|---|---|---|---|---|---|---|---|---|
| **K. Y. 4981** | Chittirai ... | 1 | Ap | 11 | 31 | 3 | My | 11 | Vaisakha ... | 1 | My | 9 | 16 | 24 | ... | 11 | 47 | 11 | 24 | 53 | 27 | 49 | 41 | 2 days, 4 ghat., 54 palas. | 3 „ 52 „ 24 „ |
| Sak. 1802 | Vaikasi ... | 4 | My | 12 | 32 | 7 | Je | 12 | Jyeshta ... | 2 | Je | 7 | 48 | 14 | ...○ | 50 | 28 | 13 | 23 | 27 | 57 | 21 | 31 | | |
| Vik. 1937 | Ani ... | 1 | Je | 13 | 31 | 3 | Jl | 13 | Ashada ... | 4 | Jl | 7 | 20 | 4 | ... | 28 | 25 | 15 | 22 | 0 | 86 | 53 | 21 | | |
| (S.) Vikrama | Adi ... | 4 | Jl | 14 | 32 | 7 | Au | 14 | Sravana ... | 5 | Au | 5 | 51 | 54 | 6 | 4 | 25 | 17 | 20 | 34 | 116 | 25 | 11 | | |
| K. 1055 (S. M.) | Avani ... | 1 | Au | 15 | 31 | 3 | S | 14 | Bhadrapada . | 7 | S | 4 | 23 | 44 | ... | 38 | 4 | 19 | 19 | 8 | 145 | 57 | 1 | | |
| K. 1055 (N. M.) | Purattasi . | 4 | S | 15 | 30 | 5 | O | 14 | Asvina ... | 1 | O | 3 | 55 | 34 | 4 | 9 | 16 | 21 | 17 | 41 | 175 | 28 | 51 | | |
| 20 g., 25 p. | Aippasi ... | 6 | O | 15 | 30 | 7 | N | 13 | Kartika ... | 3 | N | 2 | 27 | 25 | ... | 38 | 28 | 23 | 16 | 15 | 205 | 0 | 41 | | |
| B. S. 1286 | Kartigai ... | 1 | N | 14 | 30 | 2 | D | 13 | Margasira ... | 4 | D | 1 | 59 | 15 | 20○ | 6 | 9 | 25 | 14 | 48 | 234 | 32 | 32 | | |
| (N.) Nandana | Margali ... | 3 | D | 14 | 29 | 3 | Ja | 11 | Pausha ... | 6 | D | 31 | 31 | 5 | ... | 32 | 59 | 27 | 13 | 22 | 264 | 4 | 22 | | |
| **A. D. 1881** | Tai ... | 4 | Ja | 12 | 29 | 4 | F | 9 | Magha ... | 1 | Ja | 30 | 2 | 55 | 29 | 59 | 27 | 1 | 38 | 39 | 298 | 36 | 12 | | |
| H. 1299 | Masi ... | 5 | F | 10 | 30 | 6 | Mr | 11 | Phalguna ... | 2 | F | 28 | 34 | 45 | ... | 26 | 10 | 3 | 37 | 13 | 323 | 8 | 2 | | |
| Nov. 23 | Panguni ... | 7 | Mr | 12 | 31 | 2 | Ap | 11 | Chaitra ... | 4 | Mr | 30 | 6 | 35 | 29 | 53 | 50 | 5 | 35 | 46 | 352 | 39 | 52 | | |
| **K. Y. 4982** | Chittirai... | 3 | Ap | 12 | 31 | 5 | My | 12 | Vaisakha ... | 5 | Ap | 28 | 38 | 25 | ... | 23 | 39 | 7 | 34 | 20 | 16 | 56 | 11 | 1 day, 16 ghat., 0 palas. | 2 „ 21 „ 27 „ |
| Sak. 1803 | Vaikasi ... | 6 | My | 13 | 31 | 1 | Je | 12 | Jyeshta ... | 7 | My | 28 | 10 | 15 | 27●○ | 55 | 6 | 9 | 32 | 54 | 46 | 28 | 1 | | |
| Vik. 1938 | Ani ... | 2 | Je | 13 | 32 | 5 | Jl | 14 | Ashada ... | 1 | Je | 26 | 42 | 16 | ... | 29 | 58 | 11 | 31 | 27 | 75 | 59 | 51 | | |
| (S.) Vrisha | Adi ... | 6 | Jl | 15 | 31 | 1 | Au | 14 | Sravana ... | 3 | Jl | 26 | 13 | 56 | ... | 7 | 0 | 13 | 30 | 1 | 105 | 31 | 41 | | |
| K. 1056 (S. M.) | Avani ... | 2 | Au | 15 | 31 | 4 | S | 14 | Bhadrapada . | 4 | Au | 24 | 45 | 46 | ... | 46 | 33 | 15 | 28 | 34 | 135 | 3 | 31 | | |
| K 1056 (N. M.) | Purattasi . | 6 | S | 15 | 30 | 6 | O | 14 | Asvina ... | 6 | S | 23 | 17 | 36 | ... | 25 | 55 | 17 | 27 | 8 | 164 | 35 | 21 | | |
| 35 g., 56 p. | Aippasi ... | 7 | O | 15 | 30 | 1 | N | 13 | Kartika ... | 7 | O | 22 | 49 | 26 | 23 | 3 | 38 | 19 | 25 | 42 | 194 | 7 | 11 | | |
| B. S. 1287 | Kartigai ... | 2 | N | 14 | 30 | 3 | D | 13 | Margasira ... | 2 | N | 21 | 21 | 16 | ...○ | 38 | 40 | 21 | 24 | 15 | 223 | 39 | 1 | | |
| (N.) Vijaya | Margali ... | 4 | D | 14 | 29 | 4 | Ja | 11 | Pausha ... | 3 | D | 20 | 53 | 6 | 21 | 11 | 14 | 23 | 22 | 49 | 253 | 10 | 52 | | |
| **A. D. 1882** | Tai ... | 5 | Ja | 12 | 30 | 6 | F | 10 | Magha ... | 5 | Ja | 19 | 24 | 56 | ... | 39 | 22 | 25 | 21 | 22 | 282 | 42 | 42 | | |
| H. 1300 | Masi ... | 7 | F | 11 | 30 | 1 | Mr | 12 | Phalguna ... | 6 | F | 17 | 56 | 46 | 18 | 5 | 17 | 27 | 19 | 56 | 312 | 14 | 32 | | |
| Nov. 12 | Panguni ... | 2 | Mr | 13 | 30 | 3 | Ap | 11 | Chaitra ... | 1 | Mr | 19 | 28 | 36 | ... | 29 | 14 | 1 | 45 | 13 | 341 | 46 | 22 | | |
| **K. Y. 4983** | Chittirai... | 4 | Ap | 12 | 31 | 6 | My | 12 | Vaisakha ... | 3 | Ap | 18 | 0 | 27 | 17 | 52 | 27 | 3 | 43 | 47 | 6 | 2 | 41 | 0 day, [illegible] ghat., 8 palas. | 0 „ 50 „ 29 „ |
| Sak. 1804 | Vaikasi ... | 7 | My | 13 | 31 | 2 | Je | 12 | Jyeshta ... | 4 | My | 17 | 32 | 17 | ...● | 16 | 16 | 5 | 42 | 20 | 35 | 34 | 31 | | |
| Vik. 1939 | Ani ... | 3 | Je | 13 | 32 | 6 | Jl | 14 | Ashada ... | 6 | Je | 16 | 4 | 7 | 15 | 42 | 18 | 7 | 40 | 54 | 65 | 6 | 21 | | |
| (S.) Chitrabhanu. | Adi ... | 7 | Jl | 15 | 31 | 2 | Au | 14 | Sravana ... | 7 | Jl | 15 | 35 | 57 | ... | 12 | 2 | 9 | 39 | 27 | 94 | 38 | 11 | | |
| | | | | | | | | | | 2 | Au | 14 | 7 | 47 | 13 | 47 | 6 | 11 | 38 | 1 | 124 | 10 | 1 | | |
| K. 1057 (S. M.) | Avani ... | 3 | Au | 15 | 31 | 5 | S | 14 | Bhadrapada . | 3 | S | 12 | 39 | 37 | ... | 27 | 18 | 13 | 36 | 35 | 153 | 41 | 51 | | |
| K. 1057 (N. M.) | Purattasi . | 6 | S | 15 | 31 | 1 | O | 15 | Asvina ... | 5 | O | 12 | 11 | 27 | ... | 10 | 50 | 15 | 35 | 8 | 133 | 18 | 41 | | |
| 51 g., 27 p. | Aippasi ... | 2 | O | 16 | 30 | 3 | N | 14 | Kartika ... | 6 | N | 10 | 43 | 17 | ... | 54 | 33 | 17 | 33 | 42 | 212 | 45 | 31 | | |
| B. S. 1288 | Kartigai ... | 4 | N | 15 | 29 | 4 | D | 13 | Margasira ... | 1 | D | 10 | 15 | 8 | ... | 35 | 42 | 19 | 32 | 15 | 242 | 17 | 21 | | |
| (N.) Jaya | Margali ... | 5 | D | 14 | 30 | 6 | Ja | 12 | Pausha ... | 2 | Ja | 8 | 46 | 58 | 9 | 12 | 24 | 21 | 30 | 48 | 271 | 49 | 11 | | |
| **A. D. 1883** | Tai ... | 7 | Ja | 13 | 29 | 7 | F | 10 | Magha ... | 4 | F | 7 | 18 | 48 | ... | 43 | 50 | 23 | 29 | 22 | 301 | 21 | 2 | | |
| H. 1301 | Masi ... | 1 | F | 11 | 30 | 2 | Mr | 12 | Phalguna ... | 5 | Mr | 8 | 50 | 38 | 9 | 10 | 18 | 25 | 27 | 56 | 330 | 52 | 52 | | |
| Nov. 2 | Panguni ... | 3 | Mr | 13 | 30 | 4 | Ap | 11 | Chaitra ... | 7 | Ap | 7 | 22 | 28 | ...○ | 32 | 46 | 27 | 26 | 30 | 360 | 24 | 42 | | |
| **K. Y. 4984** | Chittirai ... | 5 | Ap | 12 | 31 | 7 | My | 12 | Vaisakha ... | 1 | My | 6 | 54 | 18 | ... | 53 | 54 | 1 | 51 | 47 | 24 | 41 | 1 | 1 day, 50 ghat., 47 palas. | 3 „ 26 „ 9 „ |
| Sak. 1805 | Vaikasi ... | 1 | My | 13 | 31 | 3 | Je | 12 | Jyeshta ... | 3 | Je | 5 | 26 | 8 | ... | 12 | 33 | 3 | 50 | 20 | 54 | 12 | 51 | | |
| Vik. 1940 | Ani ... | 4 | Je | 13 | 32 | 7 | Jl | 14 | Ashada ... | 4 | Jl | 4 | 57 | 59 | ... | 33 | 49 | 5 | 48 | 53 | 83 | 44 | 41 | | |
| (S.) Subhanu | Adi ... | 1 | Jl | 15 | 32 | 4 | Au | 15 | Sravana ... | 6 | Au | 3 | 29 | 49 | 2 | 59 | 20 | 7 | 47 | 28 | 113 | 16 | 31 | | |
| K. 1058 (S. M.) | Avani ... | 5 | Au | 16 | 31 | 7 | S | 15 | Bhadrapada . | 1 | S | 2 | 1 | 39 | 1 | 30 | 59 | 9 | 46 | 1 | 142 | 48 | 21 | | |
| K. 1058 (N. M.) | Purattasi . | 1 | S | 16 | 30 | 2 | O | 15 | Asvina ... | 2 | O | 1 | 33 | 29 | ...○ | 9 | 45 | 11 | 44 | 35 | 172 | 20 | 11 | | |
| 6 g., 59 p. | Aippasi ... | 3 | O | 16 | 30 | 4 | N | 14 | Kartika ... | 4 | O | 31 | 5 | 19 | 30 | 54 | 37 | 13 | 43 | 8 | 201 | 52 | 1 | | |
| B. S. 1289 | Kartigai ... | 5 | N | 15 | 29 | 5 | D | 13 | Margasira ... | 5 | N | 29 | 37 | 9 | ... | 41 | 29 | 15 | 41 | 42 | 231 | 23 | 51 | | |
| (N.) Manmatha. | Margali ... | 6 | D | 14 | 30 | 7 | Ja | 12 | Pausha ... | 7 | D | 29 | 8 | 59 | ... | 28 | 54 | 17 | 40 | 16 | 260 | 55 | 42 | | |
| **A. D. 1884** | Tai ... | 1 | Ja | 13 | 29 | 1 | F | 10 | Magha ... | 1 | Ja | 27 | 40 | 49 | 28 | 10 | 21 | 19 | 38 | 49 | 290 | 27 | 32 | | |
| H. 1302 | Masi ... | 2 | F | 11 | 30 | 3 | Mr | 11 | Phalguna ... | 3 | F | 26 | 12 | 39 | ... | 45 | 2 | 21 | 37 | 23 | 319 | 59 | 22 | | |
| Oct. 21. | Panguni ... | 4 | Mr | 12 | 30 | 5 | Ap | 10 | Chaitra ... | 4 | Mr | 26 | 44 | 29 | 27○ | 12 | 58 | 23 | 35 | 56 | 349 | 31 | 12 | | |
| **K. Y. 4985** | Chittirai... | 6 | Ap | 11 | 31 | 1 | My | 11 | Vaisakha ... | 6 | Ap | 25 | 16 | 20 | ... | 35 | 24 | 25 | 34 | 30 | 13 | 47 | 31 | 1 day, 1 ghat., 54 palas. | 1 „ 55 „ 11 „ |
| Sak. 1806 | Vaikasi ... | 2 | My | 12 | 32 | 5 | Je | 12 | Jyeshta ... | 7 | My | 24 | 48 | 10 | ... | 54 | 6 | 27 | 33 | 4 | 43 | 19 | 21 | | |
| Vik. 1941 | Ani ... | 6 | Je | 13 | 31 | 1 | Jl | 13 | Ashada ... | 2 | Je | 23 | 20 | 0 | ... | 11 | 19 | 1 | 58 | 21 | 72 | 51 | 11 | | |
| (S.) Tarana | Adi ... | 2 | Jl | 14 | 32 | 5 | Au | 14 | Sravana ... | 3 | Jl | 22 | 51 | 50 | ... | 28 | 39 | 3 | 56 | 54 | 102 | 23 | 1 | | |
| K. 1059 (S. M.) | Avani ... | 6 | Au | 15 | 31 | 1 | S | 14 | Bhadrapada . | 5 | Au | 21 | 23 | 40 | 20 | 51 | 53 | 5 | 55 | 28 | 131 | 54 | 51 | | |
| K. 1059 (N. M.) | Purattasi . | 2 | S | 15 | 30 | 3 | O | 14 | Asvina ... | 6 | S | 19 | 55 | 30 | ...○ | 20 | 17 | 7 | 54 | 1 | 161 | 26 | 41 | | |
| 22 g., 30 p. | Aippasi ... | 4 | O | 15 | 30 | 5 | N | 13 | Kartika ... | 1 | O | 19 | 27 | 20 | 18 | 56 | 51 | 9 | 52 | 35 | 190 | 58 | 31 | | |
| B. S. 1290 | Kartigai ... | 6 | N | 14 | 30 | 7 | D | 13 | Margasira ... | 2 | N | 17 | 59 | 10 | ... | 39 | 56 | 11 | 51 | 9 | 220 | 30 | 21 | | |
| (N.) Durmukha. | Margali ... | 1 | D | 14 | 29 | 1 | Ja | 11 | Pausha ... | 4 | D | 17 | 31 | 0 | ... | 28 | 38 | 13 | 49 | 42 | 250 | 2 | 11 | | |
| **A. D. 1885** | Tai ... | 2 | Ja | 12 | 29 | 2 | F | 9 | Magha ... | 6 | Ja | 16 | 2 | 51 | ... | 18 | 14 | 15 | 48 | 16 | 279 | 34 | 1 | | |
| H. 1303 | Masi ... | 3 | F | 10 | 30 | 4 | Mr | 11 | Phalguna ... | 7 | F | 14 | 34 | 41 | 15 | 4 | 42 | 17 | 46 | 49 | 309 | 5 | 52 | | |
| Oct. 10 | Panguni ... | 5 | Mr | 12 | 31 | 7 | Ap | 11 | Chaitra ... | 2 | Mr | 16 | 6 | 31 | ...○ | 41 | 32 | 19 | 45 | 23 | 338 | 37 | 42 | | |

## TABLE XII.—General Ephemeris A.D. 1885—A.D. 1890.

| Commencement of years in different eras. | Tamil month. | Tamil Month Beginning Week-day | Month | Day | Last Day | End Week-day | Month | Day | Lunar month. | New Moon Ending moment of New Moon Tithi. Week-day | Month | Mean Day (mean) | Ghatikas | Palas | Actual Day (actual) | Ghatikas | Palas | ☾'s Anom. Day | Ghatikas | Palas | ☉'s Anom. Day | Ghatikas | Palas | Deduct for Nakshatras | Deduct for Yogas. |
|---|---|---|---|---|---|---|---|---|---|---|---|---|---|---|---|---|---|---|---|---|---|---|---|---|---|
| **K. Y. 4986** | Chittirai ... | 1 | Ap | 12 | 31 | 3 | My | 12 | Vaisakha ... | 3 | Ap | 14 | 38 | 21 | 15 | 12 | 35 | 21 | 43 | 57 | 2 | 54 | 1 | 0 day, 13 ghat., 1 pala. | 0 ,, 24 ,, 14 ,, |
| Sak. 1807 | Vaikasi ... | 4 | My | 13 | 31 | 6 | Je | 12 | Jyeshta ... | 5 | My | 14 | 10 | 11 | ... | 35 | 45 | 23 | 42 | 30 | 32 | 25 | 51 | | |
| | | | | | | | | | | 6 | Je | 12 | 42 | 1 | ... | 54 | 42 | 25 | 41 | 4 | 61 | 57 | 41 | | |
| Vik. 1942 | Ani ... | 7 | Je | 13 | 32 | 3 | Jl | 14 | Ashada ... | 1 | Jl | 12 | 13 | 51 | ... | 11 | 26 | 0 | 6 | 21 | 91 | 29 | 31 | | |
| (S.) Parthiva | Adi ... | 4 | Jl | 15 | 31 | 6 | Au | 14 | Sravana ... | 2 | Au | 10 | 45 | 41 | ... | 28 | 46 | 2 | 4 | 51 | 121 | 1 | 21 | | |
| K. 1060 (S. M.) | Avani ... | 7 | Au | 15 | 31 | 2 | S | 14 | Bhadrapada . | 4 | S | 9 | 17 | 32 | 8○ | 49 | 20 | 4 | 3 | 28 | 150 | 33 | 11 | | |
| K. 1060 (N. M.) | Purattasi . | 3 | S | 15 | 31 | 5 | O | 15 | Asvina ... | 5 | O | 8 | 49 | 22 | ... | 15 | 13 | 6 | 2 | 2 | 180 | 5 | 1 | | |
| 38 g., 1 p. | Aippasi ... | 6 | O | 16 | 29 | 6 | N | 13 | Kartika ... | 7 | N | 7 | 21 | 12 | 6 | 48 | 2 | 8 | 0 | 35 | 209 | 36 | 51 | | |
| B. S. 1291 | Kartigai ... | 7 | N | 14 | 30 | 1 | D | 13 | Margasira ... | 1 | D | 6 | 53 | 2 | ... | 28 | 24 | 9 | 59 | 9 | 239 | 8 | 41 | | |
| (N.) Hemalamba. | Margali ... | 2 | D | 14 | 29 | 2 | Ja | 11 | Pausha ... | 3 | Ja | 5 | 24 | 52 | ... | 15 | 5 | 11 | 57 | 42 | 268 | 40 | 31 | | |
| **A. D. 1886** | Tai ... | 3 | Ja | 12 | 30 | 4 | F | 10 | Magha ... | 4 | F | 3 | 56 | 42 | 4 | 4 | 29 | 13 | 56 | 16 | 298 | 12 | 22 | | |
| H. 1304 | Masi ... | 5 | F | 11 | 30 | 6 | Mr | 12 | Phalguna ... | 6 | Mr | 5 | 28 | 32 | ... | 51 | 38 | 15 | 54 | 50 | 327 | 44 | 12 | | |
| Sep. 30 | Panguni ... | 7 | Mr | 13 | 30 | 1 | Ap | 11 | Chaitra ... | 1 | Ap | 4 | 0 | 22 | ... | 32 | 41 | 17 | 53 | 23 | 357 | 16 | 2 | | |
| **K. Y. 4987** | Chittirai ... | 2 | Ap | 12 | 31 | 4 | My | 12 | Vaisakha ... | 2 | My | 3 | 32 | 13 | 4 | 6 | 4 | 19 | 51 | 51 | 21 | 32 | 21 | 1 day, 36 ghat., 40 palas. | 2 ,, 59 ,, 53 ,, |
| Sak. 1808 | Vaikasi ... | 5 | My | 13 | 31 | 7 | Je | 12 | Jyeshta ... | 4 | Je | 2 | 4 | 3 | ... | 32 | 28 | 21 | 50 | 30 | 51 | 4 | 11 | | |
| Vik. 1943 | Ani ... | 1 | Je | 13 | 32 | 4 | Jl | 14 | Ashada ... | 5 | Jl | 1 | 35 | 53 | ... | 53 | 37 | 23 | 49 | 4 | 80 | 36 | 1 | | |
| (S.) Vyasa | Adi ... | 5 | Jl | 15 | 32 | 7 | Au | 14 | Sravana ... | 7 | Jl | 31 | 7 | 43 | ... | 11 | 53 | 25 | 47 | 38 | 110 | 7 | 51 | | |
| K. 1061 (S. M.) | Avani ... | 1 | Au | 15 | 31 | 3 | S | 14 | Bhadrapada . | 1 | Au | 29 | 39 | 33 | ... | 30 | 13 | 0 | 12 | 55 | 139 | 39 | 41 | | |
| K. 1061 (N. M.) | Purattasi . | 4 | S | 15 | 31 | 6 | O | 15 | Asvina ... | 3 | S | 28 | 11 | 23 | 27 | 50 | 29 | 2 | 11 | 28 | 169 | 11 | 31 | | |
| 53 g., 32 p. | Aippasi ... | 7 | O | 16 | 30 | 1 | N | 14 | Kartika ... | 4 | O | 27 | 43 | 13 | ... | 14 | 58 | 4 | 10 | 2 | 198 | 43 | 21 | | |
| B. S. 1292 | Kartigai ... | 2 | N | 15 | 29 | 2 | D | 13 | Margasira .. | 6 | N | 26 | 15 | 3 | 25 | 45 | 5 | 6 | 8 | 36 | 228 | 15 | 11 | | |
| (N.) Vilamba | Margali ... | 3 | D | 14 | 30 | 4 | Ja | 12 | Pausha ... | 7 | D | 25 | 46 | 53 | ... | 21 | 21 | 8 | 7 | 9 | 257 | 47 | 1 | | |
| **A. D. 1887** | Tai ... | 5 | Ja | 13 | 29 | 5 | F | 10 | Magha ... | 2 | Ja | 24 | 18 | 44 | ...○ | 3 | 46 | 10 | 5 | 43 | 287 | 18 | 51 | | |
| H. 1305 | Masi ... | 6 | F | 11 | 30 | 7 | Mr | 12 | Phalguna ... | 3 | F | 22 | 50 | 34 | ... | 49 | 39 | 12 | 4 | 16 | 316 | 50 | 41 | | |
| Sep. 19 | Panguni ... | 1 | Mr | 13 | 30 | 2 | Ap | 11 | Chaitra ... | 5 | Mr | 24 | 22 | 24 | ... | 35 | 46 | 14 | 2 | 50 | 346 | 22 | 31 | | |
| **K. Y. 4988** | Chittirai ... | 3 | Ap | 12 | 31 | 5 | My | 12 | Vaisakha ... | 6 | Ap | 22 | 54 | 14 | 23 | 18 | 18 | 16 | 1 | 24 | 10 | 38 | 51 | 0 day, 47 ghat., 47 palas. | 1 ,, 28 ,, 55 ,, |
| Sak. 1809 | Vaikasi ... | 6 | My | 13 | 32 | 1 | Je | 12 | Jyeshta ... | 1 | My | 22 | 26 | 4 | ... | 54 | 44 | 17 | 59 | 57 | 40 | 10 | 41 | | |
| Vik. 1944 | Ani ... | 2 | Je | 13 | 31 | 5 | Jl | 14 | Ashada ... | 2 | Je | 20 | 57 | 54 | 21 | 24 | 40 | 19 | 58 | 31 | 69 | 42 | 31 | | |
| (S.) Sarvajit | Adi ... | 6 | Jl | 15 | 32 | 2 | Au | 15 | Sravana ... | 4 | Jl | 20 | 29 | 44 | ...○ | 49 | 31 | 21 | 57 | 4 | 99 | 14 | 21 | | |
| K. 1062 (S. M.) | Avani ... | 3 | Au | 16 | 31 | 5 | S | 15 | Bhadrapada . | 6 | Au | 19 | 1 | 34 | ...● | 11 | 20 | 23 | 55 | 37 | 128 | 46 | 11 | | |
| K. 1062 (N. M.) | Purattasi . | 6 | S | 16 | 30 | 7 | O | 15 | Asvina ... | 7 | S | 17 | 33 | 25 | ... | 32 | 19 | 25 | 54 | 12 | 158 | 18 | 1 | | |
| 9 g., 4 p. | Aippasi ... | 1 | O | 16 | 30 | 2 | N | 14 | Kartika ... | 2 | O | 17 | 5 | 15 | 16 | 54 | 8 | 0 | 19 | 29 | 187 | 49 | 51 | | |
| B. S. 1293 | Kartigai ... | 3 | N | 15 | 29 | 3 | D | 13 | Margasira ... | 3 | N | 15 | 37 | 5 | ... | 18 | 25 | 2 | 18 | 2 | 217 | 21 | 41 | | |
| (N.) Vikarin | Margali ... | 4 | D | 14 | 30 | 5 | Ja | 12 | Pausha ... | 5 | D | 15 | 8 | 55 | 14 | 46 | 30 | 4 | 16 | 36 | 246 | 53 | 31 | | |
| **A. D. 1888** | Tai ... | 6 | Ja | 13 | 29 | 6 | F | 10 | Magha ... | 6 | Ja | 13 | 40 | 45 | ...○ | 19 | 0 | 6 | 15 | 9 | 276 | 25 | 21 | | |
| H. 1306 | Masi ... | 7 | F | 11 | 30 | 1 | Mr | 11 | Phalguna ... | 1 | F | 12 | 12 | 35 | 11 | 55 | 45 | 8 | 13 | 43 | 305 | 57 | 11 | | |
| Sep. 7 | Panguni ... | 2 | Mr | 12 | 30 | 3 | Ap | 10 | Chaitra ... | 2 | Mr | 12 | 44 | 25 | ... | 36 | 1 | 10 | 12 | 17 | 335 | 29 | 2 | | |
| | | | | | | | | | | 4 | Ap | 11 | 16 | 15 | ... | 18 | 7 | 12 | 10 | 50 | 365 | 0 | 52 | | |
| **K. Y. 4989** | Chittirai ... | 4 | Ap | 11 | 31 | 6 | My | 11 | Vaisakha ... | 5 | My | 10 | 48 | 5 | ... | 59 | 37 | 14 | 9 | 24 | 29 | 17 | 11 | 2 days, 11 ghat., 26 palas. | 4 ,, 4 ,, 35 ,, |
| Sak. 1810 | Vaikasi ... | 7 | My | 12 | 32 | 3 | Je | 12 | Jyeshta ... | 7 | Je | 9 | 19 | 55 | ... | 38 | 29 | 16 | 7 | 57 | 58 | 49 | 1 | | |
| Vik. 1945 | Ani ... | 4 | Je | 13 | 31 | 6 | Jl | 13 | Ashada ... | 1 | Jl | 8 | 51 | 46 | 9○ | 11 | 58 | 18 | 6 | 31 | 88 | 20 | 51 | | |
| (S.) Sarvadhari. | Adi ... | 7 | Jl | 14 | 32 | 3 | Au | 14 | Sravana ... | 3 | Au | 7 | 23 | 36 | ... | 41 | 51 | 20 | 5 | 5 | 117 | 52 | 41 | | |
| K. 1063 (S. M.) | Avani ... | 4 | Au | 15 | 31 | 6 | S | 14 | Bhadrapada . | 4 | S | 5 | 55 | 26 | 6 | 8 | 56 | 22 | 3 | 38 | 147 | 24 | 31 | | |
| K. 1063 (N. M.) | Purattasi . | 7 | S | 15 | 30 | 1 | O | 14 | Asvina ... | 6 | O | 5 | 27 | 16 | ... | 34 | 7 | 24 | 2 | 12 | 176 | 56 | 21 | | |
| 24 g., 35 p. | Aippasi ... | 2 | O | 15 | 30 | 3 | N | 13 | Kartika ... | 7 | N | 3 | 59 | 6 | ... | 58 | 52 | 26 | 0 | 45 | 206 | 28 | 11 | | |
| B. S. 1294 | Kartigai ... | 4 | N | 14 | 30 | 5 | D | 13 | Margasira ... | 2 | D | 3 | 30 | 56 | ... | 24 | 13 | 0 | 26 | 2 | 236 | 0 | 1 | | |
| (N.) Sarvari | Margali ... | 6 | D | 14 | 29 | 6 | Ja | 11 | Pausha ... | 4 | Ja | 2 | 2 | 46 | 1○ | 50 | 11 | 2 | 24 | 36 | 265 | 31 | 51 | | |
| **A. D. 1889** | Tai ... | 7 | Ja | 12 | 30 | 1 | F | 10 | Magha ... | 5 | Ja | 31 | 34 | 37 | ... | 20 | 3 | 4 | 23 | 10 | 295 | 3 | 41 | | |
| H. 1307 | Masi ... | 2 | F | 11 | 29 | 2 | Mr | 11 | Phalguna ... | 7 | Mr | 2 | 6 | 27 | 1 | 51 | 20 | 6 | 21 | 43 | 324 | 35 | 31 | | |
| Aug. 28. | Panguni ... | 3 | Mr | 12 | 31 | 5 | Ap | 11 | Chaitra ... | 1 | Mr | 31 | 38 | 17 | ... | 25 | 10 | 8 | 20 | 17 | 354 | 7 | 21 | | |
| **K. Y. 4990** | Chittirai ... | 6 | Ap | 12 | 31 | 1 | My | 12 | Vaisakha ... | 3 | Ap | 30 | 10 | 7 | ... | 1 | 23 | 10 | 18 | 50 | 18 | 23 | 41 | 1 day, 22 ghat., 33 palas. | 2 ,, 38 ,, 37 ,, |
| Sak. 1811 | Vaikasi ... | 2 | My | 13 | 31 | 4 | Je | 12 | Jyeshta ... | 4 | My | 29 | 41 | 57 | ... | 39 | 13 | 12 | 17 | 24 | 47 | 55 | 31 | | |
| Vik. 1946 | Ani ... | 5 | Je | 13 | 32 | 1 | Jl | 14 | Ashada ... | 6 | Je | 28 | 13 | 47 | ●○ | 17 | 29 | 14 | 15 | 58 | 77 | 27 | 21 | | |
| (S.) Virodhi | Adi ... | 2 | Jl | 15 | 31 | 4 | Au | 14 | Sravana ... | 7 | Jl | 27 | 45 | 37 | ... | 54 | 47 | 16 | 14 | 31 | 106 | 59 | 11 | | |
| K. 1064 (S. M.) | Avani ... | 5 | Au | 15 | 31 | 7 | S | 14 | Bhadrapada . | 2 | Au | 26 | 17 | 27 | ... | 30 | 14 | 18 | 13 | 4 | 136 | 31 | 1 | | |
| K. 1064 (N. M.) | Purattasi . | 1 | S | 15 | 31 | 3 | O | 15 | Asvina ... | 3 | S | 24 | 49 | 17 | 25 | 3 | 23 | 20 | 11 | 38 | 166 | 2 | 51 | | |
| 40 g., 6 p. | Aippasi ... | 4 | O | 16 | 29 | 4 | N | 13 | Kartika ... | 5 | O | 24 | 21 | 8 | ... | 34 | 18 | 22 | 10 | 12 | 195 | 34 | 41 | | |
| B. S. 1295 | Kartigai ... | 5 | N | 14 | 30 | 6 | D | 13 | Margasira ... | 6 | N | 22 | 52 | 58 | 23 | 3 | 13 | 24 | 8 | 46 | 225 | 6 | 31 | | |
| (N.) Plava | Margali ... | 7 | D | 14 | 29 | 7 | Ja | 11 | Pausha ... | 1 | D | 22 | 24 | 48 | ... | 31 | 20 | 26 | 7 | 19 | 254 | 38 | 21 | | |
| **A. D. 1890** | Tai ... | 1 | Ja | 12 | 30 | 2 | F | 10 | Magha ... | 2 | Ja | 20 | 56 | 38 | ... | 57 | 17 | 0 | 32 | 36 | 284 | 10 | 11 | | |
| H. 1308 | Masi ... | 3 | F | 11 | 30 | 4 | Mr | 12 | Phalguna ... | 4 | F | 19 | 28 | 28 | ... | 23 | 25 | 2 | 31 | 10 | 313 | 42 | 1 | | |
| Aug. 17. | Panguni ... | 5 | Mr | 13 | 30 | 6 | Ap | 11 | Chaitra ... | 6 | Mr | 21 | 0 | 18 | 20 | 50 | 0 | 4 | 29 | 44 | 343 | 13 | 51 | | |

# TABLE XII.—General Ephemeris A.D. 1890—A.D. 1895.

| Commencement of years in different eras. | Tamil month. | Tamil Month Beginning Week-day | Beginning Month | Beginning Day | Last Day | End Week-day | End Month | End Day | Lunar month. | New Moon Week-day | Month | Day (mean) | Ending moment of New Moon Tithi. Mean Ghatikas | Mean Palas | Day (actual) | Actual Ghatikas | Actual Palas | ☾'s Anom. Day | Ghatikas | Palas | ☉'s Anom. Day | Ghatikas | Palas | Deduct for Nakshatras | Deduct for Yogas. |
|---|---|---|---|---|---|---|---|---|---|---|---|---|---|---|---|---|---|---|---|---|---|---|---|---|---|
| **K. Y. 4991** | Chittirai | 7 | Ap | 12 | 31 | 2 | My | 12 | Vaisakha | 7 | Ap | 19 | 32 | 8 | ... | 17 | 45 | 6 | 28 | 13 | 7 | 30 | 10 | 0 day, 38 ghat., 40 palas. | 1 „ 9 „ 40 „ |
| Sak. 1812 | Vaikasi | 3 | My | 13 | 31 | 5 | Je | 12 | Jyeshta | 2 | My | 19 | 3 | 58 | 18○ | 47 | 33 | 8 | 26 | 46 | 37 | 2 | 0 | | |
| Vik. 1947 | Ani | 6 | Je | 13 | 32 | 2 | Jl | 14 | Ashada | 3 | Je | 17 | 35 | 48 | ...● | 20 | 13 | 10 | 25 | 19 | 66 | 33 | 50 | | |
| (S.) Vikrita | Adi | 3 | Jl | 15 | 31 | 5 | Au | 14 | Sravana | 5 | Jl | 17 | 7 | 39 | 16 | 56 | 4 | 12 | 23 | 53 | 96 | 5 | 41 | | |
| K. 1065 (S. M.) | Avani | 6 | Au | 15 | 31 | 1 | S | 14 | Bhadrapada | 6 | Au | 15 | 39 | 29 | ... | 34 | 53 | 14 | 22 | 26 | 125 | 37 | 31 | | |
| | | | | | | | | | | 1 | S | 14 | 11 | 19 | ... | 14 | 52 | 16 | 21 | 0 | 155 | 9 | 21 | | |
| K. 1065 (N. M.) | Purattasi | 2 | S | 15 | 31 | 4 | O | 15 | Asvina | 2 | O | 13 | 43 | 9 | ... | 54 | 33 | 18 | 19 | 34 | 184 | 41 | 11 | | |
| 55 g., 37 p. | Aippasi | 5 | O | 16 | 30 | 6 | N | 14 | Kartika | 4 | N | 12 | 14 | 59 | ...○ | 31 | 45 | 20 | 18 | 7 | 214 | 13 | 1 | | |
| B. S. 1296 | Kartigai | 7 | N | 15 | 29 | 7 | D | 13 | Margasira | 5 | D | 11 | 46 | 49 | 12● | 5 | 48 | 22 | 16 | 41 | 243 | 44 | 51 | | |
| (N.) Subhakrit | Margali | 1 | D | 14 | 30 | 2 | Ja | 12 | Pausha | 7 | Ja | 10 | 18 | 39 | ... | 36 | 22 | 24 | 15 | 14 | 273 | 16 | 41 | | |
| **A. D. 1891** | Tai | 3 | Ja | 13 | 29 | 3 | F | 10 | Magha | 1 | F | 8 | 50 | 29 | 9 | 3 | 34 | 26 | 13 | 48 | 302 | 48 | 31 | | |
| H. 1309 | Masi | 4 | F | 11 | 30 | 5 | Mr | 12 | Phalguna | 3 | Mr | 10 | 22 | 20 | ... | 28 | 1 | 0 | 39 | 5 | 332 | 20 | 21 | | |
| Aug. 7 | Panguni | 6 | Mr | 13 | 30 | 7 | Ap | 11 | Chaitra | 4 | Ap | 8 | 54 | 10 | ... | 50 | 52 | 2 | 37 | 39 | 361 | 52 | 11 | | |
| **K. Y. 4992** | Chittirai | 1 | Ap | 12 | 31 | 3 | My | 12 | Vaisakha | 6 | My | 8 | 26 | 0 | ...○ | 13 | 39 | 4 | 36 | 12 | 26 | 8 | 30 | 1 day, 57 ghat., 20 palas. | 3 „ 38 „ 19 „ |
| Sak. 1813 | Vaikasi | 4 | My | 13 | 32 | 7 | Je | 13 | Jyeshta | 7 | Je | 6 | 57 | 50 | ... | 37 | 45 | 6 | 34 | 46 | 55 | 40 | 20 | | |
| Vik. 1948 | Ani | 1 | Je | 14 | 31 | 3 | Jl | 14 | Ashada | 2 | Jl | 6 | 29 | 40 | ... | 5 | 1 | 8 | 33 | 19 | 85 | 12 | 10 | | |
| (S.) Khara | Adi | 4 | Jl | 15 | 32 | 7 | Au | 15 | Sravana | 4 | Au | 5 | 1 | 30 | 4 | 37 | 11 | 10 | 31 | 53 | 114 | 44 | 0 | | |
| K. 1066 (S. M.) | Avani | 1 | Au | 16 | 31 | 3 | S | 15 | Bhadrapada | 5 | S | 3 | 33 | 20 | ... | 15 | 4 | 12 | 30 | 27 | 144 | 15 | 50 | | |
| K. 1066. (N. M.) | Purattasi | 4 | S | 16 | 30 | 5 | O | 15 | Asvina | 7 | O | 3 | 5 | 10 | 2 | 57 | 46 | 14 | 29 | 0 | 173 | 47 | 40 | | |
| 11 g., 9 p. | Aippasi | 6 | O | 16 | 30 | 7 | N | 14 | Kartika | 1 | N | 1 | 37 | 1 | ...○ | 42 | 27 | 16 | 27 | 34 | 203 | 19 | 30 | | |
| B. S. 1297 | Kartigai | 1 | N | 15 | 29 | 1 | D | 13 | Margasira | 3 | D | 1 | 8 | 51 | ... | 25 | 50 | 18 | 26 | 7 | 232 | 51 | 21 | | |
| (N.) Sobhana | Margali | 2 | D | 14 | 30 | 3 | Ja | 12 | Pausha | 4 | D | 30 | 40 | 41 | 31 | 5 | 18 | 20 | 24 | 41 | 262 | 23 | 11 | | |
| **A. D. 1892** | Tai | 4 | Ja | 13 | 29 | 4 | F | 10 | Magha | 6 | Ja | 29 | 12 | 31 | ... | 39 | 36 | 22 | 23 | 15 | 291 | 55 | 1 | | |
| H. 1310 | Masi | 5 | F | 11 | 30 | 6 | Mr | 11 | Phalguna | 7 | F | 27 | 44 | 21 | 28 | 8 | 17 | 24 | 21 | 48 | 321 | 26 | 51 | | |
| July 26 | Panguni | 7 | Mr | 12 | 30 | 1 | Ap | 10 | Chaitra | 2 | Mr | 28 | 16 | 11 | ... | 32 | 15 | 26 | 20 | 22 | 350 | 58 | 41 | | |
| **K. Y. 4993** | Chittirai | 2 | Ap | 11 | 31 | 4 | My | 11 | Vaisakha | 3 | Ap | 26 | 48 | 1 | ...○ | 52 | 53 | 0 | 45 | 39 | 15 | 15 | 0 | 1 day, 8 ghat., 27 palas. | 2 „ 7 „ 21 „ |
| Sak. 1814 | Vaikasi | 5 | My | 12 | 32 | 1 | Je | 12 | Jyeshta | 5 | My | 26 | 19 | 51 | ... | 11 | 44 | 2 | 44 | 12 | 44 | 46 | 50 | | |
| Vik. 1949 | Ani | 2 | Je | 13 | 31 | 4 | Jl | 13 | Ashada | 6 | Je | 24 | 51 | 41 | ... | 31 | 52 | 4 | 42 | 46 | 74 | 18 | 40 | | |
| (S.) Nandana | Adi | 5 | Jl | 14 | 32 | 1 | Au | 14 | Sravana | 1 | Jl | 24 | 23 | 31 | 23 | 54 | 45 | 6 | 41 | 20 | 103 | 50 | 30 | | |
| K. 1067 (S. M.) | Avani | 2 | Au | 15 | 31 | 4 | S | 14 | Bhadrapada | 2 | Au | 22 | 55 | 21 | ... | 23 | 3 | 8 | 39 | 53 | 133 | 22 | 20 | | |
| K. 1067 (N. M.) | Purattasi | 5 | S | 15 | 30 | 6 | O | 14 | Asvina | 4 | S | 21 | 27 | 12 | 20 | 58 | 28 | 10 | 38 | 27 | 162 | 54 | 10 | | |
| 26 g., 40 p. | Aippasi | 7 | O | 15 | 30 | 1 | N | 13 | Kartika | 5 | O | 20 | 59 | 2 | ...○ | 40 | 51 | 12 | 37 | 0 | 192 | 26 | 0 | | |
| B. S. 1298 | Kartigai | 2 | N | 14 | 30 | 3 | D | 13 | Margasira | 7 | N | 19 | 30 | 52 | ... | 27 | 4 | 14 | 35 | 34 | 221 | 57 | 50 | | |
| (N.) Krodhin | Margali | 4 | D | 14 | 29 | 4 | Ja | 11 | Pausha | 2 | D | 19 | 2 | 42 | ... | 16 | 14 | 16 | 34 | 8 | 251 | 29 | 41 | | |
| **A. D. 1893** | Tai | 5 | Ja | 12 | 30 | 6 | F | 10 | Magha | 3 | Ja | 17 | 34 | 32 | 18 | 0 | 25 | 18 | 32 | 41 | 281 | 1 | 31 | | |
| H. 1311 | Masi | 7 | F | 11 | 29 | 7 | Mr | 11 | Phalguna | 5 | F | 16 | 6 | 23 | ... | 38 | 57 | 20 | 31 | 15 | 310 | 33 | 21 | | |
| July 15 | Panguni | 1 | Mr | 12 | 31 | 3 | Ap | 11 | Chaitra | 6 | Mr | 17 | 38 | 13 | 18 | 10 | 1 | 22 | 29 | 48 | 340 | 5 | 11 | | |
| **K. Y. 4994** | Chittirai | 4 | Ap | 12 | 31 | 6 | My | 12 | Vaisakha | 1 | Ap | 16 | 10 | 3 | ... | 34 | 41 | 24 | 28 | 22 | 4 | 21 | 30 | 0 day, 19 ghat., 14 palas. | 0 „ 36 „ 24 „ |
| Sak. 1815 | Vaikasi | 7 | My | 13 | 31 | 2 | Je | 12 | Jyeshta | 2 | My | 15 | 41 | 53 | ... | 54 | 43 | 26 | 26 | 56 | 33 | 53 | 20 | | |
| Vik. 1950 | Ani | 3 | Je | 13 | 32 | 6 | Jl | 14 | Ashada | 4 | Je | 14 | 13 | 43 | ... | 12 | 5 | 0 | 52 | 13 | 63 | 25 | 10 | | |
| | | | | | | | | | | 5 | Jl | 13 | 45 | 33 | ... | 29 | 21 | 2 | 50 | 46 | 92 | 57 | 0 | | |
| (S.) Vijaya | Adi | 7 | Jl | 15 | 31 | 2 | Au | 14 | Sravana | 7 | Au | 12 | 17 | 23 | 11 | 49 | 21 | 4 | 49 | 20 | 122 | 28 | 50 | | |
| K. 1068 (S. M.) | Avani | 3 | Au | 15 | 31 | 5 | S | 14 | Bhadrapada | 1 | S | 10 | 49 | 13 | ... | 14 | 34 | 6 | 47 | 54 | 152 | 0 | 40 | | |
| K. 1068 (N. M.) | Purattasi | 6 | S | 15 | 31 | 1 | O | 15 | Asvina | 3 | O | 10 | 21 | 3 | 9 | 46 | 49 | 8 | 46 | 27 | 181 | 32 | 30 | | |
| 42 g., 11 p. | Aippasi | 2 | O | 16 | 29 | 2 | N | 13 | Kartika | 4 | N | 8 | 52 | 53 | ... | 26 | 54 | 10 | 45 | 1 | 211 | 4 | 20 | | |
| B. S. 1299 | Kartigai | 3 | N | 14 | 30 | 4 | D | 13 | Margasira | 6 | D | 8 | 24 | 44 | ... | 13 | 42 | 12 | 43 | 34 | 240 | 36 | 10 | | |
| (N.) Visvavasu. | Margali | 5 | D | 14 | 29 | 5 | Ja | 11 | Pausha | 7 | Ja | 6 | 56 | 34 | 7 | 3 | 43 | 14 | 42 | 8 | 270 | 8 | 1 | | |
| **A. D. 1894** | Tai | 6 | Ja | 12 | 30 | 7 | F | 10 | Magha | 2 | F | 5 | 28 | 24 | ... | 51 | 38 | 16 | 40 | 42 | 299 | 39 | 51 | | |
| H. 1312 | Masi | 1 | F | 11 | 30 | 2 | Mr | 12 | Phalguna | 4 | Mr | 7 | 0 | 14 | ...○ | 33 | 24 | 18 | 39 | 15 | 329 | 11 | 41 | | |
| July 5 | Panguni | 3 | Mr | 13 | 30 | 4 | Ap | 11 | Chaitra | 5 | Ap | 5 | 32 | 4 | 6● | 7 | 23 | 20 | 37 | 49 | 358 | 43 | 31 | | |
| **K. Y. 4995** | Chittirai | 5 | Ap | 12 | 31 | 7 | My | 12 | Vaisakha | 7 | My | 5 | 3 | 54 | ... | 34 | 3 | 22 | 36 | 22 | 22 | 59 | 50 | 1 day, 43 ghat., 12 palas. | 3 „ 12 „ 4 „ |
| Sak. 1816 | Vaikasi | 1 | My | 13 | 31 | 3 | Je | 12 | Jyeshta | 1 | Je | 3 | 35 | 44 | ... | 54 | 59 | 24 | 34 | 56 | 52 | 31 | 40 | | |
| Vik. 1951 | Ani | 4 | Je | 13 | 32 | 7 | Jl | 14 | Ashada | 3 | Jl | 3 | 7 | 34 | ... | 12 | 31 | 26 | 33 | 30 | 82 | 3 | 30 | | |
| (S.) Jaya | Adi | 4 | Jl | 15 | 31 | 3 | Au | 14 | Sravana | 4 | Au | 1 | 39 | 24 | ... | 29 | 17 | 0 | 58 | 47 | 111 | 35 | 20 | | |
| K. 1069 (S. M.) | Avani | 4 | Au | 15 | 31 | 6 | S | 14 | Bhadrapada | 6 | Au | 31 | 11 | 14 | 30○ | 48 | 5 | 2 | 57 | 20 | 141 | 7 | 10 | | |
| K. 1069 (N. M.) | Purattasi | 7 | S | 15 | 31 | 2 | O | 15 | Asvina | 7 | S | 29 | 43 | 4 | ...● | 11 | 23 | 4 | 55 | 54 | 170 | 39 | 0 | | |
| 57 g., 42 p. | Aippasi | 3 | O | 16 | 30 | 4 | N | 14 | Kartika | 2 | O | 29 | 14 | 55 | 28 | 40 | 47 | 6 | 54 | 27 | 200 | 10 | 50 | | |
| B. S. 1300 | Kartigai | 5 | N | 15 | 29 | 5 | D | 13 | Margasira | 3 | N | 27 | 46 | 45 | ... | 17 | 33 | 8 | 52 | 1 | 229 | 42 | 40 | | |
| (N.) Parabhava | Margali | 6 | D | 14 | 30 | 7 | Ja | 12 | Pausha | 5 | D | 27 | 18 | 35 | ... | 1 | 2 | 10 | 51 | 35 | 259 | 14 | 30 | | |
| **A. D. 1895** | Tai | 1 | Ja | 13 | 29 | 1 | F | 10 | Magha | 6 | Ja | 25 | 50 | 25 | ... | 49 | 40 | 12 | 50 | 8 | 288 | 46 | 21 | | |
| H. 1313 | Masi | 2 | F | 11 | 30 | 3 | Mr | 12 | Phalguna | 1 | F | 24 | 22 | 15 | ...○ | 38 | 15 | 14 | 48 | 42 | 318 | 18 | 11 | | |
| June 24 | Panguni | 4 | Mr | 13 | 30 | 5 | Ap | 11 | Chaitra | 2 | Mr | 25 | 54 | 5 | 26 | 22 | 32 | 16 | 47 | 15 | 347 | 50 | 1 | | |

## TABLE XII.—General Ephemeris A.D. 1895—A.D. 1900.

| Commencement of years in different eras. | Tamil month. | Tamil Month Beginning Week-day | Beginning Month | Beginning Day | Last Day | Tamil Month End Week-day | End Month | End Day | Lunar month. | New Moon Ending moment of New Moon Tithi Week-day | Month | Day (mean) | Mean Ghatikas | Mean Palas | Day (actual) | Actual Ghatikas | Actual Palas | ☾'s Anom. Day | ☾'s Anom. Ghatikas | ☾'s Anom. Palas | ☉'s Anom. Day | ☉'s Anom. Ghatikas | ☉'s Anom. Palas | Deduct for Nakshatras | Deduct for Yogas. |
|---|---|---|---|---|---|---|---|---|---|---|---|---|---|---|---|---|---|---|---|---|---|---|---|---|---|
| **K. Y. 4996** | Chittirai... | 6 | Ap | 12 | 31 | 1 | My | 12 | Vaisakha ... | 4 | Ap | 24 | 25 | 55 | ... | 59 | 27 | 18 | 45 | 49 | 12 | 6 | 20 | 0 day, 54 ghat., 20 palas. | 1 „ 41 „ 6 „ |
| Sak. 1817 | Vaikasi ... | 2 | My | 13 | 32 | 5 | Je | 13 | Jyeshta ... | 5 | My | 23 | 57 | 45 | 24 | 29 | 4 | 20 | 44 | 23 | 41 | 38 | 10 | | |
| Vik. 1952 | Ani ... | 6 | Je | 14 | 31 | 1 | Jl | 14 | Ashada ... | 7 | Je | 22 | 29 | 36 | ... | 52 | 36 | 22 | 42 | 56 | 71 | 10 | 0 | | |
| (S.) Manmatha. | Adi ... | 2 | Jl | 15 | 32 | 5 | Au | 15 | Sravana ... | 2 | Jl | 22 | 1 | 26 | ... | 12 | 8 | 24 | 41 | 30 | 100 | 41 | 50 | | |
| K. 1070 (S. M.) | Avani ... | 6 | Au | 16 | 31 | 1 | S | 15 | Bhadrapada . | 3 | Au | 20 | 33 | 16 | ●○ | 30 | 29 | 26 | 40 | 3 | 130 | 13 | 40 | | |
| K. 1070 (N. M.) | Purattasi . | 2 | S | 16 | 30 | 3 | O | 15 | Asvina ... | 5 | S | 19 | 5 | 6 | 18 | 49 | 51 | 1 | 5 | 20 | 159 | 45 | 30 | | |
| 13 g., 14 p. | Aippasi ... | 4 | O | 16 | 30 | 5 | N | 14 | Kartika ... | 6 | O | 18 | 36 | 56 | ... | 12 | 27 | 3 | 3 | 54 | 189 | 17 | 20 | | |
| B. S. 1301 | Kartigai ... | 6 | N | 15 | 30 | 7 | D | 14 | Margasira ... | 1 | N | 17 | 8 | 46 | 16 | 39 | 48 | 5 | 2 | 28 | 218 | 49 | 10 | | |
| (N.) Plavanga. | Margali ... | 1 | D | 15 | 29 | 1 | Ja | 12 | Pausha ... | 2 | D | 16 | 40 | 37 | ... | 13 | 8 | 7 | 1 | 1 | 248 | 21 | 0 | | |
| **A. D. 1896** | Tai ... | 2 | Ja | 13 | 29 | 2 | F | 10 | Magha ... | 4 | Ja | 15 | 12 | 27 | 14 | 52 | 22 | 8 | 59 | 35 | 277 | 52 | 51 | | |
| H. 1314 | Masi ... | 3 | F | 11 | 30 | 4 | Mr | 11 | Phalguna ... | 5 | F | 13 | 44 | 17 | ...○ | 36 | 19 | 10 | 58 | 8 | 307 | 24 | 41 | | |
| June 12 | Panguni ... | 5 | Mr | 12 | 30 | 6 | Ap | 10 | Chaitra ... | 7 | Mr | 14 | 16 | 7 | ... | 22 | 13 | 12 | 56 | 42 | 336 | 56 | 31 | | |
| **K. Y. 4997** | Chittirai ... | 7 | Ap | 11 | 31 | 2 | My | 11 | Vaisakha ... | 1 | Ap | 12 | 47 | 57 | 13 | 6 | 31 | 14 | 55 | 16 | 1 | 12 | 50 | 0 day, 5 ghat., 27 palas. | 0 „ 10 „ 8 „ |
| Sak. 1818 | Vaikasi ... | 3 | My | 12 | 32 | 6 | Je | 12 | Jyeshta ... | 3 | My | 12 | 19 | 47 | ... | 45 | 51 | 16 | 53 | 49 | 30 | 44 | 40 | | |
| | | | | | | | | | | 4 | Je | 10 | 51 | 37 | 11 | 18 | 59 | 18 | 52 | 23 | 60 | 16 | 30 | | |
| Vik. 1953 | Ani ... | 7 | Je | 13 | 31 | 2 | Jl | 13 | Ashada ... | 6 | Jl | 10 | 23 | 27 | ... | 46 | 21 | 20 | 50 | 56 | 89 | 48 | 20 | | |
| (S.) Durmukha | Adi ... | 3 | Jl | 14 | 32 | 6 | Au | 14 | Sravana ... | 7 | Au | 8 | 55 | 17 | 9●○ | 9 | 53 | 22 | 49 | 30 | 119 | 20 | 10 | | |
| K. 1071 (S. M.) | Avani ... | 7 | Au | 15 | 31 | 2 | S | 14 | Bhadrapada . | 2 | S | 7 | 27 | 7 | ... | 31 | 35 | 24 | 48 | 4 | 148 | 52 | 0 | | |
| K. 1071 (N. M.) | Purattasi . | 3 | S | 15 | 30 | 4 | O | 14 | Asvina ... | 3 | O | 6 | 58 | 57 | ... | 53 | 18 | 26 | 46 | 37 | 178 | 23 | 50 | | |
| 28 g., 45 p. | Aippasi ... | 5 | O | 15 | 30 | 6 | N | 13 | Kartika ... | 5 | N | 5 | 30 | 48 | ... | 16 | 35 | 1 | 11 | 54 | 207 | 55 | 40 | | |
| B. S. 1302 | Kartigai ... | 7 | N | 14 | 30 | 1 | D | 13 | Margasira ... | 7 | D | 5 | 2 | 38 | 4 | 42 | 14 | 3 | 10 | 28 | 237 | 27 | 30 | | |
| (N.) Kilaka | Margali ... | 2 | D | 14 | 29 | 2 | Ja | 11 | Pausha ... | 1 | Ja | 3 | 34 | 28 | ... | 13 | 3 | 5 | 9 | 1 | 266 | 59 | 20 | | |
| **A. D. 1897** | Tai ... | 3 | Ja | 12 | 30 | 4 | F | 10 | Magha ... | 3 | F | 2 | 6 | 18 | 1 | 47 | 22 | 7 | 7 | 35 | 296 | 31 | 11 | | |
| H. 1315 | Masi ... | 5 | F | 11 | 29 | 5 | Mr | 11 | Phalguna ... | 4 | Mr | 3 | 38 | 8 | ... | 25 | 22 | 9 | 6 | 9 | 326 | 3 | 1 | | |
| June 2 | Panguni ... | 6 | Mr | 12 | 31 | 1 | Ap | 11 | Chaitra ... | 6 | Ap | 2 | 9 | 58 | ... | 6 | 4 | 11 | 4 | 42 | 355 | 34 | 51 | | |
| **K. Y. 4998** | Chittirai ... | 2 | Ap | 12 | 31 | 4 | My | 12 | Vaisakha ... | 7 | My | 1 | 41 | 48 | ... | 46 | 56 | 13 | 3 | 16 | 19 | 51 | 10 | 1 day, 29 ghat., 6 palas. | 2 „ 45 „ 48 „ |
| Sak. 1819 | Vaikasi ... | 5 | My | 13 | 31 | 7 | Je | 12 | Jyeshta ... | 2 | My | 31 | 13 | 38 | ... | 27 | 23 | 15 | 1 | 48 | 49 | 23 | 0 | | |
| Vik. 1954 | Ani ... | 1 | Je | 13 | 32 | 4 | Jl | 14 | Ashada ... | 3 | Je | 29 | 45 | 29 | 30 | 3 | 39 | 17 | 0 | 22 | 78 | 54 | 50 | | |
| (S.) Hemalamba. | Adi ... | 5 | Jl | 15 | 31 | 7 | Au | 14 | Sravana ... | 5 | Jl | 29 | 17 | 19 | ... | 35 | 54 | 18 | 58 | 55 | 108 | 26 | 40 | | |
| K. 1072 (S. M.) | Avani ... | 1 | Au | 15 | 31 | 3 | S | 14 | Bhadrapada . | 6 | Au | 27 | 49 | 9 | 28 | 4 | 0 | 20 | 57 | 29 | 137 | 58 | 30 | | |
| K. 1072 (N. M.) | Purattasi . | 4 | S | 15 | 31 | 6 | O | 15 | Asvina ... | 1 | S | 26 | 20 | 59 | ... | 31 | 29 | 22 | 56 | 2 | 167 | 30 | 20 | | |
| 44 g., 16 p. | Aippasi ... | 7 | O | 16 | 30 | 1 | N | 14 | Kartika ... | 2 | O | 25 | 52 | 49 | ... | 56 | 59 | 24 | 54 | 36 | 197 | 2 | 10 | | |
| B. S. 1303 | Kartigai ... | 2 | N | 15 | 29 | 2 | D | 13 | Margasira ... | 4 | N | 24 | 24 | 39 | ... | 22 | 20 | 26 | 53 | 10 | 226 | 34 | 0 | | |
| (N.) Saumya | Margali ... | 3 | D | 14 | 29 | 3 | Ja | 11 | Pausha ... | 5 | D | 23 | 56 | 29 | ...○ | 48 | 30 | 1 | 18 | 27 | 256 | 5 | 50 | | |
| **A. D. 1898** | Tai ... | 4 | Ja | 12 | 30 | 5 | F | 10 | Magha ... | 7 | Ja | 22 | 28 | 19 | ...● | 16 | 14 | 3 | 17 | 0 | 285 | 37 | 40 | | |
| H. 1316 | Masi ... | 6 | F | 11 | 30 | 7 | Mr | 12 | Phalguna ... | 2 | F | 21 | 0 | 9 | 20 | 45 | 51 | 5 | 15 | 34 | 315 | 9 | 31 | | |
| May 22 | Panguni ... | 1 | Mr | 13 | 30 | 2 | Ap | 11 | Chaitra ... | 3 | Mr | 22 | 32 | 0 | ... | 18 | 18 | 7 | 14 | 6 | 344 | 41 | 21 | | |
| **K. Y. 4999** | Chittirai ... | 3 | Ap | 12 | 31 | 5 | My | 12 | Vaisakha ... | 5 | Ap | 21 | 3 | 50 | 20 | 50 | 2 | 9 | 12 | 41 | 8 | 57 | 40 | 0 day, 40 ghat., 13 palas. | 1 „ 14 „ 50 „ |
| Sak. 1820 | Vaikasi ... | 6 | My | 13 | 31 | 1 | Je | 12 | Jyeshta ... | 6 | My | 20 | 35 | 40 | ... | 28 | 27 | 11 | 11 | 15 | 38 | 29 | 30 | | |
| Vik. 1955 | Ani ... | 2 | Je | 13 | 32 | 5 | Jl | 14 | Ashada ... | 1 | Je | 19 | 7 | 30 | ...○ | 6 | 13 | 13 | 9 | 48 | 68 | 1 | 20 | | |
| (S.) Vilamba | Adi ... | 6 | Jl | 15 | 31 | 1 | Au | 14 | Sravana ... | 2 | Jl | 18 | 39 | 20 | ... | 44 | 14 | 15 | 8 | 22 | 97 | 33 | 10 | | |
| K. 1073 (S. M.) | Avani ... | 2 | Au | 15 | 31 | 4 | S | 14 | Bhadrapada . | 4 | Au | 17 | 11 | 10 | ... | 21 | 11 | 17 | 6 | 55 | 127 | 5 | 0 | | |
| K. 1073 (N. M.) | Purattasi . | 5 | S | 15 | 31 | 7 | O | 15 | Asvina ... | 5 | S | 15 | 43 | 0 | ... | 56 | 16 | 19 | 5 | 29 | 156 | 36 | 50 | | |
| | | | | | | | | | | 7 | O | 15 | 14 | 50 | ... | 27 | 14 | 21 | 4 | 3 | 186 | 8 | 40 | | |
| 59 g., 47 p. | Aippasi ... | 1 | O | 16 | 30 | 2 | N | 14 | Kartika ... | 1 | N | 13 | 46 | 40 | ... | 59 | 32 | 23 | 2 | 36 | 215 | 40 | 30 | | |
| B. S. 1304 | Kartigai ... | 3 | N | 15 | 29 | 3 | D | 13 | Margasira ... | 3 | D | 13 | 18 | 31 | ...○ | 28 | 7 | 25 | 1 | 10 | 245 | 12 | 20 | | |
| (N.) Sadharana | Margali ... | 4 | D | 14 | 30 | 5 | Ja | 12 | Pausha ... | 4 | Ja | 11 | 50 | 21 | ... | 55 | 15 | 26 | 59 | 43 | 274 | 44 | 10 | | |
| **A. D. 1899** | Tai ... | 6 | Ja | 13 | 29 | 6 | F | 10 | Magha ... | 6 | F | 10 | 22 | 11 | ... | 21 | 18 | 1 | 25 | 0 | 304 | 16 | 1 | | |
| H. 1317 | Masi ... | 7 | F | 11 | 30 | 1 | Mr | 12 | Phalguna ... | 7 | Mr | 11 | 54 | 1 | ... | 47 | 4 | 3 | 23 | 34 | 333 | 47 | 51 | | |
| May 12 | Panguni ... | 2 | Mr | 13 | 30 | 3 | Ap | 11 | Chaitra ... | 2 | Ap | 10 | 25 | 51 | ... | 13 | 25 | 5 | 22 | 8 | 363 | 19 | 41 | | |
| **K. Y. 5000.** | Chittirai ... | 4 | Ap | 12 | 31 | 6 | My | 12 | Vaisakha ... | 3 | My | 9 | 57 | 41 | ... | 41 | 11 | 7 | 20 | 41 | 27 | 36 | 0 | 2 days, 3 ghat., 52 palas. | 3 „ 50 „ 30 „ |
| Sak. 1821 | Vaikasi ... | 7 | My | 13 | 32 | 3 | Je | 13 | Jyeshta ... | 5 | Je | 8 | 29 | 31 | ...○ | 11 | 54 | 9 | 19 | 15 | 57 | 7 | 50 | | |
| Vik. 1956 | Ani ... | 4 | Je | 14 | 31 | 6 | Jl | 14 | Ashada ... | 7 | Jl | 8 | 1 | 21 | 7 | 45 | 42 | 11 | 17 | 48 | 86 | 39 | 40 | | |
| (S.) Vikari | Adi ... | 7 | Jl | 15 | 32 | 3 | Au | 15 | Sravana ... | 1 | Au | 6 | 33 | 12 | ... | 23 | 11 | 13 | 16 | 22 | 116 | 11 | 80 | | |
| K. 1074 (S. M.) | Avani ... | 4 | Au | 16 | 31 | 6 | S | 15 | Bhadrapada . | 3 | S | 5 | 5 | 2 | ... | 3 | 24 | 15 | 14 | 56 | 145 | 43 | 20 | | |
| K. 1074 (N. M.) | Purattasi . | 7 | S | 16 | 30 | 1 | O | 15 | Asvina ... | 4 | O | 4 | 36 | 52 | ... | 44 | 13 | 17 | 13 | 29 | 175 | 15 | 10 | | |
| 15 g., 19 p. | Aippasi ... | 2 | O | 16 | 30 | 3 | N | 14 | Kartika ... | 6 | N | 3 | 8 | 42 | ... | 23 | 28 | 19 | 12 | 3 | 204 | 47 | 0 | | |
| B. S. 1305 | Kartigai ... | 4 | N | 15 | 30 | 5 | D | 14 | Margasira ... | 7 | D | 2 | 40 | 32 | ...○ | 59 | 47 | 21 | 10 | 36 | 234 | 18 | 50 | | |
| (N.) Virodhakrit. | Margali ... | 6 | D | 15 | 29 | 6 | Ja | 12 | Pausha ... | 2 | Ja | 1 | 12 | 22 | ... | 32 | 27 | 23 | 9 | 10 | 263 | 50 | 40 | | |
| **A. D. 1900** | Tai ... | 7 | Ja | 13 | 29 | 7 | F | 10 | Magha ... | 3 | Ja | 30 | 44 | 12 | 31 | 1 | 20 | 25 | 7 | 44 | 293 | 22 | 30 | | |
| H. 1318 | Masi ... | 1 | F | 11 | 30 | 2 | Mr | 12 | Phalguna ... | 5 | Mr | 1 | 16 | 3 | ... | 26 | 49 | 27 | 6 | 17 | 322 | 54 | 20 | | |
| May 1 | Panguni ... | 3 | Mr | 13 | 31 | 5 | Ap | 12 | Chaitra ... | 6 | Mr | 30 | 47 | 53 | ... | 49 | 50 | 1 | 31 | 34 | 352 | 26 | 10 | | |

# TABLE XII.—General Ephemeris A.D. 1900—A.D. 1905.

| Commencement of years in different eras. | Tamil month. | Tamil Month Beginning Week-day | Month | Day | Last Day | End Week-day | Month | Day | Lunar month. | New Moon Week-day | Month | Ending moment of New Moon Tithi. Mean Day (mean) | Ghatikas | Palas | Actual Day (actual) | Ghatikas | Palas | ☾'s Anom. Day | Ghatikas | Palas | ☉'s Anom. Day | Ghatikas | Palas | Deduct for Nakshatras | Deduct for Yogas |
|---|---|---|---|---|---|---|---|---|---|---|---|---|---|---|---|---|---|---|---|---|---|---|---|---|---|
| **K. Y. 5001** | Chittirai ... | 6 | Ap | 13 | 30 | 7 | My | 12 | Vaisakha ... | 1 | Ap | 29 | 19 | 43 | ... | 11 | 53 | 3 | 30 | 8 | 16 | 42 | 29 | 1 day, 14 ghat., 59 palas. | 2 „ 19 „ 32 „ |
| Sak. 1822 | Vaikasi ... | 1 | My | 13 | 32 | 4 | Je | 13 | Jyeshta ... | 2 | My | 28 | 51 | 33 | ... ●○ | 34 | 30 | 5 | 28 | 42 | 46 | 14 | 19 | | |
| Vik. 1957 | Ani ... | 5 | Je | 14 | 31 | 7 | Jl | 14 | Ashada ... | 4 | Je | 27 | 23 | 23 | 26 | 59 | 31 | 7 | 27 | 15 | 75 | 46 | 9 | | |
| (S.) Sarvari | Adi ... | 1 | Jl | 15 | 32 | 4 | Au | 15 | Sravana ... | 5 | Jl | 26 | 55 | 13 | ... | 28 | 45 | 9 | 25 | 49 | 105 | 18 | 0 | | |
| K. 1075 (S. M.) | Avani ... | 5 | Au | 16 | 31 | 7 | S | 15 | Bhadrapada . | 7 | Au | 25 | 27 | 3 | ... | 3 | 49 | 11 | 24 | 22 | 134 | 49 | 50 | | |
| K. 1075 (N. M.) | Purattasi ... | 1 | S | 16 | 30 | 2 | O | 15 | Asvina ... | 1 | S | 23 | 58 | 58 | ... | 44 | 43 | 13 | 22 | 56 | 164 | 21 | 40 | | |
| 30 g., 50 p. | Aippasi ... | 3 | O | 16 | 30 | 4 | N | 14 | Kartika ... | 3 | O | 23 | 30 | 43 | ... | 31 | 30 | 15 | 21 | 30 | 193 | 53 | 30 | | |
| B. S. 1306 | Kartigai ... | 5 | N | 15 | 30 | 6 | D | 14 | Margasira ... | 5 | N | 22 | 2 | 33 | ... ● | 14 | 26 | 17 | 20 | 8 | 223 | 25 | 20 | | |
| (N.) Paridhavin. | Margali ... | 7 | D | 15 | 29 | 7 | Ja | 12 | Pausha ... | 6 | D | 21 | 34 | 24 | ... | 56 | 39 | 19 | 18 | 37 | 252 | 57 | 10 | | |
| **A. D. 1901** | Tai ... | 1 | Ja | 13 | 30 | 2 | F | 11 | Magha ... | 1 | Ja | 20 | 6 | 14 | ... | 33 | 54 | 21 | 17 | 10 | 282 | 29 | 0 | | |
| H. 1319 | Masi ... | 3 | F | 12 | 29 | 3 | Mr | 12 | Phalguna ... | 2 | F. | 18 | 38 | 4 | 19 | 5 | 16 | 23 | 15 | 44 | 312 | 0 | 50 | | |
| Ap. 20 | Panguni ... | 4 | Mr | 13 | 31 | 6 | Ap | 12 | Chaitra ... | 4 | Mr | 20 | 9 | 54 | ... | 31 | 13 | 25 | 14 | 18 | 341 | 32 | 40 | | |
| **K. Y. 5002** | Chittrai ... | 7 | Ap | 13 | 31 | 2 | My | 13 | Vaisakha ... | 5 | Ap | 18 | 41 | 44 | ... | 52 | 54 | 27 | 12 | 51 | 5 | 48 | 59 | 0 day, 26 ghat., 6 palas. | 0 „ 48 „ 35 „ |
| Sak. 1823 | Vaikasi ... | 3 | My | 14 | 31 | 5 | Je | 13 | Jyeshta ... | 7 | My | 18 | 13 | 34 | ... ● | 12 | 7 | 1 | 38 | 8 | 35 | 20 | 49 | | |
| Vik.1958 | Ani ... | 6 | Je | 14 | 32 | 2 | Jl | 15 | Ashada ... | 1 | Je | 16 | 45 | 24 | ... | 29 | 5 | 3 | 36 | 42 | 64 | 52 | 39 | | |
| (S.) Plava. | Adi ... | 3 | Jl | 16 | 31 | 5 | Au | 15 | Sravana ... | 3 | Jl | 16 | 17 | 14 | 15 | 51 | 40 | 5 | 35 | 15 | 94 | 24 | 29 | | |
| | | | | | | | | | | 4 | Au | 14 | 49 | 5 | ... | 16 | 58 | 7 | 33 | 49 | 123 | 56 | 19 | | |
| K. 1076 (S. M.) | Avani ... | 6 | Au | 16 | 31 | 1 | S | 15 | Bhadrapada . | 6 | S | 13 | 20 | 55 | 12 | 48 | 47 | 9 | 32 | 23 | 153 | 28 | 9 | | |
| K. 1076 (N. M.) | Purattasi . | 2 | S | 16 | 31 | 4 | O | 16 | Asvina ... | 7 | O | 12 | 52 | 45 | ... ○ | 28 | 4 | 11 | 30 | 56 | 183 | 0 | 0 | | |
| 16 g., 21 p. | Aippasi ... | 5 | O | 17 | 30 | 6 | N | 15 | Kartika ... | 2 | N | 11 | 24 | 35 | ... ● | 13 | 38 | 13 | 29 | 30 | 212 | 31 | 50 | | |
| B. S. 1307. | Kartigai ... | 7 | N | 16 | 29 | 7 | D | 14 | Margasira . | 3 | D | 10 | 56 | 25 | 11 | 2 | 24 | 15 | 28 | 3 | 242 | 3 | 40 | | |
| (N.) Pramadin | Margali ... | 1 | D | 15 | 29 | 1 | Ja | 12 | Pausha ... | 5 | Ja | 9 | 28 | 15 | ... | 49 | 17 | 17 | 26 | 37 | 271 | 35 | 30 | | |
| **A. D. 1902** | Tai ... | 2 | Ja | 13 | 30 | 3 | F | 11 | Magha ... | 7 | F | 8 | 0 | 5 | ... | 31 | 6 | 19 | 25 | 11 | 301 | 7 | 20 | | |
| H. 1320 | Masi ... | 4 | F | 12 | 30 | 5 | Mr | 13 | Phalguna ... | 1 | Mr | 9 | 31 | 55 | 10 | 5 | 34 | 21 | 23 | 44 | 330 | 39 | 10 | | |
| Ap. 10. | Panguni ... | 6 | Mr | 14 | 30 | 7 | Ap | 12 | Chaitra ... | 3 | Ap | 8 | 3 | 45 | ... ○ | 33 | 3 | 23 | 22 | 18 | 360 | 11 | 0 | | |
| **K. Y. 5003** | Chittirai ... | 1 | Ap | 13 | 31 | 3 | My | 13 | Vaisakha ... | 4 | My | 7 | 35 | 36 | ... | 54 | 54 | 25 | 20 | 51 | 24 | 27 | 19 | 1 day, 49 ghat., 46 palas. | 3 „ 24 „ 14 „ |
| Sak. 1824 | Vaikasi ... | 4 | My | 14 | 31 | 6 | Je | 13 | Jyeshta ... | 6 | Je | 6 | 7 | 26 | ... | 13 | 1 | 27 | 19 | 25 | 53 | 59 | 9 | | |
| Vik. 1959. | Ani ... | 7 | Je | 14 | 32 | 3 | Jl | 15 | Ashada ... | 7 | Jl | 5 | 39 | 16 | ... | 29 | 48 | 1 | 44 | 42 | 83 | 30 | 59 | | |
| (S.) Subhakrit | Adi ... | 4 | Jl | 16 | 31 | 6 | Au | 15 | Sravana ... | 2 | Au | 4 | 11 | 6 | 3 | 48 | 8 | 3 | 43 | 16 | 113 | 2 | 50 | | |
| K. 1077 (S. M.) | Avani ... | 7 | Au | 16 | 31 | 2 | S | 15 | Bhadrapada. | 3 | S | 2 | 42 | 56 | ... | 10 | 31 | 5 | 41 | 49 | 142 | 34 | 40 | | |
| K. 1077 (N. M.) | Purattasi . | 3 | S | 16 | 31 | 5 | O | 16 | Asvina ... | 5 | O | 2 | 14 | 46 | 1 ○ | 39 | 18 | 7 | 40 | 23 | 172 | 6 | 30 | | |
| 1 g., 52 p. | Aippasi ... | 6 | O | 17 | 30 | 7 | N | 15 | Kartika ... | 6 | O | 31 | 46 | 36 | ... ● | 16 | 38 | 9 | 38 | 56 | 201 | 38 | 20 | | |
| B. S. 1308 | Kartigai ... | 1 | N | 16 | 29 | 1 | D | 14 | Margasira ... | 1 | N | 30 | 18 | 26 | 29 | 59 | 33 | 11 | 37 | 30 | 231 | 10 | 10 | | |
| (N.) Ananda | Margali ... | 2 | D | 15 | 30 | 3 | Ja | 13 | Pausha ... | 2 | D | 29 | 50 | 16 | ... | 48 | 38 | 13 | 36 | 4 | 260 | 42 | 0 | | |
| **A. D. 1903** | Tai ... | 4 | Ja | 14 | 29 | 4 | F | 11 | Magha ... | 4 | Ja | 28 | 22 | 7 | ... | 38 | 15 | 15 | 34 | 37 | 290 | 13 | 50 | | |
| H. 1321 | Masi ... | 5 | F | 12 | 30 | 6 | Mr | 13 | Phalguna ... | 5 | F | 26 | 53 | 57 | 27 | 23 | 18 | 17 | 33 | 11 | 319 | 45 | 40 | | |
| Mr. 30 | Panguni ... | 7 | Mr | 14 | 30 | 1 | Ap | 12 | Chaitra ... | 7 | Mr | 28 | 25 | 47 | 29 ●○ | 1 | 2 | 19 | 31 | 44 | 849 | 17 | 30 | | |
| **K. Y. 5004** | Chittirai ... | 2 | Ap | 13 | 31 | 4 | My | 13 | Vaisakha ... | 1 | Ap | 26 | 57 | 37 | 27 | 31 | 3 | 21 | 30 | 18 | 13 | 33 | 49 | 1 day, 0 ghat., 53 palas. | 1 „ 53 „ 17 „ |
| Sak. 1825 | Vaikasi ... | 5 | My | 14 | 32 | 1 | Je | 14 | Jyeshta ... | 3 | My | 26 | 29 | 27 | ... | 54 | 27 | 23 | 28 | 52 | 43 | 5 | 39 | | |
| Vik. 1960 | Ani ... | 2 | Je | 15 | 31 | 4 | Jl | 15 | Ashada ... | 5 | Je | 25 | 1 | 17 | ... | 13 | 17 | 25 | 27 | 25 | 72 | 37 | 29 | | |
| (S.) Sobhakrit | Adi ... | 5 | Jl | 16 | 32 | 1 | Au | 16 | Sravana ... | 6 | Jl | 24 | 33 | 7 | ... | 30 | 9 | 27 | 25 | 59 | 102 | 9 | 19 | | |
| K. 1078 (S. M.) | Avani ... | 2 | Au | 17 | 31 | 4 | S | 16 | Bhadrapada . | 1 | Au | 23 | 4 | 57 | 22 | 47 | 52 | 1 | 51 | 16 | 131 | 41 | 9 | | |
| K. 1078 (N. M.) | Purattasi . | 5 | S | 17 | 30 | 6 | O | 16 | Asvina ... | 2 | S | 21 | 36 | 48 | ●○ | 8 | 56 | 3 | 49 | 49 | 161 | 12 | 59 | | |
| 17 g., 24 p. | Aippasi ... | 7 | O | 17 | 30 | 1 | N | 15 | Kartika ... | 4 | O | 21 | 8 | 38 | 20 | 35 | 17 | 5 | 48 | 23 | 190 | 44 | 50 | | |
| B. S. 1309 | Kartigai ... | 2 | N | 16 | 30 | 3 | D | 15 | Margasira ... | 5 | N | 19 | 40 | 28 | ... | 8 | 30 | 7 | 46 | 57 | 220 | 16 | 40 | | |
| (N.) Rakshasa | Margali ... | 4 | D | 16 | 29 | 4 | Ja | 13 | Pausha ... | 7 | D | 19 | 12 | 18 | 18 | 48 | 54 | 9 | 45 | 30 | 249 | 48 | 30 | | |
| **A. D. 1904** | Tai ... | 5 | Ja | 14 | 29 | 5 | F | 11 | Magha ... | 1 | Ja | 17 | 44 | 8 | ... | 35 | 13 | 11 | 44 | 4 | 279 | 20 | 20 | | |
| H. 1322 | Masi ... | 6 | F | 12 | 30 | 7 | Mr | 12 | Phalguna ... | 3 | F | 16 | 15 | 58 | ... | 23 | 59 | 13 | 42 | 37 | 308 | 52 | 10 | | |
| Mr. 18 | Panguni ... | 1 | Mr | 13 | 31 | 3 | Ap | 12 | Chaitra ... | 4 | Mr | 16 | 47 | 48 | 17 ● | 10 | 28 | 15 | 41 | 11 | 338 | 24 | 0 | | |
| **K. Y. 5005** | Chittirai ... | 4 | Ap | 13 | 30 | 5 | My | 12 | Vaisakha ... | 6 | Ap | 15 | 19 | 38 | ... | 50 | 57 | 17 | 39 | 45 | 2 | 40 | 19 | 0 day, 12 ghat., 0 palas. | 0 „ 22 „ 19 „ |
| Sak. 1826 | Vaikasi ... | 6 | My | 13 | 32 | 2 | Je | 13 | Jyeshta ... | 7 | My | 14 | 51 | 28 | 15 | 24 | 6 | 19 | 38 | 18 | 32 | 12 | 9 | | |
| | | | | | | | | | | 2 | Je | 13 | 23 | 19 | ... | 50 | 25 | 21 | 36 | 52 | 61 | 43 | 59 | | |
| Vik. 1961 | Ani ... | 3 | Je | 14 | 31 | 5 | Jl | 14 | Ashada ... | 3 | Jl | 12 | 55 | 9 | 13 | 11 | 46 | 23 | 35 | 25 | 91 | 15 | 49 | | |
| (S.) Krodhi | Adi ... | 6 | Jl | 15 | 32 | 2 | Au | 15 | Sravana ... | 5 | Au | 11 | 26 | 59 | ... | 30 | 47 | 25 | 33 | 59 | 120 | 47 | 39 | | |
| K. 1079 (S. M.) | Avani ... | 3 | Au | 16 | 31 | 5 | S | 15 | Bhadrapada . | 6 | S | 9 | 58 | 49 | ... | 49 | 42 | 27 | 32 | 33 | 150 | 19 | 29 | | |
| K. 1079 (N. M.) | Purattasi . | 6 | S | 16 | 30 | 7 | O | 15 | Asvina ... | 1 | O | 9 | 30 | 39 | ... | 10 | 56 | 1 | 57 | 50 | 179 | 51 | 19 | | |
| 32g., 55 p. | Aippasi ... | 1 | O | 16 | 30 | 2 | N | 14 | Kartika ... | 3 | N | 8 | 2 | 29 | 7 | 36 | 0 | 3 | 56 | 23 | 209 | 23 | 10 | | |
| B. S. 1310 | Kartigai ... | 3 | N | 15 | 30 | 4 | D | 14 | Margasira ... | 4 | D | 7 | 34 | 19 | ... | 6 | 21 | 5 | 54 | 50 | 238 | 55 | 0 | | |
| (N.) Anala | Margali ... | 5 | D | 15 | 29 | 5 | Ja | 12 | Pausha ... | 6 | Ja | 6 | 6 | 9 | 5 | 42 | 34 | 7 | 53 | 30 | 268 | 26 | 50 | | |
| **A. D. 1905** | Tai ... | 6 | Ja | 13 | 30 | 7 | F | 11 | Magha ... | 7 | F | 4 | 37 | 59 | ○ | 24 | 1 | 9 | 52 | 4 | 297 | 58 | 40 | | |
| H. 1323 | Masi ... | 1 | F | 12 | 29 | 1 | Mr | 12 | Phalguna ... | 2 | Mr | 6 | 9 | 50 | ... | 8 | 45 | 11 | 50 | 38 | 327 | 30 | 30 | | |
| Mr. 8 | Panguni ... | 2 | Mr | 13 | 31 | 4 | Ap | 12 | Chaitra ... | 3 | Ap | 4 | 41 | 40 | ... | 52 | 44 | 13 | 49 | 11 | 357 | 2 | 20 | | |

# TABLE XII.—General Ephemeris A.D. 1905—A.D. 1910.

| Commencement of years in different eras. | Tamil month. | Tamil Month Beginning Week-day | Beginning Month | Beginning Day | Last Day | End Week-day | End Month | End Day | Lunar Month. | New Moon Ending moment of New Moon Tithi Week-day | Month | Mean Day (mean) | Mean Ghatikas | Mean Palas | Actual Day (actual) | Actual Ghatikas | Actual Palas | ☾'s Anom. Day | ☾'s Anom. Ghatikas | ☾'s Anom. Palas | ⊙'s Anom. Day | ⊙'s Anom. Ghatikas | ⊙'s Anom. Palas | Deduct for Nakshatras | Deduct for Yogas. |
|---|---|---|---|---|---|---|---|---|---|---|---|---|---|---|---|---|---|---|---|---|---|---|---|---|---|
| **K. Y. 5006** | Chittirai... | 5 | Ap | 13 | 31 | 7 | My | 13 | Vaisakha ... | 5 | My | 4 | 13 | 30 | ... | 37 | 5 | 15 | 47 | 45 | 21 | 18 | 39 | 1 day, 35 ghat., 39 palas. | 2 " 57 " 59 " |
| Sak. 1827 | Vaikasi ... | 1 | My | 14 | 31 | 3 | Je | 13 | Jyeshta ... | 6 | Je | 2 | 45 | 20 | 3 | 11 | 37 | 17 | 46 | 18 | 50 | 50 | 29 | | |
| Vik. 1962 | Ani ... | 4 | Je | 14 | 32 | 7 | Jl | 15 | Ashada ... | 1 | Jl | 2 | 17 | 10 | ... | 41 | 50 | 19 | 44 | 52 | 80 | 22 | 19 | | |
| (S.) Visvavasu | Adi ... | 1 | Jl | 16 | 31 | 3 | Au | 15 | Sravana ... | 2 | Jl | 31 | 49 | 0 | Au.1○ | 7 | 28 | 21 | 43 | 26 | 109 | 54 | 9 | | |
| K. 1080 (S.M.) | Avani ... | 4 | Au | 16 | 31 | 6 | S | 15 | Bhadrapada . | 4 | Au | 30 | 20 | 50 | ...● | 30 | 22 | 23 | 41 | 59 | 139 | 25 | 59 | | |
| K. 1080 (N.M.) | Purattasi . | 7 | S | 16 | 31 | 2 | O | 16 | Asvina ... | 5 | S | 28 | 52 | 40 | ... | 52 | 27 | 25 | 40 | 33 | 168 | 57 | 49 | | |
| 40 g., 26 p. | Aippasi ... | 3 | O | 17 | 30 | 4 | N | 15 | Kartika ... | 7 | O | 28 | 24 | 31 | ... | 14 | ... | 0 | 5 | 50 | 198 | 29 | 39 | | |
| B. S. 1311 | Kartigai ... | 5 | N | 16 | 29 | 5 | D | 14 | Margasira ... | 1 | N | 26 | 56 | 21 | ... | 40 | 11 | 2 | 4 | 24 | 228 | 1 | 29 | | |
| (N.) Pingala | Margali ... | 6 | D | 15 | 29 | 6 | Ja | 12 | Pausha ... | 3 | D | 26 | 28 | 11 | ... | 8 | 24 | 4 | 2 | 57 | 257 | 33 | 20 | | |
| **A. D. 1906** | Tai ... | 7 | Ja | 13 | 30 | 1 | F | 11 | Magha ... | 5 | Ja | 25 | 0 | 1 | 24○ | 40 | 22 | 6 | 1 | 31 | 287 | 5 | 10 | | |
| H. 1324 | Masi ... | 2 | F | 12 | 30 | 3 | Mr | 13 | Phalguna ... | 6 | F | 23 | 31 | 51 | ... | 16 | 3 | 8 | 0 | 4 | 316 | 36 | 59 | | |
| Feb. 25 | Panguni ... | 4 | Mr | 14 | 30 | 5 | Ap | 12 | Chaitra ... | 1 | Mr | 25 | 3 | 41 | 24 | 54 | 53 | 9 | 58 | 38 | 346 | 8 | 49 | | |
| **K. Y. 5007** | Chittirai... | 6 | Ap | 13 | 31 | 1 | My | 13 | Vaisakha ... | 2 | Ap | 23 | 35 | 31 | ... | 35 | 31 | 11 | 57 | 12 | 10 | 25 | 8 | 0 day, 46 ghat., 46 palas. | 1 " 27 " 1 " |
| Sak. 1828 | Vaikasi ... | 2 | My | 14 | 31 | 4 | Je | 13 | Jyeshta .. | 4 | My | 23 | 7 | 21 | ... | 14 | 55 | 13 | 55 | 45 | 39 | 56 | 59 | | |
| Vik. 1963 | Ani ... | 5 | Je | 14 | 32 | 1 | Jl | 15 | Ashada ... | 5 | Je | 21 | 39 | 12 | ... | 54 | 3 | 15 | 54 | 19 | 69 | 28 | 49 | | |
| (S.) Parabhava | Adi ... | 2 | Jl | 16 | 31 | 4 | Au | 15 | Sravana ... | 7 | Jl | 21 | 11 | 2 | ...○ | 28 | 30 | 17 | 52 | 52 | 99 | 0 | 39 | | |
| K. 1081 (S.M.) | Avani ... | 5 | Au | 16 | 32 | 1 | S | 16 | Bhadrapada . | 1 | Au | 19 | 42 | 52 | ...● | 59 | 34 | 19 | 51 | 26 | 128 | 32 | 29 | | |
| K. 1081 (N.M.) | Purattasi | 2 | S | 17 | 30 | 3 | O | 16 | Asvina ... | 3 | S | 18 | 14 | 42 | ... | 27 | 57 | 21 | 50 | 0 | 158 | 4 | 19 | | |
| 3 g., 58 p. | Aippasi ... | 4 | O | 17 | 30 | 5 | N | 15 | Kartika ... | 4 | O | 17 | 46 | 32 | ... | 54 | 33 | 23 | 48 | 33 | 187 | 36 | 9 | | |
| B. S. 1312 | Kartigai ... | 6 | N | 16 | 29 | 6 | D | 14 | Margasira ... | 6 | N | 16 | 18 | 22 | ... | 20 | 23 | 25 | 47 | 7 | 217 | 7 | 59 | | |
| (N.) Kalayukta | Margali ... | 7 | D | 15 | 30 | 1 | Ja | 13 | Pausha ... | 7 | D | 15 | 50 | 12 | ... | 46 | 19 | 0 | 12 | 24 | 246 | 39 | 49 | | |
| **A. D. 1907** | Tai ... | 2 | Ja | 14 | 29 | 2 | F | 11 | Magha ... | 2 | Ja | 14 | 22 | 2 | ..●○ | 13 | 13 | 2 | 10 | 57 | 276 | 11 | 39 | | |
| H. 1325 | Masi ... | 3 | F | 12 | 30 | 4 | Mr | 13 | Phalguna ... | 3 | F | 12 | 53 | 52 | ... | 41 | 30 | 4 | 9 | 31 | 305 | 43 | 29 | | |
| Feb. 14 | Panguni... | 5 | Mr | 14 | 30 | 6 | Ap | 12 | Chaitra ... | 5 | Mr | 14 | 25 | 43 | ... | 11 | 37 | 6 | 8 | 5 | 335 | 15 | 19 | | |
| | | | | | | | | | | 6 | Ap | 12 | 57 | 33 | ... | 48 | 55 | 8 | 6 | 38 | 364 | 47 | 10 | | |
| **K. Y. 5008** | Chittirai.. | 7 | Ap | 13 | 31 | 2 | My | 13 | Vaisakha ... | 1 | My | 12 | 29 | 23 | ... | 18 | 35 | 10 | 5 | 12 | 29 | 8 | 29 | 2 days, 10 ghat., 25 palas. | 4 " 2 " 41 " |
| Sak. 1829 | Vaikasi ... | 3 | My | 14 | 32 | 6 | Je | 14 | Jyeshta ... | 3 | Je | 11 | 1 | 13 | 10 | 55 | 14 | 12 | 3 | 45 | 58 | 35 | 19 | | |
| Vik. 1964 | Ani ... | 7 | Je | 15 | 31 | 2 | Jl | 15 | Ashada ... | 4 | Jl | 10 | 33 | 3 | ...○ | 33 | 7 | 14 | 2 | 19 | 88 | 7 | 9 | | |
| (S.) Plavanga | Adi ... | 3 | Jl | 16 | 32 | 6 | Au | 16 | Sravana ... | 6 | Au | 9 | 4 | 53 | ... | 11 | 2 | 16 | 0 | 53 | 117 | 38 | 59 | | |
| K. 1082 (S.M.) | Avani ... | 7 | Au | 17 | 31 | 2 | S | 16 | Bhadrapada . | 7 | S | 7 | 36 | 43 | ... | 47 | 49 | 17 | 59 | 26 | 147 | 10 | 49 | | |
| K. 1082 (N.M.) | Purattasi . | 3 | S | 17 | 30 | 4 | O | 16 | Asvina ... | 2 | O | 7 | 8 | 34 | ... | 22 | 34 | 19 | 58 | 0 | 176 | 42 | 39 | | |
| 19 g., 29 p. | Aippasi ... | 5 | O | 17 | 30 | 6 | N | 15 | Kartika ... | 3 | N | 5 | 40 | 24 | ... | 54 | 54 | 21 | 56 | 33 | 206 | 14 | 29 | | |
| B. S. 1313 | Kartigai... | 7 | N | 16 | 30 | 1 | D | 15 | Margasira ... | 5 | D | 5 | 12 | 14 | ... | 24 | 54 | 23 | 55 | 7 | 235 | 46 | 19 | | |
| (N.) Siddharthin. | Margali ... | 2 | D | 16 | 29 | 2 | Ja | 13 | Pausha ... | 6 | Ja | 3 | 44 | 4 | ... | 53 | 36 | 25 | 53 | 40 | 265 | 18 | 9 | | |
| **A. D. 1908** | Tai ... | 3 | Ja | 14 | 29 | 3 | F | 11 | Magha ... | 1 | F | 2 | 15 | 54 | ... | 19 | 22 | 0 | 18 | 58 | 294 | 49 | 59 | | |
| H. 1326 | Masi ... | 4 | F | 12 | 30 | 5 | Mr | 12 | Phalguna ... | 2 | Mr | 2 | 47 | 44 | ... | 44 | 47 | 2 | 17 | 31 | 324 | 21 | 50 | | |
| Feb. 4 | Panguni ... | 6 | Mr | 13 | 31 | 1 | Ap | 12 | Chaitra ... | 4 | Ap | 1 | 19 | 34 | ... | 10 | 9 | 4 | 16 | 5 | 353 | 53 | 40 | | |
| **K. Y. 5009** | Chittirai... | 2 | Ap | 13 | 31 | 4 | My | 13 | Vaisakha ... | 5 | Ap | 30 | 51 | 24 | ... | 36 | 8 | 6 | 14 | 38 | 18 | 9 | 59 | 1 day, 21 ghat., 32 palas. | 2 " 31 " 48 " |
| Sak. 1830 | Vaikasi ... | 5 | My | 14 | 31 | 7 | Je | 13 | Jyeshta ... | 7 | My | 30 | 23 | 14 | ... | 4 | 55 | 8 | 13 | 12 | 47 | 41 | 49 | | |
| Vik. 1965 | Ani ... | 1 | Je | 14 | 32 | 4 | Jl | 15 | Ashada ... | 1 | Je | 28 | 55 | 4 | ...● | 36 | 30 | 10 | 11 | 46 | 77 | 13 | 39 | | |
| (S.) Kilaka | Adi ... | 5 | Jl | 16 | 31 | 7 | Au | 15 | Sravana ... | 3 | Jl | 28 | 26 | 55 | ... | 12 | 3 | 12 | 10 | 19 | 106 | 45 | 29 | | |
| K. 1083 (S.M.) | Avani ... | 1 | Au | 16 | 31 | 3 | S | 15 | Bhadrapada . | 4 | Au | 26 | 58 | 45 | ... | 51 | 21 | 14 | 8 | 53 | 136 | 17 | 19 | | |
| K. 1083 (N.M.) | Purattasi | 4 | S | 16 | 30 | 5 | O | 15 | Asvina ... | 6 | S | 25 | 30 | 35 | ... | 32 | 46 | 16 | 7 | 26 | 165 | 49 | 9 | | |
| 50 g., 31 p. | Aippasi ... | 6 | O | 16 | 30 | 7 | N | 14 | Kartika ... | 1 | O | 25 | 2 | 25 | ... | 13 | 46 | 18 | 6 | 0 | 195 | 20 | 59 | | |
| B. S. 1314 | Kartigai ... | 1 | N | 15 | 30 | 2 | D | 14 | Margasira ... | 2 | N | 23 | 34 | 15 | ... | 52 | 21 | 20 | 4 | 34 | 224 | 52 | 49 | | |
| (N.) Raudra | Margali ... | 3 | D | 15 | 29 | 3 | Ja | 12 | Pausha ... | 4 | D | 23 | 6 | 5 | ...● | 27 | 26 | 22 | 3 | 7 | 254 | 24 | 39 | | |
| **A. D. 1909** | Tai ... | 4 | Ja | 13 | 30 | 5 | F | 11 | Magha ... | 5 | Ja | 21 | 37 | 55 | ... | 58 | 20 | 24 | 1 | 41 | 283 | 56 | 29 | | |
| H. 1327 | Masi ... | 6 | F | 12 | 30 | 7 | Mr | 13 | Phalguna ... | 7 | F | 20 | 9 | 45 | ... | 25 | 11 | 26 | 0 | 14 | 313 | 28 | 20 | | |
| Ja. 23 | Panguni... | 1 | Mr | 14 | 30 | 2 | Ap | 12 | Chaitra ... | 1 | Mr | 21 | 41 | 36 | ... | 48 | 57 | 0 | 25 | 31 | 343 | 0 | 10 | | |
| **K. Y. 5010** | Chittirai... | 3 | Ap | 13 | 31 | 5 | My | 13 | Vaisakha ... | 3 | Ap | 20 | 13 | 26 | ... | 10 | 48 | 2 | 24 | 5 | 7 | 16 | 29 | 0 day, 32 ghat., 39 palas. | 1 " 0 " 45 " |
| Sak. 1831 | Vaikasi ... | 6 | My | 14 | 31 | 1 | Je | 13 | Jyeshta ... | 4 | My | 19 | 45 | 16 | ...○ | 32 | 18 | 4 | 22 | 39 | 36 | 48 | 19 | | |
| Vik. 1966 | Ani ... | 2 | Je | 14 | 32 | 5 | Jl | 15 | Ashada ... | 6 | Je | 18 | 17 | 6 | 17● | 55 | 23 | 6 | 21 | 12 | 66 | 20 | 9 | | |
| (S.) Saumya | Adi ... | 6 | Jl | 16 | 31 | 1 | Au | 15 | Sravana ... | 7 | Jl | 17 | 48 | 56 | ... | 21 | 55 | 8 | 19 | 45 | 95 | 51 | 59 | | |
| | | | | | | | | | | 2 | Au | 16 | 20 | 46 | 15 | 53 | 57 | 10 | 18 | 19 | 125 | 23 | 49 | | |
| K. 1084 (S.M.) | Avani ... | 2 | Au | 16 | 31 | 4 | S | 15 | Bhadrapada . | 3 | S | 14 | 52 | 36 | ... | 32 | 6 | 12 | 16 | 53 | 154 | 55 | 39 | | |
| K. 1084 (N.M.) | Purattasi | 5 | S | 16 | 31 | 7 | O | 16 | Asvina ... | 5 | O | 14 | 24 | 26 | ... | 15 | 54 | 14 | 15 | 27 | 184 | 27 | 29 | | |
| 35 g., 0 p. | Aippasi ... | 1 | O | 17 | 30 | 2 | N | 15 | Kartika ... | 6 | N | 12 | 56 | 17 | 13○ | 1 | 50 | 16 | 14 | 0 | 218 | 59 | 19 | | |
| B. S. 1315 | Kartigai ... | 3 | N | 16 | 29 | 3 | D | 14 | Margasira ... | 1 | D | 12 | 28 | 7 | ... | 46 | 22 | 18 | 12 | 34 | 243 | 31 | 9 | | |
| (N.) Durmati | Margali ... | 4 | D | 15 | 29 | 4 | Ja | 12 | Pausha ... | 3 | Ja | 11 | 0 | 0 | ... | 25 | 37 | 20 | 11 | 7 | 273 | 2 | 59 | | |
| **A. D. 1910** | Tai ... | 5 | Ja | 13 | 30 | 6 | F | 11 | Magha ... | 4 | F | 9 | 31 | 47 | 10 | 1 | 12 | 22 | 9 | 41 | 302 | 34 | 50 | | |
| H. 1328 | Masi ... | 7 | F | 12 | 30 | 1 | Mr | 13 | Phalguna ... | 6 | Mr | 11 | 3 | 37 | ... | 29 | 19 | 24 | 8 | 15 | 332 | 6 | 40 | | |
| Ja. 13 | Panguni... | 2 | Mr | 14 | 30 | 3 | Ap | 12 | Chaitra ... | 7 | Ap | 9 | 35 | 27 | ... | 52 | 37 | 26 | 6 | 48 | 361 | 38 | 30 | | |

# TABLE XII.—General Ephemeris A.D. 1910—A.D. 1915.

| Commencement of years in different eras. | Tamil month. | Tamil Month Beginning Week-day | Beginning Month | Beginning Day | Last Day | Tamil Month End Week-day | End Month | End Day | Lunar month. | New Moon Week-day | Month | Ending moment of New Moon Tithi Mean Day (mean) | Mean Ghatikas | Mean Palas | Actual Day (actual) | Actual Ghatikas | Actual Palas | ☾'s Anom. Day | ☾'s Anom. Ghatikas | ☾'s Anom. Palas | ☉'s Anom. Day | ☉'s Anom. Ghatikas | ☉'s Anom. Palas | Deduct for Nakshatras | Deduct for Yogas. |
|---|---|---|---|---|---|---|---|---|---|---|---|---|---|---|---|---|---|---|---|---|---|---|---|---|---|
| **K. Y. 5011** | Chittirai... | 4 | Ap | 13 | 31 | 6 | My | 13 | Vaisakha ... | 2 | My | 9 | 7 | 17 | ...○ | 12 | 24 | 0 | 32 | 6 | 25 | 54 | 48 | 1 day, 56 ghat., 18 palas. | 3 „ 36 „ 25 „ |
| Sak. 1832 | Vaikasi ... | 7 | My | 14 | 31 | 2 | Je | 13 | Jyeshta ... | 3 | Je | 7 | 39 | 7 | ... | 30 | 45 | 2 | 30 | 40 | 55 | 26 | 38 | | |
| Vik. 1967 | Ani ... | 3 | Je | 14 | 32 | 6 | Jl | 15 | Ashada ... | 5 | Jl | 7 | 10 | 57 | 6 | 49 | 55 | 4 | 29 | 13 | 84 | 58 | 29 | | |
| (S.)Sadharana. | Adi ... | 7 | Jl | 16 | 32 | 3 | Au | 16 | Sravana ... | 6 | Au | 5 | 42 | 47 | ... | 12 | 30 | 6 | 27 | 47 | 114 | 30 | 19 | | |
| K. 1085 (S.M.) | Avani ... | 4 | Au | 17 | 31 | 6 | S | 16 | Bhadrapada. | 1 | S | 4 | 14 | 38 | 3 | 40 | 56 | 8 | 26 | 20 | 144 | 2 | 9 | | |
| K. 1085 (N. M) | Purattasi . | 7 | S | 17 | 30 | 1 | O | 16 | Asvina ... | 2 | O | 3 | 46 | 28 | ... | 16 | 42 | 10 | 24 | 54 | 173 | 33 | 59 | | |
| 6 g., 3 p. | Aippasi ... | 2 | O | 17 | 30 | 3 | N | 15 | Kartika ... | 4 | N | 2 | 18 | 18 | 1●○ | 59 | 42 | 12 | 23 | 28 | 203 | 5 | 49 | | |
| B. S. 1316 | Kartigai . | 4 | N | 16 | 29 | 4 | D | 14 | Margasira . | 5 | D | 1 | 50 | 8 | ... | 47 | 48 | 14 | 22 | 1 | 232 | 37 | 39 | | |
| (N.) Dundubhi | Margali ... | 5 | D | 15 | 30 | 6 | Ja | 13 | Pausha ... | 7 | D | 31 | 21 | 58 | ... | 36 | 35 | 16 | 20 | 35 | 262 | 9 | 29 | | |
| **A.D. 1911** | Tai ... | 7 | Ja | 14 | 29 | 7 | F | 11 | Magha ... | 1 | Ja | 29 | 53 | 48 | 30 | 21 | 25 | 18 | 19 | 8 | 291 | 41 | 19 | | |
| H. 1329. Ja. 2 | Masi ... | 1 | F | 12 | 30 | 2 | Mr | 13 | Phalguna ... | 3 | F | 28 | 25 | 38 | ... | 59 | 27 | 20 | 17 | 41 | 321 | 13 | 9 | | |
| H. 1330. D, 22 | Panguni ... | 3 | Mr | 14 | 30 | 4 | Ap | 12 | Chaitra ... | 4 | Mr | 29 | 57 | 29 | 30 | 30 | 8 | 22 | 16 | 16 | 350 | 45 | 0 | | |
| **K. Y. 5012** | Chittirai... | 5 | Ap | 13 | 31 | 7 | My | 13 | Vaisakha ... | 6 | Ap | 28 | 29 | 19 | ... | 54 | 20 | 24 | 14 | 49 | 15 | 1 | 18 | 1 day, 7 ghat., 25 palas. | 2 „ 5 „ 27 „ |
| Sak. 1833 | Vaikasi ... | 1 | My | 14 | 32 | 4 | Je | 14 | Jyeshta ... | 1 | My | 28 | 1 | 9 | ... | 13 | 46 | 26 | 13 | 23 | 44 | 33 | 8 | | |
| Vik. 1968 | Ani ... | 5 | Je | 15 | 31 | 7 | Jl | 15 | Ashada ... | 2 | Je | 26 | 32 | 59 | ... | 30 | 43 | 0 | 38 | 40 | 74 | 4 | 59 | | |
| (S). Virodhakrit. | Adi ... | 1 | Jl | 16 | 32 | 4 | Au | 16 | Sravana ... | 4 | Jl | 26 | 4 | 49 | 25 | 47 | 44 | 2 | 37 | 13 | 103 | 36 | 49 | | |
| K. 1086 (S.M.) | Avani ... | 5 | Au | 17 | 31 | 7 | S | 16 | Bhadrapada. | 5 | Au | 24 | 36 | 39 | ... | 8 | 4 | 4 | 35 | 47 | 133 | 8 | 39 | | |
| K. 1086 (N.M.) | Purattasi . | 1 | S | 17 | 30 | 2 | O | 16 | Asvina ... | 7 | S | 23 | 8 | 29 | 22 | 33 | 35 | 6 | 34 | 21 | 162 | 40 | 29 | | |
| 21 g., 34 p. | Aippasi ... | 3 | O | 17 | 30 | 4 | N | 15 | Kartika ... | 1 | O | 22 | 40 | 19 | ...● | 6 | 13 | 8 | 32 | 54 | 192 | 12 | 19 | | |
| B. S. 1317 | Kartigai... | 5 | N | 16 | 30 | 6 | D | 15 | Margasira ... | 3 | N | 21 | 12 | 9 | 20 | 46 | 41 | 10 | 31 | 28 | 221 | 44 | 9 | | |
| (N.)Rudhirodgarin | Margali ... | 7 | D | 16 | 29 | 7 | Ja | 13 | Pausha ... | 4 | D | 20 | 44 | 0 | ... | 33 | 38 | 12 | 30 | 1 | 251 | 15 | 59 | | |
| **A. D. 1912** | Tai ... | 1 | Ja | 14 | 29 | 1 | F | 11 | Magha ... | 6 | Ja | 19 | 15 | 50 | ... | 23 | 46 | 14 | 28 | 35 | 280 | 47 | 49 | | |
| H. 1331 | Masi ... | 2 | F | 12 | 30 | 3 | Mr | 12 | Phalguna ... | 7 | F | 17 | 47 | 40 | 18 | 11 | 26 | 16 | 27 | 9 | 310 | 19 | 39 | | |
| Dec. 11 | Panguni . | 4 | Mr | 13 | 31 | 6 | Ap | 12 | Chaitra ... | 2 | Mr | 18 | 19 | 30 | ...○ | 52 | 52 | 18 | 25 | 41 | 339 | 51 | 30 | | |
| **K. Y. 5013** | Chittirai... | 7 | Ap | 13 | 31 | 2 | My | 13 | Vaisakha ... | 3 | Ap | 16 | 51 | 20 | 17● | 26 | 26 | 20 | 24 | 16 | 4 | 7 | 48 | 0 day, 18 ghat., 32 palas. | 0 „ 34 „ 30 „ |
| Sak. 1834 | Vaikasi ... | 3 | My | 14 | 31 | 5 | Je | 13 | Jyeshta ... | 5 | My | 16 | 23 | 10 | ... | 52 | 47 | 22 | 22 | 49 | 33 | 39 | 38 | | |
| Vik. 1969 | Ani ... | 6 | Je | 14 | 31 | 1 | Jl | 14 | Ashada ... | 6 | Je | 14 | 55 | 0 | 15 | 13 | 34 | 24 | 21 | 22 | 63 | 11 | 28 | | |
| | | | | | | | | | | 1 | Jl | 14 | 26 | 50 | ... | 30 | 21 | 26 | 19 | 57 | 92 | 43 | 18 | | |
| (S.)Paridhavi. | Adi ... | 2 | Jl | 15 | 32 | 5 | Au | 15 | Sravana ... | 2 | Au | 12 | 58 | 40 | ... | 48 | 15 | 0 | 45 | 14 | 122 | 15 | 9 | | |
| K. 1087 (S.M.) | Avani ... | 6 | Au | 16 | 31 | 1 | S | 15 | Bhadrapada. | 4 | S | 11 | 30 | 31 | ...○ | 7 | 41 | 2 | 43 | 46 | 151 | 46 | 59 | | |
| K. 1087 (N.M.) | Purattasi . | 2 | S | 16 | 31 | 4 | O | 16 | Asvina ... | 6 | O | 11 | 2 | 21 | 10● | 31 | 26 | 4 | 42 | 21 | 181 | 18 | 49 | | |
| 37 g., 5 p. | Aippasi ... | 5 | O | 17 | 29 | 5 | N | 14 | Kartika ... | 7 | N | 9 | 34 | 11 | ... | 1 | 16 | 6 | 40 | 54 | 210 | 50 | 39 | | |
| B. S. 1318 | Kartigai . | 6 | N | 15 | 30 | 7 | D | 14 | Margasira ... | 2 | D | 9 | 6 | 1 | 8 | 38 | 10 | 8 | 39 | 27 | 240 | 22 | 29 | | |
| (N.) Raktaksha | Margali ... | 1 | D | 15 | 29 | 1 | Ja | 12 | Pausha ... | 3 | Ja | 7 | 37 | 51 | ... | 21 | 44 | 10 | 38 | 1 | 269 | 54 | 19 | | |
| **A. D. 1913** | Tai ... | 2 | Ja | 13 | 30 | 3 | F | 11 | Magha ... | 5 | F | 6 | 9 | 41 | ... | 9 | 26 | 12 | 36 | 35 | 299 | 26 | 9 | | |
| H. 1332 | Masi ... | 4 | F | 12 | 30 | 5 | Mr | 13 | Phalguna ... | 6 | Mr | 7 | 41 | 31 | ...○ | 57 | 14 | 14 | 35 | 9 | 328 | 57 | 59 | | |
| Nov. 30 | Panguni ... | 6 | Mr | 14 | 30 | 7 | Ap | 12 | Chaitra ... | 1 | Ap | 6 | 13 | 21 | ... | 40 | 42 | 16 | 33 | 43 | 358 | 29 | 50 | | |
| **K. Y. 5014** | Chittirai . | 1 | Ap | 13 | 32 | 3 | My | 13 | Vaisakha ... | 2 | My | 5 | 45 | 12 | 6 | 17 | 22 | 18 | 32 | 16 | 22 | 46 | 8 | 1 day, 42 ghat., 11 palas. | 3 „ 10 „ 9 „ |
| Sak. 1835 | Vaikasi ... | 4 | My | 14 | 31 | 6 | Je | 13 | Jyeshta ... | 4 | Je | 4 | 17 | 2 | ... | 46 | 50 | 20 | 30 | 50 | 52 | 17 | 58 | | |
| Vik. 1970 | Ani ... | 7 | Je | 14 | 32 | 3 | Jl | 15 | Ashada ... | 5 | Jl | 3 | 48 | 52 | 4 | 10 | 32 | 22 | 29 | 23 | 81 | 49 | 48 | | |
| (S.) Pramadi | Adi ... | 4 | Jl | 16 | 31 | 6 | Au | 15 | Sravana .. | 7 | Au | 2 | 20 | 42 | ... | 30 | 42 | 24 | 27 | 57 | 111 | 21 | 39 | | |
| K. 1088 (S.M.) | Avani ... | 7 | Au | 16 | 31 | 2 | S | 15 | Bhadrapada. | 1 | Au | 31 | 52 | 32 | ...○ | 49 | 51 | 26 | 26 | 30 | 140 | 53 | 29 | | |
| K. 1088 (N.M.) | Purattasi . | 3 | S | 16 | 30 | 4 | O | 15 | Asvina ... | 3 | S | 30 | 24 | 22 | ... | 10 | 6 | 0 | 51 | 48 | 170 | 25 | 19 | | |
| 52 g., 37 p. | Aippasi ... | 5 | O | 16 | 31 | 7 | N | 15 | Kartika ... | 4 | O | 29 | 56 | 12 | ... | 33 | 22 | 2 | 50 | 21 | 199 | 57 | 9 | | |
| B. S. 1319 | Kartigai ... | 1 | N | 16 | 29 | 1 | D | 14 | Margasira ... | 6 | N | 28 | 28 | 2 | ... | 1 | 10 | 4 | 48 | 55 | 229 | 28 | 59 | | |
| (N.) Krodhana | Margali ... | 2 | D | 15 | 30 | 3 | Ja | 13 | Pausha ... | 1 | D | 28 | 0 | 0 | 27 | 34 | 26 | 6 | 47 | 28 | 259 | 0 | 49 | | |
| **A. D. 1914.** | Tai ... | 4 | Ja | 14 | 29 | 4 | F | 11 | Magha ... | 2 | Ja | 26 | 31 | 43 | ... | 13 | 5 | 8 | 46 | 2 | 288 | 32 | 39 | | |
| H. 1333 | Masi ... | 5 | F | 12 | 31 | 6 | Mr | 13 | Phalguna ... | 4 | F | 25 | 3 | 33 | 24○ | 55 | 55 | 10 | 44 | 35 | 318 | 4 | 29 | | |
| Nov. 19. | Panguni ... | 7 | Mr | 14 | 30 | 1 | Ap | 12 | Chaitra ... | 5 | Mr | 26 | 35 | 23 | ... | 40 | 32 | 12 | 43 | 9 | 347 | 36 | 20 | | |
| **K. Y. 5015.** | Chittirai . | 2 | Ap | 13 | 31 | 4 | My | 13 | Vaisakha ... | 7 | Ap | 25 | 7 | 13 | ... | 23 | 49 | 14 | 41 | 42 | 11 | 52 | 38 | 0 day, 53 ghat., 18 palas. | 1 „ 39 „ 11 „ |
| Sak. 1836 | Vaikasi ... | 5 | My | 14 | 31 | 7 | Je | 13 | Jyeshta ... | 1 | My | 24 | 39 | 3 | 25 | 2 | 39 | 16 | 40 | 16 | 41 | 24 | 28 | | |
| Vik. 1971 | Ani ... | 1 | Je | 14 | 32 | 4 | Jl | 15 | Ashada ... | 3 | Je | 23 | 10 | 53 | ... | 35 | 50 | 18 | 38 | 50 | 70 | 56 | 18 | | |
| (S.) Ananda | Adi ... | 5 | Jl | 16 | 32 | 1 | Au | 16 | Sravana ... | 4 | Jl | 22 | 42 | 43 | 23 | 3 | 48 | 20 | 37 | 24 | 100 | 28 | 8 | | |
| K. 1089 (S. M.) | Avani ... | 2 | Au | 17 | 31 | 4 | S | 16 | Bhadrapada. | 6 | Au | 21 | 14 | 33 | ...●○ | 28 | 24 | 22 | 35 | 57 | 129 | 59 | 58 | | |
| K. 1089 (N. M.) | Purattasi . | 5 | S | 17 | 30 | 6 | O | 16 | Asvina ... | 7 | S | 19 | 46 | 23 | ... | 51 | 17 | 24 | 34 | 31 | 159 | 31 | 48 | | |
| 8 g., 8 p. | Aippasi ... | 7 | O | 17 | 30 | 1 | N | 15 | Kartika ... | 2 | O | 19 | 18 | 13 | ... | 14 | 8 | 26 | 33 | 4 | 189 | 3 | 38 | | |
| B. S. 1320 | Kartigai ... | 2 | N | 16 | 29 | 2 | D | 14 | Margasira ... | 3 | N | 17 | 50 | 4 | ... | 38 | 10 | 0 | 58 | 21 | 218 | 35 | 28 | | |
| (N.) Kshaya | Margali ... | 3 | D | 15 | 30 | 4 | Ja | 13 | Pausha ... | 5 | D | 17 | 21 | 54 | ... | 4 | 49 | 2 | 56 | 55 | 248 | 7 | 19 | | |
| **A. D. 1915** | Tai ... | 5 | Ja | 14 | 29 | 5 | F | 11 | Magha ... | 6 | Ja | 15 | 53 | 44 | ... | 34 | 47 | 4 | 55 | 29 | 277 | 39 | 9 | | |
| H. 1334 | Masi ... | 6 | F | 12 | 30 | 7 | Mr | 13 | Phalguna ... | 1 | F | 14 | 25 | 34 | ...● | 8 | 13 | 6 | 54 | 2 | 307 | 10 | 59 | | |
| Nov. 9. | Panguni ... | 1 | Mr | 14 | 30 | 2 | Ap | 12 | Chaitra ... | 2 | Mr | 15 | 57 | 24 | ... | 44 | 51 | 8 | 52 | 36 | 336 | 42 | 49 | | |

## TABLE XII.—General Ephemeris A.D. 1915—A.D. 1920.

| Commencement of years in different eras. | Tamil month. | Tamil Month Beginning Week-day | Month | Day | Last Day | End Week-day | Month | Day | Lunar Month. | New Moon Ending moment of New moon Tithi. Week-day | Month | Mean Day (mean) | Ghatikas | Palas | Actual Day (actual) | Ghatikas | Palas | ☾'s Anom. Day | Ghatikas | Palas | ☉'s Anom. Day | Ghatikas | Palas | Deduct for Nakshatras | Deduct for Yogas. |
|---|---|---|---|---|---|---|---|---|---|---|---|---|---|---|---|---|---|---|---|---|---|---|---|---|---|
| **K. Y. 5016** | Chittirai . | 3 | Ap | 13 | 31 | 5 | My | 13 | Vaisakha ... | 4 | Ap | 14 | 29 | 14 | ... | 24 | 1 | 10 | 51 | 9 | 0 | 59 | 8 | 0 day, 4 ghat, 25 palas. | 0 „ 8 „ 14 „ |
| | | | | | | | | | | 6 | My | 14 | 1 | 4 | ... | 4 | 11 | 12 | 49 | 43 | 30 | 30 | 58 | | |
| Sak. 1837 | Vaikasi ... | 6 | My | 14 | 32 | 2 | Je | 14 | Jyeshta ... | 7 | Je | 12 | 32 | 54 | ... | 43 | 23 | 14 | 48 | 17 | 60 | 2 | 48 | | |
| Vik. 1972 | Ani ... | 3 | Je | 15 | 31 | 5 | Jl | 15 | Ashada ... | 2 | Jl | 12 | 4 | 44 | ... | 19 | 47 | 16 | 46 | 50 | 89 | 34 | 38 | | |
| (S.) Rakshasa. | Adi ... | 6 | Jl | 16 | 32 | 2 | Au | 16 | Sravana ... | 3 | Au | 10 | 36 | 34 | ...● | 53 | 0 | 18 | 45 | 24 | 119 | 6 | 28 | | |
| K. 1090 (S.M.) | Avani ... | 3 | Au | 17 | 31 | 5 | S | 16 | Bhadrapada . | 5 | S | 9 | 8 | 25 | ... | 23 | 16 | 20 | 43 | 57 | 148 | 38 | 18 | | |
| K. 1090 (N.M.) | Purattasi... | 6 | S | 17 | 30 | 7 | O | 16 | Asvina ... | 6 | O | 8 | 40 | 15 | ... | 51 | 24 | 22 | 42 | 31 | 178 | 10 | 8 | | |
| 23 g., 39 p. | Aippasi ... | 1 | O | 17 | 30 | 2 | N | 15 | Kartika ... | 1 | N | 7 | 12 | 5 | ... | 18 | 6 | 24 | 41 | 5 | 207 | 41 | 59 | | |
| B. S. 1321. | Kartigai ... | 3 | N | 16 | 30 | 4 | D | 15 | Margasira ... | 2 | D | 6 | 43 | 55 | ... | 44 | 18 | 26 | 39 | 38 | 237 | 13 | 49 | | |
| (N.) Prabhava. | Margali ... | 5 | D | 16 | 29 | 5 | Ja | 13 | Pausha ... | 4 | Ja | 5 | 15 | 45 | ...○ | 10 | 44 | 1 | 4 | 55 | 266 | 45 | 39 | | |
| **A. D. 1916** | Tai ... | 6 | Ja | 14 | 30 | 7 | F | 12 | Magha ... | 5 | F | 3 | 47 | 35 | ...● | 38 | 3 | 3 | 3 | 29 | 296 | 17 | 29 | | |
| H. 1335. | Masi ... | 1 | F | 13 | 29 | 1 | Mr | 12 | Phalguna ... | 7 | Mr | 4 | 19 | 25 | ... | 6 | 40 | 5 | 2 | 2 | 325 | 49 | 19 | | |
| Oct. 28. | Panguni ... | 2 | Mr | 13 | 31 | 4 | Ap | 12 | Chaitra ... | 1 | Ap | 2 | 51 | 15 | ... | 37 | 9 | 7 | 0 | 36 | 355 | 21 | 9 | | |
| **K. Y. 5017** | Chittirai... | 5 | Ap | 13 | 31 | 7 | My | 13 | Vaisakha ... | 3 | My | 2 | 23 | 6 | ... | 9 | 52 | 8 | 59 | 10 | 19 | 37 | 28 | 1 day, 28 ghat., 5 palas. | 2 „ 43 „ 53 „ |
| Sak. 1838 | Vaikasi ... | 1 | My | 14 | 31 | 3 | Je | 13 | Jyeshta ... | 4 | My | 31 | 54 | 56 | ... | 44 | 55 | 10 | 57 | 43 | 49 | 9 | 18 | | |
| Vik. 1973 | Ani ... | 4 | Je | 14 | 32 | 7 | Jl | 15 | Ashada ... | 6 | Je | 30 | 26 | 46 | ...○ | 21 | 56 | 12 | 56 | 17 | 78 | 41 | 8 | | |
| (S.) Anala. | Adi ... | 1 | Jl | 16 | 31 | 3 | Au | 15 | Sravana ... | 7 | Jl | 29 | 58 | 36 | 30 | 0 | 9 | 14 | 54 | 50 | 108 | 12 | 58 | | |
| K. 1091 (S. M.) | Avani ... | 4 | Au | 16 | 31 | 6 | S | 15 | Bhadrapada . | 2 | Au | 28 | 30 | 26 | ... | 38 | 11 | 16 | 53 | 24 | 137 | 44 | 48 | | |
| K. 1091 (N. M.) | Purattasi... | 7 | S | 16 | 31 | 2 | O | 16 | Asvina ... | 4 | S | 27 | 2 | 16 | ... | 14 | 46 | 18 | 51 | 58 | 167 | 16 | 38 | | |
| 39 g., 10 p. | Aippasi ... | 3 | O | 17 | 29 | 3 | N | 14 | Kartika ... | 5 | O | 26 | 34 | 6 | ... | 49 | 3 | 20 | 50 | 31 | 196 | 48 | 29 | | |
| B. S. 1322 | Kartigai... | 4 | N | 15 | 30 | 5 | D | 14 | Margasira ... | 7 | N | 25 | 5 | 56 | ... | 20 | 50 | 22 | 49 | 5 | 226 | 20 | 19 | | |
| (N.) Vibhava | Margali ... | 6 | D | 15 | 29 | 6 | Ja | 12 | Pausha ... | 1 | D | 24 | 37 | 47 | ...○ | 50 | 12 | 24 | 47 | 38 | 255 | 52 | 9 | | |
| **A. D. 1917** | Tai ... | 7 | Ja | 13 | 30 | 1 | F | 11 | Magha ... | 3 | Ja | 23 | 9 | 37 | ...● | 17 | 25 | 26 | 46 | 12 | 285 | 23 | 59 | | |
| H. 1336 | Masi ... | 2 | F | 12 | 30 | 3 | Mr | 13 | Phalguna ... | 4 | F | 21 | 41 | 27 | ... | 42 | 56 | 1 | 11 | 29 | 314 | 55 | 49 | | |
| Oct. 17 | Panguni... | 4 | Mr | 14 | 30 | 5 | Ap | 12 | Chaitra ... | 6 | Mr | 23 | 13 | 17 | ... | 7 | 41 | 3 | 10 | 2 | 344 | 27 | 39 | | |
| **K. Y. 5018** | Chittirai... | 6 | Ap | 13 | 31 | 1 | My | 13 | Vaisakha ... | 7 | Ap | 21 | 45 | 7 | ... | 32 | 45 | 5 | 8 | 36 | 8 | 43 | 58 | 0 day, 39 ghat., 12 palas. | 1 „ 12 „ 56 „ |
| Sak. 1839 | Vaikasi ... | 2 | My | 14 | 31 | 4 | Je | 13 | Jyeshta ... | 2 | My | 21 | 16 | 57 | 20 | 59 | 19 | 7 | 7 | 10 | 38 | 15 | 48 | | |
| Vik. 1974 | Ani ... | 5 | Je | 14 | 32 | 1 | Jl | 15 | Ashada ... | 3 | Je | 19 | 48 | 47 | ●○ | 28 | 35 | 9 | 5 | 43 | 67 | 47 | 38 | | |
| (S.) Pingala | Adi ... | 2 | Jl | 16 | 31 | 4 | Au | 15 | Sravana ... | 5 | Jl | 19 | 20 | 37 | ... | 1 | 47 | 11 | 4 | 17 | 97 | 19 | 28 | | |
| K. 1092 (S.M.) | Avani ... | 5 | Au | 16 | 31 | 7 | S | 15 | Bhadrapada . | 6 | Au | 17 | 52 | 28 | ... | 39 | 26 | 13 | 2 | 51 | 126 | 51 | 18 | | |
| | | | | | | | | | | 1 | S | 16 | 24 | 18 | ... | 20 | 36 | 15 | 1 | 24 | 156 | 23 | 8 | | |
| K. 1092 (N.M.) | Purattasi... | 1 | S | 16 | 31 | 3 | O | 16 | Asvina ... | 2 | O | 15 | 56 | 8 | 16 | 2 | 48 | 16 | 59 | 58 | 185 | 54 | 58 | | |
| 54 g., 41p. | Aippasi ... | 4 | O | 17 | 30 | 5 | N | 15 | Kartika ... | 4 | N | 14 | 27 | 58 | ... | 43 | 34 | 18 | 58 | 31 | 215 | 26 | 49 | | |
| B. S. 1323 | Kartigai... | 6 | N | 16 | 29 | 6 | D | 14 | Margasira ... | 6 | D | 14 | 0 | 0 | ...○ | 21 | 2 | 20 | 57 | 5 | 244 | 58 | 39 | | |
| (N.) Sukla. | Margali ... | 7 | D | 15 | 30 | 1 | Ja | 13 | Pausha ... | 7 | Ja | 12 | 31 | 38 | ... | 54 | 20 | 22 | 55 | 39 | 274 | 30 | 29 | | |
| **A. D. 1918** | Tai ... | 2 | Ja | 14 | 29 | 2 | F | 11 | Magha ... | 2 | F | 11 | 3 | 28 | ... | 23 | 7 | 24 | 54 | 12 | 304 | 2 | 19 | | |
| H. 1337. | Masi ... | 3 | F | 12 | 30 | 4 | Mr | 13 | Phalguna ... | 3 | Mr | 12 | 35 | 18 | ... | 48 | 1 | 26 | 52 | 45 | 333 | 34 | 9 | | |
| Oct. 7. | Panguni... | 5 | Mr | 14 | 30 | 6 | Ap | 12 | Chaitra ... | 5 | Ap | 11 | 7 | 8 | ... | 10 | 56 | 1 | 18 | 3 | 363 | 5 | 59 | | |
| **K. Y. 5019** | Chittirai... | 7 | Ap | 13 | 31 | 2 | My | 13 | Vaisakha ... | 6 | My | 10 | 38 | 59 | ... | 31 | 4 | 3 | 16 | 36 | 27 | 22 | 18 | 2 days, 2 ghat., 51 palas. | 3 „ 48 „ 35 „ |
| Sak. 1840 | Vaikasi ... | 3 | My | 14 | 31 | 5 | Je | 13 | Jyeshta ... | 1 | Je | 9 | 10 | 49 | 8●○ | 52 | 36 | 5 | 15 | 10 | 56 | 54 | 8 | | |
| Vik. 1975 | Ani ... | 6 | Je | 14 | 32 | 2 | Jl | 15 | Ashada ... | 2 | Jl | 8 | 42 | 39 | ... | 16 | 44 | 7 | 13 | 44 | 86 | 25 | 58 | | |
| (S.) Kalayukta. | Adi ... | 3 | Jl | 16 | 32 | 6 | Au | 16 | Sravana ... | 4 | Au | 7 | 14 | 29 | 6 | 45 | 39 | 9 | 12 | 17 | 115 | 57 | 48 | | |
| K. 1093 (S.M.) | Avani ... | 7 | Au | 17 | 31 | 2 | S | 16 | Bhadrapada . | 5 | S | 5 | 46 | 19 | ... | 20 | 56 | 11 | 10 | 51 | 145 | 29 | 38 | | |
| K. 1093 (N.M) | Purattasi... | 3 | S | 17 | 30 | 4 | O | 16 | Asvina ... | 7 | O | 5 | 18 | 9 | ... | 2 | 30 | 13 | 9 | 24 | 175 | 1 | 28 | | |
| 10 g., 12 p. | Aippasi ... | 5 | O | 17 | 30 | 6 | N | 15 | Kartika ... | 1 | N | 3 | 49 | 59 | ... | 48 | 18 | 15 | 7 | 58 | 204 | 33 | 18 | | |
| B. S. 1324. | Kartigai... | 7 | N | 16 | 30 | 7 | D | 14 | Margasira ... | 3 | D | 3 | 21 | 49 | ...● | 34 | 31 | 17 | 6 | 32 | 234 | 5 | 8 | | |
| (N.) Pramoda | Margali ... | 1 | D | 15 | 30 | 2 | Ja | 13 | Pausha ... | 4 | Ja | 1 | 53 | 39 | 2 | 17 | 38 | 19 | 5 | 5 | 263 | 36 | 58 | | |
| **A. D. 1919** | Tai ... | 3 | Ja | 14 | 29 | 3 | F | 11 | Magha ... | 6 | Ja | 31 | 25 | 30 | ... | 55 | 11 | 21 | 3 | 39 | 293 | 8 | 49 | | |
| H. 1338 | Masi ... | 4 | F | 12 | 30 | 5 | Mr | 13 | Phalguna ... | 7 | Mr | 1 | 57 | 20 | 2 | 26 | 22 | 23 | 2 | 12 | 322 | 40 | 39 | | |
| Sep. 26. | Panguni ... | 6 | Mr | 14 | 30 | 7 | Ap | 12 | Chaitra ... | 2 | Mr | 31 | 29 | 10 | ... | 51 | 44 | 25 | 0 | 46 | 352 | 12 | 29 | | |
| **K. Y. 5020** | Chittirai... | 1 | Ap | 13 | 31 | 3 | My | 13 | Vaisakha ... | 4 | Ap | 30 | 1 | 0 | ... | 12 | 43 | 26 | 59 | 20 | 16 | 28 | 47 | 1 day, 13 ghat., 58 palas. | 2 „ 17 „ 38 „ |
| Sak. 1841 | Vaikasi ... | 4 | My | 14 | 32 | 7 | Je | 14 | Jyeshta ... | 5 | My | 29 | 32 | 50 | ...● | 31 | 6 | 1 | 24 | 37 | 46 | 0 | 37 | | |
| Vik. 1976. | Ani ... | 1 | Je | 15 | 31 | 3 | Jl | 15 | Ashada ... | 7 | Je | 28 | 4 | 40 | 27 | 49 | 14 | 3 | 23 | 10 | 75 | 32 | 28 | | |
| (S.) Siddharthi | Adi ... | 4 | Jl | 16 | 32 | 7 | Au | 16 | Sravana ... | 1 | Jl | 27 | 36 | 30 | ... | 9 | 44 | 5 | 21 | 45 | 105 | 4 | 18 | | |
| K. 1094 (S.M.) | Avani ... | 1 | Au | 17 | 31 | 3 | S | 16 | Bhadrapada . | 3 | Au | 26 | 8 | 20 | 25 | 34 | 55 | 7 | 20 | 18 | 134 | 36 | 8 | | |
| K. 1094 (N.M.) | Purattasi... | 4 | S | 17 | 30 | 5 | O | 16 | Asvina ... | 4 | S | 24 | 40 | 10 | ... | 7 | 1 | 9 | 18 | 51 | 164 | 7 | 58 | | |
| 25 g., 44 p. | Aippasi ... | 6 | O | 17 | 30 | 7 | N | 15 | Kartika ... | 6 | O | 24 | 12 | 1 | 23○ | 46 | 48 | 11 | 17 | 25 | 193 | 39 | 48 | | |
| B. S. 1325. | Kartigai ... | 1 | N | 16 | 30 | 2 | D | 15 | Margasira ... | 7 | N | 22 | 43 | 51 | ...● | 33 | 6 | 13 | 15 | 58 | 223 | 11 | 38 | | |
| (N.) Prajapati | Margali ... | 3 | D | 16 | 29 | 3 | Ja | 13 | Pausha ... | 2 | D | 22 | 15 | 41 | ... | 22 | 30 | 15 | 14 | 32 | 252 | 43 | 28 | | |
| **A. D. 1920** | Tai ... | 4 | Ja | 14 | 30 | 5 | F | 12 | Magha ... | 3 | Ja | 20 | 47 | 31 | 21 | 9 | 57 | 17 | 13 | 6 | 282 | 15 | 18 | | |
| H. 1339 | Masi ... | 6 | F | 13 | 29 | 6 | Mr | 12 | Phalguna ... | 5 | F | 19 | 19 | 21 | ... | 51 | 35 | 19 | 11 | 39 | 311 | 47 | 9 | | |
| Sep. 15. | Panguni ... | 7 | Mr | 13 | 31 | 2 | Ap | 12 | Chaitra ... | 6 | Mr | 19 | 51 | 11 | 20 | 25 | 44 | 21 | 10 | 13 | 341 | 18 | 59 | | |

# TABLE XIII.

## Sunrise from 8° to 12° N. Latitude.

| Day of solar year. | Lat. 8° Eqn. of time in seconds. | Lat. 8° ⊙'s Trop. Long. in seconds. | Lat. 8° +300″ Long. Total Corrn. in seconds. Trivandram. A.D. 1840—1920 | Lat. 9° Eqn. of time in seconds. | Lat. 9° ⊙'s Trop. Long. in seconds. | Lat. 9° +852″ Long. Total Corrn. in seconds Rameswaram. A.D. 1840—1920 | | Days of Solar months. | Lat. 10° Eqn. of time in seconds. | Lat. 10° ⊙'s Trop. Long. in seconds. | Lat. 10° +564″ Long. Total Corrn. in seconds. Madura. A.D. 1840—1920 | Lat. 11° Eqn. of time in seconds. | Lat. 11° ⊙'s Trop. Long. in seconds. | Lat. 11° +816″ Long. Total Corrn. in seconds. Tanjore. A.D. 1840—1920 | | Eng. date. | Lat. 12° Eqn. of time in seconds | Lat. 12° ⊙'s Trop. Long. in seconds. | Lat. 12° +212″ Long. Total Corrn. in seconds. Maisur. A.D 1840—1920 |
|---|---|---|---|---|---|---|---|---|---|---|---|---|---|---|---|---|---|---|---|
| 1 | —447 | +31 | +497 | —443 | +32 | +1089 | Tamil, Chittirai; Malayalam, Medam; Bengal, Vaisakha. | 1 | —439 | +34 | +841 | —435 | +36 | +1135 | April | 13 | —431 | +38 | +572 |
| 2 | 445 | 61 | 530 | 441 | 65 | 1124 | | 2 | 437 | 68 | 877 | 433 | 72 | 1173 | | 14 | 429 | 75 | 611 |
| 3 | 443 | 92 | 562 | 439 | 97 | 1158 | | 3 | 435 | 102 | 913 | 431 | 108 | 1211 | | 15 | 428 | 113 | 650 |
| 4 | 440 | 123 | 596 | 436 | 130 | 1194 | | 4 | 433 | 136 | 949 | 430 | 144 | 1248 | | 16 | 426 | 150 | 689 |
| 5 | 438 | 153 | 629 | 434 | 162 | 1228 | | 5 | 431 | 170 | 985 | 428 | 179 | 1285 | | 17 | 424 | 188 | 729 |
| 6 | 436 | 184 | 661 | 432 | 194 | 1262 | | 6 | 429 | 205 | 1022 | 426 | 215 | 1323 | | 18 | 422 | 226 | 769 |
| 7 | 434 | 215 | 694 | 430 | 227 | 1297 | | 7 | 427 | 239 | 1058 | 421 | 251 | 1361 | | 19 | 420 | 263 | 808 |
| 8 | 433 | 245 | 726 | 429 | 259 | 1333 | | 8 | 425 | 273 | 1094 | 421 | 287 | 1400 | | 20 | 418 | 301 | 847 |
| 9 | 429 | 276 | 760 | 425 | 292 | 1367 | | 9 | 422 | 307 | 1131 | 419 | 323 | 1438 | | 21 | 415 | 338 | 888 |
| 10 | 428 | 307 | 792 | 424 | 324 | 1400 | | 10 | 420 | 341 | 1168 | 416 | 359 | 1475 | | 22 | 413 | 376 | 927 |
| 11 | 425 | 337 | 806 | 421 | 356 | 1415 | | 11 | 417 | 375 | 1185 | 413 | 395 | 1493 | | 23 | 410 | 414 | 946 |
| 12 | 421 | 368 | 820 | 417 | 389 | 1431 | | 12 | 414 | 409 | 1202 | 411 | 431 | 1512 | | 24 | 407 | 451 | 966 |
| 13 | 418 | 399 | 834 | 414 | 421 | 1447 | | 13 | 411 | 443 | 1218 | 408 | 467 | 1530 | | 25 | 404 | 489 | 985 |
| 14 | 415 | 429 | 848 | 411 | 454 | 1462 | | 14 | 408 | 477 | 1235 | 405 | 503 | 1549 | | 26 | 401 | 526 | 1005 |
| 15 | 413 | 460 | 860 | 409 | 486 | 1476 | | 15 | 405 | 511 | 1252 | 401 | 538 | 1568 | | 27 | 398 | 564 | 1024 |
| 16 | 409 | 491 | 875 | 405 | 518 | 1492 | | 16 | 402 | 546 | 1269 | 398 | 574 | 1586 | | 28 | 395 | 602 | 1044 |
| 17 | 405 | 521 | 890 | 401 | 551 | 1509 | | 17 | 398 | 580 | 1287 | 395 | 610 | 1605 | | 29 | 391 | 639 | 1064 |
| 18 | 402 | 552 | 904 | 398 | 583 | 1524 | | 18 | 395 | 614 | 1303 | 391 | 646 | 1624 | | 30 | 388 | 678 | 1084 |
| 19 | 398 | 583 | 918 | 394 | 616 | 1540 | | 19 | 391 | 648 | 1321 | 388 | 682 | 1643 | May | 1 | 384 | 715 | 1104 |
| 20 | 394 | 613 | 933 | 390 | 648 | 1557 | | 20 | 387 | 682 | 1339 | 384 | 718 | 1662 | | 2 | 381 | 753 | 1124 |
| 21 | 390 | 644 | 948 | 386 | 680 | 1573 | | 21 | 383 | 716 | 1357 | 380 | 754 | 1681 | | 3 | 377 | 791 | 1144 |
| 22 | 387 | 675 | 961 | 382 | 713 | 1589 | | 22 | 379 | 750 | 1375 | 376 | 790 | 1701 | | 4 | 373 | 828 | 1165 |
| 23 | 384 | 705 | 975 | 379 | 745 | 1604 | | 23 | 375 | 784 | 1392 | 372 | 826 | 1720 | | 5 | 371 | 867 | 1183 |
| 24 | 381 | 736 | 989 | 377 | 778 | 1619 | | 24 | 373 | 818 | 1408 | 370 | 862 | 1738 | | 6 | 369 | 903 | 1202 |
| 25 | 378 | 767 | 1005 | 374 | 810 | 1634 | | 25 | 370 | 852 | 1425 | 368 | 897 | 1755 | | 7 | 367 | 941 | 1220 |
| 26 | 375 | 797 | 1016 | 372 | 842 | 1648 | | 26 | 368 | 887 | 1441 | 366 | 933 | 1772 | | 8 | 365 | 979 | 1239 |
| 27 | 372 | 828 | 1030 | 369 | 875 | 1653 | | 27 | 365 | 921 | 1458 | 364 | 969 | 1790 | | 9 | 363 | 1016 | 1257 |
| 28 | 369 | 859 | 1044 | 367 | 907 | 1678 | | 28 | 363 | 955 | 1475 | 362 | 1005 | 1807 | | 10 | 360 | 1053 | 1277 |
| 29 | 366 | 889 | 1057 | 364 | 940 | 1692 | | 29 | 360 | 989 | 1492 | 359 | 1041 | 1826 | | 11 | 358 | 1091 | 1295 |
| 30 | 363 | 920 | 1071 | 362 | 972 | 1707 | | 30 | 358 | 1024 | 1506 | 357 | 1076 | 1843 | | 12 | 356 | 1128 | 1304 |
| 31 | 360 | 931 | 1090 | 360 | 984 | 1722 | Tamil, Vaikasi; Malayalam, Edavam; Bengal, Jyeshta. | 1 | 356 | 1038 | 1522 | 354 | 1091 | 1861 | | 13 | 353 | 1144 | 1333 |
| 32 | 357 | 941 | 1099 | 357 | 996 | 1737 | | 2 | 354 | 1052 | 1537 | 352 | 1107 | 1879 | | 14 | 351 | 1161 | 1352 |
| 33 | 354 | 952 | 1112 | 355 | 1009 | 1751 | | 3 | 352 | 1065 | 1552 | 350 | 1122 | 1896 | | 15 | 348 | 1177 | 1371 |
| 34 | 351 | 963 | 1126 | 353 | 1021 | 1765 | | 4 | 350 | 1079 | 1568 | 348 | 1138 | 1914 | | 16 | 346 | 1194 | 1390 |
| 35 | 348 | 973 | 1140 | 350 | 1033 | 1780 | | 5 | 348 | 1093 | 1584 | 346 | 1153 | 1931 | | 17 | 344 | 1210 | 1408 |
| 36 | 346 | 984 | 1152 | 344 | 1045 | 1799 | | 6 | 342 | 1107 | 1604 | 340 | 1168 | 1952 | | 18 | 338 | 1227 | 1431 |
| 37 | 340 | 995 | 1169 | 338 | 1058 | 1817 | | 7 | 336 | 1121 | 1624 | 334 | 1184 | 1974 | | 19 | 332 | 1243 | 1453 |
| 38 | 334 | 1006 | 1186 | 332 | 1070 | 1835 | | 8 | 330 | 1134 | 1643 | 328 | 1199 | 1995 | | 20 | 326 | 1260 | 1476 |
| 39 | 328 | 1016 | 1203 | 326 | 1082 | 1853 | | 9 | 324 | 1148 | 1663 | 322 | 1215 | 2017 | | 21 | 320 | 1276 | 1498 |
| 40 | 321 | 1027 | 1220 | 319 | 1095 | 1873 | | 10 | 317 | 1162 | 1684 | 315 | 1230 | 2039 | | 22 | 313 | 1293 | 1522 |
| 41 | 315 | 1038 | 1237 | 313 | 1107 | 1891 | | 11 | 311 | 1176 | 1704 | 309 | 1245 | 2059 | | 23 | 307 | 1309 | 1545 |
| 42 | 308 | 1048 | 1247 | 306 | 1119 | 1885 | | 12 | 304 | 1190 | 1700 | 302 | 1261 | 2054 | | 24 | 300 | 1326 | 1541 |
| 43 | 301 | 1059 | 1225 | 299 | 1131 | 1880 | | 13 | 297 | 1203 | 1694 | 295 | 1276 | 2049 | | 25 | 294 | 1342 | 1535 |
| 44 | 295 | 1070 | 1219 | 293 | 1144 | 1873 | | 14 | 291 | 1217 | 1688 | 289 | 1282 | 2043 | | 26 | 287 | 1359 | 1531 |
| 45 | 288 | 1081 | 1213 | 286 | 1156 | 1864 | | 15 | 284 | 1231 | 1682 | 282 | 1307 | 2038 | | 27 | 280 | 1375 | 1527 |
| 46 | 281 | 1091 | 1207 | 279 | 1168 | 1862 | | 16 | 277 | 1245 | 1677 | 275 | 1322 | 2033 | | 28 | 274 | 1392 | 1521 |
| 47 | 274 | 1102 | 1202 | 272 | 1180 | 1857 | | 17 | 270 | 1259 | 1672 | 268 | 1338 | 2029 | | 29 | 267 | 1408 | 1517 |
| 48 | 267 | 1113 | 1196 | 265 | 1193 | 1851 | | 18 | 263 | 1272 | 1666 | 261 | 1353 | 2024 | | 30 | 260 | 1425 | 1513 |
| 49 | 260 | 1123 | 1190 | 258 | 1205 | 1846 | | 19 | 256 | 1286 | 1661 | 254 | 1369 | 2021 | | 31 | 253 | 1441 | 1509 |
| 50 | 252 | 1134 | 1185 | 250 | 1217 | 1841 | | 20 | 248 | 1300 | 1656 | 247 | 1384 | 2014 | June | 1 | 245 | 1458 | 1505 |
| 51 | 245 | 1145 | 1180 | 243 | 1229 | 1836 | | 21 | 241 | 1314 | 1651 | 240 | 1399 | 2009 | | 2 | 238 | 1474 | 1501 |
| 52 | 238 | 1156 | 1174 | 236 | 1242 | 1830 | | 22 | 234 | 1328 | 1646 | 232 | 1415 | 2005 | | 3 | 231 | 1491 | 1497 |
| 53 | 230 | 1166 | 1169 | 228 | 1254 | 1826 | | 23 | 226 | 1341 | 1641 | 225 | 1430 | 2000 | | 4 | 223 | 1507 | 1493 |
| 54 | 223 | 1177 | 1163 | 221 | 1266 | 1820 | | 24 | 219 | 1355 | 1636 | 217 | 1446 | 1996 | | 5 | 216 | 1524 | 1489 |
| 55 | 215 | 1188 | 1159 | 213 | 1278 | 1815 | | 25 | 211 | 1369 | 1631 | 210 | 1461 | 1991 | | 6 | 208 | 1540 | 1486 |
| 56 | 208 | 1198 | 1153 | 206 | 1291 | 1810 | | 26 | 204 | 1383 | 1626 | 202 | 1476 | 1987 | | 7 | 201 | 1557 | 1481 |
| 57 | 199 | 1209 | 1149 | 197 | 1303 | 1806 | | 27 | 195 | 1397 | 1623 | 194 | 1492 | 1984 | | 8 | 193 | 1573 | 1478 |
| 58 | 194 | 1220 | 1141 | 192 | 1315 | 1799 | | 28 | 190 | 1410 | 1615 | 190 | 1507 | 1976 | | 9 | 189 | 1590 | 1471 |
| 59 | 189 | 1231 | 1134 | 187 | 1327 | 1791 | | 29 | 186 | 1424 | 1607 | 186 | 1523 | 1968 | | 10 | 185 | 1606 | 1464 |
| 60 | 184 | 1241 | 1126 | 182 | 1340 | 1784 | | 30 | 182 | 1438 | 1598 | 182 | 1528 | 1960 | | 11 | 181 | 1623 | 1456 |
| 61 | 179 | 1252 | 1118 | 178 | 1352 | 1775 | | 31 | 178 | 1452 | 1590 | 178 | 1552 | 1952 | | 12 | 177 | 1640 | 1449 |
| 62 | 174 | 1220 | 1111 | 174 | 1339 | 1767 | | 32 | 174 | 1440 | 1582 | 174 | 1540 | 1944 | | 13 | 174 | 1629 | 1441 |

# TABLE XIII.

## Sunrise from 8° to 12° N. Latitude.

| Day of Solar year. | Lat. 8° +800" Long. Eqn. of time in seconds. | ☉'s Trop. Long. in seconds. | Total Corrn. in seconds. Trivandram. A.D. 1840—1920 | Lat. 9° +852" Long. Eqn. of time in seconds. | ☉'s Trop. Long. in seconds. | Total Corrn. in seconds. Rameswaram. A.D. 1840—1920 | Days of Solar months. | Solar month | Lat. 10° +564" Long. Eqn. of time in seconds. | ☉'s Trop. Long. in seconds. | Total Corrn. in seconds. Madura. A.D. 1840—1920 | Lat. 11° +816" Long. Eqn. of Time in seconds. | ☉'s Trop. Long. in seconds. | Total Corrn. in seconds. Tanjore. A.D. 1840—1920 | English month | Eng. date. | Lat. 12° +212" Long. Eqn. of Time in seconds. | ☉'s Trop. Long. in seconds. | Total Corrn. in seconds. Maisur. A.D. 1840—1920 |
|---|---|---|---|---|---|---|---|---|---|---|---|---|---|---|---|---|---|---|---|
| 63 | —165 | +1226 | +1107 | —165 | +1327 | +1763 | 1 | Tamil, Ani; Malayalam, Mithunam; Bengal, Ashada. | —165 | +1427 | +1578 | —165 | +1528 | +1941 | | 14 | —165 | +1617 | +1438 |
| 64 | 156 | 1214 | 1103 | 156 | 1314 | 1860 | 2 | | 156 | 1415 | 1575 | 156 | 1516 | 1938 | | 15 | 156 | 1606 | 1436 |
| 65 | 147 | 1201 | 1099 | 147 | 1302 | 1757 | 3 | | 147 | 1402 | 1571 | 147 | 1504 | 1935 | | 16 | 147 | 1595 | 1434 |
| 66 | 138 | 1188 | 1096 | 138 | 1289 | 1753 | 4 | | 138 | 1390 | 1568 | 138 | 1492 | 1932 | | 17 | 138 | 1583 | 1431 |
| 67 | 129 | 1176 | 1092 | 129 | 1277 | 1749 | 5 | | 129 | 1378 | 1565 | 129 | 1481 | 1930 | | 18 | 128 | 1572 | 1430 |
| 68 | 120 | 1163 | 1088 | 120 | 1264 | 1746 | 6 | | 120 | 1365 | 1561 | 119 | 1469 | 1928 | | 19 | 119 | 1561 | 1428 |
| 69 | 110 | 1150 | 1085 | 110 | 1252 | 1743 | 7 | | 110 | 1353 | 1559 | 110 | 1459 | 1925 | | 20 | 110 | 1550 | 1426 |
| 70 | 101 | 1137 | 1082 | 101 | 1239 | 1739 | 8 | | 101 | 1340 | 1555 | 101 | 1445 | 1922 | | 21 | 100 | 1538 | 1424 |
| 71 | 91 | 1125 | 1079 | 91 | 1227 | 1737 | 9 | | 91 | 1328 | 1553 | 91 | 1433 | 1921 | June | 22 | 91 | 1527 | 1421 |
| 72 | 82 | 1112 | 1065 | 82 | 1214 | 1728 | 10 | | 82 | 1316 | 1539 | 82 | 1421 | 1907 | | 23 | 81 | 1516 | 1407 |
| 73 | 72 | 1099 | 1053 | 72 | 1202 | 1711 | 11 | | 72 | 1303 | 1526 | 72 | 1409 | 1893 | | 24 | 72 | 1504 | 1392 |
| 74 | 63 | 1086 | 1039 | 63 | 1189 | 1697 | 12 | | 63 | 1291 | 1513 | 63 | 1397 | 1879 | | 25 | 62 | 1493 | 1378 |
| 75 | 54 | 1074 | 1026 | 54 | 1176 | 1683 | 13 | | 54 | 1278 | 1499 | 54 | 1385 | 1865 | | 26 | 54 | 1482 | 1362 |
| 76 | 43 | 1061 | 1014 | 43 | 1164 | 1671 | 14 | | 43 | 1266 | 1487 | 43 | 1373 | 1852 | | 27 | 43 | 1470 | 1349 |
| 77 | 34 | 1048 | 1001 | 34 | 1151 | 1658 | 15 | | 34 | 1254 | 1473 | 34 | 1362 | 1838 | | 28 | 34 | 1459 | 1335 |
| 78 | 25 | 1035 | 987 | 25 | 1139 | 1644 | 16 | | 25 | 1241 | 1459 | 25 | 1350 | 1824 | | 29 | 24 | 1448 | 1321 |
| 79 | 15 | 1023 | 975 | 15 | 1126 | 1631 | 17 | | 15 | 1229 | 1447 | 15 | 1338 | 1811 | | 30 | 15 | 1437 | 1306 |
| 80 | 5 | 1010 | 962 | 5 | 1114 | 1619 | 18 | | 5 | 1216 | 1434 | 5 | 1326 | 1797 | | 1 | 5 | 1425 | 1292 |
| 81 | + 4 | 997 | 949 | + 4 | 1101 | 1605 | 19 | | + 4 | 1204 | 1420 | + 4 | 1314 | 1783 | | 2 | + 4 | 1414 | 1277 |
| 82 | 14 | 985 | 936 | 14 | 1089 | 1592 | 20 | | 14 | 1192 | 1407 | 14 | 1302 | 1770 | | 3 | 14 | 1403 | 1263 |
| 83 | 23 | 972 | 923 | 23 | 1076 | 1571 | 21 | | 23 | 1179 | 1393 | 23 | 1290 | 1755 | | 4 | 23 | 1391 | 1248 |
| 84 | 33 | 959 | 910 | 33 | 1064 | 1566 | 22 | | 33 | 1167 | 1381 | 33 | 1278 | 1742 | | 5 | 33 | 1380 | 1234 |
| 85 | 43 | 946 | 898 | 43 | 1051 | 1553 | 23 | | 43 | 1154 | 1368 | 43 | 1266 | 1729 | | 6 | 43 | 1369 | 1220 |
| 86 | 52 | 934 | 884 | 52 | 1039 | 1539 | 24 | | 52 | 1142 | 1354 | 52 | 1254 | 1714 | | 7 | 52 | 1357 | 1205 |
| 87 | 61 | 921 | 871 | 61 | 1026 | 1526 | 25 | | 61 | 1130 | 1340 | 61 | 1243 | 1700 | | 8 | 61 | 1346 | 1191 |
| 88 | 71 | 908 | 858 | 71 | 1014 | 1513 | 26 | | 71 | 1117 | 1327 | 71 | 1231 | 1687 | | 9 | 71 | 1335 | 1177 |
| 89 | 81 | 895 | 846 | 81 | 1001 | 1500 | 27 | | 81 | 1105 | 1315 | 81 | 1219 | 1674 | | 10 | 81 | 1324 | 1163 |
| 90 | 90 | 883 | 832 | 90 | 988 | 1486 | 28 | | 90 | 1092 | 1301 | 90 | 1207 | 1659 | | 11 | 90 | 1312 | 1148 |
| 91 | 104 | 870 | 824 | 104 | 976 | 1478 | 29 | | 104 | 1080 | 1292 | 104 | 1196 | 1650 | | 12 | 104 | 1300 | 1138 |
| 92 | 114 | 847 | 811 | 114 | 953 | 1465 | 30 | | 114 | 1057 | 1279 | 114 | 1173 | 1637 | | 13 | 114 | 1276 | 1124 |
| 93 | 124 | 825 | 799 | 124 | 931 | 1452 | 31 | | 124 | 1034 | 1266 | 124 | 1149 | 1623 | July | 14 | 124 | 1252 | 1110 |
| 94 | 133 | 802 | 785 | 133 | 908 | 1439 | 32 | | 133 | 1012 | 1253 | 134 | 1126 | 1610 | | 15 | 134 | 1228 | 1096 |
| 95 | 143 | 780 | 773 | 143 | 885 | 1426 | 1 | Tamil, Adi; Malayalam, Karkatagam; Bengal, Sravana. | 143 | 989 | 1240 | 143 | 1103 | 1596 | | 16 | 144 | 1204 | 1082 |
| 96 | 153 | 757 | 760 | 153 | 862 | 1413 | 2 | | 153 | 966 | 1227 | 153 | 1079 | 1582 | | 17 | 153 | 1180 | 1067 |
| 97 | 162 | 735 | 747 | 162 | 840 | 1399 | 3 | | 162 | 943 | 1213 | 163 | 1056 | 1569 | | 18 | 163 | 1157 | 1054 |
| 98 | 172 | 712 | 734 | 172 | 817 | 1387 | 4 | | 172 | 920 | 1200 | 172 | 1033 | 1555 | | 19 | 173 | 1133 | 1040 |
| 99 | 182 | 690 | 721 | 182 | 794 | 1374 | 5 | | 182 | 898 | 1188 | 182 | 1010 | 1542 | | 20 | 182 | 1109 | 1025 |
| 100 | 191 | 667 | 708 | 191 | 772 | 1359 | 6 | | 191 | 875 | 1174 | 191 | 986 | 1527 | | 21 | 191 | 1085 | 1010 |
| 101 | 200 | 645 | 694 | 200 | 749 | 1347 | 7 | | 200 | 852 | 1160 | 200 | 963 | 1513 | | 22 | 201 | 1061 | 996 |
| 102 | 209 | 622 | 681 | 209 | 726 | 1333 | 8 | | 209 | 829 | 1145 | 210 | 940 | 1498 | | 23 | 210 | 1037 | 982 |
| 103 | 218 | 600 | 680 | 218 | 703 | 1331 | 9 | | 218 | 806 | 1141 | 219 | 916 | 1492 | | 24 | 219 | 1013 | 975 |
| 104 | 228 | 577 | 680 | 228 | 681 | 1329 | 10 | | 228 | 784 | 1138 | 228 | 893 | 1487 | | 25 | 229 | 989 | 970 |
| 105 | 237 | 555 | 680 | 237 | 658 | 1327 | 11 | | 237 | 761 | 1134 | 237 | 870 | 1481 | | 26 | 238 | 965 | 963 |
| 106 | 246 | 532 | 679 | 246 | 635 | 1325 | 12 | | 246 | 738 | 1130 | 246 | 846 | 1476 | | 27 | 247 | 941 | 956 |
| 107 | 253 | 510 | 676 | 253 | 613 | 1320 | 13 | | 254 | 715 | 1125 | 255 | 823 | 1470 | | 28 | 256 | 918 | 949 |
| 108 | 261 | 487 | 675 | 262 | 590 | 1318 | 14 | | 263 | 692 | 1122 | 264 | 800 | 1465 | | 29 | 265 | 894 | 943 |
| 109 | 271 | 465 | 675 | 272 | 567 | 1317 | 15 | | 272 | 670 | 1118 | 273 | 777 | 1459 | | 30 | 273 | 870 | 935 |
| 110 | 280 | 442 | 674 | 280 | 544 | 1313 | 16 | | 281 | 647 | 1114 | 281 | 753 | 1453 | | 31 | 282 | 846 | 928 |
| 111 | 288 | 420 | 672 | 289 | 522 | 1311 | 17 | | 289 | 624 | 1109 | 290 | 730 | 1447 | | 1 | 290 | 822 | 921 |
| 112 | 297 | 397 | 672 | 297 | 499 | 1308 | 18 | | 298 | 601 | 1105 | 298 | 707 | 1441 | | 2 | 299 | 798 | 914 |
| 113 | 306 | 375 | 671 | 306 | 476 | 1305 | 19 | | 306 | 578 | 1100 | 307 | 683 | 1435 | | 3 | 307 | 774 | 906 |
| 114 | 315 | 352 | 670 | 315 | 454 | 1293 | 20 | | 315 | 556 | 1096 | 316 | 660 | 1430 | | 4 | 316 | 750 | 900 |
| 115 | 323 | 330 | 668 | 323 | 431 | 1300 | 21 | | 323 | 533 | 1091 | 324 | 637 | 1423 | | 5 | 324 | 726 | 892 |
| 116 | 331 | 307 | 667 | 331 | 408 | 1296 | 22 | | 331 | 510 | 1086 | 332 | 613 | 1417 | | 6 | 332 | 702 | 884 |
| 117 | 338 | 285 | 664 | 339 | 385 | 1293 | 23 | | 339 | 487 | 1081 | 339 | 590 | 1409 | | 7 | 340 | 679 | 876 |
| 118 | 347 | 262 | 663 | 347 | 363 | 1290 | 24 | | 347 | 464 | 1077 | 348 | 567 | 1404 | August | 8 | 348 | 655 | 869 |
| 119 | 353 | 239 | 659 | 354 | 340 | 1285 | 25 | | 354 | 442 | 1071 | 355 | 544 | 1396 | | 9 | 356 | 631 | 861 |
| 120 | 356 | 217 | 656 | 357 | 317 | 1275 | 26 | | 358 | 419 | 1062 | 359 | 520 | 1386 | | | 360 | 607 | 849 |
| 121 | 359 | 194 | 659 | 360 | 295 | 1265 | 27 | | 362 | 396 | 1053 | 363 | 497 | 1375 | | 1 | 364 | 583 | 838 |
| 122 | 362 | 172 | 662 | 363 | 272 | 1260 | 28 | | 366 | 372 | 1044 | 367 | 472 | 1365 | | 2 | 369 | 560 | 827 |
| 123 | 365 | 162 | 665 | 367 | 261 | 1253 | 29 | | 369 | 359 | 1034 | 371 | 457 | 1354 | | 3 | 374 | 544 | 816 |
| 124 | 372 | 152 | 672 | 374 | 249 | 1249 | 30 | | 376 | 346 | 1028 | 379 | 443 | 1348 | | [illegible] | 380 | 529 | 807 |
| 125 | 379 | 143 | 679 | 381 | 238 | 1244 | 31 | | 383 | 333 | 1022 | 385 | 428 | 1339 | | 15 | 387 | 513 | 798 |

# TABLE XIII.

## Sunrise from 8° to 12° N. Latitude.

| Day of solar year. | Lat. 8° Eqn. of time in seconds. | Lat. 8° ☉'s Trop. Long. in seconds. | Lat. 8° +300″ Long. Total Corrn. in seconds. Trivandram. A.D. 1840—1920 | Lat. 9° Eqn. of time in seconds. | Lat. 9° ☉'s Trop. Long. in seconds. | Lat. 9° +852″ Long. Total Corrn. in seconds. Rameswaram. A.D. 1840—1920 | | Days of Solar months. | Lat. 10° Eqn. of time in seconds. | Lat. 10° ☉'s Trop. Long. in seconds. | Lat. 10° +564″ Long. Total Corrn. in seconds. Madura. A.D. 1840—1920 | Lat. 11° Eqn. of Time in seconds. | Lat. 11° ☉'s Trop. Long. in seconds. | Lat. 11° +786″ Long. Total Corrn. in seconds. Tanjore. A.D. 1840—1920 | | Eng. date. | Lat. 12° Eqn. of Time in seconds. | Lat. 12° ☉'s Trop. Long. in seconds. | Lat. 12° +212″ Long. Total Corrn. in seconds. Maisur. A.D. 1840—1920 |
|---|---|---|---|---|---|---|---|---|---|---|---|---|---|---|---|---|---|---|---|
| 126 | +385 | +133 | +685 | +387 | +227 | +1239 | *Tamil*, Avani; *Malayalam*, Chingam; *Bengal*, Bhadrapada. | 1 | +389 | +320 | +1015 | +392 | +414 | +1332 | August | 16 | +394 | +497 | +789 |
| 127 | 390 | 123 | 690 | 393 | 215 | 1245 | | 2 | 396 | 307 | 1009 | 399 | 399 | 1324 | | 17 | 401 | 481 | 780 |
| 128 | 396 | 114 | 696 | 399 | 204 | 1251 | | 3 | 402 | 295 | 1003 | 405 | 385 | 1316 | | 18 | 407 | 466 | 771 |
| 129 | 405 | 104 | 705 | 407 | 193 | 1259 | | 4 | 409 | 282 | 997 | 411 | 370 | 1307 | | 19 | 413 | 450 | 761 |
| 130 | 409 | 94 | 709 | 412 | 181 | 1264 | | 5 | 415 | 269 | 990 | 418 | 356 | 1300 | | 20 | 420 | 434 | 752 |
| 131 | 415 | 84 | 715 | 418 | 170 | 1270 | | 6 | 421 | 256 | 985 | 424 | 341 | 1291 | | 21 | 426 | 419 | 743 |
| 132 | 421 | 75 | 721 | 424 | 159 | 1276 | | 7 | 427 | 243 | 991 | 430 | 327 | 1282 | | 22 | 432 | 403 | 732 |
| 133 | 427 | 65 | 727 | 430 | 147 | 1282 | | 8 | 433 | 230 | 997 | 436 | 312 | 1287 | | 23 | 438 | 387 | 735 |
| 134 | 434 | 55 | 734 | 437 | 136 | 1289 | | 9 | 439 | 217 | 1003 | 441 | 298 | 1291 | | 24 | 444 | 372 | 738 |
| 135 | 438 | 45 | 738 | 441 | 125 | 1293 | | 10 | 444 | 204 | 1008 | 447 | 283 | 1295 | | 25 | 449 | 356 | 741 |
| 136 | 444 | 36 | 744 | 447 | 113 | 1299 | | 11 | 450 | 191 | 1014 | 453 | 269 | 1300 | | 26 | 455 | 340 | 744 |
| 137 | 449 | 26 | 749 | 452 | 102 | 1304 | | 12 | 455 | 178 | 1019 | 458 | 254 | 1304 | | 27 | 460 | 324 | 746 |
| 138 | 454 | 16 | 754 | 457 | 91 | 1309 | | 13 | 460 | 166 | 1024 | 463 | 240 | 1308 | | 28 | 465 | 309 | 748 |
| 139 | 461 | 6 | 761 | 464 | 79 | 1316 | | 14 | 466 | 153 | 1030 | 468 | 225 | 1312 | | 29 | 471 | 293 | 751 |
| 140 | 464 | 0 | 764 | 467 | 68 | 1319 | | 15 | 470 | 140 | 1034 | 473 | 211 | 1315 | | 30 | 475 | 277 | 753 |
| 141 | 469 | 0 | 769 | 472 | 57 | 1324 | | 16 | 475 | 127 | 1039 | 478 | 196 | 1319 | | 31 | 480 | 262 | 755 |
| 142 | 474 | 0 | 774 | 477 | 45 | 1329 | | 17 | 480 | 114 | 1044 | 483 | 182 | 1323 | September | 1 | 485 | 246 | 757 |
| 143 | 478 | 0 | 778 | 481 | 34 | 1333 | | 18 | 484 | 101 | 1048 | 487 | 167 | 1326 | | 2 | 490 | 230 | 759 |
| 144 | 483 | 0 | 783 | 486 | 23 | 1338 | | 19 | 489 | 88 | 1053 | 490 | 153 | 1328 | | 3 | 494 | 215 | 760 |
| 145 | 484 | 0 | 784 | 487 | 11 | 1339 | | 20 | 490 | 75 | 1054 | 492 | 138 | 1328 | | 4 | 497 | 199 | 761 |
| 146 | 486 | 0 | 786 | 488 | 0 | 1340 | | 21 | 491 | 62 | 1055 | 494 | 124 | 1329 | | 5 | 500 | 183 | 761 |
| 147 | 487 | 0 | 787 | 490 | 0 | 1342 | | 22 | 493 | 49 | 1057 | 495 | 109 | 1329 | | 6 | 502 | 167 | 760 |
| 148 | 489 | 0 | 789 | 491 | 0 | 1343 | | 23 | 494 | 37 | 1058 | 496 | 95 | 1329 | | 7 | 504 | 152 | 759 |
| 149 | 490 | 0 | 790 | 492 | 0 | 1344 | | 24 | 495 | 24 | 1059 | 497 | 80 | 1329 | | 8 | 506 | 136 | 758 |
| 150 | 491 | 0 | 791 | 494 | 0 | 1346 | | 25 | 496 | 11 | 1060 | 499 | 66 | 1329 | | 9 | 507 | 120 | 757 |
| 151 | 492 | 0 | 792 | 495 | 0 | 1347 | | 26 | 498 | 0 | 1062 | 500 | 51 | 1329 | | 10 | 508 | 105 | 755 |
| 152 | 494 | 0 | 794 | 496 | 0 | 1348 | | 27 | 499 | 0 | 1063 | 501 | 36 | 1329 | | 11 | 509 | 88 | 753 |
| 153 | 495 | 0 | 795 | 497 | 0 | 1349 | | 28 | 500 | 0 | 1064 | 502 | 35 | 1329 | | 12 | 510 | 85 | 751 |
| 154 | 496 | 0 | 796 | 498 | 0 | 1350 | | 29 | 501 | 0 | 1065 | 503 | 34 | 1329 | | 13 | 511 | 82 | 749 |
| 155 | 498 | 0 | 798 | 499 | 0 | 1351 | | 30 | 502 | 0 | 1066 | 504 | 32 | 1328 | | 14 | 512 | 80 | 748 |
| 156 | 500 | 0 | 800 | 500 | 0 | 1352 | | 31 | 503 | 0 | 1067 | 506 | 31 | 1329 | | 15 | 513 | 77 | 746 |
| 157 | 501 | 0 | 801 | 501 | 0 | 1353 | *Tamil*, Purattasi; *Malayalam*, Kanni; *Bengal*, Asvina. | 1 | 504 | 0 | 1068 | 508 | 30 | 1330 | | 16 | 514 | 74 | 744 |
| 158 | 502 | 0 | 802 | 503 | 0 | 1355 | | 2 | 506 | 0 | 1070 | 510 | 29 | 1331 | | 17 | 515 | 71 | 742 |
| 159 | 503 | 0 | 803 | 505 | 0 | 1357 | | 3 | 508 | 0 | 1072 | 512 | 28 | 1332 | | 18 | 516 | 68 | 740 |
| 160 | 504 | 0 | 804 | 507 | 0 | 1359 | | 4 | 510 | 0 | 1074 | 514 | 26 | 1332 | | 19 | 518 | 66 | 740 |
| 161 | 505 | 0 | 805 | 509 | 0 | 1361 | | 5 | 512 | 0 | 1076 | 515 | 25 | 1332 | | 20 | 519 | 63 | 738 |
| 162 | 506 | 0 | 806 | 510 | 0 | 1362 | | 6 | 513 | 0 | 1077 | 517 | 24 | 1333 | | 21 | 520 | 60 | 786 |
| 163 | 508 | 0 | 808 | 512 | 0 | 1364 | | 7 | 514 | 0 | 1078 | 518 | 23 | 1334 | | 22 | 522 | 57 | 734 |
| 164 | 509 | 0 | 809 | 513 | 0 | 1365 | | 8 | 516 | 0 | 1080 | 520 | 22 | 1335 | | 23 | 524 | 54 | 733 |
| 165 | 510 | 0 | 810 | 514 | 0 | 1366 | | 9 | 517 | 0 | 1081 | 521 | 20 | 1335 | | 24 | 525 | 52 | 731 |
| 166 | 511 | 0 | 811 | 515 | 0 | 1367 | | 10 | 517 | 0 | 1081 | 521 | 19 | 1333 | | 25 | 525 | 49 | 728 |
| 167 | 512 | 0 | 812 | 515 | 0 | 1367 | | 11 | 519 | 0 | 1083 | 522 | 18 | 1333 | | 26 | 526 | 46 | 726 |
| 168 | 513 | 0 | 813 | 516 | 0 | 1368 | | 12 | 519 | 0 | 1083 | 523 | 17 | 1333 | | 27 | 527 | 43 | 725 |
| 169 | 513 | 0 | 813 | 516 | 0 | 1368 | | 13 | 520 | 0 | 1084 | 524 | 16 | 1333 | | 28 | 527 | 40 | 722 |
| 170 | 513 | 0 | 813 | 516 | 0 | 1368 | | 14 | 520 | 0 | 1084 | 524 | 14 | 1332 | | 29 | 528 | 38 | 720 |
| 171 | 514 | 0 | 814 | 517 | 0 | 1369 | | 15 | 521 | 0 | 1085 | 524 | 13 | 1330 | | 30 | 528 | 35 | 717 |
| 172 | 515 | 0 | 815 | 518 | 0 | 1370 | | 16 | 521 | 0 | 1085 | 525 | 12 | 1330 | October | 1 | 528 | 32 | 714 |
| 173 | 515 | 0 | 815 | 518 | 0 | 1370 | | 17 | 521 | 0 | 1085 | 524 | 11 | 1328 | | 2 | 528 | 29 | 711 |
| 174 | 514 | 0 | 814 | 517 | 0 | 1369 | | 18 | 520 | 0 | 1084 | 524 | 10 | 1327 | | 3 | 528 | 26 | 708 |
| 175 | 513 | 0 | 813 | 516 | 0 | 1368 | | 19 | 520 | 0 | 1084 | 524 | 8 | 1326 | | 4 | 527 | 24 | 704 |
| 176 | 512 | 0 | 812 | 516 | 0 | 1368 | | 20 | 520 | 0 | 1084 | 523 | 7 | 1323 | | 5 | 527 | 21 | 701 |
| 177 | 511 | 0 | 811 | 515 | 0 | 1367 | | 21 | 519 | 0 | 1083 | 523 | 6 | 1322 | | 6 | 526 | 18 | 697 |
| 178 | 510 | 0 | 810 | 514 | 0 | 1366 | | 22 | 518 | 0 | 1082 | 522 | 5 | 1320 | | 7 | 525 | 15 | 694 |
| 179 | 509 | 0 | 809 | 513 | 0 | 1365 | | 23 | 517 | 0 | 1081 | 521 | 4 | 1318 | | 8 | 525 | 12 | 691 |
| 180 | 508 | 0 | 808 | 512 | 0 | 1364 | | 24 | 516 | 0 | 1080 | 520 | 2 | 1316 | | 9 | 524 | 10 | 687 |
| 181 | 507 | 0 | 807 | 511 | 0 | 1363 | | 25 | 514 | 0 | 1078 | 518 | 1 | 1312 | | 10 | 522 | 7 | 682 |
| 182 | 506 | 0 | 806 | 510 | 0 | 1362 | | 26 | 513 | 0 | 1077 | 517 | 0 | 1310 | | 11 | 521 | 4 | 678 |
| 183 | 505 | 0 | 805 | 509 | 0 | 1361 | | 27 | 512 | 0 | 1076 | 515 | 0 | 1307 | | 12 | 519 | 0 | 673 |
| 184 | 504 | 0 | 804 | 508 | 0 | 1360 | | 28 | 511 | 0 | 1075 | 514 | —1 | 1305 | | 13 | 518 | —3 | 669 |
| 185 | 503 | 0 | 803 | 506 | 0 | 1358 | | 29 | 509 | 0 | 1073 | 512 | 2 | 1302 | | 14 | 516 | 6 | 664 |
| 186 | 501 | 0 | 801 | 504 | 0 | 1356 | | 30 | 507 | 0 | 1071 | 510 | 4 | 1298 | | 15 | 514 | 9 | 659 |

# TABLE XIII.

## Sunrise from 8° to 12° N. Latitude.

| Day of Solar year | Lat. 8° +300″ Long.<br>Eqn. of time in seconds. | ☉'s Trop. Long. in seconds. | Total Corrn. in seconds. Trivandram. A.D. 1840—1920 | Lat. 9° +852″ Long.<br>Eqn. of time in seconds. | ☉'s Trop. Long. in seconds. | Total Corrn. in seconds. Rameswaram. A.D. 1840—1920 | Solar Months | Days of Solar Months. | Lat. 10° +564″ Long.<br>Eqn. of time in seconds. | ☉'s Trop. Long. in seconds. | Total Corrn. in seconds. Madura. A.D. 1840—1920 | Lat. 11° +816″ Long.<br>Eqn. of time in seconds. | ☉'s Trop. Long in seconds. | Total Corrn. in seconds. Tanjore. A.D. 1840—1920 | Eng. month | Eng. date. | Lat. 12° +212″ Long.<br>Eqn. of time in seconds. | ☉'s Trop. Long. in seconds. | Total Corrn. in seconds. Maisur. A.D. 1840—1920 |
|---|---|---|---|---|---|---|---|---|---|---|---|---|---|---|---|---|---|---|---|
| **187** | +499 | +0 | +799 | +502 | +0 | +1354 | *Tamil*, Aippasi; *Malayalam*, Tulam; *Bengal*, Kartika. | 1 | +505 | +0 | +1069 | +508 | —5 | +1295 | October | **16** | +512 | —12 | +654 |
| **188** | 497 | 0 | 797 | 500 | 0 | 1352 | | 2 | 503 | 0 | 1067 | 506 | 6 | 1292 | | **17** | 510 | 14 | 650 |
| **189** | 494 | 0 | 794 | 497 | 0 | 1349 | | 3 | 500 | 0 | 1064 | 503 | 7 | 1288 | | **18** | 507 | 17 | 644 |
| **190** | 491 | 0 | 791 | 494 | 0 | 1346 | | 4 | 497 | 0 | 1061 | 500 | 8 | 1284 | | **19** | 504 | 20 | 638 |
| **191** | 488 | 0 | 788 | 491 | 0 | 1343 | | 5 | 494 | 0 | 1058 | 497 | 10 | 1279 | | **20** | 501 | 23 | 632 |
| **192** | 485 | 0 | 785 | 488 | 0 | 1340 | | 6 | 491 | 0 | 1055 | 494 | 11 | 1275 | | **21** | 498 | 26 | 626 |
| **193** | 482 | 0 | 782 | 485 | 0 | 1337 | | 7 | 488 | 0 | 1052 | 491 | 12 | 1271 | | **22** | 495 | 29 | 619 |
| **194** | 479 | 0 | 779 | 482 | 0 | 1334 | | 8 | 485 | 0 | 1049 | 488 | 13 | 1254 | | **23** | 492 | 32 | 600 |
| **195** | 476 | 0 | 776 | 479 | 0 | 1331 | | 9 | 482 | 0 | 1036 | 485 | 14 | 1236 | | **24** | 489 | 35 | 581 |
| **196** | 473 | 0 | 773 | 476 | 0 | 1328 | | 10 | 479 | 0 | 1020 | 482 | 16 | 1219 | | **25** | 486 | 38 | 563 |
| **197** | 469 | 0 | 769 | 472 | 0 | 1324 | | 11 | 475 | 0 | 1003 | 478 | 17 | 1200 | | **26** | 482 | 41 | 543 |
| **198** | 465 | 0 | 765 | 468 | 0 | 1320 | | 12 | 471 | 0 | 987 | 474 | 18 | 1182 | | **27** | 478 | 43 | 524 |
| **199** | 462 | 0 | 762 | 465 | 0 | 1317 | | 13 | 468 | 0 | 971 | 471 | 19 | 1164 | | **28** | 475 | 46 | 505 |
| **200** | 458 | 0 | 758 | 461 | 0 | 1302 | | 14 | 464 | 0 | 954 | 467 | 20 | 1146 | | **29** | 471 | 49 | 485 |
| **201** | 454 | 0 | 754 | 457 | 0 | 1286 | | 15 | 460 | 0 | 937 | 463 | 22 | 1127 | | **30** | 467 | 52 | 465 |
| **202** | 450 | 0 | 750 | 453 | 0 | 1271 | | 16 | 456 | 0 | 920 | 459 | 23 | 1109 | | **31** | 463 | 55 | 446 |
| **203** | 445 | 0 | 745 | 448 | 0 | 1255 | | 17 | 451 | 0 | 902 | 454 | 24 | 1089 | November | **1** | 458 | 58 | 422 |
| **204** | 440 | 0 | 740 | 443 | 0 | 1238 | | 18 | 446 | 0 | 884 | 449 | 25 | 1070 | | **2** | 453 | 61 | 405 |
| **205** | 436 | 0 | 736 | 439 | 0 | 1223 | | 19 | 442 | 0 | 867 | 445 | 26 | 1051 | | **3** | 449 | 64 | 385 |
| **206** | 431 | 0 | 725 | 434 | 0 | 1207 | | 20 | 437 | 0 | 849 | 440 | 28 | 1032 | | **4** | 444 | 67 | 364 |
| **207** | 427 | 0 | 711 | 430 | 0 | 1191 | | 21 | 433 | 0 | 832 | 436 | 29 | 1013 | | **5** | 440 | 70 | 344 |
| **208** | 422 | 0 | 696 | 425 | 0 | 1175 | | 22 | 428 | 0 | 815 | 431 | 30 | 994 | | **6** | 435 | 72 | 324 |
| **209** | 417 | 0 | 681 | 420 | 0 | 1159 | | 23 | 423 | 0 | 797 | 425 | 31 | 973 | | **7** | 429 | 75 | 302 |
| **210** | 413 | 0 | 668 | 416 | 0 | 1143 | | 24 | 419 | 0 | 780 | 420 | 32 | 954 | | **8** | 424 | 78 | 281 |
| **211** | 411 | 0 | 656 | 414 | 0 | 1130 | | 25 | 417 | 0 | 765 | 418 | 34 | 937 | | **9** | 422 | 81 | 263 |
| **212** | 409 | 0 | 644 | 412 | 0 | 1117 | | 26 | 415 | 0 | 750 | 416 | 35 | 921 | | **10** | 420 | 84 | 246 |
| **213** | 407 | 0 | 632 | 410 | 0 | 1103 | | 27 | 413 | 0 | 735 | 414 | 36 | 904 | | **11** | 418 | 88 | 228 |
| **214** | 405 | 0 | 621 | 408 | 0 | 1090 | | 28 | 411 | 0 | 720 | 412 | 50 | 888 | | **12** | 416 | 104 | 210 |
| **215** | 403 | 0 | 609 | 406 | 0 | 1077 | | 29 | 409 | —10 | 705 | 410 | 65 | 871 | | **13** | 414 | 120 | 193 |
| **216** | 397 | 0 | 593 | 400 | 0 | 1059 | | 30 | 403 | 23 | 686 | 406 | 79 | 853 | | **14** | 408 | 135 | 171 |
| **217** | 391 | 0 | 577 | 394 | 0 | 1042 | *Tamil*, Kartigai; *Malayalam*, Vrischikam; *Bengal*, Margasira. | 1 | 397 | 36 | 667 | 400 | 94 | 832 | | **15** | 402 | 151 | 149 |
| **218** | 385 | 0 | 562 | 388 | 0 | 1025 | | 2 | 391 | 48 | 649 | 394 | 108 | 812 | | **16** | 396 | 166 | 128 |
| **219** | 378 | 0 | 545 | 381 | 0 | 1006 | | 3 | 384 | 61 | 629 | 387 | 123 | 790 | | **17** | 389 | 182 | 105 |
| **220** | 372 | 0 | 529 | 375 | —11 | 989 | | 4 | +378 | 74 | 610 | 381 | 137 | 770 | | **18** | 383 | 198 | 83 |
| **221** | 365 | 0 | 513 | 368 | 23 | 971 | | 5 | 371 | 87 | 590 | 374 | 152 | 748 | | **19** | 376 | 214 | 60 |
| **222** | 358 | 0 | 496 | 361 | 34 | 952 | | 6 | 366 | 100 | 572 | 367 | 166 | 727 | | **20** | 369 | 229 | 38 |
| **223** | 351 | 0 | 479 | 354 | 45 | 934 | | 7 | 357 | 113 | 549 | 360 | 181 | 704 | | **21** | 362 | 245 | 14 |
| **224** | 344 | 0 | 450 | 347 | 57 | 904 | | 8 | 350 | 126 | 519 | 353 | 195 | 674 | | **22** | 355 | 261 | — 17 |
| **225** | 337 | 0 | 420 | 340 | 68 | 875 | | 9 | 343 | 139 | 489 | 346 | 210 | 643 | | **23** | 348 | 276 | 47 |
| **226** | 330 | — 6 | 391 | 333 | 79 | 845 | | 10 | 336 | 152 | 460 | 339 | 224 | 613 | | **24** | 341 | 292 | 71 |
| **227** | 322 | 16 | 360 | 325 | 91 | 814 | | 11 | 328 | 165 | 429 | 331 | 239 | 582 | | **25** | 333 | 308 | 119 |
| **228** | 315 | 26 | 330 | 318 | 102 | 785 | | 12 | 321 | 177 | 399 | 324 | 253 | 552 | | **26** | 326 | 323 | 141 |
| **229** | 307 | 36 | 300 | 310 | 113 | 754 | | 13 | 313 | 190 | 368 | 316 | 268 | 520 | | **27** | 318 | 339 | 173 |
| **230** | 299 | 45 | 269 | 302 | 125 | 723 | | 14 | 305 | 203 | 337 | 308 | 282 | 489 | | **28** | 310 | 355 | 205 |
| **231** | 291 | 55 | 239 | 294 | 136 | 692 | | 15 | 297 | 216 | 307 | 300 | 297 | 458 | | **29** | 302 | 371 | 237 |
| **232** | 283 | 65 | 208 | 286 | 147 | 662 | | 16 | 289 | 229 | 276 | 292 | 311 | 426 | | **30** | 294 | 386 | 269 |
| **233** | 275 | 75 | 178 | 278 | 159 | 631 | | 17 | 281 | 242 | 245 | 284 | 326 | 395 | December | **1** | 286 | 402 | 301 |
| **234** | 266 | 84 | 146 | 269 | 170 | 599 | | 18 | 272 | 255 | 213 | 275 | 340 | 363 | | **2** | 277 | 418 | 334 |
| **235** | 258 | 94 | 116 | 261 | 181 | 569 | | 19 | 264 | 268 | 182 | 267 | 355 | 331 | | **3** | 269 | 433 | 366 |
| **236** | 250 | 104 | 85 | 253 | 193 | 538 | | 20 | 256 | 281 | 152 | 259 | 369 | 300 | | **4** | 261 | 449 | 398 |
| **237** | 242 | 114 | 55 | 245 | 204 | 507 | | 21 | 248 | 294 | 121 | 251 | 384 | 269 | | **5** | 253 | 465 | 430 |
| **238** | 233 | 123 | 23 | 236 | 215 | 475 | | 22 | 239 | 306 | 89 | 242 | 398 | 237 | | **6** | 244 | 480 | 462 |
| **239** | 224 | 133 | — 8 | 227 | 227 | 444 | | 23 | 230 | 319 | 57 | 233 | 413 | 204 | | **7** | 235 | 496 | 495 |
| **240** | 216 | 143 | 39 | 219 | 238 | 413 | | 24 | 222 | 332 | 26 | 225 | 427 | 173 | | **8** | 227 | 512 | 527 |
| **241** | 207 | 152 | 70 | 210 | 249 | 381 | | 25 | 213 | 345 | — 5 | 216 | 442 | 141 | | **9** | 218 | 528 | 560 |
| **242** | 199 | 162 | 103 | 201 | 261 | 350 | | 26 | 204 | 358 | 37 | 207 | 456 | 108 | | **10** | 209 | 543 | 593 |
| **243** | 198 | 172 | 122 | 199 | 272 | 325 | | 27 | 199 | 372 | 65 | 199 | 472 | 77 | | **11** | 199 | 560 | 627 |
| **244** | 190 | 194 | 155 | 190 | 295 | 293 | | 28 | 190 | 395 | 97 | 190 | 495 | 45 | | **12** | 190 | 584 | 660 |
| **245** | 180 | 217 | 187 | 180 | 317 | 260 | | 29 | 180 | 418 | 130 | 180 | 519 | — 11 | | **13** | 180 | 608 | 694 |
| **246** | 170 | 239 | 220 | 170 | 340 | 228 | | 30 | 170 | 440 | 162 | 170 | 542 | 22 | | **14** | 170 | 632 | 728 |

# TABLE XIII.

## Sunrise from 8° to 12° N Latitude.

| Day of Solar year. | Lat. 8° +300″ Long. Eqn. of time in seconds. | ⊙'s Trop. Long. in seconds. | Total Corrn. in seconds. Trivandram. A.D. 1840—1920 | Lat. 9° +852″ Long. Eqn. of time in seconds. | ⊙'s Trop. Long. in seconds. | Total Corrn. in seconds. Rameswaram. A.D. 1840—1920 | Solar Month | Days of Solar Months. | Lat. 10° +564″ Long. Eqn. of time in seconds. | ⊙'s Trop. Long. in seconds. | Total Corrn. in seconds. Madura. A.D. 1840—1920 | Lat. 11° +816″ Long. Eqn. of time in seconds. | ⊙'s Trop. Long. in seconds. | Total Corrn. in seconds. Tanjore. A.D. 1840—1920 | Eng. month | Eng. date | Lat. 12° +212″ Long. Eqn. of time in seconds. | ⊙'s Trop. Long. in seconds. | Total Corrn. in seconds. Maisur. A.D. 1840—1920 |
|---|---|---|---|---|---|---|---|---|---|---|---|---|---|---|---|---|---|---|---|
| 247 | +160 | —262 | —252 | +160 | —363 | +195 | *Tamil*, Margali; *Malayalam*, Dhanus; *Bengal*, Pausha. | 1 | +160 | —463 | —195 | +160 | —565 | —55 | December | 15 | +160 | —656 | —762 |
| 248 | 150 | 285 | 285 | 150 | 385 | 162 | | 2 | 150 | 486 | 228 | 150 | 588 | 88 | | 16 | 150 | 679 | 795 |
| 249 | 140 | 307 | 317 | 140 | 408 | 130 | | 3 | 140 | 509 | 261 | 140 | 612 | 122 | | 17 | 140 | 703 | 829 |
| 250 | 130 | 330 | 350 | 130 | 431 | 97 | | 4 | 130 | 532 | 294 | 130 | 635 | 155 | | 18 | 130 | 727 | 863 |
| 251 | 120 | 352 | 362 | 120 | 454 | 64 | | 5 | 120 | 554 | 326 | 120 | 658 | 188 | | 19 | 120 | 751 | 897 |
| 252 | 110 | 375 | 415 | 110 | 476 | 31 | | 6 | 110 | 577 | 359 | 110 | 682 | 222 | | 20 | 110 | 775 | 931 |
| 253 | 100 | 397 | 447 | 100 | 499 | — 1 | | 7 | 100 | 600 | 392 | 100 | 705 | 255 | | 21 | 100 | 799 | 965 |
| 254 | 90 | 420 | 480 | 90 | 522 | 34 | | 8 | 90 | 623 | 426 | 90 | 728 | 290 | | 22 | 90 | 823 | 998 |
| 255 | 80 | 442 | 503 | 80 | 544 | 56 | | 9 | 80 | 646 | 448 | 80 | 752 | 312 | | 23 | 80 | 847 | 1019 |
| 256 | 70 | 465 | 525 | 70 | 567 | 79 | | 10 | 70 | 668 | 471 | 70 | 775 | 334 | | 24 | 70 | 871 | 1041 |
| 257 | 60 | 487 | 548 | 60 | 590 | 102 | | 11 | 60 | 691 | 493 | 60 | 798 | 356 | | 25 | 60 | 895 | 1062 |
| 258 | 50 | 510 | 571 | 50 | 613 | 124 | | 12 | 50 | 714 | 516 | 50 | 821 | 378 | | 26 | 50 | 918 | 1083 |
| 259 | 40 | 532 | 594 | 40 | 635 | 147 | | 13 | 40 | 737 | 538 | 40 | 845 | 399 | | 27 | 40 | 942 | 1104 |
| 260 | 30 | 555 | 616 | 30 | 658 | 169 | | 14 | 30 | 760 | 560 | 30 | 868 | 421 | | 28 | 30 | 966 | 1126 |
| 261 | 20 | 577 | 639 | 20 | 681 | 192 | | 15 | 20 | 782 | 583 | 20 | 891 | 443 | | 29 | 20 | 990 | 1147 |
| 262 | 10 | 600 | 662 | 10 | 703 | 214 | | 16 | 10 | 805 | 605 | 10 | 915 | 465 | | 30 | 10 | 1014 | 1168 |
| 263 | 0 | 622 | 685 | 0 | 726 | 237 | | 17 | 0 | 828 | 628 | 0 | 938 | 487 | | 31 | 0 | 1038 | 1190 |
| 264 | —10 | 645 | 707 | —10 | 749 | 257 | | 18 | —10 | 851 | 649 | —10 | 961 | 508 | | 1 | —10 | 1062 | 1210 |
| 265 | 20 | 667 | 730 | 20 | 772 | 280 | | 19 | 20 | 874 | 676 | 20 | 985 | 529 | | 2 | 20 | 1086 | 1230 |
| 266 | 29 | 690 | 752 | 28 | 794 | 302 | | 20 | 28 | 896 | 693 | 28 | 1008 | 551 | | 3 | 28 | 1110 | 1252 |
| 267 | 38 | 712 | 773 | 37 | 817 | 324 | | 21 | 37 | 919 | 714 | 37 | 1031 | 572 | | 4 | 37 | 1134 | 1272 |
| 268 | 48 | 735 | 796 | 47 | 840 | 346 | | 22 | 47 | 942 | 737 | 47 | 1054 | 594 | | 5 | 47 | 1157 | 1293 |
| 269 | 59 | 757 | 820 | 57 | 862 | 369 | | 23 | 57 | 965 | 759 | 57 | 1078 | 615 | | 6 | 57 | 1181 | 1314 |
| 270 | 68 | 780 | 842 | 66 | 885 | 390 | | 24 | 66 | 988 | 780 | 66 | 1101 | 636 | | 7 | 66 | 1205 | 1335 |
| 271 | 78 | 802 | 864 | 77 | 908 | 414 | | 25 | 77 | 1010 | 804 | 77 | 1124 | 659 | | 8 | 77 | 1229 | 1357 |
| 272 | 88 | 825 | 887 | 87 | 931 | 437 | | 26 | 87 | 1033 | 826 | 87 | 1148 | 681 | | 9 | 87 | 1253 | 1378 |
| 273 | 98 | 847 | 910 | 97 | 953 | 459 | | 27 | 97 | 1056 | 849 | 97 | 1171 | 703 | | 10 | 97 | 1277 | 1400 |
| 274 | 104 | 870 | 929 | 104 | 976 | 479 | | 28 | 104 | 1080 | 868 | 104 | 1196 | 722 | | 11 | 104 | 1300 | 1418 |
| 275 | 111 | 883 | 948 | 111 | 988 | 498 | | 29 | 111 | 1092 | 887 | 111 | 1208 | 741 | | 12 | 111 | 1311 | 1436 |
| 276 | 119 | 895 | 969 | 118 | 1001 | 518 | *Tamil*, Tai; *Malayalam* Magaram; *Bengal*, Magha. | 1 | 118 | 1105 | 907 | 118 | 1220 | 760 | January | 13 | 118 | 1323 | 1455 |
| 277 | 126 | 908 | 988 | 126 | 1014 | 539 | | 2 | 125 | 1117 | 926 | 125 | 1232 | 778 | | 14 | 125 | 1334 | 1473 |
| 278 | 134 | 921 | 1010 | 133 | 1026 | 558 | | 3 | 133 | 1130 | 947 | 132 | 1246 | 798 | | 15 | 132 | 1345 | 1491 |
| 279 | 141 | 934 | 1029 | 140 | 1039 | 577 | | 4 | 140 | 1142 | 966 | 139 | 1255 | 816 | | 16 | 139 | 1356 | 1509 |
| 280 | 149 | 946 | 1050 | 148 | 1051 | 598 | | 5 | 147 | 1154 | 985 | 146 | 1267 | 835 | | 17 | 146 | 1368 | 1528 |
| 281 | 156 | 959 | 1070 | 155 | 1064 | 617 | | 6 | 154 | 1167 | 1006 | 154 | 1279 | 855 | | 18 | 153 | 1379 | 1546 |
| 282 | 164 | 972 | 1090 | 162 | 1076 | 637 | | 7 | 162 | 1179 | 1025 | 161 | 1291 | 874 | | 19 | 160 | 1390 | 1564 |
| 283 | 171 | 985 | 1110 | 170 | 1089 | 657 | | 8 | 169 | 1192 | 1045 | 168 | 1303 | 893 | | 20 | 167 | 1402 | 1583 |
| 284 | 179 | 997 | 1131 | 177 | 1101 | 677 | | 9 | 176 | 1204 | 1064 | 175 | 1315 | 911 | | 21 | 174 | 1413 | 1602 |
| 285 | 186 | 1010 | 1127 | 184 | 1114 | 672 | | 10 | 183 | 1216 | 1057 | 182 | 1329 | 903 | | 22 | 181 | [illegible] | 1592 |
| 286 | 194 | 1023 | 1125 | 192 | 1126 | 667 | | 11 | 191 | 1229 | 1051 | 189 | 1339 | 894 | | 23 | 188 | 1436 | 1583 |
| 287 | 201 | 1035 | 1121 | 199 | 1139 | 662 | | 12 | 198 | 1241 | 1045 | 196 | 1351 | 886 | | 24 | 195 | 1447 | 1573 |
| 288 | 209 | 1048 | 1118 | 206 | 1151 | 657 | | 13 | 205 | 1254 | 1038 | 204 | 1363 | 878 | | 25 | 202 | 1458 | 1564 |
| 289 | 216 | 1061 | 1114 | 214 | 1164 | 653 | | 14 | 212 | 1266 | 1031 | 211 | 1374 | 870 | | 26 | 209 | 1469 | 1554 |
| 290 | 224 | 1074 | 1112 | 221 | 1176 | 647 | | 15 | 220 | 1278 | 1025 | 218 | 1386 | 862 | | 27 | 216 | 1481 | 1545 |
| 291 | 231 | 1086 | 1108 | 228 | 1189 | 642 | | 16 | 227 | 1291 | 1018 | 225 | 1398 | 853 | | 28 | 223 | 1492 | 1535 |
| 292 | 239 | 1099 | 1105 | 236 | 1202 | 638 | | 17 | 234 | 1303 | 1012 | 232 | 1410 | 845 | | 29 | 230 | 1503 | 1526 |
| 293 | 246 | 1112 | 1102 | 243 | 1214 | 633 | | 18 | 241 | 1316 | 1005 | 239 | 1422 | 836 | | 30 | 237 | 1515 | 1516 |
| 294 | 254 | 1125 | 1099 | 250 | 1227 | 627 | | 19 | 249 | 1328 | 999 | 246 | 1434 | 828 | | 31 | 244 | 1526 | 1507 |
| 295 | 261 | 1137 | 1095 | 258 | 1239 | 623 | | 20 | 256 | 1340 | 992 | 254 | 1446 | 821 | February | 1 | 251 | 1537 | 1497 |
| 296 | 269 | 1150 | 1092 | 265 | 1252 | 618 | | 21 | 263 | 1353 | 985 | 261 | 1458 | 812 | | 2 | 258 | 1549 | 1488 |
| 297 | 276 | 1163 | 1089 | 272 | 1264 | 613 | | 22 | 270 | 1365 | 979 | 268 | 1470 | 804 | | 3 | 265 | 1560 | 1478 |
| 298 | 284 | 1176 | 1086 | 280 | 1277 | 608 | | 23 | 278 | 1378 | 973 | 275 | 1482 | 795 | | 4 | 273 | 1571 | 1470 |
| 299 | 291 | 1188 | 1082 | 287 | 1289 | 603 | | 24 | 285 | 1390 | 966 | 282 | 1493 | 787 | | 5 | 281 | 1582 | 1461 |
| 300 | 299 | 1201 | 1080 | 294 | 1302 | 598 | | 25 | 292 | 1402 | 959 | 289 | 1505 | 779 | | 6 | 288 | 1594 | 1452 |
| 301 | 306 | 1214 | 1076 | 302 | 1314 | 594 | | 26 | 299 | 1415 | 952 | 296 | 1517 | 770 | | 7 | 295 | 1605 | 1442 |
| 302 | 314 | 1226 | 1073 | 309 | 1327 | 588 | | 27 | 306 | 1427 | 946 | 304 | 1529 | 763 | | 8 | 303 | 1616 | 1434 |
| 303 | 321 | 1239 | 1069 | 316 | 1339 | 583 | | 28 | 314 | 1440 | 940 | 311 | 1541 | 754 | | 9 | 310 | 1628 | 1424 |
| 304 | 325 | 1252 | 1063 | 323 | 1352 | 578 | | 29 | 321 | 1452 | 933 | 318 | 1552 | 746 | | 10 | 317 | 1640 | 1415 |
| 305 | 328 | 1241 | 1055 | 326 | 1340 | 569 | | 30 | 324 | 1438 | 922 | 321 | 1537 | 734 | | 11 | 320 | 1623 | 1401 |

# TABLE XIII.

## Sunrise from 8° to 12° N. Latitude.

| Day of Solar year. | Lat. 8° +800″ Long. Eqn. of time in seconds. | ☉'s Trop. Long. in seconds. | Total Corrn. in seconds. **Trivandram.** A.D. 1840—1920 | Lat. 9° +852″ Long. Eqn. of time in seconds. | ☉'s Trop. Long. in seconds. | Total Corrn. in seconds. **Rameswaram.** A.D. 1840—1920 | Days of Solar months. | Lat. 10° +564″ Long. Eqn. of time in seconds. | ☉'s Trop. Long. in seconds. | Total Corrn. in seconds. **Madura.** A.D. 1840—1920 | Lat. 11° +816″ Long. Eqn. of time in seconds. | ☉'s Trop. Long. in seconds. | Total Corrn. in seconds. **Tanjore.** A.D. 1840—1920 | Eng. date. | Lat. 12° +212″ Long. Eqn. of time in seconds. | ☉'s Trop. Long. in seconds. | Total Corrn. in seconds. **Maisur.** A.D. 1840—1920 |
|---|---|---|---|---|---|---|---|---|---|---|---|---|---|---|---|---|---|
| **306** | —332 | —1231 | —1048 | —330 | —1327 | —560 | 1 | —327 | —1424 | —911 | —324 | —1521 | —721 | **12** | —323 | —1607 | —1388 |
| **307** | 335 | 1220 | 1041 | 333 | 1315 | 551 | 2 | 331 | 1411 | 902 | 328 | 1506 | 710 | **13** | 327 | 1590 | 1375 |
| **308** | 339 | 1209 | 1034 | 334 | 1303 | 543 | 3 | 334 | 1397 | 891 | 331 | 1490 | 697 | **14** | 330 | 1574 | 1362 |
| **309** | 342 | 1198 | 1026 | 340 | 1291 | 533 | 4 | 337 | 1383 | 880 | 334 | 1475 | 685 | **15** | 333 | 1557 | 1348 |
| **310** | 346 | 1188 | 1019 | 344 | 1278 | 525 | 5 | 341 | 1369 | 870 | 338 | 1460 | 674 | **16** | 337 | 1541 | 1336 |
| **311** | 349 | 1177 | 1012 | 347 | 1266 | 516 | 6 | 344 | 1355 | 859 | 341 | 1444 | 661 | **17** | 340 | 1524 | 1322 |
| **312** | 353 | 1166 | 1005 | 351 | 1254 | 509 | 7 | 347 | 1342 | 849 | 344 | 1429 | 649 | **18** | 343 | 1508 | 1309 |
| **313** | 356 | 1156 | 997 | 354 | 1242 | 498 | 8 | 351 | 1328 | 839 | 348 | 1413 | 637 | **19** | 347 | 1491 | 1296 |
| **314** | 360 | 1145 | 990 | 358 | 1229 | 490 | 9 | 354 | 1314 | 828 | 351 | 1398 | 625 | **20** | 350 | 1475 | 1283 |
| **315** | 363 | 1134 | 983 | 361 | 1217 | 481 | 10 | 357 | 1300 | 817 | 354 | 1383 | 614 | **21** | 353 | 1458 | 1268 |
| **316** | 367 | 1123 | 956 | 365 | 1205 | 453 | 11 | 361 | 1281 | 787 | 358 | 1367 | 582 | **22** | 357 | 1442 | 1245 |
| **317** | 370 | 1113 | 929 | 368 | 1193 | 423 | 12 | 364 | 1273 | 756 | 361 | 1352 | 549 | **23** | 360 | 1455 | 1201 |
| **318** | 374 | 1102 | 902 | 372 | 1180 | 395 | 13 | 367 | 1259 | 725 | 364 | 1336 | 516 | **24** | 363 | 1409 | 1166 |
| **319** | 377 | 1091 | 874 | 375 | 1168 | 360 | 14 | 371 | 1245 | 685 | 368 | 1321 | 484 | **25** | 367 | 1392 | 1133 |
| **320** | 381 | 1081 | 847 | 379 | 1156 | 337 | 15 | 374 | 1231 | 674 | 371 | 1306 | 451 | **26** | 370 | 1376 | 1098 |
| **321** | 384 | 1070 | 820 | 382 | 1144 | 308 | 16 | 377 | 1217 | 633 | 374 | 1290 | 419 | **27** | 373 | 1359 | 1063 |
| **322** | 388 | 1059 | 793 | 386 | 1131 | 279 | 17 | 381 | 1204 | 602 | 378 | 1275 | 387 | **28** | 377 | 1343 | 1030 |
| **323** | 391 | 1048 | 766 | 389 | 1119 | 250 | 18 | 384 | 1190 | 571 | 381 | 1259 | 354 | **1** | 380 | 1326 | 995 |
| **324** | 395 | 1038 | 739 | 393 | 1107 | 221 | 19 | 387 | 1176 | 540 | 384 | 1244 | 321 | **2** | 383 | 1310 | 961 |
| **325** | 398 | 1027 | 711 | 396 | 1095 | 192 | 20 | 391 | 1162 | 510 | 388 | 1229 | 289 | **3** | 387 | 1293 | 927 |
| **326** | 402 | 1016 | 685 | 400 | 1082 | 164 | 21 | 394 | 1148 | 479 | 391 | 1213 | 256 | **4** | 390 | 1277 | 892 |
| **327** | 405 | 1006 | 657 | 403 | 1070 | 134 | 22 | 397 | 1135 | 448 | 394 | 1198 | 223 | **5** | 393 | 1260 | 858 |
| **328** | 409 | 995 | 630 | 407 | 1058 | 106 | 23 | 401 | 1121 | 418 | 398 | 1182 | 191 | **6** | 397 | 1244 | 824 |
| **329** | 412 | 984 | 603 | 410 | 1045 | 76 | 24 | 404 | 1107 | 287 | 401 | 1167 | 158 | **7** | 400 | 1227 | 790 |
| **330** | 415 | 973 | 575 | 414 | 1033 | 48 | 25 | 407 | 1093 | 356 | 404 | 1152 | 125 | **8** | 403 | 1211 | 755 |
| **331** | 418 | 963 | 547 | 417 | 1021 | 19 | 26 | 411 | 1079 | 326 | 408 | 1136 | 94 | **9** | 406 | 1194 | 720 |
| **332** | 422 | 962 | 521 | 419 | 1009 | + 12 | 27 | 414 | 1066 | 294 | 411 | 1121 | 61 | **10** | 408 | 1178 | 685 |
| **333** | 425 | 941 | 493 | 422 | 996 | 41 | 28 | 417 | 1052 | 263 | 414 | 1105 | 28 | **11** | 411 | 1161 | 650 |
| **334** | 429 | 931 | 465 | 425 | 984 | 71 | 29 | 421 | 1038 | 238 | 418 | 1090 | + 4 | **12** | 414 | 1145 | 616 |
| **335** | 432 | 920 | 439 | 428 | 972 | 100 | 1 | 424 | 1024 | 202 | 420 | 1076 | 88 | **13** | 417 | 1128 | 581 |
| **336** | 432 | 889 | 408 | 428 | 940 | 132 | 2 | 424 | 990 | 168 | 420 | 1040 | 74 | **14** | 417 | 1090 | 543 |
| **337** | 433 | 859 | 378 | 429 | 907 | 164 | 3 | 425 | 956 | 135 | 421 | 1004 | 109 | **15** | 418 | 1053 | 507 |
| **338** | 433 | 828 | 347 | 429 | 875 | 196 | 4 | 425 | 922 | 101 | 421 | 968 | 145 | **16** | 418 | 1015 | 469 |
| **339** | 434 | 797 | 318 | 430 | 842 | 228 | 5 | 426 | 888 | 68 | 422 | 932 | 180 | **17** | 419 | 978 | 433 |
| **340** | 434 | 766 | 287 | 430 | 810 | 260 | 6 | 426 | 854 | 34 | 422 | 896 | 216 | **18** | 419 | 940 | 395 |
| **341** | 435 | 736 | 258 | 431 | 778 | 291 | 7 | 427 | 820 | 1 | 423 | 861 | 250 | **19** | 420 | 902 | 358 |
| **342** | 435 | 705 | 227 | 431 | 745 | 324 | 8 | 427 | 785 | + 34 | 423 | 825 | 286 | **20** | 420 | 865 | 321 |
| **343** | 436 | 675 | 197 | 432 | 713 | 355 | 9 | 428 | 751 | 67 | 424 | 789 | 321 | **21** | 421 | 827 | 284 |
| **344** | 436 | 640 | 167 | 432 | 680 | 388 | 10 | 428 | 717 | 101 | 424 | 753 | 356 | **22** | 421 | 790 | 247 |
| **345** | 437 | 613 | 137 | 433 | 648 | 419 | 11 | 429 | 683 | 135 | 425 | 717 | 391 | **23** | 422 | 752 | 210 |
| **346** | 437 | 583 | 106 | 433 | 616 | 451 | 12 | 429 | 649 | 169 | 425 | 681 | 427 | **24** | 422 | 714 | 172 |
| **347** | 438 | 552 | 77 | 434 | 583 | 483 | 13 | 430 | 615 | 202 | 426 | 645 | 462 | **25** | 423 | 677 | 136 |
| **348** | 438 | 521 | 46 | 434 | 551 | 515 | 14 | 430 | 581 | 236 | 426 | 609 | 498 | **26** | 423 | 639 | 98 |
| **349** | 439 | 491 | 16 | 435 | 518 | 547 | 15 | 431 | 547 | 269 | 427 | 573 | 533 | **27** | 424 | 602 | 62 |
| **350** | 439 | 460 | + 14 | 435 | 486 | 579 | 16 | 431 | 513 | 303 | 427 | 537 | 568 | **28** | 424 | 564 | 24 |
| **351** | 440 | 429 | 44 | 436 | 454 | 610 | 17 | 432 | 479 | 337 | 428 | 502 | 603 | **29** | 425 | 526 | + 13 |
| **352** | 440 | 399 | 75 | 436 | 421 | 643 | 18 | 432 | 444 | 371 | 428 | 466 | 639 | **30** | 425 | 489 | 50 |
| **353** | 441 | 368 | 104 | 437 | 389 | 674 | 19 | 433 | 410 | 404 | 429 | 430 | 674 | **31** | 426 | 451 | 87 |
| **354** | 441 | 337 | 135 | 437 | 356 | 707 | 20 | 433 | 376 | 438 | 429 | 394 | 710 | **1** | 426 | 414 | 124 |
| **355** | 442 | 307 | 165 | 438 | 324 | 738 | 21 | 434 | 342 | 471 | 430 | 358 | 745 | **2** | 427 | 376 | 161 |
| **356** | 442 | 276 | 195 | 438 | 292 | 770 | 22 | 434 | 308 | 505 | 430 | 322 | 781 | **3** | 427 | 338 | 199 |
| **357** | 443 | 245 | 225 | 439 | 259 | 802 | 23 | 435 | 274 | 538 | 431 | 286 | 816 | **4** | 428 | 301 | 235 |
| **358** | 443 | 215 | 256 | 439 | 227 | 834 | 24 | 435 | 240 | 572 | 431 | 250 | 852 | **5** | 428 | 263 | 273 |
| **359** | 444 | 184 | 285 | 440 | 194 | 866 | 25 | 436 | 206 | 605 | 432 | 214 | 887 | **6** | 429 | 226 | 309 |
| **360** | 444 | 153 | 316 | 440 | 162 | 898 | 26 | 436 | 172 | 639 | 432 | 178 | 922 | **7** | 429 | 188 | 347 |
| **361** | 445 | 123 | 346 | 441 | 130 | 929 | 27 | 437 | 138 | 673 | 433 | 143 | 957 | **8** | 430 | 150 | 384 |
| **362** | 445 | 92 | 376 | 441 | 97 | 962 | 28 | 437 | 103 | 707 | 433 | 107 | 993 | **9** | 430 | 113 | 421 |
| **363** | 446 | 61 | 406 | 442 | 65 | 993 | 29 | 438 | 69 | 740 | 434 | 71 | 1028 | **10** | 431 | 75 | 459 |
| **364** | 446 | 31 | 437 | 442 | 32 | 1026 | 30 | 438 | 35 | 774 | 434 | 36 | 1064 | **11** | 431 | 38 | 496 |
| **365** | 447 | 0 | 466 | 443 | 0 | 1057 | 31 | 439 | 0 | 807 | 435 | 0 | 1099 | **12** | 432 | 0 | 533 |

Days of Solar months 1–29 (days 306–334): *Tamil*, Masi; *Malayalam*, Kumbam; *Bengal*, Phalguna.

Days of Solar months 1–31 (days 335–365): *Tamil*, Panguni; *Malayalam*, Meenam; *Bengal*, Chaitra.

Eng. date: February (12–28), March (1–31), April (1–12).

# TABLE XIII.

## Sunrise from 13° to 17° N. Latitude.

| Day of Solar year. | Lat. 13° +1072" Long. Eqn. of time in seconds. | ☉'s Trop. Long. in seconds. | Total corrn. in seconds. Madras. A.D. 1840—1920 | Lat. 14° +1008" Long. Eqn. of time in seconds. | ☉'s Trop. Long. in seconds. | Total Corrn. in seconds. Nellore. A.D. 1840—1920 | Solar month | Days of Solar months. | Lat. 15° +168" Long. Eqn. of time in seconds. | ☉'s Trop. Long. in seconds. | Total Corrn. in seconds. Vijianagaram. A.D. 1840—1920 | Lat. 16° —300" Long. Eqn. of Time in seconds. | ☉'s Trop. Long. in seconds. | Total corrn. in seconds. Belgaum. A.D. 1840—1920 | Month | Eng. date. | Lat. 17° +648" Long. Eqn. of Time in seconds. | ☉'s Trop. Long. in seconds. | Total Corrn. in seconds. Hyderabad. A.D. 1840—1920 |
|---|---|---|---|---|---|---|---|---|---|---|---|---|---|---|---|---|---|---|---|
| **1** | —428 | +39 | +1464 | —424 | +41 | +1445 | *Tamil*, Chittirai; *Malayalam* Medam; *Bengal*, Vaisakha. | 1 | —420 | +43 | +651 | —417 | +45 | +212 | April | **13** | —413 | +46 | +1208 |
| **2** | 426 | 78 | 1505 | 422 | 82 | 1488 | | 2 | 418 | 86 | 697 | 415 | 89 | 259 | | **14** | 411 | 93 | 1256 |
| **3** | 424 | 118 | 1546 | 420 | 123 | 1531 | | 3 | 416 | 129 | 741 | 413 | 133 | 305 | | **15** | 409 | 139 | 1305 |
| **4** | 423 | 157 | 1586 | 419 | 164 | 1573 | | 4 | 415 | 172 | 785 | 411 | 178 | 352 | | **16** | 408 | 186 | 1354 |
| **5** | 421 | 196 | 1627 | 417 | 205 | 1616 | | 5 | 413 | 215 | 830 | 410 | 222 | 397 | | **17** | 406 | 232 | 1401 |
| **6** | 419 | 235 | 1669 | 415 | 246 | 1659 | | 6 | 411 | 258 | 875 | 407 | 267 | 445 | | **18** | 404 | 278 | 1450 |
| **7** | 417 | 274 | 1705 | 413 | 287 | 1702 | | 7 | 409 | 301 | 920 | 406 | 311 | 490 | | **19** | 402 | 325 | 1498 |
| **8** | 414 | 314 | 1752 | 410 | 328 | 1746 | | 8 | 407 | 344 | 965 | 403 | 351 | 538 | | **20** | 399 | 371 | 1547 |
| **9** | 412 | 351 | 1783 | 406 | 369 | 1791 | | 9 | 404 | 387 | 1011 | 401 | 395 | 584 | | **21** | 397 | 418 | 1596 |
| **10** | 409 | 388 | 1839 | 405 | 410 | 1835 | | 10 | 402 | 430 | 1056 | 398 | 440 | 638 | | **22** | 395 | 464 | 1645 |
| **11** | 407 | 428 | 1859 | 403 | 451 | 1856 | | 11 | 399 | 473 | 1082 | 396 | 484 | 662 | | **23** | 392 | 510 | 1672 |
| **12** | 404 | 467 | 1880 | 400 | 492 | 1879 | | 12 | 396 | 516 | 1106 | 393 | 529 | 688 | | **24** | 389 | 557 | 1688 |
| **13** | 401 | 506 | 1901 | 397 | 533 | 1901 | | 13 | 394 | 559 | 1128 | 391 | 573 | 712 | | **25** | 387 | 603 | 1724 |
| **14** | 398 | 545 | 1922 | 394 | 574 | 1923 | | 14 | 391 | 602 | 1152 | 387 | 618 | 738 | | **26** | 384 | 650 | 1751 |
| **15** | 395 | 584 | 1943 | 391 | 615 | 1945 | | 15 | 388 | 645 | 1176 | 384 | 662 | 763 | | **27** | 381 | 696 | 1777 |
| **16** | 392 | 624 | 1965 | 388 | 656 | 1968 | | 16 | 384 | 688 | 1201 | 381 | 707 | 789 | | **28** | 378 | 743 | 1801 |
| **17** | 388 | 663 | 1987 | 385 | 697 | 1990 | | 17 | 381 | 731 | 1235 | 378 | 751 | 814 | | **29** | 374 | 790 | 1832 |
| **18** | 385 | 702 | 2008 | 381 | 738 | 2013 | | 18 | 378 | 774 | 1248 | 375 | 796 | 839 | | **30** | 371 | 836 | 1859 |
| **19** | 382 | 741 | 2029 | 378 | 779 | 2036 | | 19 | 374 | 817 | 1273 | 371 | 840 | 866 | May | **1** | 368 | 883 | 1885 |
| **20** | 377 | 780 | 2052 | 374 | 820 | 2059 | | 20 | 370 | 860 | 1298 | 367 | 885 | 892 | | **2** | 364 | 929 | 1913 |
| **21** | 374 | 820 | 2073 | 370 | 861 | 2082 | | 21 | 367 | 903 | 1322 | 364 | 929 | 917 | | **3** | 360 | 973 | 1941 |
| **22** | 370 | 859 | 2095 | 366 | 902 | 2106 | | 22 | 363 | 946 | 1351 | 360 | 974 | 944 | | **4** | 358 | 1019 | 1966 |
| **23** | 368 | 898 | 2115 | 364 | 943 | 2127 | | 23 | 361 | 989 | 1369 | 358 | 1018 | 968 | | **5** | 356 | 1066 | 1992 |
| **24** | 366 | 937 | 2135 | 362 | 984 | 2148 | | 24 | 359 | 1032 | 1392 | 356 | 1063 | 992 | | **6** | 354 | 1113 | 2018 |
| **25** | 364 | 976 | 2155 | 360 | 1025 | 2166 | | 26 | 357 | 1075 | 1415 | 354 | 1107 | 1016 | | **7** | 352 | 1159 | 2043 |
| **26** | 362 | 1016 | 2176 | 358 | 1066 | 2191 | | 26 | 355 | 1118 | 1438 | 352 | 1152 | 1041 | | **8** | 350 | 1206 | 2061 |
| **27** | 360 | 1055 | 2196 | 356 | 1107 | 2212 | | 27 | 353 | 1161 | 1461 | 350 | 1196 | 1065 | | **9** | 348 | 1252 | 2095 |
| **28** | 358 | 1094 | 2216 | 354 | 1148 | 2233 | | 28 | 351 | 1204 | 1483 | 349 | 1241 | 1088 | | **10** | 347 | 1298 | 2120 |
| **29** | 356 | 1123 | 2236 | 352 | 1189 | 2255 | | 29 | 349 | 1247 | 1506 | 348 | 1285 | 1114 | | **11** | 346 | 1345 | 2144 |
| **30** | 354 | 1176 | 2256 | 350 | 1232 | 2276 | | 30 | 347 | 1290 | 1529 | 347 | 1336 | 1135 | | **12** | 345 | 1392 | 2169 |
| **31** | 352 | 1194 | 2276 | 348 | 1251 | 2297 | *Tamil*, Vaikasi; *Malayalam*, Edavam; *Bengal*, Jyeshta. | 1 | 346 | 1313 | 1545 | 346 | 1358 | 1158 | | **13** | 344 | 1416 | 2194 |
| **32** | 350 | 1212 | 2296 | 347 | 1271 | 2318 | | 2 | 345 | 1334 | 1573 | 345 | 1381 | 1182 | | **14** | 343 | 1439 | 2218 |
| **33** | 348 | 1230 | 2316 | 346 | 1290 | 2338 | | 3 | 344 | 1354 | 1594 | 344 | 1403 | 1205 | | **15** | 342 | 1463 | 2219 |
| **34** | 347 | 1248 | 2335 | 345 | 1309 | 2358 | | 4 | 343 | 1375 | 1616 | 340 | 1425 | 1231 | | **16** | 338 | 1487 | 2247 |
| **35** | 341 | 1266 | 2359 | 339 | 1328 | 2383 | | 5 | 337 | 1396 | 1643 | 335 | 1447 | 1258 | | **17** | 332 | 1510 | 2277 |
| **36** | 335 | 1285 | 2384 | 333 | 1348 | 2409 | | 6 | 331 | 1417 | 1670 | 329 | 1470 | 1287 | | **18** | 326 | 1531 | 2307 |
| **37** | 329 | 1303 | 2408 | 327 | 1367 | 2434 | | 7 | 325 | 1438 | 1697 | 323 | 1492 | 1315 | | **19** | 321 | 1558 | 2359 |
| **38** | 324 | 1321 | 2431 | 322 | 1386 | 2458 | | 8 | 320 | 1458 | 1722 | 317 | 1514 | 1343 | | **20** | 315 | 1582 | 2389 |
| **39** | 317 | 1339 | 2456 | 316 | 1406 | 2484 | | 9 | 313 | 1479 | 1750 | 311 | 1537 | 1372 | | **21** | 309 | 1605 | 2418 |
| **40** | 311 | 1357 | 2480 | 309 | 1425 | 2510 | | 10 | 307 | 1500 | 1777 | 305 | 1559 | 1400 | | **22** | 303 | 1629 | 2448 |
| **41** | 305 | 1375 | 2503 | 303 | 1444 | 2537 | | 11 | 301 | 1521 | 1808 | 299 | 1581 | 1429 | | **23** | 297 | 1653 | 2479 |
| **42** | 298 | 1393 | 2499 | 296 | 1464 | 2534 | | 12 | 294 | 1542 | 1800 | 292 | 1604 | 1427 | | **24** | 290 | 1676 | 2478 |
| **43** | 292 | 1411 | 2494 | 290 | 1483 | 2530 | | 13 | 288 | 1562 | 1797 | 286 | 1626 | 1424 | | **25** | 284 | 1700 | 2475 |
| **44** | 285 | 1429 | 2491 | 283 | 1502 | 2527 | | 14 | 282 | 1583 | 1793 | 280 | 1648 | 1421 | | **26** | 278 | 1724 | 2473 |
| **45** | 278 | 1447 | 2487 | 277 | 1521 | 2523 | | 15 | 275 | 1604 | 1791 | 273 | 1670 | 1419 | | **27** | 271 | 1747 | 2471 |
| **46** | 272 | 1466 | 2482 | 270 | 1541 | 2519 | | 16 | 268 | 1625 | 1788 | 267 | 1693 | 1416 | | **28** | 265 | 1771 | 2469 |
| **47** | 265 | 1484 | 2478 | 263 | 1560 | 2516 | | 17 | 261 | 1646 | 1785 | 259 | 1715 | 1416 | | **29** | 258 | 1795 | 2468 |
| **48** | 258 | 1502 | 2474 | 256 | 1579 | 2513 | | 18 | 254 | 1666 | 1783 | 253 | 1737 | 1413 | | **30** | 251 | 1819 | 2466 |
| **49** | 251 | 1520 | 2471 | 249 | 1599 | 2510 | | 19 | 248 | 1687 | 1779 | 246 | 1760 | 1411 | | **31** | 244 | 1842 | 2464 |
| **50** | 244 | 1538 | 2467 | 242 | 1618 | 2507 | | 20 | 241 | 1708 | 1777 | 239 | 1782 | 1409 | June | **1** | 237 | 1866 | 2463 |
| **51** | 237 | 1556 | 2463 | 235 | 1637 | 2504 | | 21 | 234 | 1729 | 1774 | 232 | 1804 | 1407 | | **2** | 230 | 1890 | 2462 |
| **52** | 229 | 1574 | 2460 | 228 | 1657 | 2501 | | 22 | 226 | 1750 | 1772 | 225 | 1827 | 1405 | | **3** | 223 | 1913 | 2461 |
| **53** | 222 | 1592 | 2457 | 220 | 1676 | 2499 | | 23 | 219 | 1770 | 1770 | 217 | 1819 | 1404 | | **4** | 216 | 1937 | 2459 |
| **54** | 214 | 1610 | 2454 | 213 | 1695 | 2496 | | 24 | 212 | 1791 | 1767 | 210 | 1871 | 1402 | | **5** | 209 | 1961 | 2458 |
| **55** | 207 | 1628 | 2450 | 206 | 1714 | 2493 | | 25 | 204 | 1812 | 1766 | 203 | 1893 | 1400 | | **6** | 201 | 1984 | 2457 |
| **56** | 200 | 1647 | 2446 | 198 | 1734 | 2490 | | 26 | 197 | 1833 | 1763 | 196 | 1916 | 1398 | | **7** | 194 | 2008 | 2456 |
| **57** | 192 | 1665 | 2443 | 190 | 1753 | 2488 | | 27 | 189 | 1854 | 1761 | 188 | 1938 | 1398 | | **8** | 186 | 2032 | 2456 |
| **58** | 188 | 1683 | 2436 | 186 | 1772 | 2482 | | 28 | 184 | 1874 | 1757 | 183 | 1960 | 1391 | | **9** | 182 | 2056 | 2451 |
| **59** | 184 | 1701 | 2430 | 182 | 1792 | 2476 | | 29 | 180 | 1895 | 1751 | 179 | 1983 | 1389 | | **10** | 178 | 2079 | 2447 |
| **60** | 180 | 1719 | 2423 | 178 | 1811 | 2470 | | 30 | 176 | 1916 | 1746 | 175 | 2005 | 1384 | | **11** | 174 | 2103 | 2442 |
| **61** | 176 | 1736 | 2416 | 174 | 1832 | 2464 | | 31 | 172 | 1936 | 1740 | 171 | 2028 | 1379 | | **12** | 171 | 2128 | 2137 |
| **62** | 172 | 1725 | 2409 | 170 | 1822 | 2458 | | 32 | 168 | 1926 | 1734 | 167 | 2019 | 1374 | | **13** | 167 | 2120 | 2433 |

# TABLE XIII.

## Sunrise from 13° to 17° N. Latitude.

| Day of Solar year. | Lat. 13° +1072″ Long. Eqn. of time in seconds. | ☉'s Trop. Long. in seconds. | Total Corrn. in seconds. Madras. A.D. 1840—1920 | Lat. 14° +1008″ Long. Eqn. of time in seconds. | ☉'s Trop. Long. in seconds. | Total Corrn. in seconds. Nellore. A.D. 1840—1920 | Days of Solar months. | Lat. 15° +168″ Long. Eqn. of time in seconds. | ☉'s Trop. Long. in seconds. | Total Corrn. in seconds. Vijianagaram. A.D. 1840—1920 | Lat. 16° —300″ Long. Eqn. of time in seconds. | ☉'s Trop. Long. in seconds. | Total Corrn. in seconds. Belgaum. A.D. 1840—1920 | Eng. date. | Lat. 17° +648″ Long. Eqn. of time in seconds. | ☉'s Trop. Long. in seconds. | Total Corrn. in seconds. Hyderabad. A.D. 1840—1920 |
|---|---|---|---|---|---|---|---|---|---|---|---|---|---|---|---|---|---|
| 63 | —164 | +1714 | +2406 | —164 | +1812 | +2454 | 1 (Tamil, Ani; Malayalam, Mithunum; Bengal, Ashada.) | —164 | +1917 | +1729 | —163 | +2010 | +1369 | 14 (June) | —163 | +2111 | +2428 |
| 64 | 155 | 1704 | 2405 | 155 | 1802 | 2453 | 2 | 155 | 1907 | 1728 | 154 | 2001 | 1369 | 15 | 154 | 2103 | 2429 |
| 65 | 146 | 1693 | 2403 | 146 | 1792 | 2452 | 3 | 146 | 1898 | 1728 | 145 | 1992 | 1369 | 16 | 145 | 2094 | 2429 |
| 66 | 137 | 1682 | 2401 | 137 | 1781 | 2450 | 4 | 137 | 1888 | 1727 | 136 | 1983 | 1369 | 17 | 136 | 2086 | 2430 |
| 67 | 128 | 1671 | 2399 | 128 | 1771 | 2449 | 5 | 127 | 1878 | 1727 | 126 | 1975 | 1371 | 18 | 127 | 2078 | 2431 |
| 68 | 119 | 1660 | 2397 | 119 | 1761 | 2448 | 6 | 118 | 1869 | 1727 | 118 | 1966 | 1370 | 19 | 118 | 2069 | 2431 |
| 69 | 109 | 1650 | 2397 | 109 | 1751 | 2448 | 7 | 109 | 1859 | 1626 | 109 | 1957 | 1370 | 20 | 108 | 2060 | 2433 |
| 70 | 100 | 1639 | 2395 | 100 | 1741 | 2447 | 8 | 100 | 1850 | 1727 | 99 | 1948 | 1371 | 21 | 99 | 2052 | 2432 |
| 71 | 91 | 1628 | 2393 | 91 | 1731 | 2445 | 9 | 90 | 1840 | 1726 | 90 | 1939 | 1370 | 22 | 90 | 2044 | 2434 |
| 72 | 81 | 1617 | 2379 | 81 | 1721 | 2430 | 10 | 81 | 1830 | 1709 | 81 | 1930 | 1353 | 23 | 80 | 2036 | 2417 |
| 73 | 72 | 1606 | 2363 | 72 | 1711 | 2414 | 11 | 71 | 1821 | 1694 | 71 | 1921 | 1337 | 24 | 71 | 2027 | 2400 |
| 74 | 62 | 1596 | 2349 | 62 | 1701 | 2399 | 12 | 62 | 1811 | 1677 | 62 | 1912 | 1319 | 25 | 62 | 2019 | 2382 |
| 75 | 54 | 1585 | 2332 | 54 | 1691 | 2382 | 13 | 54 | 1802 | 1660 | 54 | 1903 | 1301 | 26 | 54 | 2010 | 2363 |
| 76 | 43 | 1574 | 2319 | 43 | 1680 | 2368 | 14 | 43 | 1792 | 1645 | 43 | 1894 | 1286 | 27 | 43 | 2002 | 2347 |
| 77 | 34 | 1563 | 2304 | 33 | 1670 | 2353 | 15 | 33 | 1782 | 1630 | 33 | 1886 | 1270 | 28 | 33 | 1994 | 2331 |
| 78 | 24 | 1552 | 2289 | 24 | 1660 | 2337 | 16 | 24 | 1773 | 1613 | 24 | 1877 | 1253 | 29 | 24 | 1985 | 2313 |
| 79 | 15 | 1542 | 2274 | 15 | 1650 | 2321 | 17 | 15 | 1763 | 1597 | 15 | 1868 | 1235 | 30 | 15 | 1977 | 2295 |
| 80 | 5 | 1531 | 2259 | 5 | 1640 | 2306 | 18 | 5 | 1754 | 1581 | 5 | 1859 | 1219 | 1 | 5 | 1968 | 2279 |
| 81 | + 4 | 1520 | 2244 | + 4 | 1630 | 2290 | 19 | + 4 | 1744 | 1565 | + 4 | 1850 | 1202 | 2 | + 4 | 1960 | 2261 |
| 82 | 14 | 1509 | 2230 | 14 | 1620 | 2275 | 20 | 14 | 1734 | 1549 | 14 | 1841 | 1186 | 3 | 14 | 1952 | 2244 |
| 83 | 23 | 1498 | 2214 | 23 | 1610 | 2259 | 21 | 23 | 1725 | 1533 | 23 | 1832 | 1169 | 4 | 23 | 1943 | 2227 |
| 84 | 33 | 1488 | 2200 | 33 | 1600 | 2244 | 22 | 33 | 1715 | 1517 | 33 | 1823 | 1152 | 5 | 33 | 1935 | 2220 |
| 85 | 43 | 1477 | 2185 | 43 | 1590 | 2229 | 23 | 42 | 1706 | 1501 | 42 | 1814 | 1135 | 6 | 42 | 1926 | 2192 |
| 86 | 52 | 1466 | 2170 | 52 | 1579 | 2213 | 24 | 52 | 1696 | 1485 | 52 | 1805 | 1119 | 7 | 51 | 1918 | 2174 |
| 87 | 61 | 1455 | 2155 | 61 | 1569 | 2197 | 25 | 61 | 1686 | 1469 | 61 | 1797 | 1102 | 8 | 61 | 1910 | 2158 |
| 88 | 71 | 1444 | 2140 | 71 | 1559 | 2182 | 26 | 71 | 1677 | 1453 | 70 | 1788 | 1085 | 9 | 70 | 1901 | 2140 |
| 89 | 80 | 1434 | 2125 | 80 | 1549 | 2166 | 27 | 80 | 1667 | 1437 | 80 | 1779 | 1068 | 10 | 80 | 1893 | 2123 |
| 90 | 90 | 1423 | 2110 | 89 | 1539 | 2150 | 28 | 89 | 1658 | 1420 | 89 | 1770 | 1051 | 11 | 89 | 1884 | 2106 |
| 91 | 105 | 1412 | 2101 | 105 | 1528 | 2141 | 29 | 105 | 1648 | 1411 | 105 | 1760 | 1041 | 12 | 106 | 1876 | 2096 |
| 92 | 115 | 1388 | 2087 | 115 | 1503 | 2126 | 30 | 115 | 1622 | 1395 | 116 | 1734 | 1026 | 13 | 116 | 1849 | 2079 |
| 93 | 124 | 1363 | 2071 | 125 | 1478 | 2111 | 31 | 125 | 1597 | 1380 | 125 | 1708 | 1109 | 14 | 126 | 1823 | 2063 |
| 94 | 134 | 1339 | 2063 | 134 | 1453 | 2095 | 32 | 135 | 1571 | 1364 | 135 | 1681 | 992 | 15 | 135 | 1796 | 2045 |
| 95 | 144 | 1314 | 2042 | 144 | 1428 | 2080 | 1 (Tamil, Adi; Malayalam, Karkatagam; Bengal, Sravana.) | 144 | 1546 | 1348 | 145 | 1655 | 976 | 16 (July) | 145 | 1769 | 2028 |
| 96 | 154 | 1290 | 2028 | 154 | 1403 | 2065 | 2 | 154 | 1520 | 1332 | 155 | 1629 | 960 | 17 | 155 | 1742 | 2011 |
| 97 | 163 | 1266 | 2013 | 164 | 1378 | 2050 | 3 | 164 | 1495 | 1317 | 164 | 1603 | 943 | 18 | 165 | 1716 | 1995 |
| 98 | 173 | 1241 | 1998 | 173 | 1353 | 2034 | 4 | 174 | 1469 | 1301 | 174 | 1577 | 927 | 19 | 174 | 1689 | 1977 |
| 99 | 183 | 1217 | 1984 | 183 | 1328 | 2019 | 5 | 183 | 1444 | 1285 | 184 | 1550 | 910 | 20 | 184 | 1662 | 1960 |
| 100 | 192 | 1192 | 1968 | 192 | 1303 | 2003 | 6 | 193 | 1418 | 1269 | 193 | 1524 | 893 | 21 | 193 | 1636 | 1943 |
| 101 | 201 | 1168 | 1953 | 202 | 1278 | 1988 | 7 | 202 | 1393 | 1253 | 202 | 1498 | 876 | 22 | 203 | 1609 | 1926 |
| 102 | 210 | 1144 | 1938 | 211 | 1253 | 1971 | 8 | 211 | 1367 | 1235 | 212 | 1472 | 860 | 23 | 212 | 1582 | 1908 |
| 103 | 220 | 1119 | 1931 | 220 | 1228 | 1961 | 9 | 221 | 1342 | 1224 | 221 | 1446 | 847 | 24 | 222 | 1556 | 1895 |
| 104 | 229 | 1095 | 1922 | 230 | 1203 | 1953 | 10 | 230 | 1316 | 1213 | 231 | 1419 | 836 | 25 | 231 | 1539 | 1881 |
| 105 | 238 | 1070 | 1914 | 239 | 1178 | 1943 | 11 | 239 | 1291 | 1202 | 240 | 1393 | 823 | 26 | 240 | 1502 | 1866 |
| 106 | 247 | 1046 | 1906 | 248 | 1153 | 1933 | 12 | 248 | 1265 | 1191 | 249 | 1367 | 810 | 27 | 249 | 1475 | 1852 |
| 107 | 256 | 1022 | 1897 | 257 | 1128 | 1923 | 13 | 257 | 1240 | 1179 | 258 | 1341 | 797 | 28 | 258 | 1449 | 1838 |
| 108 | 266 | 997 | 1890 | 266 | 1103 | 1914 | 14 | 266 | 1214 | 1168 | 267 | 1315 | 785 | 29 | 267 | 1422 | 1824 |
| 109 | 274 | 973 | 1881 | 274 | 1078 | 1903 | 15 | 275 | 1189 | 1157 | 276 | 1288 | 772 | 30 | 276 | 1395 | 1810 |
| 110 | 283 | 948 | 1873 | 283 | 1053 | 1893 | 16 | 284 | 1163 | 1146 | 285 | 1262 | 759 | 31 | 285 | 1369 | 1795 |
| 111 | 291 | 924 | 1863 | 292 | 1028 | 1884 | 17 | 292 | 1138 | 1138 | 293 | 1236 | 746 | 1 (August) | 294 | 1342 | 1781 |
| 112 | 300 | 900 | 1855 | 300 | 1003 | 1873 | 18 | 301 | 1112 | 1122 | 301 | 1210 | 732 | 2 | 302 | 1315 | 1766 |
| 113 | 308 | 875 | 1846 | 309 | 978 | 1863 | 19 | 309 | 1087 | 1110 | 310 | 1184 | 719 | 3 | 311 | 1289 | 1752 |
| 114 | 317 | 857 | 1837 | 318 | 953 | 1854 | 20 | 318 | 1061 | 1098 | 319 | 1157 | 707 | 4 | 320 | 1262 | 1738 |
| 115 | 325 | 826 | 1828 | 326 | 928 | 1843 | 21 | 326 | 1086 | 1086 | 327 | 1131 | 693 | 5 | 328 | 1235 | 1722 |
| 116 | 333 | 802 | 1819 | 334 | 903 | 1832 | 22 | 334 | 1010 | 1074 | 335 | 1105 | 679 | 6 | 336 | 1208 | 1707 |
| 117 | 341 | 778 | 1809 | 342 | 878 | 1821 | 23 | 342 | 985 | 1061 | 343 | 1079 | 665 | 7 | 344 | 1182 | 1692 |
| 118 | 349 | 753 | 1800 | 350 | 853 | 1811 | 24 | 351 | 959 | 1050 | 352 | 1053 | 658 | 8 | 352 | 1155 | 1677 |
| 119 | 356 | 729 | 1790 | 358 | 828 | 1797 | 25 | 356 | 934 | 1035 | 358 | 1026 | 637 | 9 | 368 | 1128 | 1670 |
| 120 | 360 | 704 | 1777 | 360 | 803 | 1783 | 26 | 361 | 908 | 1020 | 363 | 1000 | 620 | 10 | 364 | 1102 | 1642 |
| 121 | 364 | 680 | 1755 | 365 | 778 | 1770 | 27 | 366 | 883 | 1004 | 369 | 974 | 605 | 11 | 370 | 1075 | 1625 |
| 122 | 368 | 656 | 1750 | 370 | 752 | 1756 | 28 | 372 | 856 | 990 | 375 | 948 | 589 | 12 | 377 | 1048 | 1609 |
| 123 | 375 | 639 | 1740 | 377 | 733 | 1744 | 29 | 379 | 835 | 977 | 382 | 926 | 574 | 13 | 384 | 1025 | 1593 |
| 124 | 382 | 621 | 1729 | 385 | 715 | 1734 | 30 | 387 | 815 | 964 | 389 | 905 | 560 | 14 | 391 | 1002 | 1577 |
| | | | | 391 | 696 | 1721 | 31 | 394 | 795 | 951 | 396 | 883 | 545 | 15 | 398 | 978 | 1559 |

# TABLE XIII.

## Sunrise from 13° to 17° N. Latitude.

| Day of Solar year. | Lat. 13° +1072″ Long. Eqn. of time in seconds. | ⊙'s Trop. Long. in seconds. | Total Corrn. in seconds. Madras. A.D. 1840—1920 | Lat. 14° +1008″ Long. Eqn. of time in seconds. | ⊙'s Trop. Long. in seconds. | Total Corrn. in seconds. Nellore. A.D. 1840—1920 | Solar month | Days of solar months. | Lat. 15° +168″ Long. Eqn. of time in seconds. | ⊙'s Trop. Long. in seconds. | Total Corrn. in seconds. Vizianagaram. A.D. 1840—1920 | Lat. 16° —300″ Long. Eqn. of Time in seconds. | ⊙'s Trop. Long. in seconds. | Total Corrn. in seconds. Belgaum. A.D. 1840—1920 | Eng. month | Eng. date. | Lat 17° +648″ Long. Eqn. of Time in seconds. | ⊙'s Trop. Long. in seconds. | Total Corrn. in seconds. Hyderabad. A.D. 1840—1920 |
|---|---|---|---|---|---|---|---|---|---|---|---|---|---|---|---|---|---|---|---|
| 126 | +396 | +587 | +1709 | +398 | +677 | +1709 | Tamil, Avani; Malayalam, Chingam; Bengal, Bhadrapada. | 1 | +401 | +775 | +938 | +403 | +861 | +530 | August | 16 | +405 | +955 | +1544 |
| 127 | 403 | 569 | 1698 | 405 | 658 | 1697 | | 2 | 408 | 754 | 924 | 410 | 839 | 515 | | 17 | 412 | 982 | 1528 |
| 128 | 409 | 552 | 1687 | 412 | 640 | 1686 | | 3 | 414 | 734 | 910 | 416 | 818 | 500 | | 18 | 419 | 909 | 1512 |
| 129 | 416 | 535 | 1677 | 418 | 621 | 1673 | | 4 | 421 | 714 | 897 | 423 | 796 | 485 | | 19 | 425 | 886 | 1495 |
| 130 | 422 | 518 | 1666 | 425 | 602 | 1661 | | 5 | 427 | 694 | 883 | 430 | 774 | 470 | | 20 | 432 | 862 | 1478 |
| 131 | 428 | 500 | 1654 | 431 | 584 | 1649 | | 6 | 433 | 673 | 868 | 436 | 753 | 445 | | 21 | 438 | 839 | 1461 |
| 132 | 434 | 483 | 1642 | 437 | 565 | 1637 | | 7 | 439 | 653 | 855 | 442 | 731 | 438 | | 22 | 444 | 816 | 1444 |
| 133 | 440 | 466 | 1644 | 443 | 546 | 1637 | | 8 | 446 | 633 | 854 | 448 | 709 | 434 | | 23 | 451 | 793 | 1440 |
| 134 | 416 | 448 | 1645 | 449 | 528 | 1637 | | 9 | 451 | 612 | 851 | 454 | 688 | 431 | | 24 | 457 | 770 | 1434 |
| 135 | 452 | 431 | 1647 | 454 | 509 | 1635 | | 10 | 457 | 592 | 849 | 460 | 666 | 427 | | 25 | 462 | 746 | 1428 |
| 136 | 457 | 414 | 1647 | 460 | 490 | 1635 | | 11 | 463 | 572 | 847 | 465 | 644 | 423 | | 26 | 468 | 723 | 1423 |
| 137 | 462 | 396 | 1648 | 465 | 471 | 1634 | | 12 | 468 | 551 | 844 | 471 | 622 | 419 | | 27 | 474 | 700 | 1417 |
| 138 | 468 | 379 | 1650 | 471 | 453 | 1634 | | 13 | 474 | 531 | 842 | 476 | 601 | 415 | | 28 | 479 | 677 | 1411 |
| 139 | 473 | 362 | 1650 | 476 | 434 | 1633 | | 14 | 479 | 511 | 839 | 482 | 579 | 411 | | 29 | 485 | 654 | 1406 |
| 140 | 478 | 345 | 1651 | 481 | 415 | 1631 | | 15 | 484 | 491 | 836 | 487 | 557 | 407 | | 30 | 489 | 630 | 1399 |
| 141 | 483 | 327 | 1651 | 486 | 397 | 1630 | | 16 | 489 | 470 | 833 | 492 | 536 | 402 | | 31 | 494 | 607 | 1392 |
| 142 | 488 | 310 | 1652 | 491 | 378 | 1629 | | 17 | 494 | 450 | 830 | 496 | 514 | 397 | September | 1 | 499 | 584 | 1386 |
| 143 | 492 | 293 | 1652 | 495 | 359 | 1627 | | 18 | 498 | 430 | 826 | 501 | 492 | 392 | | 2 | 504 | 561 | 1380 |
| 144 | 497 | 275 | 1652 | 500 | 341 | 1626 | | 19 | 503 | 409 | 823 | 506 | 471 | 388 | | 3 | 509 | 538 | 1373 |
| 145 | 501 | 258 | 1652 | 504 | 322 | 1623 | | 20 | 507 | 389 | 819 | 510 | 449 | 382 | | 4 | 514 | 513 | 1367 |
| 146 | 505 | 241 | 1651 | 508 | 303 | 1621 | | 21 | 511 | 369 | 815 | 514 | 427 | 371 | | 5 | 517 | 491 | 1359 |
| 147 | 509 | 223 | 1651 | 512 | 284 | 1619 | | 22 | 516 | 348 | 812 | 518 | 405 | 371 | | 6 | 521 | 468 | 1351 |
| 148 | 513 | 206 | 1651 | 517 | 266 | 1618 | | 23 | 520 | 328 | 808 | 523 | 384 | 367 | | 7 | 526 | 445 | 1345 |
| 149 | 517 | 189 | 1650 | 520 | 247 | 1615 | | 24 | 523 | 308 | 803 | 526 | 362 | 360 | | 8 | 529 | 422 | 1337 |
| 150 | 520 | 172 | 1649 | 524 | 228 | 1612 | | 25 | 527 | 288 | 799 | 530 | 340 | 355 | | 9 | 533 | 398 | 1330 |
| 151 | 523 | 154 | 1647 | 527 | 210 | 1609 | | 26 | 530 | 267 | 794 | 533 | 319 | 348 | | 10 | 536 | 375 | 1321 |
| 152 | 501 | 136 | 1621 | 505 | 192 | 1581 | | 27 | 508 | 248 | 764 | 512 | 296 | 318 | | 11 | 516 | 352 | 1290 |
| 153 | 504 | 132 | 1620 | 508 | 186 | 1578 | | 28 | 512 | 240 | 760 | 515 | 286 | 311 | | 12 | 519 | 341 | 1282 |
| 154 | 507 | 127 | 1618 | 511 | 180 | 1575 | | 29 | 515 | 232 | 755 | 518 | 277 | 305 | | 13 | 522 | 329 | 1273 |
| 155 | 510 | 123 | 1617 | 514 | 173 | 1571 | | 30 | 518 | 224 | 750 | 527 | 267 | 304 | | 14 | 525 | 318 | 1265 |
| 156 | 512 | 118 | 1614 | 516 | 167 | 1567 | | 31 | 520 | 216 | 744 | 523 | 258 | 291 | | 15 | 527 | 307 | 1256 |
| 157 | 514 | 114 | 1612 | 518 | 161 | 1563 | Tamil, Purattasi; Malayalam, Kanni; Bengal, Asvina. | 1 | 522 | 208 | 738 | 525 | 248 | 283 | | 16 | 529 | 295 | 1246 |
| 158 | 517 | 110 | 1611 | 521 | 155 | 1560 | | 2 | 525 | 200 | 733 | 528 | 239 | 277 | | 17 | 532 | 284 | 1238 |
| 159 | 519 | 105 | 1608 | 523 | 149 | 1556 | | 3 | 527 | 192 | 727 | 530 | 229 | 269 | | 18 | 534 | 273 | 1229 |
| 160 | 521 | 101 | 1606 | 525 | 142 | 1551 | | 4 | 529 | 184 | 721 | 532 | 220 | 262 | | 19 | 536 | 262 | 1220 |
| 161 | 523 | 96 | 1603 | 527 | 136 | 1547 | | 5 | 531 | 176 | 715 | 534 | 210 | 254 | | 20 | 538 | 250 | 1210 |
| 162 | 524 | 92 | 1600 | 528 | 130 | 1542 | | 6 | 532 | 168 | 708 | 535 | 201 | 246 | | 21 | 539 | 239 | 1200 |
| 163 | 525 | 88 | 1597 | 529 | 124 | 1537 | | 7 | 533 | 160 | 701 | 537 | 191 | 237 | | 22 | 541 | 228 | 1189 |
| 164 | 527 | 83 | 1595 | 531 | 118 | 1533 | | 8 | 535 | 152 | 695 | 538 | 182 | 228 | | 23 | 543 | 216 | 1179 |
| 165 | 528 | 79 | 1591 | 532 | 111 | 1527 | | 9 | 536 | 144 | 687 | 539 | 172 | 219 | | 24 | 548 | 205 | 1168 |
| 166 | 528 | 74 | 1588 | 532 | 105 | 1521 | | 10 | 536 | 136 | 679 | 540 | 163 | 210 | | 25 | 544 | 194 | 1157 |
| 167 | 529 | 70 | 1584 | 533 | 99 | 1515 | | 11 | 536 | 128 | 671 | 541 | 153 | 201 | | 26 | 545 | 182 | 1146 |
| 168 | 530 | 66 | 1580 | 534 | 93 | 1510 | | 12 | 538 | 120 | 665 | 542 | 144 | 193 | | 27 | 546 | 171 | 1136 |
| 169 | 531 | 61 | 1576 | 535 | 87 | 1505 | | 13 | 539 | 112 | 657 | 542 | 134 | 183 | | 28 | 547 | 160 | 1125 |
| 170 | 531 | 57 | 1572 | 535 | 80 | 1498 | | 14 | 539 | 104 | 649 | 543 | 125 | 174 | | 29 | 547 | 149 | 1113 |
| 171 | 531 | 52 | 1567 | 536 | 74 | 1493 | | 15 | 540 | 96 | 642 | 543 | 115 | 164 | | 30 | 547 | 137 | 1101 |
| 172 | 531 | 48 | 1563 | 536 | 68 | 1486 | | 16 | 540 | 88 | 633 | 543 | 106 | 154 | October | 1 | 547 | 126 | 1090 |
| 173 | 531 | 44 | 1558 | 536 | 62 | 1480 | | 17 | 540 | 80 | 625 | 543 | 96 | 144 | | 2 | 547 | 115 | 1078 |
| 174 | 531 | 39 | 1554 | 535 | 56 | 1473 | | 18 | 539 | 72 | 616 | 543 | 87 | 134 | | 3 | 547 | 103 | 1066 |
| 175 | 531 | 35 | 1549 | 535 | 49 | 1466 | | 19 | 539 | 64 | 608 | 542 | 77 | 123 | | 4 | 547 | 92 | 1055 |
| 176 | 531 | 30 | 1545 | 535 | 43 | 1460 | | 20 | 539 | 56 | 599 | 542 | 68 | 113 | | 5 | 546 | 81 | 1042 |
| 177 | 530 | 26 | 1539 | 534 | 37 | 1452 | | 21 | 538 | 48 | 590 | 541 | 58 | 102 | | 6 | 545 | 69 | 1029 |
| 178 | 529 | 22 | 1534 | 533 | 31 | 1445 | | 22 | 537 | 40 | 581 | 540 | 49 | 92 | | 7 | 544 | 58 | 1017 |
| 179 | 528 | 17 | 1528 | 532 | 25 | 1438 | | 23 | 536 | 32 | 571 | 539 | 39 | 81 | | 8 | 543 | 47 | 1004 |
| 180 | 527 | 13 | 1523 | 531 | 18 | 1430 | | 24 | 535 | 24 | 562 | 538 | 30 | 70 | | 9 | 543 | 36 | 992 |
| 181 | 525 | 8 | 1516 | 529 | 12 | 1422 | | 25 | 533 | 16 | 552 | 536 | 20 | 58 | | 10 | 541 | 24 | 978 |
| 182 | 524 | 4 | 1511 | 528 | 8 | 1414 | | 26 | 532 | 8 | 542 | 535 | 11 | 47 | | 11 | 540 | 13 | 966 |
| 183 | 522 | 0 | 1504 | 526 | 0 | 1406 | | 27 | 530 | 0 | 532 | 534 | 0 | 36 | | 12 | 538 | 0 | 952 |
| 184 | 521 | — 4 | 1499 | 525 | — 6 | 1399 | | 28 | 529 | — 8 | 523 | 533 | —10 | 25 | | 13 | 537 | —12 | 939 |
| 185 | 519 | 9 | 1492 | 523 | 13 | 1390 | | 29 | 527 | 17 | 512 | 531 | 20 | 13 | | 14 | 535 | 23 | 926 |
| 186 | 517 | 13 | 1486 | 521 | 19 | 1382 | | 30 | 525 | 25 | 502 | 529 | 30 | 1 | | 15 | 533 | 35 | 912 |

# TABLE XIII.

## Sunrise from 13° to 17° N. Latitude.

| Day of Solar year. | Lat. 13° +1072" Long. Eqn. of time in seconds. | ⊙'s Trop. Long. in seconds. | Total Corrn. in seconds. Madras. A.D. 1840–1920 | Lat. 14° +1008" Long. Eqn. of time in seconds. | ⊙'s Trop. Long. in seconds. | Total Corrn. in seconds. Nellore. A.D. 1840–1920 | Solar months | Days of solar months. | Lat. 15° +168" Long. Eqn. of time in seconds. | ⊙'s Trop. Long. in seconds. | Total Corrn. in seconds. Vizianagaram. A.D. 1840–1920 | Lat. 16° −300" Long. Eqn. of Time in seconds. | ⊙'s Trop. Long. in seconds. | Total Corrn. in seconds. Belgaum. A.D. 1840–1920 | Month | Eng. date. | Lat. 17° +648" Long. Eqn. of Time in seconds. | ⊙'s Trop. Long. in seconds. | Total Corrn. in seconds. Hyderabad. A.D. 1840–1920 |
|---|---|---|---|---|---|---|---|---|---|---|---|---|---|---|---|---|---|---|---|
| 187 | +515 | −18 | +1479 | +519 | −26 | +1373 | Tamil, Aippasi; *Malayalam*, Tulam; *Bengal*, Kartika. | 1 | +523 | −33 | +492 | +527 | −40 | −11 | | 16 | +531 | −47 | +898 |
| 188 | 513 | 22 | 1473 | 517 | 32 | 1365 | | 2 | 521 | 41 | 482 | 525 | 49 | 22 | | 17 | 529 | 58 | 885 |
| 189 | 510 | 27 | 1465 | 514 | 38 | 1356 | | 3 | 518 | 50 | 470 | 522 | 59 | 35 | | 18 | 526 | 70 | 870 |
| 190 | 507 | 31 | 1458 | 511 | 45 | 1346 | | 4 | 515 | 58 | 459 | 519 | 69 | 48 | | 19 | 523 | 82 | 855 |
| 191 | 504 | 36 | 1450 | 508 | 51 | 1337 | | 5 | 512 | 66 | 448 | 516 | 79 | 61 | | 20 | 520 | 94 | 840 |
| 192 | 501 | 40 | 1443 | 505 | 58 | 1327 | | 6 | 509 | 75 | 436 | 513 | 89 | 74 | | 21 | 517 | 105 | 826 |
| 193 | 498 | 45 | 1434 | 502 | 64 | 1318 | | 7 | 506 | 83 | 426 | 510 | 99 | 86 | | 22 | 514 | 117 | 810 |
| 194 | 495 | 49 | 1414 | 499 | 70 | 1296 | | 8 | 503 | 91 | 403 | 507 | 109 | 111 | October | 23 | 511 | 129 | 784 |
| 195 | 492 | 54 | 1393 | 496 | 77 | 1275 | | 9 | 500 | 99 | 379 | 504 | 119 | 135 | | 24 | 508 | 140 | 758 |
| 196 | 488 | 58 | 1372 | 492 | 83 | 1252 | | 10 | 496 | 108 | 355 | 500 | 129 | 161 | | 25 | 504 | 152 | 730 |
| 197 | 484 | 63 | 1351 | 488 | 90 | 1229 | | 11 | 492 | 116 | 331 | 496 | 139 | 187 | | 26 | 500 | 164 | 703 |
| 198 | 480 | 67 | 1330 | 484 | 96 | 1207 | | 12 | 488 | 124 | 307 | 492 | 148 | 212 | | 27 | 496 | 175 | 676 |
| 199 | 476 | 72 | 1308 | 480 | 102 | 1184 | | 13 | 484 | 133 | 282 | 488 | 158 | 238 | | 28 | 492 | 187 | 649 |
| 200 | 472 | 76 | 1287 | 476 | 109 | 1161 | | 14 | 479 | 141 | 257 | 483 | 163 | 265 | | 29 | 487 | 199 | 621 |
| 201 | 468 | 81 | 1266 | 472 | 115 | 1138 | | 15 | 475 | 149 | 233 | 479 | 178 | 291 | | 30 | 483 | 211 | 593 |
| 202 | 464 | 85 | 1244 | 468 | 122 | 1116 | | 16 | 471 | 158 | 208 | 475 | 188 | 316 | | 31 | 479 | 222 | 566 |
| 203 | 459 | 90 | 1222 | 463 | 128 | 1092 | | 17 | 466 | 166 | 183 | 470 | 198 | 343 | | 1 | 474 | 234 | 538 |
| 204 | 454 | 94 | 1200 | 458 | 134 | 1068 | | 18 | 461 | 174 | 158 | 465 | 208 | 370 | | 2 | 469 | 246 | 510 |
| 205 | 450 | 99 | 1178 | 454 | 141 | 1046 | | 19 | 456 | 183 | 132 | 460 | 218 | 396 | | 3 | 464 | 257 | 482 |
| 206 | 445 | 103 | 1156 | 449 | 147 | 1022 | | 20 | 451 | 191 | 112 | 455 | 228 | 423 | | 4 | 459 | 269 | 453 |
| 207 | 441 | 108 | 1135 | 445 | 154 | 999 | | 21 | 447 | 199 | 83 | 451 | 238 | 449 | | 5 | 455 | 281 | 426 |
| 208 | 436 | 112 | 1113 | 440 | 160 | 976 | | 22 | 442 | 207 | 58 | 446 | 247 | 475 | | 6 | 450 | 292 | 398 |
| 209 | 430 | 117 | 1089 | 434 | 166 | 951 | | 23 | 436 | 216 | 31 | 440 | 257 | 503 | | 7 | 444 | 304 | 369 |
| 210 | 425 | 121 | 1067 | 429 | 173 | 927 | | 24 | 431 | 224 | 6 | 433 | 267 | 532 | | 8 | 439 | 316 | 341 |
| 211 | 417 | 126 | 1042 | 423 | 179 | 902 | | 25 | 425 | 232 | − 20 | 429 | 277 | 558 | | 9 | 433 | 328 | 311 |
| 212 | 411 | 130 | 1018 | 417 | 186 | 878 | | 26 | 419 | 241 | 47 | 423 | 287 | 585 | | 10 | 427 | 339 | 282 |
| 213 | 428 | 136 | 1018 | 431 | 192 | 873 | | 27 | 433 | 248 | 53 | 436 | 296 | 594 | | 11 | 438 | 352 | 270 |
| 214 | 422 | 153 | 995 | 425 | 211 | 848 | | 28 | 427 | 268 | 79 | 430 | 318 | 622 | | 12 | 432 | 375 | 241 |
| 215 | 415 | 171 | 970 | 418 | 229 | 823 | | 29 | 420 | 289 | 107 | 423 | 339 | 650 | | 13 | 425 | 398 | 211 |
| 216 | 408 | 188 | 946 | 411 | 248 | 797 | | 30 | 413 | 309 | 134 | 416 | 361 | 679 | | 14 | 418 | 422 | 180 |
| 217 | 401 | 205 | 922 | 404 | 267 | 771 | Tamil, Kartikai; *Malayalam*, Vrischikam; *Bengal*, Margasira. | 1 | 406 | 329 | 161 | 409 | 383 | 708 | November | 15 | 411 | 445 | 150 |
| 218 | 394 | 222 | 906 | 397 | 285 | 746 | | 2 | 399 | 349 | 188 | 402 | 404 | 736 | | 16 | 404 | 468 | 120 |
| 219 | 387 | 240 | 873 | 390 | 304 | 720 | | 3 | 392 | 370 | 216 | 395 | 426 | 765 | | 17 | 397 | 491 | 90 |
| 220 | 380 | 257 | 849 | 383 | 323 | 692 | | 4 | 385 | 390 | 243 | 388 | 448 | 794 | | 18 | 390 | 514 | 60 |
| 221 | 373 | 274 | 825 | 376 | 342 | 668 | | 5 | 378 | 410 | 270 | 381 | 470 | 823 | | 19 | 383 | 538 | 29 |
| 222 | 366 | 292 | 800 | 369 | 360 | 643 | | 6 | 371 | 431 | 298 | 374 | 491 | 851 | | 20 | 376 | 561 | − 1 |
| 223 | 359 | 309 | 775 | 362 | 379 | 618 | | 7 | 364 | 451 | 324 | 367 | 513 | 881 | | 21 | 369 | 584 | 31 |
| 224 | 352 | 326 | 744 | 354 | 398 | 585 | | 8 | 356 | 471 | 357 | 359 | 535 | 915 | | 22 | 361 | 607 | 66 |
| 225 | 345 | 344 | 712 | 346 | 416 | 552 | | 9 | 348 | 492 | 391 | 351 | 556 | 949 | | 23 | 353 | 630 | 100 |
| 226 | 338 | 361 | 681 | 338 | 435 | 519 | | 10 | 340 | 512 | 424 | 343 | 578 | 984 | | 24 | 345 | 654 | 135 |
| 227 | 330 | 378 | 648 | 330 | 454 | 486 | | 11 | 332 | 532 | 458 | 335 | 600 | 1018 | | 25 | 337 | 677 | 170 |
| 228 | 323 | 395 | 617 | 322 | 472 | 453 | | 12 | 324 | 552 | 491 | 327 | 621 | 1052 | | 26 | 329 | 700 | 204 |
| 229 | 315 | 413 | 585 | 314 | 491 | 420 | | 13 | 316 | 573 | 525 | 319 | 643 | 1086 | | 27 | 321 | 723 | 239 |
| 230 | 307 | 430 | 552 | 306 | 510 | 387 | | 14 | 307 | 593 | 559 | 310 | 665 | 1121 | | 28 | 312 | 746 | 275 |
| 231 | 299 | 447 | 520 | 298 | 529 | 354 | | 15 | 299 | 613 | 593 | 302 | 687 | 1156 | | 29 | 304 | 770 | 310 |
| 232 | 291 | 465 | 487 | 290 | 547 | 321 | | 16 | 291 | 634 | 626 | 294 | 708 | 1190 | | 30 | 296 | 793 | 344 |
| 233 | 283 | 482 | 455 | 282 | 566 | 288 | | 17 | 283 | 654 | 660 | 286 | 730 | 1224 | | 1 | 288 | 816 | 379 |
| 234 | 274 | 499 | 422 | 278 | 585 | 254 | | 18 | 274 | 674 | 694 | 277 | 752 | 1259 | | 2 | 279 | 839 | 415 |
| 235 | 266 | 517 | 389 | 265 | 603 | 221 | | 19 | 265 | 695 | 729 | 268 | 773 | 1294 | | 3 | 270 | 862 | 450 |
| 236 | 258 | 534 | 357 | 257 | 622 | 188 | | 20 | 256 | 715 | 763 | 259 | 795 | 1330 | | 4 | 261 | 886 | 486 |
| 237 | 250 | 551 | 324 | 249 | 641 | 155 | | 21 | 247 | 735 | 798 | 250 | 817 | 1365 | December | 5 | 252 | 909 | 522 |
| 238 | 241 | 568 | 291 | 240 | 659 | 121 | | 22 | 238 | 755 | 832 | 241 | 838 | 1400 | | 6 | 243 | 932 | 557 |
| 239 | 232 | 586 | 258 | 231 | 678 | 87 | | 23 | 229 | 776 | 867 | 232 | 860 | 1435 | | 7 | 234 | 955 | 593 |
| 240 | 224 | 603 | 225 | 223 | 699 | 54 | | 24 | 220 | 796 | 901 | 223 | 882 | 1470 | | 8 | 225 | 978 | 629 |
| 241 | 215 | 620 | 192 | 214 | 716 | 20 | | 25 | 211 | 816 | 936 | 213 | 904 | 1507 | | 9 | 215 | 1002 | 666 |
| 242 | 206 | 638 | 158 | 205 | 734 | − 14 | | 26 | 202 | 837 | 970 | 204 | 925 | 1542 | | 10 | 206 | 1025 | 701 |
| 243 | 200 | 656 | 128 | 200 | 752 | 44 | | 27 | 201 | 856 | 997 | 201 | 948 | 1571 | | 11 | 202 | 1048 | 732 |
| 244 | 191 | 680 | 95 | 191 | 777 | 78 | | 28 | 192 | 881 | 1031 | 192 | 974 | 1606 | | 12 | 193 | 1075 | 768 |
| 245 | 181 | 705 | 60 | 181 | 802 | 113 | | 29 | 182 | 907 | 1067 | 182 | 1000 | 1642 | | 13 | 183 | 1101 | 804 |
| 246 | 171 | 729 | 26 | 171 | 827 | 148 | | 30 | 172 | 932 | 1102 | 172 | 1027 | 1679 | | 14 | 173 | 1128 | 841 |

# TABLE XIII.

## Sunrise from 13° to 17° N. Latitude.

| Day of Solar year. | Lat. 13° +1072" Long. Eqn. of time in seconds. | ☉'s Trop. Long. in seconds. | Total Corrn. in seconds. Madras. A.D. 1840—1920 | Lat. 14° +1008" Long. Eqn. of time in seconds. | ☉'s Trop. Long. in seconds. | Total Corrn. in seconds. Nellore. A.D. 1840—1920 | Days of solar months. | Lat. 15° +168" Long. Eqn. of time in seconds. | ☉'s Trop. Long. in seconds. | Total Corrn. in seconds. Vizianagaram. A.D. 840—1920 | Lat. 16° —300" Long. Eqn. of time in seconds. | ☉'s Trop. Long. in seconds. | Total Corrn. in seconds. Belgaum. A.D. 1840—1920 | Eng. date. | Lat. 17° +648" Long. Eqn. of time in seconds. | ☉'s Trop. Long. in seconds. | Total Corrn. in seconds. Hyderabad. A.D. 1840—1920 |
|---|---|---|---|---|---|---|---|---|---|---|---|---|---|---|---|---|---|
| 247 | +161 | —754 | — 9 | +161 | —852 | —183 | 1 (Tamil, Margali; Malayalam, Dhanus; Bengal, Pausha.) | +162 | —958 | —1138 | +162 | —1053 | —1715 | 15 (December.) | +163 | —1155 | —878 |
| 248 | 151 | 778 | 43 | 151 | 877 | 218 | 2 | 152 | 983 | 1173 | 152 | 1079 | 1751 | 16 | 153 | 1181 | 914 |
| 249 | 141 | 802 | 77 | 141 | 902 | 253 | 3 | 142 | 1009 | 1209 | 142 | 1105 | 1787 | 17 | 143 | 1208 | 951 |
| 250 | 131 | 827 | 112 | 131 | 927 | 288 | 4 | 132 | 1034 | 1244 | 132 | 1131 | 1823 | 18 | 133 | 1235 | 988 |
| 251 | 121 | 851 | 146 | 121 | 952 | 323 | 5 | 122 | 1060 | 1280 | 122 | 1158 | 1860 | 19 | 123 | 1262 | 1025 |
| 252 | 111 | 876 | 181 | 111 | 977 | 358 | 6 | 112 | 1085 | 1315 | 112 | 1184 | 1896 | 20 | 113 | 1288 | 1061 |
| 253 | 101 | 900 | 215 | 101 | 1002 | 393 | 7 | 102 | 1111 | 1351 | 102 | 1210 | 1932 | 21 | 103 | 1315 | 1098 |
| 254 | 91 | 924 | 249 | 91 | 1027 | 429 | 8 | 92 | 1136 | 1388 | 92 | 1236 | 1968 | 22 | 93 | 1342 | 1135 |
| 255 | 81 | 949 | 270 | 81 | 1052 | 449 | 9 | 82 | 1162 | 1408 | 82 | 1262 | 1987 | 23 | 83 | 1368 | 1158 |
| 256 | 71 | 973 | 291 | 71 | 1077 | 469 | 10 | 72 | 1187 | 1427 | 72 | 1289 | 2006 | 24 | 73 | 1395 | 1172 |
| 257 | 61 | 998 | 311 | 61 | 1102 | 489 | 11 | 62 | 1213 | 1447 | 62 | 1315 | 2025 | 25 | 63 | 1422 | 1190 |
| 258 | 51 | 1022 | 332 | 51 | 1127 | 509 | 12 | 52 | 1238 | 1466 | 52 | 1341 | 2044 | 26 | 53 | 1448 | 1209 |
| 259 | 41 | 1046 | 353 | 41 | 1152 | 529 | 13 | 42 | 1264 | 1486 | 42 | 1367 | 2062 | 27 | 43 | 1475 | 1227 |
| 260 | 31 | 1071 | 374 | 31 | 1177 | 550 | 14 | 32 | 1289 | 1506 | 32 | 1393 | 2081 | 28 | 33 | 1502 | 1245 |
| 261 | 21 | 1095 | 395 | 21 | 1202 | 570 | 15 | 22 | 1315 | 1525 | 22 | 1420 | 2100 | 29 | 23 | 1529 | 1264 |
| 262 | 11 | 1120 | 415 | 11 | 1227 | 590 | 16 | 12 | 1340 | 1545 | 12 | 1446 | 2119 | 30 | 13 | 1555 | 1282 |
| 263 | 2 | 1144 | 436 | 2 | 1252 | 610 | 17 | 2 | 1366 | 1564 | 2 | 1472 | 2138 | 31 | 2 | 1582 | 1301 |
| 264 | —10 | 1168 | 457 | —10 | 1277 | 630 | 18 | —10 | 1391 | 1586 | —10 | 1498 | 2159 | 1 (January.) | —10 | 1609 | 1323 |
| 265 | 20 | 1193 | 478 | 20 | 1302 | 650 | 19 | 20 | 1417 | 1606 | 20 | 1524 | 2178 | 2 | 20 | 1635 | 1341 |
| 266 | 29 | 1217 | 499 | 29 | 1327 | 670 | 20 | 30 | 1442 | 1625 | 30 | 1551 | 2197 | 3 | 31 | 1662 | 1362 |
| 267 | 38 | 1242 | 518 | 38 | 1352 | 689 | 21 | 39 | 1468 | 1644 | 39 | 1577 | 2215 | 4 | 40 | 1689 | 1377 |
| 268 | 48 | 1266 | 539 | 48 | 1377 | 709 | 22 | 49 | 1493 | 1663 | 49 | 1603 | 2234 | 5 | 50 | 1715 | 1396 |
| 269 | 58 | 1290 | 560 | 58 | 1402 | 729 | 23 | 59 | 1519 | 1683 | 59 | 1629 | 2252 | 6 | 60 | 1742 | 1414 |
| 270 | 67 | 1315 | 580 | 67 | 1427 | 749 | 24 | 68 | 1544 | 1702 | 68 | 1655 | 2270 | 7 | 69 | 1769 | 1431 |
| 271 | 78 | 1339 | 602 | 78 | 1452 | 770 | 25 | 79 | 1570 | 1722 | 79 | 1682 | 2290 | 8 | 80 | 1796 | 1451 |
| 272 | 88 | 1364 | 622 | 88 | 1477 | 790 | 26 | 89 | 1595 | 1742 | 89 | 1708 | 2309 | 9 | 90 | 1822 | 1469 |
| 273 | 98 | 1388 | 643 | 98 | 1502 | 810 | 27 | 99 | 1621 | 1761 | 99 | 1734 | 2328 | 10 | 100 | 1849 | 1488 |
| 274 | 108 | 1412 | 659 | 103 | 1528 | 825 | 28 | 108 | 1648 | 1775 | 108 | 1760 | 2341 | 11 | 102 | 1876 | 1498 |
| 275 | 110 | 1423 | 677 | 110 | 1538 | 842 | 29 | 109 | 1658 | 1791 | 110 | 1769 | 2357 | 12 | 109 | 1884 | 1513 |
| 276 | 117 | 1434 | 695 | 117 | 1548 | 859 | 1 (Tamil, Tai; Malayalam, Magaram; Bengal, Magha.) | 116 | 1667 | 1807 | 117 | 1778 | 2373 | 13 | 116 | 1893 | 1529 |
| 277 | 124 | 1444 | 712 | 124 | 1558 | 876 | 2 | 122 | 1677 | 1823 | 124 | 1787 | 2389 | 14 | 123 | 1901 | 1544 |
| 278 | 131 | 1455 | 730 | 131 | 1568 | 893 | 3 | 129 | 1686 | 1839 | 131 | 1796 | 2405 | 15 | 130 | 1910 | 1560 |
| 279 | 138 | 1466 | 748 | 139 | 1578 | 911 | 4 | 135 | 1696 | 1855 | 138 | 1804 | 2420 | 16 | 137 | 1918 | 1575 |
| 280 | 145 | 1477 | 766 | 146 | 1589 | 929 | 5 | 142 | 1706 | 1872 | 145 | 1813 | 2436 | 17 | 144 | 1926 | 1590 |
| 281 | 152 | 1488 | 784 | 153 | 1599 | 946 | 6 | 148 | 1715 | 1887 | 152 | 1822 | 2452 | 18 | 151 | 1935 | 1606 |
| 282 | 159 | 1498 | 801 | 160 | 1609 | 963 | 7 | 155 | 1725 | 1904 | 159 | 1831 | 2468 | 19 | 158 | 1943 | 1621 |
| 283 | 166 | 1509 | 819 | 167 | 1619 | 980 | 8 | 161 | 1734 | 1919 | 166 | 1840 | 2484 | 20 | 165 | 1952 | 1637 |
| 284 | 173 | 1520 | 837 | 175 | 1629 | 999 | 9 | 168 | 1744 | 1936 | 173 | 1849 | 2501 | 21 | 172 | 1960 | 1652 |
| 285 | 180 | 1531 | 826 | 182 | 1639 | 987 | 10 | 174 | 1754 | 1921 | 180 | 1858 | 2486 | 22 | 179 | 1968 | 1635 |
| 286 | 187 | 1542 | 815 | 189 | 1649 | 974 | 11 | 181 | 1763 | 1907 | 187 | 1867 | 2470 | 23 | 186 | 1979 | 1619 |
| 287 | 194 | 1552 | 804 | 196 | 1659 | 960 | 12 | 187 | 1773 | 1893 | 194 | 1876 | 2455 | 24 | 193 | 1985 | 1602 |
| 288 | 201 | 1563 | 793 | 203 | 1669 | 950 | 13 | 194 | 1782 | 1879 | 201 | 1885 | 2440 | 25 | 200 | 1994 | 1585 |
| 289 | 208 | 1574 | 781 | 211 | 1679 | 938 | 14 | 200 | 1792 | 1864 | 208 | 1893 | 2424 | 26 | 207 | 2002 | 1568 |
| 290 | 215 | 1585 | 770 | 218 | 1690 | 928 | 15 | 207 | 1802 | 1850 | 215 | 1902 | 2409 | 27 | 214 | 2010 | 1552 |
| 291 | 222 | 1596 | 759 | 225 | 1700 | 914 | 16 | 213 | 1811 | 1835 | 222 | 1911 | 2394 | 28 | 221 | 2019 | 1535 |
| 292 | 229 | 1606 | 748 | 232 | 1710 | 902 | 17 | 220 | 1821 | 1822 | 229 | 1920 | 2379 | 29 | 228 | 2027 | 1518 |
| 293 | 236 | 1617 | 737 | 239 | 1720 | 889 | 18 | 227 | 1830 | 1808 | 236 | 1929 | 2363 | 30 | 235 | 2036 | 1502 |
| 294 | 243 | 1628 | 726 | 247 | 1730 | 878 | 19 | 235 | 1840 | 1795 | 243 | 1938 | 2348 | 31 | 242 | 2044 | 1485 |
| 295 | 250 | 1639 | 715 | 254 | 1740 | 866 | 20 | 242 | 1850 | 1781 | 249 | 1947 | 2332 | 1 (February.) | 248 | 2052 | 1467 |
| 296 | 257 | 1650 | 704 | 261 | 1750 | 853 | 21 | 249 | 1859 | 1767 | 256 | 1956 | 2316 | 2 | 255 | 2061 | 1451 |
| 297 | 264 | 1660 | 693 | 268 | 1760 | 841 | 22 | 257 | 1869 | 1755 | 262 | 1965 | 2300 | 3 | 261 | 2069 | 1433 |
| 298 | 271 | 1671 | 682 | 275 | 1770 | 829 | 23 | 264 | 1878 | 1741 | 269 | 1974 | 2285 | 4 | 268 | 2078 | 1416 |
| 299 | 278 | 1682 | 670 | 283 | 1780 | 817 | 24 | 272 | 1888 | 1728 | 276 | 1982 | 2269 | 5 | 274 | 2086 | 1398 |
| 300 | 285 | 1693 | 659 | 290 | 1791 | 805 | 25 | 279 | 1898 | 1714 | 282 | 1991 | 2253 | 6 | 280 | 2094 | 1381 |
| 301 | 292 | 1704 | 648 | 297 | 1801 | 793 | 26 | 287 | 1907 | 1701 | 289 | 2000 | 2238 | 7 | 286 | 2103 | 1346 |
| 302 | 299 | 1714 | 637 | 304 | 1811 | 781 | 27 | 294 | 1917 | 1688 | 295 | 2009 | 2222 | 8 | 293 | 2111 | 1363 |
| 303 | 307 | 1725 | 627 | 311 | 1821 | 768 | 28 | 302 | 1926 | 1675 | 302 | 2018 | 2206 | 9 | 300 | 2120 | 1330 |
| 304 | 315 | 1736 | 617 | 319 | 1832 | 757 | 29 | 310 | 1936 | 1662 | 308 | 2028 | 2190 | 10 | 306 | 2128 | 1312 |
| 305 | 318 | 1718 | 602 | 322 | 1813 | 741 | 30 | 313 | 1915 | 1644 | 311 | 2006 | 2171 | 11 | 309 | 2104 | 1291 |

# TABLE XIII.

## Sunrise from 13° to 17° N. Latitude.

| Day of Solar year. | Lat. 13° +1072″ Long. Eqn. of time in seconds. | ⊙'s Trop. Long. in seconds. | Total Corrn. in seconds. Madras. A.D. 1840—1920 | Lat. 14° +1008″ Long. Eqn. of time in seconds. | ⊙'s Trop. Long. in seconds. | Total Corrn. in seconds. Nellore. A.D. 1840—1920 | Days of solar months. | Lat. 15° +168″ Long. Eqn. of time in seconds. | ⊙'s Trop. Long. in seconds. | Total Corrn. in seconds. Vizianagaram. A.D. 1840—1920 | Lat. 16° —300″ Long. Eqn. of time in seconds. | ⊙'s Trop. Long. in seconds. | Total Corrn. in seconds. Belgaum. A.D. 1840—1920 | Eng. date. | Lat. 17° +648″ Long. Eqn. of time in seconds. | ⊙'s Trop. Long. in seconds. | Total Corrn. in seconds. Hyderabad. A.D. 1840—1920 |
|---|---|---|---|---|---|---|---|---|---|---|---|---|---|---|---|---|---|
| 306 | —321 | —1700 | —587 | —325 | —1793 | —724 | 1 Tamil, Masi; Malayalam, Kumbam; Bengal, Phalguna. | —316 | —1894 | —1626 | —314 | —1983 | —2151 | 12 February | —312 | —2081 | —1271 |
| 307 | 324 | 1682 | 572 | 328 | 1774 | 708 | 2 | 319 | 1874 | 1604 | 317 | 1961 | 2132 | 13 | 315 | 2057 | 1250 |
| 308 | 327 | 1664 | 557 | 331 | 1755 | 692 | 3 | 322 | 1853 | 1591 | 320 | 1939 | 2113 | 14 | 318 | 2033 | 1229 |
| 309 | 330 | 1645 | 541 | 334 | 1735 | 675 | 4 | 325 | 1832 | 1571 | 323 | 1916 | 2093 | 15 | 321 | 2009 | 1208 |
| 310 | 333 | 1627 | 526 | 337 | 1716 | 659 | 5 | 328 | 1811 | 1555 | 326 | 1894 | 2074 | 16 | 324 | 1986 | 1188 |
| 311 | 336 | 1609 | 511 | 340 | 1697 | 643 | 6 | 331 | 1790 | 1539 | 329 | 1872 | 2055 | 17 | 327 | 1962 | 1167 |
| 312 | 339 | 1591 | 496 | 343 | 1678 | 727 | 7 | 334 | 1770 | 1520 | 332 | 1850 | 2036 | 18 | 330 | 1938 | 1146 |
| 313 | 342 | 1573 | 481 | 346 | 1658 | 610 | 8 | 337 | 1749 | 1501 | 334 | 1827 | 2015 | 19 | 333 | 1915 | 1126 |
| 314 | 345 | 1555 | 466 | 349 | 1639 | 594 | 9 | 340 | 1728 | 1548 | 337 | 1805 | 1996 | 20 | 336 | 1891 | 1105 |
| 315 | 348 | 1537 | 452 | 352 | 1620 | 576 | 10 | 343 | 1707 | 1467 | 340 | 1783 | 1976 | 21 | 339 | 1867 | 1083 |
| 316 | 351 | 1519 | 416 | 355 | 1600 | 538 | 11 | 346 | 1686 | 1427 | 343 | 1760 | 1934 | 22 | 342 | 1844 | 1040 |
| 317 | 354 | 1501 | 380 | 358 | 1581 | 500 | 12 | 349 | 1666 | 1387 | 346 | 1738 | 1893 | 23 | 345 | 1820 | 996 |
| 318 | 357 | 1483 | 343 | 361 | 1562 | 462 | 13 | 352 | 1645 | 1347 | 349 | 716 | 1851 | 24 | 348 | 1796 | 958 |
| 319 | 360 | 1464 | 307 | 364 | 1542 | 424 | 14 | 355 | 1624 | 1307 | 352 | 1698 | 1810 | 25 | 351 | 1772 | 949 |
| 320 | 363 | 1446 | 271 | 367 | 1523 | 386 | 15 | 358 | 1603 | 1267 | 355 | 1671 | 1768 | 26 | 354 | 1749 | 865 |
| 321 | 366 | 1428 | 235 | 370 | 1504 | 348 | 16 | 361 | 1582 | 1227 | 358 | 1649 | 1727 | 27 | 357 | 1725 | 823 |
| 322 | 369 | 1410 | 199 | 373 | 1485 | 310 | 17 | 364 | 1562 | 1187 | 361 | 1627 | 1685 | 28 | 360 | 1701 | 779 |
| 323 | 372 | 1392 | 162 | 376 | 1465 | 272 | 18 | 367 | 1541 | 1147 | 364 | 1604 | 1644 | 1 | 364 | 1678 | 737 |
| 324 | 375 | 1374 | 126 | 380 | 1446 | 235 | 19 | 370 | 1520 | 1107 | 367 | 1582 | 1602 | 2 | 367 | 1654 | 694 |
| 325 | 378 | 1356 | 90 | 383 | 1427 | 197 | 20 | 373 | 1499 | 1067 | 370 | 1560 | 1561 | 3 | 370 | 1630 | 650 |
| 326 | 381 | 1338 | 54 | 386 | 1407 | 159 | 21 | 376 | 1478 | 1027 | 373 | 1537 | 1539 | 4 | 374 | 1607 | 608 |
| 327 | 384 | 1320 | 18 | 389 | 1388 | 121 | 22 | 379 | 1458 | 987 | 376 | 1515 | 1478 | 5 | 377 | 1583 | 564 |
| 328 | 387 | 1302 | + 19 | 392 | 1369 | 83 | 23 | 382 | 1437 | 947 | 379 | 1493 | 1436 | 6 | 380 | 1559 | 521 |
| 329 | 390 | 1283 | 57 | 395 | 1349 | 45 | 24 | 385 | 1416 | 907 | 382 | 1470 | 1395 | 7 | 383 | 1535 | 477 |
| 330 | 393 | 1265 | 91 | 398 | 1330 | 7 | 25 | 388 | 1395 | 867 | 385 | 1448 | 1353 | 8 | 386 | 1512 | 434 |
| 331 | 396 | 1247 | 127 | 401 | 1311 | + 31 | 26 | 391 | 1374 | 827 | 388 | 1426 | 1312 | 9 | 389 | 1488 | 391 |
| 332 | 400 | 1229 | 162 | 404 | 1292 | 69 | 27 | 394 | 1354 | 788 | 391 | 1404 | 1270 | 10 | 392 | 1464 | 347 |
| 333 | 404 | 1211 | 198 | 406 | 1272 | 108 | 28 | 398 | 1333 | 748 | 395 | 1381 | 1230 | 11 | 395 | 1441 | 304 |
| 334 | 408 | 1193 | 235 | 408 | 1253 | 147 | 29 | 402 | 1312 | 710 | 399 | 1359 | 1179 | 12 | 397 | 1417 | 259 |
| 335 | 413 | 1176 | 267 | 410 | 1232 | 186 | 1 Tamil, Panguni; Malayalam, Meenam; Bengal, Chaitra. | 406 | 1292 | 670 | 402 | 1336 | 1148 | 13 March | 399 | 1392 | 215 |
| 336 | 413 | 1137 | 302 | 411 | 1191 | 224 | 2 | 407 | 1249 | 628 | 403 | 1291 | 1104 | 14 | 400 | 1346 | 170 |
| 337 | 414 | 1098 | 338 | 411 | 1150 | 261 | 3 | 407 | 1206 | 585 | 403 | 1247 | 1060 | 15 | 400 | 1299 | 123 |
| 338 | 414 | 1058 | 374 | 412 | 1109 | 299 | 4 | 408 | 1163 | 543 | 404 | 1202 | 1016 | 16 | 401 | 1253 | 78 |
| 339 | 415 | 1019 | 410 | 412 | 1068 | 336 | 5 | 408 | 1120 | 500 | 404 | 1158 | 981 | 17 | 401 | 1206 | 31 |
| 340 | 415 | 980 | 445 | 413 | 1027 | 374 | 6 | 409 | 1077 | 458 | 405 | 1113 | 928 | 18 | 402 | 1160 | + 0 |
| 341 | 416 | 941 | 481 | 413 | 986 | 411 | 7 | 409 | 1034 | 415 | 405 | 1069 | 884 | 19 | 402 | 1114 | 60 |
| 342 | 416 | 902 | 518 | 414 | 945 | 451 | 8 | 410 | 991 | 373 | 406 | 1024 | 840 | 20 | 403 | 1067 | 106 |
| 343 | 417 | 862 | 554 | 414 | 904 | 488 | 9 | 410 | 948 | 330 | 406 | 980 | 795 | 21 | 403 | 1021 | 143 |
| 344 | 417 | 823 | 591 | 415 | 863 | 530 | 10 | 411 | 905 | 286 | 407 | 935 | 752 | 22 | 404 | 974 | 198 |
| 345 | 418 | 784 | 630 | 416 | 822 | 570 | 11 | 411 | 862 | 243 | 407 | 891 | 707 | 23 | 404 | 928 | 244 |
| 346 | 418 | 745 | 667 | 416 | 781 | 609 | 12 | 412 | 819 | 201 | 408 | 846 | 663 | 24 | 405 | 882 | 289 |
| 347 | 419 | 706 | 705 | 417 | 740 | 649 | 13 | 412 | 776 | 158 | 408 | 802 | 619 | 25 | 405 | 835 | 336 |
| 348 | 419 | 666 | 743 | 417 | 699 | 688 | 14 | 413 | 733 | 116 | 409 | 757 | 576 | 26 | 406 | 789 | 381 |
| 349 | 420 | 627 | 781 | 418 | 658 | 728 | 15 | 413 | 690 | 73 | 409 | 713 | 531 | 27 | 406 | 742 | 428 |
| 350 | 420 | 588 | 818 | 418 | 617 | 767 | 16 | 414 | 647 | 31 | 410 | 668 | 488 | 28 | 407 | 696 | 473 |
| 351 | 421 | 549 | 858 | 419 | 576 | 809 | 17 | 414 | 604 | + 12 | 410 | 624 | 443 | 29 | 407 | 650 | 519 |
| 352 | 421 | 510 | 895 | 419 | 535 | 848 | 18 | 415 | 561 | 54 | 411 | 579 | 400 | 30 | 408 | 603 | 555 |
| 353 | 422 | 470 | 936 | 420 | 494 | 890 | 19 | 416 | 518 | 96 | 411 | 535 | 360 | 31 | 408 | 557 | 611 |
| 354 | 422 | 431 | 971 | 420 | 453 | 929 | 20 | 416 | 475 | 139 | 412 | 490 | 317 | 1 April | 409 | 510 | 657 |
| 355 | 423 | 392 | 1011 | 421 | 412 | 973 | 21 | 417 | 432 | 181 | 413 | 446 | 273 | 2 | 409 | 464 | 703 |
| 356 | 423 | 353 | 1051 | 422 | 371 | 1015 | 22 | 418 | 389 | 223 | 413 | 401 | 229 | 3 | 410 | 418 | 748 |
| 357 | 424 | 314 | 1093 | 422 | 330 | 1058 | 23 | 418 | 346 | 266 | 414 | 357 | 185 | 4 | 410 | 371 | 795 |
| 358 | 424 | 274 | 1132 | 423 | 289 | 1100 | 24 | 419 | 303 | 308 | 414 | 312 | 141 | 5 | 411 | 325 | 840 |
| 359 | 425 | 235 | 1172 | 424 | 248 | 1142 | 25 | 419 | 260 | 351 | 415 | 268 | 97 | 6 | 412 | 278 | 886 |
| 360 | 426 | 196 | 1211 | 424 | 207 | 1185 | 26 | 420 | 217 | 393 | 416 | 223 | 54 | 7 | 413 | 232 | 931 |
| 361 | 427 | 157 | 1255 | 425 | 166 | 1227 | 27 | 421 | 174 | 435 | 417 | 179 | 10 | 8 | 414 | 186 | 977 |
| 362 | 428 | 118 | 1297 | 426 | 125 | 1271 | 28 | 422 | 131 | 477 | 418 | 134 | + 33 | 9 | 415 | 139 | 1023 |
| 363 | 429 | 79 | 1337 | 427 | 84 | 1313 | 29 | 423 | 88 | 519 | 419 | 89 | 77 | 10 | 416 | 92 | 1068 |
| 364 | 430 | 40 | 1379 | 428 | 41 | 1357 | 30 | 424 | 43 | 561 | 420 | 49 | 120 | 11 | 417 | 46 | 1114 |
| 365 | 432 | 0 | 1420 | 429 | 0 | 1399 | 31 | 425 | 0 | 603 | 421 | 10 | 164 | 12 | 418 | 00 | 1159 |

# TABLE XIII.

## Sunrise from 18° to 22° N. Latitude.

| Day of Solar year. | Lat. 18° Eqn. of time in seconds. | ⊙'s Trop. Long. in seconds. | −428″ Long. Total Corrn. in seconds. Satara. A.D. 1840—1920 | Lat. 19° Eqn. of time in seconds. | ⊙'s Trop. Long. in seconds. | −460″ Long. Total Corrn. in seconds. Poona. A.D. 1840—1920 | Days of solar months. | Lat. 20° Eqn. of time in seconds. | ⊙'s Trop. Long. in seconds. | −476″ Long. Total Corrn. in seconds. Nasik. A.D. 1840—1920 | Lat. 21° Eqn. of Time in seconds. | ⊙'s Trop. Long. in seconds. | +800″ Long. Total Corrn. in seconds. Nagpur. A.D. 1840—1920 | Eng. date. | Lat. 22° Eqn. of Time in seconds. | ⊙'s Trop. Long. in seconds. | −616″ Long. Total Corrn. in seconds. Baroda. A.D. 1840—1920 |
|---|---|---|---|---|---|---|---|---|---|---|---|---|---|---|---|---|---|
| | | | | | | | Tamil, Chitirai; Malayalam, Medam; Bengal, Vaisakha. | | | | | | | April | | | |
| 1 | −409 | + 48 | +177 | −405 | + 50 | +185 | 1 | −401 | + 52 | +222 | −397 | + 54 | +1534 | 13 | −393 | + 56 | +160 |
| 2 | 407 | 96 | 227 | 403 | 100 | 237 | 2 | 399 | 104 | 277 | 395 | 107 | 1590 | 14 | 391 | 111 | 218 |
| 3 | 405 | 144 | 277 | 401 | 150 | 289 | 3 | 398 | 156 | 330 | 393 | 161 | 1646 | 15 | 389 | 167 | 276 |
| 4 | 404 | 193 | 327 | 400 | 200 | 340 | 4 | 396 | 208 | 384 | 392 | 215 | 1701 | 16 | 388 | 223 | 332 |
| 5 | 402 | 241 | 377 | 398 | 250 | 392 | 5 | 394 | 259 | 437 | 390 | 269 | 1757 | 17 | 386 | 278 | 390 |
| 6 | 400 | 289 | 427 | 396 | 300 | 444 | 6 | 392 | 311 | 491 | 388 | 323 | 1813 | 18 | 384 | 334 | 448 |
| 7 | 398 | 338 | 468 | 394 | 350 | 496 | 7 | 390 | 363 | 545 | 386 | 377 | 1869 | 19 | 382 | 390 | 505 |
| 8 | 396 | 386 | 528 | 392 | 400 | 548 | 8 | 388 | 425 | 599 | 384 | 431 | 1925 | 20 | 380 | 445 | 563 |
| 9 | 393 | 434 | 532 | 390 | 450 | 600 | 9 | 386 | 477 | 653 | 382 | 485 | 1981 | 21 | 375 | 501 | 621 |
| 10 | 391 | 483 | 629 | 387 | 500 | 653 | 10 | 383 | 529 | 697 | 379 | 539 | 2037 | 22 | 375 | 557 | 681 |
| 11 | 388 | 531 | 657 | 385 | 550 | 682 | 11 | 381 | 580 | 727 | 377 | 592 | 2069 | 23 | 378 | 612 | 714 |
| 12 | 386 | 579 | 684 | 382 | 600 | 711 | 12 | 378 | 632 | 758 | 374 | 646 | 2101 | 24 | 371 | 668 | 747 |
| 13 | 383 | 627 | 713 | 380 | 650 | 740 | 13 | 376 | 684 | 788 | 372 | 700 | 2133 | 25 | 368 | 724 | 782 |
| 14 | 380 | 676 | 741 | 377 | 700 | 770 | 14 | 373 | 736 | 819 | 369 | 754 | 2166 | 26 | 365 | 779 | 816 |
| 15 | 377 | 724 | 769 | 374 | 750 | 800 | 15 | 370 | 788 | 850 | 366 | 808 | 2198 | 27 | 362 | 835 | 850 |
| 16 | 374 | 772 | 797 | 371 | 800 | 829 | 16 | 367 | 840 | 882 | 363 | 862 | 2231 | 28 | 359 | 891 | 884 |
| 17 | 371 | 821 | 825 | 367 | 850 | 860 | 17 | 364 | 892 | 913 | 360 | 916 | 2264 | 29 | 356 | 946 | 918 |
| 18 | 367 | 869 | 855 | 364 | 900 | 890 | 18 | 361 | 944 | 944 | 357 | 970 | 2297 | 30 | 353 | 1002 | 953 |
| 19 | 364 | 917 | 883 | 361 | 950 | 919 | 19 | 357 | 996 | 976 | 353 | 1024 | 2380 | 1 | 350 | 1058 | 987 |
| 20 | 360 | 966 | 902 | 357 | 1000 | 950 | 20 | 354 | 1048 | 1007 | 350 | 1078 | 2363 | 2 | 346 | 1114 | 1022 |
| 21 | 357 | 1014 | 940 | 354 | 1050 | 980 | 21 | 350 | 1099 | 1039 | 346 | 1131 | 2397 | 3 | 343 | 1169 | 1056 |
| 22 | 353 | 1062 | 969 | 350 | 1100 | 1010 | 22 | 346 | 1152 | 1071 | 343 | 1185 | 2429 | 4 | 339 | 1225 | 1091 |
| 23 | 349 | 1110 | 999 | 346 | 1150 | 1041 | 23 | 342 | 1204 | 1103 | 339 | 1239 | 2463 | 5 | 335 | 1281 | 1127 |
| 24 | 346 | 1159 | 1027 | 342 | 1200 | 1072 | 24 | 339 | 1256 | 1134 | 335 | 1293 | 2497 | 6 | 332 | 1336 | 1161 |
| 25 | 341 | 1207 | 1057 | 338 | 1250 | 1102 | 25 | 335 | 1307 | 1166 | 331 | 1347 | 2530 | 7 | 328 | 1392 | 1196 |
| 26 | 337 | 1255 | 1086 | 334 | 1300 | 1133 | 26 | 331 | 1359 | 1199 | 327 | 1401 | 2964 | 8 | 324 | 1448 | 1231 |
| 27 | 333 | 1304 | 1115 | 330 | 1350 | 1164 | 27 | 327 | 1411 | 1231 | 323 | 1455 | 2598 | 9 | 320 | 1503 | 1266 |
| 28 | 329 | 1352 | 1145 | 326 | 1400 | 1195 | 28 | 323 | 1463 | 1263 | 319 | 1509 | 2632 | 10 | 316 | 1559 | 1302 |
| 29 | 324 | 1400 | 1175 | 322 | 1450 | 1225 | 29 | 318 | 1515 | 1296 | 315 | 1563 | 2665 | 11 | 312 | 1615 | 1337 |
| 30 | 357 | 1448 | 1167 | 364 | 1500 | 1220 | 30 | 352 | 1556 | 1290 | 349 | 1616 | 2661 | 12 | 346 | 1672 | 1334 |
| | | | | | | | Tamil, Vaikasi; Malayalam, Edavam; Bengal, Jyeshta. | | | | | | | May | | | |
| 31 | 352 | 1473 | 1197 | 349 | 1527 | 1252 | 1 | 347 | 1584 | 1323 | 344 | 1646 | 2696 | 13 | 342 | 1703 | 1369 |
| 32 | 346 | 1498 | 1228 | 344 | 1553 | 1283 | 2 | 342 | 1612 | 1356 | 339 | 1675 | 2730 | 14 | 336 | 1734 | 1406 |
| 33 | 341 | 1524 | 1259 | 339 | 1580 | 1315 | 3 | 336 | 1640 | 1390 | 334 | 1705 | 2765 | 15 | 331 | 1766 | 1442 |
| 34 | 336 | 1549 | 1289 | 333 | 1607 | 1348 | 4 | 331 | 1668 | 1423 | 328 | 1735 | 2801 | 16 | 326 | 1797 | 1479 |
| 35 | 330 | 1574 | 1320 | 328 | 1634 | 1379 | 5 | 325 | 1696 | 1457 | 323 | 1764 | 2835 | 17 | 320 | 1828 | 1516 |
| 36 | 324 | 1599 | 1351 | 322 | 1660 | 1412 | 6 | 320 | 1725 | 1491 | 317 | 1794 | 2871 | 18 | 315 | 1859 | 1552 |
| 37 | 319 | 1624 | 1381 | 316 | 1687 | 1445 | 7 | 314 | 1753 | 1525 | 312 | 1824 | 2906 | 19 | 309 | 1890 | 1589 |
| 38 | 313 | 1650 | 1413 | 311 | 1714 | 1477 | 8 | 309 | 1781 | 1558 | 306 | 1854 | 2942 | 20 | 304 | 1922 | 1626 |
| 39 | 307 | 1675 | 1444 | 305 | 1740 | 1509 | 9 | 303 | 1809 | 1592 | 300 | 1883 | 2977 | 21 | 298 | 1953 | 1663 |
| 40 | 301 | 1700 | 1475 | 299 | 1767 | 1542 | 10 | 296 | 1837 | 1627 | 294 | 1913 | 3013 | 22 | 292 | 1984 | 1700 |
| 41 | 295 | 1725 | 1505 | 293 | 1794 | 1575 | 11 | 291 | 1865 | 1661 | 288 | 1943 | 3048 | 23 | 286 | 2015 | 1738 |
| 42 | 288 | 1750 | 1504 | 286 | 1820 | 1575 | 12 | 284 | 1893 | 1662 | 282 | 1972 | 3048 | 24 | 280 | 2046 | 1739 |
| 43 | 282 | 1776 | 1503 | 280 | 1847 | 1574 | 13 | 278 | 1921 | 1661 | 276 | 2002 | 3048 | 25 | 274 | 2078 | 1740 |
| 44 | 276 | 1801 | 1501 | 274 | 1874 | 1572 | 14 | 272 | 1949 | 1661 | 270 | 2032 | 3048 | 26 | 268 | 2109 | 1740 |
| 45 | 270 | 1826 | 1499 | 267 | 1900 | 1573 | 15 | 265 | 1977 | 1661 | 263 | 2061 | 3049 | 27 | 261 | 2140 | 1742 |
| 46 | 263 | 1851 | 1499 | 261 | 1927 | 1571 | 16 | 259 | 2006 | 1661 | 257 | 2091 | 3049 | 28 | 255 | 2171 | 1743 |
| 47 | 256 | 1876 | 1498 | 254 | 1954 | 1571 | 17 | 252 | 2034 | 1662 | 250 | 2121 | 3051 | 29 | 248 | 2202 | 1745 |
| 48 | 249 | 1902 | 1498 | 247 | 1981 | 1571 | 18 | 246 | 2062 | 1662 | 244 | 2151 | 3051 | 30 | 242 | 2234 | 1746 |
| 49 | 243 | 1927 | 1496 | 241 | 2007 | 1570 | 19 | 239 | 2090 | 1662 | 237 | 2180 | 3052 | 31 | 236 | 2265 | 1746 |
| 50 | 236 | 1952 | 1496 | 234 | 2084 | 1570 | 20 | 332 | 2118 | 1662 | 230 | 2210 | 3053 | 1 | 229 | 2296 | 1748 |
| | | | | | | | | | | | | | | June | | | |
| 51 | 229 | 1977 | 1495 | 227 | 2061 | 1570 | 21 | 226 | 2146 | 1662 | 224 | 2240 | 3053 | 2 | 222 | 2327 | 1750 |
| 52 | 222 | 2002 | 1494 | 220 | 2087 | 1570 | 22 | 219 | 2174 | 1663 | 217 | 2269 | 3054 | 3 | 215 | 2358 | 1752 |
| 53 | 214 | 2028 | 1495 | 213 | 2114 | 1570 | 23 | 211 | 2202 | 1664 | 210 | 2299 | 3055 | 4 | 208 | 2390 | 1754 |
| 54 | 207 | 2053 | 1494 | 206 | 2141 | 1570 | 24 | 204 | 2230 | 1665 | 203 | 2329 | 3056 | 5 | 201 | 2421 | 1755 |
| 55 | 200 | 2078 | 1494 | 199 | 2167 | 1568 | 25 | 197 | 2258 | 1665 | 196 | 2358 | 3057 | 6 | 194 | 2452 | 1757 |
| 56 | 193 | 2103 | 1493 | 192 | 2194 | 1569 | 26 | 190 | 2287 | 1666 | 189 | 2388 | 3058 | 7 | 187 | 2483 | 1759 |
| 57 | 185 | 2128 | 1493 | 184 | 2221 | 1570 | 27 | 183 | 2315 | 1667 | 181 | 2418 | 3061 | 8 | 180 | 2514 | 1761 |
| 58 | 178 | 2154 | 1493 | 177 | 2248 | 1570 | 28 | 175 | 2343 | 1668 | 174 | 2448 | 3062 | 9 | 173 | 2546 | 1763 |
| 59 | 170 | 2179 | 1493 | 169 | 2274 | 1571 | 29 | 168 | 2371 | 1669 | 167 | 2477 | 3063 | 10 | 165 | 2577 | 1765 |
| 60 | 163 | 2204 | 1493 | 162 | 2301 | 1571 | 30 | 161 | 2399 | 1669 | 159 | 2507 | 3065 | 11 | 158 | 2608 | 1767 |
| 61 | 180 | 2228 | 1468 | 179 | 2328 | 1547 | 31 | 179 | 2428 | 1645 | 179 | 2536 | 3039 | 12 | 178 | 2640 | 1742 |
| 62 | 171 | 2220 | 1487 | 171 | 2321 | 1548 | 32 | 170 | 2422 | 1648 | 170 | 2560 | 3042 | 13 | 169 | 2635 | 1746 |

# TABLE XIII.

## Sunrise from 18° to 22° N. Latitude.

| Day of Solar year. | Lat. 18° Eqn. of time in seconds. | Lat. 18° ⊙'s Trop. Long. in seconds. | Lat. 18° −428″ Long. Total Corrn. in seconds. Satara. A.D. 1840—1920 | Lat. 19° Eqn. of time in seconds. | Lat. 19° ⊙'s Trop. Long. in seconds. | Lat. 19° −460″ Long. Total Corrn. in seconds. Poona. A.D. 1840—1920 | Days of solar months. | Lat. 20° Eqn. of time in seconds. | Lat. 20° ⊙'s Trop. Long. in seconds. | Lat. 20° −476″ Long. Total Corrn. in seconds. Nasik. A.D. 1840—1920 | Lat. 21° Eqn. of time in seconds. | Lat. 21° ⊙'s Trop. Long. in seconds. | Lat. 21° +800″ Long. Total Corrn. in seconds. Nagpur. A.D. 1840—1920 | Eng. date. | Lat. 22° Eqn. of time in seconds. | Lat. 22° ⊙'s Trop. Long. in seconds. | Lat. 22° −616″ Long. Total Corrn. in seconds. Baroda. A.D. 1840—1920 |
|---|---|---|---|---|---|---|---|---|---|---|---|---|---|---|---|---|---|
| 63 | −162 | +2213 | +1471 | −162 | +2384 | +1550 | 1 (Tamil, Ani; Malayalam, Mithunam; Bengal, Ashada.) | −162 | +2415 | +1649 | −161 | +2524 | +3045 | 14 | −161 | +2630 | +1749 |
| 64 | 153 | 2205 | 1472 | 153 | 2306 | 1552 | 2 | 153 | 2409 | 1652 | 152 | 2518 | 3048 | 15 | 152 | 2624 | 1752 |
| 65 | 145 | 2198 | 1473 | 144 | 2300 | 1554 | 3 | 144 | 2402 | 1654 | 144 | 2512 | 3050 | 16 | 143 | 2619 | 1756 |
| 66 | 135 | 2190 | 1475 | 135 | 2292 | 1555 | 4 | 135 | 2396 | 1657 | 135 | 2500 | 3053 | 17 | 134 | 2614 | 1760 |
| 67 | 126 | 2182 | 1476 | 126 | 2285 | 1557 | 5 | 126 | 2390 | 1660 | 126 | 2501 | 3057 | 18 | 125 | 2609 | 1764 |
| 68 | 117 | 2175 | 1478 | 117 | 2278 | 1559 | 6 | 117 | 2383 | 1662 | 117 | 2495 | 3060 | 19 | 116 | 2604 | 1768 |
| 69 | 108 | 2167 | 1479 | 108 | 2271 | 1561 | 7 | 107 | 2377 | 1666 | 107 | 2489 | 3064 | 20 | 107 | 2598 | 1771 |
| 70 | 99 | 2160 | 1481 | 99 | 2264 | 1563 | 8 | 98 | 2370 | 1668 | 98 | 2483 | 3067 | 21 | 98 | 2593 | 1775 |
| 71 | 90 | 2152 | 1482 | 89 | 2257 | 1567 | 9 | 89 | 2364 | 1671 | 89 | 2477 | 3071 | 22 (June) | 89 | 2588 | 1779 |
| 72 | 80 | 2144 | 1464 | 80 | 2250 | 1548 | 10 | 80 | 2358 | 1651 | 80 | 2471 | 3051 | 23 | 79 | 2583 | 1759 |
| 73 | 71 | 2137 | 1446 | 71 | 2243 | 1529 | 11 | 71 | 2351 | 1632 | 70 | 2465 | 3032 | 24 | 70 | 2578 | 1738 |
| 74 | 62 | 2129 | 1427 | 61 | 2236 | 1517 | 12 | 61 | 2345 | 1613 | 61 | 2459 | 3011 | 25 | 61 | 2572 | 1718 |
| 75 | 53 | 2122 | 1409 | 53 | 2227 | 1491 | 13 | 53 | 2338 | 1593 | 53 | 2453 | 2990 | 26 | 53 | 2567 | 1696 |
| 76 | 43 | 2114 | 1391 | 42 | 2221 | 1474 | 14 | 42 | 2332 | 1575 | 42 | 2447 | 2972 | 27 | 42 | 2562 | 1677 |
| 77 | 33 | 2106 | 1374 | 33 | 2214 | 1455 | 15 | 33 | 2326 | 1555 | 33 | 2442 | 2952 | 28 | 33 | 2557 | 1656 |
| 78 | 24 | 2099 | 1355 | 24 | 2207 | 1436 | 16 | 24 | 2319 | 1536 | 24 | 2436 | 2932 | 29 | 24 | 2552 | 1635 |
| 79 | 15 | 2091 | 1337 | 15 | 2200 | 1417 | 17 | 15 | 2313 | 1516 | 15 | 2430 | 2911 | 30 | 15 | 2546 | 1615 |
| 80 | 5 | 2084 | 1319 | 5 | 2193 | 1399 | 18 | 5 | 2306 | 1498 | 5 | 2424 | 2892 | 1 | 5 | 2541 | 1595 |
| 81 | + 4 | 2076 | 1301 | + 4 | 2186 | 1380 | 19 | + 4 | 2300 | 1478 | + 4 | 2418 | 2872 | 2 | + 4 | 2536 | 1574 |
| 82 | 14 | 2086 | 1283 | 13 | 2179 | 1361 | 20 | 13 | 2294 | 1458 | 13 | 2412 | 2852 | 3 | 13 | 2531 | 1553 |
| 83 | 23 | 2061 | 1265 | 23 | 2172 | 1348 | 21 | 23 | 2287 | 1440 | 23 | 2406 | 2833 | 4 | 23 | 2526 | 1533 |
| 84 | 33 | 2053 | 1247 | 32 | 2165 | 1324 | 22 | 32 | 2281 | 1420 | 32 | 2400 | 2812 | 5 | 32 | 2520 | 1513 |
| 85 | 42 | 2046 | 1229 | 42 | 2158 | 1306 | 23 | 42 | 2274 | 1402 | 42 | 2394 | 2793 | 6 | 42 | 2515 | 1493 |
| 86 | 51 | 2038 | 1210 | 51 | 2150 | 1287 | 24 | 51 | 2268 | 1382 | 51 | 2388 | 2773 | 7 | 51 | 2510 | 1472 |
| 87 | 60 | 2030 | 1192 | 60 | 2143 | 1268 | 25 | 60 | 2262 | 1362 | 60 | 2383 | 2753 | 8 | 60 | 2505 | 1451 |
| 88 | 70 | 2023 | 1174 | 70 | 2136 | 1250 | 26 | 70 | 2255 | 1344 | 69 | 2377 | 2733 | 9 | 69 | 2500 | 1430 |
| 89 | 79 | 2015 | 1156 | 79 | 2129 | 1239 | 27 | 79 | 2249 | 1324 | 79 | 2371 | 2713 | 10 | 79 | 2494 | 1411 |
| 90 | 88 | 2008 | 1137 | 88 | 2122 | 1212 | 28 | 88 | 2242 | 1305 | 88 | 2365 | 2693 | 11 | 88 | 2489 | 1390 |
| 91 | 106 | 2000 | 1128 | 106 | 2116 | 1202 | 29 | 106 | 2236 | 1294 | 107 | 2360 | 2683 | 12 | 107 | 2484 | 1379 |
| 92 | 116 | 1972 | 1110 | 116 | 2088 | 1184 | 30 | 117 | 2207 | 1276 | 117 | 2331 | 2664 | 13 | 117 | 2454 | 1359 |
| 93 | 126 | 1945 | 1093 | 126 | 2060 | 1166 | 31 | 126 | 2179 | 1257 | 127 | 2302 | 2645 | 14 | 127 | 2424 | 1339 |
| 94 | 136 | 1917 | 1075 | 136 | 2038 | 1148 | 32 | 136 | 2150 | 1238 | 137 | 2272 | 2625 | 15 | 137 | 2395 | 1320 |
| 95 | 146 | 1890 | 1058 | 146 | 2004 | 1130 | 1 (Tamil, Adi; Malayalam, Karkatagam; Bengal, Sravana.) | 146 | 2122 | 1220 | 146 | 2243 | 2605 | 16 (July) | 147 | 2365 | 1300 |
| 96 | 155 | 1862 | 1039 | 156 | 1976 | 1112 | 2 | 156 | 2093 | 1201 | 156 | 2214 | 2586 | 17 | 157 | 2335 | 1280 |
| 97 | 165 | 1835 | 1022 | 166 | 1948 | 1094 | 3 | 166 | 2064 | 1182 | 166 | 2185 | 2567 | 18 | 167 | 2305 | 1260 |
| 98 | 175 | 1807 | 1004 | 175 | 1920 | 1075 | 4 | 176 | 2036 | 1164 | 176 | 2156 | 2548 | 19 | 177 | 2275 | 1240 |
| 99 | 185 | 1780 | 987 | 185 | 1892 | 1057 | 5 | 186 | 2007 | 1145 | 186 | 2126 | 2528 | 20 | 187 | 2246 | 1221 |
| 100 | 194 | 1752 | 968 | 194 | 1864 | 1038 | 6 | 195 | 1979 | 1126 | 195 | 2097 | 2508 | 21 | 196 | 2216 | 1200 |
| 101 | 203 | 1725 | 950 | 204 | 1836 | 1020 | 7 | 204 | 1950 | 1106 | 205 | 2068 | 2489 | 22 | 205 | 2186 | 1179 |
| 102 | 213 | 1697 | 933 | 263 | 1808 | 1001 | 8 | 214 | 1921 | 1086 | 214 | 2039 | 2470 | 23 | 215 | 2156 | 1159 |
| 103 | 222 | 1670 | 917 | 223 | 1780 | 985 | 9 | 223 | 1893 | 1067 | 224 | 2010 | 2447 | 24 | 224 | 2126 | 1137 |
| 104 | 232 | 1642 | 903 | 232 | 1752 | 967 | 10 | 233 | 1864 | 1050 | 233 | 1980 | 2430 | 25 | 234 | 2097 | 1116 |
| 105 | 241 | 1615 | 887 | 242 | 1724 | 951 | 11 | 242 | 1836 | 1021 | 243 | 1951 | 2411 | 26 | 243 | 2067 | 1094 |
| 106 | 250 | 1587 | 871 | 251 | 1696 | 934 | 12 | 251 | 1867 | 1012 | 252 | 1922 | 2391 | 27 | 253 | 2037 | 1073 |
| 107 | 259 | 1560 | 855 | 260 | 1668 | 916 | 13 | 260 | 1778 | 993 | 261 | 1893 | 2370 | 28 | 261 | 2007 | 1050 |
| 108 | 268 | 1582 | 840 | 269 | 1640 | 899 | 14 | 269 | 1750 | 975 | 270 | 1864 | 2350 | 29 | 271 | 1977 | 1030 |
| 109 | 277 | 1505 | 825 | 278 | 1620 | 882 | 15 | 278 | 1721 | 956 | 279 | 1834 | 2330 | 30 | 280 | 1948 | 1008 |
| 110 | 286 | 1477 | 808 | 287 | 1584 | 865 | 16 | 287 | 1693 | 937 | 288 | 1805 | 2310 | 31 | 289 | 1918 | 986 |
| 111 | 294 | 1450 | 792 | 295 | 1556 | 846 | 17 | 296 | 1664 | 919 | 296 | 1776 | 2288 | 1 | 297 | 1888 | 963 |
| 112 | 303 | 1422 | 778 | 304 | 1528 | 829 | 18 | 305 | 1635 | 900 | 305 | 1747 | 2267 | 2 | 306 | 1858 | 941 |
| 113 | 312 | 1395 | 762 | 312 | 1500 | 811 | 19 | 313 | 1607 | 880 | 314 | 1718 | 2248 | 3 | 315 | 1828 | 919 |
| 114 | 321 | 1367 | 745 | 321 | 1472 | 793 | 20 | 322 | 1578 | 862 | 323 | 1688 | 2227 | 4 | 324 | 1799 | 897 |
| 115 | 329 | 1840 | 728 | 329 | 1444 | 775 | 21 | 330 | 1550 | 842 | 331 | 1659 | 2206 | 5 | 332 | 1769 | 874 |
| 116 | 337 | 1312 | 711 | 338 | 1416 | 758 | 22 | 339 | 1521 | 823 | 339 | 1630 | 2185 | 6 | 340 | 1739 | 851 |
| 117 | 345 | 1285 | 694 | 346 | 1388 | 739 | 23 | 346 | 1492 | 802 | 347 | 1601 | 2163 | 7 | 348 | 1709 | 828 |
| 118 | 353 | 1257 | 678 | 354 | 1360 | 721 | 24 | 354 | 1464 | 784 | 356 | 1572 | 2143 | 8 (August) | 357 | 1679 | 807 |
| 119 | 361 | 1230 | 661 | 362 | 1332 | 703 | 25 | 362 | 1435 | 763 | 363 | 1542 | 2121 | 9 | 364 | 1650 | 783 |
| 120 | 369 | 1202 | 644 | 369 | 1304 | 684 | 26 | 370 | 1407 | 743 | 371 | 1513 | 2100 | 10 | 372 | 1620 | 760 |
| 121 | 371 | 1175 | 622 | 373 | 1276 | 661 | 27 | 375 | 1378 | 721 | 377 | 1484 | 2076 | 11 | 379 | 1590 | 736 |
| 122 | 379 | 1148 | 605 | 381 | 1248 | 643 | 28 | 383 | 1348 | 701 | 385 | 1456 | 2055 | 12 | 387 | 1560 | 713 |
| 123 | 386 | 1123 | 587 | 388 | 1222 | 624 | 29 | 390 | 1320 | 680 | 393 | 1427 | 2034 | 13 | 395 | 1529 | 690 |
| 124 | 394 | 1099 | 571 | 396 | 1195 | 605 | 30 | 398 | 1293 | 661 | 401 | 1397 | 2012 | 14 | 403 | 1498 | 667 |
| 125 | 400 | 1074 | 552 | 403 | 1169 | 586 | 31 | 405 | 1265 | 640 | 407 | 1368 | 1989 | 15 | 410 | 1467 | 648 |

# TABLE XIII.

## Sunrise from 18° to 22° N. Latitude.

| Day of Solar year. | Lat. 18° —428″ Long. Eqn. of time in seconds. | ⊙'s Trop. Long. in seconds. | Total Corrn. in seconds. Satara. A.D. 1840—1920 | Lat. 19° —460″ Long. Eqn. of time in seconds. | ⊙'s Trop. Long. in seconds. | Total Corrn. in seconds. Poona. A.D. 1840—1920 | Solar month | Days of solar months. | Lat. 20° —476″ Long. Eqn. of time in seconds. | ⊙'s Trop. Long. in seconds. | Total Corrn. in seconds. Nasik. A.D. 1840—1920 | Lat. 21° +800″ Long. Eqn. of Time in seconds. | ⊙'s Trop. Long. in seconds. | Total Corrn. in seconds. Nagpur. A.D. 1840—1920 | Eng. month | Eng. date. | Lat. 22° —616″ Long. Eqn. of Time in seconds. | ⊙'s Trop. Long. in seconds. | Total Corrn. in seconds. Baroda. A.D. 1840—1920 |
|---|---|---|---|---|---|---|---|---|---|---|---|---|---|---|---|---|---|---|---|
| **126** | +408 | +1049 | +535 | +410 | +1143 | +567 | *Tamil*, Avani; *Malayalam*, Chingam; *Bengal*, Bhadrapada. | 1 | +412 | +1237 | +619 | +415 | +1339 | +1968 | August | **16** | +417 | +1436 | +619 |
| **127** | 415 | 1024 | 517 | 417 | 1116 | 547 | | 2 | 419 | 1209 | 598 | 422 | 1309 | 1945 | | **17** | 424 | 1405 | 595 |
| **128** | 421 | 1000 | 499 | 424 | 1090 | 528 | | 3 | 426 | 1182 | 578 | 429 | 1280 | 1923 | | **18** | 431 | 1375 | 572 |
| **129** | 428 | 975 | 481 | 430 | 1064 | 508 | | 4 | 433 | 1154 | 557 | 435 | 1251 | 1900 | | **19** | 438 | 1344 | 548 |
| **130** | 435 | 950 | 463 | 437 | 1038 | 489 | | 5 | 440 | 1126 | 526 | 412 | 1222 | 1878 | | **20** | 445 | 1313 | 524 |
| **131** | 441 | 926 | 445 | 443 | 1011 | 468 | | 6 | 446 | 1099 | 515 | 447 | 1192 | 1853 | | **21** | 451 | 1282 | 499 |
| **132** | 447 | 901 | 427 | 450 | 985 | 450 | | 7 | 452 | 1071 | 492 | 455 | 1162 | 1831 | | **22** | 457 | 1251 | 473 |
| **133** | 453 | 876 | 420 | 456 | 959 | 441 | | 8 | 459 | 1043 | 482 | 461 | 1134 | 1818 | | **23** | 464 | 1220 | 460 |
| **134** | 459 | 852 | 413 | 462 | 932 | 432 | | 9 | 464 | 1016 | 471 | 467 | 1104 | 1806 | | **24** | 470 | 1189 | 445 |
| **135** | 465 | 827 | 405 | 468 | 906 | 424 | | 10 | 470 | 988 | 460 | 473 | 1075 | 1793 | | **25** | 476 | 1158 | 481 |
| **136** | 471 | 802 | 398 | 474 | 880 | 415 | | 11 | 476 | 960 | 450 | 479 | 1046 | 1781 | | **26** | 482 | 1127 | 416 |
| **137** | 476 | 777 | 390 | 479 | 853 | 405 | | 12 | 482 | 932 | 439 | 485 | 1016 | 1768 | | **27** | 487 | 1096 | 401 |
| **138** | 483 | 753 | 389 | 485 | 827 | 396 | | 13 | 487 | 905 | 427 | 490 | 987 | 1754 | | **28** | 493 | 1066 | 387 |
| **139** | 487 | 728 | 375 | 490 | 801 | 386 | | 14 | 493 | 877 | 417 | 496 | 958 | 1742 | | **29** | 499 | 1035 | 372 |
| **140** | 492 | 703 | 366 | 495 | 775 | 377 | | 15 | 498 | 849 | 405 | 501 | 929 | 1728 | | **30** | 504 | 1004 | 357 |
| **141** | 497 | 679 | 358 | 500 | 748 | 367 | | 16 | 503 | 822 | 394 | 506 | 899 | 1715 | | **31** | 509 | 973 | 341 |
| **142** | 502 | 654 | 350 | 505 | 722 | 357 | | 17 | 508 | 794 | 382 | 511 | 870 | 1701 | | **1** | 514 | 942 | 326 |
| **143** | 507 | 629 | 342 | 510 | 696 | 347 | | 18 | 513 | 766 | 370 | 516 | 841 | 1687 | | **2** | 519 | 911 | 311 |
| **144** | 511 | 605 | 333 | 514 | 669 | 336 | | 19 | 517 | 739 | 358 | 520 | 811 | 1673 | | **3** | 523 | 880 | 294 |
| **145** | 516 | 580 | 324 | 519 | 643 | 327 | | 20 | 522 | 711 | 346 | 525 | 782 | 1659 | | **4** | 528 | 849 | 279 |
| **146** | 520 | 555 | 315 | 523 | 617 | 316 | | 21 | 526 | 683 | 334 | 529 | 753 | 1645 | | **5** | 532 | 818 | 262 |
| **147** | 524 | 530 | 306 | 527 | 590 | 305 | | 22 | 530 | 655 | 321 | 534 | 723 | 1631 | | **6** | 536 | 787 | 246 |
| **148** | 529 | 506 | 298 | 532 | 564 | 295 | | 23 | 535 | 628 | 309 | 538 | 694 | 1616 | | **7** | 540 | 757 | 230 |
| **149** | 532 | 481 | 288 | 536 | 538 | 284 | | 24 | 539 | 600 | 297 | 542 | 665 | 1602 | | **8** | 544 | 726 | 213 |
| **150** | 536 | 456 | 278 | 539 | 512 | 273 | | 25 | 542 | 572 | 283 | 545 | 636 | 1586 | | **9** | 548 | 696 | 197 |
| **151** | 539 | 432 | 268 | 542 | 485 | 261 | | 26 | 545 | 545 | 270 | 548 | 606 | 1571 | | **10** | 552 | 664 | 180 |
| **152** | 519 | 408 | 235 | 523 | 460 | 227 | | 27 | 527 | 516 | 235 | 531 | 576 | 1535 | | **11** | 535 | 632 | 148 |
| **153** | 522 | 395 | 225 | 526 | 445 | 215 | | 28 | 530 | 499 | 221 | 534 | 557 | 1519 | | **12** | 538 | 612 | 126 |
| **154** | 526 | 382 | 216 | 529 | 430 | 203 | | 29 | 533 | 483 | 208 | 537 | 539 | 1504 | | **13** | 541 | 591 | 108 |
| **155** | 529 | 368 | 205 | 532 | 416 | 192 | | 30 | 536 | 466 | 191 | 540 | 520 | 1488 | | **14** | 544 | 571 | 91 |
| **156** | 531 | 355 | 194 | 534 | 401 | 179 | | 31 | 538 | 450 | 180 | 542 | 502 | 1472 | | **15** | 546 | 550 | 72 |
| **157** | 533 | 342 | 183 | 536 | 386 | 166 | *Tamil*, Purattasi, *Malayalam*, Kanni; *Bengal*, Asvina. | 1 | 540 | 433 | 165 | 544 | 483 | 1455 | September | **16** | 548 | 530 | 54 |
| **158** | 536 | 329 | 173 | 538 | 371 | 153 | | 2 | 542 | 416 | 150 | 546 | 464 | 1438 | | **17** | 550 | 510 | 36 |
| **159** | 538 | 316 | 162 | 540 | 356 | 140 | | 3 | 544 | 400 | 136 | 548 | 446 | 1422 | | **18** | 552 | 489 | 17 |
| **160** | 540 | 302 | 150 | 542 | 342 | 128 | | 4 | 546 | 383 | 121 | 550 | 427 | 1405 | | **19** | 554 | 409 | — 1 |
| **161** | 542 | 289 | 134 | 544 | 327 | 115 | | 5 | 548 | 367 | 107 | 552 | 409 | 1389 | | **20** | 556 | 448 | 20 |
| **162** | 543 | 276 | 127 | 546 | 312 | 102 | | 6 | 549 | 350 | 91 | 553 | 390 | 1371 | | **21** | 557 | 428 | 39 |
| **163** | 545 | 263 | 117 | 547 | 297 | 87 | | 7 | 551 | 333 | 75 | 555 | 371 | 1355 | | **22** | 559 | 408 | 57 |
| **164** | 547 | 250 | 105 | 548 | 282 | 73 | | 8 | 553 | 317 | 60 | 557 | 353 | 1338 | | **23** | 561 | 387 | 76 |
| **165** | 548 | 236 | 93 | 550 | 268 | 59 | | 9 | 554 | 300 | 44 | 558 | 334 | 1320 | | **24** | 562 | 367 | 96 |
| **166** | 548 | 223 | 79 | 551 | 253 | 45 | | 10 | 554 | 284 | 26 | 558 | 316 | 1300 | | **25** | 562 | 346 | 117 |
| **167** | 549 | 210 | 67 | 552 | 238 | 31 | | 11 | 555 | 267 | 10 | 559 | 297 | 1282 | | **26** | 563 | 326 | 164 |
| **168** | 550 | 197 | 54 | 553 | 223 | 17 | | 12 | 556 | 250 | 6 | 560 | 278 | 1264 | | **27** | 564 | 306 | 157 |
| **169** | 551 | 184 | 41 | 554 | 208 | 2 | | 13 | 557 | 234 | 22 | 561 | 260 | 1246 | | **28** | 565 | 285 | 178 |
| **170** | 551 | 170 | 28 | 554 | 194 | — 13 | | 14 | 557 | 217 | 39 | 561 | 241 | 1227 | | **29** | 565 | 265 | 199 |
| **171** | 551 | 157 | 14 | 554 | 179 | 28 | | 15 | 557 | 201 | 57 | 561 | 223 | 1207 | | **30** | 565 | 244 | 220 |
| **172** | 552 | 144 | 2 | 554 | 164 | 44 | | 16 | 557 | 184 | 74 | 561 | 204 | 1188 | | **1** | 565 | 224 | 241 |
| **173** | 551 | 131 | 13 | 554 | 149 | 59 | | 17 | 557 | 167 | 91 | 561 | 185 | 1169 | | **2** | 565 | 204 | 262 |
| **174** | 551 | 118 | 27 | 554 | 134 | 74 | | 18 | 557 | 151 | 108 | 561 | 167 | 1150 | | **3** | 565 | 183 | 283 |
| **175** | 551 | 104 | 40 | 554 | 120 | 90 | | 19 | 557 | 131 | 125 | 561 | 148 | 1131 | | **4** | 565 | 163 | 304 |
| **176** | 550 | 91 | 55 | 554 | 105 | 105 | | 20 | 557 | 118 | 148 | 561 | 130 | 1111 | | **5** | 565 | 142 | 325 |
| **177** | 549 | 78 | 69 | 553 | 90 | 121 | | 21 | 557 | 101 | 160 | 560 | 111 | 1091 | October | **6** | 564 | 122 | 347 |
| **178** | 548 | 65 | 84 | 552 | 75 | 137 | | 22 | 555 | 84 | 179 | 559 | 92 | 1071 | | **7** | 563 | 102 | 369 |
| **179** | 548 | 52 | 98 | 551 | 60 | 154 | | 23 | 554 | 68 | 197 | 558 | 74 | 1051 | | **8** | 562 | 81 | 392 |
| **180** | 547 | 38 | 112 | 550 | 46 | 170 | | 24 | 553 | 51 | 215 | 557 | 55 | 1031 | | **9** | 561 | 61 | 414 |
| **181** | 545 | 25 | 128 | 548 | 31 | 187 | | 25 | 551 | 35 | 235 | 555 | 37 | 1009 | | **10** | 559 | 40 | 437 |
| **182** | 544 | 12 | 142 | 547 | 16 | 204 | | 26 | 550 | 18 | 253 | 554 | 18 | 989 | | **11** | 558 | 20 | 459 |
| **183** | 542 | 0 | 158 | 545 | 0 | 221 | | 27 | 549 | 0 | 271 | 554 | 0 | 970 | | **12** | 558 | 0 | 480 |
| **184** | 541 | —14 | 173 | 544 | —15 | 237 | | 28 | 547 | —17 | 290 | 552 | — 19 | 949 | | **13** | 556 | —21 | 503 |
| **185** | 539 | 27 | 188 | 542 | 31 | 255 | | 29 | 545 | 34 | 309 | 550 | 38 | 928 | | **14** | 554 | 42 | 526 |
| **186** | 537 | 41 | 204 | 540 | 46 | 272 | | 30 | 543 | 52 | 319 | 548 | 58 | 906 | | **15** | 552 | 63 | 549 |

# TABLE XIII.

## Sunrise from 18° to 22° N. Latitude.

| Day of Solar year. | Lat. 18° —428″ Long. Eqn. of time in seconds. | ☉'s Trop. Long. in seconds. | Total Corrn. in seconds. Satara. A.D. 1840—1920 | Lat. 19° —460″ Long. Eqn. of time in seconds. | ☉'s Trop. Long. in seconds. | Total Corrn. in seconds. Poona. A.D. 1840—1920 | Days of solar months. | Lat. 20° —476″ Long. Eqn. of time in seconds. | ☉'s Trop. Long. in seconds. | Total Corrn. in seconds. Nasik. A.D. 1840—1920 | Lat. 21° +800″ Long. Eqn. of Time in Seconds. | ☉'s Trop. Long. in seconds. | Total Corrn. in seconds. Nagpur. A.D. 1840—1920 | Eng. date. | Lat. 22° —616″ Long. Eqn. of Time in seconds. | ☉'s Trop. Long. in seconds. | Total Corrn. in seconds. Baroda. A.D. 1840—1920 |
|---|---|---|---|---|---|---|---|---|---|---|---|---|---|---|---|---|---|
| 187 | +535 | —54 | —219 | +538 | —61 | —289 | 1 | +541 | —69 | —348 | +546 | — 77 | +885 | 16 | +550 | — 84 | —572 |
| 188 | 533 | 68 | 235 | 536 | 76 | 306 | 2 | 539 | 86 | 367 | 544 | 96 | 864 | 17 | 548 | 105 | 595 |
| 189 | 530 | 82 | 252 | 533 | 92 | 325 | 3 | 536 | 103 | 387 | 541 | 115 | 842 | 18 | 545 | 127 | 620 |
| 190 | 527 | 95 | 268 | 530 | 107 | 343 | 4 | 533 | 120 | 407 | 538 | 134 | 820 | 19 | 542 | 148 | 644 |
| 191 | 524 | 109 | 285 | 527 | 122 | 361 | 5 | 530 | 138 | 428 | 535 | 154 | 797 | 20 | 539 | 169 | 668 |
| 192 | 521 | 122 | 301 | 524 | 138 | 380 | 6 | 527 | 155 | 448 | 532 | 173 | 775 | 21 | 536 | 190 | 692 |
| 193 | 518 | 136 | 318 | 521 | 153 | 399 | 7 | 524 | 172 | 468 | 529 | 192 | 753 | 22 | 533 | 211 | 715 |
| 194 | 515 | 150 | 346 | 518 | 168 | 428 | 8 | 521 | 189 | 499 | 526 | 211 | 721 | 23 | 530 | 232 | 749 |
| 195 | 512 | 163 | 373 | 515 | 184 | 458 | 9 | 518 | 206 | 532 | 523 | 230 | 688 | 24 | 527 | 253 | 788 |
| 196 | 508 | 177 | 402 | 511 | 199 | 488 | 10 | 514 | 224 | 561 | 519 | 250 | 655 | 25 | 523 | 274 | 818 |
| 197 | 504 | 190 | 431 | 507 | 214 | 518 | 11 | 510 | 241 | 593 | 515 | 269 | 622 | 26 | 519 | 295 | 853 |
| 198 | 500 | 204 | 459 | 503 | 229 | 548 | 12 | 506 | 258 | 624 | 511 | 288 | 589 | 27 | 515 | 316 | 887 |
| 199 | 496 | 218 | 488 | 499 | 245 | 579 | 13 | 502 | 275 | 656 | 507 | 307 | 555 | 28 | 511 | 338 | 922 |
| 200 | 491 | 231 | 518 | 494 | 260 | 610 | 14 | 497 | 292 | 689 | 502 | 326 | 521 | 29 | 506 | 359 | 958 |
| 201 | 487 | 245 | 547 | 490 | 275 | 640 | 15 | 493 | 310 | 721 | 498 | 346 | 488 | 30 | 502 | 380 | 993 |
| 202 | 483 | 258 | 575 | 486 | 291 | 677 | 16 | 489 | 327 | 752 | 494 | 365 | 454 | 31 | 498 | 401 | 1028 |
| 203 | 478 | 272 | 606 | 481 | 306 | 702 | 17 | 484 | 344 | 785 | 490 | 384 | 411 | 1 | 493 | 422 | 1064 |
| 204 | 473 | 286 | 635 | 476 | 321 | 733 | 18 | 479 | 361 | 818 | 484 | 403 | 386 | 2 | 488 | 443 | 1100 |
| 205 | 468 | 299 | 664 | 471 | 337 | 765 | 19 | 474 | 378 | 850 | 479 | 422 | 351 | 3 | 483 | 464 | 1136 |
| 206 | 463 | 313 | 694 | 466 | 352 | 796 | 20 | 469 | 396 | 883 | 474 | 442 | 317 | 4 | 478 | 485 | 1172 |
| 207 | 459 | 326 | 723 | 462 | 367 | 825 | 21 | 465 | 413 | 915 | 470 | 461 | 284 | 5 | 474 | 506 | 1207 |
| 208 | 454 | 340 | 752 | 457 | 382 | 857 | 22 | 460 | 430 | 947 | 465 | 480 | 250 | 6 | 469 | 527 | 1242 |
| 209 | 448 | 354 | 783 | 451 | 398 | 890 | 23 | 454 | 447 | 981 | 459 | 499 | 214 | 7 | 463 | 549 | 1279 |
| 210 | 443 | 367 | 813 | 446 | 413 | 921 | 24 | 449 | 464 | 1010 | 454 | 518 | 180 | 8 | 458 | 570 | 1315 |
| 211 | 437 | 381 | 844 | 440 | 428 | 953 | 25 | 443 | 482 | 1018 | 448 | 538 | 145 | 9 | 452 | 591 | 1352 |
| 212 | 431 | 394 | 874 | 434 | 444 | 984 | 26 | 437 | 499 | 1081 | 442 | 557 | 109 | 10 | 446 | 612 | 1389 |
| 213 | 441 | 408 | 889 | 443 | 460 | 1003 | 27 | 446 | 516 | 1100 | 449 | 576 | 87 | 11 | 451 | 632 | 1415 |
| 214 | 435 | 433 | 920 | 437 | 486 | 1035 | 28 | 440 | 544 | 1134 | 443 | 605 | 52 | 12 | 445 | 663 | 1451 |
| 2:5 | 428 | 457 | 951 | 430 | 513 | 1069 | 29 | 433 | 571 | 1168 | 436 | 635 | 15 | 13 | 438 | 694 | 1490 |
| 216 | 421 | 482 | 983 | 423 | 539 | 1102 | 30 | 426 | 599 | 1203 | 429 | 664 | — 21 | 14 | 481 | 725 | 1528 |
| 217 | 414 | 507 | 1015 | 416 | 565 | 1135 | 1 | 419 | 627 | 1238 | 422 | 693 | 57 | 15 | 424 | 756 | 1566 |
| 218 | 407 | 531 | 1046 | 409 | 591 | 1168 | 2 | 412 | 654 | 1272 | 415 | 722 | 93 | 16 | 417 | 786 | 1603 |
| 219 | 400 | 556 | 1078 | 402 | 618 | 1202 | 3 | 405 | 682 | 1307 | 408 | 752 | 130 | 17 | 409 | 817 | 1642 |
| 220 | 393 | 581 | 1110 | 395 | 644 | 1235 | 4 | 398 | 710 | 1342 | 401 | 781 | 166 | 18 | 402 | 848 | 1680 |
| 221 | 386 | 606 | 1142 | 388 | 670 | 1268 | 5 | 391 | 738 | 1377 | 394 | 810 | 202 | 19 | 394 | 879 | 1719 |
| 222 | 379 | 680 | 1173 | 381 | 697 | 1302 | 6 | 384 | 765 | 1411 | 387 | 840 | 239 | 20 | 386 | 910 | 1758 |
| 223 | 372 | 655 | 1204 | 374 | 723 | 1334 | 7 | 377 | 793 | 1447 | 380 | 809 | 276 | 21 | 378 | 941 | 1798 |
| 224 | 364 | 680 | 1239 | 366 | 749 | 1370 | 8 | 369 | 821 | 1484 | 372 | 898 | 313 | 22 | 370 | 972 | 1836 |
| 225 | -356 | 704 | 1275 | 358 | 776 | 1406 | 9 | 361 | 848 | 1520 | 364 | 928 | 350 | 23 | 362 | 1003 | 1874 |
| 226 | 348 | 729 | 1310 | 350 | 802 | 1442 | 10 | 353 | 876 | 1557 | 356 | 957 | 388 | 24 | 354 | 1034 | 1911 |
| 227 | 340 | 754 | 1346 | 342 | 828 | 1478 | 11 | 345 | 904 | 1593 | 348 | 986 | 425 | 25 | 345 | 1065 | 1950 |
| 228 | 332 | 778 | 1381 | 334 | 854 | 1514 | 12 | 337 | 931 | 1630 | 340 | 1015 | 462 | 26 | 337 | 1095 | 1988 |
| 229 | 324 | 803 | 1417 | 326 | 881 | 1550 | 13 | 329 | 959 | 1667 | 332 | 1045 | 499 | 27 | 329 | 1126 | 2026 |
| 230 | 315 | 828 | 1453 | 317 | 907 | 1587 | 14 | 320 | 987 | 1704 | 323 | 1074 | 537 | 28 | 320 | 1157 | 2065 |
| 231 | 307 | 853 | 1489 | 309 | 933 | 1623 | 15 | 312 | 1015 | 1741 | 315 | 1103 | 575 | 29 | 311 | 1188 | 2103 |
| 232 | 299 | 877 | 1524 | 301 | 960 | 1659 | 16 | 304 | 1042 | 1777 | 307 | 1133 | 612 | 30 | 303 | 1219 | 2141 |
| 233 | 291 | 902 | 1560 | 293 | 986 | 1695 | 17 | 296 | 1070 | 1814 | 298 | 1162 | 649 | 1 | 294 | 1250 | 2180 |
| 234 | 282 | 927 | 1626 | 284 | 1012 | 1732 | 18 | 287 | 1098 | 1852 | 290 | 1191 | 687 | 2 | 284 | 1280 | 2220 |
| 235 | 273 | 951 | 1633 | 275 | 1039 | 1769 | 19 | 278 | 1125 | 1889 | 281 | 1221 | 725 | 3 | 275 | 1312 | 2259 |
| 236 | 264 | 976 | 1669 | 266 | 1065 | 1806 | 20 | 269 | 1153 | 1927 | 272 | 1250 | 764 | 4 | 266 | 1343 | 2297 |
| 237 | 255 | 1001 | 1706 | 257 | 1091 | 1843 | 21 | 260 | 1181 | 1964 | 263 | 1279 | 852 | 5 | 257 | 1374 | 2336 |
| 238 | 246 | 1025 | 1742 | 248 | 1117 | 1880 | 22 | 251 | 1208 | 2002 | 254 | 1308 | 840 | 6 | 248 | 1404 | 2375 |
| 239 | 237 | 1050 | 1779 | 239 | 1144 | 1917 | 23 | 242 | 1236 | 2040 | 245 | 1338 | 878 | 7 | 238 | 1435 | 2415 |
| 240 | 228 | 1075 | 1815 | 230 | 1170 | 1954 | 24 | 233 | 1264 | 2077 | 236 | 1367 | 916 | 8 | 229 | 1466 | 2454 |
| 241 | 218 | 1100 | 1858 | 220 | 1196 | 1992 | 25 | 223 | 1292 | 2116 | 226 | 1396 | 956 | 9 | 219 | 1497 | 2493 |
| 242 | 209 | 1124 | 1889 | 211 | 1223 | 2029 | 26 | 214 | 1319 | 2153 | 217 | 1426 | 994 | 10 | 210 | 1528 | 2532 |
| 243 | 202 | 1148 | 1924 | 203 | 1248 | 2065 | 27 | 203 | 1348 | 2193 | 203 | 1456 | 1027 | 11 | 204 | 1560 | 2568 |
| 244 | 193 | 1176 | 1960 | 194 | 1276 | 2102 | 28 | 194 | 1377 | 2231 | 194 | 1485 | 1075 | 12 | 195 | 1590 | 2607 |
| 245 | 183 | 1203 | 1998 | 184 | 1304 | 2140 | 29 | 184 | 1405 | 2269 | 184 | 1514 | 1114 | 13 | 185 | 1620 | 2647 |
| 246 | 173 | 1230 | 2035 | 174 | 1332 | 2178 | 30 | 174 | 1434 | 2308 | 174 | 1544 | 1154 | 14 | 175 | 1649 | 2686 |

Days of solar months 1–30 (days 187–216): *Tamil*, Aippasi ; *Malayalam*, Tulam ; *Bengal*, Kartika.

Days of solar months 1–30 (days 217–246): *Tamil*, Kartigai ; *Malayalaw*, Vrischikam ; *Bengal*, Margasira.

Eng. date: October (16–31); November (1–30); December (1–14).

## TABLE XIII.

### Sunrise from 18° to 22° N. Latitude.

| Day of Solar year. | Lat. 18° −428″ Long. Eqn. of time in seconds. | ⊙'s Trop. Long. in seconds. | Total Corrn. in seconds. Satara. A.D. 1840—1920 | Lat. 19° −460″ Long. Eqn. of time in seconds. | ⊙'s Trop. Long. in seconds. | Total Corrn. in seconds. Poona. A.D. 1840—1920 | Solar month | Days of solar months. | Lat. 20° −476″ Long. Eqn. of time in seconds. | ⊙'s Trop. Long. in seconds. | Total Corrn. in seconds. Nasik. A.D. 1840—1920 | Lat. 21° +800″ Long. Eqn. of time in seconds. | ⊙'s Trop. Long. in seconds. | Total Corrn. in seconds. Nagpur. A.D. 1840—1920 | Eng. month | Eng. date. | Lat. 22° −616″ Long. Eqn. of time in seconds. | ⊙'s Trop. Long. in seconds. | Total Corrn. in seconds. Baroda. A.D. 1840—1920 |
|---|---|---|---|---|---|---|---|---|---|---|---|---|---|---|---|---|---|---|---|
| **247** | +163 | −1258 | −2073 | +164 | −1360 | −2216 | *Tamil*, Margali; *Malayalam*, Dhanus; *Bengal*, Pausha. | 1 | +164 | −1462 | −2346 | +164 | −1573 | −1198 | December | **15** | +165 | −1679 | −2726 |
| **248** | 153 | 1285 | 2110 | 154 | 1388 | 2254 | | 2 | 154 | 1491 | 2385 | 154 | 1602 | 1232 | | **16** | 155 | 1709 | 2766 |
| **249** | 143 | 1313 | 2148 | 144 | 1416 | 2292 | | 3 | 144 | 1520 | 2424 | 144 | 1631 | 1271 | | **17** | 145 | 1739 | 2806 |
| **250** | 133 | 1340 | 2185 | 134 | 1444 | 2330 | | 4 | 134 | 1548 | 2462 | 134 | 1660 | 1310 | | **18** | 135 | 1769 | 2846 |
| **251** | 123 | 1368 | 2223 | 124 | 1472 | 2368 | | 5 | 124 | 1577 | 2501 | 124 | 1690 | 1350 | | **19** | 125 | 1798 | 2885 |
| **252** | 113 | 1395 | 2260 | 114 | 1500 | 2406 | | 6 | 114 | 1605 | 2539 | 114 | 1719 | 1389 | | **20** | 115 | 1828 | 2925 |
| **253** | 103 | 1423 | 2298 | 104 | 1528 | 2444 | | 7 | 104 | 1634 | 2598 | 104 | 1748 | 1428 | | **21** | 105 | 1858 | 2965 |
| **254** | 93 | 1450 | 2335 | 94 | 1556 | 2482 | | 8 | 94 | 1663 | 2618 | 94 | 1777 | 1466 | | **22** | 95 | 1888 | 3005 |
| **255** | 83 | 1478 | 2353 | 84 | 1584 | 2499 | | 9 | 84 | 1691 | 2634 | 84 | 1806 | 1482 | | **23** | 85 | 1918 | 3020 |
| **256** | 73 | 1505 | 2370 | 74 | 1612 | 2522 | | 10 | 74 | 1720 | 2651 | 74 | 1836 | 1498 | | **24** | 75 | 1947 | 3035 |
| **257** | 63 | 1533 | 2388 | 64 | 1640 | 2533 | | 11 | 64 | 1748 | 2667 | 64 | 1865 | 1514 | | **25** | 65 | 1977 | 3051 |
| **258** | 53 | 1560 | 2405 | 54 | 1668 | 2550 | | 12 | 54 | 1777 | 2684 | 54 | 1894 | 1530 | | **26** | 55 | 2007 | 3066 |
| **259** | 43 | 1588 | 2423 | 44 | 1696 | 2567 | | 13 | 44 | 1806 | 2700 | 44 | 1923 | 1545 | | **27** | 44 | 2037 | 3082 |
| **260** | 33 | 1615 | 2441 | 34 | 1724 | 2585 | | 14 | 34 | 1834 | 2716 | 34 | 1952 | 1561 | | **28** | 34 | 2067 | 3097 |
| **261** | 23 | 1643 | 2458 | 24 | 1752 | 2602 | | 15 | 24 | 1863 | 2733 | 24 | 1982 | 1577 | | **29** | 24 | 2096 | 3112 |
| **262** | 13 | 1670 | 2476 | 14 | 1780 | 2619 | | 16 | 14 | 1891 | 2749 | 14 | 2011 | 1593 | | **30** | 14 | 2126 | 3128 |
| **263** | 2 | 1698 | 2493 | 2 | 1808 | 2636 | | 17 | 2 | 1920 | 2766 | 2 | 2040 | 1609 | | **31** | 2 | 2156 | 3143 |
| **264** | −10 | 1725 | 2515 | −10 | 1836 | 2659 | | 18 | −10 | 1949 | 2788 | −10 | 2069 | 1631 | January | **1** | −11 | 2186 | 3164 |
| **265** | 20 | 1753 | 2533 | 20 | 1864 | 2676 | | 19 | 20 | 1977 | 2804 | 20 | 2098 | 1647 | | **2** | 20 | 2216 | 3179 |
| **266** | 31 | 1780 | 2550 | 32 | 1892 | 2693 | | 20 | 32 | 2006 | 2821 | 32 | 2128 | 1663 | | **3** | 32 | 2245 | 3194 |
| **267** | 40 | 1808 | 2567 | 41 | 1920 | 2709 | | 21 | 41 | 2034 | 2835 | 41 | 2157 | 1678 | | **4** | 42 | 2275 | 3211 |
| **268** | 50 | 1835 | 2584 | 51 | 1948 | 2726 | | 22 | 51 | 2063 | 2853 | 51 | 2186 | 1694 | | **5** | 52 | 2305 | 3225 |
| **269** | 60 | 1863 | 2602 | 61 | 1976 | 2743 | | 23 | 61 | 2092 | 2869 | 61 | 2215 | 1709 | | **6** | 63 | 2335 | 3241 |
| **270** | 69 | 1890 | 2619 | 70 | 2004 | 2760 | | 24 | 70 | 2120 | 2881 | 70 | 2244 | 1724 | | **7** | 73 | 2365 | 3256 |
| **271** | 80 | 1918 | 2637 | 81 | 2032 | 2778 | | 25 | 81 | 2149 | 2902 | 81 | 2274 | 1741 | | **8** | 84 | 2394 | 3272 |
| **272** | 90 | 1945 | 2655 | 91 | 2060 | 2795 | | 26 | 91 | 2177 | 2918 | 91 | 2303 | 1757 | | **9** | 94 | 2424 | 3288 |
| **273** | 100 | 1973 | 2672 | 101 | 2088 | 2812 | | 27 | 101 | 2206 | 2935 | 101 | 2332 | 1773 | | **10** | 104 | 2454 | 3303 |
| **274** | 102 | 2000 | 2682 | 102 | 2116 | 2820 | | 28 | 104 | 2236 | 2941 | 101 | 2360 | 1779 | | **11** | 101 | 2484 | 3305 |
| **275** | 109 | 2008 | 2697 | 109 | 2123 | 2834 | | 29 | 108 | 2242 | 2954 | 107 | 2366 | 1791 | | **12** | 107 | 2489 | 3216 |
| **276** | 115 | 2015 | 2710 | 115 | 2130 | 2847 | *Tamil*, Tai; *Malayalam*, Magaram; *Bengal*, Magha. | 1 | 114 | 2249 | 2967 | 114 | 2372 | 1804 | | **13** | 114 | 2494 | 3328 |
| **277** | 122 | 2023 | 2725 | 122 | 2137 | 2861 | | 2 | 121 | 2255 | 2980 | 120 | 2378 | 1816 | | **14** | 120 | 2500 | 3340 |
| **278** | 129 | 2030 | 2739 | 129 | 2144 | 2875 | | 3 | 128 | 2262 | 2994 | 127 | 2384 | 1829 | | **15** | 127 | 2505 | 3352 |
| **279** | 135 | 2038 | 2753 | 135 | 2151 | 2888 | | 4 | 134 | 2268 | 3006 | 133 | 2389 | 1841 | | **16** | 133 | 2510 | 3363 |
| **280** | 142 | 2046 | 2768 | 142 | 2159 | 2903 | | 5 | 141 | 2274 | 3019 | 140 | 2395 | 1853 | | **17** | 140 | 2515 | 3375 |
| **281** | 149 | 2053 | 2782 | 149 | 2166 | 2917 | | 6 | 148 | 2281 | 3033 | 146 | 2401 | 1865 | | **18** | 146 | 2520 | 3386 |
| **282** | 155 | 2061 | 2796 | 155 | 2173 | 2940 | | 7 | 154 | 2287 | 3045 | 153 | 2407 | 1878 | | **19** | 153 | 2526 | 3399 |
| **283** | 162 | 2068 | 2810 | 162 | 2180 | 2944 | | 8 | 161 | 2294 | 3059 | 159 | 2413 | 1890 | | **20** | 159 | 2531 | 3410 |
| **284** | 169 | 2076 | 2825 | 169 | 2187 | 2957 | | 9 | 168 | 2300 | 3072 | 166 | 2419 | 1902 | | **21** | 166 | 2536 | 3422 |
| **285** | 175 | 2084 | 2806 | 175 | 2194 | 2936 | | 10 | 174 | 2306 | 3050 | 172 | 2425 | 1878 | | **22** | 172 | 2541 | 3397 |
| **286** | 182 | 2091 | 2788 | 182 | 2201 | 2917 | | 11 | 181 | 2313 | 3029 | 179 | 2431 | 1856 | | **23** | 179 | 2546 | 3373 |
| **287** | 189 | 2099 | 2769 | 189 | 2208 | 2897 | | 12 | 188 | 2318 | 3008 | 185 | 2437 | 1832 | | **24** | 185 | 2552 | 3347 |
| **288** | 195 | 2106 | 2750 | 195 | 2215 | 2876 | | 13 | 194 | 2326 | 2986 | 192 | 2443 | 1809 | | **25** | 192 | 2557 | 3323 |
| **289** | 202 | 2114 | 2732 | 202 | 2222 | 2856 | | 14 | 201 | 2332 | 2964 | 198 | 2448 | 1786 | | **26** | 198 | 2562 | 3298 |
| **290** | 209 | 2122 | 2714 | 209 | 2230 | 2837 | | 15 | 208 | 2338 | 2943 | 205 | 2454 | 1763 | | **27** | 205 | 2567 | 3274 |
| **291** | 215 | 2129 | 3695 | 215 | 2237 | 2816 | | 16 | 214 | 2345 | 2921 | 211 | 2460 | 1739 | | **28** | 211 | 2572 | 3249 |
| **292** | 222 | 2137 | 2676 | 222 | 2244 | 2796 | | 17 | 221 | 2351 | 2930 | 218 | 2466 | 1716 | | **29** | 218 | 2578 | 3224 |
| **293** | 229 | 2144 | 2658 | 229 | 2251 | 2777 | | 18 | 228 | 2358 | 2879 | 224 | 2472 | 1693 | | **30** | 224 | 2583 | 3199 |
| **294** | 235 | 2152 | 2639 | 235 | 2258 | 2756 | | 19 | 234 | 2364 | 2857 | 231 | 2474 | 1670 | | **31** | 231 | 2588 | 3175 |
| **295** | 242 | 2160 | 2621 | 242 | 2265 | 2736 | | 20 | 241 | 2370 | 2836 | 238 | 2484 | 1647 | February | **1** | 237 | 2593 | 3150 |
| **296** | 249 | 2167 | 2608 | 249 | 2272 | 2717 | | 21 | 248 | 2377 | 2815 | 245 | 2490 | 1625 | | **2** | 244 | 2598 | 3126 |
| **297** | 255 | 2175 | 2583 | 255 | 2279 | 2696 | | 22 | 254 | 2383 | 2793 | 252 | 2496 | 1602 | | **3** | 250 | 2604 | 3100 |
| **298** | 262 | 2182 | 2565 | 262 | 2286 | 2676 | | 23 | 261 | 2390 | 2772 | 259 | 2502 | 1579 | | **4** | 257 | 2609 | 3076 |
| **299** | 269 | 2190 | 2537 | 269 | 2295 | 2656 | | 24 | 268 | 2396 | 2750 | 266 | 2508 | 1556 | | **5** | 263 | 2614 | 3051 |
| **300** | 276 | 2198 | 2529 | 275 | 2301 | 2636 | | 25 | 274 | 2402 | 2728 | 272 | 2513 | 1533 | | **6** | 270 | 2619 | 3027 |
| **301** | 283 | 2205 | 2519 | 282 | 2308 | 2616 | | 26 | 281 | 2409 | 2707 | 278 | 2519 | 1509 | | **7** | 276 | 2624 | 3002 |
| **302** | 290 | 2213 | 2492 | 289 | 2315 | 2596 | | 27 | 288 | 2415 | 2686 | 284 | 2525 | 1485 | | **8** | 283 | 2630 | 2977 |
| **303** | 297 | 2220 | 2474 | 295 | 2322 | 2576 | | 28 | 294 | 2422 | 2664 | 290 | 2531 | 1462 | | **9** | 289 | 2635 | 2952 |
| **304** | 304 | 2228 | 2456 | 302 | 2328 | 2556 | | 29 | 300 | 2428 | 2642 | 297 | 2536 | 1439 | | **10** | 295 | 2640 | 2927 |
| **305** | 307 | 2208 | 2434 | 305 | 2301 | 2532 | | 30 | 303 | 2400 | 2617 | 300 | 2506 | 1412 | | **11** | 298 | 2609 | 2899 |

# TABLE XIII.

## Sunrise from 18° to 22° N. Latitude.

| Day of Solar year. | Lat. 18° —428" Long. Eqn. of time in seconds. | ☉'s Trop. Long. in seconds. | Total Corrn. in seconds. Satara. A.D. 1840—1920 | Lat. 19° —460" Long. Eqn. of time in seconds. | ☉'s Trop. Long. in seconds. | Total Corrn. in seconds. Poona. A.D. 1840—1920 | Days of solar months. | Lat. 20° —470" Long. Eqn. of time in seconds. | ☉'s Trop. Long. in seconds. | Total Corrn. in seconds. Nasik. A.D. 1840—1920 | Lat. 21° +800" Long. Eqn. of time in seconds. | ☉'s Trop. Long. in seconds. | Total Corrn. in seconds. Nagpur. A.D. 1840—1920 | Eng. date. | Lat. 22° —616" Long. Eqn. of time in seconds. | ☉'s Trop. Long. in seconds. | Total Corrn. in seconds. Baroda. A.D. 1840—1920 |
|---|---|---|---|---|---|---|---|---|---|---|---|---|---|---|---|---|---|
| 306 | —310 | —2178 | —2382 | —308 | —2275 | —2509 | 1 | —306 | —2372 | —2592 | —303 | —2477 | —1386 | 12 | —301 | —2578 | —2871 |
| 307 | 313 | 2152 | 2389 | 311 | 2268 | 2485 | 2 | 309 | 2344 | 2567 | 306 | 2447 | 1359 | 13 | 304 | 2546 | 2842 |
| 308 | 316 | 2127 | 2367 | 314 | 2221 | 2461 | 3 | 312 | 2316 | 2542 | 309 | 2417 | 1332 | 14 | 307 | 2515 | 2814 |
| 309 | 319 | 2102 | 2345 | 317 | 2194 | 2437 | 4 | 315 | 2287 | 2516 | 312 | 2388 | 1305 | 15 | 310 | 2484 | 2786 |
| 310 | 322 | 2077 | 2323 | 320 | 2168 | 2414 | 5 | 318 | 2269 | 2491 | 315 | 2358 | 1279 | 16 | 313 | 2453 | 2758 |
| 311 | 325 | 2052 | 2301 | 323 | 2141 | 2390 | 6 | 321 | 2231 | 2466 | 318 | 2328 | 1252 | 17 | 316 | 2422 | 2730 |
| 312 | 328 | 2026 | 2278 | 326 | 2114 | 2366 | 7 | 324 | 2203 | 2441 | 321 | 2298 | 1225 | 18 | 319 | 2390 | 2701 |
| 313 | 331 | 2001 | 2256 | 329 | 2088 | 2343 | 8 | 327 | 2175 | 2416 | 324 | 2269 | 1199 | 19 | 322 | 2359 | 2673 |
| 314 | 334 | 1976 | 2234 | 332 | 2061 | 2319 | 9 | 330 | 2147 | 2391 | 327 | 2239 | 1172 | 20 | 325 | 2328 | 2645 |
| 315 | 337 | 1951 | 2213 | 335 | 2034 | 2295 | 10 | 333 | 2119 | 2365 | 330 | 2209 | 1146 | 21 | 328 | 2297 | 2616 |
| 316 | 340 | 1926 | 2170 | 338 | 2008 | 2250 | 11 | 336 | 2091 | 2316 | 333 | 2180 | 1095 | 22 | 331 | 2266 | 2563 |
| 317 | 343 | 1900 | 2128 | 341 | 1981 | 2207 | 12 | 339 | 2063 | 2267 | 336 | 2150 | 1044 | 23 | 334 | 2224 | 2511 |
| 318 | 346 | 1875 | 2087 | 344 | 1954 | 2164 | 13 | 342 | 2035 | 2218 | 339 | 2120 | 993 | 24 | 337 | 2203 | 2458 |
| 319 | 349 | 1850 | 2044 | 347 | 1927 | 2119 | 14 | 345 | 2006 | 2169 | 342 | 2090 | 942 | 25 | 340 | 2172 | 2405 |
| 320 | 352 | 1825 | 2002 | 350 | 1901 | 2076 | 15 | 348 | 1978 | 2120 | 345 | 2061 | 891 | 26 | 343 | 2141 | 2352 |
| 321 | 355 | 1808 | 1959 | 353 | 1874 | 2031 | 16 | 351 | 1950 | 2073 | 348 | 2031 | 841 | 27 | 346 | 2110 | 2300 |
| 322 | 358 | 1774 | 1916 | 356 | 1847 | 1986 | 17 | 354 | 1922 | 2023 | 351 | 2001 | 790 | 28 | 349 | 2078 | 2247 |
| 323 | 361 | 1749 | 1837 | 359 | 1821 | 1941 | 18 | 357 | 1894 | 1974 | 354 | 1972 | 739 | 1 | 352 | 2047 | 2194 |
| 324 | 364 | 1724 | 1829 | 362 | 1794 | 1872 | 19 | 360 | 1866 | 1925 | 357 | 1942 | 688 | 2 | 355 | 2016 | 2142 |
| 325 | 367 | 1699 | 1786 | 365 | 1767 | 1825 | 20 | 363 | 1838 | 1876 | 360 | 1912 | 637 | 3 | 358 | 1985 | 2089 |
| 326 | 370 | 1674 | 1741 | 368 | 1741 | 1778 | 21 | 366 | 1810 | 1827 | 363 | 1883 | 588 | 4 | 361 | 1954 | 2036 |
| 327 | 373 | 1648 | 1697 | 371 | 1714 | 1731 | 22 | 369 | 1782 | 1778 | 366 | 1853 | 535 | 5 | 363 | 1922 | 1983 |
| 328 | 376 | 1623 | 1652 | 374 | 1687 | 1684 | 23 | 872 | 1754 | 1729 | 369 | 1823 | 484 | 6 | 365 | 1891 | 1929 |
| 329 | 379 | 1698 | 1607 | 377 | 1660 | 1637 | 24 | 375 | 1725 | 1680 | 371 | 1793 | 430 | 7 | 567 | 1860 | 1875 |
| 330 | 382 | 1573 | 1561 | 380 | 1634 | 1590 | 25 | 377 | 1697 | 1632 | 373 | 1764 | 380 | 8 | 369 | 1829 | 1821 |
| 331 | 385 | 1548 | 1516 | 383 | 1607 | 1543 | 26 | 379 | 1669 | 1581 | 375 | 1734 | 329 | 9 | 371 | 1798 | 1768 |
| 332 | 388 | 1522 | 1471 | 385 | 1580 | 1495 | 27 | 381 | 1641 | 1531 | 377 | 1704 | 277 | 10 | 373 | 1766 | 1714 |
| 333 | 391 | 1497 | 1426 | 387 | 1554 | 1447 | 28 | 383 | 1613 | 1481 | 379 | 1675 | 225 | 11 | 375 | 1735 | 1660 |
| 334 | 392 | 1472 | 1379 | 389 | 1527 | 1399 | 29 | 385 | 1585 | 1431 | 381 | 1645 | 173 | 12 | 377 | 1704 | 1607 |
| 335 | 395 | 1448 | 1305 | 391 | 1500 | 1351 | 1 | 387 | 1556 | 1381 | 383 | 1616 | 121 | 13 | 379 | 1672 | 1553 |
| 336 | 395 | 1400 | 1257 | 391 | 1450 | 1302 | 2 | 387 | 1504 | 1329 | 383 | 1562 | 67 | 14 | 379 | 1616 | 1497 |
| 337 | 396 | 1351 | 1209 | 392 | 1400 | 1252 | 3 | 388 | 1452 | 1278 | 384 | 1508 | 14 | 15 | 380 | 1561 | 1443 |
| 338 | 396 | 1303 | 1161 | 392 | 1350 | 1202 | 4 | 388 | 1400 | 1226 | 384 | 1454 | + 40 | 16 | 380 | 1505 | 1387 |
| 339 | 397 | 1255 | 1084 | 393 | 1300 | 1153 | 5 | 389 | 1348 | 1175 | 385 | 1400 | 93 | 17 | 381 | 1449 | 1332 |
| 340 | 397 | 1206 | 1065 | 393 | 1250 | 1103 | 6 | 389 | 1296 | 1128 | 385 | 1346 | 147 | 18 | 381 | 1393 | 1276 |
| 341 | 398 | 1158 | 1018 | 394 | 1200 | 1054 | 7 | 390 | 1246 | 1073 | 386 | 1293 | 199 | 19 | 382 | 1338 | 1222 |
| 342 | 398 | 1110 | 970 | 394 | 1150 | 1004 | 8 | 390 | 1193 | 1021 | 386 | 1239 | 253 | 20 | 382 | 1282 | 1166 |
| 343 | 399 | 1026 | 923 | 395 | 1100 | 955 | 9 | 391 | 1141 | 970 | 387 | 1185 | 306 | 21 | 383 | 1226 | 1112 |
| 344 | 399 | 1013 | 875 | 395 | 1050 | 905 | 10 | 391 | 1089 | 918 | 387 | 1131 | 360 | 22 | 383 | 1171 | 1056 |
| 345 | 400 | 965 | 828 | 396 | 1000 | 856 | 11 | 392 | 1037 | 868 | 388 | 1077 | 412 | 23 | 384 | 1115 | 1000 |
| 346 | 400 | 917 | 780 | 396 | 950 | 806 | 12 | 392 | 985 | 816 | 388 | 1023 | 466 | 24 | 384 | 1059 | 944 |
| 347 | 401 | 868 | 733 | 397 | 900 | 757 | 13 | 393 | 938 | 765 | 389 | 969 | 518 | 25 | 385 | 1004 | 890 |
| 348 | 401 | 820 | 685 | 397 | 850 | 707 | 14 | 393 | 881 | 713 | 389 | 915 | 572 | 26 | 385 | 948 | 834 |
| 349 | 402 | 772 | 637 | 398 | 800 | 658 | 15 | 394 | 829 | 662 | 390 | 861 | 625 | 27 | 386 | 592 | 779 |
| 350 | 402 | 723 | 589 | 398 | 750 | 608 | 16 | 394 | 779 | 611 | 390 | 807 | 670 | 28 | 386 | 836 | 724 |
| 351 | 403 | 675 | 542 | 399 | 700 | 559 | 17 | 395 | 726 | 560 | 391 | 754 | 732 | 29 | 387 | 781 | 669 |
| 352 | 403 | 627 | 493 | 399 | 650 | 509 | 18 | 395 | 674 | 508 | 391 | 700 | 786 | 30 | 387 | 725 | 613 |
| 353 | 404 | 579 | 446 | 400 | 600 | 460 | 19 | 396 | 622 | 447 | 392 | 646 | 839 | 31 | 388 | 668 | 559 |
| 354 | 404 | 530 | 398 | 400 | 550 | 410 | 20 | 396 | 570 | 395 | 392 | 592 | 893 | 1 | 388 | 614 | 503 |
| 355 | 405 | 482 | 350 | 401 | 500 | 361 | 21 | 397 | 518 | 344 | 393 | 538 | 946 | 2 | 389 | 558 | 448 |
| 356 | 405 | 434 | 302 | 401 | 450 | 311 | 22 | 397 | 466 | 293 | 393 | 484 | 999 | 3 | 389 | 502 | 393 |
| 357 | 406 | 385 | 255 | 402 | 400 | 262 | 23 | 398 | 414 | 242 | 394 | 430 | 1052 | 4 | 390 | 447 | 338 |
| 358 | 406 | 337 | 207 | 402 | 350 | 212 | 24 | 398 | 362 | 190 | 394 | 376 | 1106 | 5 | 390 | 391 | 282 |
| 359 | 407 | 259 | 159 | 403 | 300 | 163 | 25 | 399 | 310 | 139 | 395 | 322 | 1159 | 6 | 391 | 335 | 228 |
| 360 | 407 | 240 | 111 | 403 | 250 | 113 | 26 | 399 | 258 | 87 | 395 | 268 | 1213 | 7 | 391 | 279 | 172 |
| 361 | 408 | 192 | 64 | 404 | 200 | 64 | 27 | 400 | 207 | 36 | 396 | 215 | 1266 | 8 | 392 | 224 | 117 |
| 362 | 408 | 144 | 15 | 404 | 150 | 14 | 28 | 400 | 155 | + 16 | 396 | 161 | 1320 | 9 | 392 | 168 | 62 |
| 363 | 409 | 96 | + 32 | 405 | 100 | + 35 | 29 | 401 | 108 | 67 | 397 | 107 | 1373 | 10 | 393 | 173 | 7 |
| 364 | 409 | 48 | 80 | 405 | 50 | 85 | 30 | 401 | 51 | 119 | 397 | 53 | 1427 | 11 | 393 | 57 | + 49 |
| 365 | 409 | 0 | 129 | 405 | 0 | 136 | 31 | 401 | 0 | 171 | 397 | 0 | 1481 | 12 | 393 | 0 | 105 |

Days of solar months (days 1–29): *Tamil*, Masi; *Malayalam*, Kumbam; *Bengal*, Phalguna.

Days of solar months (days 1–31): *Tamil*, Panguni; *Malayalam*, Meenam; *Bengal*, Chaitra.

Eng. date: February (12–28), March (1–31), April (1–12).

# TABLE XIII.

## Sunrise from 23° to 27° N. Latitude.

| Day of Solar year. | Lat. 23° −20″ Long. Eqn. of time in seconds. | ☉'s Trop. Long. in seconds. | Total Corrn. in seconds. Indore. A.D. 1840—1920 | Lat. 24° −876″ Long. Eqn. of time in seconds. | ☉'s Trop. Long. in seconds. | Total Corrn. in seconds. Anhilwad. A.D. 1840—1920 | Days of solar months. | Lat. 25° +1736″ Long. Eqn. of time in seconds. | ☉'s Trop. Long. in seconds. | Total Corrn. in seconds. Benares. A.D. 1840—1920 | Lat. 26° +2264″ Long. Eqn. of Time in seconds. | ☉'s Trop. Long. in seconds. | Total Corrn. in seconds. Patna. A.D. 1840—1920 | Eng. date. | Lat. 27° +1544″ Long. Eqn. of Time in seconds. | ☉'s Trop. Long. in seconds. | Total Corrn. in seconds. Oudh. A.D. 1840—1920 |
|---|---|---|---|---|---|---|---|---|---|---|---|---|---|---|---|---|---|
| 1 | −389 | + 58 | + 801 | −384 | + 60 | − 8 | 1 (*Tamil*, Chittirai; *Malayalam*; Medam; *Bengal*, Vaisakha.) | −380 | + 62 | +2652 | −376 | + 64 | +3230 | 13 (April) | −372 | + 66 | +2556 |
| 2 | 387 | 115 | 860 | 383 | 119 | + 52 | 2 | 379 | 123 | 2714 | 374 | 128 | 3296 | 14 | 370 | 132 | 2624 |
| 3 | 385 | 173 | 920 | 381 | 179 | 114 | 3 | 377 | 185 | 2778 | 373 | 192 | 3361 | 15 | 368 | 198 | 2692 |
| 4 | 384 | 230 | 978 | 380 | 238 | 174 | 4 | 376 | 247 | 2841 | 372 | 256 | 3426 | 16 | 367 | 264 | 2759 |
| 5 | 382 | 288 | 1038 | 378 | 298 | 236 | 5 | 374 | 308 | 2904 | 370 | 319 | 3491 | 17 | 365 | 329 | 2827 |
| 6 | 380 | 346 | 1098 | 376 | 358 | 298 | 6 | 372 | 370 | 2968 | 368 | 383 | 3557 | 18 | 364 | 395 | 2894 |
| 7 | 379 | 403 | 1156 | 374 | 417 | 359 | 7 | 370 | 432 | 3032 | 366 | 447 | 3624 | 19 | 362 | 461 | 2962 |
| 8 | 376 | 461 | 1217 | 372 | 477 | 421 | 8 | 368 | 494 | 3096 | 364 | 511 | 3689 | 20 | 359 | 527 | 3031 |
| 9 | 374 | 518 | 1223 | 370 | 536 | 482 | 9 | 365 | 555 | 3160 | 361 | 575 | 3756 | 21 | 357 | 593 | 3099 |
| 10 | 372 | 576 | 1336 | 368 | 596 | 544 | 10 | 363 | 617 | 3225 | 359 | 639 | 3821 | 22 | 355 | 659 | 3165 |
| 11 | 369 | 634 | 1373 | 365 | 656 | 582 | 11 | 361 | 679 | 3263 | 357 | 703 | 3861 | 23 | 353 | 725 | 3206 |
| 12 | 367 | 691 | 1408 | 363 | 715 | 618 | 12 | 358 | 740 | 3302 | 355 | 767 | 3901 | 24 | 350 | 791 | 3248 |
| 13 | 364 | 749 | 1443 | 360 | 775 | 656 | 13 | 356 | 802 | 3340 | 352 | 831 | 3941 | 25 | 348 | 857 | 3290 |
| 14 | 362 | 806 | 1477 | 357 | 834 | 693 | 14 | 353 | 864 | 3380 | 349 | 895 | 3982 | 26 | 345 | 923 | 3332 |
| 15 | 359 | 864 | 1513 | 355 | 894 | 729 | 15 | 350 | 925 | 3418 | 347 | 958 | 4021 | 27 | 342 | 988 | 3374 |
| 16 | 356 | 922 | 1549 | 352 | 954 | 767 | 16 | 348 | 987 | 3456 | 344 | 1022 | 4062 | 28 | 340 | 1054 | 3415 |
| 17 | 353 | 979 | 1585 | 349 | 1013 | 804 | 17 | 345 | 1049 | 3495 | 341 | 1086 | 4103 | 29 | 338 | 1120 | 3456 |
| 18 | 350 | 1037 | 1620 | 346 | 1073 | 842 | 18 | 342 | 1111 | 3535 | 338 | 1150 | 4144 | 30 | 336 | 1186 | 3498 |
| 19 | 346 | 1094 | 1658 | 343 | 1132 | 879 | 19 | 338 | 1172 | 3574 | 336 | 1214 | 4183 | 1 (May) | 334 | 1252 | 3539 |
| 20 | 343 | 1152 | 1694 | 339 | 1192 | 917 | 20 | 336 | 1234 | 3612 | 334 | 1278 | 4223 | 2 | 332 | 1318 | 3580 |
| 21 | 340 | 1210 | 1730 | 335 | 1252 | 956 | 21 | 334 | 1296 | 3650 | 332 | 1342 | 4263 | 3 | 330 | 1384 | 3621 |
| 22 | 338 | 1267 | 1765 | 333 | 1311 | 994 | 22 | 332 | 1357 | 3688 | 330 | 1406 | 4302 | 4 | 328 | 1450 | 3662 |
| 23 | 336 | 1325 | 1799 | 331 | 1371 | 1028 | 23 | 330 | 1419 | 3716 | 328 | 1470 | 4342 | 5 | 326 | 1516 | 3704 |
| 24 | 334 | 1382 | 1833 | 329 | 1430 | 1065 | 24 | 328 | 1481 | 3764 | 326 | 1534 | 4382 | 6 | 324 | 1582 | 3745 |
| 25 | 332 | 1440 | 1868 | 326 | 1490 | 1102 | 25 | 326 | 1542 | 3802 | 324 | 1597 | 4421 | 7 | 322 | 1648 | 3786 |
| 26 | 330 | 1498 | 1903 | 324 | 1550 | 1129 | 26 | 324 | 1604 | 3840 | 322 | 1661 | 4461 | 8 | 320 | 1714 | 8827 |
| 27 | 328 | 1555 | 1938 | 322 | 1609 | 1185 | 27 | 322 | 1666 | 3878 | 320 | 1725 | 4501 | 9 | 318 | 1780 | 8868 |
| 28 | 326 | 1613 | 1973 | 320 | 1669 | 1211 | 28 | 320 | 1728 | 3916 | 318 | 1789 | 4541 | 10 | 316 | 1846 | 3920 |
| 29 | 324 | 1617 | 2008 | 318 | 1728 | 1248 | 29 | 318 | 1789 | 3955 | 316 | 1853 | 4590 | 11 | 314 | 1912 | 3961 |
| 30 | 322 | 1728 | 2043 | 316 | 1788 | 1284 | 30 | 316 | 1852 | 3992 | 314 | 1916 | 4620 | 12 | 312 | 1976 | 3992 |
| 31 | 320 | 1762 | 2078 | 314 | 1823 | 1321 | 1 (*Tamil*, Vaikasi; *Malayalam*, Elavam; *Bengal*, Jyeshta.) | 314 | 1888 | 4030 | 312 | 1954 | 4660 | 13 | 310 | 2015 | 4033 |
| 32 | 318 | 1795 | 2113 | 312 | 1857 | 1358 | 2 | 312 | 1924 | 4068 | 310 | 1932 | 4699 | 14 | 308 | 2054 | 4074 |
| 33 | 316 | 1827 | 2147 | 310 | 1892 | 1393 | 3 | 310 | 1960 | 4107 | 308 | 2029 | 4739 | 15 | 306 | 2094 | 4116 |
| 34 | 314 | 1859 | 2185 | 308 | 1926 | 1430 | 4 | 308 | 1997 | 4144 | 306 | 2067 | 4779 | 16 | 304 | 2133 | 4157 |
| 35 | 312 | 1892 | 2226 | 305 | 1960 | 1467 | 5 | 306 | 2032 | 4182 | 304 | 2104 | 4818 | 17 | 302 | 2172 | 4198 |
| 36 | 310 | 1925 | 2251 | 303 | 1995 | 1503 | 6 | 304 | 2068 | 4220 | 302 | 2142 | 4858 | 18 | 300 | 2211 | 4239 |
| 37 | 307 | 1958 | 2287 | 301 | 2029 | 1540 | 7 | 302 | 2104 | 4258 | 299 | 2180 | 4899 | 19 | 297 | 2250 | 4281 |
| 38 | 302 | 1990 | 2324 | 299 | 2064 | 1577 | 8 | 297 | 2141 | 4299 | 294 | 2218 | 4942 | 20 | 292 | 2290 | 4326 |
| 39 | 296 | 2024 | 2363 | 293 | 2098 | 1617 | 9 | 291 | 2176 | 4341 | 288 | 2255 | 4985 | 21 | 286 | 2329 | 4371 |
| 40 | 290 | 2057 | 2402 | 287 | 2132 | 1657 | 10 | 285 | 2212 | 4383 | 283 | 2293 | 5028 | 22 | 280 | 2368 | 4416 |
| 41 | 284 | 2090 | 2440 | 282 | 2167 | 1698 | 11 | 279 | 2248 | 4425 | 277 | 2331 | 5171 | 23 | 275 | 2407 | 4461 |
| 42 | 278 | 2123 | 2442 | 275 | 2201 | 1701 | 12 | 273 | 2284 | 4429 | 271 | 2368 | 5075 | 24 | 269 | 2446 | 4465 |
| 43 | 272 | 2155 | 2443 | 269 | 2235 | 1703 | 13 | 267 | 2320 | 4432 | 265 | 2406 | 5078 | 25 | 263 | 2486 | 4470 |
| 44 | 266 | 2187 | 2445 | 263 | 2270 | 1705 | 14 | 261 | 2356 | 4435 | 259 | 2444 | 5082 | 26 | 257 | 2525 | 4474 |
| 45 | 259 | 2220 | 2447 | 257 | 2304 | 1707 | 15 | 255 | 2392 | 4438 | 253 | 2481 | 5085 | 27 | 251 | 2564 | 4479 |
| 46 | 253 | 2253 | 2449 | 251 | 2339 | 1710 | 16 | 249 | 2428 | 4440 | 247 | 2519 | 5089 | 28 | 245 | 2603 | 4483 |
| 47 | 246 | 2286 | 2452 | 244 | 2373 | 1713 | 17 | 243 | 2464 | 4444 | 240 | 2557 | 5094 | 29 | 239 | 2642 | 4487 |
| 48 | 240 | 2318 | 2453 | 238 | 2407 | 1716 | 18 | 236 | 2500 | 4447 | 234 | 2595 | 5097 | 30 | 232 | 2682 | 4493 |
| 49 | 234 | 2352 | 2455 | 232 | 2442 | 1717 | 19 | 230 | 2537 | 4450 | 228 | 2632 | 5101 | 31 | 226 | 2721 | 4497 |
| 50 | 227 | 2385 | 2457 | 225 | 2476 | 1720 | 20 | 223 | 2572 | 4453 | 221 | 2670 | 5105 | 1 (June) | 220 | 2760 | 4502 |
| 51 | 220 | 2418 | 2460 | 218 | 2511 | 1723 | 21 | 217 | 2608 | 4456 | 215 | 2708 | 5109 | 2 | 213 | 2799 | 4507 |
| 52 | 214 | 2451 | 2462 | 212 | 2545 | 1725 | 22 | 210 | 2644 | 4460 | 208 | 2745 | 5114 | 3 | 207 | 2838 | 4511 |
| 53 | 207 | 2483 | 2464 | 205 | 2579 | 1728 | 23 | 203 | 2681 | 4464 | 201 | 2783 | 5117 | 4 | 200 | 2878 | 4517 |
| 54 | 200 | 2515 | 2466 | 198 | 2614 | 1731 | 24 | 197 | 2716 | 4467 | 195 | 2821 | 5122 | 5 | 193 | 2917 | 4522 |
| 55 | 193 | 2548 | 2469 | 191 | 2648 | 1734 | 25 | 190 | 2752 | 4471 | 188 | 2858 | 5126 | 6 | 186 | 2956 | 4528 |
| 56 | 186 | 2581 | 2472 | 184 | 2682 | 1737 | 26 | 183 | 2788 | 4474 | 181 | 2896 | 5131 | 7 | 180 | 2995 | 4532 |
| 57 | 178 | 2614 | 2476 | 177 | 2717 | 1741 | 27 | 175 | 2824 | 4479 | 174 | 2934 | 5136 | 8 | 173 | 3034 | 4537 |
| 58 | 174 | 2646 | 2475 | 174 | 2752 | 1740 | 28 | 171 | 2860 | 4480 | 171 | 2972 | 5136 | 9 | 170 | 3074 | 4539 |
| 59 | 171 | 2679 | 2474 | 171 | 2786 | 1739 | 29 | 168 | 2896 | 4480 | 168 | 3009 | 5137 | 10 | 167 | 3113 | 4540 |
| 60 | 167 | 2712 | 2473 | 167 | 2820 | 1739 | 30 | 165 | 2932 | 4480 | 165 | 3047 | 5137 | 11 | 164 | 3152 | 4542 |
| 61 | 163 | 2744 | 2473 | 163 | 2856 | 1739 | 31 | 162 | 2968 | 4480 | 162 | 3084 | 5138 | 12 | 161 | 3192 | 4543 |
| 62 | 159 | 2740 | 2472 | 159 | 2852 | 1739 | 32 | 159 | 2965 | 4480 | 159 | 3082 | 5139 | 13 | 158 | 3190 | 4544 |

# TABLE XIII.

## Sunrise from 23° to 27° N. Latitude.

| Day of Solar year. | Lat. 23° −20″ Long. Eqn. of time in seconds. | ☉'s Trop. Long. in seconds. | Total Corrn. in seconds. Indore. A.D. 1840—1920 | Lat. 24° −876″ Long. Eqn. of time in seconds. | ☉'s Trop. Long. in seconds. | Total Corrn. in seconds. Anhilwad. A.D. 1840—1920 | Days of solar months. | Lat. 25° +1736″ Long. Eqn. of time in seconds. | ☉'s Trop. Long. in seconds. | Total Corrn. in seconds. Benares. A.D. 1840—1920 | Lat. 26° +2264″ Long. Eqn. of Time in seconds. | ☉'s Trop. Long. in seconds. | Total Corrn. in seconds. Patna. A.D. 1840—1920 | Eng. date. | Lat. 27° +1544″ Long. Eqn. of Time in seconds. | ☉'s Trop. Long. in seconds. | Total Corrn. in seconds. Oudh. A.D. 1840—1920 |
|---|---|---|---|---|---|---|---|---|---|---|---|---|---|---|---|---|---|
| 63 | −155 | +2735 | +2472 | −155 | +2848 | +1739 | 1 (Tamil, Ani; Malayalam, Mithunam; Bengal, Ashada.) | −156 | +2962 | +4480 | −156 | +3079 | +5139 | 14 (June) | −155 | +3189 | +4546 |
| 64 | 151 | 2731 | 2472 | 151 | 2844 | 1739 | 2 | 150 | 2959 | 4483 | 150 | 3077 | 5143 | 15 | 150 | 3187 | 4549 |
| 65 | 143 | 2726 | 2475 | 142 | 2840 | 1744 | 3 | 142 | 2956 | 4488 | 141 | 3074 | 5149 | 16 | 147 | 3186 | 4551 |
| 66 | 134 | 2722 | 2480 | 133 | 2837 | 1749 | 4 | 133 | 2952 | 4493 | 133 | 3072 | 5155 | 17 | 132 | 3184 | 4564 |
| 67 | 125 | 2718 | 2485 | 125 | 2833 | 1754 | 5 | 124 | 2949 | 4499 | 124 | 3070 | 5162 | 18 | 123 | 3182 | 4571 |
| 68 | 116 | 2713 | 2489 | 116 | 2830 | 1759 | 6 | 115 | 2946 | 4505 | 115 | 3067 | 5168 | 19 | 115 | 3181 | 4578 |
| 69 | 107 | 2709 | 2494 | 106 | 2825 | 1765 | 7 | 106 | 2943 | 4511 | 106 | 3065 | 5175 | 20 | 105 | 3179 | 4586 |
| 70 | 98 | 2704 | 2498 | 97 | 2821 | 1770 | 8 | 97 | 2940 | 4517 | 97 | 3062 | 5181 | 21 | 97 | 3178 | 4593 |
| 71 | 88 | 2700 | 2504 | 88 | 2817 | 1776 | 9 | 88 | 2937 | 4524 | 88 | 3060 | 5188 | 22 | 87 | 3176 | 4601 |
| 72 | 79 | 2606 | 2482 | 79 | 2813 | 1754 | 10 | 79 | 2934 | 4501 | 79 | 3058 | 5164 | 23 | 78 | 3174 | 4577 |
| 73 | 70 | 2691 | 2461 | 70 | 2809 | 1722 | 11 | 70 | 2931 | 4478 | 70 | 3055 | 5141 | 24 | 69 | 3173 | 4552 |
| 74 | 61 | 2686 | 2439 | 61 | 2805 | 1710 | 12 | 60 | 2928 | 4456 | 60 | 3053 | 5118 | 25 | 60 | 3171 | 4528 |
| 75 | 53 | 2682 | 2417 | 53 | 2801 | 1687 | 13 | 52 | 2925 | 4432 | 52 | 3050 | 5094 | 26 | 52 | 3170 | 4503 |
| 76 | 42 | 2678 | 2397 | 42 | 2797 | 1666 | 14 | 42 | 2921 | 4411 | 42 | 3048 | 5071 | 27 | 41 | 3168 | 4480 |
| 77 | 33 | 2674 | 2375 | 33 | 2794 | 1644 | 15 | 33 | 2918 | 4388 | 32 | 3046 | 5049 | 28 | 32 | 3166 | 4456 |
| 78 | 24 | 2669 | 2354 | 24 | 2790 | 1622 | 16 | 24 | 2915 | 4365 | 24 | 3043 | 5024 | 29 | 24 | 3165 | 4431 |
| 79 | 15 | 2665 | 2333 | 15 | 2786 | 1600 | 17 | 14 | 2912 | 4343 | 14 | 3041 | 5002 | 30 | 14 | 3163 | 4408 |
| 80 | 5 | 2660 | 2312 | 5 | 2782 | 1579 | 18 | 5 | 2909 | 4320 | 5 | 3038 | 4978 | 1 (July) | 5 | 3162 | 4383 |
| 81 | − 4 | 2656 | 2290 | + 4 | 2778 | 1557 | 19 | + 4 | 2906 | 4297 | + 4 | 3036 | 4955 | 2 | + 4 | 3160 | 4359 |
| 82 | 13 | 2652 | 2268 | 13 | 2774 | 1535 | 20 | 13 | 2903 | 4274 | 13 | 3034 | 4931 | 3 | 13 | 3158 | 4335 |
| 83 | 23 | 2647 | 2248 | 23 | 2770 | 1514 | 21 | 22 | 2900 | 4251 | 22 | 3031 | 4908 | 4 | 22 | 3157 | 4310 |
| 84 | 32 | 2643 | 2226 | 32 | 2766 | 1492 | 22 | 32 | 2897 | 4229 | 32 | 3029 | 4885 | 5 | 32 | 3155 | 4287 |
| 85 | 42 | 2638 | 2206 | 42 | 2762 | 1471 | 23 | 41 | 2894 | 4206 | 41 | 3026 | 4862 | 6 | 41 | 3154 | 4263 |
| 86 | 51 | 2634 | 2184 | 50 | 2758 | 1447 | 24 | 50 | 2890 | 4183 | 50 | 3024 | 4838 | 7 | 50 | 3152 | 4238 |
| 87 | 60 | 2630 | 2162 | 59 | 2755 | 1425 | 25 | 59 | 2887 | 4161 | 59 | 3022 | 4815 | 8 | 59 | 3150 | 4214 |
| 88 | 69 | 2625 | 2141 | 69 | 2751 | 1404 | 26 | 69 | 2884 | 4139 | 68 | 3019 | 4791 | 9 | 68 | 3149 | 4190 |
| 89 | 78 | 2621 | 2119 | 78 | 2747 | 1382 | 27 | 78 | 2881 | 4116 | 78 | 3017 | 4769 | 10 | 77 | 3147 | 4166 |
| 90 | 87 | 2616 | 2098 | 87 | 2743 | 1360 | 28 | 87 | 2878 | 4093 | 87 | 3014 | 4745 | 11 | 86 | 3146 | 4141 |
| 91 | 107 | 2612 | 2087 | 107 | 2740 | 1349 | 29 | 108 | 2876 | 4082 | 108 | 3012 | 4734 | 12 | 108 | 3144 | 4130 |
| 92 | 117 | 2581 | 2066 | 118 | 2709 | 1329 | 30 | 118 | 2844 | 4060 | 118 | 2979 | 4711 | 13 | 119 | 3111 | 4100 |
| 93 | 127 | 2551 | 2046 | 128 | 2678 | 1308 | 31 | 128 | 2812 | 4038 | 128 | 2947 | 4689 | 14 | 129 | 3077 | 4084 |
| 94 | 137 | 2520 | 2025 | 138 | 2647 | 1287 | 32 | 138 | 2780 | 4016 | 138 | 2914 | 4666 | 15 | 139 | 3044 | 4061 |
| 95 | 147 | 2490 | 2005 | 148 | 2616 | 1266 | 1 (Tamil, Adi; Malayalam, Karkatagam; Bengal, Sravana.) | 148 | 2748 | 3994 | 148 | 2882 | 4644 | 16 | 149 | 3011 | 4038 |
| 96 | 157 | 2459 | 1984 | 158 | 2584 | 1245 | 2 | 158 | 2717 | 3972 | 158 | 2849 | 4621 | 17 | 159 | 2977 | 4014 |
| 97 | 167 | 2428 | 1963 | 168 | 2553 | 1223 | 3 | 168 | 2685 | 3951 | 168 | 2817 | 4599 | 18 | 169 | 2944 | 3991 |
| 98 | 177 | 2398 | 1943 | 177 | 2522 | 1201 | 4 | 178 | 2653 | 3929 | 178 | 2784 | 4576 | 19 | 179 | 2911 | 3968 |
| 99 | 187 | 2368 | 1922 | 187 | 2491 | 1180 | 5 | 188 | 2621 | 3907 | 188 | 2752 | 4554 | 20 | 189 | 2878 | 3945 |
| 100 | 196 | 2337 | 1901 | 197 | 2460 | 1159 | 6 | 197 | 2589 | 3884 | 198 | 2719 | 4531 | 21 | 198 | 2844 | 3920 |
| 101 | 206 | 2306 | 1880 | 206 | 2429 | 1187 | 7 | 207 | 2557 | 3862 | 207 | 2687 | 4508 | 22 | 208 | 2811 | 3897 |
| 102 | 215 | 2275 | 1859 | 216 | 2398 | 1116 | 8 | 217 | 2525 | 3841 | 217 | 2654 | 4485 | 23 | 218 | 2778 | 3874 |
| 103 | 225 | 2245 | 1836 | 226 | 2367 | 1092 | 9 | 226 | 2493 | 3814 | 227 | 2622 | 4457 | 24 | 227 | 2744 | 3844 |
| 104 | 235 | 2214 | 1814 | 235 | 2336 | 1067 | 10 | 236 | 2461 | 3788 | 236 | 2589 | 4429 | 25 | 237 | 2711 | 3815 |
| 105 | 244 | 2184 | 1790 | 244 | 2305 | 1041 | 11 | 245 | 2429 | 3761 | 246 | 2557 | 4401 | 26 | 246 | 2678 | 3784 |
| 106 | 253 | 2153 | 1767 | 254 | 2273 | 1017 | 12 | 254 | 2397 | 3734 | 255 | 2524 | 4372 | 27 | 256 | 2644 | 3755 |
| 107 | 262 | 2122 | 1743 | 263 | 2242 | 1011 | 13 | 263 | 2366 | 3707 | 264 | 2492 | 4343 | 28 | 265 | 2611 | 3725 |
| 108 | 271 | 2092 | 1720 | 272 | 2211 | 974 | 14 | 273 | 2334 | 3682 | 273 | 2459 | 4315 | 29 | 274 | 2578 | 3695 |
| 109 | 280 | 2061 | 1696 | 281 | 2180 | 941 | 15 | 282 | 2302 | 3655 | 283 | 2427 | 4287 | 30 | 283 | 2545 | 3665 |
| 110 | 291 | 2031 | 1673 | 290 | 2149 | 916 | 16 | 291 | 2270 | 3628 | 292 | 2394 | 4258 | 31 | 292 | 2511 | 3634 |
| 111 | 298 | 2000 | 1649 | 299 | 2118 | 890 | 17 | 299 | 2238 | 3600 | 300 | 2362 | 4229 | 1 (August) | 301 | 2478 | 3604 |
| 112 | 307 | 1969 | 1626 | 308 | 2087 | 865 | 18 | 308 | 2206 | 3573 | 309 | 2329 | 4200 | 2 | 310 | 2445 | 3574 |
| 113 | 315 | 1939 | 1601 | 316 | 2056 | 839 | 19 | 317 | 2174 | 3546 | 318 | 2297 | 4171 | 3 | 319 | 2411 | 3544 |
| 114 | 324 | 1908 | 1578 | 325 | 2025 | 813 | 20 | 326 | 2142 | 3519 | 327 | 2264 | 4143 | 4 | 328 | 2378 | 3514 |
| 115 | 333 | 1878 | 1554 | 333 | 1994 | 787 | 21 | 334 | 2110 | 3491 | 335 | 2232 | 4113 | 5 | 336 | 2345 | 3482 |
| 116 | 341 | 1847 | 1530 | 342 | 1963 | 762 | 22 | 343 | 2078 | 3464 | 344 | 2199 | 4084 | 6 | 344 | 2311 | 3451 |
| 117 | 349 | 1816 | 1505 | 350 | 1931 | 735 | 23 | 351 | 2047 | 3436 | 352 | 2167 | 4054 | 7 | 353 | 2278 | 3421 |
| 118 | 358 | 1786 | 1482 | 358 | 1900 | 709 | 24 | 359 | 2015 | 3409 | 360 | 2134 | 4025 | 8 | 361 | 2245 | 3390 |
| 119 | 365 | 1755 | 1456 | 366 | 1869 | 683 | 25 | 367 | 1983 | 3381 | 368 | 2102 | 3995 | 9 | 369 | 2212 | 3359 |
| 120 | 373 | 1725 | 1432 | 374 | 1838 | 657 | 26 | 375 | 1951 | 3353 | 376 | 2069 | 3965 | 10 | 377 | 2178 | 3327 |
| 121 | 382 | 1694 | 1408 | 385 | 1807 | 633 | 27 | 387 | 1919 | 3329 | 389 | 2037 | 3941 | 11 | 391 | 2145 | 3305 |
| 122 | 390 | 1664 | 1384 | 393 | 1776 | 607 | 28 | 395 | 1888 | 3301 | 397 | 2004 | 3949 | 12 | 400 | 2112 | 3272 |
| 123 | 397 | 1631 | 1358 | 400 | 1742 | 580 | 29 | 402 | 1852 | 3272 | 404 | 1966 | 3918 | 13 | 407 | 2073 | 3240 |
| 124 | 405 | 1599 | 1334 | 408 | 1708 | 553 | 30 | 410 | 1816 | 3244 | 412 | 1929 | 3888 | 14 | 415 | 2034 | 3206 |
| 125 | 412 | 1566 | 1308 | 415 | 1673 | 526 | 31 | 417 | 1780 | 3215 | 419 | 1891 | 3858 | 15 | 422 | 1994 | 3179 |

# TABLE XIII.

## Sunrise from 23° to 27° N. Latitude.

| Day of Solar year. | Lat. 23° −20″ Long. Eqn. of time in seconds. | ☉'s Trop. Long. in seconds. | Total Corrn. in seconds. Indore. A.D. 1840—1920 | Lat. 24° −876″ Long. Eqn. of time in seconds. | ☉'s Trop. Long. in seconds. | Total Corrn. in seconds. Anhilwad. A.D. 1840—1920 | Days of solar months. | Lat. 25° +1736″ Long. Eqn. of time in seconds. | ☉'s Trop. Long. in seconds. | Total Corrn. in seconds. Benares. A.D. 1840—1920 | Lat. 26° +2264″ Long. Eqn. of Time in seconds. | ☉'s Trop. Long. in seconds. | Total Corrn. in seconds. Patna. A.D. 1840—1920 | Eng. date. | Lat. 27° +1544″ Long. Eqn. of Time in seconds. | ☉'s Trop. Long. in seconds. | Total Corrn. in seconds. Oudh. A.D. 1840—1920 |
|---|---|---|---|---|---|---|---|---|---|---|---|---|---|---|---|---|---|
| 126 | +420 | +1534 | +1284 | +422 | +1639 | +499 | 1 (Tamil, Avani; Malayalam, Chingam; Bengal, Bhadrapada.) | +424 | +1744 | +3186 | +426 | +1853 | +3827 | 16 (August) | +429 | +1955 | +3144 |
| 127 | 427 | 1501 | 1258 | 429 | 1604 | 471 | 2 | 431 | 1708 | 3157 | 433 | 1815 | 3796 | 17 | 436 | 1916 | 3112 |
| 128 | 434 | 1469 | 1233 | 436 | 1570 | 444 | 3 | 438 | 1673 | 3129 | 440 | 1778 | 3765 | 18 | 443 | 1877 | 3080 |
| 129 | 441 | 1436 | 1207 | 443 | 1536 | 417 | 4 | 445 | 1637 | 3100 | 447 | 1740 | 3735 | 19 | 450 | 1838 | 3048 |
| 130 | 448 | 1404 | 1182 | 450 | 1502 | 390 | 5 | 452 | 1601 | 3071 | 454 | 1702 | 3704 | 20 | 457 | 1798 | 3015 |
| 131 | 454 | 1371 | 1155 | 456 | 1467 | 361 | 6 | 458 | 1565 | 3041 | 460 | 1665 | 3672 | 21 | 463 | 1759 | 2982 |
| 132 | 460 | 1339 | 1128 | 462 | 1433 | 334 | 7 | 464 | 1529 | 3012 | 466 | 1627 | 3641 | 22 | 469 | 1720 | 2949 |
| 133 | 467 | 1306 | 1113 | 469 | 1399 | 317 | 8 | 471 | 1493 | 2993 | 473 | 1589 | 3609 | 23 | 476 | 1681 | 2926 |
| 134 | 473 | 1274 | 1097 | 475 | 1364 | 299 | 9 | 477 | 1457 | 2973 | 479 | 1552 | 3559 | 24 | 482 | 1642 | 2902 |
| 135 | 479 | 1241 | 1080 | 481 | 1330 | 281 | 10 | 483 | 1421 | 2952 | 485 | 1514 | 3537 | 25 | 488 | 1602 | 2877 |
| 136 | 485 | 1209 | 1064 | 487 | 1296 | 263 | 11 | 489 | 1385 | 2932 | 491 | 1476 | 3515 | 26 | 494 | 1563 | 2853 |
| 137 | 490 | 1176 | 1047 | 493 | 1261 | 244 | 12 | 495 | 1349 | 2912 | 497 | 1438 | 3492 | 27 | 500 | 1524 | 2829 |
| 138 | 496 | 1144 | 1031 | 499 | 1227 | 226 | 13 | 501 | 1314 | 2892 | 503 | 1401 | 3470 | 28 | 506 | 1485 | 2805 |
| 139 | 502 | 1111 | 1015 | 505 | 1193 | 208 | 14 | 507 | 1278 | 2872 | 509 | 1363 | 3448 | 29 | 512 | 1446 | 2781 |
| 140 | 507 | 1079 | 1007 | 510 | 1159 | 189 | 15 | 512 | 1242 | 2850 | 514 | 1325 | 3425 | 30 | 517 | 1406 | 2755 |
| 141 | 512 | 1046 | 980 | 515 | 1124 | 170 | 16 | 517 | 1206 | 2829 | 519 | 1288 | 3402 | 31 | 522 | 1367 | 2730 |
| 142 | 517 | 1014 | 963 | 520 | 1090 | 141 | 17 | 522 | 1170 | 2808 | 524 | 1250 | 3379 | 1 (September) | 527 | 1328 | 2705 |
| 143 | 522 | 981 | 946 | 525 | 1056 | 132 | 18 | 527 | 1134 | 2787 | 529 | 1212 | 3356 | 2 | 532 | 1289 | 2680 |
| 144 | 527 | 949 | 929 | 530 | 1021 | 113 | 19 | 532 | 1098 | 2766 | 534 | 1175 | 3333 | 3 | 537 | 1250 | 2655 |
| 145 | 531 | 916 | 910 | 535 | 987 | 94 | 20 | 537 | 1062 | 2744 | 539 | 1137 | 3310 | 4 | 542 | 1210 | 2629 |
| 146 | 536 | 884 | 893 | 539 | 953 | 74 | 21 | 541 | 1026 | 2722 | 543 | 1099 | 3286 | 5 | 546 | 1171 | 2603 |
| 147 | 540 | 851 | 875 | 542 | 918 | 52 | 22 | 545 | 990 | 2700 | 547 | 1061 | 3261 | 6 | 550 | 1132 | 2577 |
| 148 | 544 | 819 | 857 | 546 | 884 | 32 | 23 | 546 | 955 | 2675 | 552 | 1024 | 3238 | 7 | 552 | 1093 | 2549 |
| 149 | 546 | 786 | 837 | 548 | 850 | 10 | 24 | 547 | 919 | 2650 | 554 | 986 | 3212 | 8 | 554 | 1054 | 2521 |
| 150 | 547 | 754 | 815 | 550 | 816 | — 12 | 25 | 548 | 883 | 2624 | 555 | 948 | 3185 | 9 | 556 | 1014 | 2492 |
| 151 | 548 | 721 | 794 | 551 | 781 | 35 | 26 | 550 | 847 | 2600 | 556 | 911 | 3158 | 10 | 558 | 975 | 2464 |
| 152 | 549 | 688 | 773 | 552 | 748 | 58 | 27 | 551 | 812 | 2575 | 558 | 872 | 3132 | 11 | 560 | 936 | 2436 |
| 153 | 550 | 666 | 752 | 553 | 724 | 81 | 28 | 553 | 786 | 2551 | 559 | 844 | 3105 | 12 | 562 | 906 | 2408 |
| 154 | 551 | 644 | 731 | 555 | 700 | 103 | 29 | 554 | 760 | 2526 | 560 | 816 | 3078 | 13 | 564 | 876 | 2380 |
| 155 | 552 | 621 | 709 | 556 | 676 | 126 | 30 | 555 | 733 | 2500 | 561 | 788 | 3051 | 14 | 566 | 845 | 2341 |
| 156 | 553 | 599 | 688 | 557 | 652 | 149 | 31 | 556 | 707 | 2475 | 563 | 760 | 3025 | 15 | 568 | 815 | 2328 |
| 157 | 554 | 577 | 667 | 558 | 627 | 175 | 1 (Tamil, Purattasi; Malayalam, Kanni. Bengal, Asvina.) | 558 | 681 | 2451 | 565 | 731 | 2998 | 16 | 570 | 785 | 2295 |
| 158 | 555 | 555 | 646 | 560 | 603 | 195 | 2 | 560 | 655 | 2427 | 567 | 703 | 2972 | 17 | 572 | 755 | 2267 |
| 159 | 557 | 533 | 626 | 562 | 579 | 217 | 3 | 562 | 629 | 2403 | 570 | 675 | 2947 | 18 | 574 | 725 | 2239 |
| 160 | 559 | 510 | 605 | 564 | 555 | 239 | 4 | 564 | 602 | 2380 | 572 | 647 | 2927 | 19 | 576 | 694 | 2210 |
| 161 | 561 | 488 | 585 | 566 | 531 | 359 | 5 | 566 | 576 | 2354 | 574 | 619 | 2895 | 20 | 578 | 664 | 2182 |
| 162 | 562 | 466 | 564 | 567 | 507 | 334 | 6 | 568 | 550 | 2330 | 575 | 591 | 2868 | 21 | 579 | 634 | 2153 |
| 163 | 564 | 444 | 544 | 569 | 483 | 307 | 7 | 570 | 524 | 2306 | 577 | 562 | 2841 | 22 | 581 | 604 | 2125 |
| 164 | 566 | 422 | 523 | 571 | 459 | 330 | 8 | 572 | 498 | 2281 | 579 | 535 | 2814 | 23 | 583 | 574 | 2096 |
| 165 | 567 | 399 | 501 | 572 | 435 | 354 | 9 | 574 | 471 | 2256 | 580 | 507 | 2786 | 24 | 584 | 543 | 2066 |
| 166 | 567 | 377 | 478 | 572 | 411 | 379 | 10 | 576 | 445 | 2231 | 580 | 479 | 2756 | 25 | 584 | 512 | 2034 |
| 167 | 568 | 355 | 456 | 573 | 386 | 403 | 11 | 577 | 419 | 2205 | 581 | 450 | 2729 | 26 | 585 | 483 | 2004 |
| 168 | 569 | 333 | 435 | 574 | 362 | 426 | 12 | 578 | 393 | 2179 | 582 | 422 | 2701 | 27 | 586 | 453 | 1974 |
| 169 | 570 | 311 | 413 | 575 | 338 | 450 | 13 | 579 | 367 | 2152 | 583 | 394 | 2672 | 28 | 587 | 423 | 1944 |
| 170 | 570 | 288 | 390 | 575 | 314 | 475 | 14 | 579 | 340 | 2125 | 583 | 366 | 2643 | 29 | 587 | 392 | 1913 |
| 171 | 570 | 266 | 367 | 575 | 290 | 500 | 15 | 579 | 314 | 2098 | 583 | 338 | 2614 | 30 | 587 | 362 | 1881 |
| 172 | 570 | 244 | 344 | 575 | 266 | 525 | 16 | 579 | 288 | 2071 | 583 | 310 | 2585 | 1 (October) | 587 | 332 | 1850 |
| 173 | 570 | 222 | 321 | 575 | 242 | 550 | 17 | 579 | 262 | 2044 | 583 | 282 | 2556 | 2 | 587 | 302 | 1819 |
| 174 | 570 | 200 | 298 | 575 | 218 | 575 | 18 | 579 | 236 | 2017 | 583 | 254 | 2527 | 3 | 587 | 272 | 1788 |
| 175 | 570 | 177 | 275 | 575 | 194 | 600 | 19 | 579 | 209 | 1990 | 583 | 226 | 2498 | 4 | 587 | 241 | 1757 |
| 176 | 570 | 155 | 252 | 575 | 170 | 625 | 20 | 579 | 183 | 1963 | 583 | 198 | 2469 | 5 | 587 | 211 | 1725 |
| 177 | 569 | 133 | 228 | 574 | 145 | 651 | 21 | 578 | 157 | 1935 | 582 | 169 | 2439 | 6 | 586 | 181 | 1693 |
| 178 | 568 | 111 | 205 | 573 | 121 | 676 | 22 | 577 | 131 | 1907 | 581 | 141 | 2409 | 7 | 585 | 151 | 1661 |
| 179 | 567 | 89 | 181 | 572 | 97 | 702 | 23 | 576 | 105 | 1878 | 580 | 113 | 2378 | 8 | 584 | 121 | 1629 |
| 180 | 566 | 66 | 151 | 571 | 73 | 728 | 24 | 575 | 78 | 1850 | 579 | 85 | 2348 | 9 | 583 | 90 | 1597 |
| 181 | 564 | 44 | 182 | 569 | 49 | 755 | 25 | 573 | 52 | 1821 | 577 | 57 | 2317 | 10 | 581 | 60 | 1563 |
| 182 | 563 | 22 | 108 | 568 | 25 | 781 | 26 | 572 | 26 | 1793 | 576 | 29 | 2287 | 11 | 580 | 30 | 1531 |
| 183 | 562 | 0 | 84 | 566 | 0 | 808 | 27 | 570 | 0 | 1764 | 575 | 0 | 2257 | 12 | 579 | 0 | 1499 |
| 184 | 560 | — 23 | 59 | 564 | — 25 | 835 | 28 | 568 | — 27 | 1735 | 573 | — 29 | 2226 | 13 | 577 | — 31 | 1466 |
| 185 | 558 | 46 | 84 | 562 | 50 | 862 | 29 | 566 | 54 | 1706 | 571 | 58 | 2195 | 14 | 575 | 62 | 1433 |
| 186 | 556 | 69 | 9 | 560 | 75 | 889 | 30 | 564 | 81 | 1677 | 569 | 87 | 2164 | 15 | 573 | 94 | 1399 |

# TABLE XIII.

## Sunrise from 23° to 27° N. Latitude.

| Day of Solar year. | Lat. 23° Eqn. of time in seconds. | Lat. 23° ⊙'s Trop. Long. in seconds. | Lat. 23° −20″ Long. Total Corrn. in seconds. Indore. A.D. 1840—1920 | Lat. 24° Eqn. of time in seconds. | Lat. 24° ⊙'s Trop. Long. in seconds. | Lat. 24° −876″ Long. Total Corrn. in seconds. Anhilwad. A.D. 1840—1920 | Days of solar months. | Lat 25° Eqn. of time in seconds. | Lat 25° ⊙'s Trop. Long. in seconds. | Lat 25° +1736″ Long. Total Corrn. in seconds. Benares. A.D. 1840—1920 | Lat. 26° Eqn. of Time in seconds. | Lat. 26° ⊙'s Trop. Long. in seconds. | Lat. 26° +2264″ Long. Total Corrn. in seconds. Patna. A.D. 1840—1920 | Eng. date. | Lat. 27° Eqn. of Time in seconds. | Lat. 27° ⊙'s Trop. Long. in seconds. | Lat. 27° +1544″ Long. Total Corrn. in seconds. Oudh. A.D. 1840—1920 |
|---|---|---|---|---|---|---|---|---|---|---|---|---|---|---|---|---|---|
| 187 | +554 | −92 | −24 | +558 | −100 | −916 | 1 (*Tamil*, Aippasi; *Malayalam*, Tulam; *Bengal*, Kartika.) | +562 | −108 | +1648 | +567 | −116 | +2133 | 16 (October) | +571 | −125 | +1366 |
| 188 | 551 | 114 | 41 | 555 | 124 | 943 | 2 | 559 | 135 | 1618 | 564 | 145 | 2101 | 17 | 568 | 156 | 1332 |
| 189 | 548 | 137 | 67 | 552 | 149 | 971 | 3 | 556 | 163 | 1587 | 561 | 175 | 2068 | 18 | 565 | 187 | 1298 |
| 190 | 545 | 160 | 93 | 549 | 174 | 999 | 4 | 553 | 190 | 1557 | 558 | 204 | 2036 | 19 | 562 | 218 | 1264 |
| 191 | 542 | 183 | 119 | 546 | 199 | 1027 | 5 | 550 | 217 | 1527 | 555 | 232 | 2004 | 20 | 559 | 250 | 1229 |
| 192 | 539 | 206 | 145 | 543 | 224 | 1055 | 6 | 547 | 244 | 1497 | 552 | 262 | 1972 | 21 | 556 | 281 | 1195 |
| 193 | 536 | 229 | 172 | 540 | 249 | 1084 | 7 | 543 | 271 | 1467 | 548 | 291 | 1940 | 22 | 552 | 312 | 1160 |
| 194 | 533 | 252 | 208 | 537 | 274 | 1121 | 8 | 540 | 298 | 1428 | 545 | 320 | 1899 | 23 | 549 | 343 | 1118 |
| 195 | 530 | 275 | 243 | 584 | 299 | 1159 | 9 | 536 | 325 | 1388 | 541 | 349 | 1858 | 24 | 545 | 374 | 1075 |
| 196 | 526 | 298 | 279 | 530 | 324 | 1197 | 10 | 532 | 352 | 1348 | 537 | 378 | 1816 | 25 | 541 | 406 | 1031 |
| 197 | 522 | 321 | 316 | 526 | 349 | 1235 | 11 | 528 | 379 | 1308 | 533 | 407 | 1774 | 26 | 537 | 437 | 988 |
| 198 | 518 | 343 | 352 | 522 | 373 | 1273 | 12 | 524 | 406 | 1269 | 529 | 436 | 1733 | 27 | 533 | 468 | 945 |
| 199 | 514 | 366 | 389 | 518 | 398 | 1312 | 13 | 520 | 484 | 1229 | 525 | 466 | 1691 | 28 | 529 | 499 | 902 |
| 200 | 509 | 389 | 426 | 513 | 423 | 1351 | 14 | 515 | 461 | 1188 | 520 | 495 | 1648 | 29 | 524 | 530 | 858 |
| 201 | 505 | 412 | 463 | 509 | 448 | 1389 | 15 | 511 | 488 | 1148 | 515 | 521 | 1605 | 30 | 519 | 562 | 813 |
| 202 | 501 | 435 | 499 | 505 | 473 | 1428 | 16 | 507 | 515 | 1108 | 511 | 553 | 1564 | 31 | 515 | 593 | 770 |
| 203 | 496 | 458 | 537 | 500 | 498 | 1467 | 17 | 502 | 542 | 1067 | 506 | 582 | 1521 | 1 | 510 | 624 | 726 |
| 204 | 491 | 481 | 574 | 495 | 523 | 1506 | 18 | 497 | 569 | 1026 | 500 | 611 | 1457 | 2 | 504 | 655 | 681 |
| 205 | 486 | 504 | 612 | 490 | 548 | 1546 | 19 | 492 | 596 | 985 | 495 | 640 | 1435 | 3 | 499 | 686 | 637 |
| 206 | 481 | 527 | 649 | 485 | 573 | 1585 | 20 | 487 | 623 | 944 | 490 | 669 | 1392 | 4 | 494 | 718 | 592 |
| 207 | 477 | 550 | 686 | 481 | 598 | 1623 | 21 | 483 | 650 | 904 | 486 | 698 | 1350 | 5 | 490 | 749 | 549 |
| 208 | 472 | 572 | 728 | 476 | 622 | 1662 | 22 | 478 | 677 | 864 | 480 | 727 | 1307 | 6 | 484 | 780 | 504 |
| 209 | 466 | 595 | 762 | 470 | 647 | 1703 | 23 | 472 | 705 | 822 | 474 | 757 | 1263 | 7 | 478 | 811 | 459 |
| 210 | 461 | 618 | 799 | 465 | 672 | 1742 | 24 | 467 | 732 | 781 | 468 | 786 | 1219 | 8 | 472 | 842 | 414 |
| 211 | 455 | 641 | 838 | 459 | 697 | 1782 | 25 | 461 | 759 | 739 | 462 | 815 | 1175 | 9 | 466 | 874 | 358 |
| 212 | 450 | 664 | 875 | 454 | 722 | 1822 | 26 | 455 | 786 | 697 | 458 | 844 | 1134 | 10 | 462 | 905 | 325 |
| 213 | 446 | 688 | 912 | 450 | 748 | 1860 | 27 | 452 | 812 | 658 | 455 | 872 | 1093 | 11 | 458 | 936 | 282 |
| 214 | 445 | 721 | 945 | 447 | 782 | 1897 | 28 | 449 | 848 | 613 | 452 | 910 | 1052 | 12 | 455 | 975 | 240 |
| 215 | 441 | 758 | 982 | 444 | 817 | 1935 | 29 | 446 | 884 | 580 | 449 | 947 | 1012 | 13 | 452 | 1014 | 198 |
| 216 | 434 | 785 | 1021 | 437 | 851 | 1976 | 30 | 439 | 920 | 537 | 442 | 985 | 967 | 14 | 445 | 1054 | 151 |
| 217 | 427 | 818 | 1061 | 430 | 885 | 2017 | 1 (*Tamil*, Kartigai; *Malayalam*, Vrischikam; *Bengal*, Margasira.) | 432 | 956 | 494 | 435 | 1023 | 922 | 15 (November) | 438 | 1093 | 105 |
| 218 | 420 | 850 | 1100 | 423 | 919 | 2058 | 2 | 425 | 991 | 452 | 428 | 1060 | 878 | 16 | 431 | 1132 | 59 |
| 219 | 412 | 883 | 1141 | 415 | 954 | 2101 | 3 | 417 | 1027 | 408 | 420 | 1098 | 832 | 17 | 423 | 1171 | 12 |
| 220 | 405 | 915 | 1180 | 408 | 988 | 2142 | 4 | 410 | 1063 | 365 | 413 | 1136 | 787 | 18 | 416 | 1210 | −34 |
| 221 | 397 | 948 | 1221 | 400 | 1022 | 2184 | 5 | 402 | 1099 | 321 | 405 | 1174 | 741 | 19 | 408 | 1250 | 82 |
| 222 | 389 | 980 | 1261 | 392 | 1057 | 2227 | 6 | 394 | 1135 | 277 | 397 | 1211 | 696 | 20 | 400 | 1289 | 129 |
| 223 | 381 | 1013 | 1303 | 384 | 1091 | 2268 | 7 | 386 | 1171 | 234 | 389 | 1249 | 649 | 21 | 392 | 1328 | 176 |
| 224 | 373 | 1045 | 1342 | 376 | 1125 | 2307 | 8 | 378 | 1207 | 194 | 381 | 1287 | 609 | 22 | 384 | 1367 | 217 |
| 225 | 365 | 1078 | 1380 | 368 | 1160 | 2346 | 9 | 370 | 1243 | 154 | 373 | 1324 | 568 | 23 | 376 | 1406 | 259 |
| 226 | 357 | 1110 | 1419 | 360 | 1194 | 2885 | 10 | 362 | 1279 | 114 | 365 | 1362 | 528 | 24 | 368 | 1446 | 300 |
| 227 | 348 | 1143 | 1458 | 351 | 1228 | 2425 | 11 | 353 | 1315 | 73 | 356 | 1400 | 486 | 25 | 359 | 1485 | 342 |
| 228 | 340 | 1175 | 1497 | 343 | 1262 | 2464 | 12 | 345 | 1350 | 34 | 348 | 1437 | 446 | 26 | 351 | 1524 | 383 |
| 229 | 332 | 1208 | 1536 | 335 | 1297 | 2504 | 13 | 337 | 1386 | −6 | 340 | 1475 | 405 | 27 | 343 | 1563 | 425 |
| 230 | 323 | 1240 | 1575 | 326 | 1331 | 2544 | 14 | 328 | 1422 | 47 | 331 | 1513 | 364 | 28 | 334 | 1602 | 663 |
| 231 | 314 | 1273 | 1615 | 317 | 1365 | 2584 | 15 | 319 | 1458 | 88 | 322 | 1551 | 322 | 29 | 325 | 1642 | 509 |
| 232 | 306 | 1305 | 1653 | 309 | 1400 | 2623 | 16 | 311 | 1494 | 128 | 314 | 1588 | 282 | 30 | 317 | 1681 | 551 |
| 233 | 297 | 1338 | 1693 | 300 | 1434 | 2663 | 17 | 302 | 1530 | 169 | 305 | 1626 | 240 | 1 | 308 | 1720 | 593 |
| 234 | 287 | 1370 | 1734 | 290 | 1468 | 2704 | 18 | 292 | 1566 | 211 | 295 | 1664 | 198 | 2 | 298 | 1759 | 636 |
| 235 | 278 | 1403 | 1773 | 281 | 1503 | 2744 | 19 | 282 | 1602 | 253 | 286 | 1701 | 156 | 3 | 289 | 1798 | 679 |
| 236 | 269 | 1435 | 1813 | 272 | 1537 | 2784 | 20 | 274 | 1638 | 293 | 277 | 1739 | 115 | 4 | 280 | 1838 | 721 |
| 237 | 260 | 1468 | 1852 | 263 | 1571 | 2824 | 21 | 265 | 1674 | 334 | 268 | 1777 | 73 | 5 (December) | 271 | 1877 | 763 |
| 238 | 251 | 1500 | 1892 | 254 | 1605 | 2864 | 22 | 256 | 1709 | 374 | 259 | 1814 | 32 | 6 | 262 | 1916 | 806 |
| 239 | 241 | 1533 | 1933 | 245 | 1640 | 2905 | 23 | 246 | 1745 | 416 | 249 | 1852 | −11 | 7 | 252 | 1955 | 849 |
| 240 | 232 | 1565 | 1972 | 236 | 1674 | 2945 | 24 | 237 | 1781 | 457 | 240 | 1890 | 52 | 8 | 243 | 1994 | 891 |
| 241 | 222 | 1598 | 2013 | 226 | 1708 | 2986 | 25 | 227 | 1817 | 499 | 230 | 1928 | 92 | 9 | 233 | 2034 | 934 |
| 242 | 213 | 1630 | 2052 | 217 | 1743 | 3026 | 26 | 218 | 1853 | 540 | 221 | 1965 | 136 | 10 | 224 | 2073 | 977 |
| 243 | 205 | 1664 | 2091 | 205 | 1776 | 3069 | 27 | 206 | 1888 | 584 | 206 | 2004 | 184 | 11 | 207 | 2112 | 1027 |
| 244 | 196 | 1695 | 2131 | 196 | 1807 | 3109 | 28 | 197 | 1920 | 625 | 197 | 2036 | 225 | 12 | 198 | 2145 | 1069 |
| 245 | 186 | 1725 | 2171 | 186 | 1838 | 3150 | 29 | 187 | 1952 | 667 | 187 | 2069 | 268 | 13 | 188 | 2179 | 1113 |
| 246 | 176 | 1756 | 2212 | 176 | 1869 | 3191 | 30 | 177 | 1984 | 709 | 177 | 2101 | 310 | 14 | 178 | 2212 | 1156 |

# TABLE XIII.

## Sunrise from 23° to 27° N. Latitude.

| Day of Solar year. | Lat. 23° −20″ Long. Eqn. of time in seconds. | ☉'s Trop. Long. in seconds. | Total Corrn. in seconds. Indore. A.D. 1840—1920 | Lat. 24° −876″ Long. Eqn. of time in seconds. | ☉'s Trop. Long. in seconds. | Total Corrn. in seconds. Anhilwad. A.D. 1840—1920 | Solar months | Days of solar months. | Lat. 25° +1736″ Long. Eqn. of time in seconds. | ☉'s Trop. Long. in seconds. | Total Corrn. in seconds. Benares. A.D. 1840—1920 | Lat. 26° +2264″ Long. Eqn. of time in seconds. | ☉'s Trop. Long. in seconds. | Total Corrn. in seconds. Patna. A.D. 1840—1920 | Eng. month | Eng. date. | Lat. 27° +1544″ Long. Eqn. of time in seconds. | ☉'s Trop. Long. in seconds. | Total Corrn. in seconds. Oudh. A.D. 1840—1920 |
|---|---|---|---|---|---|---|---|---|---|---|---|---|---|---|---|---|---|---|---|
| **247** | +166 | −1786 | −2252 | +166 | −1900 | −3232 | Tamil, Margali; Malayalam, Dhanus; Bengal, Pausha. | 1 | +167 | −2016 | −751 | +167 | −2134 | −353 | December | **15** | +168 | −2245 | −1199 |
| **248** | 156 | 1817 | 2293 | 156 | 1931 | 3273 | | 2 | 157 | 2047 | 792 | 157 | 2166 | 385 | | **16** | 158 | 2278 | 1242 |
| **249** | 146 | 1848 | 2334 | 146 | 1963 | 3315 | | 3 | 147 | 2079 | 834 | 147 | 2199 | 438 | | **17** | 148 | 2312 | 1286 |
| **250** | 136 | 1878 | 2374 | 136 | 1994 | 3356 | | 4 | 137 | 2111 | 876 | 137 | 2231 | 480 | | **18** | 138 | 2345 | 1329 |
| **251** | 126 | 1909 | 2415 | 126 | 2025 | 3397 | | 5 | 127 | 2143 | 918 | 127 | 2264 | 523 | | **19** | 128 | 2378 | 1372 |
| **252** | 116 | 1939 | 2455 | 116 | 2056 | 3438 | | 6 | 117 | 2175 | 960 | 117 | 2296 | 565 | | **20** | 118 | 2412 | 1416 |
| **253** | 106 | 1970 | 2496 | 106 | 2087 | 3479 | | 7 | 107 | 2207 | 1002 | 107 | 2329 | 608 | | **21** | 108 | 2445 | 1459 |
| **254** | 96 | 2001 | 2536 | 96 | 2118 | 3520 | | 8 | 97 | 2239 | 1043 | 97 | 2361 | 651 | | **22** | 98 | 2478 | 1502 |
| **255** | 86 | 2031 | 2550 | 86 | 2149 | 3534 | | 9 | 87 | 2271 | 1056 | 87 | 2394 | 663 | | **23** | 88 | 2512 | 1514 |
| **256** | 76 | 2062 | 2565 | 76 | 2180 | 3548 | | 10 | 77 | 2303 | 1069 | 77 | 2426 | 676 | | **24** | 78 | 2545 | 1525 |
| **257** | 66 | 2092 | 2579 | 66 | 2211 | 3562 | | 11 | 67 | 2335 | 1082 | 67 | 2459 | 688 | | **25** | 68 | 2578 | 1537 |
| **258** | 56 | 2123 | 2594 | 56 | 2242 | 3576 | | 12 | 57 | 2366 | 1095 | 57 | 2491 | 701 | | **26** | 58 | 2612 | 1548 |
| **259** | 45 | 2154 | 2609 | 45 | 2274 | 3590 | | 13 | 46 | 2398 | 1109 | 46 | 2524 | 714 | | **27** | 47 | 2645 | 1561 |
| **260** | 35 | 2184 | 2623 | 35 | 2305 | 3604 | | 14 | 36 | 2430 | 1123 | 36 | 2556 | 726 | | **28** | 37 | 2678 | 1573 |
| **261** | 25 | 2215 | 2638 | 25 | 2336 | 3618 | | 15 | 26 | 2462 | 1136 | 26 | 2589 | 739 | | **29** | 27 | 2711 | 1584 |
| **262** | 15 | 2245 | 2652 | 15 | 2367 | 3632 | | 16 | 16 | 2494 | 1149 | 16 | 2621 | 751 | | **30** | 17 | 2745 | 1596 |
| **263** | 2 | 2276 | 2667 | 2 | 2398 | 3646 | | 17 | 2 | 2526 | 1162 | 2 | 2654 | 764 | | **31** | 2 | 2778 | 1607 |
| **264** | −11 | 2307 | 2689 | −11 | 2429 | 3666 | | 18 | −11 | 2558 | 1185 | −11 | 2686 | 786 | January | **1** | −11 | 2811 | 1631 |
| **265** | 20 | 2337 | 2703 | 21 | 2460 | 3680 | | 19 | 21 | 2590 | 1198 | 21 | 2719 | 798 | | **2** | 21 | 2845 | 1643 |
| **266** | 33 | 2368 | 2718 | 33 | 2491 | 3696 | | 20 | 34 | 2622 | 1211 | 34 | 2751 | 811 | | **3** | 35 | 2878 | 1654 |
| **267** | 43 | 2398 | 2732 | 43 | 2522 | 3710 | | 21 | 44 | 2654 | 1224 | 44 | 2784 | 823 | | **4** | 45 | 2911 | 1666 |
| **268** | 53 | 2429 | 2747 | 53 | 2553 | 3724 | | 22 | 54 | 2685 | 1237 | 54 | 2816 | 836 | | **5** | 55 | 2944 | 1677 |
| **269** | 64 | 2460 | 2762 | 64 | 2585 | 3738 | | 23 | 65 | 2717 | 1251 | 65 | 2849 | 849 | | **6** | 66 | 2978 | 1690 |
| **270** | 74 | 2490 | 2776 | 74 | 2616 | 3752 | | 24 | 75 | 2749 | 1265 | 75 | 2881 | 861 | | **7** | 76 | 3011 | 1702 |
| **271** | 85 | 2521 | 2792 | 85 | 2647 | 3767 | | 25 | 86 | 2781 | 1279 | 86 | 2914 | 875 | | **8** | 87 | 3044 | 1714 |
| **272** | 95 | 2551 | 2806 | 95 | 2678 | 3781 | | 26 | 96 | 2813 | 1292 | 96 | 2946 | 887 | | **9** | 97 | 3078 | 1726 |
| **273** | 100 | 2582 | 2816 | 100 | 2709 | 3790 | | 27 | 100 | 2845 | 1299 | 100 | 2979 | 894 | | **10** | 100 | 3111 | 1730 |
| **274** | 101 | 2612 | 2821 | 101 | 2740 | 3795 | | 28 | 101 | 2876 | 1303 | 101 | 3012 | 897 | | **11** | 101 | 3144 | 1733 |
| **275** | 107 | 2616 | 2831 | 107 | 2744 | 3805 | | 29 | 107 | 2879 | 1312 | 107 | 3014 | 905 | | **12** | 107 | 3146 | 1741 |
| **276** | 113 | 2621 | 2842 | 113 | 2748 | 3815 | Tamil, Tai; Malayalam, Magaram; Bengal, Magha. | 1 | 113 | 2882 | 1321 | 113 | 3017 | 914 | | **13** | 113 | 3147 | 1748 |
| **277** | 120 | 2625 | 2853 | 120 | 2752 | 3826 | | 2 | 119 | 2885 | 1330 | 119 | 3019 | 922 | | **14** | 119 | 3149 | 1756 |
| **278** | 126 | 2630 | 2864 | 126 | 2756 | 3836 | | 3 | 126 | 2888 | 1340 | 125 | 3022 | 931 | | **15** | 125 | 3150 | 1763 |
| **279** | 132 | 2634 | 2874 | 132 | 2759 | 3846 | | 4 | 132 | 2891 | 1349 | 131 | 3024 | 939 | | **16** | 131 | 3152 | 1771 |
| **280** | 139 | 2638 | 2885 | 139 | 2763 | 3856 | | 5 | 138 | 2895 | 1359 | 137 | 3026 | 947 | | **17** | 137 | 3154 | 1779 |
| **281** | 145 | 2643 | 2896 | 145 | 2767 | 3866 | | 6 | 144 | 2898 | 1368 | 143 | 3029 | 956 | | **18** | 143 | 3155 | 1786 |
| **282** | 151 | 2647 | 2906 | 151 | 2771 | 3876 | | 7 | 151 | 2901 | 1378 | 149 | 3031 | 964 | | **19** | 149 | 3157 | 1794 |
| **283** | 158 | 2652 | 2918 | 158 | 2775 | 3886 | | 8 | 157 | 2904 | 1387 | 155 | 3034 | 973 | | **20** | 155 | 3158 | 1801 |
| **284** | 164 | 2656 | 2928 | 164 | 2779 | 3896 | | 9 | 163 | 2907 | 1395 | 161 | 3036 | 961 | | **21** | 161 | 3160 | 1809 |
| **285** | 170 | 2660 | 2901 | 170 | 2783 | 3868 | | 10 | 169 | 2910 | 1365 | 167 | 3038 | 949 | | **22** | 167 | 3162 | 1776 |
| **286** | 177 | 2665 | 2875 | 177 | 2787 | 3840 | | 11 | 176 | 2913 | 1336 | 173 | 3041 | 918 | | **23** | 173 | 3163 | 1743 |
| **287** | 183 | 2669 | 2849 | 183 | 2791 | 3812 | | 12 | 182 | 2916 | 1306 | 179 | 3043 | 886 | | **24** | 179 | 3165 | 1709 |
| **288** | 189 | 2674 | 2822 | 189 | 2795 | 3783 | | 13 | 188 | 2919 | 1276 | 185 | 3046 | 854 | | **25** | 185 | 3166 | 1676 |
| **289** | 196 | 2678 | 2796 | 196 | 2798 | 3756 | | 14 | 194 | 2922 | 1246 | 191 | 3048 | 822 | | **26** | 191 | 3168 | 1643 |
| **290** | 202 | 2682 | 2769 | 202 | 2802 | 3728 | | 15 | 201 | 2926 | 1217 | 197 | 3050 | 791 | | **27** | 197 | 3170 | 1610 |
| **291** | 208 | 2687 | 2742 | 208 | 2806 | 3699 | | 16 | 207 | 2929 | 1187 | 203 | 3053 | 759 | | **28** | 203 | 3171 | 1577 |
| **292** | 215 | 2691 | 2717 | 215 | 2810 | 3672 | | 17 | 213 | 2932 | 1157 | 209 | 3055 | 727 | | **29** | 209 | 3173 | 1543 |
| **293** | 221 | 2696 | 2690 | 221 | 2814 | 3643 | | 18 | 219 | 2935 | 1127 | 215 | 3058 | 696 | | **30** | 215 | 3174 | 1510 |
| **294** | 227 | 2700 | 2663 | 227 | 2818 | 3615 | | 19 | 226 | 2938 | 1098 | 221 | 3060 | 664 | | **31** | 221 | 3176 | 1477 |
| **295** | 234 | 2704 | 2637 | 234 | 2822 | 3588 | | 20 | 232 | 2941 | 1068 | 227 | 3062 | 632 | February | **1** | 227 | 3178 | 1444 |
| **296** | 240 | 2709 | 2610 | 240 | 2826 | 3559 | | 21 | 238 | 2944 | 1038 | 233 | 3065 | 601 | | **2** | 233 | 3179 | 1411 |
| **297** | 246 | 2713 | 2584 | 246 | 2830 | 3531 | | 22 | 244 | 2947 | 1008 | 239 | 3067 | 569 | | **3** | 239 | 3181 | 1377 |
| **298** | 253 | 2718 | 2558 | 253 | 2834 | 3503 | | 23 | 251 | 2950 | 979 | 245 | 3070 | 537 | | **4** | 245 | 3182 | 1344 |
| **299** | 259 | 2722 | 2531 | 259 | 2838 | 3475 | | 24 | 257 | 2953 | 949 | 251 | 3072 | 505 | | **5** | 251 | 3184 | 1311 |
| **300** | 265 | 2726 | 2504 | 265 | 2841 | 3447 | | 25 | 263 | 2957 | 919 | 258 | 3074 | 475 | | **6** | 257 | 3186 | 1278 |
| **301** | 272 | 2731 | 2488 | 272 | 2845 | 3419 | | 26 | 269 | 2960 | 889 | 265 | 3077 | 444 | | **7** | 263 | 3187 | 1245 |
| **302** | 278 | 2735 | 2452 | 278 | 2849 | 3391 | | 27 | 276 | 2963 | 860 | 272 | 3079 | 413 | | **8** | 270 | 3189 | 1212 |
| **303** | 284 | 2740 | 2425 | 284 | 2853 | 3362 | | 28 | 282 | 2966 | 830 | 279 | 3082 | 383 | | **9** | 277 | 3190 | 1180 |
| **304** | 292 | 2744 | 2400 | 290 | 2856 | 3334 | | 29 | 288 | 2968 | 800 | 285 | 3084 | 351 | | **10** | 283 | 3192 | 1147 |
| **305** | 294 | 2711 | 2369 | 292 | 2822 | 3302 | | 30 | 290 | 2932 | 766 | 287 | 3046 | 315 | | **11** | 285 | 3153 | 1110 |

# TABLE XIII.

## Sunrise from 23° to 27° N. Latitude.

| Day of Solar year. | Lat. 23° −20″ Long. Eqn. of time in seconds. | ⊙'s Trop. Long. in seconds. | Total Corrn. in seconds. Indore. A.D. 1840—1920 | Lat. 24° −876″ Long. Eqn. of Time in seconds. | ⊙'s Trop. Long. in seconds. | Total Corrn. in seconds. Anhilwad. A.D. 1840—1920 | Solar months | Days of solar months. | Lat. 25° +1736″ Long. Eqn. of time in seconds. | ⊙'s Trop. Long. in seconds. | Total Corrn. in seconds. Benares. A.D. 1840—1920 | Lat. 26° +2264″ Long. Eqn. of Time in seconds. | ⊙'s Trop. Long. in seconds. | Total Corrn. in seconds. Patna A.D. 1840—1920 | Month | Eng. date | Lat. 27° +1544″ Long. Eqn. of Time in seconds. | ⊙'s Trop. Long. in seconds. | Total Corrn. in seconds. Oudh. A.D. 1840—1920 |
|---|---|---|---|---|---|---|---|---|---|---|---|---|---|---|---|---|---|---|---|
| **306** | −297 | −2678 | −2339 | −295 | −2787 | −3270 | Bengal, Phalguna; Malayalam, Kumbam; Tamil, Masi. | 1 | −293 | −2896 | −733 | −290 | −3009 | −281 | | **12** | −288 | −3114 | −1074 |
| **307** | 300 | 2646 | 2310 | 297 | 2753 | 3238 | | 2 | 295 | 2860 | 699 | 292 | 2971 | 245 | | **13** | 290 | 3074 | 1036 |
| **308** | 303 | 2613 | 2280 | 300 | 2718 | 3206 | | 3 | 298 | 2824 | 666 | 295 | 2933 | 210 | | **14** | 293 | 3035 | 1000 |
| **309** | 305 | 2580 | 2249 | 302 | 2684 | 3174 | | 4 | 300 | 2788 | 632 | 297 | 2895 | 173 | | **15** | 295 | 2996 | 963 |
| **310** | 308 | 2547 | 2219 | 305 | 2650 | 3143 | | 5 | 303 | 2752 | 599 | 300 | 2858 | 140 | | **16** | 298 | 2957 | 927 |
| **311** | 311 | 2514 | 2189 | 307 | 2615 | 3110 | | 6 | 305 | 2716 | 565 | 302 | 2820 | 104 | | **17** | 300 | 2918 | 890 |
| **312** | 314 | 2482 | 2160 | 310 | 2581 | 3081 | | 7 | 308 | 2680 | 532 | 305 | 2782 | 69 | February | **18** | 303 | 2878 | 853 |
| **313** | 316 | 2449 | 2129 | 312 | 2546 | 3046 | | 8 | 310 | 2644 | 498 | 307 | 2745 | 34 | | **19** | 305 | 2839 | 816 |
| **314** | 319 | 2416 | 2099 | 315 | 2512 | 3015 | | 9 | 313 | 2608 | 465 | 310 | 2707 | + 1 | | **20** | 308 | 2800 | 780 |
| **315** | 322 | 2383 | 2070 | 317 | 2478 | 2981 | | 10 | 315 | 2572 | 431 | 312 | 2669 | 36 | | **21** | 310 | 2761 | 742 |
| **316** | 325 | 2350 | 2015 | 320 | 2443 | 2924 | | 11 | 318 | 2536 | 372 | 315 | 2632 | 97 | | **22** | 313 | 2722 | 679 |
| **317** | 327 | 2318 | 1960 | 322 | 2409 | 2867 | | 12 | 320 | 2500 | 318 | 317 | 2594 | 159 | | **23** | 315 | 2682 | 615 |
| **318** | 330 | 2285 | 1905 | 325 | 2374 | 2810 | | 13 | 323 | 2464 | 254 | 320 | 2556 | 220 | | **24** | 318 | 2643 | 552 |
| **319** | 333 | 2252 | 1851 | 327 | 2340 | 2753 | | 14 | 325 | 2428 | 194 | 322 | 2518 | 282 | | **25** | 320 | 2604 | 488 |
| **320** | 336 | 2219 | 1796 | 330 | 2306 | 2696 | | 15 | 328 | 2392 | 135 | 325 | 2481 | 343 | | **26** | 323 | 2565 | 425 |
| **321** | 338 | 2196 | 1740 | 332 | 2271 | 2638 | | 16 | 330 | 2356 | 76 | 327 | 2443 | 404 | | **27** | 325 | 2526 | 360 |
| **322** | 341 | 2154 | 1686 | 335 | 2237 | 2582 | | 17 | 333 | 2320 | 17 | 330 | 2405 | 465 | | **28** | 328 | 2486 | 299 |
| **323** | 344 | 2121 | 1631 | 337 | 2202 | 2524 | | 18 | 335 | 2284 | + 43 | 332 | 2368 | 527 | | **1** | 330 | 2447 | 235 |
| **324** | 347 | 2088 | 1577 | 340 | 2168 | 2468 | | 19 | 338 | 2248 | 101 | 335 | 2330 | 588 | | **2** | 333 | 2408 | 172 |
| **325** | 349 | 2055 | 1521 | 342 | 2134 | 2410 | | 20 | 340 | 2212 | 161 | 337 | 2292 | 650 | | **3** | 335 | 2369 | 108 |
| **326** | 352 | 2022 | 1466 | 345 | 2099 | 2353 | | 21 | 343 | 2176 | 220 | 340 | 2255 | 711 | | **4** | 338 | 2330 | 45 |
| **327** | 355 | 1990 | 1412 | 347 | 2065 | 2296 | | 22 | 345 | 2140 | 279 | 342 | 2217 | 773 | | **5** | 340 | 2290 | + 19 |
| **328** | 358 | 1957 | 1357 | 350 | 2030 | 2239 | | 23 | 348 | 2104 | 338 | 345 | 2179 | 834 | | **6** | 343 | 2251 | 82 |
| **329** | 360 | 1924 | 1302 | 353 | 1996 | 2183 | | 24 | 350 | 2068 | 398 | 347 | 2141 | 896 | | **7** | 345 | 2212 | 146 |
| **330** | 363 | 1891 | 1247 | 356 | 1962 | 2126 | | 25 | 353 | 2032 | 457 | 350 | 2104 | 957 | | **8** | 348 | 2173 | 209 |
| **331** | 366 | 1858 | 1192 | 359 | 1927 | 2069 | | 26 | 355 | 1996 | 516 | 352 | 2066 | 1018 | | **9** | 350 | 2134 | 272 |
| **332** | 369 | 1826 | 1138 | 362 | 1893 | 2013 | | 27 | 358 | 1960 | 575 | 355 | 2028 | 1080 | | **10** | 353 | 2094 | 335 |
| **333** | 371 | 1793 | 1082 | 365 | 1858 | 1956 | | 28 | 361 | 1924 | 634 | 357 | 1991 | 1141 | | **11** | 355 | 2055 | 399 |
| **334** | 374 | 1760 | 1028 | 368 | 1824 | 1900 | | 29 | 364 | 1888 | 702 | 360 | 1953 | 1202 | | **12** | 357 | 2016 | 463 |
| **335** | 376 | 1728 | 972 | 371 | 1788 | 1843 | Bengal, Chaitra; Malayalam, Meenam; Tamil, Panguni. | 1 | 367 | 1852 | 751 | 363 | 1916 | 1263 | | **13** | 359 | 1976 | 527 |
| **336** | 376 | 1670 | 914 | 371 | 1728 | 1783 | | 2 | 367 | 1790 | 813 | 363 | 1852 | 1327 | | **14** | 359 | 1910 | 593 |
| **337** | 377 | 1613 | 858 | 372 | 1669 | 1725 | | 3 | 368 | 1729 | 873 | 364 | 1788 | 1390 | March | **15** | 360 | 1844 | 658 |
| **338** | 377 | 1555 | 800 | 372 | 1609 | 1665 | | 4 | 368 | 1667 | 935 | 364 | 1724 | 1454 | | **16** | 360 | 1778 | 724 |
| **339** | 378 | 1498 | 744 | 373 | 1550 | 1607 | | 5 | 369 | 1605 | 996 | 365 | 1660 | 1517 | | **17** | 361 | 1712 | 789 |
| **340** | 378 | 1440 | 680 | 373 | 1490 | 1547 | | 6 | 369 | 1543 | 1058 | 365 | 1596 | 1571 | | **18** | 361 | 1646 | 855 |
| **341** | 379 | 1382 | 629 | 374 | 1430 | 1488 | | 7 | 370 | 1482 | 1118 | 366 | 1533 | 1643 | | **19** | 362 | 1581 | 919 |
| **342** | 379 | 1325 | 572 | 374 | 1371 | 1429 | | 8 | 370 | 1420 | 1180 | 366 | 1469 | 1707 | | **20** | 362 | 1515 | 985 |
| **343** | 380 | 1267 | 515 | 375 | 1311 | 1370 | | 9 | 371 | 1358 | 1240 | 367 | 1405 | 1770 | | **21** | 363 | 1449 | 1050 |
| **344** | 380 | 1210 | 456 | 375 | 1252 | 1311 | | 10 | 371 | 1297 | 1300 | 367 | 1341 | 1830 | | **22** | 363 | 1383 | 1116 |
| **345** | 381 | 1152 | 401 | 376 | 1192 | 1252 | | 11 | 372 | 1235 | 1364 | 368 | 1277 | 1896 | | **23** | 364 | 1317 | 1180 |
| **346** | 381 | 1094 | 343 | 376 | 1132 | 1192 | | 12 | 372 | 1173 | 1426 | 368 | 1213 | 1960 | | **24** | 364 | 1251 | 1246 |
| **347** | 382 | 1037 | 287 | 377 | 1073 | 1134 | | 13 | 373 | 1112 | 1486 | 369 | 1149 | 2003 | | **25** | 365 | 1185 | 1311 |
| **348** | 382 | 979 | 229 | 377 | 1013 | 1074 | | 14 | 373 | 1050 | 1548 | 369 | 1085 | 2087 | | **26** | 365 | 1119 | 1377 |
| **349** | 383 | 922 | 173 | 378 | 954 | 1016 | | 15 | 374 | 988 | 1609 | 370 | 1021 | 2150 | | **27** | 366 | 1053 | 1442 |
| **350** | 383 | 864 | 115 | 378 | 894 | 956 | | 16 | 374 | 926 | 1670 | 370 | 957 | 2213 | | **28** | 366 | 987 | 1507 |
| **351** | 384 | 806 | 58 | 379 | 834 | 897 | | 17 | 375 | 865 | 1731 | 371 | 894 | 2276 | | **29** | 367 | 922 | 1572 |
| **352** | 384 | 749 | 1 | 379 | 775 | 838 | | 18 | 375 | 803 | 1793 | 371 | 829 | 2340 | | **30** | 367 | 856 | 1638 |
| **353** | 385 | 691 | + 56 | 380 | 715 | 779 | | 19 | 376 | 741 | 1854 | 372 | 766 | 2403 | | **31** | 368 | 790 | 1703 |
| **354** | 385 | 634 | 113 | 380 | 656 | 720 | | 20 | 376 | 670 | 1945 | 372 | 702 | 2467 | | **1** | 368 | 724 | 1769 |
| **355** | 386 | 576 | 170 | 381 | 596 | 661 | | 21 | 377 | 618 | 1976 | 373 | 638 | 2530 | | **2** | 369 | 658 | 1834 |
| **356** | 386 | 518 | 225 | 381 | 536 | 601 | | 22 | 377 | 556 | 2038 | 373 | 574 | 2594 | | **3** | 369 | 592 | 1900 |
| **357** | 387 | 461 | 284 | 382 | 477 | 543 | | 23 | 378 | 495 | 2098 | 374 | 510 | 2657 | | **4** | 370 | 526 | 1965 |
| **358** | 387 | 403 | 342 | 382 | 417 | 483 | | 24 | 378 | 433 | 2160 | 374 | 446 | 2721 | | **5** | 370 | 460 | 2031 |
| **359** | 388 | 346 | 398 | 383 | 358 | 425 | | 25 | 379 | 371 | 2221 | 375 | 382 | 2784 | | **6** | 371 | 394 | 2096 |
| **360** | 388 | 288 | 456 | 383 | 298 | 365 | | 26 | 379 | 309 | 2282 | 375 | 318 | 2847 | | **7** | 371 | 328 | 2161 |
| **361** | 389 | 230 | 513 | 384 | 238 | 306 | | 27 | 380 | 248 | 2243 | 376 | 255 | 2910 | April | **8** | 372 | 263 | 2226 |
| **362** | 389 | 173 | 570 | 384 | 179 | 247 | | 28 | 380 | 186 | 2405 | 376 | 191 | 2974 | | **9** | 372 | 197 | 2292 |
| **363** | 390 | 115 | 627 | 385 | 119 | 188 | | 29 | 381 | 125 | 2466 | 377 | 127 | 3037 | | **10** | 373 | 131 | 2357 |
| **364** | 390 | 58 | 684 | 385 | 60 | 129 | | 30 | 381 | 63 | 2527 | 377 | 64 | 3101 | | **11** | 373 | 65 | 2423 |
| **365** | 391 | 0 | 741 | 386 | 0 | 70 | | 31 | 382 | 0 | 2588 | 378 | 0 | 3164 | | **12** | 374 | 0 | 2488 |

# TABLE XIII.

## Sunrise from 28° to 32° N. Latitude.

| Day of Solar year. | Lat. 28° +460″ Long. Eqn. of time in seconds. | ☉'s Trop. Long. in seconds. | Total Corrn. in seconds. Mathura. A.D. 1840—1920 | Lat. 29° +352″ Long. Eqn. of time in seconds. | ☉'s Trop. Long. in seconds. | Total Corrn. in seconds. Delhi. A.D. 1840—1920. | Days of solar months. | Lat. 30° +152″ Long. Eqn. of time in seconds. | ☉'s Trop. Long. in seconds. | Total Corrn. in seconds. Patiala. A.D. 1840—1920 | Lat. 31° +332″ Long. Eqn. of time in seconds. | ☉'s Trop. Long. in seconds. | Total Corrn. in seconds. Simla. A.D. 1840—1920 | Eng. date. | Lat. 32° −348″ Long. Eqn. of time in seconds. | ☉'s Trop. Long. in seconds. | Total Corrn. in seconds. Lahore. A.D. 1840—1920 |
|---|---|---|---|---|---|---|---|---|---|---|---|---|---|---|---|---|---|
| 1 | −367 | + 68 | +1523 | −362 | + 70 | +1462 | *Tamil*, Chittirai; *Malayalam*, Medam; *Bengal*, Vaisakha. 1 | −357 | + 72 | +1315 | −353 | + 75 | +1544 | April 13 | −348 | + 77 | + 917 |
| 2 | 364 | 136 | 1594 | 360 | 140 | 1534 | 2 | 355 | 145 | 1390 | 351 | 149 | 1620 | 14 | 346 | 154 | 998 |
| 3 | 363 | 204 | 1663 | 359 | 210 | 1605 | 3 | 354 | 217 | 1463 | 350 | 224 | 1696 | 15 | 345 | 230 | 1073 |
| 4 | 362 | 272 | 1732 | 358 | 280 | 1676 | 4 | 353 | 290 | 1537 | 348 | 298 | 1772 | 16 | 344 | 307 | 1151 |
| 5 | 360 | 341 | 1803 | 356 | 351 | 1748 | 5 | 351 | 362 | 1611 | 347 | 373 | 1848 | 17 | 342 | 384 | 1230 |
| 6 | 359 | 409 | 1872 | 354 | 421 | 1820 | 6 | 349 | 434 | 1685 | 345 | 447 | 1924 | 18 | 340 | 461 | 1309 |
| 7 | 357 | 477 | 1942 | 352 | 491 | 1893 | 7 | 348 | 507 | 1759 | 343 | 522 | 2001 | 19 | 339 | 538 | 1387 |
| 8 | 355 | 545 | 2012 | 351 | 561 | 1964 | 8 | 346 | 580 | 1833 | 341 | 596 | 2077 | 20 | 337 | 614 | 1465 |
| 9 | 353 | 613 | 2082 | 348 | 631 | 2037 | 9 | 344 | 652 | 1907 | 339 | 671 | 2154 | 21 | 335 | 691 | 1544 |
| 10 | 351 | 681 | 2153 | 346 | 701 | 2110 | 10 | 342 | 724 | 1982 | 337 | 745 | 2231 | 22 | 333 | 768 | 1623 |
| 11 | 348 | 749 | 2197 | 344 | 771 | 2155 | 11 | 339 | 796 | 2029 | 335 | 820 | 2280 | 23 | 330 | 845 | 1674 |
| 12 | 346 | 817 | 2240 | 342 | 841 | 2199 | 12 | 337 | 869 | 2076 | 333 | 894 | 2328 | 24 | 328 | 922 | 1725 |
| 13 | 344 | 885 | 2283 | 339 | 911 | 2245 | 13 | 335 | 941 | 2122 | 331 | 969 | 2374 | 25 | 326 | 998 | 1775 |
| 14 | 341 | 953 | 2327 | 337 | 981 | 2290 | 14 | 332 | 1014 | 2170 | 328 | 1043 | 2426 | 26 | 324 | 1075 | 1866 |
| 15 | 338 | 1022 | 2370 | 334 | 1051 | 2335 | 15 | 329 | 1086 | 2217 | 325 | 1118 | 2476 | 27 | 321 | 1152 | 1877 |
| 16 | 335 | 1090 | 2414 | 331 | 1122 | 2381 | 16 | 327 | 1158 | 2264 | 323 | 1192 | 2525 | 28 | 318 | 1229 | 1929 |
| 17 | 333 | 1158 | 2457 | 328 | 1192 | 2427 | 17 | 324 | 1231 | 2311 | 320 | 1267 | 2574 | 29 | 316 | 1306 | 1979 |
| 18 | 330 | 1226 | 2501 | 324 | 1262 | 2474 | 18 | 322 | 1303 | 2358 | 318 | 1341 | 2623 | 30 | 314 | 1382 | 2030 |
| 19 | 327 | 1294 | 2545 | 322 | 1332 | 2518 | 19 | 320 | 1376 | 2404 | 316 | 1416 | 2671 | May 1 | 312 | 1459 | 2080 |
| 20 | 325 | 1362 | 2588 | 320 | 1402 | 2563 | 20 | 318 | 1448 | 2451 | 314 | 1490 | 2720 | 2 | 310 | 1536 | 2131 |
| 21 | 323 | 1430 | 2631 | 318 | 1472 | 2608 | 21 | 316 | 1520 | 2497 | 312 | 1565 | 2769 | 3 | 308 | 1613 | 2181 |
| 22 | 321 | 1498 | 2674 | 316 | 1542 | 2652 | 22 | 314 | 1593 | 2544 | 310 | 1639 | 2817 | 4 | 306 | 1690 | 2232 |
| 23 | 319 | 1566 | 2717 | 314 | 1612 | 2697 | 23 | 312 | 1665 | 2590 | 308 | 1714 | 2866 | 5 | 304 | 1766 | 2282 |
| 24 | 317 | 1634 | 2760 | 312 | 1682 | 2742 | 24 | 310 | 1738 | 2637 | 306 | 1788 | 2914 | 6 | 302 | 1843 | 2333 |
| 25 | 315 | 1703 | 2802 | 310 | 1752 | 2786 | 25 | 308 | 1810 | 2683 | 304 | 1863 | 2963 | 7 | 300 | 1920 | 2383 |
| 26 | 313 | 1771 | 2845 | 308 | 1822 | 2831 | 26 | 306 | 1882 | 2740 | 302 | 1937 | 3012 | 8 | 298 | 1997 | 2434 |
| 27 | 311 | 1839 | 2888 | 306 | 1893 | 2876 | 27 | 304 | 1955 | 2776 | 300 | 2012 | 3060 | 9 | 296 | 2074 | 2484 |
| 28 | 309 | 1907 | 2931 | 304 | 1963 | 2921 | 28 | 302 | 2027 | 2823 | 298 | 2086 | 3117 | 10 | 294 | 2150 | 2535 |
| 29 | 307 | 1975 | 2974 | 302 | 2033 | 2965 | 29 | 300 | 2099 | 2869 | 296 | 2161 | 3157 | 11 | 292 | 2227 | 2585 |
| 30 | 305 | 2044 | 3017 | 300 | 2104 | 3010 | 30 | 298 | 2172 | 2916 | 294 | 2236 | 3206 | 12 | 290 | 2304 | 2636 |
| 31 | 303 | 2085 | 3060 | 298 | 2147 | 3055 | *Tamil*, Vaikasi; *Malayalam*, Edavam; *Bengal*, Jyeshta. 1 | 296 | 2216 | 2962 | 292 | 2283 | 3255 | 13 | 288 | 2352 | 2686 |
| 32 | 301 | 2126 | 3103 | 296 | 2189 | 3099 | 2 | 294 | 2261 | 3009 | 290 | 2329 | 3303 | 14 | 286 | 2401 | 2737 |
| 33 | 299 | 2167 | 3146 | 294 | 2232 | 3144 | 3 | 292 | 2305 | 3055 | 288 | 2373 | 3352 | 15 | 284 | 2449 | 2788 |
| 34 | 297 | 2208 | 3189 | 292 | 2275 | 3189 | 4 | 290 | 2350 | 4002 | 286 | 2422 | 3400 | 16 | 282 | 2498 | 2838 |
| 35 | 295 | 2248 | 3231 | 290 | 2317 | 3233 | 5 | 288 | 2394 | 3148 | 284 | 2469 | 3449 | 17 | 280 | 2546 | 2888 |
| 36 | 293 | 2289 | 3274 | 288 | 2360 | 3278 | 6 | 286 | 2439 | 3195 | 282 | 2516 | 3498 | 18 | 278 | 2595 | 2939 |
| 37 | 291 | 2330 | 3317 | 286 | 2403 | 3301 | 7 | 284 | 2483 | 3241 | 280 | 2562 | 3546 | 19 | 276 | 2648 | 2989 |
| 38 | 289 | 2371 | 3360 | 284 | 2446 | 3368 | 8 | 282 | 2528 | 3288 | 278 | 2609 | 3593 | 20 | 274 | 2692 | 3040 |
| 39 | 284 | 2412 | 3406 | 281 | 2488 | 3413 | 9 | 278 | 2572 | 3336 | 275 | 2655 | 3644 | 21 | 272 | 2740 | 3090 |
| 40 | 278 | 2453 | 3453 | 275 | 2531 | 3462 | 10 | 273 | 2617 | 3386 | 270 | 2702 | 3694 | 22 | 267 | 2789 | 3144 |
| 41 | 273 | 2494 | 3499 | 270 | 2574 | 3510 | 11 | 267 | 2661 | 3437 | 264 | 2749 | 3748 | 23 | 262 | 2837 | 3198 |
| 42 | 266 | 2535 | 3505 | 264 | 2616 | 3516 | 12 | 261 | 2706 | 3444 | 258 | 2795 | 3755 | 24 | 256 | 2886 | 3206 |
| 43 | 261 | 2576 | 3509 | 258 | 2659 | 3522 | 13 | 256 | 2750 | 3449 | 253 | 2842 | 3762 | 25 | 250 | 2934 | 3215 |
| 44 | 255 | 2617 | 3514 | 252 | 2702 | 3528 | 14 | 250 | 2795 | 3456 | 247 | 2888 | 3769 | 26 | 245 | 2988 | 3222 |
| 45 | 249 | 2657 | 3519 | 246 | 2744 | 3533 | 15 | 244 | 2839 | 3463 | 241 | 2935 | 3777 | 27 | 239 | 3031 | 3230 |
| 46 | 243 | 2698 | 3523 | 241 | 2787 | 3538 | 16 | 238 | 2884 | 3469 | 236 | 2982 | 3783 | 28 | 233 | 3080 | 3238 |
| 47 | 237 | 2739 | 3528 | 234 | 2830 | 3545 | 17 | 232 | 2928 | 3476 | 229 | 3028 | 3792 | 29 | 227 | 3128 | 3247 |
| 48 | 231 | 2780 | 3533 | 228 | 2873 | 3551 | 18 | 226 | 2973 | 3483 | 223 | 3075 | 3799 | 30 | 221 | 3177 | 3255 |
| 49 | 224 | 2821 | 3539 | 222 | 2915 | 3557 | 19 | 220 | 3017 | 3490 | 217 | 3121 | 3807 | 31 | 215 | 3225 | 3263 |
| 50 | 218 | 2862 | 3544 | 216 | 2958 | 3563 | 20 | 214 | 3062 | 3496 | 211 | 3168 | 3814 | June 1 | 209 | 3274 | 3272 |
| 51 | 211 | 2903 | 3550 | 209 | 3001 | 3570 | 21 | 207 | 3106 | 3504 | 205 | 3215 | 3822 | 2 | 203 | 3322 | 3280 |
| 52 | 205 | 2944 | 3555 | 203 | 3043 | 3576 | 22 | 201 | 3151 | 3511 | 199 | 3261 | 3829 | 3 | 197 | 3371 | 3288 |
| 53 | 198 | 2985 | 3561 | 196 | 3086 | 3582 | 23 | 194 | 3195 | 3518 | 192 | 3308 | 3838 | 4 | 190 | 3419 | 3298 |
| 54 | 192 | 3026 | 3566 | 190 | 3129 | 3588 | 24 | 188 | 3240 | 3525 | 186 | 3354 | 3845 | 5 | 184 | 3468 | 3306 |
| 55 | 185 | 3066 | 3572 | 183 | 3171 | 3595 | 25 | 181 | 3284 | 3533 | 179 | 3401 | 3854 | 6 | 178 | 3516 | 3314 |
| 56 | 182 | 3107 | 3573 | 181 | 3214 | 3597 | 26 | 179 | 3329 | 3535 | 176 | 3448 | 3858 | 7 | 174 | 3565 | 3320 |
| 57 | 178 | 3148 | 3576 | 177 | 3257 | 3601 | 27 | 175 | 3373 | 3540 | 173 | 3494 | 3863 | 8 | 170 | 3613 | 3327 |
| 58 | 174 | 3189 | 3579 | 173 | 3300 | 3605 | 28 | 171 | 3418 | 3545 | 170 | 3541 | 3867 | 9 | 166 | 3662 | 3333 |
| 59 | 170 | 3230 | 3582 | 169 | 3342 | 3609 | 29 | 167 | 3462 | 3550 | 167 | 3587 | 3870 | 10 | 162 | 3710 | 3339 |
| 60 | 166 | 3271 | 3585 | 165 | 3385 | 3612 | 30 | 163 | 3507 | 3554 | 164 | 3634 | 3876 | 11 | 158 | 3759 | 3346 |
| 61 | 162 | 3312 | 3588 | 161 | 3428 | 3616 | 31 | 159 | 3552 | 3559 | 161 | 3680 | 3881 | 12 | 154 | 3808 | 3352 |
| 62 | 158 | 3311 | 3591 | 157 | 3428 | 3620 | 32 | 155 | 3553 | 3564 | 158 | 3681 | 3885 | 13 | 150 | 3810 | 3358 |

# TABLE XIII.

## Sunrise from 28° to 32° N. Latitude.

| Day of Solar year | Lat. 28° +460″ Long. Eqn. of time in seconds. | ⊙'s Trop. Long. in seconds. | Total Corrn. in seconds. Mathura. A.D. 1840—1920 | Lat. 29° +352″ Long. Eqn. of time in seconds. | ⊙'s Trop. Long. in seconds. | Total Corrn. in seconds. Delhi. A.D. 1840—1920 | Days of solar months. | Lat. 30° +152″ Long. Eqn. of time in seconds. | ⊙'s Trop. Long. in seconds. | Total Corrn. in seconds. Patiala. A.D. 1840—1920 | Lat. 31° +332″ Long. Eqn. of time in seconds. | ⊙'s Trop. Long. in seconds. | Total Corrn. in seconds. Simla. A.D. 1840—1920 | Eng. date. | Lat. 32° —348″ Long. Eqn. of time in seconds. | ⊙'s Trop. Long. in seconds. | Total Corrn. in seconds. Lahore. A.D. 1840—1920 |
|---|---|---|---|---|---|---|---|---|---|---|---|---|---|---|---|---|---|
| | | | | | | | Tamil, Ani; *Malayalam*, Mithunam; *Bengal*, Ashada. | | | | | | | | | | |
| 63 | —154 | +3310 | +3594 | —153 | +3428 | +3624 | 1 | —151 | +3553 | +3567 | —153 | +3683 | +3890 | June 14 | —146 | +3813 | +3365 |
| 64 | 150 | 3309 | 3597 | 149 | 3428 | 3628 | 2 | 147 | 3554 | 3573 | 146 | 3684 | 3899 | 15 | 142 | 3815 | 3371 |
| 65 | 141 | 3308 | 3605 | 140 | 3427 | 3637 | 3 | 140 | 3555 | 3581 | 139 | 3686 | 3909 | 16 | 139 | 3817 | 3376 |
| 66 | 132 | 3306 | 3612 | 131 | 3427 | 3646 | 4 | 131 | 3555 | 3590 | 131 | 3687 | 3918 | 17 | 130 | 3819 | 3387 |
| 67 | 123 | 3305 | 3620 | 123 | 3427 | 3654 | 5 | 122 | 3556 | 3600 | 122 | 3689 | 3929 | 18 | 121 | 3822 | 3399 |
| 68 | 114 | 3304 | 3628 | 114 | 3427 | 3663 | 6 | 113 | 3557 | 3609 | 113 | 3690 | 3939 | 19 | 113 | 3824 | 3409 |
| 69 | 105 | 3303 | 3636 | 105 | 3427 | 3672 | 7 | 104 | 3558 | 3619 | 104 | 3692 | 3950 | 20 | 104 | 3826 | 3420 |
| 70 | 96 | 3302 | 3644 | 96 | 3427 | 3681 | 8 | 96 | 3558 | 3628 | 95 | 3693 | 3960 | 21 | 95 | 3829 | 3432 |
| 71 | 87 | 3301 | 3653 | 87 | 3427 | 3690 | 9 | 87 | 3559 | 3637 | 86 | 3695 | 3970 | 22 | 86 | 3831 | 3442 |
| 72 | 78 | 3300 | 3628 | 78 | 3427 | 3663 | 10 | 78 | 3560 | 3610 | 77 | 3696 | 3943 | 23 | 77 | 3833 | 3414 |
| 73 | 69 | 3299 | 3603 | 69 | 3426 | 3638 | 11 | 69 | 3560 | 3584 | 69 | 3698 | 3915 | 24 | 68 | 3836 | 3386 |
| 74 | 60 | 3298 | 3578 | 60 | 3426 | 3612 | 12 | 59 | 3561 | 3558 | 59 | 3699 | 3888 | 25 | 59 | 3838 | 3358 |
| 75 | 52 | 3297 | 3552 | 52 | 3426 | 3585 | 13 | 52 | 3562 | 3530 | 51 | 3701 | 3860 | 26 | 51 | 3840 | 3329 |
| 76 | 41 | 3295 | 3529 | 41 | 3426 | 3561 | 14 | 41 | 3562 | 3505 | 41 | 3702 | 3834 | 27 | 41 | 3842 | 3302 |
| 77 | 32 | 3294 | 3505 | 32 | 3426 | 3536 | 15 | 32 | 3563 | 3479 | 32 | 3704 | 3807 | 28 | 32 | 3845 | 3274 |
| 78 | 23 | 3293 | 3480 | 23 | 3426 | 3510 | 16 | 23 | 3564 | 3452 | 23 | 3705 | 3780 | 29 | 23 | 3847 | 3246 |
| 79 | 14 | 3292 | 3455 | 14 | 3426 | 3484 | 17 | 14 | 3565 | 3426 | 14 | 3707 | 3752 | 30 | 14 | 3849 | 3218 |
| 80 | 5 | 3291 | 3430 | 5 | 3425 | 3459 | 18 | 5 | 3565 | 3399 | 5 | 3708 | 3725 | July 1 | 5 | 3852 | 3190 |
| 81 | + 4 | 3290 | 3405 | + 4 | 3425 | 3433 | 19 | + 4 | 3566 | 3373 | + 4 | 3710 | 3698 | 2 | + 4 | 3854 | 3164 |
| 82 | 13 | 3289 | 3380 | 13 | 3425 | 3407 | 20 | 13 | 3567 | 3346 | 13 | 3711 | 3672 | 3 | 13 | 3856 | 3134 |
| 83 | 22 | 3288 | 3355 | 22 | 3425 | 3382 | 21 | 22 | 3567 | 3320 | 22 | 3713 | 3644 | 4 | 22 | 3859 | 3106 |
| 84 | 32 | 3287 | 3331 | 32 | 3425 | 3357 | 22 | 31 | 3568 | 3293 | 31 | 3714 | 3616 | 5 | 31 | 3861 | 3078 |
| 85 | 41 | 3286 | 3306 | 41 | 3425 | 3331 | 23 | 41 | 3569 | 3268 | 41 | 3716 | 3590 | 6 | 41 | 3863 | 3051 |
| 86 | 50 | 3284 | 3281 | 50 | 3425 | 3305 | 24 | 50 | 3569 | 3241 | 49 | 3717 | 3562 | 7 | 49 | 3865 | 3022 |
| 87 | 59 | 3283 | 3257 | 59 | 3425 | 3280 | 25 | 58 | 3570 | 3214 | 58 | 3719 | 3535 | 8 | 58 | 3868 | 2994 |
| 88 | 68 | 3282 | 3232 | 68 | 3424 | 3254 | 26 | 68 | 3570 | 3188 | 67 | 3720 | 3508 | 9 | 67 | 3870 | 2966 |
| 89 | 77 | 3281 | 3207 | 77 | 3424 | 3228 | 27 | 77 | 3571 | 3162 | 76 | 3722 | 3480 | 10 | 76 | 3872 | 2938 |
| 90 | 86 | 3280 | 3182 | 86 | 3424 | 3203 | 28 | 85 | 3572 | 3134 | 85 | 3723 | 3453 | 11 | 85 | 3875 | 2910 |
| 91 | 109 | 3280 | 3171 | 109 | 3424 | 3191 | 29 | 109 | 3572 | 3123 | 110 | 3724 | 3442 | 12 | 110 | 3876 | 2898 |
| 92 | 119 | 3246 | 3147 | 119 | 3389 | 3166 | 30 | 120 | 3536 | 3098 | 120 | 3688 | 3416 | 13 | 120 | 3839 | 2871 |
| 93 | 129 | 3212 | 3123 | 129 | 3355 | 3142 | 31 | 130 | 3501 | 3073 | 130 | 3652 | 3390 | 14 | 131 | 3802 | 2845 |
| 94 | 139 | 3178 | 3099 | 140 | 3320 | 3118 | 32 | 140 | 3465 | 3047 | 140 | 3615 | 3363 | 15 | 141 | 3765 | 2818 |
| | | | | | | | Tamil, Adi; *Malayalam*, Karkatagam; *Bengal*, Sravana. | | | | | | | | | | |
| 95 | 149 | 3144 | 3075 | 150 | 3285 | 3093 | 1 | 150 | 3430 | 3022 | 150 | 3579 | 3337 | 16 | 151 | 3728 | 2791 |
| 96 | 159 | 3110 | 3051 | 160 | 3250 | 3068 | 2 | 160 | 3394 | 2996 | 161 | 3543 | 3312 | 17 | 161 | 3691 | 2764 |
| 97 | 169 | 3077 | 3028 | 170 | 3216 | 3044 | 3 | 170 | 3359 | 2971 | 171 | 3507 | 3286 | 18 | 171 | 3654 | 2737 |
| 98 | 179 | 3043 | 3004 | 180 | 3181 | 3019 | 4 | 180 | 3323 | 2945 | 181 | 3471 | 3260 | 19 | 181 | 3617 | 2710 |
| 99 | 189 | 3009 | 2980 | 190 | 3146 | 2994 | 5 | 191 | 3288 | 2921 | 191 | 3434 | 3233 | 20 | 192 | 3580 | 2684 |
| 100 | 199 | 2975 | 2956 | 199 | 3112 | 2969 | 6 | 200 | 3252 | 2894 | 201 | 3398 | 3207 | 21 | 201 | 3543 | 2656 |
| 101 | 209 | 2941 | 2932 | 209 | 3077 | 2944 | 7 | 210 | 3217 | 2869 | 210 | 3362 | 3180 | 22 | 211 | 3508 | 2629 |
| 102 | 218 | 2907 | 2906 | 219 | 3042 | 2919 | 8 | 220 | 3181 | 2844 | 220 | 3326 | 3152 | 23 | 221 | 3469 | 2601 |
| 103 | 228 | 2873 | 2875 | 229 | 3008 | 2886 | 9 | 229 | 3146 | 2808 | 230 | 3290 | 3115 | 24 | 231 | 3432 | 2562 |
| 104 | 238 | 2839 | 2844 | 238 | 2973 | 2852 | 10 | 239 | 3110 | 2774 | 240 | 3253 | 3078 | 25 | 241 | 3395 | 2523 |
| 105 | 247 | 2805 | 2812 | 248 | 2938 | 2820 | 11 | 249 | 3075 | 2739 | 249 | 3217 | 3041 | 26 | 250 | 3358 | 2484 |
| 106 | 257 | 2771 | 2781 | 257 | 2903 | 2786 | 12 | 258 | 3039 | 2703 | 259 | 3181 | 3004 | 27 | 260 | 3321 | 2445 |
| 107 | 266 | 2738 | 2749 | 266 | 2869 | 2752 | 13 | 267 | 3004 | 2667 | 268 | 3145 | 2966 | 28 | 269 | 3284 | 2405 |
| 108 | 275 | 2704 | 2718 | 276 | 2834 | 2719 | 14 | 277 | 2968 | 2633 | 277 | 3109 | 2928 | 29 | 278 | 3247 | 2363 |
| 109 | 284 | 2670 | 2686 | 285 | 2799 | 2685 | 15 | 286 | 2933 | 2597 | 287 | 3072 | 2891 | 30 | 287 | 3210 | 2325 |
| 110 | 293 | 2636 | 2654 | 294 | 2765 | 2652 | 16 | 295 | 2897 | 2561 | 296 | 3036 | 2854 | 31 | 296 | 3173 | 2286 |
| 111 | 302 | 2602 | 2622 | 303 | 2730 | 2618 | 17 | 304 | 2862 | 2526 | 304 | 3000 | 2815 | August 1 | 305 | 3136 | 2246 |
| 112 | 311 | 2568 | 2590 | 312 | 2695 | 2584 | 18 | 313 | 2826 | 2490 | 314 | 2964 | 2778 | 2 | 314 | 3099 | 2206 |
| 113 | 320 | 2534 | 2558 | 320 | 2661 | 2549 | 19 | 321 | 2791 | 2453 | 322 | 2928 | 2739 | 3 | 323 | 3062 | 2166 |
| 114 | 329 | 2500 | 2526 | 330 | 2626 | 2516 | 20 | 330 | 2755 | 2418 | 331 | 2891 | 2701 | 4 | 332 | 3025 | 2126 |
| 115 | 337 | 2466 | 2493 | 338 | 2591 | 2482 | 21 | 339 | 2720 | 2382 | 340 | 2855 | 2664 | 5 | 341 | 2988 | 2087 |
| 116 | 345 | 2432 | 2460 | 346 | 2556 | 2447 | 22 | 347 | 2684 | 2345 | 348 | 2819 | 2625 | 6 | 349 | 2951 | 2046 |
| 117 | 353 | 2399 | 2437 | 354 | 2522 | 2412 | 23 | 355 | 2649 | 2308 | 357 | 2783 | 2587 | 7 | 358 | 2914 | 2006 |
| 118 | 362 | 2365 | 2396 | 363 | 2487 | 2378 | 24 | 364 | 2613 | 2273 | 365 | 2747 | 2548 | 8 | 366 | 2877 | 1965 |
| 119 | 370 | 2331 | 2363 | 371 | 2452 | 2343 | 25 | 372 | 2578 | 2236 | 373 | 2710 | 2509 | 9 | 374 | 2840 | 1924 |
| 120 | 378 | 2297 | 2330 | 379 | 2418 | 2309 | 26 | 380 | 2542 | 2199 | 381 | 2674 | 2471 | 10 | 382 | 2803 | 1884 |
| 121 | 394 | 2263 | 2305 | 396 | 2383 | 2283 | 27 | 399 | 2507 | 2175 | 402 | 2638 | 2445 | 11 | 390 | 2766 | 1843 |
| 122 | 402 | 2228 | 2272 | 405 | 2348 | 2249 | 28 | 407 | 2472 | 2137 | 410 | 2600 | 2406 | 12 | 413 | 2728 | 1817 |
| 123 | 409 | 2187 | 2238 | 413 | 2305 | 2214 | 29 | 415 | 2427 | 2100 | 418 | 2553 | 2367 | 13 | 421 | 2679 | 1776 |
| 124 | 417 | 2146 | 2205 | 421 | 2262 | 2179 | 30 | 423 | 2383 | 2064 | 426 | 2506 | 2328 | 14 | 429 | 2630 | 1735 |
| 125 | 424 | 2105 | 2171 | 428 | 2220 | 2144 | 31 | 430 | 2338 | 2026 | 433 | 2460 | 2289 | 15 | 436 | 2582 | 1694 |

# TABLE XIII.

## Sunrise from 28° to 32° N. Latitude.

| Day of Solar year. | Lat. 28° Eqn. of time in seconds. | ☉'s Trop. Long. in seconds. | +460″ Long. Total Corrn. in seconds. Mathura. A.D. 1840—1920 | Lat. 29° Eqn. of time in seconds. | ☉'s Trop. Long. in seconds. | +352″ Long. Total Corrn. in seconds. Delhi. A.D. 1840—1920 | Days of solar months. | Lat. 30° Eqn. of time in seconds. | ☉'s Trop. Long. in seconds. | +152″ Long. Total Corrn. in seconds. Patiala. A.D. 1840—1920 | Lat. 31° Eqn. of time in seconds. | ☉'s Trop. Long. in seconds. | +332″ Long. Total Corrn. in seconds. Simla. A.D. 1840—1920 | Eng. date. | Lat. 32° Eqn. of time in seconds. | ☉'s Trop. Long. in seconds. | —348″ Long. Total Corrn. in seconds. Lahore. A.D. 1840—1920 |
|---|---|---|---|---|---|---|---|---|---|---|---|---|---|---|---|---|---|
| | | | | | | | *Tamil*, Avani; *Malayalam*, Chingam; *Bengal*, Bhadrapada. | | | | | | | August | | | |
| 126 | +433 | +2064 | +2139 | +436 | +2177 | +2109 | 1 | +438 | +2298 | +1989 | +441 | +2413 | +2250 | 16 | +444 | +2533 | +1653 |
| 127 | 440 | 2023 | 2105 | 444 | 2134 | 2074 | 2 | 445 | 2248 | 1952 | 449 | 2366 | 2211 | 17 | 452 | 2484 | 1612 |
| 128 | 447 | 1983 | 2072 | 451 | 2091 | 2038 | 3 | 453 | 2204 | 1915 | 456 | 2319 | 2171 | 18 | 459 | 2435 | 1570 |
| 129 | 454 | 1942 | 2038 | 458 | 2048 | 2002 | 4 | 460 | 2159 | 1877 | 463 | 2272 | 2131 | 19 | 466 | 2386 | 1528 |
| 130 | 461 | 1901 | 2004 | 465 | 2006 | 1967 | 5 | 467 | 2114 | 1839 | 470 | 2226 | 2092 | 20 | 473 | 2338 | 1487 |
| 131 | 467 | 1860 | 1969 | 471 | 1963 | 1930 | 6 | 474 | 2070 | 1802 | 477 | 2179 | 2052 | 21 | 480 | 2289 | 1446 |
| 132 | 473 | 1819 | 1933 | 477 | 1920 | 1893 | 7 | 481 | 2025 | 1765 | 484 | 2132 | 2012 | 22 | 487 | 2240 | 1403 |
| 133 | 480 | 1778 | 1908 | 484 | 1877 | 1856 | 8 | 488 | 1980 | 1735 | 491 | 2085 | 1980 | 23 | 494 | 2191 | 1369 |
| 134 | 486 | 1737 | 1882 | 490 | 1834 | 1837 | 9 | 494 | 1936 | 1705 | 497 | 2038 | 1948 | 24 | 500 | 2142 | 1334 |
| 135 | 492 | 1696 | 1855 | 496 | 1792 | 1809 | 10 | 500 | 1891 | 1674 | 503 | 1992 | 1915 | 25 | 506 | 2094 | 1300 |
| 136 | 498 | 1655 | 1829 | 502 | 1749 | 1781 | 11 | 506 | 1846 | 1644 | 509 | 1945 | 1883 | 26 | 512 | 2045 | 1265 |
| 137 | 504 | 1614 | 1803 | 508 | 1706 | 1752 | 12 | 512 | 1801 | 1613 | 515 | 1898 | 1850 | 27 | 518 | 1996 | 1230 |
| 138 | 510 | 1574 | 1777 | 514 | 1663 | 1724 | 13 | 518 | 1757 | 1583 | 521 | 1851 | 1817 | 28 | 524 | 1947 | 1195 |
| 139 | 516 | 1533 | 1751 | 520 | 1620 | 1696 | 14 | 524 | 1712 | 1552 | 527 | 1804 | 1785 | 29 | 530 | 1898 | 1160 |
| 140 | 521 | 1492 | 1723 | 525 | 1578 | 1667 | 15 | 529 | 1667 | 1521 | 532 | 1758 | 1751 | 30 | 535 | 1850 | 1125 |
| 141 | 526 | 1451 | 1696 | 530 | 1535 | 1637 | 16 | 534 | 1624 | 1489 | 538 | 1711 | 1719 | 31 | 541 | 1801 | 1090 |
| | | | | | | | | | | | | | | September | | | |
| 142 | 531 | 1410 | 1669 | 535 | 1492 | 1608 | 17 | 539 | 1578 | 1458 | 543 | 1664 | 1685 | 1 | 546 | 1752 | 1054 |
| 143 | 536 | 1369 | 1642 | 540 | 1449 | 1579 | 18 | 544 | 1533 | 1426 | 548 | 1617 | 1651 | 2 | 551 | 1703 | 1018 |
| 144 | 541 | 1328 | 1615 | 545 | 1406 | 1549 | 19 | 549 | 1489 | 1395 | 553 | 1570 | 1618 | 3 | 556 | 1654 | 982 |
| 145 | 546 | 1287 | 1587 | 550 | 1364 | 1520 | 20 | 554 | 1444 | 1363 | 558 | 1524 | 1584 | 4 | 561 | 1606 | 947 |
| 146 | 550 | 1246 | 1559 | 554 | 1321 | 1490 | 21 | 558 | 1399 | 1331 | 562 | 1477 | 1550 | 5 | 566 | 1557 | 911 |
| 147 | 554 | 1205 | 1531 | 558 | 1278 | 1459 | 22 | 562 | 1354 | 1298 | 566 | 1430 | 1515 | 6 | 571 | 1508 | 875 |
| 148 | 556 | 1165 | 1501 | 560 | 1235 | 1427 | 23 | 564 | 1310 | 1264 | 568 | 1383 | 1478 | 7 | 574 | 1459 | 837 |
| 149 | 558 | 1124 | 1471 | 562 | 1192 | 1395 | 24 | 566 | 1265 | 1229 | 570 | 1336 | 1442 | 8 | 576 | 1410 | 798 |
| 150 | 560 | 1083 | 1440 | 564 | 1150 | 1363 | 25 | 568 | 1220 | 1195 | 572 | 1290 | 1405 | 9 | 578 | 1362 | 760 |
| 151 | 562 | 1042 | 1410 | 566 | 1107 | 1330 | 26 | 570 | 1176 | 1160 | 574 | 1243 | 1369 | 10 | 580 | 1313 | 721 |
| 152 | 564 | 1000 | 1380 | 568 | 1064 | 1298 | 27 | 572 | 1132 | 1126 | 576 | 1196 | 1332 | 11 | 582 | 1264 | 682 |
| 153 | 566 | 968 | 1350 | 570 | 1020 | 1265 | 28 | 574 | 1095 | 1091 | 578 | 1157 | 1295 | 12 | 584 | 1223 | 643 |
| 154 | 568 | 936 | 1320 | 572 | 995 | 1233 | 29 | 576 | 1059 | 1057 | 580 | 1119 | 1259 | 13 | 586 | 1182 | 604 |
| 155 | 570 | 903 | 1289 | 574 | 961 | 1191 | 30 | 578 | 1022 | 1022 | 582 | 1080 | 1222 | 14 | 588 | 1142 | 566 |
| 156 | 572 | 871 | 1259 | 576 | 927 | 1169 | 31 | 580 | 986 | 988 | 585 | 1042 | 1187 | 15 | 590 | 1101 | 527 |
| | | | | | | | *Tamil*, Purattasi; *Malayalam*, Kanni; *Bengal*, Asvina. | | | | | | | | | | |
| 157 | 574 | 839 | 1229 | 578 | 892 | 1136 | 1 | 583 | 949 | 954 | 587 | 1003 | 1150 | 16 | 592 | 1060 | 488 |
| 158 | 577 | 807 | 1200 | 581 | 858 | 1105 | 2 | 586 | 913 | 921 | 590 | 964 | 1114 | 17 | 595 | 1019 | 450 |
| 159 | 579 | 775 | 1170 | 583 | 824 | 1073 | 3 | 588 | 876 | 886 | 592 | 926 | 1078 | 18 | 597 | 978 | 411 |
| 160 | 581 | 742 | 1139 | 585 | 790 | 1041 | 4 | 590 | 840 | 852 | 594 | 887 | 1041 | 19 | 599 | 938 | 373 |
| 161 | 583 | 710 | 1109 | 587 | 755 | 1008 | 5 | 592 | 803 | 817 | 596 | 849 | 1005 | 20 | 601 | 897 | 334 |
| 162 | 584 | 678 | 1078 | 589 | 721 | 976 | 6 | 593 | 767 | 782 | 597 | 810 | 967 | 21 | 602 | 856 | 294 |
| 163 | 586 | 646 | 1046 | 590 | 687 | 942 | 7 | 595 | 730 | 747 | 599 | 771 | 931 | 22 | 604 | 815 | 256 |
| 164 | 588 | 614 | 1015 | 592 | 652 | 909 | 8 | 597 | 694 | 711 | 601 | 733 | 893 | 23 | 606 | 774 | 216 |
| 165 | 589 | 581 | 982 | 593 | 618 | 874 | 9 | 598 | 657 | 675 | 602 | 694 | 854 | 24 | 607 | 734 | 175 |
| 166 | 589 | 549 | 949 | 593 | 584 | 839 | 10 | 598 | 621 | 637 | 602 | 656 | 814 | 25 | 607 | 693 | 133 |
| 167 | 590 | 517 | 917 | 594 | 549 | 804 | 11 | 599 | 584 | 600 | 603 | 617 | 775 | 26 | 608 | 652 | 92 |
| 168 | 591 | 485 | 885 | 595 | 515 | 770 | 12 | 600 | 548 | 564 | 604 | 578 | 737 | 27 | 609 | 611 | 51 |
| 169 | 592 | 453 | 852 | 596 | 481 | 735 | 13 | 601 | 511 | 527 | 605 | 540 | 698 | 28 | 610 | 570 | 9 |
| 170 | 592 | 420 | 819 | 596 | 447 | 700 | 14 | 601 | 475 | 489 | 605 | 501 | 658 | 29 | 610 | 530 | — 33 |
| 171 | 592 | 388 | 786 | 596 | 412 | 664 | 15 | 601 | 438 | 451 | 605 | 463 | 618 | 30 | 610 | 489 | 75 |
| | | | | | | | | | | | | | | October | | | |
| 172 | 592 | 356 | 752 | 596 | 378 | 629 | 16 | 601 | 402 | 414 | 605 | 424 | 578 | 1 | 610 | 448 | 117 |
| 173 | 592 | 324 | 709 | 596 | 343 | 593 | 17 | 601 | 365 | 376 | 605 | 385 | 538 | 2 | 610 | 407 | 159 |
| 174 | 592 | 292 | 686 | 596 | 309 | 558 | 18 | 601 | 329 | 338 | 605 | 347 | 498 | 3 | 610 | 366 | 201 |
| 175 | 592 | 259 | 652 | 596 | 275 | 522 | 19 | 601 | 292 | 301 | 605 | 308 | 458 | 4 | 610 | 326 | 243 |
| 176 | 592 | 227 | 619 | 596 | 241 | 487 | 20 | 601 | 256 | 263 | 605 | 270 | 418 | 5 | 610 | 285 | 285 |
| 177 | 591 | 195 | 585 | 595 | 206 | 450 | 21 | 600 | 219 | 224 | 604 | 231 | 377 | 6 | 609 | 244 | 328 |
| 178 | 590 | 163 | 557 | 594 | 172 | 414 | 22 | 599 | 183 | 186 | 603 | 192 | 337 | 7 | 608 | 203 | 371 |
| 179 | 589 | 131 | 516 | 593 | 138 | 377 | 23 | 598 | 146 | 147 | 602 | 154 | 296 | 8 | 607 | 162 | 415 |
| 180 | 588 | 98 | 482 | 592 | 104 | 341 | 24 | 597 | 110 | 108 | 601 | 115 | 255 | 9 | 606 | 122 | 458 |
| 181 | 586 | 66 | 447 | 590 | 69 | 303 | 25 | 595 | 73 | 68 | 599 | 77 | 213 | 10 | 604 | 81 | 502 |
| 182 | 585 | 34 | 412 | 589 | 35 | 267 | 26 | 594 | 37 | 80 | 598 | 38 | 172 | 11 | 603 | 40 | 545 |
| 183 | 584 | 0 | 378 | 588 | 0 | 230 | 27 | 593 | 0 | — 9 | 598 | 0 | 132 | 12 | 603 | 0 | 587 |
| 184 | 582 | — 33 | 343 | 586 | — 35 | 193 | 28 | 591 | — 38 | 49 | 596 | — 40 | 90 | 13 | 601 | — 42 | 631 |
| 185 | 580 | 67 | 307 | 584 | 71 | 155 | 29 | 589 | 75 | 88 | 594 | 80 | 48 | 14 | 599 | 84 | 675 |
| 186 | 578 | 100 | 272 | 582 | 106 | 118 | 30 | 587 | 113 | 128 | 592 | 120 | 6 | 15 | 597 | 126 | 719 |

# TABLE XIII.

## Sunrise from 28° to 32° N. Latitude.

| Day of Solar year. | Lat. 28° +460″ Long. Eqn. of time in seconds. | ☉'s Trop. Long. in seconds. | Total Corrn. in seconds. Mathura. A.D. 1840—1920 | Lat. 29° +352″ Long. Eqn. of time in seconds. | ☉'s Trop. Long. in seconds. | Total Corrn. in seconds. Delhi. A.D. 1840—1920 | Days of solar months. | Lat. 30° +152″ Long. Eqn. of time in seconds. | ☉'s Trop. Long. in seconds. | Total Corrn. in seconds. Patiala. A.D. 1840—1920 | Lat. 31° +332″ Long. Eqn. of time in seconds. | ☉'s Trop. Long. in seconds. | Total Corrn. in seconds. Simla. A.D. 1840—1920 | Eng. date | Lat. 32° —348″ Long. Eqn. of time in seconds. | ☉'s Trop. Long. in seconds. | Total Corrn. in seconds. Lahore. A.D. 1840—1920 |
|---|---|---|---|---|---|---|---|---|---|---|---|---|---|---|---|---|---|
| 187 | +576 | —183 | +237 | +580 | —142 | + 80 | 1 (Tamil, Aippasi; Malayalam, Tulam; Bengal, Kartika.) | +585 | —151 | —168 | +590 | —160 | — 36 | 16 October | +595 | —168 | —763 |
| 188 | 573 | 166 | 201 | 577 | 177 | 42 | 2 | 582 | 188 | 208 | 587 | 199 | 78 | 17 | 592 | 210 | 808 |
| 189 | 570 | 200 | 164 | 574 | 213 | 3 | 3 | 579 | 226 | 249 | 584 | 239 | 121 | 18 | 589 | 253 | 854 |
| 190 | 567 | 233 | 128 | 571 | 248 | — 35 | 4 | 576 | 264 | 290 | 581 | 279 | 164 | 19 | 586 | 295 | 899 |
| 191 | 564 | 266 | 92 | 568 | 284 | 74 | 5 | 573 | 302 | 331 | 578 | 319 | 207 | 20 | 583 | 337 | 944 |
| 192 | 561 | 300 | 55 | 565 | 319 | 112 | 6 | 570 | 339 | 371 | 575 | 359 | 250 | 21 | 580 | 379 | 989 |
| 193 | 557 | 333 | 17 | 561 | 355 | 151 | 7 | 566 | 377 | 414 | 571 | 399 | 293 | 22 | 576 | 421 | 1036 |
| 194 | 554 | 366 | — 27 | 558 | 390 | 197 | 8 | 563 | 415 | 462 | 568 | 439 | 343 | 23 | 573 | 463 | 1088 |
| 195 | 550 | 400 | 72 | 554 | 426 | 244 | 9 | 559 | 452 | 510 | 564 | 479 | 394 | 24 | 569 | 505 | 1141 |
| 196 | 546 | 433 | 117 | 550 | 461 | 290 | 10 | 555 | 490 | 559 | 560 | 519 | 444 | 25 | 565 | 547 | 1193 |
| 197 | 542 | 466 | 162 | 546 | 497 | 337 | 11 | 551 | 528 | 608 | 556 | 559 | 495 | 26 | 561 | 589 | 1246 |
| 198 | 538 | 499 | 206 | 542 | 532 | 384 | 12 | 547 | 565 | 656 | 552 | 598 | 546 | 27 | 557 | 631 | 1299 |
| 199 | 534 | 533 | 251 | 538 | 568 | 431 | 13 | 543 | 603 | 705 | 548 | 638 | 597 | 28 | 553 | 674 | 1352 |
| 200 | 529 | 566 | 297 | 533 | 603 | 499 | 14 | 538 | 641 | 755 | 543 | 678 | 649 | 29 | 548 | 716 | 1406 |
| 201 | 524 | 599 | 343 | 528 | 639 | 520 | 15 | 533 | 679 | 805 | 538 | 718 | 700 | 30 | 543 | 758 | 1459 |
| 202 | 520 | 633 | 388 | 524 | 674 | 573 | 16 | 529 | 716 | 858 | 534 | 758 | 751 | 31 | 539 | 800 | 1512 |
| 203 | 515 | 666 | 434 | 519 | 710 | 621 | 17 | 524 | 754 | 903 | 529 | 798 | 808 | 1 November | 534 | 842 | 1566 |
| 204 | 509 | 699 | 481 | 513 | 745 | 670 | 18 | 518 | 792 | 954 | 523 | 836 | 856 | 2 | 528 | 884 | 1621 |
| 205 | 504 | 733 | 527 | 508 | 781 | 718 | 19 | 513 | 829 | 1008 | 518 | 878 | 908 | 3 | 523 | 926 | 1675 |
| 206 | 499 | 766 | 573 | 503 | 816 | 765 | 20 | 508 | 867 | 1053 | 513 | 918 | 959 | 4 | 518 | 968 | 1728 |
| 207 | 495 | 799 | 618 | 499 | 852 | 812 | 21 | 504 | 905 | 1102 | 509 | 958 | 1010 | 5 | 514 | 1010 | 1781 |
| 208 | 489 | 832 | 664 | 493 | 887 | 861 | 22 | 498 | 942 | 1152 | 503 | 997 | 1063 | 6 | 508 | 1052 | 1836 |
| 209 | 483 | 866 | 711 | 487 | 923 | 910 | 23 | 492 | 980 | 1203 | 497 | 1037 | 1116 | 7 | 502 | 1095 | 1891 |
| 210 | 477 | 899 | 758 | 481 | 958 | 959 | 24 | 486 | 1018 | 1254 | 491 | 1077 | 1169 | 8 | 496 | 1137 | 1946 |
| 211 | 471 | 932 | 805 | 477 | 994 | 1005 | 25 | 480 | 1056 | 1305 | 485 | 1117 | 1221 | 9 | 490 | 1179 | 1960 |
| 212 | 468 | 966 | 849 | 473 | 1029 | 1052 | 26 | 474 | 1093 | 1355 | 479 | 1157 | 1275 | 10 | 483 | 1221 | 2056 |
| 213 | 465 | 1000 | 893 | 469 | 1064 | 1099 | 27 | 473 | 1132 | 1401 | 478 | 1196 | 1322 | 11 | 481 | 1264 | 2107 |
| 214 | 462 | 1041 | 937 | 465 | 1107 | 1136 | 28 | 467 | 1177 | 1452 | 471 | 1243 | 1376 | 12 | 474 | 1313 | 2163 |
| 215 | 455 | 1082 | 985 | 458 | 1150 | 1196 | 29 | 460 | 1221 | 1503 | 464 | 1290 | 1430 | 13 | 467 | 1362 | 2219 |
| 216 | 448 | 1123 | 1033 | 451 | 1192 | 1245 | 30 | 453 | 1266 | 1555 | 457 | 1336 | 1483 | 14 | 460 | 1410 | 2274 |
| 217 | 441 | 1164 | 1081 | 444 | 1235 | 1295 | 1 (Tamil, Kartigai; Malayalam, Vrischikam; Bengal, Margasira.) | 446 | 1311 | 1607 | 450 | 1383 | 1537 | 15 | 453 | 1459 | 2330 |
| 218 | 434 | 1204 | 1128 | 437 | 1276 | 1345 | 2 | 439 | 1355 | 1658 | 443 | 1430 | 1591 | 16 | 446 | 1508 | 2386 |
| 219 | 426 | 1245 | 1177 | 429 | 1321 | 1396 | 3 | 431 | 1400 | 1711 | 435 | 1477 | 1646 | 17 | 438 | 1557 | 2443 |
| 220 | 419 | 1286 | 1225 | 422 | 1364 | 1446 | 4 | 424 | 1445 | 1763 | 428 | 1524 | 1700 | 18 | 431 | 1606 | 2499 |
| 221 | 411 | 1327 | 1315 | 414 | 1406 | 1496 | 5 | 416 | 1490 | 1816 | 420 | 1570 | 1758 | 19 | 423 | 1654 | 2555 |
| 222 | 403 | 1368 | 1365 | 406 | 1449 | 1567 | 6 | 408 | 1534 | 1868 | 412 | 1617 | 1809 | 20 | 415 | 1703 | 2612 |
| 223 | 395 | 1409 | 1373 | 398 | 1492 | 1598 | 7 | 400 | 1579 | 1920 | 404 | 1664 | 1864 | 21 | 407 | 1752 | 2669 |
| 224 | 387 | 1450 | 1407 | 390 | 1535 | 1641 | 8 | 392 | 1624 | 1963 | 396 | 1711 | 1908 | 22 | 399 | 1801 | 2714 |
| 225 | 379 | 1491 | 1457 | 382 | 1578 | 1683 | 9 | 384 | 1668 | 2007 | 388 | 1758 | 1952 | 23 | 391 | 1850 | 2759 |
| 226 | 371 | 1532 | 1499 | 374 | 1620 | 1726 | 10 | 376 | 1713 | 2050 | 380 | 1804 | 1997 | 24 | 383 | 1898 | 2804 |
| 227 | 362 | 1573 | 1462 | 365 | 1668 | 1770 | 11 | 367 | 1758 | 2095 | 371 | 1851 | 2042 | 25 | 374 | 1947 | 2850 |
| 228 | 354 | 1613 | 1583 | 357 | 1706 | 1812 | 12 | 359 | 1802 | 2138 | 363 | 1898 | 2087 | 26 | 366 | 1996 | 2895 |
| 229 | 345 | 1654 | 1626 | 348 | 1749 | 1856 | 13 | 350 | 1847 | 2183 | 354 | 1945 | 2131 | 27 | 357 | 2045 | 2941 |
| 230 | 336 | 1695 | 1669 | 339 | 1792 | 1900 | 14 | 341 | 1892 | 2227 | 345 | 1992 | 2176 | 28 | 348 | 2094 | 2987 |
| 231 | 327 | 1736 | 1712 | 330 | 1834 | 1944 | 15 | 332 | 1937 | 2272 | 336 | 2038 | 2222 | 29 | 339 | 2142 | 3038 |
| 232 | 318 | 1777 | 1735 | 321 | 1877 | 1987 | 16 | 323 | 1981 | 2316 | 327 | 2085 | 2267 | 30 | 330 | 2191 | 3079 |
| 233 | 309 | [illegible] | 1798 | 312 | 1920 | 2031 | 17 | 314 | 2026 | 2361 | 318 | 2132 | 2312 | 1 December | 321 | 2240 | 3125 |
| 234 | 299 | [illegible] | 1842 | 302 | 1968 | 2076 | 18 | 304 | 2071 | 2406 | 308 | 2179 | 2358 | 2 | 311 | 2289 | 3172 |
| 235 | 290 | 1900 | 1885 | 293 | 2006 | 2119 | 19 | 295 | 2115 | 2451 | 299 | 2226 | 2403 | 3 | 302 | 2338 | 3218 |
| 236 | 281 | 1941 | 1928 | 284 | 2048 | 2159 | 20 | 286 | 2160 | 2495 | 290 | 2272 | 2449 | 4 | 293 | 2386 | 3264 |
| 237 | 272 | 1982 | 1971 | 275 | 2091 | 2207 | 21 | 277 | 2205 | 2540 | 281 | 2319 | 2494 | 5 | 284 | 2435 | 3310 |
| 238 | 262 | 2022 | 2014 | 265 | 2134 | 2251 | 22 | 267 | 2249 | 2585 | 271 | 2366 | 2540 | 6 | 274 | 2484 | 3357 |
| 239 | 252 | 2063 | 2058 | 255 | 2177 | 2296 | 23 | 257 | 2294 | 2631 | 261 | 2413 | 2586 | 7 | 264 | 2533 | 3404 |
| 240 | 243 | 2104 | 2101 | 246 | 2220 | 2340 | 24 | 248 | 2339 | 2675 | 252 | 2460 | 2631 | 8 | 255 | 2582 | 3456 |
| 241 | 233 | 2145 | 2165 | 236 | 2262 | 2385 | 25 | 238 | 2384 | 2721 | 242 | 2506 | 2678 | 9 | 244 | 2630 | 3498 |
| 242 | 223 | 2186 | 2189 | 226 | 2305 | 2429 | 26 | 228 | 2428 | 2766 | 232 | 2553 | 2724 | 10 | 234 | 2679 | 3545 |
| 243 | 207 | 2228 | 2239 | 208 | 2348 | 2482 | 27 | 208 | 2472 | 2822 | 209 | 2600 | 2783 | 11 | 210 | 2728 | 3606 |
| 244 | 198 | 2262 | 2282 | 199 | 2383 | 2526 | 28 | 199 | 2507 | 2866 | 200 | 2636 | 2828 | 12 | 201 | 2765 | 3652 |
| 245 | 188 | 2296 | 2326 | 189 | 2417 | 2570 | 29 | 189 | 2543 | 2912 | 190 | 2672 | 2867 | 13 | 191 | 2802 | 3699 |
| 246 | 178 | 2330 | 2370 | 179 | 2452 | 2615 | 30 | 179 | 2578 | 2957 | 180 | 2709 | 2921 | 14 | 181 | 2839 | 3746 |

# TABLE XIII.

## Sunrise from 28° to 32° N. Latitude.

Solar months (days 1–29): Tamil, Margali; Malayalam Dhanus; Bengal Pausha. — English dates: December, January.

Solar months (days 1–30): Tamil, Tai; Malayalam, Magaram; Bengal, Magha. — English dates: January, February.

| | Lat. 28° | +460″ Long. | | Lat. 29° | +352″ Long. | | | Lat. 30° | +152″ Long. | | Lat. 31° | +332″ Long. | | | Lat. 32° | −348″ Long. | |
|---|---|---|---|---|---|---|---|---|---|---|---|---|---|---|---|---|---|
| **Day of Solar year.** | Eqn. of time in seconds. | ☉'s Trop. Long. in seconds. | Total Corrn. in seconds. **Mathura.** A.D. 1840—1920 | Eqn. of time in seconds. | ☉'s Trop. Long. in seconds. | Total Corrn. in seconds. **Delhi.** A.D. 1840—1920 | Days of solar months. | Eqn. of time in seconds. | ☉'s Trop. Long. in seconds. | Total Corrn. in seconds. **Patiala.** A.D. 1840—1920 | Eqn. of time in seconds. | ☉'s Trop. Long. in seconds. | Total Corrn. in seconds. **Simla.** A.D. 1840—1920 | **Eng. date.** | Eqn. of time in seconds. | ☉'s Trop. Long. in seconds. | Total Corrn. in seconds. **Lahore.** A.D. 1840—1920 |
| **247** | +168 | −2364 | −2414 | +169 | −2487 | −2660 | 1 | +169 | −2614 | −3003 | +170 | −2745 | −2962 | **15** | +171 | −2876 | −3793 |
| **248** | 158 | 2397 | 2457 | 159 | 2521 | 2704 | 2 | 159 | 2649 | 3048 | 160 | 2781 | 3013 | **16** | 161 | 2913 | 3840 |
| **249** | 148 | 2431 | 2501 | 149 | 2556 | 2749 | 3 | 149 | 2685 | 3094 | 150 | 2817 | 3059 | **17** | 151 | 2950 | 3887 |
| **250** | 138 | 2465 | 2545 | 139 | 2591 | 2794 | 4 | 139 | 2720 | 3139 | 140 | 2853 | 3105 | **18** | 141 | 2987 | 3934 |
| **251** | 128 | 2499 | 2589 | 129 | 2626 | 2839 | 5 | 129 | 2756 | 3185 | 130 | 2890 | 3152 | **19** | 131 | 3024 | 3981 |
| **252** | 118 | 2533 | 2633 | 119 | 2660 | 2888 | 6 | 119 | 2791 | 3230 | 120 | 2926 | 3198 | **20** | 121 | 3061 | 4028 |
| **253** | 108 | 2567 | 2677 | 109 | 2695 | 2929 | 7 | 109 | 2827 | 3276 | 110 | 2962 | 3244 | **21** | 111 | 3098 | 4075 |
| **254** | 98 | 2601 | 2722 | 99 | 2730 | 2973 | 8 | 99 | 2862 | 3321 | 100 | 2998 | 3292 | **22** | 101 | 3135 | 4123 |
| **255** | 88 | 2635 | 2733 | 89 | 2764 | 2983 | 9 | 89 | 2898 | 3330 | 90 | 3034 | 3300 | **23** | 91 | 3172 | 4131 |
| **256** | 78 | 2669 | 2744 | 79 | 2799 | 2993 | 10 | 79 | 2933 | 3340 | 80 | 3071 | 3309 | **24** | 81 | 3209 | 4136 |
| **257** | 68 | 2703 | 2775 | 69 | 2834 | 3003 | 11 | 69 | 2969 | 3349 | 70 | 3107 | 3317 | **25** | 71 | 3246 | 4146 |
| **258** | 58 | 2736 | 2766 | 59 | 2868 | 3013 | 12 | 59 | 3004 | 3358 | 60 | 3143 | 3326 | **26** | 61 | 3283 | 4154 |
| **259** | 47 | 2770 | 2778 | 48 | 2903 | 3025 | 13 | 48 | 3040 | 3368 | 49 | 3179 | 3335 | **27** | 50 | 3320 | 4162 |
| **260** | 37 | 2804 | 2790 | 38 | 2938 | 3035 | 14 | 38 | 3075 | 3378 | 39 | 3215 | 3344 | **28** | 40 | 3357 | 4170 |
| **261** | 27 | 2838 | 2801 | 28 | 2973 | 3045 | 15 | 28 | 3111 | 3387 | 29 | 3252 | 3352 | **29** | 30 | 3394 | 4178 |
| **262** | 17 | 2872 | 2812 | 18 | 3007 | 3055 | 16 | 18 | 3146 | 3396 | 19 | 3288 | 3361 | **30** | 20 | 3431 | 4186 |
| **263** | 2 | 2906 | 2828 | 2 | 3042 | 3071 | 17 | 2 | 3182 | 3412 | 2 | 3324 | 3376 | **31** | 4 | 3468 | 4199 |
| **264** | −11 | 2940 | 2842 | −11 | 3077 | 3084 | 18 | −11 | 3217 | 3424 | −11 | 3360 | 3388 | **1** | −12 | 3505 | 4213 |
| **265** | 21 | 2974 | 2853 | 21 | 3111 | 3094 | 19 | 21 | 3253 | 3433 | 21 | 3396 | 3396 | **2** | 28 | 3542 | 4226 |
| **266** | 35 | 3008 | 2868 | 36 | 3146 | 3110 | 20 | 36 | 3288 | 3448 | 37 | 3433 | 3412 | **3** | 38 | 3579 | 4234 |
| **267** | 45 | 3042 | 2879 | 46 | 3181 | 3120 | 21 | 46 | 3324 | 3457 | 47 | 3469 | 3421 | **4** | 48 | 3616 | 4242 |
| **268** | 55 | 3075 | 2890 | 56 | 3215 | 3130 | 22 | 56 | 3359 | 3466 | 57 | 3505 | 3429 | **5** | 58 | 3653 | 4250 |
| **269** | 66 | 3109 | 2902 | 67 | 3250 | 3141 | 23 | 67 | 3395 | 3476 | 68 | 3541 | 3439 | **6** | 69 | 3690 | 4258 |
| **270** | 76 | 3143 | 2914 | 77 | 3285 | 3151 | 24 | 77 | 3430 | 3486 | 78 | 3577 | 3447 | **7** | 79 | 3727 | 4266 |
| **271** | 87 | 3177 | 2926 | 88 | 3320 | 3162 | 25 | 88 | 3466 | 3496 | 89 | 3614 | 3457 | **8** | 90 | 3764 | 4275 |
| **272** | 97 | 3211 | 2937 | 98 | 3354 | 3172 | 26 | 98 | 3501 | 3505 | 99 | 3650 | 3465 | **9** | 100 | 3801 | 4283 |
| **273** | 107 | 3245 | 2948 | 108 | 3390 | 3182 | 27 | 108 | 3537 | 3515 | 109 | 3686 | 3474 | **10** | 110 | 3838 | 4290 |
| **274** | 100 | 3280 | 2942 | 100 | 3424 | 3173 | 28 | 100 | 3572 | 3506 | 98 | 3724 | 3460 | **11** | 98 | 3876 | 4275 |
| **275** | 106 | 3281 | 2949 | 106 | 3424 | 3181 | 29 | 106 | 3571 | 3511 | 104 | 3722 | 3464 | **12** | 104 | 3874 | 4280 |
| **276** | 112 | 3282 | 2956 | 112 | 3424 | 3187 | 1 | 112 | 3571 | 3517 | 110 | 3721 | 3469 | **13** | 110 | 3871 | 4283 |
| **277** | 118 | 3283 | 2963 | 118 | 3424 | 3193 | 2 | 118 | 3570 | 3522 | 116 | 3719 | 3473 | **14** | 116 | 3869 | 4287 |
| **278** | 124 | 3284 | 2970 | 124 | 3424 | 3199 | 3 | 124 | 3569 | 3526 | 122 | 3718 | 3478 | **15** | 122 | 3867 | 4291 |
| **279** | 130 | 3285 | 2977 | 130 | 3425 | 3205 | 4 | 130 | 3568 | 3532 | 128 | 3716 | 3482 | **16** | 128 | 3864 | 4294 |
| **280** | 136 | 3287 | 2985 | 136 | 3425 | 3211 | 5 | 136 | 3568 | 3537 | 134 | 3715 | 3487 | **17** | 134 | 3862 | 4298 |
| **281** | 142 | 3288 | 2992 | 142 | 3425 | 3217 | 6 | 142 | 3567 | 3543 | 140 | 3713 | 3491 | **18** | 140 | 3860 | 4302 |
| **282** | 148 | 3289 | 2999 | 148 | 3425 | 3224 | 7 | 148 | 3566 | 3548 | 146 | 3712 | 3496 | **19** | 146 | 3858 | 4306 |
| **283** | 154 | 3290 | 3006 | 154 | 3425 | 3230 | 8 | 154 | 3566 | 3554 | 152 | 3710 | 3500 | **20** | 152 | 3855 | 4309 |
| **284** | 160 | 3291 | 2992 | 160 | 3425 | 3236 | 9 | 160 | 3565 | 3560 | 158 | 3709 | 3506 | **21** | 158 | 3853 | 4314 |
| **285** | 166 | 3292 | 2977 | 166 | 3425 | 3199 | 10 | 166 | 3564 | 3523 | 164 | 3707 | 3465 | **22** | 164 | 3850 | 4271 |
| **286** | 172 | 3293 | 2942 | 172 | 3426 | 3163 | 11 | 172 | 3564 | 3483 | 170 | 3706 | 3425 | **23** | 170 | 3848 | 4229 |
| **287** | 178 | 3294 | 2907 | 178 | 3426 | 3126 | 12 | 178 | 3563 | 3444 | 176 | 3704 | 3384 | **24** | 176 | 3846 | 4186 |
| **288** | 184 | 3295 | 2872 | 184 | 3426 | 3089 | 13 | 184 | 3562 | 3406 | 182 | 3703 | 3344 | **25** | 182 | 3844 | 4144 |
| **289** | 190 | 3296 | 2837 | 190 | 3426 | 3052 | 14 | 190 | 3561 | 3367 | 188 | 3701 | 3303 | **26** | 188 | 3841 | 4101 |
| **290** | 196 | 3298 | 2803 | 196 | 3426 | 3016 | 15 | 196 | 3561 | 3329 | 194 | 3700 | 3262 | **27** | 194 | 3839 | 4059 |
| **291** | 202 | 3299 | 2768 | 202 | 3426 | 2979 | 16 | 202 | 3560 | 3290 | 200 | 3698 | 3222 | **28** | 200 | 3837 | 4016 |
| **292** | 208 | 3300 | 2733 | 208 | 3426 | 2942 | 17 | 208 | 3559 | 3252 | 206 | 3697 | 3181 | **29** | 206 | 3835 | 3974 |
| **293** | 214 | 3301 | 2698 | 214 | 3426 | 2906 | 18 | 214 | 3559 | 3213 | 212 | 3695 | 3141 | **30** | 212 | 3832 | 3931 |
| **294** | 220 | 3302 | 2663 | 220 | 3426 | 2869 | 19 | 220 | 3558 | 3175 | 218 | 3694 | 3100 | **31** | 218 | 3830 | 3889 |
| **295** | 226 | 3303 | 2628 | 226 | 3427 | 2832 | 20 | 226 | 3557 | 3136 | 224 | 3692 | 3059 | **1** | 223 | 3828 | 3845 |
| **296** | 232 | 3304 | 2593 | 232 | 3427 | 2796 | 21 | 232 | 3557 | 3098 | 230 | 3691 | 2999 | **2** | 229 | 3825 | 3803 |
| **297** | 238 | 3305 | 2558 | 238 | 3427 | 2759 | 22 | 238 | 3556 | 3059 | 235 | 3689 | 2977 | **3** | 234 | 3823 | 3759 |
| **298** | 244 | 3306 | 2523 | 244 | 3427 | 2722 | 23 | 244 | 3555 | 3021 | 240 | 3688 | 2936 | **4** | 239 | 3821 | 3716 |
| **299** | 250 | 3307 | 2488 | 250 | 3427 | 2685 | 24 | 250 | 3554 | 2992 | 245 | 3686 | 2894 | **5** | 244 | 3818 | 3672 |
| **300** | 256 | 3309 | 2454 | 256 | 3427 | 2649 | 25 | 255 | 3554 | 2943 | 250 | 3685 | 2852 | **6** | 249 | 3816 | 3631 |
| **301** | 262 | 3310 | 2419 | 262 | 3427 | 2612 | 26 | 260 | 3553 | 2903 | 255 | 3683 | 2807 | **7** | 254 | 3814 | 3585 |
| **302** | 268 | 3311 | 2384 | 268 | 3428 | 2575 | 27 | 265 | 3552 | 2864 | 260 | 3682 | 2769 | **8** | 260 | 3812 | 3543 |
| **303** | 274 | 3312 | 2349 | 273 | 3428 | 2538 | 28 | 270 | 3552 | 2824 | 266 | 3680 | 2729 | **9** | 265 | 3809 | 3499 |
| **304** | 280 | 3312 | 2314 | 278 | 3428 | 2500 | 29 | 276 | 3552 | 2786 | 272 | 3680 | 2688 | **10** | 270 | 3808 | 3456 |
| **305** | 282 | 3271 | 2275 | 280 | 3385 | 2459 | 30 | 278 | 3507 | 2743 | 274 | 3633 | 2643 | **11** | 272 | 3759 | 3419 |

# TABLE XIII.

## Sunrise from 28° to 32° N. Latitude.

| Day of Solar year. | Lat. 28° +460″ Long. Eqn. of time in seconds. | ☉'s Trop. Long. in seconds. | Total Corrn. in seconds. Mathura. A.D. 1840—1920 | Lat. 29° +352″ Long. Eqn. of time in seconds. | ☉'s Trop. Long. in seconds. | Total Corrn. in seconds. Delhi. A.D. 1840—1920 | Days of solar months. | Lat. 30° +152″ Long. Eqn. of time in seconds. | ☉'s Trop. Long. in seconds. | Total Corrn. in seconds. Patiala. A.D. 1840—1920 | Lat. 31° +332″ Long. Eqn. of time in seconds. | ☉'s Trop. Long. in seconds. | Total Corrn. in seconds. Simla. A.D. 1840—1920 | Eng. date. | Lat. 32° —348″ Long. Eqn. of time in seconds. | ☉'s Trop. Long. in seconds. | Total Corrn. in seconds. Lahore. A.D. 1840—1920 |
|---|---|---|---|---|---|---|---|---|---|---|---|---|---|---|---|---|---|
| **306** | —285 | —3230 | —2237 | —283 | —3343 | —2420 | 1 | —280 | —3463 | —2701 | —276 | —3587 | —2599 | **12** | —274 | —3711 | —3368 |
| **307** | 287 | 3189 | 2198 | 285 | 3300 | 2379 | 2 | 283 | 3418 | 2659 | 279 | 3540 | 2556 | **13** | 276 | 3662 | 3316 |
| **308** | 290 | 3148 | 2160 | 288 | 3257 | 2339 | 3 | 285 | 3371 | 2617 | 281 | 3494 | 2511 | **14** | 278 | 3614 | 3270 |
| **309** | 292 | 3107 | 2121 | 290 | 3214 | 2298 | 4 | 287 | 3329 | 2574 | 283 | 3447 | 2466 | **15** | 281 | 3565 | 3224 |
| **310** | 295 | 3067 | 2084 | 293 | 3172 | 2264 | 5 | 290 | 3285 | 2533 | 286 | 3400 | 2422 | **16** | 283 | 3517 | 3178 |
| **311** | 297 | 3026 | 2045 | 295 | 3129 | 2218 | 6 | 292 | 3240 | 2490 | 288 | 3354 | 2378 | **17** | 285 | 3468 | 3131 |
| **312** | 300 | 2985 | 2007 | 298 | 3086 | 2168 | 7 | 294 | 3196 | 2448 | 290 | 3307 | 2333 | **18** | 287 | 3420 | 3085 |
| **313** | 302 | 2944 | 1968 | 300 | 3044 | 2138 | 8 | 297 | 3151 | 2406 | 293 | 3261 | 2290 | **19** | 289 | 3371 | 3038 |
| **314** | 305 | 2903 | 1930 | 303 | 3001 | 2098 | 9 | 299 | 3107 | 2364 | 295 | 3214 | 2245 | **20** | 292 | 3323 | 2993 |
| **315** | 307 | 2862 | 1891 | 305 | 2958 | 2057 | 10 | 301 | 3062 | 2321 | 297 | 3167 | 2201 | **21** | 294 | 3274 | 2946 |
| **316** | 310 | 2821 | 1826 | 308 | 2916 | 1990 | 11 | 304 | 3018 | 2252 | 300 | 3121 | 2129 | **22** | 296 | 3226 | 2871 |
| **317** | 312 | 2780 | 1760 | 310 | 2873 | 1922 | 12 | 306 | 2973 | 2181 | 302 | 3074 | 2057 | **23** | 298 | 3177 | 2796 |
| **318** | 315 | 2739 | 1695 | 313 | 2830 | 1855 | 13 | 308 | 2929 | 2111 | 304 | 3028 | 1984 | **24** | 300 | 3129 | 2722 |
| **319** | 317 | 2698 | 1629 | 315 | 2787 | 1787 | 14 | 311 | 2884 | 2041 | 307 | 2981 | 1913 | **25** | 303 | 3080 | 2648 |
| **320** | 320 | 2658 | 1567 | 318 | 2745 | 1719 | 15 | 313 | 2840 | 1971 | 309 | 2934 | 1840 | **26** | 305 | 3032 | 2573 |
| **321** | 322 | 2617 | 1497 | 320 | 2702 | 1651 | 16 | 315 | 2795 | 1901 | 311 | 2888 | 1768 | **27** | 307 | 2983 | 2498 |
| **322** | 325 | 2576 | 1432 | 323 | 2659 | 1584 | 17 | 318 | 2751 | 1831 | 314 | 2841 | 1696 | **28** | 309 | 2935 | 2423 |
| **323** | 327 | 2535 | 1366 | 325 | 2617 | 1515 | 18 | 320 | 2706 | 1761 | 316 | 2795 | 1624 | **1** | 311 | 2866 | 2349 |
| **324** | 330 | 2494 | 1301 | 328 | 2574 | 1449 | 19 | 322 | 2662 | 1690 | 318 | 2748 | 1551 | **2** | 314 | 2838 | 2275 |
| **325** | 332 | 2453 | 1235 | 330 | 2531 | 1381 | 20 | 325 | 2617 | 1621 | 321 | 2701 | 1480 | **3** | 316 | 2789 | 2200 |
| **326** | 335 | 2412 | 1170 | 332 | 2489 | 1313 | 21 | 327 | 2573 | 1551 | 323 | 2655 | 1407 | **4** | 318 | 2741 | 2125 |
| **327** | 337 | 2371 | 1104 | 334 | 2446 | 1245 | 22 | 329 | 2528 | 1480 | 325 | 2608 | 1335· | **5** | 820 | 2692 | 2050 |
| **328** | 340 | 2330 | 1039 | 336 | 2403 | 1177 | 23 | 331 | 2484 | 1410 | 327 | 2562 | 1264 | **6** | 322 | 2641 | 1976 |
| **329** | 342 | 2289 | 973 | 338 | 2360 | 1109 | 24 | 333 | 2439 | 1339 | 329 | 2515 | 1190 | **7** | 324 | 2595 | 1901 |
| **330** | 345 | 2249 | 908 | 340 | 2318 | 1040 | 25 | 335 | 2395 | 1269 | 331 | 2468 | 1117 | **8** | 326 | 2547 | 1826 |
| **331** | 347 | 2208 | 841 | 342 | 2275 | 972 | 26 | 337 | 2350 | 1199 | 333 | 2422 | 1045 | **9** | 328 | 2498 | 1851 |
| **332** | 349 | 2167 | 775 | 344 | 2232 | 904 | 27 | 339 | 2306 | 1128 | 335 | 2375 | 972 | **10** | 330 | 2450 | 1676 |
| **333** | 351 | 2126 | 709 | 346 | 2190 | 836 | 28 | 341 | 2261 | 1058 | 337 | 2329 | 900 | **11** | 332 | 2401 | 1602 |
| **334** | 353 | 2085 | 643 | 348 | 2147 | 768 | 29 | 343 | 2217 | 987 | 339 | 2282 | 827 | **12** | 334 | 2353 | 1527 |
| **335** | 355 | 2044 | 577 | 350 | 2104 | 700 | 1 | 345 | 2172 | 917 | 341 | 2236 | 755 | **13** | 336 | 2304 | 1452 |
| **336** | 355 | 1976 | 509 | 350 | 2034 | 630 | 2 | 345 | 2100 | 845 | 341 | 2161 | 680 | **14** | 336 | 2227 | 1375 |
| **337** | 356 | 1908 | 442 | 351 | 1964 | 561 | 3 | 346 | 2027 | 773 | 342 | 2087 | 607 | **15** | 337 | 2150 | 1299 |
| **338** | 356 | 1840 | 374 | 351 | 1894 | 491 | 4 | 346 | 1955 | 701 | 342 | 2012 | 532 | **16** | 337 | 2074 | 1223 |
| **339** | 357 | 1772 | 307 | 352 | 1824 | 422 | 5 | 347 | 1882 | 629 | 343 | 1938 | 459 | **17** | 338 | 1997 | 1147 |
| **340** | 357 | 1704 | 239 | 352 | 1753 | 351 | 6 | 347 | 1810 | 557 | 343 | 1863 | 384 | **18** | 338 | 1920 | 1070 |
| **341** | 358 | 1635 | 171 | 353 | 1683 | 282 | 7 | 348 | 1738 | 486 | 344 | 1789 | 311 | **19** | 339 | 1843 | 994 |
| **342** | 358 | 1567 | 103 | 353 | 1613 | 212 | 8 | 348 | 1665 | 413 | 344 | 1714 | 236 | **20** | 339 | 1766 | 917 |
| **343** | 359 | 1499 | 36 | 354 | 1543 | 143 | 9 | 349 | 1593 | 342 | 345 | 1640 | 153 | **21** | 340 | 1690 | 842 |
| **344** | 359 | 1431 | + 32 | 354 | 1473 | 73 | 10 | 349 | 1520 | 269 | 345 | 1565 | 88 | **22** | 340 | 1613 | 765 |
| **345** | 360 | 1363 | 100 | 355 | 1403 | 3 | 11 | 350 | 1448 | 198 | 346 | 1491 | 14 | **23** | 341 | 1536 | 689 |
| **346** | 360 | 1295 | 168 | 355 | 1333 | + 66 | 12 | 350 | 1376 | 126 | 346 | 1416 | + 61 | **24** | 341 | 1459 | 612 |
| **347** | 361 | 1227 | 235 | 356 | 1263 | 135 | 13 | 351 | 1303 | 54 | 347 | 1342 | 134 | **25** | 342 | 1382 | 536 |
| **348** | 361 | 1159 | 303 | 356 | 1193 | 205 | 14 | 351 | 1231 | + 18 | 347 | 1267 | 209 | **26** | 342 | 1306 | 460 |
| **349** | 362 | 1091 | 370 | 357 | 1123 | 274 | 15 | 352 | 1158 | 90 | 348 | 1193 | 282 | **27** | 343 | 1229 | 384 |
| **350** | 362 | 1023 | 439 | 357 | 1052 | 346 | 16 | 352 | 1086 | 162 | 348 | 1118 | 357 | **28** | 343 | 1152 | 307 |
| **351** | 363 | 954 | 506 | 357 | 982 | 415 | 17 | 352 | 1014 | 234 | 349 | 1044 | 430 | **29** | 344 | 1175 | 231 |
| **352** | 363 | 886 | 574 | 358 | 912 | 485 | 18 | 353 | 941 | 306 | 349 | 969 | 505 | **30** | 344 | 998 | 154 |
| **353** | 364 | 818 | 641 | 358 | 842 | 554 | 19 | 353 | 869 | 379 | 349 | 895 | 579 | **31** | 344 | 922 | 78 |
| **354** | 364 | 750 | 709 | 358 | 772 | 626 | 20 | 353 | 796 | 451 | 350 | 820 | 653 | **1** | 345 | 845 | 2 |
| **355** | 365 | 682 | 776 | 359 | 702 | 694 | 21 | 354 | 724 | 522 | 350 | 746 | 727 | **2** | 345 | 768 | + 75 |
| **356** | 365 | 614 | 844 | 359 | 632 | 765 | 22 | 354 | 652 | 594 | 350 | 671 | 802 | **3** | 345 | 691 | 152 |
| **357** | 366 | 546 | 911 | 360 | 562 | 833 | 23 | 354 | 579 | 667 | 351 | 597 | 875 | **4** | 346 | 614 | 228 |
| **358** | 366 | 478 | 979 | 360 | 492 | 903 | 24 | 355 | 507 | 738 | 351 | 522 | 950 | **5** | 346 | 538 | 304 |
| **359** | 367 | 410 | 1046 | 360 | 422 | 973 | 25 | 355 | 434 | 821 | 351 | 448 | 1024 | **6** | 346 | 461 | 381 |
| **360** | 367 | 342 | 1115 | 361 | 351 | 1042 | 26 | 355 | 362 | 876 | 352 | 373 | 1098 | **7** | 347 | 384 | 457 |
| **361** | 368 | 273 | 1182 | 361 | 281 | 1113 | 27 | 356 | 290 | 954 | 352 | 299 | 1172 | **8** | 347 | 307 | 534 |
| **362** | 368 | 205 | 1250 | 361 | 211 | 1183 | 28 | 356 | 317 | 1027 | 352 | 224 | 1247 | **9** | 347 | 230 | 611 |
| **363** | 368 | 137 | 1318 | 362 | 141 | 1252 | 29 | 356 | 145 | 1099 | 353 | 150 | 1320 | **10** | 348 | 154 | 686 |
| **364** | 367 | 69 | 1387 | 362 | 71 | 1322 | 30 | 357 | 72 | 1171 | 353 | 75 | 1395 | **11** | 348 | 77 | 763 |
| **365** | 367 | 0 | 1455 | 362 | 0 | 1392 | 31 | 357 | 0 | 1243 | 353 | 0 | 1469 | **12** | 348 | 0 | 840 |

Days of solar months 1–29 (days 306–334): *Tamil*, Masi; *Malayalam*, Kumbam, *Bengal*, Phalguna.

Days of solar months 1–31 (days 335–365): *Tamil*, Panguni; *Malayalam*, Meenam; *Bengal*, Chaitra.

Eng. date: February (12–28); March (1–31); April (1–12).

# TABLE XIII.

## Sunrise from 33° to 35° N. Latitude; also for BOMBAY and CALCUTTA.

| Day of Solar year. | Lat. 33° —480″ Long. Eqn. of time in seconds. | ☉'s Trop. Long. in seconds. | Total Corrn. in seconds. Jehlum. A.D. 1840—1920 | Lat. 34° —228″ Long. Eqn. of time in seconds. | ☉'s Trop. Long. in seconds. | Total Corrn. in seconds. Srinagar. A.D. 1840—1920 | Days of solar months. | Lat. 35° —1592″ Long. Eqn. of time in seconds. | ☉'s Trop. Long. in seconds. | Total Corrn. in seconds. Kabul. A.D. 1840—1920 | Lat. 19° —692″ Long. Eqn. of time in seconds. | ☉'s Trop. Long. in seconds. | Total Corrn. in seconds. Bombay. A.D. 1840—1920 | Eng. date. | Lat. 22·5° +3020″ Long. Eqn. of time in seconds. | ☉'s Trop. Long. in seconds. | Total Corrn. in seconds. Calcutta. A.D. 1840—1920 |
|---|---|---|---|---|---|---|---|---|---|---|---|---|---|---|---|---|---|
| 1 | —343 | + 79 | + 840 | —338 | + 82 | +1148 | 1 | — 333 | + 84 | —161 | —405 | + 50 | — 47 | 13 | —391 | + 57 | +3891 |
| 2 | 341 | 158 | 921 | 336 | 163 | 1231 | 2 | 331 | 168 | 75 | 403 | 100 | + 5 | 14 | 389 | 113 | 3877 |
| 3 | 340 | 238 | 1002 | 335 | 245 | 1314 | 3 | 330 | 252 | + 10 | 401 | 150 | 57 | 15 | 387 | 170 | 3936 |
| 4 | 339 | 317 | 1082 | 334 | 326 | 1396 | 4 | 329 | 336 | 95 | 400 | 200 | 108 | 16 | 386 | 227 | 3993 |
| 5 | 337 | 396 | 1163 | 332 | 408 | 1480 | 5 | 328 | 420 | 180 | 398 | 250 | 160 | 17 | 384 | 283 | 4052 |
| 6 | 336 | 475 | 1243 | 331 | 490 | 1563 | 6 | 327 | 504 | 265 | 396 | 300 | 212 | 18 | 382 | 340 | 4111 |
| 7 | 334 | 554 | 1324 | 330 | 571 | 1645 | 7 | 325 | 588 | 351 | 394 | 350 | 264 | 19 | 380 | 396 | 4170 |
| 8 | 332 | 634 | 1406 | 328 | 653 | 1729 | 8 | 323 | 672 | 437 | 392 | 400 | 316 | 20 | 378 | 453 | 4229 |
| 9 | 330 | 713 | 1487 | 326 | 734 | 1812 | 9 | 321 | 756 | 523 | 390 | 450 | 368 | 21 | 375 | 510 | 4289 |
| 10 | 328 | 792 | 1568 | 324 | 816 | 1896 | 10 | 319 | 840 | 609 | 387 | 500 | 421 | 22 | 373 | 567 | 4347 |
| 11 | 326 | 871 | 1620 | 321 | 898 | 1951 | 11 | 317 | 924 | 665 | 385 | 550 | 450 | 23 | 371 | 623 | 4381 |
| 12 | 324 | 950 | 1673 | 319 | 979 | 2005 | 12 | 315 | 1008 | 721 | 382 | 600 | 479 | 24 | 369 | 679 | 4415 |
| 13 | 322 | 1030 | 1725 | 317 | 1061 | 2059 | 13 | 313 | 1092 | 778 | 380 | 650 | 508 | 25 | 366 | 736 | 4450 |
| 14 | 319 | 1109 | 1778 | 315 | 1142 | 2112 | 14 | 310 | 1176 | 835 | 377 | 700 | 538 | 26 | 363 | 792 | 4485 |
| 15 | 316 | 1188 | 1832 | 312 | 1224 | 2167 | 15 | 308 | 1260 | 891 | 374 | 750 | 568 | 27 | 360 | 850 | 4520 |
| 16 | 314 | 1267 | 1884 | 310 | 1306 | 2221 | 16 | 306 | 1344 | 947 | 371 | 800 | 597 | 28 | 357 | 906 | 4555 |
| 17 | 312 | 1346 | 1936 | 308 | 1387 | 2275 | 17 | 304 | 1428 | 1003 | 367 | 850 | 628 | 29 | 354 | 962 | 4590 |
| 18 | 310 | 1426 | 1989 | 306 | 1469 | 2329 | 18 | 302 | 1512 | 1059 | 364 | 900 | 658 | 30 | 352 | 1019 | 4624 |
| 19 | 308 | 1505 | 2041 | 304 | 1550 | 2383 | 19 | 300 | 1596 | 1116 | 361 | 950 | 687 | 1 | 348 | 1072 | 4660 |
| 20 | 306 | 1584 | 2093 | 302 | 1632 | 2437 | 20 | 298 | 1680 | 1172 | 357 | 1000 | 718 | 2 | 344 | 1133 | 4696 |
| 21 | 304 | 1663 | 2145 | 300 | 1714 | 2491 | 21 | 296 | 1764 | 1228 | 354 | 1050 | 748 | 3 | 341 | 1190 | 4731 |
| 22 | 302 | 1742 | 2198 | 298 | 1795 | 2544 | 22 | 294 | 1848 | 1284 | 350 | 1100 | 778 | 4 | 338 | 1246 | 4766 |
| 23 | 300 | 1822 | 2250 | 296 | 1877 | 2598 | 23 | 292 | 1932 | 1340 | 346 | 1150 | 809 | 5 | 335 | 1303 | 4801 |
| 24 | 298 | 1901 | 2302 | 294 | 1958 | 2652 | 24 | 290 | 2016 | 1397 | 342 | 1200 | 840 | 6 | 333 | 1359 | 4835 |
| 25 | 296 | 1980 | 2355 | 292 | 2040 | 2706 | 25 | 288 | 2100 | 1453 | 338 | 1250 | 870 | 7 | 330 | 1416 | 4870 |
| 26 | 294 | 2059 | 2407 | 290 | 2122 | 2760 | 26 | 286 | 2184 | 1509 | 334 | 1300 | 901 | 8 | 327 | 1473 | 4904 |
| 27 | 292 | 2138 | 2459 | 288 | 2203 | 2814 | 27 | 284 | 2268 | 1565 | 330 | 1350 | 932 | 9 | 324 | 1530 | 4940 |
| 28 | 290 | 2218 | 2512 | 286 | 2285 | 2868 | 28 | 282 | 2352 | 1621 | 326 | 1400 | 963 | 10 | 321 | 1587 | 4975 |
| 29 | 288 | 2297 | 2564 | 284 | 2366 | 2921 | 29 | 280 | 2436 | 1678 | 322 | 1450 | 993 | 11 | 318 | 1644 | 5010 |
| 30 | 286 | 2376 | 2616 | 282 | 2448 | 2975 | 30 | 278 | 2520 | 1734 | 354 | 1500 | 988 | 12 | 334 | 1700 | 5026 |
| 31 | 284 | 2426 | 2669 | 280 | 2500 | 3029 | 1 | 276 | 2574 | 1790 | 349 | 1527 | 1020 | 13 | 331 | 1732 | 5061 |
| 32 | 282 | 2477 | 2721 | 278 | 2552 | 3083 | 2 | 274 | 2628 | 1846 | 344 | 1553 | 1051 | 14 | 327 | 1764 | 5097 |
| 33 | 280 | 2527 | 2773 | 276 | 2604 | 3137 | 3 | 272 | 2683 | 1902 | 339 | 1580 | 1083 | 15 | 324 | 1796 | 5132 |
| 34 | 278 | 2577 | 2826 | 274 | 2655 | 3201 | 4 | 270 | 2737 | 1959 | 323 | 1607 | 1116 | 16 | 320 | 1828 | 5168 |
| 35 | 276 | 2628 | 2878 | 272 | 2707 | 3245 | 5 | 268 | 2791 | 2015 | 328 | 1634 | 1147 | 17 | 316 | 1860 | 5204 |
| 36 | 274 | 2678 | 2930 | 270 | 2759 | 3299 | 6 | 266 | 2845 | 2071 | 322 | 1660 | 1180 | 18 | 313 | 1892 | 5239 |
| 37 | 272 | 2728 | 2983 | 268 | 2811 | 3352 | 7 | 264 | 2899 | 2127 | 316 | 1687 | 1213 | 19 | 308 | 1924 | 5276 |
| 38 | 270 | 2779 | 3035 | 266 | 2863 | 3406 | 8 | 262 | 2953 | 2183 | 311 | 1714 | 1245 | 20 | 303 | 1956 | 5313 |
| 39 | 269 | 2829 | 3086 | 264 | 2915 | 3460 | 9 | 260 | 3008 | 2240 | 305 | 1740 | 1277 | 21 | 297 | 1988 | 5351 |
| 40 | 264 | 2879 | 3142 | 261 | 2967 | 3515 | 10 | 258 | 3062 | 2296 | 299 | 1767 | 1310 | 22 | 291 | 2020 | 5389 |
| 41 | 260 | 2929 | 3196 | 258 | 3019 | 3570 | 11 | 256 | 3116 | 2352 | 293 | 1794 | 1343 | 23 | 285 | 2052 | 5427 |
| 42 | 254 | 2980 | 3205 | 252 | 3070 | 3580 | 12 | 250 | 3170 | 2363 | 286 | 1820 | 1343 | 24 | 279 | 2084 | 5428 |
| 43 | 247 | 3030 | 3215 | 245 | 3122 | 3591 | 13 | 243 | 3224 | 2375 | 280 | 1847 | 1342 | 25 | 273 | 2116 | 5429 |
| 44 | 242 | 3080 | 3224 | 240 | 3174 | 3600 | 14 | 237 | 3279 | 2385 | 274 | 1874 | 1340 | 26 | 267 | 2148 | 5430 |
| 45 | 236 | 3131 | 3233 | 234 | 3226 | 3610 | 15 | 231 | 3333 | 2396 | 267 | 1900 | 1341 | 27 | 260 | 2180 | 5432 |
| 46 | 231 | 3181 | 3241 | 228 | 3278 | 3620 | 16 | 226 | 3387 | 2406 | 261 | 1927 | 1339 | 28 | 254 | 2211 | 5434 |
| 47 | 224 | 3231 | 3251 | 222 | 3330 | 3630 | 17 | 219 | 3441 | 2418 | 254 | 1954 | 1339 | 29 | 247 | 2244 | 5436 |
| 48 | 218 | 3282 | 3260 | 216 | 3382 | 3640 | 18 | 213 | 3495 | 2429 | 247 | 1981 | 1339 | 30 | 241 | 2276 | 5437 |
| 49 | 212 | 3332 | 3270 | 210 | 3433 | 3650 | 19 | 208 | 3550 | 2438 | 241 | 2007 | 1338 | 31 | 235 | 2308 | 5438 |
| 50 | 207 | 3382 | 3278 | 205 | 3485 | 3659 | 20 | 203 | 3604 | 2448 | 234 | 2034 | 1338 | 1 | 228 | 2340 | 5440 |
| 51 | 201 | 3433 | 3287 | 199 | 3537 | 3669 | 21 | 197 | 3658 | 2459 | 227 | 2061 | 1338 | 2 | 221 | 2372 | 5443 |
| 52 | 195 | 3483 | 3296 | 193 | 3589 | 3679 | 22 | 191 | 3712 | 2470 | 220 | 2087 | 1338 | 3 | 214 | 2404 | 5445 |
| 53 | 188 | 3533 | 3306 | 186 | 3641 | 3690 | 23 | 184 | 3766 | 2482 | 213 | 2114 | 1338 | 4 | 207 | 2436 | 5447 |
| 54 | 182 | 3584 | 3316 | 180 | 3693 | 3700 | 24 | 178 | 3821 | 2492 | 206 | 2141 | 1338 | 5 | 200 | 2468 | 5449 |
| 55 | 176 | 3634 | 3325 | 177 | 3745 | 3707 | 25 | 176 | 3875 | 2499 | 199 | 2167 | 1336 | 6 | 193 | 2500 | 5451 |
| 56 | 172 | 3684 | 3332 | 174 | 3797 | 3714 | 26 | 172 | 3929 | 2508 | 192 | 2194 | 1337 | 7 | 186 | 2532 | 5454 |
| 57 | 169 | 3735 | 3338 | 171 | 3848 | 3721 | 27 | 168 | 3983 | 2517 | 184 | 2221 | 1338 | 8 | 179 | 2564 | 5456 |
| 58 | 166 | 3785 | 3344 | 168 | 3900 | 3728 | 28 | 164 | 4037 | 2526 | 177 | 2248 | 1338 | 9 | 174 | 2596 | 5456 |
| 59 | 163 | 3835 | 3351 | 165 | 3952 | 3735 | 29 | 160 | 4092 | 2534 | 169 | 2274 | 1339 | 10 | 168 | 2628 | 5457 |
| 60 | 160 | 3886 | 3357 | 162 | 4004 | 3742 | 30 | 156 | 4146 | 2543 | 162 | 2301 | 1339 | 11 | 162 | 2660 | 5458 |
| 61 | 157 | 3936 | 3363 | 159 | 4056 | 3749 | 31 | 152 | 4200 | 2552 | 179 | 2328 | 1315 | 12 | 171 | 2692 | 5445 |
| 62 | 154 | 3939 | 3369 | 156 | 4060 | 3756 | 32 | 148 | 4205 | 2561 | 171 | 2321 | 1316 | 13 | 164 | 2687 | 5447 |

Days of solar months 1–30 (days 1–30 of solar year): Tamil, Chittirai; Malayalam, Medam; Bengal, Vaisakha.

Days of solar months 1–32 (days 31–62 of solar year): Tamil, Vaikasi; Malayalam, Edavam; Bengal, Jyeshta.

Eng. date: April (13–30), May (1–31), June (1–13).

# TABLE XIII.

## Sunrise from 33° to 35° N. Latitude; also for BOMBAY and CALCUTTA.

| Day of Solar year. | Lat. 33°—480″ Long. Eqn. of time in seconds. | ☉'s Trop. Long. in seconds. | Total Corrn. in seconds. Jhelum. A.D. 1840—1920 | Lat. 34°—228″ Long. Eqn. of time in seconds. | ☉'s Trop. Long in seconds. | Total Corrn. in seconds. Srinagar. A.D. 1840—1920 | Days of solar months. | Lat. 35°—1592″ Long. Eqn. of time. in seconds. | ☉'s Trop. Long. in seconds. | Total Corrn. in seconds. Kabul. A.D 1840—1920 | Lat. 19°—692″ Long. Eqn. of time in seconds. | ☉'s Trop. Long. in seconds. | Total Corrn. in seconds. Bombay. A.D. 1840—1920 | Eng. date. | Lat. 22·5°+3020″ Long. Eqn. of time in seconds. | ☉'s Trop. Long. in seconds. | Total Corrn. in seconds. Calcutta. A.D. 1840—1920 |
|---|---|---|---|---|---|---|---|---|---|---|---|---|---|---|---|---|---|
| | | | | | | | *Tamil,* Ani; *Malayalam,* Mithunam; *Bengal,* Ashada. | | | | | | | June | | | |
| **63** | —151 | +3942 | +3375 | —153 | +4064 | +3763 | 1 | —144 | +4210 | +2570 | —162 | +2384 | +1318 | **14** | —159 | +2682 | +5447 |
| **64** | 148 | 3946 | 3382 | 150 | 4068 | 3778 | 2 | 140 | 4214 | 2578 | 153 | 2306 | 1320 | **15** | 151 | 2677 | 5450 |
| **65** | 138 | 3949 | 3395 | 138 | 4072 | 3786 | 3 | 137 | 4219 | 2589 | 144 | 2300 | 1322 | **16** | 143 | 2672 | 5453 |
| **66** | 130 | 3952 | 3406 | 130 | 4076 | 3798 | 4 | 130 | 4224 | 2598 | 135 | 2292 | 1323 | **17** | 134 | 2668 | 5457 |
| **67** | 121 | 3955 | 3418 | 120 | 4080 | 3812 | 5 | 120 | 4229 | 2613 | 126 | 2285 | 1325 | **18** | 125 | 2663 | 5462 |
| **68** | 112 | 3958 | 3430 | 112 | 4084 | 3824 | 6 | 112 | 4234 | 2626 | 117 | 2278 | 1327 | **19** | 116 | 2658 | 5466 |
| **69** | 103 | 3962 | 3443 | 103 | 4088 | 3837 | 7 | 103 | 4238 | 2639 | 108 | 2271 | 1329 | **20** | 107 | 2653 | 5470 |
| **70** | 95 | 3965 | 3454 | 94 | 4092 | 3850 | 8 | 94 | 4243 | 2653 | 99 | 2264 | 1331 | **21** | 96 | 2648 | 5476 |
| **71** | 86 | 3968 | 3466 | 85 | 4096 | 3863 | 9 | 85 | 4248 | 2667 | 89 | 2257 | 1335 | **22** | 88 | 2644 | 5480 |
| **72** | 77 | 3971 | 3437 | 76 | 4100 | 3833 | 10 | 76 | 4253 | 2636 | 80 | 2250 | 1316 | **23** | 79 | 2639 | 5458 |
| **73** | 68 | 3974 | 3408 | 68 | 4104 | 3803 | 11 | 68 | 4258 | 2605 | 71 | 2243 | 1297 | **24** | 70 | 2634 | 5437 |
| **74** | 58 | 3978 | 3380 | 58 | 4108 | 3774 | 12 | 58 | 4262 | 2575 | 61 | 2236 | 1285 | **25** | 61 | 2629 | 5416 |
| **75** | 51 | 3981 | 3343 | 50 | 4112 | 3743 | 13 | 50 | 4267 | 2544 | 53 | 2227 | 1259 | **26** | 53 | 2624 | 5394 |
| **76** | 41 | 3984 | 3321 | 41 | 4116 | 3713 | 14 | 41 | 4272 | 2514 | 42 | 2221 | 1242 | **27** | 42 | 2620 | 5375 |
| **77** | 32 | 3987 | 3292 | 32 | 4120 | 3684 | 15 | 32 | 4277 | 2483 | 33 | 2214 | 1223 | **28** | 33 | 2615 | 5353 |
| **78** | 23 | 3990 | 3263 | 23 | 4124 | 3654 | 16 | 23 | 4282 | 2453 | 24 | 2207 | 1204 | **29** | 24 | 2610 | 5332 |
| **79** | 14 | 3994 | 3234 | 14 | 4128 | 3624 | 17 | 14 | 4286 | 2422 | 15 | 2200 | 1185 | **30** | 15 | 2605 | 5312 |
| | | | | | | | | | | | | | | July | | | |
| **80** | 5 | 3997 | 3206 | 5 | 4132 | 3595 | 18 | 5 | 4291 | 2392 | 5 | 2193 | 1167 | **1** | 5 | 2600 | 5291 |
| **81** | + 4 | 4000 | 3177 | + 4 | 4136 | 3565 | 19 | + 4 | 4296 | 2361 | + 4 | 2186 | 1148 | **2** | + 4 | 2596 | 5270 |
| **82** | 13 | 4003 | 3148 | 13 | 4140 | 3535 | 20 | 13 | 4301 | 2331 | 13 | 2179 | 1129 | **3** | 13 | 2591 | 5248 |
| **83** | 22 | 4006 | 3119 | 22 | 4144 | 3505 | 21 | 22 | 4306 | 2300 | 23 | 2172 | 1111 | **4** | 23 | 2586 | 5288 |
| **84** | 30 | 4010 | 3089 | 30 | 4148 | 3475 | 22 | 30 | 4310 | 2266 | 32 | 2165 | 1092 | **5** | 32 | 2581 | 5207 |
| **85** | 40 | 4013 | 3061 | 40 | 4152 | 3446 | 23 | 40 | 4315 | 2236 | 42 | 2158 | 1074 | **6** | 42 | 2576 | 5187 |
| **86** | 49 | 4016 | 3032 | 48 | 4156 | 3415 | 24 | 48 | 4320 | 2203 | 51 | 2150 | 1055 | **7** | 51 | 2571 | 5166 |
| **87** | 57 | 4019 | 3002 | 57 | 4160 | 3386 | 25 | 57 | 4325 | 2174 | 60 | 2143 | 1036 | **8** | 60 | 2567 | 5144 |
| **88** | 67 | 4022 | 2974 | 66 | 4164 | 3356 | 26 | 66 | 4330 | 2145 | 70 | 2136 | 1026 | **9** | 69 | 2562 | 5123 |
| **89** | 76 | 4026 | 2945 | 75 | 4168 | 3326 | 27 | 75 | 4334 | 2116 | 79 | 2129 | 1007 | **10** | 78 | 2557 | 5102 |
| **90** | 84 | 4029 | 2915 | 84 | 4172 | 3296 | 28 | 84 | 4339 | 2086 | 88 | 2122 | 980 | **11** | 87 | 2552 | 5081 |
| **91** | 110 | 4032 | 2903 | 111 | 4176 | 3285 | 29 | 111 | 4344 | 2073 | 106 | 2116 | 970 | **12** | 107 | 2548 | 5071 |
| **92** | 121 | 3994 | 2876 | 121 | 4137 | 3256 | 30 | 121 | 4304 | 2044 | 116 | 2088 | 952 | **13** | 117 | 2517 | 5050 |
| **93** | 131 | 3956 | 2848 | 132 | 4099 | 3228 | 31 | 132 | 4265 | 2015 | 126 | 2060 | 934 | **14** | 127 | 2487 | 5030 |
| **94** | 141 | 3918 | 2820 | 141 | 4060 | 3199 | 32 | 142 | 4225 | 1986 | 136 | 2038 | 916 | **15** | 137 | 2457 | 5010 |
| | | | | | | | *Tamil,* Adi; *Malayalam,* Karkatagam; *Bengal,* Sravana. | | | | | | | | | | |
| **95** | 151 | 3880 | 2792 | 151 | 4021 | 3170 | 1 | 152 | 4186 | 1956 | 146 | 2004 | 898 | **16** | 147 | 2427 | 4990 |
| **96** | 162 | 3842 | 2766 | 162 | 3982 | 3142 | 2 | 163 | 4147 | 1928 | 156 | 1976 | 880 | **17** | 157 | 2397 | 4970 |
| **97** | 172 | 3804 | 2738 | 172 | 3944 | 3113 | 3 | 173 | 4107 | 1898 | 166 | 1948 | 862 | **18** | 167 | 2366 | 4949 |
| **98** | 182 | 3766 | 2710 | 182 | 3905 | 3085 | 4 | 183 | 4068 | 1869 | 175 | 1920 | 843 | **19** | 177 | 2336 | 4929 |
| **99** | 192 | 3728 | 2662 | 193 | 3866 | 3057 | 5 | 193 | 4028 | 1839 | 185 | 1892 | 825 | **20** | 187 | 2307 | 4909 |
| **100** | 202 | 3691 | 2654 | 202 | 3828 | 3027 | 6 | 203 | 3989 | 1810 | 194 | 1864 | 806 | **21** | 196 | 2276 | 4888 |
| **101** | 211 | 3653 | 2625 | 212 | 3789 | 2999 | 7 | 212 | 3949 | 1779 | 204 | 1836 | 788 | **22** | 205 | 2246 | 4867 |
| **102** | 221 | 3615 | 2597 | 222 | 3750 | 2970 | 8 | 222 | 3910 | 1750 | 263 | 1808 | 769 | **23** | 215 | 2215 | 4847 |
| **103** | 231 | 3577 | 2557 | 232 | 3711 | 2928 | 9 | 232 | 3870 | 1706 | 223 | 1780 | 753 | **24** | 224 | 2185 | 4824 |
| **104** | 242 | 3539 | 2517 | 243 | 3673 | 2887 | 10 | 244 | 3831 | 1663 | 232 | 1752 | 735 | **25** | 235 | 2155 | 4803 |
| **105** | 251 | 3501 | 2476 | 252 | 3634 | 2844 | 11 | 253 | 3791 | 1618 | 242 | 1724 | 719 | **26** | 244 | 2125 | 4780 |
| **106** | 261 | 3463 | 2435 | 262 | 3595 | 2802 | 12 | 263 | 3752 | 1573 | 251 | 1696 | 702 | **27** | 253 | 2095 | 4758 |
| **107** | 270 | 3425 | 2394 | 271 | 3557 | 2759 | 13 | 272 | 3712 | 1528 | 260 | 1668 | 684 | **28** | 261 | 2064 | 4735 |
| **108** | 278 | 3387 | 2352 | 279 | 3518 | 2715 | 14 | 280 | 3673 | 1482 | 269 | 1640 | 667 | **29** | 271 | 2034 | 4714 |
| **109** | 288 | 3349 | 2311 | 288 | 3479 | 2672 | 15 | 289 | 3633 | 1436 | 278 | 1620 | 650 | **30** | 280 | 2004 | 4692 |
| **110** | 297 | 3311 | 2270 | 297 | 3440 | 2629 | 16 | 298 | 3594 | 1391 | 287 | 1584 | 633 | **31** | 290 | 1974 | 4671 |
| | | | | | | | | | | | | | | August | | | |
| **111** | 305 | 3273 | 2227 | 305 | 3402 | 2585 | 17 | 306 | 3554 | 1344 | 295 | 1556 | 614 | **1** | 297 | 1944 | 4647 |
| **112** | 315 | 3235 | 2187 | 316 | 3363 | 2544 | 18 | 316 | 3515 | 1300 | 304 | 1528 | 587 | **2** | 306 | 1913 | 4625 |
| **113** | 324 | 3197 | 2146 | 325 | 3324 | 2501 | 19 | 326 | 3475 | 1257 | 312 | 1500 | 579 | **3** | 315 | 1883 | 4603 |
| **114** | 333 | 3159 | 2104 | 334 | 3286 | 2468 | 20 | 335 | 3436 | 1210 | 321 | 1472 | 561 | **4** | 324 | 1853 | 4581 |
| **115** | 342 | 3121 | 2163 | 343 | 3247 | 2415 | 21 | 344 | 3396 | 1165 | 329 | 1444 | 543 | **5** | 332 | 1823 | 4558 |
| **116** | 350 | 3084 | 2020 | 351 | 3208 | 2371 | 22 | 352 | 3357 | 1118 | 338 | 1416 | 526 | **6** | 340 | 1793 | 4535 |
| **117** | 359 | 3046 | 1979 | 360 | 3169 | 2328 | 23 | 361 | 3317 | 1073 | 346 | 1388 | 507 | **7** | 348 | 1762 | 4512 |
| **118** | 367 | 3008 | 1937 | 368 | 3131 | 2284 | 24 | 369 | 3278 | 1027 | 354 | 1360 | 489 | **8** | 357 | 1732 | 4490 |
| **119** | 375 | 2970 | 1894 | 376 | 3092 | 2240 | 25 | 377 | 3238 | 980 | 362 | 1332 | 471 | **9** | 364 | 1702 | 4466 |
| **120** | 383 | 2932 | 1852 | 384 | 3053 | 2196 | 26 | 385 | 3199 | 934 | 369 | 1304 | 452 | **10** | 372 | 1672 | 4443 |
| **121** | 408 | 2894 | 1898 | 411 | 3015 | 2171 | 27 | 414 | 3159 | 908 | 373 | 1276 | 429 | **11** | 381 | 1642 | 4421 |
| **122** | 416 | 2856 | 1784 | 419 | 2976 | 2127 | 28 | 422 | 3120 | 862 | 381 | 1248 | 411 | **12** | 388 | 1612 | 4397 |
| **123** | 424 | 2806 | 1742 | 427 | 2924 | 2083 | 29 | 430 | 3066 | 816 | 388 | 1222 | 392 | **13** | 396 | 1580 | 4374 |
| **124** | 432 | 2755 | 1699 | 435 | 2872 | 2039 | 30 | 438 | 3011 | 769 | 396 | 1195 | 373 | **14** | 404 | 1548 | 4351 |
| **125** | 439 | 2705 | 1656 | 442 | 2820 | 1994 | 31 | 445 | 2957 | 722 | 403 | 1169 | 354 | **15** | 411 | 1516 | 4327 |

# TABLE XIII.

## Sunrise from 33° to 35° N. Latitude; also for BOMBAY and CALCUTTA.

| Day of Solar year | Lat. 33°—480″ Long.: Eqn. of time in seconds. | ⊙'s Trop. Long. in seconds. | Total Corrn. in seconds. Jehlum. A.D. 1840—1920 | Lat. 34°—228″ Long.: Eqn. of time in seconds. | ⊙'s Trop. Long. in seconds. | Total Corrn. in seconds. Srinagar. A.D. 1840—1920 | Days of solar months | Lat. 35°—1592″ Long.: Eqn. of time in seconds. | ⊙'s Trop. Long. in seconds. | Total Corrn. in seconds. Kabul. A.D. 1840—1920 | Lat. 19°—692″ Long.: Eqn. of time in seconds. | ⊙'s Trop. Long. in seconds. | Total Corrn. in seconds. Bombay. A.D. 1840—1920 | Eng. date | Lat. 22·5°+3020″ Long.: Eqn. of time in seconds. | ⊙'s Trop. Long. in seconds. | Total Corrn. in seconds. Calcutta. A.D. 1840—1920 |
|---|---|---|---|---|---|---|---|---|---|---|---|---|---|---|---|---|---|
| **126** | +447 | +2654 | +1613 | +450 | +2768 | +1950 | 1 (*Tamil*, Avani; *Malayalam*, Chingam; *Bengal*, Bhadrapada.) | +453 | +2902 | +675 | +410 | +1148 | +335 | **16** (August.) | +419 | +1485 | +4304 |
| **127** | 455 | 2604 | 1571 | 458 | 2716 | 1906 | 2 | 461 | 2848 | 629 | 417 | 1116 | 315 | **17** | 425 | 1454 | 4276 |
| **128** | 462 | 2554 | 1528 | 465 | 2664 | 1861 | 3 | 468 | 2794 | 582 | 424 | 1090 | 296 | **18** | 432 | 1423 | 4252 |
| **129** | 469 | 2503 | 1484 | 472 | 2612 | 1816 | 4 | 475 | 2739 | 534 | 430 | 1064 | 276 | **19** | 439 | 1392 | 4227 |
| **130** | 476 | 2453 | 1441 | 479 | 2560 | 1771 | 5 | 482 | 2685 | 487 | 437 | 1038 | 257 | **20** | 446 | 1361 | 4201 |
| **131** | 483 | 2402 | 1397 | 486 | 2508 | 1726 | 6 | 489 | 2630 | 439 | 443 | 1011 | 236 | **21** | 452 | 1330 | 4176 |
| **132** | 490 | 2352 | 1354 | 493 | 2456 | 1681 | 7 | 496 | 2576 | 392 | 450 | 985 | 218 | **22** | 458 | 1299 | 4150 |
| **133** | 497 | 2302 | 1318 | 500 | 2404 | 1642 | 8 | 503 | 2522 | 351 | 456 | 959 | 209 | **23** | 465 | 1268 | 4136 |
| **134** | 503 | 2251 | 1280 | 506 | 2352 | 1603 | 9 | 509 | 2467 | 309 | 462 | 932 | 200 | **24** | 471 | 1237 | 4120 |
| **135** | 509 | 2201 | 1243 | 512 | 2300 | 1563 | 10 | 515 | 2413 | 267 | 468 | 906 | 192 | **25** | 477 | 1206 | 4104 |
| **136** | 515 | 2150 | 1206 | 518 | 2248 | 1523 | 11 | 521 | 2358 | 225 | 474 | 880 | 183 | **26** | 483 | 1175 | 4068 |
| **137** | 521 | 2100 | 1168 | 524 | 2196 | 1484 | 12 | 527 | 2304 | 183 | 479 | 853 | 173 | **27** | 488 | 1144 | 4072 |
| **138** | 527 | 2050 | 1131 | 530 | 2144 | 1444 | 13 | 533 | 2250 | 141 | 485 | 827 | 164 | **28** | 493 | 1113 | 4055 |
| **139** | 533 | 1999 | 1093 | 536 | 2092 | 1404 | 14 | 539 | 2195 | 99 | 490 | 801 | 154 | **29** | 500 | 1082 | 4041 |
| **140** | 538 | 1949 | 1055 | 541 | 2040 | 1364 | 15 | 544 | 2141 | 56 | 495 | 775 | 145 | **30** | 505 | 1051 | 4028 |
| **141** | 544 | 1898 | 1018 | 547 | 1988 | 1324 | 16 | 550 | 2086 | 14 | 500 | 748 | 135 | **31** | 510 | 1020 | 4007 |
| **142** | 549 | 1848 | 979 | 552 | 1936 | 1283 | 17 | 555 | 2032 | — 29 | 505 | 722 | 125 | **1** (September.) | 515 | 989 | 3990 |
| **143** | 554 | 1798 | 941 | 557 | 1884 | 1242 | 18 | 560 | 1978 | 72 | 510 | 696 | 115 | **2** | 520 | 958 | 3974 |
| **144** | 559 | 1747 | 903 | 562 | 1832 | 1202 | 19 | 565 | 1923 | 115 | 514 | 669 | 104 | **3** | 525 | 927 | 3956 |
| **145** | 564 | 1697 | 864 | 567 | 1780 | 1161 | 20 | 570 | 1869 | 158 | 519 | 643 | 95 | **4** | 529 | 896 | 3939 |
| **146** | 569 | 1646 | 826 | 573 | 1728 | 1121 | 21 | 576 | 1814 | 200 | 523 | 617 | 84 | **5** | 533 | 865 | 3921 |
| **147** | 576 | 1596 | 790 | 580 | 1676 | 1083 | 22 | 585 | 1760 | 239 | 527 | 590 | 73 | **6** | 537 | 831 | 3904 |
| **148** | 578 | 1546 | 748 | 586 | 1624 | 1045 | 23 | 591 | 1706 | 281 | 532 | 564 | 63 | **7** | 541 | 800 | 3886 |
| **149** | 580 | 1495 | 707 | 592 | 1572 | 1003 | 24 | 596 | 1651 | 324 | 536 | 538 | 52 | **8** | 543 | 768 | 3866 |
| **150** | 582 | 1445 | 666 | 595 | 1520 | 961 | 25 | 601 | 1597 | 367 | 539 | 512 | 41 | **9** | 546 | 735 | 3847 |
| **151** | 584 | 1394 | 624 | 596 | 1468 | 916 | 26 | 602 | 1542 | 414 | 542 | 485 | 29 | **10** | 549 | 704 | 3829 |
| **152** | 586 | 1344 | 583 | 597 | 1416 | 871 | 27 | 603 | 1488 | 461 | 523 | 460 | — 5 | **11** | 540 | 672 | 3798 |
| **153** | 588 | 1301 | 541 | 598 | 1370 | 827 | 28 | 604 | 1440 | 508 | 526 | 445 | 17 | **12** | 542 | 651 | 3779 |
| **154** | 590 | 1257 | 500 | 599 | 1325 | 782 | 29 | 605 | 1392 | 555 | 529 | 430 | 29 | **13** | 545 | 629 | 3760 |
| **155** | 592 | 1214 | 459 | 600 | 1279 | 737 | 30 | 606 | 1344 | 602 | 532 | 416 | 40 | **14** | 548 | 607 | 3736 |
| **156** | 594 | 1171 | 417 | 601 | 1233 | 693 | 31 | 607 | 1296 | 649 | 534 | 401 | 53 | **15** | 550 | 585 | 3721 |
| **157** | 596 | 1127 | 376 | 602 | 1188 | 648 | 1 (*Tamil*, Purattasi; *Malayalam*, Kanni; *Bengal*, Asvina.) | 608 | 1248 | 696 | 536 | 386 | 66 | **16** | 552 | 564 | 3701 |
| **158** | 599 | 1084 | 336 | 604 | 1142 | 604 | 2 | 609 | 1200 | 743 | 538 | 371 | 79 | **17** | 554 | 542 | 3672 |
| **159** | 601 | 1040 | 294 | 606 | 1096 | 561 | 3 | 610 | 1152 | 790 | 540 | 356 | 92 | **18** | 556 | 521 | 3662 |
| **160** | 603 | 997 | 253 | 608 | 1051 | 517 | 4 | 612 | 1104 | 836 | 542 | 342 | 104 | **19** | 558 | 498 | 3642 |
| **161** | 605 | 954 | 212 | 610 | 1005 | 473 | 5 | 614 | 1056 | 882 | 544 | 327 | 120 | **20** | 560 | 477 | 3622 |
| **162** | 606 | 910 | 169 | 611 | 959 | 429 | 6 | 615 | 1008 | 929 | 546 | 312 | 130 | **21** | 561 | 455 | 3602 |
| **163** | 608 | 867 | 128 | 613 | 913 | 385 | 7 | 617 | 960 | 975 | 547 | 297 | 145 | **22** | 563 | 434 | 3583 |
| **164** | 610 | 824 | 85 | 615 | 868 | 340 | 8 | 619 | 912 | 1023 | 548 | 282 | 159 | **23** | 565 | 412 | 3563 |
| **165** | 611 | 780 | 41 | 616 | 822 | 294 | 9 | 620 | 864 | 1071 | 550 | 268 | 173 | **24** | 566 | 390 | 3542 |
| **166** | 611 | 737 | — 3 | 616 | 776 | 246 | 10 | 620 | 816 | 1121 | 551 | 253 | 187 | **25** | 566 | 368 | 3519 |
| **167** | 612 | 694 | 47 | 617 | 731 | 200 | 11 | 621 | 768 | 1169 | 552 | 238 | 201 | **26** | 567 | 347 | 3496 |
| **168** | 613 | 650 | 91 | 618 | 685 | 154 | 12 | 622 | 720 | 1218 | 553 | 223 | 215 | **27** | 567 | 325 | 3475 |
| **169** | 614 | 607 | 135 | 619 | 639 | 108 | 13 | 623 | 672 | 1267 | 554 | 208 | 230 | **28** | 568 | 303 | 3454 |
| **170** | 614 | 564 | 180 | 619 | 594 | 61 | 14 | 623 | 624 | 1316 | 554 | 194 | 245 | **29** | 569 | 281 | 3408 |
| **171** | 614 | 520 | 224 | 619 | 548 | 13 | 15 | 623 | 576 | 1366 | 554 | 179 | 260 | **30** | 569 | 260 | 3410 |
| **172** | 614 | 477 | 269 | 619 | 502 | — 34 | 16 | 623 | 528 | 1415 | 454 | 164 | 276 | **1** (October.) | 570 | 238 | 3389 |
| **173** | 614 | 433 | 314 | 619 | 457 | 81 | 17 | 623 | 480 | 1465 | 554 | 149 | 291 | **2** | 569 | 217 | 3365 |
| **174** | 614 | 390 | 359 | 619 | 411 | 128 | 18 | 623 | 432 | 1515 | 554 | 134 | 306 | **3** | 569 | 195 | 3343 |
| **175** | 614 | 347 | 404 | 619 | 365 | 175 | 19 | 623 | 384 | 1564 | 554 | 120 | 322 | **4** | 569 | 173 | 3321 |
| **176** | 614 | 303 | 448 | 619 | 320 | 223 | 20 | 623 | 336 | 1614 | 554 | 105 | 337 | **5** | 569 | 151 | 3298 |
| **177** | 613 | 260 | 494 | 618 | 274 | 271 | 21 | 622 | 288 | 1664 | 553 | 90 | 353 | **6** | 568 | 129 | 3275 |
| **178** | 612 | 217 | 540 | 617 | 228 | 319 | 22 | 621 | 240 | 1715 | 552 | 75 | 369 | **7** | 567 | 108 | 3251 |
| **179** | 611 | 173 | 586 | 616 | 183 | 367 | 23 | 620 | 192 | 1766 | 551 | 60 | 386 | **8** | 566 | 86 | 3238 |
| **180** | 610 | 130 | 632 | 615 | 137 | 415 | 24 | 619 | 144 | 1816 | 550 | 46 | 402 | **9** | 565 | 64 | 3205 |
| **181** | 608 | 87 | 678 | 613 | 91 | 465 | 25 | 617 | 96 | 1868 | 548 | 31 | 419 | **10** | 563 | 42 | 3180 |
| **182** | 607 | 43 | 724 | 612 | 46 | 513 | 26 | 616 | 48 | 1918 | 547 | 16 | 436 | **11** | 562 | 21 | 3147 |
| **183** | 607 | 0 | 769 | 612 | 0 | 562 | 27 | 616 | 0 | 1968 | 545 | 0 | 453 | **12** | 561 | 0 | 3133 |
| **184** | 606 | — 45 | 815 | 611 | — 47 | 608 | 28 | 616 | — 50 | 2018 | 544 | — 15 | 469 | **13** | 559 | — 22 | 3124 |
| **185** | 604 | 90 | 862 | 609 | 94 | 657 | 29 | 614 | 99 | 2069 | 542 | 31 | 487 | **14** | 557 | 44 | 3085 |
| **186** | 602 | 134 | 908 | 607 | 142 | 707 | 30 | 612 | 149 | 2121 | 540 | 46 | 504 | **15** | 555 | 67 | 3060 |

# TABLE XIII.

## Sunrise from 33° to 35° N. Latitude; also for BOMBAY and CALCUTTA.

| Day of Solar year. | Lat. 33°—480″ Long. Eqn. of time in seconds. | ☉'s Trop. Long. in seconds. | Total Corrn. in seconds. Jehlum. A.D. 1840—1920 | Lat. 34°—228″ Long. Eqn. of time in seconds. | ☉'s Trop. Long. in seconds. | Total Corrn. in seconds. Srinagar. A.D. 1840—1920 | Solar months | Days of solar months. | Lat. 35°—1592″ Long. Eqn. of time in seconds. | ☉'s Trop. Long. in seconds. | Total Corrn. in seconds. Kabul. A.D. 1840—1920 | Lat. 19°—692″ Long. Eqn. of time in seconds. | ☉'s Trop. Long. in seconds. | Total Corrn. in seconds. Bombay. A.D. 1840—1920 | Eng. month | Eng. date. | Lat. 22·5°+3020″ Long. Eqn. of time in seconds. | ☉'s Trop. Long. in seconds. | Total Corrn. in seconds. Calcutta. A.D. 1840—1920 |
|---|---|---|---|---|---|---|---|---|---|---|---|---|---|---|---|---|---|---|---|
| **187** | +600 | —179 | — 955 | +605 | — 189 | — 756 | *Tamil,* Aippasi; *Malayalam,* Tulam; *Bengal,* Kartika. | 1 | +610 | —198 | —2173 | +538 | — 61 | —511 | October | **16** | +552 | — 89 | +3035 |
| **188** | 597 | 224 | 1003 | 602 | 236 | 806 | | 2 | 607 | 248 | 2225 | 536 | 76 | 538 | | **17** | 551 | 112 | 3011 |
| **189** | 604 | 269 | 1041 | 609 | 283 | 846 | | 3 | 614 | 298 | 2268 | 533 | 92 | 557 | | **18** | 548 | 134 | 2986 |
| **190** | 601 | 314 | 1091 | 606 | 330 | 896 | | 4 | 611 | 347 | 2310 | 530 | 107 | 575 | | **19** | 545 | 156 | 2961 |
| **191** | 588 | 358 | 1146 | 593 | 378 | 953 | | 5 | 598 | 397 | 2383 | 527 | 122 | 593 | | **20** | 542 | 179 | 2935 |
| **192** | 585 | 403 | 1194 | 590 | 425 | 1007 | | 6 | 595 | 446 | 2435 | 524 | 138 | 612 | | **21** | 538 | 201 | 2919 |
| **193** | 581 | 448 | 1243 | 586 | 472 | 1058 | | 7 | 591 | 496 | 2489 | 521 | 153 | 631 | | **22** | 535 | 224 | 2888 |
| **194** | 578 | 493 | 1296 | 583 | 519 | 1113 | | 8 | 588 | 546 | 2546 | 518 | 168 | 660 | | **23** | 532 | 246 | 2848 |
| **195** | 574 | 538 | 1351 | 579 | 566 | 1169 | | 9 | 584 | 595 | 2605 | 515 | 184 | 690 | | **24** | 529 | 268 | 2814 |
| **196** | 570 | 582 | 1405 | 575 | 614 | 1225 | | 10 | 580 | 645 | 2663 | 511 | 199 | 720 | | **25** | 525 | 291 | 2778 |
| **197** | 566 | 627 | 1460 | 571 | 661 | 1281 | | 11 | 576 | 694 | 2722 | 507 | 214 | 750 | | **26** | 521 | 318 | 2741 |
| **198** | 562 | 672 | 1514 | 567 | 708 | 1337 | | 12 | 572 | 744 | 2780 | 503 | 229 | 780 | | **27** | 517 | 336 | 2696 |
| **199** | 558 | 717 | 1568 | 563 | 755 | 1393 | | 13 | 568 | 794 | 2838 | 499 | 245 | 811 | | **28** | 513 | 358 | 2670 |
| **200** | 553 | 762 | 1624 | 558 | 802 | 1450 | | 14 | 563 | 843 | 2898 | 494 | 260 | 842 | | **29** | 508 | 380 | 2633 |
| **201** | 548 | 806 | 1679 | 553 | 850 | 1507 | | 15 | 558 | 893 | 2957 | 490 | 275 | 872 | | **30** | 503 | 403 | 2596 |
| **202** | 544 | 851 | 1734 | 549 | 897 | 1563 | | 16 | 553 | 942 | 3017 | 486 | 291 | 909 | | **31** | 499 | 425 | 2560 |
| **203** | 539 | 896 | 1789 | 544 | 944 | 1620 | | 17 | 549 | 992 | 3075 | 481 | 306 | 934 | November | **1** | 494 | 448 | 2523 |
| **204** | 533 | 941 | 1845 | 538 | 991 | 1678 | | 18 | 543 | 1042 | 3135 | 476 | 321 | 965 | | **2** | 489 | 455 | 2486 |
| **205** | 528 | 986 | 1901 | 533 | 1038 | 1735 | | 19 | 538 | 1091 | 3195 | 471 | 337 | 997 | | **3** | 484 | 492 | 2449 |
| **206** | 523 | 1030 | 1956 | 528 | 1086 | 1792 | | 20 | 533 | 1141 | 3254 | 466 | 352 | 1028 | | **4** | 479 | 515 | 2412 |
| **207** | 519 | 1075 | 2011 | 524 | 1133 | 1848 | | 21 | 529 | 1191 | 3313 | 462 | 367 | 1057 | | **5** | 475 | 537 | 2376 |
| **208** | 513 | 1120 | 2067 | 518 | 1180 | 1906 | | 22 | 523 | 1240 | 3373 | 457 | 382 | 1089 | | **6** | 469 | 560 | 2338 |
| **209** | 507 | 1165 | 2123 | 512 | 1227 | 1964 | | 23 | 517 | 1290 | 3433 | 451 | 398 | 1122 | | **7** | 463 | 582 | 2300 |
| **210** | 501 | 1210 | 2180 | 506 | 1274 | 2022 | | 24 | 511 | 1339 | 3494 | 446 | 413 | 1153 | | **8** | 458 | 604 | 2263 |
| **211** | 495 | 1254 | 2236 | 500 | 1322 | 2080 | | 25 | 505 | 1389 | 3554 | 440 | 428 | 1185 | | **9** | 452 | 627 | 2225 |
| **212** | 488 | 1299 | 2294 | 493 | 1369 | 2139 | | 26 | 498 | 1438 | 3616 | 434 | 444 | 1216 | | **10** | 450 | 649 | 2191 |
| **213** | 484 | 1344 | 2348 | 487 | 1416 | 2197 | | 27 | 490 | 1488 | 3678 | 443 | 460 | 1235 | | **11** | 443 | 672 | 2157 |
| **214** | 477 | 1394 | 2405 | 480 | 1468 | 2256 | | 28 | 483 | 1542 | 3739 | 437 | 486 | 1267 | | **12** | 445 | 704 | 2122 |
| **215** | 470 | 1445 | 2463 | 473 | 1520 | 2315 | | 29 | 476 | 1597 | 3801 | 430 | 513 | 1301 | | **13** | 440 | 735 | 2186 |
| **216** | 463 | 1495 | 2520 | 466 | 1572 | 2374 | | 30 | 469 | 1651 | 3862 | 423 | 539 | 1334 | | **14** | 433 | 767 | 2046 |
| **217** | 456 | 1546 | 2578 | 459 | 1624 | 2433 | *Tamil,* Kartigai; *Malayalam,* Vrischikam; *Bengal,* Margasira. | 1 | 462 | 1706 | 3924 | 416 | 565 | 1367 | | **15** | 426 | 800 | 2007 |
| **218** | 449 | 1596 | 2635 | 452 | 1676 | 2492 | | 2 | 455 | 1760 | 3975 | 409 | 591 | 1400 | | **16** | 419 | 831 | 1969 |
| **219** | 441 | 1646 | 2693 | 444 | 1728 | 2552 | | 3 | 447 | 1814 | 4047 | 402 | 618 | 1434 | | **17** | 412 | 863 | 1930 |
| **220** | 434 | 1697 | 2751 | 437 | 1780 | 2611 | | 4 | 440 | 1869 | 4109 | 395 | 644 | 1467 | | **18** | 405 | 895 | 1891 |
| **221** | 426 | 1747 | 2809 | 429 | 1832 | 2671 | | 5 | 432 | 1923 | 4171 | 388 | 670 | 1500 | | **19** | 397 | 927 | 1850 |
| **222** | 418 | 1798 | 2868 | 421 | 1884 | 2731 | | 6 | 424 | 1978 | 4234 | 381 | 697 | 1534 | | **20** | 389 | 959 | 1811 |
| **223** | 410 | 1848 | 2926 | 413 | 1936 | 2791 | | 7 | 416 | 2032 | 4296 | 374 | 723 | 1566 | | **21** | 382 | 991 | 1778 |
| **224** | 402 | 1898 | 2972 | 405 | 1988 | 2838 | | 8 | 408 | 2086 | 4343 | 366 | 749 | 1602 | | **22** | 374 | 1023 | 1734 |
| **225** | 394 | 1949 | 3018 | 397 | 2040 | 2884 | | 9 | 400 | 2141 | 4391 | 358 | 776 | 1638 | | **23** | 366 | 1055 | 1695 |
| **226** | 386 | 1999 | 3064 | 389 | 2092 | 2931 | | 10 | 392 | 2195 | 4438 | 350 | 802 | 1674 | | **24** | 358 | 1087 | 1657 |
| **227** | 377 | 2050 | 3111 | 380 | 2144 | 2979 | | 11 | 383 | 2250 | 4487 | 342 | 828 | 1710 | | **25** | 350 | 1119 | 1629 |
| **228** | 369 | 2100 | 3157 | 372 | 2196 | 3025 | | 12 | 375 | 2304 | 4534 | 334 | 854 | 1746 | | **26** | 341 | 1151 | 1599 |
| **229** | 360 | 2150 | 3204 | 363 | 2248 | 3073 | | 13 | 366 | 2358 | 4583 | 326 | 881 | 1782 | | **27** | 333 | 1183 | 1560 |
| **230** | 351 | 2201 | 3250 | 354 | 2300 | 3121 | | 14 | 387 | 2413 | 4631 | 317 | 907 | 1819 | | **28** | 324 | 1215 | 1521 |
| **231** | 342 | 2251 | 3297 | 345 | 2352 | 3169 | | 15 | 348 | 2467 | 4680 | 309 | 933 | 1855 | | **29** | 316 | 1247 | 1483 |
| **232** | 333 | 2302 | 3344 | 336 | 2404 | 3216 | | 16 | 339 | 2522 | 4728 | 301 | 960 | 1891 | | **30** | 307 | 1279 | 1444 |
| **233** | 324 | 2352 | 3391 | 327 | 2456 | 3264 | | 17 | 330 | 2576 | 4777 | 293 | 986 | 1927 | December | **1** | 299 | 1311 | 1405 |
| **234** | 314 | 2402 | 3439 | 317 | 2508 | 3313 | | 18 | 320 | 2630 | 4826 | 284 | 1012 | 1964 | | **2** | 290 | 1343 | 1366 |
| **235** | 305 | 2453 | 3486 | 308 | 2560 | 3360 | | 19 | 311 | 2685 | 4875 | 275 | 1039 | 2001 | | **3** | 281 | 1374 | 1327 |
| **236** | 296 | 2508 | 3533 | 299 | 2612 | 3408 | | 20 | 302 | 2739 | 4923 | 266 | 1065 | 2038 | | **4** | 272 | 1407 | 1288 |
| **237** | 287 | 2554 | 3580 | 290 | 2664 | 3456 | | 21 | 293 | 2794 | 4972 | 257 | 1091 | 2075 | | **5** | 263 | 1439 | 1249 |
| **238** | 277 | 2604 | 3628 | 280 | 2716 | 3505 | | 22 | 283 | 2848 | 5021 | 248 | 1117 | 2112 | | **6** | 254 | 1470 | 1209 |
| **239** | 267 | 2654 | 3676 | 270 | 2768 | 3553 | | 23 | 273 | 2902 | 5071 | 239 | 1144 | 2149 | | **7** | 245 | 1502 | 1170 |
| **240** | 258 | 2705 | 3723 | 261 | 2820 | 3601 | | 24 | 264 | 2957 | 5119 | 230 | 1170 | 2186 | | **8** | 236 | 1534 | 1131 |
| **241** | 246 | 2755 | 3773 | 248 | 2872 | 3653 | | 25 | 251 | 3011 | 5172 | 220 | 1196 | 2224 | | **9** | 226 | 1567 | 1091 |
| **242** | 236 | 2806 | 3821 | 238 | 2924 | 3701 | | 26 | 241 | 3066 | 5221 | 211 | 1223 | 2261 | | **10** | 216 | 1598 | 1051 |
| **243** | 211 | 2856 | 3884 | 212 | 2976 | 3766 | | 27 | 213 | 3120 | 5287 | 203 | 1248 | 2297 | | **11** | 205 | 1630 | 1009 |
| **244** | 202 | 2894 | 3931 | 203 | 3015 | 3814 | | 28 | 204 | 3159 | 5337 | 194 | 1276 | 2334 | | **12** | 196 | 1660 | 970 |
| **245** | 192 | 2932 | 3979 | 193 | 3053 | 3863 | | 29 | 194 | 3199 | 5387 | 184 | 1304 | 2372 | | **13** | 186 | 1691 | 930 |
| **246** | 182 | 2970 | 4026 | 184 | 3092 | 3911 | | 30 | 184 | 3238 | 5436 | 174 | 1332 | 2410 | | **14** | 176 | 1721 | 890 |

# TABLE XIII.

## Sunrise from 33° to 35° N. Latitude; also for BOMBAY and CALCUTTA.

Days of solar months 1–29 (Days 247–275): Tamil, Margali; *Malayalam*, Dhanus; *Bengal*, Pausha.
Days of solar months 1–30 (Days 276–305): Tamil, Tai; *Malayalam*, Magaram; *Bengal*, Magha.
Eng. date: December (15–31), January (1–31), February (1–11).

| | Lat. 33° —480″ Long. | | | Lat. 34°—228″ Long. | | | | Lat. 35° —1592″ Long. | | | Lat. 19°—692″ Long. | | | | Lat. 22·5° +3020″ Long. | | |
|---|---|---|---|---|---|---|---|---|---|---|---|---|---|---|---|---|---|
| Day of Solar year. | Eqn. of time in seconds. | ☉'s Trop. Long. in seconds. | Total Corrn. in seconds. Jehlum. A.D. 1840—1920 | Eqn. of time in seconds. | ☉'s Trop. Long. in seconds. | Total Corrn. in seconds. Srinagar. A.D. 1840—1920 | Days of solar months. | Eqn. of time in seconds. | ☉'s Trop. Long. in seconds. | Total Corrn. in seconds. Kabul. A.D. 1840—1920 | Eqn. of time in seconds. | ☉'s Trop. Long. in seconds. | Total Corrn. in seconds. Bombay. A.D. 1840—1920 | Eng. date. | Eqn. of time in seconds. | ☉'s Trop. Long. in seconds. | Total Corrn. in seconds. Calcutta. A.D. 1840—1920 |
| **247** | +172 | —3008 | —4074 | +173 | —3131 | —3960 | 1 | +174 | —3278 | —5486 | +164 | —1360 | —2448 | **15** | +166 | —1741 | +850 |
| **248** | 162 | 3046 | 4122 | 163 | 3169 | 4009 | 2 | 164 | 3317 | 5535 | 154 | 1388 | 2486 | **16** | 156 | 1762 | 809 |
| **249** | 152 | 3084 | 4170 | 153 | 3208 | 4057 | 3 | 154 | 3357 | 5585 | 144 | 1416 | 2524 | **17** | 146 | 1793 | 769 |
| **250** | 142 | 3121 | 4218 | 143 | 3247 | 4106 | 4 | 144 | 3396 | 5634 | 134 | 1444 | 2562 | **18** | 135 | 1823 | 728 |
| **251** | 132 | 3159 | 4266 | 133 | 3286 | 4155 | 5 | 134 | 3436 | 5683 | 124 | 1472 | 2600 | **19** | 125 | 1853 | 688 |
| **252** | 122 | 3197 | 4314 | 123 | 3324 | 4204 | 6 | 124 | 3475 | 5733 | 114 | 1500 | 2638 | **20** | 115 | 1883 | 648 |
| **253** | 112 | 3235 | 4362 | 113 | 3363 | 4252 | 7 | 114 | 3515 | 5782 | 104 | 1528 | 2676 | **21** | 105 | 1914 | 607 |
| **254** | 102 | 3273 | 4410 | 103 | 3402 | 4301 | 8 | 104 | 3554 | 5832 | 94 | 1556 | 2714 | **22** | 95 | 1944 | 567 |
| **255** | 92 | 3311 | 4417 | 93 | 3440 | 4307 | 9 | 94 | 3594 | 5837 | 84 | 1584 | 2731 | **23** | 85 | 1974 | 553 |
| **256** | 82 | 3349 | 4424 | 83 | 3479 | 4313 | 10 | 84 | 3633 | 5842 | 74 | 1612 | 2754 | **24** | 75 | 2004 | 538 |
| **257** | 72 | 3387 | 4430 | 73 | 3518 | 4319 | 11 | 74 | 3673 | 5848 | 64 | 1640 | 2765 | **25** | 65 | 2034 | 525 |
| **258** | 62 | 3425 | 4437 | 63 | 3557 | 4325 | 12 | 64 | 3712 | 5853 | 54 | 1668 | 2782 | **26** | 55 | 2065 | 508 |
| **259** | 51 | 3463 | 4445 | 52 | 3595 | 4332 | 13 | 53 | 3752 | 5859 | 44 | 1696 | 2799 | **27** | 44 | 2095 | 492 |
| **260** | 41 | 3501 | 4452 | 42 | 3634 | 4338 | 14 | 43 | 3791 | 5864 | 34 | 1724 | 2817 | **28** | 34 | 2125 | 478 |
| **261** | 31 | 3539 | 4459 | 32 | 3673 | 4344 | 15 | 33 | 3831 | 5869 | 24 | 1752 | 2834 | **29** | 24 | 2155 | 463 |
| **262** | 21 | 3577 | 4465 | 22 | 3711 | 4350 | 16 | 23 | 3870 | 5875 | 14 | 1780 | 2851 | **30** | 14 | 2185 | 448 |
| **263** | 11 | 3615 | 4472 | 12 | 3750 | 4356 | 17 | 13 | 3910 | 5880 | 2 | 1808 | 2868 | **31** | 2 | 2216 | 431 |
| **264** | — 19 | 3653 | 4499 | — 20 | 3789 | 4384 | 18 | — 21 | 3949 | 5909 | — 10 | 1836 | 2891 | **1** | — 11 | 2246 | 414 |
| **265** | 29 | 3691 | 4506 | 30 | 3828 | 4390 | 19 | 31 | 3989 | 5914 | 20 | 1864 | 2908 | **2** | 21 | 2276 | 399 |
| **266** | 39 | 3728 | 4513 | 40 | 3866 | 4396 | 20 | 41 | 4028 | 5919 | 32 | 1892 | 2925 | **3** | 32 | 2306 | 383 |
| **267** | 49 | 3766 | 4519 | 50 | 3905 | 4402 | 21 | 51 | 4068 | 5925 | 41 | 1920 | 2941 | **4** | 42 | 2336 | 368 |
| **268** | 59 | 3804 | 4526 | 60 | 3944 | 4408 | 22 | 61 | 4107 | 5930 | 51 | 1948 | 2958 | **5** | 52 | 2367 | 353 |
| **269** | 70 | 3842 | 4534 | 71 | 3982 | 4415 | 23 | 72 | 4147 | 5936 | 61 | 1976 | 2975 | **6** | 63 | 2397 | 337 |
| **270** | 80 | 3880 | 4541 | 81 | 4021 | 4421 | 24 | 82 | 4186 | 5941 | 70 | 2004 | 2992 | **7** | 73 | 2427 | 323 |
| **271** | 91 | 3918 | 4549 | 92 | 4060 | 4428 | 25 | 93 | 4225 | 5947 | 81 | 2032 | 3010 | **8** | 84 | 2457 | 307 |
| **272** | 92 | 3956 | 4546 | 93 | 4099 | 4425 | 26 | 94 | 4265 | 5944 | 91 | 2060 | 3027 | **9** | 94 | 2487 | 292 |
| **273** | 93 | 3994 | 4544 | 94 | 4137 | 4422 | 27 | 95 | 4304 | 5940 | 101 | 2088 | 3044 | **10** | 100 | 2518 | 281 |
| **274** | 98 | 4032 | 4546 | 97 | 4176 | 4421 | 28 | 97 | 4344 | 5937 | 102 | 2116 | 3052 | **11** | 102 | 2548 | 274 |
| **275** | 103 | 4029 | 4548 | 102 | 4172 | 4422 | 29 | 102 | 4339 | 5937 | 109 | 2123 | 3066 | **12** | 107 | 2552 | 265 |
| **276** | 109 | 4026 | 4551 | 108 | 4168 | 4424 | 1 | 107 | 4334 | 5937 | 115 | 2130 | 3079 | **13** | 114 | 2557 | 253 |
| **277** | 115 | 4022 | 4553 | 114 | 4164 | 4426 | 2 | 113 | 4330 | 5939 | 122 | 2137 | 3093 | **14** | 120 | 2561 | 243 |
| **278** | 120 | 4019 | 4555 | 119 | 4160 | 4427 | 3 | 119 | 4325 | 5940 | 129 | 2144 | 3107 | **15** | 127 | 2567 | 231 |
| **279** | 126 | 4016 | 4558 | 125 | 4156 | 4429 | 4 | 124 | 4320 | 5940 | 135 | 2151 | 3120 | **16** | 133 | 2572 | 221 |
| **280** | 132 | 4013 | 4561 | 131 | 4152 | 4431 | 5 | 130 | 4315 | 5941 | 142 | 2159 | 3135 | **17** | 140 | 2576 | 210 |
| **281** | 137 | 4010 | 4563 | 136 | 4148 | 4432 | 6 | 136 | 4310 | 5942 | 149 | 2166 | 3149 | **18** | 146 | 2581 | 199 |
| **282** | 143 | 4006 | 4565 | 142 | 4144 | 4434 | 7 | 141 | 4306 | 5943 | 155 | 2173 | 3172 | **19** | 152 | 2586 | 188 |
| **283** | 149 | 4003 | 4568 | 148 | 4140 | 4436 | 8 | 146 | 4301 | 5943 | 162 | 2180 | 3176 | **20** | 159 | 2591 | 176 |
| **284** | 154 | 4000 | 4570 | 153 | 4136 | 4437 | 9 | 152 | 4296 | 5944 | 169 | 2187 | 3189 | **21** | 165 | 2595 | 166 |
| **285** | 160 | 3997 | 4526 | 159 | 4132 | 4391 | 10 | 158 | 4291 | 5896 | 175 | 2194 | 3168 | **22** | 171 | 2600 | 190 |
| **286** | 166 | 3994 | 4481 | 165 | 4128 | 4345 | 11 | 163 | 4286 | 5847 | 182 | 2201 | 3149 | **23** | 178 | 2605 | 212 |
| **287** | 171 | 3990 | 4436 | 170 | 4124 | 4298 | 12 | 168 | 4282 | 5797 | 189 | 2208 | 3129 | **24** | 184 | 2610 | 240 |
| **288** | 177 | 3987 | 4392 | 176 | 4120 | 4252 | 13 | 174 | 4277 | 5749 | 195 | 2215 | 3108 | **25** | 190 | 2615 | 266 |
| **289** | 183 | 3984 | 4347 | 182 | 4116 | 4201 | 14 | 180 | 4272 | 5701 | 202 | 2222 | 3088 | **26** | 197 | 2620 | 291 |
| **290** | 188 | 3981 | 4302 | 187 | 4112 | 4160 | 15 | 185 | 4267 | 5652 | 209 | 2230 | 3069 | **27** | 204 | 2624 | 316 |
| **291** | 194 | 3978 | 4258 | 193 | 4108 | 4114 | 16 | 191 | 4262 | 5604 | 215 | 2237 | 3048 | **28** | 210 | 2629 | 352 |
| **292** | 200 | 3974 | 4213 | 199 | 4104 | 4068 | 17 | 197 | 4258 | 5555 | 222 | 2244 | 3028 | **29** | 216 | 2634 | 368 |
| **293** | 205 | 3971 | 4168 | 204 | 4100 | 4021 | 18 | 202 | 4253 | 5506 | 229 | 2251 | 3009 | **30** | 222 | 2639 | 394 |
| **294** | 211 | 3968 | 4124 | 210 | 4096 | 3975 | 19 | 207 | 4248 | 5457 | 235 | 2258 | 2988 | **31** | 229 | 2644 | 419 |
| **295** | 216 | 3965 | 4078 | 216 | 4092 | 3929 | 20 | 213 | 4242 | 5409 | 242 | 2265 | 2968 | **1** | 236 | 2648 | 444 |
| **296** | 222 | 3962 | 4034 | 221 | 4088 | 3882 | 21 | 219 | 4238 | 5361 | 249 | 2272 | 2949 | **2** | 242 | 2653 | 470 |
| **297** | 228 | 3958 | 3990 | 227 | 4084 | 3837 | 22 | 224 | 4234 | 5311 | 255 | 2279 | 2928 | **3** | 248 | 2657 | 496 |
| **298** | 233 | 3955 | 3944 | 233 | 4080 | 3791 | 23 | 229 | 4229 | 5262 | 262 | 2286 | 2908 | **4** | 255 | 2662 | 521 |
| **299** | 239 | 3952 | 3900 | 238 | 4076 | 3744 | 24 | 235 | 4224 | 5214 | 269 | 2295 | 2888 | **5** | 261 | 2666 | 547 |
| **300** | 245 | 3949 | 3856 | 244 | 4072 | 3698 | 25 | 241 | 4219 | 5166 | 275 | 2301 | 2868 | **6** | 268 | 2670 | 572 |
| **301** | 250 | 3946 | 3810 | 249 | 4068 | 3651 | 26 | 246 | 4214 | 5111 | 282 | 2308 | 2848 | **7** | 274 | 2675 | 598 |
| **302** | 256 | 3942 | 3766 | 254 | 4064 | 3604 | 27 | 251 | 4210 | 5067 | 289 | 2315 | 2828 | **8** | 280 | 2680 | 624 |
| **303** | 262 | 3939 | 3722 | 259 | 4060 | 3557 | 28 | 256 | 4205 | 5018 | 295 | 2322 | 2808 | **9** | 286 | 2685 | 650 |
| **304** | 267 | 3936 | 3676 | 264 | 4056 | 3511 | 29 | 261 | 4200 | 4969 | 302 | 2328 | 2788 | **10** | 293 | 2689 | 675 |
| **305** | 269 | 3886 | 3628 | 266 | 4004 | 3461 | 30 | 263 | 4146 | 4917 | 305 | 2301 | 2764 | **11** | 296 | 2659 | 701 |

# TABLE XIII.

## Sunrise from 33° to 35° N. Latitude; also for BOMBAY and CALCUTTA.

| | Lat. 33°—480″ Long. | | | Lat. 34°—228″ Long. | | | | Lat. 35°—1592″ Long. | | | Lat. 19°—692″ Long. | | | | Lat. 22·5°+3020″ Long. | | |
|---|---|---|---|---|---|---|---|---|---|---|---|---|---|---|---|---|---|
| **Day of Solar year.** | Eqn. of time in seconds. | ⊙'s Trop. Long. in seconds. | Total Corrn. in seconds. **Jehlum.** A.D. 1840—1920 | Eqn. of time in seconds. | ⊙'s Trop. Long. in seconds. | Total Corrn. in seconds. **Srinagar.** A.D 1840—1920 | Days of solar months. | Eqn. of time in seconds. | ⊙'s Trop. Long. in seconds. | Total Corrn. in seconds. **Kabul.** A.D. 1840—1920 | Eqn. of time in seconds. | ⊙'s Trop. Long. in seconds. | Total Corrn. in seconds. **Bombay.** A.D. 1840—1920 | **Eng. date.** | Eqn. of time in seconds. | ⊙'s Trop. Long. in seconds. | Total Corrn. in seconds. **Calcutta.** A.D. 1840—1920 |
| | | | | | | | *Bengal*, Phalguna; *Malayalam*, Kumbam; *Tamil*, Masi. | | | | | | | | | | |
| **306** | —271 | —3835 | —3580 | —268 | —3952 | —3411 | 1 | —265 | —4092 | —4865 | —308 | —2275 | —2741 | **February 12** | —299 | —2628 | +738 |
| **307** | 273 | 3785 | 3532 | 270 | 3900 | 3361 | 2 | 267 | 4037 | 4812 | 311 | 2263 | 2717 | **13** | 302 | 2596 | 762 |
| **308** | 275 | 3735 | 3483 | 272 | 3848 | 3311 | 3 | 269 | 3983 | 4760 | 314 | 2221 | 2693 | **14** | 305 | 2564 | 791 |
| **309** | 277 | 3684 | 3435 | 274 | 3797 | 3261 | 4 | 271 | 3929 | 4708 | 317 | 2194 | 2669 | **15** | 308 | 2532 | 820 |
| **310** | 280 | 3634 | 3388 | 277 | 3745 | 3212 | 5 | 273 | 3875 | 4656 | 320 | 2168 | 2646 | **16** | 310 | 2500 | 850 |
| **311** | 282 | 3584 | 3339 | 279 | 3693 | 3162 | 6 | 275 | 3821 | 4604 | 323 | 2141 | 2622 | **17** | 313 | 2468 | 879 |
| **312** | 284 | 3533 | 3291 | 281 | 3641 | 3113 | 7 | 277 | 3766 | 4552 | 326 | 2114 | 2598 | **18** | 316 | 2436 | 908 |
| **313** | 286 | 3483 | 3243 | 283 | 3589 | 3063 | 8 | 279 | 3712 | 4499 | 329 | 2088 | 2575 | **19** | 319 | 2404 | 937 |
| **314** | 288 | 3433 | 3194 | 285 | 3537 | 3013 | 9 | 281 | 3658 | 4447 | 332 | 2061 | 2551 | **20** | 322 | 2372 | 966 |
| **315** | 290 | 3382 | 3146 | 287 | 3485 | 2963 | 10 | 283 | 3604 | 4395 | 335 | 2034 | 2527 | **21** | 325 | 2340 | 995 |
| **316** | 293 | 3332 | 3070 | 289 | 3433 | 2883 | 11 | 285 | 3550 | 4313 | 338 | 2008 | 2482 | **22** | 328 | 2308 | 1049 |
| **317** | 295 | 3282 | 2993 | 291 | 3382 | 2804 | 12 | 287 | 3495 | 4231 | 341 | 1981 | 2439 | **23** | 331 | 2276 | 1102 |
| **318** | 297 | 3231 | 2915 | 293 | 3330 | 2724 | 13 | 289 | 3441 | 4149 | 344 | 1954 | 2396 | **24** | 334 | 2244 | 1156 |
| **319** | 299 | 3181 | 2838 | 295 | 3278 | 2645 | 14 | 291 | 3387 | 4067 | 347 | 1927 | 2351 | **25** | 337 | 2212 | 1210 |
| **320** | 301 | 3131 | 2761 | 297 | 3226 | 2585 | 15 | 293 | 3333 | 3985 | 350 | 1901 | 2308 | **26** | 340 | 2180 | 1264 |
| **321** | 303 | 3080 | 2684 | 299 | 3174 | 2485 | 16 | 295 | 3279 | 3903 | 353 | 1874 | 2263 | **27** | 342 | 2148 | 1318 |
| **322** | 306 | 3030 | 2608 | 301 | 3122 | 2406 | 17 | 297 | 3224 | 3821 | 356 | 1847 | 2218 | **28** | 345 | 2116 | 1372 |
| **323** | 308 | 2980 | 2530 | 303 | 3070 | 2326 | 18 | 299 | 3170 | 3739 | 359 | 1821 | 2173 | **March 1** | 348 | 2084 | 1426 |
| **324** | 310 | 2929 | 2453 | 305 | 3019 | 2247 | 19 | 301 | 3116 | 3657 | 362 | 1794 | 2104 | **2** | 351 | 2052 | 1479 |
| **325** | 312 | 2879 | 2376 | 307 | 2967 | 2167 | 20 | 303 | 3062 | 3575 | 365 | 1767 | 2057 | **3** | 353 | 2020 | 1534 |
| **326** | 314 | 2829 | 2299 | 309 | 2915 | 2087 | 21 | 305 | 3008 | 3493 | 368 | 1741 | 2010 | **4** | 356 | 1988 | 1588 |
| **327** | 316 | 2779 | 2222 | 311 | 2863 | 2008 | 22 | 307 | 2953 | 3411 | 371 | 1714 | 1963 | **5** | 359 | 1956 | 1641 |
| **328** | 318 | 2728 | 2144 | 313 | 2811 | 1928 | 23 | 309 | 2899 | 3329 | 374 | 1687 | 1916 | **6** | 361 | 1924 | 1696 |
| **329** | 320 | 2678 | 2067 | 315 | 2759 | 1849 | 24 | 311 | 2845 | 3247 | 377 | 1660 | 1869 | **7** | 364 | 1892 | 1749 |
| **330** | 322 | 2628 | 1990 | 317 | 2707 | 1769 | 25 | 313 | 2791 | 3165 | 380 | 1634 | 1822 | **8** | 366 | 1860 | 1804 |
| **331** | 324 | 2577 | 1913 | 319 | 2655 | 1689 | 26 | 315 | 2737 | 3083 | 383 | 1607 | 1775 | **9** | 369 | 1828 | 1858 |
| **332** | 326 | 2527 | 1836 | 321 | 2604 | 1610 | 27 | 317 | 2683 | 3001 | 385 | 1580 | 1727 | **10** | 371 | 1796 | 1912 |
| **333** | 328 | 2477 | 1758 | 323 | 2552 | 1530 | 28 | 319 | 2628 | 2919 | 387 | 1554 | 1679 | **11** | 373 | 1764 | 2068 |
| **334** | 330 | 2426 | 1681 | 325 | 2500 | 1451 | 29 | 321 | 2574 | 2837 | 389 | 1527 | 1631 | **12** | 375 | 1732 | 2021 |
| | | | | | | | *Bengal*, Chaitra; *Malayalam*, Meenam; *Tamil*, Panguni. | | | | | | | | | | |
| **335** | 332 | 2376 | 1604 | 327 | 2448 | 1371 | 1 | 323 | 2520 | 2755 | 391 | 1500 | 1583 | **13** | 377 | 1700 | 2076 |
| **336** | 332 | 2297 | 1525 | 327 | 2366 | 1289 | 2 | 323 | 2436 | 2671 | 391 | 1450 | 1534 | **14** | 377 | 1643 | 2133 |
| **337** | 332 | 2218 | 1446 | 327 | 2285 | 1208 | 3 | 323 | 2352 | 2587 | 392 | 1400 | 1484 | **15** | 378 | 1587 | 2188 |
| **338** | 333 | 2138 | 1367 | 328 | 2203 | 1127 | 4 | 324 | 2268 | 2504 | 392 | 1350 | 1434 | **16** | 378 | 1530 | 2300 |
| **339** | 333 | 2059 | 1288 | 328 | 2122 | 1046 | 5 | 324 | 2184 | 2420 | 393 | 1300 | 1385 | **17** | 380 | 1473 | 2357 |
| **340** | 333 | 1980 | 1209 | 328 | 2040 | 964 | 6 | 324 | 2100 | 2336 | 393 | 1250 | 1335 | **18** | 380 | 1416 | 2412 |
| **341** | 334 | 1901 | 1131 | 329 | 1958 | 883 | 7 | 325 | 2016 | 2253 | 394 | 1200 | 1286 | **19** | 381 | 1360 | 2468 |
| **342** | 334 | 1822 | 1052 | 329 | 1877 | 802 | 8 | 325 | 1932 | 2169 | 394 | 1150 | 1236 | **20** | 381 | 1303 | 2524 |
| **343** | 334 | 1742 | 972 | 329 | 1795 | 720 | 9 | 325 | 1848 | 2085 | 395 | 1100 | 1187 | **21** | 382 | 1246 | 2581 |
| **344** | 335 | 1663 | 894 | 330 | 1714 | 640 | 10 | 326 | 1764 | 2002 | 395 | 1050 | 1137 | **22** | 382 | 1190 | 2638 |
| **345** | 335 | 1584 | 815 | 330 | 1632 | 558 | 11 | 326 | 1680 | 1918 | 396 | 1000 | 1088 | **23** | 383 | 1133 | 2637 |
| **346** | 335 | 1505 | 736 | 330 | 1550 | 476 | 12 | 326 | 1596 | 1834 | 396 | 950 | 1038 | **24** | 383 | 1076 | 2694 |
| **347** | 336 | 1426 | 658 | 331 | 1469 | 396 | 13 | 327 | 1512 | 1751 | 397 | 900 | 989 | **25** | 384 | 1020 | 2749 |
| **348** | 336 | 1346 | 578 | 331 | 1387 | 314 | 14 | 327 | 1423 | 1667 | 397 | 850 | 939 | **26** | 384 | 963 | 2806 |
| **349** | 336 | 1267 | 499 | 331 | 1306 | 233 | 15 | 327 | 1344 | 1583 | 398 | 800 | 890 | **27** | 385 | 907 | 2862 |
| **350** | 337 | 1188 | 421 | 332 | 1224 | 152 | 16 | 328 | 1260 | 1500 | 398 | 750 | 840 | **28** | 385 | 850 | 2918 |
| **351** | 337 | 1109 | 342 | 332 | 1142 | 70 | 17 | 328 | 1176 | 1416 | 399 | 700 | 791 | **29** | 386 | 793 | 2974 |
| **352** | 337 | 1030 | 263 | 332 | 1061 | 11 | 18 | 328 | 1092 | 1332 | 399 | 650 | 741 | **30** | 386 | 737 | 3030 |
| **353** | 338 | 950 | 184 | 333 | 979 | +92 | 19 | 329 | 1008 | 1249 | 400 | 600 | 692 | **31** | 387 | 679 | 3086 |
| **354** | 338 | 871 | 105 | 333 | 898 | 178 | 20 | 329 | 924 | 1165 | 400 | 550 | 642 | **April 1** | 387 | 624 | 3143 |
| **355** | 338 | 792 | 26 | 333 | 816 | 255 | 21 | 329 | 840 | 1081 | 401 | 500 | 593 | **2** | 388 | 567 | 3199 |
| **356** | 339 | 713 | +52 | 334 | 734 | 336 | 22 | 330 | 756 | 998 | 401 | 450 | 543 | **3** | 388 | 510 | 3255 |
| **357** | 339 | 634 | 131 | 334 | 653 | 417 | 23 | 330 | 672 | 914 | 402 | 400 | 494 | **4** | 389 | 454 | 3310 |
| **358** | 339 | 554 | 201 | 334 | 571 | 499 | 24 | 330 | 588 | 830 | 402 | 350 | 444 | **5** | 389 | 397 | 3367 |
| **359** | 340 | 475 | 289 | 335 | 490 | 579 | 25 | 331 | 504 | 747 | 403 | 300 | 395 | **6** | 390 | 340 | 3422 |
| **360** | 340 | 396 | 368 | 335 | 408 | 661 | 26 | 331 | 420 | 663 | 403 | 250 | 345 | **7** | 390 | 283 | 3480 |
| **361** | 340 | 317 | 447 | 335 | 326 | 743 | 27 | 331 | 336 | 579 | 404 | 200 | 296 | **8** | 391 | 227 | 3535 |
| **362** | 341 | 238 | 525 | 336 | 245 | 823 | 28 | 332 | 252 | 496 | 404 | 150 | 246 | **9** | 391 | 171 | 3591 |
| **363** | 341 | 158 | 605 | 336 | 163 | 905 | 29 | 332 | 168 | 412 | 405 | 100 | 197 | **10** | 392 | 114 | 3647 |
| **364** | 341 | 79 | 684 | 336 | 82 | 986 | 30 | 332 | 84 | 328 | 405 | 50 | 147 | **11** | 392 | 57 | 3700 |
| **365** | 342 | 0 | 762 | 337 | 0 | 1067 | 31 | 333 | 0 | 245 | 405 | 0 | 97 | **12** | 392 | 0 | 3761 |

# TABLE XIII—A.

## Latitudes and Longitudes of Chief places in British India and the Feudatory States.

**(Longitudes are expressed by the difference in Seconds of time as compared with Ujjain.)**

| Serial No. | District. | Town. | Latitude in degrees & decimals of a degree. | Time difference in Seconds. |
|---|---|---|---|---|
| | | **BRITISH TERRITORY.** | | |
| | | **Aden Province.** | | |
| 1 | Aden ... | Aden | 12·7 | —7348 |
| | | **Ajmere—Merwara Province.** | | |
| 2 | Ajmere ... | Ajmere | 26·5 | — 280 |
| | | Nazirabad | 26·3 | — 256 |
| 3 | Merwara ... | Beawar (Nayanagar) | 26·0 | — 352 |
| | | **Baluchistan Province.** | | |
| 4 | Quetta—Pesbin | Quetta | 30·2 | —2104 |
| | | **Bengal Province.** | | |
| 5 | Balasore ... | Balasore | 21·5 | +2676 |
| 6 | Bankura ... | Bankura | 23·2 | +2708 |
| 7 | Bhagalpur ... | Bhagalpur | 25·2 | +2692 |
| 8 | Burdwan ... | Burdwan | 23·2 | +2896 |
| | | Howrah | 22·5 | +3016 |
| 9 | Champaram ... | Bettiah | 26·5 | +2092 |
| 10 | Cuttack ... | Cuttack | 20·5 | +2420 |
| 11 | Darbhanga ... | Darbhanga | 26·2 | +2428 |
| 12 | Gaya ... | Gaya | 24·5 | +2216 |
| 13 | Hooghly ... | Hoogly and Chinsura | 23·0 | +3028 |
| | | Serampore | 22·7 | +3016 |
| 14 | Midnapore ... | Midnapore | 22·5 | +2768 |
| 15 | Monghyr ... | Monghyr | 25·3 | +2564 |
| 16 | Murshidabad . | Berhampore | 24·0 | +2996 |
| 17 | Muzaffarpur ... | Hajipur | 25·7 | +2260 |
| | | Muzaffarpur | 26·0 | +2308 |
| 18 | Nadia ... | Krishnagar | 23·5 | +3056 |
| | | Santipur | 23·2 | +3040 |
| 19 | Twenty-four Parganas ... | Calcutta<br>Cossipur-Chitpur<br>Garden Reach<br>Maniktala | 22·5 | +3020 |
| | | Naihati | 23·0 | +3032 |
| | | Baranagar | 22·7 | +3020 |
| | | Bhatpara | 23·0 | +3032 |
| 20 | Patna ... | Patna | 25·7 | +2252 |
| | | Bihar | 25·2 | +2336 |
| | | Dinapore | 25·7 | +2224 |
| 21 | Puri ... | Puri | 19·8 | +2408 |
| 22 | Ranchi ... | Ranchi | 23·3 | +2292 |
| 23 | Saran ... | Chapra | 25·8 | +2148 |
| 24 | Shahabad ... | Sasaram | 25·0 | +1976 |
| | | Arrah | 25·5 | +2132 |
| | | **Berar Province** | | |
| 25 | Akola ... | Akola | 20·7 | + 300 |
| 26 | Amraoti ... | Amraoti | 21·0 | + 480 |
| 27 | Ellichpur ... | Ellichpur | 21·2 | + 424 |
| | | **Bombay Province.** | | |
| 28 | Ahmadabad ... | Ahmadabad | 23·0 | — 768 |
| 29 | Ahmadnagar ... | Ahmadnagar | 19·0 | — 208 |
| 30 | Belgaum ... | Belgaum | 16·0 | — 304 |
| 31 | Bijapur ... | Bijapur | 16·8 | — 16 |
| 32 | Bombay ... | Bombay | 19·0 | — 692 |
| 33 | Broach ... | Broach | 21·7 | — 672 |
| 34 | Dharwar ... | Hubli | 15·3 | — 152 |
| | | Dharwar | 15·5 | — 184 |
| | | Gadag | 15·5 | — 36 |
| 35 | Kaira ... | Nadiad | 22·7 | — 700 |
| 36 | Khandesh ... | Dhulia | 21·0 | — 240 |
| | | **Bombay Province—(continued).** | | |
| 37 | Nasik ... | Nasik | 20·0 | — 480 |
| 38 | Panch Mahal ... | Godhra | 22·7 | — 520 |
| 39 | Poona ... | Poona | 18·5 | — 464 |
| 40 | Sholapur ... | Sholapur | 17·7 | + 28 |
| | | Pandharpur | 17·7 | — 84 |
| | | Barsi | 18·2 | — 24 |
| 41 | Satara ... | Satara | 17·7 | — 428 |
| 42 | Surat ... | Surat | 12·2 | — 708 |
| 43 | Thana ... | Bandra | 19·0 | — 708 |
| | | **Burma Province.** | | |
| 44 | Akyab ... | Akyab | 20·2 | +4112 |
| 45 | Amherst ... | Moulmein | 16·5 | +5244 |
| 46 | Bassein ... | Bassein | 16·7 | +4556 |
| 47 | Henzada ... | Henzada | 17·7 | +4732 |
| 48 | Mandalay ... | Mandalay | 22·0 | +4876 |
| 49 | Prome ... | Prome | 18·8 | +4664 |
| 50 | Rangoon ... | Rangoon | 16·7 | +4896 |
| 51 | Tavoy ... | Tavoy | 14·1 | +5380 |
| | | **Central Provinces.** | | |
| 52 | Jubbalpore ... | Jubbalpore | 23·2 | +1000 |
| 53 | Nagpur ... | Kamptee | 21·2 | + 820 |
| | | Nagpur | 21·2 | + 800 |
| 54 | Nimar ... | Burhanpur | 21·3 | + 108 |
| 55 | Raipur ... | Raipur | 21·2 | +1408 |
| 56 | Saugor ... | Saugor | 23·8 | + 712 |
| | | **Eastern Bengal & Assam Province.** | | |
| 57 | Chittagong ... | Chittagong | 23·3 | +3852 |
| 58 | Decca ... | Decca<br>Narayanganj | 23·7 | +3508 |
| 59 | Manipur ... | Imphal | 24·8 | +4360 |
| 60 | Pabna ... | Sirajganj | 24·5 | +3352 |
| 61 | Rajshahi ... | Rampur-Bolia | 24·3 | +3076 |
| | | **Madras Province.** | | |
| 62 | Bellary ... | Adoni | 15·7 | + 360 |
| | | Bellary | 15·2 | + 256 |
| 63 | Chingleput ... | Conjeeveram | 12·8 | + 940 |
| 64 | Coimbatore ... | Coimbatore | 11·0 | + 284 |
| 65 | Ganjam ... | Berhampore | 19·3 | +2164 |
| 66 | Godaveri ... | Cocanada | 17·0 | +1548 |
| | | Rajahmundry | 17·0 | +1436 |
| 67 | Kistna ... | Bezwada | 16·5 | +1160 |
| | | Ellore | 16·7 | +1280 |
| | | Masulipatam | 16·2 | +1284 |
| 68 | Guntur ... | Guntur | 16·3 | +1124 |
| 69 | Kurnool ... | Kurnool | 15·8 | + 548 |
| 70 | Madras ... | Madras | 13·0 | +1072 |
| 71 | Madura ... | Madura | 10·0 | + 560 |
| | | Bodinayakkanur | 10·0 | + 376 |
| | | Dindigul | 10·3 | + 528 |
| 72 | Malabar ... | Calicut | 11·2 | + 0 |
| | | Cannanore | 11·8 | — 100 |
| | | Palghat | 10·8 | + 208 |
| | | Tellicherry | 11·7 | — 72 |
| 73 | Nellore ... | Nellore | 14·5 | +1008 |
| 74 | North Arcot ... | Gudiyattam | 13·0 | + 744 |
| | | Vellore | 12·9 | + 808 |
| 75 | Salem ... | Salem | 11·7 | + 572 |
| 76 | South Arcot ... | Cuddalore | 11·7 | + 956 |
| 77 | South Canara . | Mangalore | 12·8 | — 224 |
| 78 | Tanjore ... | Kumbakonam | 11·0 | + 860 |
| | | Negapatam | 10·7 | + 976 |

## TABLE XIII—A.

### Latitudes and Longitudes of Chief places in British India and the Feudatory States.

**(Longitudes are expressed by the difference in Seconds of time as compared with Ujjain.)**

| Serial No. | District. | Town. | Latitude in degrees & decimals of a degree. | Time difference in Seconds. |
|---|---|---|---|---|
| | **Madras Province—(*continued*).** | | | |
| 78 | Tanjore—*contd* ... | Tanjore. | 10·7 | + 804 |
| | | Mannargudi | 10·7 | + 880 |
| | | Mayavaram | 11·1 | + 928 |
| 79 | Tinnevelly ... | Palamcottah | 8·7 | + 472 |
| | | Rajapalaiyam | 9·5 | + 424 |
| | | Srivilliputtur | 9·5 | + 440 |
| | | Tinnevelly | 8·7 | + 456 |
| | | Tiruchendur | 8·5 | + 560 |
| | | Tuticorin | 8·8 | + 568 |
| 80 | Trichinopoly ... | { Trichinopoly<br>Srirangam } | 10·8 | + 700 |
| 81 | Vizagapatam. | Vizagapatam | 17·7 | +1804 |
| | | Vizianagaram | 18·2 | +1832 |
| | **North West Frontier Province.** | | | |
| 82 | Dera Ismail Khan } | Dera Ismail Khan | 31·8 | —1168 |
| 83 | Kohat ... | Kohat | 33·6 | —1044 |
| 84 | Peshawar ... | Peshawar | 34·0 | —1008 |
| | **Punjab Province.** | | | |
| 85 | Amritsar ... | Amritsar | 31·7 | — 216 |
| 86 | Delhi ... | Delhi | 28·7 | + 352 |
| 87 | Dera Ghazi Khan | Dera Ghazi Khan | 30·0 | —1200 |
| 88 | Ferozepore ... | Ferozepore | 31·0 | — 280 |
| 89 | Gurdaspur ... | Batala | 30·8 | — 140 |
| 90 | Gujranwala ... | Gujranwala | 32·2 | — 384 |
| 91 | Gurgaon ... | Rewari | 28·2 | + 204 |
| 92 | Hissar ... | Bhiwani | 28·8 | + 84 |
| 93 | Jhang ... | Jhang-Maghiana | 31·3 | — 828 |
| 94 | Karnal ... | { Karnal<br>Panipat | 29·7<br>29·4 } | + 288 |
| 95 | Jullundur ... | Jullundur | 31·3 | — 48 |
| 96 | Lahore ... | Lahore | 31·6 | — 348 |
| | | Kasur | 31·1 | — 316 |
| 97 | Ludhiana ... | Ludhiana | 31·0 | + 20 |
| 98 | Multan ... | Multan | 30·2 | —1024 |
| 99 | Rawalpandi ... | Rawalpandi | 33·6 | — 640 |
| 100 | Rohtak ... | Rohtak | 28·9 | + 192 |
| 101 | Sialkot ... | Sialkot | 32·5 | — 300 |
| | **Sindh Province.** | | | |
| 102 | Hyderabad ... | Hyderabad | 25·3 | —1768 |
| 103 | Karachi ... | Karachi | 24·8 | —2092 |
| 104 | Sukkur ... | Shikarpur | 28·0 | —1708 |
| | | Sukkur | 27·7 | —1652 |
| | **United Provinces of Agra and Oudh.** | | | |
| 105 | Harodi ... | Shahabad | 27·7 | +1000 |
| 106 | Sitapur ... | Sitapur | 27·6 | +1172 |
| | **NATIVE STATES.** | | | |
| | **Baroda State.** | | | |
| 107 | Baroda ... | Baroda | 22·3 | — 608 |
| 108 | Kadi ... | Patan (Anhilwad) | 23·8 | — 868 |
| 109 | Navasan ... | Navasan | 21·0 | — 684 |
| | **Bombay Native States.** | | | |
| 110 | Cambay ... | Cambay | 22·3 | — 748 |
| 111 | Cutch ... | Bhuj | 23·2 | —1436 |
| 112 | Kathiawar ... | Bhaunagar | 21·7 | — 860 |
| | | Dhoraji | 21·7 | —1240 |
| | **Bombay Native States—(*continued*).** | | | |
| 112 | Kathiawar—*contd*. | Junagarah | 21·5 | —1244 |
| | | Navanagar | 22·4 | —1324 |
| | | Porabandar | 21·6 | —1436 |
| | | Rajkot | 22·3 | —1188 |
| | | Wadhwan | 22·6 | — 972 |
| 113 | Kolhapur ... | Kolhapur | 16·6 | — 364 |
| | **Central India Agency.** | | | |
| 114 | Rewah ... | Rewah | 24·5 | +1324 |
| 115 | Bhopal ... | Bhopal | 23·2 | + 392 |
| 116 | Datia ... | Datia | 25·6 | + 644 |
| 117 | Gwalior ... | Lashkar | 26·2 | + 572 |
| | | Ujjain | 23·1 | + 0 |
| 118 | Indore ... | Indore | 22·7 | + 28 |
| 119 | Iaora ... | Iaora | 23·6 | — 156 |
| 120 | Mandasar in Gwalior } | Mandasar | 24·1 | — 168 |
| 121 | Ratlam ... | Ratlam | 23·3 | — 176 |
| | **Hyderabad State.** | | | |
| 122 | Aurangabad ... | { Aurangabad<br>Jalna } | 19·8 | — 108 |
| 123 | Gulbarga ... | Gulbarga | 17·3 | + 256 |
| 124 | Hyderabad ... | Hyderabad | 17·3 | + 640 |
| 125 | Raichur ... | Raichur | 16·2 | + 376 |
| | **Kashmir State.** | | | |
| 126 | Jammu ... | Jammu | 32·7 | — 208 |
| 127 | Kashmir ... | Srinagar | 34·8 | — 228 |
| | **Madras Native States.** | | | |
| 128 | Cochin ... | Ernakulam | 10·0 | + 120 |
| | | Mattancheri | 10·0 | + 112 |
| 129 | Pudukkottai ... | Pudukkottai | 10·3 | + 728 |
| 130 | Travancore ... | Alleppey | 9·5 | + 132 |
| | | Nagercoil | 8·1 | + 400 |
| | | Trivandrum | 8·5 | + 280 |
| | **Mysore State.** | | | |
| 131 | Bangalore ... | Bangalore | 13·0 | + 432 |
| 132 | Kolar ... | Kolar | 13·1 | + 564 |
| 133 | Mysore ... | Mysore | 12·3 | + 212 |
| | **Punjab Native States.** | | | |
| 134 | Malar Kotla ... | Malar Kotla | 30·5 | + 48 |
| 135 | Patiala ... | Patiala | 30·3 | + 164 |
| | **Rajputana Agency.** | | | |
| 136 | Alwar ... | Alwar | 27·6 | + 196 |
| 137 | Bharatpur ... | Bharatpur | 27·2 | + 412 |
| 138 | Bikaner ... | Bikaner | 28·0 | — 596 |
| 139 | Jaipur ... | Sikar | 27·6 | — 156 |
| | | Jaipur | 26·9 | + 12 |
| 140 | Karauli ... | Karauli | 26·5 | + 300 |
| 141 | Kotah ... | Kotah | 25·1 | + 16 |
| 142 | Jodhpur ... | Jodhpur | 26·3 | — 664 |
| 143 | Mewar or Udaipur } ... | Udaipur | 24·6 | — 500 |
| 144 | Tonk ... | Tonk | 26·1 | + 4 |
| | **United Provinces Native States.** | | | |
| 145 | Rampur .. | Rampur | 28·8 | + 780 |

# TABLE XIV.

**Jupiter's Cycle of 60 years—Northern System, showing suppressed years. A.D. 250 to A.D. 2000.**

| 0 | Vijaya ... | 2 80 | 3 39 | 3 98 | 4 58 | 5 17 | 5 76 | 6 36 | 0 | 6 95 | 7 54 | 8 14 | 8 73 | 9 32 | 9 92 | 10 51 | 11 10 |
|---|---|---|---|---|---|---|---|---|---|---|---|---|---|---|---|---|---|
| 1 | Jaya ... | 81 | 40 | 99 | 59 | 18 | 77 | 37 | 1 | 96 | 55 | 15 | 74 | 33 | 93 | 52 | 11 |
| 2 | Manmatha ... | 82 | 41 | 4 00 | 60 | 19 | 78 | 38 | 2 | 97 | 56 | 16 | 75 | 34 | 94 | 53 | 12 |
| 3 | Durmukha ... | 83 | 42 | 01 | 61 | 20 | 79 | 39 | 3 | 98 | 57 | 17 | 76 | 35 | ... | 54 | 18 |
| 4 | Hemalamba ... | 84 | 43 | 02 | 62 | 21 | 80 | 40 | 4 | 99 | 58 | 18 | 77 | 36 | 95 | 55 | 14 |
| 5 | Vilamba ... | 85 | 44 | 03 | 63 | 22 | 81 | 41 | 5 | 7 00 | 59 | 19 | 78 | 37 | 96 | 56 | 15 |
| 6 | Vikarin ... | 86 | 45 | 04 | 64 | 23 | 82 | 42 | 6 | 01 | 60 | 20 | 79 | 38 | 97 | 57 | 16 |
| 7 | Sarvari ... | 87 | 46 | 05 | 65 | 24 | 83 | 43 | 7 | 02 | 61 | 21 | 80 | 39 | 98 | 58 | 17 |
| 8 | Plava ... | 88 | 47 | 06 | 66 | 25 | 84 | 44 | 8 | 03 | 62 | 22 | 81 | 40 | 99 | 59 | 18 |
| 9 | Subhakrit ... | 89 | 48 | 07 | 67 | 26 | 85 | 45 | 9 | 04 | 63 | 23 | 82 | 41 | 10 00 | 60 | 19 |
| 10 | Sobhana ... | 90 | 49 | 08 | 68 | 27 | 86 | 46 | 10 | 05 | 64 | ... | 83 | 42 | 01 | 61 | 20 |
| 11 | Krodhin ... | 91 | 50 | 09 | 69 | 28 | 87 | 47 | 11 | 06 | 65 | 24 | 84 | 43 | 02 | 62 | 21 |
| 12 | Visvavasu ... | 92 | 51 | 10 | 70 | 29 | 88 | 48 | 12 | 07 | 66 | 25 | 85 | 44 | 03 | 63 | 22 |
| 13 | Parabhava ... | 93 | 52 | 11 | 71 | 30 | 89 | 49 | 13 | 08 | 67 | 26 | 86 | 45 | 04 | 64 | 23 |
| 14 | Plavanga ... | 94 | 53 | 12 | 72 | 31 | 90 | 50 | 14 | 09 | 68 | 27 | 87 | 46 | 05 | 65 | 24 |
| 15 | Kilaka ... | 95 | 54 | 13 | 73 | 32 | 91 | 51 | 15 | 10 | 69 | 28 | 88 | 47 | 06 | 66 | 25 |
| 16 | Saumya ... | 96 | 55 | 14 | 74 | 33 | 92 | 52 | 16 | 11 | 70 | 29 | 89 | 48 | 07 | 67 | 26 |
| 17 | Sadharana ... | 97 | 56 | 15 | 75 | 34 | 93 | 53 | 17 | 12 | 71 | 30 | 90 | 49 | 08 | 68 | 27 |
| 18 | Virodhakrit ... | 98 | 57 | 16 | 76 | 35 | 94 | ... | 18 | 13 | 72 | 31 | 91 | 50 | 09 | 69 | 28 |
| 19 | Paridhavin ... | 99 | 58 | 17 | 77 | 36 | 95 | 54 | 19 | 14 | 73 | 32 | 92 | 51 | 10 | 70 | 29 |
| 20 | Pramadin ... | 3 00 | 59 | 18 | 78 | 37 | 96 | 55 | 20 | 15 | 74 | 33 | 93 | 52 | 11 | 71 | 30 |
| 21 | Ananda ... | 01 | 60 | 19 | 79 | 38 | 97 | 56 | 21 | 16 | 75 | 34 | 94 | 53 | 12 | 72 | 31 |
| 22 | Rakshasa ... | 02 | 61 | 20 | 80 | 39 | 98 | 57 | 22 | 17 | 76 | 35 | 95 | 54 | 13 | 73 | 32 |
| 23 | Anala ... | 03 | 62 | 21 | 81 | 40 | 99 | 58 | 23 | 18 | 77 | 36 | 96 | 55 | 14 | 74 | 33 |
| 24 | Pingala ... | 04 | 63 | 22 | 82 | 41 | 6 00 | 59 | 24 | 19 | 78 | 37 | 97 | 56 | 15 | 75 | 34 |
| 25 | Kalayukta ... | 05 | 64 | 23 | ... | 42 | 01 | 60 | 25 | 20 | 79 | 38 | 98 | 57 | 16 | 76 | 35 |
| 26 | Siddharthin ... | 06 | 65 | 24 | 83 | 43 | 02 | 61 | 26 | 21 | 80 | 39 | 99 | 58 | 17 | 77 | 36 |
| 27 | Raudra ... | 07 | 66 | 25 | 84 | 44 | 03 | 62 | 27 | 22 | 81 | 40 | 9 00 | 59 | 18 | 78 | 37 |
| 28 | Durmati ... | 08 | 67 | 26 | 85 | 45 | 04 | 63 | 28 | 23 | 82 | 41 | 01 | 60 | 19 | 79 | 38 |
| 29 | Dundubhi ... | 09 | 68 | 27 | 86 | 46 | 05 | 64 | 29 | 24 | 83 | 42 | 02 | 61 | 20 | ... | 39 |
| 30 | Rudhirodgarin . | 10 | 69 | 28 | 87 | 47 | 06 | 65 | 30 | 25 | 84 | 43 | 03 | 62 | 21 | 80 | 40 |
| 31 | Raktaksha ... | 11 | 70 | 29 | 88 | 48 | 07 | 66 | 31 | 26 | 85 | 44 | 04 | 63 | 22 | 81 | 41 |
| 32 | Krodhana ... | ... | 71 | 30 | 89 | 49 | 08 | 67 | 32 | 27 | 86 | 45 | 05 | 64 | 23 | 82 | 42 |
| 33 | Kshaya ... | 12 | 72 | 31 | 90 | 50 | 09 | 68 | 33 | 28 | 87 | 46 | 06 | 65 | 24 | 83 | 43 |
| 34 | Prabhava ... | 13 | 73 | 32 | 91 | 51 | 10 | 69 | 34 | 29 | 88 | 47 | 07 | 66 | 25 | 84 | 44 |
| 35 | Vibhava ... | 14 | 74 | 33 | 92 | 52 | 11 | 70 | 35 | 30 | 89 | 48 | 08 | 67 | 26 | 85 | 45 |
| 36 | Sukla ... | 15 | 75 | 34 | 93 | 53 | 12 | 71 | 36 | 31 | 90 | 49 | ... | 68 | 27 | 86 | 46 |
| 37 | Pramoda ... | 16 | 76 | 35 | 94 | 54 | 13 | 72 | 37 | 32 | 91 | 50 | 09 | 69 | 28 | 87 | 47 |
| 38 | Prajapati ... | 17 | 77 | 36 | 95 | 55 | 14 | 73 | 38 | 33 | 92 | 51 | 10 | 70 | 29 | 88 | 48 |
| 39 | Angiras ... | 18 | 78 | 37 | 96 | 56 | 15 | 74 | 39 | 34 | 93 | 52 | 11 | 71 | 30 | 89 | 49 |
| 40 | Srimukha ... | 19 | 79 | 38 | 97 | 57 | 16 | 75 | 40 | 35 | 94 | 53 | 12 | 72 | 31 | 90 | 50 |
| 41 | Bhava ... | 20 | 80 | 39 | 98 | 58 | 17 | 76 | 41 | 36 | 95 | 54 | 13 | 73 | 32 | 91 | 51 |
| 42 | Yuvan ... | 21 | 81 | 40 | 99 | 59 | 18 | 77 | 42 | 37 | 96 | 55 | 14 | 74 | 35 | 92 | 52 |
| 43 | Dhatri ... | 22 | 82 | 41 | 5 00 | 60 | 19 | 78 | 43 | 38 | 97 | 56 | 15 | 75 | 34 | 93 | 53 |
| 44 | Isvara ... | 23 | 83 | 42 | 01 | 61 | 20 | 79 | 44 | ... | 98 | 57 | 16 | 76 | 35 | 94 | 54 |
| 45 | Bahudhanya ... | 24 | 84 | 43 | 02 | 62 | 21 | 80 | 45 | 39 | 99 | 58 | 17 | 77 | 36 | 95 | 55 |
| 46 | Pramathin ... | 25 | 85 | 44 | 03 | 63 | 22 | 81 | 46 | 40 | 8 00 | 59 | 18 | 78 | 37 | 96 | 56 |
| 47 | Vikrama ... | 26 | 86 | 45 | 04 | 64 | 23 | 82 | 47 | 41 | 01 | 60 | 19 | 79 | 38 | 97 | 57 |
| 48 | Vrisha ... | 27 | 87 | 46 | 05 | 65 | 24 | 83 | 48 | 42 | 02 | 61 | 20 | 80 | 39 | 98 | 58 |
| 49 | Chitrabhanu ... | 28 | 88 | 47 | 06 | 66 | 25 | 84 | 49 | 43 | 03 | 62 | 21 | 81 | 40 | 99 | 59 |
| 50 | Subhanu ... | 29 | 89 | 48 | 07 | 67 | 26 | 85 | 50 | 44 | 04 | 63 | 22 | 82 | 41 | 11 00 | 60 |
| 51 | Tarana ... | 30 | 90 | 49 | 08 | ... | 27 | 86 | 51 | 45 | 05 | 64 | 23 | 83 | 42 | 01 | 61 |
| 52 | Parthiva ... | 31 | 91 | 50 | 09 | 68 | 28 | 87 | 52 | 46 | 06 | 65 | 24 | 84 | 43 | 02 | 62 |
| 53 | Vyaya ... | 32 | 92 | 51 | 10 | 69 | 29 | 88 | 53 | 47 | 07 | 66 | 25 | 85 | 44 | 03 | 63 |
| 54 | Sarvajit ... | 33 | 93 | 52 | 11 | 70 | 30 | 89 | 54 | 48 | 08 | 67 | 26 | 86 | 45 | 04 | 64 |
| 55 | Sarvadharin ... | 34 | 94 | 53 | 12 | 71 | 31 | 90 | 55 | 49 | 09 | 68 | 27 | 87 | 46 | 05 | ... |
| 56 | Virodhin ... | 35 | 95 | 54 | 13 | 72 | 32 | 91 | 56 | 50 | 10 | 69 | 28 | 88 | 47 | 06 | 65 |
| 57 | Vikrita ... | 36 | 96 | 55 | 14 | 73 | 33 | 92 | 57 | 51 | 11 | 70 | 29 | 89 | 48 | 07 | 66 |
| 58 | Khara ... | 37 | 97 | 56 | 15 | 74 | 34 | 93 | 58 | 52 | 12 | 71 | 30 | 90 | 49 | 08 | 67 |
| 59 | Nandana ... | 38 | ... | 57 | 16 | 75 | 35 | 94 | 59 | 53 | 13 | 72 | 31 | 91 | 50 | 09 | 68 |

# TABLE XIV.

**Jupiter's Cycle of 60 years—Northern System, showing suppressed years. A.D. 280 to A.D. 2000.**

| | | | | | | | | |
|---|---|---|---|---|---|---|---|---|
| 0 | Vijaya ... | 11 69 | 12 29 | 12 88 | 13 47 | 14 07 | 14 66 | 15 25 |
| 1 | Jaya ... | 70 | 30 | 89 | 48 | 08 | 67 | 26 |
| 2 | Manmatha ... | 71 | 31 | 90 | 49 | 09 | 68 | 27 |
| 3 | Durmukha ... | 72 | 32 | 91 | 50 | 10 | 69 | 28 |
| 4 | Hemalamba ... | 73 | 33 | 92 | 51 | 11 | 70 | 29 |
| 5 | Vilamba ... | 74 | 34 | 93 | 52 | 12 | 71 | 30 |
| 6 | Vikarin ... | 75 | 35 | 94 | 53 | 13 | 72 | 31 |
| 7 | Sarvari ... | 76 | 36 | 95 | 54 | 14 | 73 | 32 |
| 8 | Plava ... | 77 | 37 | 96 | 55 | 15 | 74 | 33 |
| 9 | Subhakrit ... | 78 | 38 | 97 | 56 | 16 | 75 | 34 |
| 10 | Sobhana ... | 79 | 39 | 98 | 57 | 17 | 76 | 35 |
| 11 | Krodhin ... | 80 | 40 | 99 | 58 | 18 | 77 | 36 |
| 12 | Visvavasu ... | 81 | 41 | 13 00 | 59 | 19 | 78 | 37 |
| 13 | Parabhava ... | 82 | 42 | 01 | 60 | 20 | 79 | 38 |
| 14 | Plavanga ... | 83 | 43 | 02 | 61 | ... | 80 | 39 |
| 15 | Kilaka ... | 84 | 44 | 03 | 62 | 21 | 81 | 40 |
| 16 | Saumya ... | 85 | 45 | 04 | 63 | 22 | 82 | 41 |
| 17 | Sadharana ... | 86 | 46 | 05 | 64 | 23 | 83 | 42 |
| 18 | Virodhakrit ... | 87 | 47 | 06 | 65 | 24 | 84 | 43 |
| 19 | Paridhavin ... | 88 | 48 | 07 | 66 | 25 | 85 | 44 |
| 20 | Pramadin ... | 89 | 49 | 08 | 67 | 26 | 86 | 45 |
| 21 | Ananda ... | 90 | 50 | 09 | 68 | 27 | 87 | 46 |
| 22 | Rakshasa ... | 91 | ... | 10 | 69 | 28 | 88 | 47 |
| 23 | Anala ... | 92 | 51 | 11 | 70 | 29 | 89 | 48 |
| 24 | Pingala ... | 93 | 52 | 12 | 71 | 30 | 90 | 49 |
| 25 | Kalayukta ... | 94 | 53 | 13 | 72 | 31 | 91 | 50 |
| 26 | Siddharthin ... | 95 | 54 | 14 | 73 | 32 | 92 | 51 |
| 27 | Raudra ... | 96 | 55 | 15 | 74 | 33 | 93 | 52 |
| 28 | Durmati ... | 97 | 56 | 16 | 75 | 34 | 94 | 53 |
| 29 | Dundubhi ... | 98 | 57 | 17 | 76 | 35 | 95 | 54 |
| 30 | Rudhirodgarin . | 99 | 58 | 18 | 77 | 36 | 96 | 55 |
| 31 | Raktaksha ... | 12 00 | 59 | 19 | 78 | 37 | 97 | 56 |
| 32 | Krodhana ... | 01 | 60 | 20 | 79 | 38 | 98 | 57 |
| 33 | Kshaya ... | 02 | 61 | 21 | 80 | 39 | 99 | 58 |
| 34 | Prabhava ... | 03 | 62 | 22 | 81 | 40 | 15 00 | 59 |
| 35 | Vibhava ... | 04 | 63 | 23 | 82 | 41 | 01 | 60 |
| 36 | Sukla ... | 05 | 64 | 24 | 83 | 42 | 02 | 61 |
| 37 | Pramoda ... | 06 | 65 | 25 | 84 | 43 | 03 | 62 |
| 38 | Prajapati ... | 07 | 66 | 26 | 85 | 44 | 04 | 63 |
| 39 | Angiras ... | 08 | 67 | 27 | 86 | 45 | 05 | 64 |
| 40 | Srimukha ... | 09 | 68 | 28 | 87 | 46 | 06 | 65 |
| 41 | Bhava ... | 10 | 69 | 29 | 88 | 47 | 07 | 66 |
| 42 | Yuvan ... | 11 | 70 | 30 | 89 | 48 | 08 | 67 |
| 43 | Dhatri ... | 12 | 71 | 31 | 90 | 49 | 09 | 68 |
| 44 | Isvara ... | 13 | 72 | 32 | 91 | 50 | 10 | 69 |
| 45 | Bahudhanya ... | 14 | 73 | 33 | 92 | 51 | 11 | 70 |
| 46 | Pramathin ... | 15 | 74 | 34 | 93 | 52 | 12 | 71 |
| 47 | Vikrama ... | 16 | 75 | 35 | 94 | 53 | 13 | 72 |
| 48 | Vrisha ... | 17 | 76 | ... | 95 | 54 | 14 | 73 |
| 49 | Chitrabhanu ... | 18 | 77 | 36 | 96 | 55 | ... | 74 |
| 50 | Subhanu ... | 19 | 78 | 37 | 97 | 56 | 15 | 75 |
| 51 | Tarana ... | 20 | 79 | 38 | 98 | 57 | 16 | 76 |
| 52 | Parthiva ... | 21 | 80 | 39 | 99 | 58 | 17 | 77 |
| 53 | Vyaya ... | 22 | 81 | 40 | 14 00 | 59 | 18 | 78 |
| 54 | Sarvajit ... | 23 | 82 | 41 | 01 | 60 | 19 | 79 |
| 55 | Sarvadharin ... | 24 | 83 | 42 | 02 | 61 | 20 | 80 |
| 56 | Virodhin ... | 25 | 84 | 43 | 03 | 62 | 21 | 81 |
| 57 | Vikrita ... | 26 | 85 | 44 | 04 | 63 | 22 | 82 |
| 58 | Khara ... | 27 | 86 | 45 | 05 | 64 | 23 | 83 |
| 59 | Nandana ... | 28 | 87 | 46 | 06 | 65 | 24 | 84 |

| | | | | | | | |
|---|---|---|---|---|---|---|---|
| 0 | 15 85 | 16 44 | 17 03 | 17 63 | 18 22 | 18 81 | 19 41 |
| 1 | 86 | 45 | 04 | 64 | 23 | 82 | ... |
| 2 | 87 | 46 | 05 | 65 | 24 | 83 | 42 |
| 3 | 88 | 47 | 06 | 66 | 25 | 84 | 43 |
| 4 | 89 | 48 | 07 | 67 | 26 | 85 | 44 |
| 5 | 90 | 49 | 08 | 68 | 27 | 86 | 45 |
| 6 | 91 | 50 | 09 | 69 | 28 | 87 | 46 |
| 7 | 92 | 51 | 10 | 70 | 29 | 88 | 47 |
| 8 | 93 | 52 | 11 | 71 | 30 | 89 | 48 |
| 9 | 94 | 53 | 12 | ... | 31 | 90 | 49 |
| 10 | 95 | 54 | 13 | 72 | 32 | 91 | 50 |
| 11 | 96 | 55 | 14 | 73 | 33 | 92 | 51 |
| 12 | 97 | 56 | 15 | 74 | 34 | 93 | 52 |
| 13 | 98 | 57 | 16 | 75 | 35 | 94 | 53 |
| 14 | 99 | 58 | 17 | 76 | 36 | 95 | 54 |
| 15 | 16 00 | 59 | 18 | 77 | 37 | 96 | 55 |
| 16 | ... | 60 | 19 | 78 | 38 | 97 | 56 |
| 17 | 01 | 61 | 20 | 79 | 39 | 98 | 57 |
| 18 | 02 | 62 | 21 | 80 | 40 | 99 | 58 |
| 19 | 03 | 63 | 22 | 81 | 41 | 19 00 | 59 |
| 20 | 04 | 64 | 23 | 82 | 42 | 01 | 60 |
| 21 | 05 | 65 | 24 | 83 | 43 | 02 | 61 |
| 22 | 06 | 66 | 25 | 84 | 44 | 03 | 62 |
| 23 | 07 | 67 | 26 | 85 | 45 | 04 | 63 |
| 24 | 08 | 68 | 27 | 86 | 46 | 05 | 64 |
| 25 | 09 | 69 | 28 | 87 | 47 | 06 | 65 |
| 26 | 10 | 70 | 29 | 88 | 48 | 07 | 66 |
| 27 | 11 | 71 | 30 | 89 | 49 | 08 | 67 |
| 28 | 12 | 72 | 31 | 90 | 50 | 09 | 68 |
| 29 | 13 | 73 | 32 | 91 | 51 | 10 | 69 |
| 30 | 14 | 74 | 33 | 92 | 52 | 11 | 70 |
| 31 | 15 | 75 | 34 | 93 | 53 | 12 | 71 |
| 32 | 16 | 76 | 35 | 94 | 54 | 13 | 72 |
| 33 | 17 | 77 | 36 | 95 | 55 | 14 | 73 |
| 34 | 18 | 78 | 37 | 96 | 56 | 15 | 74 |
| 35 | 19 | 79 | 38 | 97 | ... | 16 | 75 |
| 36 | 20 | 80 | 39 | 98 | 57 | 17 | 76 |
| 37 | 21 | 81 | 40 | 99 | 58 | 18 | 77 |
| 38 | 22 | 82 | 41 | 18 00 | 59 | 19 | 78 |
| 39 | 23 | 83 | 42 | 01 | 60 | 20 | 79 |
| 40 | 24 | 84 | 43 | 02 | 61 | 21 | 80 |
| 41 | 25 | 85 | 44 | 03 | 62 | 22 | 81 |
| 42 | 26 | ... | 45 | 04 | 63 | 23 | 82 |
| 43 | 27 | 86 | 46 | 05 | 64 | 24 | 83 |
| 44 | 28 | 87 | 47 | 06 | 65 | 25 | 84 |
| 45 | 29 | 88 | 48 | 07 | 66 | 26 | 85 |
| 46 | 30 | 89 | 49 | 08 | 67 | 27 | 86 |
| 47 | 31 | 90 | 50 | 09 | 68 | 28 | 87 |
| 48 | 32 | 91 | 51 | 10 | 69 | 29 | 88 |
| 49 | 33 | 92 | 52 | 11 | 70 | 30 | 89 |
| 50 | 34 | 93 | 53 | 12 | 71 | 31 | 90 |
| 51 | 35 | 94 | 54 | 13 | 72 | 32 | 91 |
| 52 | 36 | 95 | 55 | 14 | 73 | 33 | 92 |
| 53 | 37 | 96 | 56 | 15 | 74 | 34 | 93 |
| 54 | 38 | 97 | 57 | 16 | 75 | 35 | 94 |
| 55 | 39 | 98 | 58 | 17 | 76 | 36 | 95 |
| 56 | 40 | 99 | 59 | 18 | 77 | 37 | 96 |
| 57 | 41 | 17 00 | 60 | 19 | 78 | 38 | 97 |
| 58 | 42 | 01 | 61 | 20 | 79 | 39 | 98 |
| 59 | 43 | 02 | 62 | 21 | 80 | 40 | 99 |

# TABLE XV.—(See Explanation on page 197.)

## Correspondence of Hijra (Muhammadan) and A.D. Years, A.D. 622 to A.D. 965.

| | A.D. | H. | A.D. | H. | A.D. | H. | A.D. | H. | A.D. | H. | A.D. | H. | | A.D. | H. | A.D. | H. | A.D. | H. | A.D. | H. | A.D. | H. |
|---|---|---|---|---|---|---|---|---|---|---|---|---|---|---|---|---|---|---|---|---|---|---|---|
| Dec | ... | | 6 40* | 0 20 | 6 72 | 0 53 | 7 05 | 0 87 | 7 37 | 1 20 | 7 70 | 1 54 | Dec | 8 02 | 1 87 | 8 35 | 2 21 | 8 68 | 2 55 | 9 00 | 2 88 | 9 33 | 3 22 |
| Dec | ... | | 41 | 21 | 73 | 54 | 06 | 88 | (5) 38 | 21 | 71 | 55 | Dec | 03* | 88 | 36 | 22 | 69 | 56 | ·63 01* | 89 | 34 | 23 |
| Dec | ... | | ... | ... | 74 | 55 | 07 | 89 | 39 | 22 | 72 | 56 | Dec | 04 | 89 | 37 | 23 | ... | ... | 02 | 90 | ... | ... |
| Nov | ... | | 42 | 22 | 75 | 56 | 08 | 90 | 40 | 23 | 73 | 57 | Nov | 05 | 90 | 38 | 24 | 70 | 57 | 03 | 91 | 35 | 24 |
| Nbv | ... | | 43 | 23 | 76* | 57 | (4) 09 | 91 | 41 | 24 | 74* | 58 | Nov | 06 | 91 | 39 | 25 | 71 | 58 | 04 | 92 | 36 | 25 |
| Nov | ... | | 44 | 24 | 77 | 58 | ... | ... | 42 | 25 | ... | ... | Nov | 07 | 92 | ... | ... | ·61 72* | 59 | 05 | 93 | 37 | 26 |
| Oct | ... | | 45 | 25 | 78 | 59 | 10 | 92 | 43 | 26 | 75 | 59 | Oct | 08 | 93 | 40 | 26 | 73 | 60 | 06 | 94 | 38* | 27 |
| Oct | ... | | 46 | 26 | 79 | 60 | 11 | 93 | 44 | 27 | 76 | 60 | Oct | 09 | 94 | 41 | 27 | 74 | 61 | 07 | 95 | 39 | 28 |
| Oct | ... | | 47* | 27 | (3) 80 | 61 | 12 | 94 | 45* | 28 | 77 | 61 | Oct | 10 | 95 | 42 | 28 | 75 | 62 | ... | ... | 40 | 29 |
| Sep | ... | | 48 | 28 | 81 | 62 | 13 | 95 | 46 | 29 | 78 | 62 | Sep | 11 | 96 | ·60 43* | 29 | 76 | 63 | 08 | 96 | 41 | 30 |
| Sep | ... | | 49 | 29 | 82 | 63 | 14 | 96 | 47 | 30 | 79 | 63 | Sep | 12 | 97 | 44 | 30 | 77 | 64 | 09* | 97 | (12) 42 | 31 |
| Sep | ... | | 50 | 30 | ... | ... | 15 | 97 | ... | ... | 80 | 64 | Sep | 13 | 98 | 45 | 31 | 78 | 65 | 10 | 98 | 43 | 32 |
| Aug | ... | | (2) 51 | 31 | 83 | 64 | 16* | 98 | 48 | 31 | 81 | 65 | Aug | ·59 14* | 99 | 46 | 32 | 79 | 66 | 11 | 99 | 44 | 33 |
| Aug | ... | | 52 | 32 | 84 | 65 | 17 | 99 | 49 | 32 | 82 | 66 | Aug | 15 | 2 00 | 47 | 33 | 80* | 67 | 12 | 3 00 | 45 | 34 |
| Aug | ... | | 53 | 33 | 85 | 66 | 18 | 1 00 | 50 | 33 | 83 | 67 | Aug | ... | ... | 48 | 34 | 81 | 68 | (11) 13 | 01 | 46 | 35 |
| July | 6 22 | 0 1 | 54 | 34 | 86 | 67 | 19 | 01 | 51 | 34 | 84 | 68 | July | 16 | 01 | 49 | 35 | 82 | 69 | 14 | 02 | 47 | 36 |
| July | 23 | 2 | 55 | 35 | 87* | 68 | 20 | 02 | 52 | 35 | ·58 85* | 69 | July | 17 | 02 | 50 | 36 | 83 | 70 | 15 | 03 | 48 | 37 |
| July | ... | ... | ... | ... | 88 | 69 | 21 | 03 | 53 | 36 | 86 | 70 | July | 18 | 03 | 51* | 37 | ... | ... | 16 | 04 | ... | ... |
| June | 24 | 3 | 56 | 36 | 89 | 70 | 22 | 04 | 54 | 37 | 87 | 71 | June | 19 | 04 | 52 | 38 | (10) 84 | 71 | 17 | 05 | 49* | 38 |
| June | 25 | 4 | 57 | 37 | 90 | 71 | 23 | 05 | 55 | 38 | 88 | 72 | June | 20 | 05 | 53 | 39 | 85 | 72 | 18 | 06 | 50 | 39 |
| June | 26 | 5 | 58* | 38 | 91 | 72 | ... | ... | ·56 56* | 39 | ... | ... | June | 21 | 06 | 54 | 40 | 86 | 73 | 19 | 07 | 51 | 40 |
| May | 27 | 6 | 59 | 39 | 92 | 73 | 24 | 06 | 57 | 40 | 89 | 73 | May | 22* | 07 | (9) 55 | 41 | 87 | 74 | 20* | 08 | 52 | 41 |
| May | 28 | 7 | 60 | 40 | 93 | 74 | 25 | 07 | 58 | 41 | 90 | 74 | May | 23 | 08 | 56 | 42 | 88 | 75 | 21 | 09 | 53 | 42 |
| May | ... | ... | 61 | 41 | 94 | 75 | 26 | 08 | 59 | 42 | 91 | 75 | May | 24 | 09 | ... | ... | 89 | 76 | 22 | 10 | 54 | 43 |
| Apr | 29* | 8 | 62 | 42 | 95 | 76 | ·55 27* | 09 | 60 | 43 | 92 | 76 | Apr | 25 | 10 | 57 | 43 | 90 | 77 | 23 | 11 | 55 | 44 |
| Apr | 30 | 9 | 63 | 43 | 96 | 77 | 28 | 10 | 61 | 44 | 93* | 77 | Apr | (8) 26 | 11 | 58 | 44 | 91* | 78 | 24 | 12 | 56 | 45 |
| Apr | 31 | 10 | 64 | 44 | ... | ... | 29 | 11 | 62 | 45 | 94 | 78 | Apr | 27 | 12 | 59 | 45 | 92 | 79 | ... | ... | 57 | 46 |
| Mar | 32 | 11 | 65 | 45 | 97 | 78 | 30 | 12 | 63 | 46 | 95 | 79 | Mar | 28 | 13 | 60 | 46 | 93 | 80 | 25 | 13 | 58 | 47 |
| Mar | 33 | 12 | 66 | 46 | ·53 98* | 79 | 31 | 13 | 64* | 47 | 96 | 80 | Mar | 29 | 14 | 61 | 47 | 94 | 81 | 26 | 14 | 59 | 48 |
| Mar | 34 | 13 | 67 | 47 | 99 | 80 | 32 | 14 | ... | ... | (7) 97 | 81 | Mar | ... | ... | 62* | 48 | 95 | 82 | 27 | 15 | ·66 60* | 49 |
| Feb | 35 | 14 | 68 | 48 | 7 00 | 81 | 33 | 15 | 65 | 48 | 98 | 82 | Feb | 30 | 15 | 63 | 49 | 96 | 83 | 28 | 16 | 61 | 50 |
| Feb | 36 | 15 | ·52 69* | 49 | 01 | 82 | 34 | 16 | 66 | 49 | 99 | 83 | Feb | 31 | 16 | 64 | 50 | 97 | 84 | 29 | 17 | 62 | 51 |
| Feb | 37 | 16 | ... | ... | 02 | 83 | ... | ... | 67 | 50 | 8 00 | 84 | Feb | 32 | 17 | 65 | 51 | ... | ... | 30 | 18 | ... | ... |
| Jan | 38 | 17 | 70 | 50 | 03 | 84 | 35* | 17 | (6) 68 | 51 | 01 | 85 | Jan | 33* | 18 | 66 | 52 | 98 | 85 | ·64 31* | 19 | 63 | 52 |
| Jan | 39 | 18 | 71 | 51 | 04 | 85 | 36 | 18 | 69 | 52 | 02 | 86 | Jan | 34 | 19 | 67 | 53 | 99 | 86 | 32 | 20 | 64 | 53 |
| Jan | ·51 40 | 19 | 72 | 52 | 05* | 86 | 37 | 19 | 70 | 53 | ... | ... | Jan | 35 | 20 | 68 | 54 | 9 00 | 87 | 33 | 21 | 65 | 54 |

# TABLE XV.—(See Explanation on page 197.)

## Correspondence of Hijra (Muhammadan) and A.D. Years, A.D. 965 to A.D. 1323.

| | A.D. | H. | A.D. | H. | A.D. | H. | A.D. | H. | A.D. | H. | A.D. | H. |
|---|---|---|---|---|---|---|---|---|---|---|---|---|
| Dec | 9 65 | 3 55 | 9 98 | 3 89 | 10 30 | 4 22 | 10 63 | 4 56 | 10 95 | 4 89 | 11 28 | 5 23 |
| Dec | 66 | 56 | 99 | 90 | 31 | 23 | 64 | 57 | 96 | 90 | 29 | 24 |
| Dec | 67* | 57 | 10 00 (14) | 91 | 32 | 24 | 65* | 58 | 97 | 91 | 30 | 25 |
| Nov | 68 | 58 | 01 | 92 | 33 | 25 | 66 | 59 | 98 | 92 | 31 | 26 |
| Nov | 69 | 59 | 02 | 93 | 34 | 26 | 67 | 60 | 99 | 93 | 32 | 27 |
| Nov | 70 | 60 | ... | ... | 35 | 27 | ... | ... | 11 00 | 94 | 33 | 28 |
| Oct | (13)71 | 61 | 03 | 94 | 36* | 28 | 68 | 61 | 01 | 95 | ·72 34* | 29 |
| Oct | 72 | 62 | 04 | 95 | 37 | 29 | 69 | 62 | 02 | 96 | 35 | 30 |
| Oct | 73 | 63 | 05 | 96 | 38 | 30 | 70 | 63 | 03 | 97 | ... | ... |
| Sep | 74 | 64 | 06 | 97 | 39 | 31 | 71 | 64 | 04 | 98 | 36 | 31 |
| Sep | 75 | 65 | 07* | 98 | 40 | 32 | 72 | 65 | ·71 05* | 99 | 37 | 32 |
| Sep | ... | ... | 08 | 99 | ... | ... | 73 | 66 | 06 | 5 00 | 28 | 33 |
| Aug | 76 | 66 | 09 | 4 00 | 41 | 33 | 74 | 67 | 07 | 01 | 39 | 34 |
| Aug | 77 | 67 | 10 | 01 | 42 | 34 | 75 | 68 | 08 | 02 | 40 | 35 |
| Aug | 78* | 68 | 11 | 02 | 43 | 35 | ·70 76* | 69 | ... | ... | 41 | 36 |
| July | 79 | 69 | 12 | 03 | 44 | 36 | 77 | 70 | 09 | 03 | 42* | 37 |
| July | 80 | 70 | 13 | 04 | 45 | 37 | 78 | 71 | 10 | 04 | 43 | 38 |
| July | 81 | 71 | 14 | 05 | 46 | 38 | 79 | 72 | 11 | 05 | 44 | 39 |
| June | 82 | 72 | 15 | 06 | ·69 47* | 39 | 80 | 73 | 12 | 06 | 45 | 40 |
| June | 83 | 73 | 16 | 07 | 48 | 40 | 81 | 74 | 13* | 07 | (19)46 | 41 |
| June | 84 | 74 | ... | ... | 49 | 41 | 82 | 75 | 14 | 08 | 47 | 42 |
| May | 85 | 75 | 17 | 08 | 50 | 42 | 83 | 76 | 15 | 09 | 48 | 43 |
| May | 86 | 76 | ·67 18* | 09 | 51 | 43 | 84* | 77 | 16 | 10 | 49 | 44 |
| May | 87 | 77 | 19 | 10 | 52 | 44 | ... | ... | (18)17 | 11 | ... | ... |
| Apr | 88 | 78 | 20 | 11 | 53 | 45 | 85 | 78 | 18 | 12 | 50 | 45 |
| Apr | ·67 89* | 79 | 21 | 12 | 54 | 46 | 86 | 79 | 19 | 13 | 51 | 46 |
| Apr | ... | ... | 22 | 13 | 55 | 47 | 87 | 80 | 20 | 14 | 52 | 47 |
| Mar | 90 | 80 | 23 | 14 | 56* | 48 | (17)88 | 81 | 21 | 15 | 53* | 48 |
| Mar | 91 | 81 | 24 | 15 | 57 | 49 | 89 | 82 | 22 | 16 | 54 | 49 |
| Mar | 92 | 82 | 25 | 16 | ... | ... | 90 | 83 | 23 | 17 | 55 | 50 |
| Feb | 93 | 83 | 26* | 17 | 58 | 50 | 91 | 84 | 24* | 18 | 56 | 51 |
| Feb | 94 | 84 | 27 | 18 | (16)59 | 51 | 92 | 85 | 25 | 19 | 57 | 52 |
| Feb | 95 | 85 | ... | ... | 60 | 52 | 93 | 86 | ... | ... | 58 | 53 |
| Jan | 96 | 86 | 28 | 19 | 61 | 53 | 94 | 87 | 26 | 20 | 59 | 54 |
| Jan | 97* | 87 | 29 | 20 | 62 | 54 | 95* | 88 | 27 | 21 | 60 | 55 |
| Jan | 98 | 88 | (15)30 | 21 | 63 | 55 | ... | ... | 28 | 22 | ... | ... |

| | A.D. | H. | A.D. | H. | A.D. | H. | A.D. | H. | A.D. | H. |
|---|---|---|---|---|---|---|---|---|---|---|
| Dec | 11 60 | 5 56 | 11 93 | 5 90 | 12 26 | 6 24 | 12 58 | 6 57 | 12 91 | 6 91 |
| Dec | 61 | 57 | 94 | 91 | 27 | 25 | 59 | 58 | (24)92 | 92 |
| Dec | 62 | 58 | 95 | 92 | ... | ... | 60 | 59 | 93 | 93 |
| Nov | ·78 68* | 59 | 96 | 93 | 28 | 26 | 61 | 60 | 94 | 94 |
| Nov | 64 | 60 | 97 | 94 | 29* | 27 | (23)62 | 61 | 95 | 95 |
| Nov | 65 | 61 | 98 | 95 | 30 | 28 | 63 | 62 | ... | ... |
| Oct | 66 | 62 | 99 | 96 | 31 | 29 | 64 | 63 | 96 | 96 |
| Oct | 67 | 63 | 12 00* | 97 | 32 | 30 | 65 | 64 | 97 | 97 |
| Oct | 68 | 64 | 01 | 98 | (22)33 | 31 | 66 | 65 | 98* | 98 |
| Sep | 69 | 65 | 02 | 99 | 34 | 32 | 67 | 66 | 99 | 99 |
| Sep | 70 | 66 | 03 | 6 00 | 35 | 33 | 68 | 67 | 13 00 | 7 00 |
| Sep | 71* | 67 | ... | ... | 36 | 34 | ... | ... | 01 | 01 |
| Aug | 72 | 68 | (21)04 | 01 | 37 | 35 | 69* | 68 | 02 | 02 |
| Aug | 73 | 69 | 05 | 02 | 38 | 36 | 70 | 69 | 03 | 03 |
| Aug | 74 | 70 | 06 | 03 | 39 | 37 | 71 | 70 | 04 | 04 |
| July | (20)75 | 71 | 07 | 04 | 40* | 38 | 72 | 71 | 05 | 05 |
| July | 76 | 72 | 08 | 05 | 41 | 39 | 73 | 72 | 06 | 06 |
| July | ... | ... | 09 | 06 | 42 | 40 | 74 | 73 | 07 | 07 |
| June | 77 | 73 | 10 | 07 | 43 | 41 | 75 | 74 | 08 | 08 |
| June | 78 | 74 | 11* | 08 | 44 | 42 | 76 | 75 | ·72 09* | 09 |
| June | 79 | 75 | 12 | 09 | ... | ... | 77 | 76 | ... | ... |
| May | 80 | 76 | 13 | 10 | 45 | 43 | 78 | 77 | 10 | 10 |
| May | 81 | 77 | 14 | 11 | 46 | 44 | 79 | 78 | 11 | 11 |
| May | 82* | 78 | 15 | 12 | 47 | 45 | ·78 80* | 79 | 12 | 12 |
| Apr | 83 | 79 | 16 | 13 | 48 | 46 | 81 | 80 | 13 | 13 |
| Apr | 84 | 80 | 17 | 14 | 49 | 47 | 82 | 81 | 14 | 14 |
| Apr | 85 | 81 | ... | ... | 50 | 48 | 83 | 82 | 15 | 15 |
| Mar | 86 | 82 | 18 | 15 | ·77 51* | 49 | 84 | 83 | 16 | 16 |
| Mar | 87 | 83 | 19 | 16 | 52 | 50 | 85 | 84 | 17* | 17 |
| Mar | 88 | 84 | 20 | 17 | 53 | 51 | ... | ... | 18 | 18 |
| Feb | 89 | 85 | 21 | 18 | 54 | 52 | 86 | 85 | 19 | 19 |
| Feb | 90 | 86 | ·76 22* | 19 | 55 | 53 | 87 | 86 | 20 | 20 |
| Feb | ... | ... | 23 | 20 | ... | ... | 88* | 87 | ... | ... |
| Jan | 91 | 87 | 24 | 21 | 56 | 54 | 89 | 88 | (25)21 | 21 |
| Jan | 92 | 88 | 25 | 22 | 57 | 55 | 90 | 89 | 22 | 22 |
| Jan | ·74 93* | 89 | 26 | 23 | 58 | 56 | 91 | 90 | 23 | 23 |

## TABLE XV.—(See Explanation on page 197.)

### Correspondence of Hijra (Muhammadan) and A.D. Years, A.D. 1323 to A.D. 1681.

| | A.D. | H. | A.D. | H. | A.D. | H. | A.D. | H. | A.D. | H. | A.D. | H. | | A.D. | H. | A.D. | H. | A.D. | H. | A.D. | H. | A.D. | H. |
|---|---|---|---|---|---|---|---|---|---|---|---|---|---|---|---|---|---|---|---|---|---|---|---|
| Dec | 1323 | 724 | 1356 | 758 | 1388 | 791 | 1421 | 825 | 1454* | 859 | 1486 | 892 | Dec | 1519 | 926 | 1551 | 959 | 1584 | 993 | 1616 | 1026 | 1649 | 1060 |
| Dec | 24 | 25 | 57 | 59 | 89 | 92 | 22 | 26 | ·85 55 | 60 | 87 | 93 | Dec | ·20 | 27 | 52 | 60 | 85 | 94 | 17 | 27 | 50 | 61 |
| Dec | 25 | 26 | 58 | 60 | 90 | 93 | 23 | 27 | ... | ... | 88 | 94 | Dec | 21* | 28 | (33)53 | 61 | 86 | 95 | 18* | 28 | 51 | 62 |
| Nov | 26 | 27 | 59 | 61 | 91 | 94 | 24 | 28 | 56 | 61 | 89 | 95 | Nov | 22 | 29 | 54 | 62 | 87 | 96 | 19 | 29 | 52 | 63 |
| Nov | 27* | 28 | 60 | 62 | 92 | 95 | ·84 25* | 29 | 57 | 62 | 90 | 96 | Nov | 23 | 30 | 55 | 63 | 88 | 97 | 20 | 30 | 53 | 64 |
| Nov | 28 | 29 | ... | ... | 93 | 96 | 26 | 30 | 58 | 63 | 91* | 97 | Nov | ... | ... | 56 | 64 | ... | ... | 21 | 31 | 54 | 65 |
| Oct | 29 | 30 | 61 | 63 | 94 | 97 | 27 | 31 | 59 | 64 | 92 | 98 | Oct | (32)24 | 31 | 57 | 65 | 89* | 98 | 22 | 32 | 55 | 66 |
| Oct | 30 | 31 | 62 | 64 | 95 | 98 | 28 | 32 | 60 | 65 | 93 | 99 | Oct | 25 | 32 | 58 | 66 | 90 | 99 | 23 | 33 | 56 | 67 |
| Oct | 31 | 32 | 63 | 65 | ·83 96* | 99 | ... | ... | 61 | 66 | 94 | 900 | Oct | 26 | 33 | 59 | 67 | 91 | 1000 | 24 | 34 | ... | ... |
| Sep | 32 | 33 | 64 | 66 | 87 | 800 | 29 | 33 | 62* | 67 | (31)95 | 01 | Sep | 27 | 34 | 60* | 68 | 92 | 01 | 25 | 35 | 57 | 68 |
| Sep | 33 | 34 | 65 | 67 | 98 | 01 | 30 | 34 | 63 | 68 | 96 | 02 | Sep | 28 | 35 | 61 | 69 | 93 | 02 | 26 | 36 | ·93 58* | 69 |
| Sep | 34 | 35 | 66 | 68 | 99 | 02 | 31 | 35 | 64 | 69 | ... | ... | Sep | 29 | 36 | ... | ... | 94 | 03 | 27 | 37 | 59 | 70 |
| Aug | 35 | 36 | ·82 67* | 69 | 1400 | 08 | 32 | 36 | 65 | 70 | 97 | 03 | Aug | 30 | 37 | 62 | 70 | 95 | 04 | 28 | 38 | 60 | 71 |
| Aug | 36 | 37 | 68 | 70 | 01 | 04 | 33* | 37 | (30)66 | 71 | 98 | 04 | Aug | 31* | 38 | 63 | 71 | 96 | 05 | ·92 29* | 39 | 61 | 72 |
| Aug | ... | ... | 69 | 71 | 02 | 05 | 34 | 38 | 67 | 72 | 99 | 05 | Aug | 32 | 39 | 64 | 72 | 97 | 06 | ... | ... | 62 | 73 |
| July | 37 | 38 | 70 | 72 | 03 | 06 | 35 | 39 | 68 | 73 | 1500 | 06 | July | 33 | 40 | 65 | 73 | 98 | 07 | 30 | 40 | 63 | 74 |
| July | ·81 38* | 39 | 71 | 73 | 04* | 07 | 36 | 40 | 69 | 74 | 01 | 07 | July | 34 | 41 | 66 | 74 | 99 | 08 | 31 | 41 | 64 | 75 |
| July | 39 | 40 | 72 | 74 | ... | ... | (29)37 | 41 | ... | ... | 02* | 08 | July | ... | ... | 67 | 75 | 1600* | 09 | 32 | 42 | 65 | 76 |
| June | 40 | 41 | 73 | 75 | 05 | 08 | 38 | 42 | 70 | 75 | 03 | 09 | June | 35 | 42 | 68 | 76 | ·91 01 | 10 | 33 | 43 | 66* | 77 |
| June | 41 | 42 | 74 | 76 | 06 | 09 | 39 | 43 | 71 | 76 | 04 | 10 | June | 36 | 43 | 69 | 77 | 02 | 11 | 34 | 44 | 67 | 78 |
| June | 42 | 43 | 75* | 77 | 07 | 10 | 40 | 44 | 72 | 77 | 05 | 11 | June | 37 | 44 | 70 | 78 | 03 | 12 | 35 | 45 | 68 | 79 |
| May | 43 | 44 | 76 | 78 | (28)08 | 11 | 41 | 45 | 73* | 78 | 06 | 12 | May | 38 | 45 | ·90 71* | 79 | 04 | 13 | 36 | 46 | 69 | 80 |
| May | 44 | 45 | ... | ... | 09 | 12 | 42 | 46 | 74 | 79 | 07 | 13 | May | 39 | 46 | 72 | 80 | 05 | 14 | 37* | 47 | (37)70 | 81 |
| May | 45 | 46 | 77 | 79 | 10 | 13 | 43 | 47 | 75 | 80 | 08 | 14 | May | 40 | 47 | 73 | 81 | ... | ... | 38 | 48 | ... | ... |
| Apr | 46* | 47 | 78 | 80 | 11 | 14 | 44* | 48 | 76 | 81 | 09 | 15 | Apr | 41 | 48 | 74 | 82 | 06 | 15 | 39 | 49 | 71 | 82 |
| Apr | 47 | 48 | (27)79 | 81 | 12 | 15 | 45 | 49 | 77 | 82 | 10 | 16 | Apr | ·88 42* | 49 | 75 | 83 | 07 | 16 | 40 | 50 | 72 | 83 |
| Apr | 48 | 49 | 80 | 82 | 13 | 16 | ... | ... | 78 | 83 | ... | ... | Apr | 43 | 50 | ... | ... | 08* | 17 | (36)41 | 51 | 73 | 84 |
| Mar | 49 | 50 | 81 | 83 | 14 | 17 | 46 | 50 | 79 | 84 | 11 | 17 | Mar | 44 | 51 | 76 | 84 | 09 | 18 | 42 | 52 | 74 | 85 |
| Mar | (26)50 | 51 | 82 | 84 | 15* | 18 | 47 | 51 | 80 | 85 | 12 | 18 | Mar | 45 | 52 | 77 | 85 | 10 | 19 | 43 | 53 | 75 | 86 |
| Mar | ... | ... | 83 | 85 | 16 | 19 | 48 | 52 | 81 | 86 | ·87 13* | 19 | Mar | 46 | 53 | 78 | 86 | 11 | 20 | ... | ... | 76 | 87 |
| Feb | 51 | 52 | 84 | 86 | 17 | 20 | 49 | 53 | 82 | 87 | 14 | 20 | Feb | 47 | 54 | 79* | 87 | (35)12 | 21 | 44 | 54 | 77* | 88 |
| Feb | 52 | 53 | 85 | 87 | 18 | 21 | 50 | 54 | 83 | 88 | 15 | 21 | Feb | 48 | 55 | 80 | 88 | 13 | 22 | 45 | 55 | 78 | 89 |
| Feb | 53 | 54 | 86* | 88 | ... | ... | 51 | 55 | ... | ... | 16 | 22 | Feb | ... | ... | 81 | 89 | 14 | 23 | 46 | 56 | 79 | 90 |
| Jan | 54 | 55 | 87 | 89 | 19 | 22 | 52 | 56 | ·86 84* | 89 | 17 | 23 | Jan | 49 | 56 | 82 | 90 | 15 | 24 | 47 | 57 | 80 | 91 |
| Jan | 55 | 56 | 88 | 90 | 20 | 23 | 53 | 57 | 85 | 90 | 18 | 24 | Jan | 50* | 57 | (34)83 | 91 | 16 | 25 | 48* | 58 | 81 | 92 |
| Jan | 56* | 57 | ... | ... | 21 | 24 | 54 | 58 | 86 | 91 | 19 | 25 | Jan | 51 | 58 | 84 | 92 | ... | ... | 49 | 59 | ... | ... |

# TABLE XV.—(See Explanation on page 197.)

## Correspondence of Hijra (Muhammadan) and A.D. Years, A.D. 1681 to A.D. 2000.

| | A.D. | H. | A.D. | H. | A.D. | H. | A.D. | H. | A.D. | H. | | A.D. | H. | A.D. | H. | A.D. | H. | A.D. | H. | A.D. | H. |
|---|---|---|---|---|---|---|---|---|---|---|---|---|---|---|---|---|---|---|---|---|---|
| Dec | 1681 | 1093 | 1714 | 1127 | 1747 | 1161 | 1780 | 1195 | 1813 | 1229 | Dec | 1845 | 1262 | 1878 | 1296 | 1911 | 1330 | 1943 | 1363 | 1976 | 1397 |
| Dec | 82 | 94 | 15 | 28 | 48 | 62 | 81 | 96 | 14 | 30 | Dec | 46 | 63 | 79 | 97 | 12 | 31 | 44 | 64 | 77 | 98 |
| Dec | 83 | 95 | ·95 16* | 29 | ... | ... | 82* | 97 | (42)15 | 31 | Dec | 47 | 64 | 80* | 98 | ... | ... | 45 | 65 | ·06/1d 78* | 99 |
| Nov | 84 | 96 | 17 | 30 | 49 | 63 | 83 | 98 | 16 | 32 | Nov | 48 | 65 | 81 | 99 | 13 | 32 | 46 | 66 | 79 | 1400 |
| Nov | 85 | 97 | 18 | 31 | 50 | 64 | 84 | 99 | 17 | 33 | Nov | 49 | 66 | 82 | 1300 | 14 | 33 | 47 | 67 | 80 | 01 |
| Nov | 86 | 98 | 19 | 32 | 51 | 65 | 85 | 1200 | ... | ... | Nov | 50 | 67 | 83 | 01 | 15 | 34 | 48 | 68 | ... | ... |
| Nov | ... | ... | ... | ... | 52 | 66 | ... | ... | ... | ... | Nov | ... | ... | ... | ... | ... | ... | ... | ... | ... | ... |
| Oct | ·94 87* | 99 | 20 | 33 | 53* | 67 | (41)86 | 01 | 18 | 34 | Oct | 51* | 68 | 84 | 02 | 16 | 35 | ·05/1d 49* | 69 | 81 | 02 |
| Oct | 88 | 1100 | 21 | 34 | 54 | 68 | 87 | 02 | 19 | 35 | Oct | 52 | 69 | 85 | 03 | 17 | 36 | 50 | 70 | 82 | 03 |
| Oct | 89 | 01 | 22 | 35 | 55 | 69 | 88 | 03 | 20 | 36 | Oct | 53 | 70 | ... | ... | 18 | 37 | 51 | 71 | 83 | 04 |
| Sep | 90 | 02 | 23 | 36 | 56 | 70 | 89 | 04 | 21 | 37 | Sep | 54 | 71 | 86 | 04 | 19 | 38 | 52 | 72 | 84 | 05 |
| Sep | 91 | 03 | 24* | 37 | (40)57 | 71 | 90 | 05 | 22* | 38 | Sep | 55 | 72 | 87 | 05 | ·04/1d 20* | 39 | 53 | 73 | 85 | 06 |
| Sep | 92 | 04 | ... | ... | 58 | 72 | ... | ... | 23 | 39 | Sep | 56 | 73 | 88 | 06 | 21 | 40 | ... | ... | 86* | 07 |
| Aug | 93 | 05 | 25 | 38 | 59 | 73 | 91 | 06 | 24 | 40 | Aug | 57 | 74 | 89 | 07 | 22 | 41 | 54 | 74 | 87 | 08 |
| Aug | 94 | 06 | 26 | 39 | 60 | 74 | 92 | 07 | 25 | 41 | Aug | 58 | 75 | 90 | 08 | 23 | 42 | 55 | 75 | 88 | 09 |
| Aug | 95* | 07 | 27 | 40 | 61 | 75 | 93* | 08 | 26 | 42 | Aug | ... | ... | ·02/1d 91* | 09 | 24 | 43 | 56 | 76 | 89 | 10 |
| July | 96 | 08 | (39)28 | 41 | 62 | 76 | 94 | 09 | 27 | 43 | July | 59 | 76 | 92 | 10 | 25 | 44 | 57* | 77 | (48)90 | 11 |
| July | 97 | 09 | 29 | 42 | 63 | 77 | 95 | 10 | 28 | 44 | July | 60 | 77 | 93 | 11 | 26 | 45 | 58 | 78 | 91 | 12 |
| July | ... | ... | 30 | 43 | ... | ... | 96 | 11 | ... | ... | July | 61 | 78 | 94 | 12 | 27 | 46 | 59 | 79 | 92 | 13 |
| June | 98 | 10 | 31 | 44 | 64* | 78 | 97 | 12 | 29 | 45 | June | ·01/1d 62* | 79 | 95 | 13 | 28* | 47 | 60 | 80 | 93 | 14 |
| June | (38)99 | 11 | 32 | 45 | 65 | 79 | 98 | 13 | 30 | 46 | June | 63 | 80 | 96 | 14 | 29 | 48 | (47)61 | 81 | 94 | 15 |
| June | 1700 | 12 | 33 | 46 | 66 | 80 | 99 | 14 | 31 | 47 | June | 64 | 81 | 97 | 15 | ... | ... | 62 | 82 | ... | ... |
| May | 01 | 13 | 34 | 47 | 67 | 81 | 1800 | 15 | 32 | 48 | May | 65 | 82 | 98 | 16 | 30 | 49 | 63 | 83 | 95 | 16 |
| May | 02 | 14 | 35* | 48 | 68 | 82 | 01 | 16 | ·00/1d 33* | 49 | May | 66 | 83 | 99* | 17 | 31 | 50 | 64 | 84 | 96 | 17 |
| May | 03 | 15 | 36 | 49 | 69 | 83 | 02 | 17 | 34 | 50 | May | 67 | 84 | 1900 | 18 | (46)32 | 51 | 65 | 85 | 97* | 18 |
| Apr | 04 | 16 | 37 | 50 | 70 | 84 | 03 | 18 | 35 | 51 | Apr | 68 | 85 | 01 | 19 | 33 | 52 | 66 | 86 | 98 | 19 |
| Apr | 05 | 17 | 38 | 51 | 71 | 85 | ·99 04* | 19 | 36 | 52 | Apr | 69 | 86 | 02 | 20 | 34 | 53 | 67 | 87 | 99 | 20 |
| Apr | 06* | 18 | ... | ... | 72 | 86 | 05 | 20 | 37 | 53 | Apr | 70* | 87 | ... | ... | 35 | 54 | ... | ... | 2000 | 21 |
| Mar | 07 | 19 | 39 | 52 | 73 | 87 | 06 | 21 | 38 | 54 | Mar | 71 | 88 | (45)03 | 21 | 36 | 55 | 68* | 88 | ... | ... |
| Mar | 08 | 20 | 40 | 53 | 74 | 88 | 07 | 22 | 39 | 55 | Mar | 72 | 89 | 04 | 22 | 37 | 56 | 69 | 89 | ... | ... |
| Mar | 09 | 21 | 41 | 54 | ·98 75* | 89 | ... | ... | 40 | 56 | Mar | 73 | 90 | 05 | 23 | 38 | 57 | 70 | 90 | ... | ... |
| Feb | 10 | 22 | 42 | 55 | 76 | 90 | 08 | 23 | 41* | 57 | Feb | (44)74 | 91 | 06 | 24 | 39* | 58 | 71 | 91 | ... | ... |
| Feb | 11 | 23 | 43 | 56 | 77 | 91 | 09 | 24 | 42 | 58 | Feb | 75 | 92 | 07 | 25 | 40 | 59 | 72 | 92 | ... | ... |
| Feb | ... | ... | 44 | 57 | ... | ... | 10 | 25 | 43 | 59 | Feb | ... | ... | 08 | 26 | ... | ... | 73 | 93 | ... | ... |
| Jan | 12 | 24 | 45 | 58 | 78 | 92 | 11 | 26 | 44 | 60 | Jan | 76 | 93 | 09 | 27 | 41 | 60 | 74 | 94 | ... | ... |
| Jan | 13 | 25 | ·97 46* | 59 | 79 | 93 | 12* | 27 | (43)45 | 61 | Jan | 77 | 94 | 10* | 28 | 42 | 61 | 75 | 95 | ... | ... |
| Jan | 14 | 26 | 47 | 60 | 80 | 94 | 13 | 28 | ... | ... | Jan | 78 | 95 | 11 | 29 | 43 | 62 | 76 | 96 | ... | ... |

# TABLE XV.—Explanation.

## Correspondence of Hijra (Muhammadan) and A.D. Years, A.D. 622 to A.D. 2000.

1. This table shows the A.D. year and month when each Hijra year begins.

2. Hijra years are divided into cycles of 30 each. From 622 A.D. to 2000 A.D. there are 48 such cycles, the first cycle being an incomplete one of 29 years. The cycles are indicated by the figures between brackets.

3. The actual date of the A.D. month marking the commencement of each Hijra year can be obtained from Table X by adding 1, 2 or 3 days, according to the following rules, to the date of mean new-moon shown in that table.

(*a*) Subject to the exceptions noted in Rule (*c*) below, 2 days are to be added to the date of mean new-moon when the fraction of day shown in Table X against the mean new-moon does not exceed the decimal figure printed in italic under each cycle in this table. When this figure is exceeded, add 3 days.

(*b*) The figure printed in italic may be called the 'argument of the crescent' and the quantity to be added to the date of mean new-moon, whether 1, 2 or 3 days, may be called the 'equation of the crescent'. In cycles (1) to (41), the highest argument for an equation of 2 days is shown against the year in which that argument occurs. For instance, in the 30th cycle, the maximum argument for an equation of 2 days is ·86 which is entered against A.D. 1484, H. 889. This entry means that in the 30th cycle, whenever the fraction of day set against a new moon in Table X, is ·86 (as in January 1484 A.D.) or less, 2 days are to be added to the date of mean new moon in order to get the commencement of the corresponding Hijra year. Otherwise, that is, if the fraction of day exceeds ·86, 3 days should be added. Thus, taking 10 successive years of the 30th cycle, the dates of commencement of the corresponding Hijra years are ascertained as follows from this Table and Table X :—

| A.D. year. | Month. | Day and fraction of day. | Hijra year. | Commencement of Hijra year. |
|---|---|---|---|---|
| 1480 | March | 11·39 | 885 | 13th March |
| 1481 | February | 28·76 | 886 | 2nd March |
| 1482 | February | 18·13 | 887 | 20th February |
| 1483 | February | 7·50 | 888 | 9th February |
| 1484 | January | 27·86 | 889 | 29th January |
| 1485 | January | 16·23 | 890 | 18th January |
| 1486 | January | 5·60 | 891 | 7th January |
| 1486 | December | 25·96 | 892 | 28th December |
| 1487 | December | 15·33 | 893 | 17th December |
| 1488 | December | 3·70 | 894 | 5th December |

(*c*) From the 42nd to the 48th cycles (A.D. 1815—2000), the additions to be made to mean new moon dates in order to get the commencement of Hijra years are 1 day and 2 days respectively, according as the limit shown in this table is not, or is, exceeded. Thus in the 42nd cycle, 1 day has to be added only when the fraction of day of mean new moon is ·00, *i.e.*, in the year 1833, when new moon fell on May 19·00. The corresponding Hijra year 1249, therefore, commenced on 20th May 1833. In all other years of this cycle the number of days to be added to mean new moon in order to get the commencement of Hijra year is 2. Similarly in 1862, mean new-moon fell on June 27·01 and the Hijra year 1279 commenced on 28th June. In A.D. 1978, the Hijra year will commence on Dec. 1, because new-moon will fall on Nov. 30·06, and ·06, is the argument of the crescent for that cycle.

4. The 2nd, 5th, 8th, 10th, 13th, 15th, 19th, 21st, 24th, 27th and 29th years in each cycle are leap years according to the present table, that is, they will be found to contain 355 days instead of 354. In certain countries, however, the 7th, 18th and 26th years of each cycle are reckoned as leap years instead of the 8th, 19th and 27th. Under the former reckoning, the 8th, 19th and 27th years following as they do, certain leap years, would begin 1 day later than they do according to the present table. For the convenience of reference, the 8th, 19th and 27th years of each cycle are marked by an asterisk. The reader will understand that the date of commencement of such Hijra years, as determined by the rules above laid down, is liable to be increased by 1 day in those countries where the 7th, 18th and 26th years of each cycle are regarded as leap years.

# TABLE XVI.

## Modern Equations for Lunar Inequalities.

### A.—Equation for Lunar Variation.

| Tithis | 2 × (☾'s—☉'s Long.) | d. | g. | p. | d. | g. | p. | Eqn. g. | Eqn. p. |
|---|---|---|---|---|---|---|---|---|---|
| **1** | 24° | 1 | 50 | 13 | 29 | 23 | 30 | −1 | 11 |
| **2** | 48° | 3 | 40 | 26 | 31 | 13 | 43 | −2 | 11 |
| **3** | 72° | 5 | 30 | 39 | 33 | 3 | 56 | −2 | 47 |
| **4** | 96° | 7 | 20 | 52 | 34 | 54 | 9 | −2 | 55 |
| **5** | 120° | 9 | 11 | 6 | 36 | 44 | 23 | −2 | 32 |
| **6** | 144° | 11 | 1 | 19 | 38 | 34 | 36 | −1 | 43 |
| **7** | 168° | 12 | 51 | 32 | 40 | 24 | 49 | −0 | 37 |
| **8** | 192° | 14 | 41 | 45 | 42 | 15 | 2 | +0 | 37 |
| **9** | 216° | 16 | 31 | 58 | 44 | 5 | 15 | +1 | 43 |
| **10** | 240° | 18 | 22 | 11 | 45 | 55 | 28 | +2 | 32 |
| **11** | 264° | 20 | 12 | 24 | 47 | 45 | 41 | +2 | 55 |
| **12** | 288° | 22 | 2 | 37 | 49 | 35 | 54 | +2 | 47 |
| **13** | 312° | 23 | 52 | 50 | 51 | 26 | 7 | +2 | 11 |
| **14** | 336° | 25 | 43 | 3 | 53 | 16 | 20 | +1 | 11 |
| **15** | 360° | 27 | 33 | 17 | 55 | 6 | 33 | +0 | 0 |

## TABLE XVI.—Modern Equations for

### B.—☾'s Anomaly in days, ghatikas and palas and corresponding Equation of the

| Eqn. Palas. | 0gh. | 1 | 2. | 3 | 4 | 5 | 6 | 7 | 8 | 9 | Eqn. Palas. | 10gh. | 11 | 12 | 13 | 14 | 15 | 16 | 17 | 18 | 19 |
|---|---|---|---|---|---|---|---|---|---|---|---|---|---|---|---|---|---|---|---|---|---|
| | 0 | ← Days of Anom. → | | | | | 0 | I | I | I | | I | I | I | I | II | II | II | II | II | II |
| | g.p. | g.p. | g.p. | g.p. | g.p. | g.p. | g.p. | g.p. | g.p. | g.p. | | g.p. | g.p. | g.p. | g.p. | g.p. | g.p. | g.p. | g.p. | g.p. | g.p. |
| 0 | 0, 6 | 8,30 | 17, 0 | 25,34 | 34, 8 | 42,43 | 51,22 | 0,10 | 8,53 | 17,41 | 0 | 26,39 | 35,36 | 44,52 | 54,12 | 3,32 | 13,16 | 23, 8 | 33, 5 | 43,25 | 54, 3 |
| 1 | 0, 9 | 8,39 | 17,13 | 25,43 | 34,17 | 42,52 | 51,31 | 0,14 | 8,58 | 17,15 | 1 | 26,43 | 35,45 | 45, 1 | 54,17 | 3,46 | 13,25 | 23,13 | 33,14 | 43,34 | 54,13 |
| 2 | 0,14 | 8,48 | 17,18 | 25,52 | 34,27 | 43, 1 | 51,40 | 0,23 | 9, 7 | 17,55 | 2 | 26,52 | 35,54 | 45, 5 | 54,26 | 3,55 | 13,34 | 23,26 | 33,28 | 43,43 | 54,22 |
| 3 | 0,23 | 8,57 | 17,27 | 26, 1 | 34,31 | 43,10 | 51,45 | 0,33 | 9,16 | 18, 4 | 3 | 27, 1 | 36, 4 | 45,15 | 54,40 | 4, 4 | 13,43 | 23,36 | 33,37 | 43,52 | 54,35 |
| 4 | 0,32 | 9, 6 | 17,36 | 26,11 | 34,41 | 43,19 | 51,54 | 0,42 | 9,25 | 18,14 | 4 | 27,15 | 36,13 | 45,28 | 54,49 | 4,13 | 13,52 | 23,45 | 33,46 | 44, 6 | 54,45 |
| 5 | 0,41 | 9,11 | 17,46 | 26,20 | 34,50 | 43,29 | 52, 7 | 0,51 | 9,35 | 18,23 | 5 | 27,20 | 36,22 | 45,38 | 54,53 | 4,23 | 14, 1 | 23,54 | 33,55 | 44,15 | 54,54 |
| 6 | 0,50 | 9,20 | 17,55 | 26,29 | 34,59 | 43,38 | 52,17 | 1, 0 | 9,44 | 18,32 | 6 | 27,29 | 36,31 | 45,47 | 55, 2 | 4,32 | 14,15 | 24, 3 | 34, 1 | 44,29 | 55, 8 |
| 7 | 1, 0 | 9,30 | 17,59 | 26,34 | 35, 8 | 43,47 | 52,21 | 1, 9 | 9,53 | 18,41 | 7 | 27,38 | 36,40 | 45,52 | 55,16 | 4,46 | 14,20 | 24,17 | 34,19 | 44,34 | 55,17 |
| 8 | 1, 9 | 9,39 | 18, 8 | 26,43 | 35,17 | 43,52 | 52,30 | 1,19 | 9,58 | 18,50 | 8 | 27,48 | 36,49 | 46, 5 | 55,25 | 4,50 | 14,29 | 24,26 | 34,23 | 44,47 | 55,30 |
| 9 | 1,18 | 9,48 | 18,18 | 26,52 | 35,26 | 44, 1 | 52,40 | 1,28 | 10, 7 | 18,59 | 9 | 28, 1 | 36,59 | 46,14 | 55,35 | 5, 0 | 14,43 | 24,35 | 34,37 | 44,57 | 55,40 |
| 10 | 1,27 | 9,53 | 18,27 | 27, 1 | 35,31 | 44,10 | 52,49 | 1,37 | 10,16 | 19, 9 | 10 | 28, 6 | 37, 8 | 46,24 | 55,44 | 5,13 | 14,52 | 24,49 | 34,46 | 45,10 | 55,52 |
| 11 | 1,32 | 10, 2 | 18,36 | 27,11 | 35,40 | 44,19 | 52,58 | 1,46 | 10,25 | 19,13 | 11 | 28,15 | 37,17 | 46,33 | 55,53 | 5,23 | 14,56 | 24,54 | 35, 0 | 45,15 | 56, 3 |
| 12 | 1,41 | 10,11 | 18,45 | 27,20 | 35,49 | 44,28 | 53, 7 | 1,55 | 10,34 | 19,23 | 12 | 28,24 | 37,26 | 46,42 | 56, 2 | 5,32 | 15,10 | 25, 3 | 35, 9 | 45,29 | 56,12 |
| 13 | 1,50 | 10,20 | 18,50 | 27,29 | 35,59 | 44,33 | 53,17 | 2, 5 | 10,43 | 19,31 | 13 | 28,33 | 37,35 | 46,51 | 56,11 | 5,41 | 15,20 | 25,16 | 35,14 | 45,38 | 56,26 |
| 14 | 2, 0 | 10,29 | 18,59 | 27,38 | 36, 8 | 44,42 | 53,25 | 2, 9 | 10,53 | 19,46 | 14 | 28,43 | 37,44 | 47, 0 | 56,20 | 5,50 | 15,29 | 25,21 | 35,27 | 45,52 | 56,35 |
| 15 | 2, 9 | 10,38 | 19, 8 | 27,47 | 36,17 | 44,51 | 53,35 | 2,18 | 11, 2 | 19,54 | 15 | 28,52 | 37,54 | 47, 9 | 56,30 | 5,59 | 15,38 | 25,35 | 35,32 | 46, 1 | 56,44 |
| 16 | 2,18 | 10,48 | 19,17 | 27,52 | 36,26 | 45, 0 | 53,39 | 2,28 | 11,11 | 19,59 | 16 | 29, 1 | 38, 3 | 47,19 | 56,39 | 6, 8 | 15,47 | 25,44 | 35,46 | 46,10 | 56,58 |
| 17 | 2,22 | 10,57 | 19,26 | 28, 1 | 36,35 | 45,10 | 53,49 | 2,37 | 11,20 | 20, 8 | 17 | 29,10 | 38,12 | 47,28 | 56,48 | 6,18 | 15,56 | 25,53 | 35,50 | 46,20 | 57, 7 |
| 18 | 2,32 | 11, 6 | 19,36 | 28,10 | 36,44 | 45,19 | 53,58 | 2,46 | 11,29 | 20,18 | 18 | 29,19 | 38,21 | 47,37 | 56,57 | 6,27 | 16, 6 | 26, 7 | 36, 4 | 46,33 | 57,21 |
| 19 | 2,41 | 11,10 | 19,45 | 28,19 | 36,49 | 45,28 | 54, 7 | 2,55 | 11,38 | 20,27 | 19 | 29,29 | 38,30 | 47,46 | 57, 7 | 6,36 | 16,19 | 26,11 | 36,13 | 46,42 | 57,30 |
| 20 | 2,50 | 11,20 | 19,54 | 28,24 | 36,58 | 45,37 | 54,16 | 3, 4 | 11,48 | 20,36 | 20 | 29,38 | 38,39 | 47,55 | 57,16 | 6,50 | 16,24 | 26,26 | 36,27 | 46,52 | 57,39 |
| 21 | 2,59 | 11,29 | 20, 3 | 28,38 | 37, 7 | 45,46 | 54,25 | 3,13 | 11,57 | 20,45 | 21 | 29,47 | 38,49 | 48, 5 | 57,25 | 6,59 | 16,38 | 26,35 | 36,36 | 47, 1 | 57,53 |
| 22 | 3, 8 | 11,38 | 20, 8 | 28,42 | 37,17 | 45,51 | 54,35 | 3,23 | 12, 1 | 20,54 | 22 | 29,56 | 38,58 | 48,14 | 57,34 | 7, 8 | 16,47 | 26,44 | 36,45 | 47,15 | 58, 7 |
| 23 | 3,17 | 11,43 | 20,17 | 28,51 | 37,26 | 46, 0 | 54,44 | 3,27 | 12,11 | 21, 4 | 23 | 30, 5 | 39, 7 | 48,23 | 57,43 | 7,17 | 16,56 | 26,53 | 36,54 | 47,24 | 58,11 |
| 24 | 3,27 | 11,52 | 20,26 | 29, 1 | 37,35 | 46, 9 | 54,53 | 3,36 | 12,20 | 21,13 | 24 | 30,14 | 39,16 | 48,32 | 57,52 | 7,26 | 17, 5 | 27, 2 | 37, 8 | 47,37 | 58,25 |
| 25 | 3,31 | 12, 1 | 20,36 | 29,10 | 37,44 | 46,19 | 55, 2 | 3,45 | 12,29 | 21,17 | 25 | 30,24 | 39,26 | 48,41 | 58, 2 | 7,36 | 17,14 | 27,16 | 37,17 | 47,42 | 58,34 |
| 26 | 3,40 | 12,10 | 20,45 | 29,19 | 37,53 | 46,28 | 55,11 | 3,59 | 12,38 | 21,26 | 26 | 30,33 | 39,30 | 48,50 | 58,11 | 7,45 | 17,23 | 27,25 | 37,27 | 47,56 | 58,48 |
| 27 | 3,50 | 12,20 | 20,54 | 29,28 | 38, 2 | 46,37 | 55,20 | 4, 8 | 12,48 | 21,36 | 27 | 30,42 | 39,39 | 49, 0 | 58,25 | 7,54 | 17,33 | 27,34 | 37,36 | 48, 5 | 59, 2 |
| 28 | 3,59 | 12,29 | 21, 3 | 29,37 | 38, 7 | 46,46 | 55,30 | 4,18 | 12,57 | 21,49 | 28 | 30,47 | 39,53 | 49,14 | 58,29 | 8, 3 | 17,46 | 27,44 | 37,50 | 48,19 | 59, 6 |
| 29 | 4, 3 | 12,38 | 21,12 | 29,42 | 38,16 | 46,55 | 55,34 | 4,22 | 13, 6 | 21,54 | 29 | 30,56 | 40, 2 | 49,18 | 58,38 | 8,17 | 17,51 | 27,53 | 37,58 | 48,23 | 59,20 |
| 30 | 4,13 | 12,47 | 21,21 | 29,51 | 38,26 | 47, 4 | 55,43 | 4,31 | 13,15 | 22, 3 | 30 | 31, 5 | 40,12 | 49,27 | 58,47 | 8,26 | 18, 5 | 28, 6 | 38,13 | 48,37 | 59,29 |
| 31 | 4,22 | 12,56 | 21,26 | 30, 0 | 38,34 | 47, 9 | 55,52 | 4,41 | 13,20 | 22,12 | 31 | 31,14 | 40,16 | 49,36 | 59, 1 | 8,35 | 18,14 | 28,16 | 38,17 | 48,47 | 59,43 |
| 32 | 4,31 | 13, 1 | 21,35 | 30, 9 | 38,39 | 47,18 | 56, 2 | 4,50 | 13,29 | 22,21 | 32 | 31,23 | 40,25 | 49,50 | 59, 6 | 8,45 | 18,23 | 28,25 | 38,31 | 49, 0 | 59,53 |
| 33 | 4,40 | 13,10 | 21,44 | 30,19 | 38,49 | 47,27 | 56,11 | 4,59 | 13,43 | 22,31 | 33 | 31,33 | 40,39 | 49,59 | 59,15 | 8,54 | 18,37 | 28,34 | 38,40 | 49, 9 | 3 0, 2 |
| 34 | 4 49 | 13,19 | 21,53 | 30,23 | 38,58 | 47,37 | 56,20 | 5, 8 | 13,52 | 22,40 | 34 | 31,42 | 40,48 | 50, 4 | 59,29 | 9, 3 | 18,41 | 28,43 | 38,54 | 49,19 | 0,15 |
| 35 | 4,59 | 13,28 | 21,58 | 30,32 | 39, 7 | 47,46 | 56,29 | 5 18 | 13,56 | 22,49 | 35 | 31,51 | 40,53 | 50,13 | 59,38 | 9,12 | 18,56 | 28,57 | 38,59 | 49,28 | 0,29 |
| 36 | 5, 8 | 13,38 | 22, 7 | 30,41 | 39,16 | 47,55 | 56,38 | 5,27 | 14, 6 | 22,58 | 36 | 32, 0 | 41, 2 | 50,27 | 59,47 | 9,21 | 19, 4 | 29, 6 | 39,12 | 49,42 | 0,38 |
| 37 | 5,17 | 13,42 | 22,16 | 30,51 | 39,25 | 48, 4 | 56,48 | 5,31 | 14,15 | 23, 8 | 37 | 32, 5 | 41,11 | 50,36 | 59,52 | 9,30 | 19,14 | 29,15 | 39,22 | 49,51 | 0,48 |
| 38 | 5,22 | 13,51 | 22,26 | 31, 0 | 39,34 | 48,13 | 56,52 | 5,40 | 14,24 | 23,12 | 38 | 32,19 | 41,20 | 50,41 | 2 0, 5 | 9,44 | 19,28 | 29,24 | 39,35 | 50, 0 | 0,57 |
| 39 | 5,31 | 14, 0 | 22,35 | 31, 9 | 39,44 | 48,22 | 57, 2 | 5,50 | 14,33 | 23,21 | 39 | 32,28 | 41,34 | 50,50 | 0,15 | 9,53 | 19,32 | 29,34 | 39,44 | 50, 9 | 1,10 |
| 40 | 5,40 | 14,10 | 22,44 | 31,19 | 39,53 | 48,27 | 57,10 | 5,59 | 14,42 | 23,30 | 40 | 32,37 | 41,39 | 51, 4 | 0,24 | 9,58 | 19,46 | 29,47 | 39,54 | 50,23 | 1,24 |
| 41 | 5,49 | 14,19 | 22,53 | 31,28 | 39,57 | 48,36 | 57,20 | 6, 8 | 14,51 | 23,39 | 41 | 32,41 | 41,48 | 51,13 | 0,33 | 10,12 | 19,55 | 29,52 | 40, 3 | 50,32 | 1,29 |
| 42 | 5,58 | 14,28 | 23, 2 | 31,37 | 40, 7 | 48,45 | 57,33 | 6,17 | 15, 1 | 23,54 | 42 | 32,51 | 41,57 | 51,22 | 0,42 | 10,21 | 20, 4 | 30, 6 | 40,17 | 50,46 | 1,43 |
| 43 | 6, 3 | 14,37 | 23,11 | 31,41 | 40,16 | 48,55 | 57,38 | 6,26 | 15,10 | 23,58 | 43 | 33, 4 | 42,11 | 51,31 | 0,52 | 10,30 | 20,18 | 30,15 | 40,26 | 50,50 | 1,56 |
| 44 | 6,12 | 14,46 | 23,16 | 31,51 | 40,25 | 49, 4 | 57,47 | 6,35 | 15,14 | 24, 7 | 44 | 33,14 | 42,15 | 51,40 | 1, 0 | 13,39 | 20,23 | 30,24 | 40,35 | 51, 4 | 2, 6 |
| 45 | 6,21 | 14,56 | 23,26 | 32, 0 | 40,34 | 49,13 | 57,57 | 6,45 | 15,24 | 24,16 | 45 | 33,23 | 42,24 | 51,50 | 1,15 | 10,48 | 20,36 | 30,38 | 40,44 | 51,18 | 2,19 |
| 46 | 6,30 | 15, 0 | 23,34 | 32, 9 | 40,48 | 49,22 | 58, 6 | 6,49 | 15,38 | 24,26 | 46 | 33,27 | 42,34 | 51,59 | 1,19 | 10,58 | 20,46 | 30,43 | 40,58 | 51,27 | 2,24 |
| 47 | 6,40 | 15, 9 | 23,44 | 32,18 | 40,52 | 49,32 | 58,15 | 6,58 | 15,46 | 24,35 | 47 | 33,36 | 42,43 | 52, 8 | 1,28 | 11,11 | 20,55 | 30,56 | 41, 7 | 51,36 | 2,38 |
| 48 | 6,49 | 15,18 | 23,53 | 32,27 | 41, 2 | 49,41 | 58,24 | 7,12 | 15,56 | 24,44 | 48 | 33,46 | 42,52 | 52,17 | 1,38 | 11,21 | 21, 4 | 31, 6 | 41,21 | 51,50 | 2,51 |
| 49 | 6,58 | 15,28 | 24, 2 | 32,36 | 41,11 | 49,45 | 58,33 | 7,21 | 16, 0 | 24,53 | 49 | 33.59 | 43, 2 | 52,26 | 1,51 | 11,25 | 21,13 | 31,15 | 41,25 | 51,59 | 3, 1 |
| 50 | 7, 7 | 15,32 | 24,11 | 32,46 | 41,15 | 49,54 | 58,43 | 7,26 | 16, 9 | 25, 2 | 50 | 34, 4 | 43,11 | 52,35 | 1,56 | 11,39 | 21,27 | 31,24 | 41,39 | 52,13 | 3,10 |
| 51 | 7,16 | 15,41 | 24,21 | 32,55 | 41,24 | 50, 4 | 58,52 | 7,35 | 16,19 | 25,11 | 51 | 34,13 | 43,20 | 52,45 | 2,10 | 11,48 | 21,36 | 31,33 | 41,49 | 52,22 | 3,19 |
| 52 | 7,21 | 15,51 | 24,30 | 32,59 | 41,34 | 50,13 | 58,56 | 7,44 | 16,28 | 25,16 | 52 | 34,22 | 43,34 | 52,54 | 2,19 | 11,57 | 21,45 | 31,47 | 42, 2 | 52,32 | 3,33 |
| 53 | 7,30 | 16, 0 | 24,34 | 33, 9 | 41,43 | 50,22 | 59, 5 | 7,53 | 16,37 | 25,25 | 53 | 34,32 | 43,43 | 53, 3 | 2,28 | 12, 7 | 21,54 | 31,56 | 42, 7 | 52,45 | 3,46 |
| 54 | 7,39 | 16, 9 | 24,43 | 33,18 | 41,52 | 50,31 | 59,15 | 8, 3 | 16,46 | 25,34 | 54 | 34,46 | 43,52 | 53,12 | 2,37 | 12,16 | 22, 4 | 32, 5 | 42,21 | 52,54 | 3,56 |
| 55 | 7,48 | 16,18 | 24,58 | 33,27 | 42, 1 | 50,40 | 59,19 | 8,12 | 16,56 | 25,44 | 55 | 34,50 | 44, 1 | 53,21 | 2,46 | 12,25 | 22,17 | 32,14 | 42,30 | 53, 8 | 4, 5 |
| 56 | 7,53 | 16,28 | 25, 2 | 33,31 | 42,10 | 50,50 | 59,33 | 8,21 | 17, 5 | 25,53 | 56 | 34,59 | 44,10 | 53,30 | 2.55 | 12,34 | 22,22 | 32,24 | 42,44 | 53,17 | 4,19 |
| 57 | 8, 2 | 16,37 | 25,11 | 33,41 | 42,20 | 50,59 | 59,42 | 8,30 | 17,14 | 26, 7 | 57 | 35, 9 | 44,24 | 53,40 | 3, 5 | 12,48 | 22,36 | 32,37 | 42,53 | 53,27 | 4,28 |
| 58 | 8,11 | 16,46 | 25,16 | 33,50 | 42,29 | 51, 3 | 59,51 | 8,39 | 17,18 | 26,16 | 58 | 35.18 | 44,29 | 53,49 | 3,18 | 12,53 | 22,45 | 32,47 | 43, 2 | 53,40 | 4,42 |
| 59 | 8,20 | 16,50 | 25,25 | 33,59 | 42,34 | 51,12 | 1 0, 0 | 8,44 | 17,28 | 26,29 | 59 | 35,27 | 44,38 | 53,58 | 3,23 | 13, 2 | 22,54 | 32,56 | 43,11 | 53,50 | 4,51 |

## Lunar Inequalities.

**centre in ghatikas and palas.**

| 20gh. III g.p. | 21 III g.p. | 22 III g.p. | 23 III g.p. | 24 III g.p. | Eqn. Palas. | 25gh. IV g.p. | 26 IV g.p. | 27 IV g.p. | 28 IV g.p. | 29 V g.p. | 30 V g.p. |
|---|---|---|---|---|---|---|---|---|---|---|---|
| 5, 0 | 16,24 | 28,11 | 40,32 | 53,45 | **0** | 7,45 | 22,41 | 39,13 | 57,45 | 19,38 | 48,39 |
| 5,14 | 16,33 | 28,21 | 40,49 | 53,59 | **1** | 7,55 | 23, 0 | 39,32 | 57,58 | 20, 1 | 49,20 |
| 5,23 | 16,47 | 28,34 | 41, 8 | 54, 8 | **2** | 8, 9 | 23,18 | 39,45 | 58,21 | 20,29 | 49,48 |
| 5,37 | 16,56 | 28,48 | 41.17 | 54,22 | **3** | 8,27 | 23,32 | 40. 4 | 58,44 | 20,51 | 50.29 |
| 5,46 | 17,10 | 29, 1 | 41,26 | 54,36 | **4** | 8,41 | 23,46 | 40,23 | 59, 3 | 21,14 | 51,10 |
| 5,54 | 17,24 | 29,11 | 41,40 | 54,54 | **5** | 8,50 | 24, 4 | 40,45 | 59,21 | 21,38 | 51,47 |
| 6, 9 | 17,33 | 29,25 | 41.49 | 55, 8 | **6** | 9, 8 | 24,18 | 41, 4 | 59,44 | 22, 5 | 52,28 |
| 6,18 | 17,42 | 29,39 | 42, 3 | 55,17 | **7** | 9,22 | 24,36 | 41,17 | 5 0. 3 | 22,28 | 52,56 |
| 6,27 | 17,56 | 29,52 | 42,12 | 55,31 | **8** | 9,40 | 24,50 | 41,36 | 0,25 | 22,51 | 53,37 |
| 6,41 | 18,10 | 30, 2 | 42,26 | 55,45 | **9** | 9,54 | 25, 4 | 41,59 | 0,48 | 23,19 | 54,18 |
| 6,50 | 18,19 | 30,11 | 42,40 | 55,58 | **10** | 10, 8 | 25,22 | 42,17 | 1, 2 | 23,41 | 55, 0 |
| 7, 4 | 18,28 | 30,25 | 42,54 | 56,12 | **11** | 10,22 | 25,40 | 42,31 | 1,25 | 24, 9 | 55,23 |
| 7,13 | 18,41 | 30,39 | 43, 7 | 56,26 | **12** | 10,36 | 25,59 | 42,49 | 1,48 | 24,37 | 56, 4 |
| 7,22 | 18,56 | 30,52 | 43,16 | 56,40 | **13** | 10,54 | 26, 8 | 43, 7 | 2,11 | 24,55 | 56,45 |
| 7,36 | 19,10 | 31, 2 | 43,30 | 56,54 | **14** | 11, 3 | 26,26 | 43,26 | 2,25 | 25,18 | 57,27 |
| 7,45 | 19,19 | 31,15 | 43,44 | 57, 8 | **15** | 11,22 | 26,45 | 43,44 | 2,48 | 25,45 | 58, 8 |
| 7,59 | 19,28 | 31,29 | 43,53 | 57,17 | **16** | 11,35 | 27, 3 | 43,58 | 3,11 | 26,13 | 58,36 |
| 8, 8 | 19,42 | 31,43 | 44, 7 | 57,30 | **17** | 11,49 | 27,12 | 44,17 | 3,34 | 26,41 | 59,17 |
| 8,22 | 19,56 | 31,56 | 44,20 | 57,44 | **18** | 12, 8 | 27,31 | 44,35 | 3,52 | 27,13 | 59,54 |
| 8,36 | 20, 5 | 32, 7 | 44,34 | 57,55 | **19** | 12,17 | 27,49 | 44,58 | 4,10 | 27,40 | 6 0,35 |
| 8,45 | 20,14 | 32,19 | 44,48 | 58, 4 | **20** | 12,35 | 28, 7 | 45,12 | 4,33 | 28,12 | 1, 3 |
| 8,59 | 20,28 | 32,33 | 45, 2 | 58,25 | **21** | 12,49 | 28,21 | 45,30 | 4,52 | 28,35 | 1,39 |
| 9, 8 | 20,41 | 32,47 | 45,16 | 58,39 | **22** | 13, 3 | 28,35 | 45,48 | 5,15 | 29, 7 | 2,39 |
| 9,17 | 20,51 | 32,56 | 45,29 | 58,53 | **23** | 13,16 | 28,53 | 46, 7 | 5,28 | 29,40 | 3,34 |
| 9,26 | 21, 0 | 33,10 | 45,44 | 59, 7 | **24** | 13,30 | 29, 7 | 46,30 | 5,51 | 30,12 | 4,29 |
| 9,40 | 21,14 | 33,19 | 45,57 | 59,16 | **25** | 13,48 | 29,25 | 46,39 | 6,14 | 30,44 | 5,11 |
| 9,45 | 21,28 | 33,33 | 46, 6 | 59,34 | **26** | 14, 2 | 29,39 | 47, 2 | 6,37 | 31, 7 | 6, 6 |
| 9,58 | 21,36 | 33,47 | 46,20 | 59,48 | **27** | 14,16 | 29,57 | 47,20 | 7, 0 | 31,34 | 7, 1 |
| 10,12 | 21,46 | 33,56 | 46,34 | 4 0, 2 | **28** | 14,30 | 30,11 | 47,38 | 7,14 | 32, 7 | 7,56 |
| 10,26 | 21,59 | 34,10 | 46,52 | 0,11 | **29** | 14,43 | 30,30 | 47,52 | 7,37 | 32,39 | 8,37 |
| 10,35 | 22,13 | 34,24 | 47, 6 | 0,25 | **30** | 15, 1 | 30,48 | 48,10 | 8, 0 | 33,11 | 9,32 |
| 10,44 | 22,27 | 34,37 | 47,15 | 0,39 | **31** | 15,16 | 30,57 | 48,29 | 8,23 | 33,29 | 10,32 |
| 10,58 | 22,36 | 34,47 | 47,29 | 0,57 | **32** | 15,25 | 31,16 | 48,52 | 8,42 | 34, 2 | 11,27 |
| 11,12 | 22,46 | 35, 0 | 47,42 | 1, 6 | **33** | 15,43 | 31,34 | 49,10 | 9, 9 | 34,34 | 12,27 |
| 11,21 | 22,59 | 35,14 | 47,56 | 1,20 | **34** | 15.57 | 31,53 | 49,24 | 9,32 | 35, 6 | 13, 8 |
| 11,30 | 23,13 | 35,28 | 48, 6 | 1,34 | **35** | 16,16 | 32, 2 | 49,42 | 9,59 | 35,20 | 14, 4 |
| 11,44 | 23,22 | 35,42 | 48,24 | 1,52 | **36** | 16,29 | 32,20 | 50, 1 | 10,18 | 35,56 | 14,59 |
| 11,58 | 23,31 | 35,51 | 48,37 | 2, 6 | **37** | 16,38 | 32.38 | 50,19 | 10,32 | 36,20 | 15,54 |
| 12, 7 | 23,45 | 36, 0 | 48,51 | 2,20 | **38** | 16,57 | 32,57 | 50,33 | 10,55 | 37, 1 | 16,35 |
| 12,21 | 23,59 | 36,14 | 49, 5 | 2,33 | **39** | 17,11 | 33,15 | 50,51 | 11,18 | 37,33 | 17,30 |
| 12,30 | 24, 8 | 36,27 | 49,15 | 2,52 | **40** | 17,29 | 33,24 | 51,14 | 11,36 | 37,56 | 18,21 |
| 12,44 | 24,22 | 36,37 | 49,28 | 3, 6 | **41** | 17,38 | 33,43 | 51,33 | 11,59 | 38,28 | 18,53 |
| 12,53 | 24,31 | 36,50 | 49,42 | 3,20 | **42** | 17,52 | 34, 1 | 51,51 | 12,22 | 39. 0 | 21, 1 |
| 13, 7 | 24,45 | 37, 4 | 50, 0 | 3,33 | **43** | 18,10 | 34,15 | 52, 9 | 12,49 | 39,32 | 22, 6 |
| 13,16 | 24,54 | 37,17 | 50,10 | 3,47 | **44** | 18,20 | 34,29 | 52,23 | 13,12 | 39,50 | 23,37 |
| 13,30 | 25, 8 | 37,32 | 50,23 | 4, 6 | **45** | 18,42 | 34,47 | 52,46 | 13,40 | 40,23 | 25, 9 |
| 13,39 | 25,17 | 37,41 | 50,37 | 4,19 | **46** | 18,56 | 35, 1 | 53, 9 | 13,58 | 40,55 | 26,46 |
| 13,53 | 25,30 | 37,55 | 50,51 | 4,28 | **47** | 19,10 | 35,19 | 53,23 | 14,21 | 41,27 | 27,50 |
| 14, 6 | 25,40 | 38, 9 | 51, 5 | 4,47 | **48** | 19,28 | 35,33 | 53,41 | 14,49 | 41,59 | 29,27 |
| 14,16 | 25,54 | 38,22 | 51,14 | 5, 0 | **49** | 19,47 | 35,51 | 54, 4 | 15,16 | 42,17 | 31, 7 |
| 14,29 | 26, 8 | 38,32 | 51,32 | 5,19 | **50** | 20, 1 | 36, 9 | 54,27 | 15,35 | 42,50 | 32,53 |
| 14,38 | 26,17 | 38,45 | 51,46 | 5,29 | **51** | 20,19 | 36,28 | 54,45 | 15,58 | 43,22 | 34,25 |
| 14,53 | 26,26 | 38,54 | 52, 0 | 5,42 | **52** | 20,33 | 36,51 | 54,59 | 16,25 | 43,54 | 35,11 |
| 15, 3 | 26,40 | 39, 8 | 52, 9 | 6, 0 | **53** | 20,51 | 37, 5 | 55,18 | 16,53 | 44,26 | 36, 6 |
| 15,15 | 26,53 | 39,22 | 52,23 | 6,13 | **54** | 21, 9 | 37,23 | 55,36 | 17,20 | 44,49 | 40,32 |
| 15,29 | 27, 7 | 39,31 | 52,36 | 6,25 | **55** | 21,19 | 37,42 | 55,59 | 17,39 | 45,30 | 45, 0 |
| 15,38 | 27,16 | 39,45 | 52,51 | 6,41 | **56** | 21,37 | 38, 0 | 56,13 | 18, 2 | 46,12 | 50, 0 |
| 15,52 | 27,30 | 39,59 | 53, 9 | 6,55 | **57** | 21,55 | 38,19 | 56,36 | 18,24 | 46,53 | 53,19 |
| 16, 1 | 27,44 | 40,12 | 53,18 | 7,15 | **58** | 22,14 | 38,32 | 56,59 | 18,52 | 47,20 | ... |
| 16,11 | 27,58 | 40,22 | 53,32 | 7,28 | **59** | 22,23 | 38,55 | 57,22 | 19,11 | 47,57 | .. |

**C.—Eqn., for evection in gh., and palas.**

Argument: 2 × (☾'s – ☉'s Longitude) – ☾'s Anomaly.

| Eqn. Palas. | 0gh. 0 g.p. | 1 Days g.p. | 2 I g.p. | 3 II g.p. | 4 II g.p. | 5 III g.p. | 6 V g.p. |
|---|---|---|---|---|---|---|---|
| **0** | 0, 0 | 40,11 | 20,59 | 4, 9 | 51,23 | 46, 2 | 5 0,21 |
| **1** | 0,41 | 40,39 | 21,45 | 5, 0 | 52,13 | 47,11 | 2. 6 |
| **2** | 1, 9 | 41,24 | 22,31 | 5,36 | 52.59 | 48,15 | 3,15 |
| **3** | 1,55 | 42, 6 | 23,17 | 6,22 | 53,50 | 49,19 | 4,56 |
| **4** | 2,36 | 42,52 | 23,44 | 7, 8 | 54,45 | 50, 5 | 6,42 |
| **5** | 3,17 | 43,19 | 24,30 | 7,59 | 55,26 | 51,14 | 8,23 |
| **6** | 4, 4 | 44, 5 | 25,16 | 8,49 | 55,59 | 52,23 | 9,59 |
| **7** | 4,31 | 44,47 | 26, 7 | 9,21 | 57,12 | 53,27 | 11,31 |
| **8** | 5,17 | 45,32 | 26,24 | 10, 7 | 58, 7 | 54.13 | 13,40 |
| **9** | 5,58 | 46,14 | 27,16 | 10,58 | 59, 6 | 55,22 | 15,49 |
| **10** | 6,40 | 46,42 | 27,42 | 11,48 | 59.43 | 56,31 | 17,53 |
| **11** | 7, 7 | 47,27 | 28,47 | 12,25 | 3 0,39 | 57,39 | 19,14 |
| **12** | 7,53 | 48, 9 | 29,33 | 13,11 | 1,34 | 58,49 | 21,14 |
| **13** | 8,39 | 48,54 | 30, 1 | 13,57 | 2,29 | 59,34 | 23 19 |
| **14** | 9,20 | 49.22 | 30,47 | 14,47 | 3, 9 | 4 1, 6 | 25,18 |
| **15** | 10, 2 | 50, 8 | 31,33 | 15,33 | 4, 1 | 1,57 | 27,13 |
| **16** | 10,29 | 50,50 | 32,19 | 16,10 | 4,56 | 3,10 | 28,49 |
| **17** | 11,15 | 51,40 | 32,51 | 16,56 | 5,51 | 3,56 | 31,16 |
| **18** | 11,56 | 52,17 | 33,37 | 17,47 | 6,51 | 5,15 | 33,43 |
| **19** | 12,42 | 52,44 | 34,23 | 18. 1 | 7,32 | 6,28 | 36,10 |
| **20** | 13, 5 | 53,30 | 35, 9 | 19,14 | 8,22 | 7,42 | 37,51 |
| **21** | 13,51 | 54,16 | 35,54 | 20, 0 | 9,27 | 8,55 | 40,18 |
| **22** | 14,37 | 55, 2 | 36,27 | 20,50 | 10,18 | 9,45 | 42,50 |
| **23** | 15,18 | 55,29 | 37,13 | 21,41 | 10,58 | 10,59 | 44,26 |
| **24** | 16, 4 | 56,15 | 37,59 | 22,31 | 11,53 | 12,12 | 50,14 |
| **25** | 16,27 | 56,57 | 38,44 | 23, 8 | 12,53 | 13,26 | 50,47 |
| **26** | 17,13 | 57,43 | 39,18 | 23,54 | 13,53 | 14,16 | 53,55 |
| **27** | 17,59 | 58,29 | 40, 2 | 24,44 | 14,52 | 15,30 | 57, 4 |
| **28** | 18,45 | 58,52 | 40,49 | 25,39 | 15,34 | 16,43 | 6 0,12 |
| **29** | 19,13 | 59,38 | 41,34 | 26,16 | 16,29 | 18,20 | 1,39 |
| **30** | 19,58 | 1 0,28 | 42,21 | 27, 2 | 17,29 | 19,15 | 7, 1 |
| **31** | 20, 4 | 1, 9 | 42,52 | 27,53 | 18,29 | 20,10 | 11,36 |
| **32** | 21,26 | 1,37 | 43,20 | 28,44 | 19,10 | 21,38 | 16,12 |
| **33** | 22, 8 | 2,23 | 44,24 | 29.34 | 20, 5 | 22,55 | 18,53 |
| **34** | 22,35 | 3, 4 | 45,10 | 30,10 | 21, 5 | 24,18 | 36,22 |
| **35** | 23,20 | 3,45 | 45,42 | 30,57 | 22, 4 | 25, 8 | 36, 6 |
| **36** | 24, 3 | 4,32 | 46,28 | 31,47 | 23, 4 | 26,31 | 53,19 |
| **37** | 24,43 | 4,59 | 47,14 | 32,37 | 23,42 | 27,54 | ... |
| **38** | 25,11 | 5,45 | 48,41 | 33,14 | 24,41 | 29,16 | ... |
| **39** | 25,57 | 6,31 | 48,50 | 34, 5 | 25,40 | 30,39 | ... |
| **40** | 26,43 | 7,18 | 49,23 | 35, 0 | 26,40 | 31,34 | ... |
| **41** | 27,24 | 7,40 | 50, 9 | 35,46 | 27,26 | 31,57 | ... |
| **42** | 28, 6 | 8,26 | 50,55 | 36,41 | 28,21 | 34,20 | ... |
| **43** | 28,32 | 8,53 | 51,45 | 37,18 | 29,26 | 35,33 | ... |
| **44** | 29,19 | 9,58 | 52,17 | 38, 8 | 30,30 | 36,33 | ... |
| **45** | 30, 0 | 10,43 | 53, 3 | 39, 4 | 31,34 | 38, 5 | ... |
| **46** | 30,46 | 11,12 | 53,49 | 39,58 | 32,15 | 39,36 | ... |
| **47** | 31,14 | 11,57 | 54,35 | 40,35 | 33,15 | 41, 9 | ... |
| **48** | 32, 0 | 12,43 | 55,26 | 41,21 | 34,19 | 42,40 | ... |
| **49** | 32,41 | 13,29 | 55,58 | 42,16 | 35,23 | 43,40 | ... |
| **50** | 33,22 | 13,57 | 56,44 | 43, 7 | 36, 9 | 45,12 | ... |
| **51** | 34, 9 | 14,42 | 57,30 | 44, 1 | 37, 9 | 46,44 | ... |
| **52** | 34,27 | 15,28 | 58,15 | 44,38 | 38, 9 | 48,52 | ... |
| **53** | 35,17 | 16,14 | 58,48 | 45,30 | 39,13 | 49,15 | ... |
| **54** | 36, 3 | 17, 0 | 59, 2 | 46,25 | 40,17 | 50,47 | ... |
| **55** | 36,49 | 17,28 | 2 0,38 | 47,15 | 41, 3 | 52,46 | ... |
| **56** | 37,12 | 18,14 | 1,14 | 47,56 | 42, 3 | 53,55 | ... |
| **57** | 37,59 | 18,59 | 2, 1 | 48,46 | 43, 3 | 55,41 | ... |
| **58** | 38,44 | 19,46 | 2,37 | 49,37 | 43,53 | 56,59 | ... |
| **59** | 39,30 | 20,13 | 3,23 | 50,32 | 44,58 | 58,35 | ... |

# TABLE XVII.—MARS.

## Data for calculating Latitude and geocentric Longitude of Mars from B.C. 1 to A.D. 2000.

### Mars' mean motion for Centuries.

| A.D. | Kaliyuga. | Mean long. at commencement of Solar Year. | Longitude of apsis. | Longitude of node. |
|---|---|---|---|---|
| | | Deg. | Deg. | Deg. |
| **1 B.C.** | 3101 K.Y. | 254·7442 | 130·0274 | 40·0866 |
| **100 A.D.** | 3201 | 315·2712 | 130·0291 | 40·0848 |
| **200** | 3301 | 15·7982 | 130·0308 | 40·0830 |
| **300** | 3401 | 76·3252 | 130·0325 | 40·0813 |
| **400** | 3501 | 136·8522 | 130·0342 | 40·0795 |
| **500** | 3601 | 197·3793 | 130·0359 | 40·0777 |
| **600** | 3701 | 257·9063 | 130·0376 | 40·0759 |
| **700** | 3801 | 318·4333 | 130·0393 | 40·0741 |
| **800** | 3901 | 18·9603 | 130·0410 | 40·0723 |
| **900** | 4001 | 79·4873 | 130·0427 | 40·0706 |

| A.D. | Kaliyuga. | Mean long. at commencement of Solar Year. | Longitude of apsis. | Longitude of node. |
|---|---|---|---|---|
| | | Deg. | Deg. | Deg. |
| **1000 A.D.** | 4101 K.Y. | 140·0144 | 130·0444 | 40·0688 |
| **1100** | 4201 | 200·5414 | 130·0461 | 40·0670 |
| **1200** | 4301 | 261·0684 | 130·0478 | 40·0652 |
| **1300** | 4401 | 321·5954 | 130·0495 | 40·0634 |
| **1400** | 4501 | 22·1224 | 130·0512 | 40·0616 |
| **1500** | 4601 | 82·6495 | 130·0529 | 40·0599 |
| **1600** | 4701 | 143·1765 | 130·0546 | 40·0581 |
| **1700** | 4801 | 203·7035 | 130·0563 | 40·0563 |
| **1800** | 4901 | 264·2305 | 130·0580 | 40·0545 |
| **1900** | 5001 | 324·7575 | 130·0597 | 40·0527 |
| **2000** | 5101 | 25·2846 | 130·0614 | 40·0510 |

### Mars' mean motion for Odd Years.

| Yrs. | Deg. | Yrs. | Deg. | Yrs. | Deg. | Yrs. | Deg. | Yrs. | Deg. | Yrs. | Deg. | Yrs. | Deg. | Yrs. | Deg. | Yrs. | Deg. | Yrs. | Deg. |
|---|---|---|---|---|---|---|---|---|---|---|---|---|---|---|---|---|---|---|---|
| **1** | 191·40 | **11** | 305·46 | **21** | 59·51 | **31** | 173·56 | **41** | 287·62 | **51** | 41·67 | **61** | 155·72 | **71** | 269·77 | **81** | 23·83 | **91** | 137·88 |
| **2** | 22·81 | **12** | 136·86 | **22** | 250·92 | **32** | 4·97 | **42** | 119·02 | **52** | 233·07 | **62** | 347·13 | **72** | 101·18 | **82** | 215·23 | **92** | 329·29 |
| **3** | 214·22 | **13** | 328·27 | **23** | 82·32 | **33** | 196·37 | **43** | 310·43 | **53** | 64·48 | **63** | 178·53 | **73** | 292·58 | **83** | 46·64 | **93** | 160·69 |
| **4** | 45·62 | **14** | 159·67 | **24** | 273·73 | **34** | 27·77 | **44** | 141·83 | **54** | 255·88 | **64** | 9·94 | **74** | 123·99 | **84** | 238·04 | **94** | 352·09 |
| **5** | 237·03 | **15** | 351·08 | **25** | 105·13 | **35** | 219·18 | **45** | 333·24 | **55** | 87·29 | **65** | 201·34 | **75** | 315·39 | **85** | 69·45 | **95** | 183·50 |
| **6** | 68·43 | **16** | 182·48 | **26** | 296·54 | **36** | 50·59 | **46** | 164·64 | **56** | 278·69 | **66** | 32·75 | **76** | 146·80 | **86** | 260·85 | **96** | 14·91 |
| **7** | 257·84 | **17** | 13·89 | **27** | 127·94 | **37** | 241·99 | **47** | 356·05 | **57** | 110·10 | **67** | 224·15 | **77** | 338·21 | **87** | 92·26 | **97** | 206·31 |
| **8** | 91·24 | **18** | 205·29 | **28** | 319·35 | **38** | 73·40 | **48** | 187·45 | **58** | 301·51 | **68** | 55·56 | **78** | 169·61 | **88** | 283·66 | **98** | 37·72 |
| **9** | 282·65 | **19** | 36·70 | **29** | 150·75 | **39** | 264·80 | **49** | 18·86 | **59** | 132·91 | **69** | 246·96 | **79** | 1·02 | **89** | 115·07 | **99** | 229·12 |
| **10** | 114·05 | **20** | 228·10 | **30** | 342·16 | **40** | 96·21 | **50** | 210·26 | **60** | 324·32 | **70** | 78·37 | **80** | 192·42 | **90** | 306·47 | **100** | 60·53 |

### Mars' mean motion for Days.

| Days. | 0 | 1 | 2 | 3 | 4 | 5 | 6 | 7 | 8 | 9 |
|---|---|---|---|---|---|---|---|---|---|---|
| | | Degrees. | | | | Degrees. | | | Degrees. | |
| **0** | ... | 0·52 | 1·05 | 1·57 | 2·10 | 2·62 | 3·14 | 3·67 | 4·19 | 4·72 |
| **1** | 5·24 | 5·76 | 6·29 | 6·81 | 7·33 | 7·86 | 8·38 | 8·91 | 9·43 | 9·96 |
| **2** | 10·48 | 11·00 | 11·52 | 12·05 | 12·58 | 13·10 | 13·62 | 14·15 | 14·67 | 15·19 |
| **3** | 15·72 | 16·24 | 16·77 | 17·29 | 17·82 | 18·34 | 18·86 | 19·39 | 19·91 | 20·44 |
| **4** | 20·96 | 21·48 | 22·01 | 22·53 | 23·06 | 23·58 | 24·11 | 24·63 | 25·15 | 25·68 |
| **5** | 26·20 | 26·73 | 27·25 | 27·77 | 28·30 | 28·82 | 29·35 | 29·87 | 30·39 | 30·92 |
| **6** | 31·44 | 31·97 | 32·49 | 33·01 | 33·54 | 34·07 | 34·59 | 35·11 | 35·63 | 36·16 |
| **7** | 36·68 | 37·21 | 37·73 | 38·25 | 38·78 | 39·30 | 39·83 | 40·35 | 40·87 | 41·40 |
| **8** | 41·92 | 42·45 | 42·97 | 43·49 | 44·02 | 44·54 | 45·07 | 45·59 | 46·11 | 46·64 |
| **9** | 47·16 | 47·69 | 48·21 | 48·73 | 49·26 | 49·78 | 50·31 | 50·83 | 51·35 | 51·88 |
| **10** | 52·40 | 52·93 | 53·45 | 53·97 | 54·50 | 55·02 | 55·55 | 56·07 | 56·59 | 57·12 |
| **11** | 57·64 | 58·17 | 58·70 | 59·22 | 59·74 | 60·26 | 60·79 | 61·31 | 61·84 | 62·36 |
| **12** | 62·88 | 63·41 | 63·93 | 64·46 | 64·98 | 65·50 | 66·03 | 66·55 | 67·08 | 67·60 |
| **13** | 68·12 | 68·65 | 69·17 | 69·70 | 70·22 | 70·74 | 71·27 | 71·79 | 72·32 | 72·84 |
| **14** | 73·36 | 73·89 | 74·41 | 74·94 | 75·46 | 75·98 | 76·51 | 77·03 | 77·56 | 78·08 |
| **15** | 78·60 | 79·13 | 79·65 | 80·18 | 80·70 | 81·22 | 81·75 | 82·27 | 82·80 | 83·32 |
| **16** | 83·84 | 84·37 | 84·89 | 85·42 | 85·94 | 86·46 | 86·99 | 87·51 | 88·04 | 88·56 |
| **17** | 89·09 | 89·61 | 90·13 | 90·66 | 91·88 | 91·71 | 92·23 | 92·75 | 93·28 | 93·80 |
| **18** | 94·33 | 94·85 | 95·37 | 95·90 | 96·42 | 95·95 | 97·47 | 97·99 | 98·52 | 99·04 |
| **19** | 99·57 | 100·09 | 100·61 | 101·14 | 101·66 | 102·19 | 102·71 | 103·23 | 103·76 | 104·28 |
| **20** | 104·81 | 105·33 | 105·85 | 106·38 | 106·90 | 107·43 | 107·95 | 108·47 | 109·00 | 109·52 |
| **21** | 110·05 | 110·57 | 111·09 | 111·62 | 112·14 | 112·67 | 113·19 | 113·71 | 114·24 | 114·77 |
| **22** | 115·29 | 115·81 | 116·33 | 116·86 | 117·38 | 117·91 | 118·43 | 118·96 | 119·46 | 120·00 |
| **23** | 120·53 | 121·05 | 121·58 | 122·10 | 122·62 | 123·15 | 123·67 | 124·20 | 124·72 | 125·25 |
| **24** | 125·77 | 126·29 | 126·82 | 127·34 | 127·86 | 128·39 | 126·91 | 129·44 | 129·96 | 130·48 |
| **25** | 131·01 | 131·53 | 132·06 | 132·55 | 133·10 | 133·63 | 134·15 | 134·68 | 135·20 | 135·72 |
| **26** | 136·25 | 136·77 | 137·30 | 137·82 | 138·34 | 138·87 | 139·39 | 139·92 | 140·44 | 140·96 |
| **27** | 141·49 | 142·01 | 142·54 | 143·06 | 143·58 | 144·11 | 144·63 | 145·16 | 145·68 | 146·20 |
| **28** | 146·73 | 147·25 | 147·78 | 148·30 | 148·82 | 149·35 | 149·87 | 150·40 | 150·92 | 151·45 |
| **29** | 151·97 | 152·49 | 153·02 | 153·54 | 154·07 | 154·59 | 155·11 | 155·64 | 156·16 | 156·69 |
| **30** | 157·21 | 157·73 | 158·26 | 158·78 | 159·31 | 159·83 | 160·35 | 160·88 | 161·40 | 161·93 |
| **31** | 162·45 | 162·97 | 163·50 | 164·02 | 164·55 | 165·07 | 165·59 | 166·12 | 166·64 | 167·17 |
| **32** | 167·69 | 168·21 | 168·74 | 169·26 | 169·79 | 170·31 | 170·83 | 171·36 | 171·88 | 172·41 |
| **33** | 172·93 | 173·45 | 173·98 | 174·50 | 175·03 | 175·55 | 176·07 | 176·60 | 177·12 | 177·65 |
| **34** | 178·17 | 178·69 | 179·22 | 179·74 | 180·27 | 180·79 | 181·31 | 181·84 | 182·36 | 182·89 |
| **35** | 183·41 | 183·94 | 184·46 | 184·98 | 185·51 | 186·03 | 186·56 | 187·08 | 187·60 | 188·13 |
| **36** | 188·65 | 189·18 | 189·70 | 190·22 | 190·75 | 191·27 | ... | ... | ... | ... |

### Mars' Annual Eqn. in Deg.

| | + | + | | — |
|---|---|---|---|---|
| **0** | 0·0 | 180·0 | 180·0 | 360·0 |
| **1** | 2·5 | 179·5 | 180·5 | 357·5 |
| **2** | 5·1 | 178·9 | 181·1 | 354·9 |
| **3** | 7·6 | 178·4 | 181·6 | 352·4 |
| **4** | 10·1 | 177·8 | 182·2 | 349·9 |
| **5** | 12·7 | 177·3 | 182·7 | 347·3 |
| **6** | 15·2 | 176·8 | 183·2 | 344·8 |
| **7** | 17·8 | 176·2 | 183·8 | 342·2 |
| **8** | 20·4 | 175·7 | 184·3 | 339·6 |
| **9** | 23·0 | 175·1 | 184·9 | 337·0 |
| **10** | 25·5 | 174·5 | 185·5 | 334·5 |
| **11** | 28·1 | 173·9 | 186·1 | 331·9 |
| **12** | 30·7 | 173·4 | 186·6 | 329·3 |
| **13** | 33·3 | 172·8 | 187·2 | 326·7 |
| **14** | 35·9 | 172·2 | 187·8 | 324·1 |
| **15** | 38·5 | 171·5 | 188·5 | 321·5 |
| **16** | 41·2 | 170·9 | 189·1 | 318·8 |
| **17** | 43·9 | 170·3 | 189·7 | 316·1 |
| **18** | 46·5 | 169·6 | 190·4 | 313·5 |
| **19** | 49·2 | 169·0 | 191·0 | 310·8 |
| **20** | 51·8 | 168·3 | 191·7 | 308·2 |
| **21** | 54·6 | 167·5 | 192·5 | 305·4 |
| **22** | 57·3 | 166·8 | 193·2 | 302·7 |
| **23** | 60·2 | 166·0 | 194·0 | 299·8 |
| **24** | 63·0 | 165·3 | 194·7 | 297·0 |
| **25** | 65·8 | 164·5 | 195·5 | 294·2 |
| **26** | 68·7 | 163·6 | 196·4 | 291·3 |
| **27** | 71·6 | 162·6 | 197·4 | 288·4 |
| **28** | 74·6 | 161·7 | 198·3 | 285·4 |
| **29** | 77·7 | 160·7 | 199·3 | 282·3 |
| **30** | 80·8 | 159·6 | 200·4 | 279·2 |
| **31** | 84·0 | 158·5 | 201·5 | 276·0 |
| **32** | 87·3 | 157·3 | 202·7 | 272·7 |
| **33** | 90·7 | 155·9 | 204·1 | 269·3 |
| **34** | 94·2 | 154·5 | 205·5 | 265·8 |
| **35** | 97·9 | 152·8 | 207·2 | 262·1 |
| **36** | 101·7 | 151·0 | 209·0 | 258·3 |
| **37** | 106·0 | 148·7 | 211·3 | 254·0 |
| **38** | 110·6 | 146·2 | 213·8 | 249·4 |
| **39** | 116·1 | 142·7 | 217·3 | 243·9 |
| **40** | 124·2 | 136·5 | 223·5 | 235·8 |
| **40·15** | 135·0 | ... | 225·0 | ... |

## TABLE XVII.—*contd.*

### Mars' Anomaly.

| + | ·0 | ·1 | ·2 | ·3 | ·4 | ·5 | ·6 | ·7 | ·8 | ·9 |
|---|---|---|---|---|---|---|---|---|---|---|
| | Degrees. | | | | | | | | | |
| 0 | 0·0 | 0·5 | 0·9 | 1·4 | 1·9 | 2·4 | 2·9 | 3·4 | 3·9 | 4·4 |
| 1 | 4·8 | 5·3 | 5·8 | 6·3 | 6·7 | 7·2 | 7·7 | 8·2 | 8·7 | 9·2 |
| 2 | 9·7 | 10·2 | 10·7 | 11·2 | 11·7 | 12·2 | 12·7 | 13·2 | 13·7 | 14·2 |
| 3 | 14·7 | 15·2 | 15·7 | 16·2 | 16·7 | 17·3 | 17·8 | 18·3 | 18·8 | 19·4 |
| 4 | 19·9 | 20·4 | 20·9 | 21·4 | 22·0 | 22·5 | 23·0 | 23·6 | 24·1 | 24·7 |
| 5 | 25·2 | 25·7 | 26·3 | 26·9 | 27·4 | 28·0 | 28·6 | 29·1 | 29·7 | 30·2 |
| 6 | 30·8 | 31·4 | 32·0 | 32·6 | 33·2 | 33·7 | 34·4 | 35·0 | 35·6 | 36·2 |
| 7 | 36·9 | 37·5 | 38·1 | 38·7 | 39·4 | 40·1 | 40·7 | 41·4 | 42·1 | 42·7 |
| 8 | 43·4 | 44·1 | 44·8 | 45·5 | 46·3 | 47·0 | 47·7 | 48·5 | 49·2 | 50·0 |
| 9 | 50·8 | 51·6 | 52·4 | 53·2 | 54·2 | 55·0 | 55·9 | 56·8 | 57·7 | 58·7 |
| 10 | 59·7 | 60·7 | 61·8 | 62·9 | 64·0 | 65·2 | 66·5 | 67·7 | 69·2 | 70·6 |
| 11 | 72·2 | 74·1 | 76·1 | 78·4 | 81·2 | 85·7 | ... | ... | ... | ... |

**Maximum Equation 11·53** for Anomaly of 90·00.

| + | ·9 | ·8 | ·7 | ·6 | ·5 | ·4 | ·3 | ·2 | ·1 | ·0 |
|---|---|---|---|---|---|---|---|---|---|---|
| | Degrees. | | | | | | | | | |
| 11 | ... | ... | ... | ... | 94·2 | 98·7 | 101·6 | 103·9 | 105·9 | 107·8 |
| 10 | 109·4 | 110·8 | 112·3 | 113·5 | 114·8 | 116·0 | 117·1 | 118·2 | 119·3 | 120·3 |
| 9 | 121·3 | 122·3 | 123·2 | 124·1 | 125·0 | 125·8 | 126·8 | 127·6 | 128·4 | 129·2 |
| 8 | 130·0 | 130·8 | 131·5 | 132·3 | 133·0 | 133·7 | 134·5 | 135·2 | 135·9 | 136·6 |
| 7 | 137·3 | 137·9 | 138·6 | 139·3 | 139·9 | 140·6 | 141·3 | 141·9 | 142·5 | 143·1 |
| 6 | 143·8 | 144·4 | 145·0 | 145·6 | 146·3 | 146·8 | 147·4 | 148·0 | 148·6 | 149·2 |
| 5 | 149·8 | 150·3 | 150·9 | 151·4 | 152·0 | 152·6 | 153·1 | 153·7 | 154·3 | 154·8 |
| 4 | 155·3 | 155·9 | 156·4 | 157·0 | 157·5 | 158·0 | 158·6 | 159·1 | 159·6 | 160·1 |
| 3 | 160·6 | 161·2 | 161·7 | 162·2 | 162·7 | 163·3 | 163·8 | 164·3 | 164·8 | 165·3 |
| 2 | 165·8 | 166·3 | 166·8 | 167·3 | 167·8 | 168·3 | 168·8 | 169·3 | 169·8 | 170·3 |
| 1 | 170·8 | 171·3 | 171·8 | 172·3 | 172·8 | 173·3 | 173·7 | 174·2 | 174·7 | 175·2 |
| 0 | 175·6 | 176·1 | 176·6 | 177·1 | 177·6 | 178·1 | 178·6 | 179·1 | 179·5 | 180·0 |

| — | ·0 | ·1 | ·2 | ·3 | ·4 | ·5 | ·6 | ·7 | ·8 | ·9 |
|---|---|---|---|---|---|---|---|---|---|---|
| | Degrees. | | | | | | | | | |
| 0 | 180·0 | 180·5 | 180·9 | 181·4 | 181·9 | 182·4 | 182·9 | 183·4 | 183·9 | 184·4 |
| 1 | 184·8 | 185·3 | 185·8 | 186·3 | 186·7 | 187·2 | 187·7 | 188·2 | 188·7 | 189·2 |
| 2 | 189·7 | 190·2 | 190·7 | 191·2 | 191·7 | 192·2 | 192·7 | 193·2 | 193·7 | 194·2 |
| 3 | 194·7 | 195·2 | 195·7 | 196·2 | 196·7 | 197·3 | 197·8 | 198·3 | 198·8 | 199·4 |
| 4 | 199·9 | 200·4 | 200·9 | 201·4 | 202·0 | 202·5 | 203·0 | 203·6 | 204·1 | 204·7 |
| 5 | 205·2 | 205·7 | 206·3 | 206·9 | 207·4 | 208·0 | 208·6 | 209·1 | 209·7 | 210·2 |
| 6 | 210·8 | 211·4 | 212·0 | 212·6 | 213·2 | 213·7 | 214·4 | 215·0 | 215·6 | 216·2 |
| 7 | 216·9 | 217·5 | 218·1 | 218·7 | 219·4 | 220·1 | 220·7 | 221·4 | 222·1 | 222·7 |
| 8 | 223·4 | 224·1 | 224·8 | 225·5 | 226·3 | 227·9 | 227·7 | 229·5 | 229·2 | 230·0 |
| 9 | 230·8 | 231·6 | 232·4 | 233·2 | 234·2 | 235·0 | 235·9 | 236·8 | 237·7 | 238·7 |
| 10 | 239·7 | 240·7 | 241·8 | 242·9 | 244·0 | 245·2 | 246·5 | 247·7 | 249·2 | 250·6 |
| 11 | 252·2 | 254·1 | 256·1 | 258·4 | 261·2 | 265·7 | ... | ... | ... | ... |

**Maximum Equation 11·53** for Anomaly of 270·00.

| — | ·9 | ·8 | ·7 | ·6 | ·5 | ·4 | ·3 | ·2 | ·1 | ·0 |
|---|---|---|---|---|---|---|---|---|---|---|
| | Degrees. | | | | | | | | | |
| 11 | ... | ... | ... | ... | 274·2 | 278·7 | 281·6 | 283·9 | 285·9 | 287·8 |
| 10 | 289·4 | 290·8 | 292·3 | 293·5 | 294·8 | 296·0 | 297·1 | 298·2 | 299·3 | 300·3 |
| 9 | 301·3 | 302·3 | 303·2 | 304·1 | 305·0 | 305·8 | 306·8 | 307·6 | 308·4 | 309·2 |
| 8 | 310·0 | 310·8 | 311·5 | 312·3 | 313·0 | 313·7 | 314·5 | 315·2 | 315·9 | 316·6 |
| 7 | 317·3 | 317·9 | 318·6 | 319·3 | 319·9 | 320·6 | 321·3 | 321·9 | 322·5 | 323·1 |
| 6 | 323·8 | 324·4 | 325·0 | 325·6 | 326·3 | 326·8 | 327·4 | 328·0 | 328·6 | 329·2 |
| 5 | 329·8 | 330·3 | 330·9 | 331·4 | 332·0 | 332·6 | 333·1 | 333·7 | 334·3 | 334·8 |
| 4 | 335·3 | 335·9 | 336·4 | 337·0 | 337·5 | 338·0 | 338·6 | 339·1 | 339·6 | 340·1 |
| 3 | 340·6 | 341·2 | 341·7 | 342·2 | 342·7 | 343·3 | 343·8 | 344·3 | 344·8 | 345·3 |
| 2 | 345·8 | 346·3 | 346·8 | 347·3 | 347·8 | 348·3 | 348·8 | 349·3 | 349·8 | 350·3 |
| 1 | 350·8 | 351·3 | 351·8 | 352·3 | 352·8 | 353·3 | 353·7 | 354·2 | 354·7 | 355·2 |
| 0 | 355·6 | 356·1 | 356·6 | 357·1 | 357·6 | 358·1 | 355·6 | 359·1 | 359·5 | 360·0 |

## MERCURY.

### Data for calculating Latitude and geocentric Longitude of Mercury from B. C. 1 to A. D. 2000.

### Mercury's mean motion for Centuries.

| A.D. | Kaliyuga. | Mean Long. at commencement of Solar Year. | Longitude of apsis. | Longitude of node. | A.D. | Kaliyuga. | Mean Long. at commencement of Solar Year. | Longitude of apsis. | Longitude of node. |
|---|---|---|---|---|---|---|---|---|---|
| | | Deg. | Deg. | Deg. | | | Deg. | Deg. | Deg. |
| 1 B.C. | 3101 K.Y. | 209·3843 | 220·4118 | 20·7580 | 1000 A.D. | 4101 K.Y. | 249·5645 | 220·4425 | 20·7173 |
| 100 A.D. | 3201 | 285·4023 | 220·4149 | 20·7539 | 1100 | 4201 | 325·5825 | 220·4455 | 20·7133 |
| 200 | 3301 | 1·4203 | 220·4180 | 20·7499 | 1200 | 4301 | 41·6005 | 220·4486 | 20·7092 |
| 300 | 3401 | 77·4383 | 220·4210 | 20·7458 | 1300 | 4401 | 117·6185 | 220·4517 | 20·7051 |
| 400 | 3501 | 153·4563 | 220·4241 | 20·7417 | 1400 | 4501 | 193·6365 | 220·4547 | 20·7011 |
| 500 | 3601 | 229·4744 | 220·4272 | 20·7377 | 1500 | 4601 | 269·6546 | 220·4578 | 20·6970 |
| 600 | 3701 | 305·4924 | 220·4302 | 20·7336 | 1600 | 4701 | 345·6726 | 220·4609 | 20·6929 |
| 700 | 3801 | 21·5104 | 220·4333 | 20·7295 | 1700 | 4801 | 61·6906 | 220·4639 | 20·6889 |
| 800 | 3901 | 97·5284 | 220·4363 | 20·7255 | 1800 | 4901 | 137·7086 | 220·4670 | 20·6848 |
| 900 | 4001 | 173·5464 | 220·4394 | 20·7214 | 1900 | 5001 | 213·7266 | 220·4701 | 20·6807 |
| | | | | | 2000 | 5101 | 289·7447 | 220·4731 | 20·6767 |

### Mercury's mean motion for Odd Years.

| Yrs. | Deg. | Yrs. | Deg. | Yrs. | Deg. | Yrs. | Deg. | Yrs. | Deg. | Yrs. | Deg. | Yrs. | Deg. | Yrs. | Deg. | Yrs. | Deg. | Yrs. | Deg. |
|---|---|---|---|---|---|---|---|---|---|---|---|---|---|---|---|---|---|---|---|
| 1 | 54·76 | 11 | 242·36 | 21 | 69·96 | 31 | 257·57 | 41 | 85·17 | 51 | 272·77 | 61 | 100·37 | 71 | 287·97 | 81 | 115·57 | 91 | 303·18 |
| 2 | 109·52 | 12 | 297·12 | 22 | 124·72 | 32 | 312·33 | 42 | 139·93 | 52 | 327·53 | 62 | 155·13 | 72 | 342·73 | 82 | 170·33 | 92 | 357·94 |
| 3 | 164·28 | 13 | 351·88 | 23 | 179·48 | 33 | 7·09 | 43 | 194·69 | 53 | 22·29 | 63 | 209·89 | 73 | 37·49 | 83 | 225·09 | 93 | 52·70 |
| 4 | 219·04 | 14 | 46·64 | 24 | 234·24 | 34 | 61·85 | 44 | 249·45 | 54 | 77·05 | 64 | 264·65 | 74 | 92·25 | 84 | 279·85 | 94 | 107·46 |
| 5 | 273·80 | 15 | 101·40 | 25 | 289·00 | 35 | 116·61 | 45 | 304·21 | 55 | 131·81 | 65 | 319·41 | 75 | 147·01 | 85 | 334·61 | 95 | 162·22 |
| 6 | 328·56 | 16 | 156·16 | 26 | 343·76 | 36 | 171·37 | 46 | 358·97 | 56 | 186·57 | 66 | 14·17 | 76 | 201·77 | 86 | 29·37 | 96 | 216·98 |
| 7 | 23·32 | 17 | 210·92 | 27 | 38·52 | 37 | 226·13 | 47 | 53·73 | 57 | 241·33 | 67 | 68·93 | 77 | 256·53 | 87 | 84·13 | 97 | 271·74 |
| 8 | 78·08 | 18 | 265·68 | 28 | 93·28 | 38 | 280·89 | 48 | 108·49 | 58 | 296·09 | 68 | 123·69 | 78 | 311·29 | 88 | 138·89 | 98 | 326·50 |
| 9 | 132·84 | 19 | 320·44 | 29 | 148·04 | 39 | 335·65 | 49 | 163·25 | 59 | 350·85 | 69 | 178·45 | 79 | 6·05 | 89 | 193·66 | 99 | 21·26 |
| 10 | 187·60 | 20 | 15·20 | 30 | 202·80 | 40 | 30·41 | 50 | 218·01 | 60 | 45·61 | 70 | 233·21 | 80 | 60·81 | 90 | 248·42 | 100 | 76·02 |

## TABLE XVII.—*contd.*

### Mercury's mean motion for Days.

| Days. | 0 | 1 | 2 | 3 | 4 | 5 | 6 | 7 | 8 | 9 |
|---|---|---|---|---|---|---|---|---|---|---|
| | Degrees. | | | | | | | | | |
| 0 | ... | 4·09 | 8·18 | 12·28 | 16·37 | 20·46 | 24·55 | 28·65 | 32·74 | 36·83 |
| 1 | 40·92 | 45·02 | 49·11 | 53·20 | 57·29 | 61·38 | 65·48 | 69·57 | 73·66 | 77·75 |
| 2 | 81·87 | 85·94 | 90·03 | 94·12 | 98·22 | 102·31 | 106·40 | 110·49 | 114·59 | 118·68 |
| 3 | 122·77 | 126·86 | 130·95 | 135·05 | 139·14 | 143·23 | 147·32 | 151·42 | 155·51 | 159·60 |
| 4 | 163.69 | 167·79 | 171·88 | 175·97 | 180·06 | 184·15 | 188·25 | 192·34 | 196·43 | 200·52 |
| 5 | 204·62 | 208·71 | 212·80 | 216·89 | 220·99 | 225·08 | 229·17 | 233·26 | 237·36 | 241·45 |
| 6 | 245·54 | 249·63 | 253·72 | 257·82 | 261·91 | 266·00 | 270·09 | 274·19 | 278·28 | 282·37 |
| 7 | 286·46 | 290·56 | 294·65 | 298·74 | 302·83 | 306·92 | 311·02 | 315·11 | 319·20 | 323·29 |
| 8 | 327·39 | 331·48 | 335·57 | 339·66 | 343·76 | 347·85 | 351·94 | 356·03 | 0·13 | 4·22 |
| 9 | 8·31 | 12·40 | 16·49 | 20·59 | 24·68 | 28·77 | 32·86 | 36·96 | 41·05 | 45·14 |
| 10 | 49·23 | 53·33 | 57·42 | 61·51 | 65·60 | 69·69 | 73·79 | 77·88 | 81·97 | 86·06 |
| 11 | 90·16 | 94·25 | 98·34 | 102·43 | 106·53 | 118·62 | 114·71 | 118·80 | 122·90 | 126·99 |
| 12 | 131·08 | 135·17 | 139·26 | 143·36 | 147·45 | 151·54 | 155·63 | 159·73 | 163·82 | 167·91 |
| 13 | 172·00 | 176·10 | 180·19 | 184·28 | 188·37 | 192·47 | 196·56 | 200·65 | 204·74 | 208·83 |
| 14 | 212·93 | 217·02 | 221·11 | 225·20 | 229·30 | 233·39 | 237·48 | 247·57 | 245·67 | 249·76 |
| 15 | 253·85 | 257·94 | 262·03 | 266·13 | 270·22 | 274·31 | 278·40 | 282·50 | 286·59 | 290·68 |
| 16 | 294·77 | 298·87 | 302·96 | 307·05 | 311·14 | 315·24 | 319·33 | 323·42 | 327·51 | 331·60 |
| 17 | 335·70 | 339·79 | 343·88 | 347·97 | 352·07 | 356·16 | 0·25 | 4·34 | 8·44 | 12·53 |
| 18 | 16·62 | 20·71 | 24·80 | 28·90 | 32·99 | 37·08 | 41·17 | 45·27 | 49·36 | 53·45 |
| 19 | 57·54 | 61·64 | 65·73 | 69·82 | 73·91 | 78·01 | 82·10 | 86·19 | 90·28 | 94·37 |
| 20 | 98·47 | 102·56 | 106·65 | 110·74 | 114·84 | 118·93 | 123·02 | 127·11 | 131·21 | 135·30 |
| 21 | 139·39 | 143·48 | 147·58 | 151·67 | 155·76 | 159·85 | 163·94 | 168·04 | 172·98 | 176·22 |
| 22 | 180·31 | 184·41 | 188·50 | 192·59 | 196·68 | 200·78 | 204·87 | 208·96 | 213·05 | 217·15 |
| 23 | 221·24 | 225·33 | 228·42 | 233·51 | 237·61 | 241·70 | 245·80 | 249·88 | 253·98 | 258·07 |
| 24 | 262·16 | 266·25 | 270·35 | 274·44 | 278·53 | 282·62 | 286·71 | 290·80 | 291·90 | 298·99 |
| 25 | 303·08 | 307·18 | 311·27 | 315·36 | 319·45 | 323·55 | 327·64 | 331·73 | 335·82 | 339·92 |
| 26 | 344·01 | 348·10 | 352·19 | 356·28 | 0·38 | 4·47 | 8·56 | 12·65 | 16·75 | 20·84 |
| 27 | 24·93 | 29·02 | 33·12 | 37·21 | 41·30 | 45·39 | 49·49 | 53·58 | 57·67 | 61·76 |
| 28 | 65·85 | 69·95 | 74·04 | 78·13 | 82·22 | 86·32 | 90·41 | 94·50 | 98·59 | 102·69 |
| 29 | 106·78 | 110·87 | 114·96 | 119·05 | 123·15 | 127·24 | 131·33 | 135·42 | 139·52 | 143·61 |
| 30 | 147·70 | 151·79 | 155·89 | 159·98 | 164·07 | 168·16 | 172·26 | 176·35 | 180·44 | 184·53 |
| 31 | 188·62 | 192·72 | 196·81 | 200·90 | 204·99 | 209·09 | 213·18 | 217·27 | 221·36 | 225·46 |
| 32 | 229·55 | 233·64 | 237·73 | 241·82 | 245·92 | 250·00 | 254·10 | 258·19 | 262·29 | 266·38 |
| 33 | 270·47 | 274·56 | 278·66 | 282·75 | 286·84 | 290·93 | 295·03 | 299·12 | 303·21 | 307·30 |
| 34 | 311·39 | 315·49 | 319·58 | 323·67 | 327·76 | 331·86 | 335·95 | 340·04 | 344·13 | 348·23 |
| 35 | 352·32 | 356·41 | 0·50 | 4·60 | 8·69 | 12·78 | 16·87 | 20·96 | 25·06 | 29·15 |
| 36 | 33·24 | 37·33 | 41·43 | 45·52 | 49·61 | 53·70 | ... | ... | ... | ... |

### Mercury's Annual Equation.

| | + | — |
|---|---|---|
| | Degrees. | |
| 0 | 0·0 | 360·0 |
| 1 | 3·7 | 356·3 |
| 2 | 7·4 | 352·6 |
| 3 | 11·2 | 348·8 |
| 4 | 14·9 | 345·1 |
| 5 | 18·7 | 341·3 |
| 6 | 22·5 | 337·5 |
| 7 | 26·4 | 333·6 |
| 8 | 30·2 | 329·8 |
| 9 | 34·7 | 325·3 |
| 10 | 38·2 | 321·8 |
| 11 | 42·3 | 317·7 |
| 12 | 46·5 | 313·5 |
| 13 | 50·7 | 309·3 |
| 14 | 55·2 | 304·8 |
| 15 | 59·8 | 300·2 |
| 16 | 64·6 | 295·4 |
| 17 | 69·8 | 290·2 |
| 18 | 75·4 | 284·6 |
| 19 | 81·6 | 278·4 |
| 20 | 88·7 | 271·3 |
| 21 | 98·6 | 261·4 |
| 21·5 | 112·2 | 247·8 |
| | + | — |
| 21 | 123·7 | 236·3 |
| 20 | 131·5 | 228·5 |
| 19 | 136·6 | 223·4 |
| 18 | 140·7 | 219·3 |
| 17 | 144·3 | 215·7 |
| 16 | 147·4 | 212·6 |
| 15 | 150·3 | 209·7 |
| 14 | 152·9 | 207·1 |
| 13 | 155·3 | 204·7 |
| 12 | 157·6 | 202·4 |
| 11 | 159·8 | 200·2 |
| 10 | 161·9 | 198·1 |
| 9 | 163·9 | 196·1 |
| 8 | 165·8 | 194·2 |
| 7 | 167·7 | 192·3 |
| 6 | 169·5 | 190·5 |
| 5 | 171·3 | 188·7 |
| 4 | 173·1 | 186·9 |
| 3 | 174·8 | 185·2 |
| 2 | 176·6 | 183·4 |
| 1 | 178·3 | 181·7 |
| 0 | 180·0 | 180·0 |

### Mercury's Anomaly.

| | + | + | — | — |
|---|---|---|---|---|
| | Degrees. | | | |
| 0·0 | 0·0 | 180·0 | 180·0 | 360·0 |
| 0·1 | 1·2 | 178·8 | 181·2 | 358·8 |
| 0·2 | 2·4 | 177·6 | 182·4 | 357·6 |
| 0·3 | 3·6 | 176·4 | 183·6 | 356·4 |
| 0·4 | 4·9 | 175·1 | 184·9 | 355·1 |
| 0·5 | 6·0 | 174·0 | 186·0 | 354·0 |
| 0·6 | 7·2 | 172·8 | 187·2 | 352·8 |
| 0·7 | 8·5 | 171·5 | 188·5 | 351·5 |
| 0·8 | 9·7 | 170·3 | 189·7 | 350·3 |
| 0·9 | 11·0 | 169·0 | 191·0 | 349·0 |
| 1·0 | 12·2 | 167·8 | 192·2 | 347·8 |
| 1·1 | 13·5 | 166·5 | 193·5 | 346·5 |
| 1·2 | 14·7 | 165·3 | 194·7 | 345·3 |
| 1·3 | 16·1 | 163·9 | 196·1 | 343·9 |
| 1·4 | 17·4 | 162·6 | 197·4 | 342·6 |
| 1·5 | 18·7 | 161·3 | 198·7 | 341·3 |
| 1·6 | 20·1 | 159·9 | 200·1 | 339·9 |
| 1·7 | 21·5 | 158·5 | 201·5 | 338·5 |
| 1·8 | 22·7 | 157·3 | 202·7 | 337·3 |
| 1·9 | 24·2 | 155·8 | 204·2 | 335·8 |
| 2·0 | 25·6 | 154·4 | 205·6 | 334·4 |
| 2·1 | 27·0 | 153·0 | 207·0 | 333·0 |
| 2·2 | 28·5 | 151·5 | 208·5 | 331·5 |
| 2·3 | 29·9 | 150·1 | 209·9 | 330·1 |
| 2·4 | 31·4 | 148·6 | 211·4 | 328·6 |
| 2·5 | 33·0 | 147·0 | 213·0 | 327·0 |
| 2·6 | 34·5 | 145·5 | 214·5 | 325·5 |
| 2·7 | 36·0 | 144·0 | 216·0 | 324·0 |
| 2·8 | 37·7 | 142·3 | 217·7 | 322·3 |
| 2·9 | 39·4 | 140·6 | 219·4 | 320·6 |
| 3·0 | 41·2 | 138·8 | 221·2 | 318·8 |
| 3·1 | 42·9 | 137·1 | 222·9 | 317·1 |
| 3·2 | 44·7 | 135·3 | 224·7 | 315·3 |
| 3·3 | 46·6 | 133·4 | 226·6 | 313·4 |
| 3·4 | 48·5 | 131·5 | 228·5 | 311·5 |
| 3·5 | 50·6 | 129·4 | 230·6 | 309·4 |
| 3·6 | 52·7 | 127·3 | 232·7 | 307·3 |
| 3·7 | 55·0 | 125·0 | 235·0 | 305·0 |
| 3·8 | 57·5 | 122·5 | 237·5 | 302·5 |
| 3·9 | 60·0 | 120·0 | 240·0 | 300·0 |
| 4·0 | 63·0 | 117·0 | 243·0 | 297·0 |
| 4·1 | 66·0 | 114·0 | 246·0 | 294·0 |
| 4·2 | 69·7 | 110·2 | 249·7 | 290·2 |
| 4·3 | 74·0 | 106·0 | 254·0 | 286·0 |
| 4·4 | 80·5 | 99·5 | 260·5 | 279·5 |
| 4·5 | 90·0 | 90·0 | 270·0 | 270·0 |

## JUPITER.

### Data for calculating Latitude and geocentric Longitude of Jupiter from B.C. 1 to A.D. 2000.

#### Jupiter's mean motion for Centuries.

| A.D. | Kaliyuga. | Mean long. at commencement of Solar Year. | Longitude of apsis. | Longitude of node. | A.D. | Kaliyuga. | Mean long. at commencement of Solar Year. | Longitude of apsis. | Longitude of node. |
|---|---|---|---|---|---|---|---|---|---|
| | | Deg. | Deg. | Deg. | | | Deg. | Deg. | Deg |
| 1 B.C. | 3101 K.Y. | 163·1655 | 171·2250 | 79·6967 | 1000 A.D. | 4101 K.Y. | 270·6117 | 171·3001 | 79·6822 |
| 100 A.D. | 3201 | 317·9101 | 171·2325 | 79·6952 | 1100 | 4201 | 65·3563 | 171·3076 | 79·6807 |
| 200 | 3301 | 112·6547 | 171·2400 | 79·6938 | 1200 | 4301 | 220·1009 | 171·3151 | 79·6793 |
| 300 | 3401 | 267·3993 | 171·2475 | 79·6923 | 1300 | 4401 | 14·8455 | 171·3226 | 79·6778 |
| 400 | 3501 | 62·1439 | 171·2550 | 79·6909 | 1400 | 4501 | 169·5901 | 171·3301 | 79·6764 |
| 500 | 3601 | 216·8885 | 171·2625 | 79·6894 | 1500 | 4601 | 324·3347 | 171·3376 | 79·6749 |
| 600 | 3701 | 11·6331 | 171·2700 | 79·6880 | 1600 | 4701 | 119·0793 | 171·3451 | 79·6735 |
| 700 | 3801 | 166·3778 | 171·2776 | 79·6865 | 1700 | 4801 | 273·8239 | 171·3526 | 79·6720 |
| 800 | 3901 | 321·1225 | 171·2851 | 79·6851 | 1800 | 4901 | 68·5665 | 171·3601 | 79·6706 |
| 900 | 4001 | 115·8671 | 171·2926 | 79·6836 | 1900 | 5001 | 223·3130 | 171·3676 | 79·6691 |

## TABLE XVII.—*contd.*

### Jupiter's mean motion for Odd Years.

| Yrs. | Deg. | Yrs. | Deg. | Yrs. | Deg. | Yrs. | Deg. | Yrs. | Deg. | Yrs. | Deg. | Yrs. | Deg. | Yrs. | Deg. | Yrs. | Deg. | Yrs. | Deg. |
|---|---|---|---|---|---|---|---|---|---|---|---|---|---|---|---|---|---|---|---|
| 1 | 30·35 | 11 | 333·82 | 21 | 277·30 | 31 | 220·77 | 41 | 164·24 | 51 | 107·72 | 61 | 51·19 | 71 | 354·67 | 81 | 298·14 | 91 | 241·62 |
| 2 | 60·69 | 12 | 4·17 | 22 | 307·64 | 32 | 251·12 | 42 | 194·59 | 52 | 138·07 | 62 | 81·54 | 72 | 25·02 | 82 | 328·49 | 92 | 271·96 |
| 3 | 91·04 | 13 | 34·52 | 23 | 337·99 | 33 | 281·46 | 43 | 224·94 | 53 | 168·41 | 63 | 111·89 | 73 | 55·36 | 83 | 358·84 | 93 | 302·31 |
| 4 | 121·39 | 14 | 64·86 | 24 | 8·34 | 34 | 311·81 | 44 | 255·29 | 54 | 198·76 | 64 | 142·24 | 74 | 85·71 | 84 | 29·18 | 94 | 332·66 |
| 5 | 151·74 | 15 | 95·21 | 25 | 38·69 | 35 | 342·16 | 45 | 285·63 | 55 | 229·11 | 65 | 172·58 | 75 | 116·06 | 85 | 59·53 | 95 | 3·01 |
| 6 | 182·08 | 16 | 125·56 | 26 | 69·03 | 36 | 12·51 | 46 | 315·98 | 56 | 259·46 | 66 | 202·93 | 76 | 146·41 | 86 | 89·88 | 96 | 33·35 |
| 7 | 212·43 | 17 | 155·91 | 27 | 99·38 | 37 | 42·85 | 47 | 346·33 | 57 | 289·80 | 67 | 233·28 | 77 | 176·75 | 87 | 120·23 | 97 | 63·70 |
| 8 | 242·78 | 18 | 186·25 | 28 | 129·73 | 38 | 73·20 | 48 | 16·68 | 58 | 320·15 | 68 | 263·63 | 78 | 207·10 | 88 | 150·57 | 98 | 94·05 |
| 9 | 273·13 | 19 | 216·60 | 29 | 160·07 | 39 | 103·55 | 49 | 47·02 | 59 | 350·50 | 69 | 293·97 | 79 | 237·45 | 89 | 180·92 | 99 | 124·40 |
| 10 | 303·47 | 20 | 246·95 | 30 | 190·42 | 40 | 133·90 | 50 | 77·37 | 60 | 20·85 | 70 | 324·32 | 80 | 267·80 | 90 | 211·27 | 100 | 154·74 |

### Jupiter's mean motion for Days.

| Days. | 0 | 1 | 2 | 3 | 4 | 5 | 6 | 7 | 8 | 9 |
|---|---|---|---|---|---|---|---|---|---|---|
| | Degrees. | | | | | | | | | |
| 0 | ... | 0·80 | 0·17 | 0·25 | 0·33 | 0·41 | 0·50 | 0·58 | 0·66 | 0·75 |
| 1 | 0·83 | 0·91 | 1·00 | 1·08 | 1·16 | 1·25 | 1·33 | 1·41 | 1·50 | 1·58 |
| 2 | 1·66 | 1·74 | 1·83 | 1·91 | 1·99 | 2·07 | 2·16 | 2·24 | 2·33 | 2·41 |
| 3 | 2·49 | 2·58 | 2·66 | 2·74 | 2·82 | 2·91 | 2·99 | 3·07 | 3·16 | 3·24 |
| 4 | 3·32 | 3·41 | 3·49 | 3·57 | 3·66 | 3·74 | 3·82 | 3·90 | 3·99 | 4·07 |
| 5 | 4·15 | 4·24 | 4·32 | 4·40 | 4·49 | 4·57 | 4·65 | 4·74 | 4·82 | 4·90 |
| 6 | 4·99 | 5·07 | 5·15 | 5.23 | 5·32 | 5·40 | 5·48 | 5·57 | 5·65 | 5·73 |
| 7 | 5·82 | 5·90 | 5·98 | 6·07 | 6·15 | 6·23 | 6·31 | 6·40 | 6·48 | 6·56 |
| 8 | 6·65 | 6·73 | 6·81 | 6·90 | 6·98 | 7·06 | 7·15 | 7·23 | 7·31 | 7·39 |
| 9 | 7·48 | 7·56 | 7·64 | 7·73 | 7·81 | 7·89 | 7·98 | 8·06 | 8·14 | 8·23 |
| 10 | 8·31 | 8·39 | 8·47 | 8·56 | 8·64 | 8·72 | 8·81 | 8·89 | 8·97 | 9·06 |
| 11 | 9·14 | 9·22 | 9·31 | 9·39 | 9·47 | 9·55 | 9·64 | 9·72 | 9·80 | 9·89 |
| 12 | 9·97 | 10·05 | 10·14 | 10·22 | 10·30 | 10·39 | 10·47 | 10·55 | 10·64 | 10·72 |
| 13 | 10·80 | 10·88 | 10·97 | 11·05 | 11·13 | 11·22 | 11·30 | 11·38 | 11·47 | 11·55 |
| 14 | 11·63 | 11·72 | 11·80 | 11·88 | 11·96 | 12·05 | 12·13 | 12·21 | 12·30 | 12·38 |
| 15 | 12·46 | 12·55 | 12 63 | 12·71 | 12·80 | 12·88 | 12·96 | 13·04 | 13·13 | 13·21 |
| 16 | 13·29 | 13·38 | 13·46 | 13·54 | 13·63 | 13·71 | 13·79 | 13·88 | 13·96 | 14·04 |
| 17 | 14·13 | 14·21 | 14·29 | 14·37 | 14·46 | 14·54 | 14·62 | 14·71 | 14·79 | 14·87 |
| 18 | 14·96 | 15·04 | 15·12 | 15·21 | 15·29 | 15·37 | 15·45 | 15·54 | 15·62 | 15·70 |

| Days. | 0 | 1 | 2 | 3 | 4 | 5 | 6 | 7 | 8 | 9 |
|---|---|---|---|---|---|---|---|---|---|---|
| | Degrees. | | | | | | | | | |
| 19 | 15·79 | 15·87 | 15·95 | 16·04 | 16·12 | 16·20 | 16·29 | 16·37 | 16·45 | 16·53 |
| 20 | 16·62 | 16·70 | 16·78 | 16·87 | 16·95 | 17·03 | 17·12 | 17·20 | 17·28 | 17·37 |
| 21 | 17·45 | 17·53 | 17·61 | 17·70 | 17·78 | 17·86 | 17·95 | 18·03 | 18·11 | 18·20 |
| 22 | 18·28 | 18·36 | 18·45 | 18·53 | 18·61 | 18·70 | 18·78 | 18·86 | 18·94 | 19·03 |
| 23 | 19·11 | 19·19 | 19·28 | 19·36 | 19·44 | 19·53 | 19·61 | 19·69 | 19·78 | 19·86 |
| 24 | 19·94 | 20·02 | 20·11 | 20·19 | 20·27 | 20·36 | 20·44 | 20·52 | 20·61 | 20·69 |
| 25 | 20·77 | 20·86 | 20·94 | 21·02 | 21·11 | 21·19 | 21·27 | 21·35 | 21·44 | 21·52 |
| 26 | 21·60 | 21·69 | 21·77 | 21·85 | 21·94 | 22·02 | 22·10 | 22·18 | 22·27 | 22·35 |
| 27 | 22·43 | 22·52 | 22·60 | 22·68 | 22·77 | 22·85 | 22·93 | 23·02 | 23·10 | 23·18 |
| 28 | 23·27 | 23·35 | 23·43 | 23·51 | 23·60 | 23·68 | 23·76 | 23·85 | 23·93 | 24·01 |
| 29 | 24·10 | 24·18 | 24·26 | 24·35 | 24·43 | 24·51 | 24·59 | 24·68 | 24·76 | 24·84 |
| 30 | 24·93 | 25·01 | 25·09 | 25·18 | 25·26 | 25·34 | 25·43 | 25·51 | 25·59 | 25·67 |
| 31 | 25·76 | 25·84 | 25·92 | 26·01 | 26·09 | 26·17 | 26·26 | 26·34 | 26·42 | 26·51 |
| 32 | 26·59 | 26·67 | 26·75 | 26·84 | 26·92 | 27·00 | 27·09 | 27·17 | 27·25 | 27·34 |
| 33 | 27·42 | 27·50 | 27·59 | 27·67 | 27·75 | 27·84 | 27·92 | 28·00 | 28·08 | 28·17 |
| 34 | 28·25 | 28·33 | 28·42 | 28·50 | 28·58 | 28·67 | 28·75 | 28·83 | 28·92 | 29·00 |
| 35 | 29·08 | 29·16 | 29·25 | 29·33 | 29·41 | 29·50 | 29·58 | 29·66 | 29·75 | 29·83 |
| 36 | 29·91 | 30·00 | 30·08 | 30·16 | 30·24 | 30·33 | ... | ... | ... | ... |

### Jupiter's Annual Equation.

| + | ·0 | ·1 | ·2 | ·3 | ·4 | ·5 | ·6 | ·7 | ·8 | ·9 |
|---|---|---|---|---|---|---|---|---|---|---|
| | Degrees. | | | | | | | | | |
| 0 | 0·0 | 0·6 | 1·2 | 1·8 | 2·4 | 3·1 | 3·7 | 4·3 | 4·9 | 5·5 |
| 1 | 6·1 | 6·7 | 7·4 | 8·0 | 8·6 | 9·2 | 9·8 | 10·4 | 11·1 | 11·7 |
| 2 | 12·2 | 12·9 | 13·5 | 14·1 | 14·7 | 15·4 | 16·0 | 16·6 | 17·2 | 17·9 |
| 3 | 18·5 | 19·1 | 19·7 | 20·4 | 21·0 | 21·6 | 22·2 | 22·9 | 23·5 | 24·2 |
| 4 | 24·8 | 25·4 | 26·1 | 26·7 | 27·4 | 28·0 | 28·6 | 29·2 | 29·9 | 30·6 |
| 5 | 31·2 | 31·9 | 32·6 | 33·2 | 33·9 | 34·5 | 35·2 | 35·9 | 36·6 | 37·2 |
| 6 | 37·9 | 38·6 | 39·3 | 40·0 | 40·7 | 41·4 | 42·0 | 42·8 | 43·5 | 44·2 |
| 7 | 44·9 | 45·7 | 46·4 | 47·2 | 47·9 | 48·6 | 49·4 | 50·2 | 50·9 | 51·7 |
| 8 | 52·4 | 53·2 | 54·0 | 54·8 | 55·7 | 56·4 | 57·3 | 58·2 | 59·0 | 59·8 |
| 9 | 60·7 | 61·6 | 62·5 | 63·4 | 64·4 | 65·3 | 66·3 | 67·2 | 68·3 | 69·4 |
| 10 | 70·4 | 71·5 | 72·7 | 73·9 | 75·0 | 76·4 | 77·6 | 79·0 | 80·5 | 82·0 |
| 11 | 83·7 | 85·5 | 87·3 | 90·0 | 93·0 | 97·5 | ... | ... | ... | ... |

| — | ·0 | ·1 | ·2 | ·3 | ·4 | ·5 | ·6 | ·7 | ·8 | ·9 |
|---|---|---|---|---|---|---|---|---|---|---|
| | Degrees. | | | | | | | | | |
| 0 | 180·0 | 180·4 | 180·8 | 181·3 | 181·7 | 182·1 | 182·5 | 182·9 | 183·3 | 183·7 |
| 1 | 184·1 | 184·5 | 185·0 | 185·4 | 185·8 | 186·2 | 186·6 | 187·0 | 187·5 | 187·9 |
| 2 | 188·3 | 188·7 | 189·1 | 189·6 | 190·0 | 190·4 | 190·8 | 191·3 | 191·7 | 192·1 |
| 3 | 192·5 | 193·0 | 193·4 | 193·8 | 194·3 | 194·7 | 195·1 | 195·5 | 196·0 | 196·4 |
| 4 | 196·9 | 197·3 | 197·8 | 198·2 | 198·6 | 199·1 | 199·6 | 200·0 | 200·5 | 200·9 |
| 5 | 201·4 | 201·8 | 202·3 | 202·8 | 203·2 | 203·7 | 204·2 | 204·7 | 205·2 | 205·6 |
| 6 | 206·1 | 206·6 | 207·1 | 207·6 | 208·1 | 208·6 | 209·1 | 209·6 | 210·1 | 210·7 |
| 7 | 211·2 | 211·7 | 212·3 | 212·8 | 213·4 | 213·9 | 214·5 | 215·1 | 215·6 | 216·2 |
| 8 | 216·8 | 217·4 | 217·9 | 218·6 | 219·2 | 219·8 | 220·4 | 221·1 | 221·8 | 222·4 |
| 9 | 223·2 | 223·8 | 224·6 | 225·3 | 226·0 | 226·8 | 227·6 | 228·3 | 229·2 | 229·9 |
| 10 | 230·9 | 231·8 | 232·8 | 233·8 | 234·8 | 236·0 | 237·0 | 238·1 | 239·4 | 240·8 |
| 11 | 242·3 | 243·8 | 245·8 | 247·8 | 250·5 | 255·0 | ... | ... | ... | ... |

**Maximum Equation 11·54** for Anomaly of 103·00.

| + | ·9 | ·8 | ·7 | ·6 | ·5 | ·4 | ·3 | ·2 | ·1 | ·0 |
|---|---|---|---|---|---|---|---|---|---|---|
| | Degrees. | | | | | | | | | |
| 11 | ... | ... | ... | ... | 105·0 | 109·5 | 112·2 | 114·2 | 116·2 | 117·7 |
| 10 | 119·2 | 120·6 | 121·9 | 123·0 | 124·0 | 125·2 | 126·2 | 127·2 | 128·2 | 129·1 |
| 9 | 130·1 | 130·8 | 131·7 | 132·4 | 133·2 | 134·0 | 134·7 | 135·4 | 136·2 | 136·8 |
| 8 | 137·6 | 138·2 | 138·9 | 139·6 | 140·2 | 140·8 | 141·4 | 142·1 | 142·6 | 143·2 |
| 7 | 143·8 | 144·4 | 144·9 | 145·5 | 146·1 | 146·6 | 147·2 | 147·7 | 148·3 | 148·8 |
| 6 | 149·3 | 149·9 | 150·4 | 150·9 | 151·4 | 151·9 | 152·4 | 152·9 | 153·4 | 153·9 |
| 5 | 154·4 | 154·8 | 155·3 | 155·8 | 156·3 | 156·8 | 157·2 | 157·7 | 158·2 | 158·6 |
| 4 | 159·1 | 159·5 | 160·0 | 160·4 | 160·9 | 161·4 | 161·8 | 162·2 | 162·7 | 163·1 |
| 3 | 163·6 | 164·0 | 164·5 | 164·9 | 165·3 | 165·7 | 166·2 | 166·6 | 167·0 | 167·5 |
| 2 | 167·9 | 168·3 | 168·7 | 169·2 | 169·6 | 170·0 | 170·4 | 170·9 | 171·3 | 171·7 |
| 1 | 172·1 | 172·5 | 173·0 | 173·4 | 173·8 | 174·2 | 174·6 | 175·0 | 175·5 | 175·9 |
| 0 | 176·3 | 176·7 | 177·1 | 177·5 | 177·9 | 178·3 | 178·7 | 179·2 | 179·6 | 180·0 |

**Maximum Equation 11·54** for Anomaly of 257·00

| — | ·9 | ·8 | ·7 | ·6 | ·5 | ·4 | ·3 | ·2 | ·1 | ·0 |
|---|---|---|---|---|---|---|---|---|---|---|
| | Degrees. | | | | | | | | | |
| 11 | ... | ... | ... | ... | 262·5 | 267·0 | 270·0 | 272·5 | 274·5 | 276·3 |
| 10 | 278·0 | 279·5 | 281·0 | 282·4 | 283·6 | 285·0 | 286·4 | 287·3 | 288·5 | 289·6 |
| 9 | 290·6 | 291·7 | 292·8 | 293·7 | 294·7 | 295·6 | 296·6 | 297·5 | 298·4 | 299·3 |
| 8 | 300·2 | 301·0 | 301·8 | 302·7 | 303·6 | 304·3 | 305·2 | 306·0 | 306·8 | 307·6 |
| 7 | 308·3 | 309·1 | 309·8 | 310·6 | 311·4 | 312·1 | 312·8 | 313·6 | 314·3 | 315·1 |
| 6 | 315·8 | 316·5 | 317·2 | 318·0 | 318·6 | 319·3 | 320·0 | 320·7 | 321·4 | 322·1 |
| 5 | 322·8 | 323·4 | 324·1 | 324·8 | 325·5 | 326·1 | 326·8 | 327·4 | 328·1 | 328·8 |
| 4 | 329·4 | 330·1 | 330·8 | 331·4 | 332·0 | 332·6 | 333·3 | 333·9 | 334·6 | 335·2 |
| 3 | 335·8 | 336·5 | 337·1 | 337·8 | 338·4 | 339·0 | 339·6 | 340·3 | 340·9 | 341·5 |
| 2 | 342·1 | 342·8 | 343·4 | 344·0 | 344·6 | 345·3 | 345·9 | 346·5 | 347·1 | 347·8 |
| 1 | 348·3 | 348·9 | 349·6 | 350·2 | 350·8 | 351·4 | 352·0 | 352·6 | 353·3 | 353·9 |
| 0 | 354·5 | 355·1 | 355·7 | 356·3 | 356·9 | 357·6 | 358·2 | 358·8 | 359·4 | 360·0 |

## TABLE XVII.—*contd.*

### Jupiter's Anomaly.

| + | ·0 | ·1 | ·2 | ·3 | ·4 | ·5 | ·6 | ·7 | ·8 | ·9 |
|---|---|---|---|---|---|---|---|---|---|---|
| | Degrees. | | | | | | | | | |
| 0 | 0·0 | 1·1 | 2·2 | 3·2 | 4·4 | 5·5 | 6·6 | 7·7 | 8·8 | 9·9 |
| 1 | 11·0 | 12·2 | 13·3 | 14·5 | 15·6 | 16·7 | 17·9 | 19·1 | 20·2 | 21·5 |
| 2 | 22·6 | 23·9 | 25·1 | 26·4 | 27·6 | 28·9 | 30·2 | 31·5 | 32·9 | 34·1 |
| 3 | 35·5 | 37·0 | 38·4 | 39·9 | 41·4 | 42·9 | 44·5 | 46·1 | 47·7 | 49·5 |
| 4 | 51·2 | 53·1 | 55·0 | 57·1 | 59·2 | 61·7 | 64·0 | 67·0 | 70·0 | 74·0 |
| 5 | 77·5 | 90·0 | ... | ... | ... | ... | ... | ... | ... | ... |

**Maximum Equation 5·0994** for Anomaly of 90·00

| + | ·9 | ·8 | ·7 | ·6 | ·5 | ·4 | ·3 | ·2 | ·1 | ·0 |
|---|---|---|---|---|---|---|---|---|---|---|
| | Degrees. | | | | | | | | | |
| 5 | ... | ... | ... | ... | ... | ... | ... | ... | 90·0 | 102·5 |
| 4 | 106·0 | 110·0 | 113·0 | 116·0 | 118·3 | 120·8 | 122·9 | 125·0 | 126·9 | 128·8 |
| 3 | 130·5 | 132·3 | 133·9 | 135·5 | 137·1 | 138·6 | 140·1 | 141·6 | 143·0 | 144·5 |
| 2 | 145·9 | 147·1 | 148·5 | 149·8 | 151·1 | 152·4 | 153·6 | 154·9 | 156·1 | 157·4 |
| 1 | 158·5 | 159·8 | 160·9 | 162·1 | 163·3 | 164·4 | 165·5 | 166·7 | 167·8 | 169·0 |
| 0 | 170·1 | 171·2 | 172·3 | 173·4 | 174·5 | 175·6 | 176·8 | 177·8 | 178·9 | 180·0 |

| — | ·0 | ·1 | ·2 | ·3 | ·4 | ·5 | ·6 | ·7 | ·8 | ·9 |
|---|---|---|---|---|---|---|---|---|---|---|
| | Degrees. | | | | | | | | | |
| 0 | 180·0 | 181·1 | 182·2 | 183·2 | 184·4 | 185·5 | 186·6 | 187·7 | 188·8 | 189·9 |
| 1 | 191·0 | 192·2 | 193·3 | 194·5 | 195·6 | 196·7 | 197·9 | 199·0 | 200·2 | 201·5 |
| 2 | 202·6 | 203·9 | 205·1 | 206·4 | 207·6 | 208·9 | 210·2 | 211·5 | 212·9 | 214·1 |
| 3 | 215·5 | 217·0 | 218·4 | 219·9 | 221·4 | 222·9 | 224·5 | 226·1 | 227·7 | 229·5 |
| 4 | 231·2 | 233·1 | 235·0 | 237·1 | 239·2 | 241·7 | 244·0 | 247·0 | 250·2 | 254·0 |
| 5 | 257·5 | 270·0 | ... | ... | ... | ... | ... | ... | ... | ... |

**Maximum Equation 5·0994** for Anomaly of 270·00.

| — | ·9 | ·8 | ·7 | ·6 | ·5 | ·4 | ·3 | ·2 | ·1 | ·0 |
|---|---|---|---|---|---|---|---|---|---|---|
| | Degrees. | | | | | | | | | |
| 5 | ... | ... | ... | ... | ... | ... | ... | ... | 270·0 | 282·5 |
| 4 | 286·0 | 290·0 | 293·0 | 296·0 | 298·3 | 300·8 | 302·9 | 305·0 | 306·9 | 308·8 |
| 3 | 310·5 | 312·3 | 313·9 | 315·5 | 317·1 | 318·6 | 320·1 | 321·6 | 323·0 | 324·5 |
| 2 | 325·9 | 327·1 | 328·5 | 329·8 | 331·1 | 332·4 | 333·6 | 334·9 | 336·1 | 337·1 |
| 1 | 338·5 | 339·8 | 340·9 | 342·1 | 343·3 | 344·4 | 345·5 | 346·7 | 347·8 | 349·0 |
| 0 | 350·1 | 351·2 | 352·3 | 353·4 | 354·5 | 355·6 | 356·8 | 357·8 | 358·9 | 360·0 |

## VENUS.

### Data for calculating Latitude and geocentric Longitude of Venus from B.C. 1 to A.D. 2000.

### Venus' mean motion for Centuries.

| A.D. | Kaliyuga. | Mean long. at commencement of Solar Year. | Longitude of apsis. | Longitude of node. |
|---|---|---|---|---|
| | | Deg. | Deg. | Deg. |
| 1 B.C. | 3101 K.Y. | 303·2772 | 79·7837 | 59·8039 |
| 100 A.D. | 3201 | 142·2624 | 79·7882 | 59·7964 |
| 200 | 3301 | 341·2475 | 79·7926 | 59·7889 |
| 300 | 3401 | 180·2327 | 79·7971 | 59·7814 |
| 400 | 3501 | 19·2178 | 79·8016 | 59·7738 |
| 500 | 3601 | 218·2030 | 79·8060 | 59·7663 |
| 600 | 3701 | 57·1882 | 79·8105 | 59·7588 |
| 700 | 3801 | 256·1733 | 79·8149 | 59·7513 |
| 800 | 3901 | 95·1585 | 79·8194 | 59·7438 |
| 900 | 4001 | 294·1436 | 79·8238 | 59·7363 |
| 1000 A.D. | 4101 K.Y. | 133·1288 | 79·8283 | 59·7288 |
| 1100 | 4201 | 332·1140 | 79·8328 | 59·7213 |
| 1200 | 4301 | 171·0991 | 79·8372 | 59·7138 |
| 1300 | 4401 | 10·0843 | 79·8417 | 59·7063 |
| 1400 | 4501 | 209·0694 | 79·8461 | 59·6988 |
| 1500 | 4601 | 48·0546 | 79·8506 | 59·6913 |
| 1600 | 4701 | 247·0398 | 79·8550 | 59·6838 |
| 1700 | 4801 | 86·0249 | 79·8595 | 59·6763 |
| 1800 | 4901 | 285·0101 | 79·8639 | 59·6687 |
| 1900 | 5001 | 123·9952 | 79·8684 | 59·6612 |
| 2000 | 5101 | 322·9804 | 79·8729 | 59·6537 |

### Venus' mean motion for Odd Years.

| Yrs. | Deg. | Yrs. | Deg. | Yrs. | Deg. | Yrs. | Deg. | Yrs. | Deg. | Yrs. | Deg. | Yrs. | Deg. | Yrs. | Deg. | Yrs. | Deg. | Yrs. | Deg. |
|---|---|---|---|---|---|---|---|---|---|---|---|---|---|---|---|---|---|---|---|
| 1 | 225·19 | 11 | 317·09 | 21 | 48·99 | 31 | 140·88 | 41 | 232·78 | 51 | 324·68 | 61 | 56·58 | 71 | 148·48 | 81 | 240·38 | 91 | 332·28 |
| 2 | 90·38 | 12 | 182·28 | 22 | 274·18 | 32 | 6·07 | 42 | 97·97 | 52 | 189·87 | 62 | 281·77 | 72 | 13·67 | 82 | 105·57 | 92 | 197·47 |
| 3 | 315·57 | 13 | 47·47 | 23 | 139·37 | 33 | 231·26 | 43 | 323·16 | 53 | 55·06 | 63 | 146·96 | 73 | 238·86 | 83 | 330·76 | 93 | 62·66 |
| 4 | 180·76 | 14 | 272·66 | 24 | 4·56 | 34 | 96·45 | 44 | 188·35 | 54 | 280·25 | 64 | 12·15 | 74 | 104·05 | 84 | 195·95 | 94 | 287·85 |
| 5 | 45·95 | 15 | 137·85 | 25 | 229·75 | 35 | 321·64 | 45 | 53·54 | 55 | 145·44 | 65 | 237·34 | 75 | 329·24 | 85 | 61·14 | 95 | 153·04 |
| 6 | 271·14 | 16 | 3·04 | 26 | 94·94 | 36 | 186·83 | 46 | 278·73 | 56 | 10·63 | 66 | 102·53 | 76 | 194·43 | 86 | 286·33 | 96 | 18·23 |
| 7 | 136·33 | 17 | 228·23 | 27 | 320·13 | 37 | 52·02 | 47 | 143·92 | 57 | 235·82 | 67 | 327·72 | 77 | 59·62 | 87 | 151·52 | 97 | 243·42 |
| 8 | 1·52 | 18 | 93·42 | 28 | 185·32 | 38 | 277·21 | 48 | 9·11 | 58 | 101·01 | 68 | 192·91 | 78 | 284·81 | 88 | 16·71 | 98 | 108·60 |
| 9 | 226·71 | 19 | 318·61 | 29 | 50·51 | 39 | 142·40 | 49 | 234·30 | 59 | 326·20 | 69 | 58·10 | 79 | 150·00 | 89 | 241·90 | 99 | 333·79 |
| 10 | 91·90 | 20 | 183·80 | 30 | 275·70 | 40 | 7·59 | 50 | 99·49 | 60 | 191·39 | 70 | 283·29 | 80 | 15·19 | 90 | 107·09 | 100 | 198·98 |

## TABLE XVII.—*contd.*

### Venus' mean motion for Days.

Degrees.

| Days. | 0 | 1 | 2 | 3 | 4 | 5 | 6 | 7 | 8 | 9 |
|---|---|---|---|---|---|---|---|---|---|---|
| 0 | ... | 1·60 | 3·20 | 4·81 | 6·41 | 8·01 | 9·61 | 11·21 | 12·82 | 14·42 |
| 1 | 16·02 | 17·62 | 19·23 | 20·83 | 22·43 | 24·03 | 25·63 | 27·24 | 28·84 | 30·44 |
| 2 | 32·04 | 33·64 | 35·25 | 36·65 | 38·45 | 40·05 | 41·65 | 43·26 | 44·86 | 46·46 |
| 3 | 48·06 | 49·67 | 51·27 | 52·87 | 54·47 | 56·07 | 57·68 | 59·28 | 60·88 | 62·48 |
| 4 | 64·08 | 65·68 | 67·29 | 68·89 | 70·49 | 72·09 | 73·70 | 75·30 | 76.90 | 78·50 |
| 5 | 80·11 | 81·71 | 83·31 | 84·91 | 86·51 | 88·12 | 89·72 | 91·32 | 92·32 | 94·53 |
| 6 | 96·13 | 97·73 | 99·33 | 100·93 | 102·54 | 104·14 | 105·74 | 107·34 | 108·94 | 110·55 |
| 7 | 112·15 | 113·75 | 115·35 | 116·95 | 118·56 | 120·16 | 121·76 | 123·36 | 124·97 | 126·57 |
| 8 | 128·70 | 129·77 | 131·37 | 132·98 | 134·58 | 136·18 | 137·78 | 139·38 | 140·99 | 142·59 |
| 9 | 144·19 | 145·79 | 147·40 | 149·00 | 150·60 | 152·20 | 153·80 | 155·41 | 157·01 | 158·61 |
| 10 | 160·21 | 161·81 | 163·42 | 165·02 | 166·62 | 168·22 | 169·83 | 171·43 | 173·03 | 174·63 |
| 11 | 176·23 | 177·84 | 179·44 | 181·04 | 182·64 | 182·24 | 185·85 | 187·45 | 189·05 | 190·65 |
| 12 | 192·25 | 193·86 | 195·46 | 197·06 | 198·66 | 200·27 | 201·87 | 203·47 | 205·07 | 206·67 |
| 13 | 208·28 | 209·88 | 211·48 | 213·08 | 214·69 | 216·29 | 217·89 | 219·49 | 221·09 | 222·70 |
| 14 | 224·30 | 225·90 | 227·50 | 229·10 | 230·71 | 232·31 | 233·91 | 235·51 | 237·11 | 238·72 |
| 15 | 240·32 | 241·92 | 243·52 | 245·13 | 246·73 | 248·33 | 249·93 | 251·53 | 253·14 | 254·74 |
| 16 | 256·34 | 257·94 | 259·54 | 261·15 | 262·75 | 264·35 | 265·95 | 267·56 | 269·16 | 270·76 |
| 17 | 272·56 | 273·96 | 275·57 | 277·17 | 278·77 | 280·37 | 281·97 | 283·58 | 285·18 | 286·78 |
| 18 | 288·38 | 289·98 | 291·59 | 293·19 | 294·79 | 296·39 | 297·99 | 299·60 | 301·20 | 302·80 |
| 19 | 304·40 | 306·01 | 307·61 | 309·21 | 310·81 | 312·41 | 314·02 | 315·62 | 317·22 | 318·82 |
| 20 | 320·43 | 322·03 | 323·63 | 325·23 | 326·83 | 328·44 | 330·04 | 331·64 | 333·24 | 334·84 |
| 21 | 336·45 | 338·05 | 339·65 | 341·25 | 342·86 | 344·46 | 346·06 | 347·66 | 349·26 | 350·87 |
| 22 | 352·47 | 354·07 | 355·67 | 357·27 | 358·88 | 0·48 | 2·08 | 3·68 | 5·29 | 6·89 |
| 23 | 8·49 | 10·09 | 11·69 | 13·30 | 14·90 | 16·50 | 18·10 | 19·70 | 21·31 | 22·91 |
| 24 | 24·51 | 26·11 | 27·72 | 29·32 | 30·92 | 32·52 | 34·12 | 35·73 | 37·33 | 38·93 |
| 25 | 40·53 | 42·13 | 43·74 | 45·34 | 46·94 | 48·54 | 50·15 | 51·75 | 53·35 | 54·95 |
| 26 | 56·55 | 58·16 | 59·76 | 61·36 | 62·96 | 64·56 | 66·17 | 67·77 | 69·37 | 70·97 |
| 27 | 72·57 | 74·17 | 75·78 | 77·38 | 78·98 | 80·59 | 82·19 | 83·79 | 85·39 | 86·99 |
| 28 | 88·60 | 90·20 | 91·80 | 93·40 | 95·00 | 96·61 | 98·21 | 99·81 | 101·41 | 103·02 |
| 29 | 104·62 | 106·22 | 107·82 | 109·42 | 111·03 | 112·63 | 114·23 | 115·83 | 117·43 | 119·04 |
| 30 | 120·64 | 122·24 | 123·84 | 125·45 | 127·05 | 128·65 | 130·25 | 131·85 | 133·46 | 135·06 |
| 31 | 136·66 | 138·26 | 139·86 | 141·47 | 143·07 | 144·67 | 146·27 | 147·87 | 149·48 | 151·08 |
| 32 | 152·68 | 154·28 | 155·89 | 157·49 | 159·09 | 160·69 | 162·29 | 163·90 | 165·50 | 167·10 |
| 33 | 168·70 | 170·30 | 171·91 | 173·51 | 175·11 | 176·71 | 178·32 | 179·92 | 181·52 | 183·12 |
| 34 | 184·72 | 186·33 | 187·93 | 189·53 | 191·13 | 192·73 | 194·34 | 195·94 | 197·54 | 199·14 |
| 35 | 200·75 | 202·35 | 203·95 | 205·55 | 207·15 | 208·76 | 210·36 | 211·96 | 213·56 | 215·16 |
| 36 | 216·77 | 218·37 | 219·97 | 221·57 | 223·18 | 224·78 | ... | ... | ... | ... |

### Venus' Annual Equation.

Degrees.

| | + | + | − | − |
|---|---|---|---|---|
| 0 | 0·0 | 180·0 | 180·0 | 360·0 |
| 1 | 2·4 | 179·6 | 180·4 | 357·6 |
| 2 | 4·7 | 179·2 | 180·8 | 355·3 |
| 3 | 7·1 | 178·8 | 181·2 | 352·9 |
| 4 | 9·5 | 178·5 | 181·5 | 350·5 |
| 5 | 11·9 | 178·1 | 181·9 | 348·1 |
| 6 | 14·3 | 177·7 | 182·3 | 345·7 |
| 7 | 16·7 | 177·3 | 182·7 | 343·3 |
| 8 | 19·1 | 176·9 | 183·1 | 340·9 |
| 9 | 21·4 | 176·6 | 183·4 | 338·6 |
| 10 | 23·9 | 176·2 | 183·8 | 336·1 |
| 11 | 26·3 | 175·8 | 184·2 | 333·7 |
| 12 | 28·7 | 175·3 | 184·7 | 331·3 |
| 13 | 31·1 | 174·9 | 185·1 | 328·9 |
| 14 | 33·5 | 174·5 | 185·5 | 326·5 |
| 15 | 35·9 | 174·1 | 185·9 | 324·1 |
| 16 | 38·4 | 173·7 | 186·3 | 321·6 |
| 17 | 40·8 | 173·2 | 186·8 | 319·2 |
| 18 | 43·2 | 172·8 | 187·2 | 316·8 |
| 19 | 45·7 | 172·4 | 187·6 | 314·3 |
| 20 | 48·2 | 171·9 | 188·1 | 311·8 |
| 21 | 50·6 | 171·4 | 188·6 | 309·4 |
| 22 | 53·1 | 170·9 | 189·1 | 306·9 |
| 23 | 55·6 | 170·4 | 189·6 | 304·4 |
| 24 | 58·2 | 169·9 | 190·1 | 301·8 |
| 25 | 60·7 | 169·4 | 190·6 | 299·3 |
| 26 | 63·3 | 168·9 | 191·1 | 296·7 |
| 27 | 65·9 | 168·3 | 191·7 | 294·1 |
| 28 | 68·4 | 167·7 | 192·3 | 291·6 |
| 29 | 71·1 | 167·1 | 192·9 | 288·9 |
| 30 | 73·8 | 166·4 | 193·6 | 286·2 |
| 31 | 76·4 | 165·8 | 194·2 | 283·6 |
| 32 | 79·1 | 165·1 | 194·9 | 280·9 |
| 33 | 81·9 | 164·3 | 195·7 | 278·1 |
| 34 | 84·7 | 163·5 | 196·5 | 275·3 |
| 35 | 87·5 | 162·7 | 197·3 | 272·5 |
| 36 | 90·4 | 161·9 | 198·1 | 269·6 |
| 37 | 93·4 | 161·0 | 199·0 | 266·6 |
| 38 | 96·4 | 159·9 | 200·1 | 263·6 |
| 39 | 99·6 | 158·8 | 201·2 | 260·4 |
| 40 | 102·8 | 157·7 | 202·3 | 257·2 |
| 41 | 106·2 | 156·2 | 203·8 | 253·8 |
| 42 | 109·8 | 154·6 | 205·4 | 250·2 |
| 43 | 113·6 | 152·8 | 207·2 | 246·4 |
| 44 | 117·9 | 150·6 | 209·4 | 242·1 |
| 45 | 122·9 | 147·8 | 212·2 | 237·1 |
| 46 | 129·7 | 142·8 | 217·2 | 230·3 |
| 46·38 | 135·0 | ... | 225·0 | ... |

### Venus' Anomaly.

Degrees.

| | + | − |
|---|---|---|
| 0·0 | 0·0 | 180·0 |
| 0·1 | 3·0 | 183·0 |
| 0·2 | 6·0 | 186·0 |
| 0·3 | 9·2 | 189·2 |
| 0·4 | 10·2 | 190·2 |
| 0·5 | 15·5 | 195·5 |
| 0·6 | 18·7 | 198·7 |
| 0·7 | 22·2 | 202·2 |
| 0·8 | 25·7 | 205·7 |
| 0·9 | 29·5 | 209·5 |
| 1·0 | 33·2 | 213·2 |
| 1·1 | 37·2 | 217·2 |
| 1·2 | 41·5 | 221·5 |
| 1·3 | 45·2 | 225·2 |
| 1·4 | 51·5 | 231·5 |
| 1·5 | 57·7 | 237·5 |
| 1·6 | 64·7 | 244·7 |
| 1·7 | 75·0 | 255·0 |
| 1·75 | 90·0 | 270·0 |
| | + | − |
| 1·7 | 105·0 | 285·0 |
| 1·6 | 115·3 | 295·3 |
| 1·5 | 122·3 | 302·2 |
| 1·4 | 128·5 | 308·5 |
| 1·3 | 134·8 | 314·8 |
| 1·2 | 138·5 | 318·5 |
| 1·1 | 142·8 | 322·8 |
| 1·0 | 146·8 | 326·8 |
| 0·9 | 150·5 | 330·5 |
| 0·8 | 154·3 | 334·3 |
| 0·7 | 157·8 | 337·8 |
| 0·6 | 161·3 | 341·3 |
| 0·5 | 164·5 | 344·5 |
| 0·4 | 169·8 | 349·8 |
| 0·3 | 170·8 | 350·8 |
| 0·2 | 174·0 | 354·0 |
| 0·1 | 177·0 | 357·0 |
| 0·0 | 180·0 | 360·0 |

## SATURN.

### Data for calculating Latitude and geocentric Longitude of Saturn from B.C. 1 to A.D. 2000.

### Saturn's mean motion for Centuries.

| A.D. | Kaliyuga. | Mean Long at commencement of Solar Year. | Longitude of apsis. | Longitude of node. | A.D. | Kaliyuga | Mean Long. at commencement of Solar Year. | Longitude of aapsis. | Longitude of node. |
|---|---|---|---|---|---|---|---|---|---|
| | | Deg. | Deg | Deg. | | | Deg. | Deg. | Deg. |
| 1 B.C. | 3101 K.Y. | 70·5378 | 236·6194 | 100·4714 | 1000 A.D. | 4101 K.Y. | 49·6072 | 236·6226 | 100·4162 |
| 100 A.D. | 3201 | 212·4447 | 236·6197 | 100·4659 | 1100 | 4201 | 191·5141 | 236·6229 | 100·4107 |
| 200 | 3301 | 354·3517 | 236·6200 | 100·4604 | 1200 | 4301 | 333·4211 | 236·6233 | 100·4052 |
| 300 | 3401 | 136·2586 | 236·6203 | 100·4548 | 1300 | 4401 | 115·3280 | 236·6236 | 100·3997 |
| 400 | 3501 | 278·1655 | 236·6207 | 100·4493 | 1400 | 4501 | 257·2349 | 236·6239 | 100·3941 |
| 500 | 3601 | 60·0725 | 236·6210 | 100·4438 | 1500 | 4601 | 39·1419 | 236·6242 | 100·3886 |
| 600 | 3701 | 201·9794 | 236·6213 | 100·4383 | 1600 | 4701 | 181·0488 | 236·6246 | 100·3831 |
| 700 | 3801 | 343·8864 | 236·6216 | 100·4328 | 1700 | 4801 | 322·9558 | 236·6249 | 100·3776 |
| 800 | 3901 | 125·7933 | 236·6220 | 100·4272 | 1800 | 4901 | 104·8627 | 236·6252 | 100·3721 |
| 900 | 4001 | 267·7002 | 236·6223 | 100·4217 | 1900 | 5001 | 246·7696 | 236·6255 | 100·3665 |
| | | | | | 2000 | 5101 | 28·6766 | 236·6259 | 100·3610 |

## TABLE XVII.—*contd.*

### Saturn's mean motion for Odd Years.

| Yrs. | Deg. | Yrs. | Deg. | Yrs. | Deg. | Yrs. | Deg. | Yrs. | Deg. | Yrs. | Deg. | Yrs. | Deg. | Yrs. | Deg. | Yrs. | Deg. | Yrs. | Deg. |
|---|---|---|---|---|---|---|---|---|---|---|---|---|---|---|---|---|---|---|---|
| 1 | 12·22 | 11 | 134·41 | 21 | 256·60 | 31 | 18·79 | 41 | 140·98 | 51 | 263·17 | 61 | 25·36 | 71 | 147·55 | 81 | 269·74 | 91 | 31·93 |
| 2 | 24·44 | 12 | 146·63 | 22 | 268·82 | 32 | 31·01 | 42 | 153·20 | 52 | 275·39 | 62 | 37·58 | 72 | 159·77 | 82 | 281·96 | 92 | 44·15 |
| 3 | 36·66 | 13 | 158·85 | 23 | 281·04 | 33 | 43·23 | 43 | 165·42 | 53 | 287·61 | 63 | 49·80 | 73 | 171·99 | 83 | 294·18 | 93 | 56·37 |
| 4 | 48·88 | 14 | 171·07 | 24 | 293·26 | 34 | 55·45 | 44 | 177·64 | 54 | 299·83 | 64 | 62·02 | 74 | 184·21 | 84 | 306·40 | 94 | 68·59 |
| 5 | 61·09 | 15 | 183·29 | 25 | 305·48 | 35 | 67·67 | 45 | 189·86 | 55 | 312·05 | 65 | 74·24 | 75 | 196·43 | 85 | 318·62 | 95 | 80·81 |
| 6 | 73·31 | 16 | 195·50 | 26 | 317·70 | 36 | 79·89 | 46 | 202·08 | 56 | 324·27 | 66 | 86·46 | 76 | 208·65 | 86 | 330·84 | 96 | 93·03 |
| 7 | 85·53 | 17 | 207·72 | 27 | 329·91 | 37 | 92·10 | 47 | 214·30 | 57 | 336·49 | 67 | 98·68 | 77 | 220·87 | 87 | 343·06 | 97 | 105·25 |
| 8 | 97·75 | 18 | 219·94 | 28 | 342·13 | 38 | 104·32 | 48 | 226·51 | 58 | 348·71 | 68 | 110·90 | 78 | 233·09 | 88 | 355·28 | 98 | 117·47 |
| 9 | 109·97 | 19 | 232·16 | 29 | 354·35 | 39 | 116·54 | 49 | 238·73 | 59 | 0·92 | 69 | 123·12 | 79 | 245·31 | 89 | 7·50 | 99 | 129·69 |
| 10 | 122·19 | 20 | 244·38 | 30 | 6·57 | 40 | 128·76 | 50 | 250·95 | 60 | 13·14 | 70 | 135·33 | 80 | 257·52 | 90 | 19·72 | 100 | 141·91 |

### Saturn's mean motion for Days.

Degrees.

| Days. | 0 | 1 | 2 | 3 | 4 | 5 | 6 | 7 | 8 | 9 |
|---|---|---|---|---|---|---|---|---|---|---|
| 0 | ... | 0·03 | 0·07 | 0·10 | 0·13 | 0·17 | 0·20 | 0·23 | 0·27 | 0·30 |
| 1 | 0·33 | 0·37 | 0·40 | 0·43 | 0·47 | 0·50 | 0·53 | 0·57 | 0·60 | 0·61 |
| 2 | 0·67 | 0·70 | 0·74 | 0·77 | 0·80 | 0·84 | 0·87 | 0·90 | 0·94 | 0·97 |
| 3 | 1·00 | 1·04 | 1·07 | 1·10 | 1·14 | 1·17 | 1·20 | 1·24 | 1·27 | 1·30 |
| 4 | 1·34 | 1·37 | 1·40 | 1·44 | 1·47 | 1·51 | 1·54 | 1·57 | 1·61 | 1·64 |
| 5 | 1·67 | 1·71 | 1·74 | 1·77 | 1·81 | 1·84 | 1·87 | 1·91 | 1·94 | 1·27 |
| 6 | 2·01 | 2·04 | 2·07 | 2·11 | 2·14 | 2·17 | 2·21 | 2·24 | 2·27 | 2·31 |
| 7 | 2·34 | 2·38 | 2·41 | 2·44 | 2·48 | 2·51 | 2·54 | 2·58 | 2·61 | 2·64 |
| 8 | 2·68 | 2·71 | 2·74 | 2·78 | 2·81 | 2·84 | 2·88 | 2·91 | 2·94 | 2·98 |
| 9 | 3·01 | 3·04 | 3·08 | 3·11 | 3·14 | 3·18 | 3·21 | 3·25 | 3·28 | 3·31 |
| 10 | 3·35 | 3·38 | 3·41 | 3·45 | 3·48 | 3·51 | 3·55 | 3·58 | 3·61 | 3·65 |
| 11 | 3·68 | 3·71 | 3·75 | 3·78 | 3·81 | 3·85 | 3·88 | 3·91 | 3·95 | 3·98 |
| 12 | 4·01 | 4·05 | 4·08 | 4·12 | 4·15 | 4·18 | 4·22 | 4·25 | 4·28 | 4·32 |
| 13 | 4·35 | 4·38 | 4·42 | 4·45 | 4·48 | 4·52 | 4·55 | 4·58 | 4·62 | 4·65 |
| 14 | 4·68 | 4·72 | 4·75 | 4·78 | 4·82 | 4·85 | 4·88 | 4·92 | 4·95 | 4·99 |
| 15 | 5·02 | 5·05 | 5·09 | 5·12 | 5·15 | 5·19 | 5·22 | 5·25 | 5·29 | 5·32 |
| 16 | 5·35 | 5·39 | 5·42 | 5·45 | 5·49 | 5·52 | 5·55 | 5·59 | 5·62 | 5·65 |
| 17 | 5·69 | 5·72 | 5·75 | 5·79 | 5·82 | 5·85 | 5·89 | 5·92 | 5·96 | 5·99 |
| 18 | 6·02 | 6·06 | 6·09 | 6·12 | 6·16 | 6·19 | 6·22 | 6·26 | 6·29 | 6·32 |
| 19 | 6·36 | 6·39 | 6·42 | 6·46 | 6·49 | 6·52 | 6·56 | 6·59 | 6·62 | 6·66 |
| 20 | 6·69 | 6·72 | 6·76 | 6·79 | 6·83 | 6·86 | 6·89 | 6·93 | 6·96 | 6·99 |
| 21 | 7·03 | 7·06 | 7·09 | 7·13 | 7·16 | 7·19 | 7·23 | 7·26 | 7·29 | 7·33 |
| 22 | 7·36 | 7·39 | 7·43 | 7·46 | 7·49 | 7·53 | 7·56 | 7·59 | 7·63 | 7·36 |
| 23 | 7·70 | 7·73 | 7·76 | 7·80 | 7·93 | 7·86 | 7·90 | 7·93 | 7·96 | 8·00 |
| 24 | 8·03 | 8·06 | 8·10 | 8·13 | 8·16 | 8·20 | 8·23 | 8·26 | 8·29 | 8·33 |
| 25 | 8·36 | 8·40 | 8·43 | 8·46 | 8·50 | 8·53 | 8·57 | 8·60 | 8·63 | 8·67 |
| 26 | 8·70 | 8·73 | 8·77 | 8·80 | 8·83 | 8·87 | 8·90 | 8·93 | 8·97 | 9·00 |
| 27 | 9·03 | 9·07 | 9·10 | 9·13 | 9·17 | 9·20 | 9·23 | 9·27 | 9·30 | 9·33 |
| 28 | 9·37 | 9·40 | 9·43 | 9·47 | 9·50 | 9·54 | 9·57 | 9·60 | 9·64 | 9·67 |
| 29 | 9·70 | 9·74 | 9·77 | 9·80 | 9·84 | 9·87 | 9·90 | 9·94 | 9·97 | 10·00 |
| 30 | 10·04 | 10·07 | 10·10 | 10·14 | 10·17 | 10·20 | 10·24 | 10·27 | 10·31 | 10·34 |
| 31 | 10·37 | 10·41 | 10·44 | 10·47 | 10·51 | 10·54 | 10·57 | 10·61 | 10·64 | 10·67 |
| 32 | 10·71 | 10·74 | 10·77 | 10·81 | 10·84 | 10·87 | 10·91 | 10·94 | 10·97 | 11·01 |
| 33 | 11·04 | 11·07 | 11·11 | 11·14 | 11·18 | 11·21 | 11·24 | 11·28 | 11·31 | 11·34 |
| 34 | 11·38 | 11·41 | 11·44 | 11·48 | 11·51 | 11·54 | 11·58 | 11·61 | 11·64 | 11·68 |
| 35 | 11·71 | 11·74 | 11·78 | 11·81 | 11·84 | 11·88 | 11·91 | 11·94 | 11·98 | 12·01 |
| 36 | 12·05 | 12·08 | 12·11 | 12·15 | 12·18 | 12·21 | ... | ... | ... | ... |

### Saturn's Annual Equation.

Degrees.

| + | ·0 | ·1 | ·2 | ·3 | ·4 | ·5 | ·6 | ·7 | ·8 | ·9 |
|---|---|---|---|---|---|---|---|---|---|---|
| 0 | 0·0 | 1·0 | 2·0 | 3·1 | 4·1 | 5·1 | 6·1 | 7·1 | 8·2 | 9·2 |
| 1 | 10·2 | 11·2 | 12·2 | 13·3 | 14·4 | 15·4 | 16·4 | 17·5 | 18·5 | 19·6 |
| 2 | 20·6 | 21·7 | 22·7 | 23·8 | 24·9 | 26·0 | 27·1 | 28·2 | 29·2 | 30·4 |
| 3 | 31·5 | 32·6 | 33·7 | 34·9 | 36·1 | 37·2 | 38·4 | 39·6 | 40·8 | 42·0 |
| 4 | 43·2 | 44·5 | 45·9 | 47·1 | 48·5 | 49·7 | 51·1 | 52·5 | 54·0 | 55·5 |
| 5 | 57·0 | 58·5 | 60·1 | 61·7 | 63·5 | 65·4 | 67·2 | 69·2 | 71·5 | 73·7 |
| 6 | 76·2 | 79·3 | 82·7 | 87·5 | 97·5 | ... | ... | ... | ... | ... |

**Maximum Equation 6·3767** for Anomaly of 97·50.

Degrees.

| + | ·9 | ·8 | ·7 | ·6 | ·5 | ·4 | ·3 | ·2 | ·1 | ·0 |
|---|---|---|---|---|---|---|---|---|---|---|
| 6 | ... | ... | ... | ... | ... | 97·5 | 105·0 | 109·2 | 112·7 | 115·7 |
| 5 | 117·7 | 120·0 | 121·7 | 123·7 | 125·4 | 127·0 | 128·7 | 130·2 | 131·5 | 132·7 |
| 4 | 134·1 | 135·4 | 136·6 | 137·9 | 139·0 | 140·2 | 141·3 | 142·5 | 143·5 | 144·6 |
| 3 | 145·6 | 146·7 | 147·7 | 148·7 | 149·7 | 150·3 | 151·6 | 152·5 | 153·5 | 154·4 |
| 2 | 155·3 | 156·2 | 157·2 | 158·1 | 158·9 | 159·2 | 160·7 | 161·6 | 162·5 | 163·3 |
| 1 | 164·2 | 165·1 | 165·9 | 166·7 | 167·6 | 168·4 | 169·2 | 170·1 | 170·9 | 171·7 |
| 0 | 172·6 | 173·4 | 174·2 | 175·1 | 175·9 | 176·7 | 177·5 | 178·3 | 179·2 | 180·0 |

Degrees.

| — | ·0 | ·1 | ·2 | ·3 | ·4 | ·5 | ·6 | ·7 | ·8 | ·9 |
|---|---|---|---|---|---|---|---|---|---|---|
| 0 | 180·0 | 180·8 | 181·7 | 182·5 | 183·3 | 184·1 | 184·9 | 185·8 | 186·6 | 187·4 |
| 1 | 188·3 | 189·1 | 189·9 | 190·8 | 191·6 | 192·4 | 193·3 | 194·1 | 194·9 | 195·8 |
| 2 | 196·7 | 197·5 | 198·4 | 199·4 | 200·8 | 201·1 | 201·9 | 202·8 | 203·8 | 204·7 |
| 3 | 205·6 | 206·5 | 207·5 | 208·4 | 209·4 | 210·3 | 211·3 | 212·3 | 213·3 | 214·4 |
| 4 | 215·4 | 216·5 | 217·5 | 218·7 | 219·8 | 221·0 | 222·1 | 223·4 | 224·6 | 225·9 |
| 5 | 227·3 | 228·5 | 229·8 | 231·3 | 233·0 | 234·6 | 236·3 | 238·3 | 240·0 | 242·3 |
| 6 | 244·3 | 247·3 | 250·8 | 255·0 | 262·5 | ... | ... | ... | ... | ... |

**Maximum Equation 6·3767** for Anomaly of 262·50.

Degrees.

| — | ·9 | ·8 | ·7 | ·6 | ·5 | ·4 | ·3 | ·2 | ·1 | ·0 |
|---|---|---|---|---|---|---|---|---|---|---|
| 6 | ... | ... | ... | ... | ... | 262·5 | 272·5 | 277·3 | 280·8 | 283·8 |
| 5 | 286·3 | 288·5 | 290·8 | 292·8 | 294·6 | 296·5 | 298·3 | 299·9 | 301·5 | 303·0 |
| 4 | 304·5 | 306·0 | 307·5 | 308·9 | 310·3 | 311·5 | 312·9 | 314·1 | 315·5 | 316·8 |
| 3 | 318·0 | 319·2 | 320·4 | 321·6 | 322·8 | 323.9 | 325·1 | 326·3 | 327·4 | 328·5 |
| 2 | 329·6 | 330·3 | 331·8 | 332·9 | 334·0 | 335·1 | 336·2 | 337·3 | 338·3 | 339·4 |
| 1 | 340·4 | 341·5 | 342·5 | 343·6 | 344·6 | 345·6 | 346·7 | 347·8 | 348·8 | 349·8 |
| 0 | 350·8 | 351·8 | 352·9 | 353·9 | 354·9 | 355·9 | 356·9 | 358·0 | 359·0 | 360·0 |

### Saturn's Anomaly.

Degrees.

| + | ·0 | ·1 | ·2 | ·3 | ·4 | ·5 | ·6 | ·7 | ·8 | ·9 |
|---|---|---|---|---|---|---|---|---|---|---|
| 0 | 0·0 | 0·7 | 1·5 | 2·2 | 2·9 | 3·7 | 4·4 | 5·2 | 5·9 | 6·7 |
| 1 | 7·4 | 8·1 | 8·9 | 9·6 | 10·4 | 11·2 | 11·9 | 12·7 | 13·4 | 14·2 |
| 2 | 14·9 | 15·7 | 16·5 | 17·2 | 18·1 | 18·8 | 19·6 | 20·4 | 21·2 | 22·0 |
| 3 | 22·8 | 23·7 | 24·5 | 25·3 | 26·1 | 26·9 | 27·8 | 28·7 | 29·5 | 30·3 |
| 4 | 31·2 | 32·1 | 33·0 | 33·9 | 34·8 | 35·7 | 36·7 | 37·6 | 38·5 | 39·5 |
| 5 | 40·5 | 41·5 | 42·5 | 43·5 | 44·6 | 45·6 | 46·7 | 47·9 | 49·0 | 50·1 |
| 6 | 51·4 | 52·5 | 53·9 | 55·1 | 56·5 | 57·9 | 59·2 | 60·9 | 62·5 | 64·1 |
| 7 | 66·0 | 67·7 | 70·0 | 72·2 | 75·0 | 78·2 | 83·0 | 90·0 | | |

**Maximum Equation 7·6586** for Anomaly of 90·00.

Degrees.

| + | ·9 | ·8 | ·7 | ·6 | ·5 | ·4 | ·3 | ·2 | ·1 | ·0 |
|---|---|---|---|---|---|---|---|---|---|---|
| 7 | ... | ... | 90·0 | 97·0 | 101·8 | 105·0 | 107·8 | 110·0 | 112·3 | 114·0 |
| 6 | 115·9 | 117·5 | 119·1 | 120·8 | 122·1 | 123·5 | 124·9 | 126·1 | 127·5 | 128·6 |
| 5 | 129·9 | 131·0 | 132·1 | 133·3 | 134·4 | 135·4 | 136·5 | 137·5 | 138·5 | 139·5 |
| 4 | 140·5 | 141·5 | 142·4 | 143·3 | 144·3 | 145·2 | 146·1 | 147·0 | 147·9 | 148·8 |
| 3 | 149·7 | 150·5 | 151·3 | 152·2 | 153·1 | 153·9 | 154·7 | 155·5 | 156·3 | 157·2 |
| 2 | 158·0 | 158·8 | 159·6 | 160·4 | 161·2 | 161·9 | 162·8 | 163·5 | 164·3 | 165·1 |
| 1 | 165·8 | 166·6 | 167·3 | 168·1 | 168·8 | 169·6 | 170·4 | 171·1 | 171·9 | 172·6 |
| 0 | 173·3 | 174·1 | 174·8 | 175·6 | 176·3 | 177·1 | 177·8 | 178·5 | 179·3 | 180·0 |

Degrees.

| — | ·0 | ·1 | ·2 | ·3 | ·4 | ·5 | ·6 | ·7 | ·8 | ·9 |
|---|---|---|---|---|---|---|---|---|---|---|
| 0 | 180·0 | 180·7 | 181·5 | 182·2 | 182·9 | 183·7 | 184·4 | 185·2 | 185·9 | 186·7 |
| 1 | 187·4 | 188·1 | 188·9 | 189·6 | 190·4 | 191·2 | 191·9 | 192·7 | 193·4 | 194·2 |
| 2 | 194·9 | 195·7 | 196·5 | 197·2 | 198·1 | 198·8 | 199·6 | 200·4 | 201·2 | 202·0 |
| 3 | 202·8 | 203·7 | 204·5 | 205·3 | 206·1 | 206·9 | 207·8 | 208·7 | 209·5 | 210·3 |
| 4 | 211·2 | 212·1 | 213·0 | 213·9 | 214·8 | 215·7 | 216·7 | 217·6 | 218·5 | 219·5 |
| 5 | 220·5 | 221·5 | 222·5 | 223·5 | 224·6 | 225·6 | 226·7 | 227·9 | 229·0 | 230·1 |
| 6 | 231·4 | 232·5 | 233·9 | 235·1 | 236·5 | 237·9 | 239·2 | 240·9 | 242·5 | 244·1 |
| 7 | 246·0 | 247·7 | 250·0 | 252·2 | 255·0 | 258·2 | 263·0 | 270·0 | ... | ... |

**Maximum Equation 7·6586** for Anomaly of 270·00.

Degrees.

| — | ·9 | ·8 | ·7 | ·6 | ·5 | ·4 | ·3 | ·2 | ·1 | ·0 |
|---|---|---|---|---|---|---|---|---|---|---|
| 7 | ... | ... | 270·0 | 277·0 | 281·8 | 285·0 | 287·8 | 290·0 | 292·8 | 294·0 |
| 6 | 295·9 | 297·5 | 299·1 | 300·8 | 302·1 | 303·5 | 304·9 | 306·1 | 307·5 | 308·6 |
| 5 | 309·9 | 311·0 | 312·1 | 313·3 | 314·4 | 315·4 | 316·5 | 317·5 | 318·5 | 319·5 |
| 4 | 320·5 | 321·5 | 322·4 | 323·3 | 324·3 | 325·2 | 326·1 | 327·0 | 327·9 | 328·8 |
| 3 | 329·7 | 330·5 | 331·3 | 332·2 | 333·1 | 333·9 | 334·7 | 335·5 | 336·3 | 337·2 |
| 2 | 338·0 | 338·8 | 339·6 | 340·4 | 341·2 | 341·9 | 342·8 | 343·5 | 344·3 | 345·1 |
| 1 | 345·8 | 346·6 | 347·3 | 348·1 | 348·8 | 349·6 | 350·4 | 351·1 | 351·9 | 352·6 |
| 0 | 353·3 | 354·1 | 354·8 | 355·6 | 356·3 | 357·1 | 357·8 | 358·5 | 359·3 | 360·0 |

## TABLE XVII. A.

### Sun's mean Longitude and Equation for every complete day of Hindu Solar year.

| Day | ☉'s mean longitude. Degrees. | Eqn. | Day | ☉'s mean longitude. Degrees. | Eqn. | Day | ☉'s mean longitude. Degrees. | Eqn. | Day | ☉'s mean longitude. Degrees. | Eqn. | Day | ☉'s mean longitude. Degrees. | Eqn. |
|---|---|---|---|---|---|---|---|---|---|---|---|---|---|---|
| **0** | 357·8606 | +2·1378 | **61** | 57·9823 | +·7267 | **122** | 118·1041 | —1·4333 | **183** | 178·2258 | —2·1367 | **244** | 238·3476 | —·7203 |
| **1** | 358·8462 | 2·1333 | **62** | 58·9679 | ·6919 | **123** | 119·0897 | 1·4586 | **184** | 179·2114 | 2·1305 | **245** | 239·3332 | ·6833 |
| **2** | 359·8318 | 2·1228 | **63** | 59·9535 | ·6561 | **124** | 120·0753 | 1·4883 | **185** | 180·1970 | 2·1219 | **246** | 240·3188 | ·6469 |
| **3** | 0·8174 | 2·1133 | **64** | 60·9391 | ·6192 | **125** | 121·0609 | 1·5133 | **186** | 181·1827 | 2·1133 | **247** | 241·3044 | ·6097 |
| **4** | 1·8030 | 2·1036 | **65** | 61·9247 | ·5847 | **126** | 122·0465 | 1·5405 | **187** | 182·1683 | 2·1061 | **248** | 242·2190 | ·5739 |
| **5** | 2·7886 | 2·0967 | **66** | 62·9103 | ·5480 | **127** | 123·0321 | 1·5667 | **188** | 183·1539 | 2·0964 | **249** | 243·2756 | ·5369 |
| **6** | 3·7742 | 2·0858 | **67** | 63·8959 | ·5111 | **128** | 124·0177 | 1·5919 | **189** | 184·1395 | 2·0853 | **250** | 244·2612 | ·5000 |
| **7** | 4·7598 | 2·0767 | **68** | 64·8815 | ·4739 | **129** | 125·0037 | 1·6167 | **190** | 185·1251 | 2·0753 | **251** | 245·2468 | ·4636 |
| **8** | 5·7454 | 2·0647 | **69** | 65·8671 | ·4369 | **130** | 125·8889 | 1·6433 | **191** | 186·1107 | 2·0642 | **252** | 246·3324 | ·4253 |
| **9** | 6·7310 | 2·0514 | **70** | 66·8527 | ·4000 | **131** | 126·9745 | 1·6667 | **192** | 187·0963 | 2·0519 | **253** | 247·2180 | ·3883 |
| **10** | 7·7166 | 2·0394 | **71** | 67·8383 | ·3622 | **132** | 127·9601 | 1·6894 | **193** | 188·0819 | 2·0383 | **254** | 248·2036 | ·3500 |
| **11** | 8·7022 | 2·0267 | **72** | 68·8239 | ·3192 | **133** | 128·9457 | 1·7133 | **194** | 189·0675 | 2·0261 | **255** | 249·1892 | ·3122 |
| **12** | 9·6878 | 2·0122 | **73** | 69·8096 | ·2869 | **134** | 129·9313 | 1·7358 | **195** | 190·0531 | 2·0125 | **256** | 250·1748 | ·2753 |
| **13** | 10·6734 | 1·9986 | **74** | 70·7952 | ·2486 | **135** | 130·9169 | 1·7572 | **196** | 191·0387 | 1·9978 | **257** | 251·1604 | ·2369 |
| **14** | 11·6590 | 1·9833 | **75** | 71·7808 | ·2167 | **136** | 131·9025 | 1·7789 | **197** | 192·0243 | 1·9819 | **258** | 252·1460 | ·1992 |
| **15** | 12·6446 | 1·9667 | **76** | 72·7664 | ·1717 | **137** | 132·8881 | 1·8000 | **198** | 193·0099 | 1·9661 | **259** | 253·1316 | ·1597 |
| **16** | 13·6302 | 1·9500 | **77** | 73·7520 | ·1333 | **138** | 133·8737 | 1·8217 | **199** | 193·9955 | 1·9514 | **260** | 254·1172 | ·1228 |
| **17** | 14·6158 | 1·9333 | **78** | 74·7376 | ·0975 | **139** | 134·8593 | 1·8419 | **200** | 194·9811 | 1·9333 | **261** | 255·1029 | ·0847 |
| **18** | 15·6014 | 1·9167 | **79** | 75·7232 | ·0600 | **140** | 135·8449 | 1·8597 | **201** | 195·9667 | 1·9161 | **262** | 256·0885 | ·0469 |
| **19** | 16·5870 | 1·9000 | **80** | 76·7088 | ·0203 | **141** | 136·8305 | 1·8789 | **202** | 196·9523 | 1·9008 | **263** | 257·0741 | ·0086 |
| **20** | 17·5726 | 1·8833 | **81** | 77·6944 | —·0167 | **142** | 137·8161 | 1·8980 | **203** | 197·9379 | 1·8808 | **264** | 258·0597 | +·0394 |
| **21** | 18·5582 | 1·8622 | **82** | 78·6800 | ·0550 | **143** | 138·8017 | 1·9167 | **204** | 198·9235 | 1·8600 | **265** | 259·0453 | ·0764 |
| **22** | 19·5438 | 1·8419 | **83** | 79·6656 | ·0933 | **144** | 139·7873 | 1·9333 | **205** | 199·9091 | 1·8417 | **266** | 260·0309 | ·1111 |
| **23** | 20·5294 | 1·8217 | **84** | 80·6512 | ·1314 | **145** | 140·7729 | 1·9500 | **206** | 200·8947 | 1·8222 | **267** | 261·0165 | ·1469 |
| **24** | 21·5150 | 1·8025 | **85** | 81·6368 | ·1692 | **146** | 141·7585 | 1·9667 | **207** | 201·8803 | 1·8014 | **268** | 262·0021 | ·1847 |
| **25** | 22·5006 | 1·7814 | **86** | 82·6224 | ·2072 | **147** | 142·7441 | 1·9825 | **208** | 202·8659 | 1·7805 | **269** | 262·9877 | ·2239 |
| **26** | 23·4862 | 1·7600 | **87** | 83·6080 | ·2444 | **148** | 143·7298 | 1·9980 | **209** | 203·8515 | 1·7575 | **270** | 263·9733 | ·2586 |
| **27** | 24·4718 | 1·7383 | **88** | 84·5936 | ·2833 | **149** | 144·7154 | 2·0122 | **210** | 204·8371 | 1·7369 | **271** | 264·9589 | ·2992 |
| **28** | 25·4574 | 1·7167 | **89** | 85·5792 | ·3203 | **150** | 145·7010 | 2·0253 | **211** | 205·8227 | 1·7139 | **272** | 265·9445 | ·3369 |
| **29** | 26·4430 | 1·6933 | **90** | 86·5648 | ·3572 | **151** | 146·6866 | 2·0369 | **212** | 206·8083 | 1·6894 | **273** | 266·9301 | ·3739 |
| **30** | 27·4286 | 1·6680 | **91** | 87·5504 | ·3955 | **152** | 147·6722 | 2·0492 | **213** | 207·7939 | 1·6664 | **274** | 267·9157 | ·4122 |
| **31** | 28·4142 | 1·6455 | **92** | 88·5360 | ·4333 | **153** | 148·6578 | 2·0611 | **214** | 208·7795 | 1·6433 | **275** | 268·9013 | ·4500 |
| **32** | 29·3998 | 1·6192 | **93** | 89·5216 | ·4692 | **154** | 149·6434 | 2·0739 | **215** | 209·7651 | 1.6178 | **276** | 269·8869 | ·4869 |
| **33** | 30·3854 | 1·5955 | **94** | 90·5072 | ·5061 | **155** | 150·6290 | 2·0858 | **216** | 210·7507 | 1·5922 | **277** | 270·8725 | ·5253 |
| **34** | 31·3710 | 1·5692 | **95** | 91·4928 | ·5433 | **156** | 151·6146 | 2·0955 | **217** | 211·7363 | 1·5667 | **278** | 271·8581 | ·5611 |
| **35** | 32·3567 | 1·5455 | **96** | 92·4784 | ·5789 | **157** | 152·6002 | 2·1036 | **218** | 212·7219 | 1·5411 | **279** | 272·8437 | [illegible] |
| **36** | 33·3423 | 1·5167 | **97** | 93·4640 | ·6167 | **158** | 153·5858 | 2·1136 | **219** | 213·7105 | 1·5130 | **280** | 278·8293 | ·6347 |
| **37** | 34·3279 | 1·4914 | **98** | 94·4496 | ·6525 | **159** | 154·5714 | 2·1228 | **220** | 214·6931 | 1·4869 | **281** | 274·8149 | ·6692 |
| **38** | 35·3135 | 1·4633 | **99** | 95·4352 | ·6894 | **160** | 155·5570 | 2·1314 | **221** | 215·6787 | 1·4589 | **282** | 275·8005 | ·7061 |
| **39** | 36·2991 | 1·4358 | **100** | 96·4208 | ·7228 | **161** | 156·5426 | 2·1383 | **222** | 216·6643 | 1·4308 | **283** | 276·7861 | ·7419 |
| **40** | 37·2847 | 1·4061 | **101** | 97·4064 | ·7586 | **162** | 157·5282 | 2·1438 | **223** | 217·6499 | 1·4028 | **284** | 277·7717 | ·7764 |
| **41** | 38·2703 | 1·3778 | **102** | 98·3920 | ·7944 | **163** | 158·5138 | 2·1494 | **224** | 218·6356 | 1·3733 | **285** | 278·7573 | ·8122 |
| **42** | 39·2559 | 1·3480 | **103** | 99·3776 | ·8289 | **164** | 159·4994 | 2·1567 | **225** | 219·6212 | 1·3439 | **286** | 279·7429 | ·8469 |
| **43** | 40·2415 | 1·3180 | **104** | 100·3632 | ·8633 | **165** | 160·4850 | 2·1603 | **226** | 220·6068 | 1·3144 | **287** | 280·7285 | ·8808 |
| **44** | 41·2271 | 1·2894 | **105** | 101·3488 | ·8986 | **166** | 161·4706 | 2·1617 | **227** | 221·5924 | 1·2828 | **288** | 281·7141 | ·9147 |
| **45** | 42·2127 | 1·2586 | **106** | 102·3344 | ·9333 | **167** | 162·4562 | 2·1667 | **228** | 222·5780 | 1·2533 | **289** | 282·6997 | ·9500 |
| **46** | 43·1983 | 1·2197 | **107** | 103·3200 | ·9667 | **168** | 163·4418 | 2·1692 | **229** | 223·5636 | 1·2228 | **290** | 283·6853 | ·9833 |
| **47** | 44·1839 | 1·1967 | **108** | 104·3056 | 1·0009 | **169** | 164·4274 | 2·1728 | **230** | 224·5492 | 1·1905 | **291** | 284·6709 | 1·0192 |
| **48** | 45·1695 | 1·1647 | **109** | 105·2912 | 1·0333 | **170** | 165·4130 | 2·1742 | **231** | 225·5348 | 1·1589 | **292** | 285·6565 | 1·0508 |
| **49** | 46·1551 | 1·1347 | **110** | 106·2760 | 1·0667 | **171** | 166·3986 | 2·1755 | **232** | 226·5204 | 1·1283 | **293** | 286·6421 | 1·0850 |
| **50** | 47·1407 | 1·1014 | **111** | 107·2625 | 1·1000 | **172** | 167·3842 | 2·1769 | **233** | 227·5060 | 1·0967 | **294** | 287·6277 | 1·1167 |
| **51** | 48·1263 | 1·0703 | **112** | 108·2481 | 1·1308 | **173** | 168·3698 | 2·1755 | **234** | 228·4916 | 1·0614 | **295** | 288·6133 | 1·1483 |
| **52** | 49·1119 | 1·0369 | **113** | 109·2337 | 1·1622 | **174** | 169·3554 | 2·1742 | **235** | 229·4772 | 1·0283 | **296** | 289·5984 | 1·1814 |
| **53** | 50·0975 | 1·0025 | **114** | 110·2193 | 1·1955 | **175** | 170·3410 | 2·1728 | **236** | 230·4628 | ·9953 | **297** | 290·5845 | 1·2130 |
| **54** | 51·0831 | ·9703 | **115** | 111·2049 | 1·2253 | **176** | 171·3266 | 2·1714 | **237** | 231·4484 | ·9622 | **298** | 291·5701 | 1·2439 |
| **55** | 52·0687 | ·9358 | **116** | 112·1905 | 1·2561 | **177** | 172·3122 | 2·1678 | **238** | 232·4340 | ·9280 | **299** | 292·5558 | 1·2747 |
| **56** | 53·0543 | ·9025 | **117** | 113·1761 | 1·2858 | **178** | 173·2978 | 2·1642 | **239** | 233·4196 | ·8928 | **300** | 293·5414 | 1·3086 |
| **57** | 54·0399 | ·8667 | **118** | 114·1617 | 1·3167 | **179** | 174·2834 | 2·1597 | **240** | 234·4052 | ·8597 | **301** | 294·5270 | 1·3322 |
| **58** | 55·0955 | ·8333 | **119** | 115·1473 | 1·3455 | **180** | 175·2690 | 2·1561 | **241** | 235·3908 | ·8228 | **302** | 295·5126 | 1·3622 |
| **59** | 56·0111 | ·7964 | **120** | 116·1329 | 1·3739 | **181** | 176·2546 | 2·1489 | **242** | 236·3764 | ·7894 | **303** | 296·4982 | 1·3903 |
| **60** | 56·9967 | ·7622 | **121** | 117·1185 | 1·4025 | **182** | 177·2402 | 2·1439 | **243** | 237·3620 | ·7536 | **304** | 297·4838 | 1·4197 |

**Tab. XVII. A.**- *contd.*

| Day | ☉'s mean longitude. Degrees. | Eqn. |
|---|---|---|
| 305 | 298·4694 | +1·4492 |
| 306 | 299·4550 | 1·4747 |
| 307 | 300·4406 | 1·5042 |
| 308 | 301·4262 | 1·5297 |
| 309 | 302·4118 | 1·5567 |
| 310 | 303·3974 | 1·5833 |
| 311 | 304·3830 | 1·6089 |
| 312 | 305·3686 | 1·6344 |
| 313 | 306·3542 | 1·6589 |
| 314 | 307·3398 | 1·6819 |
| 315 | 308·3254 | 1·7050 |
| 316 | 309·3110 | 1·7280 |
| 317 | 310·2966 | 1·7497 |
| 318 | 311·2822 | 1·7728 |
| 319 | 312·2678 | 1·7922 |
| 320 | 313·2534 | 1·8153 |
| 321 | 314·2390 | 1·8347 |
| 322 | 315·2226 | 1·8530 |
| 323 | 316·2102 | 1·8725 |
| 324 | 317·1958 | 1·8917 |
| 325 | 318·1814 | 1·9111 |
| 326 | 319·1670 | 1·9269 |
| 327 | 320·1526 | 1·6453 |
| 328 | 321·1382 | 1·9611 |
| 329 | 322·1238 | 1·9758 |
| 330 | 323·1094 | 1·9917 |
| 331 | 324·0950 | 2·0064 |
| 332 | 325·0806 | 2·0211 |
| 333 | 326·0662 | 2·0347 |
| 334 | 327·0518 | 2·0483 |
| 335 | 328·0370 | 2·0597 |
| 336 | 329·0231 | 2·0708 |
| 337 | 330·0087 | 2·0819 |
| 338 | 330·9943 | 2·0930 |
| 339 | 331·9799 | 2·1017 |
| 340 | 332·9655 | 2·1114 |
| 341 | 333·9511 | 2·1211 |
| 342 | 334·9367 | 2·1283 |
| 343 | 335·9223 | 2·1369 |
| 344 | 336·9079 | 2·1430 |
| 345 | 337·8935 | 2·1480 |
| 346 | 338·8791 | 2·1542 |
| 347 | 339·8647 | 2·1592 |
| 348 | 340·8503 | 2·1628 |
| 349 | 341·8359 | 2·1653 |
| 350 | 342·8215 | 2·1689 |
| 351 | 343·8071 | 2·1708 |
| 352 | 344·7927 | 2·1717 |
| 353 | 345·7783 | 2·1730 |
| 354 | 346·7639 | 2·1755 |
| 355 | 347·7495 | 2·1764 |
| 356 | 348·7351 | 2·1739 |
| 357 | 349·7207 | 2·1714 |
| 358 | 350·7063 | 2·1700 |
| 359 | 351·6919 | 2·1686 |
| 360 | 352·6775 | 2·1650 |
| 361 | 353·6631 | 2·1625 |
| 362 | 354·6487 | 2·1589 |
| 363 | 355·6343 | 2·1528 |
| 364 | 356·6199 | 2·1467 |
| 365 | 357·6055 | 2·1405 |

# TABLE XVII B.

## RAHU.

### Mean place of Rahu for centuries, odd years, days and fractions of days.

#### Mean place of Rahu for Centuries.

| | K.Y. | Deg. | A.D. | K.Y. | Deg. | A.D. | K.Y. | Deg. |
|---|---|---|---|---|---|---|---|---|
| 1 B.C. | 3101 | 285·2370 | 700 | 3801 | 58·2280 | 1400 | 4501 | 191·2191 |
| 100 A.D. | 3201 | 149·9500 | 800 | 3901 | 282·9410 | 1500 | 4601 | 55·9321 |
| 200 | 3301 | 14·6630 | 900 | 4001 | 147·6540 | 1600 | 4701 | 280·6451 |
| 300 | 3401 | 239·3760 | 1000 | 4101 | 12·3670 | 1700 | 4801 | 145·3581 |
| 400 | 3501 | 104·0890 | 1100 | 4201 | 237·0800 | 1800 | 4901 | 10·0711 |
| 500 | 3601 | 328·8020 | 1200 | 4301 | 101·7931 | 1900 | 5001 | 234·7841 |
| 600 | 3701 | 193·5150 | 1300 | 4401 | 326·5061 | 2000 | 5101 | 99·4971 |

#### Mean place of Rahu for Odd Years.

| Yrs. | Deg. | Yrs. | Deg. | Yrs. | Deg. | Yrs. | Deg. | Yrs. | Deg. | Yrs. | Deg. | Yrs. | Deg. |
|---|---|---|---|---|---|---|---|---|---|---|---|---|---|
| 1 | 19·3529 | 18 | 348·3517 | 35 | 317·3504 | 52 | 286·3492 | 69 | 255·3480 | 86 | 224·3468 | 100 | 135·2870 |
| 2 | 38·7057 | 19 | 7·7045 | 36 | 336·7033 | 53 | 305·7021 | 70 | 274·7009 | 87 | 243·6997 | 200 | 270·5740 |
| 3 | 58·0586 | 20 | 27·0574 | 37 | 356·0562 | 54 | 325·0550 | 71 | 294·0538 | 88 | 263·0525 | 300 | 45·8610 |
| 4 | 77·4115 | 21 | 46·4103 | 38 | 15·4091 | 55 | 344·4078 | 72 | 313·4066 | 89 | 282·4054 | 400 | 181·1480 |
| 5 | 96·7643 | 22 | 65·7631 | 39 | 34·7619 | 56 | 3·7607 | 73 | 332·7595 | 90 | 301·7583 | 500 | 316·4350 |
| 6 | 116·1172 | 23 | 85·1160 | 40 | 54·1148 | 57 | 23·1136 | 74 | 352·1124 | 91 | 321·1112 | 600 | 91·7220 |
| 7 | 135·4701 | 24 | 104·4689 | 41 | 73·4677 | 58 | 42·4664 | 75 | 11·4652 | 92 | 340·4640 | 700 | 227·0090 |
| 8 | 154·8229 | 25 | 123·8217 | 42 | 92·8205 | 59 | 61·8193 | 76 | 30·8181 | 93 | 359·8169 | 800 | 2·2960 |
| 9 | 174·1758 | 26 | 143·1746 | 43 | 112·1734 | 60 | 81·1722 | 77 | 50·1710 | 94 | 19·1698 | 900 | 137·5830 |
| 10 | 193·5287 | 27 | 162·5275 | 44 | 131·5263 | 61 | 100·5251 | 78 | 69·5238 | 95 | 38·5226 | 1000 | 272·8700 |
| 11 | 212·8816 | 28 | 181·8803 | 45 | 150·8791 | 62 | 119·8779 | 79 | 88·8767 | 96 | 57·8755 | 2000 | 185·7400 |
| 12 | 232·2344 | 29 | 201·2332 | 46 | 170·2320 | 63 | 139·2308 | 80 | 108·2296 | 97 | 77·2284 | 3000 | 98·6098 |
| 13 | 251·5873 | 30 | 220·5861 | 47 | 189·5849 | 64 | 158·5837 | 81 | 127·5825 | 98 | 96·5812 | 4000 | 11·4798 |
| 14 | 270·9402 | 31 | 239·9390 | 48 | 208·9377 | 65 | 177·9365 | 82 | 146·9353 | 99 | 115·9341 | 5000 | 284·3497 |
| 15 | 290·2930 | 32 | 259·2918 | 49 | 228·2906 | 66 | 197·2894 | 83 | 166·2882 | ... | ... | ... | ... |
| 16 | 309·6459 | 33 | 278·6447 | 50 | 247·6435 | 67 | 216·6423 | 84 | 185·6411 | ... | ... | ... | ... |
| 17 | 328·9988 | 34 | 297·9976 | 51 | 266·9964 | 68 | 235·9951 | 85 | 204·9939 | ... | ... | ... | ... |

#### Mean place of Rahu at Ujjain Sunrise on 0 day of each Solar Year.

#### A.D. 1840 to A.D. 1920.

| A.D. | Deg. | A.D. | Deg. | A.D. | Deg. | A.D. | Deg. | A.D. | Deg. | A.D. | Deg. |
|---|---|---|---|---|---|---|---|---|---|---|---|
| 1840 | 315·9563 | 1854 | 45·0161 | 1868 | 134·0759 | 1882 | 223·1357 | 1896 | 312·1956 | 1910 | 41·2554 |
| 1841 | 296·6034 | 1855 | 25·6632 | 1869 | 114·7231 | 1883 | 203·7829 | 1897 | 292·8427 | 1911 | 21·9025 |
| 1842 | 277·2505 | 1856 | 6·3104 | 1870 | 95·3702 | 1884 | 184·4300 | 1898 | 273·4898 | 1912 | 2·5496 |
| 1843 | 257·8977 | 1857 | 346·9575 | 1871 | 76·0173 | 1885 | 165·0771 | 1899 | 254·1370 | 1913 | 343·1968 |
| 1844 | 238·5448 | 1858 | 327·6046 | 1872 | 56·6644 | 1886 | 145·7243 | 1900 | 234·7841 | 1914 | 323·8439 |
| 1845 | 219·1919 | 1859 | 308·2518 | 1873 | 37·3116 | 1887 | 126·3714 | 1901 | 215·4312 | 1915 | 304·4910 |
| 1846 | 199·8391 | 1860 | 288·8989 | 1874 | 17·9587 | 1888 | 107·0185 | 1902 | 196·0783 | 1916 | 285·1382 |
| 1847 | 180·4862 | 1861 | 269·5460 | 1875 | 358·6058 | 1889 | 87·6657 | 1903 | 176·7255 | 1917 | 265·7853 |
| 1848 | 161·1333 | 1862 | 250·1931 | 1876 | 339·2530 | 1890 | 68·3128 | 1904 | 157·3726 | 1918 | 246·4324 |
| 1849 | 141·7805 | 1863 | 230·8403 | 1877 | 319·9001 | 1891 | 48·9599 | 1905 | 138·0197 | 1919 | 227·0795 |
| 1850 | 122·4276 | 1864 | 211·4874 | 1878 | 300·5472 | 1892 | 29·6070 | 1906 | 118·6669 | ... | ... |
| 1851 | 103·0747 | 1865 | 192·1345 | 1879 | 281·1944 | 1893 | 10·2542 | 1907 | 99·3140 | ... | ... |
| 1852 | 83·7218 | 1866 | 172·7817 | 1880 | 261·8415 | 1894 | 350·9013 | 1908 | 79·9611 | ... | ... |
| 1853 | 64·3690 | 1867 | 153·4288 | 1881 | 242·4886 | 1895 | 331·5484 | 1909 | 60·6083 | ... | ... |

#### Mean place of Rahu for every 10 days and for fractions of a Day.

| Day. | Deg. | Days. | Deg. | Days. | Deg. | Days. | Deg. | Days. | Deg. | Days. | Deg. |
|---|---|---|---|---|---|---|---|---|---|---|---|
| ·1 | ·0053 | 1 | ·0530 | 10 | ·5298 | 100 | 5·2984 | 190 | 10·0669 | 280 | 14·8355 |
| ·2 | ·0106 | 2 | ·1060 | 20 | 1·0597 | 110 | 5·8282 | 200 | 10·5968 | 290 | 15·3654 |
| ·3 | ·0159 | 3 | ·1589 | 30 | 1·5895 | 120 | 6·3581 | 210 | 11·1266 | 300 | 15·8952 |
| ·4 | ·0212 | 4 | ·2119 | 40 | 2·6194 | 130 | 6·8879 | 220 | 11·6565 | 310 | 16·4250 |
| ·5 | ·0265 | 5 | ·2649 | 50 | 2·6492 | 140 | 7·4177 | 230 | 12·1863 | 320 | 16·9549 |
| ·6 | ·0318 | 6 | ·3179 | 60 | 3·1790 | 150 | 7·9476 | 240 | 12·7162 | 330 | 17·4847 |
| ·7 | ·0371 | 7 | ·3709 | 70 | 3·7089 | 160 | 8·4774 | 250 | 13·2460 | 340 | 18·0145 |
| ·8 | ·0424 | 8 | ·4239 | 80 | 4·2387 | 170 | 9·0073 | 260 | 13·7758 | 350 | 18·5444 |
| ·9 | ·0477 | 9 | ·4768 | 90 | 4·7686 | 180 | 9·5371 | 270 | 14·3057 | 360 | 19·0742 |

## TABLE XVII—C.

Decimal parts of ⊙'s L. for one day.

| Day. | Degree. | Day. | Degree. |
|---|---|---|---|
| ·01 | ·0099 | ·51 | ·5027 |
| ·02 | ·0197 | ·52 | ·5125 |
| ·03 | ·0296 | ·53 | ·5224 |
| ·04 | ·0394 | ·54 | ·5322 |
| ·05 | ·0493 | ·55 | ·5421 |
| ·06 | ·0591 | ·56 | ·5519 |
| ·07 | ·0690 | ·57 | ·5618 |
| ·08 | ·0788 | ·58 | ·5716 |
| ·09 | ·0887 | ·59 | ·5815 |
| ·10 | ·0986 | ·60 | ·5914 |
| ·11 | ·1084 | ·61 | ·6012 |
| ·12 | ·1183 | ·62 | ·6111 |
| ·13 | ·1281 | ·63 | ·6209 |
| ·14 | ·1380 | ·64 | ·6308 |
| ·15 | ·1478 | ·65 | ·6406 |
| ·16 | ·1577 | ·66 | 6505 |
| ·17 | ·1675 | ·67 | ·6603 |
| ·18 | ·1774 | ·68 | ·6702 |
| ·19 | ·1873 | ·69 | ·6801 |
| ·20 | ·1971 | ·70 | ·6899 |
| ·21 | ·2070 | ·71 | ·6998 |
| ·22 | ·2168 | ·72 | ·7096 |
| ·23 | ·2267 | ·73 | ·7195 |
| ·24 | ·2365 | ·74 | ·7293 |
| ·25 | ·2464 | ·75 | ·7392 |
| ·26 | ·2563 | ·76 | ·7491 |
| ·27 | ·2661 | ·77 | ·7589 |
| ·28 | ·2760 | ·78 | ·7688 |
| ·29 | ·2858 | ·79 | ·7786 |
| ·30 | ·2957 | ·80 | ·7885 |
| ·31 | ·3055 | ·81 | ·7983 |
| ·32 | ·3154 | ·82 | ·8082 |
| ·33 | ·3252 | ·83 | ·8180 |
| ·34 | ·3351 | ·84 | ·8279 |
| ·35 | ·3450 | ·85 | ·8378 |
| ·36 | ·3548 | ·86 | ·8476 |
| ·37 | ·3647 | ·87 | ·8575 |
| ·38 | ·3745 | ·88 | ·8673 |
| ·39 | ·3844 | ·89 | ·8772 |
| ·40 | ·3942 | ·90 | ·8870 |
| ·41 | ·4041 | ·91 | ·8969 |
| ·42 | ·4139 | ·92 | ·9067 |
| ·43 | ·4238 | ·93 | ·9166 |
| ·44 | ·4337 | ·94 | ·9265 |
| ·45 | ·4435 | ·95 | ·9363 |
| ·46 | ·4534 | ·96 | ·9462 |
| ·47 | .4632 | ·97 | ·9560 |
| ·48 | ·4731 | ·98 | ·9659 |
| ·49 | ·4829 | ·99 | ·9757 |
| ·50 | ·4928 | 1·00 | ·9856 |

## TABLE XVII—D.

Decimal parts of increase of ☾'s L. for one nakshatra.

| | Degrees. | | Degrees. |
|---|---|---|---|
| ·01 | ·1333 | ·51 | 6·8000 |
| ·02 | ·2666 | ·52 | 6·9333 |
| ·03 | ·4000 | ·53 | 7·0666 |
| ·04 | ·5333 | ·54 | 7·2000 |
| ·05 | ·6666 | ·55 | 7·3333 |
| ·06 | ·8000 | ·56 | 7·4666 |
| ·07 | ·9333 | ·57 | 7·6000 |
| ·08 | 1·0666 | ·58 | 7·7333 |
| ·09 | 1·2000 | ·59 | 7·8666 |
| ·10 | 1·3333 | ·60 | 8·0000 |
| ·11 | 1·4666 | ·61 | 8·1333 |
| ·12 | 1·6000 | ·62 | 8·2666 |
| ·13 | 1·7333 | ·63 | 8·4000 |
| ·14 | 1·8666 | ·64 | 8·5333 |
| ·15 | 2·0000 | ·65 | 8·6666 |
| ·16 | 2·1333 | ·66 | 8·8000 |
| ·17 | 2·2666 | ·67 | 8·9333 |
| ·18 | 2·4000 | ·68 | 9·0666 |
| ·19 | 2·5333 | ·69 | 9·2000 |
| ·20 | 2·6666 | ·70 | 9·3333 |
| ·21 | 2·8000 | ·71 | 9·4666 |
| ·22 | 2·9333 | ·72 | 9·6000 |
| ·23 | 3·0666 | ·73 | 9·7333 |
| ·24 | 3·2000 | ·74 | 9·8666 |
| ·25 | 3·3333 | ·75 | 10·0000 |
| ·26 | 3·4666 | ·76 | 10·1333 |
| ·27 | 3·6000 | ·77 | 10·2666 |
| ·28 | 2·7333 | ·78 | 10·4000 |
| ·29 | 3·8666 | ·79 | 10·5333 |
| ·30 | 4·0000 | ·80 | 10·6666 |
| ·31 | 4·1333 | ·81 | 10·8000 |
| ·32 | 4·2666 | ·82 | 10·9333 |
| ·33 | 4·4000 | ·83 | 11·0666 |
| ·34 | 4·5333 | ·84 | 11·2000 |
| ·35 | 4·6666 | ·85 | 11·3333 |
| ·36 | 4·8000 | ·86 | 11·4666 |
| ·37 | 4·9333 | ·87 | 11·6000 |
| ·38 | 5·0666 | ·88 | 11·7333 |
| ·39 | 5·2000 | ·89 | 11·8666 |
| ·40 | 5·3333 | ·90 | 12·0000 |
| ·41 | 5·4666 | ·91 | 12·1333 |
| ·42 | 5·6000 | ·92 | 12·2666 |
| ·43 | 5·7333 | ·93 | 12·4000 |
| ·44 | 5·8666 | ·94 | 12·5333 |
| ·45 | 6·0000 | ·95 | 12·6666 |
| ·46 | 6·1333 | ·96 | 12·8000 |
| ·47 | 6·2666 | ·97 | 12·9333 |
| ·48 | 6·4000 | ·98 | 13·0666 |
| ·49 | 6·5333 | ·99 | 13·2000 |
| ·50 | 6·6666 | 1·00 | 13·3333 |

## TABLE XVII—E.

Eqn. in minutes of a degree for ascertaining the latitude of a planet.

| Degree. | Mar. | Mer. | Jup. | Ven. | Sat. | Degree. |
|---|---|---|---|---|---|---|
| 0·00 | 5682 | 4708 | 4107 | 5940 | 3811 | 360·00 |
| 3·75 | 5678 | 4706 | 4107 | 5936 | 3810 | 356·25 |
| 7·50 | 5667 | 4699 | 4104 | 5925 | 3809 | 352·50 |
| 11·25 | 5651 | 4688 | 4099 | 5909 | 3806 | 348·75 |
| 15·00 | 5629 | 4674 | 4092 | 5886 | 3801 | 345·00 |
| 18·75 | 5601 | 4656 | 4083 | 5857 | 3795 | 341·25 |
| 22·50 | 5567 | 4634 | 4071 | 5822 | 3788 | 337·50 |
| 26·25 | 5528 | 4608 | 4056 | 5781 | 3779 | 333·75 |
| 30·00 | 5484 | 4578 | 4040 | 5734 | 3769 | 330·00 |
| 33·75 | 5431 | 4545 | 4021 | 5681 | 3758 | 326·25 |
| 37·50 | 5379 | 4507 | 3999 | 5622 | 3745 | 322·50 |
| 41·25 | 5318 | 4467 | 3976 | 5558 | 3731 | 318·75 |
| 45·00 | 5252 | 4423 | 3951 | 5487 | 3716 | 315·00 |
| 48·75 | 5181 | 4375 | 3922 | 5412 | 3699 | 311·25 |
| 52·50 | 5104 | 4324 | 3892 | 5331 | 3682 | 307·50 |
| 56·25 | 5023 | 4270 | 3860 | 5244 | 3663 | 303·75 |
| 60·00 | 4937 | 4213 | 3827 | 5152 | 3643 | 300·00 |
| 63·75 | 4846 | 4153 | 3791 | 5055 | 3623 | 296·25 |
| 67·50 | 4751 | 4090 | 3754 | 4953 | 3601 | 292·50 |
| 71·25 | 4651 | 4025 | 3716 | 4846 | 3579 | 288·75 |
| 75·00 | 4547 | 3957 | 3676 | 4734 | 3556 | 285·00 |
| 78·75 | 4429 | 3836 | 3635 | 4617 | 3534 | 281·25 |
| 82·50 | 4327 | 3813 | 3593 | 4496 | 3509 | 277·50 |
| 86·25 | 4210 | 3738 | 3550 | 4371 | 3484 | 273·75 |
| 90·00 | 4090 | 3662 | 3506 | 4241 | 3459 | 270·00 |
| 93·75 | 3966 | 3584 | 3461 | 4107 | 3434 | 266·25 |
| 97·50 | 3839 | 3504 | 3417 | 3970 | 3408 | 262·50 |
| 101·25 | 3708 | 3423 | 3372 | 3828 | 3384 | 258·75 |
| 105·00 | 3575 | 3341 | 3326 | 3684 | 3359 | 255·00 |
| 108·75 | 3410 | 3259 | 3282 | 3536 | 3335 | 251·25 |
| 112·50 | 3303 | 3177 | 3238 | 3385 | 3313 | 247·50 |
| 116·25 | 3162 | 3095 | 3195 | 3231 | 3289 | 243·75 |
| 120·00 | 3019 | 3012 | 3152 | 3075 | 3265 | 240·00 |
| 123·75 | 2875 | 2930 | 3110 | 2916 | 3242 | 236·25 |
| 127·50 | 2730 | 2851 | 3070 | 2756 | 3221 | 232·50 |
| 131·25 | 2585 | 2773 | 3032 | 2594 | 3200 | 228·75 |
| 135·00 | 2440 | 2697 | 2995 | 2432 | 3181 | 225·00 |
| 138·75 | 2295 | 2624 | 2961 | 2269 | 3162 | 221·25 |
| 142·50 | 2152 | 2554 | 2928 | 2106 | 3146 | 217·50 |
| 146·25 | 2011 | 2487 | 2898 | 1945 | 3131 | 213·75 |
| 150·00 | 1874 | 2426 | 2871 | 1786 | 3117 | 210·00 |
| 153·75 | 1743 | 2370 | 2847 | 1630 | 3104 | 206·25 |
| 157·50 | 1618 | 2319 | 2825 | 1481 | 3092 | 202·50 |
| 161·25 | 1503 | 2275 | 2807 | 1341 | 3083 | 198·75 |
| 165·00 | 1402 | 2238 | 2798 | 1213 | 3077 | 195·00 |
| 168·75 | 1317 | 2208 | 2782 | 1101 | 3072 | 191·25 |
| 172·50 | 1251 | 2186 | 2774 | 1013 | 3068 | 187·50 |
| 176·25 | 1209 | 2173 | 2770 | 956 | 3066 | 183·75 |
| 180·00 | 1194 | 2168 | 2769 | 936 | 3065 | 180·00 |

In applying the above equations the following data, giving the inclination of the moon's orbit to the ecliptic and the greatest apparent latitude of the other planets have to be remembered :—

| | | | | | |
|---|---|---|---|---|---|
| Inclination of ☾'s orbit | | | 4° 30′, | Sine | 270′ |
| Greatest latitude | | Mars | 1° 30′ | „ | 90′ |
| „ | „ | Mercury | 2° 0′ | „ | 120′ |
| „ | „ | Jupiter | 1° 16′ | „ | 78′ |
| „ | „ | Venus | 2° 16′ | „ | 138′ |
| „ | „ | Saturn | 2° 10′ | „ | 132′ |

## TABLE XVII.—F.

Table of Hindu Sines.

| Degree. + | Sine. | Degree. — | Diff. + |
|---|---|---|---|
| 0·00 | 000 | 180·00 | ... |
| 3·75 | 225 | 183·75 | 225 |
| 7·50 | 449 | 187·50 | 224 |
| 11·25 | 671 | 191·25 | 222 |
| 15·00 | 890 | 195·00 | 219 |
| 18·75 | 1105 | 198·75 | 215 |
| 22·50 | 1315 | 202·50 | 210 |
| 26·25 | 1520 | 206·25 | 205 |
| 30·00 | 1719 | 210·00 | 199 |
| 33·75 | 1910 | 213·75 | 191 |
| 37·50 | 2093 | 217·50 | 183 |
| 41·25 | 2267 | 221·25 | 174 |
| 45·00 | 2431 | 225·00 | 164 |
| 48·75 | 2585 | 228·75 | 154 |
| 52·50 | 2728 | 232·50 | 143 |
| 56·25 | 2859 | 236·25 | 131 |
| 60·00 | 2978 | 240·00 | 119 |
| 63·75 | 3084 | 243·75 | 106 |
| 67·50 | 3177 | 247·50 | 93 |
| 71·25 | 3256 | 251·25 | 79 |
| 75·00 | 3321 | 255·00 | 65 |
| 78·75 | 3372 | 258·75 | 51 |
| 82·50 | 3409 | 262·50 | 37 |
| 86·25 | 3431 | 266·25 | 22 |
| 90·00 | 3438 | 270·00 | 7 |
| | | | — |
| 93·75 | 3431 | 273·75 | 7 |
| 97·50 | 3409 | 277·50 | 22 |
| 101·25 | 3372 | 281·25 | 37 |
| 105·00 | 3321 | 285·00 | 51 |
| 108·75 | 3256 | 288·75 | 65 |
| 112·50 | 3177 | 292·50 | 79 |
| 116·25 | 3084 | 296·25 | 93 |
| 120·00 | 2978 | 300·00 | 106 |
| 123·75 | 2859 | 303·75 | 119 |
| 127·50 | 2728 | 307·50 | 131 |
| 131·25 | 2585 | 311·25 | 143 |
| 135·00 | 2431 | 315·00 | 154 |
| 138·75 | 2267 | 318·75 | 164 |
| 142·50 | 2093 | 322·50 | 174 |
| 146·25 | 1910 | 326·25 | 183 |
| 150·00 | 1719 | 330·00 | 191 |
| 153·75 | 1520 | 333·75 | 199 |
| 157·50 | 1315 | 337·50 | 205 |
| 161·25 | 1105 | 341·25 | 210 |
| 165·00 | 890 | 345·00 | 215 |
| 168·75 | 671 | 348·75 | 219 |
| 172·50 | 449 | 352·50 | 222 |
| 176·25 | 225 | 356·25 | 224 |
| 180·00 | 000 | 360·00 | 225 |

## TABLE XVII.—G.

**Decimal parts of 3·75 degrees [Example 3·2°=·853 of 3·75°].**

| | | | | | | | | | | | | | | | |
|---|---|---|---|---|---|---|---|---|---|---|---|---|---|---|---|
| 0·1 | ·026666 | 0·6 | ·160000 | 1·1 | ·293333 | 1·6 | ·426666 | 2·1 | ·560000 | 2·6 | ·693333 | 3·1 | ·826666 | 3·6 | ·960000 |
| 0·2 | ·053333 | 0·7 | ·186666 | 1·2 | ·320000 | 1·7 | ·453333 | 2·2 | ·586666 | 2·7 | ·720000 | 3·2 | ·853333 | 3·7 | ·986666 |
| 0·3 | ·080000 | 0·8 | ·213333 | 1·3 | ·346666 | 1·8 | ·480000 | 2·3 | ·613333 | 2·8 | ·746666 | 3·3 | ·880000 | | |
| 0·4 | ·106666 | 0·9 | ·240000 | 1·4 | ·373333 | 1·9 | ·506666 | 2·4 | ·640000 | 2·9 | ·773333 | 3·4 | ·906666 | | |
| 0·5 | ·133333 | 1·0 | ·266666 | 1·5 | ·400000 | 2·0 | ·533333 | 2·5 | ·666666 | 3·0 | ·800000 | 3·5 | ·933333 | | |

## Geocentric (Indian) Longitudes of Sun and Major planets at Ujjain

| | Day of S. Yr. | 0 | 10 | 20 | 30 | 40 | 50 | 60 | 70 | 80 | 90 | 100 | 110 | 120 | 130 | 140 | 150 | 160 | 170 |
|---|---|---|---|---|---|---|---|---|---|---|---|---|---|---|---|---|---|---|---|
| | ⊙'s Mean L. | 357·0 | 6·9 | 16·7 | 26·6 | 36·4 | 46·3 | 56·1 | 66·0 | 75·9 | 85·7 | 95·6 | 105·4 | 115·3 | 125·1 | 135·0 | 144·8 | 154·7 | 164·6 |
| | Corrn. for do. | +2·1 | +2·0 | +1·9 | +1·7 | +1·4 | +1·1 | +0·8 | +0·4 | +0·0 | −0·3 | −0·7 | −1·1 | −1·4 | −1·6 | −1·9 | −2·0 | −2·1 | −2·2 |
| | English date. | Ap 11 | Ap 21 | My 1 | My 11 | My 21 | My 31 | Je 10 | Je 20 | Je 30 | Jl 10 | Jl 20 | Jl 30 | Au 9 | Au 19 | Au 29 | S 8 | S 18 | S 28 |
| | Tamil month. | Chit | Chit | Chit | Chit | Vaik | Vaik | Vaik | Ani | Ani | Ani | Adi | Adi | Adi | Avan | Avan | Avan | Pur | Pur |
| | Do. date. | 1 | 11 | 21 | 31 | 10 | 20 | 30 | 9 | 19 | 29 | 7 | 17 | 27 | 6 | 16 | 26 | 5 | 15 |
| 1840 | Mars ... | 5·0 | 12·6 | 20·0 | 27·2 | 34·5 | 41·7 | 48·8 | 55·7 | 62·5 | 69·4 | 76·1 | 82·9 | 89·3 | 95·8 | 102·3 | 108·6 | 114·9 | 121·0 |
| ⊙'s L. | Merc. ... | 0·0 | 352·6 | 359·4 | 3·2 | 18·4 | 36·1 | 55·0 | 73·8 | 92·0 | 108·5 | 121·1 | 126·7 | 122·4 | 115·9 | 115·5 | 128·6 | 149·8 | 160·8 |
| +·8 | Jup. ... | 204·8 | 203·5 | 202·2 | 200·9 | 199·8 | 198·5 | 197·7 | 197·0 | 196·7 | 196·3 | 196·4 | 197·4 | 198·3 | 199·5 | 200·9 | 202·5 | 204·3 | 206·3 |
| | Ven. ... | 332·1 | 344·4 | 356·4 | 8·8 | 21·0 | 33·2 | 45·5 | 57·7 | 69·6 | 82·3 | 94·5 | 106·8 | 119·0 | 131·3 | 143·1 | 155·9 | 168·4 | 180·6 |
| | Sat. ... | 239·1 | 238·6 | 238·1 | 237·7 | 237·0 | 236·3 | 235·5 | 234·8 | 234·1 | 233·4 | 233·0 | 232·6 | 232·3 | 232·4 | 232·4 | 232·5 | 233·1 | 233·7 |
| 1841 | Mars ... | 187·5 | 183·9 | 180·7 | 177·8 | 176·8 | 177·1 | 178·4 | 181·0 | 184·1 | 188·0 | 192·6 | 197·6 | 203·0 | 208·8 | 214·7 | 221·1 | 227·6 | 234·4 |
| ⊙'s L. | Merc. ... | 333·8 | 343·0 | 357·2 | 14·7 | 33·5 | 52·6 | 70·8 | 85·9 | 101·0 | 108·3 | 105·4 | 98·2 | 99·3 | 109·2 | 123·8 | 140·6 | 158·4 | 176·0 |
| +·5 | Jup. ... | 238·2 | 237·8 | 237·0 | 236·1 | 235·1 | 233·8 | 232·6 | 231·2 | 230·0 | 228·9 | 228·3 | 227·7 | 227·7 | 228·0 | 228·6 | 229·3 | 230·4 | 231·8 |
| | Ven. ... | 39·1 | 41·9 | 41·5 | 36·9 | 30·9 | 26·1 | 23·8 | 27·7 | 33·5 | 41·1 | 49·9 | 59·4 | 69·7 | 80·5 | 91·5 | 102·8 | 114·5 | 126·4 |
| | Sat. ... | 250·5 | 250·2 | 249·9 | 249·5 | 248·5 | 247·5 | 246·5 | 246·1 | 245·7 | 245·3 | 245·4 | 245·5 | 245·6 | 244·8 | 244·0 | 243·2 | 243·6 | 244·0 |
| 1842 | Mars ... | 19·7 | 26·9 | 33·9 | 41·0 | 47·9 | 55·0 | 61·8 | 68·4 | 75·3 | 81·9 | 88·6 | 95·0 | 101·5 | 107·5 | 114·3 | 120·9 | 127·0 | 133·3 |
| ⊙'s L. | Merc. ... | 335·0 | 351·8 | 10·8 | 29·1 | 47·9 | 65·4 | 79·9 | 88·8 | 88·2 | 80·8 | 79·8 | 88·1 | 102·0 | 118·6 | 136·7 | 154·5 | 171·7 | 187·2 |
| +·3 | Jup. ... | 269·6 | 270·3 | 270·5 | 270·7 | 270·5 | 269·8 | 268·9 | 267·7 | 266·5 | 265·2 | 263·9 | 262·7 | 261·7 | 260·9 | 260·4 | 260·2 | 260·2 | 260·6 |
| | Ven. ... | 8·6 | 21·0 | 33·3 | 45·5 | 57·7 | 70·1 | 82·0 | 94·1 | 105·9 | 117·9 | 129·7 | 141·7 | 153·4 | 164·8 | 176·7 | 187·6 | 198·7 | 209·5 |
| | Sat. ... | 261·3 | 260·9 | 260·5 | 260·0 | 259·8 | 259·6 | 259·4 | 258·7 | 258·0 | 257·3 | 256·6 | 255·9 | 255·1 | 254·8 | 254·5 | 254·2 | 254·5 | 254·8 |
| 1843 | Mars ... | 237·0 | 239·1 | 240·3 | 240·3 | 239·3 | 237·3 | 234·4 | 231·5 | 229·4 | 228·4 | 228·5 | 230·3 | 233·0 | 236·7 | 241·1 | 246·6 | 251·5 | 257·4 |
| ⊙'s L. | Merc. ... | 347·0 | 6·7 | 24·5 | 42·3 | 57·7 | 68·2 | 70·4 | 63·2 | 60·1 | 67·0 | 80·0 | 96·3 | 114·4 | 132·6 | 150·3 | 166·8 | 179·7 | 186·8 |
| +·0 | Jup. ... | 298·9 | 300·6 | 302·0 | 303·1 | 304·2 | 304·7 | 305·2 | 305·3 | 304·8 | 304·2 | 302·9 | 301·4 | 300·7 | 299·3 | 298·0 | 296·9 | 295·9 | 295·2 |
| | Ven. ... | 315·9 | 327·5 | 339·4 | 351·3 | 3·3 | 15·2 | 27·0 | 39·0 | 51·1 | 63·3 | 75·4 | 87·3 | 99·4 | 111·8 | 124·0 | 136·4 | 148·3 | 161·2 |
| | Sat. ... | 272·2 | 272·3 | 272·4 | 272·5 | 272·1 | 271·7 | 271·3 | 270·6 | 269·9 | 269·2 | 268·5 | 267·8 | 267·0 | 266·5 | 266·0 | 265·5 | 265·5 | 265·5 |
| 1844 | Mars ... | 34·7 | 41·5 | 48·3 | 55·0 | 61·8 | 68·4 | 75·0 | 81·5 | 88·1 | 94·6 | 100·0 | 107·5 | 113·9 | 120·3 | 126·6 | 132·9 | 139·4 | 145·7 |
| ⊙'s L. | Merc. ... | 3·1 | 21·0 | 36·8 | 48·3 | 51·3 | 45·0 | 40·8 | 46·4 | 59·0 | 75·4 | 93·5 | 112·1 | 130·2 | 147·0 | 161·2 | 169·7 | 169·3 | 161·4 |
| +·7 | Jup. ... | 327·3 | 329·5 | 331·5 | 333·6 | 335·4 | 337·1 | 338·6 | 339·7 | 340·5 | 341·0 | 341·2 | 341·1 | 340·5 | 339·6 | 338·5 | 337·4 | 336·1 | 334·7 |
| | Ven. ... | 42·4 | 53·4 | 64·1 | 74·6 | 83·9 | 93·0 | 100·1 | 105·7 | 108·3 | 107·1 | 102·1 | 95·1 | 91·5 | 91·4 | 94·5 | 100·5 | 108·4 | 117·0 |
| | Sat. ... | 283·1 | 283·4 | 283·7 | 284·0 | 283·8 | 283·6 | 283·3 | 282·7 | 282·0 | 281·4 | 280·6 | 279·9 | 279·1 | 278·5 | 277·9 | 277·2 | 277·0 | 276·8 |
| 1845 | Mars ... | 267·9 | 273·6 | 279·4 | 284·6 | 289·8 | 294·5 | 298·8 | 302·5 | 305·1 | 307·3 | 307·7 | 306·4 | 304·7 | 301·9 | 299·4 | 298·2 | 297·8 | 299·1 |
| ⊙'s L. | Merc. ... | 14·4 | 25·6 | 32·5 | 27·6 | 21·5 | 25·2 | 36·8 | 52·7 | 70·6 | 89·4 | 107·9 | 125·4 | 140·7 | 151·3 | 153·6 | 146·8 | 142·5 | 147·4 |
| +·5 | Jup. ... | 354·9 | 357·3 | 359·7 | 2·1 | 4·5 | 6·5 | 8·2 | 10·6 | 12·4 | 14·0 | 15·4 | 16·5 | 17·3 | 17·6 | 17·7 | 17·4 | 16·8 | 15·8 |
| | Ven. ... | 350·2 | 2·7 | 15·0 | 27·5 | 39·7 | 51·9 | 64·2 | 76·8 | 88·6 | 100·7 | 113·3 | 125·4 | 137·5 | 149·5 | 161·8 | 174·0 | 186·1 | 198·1 |
| | Sat. ... | 293·8 | 294·4 | 294·9 | 295·4 | 295·4 | 295·4 | 295·4 | 294·8 | 294·3 | 293·8 | 293·0 | 292·3 | 291·6 | 290·8 | 290·1 | 289·4 | 289·1 | 288·7 |
| 1846 | Mars ... | 49·0 | 55·4 | 61·8 | 68·2 | 74·5 | 80·9 | 87·3 | 93·5 | 100·0 | 106·4 | 112·7 | 118·9 | 125·4 | 131·7 | 138·1 | 144·4 | 150·9 | 157·2 |
| ⊙'s L. | Merc. ... | 12·9 | 10·3 | 3·1 | 4·4 | 14·4 | 29·5 | 47·4 | 66·5 | 85·0 | 103·0 | 119·3 | 131·6 | 136·9 | 132·0 | 125·7 | 128·3 | 139·0 | 153·7 |
| ·2 | Jup. ... | 22·5 | 24·8 | 27·1 | 29·5 | 31·9 | 34·3 | 36·7 | 39·0 | 41·1 | 43·4 | 45·4 | 47·2 | 48·9 | 51·0 | 51·6 | 52·6 | 53·3 | 53·6 |
| | Ven. ... | 317·7 | 324·6 | 332·8 | 341·9 | 352·0 | 2·6 | 13·2 | 24·4 | 35·7 | 47·4 | 59·1 | 70·7 | 82·5 | 94·4 | 106·4 | 118·8 | 130·9 | 143·2 |
| | Sat. ... | 304·6 | 305·4 | 306·2 | 306·9 | 307·2 | 307·4 | 307·6 | 307·3 | 307·0 | 306·7 | 306·0 | 305·3 | 304·6 | 303·8 | 303·0 | 302·2 | 301·6 | 301·0 |
| 1847 | Mars ... | 289·5 | 296·7 | 303·8 | 310·3 | 317·8 | 324·9 | 331·6 | 338·2 | 344·8 | 351·0 | 356·7 | 2·1 | 8·1 | 12·5 | 16·0 | 18·4 | 19·8 | 20·0 |
| ⊙'s L. | Merc. ... | 345·6 | 344·3 | 352·6 | 6·9 | 24·2 | 42·9 | 61·8 | 79·3 | 97·3 | 111·1 | 118·5 | 116·2 | 108·9 | 109·3 | 118·5 | 132·6 | 149·2 | 167·0 |
| +·0 | Jup. ... | 50·3 | 52·2 | 54·2 | 56·5 | 58·7 | 60·0 | 63·3 | 65·7 | 67·9 | 70·2 | 72·5 | 73·9 | 76·9 | 79·0 | 80·9 | 82·7 | 84·1 | 85·3 |
| | Ven. ... | 26·8 | 38·9 | 50·8 | 62·8 | 74·6 | 86·2 | 97·6 | 108·9 | 120·0 | 130·6 | 140·9 | 150·3 | 159·2 | 166·7 | 172·5 | 175·5 | 175·4 | 170·6 |
| | Sat. ... | 315·7 | 316·6 | 317·5 | 318·4 | 318·9 | 319·4 | 319·8 | 319·7 | 319·6 | 319·5 | 319·0 | 318·5 | 317·9 | 317·1 | 316·3 | 315·6 | 314·8 | 314·1 |
| 1848 | Mars ... | 64·4 | 70·2 | 76·1 | 82·1 | 88·0 | 94·2 | 100·3 | 106·4 | 112·9 | 118·9 | 124·9 | 131·3 | 137·6 | 144·0 | 150·4 | 156·8 | 163·2 | 169·7 |
| ⊙'s L. | Merc. ... | 332·5 | 345·7 | 2·7 | 21·4 | 40·2 | 58·2 | 76·5 | 91·0 | 99·4 | 99·1 | 91·4 | 90·3 | 98·7 | 112·6 | 129·4 | 147·0 | 164·9 | 182·1 |
| +·7 | Jup. ... | 79·1 | 80·3 | 81·7 | 83·4 | 85·2 | 87·2 | 89·3 | 91·5 | 93·6 | 95·9 | 98·1 | 100·5 | 102·8 | 105·0 | 107·2 | 109·3 | 111·3 | 113·1 |
| | Ven. ... | 332·6 | 344·9 | 357·1 | 9·4 | 21·5 | 33·7 | 45·9 | 58·2 | 70·5 | 82·8 | 94·9 | 106·9 | 119·5 | 131·8 | 144·2 | 156·4 | 168·9 | 181·1 |
| | Sat. ... | 326·9 | 327·9 | 328·9 | 330·0 | 330·8 | 331·5 | 332·2 | 332·0 | 331·8 | 331·5 | 331·4 | 331·4 | 331·4 | 330·7 | 330·0 | 329·2 | 328·4 | 327·6 |
| 1849 | Mars ... | 308·9 | 316·4 | 324·1 | 331·7 | 339·3 | 346·8 | 354·0 | 1·4 | 8·4 | 15·5 | 22·3 | 29·0 | 35·5 | 41·5 | 47·4 | 52·9 | 58·1 | 62·7 |
| ⊙'s L. | Merc. ... | 341·5 | 358·1 | 16·9 | 35·7 | 53·4 | 68·9 | 79·6 | 81·1 | 74·0 | 71·1 | 77·9 | 90·7 | 107·2 | 125·0 | 143·2 | 160·6 | 176·7 | 189·6 |
| +·4 | Jup. ... | 109·4 | 109·7 | 110·3 | 111·2 | 112·2 | 113·5 | 115·1 | 116·9 | 118·6 | 120·7 | 122·8 | 124·9 | 127·0 | 129·3 | 131·5 | 133·8 | 135·9 | 138·0 |
| | Ven. ... | 37·4 | 40·2 | 38·4 | 33·5 | 26·2 | 23·6 | 23·8 | 26·8 | 32·8 | 40·8 | 49·7 | 59·5 | 70·0 | 80·8 | 91·8 | 103·3 | 115·0 | 126·7 |
| | Sat. ... | 338·1 | 339·3 | 340·5 | 341·7 | 342·6 | 343·5 | 344·4 | 344·8 | 345·2 | 345·5 | 345·3 | 345·1 | 345·0 | 344·5 | 343·9 | 343·3 | 342·5 | 341·7 |

| Mesh. | Vrsh. | Mith. | Kat. | Sim. | Kan. | Tul. | Vrsch. | Dhan. | Mak. | Kum. | Min. | Asvn. | Bhar. | Krit. | Rohi. | Mrig. | Ardh. | Punar. | Push. | Asles. |
|---|---|---|---|---|---|---|---|---|---|---|---|---|---|---|---|---|---|---|---|---|
| 30 | 60 | 90 | 120 | 150 | 180 | 210 | 240 | 270 | 300 | 330 | 360 | 13·3 | 26·7 | 40·0 | 53·3 | 66·7 | 80·0 | 93·3 | 106·7 | 120·0 |

## mean Sunrise for every ten days from A.D. 1840 to A.D. 1849.

| 180 | 190 | 200 | 210 | 220 | 230 | 240 | 250 | 260 | | 270 | 280 | 290 | 300 | 310 | 320 | 330 | 340 | 350 | 360 |
|---|---|---|---|---|---|---|---|---|---|---|---|---|---|---|---|---|---|---|---|
| 174·4 | 184·3 | 194·1 | 204·0 | 213·8 | 223·7 | 233·5 | 243·4 | 253·3 | | 263·1 | 273·0 | 282·8 | 292·7 | 302·5 | 312·4 | 322·2 | 332·1 | 342·0 | 351·8 |
| −2·2 | −2·1 | −1·9 | −1·7 | −1·5 | −1·2 | −0·9 | −0·5 | −0·1 | | +0·8 | +0·6 | +1·0 | +1·3 | +1·6 | +1·8 | +2·0 | +2·1 | +2·2 | +2·2 |
| O 8 | O 18 | O 28 | N 7 | N 17 | N 27 | D 7 | D 17 | D 27 | | Ja 6 | Ja 16 | Ja 26 | F 5 | F 15 | F 25 | Mr 6,7 | 16,17 | 26,27 | Ap 5,6 |
| Pur 25 | Aipp 4 | Aipp 14 | Aipp 24 | Kart 4 | Kart 14 | Kart 24 | Marg 5 | Marg 15 | | Marg 25 | Tai 5 | Tai 15 | Tai 25 | Masi 6 | Masi 16 | Masi 26 | Pang 6 | Pang 16 | Pang 26 |
| 127·0 | 133·2 | 189·0 | 144·9 | 150·4 | 156·0 | 161·5 | 166·8 | 171·8 | 41 | 176·6 | 180·8 | 184·8 | 188·3 | 191·1 | 193·1 | 193·8 | 193·9 | 192·0 | 189·0 |
| 177·9 | 195·4 | 211·6 | 225·3 | 233·7 | 233·1 | 225·4 | 223·1 | 229·9 | | 242·9 | 259·0 | 276·4 | 294·4 | 311·4 | 327·3 | 339·4 | 344·3 | 339·3 | 334·2 |
| 208·0 | 210·2 | 212·3 | 215·3 | 216·8 | 218·9 | 221·2 | 223·4 | 225·6 | | 227·7 | 229·6 | 231·4 | 233·2 | 234·7 | 236·0 | 237·1 | 237·8 | 238·3 | 238·4 |
| 193·1 | 205·7 | 217·9 | 230·4 | 242·7 | 255·1 | 267·4 | 279·5 | 292·0 | | 308·9 | 315·6 | 327·5 | 339·2 | 350·3 | 0·9 | 11·1 | 20·6 | 29·3 | 35·8 |
| 234·2 | 235·1 | 236·0 | 237·0 | 238·1 | 239·2 | 240·4 | 241·7 | 243·0 | | 244·2 | 245·3 | 246·4 | 247·5 | 248·4 | 249·3 | 250·1 | 250·4 | 250·7 | 251·1 |
| 241·2 | 248·3 | 255·7 | 262·9 | 270·4 | 277·8 | 285·2 | 293·0 | 300·6 | 42 | 308·5 | 316·0 | 323·6 | 331·2 | 338·8 | 346·5 | 354·0 | 1·3 | 8·9 | 16·0 |
| 192·8 | 207·3 | 216·8 | 218·4 | 210·8 | 207·1 | 211·2 | 224·2 | 239·7 | | 256·9 | 274·6 | 291·8 | 307·6 | 320·3 | 325·6 | 322·1 | 315·3 | 317·3 | 327·8 |
| 233·3 | 235·2 | 237·1 | 239·1 | 241·2 | 243·4 | 245·7 | 248·0 | 250·2 | | 252·5 | 254·8 | 257·0 | 259·3 | 261·3 | 268·1 | 265·0 | 266·5 | 268·1 | 269·3 |
| 138·4 | 150·1 | 162·3 | 174·6 | 187·0 | 199·3 | 211·7 | 224·1 | 236·7 | | 249·3 | 262·0 | 274·7 | 287·1 | 299·8 | 312·3 | 324·9 | 337·5 | 349·8 | 2·2 |
| 244·4 | 245·3 | 246·2 | 247·0 | 248·1 | 249·2 | 250·2 | 251·4 | 252·6 | | 253·7 | 254·8 | 255·9 | 257·1 | 258·0 | 258·9 | 259·7 | 260·8 | 260·9 | 261·4 |
| 140·4 | 145·7 | 151·8 | 157·8 | 163·8 | 169·9 | 175·7 | 181·7 | 187·5 | 43 | 193·2 | 198·3 | 204·3 | 209·7 | 214·8 | 219·7 | 224·3 | 228·6 | 232·5 | 235·7 |
| 198·9 | 203·3 | 197·8 | 191·5 | 194·0 | 204·7 | 219·1 | 235·7 | 253·3 | | 270·7 | 287·0 | 300·3 | 308·2 | 306·6 | 298·9 | 298·0 | 306·8 | 320·9 | 338·0 |
| 261·7 | 262·8 | 264·3 | 265·8 | 267·4 | 269·3 | 271·5 | 273·6 | 275·9 | | 278·2 | 280·4 | 282·8 | 285·2 | 287·5 | 289·7 | 292·0 | 294·2 | 296·2 | 298·1 |
| 219·5 | 229·2 | 237·5 | 244·8 | 250·0 | 251·8 | 250·1 | 244·1 | 240·9 | | 236·6 | 237·7 | 242·6 | 249·6 | 257·6 | 266·2 | 277·2 | 288·0 | 299·0 | 310·2 |
| 255·0 | 255·7 | 256·4 | 257·0 | 258·0 | 259·0 | 260·1 | 261·2 | 262·3 | | 263·5 | 264·7 | 265·9 | 267·0 | 268·0 | 269·0 | 270·1 | 270·8 | 271·5 | 272·1 |
| 263·6 | 270·2 | 277·7 | 283·6 | 290·5 | 297·6 | 304·9 | 312·0 | 319·3 | 44 | 326·6 | 333·9 | 341·2 | 348·2 | 355·5 | 2·5 | 9·6 | 16·6 | 23·6 | 30·6 |
| 184·5 | 176·7 | 176·4 | 185·0 | 198·9 | 215·0 | 232·1 | 249·4 | 266·4 | | 280·3 | 290·1 | 291·1 | 283·7 | 280·3 | 286·4 | 299·1 | 315·6 | 333·6 | 352·6 |
| 294·7 | 294·8 | 295·2 | 295·8 | 296·7 | 297·9 | 299·5 | 301·3 | 303·1 | | 305·1 | 307·3 | 309·5 | 311·8 | 314·1 | 316·5 | 318·9 | 321·4 | 323·8 | 326·1 |
| 174·3 | 186·2 | 198·9 | 211·2 | 223·6 | 236·2 | 248·7 | 261·6 | 274·2 | | 286·6 | 299·2 | 311·5 | 322·9 | 336·0 | 348·5 | 0·3 | 12·2 | 24·1 | 35·6 |
| 265·5 | 266·1 | 266·7 | 267·2 | 268·1 | 269·0 | 269·9 | 270·9 | 272·0 | | 273·1 | 274·4 | 275·6 | 276·9 | 277·9 | 279·0 | 280·2 | 281·0 | 281·8 | 282·7 |
| 152·1 | 158·3 | 164·5 | 170·9 | 177·4 | 183·6 | 189·9 | 196·1 | 202·5 | 45 | 208·8 | 215·1 | 221·4 | 227·8 | 234·0 | 240·2 | 246·5 | 252·8 | 258·8 | 264·8 |
| 159·5 | 166·8 | 179·9 | 195·7 | 213·0 | 230·2 | 246·6 | 261·4 | 272·2 | | 275·2 | 268·6 | 263·7 | 267·9 | 279·6 | 295·3 | 312·9 | 331·2 | 349·4 | 6·3 |
| 333·5 | 332·3 | 331·5 | 330·9 | 330·7 | 330·9 | 331·4 | 332·2 | 333·3 | | 334·7 | 336·2 | 338·1 | 339·9 | 342·2 | 344·4 | 346·7 | 349·0 | 351·4 | 353·8 |
| 127·2 | 137·7 | 148·7 | 159·8 | 171·6 | 183·4 | 195·1 | 207·5 | 219·8 | | 232·0 | 244·3 | 256·7 | 269·3 | 281·8 | 294·4 | 306·8 | 319·1 | 331·6 | 344·0 |
| 276·7 | 277·0 | 277·4 | 277·7 | 278·5 | 279·4 | 280·2 | 281·3 | 282·3 | | 283·4 | 284·7 | 285·9 | 287·1 | 288·3 | 289·5 | 290·6 | 291·5 | 292·4 | 293·4 |
| 301·4 | 304·8 | 308·9 | 313·4 | 318·7 | 324·1 | 330·0 | 335·9 | 341·9 | 46 | 348·1 | 354·4 | 0·6 | 7·1 | 13·6 | 20·0 | 26·5 | 32·9 | 39·4 | 45·7 |
| 159·3 | 174·9 | 192·0 | 209·3 | 226·3 | 241·8 | 253·9 | 259·4 | 255·3 | | 247·9 | 249·5 | 259·4 | 273·8 | 290·8 | 308·8 | 326·9 | 344·4 | 359·6 | 10·4 |
| 14·1 | 13·4 | 12·2 | 10·9 | 10·0 | 9·6 | 8·4 | 7·5 | 7·4 | | 7·4 | 7·9 | 8·8 | 10·1 | 11·5 | 13·2 | 15·1 | 17·1 | 19·1 | 21·3 |
| 210·3 | 222·1 | 233·9 | 245·7 | 257·4 | 268·9 | 279·9 | 290·6 | 300·5 | | 310·0 | 317·9 | 324·3 | 328·1 | 328·2 | 324·1 | 317·9 | 313·1 | 312·5 | 315·4 |
| 288·3 | 288·5 | 288·7 | 288·8 | 289·4 | 290·0 | 290·7 | 291·7 | 292·7 | | 293·8 | 294·9 | 296·6 | 297·3 | 298·5 | 299·7 | 300·9 | 302·0 | 303·0 | 304·0 |
| 163·7 | 170·2 | 176·5 | 183·3 | 189·7 | 196·4 | 203·1 | 209·7 | 216·3 | 47 | 223·1 | 230·0 | 236·9 | 243·6 | 250·7 | 257·6 | 264·6 | 271·8 | 278·8 | 286·0 |
| 170·7 | 188·3 | 205·5 | 221·6 | 235·1 | 242·7 | 241·6 | 234·2 | 231·9 | | 239·8 | 253·1 | 266·8 | 284·3 | 302·1 | 319·2 | 336·3 | 350·0 | 357·5 | 349·1 |
| 53·7 | 53·1 | 52·0 | 51·2 | 50·2 | 48·8 | 47·5 | 46·3 | 45·1 | | 44·0 | 43·4 | 43·0 | 43·0 | 43·4 | 44·0 | 45·0 | 46·2 | 47·1 | 49·4 |
| 155·5 | 167·8 | 180·2 | 192·7 | 205·2 | 217·8 | 230·4 | 243·0 | 255·8 | | 268·7 | 281·0 | 293·5 | 306·3 | 318·7 | 331·2 | 343·9 | 356·2 | 8·5 | 20·1 |
| 300·5 | 300·4 | 300·3 | 300·3 | 300·7 | 301·2 | 301·7 | 302·6 | 303·5 | | 304·5 | 305·6 | 306·7 | 307·8 | 209·1 | 310·4 | 311·7 | 312·9 | 314·0 | 315·2 |
| 18·6 | 16·6 | 13·8 | 11·1 | 9·4 | 8·9 | 9·5 | 11·3 | 13·7 | 48 | 16·9 | 20·6 | 24·9 | 29·5 | 34·5 | 39·5 | 45·1 | 50·5 | 56·0 | 61·9 |
| 184·6 | 201·2 | 215·7 | 225·7 | 227·2 | 220·5 | 215·8 | 221·1 | 232·6 | | 248·0 | 265·4 | 283·0 | 300·7 | 316·6 | 329·6 | 336·2 | 332·9 | 325·9 | 327·3 |
| 86·5 | 87·2 | 87·5 | 87·6 | 87·5 | 86·9 | 86·0 | 84·8 | 83·6 | | 82·3 | 81·0 | 79·8 | 79·1 | 77·9 | 77·4 | 77·2 | 77·3 | 77·7 | 78·6 |
| 164·7 | 160·2 | 159·6 | 163·1 | 168·3 | 176·8 | 185·9 | 195·6 | 206·3 | | 217·4 | 228·9 | 240·4 | 252·5 | 264·5 | 276·5 | 288·7 | 300·8 | 313·0 | 325·3 |
| 313·4 | 313·1 | 312·8 | 312·5 | 312·7 | 313·0 | 313·3 | 314·0 | 314·7 | | 315·5 | 316·6 | 317·7 | 318·8 | 320·0 | 321·2 | 322·5 | 323·5 | 324·5 | 325·6 |
| 176·4 | 182·9 | 189·7 | 196·5 | 203·3 | 210·1 | 216·9 | 223·9 | 231·1 | 49 | 238·1 | 245·1 | 252·6 | 259·9 | 267·3 | 274·8 | 282·3 | 289·9 | 297·4 | 305·1 |
| 197·1 | 208·5 | 212·0 | 206·6 | 200·2 | 203·3 | 214·1 | 229·0 | 245·8 | | 263·3 | 280·7 | 297·1 | 310·4 | 318·1 | 316·5 | 308·8 | 308·9 | 318·1 | 332·3 |
| 114·9 | 116·5 | 117·0 | 118·2 | 119·6 | 120·0 | 120·1 | 119·8 | 119·3 | | 118·5 | 117·5 | 116·2 | 114·9 | 113·6 | 112·5 | 111·3 | 110·5 | 109·8 | 109·5 |
| 193·8 | 206·2 | 218·6 | 230·9 | 243·2 | 255·6 | 267·7 | 280·2 | 292·2 | | 304·2 | 316·2 | 328·0 | 339·3 | 350·2 | 1·0 | 11·0 | 18·9 | 25·6 | 34·1 |
| 326·8 | 326·2 | 325·7 | 325·2 | 325·2 | 325·3 | 325·3 | 325·6 | 325·9 | | 326·2 | 327·5 | 328·8 | 330·1 | 331·3 | 332·5 | 333·7 | 335·0 | 336·3 | 337·6 |
| 66·7 | 70·2 | 72·6 | 74·0 | 74·4 | 73·4 | 71·4 | 67·8 | 64·1 | 50 | 61·1 | 58·9 | 58·2 | 58·5 | 60·0 | 62·4 | 65·4 | 69·3 | 73·5 | 77·9 |
| 196·1 | 193·3 | 185·7 | 185·9 | 194·7 | 208·4 | 224·7 | 242·2 | 259·5 | | 276·0 | 290·3 | 299·8 | 300·7 | 293·2 | 290·0 | 296·5 | 309·7 | 326·3 | 344·4 |
| 140·2 | 142·1 | 144·0 | 145·7 | 147·3 | 148·5 | 149·6 | 150·3 | 150·8 | | 151·0 | 150·9 | 150·4 | 149·4 | 148·6 | 147·3 | 146·0 | 144·7 | 143·5 | 142·4 |
| 138·7 | 150·8 | 162·8 | 175·0 | 187·5 | 199·8 | 212·2 | 224·8 | 237·4 | | 249·8 | 262·5 | 275·0 | 287·8 | 300·3 | 313·0 | 325·4 | 338·0 | 350·5 | 2·9 |
| 340·9 | 340·2 | 339·5 | 338·9 | 338·6 | 338·3 | 338·1 | 338·4 | 338·8 | | 339·2 | 340·0 | 340·8 | 341·6 | 342·7 | 343·9 | 345·1 | 346·4 | 347·7 | 349·0 |

| Mag. | P. Phal. | U. Phal. | Hasta. | Chit. | Svati. | Visa | Anur. | Jyesh. | Mula. | P. Ash. | U. Ash. | Srav. | Dan. | Satab. | P. Bhad. | U. Bhad. | Revati. |
|---|---|---|---|---|---|---|---|---|---|---|---|---|---|---|---|---|---|
| 133·3 | 146·7 | 160·0 | 173·3 | 186·7 | 200·0 | 213·3 | 226·7 | 240·0 | 253·3 | 266·7 | 280·0 | 293·3 | 306·7 | 320·0 | 333·3 | 346·7 | 360·0 |

## Geocentric (Indien) Longitudes of Sun and Major planets at Ujjain

| | Day of S. Yr. | 0 | 10 | 20 | 30 | 40 | 50 | 60 | 70 | 80 | 90 | 100 | 110 | 120 | 130 | 140 | 150 | 160 | 170 |
|---|---|---|---|---|---|---|---|---|---|---|---|---|---|---|---|---|---|---|---|
| | ☉'s Mean L. | 357.0 | 6.9 | 16.7 | 26.6 | 36.4 | 46.3 | 56.1 | 66.0 | 75.9 | 85.7 | 95.6 | 105.4 | 115.3 | 125.1 | 135.0 | 144.8 | 154.7 | 164.6 |
| | Corrn. for do. | +2.1 | +2.0 | +1.9 | +1.7 | +1.4 | +1.1 | +0.8 | +0.4 | +0.0 | −0.3 | −0.7 | −1.1 | −1.4 | −1.6 | −1.9 | −2.0 | −2.1 | −2.2 |
| | English date. | Ap 11 | Ap 21 | My 1 | My 11 | My 21 | My 31 | Je 10 | Je 20 | Je 30 | Jl 10 | Jl 20 | Jl 30 | Au 9 | Au19 | Au 29 | S 8 | S 18 | S 28 |
| | Tamil month. | Chit | Chit | Chit | Chit | Vaik | Vaik | Vaik | Ani | Ani | Ani | Adi | Adi | Adi | Avan | Avan | Avan | Pur | Pur |
| | Do. Date. | 1 | 11 | 21 | 31 | 10 | 20 | 30 | 9 | 19 | 29 | 7 | 17 | 27 | 6 | 16 | 26 | 5 | 15 |
| 1850 | Mars ... | 80.3 | 85.5 | 90.8 | 96.1 | 101.6 | 107.5 | 113.2 | 118.9 | 125.0 | 130.7 | 137.2 | 143.4 | 149.7 | 156.2 | 162.4 | 168.9 | 175.6 | 182.0 |
| ☉'s L. | Merc. ... | 358.8 | 12.0 | 30.5 | 46.7 | 58.5 | 62.8 | 58.6 | 51.8 | 56.4 | 68.7 | 34.9 | 102.7 | 121.1 | 139.0 | 155.9 | 170.0 | 179.0 | 179.0 |
| +.2 | Jup. ... | 141.8 | 140.9 | 140.5 | 140.3 | 140.4 | 140.8 | 141.8 | 142.7 | 144.0 | 145.5 | 147.1 | 149.1 | 151.1 | 153.1 | 155.2 | 157.4 | 159.6 | 161.7 |
| | Ven. ... | 9.1 | 21.5 | 33.7 | 46.0 | 58.4 | 70.6 | 82.4 | 94.6 | 106.4 | 118.4 | 130.4 | 141.9 | 153.6 | 165.3 | 176.9 | 187.9 | 198.8 | 209.4 |
| | Sat. ... | 349.7 | 351.0 | 352.3 | 353.5 | 354.5 | 355.5 | 356.6 | 357.3 | 358.0 | 358.7 | 358.8 | 358.9 | 359.1 | 358.6 | 358.1 | 357.7 | 357.0 | 356.3 |
| 1851 | Mars ... | 326.1 | 333.9 | 341.6 | 849.3 | 356.9 | 4.3 | 12.0 | 19.6 | 26.4 | 33.6 | 40.1 | 47.3 | 54.1 | 60.7 | 66.9 | 73.2 | 79.1 | 84.7 |
| ☉'s L. | Merc. ... | 8.1 | 24.0 | 37.0 | 43.4 | 39.2 | 32.9 | 34.6 | 46.6 | 62.1 | 79.8 | 98.6 | 116.8 | 134.4 | 149.6 | 160.6 | 163.5 | 156.9 | 152.0 |
| —1 | Jup. ... | 175.3 | 175.1 | 175.0 | 174.8 | 174.7 | 174.9 | 175.2 | 175.7 | 176.3 | 176.9 | 177.8 | 178.8 | 179.9 | 181.3 | 182.6 | 184.0 | 185.5 | 187.0 |
| | Ven. ... | 316.2 | 328.1 | 339.9 | 351.8 | 3.6 | 15.8 | 27.5 | 39.5 | 51.6 | 63.8 | 75.9 | 88.0 | 100.1 | 112.3 | 124.5 | 136.9 | 149.3 | 161.4 |
| | Sat. ... | 1.4 | 2.7 | 4.0 | 5.3 | 6.6 | 7.9 | 9.1 | 10.0 | 10.8 | 11.6 | 12.1 | 12.5 | 12.9 | 12.7 | 12.4 | 12.1 | 11.6 | 11.0 |
| 1852 | Mars ... | 100.3 | 103.9 | 108.0 | 112.5 | 117.2 | 122.2 | 127.5 | 133.2 | 138.9 | 144.8 | 150.7 | 156.8 | 163.0 | 169.4 | 175.9 | 182.4 | 189.1 | 195.8 |
| ☉'s L. | Merc. ... | 16.8 | 23.9 | 21.1 | 15.1 | 15.3 | 25.7 | 40.9 | 58.7 | 77.3 | 96.1 | 113.9 | 129.9 | 142.1 | 146.8 | 141.5 | 135.5 | 138.5 | 149.2 |
| +.7 | Jup. ... | 209.6 | 208.5 | 207.2 | 205.9 | 204.6 | 203.5 | 202.4 | 201.5 | 201.0 | 200.8 | 200.9 | 201.3 | 202.0 | 203.2 | 204.6 | 206.0 | 207.6 | 209.3 |
| | Ven. ... | 42.8 | 54.0 | 64.5 | 74.6 | 83.9 | 92.4 | 99.3 | 104.4 | 106.4 | 104.2 | 98.3 | 92.0 | 89.0 | 89.4 | 93.3 | 99.7 | 108.0 | 117.0 |
| | Sat. ... | 13.7 | 15.0 | 16.3 | 17.6 | 18.9 | 20.2 | 21.6 | 22.6 | 23.6 | 24.6 | 25.3 | 26.0 | 26.7 | 26.8 | 26.9 | 27.0 | 26.6 | 26.1 |
| 1853 | Mars ... | 342.9 | 350.7 | 358.5 | 6. | 13.6 | 21.1 | 28.3 | 35.6 | 42.7 | 49.8 | 56.6 | 63.3 | 70.1 | 76.6 | 83.0 | 89.3 | 95.5 | 101.6 |
| ☉'s L. | Merc. ... | 3.5 | 356.2 | 355.1 | 3.6 | 17.9 | 35.4 | 54.0 | 73.1 | 91.2 | 108.1 | 121.8 | 128.7 | 126.3 | 118.9 | 119.5 | 128.6 | 142.9 | 159.5 |
| +.4 | Jup. ... | 242.8 | 242.4 | 241.9 | 241.1 | 240.1 | 238.8 | 237.5 | 237.0 | 235.4 | 233.9 | 233.1 | 232.4 | 232.2 | 232.1 | 232.5 | 233.4 | 234.2 | 235.4 |
| | Ven. ... | 350.7 | 3.1 | 15.5 | 28.0 | 40.2 | 52.6 | 64.7 | 77.1 | 89.1 | 101.4 | 113.6 | 125.9 | 138.0 | 150.2 | 162.4 | 174.3 | 186.4 | 198.4 |
| | Sat. ... | 26.4 | 27.5 | 28.6 | 29.8 | 31.1 | 32.5 | 33.9 | 35.1 | 36.3 | 37.5 | 38.5 | 39.4 | 40.3 | 40.7 | 41.1 | 41.4 | 41.2 | 41.0 |
| 1854 | Mars ... | 126.8 | 127.4 | 129.6 | 132.1 | 135.4 | 139.5 | 144.0 | 148.9 | 154.0 | 159.5 | 165.2 | 171.1 | 177.3 | 183.6 | 190.0 | 196.6 | 203.3 | 210.1 |
| ☉'s L. | Merc. ... | 335.6 | 342.3 | 355.3 | 12.0 | 30.6 | 49.6 | 68.2 | 85.8 | 100.5 | 109.8 | 109.8 | 102.3 | 100.5 | 108.1 | 121.6 | 137.9 | 155.4 | 173.3 |
| +.1 | Jup. ... | 273.7 | 274.4 | 275.1 | 275.6 | 275.6 | 275.2 | 274.5 | 273.5 | 272.3 | 271.0 | 269.7 | 268.5 | 267.2 | 266.3 | 265.7 | 265.1 | 265.0 | 265.4 |
| | Ven. ... | 316.8 | 323.9 | 332.3 | 341.8 | 351.9 | 2.7 | 13.5 | 24.7 | 36.0 | 47.8 | 59.5 | 71.2 | 83.2 | 94.9 | 107.1 | 119.3 | 131.4 | 143.7 |
| | Sat. ... | 38.5 | 39.8 | 41.1 | 42.3 | 43.7 | 45.1 | 46.5 | 47.8 | 49.1 | 50.5 | 51.5 | 52.6 | 53.6 | 54.3 | 55.0 | 55.7 | 55.8 | 55.9 |
| 1855 | Mars ... | 358.4 | 6.1 | 13.4 | 20.9 | 28.2 | 35.5 | 42.5 | 49.5 | 56.6 | 63.6 | 70.4 | 77.0 | 83.7 | 90.1 | 96.6 | 102.8 | 109.3 | 115.3 |
| ☉'s L. | Merc. ... | 332.9 | 349.1 | 7.2 | 26.2 | 45.2 | 63.0 | 78.8 | 89.5 | 92.4 | 85.6 | 81.7 | 87.4 | 100.1 | 116.0 | 133.9 | 151.7 | 169.2 | 185.4 |
| —1 | Jup. ... | 302.7 | 304.4 | 306.1 | 307.5 | 308.7 | 309.4 | 310.1 | 310.2 | 310.0 | 309.6 | 308.8 | 307.8 | 306.5 | 305.2 | 303.8 | 302.5 | 301.4 | 300.5 |
| | Ven. ... | 27.2 | 39.3 | 51.3 | 63.3 | 74.8 | 86.5 | 97.9 | 109.8 | 121.3 | 132.5 | 141.8 | 150.2 | 158.9 | 166.0 | 171.0 | 173.6 | 172.1 | 166.7 |
| | Sat. ... | 51.3 | 52.4 | 53.6 | 54.8 | 56.1 | 57.4 | 58.7 | 60.0 | 61.3 | 62.7 | 63.9 | 65.0 | 66.3 | 67.3 | 68.2 | 69.1 | 69.4 | 69.8 |
| 1856 | Mars ... | 166.0 | 163.3 | 161.4 | 160.9 | 161.5 | 163.5 | 166.3 | 169.5 | 173.8 | 178.5 | 183.6 | 189.1 | 194.7 | 200.8 | 207.0 | 213.4 | 220.1 | 227.0 |
| ☉'s L. | Merc. ... | 346.2 | 4.7 | 23.5 | 41.7 | 57.8 | 69.5 | 73.8 | 68.2 | 62.8 | 68.7 | 79.6 | 95.5 | 113.4 | 131.6 | 149.6 | 166.0 | 179.9 | 188.5 |
| +.6 | Jup. ... | 330.9 | 333.2 | 335.2 | 337.4 | 339.4 | 341.1 | 342.8 | 344.0 | 345.0 | 345.3 | 345.6 | 346.4 | 346.1 | 345.4 | 344.5 | 343.3 | 342.0 | 340.8 |
| | Ven. ... | 333.1 | 345.4 | 357.8 | 10.0 | 22.2 | 34.2 | 46.6 | 58.9 | 71.0 | 83.3 | 95.5 | 107.8 | 120.0 | 132.3 | 144.6 | 156.9 | 169.4 | 181.6 |
| | Sat. ... | 64.4 | 65.5 | 66.5 | 67.5 | 68.8 | 70.1 | 71.3 | 72.6 | 73.9 | 75.2 | 76.5 | 77.8 | 79.2 | 80.2 | 81.3 | 82.2 | 82.9 | 83.5 |
| 1857 | Mars ... | 14.0 | 21.3 | 28.4 | 35.7 | 42.7 | 49.6 | 56.7 | 63.5 | 70.4 | 77.0 | 83.6 | 90.1 | 96.5 | 103.1 | 109.5 | 115.9 | 122.2 | 128.3 |
| ☉'s L. | Merc. ... | 0.8 | 18.7 | 35.0 | 48.2 | 54.3 | 50.4 | 43.7 | 46.6 | 57.6 | 72.9 | 90.7 | 109.3 | 127.4 | 144.8 | 160.0 | 170.3 | 172.8 | 166.0 |
| +.4 | Jup. ... | 358.5 | 0.9 | 3.3 | 5.7 | 8.0 | 10.3 | 12.5 | 14.6 | 16.3 | 18.1 | 19.7 | 20.9 | 21.9 | 22.5 | 22.7 | 22.7 | 22.3 | 21.6 |
| | Ven. ... | 35.7 | 37.6 | 35.1 | 29.0 | 23.2 | 20.2 | 21.5 | 25.5 | 32.1 | 40.3 | 49.7 | 59.4 | 69.9 | 80.9 | 92.1 | 103.6 | 115.3 | 127.2 |
| | Sat. ... | 77.8 | 78.5 | 79.2 | 80.1 | 81.2 | 82.4 | 83.5 | 84.4 | 86.1 | 87.5 | 88.8 | 90.1 | 91.4 | 92.6 | 93.7 | 94.9 | 95.7 | 96.6 |
| 1858 | Mars ... | 220.2 | 219.6 | 217.9 | 215.2 | 211.7 | 209.0 | 206.8 | 206.2 | 206.4 | 208.2 | 210.9 | 214.3 | 218.8 | 223.6 | 229.0 | 234.8 | 240.7 | 247.2 |
| ☉'s L. | Merc. ... | 12.6 | 26.6 | 34.6 | 32.8 | 25.4 | 25.8 | 35.6 | 50.5 | 67.3 | 86.5 | 105.0 | 122.8 | 139.0 | 151.2 | 156.6 | 151.8 | 145.2 | 147.4 |
| +.1 | Jup. ... | 26.1 | 28.4 | 30.8 | 33.2 | 35.6 | 38.0 | 40.3 | 42.6 | 44.9 | 47.2 | 49.2 | 51.1 | 52.9 | 54.5 | 55.8 | 56.9 | 57.7 | 58.2 |
| | Ven. ... | 9.6 | 22.0 | 34.2 | 46.8 | 58.9 | 71.1 | 82.9 | 95.0 | 106.9 | 118.7 | 130.7 | 142.4 | 153.9 | 165.6 | 177.0 | 188.2 | 198.9 | 209.3 |
| | Sat. ... | 91.8 | 92.3 | 92.8 | 93.3 | 94.4 | 95.4 | 96.4 | 97.6 | 98.8 | 100.1 | 101.4 | 102.7 | 104.1 | 105.4 | 106.6 | 107.8 | 108.8 | 109.7 |
| | Eng. date. | Ap 12 | Ap 22 | My 2 | My 12 | My 22 | Je 1 | Je 11 | Je 21 | Jl 1 | Jl 11 | Jl 21 | Jl 31 | Au 10 | Au 20 | Au 30 | S 9 | S 19 | S 29 |
| 1859 | Mars ... | 29.0 | 35.8 | 43.0 | 49.8 | 56.6 | 63.3 | 70.0 | 76.6 | 83.3 | 89.8 | 96.4 | 102.7 | 109.2 | 115.6 | 122.2 | 128.5 | 134.3 | 141.1 |
| ☉'s L. | Merc. ... | 14.6 | 13.6 | 6.9 | 6.1 | 14.7 | 29.1 | 45.9 | 65.2 | 83.7 | 102.2 | 118.9 | 132.2 | 139.0 | 136.2 | 128.9 | 129.4 | 138.8 | 153.0 |
| +.8 | Jup. ... | 54.2 | 56.0 | 58.1 | 60.2 | 62.3 | 64.6 | 66.9 | 69.3 | 71.7 | 74.0 | 76.2 | 78.5 | 80.7 | 82.7 | 84.6 | 86.4 | 88.0 | 89.3 |
| | Ven ... | 317.7 | 329.6 | 341.4 | 353.4 | 5.3 | 17.2 | 29.2 | 41.2 | 53.1 | 65.4 | 77.7 | 89.9 | 101.8 | 114.1 | 126.4 | 138.6 | 151.0 | 163.4 |
| | Sat. ... | 105.6 | 105.9 | 106.1 | 106.5 | 107.2 | 108.0 | 108.8 | 110.0 | 111.2 | 112.4 | 113.7 | 115.0 | 116.2 | 117.5 | 118.8 | 120.0 | 121.2 | 122.3 |

| Mesh. | Vrsh. | Mith. | Kat. | Sim. | Kan. | Tul. | Vrsch. | Dhan. | Mak. | Kum. | Min. | Asvn. | Bhar. | Krit. | Rohi. | Mrig. | Ardh. | Punar. | Push. | Asles. |
|---|---|---|---|---|---|---|---|---|---|---|---|---|---|---|---|---|---|---|---|---|
| 30 | 60 | 90 | 120 | 150 | 180 | 210 | 240 | 270 | 300 | 330 | 360 | 13.3 | 26.7 | 40.0 | 53.3 | 66.7 | 80.0 | 93.3 | 106.7 | 120.0 |

## mean Sunrise for every ten days from A.D. 1850 to A.D. 1859.

| | 180 | 190 | 200 | 210 | 220 | 230 | 240 | 250 | 260 | 270 | 280 | 290 | 300 | 310 | 320 | 330 | 340 | 350 | 360 |
|---|---|---|---|---|---|---|---|---|---|---|---|---|---|---|---|---|---|---|---|
| | 174·4 | 184·3 | 194·1 | 204·0 | 213·8 | 223·7 | 233·5 | 243·4 | 253·3 | 263·1 | 273·0 | 282·8 | 292·7 | 302·5 | 312·4 | 322·2 | 332·1 | 342·0 | 351·8 |
| | −2·2 | −2·1 | −1·9 | −1·7 | −1·5 | −1·2 | −0·9 | −0·5 | −0·1 | +0·3 | +0·6 | +1·0 | +1·3 | +1·6 | +1·8 | +2·0 | +2·1 | +2·2 | +2·2 |
| | O 8 | O 18 | O 28 | N 7 | N 17 | N 27 | D 7 | D 17 | D 27 | Ja 6 | Ja 16 | Ja 26 | F 5 | F 15 | F 25 | Mr 6,7 | 16, 17 | 26, 27 | Ap 5,6 |
| | Pur 25 | Aipp 4 | Aipp 14 | Aipp 24 | Kart 4 | Kart 14 | Kart 24 | Marg 5 | Marg 15 | Marg 25 | Tai 5 | Tai 15 | Tai 25 | Masi 6 | Masi 16 | Masi 26 | Pang 6 | Pang 16 | Pang 26 |
| **51** | 188·7 | 195·6 | 202·5 | 209·3 | 216·4 | 223·4 | 230·8 | 238·0 | 245·3 | 252·7 | 260·3 | 267·9 | 275·5 | 282·9 | 290·9 | 298·7 | 306·6 | 314·3 | 322·1 |
| | 171·2 | 168·8 | 175·6 | 188·3 | 204·1 | 221·3 | 238·4 | 255·4 | 270·3 | 281·1 | 284·8 | 278·5 | 273·1 | 276·9 | 288·4 | 303·9 | 321·8 | 340·3 | 358·6 |
| | 164·0 | 166·2 | 168·3 | 170·4 | 172·3 | 174·0 | 175·7 | 177·3 | 178·5 | 179·6 | 180·4 | 180·9 | 181·1 | 180·7 | 180·2 | 179·5 | 178·6 | 177·4 | 176·1 |
| | 219·2 | 228·9 | 237·0 | 243·9 | 248·3 | 247·7 | 246·4 | 239·9 | 234·9 | 233·8 | 236·0 | 241·3 | 248·7 | 256·5 | 266·9 | 277·3 | 288·1 | 299·2 | 310·5 |
| | 355·5 | 354·7 | 353·9 | 353·1 | 352·6 | 352·1 | 351·7 | 351·8 | 351·9 | 351·9 | 352·5 | 353·1 | 353·8 | 354·8 | 355·9 | 357·0 | 358·2 | 359·5 | 0·8 |
| **52** | 90·3 | 95·3 | 99·9 | 104·2 | 107·9 | 110·6 | 112·9 | 113·7 | 113·7 | 112·3 | 109·7 | 106·4 | 102·3 | 98·7 | 96·1 | 95·0 | 95·0 | 96·5 | 98·2 |
| | 156·4 | 168·0 | 183·2 | 200·2 | 217·6 | 234·5 | 250·1 | 262·4 | 268·2 | 264·7 | 257·3 | 258·3 | 268·0 | 282·3 | 299·5 | 317·5 | 335·9 | 353·5 | 9·2 |
| | 189·1 | 190·2 | 191·7 | 193·4 | 194·9 | 196·5 | 198·1 | 199·3 | 200·6 | 201·9 | 202·9 | 204·0 | 204·8 | 205·4 | 205·9 | 206·3 | 206·6 | 206·5 | 206·5 |
| | 174·2 | 186·7 | 199·2 | 211·7 | 224·1 | 236·9 | 249·4 | 262·1 | 274·7 | 287·1 | 299·7 | 312·0 | 324·4 | 336·6 | 348·8 | 0·8 | 12·5 | 24·4 | 35·7 |
| | 10·4 | 9·6 | 8·7 | 7·8 | 7·2 | 6·6 | 6·0 | 5·8 | 5·6 | 5·4 | 5·7 | 6·1 | 6·5 | 7·4 | 8·3 | 9·3 | 10·4 | 11·5 | 12·7 |
| **53** | 202·7 | 209·7 | 216·6 | 223·9 | 231·0 | 238·4 | 245·8 | 253·2 | 260·8 | 268·4 | 276·2 | 284·0 | 291·9 | 299·7 | 307·7 | 315·6 | 323·4 | 331·3 | 339·0 |
| | 164·0 | 180·9 | 198·4 | 215·5 | 231·3 | 244·7 | 252·0 | 250·2 | 242·1 | 240·2 | 249·7 | 263·1 | 279·5 | 297·6 | 315·4 | 333·0 | 348·7 | 0·7 | 5·3 |
| | 211·3 | 213·4 | 215·5 | 217·7 | 219·9 | 222·1 | 224·3 | 226·6 | 228·8 | 231·0 | 233·0 | 235·0 | 236·7 | 238·3 | 240·8 | 240·8 | 241·9 | 242·6 | 242·9 |
| | 127·2 | 138·1 | 149·1 | 160·2 | 171·9 | 183·9 | 195·9 | 208·1 | 220·4 | 232·6 | 244·9 | 257·5 | 269·9 | 282·4 | 295·0 | 307·4 | 319·6 | 332·2 | 344·6 |
| | 25·6 | 24·8 | 24·0 | 23·2 | 22·4 | 21·6 | 20·9 | 20·4 | 20·0 | 19·6 | 19·7 | 19·8 | 19·9 | 20·7 | 21·5 | 22·3 | 23·3 | 24·3 | 25·3 |
| **54** | 107·4 | 113·2 | 118·5 | 123·7 | 128·5 | 133·2 | 137·2 | 140·8 | 143·4 | 145·5 | 146·2 | 146·0 | 144·1 | 141·0 | 137·4 | 133·7 | 130·4 | 128·1 | 127·0 |
| | 177·2 | 194·6 | 211·1 | 225·4 | 234·9 | 236·0 | 228·7 | 224·9 | 220·4 | 242·6 | 258·2 | 275·5 | 293·6 | 310·8 | 327·2 | 340·2 | 346·6 | 343·2 | 336·1 |
| | 236·9 | 238·7 | 240·7 | 242·5 | 244·6 | 246·7 | 248·8 | 251·1 | 253·4 | 255·7 | 258·0 | 260·2 | 262·5 | 264·6 | 266·6 | 268·6 | 270·4 | 271·9 | 273·3 |
| | 209·6 | 222·4 | 234·4 | 246·0 | 257·7 | 268·8 | 278·8 | 290·6 | 300·2 | 309·3 | 317·0 | 322·8 | 325·8 | 324·7 | 320·3 | 315·3 | 310·8 | 310·7 | 314·0 |
| | 40·8 | 40·2 | 39·6 | 38·9 | 38·1 | 37·3 | 36·4 | 35·7 | 35·0 | 34·4 | 34·2 | 34·0 | 33·8 | 34·2 | 34·6 | 35·1 | 36·0 | 37·0 | 38·0 |
| **55** | 217·1 | 224·1 | 231·2 | 238·6 | 246·1 | 253·3 | 261·1 | 268·7 | 276·3 | 284·3 | 292·0 | 299·9 | 307·9 | 315·7 | 323·5 | 331·4 | 339·1 | 346·9 | 354·5 |
| | 190·3 | 205·6 | 217·0 | 221·2 | 215·8 | 209·2 | 212·0 | 222·5 | 237·2 | 254·2 | 271·8 | 289·2 | 305·6 | 319·5 | 327·9 | 327·0 | 319·3 | 318·3 | 326·6 |
| | 265·9 | 266·7 | 267·9 | 269·4 | 271·0 | 272·8 | 274·9 | 277·1 | 279·2 | 281·5 | 283·8 | 286·2 | 288·4 | 290.7 | 293·1 | 295·4 | 297·6 | 299·6 | 301·8 |
| | 156·0 | 168·3 | 180·7 | 193·3 | 205·7 | 218·5 | 230·9 | 243·6 | 256·3 | 268·8 | 281·7 | 294·2 | 306·8 | 319·2 | 331·7 | 344·2 | 356·7 | 9·0 | 21·2 |
| | 56·0 | 55·5 | 55·1 | 54·5 | 53·5 | 52·6 | 51·7 | 51·1 | 50·5 | 49·8 | 49·4 | 48·9 | 48·6 | 48·8 | 48·9 | 49·0 | 49·7 | 50·5 | 51·1 |
| **56** | 121·5 | 127·2 | 133·0 | 138·9 | 144·3 | 149·9 | 154·1 | 159·5 | 164·2 | 168·5 | 172·4 | 175·3 | 177·6 | 179·0 | 179·3 | 178·0 | 175·2 | 172·0 | 168·6 |
| | 198·3 | 205·3 | 202·6 | 195·1 | 194·6 | 203·2 | 217·0 | 233·1 | 249·3 | 268·0 | 284·7 | 299·0 | 309·0 | 310·6 | 303·4 | 299·8 | 306·2 | 318·9 | 335·3 |
| | 299·9 | 299·6 | 299·9 | 300·4 | 301·0 | 302·0 | 303·3 | 305·0 | 306·9 | 308·7 | 310·7 | 313·0 | 315·3 | 317·6 | 320·0 | 322·4 | 324·9 | 327·3 | 329·6 |
| | 160·2 | 157·3 | 157·7 | 161·8 | 168·1 | 176·2 | 185·6 | 195·7 | 206·4 | 217·6 | 226·0 | 240·8 | 252·9 | 264·8 | 277·0 | 289·2 | 301·5 | 313·5 | 325·6 |
| | 70·2 | 69·9 | 69·7 | 69·5 | 68·6 | 67·6 | 66·8 | 66·2 | 65·7 | 65·1 | 64·4 | 63·8 | 63·1 | 63·0 | 62·8 | 62·7 | 63·1 | 63·6 | 63·9 |
| **57** | 233·9 | 241·1 | 248·3 | 255·6 | 263·0 | 270·8 | 278·2 | 285·9 | 293·6 | 301·3 | 309·1 | 317·0 | 324·7 | 332·6 | 340·1 | 347·8 | 355·3 | 2·7 | 10·3 |
| | 188·0 | 180·5 | 178·1 | 185·2 | 198·1 | 214·0 | 231·1 | 248·6 | 265·1 | 280·0 | 291·0 | 294·0 | 287·7 | 282·6 | 287·0 | 298·7 | 314·5 | 332·6 | 351·0 |
| | 339·5 | 338·2 | 337·1 | 336·3 | 335·9 | 335·7 | 336·1 | 336·6 | 337·5 | 338·7 | 340·1 | 341·9 | 343·8 | 345·9 | 348·0 | 350·2 | 352·6 | 355·0 | 357·4 |
| | 194·3 | 206·5 | 218·9 | 231·4 | 243·5 | 255·6 | 268·4 | 270·5 | 292·7 | 304·5 | 316·5 | 328·3 | 339·6 | 350·5 | 1·1 | 9·9 | 19·6 | 27·9 | 33·[illegible] |
| | 84·2 | 84·3 | 84·5 | 84·5 | 84·0 | 83·4 | 82·8 | 82·0 | 81·2 | 80·5 | 79·7 | 79·0 | 78·1 | 77·8 | 77·4 | 77·0 | 77·3 | 77·6 | 77·8 |
| **58** | 134·5 | 140·6 | 146·7 | 152·8 | 158·6 | 164·6 | 170·2 | 175·8 | 181·4 | 186·8 | 192·1 | 197·2 | 202·0 | 206·3 | 210·3 | 213·8 | 216·8 | 218·8 | 219·8 |
| | 161·5 | 166·2 | 177·9 | 193·2 | 210·5 | 227·5 | 244·3 | 259·9 | 272·3 | 277·7 | 273·9 | 266·4 | 268·1 | 278·5 | 292·7 | 310·1 | 328·2 | 346·2 | 4·[illegible] |
| | 20·6 | 19·4 | 18·2 | 16·8 | 15·4 | 14·3 | 13·3 | 12·6 | 11·7 | 11·8 | 12·6 | 13·3 | 14·3 | 15·6 | 17·2 | 18·9 | 20·8 | 22·8 | 25·0 |
| | 139·2 | 151·3 | 163·3 | 175·5 | 188·0 | 200·3 | 212·7 | 225·5 | 237·9 | 250·3 | 263·0 | 275·7 | 288·3 | 300·8 | 312·0 | 325·9 | 338·5 | 351·0 | 3·4 |
| | 97·4 | 97·8 | 98·1 | 98·4 | 98·2 | 98·1 | 97·8 | 97·2 | 96·5 | 95·8 | 95·0 | 94·1 | 93·2 | 92·9 | 92·6 | 92·3 | 91·9 | 91·5 | 91·0 |
| **59** | 253·7 | 260·7 | 266·6 | 274·6 | 282·0 | 289·2 | 296·7 | 304·2 | 311·7 | 319·3 | 326·6 | 334·0 | 341·0 | 348·7 | 356·2 | 3·3 | 10·7 | 17·6 | 24·7 |
| | 157·9 | 172·4 | 189·2 | 206·7 | 223·7 | 239·7 | 253·2 | 260·9 | 259·6 | 251·9 | 250·3 | 258·2 | 271·7 | 288·1 | 306·0 | 324·1 | 342·0 | 358·1 | 10·6 |
| | 58·3 | 57·9 | 57·9 | 56·2 | 55·4 | 55·2 | 52·8 | 51·4 | 50·3 | 49·1 | 48·3 | 47·9 | 47·6 | 48·0 | 48·4 | 49·2 | 50·3 | 51·6 | 53·2 |
| | 219·1 | 228·6 | 236·1 | 242·6 | 245·9 | 246·1 | 242·2 | 236·4 | 231·5 | 231·1 | 234·3 | 240·1 | 248·2 | 257·2 | 267·0 | 277·3 | 288·4 | 299·7 | 310·8 |
| | 110·8 | 111·5 | 112·1 | 112·8 | 112·8 | 112·9 | 112·9 | 112·4 | 111·8 | 111·3 | 110·5 | 109·8 | 109·0 | 108·2 | 107·4 | 106·7 | 106·4 | 106·1 | 105·6 |
| | O 9 | O 19 | O 29 | N 8 | N 18 | N 28 | D 8 | D 18 | D 28 | Ja 7 | Ja 17 | Ja 27 | F 6 | F 16 | F 26 | Mr 7 | Mr 17 | Mr 27 | Ap 6 |
| **60** | 147·2 | 153·6 | 159·7 | 166·1 | 172·2 | 178·3 | 184·5 | 190·7 | 196·8 | 202·9 | 209·1 | 215·0 | 221·1 | 226·9 | 232·9 | 238·5 | 244·1 | 249·6 | 254·8 |
| | 169·5 | 187·2 | 204·5 | 221·0 | 235·0 | 244·2 | 244·8 | 237·2 | 234·1 | 239·9 | 252·5 | 268·3 | 285·9 | 303·9 | 321·4 | 337·6 | 350·8 | 357·1 | 353·4 |
| | 90·5 | 91·3 | 91·9 | 92·1 | 92·0 | 91·6 | 90·8 | 89·7 | 88·5 | 87·2 | 85·9 | 84·7 | 83·5 | 82·6 | 82·0 | 81·5 | 81·2 | 81·1 | 82·5 |
| | 176·1 | 188·4 | 200·9 | 213·4 | 226·0 | 238·6 | 251·0 | 264·0 | 276·1 | 288·9 | 301·5 | 313·8 | 326·1 | 338·1 | 350·3 | 2·3 | 14·0 | 25·9 | 37·4 |
| | 123·5 | 124·2 | 125·0 | 125·8 | 126·1 | 126·3 | 126·7 | 126·4 | 126·2 | 125·8 | 125·1 | 124·3 | 123·7 | 122·9 | 122·2 | 121·3 | 120·7 | 120·2 | 119·6 |

| Mag. | P. Phal. | U. Phal. | Hasta. | Chit. | Svati. | Visa. | Anur. | Jyesh. | Mula. | P. Ash. | U. Ash. | Srav. | Dan. | Satab. | P. Bhad. | U. Bhad. | Revati. |
|---|---|---|---|---|---|---|---|---|---|---|---|---|---|---|---|---|---|
| 133·3 | 146·7 | 160·0 | 173·3 | 186·7 | 200·0 | 213·3 | 226·7 | 240·0 | 253·3 | 266·7 | 280·0 | 293·3 | 306·7 | 320·0 | 333·3 | 346·7 | 360·0 |

## Geocentric (Indian) Longitudes of Sun and Major planets at Ujjain

| Day of S. Yr. | | 0 | 10 | 20 | 30 | 40 | 50 | 60 | 70 | 80 | 90 | 100 | 110 | 120 | 130 | 140 | 150 | 160 | 170 |
|---|---|---|---|---|---|---|---|---|---|---|---|---|---|---|---|---|---|---|---|
| ⊙'s Mean Long. | | 357·0 | 6·9 | 16·7 | 26·6 | 36·4 | 46·3 | 56·1 | 66·0 | 75·9 | 85·7 | 95·6 | 105·4 | 115·3 | 125·1 | 135·0 | 144·8 | 154·7 | 164·6 |
| Corrn. for do. | | +2·1 | +2·0 | +1·9 | +1·7 | +1·4 | +1·1 | +0·8 | +0·4 | +0·0 | −0·3 | −0·7 | −1·1 | −1·4 | −1·6 | −1·9 | −2·0 | −2·1 | −2·2 |
| English date. | | Ap 11 | Ap 21 | My 1 | My 11 | My 21 | My 31 | Je 10 | Jl 20 | Je 30 | Jl 10 | Jl 20 | Jl 30 | Au 9 | Au 19 | Au 29 | S 8 | S 18 | S 28 |
| Tamil month. | | Chit | Chit | Chit | Chit | Vaik | Vaik | Vaik | Ani | Ani | Ani | Adi | Adi | Adi | Avan | Avan | Avan | Pur | Pur |
| Do. date. | | 1 | 11 | 21 | 31 | 10 | 20 | 30 | 9 | 19 | 29 | 7 | 17 | 27 | 6 | 16 | 26 | 5 | 15 |
| 1860 | Mars ... | 257·4 | 262·2 | 266·6 | 270·4 | 274·1 | 276·6 | 278·3 | 279·0 | 278·2 | 276·5 | 274·0 | 271·4 | 269·5 | 269·0 | 269·7 | 271·6 | 274·6 | 278·8 |
| ⊙'s L. | Merc. ... | 348·8 | 346·4 | 353·2 | 6·5 | 23·2 | 41·9 | 60·8 | 79·4 | 96·7 | 111·4 | 120·3 | 120·1 | 112·6 | 110·8 | 118·7 | 132·0 | 148·4 | 166·0 |
| +·6 | Jup. ... | 83·0 | 84·1 | 85·5 | 87·0 | 88·6 | 90·7 | 92·7 | 94·9 | 97·0 | 99·3 | 101·6 | 103·7 | 106·0 | 108·3 | 109·9 | 112·7 | 114·6 | 116·5 |
| | Ven. ... | 42·9 | 54·1 | 64·4 | 74·5 | 83·6 | 92·1 | 98·2 | 102·9 | 103·8 | 100·7 | 94·4 | 88·3 | 86·5 | 87·7 | 92·4 | 99·2 | 107·7 | 117·0 |
| | Sat. ... | 119·4 | 119·5 | 119·6 | 119·5 | 119·5 | 120·1 | 120·6 | 121·2 | 122·3 | 123·3 | 124·3 | 125·5 | 126·7 | 127·9 | 129·2 | 130·5 | 131·8 | 133·0 |
| 1861 | Mars ... | 43·2 | 49·8 | 55·4 | 62·9 | 69·5 | 75·9 | 82·4 | 88·7 | 95·3 | 101·7 | 108·0 | 114·5 | 120·7 | 127·1 | 133·7 | 139·8 | 146·2 | 152·6 |
| ⊙'s L. | Merc. ... | 332·1 | 344·0 | 0·2 | 18·3 | 37·2 | 56·4 | 74·2 | 89·7 | 100·4 | 102·8 | 96·1 | 92·1 | 98·4 | 110·5 | 126·7 | 144·2 | 162·1 | 179·3 |
| +·3 | Jup. ... | 113·6 | 113·8 | 114·2 | 114·9 | 115·9 | 117·1 | 118·5 | 120·8 | 122·1 | 123·9 | 126·0 | 128·1 | 130·3 | 132·5 | 134·6 | 136·8 | 139·1 | 141·3 |
| | Ven. ... | 351·2 | 3·7 | 16·0 | 28·5 | 40·7 | 53·1 | 65·2 | 77·6 | 89·6 | 101·9 | 114·3 | 126·4 | 138·5 | 150·7 | 162·7 | 174·8 | 186·9 | 199·1 |
| | Sat. ... | 133·4 | 133·1 | 132·9 | 132·6 | 132·8 | 133·2 | 133·5 | 134·4 | 135·2 | 136·0 | 137·1 | 138·2 | 139·3 | 140·5 | 141·7 | 143·0 | 144·3 | 145·6 |
| 1862 | Mars ... | 281·6 | 288·2 | 294·9 | 301·5 | 308·0 | 314·4 | 320·7 | 326·7 | 232·3 | 237·6 | 342·4 | 346·4 | 349·5 | 351·5 | 352·4 | 351·9 | 349·9 | 347·1 |
| ⊙'s L. | Merc. ... | 337·5 | 355·3 | 14·0 | 32·9 | 51·2 | 67·5 | 79·7 | 84·6 | 79·4 | 73·6 | 77·8 | 89·3 | 104·6 | 122·2 | 140·3 | 158·1 | 174·7 | 188·7 |
| +·1 | Jup. ... | 146·5 | 145·5 | 144·9 | 144·5 | 144·5 | 144·8 | 145·4 | 146·4 | 147·7 | 149·0 | 150·6 | 152·3 | 154·3 | 156·2 | 158·3 | 160·5 | 162·6 | 164·8 |
| | Ven. ... | 316·6 | 324·1 | 332·7 | 342·6 | 352·7 | 3·4 | 14·5 | 25·7 | 37·2 | 48·8 | 60·6 | 72·4 | 84·3 | 96·1 | 108·3 | 120·5 | 132·6 | 144·9 |
| | Sat. ... | 147·0 | 146·5 | 146·1 | 145·5 | 145·6 | 145·6 | 145·7 | 146·4 | 147·0 | 147·6 | 148·5 | 149·5 | 150·4 | 151·6 | 152·8 | 154·0 | 155·2 | 156·4 |
| Eng. date. | | Ap 12 | Ap 22 | My 2 | My 12 | My 22 | Je 1 | Je 11 | Je 21 | Jl 1 | Jl 11 | Jl 21 | Jl 31 | Au 10 | Au 20 | Au 30 | S 9 | S 19 | S 29 |
| 1863 | Mars ... | 58·4 | 64·5 | 70·6 | 76·7 | 82·9 | 89·1 | 95·3 | 101·6 | 107·8 | 114·1 | 120·3 | 126·7 | 133·0 | 139·3 | 145·7 | 152·1 | 158·5 | 165·2 |
| ⊙'s L. | Merc. ... | 352·8 | 11·5 | 29·5 | 46·3 | 59·4 | 65·5 | 61·4 | 54·9 | 57·3 | 68·7 | 84·1 | 101·7 | 120·0 | 138·2 | 155·3 | 170·0 | 180·2 | 182·1 |
| +·8 | Jup. ... | 179·9 | 178·6 | 177·4 | 176·2 | 175·4 | 174·7 | 174·5 | 174·6 | 174·8 | 175·5 | 176·4 | 177·6 | 179·1 | 180·7 | 182·5 | 184·4 | 186·4 | 188·4 |
| | Ven. ... | 29·0 | 40·9 | 52·9 | 64·8 | 75·9 | 87·9 | 99·4 | 110·5 | 121·5 | 131·6 | 141·6 | 150·9 | 159·2 | 165·5 | 169·9 | 171·1 | 167·8 | 161·6 |
| | Sat. ... | 160·3 | 159·7 | 159·0 | 158·4 | 158·2 | 158·2 | 157·9 | 158·3 | 158·8 | 159·1 | 160·0 | 160·8 | 161·6 | 162·7 | 163·9 | 164·9 | 166·1 | 167·3 |
| Eng. date. | | Ap 11 | Ap 21 | My 1 | My 11 | My 21 | My 31 | Je 10 | Je 20 | Je 30 | Jl 10 | Jl 20 | Jl 30 | Au 9 | Au 19 | Au 29 | S 8 | S 18 | S 28 |
| 1864 | Mars ... | 301·6 | 309·1 | 316·7 | 324·0 | 331·6 | 338·0 | 346·1 | 353·3 | 0·2 | 7·2 | 13·7 | 20·0 | 26·0 | 31·9 | 37·3 | 42·2 | 46·4 | 50·1 |
| ⊙'s L. | Merc. ... | 6·7 | 24·0 | 37·8 | 45·8 | 44·0 | 36·5 | 37·0 | 46·8 | 61·7 | 79·0 | 97·5 | 115·9 | 133·5 | 149·4 | 161·5 | 166·3 | 161·4 | 154·7 |
| +·6 | Jup. ... | 214·0 | 213·0 | 211·8 | 210·5 | 209·2 | 208·0 | 206·9 | 206·8 | 205·4 | 204·9 | 204·9 | 205·2 | 205·9 | 206·8 | 208·1 | 209·5 | 211·1 | 212·9 |
| | Ven. ... | 333·7 | 346·0 | 358·3 | 10·5 | 23·2 | 35·0 | 47·3 | 58·3 | 71·6 | 83·7 | 96·1 | 108·8 | 120·8 | 132·9 | 145·3 | 157·5 | 170·0 | 182·2 |
| | Sat. ... | 173·5 | 172·7 | 171·9 | 171·2 | 170·8 | 170·6 | 170·1 | 170·2 | 170·4 | 170·5 | 171·2 | 171·8 | 172·6 | 173·7 | 174·9 | 176·1 | 177·1 | 178·1 |
| 1865 | Mars ... | 73·9 | 79·3 | 84·8 | 90·5 | 96·1 | 102·0 | 108·0 | 113·9 | 119·7 | 126·0 | 132·1 | 138·5 | 144·8 | 151·2 | 157·5 | 164·0 | 170·5 | 177·0 |
| ⊙'s L. | Merc. ... | 16·2 | 25·5 | 25·8 | 18·4 | 18·6 | 24·6 | 38·6 | 56·0 | 74·3 | 93·1 | 111·3 | 128·0 | 141·7 | 148·8 | 146·5 | 138·9 | 138·9 | 147·6 |
| +·3 | Jup. ... | 246·9 | 246·8 | 246·3 | 245·7 | 244·8 | 243·6 | 242·3 | 241·0 | 239·8 | 238·6 | 237·5 | 236·8 | 236·4 | 236·4 | 236·7 | 237·2 | 238·1 | 239·4 |
| | Ven. ... | 33·9 | 35·2 | 32·0 | 25·2 | 19·8 | 18·0 | 20·1 | 24·8 | 31·7 | 40·1 | 49·7 | 59·6 | 70·4 | 81·2 | 92·5 | 104·0 | 115·9 | 127·6 |
| | Sat. ... | 186·3 | 185·5 | 184·8 | 184·0 | 183·4 | 182·9 | 182·2 | 182·2 | 182·1 | 182·1 | 182·5 | 183·0 | 183·4 | 184·3 | 185·1 | 186·1 | 187·2 | 188·3 |
| 1866 | Mars ... | 319·4 | 327·7 | 334·9 | 342·5 | 350·3 | 357·7 | 5·2 | 12·4 | 19·8 | 26·7 | 33·7 | 40·7 | 47·3 | 53·8 | 60·0 | 66·0 | 71·7 | 77·1 |
| ⊙'s L. | Merc. ... | 3·2 | 1·0 | 357·0 | 3·0 | 15·8 | 32·7 | 51·0 | 69·0 | 88·7 | 106·0 | 120·7 | 130·1 | 130·7 | 123·1 | 120·7 | 127·6 | 140·8 | 157·0 |
| +·0 | Jup. ... | 277·7 | 278·7 | 279·5 | 280·0 | 280·1 | 279·8 | 279·2 | 278·3 | 277·1 | 275·9 | 274·7 | 273·3 | 272·1 | 271·0 | 270·2 | 269·6 | 269·5 | 269·6 |
| | Ven. ... | 10·1 | 22·4 | 34·7 | 47·0 | 59·3 | 71·5 | 83·4 | 95·4 | 107·2 | 119·2 | 131·2 | 142·7 | 154·4 | 165·9 | 177·3 | 188·3 | 199·0 | 209·4 |
| | Sat. ... | 198·7 | 197·9 | 197·2 | 196·3 | 195·6 | 194·9 | 194·3 | 194·0 | 193·6 | 193·4 | 193·6 | 193·9 | 194·2 | 194·9 | 195·6 | 196·3 | 197·3 | 198·4 |
| Eng. date. | | Ap 12 | Ap 22 | My 2 | My 12 | My 22 | Je 1 | Je 11 | Je 21 | Jl 1 | Jl 11 | Jl 21 | Jl 31 | Au 10 | Au 20 | Au 30 | S 9 | S 19 | S 29 |
| 1867 | Mars ... | 92·1 | 96·4 | 100·9 | 105·9 | 111·1 | 116·4 | 121·9 | 127·5 | 133·6 | 139·3 | 145·4 | 151·5 | 157·9 | 164·2 | 170·7 | 177·1 | 183·9 | 190·5 |
| ⊙'s L. | Merc. ... | 338·4 | 343·0 | 354·9 | 11·3 | 29·6 | 48·8 | 67·4 | 85·3 | 100·4 | 111·3 | 113·4 | 106·4 | 102·9 | 108·6 | 121·3 | 137·2 | 154·7 | 172·4 |
| +·8 | Jup. ... | 306·8 | 308·8 | 310·4 | 311·8 | 313·1 | 314·1 | 314·7 | 315·0 | 314·9 | 314·5 | 313·8 | 312·8 | 311·6 | 310·3 | 308·9 | 307·6 | 306·3 | 305·0 |
| | Ven. ... | 318·3 | 330·0 | 342·0 | 353·8 | 5·9 | 17·8 | 29·8 | 41·8 | 53·7 | 66·0 | 78·9 | 91·6 | 102·4 | 115·7 | 127·0 | 139·4 | 151·8 | 164·0 |
| | Sat. ... | 210·8 | 210·0 | 209·2 | 208·5 | 207·7 | 207·0 | 206·3 | 205·8 | 205·4 | 204·9 | 205·0 | 205·0 | 205·1 | 205·4 | 205·6 | 205·8 | 207·1 | 208·5 |
| Eng. date. | | Ap 11 | Ap 21 | My 1 | My 11 | My 21 | My 31 | Je 10 | Je 20 | Je 30 | Jl 10 | Jl 20 | Jl 30 | Au 9 | Au 19 | Au 29 | S 8 | S 18 | S 28 |
| 1868 | Mars ... | 336·8 | 344·4 | 352·3 | 359·8 | 7·3 | 14·9 | 22·2 | 29·5 | 36·8 | 43·8 | 50·8 | 57·5 | 64·4 | 70·8 | 77·3 | 83·5 | 89·6 | 95·6 |
| ⊙'s L. | Merc. ... | 333·2 | 348·3 | 6·2 | 25·2 | 44·1 | 62·4 | 78·5 | 90·7 | 95·3 | 90·1 | 84·3 | 88·3 | 99·9 | 115·4 | 132·8 | 150·7 | 168·4 | 184·8 |
| +·5 | Jup. ... | 384·9 | 337·2 | 339·3 | 341·6 | 343·5 | 345·4 | 347·0 | 348·5 | 349·6 | 350·5 | 351·1 | 351·3 | 351·2 | 350·6 | 349·7 | 348·7 | 347·5 | 346·2 |
| | Ven. ... | 43·2 | 53·9 | 64·5 | 74·4 | 83·4 | 91·4 | 97·5 | 101·0 | 101·7 | 97·3 | 91·0 | 85·7 | 83·8 | 87·1 | 91·5 | 98·9 | 107·4 | 117·3 |
| | Sat. ... | 222·3 | 221·7 | 221·0 | 220·5 | 219·7 | 219·0 | 218·3 | 217·6 | 217·0 | 216·4 | 216·3 | 216·1 | 216·0 | 216·3 | 216·7 | 217·1 | 218·0 | 218·8 |
| 1869 | Mars ... | 115·2 | 117·3 | 120·2 | 123·5 | 127·8 | 132·1 | 137·0 | 142·3 | 147·6 | 153·4 | 159·2 | 165·6 | 171·5 | 177·8 | 184·3 | 190·8 | 197·4 | 204·3 |
| ⊙'s L. | Merc. ... | 344·2 | 1·8 | 20·7 | 39·1 | 56·0 | 69·8 | 76·4 | 73·0 | 66·0 | 68·2 | 78·4 | 93·2 | 111·7 | 128·9 | 146·9 | 164·0 | 179·0 | 189·1 |
| +·3 | Jup. ... | 2·4 | 4·8 | 7·2 | 9·7 | 12·8 | 14·3 | 16·5 | 18·6 | 20·4 | 22·4 | 23·9 | 25·2 | 26·5 | 27·2 | 27·6 | 27·4 | 27·0 | 26·9 |
| | Ven. ... | 351·8 | 4·3 | 16·8 | 29·1 | 41·2 | 53·7 | 65·8 | 78·1 | 90·2 | 102·5 | 114·7 | 127·0 | 139·1 | 151·1 | 163·4 | 175·4 | 187·5 | 199·5 |
| | Sat. ... | 234·0 | 233·4 | 232·9 | 232·3 | 231·4 | 230·6 | 229·8 | 229·2 | 228·6 | 227·9 | 227·6 | 227·2 | 227·0 | 227·1 | 227·3 | 227·6 | 228·2 | 228·9 |

| Mesh. | Vrsh. | Mith. | Kat. | Sim. | Kan.. | Tul. | Vrsch. | Dhan. | Mak. | Kum. | Min. | Asvn. | Bhar. | Krit. | Rohi. | Mrig. | Ardh. | Punar. | Push. | Asles. |
|---|---|---|---|---|---|---|---|---|---|---|---|---|---|---|---|---|---|---|---|---|
| 30 | 60 | 90 | 120 | 150 | 180 | 210 | 240 | 270 | 300 | 330 | 360 | 13·3 | 26·7 | 40·0 | 53·3 | 66·7 | 80·0 | 93·3 | 106·7 | 120·0 |

**mean Sunrise for every ten days from A.D. 1860 to A.D. 1869.**

| 180 | 190 | 200 | 210 | 220 | 230 | 240 | 250 | 260 | | 270 | 280 | 290 | 300 | 310 | 320 | 330 | 340 | 350 | 360 |
|---|---|---|---|---|---|---|---|---|---|---|---|---|---|---|---|---|---|---|---|
| 174·4 | 184·3 | 194·1 | 204·0 | 213·8 | 223·7 | 233·5 | 243·4 | 253·3 | | 263·1 | 273·0 | 282·8 | 292·7 | 302·5 | 312·4 | 322·2 | 332·1 | 342·0 | 351·8 |
| −2·2 | −2·1 | −1·9 | −1·7 | −1·5 | −1·2 | −0·9 | −0·5 | −0·1 | | +0·3 | +0·6 | +1·0 | +1·3 | +1·6 | +1·8 | +2·0 | +2·1 | +2·2 | +2·2 |
| O 8 | O 18 | O 28 | N 7 | N 17 | N 27 | D 7 | D 17 | D 27 | | Ja 6 | Ja 16 | Ja 26 | F 5 | F 15 | F 25 | Mr 7 | Mr 17 | Mr 27 | Ap 6 |
| Pur 25 | Aipp 4 | Aipp 14 | Aipp 24 | Kart 4 | Kart 14 | Kart 24 | Marg 5 | Marg 15 | | Marg 25 | Tai 5 | Tai 15 | Tai 25 | Masi 6 | Masi 16 | Masi 26 | Pang 6 | Pang 16 | Pang 26 |
| 283·3 | 288·7 | 294·3 | 300·3 | 306·5 | 312·9 | 319·5 | 325·9 | 332·7 | **61** | 339·3 | 346·2 | 353·0 | 359·8 | 6·4 | 13·3 | 19·9 | 26·7 | 33·3 | 39·7 |
| 183·6 | 200·6 | 215·4 | 226·5 | 229·9 | 224·5 | 218·2 | 221·5 | 232·2 | | 247·3 | 264·4 | 282·0 | 299·6 | 316·2 | 330·0 | 338·0 | 337·3 | 329·7 | 328·7 |
| 118·4 | 120·6 | 121·4 | 122·6 | 123·5 | 124·2 | 124·3 | 124·3 | 123·7 | | 123·1 | 122·5 | 121·1 | 119·8 | 118·5 | 117·2 | 116·0 | 115·1 | 114·4 | 113·3 |
| 126·3 | 138·0 | 149·2 | 160·3 | 172·4 | 184·2 | 196·0 | 208·2 | 220·5 | | 232·9 | 245·4 | 257·8 | 270·4 | 282·7 | 295·3 | 307·7 | 320·1 | 332·7 | 344·9 |
| 134·1 | 135·3 | 136·3 | 137·2 | 138·2 | 139·0 | 139·9 | 139·7 | 139·6 | | 139·7 | 139·1 | 138·6 | 138·0 | 137·2 | 136·4 | 135·6 | 134·9 | 134·2 | 133·5 |
| 159·2 | 165·4 | 171·8 | 178·3 | 184·7 | 191·3 | 197·8 | 204·3 | 210·9 | **62** | 217·5 | 224·2 | 230·7 | 237·4 | 244·0 | 250·9 | 257·7 | 264·5 | 271·3 | 278·5 |
| 195·5 | 208·1 | 215·3 | 211·2 | 203·7 | 203·9 | 212·8 | 226·5 | 243·2 | | 260·6 | 278·2 | 294·7 | 309·4 | 319·1 | 320·5 | 312·1 | 309·6 | 316·7 | 329·6 |
| 143·5 | 145·5 | 147·3 | 149·2 | 150·8 | 152·1 | 153·3 | 154·2 | 154·7 | | 155·1 | 155·1 | 154·7 | 154·0 | 153·1 | 151·9 | 150·9 | 149·8 | 148·1 | 147·0 |
| 211·1 | 222·9 | 234·7 | 246·5 | 257·7 | 268·9 | 279·9 | 290·3 | 300·1 | | 308·8 | 315·9 | 321·3 | 323·1 | 321·3 | 315·5 | 310·2 | 307·5 | 308·9 | 312·0 |
| 146·9 | 147·9 | 149·0 | 150·0 | 150·7 | 151·5 | 152·1 | 152·3 | 152·6 | | 152·8 | 152·6 | 152·3 | 151·9 | 151·2 | 150·4 | 149·7 | 148·9 | 148·2 | 147·4 |
| 344·6 | 343·0 | 342·1 | 342·6 | 345·4 | 347·8 | 350·5 | 354·3 | 358·7 | **63** | 3·6 | 8·4 | 14·0 | 19·4 | 24·9 | 30·8 | 36·7 | 42·9 | 48·3 | 54·7 |
| 197·7 | 197·4 | 189·7 | 187·1 | 193·3 | 206·6 | 222·0 | 239·4 | 256·9 | | 273·6 | 288·8 | 300·1 | 303·8 | 297·8 | 292·5 | 296·2 | 307·8 | 323·7 | 341·5 |
| 167·1 | 169·2 | 171·3 | 173·5 | 175·4 | 177·4 | 179·1 | 180·6 | 181·9 | | 183·2 | 184·1 | 184·8 | 185·1 | 185·1 | 184·7 | 184·0 | 183·1 | 182·0 | 180·8 |
| 157·4 | 169·5 | 181·9 | 194·4 | 206·9 | 219·7 | 232·0 | 244·8 | 257·6 | | 270·1 | 282·7 | 295·4 | 308·0 | 320·4 | 332·9 | 345·6 | 357·9 | 10·2 | 22·2 |
| 157·7 | 158·9 | 160·1 | 161·3 | 162·0 | 163·0 | 163·9 | 164·3 | 164·8 | | 165·2 | 165·0 | 164·9 | 164·8 | 164·3 | 163·7 | 163·1 | 162·3 | 161·6 | 160·8 |
| O 9 | O 19 | O 29 | N 8 | N 18 | N 28 | D 8 | D 18 | D 28 | | Ja 7 | Ja 17 | Ja 27 | F 6 | F 16 | F 26 | Mr 7 | Mr 17 | Mr 27 | Ap 6 |
| 171·6 | 178·2 | 184·7 | 191·3 | 198·1 | 204·8 | 211·6 | 218·6 | 225·4 | **64** | 232·5 | 239·5 | 246·6 | 253·8 | 261·0 | 268·8 | 275·7 | 283·1 | 290·4 | 297·9 |
| 175·1 | 170·8 | 176·0 | 187·8 | 203·1 | 219·1 | 237·5 | 254·4 | 269·9 | | 282·0 | 287·3 | 282·7 | 275·9 | 278·0 | 288·4 | 303·2 | 320·8 | 339·2 | 357·7 |
| 190·6 | 192·7 | 194·9 | 197·2 | 199·4 | 201·5 | 203·6 | 205·6 | 207·4 | | 209·3 | 210·9 | 212·3 | 213·5 | 214·5 | 215·0 | 215·6 | 215·5 | 215·1 | 214·5 |
| 156·3 | 154·6 | 156·2 | 161·1 | 168·2 | 177·0 | 186·5 | 196·9 | 207·8 | | 218·9 | 230·6 | 242·3 | 254·4 | 266·9 | 278·7 | 290·7 | 303·0 | 315·1 | 327·5 |
| 168·6 | 169·8 | 171·0 | 172·3 | 173·2 | 174·2 | 175·1 | 175·7 | 176·2 | | 176·9 | 177·0 | 177·2 | 177·3 | 176·8 | 176·4 | 175·9 | 175·2 | 174·5 | 173·7 |
| O 8 | O 18 | O 28 | N 7 | N 17 | N 27 | D 7 | D 17 | D 27 | | Ja 6 | Ja 16 | Ja 26 | F 5 | F 15 | F 25 | Mr 7 | Mr 17 | Mr 27 | Ap 6 |
| 52·6 | 54·4 | 55·0 | 54·0 | 52·1 | 49·1 | 45·6 | 42·9 | 41·2 | **65** | 40·7 | 41·3 | 42·9 | 45·5 | 48·5 | 52·2 | 56·5 | 61·2 | 66·2 | 71·2 |
| 157·5 | 168·0 | 182·5 | 199·2 | 216·6 | 223·8 | 249·6 | 262·9 | 270·4 | | 268·9 | 260·7 | 259·9 | 268·3 | 282·0 | 298·6 | 316·8 | 335·0 | 353·0 | 9·0 |
| 214·9 | 216·8 | 218·8 | 221·1 | 223·3 | 225·5 | 227·7 | 230·0 | 232·2 | | 234·4 | 236·5 | 238·5 | 240·2 | 242·3 | 243·3 | 244·7 | 245·8 | 246·5 | 246·9 |
| 194·9 | 207·3 | 219·5 | 231·8 | 244·1 | 256·5 | 268·7 | 281·1 | 293.1 | | 304·9 | 316·9 | 328·5 | 340·0 | 350·8 | 1·3 | 10·5 | 19·2 | 27·1 | 32·4 |
| 179·1 | 180·3 | 181·5 | 182·7 | 183·8 | 184·9 | 185·9 | 186·7 | 187·4 | | 188·3 | 188·5 | 188·7 | 189·0 | 188·9 | 188·8 | 188·6 | 188·0 | 187·1 | 186·5 |
| 183·8 | 190·5 | 197·3 | 204·3 | 211·1 | 218·1 | 225·1 | 232·3 | 239·6 | **66** | 246·8 | 254·3 | 261·6 | 269·3 | 276·8 | 284·4 | 292·1 | 300·0 | 307·7 | 315·4 |
| 161·6 | 178·2 | 195·7 | 213·0 | 229·4 | 243·5 | 252·8 | 254·1 | 246·7 | | 242·9 | 248·7 | 261·2 | 277·0 | 294·4 | 312·6 | 330·2 | 347·1 | 0·5 | 7·7 |
| 240·8 | 242·4 | 244·1 | 246·0 | 248·0 | 250·1 | 252·4 | 254·7 | 257·0 | | 259·3 | 261·6 | 263·9 | 266·2 | 268·3 | 270·3 | 272·4 | 274·2 | 275·8 | 277·3 |
| 139·8 | 151·7 | 163·9 | 176·1 | 188·4 | 200·9 | 213·3 | 226·0 | 238·3 | | 251·0 | 263·6 | 276·3 | 289·0 | 301·4 | 314·1 | 326·7 | 339·1 | 351·6 | 4·0 |
| 189·2 | 190·6 | 191·8 | 192·9 | 194·0 | 195·2 | 196·3 | 197·2 | 198·0 | | 199·0 | 199·6 | 200·1 | 200·6 | 200·6 | 200·7 | 200·6 | 200·1 | 199·1 | 198·9 |
| 82·4 | 86·9 | 90·8 | 94·4 | 97·2 | 99·0 | 99·8 | 99·3 | 97·7 | **67** | 94·8 | 90·9 | 87·0 | 83·9 | 81·8 | 81·2 | 82·3 | 83·8 | 86·5 | 89·9 |
| 174·3 | 191·9 | 208·5 | 223·9 | 235·1 | 239·0 | 233·5 | 227·2 | 230·1 | | 240·7 | 255·6 | 272·8 | 290·7 | 308·3 | 325·0 | 339·4 | 348·2 | 347·9 | 340·3 |
| 270·0 | 271·0 | 272·1 | 273·4 | 275·0 | 276·8 | 278·6 | 280·8 | 282·9 | | 285·2 | 287·4 | 289·7 | 292·1 | 294·4 | 296·8 | 299·2 | 301·5 | 303·6 | 305·8 |
| 219·0 | 227·9 | 235·2 | 242·1 | 244·0 | 243·2 | 238·5 | 232·2 | 228·7 | | 229·2 | 232·9 | 239·2 | 247·6 | 256·7 | 266·9 | 277·3 | 288·2 | 299·7 | 311·1 |
| 199·4 | 200·6 | 201·8 | 202·9 | 204·0 | 205·1 | 206·3 | 207·2 | 208·4 | | 209·5 | 210·0 | 210·6 | 211·2 | 211·0 | 210·9 | 210·7 | 210·9 | 211·0 | 211·1 |
| O 9 | O 19 | O 29 | N 8 | N 18 | N 28 | D 8 | D 18 | D 28 | | Ja 7 | Ja 17 | Ja 27 | F 6 | F 16 | F 26 | Mr 7 | Mr 17 | Mr 27 | Ap 6 |
| 197·2 | 204·2 | 211·2 | 218·3 | 225·3 | 232·7 | 240·0 | 247·4 | 254·8 | **68** | 262·5 | 270·1 | 277·9 | 285·5 | 293·4 | 301·2 | 309·2 | 317·1 | 324·9 | 332·7 |
| 189·4 | 205·5 | 217·7 | 223·6 | 219·8 | 212·3 | 213·2 | 222·9 | 236·6 | | 253·3 | 270·9 | 288·3 | 305·3 | 319·6 | 329·6 | 330·7 | 323·2 | 320·2 | 327·1 |
| 304·7 | 304·5 | 304·5 | 304·9 | 305·5 | 306·5 | 307·7 | 309·3 | 311·1 | | 313·0 | 314·8 | 317·0 | 319·3 | 321·6 | 324·0 | 326·5 | 328·8 | 331·1 | 333·7 |
| 176·7 | 189·0 | 201·7 | 214·0 | 226·7 | 239·2 | 251·8 | 264·4 | 277·1 | | 289·5 | 301·9 | 314·3 | 326·7 | 339·7 | 350·9 | 2·7 | 14·6 | 26·3 | 37·8 |
| 209·5 | 210·6 | 211·7 | 212·8 | 214·1 | 215·3 | 216·5 | 217·4 | 218·6 | | 219·6 | 220·5 | 221·3 | 222·1 | 222·4 | 222·8 | 223·1 | 222·9 | 222·8 | 222·6 |
| O 8 | O 18 | O 28 | N 7 | N 17 | N 27 | D 7 | D 17 | D 27 | | Ja 6 | Ja 16 | Ja 26 | F 5 | F 15 | F 25 | Mr 7 | Mr 17 | Mr 27 | Ap 6 |
| 101·2 | 106·8 | 111·9 | 116·8 | 121·3 | 125·4 | 128·7 | 131·4 | 133·2 | **69** | 133·9 | 133·5 | 131·5 | 128·4 | 124·9 | 121·2 | 117·9 | 115·4 | 114·5 | 114·7 |
| 198·4 | 207·0 | 206·1 | 198·5 | 196·4 | 203·5 | 216·3 | 232·1 | 249·6 | | 267·0 | 283·9 | 299·1 | 310·2 | 313·7 | 307·4 | 301·6 | 306·6 | 318·5 | 334·5 |
| 344·7 | 343·5 | 342·5 | 341·6 | 340·9 | 340·8 | 340·9 | 341·2 | 342·2 | | 343·0 | 344·5 | 346·1 | 347·9 | 349·9 | 352·1 | 354·3 | 356·5 | 358·8 | 1·2 |
| 127·2 | 138·3 | 149·3 | 161·0 | 172·7 | 184·7 | 196·7 | 208·9 | 221·3 | | 233·4 | 245·8 | 258·3 | 270·9 | 283·4 | 296·0 | 308·4 | 320·8 | 333·2 | 345·6 |
| 219·6 | 220·6 | 221·6 | 222·7 | 223·9 | 225·1 | 226·2 | 227·3 | 228·3 | | 229·5 | 230·4 | 231·2 | 232·2 | 232·8 | 233·3 | 233·9 | 233·9 | 234·0 | 234·0 |
| 211·1 | 218·2 | 225·3 | 232·8 | 239·9 | 247·2 | 255·0 | 262·6 | 270·2 | **70** | 277·9 | 285·7 | 293·6 | 301·4 | 309·4 | 317·3 | 325·0 | 332·9 | 340·7 | 348·4 |
| 191·6 | 185·0 | 180·2 | 184·7 | 196·3 | 211·6 | 228·5 | 245·8 | 262·8 | | 278·5 | 291·1 | 296·7 | 292·7 | 294.3 | 296·1 | 306·4 | 321·1 | 338·0 | 348·2 |
| 26·0 | 24·9 | 23·6 | 22·3 | 21·0 | 19·9 | 18·8 | 18·1 | 17·5 | | 17·3 | 17·5 | 18·0 | 19·0 | 20·1 | 21·5 | 23·5 | 25·0 | 26·9 | 29·1 |
| 211·4 | 223·3 | 235·1 | 246·9 | 258·2 | 269·3 | 280·1 | 290·1 | 299·9 | | 308·4 | 314·5 | 319·6 | 320·8 | 317·7 | 311·6 | 306·6 | 304·9 | 306·9 | 311·8 |
| 229·4 | 230·4 | 231.4 | 232·4 | 233·5 | 234·6 | 235·8 | 236·9 | 238·0 | | 239·2 | 240·3 | 241·3 | 242·3 | 243·0 | 243·8 | 244·4 | 244·7 | 244·9 | 245·2 |

| Mag. | P. Phal. | U. Phal. | Hasta. | Chit. | Svati. | Visa. | Anur. | Jyesh. | Mula. | P. Ash. | U. Ash. | Srav. | Dan. | Satab. | P.Bhad. | U. Bhad. | Revati. |
|---|---|---|---|---|---|---|---|---|---|---|---|---|---|---|---|---|---|
| 133·3 | 146·7 | 160·0 | 173·3 | 186·7 | 200·0 | 213·3 | 226·7 | 240·0 | 253·3 | 266·7 | 280·0 | 293·3 | 306·7 | 320·0 | 333·3 | 346·7 | 360·0 |

## Geocentric (Indian) Longitudes of Sun and Major planets at Ujjain

| Day of S. Yr. | | 0 | 10 | 20 | 30 | 40 | 50 | 60 | 70 | 80 | 90 | 100 | 110 | 120 | 130 | 140 | 150 | 160 | 170 |
|---|---|---|---|---|---|---|---|---|---|---|---|---|---|---|---|---|---|---|---|
| ☉'s Mean L. | | 357·0 | 6·9 | 16·7 | 26·6 | 36·4 | 46·3 | 56·1 | 66·0 | 75·9 | 85·7 | 95·6 | 105·4 | 115·3 | 125·1 | 135·0 | 144·8 | 154·7 | 164·6 |
| Corrn. for do. | | +2·1 | +2·0 | +1·9 | +1·7 | +1·4 | +1·1 | +0·8 | +0·4 | +0·0 | −0·3 | −0·7 | −1·1 | −1·4 | −1·6 | −1·9 | −2·0 | −2·1 | −2·2 |
| English date. | | Ap 11 | Ap 21 | My 1 | My 11 | My 21 | My 31 | Je 10 | Je 20 | Je 30 | Jl 10 | Jl 20 | Jl 30 | Au 9 | Au 19 | Au 29 | S 8 | S 18 | S 28 |
| Tamil month. | | Chit | Chit | Chit | Chit | Vaik | Vaik | Vaik | Ani | Ani | Ani | Adi | Adi | Adi | Avan | Avan | Avan | Pura | Pura |
| Do. date. | | 1 | 11 | 21 | 31 | 10 | 20 | 30 | 9 | 19 | 29 | 7 | 17 | 27 | 6 | 16 | 26 | 5 | 15 |
| 1870 | Mars ... | 352·3 | 0·0 | 7·7 | 15·1 | 22·5 | 29·8 | 37·1 | 44·2 | 51·4 | 58·2 | 65·0 | 71·8 | 78·4 | 85·0 | 91·4 | 97·7 | 104·0 | 110·0 |
| ☉'s L. | Merc. ... | 357·5 | 14·8 | 33·3 | 47·7 | 56·5 | 55·7 | 48·2 | 47·5 | 56·4 | 71·0 | 88·2 | 106·3 | 124·7 | 142·4 | 158·3 | 170·4 | 175·7 | 171·0 |
| +·0 | Jup. ... | 30·0 | 32·2 | 34·5 | 36·9 | 39·2 | 41·6 | 44·0 | 16·3 | 48·8 | 51·1 | 53·1 | 55·1 | 56·9 | 58·6 | 59·9 | 61·1 | 62·1 | 62·8 |
| | Ven. ... | 314·1 | 322·9 | 332·0 | 341·9 | 352·2 | 3·0 | 14·0 | 25·4 | 36·9 | 48·7 | 60·1 | 72·3 | 84·1 | 96·0 | 108·2 | 120·4 | 132·5 | 144·8 |
| | Sat. ... | 245·2 | 245·0 | 244·7 | 244·6 | 243·8 | 242·9 | 242·0 | 241·2 | 240·5 | 239·8 | 239·3 | 238·9 | 238·3 | 238·3 | 238·2 | 238·2 | 238·6 | 238·9 |
| Eng. date. | | Ap 12 | Ap 22 | My 2 | My 12 | My 22 | Je 1 | Je 11 | Je 21 | Jl 1 | Jl 11 | Jl 21 | Jl 31 | Au 10 | Au 20 | Au 30 | S 9 | S 19 | S 29 |
| 1871 | Mars ... | 148·6 | 147·1 | 147·0 | 148·2 | 150·1 | 152·9 | 156·7 | 160·8 | 165·6 | 170·6 | 176·0 | 181·8 | 187·6 | 193·8 | 200·2 | 206·7 | 213·4 | 220·4 |
| ☉'s L. | Merc. ... | 12·1 | 27·1 | 36·5 | 36·8 | 29·2 | 27·7 | 35·7 | 50·0 | 66·9 | 85·5 | 103·9 | 122·0 | 138·7 | 151·8 | 158·6 | 155·9 | 148·4 | 148·8 |
| +·7 | Jup. ... | 58·0 | 59·9 | 61·8 | 63·8 | 66·0 | 68·1 | 70·4 | 72·8 | 75·2 | 77·5 | 79·8 | 82·0 | 84·3 | 86·4 | 88·3 | 90·2 | 91·9 | 93·3 |
| | Ven. ... | 29·5 | 41·5 | 53·2 | 65·3 | 77·0 | 88·4 | 99·8 | 110·8 | 121·3 | 131·0 | 141·5 | 150·6 | 158·5 | 164·4 | 168·0 | 168·6 | 164·3 | 158·1 |
| | Sat. ... | 256·3 | 256·0 | 255·8 | 255·6 | 255·4 | 255·1 | 254·9 | 253·7 | 252·6 | 251·4 | 250·8 | 250·3 | 249·7 | 249·5 | 249·4 | 249·2 | 249·5 | 249·7 |
| Eng. date. | | Ap 11 | Ap 21 | My 1 | My 11 | My 21 | My 31 | Je 10 | Je 20 | Je 30 | Jl 10 | Jl 20 | Jl 30 | Au 9 | Au 19 | Au 29 | S 8 | S 18 | S 28 |
| 1872 | Mars ... | 8·1 | 15·5 | 23·0 | 30·2 | 37·5 | 44·5 | 51·5 | 58·4 | 65·3 | 72·1 | 78·8 | 85·4 | 92·0 | 98·4 | 104·8 | 111·1 | 117·5 | 123·6 |
| ☉'s L. | Merc. ... | 14·2 | 18·3 | 11·8 | 8·0 | 14·1 | 27·4 | 43·9 | 62·4 | 81·2 | 99·6 | 116·8 | 131·3 | 140·7 | 140·6 | 133·0 | 130·7 | 138·1 | 151·1 |
| +·5 | Jup. ... | 87·1 | 88·1 | 89·3 | 90·7 | 92·4 | 94·3 | 96·2 | 98·3 | 100·4 | 102·6 | 104·8 | 107·0 | 109·3 | 111·7 | 113·9 | 116·1 | 118·2 | 120·1 |
| | Ven. ... | 334·2 | 346·5 | 358·7 | 11·0 | 23·1 | 35·3 | 47·6 | 59·8 | 72·1 | 84·4 | 96·6 | 108·8 | 121·3 | 133·4 | 145·8 | 158·0 | 170·5 | 182·9 |
| | Sat. ... | 267·2 | 267·2 | 267·3 | 267·2 | 266·7 | 266·0 | 265·6 | 264·9 | 264·2 | 263·5 | 262·8 | 262·2 | 261·4 | 261·0 | 260·5 | 260·2 | 260·3 | 260·5 |
| 1873 | Mars ... | 199·5 | 196·6 | 193·3 | 190·0 | 187·7 | 186·7 | 186·8 | 188·3 | 190·9 | 194·2 | 198·2 | 202·7 | 208·0 | 213·5 | 219·5 | 225·5 | 231·9 | 238·7 |
| ☉'s L. | Merc. ... | 354·8 | 348·9 | 252·8 | 4·9 | 20·7 | 38·9 | 57·9 | 76·8 | 94·4 | 110·1 | 121·1 | 123·9 | 117·4 | 112·8 | 118·1 | 130·0 | 145·9 | 163·0 |
| +·2 | Jup. ... | 118·8 | 118·2 | 118·3 | 118·9 | 119·7 | 120·8 | 122·2 | 123·8 | 125·5 | 127·3 | 129·3 | 131·5 | 133·5 | 135·8 | 138·0 | 140·0 | 142·4 | 144·5 |
| | Ven. ... | 32·2 | 32·3 | 28·3 | 22·0 | 16·7 | 15·7 | 18·6 | 24·1 | 31·2 | 40·0 | 49·6 | 59·9 | 70·4 | 81·4 | 92·8 | 104·3 | 116·2 | 128·1 |
| | Sat. .. | 278·0 | 278·3 | 278·5 | 278·7 | 278·3 | 278·0 | 277·7 | 277·1 | 276·4 | 275·7 | 274·3 | 273·0 | 271·5 | 271·6 | 271·7 | 271·7 | 271·7 | 271·6 |
| 1874 | Mars ... | 22·8 | 29·8 | 36·8 | 43·9 | 50·8 | 57·7 | 64·4 | 71·1 | 78·1 | 84·4 | 91·0 | 97·5 | 103·9 | 110·4 | 116·8 | 123·1 | 129·5 | 135·8 |
| ☉'s L. | Merc. ... | 332·7 | 342·6 | 357·8 | 15·7 | 34·4 | 53·4 | 71·4 | 87·2 | 100·8 | 106·2 | 101·3 | 95·1 | 98·3 | 109·1 | 124·3 | 141·5 | 159·4 | 176·9 |
| +·0 | Jup. ... | 150·9 | 149·9 | 149·2 | 148·7 | 148·6 | 148·7 | 149·4 | 150·2 | 151·3 | 152·5 | 154·0 | 156·0 | 157·7 | 159·5 | 161·6 | 163·8 | 165·9 | 168·1 |
| | Ven. ... | 10·7 | 23·0 | 35·3 | 47·6 | 59·9 | 72·0 | 84·0 | 96·1 | 107·7 | 119·6 | 131·6 | 143.1 | 154·8 | 166·5 | 177·5 | 188·5 | 199·0 | 209·2 |
| | Sat. ... | 288·8 | 289·2 | 289·7 | 290·1 | 290·0 | 289·8 | 289·8 | 289·3 | 288·7 | 288·2 | 287·4 | 286·7 | 285·8 | 285·1 | 284·4 | 283·8 | 283·5 | 283·4 |
| Eng. date. | | Ap 12 | Ap 22 | My 2 | My 12 | My 22 | Je 1 | Je 11 | Je 21 | Jl 1 | Jl 11 | Jl 21 | Jl 31 | Au 10 | Au 20 | Au 30 | S 9 | S 19 | S 29 |
| 1875 | Mars ... | 245·1 | 248·5 | 251·2 | 252·7 | 253·1 | 252·7 | 251·3 | 248·7 | 246·4 | 243·6 | 242·1 | 243·4 | 243·7 | 246·1 | 249·7 | 253·8 | 258·9 | 264·2 |
| ☉'s L. | Merc. ... | 337·1 | 354·3 | 13·0 | 31·5 | 50·2 | 67·1 | 80·8 | 87·4 | 83·8 | 76·9 | 78·7 | 89·1 | 104·0 | 121·2 | 139·5 | 157·3 | 174·1 | 188·7 |
| +·7 | Jup. ... | 184·7 | 183·3 | 182·1 | 180·9 | 180·0 | 179·3 | 178·8 | 178·7 | 178·9 | 179·4 | 180·2 | 181·3 | 182·6 | 184·1 | 185·8 | 187·9 | 189·8 | 191·7 |
| | Ven. ... | 318·6 | 330·3 | 342·3 | 354·3 | 6·2 | 18·1 | 30·2 | 42·3 | 54·0 | 66·5 | 78·7 | 90·7 | 102·9 | 115·2 | 127·5 | 139·9 | 152·3 | 164·7 |
| | Sat. ... | 299·7 | 300·4 | 301·0 | 301·6 | 301·8 | 301·8 | 301·9 | 301·4 | 301·0 | 300·6 | 299·9 | 299·2 | 298·5 | 297·7 | 297·0 | 296·2 | 295·7 | 295·3 |
| Eng. date. | | Ap 11 | Ap 21 | My 1 | My 11 | My 21 | My 31 | Je 10 | Je 20 | Je 30 | Jl 10 | Jl 20 | Jl 30 | Au 9 | Au 19 | Au 29 | S 8 | S 18 | S 28 |
| 1876 | Mars ... | 27·6 | 44·4 | 51·1 | 57·9 | 64·4 | 71·0 | 77·6 | 84·2 | 90·6 | 97·1 | 103·5 | 109·9 | 116·3 | 122·6 | 129·1 | 135·3 | 141·7 | 148·0 |
| ☉'s L. | Merc. ... | 349·8 | 7·4 | 27·1 | 44·7 | 58·9 | 67·7 | 66·6 | 59·0 | 58·6 | 67·8 | 81·9 | 98·9 | 117·2 | 135·3 | 152·8 | 168·5 | 180·4 | 185·1 |
| +·5 | Jup. ... | 218·7 | 217·8 | 216·6 | 215·3 | 214·1 | 212·8 | 211·6 | 210·6 | 209·9 | 209·4 | 209·2 | 209·4 | 209·9 | 210·8 | 211·8 | 213·8 | 214·6 | 216·4 |
| | Ven. ... | 43·4 | 54·4 | 64·7 | 74·2 | 83·1 | 90·6 | 96·6 | 99·3 | 98·8 | 93·8 | 87·4 | 82·5 | 81·8 | 84·8 | 90·9 | 98·5 | 107·4 | 117·0 |
| | Sat. ... | 310·6 | 311·6 | 312·5 | 313·6 | 313·8 | 313·8 | 314·1 | 313·9 | 313·8 | 313·5 | 312·9 | 312·2 | 311·6 | 310·8 | 310·1 | 309·2 | 308·6 | 308·1 |
| 1877 | Mars ... | 272·8 | 279·0 | 285·3 | 291·1 | 296·9 | 302·2 | 307·5 | 312·1 | 316·5 | 319·8 | 322·4 | 323·8 | 323·3 | 322·1 | 319·8 | 317·2 | 314·9 | 314·0 |
| ☉'s L. | Merc. ... | 3·5 | 21·6 | 36·8 | 47·1 | 48·5 | 41.1 | 38·5 | 45·9 | 59·4 | 76·3 | 94·3 | 113·0 | 130·9 | 147·5 | 160·9 | 168·3 | 166·1 | 158·4 |
| +·2 | Jup. ... | 251·3 | 251·5 | 251·1 | 250·6 | 249·8 | 248·7 | 247·4 | 246·1 | 244·7 | 243·5 | 242·5 | 241·7 | 241·0 | 240·8 | 241·1 | 241·4 | 242·1 | 243·1 |
| | Ven. ... | 352·5 | 5·0 | 17·3 | 29·4 | 42·0 | 54·2 | 66·3 | 78·6 | 90·8 | 102·8 | 115·2 | 127·5 | 139·6 | 151·6 | 163·8 | 175·7 | 187·8 | 200·0 |
| | Sat. ... | 321·7 | 322·6 | 323·6 | 324·5 | 325·2 | 325·8 | 326·5 | 326·5 | 326·6 | 326·6 | 326·1 | 325·7 | 325·1 | 324·4 | 323·7 | 322·9 | 322·1 | 321·4 |
| 1878 | Mars ... | 52·1 | 58·4 | 64·9 | 71·0 | 77·2 | 83·5 | 89·9 | 96·1 | 102·6 | 108·7 | 115·0 | 121·4 | 127·8 | 134·2 | 140·5 | 146·9 | 153·2 | 159·6 |
| ☉'s L. | Marc. ... | 14·7 | 26·2 | 29·6 | 23·4 | 18·8 | 24·2 | 36·9 | 53·3 | 71·5 | 90·2 | 108·7 | 125·8 | 140·6 | 150·0 | 150·9 | 143·7 | 140·5 | 146·9 |
| −·1 | Jup. ... | 281·8 | 283·0 | 283·8 | 284·5 | 284·8 | 284·8 | 284·4 | 283·6 | 282·6 | 281·5 | 280·1 | 278·7 | 277·5 | 276·4 | 275·4 | 274·6 | 274·2 | 274·2 |
| | Ven. ... | 314·0 | 322·7 | 331·7 | 341·2 | 352·5 | 3·2 | 14·3 | 25·7 | 37·2 | 49·4 | 60·9 | 72·6 | 84·5 | 96·6 | 108·7 | 120·9 | 133·0 | 145·3 |
| | Sat. ... | 332·9 | 334·0 | 335·0 | 336·2 | 337·0 | 337·6 | 338·5 | 338·9 | 339·2 | 339·6 | 339·3 | 339·1 | 338·8 | 338·2 | 337·5 | 336·8 | 336·0 | 335·3 |
| Eng. date. | | Ap 12 | Ap 22 | My 2 | My 12 | My 22 | Je 1 | Je 11 | Je 21 | Jl 1 | Jl 11 | Jl 21 | Jl 31 | Au 10 | Au 20 | Au 30 | S 9 | S 19 | S 29 |
| 1879 | Mars ... | 294·5 | 301·5 | 308·7 | 316·1 | 323·4 | 330·5 | 337·4 | 344·2 | 350·9 | 357·5 | 3·6 | 9·5 | 14·9 | 20·1 | 24·4 | 28·0 | 30·9 | 32·3 |
| ☉'s L. | Marc. ... | 10·8 | 5·5 | 359·8 | 4·0 | 16·0 | 31·9 | 50·5 | 69·1 | 87·8 | 105·5 | 120·8 | 131·7 | 134·0 | 127·2 | 123·6 | 128·3 | 140·4 | 156·1 |
| +·7 | Jup. ... | 310·8 | 312·7 | 314·5 | 316·0 | 317·4 | 318·5 | 319·4 | 319·9 | 319·9 | 319·7 | 319·3 | 318·4 | 317·2 | 316·0 | 314·6 | 313·2 | 312·1 | 310·7 |
| | Ven. ... | 29·9 | 41·9 | 53·7 | 65·7 | 77·2 | 88·8 | 100·0 | 111·2 | 121·7 | 131·9 | 141·5 | 150·2 | 157·5 | 163·2 | 165·4 | 165·2 | 160·5 | 154·7 |
| | Sat. ... | 344·4 | 345·6 | 346·8 | 348·1 | 349·1 | 349·9 | 351·0 | 351·5 | 352·1 | 352·6 | 353·0 | 353·3 | 353·7 | 352·7 | 352·0 | 351·1 | 350·3 | 349·6 |

| Mesh. | Vrsh. | Mith. | Kat. | Sim. | Kan. | Tul. | Vrsch. | Dhan. | Mak. | Kum. | Min. | Asvn. | Bhar. | Krit. | Rohi. | Mrig. | Ardh. | Punar. | Push. | Asles. |
|---|---|---|---|---|---|---|---|---|---|---|---|---|---|---|---|---|---|---|---|---|
| 30 | 60 | 90 | 120 | 150 | 180 | 210 | 240 | 270 | 300 | 330 | 360 | 13·3 | 26·7 | 40·0 | 53·3 | 66·7 | 80·0 | 93·3 | 106·7 | 120·0 |

**mean Sunrise for every ten days from A.D. 1870 to A.D. 1879.**

| 180 | 190 | 200 | 210 | 220 | 230 | 240 | 250 | 260 | | 270 | 280 | 290 | 300 | 310 | 320 | 330 | 340 | 350 | 360 |
|---|---|---|---|---|---|---|---|---|---|---|---|---|---|---|---|---|---|---|---|
| 174·4 | 184·3 | 194·1 | 204·0 | 213·8 | 223·7 | 233·5 | 243·4 | 253·3 | | 263·1 | 273·0 | 282·8 | 292·7 | 302·5 | 312·4 | 322·2 | 332·1 | 342·0 | 351·8 |
| −2·2 | −2·1 | −1·9 | −1·7 | −1·5 | −1·2 | −0·9 | −0·5 | −0·1 | | +0·3 | +0·6 | +1·0 | +1·3 | +1·6 | +1·8 | +2·0 | +2·1 | +2·2 | +2·2 |
| O 8 | O 18 | O 28 | N 7 | N 17 | N 27 | D 7 | D 17 | D 27 | | Ja 6 | Ja 16 | Ja 26 | F 5 | F 15 | F 25 | Mr 7 | Mr 17 | Mr 27 | Ap 6 |
| Pur 25 | Aipp 4 | Aipp 14 | Aipp 24 | Kart 4 | Kart 14 | Kart 24 | Marg 5 | Marg 15 | | Marg 25 | Tai 5 | Tai 15 | Tai 25 | Masi 6 | Masi 16 | Masi 26 | Pang 6 | Pang 16 | Pang 26 |
| 116·2 | 121·9 | 127·5 | 133·2 | 138·4 | 143·7 | 148·3 | 153·0 | 156·8 | **71** | 160·5 | 163·2 | 165·0 | 165·9 | 165·6 | 163·7 | 160·8 | 157·0 | 153·0 | 149·7 |
| 164·4 | 166·6 | 176·5 | 191·0 | 207·4 | 224·8 | 241·9 | 258·1 | 271·6 | | 279·4 | 278·3 | 270·7 | 267·2 | 277·1 | 290·8 | 307·3 | 323·2 | 342·0 | 0·1 |
| 63·0 | 63·0 | 62·5 | 61·8 | 60·8 | 59·6 | 58·3 | 57·0 | 55·8 | | 54·5 | 53·6 | 52·9 | 52·5 | 52·5 | 52·9 | 53·6 | 54·5 | 55·7 | 57·1 |
| 157·3 | 169·4 | 181·8 | 194·5 | 206·8 | 219·6 | 232·1 | 245·7 | 257·3 | | 270·0 | 282·8 | 295·3 | 307·9 | 320·3 | 333·0 | 345·3 | 357·8 | 10·1 | 22·1 |
| 239·4 | 240·4 | 241·5 | 242·4 | 243·4 | 244·4 | 245·6 | 246·8 | 248·0 | | 249·1 | 250·2 | 251·2 | 252·4 | 253·2 | 253·9 | 254·8 | 255·2 | 255·7 | 256·2 |
| O 9 | O 19 | O 29 | N 8 | N 18 | N 28 | D 8 | D 18 | D 28 | | Ja 7 | Ja 17 | Ja 27 | F 6 | F 16 | F 26 | Mr 7 | Mr 17 | Mr 27 | Ap 6 |
| 227·3 | 234·3 | 241·6 | 249·1 | 256·4 | 264·0 | 271·6 | 279·2 | 287·2 | **72** | 294·9 | 302·9 | 310·6 | 318·4 | 326·1 | 334·0 | 341·6 | 349·3 | 356·8 | 4·5 |
| 157·7 | 171·8 | 188·2 | 205·5 | 222·8 | 239·3 | 253·2 | 262·4 | 262·9 | | 255·5 | 252·2 | 258·4 | 271·2 | 287·3 | 305·0 | 323·1 | 341·0 | 357·8 | 11·2 |
| 94·6 | 95·5 | 96·1 | 96·5 | 96·6 | 96·2 | 95·7 | 94·7 | 93·5 | | 92·3 | 91·0 | 89·6 | 88·4 | 87·4 | 86·7 | 86·2 | 86·0 | 86·0 | 86·8 |
| 153·4 | 152·3 | 154·5 | 160·2 | 167·7 | 176·7 | 186·4 | 196·9 | 208·0 | | 219·2 | 230·9 | 242·6 | 254·7 | 266·9 | 279·0 | 291·2 | 303·3 | 315·6 | 328·0 |
| 250·0 | 250·8 | 251·5 | 252·3 | 253·4 | 254·5 | 255·6 | 256·7 | 257·8 | | 259·0 | 260·1 | 261·2 | 262·4 | 263·3 | 264·3 | 265·2 | 265·9 | 266·5 | 267·1 |
| O 8 | O 18 | O 28 | N 7 | N 17 | N 27 | D 7 | D 17 | D 27 | | Ja 6 | Ja 16 | Ja 26 | F 5 | F 15 | F 25 | Mr 7 | Mr 17 | Mr 27 | Ap 6 |
| 129·7 | 135·7 | 141·8 | 147·5 | 153·3 | 158·9 | 164·7 | 170·1 | 175·2 | **73** | 180·2 | 185·0 | 189·5 | 193·4 | 197·0 | 199·7 | 201·7 | 202·6 | 202·5 | 200·9 |
| 167·2 | 184·6 | 202·0 | 218·6 | 233·6 | 244·5 | 248·0 | 241·9 | 236·3 | | 239·6 | 250·7 | 265·8 | 283·1 | 301·2 | 318·9 | 334·7 | 350·1 | 358·8 | 358·5 |
| 121·9 | 123·6 | 125·1 | 127·3 | 126·5 | 128·1 | 128·6 | 128·7 | 128·3 | | 127·8 | 127·1 | 126·1 | 124·8 | 123·5 | 122·1 | 120·9 | 119·9 | 119·0 | 118·4 |
| 195·4 | 207·8 | 220·8 | 232·3 | 244·6 | 257·0 | 269·3 | 281·4 | 293·5 | | 305·6 | 317·4 | 328·8 | 339·9 | 350·6 | 1·2 | 10·6 | 19·1 | 26·6 | 31·1 |
| 260·5 | 261·2 | 261·8 | 262·4 | 263·3 | 264·3 | 265·3 | 266·6 | 267·7 | | 268·8 | 270·0 | 271·2 | 272·3 | 273·4 | 274·4 | 275·4 | 276·2 | 276·9 | 277·7 |
| 245·4 | 252·5 | 259·6 | 266·8 | 274·3 | 281·8 | 289·3 | 296·9 | 304·4 | **74** | 312·1 | 319·6 | 327·3 | 334·8 | 342·3 | 349·8 | 357·2 | 4·7 | 11·8 | 19·0 |
| 180·8 | 197·9 | 213·8 | 226·2 | 232·2 | 228·8 | 221·6 | 222·0 | 231·0 | | 245·0 | 261·7 | 279·3 | 297·9 | 314·0 | 328·8 | 339·0 | 341·1 | 334·3 | 331·6 |
| 146·7 | 148·8 | 150·7 | 152·6 | 154·3 | 155·7 | 157·2 | 158·2 | 158·7 | | 159·2 | 159·3 | 159·0 | 158·5 | 157·8 | 156·5 | 155·3 | 154·1 | 152·6 | 151·7 |
| 139·5 | 152·2 | 164·2 | 176·6 | 188·9 | 201·4 | 213·8 | 226·4 | 238·8 | | 251·5 | 264·1 | 276·7 | 289·5 | 302·0 | 314·1 | 326·6 | 339·2 | 351·7 | 4·4 |
| 271·5 | 271·9 | 272·4 | 272·8 | 273·7 | 274·5 | 275·4 | 276·5 | 277·6 | | 278·8 | 280·0 | 281·2 | 282·3 | 283·5 | 284·7 | 285·8 | 286·7 | 287·7 | 288·5 |
| 142·1 | 148·2 | 154·3 | 160·5 | 166·5 | 172·6 | 178·7 | 184·6 | 190·5 | **75** | 196·5 | 202·4 | 208·0 | 213·6 | 219·1 | 224·3 | 229·5 | 234·2 | 238·6 | 242·8 |
| 193·4 | 207·2 | 215·8 | 215·5 | 207·7 | 205·2 | 211·9 | 224·7 | 240·6 | | 257·8 | 275·6 | 292·7 | 307·9 | 319·4 | 323·8 | 318·3 | 312·4 | 316·3 | 328·0 |
| 170·2 | 172·4 | 174·6 | 176·8 | 178·8 | 180·7 | 182·5 | 184·2 | 185·7 | | 186·9 | 187·9 | 188·6 | 189·2 | 189·3 | 189·0 | 188·7 | 187·8 | 186·8 | 185·5 |
| 218·8 | 227·4 | 234·4 | 239·7 | 241·4 | 239·8 | 233·4 | 228·2 | 225·9 | | 227·2 | 234·2 | 238·8 | 247·4 | 256·7 | 266·9 | 277·6 | 288·7 | 300·2 | 311·5 |
| 283·0 | 283·2 | 283·3 | 283·5 | 284·3 | 285·1 | 285·7 | 286·7 | 287·8 | | 288·9 | 290·1 | 291·3 | 292·4 | 293·6 | 294·7 | 296·0 | 297·1 | 298·2 | 299·3 |
| O 9 | O 19 | O 29 | N 8 | N 18 | N 28 | D 8 | D 18 | D 28 | | Ja 7 | Ja 17 | Ja 27 | F 6 | F 16 | F 26 | Mr 7 | Mr 17 | Mr 27 | Ap 6 |
| 270·3 | 276·4 | 282·8 | 289·6 | 296·3 | 303·3 | 310·1 | 317·1 | 324·3 | **76** | 331·4 | 338·5 | 345·5 | 352·7 | 359·8 | 6·7 | 13·6 | 20·5 | 27·4 | 34·3 |
| 198·9 | 200·6 | 193·6 | 189·2 | 194·4 | 206·0 | 221·4 | 238·4 | 255·9 | | 272·8 | 288·6 | 301·1 | 306·4 | 302·2 | 295·5 | 297·4 | 307·8 | 322·3 | 340·6 |
| 193·9 | 196·0 | 198·2 | 200·4 | 202·6 | 204·7 | 206·9 | 209·0 | 211·0 | | 212·9 | 214·6 | 216·1 | 217·3 | 218·3 | 219·2 | 219·7 | 219·9 | 219·5 | 219·1 |
| 177·2 | 189·5 | 202·0 | 214·5 | 227·2 | 239·8 | 252·1 | 265·0 | 277·5 | | 290·0 | 302·4 | 314·8 | 327·0 | 339·2 | 351·4 | 3·2 | 14·9 | 26·6 | 37·9 |
| 294·8 | 294·8 | 294·9 | 294·9 | 295·5 | 296·0 | 296·6 | 297·6 | 298·7 | | 299·6 | 300·8 | 302·0 | 303·1 | 304·3 | 305·5 | 306·7 | 307·9 | 309·0 | 310·1 |
| O 8 | O 18 | O 28 | N 7 | N 17 | N 27 | D 7 | D 17 | D 27 | | Ja 6 | Ja 16 | Ja 26 | F 5 | F 15 | F 25 | Mr 7 | Mr 17 | Mr 27 | Ap 6 |
| 154·4 | 160·6 | 167·2 | 173·5 | 179·8 | 186·3 | 192·6 | 199·0 | 205·5 | **77** | 211·9 | 218·4 | 224·7 | 231·1 | 237·5 | 244·1 | 250·5 | 256·9 | 263·4 | 269·6 |
| 180·1 | 173·7 | 176·0 | 186·4 | 200·9 | 217·5 | 234·8 | 252·0 | 268·2 | | 281·5 | 289·3 | 287·4 | 279·7 | 278·5 | 287·2 | 301·2 | 318·1 | 336·4 | 355·0 |
| 218·2 | 220·2 | 222·3 | 224·4 | 226·6 | 228·8 | 231·1 | 233·4 | 235·5 | | 237·8 | 40·0 | 242·0 | 243·8 | 245·6 | 247·2 | 248·6 | 249·8 | 250·7 | 251·3 |
| 127·6 | 138·5 | 149·7 | 161·4 | 173·1 | 185·0 | 197·1 | 209·5 | 221·9 | | 234·2 | 246·5 | 258·9 | 271·5 | 284·0 | 296·6 | 309·0 | 321·4 | 333·8 | 346·2 |
| 307·4 | 307·1 | 306·9 | 306·7 | 307·0 | 307·5 | 307·8 | 308·7 | 309·5 | | 310·4 | 311·5 | 312·6 | 313·7 | 314·9 | 316·1 | 317·4 | 318·7 | 320·0 | 321·2 |
| 314·4 | 315·9 | 318·8 | 322·4 | 326·8 | 331·2 | 336·4 | 341·1 | 347·3 | **78** | 353·3 | 359·0 | 5·2 | 11·3 | 17·5 | 23·7 | 29·9 | 36·0 | 42·6 | 48·9 |
| 158·1 | 166·5 | 180·4 | 196·6 | 213·9 | 231·1 | 247·3 | 261·8 | 271·3 | | 272·6 | 265·2 | 261·6 | 267·4 | 279·9 | 296·0 | 313·8 | 332·1 | 350·3 | 7·2 |
| 244·5 | 246·0 | 247·7 | 249·5 | 251·7 | 253·6 | 255·9 | 258·2 | 260·5 | | 262·9 | 265·2 | 267·5 | 269·8 | 272·0 | 274·0 | 276·1 | 278·0 | 279·8 | 281·4 |
| 211·8 | 224·0 | 235·4 | 247·2 | 258·3 | 269·4 | 279·0 | 290·0 | 299·6 | | 307·5 | 314·0 | 317·5 | 317·9 | 313·7 | 307·9 | 305·2 | 302·1 | 304·2 | 310·9 |
| 320·6 | 320·2 | 319·9 | 319·5 | 319·5 | 319·7 | 319·7 | 320·3 | 321·0 | | 321·7 | 322·7 | 323·8 | 324·8 | 326·0 | 327·2 | 328·5 | 329·8 | 331·1 | 332·3 |
| 166·2 | 172·7 | 179·2 | 185·8 | 192·3 | 199·1 | 205·8 | 212·5 | 219·4 | **79** | 226·1 | 233·0 | 240·0 | 246·9 | 254·0 | 261·0 | 268·3 | 275·5 | 282·6 | 289·9 |
| 159·7 | 175·6 | 192·9 | 210·1 | 226·9 | 242·2 | 253·1 | 257·0 | 251·5 | | 245·4 | 248·5 | 259·2 | 274·4 | 291·7 | 309·8 | 327·7 | 344·8 | 359·6 | 9·1 |
| 274·6 | 275·2 | 276·1 | 277·4 | 278·9 | 280·5 | 282·5 | 284·5 | 286·6 | | 288·8 | 291·0 | 293·4 | 295·8 | 298·1 | 300·5 | 302·9 | 305·2 | 307·5 | 309·6 |
| 157·6 | 169·9 | 182·3 | 195·0 | 207·4 | 220·1 | 232·6 | 245·2 | 258·0 | | 270·9 | 283·3 | 295·8 | 308·4 | 320·8 | 333·3 | 346·0 | 358·3 | 10·6 | 22·6 |
| 334·5 | 333·8 | 333·2 | 332·5 | 332·4 | 332·2 | 332·2 | 332·6 | 333·1 | | 333·5 | 334·5 | 335·4 | 336·3 | 337·4 | 338·5 | 339·7 | 341·0 | 342·3 | 343·5 |
| O 9 | O 19 | O 29 | N 8 | N 18 | N 28 | D 8 | D 18 | D 28 | | Ja 7 | Ja 17 | Ja 27 | F 6 | F 16 | F 26 | Mr 7 | Mr 17 | Mr 27 | Ap 6 |
| 23·0 | 31·6 | 29·6 | 26·8 | 24·2 | 21·9 | 20·7 | 20·8 | 22·1 | **80** | 23·8 | 27·0 | 30·4 | 34·3 | 37·8 | 43·6 | 48·7 | 53·8 | 59·2 | 64·5 |
| 173·3 | 191·0 | 207·9 | 223·6 | 235·7 | 241·4 | 237·4 | 230·2 | 231·2 | | 240·8 | 255·1 | 271·9 | 289·6 | 307·6 | 324·7 | 339·5 | 348·7 | 351·7 | 346·9 |
| 310·2 | 309·5 | 308·8 | 308·6 | 309·0 | 309·8 | 310·9 | 312·3 | 313·9 | | 315·8 | 317·8 | 319·9 | 322·2 | 324·5 | 325·8 | 329·3 | 331·7 | 334·1 | 336·5 |
| 150·0 | 150·1 | 153·3 | 159·4 | 167·3 | 176·5 | 186·2 | 197·0 | 208·0 | | 219·6 | 231·3 | 243·3 | 255·3 | 267·4 | 279·9 | 291·8 | 303·9 | 316·2 | 328·6 |
| 348·7 | 348·0 | 347·6 | 346·5 | 346·2 | 345·8 | 345·5 | 345·7 | 345·8 | | 346·1 | 346·8 | 347·6 | 348·2 | 349·4 | 350·5 | 351·5 | 352·7 | 354·1 | 355·4 |

| Mag. | P.Phal. | U.Phal. | Hast. | Chit. | Svati. | Visa. | Anur. | Jyesh. | Mula. | P.Ash. | U.Ash. | Srav. | Dan. | Satab. | P.Bhad. | U.Bhad. | Revati. |
|---|---|---|---|---|---|---|---|---|---|---|---|---|---|---|---|---|---|
| 133·3 | 146·7 | 160·0 | 173·3 | 186·7 | 200·0 | 213·3 | 226·7 | 240·0 | 253·3 | 266·7 | 280·0 | 293·3 | 306·7 | 320·0 | 333·3 | 346·7 | 360·0 |

## Geocentric (Indian) Longitudes of Sun and Major planets at Ujjain

| Day of S. Yr. | | 0 | 10 | 20 | 30 | 40 | 50 | 60 | 70 | 80 | 90 | 100 | 110 | 120 | 130 | 140 | 150 | 160 | 170 |
|---|---|---|---|---|---|---|---|---|---|---|---|---|---|---|---|---|---|---|---|
| ☉'s Mean L. | | 357·0 | 6·9 | 16·7 | 26·6 | 36·4 | 46·3 | 56·1 | 66·0 | 75·9 | 85·7 | 95·6 | 105·4 | 115·3 | 125·1 | 135·0 | 144·8 | 154·7 | 164·6 |
| Corrn. for do. | | +2·1 | +2·0 | +1·9 | +1·7 | +1·4 | +1·1 | +0·8 | +0·4 | +0·0 | −0·3 | −0·7 | −1·1 | −1·4 | −1·6 | −1·9 | −2·0 | −2·1 | −2·2 |
| English date. | | Ap 11 | Ap 21 | My 1 | My 11 | My 21 | My 31 | Je 10 | Je 20 | Je 30 | Jl 10 | Jl 20 | Jl 30 | Au 9 | Au 19 | Au 29 | S 8 | S 18 | S 28 |
| Tamil month. | | Chit | Chit | Chit | Chit | Vaik | Vaik | Vaik | Ani | Ani | Ani | Adi | Adi | Adi | Avan | Avan | Avan | Pur | Pur |
| Do. date. | | 1 | 11 | 21 | 31 | 10 | 20 | 30 | 9 | 19 | 29 | 7 | 17 | 27 | 6 | 16 | 26 | 5 | 15 |
| 1880 | Mars ... | 67·5 | 73·2 | 79·1 | 84·9 | 90·9 | 96·9 | 103·0 | 108·9 | 115·1 | 121·3 | 127·4 | 133·8 | 140·1 | 146·4 | 152·8 | 159·2 | 165·7 | 172·2 |
| ☉'s L. | Merc. ... | 341·5 | 343·2 | 353·7 | 8·8 | 26·7 | 45·7 | 64·5 | 83·1 | 99·1 | 111·5 | 116·6 | 111·5 | 105·6 | 108·6 | 119·7 | 135·6 | 151·8 | 169·7 |
| +·4 | Jup. ... | 338·6 | 340·9 | 343·2 | 345·4 | 347·5 | 349·5 | 351·2 | 352·7 | 353·9 | 355·0 | 355·7 | 356·0 | 356·1 | 355·6 | 355·1 | 354·2 | 353·2 | 351·8 |
| | Ven. ... | 334·7 | 346·9 | 359·2 | 11·5 | 23·6 | 36·0 | 48·1 | 60·5 | 72·6 | 84·9 | 97·2 | 109·3 | 121·8 | 133·9 | 146·2 | 158·5 | 171·0 | 183·4 |
| | Sat. ... | 354·9 | 356·5 | 358·2 | 359·9 | 1·1 | 2·2 | 3·3 | 4·0 | 4·8 | 5·6 | 5·8 | 6·1 | 6·4 | 6·2 | 5·9 | 5·7 | 5·3 | 4·8 |
| 1881 | Mars ... | 312·7 | 320·2 | 328·0 | 335·6 | 343·2 | 350·6 | 358·0 | 5·4 | 12·5 | 19·7 | 26·5 | 33·3 | 39·9 | 46·1 | 52·1 | 58·0 | 63·4 | 68·3 |
| ☉'s L. | Merc. ... | 332·1 | 346·3 | 3·6 | 22·3 | 41·3 | 59·8 | 76·7 | 90·6 | 97·8 | 95·8 | 88·0 | 88·9 | 98·6 | 113·1 | 130·1 | 148·1 | 165·8 | 182·8 |
| +·2 | Jup. ... | 6·2 | 8·6 | 11·0 | 13·4 | 15·8 | 18·2 | 20·5 | 22·6 | 24·6 | 26·5 | 28·2 | 29·7 | 30·9 | 31·8 | 32·4 | 32·7 | 32·6 | 32·3 |
| | Ven. ... | 30·3 | 29·3 | 24·3 | 17·2 | 13·9 | 13·9 | 17·4 | 23·5 | 31·0 | 39·9 | 49·8 | 60·1 | 70·6 | 82·2 | 93·4 | 104·9 | 116·8 | 128·5 |
| | Sat. ... | 7·9 | 9·2 | 10·5 | 11·9 | 13·2 | 14·4 | 15·6 | 16·6 | 17·5 | 18·6 | 19·1 | 19·4 | 20·3 | 20·4 | 20·4 | 20·5 | 19·9 | 19·4 |
| 1882 | Mars ... | 84·3 | 89·0 | 94·1 | 99·3 | 104·7 | 110·3 | 116·0 | 121·8 | 127·7 | 133·7 | 139·8 | 146·0 | 152·4 | 158·6 | 165·0 | 171·6 | 178·1 | 184·8 |
| ☉'s L. | Merc. ... | 340·7 | 358·9 | 17·8 | 36·4 | 53·9 | 68·9 | 78·1 | 77·9 | 70·5 | 69·2 | 77·4 | 91·2 | 108·0 | 125·9 | 144·1 | 161·4 | 177·2 | 189·2 |
| −·1 | Jup. ... | 33·8 | 36·0 | 38·3 | 40·6 | 42·9 | 45·3 | 47·8 | 50·1 | 52·8 | 54·6 | 56·9 | 58·9 | 60·9 | 62·6 | 64·1 | 65·4 | 66·5 | 67·2 |
| | Ven. ... | 11·2 | 23·6 | 35·8 | 48·3 | 60·0 | 72·4 | 84·5 | 96·4 | 108·3 | 120·1 | 132·1 | 143·4 | 155·1 | 166·8 | 177·6 | 188·6 | 199·3 | 203·9 |
| | Sat. ... | 20·0 | 21·3 | 22·6 | 24·0 | 25·4 | 26·8 | 28·1 | 29·3 | 30·4 | 31·5 | 32·3 | 33·1 | 33·9 | 34·2 | 34·5 | 34·8 | 34·5 | 34·2 |
| Eng. Date. | | Ap 12 | Ap 22 | My 2 | My 12 | My 22 | Je 1 | Je 11 | Je 21 | Jl 1 | Jl 11 | Jl 21 | Jl 31 | Au 10 | Au 20 | Au 30 | S 9 | S 19 | S 29 |
| 1883 | Mars ... | 330·4 | 338·1 | 345·9 | 353·5 | 1·1 | 8·7 | 16·1 | 23·4 | 30·6 | 37·7 | 44·8 | 51·5 | 58·2 | 64·7 | 71·1 | 77·4 | 83·3 | 89·2 |
| ☉'s L. | Merc. ... | 356·5 | 15·0 | 32·7 | 48·0 | 58·2 | 59·5 | 51·9 | 49·6 | 56·9 | 70·6 | 87·2 | 105·3 | 123·7 | 141·6 | 157·9 | 171·0 | 177·8 | 175·1 |
| +·6 | Jup. ... | 62·1 | 63·8 | 65·6 | 67·5 | 69·7 | 71·8 | 74·1 | 76·4 | 78·7 | 81·0 | 83·3 | 85·6 | 87·9 | 89·9 | 91·8 | 93·8 | 95·6 | 97·6 |
| | Ven. ... | 318·9 | 330·8 | 342·8 | 354·8 | 6·7 | 18·6 | 30·6 | 42·8 | 54·9 | 67·0 | 79·9 | 92·4 | 103·4 | 115·7 | 128·0 | 140·4 | 153·8 | 165·2 |
| | Sat. ... | 32·7 | 34·0 | 35·3 | 36·6 | 37·9 | 39·3 | 40·7 | 41·9 | 43·1 | 44·4 | 45·4 | 46·4 | 47·4 | 48·0 | 48·6 | 49·1 | 49·1 | 49·0 |
| Eng. Date. | | Ap 11 | Ap 21 | My 1 | My 11 | My 21 | My 31 | Je 10 | Je 20 | Je 30 | Jl 10 | Jl 20 | Jl 30 | Au 9 | Au 19 | Au 29 | S 8 | S 18 | S 28 |
| 1884 | Mars ... | 105·4 | 107·6 | 111·9 | 115·9 | 120·6 | 125·4 | 130·8 | 136·3 | 141·8 | 147·6 | 153·5 | 159·7 | 165·8 | 172·2 | 178·7 | 185·1 | 191·9 | 198·7 |
| ☉'s L. | Merc. ... | 9·9 | 25·7 | 37·3 | 40·8 | 34·3 | 30·1 | 35·4 | 48·0 | 64·3 | 82·6 | 101·3 | 119·5 | 136·5 | 150·9 | 160·2 | 160·3 | 152·9 | 150·0 |
| +·4 | Jup. ... | 91·3 | 92·1 | 93·2 | 94·5 | 96·1 | 97·9 | 99·7 | 101·7 | 103·8 | 106·0 | 108·2 | 110·4 | 112·8 | 115·0 | 117·3 | 119·5 | 121·6 | 123·6 |
| | Ven. ... | 43·7 | 54·6 | 64·6 | 74·1 | 82·9 | 89·9 | 94·9 | 97·2 | 95·5 | 89·9 | 84·2 | 79·6 | 81·6 | 83·8 | 90·2 | 98·0 | 107·1 | 117·4 |
| | Sat. ... | 45·5 | 46·7 | 47·9 | 49·1 | 50·5 | 51·9 | 53·3 | 54·6 | 55·9 | 57·2 | 58·3 | 59·4 | 60·5 | 61·4 | 62·2 | 63·0 | 63·3 | 63·5 |
| 1885 | Mars ... | 346·3 | 354·1 | 1·7 | 9·2 | 16·8 | 24·0 | 31·4 | 38·8 | 45·6 | 52·8 | 59·6 | 66·5 | 73·3 | 79·5 | 86·1 | 92·4 | 98·5 | 104· |
| ☉'s L. | Merc. ... | 16·1 | 21·6 | 17·1 | 11·0 | 14·0 | 25·6 | 41·3 | 59·5 | 78·3 | 96·9 | 114·6 | 130·1 | 141·3 | 144·3 | 137·8 | 132·9 | 137·5 | 149·6 |
| +·1 | Jup. ... | 122·5 | 122·3 | 122·5 | 122·8 | 123·5 | 124·5 | 125·8 | 127·2 | 128·9 | 130·8 | 132·8 | 134·8 | 136·9 | 139·1 | 141·2 | 143·4 | 145·6 | 147·8 |
| | Ven. ... | 353·0 | 5·5 | 17·8 | 30·1 | 42·4 | 53·9 | 66·8 | 79·1 | 91·3 | 103·5 | 115·9 | 128·0 | 140·1 | 152·1 | 164·2 | 176·2 | 188·3 | 200·3 |
| | Sat. ... | 58·6 | 59·6 | 60·7 | 61·8 | 63·1 | 64·4 | 65·7 | 67·0 | 68·4 | 69·8 | 71·2 | 72·5 | 73·8 | 74·7 | 75·6 | 76·5 | 77·1 | 77·7 |
| 1886 | Mars ... | 134·0 | 134·0 | 135·3 | 137·3 | 140·0 | 143·7 | 148·0 | 152·7 | 157·6 | 163·0 | 168·6 | 174·6 | 180·5 | 186·8 | 193·2 | 199·9 | 206·5 | 213·4 |
| ☉'s L. | Merc. ... | 0·1 | 352·7 | 353·6 | 3·3 | 18·5 | 36·2 | 55·1 | 73·9 | 92·1 | 108·6 | 121·2 | 126·8 | 122·5 | 116·0 | 115·6 | 128·7 | 143·4 | 160·4 |
| −·1 | Jup. ... | 155·6 | 154·5 | 153·7 | 152·9 | 152·8 | 152·9 | 153·2 | 154·1 | 155·0 | 156·0 | 157·5 | 159·2 | 161·0 | 162·9 | 165·0 | 167·0 | 169·1 | 171·3 |
| | Ven. ... | 323·8 | 322·3 | 331·7 | 342·0 | 353·1 | 3·7 | 14·7 | 26·1 | 37·8 | 49·6 | 61·3 | 73·2 | 85·1 | 97·1 | 109·3 | 121·5 | 133·6 | 145·9 |
| | Sat. ... | 71·9 | 72·8 | 73·7 | 74·6 | 75·8 | 77·0 | 78·1 | 79·4 | 80·7 | 82·1 | 83·5 | 84·8 | 86·1 | 87·2 | 88·3 | 89·4 | 90·3 | 91·1 |
| Eng. Date. | | Ap 12 | Ap 22 | My 2 | My 12 | My 22 | Je 1 | Je 11 | Je 21 | Jl 1 | Jl 11 | Jl 21 | Jl 31 | Au 10 | Au 20 | Au 30 | S 9 | S 19 | S 29 |
| 1887 | Mars ... | 2·4 | 9·7 | 17·3 | 24·6 | 31·9 | 39·0 | 46·1 | 53·2 | 60·2 | 66·9 | 73·8 | 80·8 | 87·0 | 93·5 | 99·9 | 106·1 | 112·4 | 118·8 |
| ☉'s L. | Merc. ... | 333·9 | 343·1 | 357·3 | 14·8 | 33·6 | 52·7 | 70·9 | 86·0 | 101·1 | 108·4 | 105·5 | 98·3 | 99·4 | 109·3 | 123·9 | 140·7 | 158·5 | 176·1 |
| +·6 | Jup. ... | 189·4 | 188·1 | 186·8 | 185·5 | 184·5 | 183·6 | 183·1 | 182·9 | 183·0 | 183·3 | 184·0 | 184·9 | 186·2 | 187·6 | 189·4 | 191·2 | 193·2 | 195·1 |
| | Ven. ... | 30·4 | 42·3 | 54·4 | 66·2 | 77·3 | 89·1 | 100·3 | 111·3 | 121·8 | 131·8 | 141·3 | 149·7 | 156·6 | 161·5 | 163·9 | 162·1 | 156·6 | 150·2 |
| | Sat. ... | 85·4 | 86·1 | 86·8 | 87·5 | 88·5 | 89·6 | 90·7 | 92·0 | 93·3 | 94·6 | 95·9 | 97·2 | 98·5 | 99·8 | 101·0 | 102·2 | 103·2 | 104·2 |
| Eng. Date. | | Ap 11 | Ap 21 | My 1 | My 11 | My 21 | My 31 | Je 10 | Je 20 | Je 30 | Jl 10 | Jl 20 | Jl 30 | Au 9 | Au 19 | Au 29 | S 8 | S 18 | S 28 |
| 1888 | Mars ... | 178·0 | 174·2 | 170·2 | 168·3 | 169·6 | 170·5 | 173·9 | 175·7 | 179·4 | 183·4 | 188·3 | 193·6 | 199·2 | 205·0 | 211·1 | 217·4 | 224·2 | 230·8 |
| ☉'s L. | Merc. ... | 335·1 | 351·9 | 10·9 | 29·2 | 48·0 | 65·5 | 80·0 | 88·9 | 88·3 | 80·9 | 79·9 | 88·2 | 102·1 | 118·7 | 136·8 | 154·6 | 171·8 | 187·3 |
| +·3 | Jup. ... | 223·3 | 222·4 | 221·3 | 220·6 | 218·9 | 217·5 | 216·3 | 215·2 | 214·4 | 213·6 | 213·3 | 213·4 | 213·8 | 214·6 | 215·7 | 216·3 | 218·2 | 219·9 |
| | Ven. ... | 335·2 | 347·5 | 359·7 | 12·2 | 24·2 | 36·5 | 48·8 | 60·0 | 73·1 | 85·6 | 97·7 | 110·0 | 122·4 | 134·4 | 146·9 | 159·0 | 171·5 | 183·9 |
| | Sat. ... | 99·2 | 99·6 | 100·0 | 100·5 | 101·3 | 102·2 | 103·1 | 104·3 | 105·5 | 106·8 | 108·1 | 109·4 | 110·6 | 112·0 | 113·3 | 114·6 | 115·7 | 116·8 |
| 1889 | Mars ... | 16·9 | 24·2 | 31·4 | 38·6 | 45·6 | 52·4 | 59·4 | 66·2 | 73·0 | 79·6 | 86·4 | 92·8 | 99·4 | 105·8 | 112·2 | 118·7 | 124·8 | 131·1 |
| ☉'s L. | Merc. ... | 347·1 | 6·8 | 24·6 | 42·4 | 57·8 | 68·3 | 70·5 | 63·3 | 60·2 | 67·1 | 80·1 | 96·4 | 114·5 | 132·7 | 150·4 | 166·9 | 179·8 | 186·9 |
| +·1 | Jup. ... | 255·7 | 255·9 | 255·7 | 255·3 | 254·7 | 253·7 | 252·5 | 251·2 | 249·9 | 248·6 | 247·4 | 246·4 | 245·7 | 245·5 | 245·4 | 245·6 | 246·2 | 247·2 |
| | Ven. ... | 28·2 | 25·8 | 20·2 | 14·1 | 11·0 | 12·0 | 16·2 | 22·6 | 30·7 | 40·0 | 51·2 | 60·2 | 71·1 | 82·1 | 93·7 | 105·2 | 117·1 | 129·0 |
| | Sat. ... | 113·2 | 113·4 | 113·4 | 113·4 | 114·1 | 114·8 | 115·5 | 116·5 | 117·5 | 118·3 | 119·8 | 121·1 | 122·4 | 123·7 | 125·0 | 126·4 | 127·6 | 128·7 |

| Mesh. | Vrsh. | Mith. | Kat. | Sim. | Kan. | Tul. | Vrsch. | Dhan. | Mak. | Kum. | Min. | Asvn. | Bhar. | Krit. | Rohi. | Mrig. | Ardh. | Punar. | Push. | Asles. |
|---|---|---|---|---|---|---|---|---|---|---|---|---|---|---|---|---|---|---|---|---|
| 30 | 60 | 90 | 120 | 150 | 180 | 210 | 240 | 270 | 300 | 330 | 360 | 13·3 | 26·7 | 40·0 | 53·3 | 66·7 | 80·0 | 93·3 | 106·7 | 120·0 |

**mean Sunrise for every ten days from A.D. 1880 to A.D. 1889.**

| 180 | 190 | 200 | 210 | 220 | 230 | 240 | 250 | 260 | | 270 | 280 | 290 | 300 | 310 | 320 | 330 | 340 | 350 | 360 |
|---|---|---|---|---|---|---|---|---|---|---|---|---|---|---|---|---|---|---|---|
| 174·4 | 184·3 | 194·1 | 204·0 | 213·8 | 223·7 | 233·5 | 243·4 | 253·3 | | 263·1 | 273·0 | 282·8 | 292·7 | 302·5 | 312·4 | 322·2 | 332·1 | 342·0 | 351·8 |
| −2·2 | −2·1 | −1·9 | −1·7 | −1·5 | −1·2 | −0·9 | −0·5 | −0·1 | | +0·3 | +0·6 | +1·0 | +1·3 | +1·6 | +1·8 | +2·0 | +2·1 | +2·2 | +2·2 |
| O 8 | O 18 | O 28 | N 7 | N 17 | N 27 | D 7 | D 17 | D 27 | | Ja 6 | Ja 16 | Ja 26 | F 5 | F 15 | F 25 | Mr 7 | Mr 17 | Mr 27 | Ap 6 |
| Pur 25 | Aipp 4 | Aipp 14 | Aipp 24 | Kart 4 | Kart 14 | Kart 24 | Marg 5 | Marg 15 | | Marg 25 | Tai 5 | Tai 15 | Tai 25 | Masi 6 | Masi 16 | Masi 26 | Pang 6 | Pang 16 | Pang 26 |
| 179·0 | 185·7 | 192·2 | 199·0 | 205·9 | 212·8 | 219·9 | 226·9 | 233·8 | **81** | 241·1 | 248·4 | 255·8 | 263·1 | 270·7 | 278·2 | 285·7 | 293·3 | 300·9 | 308·7 |
| 187·1 | 203·3 | 216·7 | 224·9 | 224·0 | 216·4 | 214·3 | 221·5 | 234·6 | | 250·6 | 268·1 | 285·6 | 302·7 | 318·2 | 329·8 | 334·3 | 328·1 | 322·8 | 326·8 |
| 350·5 | 349·2 | 348·0 | 347·0 | 246·2 | 345·6 | 344·6 | 345·8 | 346·6 | | 347·5 | 349·5 | 350·2 | 351·9 | 353·8 | 356·0 | 358·1 | 0·3 | 2·7 | 5·1 |
| 195·7 | 208·1 | 220·5 | 232·8 | 245·1 | 257·3 | 269·6 | 281·9 | 294·0 | | 305·8 | 317·7 | 328·9 | 340·0 | 350·7 | 0·9 | 10·1 | 18·6 | 25·1 | 29·3 |
| 4·5 | 3·3 | 2·2 | 1·0 | 0·4 | 359·9 | 359·3 | 259·3 | 359·2 | | 359·1 | 359·6 | 0·2 | 0·6 | 1·6 | 2·5 | 3·5 | 4·7 | 5·9 | 7·2 |
| 72·9 | 76·9 | 80·0 | 82·5 | 83·9 | 84·2 | 83·0 | 81·6 | 77·1 | **82** | 73·4 | 70·0 | 68·1 | 66·9 | 67·1 | 68·6 | 70·8 | 74·2 | 77·9 | 82·0 |
| 197·6 | 207·6 | 209·8 | 203·1 | 198·0 | 202·9 | 214·4 | 229·7 | 246·8 | | 264·4 | 281·5 | 297·4 | 310·3 | 316·2 | 312·7 | 305·5 | 307·1 | 317·2 | 331·8 |
| 31·5 | 30·5 | 29·2 | 27·8 | 26·4 | 25·1 | 24·1 | 23·2 | 22·6 | | 22·2 | 22·2 | 22·6 | 23·3 | 24·3 | 25·6 | 27·1 | 28·8 | 30·7 | 32·8 |
| 140·7 | 152·8 | 164·8 | 177·2 | 189·5 | 202·0 | 214·4 | 227·0 | 239·6 | | 252·1 | 234·7 | 277·5 | 290·1 | 302·7 | 315·2 | 327·8 | 340·2 | 352·3 | 4·7 |
| 18·7 | 17·9 | 17·2 | 16·3 | 15·5 | 14·7 | 13·8 | 13·6 | 13·2 | | 13·0 | 13·3 | 13·5 | 13·8 | 14·5 | 15·3 | 16·0 | 17·1 | 18·3 | 19·5 |
| 191·5 | 198·4 | 205·3 | 212·2 | 219·4 | 226·4 | 233·4 | 241·0 | 248·4 | **83** | 256·2 | 263·4 | 270·9 | 278·7 | 286·4 | 294·2 | 301·9 | 309·9 | 317·8 | 325·6 |
| 194·3 | 189·7 | 183·9 | 184·9 | 194·8 | 209·2 | 225·6 | 243·1 | 260·2 | | 276·5 | 290·4 | 298·3 | 297·6 | 289·8 | 286·9 | 293·3 | 310·2 | 327·1 | 345·3 |
| 67·7 | 67·8 | 67·5 | 67·0 | 67·2 | 65·1 | 63·8 | 62·5 | 61·2 | | 59·9 | 58·8 | 58·0 | 57·4 | 57·2 | 57·5 | 58·0 | 58·7 | 59·7 | 61·1 |
| 218·5 | 226·9 | 233·3 | 238·9 | 240·4 | 235·6 | 229·1 | 224·2 | 223·4 | | 225·5 | 230·8 | 238·1 | 246·9 | 256·4 | 266·8 | 277·7 | 289·0 | 300·3 | 312·2 |
| 33·9 | 33·2 | 32·4 | 31·6 | 30·8 | 30·0 | 29·2 | 28·6 | 28·0 | | 27·4 | 27·3 | 27·3 | 27·3 | 27·8 | 28·3 | 28·9 | 30·0 | 31·0 | 32·0 |
| O 9 | O 19 | O 29 | N 8 | N 18 | N 28 | D 8 | D 18 | D 28 | | Ja 7 | Ja 17 | Ja 27 | F 6 | F 16 | F 26 | Mr 7 | Mr 17 | Mr 27 | Ap 6 |
| 94·7 | 99·9 | 104·7 | 109·3 | 113·2 | 116·5 | 119·7 | 120·9 | 121·2 | **84** | 120·5 | 118·9 | 115·7 | 111·8 | 108·2 | 105·4 | 102·6 | 101·9 | 102·5 | 104·2 |
| 167·5 | 167·7 | 177·3 | 190·2 | 206·6 | 223·8 | 240·9 | 257·5 | 271·5 | | 281·0 | 281·8 | 274·2 | 270·9 | 277·3 | 290·2 | 306·5 | 324·4 | 343·0 | 1·1 |
| 98·5 | 99·7 | 100·4 | 101·0 | 101·0 | 100·9 | 100·4 | 99·5 | 98·5 | | 97·3 | 96·1 | 94·2 | 93·5 | 92·4 | 91·6 | 90·8 | 90·5 | 90·5 | 90·9 |
| 177·7 | 190·2 | 202·6 | 215·0 | 227·7 | 240·3 | 252·8 | 265·5 | 278·0 | | 290·5 | 302·9 | 315·2 | 327·5 | 339·5 | 351·7 | 3·5 | 15·4 | 27·1 | 38·1 |
| 49·0 | 48·4 | 47·8 | 47·3 | 46·5 | 45·7 | 44·8 | 44·1 | 43·4 | | 42·6 | 42·3 | 42·0 | 41·7 | 42·1 | 42·5 | 42·8 | 43·5 | 44·2 | 44·9 |
| O 8 | O 18 | O 28 | N 7 | N 17 | N 27 | D 7 | D 17 | D 27 | | Ja 6 | Ja 16 | Ja 26 | F 5 | F 15 | F 25 | Mr 7 | Mr 17 | Mr 27 | Ap 6 |
| 205·7 | 212·5 | 219·5 | 226·9 | 234·7 | 241·4 | 248·5 | 256·3 | 264·0 | **85** | 271·6 | 279·5 | 287·3 | 295·1 | 303·1 | 310·9 | 318·8 | 326·8 | 334·6 | 342·5 |
| 157·0 | 169·9 | 185·7 | 202·9 | 220·2 | 236·9 | 251·7 | 262·5 | 266·2 | | 260·1 | 254·7 | 258·2 | 269·4 | 284·7 | 302·2 | 320·4 | 338·5 | 355·8 | 10·5 |
| 125·4 | 127·2 | 128·7 | 130·1 | 131·2 | 132·1 | 132·6 | 132·8 | 132·8 | | 132·4 | 131·7 | 131·1 | 129·5 | 128·2 | 126·8 | 125·5 | 124·4 | 123·4 | 122·7 |
| 127·7 | 139·1 | 150·0 | 161·7 | 173·4 | 185·6 | 197·6 | 209·2 | 222·2 | | 234·5 | 247·0 | 259·4 | 272·0 | 284·5 | 297·1 | 309·3 | 321·9 | 334·3 | 346·7 |
| 63·8 | 63·6 | 63·4 | 63·2 | 62·4 | 61·5 | 60·6 | 59·7 | 58·9 | | 58·1 | 57·5 | 57·0 | 56·5 | 56·4 | 56·4 | 56·4 | 56·8 | 57·2 | 57·7 |
| 110·5 | 116·3 | 121·7 | 127·2 | 132·3 | 137·0 | 141·3 | 145·2 | 148·6 | **86** | 151·0 | 152·9 | 153·3 | 152·5 | 150·4 | 147·0 | 143·2 | 139·3 | 136·3 | 134·4 |
| 164·6 | 181·7 | 199·2 | 216·3 | 231·9 | 244·3 | 250·1 | 246·8 | 239·5 | | 240·0 | 249·4 | 263·5 | 280·3 | 298·2 | 316·2 | 333·4 | 348·9 | 359·8 | 2·4 |
| 150·0 | 152·1 | 154·2 | 156·1 | 157·7 | 159·3 | 160·6 | 161·7 | 162·5 | | 163·2 | 163·5 | 163·5 | 163·0 | 162·2 | 161·3 | 160·2 | 158·8 | 157·6 | 156·3 |
| 212·1 | 224·1 | 235·7 | 247·3 | 258·4 | 269·5 | 280·1 | 289·7 | 298·8 | | 306·5 | 312·8 | 315·6 | 316·8 | 309·6 | 308·1 | 299·6 | 300·2 | 303·9 | 310·0 |
| 78·3 | 78·2 | 78·1 | 78·0 | 77·4 | 76·8 | 76·2 | 75·5 | 74·7 | | 73·9 | 73·1 | 72·3 | 71·6 | 71·3 | 71·0 | 70·7 | 71·0 | 71·3 | 71·6 |
| 220·2 | 227·0 | 234·7 | 241·9 | 249·2 | 256·4 | 264·0 | 272·1 | 279·9 | **87** | 288·0 | 295·5 | 303·2 | 311·2 | 318·9 | 326·9 | 334·8 | 342·3 | 350·1 | 357·7 |
| 178·0 | 195·5 | 211·7 | 225·4 | 233·8 | 233·2 | 225·5 | 223·2 | 230·0 | | 243·0 | 259·1 | 276·5 | 294·5 | 311·5 | 327·4 | 339·5 | 344·4 | 339·4 | 334·3 |
| 173·5 | 175·7 | 177·8 | 180·0 | 182·1 | 184·0 | 185·9 | 187·7 | 189·2 | | 190·6 | 191·7 | 192·6 | 193·3 | 193·4 | 193·4 | 193·0 | 192·4 | 191·4 | 190·2 |
| 158·2 | 170·5 | 182·9 | 195·6 | 208·1 | 220·7 | 233·2 | 246·0 | 258·6 | | 271·3 | 283·9 | 296·4 | 309·0 | 321·4 | 334·1 | 346·4 | 358·7 | 11·0 | 23·2 |
| 91·9 | 92·2 | 92·4 | 92·6 | 92·3 | 92·0 | 91·6 | 90·9 | 90·1 | | 89·3 | 88·5 | 87·7 | 86·9 | 86·3 | 85·7 | 85·2 | 85·2 | 85·2 | 85·2 |
| O 9 | O 19 | O 29 | N 8 | N 18 | N 28 | D 8 | D 18 | D 28 | | Ja 7 | Ja 17 | Ja 27 | F 6 | F 16 | F 26 | Mr 7 | Mr 17 | Mr 27 | Ap 6 |
| 124·8 | 130·8 | 136·6 | 142·3 | 147·9 | 153·5 | 158·6 | 163·7 | 168·6 | **88** | 173·5 | 177·2 | 180·9 | 183·8 | 186·9 | 187·2 | 186·2 | 185·8 | 183·6 | 179·5 |
| 192·9 | 207·4 | 216·9 | 218·5 | 210·9 | 207·2 | 211·3 | 224·3 | 239·8 | | 257·0 | 274·7 | 291·9 | 307·7 | 320·4 | 325·7 | 322·2 | 315·4 | 317·4 | 327·9 |
| 197·2 | 199·4 | 201·4 | 203·7 | 205·9 | 208·1 | 210·2 | 212·4 | 214·4 | | 216·3 | 218·0 | 219·6 | 221·1 | 222·2 | 223·0 | 223·8 | 224·1 | 224·1 | 223·7 |
| 147·3 | 148·0 | 152·0 | 158·7 | 167·0 | 176·2 | 186·5 | 197·2 | 208·3 | | 219·7 | 231·6 | 243·5 | 255·6 | 267·9 | 280·1 | 292·3 | 304·4 | 316·7 | 329·1 |
| 105·1 | 105·6 | 106·1 | 106·5 | 106·5 | 106·4 | 106·3 | 105·8 | 105·2 | | 104·6 | 103·8 | 103·0 | 102·1 | 101·3 | 100·6 | 99·9 | 99·6 | 99·3 | 99·1 |
| O 8 | O 18 | O 28 | N 7 | N 17 | N 27 | D 7 | D 17 | D 27 | | Ja 6 | Ja 16 | Ja 26 | F 5 | F 15 | F 25 | Mr 7 | Mr 17 | Mr 27 | Ap 6 |
| 238·0 | 245·0 | 252·1 | 259·5 | 267·0 | 274·5 | 282·0 | 289·7 | 297·4 | **89** | 305·2 | 312·8 | 320·5 | 328·2 | 336·0 | 343·4 | 351·1 | 358·5 | 5·9 | 13·3 |
| 199·0 | 203·4 | 197·9 | 191·6 | 194·1 | 204·8 | 219·2 | 235·8 | 253·4 | | 270·8 | 287·1 | 300·4 | 308·3 | 306·7 | 299·0 | 298·1 | 306·8 | 321·0 | 338·1 |
| 221·6 | 223·6 | 225·6 | 227·7 | 229·9 | 232·1 | 234·4 | 236·6 | 238·9 | | 241·1 | 243·3 | 245·5 | 247·4 | 249·2 | 250·3 | 252·4 | 253·5 | 254·5 | 255·5 |
| 196·2 | 208·6 | 221·0 | 233·3 | 245·4 | 257·8 | 270·0 | 282·2 | 294·3 | | 306·0 | 318·0 | 329·2 | 340·0 | 350·8 | 0·6 | 9·8 | 17·9 | 24·2 | 27·6 |
| 117·8 | 118·5 | 119·2 | 119·9 | 120·1 | 120·3 | 120·5 | 120·1 | 119·7 | | 119·4 | 118·6 | 117·8 | 117·0 | 116·2 | 115·4 | 114·7 | 114·2 | 113·7 | 113·2 |
| 137·3 | 143·3 | 149·4 | 155·6 | 161·6 | 167·4 | 173·2 | 179·1 | 184·7 | **90** | 190·3 | 195·9 | 201·2 | 206·2 | 211·3 | 215·7 | 219·9 | 223·5 | 226·6 | 229·1 |
| 184·6 | 176·8 | 176·5 | 185·1 | 199·0 | 215·1 | 232·2 | 249·5 | 266·5 | | 280·4 | 290·2 | 291·2 | 283·8 | 280·4 | 286·5 | 299·1 | 315·7 | 333·7 | 352·7 |
| 248·4 | 250·4 | 251·5 | 253·3 | 255·2 | 257·4 | 259·5 | 261·8 | 264·1 | | 266·4 | 268·6 | 271·0 | 273·4 | 275·5 | 277·6 | 279·7 | 281·7 | 283·6 | 285·6 |
| 141·2 | 153·1 | 165·3 | 177·7 | 190·0 | 202·5 | 214·9 | 227·6 | 240·1 | | 252·8 | 265·4 | 278·0 | 290·6 | 303·2 | 315·8 | 328·3 | 340·7 | 353·2 | 5·6 |
| 129·8 | 131·0 | 132·3 | 133·6 | 133·7 | 133·8 | 133·9 | 133·8 | 133·7 | | 133·5 | 132·9 | 132·3 | 131·6 | 130·9 | 130·1 | 129·3 | 128·6 | 127·9 | 127·2 |

| Mag. | P.Phal. | U.Phal. | Hast. | Chit. | Svati. | Visa. | Anur. | Jyesh. | Mula. | P.Ash. | U.Ash. | Srav. | Dan. | Satab. | P.Bhad. | U.Bhad. | Revati. |
|---|---|---|---|---|---|---|---|---|---|---|---|---|---|---|---|---|---|
| 133·3 | 146·7 | 160·0 | 173·3 | 186·7 | 200·0 | 213·3 | 226·7 | 240·0 | 253·3 | 266·7 | 280·0 | 293·3 | 306·7 | 320·0 | 333·3 | 346·7 | 360·0 |

## Geocentric (Indian) Longitudes of sun and Major planets at Ujjain

| Day of S. Yr. | 0 | 10 | 20 | 30 | 40 | 50 | 60 | 70 | 80 | 90 | 100 | 110 | 120 | 130 | 140 | 150 | 160 | 170 |
|---|---|---|---|---|---|---|---|---|---|---|---|---|---|---|---|---|---|---|
| ⊙'s mean L. | 357·0 | 6·9 | 16·7 | 26·6 | 36·4 | 46·3 | 56·1 | 66·0 | 75·9 | 85·7 | 95·6 | 105·4 | 115·3 | 125·1 | 135·0 | 144·8 | 154·7 | 164·6 |
| Corrn. for do. | +2·1 | +2·0 | +1·9 | +1·7 | +1·4 | +1·1 | +0·8 | +0·4 | +0·0 | −0·3 | −0·7 | −1·1 | −1·4 | −1·6 | −1·9 | −2·0 | −2·1 | −2·2 |
| English date. | Ap 12 | Ap 22 | My 2 | My 12 | My 22 | Je 1 | Je 11 | Je 21 | Jl 1 | Jl 11 | Jl 21 | Jl 31 | Au 10 | Au 20 | Au 30 | S 9 | S 19 | S 29 |
| Tamil month. | Chit | Chit | Chit | Chit | Vaik | Vaik | Vaik | Ani | Ani | Ani | Adi | Adi | Adi | Avan | Avan | Avan | Pur | Pur |
| Do. date. | 1 | 11 | 21 | 31 | 10 | 20 | 30 | 9 | 19 | 29 | 7 | 17 | 27 | 6 | 16 | 26 | 5 | 15 |
| 1890 Mars ... | 230·0 | 231·0 | 230·7 | 229·6 | 227·1 | 224·3 | 221·2 | 219·1 | 218·1 | 218·6 | 220·2 | 222·9 | 226·5 | 230·8 | 235·8 | 241·2 | 247·3 | 253·2 |
| ⊙'s L. Merc. ... | 3·2 | 21·1 | 36·9 | 48·4 | 51·4 | 45·1 | 40·9 | 46·5 | 59·1 | 75·5 | 93·6 | 112·2 | 130·3 | 147·1 | 161·3 | 169·8 | 169·4 | 161·5 |
| +·8 Jup. ... | 286·1 | 287·3 | 288·4 | 289·1 | 289·5 | 289·6 | 289·3 | 288·7 | 287·8 | 286·8 | 285·4 | 284·0 | 282·7 | 281·9 | 280·7 | 279·6 | 279·1 | 279·0 |
| Ven. ... | 13·1 | 25·5 | 37·5 | 50·0 | 62·1 | 74·3 | 86·2 | 98·3 | 110·3 | 121·9 | 133·5 | 145·1 | 156·6 | 167·9 | 179·3 | 189·9 | 200·2 | 209·8 |
| Sat. ... | 127·0 | 126·6 | 126·5 | 126·6 | 127·0 | 127·4 | 127·9 | 128·8 | 129·7 | 130·6 | 131·7 | 132·8 | 134·0 | 135·2 | 136·5 | 137·8 | 139·1 | 140·4 |
| 1891 Mars ... | 31·1 | 38·8 | 45·8 | 52·5 | 59·3 | 66·1 | 72·7 | 79·3 | 85·7 | 92·3 | 98·8 | 105·2 | 111·7 | 118·2 | 124·6 | 130·8 | 137·1 | 143·5 |
| ⊙'s L. Merc. ... | 14·5 | 25·7 | 32·6 | 27·7 | 21·6 | 25·3 | 36·9 | 52·8 | 70·7 | 89·5 | 108·0 | 125·0 | 140·8 | 151·4 | 153·7 | 146·9 | 142·6 | 147·5 |
| +·6 Jup. ... | 314·7 | 316·7 | 318·5 | 320·2 | 321·6 | 322·9 | 324·1 | 324·6 | 324·9 | 324·9 | 324·5 | 323·8 | 322·8 | 321·5 | 320·3 | 319·0 | 317·6 | 316·4 |
| Ven. ... | 319·5 | 331·4 | 343·2 | 355·2 | 7·3 | 19·2 | 31·3 | 43·4 | 55·5 | 67·6 | 79·9 | 91·8 | 103·6 | 116·3 | 128·6 | 141·0 | 153·4 | 165·8 |
| Sat. ... | 140·8 | 140·4 | 140·0 | 139·6 | 139·8 | 140·0 | 140·2 | 140·8 | 141·5 | 142·2 | 143·2 | 144·3 | 145·4 | 146·6 | 147·8 | 149·1 | 150·4 | 151·6 |
| Eng. Date. | Ap 11 | Ap 21 | My 1 | My 11 | My 21 | My 31 | Je 10 | Je 20 | Je 30 | Jl 10 | Jl 20 | Jl 30 | Au 9 | Au 19 | Au 29 | S 8 | S 18 | S 28 |
| 1892 Mars ... | 263·3 | 268·6 | 273·7 | 278·7 | 283·0 | 287·1 | 290·3 | 294·2 | 294·2 | 294·7 | 293·5 | 291·0 | 288·7 | 286·2 | 284·6 | 284·6 | 286·4 | 289·1 |
| ⊙'s L. Merc. ... | 13·0 | 10·4 | 3·2 | 4·5 | 14·5 | 29·6 | 47·5 | 66·6 | 85·1 | 103·1 | 119·4 | 131·7 | 137·0 | 132·1 | 125·8 | 128·4 | 139·1 | 153·8 |
| +·3 Jup. ... | 342·4 | 344·8 | 347·1 | 349·3 | 351·5 | 353·5 | 355·4 | 357·1 | 358·5 | 359·8 | 0·3 | 0·8 | 1·3 | 1·2 | 0·7 | 0·0 | 359·0 | 357·6 |
| Ven. ... | 44·0 | 54·6 | 64·7 | 73·8 | 82·2 | 88·9 | 93·6 | 94·8 | 91·9 | 85·7 | 80·1 | 77·2 | 78·5 | 82·7 | 89·3 | 97·9 | 107·2 | 117·1 |
| Sat. ... | 154·3 | 153·7 | 153·2 | 152·7 | 152·6 | 152·5 | 152·5 | 152·9 | 153·3 | 153·8 | 154·8 | 155·7 | 156·6 | 157·7 | 158·8 | 160·0 | 161·2 | 162·5 |
| 1893 Mars ... | 46·3 | 52·8 | 59·3 | 65·8 | 72·1 | 78·5 | 85·0 | 91·4 | 97·8 | 104·1 | 110·5 | 116·9 | 123·3 | 129·5 | 136·0 | 142·3 | 148·9 | 155·0 |
| ⊙'s L. Merc. ... | 345·7 | 344·4 | 352·7 | 7·0 | 24·3 | 43·0 | 61·9 | 79·4 | 97·4 | 111·2 | 118·6 | 116·3 | 109·0 | 109·4 | 118·6 | 132·7 | 149·3 | 167·1 |
| +·1 Jup. ... | 9·9 | 12·3 | 14·7 | 17·2 | 19·5 | 21·9 | 24·4 | 26·5 | 28·5 | 30·6 | 32·4 | 34·0 | 35·3 | 36·4 | 37·0 | 37·6 | 37·7 | 37·5 |
| Ven. ... | 358·6 | 6·1 | 18·4 | 30·7 | 43·0 | 55·3 | 67·5 | 79·7 | 91·9 | 104·1 | 116·5 | 128·6 | 140·6 | 152·7 | 164·6 | 176·8 | 188·9 | 200·7 |
| Sat. ... | 167·7 | 166·9 | 166·2 | 165·7 | 165·2 | 164·2 | 164·7 | 164·9 | 165·1 | 165·4 | 166·2 | 166·9 | 167·6 | 168·6 | 169·6 | 170·7 | 171·9 | 173·1 |
| Eng. Date. | Ap 12 | Ap 22 | My 2 | My 12 | My 22 | Je 1 | Je 11 | Je 21 | Jl 1 | Jl 11 | Jl 21 | Jl 31 | Au 10 | Au 20 | Au 30 | S 9 | S 19 | S 29 |
| 1894 Mars ... | 286·5 | 293·9 | 300·5 | 307·4 | 314·1 | 321·0 | 327·8 | 333·9 | 340·1 | 345·7 | 351·4 | 356·1 | 0·5 | 4·0 | 6·3 | 7·7 | 7·5 | 6·1 |
| ⊙'s L. Merc. ... | 332·6 | 345·8 | 2·8 | 21·5 | 40·3 | 58·3 | 76·6 | 91·1 | 99·5 | 99·2 | 91·5 | 90·4 | 98·8 | 112·7 | 129·5 | 147·1 | 165·0 | 182·2 |
| +·8 Jup. ... | 37·8 | 40·0 | 42·3 | 44·6 | 46·9 | 49·3 | 51·6 | 54·0 | 56·3 | 58·6 | 60·9 | 62·9 | 64·8 | 66·6 | 68·3 | 69·8 | 70·9 | 71·8 |
| Ven. ... | 314·3 | 323·0 | 332·8 | 343·2 | 353·8 | 5·0 | 16·4 | 27·6 | 39·1 | 51·3 | 63·0 | 74·8 | 86·8 | 98·7 | 110·9 | 123·2 | 135·5 | 147·6 |
| Sat. ... | 180·4 | 180·0 | 179·6 | 179·1 | 178·4 | 177·6 | 176·8 | 176·8 | 176·8 | 176·8 | 177·4 | 178·0 | 178·5 | 179·4 | 180·4 | 181·4 | 182·5 | 183·6 |
| 1895 Mars ... | 61·6 | 67·5 | 73·6 | 79·6 | 85·8 | 91·8 | 98·0 | 104·2 | 110·3 | 116·7 | 122·7 | 129·0 | 135·4 | 141·7 | 148·1 | 154·6 | 161·0 | 167·6 |
| ⊙'s L. Merc. ... | 341·6 | 358·2 | 17·0 | 35·8 | 53·6 | 69·0 | 79·7 | 81·2 | 74·1 | 71·2 | 78·0 | 90·8 | 107·3 | 125·1 | 143·3 | 160·7 | 176·8 | 189·7 |
| +·5 Jup ... | 65·9 | 67·4 | 69·3 | 71·1 | 73·2 | 75·3 | 77·6 | 79·9 | 82·1 | 84·4 | 86·7 | 89·0 | 91·2 | 93·4 | 95·4 | 97·4 | 99·2 | 100·8 |
| Ven. ... | 30·9 | 42·7 | 54·9 | 66·3 | 78·0 | 89·4 | 100·4 | 111·4 | 121·7 | 131·7 | 141·0 | 149·2 | 155·8 | 160·2 | 161·3 | 158·4 | 152·1 | 147·1 |
| Sat ... | 192·9 | 192·1 | 191·3 | 190·6 | 190·0 | 189·4 | 188·8 | 188·6 | 188·4 | 188·1 | 188·4 | 188·7 | 189·1 | 189·8 | 190·7 | 191·6 | 192·2 | 193·7 |
| Eng. Date. | Ap 11 | Ap 21 | My 1 | My 11 | My 21 | My 31 | Je 10 | Je 20 | Je 30 | Jl 10 | Jl 20 | Jl 30 | Au 9 | Au 19 | Au 29 | S 8 | S 18 | S 28 |
| 1896 Mars ... | 305·5 | 313·0 | 320·6 | 328·2 | 335·8 | 343·1 | 350·5 | 357·8 | 4·8 | 11·9 | 18·4 | 25·0 | 31·3 | 37·3 | 43·2 | 49·1 | 53·2 | 57·4 |
| ⊙'s L. Merc. ... | 353·9 | 12·1 | 30·6 | 46·8 | 58·6 | 62·9 | 58·7 | 51·9 | 56·5 | 68·8 | 85·0 | 102·8 | 121·2 | 139·1 | 156·0 | 170·1 | 179·1 | 179·1 |
| +·3 Jup. ... | 95·3 | 96·0 | 97·0 | 98·3 | 99·8 | 101·4 | 103·2 | 105·2 | 107·3 | 109·4 | 111·6 | 113·8 | 115·9 | 118·3 | 120·5 | 122·8 | 124·9 | 127·0 |
| Ven. ... | 335·8 | 348·1 | 0·3 | 12·8 | 24·8 | 37·1 | 49·2 | 61·6 | 73·8 | 86·2 | 98·3 | 110·6 | 122·8 | 135·0 | 147·5 | 159·6 | 172·1 | 184·5 |
| Sat. ... | 205·2 | 203·7 | 202·1 | 200·6 | 199·9 | 199·2 | 198·4 | 198·1 | 197·7 | 197·4 | 197·6 | 197·7 | 198·0 | 198·6 | 199·3 | 200·0 | 201·0 | 202·1 |
| 1897 Mars ... | 78·3 | 82·7 | 87·9 | 93·4 | 99·0 | 104·7 | 110·7 | 116·6 | 122·6 | 128·3 | 134·6 | 141·1 | 147·3 | 153·6 | 160·1 | 166·5 | 173·0 | 179·6 |
| ⊙'s L. Merc. ... | 8·2 | 24·1 | 37·1 | 43·5 | 39·3 | 33·0 | 34·7 | 46·7 | 62·2 | 79·9 | 98·7 | 116·9 | 134·5 | 149·7 | 160·7 | 163·6 | 157·0 | 152·1 |
| +·0 Jup. ... | 126·8 | 126·6 | 126·6 | 126·8 | 127·3 | 128·3 | 129·4 | 130·8 | 132·5 | 134·1 | 136·0 | 138·1 | 140·9 | 142·3 | 144·5 | 146·7 | 148·9 | 151·0 |
| Ven. ... | 15·7 | 10·5 | 3·9 | 1·0 | 1·3 | 4·9 | 11·3 | 19·3 | 28·6 | 38·8 | 49·4 | 60·6 | 72·1 | 83·4 | 95·4 | 107·5 | 119·6 | 131·5 |
| Sat. ... | 217·0 | 216·4 | 215·7 | 215·0 | 214·2 | 213·5 | 212·7 | 212·1 | 211·6 | 211·1 | 211·1 | 211·0 | 210·9 | 211·3 | 211·8 | 212·1 | 213·0 | 213·8 |
| Eng. Date. | Ap 12 | Ap 22 | My 2 | My 12 | My 22 | Je 1 | Je 11 | Je 21 | Jl 1 | Jl 11 | Jl 21 | Jl 31 | Au 10 | Au 20 | Au 30 | S 9 | S 19 | S 29 |
| 1898 Mars ... | 323·7 | 331·5 | 339·2 | 346·9 | 354·6 | 2·0 | 9·5 | 16·9 | 24·1 | 31·2 | 38·1 | 44·9 | 51·4 | 58·1 | 64·3 | 70·5 | 76·4 | 82·0 |
| ⊙'s L. Merc. ... | 16·9 | 24·0 | 21·2 | 15·2 | 15·4 | 25·8 | 41·0 | 58·8 | 77·4 | 96·2 | 114·0 | 130·0 | 142·2 | 146·9 | 141·6 | 135·6 | 138·6 | 149·3 |
| +·8 Jup. ... | 160·1 | 158·9 | 158·1 | 157·4 | 156·9 | 156·9 | 157·1 | 157·8 | 158·8 | 159·8 | 161·2 | 162·8 | 164·4 | 166·4 | 168·3 | 170·4 | 172·5 | 174·6 |
| Ven. ... | 11·7 | 24·1 | 36·7 | 49·0 | 61·3 | 73·9 | 86·3 | 98·5 | 110·5 | 123·3 | 135·4 | 147·3 | 159·2 | 171·0 | 182·1 | 192·9 | 203·6 | 213·2 |
| Sat. ... | 228·6 | 228·0 | 227·3 | 226·8 | 226·0 | 225·3 | 224·6 | 224·0 | 223·5 | 222·8 | 222·4 | 222·1 | 221·8 | 222·1 | 222·3 | 222·6 | 223·4 | 224·1 |
| 1899 Mars ... | 96·4 | 100·3 | 104·6 | 110·2 | 114·3 | 119·5 | 124·9 | 130·5 | 136·4 | 142·3 | 148·1 | 154·4 | 160·6 | 166·9 | 173·3 | 179·7 | 186·6 | 193·2 |
| ⊙'s L. Merc. ... | 3·6 | 356·3 | 355·2 | 3·7 | 18·0 | 35·5 | 54·1 | 73·2 | 91·3 | 108·2 | 121·9 | 128·8 | 126·4 | 119·0 | 119·6 | 128·7 | 143·0 | 159·6 |
| +·5 Jup. ... | 194·1 | 192·8 | 191·5 | 190·3 | 189·2 | 188·2 | 187·5 | 187·2 | 187·1 | 187·2 | 187·8 | 188·7 | 189·8 | 191·2 | 192·8 | 194·5 | 196·4 | 198·4 |
| Ven. ... | 319·8 | 331·9 | 343·7 | 355·7 | 7·6 | 19·7 | 31·7 | 43·9 | 55·5 | 68·1 | 80·4 | 92·3 | 104·7 | 116·8 | 129·3 | 141·5 | 153·7 | 166·3 |
| Sat. ... | 239·9 | 239·5 | 239·1 | 238·6 | 237·8 | 237·0 | 236·3 | 235·7 | 235·0 | 234·2 | 233·8 | 233·3 | 233·0 | 233·0 | 233·1 | 233·1 | 233·7 | 234·2 |

| Mesh. | Vrsh. | Mith. | Kat. | Sim. | Kan. | Tul. | Vrsch. | Dhan. | Mak. | Kum. | Min. | Asvn. | Bhar. | Krit. | Rohi. | Mrig. | Ardh. | Punar. | Push. | Asles. |
|---|---|---|---|---|---|---|---|---|---|---|---|---|---|---|---|---|---|---|---|---|
| 30 | 60 | 90 | 120 | 150 | 180 | 210 | 240 | 270 | 300 | 330 | 360 | 13·3 | 26·7 | 40·0 | 53·3 | 66·7 | 80·0 | 93·3 | 106·7 | 120·0 |

**mean Sunrise for every ten days from A.D. 1890 to A.D. 1899.**

| | 180 | 190 | 200 | 210 | 220 | 230 | 240 | 250 | 260 | 270 | 280 | 290 | 300 | 310 | 320 | 330 | 340 | 350 | 360 |
|---|---|---|---|---|---|---|---|---|---|---|---|---|---|---|---|---|---|---|---|
| | 174·4 | 184·3 | 194·1 | 204·0 | 213·8 | 223·7 | 233·5 | 243·4 | 253·3 | 263·1 | 273·0 | 282·8 | 292·7 | 302·5 | 312·4 | 322·2 | 332·1 | 342·0 | 351·8 |
| | −2·2 | −2·1 | −1·9 | −1·7 | −1·5 | −1·2 | −0·9 | −0·5 | −0·1 | +0·3 | +0·6 | +1·0 | +1·3 | +1·6 | +1·8 | +2·0 | +2·1 | +2·2 | +2·2 |
| | O 9 | O 19 | O 29 | N 8 | N 18 | N 28 | D 8 | D 18 | D 28 | Ja 7 | Ja 17 | Ja 27 | F 6 | F 16 | F 26 | Mr 7,8 | 17,18 | 27,28 | Ap 6,7 |
| | Pur | Aipp | Aipp | Aipp | Kart | Kart | Kart | Marg | Marg | Marg | Tai | Tai | Tai | Masi | Masi | Masi | Pang | Pang | Pang |
| | 25 | 4 | 14 | 24 | 4 | 14 | 24 | 5 | 15 | 25 | 5 | 15 | 25 | 6 | 16 | 26 | 6 | 16 | 26 |
| **91** | 259·6 | 266·3 | 273·1 | 280·2 | 287·3 | 294·5 | 302·0 | 309·1 | 316·6 | 324·0 | 331·2 | 338·6 | 345·8 | 353·1 | 0·3 | 7·5 | 14·6 | 21·6 | 28·6 |
| | 159·6 | 166·9 | 180·0 | 195·8 | 213·1 | 230·3 | 246·7 | 261·5 | 272·3 | 275·3 | 268·7 | 263·8 | 268·0 | 279·7 | 295·4 | 313·0 | 331·3 | 349·5 | 6·4 |
| | 279·2 | 280·0 | 280·6 | 281·7 | 283·1 | 284·6 | 286·4 | 288·5 | 290·6 | 292·8 | 295·0 | 297·4 | 299·8 | 302·1 | 304·5 | 306·9 | 309·2 | 311·6 | 313·7 |
| | 219·0 | 226·6 | 232·6 | 235·8 | 235·3 | 231·3 | 225·1 | 221·1 | 220·9 | 224·4 | 230·5 | 238·4 | 247·5 | 257·5 | 267·9 | 279·0 | 290·3 | 301·8 | 313·5 |
| | 141·6 | 142·7 | 143·7 | 144·7 | 145·3 | 145·9 | 146·5 | 146·6 | 146·7 | 146·8 | 146·4 | 146·0 | 145·5 | 144·8 | 144·0 | 143·2 | 142·4 | 141·6 | 140·9 |
| **92** | 149·8 | 156·2 | 162·3 | 168·7 | 174·9 | 181·2 | 187·5 | 193·7 | 199·9 | 206·2 | 212·3 | 218·6 | 224·6 | 230·9 | 237·0 | 242·9 | 249·0 | 254·7 | 260·6 |
| | 159·4 | 175·0 | 192·1 | 209·4 | 226·4 | 241·9 | 254·0 | 259·5 | 255·4 | 248·0 | 249·6 | 259·5 | 273·9 | 290·9 | 308·9 | 327·0 | 344·5 | 359·7 | 10·5 |
| | 315·8 | 314·4 | 314·4 | 314·3 | 314·7 | 315·4 | 316·3 | 317·6 | 319·2 | 321·0 | 322·8 | 324·8 | 327·0 | 329·3 | 331·7 | 334·1 | 336·5 | 338·9 | 341·3 |
| | 178·3 | 190·8 | 203·2 | 215·8 | 228·3 | 240·9 | 253·4 | 266·1 | 278·6 | 291·1 | 303·5 | 315·8 | 327·9 | 340·3 | 352·3 | 3·9 | 15·8 | 27·5 | 38·4 |
| | 152·8 | 153·9 | 155·0 | 156·1 | 157·0 | 157·8 | 158·6 | 158·9 | 159·2 | 159·5 | 159·3 | 159·0 | 158·8 | 158·2 | 157·5 | 156·8 | 156·1 | 155·3 | 154·6 |
| | O 8 | O 18 | O 28 | N 7 | N 17 | N 27 | D 7 | D 17 | D 27 | Ja 6 | Ja 16 | Ja 26 | F 5 | F 15 | F 25 | Mr 7 | Mr 17 | Mr 27 | Ap 6 |
| **93** | 292·6 | 296·9 | 301·7 | 307·1 | 312·8 | 318·7 | 325·0 | 331·1 | 337·5 | 344·0 | 350·6 | 357·0 | 3·7 | 10·2 | 16·9 | 23·4 | 29·9 | 36·4 | 43·1 |
| | 170·8 | 188·4 | 205·6 | 221·7 | 235·2 | 242·8 | 241·7 | 234·3 | 232·0 | 239·9 | 253·2 | 266·9 | 284·4 | 302·2 | 320·0 | 336·4 | 350·1 | 357·6 | 349·2 |
| | 356·4 | 354·3 | 353·7 | 352·5 | 351·1 | 351·0 | 350·8 | 350·8 | 351·4 | 352·1 | 353·2 | 354·5 | 356·1 | 358·0 | 359·8 | 2·0 | 4·1 | 6·5 | 8·8 |
| | 127·8 | 139·1 | 150·3 | 162·0 | 173·9 | 185·9 | 198·1 | 210·3 | 222·7 | 235·0 | 247·5 | 259·9 | 272·5 | 285·0 | 297·6 | 309·8 | 322·4 | 334·8 | 346·2 |
| | 163·8 | 165·0 | 166·1 | 167·2 | 168·2 | 169·2 | 170·1 | 170·6 | 171·2 | 171·7 | 171·7 | 171·7 | 171·7 | 171·2 | 170·7 | 170·1 | 169·4 | 168·7 | 167·9 |
| **94** | 161·8 | 168·1 | 174·4 | 181·1 | 187·5 | 194·0 | 200·6 | 207·2 | 213·9 | 220·6 | 227·4 | 234·0 | 241·0 | 247·7 | 254·4 | 261·5 | 268·5 | 275·9 | 282·4 |
| | 184·7 | 201·3 | 215·3 | 225·8 | 227·3 | 220·6 | 215·9 | 221·2 | 232·7 | 248·1 | 265·5 | 283·1 | 300·8 | 316·7 | 329·7 | 336·3 | 333·0 | 326·0 | 327·4 |
| | 36·8 | 35·8 | 34·8 | 33·6 | 32·3 | 30·9 | 29·7 | 28·6 | 27·8 | 27·3 | 27·1 | 27·4 | 27·8 | 28·8 | 30·0 | 31·4 | 32·9 | 34·7 | 36·6 |
| | 212·5 | 224·6 | 236·1 | 247·5 | 258·6 | 269·6 | 279·9 | 289·7 | 298·4 | 305·5 | 311·0 | 312·9 | 311·1 | 304·7 | 299·7 | 298·6 | 298·2 | 302·6 | 309·3 |
| | 174·3 | 175·5 | 176·7 | 178·0 | 179·0 | 180·0 | 181·0 | 181·7 | 182·4 | 183·1 | 183·4 | 183·7 | 183·9 | 183·6 | 183·3 | 183·0 | 182·3 | 181·6 | 180·8 |
| | O 9 | O 19 | O 29 | N 8 | N 18 | N 28 | D 8 | D 18 | D 28 | Ja 7 | Ja 17 | Ja 27 | F 6 | F 16 | F 26 | Mr 7,8 | 17,18 | 27,28 | Ap 6,7 |
| **95** | 3·8 | 1·0 | 358·6 | 357·4 | 357·0 | 358·2 | 0·1 | 3·1 | 6·7 | 10·4 | 15·0 | 19·8 | 24·9 | 30·1 | 35·5 | 41·2 | 46·8 | 52·9 | 58·5 |
| | 197·2 | 208·6 | 212·1 | 206·7 | 200·3 | 203·4 | 214·2 | 229·1 | 245·9 | 263·4 | 280·8 | 297·2 | 310·5 | 318·2 | 316·6 | 308·9 | 309·0 | 318·2 | 332·4 |
| | 72·4 | 72·6 | 72·3 | 72·3 | 71·5 | 70·3 | 68·0 | 67·7 | 66·3 | 65·1 | 64·0 | 63·0 | 62·3 | 62·1 | 62·1 | 62·5 | 63·0 | 64·0 | 65·2 |
| | 159·9 | 172·4 | 184·8 | 197·3 | 209·9 | 222·0 | 235·1 | 247·7 | 260·4 | 273·0 | 285·6 | 298·2 | 310·9 | 323·1 | 336·0 | 348·3 | 0·5 | 12·7 | 24·9 |
| | 184·8 | 186·0 | 187·2 | 188·4 | 189·5 | 190·6 | 191·7 | 192·6 | 193·4 | 194·2 | 194·6 | 195·0 | 195·3 | 195·2 | 195·1 | 195·0 | 194·5 | 194·0 | 193·4 |
| **96** | 174·2 | 180·6 | 187·3 | 194·0 | 200·8 | 207·6 | 214·5 | 221·4 | 228·4 | 235·4 | 242·5 | 249·8 | 257·4 | 264·3 | 271·8 | 279·1 | 286·6 | 294·2 | 301·6 |
| | 196·2 | 193·4 | 185·8 | 186·0 | 194·8 | 208·5 | 224·8 | 242·3 | 259·6 | 276·1 | 290·4 | 299·9 | 300·8 | 293·3 | 290·1 | 296·6 | 309·8 | 326·4 | 344·5 |
| | 102·4 | 103·7 | 104·5 | 105·3 | 105·8 | 105·8 | 105·4 | 104·7 | 103·8 | 102·7 | 101·4 | 99·9 | 98·6 | 97·1 | 96·4 | 95·7 | 95·3 | 95·1 | 95·4 |
| | 144·6 | 146·3 | 151·1 | 158·0 | 166·2 | 176·1 | 186·2 | 197·4 | 208·7 | 220·0 | 232·1 | 243·8 | 255·9 | 268·2 | 280·4 | 292·7 | 304·9 | 317·2 | 329·8 |
| | 194·8 | 195·9 | 197·0 | 198·2 | 199·4 | 200·6 | 201·8 | 202·7 | 203·6 | 204·6 | 205·2 | 205·8 | 206·3 | 206·4 | 206·5 | 206·5 | 206·1 | 205·7 | 205·3 |
| | O 8 | O 18 | O 28 | N 7 | N 17 | N 27 | D 7 | D 17 | D 27 | Ja 6 | Ja 16 | Ja 26 | F 5 | F 15 | F 25 | Mr 7 | Mr 17 | Mr 27 | Ap 6 |
| **97** | 60·9 | 63·7 | 65·3 | 65·9 | 65·2 | 63·4 | 60·3 | 56·9 | 53·9 | 51·7 | 50·5 | 51·0 | 52·4 | 54·4 | 57·6 | 61·2 | 65·5 | 70·0 | 74·8 |
| | 171·3 | 168·9 | 175·7 | 188·4 | 204·2 | 221·4 | 238·5 | 255·5 | 270·4 | 281·2 | 284·7 | 278·5 | 273·2 | 277·0 | 288·5 | 304·0 | 321·9 | 340·4 | 358·7 |
| | 128·9 | 130·6 | 132·2 | 133·7 | 135·0 | 135·9 | 136·6 | 137·0 | 137·0 | 136·6 | 136·1 | 135·2 | 134·4 | 132·9 | 131·6 | 130·3 | 129·1 | 128·0 | 127·2 |
| | 196·8 | 209·2 | 221·6 | 233·9 | 246·0 | 258·4 | 270·6 | 282·6 | 294·1 | 305·4 | 316·4 | 327·2 | 337·9 | 347·6 | 356·8 | 5·2 | 11·9 | 16·0 | 17·0 |
| | 203·2 | 204·4 | 205·6 | 206·7 | 207·9 | 209·1 | 210·3 | 211·5 | 212·8 | 214·1 | 214·6 | 215·2 | 215·7 | 216·0 | 216·2 | 216·6 | 216·6 | 216·7 | 216·8 |
| **98** | 186·5 | 193·2 | 199·9 | 206·9 | 213·8 | 220·9 | 228·0 | 235·3 | 242·6 | 250·0 | 257·3 | 264·9 | 272·5 | 280·2 | 287·7 | 295·7 | 303·4 | 311·1 | 319·0 |
| | 156·5 | 168·1 | 183·3 | 200·3 | 217·7 | 234·6 | 250·2 | 262·5 | 268·3 | 264·5 | 257·4 | 258·4 | 268·1 | 282·4 | 299·6 | 317·6 | 336·0 | 353·6 | 9·3 |
| | 153·3 | 155·4 | 157·3 | 159·2 | 161·1 | 162·7 | 164·3 | 165·4 | 166·3 | 167·0 | 167·5 | 167·6 | 167·3 | 166·6 | 165·7 | 164·7 | 163·6 | 162·7 | 160·9 |
| | 143·7 | 155·8 | 168·0 | 180·2 | 192·3 | 204·6 | 216·8 | 229·0 | 241·3 | 253·3 | 265·4 | 277·7 | 289·7 | 302·2 | 314·7 | 327·4 | 339·4 | 351·5 | 4·3 |
| | 214·8 | 215·9 | 217·0 | 218·0 | 219·2 | 220·4 | 221·5 | 222·6 | 223·8 | 224·9 | 225·8 | 226·6 | 227·4 | 227·8 | 228·3 | 228·8 | 228·8 | 228·7 | 228·8 |
| | O 9 | O 19 | O 29 | N 8 | N 18 | N 28 | D 8 | D 18 | D 28 | Ja 7 | Ja 17 | Ja 27 | F 6 | F 16 | F 26 | Mr 8 | Mr 18 | Mr 28 | Ap 7 |
| **99** | 87·2 | 91·0 | 96·4 | 100·3 | 103·5 | 106·0 | 107·5 | 107·7 | 106·8 | 104·7 | 101·4 | 97·6 | 93·8 | 90·9 | 89·1 | 88·1 | 90·0 | 92·0 | 94·9 |
| | 164·1 | 181·0 | 198·5 | 215·6 | 231·4 | 244·8 | 252·1 | 250·3 | 242·2 | 240·3 | 249·8 | 263·2 | 279·6 | 297·7 | 315·5 | 333·1 | 348·1 | 0·8 | 5·4 |
| | 176·8 | 179·1 | 181·2 | 183·7 | 185·6 | 187·5 | 189·5 | 191·2 | 192·8 | 194·1 | 195·5 | 196·4 | 197·3 | 197·6 | 197·6 | 197·4 | 196·8 | 195·8 | 194·7 |
| | 222·4 | 230·4 | 236·2 | 239·4 | 239·1 | 234·6 | 227·7 | 223·3 | 222·5 | 225·0 | 230·5 | 238·0 | 246·7 | 256·2 | 266·7 | 277·4 | 288·5 | 300·0 | 311·5 |
| | 224·9 | 225·9 | 226·9 | 228·0 | 229·1 | 230·2 | 231·4 | 232·6 | 233·8 | 234·9 | 235·8 | 236·8 | 237·7 | 238·4 | 239·0 | 239·6 | 239·7 | 239·9 | 240·0 |
| **1900** | 200·1 | 207·0 | 214·0 | 221·2 | 228·3 | 235·6 | 243·0 | 250·5 | 258·0 | 265·5 | 273·3 | 281·0 | 288·9 | 296·8 | 304·6 | 312·5 | 320·5 | 328·3 | 336·1 |
| | 177·3 | 194·7 | 211·2 | 225·8 | 235·0 | 236·1 | 228·8 | 225·0 | 230·5 | 242·7 | 258·3 | 275·6 | 293·7 | 310·9 | 327·3 | 340·3 | 346·7 | 343·3 | 336·2 |
| | 200·2 | 202·4 | 204·6 | 206·8 | 209·0 | 211·2 | 213·5 | 215·7 | 217·7 | 219·6 | 221·4 | 223·2 | 224·7 | 225·9 | 227·0 | 228·0 | 228·7 | 228·4 | 228·3 |
| | 178·8 | 191·3 | 203·7 | 216·3 | 228·8 | 241·4 | 254·1 | 266·5 | 279·2 | 291·6 | 304·0 | 316·1 | 328·4 | 340·6 | 352·6 | 4·2 | 16·1 | 27·6 | 38·7 |
| | 234·8 | 235·7 | 236·7 | 237·7 | 238·8 | 239·9 | 241·1 | 242·3 | 243·5 | 244·6 | 245·7 | 246·7 | 247·7 | 248·5 | 249·2 | 250·1 | 250·4 | 250·8 | 251·4 |

| Mag. | P.Phal. | U.Phal. | Hast. | Chit. | Svati. | Visa. | Anur. | Jyesh. | Mula. | P.Ash. | U.Ash. | Srav. | Dan. | Satab. | P.Bhad. | U.Bhad. | Revati |
|---|---|---|---|---|---|---|---|---|---|---|---|---|---|---|---|---|---|
| 133·3 | 146·7 | 160·0 | 173·3 | 186·7 | 200·0 | 213·3 | 226·7 | 240·0 | 253·3 | 266·7 | 280·0 | 293·3 | 306·7 | 320·0 | 333·3 | 346·7 | 360·0 |

## Geocentric (Indian) Longitudes of Sun and Major planets at Ujjain

| | | 0 | 10 | 20 | 30 | 40 | 50 | 60 | 70 | 80 | 90 | 100 | 110 | 120 | 130 | 140 | 150 | 160 | 170 |
|---|---|---|---|---|---|---|---|---|---|---|---|---|---|---|---|---|---|---|---|
| Day of S. Yr. | | 0 | 10 | 20 | 30 | 40 | 50 | 60 | 70 | 80 | 90 | 100 | 110 | 120 | 130 | 140 | 150 | 160 | 170 |
| ☉'s Mean L. | | 357·0 | 6·9 | 16·7 | 26·6 | 36·4 | 46·3 | 56·1 | 66·0 | 75·9 | 85·7 | 95·6 | 105·4 | 115·3 | 125·1 | 135·0 | 144·8 | 154·7 | 164·6 |
| Corrn. for do. | | +2·1 | +2·0 | +1·9 | +1·7 | +1·4 | +1·1 | +0·8 | +0·4 | +0·0 | −0·3 | −0·7 | −1·1 | −1·4 | −1·6 | −1·9 | −2·0 | −2·1 | −2·2 |
| English. date. | | Ap 12 | Ap 22 | My 2 | My 12 | My 22 | Je 1. | Je 11 | Je 21 | Jl 1 | Jl 11 | Jl 21 | Jl 31 | Au 10 | Au 20 | Au 30 | S 9 | S 19 | S 29 |
| Tamil month. | | Chit | Chit | Chit | Chit | Vaik | Vaik | Vaik | Ani | Ani | Ani | Adi | Adi | Adi | Avan | Avan | Avan | Pur | Pur |
| Do. date. | | 1 | 11 | 21 | 31 | 10 | 20 | 30 | 9 | 19 | 29 | 7 | 17 | 27 | 6 | 16 | 26 | 5 | 15 |
| 1900 | Mars ... | 340·1 | 347·8 | 355·5 | 3·2 | 10·8 | 18·2 | 25·4 | 32·7 | 39·8 | 47·0 | 53·9 | 60·9 | 67·3 | 73·8 | 80·4 | 86·6 | 92·8 | 98·9 |
| ☉'s L. | Merc. ... | 335·7 | 342·4 | 355·5 | 12·1 | 30·7 | 49·7 | 68·3 | 85·9 | 100·6 | 109·9 | 109·9 | 102·4 | 100·6 | 108·2 | 121·7 | 138·0 | 155·5 | 173·4 |
| +·2 | Jup. ... | 227·8 | 227·0 | 226·1 | 225·0 | 223·7 | 222·3 | 221·0 | 219·9 | 218·9 | 218·1 | 217·6 | 217·5 | 217·8 | 218·4 | 219·3 | 220·3 | 221·6 | 223·1 |
| | Ven. ... | 44·1 | 54·5 | 64·5 | 73·7 | 81·7 | 88·0 | 92·0 | 92·5 | 88·6 | 82·3 | 76·7 | 74·9 | 77·0 | 81·8 | 89·0 | 97·4 | 106·9 | 116·9 |
| | Sat. ... | 251·1 | 250·8 | 250·6 | 250·3 | 249·6 | 248·8 | 248·2 | 247·4 | 246·7 | 246·0 | 245·4 | 244·9 | 244·3 | 244·2 | 244·0 | 243·9 | 244·3 | 244·8 |
| 1901 | Mars ... | 121·4 | 122·8 | 124·9 | 128·0 | 131·8 | 136·0 | 140·7 | 145·7 | 151·0 | 156·5 | 162·2 | 168·4 | 174·4 | 180·7 | 187·2 | 193·8 | 200·4 | 207·2 |
| ☉'s L. | Merc. ... | 333·0 | 349·2 | 7·3 | 26·3 | 45·3 | 63·1 | 78·9 | 89·6 | 92·5 | 85·7 | 81·8 | 87·5 | 100·2 | 116·1 | 134·0 | 151·0 | 169·3 | 185·5 |
| +·0 | Jup. ... | 259·9 | 260·3 | 260·3 | 260·2 | 259·9 | 259·4 | 257·6 | 256·1 | 254·9 | 253·6 | 252·5 | 251·4 | 250·6 | 250·0 | 249·9 | 250·0 | 250·4 | 251·2 |
| | Ven. ... | 354·1 | 6·5 | 18·9 | 31·2 | 43·6 | 55·8 | 68·1 | 80·2 | 92·4 | 104·6 | 117·0 | 128·9 | 141·2 | 153·2 | 165·2 | 177·1 | 189·4 | 201·2 |
| | Sat. ... | 262·1 | 262·1 | 262·0 | 261·9 | 261·3 | 260·8 | 260·2 | 259·5 | 258·9 | 258·1 | 257·0 | 255·9 | 254·9 | 255·0 | 255·0 | 255·1 | 255·2 | 255·4 |
| Eng. Date. | | Ap 13 | Ap 23 | My 3 | My 13 | My 23 | Je 2 | Je 12 | Je 22 | Jl 2 | Jl 12 | Jl 22 | Au 1 | Au 11 | Au 21 | Au 31 | S 10 | S 20 | S 30 |
| 1902 | Mars ... | 356·4 | 3·9 | 11·5 | 18·9 | 26·2 | 33·6 | 40·7 | 47·8 | 54·9 | 61·8 | 68·5 | 75·2 | 81·9 | 88·4 | 94·8 | 101·0 | 107·3 | 113·4 |
| ☉'s L. | Merc. ... | 346·3 | 4·8 | 23·6 | 41·8 | 57·9 | 69·6 | 73·9 | 68·3 | 62·9 | 68·8 | 79·7 | 95·6 | 113·5 | 131·7 | 149·7 | 166·1 | 180·0 | 188·6 |
| +·7 | Jup. ... | 289·9 | 291·3 | 292·4 | 293·4 | 294·1 | 294·5 | 294·2 | 293·0 | 293·0 | 291·8 | 290·7 | 289·5 | 288·1 | 286·8 | 285·7 | 284·7 | 284·0 | 283·7 |
| | Ven. ... | 313·5 | 322·5 | 332·4 | 342·9 | 353·8 | 4·9 | 16·3 | 27·8 | 39·5 | 51·5 | 63·2 | 74·8 | 87·0 | 98·9 | 111·1 | 123·3 | 135·5 | 147·8 |
| | Sat. ... | 273·0 | 273·1 | 273·3 | 273·3 | 273·0 | 272·6 | 272·2 | 271·5 | 270·7 | 270·1 | 269·3 | 268·6 | 267·7 | 267·2 | 266·8 | 266·3 | 266·3 | 266·2 |
| 1903 | Mars ... | 157·8 | 155·4 | 154·7 | 154·7 | 156·2 | 158·5 | 161·4 | 165·4 | 169·9 | 174·8 | 179·9 | 185·6 | 191·3 | 197·6 | 203·9 | 210·2 | 217·0 | 223·9 |
| ☉'s L. | Merc. ... | 0·9 | 18·8 | 35·1 | 48·3 | 54·4 | 50·5 | 43·8 | 46·7 | 57·7 | 73·0 | 90·8 | 109·4 | 127·5 | 144·9 | 160·1 | 170·4 | 172·9 | 166·1 |
| +·5 | Jup. ... | 318·4 | 320·4 | 322·4 | 324·3 | 325·9 | 327·3 | 328·4 | 329·2 | 329·5 | 329·7 | 329·7 | 329·2 | 328·3 | 327·0 | 325·8 | 324·5 | 323·1 | 321·9 |
| | Ven. ... | 31·3 | 43·6 | 55·2 | 66·9 | 78·6 | 90·0 | 100·8 | 111·6 | 121·9 | 131·7 | 140·8 | 148·8 | 155·0 | 158·2 | 158·9 | 155·1 | 148·5 | 143·3 |
| | Sat. ... | 283·8 | 284·1 | 284·5 | 284·8 | 284·7 | 284·5 | 284·2 | 283·6 | 282·9 | 282·3 | 281·5 | 281·8 | 280·1 | 279·4 | 278·8 | 278·1 | 277·9 | 277·8 |
| Eng. Date. | | Ap 12 | Ap 22 | My 2 | My 12 | My 22 | Je 1 | Je 11 | Je 21 | Jl 1 | Jl 11 | Jl 21 | Jl 31 | Au 10 | Au 20 | Au 30 | S 9 | S 19 | S 29 |
| 1904 | Mars ... | 11·2 | 18·6 | 25·8 | 33·1 | 40·3 | 47·3 | 54·2 | 61·2 | 67·9 | 74·5 | 81·4 | 88·0 | 94·5 | 101·1 | 107·4 | 113·7 | 120·0 | 126·3 |
| ☉'s L. | Merc. ... | 12·7 | 26·7 | 34·7 | 32·9 | 25·5 | 25·9 | 35·7 | 50·6 | 67·9 | 86·6 | 105·1 | 122·9 | 139·1 | 151·3 | 156·7 | 151·9 | 145·3 | 147·5 |
| +·2 | Jup. ... | 346·1 | 348·5 | 350·9 | 353·1 | 355·3 | 357·4 | 359·4 | 1·1 | 2·7 | 4·1 | 5·2 | 5·9 | 6·3 | 6·3 | 5·9 | 5·3 | 4·4 | 3·2 |
| | Ven. ... | 336·3 | 348·4 | 0·7 | 13·1 | 25·3 | 37·6 | 49·3 | 62·1 | 74·3 | 86·7 | 98·8 | 111·1 | 123·3 | 135·5 | 148·0 | 160·1 | 172·6 | 185·0 |
| | Sat. ... | 294·6 | 295·2 | 295·7 | 296·2 | 296·3 | 296·2 | 296·3 | 296·1 | 296·0 | 294·8 | 293·4 | 293·0 | 292·6 | 291·8 | 291·1 | 290·4 | 290·0 | 289·5 |
| 1905 | Mars ... | 211·5 | 209·4 | 206·9 | 203·7 | 200·4 | 198·2 | 197·0 | 198·1 | 198·8 | 201·3 | 204·6 | 208·8 | 213·6 | 218·7 | 224·5 | 230·5 | 236·5 | 243·2 |
| ☉'s L. | Merc. ... | 14·8 | 15·4 | 8·1 | 5·1 | 13·5 | 27·6 | 44·8 | 63·3 | 82·2 | 100·5 | 117·5 | 131·2 | 139·2 | 137·4 | 129·7 | 129·2 | 137·8 | 151·7 |
| +·0 | Jup. ... | 13·7 | 16·0 | 18·4 | 20·8 | 23·3 | 25·7 | 28·0 | 30·2 | 32·4 | 34·2 | 36·3 | 37·9 | 39·4 | 40·5 | 41·4 | 42·2 | 42·5 | 42·3 |
| | Ven. ... | 23·1 | 18·9 | 12·8 | 8·1 | 6·4 | 8·9 | 14·2 | 21·7 | 30·2 | 39·8 | 50·0 | 60·7 | 71·6 | 83·0 | 94·6 | 106·3 | 118·0 | 130·1 |
| | Sat. ... | 305·6 | 306·3 | 307·1 | 307·7 | 308·0 | 308·2 | 308·5 | 308·2 | 308·0 | 307·6 | 307·0 | 306·3 | 305·6 | 304·8 | 304·1 | 303·2 | 302·8 | 302·3 |
| Eng. Date. | | Ap 13 | Ap 23 | My 3 | My 13 | My 23 | Je 2 | Je 12 | Je 22 | Jl 2 | Jl 12 | Jl 22 | Au 1 | Au 11 | Au 21 | Au 31 | S 10 | S 20 | S 30 |
| 1906 | Mars ... | 26·5 | 33·4 | 40·4 | 47·3 | 54·4 | 61·0 | 67·8 | 74·7 | 81·1 | 87·6 | 94·2 | 100·6 | 107·2 | 113·5 | 120·0 | 126·3 | 132·5 | 138·9 |
| ☉'s L. | Merc ... | 348·9 | 346·5 | 353·3 | 6·6 | 23·3 | 42·0 | 60·9 | 79·5 | 96·8 | 111·5 | 120·4 | 120·2 | 112·7 | 110·9 | 118·8 | 132·1 | 148·5 | 166·1 |
| +·7 | Jup. ... | 41·6 | 43·8 | 45·9 | 48·2 | 50·5 | 52·9 | 55·2 | 57·5 | 59·9 | 62·2 | 64·5 | 66·5 | 68·5 | 70·2 | 71·9 | 73·6 | 75·0 | 76·0 |
| | Ven. ... | 14·2 | 26·6 | 38·8 | 51·1 | 63·2 | 75·2 | 87·2 | 99·0 | 110·8 | 122·6 | 134·3 | 146·0 | 157·5 | 168·4 | 179·4 | 190·0 | 200·1 | 209·3 |
| | Sat. ... | 316·6 | 317·5 | 318·3 | 319·3 | 319·7 | 320·2 | 320·7 | 320·6 | 320·6 | 320·5 | 320·0 | 319·4 | 318·8 | 318·0 | 317·3 | 316·5 | 315·8 | 315·1 |
| 1907 | Mars ... | 252·3 | 256·3 | 259·9 | 263·1 | 265·3 | 266·9 | 267·5 | 266·0 | 263·7 | 261·8 | 259·0 | 257·5 | 256·4 | 257·8 | 259·7 | 263·0 | 267·0 | 271·5 |
| ☉'s L. | Merc. ... | 332·2 | 344·1 | 0·3 | 18·4 | 37·3 | 56·5 | 74·3 | 89·8 | 100·5 | 102·9 | 96·2 | 92·2 | 98·5 | 110·6 | 126·8 | 144·3 | 162·2 | 179·4 |
| +·4 | Jup. ... | 69·7 | 71·2 | 72·8 | 74·7 | 76·7 | 78·8 | 81·0 | 83·2 | 85·6 | 87·9 | 90·1 | 92·5 | 94·7 | 97·0 | 99·0 | 101·0 | 102·8 | 104·6 |
| | Ven. ... | 320·1 | 332·1 | 344·2 | 356·2 | 8·1 | 20·2 | 32·3 | 44·5 | 57·2 | 68·7 | 80·7 | 92·6 | 105·2 | 117·3 | 129·8 | 142·0 | 154·2 | 166·8 |
| | Sat. ... | 327·7 | 328·7 | 329·8 | 320·8 | 331·6 | 332·3 | 333·0 | 333·1 | 333·3 | 333·5 | 333·2 | 332·8 | 332·5 | 331·9 | 331·0 | 330·2 | 329·4 | 328·7 |
| Eng. Date. | | Ap 12 | Ap 22 | My 2 | My 12 | My 22 | Je 1 | Je 11 | Je 21 | Jl 1 | Jl 11 | Jl 21 | Jl 31 | Au 10 | Au 20 | Au 30 | S 9 | S 19 | S 29 |
| 1908 | Mars ... | 40·5 | 47·3 | 54·0 | 60·5 | 67·0 | 73·6 | 80·0 | 86·6 | 92·9 | 99·4 | 105·8 | 112·2 | 118·7 | 125·0 | 131·4 | 137·8 | 144·1 | 150·5 |
| ☉'s L. | Merc. ... | 337·6 | 355·4 | 14·1 | 33·0 | 51·3 | 67·6 | 79·8 | 84·7 | 79·5 | 73·7 | 77·9 | 89·4 | 104·7 | 122·3 | 140·4 | 158·2 | 174·8 | 188·8 |
| +·2 | Jup. ... | 99·5 | 100·1 | 101·0 | 102·0 | 103·4 | 104·9 | 106·8 | 108·8 | 110·7 | 112·8 | 115·0 | 117·2 | 119·4 | 121·6 | 123·8 | 126·1 | 128·2 | 130·3 |
| | Ven. ... | 44·3 | 54·9 | 64·2 | 73·3 | 80·9 | 87·2 | 90·0 | 89·8 | 85·1 | 78·7 | 73·7 | 72·7 | 75·6 | 81·2 | 88·6 | 97·4 | 107·1 | 117·5 |
| | Sat. ... | 339·1 | 340·3 | 341·4 | 342·6 | 343·5 | 344·3 | 345·3 | 345·7 | 346·2 | 346·7 | 346·5 | 346·4 | 346·3 | 345·7 | 345·0 | 344·4 | 343·6 | 342·9 |
| 1909 | Mars ... | 277·6 | 284·1 | 290·7 | 297·1 | 303·3 | 309·4 | 315·0 | 320·6 | 325·7 | 330·3 | 334·4 | 337·3 | 339·2 | 339·9 | 339·1 | 337·1 | 334·4 | 332·2 |
| ☉'s L. | Merc. ... | 350·8 | 9·4 | 27·8 | 44·9 | 58·7 | 65·8 | 62·6 | 55·5 | 57·1 | 67·4 | 82·7 | 99·7 | 118·1 | 136·3 | 153·6 | 168·5 | 179·6 | 182·8 |
| —·1 | Jup. ... | 131·4 | 130·9 | 130·9 | 130·8 | 131·3 | 132·1 | 133·2 | 134·5 | 136·0 | 137·6 | 139·4 | 141·4 | 143·4 | 145·5 | 147·6 | 149·8 | 152·0 | 154·2 |
| | Ven. ... | 354·6 | 7·1 | 19·4 | 31·7 | 44·0 | 56·3 | 68·6 | 80·7 | 92·9 | 105·1 | 117·5 | 129·4 | 142·2 | 153·7 | 165·6 | 177·6 | 189·7 | 201·5 |
| | Sat. ... | 350·5 | 351·8 | 353·0 | 354·3 | 355·4 | 356·4 | 357·5 | 358·2 | 358·8 | 359·6 | 359·7 | 359·9 | 0·1 | 359·8 | 359·4 | 358·9 | 358·2 | 357·5 |

| Mesh. | Vrsh. | Mith. | Kat. | Sim. | Kan. | Tul. | Vrsch. | Dhan. | Mak. | Kum. | Min. | Asvn. | Bhar. | Krit. | Rohi. | Mrig. | Ardh. | Punar. | Push. | Asles. |
|---|---|---|---|---|---|---|---|---|---|---|---|---|---|---|---|---|---|---|---|---|
| 30 | 60 | 90 | 120 | 150 | 180 | 210 | 240 | 270 | 300 | 330 | 360 | 13·0 | 26·7 | 40·0 | 53·3 | 66·7 | 80·0 | 93·3 | 106·7 | 120·0 |

## mean Sunrise for every ten days from A.D. 1900 to A.D. 1909.

| Year | 180 | 190 | 200 | 210 | 220 | 230 | 240 | 250 | 260 |
|---|---|---|---|---|---|---|---|---|---|
| | 174·4 | 184·3 | 194·1 | 204·0 | 213·8 | 223·7 | 233·5 | 243·4 | 253·3 |
| | −2·2 | −2·1 | −1·9 | −1·7 | −1·5 | −1·2 | −0·9 | −0·5 | −0·1 |
| | O 9 | O 19 | O 29 | N 8 | N 18 | N 28 | D 8 | D 18 | D 28 |
| | Pur | Aipp | Aipp | Aipp | Kart | Kart | Kart | Marg | Marg |
| | 25 | 4 | 14 | 24 | 4 | 14 | 24 | 5 | 15 |
| 01 | 104·7 | 110·2 | 115·6 | 120·6 | 125·3 | 129·7 | 133·6 | 136·5 | 139·0 |
| | 190·4 | 205·7 | 217·1 | 220·3 | 215·9 | 209·3 | 212·1 | 222·6 | 237·3 |
| | 225·0 | 226·9 | 228·9 | 231·1 | 233·3 | 235·4 | 237·7 | 240·0 | 242·3 |
| | 127·9 | 139·4 | 150·6 | 162·5 | 174·4 | 186·4 | 198·6 | 210·8 | 223·0 |
| | 245·1 | 246·0 | 246·8 | 247·7 | 248·7 | 249·7 | 250·8 | 252·0 | 253·2 |
| 02 | 214·0 | 221·1 | 228·3 | 235·6 | 243·0 | 250·5 | 258·1 | 265·7 | 273·3 |
| | 198·4 | 205·4 | 202·7 | 195·2 | 194·7 | 203·3 | 217·1 | 233·2 | 249·4 |
| | 252·4 | 253·7 | 255·7 | 257·2 | 258·8 | 260·9 | 263·1 | 265·3 | 267·5 |
| | 213·0 | 224·0 | 236·4 | 247·8 | 258·9 | 268·4 | 279·6 | 289·4 | 297·9 |
| | 255·6 | 256·2 | 256·9 | 257·6 | 258·6 | 259·7 | 260·7 | 261·8 | 262·9 |
| | O 10 | O 20 | O 30 | N 9 | N 19 | N 29 | D 9 | D 19 | D 29 |
| 03 | 119·6 | 125·4 | 131·2 | 136·7 | 142·2 | 147·5 | 152·5 | 157·1 | 161·4 |
| | 188·1 | 180·6 | 178·2 | 185·3 | 198·2 | 214·1 | 231·2 | 248·7 | 265·2 |
| | 283·7 | 284·1 | 284·8 | 285·9 | 287·1 | 288·5 | 290·3 | 292·2 | 294·4 |
| | 160·1 | 172·6 | 185·1 | 197·5 | 210·1 | 222·6 | 235·3 | 247·9 | 260·7 |
| | 266·3 | 266·8 | 267·4 | 268·0 | 268·9 | 269·9 | 270·8 | 271·9 | 273·0 |
| 04 | 230·9 | 238·0 | 245·2 | 252·6 | 259·9 | 267·5 | 275·1 | 282·8 | 290·6 |
| | 161·6 | 166·3 | 178·0 | 193·3 | 210·6 | 226·6 | 244·4 | 260·0 | 272·4 |
| | 320·8 | 320·0 | 319·7 | 319·4 | 319·3 | 319·8 | 320·7 | 322·0 | 323·4 |
| | 142·4 | 145·7 | 150·1 | 157·6 | 166·5 | 176·3 | 186·6 | 197·6 | 208·9 |
| | 277·5 | 277·9 | 278·2 | 278·6 | 279·4 | 280·1 | 281·0 | 282·1 | 283·2 |
| | O 9 | O 19 | O 29 | N 8 | N 18 | N 28 | D 8 | D 18 | D 28 |
| 05 | 132·3 | 138·4 | 143·9 | 150·4 | 156·2 | 161·9 | 167·6 | 172·3 | 178·5 |
| | 158·0 | 172·5 | 189·3 | 206·8 | 223·8 | 236·8 | 253·3 | 261·0 | 259·7 |
| | 2·0 | 0·6 | 359·3 | 358·0 | 357·0 | 356·3 | 355·8 | 355·8 | 356·1 |
| | 197·3 | 209·7 | 221·9 | 234·2 | 246·5 | 258·7 | 270·9 | 283·1 | 295·2 |
| | 289·2 | 289·3 | 289·4 | 289·5 | 290·2 | 290·8 | 291·4 | 292·4 | 293·5 |
| 06 | 249·8 | 256·8 | 263·9 | 270·9 | 278·4 | 285·8 | 293·3 | 300·8 | 308·3 |
| | 168·0 | 185·5 | 202·8 | 219·5 | 233·9 | 243·8 | 245·6 | 238·4 | 234·1 |
| | 41·9 | 41·1 | 40·1 | 38·9 | 37·5 | 35·9 | 34·9 | 35·7 | 32·9 |
| | 142·1 | 154·2 | 166·4 | 178·8 | 191·1 | 203·6 | 216·3 | 228·7 | 241·2 |
| | 301·7 | 301·6 | 301·4 | 301·3 | 301·7 | 302·2 | 302·5 | 303·4 | 304·2 |
| | O 10 | O 20 | O 30 | N 9 | N 19 | N 29 | D 9 | D 19 | D 29 |
| 07 | 145·0 | 151·3 | 157·5 | 163·6 | 169·9 | 175·9 | 182·0 | 188·1 | 194·2 |
| | 183·7 | 200·7 | 215·5 | 226·6 | 230·0 | 224·6 | 218·3 | 221·6 | 232·3 |
| | 76·7 | 77·0 | 77·1 | 76·8 | 76·2 | 75·3 | 74·2 | 72·7 | 71·5 |
| | 217·9 | 224·7 | 229·2 | 230·8 | 228·3 | 222·7 | 217·3 | 215·1 | 217·1 |
| | 314·3 | 314·0 | 313·6 | 313·4 | 313·6 | 313·9 | 314·2 | 314·9 | 315·7 |
| 08 | 277·1 | 282·7 | 288·8 | 295·2 | 301·6 | 308·3 | 315·0 | 321·8 | 328·7 |
| | 195·6 | 208·2 | 215·4 | 211·3 | 203·8 | 204·8 | 212·8 | 226·6 | 243·3 |
| | 106·1 | 107·6 | 108·6 | 109·3 | 109·8 | 109·9 | 109·5 | 109·1 | 108·4 |
| | 179·3 | 191·8 | 204·2 | 216·8 | 229·3 | 242·1 | 254·6 | 266·9 | 279·5 |
| | 328·0 | 327·4 | 326·9 | 326·3 | 326·4 | 326·4 | 326·4 | 326·9 | 327·5 |
| | O 9 | O 19 | O 29 | N 8 | N 18 | N 28 | D 8 | D 18 | D 28 |
| 09 | 156·8 | 163·3 | 169·7 | 176·0 | 182·5 | 188·9 | 195·3 | 201·8 | 208·2 |
| | 197·8 | 197·5 | 189·8 | 187·2 | 193·9 | 206·7 | 222·1 | 239·5 | 257·0 |
| | 132·3 | 134·0 | 135·7 | 137·3 | 138·7 | 139·8 | 140·6 | 141·1 | 141·2 |
| | 128·3 | 139·6 | 151·2 | 162·9 | 174·6 | 186·8 | 199·1 | 211·4 | 223·6 |
| | 342·2 | 341·4 | 340·6 | 339·9 | 339·6 | 339·4 | 339·1 | 339·5 | 339·8 |
| 10 | 330·5 | 330·3 | 331·6 | 332·8 | 336·0 | 339·5 | 343·8 | 348·7 | 353·5 |
| | 175·6 | 171·2 | 175·1 | 186·5 | 201·5 | 218·4 | 235·6 | 252·8 | 268·6 |
| | 156·4 | 158·6 | 160·6 | 162·6 | 164·5 | 166·1 | 167·6 | 168·9 | 169·9 |
| | 213·5 | 225·5 | 236·7 | 248·1 | 259·0 | 269·5 | 279·5 | 288·9 | 297·0 |
| | 356·7 | 355·9 | 355·2 | 354·3 | 353·8 | 353·4 | 352·8 | 352·9 | 352·9 |

| Year | 270 | 280 | 290 | 300 | 310 | 320 | 330 | 340 | 350 | 360 |
|---|---|---|---|---|---|---|---|---|---|---|
| | 263·1 | 273·0 | 282·8 | 292·7 | 302·5 | 312·4 | 322·2 | 332·1 | 342·0 | 351·8 |
| | +0·3 | +0·6 | +1·0 | +1·3 | +1·6 | +1·8 | +2·0 | +2·1 | +2·2 | +2·2 |
| | Ja 7 | Ja 17 | Ja 27 | F 6 | F 16 | F 26 | Mr 8 | Mr 18 | Mr 28 | Ap 7 |
| | Marg | Tai | Tai | Tai | Masi | Masi | Masi | Pang | Pang | Pang |
| | 25 | 5 | 15 | 25 | 6 | 16 | 26 | 6 | 16 | 26 |
| 01 | 140·5 | 140·7 | 139·5 | 137·4 | 134·1 | 130·4 | 126·6 | 123·7 | 121·7 | 121·2 |
| | 254·3 | 271·9 | 289·3 | 305·7 | 319·6 | 328·0 | 327·1 | 319·4 | 318·4 | 326·7 |
| | 244·5 | 246·7 | 248·9 | 250·9 | 252·8 | 254·6 | 256·2 | 257·6 | 258·8 | 259·7 |
| | 235·3 | 248·1 | 260·4 | 273·0 | 285·5 | 298·1 | 310·5 | 322·9 | 335·5 | 347·9 |
| | 254·3 | 255·4 | 256·5 | 257·7 | 258·5 | 259·4 | 260·3 | 260·9 | 261·4 | 262·0 |
| 02 | 281·2 | 289·0 | 296·7 | 304·7 | 312·6 | 320·6 | 328·4 | 336·3 | 343·9 | 351·6 |
| | 268·1 | 284·8 | 299·1 | 309·1 | 310·7 | 303·5 | 299·9 | 306·3 | 319·0 | 335·4 |
| | 269·8 | 272·2 | 274·4 | 276·7 | 279·0 | 281·3 | 283·4 | 285·4 | 287·4 | 289·1 |
| | 304·6 | 309·3 | 310·4 | 307·1 | 301·0 | 296·0 | 295·5 | 296·5 | 301·5 | 308·6 |
| | 264·1 | 265·3 | 266·5 | 267·7 | 268·7 | 269·6 | 270·6 | 271·3 | 272·1 | 272·7 |
| | Ja 8 | Ja 18 | Ja 28 | F 7 | F 17 | F 27 | Mr 8,9 | 18, 19 | 28,29, | Ap 7,8 |
| 03 | 165·2 | 168·7 | 171·1 | 172·8 | 173·2 | 172·5 | 170·4 | 167·0 | 163·2 | 160·1 |
| | 280·1 | 291·1 | 294·1 | 287·8 | 282·7 | 287·1 | 298·8 | 314·6 | 332·7 | 351·1 |
| | 296·5 | 298·7 | 301·0 | 303·3 | 305·7 | 308·1 | 310·5 | 312·9 | 315·3 | 317·5 |
| | 273·2 | 285·8 | 298·5 | 311·1 | 323·5 | 336·2 | 348·3 | 0·6 | 12·9 | 24·9 |
| | 274·1 | 275·3 | 276·5 | 277·7 | 278·8 | 279·8 | 281·0 | 281·9 | 282·7 | 283·5 |
| 04 | 298·4 | 306·3 | 314·1 | 321·8 | 329·5 | 337·3 | 344·9 | 352·4 | 0·1 | 7·5 |
| | 277·8 | 274·0 | 266·5 | 268·2 | 278·6 | 292·8 | 310·2 | 328·3 | 346·3 | 4·5 |
| | 324·9 | 326·7 | 328·6 | 330·8 | 333·1 | 335·4 | 337·8 | 340·2 | 342·6 | 345·0 |
| | 220·4 | 232·5 | 244·5 | 256·7 | 268·8 | 281·0 | 293·0 | 305·5 | 317·8 | 330·2 |
| | 284·2 | 285·4 | 286·6 | 287·7 | 288·9 | 290·1 | 291·3 | 292·2 | 293·2 | 294·2 |
| | Ja 7 | Ja 17 | Ja 27 | F 6 | F 16 | F 26 | Mr 8 | Mr 18 | Mr 28 | Ap 7 |
| 05 | 183·8 | 188·8 | 193·6 | 197·9 | 201·9 | 205·6 | 208·6 | 210·8 | 211·8 | 211·8 |
| | 252·0 | 250·4 | 258·3 | 271·8 | 288·2 | 306·1 | 324·2 | 342·1 | 358·2 | 10·7 |
| | 356·6 | 357·5 | 358·8 | 0·2 | 2·0 | 3·8 | 5·9 | 8·0 | 10·3 | 12·5 |
| | 306·9 | 318·5 | 329·7 | 340·4 | 350·9 | 0·5 | 8·9 | 16·2 | 21·7 | 23·3 |
| | 294·5 | 295·7 | 296·9 | 298·1 | 299·3 | 300·5 | 301·7 | 302·8 | 303·8 | 305·0 |
| 06 | 315·9 | 323·4 | 330·9 | 338·3 | 345·8 | 353·2 | 0·9 | 7·8 | 15·1 | 22·1 |
| | 239·2 | 251·2 | 266·7 | 284·1 | 301·9 | 319·8 | 336·4 | 349·8 | 357·1 | 354·8 |
| | 32·1 | 31·8 | 31·9 | 32·2 | 33·0 | 34·0 | 35·2 | 36·7 | 38·4 | 40·3 |
| | 253·9 | 266·6 | 279·1 | 291·9 | 304·4 | 316·9 | 329·4 | 342·0 | 354·5 | 6·7 |
| | 305·2 | 306·4 | 307·6 | 308·7 | 309·9 | 311·1 | 312·3 | 313·5 | 314·7 | 316·0 |
| | Ja 8 | Ja 18 | Ja 28 | F 7 | F 17 | F 27 | Mr 8,9 | 18,19 | 28,29 | Ap 7,8 |
| 07 | 200·2 | 206·2 | 212·2 | 217·9 | 223·6 | 229·3 | 234·7 | 240·0 | 245·2 | 250·1 |
| | 247·4 | 261·5 | 282·1 | 299·7 | 316·3 | 330·1 | 338·1 | 337·4 | 329·8 | 328·8 |
| | 70·4 | 69·1 | 68·1 | 67·2 | 66·8 | 66·8 | 66·9 | 67·3 | 68·1 | 68·9 |
| | 221·8 | 229·0 | 237·7 | 247·2 | 257·5 | 268·4 | 279·5 | 290·8 | 302·5 | 314·2 |
| | 316·4 | 317·4 | 318·5 | 319·6 | 320·9 | 322·2 | 323·5 | 324·7 | 325·9 | 327·2 |
| 08 | 335·7 | 342·7 | 349·5 | 356·5 | 3·4 | 10·2 | 17·0 | 23·8 | 30·5 | 37·2 |
| | 260·7 | 278·3 | 294·8 | 309·5 | 319·2 | 320·6 | 312·2 | 309·7 | 316·8 | 329·7 |
| | 107·4 | 106·3 | 105·1 | 103·6 | 102·5 | 101·4 | 100·4 | 99·8 | 99·5 | 99·6 |
| | 291·9 | 304·5 | 316·6 | 328·9 | 341·1 | 353·1 | 4·7 | 16·4 | 27·9 | 38·8 |
| | 328·0 | 328·9 | 329·9 | 330·9 | 331·1 | 332·3 | 334·5 | 335·8 | 337·1 | 338·4 |
| | Ja 7 | Ja 17 | Ja 27 | F 6 | F 16 | F 26 | Mr 8 | Mr 18 | Mr 28 | Ap 7 |
| 09 | 215·0 | 221·3 | 227·9 | 234·5 | 241·2 | 247·7 | 254·5 | 261·0 | 267·6 | 274·3 |
| | 273·7 | 288·9 | 300·2 | 303·9 | 297·9 | 292·6 | 296·3 | 307·9 | 323·8 | 341·6 |
| | 141·1 | 140·7 | 139·9 | 138·7 | 137·6 | 136·4 | 135·1 | 133·8 | 132·7 | 131·8 |
| | 236·0 | 248·7 | 261·0 | 273·7 | 286·1 | 298·7 | 311·1 | 323·5 | 336·1 | 348·3 |
| | 340·1 | 340·9 | 341·8 | 342·6 | 343·7 | 344·8 | 346·0 | 347·3 | 348·6 | 349·8 |
| 10 | 358·7 | 4·2 | 9·8 | 15·7 | 21·4 | 27·5 | 33·6 | 39·7 | 45·8 | 51·9 |
| | 281·1 | 288·1 | 283·9 | 272·7 | 277·6 | 287·1 | 301·6 | 318·9 | 337·0 | 355·8 |
| | 170·9 | 171·4 | 171·5 | 171·4 | 170·9 | 170·2 | 169·1 | 168·0 | 166·7 | 165·5 |
| | 303·5 | 306·9 | 307·2 | 303·4 | 297·0 | 292·4 | 291·8 | 294·8 | 300·7 | 308·2 |
| | 352·9 | 353·6 | 354·2 | 354·8 | 355·8 | 356·9 | 358·0 | 359·2 | 0·4 | 1·6 |

| Mag. | P.Phal. | U.Phal. | Hast. | Chit. | Svati. | Visa. | Anur. | Jyesh. | Mula. | P.Ash. | U.Ash. | Srav. | Dan. | Satab. | P.Bhad. | U.Bhad. | Revati. |
|---|---|---|---|---|---|---|---|---|---|---|---|---|---|---|---|---|---|
| 133·3 | 146·7 | 160·0 | 173·3 | 186·7 | 200·0 | 213·3 | 226·7 | 240·0 | 253·3 | 266·7 | 280·0 | 293·3 | 306·7 | 320·0 | 333·3 | 346·7 | 360·0 |

## Geocentric (Indian) Longitudes of Sun and Major planets at Ujjain

| Day of S. Yr. | | 0 | 10 | 20 | 30 | 40 | 50 | 60 | 70 | 80 | 90 | 100 | 110 | 120 | 130 | 140 | 150 | 160 | 170 |
|---|---|---|---|---|---|---|---|---|---|---|---|---|---|---|---|---|---|---|---|
| ⊙'s Mean L. | | 357·0 | 6·9 | 16·7 | 26·6 | 36·4 | 46·3 | 56·1 | 66·0 | 75·9 | 85·7 | 95·6 | 105·4 | 115·3 | 125·1 | 135·0 | 144·8 | 154·7 | 164·6 |
| Corrn. for do. | | +2·1 | +2·0 | +1·9 | +1·7 | +1·4 | +1·1 | +0·8 | +0·4 | +0·0 | −0·3 | −0·7 | −1·1 | −1·4 | −1·6 | −1·9 | −2·0 | −2·1 | −2·2 |
| English date. | | Ap 13 | Ap 23 | My 3 | My 13 | My 23 | Je 2 | Je 12 | Je 22 | Jl 2 | Jl 12 | Jl 22 | Au 1 | Au 11 | Au 21 | Au 31 | S 10 | S 20 | S 30 |
| Tamil month. | | Chit | Chit | Chit | Chit | Vaik | Vaik | Vaik | Ani | Ani | Ani | Adi | Adi | Adi | Avan | Avan | Avan | Pur | Pur |
| Do. date. | | 1 | 11 | 21 | 31 | 10 | 20 | 30 | 9 | 19 | 29 | 7 | 17 | 27 | 6 | 16 | 26 | 5 | 15 |
| 1910 | Mars ... | 55·6 | 61·8 | 68·0 | 74·3 | 80·5 | 86·8 | 93·0 | 99·7 | 105·6 | 111·9 | 118·2 | 124·4 | 130·8 | 137·1 | 143·6 | 149·9 | 156·3 | 162·9 |
| ⊙'s L. | Merc. ... | 6·8 | 24·1 | 37·9 | 45·9 | 44·1 | 36·6 | 37·1 | 46·9 | 61·8 | 79·1 | 97·4 | 116·0 | 133·6 | 149·5 | 161·6 | 166·4 | 161·5 | 154·8 |
| +·6 | Jup. ... | 164·8 | 163·7 | 162·5 | 161·6 | 161·1 | 161·0 | 161·1 | 161·3 | 162·3 | 163·4 | 164·6 | 166·1 | 167·9 | 169·7 | 171·6 | 173·7 | 175·7 | 177·9 |
| | Ven. ... | 313·6 | 322·7 | 332.9 | 343·6 | 354·3 | 5·4 | 17·0 | 28·7 | 40·2 | 52·2 | 64·0 | 75·7 | 87·7 | 99·8 | 112·0 | 124·2 | 136·6 | 148·5 |
| | Sat. ... | 2·3 | 3·6 | 4·9 | 6·3 | 7·5 | 8·6 | 9·9 | 10·9 | 11·8 | 12·7 | 13·1 | 13·6 | 13·9 | 13·7 | 13·6 | 13·5 | 12·9 | 12·2 |
| 1911 | Mars ... | 298·4 | 305·7 | 313·0 | 320·5 | 327·7 | 335·1 | 342·2 | 349·1 | 356·2 | 2·9 | 9·3 | 15·4 | 21·3 | 26·9 | 31·9 | 36·4 | 39·7 | 42·9 |
| ⊙'s L. | Merc. ... | 16·3 | 25·6 | 25·9 | 18·5 | 18·7 | 24·7 | 38·7 | 56·1 | 74·4 | 93·2 | 111·4 | 128·1 | 141·8 | 148·9 | 146·6 | 139·0 | 139·0 | 147·7 |
| +·4 | Jup. ... | 198·8 | 197·6 | 197·0 | 194·9 | 193·8 | 192·8 | 191·9 | 191·4 | 191·4 | 191·4 | 191·7 | 192·6 | 193·5 | 194·9 | 196·3 | 198·0 | 199·7 | 201·7 |
| | Ven. ... | 31·8 | 43·6 | 55·7 | 67·2 | 78·9 | 90·3 | 100·9 | 111·7 | 122·0 | 131·6 | 140·8 | 147·9 | 153·7 | 156·6 | 156·0 | 151·2 | 145·1 | 140·2 |
| | Sat. ... | 14·3 | 15·7 | 17·1 | 18·5 | 19·8 | 21·1 | 22·4 | 23·7 | 25·2 | 26·5 | 26·9 | 27·2 | 27·6 | 27·7 | 27·9 | 28·0 | 27·5 | 27·1 |
| Eng. Date. | | Ap 12 | Ap 22 | My 2 | My 12 | My 22 | Je 1 | Je 11 | Je 21 | Jl 1 | Jl 11 | Jl 21 | Jl 31 | Au 10 | Au 20 | Au 30 | S 9 | S 19 | S 29 |
| 1912 | Mars ... | 70·9 | 76·5 | 82·1 | 87·9 | 93·7 | 99·5 | 105·6 | 111·5 | 117·7 | 123·8 | 130·1 | 136·3 | 142·4 | 148·9 | 155·4 | 161·8 | 168·3 | 174·8 |
| ⊙'s L. | Merc. ... | 8·3 | 1·1 | 357.1 | 3·1 | 15·9 | 32·8 | 51·1 | 69·1 | 88·8 | 106·1 | 120·8 | 130·2 | 130·8 | 123·2 | 120·8 | 127·7 | 140·9 | 157·1 |
| +·1 | Jup. ... | 232·4 | 231·0 | 230·6 | 229·9 | 228·6 | 227·3 | 225·9 | 224·8 | 223·7 | 222·9 | 222·2 | 222·0 | 222·0 | 222·4 | 223·3 | 224·4 | 225·4 | 226·8 |
| | Ven. ... | 336·8 | 349·1 | 1·4 | 13·6 | 25·8 | 37·1 | 50·4 | 62·6 | 74·8 | 87·2 | 99·3 | 111·6 | 124·6 | 136·2 | 148·5 | 160·6 | 173·1 | 185·5 |
| | Sat. ... | 26·9 | 28·2 | 29·5 | 30·8 | 32·1 | 33·4 | 34·8 | 36·1 | 37·3 | 38·5 | 39·4 | 40·2 | 41·2 | 41·6 | 42·1 | 42·5 | 42·4 | 42·3 |
| 1913 | Mars ... | 316·3 | 323·9 | 331·7 | 339·3 | 347·0 | 354·5 | 1·9 | 9·3 | 16·4 | 23·6 | 31·0 | 37·3 | 43·8 | 50·2 | 56·5 | 62·4 | 68·0 | 73·2 |
| ⊙'s L. | Merc. ... | 338·9 | 342·2 | 353·7 | 9·3 | 27·6 | 46·7 | 65·4 | 83·7 | 98·4 | 110·8 | 114·1 | 106·3 | 103·2 | 107·7 | 119·8 | 135·3 | 152·9 | 170·6 |
| −·1 | Jup. ... | 264·0 | 264·6 | 264·8 | 264·7 | 264·4 | 263·6 | 262·6 | 261·4 | 260·1 | 258·8 | 257·5 | 256·3 | 255·5 | 254·7 | 254·4 | 254·4 | 254·6 | 255·3 |
| | Ven. ... | 19·4 | 14·7 | 9·1 | 4·3 | 4·2 | 7·6 | 13·5 | 21·2 | 30·1 | 39·9 | 50·3 | 60·8 | 71·9 | 83·2 | 94·9 | 106·6 | 118·5 | 130·4 |
| | Sat. ... | 39·5 | 40·7 | 41·9 | 43·1 | 44·5 | 45·9 | 47·3 | 48·6 | 49·9 | 51·3 | 52·3 | 53·4 | 54·4 | 55·2 | 55·9 | 56·4 | 56·7 | 56·9 |
| Eng. Date. | | Ap 13 | Ap 23 | My 3 | My 13 | My 23 | Je 2 | Je 12 | Je 22 | Jl 2 | Jl 12 | Jl 22 | Au 1 | Au 11 | Au 21 | Au 31 | S 10 | S 20 | S 30 |
| 1914 | Mars ... | 88·7 | 93·0 | 97·9 | 103·0 | 108·4 | 114·1 | 119·4 | 125·1 | 131·0 | 137·0 | 143·0 | 149·1 | 155·5 | 161·7 | 168·3 | 174·8 | 181·4 | 188·0 |
| ⊙'s L. | Merc. ... | 333·3 | 348·4 | 6·3 | 25·3 | 44·2 | 62·5 | 78·6 | 90·8 | 95·4 | 90·2 | 84·4 | 88·4 | 100·0 | 115·5 | 132·9 | 150·8 | 168·5 | 184·9 |
| +·6 | Jup. ... | 293·9 | 295·4 | 296·7 | 297·7 | 298·9 | 299·0 | 299.2 | 298·6 | 298·1 | 297·2 | 296·2 | 294·9 | 293·6 | 292·4 | 231·2 | 290·1 | 289·2 | 288·7 |
| | Ven. ... | 14·7 | 27·1 | 39·5 | 51·6 | 63·7 | 75·7 | 87·6 | 99·5 | 111·1 | 123·1 | 134·7 | 146·3 | 157·6 | 168·9 | 179·7 | 190·1 | 199·8 | 209·1 |
| | Sat. ... | 52·4 | 53·5 | 54·8 | 55·9 | 57·2 | 58·5 | 59·9 | 61·2 | 62·5 | 63·9 | 64·8 | 65·6 | 66·6 | 67·9 | 69·1 | 70·3 | 70·7 | 71·2 |
| 1915 | Mars ... | 333·7 | 341·6 | 349·3 | 356·8 | 4·6 | 12·1 | 19·4 | 26·8 | 34·0 | 41·1 | 48·0 | 54·8 | 61·6 | 68·1 | 74·5 | 80·8 | 86·8 | 92·7 |
| ⊙'s L. | Merc. ... | 344·3 | 1·9 | 20·8 | 39·2 | 56·1 | 69·9 | 76·5 | 73·1 | 66·1 | 68·3 | 78·5 | 93·3 | 111·8 | 129·0 | 147·0 | 164·1 | 179·1 | 189·2 |
| +·4 | Jup. ... | 322·2 | 324·4 | 326·4 | 328·4 | 330·0 | 331·6 | 332·8 | 333·7 | 334·4 | 334·8 | 334·9 | 334·4 | 333·7 | 332·7 | 331·5 | 330·2 | 328·8 | 327·4 |
| | Ven. ... | 320·7 | 332·5 | 344·6 | 356·8 | 8·5 | 20·8 | 32·9 | 44·9 | 57·0 | 69·1 | 81·3 | 93·4 | 105·7 | 118·0 | 130·4 | 142·6 | 155·0 | 167·4 |
| | Sat. ... | 65·7 | 66·7 | 67·6 | 68·6 | 69·8 | 71·1 | 72·3 | 73·7 | 75·1 | 76·4 | 77·7 | 79·0 | 80·3 | 81·3 | 82·4 | 83·5 | 84·2 | 84·8 |
| Eng. Date. | | Ap 12 | Ap 22 | My 2 | My 12 | My 22 | Je 1 | Je 11 | Je 21 | Jl 1 | Jl 11 | Jl 21 | Jl 31 | Au 10 | Au 20 | Au 30 | S 9 | S 19 | S 29 |
| 1916 | Mars ... | 110·8 | 113·2 | 116·4 | 120·1 | 124·4 | 129·2 | 134·0 | 139·3 | 145·3 | 151·0 | 156·7 | 162·6 | 168·7 | 175·1 | 181·6 | 188·2 | 194·7 | 201·6 |
| ⊙'s L. | Merc. ... | 357·6 | 14·9 | 33·4 | 47·8 | 56·6 | 55·8 | 48·3 | 47·6 | 56·5 | 71·1 | 88·3 | 106·4 | 124·8 | 142·5 | 158·4 | 170·5 | 175·8 | 171·1 |
| +·1 | Jup. ... | 349·9 | 352·3 | 354·7 | 357·1 | 359·3 | 1·4 | 3·5 | 5·3 | 7·0 | 8·5 | 9·6 | 10·5 | 11·1 | 11·3 | 11·1 | 10·6 | 10·0 | 8·9 |
| | Ven. ... | 34·4 | 54·8 | 64·3 | 73·0 | 80·0 | 85·5 | 88·4 | 86·7 | 82·9 | 75·3 | 70·7 | 71·1 | 74·5 | 80·3 | 88·3 | 97·1 | 107·2 | 117·6 |
| | Sat. ... | 79·2 | 79·9 | 80·7 | 81·3 | 82·4 | 83·6 | 84·8 | 86·1 | 87·4 | 88·7 | 90·1 | 91·5 | 92·8 | 94·0 | 95·1 | 96·4 | 97·3 | 98·1 |
| Eng. Date. | | Ap 13 | Ap 23 | My 3 | My 13 | My 23 | Je 2 | Je 12 | Je 22 | Jl 2 | Jl 12 | Jl 22 | Au 1 | Au 11 | Au 21 | Au 31 | S 10 | S 20 | S 30 |
| 1917 | Mars ... | 350·4 | 358·1 | 5·7 | 13·1 | 20·6 | 28·0 | 35·2 | 42·3 | 49·5 | 56·3 | 63·3 | 70·0 | 76·5 | 83·1 | 89·5 | 96·0 | 102·1 | 108·1 |
| ⊙'s L. | Merc. ... | 12·2 | 27·2 | 36·6 | 36·9 | 29·3 | 27·8 | 35·8 | 50·1 | 67·0 | 85·6 | 104·0 | 122·1 | 138·8 | 151·9 | 158·7 | 156·0 | 148·5 | 148·9 |
| +·8 | Jup. ... | 17·7 | 20·1 | 22·5 | 24·9 | 27·3 | 29·7 | 32·0 | 34·4 | 36·5 | 38·6 | 40·5 | 42·2 | 43·8 | 45·0 | 46·1 | 46·8 | 47·2 | 47·3 |
| | Ven. ... | 356·5 | 8·8 | 21·2 | 33·5 | 45·9 | 58·1 | 70·2 | 82·5 | 94·4 | 106·9 | 119·1 | 131·2 | 143.3 | 155·3 | 167·3 | 179·1 | 191·4 | 203·0 |
| | Sat. ... | 92·9 | 93·4 | 93·9 | 94·5 | 95·4 | 96·3 | 97·4 | 98·6 | 99·8 | 101·1 | 102·4 | 103·7 | 105·1 | 106·3 | 107·6 | 108·9 | 110·0 | 111·0 |
| 1918 | Mars ... | 141·7 | 141·1 | 141·3 | 143·1 | 145·5 | 148·9 | 152·7 | 157·2 | 162·0 | 167·3 | 172·9 | 178·5 | 184·5 | 190·7 | 197·2 | 203·8 | 210·4 | 217·4 |
| ⊙'s L. | Merc. ... | 14·3 | 18·4 | 11·9 | 8·1 | 14·2 | 27·5 | 44·0 | 62·5 | 81·3 | 99·7 | 116·9 | 131·4 | 140·8 | 140·7 | 133·1 | 130·8 | 138·2 | 151·2 |
| +·6 | Jup. ... | 45·3 | 47·5 | 49·6 | 51·8 | 53·7 | 56·4 | 58·7 | 61·0 | 63·4 | 65·7 | 68·0 | 70·2 | 72·2 | 73·9 | 75·8 | 77·5 | 78·9 | 80·0 |
| | Ven. ... | 313·1 | 322·8 | 332·8 | 343·7 | 354·6 | 5·9 | 17·3 | 28·9 | 40·7 | 52·6 | 64·6 | 76·2 | 88·3 | 100·3 | 112·5 | 124·7 | 136·9 | 149·2 |
| | Sat. ... | 106·6 | 106·9 | 107·2 | 107·6 | 108·3 | 109·1 | 109·8 | 110·9 | 112·1 | 113·2 | 114·5 | 116·8 | 117·0 | 118·3 | 119·6 | 121·0 | 122·2 | 123·3 |
| 1919 | Mars ... | 5·5 | 13·0 | 20·4 | 27·6 | 34·8 | 41·9 | 48·9 | 56·1 | 62·8 | 69·8 | 76·4 | 82·9 | 89·7 | 96·0 | 102·7 | 109·0 | 115·2 | 121·3 |
| ⊙'s L. | Merc. ... | 354·9 | 349·0 | 352·9 | 5·0 | 20·8 | 39·0 | 58·0 | 76·9 | 94·5 | 110·2 | 121·2 | 124·0 | 117·5 | 112·9 | 118·2 | 130·1 | 146·0 | 163·1 |
| +·3 | Jup. ... | 73·9 | 75·2 | 76·9 | 78·6 | 80·5 | 82·5 | 84·6 | 86·8 | 89·1 | 91·4 | 93·6 | 95·9 | 98·3 | 100·5 | 102·6 | 104·7 | 106·7 | 108·4 |
| | Ven. ... | 32·3 | 44·2 | 56·0 | 67·7 | 79·2 | 90·4 | 101·4 | 112·0 | 121·9 | 131·3 | 139·8 | 147·0 | 152·0 | 154·3 | 152·7 | 147·1 | 139·6 | 137·6 |
| | Sat. ... | 119·5 | 120·6 | 120·5 | 120·6 | 121·0 | 121·6 | 122·1 | 122·8 | 123·6 | 124·3 | 125·4 | 126·5 | 127·7 | 129·0 | 130·3 | 131·5 | 132·8 | 134·1 |

| Mesh. | Vrsh. | Mith. | Kat. | Sim. | Kan. | Tul. | Vrsch. | Dhan. | Mak. | Kum. | Min. | Asvn. | Bhar. | Krit. | Rohi. | Mrig. | Ardh. | Punar. | Push. | Asles. |
|---|---|---|---|---|---|---|---|---|---|---|---|---|---|---|---|---|---|---|---|---|
| 30 | 60 | 90 | 120 | 150 | 180 | 210 | 240 | 270 | 300 | 330 | 360 | 13·3 | 26·7 | 40·0 | 53·3 | 66·7 | 80·0 | 93·3 | 106·7 | 120·0 |

## ean Sunrise for every ten days from A.D. 1910 to A.D. 1919.

| 180 | 190 | 200 | 210 | 220 | 230 | 240 | 250 | 260 | | 270 | 280 | 290 | 300 | 310 | 320 | 330 | 340 | 350 | 360 |
|---|---|---|---|---|---|---|---|---|---|---|---|---|---|---|---|---|---|---|---|
| 174·4 | 184·3 | 194·1 | 204·0 | 213·8 | 223·7 | 233·5 | 243·4 | 253·3 | | 263·1 | 273·0 | 282·8 | 292·7 | 302·5 | 312·4 | 322·2 | 332·1 | 342·0 | 351·8 |
| —2·2 | —2·1 | —1·9 | —1·7 | —1·5 | —1·2 | —0·9 | —0·5 | —0·1 | | +0·3 | +0·6 | +1·0 | +1·3 | +1·6 | +1·8 | +2·0 | +2·1 | +2·2 | +2·2 |
| O 10 | O 20 | O 30 | N 9 | N 19 | N 29 | D 9 | D 19 | D 29 | | Ja 8 | Ja 18 | Ja 28 | F 7 | F 17 | F 27 | Mr 8,9 | 18,19 | 28,29 | Ap 7,8 |
| Pur | Aipp | Aipp | Aipp | Kart | Kart | Kart | Marg | Marg | | Marg | Tai | Tai | Tai | Masi | Masi | Masi | Pang | Pang | Pang |
| 25 | 4 | 14 | 24 | 4 | 14 | 24 | 5 | 15 | | 25 | 5 | 15 | 25 | 6 | 16 | 26 | 6 | 16 | 26 |
| 169·4 | 176·0 | 182·5 | 189·0 | 195·7 | 202·4 | 209·3 | 216·1 | 223·0 | **11** | 229·9 | 236·8 | 243·9 | 250·9 | 258·2 | 265·3 | 272·6 | 280·0 | 287·2 | 294·5 |
| 157·6 | 168·1 | 182·6 | 199·3 | 216·7 | 233·9 | 249·7 | 263·0 | 270·5 | | 269·0 | 260·8 | 260·0 | 268·4 | 282·1 | 298·7 | 316·9 | 335·1 | 353·1 | 9·1 |
| 80·1 | 182·3 | 184·4 | 186·6 | 188·8 | 190·9 | 192·8 | 194·6 | 196·3 | | 197·9 | 199·4 | 200·3 | 201·2 | 201·6 | 201·7 | 201·6 | 201·3 | 200·5 | 199·5 |
| 61·0 | 173·5 | 186·0 | 198·6 | 211·0 | 223·7 | 236·2 | 248·9 | 261·5 | | 274·1 | 286·8 | 299·3 | 311·9 | 324·3 | 336·9 | 349·2 | 1·5 | 13·8 | 25·8 |
| 11·5 | 10·6 | 9·8 | 9·0 | 8·3 | 7·7 | 7·1 | 6·8 | 6·6 | | 6·4 | 6·8 | 7·1 | 7·6 | 8·4 | 9·3 | 10·2 | 11·4 | 12·6 | 13·8 |
| 44·8 | 45·7 | 44·9 | 42·7 | 40·1 | 37·0 | 34·5 | 32·8 | 32·0 | **12** | 32·7 | 34·4 | 37·0 | 40·3 | 44·0 | 48·1 | 52·1 | 57·6 | 62·9 | 68·1 |
| 61·7 | 178·3 | 195·8 | 213·8 | 229·5 | 243·6 | 252·9 | 254·2 | 246·8 | | 243·0 | 248·8 | 261·3 | 277·1 | 294·5 | 312·7 | 330·3 | 347·2 | 0·6 | 7·8 |
| 03·8 | 205·8 | 207·9 | 210·2 | 212·4 | 214·6 | 216·8 | 219·0 | 221·1 | | 223·2 | 225·0 | 226·7 | 228·3 | 229·8 | 230·9 | 231·8 | 232·4 | 232·7 | 232·7 |
| 139·9 | 142·4 | 149·4 | 157·1 | 166·2 | 176·2 | 186·5 | 197·7 | 209·2 | | 220·7 | 232·2 | 244·8 | 257·0 | 269·1 | 281·5 | 293·5 | 305·8 | 318·2 | 330·7 |
| 26·7 | 25·9 | 25·1 | 24·4 | 23·6 | 22·9 | 22·1 | 21·6 | 21·1 | | 20·7 | 20·8 | 20·9 | 20·9 | 21·6 | 22·2 | 23·0 | 24·0 | 25·2 | 26·2 |
| 9 | O 19 | O 29 | N 8 | N 18 | N 28 | D 8 | D 18 | D 28 | | Ja 7 | Ja 17 | Ja 27 | F 6 | F 16 | F 26 | Mr 8 | Mr 18 | Mr 28 | Ap 7 |
| 181·4 | 188·2 | 195·0 | 201·8 | 208·6 | 216·1 | 223·1 | 230·2 | 237·4 | **13** | 244·5 | 251·9 | 259·3 | 266·8 | 274·3 | 281·9 | 289·5 | 297·1 | 304·9 | 312·6 |
| 174·4 | 192·0 | 208·6 | 224·0 | 235·2 | 239·1 | 233·6 | 227·3 | 230·2 | | 240·8 | 255·7 | 272·9 | 290·8 | 308·4 | 325·1 | 339·5 | 348·3 | 348·0 | 340·4 |
| 228·6 | 230·4 | 232·4 | 234·5 | 236·7 | 238·9 | 241·1 | 243·3 | 245·7 | | 247·9 | 250·1 | 252·3 | 254·3 | 256·3 | 258·2 | 259·9 | 261·5 | 262·8 | 263·6 |
| 197·8 | 210·1 | 222·4 | 234·7 | 247·0 | 259·4 | 271·4 | 283·5 | 295·4 | | 307·2 | 319·4 | 329·8 | 340·5 | 350·6 | 0·2 | 8·4 | 14·9 | 19·6 | 20·6 |
| 42·1 | 41·2 | 40·2 | 39·2 | 38·6 | 38·1 | 37·6 | 36·9 | 36·3 | | 35·5 | 35·3 | 35·2 | 34·9 | 35·4 | 35·8 | 36·2 | 37·0 | 37·9 | 38·9 |
| 78·1 | 82·5 | 86·3 | 89·4 | 91·7 | 92·7 | 92·9 | 91·4 | 88·8 | **14** | 85·5 | 81·7 | 78·5 | 76·0 | 74·9 | 75·3 | 76·9 | 79·3 | 82·5 | 86·1 |
| 188·0 | 204·2 | 216·9 | 223·5 | 221·1 | 213·2 | 212·8 | 221·3 | 234·8 | | 251·4 | 269·1 | 286·7 | 303·6 | 318·5 | 328·9 | 331·2 | 324·4 | 320·5 | 326·2 |
| 256·2 | 257·6 | 259·2 | 260·8 | 262·5 | 264·5 | 266·6 | 268·0 | 271·0 | | 273·3 | 275·6 | 278·0 | 280·3 | 282·7 | 284·9 | 287·1 | 289·2 | 291·2 | 293·2 |
| 42·4 | 154 7 | 166·9 | 179·4 | 191·7 | 204·1 | 216·8 | 229·2 | 241·7 | | 254·4 | 267·1 | 279·6 | 292·4 | 305·3 | 318·2 | 329·9 | 342·5 | 355·0 | 7·2 |
| 57·0 | 56·6 | 56·1 | 55·7 | 55·9 | 54·1 | 53·3 | 52·5 | 51·8 | | 51·0 | 50·6 | 50·1 | 49·7 | 49·7 | 49·8 | 49·8 | 50·5 | 51·2 | 51·9 |
| O 10 | O 20 | O 30 | N 9 | N 19 | N 29 | D 9 | D 19 | D 29 | | Ja 8 | Ja 18 | Ja 28 | F 7 | F 17 | F 27 | Mr 8,9 | 18,19 | 28,29 | Ap 7,8 |
| 94·9 | 201·7 | 208·6 | 215·6 | 222·8 | 230·1 | 237·4 | 244·7 | 252·1 | **15** | 259·7 | 267·3 | 275·1 | 282·7 | 290·4 | 298·3 | 306·2 | 314·1 | 321·9 | 329·8 |
| 198·5 | 207·1 | 206·2 | 198·6 | 196·5 | 203·6 | 216·4 | 232·2 | 249·7 | | 267·1 | 284·0 | 299·2 | 310·3 | 313·8 | 307·5 | 301·7 | 306·7 | 318·6 | 334·6 |
| 2 3·5 | 288·8 | 289·3 | 290·1 | 291·3 | 292·6 | 294·2 | 295·9 | 298·2 | | 300·3 | 302·4 | 304·8 | 307·1 | 309·4 | 311·8 | 314·3 | 316·7 | 319·1 | 321·4 |
| 21[illegible]·1 | 223·8 | 227·5 | 228·7 | 224·6 | 218·5 | 213·6 | 212·6 | 215·4 | | 220·9 | 228·3 | 237·2 | 247·1 | 257·6 | 268·5 | 279·6 | 291·1 | 302·8 | 314·7 |
| 71·5 | 71·3 | 71·2 | 71·0 | 70·3 | 69·5 | 68·9 | 68·1 | 67·4 | | 66·5 | 65·9 | 65·2 | 64·5 | 64·4 | 64·1 | 64·0 | 64·4 | 64·9 | 65·2 |
| 98·4 | 103·8 | 108·9 | 113·4 | 117·9 | 121·8 | 124·9 | 127·1 | 128·4 | **16** | 128·6 | 127·5 | 125·3 | 122·0 | 118·0 | 114·2 | 111·4 | 109·7 | 109·2 | 110·1 |
| 191·7 | 185·1 | 180·3 | 184·8 | 196·4 | 211·7 | 228·6 | 245·9 | 262·9 | | 278·6 | 291·2 | 296·8 | 292·8 | 294·4 | 296·2 | 306·5 | 321·2 | 338·1 | 343·3 |
| 323·3 | 325·6 | 324·9 | 324·4 | 324·3 | 324·7 | 325·3 | 326·3 | 327·6 | | 329·2 | 329·8 | 332·7 | 334·8 | 337·0 | 339·2 | 341·6 | 344·0 | 346·4 | 348·8 |
| 7[illegible]9 | 192·4 | 205·0 | 217·5 | 229·9 | 242·5 | 255·5 | 267·7 | 280·1 | | 292·6 | 304·9 | 317·2 | 329·3 | 341·5 | 353·5 | 5·1 | 17·0 | 2[illegible]·1 | 39·2 |
| 85·4 | 85·5 | 85·7 | 85·8 | 85·3 | 84·9 | 84·4 | 83·6 | 82·8 | | 82·0 | 81·1 | 80·3 | 79·4 | 79·1 | 78·7 | 78·4 | 78·5 | 78·5 | 78·[illegible] |
| O 9 | O 19 | O 29 | N 8 | N 18 | N 28 | D 8 | D 18 | D 28 | | Ja 7 | Ja 17 | Ja 27 | F 6 | F 16 | F 26 | Mr 8 | Mr 18 | Mr 28 | Ap 7 |
| 208·5 | 215·6 | 222·6 | 229·9 | 237·1 | 244·6 | 252·0 | 259·6 | 267·2 | **17** | 275·0 | 282·9 | 290·6 | 298·6 | 306·4 | 314·1 | 322·1 | 330·0 | 338·0 | 345·6 |
| 164·5 | 166·7 | 176·6 | 191·1 | 207·5 | 224·9 | 242·0 | 258·2 | 271·7 | | 279·5 | 278·4 | 270·8 | 267·3 | 277·2 | 290·9 | 307·4 | 323·3 | 342·1 | 0·5 |
| 7·8 | 6·4 | 5·0 | 3·8 | 2·6 | 1·7 | 1·0 | 0·9 | 0·9 | | 1·4 | 2·1 | 3·2 | 4·4 | 6·1 | 7·9 | 9·7 | 11·9 | 14·1 | 16·4 |
| 128·6 | 139·9 | 151·5 | 163·2 | 175·3 | 187·3 | 199·5 | 211·7 | 224·0 | | 236·5 | 249·2 | 261·5 | 274·2 | 287·0 | 299·0 | 311·6 | 324·1 | 336·6 | 349·0 |
| 98·9 | 99·3 | 99·7 | 100·1 | 99·8 | 99·6 | 99·4 | 98·7 | 98·0 | | 97·4 | 96·6 | 96·9 | 95·0 | 94·3 | 93·7 | 93·0 | 92·9 | 92·7 | 92·5 |
| O 10 | O 20 | O 30 | N 9 | N 19 | N 29 | D 9 | D 19 | D 29 | | Ja 8 | Ja 18 | Ja 28 | F 7 | F 17 | F 27 | Mr 8,9 | 18,19 | 28,29 | Ap 7,8 |
| 114·1 | 119·9 | 125·6 | 131·0 | 136·1 | 141·2 | 145·8 | 149·9 | 153·6 | **18** | 156·4 | 158·8 | 159·9 | 160·2 | 158·6 | 155·8 | 152·6 | 148·4 | 145·2 | 142·4 |
| 157·8 | 171·9 | 188·3 | 205·6 | 222·9 | 239·4 | 253·3 | 262·5 | 263·0 | | 255·6 | 252·3 | 258·5 | 271·3 | 287·4 | 305·1 | 323·2 | 341·1 | 357·9 | 11·3 |
| 47·0 | 46·4 | 45·5 | 44·3 | 43·1 | 36·7 | 40·3 | 39·2 | 38·1 | | 37·3 | 36·9 | 36·7 | 37·0 | 37·6 | 38·5 | 39·6 | 40·9 | 42·5 | 44·3 |
| 215·0 | 226·7 | 237·8 | 249·2 | 260·1 | 270·5 | 280·2 | 289·2 | 296·8 | | 302·2 | 304·8 | 303·7 | 298·4 | 292·1 | 289·1 | 290·3 | 294·1 | 300·8 | 308·7 |
| 111·9 | 112·6 | 113·2 | 113·8 | 113·9 | 113·9 | 114·0 | 113·9 | 113·1 | | 112·5 | 111·7 | 110·9 | 110·2 | 109·4 | 108·6 | 107·8 | 107·4 | 106·9 | 106·6 |
| 224·4 | 231·4 | 238·6 | 246·0 | 253·5 | 260·9 | 268·7 | 276·3 | 283·9 | **19** | 291·8 | 299·6 | 307·5 | 315·0 | 323·2 | 331·0 | 338·8 | 346·6 | 354·2 | 1·4 |
| 167·3 | 184·7 | 202·1 | 218·7 | 233·7 | 244·6 | 248·1 | 242·0 | 236·4 | | 239·7 | 250·8 | 265·9 | 283·2 | 301·3 | 319·0 | 334·8 | 350·2 | 358·9 | 358·6 |
| 80·9 | 81·1 | 81·8 | 81·6 | 81·2 | 80·3 | 79·0 | 78·1 | 76·8 | | 75·6 | 74·3 | 73·2 | 72·2 | 71·6 | 71·3 | 71·3 | 71·5 | 72·3 | 73·3 |
| 161·5 | 174·0 | 186·4 | 198·7 | 211·5 | 224·1 | 236·7 | 249·4 | 262·0 | | 274·7 | 287·3 | 299·6 | 312·4 | 325·0 | 327·8 | 349·7 | 2·0 | 14·3 | 26·2 |
| 124·4 | 125·1 | 125·9 | 126·7 | 127·1 | 127·3 | 127·7 | 127·7 | 127·8 | | 127·9 | 126·3 | 125·6 | 124·9 | 124·0 | 123·2 | 122·4 | 122·4 | 120·5 | 119·6 |
| 127·5 | 133·5 | 139·5 | 145·2 | 151·0 | 156·5 | 162·0 | 167·1 | 172·2 | **20** | 177·2 | 181·7 | 185·6 | 189·3 | 192·3 | 194·4 | 195·7 | 195·6 | 194·3 | 191·7 |
| 180·9 | 198·0 | 213·9 | 226·3 | 232·3 | 228·9 | 221·7 | 222·1 | 231·1 | | 245·1 | 261·8 | 279·4 | 298·0 | 314·1 | 328·9 | 339·1 | 341·2 | 334·4 | 331·7 |
| 110·0 | 111·3 | 112·5 | 113·4 | 114·1 | 114·4 | 114·4 | 114·1 | 113·3 | | 112·4 | 111·3 | 109·9 | 108·6 | 107·4 | 106·2 | 105·2 | 104·4 | 104·0 | 103·9 |
| 138·2 | 141·9 | 149·5 | 156·8 | 165·8 | 176·1 | 186·9 | 198·0 | 209·3 | | 221·0 | 233·0 | 245·3 | 257·5 | 269·6 | 282·0 | 294·1 | 306·8 | 318·8 | 331·2 |
| 135·4 | 136·3 | 137·4 | 138·4 | 139·2 | 139·9 | 140·6 | 140·9 | 141·1 | | 141·3 | 140·8 | 140·4 | 140·1 | 139·4 | 138·6 | 137·9 | 137·1 | 136·4 | 135·6 |

| Mag. | P.Phal. | U.Phal. | Hasta. | Chit. | Svati. | Visa. | Anur. | Jyesh. | Mula. | P.Ash. | U.Ash. | Srav. | Dan. | Satab. | P.Bhad. | U.Bhad. | Revati. |
|---|---|---|---|---|---|---|---|---|---|---|---|---|---|---|---|---|---|
| 133·3 | 146·7 | 160·0 | 173·3 | 186·7 | 200·0 | 213·3 | 226·7 | 240·0 | 253·3 | 266·7 | 280·0 | 293·3 | 306·7 | 320·0 | 333·3 | 346·7 | 360·0 |

# TABLE XIX :—Ghatikas and Palas (Naligais and Vinadis) as Fractions of a day.

**N.B.**—The same table will serve for minutes and seconds as fractions of an **Hour** or a **Degree**.

| Palas. | Ghatikas. 0 | 1 | 2 | 3 | 4 | 5 | 6 | 7 | 8 | 9 | 10 | 11 | 12 | 13 | 14 | 15 | 16 | 17 | 18 | 19 |
|---|---|---|---|---|---|---|---|---|---|---|---|---|---|---|---|---|---|---|---|---|
| 0 | ... | ·0167 | ·0333 | ·0500 | ·0667 | ·0833 | ·1000 | ·1167 | ·1333 | ·1500 | ·1667 | ·1833 | ·2000 | ·2167 | ·2333 | ·2500 | ·2667 | ·2833 | ·3000 | ·3167 |
| 1 | ·0003 | ·0169 | ·0336 | ·0503 | ·0669 | ·0836 | ·1003 | ·1169 | ·1336 | ·1503 | ·1669 | ·1836 | ·2003 | ·2169 | ·2336 | ·2503 | ·2669 | ·2836 | ·3003 | ·3169 |
| 2 | ·0005 | ·0172 | ·0339 | ·0505 | ·0672 | ·0839 | ·1005 | ·1172 | ·1339 | ·1505 | ·1672 | ·1839 | ·2005 | ·2172 | ·2339 | ·2505 | ·2672 | ·2839 | ·3005 | ·3172 |
| 3 | ·0008 | ·0175 | ·0342 | ·0508 | ·0675 | ·0842 | ·1008 | ·1175 | ·1342 | ·1508 | ·1675 | ·1842 | ·2008 | ·2175 | ·2342 | ·2508 | ·2675 | ·2842 | ·3008 | ·3175 |
| 4 | ·0011 | ·0178 | ·0344 | ·0511 | ·0678 | ·0844 | ·1011 | ·1178 | ·1344 | ·1511 | ·1678 | ·1844 | ·2011 | ·2178 | ·2344 | ·2511 | ·7678 | ·2844 | ·3011 | ·3178 |
| 5 | ·0014 | ·0180 | ·0347 | ·0514 | ·0680 | ·0847 | ·1014 | ·1180 | ·1347 | ·1514 | ·1680 | ·1847 | ·2014 | ·2180 | ·2347 | ·2514 | ·2680 | ·2847 | ·3014 | ·3180 |
| 6 | ·0017 | ·0183 | ·0350 | ·0517 | ·0683 | ·0850 | ·1017 | ·1183 | ·1350 | ·1517 | ·1683 | ·1850 | ·2017 | ·2183 | ·2350 | ·2517 | ·2683 | ·2850 | ·3017 | ·3183 |
| 7 | ·0019 | ·0186 | ·0353 | ·0519 | ·0686 | ·0853 | ·1019 | ·1186 | ·1353 | ·1519 | ·1686 | ·1853 | ·2019 | ·2186 | ·2353 | ·2519 | ·2686 | ·2853 | ·3019 | ·3186 |
| 8 | ·0022 | ·0189 | ·0355 | ·0522 | ·0689 | ·0855 | ·1022 | ·1189 | ·1355 | ·1522 | ·1689 | ·1855 | ·2022 | ·2189 | ·2355 | ·2522 | ·2689 | ·2855 | ·3022 | ·3189 |
| 9 | ·0025 | ·0192 | ·0358 | ·0525 | ·0692 | ·0858 | ·1025 | ·1192 | ·1358 | ·1525 | ·1692 | ·1858 | ·2025 | ·2192 | ·2358 | ·2525 | ·2692 | ·2858 | ·3025 | ·3192 |
| 10 | ·0028 | ·0194 | ·0361 | ·0528 | ·0694 | ·0861 | ·1028 | ·1194 | ·1361 | ·1528 | ·1694 | ·1861 | ·2028 | ·2194 | ·2361 | ·2528 | ·2694 | ·2861 | ·3028 | ·3194 |
| 11 | ·0030 | ·0197 | ·0364 | ·0530 | ·0697 | ·0864 | ·1030 | ·1197 | ·1364 | ·1530 | ·1697 | ·1864 | ·2030 | ·2197 | ·2364 | ·2530 | ·2697 | ·2864 | ·3030 | ·3197 |
| 12 | ·0033 | ·0200 | ·0367 | ·0533 | ·0700 | ·0867 | ·1033 | ·1200 | ·1367 | ·1533 | ·1700 | ·1867 | ·2033 | ·2200 | ·2367 | ·2533 | ·2700 | ·2867 | ·3033 | ·3200 |
| 13 | ·0036 | ·0203 | ·0369 | ·0536 | ·0703 | ·0869 | ·1036 | ·1203 | ·1369 | ·1536 | ·1702 | ·1869 | ·2036 | ·2203 | ·2369 | ·2536 | ·2703 | ·2869 | ·3036 | ·3203 |
| 14 | ·0039 | ·0205 | ·0372 | ·0539 | ·0705 | ·0872 | ·1039 | ·1205 | ·1372 | ·1539 | ·1705 | ·1872 | ·2039 | ·2205 | ·2372 | ·2539 | ·2705 | ·2872 | ·3039 | ·3205 |
| 15 | ·0042 | ·0208 | ·0375 | ·0542 | ·0708 | ·0875 | ·1042 | ·1208 | ·1375 | ·1542 | ·1708 | ·1875 | ·2042 | ·2208 | ·2375 | ·2542 | ·2708 | ·2875 | ·3042 | ·3208 |
| 16 | ·0044 | ·0211 | ·0378 | ·0544 | ·0711 | ·0878 | ·1044 | ·1211 | ·1378 | ·1544 | ·1711 | ·1878 | ·2044 | ·2211 | ·2378 | ·2544 | ·2711 | ·2878 | ·3044 | ·3211 |
| 17 | ·0047 | ·0214 | ·0380 | ·0547 | ·0714 | ·0880 | ·1047 | ·1214 | ·1380 | ·1547 | ·1714 | ·1880 | ·2047 | ·2214 | ·2380 | ·2547 | ·2714 | ·2880 | ·3047 | ·3214 |
| 18 | ·0050 | ·0217 | ·0383 | ·0550 | ·0717 | ·0883 | ·1050 | ·1217 | ·1383 | ·1550 | ·1717 | ·1883 | ·2050 | ·2217 | ·2383 | ·2550 | ·2717 | ·2883 | ·3050 | ·3217 |
| 19 | ·0053 | ·0219 | ·0386 | ·0553 | ·0719 | ·0886 | ·1053 | ·1219 | ·1386 | ·1553 | ·1719 | ·1886 | ·2053 | ·2219 | ·2386 | ·2553 | ·2719 | ·2886 | ·3053 | ·3219 |
| 20 | ·0055 | ·0222 | ·0389 | ·0555 | ·0722 | ·0889 | ·1055 | ·1222 | ·1389 | ·1555 | ·1722 | ·1889 | ·2055 | ·2222 | ·2389 | ·2555 | ·2722 | ·2889 | ·3055 | ·3222 |
| 21 | ·0058 | ·0225 | ·0392 | ·0558 | ·0725 | ·0892 | ·1058 | ·1225 | ·1392 | ·1558 | ·1725 | ·1892 | ·2058 | ·2225 | ·2392 | ·2558 | ·2725 | ·2892 | ·3058 | ·3225 |
| 22 | ·0061 | ·0228 | ·0394 | ·0561 | ·0728 | ·0894 | ·1061 | ·1228 | ·1394 | ·1561 | ·1728 | ·1894 | ·2061 | ·2228 | ·2394 | ·2561 | ·2728 | ·2894 | ·3061 | ·3228 |
| 23 | ·0064 | ·0230 | ·0397 | ·0564 | ·0730 | ·0897 | ·1064 | ·1230 | ·1397 | ·1564 | ·1730 | ·1897 | ·2064 | ·2230 | ·2397 | ·2564 | ·2730 | ·2897 | ·3064 | ·3230 |
| 24 | ·0067 | ·0233 | ·0400 | ·0567 | ·0733 | ·0900 | ·1067 | ·1233 | ·1400 | ·1567 | ·1733 | ·1900 | ·2067 | ·2233 | ·2400 | ·2567 | ·2733 | ·2900 | ·3067 | ·3233 |
| 25 | ·0069 | ·0236 | ·0403 | ·0569 | ·0736 | ·0903 | ·1069 | ·1236 | ·1403 | ·1569 | ·1736 | ·1903 | ·2069 | ·2236 | ·2403 | ·2569 | ·2736 | ·2903 | ·3069 | ·3236 |
| 26 | ·0072 | ·0239 | ·0405 | ·0572 | ·0739 | ·0905 | ·1072 | ·1239 | ·1405 | ·1572 | ·1739 | ·1905 | ·3072 | ·2239 | ·2405 | ·2572 | ·2739 | ·2905 | ·3072 | ·3239 |
| 27 | ·0075 | ·0242 | ·0408 | ·0575 | ·0742 | ·0908 | ·1075 | ·1242 | ·1408 | ·1575 | ·1742 | ·1908 | ·2075 | ·2242 | ·2408 | ·2575 | ·2742 | ·2908 | ·3075 | ·3242 |
| 28 | ·0078 | ·0244 | ·0411 | ·0578 | ·0744 | ·0911 | ·1078 | ·1244 | ·1411 | ·1578 | ·1744 | ·1911 | ·2078 | ·2244 | ·2411 | ·2578 | ·2744 | ·2911 | ·3078 | ·3244 |
| 29 | ·0080 | ·0247 | ·0414 | ·0580 | ·0747 | ·0914 | ·1080 | ·1247 | ·1414 | ·1580 | ·1747 | ·1914 | ·2080 | ·2247 | ·2414 | ·2580 | ·2747 | ·2914 | ·3080 | ·3247 |
| 30 | ·0083 | ·0250 | ·0417 | ·0583 | ·0750 | ·0917 | ·1083 | ·1250 | ·1417 | ·1583 | ·1750 | ·1917 | ·2083 | ·2250 | ·2417 | ·2583 | ·2750 | ·2917 | ·3083 | ·3250 |
| 31 | ·0086 | ·0253 | ·0419 | ·0586 | ·0753 | ·0919 | ·1086 | ·1253 | ·1419 | ·1586 | ·1753 | ·1919 | ·2086 | ·2253 | ·2419 | ·2586 | ·2753 | ·2919 | ·3086 | ·3253 |
| 32 | ·0089 | ·0255 | ·0422 | ·0589 | ·0755 | ·0922 | ·1089 | ·1255 | ·1422 | ·1589 | ·1755 | ·1922 | ·2089 | ·2255 | ·2422 | ·2589 | ·2755 | ·2922 | ·3089 | ·3255 |
| 33 | ·0092 | ·0258 | ·0425 | ·0592 | ·0758 | ·0925 | ·1092 | ·1258 | ·1425 | ·1592 | ·1758 | ·1925 | ·2092 | ·2258 | ·2425 | ·2592 | ·2758 | ·2925 | ·3092 | ·3258 |
| 34 | ·0094 | ·0261 | ·0428 | ·0594 | ·0761 | ·0928 | ·1094 | ·1261 | ·1428 | ·1594 | ·1761 | ·1928 | ·2094 | ·2261 | ·2428 | ·2594 | ·2761 | ·2928 | ·3094 | ·3261 |
| 35 | ·0097 | ·0264 | ·0430 | ·0597 | ·0764 | ·0930 | ·1097 | ·1264 | ·1430 | ·1597 | ·1764 | ·1930 | ·2097 | ·2264 | ·2430 | ·2597 | ·2764 | ·2930 | ·3097 | ·3264 |
| 36 | ·0100 | ·0267 | ·0433 | ·0600 | ·0767 | ·0933 | ·1100 | ·1267 | ·1433 | ·1600 | ·1767 | ·1933 | ·2100 | ·2267 | ·2433 | ·2600 | ·2767 | ·2933 | ·3100 | ·3267 |
| 37 | ·0103 | ·0269 | ·0436 | ·0603 | ·0769 | ·0936 | ·1103 | ·1269 | ·1436 | ·1603 | ·1769 | ·1936 | ·2103 | ·2269 | ·2436 | ·2603 | ·2769 | ·2936 | ·3103 | ·3269 |
| 38 | ·0105 | ·0272 | ·0439 | ·0605 | ·0772 | ·0939 | ·1105 | ·1272 | ·1439 | ·1605 | ·1772 | ·1939 | ·2105 | ·2272 | ·2439 | ·2605 | ·2772 | ·2939 | ·3105 | ·3272 |
| 39 | ·0108 | ·0275 | ·0442 | ·0608 | ·0775 | ·0942 | ·1108 | ·1275 | ·1442 | ·1608 | ·1775 | ·1942 | ·2108 | ·2275 | ·2442 | ·2608 | ·2775 | ·2942 | ·3108 | ·3275 |
| 40 | ·0111 | ·0278 | ·0444 | ·0611 | ·0778 | ·0944 | ·1111 | ·1278 | ·1444 | ·1611 | ·1778 | ·1944 | ·2111 | ·2278 | ·2444 | ·2611 | ·2778 | ·2944 | ·3111 | ·3278 |
| 41 | ·0114 | ·0280 | ·0447 | ·0614 | ·0780 | ·0947 | ·1114 | ·1280 | ·1447 | ·1614 | ·1780 | ·1947 | ·2114 | ·2280 | ·2447 | ·2614 | ·2780 | ·2947 | ·3114 | ·3280 |
| 42 | ·0117 | ·0283 | ·0450 | ·0617 | ·0783 | ·0950 | ·1117 | ·1283 | ·1450 | ·1617 | ·1783 | ·1950 | ·2117 | ·2283 | ·2450 | ·2617 | ·2783 | ·2950 | ·3117 | ·3283 |
| 43 | ·0119 | ·0286 | ·0453 | ·0619 | ·0786 | ·0953 | ·1119 | ·1286 | ·1453 | ·1619 | ·1786 | ·1953 | ·2119 | ·2286 | ·2453 | ·2619 | ·2786 | ·2953 | ·3119 | ·3286 |
| 44 | ·0122 | ·0289 | ·0455 | ·0622 | ·0789 | ·0955 | ·1122 | ·1289 | ·1455 | ·1622 | ·1789 | ·1955 | ·2122 | ·2289 | ·2455 | ·2622 | ·2789 | ·2955 | ·3122 | ·3289 |
| 45 | ·0125 | ·0292 | ·0458 | ·0625 | ·0792 | ·0958 | ·1125 | ·1292 | ·1458 | ·1625 | ·1792 | ·1958 | ·2125 | ·2292 | ·2458 | ·2625 | ·2792 | ·2958 | ·3125 | ·3292 |
| 46 | ·0128 | ·0294 | ·0461 | ·0628 | ·0794 | ·0961 | ·1128 | ·1294 | ·1461 | ·1628 | ·1794 | ·1961 | ·2128 | ·2294 | ·2461 | ·2628 | ·2794 | ·2961 | ·3128 | ·3294 |
| 47 | ·0130 | ·0297 | ·0464 | ·0630 | ·0797 | ·0964 | ·1130 | ·1297 | ·1464 | ·1630 | ·1797 | ·1964 | ·2130 | ·2297 | ·2464 | ·2630 | ·2797 | ·2964 | ·3130 | ·3297 |
| 48 | ·0133 | ·0300 | ·0467 | ·0633 | ·0800 | ·0967 | ·1133 | ·1300 | ·1467 | ·1633 | ·1800 | ·1967 | ·2133 | ·2300 | ·2467 | ·2633 | ·2800 | ·2967 | ·3133 | ·3300 |
| 49 | ·0136 | ·0303 | ·0469 | ·0636 | ·0803 | ·0969 | ·1136 | ·1303 | ·1470 | ·1636 | ·1803 | ·1969 | ·2136 | ·2303 | ·2469 | ·2636 | ·2803 | ·2969 | ·3136 | ·3303 |
| 50 | ·0139 | ·0305 | ·0472 | ·0639 | ·0805 | ·0972 | ·1139 | ·1305 | ·1472 | ·1639 | ·1805 | ·1972 | ·2139 | ·2305 | ·2472 | ·2639 | ·2805 | ·2972 | ·3139 | ·3305 |
| 51 | ·0142 | ·0308 | ·0475 | ·0642 | ·0808 | ·0975 | ·1142 | ·1308 | ·1475 | ·1642 | ·1808 | ·1975 | ·2142 | ·2308 | ·2475 | ·2642 | ·2808 | ·2975 | ·3142 | ·3308 |
| 52 | ·0144 | ·0311 | ·0478 | ·0644 | ·0811 | ·0978 | ·1144 | ·1311 | ·1478 | ·1644 | ·1811 | ·1978 | ·2144 | ·2311 | ·2478 | ·2644 | ·2811 | ·2978 | ·3144 | ·3311 |
| 53 | ·0147 | ·0314 | ·0480 | ·0647 | ·0814 | ·0980 | ·1147 | ·1314 | ·1480 | ·1647 | ·1814 | ·1980 | ·2147 | ·2314 | ·2480 | ·2647 | ·2814 | ·2980 | ·3147 | ·3314 |
| 54 | ·0150 | ·0317 | ·0483 | ·0650 | ·0817 | ·0983 | ·1150 | ·1317 | ·1483 | ·1650 | ·1817 | ·1983 | ·2150 | ·2317 | ·2483 | ·2650 | ·2817 | ·2983 | ·3150 | ·3317 |
| 55 | ·0153 | ·0319 | ·0486 | ·0653 | ·0819 | ·0986 | ·1153 | ·1319 | ·1486 | ·1653 | ·1819 | ·1986 | ·2153 | ·2319 | ·2486 | ·2653 | ·2819 | ·2986 | ·3153 | ·3319 |
| 56 | ·0155 | ·0322 | ·0489 | ·0655 | ·0822 | ·0989 | ·1155 | ·1322 | ·1489 | ·1655 | ·1822 | ·1989 | ·2155 | ·2322 | ·2489 | ·2655 | ·2822 | ·2989 | ·3155 | ·3322 |
| 57 | ·0158 | ·0325 | ·0492 | ·0658 | ·0825 | ·0992 | ·1158 | ·1325 | ·1492 | ·1658 | ·1825 | ·1992 | ·2158 | ·2325 | ·2492 | ·2658 | ·2825 | ·2992 | ·3158 | ·3325 |
| 58 | ·0161 | ·0328 | ·0494 | ·0661 | ·0828 | ·0994 | ·1161 | ·1328 | ·1494 | ·1661 | ·1828 | ·1994 | ·2161 | ·2328 | ·2494 | ·2661 | ·2828 | ·2994 | ·3161 | ·3328 |
| 59 | ·0164 | ·0330 | ·0497 | ·0664 | ·0830 | ·0997 | ·1164 | ·1330 | ·1497 | ·1664 | ·1830 | ·1997 | ·2164 | ·2330 | ·2497 | ·2664 | ·2830 | ·2997 | ·3164 | ·3330 |

**TABLE XIX:—Ghatikas and Palas (Naligais and Vinadis) as Fractions of a day.**—*contd.*

**N.B.**—The same table will serve for minutes and seconds as fractions of an **Hour** or a **Degree**.

| Ghatikas. | | | | | Palas. | Ghatikas. | | | | | Palas. | Ghatikas. | | | | | Palas. | Ghatikas. | | | | |
|---|---|---|---|---|---|---|---|---|---|---|---|---|---|---|---|---|---|---|---|---|---|---|
| **20** | **21** | **22** | **23** | **24** | | **25** | **26** | **27** | **28** | **29** | | **30** | **31** | **32** | **33** | **34** | | **35** | **36** | **37** | **38** | **39** |
| ·3333 | ·3500 | ·3667 | ·3833 | ·4000 | **0** | ·4167 | ·4333 | ·4500 | ·4667 | ·4833 | **0** | ·5000 | ·5167 | ·5333 | ·5500 | ·5667 | **0** | ·5833 | ·6000 | ·6167 | ·6333 | ·6500 |
| ·3336 | ·3503 | ·3669 | ·3836 | ·4003 | **1** | ·4169 | ·4336 | ·4503 | ·4669 | ·4836 | **1** | ·5003 | ·5169 | ·5336 | ·5503 | ·5669 | **1** | ·5836 | ·6003 | ·6169 | ·6336 | ·6503 |
| ·3339 | ·3505 | ·3672 | ·3839 | ·4005 | **2** | ·4172 | ·4339 | ·4505 | ·4672 | ·4839 | **2** | ·5005 | ·5172 | ·5339 | ·5505 | ·5672 | **2** | ·5839 | ·6005 | ·6172 | ·6339 | ·6505 |
| ·3342 | ·3508 | ·3675 | ·3842 | ·4008 | **3** | ·4175 | ·4342 | ·4508 | ·4675 | ·4842 | **3** | ·5008 | ·5175 | ·5342 | ·5508 | ·5675 | **3** | ·5842 | ·6008 | ·6175 | ·6342 | ·6508 |
| ·3344 | ·3511 | ·3678 | ·3844 | ·4011 | **4** | ·4178 | ·4344 | ·4511 | ·4678 | ·4844 | **4** | ·5011 | ·5178 | ·5344 | ·5511 | ·5678 | **4** | ·5844 | ·6011 | ·6178 | ·6344 | ·6511 |
| ·3347 | ·3514 | ·3680 | ·3847 | ·4014 | **5** | ·4180 | ·4347 | ·4514 | ·4680 | ·4847 | **5** | ·5014 | ·5180 | ·5347 | ·5514 | ·5680 | **5** | ·5847 | ·6014 | ·6180 | ·6347 | ·6514 |
| ·3350 | ·3517 | ·3683 | ·3850 | ·4017 | **6** | ·4183 | ·4350 | ·4517 | ·4683 | ·4850 | **6** | ·5017 | ·5183 | ·5350 | ·5517 | ·5683 | **6** | ·5850 | ·6017 | ·6183 | ·6350 | ·6517 |
| ·3353 | ·3519 | ·3686 | ·3853 | ·4019 | **7** | ·4186 | ·4353 | ·4519 | ·4686 | ·4853 | **7** | ·5019 | ·5186 | ·5353 | ·5519 | ·5686 | **7** | ·5853 | ·6019 | ·6186 | ·6353 | ·6519 |
| ·3355 | ·3522 | ·3689 | ·3855 | ·4022 | **8** | ·4189 | ·4355 | ·4522 | ·4689 | ·4855 | **8** | ·5022 | ·5189 | ·5355 | ·5522 | ·5689 | **8** | ·5855 | ·6022 | ·6189 | ·6355 | ·6522 |
| ·3358 | ·3525 | ·3692 | ·3858 | ·4025 | **9** | ·4192 | ·4358 | ·4525 | ·4692 | ·4858 | **9** | ·5025 | ·5192 | ·5358 | ·5525 | ·5692 | **9** | ·5858 | ·6025 | ·6192 | ·6358 | ·6525 |
| ·3361 | ·3528 | ·3694 | ·3861 | ·4028 | **10** | ·4194 | ·4361 | ·4528 | ·4694 | ·4861 | **10** | ·5028 | ·5194 | ·5361 | ·5528 | ·5694 | **10** | ·5861 | ·6028 | ·6194 | ·6361 | ·6528 |
| ·3364 | ·3530 | ·3697 | ·3864 | ·4030 | **11** | ·4197 | ·4364 | ·4530 | ·4697 | ·4864 | **11** | ·5030 | ·5197 | ·5364 | ·5530 | ·5697 | **11** | ·5864 | ·6030 | ·6197 | ·6364 | ·6530 |
| ·3367 | ·3533 | ·3700 | ·3867 | ·4033 | **12** | ·4200 | ·4367 | ·4533 | ·4700 | ·4867 | **12** | ·5033 | ·5200 | ·5367 | ·5533 | ·5700 | **12** | ·5867 | ·6033 | ·6200 | ·6367 | ·6533 |
| ·3369 | ·3536 | ·3703 | ·3869 | ·4036 | **13** | ·4203 | ·4369 | ·4536 | ·4703 | ·4869 | **13** | ·5036 | ·5203 | ·5369 | ·5536 | ·5703 | **13** | ·5869 | ·6036 | ·6203 | ·6369 | ·6536 |
| ·3372 | ·3539 | ·3705 | ·3872 | ·4039 | **14** | ·4205 | ·4372 | ·4539 | ·4705 | ·4872 | **14** | ·5039 | ·5205 | ·5372 | ·5539 | ·5705 | **14** | ·5872 | ·6039 | ·6205 | ·6372 | ·6539 |
| ·3375 | ·3542 | ·2708 | ·3875 | ·4042 | **15** | ·4208 | ·4375 | ·4542 | ·4708 | ·4875 | **15** | ·5042 | ·5208 | ·5375 | ·5542 | ·5708 | **15** | ·5875 | ·6042 | ·6208 | ·6375 | ·6542 |
| ·3378 | ·3544 | ·3711 | ·3878 | ·4044 | **16** | ·4211 | ·4378 | ·4544 | ·4711 | ·4878 | **16** | ·5044 | ·5211 | ·5378 | ·5544 | ·5711 | **16** | ·5878 | ·6044 | ·6211 | ·6378 | ·6544 |
| ·3380 | ·3547 | ·3714 | ·3880 | ·4047 | **17** | ·4214 | ·4380 | ·4547 | ·4714 | ·4880 | **17** | ·5047 | ·5214 | ·5380 | ·5547 | ·5714 | **17** | ·5880 | ·6047 | ·6214 | ·6380 | ·6547 |
| ·3383 | ·3550 | ·3717 | ·3883 | ·4050 | **18** | ·4217 | ·4383 | ·4550 | ·4717 | ·4883 | **18** | ·5050 | ·5217 | ·5383 | ·5550 | ·5717 | **18** | ·5883 | ·6050 | ·6217 | ·6383 | ·6550 |
| ·3386 | ·3553 | ·3719 | ·3886 | ·4053 | **19** | ·4219 | ·4386 | ·4553 | ·4719 | ·4886 | **19** | ·5053 | ·5219 | ·5386 | ·5553 | ·5719 | **19** | ·5886 | ·6053 | ·6219 | ·6386 | ·6553 |
| ·3389 | ·3555 | ·3722 | ·3889 | ·4055 | **20** | ·4222 | ·4389 | ·4555 | ·4722 | ·4889 | **20** | ·5055 | ·5222 | ·5389 | ·5555 | ·5722 | **20** | ·5889 | ·6055 | ·6222 | ·6389 | ·6555 |
| ·3392 | ·3558 | ·3725 | ·3892 | ·4058 | **21** | ·4225 | ·4392 | ·4558 | ·4725 | ·4892 | **21** | ·5058 | ·5225 | ·5392 | ·5558 | ·5725 | **21** | ·5892 | ·6058 | ·6225 | ·6392 | ·6558 |
| ·3394 | ·3561 | ·3728 | ·3894 | ·4061 | **22** | ·4228 | ·4394 | ·4561 | ·4728 | ·4894 | **22** | ·5061 | ·5228 | ·5394 | ·5561 | ·5728 | **22** | ·5894 | ·6061 | ·6228 | ·6394 | ·6561 |
| ·3397 | ·3564 | ·3730 | ·3897 | ·4064 | **23** | ·4230 | ·4397 | ·4564 | ·4730 | ·4897 | **23** | ·5064 | ·5230 | ·5397 | ·5564 | ·5730 | **23** | ·5897 | ·6064 | ·6230 | ·6397 | ·6564 |
| ·3400 | ·3567 | ·3733 | ·3900 | ·4067 | **24** | ·4233 | ·4400 | ·4567 | ·4733 | ·4900 | **24** | ·5067 | ·5233 | ·5400 | ·5567 | ·5733 | **24** | ·5900 | ·6067 | ·6233 | ·6400 | ·6567 |
| ·3403 | ·3569 | ·3736 | ·3903 | ·4069 | **25** | ·4236 | ·4403 | ·4569 | ·4736 | ·4903 | **25** | ·5069 | ·5236 | ·5403 | ·5569 | ·5736 | **25** | ·5903 | ·6069 | ·6236 | ·6403 | ·6569 |
| ·3405 | ·3572 | ·3739 | ·3905 | ·4072 | **26** | ·4239 | ·4405 | ·4572 | ·4739 | ·4905 | **26** | ·5072 | ·5239 | ·5405 | ·5572 | ·5739 | **26** | ·5905 | ·6072 | ·6239 | ·6405 | ·6572 |
| ·3408 | ·3575 | ·3742 | ·3908 | ·4075 | **27** | ·4242 | ·4408 | ·4575 | ·4742 | ·4908 | **27** | ·5075 | ·5242 | ·5408 | ·5575 | ·5742 | **27** | ·5908 | ·6075 | ·6242 | ·6408 | ·6575 |
| ·3411 | ·3578 | ·3744 | ·3911 | ·4078 | **28** | ·4244 | ·4411 | ·4578 | ·4744 | ·4911 | **28** | ·5078 | ·5244 | ·5411 | ·5578 | ·5744 | **28** | ·5911 | ·6078 | ·6244 | ·6411 | ·6578 |
| ·3414 | ·3580 | ·3747 | ·3914 | ·4080 | **29** | ·4247 | ·4414 | ·4580 | ·4747 | ·4914 | **29** | ·5080 | ·5247 | ·5414 | ·5580 | ·5747 | **29** | ·5914 | ·6080 | ·6247 | ·6414 | ·6580 |
| ·3417 | ·3583 | ·3750 | ·3917 | ·4083 | **30** | ·4250 | ·4417 | ·4583 | ·4750 | ·4917 | **30** | ·5083 | ·5250 | ·5417 | ·5583 | ·5750 | **30** | ·5917 | ·6083 | ·6250 | [illegible] | ·6583 |
| ·3419 | ·3586 | ·3753 | ·3919 | ·4086 | **31** | ·4253 | ·4419 | ·4586 | ·4753 | ·4919 | **31** | ·5086 | ·5253 | ·5419 | ·5586 | ·5753 | **31** | ·5919 | ·6086 | ·6253 | ·6419 | ·6586 |
| ·3422 | ·3589 | ·3755 | ·3922 | ·4089 | **32** | ·4255 | ·4422 | ·4589 | ·4755 | ·4922 | **32** | ·5089 | ·5255 | ·5422 | ·5589 | ·5755 | **32** | ·5922 | ·6089 | ·6255 | ·6422 | ·6589 |
| ·3425 | ·3592 | ·3758 | ·3925 | ·4092 | **33** | ·4258 | ·4425 | ·4592 | ·4758 | ·4925 | **33** | ·5092 | ·5258 | ·5425 | ·5592 | ·5758 | **33** | ·5925 | ·6092 | ·6258 | ·6425 | ·6592 |
| ·3428 | ·3594 | ·3761 | ·3928 | ·4094 | **34** | ·4261 | ·4428 | ·4594 | ·4761 | ·4928 | **34** | ·5094 | ·5261 | ·5428 | ·5594 | ·5761 | **34** | ·5928 | ·6094 | ·6261 | ·6428 | ·6594 |
| ·3430 | ·3597 | ·3764 | ·3930 | ·4097 | **35** | ·4264 | ·4431 | ·4597 | ·4764 | ·4930 | **35** | ·5097 | ·5264 | ·5431 | ·5597 | ·5764 | **35** | ·5930 | ·6097 | ·6264 | ·6431 | ·6597 |
| ·3433 | ·3600 | ·3767 | ·3933 | ·4100 | **36** | ·4267 | ·4433 | ·4600 | ·4767 | ·4933 | **36** | ·5100 | ·5267 | ·5433 | ·5600 | ·5767 | **36** | ·5933 | ·6100 | ·6267 | ·6433 | ·6600 |
| ·3436 | ·3603 | ·3769 | ·3936 | ·4103 | **37** | ·4269 | ·4436 | ·4603 | ·4769 | ·4936 | **37** | ·5103 | ·5269 | ·5436 | ·5603 | ·5769 | **37** | ·5936 | ·6103 | ·6269 | ·6436 | ·6608 |
| ·3439 | ·3605 | ·3772 | ·3939 | ·4105 | **38** | ·4272 | ·4439 | ·4605 | ·4772 | ·4939 | **38** | ·5105 | ·5272 | ·5439 | ·5605 | ·5772 | **38** | ·5939 | ·6105 | ·6272 | ·6439 | ·6605 |
| ·3442 | ·3608 | ·3775 | ·3942 | ·4108 | **39** | ·4275 | ·4442 | ·4608 | ·4775 | ·4942 | **39** | ·5108 | ·5275 | ·5442 | ·5608 | ·5775 | **39** | ·5942 | ·6108 | ·6275 | ·6442 | ·6608 |
| ·3444 | ·3611 | ·3778 | ·3944 | ·4111 | **40** | ·4278 | ·4444 | ·4611 | ·4778 | ·4944 | **40** | ·5111 | ·5278 | ·5444 | ·5611 | ·5778 | **40** | ·5944 | ·6111 | ·6278 | ·6444 | ·6611 |
| ·3447 | ·3614 | ·3780 | ·3947 | ·4114 | **41** | ·4280 | ·4447 | ·4614 | ·4780 | ·4947 | **41** | ·5114 | ·5280 | ·5447 | ·5614 | ·5780 | **41** | ·5947 | ·6114 | ·6280 | ·6447 | ·6614 |
| ·3450 | ·3617 | ·3783 | ·3950 | ·4117 | **42** | ·4283 | ·4450 | ·4617 | ·4783 | ·4950 | **42** | ·5117 | ·5283 | ·5450 | ·5617 | ·5783 | **42** | ·5950 | ·6117 | ·6283 | ·6450 | ·6617 |
| ·3453 | ·3619 | ·3786 | ·3953 | ·4119 | **43** | ·4286 | ·4453 | ·4619 | ·4786 | ·4953 | **43** | ·5119 | ·5286 | ·5453 | ·5619 | ·5786 | **43** | ·5953 | ·6119 | ·6286 | ·6453 | ·6620 |
| ·3455 | ·3622 | ·3789 | ·3955 | ·4122 | **44** | ·4289 | ·4455 | ·4622 | ·4789 | ·4955 | **44** | ·5122 | ·5289 | ·5455 | ·5622 | ·5789 | **44** | ·5955 | ·6122 | ·6289 | ·6455 | ·6622 |
| ·3458 | ·3625 | ·3792 | ·3958 | ·4125 | **45** | ·4292 | ·4458 | ·4625 | ·4792 | ·4958 | **45** | ·5125 | ·5292 | ·5458 | ·5625 | ·5792 | **45** | ·5958 | ·6125 | ·6292 | ·6458 | ·6625 |
| ·3461 | ·3628 | ·3794 | ·3961 | ·4128 | **46** | ·4294 | ·4461 | ·4628 | ·4794 | ·4961 | **46** | ·5128 | ·5294 | ·5461 | ·5628 | ·5794 | **46** | ·5961 | ·6128 | ·6294 | ·6461 | ·6628 |
| ·3464 | ·3630 | ·3797 | ·3964 | ·4130 | **47** | ·4297 | ·4464 | ·4630 | ·4797 | ·4964 | **47** | ·5130 | ·5297 | ·5464 | ·5630 | ·5797 | **47** | ·5964 | ·6130 | ·6297 | ·6464 | ·6630 |
| ·3467 | ·3633 | ·3800 | ·3967 | ·4133 | **48** | ·4300 | ·4467 | ·4633 | ·4800 | ·4967 | **48** | ·5133 | ·5300 | ·5467 | ·5633 | ·5800 | **48** | ·5967 | ·6133 | ·6300 | ·6467 | ·6633 |
| ·3469 | ·3636 | ·3803 | ·3969 | ·4136 | **49** | ·4303 | ·4469 | ·4636 | ·4803 | ·4969 | **49** | ·5136 | ·5303 | ·5469 | ·5636 | ·5803 | **49** | ·5969 | ·6136 | ·6303 | ·6469 | ·6636 |
| ·3472 | ·3639 | ·3805 | ·3972 | ·4139 | **50** | ·4305 | ·4472 | ·4639 | ·4805 | ·4972 | **50** | ·5139 | ·5305 | ·5472 | ·5639 | ·5805 | **50** | ·5972 | ·6139 | ·6305 | ·6472 | ·6639 |
| ·3475 | ·3642 | ·3808 | ·3975 | ·4142 | **51** | ·4308 | ·4475 | ·4642 | ·4808 | ·4975 | **51** | ·5142 | ·5308 | ·5475 | ·5642 | ·5808 | **51** | ·5975 | ·6142 | ·6308 | ·6475 | ·6642 |
| ·3478 | ·3644 | ·3811 | ·3978 | ·4144 | **52** | ·4311 | ·4478 | ·4644 | ·4811 | ·4978 | **52** | ·5144 | ·5311 | ·5478 | ·5644 | ·5811 | **52** | ·5978 | ·6144 | ·6311 | ·6478 | ·6644 |
| ·3480 | ·3647 | ·3814 | ·3980 | ·4147 | **53** | ·4314 | ·4480 | ·4647 | ·4814 | ·4980 | **53** | ·5147 | ·5314 | ·5480 | ·5647 | ·5814 | **53** | ·5980 | ·6147 | ·6314 | ·6480 | ·6647 |
| ·3483 | ·3650 | ·3817 | ·3983 | ·4150 | **54** | ·4317 | ·4483 | ·4650 | ·4817 | ·4983 | **54** | ·5150 | ·5317 | ·5483 | ·5650 | ·5817 | **54** | ·5983 | ·6150 | ·6317 | ·6483 | ·6650 |
| ·3486 | ·3653 | ·3819 | ·3986 | ·4153 | **55** | ·4319 | ·4486 | ·4653 | ·4819 | ·4986 | **55** | ·5153 | ·5319 | ·5486 | ·5653 | ·5819 | **55** | ·5986 | ·6153 | ·6319 | ·6486 | ·6653 |
| ·3489 | ·3655 | ·3822 | ·3989 | ·4155 | **56** | ·4322 | ·4489 | ·4655 | ·4822 | ·4989 | **56** | ·5155 | ·5322 | ·5489 | ·5655 | ·5822 | **56** | ·5989 | ·6155 | ·6322 | ·6489 | ·6655 |
| ·3492 | ·3658 | ·3825 | ·3992 | ·4158 | **57** | ·4325 | ·4492 | ·4658 | ·4825 | ·4992 | **57** | ·5158 | ·5325 | ·5492 | ·5658 | ·5825 | **57** | ·5992 | ·6158 | ·6325 | ·6492 | ·6658 |
| ·3494 | ·3661 | ·3828 | ·3994 | ·4161 | **58** | ·4328 | ·4494 | ·4661 | ·4828 | ·4994 | **58** | ·5161 | ·5328 | ·5494 | ·5661 | ·5828 | **58** | ·5994 | ·6161 | ·6328 | ·6494 | ·6661 |
| ·3497 | ·3664 | ·3830 | ·3997 | ·4164 | **59** | ·4330 | ·4497 | ·4664 | ·4830 | ·4997 | **59** | ·5164 | ·5330 | ·5497 | ·5664 | ·5830 | **59** | ·5997 | ·6164 | ·6330 | ·6497 | ·6664 |

## TABLE XIX:—Ghatikas and Palas (Naligais and Vinadis) as Fractions of a day.—*contd.*

**N.B.**—The same table will serve for minutes and seconds as fractions of an **Hour** or a **Degree**.

| Palas. | Ghatikas. | | | | | | | | | | | | | | | | | | | |
|---|---|---|---|---|---|---|---|---|---|---|---|---|---|---|---|---|---|---|---|---|
| | **40** | **41** | **42** | **43** | **44** | **45** | **46** | **47** | **48** | **49** | **50** | **51** | **52** | **53** | **54** | **55** | **56** | **57** | **58** | **59** |
| **0** | ·6667 | ·6833 | ·7000 | ·7167 | ·7333 | ·7500 | ·7667 | ·7833 | ·8000 | ·8167 | ·8333 | ·8500 | ·8667 | ·8833 | ·9000 | ·9167 | ·9333 | ·9500 | ·9667 | ·9833 |
| **1** | ·6669 | ·6836 | ·7003 | ·7169 | ·7336 | ·7503 | ·7669 | ·7836 | ·8003 | ·8169 | ·8336 | ·8503 | ·8669 | ·8836 | ·9003 | ·9169 | ·9336 | ·9503 | ·9669 | ·9836 |
| **2** | ·6672 | ·6839 | ·7006 | ·7172 | ·7339 | ·7505 | ·7672 | ·7839 | ·8005 | ·8172 | ·8339 | ·8505 | ·8672 | ·8839 | ·9005 | ·9172 | ·9339 | ·9505 | ·9672 | ·9839 |
| **3** | ·6675 | ·6842 | ·7008 | ·7175 | ·7342 | ·7508 | ·7675 | ·7842 | ·8008 | ·8175 | ·8342 | ·8508 | ·8675 | ·8842 | ·9008 | ·9175 | ·9342 | ·9508 | ·9675 | ·9842 |
| **4** | ·6678 | ·6844 | ·7011 | ·7178 | ·7344 | ·7511 | ·7678 | ·7844 | ·8011 | ·8178 | ·8344 | ·8511 | ·8678 | ·8844 | ·9011 | ·9178 | ·9344 | ·9511 | ·9678 | ·9844 |
| **5** | ·6680 | ·6847 | ·7014 | ·7180 | ·7347 | ·7514 | ·7680 | ·7847 | ·8014 | ·8180 | ·8347 | ·8514 | ·8680 | ·8847 | ·9014 | ·9180 | ·9347 | ·9514 | ·9680 | ·9847 |
| **6** | ·6683 | ·6850 | ·7017 | ·7183 | ·7350 | ·7517 | ·7683 | ·7850 | ·8017 | ·8183 | ·8350 | ·8517 | ·8683 | ·8850 | ·9017 | ·9183 | ·9350 | ·9517 | ·9683 | ·9850 |
| **7** | ·6686 | ·6853 | ·7019 | ·7186 | ·7353 | ·7519 | ·7686 | ·7853 | ·8019 | ·8186 | ·8353 | ·8519 | ·8686 | ·8853 | ·9019 | ·9186 | ·9353 | ·9519 | ·9686 | ·9853 |
| **8** | ·6689 | ·6855 | ·7022 | ·7189 | ·7355 | ·7522 | ·7689 | ·7855 | ·8022 | ·8189 | ·8355 | ·8522 | ·8689 | ·8855 | ·9022 | ·9189 | ·9355 | ·9522 | ·9689 | ·9855 |
| **9** | ·6692 | ·6858 | ·7025 | ·7192 | ·7358 | ·7525 | ·7692 | ·7858 | ·8025 | ·8192 | ·8358 | ·8525 | ·8692 | ·8858 | ·9025 | ·9192 | ·9358 | ·9525 | ·9692 | ·9858 |
| **10** | ·6694 | ·6861 | ·7028 | ·7194 | ·7361 | ·7528 | ·7694 | ·7861 | ·8028 | ·8194 | ·8361 | ·8528 | ·8694 | ·8861 | ·9028 | ·9194 | ·9361 | ·9528 | ·9694 | ·9861 |
| **11** | ·6697 | ·6864 | ·7030 | ·7197 | ·7364 | ·7530 | ·7697 | ·7864 | ·8030 | ·8197 | ·8364 | ·8530 | ·8697 | ·8864 | ·9030 | ·9197 | ·9364 | ·9530 | ·9697 | ·9864 |
| **12** | ·6700 | ·6867 | ·7033 | ·7200 | ·7367 | ·7533 | ·7700 | ·7867 | ·8033 | ·8200 | ·8367 | ·8533 | ·8700 | ·8867 | ·9033 | ·9200 | ·9367 | ·9533 | ·9700 | ·9867 |
| **13** | ·6703 | ·6869 | ·7036 | ·7203 | ·7369 | ·7536 | ·7703 | ·7869 | ·8036 | ·8203 | ·8369 | ·8536 | ·8703 | ·8869 | ·9036 | ·9203 | ·9369 | ·9536 | ·9703 | ·9869 |
| **14** | ·6705 | ·6872 | ·7039 | ·7205 | ·7372 | ·7539 | ·7705 | ·7872 | ·8039 | ·8205 | ·8372 | ·8539 | ·8705 | ·8872 | ·9039 | ·9205 | ·9372 | ·9539 | ·9705 | ·9872 |
| **15** | ·6708 | ·6875 | ·7042 | ·7208 | ·7375 | ·7542 | ·7708 | ·7875 | ·8042 | ·8208 | ·8375 | ·8542 | ·8708 | ·8875 | ·9042 | ·9208 | ·9375 | ·9542 | ·9708 | ·9875 |
| **16** | ·6711 | ·6878 | ·7044 | ·7211 | ·7378 | ·7544 | ·7711 | ·7878 | ·8044 | ·8211 | ·8378 | ·8544 | ·8711 | ·8878 | ·9044 | ·9211 | ·9378 | ·9544 | ·9711 | ·9878 |
| **17** | ·6714 | ·6880 | ·7047 | ·7214 | ·7380 | ·7547 | ·7714 | ·7880 | ·8047 | ·8214 | ·8380 | ·8547 | ·8714 | ·8880 | ·9047 | ·9214 | ·9380 | ·9547 | ·9714 | ·9880 |
| **18** | ·6717 | ·6883 | ·7050 | ·7217 | ·7383 | ·7550 | ·7717 | ·7883 | ·8050 | ·8217 | ·8383 | ·8550 | ·8717 | ·8883 | ·9050 | ·9217 | ·9383 | ·9550 | ·9717 | ·9883 |
| **19** | ·6719 | ·6886 | ·7053 | ·7219 | ·7386 | ·7553 | ·7719 | ·7886 | ·8053 | ·8219 | ·8386 | ·8553 | ·8719 | ·8886 | ·9053 | ·9219 | ·9386 | ·9553 | ·9719 | ·9886 |
| **20** | ·6722 | ·6889 | ·7055 | ·7222 | ·7389 | ·7555 | ·7722 | ·7889 | ·8055 | ·8222 | ·8389 | ·8555 | ·8722 | ·8889 | ·9055 | ·9222 | ·9389 | ·9555 | ·9722 | ·9889 |
| **21** | ·6725 | ·6892 | ·7058 | ·7225 | ·7392 | ·7558 | ·7725 | ·7892 | ·8058 | ·8225 | ·8392 | ·8558 | ·8725 | ·8892 | ·9058 | ·9225 | ·9392 | ·9558 | ·9725 | ·9892 |
| **22** | ·6728 | ·6894 | ·7061 | ·7228 | ·7394 | ·7561 | ·7728 | ·7894 | ·8061 | ·8228 | ·8394 | ·8561 | ·8728 | ·8894 | ·9061 | ·9228 | ·9394 | ·9561 | ·9728 | ·9894 |
| **23** | ·6730 | ·6897 | ·7064 | ·7230 | ·7397 | ·7564 | ·7730 | ·7897 | ·8064 | ·8230 | ·8397 | ·8564 | ·8730 | ·8897 | ·9064 | ·9230 | ·9397 | ·9564 | ·9730 | ·9897 |
| **24** | ·6733 | ·6900 | ·7067 | ·7233 | ·7400 | ·7567 | ·7733 | ·7900 | ·8067 | ·8233 | ·8400 | ·8567 | ·8733 | ·8900 | ·9067 | ·9233 | ·9400 | ·9567 | ·9733 | ·9900 |
| **25** | ·6736 | ·6903 | ·7069 | ·7236 | ·7403 | ·7569 | ·7736 | ·7903 | ·8069 | ·8236 | ·8403 | ·8569 | 8736 | ·8903 | ·9069 | ·9236 | ·9403 | ·9569 | ·9736 | ·9903 |
| **26** | ·6739 | ·6905 | ·7072 | ·7239 | ·7405 | ·7572 | ·7739 | ·7905 | ·8072 | ·8239 | ·8405 | ·8572 | ·8739 | ·8905 | ·9072 | ·9239 | ·9405 | ·9572 | ·9739 | ·9905 |
| **27** | ·6742 | ·6908 | ·7075 | ·7242 | ·7408 | ·7575 | ·7742 | ·7908 | ·8075 | ·8242 | ·8408 | ·8575 | ·8742 | ·8908 | ·9075 | ·9242 | ·9408 | ·9575 | ·9742 | ·9908 |
| **28** | ·6744 | ·6911 | ·7078 | ·7244 | ·7411 | ·7578 | ·7744 | ·7911 | ·8078 | ·8244 | ·8411 | ·8578 | ·8744 | ·8911 | ·9078 | ·9244 | ·9411 | ·9578 | ·9744 | ·9911 |
| **29** | ·6747 | ·6914 | ·7080 | ·7247 | ·7414 | ·7580 | ·7747 | ·7914 | ·8080 | ·8247 | ·8414 | ·8580 | ·8747 | ·8914 | ·9080 | ·9247 | ·9414 | ·9580 | ·9747 | ·9914 |
| **30** | ·6750 | ·6917 | ·7083 | ·7250 | ·7417 | ·7583 | ·7750 | ·7917 | ·8083 | ·8250 | ·8417 | ·8583 | ·8750 | ·8917 | ·9083 | ·9250 | ·9417 | ·9583 | ·9750 | ·9917 |
| **31** | ·6753 | ·6919 | ·7086 | ·7253 | ·7419 | ·7586 | ·7753 | ·7919 | ·8086 | ·8253 | ·8419 | ·8586 | ·8753 | ·8919 | ·9086 | ·9253 | ·9419 | ·9586 | ·9753 | ·9919 |
| **32** | ·6755 | ·6922 | ·7089 | ·7255 | ·7422 | ·7589 | ·7755 | ·7922 | ·8089 | ·8255 | ·8422 | ·8589 | ·8755 | ·8922 | ·9089 | ·9255 | ·9422 | ·9589 | ·9755 | ·9922 |
| **33** | ·6758 | ·6925 | ·7092 | ·7258 | ·7425 | ·7592 | ·7758 | ·7925 | ·8092 | ·8258 | ·8425 | ·8592 | ·8758 | ·8925 | ·9092 | ·9258 | ·9425 | ·9592 | ·9758 | ·9925 |
| **34** | ·6761 | ·6928 | ·7094 | ·7261 | ·7428 | ·7594 | ·7761 | ·7928 | ·8094 | ·8261 | ·8428 | ·8594 | ·8761 | ·8928 | ·9094 | ·9261 | ·9428 | ·9594 | ·9761 | ·9928 |
| **35** | ·6764 | ·6930 | ·7097 | ·7264 | ·7430 | ·7597 | ·7764 | ·7930 | ·8097 | ·8264 | ·8430 | ·8597 | ·8764 | ·8930 | ·9097 | ·9264 | ·9430 | ·9597 | ·9764 | ·9930 |
| **36** | ·6767 | ·6933 | ·7100 | ·7267 | ·7433 | ·7600 | ·7767 | ·7933 | ·8100 | ·8267 | ·8433 | ·8600 | ·8767 | ·8933 | ·9100 | ·9267 | ·9433 | ·9600 | ·9767 | ·9933 |
| **37** | ·6769 | ·6936 | ·7103 | ·7269 | ·7436 | ·7603 | ·7769 | ·7936 | ·8103 | ·8269 | ·8436 | ·8603 | ·8769 | ·8936 | ·9103 | ·9269 | ·9436 | ·9603 | ·9769 | ·9936 |
| **38** | ·6772 | ·6939 | ·7105 | ·7272 | ·7439 | ·7605 | ·7772 | ·7939 | ·8105 | ·8272 | ·8439 | ·8605 | ·8772 | ·8939 | ·9105 | ·9272 | ·9439 | ·9605 | ·9772 | ·9939 |
| **39** | ·6775 | ·6942 | ·7108 | ·7275 | ·7442 | ·7608 | ·7775 | ·7942 | ·8108 | ·8275 | ·8442 | ·8608 | ·8775 | ·8942 | ·9108 | ·9275 | ·9442 | ·9608 | ·9775 | ·9942 |
| **40** | ·6778 | ·6944 | ·7111 | ·7278 | ·7444 | ·7611 | ·7778 | ·7944 | ·8111 | ·8278 | ·8444 | ·8611 | ·8778 | ·8944 | ·9111 | ·9278 | ·9444 | ·9611 | ·9778 | ·9944 |
| **41** | ·6780 | ·6947 | ·7114 | ·7280 | ·7447 | ·7614 | ·7780 | ·7947 | ·8114 | ·8280 | ·8447 | ·8614 | ·8780 | ·8947 | ·9114 | ·9280 | ·9447 | ·9614 | ·9780 | ·9947 |
| **42** | ·6783 | ·6950 | ·7117 | ·7283 | ·7450 | ·7617 | ·7783 | ·7950 | ·8117 | ·8283 | ·8450 | ·8617 | ·8783 | ·8950 | ·9117 | ·9283 | ·9450 | ·9617 | ·9783 | ·9950 |
| **43** | ·6786 | ·6953 | ·7119 | ·7286 | ·7453 | ·7619 | ·7786 | ·7953 | ·8119 | ·8286 | ·8453 | ·8619 | ·8786 | ·8953 | ·9119 | ·9286 | ·9453 | ·9619 | ·9786 | ·9953 |
| **44** | ·6789 | ·6956 | ·7122 | ·7289 | ·7455 | ·7622 | ·7789 | ·7955 | ·8122 | ·8289 | ·8455 | ·8622 | ·8789 | ·8955 | ·9122 | ·9289 | ·9455 | ·9622 | ·9789 | ·9955 |
| **45** | ·6792 | ·6958 | ·7125 | ·7292 | ·7458 | ·7625 | ·7792 | ·7958 | ·8125 | ·8292 | ·8458 | ·8625 | ·8792 | ·8958 | ·9125 | ·9292 | ·9458 | ·9625 | ·9792 | ·9958 |
| **46** | ·6794 | ·6961 | ·7128 | ·7294 | ·7461 | ·7628 | ·7794 | ·7961 | ·8128 | ·8294 | ·8461 | ·8628 | ·8794 | ·8961 | ·9128 | ·9294 | ·9461 | ·9628 | ·9794 | ·9961 |
| **47** | ·6797 | ·6964 | ·7130 | ·7297 | ·7464 | ·7630 | ·7797 | ·7964 | ·8130 | ·8297 | ·8464 | ·8630 | ·8797 | ·8964 | ·9130 | ·9297 | ·9464 | ·9630 | ·9797 | ·9964 |
| **48** | ·6800 | ·6967 | ·7133 | ·7300 | ·7467 | ·7633 | ·7800 | ·7967 | ·8133 | ·8300 | ·8467 | ·8633 | ·8800 | ·8967 | ·9133 | ·9300 | ·9467 | ·9633 | ·9800 | ·9967 |
| **49** | ·6803 | ·6969 | ·7136 | ·7303 | ·7469 | ·7636 | ·7803 | ·7969 | ·8136 | ·8303 | ·8469 | ·8636 | ·8803 | ·8969 | ·9136 | ·9303 | ·9469 | ·9636 | ·9803 | ·9969 |
| **50** | ·6805 | ·6972 | ·7139 | ·7305 | ·7472 | ·7639 | ·7805 | ·7972 | ·8139 | ·8305 | ·8472 | ·8639 | ·8805 | ·8972 | ·9139 | ·9305 | ·9472 | ·9639 | ·9805 | ·9972 |
| **51** | ·6808 | ·6975 | ·7142 | ·7308 | ·7475 | ·7642 | ·7808 | ·7975 | ·8142 | ·8308 | ·8475 | ·8642 | ·8808 | ·8975 | ·9142 | ·9308 | ·9475 | ·9642 | ·9808 | ·9975 |
| **52** | ·6811 | ·6978 | ·7144 | ·7311 | ·7478 | ·7644 | ·7811 | ·7978 | ·8144 | ·8311 | ·8478 | ·8644 | ·8811 | ·8978 | ·9144 | ·9311 | ·9478 | ·9644 | ·9811 | ·9978 |
| **53** | ·6814 | ·6980 | ·7147 | ·7314 | ·7480 | ·7647 | ·7814 | ·7980 | ·8147 | ·8314 | ·8480 | ·8647 | ·8814 | ·8980 | ·9147 | ·9314 | ·9480 | ·9647 | ·9814 | ·9980 |
| **54** | ·6817 | ·6983 | ·7150 | ·7317 | ·7483 | ·7650 | ·7817 | ·7983 | ·8150 | ·8317 | ·8483 | ·8650 | ·8817 | ·8983 | ·9150 | ·9317 | ·9483 | ·9650 | ·9817 | ·9983 |
| **55** | ·6819 | ·6986 | ·7153 | ·7319 | ·7486 | ·7653 | ·7819 | ·7986 | ·8153 | ·8319 | ·8486 | ·8653 | ·8819 | ·8986 | ·9153 | ·9319 | ·9486 | ·9653 | ·9819 | ·9986 |
| **56** | ·6822 | ·6989 | ·7155 | ·7322 | ·7489 | ·7655 | ·7822 | ·7989 | ·8155 | ·8322 | ·8489 | ·8655 | ·8822 | ·8989 | ·9155 | ·9322 | ·9489 | ·9655 | ·9822 | ·9989 |
| **57** | ·6825 | ·6992 | ·7158 | ·7325 | ·7492 | ·7658 | ·7825 | ·7992 | ·8158 | ·8325 | ·8492 | ·8658 | ·8825 | ·8992 | ·9158 | ·9325 | ·9492 | ·9658 | ·9825 | ·9992 |
| **58** | ·6828 | ·6994 | ·7161 | ·7328 | ·7494 | ·7661 | ·7828 | ·7994 | ·8161 | ·8328 | ·8494 | ·8661 | ·8828 | ·8994 | ·9161 | ·9328 | ·9494 | ·9661 | ·9828 | ·9994 |
| **59** | ·6830 | ·6997 | ·7164 | ·7330 | ·7497 | ·7664 | ·7830 | ·7997 | ·8164 | ·8330 | ·8497 | ·8664 | ·8830 | ·8997 | ·9163 | ·9330 | ·9497 | ·9664 | ·9830 | ·9997 |

# TABLE XX:—Hours and Minutes as Fractions of a Day.

| Minutes. | Hours. 0 | 1 | 2 | 3 | Minutes. | Hours. 4 | 5 | 6 | 7 | Minutes. | Hours. 8 | 9 | 10 | 11 | Minutes. | Seconds of hours as fractions of a day. |
|---|---|---|---|---|---|---|---|---|---|---|---|---|---|---|---|---|
| 0 | ... | ·0416666 | ·0833333 | ·1250000 | 0 | ·1666666 | ·2083333 | ·2500000 | ·2916666 | 0 | ·3333333 | ·3750000 | ·4166666 | ·4583333 | 0 | ... |
| 1 | ·0006944 | ·0423611 | ·0840277 | ·1256944 | 1 | ·1673611 | ·2090277 | ·2506944 | ·2923611 | 1 | ·3340277 | ·3756944 | ·4173611 | ·4590277 | 1 | ·00001 |
| 2 | ·0013888 | ·0430555 | ·0847222 | ·1263888 | 2 | ·1680555 | ·2097222 | ·2513888 | ·2930555 | 2 | ·3347222 | ·3763888 | ·4180555 | ·4597222 | 2 | ·00002 |
| 3 | ·0020833 | ·0437500 | ·0854166 | ·1270833 | 3 | ·1687500 | ·2104166 | ·2520833 | ·2937500 | 3 | ·3354166 | ·3770833 | ·4187500 | ·4604166 | 3 | ·00003 |
| 4 | ·0027777 | ·0444444 | ·0861111 | ·1277777 | 4 | ·1694444 | ·2111111 | ·2527777 | ·2944444 | 4 | ·3361111 | ·3777777 | ·4194444 | ·4611111 | 4 | ·00005 |
| 5 | ·0034722 | ·0451388 | ·0868055 | ·1284722 | 5 | ·1701388 | ·2118055 | ·2534722 | ·2951388 | 5 | ·3368055 | ·3784722 | ·4201388 | ·4618055 | 5 | ·00006 |
| 6 | ·0041666 | ·0458333 | ·0875000 | ·1291666 | 6 | ·1708333 | ·2125000 | ·2541666 | ·2958333 | 6 | ·3375000 | ·3791666 | ·4208333 | ·4625000 | 6 | ·00007 |
| 7 | ·0048611 | ·0465277 | ·0881944 | ·1298611 | 7 | ·1715277 | ·2131944 | ·2548611 | ·2965277 | 7 | ·3381944 | ·3798611 | ·4215277 | ·4631944 | 7 | ·00008 |
| 8 | ·0055555 | ·0472222 | ·0888888 | ·1305555 | 8 | ·1722222 | ·2138888 | ·2555555 | ·2972222 | 8 | ·3388888 | ·3805555 | ·4222222 | ·4638888 | 8 | ·00009 |
| 9 | ·0062500 | ·0479166 | ·0895833 | ·1312500 | 9 | ·1729166 | ·2145833 | ·2562500 | ·2979166 | 9 | ·3395833 | ·3812500 | ·4229166 | ·4645833 | 9 | ·00010 |
| 10 | ·0069444 | ·0486111 | ·0902777 | ·1319444 | 10 | ·1736111 | ·2152777 | ·2569444 | ·2986111 | 10 | ·3402777 | ·3819444 | ·4236111 | ·4652777 | 10 | ·00012 |
| 11 | ·0076388 | ·0493055 | ·0909722 | ·1326388 | 11 | ·1743055 | ·2159722 | ·2576388 | ·2993055 | 11 | ·3409722 | ·3826388 | ·4243055 | ·4659722 | 11 | ·00013 |
| 12 | ·0083333 | ·0500000 | ·0916666 | ·1333333 | 12 | ·1750000 | ·2166666 | ·2583333 | ·3000000 | 12 | ·3416666 | ·3833333 | ·4250000 | ·4666666 | 12 | ·00014 |
| 13 | ·0090277 | ·0506944 | ·0923611 | ·1340277 | 13 | ·1756944 | ·2173611 | ·2590277 | ·3006944 | 13 | ·3423611 | ·3840277 | ·4256944 | ·4673611 | 13 | ·00015 |
| 14 | ·0097222 | ·0513888 | ·0930555 | ·1347222 | 14 | ·1763888 | ·2180555 | ·2597222 | ·3013888 | 14 | ·3430555 | ·3847222 | ·4263888 | ·4680555 | 14 | ·00016 |
| 15 | ·0104166 | ·0520833 | ·0937500 | ·1354166 | 15 | ·1770833 | ·2187500 | ·2604166 | ·3020833 | 15 | ·3437500 | ·3854166 | ·4270833 | ·4687500 | 15 | ·00017 |
| 16 | ·0111111 | ·0527777 | ·0944444 | ·1361111 | 16 | ·1777777 | ·2194444 | ·2611111 | ·3027777 | 16 | ·3444444 | ·3861111 | ·4277777 | ·4694444 | 16 | ·00018 |
| 17 | ·0118055 | ·0534722 | ·0951388 | ·1368055 | 17 | ·1784722 | ·2201388 | ·2618055 | ·3034722 | 17 | ·3451388 | ·3868055 | ·4284722 | ·4701388 | 17 | ·00020 |
| 18 | ·0125000 | ·0541666 | ·0958333 | ·1375000 | 18 | ·1791666 | ·2208333 | ·2625000 | ·3041666 | 18 | ·3458333 | ·3875000 | ·4291666 | ·4708333 | 18 | ·00021 |
| 19 | ·0131944 | ·0548611 | ·0965277 | ·1381944 | 19 | ·1798611 | ·2215277 | ·2631944 | ·3048611 | 19 | ·3465277 | ·3881944 | ·4298611 | ·4715277 | 19 | ·00022 |
| 20 | ·0138888 | ·0555555 | ·0972222 | ·1388888 | 20 | ·1805555 | ·2222222 | ·2638888 | ·3055555 | 20 | ·3472222 | ·3888888 | ·4305555 | ·4722222 | 20 | ·00023 |
| 21 | ·0145833 | ·0562500 | ·0979166 | ·1395833 | 21 | ·1812500 | ·2229166 | ·2645833 | ·3062500 | 21 | ·3479166 | ·3895833 | ·4312500 | ·4729166 | 21 | ·00024 |
| 22 | ·0152777 | ·0569444 | ·0986111 | ·1402777 | 22 | ·1819444 | ·2236111 | ·2652777 | ·3069444 | 22 | ·3486111 | ·3902777 | ·4319444 | ·4736111 | 22 | ·00025 |
| 23 | ·0159722 | ·0576388 | ·0993055 | ·1409722 | 23 | ·1826388 | ·2243055 | ·2659722 | ·3076388 | 23 | ·3493055 | ·3909722 | ·4326388 | ·4743055 | 23 | ·00027 |
| 24 | ·0166666 | ·0583333 | ·1000000 | ·1416666 | 24 | ·1833333 | ·2250000 | ·2666666 | ·3083333 | 24 | ·3500000 | ·3916666 | ·4333333 | ·4750000 | 24 | ·00028 |
| 25 | ·0173611 | ·0590277 | ·1006944 | ·1423611 | 25 | ·1840277 | ·2256944 | ·2673611 | ·3090277 | 25 | ·3506944 | ·3923611 | ·4340277 | ·4756944 | 25 | ·00029 |
| 26 | ·0180555 | ·0597222 | ·1013888 | ·1430555 | 26 | ·1847222 | ·2263888 | ·2680555 | ·3097222 | 26 | ·3513888 | ·3930555 | ·4347222 | ·4763888 | 26 | ·00030 |
| 27 | ·0187500 | ·0604166 | ·1020833 | ·1437500 | 27 | ·1854166 | ·2270833 | ·2687500 | ·3104166 | 27 | ·3520833 | ·3937500 | ·4354166 | ·4770833 | 27 | ·00031 |
| 28 | ·0194444 | ·0611111 | ·1027777 | ·1444444 | 28 | ·1861111 | ·2277777 | ·2694444 | ·3111111 | 28 | ·3527777 | ·3944444 | ·4361111 | ·4777777 | 28 | ·00032 |
| 29 | ·0201388 | ·0618055 | ·1034722 | ·1451388 | 29 | ·1868055 | ·2284722 | ·2701388 | ·3118055 | 29 | ·3534722 | ·3951388 | ·4368055 | ·4784722 | 29 | ·00034 |
| 30 | ·0208333 | ·0625000 | ·1041666 | ·1458333 | 30 | ·1875000 | ·2291666 | ·2708333 | ·3125000 | 30 | ·3541666 | ·3958333 | ·4375000 | ·4791666 | 30 | ·00035 |
| 31 | ·0215277 | ·0631944 | ·1048611 | ·1465277 | 31 | ·1881944 | ·2298611 | ·2715277 | ·3131944 | 31 | ·3548611 | ·3965277 | ·4381944 | ·4798611 | 31 | ·00036 |
| 32 | ·0222222 | ·0638888 | ·1055555 | ·1472222 | 32 | ·1888888 | ·2305555 | ·2722222 | ·3138888 | 32 | ·3555555 | ·3972222 | ·4388888 | ·4805555 | 32 | ·00037 |
| 33 | ·0229166 | ·0645833 | ·1062500 | ·1479166 | 33 | ·1895833 | ·2312500 | ·2729166 | ·3145833 | 33 | ·3562500 | ·3979166 | ·4395833 | ·4812500 | 33 | ·00038 |
| 34 | ·0236111 | ·0652777 | ·1069444 | ·1486111 | 34 | ·1902777 | ·2319444 | ·2736111 | ·3152777 | 34 | ·3569444 | ·3986111 | ·4402777 | ·4819444 | 34 | ·00039 |
| 35 | ·0243055 | ·0659722 | ·1076388 | ·1493055 | 35 | ·1909722 | ·2326388 | ·2743055 | ·3159722 | 35 | ·3576388 | ·3993055 | ·4409722 | ·4826388 | 35 | ·00040 |
| 36 | ·0250000 | ·0666666 | ·1083333 | ·1500000 | 36 | ·1916666 | ·2333333 | ·2750000 | ·3166666 | 36 | ·3583333 | ·4000000 | ·4416666 | ·4833333 | 36 | ·00042 |
| 37 | ·0256944 | ·0673611 | ·1090277 | ·1506944 | 37 | ·1923611 | ·2340277 | ·2756944 | ·3173611 | 37 | ·3590277 | ·4006944 | ·4423611 | ·4840277 | 37 | ·00043 |
| 38 | ·0263888 | ·0680555 | ·1097222 | ·1513888 | 38 | ·1930555 | ·2347222 | ·2763888 | ·3180555 | 38 | ·3597222 | ·4013888 | ·4430555 | ·4847222 | 38 | ·00044 |
| 39 | ·0270833 | ·0687500 | ·1104166 | ·1520833 | 39 | ·1937500 | ·2354166 | ·2770833 | ·3187500 | 39 | ·3604166 | ·4020833 | ·4437500 | ·4854166 | 39 | ·00045 |
| 40 | ·0277777 | ·0694444 | ·1111111 | ·1527777 | 40 | ·1944444 | ·2361111 | ·2777777 | ·3194444 | 40 | ·3611111 | ·4027777 | ·4444444 | ·4861111 | 40 | ·00046 |
| 41 | ·0284722 | ·0701388 | ·1118055 | ·1534722 | 41 | ·1951388 | ·2368055 | ·2784722 | ·3201388 | 41 | ·3618055 | ·4034722 | ·4451388 | ·4868055 | 41 | ·00047 |
| 42 | ·0291666 | ·0708333 | ·1125000 | ·1541666 | 42 | ·1958333 | ·2375000 | ·2791666 | ·3208333 | 42 | ·3625000 | ·4041666 | ·4458333 | ·4875000 | 42 | ·00049 |
| 43 | ·0298611 | ·0715277 | ·1131944 | ·1548611 | 43 | ·1965277 | ·2381944 | ·2798611 | ·3215277 | 43 | ·3631944 | ·4048611 | ·4465277 | ·4881944 | 43 | ·00050 |
| 44 | ·0305555 | ·0722222 | ·1138888 | ·1555555 | 44 | ·1972222 | ·2388888 | ·2805555 | ·3222222 | 44 | ·3638888 | ·4055555 | ·4472222 | ·4888888 | 44 | ·00051 |
| 45 | ·0312500 | ·0729166 | ·1145833 | ·1562500 | 45 | ·1979166 | ·2395833 | ·2812500 | ·3229166 | 45 | ·3645833 | ·4062500 | ·4479166 | ·4895833 | 45 | ·00052 |
| 46 | ·0319444 | ·0736111 | ·1152777 | ·1569444 | 46 | ·1986111 | ·2402777 | ·2819444 | ·3236111 | 46 | ·3652777 | ·4069444 | ·4486111 | ·4902777 | 46 | ·00053 |
| 47 | ·0326388 | ·0743055 | ·1159722 | ·1576388 | 47 | ·1993055 | ·2409722 | ·2826388 | ·3243055 | 47 | ·3659722 | ·4076388 | ·4493055 | ·4909722 | 47 | ·00054 |
| 48 | ·0333333 | ·0750000 | ·1166666 | ·1583333 | 48 | ·2000000 | ·2416666 | ·2833333 | ·3250000 | 48 | ·3666666 | ·4083333 | ·4500000 | ·4916666 | 48 | ·00055 |
| 49 | ·0340277 | ·0756944 | ·1173611 | ·1590277 | 49 | ·2006944 | ·2423611 | ·2840277 | ·3256944 | 49 | ·3673611 | ·4090277 | ·4506944 | ·4923611 | 49 | ·00057 |
| 50 | ·0347222 | ·0763888 | ·1180555 | ·1597222 | 50 | ·2013888 | ·2430555 | ·2847222 | ·3263888 | 50 | ·3680555 | ·4097222 | ·4513888 | ·4930555 | 50 | ·00058 |
| 51 | ·0354166 | ·0770833 | ·1187500 | ·1604166 | 51 | ·2020833 | ·2437500 | ·2854166 | ·3270833 | 51 | ·3687500 | ·4104166 | ·4520833 | ·4937500 | 51 | ·00059 |
| 52 | ·0361111 | ·0777777 | ·1194444 | ·1611111 | 52 | ·2027777 | ·2444444 | ·2861111 | ·3277777 | 52 | ·3694444 | ·4111111 | ·4527777 | ·4944444 | 52 | ·00060 |
| 53 | ·0368055 | ·0784722 | ·1201388 | ·1618055 | 53 | ·2034722 | ·2451388 | ·2868055 | ·3284722 | 53 | ·3701388 | ·4118055 | ·4534722 | ·4951388 | 53 | ·00061 |
| 54 | ·0375000 | ·0791666 | ·1208333 | ·1625000 | 54 | ·2041666 | ·2458333 | ·2875000 | ·3291666 | 54 | ·3708333 | ·4125000 | ·4541666 | ·4958333 | 54 | ·00062 |
| 55 | ·0381944 | ·0798611 | ·1215277 | ·1631944 | 55 | ·2048611 | ·2465277 | ·2881944 | ·3298611 | 55 | ·3715277 | ·4131944 | ·4548611 | ·4965277 | 55 | ·00064 |
| 56 | ·0388888 | ·0805555 | ·1222222 | ·1638888 | 56 | ·2055555 | ·2472222 | ·2888888 | ·3305555 | 56 | ·3722222 | ·4138888 | ·4555555 | ·4972222 | 56 | ·00065 |
| 57 | ·0395833 | ·0812500 | ·1229166 | ·1645833 | 57 | ·2062500 | ·2479166 | ·2895833 | ·3312500 | 57 | ·3729166 | ·4145833 | ·4562500 | ·4979166 | 57 | ·00066 |
| 58 | ·0402777 | ·0819444 | ·1236111 | ·1652777 | 58 | ·2069444 | ·2486111 | ·2202777 | ·3319444 | 58 | ·3736111 | ·4152777 | ·4569144 | ·4986111 | 58 | ·00067 |
| 59 | ·0409722 | ·0826388 | ·1243055 | ·1659722 | 59 | ·2076388 | ·2493055 | ·2909722 | ·3326388 | 59 | ·3743055 | ·4159722 | ·4576388 | ·4993055 | 59 | ·00068 |

## TABLE XX:— Hours and Minutes as Fractions of a Day—*contd.*

| Minutes. | Hours. 12 | 13 | 14 | 15 | Minutes. | Hours. 16 | 17 | 18 | 19 | Minutes. | Hours. 20 | 21 | 22 | 23 | Minutes. | Seconds of hours as fractions of a day. |
|---|---|---|---|---|---|---|---|---|---|---|---|---|---|---|---|---|
| 0 | ·5000000 | ·5416666 | ·5833333 | ·6250000 | 0 | ·6666666 | ·7083333 | ·7500000 | ·7916666 | 0 | ·8333333 | ·8750000 | ·9166666 | ·9583333 | 0 | ... |
| 1 | ·5006944 | ·5423611 | ·5840277 | ·6256944 | 1 | ·6673611 | ·7090277 | ·7506944 | ·7923611 | 1 | ·8340277 | ·8756944 | ·9173611 | ·9590277 | 1 | ·00001 |
| 2 | ·5013888 | ·5430555 | ·5847222 | ·6263888 | 2 | ·6680555 | ·7097222 | ·7513888 | ·7930555 | 2 | ·8347222 | ·8763888 | ·9180555 | ·9597222 | 2 | ·00002 |
| 3 | ·5020833 | ·5437500 | ·5854166 | ·6270833 | 3 | ·6687500 | ·7104166 | ·7520833 | ·7937500 | 3 | ·8354166 | ·8770833 | ·9187500 | ·9604166 | 3 | ·00003 |
| 4 | ·5027777 | ·5444444 | ·5861111 | ·6277777 | 4 | ·6694444 | ·7111111 | ·7527777 | ·7944444 | 4 | ·8361111 | ·8777777 | ·9194444 | ·9611111 | 4 | ·00005 |
| 5 | ·5034722 | ·5451388 | ·5868055 | ·6284722 | 5 | ·6701388 | ·7118055 | ·7534722 | ·7951388 | 5 | ·8368055 | ·8784722 | ·9201388 | ·9618055 | 5 | ·00006 |
| 6 | ·5041666 | ·5458333 | ·5875000 | ·6291666 | 6 | ·6708333 | ·7125000 | ·7541666 | ·7958333 | 6 | ·8375000 | ·8791666 | ·9208333 | ·9625000 | 6 | ·00007 |
| 7 | ·5048611 | ·5465277 | ·5881944 | ·6298611 | 7 | ·6715277 | ·7131944 | ·7548611 | ·7965277 | 7 | ·8381944 | ·8798611 | ·9215277 | ·9631944 | 7 | ·00008 |
| 8 | ·5055555 | ·5472222 | ·5888888 | ·6305555 | 8 | ·6722222 | ·7138888 | ·7555555 | ·7972222 | 8 | ·8388888 | ·8805555 | ·9222222 | ·9638888 | 8 | ·00009 |
| 9 | ·5062500 | ·5479166 | ·5895833 | ·6312500 | 9 | ·6729166 | ·7145833 | ·7562500 | ·7979166 | 9 | ·8395833 | ·8812500 | ·9229166 | ·9645833 | 9 | ·00010 |
| 10 | ·5069444 | ·5486111 | ·5902777 | ·6319444 | 10 | ·6736111 | ·7152777 | ·7569444 | ·7986111 | 10 | ·8402777 | ·8819444 | ·9236111 | ·9652777 | 10 | ·00012 |
| 11 | ·5076388 | ·5493055 | ·5909722 | ·6326388 | 11 | ·6743055 | ·7159722 | ·7576388 | ·7993055 | 11 | ·8409722 | ·8826388 | ·9243055 | ·9659722 | 11 | ·00013 |
| 12 | ·5083333 | ·5500000 | ·5916666 | ·6333333 | 12 | ·6750000 | ·7166666 | ·7583333 | ·8000000 | 12 | ·8416666 | ·8833333 | ·9250000 | ·9666666 | 12 | ·00014 |
| 13 | ·5090277 | ·5506944 | ·5923611 | ·6340277 | 13 | ·6756944 | ·7173611 | ·7590277 | ·8006944 | 13 | ·8423611 | ·8840277 | ·9256944 | ·9673611 | 13 | ·00015 |
| 14 | ·5097222 | ·5513888 | ·5930555 | ·6347222 | 14 | ·6763888 | ·7180555 | ·7597222 | ·8013888 | 14 | ·8430555 | ·8847222 | ·9263888 | ·9680555 | 14 | ·00016 |
| 15 | ·5104166 | ·5520833 | ·5937500 | ·6354166 | 15 | ·6770833 | ·7187500 | ·7604166 | ·8020833 | 15 | ·8437500 | ·8854166 | ·9270833 | ·9687500 | 15 | ·00017 |
| 16 | ·5111111 | ·5527777 | ·5944444 | ·6361111 | 16 | ·6777777 | ·7194444 | ·7611111 | ·8027777 | 16 | ·8444444 | ·8861111 | ·9277777 | ·9694444 | 16 | ·00018 |
| 17 | ·5118055 | ·5534722 | ·5951388 | ·6368055 | 17 | ·6784722 | ·7201388 | ·7618055 | ·8034722 | 17 | ·8451388 | ·8868055 | ·9284722 | ·9701388 | 17 | ·00020 |
| 18 | ·5125000 | ·5541666 | ·5958333 | ·6375000 | 18 | ·6791666 | ·7208333 | ·7625000 | ·8041666 | 18 | ·8458333 | ·8875000 | ·9291666 | ·9708333 | 18 | ·0002 |
| 19 | ·5131944 | ·5548611 | ·5965277 | ·6381944 | 19 | ·6798611 | ·7215277 | ·7631944 | ·8048611 | 19 | ·8465277 | ·8881944 | ·9298611 | ·9715277 | 19 | ·00022 |
| 20 | ·5138888 | ·5555555 | ·5972222 | ·6388888 | 20 | ·6805555 | ·7222222 | ·7638888 | ·8055555 | 20 | ·8472222 | ·8888888 | ·9305555 | ·9722222 | 20 | ·00023 |
| 21 | ·5145833 | ·5562500 | ·5979166 | ·6395833 | 21 | ·6812500 | ·7229166 | ·7645833 | ·8062500 | 21 | ·8479166 | ·8895833 | ·9312500 | ·9729166 | 21 | ·00024 |
| 22 | ·5152777 | ·5569444 | ·5986111 | ·6402777 | 22 | ·6819444 | ·7236111 | ·7652777 | ·8069444 | 22 | ·8486111 | ·8902777 | ·9319444 | ·9736111 | 22 | ·00025 |
| 23 | ·5159722 | ·5576388 | ·5993055 | ·6409722 | 23 | ·6826388 | ·7243055 | ·7659722 | ·8076388 | 23 | ·8493055 | ·8909722 | ·9326388 | ·9743055 | 23 | ·00027 |
| 24 | ·5166666 | ·5583333 | ·6000000 | ·6416666 | 24 | ·6833333 | ·7250000 | ·7666666 | ·8083333 | 24 | ·8500000 | ·8916666 | ·9333333 | ·9750000 | 24 | ·00028 |
| 25 | ·5173611 | ·5590277 | ·6006944 | ·6423611 | 25 | ·6840277 | ·7256944 | ·7673611 | ·8090277 | 25 | ·8506944 | ·8923611 | ·9340277 | ·9756944 | 25 | ·00029 |
| 26 | ·5180555 | ·5597222 | ·6013888 | ·6430555 | 26 | ·6847222 | ·7263888 | ·7680555 | ·8097222 | 26 | ·8513888 | ·8930555 | ·9347222 | ·9763888 | 26 | ·00030 |
| 27 | ·5187500 | ·5604166 | ·6020833 | ·6437500 | 27 | ·6854166 | ·7270833 | ·7687500 | ·8104166 | 27 | ·8520833 | ·8937500 | ·9354166 | ·9770833 | 27 | ·00031 |
| 28 | ·5194444 | ·5611111 | ·6027777 | ·6444444 | 28 | ·6861111 | ·7277777 | ·7694444 | ·8111111 | 28 | ·8527777 | ·8944444 | ·9361111 | ·9777777 | 28 | ·00032 |
| 29 | ·5201388 | ·5618055 | ·6034722 | ·6451388 | 29 | ·6868055 | ·7284722 | ·7701388 | ·8118055 | 29 | ·8534722 | ·8951388 | ·9368055 | ·9784722 | 29 | ·00034 |
| 30 | ·5208333 | ·5625000 | ·6041666 | ·6458333 | 30 | ·6875000 | ·7291666 | ·7708333 | ·8125000 | 30 | ·8541666 | ·8958333 | ·9375000 | ·9791666 | 30 | ·00035 |
| 31 | ·5215277 | ·5631944 | ·6048611 | ·6465277 | 31 | ·6881944 | ·7298611 | ·7715277 | ·8131944 | 31 | ·8548611 | ·8965277 | ·9381944 | ·9798611 | 31 | ·00036 |
| 32 | ·5222222 | ·5638888 | ·6055555 | ·6472222 | 32 | ·6888888 | ·7305555 | ·7722222 | ·8138888 | 32 | ·8555555 | ·8972222 | ·9388888 | ·9805555 | 32 | ·00037 |
| 33 | ·5229166 | ·5645833 | ·6062500 | ·6479166 | 33 | ·6895833 | ·7312500 | ·7729166 | ·8145833 | 33 | ·8562500 | ·8979166 | ·9395833 | ·9812500 | 33 | ·00038 |
| 34 | ·5236111 | ·5652777 | ·6069444 | ·6486111 | 34 | ·6902777 | ·7319444 | ·7736111 | ·8152777 | 34 | ·8569444 | ·8986111 | ·9402777 | ·9819444 | 34 | ·00039 |
| 35 | ·5243055 | ·5659722 | ·6076388 | ·6493055 | 35 | ·6909722 | ·7326388 | ·7743055 | ·8159722 | 35 | ·8576388 | ·8993055 | ·9409722 | ·9826388 | 35 | ·00040 |
| 36 | ·5250000 | ·5666666 | ·6083333 | ·6500000 | 36 | ·6916666 | ·7333333 | ·7750000 | ·8166666 | 36 | ·8583333 | ·9000000 | ·9416666 | ·9833333 | 36 | ·00042 |
| 37 | ·5256944 | ·5673611 | ·6090277 | ·6506944 | 37 | ·6923611 | ·7340277 | ·7756945 | ·8173611 | 37 | ·8590277 | ·9006944 | ·9423611 | ·9840277 | 37 | ·00043 |
| 38 | ·5263888 | ·5680555 | ·6097222 | ·6513888 | 38 | ·6930555 | ·7347222 | ·7763888 | ·8180555 | 38 | ·8597222 | ·9013888 | ·9430555 | ·9847222 | 38 | ·00044 |
| 39 | ·5270833 | ·5687500 | ·6104166 | ·6520833 | 39 | ·6937500 | ·7354166 | ·7770833 | ·8187500 | 39 | ·8604166 | ·9020833 | ·9437500 | ·9854166 | 39 | ·00045 |
| 40 | ·5277777 | ·5694444 | ·6111111 | ·6527777 | 40 | ·6944444 | ·7361111 | ·7777777 | ·8194444 | 40 | ·8611111 | ·9027777 | ·9444444 | ·9861111 | 40 | ·00046 |
| 41 | ·5284722 | ·5701388 | ·6118055 | ·6534722 | 41 | ·6951388 | ·7368055 | ·7784722 | ·8201388 | 41 | ·8618055 | ·9034722 | ·9451388 | ·9868055 | 41 | ·00047 |
| 42 | ·5291666 | ·5708333 | ·6125000 | ·6541666 | 42 | ·6958333 | ·7375000 | ·7791666 | ·8208333 | 42 | ·8625000 | ·9041666 | ·9458333 | ·9875000 | 42 | ·00049 |
| 43 | ·5298611 | ·5715277 | ·6131944 | ·6548611 | 43 | ·6965277 | ·7381944 | ·7798611 | ·8215277 | 43 | ·8631944 | ·9048611 | ·9465277 | ·9881944 | 43 | ·00050 |
| 44 | ·5305555 | ·5722222 | ·6138888 | ·6555555 | 44 | ·6972222 | ·7388888 | ·7805555 | ·8222222 | 44 | ·8638888 | ·9055555 | ·9472222 | ·9888888 | 44 | ·00051 |
| 45 | ·5312500 | ·5729166 | ·6145833 | ·6562500 | 45 | ·6979166 | ·7395833 | ·7812500 | ·8229166 | 45 | ·8645833 | ·9062500 | ·9479166 | ·9895833 | 45 | ·0005 |
| 46 | ·5319444 | ·5736111 | ·6152777 | ·6569444 | 46 | ·6986111 | ·7402777 | ·7819444 | ·8236111 | 46 | ·8652777 | ·9069444 | ·9486111 | ·9902777 | 46 | ·0005 |
| 47 | ·5326388 | ·5743055 | ·6159722 | ·6576388 | 47 | ·6993055 | ·7409722 | ·7826388 | ·8243055 | 47 | ·8659722 | ·9076388 | ·9493055 | ·9909722 | 47 | ·0005 |
| 48 | ·5333333 | ·5750000 | ·6166666 | ·6583333 | 48 | ·7000000 | ·7416666 | ·7833333 | ·8250000 | 48 | ·8666666 | ·9083333 | ·9500000 | ·9916666 | 48 | ·0005 |
| 49 | ·5340277 | ·5756944 | ·6173611 | ·6590277 | 49 | ·7006944 | ·7423611 | ·7840277 | ·8256944 | 49 | ·8673611 | ·9090277 | ·9506944 | ·9923611 | 49 | ·0005 |
| 50 | ·5347222 | ·5763888 | ·6180555 | ·6597222 | 50 | ·7013888 | ·7430555 | ·7847222 | ·8263888 | 50 | ·8680555 | ·9097222 | ·9513888 | ·9930555 | 50 | ·0005 |
| 51 | ·5354166 | ·5770833 | ·6187500 | ·6604166 | 51 | ·7020833 | ·7437500 | ·7854166 | ·8270833 | 51 | ·8687500 | ·9104166 | ·9520833 | ·9937500 | 51 | ·0005 |
| 52 | ·5361111 | ·5777777 | ·6194444 | ·6611111 | 52 | ·7027777 | ·7444444 | ·7861111 | ·8277777 | 52 | ·8694444 | ·9111111 | ·9527777 | ·9944444 | 52 | ·0006 |
| 53 | ·5368055 | ·5784722 | ·6201388 | ·6618055 | 53 | ·7034722 | ·7451388 | ·7868055 | ·8284722 | 53 | ·8701388 | ·9118055 | ·9534722 | ·9951388 | 53 | ·000 |
| 54 | ·5375000 | ·5791666 | ·6208333 | ·6625000 | 54 | ·7041666 | ·7458333 | ·7875000 | ·8291666 | 54 | ·8708333 | ·9125000 | ·9541666 | ·9958333 | 54 | ·0006 |
| 55 | ·5381944 | ·5798611 | ·6215277 | ·6631944 | 55 | ·7048611 | ·7465277 | ·7881944 | ·8298611 | 55 | ·8715277 | ·9131944 | ·9548611 | ·9965277 | 55 | ·0006 |
| 56 | ·5388888 | ·5805555 | ·6222222 | ·6638888 | 56 | ·7055555 | ·7472222 | ·7888888 | ·8305555 | 56 | ·8722222 | ·9138888 | ·9555555 | ·9972222 | 56 | ·0006 |
| 57 | ·5395833 | ·5812500 | ·6229166 | ·6645833 | 57 | ·7062500 | ·7479166 | ·7895833 | ·8312500 | 57 | ·8729166 | ·9145833 | ·9562500 | ·9979166 | 57 | ·0006 |
| 58 | ·5402777 | ·5819444 | ·6236111 | ·6652777 | 58 | ·7069444 | ·7486111 | ·7902777 | ·8319444 | 58 | ·8736111 | ·9152777 | ·9569444 | ·9986111 | 58 | ·0006 |
| 59 | ·5409722 | ·5826388 | ·6243055 | ·6659722 | 59 | ·7076388 | ·7493055 | ·7909722 | ·8326388 | 59 | ·8743055 | ·9159722 | ·9576388 | ·9993055 | 59 | ·0006 |

## TABLE XXI.—Navamsas.

| Nav. | Long. | Nav. Sign. | Nav. | Long. | Nav. Sign. | Nav. | Long. | Nav. Sign. | Nav. | Long. | Nav. Sign. | Nav. | Long. | Nav. Sign. | Nav. | Long. | Nav. Sign. |
|---|---|---|---|---|---|---|---|---|---|---|---|---|---|---|---|---|---|
| 1 | 3·3 | I | 1 | 63·3 | VII | 1 | 123·3 | I | 1 | 183·3 | VII | 1 | 243·3 | I | 1 | 303·3 | VII |
| 2 | 6·6 | II | 2 | 66·6 | VIII | 2 | 126·6 | II | 2 | 186·6 | VIII | 2 | 246·6 | II | 2 | 306·6 | VIII |
| 3 | 10·0 | III | 3 | 70·0 | IX | 3 | 130·0 | III | 3 | 190·0 | IX | 3 | 250·0 | III | 3 | 310·0 | IX |
| 4 | 13·3 | IV | 4 | 73·3 | X | 4 | 133·3 | IV | 4 | 193·3 | X | 4 | 253·3 | IV | 4 | 313·3 | X |
| 5 | 16·6 | V | 5 | 76·6 | XI | 5 | 136·6 | V | 5 | 196·6 | XI | 5 | 256·6 | V | 5 | 316·6 | XI |
| 6 | 20·0 | VI | 6 | 80·0 | XII | 6 | 140·0 | VI | 6 | 200·0 | XII | 6 | 260·0 | VI | 6 | 320·0 | XII |
| 7 | 23·3 | VII | 7 | 83·3 | I | 7 | 143·3 | VII | 7 | 203·3 | I | 7 | 263·3 | VII | 7 | 323·3 | I |
| 8 | 26·6 | VIII | 8 | 86·6 | II | 8 | 146·6 | VIII | 8 | 206·6 | II | 8 | 266·6 | VIII | 8 | 326·6 | II |
| 9 | 30·0 | IX | 9 | 90·0 | III | 9 | 150·0 | IX | 9 | 210·0 | III | 9 | 270·0 | IX | 9 | 330·0 | III |
| 1 | 33·3 | X | 1 | 93·3 | IV | 1 | 153·3 | X | 1 | 213·3 | IV | 1 | 273·3 | X | 1 | 333·3 | IV |
| 2 | 36·6 | XI | 2 | 96·6 | V | 2 | 156·6 | XI | 2 | 216·6 | V | 2 | 276·6 | XI | 2 | 336·6 | V |
| 3 | 40·0 | XII | 3 | 100·0 | VI | 3 | 160·0 | XII | 3 | 220·0 | VI | 3 | 280·0 | XII | 3 | 340·0 | VI |
| 4 | 43·3 | I | 4 | 103·3 | VII | 4 | 163·3 | I | 4 | 223·3 | VII | 4 | 283·3 | I | 4 | 343·3 | VII |
| 5 | 46·6 | II | 5 | 106·6 | VIII | 5 | 166·6 | II | 5 | 226·6 | VIII | 5 | 286·6 | II | 5 | 346·6 | VIII |
| 6 | 50·0 | III | 6 | 110·0 | IX | 6 | 170·0 | III | 6 | 230·0 | IX | 6 | 290·0 | III | 6 | 350·0 | IX |
| 7 | 53·3 | IV | 7 | 113·3 | X | 7 | 173·3 | IV | 7 | 233·3 | X | 7 | 293·3 | IV | 7 | 353·3 | X |
| 8 | 56·6 | V | 8 | 116·6 | XI | 8 | 176·6 | V | 8 | 236·6 | XI | 8 | 296·6 | V | 8 | 356·6 | XI |
| 9 | 60·0 | VI | 9 | 120·0 | XII | 9 | 180·0 | VI | 9 | 240·0 | XII | 9 | 300·0 | VI | 9 | 360·0 | XII |

## TABLE XXII.—Ahargana.

### CENTURIES.

| A.D. | K.Y. | Days. | A.D. | K.Y. | Days. | A.D. | K.Y. | Days. | A.D. | K.Y. | Days. | A.D. | K.Y. | Days. |
|---|---|---|---|---|---|---|---|---|---|---|---|---|---|---|
| **101 B.C.** | 3001 | 1096141·28 | **400** | 3501 | 1278770·66 | **900** | 4001 | 1461400·03 | **1400** | 4501 | 1644029·41 | **1900** | 5001 | 1826658·79 |
| **1 B.C.** | 3101 | 1132667·15 | **500** | 3601 | 1315296·53 | **1000** | 4101 | 1497925·91 | **1500** | 4601 | 1680555·29 | **2000** | 5101 | 1863184·67 |
| **100 A.D.** | 3201 | 1169193·03 | **600** | 3701 | 1351822·41 | **1100** | 4201 | 1534451·78 | **1600** | 4701 | 1717081·16 | **3000** | 5201 | 2228443·42 |
| **200** | 3301 | 1205718·90 | **700** | 3801 | 1388348·28 | **1200** | 4301 | 1570977·66 | **1700** | 4801 | 1753607·04 | **4000** | 5301 | 2593702·18 |
| **300** | 3401 | 1242244·78 | **800** | 3901 | 1424874·16 | **1300** | 4401 | 1607503·54 | **1800** | 4901 | 1790132·91 | **5000** | 5401 | 2958960·94 |

*N.B.*—For centuries prior to **1 B.C.** (3101 K.Y.), subtract from the ahargana for **1** B.C. the ahargana for 100, 200......3000 years as the case may be according to last column of next table. Supposing we want the ahargana for Rama's birthday (according to Bentley), *viz.*, the 33rd day of Solar Year 962 B.C. (*i.e.*, 39th year of century beginning with B.C. 1001),

| | | | Days. |
|---|---|---|---|
| We have B.C. 1 | Ahargana | ... ... ... | 1132667·15 |
| Deduct for 1000 years | Do. | ... ... ... | —365258·76 |
| | | | 767408·39 |
| Add for 39 years | Do. | ... ... ... | +14245·09 |
| Do. 33 days | Do. | ... ... ... | +33·00 |
| B.C. 962, 33rd day of Solar Year, | | | 781686·48 Ahargana required. |

### ODD YEARS.

| Yrs. | Days. | Yrs. | Days. | Yrs. | Days. | Yrs. | Days. | Yrs. | Days. | Yrs. | Days. |
|---|---|---|---|---|---|---|---|---|---|---|---|
| **1** | 365·26 | **21** | 7670·48 | **41** | 14975·61 | **61** | 22281·78 | **81** | 29585·95 | **200** | 73051·75 |
| **2** | 730·52 | **22** | 8035·69 | **42** | 15340·87 | **62** | 22646·04 | **82** | 29951·22 | **300** | 109577·63 |
| **3** | 1095·78 | **23** | 8400·95 | **43** | 15706·13 | **63** | 23011·30 | **83** | 30316·48 | **400** | 146103·50 |
| **4** | 1461·03 | **24** | 8766·21 | **44** | 16071·38 | **64** | 23376·56 | **84** | 30681·73 | **500** | 182629·38 |
| **5** | 1826·29 | **25** | 9131·47 | **45** | 16436·64 | **65** | 23741·82 | **85** | 31046·99 | **600** | 219155·25 |
| **6** | 2191·55 | **26** | 9496·73 | **46** | 16801·90 | **66** | 24107·08 | **86** | 31412·25 | **700** | 255681·13 |
| **7** | 2556·81 | **27** | 9861·99 | **47** | 17167·16 | **67** | 24472·34 | **87** | 31777·51 | **800** | 292207·00 |
| **8** | 2922·07 | **28** | 10227·24 | **48** | 17532·42 | **68** | 24837·59 | **88** | 32142·77 | **900** | 328732·88 |
| **9** | 3287·33 | **29** | 10592·50 | **49** | 17897·68 | **69** | 25202·85 | **89** | 32508·03 | **1000** | 365258·76 |
| **10** | 3652·59 | **30** | 10957·76 | **50** | 18262·94 | **70** | 25568·11 | **90** | 32873·29 | **2000** | 730517·51 |
| **11** | 4017·85 | **31** | 11323·02 | **51** | 18628·20 | **71** | 25933·37 | **91** | 33238·55 | **3000** | 1095776·27 |
| **12** | 4383·10 | **32** | 11688·28 | **52** | 18993·45 | **72** | 26298·63 | **92** | 33603·80 | **4000** | 1461035·03 |
| **13** | 4748·36 | **33** | 12053·54 | **53** | 19358·71 | **73** | 26663·89 | **93** | 33969·06 | **5000** | 1826293·78 |
| **14** | 5113·62 | **34** | 12418·80 | **54** | 19723·97 | **74** | 27029·15 | **94** | 34334·32 | | |
| **15** | 5478·88 | **35** | 12784·06 | **55** | 20089·23 | **75** | 27394·41 | **95** | 34699·58 | | |
| **16** | 5844·14 | **36** | 13149·31 | **56** | 20454·49 | **76** | 27759·66 | **96** | 35064·84 | | |
| **17** | 6209·40 | **37** | 13514·57 | **57** | 20819·75 | **77** | 28124·92 | **97** | 35430·10 | | |
| **18** | 6574·66 | **38** | 13879·83 | **58** | 21185·01 | **78** | 28490·18 | **98** | 35795·36 | | |
| **19** | 6939·92 | **39** | 14245·09 | **59** | 21550·27 | **79** | 28855·44 | **99** | 36160·62 | | |
| **20** | 7305·17 | **40** | 14610·35 | **60** | 21915·52 | **80** | 29220·70 | **100** | 36525·88 | | |

## Models for Compendious Working of Tithis, Nakshatras and Yogas.

These models are chiefly intended to illustrate the use of the Eye-Table, which is a companion to Tables VIII IX, X, XI, XII and XVII for finding longitudes, rasis, solar dates and, lastly and chiefly, the ending moments of an Tithi, Nakshatra, or Yoga.

### Table XII and Eye-Table.—Lunar Tithis (A.D. 1840 –A.D. 1920).

*Required the ending moment of (Lunar) Ashada Sukla Ashtami, Saka 1832, Vik. 1967.*

| | | d. | gh. | p. | | gh. | ☉'s Anom. d. | ☉'s Anom. gh. | ☾'s Anom. d. | ☾'s Anom. gh. | |
|---|---|---|---|---|---|---|---|---|---|---|---|
| Mean ending moment of Ashadha New Moon Saka 1832. (Table XII) A.D. 1910. ... ... | Jl. | 7 | 10 | 57 | | | 84 | 58 | 4 | 29 | |
| Collective duration of 8 tithis (Eye-Table) | | 7 | 52 | 29 | ☉'s Eqn. (Eye-Tab.) | —2 | 7 | 52 | 7 | 52 | |
| Ending moment of mean Tithi ... | Jl. | 15 | 3 | 26 | ☾'s Eq. (Eye-Tab.) | —9 | 92 | 50 | 12 | 21 | |
| Sum of ☉'s and ☾'s Equations ... | | | —11 | | | | | | | —2 | (☉'s eqn.) |
| | Jl. | 14 | 52 | | | —11 | | | 12 | 19 | |

The ending moment of the tithi (Lanka time) was, therefore, 52 ghatikas after Lanka sunrise on 14 July 1910. If we want local time, we should use Table XIII.

### Tables XII, XI and IX-*l*:—Nakshatras and Yogas (A.D. 1840—A.D. 1920). Eye-Table not used.

Required the ending moment of *Krittika Nakshatra for the month of July 1910.*

Since Ashadha New Moon fell on 7 July 1910 and Krittika Nakshatra in Ashadha ends about 25 days from New moon (Table XI) which would take us into August, we take the previous New moon.

| | d. | gh. | p. | | d. | gh. | p. | ☾'s Anom. d. | gh. | p. | |
|---|---|---|---|---|---|---|---|---|---|---|---|
| (Tab. XII) Jyeshtha New Moon ... A. D. 1910. | | | | Je. | 7 | 39 | 7 | 2 | 30 | 40 | |
| (Tab. XI-A) Longest interval from Jyesh. New Moon to Krit. (No. 3) Nak. ... | 28 | 18 | 39 | | | | | | | | |
| (Tab. XI-A) Deduct for Nak. in S.Y. A.D. 1910 ... | 1 | 56 | 18 | | 26 | 22 | 21 | 26 | 22 | 21 | |
| | 26 | 22 | 21 | Je. | 34 | 1 | 28 | 28 | 53 | 1 | |
| (Tab. IX-*l*) ☾'s Nak. Eqn. for An. of 1 d. 19 gh. 44 p. | | | | | | —6 | 20 | 27 | 33 | 17 | Anom. month |
| | | | | 33 Je. *i.e.*, Jl. | 3, | 55 | 8 | 1 | 19 | 44 | |

The ending moment of Krittika Nakshatra in July 1910, was 55 ghatikas after Lanka sunrise on 3 July 1910. If we want local time, we should use Table XIII.

For Yogas proceed similarly, but add together ☉'s and ☾'s Yoga Eqns. by Tab. IX-*i*.

### Table X and Eye-Table.—Lunar Tithis (ANY YEAR IN TAB. X.)

This method will be of immense service to Epigraphists who will require the English equivalents of thousands of tithis ranging from 1 B.C. to modern times. With Table X and the Eye-Table the operation can be performed in less than a minute. Thus, *Required the ending moment of Ashada Sukla 12, AD. 484, K.Y. 3585.*

| | Week-day. | Month and day. | Fraction of day. | | | ☉'s Anom. | ☾'s Anom. | |
|---|---|---|---|---|---|---|---|---|
| | | | | | | 24·39 | 4·49 | |
| Mean ending moment of Ashada New moon Tithi. A.D. 484 (Table X). | 7 | Je. 9 | ·67 | ☉'s Eqn. (Eye-Tab.) | —·04. | 59·06 | 3·95 | |
| | | | | | | 83·45 | 8·44 | |
| Collective duration of 12 tithis (Eye-Tab.) | 11 +1* | 11 | ·81 | ☾'s Eqn. (Eye-Tab.) | +·41 | 11·81 | 11·81 | |
| | 19 | 21 | ·48 | | +·37 | 95·26 | 20·25 | |
| Sum of ☉'s and ☾'s Equations | ... | ... | +·37 | | | | —·04 | (☉'s Eqn.) |
| (19 – 14)= | 5 | 21 | ·85 | | | | 20·21 | |

**Answer.**—*Thursday, 21 June A.D. 484, at ·85 (i.e., by Eye-Table, 51 ghatikas or 20½ hours) after sunrise (Lanka time).* The absolutely correct time differs from the above by only 11 palas, as we may see at p. 91, Sec. **259** of the text.

* When the sum of the several fractions of a day in the column headed *fraction of day* exceeds unity, we should add 1 to the "*Week-day*" column.

### TABLES XVII, XI, X, VIII, V, III, AND EYE-TABLE.—NAKSHATRAS AND YOGAS

THREE ALTERNATIVE METHODS TO CHOOSE FROM. (ANY YEAR IN TAB. X). Table IX not used.

***Required the Nakshatra current at the ending moment of above Tithi.***

| | |
|---|---|
| By Table VIII (last col.) the sun's long. for Nak. corresponding to 95 days of the solar year is, | 7·5051 |
| By Table V (last col.) ☉'s long. for ·26 day is ... | ·02 |
| ☉'s long. corresponding to 95·26 days ... | 7·53 |
| Add Tithi equivalent in days ... ... ... | 11·81 |
| | 19·34 |

By Table III, at 19·34 days of lunation space, the Nakshatra current is the 18th, *i.e.*, Jyeshtha.

This, then, was the Nakshatra *current* at the ending moment of Ash. Suk. 12, A.D. 484.

SECOND METHOD

We may calculate directly from Tab. XVII-A and C (pp. 207, 209), ☉'s long. for 95·26 days of Solar year, *i.e.*, 91·49°–·54° + ·26°=91·21°. Add ☾'s–☉'s long. *i.e.*, No. of tithi × 12°=12 × 12°=144°

Total ... 235·21°

This, by Eye-Tab. is the long. of *Nakshatra* Jyeshtha.

THIRD METHOD BY TABLE XI, FOR ENDING MOMENT OF NAKSHATRA.

Annual correction for Nakshatras for A.D. 484 is that for 24·39, *i.e.*, 1·79+·03=1·82 (Tab. X and XI).

The ending moment of No. 18 Nak. in Lunar Ashada was 13·95 – 1·82=12·13 days from the new moon, *i.e.*, ·32 of a day after the mean tithi, whose ending moment is 11·81 days from New moon.

The ☾'s anom. then was ·32 more than that of the tithi, *i.e.*, 20·57, for which the Nak. eqn. is +·38 (Eye-Tab.).

The ending moment of the Nak. No. 18, *i.e.*, Jyeshtha, was, therefore, ·48+·32+·38=1·18 or ·18 of a day after sunrise on the next day after the 21st June.

*Yoga current at ending moment of above tithi.*

| | |
|---|---|
| 2×☉'s Longitude as in 1st column | 182·42. |
| Add. ☾'s–☉'s long.=No. of Tithi × 12°= | 144° |
| Total ... | 326·42° |

This, by Eye-Tab., is the long. of *Yoga* Brahman.

[*N. B.*—Readers who have to use this Eye-Table frequently are advised either to get it mounted on cloth or to get a mounted spare copy, *price* four annas.]

# EYE-TABLE,

**SANKRANTIS**, *i. e.*, (1) **MOMENTS OF COMMENCEMENT OF** *SOLAR* **MONTHS**, (2) **LATEST POSSIBLE MOMENT**

| | | | |
|---|---|---|---|---|
| **Order of Signs of Zodiac or Rasis** ... ... ... ... | **1** | **2** | **3** | **4** |
| **Bengal Solar months** ... ... ... ... | **Vaisakha** | **Jyeshtha** | **Ashadha** | **Sravana** |
| **Malayalam Solar months ; also names of Rasis** ... ... ... | { **Mesha** / **Medam** | { **Vrishabha** / **Edavam** | **Mithunam** | **Karkadam** |
| **Degree of Longitude at planet's entry into each sign** ... ... | **0°** | **30°** | **60°** | **90°** |
| **Tamil Solar months** ... ... .. ... | **Chittirai** | **Vaikasi** | **Ani** | **Adi** |
| Moment of Sankranti in days of Solar Year, and decimals of a day ... | 0·0 | 30·9353 | 62·3555 | 94·0003 |
| Do. do. Ghatikas and Palas ... ... | 0, 0, 0 | 30, 56, 7 | 62, 21, 20 | 94, 0, 1 |
| English date corresponding to each Sankranti (A.D. 1911) ... ... | Ap. 13 | May 14 | June 14 | July 16 |
| Lunar Month : each month commences *before* Sankranti noted in *next* column. ... | (*1*) *Vaisakha* | (*2*) *Jyeshtha* | (*3*) *Ashadha* | (*4*) *Sravana* (*5*] |

**Tithi equivalents in days, deci**

| **Tithis** ... ... | 1 | 2 | 3 | 4 | 5 | 6 | 7 | 8 | 9 | 10 | 11 | 12 | 13 | 14 | 15 |
|---|---|---|---|---|---|---|---|---|---|---|---|---|---|---|---|
| **Tithi equivalent in days** ... ... | ·98 | 1·97 | 2·95 | 3·94 | 4·92 | 5·91 | 6·89 | 7·87 | 8·86 | 9·84 | 10·83 | 11·81 | 12·80 | 13·78 | 14·76 |

| | 1 | 2 | 3 | 4 | 5 | 6 | 7 |
|---|---|---|---|---|---|---|---|
| **Do. do. Ghatikas, and Palas** ... | 0, 59, 4 | 1, 58, 7 | 2, 57, 11 | 3, 56, 15 | 4, 55, 18 | 5, 54, 22 | 6, 53, 26 |
| | **16** | **17** | **18** | **19** | **20** | **21** | **22** |
| **Do. do. do.** ... | 15, 44, 59 | 16, 44, 2 | 17, 43, 6 | 18, 42, 10 | 19, 41, 14 | 20, 40, 17 | 21, 39, 21 |

**Order, Names and Longitudes of**

| | | | | | | |
|---|---|---|---|---|---|---|
| **Nakshatras** ... ... ... ... | **Asvini** | **Bharani** | **Krittika** | **Rohini** | **Mrigasira** | **Ardra** |
| *YOGAS* ... ... ... ... | *Vishkambha* 1 | *Priti* 2 | *Ayushmat* 3 | *Saubhagya* 4 | *Sobhana* 5 | *Atiganda* 6 |
| **Longitude at Planet's entry into each Nakshatra** ; *also longitude, at commencement, of each Yoga.* | 0° | 13° 20′ | 26° 40′ | 40° 0′ | 53° 20′ | 66° 40′ |
| **Nakshatras** ... ... ... ... | **Svati** | **Visakha** | **Anuradha** | **Jyeshtha** | **Mula** | **Pur. Ashad** |
| *YOGAS* ... ... ... ... | *Vajra* 15 | *Siddhi* 16 | *Vyatipata* 17 | *Variyas* 18 | *Parigha* 19 | *Siva* 20 |
| **Longitude at Planet's entry into each Nakshatra** ; *also longitude, at commencement, of each Yoga.* | 186° 40′ | 200° 0′ | 213° 0′ | 226° 40′ | 240° 0′ | 253° 20′ |

(Surya Siddhanta.) **Moon's Equation of the Centre and Moon's Anomaly in Days and Decimals**

| | ·00 | ·01 | ·02 | ·03 | ·04 | ·05 | ·06 | ·07 | ·08 | ·09 | ·10 | ·11 | ·12 | ·13 | ·14 | ·15 | ·16 | ·17 | ·18 | ·19 | ·20 | ·21 |
|---|---|---|---|---|---|---|---|---|---|---|---|---|---|---|---|---|---|---|---|---|---|---|
| * – | ·00 | ·11 | ·23 | ·34 | ·46 | ·58 | ·69 | ·81 | ·93 | 1·04 | 1·16 | 1·28 | 1·40 | 1·52 | 1·64 | 1·77 | 1·89 | 2·02 | 2·14 | 2·27 | 2·40 | 2·53 |
| * – | 13·78 | 13·68 | 13·59 | 13·49 | 13·40 | 13·30 | 13·20 | 13·11 | 13·01 | 12·91 | 12·81 | 12·71 | 12·61 | 12·51 | 12·41 | 12·31 | 12·20 | 12·10 | 11·99 | 11·88 | 11·77 | 11·66 |
| *+ | 13·78 | 13·87 | 13·97 | 14·06 | 14·16 | 14·25 | 14·35 | 14·45 | 14·54 | 14·64 | 14·74 | 14·84 | 14·94 | 15·04 | 15·14 | 15·24 | 15·35 | 15·45 | 15·56 | 15·67 | 15·78 | 15·89 |
| *+ | 27·55 | 27·44 | 27·32 | 27·21 | 27·09 | 26·98 | 26·86 | 26·74 | 26·63 | 26·51 | 26·39 | 26·27 | 26·15 | 26·03 | 25·91 | 25·79 | 25·66 | 25·54 | 25·41 | 25·28 | 25·15 | 25·02 |
| **Equation for Nakshatras** ... | ·04 | ·05 | ·06 | ·06 | ·07 | ·08 | ·09 | ·10 | ·11 | ·12 | ·13 | ·14 | ·15 | ·16 | ·17 | ·18 | ·18 | ·19 | | | | |
| **Equation for Yogas** ... | ·03 | ·04 | ·05 | ·06 | ·06 | ·08 | ·08 | ·09 | ·10 | ·11 | ·12 | ·13 | ·14 | ·15 | ·15 | ·16 | ·17 | ·18 | | | | |

(Surya Siddhanta.) **Sun's equation of the Centre and Sun's Anomaly in days of the Solar year and decimals**

| +·18 | +·175 | +·17 | +·16 | +·15 | +·14 | +·13 | +·12 | +·11 | +·10 | +·09 | +·08 | +·07 | +·06 | +·05 | +·04 | +·03 | +·02 | +·01 | ·00 | –·01 |
|---|---|---|---|---|---|---|---|---|---|---|---|---|---|---|---|---|---|---|---|---|
| 358 | 0 | 7 | 16 | 23 | 28 | 33 | 38 | 42 | 46 | 50 | 54 | 57 | 61 | 64 | 68 | 71 | 74 | 77 | 80·5 | 84 |
| 354 | 343 | 334 | 327 | 321 | 315 | 310 | 305·5 | 301 | 297 | 293 | 289 | 286 | 283 | 279 | 276 | 273 | 269·5 | 266 | 263 | 260 |

(Surya Siddhanta.) **Moon's Equation of the Centre in Ghatikas and Moon's Anoma**

| **Ghatikas** ... ... | | 0 | 1 | 2 | 3 | 4 | 5 | 6 | 7 | 8 | 9 | 10 |
|---|---|---|---|---|---|---|---|---|---|---|---|---|
| **Days & Ghat. (neg. eqn.)** ... ... | *– | ... | 0, 11 | 0, 23 | 0, 34 | 0, 46 | 0, 58 | 1, 10 | 1, 22 | 1, 34 | 1, 46 | 1, 59 |
| **Do. do.** ... ... | *– | 13, 47 | 13, 37 | 13, 28 | 13, 18 | 13, 8 | 12, 59 | 12, 49 | 12, 39 | 12, 29 | 12, 19 | 12, 8 |
| **Do. (pos. eqn.)** ... ... | *+ | 13, 47 | 13, 56 | 14, 6 | 14, 15 | 14, 25 | 14, 35 | 14, 44 | 14, 54 | 15, 4 | 15, 15 | 15, 25 |
| **Do. do.** ... ... | *+ | 27, 33 | 27, 22 | 27, 10 | 26, 59 | 26, 47 | 26, 35 | 26, 23 | 26, 12 | 26, 0 | 25, 47 | 25, 35 |

(Surya Siddhanta.) **Sun's Equation of the Centre in Ghatikas and Sun's Anomaly in Days.** (The

| **Ghatikas** ... ... ... | +11 | +10½ | +10 | +9 | +8 | +7 | +6 | +5 | +4 | +3 | +[illegible] |
|---|---|---|---|---|---|---|---|---|---|---|---|
| **Days** ... ... ... | 359 | 1 | 11 | 23 | 32 | 39½ | 46 | 53 | 59 | 64 | 7[illegible] |
| **Do.** ... ... ... | 354 | 343 | 333 | 321 | 312 | 304 | 297 | 291 | 285 | 279 | 27[illegible] |

29·53059 d. (1 Syn. mo.)=29 d., 12 hrs. ; 44 min., 3½ sec. ; 27·5546 d.=27 d. 13 hrs., 18 min., 37 sec. **Hours as Decimal Fractions of a D**

| | 1 | 2 | 3 | 4 | 5 | 6 | 7 | 8 | 9 | 10 | 11 | 12 | 13 |
|---|---|---|---|---|---|---|---|---|---|---|---|---|---|
| **Fraction of a day=hours** ... ... | ·04 | ·08 | ·12 | ·17 | ·21 | ·25 | ·29 | ·33 | ·37 | ·42 | ·46 | ·50 | ·54 |

29·5306 d. (1 Syn. mo.)=29 d., 31 gh., 50 p. ; 27·5546 d. (1 an. mo.)=27 d., 33 gh., 17 p. **Ghatikas as Decimal**

| | 1 | 2 | 3 | 4 | 5 | 6 | 7 | 8 | 9 | 10 | 11 | 12 | 13 |
|---|---|---|---|---|---|---|---|---|---|---|---|---|---|
| **Do. do. =Ghatikas** ... | ·02 | ·03 | ·05 | ·07 | ·08 | ·10 | ·12 | ·13 | ·15 | ·17 | ·18 | ·20 | ·22 |
| | **31** | **32** | **33** | **34** | **35** | **36** | **37** | **38** | **39** | **40** | **41** | **42** | **43** |
| **Do. do. = do.** ... | ·52 | ·53 | ·55 | ·57 | ·58 | ·60 | ·62 | ·63 | ·65 | ·67 | ·68 | ·70 | ·72 |

**.ATEST POSSIBLE MOMENTS OF COMMENCEMENT OF *LUNAR* MONTHS (Surya Siddhanta), (3) Longitudes of Rasis.**

| 3 Ashadha | 4 Sravana | 5 Bhadrapada | 6 Asvina | 7 Karttika | 8 Margasira | 9 Pausha | 10 Magha | 11 Phalguna | 12 Chaitra | |
|---|---|---|---|---|---|---|---|---|---|---|
| Mithunam | Karkadam | Simham | Kanni | Tulam | Vrischikam | Dhanus | Makaram | Kumbham | Minam | |
| 60° | 90° | 120° | 150° | 180° | 210° | 240° | 270° | 300° | 330° | |
| Ani | Adi | Avani | Purattasi | Aippasi | Kartigai | Margali | Tai | Masi | Panguni | |
| 62·3555 | 94·0003 | 125·4755 | 156·4912 | 186·9355 | 216·8289 | 246·3192 | 275·6369 | 305·0850 | 334·9053 | 365·2588 |
| 62, 21, 20 | 94, 0, 1 | 125, 28, 32 | 156, 29, 39 | 186,56, 8 | 216, 49, 44 | 246, 19, 9 | 275, 38, 13 | 305, 5, 6, | 334, 54, 19 | 365, 15, 32 |
| June 14 | July 16 | Aug. 16 | Sep. 16 | Oct. 17 | Nov. 16 | Dec. 15 | Jan. 14 | Feb. 12 | Mar. 13 | |

*(3) Ashadha (4) Sravana (5) Bhadrapada (6) Asvina (7) Karttika (8) Margasira (9) Pausha (10) Magha (11) Phalguna (12) Chaitra (13) Chaitra*

**thi equivalents in days, decimals of a day, Ghatikas and Palas.**

| 11 | 12 | 13 | 14 | 15 | 16 | 17 | 18 | 19 | 20 | 21 | 22 | 23 | 24 | 25 | 26 | 27 | 28 | 29 | 30 |
|---|---|---|---|---|---|---|---|---|---|---|---|---|---|---|---|---|---|---|---|
| 10·83 | 11·81 | 12·80 | 13·78 | 14·76 | 15·75 | 16·73 | 17·72 | 18·70 | 19·69 | 20·67 | 21·65 | 22·64 | 23·62 | 24·61 | 25·59 | 26·58 | 27·56 | 28·55 | 29·53 |

(Increase of Anomaly).

| 28 | 29 | 30 |
|---|---|---|
| ·01 | 1·00 | 1·98 |

| 5 | 6 | 7 | 8 | 9 | 10 | 11 | 12 | 13 | 14 | 15 | 30 tithis-1 Anom. mo. |
|---|---|---|---|---|---|---|---|---|---|---|---|
| 55, 18 | 5, 54, 22 | 6, 53, 26 | 7, 52, 29 | 8, 51, 33 | 9, 50, 37 | 10, 49, 40 | 11, 48, 44 | 12, 47, 48 | 13, 46, 51 | 14, 45, 55 | 29, 31, 50, 27, 13, 17, |

| 20 | 21 | 22 | 23 | 24 | 25 | 26 | 27 | 28 | 29 | 30 |
|---|---|---|---|---|---|---|---|---|---|---|
| 41, 14 | 20, 40, 17 | 21, 39, 21 | 22, 38, 24 | 23, 37, 28 | 24, 36, 32 | 25, 35, 36 | 26, 34, 39 | 27, 33, 43 | 28, 32, 46 | 29, 31, 50 |

(Increase of anomaly)

| 28 | 29 | 30 |
|---|---|---|
| 0, 0, 26 | 0, 59, 29 | 1, 58, 33 |

**. Names and Longitudes of Nakshatras and *YOGAS*.**

| ini | Mrigasira | Ardra | Punarvasu | Pushya | Aslesha | Magha | Pur. Phalguni | Utt.Phalguni | Hasta | Chitra |
|---|---|---|---|---|---|---|---|---|---|---|
| *gya* | *Sobhana* | *Atiganda* | *Sukarman* | *Dhriti* | *Sula* | *Ganda* | *Vriddhi* | *Dhruva* | *Vyaghata* | *Harshana* |
| | 5 | 6 | 7 | 8 | 9 | 10 | 11 | 12 | 13 | 14 |
| 0′ | 53° 20′ | 66° 40′ | 80° 0′ | 93° 20′ | 106° 40 | 120° 0′ | 133° 20′ | 146° 40′ | 160° 0′ | 173° 20′ |

| shtha | Mula | Pur. Ashada | Utt. Ashada | Sravana | Sravishta | Satabhishaj | Pur. Bhadr. | Utt. Bhadr. | Revati |
|---|---|---|---|---|---|---|---|---|---|
| *riyas* | *Parigha* | *Siva* | *Siddha* | *Sadhya* | *Subha* | *Sukla* | *Brahman* | *Indra* | *Vaidhriti* |
| 18 | 19 | 20 | 21 | 22 | 23 | 24 | 25 | 26 | 27 |
| ° 40′ | 240° 0′ | 253° 20′ | 266° 40′ | 280° 0′ | 293° 20′ | 306° 40′ | 320° 0′ | 333° 20′ | 346° 40′ |

**maly in Days and Decimals of a Day.** (*The signs – and + are to be applied to the equation in heavy type and not to the anomaly.)

| 6 | ·17 | ·18 | ·19 | ·20 | ·21 | ·22 | ·23 | ·24 | ·25 | ·26 | ·27 | ·28 | ·29 | ·30 | ·31 | ·32 | ·33 | ·34 | ·35 | ·36 | ·37 | ·38 | ·39 | ·40 | ·41 |
|---|---|---|---|---|---|---|---|---|---|---|---|---|---|---|---|---|---|---|---|---|---|---|---|---|---|
| 9 | 2·02 | 2·14 | 2·27 | 2·40 | 2·53 | 2·67 | 2·80 | 2·94 | 3·08 | 3·23 | 3·38 | 3·53 | 3·69 | 3·85 | 4·01 | 4·19 | 4·37 | 4·56 | 4·76 | 4·98 | 5·21 | 5·48 | 5·78 | 6·14 | 6·70 |
| 0 | 12·10 | 11·99 | 11·88 | 11·77 | 11·66 | 11·55 | 11·43 | 11·31 | 11·19 | 11·07 | 10·94 | 10·81 | 10·67 | 10·53 | 10·38 | 10·23 | 10·07 | 9·90 | 9·71 | 9·51 | 9·30 | 9·06 | 8·78 | 8·43 | 7·90 |
| 5 | 15·45 | 15·56 | 15·67 | 15·78 | 15·89 | 16·00 | 16·12 | 16·24 | 16·36 | 16·49 | 16·61 | 16·75 | 16·88 | 17·03 | 17·17 | 17·33 | 17·49 | 17·66 | 17·84 | 18·03 | 18·25 | 18·49 | 18·77 | 19·12 | 19·65 |
| 6 | 25·54 | 25·41 | 25·28 | 25·15 | 25·02 | 24·89 | 24·75 | 24·61 | 24·47 | 24·32 | 24·18 | 24·02 | 23·87 | 23·70 | 23·54 | 23·36 | 23·18 | 22·99 | 22·79 | 22·58 | 22·34 | 22·08 | 21·73 | 21·41 | 20·86 |
| 5 | ·16 | ·17 | ·18 | ·18 | ·19 | ·20 | ·21 | ·22 | ·23 | ·24 | ·25 | ·26 | ·27 | ·28 | ·29 | ·30 | ·30 | ·31 | ·32 | ·33 | ·34 | ·35 | ·36 | ·37 | ·38 |
| 4 | ·15 | ·15 | ·16 | ·17 | ·18 | ·19 | ·20 | ·21 | ·21 | ·22 | ·23 | ·24 | ·25 | ·26 | ·27 | ·27 | ·28 | ·29 | ·30 | ·31 | ·32 | ·33 | ·34 | ·34 | ·35 |

**he Solar year and decimals of a day.** (The anomalies are entered in ordinary type and the equations, with appropriate signs, in heavy type.)

| +·03 | +·02 | +·01 | ·00 | –·01 | –·02 | –·03 | –·04 | –·05 | –·06 | –·07 | –·08 | –·09 | –·10 | –·11 | –·12 | –·13 | –·14 | –·15 | –·16 | –·17 | –·175 | –·18 |
|---|---|---|---|---|---|---|---|---|---|---|---|---|---|---|---|---|---|---|---|---|---|---|
| 71 | 74 | 77 | 80·5 | 84 | 87 | 90 | 93·5 | 97 | 100 | 103 | 107 | 111 | 115 | 119 | 123 | 128 | 132 | 138 | 145 | 154 | 160 | 168 |
| 273 | 269·5 | 266 | 263 | 260 | 257 | 254 | 250 | 247 | 244 | 240 | 237 | 233 | 229 | 225 | 221 | 216 | 211 | 206 | 199 | 190 | 183 | 176 |

**hatikas and Moon's Anomaly in Days and Ghatikas.** (* The signs – and + are to be applied to the equation and not to the anomaly.)

| 7 | 8 | 9 | 10 | 11 | 12 | 13 | 14 | 15 | 16 | 17 | 18 | 19 | 20 | 21 | 22 | 23 | 24 | 25 |
|---|---|---|---|---|---|---|---|---|---|---|---|---|---|---|---|---|---|---|
| 1, 22 | 1, 34 | 1, 46 | 1, 59 | 2, 11 | 2, 24 | 2, 37 | 2, 51 | 3, 5 | 3, 20 | 3, 35 | 3, 51 | 4, 8 | 4, 26 | 4, 46 | 5, 8 | 5, 34 | 6, 8 | 7, 18 |
| 12, 39 | 12, 29 | 12, 19 | 12, 8 | 11, 58 | 11, 46 | 11, 35 | 11, 24 | 11, 12 | 10, 59 | 10, 46 | 10, 32 | 10, 17 | 10, 1 | 9, 48 | 9, 22 | 8, 59 | 8, 26 | 7, 20 |
| 14, 54 | 15, 4 | 15, 15 | 15, 25 | 15, 36 | 15, 47 | 15, 58 | 16, 10 | 16, 22 | 16, 34 | 16, 47 | 17, 2 | 17, 16 | 17, 33 | 17, 50 | 18, 11 | 18, 35 | 19, 7 | 20, 15 |
| 26, 12 | 26, 0 | 25, 47 | 25, 35 | 25, 22 | 25, 9 | 24, 56 | 24, 42 | 24, 28 | 24, 14 | 23, 58 | 23, 42 | 23, 25 | 23, 7 | 22, 47 | 22, 25 | 21, 59 | 21, 25 | 20, 17 |

**Anomaly in Days.** (The anomalies are entered in ordinary type and the equations, with appropriate signs, in heavy type. *Reverse Sign for Yogas.*)

| +5 | +4 | +3 | +2 | +1 | 0 | –1 | –2 | –3 | –4 | –5 | –6 | –7 | –8 | –9 | –10 | –10½ | –11 |
|---|---|---|---|---|---|---|---|---|---|---|---|---|---|---|---|---|---|
| 53 | 59 | 64 | 70 | 75 | 80 | 86 | 91 | 97 | 102 | 108 | 115 | 122 | 129 | 138 | 150 | 160 | 167 |
| 291 | 285 | 279 | 274 | 269 | 263 | 258 | 252 | 247 | 241 | 235 | 229 | 222 | 215 | 205 | 193 | 183 | 177 |

**as Decimal Fractions of a Day.** 365·25876 d. (1 Indian Sol. Year)=365 d., 6 hrs., 12 min., 37 sec.

| 9 | 10 | 11 | 12 | 13 | 14 | 15 | 16 | 17 | 18 | 19 | 20 | 21 | 22 | 23 | Minutes 14 | 29 | 44 |
|---|---|---|---|---|---|---|---|---|---|---|---|---|---|---|---|---|---|
| 7 | ·42 | ·46 | ·50 | ·54 | ·58 | ·62 | ·67 | ·71 | ·75 | ·79 | ·83 | ·87 | ·92 | ·96 | ·01 | ·02 | ·03 |

.., 17 p. **Ghatikas as Decimal Fractions of a Day.** 365·25876 d. (1 Indian Sol. Year)=365 d., 15 gh., 32 p.

| 9 | 10 | 11 | 12 | 13 | 14 | 15 | 16 | 17 | 18 | 19 | 20 | 21 | 22 | 23 | 24 | 25 | 26 | 27 | 28 | 29 | 30 |
|---|---|---|---|---|---|---|---|---|---|---|---|---|---|---|---|---|---|---|---|---|---|
| .5 | ·17 | ·18 | ·20 | ·22 | ·23 | ·25 | ·27 | ·28 | ·30 | ·32 | ·33 | ·35 | ·37 | ·38 | ·40 | ·42 | ·43 | ·45 | ·47 | ·48 | ·50 |

| 9 | 40 | 41 | 42 | 43 | 44 | 45 | 46 | 47 | 48 | 49 | 50 | 51 | 52 | 53 | 54 | 55 | 56 | 57 | 58 | 59 | |
|---|---|---|---|---|---|---|---|---|---|---|---|---|---|---|---|---|---|---|---|---|---|
| 36 | ·67 | ·68 | ·70 | ·72 | ·73 | ·75 | ·77 | ·78 | ·80 | ·82 | ·83 | ·85 | ·87 | ·88 | ·90 | ·92 | ·93 | ·95 | ·97 | ·98 | ... |

# INDEX.

*N.B.*—References to the Text are by Section in *Arabic* numerals, not by page, unless the page is specially indicated; similarly, references to the Tables are by the *Roman number* of the Table (and occasionally by page, which is then specially indicated). Thus "215" means "Section **215** of the Text" and "II" means "Table II". Arabic numerals placed between brackets (except when a page of the Text is indicated), refer to the numbers of sub-paragraphs or sub-entries. Thus "112 (16)", "XIII-A (82)", "XIV (57)," "I (15)". The names of (12) signs of the Zodiac, (15) Tithis, (27) Nakshatras, (27) Yogas, (8) Karanas, (60) years in Jupiter's Cycle (Northern and Southern Systems), places whose latitude and longitude are given in Tables XIII and XIII-A, festivals noted under tithis, (21) eras noted under Sec. **112** etc., etc., have all been carefully indexed by references to Section, Table and sub-number.

**A**



**A**—(*contd.*)

**A—(contd.)**



**B**



**B—(contd.)**

## C



## C—(*contd.*)

## G



## H



## H—(contd.)



## I



## J

J—(*contd.*)



K



K—(*contd.*)



L

**L—(contd.)**



**M—(contd.).**

### N



### O



### P

**P—(contd.)**



**Q**



**R**



**S**

S—(*contd.*)



S—(*contd.*)

**S—(contd.)**



**T**



**U**



**V**